BIG IDEAS MATH®

ALGEBRA 1

A Bridge to Success

TEACHING EDITION

Ron Larson
Laurie Boswell

BIG IDEAS LEARNING®

Erie, Pennsylvania
BigIdeasLearning.com

Big Ideas Learning, LLC
1762 Norcross Road
Erie, PA 16510-3838
USA

For product information and customer support, contact Big Ideas Learning at **1-877-552-7766** or visit us at ***BigIdeasLearning.com***.

Cover Image
Nithon/Shutterstock.com

Printed in the U.S.A.

ISBN 13: 978-1-68033-115-8
ISBN 10: 1-68033-115-9

2 3 4 5 6 7 8 9 10 WEB 18 17 16 15

Authors

Ron Larson, Ph.D., is well known as the lead author of a comprehensive program for mathematics that spans middle school, high school, and college courses. He holds the distinction of Professor Emeritus from Penn State Erie, The Behrend College, where he taught for nearly 40 years. He received his Ph.D. in mathematics from the University of Colorado. Dr. Larson's numerous professional activities keep him actively involved in the mathematics education community and allow him to fully understand the needs of students, teachers, supervisors, and administrators.

Ron Larson

Laurie Boswell, Ed.D., is the Head of School and a mathematics teacher at the Riverside School in Lyndonville, Vermont. Dr. Boswell is a recipient of the Presidential Award for Excellence in Mathematics Teaching and has taught mathematics to students at all levels, from elementary through college. Dr. Boswell was a Tandy Technology Scholar and served on the NCTM Board of Directors from 2002 to 2005. She currently serves on the board of NCSM and is a popular national speaker.

Laurie Boswell

Dr. Ron Larson and **Dr. Laurie Boswell** began writing together in 1992. Since that time, they have authored over two dozen textbooks. In their collaboration, Ron is primarily responsible for the student edition while Laurie is primarily responsible for the teaching edition.

For the Student

Welcome to *Big Ideas Math Algebra 1*. From start to finish, this program was designed with you, the learner, in mind.

As you work through the chapters in your Algebra 1 course, you will be encouraged to think and to make conjectures while you persevere through challenging problems and exercises. You will make errors—and that is ok! Learning and understanding occur when you make errors and push through mental roadblocks to comprehend and solve new and challenging problems.

In this program, you will also be required to explain your thinking and your analysis of diverse problems and exercises. You will master content through engaging explorations that will provide deeper understanding, concise stepped-out examples, and rich thought-provoking exercises. Being actively involved in learning will help you develop mathematical reasoning and use it to solve math problems and work through other everyday challenges.

We wish you the best of luck as you explore Algebra 1. We are excited to be a part of your preparation for the challenges you will face in the remainder of your high school career and beyond.

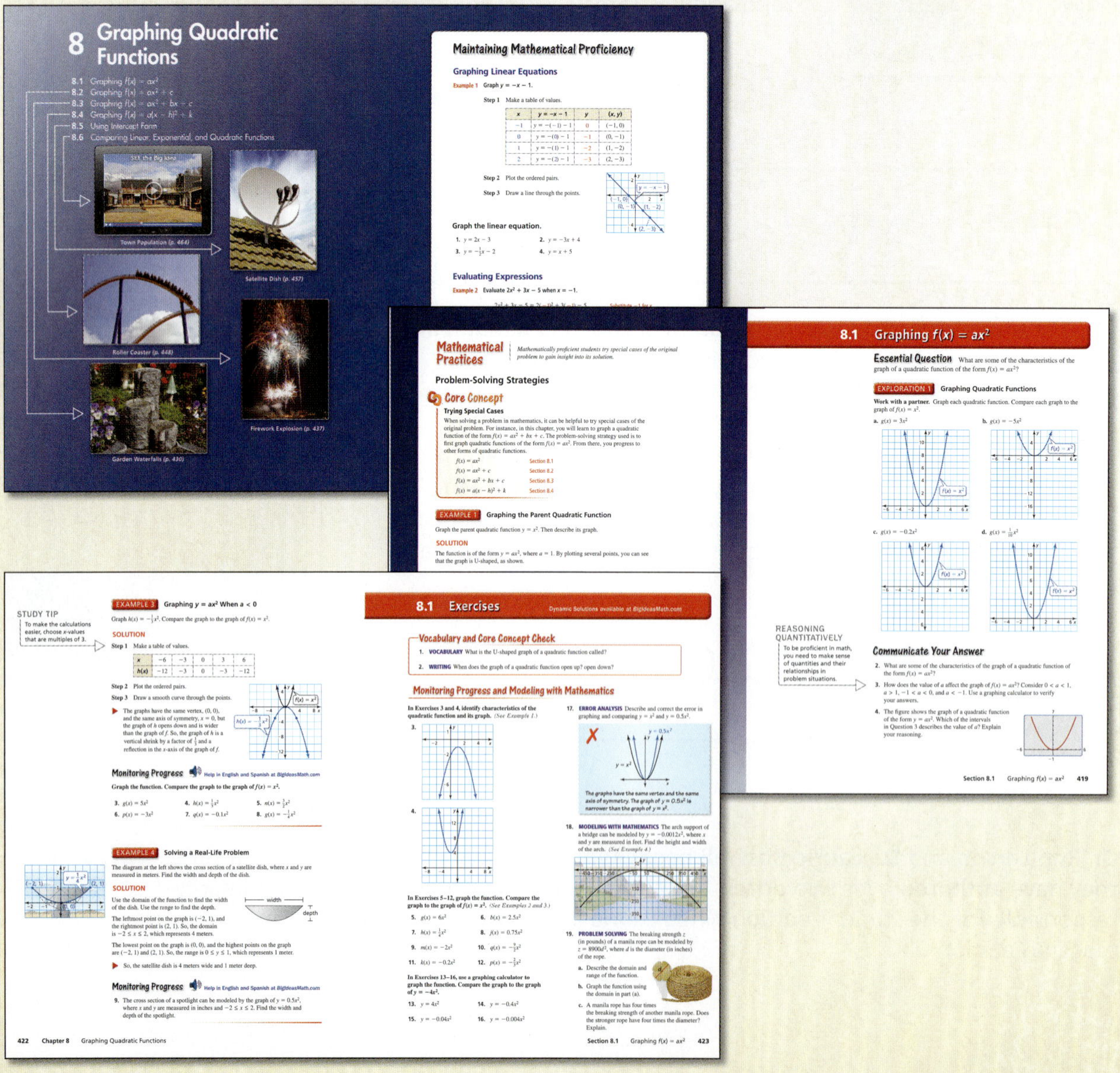

Big Ideas Math High School Research

Big Ideas Math Algebra 1, *Geometry*, and *Algebra 2* is a research-based program providing a rigorous, focused, and coherent curriculum for high school students. Ron Larson and Laurie Boswell utilized their expertise as well as the body of knowledge collected by additional expert mathematicians and researchers to develop each course.

The pedagogical approach to this program follows the best practices outlined in the most prominent and widely-accepted educational research and standards, including:

Achieve, ACT, and The College Board

Adding It Up: Helping Children Learn Mathematics
National Research Council ©2001

Common Core State Standards for Mathematics
National Governors Association Center for Best Practices and the Council of Chief State School Officers ©2010

Curriculum Focal Points and the *Principles and Standards for School Mathematics* ©2000
National Council of Teachers of Mathematics (NCTM)

Project Based Learning
The Buck Institute

Rigor/Relevance Framework™
International Center for Leadership in Education

Universal Design for Learning Guidelines
CAST ©2011

Big Ideas Math would like to express our gratitude to the mathematics education and instruction experts who served as consultants during the writing of *Big Ideas Math Algebra 1*, *Geometry*, and *Algebra 2*. Their input was an invaluable asset during the development of this program.

Kristen Karbon
Curriculum and Assessment Coordinator
Troy School District
Troy, Michigan

Jean Carwin
Math Specialist/TOSA
Snohomish School District
Snohomish, Washington

Carolyn Briles
Performance Tasks Consultant
Mathematics Teacher, Loudoun County Public Schools
Leesburg, Virginia

Bonnie Spence
Differentiated Instruction Consultant
Mathematics Lecturer, The University of Montana
Missoula, Montana

Connie Schrock, Ph.D.
Performance Tasks Consultant
Mathematics Professor, Emporia State University
Emporia, Kansas

We would also like to thank all of our reviewers who took the time to provide feedback during the final development phases. For a complete list of the *Big Ideas Math* program reviewers, please visit *www.BigIdeasLearning.com*.

Mathematical Practices

Make sense of problems and persevere in solving them.

- *Essential Questions* help students focus on core concepts as they analyze and work through each *Exploration*.
- Section opening *Explorations* allow students to struggle with new mathematical concepts and explain their reasoning in the *Communicate Your Answer* questions.

Reason abstractly and quantitatively.

- *Reasoning*, *Critical Thinking*, *Abstract Reasoning*, and *Problem Solving* exercises challenge students to apply their acquired knowledge and reasoning skills to solve each problem.
- *Thought Provoking* exercises test the reasoning skills of students as they analyze and interpret perplexing scenarios.

Construct viable arguments and critique the reasoning of others.

- Students must justify their responses to each *Essential Question* in the *Communicate Your Answer* questions at the end of each *Exploration* set.
- Students are asked to construct arguments and critique the reasoning of others in specialized exercises, including *Making an Argument*, *How Do You See It?*, *Drawing Conclusions*, *Reasoning*, *Error Analysis*, *Problem Solving*, and *Writing*.

Model with mathematics.

- Real-life scenarios are utilized in *Explorations*, *Examples*, *Exercises*, and *Assessments* so students have opportunities to apply the mathematical concepts they have learned to realistic situations.
- *Modeling with Mathematics* exercises allow students to interpret a problem in the context of a real-life situation, often utilizing tables, graphs, visual representations, and formulas.

Use appropriate tools strategically.

- Students are provided opportunities for selecting and utilizing the appropriate mathematical tool in *Using Tools* exercises. Students work with graphing calculators, dynamic geometry software, models, and more.
- A variety of tool papers and manipulatives are available for students to use in problems as strategically appropriate.

Attend to precision.

- *Vocabulary and Core Concept Check* exercises require students to use clear, precise mathematical language in their solutions and explanations.
- The many opportunities for cooperative learning in this program, including working with partners for each *Exploration*, support precise, explicit mathematical communication.

Look for and make use of structure.

- *Using Structure* exercises provide students with the opportunity to explore patterns and structure in mathematics.
- Students analyze structure in problems through *Justifying Steps* and *Analyzing Equations* exercises.

Look for and express regularity in repeated reasoning.

- Students are continually encouraged to evaluate the reasonableness of their solutions and their steps in the problem-solving process.
- Stepped-out *Examples* encourage students to maintain oversight of their problem-solving process and pay attention to the relevant details in each step.

Go to *BigIdeasLearning.com* for more information on the Mathematical Practices.

Mathematical Content

Chapter Coverage

1 2 3 4 5 6 7 8 9 10 11

Number and Quantity

- The Real Number System
- Quantities

1 2 3 4 5 6 7 8 9 10 11

Algebra

- Seeing Structure in Expressions
- Arithmetic with Polynomials and Rational Expressions
- Creating Equations
- Reasoning with Equations and Inequalities

1 2 3 4 5 6 7 8 9 10 11

Functions

- Interpreting Functions
- Building Functions
- Linear, Quadratic, and Exponential Models

1 2 3 4 5 6 7 8 9 10 11

Statistics and Probability

- Interpreting Categorical and Quantitative Data

1 Solving Linear Equations

Laurie's Notes

See the Big Idea
Learn how boat navigators use dead reckoning to calculate their distance covered in a single direction.

Solving Linear Inequalities 2

Laurie's Notes

See the Big Idea
Determine how designers decide on the number of electrical circuits needed in a house.

3 Graphing Linear Functions

Laurie's Notes

See the Big Idea
Discover why unlike almost any other natural phenomenon, light travels at a constant speed.

Writing Linear Functions 4

Laurie's Notes

See the Big Idea
Explore wind power and discover where the future of wind power will take us.

5 Solving Systems of Linear Equations

Laurie's Notes

See the Big Idea
Learn how fisheries manage their complex ecosystems.

6 Exponential Functions and Sequences

Laurie's Notes

See the Big Idea
Explore the variety of recursive sequences in language, art, music, nature, and games.

7 Polynomial Equations and Factoring

Laurie's Notes

See the Big Idea

Explore whether seagulls and crows use the optimal height while dropping hard-shelled food to crack it open.

8 Graphing Quadratic Functions

Laurie's Notes

See the Big Idea
Investigate the link between population growth and the classic exponential pay doubling application.

9 Solving Quadratic Equations

Laurie's Notes

See the Big Idea
Explore the Parthenon and investigate how the use of the golden rectangle has evolved since its discovery.

Radical Functions and Equations 10

Laurie's Notes

See the Big Idea
Explore how a tsunami's speed changes with depth and explore other factors in determining tsunami danger zones.

11 Data Analysis and Displays

Laurie's Notes

See the Big Idea
Foray into the fashion differences between men and women.

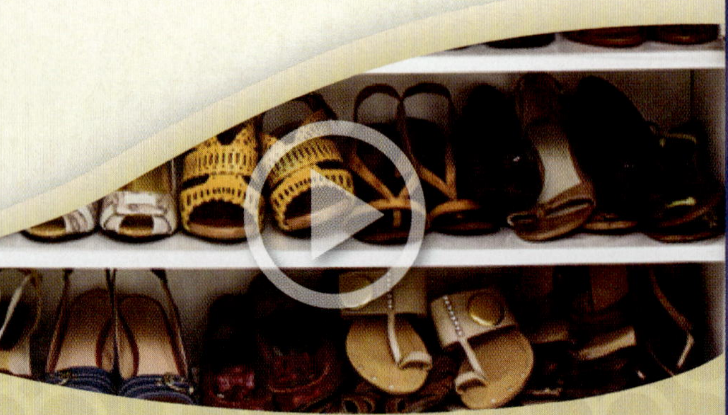

How to Use Your Math Book

Get ready for each chapter by **Maintaining Mathematical Proficiency** and reviewing the **Mathematical Practices**. Begin each section by working through the **EXPLORATIONS** to **Communicate Your Answer** to the **Essential Question**. Each **Lesson** will explain **What You Will Learn** through **EXAMPLES**, **Core Concepts**, and **Core Vocabulary**.

Answer the **Monitoring Progress** questions as you work through each lesson. Look for STUDY TIPS, COMMON ERRORS, and suggestions for looking at a problem ANOTHER WAY throughout the lessons. We will also provide you with guidance for accurate mathematical READING and concept details you should REMEMBER.

Sharpen your newly acquired skills with **Exercises** at the end of every section. Halfway through each chapter you will be asked **What Did You Learn?** and you can use the Mid-Chapter **Quiz** to check your progress. You can also use the **Chapter Review** and **Chapter Test** to review and assess yourself after you have completed a chapter.

Apply what you learned in each chapter to a **Performance Task** and build your confidence for taking standardized tests with each chapter's **Cumulative Assessment**.

For extra practice in any chapter, use your *Online Resources*, *Skills Review Handbook*, or your *Student Journal*.

Program Overview

Program Philosophy: Rigor and Balance with Real-Life Applications

The ***Big Ideas Math***® program balances conceptual understanding with procedural fluency. Real-life applications help turn mathematical learning into an engaging and meaningful way to see and explore the real world.

Essential Question How can you use function notation to represent a function?

The notation $f(x)$, called **function notation**, is another name for y. This notation is read as "the value of f at x" or "f of x." The parentheses do not imply multiplication. You can use letters other than f to name a function. The letters g, h, j, and k are often used to name functions.

EXPLORATION 1 **Matching Functions with Their Graphs**

Work with a partner. Match each function with its graph.

a. $f(x) = 2x - 3$ **b.** $g(x) = -x + 2$

c. $h(x) = x^2 - 1$ **d.** $j(x) = 2x^2 - 3$

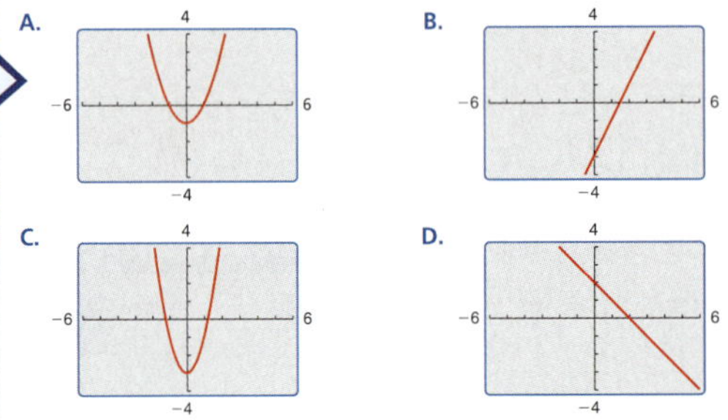

Explorations and guiding *Essential Questions* encourage **conceptual understanding**.

3.3 Lesson

Core Vocabulary

function notation, *p. 122*

Previous

linear function

quadrant

What You Will Learn

- Use function notation to evaluate and interpret func
- Use function notation to solve and graph functions.
- Solve real-life problems using function notation.

Using Function Notation to Evaluate and Interpret

You know that a linear function can be written in the form $y = mx + b$. By naming a linear function f, you can also write the function using **function notation**.

$f(x) = mx + b$ Function notation

The notation $f(x)$ is another name for y. If f is a function, and x is in its domain, then $f(x)$ represents the output of f corresponding to the input x. You can use letters other than f to name a function, such as g or h.

READING

The notation $f(x)$ is read as "the value of f at x" or "f of x." It does not mean "f times x."

EXAMPLE 1 **Evaluating a Function**

Evaluate $f(x) = -4x + 7$ when $x = 2$ and $x = -2$.

SOLUTION

$f(x) = -4x + 7$	Write the function.	$f(x) = -4x + 7$
$f(2) = -4(2) + 7$	Substitute for x.	$f(-2) = -4(-2) + 7$
$= -8 + 7$	Multiply.	$= 8 + 7$
$= -1$	Add.	$= 15$

► When $x = 2$, $f(x) = -1$, and when $x = -2$, $f(x) = 15$.

Direct instruction lessons allow for **procedural fluency** and provide the opportunity to use clear, precise mathematical language.

Real-life applications provide students with opportunities to connect classroom lessons to realistic scenarios.

Solving Real-Life Problems

EXAMPLE 5 **Modeling with Mathematics**

The graph shows the number of miles a helicopter is from its destination after x hours on its first flight. On its second flight, the helicopter travels 50 miles farther and increases its speed by 25 miles per hour. The function $f(x) = 350 - 125x$ represents the second flight, where $f(x)$ is the number of miles the helicopter is from its destination after x hours. Which flight takes less time? Explain.

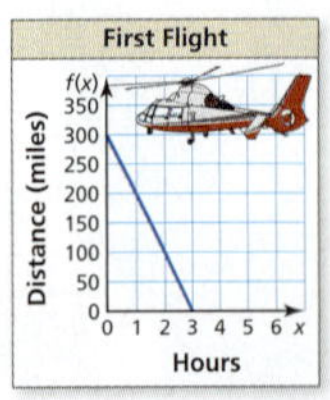

SOLUTION

1. **Understand the Problem** You are given a graph of the first flight and an equation of the second flight. You are asked to compare the flight times to determine which flight takes less time.
2. **Make a Plan** Graph the function that represents the second flight. Compare the graph to the graph of the first flight. The x-value that corresponds to $f(x) = 0$ represents the flight time.
3. **Solve the Problem** Graph $f(x) = 350 - 125x$.

Step 1 Make an input-output table to find the ordered pairs.

x	0	1	2	3
$f(x)$	350	225	100	−25

Step 2 Plot the ordered pairs.

Chapter openers focused on **Maintaining Mathematical Proficiency** promote the development of the habits of mind mathematically proficient students demonstrate.

Maintaining Mathematical Proficiency

Using Order of Operations

Example 1 Evaluate $10^2 \div (30 \div 3) - 4(3 - 9) + 5^1$.

First:	Parentheses	$10^2 \div (30 \div 3) - 4(3 - 9) + 5^1 = 10^2 \div 10 - 4(-6) + 5^1$
Second:	Exponents	$= 100 \div 10 - 4(-6) + 5$
Third:	Multiplication and Division (from left to right)	$= 10 + 24 + 5$
Fourth:	Addition and Subtraction (from left to right)	$= 39$

Evaluate the expression.

1. $12\left(\frac{14}{2}\right) - 3^3 + 15 - 9^2$ **2.** $5^2 \cdot 8 \div 2^2 + 20 \cdot 3 - 4$ **3.** $-7 + 16 \div 2^4 + (10 - 4^2)$

Finding Square Roots

Example 2 Find $-\sqrt{81}$.

▶ $-\sqrt{81}$ represents the negative square root. Because $9^2 = 81$, $-\sqrt{81} = -\sqrt{9^2} = -9$.

Find the square root(s).

4. $\sqrt{64}$ **5.** $-\sqrt{4}$ **6.** $-\sqrt{25}$ **7.** $\pm\sqrt{121}$

Writing Equations for Arithmetic Sequences

Write an equation for the *n*th term of the arithmetic sequence 5, 15, 25, 35, ...

The first term is 5, and the common difference is 10.

$a_n = a_1 + (n - 1)d$ Equation for an arithmetic sequence

$a_n = 5 + (n - 1)(10)$ Substitute 5 for a_1 and 10 for d.

$a_n = 10n - 5$ Simplify.

n equation for the *n*th term of the arithmetic sequence.

, 16, 18, . . . **9.** 6, 3, 0, −3, . . . **10.** 22, 15, 8, 1, . . .

RACT REASONING Recall that a perfect square is a number with integers as its square Is the product of two perfect squares always a perfect square? Is the quotient of two t squares always a perfect square? Explain your reasoning.

The Mathematical Practices are woven into every chapter, including a full page dedicated to mastering one of the **Mathematical Practices**.

Mathematical Practices

Mathematically proficient students look closely to find a pattern.

Problem-Solving Strategies

Core Concept

Finding a Pattern

When solving a real-life problem, look for a pattern in the data. The pattern could include repeating items, numbers, or events. After you find the pattern, describe it and use it to solve the problem.

EXAMPLE 1 Using a Problem-Solving Strategy

The volumes of seven chambers of a chambered nautilus are given. Find the volume of Chamber 10.

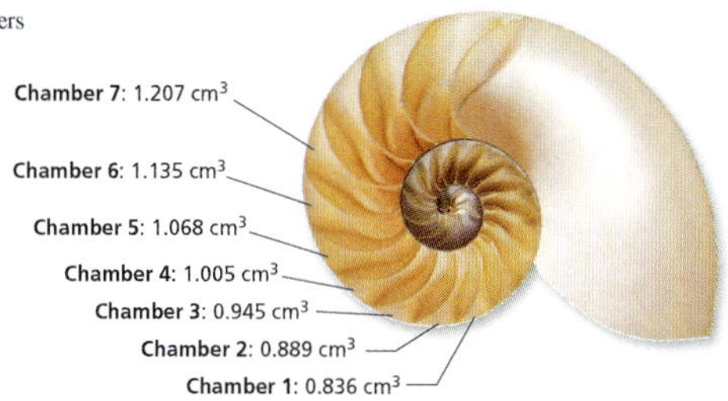

SOLUTION

To find a pattern, try dividing each volume by the volume of the previous chamber.

$\frac{0.889}{0.836} \approx 1.063$ $\frac{0.945}{0.889} \approx 1.063$ $\frac{1.005}{0.945} \approx 1.063$

$\frac{1.068}{1.005} \approx 1.063$ $\frac{1.135}{1.068} \approx 1.063$ $\frac{1.207}{1.135} \approx 1.063$

From this, you can see that the volume of each chamber is about 6.3% greater than the volume of the previous chamber. To find the volume of Chamber 10, multiply the volume of Chamber 7 by 1.063 three times.

$1.207(1.063) \approx \mathbf{1.283}$ (volume of Chamber 8) $1.283(1.063) \approx \mathbf{1.364}$ (volume of Chamber 9) $1.364(1.063) \approx \mathbf{1.450}$ (volume of Chamber 10)

▶ The volume of Chamber 10 is about 1.450 cubic centimeters.

Monitoring Progress

1. A rabbit population over 8 consecutive years is given by 50, 80, 128, 205, 328, 524, 839, 1342. Find the population in the tenth year.
2. The sums of the numbers in the first eight rows of Pascal's Triangle are 1, 2, 4, 8, 16, 32, 64, 128. Find the sum of the numbers in the tenth row.

Monitoring Progress problems allow students to practice and sharpen their skills as they work toward mathematical understanding.

Personalized Learning

The ***Big Ideas Math*** program offers teachers and students many ways to personalize and enrich the learning experience of all levels of learners.

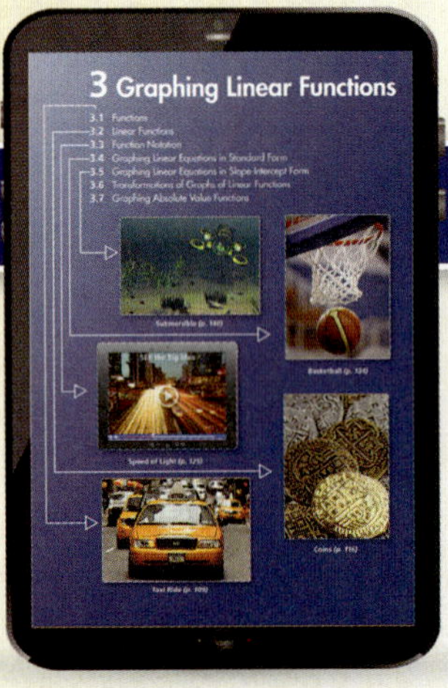

Dynamic Student Edition

This unique tool, available online or as an eBook App, provides students with embedded 21st century learning resources. Students have the opportunity to interact with the underlying mathematics in a number of ways, including engaging tutorials, interactive manipulatives, flashcards, vocabulary support, and games that enhance the learning experience and promote mathematical understanding.

Dynamic Assessment & Progress Monitoring Tool

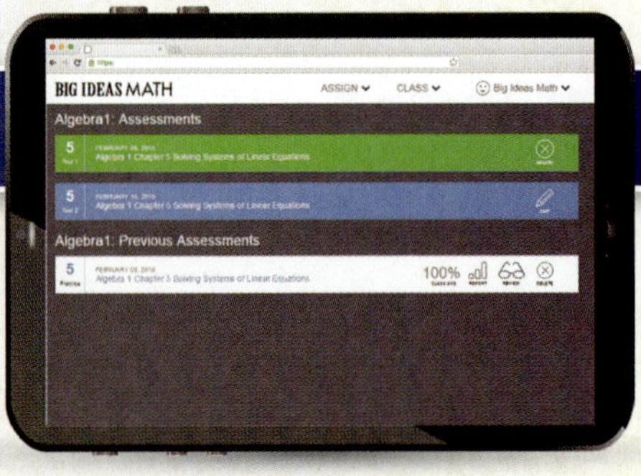

This tool allows teachers to provide customizable homework directly related to the ***Big Ideas Math*** program. Question types include technology enhanced items such as drag and drop, graphing, point plotting, multiple select, and fill in the blank using math expressions. Assignments are automatically scored and students have access to immediate remediation on homework questions.

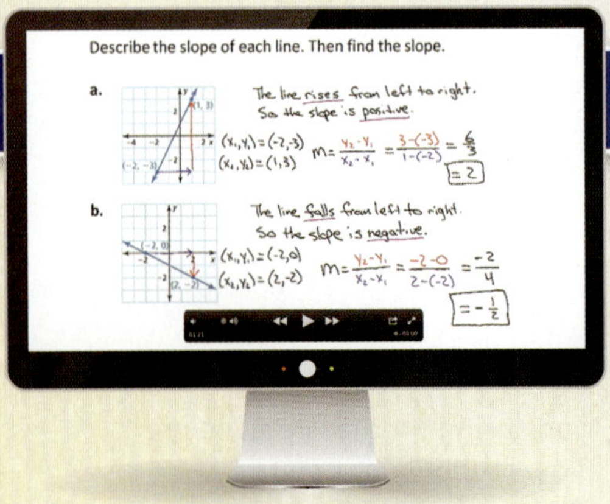

Online Lesson Tutorials

Two- to three-minute lesson tutorial videos provide video and audio support for every example in the textbook. These are valuable for students who miss a class, need a second explanation, or need extra assistance with a homework assignment. Parents can also utilize the tutorials to stay connected or to provide additional help at home.

Differentiated Instruction

Through print and digital resources, the ***Big Ideas Math*** program completely supports the 3-Tier Response to Intervention model. Using research-based strategies, teachers can reach, challenge, and motivate each student with high-quality instruction targeted to individual needs.

Customized Learning Intervention (Tier 3)

- ***Big Ideas Math*** Middle School program at *BigIdeasMath.com*

Strategic Intervention (Tier 2)

- Lesson Tutorials
- Game Closet
- Skills Review Handbook
- Differentiated Instruction
- Dynamic Assesssment and Progress Monitoring Tool

Daily Intervention (Tier 1)

- Student Journal
- Vocabulary Support
- Lesson Tutorials
- Communicate Your Answer
- Monitoring Progress
- Maintaining Mathematical Proficiency
- Dynamic Assessment and Progress Monitoring Tool

Big Ideas Math Dynamic Technology

The Dynamic Assessment and Progress Monitoring Tool

The **Dynamic Assessment and Progress Monitoring Tool** allows teachers to track and evaluate their students' advancement through the curriculum. Developed exclusively for ***Big Ideas Math***, this technology provides teachers and students an intuitive and state-of-the-art tool to help students effectively learn mathematics. Built for ease of use, the tool is available on a wide range of devices.

Homework and Assessment

- Includes multiple assignments available for each chapter that are customizable
- Allows you to assign homework and assessments for the entire class or a select group of students
- Offers progress monitoring assessments for an adaptive testing experience

Direct Ties to Remediation

- Includes direct links to Lesson Tutorial Videos and relevant lesson sections

All-In-One Reporting

- Offers real-time reporting at both the class and student level
- Tracks progress through Item Analysis, Standard Analysis, or Remediation reports

Assessment Delivery

- Provides embedded tools for students
- Includes auto-scored technology enhanced items such as drag and drop, graphing, point plotting, multiple select, and fill in the blank using math expressions
- Allows you to include reminders or notes to students

Intuitive Design

- Operates on a wide range of devices with large and clear icons for visibility
- Allows for multiple reporting views through toggle options
- Includes intelligent presets and easy navigation

Dynamic Student Edition

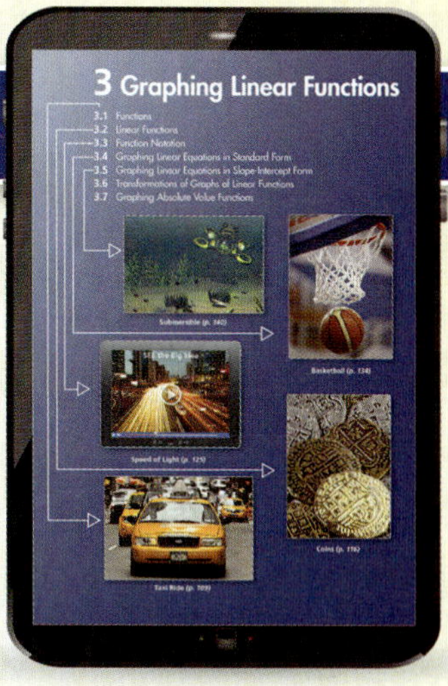

Through the **Dynamic Student Edition eBook App**, students not only have access to the complete textbook, but they can also explore robust interactive digital resources embedded within each lesson. There is audio support for the text and Lesson Tutorial Videos in both English and Spanish. Interactive investigations, direct links to remediation, and additional resources are linked right to the lesson.

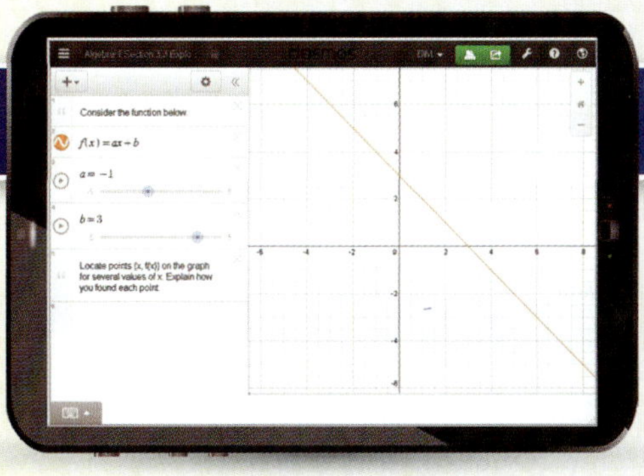

Dynamic Investigations

Dynamic Investigations in the ***Big Ideas Math*** program are powered by Desmos® and GeoGebra®. Teachers and students can integrate these investigations into their discovery learning to interact with the Explorations in the Student Edition.

Real-Life STEM Videos

Science - Technology - Engineering - Mathematics

Every chapter in the ***Big Ideas Math*** program contains a Real-Life STEM Video allowing students to further engage with mathematical concepts. Students learn about the speed of light, natural disasters, solar power, and more!

Teaching Support

Lesson Planning Support

Online Lesson Plans

Complete, editable Lesson Plans are included for every lesson in the program.

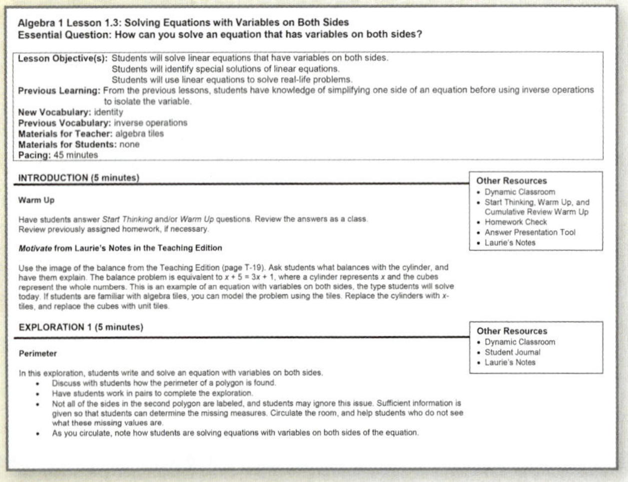
Algebra 1 Lesson 1.3: Solving Equations with Variables on Both Sides
Essential Question: How can you solve an equation that has variables on both sides?

Lesson Objective(s): Students will solve linear equations that have variables on both sides.
Students will identify special solutions of linear equations.
Students will use linear equations to solve real-life problems.
Previous Learning: From the previous lessons, students have knowledge of simplifying one side of an equation before using inverse operations to isolate the variable.
New Vocabulary: identity
Previous Vocabulary: inverse operations
Materials for Teacher: algebra tiles
Materials for Students: none
Pacing: 45 minutes

INTRODUCTION (5 minutes)

Warm Up

Have students answer *Start Thinking* and/or *Warm Up* questions. Review the answers as a class. Review previously assigned homework, if necessary.

***Motivate* from Laurie's Notes in the Teaching Edition**

Use the image of the balance from the Teaching Edition (page T-19). Ask students what balances with the cylinder, and have them explain. The balance problem is equivalent to $x + 5 = 3x + 1$, where a cylinder represents x and the cubes represent the whole numbers. This is an example of an equation with variables on both sides, the type students will solve today. If students are familiar with algebra tiles, you can model the problem using the tiles. Replace the cylinders with x-tiles, and replace the cubes with unit tiles.

Other Resources
- Dynamic Classroom
- Start Thinking, Warm Up, and Cumulative Review Warm Up
- Homework Check
- Answer Presentation Tool
- Laurie's Notes

EXPLORATION 1 (5 minutes)

Perimeter

In this exploration, students write and solve an equation with variables on both sides.
- Discuss with students how the perimeter of a polygon is found.
- Have students work in pairs to complete the exploration.
- Not all of the sides in the second polygon are labeled, and students may ignore this issue. Sufficient information is given so that students can determine the missing measures. Circulate the room, and help students who do not see what these missing values are.
- As you circulate, note how students are solving equations with variables on both sides of the equation.

Other Resources
- Dynamic Classroom
- Student Journal
- Laurie's Notes

Laurie's Notes

Laurie Boswell provides comprehensive teaching support for every student page in the Teaching Edition.

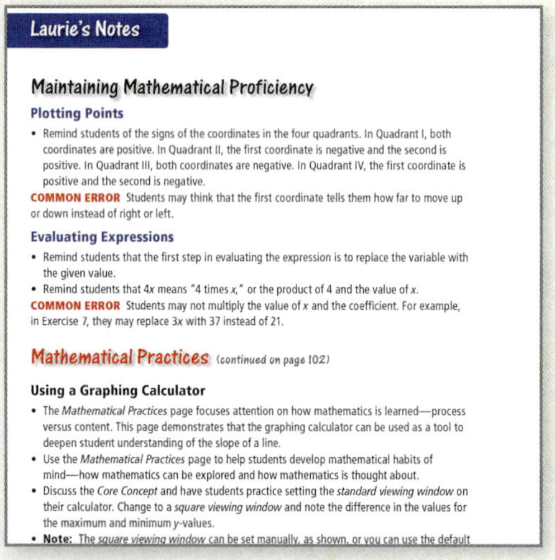
Laurie's Notes

Maintaining Mathematical Proficiency

Plotting Points
- Remind students of the signs of the coordinates in the four quadrants. In Quadrant I, both coordinates are positive. In Quadrant II, the first coordinate is negative and the second is positive. In Quadrant III, both coordinates are negative. In Quadrant IV, the first coordinate is positive and the second is negative.

COMMON ERROR Students may think that the first coordinate tells them how far to move up or down instead of right or left.

Evaluating Expressions
- Remind students that the first step in evaluating the expression is to replace the variable with the given value.
- Remind students that $4x$ means "4 times x," or the product of 4 and the value of x.

COMMON ERROR Students may not multiply the value of x and the coefficient. For example, in Exercise 7, they may replace $3x$ with 37 instead of 21.

Mathematical Practices *(continued on page 102)*

Using a Graphing Calculator
- The *Mathematical Practices* page focuses attention on how mathematics is learned—process versus content. This page demonstrates that the graphing calculator can be used as a tool to deepen student understanding of the slope of a line.
- Use the *Mathematical Practices* page to help students develop mathematical habits of mind—how mathematics can be explored and how mathematics is thought about.
- Discuss the *Core Concept* and have students practice setting the *standard viewing window* on their calculator. Change to a *square viewing window* and note the difference in the values for the maximum and minimum y-values.
- **Note:** The square viewing window can be set manually, as shown, or you can use the default

Lesson Presentation Support

Dynamic Classroom

This online tool contains a collection of interactive resources for presenting lessons.

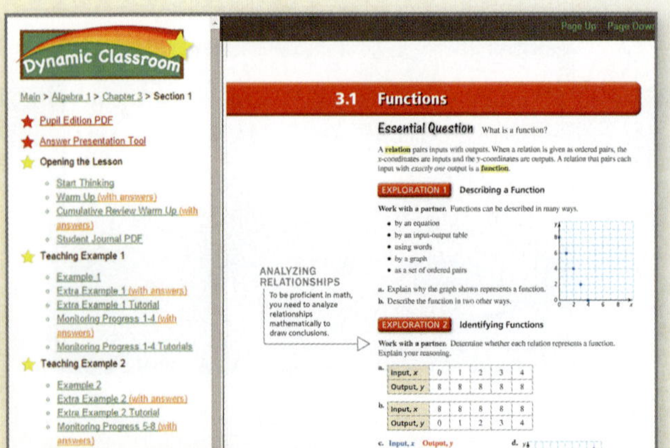

Interactive Whiteboard Lesson Library

Standard or customizable lessons are provided for SMART®, Promethean®, and Mimeo® interactive whiteboards.

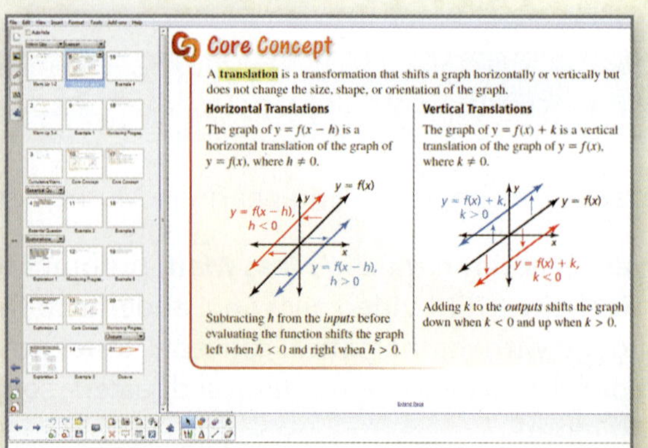

Program Resources

Print

Student Edition

Teaching Edition

- Laurie's Notes

Student Journal

Available in English and Spanish

Resources by Chapter

- Start Thinking
- Warm-Up
- Cumulative Review Warm-Up
- Practice A and B
- Enrichment and Extension
- Puzzle Time
- Family Communication Letters
  Available in English and Spanish

Assessment Book

- Performance Tasks
- Prerequisite Skills Tests with Item Analysis
- Cumulative Test
- Quizzes
- Chapter Tests
- Alternative Assessments with Scoring Rubrics
- Pre-Course Test with Item Analysis
- Post-Course Test with Item Analysis

Technology

Student Edition

With complete English and Spanish audio

- Dynamic eBook App
  - Lesson Tutorial Videos
  - Dynamic Investigations
- Online Home Edition
  - Multi-Language Glossary
  - Skills Review Handbook
- Dynamic Solutions Tool

Dynamic Classroom

- Interactive Manipulatives
- Answer Presentation Tool
- Extra Examples
- Mini-Assessment

Dynamic Teaching Tools

- Interactive Whiteboard Lesson Library
  - Includes standard and customizable lessons.
  - Compatible with SMART®, Promethean®, and Mimio® technology
- ExamView
- Real-Life STEM Videos
- Editable Online Resources
  - Lesson Plans
  - Pacing Guides
  - Assessment Book
  - Resources by Chapter

Dynamic Assessment and Progress Monitoring Tool

- Homework and Assessment Creation
- Progress Monitoring
- Direct Ties to Remediation
- All-In-One Reporting
- Online Chat Tutor

Learning Progression
Algebra 1, Geometry, and Algebra 2

Domain	Algebra 1	Geometry	Algebra 2
Number and Quantity			
Quantities	Reason quantitatively and use units. *Chapters 1, 3–10*		
Real Numbers	Use properties of rational exponents, rational numbers, and irrational numbers. *Chapters 6, 9, 10*		
Complex Numbers			Perform arithmetic with complex numbers and use complex numbers in quadratic identities and equations. *Chapters 3, 4* Use complex numbers in polynomial identities and equations. *Chapters 3, 4*
Algebra			
Structure	Interpret the structure of linear, quadratic, and exponential expressions. Write equivalent expressions. *Chapters 1, 3, 6–9*		Interpret the structure of polynomial and rational expressions. Write equivalent expressions. *Chapters 3–9*
Arithmetic with Polynomials and Rational Expressions	Perform arithmetic on linear and quadratic polynomials. *Chapters 7–9*		Perform arithmetic on polynomials beyond quadratics. *Chapters 4, 7* Rewrite rational expressions. *Chapters 4, 7*
Creating Equations	Create linear, quadratic, and exponential equations. *Chapters 1–10*		Create equations of all available types. *Chapters 1–7, 9*
Reasoning with Equations and Inequalities	Solve linear and absolute value equations and inequalities, and quadratic and exponential equations. *Chapters 1–4, 6–10* Solve linear systems of equations and inequalities, and nonlinear systems of equations. *Chapters 5, 9*		Solve polynomial, radical, exponential, logarithmic, and rational equations, and quadratic and radical inequalities. *Chapters 3–7*

Domain	Algebra 1	Geometry	Algebra 2
Functions			
Interpreting Functions	Interpret and analyze linear, quadratic, exponential, absolute value, step, and piecewise functions. *Chapters 3, 4, 6, 8–10*		Interpret and analyze polynomial, radical, exponential, logarithmic, and rational functions. *Chapters 4–7*
Building Functions	Build linear, quadratic, and exponential models. *Chapters 4, 6, 8* Build linear, quadratic, exponential, and absolute value functions from existing functions. *Chapters 3, 6, 8, 10*		Build models from all types of functions studied. *Chapters 1–8* Build polynomial, radical, exponential, logarithmic, and rational functions from existing functions. *Chapters 1, 2, 4–7*
Trigonometric Functions			Extend the domain of trigonometric functions. Model with trigonometric functions. Prove and use trigonometric identities. *Chapter 9*
Geometry			
Congruence		Understand transformations and congruence, prove geometric theorems, and make constructions. *Chapters 1–7*	
Similarity and Trigonometry		Understand similarity and prove theorems involving similarity. *Chapters 8, 9* Use trigonometry with right triangles. Chapter 9	Use trigonometry with general triangles. *Chapter 9*
Circles		Use theorems about circles. Find arc lengths and areas of sectors. *Chapters 10, 11*	
Measurement		Understand and use volume formulas. Relate two-dimensional and three-dimensional objects. *Chapter 11*	
Statistics and Probability			
Interpreting Data	Summarize, represent, and interpret data using measures of center and spread. *Chapter 11*		Fit data to a normal distribution using the mean and standard deviation. *Chapter 11*
Making Inferences			Make inferences and justify conclusions. *Chapter 11*
Probability		Use conditional probability and find the probabilities of compound events. *Chapter 12*	Use conditional probability and find the probabilities of compound events. *Chapter 10*

Pacing Guide

Chapters 1–11 160 Days

Chapter 1 (12 Days)

Chapter Opener/ Mathematical Practices	1 Day
Section 1	1 Day
Section 2	2 Days
Section 3	1 Day
Quiz	1 Day
Section 4	2 Days
Section 5	2 Days
Chapter Review/Chapter Tests	2 Days

Year-to-Date 12 Days

Chapter 2 (12 Days)

Chapter Opener/ Mathematical Practices	1 Day
Section 1	1 Day
Section 2	1 Day
Section 3	1 Day
Section 4	1 Day
Quiz	1 Day
Section 5	2 Days
Section 6	2 Days
Chapter Review/Chapter Tests	2 Days

Year-to-Date 24 Days

Chapter 3 (14 Days)

Chapter Opener/ Mathematical Practices	1 Day
Section 1	1 Day
Section 2	2 Days
Section 3	1 Day
Quiz	1 Day
Section 4	1 Day
Section 5	1 Day
Section 6	2 Days
Section 7	2 Days
Chapter Review/Chapter Tests	2 Days

Year-to-Date 38 Days

Chapter 4 (14 Days)

Chapter Opener/ Mathematical Practices	1 Day
Section 1	1 Day
Section 2	1 Day
Section 3	1 Day
Quiz	1 Day
Section 4	1 Day
Section 5	2 Days
Section 6	2 Days
Section 7	2 Days
Chapter Review/Chapter Tests	2 Days

Year-to-Date 52 Days

Chapter 5 (15 Days)

Chapter Opener/ Mathematical Practices	1 Day
Section 1	1 Day
Section 2	1 Day
Section 3	2 Days
Section 4	1 Day
Quiz	1 Day
Section 5	2 Days
Section 6	2 Days
Section 7	2 Days
Chapter Review/Chapter Tests	2 Days

Year-to-Date 67 Days

Chapter 6 (17 Days)

Chapter Opener/ Mathematical Practices	1 Day
Section 1	1 Day
Section 2	2 Days
Section 3	2 Days
Section 4	2 Days
Quiz	1 Day
Section 5	2 Days
Section 6	2 Days
Section 7	2 Days
Chapter Review/Chapter Tests	2 Days

Year-to-Date 84 Days

Chapter 7 (19 Days)

Chapter Opener/ Mathematical Practices	1 Day
Section 1	1 Day
Section 2	2 Days
Section 3	2 Days
Section 4	2 Days
Quiz	1 Day
Section 5	2 Days
Section 6	2 Days
Section 7	2 Days
Section 8	2 Days
Chapter Review/Chapter Tests	2 Days

Year-to-Date 103 Days

Chapter 8 (16 Days)

Chapter Opener/ Mathematical Practices	1 Day
Section 1	2 Days
Section 2	2 Days
Section 3	2 Days
Quiz	1 Day
Section 4	2 Days
Section 5	2 Days
Section 6	2 Days
Chapter Review/Chapter Tests	2 Days

Year-to-Date 119 Days

Chapter 9 (16 Days)

Chapter Opener/ Mathematical Practices	1 Day
Section 1	2 Days
Section 2	2 Days
Section 3	2 Days
Quiz	1 Day
Section 4	2 Days
Section 5	2 Days
Section 6	2 Days
Chapter Review/Chapter Tests	2 Days

Year-to-Date 135 Days

Chapter 10 (11 Days)

Chapter Opener/ Mathematical Practices	1 Day
Section 1	2 Days
Section 2	2 Days
Section 3	2 Days
Section 4	2 Days
Chapter Review/Chapter Tests	2 Days

Year-to-Date 146 Days

Chapter 11 (14 Days)

Chapter Opener/ Mathematical Practices	1 Day
Section 1	2 Days
Section 2	2 Days
Section 3	2 Days
Quiz	1 Day
Section 4	2 Days
Section 5	2 Days
Chapter Review/Chapter Tests	2 Days

Year-to-Date 160 Days

Chapter 1 Pacing Guide

Chapter 1 Pacing Guide	
Chapter Opener/ Mathematical Practices	1 Day
Section 1	1 Day
Section 2	2 Days
Section 3	1 Day
Quiz	1 Day
Section 4	2 Days
Section 5	2 Days
Chapter Review/ Chapter Tests	2 Days
Total Chapter 1	12 Days
Year-to-Date	12 Days

1 Solving Linear Equations

Density of Pyrite *(p. 41)*

Cheerleading Competition *(p. 29)*

Boat *(p. 22)*

Biking *(p. 14)*

Average Speed *(p. 6)*

Laurie's Notes

Chapter Summary

Welcome to a new school year, and for many students, a new school. There is always great excitement, and students are anxious to start anew. As teachers, we need to capitalize on the opportunity, establishing norms and routines for student discourse and classroom climate.

- In this book, students are expected to work together on explorations, to make conjectures, to construct viable arguments, and to critique the reasoning of others. Take time in this first chapter to make explicit what classroom productive dialogue sounds like. Listen for students explaining their thinking, not just their process.
- Chapter 1 presents the foundational skills related to solving linear equations and the connected skills of solving absolute value equations and rewriting equations and formulas.
- Most students will have prior experience with the properties of equality and techniques presented in the first three sections. It will sound familiar that whatever operation is performed on one side of the equation, the same operation must be performed on the other side of the equation to keep equality, or balance.
- The fourth section of the chapter applies the techniques of equation solving to the context of absolute value equations. Understanding absolute value as a function concept and not simply two vertical lines can be challenging for students.
- Solving literal equations in the last section requires students to see the structure of equations and perform operations on variable terms (i.e., $4x$) as they would perform operations on constants (i.e., 4).
- Essential to success in this chapter is accuracy in computation. Feedback to students should distinguish between an error in computation and a process error.

What Your Students Have Learned

Middle School
- Solve multi-step real-life problems involving operations with rational numbers.
- Understand the absolute value of a rational number as its distance from 0 on the number line.
- Evaluate expressions using Order of Operations.
- Use variables to represent quantities in real-life problems.
- Solve linear equations using the Distributive Property and combining like terms.

What Your Students Will Learn

Algebra 1
- Solve multi-step linear equations in one variable that have variables on both sides using inverse operations.
- Use unit analysis to understand and model real-life problems.
- Identify equations with infinitely many solutions or no solution.
- Solve absolute value equations including equations involving two absolute values.
- Identify extraneous solutions of absolute value equations.
- Solve literal equations for one variable in terms of the other variables.

Dynamic Teaching Tools

- Dynamic Assessment & Progress Monitoring Tool
- Lesson Planning Tool
- Interactive Whiteboard Lesson Library
- Dynamic Classroom with Dynamic Investigations
- Real-Life STEM Videos

Scaffolding in the Classroom

Graphic Organizers: Word Magnet

A Word Magnet can be used to organize information associated with a vocabulary word or term. Students write the word or term inside the magnet. Students write associated information on the blank lines that "radiate" from the magnet. Associated information can include, but is not limited to: other vocabulary words or terms, definitions, formulas, procedures, examples, and visuals. This type of organizer serves as a good summary tool because any information related to a topic can be included.

Questioning in the Classroom

But why?

Ask questions that require critical thinking so that follow-up questions may be asked. Avoid questions having *yes* or *no* answers. If this cannot be avoided, always follow with *why?*

Laurie's Notes

Maintaining Mathematical Proficiency

Adding and Subtracting Integers

- Remind students how to add integers with the same sign. They should add the absolute values of the integers and then use the common sign.
- Remind students how to add integers with different signs. They should subtract the lesser absolute value from the greater absolute value and use the sign of the integer with the greater absolute value.

COMMON ERROR Students may ignore the signs and just add the integers.

Multiplying and Dividing Integers

- Remind students that the product or quotient of two integers with the same sign is always positive.
- Remind students that the product or quotient of two integers with different signs is always negative.

COMMON ERROR When both factors are negative, students may write the product as negative.

Mathematical Practices *(continued on page 2)*

Specifying Units of Measure

- The *Mathematical Practices* page focuses attention on how mathematics is learned—process versus content. This page demonstrates that focusing on unit analysis leads to labeling answers with correct units of measures, providing an entry point into solving problems.

If students need help...	*If students got it...*
Student Journal • Maintaining Mathematical Proficiency	Game Closet at *BigIdeasMath.com*
Lesson Tutorials	Start the *next* Section
Skills Review Handbook	

Maintaining Mathematical Proficiency

Adding and Subtracting Integers

Example 1 Evaluate $4 + (-12)$.

$|-12| > |4|$. So, subtract $|4|$ from $|-12|$.

$4 + (-12) = -8$

Use the sign of -12.

Example 2 Evaluate $-7 - (-16)$.

$-7 - (-16) = -7 + 16$ Add the opposite of -16.

$= 9$ Add.

Add or subtract.

1. $-5 + (-2)$ **2.** $0 + (-13)$ **3.** $-6 + 14$

4. $19 - (-13)$ **5.** $-1 - 6$ **6.** $-5 - (-7)$

7. $17 + 5$ **8.** $8 + (-3)$ **9.** $11 - 15$

Multiplying and Dividing Integers

Example 3 Evaluate $-3 \cdot (-5)$.

The integers have the same sign.

$-3 \cdot (-5) = 15$

The product is positive.

Example 4 Evaluate $15 \div (-3)$.

The integers have different signs.

$15 \div (-3) = -5$

The quotient is negative.

Multiply or divide.

10. $-3(8)$ **11.** $-7 \cdot (-9)$ **12.** $4 \cdot (-7)$

13. $-24 \div (-6)$ **14.** $-16 \div 2$ **15.** $12 \div (-3)$

16. $6 \cdot 8$ **17.** $36 \div 6$ **18.** $-3(-4)$

19. **ABSTRACT REASONING** Summarize the rules for (a) adding integers, (b) subtracting integers, (c) multiplying integers, and (d) dividing integers. Give an example of each.

What Your Students Have Learned

- Add integers with the same sign or with different signs using absolute value.
- Subtract integers by adding the opposite.
- Multiply integers with the same sign or with different signs.
- Divide integers with the same sign or with different signs and recognize the quotient as a rational number.

ANSWERS

1. -7
2. -13
3. 8
4. 32
5. -7
6. 2
7. 22
8. 5
9. -4
10. -24
11. 63
12. -28
13. 4
14. -8
15. -4
16. 48
17. 6
18. 12
19. a. If the signs are the same, add the absolute values and attach the sign. If the signs are different, subtract the absolute values and attach the sign of the number with the greatest absolute value; *Sample answer:* $-6 + 2 = -4$

19b–d. See Additional Answers.

Vocabulary Review

Have students make Information Frames for the following words.

- Integer
- Opposite
- Absolute value

Mathematical Practices

Mathematically proficient students carefully specify units of measure.

Specifying Units of Measure

Core Concept

Operations and Unit Analysis

Addition and Subtraction

When you add or subtract quantities, they must have the same units of measure. The sum or difference will have the *same* unit of measure.

Example

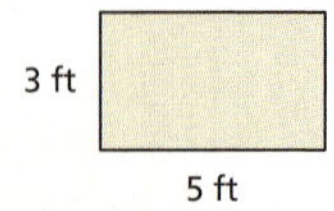

Perimeter of rectangle

= (3 ft) + (5 ft) + (3 ft) + (5 ft)

= 16 feet ← When you add **feet**, you get **feet**.

Multiplication and Division

When you multiply or divide quantities, the product or quotient will have a *different* unit of measure.

Example Area of rectangle = (3 ft) × (5 ft)

= 15 square feet ← When you multiply **feet,** you get feet squared, or **square feet.**

EXAMPLE 1 Specifying Units of Measure

You work 8 hours and earn \$72. What is your hourly wage?

SOLUTION

dollars per hour		dollars per hour
Hourly wage (\$ per h)	=	\$72 ÷ 8 h
	=	\$9 per hour

← The units on each side of the equation balance. Both are specified in dollars per hour.

▶ Your hourly wage is \$9 per hour.

Monitoring Progress

Solve the problem and specify the units of measure.

1. The population of the United States was about 280 million in 2000 and about 310 million in 2010. What was the annual rate of change in population from 2000 to 2010?
2. You drive 240 miles and use 8 gallons of gasoline. What was your car's gas mileage (in miles per gallon)?
3. A bathtub is in the shape of a rectangular prism. Its dimensions are 5 feet by 3 feet by 18 inches. The bathtub is three-fourths full of water and drains at a rate of 1 cubic foot per minute. About how long does it take for all the water to drain?

Laurie's Notes — Mathematical Practices (continued from page T-1)

Use this page to help students develop mathematical habits of mind—how mathematics can be explored and how mathematics is thought about.

- Allow students time to read through the *Core Concept* and example.
- ? Ask probing questions to assess students' understanding of the units of measure being the same on each side of the equation.
- Students could work with partners or in groups on *Monitoring Progress*. Allow private think time before dialogue begins.

MONITORING PROGRESS ANSWERS

1. about 3 million per year
2. 30 mi/gal
3. about 17 min

Laurie's Notes

Overview of Section 1.1

Introduction

- Students will have prior experience solving simple equations and will recall the basic concept that you must perform the same operation on both sides of an equation when solving for the variable. What will be apparent as you begin this first lesson of the year will be the recall and understanding of basic computational skills. As you progress through the first few sections, it will be important to assess understanding of equation solving and accuracy of computation. If an answer is incorrect, was the process correct and the error in the computation? Was the computation done correctly but the process flawed? Feedback to students should distinguish between these two possibilities.

Resources

- If students are familiar with algebra tiles (they were used to develop understanding of operations with integers), use the tiles to model one-step equation solving.
- Search online for applets that model one-step equation solving.

Common Misconceptions

- Students may think of the equation sign as a right-pointing arrow and not as the statement that two quantities are equal. For that reason, they think of $x - 8 = -17$ as different from $-17 = x - 8$.

Formative Assessment Tips

- **Thumbs Up:** This technique asks students to indicate the extent to which they understand a concept, procedure, or even the directions for an activity.

I get it. *I don't get it.* *I'm not sure.*

- Use this technique to assess students' understanding of the directions for Exploration 1 or why you divide by π and not -2 in Example 2(b).

Pacing Suggestion

- If you believe your students are secure with one-step equation solving, be selective in the problems worked in class. Explorations 1–3 provide a context for establishing classroom norms for partner work. If you believe students are secure in this, you might start with the formal lesson.

Dynamic Teaching Tools

- Dynamic Assessment & Progress Monitoring Tool
- Lesson Planning Tool
- Interactive Whiteboard Lesson Library
- Dynamic Classroom with Dynamic Investigations

What Your Students Will Learn

- Solve linear equations in one variable using addition, subtraction, multiplication, division, and the properties of equality.

Laurie's Notes

Exploration

Motivate

? "Who is considered the fastest runner?" Students will likely say Usain Bolt, an Olympic sprinter from Jamaica.

? "What is considered the fastest bird?" peregrine falcon when in its hunting dive

- **Extension:** Share a short video of a falcon diving at over 150 miles per hour. Search the Internet for "fastest bird."
- Discuss the distance formula ($d = rt$) and explain that in the lesson they will determine the speed of Usain Bolt in the 2012 Olympics. The formula involves three variables and knowing two out of the three, you can solve for the third.

Exploration 1

- From the first day, you want to establish the classroom culture of students working together and having discussions about mathematics. In working through the explorations, students are expected to talk with one another and share their thinking about the mathematics they are doing.
- **Use Appropriate Tools Strategically:** Assess students' knowledge of using a protractor.
- **Alternate Approach:** If you believe the majority of your students know that the sum of the angles of a quadrilateral is 360°, you could ask students to use straightedges to draw random quadrilaterals. Exchange drawings with partners. The partners measure the four angles and compute the sum. Compare results throughout the class.

Exploration 2

- Have a discussion with the class about the vocabulary: *conjecture*, *rule*, and *theorem*.
- If the *Alternate Approach* is used in Exploration 1, students could use the evidence of the class's work to write a conjecture.

Exploration 3

- As time permits, the focus here should be on the writing of the equation.

? "Describe the process for solving each equation." Add the values of the three given angles. Subtract this sum from each side of the equation.

Communicate Your Answer

- Have students share a context of a real-life example with their partners.
- Question 5 is an informal proof that the sum of the interior angles of a quadrilateral equals 360°. Check to see that students understand what a conjecture is and how one might affirm or disprove a conjecture.

Connecting to Next Step

- Time is always used on Day 1 to check rosters, assign books, discuss course expectations, and so on. If you feel students are comfortable with simple one-step equation solving, assign exercises from Exercises 5–30 on page 8.

1.1 Solving Simple Equations

Essential Question How can you use simple equations to solve real-life problems?

EXPLORATION 1 **Measuring Angles**

Work with a partner. Use a protractor to measure the angles of each quadrilateral. Copy and complete the table to organize your results. (The notation $m\angle A$ denotes the measure of angle A.) How precise are your measurements?

a.

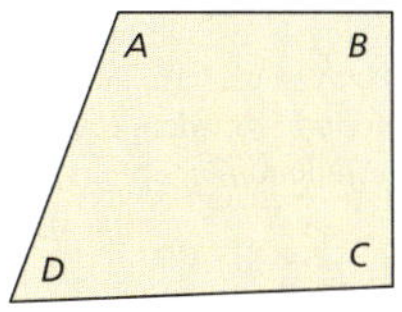

b.

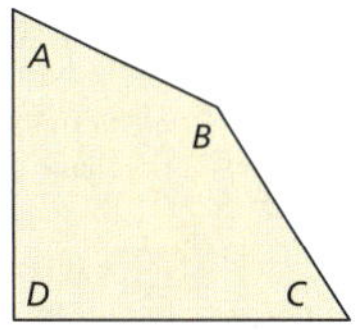

c.

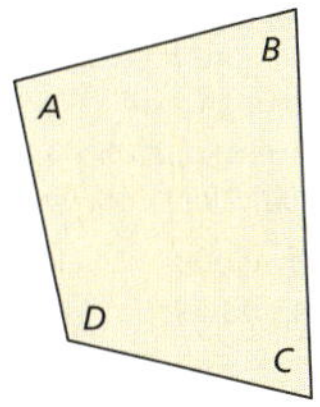

Quadrilateral	$m\angle A$ (degrees)	$m\angle B$ (degrees)	$m\angle C$ (degrees)	$m\angle D$ (degrees)	$m\angle A + m\angle B + m\angle C + m\angle D$
a.					
b.					
c.					

UNDERSTANDING MATHEMATICAL TERMS

A **conjecture** is an unproven statement about a general mathematical concept. After the statement is proven, it is called a **rule** or a **theorem**.

EXPLORATION 2 **Making a Conjecture**

Work with a partner. Use the completed table in Exploration 1 to write a conjecture about the sum of the angle measures of a quadrilateral. Draw three quadrilaterals that are different from those in Exploration 1 and use them to justify your conjecture.

EXPLORATION 3 **Applying Your Conjecture**

Work with a partner. Use the conjecture you wrote in Exploration 2 to write an equation for each quadrilateral. Then solve the equation to find the value of x. Use a protractor to check the reasonableness of your answer.

a.

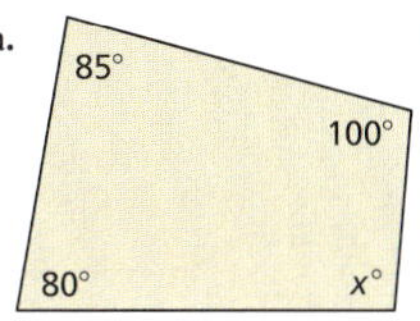

b.

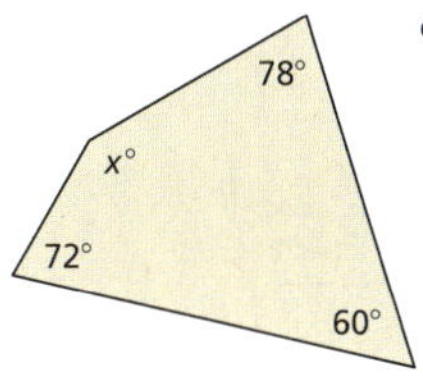

c.

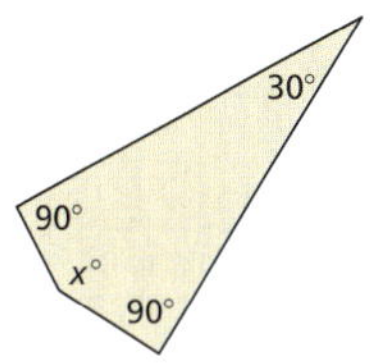

Communicate Your Answer

4. How can you use simple equations to solve real-life problems?
5. Draw your own quadrilateral and cut it out. Tear off the four corners of the quadrilateral and rearrange them to affirm the conjecture you wrote in Exploration 2. Explain how this affirms the conjecture.

Dynamic Teaching Tools

- Dynamic Assessment & Progress Monitoring Tool
- Lesson Planning Tool
- Interactive Whiteboard Lesson Library
- Dynamic Classroom with Dynamic Investigations

ANSWERS

1. **a.** 110; 90; 92; 68; 360°
   b. 65; 147; 58; 90; 360°
   c. 91; 79; 75; 115; 360°
   Answers will vary.
2. equals 360°
   Sample answer:

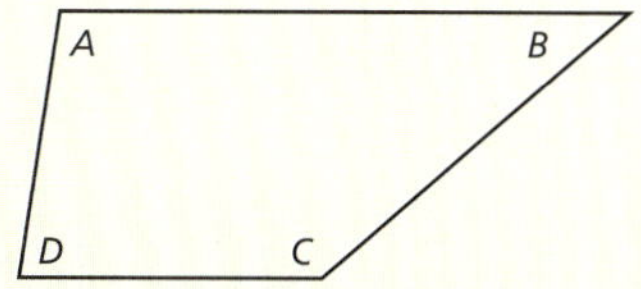

$100 + 40 + 140 + 80 = 360$

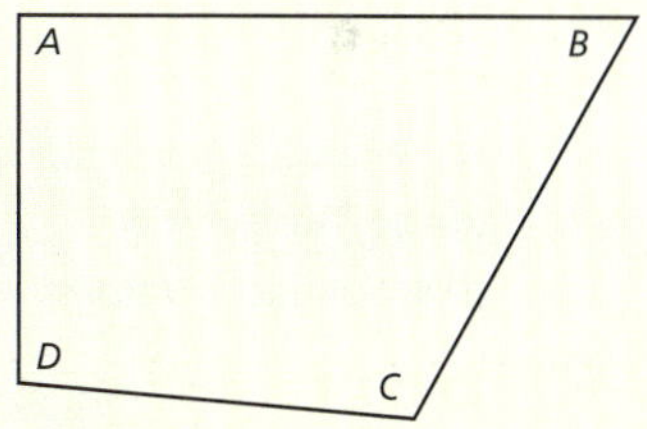

$90 + 60 + 115 + 95 = 360$

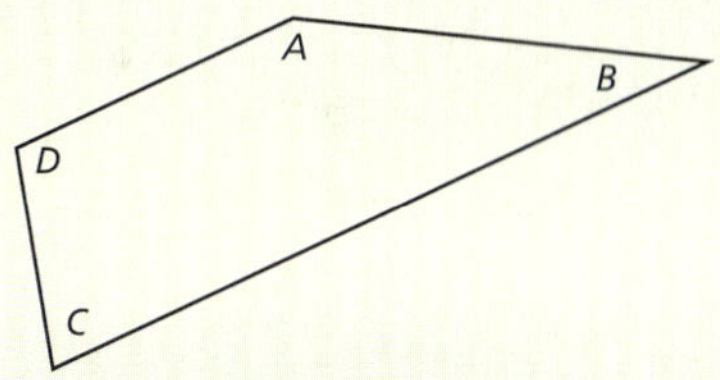

$150 + 30 + 75 + 105 = 360$

Divide the quadrilateral into two triangles.

The sum of the angle measures of a triangle is 180°, so the sum of the angle measures of a quadrilateral is $2(180°) = 360°$.

3. **a.** $85 + 100 + x + 80 = 360; x = 95$
   b. $x + 78 + 60 + 72 = 360; x = 150$
   c. $90 + 30 + 90 + x = 360; x = 150$
4. Simple equations can relate parts of geometric shapes and can be used to find missing parts.
5. The corners can be arranged so the angles complete a full circle, which is 360°.

Differentiated Instruction

Kinesthetic

Ask two students to assist you at the board or overhead when solving equations. Assign one student to the left side of the equation and the other student to the right side. Each student is responsible for performing the operations on his or her side. Emphasize that to keep the equality, both students must perform the same operation at each step.

Extra Example 1

Solve each equation. Justify each step. Check your answer.

a. $d - \frac{1}{4} = -\frac{1}{2}$ $d = -\frac{1}{4}$

b. $m + 4.8 = 9.2$ $m = 4.4$

MONITORING PROGRESS ANSWERS

1. $n = -10$; Subtract 3 from each side.
2. $g = -\frac{1}{3}$; Add $\frac{1}{3}$ to each side.
3. $p = -10.4$; Subtract 3.9 from each side.

1.1 Lesson

Core Vocabulary

conjecture, *p. 3*
rule, *p. 3*
theorem, *p. 3*
equation, *p. 4*
linear equation in one variable, *p. 4*
solution, *p. 4*
inverse operations, *p. 4*
equivalent equations, *p. 4*

Previous
expression

What You Will Learn

- Solve linear equations using addition and subtraction.
- Solve linear equations using multiplication and division.
- Use linear equations to solve real-life problems.

Solving Linear Equations by Adding or Subtracting

An **equation** is a statement that two expressions are equal. A **linear equation in one variable** is an equation that can be written in the form $ax + b = 0$, where a and b are constants and $a \neq 0$. A **solution** of an equation is a value that makes the equation true.

Inverse operations are two operations that undo each other, such as addition and subtraction. When you perform the same inverse operation on each side of an equation, you produce an equivalent equation. **Equivalent equations** are equations that have the same solution(s).

Core Concept

Addition Property of Equality

Words Adding the same number to each side of an equation produces an equivalent equation.

Algebra If $a = b$, then $a + c = b + c$.

Subtraction Property of Equality

Words Subtracting the same number from each side of an equation produces an equivalent equation.

Algebra If $a = b$, then $a - c = b - c$.

EXAMPLE 1 Solving Equations by Addition or Subtraction

Solve each equation. Justify each step. Check your answer.

a. $x - 3 = -5$ **b.** $0.9 = y + 2.8$

SOLUTION

a. $x - 3 = -5$ Write the equation.

Addition Property of Equality → $+3 \quad +3$ Add 3 to each side.

$x = -2$ Simplify.

▶ The solution is $x = -2$.

Check

$x - 3 = -5$

$-2 - 3 \stackrel{?}{=} -5$

$-5 = -5$ ✓

b. $0.9 = y + 2.8$ Write the equation.

Subtraction Property of Equality → $-2.8 \quad -2.8$ Subtract 2.8 from each side.

$-1.9 = y$ Simplify.

▶ The solution is $y = -1.9$.

Check

$0.9 = y + 2.8$

$0.9 \stackrel{?}{=} -1.9 + 2.8$

$0.9 = 0.9$ ✓

Monitoring Progress

Help in English and Spanish at *BigIdeasMath.com*

Solve the equation. Justify each step. Check your solution.

1. $n + 3 = -7$
2. $g - \frac{1}{3} = -\frac{2}{3}$
3. $-6.5 = p + 3.9$

Laurie's Notes Teacher Actions

- In Examples 1 and 2, students need to understand that solving an equation is a process of reasoning. In the process of learning to solve equations, students learn "if … then" statements as stated in the properties of equality such as if $x = y$, then $x + 4 = y + 4$.
- *Actively listen* as you probe student understanding of the vocabulary and *Core Concept*.

Solving Linear Equations by Multiplying or Dividing

Core Concept

Multiplication Property of Equality

Words Multiplying each side of an equation by the same nonzero number produces an equivalent equation.

Algebra If $a = b$, then $a \cdot c = b \cdot c$, $c \neq 0$.

Division Property of Equality

Words Dividing each side of an equation by the same nonzero number produces an equivalent equation.

Algebra If $a = b$, then $a \div c = b \div c$, $c \neq 0$.

REMEMBER

Multiplication and division are inverse operations.

EXAMPLE 2 Solving Equations by Multiplication or Division

Solve each equation. Justify each step. Check your answer.

a. $-\frac{n}{5} = -3$ **b.** $\pi x = -2\pi$ **c.** $1.3z = 5.2$

SOLUTION

a.

$-\frac{n}{5} = -3$ Write the equation.

Multiplication Property of Equality → $-5 \cdot \left(-\frac{n}{5}\right) = -5 \cdot (-3)$ Multiply each side by -5.

$n = 15$ Simplify.

▶ The solution is $n = 15$.

Check

$-\frac{n}{5} = -3$

$-\frac{15}{5} \stackrel{?}{=} -3$

$-3 = -3$ ✓

b.

$\pi x = -2\pi$ Write the equation.

Division Property of Equality → $\frac{\pi x}{\pi} = \frac{-2\pi}{\pi}$ Divide each side by π.

$x = -2$ Simplify.

▶ The solution is $x = -2$.

Check

$\pi x = -2\pi$

$\pi(-2) \stackrel{?}{=} -2\pi$

$-2\pi = -2\pi$ ✓

c.

$1.3z = 5.2$ Write the equation.

Division Property of Equality → $\frac{1.3z}{1.3} = \frac{5.2}{1.3}$ Divide each side by 1.3.

$z = 4$ Simplify.

▶ The solution is $z = 4$.

Check

$1.3z = 5.2$

$1.3(4) \stackrel{?}{=} 5.2$

$5.2 = 5.2$ ✓

Monitoring Progress

Help in English and Spanish at *BigIdeasMath.com*

Solve the equation. Justify each step. Check your solution.

4. $\frac{y}{3} = -6$ **5.** $9\pi = \pi x$ **6.** $0.05w = 1.4$

English Language Learners

Vocabulary

In this section, students will use inverse (or opposite) operations to solve equations. Students will use addition to solve a subtraction equation and use subtraction to solve an addition equation. Review the pairs of words listed below. Then give students one word of a pair, and ask them to provide the opposite word.

Examples:

product, quotient	add, subtract
multiply, divide	plus, minus
positive, negative	sum, difference
odd, even	

Extra Example 2

Solve each equation. Justify each step. Check your answer.

a. $\frac{m}{5} = -4$ $m = -20$

b. $3p = -\frac{2}{3}$ $p = -\frac{2}{9}$

c. $2.5r = 75$ $r = 30$

MONITORING PROGRESS ANSWERS

4. $y = -18$; Multiply each side by 3.

5. $x = 9$; Divide each side by π.

6. $w = 28$; Divide each side by 0.05.

Laurie's Notes Teacher Actions

? "Why is $c \neq 0$ in the Multiplication and Division Properties of Equality?" *Sample answer:* When multiplying, if $c = 0$, then $a \cdot c = b \cdot c$ is true when $a \neq b$. Division by 0 is undefined.

COMMON ERROR Students often view π as a variable and not a constant. Be sure to model a problem with π as the coefficient.

? **Check for Understanding:** Write $-\frac{n}{5} = \frac{-n}{5} = \frac{n}{-5}$ on the board. "Are all three expressions equivalent?" yes

Extra Example 3

You clean a community park for 6.5 hours. You earn \$42.25. Write and solve an equation to find how much you earn per hour. $6.5r = 42.25$; $r = 6.5$; You earn \$6.50 per hour.

MONITORING PROGRESS ANSWER

7. 38.65 sec

Solving Real-Life Problems

MODELING WITH MATHEMATICS

Mathematically proficient students routinely check that their solutions make sense in the context of a real-life problem.

Core Concept

Four-Step Approach to Problem Solving

1. **Understand the Problem** What is the unknown? What information is being given? What is being asked?
2. **Make a Plan** This plan might involve one or more of the problem-solving strategies shown on the next page.
3. **Solve the Problem** Carry out your plan. Check that each step is correct.
4. **Look Back** Examine your solution. Check that your solution makes sense in the original statement of the problem.

EXAMPLE 3 Modeling with Mathematics

In the 2012 Olympics, Usain Bolt won the 200-meter dash with a time of 19.32 seconds. Write and solve an equation to find his average speed to the nearest hundredth of a meter per second.

REMEMBER

The formula that relates distance d, rate or speed r, and time t is

$d = rt$.

SOLUTION

1. **Understand the Problem** You know the winning time and the distance of the race. You are asked to find the average speed to the nearest hundredth of a meter per second.
2. **Make a Plan** Use the Distance Formula to write an equation that represents the problem. Then solve the equation.
3. **Solve the Problem**

$d = r \cdot t$ Write the Distance Formula.

$200 = r \cdot 19.32$ Substitute 200 for d and 19.32 for t.

$\frac{200}{19.32} = \frac{19.32r}{19.32}$ Divide each side by 19.32.

$10.35 \approx r$ Simplify.

REMEMBER

The symbol $\approx$ means "approximately equal to."

▶ Bolt's average speed was about 10.35 meters per second.

4. **Look Back** Round Bolt's average speed to 10 meters per second. At this speed, it would take

$$\frac{200 \text{ m}}{10 \text{ m/sec}} = 20 \text{ seconds}$$

to run 200 meters. Because 20 is close to 19.32, your solution is reasonable.

Monitoring Progress

Help in English and Spanish at *BigIdeasMath.com*

7. Suppose Usain Bolt ran 400 meters at the same average speed that he ran the 200 meters. How long would it take him to run 400 meters? Round your answer to the nearest hundredth of a second.

Laurie's Notes Teacher Actions

- Students may quickly reason that you divide distance by time to find the rate. Connecting the steps to the problem-solving plan should not be overlooked.
- Engage students by showing a video of the race. Search the Internet for "Usain Bolt 200m finals."
- **Attend to Precision:** Discuss with students the importance of labeling answers with appropriate units.

Core Concept

Common Problem-Solving Strategies

Use a verbal model.	Guess, check, and revise.
Draw a diagram.	Sketch a graph or number line.
Write an equation.	Make a table.
Look for a pattern.	Make a list.
Work backward.	Break the problem into parts.

EXAMPLE 4 Modeling with Mathematics

On January 22, 1943, the temperature in Spearfish, South Dakota, fell from 54°F at 9:00 A.M. to -4°F at 9:27 A.M. How many degrees did the temperature fall?

SOLUTION

1. **Understand the Problem** You know the temperature before and after the temperature fell. You are asked to find how many degrees the temperature fell.
2. **Make a Plan** Use a verbal model to write an equation that represents the problem. Then solve the equation.
3. **Solve the Problem**

Words Temperature at 9:27 A.M. = Temperature at 9:00 A.M. − Number of degrees the temperature fell

Variable Let T be the number of degrees the temperature fell.

Equation $-4 = 54 - T$

$-4 = 54 - T$	Write the equation.
$-4 - 54 = 54 - 54 - T$	Subtract 54 from each side.
$-58 = -T$	Simplify.
$58 = T$	Divide each side by -1.

▶ The temperature fell 58°F.

REMEMBER
The distance between two points on a number line is always positive.

4. **Look Back** The temperature fell from 54 degrees *above* 0 to 4 degrees *below* 0. You can use a number line to check that your solution is reasonable.

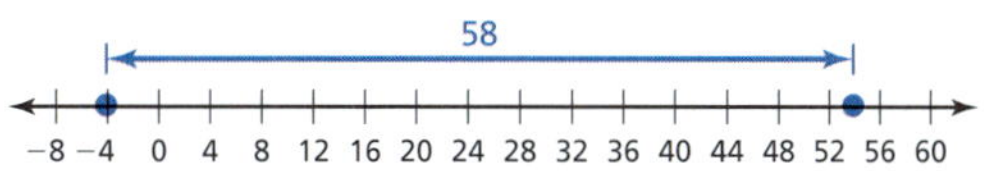

Monitoring Progress
Help in English and Spanish at *BigIdeasMath.com*

8. You thought the balance in your checking account was \$68. When your bank statement arrives, you realize that you forgot to record a check. The bank statement lists your balance as \$26. Write and solve an equation to find the amount of the check that you forgot to record.

Extra Example 4

A discounted concert ticket is \$14.50 less than the original price p. You pay \$53 for a discounted ticket. What is the original price of the ticket? \$67.50

MONITORING PROGRESS ANSWER

8. $26 = 68 - c$; \$42

Laurie's Notes Teacher Actions

- Make a classroom poster to display Common Problem-Solving Strategies.
- **Model with Mathematics:** Students may suggest that drawing a number line and plotting the two temperatures helps them to quickly see the solution.

Closure

- Describe in words how to solve a one-step equation. *Sample answer:* Use the same inverse operation on each side of the equation to produce an equivalent equation.
- Solve for x: $-13.8 = x - 4.3$ $x = -9.5$

Assignment Guide and Homework Check

ASSIGNMENT

Basic: 1–4, 5–39 odd, 47, 50, 57–65

Average: 1–4, 6–38 even, 39–45, 50, 57–65

Advanced: 1–4, 6, 10, 14–16, 20, 32–48 even, 50–65

HOMEWORK CHECK

Basic: 9, 11, 15, 27, 29

Average: 14, 16, 28, 34, 43

Advanced: 15, 16, 20, 34, 46

ANSWERS

1. $+$ and $-$; $\times$ and $\div$
2. yes; $x = -5$
3. Division Property of Equality; Divide each side by 14.
4. $x - 6 = 5$; It is the only one that involves addition and subtraction.
5. $x = 3$; Subtract 5 from each side.
6. $m = -7$; Subtract 9 from each side.
7. $y = 7$; Add 4 to each side.
8. $s = 3$; Add 2 to each side.
9. $w = -7$; Subtract 3 from each side.
10. $n = -1$; Add 6 to each side.
11. $p = -3$; Add 11 to each side.
12. $q = -4$; Subtract 4 from each side.
13. $r = 18$; Add 8 to each side.
14. $t = 4$; Subtract 5 from each side.
15. $p - 12.95 = 44$; \$56.95
16. $x + 12 = 195$; 183 points
17. $x + 100 + 120 + 100 = 360$; $x = 40$
18. $x + 150 + 77 + 48 = 360$; $x = 85$
19. $x + 76 + 92 + 122 = 360$; $x = 70$
20. $x + 60 + 115 + 85 = 360$; $x = 100$
21. $g = 4$; Divide each side by 5.
22. $q = 13$; Divide each side by 4.
23. $p = 15$; Multiply each side by 5.
24. $y = 7$; Multiply each side by 7.
25. $r = -8$; Divide each side by -8.
26. $x = -16$; Multiply each side by -2.
27. $x = 48$; Multiply each side by 6.
28. $w = -18$; Multiply each side by -3.
29. $s = -6$; Divide each side by 9.
30. $t = -49$; Multiply each side by 7.

1.1 Exercises

Dynamic Solutions available at *BigIdeasMath.com*

Vocabulary and Core Concept Check

1. **VOCABULARY** Which of the operations $+$, $-$, $\times$, and $\div$ are inverses of each other?
2. **VOCABULARY** Are the equations $-2x = 10$ and $-5x = 25$ equivalent? Explain.
3. **WRITING** Which property of equality would you use to solve the equation $14x = 56$? Explain.
4. **WHICH ONE DOESN'T BELONG?** Which expression does not belong with the other three? Explain your reasoning.

$8 = \frac{x}{2}$ $3 = x \div 4$ $x - 6 = 5$ $\frac{x}{3} = 9$

Monitoring Progress and Modeling with Mathematics

In Exercises 5–14, solve the equation. Justify each step. Check your solution. *(See Example 1.)*

5. $x + 5 = 8$
6. $m + 9 = 2$
7. $y - 4 = 3$
8. $s - 2 = 1$
9. $w + 3 = -4$
10. $n - 6 = -7$
11. $-14 = p - 11$
12. $0 = 4 + q$
13. $r + (-8) = 10$
14. $t - (-5) = 9$

15. **MODELING WITH MATHEMATICS** A discounted amusement park ticket costs \$12.95 less than the original price p. Write and solve an equation to find the original price.

16. **MODELING WITH MATHEMATICS** You and a friend are playing a board game. Your final score x is 12 points less than your friend's final score. Write and solve an equation to find your final score.

	ROUND 9	ROUND 10	FINAL SCORE
Your Friend	22	12	195
You	9	25	?

USING TOOLS **The sum of the angle measures of a quadrilateral is 360°. In Exercises 17–20, write and solve an equation to find the value of x. Use a protractor to check the reasonableness of your answer.**

17. 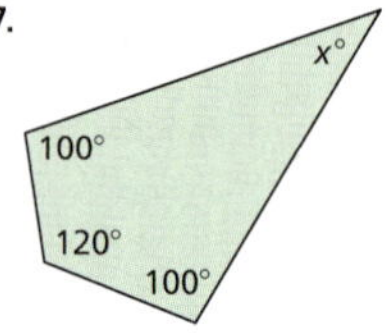

18. 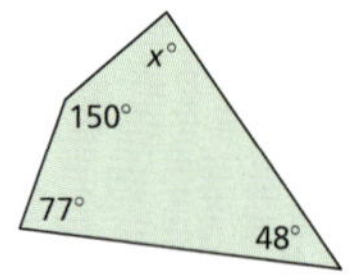

19. 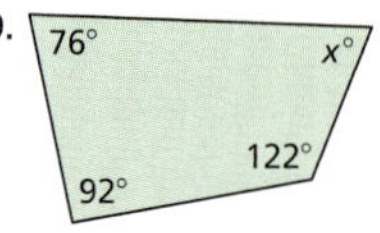

20. 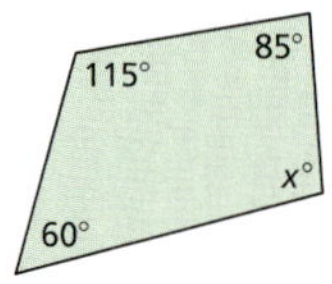

In Exercises 21–30, solve the equation. Justify each step. Check your solution. *(See Example 2.)*

21. $5g = 20$
22. $4q = 52$
23. $p \div 5 = 3$
24. $y \div 7 = 1$
25. $-8r = 64$
26. $x \div (-2) = 8$
27. $\frac{x}{6} = 8$
28. $\frac{w}{-3} = 6$
29. $-54 = 9s$
30. $-7 = \frac{t}{7}$

In Exercises 31–38, solve the equation. Check your solution.

31. $\frac{3}{2} + t = \frac{1}{2}$ **32.** $b - \frac{3}{16} = \frac{5}{16}$

33. $\frac{3}{7}m = 6$ **34.** $-\frac{2}{5}y = 4$

35. $5.2 = a - 0.4$ **36.** $f + 3\pi = 7\pi$

37. $-108\pi = 6\pi j$ **38.** $x \div (-2) = 1.4$

ERROR ANALYSIS **In Exercises 39 and 40, describe and correct the error in solving the equation.**

39. ✗

$$-0.8 + r = 12.6$$
$$r = 12.6 + (-0.8)$$
$$r = 11.8$$

40. ✗

$$-\frac{m}{3} = -4$$
$$3 \cdot \left(-\frac{m}{3}\right) = 3 \cdot (-4)$$
$$m = -12$$

41. **ANALYZING RELATIONSHIPS** A baker orders 162 eggs. Each carton contains 18 eggs. Which equation can you use to find the number x of cartons? Explain your reasoning and solve the equation.

Ⓐ $162x = 18$ Ⓑ $\frac{x}{18} = 162$

Ⓒ $18x = 162$ Ⓓ $x + 18 = 162$

MODELING WITH MATHEMATICS **In Exercises 42–44, write and solve an equation to answer the question.** *(See Examples 3 and 4.)*

42. The temperature at 5 P.M. is 20°F. The temperature at 10 P.M. is −5°F. How many degrees did the temperature fall?

43. The length of an American flag is 1.9 times its width. What is the width of the flag?

9.5 ft

44. The balance of an investment account is \$308 more than the balance 4 years ago. The current balance of the account is \$4708. What was the balance 4 years ago?

45. **REASONING** Identify the property of equality that makes Equation 1 and Equation 2 equivalent.

Equation 1	$x - \frac{1}{2} = \frac{x}{4} + 3$
Equation 2	$4x - 2 = x + 12$

46. **PROBLEM SOLVING** Tatami mats are used as a floor covering in Japan. One possible layout uses four identical rectangular mats and one square mat, as shown. The area of the square mat is half the area of one of the rectangular mats.

Total area = 81 ft²

a. Write and solve an equation to find the area of one rectangular mat.

b. The length of a rectangular mat is twice the width. Use Guess, Check, and Revise to find the dimensions of one rectangular mat.

47. **PROBLEM SOLVING** You spend \$30.40 on 4 CDs. Each CD costs the same amount and is on sale for 80% of the original price.

a. Write and solve an equation to find how much you spend on each CD.

b. The next day, the CDs are no longer on sale. You have \$25. Will you be able to buy 3 more CDs? Explain your reasoning.

48. **ANALYZING RELATIONSHIPS** As c increases, does the value of x *increase*, *decrease*, or *stay the same* for each equation? Assume c is positive.

Equation	Value of x
$x - c = 0$	
$cx = 1$	
$cx = c$	
$\frac{x}{c} = 1$	

Dynamic Teaching Tools

Dynamic Assessment & Progress Monitoring Tool

Interactive Whiteboard Lesson Library

Dynamic Classroom with Dynamic Investigations

ANSWERS

31. $t = -1$

32. $b = \frac{1}{2}$

33. $m = 14$

34. $y = -10$

35. $a = 5.6$

36. $f = 4\pi$

37. $j = -18$

38. $x = -2.8$

39. Subtract −0.8 from each side, not add; $r = 12.6 - (-0.8)$; $r = 13.4$

40. Multiply each side by −3, not 3; $-3 \cdot \left(-\frac{m}{3}\right) = -3 \cdot (-4)$; $m = 12$

41. C; Multiplying the number of eggs in each carton by the number of cartons will give the total number of eggs; 9 cartons

42. $-5 = 20 - T$; 25°F

43. $9.5 = 1.9w$; 5 ft

44. $4708 = b + 308$; \$4400

45. Multiplication Property of Equality

46. **a.** $4.5A = 81$; 18 ft²

b. 6 ft by 3 ft

47. **a.** $4p = 30.40$; \$7.60

b. no; Each CD costs \$9.50 at the regular price, so 3 CDs would cost \$28.50 which is greater than \$25.

48. increase; decrease; stay the same; increase

ANSWERS

49. **a.** 5; 10

b. −2; 9

50. **a.** 100%

b. It shows the sum of all the parts is 100%; Solve it for x; $x = 31$

51. 71; Because $\frac{1}{6}$ of the girls is 6, there are 36 girls. Because $\frac{2}{7}$ of the boys is 10, there are 35 boys; $36 + 35 = 71$

52. *Sample answer:* $5x = 25 + 5$; You and 4 friends go to a movie. The total cost of the 5 tickets is \$25 and the snacks you and your friends split cost an additional \$5. How much does each person pay?; $x = 6$; Each person pays \$6.

53. $B = 12\pi \text{ in.}^2$

54. $h = 9$ cm

55. $B = 9\pi \text{ m}^2$

56. $h = 3.5$ ft

57. **a.** 132 hits

b. no; Dividing by a larger number of at-bats decreases the value of the average.

58. $8y + 24$

59. $\frac{5}{6}x + \frac{15}{4}$

60. $5m + 15 + 5n$

61. $8p + 16q + 24$

62. 300

63. 0.02

64. 0.47

65. 298.26

49. USING STRUCTURE Use the values −2, 5, 9, and 10 to complete each statement about the equation $ax = b - 5$.

a. When $a =$ ___ and $b =$ ___, x is a positive integer.

b. When $a =$ ___ and $b =$ ___, x is a negative integer.

50. HOW DO YOU SEE IT? The circle graph shows the percents of different animals sold at a local pet store in 1 year.

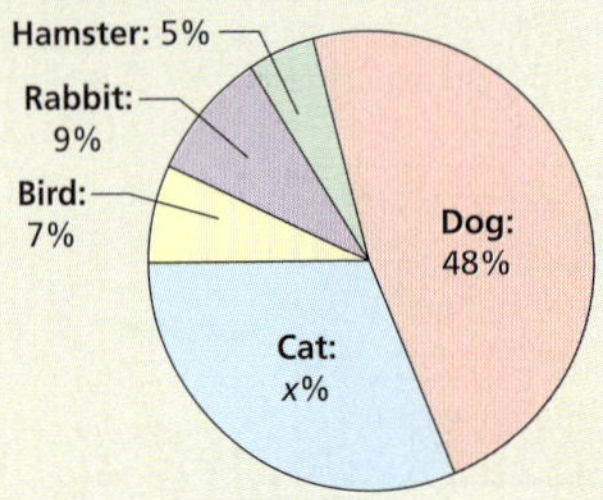

a. What percent is represented by the entire circle?

b. How does the equation $7 + 9 + 5 + 48 + x = 100$ relate to the circle graph? How can you use this equation to find the percent of cats sold?

51. REASONING One-sixth of the girls and two-sevenths of the boys in a school marching band are in the percussion section. The percussion section has 6 girls and 10 boys. How many students are in the marching band? Explain.

52. THOUGHT PROVOKING Write a real-life problem that can be modeled by an equation equivalent to the equation $5x = 30$. Then solve the equation and write the answer in the context of your real-life problem.

MATHEMATICAL CONNECTIONS In Exercises 53–56, find the height h or the area of the base B of the solid.

53.

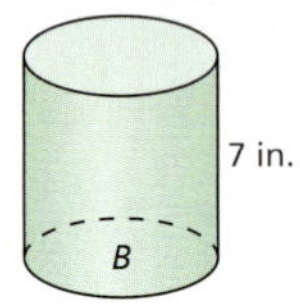

Volume = $84\pi \text{ in.}^3$

54.

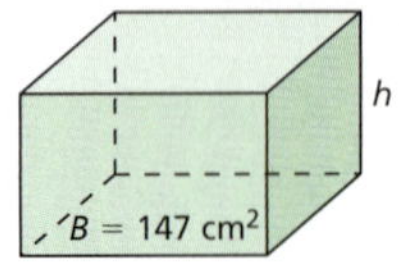

Volume = 1323 cm^3

55.

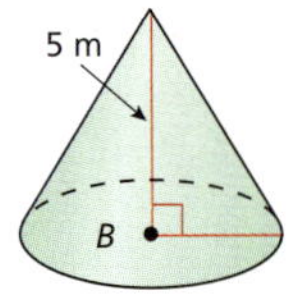

Volume = $15\pi \text{ m}^3$

56.

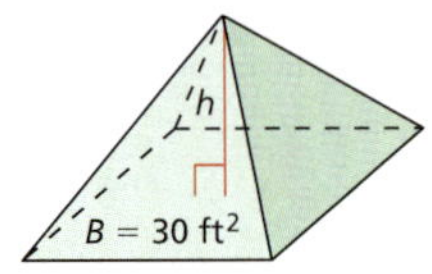

Volume = 35 ft^3

57. MAKING AN ARGUMENT In baseball, a player's batting average is calculated by dividing the number of hits by the number of at-bats. The table shows Player A's batting average and number of at-bats for three regular seasons.

Season	Batting average	At-bats
2010	.312	596
2011	.296	446
2012	.295	599

a. How many hits did Player A have in the 2011 regular season? Round your answer to the nearest whole number.

b. Player B had 33 fewer hits in the 2011 season than Player A but had a greater batting average. Your friend concludes that Player B had more at-bats in the 2011 season than Player A. Is your friend correct? Explain.

Maintaining Mathematical Proficiency

Reviewing what you learned in previous grades and lessons

Use the Distributive Property to simplify the expression. *(Skills Review Handbook)*

58. $8(y + 3)$ **59.** $\frac{5}{6}\left(x + \frac{1}{2} + 4\right)$ **60.** $5(m + 3 + n)$ **61.** $4(2p + 4q + 6)$

Copy and complete the statement. Round to the nearest hundredth, if necessary. *(Skills Review Handbook)*

62. $\frac{5 \text{ L}}{\text{min}} = \frac{\square \text{ L}}{\text{h}}$ **63.** $\frac{68 \text{ mi}}{\text{h}} \approx \frac{\square \text{ mi}}{\text{sec}}$

64. $\frac{7 \text{ gal}}{\text{min}} \approx \frac{\square \text{ qt}}{\text{sec}}$ **65.** $\frac{8 \text{ km}}{\text{min}} \approx \frac{\square \text{ mi}}{\text{h}}$

Mini-Assessment

Solve the equation.

1. $t + 17 = 3$ $t = -14$
2. $-2\pi + d = -3\pi$ $d = -\pi$
3. $-13.5 = 2.7s$ $s = -5$
4. $\frac{2}{3}j = 8$ $j = 12$
5. You earn \$9.65 per hour. This week, you earned \$308.80 before taxes. Write and solve an equation to find the number of hours you worked this week. $9.65x = 308.80$; $x = 32$; You worked 32 hours this week.

If students need help...	*If students got it...*
Resources by Chapter • Practice A and Practice B • Puzzle Time	Resources by Chapter • Enrichment and Extension • Cumulative Review
Student Journal • Practice	Start the *next* Section
Differentiating the Lesson Skills Review Handbook	

Laurie's Notes

Overview of Section 1.2

Introduction

- **Make Sense of Problems and Persevere in Solving Them** and **Reason Abstractly and Quantitatively:** To make sense of solving multi-step equations, students must understand what the symbols and operations represent. When they evaluate expressions, they follow the order of operations. When they solve equations, they undo that process by performing the inverse operations in a reverse order. For example:

Evaluate $3x - 4$ when $x = 5$.		Solve $3x - 4 = 11$.	
$3(5) - 4$		$3x = 15$	$+4$
$15 - 4$	$\times 3$	$x = 5$	$\div 3$
11	-4		

Mathematically proficient students make sense of this relationship and perform symbolic manipulations to solve for the variable.

Resources

- If students are familiar with algebra tiles, use the tiles to model multi-step equation solving.
- Search online for applets that model multi-step equation solving.

Common Misconceptions

- In solving an equation such as $5 - 2x = -8$, students see the subtraction sign and conclude that they need to "undo" the operation by adding. Ask probing questions so students recognize that the variable term is being subtracted from a constant term versus the constant term being subtracted from the variable term.
- Rewrite subtraction expressions using the related addition problem.
- $5 - 2x = -8$ can be rewritten as $5 + (-2x) = -8$.
- $24 = 5x - 2(7 - 4x)$ can be rewritten as $24 = 5x + (-2)[7 + (-4x)]$.

Formative Assessment Tips

- **Turn and Talk:** This technique allows all students in the class to have a voice. Using a 3-foot voice, students turn and talk to their partners about a problem or to discuss a question. There may be different roles that I ask partners to assume, so I refer to partner A and partner B. In discussing a procedure or explaining an answer, I might ask partner A to talk uninterrupted for a fixed period of time. Then partner B might be asked to repeat back what he/she heard or to ask a question about what has been shared.
  Example: *Turn and Talk*, so that partner A explains how solving $5 - 2x = -8$ differs from solving $2x - 5 = -8$.
- It is important to establish norms: Three-foot voices should be expected when students are doing partner work. Discuss with students the difference between *authentic listening* and being quiet while your partners are speaking.

Pacing Suggestion

- The questions students will work in the exploration require understanding of combining like terms and simplifying expressions, along with solving equations. These are foundational skills that can be worked on during the explorations and formal lesson. Do not rush!

Dynamic Teaching Tools

- Dynamic Assessment & Progress Monitoring Tool
- Lesson Planning Tool
- Interactive Whiteboard Lesson Library
- Dynamic Classroom with Dynamic Investigations

What Your Students Will Learn

- Solve multi-step linear equations in one variable using inverse operations.
- Use unit analysis to understand and model real-life problems.

Laurie's Notes

Exploration

Motivate

- Make a card for each student in your class. Write a variable term on each card. Students will walk around to find others with a card containing a like term to the one they are holding.

  Samples: $5x$, $-13x$, $5y$, $6xy$, x, $3.8x$, $\frac{1}{2}y$, $-3.8y$

? "What does it mean for terms to be 'like terms'?" The variables of the terms are the same.

Exploration 1

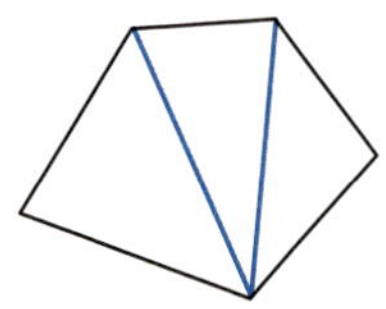

- Students may be familiar with the formula for the sum of the angles of a polygon from middle school.

? "What is the sum of the interior angles of a pentagon?" 540°

- Ask a volunteer to explain how he/she found the answer. Specifically, ask why the formula $180(n - 2)$ works.
  - **Model with Mathematics:** Draw the sketch of a pentagon with the diagonals drawn from one vertex. The model should help students make sense of the formula.
  - **Construct Viable Arguments and Critique the Reasoning of Others:** Listen for a student explanation: There are $n - 2$ triangles and the sum of the angle measures of each triangle is 180°.
  - As students are working on the questions, circulate to see that: (1) They have used the formula correctly, and (2) They are combining like terms correctly.
- **Formative Assessment Tip:** As you circulate, make note of student work that you want to highlight at the document camera or have done at the board. For instance, if a student uses the Distributive Property in part (f), make a note to ask that student to share his/her work with the class.
- **Alternate Approach:** You may want to select only 2–3 questions for students to work with partners as a warm-up before beginning the formal lesson.

Exploration 2

- This exploration reinforces the skill of measuring accurately with a protractor.
- Students may find it challenging to write expressions for each angle of the polygon. What often happens is that the expressions will involve addition, and the coefficient of the variable will be a positive integer.

Communicate Your Answer

- **Construct Viable Arguments and Critique the Reasoning of Others:** For Question 4, have students *Turn and Talk* about their explanation. Partners should listen critically to make sense of the explanation stated by their partners.
- There are different ways in which students could answer Question 5. Allow time for different methods to be shared.

Connecting to Next Step

- If a few problems from Exploration 1 were used as a warm-up, homework would be assigned based on the completion of the formal lesson.

1.2 Solving Multi-Step Equations

Essential Question How can you use multi-step equations to solve real-life problems?

EXPLORATION 1 **Solving for the Angle Measures of a Polygon**

Work with a partner. The sum S of the angle measures of a polygon with n sides can be found using the formula $S = 180(n - 2)$. Write and solve an equation to find each value of x. Justify the steps in your solution. Then find the angle measures of each polygon. How can you check the reasonableness of your answers?

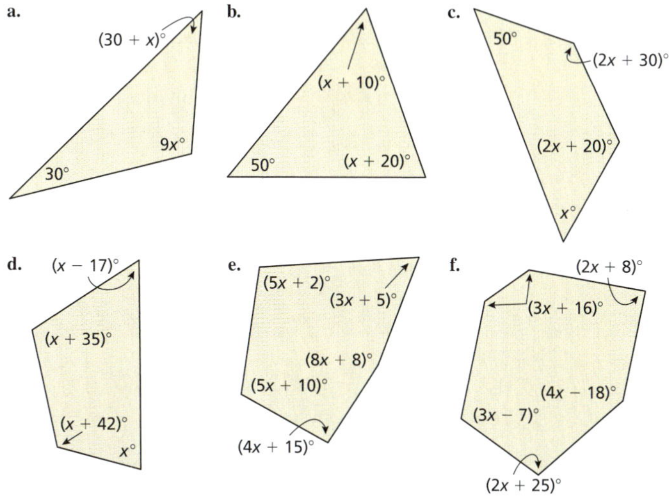

JUSTIFYING CONCLUSIONS

To be proficient in math, you need to be sure your answers make sense in the context of the problem. For instance, if you find the angle measures of a triangle, and they have a sum that is not equal to 180°, then you should check your work for mistakes.

EXPLORATION 2 **Writing a Multi-Step Equation**

Work with a partner.

a. Draw an irregular polygon.

b. Measure the angles of the polygon. Record the measurements on a separate sheet of paper.

c. Choose a value for x. Then, using this value, work backward to assign a variable expression to each angle measure, as in Exploration 1.

d. Trade polygons with your partner.

e. Solve an equation to find the angle measures of the polygon your partner drew. Do your answers seem reasonable? Explain.

Communicate Your Answer

3. How can you use multi-step equations to solve real-life problems?

4. In Exploration 1, you were given the formula for the sum S of the angle measures of a polygon with n sides. Explain why this formula works.

5. The sum of the angle measures of a polygon is 1080°. How many sides does the polygon have? Explain how you found your answer.

Dynamic Teaching Tools

- Dynamic Assessment & Progress Monitoring Tool
- Lesson Planning Tool
- Interactive Whiteboard Lesson Library
- Dynamic Classroom with Dynamic Investigations

ANSWERS

1. a. $(30 + x) + 9x + 30 = 180$; Write the equation.

$10x + 60 = 180$; Combine like terms.

$10x = 120$; Subtract 60 from each side.

$x = 12$; Divide each side by 10.

42°, 108°, 30°

b. $(x + 10) + (x + 20) + 50 = 180$; Write the equation.

$2x + 80 = 180$; Combine like terms.

$2x = 100$; Subtract 80 from each side.

$x = 50$; Divide each side by 2.

60°, 70°, 50°

c. $50 + (2x + 30) + (2x + 20) + x = 360$; Write the equation.

$5x + 100 = 360$; Combine like terms.

$5x = 260$; Subtract 100 from each side.

$x = 52$; Divide each side by 5.

50°, 134°, 124°, 52°

d. $(x - 17) + x + (x + 42) + (x + 35) = 360$; Write the equation.

$4x + 60 = 360$; Combine like terms.

$4x = 300$; Subtract 60 from each side.

$x = 75$; Divide each side by 4.

58°, 75°, 117°, 110°

e. $(5x + 2) + (3x + 5) + (8x + 8) + (4x + 15) + (5x + 10) = 540$; Write the equation.

$25x + 40 = 540$; Combine like terms.

$25x = 500$; Subtract 40 from each side.

$x = 20$; Divide each side by 25.

102°, 65°, 168°, 95°, 110°

1. f. $(3x + 16) + (3x + 16) + (2x + 8) + (4x - 18) + (2x + 25) + (3x - 7) = 720$; Write the equation.

$17x + 40 = 720$; Combine like terms.

$17x = 680$; Subtract 40 from each side.

$x = 40$; Divide each side by 17.

136°, 136°, 88°, 142°, 105°, 113°

Measure the angles with a protractor.

2. a. *Sample answer:*

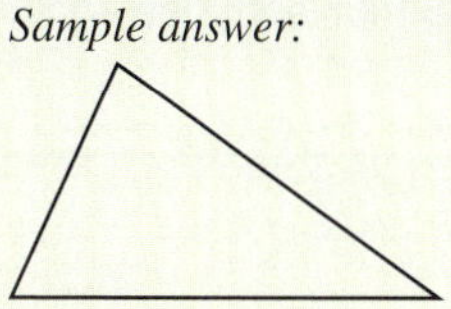

b. *Sample answer:* 80°, 35°, 65°

c. *Sample answer:* $x = 10$; $8x$, $3x + 5$, $6x + 5$

d. Check students' work.

e. *Sample answer:* $(16x + 4) + (6x + 4) + (23x + 2) + (13x + 2) = 360$, $x = 6$; 100°, 40°, 140°, 80°; yes; The angles are close to those measured with a protractor.

3. Use multi-step equations to find the values of missing parts in an object with the shape of a geometric figure.

4–5. See Additional Answers.

Differentiated Instruction

Organization

In order to minimize mistakes when solving equations, students need to keep their work organized. Provide grid paper with squares large enough so that students can write one number per square. Encourage students to use the grids for vertical and horizontal alignment.

Extra Example 1

Solve $15 = 1.5r + 3$. Check your solution. $r = 8$

Extra Example 2

Solve $5f + 4f - 8 = 19$. Check your solution. $f = 3$

MONITORING PROGRESS ANSWERS

1. $n = -3$
2. $c = -20$
3. $x = -\frac{1}{2}$

1.2 Lesson

Core Vocabulary

Previous
inverse operations
mean

What You Will Learn

- Solve multi-step linear equations using inverse operations.
- Use multi-step linear equations to solve real-life problems.
- Use unit analysis to model real-life problems.

Solving Multi-Step Linear Equations

Core Concept

Solving Multi-Step Equations

To solve a multi-step equation, simplify each side of the equation, if necessary. Then use inverse operations to isolate the variable.

EXAMPLE 1 Solving a Two-Step Equation

Solve $2.5x - 13 = 2$. Check your solution.

SOLUTION

	$2.5x - 13 = 2$	Write the equation.
Undo the subtraction.	$+13 \quad +13$	Add 13 to each side.
	$2.5x = 15$	Simplify.
Undo the multiplication.	$\frac{2.5x}{2.5} = \frac{15}{2.5}$	Divide each side by 2.5.
	$x = 6$	Simplify.

Check

$2.5x - 13 = 2$

$2.5(6) - 13 \stackrel{?}{=} 2$

$2 = 2$ ✓

▶ The solution is $x = 6$.

EXAMPLE 2 Combining Like Terms to Solve an Equation

Solve $-12 = 9x - 6x + 15$. Check your solution.

SOLUTION

	$-12 = 9x - 6x + 15$	Write the equation.
	$-12 = 3x + 15$	Combine like terms.
Undo the addition.	$-15 \quad -15$	Subtract 15 from each side.
	$-27 = 3x$	Simplify.
Undo the multiplication.	$\frac{-27}{3} = \frac{3x}{3}$	Divide each side by 3.
	$-9 = x$	Simplify.

Check

$-12 = 9x - 6x + 15$

$-12 \stackrel{?}{=} 9(-9) - 6(-9) + 15$

$-12 = -12$ ✓

▶ The solution is $x = -9$.

Monitoring Progress

Help in English and Spanish at *BigIdeasMath.com*

Solve the equation. Check your solution.

1. $-2n + 3 = 9$
2. $-21 = \frac{1}{2}c - 11$
3. $-2x - 10x + 12 = 18$

Laurie's Notes Teacher Actions

- In solving multi-step equations, the goal is to isolate the variable. Have students *Turn and Talk* to their neighbor. Partner A explains what "isolating the variable" means and how it is done. Select a partner B to summarize for the class what he/she heard.

? Write $2.5x - 13 = 2$ and $2 = 2.5x - 13$ and ask, "Are these equations solved the same way? Explain." Yes. *Sample answer:* Add 13 to each side and then divide each side by 2.5.

? Advancing Question: "Are the solutions the same for $-12 = 9x - 6x + 15$ and $-12 = 9x + 15 - 6x$? Explain." Yes. *Sample answer:* Subtract 15 from each side, combine like terms, and then divide each side by 3.

EXAMPLE 3 Using Structure to Solve a Multi-Step Equation

Solve $2(1 - x) + 3 = -8$. Check your solution.

SOLUTION

Method 1 One way to solve the equation is by using the Distributive Property.

$2(1 - x) + 3 = -8$	Write the equation.
$2(1) - 2(x) + 3 = -8$	Distributive Property
$2 - 2x + 3 = -8$	Multiply.
$-2x + 5 = -8$	Combine like terms.
$\underline{-5} \quad \underline{-5}$	Subtract 5 from each side.
$-2x = -13$	Simplify.
$\frac{-2x}{-2} = \frac{-13}{-2}$	Divide each side by -2.
$x = 6.5$	Simplify.

▶ The solution is $x = 6.5$.

Check

$2(1 - x) + 3 = -8$

$2(1 - 6.5) + 3 \stackrel{?}{=} -8$

$-8 = -8$ ✓

Method 2 Another way to solve the equation is by interpreting the expression $1 - x$ as a single quantity.

$2(1 - x) + 3 = -8$	Write the equation.
$\underline{-3} \quad \underline{-3}$	Subtract 3 from each side.
$2(1 - x) = -11$	Simplify.
$\frac{2(1 - x)}{2} = \frac{-11}{2}$	Divide each side by 2.
$1 - x = -5.5$	Simplify.
$\underline{-1} \quad \underline{-1}$	Subtract 1 from each side.
$-x = -6.5$	Simplify.
$\frac{-x}{-1} = \frac{-6.5}{-1}$	Divide each side by -1.
$x = 6.5$	Simplify.

LOOKING FOR STRUCTURE

First solve for the expression $1 - x$, and then solve for x.

▶ The solution is $x = 6.5$, which is the same solution obtained in Method 1.

Monitoring Progress

Help in English and Spanish at *BigIdeasMath.com*

Solve the equation. Check your solution.

4. $3(x + 1) + 6 = -9$

5. $15 = 5 + 4(2d - 3)$

6. $13 = -2(y - 4) + 3y$

7. $2x(5 - 3) - 3x = 5$

8. $-4(2m + 5) - 3m = 35$

9. $5(3 - x) + 2(3 - x) = 14$

English Language Learners

Visual Aids

To solve an equation, students must perform the same operation on each side of an equation to maintain equivalency from one step to the next until they reach the solution. Use a balance scale to model this concept. Explain that the equal sign (=) divides the equation into two sides, which must be kept in balance. Aligning the equal signs vertically on their papers makes it easy for students to identify the result of each operation on both sides of an equation as they check their work.

Extra Example 3

Solve $4(n - 6) - 1 = 31$. Check your solution. $n = 14$

MONITORING PROGRESS ANSWERS

4. $x = -6$

5. $d = 2.75$

6. $y = 5$

7. $x = 5$

8. $m = -5$

9. $x = 1$

Laurie's Notes Teacher Actions

- **Look For and Make Use of Structure:** Two methods are shown for solving the equation. Students might not think to divide by 2 as the second step in Method 2 because a decimal is introduced. The focus needs to be on seeing the structure—treating $(1 - x)$ as a term that is being isolated.

? Advancing Question: "Could the first step in solving the equation be dividing each side by 2? Explain." Yes. *Sample answer:* The Division Property of Equality is used to justify division as the first step. $[2(1 - x) + 3] \div 2 = -8 \div 2$

COMMON ERROR In Method 1, when simplifying $2 - 2x + 3$, students may subtract 3 from 2 and write $2x - 1$. Rewriting $2 - 2x + 3$ as $2 + (-2x) + 3$ will help.

Extra Example 4

Use the table to find the high temperature x on Wednesday so that the mean high temperature for the 5 days is 65°F.

Day	High Temperature (°F)
Monday	76
Tuesday	70
Wednesday	x
Thursday	68
Friday	55

The high temperature on Wednesday needs to be 56°F.

MONITORING PROGRESS ANSWER

10. 48 lb/in.^{2}

Solving Real-Life Problems

EXAMPLE 4 **Modeling with Mathematics**

Use the table to find the number of miles x you need to bike on Friday so that the mean number of miles biked per day is 5.

Day	Miles
Monday	3.5
Tuesday	5.5
Wednesday	0
Thursday	5
Friday	x

SOLUTION

1. **Understand the Problem** You know how many miles you biked Monday through Thursday. You are asked to find the number of miles you need to bike on Friday so that the mean number of miles biked per day is 5.
2. **Make a Plan** Use the definition of mean to write an equation that represents the problem. Then solve the equation.
3. **Solve the Problem** The mean of a data set is the sum of the data divided by the number of data values.

$$\frac{3.5 + 5.5 + 0 + 5 + x}{5} = 5 \quad \text{Write the equation.}$$

$$\frac{14 + x}{5} = 5 \quad \text{Combine like terms.}$$

$$5 \cdot \frac{14 + x}{5} = 5 \cdot 5 \quad \text{Multiply each side by 5.}$$

$$14 + x = 25 \quad \text{Simplify.}$$

$$-14 \qquad -14 \quad \text{Subtract 14 from each side.}$$

$$x = 11 \quad \text{Simplify.}$$

▶ You need to bike 11 miles on Friday.

4. **Look Back** Notice that on the days that you did bike, the values are close to the mean. Because you did not bike on Wednesday, you need to bike about twice the mean on Friday. Eleven miles is about twice the mean. So, your solution is reasonable.

Monitoring Progress

Help in English and Spanish at *BigIdeasMath.com*

10. The formula $d = \frac{1}{2}n + 26$ relates the nozzle pressure n (in pounds per square inch) of a fire hose and the maximum horizontal distance the water reaches d (in feet). How much pressure is needed to reach a fire 50 feet away?

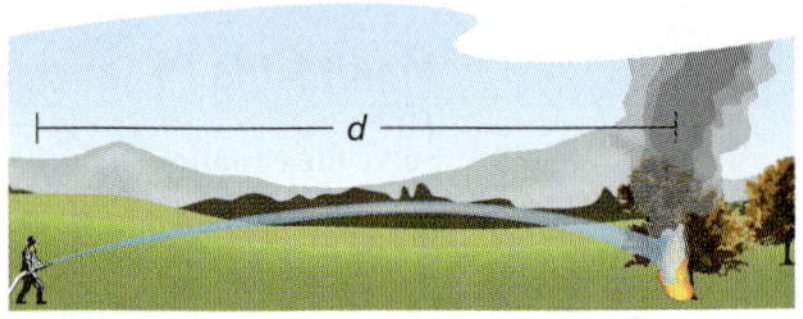

Laurie's Notes Teacher Actions

? "How do you find the mean of five numbers?" Sum the numbers and divide by 5.

- **FYI:** It may be helpful to write the third step with parentheses: $5\left(\frac{14 + x}{5}\right)$.
- **Note:** This is a classic question. When all of the data are known except for one, what is needed in order to achieve a particular average? Students often ask this in the context of wanting to know what they have to score on a test in order to achieve a certain average.

REMEMBER

When you add **miles** to **miles**, you get **miles**. But, when you divide **miles** by **days**, you get **miles per day**.

Using Unit Analysis to Model Real-Life Problems

When you write an equation to model a real-life problem, you should check that the units on each side of the equation balance. For instance, in Example 4, notice how the units balance.

miles | miles per day

$$\frac{3.5 + 5.5 + 0 + 5 + x}{5} = 5 \qquad \frac{\text{mi}}{\text{day}} = \frac{\text{mi}}{\text{day}} \ ✓$$

per | days

EXAMPLE 5 Solving a Real-Life Problem

Your school's drama club charges \$4 per person for admission to a play. The club borrowed \$400 to pay for costumes and props. After paying back the loan, the club has a profit of \$100. How many people attended the play?

SOLUTION

1. **Understand the Problem** You know how much the club charges for admission. You also know how much the club borrowed and its profit. You are asked to find how many people attended the play.
2. **Make a Plan** Use a verbal model to write an equation that represents the problem. Then solve the equation.
3. **Solve the Problem**

REMEMBER

When you multiply **dollars per person** by **people**, you get **dollars**.

Words Ticket price • Number of people who attended − Amount of loan = Profit

Variable Let x be the number of people who attended.

Equation $\frac{\$4}{\text{person}} \cdot x \text{ people} - \$400 = \$100$ $\qquad \$ = \$ \ ✓$

$4x - 400 = 100$	Write the equation.
$4x - 400 + 400 = 100 + 400$	Add 400 to each side.
$4x = 500$	Simplify.
$\frac{4x}{4} = \frac{500}{4}$	Divide each side by 4.
$x = 125$	Simplify.

▶ So, 125 people attended the play.

4. **Look Back** To check that your solution is reasonable, multiply \$4 per person by 125 people. The result is \$500. After paying back the \$400 loan, the club has \$100, which is the profit.

Monitoring Progress

Help in English and Spanish at *BigIdeasMath.com*

11. You have 96 feet of fencing to enclose a rectangular pen for your dog. To provide sufficient running space for your dog to exercise, the pen should be three times as long as it is wide. Find the dimensions of the pen.

Extra Example 5

You spent \$50 on supplies for the bake sale. Each item sold for \$2. After the sale, you had a profit of \$180. How many items did you sell? 115 items

MONITORING PROGRESS ANSWER

11. 36 ft by 12 ft

Laurie's Notes Teacher Actions

- **Attend to Precision:** Students should pay attention to the units associated with the numbers used in real-life application problems. The units help students check to see when the process used is yielding reasonable results.
- ? **Assessing Question:** "Can you explain in words (i.e., verbal model) how to determine the number of people who attended?" I remind students that if they cannot explain in words how to solve a problem, writing an equation is not going to help. *Sample answer:* 4 dollars per person times the number of people who attended, minus the 400 dollar loan is equal to the profit in dollars.

Closure

- **Exit Ticket:** Solve $8x + 9 - 4x = 25$. Check your solution. $x = 4$

Assignment Guide and Homework Check

ASSIGNMENT

Basic: 1, 2, 3–25 odd, 35, 37, 41, 47, 50, 57–65

Average: 1, 2, 4–34 even, 35–37, 39–42, 47, 49–51, 57–65

Advanced: 1, 2, 6, 10, 14–16, 18–34 even, 35–41, 44, 47, 48–56 even, 57–65

HOMEWORK CHECK

Basic: 3, 9, 13, 15, 21

Average: 10, 14, 16, 22, 37

Advanced: 6, 16, 20, 35, 44

ANSWERS

1. like terms
2. Use the Distributive Property to simplify, then solve for x; Solve for the expression $4x - 11$, then solve for x.
3. $w = 4$
4. $g = 8$
5. $q = 1$
6. $m = -3$
7. $z = -32$
8. $a = 6$
9. $h = 4$
10. $d = -16$
11. $y = 4$
12. $n = 4$
13. $v = 3$
14. $c = -2$
15. 6 min
16. $553 = 265 + 48t$; 6 h
17. $z = 3$
18. $g = -3$
19. $m = 3$
20. $h = -9$
21. $c = 5$
22. $y = -19$
23. $x = 29$
24. $r = -6$
25. $k = 45$; 45°, 90°, 45°
26. $a = 60$; 120°, 60°, 120°, 60°
27. $b = 90$; 90°, 135°, 90°, 90°, 135°
28. $x = 125$; 120°, 100°, 120°, 135°, 120°, 125°
29. $2n + 13 = 75$; $n = 31$
30. $3n - 4 = -19$; $n = -5$
31. $8 + \frac{n}{3} = -2$; $n = -30$
32. $2n + \frac{1}{2}n = 10$; $n = 4$
33. $6(n + 15) = -42$; $n = -22$
34. $4(n - 7) = 12$; $n = 10$

1.2 Exercises

Dynamic Solutions available at *BigIdeasMath.com*

Vocabulary and Core Concept Check

1. **COMPLETE THE SENTENCE** To solve the equation $2x + 3x = 20$, first combine $2x$ and $3x$ because they are ________.

2. **WRITING** Describe two ways to solve the equation $2(4x - 11) = 10$.

Monitoring Progress and Modeling with Mathematics

In Exercises 3–14, solve the equation. Check your solution. *(See Examples 1 and 2.)*

3. $3w + 7 = 19$
4. $2g - 13 = 3$
5. $11 = 12 - q$
6. $10 = 7 - m$
7. $5 = \frac{z}{-4} - 3$
8. $\frac{a}{3} + 4 = 6$
9. $\frac{h + 6}{5} = 2$
10. $\frac{d - 8}{-2} = 12$
11. $8y + 3y = 44$
12. $36 = 13n - 4n$
13. $12v + 10v + 14 = 80$
14. $6c - 8 - 2c = -16$

15. **MODELING WITH MATHEMATICS** The altitude a (in feet) of a plane t minutes after liftoff is given by $a = 3400t + 600$. How many minutes after liftoff is the plane at an altitude of 21,000 feet?

16. **MODELING WITH MATHEMATICS** A repair bill for your car is \$553. The parts cost \$265. The labor cost is \$48 per hour. Write and solve an equation to find the number of hours of labor spent repairing the car.

In Exercises 17–24, solve the equation. Check your solution. *(See Example 3.)*

17. $4(z + 5) = 32$
18. $-2(4g - 3) = 30$
19. $6 + 5(m + 1) = 26$
20. $5h + 2(11 - h) = -5$
21. $27 = 3c - 3(6 - 2c)$
22. $-3 = 12y - 5(2y - 7)$
23. $-3(3 + x) + 4(x - 6) = -4$
24. $5(r + 9) - 2(1 - r) = 1$

USING TOOLS **In Exercises 25–28, find the value of the variable. Then find the angle measures of the polygon. Use a protractor to check the reasonableness of your answer.**

25.

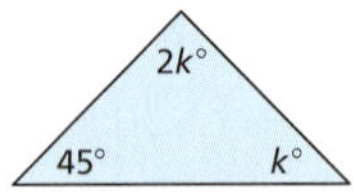

Sum of angle measures: 180°

26.

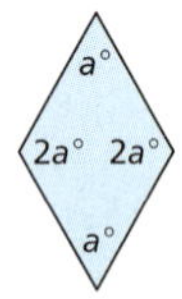

Sum of angle measures: 360°

27.

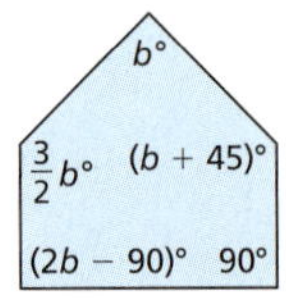

Sum of angle measures: 540°

28.

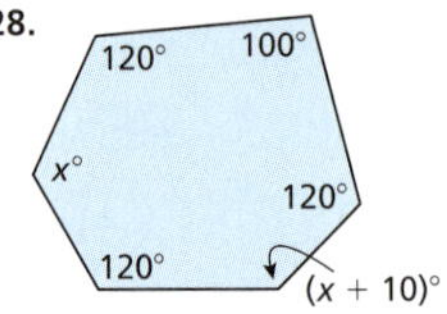

Sum of angle measures: 720°

In Exercises 29–34, write and solve an equation to find the number.

29. The sum of twice a number and 13 is 75.
30. The difference of three times a number and 4 is −19.
31. Eight plus the quotient of a number and 3 is −2.
32. The sum of twice a number and half the number is 10.
33. Six times the sum of a number and 15 is −42.
34. Four times the difference of a number and 7 is 12.

USING EQUATIONS **In Exercises 35–37, write and solve an equation to answer the question. Check that the units on each side of the equation balance.** *(See Examples 4 and 5.)*

35. During the summer, you work 30 hours per week at a gas station and earn \$8.75 per hour. You also work as a landscaper for \$11 per hour and can work as many hours as you want. You want to earn a total of \$400 per week. How many hours must you work as a landscaper?

36. The area of the surface of the swimming pool is 210 square feet. What is the length d of the deep end (in feet)?

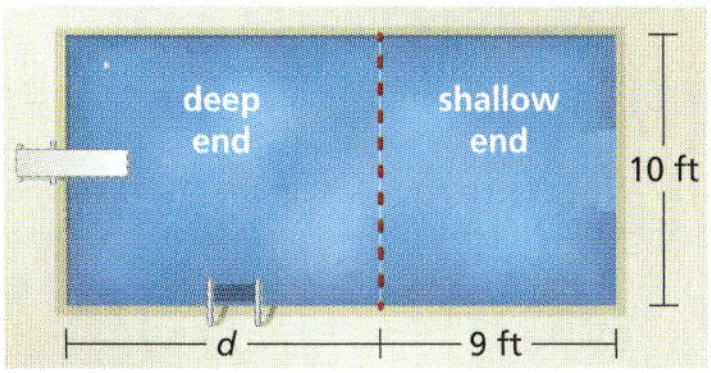

37. You order two tacos and a salad. The salad costs \$2.50. You pay 8% sales tax and leave a \$3 tip. You pay a total of \$13.80. How much does one taco cost?

JUSTIFYING STEPS **In Exercises 38 and 39, justify each step of the solution.**

38. $-\frac{1}{2}(5x - 8) - 1 = 6$ — Write the equation.

$-\frac{1}{2}(5x - 8) = 7$ — ______

$5x - 8 = -14$ — ______

$5x = -6$ — ______

$x = -\frac{6}{5}$ — ______

39. $2(x + 3) + x = -9$ — Write the equation.

$2(x) + 2(3) + x = -9$ — ______

$2x + 6 + x = -9$ — ______

$3x + 6 = -9$ — ______

$3x = -15$ — ______

$x = -5$ — ______

ERROR ANALYSIS **In Exercises 40 and 41, describe and correct the error in solving the equation.**

40. ✗

$$-2(7 - y) + 4 = -4$$
$$-14 - 2y + 4 = -4$$
$$-10 - 2y = -4$$
$$-2y = 6$$
$$y = -3$$

41. ✗

$$\frac{1}{4}(x - 2) + 4 = 12$$
$$\frac{1}{4}(x - 2) = 8$$
$$x - 2 = 2$$
$$x = 4$$

MATHEMATICAL CONNECTIONS **In Exercises 42–44, write and solve an equation to answer the question.**

42. The perimeter of the tennis court is 228 feet. What are the dimensions of the court?

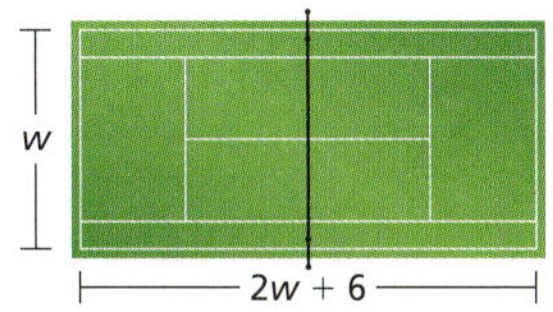

43. The perimeter of the Norwegian flag is 190 inches. What are the dimensions of the flag?

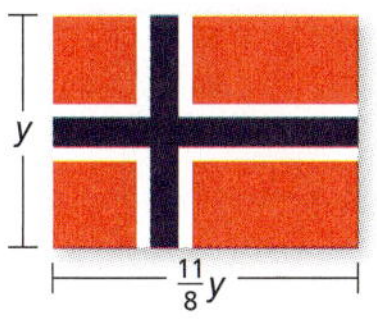

44. The perimeter of the school crossing sign is 102 inches. What is the length of each side?

Dynamic Teaching Tools

- Dynamic Assessment & Progress Monitoring Tool
- Interactive Whiteboard Lesson Library
- Dynamic Classroom with Dynamic Investigations

ANSWERS

35. $30(8.75) + 11t = 400$; 12.5 h

36. $10d + 10(9) = 210$; 12 ft

37. $1.08(2t + 2.50) + 3 = 13.80$; \$3.75

38. Add 1 to each side; Multiply each side by -2; Add 8 to each side; Divide each side by 5.

39. Distributive Property; Simplify; Combine like terms; Subtract 6 from each side; Divide each side by 3.

40. When using the Distributive Property in the second step, the second term should be positive; $-14 + 2y + 4 = -4$; $-10 + 2y = -4$; $2y = 6$; $y = 3$

41. In the third step, the right side should be 8×4, not $8 \div 4$; $x - 2 = 32$; $x = 34$

42. $2(2w + 6) + 2w = 228$, $w = 36$; 78 ft by 36 ft

43. $2y + 2 \cdot \frac{11}{8}y = 190$, $y = 40$; 55 in. by 40 in.

44. $(s + 6) + (s + 6) + s + 2s + s = 102$; $s = 15$; 21 in., 21 in., 15 in., 30 in., 15 in.

ANSWERS

45. $x = \frac{15}{16}$; *Sample answer:* method 1; There are no fractions until the last step.

46. 6 tickets

47. no; Solving the equation $0.25(d + 8) + 0.10d = 2.80$ results in the number of dimes not being a whole number.

48. See Additional Answers.

49. 16, 18, 20; The next consecutive even integers after $2n$ are $2n + 2$ and $2n + 4$. Solve the equation $2n + (2n + 2) + (2n + 4) = 54$. Then substitute the solution into the expressions for the integers.

50. **a.** greater than 20; The attendance at the second meeting is only 1 above 20, and the attendance at the other two is more than 1 below 20, so the attendance at the fourth meeting must be greater than 20 to have a mean of 20.

b. *Sample answer:* 24

c. Use the mean formula to find the attendance at the fourth meeting.

51. $x = -\frac{7}{b}$

52–65. See Additional Answers.

Mini-Assessment

Solve the equation.

1. $-5d + 4 = 12$ $d = -1.6$
2. $-2n + 9n - 8 = 27$ $n = 5$
3. $-18 = 5(2 - x) + 2$ $x = 6$
4. Use the table to find the grade g on Quiz 3 so that the mean grade for the quizzes is 90.

Quiz	Grade
Quiz 1	92
Quiz 2	89
Quiz 3	g
Quiz 4	85

94

5. A team pays $50 to reserve a banquet table. The players on the team pay $9 per person. The team spends a total of $104. How many players sit at the team table?
6 players

45. COMPARING METHODS Solve the equation $2(4 - 8x) + 6 = -1$ using (a) Method 1 from Example 3 and (b) Method 2 from Example 3. Which method do you prefer? Explain.

46. PROBLEM SOLVING An online ticket agency charges the amounts shown for basketball tickets. The total cost for an order is $220.70. How many tickets are purchased?

Charge	Amount
Ticket price	$32.50 per ticket
Convenience charge	$3.30 per ticket
Processing charge	$5.90 per order

47. MAKING AN ARGUMENT You have quarters and dimes that total $2.80. Your friend says it is possible that the number of quarters is 8 more than the number of dimes. Is your friend correct? Explain.

48. THOUGHT PROVOKING You teach a math class and assign a weight to each component of the class. You determine final grades by totaling the products of the weights and the component scores. Choose values for the remaining weights and find the necessary score on the final exam for a student to earn an A (90%) in the class, if possible. Explain your reasoning.

Component	Student's score	Weight	Score × Weight
Class Participation	92%	0.20	92% × 0.20 = 18.4%
Homework	95%		
Midterm Exam	88%		
Final Exam			
Total		1	

49. REASONING An even integer can be represented by the expression $2n$, where n is any integer. Find three consecutive even integers that have a sum of 54. Explain your reasoning.

50. HOW DO YOU SEE IT? The scatter plot shows the attendance for each meeting of a gaming club.

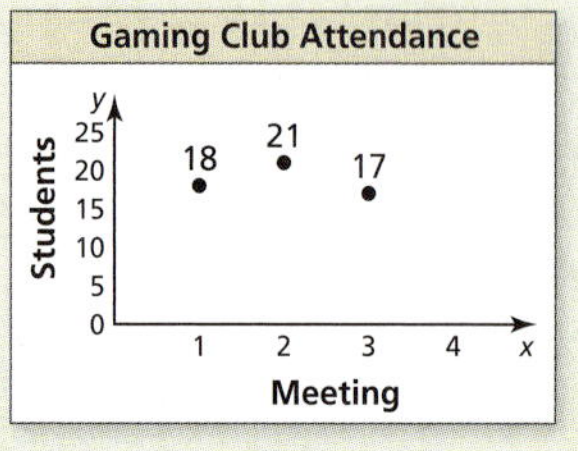

a. The mean attendance for the first four meetings is 20. Is the number of students who attended the fourth meeting greater than or less than 20? Explain.

b. Estimate the number of students who attended the fourth meeting.

c. Describe a way you can check your estimate in part (b).

REASONING In Exercises 51–56, the letters a, b, and c represent nonzero constants. Solve the equation for x.

51. $bx = -7$

52. $x + a = \frac{3}{4}$

53. $ax - b = 12.5$

54. $ax + b = c$

55. $2bx - bx = -8$

56. $cx - 4b = 5b$

Maintaining Mathematical Proficiency

Reviewing what you learned in previous grades and lessons

Simplify the expression. *(Skills Review Handbook)*

57. $4m + 5 - 3m$

58. $9 - 8b + 6b$

59. $6t + 3(1 - 2t) - 5$

Determine whether (a) $x = -1$ or (b) $x = 2$ is a solution of the equation. *(Skills Review Handbook)*

60. $x - 8 = -9$

61. $x + 1.5 = 3.5$

62. $2x - 1 = 3$

63. $3x + 4 = 1$

64. $x + 4 = 3x$

65. $-2(x - 1) = 1 - 3x$

If students need help...	*If students got it...*
Resources by Chapter • Practice A and Practice B • Puzzle Time	Resources by Chapter • Enrichment and Extension • Cumulative Review
Student Journal • Practice	Start the *next* Section
Differentiating the Lesson Skills Review Handbook	

Laurie's Notes

Overview of Section 1.3

Introduction

- Students can generally solve visual equations where objects are shown on a balance scale. (See *Motivate* next page.) Placing (or removing) the same amount of items on each side of a balance scale makes sense. Applying this to equation solving, explain that either variable terms or constant terms can be added to (or removed from) each side of the equation. A challenge for students is having confidence in what step to do first. It is helpful to model at least one equation in class where the variable terms are dealt with first and then solve the same equation again where the constant terms are dealt with first.

Example:

$5x - 4 = 3x + 12$	$5x - 4 = 3x + 12$
$2x - 4 = 12$	$5x = 3x + 16$
$2x = 16$	$2x = 16$
$x = 8$	$x = 8$

- As teachers, we tend to use a lot of familiar vocabulary when solving equations, because we have been solving equations for many years. Our vocabulary may not be familiar to students. Be sure to check for understanding when using phrases such as: "isolate the variable," "collect variable terms," and "collect constant terms."

Resources

- If students are familiar with algebra tiles, the tiles can be used to model equations with variables on both sides of the equation.
- Search online for applets that model equation solving with variables on both sides of the equation.

Formative Assessment Tips

- **I Used to Think ... But Now I Know:** This technique asks students to consider how their thinking about a concept or skill has changed from the beginning of instruction to the end of instruction. This can be done orally or in writing. It is important for students to be able to self-assess and reflect on their own learning.
- Use this technique at the end of the formal lesson. If time permits, have students discuss with one another or the whole class how their understanding developed and/or changed. They may be able to identify the example or class discussion that helped to promote the learning.

Extensions

- If you have access to a symbolic manipulator (CAS calculator or software), you may wish to model how equations can be solved using such technology.

Pacing Suggestion

- If you feel the problems in the explorations will take too long, you might have students work only one problem in each exploration before moving on to the formal lesson.

Dynamic Teaching Tools

- Dynamic Assessment & Progress Monitoring Tool
- Lesson Planning Tool
- Interactive Whiteboard Lesson Library
- Dynamic Classroom with Dynamic Investigations

What Your Students Will Learn

- Solve linear equations in one variable that have variables on both sides.
- Identify equations with infinitely many solutions or no solution.

Laurie's Notes

Exploration

Motivate

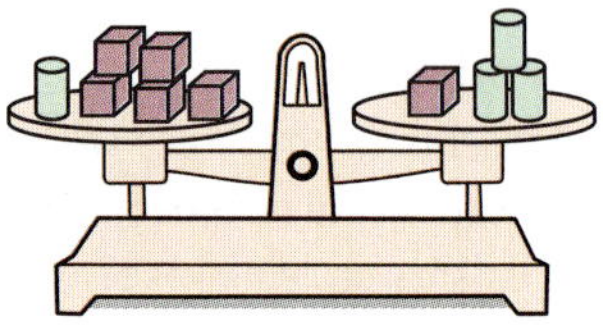

? "What balances with the cylinder? Explain." 2 cubes; Remove one cube and one cylinder from each side. 2 cylinders balance with 4 cubes, so 1 cylinder would balance with 2 cubes.

- The balance problem is equivalent to $x + 5 = 3x + 1$, where a cylinder represents x and the cubes represent the whole numbers. This is an example of an equation with variables on both sides, the type students will solve today.
- If students are familiar with algebra tiles, you can model the problem using the tiles. Replace the cylinders with x-tiles, and replace the cubes with unit tiles.

Exploration 1

- Discuss with students how the perimeter of a polygon is found.

COMMON ERROR Not all of the sides in the second polygon are labeled and students may ignore this issue. Sufficient information is given so that students can determine the missing measures.

- As you circulate, note how students are solving equations with variables on both sides of the equation.

Exploration 2

- Discuss with students the general concept of what it means to measure the attributes of a two-dimensional figure. In other words, what is the difference between the perimeter of a rectangle and the area of a rectangle?

? "What type of unit is used to measure each?" linear units for perimeter and square units for area

- **FYI:** Be sure to make it clear that the directions are saying that the perimeter and the area are not the same, but their values are equal. For example, a square that measures 4 centimeters on each edge has a perimeter of 16 centimeters and an area of 16 square centimeters. The value (16) is the same, but the units of measure are not.

? Before students begin, ask a few questions.

  - "How do you find the area of a triangle?" $\frac{1}{2}bh$ "Area of a semi-circle?" $\frac{1}{2}\pi r^2$
  - "What is a composite figure?" two or more geometric shapes that share a common border
  - "How do you find the perimeter of a composite figure?" Add the perimeters, less the shared sides.
  - "How do you find the area of a composite figure?" Add the areas of each.

- The three problems in this exploration require more simplifying of expressions before the variable term is isolated. Student understanding and success on these problems will inform your instruction during the formal lesson.

Communicate Your Answer

- In answering Question 3, you should hear students refer to collecting the variable terms on one side of the equation and the constant terms on the other side of the equation. You could probe this idea further.

? "Does it matter which you do first—isolating the variable terms or isolating the constant terms?" no

Connecting to Next Step

- If you feel students are comfortable with solving simple equations with variables on both sides of the equation, assign exercises from Exercises 3–10 on page 23.

1.3 Solving Equations with Variables on Both Sides

Essential Question How can you solve an equation that has variables on both sides?

EXPLORATION 1 **Perimeter**

Work with a partner. The two polygons have the same perimeter. Use this information to write and solve an equation involving x. Explain the process you used to find the solution. Then find the perimeter of each polygon.

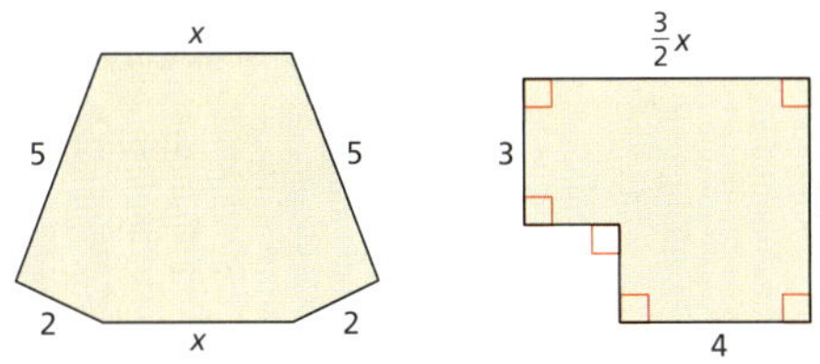

EXPLORATION 2 **Perimeter and Area**

Work with a partner.

- Each figure has the unusual property that the value of its perimeter (in feet) is equal to the value of its area (in square feet). Use this information to write an equation for each figure.
- Solve each equation for x. Explain the process you used to find the solution.
- Find the perimeter and area of each figure.

LOOKING FOR STRUCTURE

To be proficient in math, you need to visualize complex things, such as composite figures, as being made up of simpler, more manageable parts.

a.

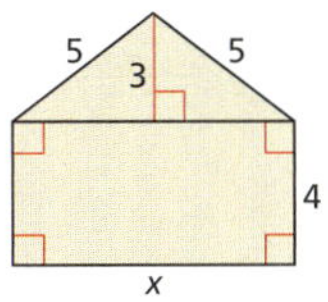

b.

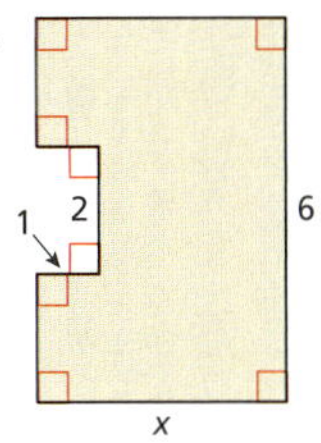

c.

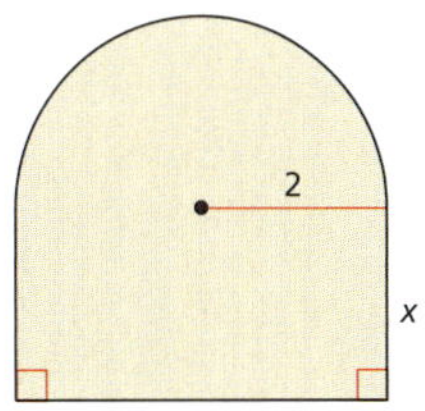

Communicate Your Answer

3. How can you solve an equation that has variables on both sides?
4. Write three equations that have the variable x on both sides. The equations should be different from those you wrote in Explorations 1 and 2. Have your partner solve the equations.

Dynamic Teaching Tools

- Dynamic Assessment & Progress Monitoring Tool
- Lesson Planning Tool
- Interactive Whiteboard Lesson Library
- Dynamic Classroom with Dynamic Investigations

ANSWERS

1. $2x + 14 = 3x + 10$; $x = 4$; Add the side lengths of each figure to get the perimeters and set them equal to each other; $P = 22$
2. **a.** $x + 18 = 4x + \frac{3}{2}x$; $x = 4$; Add the side lengths to get the perimeter. Add the area of the triangle to the area of the rectangle to get the total area. Then set the perimeter equal to the area; $P = 22$ ft, $A = 22$ ft^2

   b. $2x + 14 = 6x - 2$; $x = 4$; Add the side lengths to get the perimeter. Subtract the area of the small rectangle from the area of the large rectangle to get the total area. Then set the perimeter equal to the area; $P = 22$ ft, $A = 22$ ft^2

   c. $2\pi + 2x + 4 = 2\pi + 4x$; $x = 2$; Add the circumference of the semicircle to the remaining 3 side lengths to find the perimeter. Add the area of the semicircle to the area of the rectangle to find the total area. Then set the perimeter equal to the area; $P = 2\pi + 8$ ft, $A = 2\pi + 8$ ft^2
3. Collect the variable terms on one side of the equation and the constant terms on the other side of the equation, then solve.
4. *Sample answer:* $5x - 7 = 2x + 5$, $x = 4$; $-x + 3 = 4x + 13$, $x = -2$; $7x = 6x + 4$, $x = 4$

Differentiated Instruction

Visual

Encourage students to highlight any term that includes a variable. This makes it easy to see which terms they need to combine. Have students draw a vertical line through the equal sign. Explain that they can use inverse operations to collect all the highlighted terms on one side of the line. Then, they can combine those terms to simplify the equation and find the solution.

Extra Example 1

Solve $5p = 24 - 3p$. Check your solution. $p = 3$

Extra Example 2

Solve $2(3n - 4) = 0.5(10 - n)$. $n = 2$

MONITORING PROGRESS ANSWERS

1. $x = -2$
2. $h = \frac{3}{8}$
3. $n = 0$

1.3 Lesson

Core Vocabulary

identity, *p. 21*

Previous

inverse operations

What You Will Learn

- Solve linear equations that have variables on both sides.
- Identify special solutions of linear equations.
- Use linear equations to solve real-life problems.

Solving Equations with Variables on Both Sides

Core Concept

Solving Equations with Variables on Both Sides

To solve an equation with variables on both sides, simplify one or both sides of the equation, if necessary. Then use inverse operations to collect the variable terms on one side, collect the constant terms on the other side, and isolate the variable.

EXAMPLE 1 **Solving an Equation with Variables on Both Sides**

Solve $10 - 4x = -9x$. Check your solution.

SOLUTION

$10 - 4x = -9x$	Write the equation.
$+4x \quad +4x$	Add $4x$ to each side.
$10 = -5x$	Simplify.
$\frac{10}{-5} = \frac{-5x}{-5}$	Divide each side by -5.
$-2 = x$	Simplify.

Check

$10 - 4x = -9x$

$10 - 4(-2) \stackrel{?}{=} -9(-2)$

$18 = 18$ ✓

▶ The solution is $x = -2$.

EXAMPLE 2 **Solving an Equation with Grouping Symbols**

Solve $3(3x - 4) = \frac{1}{4}(32x + 56)$.

SOLUTION

$3(3x - 4) = \frac{1}{4}(32x + 56)$	Write the equation.
$9x - 12 = 8x + 14$	Distributive Property
$+12 \quad +12$	Add 12 to each side.
$9x = 8x + 26$	Simplify.
$-8x \quad -8x$	Subtract $8x$ from each side.
$x = 26$	Simplify.

▶ The solution is $x = 26$.

Monitoring Progress

Help in English and Spanish at *BigIdeasMath.com*

Solve the equation. Check your solution.

1. $-2x = 3x + 10$ **2.** $\frac{1}{2}(6h - 4) = -5h + 1$ **3.** $-\frac{3}{4}(8n + 12) = 3(n - 3)$

Laurie's Notes Teacher Actions

- Write the *Core Concept* and then ask students to *Turn and Talk*. Partner A explains his/her understanding of the *Core Concept*. After one minute, select a partner B to explain the *Core Concept* to the class.
- **Common Misconception:** In Example 1, students want to add $9x$ to each side so that the variable term is on the left side of the equation. Explain that you can solve for the variable on either side of the equation.
- **Monitoring Progress:** Have students check their work and solutions with their partners.

Identifying Special Solutions of Linear Equations

Core Concept

Special Solutions of Linear Equations

Equations do not always have one solution. An equation that is true for all values of the variable is an **identity** and has *infinitely many solutions*. An equation that is not true for any value of the variable has *no solution*.

REASONING

The equation $15x + 6 = 15x$ is not true because the number $15x$ cannot be equal to 6 more than itself.

EXAMPLE 3 Identifying the Number of Solutions

Solve each equation.

a. $3(5x + 2) = 15x$ **b.** $-2(4y + 1) = -8y - 2$

SOLUTION

a. $3(5x + 2) = 15x$ Write the equation.

$15x + 6 = 15x$ Distributive Property

$-15x \quad -15x$ Subtract $15x$ from each side.

$6 = 0$ ✗ Simplify.

▶ The statement $6 = 0$ is never true. So, the equation has no solution.

READING

All real numbers are solutions of an identity.

b. $-2(4y + 1) = -8y - 2$ Write the equation.

$-8y - 2 = -8y - 2$ Distributive Property

$+8y \quad +8y$ Add $8y$ to each side.

$-2 = -2$ Simplify.

▶ The statement $-2 = -2$ is always true. So, the equation is an identity and has infinitely many solutions.

Monitoring Progress

Help in English and Spanish at *BigIdeasMath.com*

Solve the equation.

4. $4(1 - p) = -4p + 4$

5. $6m - m = \frac{5}{6}(6m - 10)$

6. $10k + 7 = -3 - 10k$

7. $3(2a - 2) = 2(3a - 3)$

Concept Summary

Steps for Solving Linear Equations

Here are several steps you can use to solve a linear equation. Depending on the equation, you may not need to use some steps.

Step 1 Use the Distributive Property to remove any grouping symbols.

Step 2 Simplify the expression on each side of the equation.

Step 3 Collect the variable terms on one side of the equation and the constant terms on the other side.

Step 4 Isolate the variable.

Step 5 Check your solution.

STUDY TIP

To check an identity, you can choose several different values of the variable.

Extra Example 3

Solve each equation.

a. $4(3d - 1) = 12d$

The statement $-4 = 0$ is never true. So, the equation has no solution.

b. $21 - 7w = 7(3 - w)$

The statement $21 = 21$ is always true. So, the equation is an identity and has infinitely many solutions.

MONITORING PROGRESS ANSWERS

4. infinitely many solutions

5. no solution

6. $k = -\frac{1}{2}$

7. infinitely many solutions

Laurie's Notes Teacher Actions

- **?** Before writing the *Core Concept*, ask, "Can you think of an equation that has infinitely many solutions? No solution?" If students do not offer answers, provide equations such as $2(x + 5) = 2x + 10$ (infinitely many solutions) and $2x + 4 = 2x + 7$ (no solution).
- **Reason Abstractly and Quantitatively:** Note the *Reasoning* in the minor column. At the point of simplifying the equation in part (a) to $15x + 6 = 15x$, students should reason that there is no solution. Similarly, in part (b), $-8y - 2 = -8y - 2$ is always true, so there are infinitely many solutions.
- **?** "What does it mean to 'isolate the variable'?" Once the variable term is on one side of the equation and the constant terms are on the other, you want the coefficient of the variable to equal 1.

Extra Example 4

Paul bikes for 3 hours to reach his favorite beach. He bikes 2 miles per hour faster on the way back. So, the trip back takes only 2.5 hours. How fast did he bike going to the beach? 10 miles per hour

MONITORING PROGRESS ANSWER

8. 17.5 mi

Solving Real-Life Problems

EXAMPLE 4 **Modeling with Mathematics**

A boat leaves New Orleans and travels upstream on the Mississippi River for 4 hours. The return trip takes only 2.8 hours because the boat travels 3 miles per hour faster downstream due to the current. How far does the boat travel upstream?

SOLUTION

1. **Understand the Problem** You are given the amounts of time the boat travels and the difference in speeds for each direction. You are asked to find the distance the boat travels upstream.
2. **Make a Plan** Use the Distance Formula to write expressions that represent the problem. Because the distance the boat travels in both directions is the same, you can use the expressions to write an equation.
3. **Solve the Problem** Use the formula (distance) = (rate)(time).

Words Distance upstream = Distance downstream

Variable Let x be the speed (in miles per hour) of the boat traveling upstream.

Equation $\frac{x \text{ mi}}{1 \cancel{h}} \cdot 4 \cancel{h} = \frac{(x+3) \text{ mi}}{1 \cancel{h}} \cdot 2.8 \cancel{h}$ (mi = mi) ✓

$4x = 2.8(x + 3)$ Write the equation.

$4x = 2.8x + 8.4$ Distributive Property

$-2.8x \quad -2.8x$ Subtract 2.8x from each side.

$1.2x = 8.4$ Simplify.

$\frac{1.2x}{1.2} = \frac{8.4}{1.2}$ Divide each side by 1.2.

$x = 7$ Simplify.

▶ So, the boat travels 7 miles per hour upstream. To determine how far the boat travels upstream, multiply 7 miles per hour by 4 hours to obtain 28 miles.

4. **Look Back** To check that your solution is reasonable, use the formula for distance. Because the speed upstream is 7 miles per hour, the speed downstream is 7 + 3 = 10 miles per hour. When you substitute each speed into the Distance Formula, you get the same distance for upstream and downstream.

Upstream

$$\text{Distance} = \frac{7 \text{ mi}}{1 \cancel{h}} \cdot 4 \cancel{h} = 28 \text{ mi}$$

Downstream

$$\text{Distance} = \frac{10 \text{ mi}}{1 \cancel{h}} \cdot 2.8 \cancel{h} = 28 \text{ mi}$$

Monitoring Progress

Help in English and Spanish at *BigIdeasMath.com*

8. A boat travels upstream on the Mississippi River for 3.5 hours. The return trip only takes 2.5 hours because the boat travels 2 miles per hour faster downstream due to the current. How far does the boat travel upstream?

Laurie's Notes Teacher Actions

- **Make Sense of Problems and Persevere in Solving Them** and **Model with Mathematics:** Students must first understand that the distances are the same, but the times are different because the rates change.

? "How do you find the distance traveled when you know the rate and time?" multiply; $d = rt$

- **Attend to Precision:** Discuss with students how unit analysis (see *Mathematical Practices* on page 2) is used to verify that the equation written makes sense.

Closure

- **I Used to Think … But Now I Know:** Take time for students to reflect on their current understanding of solving equations.
- **Exit Ticket:** Solve $6 - 2x = 4x - 9$. $x = 2.5$

1.3 Exercises

Dynamic Solutions available at *BigIdeasMath.com*

Vocabulary and Core Concept Check

1. **VOCABULARY** Is the equation $-2(4 - x) = 2x + 8$ an identity? Explain your reasoning.

2. **WRITING** Describe the steps in solving the linear equation $3(3x - 8) = 4x + 6$.

Monitoring Progress and Modeling with Mathematics

In Exercises 3–16, solve the equation. Check your solution. *(See Examples 1 and 2.)*

3. $15 - 2x = 3x$
4. $26 - 4s = 9s$
5. $5p - 9 = 2p + 12$
6. $8g + 10 = 35 + 3g$
7. $5t + 16 = 6 - 5t$
8. $-3r + 10 = 15r - 8$
9. $7 + 3x - 12x = 3x + 1$
10. $w - 2 + 2w = 6 + 5w$
11. $10(g + 5) = 2(g + 9)$
12. $-9(t - 2) = 4(t - 15)$
13. $\frac{2}{3}(3x + 9) = -2(2x + 6)$
14. $2(2t + 4) = \frac{3}{4}(24 - 8t)$
15. $10(2y + 2) - y = 2(8y - 8)$
16. $2(4x + 2) = 4x - 12(x - 1)$

17. **MODELING WITH MATHEMATICS** You and your friend drive toward each other. The equation $50h = 190 - 45h$ represents the number h of hours until you and your friend meet. When will you meet?

18. **MODELING WITH MATHEMATICS** The equation $1.5r + 15 = 2.25r$ represents the number r of movies you must rent to spend the same amount at each movie store. How many movies must you rent to spend the same amount at each movie store?

Membership Fee: $15 | Membership Fee: Free

In Exercises 19–24, solve the equation. Determine whether the equation has *one solution*, *no solution*, or *infinitely many solutions*. *(See Example 3.)*

19. $3t + 4 = 12 + 3t$
20. $6d + 8 = 14 + 3d$
21. $2(h + 1) = 5h - 7$
22. $12y + 6 = 6(2y + 1)$
23. $3(4g + 6) = 2(6g + 9)$
24. $5(1 + 2m) = \frac{1}{2}(8 + 20m)$

ERROR ANALYSIS **In Exercises 25 and 26, describe and correct the error in solving the equation.**

25. ✗

$$5c - 6 = 4 - 3c$$
$$2c - 6 = 4$$
$$2c = 10$$
$$c = 5$$

26. ✗

$$6(2y + 6) = 4(9 + 3y)$$
$$12y + 36 = 36 + 12y$$
$$12y = 12y$$
$$0 = 0$$

The equation has no solution.

27. **MODELING WITH MATHEMATICS** Write and solve an equation to find the month when you would pay the same total amount for each Internet service.

	Installation fee	Price per month
Company A	$60.00	$42.95
Company B	$25.00	$49.95

Assignment Guide and Homework Check

ASSIGNMENT

Basic: 1, 2, 3–15 odd, 19–25 odd, 27–29, 34, 38, 41–44

Average: 1, 2, 4–34 even, 38, 41–44

Advanced: 1, 2, 10, 16, 18–24 even, 25–30, 32–44

HOMEWORK CHECK

Basic: 9, 15, 19, 21, 23

Average: 8, 18, 20, 22, 24

Advanced: 18, 24, 30, 32, 36

ANSWERS

1. no; Solving the equation gives a statement that is never true, not one that is always true.
2. Use the Distributive Property to get $9x - 24 = 4x + 6$. Subtract $4x$ from each side to get $5x - 24 = 6$. Add 24 to each side to get $5x = 30$. Divide each side by 5 to get $x = 6$.
3. $x = 3$
4. $s = 2$
5. $p = 7$
6. $g = 5$
7. $t = -1$
8. $r = 1$
9. $x = \frac{1}{2}$
10. $w = -4$
11. $g = -4$
12. $t = 6$
13. $x = -3$
14. $t = 1$
15. $y = -12$
16. $x = \frac{1}{2}$
17. 2 h
18. 20 movies
19. no solution
20. $d = 2$; one solution
21. $h = 3$; one solution
22. infinitely many solutions
23. infinitely many solutions
24. no solution
25. In the second step, you should add $3c$ to each side; $8c - 6 = 4$, $8c = 10$, $c = \frac{5}{4}$
26. $0 = 0$ is always true, so it is an identity; The equation has infinitely many solutions.
27. $60 + 42.95x = 25 + 49.95x$; 5th month

Dynamic Teaching Tools

- Dynamic Assessment & Progress Monitoring Tool
- Interactive Whiteboard Lesson Library
- Dynamic Classroom with Dynamic Investigations

ANSWERS

28. 50 g

29. $r = -2$

30. $r = 23$

31. $x = 10$; $S = 62.5\pi$ cm^2 or about 196.35 cm^2; $V = 62.5\pi$ cm^3 or about 196.35 cm^3

32. $x = \frac{9}{2}$; $S = 58.32\pi$ ft^2 or about 183.22 ft^2; $V = 58.32\pi$ ft^3 or about 183.22 ft^3

33. 4 sec

34. no; The cheetah would have to maintain top speed for about 21.7 seconds, which is greater than 20 seconds.

35. $a = 5$; Both sides simplify to $10x + 15$.

36. $a = 4$; Both sides simplify to $20x - 8$.

37. 11, 12

38. **a.** 6 yr

b. The left side shows the Spanish enrollment after x years, and the right side shows the French enrollment after x years; Solving the equation will indicate after how many years the enrollments will be equal.

39–44. See Additional Answers.

Mini-Assessment

Solve the equation.

1. $8a + 96 = -24a$ $a = -3$
2. $0.2(10b + 60) = -4(b - 12)$ $b = 6$
3. $3(2m + 4) = 2(3m + 6)$ infinitely many solutions
4. $4y = -2(6 - 2y)$ no solution
5. Mary drove 4 hours to reach her camping spot. Driving home, she increased her speed by 15 miles per hour, reducing her travel time by 1 hour. How fast did she drive going to the camping spot? 45 miles per hour

28. PROBLEM SOLVING One serving of granola provides 4% of the protein you need daily. You must get the remaining 48 grams of protein from other sources. How many grams of protein do you need daily?

USING STRUCTURE **In Exercises 29 and 30, find the value of r.**

29. $8(x + 6) - 10 + r = 3(x + 12) + 5x$

30. $4(x - 3) - r + 2x = 5(3x - 7) - 9x$

MATHEMATICAL CONNECTIONS **In Exercises 31 and 32, the value of the surface area of the cylinder is equal to the value of the volume of the cylinder. Find the value of x. Then find the surface area and volume of the cylinder.**

31.

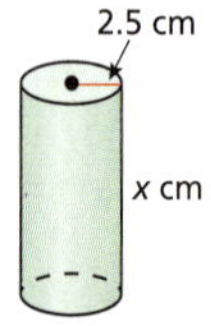

32.

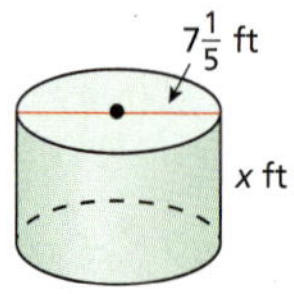

33. MODELING WITH MATHEMATICS A cheetah that is running 90 feet per second is 120 feet behind an antelope that is running 60 feet per second. How long will it take the cheetah to catch up to the antelope? *(See Example 4.)*

34. MAKING AN ARGUMENT A cheetah can run at top speed for only about 20 seconds. If an antelope is too far away for a cheetah to catch it in 20 seconds, the antelope is probably safe. Your friend claims the antelope in Exercise 33 will not be safe if the cheetah starts running 650 feet behind it. Is your friend correct? Explain.

REASONING **In Exercises 35 and 36, for what value of a is the equation an identity? Explain your reasoning.**

35. $a(2x + 3) = 9x + 15 + x$

36. $8x - 8 + 3ax = 5ax - 2a$

37. REASONING Two times the greater of two consecutive integers is 9 less than three times the lesser integer. What are the integers?

38. HOW DO YOU SEE IT? The table and the graph show information about students enrolled in Spanish and French classes at a high school.

	Students enrolled this year	Average rate of change
Spanish	355	9 fewer students each year
French	229	12 more students each year

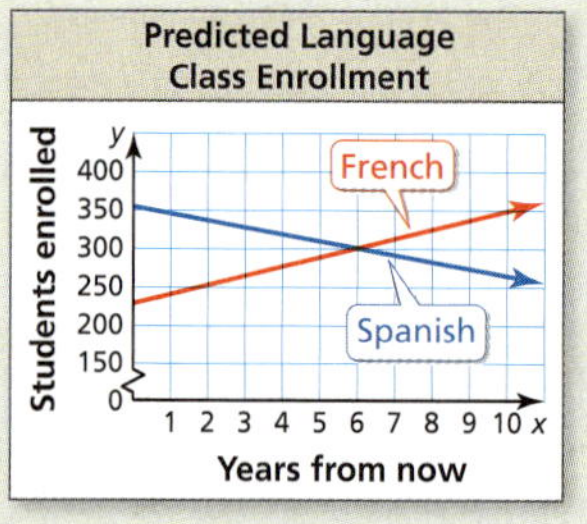

a. Use the graph to determine after how many years there will be equal enrollment in Spanish and French classes.

b. How does the equation $355 - 9x = 229 + 12x$ relate to the table and the graph? How can you use this equation to determine whether your answer in part (a) is reasonable?

39. WRITING EQUATIONS Give an example of a linear equation that has (a) no solution and (b) infinitely many solutions. Justify your answers.

40. THOUGHT PROVOKING Draw a different figure that has the same perimeter as the triangle shown. Explain why your figure has the same perimeter.

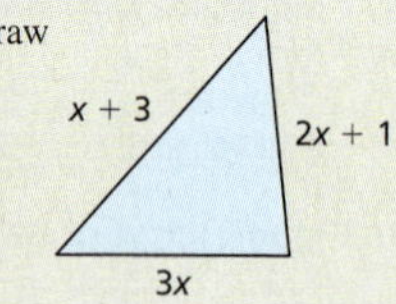

Maintaining Mathematical Proficiency

Reviewing what you learned in previous grades and lessons

Order the values from least to greatest. *(Skills Review Handbook)*

41. $9, |-4|, -4, 5, |2|$

42. $|-32|, 22, -16, -|21|, |-10|$

43. $-18, |-24|, -19, |-18|, |22|$

44. $-|-3|, |0|, -1, |2|, -2$

If students need help...	*If students got it...*
Resources by Chapter • Practice A and Practice B • Puzzle Time	Resources by Chapter • Enrichment and Extension • Cumulative Review
Student Journal • Practice	Start the *next* Section
Differentiating the Lesson Skills Review Handbook	

1.1–1.3 What Did You Learn?

Core Vocabulary

conjecture, *p. 3*
rule, *p. 3*
theorem, *p. 3*
equation, *p. 4*
linear equation in one variable, *p. 4*
solution, *p. 4*
inverse operations, *p. 4*
equivalent equations, *p. 4*
identity, *p. 21*

Core Concepts

Section 1.1

Addition Property of Equality, *p. 4*
Subtraction Property of Equality, *p. 4*
Multiplication Property of Equality, *p. 5*
Division Property of Equality, *p. 5*
Four-Step Approach to Problem Solving, *p. 6*
Common Problem-Solving Strategies, *p. 7*

Section 1.2

Solving Multi-Step Equations, *p. 12*
Unit Analysis, *p. 15*

Section 1.3

Solving Equations with Variables on Both Sides, *p. 20*
Special Solutions of Linear Equations, *p. 21*

Mathematical Practices

1. How did you make sense of the relationships between the quantities in Exercise 46 on page 9?
2. What is the limitation of the tool you used in Exercises 25–28 on page 16?
3. What definition did you use in your reasoning in Exercises 35 and 36 on page 24?

Study Skills

Completing Homework Efficiently

Before doing homework, review the Core Concepts and examples. Use the tutorials at *BigIdeasMath.com* for additional help.

Complete homework as though you are also preparing for a quiz. Memorize different types of problems, vocabulary, rules, and so on.

25

Dynamic Teaching Tools

- Dynamic Assessment & Progress Monitoring Tool
- Interactive Whiteboard Lesson Library
- Dynamic Classroom with Dynamic Investigations

ANSWERS

1. *Sample answer:* Let A represent the area of a single rectangle. Because each rectangle has the same area, and the area of the square is half the area of a rectangle, use the expression $4A + \frac{1}{2}A$ to represent the total area.
2. *Sample answer:* A limitation of a protractor is centering it exactly on each vertex point.
3. the definition of "identity"

ANSWERS

1. $x = -2$; Subtract 9 from each side.
2. $z = 12.4$; Add 3.8 to each side.
3. $r = -5$; Divide each side by -12.
4. $p = 24$; Multiply each side by $\frac{4}{3}$.
5. $m = 8$
6. $v = 5$
7. $w = \frac{1}{5}$
8. $a = -0.6$
9. $k = -9$
10. $x = 14$
11. $c = 0$
12. $n = -9$
13. $q = \frac{5}{6}$
14. $y = -2$
15. no solution
16. infinitely many solutions
17. $2 = \frac{s}{5}$; 10 sec
18. $12 + 2x = 15$; $1\frac{1}{2}$ ft
19. **a.** 3 h; The cost at studio A is $10 + 8h$ and the cost at Studio B is $16 + 6h$. To find when the costs are the same, set these two expressions equal and solve for the time.

    b. The costs will never be the same; *Sample answer:* The cost for studio B changes to $16 + 8h$, and the new equation has no solution.

1.1–1.3 Quiz

Solve the equation. Justify each step. Check your solution. *(Section 1.1)*

1. $x + 9 = 7$
2. $8.6 = z - 3.8$
3. $60 = -12r$
4. $\frac{3}{4}p = 18$

Solve the equation. Check your solution. *(Section 1.2)*

5. $2m - 3 = 13$
6. $5 = 10 - v$
7. $5 = 7w + 8w + 2$
8. $-21a + 28a - 6 = -10.2$
9. $2k - 3(2k - 3) = 45$
10. $68 = \frac{1}{5}(20x + 50) + 2$

Solve the equation. *(Section 1.3)*

11. $3c + 1 = c + 1$
12. $-8 - 5n = 64 + 3n$
13. $2(8q - 5) = 4q$
14. $9(y - 4) - 7y = 5(3y - 2)$
15. $4(g + 8) = 7 + 4g$
16. $-4(-5h - 4) = 2(10h + 8)$

17. To estimate how many miles you are from a thunderstorm, count the seconds between when you see lightning and when you hear thunder. Then divide by 5. Write and solve an equation to determine how many seconds you would count for a thunderstorm that is 2 miles away. *(Section 1.1)*

18. You want to hang three equally-sized travel posters on a wall so that the posters on the ends are each 3 feet from the end of the wall. You want the spacing between posters to be equal. Write and solve an equation to determine how much space you should leave between the posters. *(Section 1.2)*

19. You want to paint a piece of pottery at an art studio. The total cost is the cost of the piece plus an hourly studio fee. There are two studios to choose from. *(Section 1.3)*

    a. After how many hours of painting are the total costs the same at both studios? Justify your answer.

    b. Studio B increases the hourly studio fee by \$2. How does this affect your answer in part (a)? Explain.

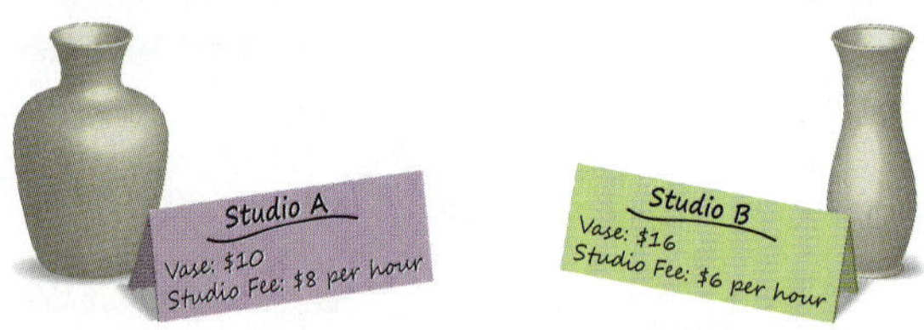

Laurie's Notes

Overview of Section 1.4

Introduction

- When absolute value is first introduced in middle school, students think of the absolute value of a number as being the distance a number is from 0. So $|4| = 4$ and $|-4| = 4$. This is the geometric definition of absolute value.
- In this lesson, students need to be secure with the algebraic definition of absolute value: $|x| = x$ when $x \geq 0$ or $|x| = -x$ when $x < 0$.
- I like to write an equation such as $|x| = 6$ on the board and then place my hand over the x and ask, "What could be hidden under my hand that would make this equation true?" Students quickly say 6 or -6. This same technique is then used for the equation $|x - 4| = 6$. I place my hand over the expression $x - 4$ and say, "Whatever is under my hand must be equal to 6 or -6." This leads to the two linear equations, $x - 4 = 6$ and $x - 4 = -6$.
- Placing my hand over the expression inside the absolute value symbols is a way of drawing attention to the input of the absolute value function. This is not language I use at this time, but it is an approach that will re-occur in high school mathematics as students study quadratics $[(x - 4)^2]$, logarithms $[\log(x - 4)]$, and trigonometry $[\sin(x - 4)]$. In each case, there is a transformation of the parent function 4 units to the right.
- This lesson provides further practice in solving multi-step equations and equations with variables on both sides.

Resources

- I have a long number line (8 feet) above one of my whiteboards. The number line is a helpful visual reference for students in this lesson.

Common Misconceptions

- Students are taught that the absolute value of a number is always positive, or 0. This is not the same as asking the question, "What value or values of x will make the equation $|x| = 4$ true?"
- Students may believe that absolute value equations must have two solutions, one positive and one negative. Address this possible misconception when doing Examples 2 and 3 in the lesson.

Formative Assessment Tips

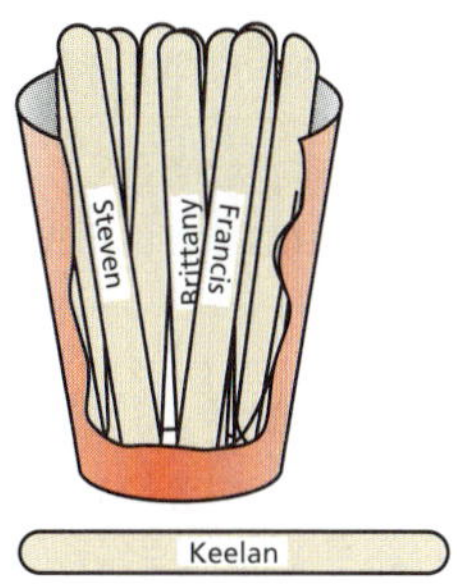

- **Popsicle Sticks:** This technique ensures that *any* student can be called on during questioning time in class. Write the name of each student on a Popsicle stick. Place the sticks in a cup (or can). When questions are posed in class during *No Hands Questioning**, each student should think and be prepared to answer. If you only call on students who raise their hands, students can then opt out of being engaged. All students think they have an equal chance of being called on when a stick is pulled randomly, so they engage more in the lesson.
- If there are certain students that you want to hear from, the cup could have an inner cylinder where *select* sticks are placed. It will appear that the process is still random, and the voices you have not heard much from will still have the opportunity to prepare and formulate their answer.

* See Section 2.4 for a description of *No Hands Questioning*.

Pacing Suggestion

- The problems students will work in the explorations will help them develop conceptual understanding of solving absolute value equations. This can be challenging for students. When students have finished the explorations, begin the formal lesson.

Dynamic Teaching Tools

- Dynamic Assessment & Progress Monitoring Tool
- Lesson Planning Tool
- Interactive Whiteboard Lesson Library
- Dynamic Classroom with Dynamic Investigations

What Your Students Will Learn

- Solve absolute value equations including equations involving two absolute values.
- Identify extraneous solutions of absolute value equations.

Laurie's Notes

Exploration

Motivate

? "Can you think of an equation that has more than one answer?" Students may suggest something like $x + 4 = 4 + x$, which has infinitely many solutions.

? "What does $|5|$ mean?" absolute value of 5 "What does $|-5|$ mean?" absolute value of -5

? "What is the solution of $|x| = 5$?" 5 or -5

For Your Information

- This exploration page models the "Rule of 3." The *Rule of 3* is the concept that ideas presented in sets of 3 are more powerful and more likely to be remembered.
- The same absolute value equation is solved using three different approaches: algebraic (symbolic), graphical, and numerical.

Exploration 1

- Students will very quickly see that $x = 1$ is a solution. Many students will then conclude that $x = -1$ is also a solution. Students incorrectly reason that if you take the absolute value of -1 and then add 2, the solution is 3. Point out to students that the addition occurs first, and then the absolute value of the sum is taken. You could also suggest that students substitute $x = -1$ and evaluate correctly to see that their reasoning is incorrect.
- Students may be uncertain of how to describe the values of $x + 2$ that make the equation true. Relating it back to the *Motivate*, say, "The quantity inside the absolute value symbols is either equal to 3 or -3." That leads to the equations $x + 2 = 3$ and $x + 2 = -3$.

? "Solving each of these linear equations, what solutions do you have?" 1 and -5

Exploration 2

- **Connection:** The first step, finding the point where $x + 2 = 0$, is a translation. When solving $|x| = 3$, you are finding the two values that are 3 units from 0 thus making 0 the midpoint. When solving $|x + 2| = 3$, you are still finding two values that are 3 units from a midpoint, but the midpoint has been translated 2 units to the left, where $x + 2 = 0$. Making the connection to the translation ($|x| = 3 \rightarrow |x + 2| = 3$) is a *Big Idea* that students will see throughout their study of algebra.

? "Using the number line, what solutions do you have?" -5 and 1

Exploration 3

- Students can use the table feature on a graphing calculator.

X	Y1	
−6	4	
−5	3	
−4	2	
−3	1	
−2	0	
−1	1	
0	2	
1	3	
2	4	

X=−5

? "Using the spreadsheet, what solutions do you have?" -5 and 1

Communicate Your Answer

- After students have shared their thoughts about the different methods for solving absolute value equations, use *Popsicle Sticks* to hear a sampling of answers.

Connecting to Next Step

- If you feel students are comfortable with solving simple absolute value equations, assign exercises from Exercises 3–18 on page 32.

1.4 Solving Absolute Value Equations

Essential Question How can you solve an absolute value equation?

EXPLORATION 1 Solving an Absolute Value Equation Algebraically

Work with a partner. Consider the absolute value equation

$|x + 2| = 3.$

a. Describe the values of $x + 2$ that make the equation true. Use your description to write two linear equations that represent the solutions of the absolute value equation.

b. Use the linear equations you wrote in part (a) to find the solutions of the absolute value equation.

c. How can you use linear equations to solve an absolute value equation?

MAKING SENSE OF PROBLEMS

To be proficient in math, you need to explain to yourself the meaning of a problem and look for entry points to its solution.

EXPLORATION 2 Solving an Absolute Value Equation Graphically

Work with a partner. Consider the absolute value equation

$|x + 2| = 3.$

a. On a real number line, locate the point for which $x + 2 = 0$.

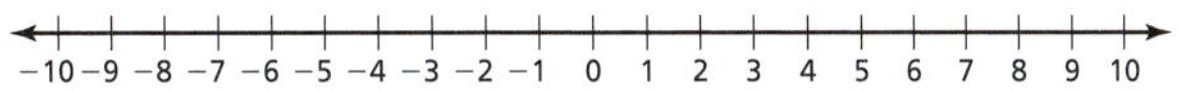

b. Locate the points that are 3 units from the point you found in part (a). What do you notice about these points?

c. How can you use a number line to solve an absolute value equation?

EXPLORATION 3 Solving an Absolute Value Equation Numerically

Work with a partner. Consider the absolute value equation

$|x + 2| = 3.$

a. Use a spreadsheet, as shown, to solve the absolute value equation.

b. Compare the solutions you found using the spreadsheet with those you found in Explorations 1 and 2. What do you notice?

c. How can you use a spreadsheet to solve an absolute value equation?

	A	B
1	x	|x + 2|
2	-6	4 (abs(A2 + 2))
3	-5	
4	-4	
5	-3	
6	-2	
7	-1	
8	0	
9	1	
10	2	

Communicate Your Answer

4. How can you solve an absolute value equation?

5. What do you like or dislike about the algebraic, graphical, and numerical methods for solving an absolute value equation? Give reasons for your answers.

Dynamic Teaching Tools

- Dynamic Assessment & Progress Monitoring Tool
- Lesson Planning Tool
- Interactive Whiteboard Lesson Library
- Dynamic Classroom with Dynamic Investigations

ANSWERS

1. **a.** The equation is true if the expression inside the absolute value symbol is 3 or −3; $x + 2 = 3; x + 2 = -3$

   b. $x = 1, x = -5$

   c. Set the expression inside the absolute value symbol equal to both the positive and negative of the constant value, then solve both equations.

2. **a.** 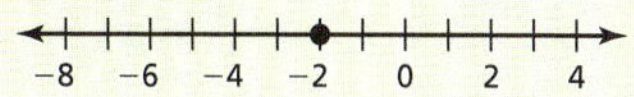

   $x = -2$

   b. 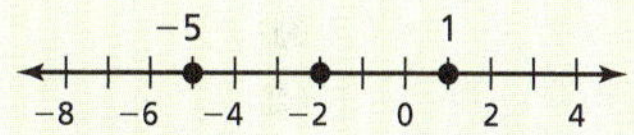

   $x = 1$ and $x = -5$; They are the solutions of the equation.

   c. Set the expression inside the absolute value symbol equal to 0 and solve. Plot the solution on a number line. Then plot the points that are the constant amount of units from that point. These last 2 points are the solutions to the original equation.

3. **a.** $x = 1, x = -5$

   b. They are the same.

   c. Have the spreadsheet calculate the value of the absolute value expression for many values of x, and find the ones that give the expected solution.

4. solving algebraic equations, graphically on a number line, or by trial and error with a spreadsheet

5. *Sample answer:* The algebraic method is favorable because it is the quickest method. The graphical method is also favorable because it helps to visualize absolute value. The numerical method is not favorable because setting up the spreadsheet is time consuming.

English Language Learners

Illustrate

Remind students that the absolute value of a number is its distance from zero on the number line. Because it represents a distance, absolute value is always nonnegative. Illustrate how to find the absolute value of a given number on a number line. Then, given an absolute value of a number other than zero, show how to find the pair of numbers with that absolute value.

Extra Example 1

Solve each equation. Graph the solutions, if possible.

a. $|x + 6| = 11$ $x = 5$ and $x = -17$

b. $|2x - 1| = -4$ no solution

MONITORING PROGRESS ANSWERS

1. $x = -10, x = 10;$

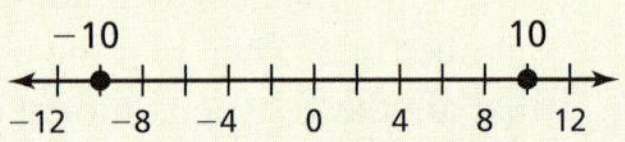

2. $x = -3, x = 5;$

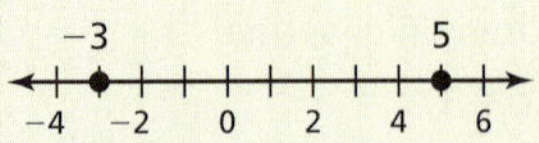

3. no solution

1.4 Lesson

Core Vocabulary

absolute value equation, *p. 28*
extraneous solution, *p. 31*

Previous
absolute value
opposite

What You Will Learn

- Solve absolute value equations.
- Solve equations involving two absolute values.
- Identify special solutions of absolute value equations.

Solving Absolute Value Equations

An **absolute value equation** is an equation that contains an absolute value expression. You can solve these types of equations by solving two related linear equations.

Core Concept

Properties of Absolute Value

Let a and b be real numbers. Then the following properties are true.

1. $|a| \geq 0$ **2.** $|-a| = |a|$

3. $|ab| = |a||b|$ **4.** $\left|\frac{a}{b}\right| = \frac{|a|}{|b|}, b \neq 0$

Solving Absolute Value Equations

To solve $|ax + b| = c$ when $c \geq 0$, solve the related linear equations

$$ax + b = c \quad or \quad ax + b = -c.$$

When $c < 0$, the absolute value equation $|ax + b| = c$ has no solution because absolute value always indicates a number that is not negative.

EXAMPLE 1 Solving Absolute Value Equations

Solve each equation. Graph the solutions, if possible.

a. $|x - 4| = 6$ **b.** $|3x + 1| = -5$

SOLUTION

a. Write the two related linear equations for $|x - 4| = 6$. Then solve.

$x - 4 = 6$ *or* $x - 4 = -6$ Write related linear equations.

$x = 10$ $x = -2$ Add 4 to each side.

▶ The solutions are $x = 10$ and $x = -2$.

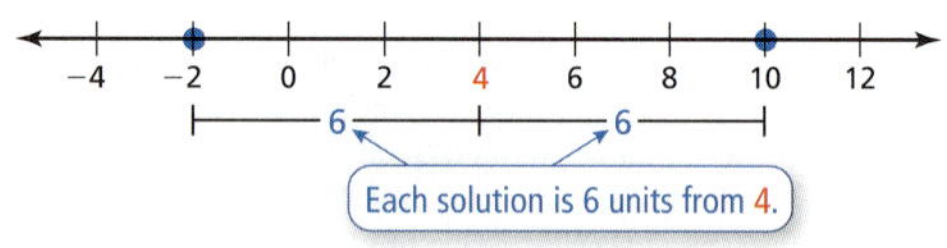

Property of Absolute Value → **b.** The absolute value of an expression must be greater than or equal to 0. The expression $|3x + 1|$ cannot equal -5.

▶ So, the equation has no solution.

Monitoring Progress

Help in English and Spanish at *BigIdeasMath.com*

Solve the equation. Graph the solutions, if possible.

1. $|x| = 10$ **2.** $|x - 1| = 4$ **3.** $|3 + x| = -3$

Laurie's Notes Teacher Actions

- Write the *Core Concept* and discuss each property. Substitute values for a and b to help students make sense of each property. Be sure to connect the algebraic definition of *absolute value* to how these equations are solved.

? "What are the two related linear equations for $|x - 4| = 6$?" $x - 4 = 6$ or $x - 4 = -6$

- **Connection:** Connect the graphical representation of the solution to transformations. The solutions to the equation $|x| = 6$ have been translated 4 units to the right.

? Check for Understanding: "How are the solutions to $|x| = 10$ and $|x + 2| = 10$ alike?" The difference between the solutions is 20 units (−10 and 10; −12 and 8).

EXAMPLE 2 Solving an Absolute Value Equation

Solve $|3x + 9| - 10 = -4$.

SOLUTION

First isolate the absolute value expression on one side of the equation.

$$|3x + 9| - 10 = -4 \quad \text{Write the equation.}$$
$$|3x + 9| = 6 \quad \text{Add 10 to each side.}$$

Now write two related linear equations for $|3x + 9| = 6$. Then solve.

$$3x + 9 = 6 \quad \textit{or} \quad 3x + 9 = -6 \quad \text{Write related linear equations.}$$
$$3x = -3 \qquad 3x = -15 \quad \text{Subtract 9 from each side.}$$
$$x = -1 \qquad x = -5 \quad \text{Divide each side by 3.}$$

▶ The solutions are $x = -1$ and $x = -5$.

ANOTHER WAY

Using the product property of absolute value, $|ab| = |a|\,|b|$, you could rewrite the equation as

$$3|x + 3| - 10 = -4$$

and then solve.

EXAMPLE 3 Writing an Absolute Value Equation

In a cheerleading competition, the minimum length of a routine is 4 minutes. The maximum length of a routine is 5 minutes. Write an absolute value equation that represents the minimum and maximum lengths.

SOLUTION

1. **Understand the Problem** You know the minimum and maximum lengths. You are asked to write an absolute value equation that represents these lengths.
2. **Make a Plan** Consider the minimum and maximum lengths as solutions to an absolute value equation. Use a number line to find the halfway point between the solutions. Then use the halfway point and the distance to each solution to write an absolute value equation.
3. **Solve the Problem**

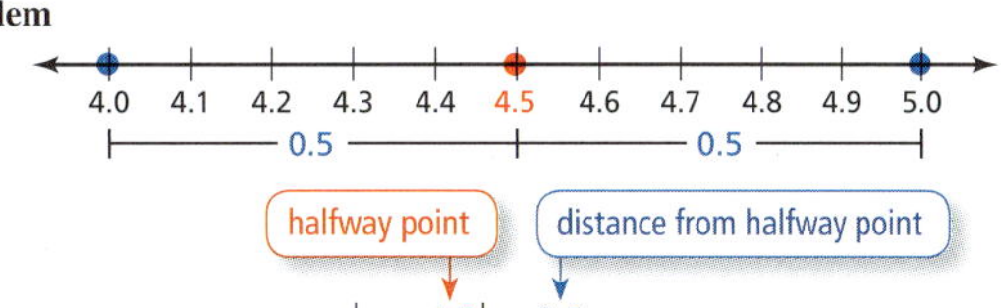

$$|x - 4.5| = 0.5$$

▶ The equation is $|x - 4.5| = 0.5$.

4. **Look Back** To check that your equation is reasonable, substitute the minimum and maximum lengths into the equation and simplify.

Minimum $|4 - 4.5| = 0.5$ ✓ **Maximum** $|5 - 4.5| = 0.5$ ✓

REASONING

Mathematically proficient students have the ability to decontextualize problem situations.

Monitoring Progress

Help in English and Spanish at *BigIdeasMath.com*

Solve the equation. Check your solutions.

4. $|x - 2| + 5 = 9$
5. $4|2x + 7| = 16$
6. $-2|5x - 1| - 3 = -11$
7. For a poetry contest, the minimum length of a poem is 16 lines. The maximum length is 32 lines. Write an absolute value equation that represents the minimum and maximum lengths.

Differentiated Instruction

Auditory

Remind students that the absolute value of an expression cannot be negative.

Extra Example 2

Solve $|3x - 6| - 9 = -3$. $x = 0$ and $x = 4$

Extra Example 3

For a writing assignment, the essay should be at least 375 words but no more than 425 words. Write an absolute value equation that represents the minimum and maximum number of words.
$|w - 400| = 25$

MONITORING PROGRESS ANSWERS

4. $x = -2, x = 6$
5. $x = -5.5, x = -1.5$
6. $x = -0.6, x = 1$
7. $|x - 24| = 8$

Laurie's Notes Teacher Actions

- **Look For and Make Use of Structure:** In Example 2, students should recognize that isolating the absolute value term is the same as isolating the variable term in the equation $2x - 10 = -4$.
- **Popsicle Sticks:** Once the absolute value term has been isolated, write the two related linear equations. Give students time to finish the example with their partners. Draw a Popsicle stick to ask for a student's solutions.
- **Reason Abstractly and Quantitatively** and **Model with Mathematics:** Pose the problem in Example 3 and give students time with their partners to consider a method of solution.
- **Monitoring Progress:** As students begin to check with their neighbors, ask volunteers to put solutions on the board.

Extra Example 4

a. Solve $|2m + 5| = |m|$.

$m = -5$ and $m = -\frac{5}{3}$

b. Solve $|5k + 12| = 2|4k - 1|$.

$k = \frac{14}{3}$ and $k = -\frac{10}{13}$

MONITORING PROGRESS ANSWERS

8. $x = -3, x = 7$

9. $x = 1.4, x = 17$

Solving Equations with Two Absolute Values

If the absolute values of two algebraic expressions are equal, then they must either be equal to each other or be opposites of each other.

Core Concept

Solving Equations with Two Absolute Values

To solve $|ax + b| = |cx + d|$, solve the related linear equations

$$ax + b = cx + d \quad or \quad ax + b = -(cx + d).$$

EXAMPLE 4 Solving Equations with Two Absolute Values

Solve (a) $|3x - 4| = |x|$ and (b) $|4x - 10| = 2|3x + 1|$.

SOLUTION

a. Write the two related linear equations for $|3x - 4| = |x|$. Then solve.

$3x - 4 = x$	*or* $3x - 4 = -x$
$\underline{-x \quad -x}$	$\underline{+x \quad +x}$
$2x - 4 = 0$	$4x - 4 = 0$
$\underline{+4 \quad +4}$	$\underline{+4 \quad +4}$
$2x = 4$	$4x = 4$
$\frac{2x}{2} = \frac{4}{2}$	$\frac{4x}{4} = \frac{4}{4}$
$x = 2$	$x = 1$

▶ The solutions are $x = 2$ and $x = 1$.

Check

$|3x - 4| = |x|$

$|3(2) - 4| \stackrel{?}{=} |2|$

$|2| \stackrel{?}{=} |2|$

$2 = 2$ ✓

$|3x - 4| = |x|$

$|3(1) - 4| \stackrel{?}{=} |1|$

$|-1| \stackrel{?}{=} |1|$

$1 = 1$ ✓

b. Write the two related linear equations for $|4x - 10| = 2|3x + 1|$. Then solve.

$4x - 10 = 2(3x + 1)$	*or* $4x - 10 = 2[-(3x + 1)]$
$4x - 10 = 6x + 2$	$4x - 10 = 2(-3x - 1)$
$\underline{-6x \quad -6x}$	$4x - 10 = -6x - 2$
$-2x - 10 = 2$	$\underline{+6x \quad +6x}$
$\underline{+10 \quad +10}$	$10x - 10 = -2$
$-2x = 12$	$\underline{+10 \quad +10}$
$\frac{-2x}{-2} = \frac{12}{-2}$	$10x = 8$
$x = -6$	$\frac{10x}{10} = \frac{8}{10}$
	$x = 0.8$

▶ The solutions are $x = -6$ and $x = 0.8$.

Monitoring Progress

Help in English and Spanish at *BigIdeasMath.com*

Solve the equation. Check your solutions.

8. $|x + 8| = |2x + 1|$ **9.** $3|x - 4| = |2x + 5|$

Laurie's Notes Teacher Actions

? "Consider the equation $|a| = |b|$. If $a = 4$, what could b equal?" 4 or −4 "If $a = -8$, what could b equal?" 8 or −8

- Write the *Core Concept* and discuss the need for parentheses around the expression $(cx + d)$.
- Students are solving equations with variables on both sides. As you circulate, you may notice students with the same solutions but the steps in different orders. As time permits, have students share different methods of solution.
- Model checking the solutions.

Identifying Special Solutions

When you solve an absolute value equation, it is possible for a solution to be *extraneous*. An **extraneous solution** is an apparent solution that must be rejected because it does not satisfy the original equation.

EXAMPLE 5 Identifying Extraneous Solutions

Solve $|2x + 12| = 4x$. Check your solutions.

SOLUTION

Write the two related linear equations for $|2x + 12| = 4x$. Then solve.

$2x + 12 = 4x$	*or*	$2x + 12 = -4x$	Write related linear equations.
$12 = 2x$		$12 = -6x$	Subtract $2x$ from each side.
$6 = x$		$-2 = x$	Solve for x.

Check the apparent solutions to see if either is extraneous.

▶ The solution is $x = 6$. Reject $x = -2$ because it is extraneous.

Check

$|2x + 12| = 4x$

$|2(6) + 12| \stackrel{?}{=} 4(6)$

$|24| \stackrel{?}{=} 24$

$24 = 24$ ✓

$|2x + 12| = 4x$

$|2(-2) + 12| \stackrel{?}{=} 4(-2)$

$|8| \stackrel{?}{=} -8$

$8 \neq -8$ ✗

When solving equations of the form $|ax + b| = |cx + d|$, it is possible that one of the related linear equations will not have a solution.

EXAMPLE 6 Solving an Equation with Two Absolute Values

Solve $|x + 5| = |x + 11|$.

SOLUTION

By equating the expression $x + 5$ and the opposite of $x + 11$, you obtain

$x + 5 = -(x + 11)$	Write related linear equation.
$x + 5 = -x - 11$	Distributive Property
$2x + 5 = -11$	Add x to each side.
$2x = -16$	Subtract 5 from each side.
$x = -8.$	Divide each side by 2.

However, by equating the expressions $x + 5$ and $x + 11$, you obtain

$x + 5 = x + 11$	Write related linear equation.
$x = x + 6$	Subtract 5 from each side.
$0 = 6$ ✗	Subtract x from each side.

which is a false statement. So, the original equation has only one solution.

▶ The solution is $x = -8$.

REMEMBER

Always check your solutions in the original equation to make sure they are not extraneous.

Monitoring Progress

Help in English and Spanish at *BigIdeasMath.com*

Solve the equation. Check your solutions.

10. $|x + 6| = 2x$

11. $|3x - 2| = x$

12. $|2 + x| = |x - 8|$

13. $|5x - 2| = |5x + 4|$

Extra Example 5

Solve $|3n + 18| = 6n$. Check your solutions. $n = 6$; Reject $n = -2$ because it is extraneous.

Extra Example 6

Solve $|2r - 4| = |2r + 10|$. $r = -1.5$

MONITORING PROGRESS ANSWERS

10. $x = 6$

11. $x = 0.5, x = 1$

12. $x = 3$

13. $x = -0.2$

Laurie's Notes Teacher Actions

- ? "What does the word 'extraneous' mean?" extra; not needed
- Define *extraneous solution*.
- Students who have been successful in solving absolute value equations will ask why they did not have extraneous solutions before. It is when there are variables on both sides of the absolute value equation that it is possible to have an extraneous solution.

Closure

- Describe in words the first step in solving each equation.
  - **a.** $|x - 4| = 8$ *Sample answer:* Write the related linear equations, in this case, $x - 4 = 8$ or $x - 4 = -8$.
  - **b.** $|x + 5| = |x - 3|$ *Sample answer:* Write the related linear equations, in this case, $x + 5 = x - 3$ or $x + 5 = -(x - 3)$.

Assignment Guide and Homework Check

ASSIGNMENT

Basic: 1, 2, 3–45 odd, 46–47, 49, 60, 62–67

Average: 1, 2, 4–44 even, 45–50, 60, 62–67

Advanced: 1, 2, 19–32, 36–44 even, 45–56, 58–60, 62–67

HOMEWORK CHECK

Basic: 19, 21, 23, 25, 37

Average: 20, 22, 24, 26, 38

Advanced: 21, 24, 26, 30, 40

1.4 Exercises

Dynamic Solutions available at *BigIdeasMath.com*

Vocabulary and Core Concept Check

1. **VOCABULARY** What is an extraneous solution?

2. **WRITING** Without calculating, how do you know that the equation $|4x - 7| = -1$ has no solution?

Monitoring Progress and Modeling with Mathematics

In Exercises 3–10, simplify the expression.

3. $|-9|$
4. $-|15|$
5. $|14| - |-14|$
6. $|-3| + |3|$
7. $-|-5 \cdot (-7)|$
8. $|-0.8 \cdot 10|$
9. $\left|\frac{27}{-3}\right|$
10. $\left|-\frac{-12}{4}\right|$

In Exercises 11–24, solve the equation. Graph the solution(s), if possible. *(See Examples 1 and 2.)*

11. $|w| = 6$
12. $|r| = -2$
13. $|y| = -18$
14. $|x| = 13$
15. $|m + 3| = 7$
16. $|q - 8| = 14$
17. $|-3d| = 15$
18. $\left|\frac{t}{2}\right| = 6$
19. $|4b - 5| = 19$
20. $|x - 1| + 5 = 2$
21. $-4|8 - 5n| = 13$
22. $-3\left|1 - \frac{2}{3}v\right| = -9$
23. $3 = -2\left|\frac{1}{4}s - 5\right| + 3$
24. $9|4p + 2| + 8 = 35$

25. **WRITING EQUATIONS** The minimum distance from Earth to the Sun is 91.4 million miles. The maximum distance is 94.5 million miles. *(See Example 3.)*
    a. Represent these two distances on a number line.
    b. Write an absolute value equation that represents the minimum and maximum distances.

26. **WRITING EQUATIONS** The shoulder heights of the shortest and tallest miniature poodles are shown.

    a. Represent these two heights on a number line.
    b. Write an absolute value equation that represents these heights.

USING STRUCTURE In Exercises 27–30, match the absolute value equation with its graph without solving the equation.

27. $|x + 2| = 4$
28. $|x - 4| = 2$
29. $|x - 2| = 4$
30. $|x + 4| = 2$

A.
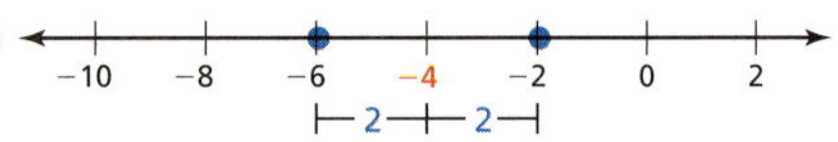

B.
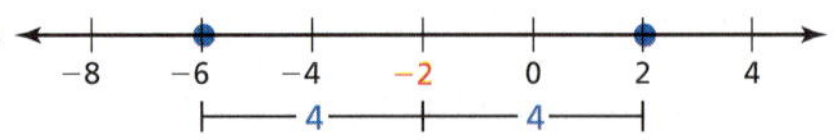

C.
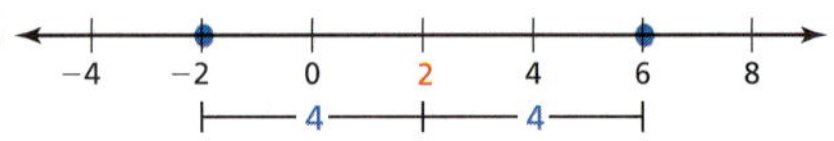

D.
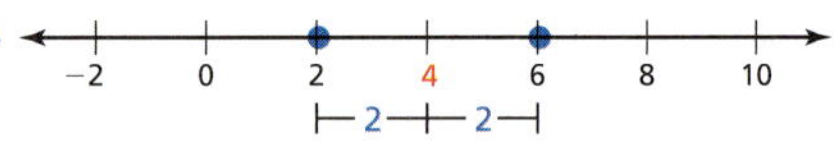

ANSWERS

1. an apparent solution that must be rejected because it does not satisfy the original equation
2. An absolute value cannot be negative.
3. 9
4. −15
5. 0
6. 6
7. −35
8. 8
9. 9
10. 3
11. $w = -6, w = 6;$

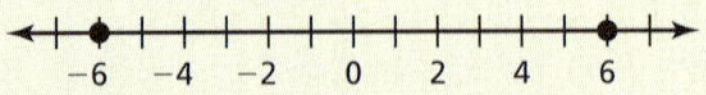

12. no solution
13. no solution
14. $x = -13, x = 13;$

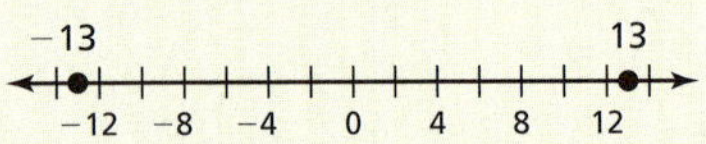

15. $m = -10, m = 4;$

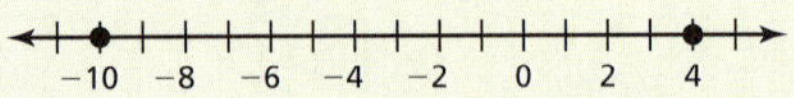

16. $q = -6, q = 22;$

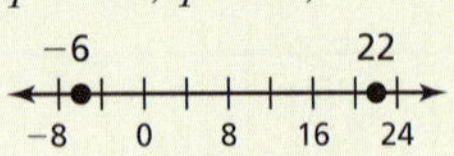

17. $d = -5, d = 5;$

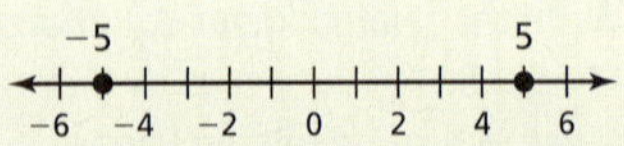

18. $t = -12, t = 12;$

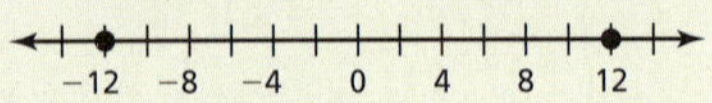

19. $b = -3.5, b = 6;$

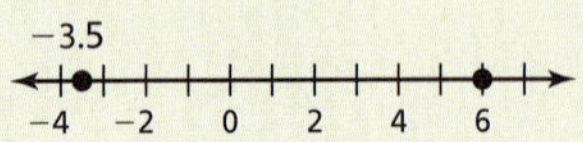

20. no solution
21. no solution
22. $v = -3, v = 6;$

23. $s = 20;$

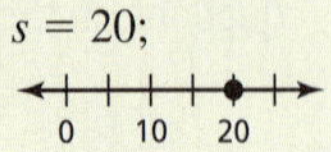

24. $p = -\frac{5}{4}, p = \frac{1}{4};$

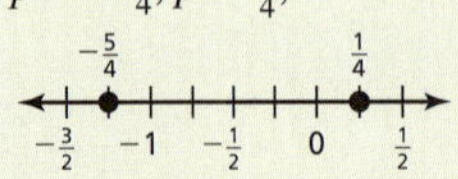

25. a.

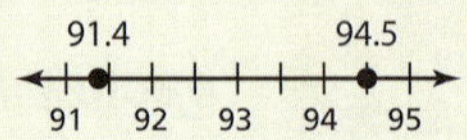

    b. $|d - 92{,}950{,}000| = 1{,}550{,}000$

26. a.

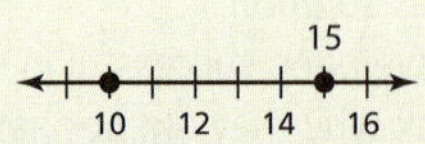

    b. $|h - 12.5| = 2.5$

27. B
28. D
29. C
30. A

In Exercises 31–34, write an absolute value equation that has the given solutions.

31. $x = 8$ and $x = 18$ **32.** $x = -6$ and $x = 10$

33. $x = 2$ and $x = 9$ **34.** $x = -10$ and $x = -5$

In Exercises 35–44, solve the equation. Check your solutions. *(See Examples 4, 5, and 6.)*

35. $|4n - 15| = |n|$ **36.** $|2c + 8| = |10c|$

37. $|2b - 9| = |b - 6|$ **38.** $|3k - 2| = 2|k + 2|$

39. $4|p - 3| = |2p + 8|$ **40.** $2|4w - 1| = 3|4w + 2|$

41. $|3h + 1| = 7h$ **42.** $|6a - 5| = 4a$

43. $|f - 6| = |f + 8|$ **44.** $|3x - 4| = |3x - 5|$

45. MODELING WITH MATHEMATICS Starting from 300 feet away, a car drives toward you. It then passes by you at a speed of 48 feet per second. The distance d (in feet) of the car from you after t seconds is given by the equation $d = |300 - 48t|$. At what times is the car 60 feet from you?

46. MAKING AN ARGUMENT Your friend says that the absolute value equation $|3x + 8| - 9 = -5$ has no solution because the constant on the right side of the equation is negative. Is your friend correct? Explain.

47. MODELING WITH MATHEMATICS You randomly survey students about year-round school. The results are shown in the graph.

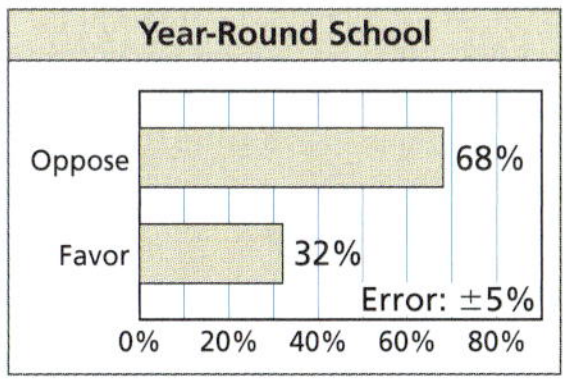

The error given in the graph means that the actual percent could be 5% more or 5% less than the percent reported by the survey.

a. Write and solve an absolute value equation to find the least and greatest percents of students who could be in favor of year-round school.

b. A classmate claims that $\frac{1}{3}$ of the student body is actually in favor of year-round school. Does this conflict with the survey data? Explain.

48. MODELING WITH MATHEMATICS The recommended weight of a soccer ball is 430 grams. The actual weight is allowed to vary by up to 20 grams.

a. Write and solve an absolute value equation to find the minimum and maximum acceptable soccer ball weights.

b. A soccer ball weighs 423 grams. Due to wear and tear, the weight of the ball decreases by 16 grams. Is the weight acceptable? Explain.

ERROR ANALYSIS In Exercises 49 and 50, describe and correct the error in solving the equation.

49.

$$|2x - 1| = -9$$

$2x - 1 = -9$ *or* $2x - 1 = -(-9)$

$2x = -8$ $\quad 2x = 10$

$x = -4$ $\quad x = 5$

The solutions are $x = -4$ and $x = 5$.

50.

$$|5x + 8| = x$$

$5x + 8 = x$ *or* $5x + 8 = -x$

$4x + 8 = 0$ $\quad 6x + 8 = 0$

$4x = -8$ $\quad 6x = -8$

$x = -2$ $\quad x = -\frac{4}{3}$

The solutions are $x = -2$ and $x = -\frac{4}{3}$.

51. ANALYZING EQUATIONS Without solving completely, place each equation into one of the three categories.

No solution	One solution	Two solutions

$|x - 2| + 6 = 0$ $\quad |x + 3| - 1 = 0$

$|x + 8| + 2 = 7$ $\quad |x - 1| + 4 = 4$

$|x - 6| - 5 = -9$ $\quad |x + 5| - 8 = -8$

Dynamic Teaching Tools

- Dynamic Assessment & Progress Monitoring Tool
- Interactive Whiteboard Lesson Library
- Dynamic Classroom with Dynamic Investigations

ANSWERS

31. $|x - 13| = 5$

32. $|x - 2| = 8$

33. $|x - 5.5| = 3.5$

34. $|x + 7.5| = 2.5$

35. $n = 3, n = 5$

36. $c = -\frac{2}{3}, c = 1$

37. $b = 3, b = 5$

38. $k = -0.4, k = 6$

39. $p = \frac{2}{3}, p = 10$

40. $w = -2, w = -0.2$

41. $h = 0.25$

42. $a = 0.5, a = 2.5$

43. $f = -1$

44. $x = 1.5$

45. 5 sec, 7.5 sec

46. no; The absolute value has to be isolated first, which makes the constant on the right positive.

47. a. $|x - 32| = 5$; 27%, 37%

b. no; $\frac{1}{3}\left(33\frac{1}{3}\%\right)$ falls within the range of possible values.

48. a. $|x - 430| = 20$; 410 g, 450 g

b. no; 407 grams is below the minimum acceptable weight.

49. The absolute value cannot be negative. So, there is no solution.

50. Both solutions are extraneous. So, there is no solution.

51. No solution: $|x - 2| + 6 = 0$, $|x - 6| - 5 = -9$

One solution: $|x - 1| + 4 = 4$, $|x + 5| - 8 = -8$

Two solutions: $|x + 8| + 2 = 7$, $|x + 3| - 1 = 0$

ANSWERS

52. b; d

53. always; Square roots of the same number have the same absolute value.

54. always; $a - b$ and $b - a$ are opposites, so they will have the same absolute values.

55. sometimes; The equation will only have two solutions if p is positive.

56. always; The solutions are $p + 4$ and $p - 4$.

57. Absolute value equations will have no solution when the absolute value is equal to a negative number, one solution when the absolute value is equal to zero, and two solutions when the absolute value is equal to a positive number; *Sample answer:* $|x + 12| = -2$ has no solution, $|x + 12| = 0$ has one solution, and $|x + 12| = 2$ has two solutions.

58. *Sample answer:* In a round of a trivia game, you lose 5 points for an incorrect answer and win 5 points for a correct answer. You have 67 points. x represents your score after your next answer.

59. $x = 1, x = -5$; Interpret $|x + 2|$ as a single quantity and solve for it, then solve the resulting absolute value equation.

60. **a.** 40%, 44%; 0%, 4%

b. Set the absolute value of x minus the percentage of the party interested in equal to 2, then solve for the minimum and maximum values.

c. Republican; 44% is within the error range of 42%.

61–67. See Additional Answers.

Mini-Assessment

Solve the equation.

1. $|x + 3| = 13$ $x = 10$ and $x = -16$
2. $|3x - 1| + 6 = 4$ no solution
3. $|6x + 8| = |10x|$ $x = 2$ and $x = -0.5$
4. $2x - 3 = |x - 2|$ $x = \frac{5}{3}$
5. To drive safely along a road, drivers need to maintain a speed of at least 50 miles per hour and at most 60 miles per hour. Write an absolute value equation that represents the minimum and maximum speeds. $|s - 55| = 5$

52. USING STRUCTURE Fill in the equation $|x - \boxed{}| = \boxed{}$ with a, b, c, or d so that the equation is graphed correctly.

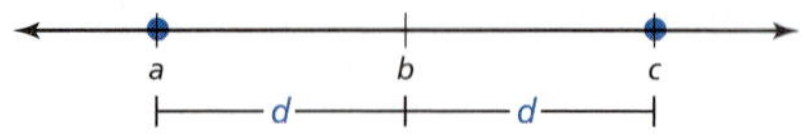

ABSTRACT REASONING **In Exercises 53–56, complete the statement with *always*, *sometimes*, or *never*. Explain your reasoning.**

53. If $x^2 = a^2$, then $|x|$ is ________ equal to $|a|$.

54. If a and b are real numbers, then $|a - b|$ is ________ equal to $|b - a|$.

55. For any real number p, the equation $|x - 4| = p$ will ________ have two solutions.

56. For any real number p, the equation $|x - p| = 4$ will ________ have two solutions.

57. WRITING Explain why absolute value equations can have no solution, one solution, or two solutions. Give an example of each case.

58. THOUGHT PROVOKING Describe a real-life situation that can be modeled by an absolute value equation with the solutions $x = 62$ and $x = 72$.

59. CRITICAL THINKING Solve the equation shown. Explain how you found your solution(s).

$$8|x + 2| - 6 = 5|x + 2| + 3$$

60. HOW DO YOU SEE IT? The circle graph shows the results of a survey of registered voters the day of an election.

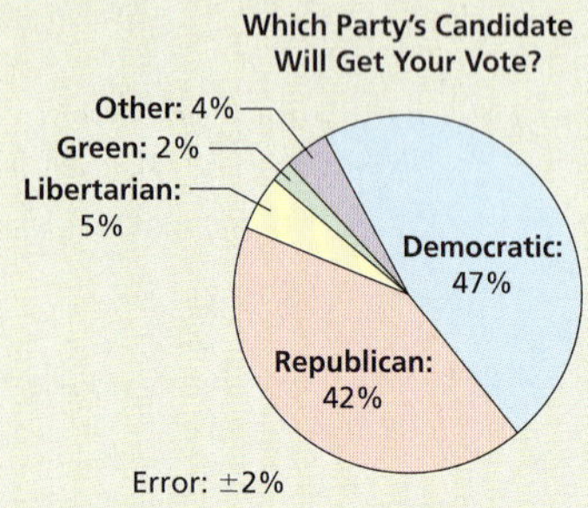

The error given in the graph means that the actual percent could be 2% more or 2% less than the percent reported by the survey.

a. What are the minimum and maximum percents of voters who could vote Republican? Green?

b. How can you use absolute value equations to represent your answers in part (a)?

c. One candidate receives 44% of the vote. Which party does the candidate belong to? Explain.

61. ABSTRACT REASONING How many solutions does the equation $a|x + b| + c = d$ have when $a > 0$ and $c = d$? when $a < 0$ and $c > d$? Explain your reasoning.

Maintaining Mathematical Proficiency

Reviewing what you learned in previous grades and lessons

Identify the property of equality that makes Equation 1 and Equation 2 equivalent. *(Section 1.1)*

62.

Equation 1	$3x + 8 = x - 1$
Equation 2	$3x + 9 = x$

63.

Equation 1	$4y = 28$
Equation 2	$y = 7$

Use a geometric formula to solve the problem. *(Skills Review Handbook)*

64. A square has an area of 81 square meters. Find the side length.

65. A circle has an area of 36π square inches. Find the radius.

66. A triangle has a height of 8 feet and an area of 48 square feet. Find the base.

67. A rectangle has a width of 4 centimeters and a perimeter of 26 centimeters. Find the length.

If students need help...	*If students got it...*
Resources by Chapter • Practice A and Practice B • Puzzle Time	Resources by Chapter • Enrichment and Extension • Cumulative Review
Student Journal • Practice	Start the *next* Section
Differentiating the Lesson Skills Review Handbook	

Laurie's Notes

Overview of Section 1.5

Introduction

- In the study of high school mathematics, students will work with equations involving more than one variable. In Algebra I, this occurs when students start to represent linear equations in two variables, x and y. This chapter presents equation-solving techniques, often referred to as symbolic manipulation, that apply to equations involving more than one variable.
- This lesson uses formulas that should be familiar to students, in order to practice the skill of rewriting equations. In addition, students will rewrite an equation in two variables, x and y, to solve for y. Rewriting an equation in function form is an important skill that you want students to be secure with now. Any weaknesses in rational number operations will surface, so it is important to differentiate where students are having a problem in this lesson. Is the challenge with using inverse operations correctly, or is it problems with rational numbers, or both?

Common Misconceptions

- When an equation involves more than one variable, students can be uncertain how to treat the variable(s) that is not being solved for. It is helpful to review vocabulary such as *coefficient, variable term*, and *constant term* before you begin this lesson.

Formative Assessment Tips

- **Whiteboarding:** Whiteboards can be used to provide individual responses, or used with small groups to encourage student collaboration and consensus on a problem or solution method. Whiteboards can be used at the beginning of class for warm-ups or throughout the lesson to elicit student responses. Unlike writing on scrap paper (individual response) or chart paper (group response), responses can be erased and modified easily. As understanding progresses, responses can reflect this growth.
- Use boards for more than quick responses. Sizable boards can be used to communicate thinking, providing evidence of how a problem was solved. When students display their whiteboards in the front of the room, classmates can critique their reasoning or method of solution.

Pacing Suggestion

- The formal lesson is quite long. You might have students complete only Exploration 1 and then move to Examples 3, 4, 7, and 8 in the formal lesson. That would leave Examples 1 and 2 (literal equations) and Examples 5 and 6 (temperature formulas) for the second day.

Dynamic Teaching Tools

- Dynamic Assessment & Progress Monitoring Tool
- Lesson Planning Tool
- Interactive Whiteboard Lesson Library
- Dynamic Classroom with Dynamic Investigations

What Your Students Will Learn

- Solve literal equations for one variable in terms of the other variables.

Laurie's Notes

Exploration

Motivate

? Distribute whiteboards and ask students to answer the first three questions.

- "I drove 2.1 hours at 65 miles per hour. How far did I go?" 136.5 miles
- "It took me 32 minutes to walk 4 kilometers. How fast was I walking on average?" $\frac{1}{8}$ kilometer per minute or 7.5 kilometers per hour
- "At 8 miles per hour, how long will it take me to bike 6 miles?" $\frac{3}{4}$ hour

? "How are these three equations alike?" They all involve distance, rate, and time.

- Write the three equations relating distance, rate, and time.

  $d = rt, r = \frac{d}{t}, t = \frac{d}{r}$
- Explain that today they will be rewriting familiar formulas and solving them for different variables, just as the familiar formula $d = rt$ can be solved for r or t.

Exploration 1

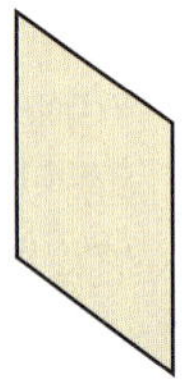

? Draw a parallelogram without a horizontal side and ask, "How do you find the area of a parallelogram?" $A = bh$

? "Are the two adjacent sides the base and height?" No, not unless the parallelogram is a rectangle.

- Circulate as students begin the exploration. Note student work in part (c).

? **Assessing Question:** "How did you solve for b (base) when you knew the area and height?" Divide each side by 5. "Can this same process be used when you do not know the area or height? Explain." Yes. Divide each side of the literal equation by h.

- Have partners share their answers to part (d). You might also ask which was easier, part (b) or part (c). The steps are the same; however, students often view working with constants as easier than working with variables.

Exploration 2

- Have students read through the directions. Check to see that students understand that they are solving the equation for a variable and then making a substitution, just as they did in part (c) of Exploration 1.
- **Differentiate:** You might assign one of the four problems to different groups or pairs of students in the class. Work could be recorded on a whiteboard, enabling presentation of the work to the whole class.
- **Teaching Tip:** The formula for the volume of a cone involves the fraction $\frac{1}{3}$. Suggest to students that they first work with the fraction before they consider the variables involved.

  $V = \frac{1}{3}Bh$

  $3V = Bh$

Communicate Your Answer

- Students should refer to solving an equation for a particular variable.
- Students should be familiar with many formulas, and not always from mathematics!

Connecting to Next Step

- If you feel students are comfortable with solving a formula for a variable, assign Exercises 28 and 29 on page 40 and Exercise 34 on page 41.

1.5 Rewriting Equations and Formulas

Essential Question How can you use a formula for one measurement to write a formula for a different measurement?

EXPLORATION 1 **Using an Area Formula**

Work with a partner.

a. Write a formula for the area A of a parallelogram.

b. Substitute the given values into the formula. Then solve the equation for b. Justify each step.

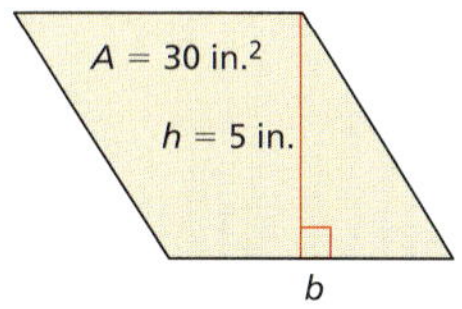

c. Solve the formula in part (a) for b without first substituting values into the formula. Justify each step.

d. Compare how you solved the equations in parts (b) and (c). How are the processes similar? How are they different?

REASONING QUANTITATIVELY

To be proficient in math, you need to consider the given units. For instance, in Exploration 1, the area A is given in square inches and the height h is given in inches. A unit analysis shows that the units for the base b are also inches, which makes sense.

EXPLORATION 2 **Using Area, Circumference, and Volume Formulas**

Work with a partner. Write the indicated formula for each figure. Then write a new formula by solving for the variable whose value is not given. Use the new formula to find the value of the variable.

a. Area A of a trapezoid

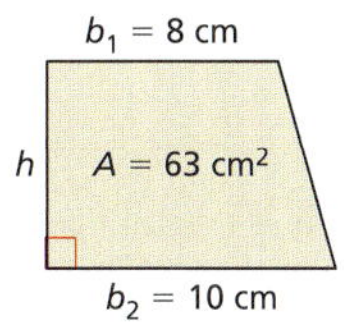

b. Circumference C of a circle

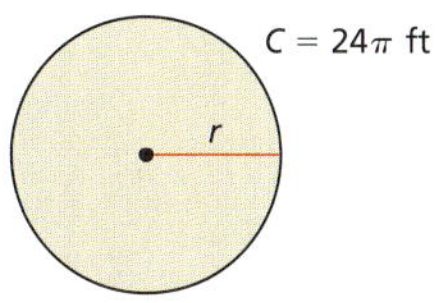

c. Volume V of a rectangular prism

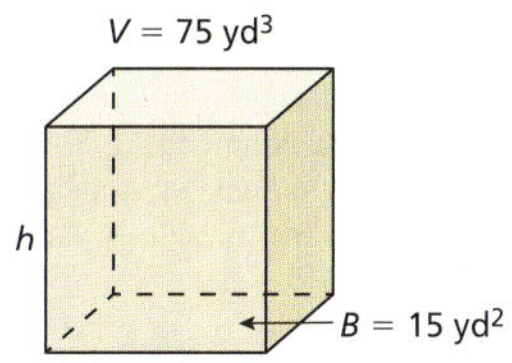

d. Volume V of a cone

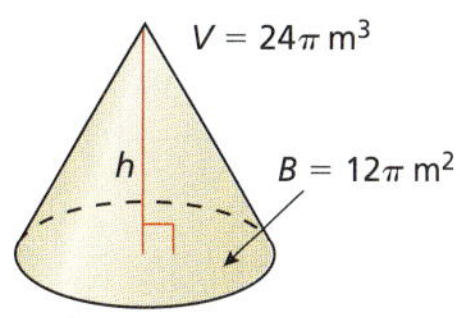

Communicate Your Answer

3. How can you use a formula for one measurement to write a formula for a different measurement? Give an example that is different from those given in Explorations 1 and 2.

Dynamic Teaching Tools
- Dynamic Assessment & Progress Monitoring Tool
- Lesson Planning Tool
- Interactive Whiteboard Lesson Library
- Dynamic Classroom with Dynamic Investigations

ANSWERS

1. a. $A = bh$

b. $30 = b \cdot 5$; Write the equation.
$6 = b$; Divide each side by 5.

c. $A = bh$; Write the formula.
$\frac{A}{h} = b$; Divide each side by h.

d. applied the same steps to solve; used variables instead of constants

2. a. $A = \frac{1}{2}h(b_1 + b_2)$; $h = \frac{2A}{b_1 + b_2}$; $h = 7$ cm

b. $C = 2\pi r$; $r = \frac{C}{2\pi}$; $r = 12$ ft

c. $V = Bh$; $h = \frac{V}{B}$; $h = 5$ yd

d. $V = \frac{1}{3}Bh$; $h = \frac{3V}{B}$; $h = 6$ m

3. Solve the formula for a different variable; *Sample answer:* (volume of a cylinder) $V = \pi r^2 h$; $h = \frac{V}{\pi r^2}$

English Language Learners

Vocabulary

Write the word *literal* on the board. Explain that in everyday speech, the *literal* meaning of a phrase is based on the strict definition of each word. Compare the literal and intended meanings of the expression *It's raining cats and dogs*. Explain that, in mathematics, a *literal* equation is an equation with two or more variables. When you solve for one variable, the solution includes the other variables.

Extra Example 1

Solve the literal equation $4x - 7y = 12$ for y. $y = \frac{4x - 12}{7}$

Extra Example 2

Solve the literal equation $3w + 4wp = a$ for w. $w = \frac{a}{3 + 4p}$

MONITORING PROGRESS ANSWERS

1. $y = 3 + \frac{1}{3}x$
2. $y = x - \frac{5}{2}$
3. $y = 5 - 2x$
4. $x = y$
5. $x = \frac{m}{2 + k}$
6. $x = \frac{y - 3}{5 - k}$

1.5 Lesson

Core Vocabulary

literal equation, *p. 36*
formula, *p. 37*

Previous
surface area

What You Will Learn

- Rewrite literal equations.
- Rewrite and use formulas for area.
- Rewrite and use other common formulas.

Rewriting Literal Equations

An equation that has two or more variables is called a **literal equation**. To rewrite a literal equation, solve for one variable in terms of the other variable(s).

EXAMPLE 1 Rewriting a Literal Equation

Solve the literal equation $3y + 4x = 9$ for y.

SOLUTION

$3y + 4x = 9$	Write the equation.
$3y + 4x - 4x = 9 - 4x$	Subtract $4x$ from each side.
$3y = 9 - 4x$	Simplify.
$\frac{3y}{3} = \frac{9 - 4x}{3}$	Divide each side by 3.
$y = 3 - \frac{4}{3}x$	Simplify.

▶ The rewritten literal equation is $y = 3 - \frac{4}{3}x$.

EXAMPLE 2 Rewriting a Literal Equation

Solve the literal equation $y = 3x + 5xz$ for x.

SOLUTION

$y = 3x + 5xz$	Write the equation.
$y = x(3 + 5z)$	Distributive Property
$\frac{y}{3 + 5z} = \frac{x(3 + 5z)}{3 + 5z}$	Divide each side by $3 + 5z$.
$\frac{y}{3 + 5z} = x$	Simplify.

▶ The rewritten literal equation is $x = \frac{y}{3 + 5z}$.

REMEMBER
Division by 0 is undefined.

In Example 2, you must assume that $z \neq -\frac{3}{5}$ in order to divide by $3 + 5z$. In general, if you have to divide by a variable or variable expression when solving a literal equation, you should assume that the variable or variable expression does not equal 0.

Monitoring Progress

Help in English and Spanish at *BigIdeasMath.com*

Solve the literal equation for *y*.

1. $3y - x = 9$
2. $2x - 2y = 5$
3. $20 = 8x + 4y$

Solve the literal equation for *x*.

4. $y = 5x - 4x$
5. $2x + kx = m$
6. $3 + 5x - kx = y$

Laurie's Notes Teacher Actions

- **Look For and Make Use of Structure:** When solving literal equations, you want students to verbalize the operations represented. For example, when solving $3y + 4x = 9$, students should understand there are two variable terms. To isolate the $3y$-term, subtract $4x$. In other words, approach solving $3y + 4x = 9$ for y in the same way as you would solve $3y + 4 = 9$ for y.
- Write the definition of a *literal equation*.

? "In Example 1, can 9 and $4x$ be combined? Explain." No. They are not like terms.

- In Example 2, use color to highlight each x: $y = 3x + 5xz$. Using the Distributive Property is not obvious to most students because they do not think of the property in a factoring context.

? **Turn and Talk:** "Why can't $z = -\frac{3}{5}$?" The denominator would be 0 and, therefore, undefined.

Rewriting and Using Formulas for Area

A **formula** shows how one variable is related to one or more other variables. A formula is a type of literal equation.

EXAMPLE 3 Rewriting a Formula for Surface Area

The formula for the surface area S of a rectangular prism is $S = 2\ell w + 2\ell h + 2wh$. Solve the formula for the length ℓ.

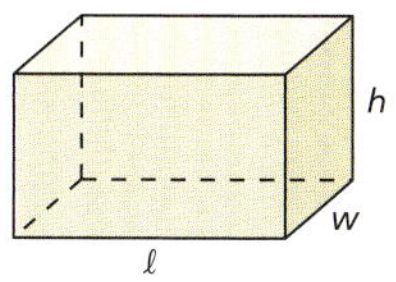

SOLUTION

$S = 2\ell w + 2\ell h + 2wh$ Write the equation.

$S - 2wh = 2\ell w + 2\ell h + 2wh - 2wh$ Subtract $2wh$ from each side.

$S - 2wh = 2\ell w + 2\ell h$ Simplify.

$S - 2wh = \ell(2w + 2h)$ Distributive Property

$\dfrac{S - 2wh}{2w + 2h} = \dfrac{\ell(2w + 2h)}{2w + 2h}$ Divide each side by $2w + 2h$.

$\dfrac{S - 2wh}{2w + 2h} = \ell$ Simplify.

▶ When you solve the formula for ℓ, you obtain $\ell = \dfrac{S - 2wh}{2w + 2h}$.

EXAMPLE 4 Using a Formula for Area

You own a rectangular lot that is 500 feet deep. It has an area of 100,000 square feet. To pay for a new water system, you are assessed \$5.50 per foot of lot frontage.

a. Find the frontage of your lot.

b. How much are you assessed for the new water system?

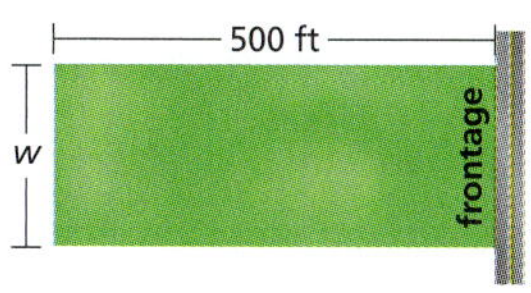

SOLUTION

a. In the formula for the area of a rectangle, let the width w represent the lot frontage.

$A = \ell w$ Write the formula for area of a rectangle.

$\dfrac{A}{\ell} = w$ Divide each side by ℓ to solve for w.

$\dfrac{100{,}000}{500} = w$ Substitute 100,000 for A and 500 for ℓ.

$200 = w$ Simplify.

▶ The frontage of your lot is 200 feet.

b. Each foot of frontage costs \$5.50, and $\dfrac{\$5.50}{1\ \cancel{\text{ft}}} \cdot 200\ \cancel{\text{ft}} = \$1100$.

▶ So, your total assessment is \$1100.

Monitoring Progress

Help in English and Spanish at *BigIdeasMath.com*

Solve the formula for the indicated variable.

7. Area of a triangle: $A = \frac{1}{2}bh$; Solve for h.

8. Surface area of a cone: $S = \pi r^2 + \pi r\ell$; Solve for ℓ.

Differentiated Instruction

Inclusion

Some students may have difficulty keeping track of the variable. Have students circle the variable for which they are solving. Remind them that they have found the answer when that variable is alone on one side of the literal equation.

Extra Example 3

The formula for the area of a circle is $A = \pi r^2$. Solve the formula for r. $r = \sqrt{\dfrac{A}{\pi}}$

Extra Example 4

A patio is in the shape of a parallelogram. Its base, which is up against the side of the house, is 13 feet. The area of the patio is 156 square feet. The height of the parallelogram represents the distance from the house to the edge of the patio. The yard is 24 yards deep from the house to the back fence.

a. Find the distance from the house to the edge of the patio. 12 feet

b. How far is it from that edge of the patio to the back fence? 60 feet, or 20 yards

MONITORING PROGRESS ANSWERS

7. $h = \dfrac{2A}{b}$

8. $\ell = \dfrac{S - \pi r^2}{\pi r}$

Laurie's Notes Teacher Actions

- Use color to highlight each ℓ: $S = 2\ell w + 2\ell h + 2hw$
- **Look For and Make Use of Structure:** Students must see the $2hw$ as a constant term that can be subtracted from each side of the equation.
- Pose the problem in Example 4. Give time for partners to work the example on whiteboards. As you circulate, observe solution methods. Determine if there are unique and/or particularly clear solutions that should be seen by the class. Either ask for those solutions to be shared, or make it appear random by placing particular Popsicle sticks in the center of the can.

Extra Example 5

Solve the distance formula $d = rt$ for r.

$r = \frac{d}{t}$

Extra Example 6

On a July day, the temperature at the peak of Mt. Everest is −16°C. Is that temperature warmer or colder than 0°F?

warmer

MONITORING PROGRESS ANSWER

9. no

Rewriting and Using Other Common Formulas

Core Concept

Common Formulas

Temperature F = degrees Fahrenheit, C = degrees Celsius

$$C = \frac{5}{9}(F - 32)$$

Simple Interest I = interest, P = principal,
r = annual interest rate (decimal form),
t = time (years)

$$I = Prt$$

Distance d = distance traveled, r = rate, t = time

$$d = rt$$

EXAMPLE 5 Rewriting the Formula for Temperature

Solve the temperature formula for F.

SOLUTION

$C = \frac{5}{9}(F - 32)$	Write the temperature formula.
$\frac{9}{5}C = F - 32$	Multiply each side by $\frac{9}{5}$.
$\frac{9}{5}C + 32 = F - 32 + 32$	Add 32 to each side.
$\frac{9}{5}C + 32 = F$	Simplify.

▶ The rewritten formula is $F = \frac{9}{5}C + 32$.

EXAMPLE 6 Using the Formula for Temperature

Which has the greater surface temperature: Mercury or Venus?

SOLUTION

Convert the Celsius temperature of Mercury to degrees Fahrenheit.

$F = \frac{9}{5}C + 32$	Write the rewritten formula from Example 5.
$= \frac{9}{5}(427) + 32$	Substitute 427 for C.
$= 800.6$	Simplify.

▶ Because 864°F is greater than 800.6°F, Venus has the greater surface temperature.

Monitoring Progress

Help in English and Spanish at *BigIdeasMath.com*

9. A fever is generally considered to be a body temperature greater than 100°F. Your friend has a temperature of 37°C. Does your friend have a fever?

Laurie's Notes Teacher Actions

- **Attend to Precision:** In the simple interest formula, students should note that both rate and time are in terms of years.
- Students may know the formula for converting Celsius to Fahrenheit. If so, that should be the common formula stated and change Example 5 to rewriting the formula to solve for Celsius.
- **Extension:** Students have likely heard about the Kelvin temperature scale. Have students research a formula for converting Kelvin to Fahrenheit or Celsius. Rewrite the formula.

EXAMPLE 7 Using the Formula for Simple Interest

You deposit \$5000 in an account that earns simple interest. After 6 months, the account earns \$162.50 in interest. What is the annual interest rate?

COMMON ERROR

The unit of t is years. Be sure to convert months to years.

$\frac{1 \text{ yr}}{12 \text{ mo}} \cdot 6 \text{ mo} = 0.5 \text{ yr}$

SOLUTION

To find the annual interest rate, solve the simple interest formula for r.

$I = Prt$ Write the simple interest formula.

$\frac{I}{Pt} = r$ Divide each side by Pt to solve for r.

$\frac{162.50}{(5000)(0.5)} = r$ Substitute 162.50 for I, 5000 for P, and 0.5 for t.

$0.065 = r$ Simplify.

▶ The annual interest rate is 0.065, or 6.5%.

EXAMPLE 8 Solving a Real-Life Problem

A truck driver averages 60 miles per hour while delivering freight to a customer. On the return trip, the driver averages 50 miles per hour due to construction. The total driving time is 6.6 hours. How long does each trip take?

SOLUTION

Step 1 Rewrite the Distance Formula to write expressions that represent the two trip times. Solving the formula $d = rt$ for t, you obtain $t = \frac{d}{r}$. So, $\frac{d}{60}$ represents the delivery time, and $\frac{d}{50}$ represents the return trip time.

Step 2 Use these expressions and the total driving time to write and solve an equation to find the distance one way.

$\frac{d}{60} + \frac{d}{50} = 6.6$ The sum of the two trip times is 6.6 hours.

$\frac{11d}{300} = 6.6$ Add the left side using the LCD.

$11d = 1980$ Multiply each side by 300 and simplify.

$d = 180$ Divide each side by 11 and simplify.

The distance one way is 180 miles.

Step 3 Use the expressions from Step 1 to find the two trip times.

▶ So, the delivery takes $180 \text{ mi} \div \frac{60 \text{ mi}}{1 \text{ h}} = 3$ hours, and the return trip takes $180 \text{ mi} \div \frac{50 \text{ mi}}{1 \text{ h}} = 3.6$ hours.

Monitoring Progress

Help in English and Spanish at *BigIdeasMath.com*

10. How much money must you deposit in a simple interest account to earn \$500 in interest in 5 years at 4% annual interest?

11. A truck driver averages 60 miles per hour while delivering freight and 45 miles per hour on the return trip. The total driving time is 7 hours. How long does each trip take?

Extra Example 7

You deposit \$3500 in an account that earns 4% simple interest. How long will you have to leave the money in the account to earn \$350 interest? 2.5 years

Extra Example 8

Dan drove 330 miles at 60 miles per hour. Ryan drove 275 miles at 55 miles per hour. How much more time did Dan drive than Ryan? Dan drove 0.5 hour longer than Ryan.

MONITORING PROGRESS ANSWERS

10. \$2500

11. delivery: 3h, return: 4h

Laurie's Notes Teacher Actions

COMMON ERROR In solving the simple interest formula for r in Example 7, students often divide by P and then divide by t. By doing so, they end up with a compound fraction on the left side of the equation.

? Turn and Talk: "Why can you divide by P and t in one step?" *Sample answer:* P and t are multiplied together and can be considered the coefficient of r.

- **Whiteboarding:** Pose the question in Example 8 and give partners time to work the example. Circulate and ask *advancing questions* to assist students in making progress.
- **Construct Viable Arguments and Critique the Reasoning of Others:** Have a pair of students share their solution.

Closure

- **Exit Ticket:** Solve $2x + 4y = 11$ for y. $y = \frac{11 - 2x}{4}$

1.5 Exercises

Dynamic Solutions available at *BigIdeasMath.com*

Vocabulary and Core Concept Check

1. **VOCABULARY** Is $9r + 16 = \frac{\pi}{5}$ a literal equation? Explain.

2. **DIFFERENT WORDS, SAME QUESTION** Which is different? Find "both" answers.

Solve $3x + 6y = 24$ for x.	Solve $24 - 3x = 6y$ for x.
Solve $6y = 24 - 3x$ for y in terms of x.	Solve $24 - 6y = 3x$ for x in terms of y.

Monitoring Progress and Modeling with Mathematics

In Exercises 3–12, solve the literal equation for *y*. *(See Example 1.)*

3. $y - 3x = 13$
4. $2x + y = 7$
5. $2y - 18x = -26$
6. $20x + 5y = 15$
7. $9x - y = 45$
8. $6 - 3y = -6$
9. $4x - 5 = 7 + 4y$
10. $16x + 9 = 9y - 2x$
11. $2 + \frac{1}{6}y = 3x + 4$
12. $11 - \frac{1}{2}y = 3 + 6x$

In Exercises 13–22, solve the literal equation for *x*. *(See Example 2.)*

13. $y = 4x + 8x$
14. $m = 10x - x$
15. $a = 2x + 6xz$
16. $y = 3bx - 7x$
17. $y = 4x + rx + 6$
18. $z = 8 + 6x - px$
19. $sx + tx = r$
20. $a = bx + cx + d$
21. $12 - 5x - 4kx = y$
22. $x - 9 + 2wx = y$

23. **MODELING WITH MATHEMATICS** The total cost C (in dollars) to participate in a ski club is given by the literal equation $C = 85x + 60$, where x is the number of ski trips you take.
   a. Solve the equation for x.
   b. How many ski trips do you take if you spend a total of \$315? \$485?

24. **MODELING WITH MATHEMATICS** The penny size of a nail indicates the length of the nail. The penny size d is given by the literal equation $d = 4n - 2$, where n is the length (in inches) of the nail.
   a. Solve the equation for n.
   b. Use the equation from part (a) to find the lengths of nails with the following penny sizes: 3, 6, and 10.

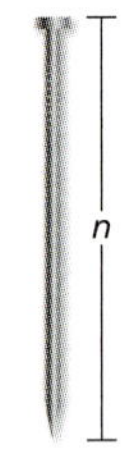

ERROR ANALYSIS In Exercises 25 and 26, describe and correct the error in solving the equation for *x*.

25. ✗
$$12 - 2x = -2(y - x)$$
$$-2x = -2(y - x) - 12$$
$$x = (y - x) + 6$$

26. ✗
$$10 = ax - 3b$$
$$10 = x(a - 3b)$$
$$\frac{10}{a - 3b} = x$$

In Exercises 27–30, solve the formula for the indicated variable. *(See Examples 3 and 5.)*

27. Profit: $P = R - C$; Solve for C.
28. Surface area of a cylinder: $S = 2\pi r^2 + 2\pi rh$; Solve for h.
29. Area of a trapezoid: $A = \frac{1}{2}h(b_1 + b_2)$; Solve for b_2.
30. Average acceleration of an object: $a = \frac{v_1 - v_0}{t}$; Solve for v_1.

Assignment Guide and Homework Check

ASSIGNMENT

Basic: 1–2, 3–33 odd, 40–41, 47–54

Average: 1–2, 4–22 even, 23–34, 39–41, 47–54

Advanced: 1–2, 10–22 even, 26–34 even, 35–37, 39–54

HOMEWORK CHECK

Basic: 7, 17, 23, 27, 31

Average: 12, 22, 30, 32, 34

Advanced: 12, 22, 35, 37, 40

ANSWERS

1. no; It only has one variable.
2. Solve $6y = 24 - 3x$ for y in terms of x; $y = 4 - \frac{1}{2}x$; $x = 8 - 2y$
3. $y = 13 + 3x$
4. $y = 7 - 2x$
5. $y = -13 + 9x$
6. $y = 3 - 4x$
7. $y = 9x - 45$
8. $y = 4$
9. $y = x - 3$
10. $y = 2x + 1$
11. $y = 18x + 12$
12. $y = 16 - 12x$
13. $x = \frac{1}{12}y$
14. $x = \frac{1}{9}m$
15. $x = \frac{a}{2 + 6z}$
16. $x = \frac{y}{3b - 7}$
17. $x = \frac{y - 6}{4 + r}$
18. $x = \frac{z - 8}{6 - p}$
19. $x = \frac{r}{s + t}$
20. $x = \frac{a - d}{b + c}$
21. $x = \frac{y - 12}{-5 - 4k}$
22. $x = \frac{y + 9}{1 + 2w}$
23. a. $x = \frac{C - 60}{85}$
   b. 3 trips; 5 trips
24. a. $n = \frac{d + 2}{4}$
   b. $1\frac{1}{4}$ in.; 2 in.; 3 in.
25. The equation is not solved for x because there is still a term with x on both sides; $x = y - x + 6$; $2x = y + 6$; $x = \frac{y + 6}{2}$
26. There is no x in $3b$ to factor out; $10 = ax - 3b$; $10 + 3b = ax$; $x = \frac{10 + 3b}{a}$
27. $C = R - P$
28. $h = \frac{S - 2\pi r^2}{2\pi r}$
29. $b_2 = \frac{2A}{h} - b_1$
30. $v_1 = at + v_0$

31. REWRITING A FORMULA A common statistic used in professional football is the quarterback rating. This rating is made up of four major factors. One factor is the completion rating given by the formula

$$R = 5\left(\frac{C}{A} - 0.3\right)$$

where C is the number of completed passes and A is the number of attempted passes. Solve the formula for C.

32. REWRITING A FORMULA Newton's law of gravitation is given by the formula

$$F = G\left(\frac{m_1 m_2}{d^2}\right)$$

where F is the force between two objects of masses m_1 and m_2, G is the gravitational constant, and d is the distance between the two objects. Solve the formula for m_1.

33. MODELING WITH MATHEMATICS The sale price S (in dollars) of an item is given by the formula $S = L - rL$, where L is the list price (in dollars) and r is the discount rate (in decimal form). *(See Examples 4 and 6.)*

a. Solve the formula for r.

b. The list price of the shirt is $30. What is the discount rate?

34. MODELING WITH MATHEMATICS The density d of a substance is given by the formula $d = \frac{m}{V}$, where m is its mass and V is its volume.

Pyrite

Density: 5.01g/cm³ Volume: 1.2 cm³

a. Solve the formula for m.

b. Find the mass of the pyrite sample.

35. PROBLEM SOLVING You deposit $2000 in an account that earns simple interest at an annual rate of 4%. How long must you leave the money in the account to earn $500 in interest? *(See Example 7.)*

36. PROBLEM SOLVING A flight averages 460 miles per hour. The return flight averages 500 miles per hour due to a tailwind. The total flying time is 4.8 hours. How long is each flight? Explain. *(See Example 8.)*

37. USING STRUCTURE An athletic facility is building an indoor track. The track is composed of a rectangle and two semicircles, as shown.

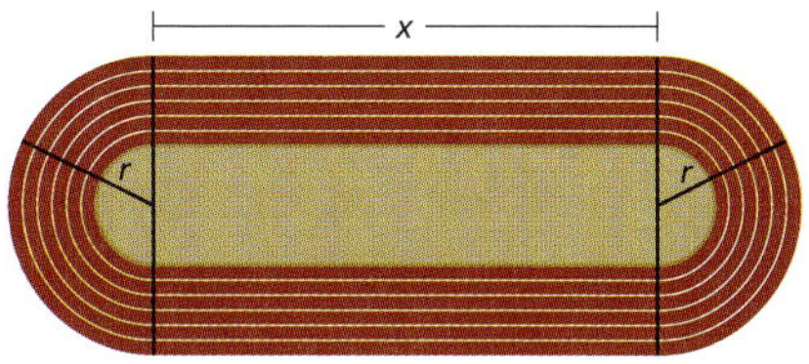

a. Write a formula for the perimeter of the indoor track.

b. Solve the formula for x.

c. The perimeter of the track is 660 feet, and r is 50 feet. Find x. Round your answer to the nearest foot.

38. MODELING WITH MATHEMATICS The distance d (in miles) you travel in a car is given by the two equations shown, where t is the time (in hours) and g is the number of gallons of gasoline the car uses.

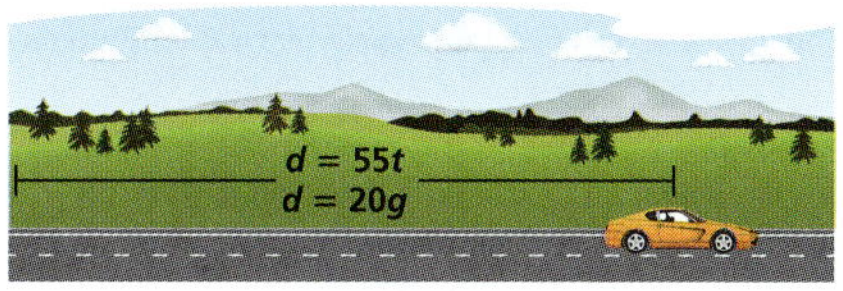

a. Write an equation that relates g and t.

b. Solve the equation for g.

c. You travel for 6 hours. How many gallons of gasoline does the car use? How far do you travel? Explain.

Dynamic Teaching Tools

- Dynamic Assessment & Progress Monitoring Tool
- Interactive Whiteboard Lesson Library
- Dynamic Classroom with Dynamic Investigations

ANSWERS

31. $C = A\left(\frac{R}{5} + 0.3\right)$

32. $m_1 = \frac{Fd^2}{m_2 G}$

33. a. $r = \frac{L - S}{L}$

b. 0.4

34. a. $m = dV$

b. 6.012 g

35. 6.25 yr

36. 2.5 h; 2.3 h; $\frac{d}{460}$ represents the original trip time, and $\frac{d}{500}$ represents the return trip time. Add these expressions and solve for the one-way distance. Substituting the distance into each of the expressions gives the time for each flight.

37. a. $P = 2x + 2\pi r$

b. $x = \frac{1}{2}P - \pi r$

c. 173 ft

38. a. $55t = 20g$

b. $g = \frac{11t}{4}$

c. 16.5 gal; 330 mi; The amount of gasoline used can be found using the formula from part (b). Either of the original formulas can be used to find the distance.

ANSWERS

39. **a.** $r = \dfrac{C}{2\pi}$

b. 1.1 ft; 1.3 ft; 1.4 ft

c. First find the radius using the formula from part (a), then substitute this into the formula for the area of a circle.

40. **a.** $S = 4\ell b + 2b^2$

b. *Sample answer:* ℓ; The formula contains terms with both b and b^2, but only one term with ℓ.

41. no; 70°F is about 21.1°C, which is greater than 20°C.

42. *Sample answer:* $h = 8$; missing base $= 2$; $A = \frac{1}{2}h(b_1 + b_2)$, $40 = \frac{1}{2} \cdot 8 \cdot (2 + 8)$

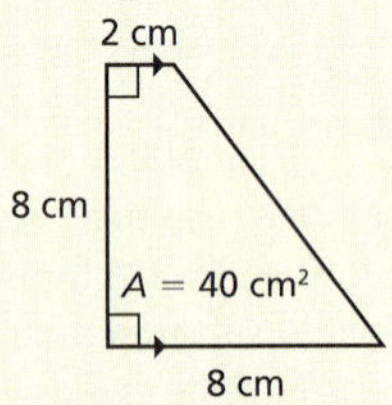

43. $A = \dfrac{5}{2}bh;\ h = \dfrac{2A}{5b}$

44. $A = 4bh;\ h = \dfrac{A}{4b}$

45. $a = \dfrac{b + c}{bx - 1}$

46. $a = \dfrac{yb}{y - xb}$

47–54. See Additional Answers.

Mini-Assessment

1. Solve $9 - c = 17d$ for c.
   $c = 9 - 17d$
2. Solve $nt + 6 - xt = s$ for t.
   $t = \dfrac{s - 6}{n - x}$
3. Solve the formula for the volume of a rectangular prism $V = \ell hw$ for w.
   $w = \dfrac{V}{\ell h}$
4. A circular trampoline has a padded rope that goes around the outside of the frame. The rope is sold for \$2.75 per foot and costs \$104.50 before tax for this trampoline. What is the diameter of the trampoline, to the nearest foot? 12 feet
5. Kayla drives 144 miles in 3 hours. Kendra drives 224 miles in 4 hours. How much faster does Kendra drive on average? Kendra drives an average of 8 miles per hour faster than Kayla.

39. MODELING WITH MATHEMATICS One type of stone formation found in Carlsbad Caverns in New Mexico is called a column. This cylindrical stone formation connects to the ceiling and the floor of a cave.

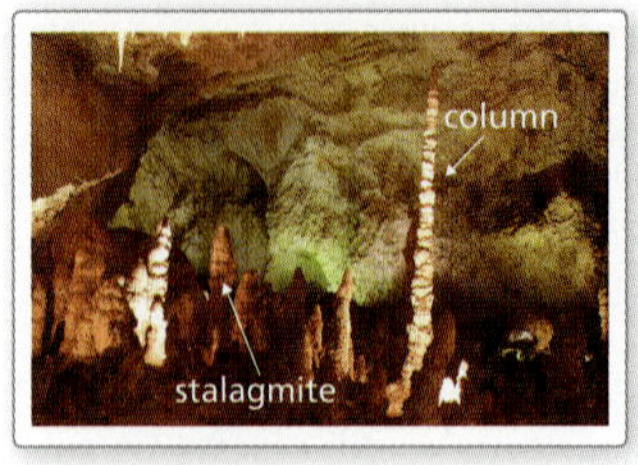

a. Rewrite the formula for the circumference of a circle, so that you can easily calculate the radius of a column given its circumference.

b. What is the radius (to the nearest tenth of a foot) of a column that has a circumference of 7 feet? 8 feet? 9 feet?

c. Explain how you can find the area of a cross section of a column when you know its circumference.

40. HOW DO YOU SEE IT? The rectangular prism shown has bases with equal side lengths.

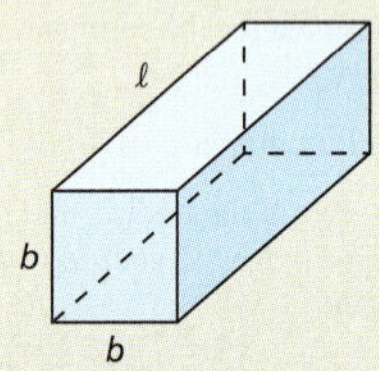

a. Use the figure to write a formula for the surface area S of the rectangular prism.

b. Your teacher asks you to rewrite the formula by solving for one of the side lengths, b or ℓ. Which side length would you choose? Explain your reasoning.

41. MAKING AN ARGUMENT Your friend claims that Thermometer A displays a greater temperature than Thermometer B. Is your friend correct? Explain your reasoning.

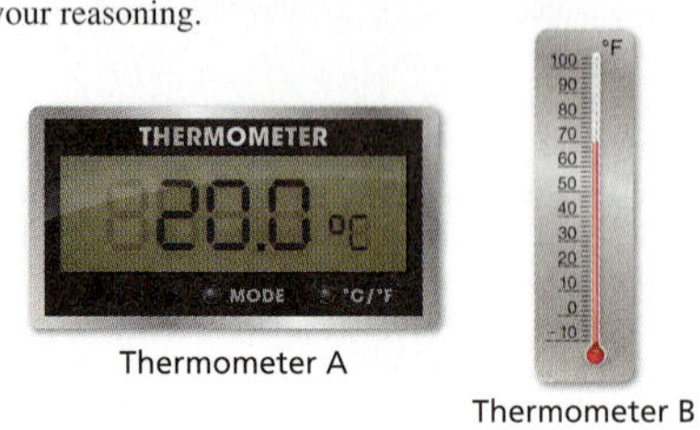

42. THOUGHT PROVOKING Give a possible value for h. Justify your answer. Draw and label the figure using your chosen value of h.

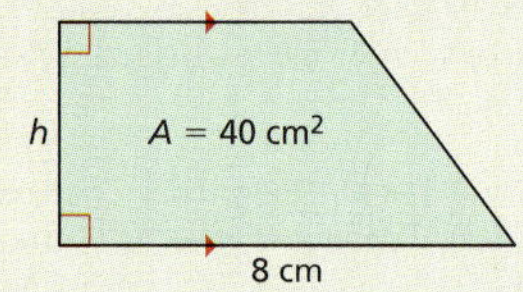

MATHEMATICAL CONNECTIONS **In Exercises 43 and 44, write a formula for the area of the regular polygon. Solve the formula for the height h.**

43.

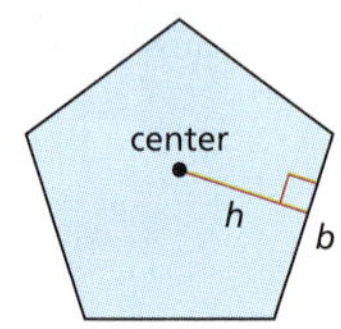

44.

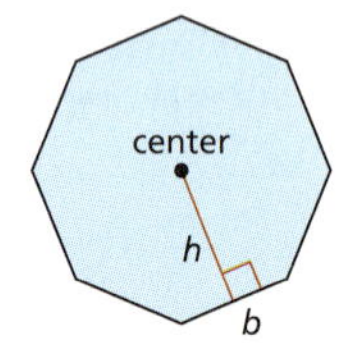

REASONING **In Exercises 45 and 46, solve the literal equation for a.**

45. $x = \dfrac{a + b + c}{ab}$

46. $y = x\left(\dfrac{ab}{a - b}\right)$

Maintaining Mathematical Proficiency
Reviewing what you learned in previous grades and lessons

Evaluate the expression. *(Skills Review Handbook)*

47. $15 - 5 + 5^2$ **48.** $18 \cdot 2 - 4^2 \div 8$ **49.** $3^3 + 12 \div 3 \cdot 5$ **50.** $2^5(5 - 6) + 9 \div 3$

Solve the equation. Graph the solutions, if possible. *(Section 1.4)*

51. $|x - 3| + 4 = 9$ **52.** $|3y - 12| - 7 = 2$ **53.** $2|2r + 4| = -16$ **54.** $-4|s + 9| = -24$

If students need help...	*If students got it...*
Resources by Chapter • Practice A and Practice B • Puzzle Time	Resources by Chapter • Enrichment and Extension • Cumulative Review
Student Journal • Practice	Start the *next* Section
Differentiating the Lesson Skills Review Handbook	

1.4–1.5 What Did You Learn?

Core Vocabulary

absolute value equation, *p. 28*
extraneous solution, *p. 31*
literal equation, *p. 36*
formula, *p. 37*

Core Concepts

Section 1.4

Properties of Absolute Value, *p. 28*
Solving Absolute Value Equations, *p. 28*
Solving Equations with Two Absolute Values, *p. 30*
Special Solutions of Absolute Value Equations, *p. 31*

Section 1.5

Rewriting Literal Equations, *p. 36*
Common Formulas, *p. 38*

Mathematical Practices

1. How did you decide whether your friend's argument in Exercise 46 on page 33 made sense?
2. How did you use the structure of the equation in Exercise 59 on page 34 to rewrite the equation?
3. What entry points did you use to answer Exercises 43 and 44 on page 42?

Performance Task

Magic of Mathematics

Have you ever watched a magician perform a number trick? You can use algebra to explain how these types of tricks work.

To explore the answers to these questions and more, go to ***BigIdeasMath.com***.

Dynamic Teaching Tools

- Dynamic Assessment & Progress Monitoring Tool
- Interactive Whiteboard Lesson Library
- Dynamic Classroom with Dynamic Investigations

ANSWERS

1. *Sample answer:* An absolute value equation must be set equal to a positive value to have a solution. Isolate the absolute value on one side of the equation to determine whether it has a solution.
2. *Sample answer:* The absolute values are the same, so treat them as a single variable.
3. *Sample answer:* the center and two adjacent vertices to form congruent triangles

ANSWERS

1. $z = -9$; Subtract 3 from each side.
2. $t = -13$; Divide each side by -0.2.
3. $n = 10$; Multiply each side by -5.
4. $y = -9$
5. $b = -5$
6. $n = 6$
7. $z = -5$
8. $x = 18$
9. $w = \frac{25}{4}$
10. $x = 10$; 110°, 50°, 20°
11. $x = 126$; 126°, 96°, 126°, 96°, 96°

1 Chapter Review

Dynamic Solutions available at *BigIdeasMath.com*

1.1 Solving Simple Equations *(pp. 3–10)*

a. Solve $x - 5 = -9$. Justify each step.

	$x - 5 = -9$	Write the equation.
Addition Property of Equality →	$+5 \quad +5$	Add 5 to each side.
	$x = -4$	Simplify.

▶ The solution is $x = -4$.

b. Solve $4x = 12$. Justify each step.

	$4x = 12$	Write the equation.
Division Property of Equality →	$\frac{4x}{4} = \frac{12}{4}$	Divide each side by 4.
	$x = 3$	Simplify.

▶ The solution is $x = 3$.

Solve the equation. Justify each step. Check your solution.

1. $z + 3 = -6$ **2.** $2.6 = -0.2t$ **3.** $-\frac{n}{5} = -2$

1.2 Solving Multi-Step Equations *(pp. 11–18)*

Solve $-6x + 23 + 2x = 15$.

$-6x + 23 + 2x = 15$	Write the equation.
$-4x + 23 = 15$	Combine like terms.
$-4x = -8$	Subtract 23 from each side.
$x = 2$	Divide each side by -4.

▶ The solution is $x = 2$.

Solve the equation. Check your solution.

4. $3y + 11 = -16$ **5.** $6 = 1 - b$ **6.** $n + 5n + 7 = 43$

7. $-4(2z + 6) - 12 = 4$ **8.** $\frac{3}{2}(x - 2) - 5 = 19$ **9.** $6 = \frac{1}{5}w + \frac{7}{5}w - 4$

Find the value of x. Then find the angle measures of the polygon.

10.

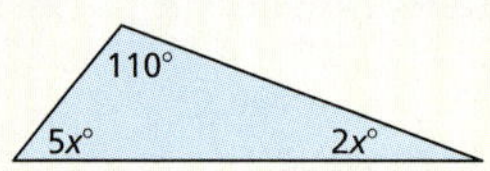

Sum of angle measures: 180°

11.

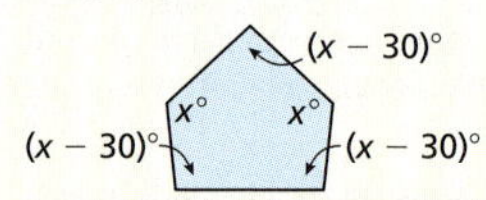

Sum of angle measures: 540°

1.3 Solving Equations with Variables on Both Sides (pp. 19–24)

Solve $2(y - 4) = -4(y + 8)$.

$2(y - 4) = -4(y + 8)$	Write the equation.
$2y - 8 = -4y - 32$	Distributive Property
$6y - 8 = -32$	Add $4y$ to each side.
$6y = -24$	Add 8 to each side.
$y = -4$	Divide each side by 6.

▶ The solution is $y = -4$.

Solve the equation.

12. $3n - 3 = 4n + 1$ **13.** $5(1 + x) = 5x + 5$ **14.** $3(n + 4) = \frac{1}{2}(6n + 4)$

1.4 Solving Absolute Value Equations (pp. 27–34)

a. Solve $|x - 5| = 3$.

$x - 5 = 3$	*or*	$x - 5 = -3$	Write related linear equations.
$+5 \quad +5$		$+5 \quad +5$	Add 5 to each side.
$x = 8$		$x = 2$	Simplify.

▶ The solutions are $x = 8$ and $x = 2$.

b. Solve $|2x + 6| = 4x$. Check your solutions.

$2x + 6 = 4x$	*or*	$2x + 6 = -4x$	Write related linear equations.
$-2x \quad -2x$		$-2x \quad -2x$	Subtract $2x$ from each side.
$6 = 2x$		$6 = -6x$	Simplify.
$\frac{6}{2} = \frac{2x}{2}$		$\frac{6}{-6} = \frac{-6x}{-6}$	Solve for x.
$3 = x$		$-1 = x$	Simplify.

Check the apparent solutions to see if either is extraneous.

▶ The solution is $x = 3$. Reject $x = -1$ because it is extraneous.

Check

$|2x + 6| = 4x$

$|2(3) + 6| \stackrel{?}{=} 4(3)$

$|12| \stackrel{?}{=} 12$

$12 = 12$ ✓

$|2x + 6| = 4x$

$|2(-1) + 6| \stackrel{?}{=} 4(-1)$

$|4| \stackrel{?}{=} -4$

$4 \neq -4$ ✗

Solve the equation. Check your solutions.

15. $|y + 3| = 17$ **16.** $-2|5w - 7| + 9 = -7$ **17.** $|x - 2| = |4 + x|$

18. The minimum sustained wind speed of a Category 1 hurricane is 74 miles per hour. The maximum sustained wind speed is 95 miles per hour. Write an absolute value equation that represents the minimum and maximum speeds.

ANSWERS

12. $n = -4$

13. infinitely many solutions

14. no solution

15. $y = 14, y = -20$

16. $w = 3, w = -\frac{1}{5}$

17. $x = -1$

18. $|v - 84.5| = 10.5$

ANSWERS

19. $y = \frac{1}{2}x - 5$

20. $y = 2x - 2$

21. $y = \dfrac{a}{9 + 3x}$

22. **a.** $h = \dfrac{3V}{B}$

b. 18 cm

23. **a.** $K = \frac{5}{9}(F - 32) + 273.15$

b. about 355.37 K

1.5 Rewriting Equations and Formulas (pp. 35–42)

a. The slope-intercept form of a linear equation is $y = mx + b$. Solve the equation for m.

$y = mx + b$	Write the equation.
$y - b = mx + b - b$	Subtract b from each side.
$y - b = mx$	Simplify.
$\dfrac{y-b}{x} = \dfrac{mx}{x}$	Divide each side by x.
$\dfrac{y-b}{x} = m$	Simplify.

▶ When you solve the equation for m, you obtain $m = \dfrac{y-b}{x}$.

b. The formula for the surface area S of a cylinder is $S = 2\pi r^2 + 2\pi rh$. Solve the formula for the height h.

$S = 2\pi r^2 + 2\pi rh$	Write the equation.
$-2\pi r^2 \quad -2\pi r^2$	Subtract $2\pi r^2$ from each side.
$S - 2\pi r^2 = 2\pi rh$	Simplify.
$\dfrac{S - 2\pi r^2}{2\pi r} = \dfrac{2\pi rh}{2\pi r}$	Divide each side by $2\pi r$.
$\dfrac{S - 2\pi r^2}{2\pi r} = h$	Simplify.

▶ When you solve the formula for h, you obtain $h = \dfrac{S - 2\pi r^2}{2\pi r}$.

Solve the literal equation for y.

19. $2x - 4y = 20$ **20.** $8x - 3 = 5 + 4y$ **21.** $a = 9y + 3yx$

22. The volume V of a pyramid is given by the formula $V = \frac{1}{3}Bh$, where B is the area of the base and h is the height.

a. Solve the formula for h.

b. Find the height h of the pyramid.

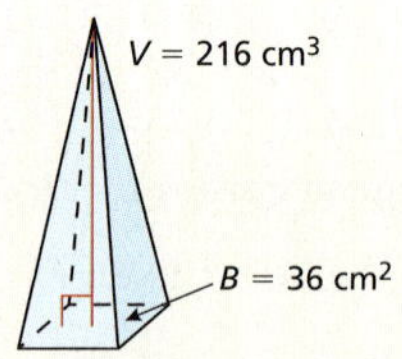

23. The formula $F = \frac{9}{5}(K - 273.15) + 32$ converts a temperature from kelvin K to degrees Fahrenheit F.

a. Solve the formula for K.

b. Convert 180°F to kelvin K. Round your answer to the nearest hundredth.

1 Chapter Test

Solve the equation. Justify each step. Check your solution.

1. $x - 7 = 15$
2. $\frac{2}{3}x + 5 = 3$
3. $11x + 1 = -1 + x$

Solve the equation.

4. $2|x - 3| - 5 = 7$
5. $|2x - 19| = 4x + 1$
6. $-2 + 5x - 7 = 3x - 9 + 2x$
7. $3(x + 4) - 1 = -7$
8. $|20 + 2x| = |4x + 4|$
9. $\frac{1}{3}(6x + 12) - 2(x - 7) = 19$

Describe the values of c for which the equation has no solution. Explain your reasoning.

10. $3x - 5 = 3x - c$
11. $|x - 7| = c$

12. A safety regulation states that the minimum height of a handrail is 30 inches. The maximum height is 38 inches. Write an absolute value equation that represents the minimum and maximum heights.

13. The perimeter P (in yards) of a soccer field is represented by the formula $P = 2\ell + 2w$, where ℓ is the length (in yards) and w is the width (in yards).
    - **a.** Solve the formula for w.
    - **b.** Find the width of the field.
    - **c.** About what percent of the field is inside the circle?

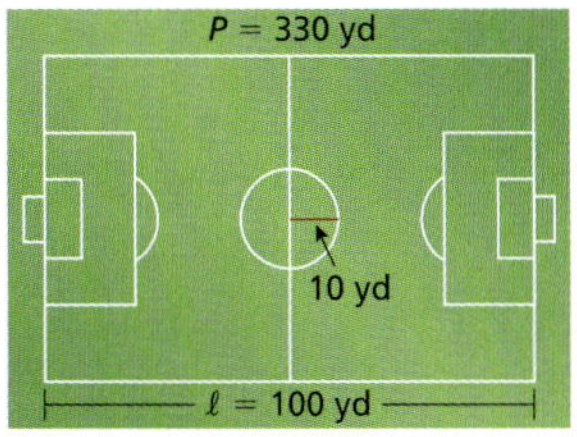

14. Your car needs new brakes. You call a dealership and a local mechanic for prices.

	Cost of parts	Labor cost per hour
Dealership	$24	$99
Local Mechanic	$45	$89

- **a.** After how many hours are the total costs the same at both places? Justify your answer.
- **b.** When do the repairs cost less at the dealership? at the local mechanic? Explain.

15. Consider the equation $|4x + 20| = 6x$. Without calculating, how do you know that $x = -2$ is an extraneous solution?

16. Your friend was solving the equation shown and was confused by the result "$-8 = -8$." Explain what this result means.

$$4(y - 2) - 2y = 6y - 8 - 4y$$
$$4y - 8 - 2y = 6y - 8 - 4y$$
$$2y - 8 = 2y - 8$$
$$-8 = -8$$

ANSWERS

1. $x = 22$; Add 7 to each side.
2. $x = -3$; Subtract 5 from each side; Multiply each side by $\frac{3}{2}$.
3. $x = -\frac{1}{5}$; Subtract x and 1 from each side; Divide each side by 10.
4. $x = -3, x = 9$
5. $x = 3$
6. infinitely many solutions
7. $x = -6$
8. $x = -4, x = 8$
9. no solution
10. $c \neq 5$; If c is 5, then the equation is an identity. For all other values of c, subtracting $3x$ from each side will give a statement that is always false.
11. $c < 0$; An absolute value cannot be negative.
12. $|h - 34| = 4$
13. **a.** $w = \dfrac{P - 2\ell}{2}$
    b. 65 yd
    c. about 4.8%
14. **a.** 2.1 h; The cost at the dealership is $24 + 99t$ and the cost at the local mechanic is $45 + 89t$. Set these two expressions equal and solve for the time.
    b. time is less than 2.1 h; time is greater than 2.1 h; Because the expressions are equal for 2.1 hours, that is the cutoff point from the dealership being less expensive to the local mechanic being less expensive.
15. It will give a negative value on the right and absolute value cannot be negative.
16. It is an identity, meaning that all real numbers are solutions of the equation.

If students need help...	*If students got it...*
Lesson Tutorials	Resources by Chapter • Enrichment and Extension • Cumulative Review
Skills Review Handbook	Performance Task
BigIdeasMath.com	Start the *next* Section

ANSWERS

1. B
2. $cx - a + b = 2b$, $x = \dfrac{a+b}{c}$, $b + a = cx$
3. a. $<$
   b. $<$
   c. $>$
   d. $<$
   e. $=$
   f. $=$
4. a. $24x + 28(5 - x) = 132$ or $24(5 - x) + 28x = 132$
   b. \$4; *Sample answer:* Switching gives a total cost of \$128, which is \$4 less than \$132.

1 Cumulative Assessment

1. A mountain biking park has 48 trails, 37.5% of which are beginner trails. The rest are divided evenly between intermediate and expert trails. How many of each kind of trail are there?

   (A) 12 beginner, 18 intermediate, 18 expert
   (B) 18 beginner, 15 intermediate, 15 expert
   (C) 18 beginner, 12 intermediate, 18 expert
   (D) 30 beginner, 9 intermediate, 9 expert

2. Which of the equations are equivalent to $cx - a = b$?

$cx - a + b = 2b$	$0 = cx - a + b$	$2cx - 2a = \dfrac{b}{2}$
$x - a = \dfrac{b}{c}$	$x = \dfrac{a+b}{c}$	$b + a = cx$

3. Let N represent the number of solutions of the equation $3(x - a) = 3x - 6$. Complete each statement with the symbol <, >, or =.

   a. When $a = 3$, N ____ 1.
   b. When $a = -3$, N ____ 1.
   c. When $a = 2$, N ____ 1.
   d. When $a = -2$, N ____ 1.
   e. When $a = x$, N ____ 1.
   f. When $a = -x$, N ____ 1.

4. You are painting your dining room white and your living room blue. You spend \$132 on 5 cans of paint. The white paint costs \$24 per can, and the blue paint costs \$28 per can.

   a. Use the numbers and symbols to write an equation that represents how many cans of each color you bought.

x	132	5	24	28	=
(	)	+	−	×	÷

   b. How much would you have saved by switching the colors of the dining room and living room? Explain.

5. Which of the equations are equivalent?

$6x + 6 = -14$ $\quad$ $8x + 6 = -2x - 14$

$5x + 3 = -7$ $\quad$ $7x + 3 = 2x - 13$

6. The perimeter of the triangle is 13 inches. What is the length of the shortest side?

Ⓐ 2 in.

Ⓑ 3 in.

Ⓒ 4 in.

Ⓓ 8 in.

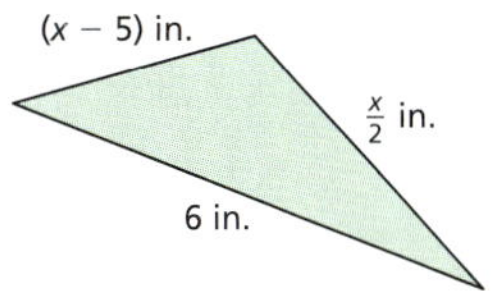

7. You pay \$45 per month for cable TV. Your friend buys a satellite TV receiver for \$99 and pays \$36 per month for satellite TV. Your friend claims that the expenses for a year of satellite TV are less than the expenses for a year of cable.

a. Write and solve an equation to determine when you and your friend will have paid the same amount for TV services.

b. Is your friend correct? Explain.

8. Place each equation into one of the four categories.

No solution	One solution	Two solutions	Infinitely many solutions

$\lvert 8x + 3\rvert = 0$	$-6 = 5x - 9$	$3x - 12 = 3(x - 4) + 1$
$-2x + 4 = 2x + 4$	$0 = \lvert x + 13\rvert + 2$	$-4(x + 4) = -4x - 16$
$12x - 2x = 10x - 8$	$9 = 3\lvert 2x - 11\rvert$	$7 - 2x = 3 - 2(x - 2)$

9. A car travels 1000 feet in 12.5 seconds. Which of the expressions do *not* represent the average speed of the car?

$80\,\dfrac{\text{second}}{\text{feet}}$ $\quad$ $80\,\dfrac{\text{feet}}{\text{second}}$

$\dfrac{80\text{ feet}}{\text{second}}$ $\quad$ $\dfrac{\text{second}}{80\text{ feet}}$

ANSWERS

5. $8x + 6 = -2x - 14$ and $5x + 3 = -7$

6. B

7. a. $45m = 99 + 36m$; month 11

b. yes; Because the expenses are equal for month 11, any time after that satellite TV will be less expensive.

8. no solution: $12x - 2x = 10x - 8$, $0 = \lvert x + 13\rvert + 2$, $3x - 12 = 3(x - 4) + 1$

one solution: $\lvert 8x + 3\rvert = 0$, $-2x + 4 = 2x + 4$, $-6 = 5x - 9$

two solutions: $9 = 3\lvert 2x - 11\rvert$

infinitely many solutions: $-4(x + 4) = -4x - 16$, $7 - 2x = 3 - 2(x - 2)$

9. $80\dfrac{\text{sec}}{\text{ft}}, \dfrac{\text{sec}}{80\text{ ft}}$

Chapter 2 Pacing Guide

Chapter Opener/ Mathematical Practices	1 Day
Section 1	1 Day
Section 2	1 Day
Section 3	1 Day
Section 4	1 Day
Quiz	1 Day
Section 5	2 Days
Section 6	2 Days
Chapter Review/ Chapter Tests	2 Days
Total Chapter 2	12 Days
Year-to-Date	24 Days

2 Solving Linear Inequalities

Camel Physiology (*p. 91*)

Mountain Plant Life (*p. 85*)

Digital Camera (*p. 70*)

Microwave Electricity (*p. 64*)

Natural Arch (*p. 59*)

Laurie's Notes

Chapter Summary

- Students have just finished a chapter on solving linear equations—simple and multi-step equations along with absolute value equations. The techniques used in solving linear equations are applied to linear inequalities in this chapter. In middle school, students solved and graphed linear inequalities, so many of the topics in this chapter should be familiar to them.
- The chapter begins with an introduction to writing and graphing inequalities. Color coding and verbal models are used to help students develop confidence in writing inequalities, a necessary skill for the chapter. The graphs are used to display and check solutions.
- The next three lessons focus on solving increasingly complex inequalities. Tools used in developing facility with these problems include symbolic manipulation, tables, and spreadsheets. Practice with real number operations is integrated throughout.
- The last two lessons of the chapter introduce compound inequalities, which are necessary in solving absolute value inequalities. Look for the helpful teaching strategies offered in these lessons.
- New formative assessment tips are offered in many of the lessons, and tips from the previous chapter are referenced throughout the notes at point of use.

What Your Students Have Learned

Middle School
- Use substitution to determine whether numbers are solutions to linear inequalities.
- Write and graph inequalities of the form $x > c$ or $x < c$.
- Recognize that inequalities have infinitely many solutions.
- Solve linear inequalities with positive coefficients and graph the solution set.

What Your Students Will Learn

Algebra 1
- Write linear inequalities and sketch graphs of linear inequalities.
- Solve multi-step linear inequalities in one variable using inverse operations, including multiplying and dividing by negative numbers.
- Write compound linear inequalities in one variable joined by the word *and* or the word *or*.
- Graph and solve compound linear inequalities in one variable.
- Solve absolute value inequalities.

Dynamic Teaching Tools
- Dynamic Assessment & Progress Monitoring Tool
- Lesson Planning Tool
- Interactive Whiteboard Lesson Library
- Dynamic Classroom with Dynamic Investigations
- Real-Life STEM Videos

Scaffolding in the Classroom

Graphic Organizers: Information Frame

An Information Frame can be used to help students organize and remember concepts. Students write the topic in the middle rectangle. Then students write related concepts in the spaces around the rectangle. Related concepts can include *Words, Numbers, Algebra, Definition, Example, Non-Example, Visual, Procedure, Details,* and *Vocabulary.* Students can place their Information Frames on note cards to use as a quick study reference.

Questioning in the Classroom

Be open to multiple answers.
Questions can frequently be interpreted differently. Allow students time to discuss and explain their point of view.

Laurie's Notes

Maintaining Mathematical Proficiency

Graphing Numbers on a Number Line

- Remind students that positive numbers are greater than 0 and are to the right of 0 on a number line. Negative numbers are less than 0 and are to the left of 0 on a number line.
- Remind students that the absolute value of a number is the distance between the number and 0 on a number line.

COMMON ERROR Students may ignore the absolute value symbols when they graph the numbers or they may find the sum or difference without first finding the absolute value.

Comparing Real Numbers

- Remind students to graph each pair of integers on a number line. On a horizontal number line, the number on the right is the greater number.

COMMON ERROR Students may ignore the negative signs when they compare the numbers.

Mathematical Practices (continued on page 52)

- The *Mathematical Practices* page focuses attention on how mathematics is learned—process versus content. This page demonstrates the use of graphing calculators as a tool in learning important mathematics.
- Use the *Mathematical Practices* page to help students develop mathematical habits of mind—how mathematics can be explored and how mathematics is thought about.
- Refer to the *Core Concept* and explain that the graphing calculators can be used to graph solutions to single-variable inequalities.
- Example 1 demonstrates the syntax needed when entering an inequality. The example shown has variables on both sides of the inequality symbol. The same technique is used when one side involves only a constant.

If students need help...	*If students got it...*
Student Journal • Maintaining Mathematical Proficiency	Game Closet at *BigIdeasMath.com*
Lesson Tutorials	Start the *next* Section
Skills Review Handbook	

Maintaining Mathematical Proficiency

Graphing Numbers on a Number Line

Example 1 Graph each number.

a. 3

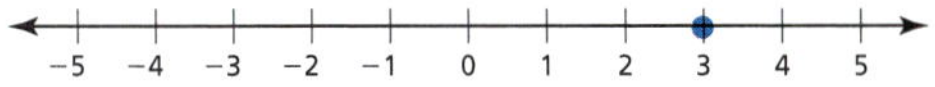

b. -5

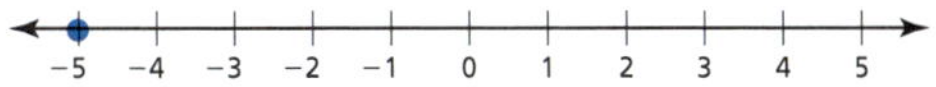

Example 2 Graph each number.

a. $|4|$

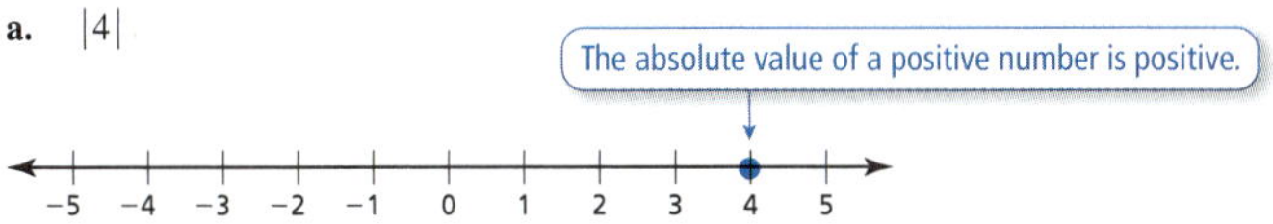

b. $|-2|$

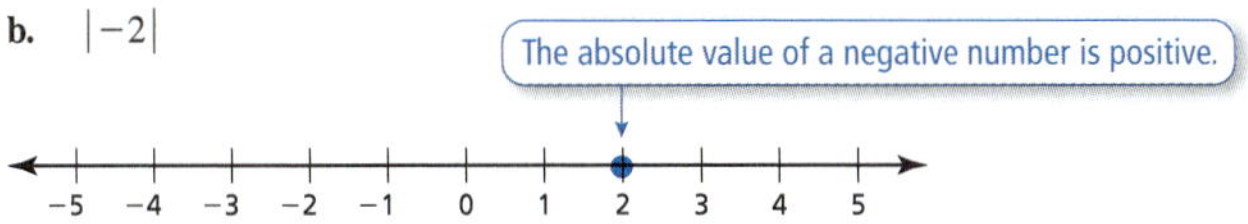

Graph the number.

1. 6
2. $|2|$
3. $|-1|$
4. $2 + |-2|$
5. $1 - |-4|$
6. $-5 + |3|$

Comparing Real Numbers

Example 3 Complete the statement -1 ▒ -5 with <, >, or =.

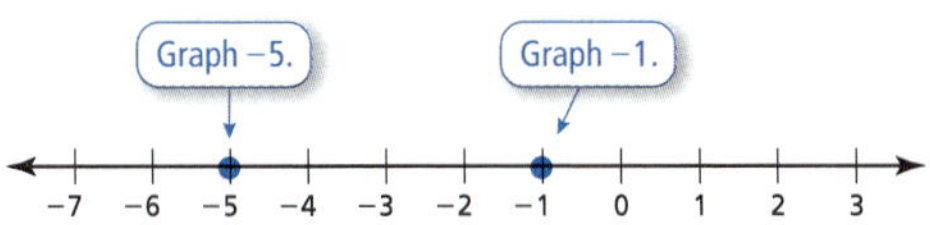

▶ -1 is to the right of -5. So, $-1 > -5$.

Complete the statement with <, >, or =.

7. 2 ▒ 9
8. -6 ▒ 5
9. -12 ▒ -4
10. -7 ▒ -13
11. $|-8|$ ▒ $|8|$
12. -10 ▒ $|-18|$
13. **ABSTRACT REASONING** A number a is to the left of a number b on the number line. How do the numbers $-a$ and $-b$ compare?

What Your Students Have Learned

- Graph numbers and numerical expressions on a number line.
- Compare numbers and numerical expressions and write statements with <, >, or =.

ANSWERS

1. 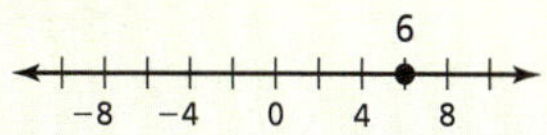
2. 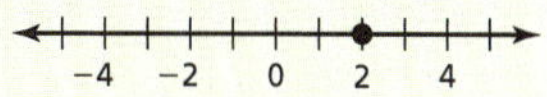
3. 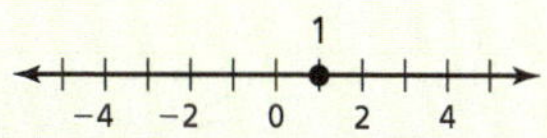
4. 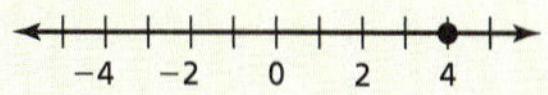
5. 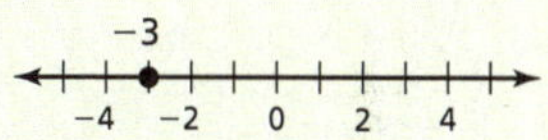
6. 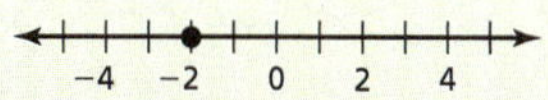
7. <
8. <
9. <
10. >
11. =
12. <
13. $-b < -a$

Vocabulary Review

Have students make an Idea and Examples Chart for the following words.

- Absolute value
- Negative number
- Positive number

MONITORING PROGRESS ANSWERS

1. $x < -4$
2. $x > 3$
3. $x \geq -2$

Mathematical Practices

Mathematically proficient students use technology tools to explore concepts.

Using a Graphing Calculator

Core Concept

Solving an Inequality in One Variable

You can use a graphing calculator to solve an inequality.

1. Enter the inequality into a graphing calculator.
2. Graph the inequality.
3. Use the graph to write the solution.

EXAMPLE 1 Using a Graphing Calculator

Use a graphing calculator to solve (a) $2x - 1 < x + 2$ and (b) $2x - 1 \leq x + 2$.

SOLUTION

a. Enter the inequality $2x - 1 < x + 2$ into a graphing calculator. Press *graph*.

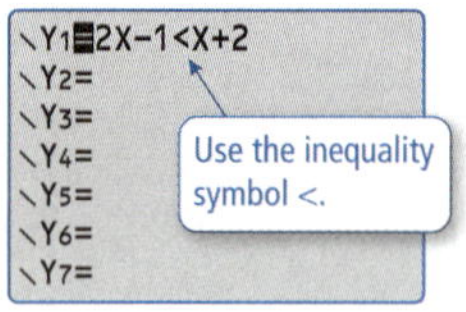

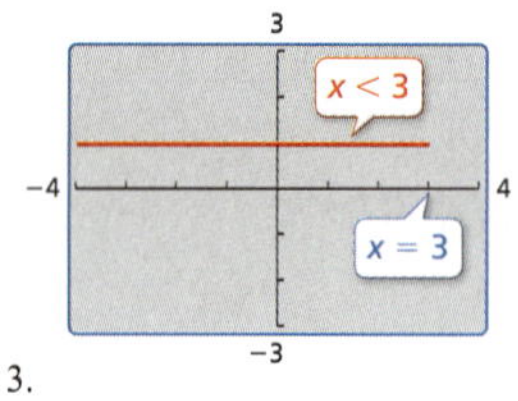

▶ The solution of the inequality is $x < 3$.

b. Enter the inequality $2x - 1 \leq x + 2$ into a graphing calculator. Press *graph*.

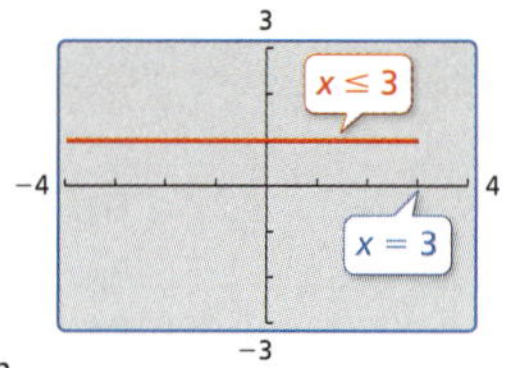

▶ The solution of the inequality is $x \leq 3$.

Notice that the graphing calculator does not distinguish between the solutions $x < 3$ and $x \leq 3$. You must distinguish between these yourself, based on the inequality symbol used in the original inequality.

Monitoring Progress

Use a graphing calculator to solve the inequality.

1. $2x + 3 < x - 1$
2. $-x - 1 > -2x + 2$
3. $\frac{1}{2}x + 1 \leq \frac{3}{2}x + 3$

Laurie's Notes Mathematical Practices (continued from page T-51)

- Explain why the graphs given in Example 1 are horizontal and not vertical: the graphing calculator graphs one dimension horizontally, similar to how the graph would be on a number line. It offsets the solution from the *x*-axis so that the solution is visible.
- Be sure to point out that the graph does not distinguish $x < 3$ and $x \leq 3$, meaning that students will not see the open circle or closed circle on the calculator screen! They will need to interpret any endpoints based on the inequality symbols used in the original problem.
- The three questions in *Monitoring Progress* should not take long. Students may enter the fractions in Question 3 as fractions using the division symbol or as decimals.

Laurie's Notes

Overview of Section 2.1

Introduction

- Students have written and graphed inequalities in middle school. The domain may have been restricted to whole numbers, followed by integers. In this lesson, students will work with real numbers, and they will also decide whether a number is a solution of an inequality.
- When students translate a verbal phrase or context into symbols, the variable may end up on the right side of the inequality symbol. *Example:* 80 points was the lowest score on the test and students write $80 < p$. Suggest to students that they rewrite the inequality so that the variable is read first (left side of inequality). In doing so, the direction of the inequality symbol must be switched; $p > 80$.
- Solutions of an inequality are represented in symbols and by a graph. An alternate way of representing solutions is to use set-builder notation.

Common Misconceptions

- Students will often make the mistake of thinking $-1 < -2$, forgetting that relationships are reversed on the negative side of 0; $-1 > -2$.

Formative Assessment Tips

- **Learning Goals Inventory:** At the beginning of a chapter or unit of study, it is important for students to identify their level of understanding or knowledge of the learning intentions (goals) for the chapter. Taking an inventory of the explicit goals for the chapter informs students what you want them to learn—they know the target. On one side of a sheet of paper, I type the assignment guide for the chapter. On the other side, I list the learning intentions for the chapter and ask students to do a self-assessment for each learning goal.

  0—no knowledge, understanding, or ability
  1—beginning to get it
  2—feeling okay
  3—very confident
- Students return to the *Learning Goals Inventory* midway through the chapter and after the chapter assessment. Students recognize their own growth in completing the inventory multiple times.

Applications

- **Model with Mathematics:** In Exercise 46, maximum load limits on roads are explored. Have students research the load limits for state and federal roads in your state. The number of axles on the vehicle will be a factor.

Pacing Suggestion

- Given the review nature of the content in this lesson, you should only spend one day on it. You could do the *Motivate* and explorations or simply present the formal lesson. Students would prefer the more active version!

Dynamic Teaching Tools

- Dynamic Assessment & Progress Monitoring Tool
- Lesson Planning Tool
- Interactive Whiteboard Lesson Library
- Dynamic Classroom with Dynamic Investigations

What Your Students Will Learn

- Write linear inequalities that represent sentences or graphs.
- Sketch graphs of linear inequalities.

Laurie's Notes

Exploration

Motivate

- **Preparation:** Write eight inequalities on index cards. Draw the matching graphs on eight strips of paper large enough to be seen by students across your room. Tape the eight graphs in different locations around your room.
- Examples of inequalities to explore: $x > 4$; $x \leq -4$; $x > -4$; $x \leq 4$; $x < -2.5$; $x \leq -2.5$; $x > 3.5$; $x \geq 3.5$
- Explain to students that they are starting a new chapter today. Express your confidence in them, knowing that they will have little difficulty with graphing inequalities.
- Select eight students at random and hand each an index card. Ask students to find their graphs and to go stand next to the graphs.

? After students have matched their cards to the graphs, ask each student to explain how he or she knows the match is correct. What features of the graph did they look for? Listen for: open circle versus closed circle, shading the correct side of the number line.

- After all of the students have made their explanations, collect their cards. Next, ask eight different students to go to one of the graphs and say aloud the inequality that is shown by the graph.

Exploration 1

? "The variable *t* is suggested to represent temperatures. Could any variable be used?" yes

? "Could the temperature in Sweden be −10°C? Explain." yes; The phrase "at least" would include −10°C and anything warmer.

- Discuss the real-life physical limitations. For instance, although the graph displays solutions to infinity, the temperatures in Sweden are unlikely to approach infinity.
- **Reason Abstractly and Quantitatively:** Learning to interpret solutions in the context of the problem is a practice you want students to develop.

Exploration 2

- Students may have forgotten the difference in notation between the closed and open circle. When both partners have forgotten what the notation means, I encourage them to check in with another group.
- Teachers should not be the only source of knowledge and information in the class. Develop the culture in your classroom of students listening to others and offering information to others.
- There may be different words students use to describe the inequalities. For part (a) students may say, "all values of *x* greater than or equal to 1" or "all *x*-values that are at least 1."
- **Attend to Precision:** Solutions should refer to a variable, meaning it would not be acceptable to give "greater than or equal to 1" as an answer to part (a).

Communicate Your Answer

- Students sometimes have difficulty thinking of a context much different from the one they have just completed. Suggest that they think of units that might be associated with each number. In Question 4(a), 3.5 might refer to inches, pieces of pizza, pounds, weeks, hours, sheets of plywood, and so on.

Connecting to Next Step

- If students are confident of their understanding of writing and graphing inequalities, you might have them record the *Core Vocabulary* and *Concept Summary* in the formal lesson and perhaps complete Example 4, an application. Students would then be ready for homework exercises.

2.1 Writing and Graphing Inequalities

Essential Question How can you use an inequality to describe a real-life statement?

EXPLORATION 1 **Writing and Graphing Inequalities**

Work with a partner. Write an inequality for each statement. Then sketch the graph of the numbers that make each inequality true.

a. Statement The temperature t in Sweden is at least $-10°C$.

Inequality

Graph

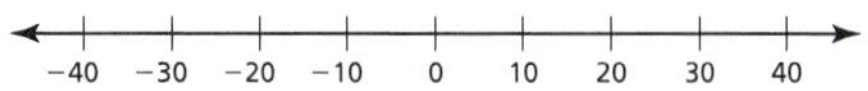

b. Statement The elevation e of Alabama is at most 2407 feet.

Inequality

Graph

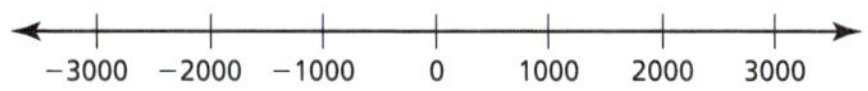

ATTENDING TO PRECISION

To be proficient in math, you need to communicate precisely. You also need to state the meanings of the symbols you use.

EXPLORATION 2 **Writing Inequalities**

Work with a partner. Write an inequality for each graph. Then, in words, describe all the values of x that make each inequality true.

a.

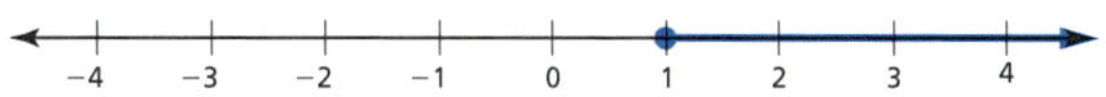

b.

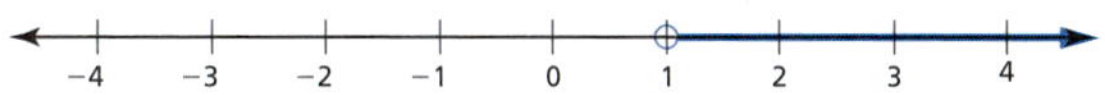

c.

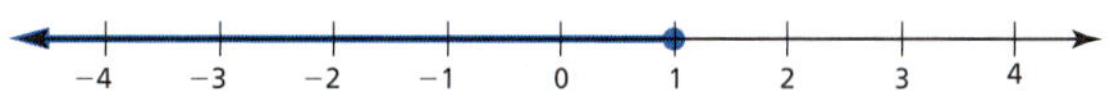

d.

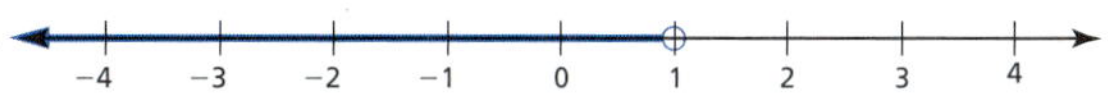

Communicate Your Answer

3. How can you use an inequality to describe a real-life statement?

4. Write a real-life statement that involves each inequality.

   a. $x < 3.5$

   b. $x \leq 6$

   c. $x > -2$

   d. $x \geq 10$

Dynamic Teaching Tools

- Dynamic Assessment & Progress Monitoring Tool
- Lesson Planning Tool
- Interactive Whiteboard Lesson Library
- Dynamic Classroom with Dynamic Investigations

ANSWERS

1. **a.** $t \geq -10°C$

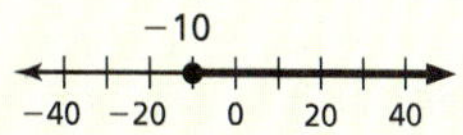

   b. $e \leq 2407$ ft

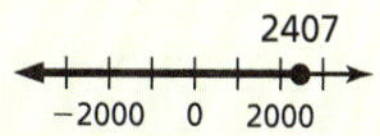

2. **a.** $x \geq 1$; one and all numbers greater than one

   b. $x > 1$; all numbers greater than one

   c. $x \leq 1$; one and all numbers less than one

   d. $x < 1$; all numbers less than one

3. An inequality can indicate the largest or smallest value for a quantity.

4. **a.** *Sample answer:* The judge's score is less than 3.5.

   b. *Sample answer:* The time in the pool was at most 6 hours.

   c. *Sample answer:* The temperature is greater than $-2°C$.

   d. *Sample answer:* The profit was at least \$10.

English Language Learners

Vocabulary and Symbols

Have students create an index card for each of the symbols <, >, ≤, and ≥. Have them write the symbol on the front of the card and the descriptive phrase on the back. Students can use the cards to help them choose the correct symbol when they write inequalities for word sentences.

Extra Example 1

Write each sentence as an inequality.

a. A number c plus 9.4 is greater than or equal to -6. $c + 9.4 \geq -6$

b. Six is greater than a number p minus 8. $6 > p - 8$

c. Ten is no more than the sum of 3 times a number y and 4. $10 \leq 3y + 4$

MONITORING PROGRESS ANSWERS

1. $b < 30.4$
2. $-\frac{7}{10} \geq 2k - 4$

2.1 Lesson

Core Vocabulary

inequality, *p. 54*
solution of an inequality, *p. 55*
solution set, *p. 55*
graph of an inequality, *p. 56*

Previous
expression

What You Will Learn

- Write linear inequalities.
- Sketch the graphs of linear inequalities.
- Write linear inequalities from graphs.

Writing Linear Inequalities

An **inequality** is a mathematical sentence that compares expressions. An inequality contains the symbol <, >, ≤, or ≥. To write an inequality, look for the following phrases to determine what inequality symbol to use.

Inequality Symbols				
Symbol	<	>	≤	≥
Key Phrases	• is less than • is fewer than	• is greater than • is more than	• is less than or equal to • is at most • is no more than	• is greater than or equal to • is at least • is no less than

EXAMPLE 1 Writing Inequalities

Write each sentence as an inequality.

a. A number w minus 3.5 is less than or equal to -2.

b. Three is less than a number n plus 5.

c. Zero is greater than or equal to twice a number x plus 1.

SOLUTION

a. A number w minus 3.5 is less than or equal to -2.

$w - 3.5$ ≤ -2

An inequality is $w - 3.5 \leq -2$.

b. Three is less than a number n plus 5.

3 < $n + 5$

An inequality is $3 < n + 5$.

READING

The inequality $3 < n + 5$ is the same as $n + 5 > 3$.

c. Zero is greater than or equal to twice a number x plus 1.

0 ≥ $2x + 1$

An inequality is $0 \geq 2x + 1$.

Monitoring Progress

Help in English and Spanish at *BigIdeasMath.com*

Write the sentence as an inequality.

1. A number b is fewer than 30.4.
2. $-\frac{7}{10}$ is at least twice a number k minus 4.

Laurie's Notes Teacher Actions

- To review the vocabulary associated with inequalities, you can do a matching activity. Write expressions (key phrases) on cards and hand them to students. Include cards that have the inequality symbols written on them. Have students sort themselves into four groups in different regions of the room.
- **Use Appropriate Tools Strategically:** Using color can be a helpful tool as students are learning to translate words into symbols.

Sketching the Graphs of Linear Inequalities

A **solution of an inequality** is a value that makes the inequality true. An inequality can have more than one solution. The set of all solutions of an inequality is called the **solution set**.

Value of x	$x + 5 \geq -2$	Is the inequality true?
-6	$-6 + 5 \overset{?}{\geq} -2$ $-1 \geq -2$ ✓	yes
-7	$-7 + 5 \overset{?}{\geq} -2$ $-2 \geq -2$ ✓	yes
-8	$-8 + 5 \overset{?}{\geq} -2$ $-3 \not\geq -2$ ✗	no

Recall that a diagonal line through an inequality symbol means the inequality is *not* true. For instance, the symbol $\not\geq$ means "is not greater than or equal to."

EXAMPLE 2 Checking Solutions

Tell whether -4 is a solution of each inequality.

a. $x + 8 < -3$

b. $-4.5x > -21$

SOLUTION

a. $x + 8 < -3$ Write the inequality.

$-4 + 8 \overset{?}{<} -3$ Substitute -4 for x.

$4 \not< -3$ ✗ Simplify.

4 is *not* less than -3.

▶ So, -4 is *not* a solution of the inequality.

b. $-4.5x > -21$ Write the inequality.

$-4.5(-4) \overset{?}{>} -21$ Substitute -4 for x.

$18 > -21$ ✓ Simplify.

18 is greater than -21.

▶ So, -4 is a solution of the inequality.

Monitoring Progress

Help in English and Spanish at *BigIdeasMath.com*

Tell whether -6 is a solution of the inequality.

3. $c + 4 < -1$

4. $10 \leq 3 - m$

5. $21 \div x \geq -3.5$

6. $4x - 25 > -2$

Laurie's Notes Teacher Actions

- **? Assessing Question:** Write a simple inequality on the board, such as $x - 5 > -2$, and ask for a solution. "How do you know 10 is a solution?" Substitute 10 in place of x and do the arithmetic to see whether the inequality is true. "How many solutions are there?" infinitely many
- Work through Example 2 as shown, noting the use of the diagonal line through the inequality symbol.
- **? Probing Question:** "Is there a symbol equivalent to $<$?" yes, $\not\geq$
- When working with inequalities in real life, there are some cases where negative values do not apply in the real-life context or where only certain values like whole numbers do apply. Students need to be aware that an inequality like $x < 4$ represents *all possible* solutions less than 4 and *not all numbers* less than 4.

Extra Example 2

Tell whether 3 is a solution of each inequality.

a. $x - 7 > -5$ 3 is a solution.

b. $-3x \leq -12.5$ 3 is *not* a solution.

MONITORING PROGRESS ANSWERS

3. yes

4. no

5. yes

6. no

Differentiated Instruction

Visual

Some students may have difficulty graphing inequalities. Have students complete this table to describe the type of circle to use and the direction to draw the arrow in the graph. Then have students graph the inequalities and exchange graphs with a classmate to check their work.

Inequality	Circle	Direction
$x < 6$	open	left
$x > -4$	open	right
$x \leq -1$	closed	left
$x \geq 1$	closed	right

Extra Example 3

Graph each inequality.

a. $x \geq 3$

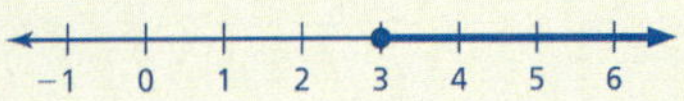

b. $-4 > y$

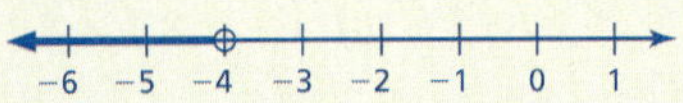

c. $x \leq -2$

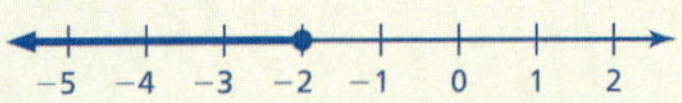

MONITORING PROGRESS ANSWERS

7. 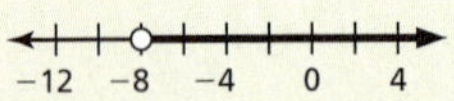

8. 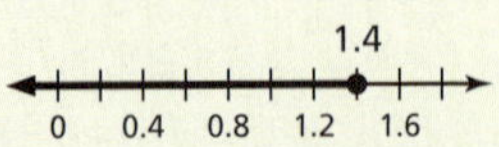

9. 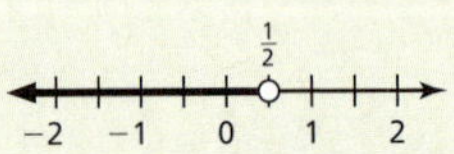

10. 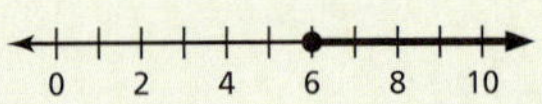

The **graph of an inequality** shows the solution set of the inequality on a number line. An open circle, ○, is used when a number is *not* a solution. A closed circle, ●, is used when a number is a solution. An arrow to the left or right shows that the graph continues in that direction.

EXAMPLE 3 Graphing Inequalities

Graph each inequality.

a. $y \leq -3$ **b.** $2 < x$ **c.** $x > 0$

SOLUTION

a. Test a number to the left of -3. $y = -4$ is a solution.

Test a number to the right of -3. $y = 0$ is *not* a solution.

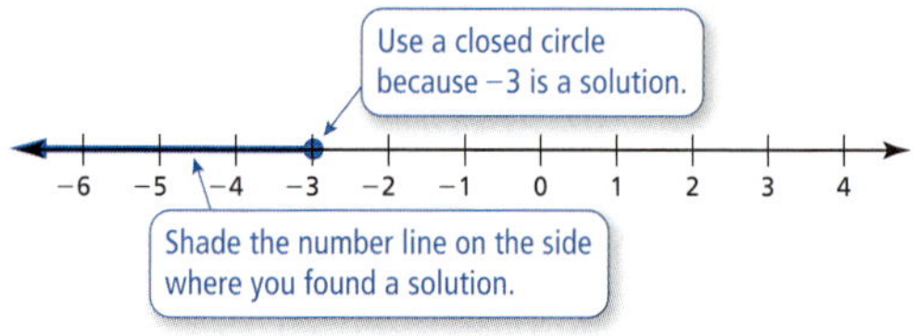

b. Test a number to the left of 2. $x = 0$ is *not* a solution.

Test a number to the right of 2. $x = 4$ is a solution.

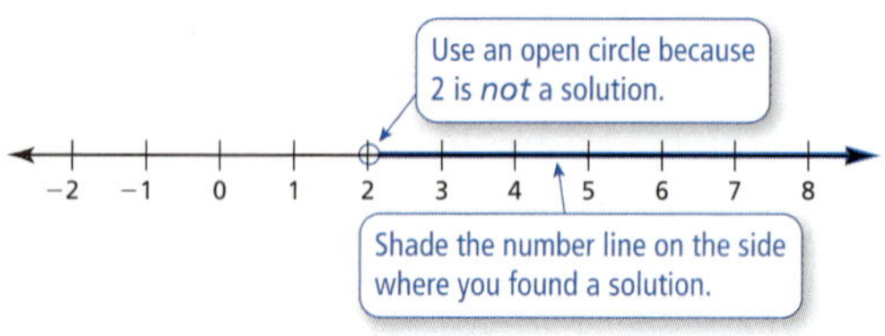

c. Just by looking at the inequality, you can see that it represents the set of all positive numbers.

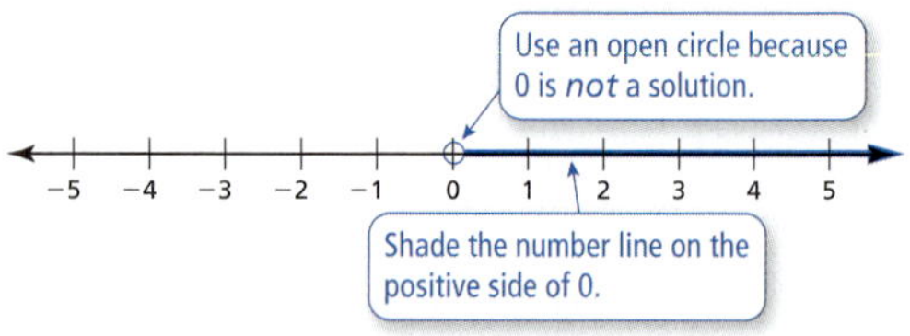

ANOTHER WAY

Another way to represent the solutions of an inequality is to use *set-builder notation*. In Example 3b, the solutions can be written as $\{x \mid x > 2\}$, which is read as "the set of all numbers x such that x is greater than 2."

Monitoring Progress

Help in English and Spanish at *BigIdeasMath.com*

Graph the inequality.

7. $b > -8$

8. $1.4 \geq g$

9. $r < \frac{1}{2}$

10. $v \geq \sqrt{36}$

Laurie's Notes Teacher Actions

- Students may well remember how to graph inequalities. A quick reminder of the open circle versus closed circle may be all that is necessary.
- The inequality symbol points in the direction of the graph when the variable is on the left side of the symbol. This is not true when the variable is on the right side.
- **Big Idea:** When looking at a graph we can read the values of the variable that make the inequality true and the values of the variable that make the inequality *not* true. Numbers *not* shaded are *not* solutions.
- Ask students to show a *Thumbs Up* indicating their understanding of graphing inequalities.
- **Monitoring Progress:** Check to make sure that students have shaded graphs in the correct direction.

Writing Linear Inequalities from Graphs

EXAMPLE 4 **Writing Inequalities from Graphs**

The graphs show the height restrictions h (in inches) for two rides at an amusement park. Write an inequality that represents the height restriction of each ride.

Ride A

Ride B

SOLUTION

Ride A

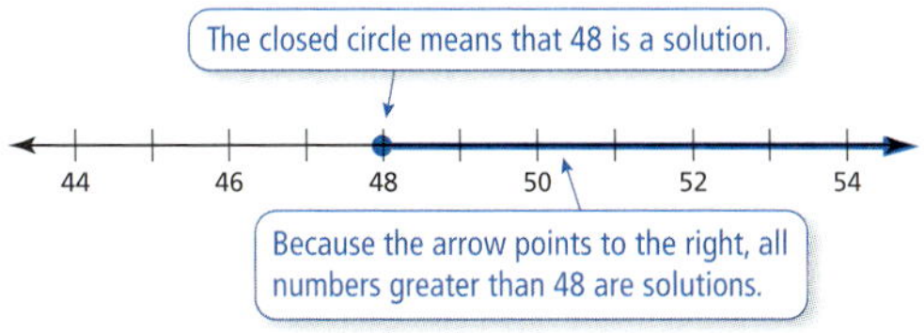

Ride B

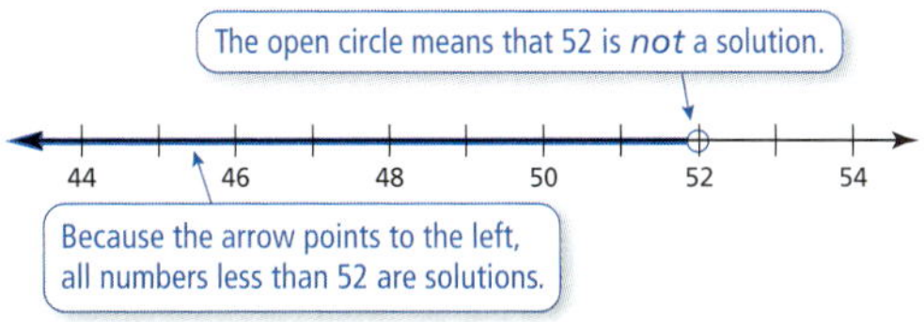

▶ So, $h \geq 48$ represents the height restriction for Ride A, and $h < 52$ represents the height restriction for Ride B.

Monitoring Progress

Help in English and Spanish at *BigIdeasMath.com*

11. Write an inequality that represents the graph.

Concept Summary

Representing Linear Inequalities

Words	Algebra	Graph
x is less than 2	$x < 2$	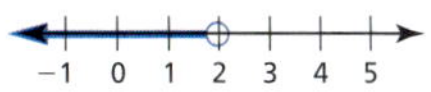
x is greater than 2	$x > 2$	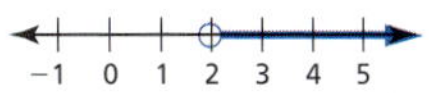
x is less than or equal to 2	$x \leq 2$	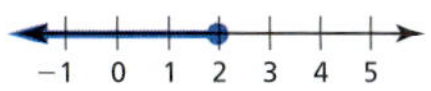
x is greater than or equal to 2	$x \geq 2$	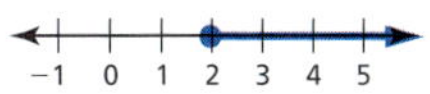

Laurie's Notes Teacher Actions

? "How many of you have been to an amusement park where there was a sign indicating a height limit or height requirement?" Answers will vary.

- **Whiteboarding:** Pose Example 4 and have students work with partners to complete it. Have students hold the whiteboards to reveal their answers.
- Have students write the *Concept Summary* in their notebooks or journals.

Closure

- Explain how to graph the inequality $x > -2$ to someone you are speaking to on the phone. *Sample answer:* On a number line, draw an open circle on -2 and shade the line to the right of -2.

Extra Example 4

The graphs show the age restrictions (in years) for two types of movie tickets. Write an inequality that represents the age restriction of each ticket.

Ticket A

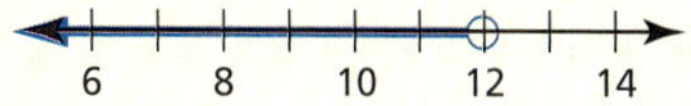

Ticket B

Ticket A: $a < 12$; Ticket B: $a \geq 62$

MONITORING PROGRESS ANSWER

11. $x \geq -6$

Assignment Guide and Homework Check

ASSIGNMENT

Basic: 1–4, 5–23 odd, 27–35 odd, 41, 45, 50, 52, 60–67

Average: 1–4, 6–46 even, 50–54 even, 60–67

Advanced: 1–4, 10–14, 22–26, 28, 34–58 even, 59–67

HOMEWORK CHECK

Basic: 9, 17, 33, 41

Average: 10, 16, 32, 44

Advanced: 12, 24, 36, 44

ANSWERS

1. inequality
2. no; If x is 5, then the value on the left simplifies to 8, which is not greater than 8.
3. Draw an open circle when a number is not part of the solution. Draw a closed circle when a number is part of the solution. Draw an arrow to the left or right to show that the graph continues in that direction.
4. w is no more than -7; $w \le -7$; $w \ge -7$
5. $x > 3$
6. $n + 7 \le 9$
7. $15 \le \frac{t}{5}$
8. $3w < 18$
9. $\frac{1}{2}y > 22$
10. $3 < s + 4$
11. $13 \ge v - 1$
12. $4 \ge \frac{x}{2.1}$
13. $w \ge 1.7$
14. $x \le 170$
15. no
16. no
17. yes
18. yes
19. yes
20. no
21. yes
22. no
23. yes
24. no
25. **a.** $h < 107$

    b. no; A height of 9 feet is equal to 108 inches, which is not less than 107 inches.

2.1 Exercises

Dynamic Solutions available at *BigIdeasMath.com*

Vocabulary and Core Concept Check

1. **COMPLETE THE SENTENCE** A mathematical sentence using the symbols <, >, ≤, or ≥ is called a(n)________.
2. **VOCABULARY** Is 5 in the solution set of $x + 3 > 8$? Explain.
3. **ATTENDING TO PRECISION** Describe how to graph an inequality.
4. **DIFFERENT WORDS, SAME QUESTION** Which is different? Write "both" inequalities.

w is greater than or equal to -7.	w is no less than -7.
w is no more than -7.	w is at least -7.

Monitoring Progress and Modeling with Mathematics

In Exercises 5–12, write the sentence as an inequality. *(See Example 1.)*

5. A number x is greater than 3.
6. A number n plus 7 is less than or equal to 9.
7. Fifteen is no more than a number t divided by 5.
8. Three times a number w is less than 18.
9. One-half of a number y is more than 22.
10. Three is less than the sum of a number s and 4.
11. Thirteen is at least the difference of a number v and 1.
12. Four is no less than the quotient of a number x and 2.1.

13. **MODELING WITH MATHEMATICS** On a fishing trip, you catch two fish. The weight of the first fish is shown. The second fish weighs at least 0.5 pound more than the first fish. Write an inequality that represents the possible weights of the second fish.

14. **MODELING WITH MATHEMATICS** There are 430 people in a wave pool. Write an inequality that represents how many more people can enter the pool.

HOURS
Monday – Friday: 10 A.M. – 6 P.M.
Saturday – Sunday: 10 A.M. – 7 P.M.
Maximum Capacity: 600

In Exercises 15–24, tell whether the value is a solution of the inequality. *(See Example 2.)*

15. $r + 4 > 8; r = 2$
16. $5 - x < 8; x = -3$
17. $3s \le 19; s = -6$
18. $17 \ge 2y; y = 7$
19. $-1 > -\frac{x}{2}; x = 3$
20. $\frac{4}{z} \ge 3; z = 2$
21. $14 \ge -2n + 4; n = -5$
22. $-5 \div (2s) < -1; s = 10$
23. $20 \le \frac{10}{2z} + 20; z = 5$
24. $\frac{3m}{6} - 2 > 3; m = 8$

25. **MODELING WITH MATHEMATICS** The tallest person who ever lived was approximately 8 feet 11 inches tall.

    a. Write an inequality that represents the heights of every other person who has ever lived.

    b. Is 9 feet a solution of the inequality? Explain.

26. **DRAWING CONCLUSIONS** The winner of a weight-lifting competition bench-pressed 400 pounds. The other competitors all bench-pressed at least 23 pounds less.

a. Write an inequality that represents the weights that the other competitors bench-pressed.

b. Was one of the other competitors able to bench-press 379 pounds? Explain.

ERROR ANALYSIS In Exercises 27 and 28, describe and correct the error in determining whether 8 is in the solution set of the inequality.

27. ✗

$$-y + 7 < -4$$
$$-8 + 7 \overset{?}{<} -4$$
$$-1 < -4$$

8 is in the solution set.

28. ✗

$$\frac{1}{2}x + 2 \leq 6$$
$$\frac{1}{2}(8) + 2 \overset{?}{\leq} 6$$
$$4 + 2 \overset{?}{\leq} 6$$
$$6 \leq 6$$

8 is *not* in the solution set.

In Exercises 29–36, graph the inequality. *(See Example 3.)*

29. $x \geq 2$

30. $z \leq 5$

31. $-1 > t$

32. $-2 < w$

33. $v \leq -4$

34. $s < 1$

35. $\frac{1}{4} < p$

36. $r \geq -|5|$

In Exercises 37–40, write and graph an inequality for the given solution set.

37. $\{x \mid x < 7\}$

38. $\{n \mid n \geq -2\}$

39. $\{z \mid 1.3 \leq z\}$

40. $\{w \mid 5.2 > w\}$

In Exercises 41–44, write an inequality that represents the graph. *(See Example 4.)*

41. 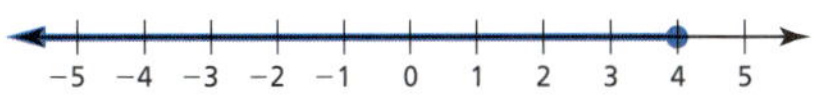

42. 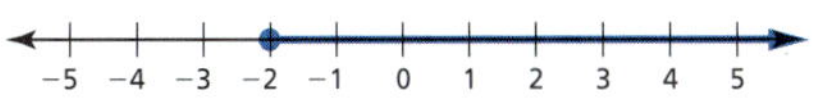

43. 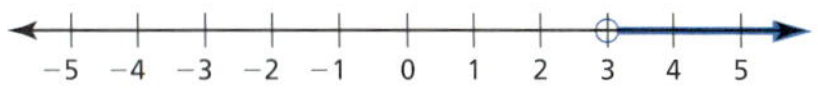

44. 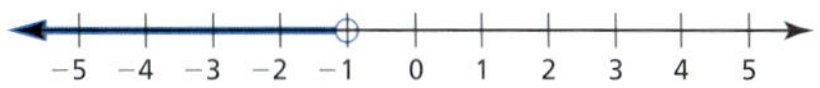

45. **ANALYZING RELATIONSHIPS** The water temperature of a swimming pool must be no less than 76°F. The temperature is currently 74°F. Which graph correctly shows how much the temperature needs to increase? Explain your reasoning.

Ⓐ 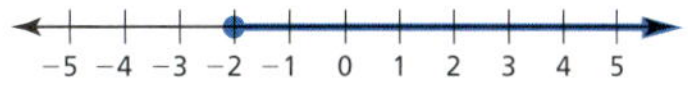

Ⓑ 

Ⓒ 

Ⓓ 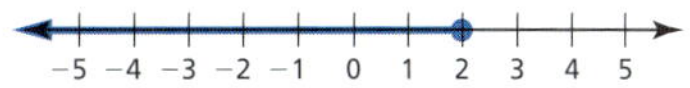

46. **MODELING WITH MATHEMATICS** According to a state law for vehicles traveling on state roads, the maximum total weight of a vehicle and its contents depends on the number of axles on the vehicle. For each type of vehicle, write and graph an inequality that represents the possible total weights w (in pounds) of the vehicle and its contents.

Maximum Total Weights

2 axles, 40,000 lb	3 axles, 60,000 lb	4 axles, 80,000 lb

47. **PROBLEM SOLVING** The Xianren Bridge is located in Guangxi Province, China. This arch is the world's longest natural arch, with a length of 400 feet. Write and graph an inequality that represents the lengths ℓ (in *inches*) of all other natural arches.

Dynamic Teaching Tools

- Dynamic Assessment & Progress Monitoring Tool
- Interactive Whiteboard Lesson Library
- Dynamic Classroom with Dynamic Investigations

ANSWERS

26. a. $x \leq 377$

b. no; A weight of 379 pounds is not a solution of the inequality in part (a).

27. Because -1 is not less than -4, the final result is not true; $-1 \nless -4$; 8 is not in the solution set.

28. Because 6 equals 6, the final inequality is true; 8 is in the solution set.

29. 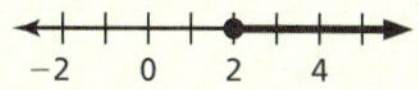

30. 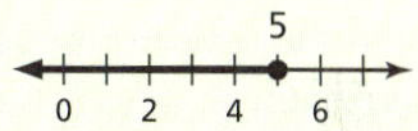

31. 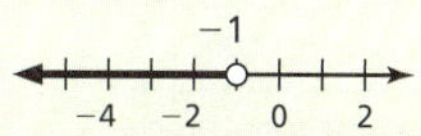

32. 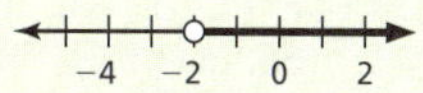

33. 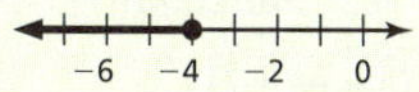

34. 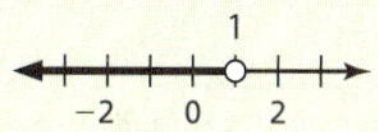

35. 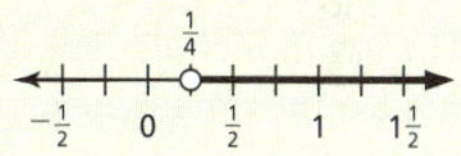

36. 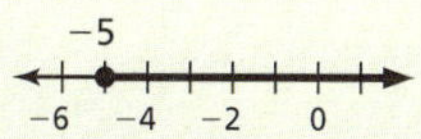

37. $x < 7$

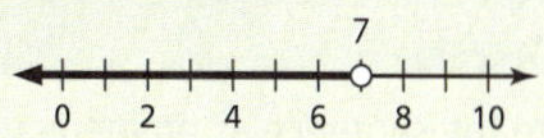

38. $n \geq -2$

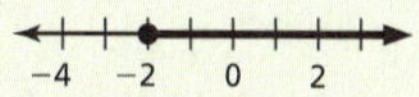

39. $1.3 \leq z$

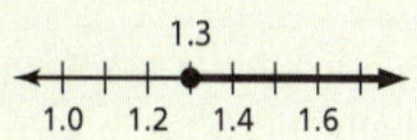

40. $5.2 > w$

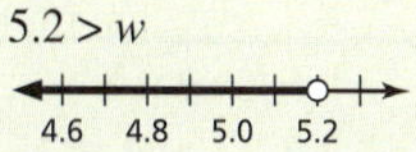

41. $x \leq 4$

42. $x \geq -2$

43. $x > 3$

44. $x < -1$

45. C; The temperature must be at least 2°F warmer, so the increase is represented by $x \geq 2$.

46. 2 axles: $w \leq 40,000$

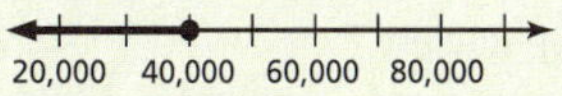

3 axles: $w \leq 60,000$

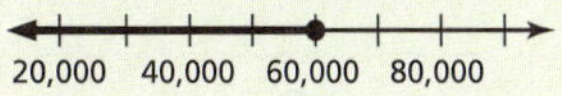

4 axles: $w \leq 80,000$

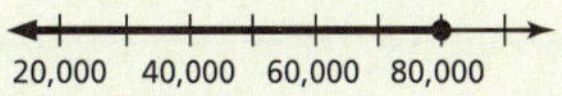

47. $\ell < 4800$

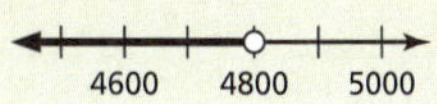

ANSWERS

48. *Sample answer:* $h \leq \frac{25}{7}$

49. *Sample answer:* You spend \$23 on admission and x dollars on snacks, and you can spend no more than \$31 total.

50. **a.** $T \geq -38.87$

b. yes; The graph shows $-38.87°C$ as the lowest possible melting point.

51. $0.90x \leq 24$; yes; Because $0.9(25) = \$22.50$, which is less than \$24, the inequality is true.

52. cousin; Because the inequality is "less than or equal to," a weight equal to the given amount is possible.

53. *Sample answer:* A temperature above the freezing point of water can be represented by $T > 0$ if the temperature is in degrees Celsius, or by $T > 32$ if the temperature is in degrees Fahrenheit.

54. $b < 17{,}000$

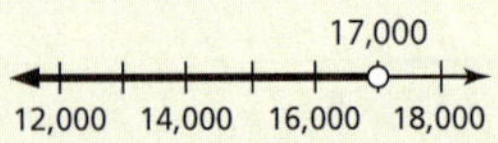

55. $x < 14$ **56.** $x \geq 1.6$

57. $x < 3$ **58.** $x > 6$

59. **a.** $r > \frac{40}{7}$ (about 5.71)

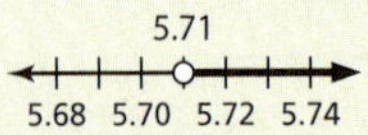

b. no; The graph includes speeds beyond the maximum speed a human can run.

60–67. See Additional Answers.

Mini-Assessment

1. Write the sentence as an inequality.
   One-half is at least 3 times a number p. $\frac{1}{2} \geq 3p$
2. Tell whether -2 is a solution of the inequality $n - 5 < 8$. -2 is a solution.
3. Graph the inequality $s > -3$.

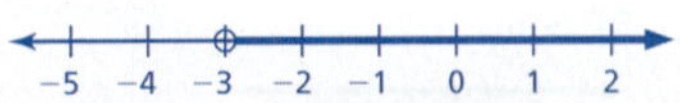

4. Write an inequality that represents the graph.

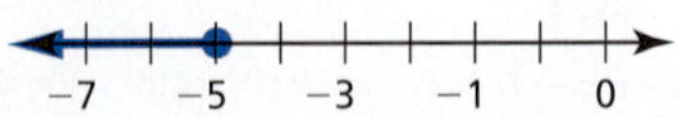

$x \leq -5$

48. **THOUGHT PROVOKING** A student works no more than 25 hours each week at a part-time job. Write an inequality that represents how many hours the student can work each day.

49. **WRITING** Describe a real-life situation modeled by the inequality $23 + x \leq 31$.

50. **HOW DO YOU SEE IT?** The graph represents the known melting points of all metallic elements (in degrees Celsius).

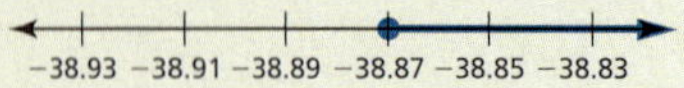

a. Write an inequality represented by the graph.

b. Is it possible for a metallic element to have a melting point of $-38.87°C$? Explain.

51. **DRAWING CONCLUSIONS** A one-way ride on a subway costs \$0.90. A monthly pass costs \$24. Write an inequality that represents how many one-way rides you can buy before it is cheaper to buy the monthly pass. Is it cheaper to pay the one-way fare for 25 rides? Explain.

Subway Prices	
One-way ride	\$0.90
Monthly pass	\$24.00

52. **MAKING AN ARGUMENT** The inequality $x \leq 1324$ represents the weights (in pounds) of all mako sharks ever caught using a rod and reel. Your friend says this means no one using a rod and reel has ever caught a mako shark that weighs 1324 pounds. Your cousin says this means someone using a rod and reel *has* caught a mako shark that weighs 1324 pounds. Who is correct? Explain your reasoning.

53. **CRITICAL THINKING** Describe a real-life situation that can be modeled by more than one inequality.

54. **MODELING WITH MATHEMATICS** In 1997, Superman's cape from the 1978 movie *Superman* was sold at an auction. The winning bid was \$17,000. Write and graph an inequality that represents the amounts all the losing bids.

MATHEMATICAL CONNECTIONS In Exercises 55–58, write an inequality that represents the missing dimension x.

55. The area is less than 42 square meters.

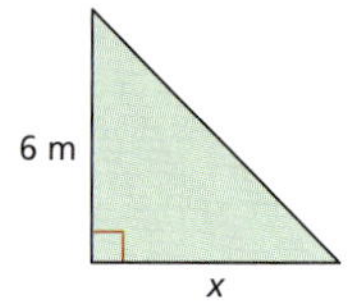

56. The area is greater than or equal to 8 square feet.

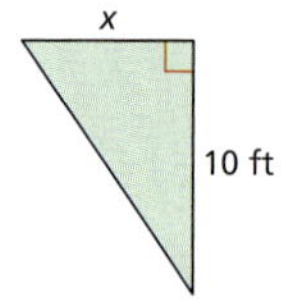

57. The area is less than 18 square centimeters.

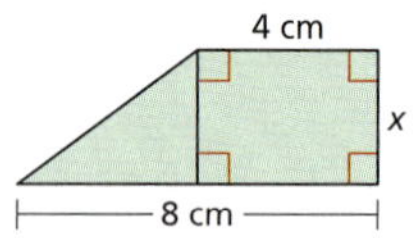

58. The area is greater than 12 square inches.

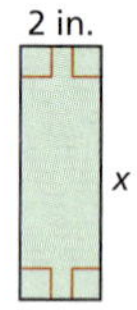

59. **WRITING** A runner finishes a 200-meter dash in 35 seconds. Let r represent any speed (in meters per second) faster than the runner's speed.

a. Write an inequality that represents r. Then graph the inequality.

b. Every point on the graph represents a speed faster than the runner's speed. Do you think every point could represent the speed of a runner? Explain.

Maintaining Mathematical Proficiency

Reviewing what you learned in previous grades and lessons

Solve the equation. Check your solution. *(Section 1.1)*

60. $x + 2 = 3$

61. $y - 9 = 5$

62. $6 = 4 + y$

63. $-12 = y - 11$

Solve the literal equation for x. *(Section 1.5)*

64. $v = x \cdot y \cdot z$

65. $s = 2r + 3x$

66. $w = 5 + 3(x - 1)$

67. $n = \frac{2x + 1}{2}$

If students need help...	*If students got it...*
Resources by Chapter • Practice A and Practice B • Puzzle Time	Resources by Chapter • Enrichment and Extension • Cumulative Review
Student Journal • Practice	Start the *next* Section
Differentiating the Lesson Skills Review Handbook	

Laurie's Notes

Overview of Section 2.2

Introduction

- In this lesson, students will write, solve, and graph inequalities that involve addition and subtraction. Students solved equations in the last chapter and inequalities in middle school. Conceptually the process is the same.
- **Look For and Make Use of Structure:** Mathematically proficient students will recognize the similarity between solving $x + 7 = -12$ and $x + 7 < -12$.

Common Misconceptions

- When students encounter an inequality involving subtraction, they may immediately think, "I need to add in order to solve this inequality." In this example: $x - (-4) > 10$, they should first simplify the inequality and write $x + 4 > 10$. To solve, they subtract 4 from each side of the inequality. Show the following examples in parallel and have students discuss the approaches.

$$\begin{array}{rr} x - (-4) > & 10 \\ \underline{+(-4)} & \underline{+(-4)} \\ x > & 6 \end{array} \qquad\qquad \begin{array}{rr} x + 4 > & 10 \\ \underline{-4} & \underline{-4} \\ x > & 6 \end{array}$$

Formative Assessment Tips

- **Think-Pair-Share:** This technique allows students to share their thinking about a problem with their partners after they have had time to consider the problem alone. Once partners have discussed the problem, small groups or the whole class should discuss the problem.
- The initial time working alone is important for students to develop their own understanding of the mathematics. "Private think time" is what I call it. Once students have engaged in the problem, sharing with partners helps confirm their understanding or perhaps the need to modify their thinking. Sharing your thinking with the whole class is more comfortable for students when they have had the chance to discuss their thinking with partners.
- **Construct Viable Arguments and Critique the Reasoning of Others:** Using this formative assessment technique allows you to check students' conceptual knowledge and their ability to construct a viable argument.

Applications

- Inequalities are more common than students might guess. Applications often involve measurement (weight, height, length). For instance, airlines have guidelines for the maximum weight of luggage, depending upon whether it is carry-on or checked luggage, or whether the travel is domestic or international.

Another Way

- Introduce the Addition or Subtraction Properties of Inequalities with a personal example. I summarize these two properties in the following way:

  I am older than you. In two years, I will still be older than you.

  Laurie's age > Student's age — if $a > b$,
  Laurie's age + 2 > Student's age + 2 — then $a + c > b + c$

  Two years ago, I was older than you.

  Laurie's age > Student's age — if $a > b$,
  Laurie's age − 2 > Student's age − 2 — then $a - c > b - c$

Pacing Suggestion

- Once students have worked the explorations, continue with the formal lesson.

Dynamic Teaching Tools

- Dynamic Assessment & Progress Monitoring Tool
- Lesson Planning Tool
- Interactive Whiteboard Lesson Library
- Dynamic Classroom with Dynamic Investigations

What Your Students Will Learn

- Solve linear inequalities in one variable by adding or subtracting.

Laurie's Notes

Exploration

Motivate

- Wear a football-related piece of clothing today if you own one.
- Set the tone by tossing a few passes in class with a small foam football. Ask a statistician to record your efforts in a table at the board. Use three columns: **C**ompleted, I**N**tercepted, and Inco**M**plete.
- I recommend **A**ttempting 10 short passes to students nearby. You may need to give permission to have a pass intercepted.

? Ask the following questions.
  - "How many passes did I attempt?" 10 Record this next to the table.
  - "Can I complete more passes than I attempt?" no
  - "Are completed passes + incomplete passes always less than or equal to attempted passes?" yes
  - "Are completed passes + incomplete passes always less than attempted passes?" No, they could be equal.

Exploration 1

- The tree diagram should be a helpful aid to students.
- Discuss students' answers and their reasoning when they have finished.
- For parts (c) and (d), point out the need to pay attention to the inequality symbol. It is possible, though unlikely, that $T = C$. In that case, the inequality $T < C$ may *not* be true, while the inequality $T \leq C$ is true.
- **Reason Abstractly and Quantitatively:** There will be some heated discussion about the inequalities, but remember to ask, "is it *possible*" versus "is it *probable*." Take time for students to explain their reasoning.

Exploration 2

- In this exploration, two of the inequalities force students to think about solving the inequality before they begin. In part (b), if $P + 100 > 250$, then it means $P > 150$. Students should decontextualize the situation and manipulate the symbols to solve the inequality.
- Answers will vary for this exploration, but suggest to students that they keep the numbers as simple as possible. For instance, one possible answer for the first question is (1, 0, 0, 0, 1), where only one pass is attempted and it is intercepted. The result is $P = -200$, and so $P < 0$.
- **Reason Abstractly and Quantitatively:** Students will need to do a little trial and error with these problems. They should ask themselves, "What happens when I increase this variable; does another variable decrease?" Students need the opportunity to practice this type of reasoning.

Communicate Your Answer

- Students should recognize the similarity between solving $x + 3 < 4$ and $x + 3 = 4$.

Connecting to Next Step

- The inequalities students have worked in the explorations have been done using mental math. In the formal lesson, they will solve inequalities symbolically and graph the solutions.

2.2 Solving Inequalities Using Addition or Subtraction

Essential Question How can you use addition or subtraction to solve an inequality?

EXPLORATION 1 **Quarterback Passing Efficiency**

Work with a partner. The National Collegiate Athletic Association (NCAA) uses the following formula to rank the passing efficiencies P of quarterbacks.

$$P = \frac{8.4Y + 100C + 330T - 200N}{A}$$

Y = total length of all completed passes (in Yards)

C = Completed passes

T = passes resulting in a Touchdown

N = iNtercepted passes

A = Attempted passes

M = incoMplete passes

MODELING WITH MATHEMATICS

To be proficient in math, you need to identify and analyze important relationships and then draw conclusions, using tools such as diagrams, flowcharts, and formulas.

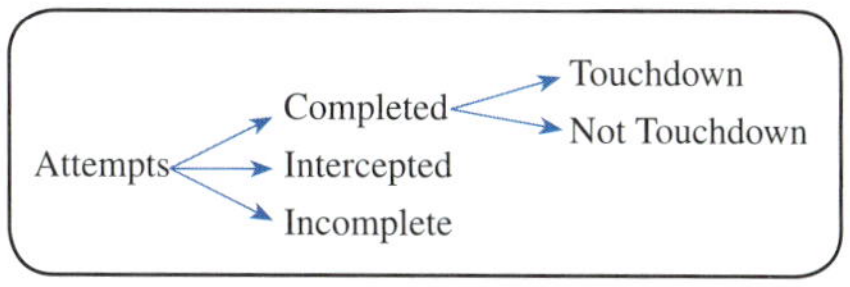

Determine whether each inequality must be true. Explain your reasoning.

a. $T < C$ **b.** $C + N \leq A$ **c.** $N < A$ **d.** $A - C \geq M$

EXPLORATION 2 **Finding Solutions of Inequalities**

Work with a partner. Use the passing efficiency formula to create a passing record that makes each inequality true. Record your results in the table. Then describe the values of P that make each inequality true.

Attempts	Completions	Yards	Touchdowns	Interceptions

a. $P < 0$

b. $P + 100 \geq 250$

c. $P - 250 > -80$

Communicate Your Answer

3. How can you use addition or subtraction to solve an inequality?

4. Solve each inequality.

a. $x + 3 < 4$ **b.** $x - 3 \geq 5$

c. $4 > x - 2$ **d.** $-2 \leq x + 1$

Dynamic Teaching Tools

- Dynamic Assessment & Progress Monitoring Tool
- Lesson Planning Tool
- Interactive Whiteboard Lesson Library
- Dynamic Classroom with Dynamic Investigations

ANSWERS

1. a. no; It is also possible for the number of touchdowns to equal the number of completed passes.

b. yes; Completed and intercepted passes are both included in attempts.

c. no; It is also possible for the number of intercepted passes to equal the number of attempts.

d. yes; After subtracting completed passes from attempts, all remaining passes could be incomplete.

2. a. *Sample answer:* 25; 5; 100; 0; 10; negative numbers

b. *Sample answer:* 10; 9; 600; 4; 0; numbers greater than or equal to 150

c. *Sample answer:* 10; 7; 300; 2; 1; numbers greater than 170

3. Add or subtract the same number to or from each side of the inequality to create an equivalent inequality.

4. a. $x < 1$

b. $x \geq 8$

c. $x < 6$

d. $x \geq -3$

Differentiated Instruction

Kinesthetic

To help students understand the difference between the solutions of equations and inequalities, write the equation $x + 5 = 10$ and the inequality $x + 5 \geq 10$ on the board. Ask two students to assist you at the board. Assign the equation to one student to solve and graph on a number line. Assign the inequality to the second student to solve and graph on a number line. Have students discuss how and why the solutions are different.

Extra Example 1

Solve $x - 5 < -3$. Graph the solution.

$x < 2$

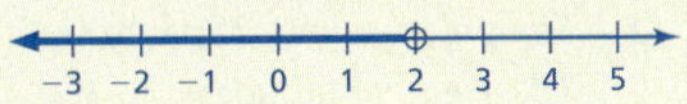

MONITORING PROGRESS ANSWERS

1. $b > -7$

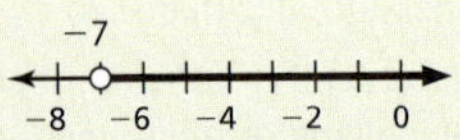

2. $m \leq 8$

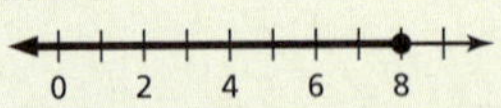

3. $y < \frac{1}{2}$

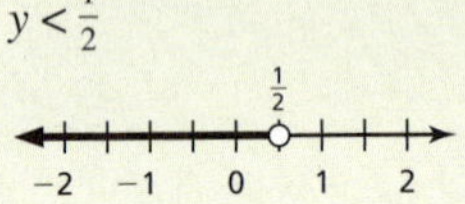

2.2 Lesson

Core Vocabulary

equivalent inequalities, *p. 62*

Previous

inequality

What You Will Learn

- Solve inequalities using addition.
- Solve inequalities using subtraction.
- Use inequalities to solve real-life problems.

Solving Inequalities Using Addition

Just as you used the properties of equality to produce equivalent equations, you can use the properties of inequality to produce equivalent inequalities. **Equivalent inequalities** are inequalities that have the same solutions.

Core Concept

Addition Property of Inequality

Words Adding the same number to each side of an inequality produces an equivalent inequality.

Numbers

$$\begin{array}{rcl} -3 & < & 2 \\ +4 & & +4 \\ \hline 1 & < & 6 \end{array} \qquad \begin{array}{rcl} -3 & \geq & -10 \\ +3 & & +3 \\ \hline 0 & \geq & -7 \end{array}$$

Algebra If $a > b$, then $a + c > b + c$. If $a \geq b$, then $a + c \geq b + c$.
If $a < b$, then $a + c < b + c$. If $a \leq b$, then $a + c \leq b + c$.

The diagram shows one way to visualize the Addition Property of Inequality when $c > 0$.

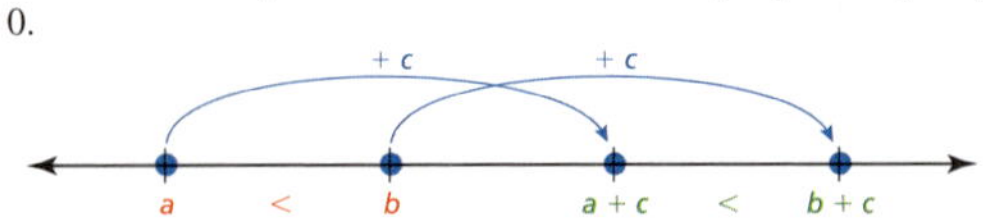

EXAMPLE 1 Solving an Inequality Using Addition

Solve $x - 6 \geq -10$. Graph the solution.

SOLUTION

$x - 6 \geq -10$	Write the inequality.
$+6 \quad +6$	Add 6 to each side.
$x \geq -4$	Simplify.

Addition Property of Inequality →

▶ The solution is $x \geq -4$.

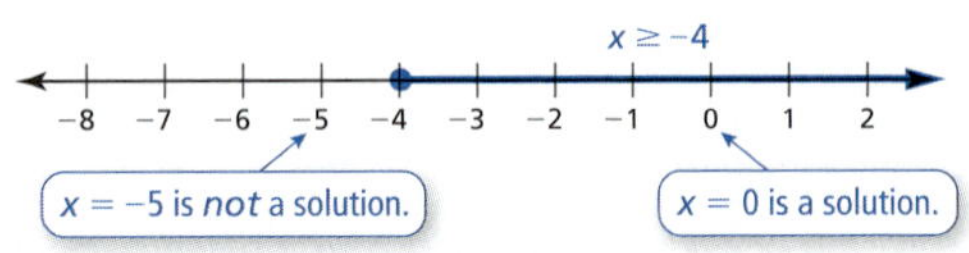

REMEMBER
To check this solution, substitute a few numbers to the left and right of −4 into the original inequality.

Monitoring Progress

Help in English and Spanish at *BigIdeasMath.com*

Solve the inequality. Graph the solution.

1. $b - 2 > -9$ **2.** $m - 3 \leq 5$ **3.** $\frac{1}{4} > y - \frac{1}{4}$

Laurie's Notes Teacher Actions

- Define *equivalent inequalities* and write the *Core Concept*.
- **Look For and Make Use of Structure:** This property should look very familiar, as it is similar to the Addition Property of Equality. Demonstrating the property using just numbers helps students make sense of the property.
- Discuss the visual representation of the property.
- **Attend to Precision:** In Example 1, remind students that they are using the Addition Property of Inequality.

COMMON ERROR In graphing the solution $x \geq -4$, students may shade in the wrong direction by thinking $-5 > -4$. Remind them that the further to the right you are on a number line, the greater the number.

Solving Inequalities Using Subtraction

Core Concept

Subtraction Property of Inequality

Words Subtracting the same number from each side of an inequality produces an equivalent inequality.

Numbers

$$\begin{array}{rcl} -3 & \le & 1 \\ -5 & & -5 \\ \hline -8 & \le & -4 \end{array} \qquad \begin{array}{rcl} 7 & > & -20 \\ -7 & & -7 \\ \hline 0 & > & -27 \end{array}$$

Algebra If $a > b$, then $a - c > b - c$. If $a \ge b$, then $a - c \ge b - c$.

If $a < b$, then $a - c < b - c$. If $a \le b$, then $a - c \le b - c$.

The diagram shows one way to visualize the Subtraction Property of Inequality when $c > 0$.

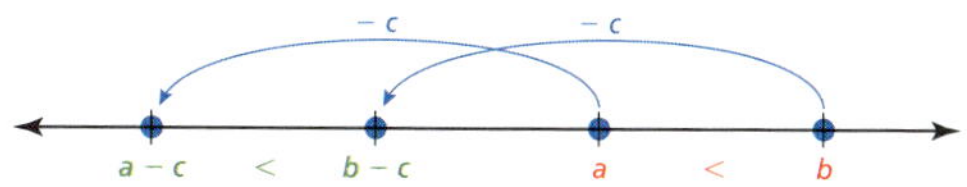

EXAMPLE 2 Solving an Inequality Using Subtraction

Solve each inequality. Graph the solution.

a. $y + 8 \le 5$ **b.** $-8 < 1.4 + m$

SOLUTION

a.

	$y + 8 \le \ \ 5$	Write the inequality.
Subtraction Property of Inequality →	$-8 \quad -8$	Subtract 8 from each side.
	$y \le -3$	Simplify.

▶ The solution is $y \le -3$.

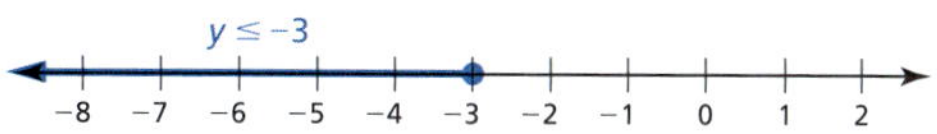

b.

	$-8 < \ \ 1.4 + m$	Write the inequality.
Subtraction Property of Inequality →	$-1.4 \quad -1.4$	Subtract 1.4 from each side.
	$-9.4 < m$	Simplify.

▶ The solution is $m > -9.4$.

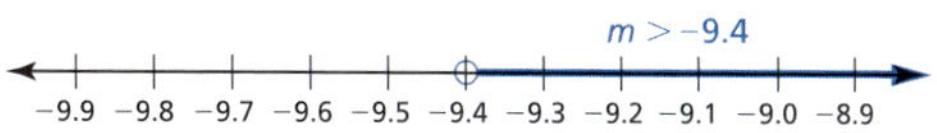

Monitoring Progress

Help in English and Spanish at *BigIdeasMath.com*

Solve the inequality. Graph the solution.

4. $k + 5 \le -3$ **5.** $\frac{5}{6} \le z + \frac{1}{6}$ **6.** $p + 0.7 > -2.3$

Extra Example 2

Solve each inequality. Graph the solution.

a. $s + 18 \ge 21$ $s \ge 3$

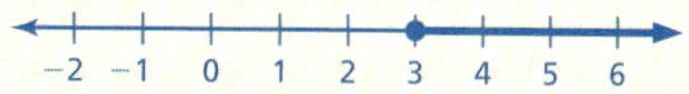

b. $-5.2 > x + 2$ $x < -7.2$

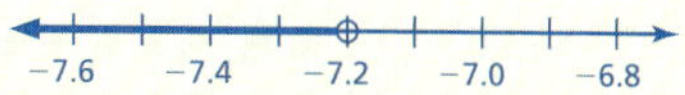

MONITORING PROGRESS ANSWERS

4. $k \le -8$

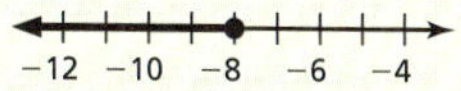

5. $z \ge \frac{2}{3}$

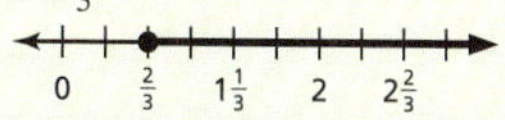

6. $p > -3$

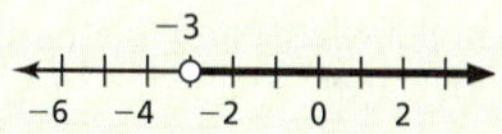

Laurie's Notes — Teacher Actions

- The Subtraction Property of Inequality is first demonstrated with numbers. In the example shown, students may need to be reminded that $-8 \le -4$.

? "Can the inequality in Example 2(b) be written as $-8 < m + 1.4$? Explain." yes; Commutative Property

- Encourage students to check their solution by selecting a value in the solution set and substituting it into the original inequality.
- **Attend to Precision:** When graphing the solution $m > -9.4$, be sure students scale the number line correctly.

Extra Example 3

An electric grill that uses 1600 watts of electricity is plugged into the circuit described in Example 3.

a. Write and solve an inequality that represents how many watts you can add to the circuit without overloading the circuit. $1600 + w < 1800$; $w < 200$; You can add up to 200 watts.

b. In addition to the electric grill, which of the appliances listed in the table in Example 3 can you plug into the circuit at the same time without overloading the circuit? clock radio

MONITORING PROGRESS ANSWER

7. no; The additional electricity must be less than 800 watts, and the toaster uses 800 watts.

Solving Real-Life Problems

EXAMPLE 3 **Modeling with Mathematics**

A circuit overloads at 1800 watts of electricity. You plug a microwave oven that uses 1100 watts of electricity into the circuit.

a. Write and solve an inequality that represents how many watts you can add to the circuit without overloading the circuit.

b. In addition to the microwave oven, which of the following appliances can you plug into the circuit at the same time without overloading the circuit?

Appliance	Watts
Clock radio	50
Blender	300
Hot plate	1200
Toaster	800

SOLUTION

1. **Understand the Problem** You know that the microwave oven uses 1100 watts out of a possible 1800 watts. You are asked to write and solve an inequality that represents how many watts you can add without overloading the circuit. You also know the numbers of watts used by four other appliances. You are asked to identify the appliances you can plug in at the same time without overloading the circuit.
2. **Make a Plan** Use a verbal model to write an inequality. Then solve the inequality and identify other appliances that you can plug into the circuit at the same time without overloading the circuit.
3. **Solve the Problem**

Words Watts used by microwave oven + Additional watts < Overload wattage

Variable Let w be the additional watts you can add to the circuit.

Inequality $1100 \quad + \quad w \quad < \quad 1800$

$1100 + w < 1800$ Write the inequality.

Subtraction Property of Inequality → $-1100 \quad\quad -1100$ Subtract 1100 from each side.

$w < 700$ Simplify.

▶ You can add up to 700 watts to the circuit, which means that you can also plug in the clock radio and the blender.

4. **Look Back** You can check that your answer is correct by adding the numbers of watts used by the microwave oven, clock radio, and blender.

$1100 + 50 + 300 = 1450$

The circuit will not overload because the total wattage is less than 1800 watts.

Monitoring Progress

Help in English and Spanish at *BigIdeasMath.com*

7. The microwave oven uses only 1000 watts of electricity. Does this allow you to have both the microwave oven and the toaster plugged into the circuit at the same time? Explain your reasoning.

Laurie's Notes Teacher Actions

- The problem posed is not difficult to solve with mental math. The point is not that you can solve the problem mentally. The practice you want to provide students is the opportunity to *read* a problem, *make sense* of the context, and *translate* words into symbols. This is a building block for future problems. I ask a volunteer to read Steps 1 and 2 in the problem-solving plan.
- **FYI:** Most modern residential circuits are 15 or 20 amps, so there is a maximum load of either (15A × 120V =) 1800 watts or (20A × 120V =) 2400 watts before the breaker trips.
- **Use Appropriate Tools Strategically:** The color coding is a tool to help students connect words and symbols.
- **Think-Pair-Share:** Partners should work Question 7 together.

Closure

- **Think-Pair-Share:** Solve and graph.

$x + 3.8 \leq -9$ $\quad x \leq -12.8$

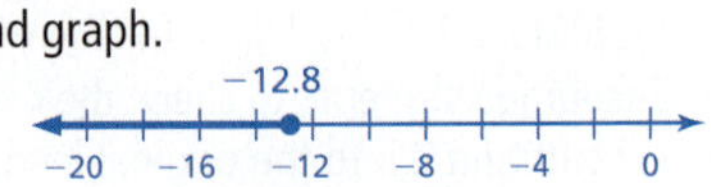

$\frac{2}{5} > x - \frac{3}{4}$ $\quad x < 1\frac{3}{20}$

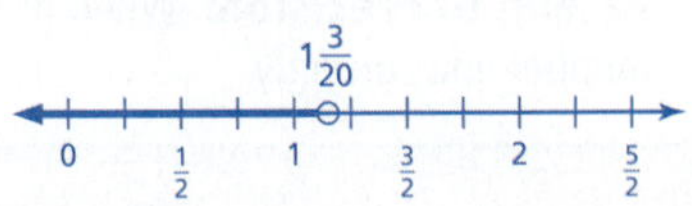

2.2 Exercises

Dynamic Solutions available at *BigIdeasMath.com*

Vocabulary and Core Concept Check

1. **VOCABULARY** Why is the inequality $x \leq 6$ equivalent to the inequality $x - 5 \leq 6 - 5$?
2. **WRITING** Compare solving equations using addition with solving inequalities using addition.

Monitoring Progress and Modeling with Mathematics

In Exercises 3–6, tell which number you would add to or subtract from each side of the inequality to solve it.

3. $k + 11 < -3$
4. $v - 2 > 14$
5. $-1 \geq b - 9$
6. $-6 \leq 17 + p$

In Exercises 7–20, solve the inequality. Graph the solution. *(See Examples 1 and 2.)*

7. $x - 4 < -5$
8. $1 \leq s - 8$
9. $6 \geq m - 1$
10. $c - 12 > -4$
11. $r + 4 < 5$
12. $-8 \leq 8 + y$
13. $9 + w > 7$
14. $15 \geq q + 3$
15. $h - (-2) \geq 10$
16. $-6 > t - (-13)$
17. $j + 9 - 3 < 8$
18. $1 - 12 + y \geq -5$
19. $10 \geq 3p - 2p - 7$
20. $18 - 5z + 6z > 3 + 6$

In Exercises 21–24, write the sentence as an inequality. Then solve the inequality.

21. A number plus 8 is greater than 11.
22. A number minus 3 is at least -5.
23. The difference of a number and 9 is fewer than 4.
24. Six is less than or equal to the sum of a number and 15.
25. **MODELING WITH MATHEMATICS** You are riding a train. Your carry-on bag can weigh no more than 50 pounds. Your bag weighs 38 pounds. *(See Example 3.)*
    a. Write and solve an inequality that represents how much weight you can add to your bag.
    b. Can you add both a 9-pound laptop and a 5-pound pair of boots to your bag without going over the weight limit? Explain.
26. **MODELING WITH MATHEMATICS** You order the hardcover book shown from a website that offers free shipping on orders of $25 or more. Write and solve an inequality that represents how much more you must spend to get free shipping.

Price: $19.76

ERROR ANALYSIS In Exercises 27 and 28, describe and correct the error in solving the inequality or graphing the solution.

27. ✗

$$-17 < x - 14$$
$$-17 + 14 < x - 14 + 14$$
$$-3 < x$$

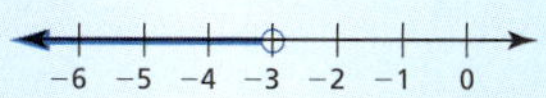

28. ✗

$$-10 + x \geq -9$$
$$-10 + 10 + x \geq -9$$
$$x \geq -9$$

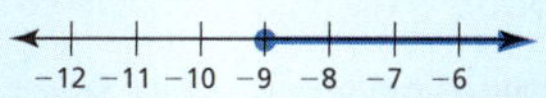

29. **PROBLEM SOLVING** An NHL hockey player has 59 goals so far in a season. What are the possible numbers of additional goals the player can score to match or break the NHL record of 92 goals in a season?

Assignment Guide and Homework Check

ASSIGNMENT

Basic: 1–6, 7–29 odd, 30, 36, 39–46

Average: 1, 2–32 even, 36–46

Advanced: 1–2, 16–20, 22, 24–26, 28–38 even, 39–46

HOMEWORK CHECK

Basic: 5, 7, 13, 25

Average: 8, 18, 22, 26

Advanced: 16, 17, 24, 25

ANSWERS

1. Subtraction Property of Inequality
2. *Sample answer:* In both cases the same number is added to each side to obtain an equivalent mathematical sentence.
3. subtract 11
4. add 2
5. add 9
6. subtract 17
7. $x < -1$

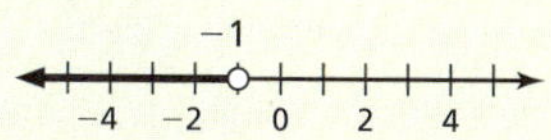

8. $s \geq 9$

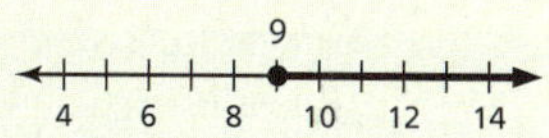

9. $m \leq 7$

10. $c > 8$

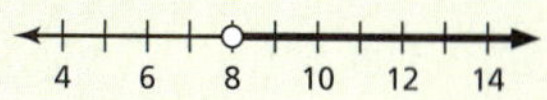

11. $r < 1$

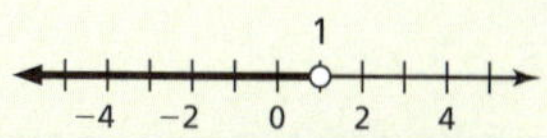

12. $y \geq -16$

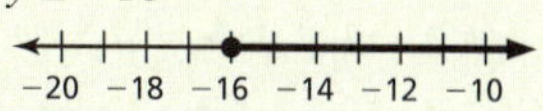

13. $w > -2$

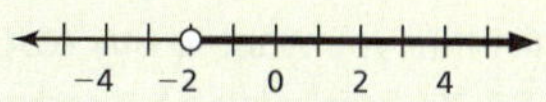

14. $q \leq 12$

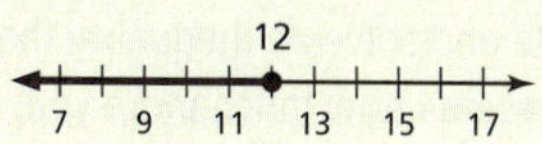

15. $h \geq 8$

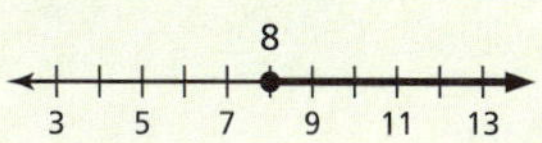

16. $t < -19$

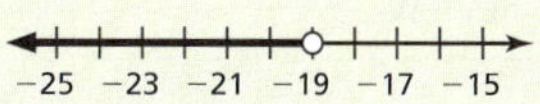

17. $j < 2$

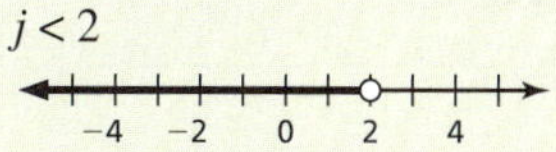

18. $y \geq 6$

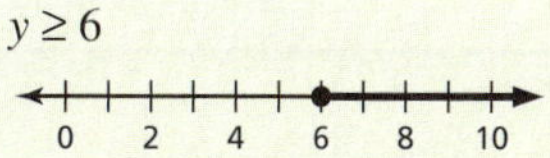

19. $p \leq 17$

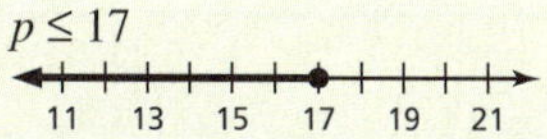

20. $z > -9$

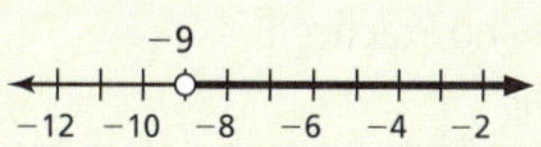

21. $n + 8 > 11; n > 3$
22. $n - 3 \geq -5; n \geq -2$
23. $n - 9 < 4; n < 13$
24. $6 \leq n + 15; n \geq -9$

25–29. See Additional Answers.

Dynamic Teaching Tools

- Dynamic Assessment & Progress Monitoring Tool
- Interactive Whiteboard Lesson Library
- Dynamic Classroom with Dynamic Investigations

ANSWERS

30. **a.** score greater than 117.4 points

b. they both are; Both scores are greater than 117.4 points.

31. A; Subtract 3 from each side; D; order of inequality reverses for opposites

32. $14.2 + 15.5 + x < 51.3$; $x < 21.6$

33. $6.4 + 4.9 + 4.1 + x \leq 18.7$; $x \leq 3.3$

34. *Sample answer:* $x + 34 \leq 50$; A shelf can safely support 50 pounds. The combined weight of items on the shelf is 34 pounds. x represents the possible weights of items that can be put on the shelf safely.

35. no; *Sample answer:* 3, 7, 8, 9, and 12; There are infinitely many solutions. Check 8 and a few numbers greater than and less than 8.

36–46. See Additional Answers.

Mini-Assessment

Solve the inequality. Graph the solution.

1. $x - 2 > -3$ $x > -1$

2. $4 \geq m - 3$ $m \leq 7$

3. $r + 5 < 3$ $r < -2$

4. $-8 + y \geq -4$ $y \geq 4$

5. A shirt that you want to buy costs \$15.50. You can use a \$5 discount coupon if you spend at least \$20. Write and solve an inequality that represents how much more you must spend to use the \$5 coupon.
$x + 15.5 \geq 20$; $x \geq 4.5$; at least \$4.50 more

30. MAKING AN ARGUMENT In an aerial ski competition, you perform two acrobatic ski jumps. The scores on the two jumps are then added together.

Ski jump	Competitor's score	Your score
1	117.1	119.5
2	119.8	

a. Describe the score that you must earn on your second jump to beat your competitor.

b. Your coach says that you will beat your competitor if you score 118.4 points. A teammate says that you only need 117.5 points. Who is correct? Explain.

31. REASONING Which of the following inequalities are equivalent to the inequality $x - b < 3$, where b is a constant? Justify your answer.

Ⓐ $x - b - 3 < 0$ Ⓑ $0 > b - x + 3$

Ⓒ $x < 3 - b$ Ⓓ $-3 < b - x$

MATHEMATICAL CONNECTIONS In Exercises 32 and 33, write and solve an inequality to find the possible values of x.

32. Perimeter < 51.3 inches

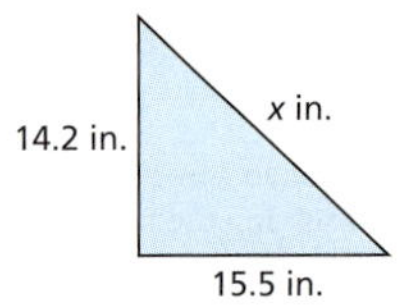

33. Perimeter ≤ 18.7 feet

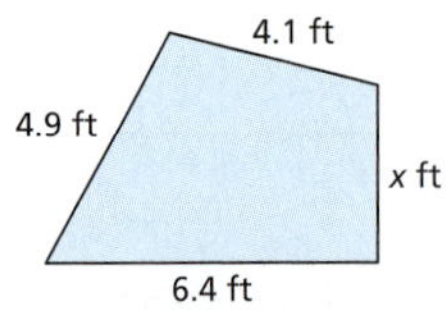

34. THOUGHT PROVOKING Write an inequality that has the solution shown in the graph. Describe a real-life situation that can be modeled by the inequality.

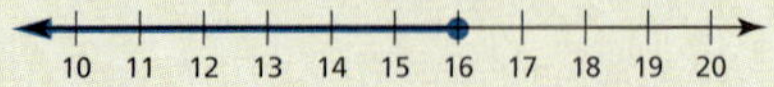

35. WRITING Is it possible to check all the numbers in the solution set of an inequality? When you solve the inequality $x - 11 \geq -3$, which numbers can you check to verify your solution? Explain your reasoning.

36. HOW DO YOU SEE IT? The diagram represents the numbers of students in a school with brown eyes, brown hair, or both.

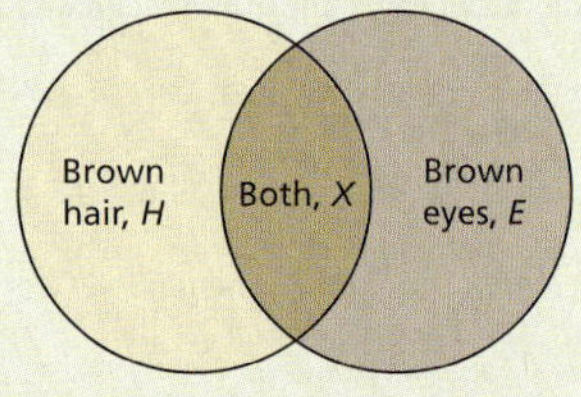

Determine whether each inequality must be true. Explain your reasoning.

a. $H \geq E$ b. $H + 10 \geq E$

c. $H \geq X$ d. $H + 10 \geq X$

e. $H > X$ f. $H + 10 > X$

37. REASONING Write and graph an inequality that represents the numbers that are *not* solutions of each inequality.

a. $x + 8 < 14$

b. $x - 12 \geq 5.7$

38. PROBLEM SOLVING Use the inequalities $c - 3 \geq d$, $b + 4 < a + 1$, and $a - 2 \leq d - 7$ to order a, b, c, and d from least to greatest.

Maintaining Mathematical Proficiency

Reviewing what you learned in previous grades and lessons

Find the product or quotient. *(Skills Review Handbook)*

39. $7 \cdot (-9)$ **40.** $-11 \cdot (-12)$ **41.** $-27 \div (-3)$ **42.** $20 \div (-5)$

Solve the equation. Check your solution. *(Section 1.1)*

43. $6x = 24$ **44.** $-3y = -18$ **45.** $\frac{s}{-8} = 13$ **46.** $\frac{n}{4} = -7.3$

If students need help...	*If students got it...*
Resources by Chapter • Practice A and Practice B • Puzzle Time	Resources by Chapter • Enrichment and Extension • Cumulative Review
Student Journal • Practice	Start the *next* Section
Differentiating the Lesson Skills Review Handbook	

Laurie's Notes

Overview of Section 2.3

Introduction

- In this lesson, students will write, solve, and graph inequalities that involve multiplication and division. Students solved equations in the last chapter and inequalities in middle school. Conceptually the process is the same.
- **Look For and Make Use of Structure:** Mathematically proficient students will recognize the similarity between solving $3x = -12$ and $3x < -12$.
- The direction of the inequality sign can change for two reasons: (1) you have multiplied or divided by a negative number or (2) you want to make the final solution more readable. Before the lesson ends, students need to identify each case. Here they both are shown in the solution of an inequality:

$12 < -4x$	
$-3 > x$	Divide by -4 and change direction of inequality symbol.
$x < -3$	Change direction of inequality symbol to read solution.

Formative Assessment Tips

- **Exit Ticket:** This technique asks students to respond to a question at the end of the lesson, activity, or learning experience. The *Exit Ticket* allows you to collect evidence of student learning. I cut scrap paper into smaller pieces so that "exit tickets" can be distributed quickly to students.
- The *Exit Ticket* is helpful in planning instruction. During the class there may be students you have not heard from. They may not have raised their hands, or they may have been less vocal when working with partners. The *Exit Ticket* helps you gauge the ability of all students to answer a particular type of question.
- Students write their names on the exit tickets, which are then collected.
- When a subset of students has difficulty with the skill addressed by the *Exit Ticket*, instruction for the following day should address this problem.

Pacing Suggestion

- There are four inequalities in part (b) of each exploration. Divide the class into four different groups, have each group do one of the four inequalities, and then share as a class.

Dynamic Teaching Tools

- Dynamic Assessment & Progress Monitoring Tool
- Lesson Planning Tool
- Interactive Whiteboard Lesson Library
- Dynamic Classroom with Dynamic Investigations

What Your Students Will Learn

- Solve linear inequalities in one variable by multiplying or dividing by positive numbers.
- Solve linear inequalities in one variable by multiplying or dividing by negative numbers.

Laurie's Notes

Exploration

Motivate

- Ask a series of questions and record students' solutions.
  - ? "What integers are solutions of $x > 4$?" 5, 6, 7, ...
  - ? "What integers are solutions of $-x > 4$, meaning what numbers have an opposite that is greater than 4?" $-5, -6, -7, \ldots$
  - ? "What integers are solutions of $x < -4$?" $-5, -6, -7, \ldots$
- Leave these three problems on the board and refer to them at the end of class.

Exploration Note

- **Look For and Express Regularity in Repeated Reasoning:** The approach in this investigation is to use a table of values to see what numbers satisfy the inequality. The inequalities involving positive coefficients behave as expected. It is the inequalities involving negative coefficients that seem not to work as expected—from the student perspective. Through repeated trials students should recognize that when the coefficient is negative, the inequality symbol is reversed.

Exploration 1

- Using the information in the table, students decide which graph represents the solution. Finally, they write the solution.
- Discuss how the solution is written: $2 < x$ or $x > 2$.
- Discuss the results with students:
  - ? "What did you find as the solution for part (a)?" $x > 2$
  - ? "Is this what you would have expected?" Likely, they will say "yes."
  - Repeat this line of questions for the inequalities in part (b).
- Do not tell students a rule at this point. Simply say that perhaps they need to try a few more inequalities to help figure out what is going on.

Exploration 2

- The format of this exploration is the same as the first.
- Discuss the results with students:
  - ? "What did you find as the solution for part (a)?" $x < -2$
  - ? "Is this what you would have expected?" Likely, they will say "no."
  - Repeat this line of questions for the inequalities in part (b).

? "What was different about these inequalities from those in Exploration 1?" The coefficients were negative.

Communicate Your Answer

- *Actively listen* as students describe how to use division to solve an inequality.

? If you began with the *Motivate*, ask "Which of the inequalities had the same solution?" $-x > 4$ and $x < -4$

? "Is this consistent with what you discovered in the explorations?" yes; Listen for comments about the negative coefficient of x and the switching of the inequality symbol.

Connecting to Next Step

- The rule for changing the direction of the inequality symbol when dividing by a negative should make sense to students. Connect the operations of multiplication and division when stating the *Core Concepts*.

2.3 Solving Inequalities Using Multiplication or Division

Essential Question

How can you use division to solve an inequality?

EXPLORATION 1 Writing a Rule

Work with a partner.

a. Copy and complete the table. Decide which graph represents the solution of the inequality $6 < 3x$. Write the solution of the inequality.

x	-1	0	1	2	3	4	5
$3x$	-3						
$6 \overset{?}{<} 3x$	No						

LOOKING FOR A PATTERN

To be proficient in math, you need to investigate relationships, observe patterns, and use your observations to write general rules.

b. Use a table to solve each inequality. Then write a rule that describes how to use division to solve the inequalities.

i. $2x < 4$ **ii.** $3 \geq 3x$ **iii.** $2x < 8$ **iv.** $6 \geq 3x$

EXPLORATION 2 Writing a Rule

Work with a partner.

a. Copy and complete the table. Decide which graph represents the solution of the inequality $6 < -3x$. Write the solution of the inequality.

x	-5	-4	-3	-2	-1	0	1
$-3x$							
$6 \overset{?}{<} -3x$							

b. Use a table to solve each inequality. Then write a rule that describes how to use division to solve the inequalities.

i. $-2x < 4$ **ii.** $3 \geq -3x$ **iii.** $-2x < 8$ **iv.** $6 \geq -3x$

Communicate Your Answer

3. How can you use division to solve an inequality?

4. Use the rules you wrote in Explorations 1(b) and 2(b) to solve each inequality.

a. $7x < -21$ **b.** $12 \leq 4x$ **c.** $10 < -5x$ **d.** $-3x \leq 0$

Dynamic Teaching Tools

- Dynamic Assessment & Progress Monitoring Tool
- Lesson Planning Tool
- Interactive Whiteboard Lesson Library
- Dynamic Classroom with Dynamic Investigations

ANSWERS

1. a. 0; 3; 6; 9; 12; 15; no; no; no; yes; yes; yes; right graph; $x > 2$

b. i. $x < 2$

ii. $x \leq 1$

iii. $x < 4$

iv. $x \leq 2$

Dividing each side of an inequality by the same positive number produces an equivalent inequality.

2. a. 15; 12; 9; 6; 3; 0; -3; yes; yes; yes; no; no; no; no; left graph; $x < -2$

b. i. $x > -2$

ii. $x \geq -1$

iii. $x > -4$

iv. $x \geq -2$

When dividing each side of an inequality by the same negative number, the direction of the inequality must be reversed to produce an equivalent inequality.

3. Divide each side of the inequality by the same number. If the number is positive, this produces an equivalent inequality. If the number is negative, the inequality must be reversed to be equivalent.

4. a. $x < -3$

b. $x \geq 3$

c. $x < -2$

d. $x \geq 0$

English Language Learners

Pair Activity

Have students work in pairs to solve inequalities. Each student solves a different inequality. Students share their work and explain their reasoning while their partners follow along.

Extra Example 1

a. Solve $-4 > \frac{x}{4}$. Graph the solution.

$x < -16$

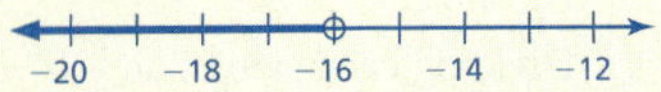

b. Solve $2n \geq -14$. Graph the solution.

$n \geq -7$

MONITORING PROGRESS ANSWERS

1. $n \geq -7$

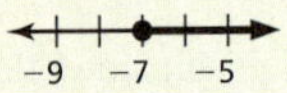

2. $w \leq -32$

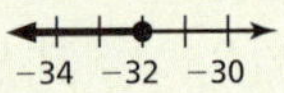

3. $b \geq 9$

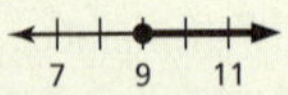

4. $q < -12$

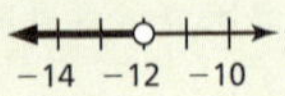

2.3 Lesson

What You Will Learn

- Solve inequalities by multiplying or dividing by *positive* numbers.
- Solve inequalities by multiplying or dividing by *negative* numbers.
- Use inequalities to solve real-life problems.

Multiplying or Dividing by Positive Numbers

Core Concept

Multiplication and Division Properties of Inequality ($c > 0$)

Words Multiplying or dividing each side of an inequality by the same *positive* number produces an equivalent inequality.

Numbers

$-6 < 8$ $\qquad$ $6 > -8$

$2 \cdot (-6) < 2 \cdot 8$ $\qquad$ $\frac{6}{2} > \frac{-8}{2}$

$-12 < 16$ $\qquad$ $3 > -4$

Algebra If $a > b$ and $c > 0$, then $ac > bc$. If $a > b$ and $c > 0$, then $\frac{a}{c} > \frac{b}{c}$.

If $a < b$ and $c > 0$, then $ac < bc$. If $a < b$ and $c > 0$, then $\frac{a}{c} < \frac{b}{c}$.

These properties are also true for $\leq$ and $\geq$.

EXAMPLE 1 Multiplying or Dividing by Positive Numbers

Solve (a) $\frac{x}{8} > -5$ and (b) $-24 \geq 3x$. Graph each solution.

SOLUTION

a. $\frac{x}{8} > -5$ $\qquad$ Write the inequality.

Multiplication Property of Inequality → $8 \cdot \frac{x}{8} > 8 \cdot (-5)$ $\qquad$ Multiply each side by 8.

$x > -40$ $\qquad$ Simplify.

▶ The solution is $x > -40$.

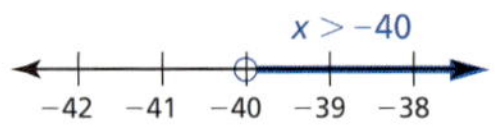

b. $-24 \geq 3x$ $\qquad$ Write the inequality.

Division Property of Inequality → $\frac{-24}{3} \geq \frac{3x}{3}$ $\qquad$ Divide each side by 3.

$-8 \geq x$ $\qquad$ Simplify.

▶ The solution is $x \leq -8$.

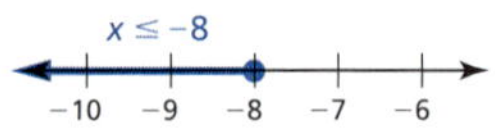

Monitoring Progress

Help in English and Spanish at *BigIdeasMath.com*

Solve the inequality. Graph the solution.

1. $\frac{n}{7} \geq -1$ $\qquad$ **2.** $-6.4 \geq \frac{1}{5}w$ $\qquad$ **3.** $4b \geq 36$ $\qquad$ **4.** $-18 > 1.5q$

Laurie's Notes Teacher Actions

- **Look For and Make Use of Structure:** These properties should look very familiar, as they are similar to the Multiplication and Division Properties of Equality. Demonstrating the property using just numbers helps students make sense of it.
- Note that the properties are restricted to multiplying and dividing by a positive number. This is very important.
- **Representation:** Multiplication is represented by the dot notation, and −5 is enclosed in parentheses for clarity only. Otherwise, students might become confused and think 5 is being subtracted.
- Division is represented differently in Questions 1 and 2. The first representation is more common in algebra. In Question 2, check to see that students multiply by the reciprocal of the coefficient (most efficient) versus dividing both sides by $\frac{1}{5}$.

Multiplying or Dividing by Negative Numbers

Core Concept

Multiplication and Division Properties of Inequality ($c < 0$)

Words When multiplying or dividing each side of an inequality by the same *negative* number, the direction of the inequality symbol must be reversed to produce an equivalent inequality.

Numbers

$-6 < 8$ $\qquad$ $6 > -8$

$-2 \cdot (-6) > -2 \cdot 8$ $\qquad$ $\frac{6}{-2} < \frac{-8}{-2}$

$12 > -16$ $\qquad$ $-3 < 4$

Algebra If $a > b$ and $c < 0$, then $ac < bc$. If $a > b$ and $c < 0$, then $\frac{a}{c} < \frac{b}{c}$.

If $a < b$ and $c < 0$, then $ac > bc$. If $a < b$ and $c < 0$, then $\frac{a}{c} > \frac{b}{c}$.

These properties are also true for $\leq$ and $\geq$.

COMMON ERROR

A negative sign in an inequality does not necessarily mean you must reverse the inequality symbol, as shown in Example 1.

Only reverse the inequality symbol when you multiply or divide each side by a negative number.

EXAMPLE 2 Multiplying or Dividing by Negative Numbers

Solve each inequality. Graph each solution.

a. $2 < \frac{y}{-3}$ $\qquad$ **b.** $-7y \leq -35$

SOLUTION

a. $2 < \frac{y}{-3}$ — Write the inequality.

Multiplication Property of Inequality → $-3 \cdot 2 > -3 \cdot \frac{y}{-3}$ — Multiply each side by -3. Reverse the inequality symbol.

$-6 > y$ — Simplify.

▶ The solution is $y < -6$.

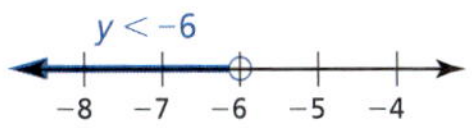

b. $-7y \leq -35$ — Write the inequality.

Division Property of Inequality → $\frac{-7y}{-7} \geq \frac{-35}{-7}$ — Divide each side by -7. Reverse the inequality symbol.

$y \geq 5$ — Simplify.

▶ The solution is $y \geq 5$.

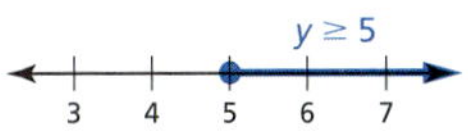

Monitoring Progress

Help in English and Spanish at *BigIdeasMath.com*

Solve the inequality. Graph the solution.

5. $\frac{p}{-4} < 7$ $\qquad$ **6.** $\frac{x}{-5} \leq -5$ $\qquad$ **7.** $1 \geq -\frac{1}{10}z$

8. $-9m > 63$ $\qquad$ **9.** $-2r \geq -22$ $\qquad$ **10.** $-0.4y \geq -12$

Extra Example 2

Solve each inequality. Graph each solution.

a. $3 > \frac{x}{-7}$ $\quad x > -21$

b. $-5m \geq -45$ $\quad m \leq 9$

MONITORING PROGRESS ANSWERS

5. $p > -28$

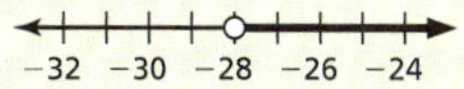

6. $x \geq 25$

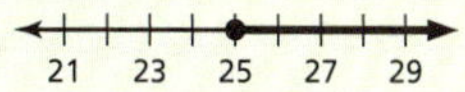

7. $z \geq -10$

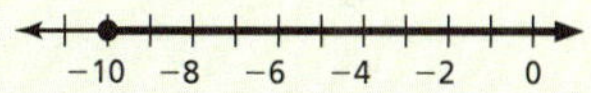

8. $m < -7$

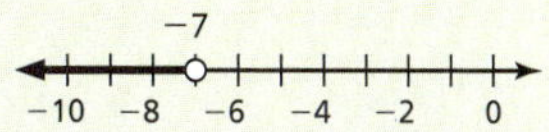

9. $r \leq 11$

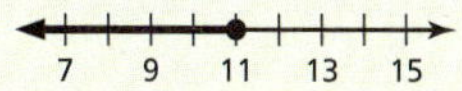

10. $y \leq 30$

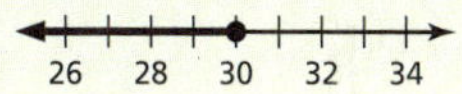

Laurie's Notes Teacher Actions

- Write the *Core Concept*. Discuss the difference between these properties and those at the beginning of the lesson.
- The short version of the properties: When you multiply or divide by a negative quantity, reverse the direction of the inequality symbol.
- After working each inequality in Example 2, check numbers that are in the solution set and numbers *not* in the solution set.

COMMON ERROR When students solve $2x < -4$, they sometimes reverse the inequality symbol because there is a negative number in the problem. The inequality symbol is reversed only when the coefficient is negative, *not* when the constant is negative.

- **Neighbor Check:** Have students work independently, and then have their neighbors check their work. Have students discuss any discrepancies.

Extra Example 3

You earn \$8.50 per hour at the sub shop. Write and solve an inequality that represents the numbers of hours you need to work to earn \$187 to buy a tablet computer. $8.5h \geq 187$; $h \geq 22$; at least 22 hours

MONITORING PROGRESS ANSWERS

11. $0.25c \leq 3.65$, $c \leq 14.6$

12. $\frac{165}{t} \leq 55$, $t \geq 3$

Solving Real-Life Problems

EXAMPLE 3 **Modeling with Mathematics**

You earn \$9.50 per hour at your summer job. Write and solve an inequality that represents the numbers of hours you need to work to buy a digital camera that costs \$247.

SOLUTION

1. **Understand the Problem** You know your hourly wage and the cost of the digital camera. You are asked to write and solve an inequality that represents the numbers of hours you need to work to buy the digital camera.
2. **Make a Plan** Use a verbal model to write an inequality. Then solve the inequality.
3. **Solve the Problem**

Words Hourly wage • Hours worked ≥ Cost of camera

Variable Let n be the number of hours worked.

Inequality $9.5 \cdot n \geq 247$

$9.5n \geq 247$ — Write the inequality.

Division Property of Inequality → $\frac{9.5n}{9.5} \geq \frac{247}{9.5}$ — Divide each side by 9.5.

$n \geq 26$ — Simplify.

▶ You need to work at least 26 hours for your gross pay to be at least \$247. If you have payroll deductions, such as Social Security taxes, you need to work more than 26 hours.

REMEMBER
Compatible numbers are numbers that are easy to compute mentally.

4. **Look Back** You can use estimation to check that your answer is reasonable.

\$247 ÷ \$9.50/h
↓ ↓
\$250 ÷ \$10/h = 25 h — Use compatible numbers.

Your hourly wage is about \$10 per hour. So, to earn about \$250, you need to work about 25 hours.

Unit Analysis Each time you set up an equation or inequality to represent a real-life problem, be sure to check that the units balance.

$$\frac{\$9.50}{\cancel{h}} \times 26\cancel{h} = \$247$$

Monitoring Progress

Help in English and Spanish at *BigIdeasMath.com*

11. You have at most \$3.65 to make copies. Each copy costs \$0.25. Write and solve an inequality that represents the numbers of copies you can make.

12. The maximum speed limit for a school bus is 55 miles per hour. Write and solve an inequality that represents the numbers of hours it takes to travel 165 miles in a school bus.

Laurie's Notes — Teacher Actions

- The problem posed is not difficult to solve with mental math. The point is not that you can solve the problem mentally. The practice you want to provide students is the opportunity to *read* a problem, *make sense* of the context, and *translate* words into symbols. This is a building block for future problems. I ask a volunteer to read Steps 1 and 2 in the problem-solving plan.
- **Use Appropriate Tools Strategically:** The color coding is a tool to help students connect words and symbols.
- **Think-Pair-Share:** Partners should work Questions 11 and 12 together.

Closure

- **I Used to Think … But Now I Know:** Take time for students to reflect on their current understanding of solving inequalities involving multiplication and division.
- **Exit Ticket:** Solve and graph.

$\frac{x}{-3} \leq -9$ $x \geq 27$

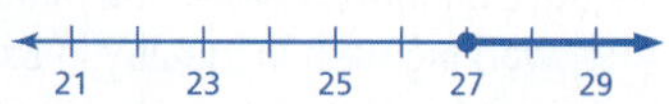

$-8 > 4x$ $x < -2$

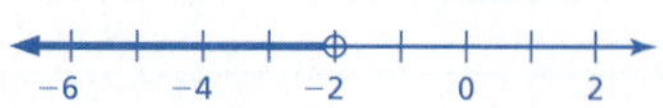

2.3 Exercises

Dynamic Solutions available at *BigIdeasMath.com*

Vocabulary and Core Concept Check

1. **WRITING** Explain how solving $2x < -8$ is different from solving $-2x < 8$.

2. **OPEN-ENDED** Write an inequality that is solved using the Division Property of Inequality where the inequality symbol needs to be reversed.

Monitoring Progress and Modeling with Mathematics

In Exercises 3–10, solve the inequality. Graph the solution. *(See Example 1.)*

3. $4x < 8$
4. $3y \leq -9$
5. $-20 \leq 10n$
6. $35 < 7t$
7. $\frac{x}{2} > -2$
8. $\frac{a}{4} < 10.2$
9. $20 \geq \frac{4}{5}w$
10. $-16 \leq \frac{8}{3}t$

In Exercises 11–18, solve the inequality. Graph the solution. *(See Example 2.)*

11. $-6t < 12$
12. $-9y > 9$
13. $-10 \geq -2z$
14. $-15 \leq -3c$
15. $\frac{n}{-3} \geq 1$
16. $\frac{w}{-5} \leq 16$
17. $-8 < -\frac{1}{4}m$
18. $-6 > -\frac{2}{3}y$

19. **MODELING WITH MATHEMATICS** You have \$12 to buy five goldfish for your new fish tank. Write and solve an inequality that represents the prices you can pay per fish. *(See Example 3.)*

20. **MODELING WITH MATHEMATICS** A weather forecaster predicts that the temperature in Antarctica will decrease 8°F each hour for the next 6 hours. Write and solve an inequality to determine how many hours it will take for the temperature to drop at least 36°F.

USING TOOLS In Exercises 21–26, solve the inequality. Use a graphing calculator to verify your answer.

21. $36 < 3y$
22. $17v \geq 51$
23. $2 \leq -\frac{2}{9}x$
24. $4 > \frac{n}{-4}$
25. $2x > \frac{3}{4}$
26. $1.1y < 4.4$

ERROR ANALYSIS In Exercises 27 and 28, describe and correct the error in solving the inequality.

27. ✗

$$-6 > \frac{2}{3}x$$
$$\frac{3}{2} \cdot (-6) < \frac{3}{2} \cdot \frac{2}{3}x$$
$$-\frac{18}{2} < x$$
$$-9 < x$$

The solution is $x > -9$.

28. ✗

$$-4y \leq -32$$
$$\frac{-4y}{-4} \leq \frac{-32}{-4}$$
$$y \leq 8$$

The solution is $y \leq 8$.

29. **ATTENDING TO PRECISION** You have \$700 to buy new carpet for your bedroom. Write and solve an inequality that represents the costs per square foot that you can pay for the new carpet. Specify the units of measure in each step.

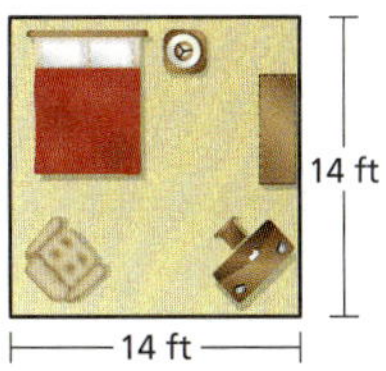

Assignment Guide and Homework Check

ASSIGNMENT

Basic: 1, 2, 3–19 odd, 27, 30, 31, 33, 40–47

Average: 1, 2, 4–30 even, 31, 35, 40–47

Advanced: 1, 2, 6–28 even, 29–31, 32–38 even, 40–47

HOMEWORK CHECK

Basic: 3, 7, 11, 15, 19

Average: 4, 8, 12, 16, 20

Advanced: 6, 8, 16, 18, 29

ANSWERS

1. When solving $2x < -8$, the inequality symbol is not reversed when dividing each side by 2. When solving $-2x < 8$, the inequality is reversed when dividing each side by -2.
2. *Sample answer:* $-5x \geq 25$
3. $x < 2$

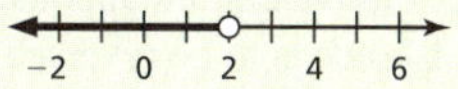

4. $y \leq -3$

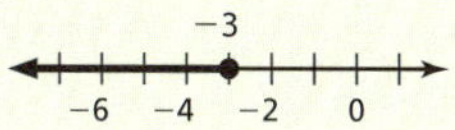

5. $n \geq -2$

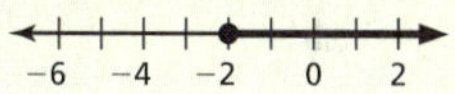

6. $t > 5$

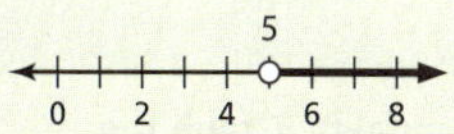

7. $x > -4$

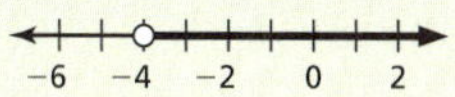

8. $a < 40.8$

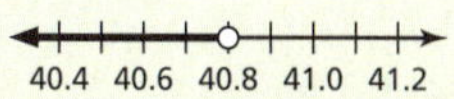

9. $w \leq 25$

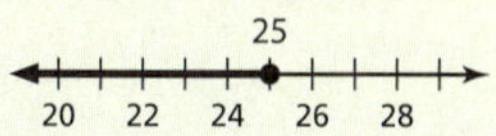

10. $t \geq -6$

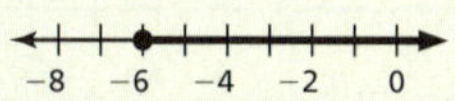

11. $t > -2$

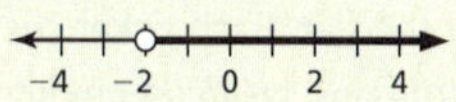

12. $y < -1$

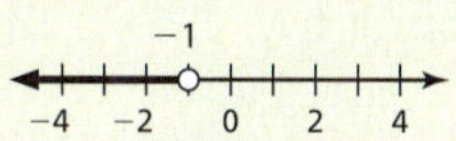

13. $z \geq 5$

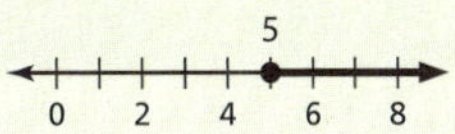

14. $c \leq 5$

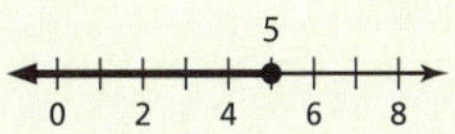

15. $n \leq -3$

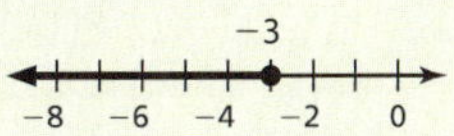

16. $w \geq -80$

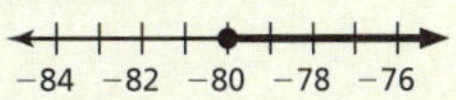

17. $m < 32$

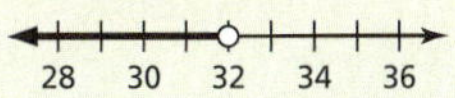

18. $y > 9$

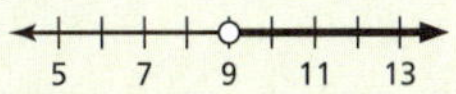

19. $5p \leq 12, p \leq 2.4$
20. $-8t \leq -36, t \geq 4.5$
21. $y > 12$
22. $v \geq 3$
23. $x \leq -9$

24–29. See Additional Answers.

Dynamic Teaching Tools

- Dynamic Assessment & Progress Monitoring Tool
- Interactive Whiteboard Lesson Library
- Dynamic Classroom with Dynamic Investigations

ANSWERS

30. **a.** C; Multiplying both sides by m gives $x < -m$.

b. A; Multiplying both sides by m gives $x > m$.

c. B; Multiplying both sides by m gives $x < m$.

d. D; Multiplying both sides by $-m$ and reversing the inequality symbol gives $x > -m$.

31. **a.** $d \le 6.3(2)$, $d \le 12.6$

b. yes; The distance traveled in 4 hours would be no more than 25.2 miles, which is less than the distance required for a marathon.

32. *Sample answer:* $x < 21$

33. more than 300 million pennies

34. no; To get the second inequality, each side of the first is multiplied by -3. To be equivalent, the inequality symbol also needs to be reversed.

35–47. See Additional Answers.

Mini-Assessment

Solve the inequality. Graph the solution.

1. $6x < -24$ $x < -4$

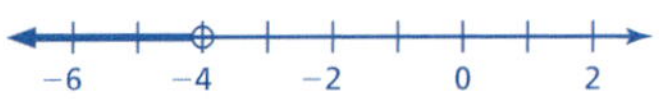

2. $\frac{s}{-4} \le -2$ $s \ge 8$

3. $18 < -\frac{2}{3}y$ $y < -27$

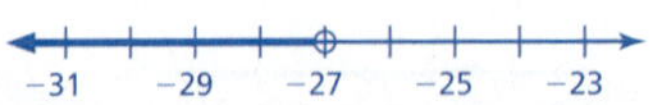

4. You have \$12 to spend on ride tickets at the fair. Each ticket costs \$1.25. Write and solve an inequality that represents the number of tickets you can buy. $1.25x \le 12$; $x \le 9.6$; at most 9 tickets

30. HOW DO YOU SEE IT? Let $m > 0$. Match each inequality with its graph. Explain your reasoning.

a. $\frac{x}{m} < -1$ **b.** $\frac{x}{m} > 1$

c. $\frac{x}{m} < 1$ **d.** $-\frac{x}{m} < 1$

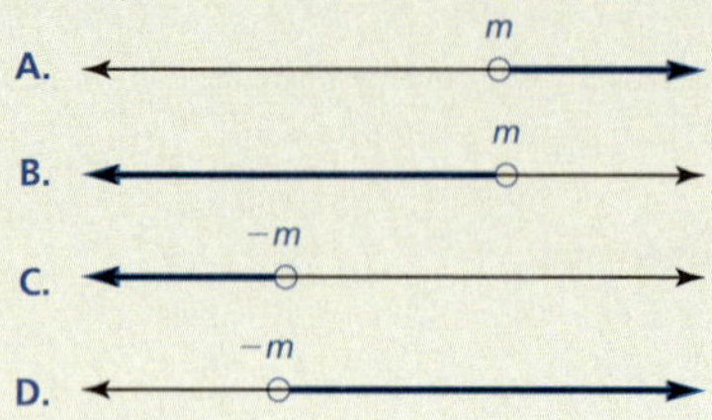

31. MAKING AN ARGUMENT You run for 2 hours at a speed no faster than 6.3 miles per hour.

a. Write and solve an inequality that represents the possible numbers of miles you run.

b. A marathon is approximately 26.2 miles. Your friend says that if you continue to run at this speed, you will not be able to complete a marathon in less than 4 hours. Is your friend correct? Explain.

32. THOUGHT PROVOKING The inequality $\frac{x}{4} \le 5$ has a solution of $x = p$. Write a second inequality that also has a solution of $x = p$.

33. PROBLEM SOLVING The U.S. Mint pays \$0.02 to produce every penny. How many pennies are produced when the U.S. Mint pays more than \$6 million in production costs?

34. REASONING Are $x \le \frac{2}{3}$ and $-3x \le -2$ equivalent? Explain your reasoning.

35. ANALYZING RELATIONSHIPS Consider the number line shown.

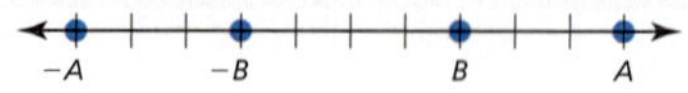

a. Write an inequality relating A and B.

b. Write an inequality relating $-A$ and $-B$.

c. Use the results from parts (a) and (b) to explain why the direction of the inequality symbol must be reversed when multiplying or dividing each side of an inequality by the same negative number.

36. REASONING Why might solving the inequality $\frac{4}{x} \ge 2$ by multiplying each side by x lead to an error? (*Hint:* Consider $x > 0$ and $x < 0$.)

37. MATHEMATICAL CONNECTIONS The radius of a circle is represented by the formula $r = \frac{C}{2\pi}$. Write and solve an inequality that represents the possible circumferences C of the circle.

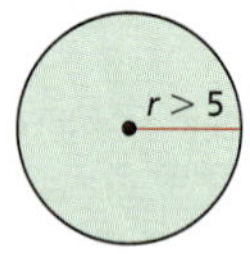

38. CRITICAL THINKING A water-skiing instructor recommends that a boat pulling a beginning skier has a speed less than 18 miles per hour. Write and solve an inequality that represents the possible distances d (in miles) that a beginner can travel in 45 minutes of practice time.

39. CRITICAL THINKING A local zoo employs 36 people to take care of the animals each day. At most, 24 of the employees work full time. Write and solve an inequality that represents the fraction of employees who work part time. Graph the solution.

Maintaining Mathematical Proficiency

Reviewing what you learned in previous grades and lessons

Solve the equation. Check your solution. *(Section 1.2 and Section 1.3)*

40. $5x + 3 = 13$

41. $\frac{1}{2}y - 8 = -10$

42. $-3n + 2 = 2n - 3$

43. $\frac{1}{2}z + 4 = \frac{5}{2}z - 8$

Tell which number is greater. *(Skills Review Handbook)*

44. 0.8, 85%

45. $\frac{16}{30}$, 50%

46. 120%, 0.12

47. 60%, $\frac{2}{3}$

If students need help...	*If students got it...*
Resources by Chapter • Practice A and Practice B • Puzzle Time	Resources by Chapter • Enrichment and Extension • Cumulative Review
Student Journal • Practice	Start the *next* Section
Differentiating the Lesson Skills Review Handbook	

Laurie's Notes

Overview of Section 2.4

Introduction

- You solve multi-step inequalities the same way you solve multi-step equations. You only need to remember to change the direction of the inequality symbol when you multiply or divide by a negative quantity. Students solved multi-step equations in the first chapter, so they should be ready to apply that knowledge to inequalities.
- Just like linear equations, inequalities may have no solution or a solution of all real numbers. These special solutions are introduced after students have solved a number of multi-step inequalities.
- Students should be encouraged to check their solutions by substituting a value from the solution set to verify that it gives a true statement. They should also substitute a value that is *not* in the solution set to verify that it gives a false statement.

Formative Assessment Tips

- **No-Hands Questioning:** Typically when you ask a question there are hands that immediately go up, often the same hands each time. Some students need a longer time to process a question and think through their response. This technique instructs students not to put their hands in the air when the question is posed. *Wait Time** is exercised.
- This technique encourages all students to be active and engaged in the lesson. Students who need additional think time are provided that opportunity. Teachers can then use *Popsicle Sticks* to call on students for a response, or they can purposely call on those students whose voice they do not hear enough.
- To be effective, the question(s) posed during *No-Hands Questioning* have to require more than a simple response.

*See Section 2.5 for a description of *Wait Time*.

Another Way

- The exploration demonstrates that the solution of a multi-step inequality can be graphed using a graphing calculator. When the solution is a ray or half-line, the *trace* key and *zoom* feature can be used to get a good approximation of the endpoint even when it is not an integer value. The calculator is a tool to be used wisely.

Pacing Suggestion

- The inequalities students will work in the exploration are similar to Examples 1 and 2 in the formal lesson. If students have been successful, you may want to begin the formal lesson with Example 3.

Dynamic Teaching Tools

- Dynamic Assessment & Progress Monitoring Tool
- Lesson Planning Tool
- Interactive Whiteboard Lesson Library
- Dynamic Classroom with Dynamic Investigations

What Your Students Will Learn

- Solve multi-step linear inequalities in one variable using inverse operations.

Laurie's Notes

Exploration

Motivate

? "Have you ever heard of or played Ultimate? Answers will vary.
- It is a sport played with a flying disc at colleges, high schools, and some middle schools. There are 10 simple rules, one of which is that there are no officials! Pretty cool.
- A regulation Ultimate field is 64 meters long and 37 meters wide, with two end zones of 18 meters each. Sometimes the length (64 meters) varies when using existing fields. Let the length be x. Draw a sketch and find the area of the field in terms of x.
  Area $= 37(x + 36) = (37x + 1332)$ m^2

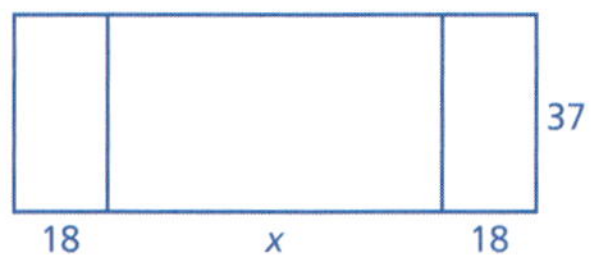

Discuss

- **Look For and Make Use of Structure:** Students have solved multi-step equations. Structurally, multi-step inequalities are solved in the same fashion. Mathematically proficient students recognize the similarity between solving $-4x - 5 = 41$ and $-4x - 5 < 41$. Students must recall that when the variable term has been isolated, if the coefficient is negative, then the inequality symbol is reversed.

Exploration 1

- This may seem like a stretch for students to solve multi-step inequalities of this type, yet the connection to multi-step equations and working with partners is generally enough for students to meet with success.
- Discuss the directions to ensure that students do not immediately use a calculator to determine the solution. The calculator is used to check their answers.
- In checking their solutions, students may find that their endpoint is correct but the shading is going in the opposite direction. Students should then be able to do a self-check to observe that they forgot to change the direction of the inequality symbol when they multiplied or divided by a negative number.
- Circulate and make note of common errors, questions, or procedures. When several groups are challenged by one of the inequalities, look for students who have solved it correctly. Ask those students to show the first step in their solution.
- **Construct Viable Arguments and Critique the Reasoning of Others:** Ask two volunteers who have their work displayed in a neat and organized fashion to share their solutions at the board for parts (e) and (f). You want to model good problem solving, as well as good mathematics. Hearing the reasoning process of classmates helps students improve their own problem-solving strategies.

? "Is it possible to have the same solutions but the steps in a different order? Explain." yes; There are different orders in which the steps might be done and still correctly solve the inequality.

Communicate Your Answer

- Ask two volunteers to share their work for Question 3.

Connecting to Next Step

- If students have been very successful with solving the inequalities in the exploration, be selective in examples you work in the formal lesson. It may be sufficient to try a few questions from *Monitoring Progress* and the real-life example.

2.4 Solving Multi-Step Inequalities

Essential Question How can you solve a multi-step inequality?

EXPLORATION 1 **Solving a Multi-Step Inequality**

Work with a partner.

- Use what you already know about solving equations and inequalities to solve each multi-step inequality. Justify each step.
- Match each inequality with its graph. Use a graphing calculator to check your answer.

JUSTIFYING STEPS
To be proficient in math, you need to justify each step in a solution and communicate your justification to others.

a. $2x + 3 \leq x + 5$ **b.** $-2x + 3 > x + 9$

c. $27 \geq 5x + 4x$ **d.** $-8x + 2x - 16 < -5x + 7x$

e. $3(x - 3) - 5x > -3x - 6$ **f.** $-5x - 6x \leq 8 - 8x - x$

A.

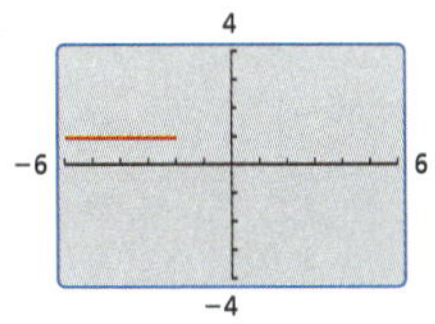

B.

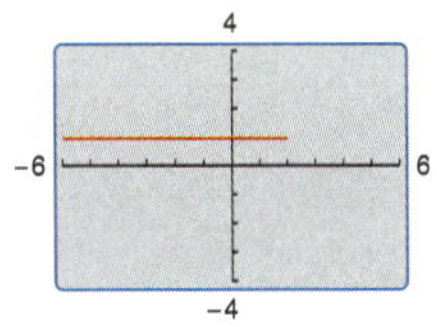

C.

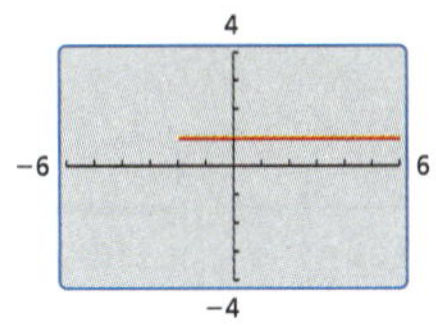

D.

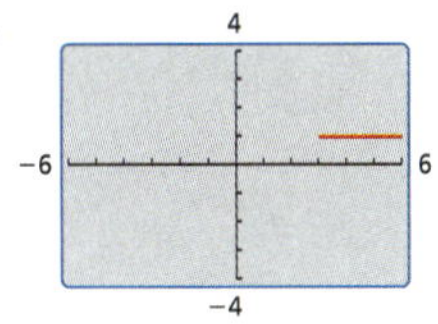

E.

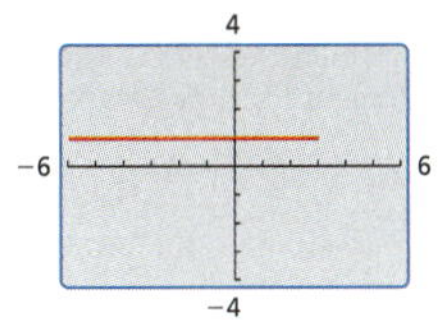

F.

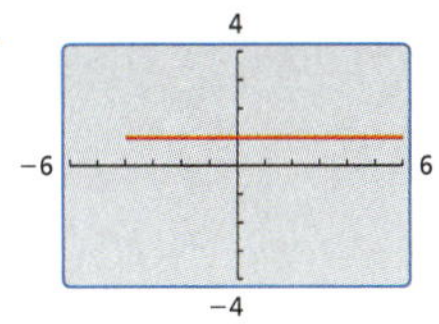

Communicate Your Answer

2. How can you solve a multi-step inequality?
3. Write two different multi-step inequalities whose solutions are represented by the graph.

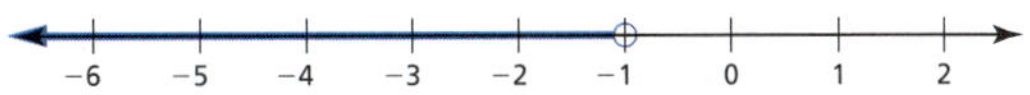

Dynamic Teaching Tools

- Dynamic Assessment & Progress Monitoring Tool
- Lesson Planning Tool
- Interactive Whiteboard Lesson Library
- Dynamic Classroom with Dynamic Investigations

ANSWERS

1. **a.** $x \leq 2$; Subtract x and 3 from each side; B
   b. $x < -2$; Subtract x and 3 from each side; Divide each side by -3; A
   c. $3 \geq x$; Divide each side by 9; E
   d. $-2 < x$; Add $6x$ to each side; Divide each side by 8; C
   e. $x > 3$; Add $3x$ and 9 to each side; D
   f. $x \geq -4$; Add $9x$ to each side; Divide each side by -2; F
2. Simplify each side, if possible, then use inverse operations to isolate the variable. Reverse the inequality symbol if multiplying or dividing by a negative number.
3. *Sample answer:* $3x + 4 < 1$, $-2x - 10 > -8$

English Language Learners

Build on Past Knowledge

Explain to students that when they solve multi-step inequalities, they will use a process similar to the one they used to solve multi-step equations. They will be using inverse operations and performing one step at a time in order to isolate the variable.

Extra Example 1

Solve each inequality. Graph each solution.

a. $\frac{x}{-2} + 4 > -1$ $x < 10$

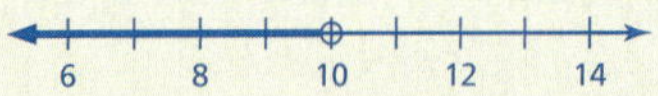

b. $3c - 2 \le -11$ $c \le -3$

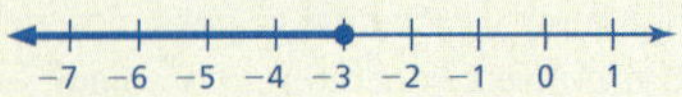

MONITORING PROGRESS ANSWERS

1. $b < 2$

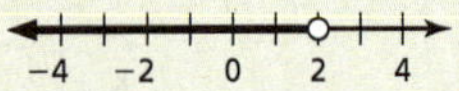

2. $c \le 4$

3. $n < -2$

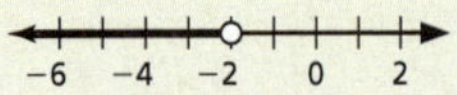

4. $v \ge -3$

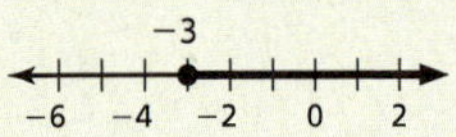

2.4 Lesson

What You Will Learn

- Solve multi-step inequalities.
- Use multi-step inequalities to solve real-life problems.

Solving Multi-Step Inequalities

To solve a multi-step inequality, simplify each side of the inequality, if necessary. Then use inverse operations to isolate the variable. Be sure to reverse the inequality symbol when multiplying or dividing by a negative number.

EXAMPLE 1 Solving Multi-Step Inequalities

Solve each inequality. Graph each solution.

a. $\frac{y}{-6} + 7 < 9$ **b.** $2v - 4 \ge 8$

SOLUTION

a. $\frac{y}{-6} + 7 < 9$ Write the inequality.

$-7 \quad -7$ Subtract 7 from each side.

$\frac{y}{-6} < 2$ Simplify.

$-6 \cdot \frac{y}{-6} > -6 \cdot 2$ Multiply each side by −6. Reverse the inequality symbol.

$y > -12$ Simplify.

▶ The solution is $y > -12$.

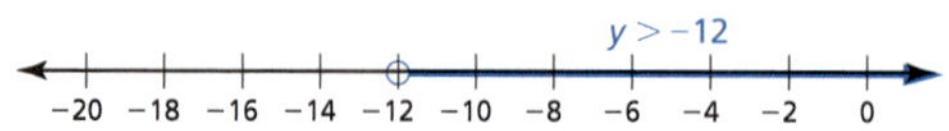

b. $2v - 4 \ge 8$ Write the inequality.

$+4 \quad +4$ Add 4 to each side.

$2v \ge 12$ Simplify.

$\frac{2v}{2} \ge \frac{12}{2}$ Divide each side by 2.

$v \ge 6$ Simplify.

▶ The solution is $v \ge 6$.

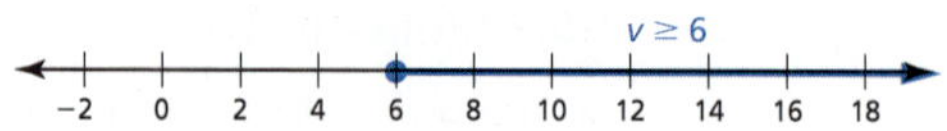

Monitoring Progress

Help in English and Spanish at *BigIdeasMath.com*

Solve the inequality. Graph the solution.

1. $4b - 1 < 7$
2. $8 - 9c \ge -28$
3. $\frac{n}{-2} + 11 > 12$
4. $6 \ge 5 - \frac{v}{3}$

Laurie's Notes Teacher Actions

? "What operations are being performed on the left side of the inequality?" division and addition

? "What is the first step in isolating the variable, meaning getting the *y*-term by itself?" Subtract 7 from each side of the inequality.

? "To solve for *y*, what is the last step?" Multiply each side by −6, and change the direction of the inequality symbol.

- Graph and check. Remember to use an open circle because the variable cannot equal −12.

COMMON ERROR In solving Question 2, students add 8 to each side. Suggest to students that they rewrite the inequality as $8 + (-9c) \ge -28$.

EXAMPLE 2 Solving an Inequality with Variables on Both Sides

Solve $6x - 5 < 2x + 11$.

SOLUTION

$6x - 5 < 2x + 11$	Write the inequality.
$+5 \qquad +5$	Add 5 to each side.
$6x < 2x + 16$	Simplify.
$-2x \qquad -2x$	Subtract $2x$ from each side.
$4x < 16$	Simplify.
$\frac{4x}{4} < \frac{16}{4}$	Divide each by 4.
$x < 4$	Simplify.

▶ The solution is $x < 4$.

When solving an inequality, if you obtain an equivalent inequality that is true, such as $-5 < 0$, the solutions of the inequality are *all real numbers*. If you obtain an equivalent inequality that is false, such as $3 \leq -2$, the inequality has *no solution*.

Graph of an inequality whose solutions are all real numbers

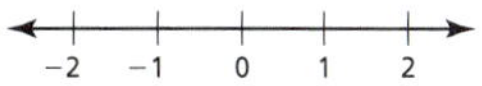

Graph of an inequality that has no solution

EXAMPLE 3 Inequalities with Special Solutions

Solve (a) $8b - 3 > 4(2b + 3)$ and (b) $2(5w - 1) \leq 7 + 10w$.

SOLUTION

a.

$8b - 3 > 4(2b + 3)$	Write the inequality.
$8b - 3 > 8b + 12$	Distributive Property
$-8b \qquad -8b$	Subtract $8b$ from each side.
$-3 > 12$ ✗	Simplify.

▶ The inequality $-3 > 12$ is false. So, there is no solution.

b.

$2(5w - 1) \leq 7 + 10w$	Write the inequality.
$10w - 2 \leq 7 + 10w$	Distributive Property
$-10w \qquad -10w$	Subtract $10w$ from each side.
$-2 \leq 7$	Simplify.

▶ The inequality $-2 \leq 7$ is true. So, all real numbers are solutions.

LOOKING FOR STRUCTURE

When the variable terms on each side of an inequality are the same, the constant terms will determine whether the inequality is true or false.

Monitoring Progress

Help in English and Spanish at *BigIdeasMath.com*

Solve the inequality.

5. $5x - 12 \leq 3x - 4$

6. $2(k - 5) < 2k + 5$

7. $-4(3n - 1) > -12n + 5.2$

8. $3(2a - 1) \geq 10a - 11$

Extra Example 2

Solve $5x - 6 \geq 3x + 4$. $x \geq 5$

Extra Example 3

a. Solve $6t - 4 > 2(3t + 5)$. The inequality $-4 > 10$ is false. So, there is no solution.

b. Solve $2(2x - 3) \leq 9 + 4x$. The inequality $-6 \leq 9$ is true. So, all real numbers are solutions.

MONITORING PROGRESS ANSWERS

5. $x \leq 4$

6. all real numbers

7. no solution

8. $a \leq 2$

Laurie's Notes Teacher Actions

- Discuss with students the side on which they want to solve for the variable. In Example 2, solving for the variable on the left gives a coefficient of 4. Solving for the variable on the right gives a coefficient of –4. To avoid a negative coefficient, solve for the variable on the side with the coefficient with the greater value.

? **Turn and Talk:** Before introducing inequalities with special solutions, have students *Turn and Talk:* "What if Example 2 had been $6x - 5 < 6x + 11$? What would the solution have been?" Students will likely say it does not make sense; there is no x left in the inequality!

? **Assessing Question:** "Does every equation have a solution?" no; Students may have forgotten about an equation such as $5x - 2 = 5x + 3$.

- **Look For and Make Use of Structure:** Work through both parts of Example 3.
- **No-Hands Questioning:** "How will you know that an inequality will have a special solution? Explain." When like terms have been combined and the variable terms are the same on both sides.
- **Make Sense of Problems and Persevere in Solving Them:** In Example 3(a), have students substitute different values for b. This does not prove that there is no solution, but it helps students make sense of the problem. In Example 3(b), have students substitute different values of w. They will find that any value works!

Extra Example 4

Your grades on 4 math tests are 79, 85, 86, and 88. You want your mean grade to be at least 85. What grade on your fifth test will give you a mean score of at least 85? a grade of at least 87

MONITORING PROGRESS ANSWER

9. a score of at least 73

Solving Real-Life Problems

EXAMPLE 4 **Modeling with Mathematics**

You need a mean score of at least 90 points to advance to the next round of the touch-screen trivia game. What scores in the fifth game will allow you to advance?

SOLUTION

REMEMBER
The mean in Example 4 is equal to the sum of the game scores divided by the number of games.

1. **Understand the Problem** You know the scores of your first four games. You are asked to find the scores in the fifth game that will allow you to advance.
2. **Make a Plan** Use the definition of the mean of a set of numbers to write an inequality. Then solve the inequality and answer the question.
3. **Solve the Problem** Let x be your score in the fifth game.

$$\frac{95 + 91 + 77 + 89 + x}{5} \geq 90 \qquad \text{Write an inequality.}$$

$$\frac{352 + x}{5} \geq 90 \qquad \text{Simplify.}$$

$$5 \cdot \frac{352 + x}{5} \geq 5 \cdot 90 \qquad \text{Multiply each side by 5.}$$

$$352 + x \geq 450 \qquad \text{Simplify.}$$

$$\underline{-352} \qquad \underline{-352} \qquad \text{Subtract 352 from each side.}$$

$$x \geq 98 \qquad \text{Simplify.}$$

▶ A score of at least 98 points will allow you to advance.

4. **Look Back** You can draw a diagram to check that your answer is reasonable. The horizontal bar graph shows the differences between the game scores and the desired mean of 90.

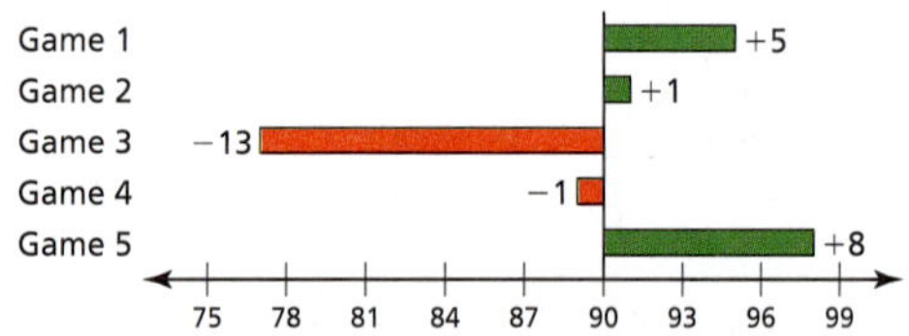

To have a mean of 90, the sum of the differences must be zero.

$$5 + 1 - 13 - 1 + 8 = 0 \quad ✓$$

Monitoring Progress

Help in English and Spanish at *BigIdeasMath.com*

9. **WHAT IF?** You need a mean score of at least 85 points to advance to the next round. What scores in the fifth game will allow you to advance?

Laurie's Notes Teacher Actions

- This is a classic problem. Students always want to know what they have to score on a test in order to have an average (mean) of ___________. This is the same type of problem.
- Set up the inequality to compute the mean. Because you want your score to be a minimum of 90, you need to set the mean greater than or equal to 90.
- **Use Appropriate Tools Strategically:** The horizontal bar graph is a helpful tool for students to judge the reasonableness of their solutions.

Closure

- **Writing:** How are these alike? How are they different?
  $3n - 4 = -25$ $\quad 3n - 4 > -25$ $\quad 3n - 4 \leq -25$
  Sample answer: They are alike because they each use the expressions $(3n - 4)$ and (-25). They are different because of the way the expressions are related: equal to, greater than, and less than or equal to.

2.4 Exercises

Dynamic Solutions available at *BigIdeasMath.com*

Vocabulary and Core Concept Check

1. **WRITING** Compare solving multi-step inequalities and solving multi-step equations.

2. **WRITING** Without solving, how can you tell that the inequality $4x + 8 \le 4x - 3$ has no solution?

Monitoring Progress and Modeling with Mathematics

In Exercises 3–6, match the inequality with its graph.

3. $7b - 4 \le 10$
4. $4p + 4 \ge 12$
5. $-6g + 2 \ge 20$
6. $3(2 - f) \le 15$

A. 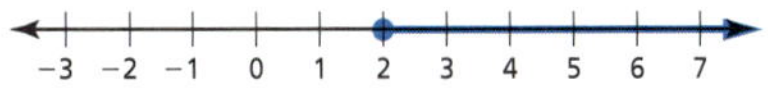

B. 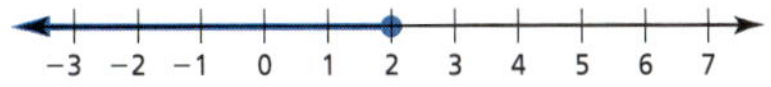

C. 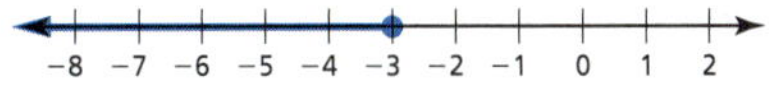

D. 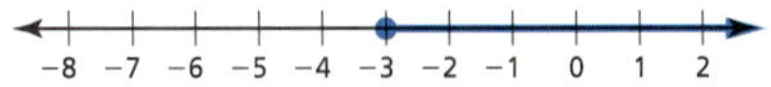

In Exercises 7–16, solve the inequality. Graph the solution. *(See Example 1.)*

7. $2x - 3 > 7$
8. $5y + 9 \le 4$
9. $-9 \le 7 - 8v$
10. $2 > -3t - 10$
11. $\frac{w}{2} + 4 > 5$
12. $1 + \frac{m}{3} \le 6$
13. $\frac{p}{-8} + 9 > 13$
14. $3 + \frac{r}{-4} \le 6$
15. $6 \ge -6(a + 2)$
16. $18 \le 3(b - 4)$

In Exercises 17–28, solve the inequality. *(See Examples 2 and 3.)*

17. $4 - 2m > 7 - 3m$
18. $8n + 2 \le 8n - 9$
19. $-2d - 2 < 3d + 8$
20. $8 + 10f > 14 - 2f$
21. $8g - 5g - 4 \le -3 + 3g$
22. $3w - 5 > 2w + w - 7$
23. $6(\ell + 3) < 3(2\ell + 6)$
24. $2(5c - 7) \ge 10(c - 3)$
25. $4\left(\frac{1}{2}t - 2\right) > 2(t - 3)$
26. $15\left(\frac{1}{3}b + 3\right) \le 6(b + 9)$
27. $9j - 6 + 6j \ge 3(5j - 2)$
28. $6h - 6 + 2h < 2(4h - 3)$

ERROR ANALYSIS **In Exercises 29 and 30, describe and correct the error in solving the inequality.**

29. ✗

$$\frac{x}{4} + 6 \ge 3$$
$$x + 6 \ge 12$$
$$x \ge 6$$

30. ✗

$$-2(1 - x) \le 2x - 7$$
$$-2 + 2x \le 2x - 7$$
$$-2 \le -7$$

All real numbers are solutions.

31. **MODELING WITH MATHEMATICS** Write and solve an inequality that represents how many \$20 bills you can withdraw from the account without going below the minimum balance. *(See Example 4.)*

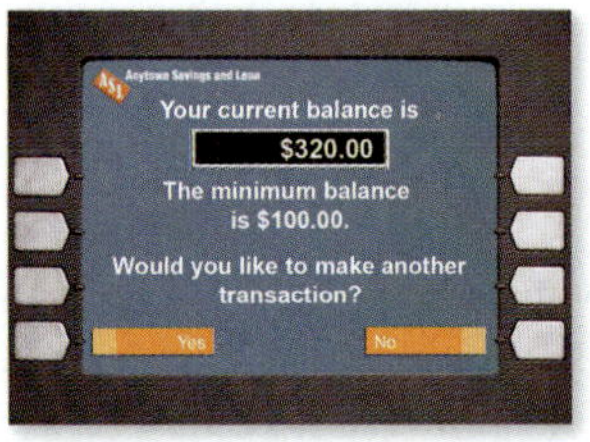

Assignment Guide and Homework Check

ASSIGNMENT

Basic: 1, 2, 3–23 odd, 29, 31, 34, 36, 41–43

Average: 1, 2, 4–30 even, 31–36, 41–43

Advanced: 1, 2, 13–16, 22–28 even, 30–43

HOMEWORK CHECK

Basic: 7, 13, 17, 19, 31

Average: 8, 14, 18, 20, 32

Advanced: 14, 16, 22, 28, 32

ANSWERS

1. *Sample answer:* The same steps can be applied when solving multi-step inequalities and multi-step equations, except that when each side of an inequality is divided by a negative number, the inequality must be reversed.
2. *Sample answer:* Because the terms with the variable are the same, they will cancel, and $8 \le -3$ is not true.
3. B
4. A
5. C
6. D
7. $x > 5$

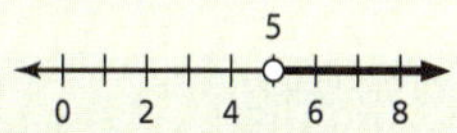

8. $y \le -1$

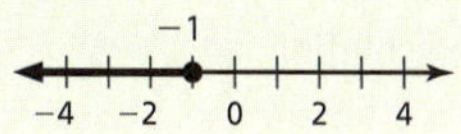

9. $v \le 2$

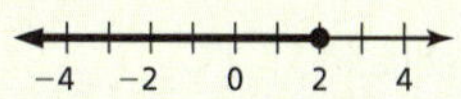

10. $t > -4$

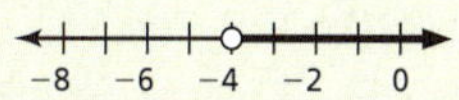

11. $w > 2$

12. $m \le 15$

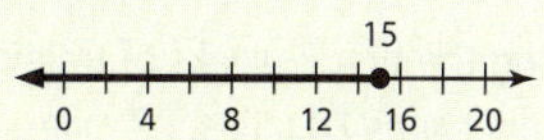

13. $p < -32$

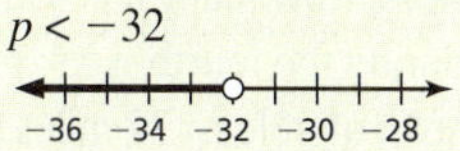

14. $r \ge -12$

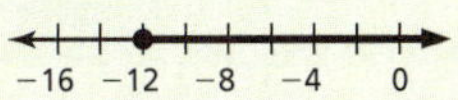

15. $a \ge -3$

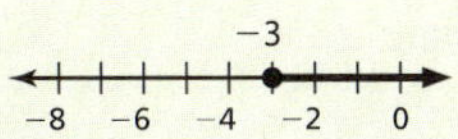

16. $b \ge 10$

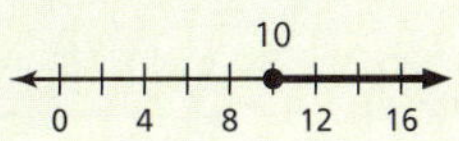

17. $m > 3$
18. no solution
19. $d > -2$
20. $f > 0.5$
21. all real numbers
22. all real numbers
23. no solution
24. all real numbers
25. no solution
26. $b \ge -9$
27. all real numbers
28. no solution

29–31. See Additional Answers.

Dynamic Teaching Tools

- Dynamic Assessment & Progress Monitoring Tool
- Interactive Whiteboard Lesson Library
- Dynamic Classroom with Dynamic Investigations

ANSWERS

32. $25h + 125 \leq 500$, $h \leq 15$

33. $12(2x - 3) > 60$; $x > 4$

34. no; For Woodland to be cheaper, its total cost needs to be less than Forest Park's total cost. Solving the inequality $20 + 55n \leq 100 + 35n$ indicates that Woodland is cheaper for no more than 4 nights, not more than 4 nights.

35. 7 stories; Using the Pythagorean Theorem, the 74-foot ladder can reach at most 70 feet. Solving the inequality $10n - 8 \leq 70$ gives $n \leq 7.8$, so the ladder cannot quite reach the 8th story.

36. **a.** \$40

b. \$3.55; \$8

c. $3.55x + 8 \leq 40$

d. no more than 9 gallons

37. $r \geq 3$

38–43. See Additional Answers.

Mini-Assessment

1. Solve the inequality $-2x + 6 > 16$. Graph the solution. $x < -5$

Solve the inequality.

2. $6n + 5 > 2n - 3$ $n > -2$
3. $10x - 3 < 2(5x + 1)$ The inequality $-3 < 2$ is true. So, all real numbers are solutions.
4. $3(4s - 3) \geq 5 + 12s$ The inequality $-9 \geq 5$ is false. So, there is no solution.
5. During the first 3 weeks of training, a runner runs 20 miles, 22 miles, and 21 miles. How many miles must she run during the fourth week to have a mean of at least 22 miles per week? at least 25 miles

32. MODELING WITH MATHEMATICS A woodworker wants to earn at least \$25 an hour making and selling cabinets. He pays \$125 for materials. Write and solve an inequality that represents how many hours the woodworker can spend building the cabinet.

33. MATHEMATICAL CONNECTIONS The area of the rectangle is greater than 60 square feet. Write and solve an inequality to find the possible values of x.

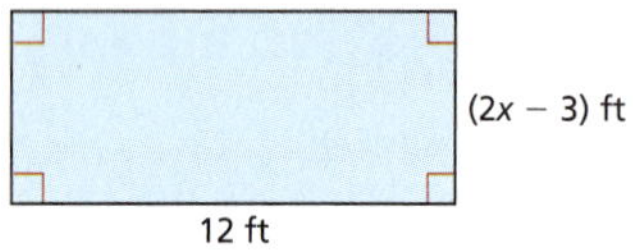

34. MAKING AN ARGUMENT Forest Park Campgrounds charges a \$100 membership fee plus \$35 per night. Woodland Campgrounds charges a \$20 membership fee plus \$55 per night. Your friend says that if you plan to camp for four or more nights, then you should choose Woodland Campgrounds. Is your friend correct? Explain.

35. PROBLEM SOLVING The height of one story of a building is about 10 feet. The bottom of the ladder on the fire truck must be at least 24 feet away from the building. How many stories can the ladder reach? Justify your answer.

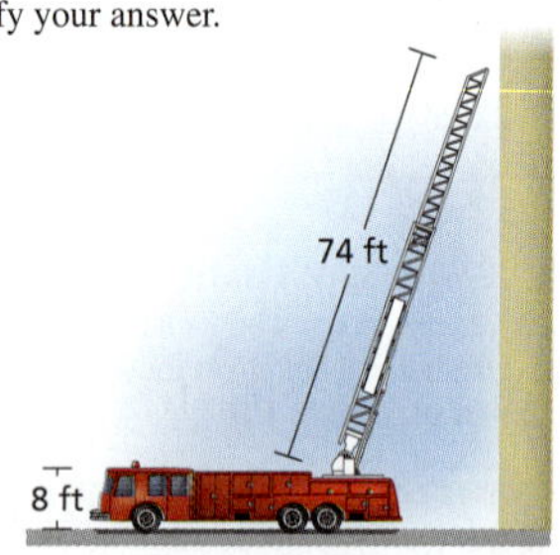

36. HOW DO YOU SEE IT? The graph shows your budget and the total cost of x gallons of gasoline and a car wash. You want to determine the possible amounts (in gallons) of gasoline you can buy within your budget.

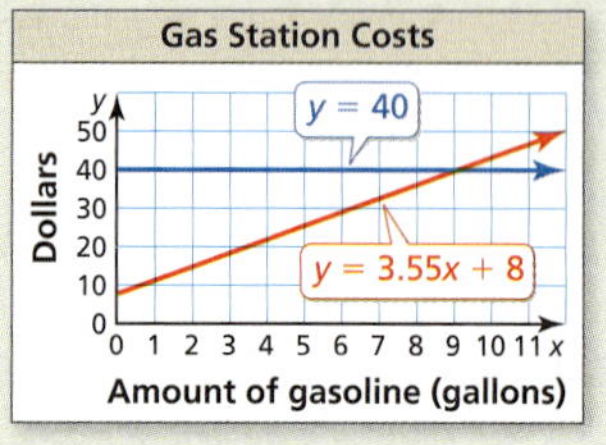

a. What is your budget?

b. How much does a gallon of gasoline cost? How much does a car wash cost?

c. Write an inequality that represents the possible amounts of gasoline you can buy.

d. Use the graph to estimate the solution of your inequality in part (c).

37. PROBLEM SOLVING For what values of r will the area of the shaded region be greater than or equal to $9(\pi - 2)$?

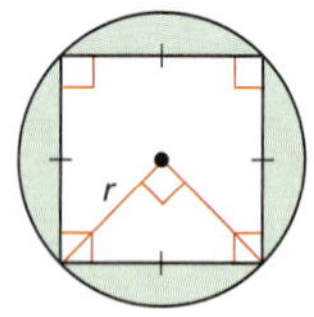

38. THOUGHT PROVOKING A runner's times (in minutes) in the four races he has completed are 25.5, 24.3, 24.8, and 23.5. The runner plans to run at least one more race and wants to have an average time less than 24 minutes. Write and solve an inequality to show how the runner can achieve his goal.

REASONING **In Exercises 39 and 40, find the value of a for which the solution of the inequality is all real numbers.**

39. $a(x + 3) < 5x + 15 - x$

40. $3x + 8 + 2ax \geq 3ax - 4a$

Maintaining Mathematical Proficiency

Reviewing what you learned in previous grades and lessons

Write the sentence as an inequality. *(Section 2.1)*

41. Six times a number y is less than or equal to 10.

42. A number p plus 7 is greater than 24.

43. The quotient of a number r and 7 is no more than 18.

If students need help...	*If students got it...*
Resources by Chapter • Practice A and Practice B • Puzzle Time	Resources by Chapter • Enrichment and Extension • Cumulative Review
Student Journal • Practice	Start the *next* Section
Differentiating the Lesson Skills Review Handbook	

2.1–2.4 What Did You Learn?

Core Vocabulary

inequality, *p. 54*
solution of an inequality, *p. 55*
solution set, *p. 55*
graph of an inequality, *p. 56*
equivalent inequalities, *p. 62*

Core Concepts

Section 2.1
Representing Linear Inequalities, *p. 57*

Section 2.2
Addition Property of Inequality, *p. 62*
Subtraction Property of Inequality, *p. 63*

Section 2.3
Multiplication and Division Properties of Inequality ($c > 0$), *p. 68*
Multiplication and Division Properties of Inequality ($c < 0$), *p. 69*

Section 2.4
Solving Multi-Step Inequalities, *p. 74*
Special Solutions of Linear Inequalities, *p. 75*

Mathematical Practices

1. Explain the meaning of the inequality symbol in your answer to Exercise 47 on page 59. How did you know which symbol to use?
2. In Exercise 30 on page 66, why is it important to check the reasonableness of your answer in part (a) before answering part (b)?
3. Explain how considering the units involved in Exercise 29 on page 71 helped you answer the question.

Study Skills

Analyzing Your Errors

Application Errors

What Happens: You can do numerical problems, but you struggle with problems that have context.

How to Avoid This Error: Do not just mimic the steps of solving an application problem. Explain out loud what the question is asking and why you are doing each step. After solving the problem, ask yourself, "Does my solution make sense?"

79

Dynamic Teaching Tools
- Dynamic Assessment & Progress Monitoring Tool
- Interactive Whiteboard Lesson Library
- Dynamic Classroom with Dynamic Investigations

ANSWERS

1. The inequality symbol means that all the other bridges are shorter; Because the Xianren Bridge is the longest, all other bridges must have a shorter length.
2. The value from part (a) is used to answer part (b), so checking its reasonableness is important to be sure that the correct result is used.
3. Showing the units in each step helps to compare the units in the final answer with the expected units.

ANSWERS

1. $z - 6 \geq 11$
2. $12 \leq -1.5w + 4$
3. $x < 0$
4. $x \geq 8$
5. $q \leq 6$

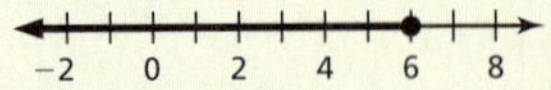

6. $z < -2$

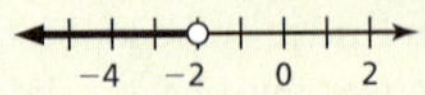

7. $y > 1$

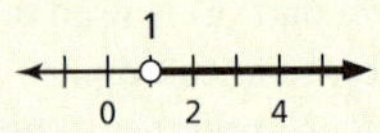

8. $p \geq 6$

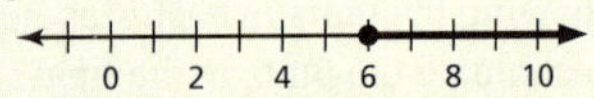

9. $w > -12$

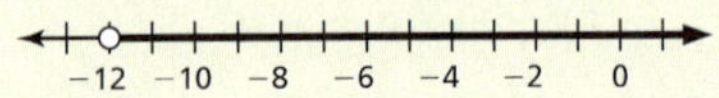

10. $x < -\frac{1}{4}$

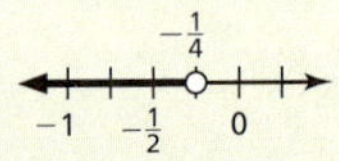

11. $y \geq 8$
12. all real numbers
13. no solution
14. a. $s \geq 100$

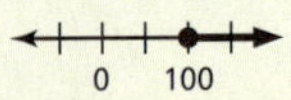

$t \geq 5$

$u \geq 10$

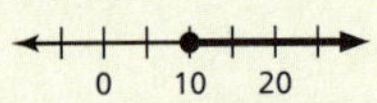

b. no; One of the requirements is to swim 100 yards, which is equal to 300 feet. If you only swim 250 feet you do not meet this requirement.

15. $V + 100 \leq 700$; $V \leq 600$
16. a. $15w \geq 120$; $w \geq 8$

b. The number of weeks is reduced; $65 + 15w \geq 120$; $w \geq 3\frac{2}{3}$

2.1–2.4 Quiz

Write the sentence as an inequality. *(Section 2.1)*

1. A number z minus 6 is greater than or equal to 11.
2. Twelve is no more than the sum of -1.5 times a number w and 4.

Write an inequality that represents the graph. *(Section 2.1)*

3. 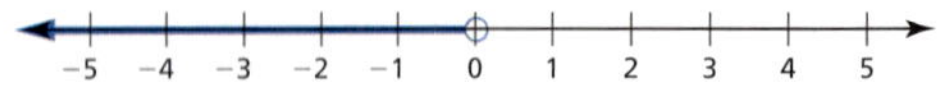

4. 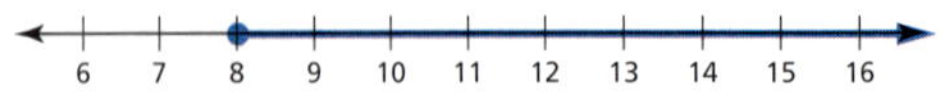

Solve the inequality. Graph the solution. *(Section 2.2 and Section 2.3)*

5. $9 + q \leq 15$
6. $z - (-7) < 5$
7. $-3 < y - 4$
8. $3p \geq 18$
9. $6 > \frac{w}{-2}$
10. $-20x > 5$

Solve the inequality. *(Section 2.4)*

11. $3y - 7 \geq 17$
12. $8(3g - 2) \leq 12(2g + 1)$
13. $6(2x - 1) \geq 3(4x + 1)$

14. Three requirements for a lifeguard training course are shown. *(Section 2.1)*
    a. Write and graph three inequalities that represent the requirements.
    b. You can swim 250 feet, tread water for 6 minutes, and swim 35 feet underwater without taking a breath. Do you satisfy the requirements of the course? Explain.

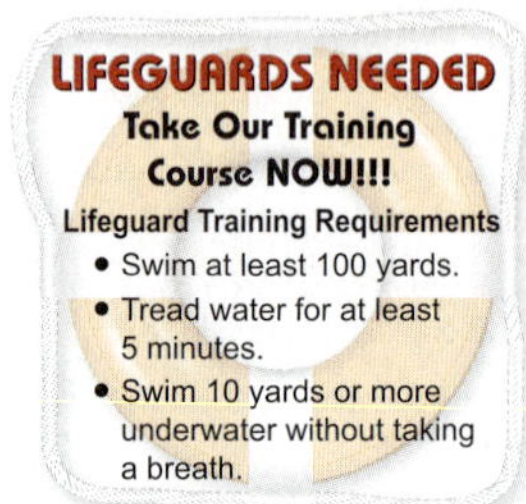

15. The maximum volume of an American white pelican's bill is about 700 cubic inches. A pelican scoops up 100 cubic inches of water. Write and solve an inequality that represents the additional volumes the pelican's bill can contain. *(Section 2.2)*

16. You save \$15 per week to purchase one of the bikes shown. *(Section 2.3 and Section 2.4)*
    a. Write and solve an inequality to find the numbers of weeks you need to save to purchase a bike.
    b. Your parents give you \$65 to help you buy the new bike. How does this affect you answer in part (a)? Use an inequality to justify your answer.

Laurie's Notes

Overview of Section 2.5

Introduction

- Compound inequalities are introduced in this lesson. The skills involved with solving and graphing the individual inequalities are familiar to students. What is new is using the conjunction "and" or "or" to combine two inequalities. It is helpful to first review what it means to form the intersection (and) or the union (or) of two sets.
- There are two ways in which an "and" compound inequality can be written, and hence, there are two strategies for solving as shown in Example 2. When reading the solutions of a compound inequality, students may need to be reminded to use the correct conjunction.

Common Misconceptions

- Students often believe that the graph of an "and" compound inequality is a segment (with open or closed endpoints) and the graph of an "or" compound inequality is two rays in opposite directions (with open or closed endpoints). There are other possibilities.

$x > 2$ *and* $x > -1$

$x > -1$ *or* $x < 3$

Formative Assessment Tips

- **Wait Time:** Wait time is the interval between a question being posed and a student (or the teacher) response. Silence can be uncomfortable in a classroom, but research has shown that increasing wait time increases class participation and answers are more detailed. For complex, higher-order thinking questions, increased wait time is necessary.
- **Reason Abstractly and Quantitatively** and **Construct Viable Arguments and Critique the Reasoning of Others:** If we want students to reason abstractly and quantitatively, and to construct viable arguments, we need to first pose complex questions. Students then need additional *Wait Time* to allow for thinking and formulation of responses. With increased participation, teachers learn more about students' progress and their learning.

Another Way

- **Teaching Tip:** To introduce graphing compound inequalities, use two transparencies or use the *copy* command on an interactive whiteboard. You want students to see that a compound inequality is composed of two parts. When joined with the word "and," you are graphing the intersection of the two parts. When joined with the word "or," you are graphing the union of the two parts. You can overlay the two parts using the transparencies or *copy* command.

Pacing Suggestion

- The explorations are similar to Example 1 in the formal lesson. If you use the *Motivate* and both explorations, you may want to omit Example 1 in the formal lesson.

Dynamic Teaching Tools

- Dynamic Assessment & Progress Monitoring Tool
- Lesson Planning Tool
- Interactive Whiteboard Lesson Library
- Dynamic Classroom with Dynamic Investigations

What Your Students Will Learn

- Write compound linear inequalities in one variable joined by the word *and* or the word *or.*
- Graph compound linear inequalities in one variable.
- Solve compound linear inequalities in one variable using inverse operations.

Laurie's Notes

Exploration

Motivate

- **FYI:** Every state has a law on the books that says something along the lines of: "A person shall not drive a motor vehicle at such a slow speed so as to impede or block the normal and reasonable forward movement of traffic."
- Some states specify a minimum speed limit for highway driving. In Florida, that minimum speed is 40 miles per hour on highways with at least four lanes. The maximum speed on a four-lane divided highway is 65 miles per hour.

? "Can you graph the legal speeds on a four-lane highway in Florida?"

- Explain to students that this is the graph of a compound inequality, the type they will look at in this lesson.

Discuss

- Have students discuss the difference between your boss saying, "You can take a 15-minute break *or* leave 15 minutes early today," versus "You can take a 15-minute break *and* leave 15 minutes early today."
- The words "and" and "or" have meaning in mathematics that students will see today.

Exploration 1

- Superimposed above each graph is a description of the interval. A definition is not necessary because students can make sense of the words.
- It is natural for students to think (and say) that the shaded part is the numbers between ____ and ____. Remind students of the directions: use two inequalities to describe the interval.
- **Whiteboarding:** Have students write their answers and share with the class.
- **Attend to Precision:** Students should check that they have used the correct inequality symbols.
- **Construct Viable Arguments and Critique the Reasoning of Others:** Ask several students to share their reasoning for part (e).

Exploration 2

- The second exploration is very similar to the first. Now that the first exploration has been discussed, it should not take long for students to write two inequalities for the intervals shown.
- **Whiteboarding:** Have students write their answers and share with the class.

Communicate Your Answer

- **Turn and Talk:** Give students time to answer Question 3 with their partners.

Connecting to Next Step

- If you feel students are comfortable using two inequalities to describe an interval, assign Exercises 3–6 on page 85. Explain that two inequalities joined by "and" or "or" are called compound inequalities. You could also have students copy the *Core Vocabulary* into their notebooks.

2.5 Solving Compound Inequalities

Essential Question

How can you use inequalities to describe intervals on the real number line?

EXPLORATION 1 Describing Intervals on the Real Number Line

Work with a partner. In parts (a)–(d), use two inequalities to describe the interval.

> **REASONING ABSTRACTLY**
> To be proficient in math, you need to create a clear representation of the problem at hand.

a. Half-Open Interval

b. Half-Open Interval

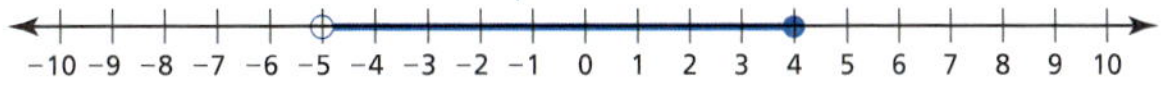

c. Closed Interval

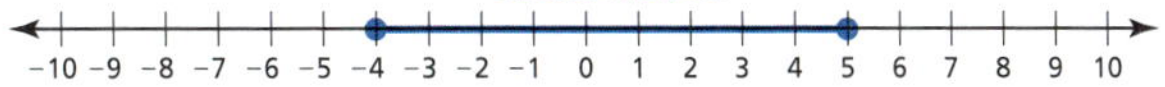

d. Open Interval

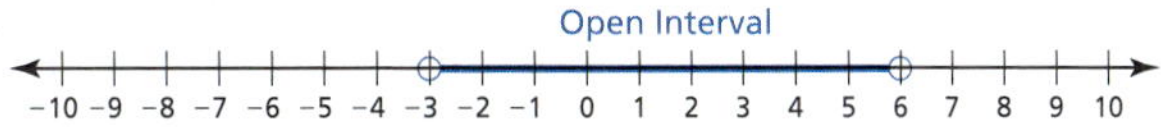

e. Do you use "and" or "or" to connect the two inequalities in parts (a)–(d)? Explain.

EXPLORATION 2 Describing Two Infinite Intervals

Work with a partner. In parts (a)–(d), use two inequalities to describe the interval.

a.

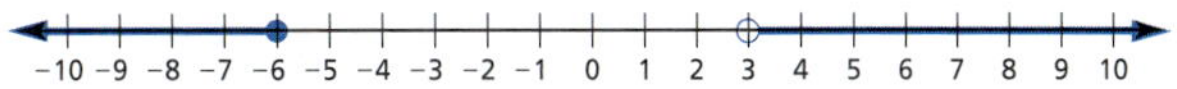

b.

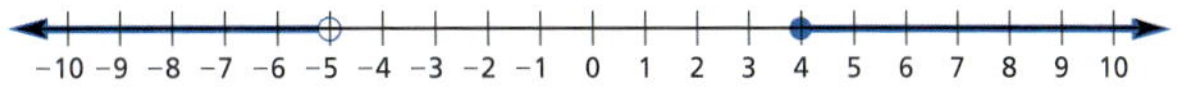

c.

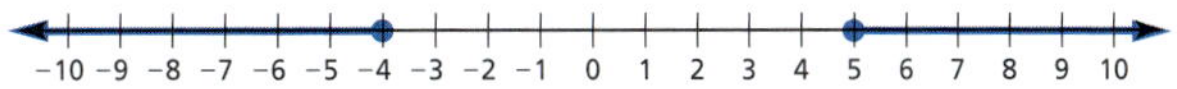

d.

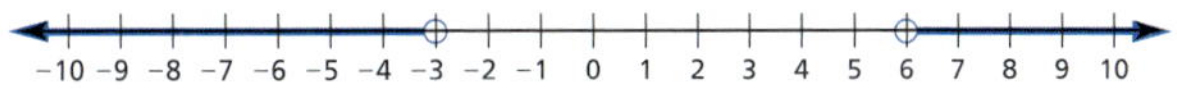

e. Do you use "and" or "or" to connect the two inequalities in parts (a)–(d)? Explain.

Communicate Your Answer

3. How can you use inequalities to describe intervals on the real number line?

Dynamic Teaching Tools

- Dynamic Assessment & Progress Monitoring Tool
- Lesson Planning Tool
- Interactive Whiteboard Lesson Library
- Dynamic Classroom with Dynamic Investigations

ANSWERS

1. **a.** $x \geq -6$ and $x < 3$

b. $x > -5$ and $x \leq 4$

c. $x \geq -4$ and $x \leq 5$

d. $x > -3$ and $x < 6$

e. and; Both inequalities need to be true for values that are in the interval.

2. **a.** $x \leq -6$ or $x > 3$

b. $x < -5$ or $x \geq 4$

c. $x \leq -4$ or $x \geq 5$

d. $x < -3$ or $x > 6$

e. or; Either inequality needs to be true for values that are in the interval.

3. Write 2 inequalities joined by "and" or "or."

Differentiated Instruction

Kinesthetic/Visual

Have pairs of students work together to graph compound inequalities. Have one student use a red pencil to graph the first part of the inequality. The other student uses a different color to graph the second part on the same number line. For a compound inequality with "and," the graph is the interval where both colors overlap.

Extra Example 1

Write each sentence as an inequality. Graph each inequality.

a. A number c is less than 5 and greater than or equal to -2. $-2 \le c < 5$

b. A number t is less than -1 or greater than or equal to 4. $t < -1$ *or* $t \ge 4$

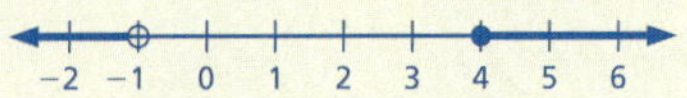

MONITORING PROGRESS ANSWERS

1. $0 < d < 10$

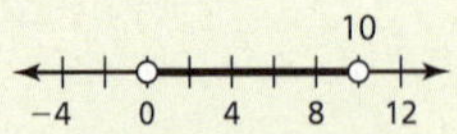

2. $a < -6$ *or* $a \ge -3$

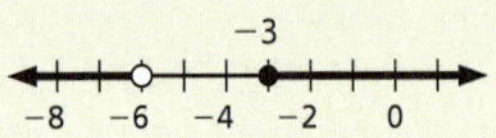

2.5 Lesson

Core Vocabulary

compound inequality, *p. 82*

What You Will Learn

- Write and graph compound inequalities.
- Solve compound inequalities.
- Use compound inequalities to solve real-life problems.

Writing and Graphing Compound Inequalities

A **compound inequality** is an inequality formed by joining two inequalities with the word "and" or the word "or."

The graph of a compound inequality with "and" is the *intersection* of the graphs of the inequalities. The graph shows numbers that are solutions of *both* inequalities.

$x \ge 2$

$x < 5$

$2 \le x$ *and* $x < 5$

$2 \le x < 5$

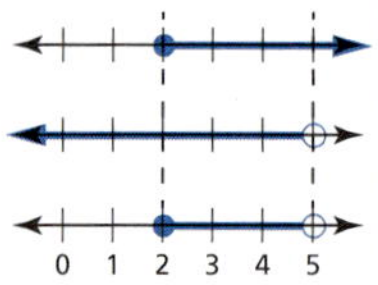

The graph of a compound inequality with "or" is the *union* of the graphs of the inequalities. The graph shows numbers that are solutions of *either* inequality.

$y \le -2$

$y > 1$

$y \le -2$ *or* $y > 1$

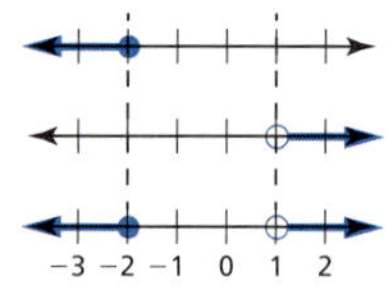

EXAMPLE 1 Writing and Graphing Compound Inequalities

Write each sentence as an inequality. Graph each inequality.

a. A number x is greater than -8 and less than or equal to 4.

b. A number y is at most 0 or at least 2.

REMEMBER

A compound inequality with "and" can be written as a single inequality. For example, you can write $x > -8$ and $x \le 4$ as $-8 < x \le 4$.

SOLUTION

a. A number x is greater than -8 and less than or equal to 4.

$x > -8$ *and* $x \le 4$

▶ An inequality is $-8 < x \le 4$.

Graph the intersection of the graphs of $x > -8$ and $x \le 4$.

b. A number y is at most 0 or at least 2.

$y \le 0$ *or* $y \ge 2$

▶ An inequality is $y \le 0$ *or* $y \ge 2$.

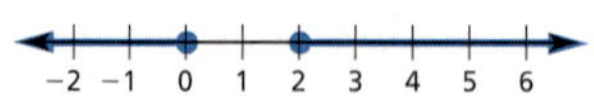

Graph the union of the graphs of $y \le 0$ and $y \ge 2$.

Monitoring Progress

Help in English and Spanish at *BigIdeasMath.com*

Write the sentence as an inequality. Graph the inequality.

1. A number d is more than 0 and less than 10.
2. A number a is fewer than -6 or no less than -3.

Laurie's Notes Teacher Actions

- Define *compound inequality* and connect to the explorations or *Motivate*.
- Review what intersection and union mean in the context of a set of numbers.
- Discuss the *Remember* note. Be sure students understand that when they encounter a compound inequality written in the form $a < x < b$, the conjunction is always "and."
- As you graph say, "We want to shade the values of x that are greater than -8 and less than or equal to 4."

? "What does the phrase 'at most' mean?" As an example, if you have at most 3 cookies, then you could have 3, 2, 1, or no cookies.

Solving Compound Inequalities

You can solve a compound inequality by solving two inequalities separately. When a compound inequality with "and" is written as a single inequality, you can solve the inequality by performing the same operation on each expression.

LOOKING FOR STRUCTURE

To be proficient in math, you need to see complicated things as single objects or as being composed of several objects.

EXAMPLE 2 Solving Compound Inequalities with "And"

Solve each inequality. Graph each solution.

a. $-4 < x - 2 < 3$ **b.** $-3 < -2x + 1 \le 9$

SOLUTION

a. Separate the compound inequality into two inequalities, then solve.

$-4 < x - 2$	*and*	$x - 2 < 3$	Write two inequalities.
$\underline{+2} \quad \underline{+2}$		$\underline{+2} \quad \underline{+2}$	Add 2 to each side.
$-2 < x$	*and*	$x < 5$	Simplify.

▶ The solution is $-2 < x < 5$.

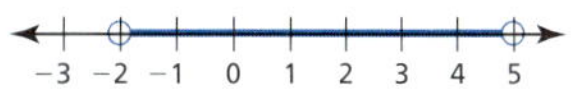

b.

$-3 < -2x + 1 \le 9$	Write the inequality.
$\underline{-1} \quad \underline{-1} \quad \underline{-1}$	Subtract 1 from each expression.
$-4 < -2x \le 8$	Simplify.
$\dfrac{-4}{-2} > \dfrac{-2x}{-2} \ge \dfrac{8}{-2}$	Divide each expression by −2. Reverse each inequality symbol.
$2 > x \ge -4$	Simplify.

▶ The solution is $-4 \le x < 2$.

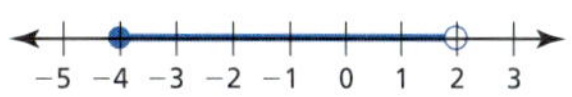

EXAMPLE 3 Solving a Compound Inequality with "Or"

Solve $3y - 5 < -8$ *or* $2y - 1 > 5$. Graph the solution.

SOLUTION

$3y - 5 < -8$	*or*	$2y - 1 > 5$	Write the inequality.
$\underline{+5} \quad \underline{+5}$		$\underline{+1} \quad \underline{+1}$	Addition Property of Inequality
$3y < -3$		$2y > 6$	Simplify.
$\dfrac{3y}{3} < \dfrac{-3}{3}$		$\dfrac{2y}{2} > \dfrac{6}{2}$	Division Property of Inequality
$y < -1$	*or*	$y > 3$	Simplify.

▶ The solution is $y < -1$ *or* $y > 3$.

Monitoring Progress Help in English and Spanish at *BigIdeasMath.com*

Solve the inequality. Graph the solution.

3. $5 \le m + 4 < 10$ **4.** $-3 < 2k - 5 < 7$

5. $4c + 3 \le -5$ *or* $c - 8 > -1$ **6.** $2p + 1 < -7$ *or* $3 - 2p \le -1$

Extra Example 2

Solve each inequality. Graph each solution.

a. $-3 < x + 1 < 5$ $-4 < x < 4$

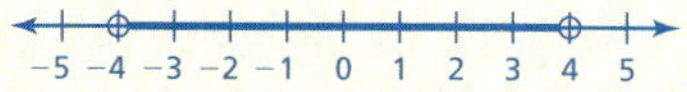

b. $-7 < -3w - 4 \le 2$ $-2 \le w < 1$

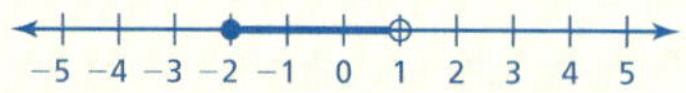

Extra Example 3

Solve $2x + 6 < -2$ *or* $4x - 5 > 3$. Graph the solution. $x < -4$ *or* $x > 2$

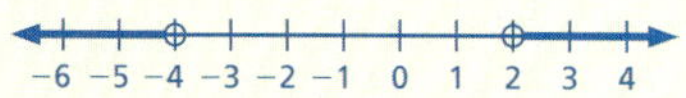

MONITORING PROGRESS ANSWERS

3. $1 \le m < 6$

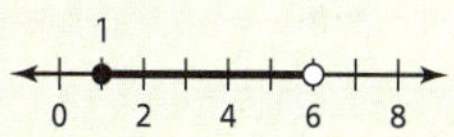

4. $1 < k < 6$

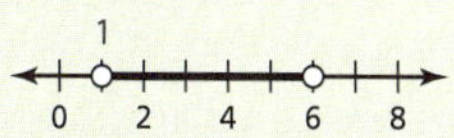

5. $c \le -2$ *or* $c > 7$

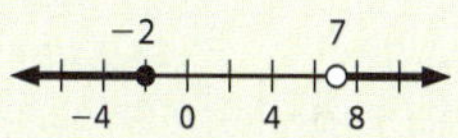

6. $p < -4$ *or* $p \ge 2$

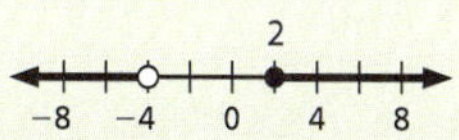

Laurie's Notes Teacher Actions

- Discuss the general process for solving compound inequalities. Explain that each inequality can be solved separately. When the compound inequality with "and" is written as a single inequality statement, perform the same operation on each of the three parts (left, middle, and right).
- Part (a) is solved by separating the inequalities, while part (b) models performing operations on all three parts of the inequality.

COMMON ERROR When dividing each of the three expressions by −2, students may forget to reverse the direction of the inequality symbols.

- Discuss why the solution $2 > x > -4$ is rewritten. By convention we read inequalities from the least value (−4) to the greatest value (2), just as we would read the graph of the solution from left to right.
- In both examples, be sure to substitute values not in the solution set to show that they do not make the inequality true.

Extra Example 4

A recommended storage temperature for chocolate is from 15°C to 20°C.

a. Write and solve a compound inequality that represents the possible storage temperatures (in degrees Fahrenheit) for chocolate. $15 \le \frac{5}{9}(F - 32) \le 20$; $59 \le F \le 68$; The storage temperature is 59°F to 68°F.

b. Describe one situation in which the surrounding temperature could be below the storage temperature range and one in which it could be above. *Sample answer:* The surrounding temperature could be below this range outdoors on a cold day. The surrounding temperature could be above this range in a car parked in the sun on a hot day.

MONITORING PROGRESS ANSWER

7. $-40 \le \frac{5}{9}(F - 32) \le 15$, $-40 \le F \le 59$

Solving Real-Life Problems

EXAMPLE 4 **Modeling with Mathematics**

Operating temperature: 0°C to 35°C

Electrical devices should operate effectively within a specified temperature range. Outside the operating temperature range, the device may fail.

a. Write and solve a compound inequality that represents the possible operating temperatures (in degrees Fahrenheit) of the smartphone.

b. Describe one situation in which the surrounding temperature could be below the operating range and one in which it could be above.

SOLUTION

1. **Understand the Problem** You know the operating temperature range in degrees Celsius. You are asked to write and solve a compound inequality that represents the possible operating temperatures (in degrees Fahrenheit) of the smartphone. Then you are asked to describe situations outside this range.

2. **Make a Plan** Write a compound inequality in degrees Celsius. Use the formula $C = \frac{5}{9}(F - 32)$ to rewrite the inequality in degrees Fahrenheit. Then solve the inequality and describe the situations.

STUDY TIP

You can also solve the inequality by first multiplying each expression by $\frac{9}{5}$.

3. **Solve the Problem** Let C be the temperature in degrees Celsius, and let F be the temperature in degrees Fahrenheit.

$0 \le C \le 35$	Write the inequality using C.
$0 \le \frac{5}{9}(F - 32) \le 35$	Substitute $\frac{5}{9}(F - 32)$ for C.
$9 \cdot 0 \le 9 \cdot \frac{5}{9}(F - 32) \le 9 \cdot 35$	Multiply each expression by 9.
$0 \le 5(F - 32) \le 315$	Simplify.
$0 \le 5F - 160 \le 315$	Distributive Property
$+160 \quad +160 \quad +160$	Add 160 to each expression.
$160 \le 5F \le 475$	Simplify.
$\frac{160}{5} \le \frac{5F}{5} \le \frac{475}{5}$	Divide each expression by 5.
$32 \le F \le 95$	Simplify.

▶ The solution is $32 \le F \le 95$. So, the operating temperature range of the smartphone is 32°F to 95°F. One situation when the surrounding temperature could be below this range is winter in Alaska. One situation when the surrounding temperature could be above this range is daytime in the Mojave Desert of the American Southwest.

4. **Look Back** You can use the formula $C = \frac{5}{9}(F - 32)$ to check that your answer is correct. Substitute 32 and 95 for F in the formula to verify that 0°C and 35°C are the minimum and maximum operating temperatures in degrees Celsius.

Monitoring Progress

Help in English and Spanish at *BigIdeasMath.com*

−40°C to 15°C

7. Write and solve a compound inequality that represents the temperature rating (in degrees Fahrenheit) of the winter boots.

Laurie's Notes Teacher Actions

? Hold up a cell phone and ask, "Have any of you ever had difficulty getting your phone to work properly because the outside temperature was too hot or too cold?" Answers will vary.

- Share the range of operating temperatures for a generic cell phone.

? **Wait Time:** Pose and give *Wait Time*, and even alert students that this is a *No-Hands Question*. "How can you determine the range of operating temperatures in degrees Fahrenheit?" Wait 8–10 seconds and solicit a response from a student whose voice is not often heard. Students may offer strategies that do not involve writing the compound inequality first and then solving for the interval of operating temperatures.

Closure

- **Exit Ticket:** Solve and graph.

a. $-1 \le 2x + 3 \le 7$ $\quad -2 \le x \le 2$

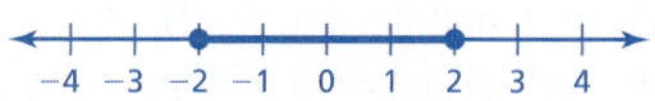

b. $4x + 1 \le -11$ *or* $3x - 4 \ge 5$ $\quad x \le -3$ *or* $x \ge 3$

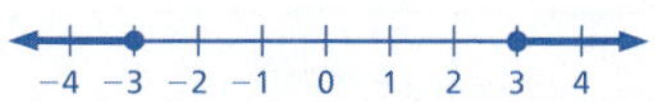

2.5 Exercises

Dynamic Solutions available at *BigIdeasMath.com*

Vocabulary and Core Concept Check

1. **WRITING** Compare the graph of $-6 \le x \le -4$ with the graph of $x \le -6$ *or* $x \ge -4$.

2. **WHICH ONE DOESN'T BELONG?** Which compound inequality does *not* belong with the other three? Explain your reasoning.

$a > 4$ *or* $a < -3$	$a < -2$ *or* $a > 8$	$a > 7$ *or* $a < -5$	$a < 6$ *or* $a > -9$

Monitoring Progress and Modeling with Mathematics

In Exercises 3–6, write a compound inequality that is represented by the graph.

3. 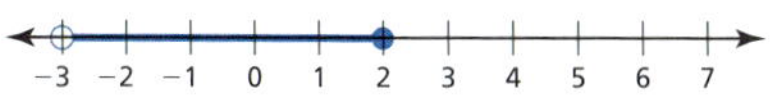

4. 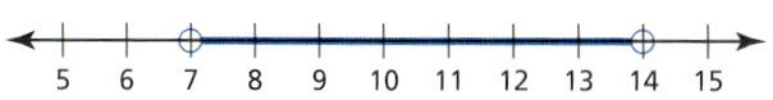

5. 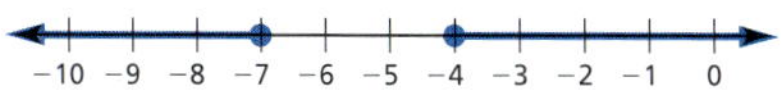

6. 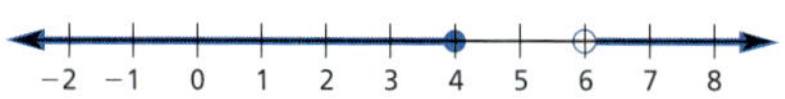

In Exercises 7–10, write the sentence as an inequality. Graph the inequality. *(See Example 1.)*

7. A number p is less than 6 and greater than 2.

8. A number n is less than or equal to -7 or greater than 12.

9. A number m is more than $-7\frac{2}{3}$ or at most -10.

10. A number r is no less than -1.5 and fewer than 9.5.

11. **MODELING WITH MATHEMATICS** Slitsnails are large mollusks that live in deep waters. They have been found in the range of elevations shown. Write and graph a compound inequality that represents this range.

12. **MODELING WITH MATHEMATICS** The life zones on Mount Rainier, a mountain in Washington, can be approximately classified by elevation, as follows.

Low-elevation forest: above 1700 feet to 2500 feet
Mid-elevation forest: above 2500 feet to 4000 feet
Subalpine: above 4000 feet to 6500 feet
Alpine: above 6500 feet to the summit

Elevation of Mount Rainier: 14,410 ft

Write a compound inequality that represents the elevation range for each type of plant life.

a. trees in the low-elevation forest zone

b. flowers in the subalpine and alpine zones

In Exercises 13–20, solve the inequality. Graph the solution. *(See Examples 2 and 3.)*

13. $6 < x + 5 \le 11$

14. $24 > -3r \ge -9$

15. $v + 8 < 3$ *or* $-8v < -40$

16. $-14 > w + 3$ *or* $3w \ge -27$

17. $2r + 3 < 7$ *or* $-r + 9 \le 2$

18. $-6 < 3n + 9 < 21$

19. $-12 < \frac{1}{2}(4x + 16) < 18$

20. $35 < 7(2 - b)$ *or* $\frac{1}{3}(15b - 12) \ge 21$

Assignment Guide and Homework Check

ASSIGNMENT

Basic: 1, 2, 3–11 odd, 12, 13–17 odd, 21, 25, 33–40

Average: 1, 2, 4–18 even, 22, 23, 24–28 even, 33–40

Advanced: 1, 2, 6, 9–12, 18–20, 22–24, 28–40

HOMEWORK CHECK

Basic: 5, 7, 13, 15, 21

Average: 8, 10, 14, 16, 23

Advanced: 9, 10, 19, 20, 23

ANSWERS

1. The graph of $-6 \le x \le -4$ shows a single segment between -6 and -4. The graph of $x \le -6$ *or* $x \ge -4$ shows two opposite rays with endpoints at -6 and -4.

2. $a < 6$ *or* $a > -9$; The graph of this inequality is the only one with overlapping rays that represent all real numbers.

3. $-3 < x \le 2$

4. $7 < x < 14$

5. $x \le -7$ *or* $x \ge -4$

6. $x \le 4$ *or* $x > 6$

7. $2 < p < 6$
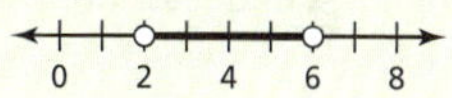

8. $n \le -7$ *or* $n > 12$

9. $m > -7\frac{2}{3}$ *or* $m \le -10$
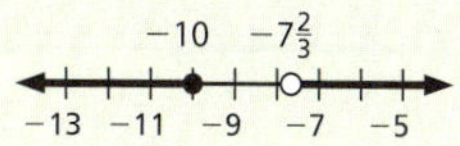

10. $-1.5 \le r < 9.5$
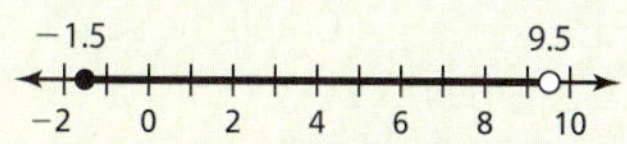

11. $-2500 \le e \le -100$
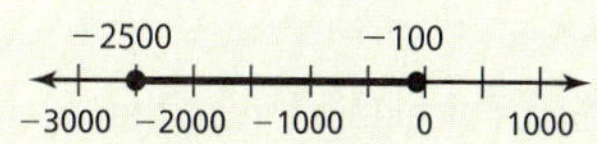

12. a. $1700 < h \le 2500$
    b. $4000 < h \le 14{,}410$

13. $1 < x \le 6$
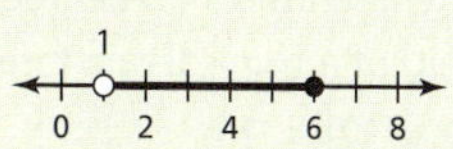

14. $-8 < r \le 3$
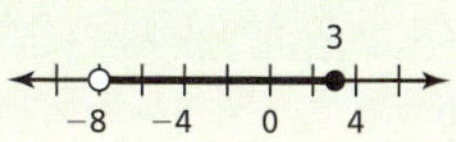

15. $v < -5$ *or* $v > 5$
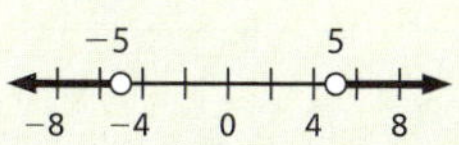

16. $w < -17$ *or* $w \ge -9$
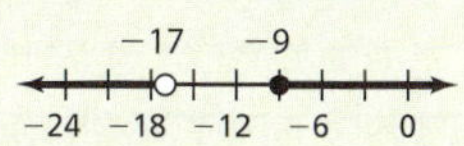

17. $r < 2$ *or* $r \ge 7$
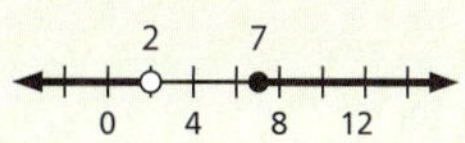

18. $-5 < n < 4$
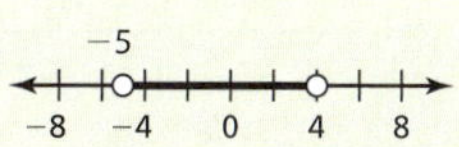

19. $-10 < x < 5$
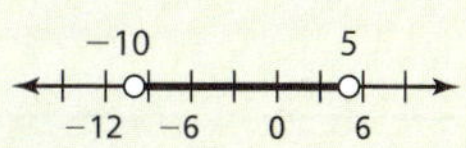

20. See Additional Answers.

Dynamic Teaching Tools

- Dynamic Assessment & Progress Monitoring Tool
- Interactive Whiteboard Lesson Library
- Dynamic Classroom with Dynamic Investigations

ANSWERS

21. In the second step, 3 should have been subtracted from 4 on the left side; $1 < -2x < 6$; $-\frac{1}{2} > x > -3$

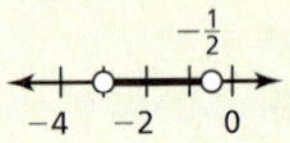

22. The graph should have 2 opposite rays, not 1 segment.

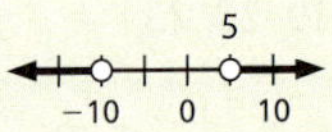

23. $-20 \le \frac{5}{9}(F - 32) \le -15$, $-4 \le F \le 5$

24–40. See Additional Answers.

ERROR ANALYSIS In Exercises 21 and 22, describe and correct the error in solving the inequality or graphing the solution.

21.

✗

$4 < -2x + 3 < 9$

$4 < -2x < 6$

$-2 > x > -3$

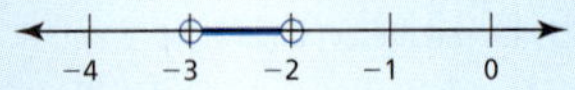

22.

✗

$x - 2 > 3$ or $x + 8 < -2$

$x > 5$ or $x < -10$

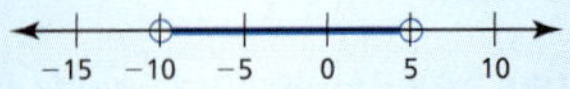

23. MODELING WITH MATHEMATICS Write and solve a compound inequality that represents the possible temperatures (in degrees Fahrenheit) of the interior of the iceberg. *(See Example 4.)*

24. PROBLEM SOLVING A ski shop sells skis with lengths ranging from 150 centimeters to 220 centimeters. The shop says the length of the skis should be about 1.16 times a skier's height (in centimeters). Write and solve a compound inequality that represents the heights of skiers the shop does *not* provide skis for.

In Exercises 25–30, solve the inequality. Graph the solution, if possible.

25. $22 < -3c + 4 < 14$

26. $2m - 1 \ge 5$ *or* $5m > -25$

27. $-y + 3 \le 8$ *and* $y + 2 > 9$

28. $x - 8 \le 4$ *or* $2x + 3 > 9$

29. $2n + 19 \le 10 + n$ *or* $-3n + 3 < -2n + 33$

30. $3x - 18 < 4x - 23$ *and* $x - 16 < -22$

31. REASONING Fill in the compound inequality $4(x - 6)$ ▭ $2(x - 10)$ *and* $5(x + 2) \ge 2(x + 8)$ with <, ≤, >, or ≥ so that the solution is only one value.

32. THOUGHT PROVOKING Write a real-life story that can be modeled by the graph.

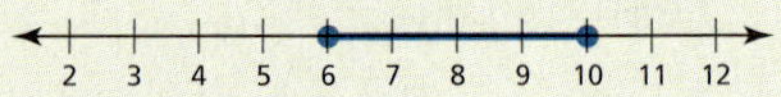

33. MAKING AN ARGUMENT The sum of the lengths of any two sides of a triangle is greater than the length of the third side. Use the triangle shown to write and solve three inequalities. Your friend claims the value of x can be 1. Is your friend correct? Explain.

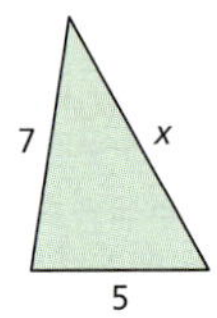

34. HOW DO YOU SEE IT? The graph shows the annual profits of a company from 2006 to 2013.

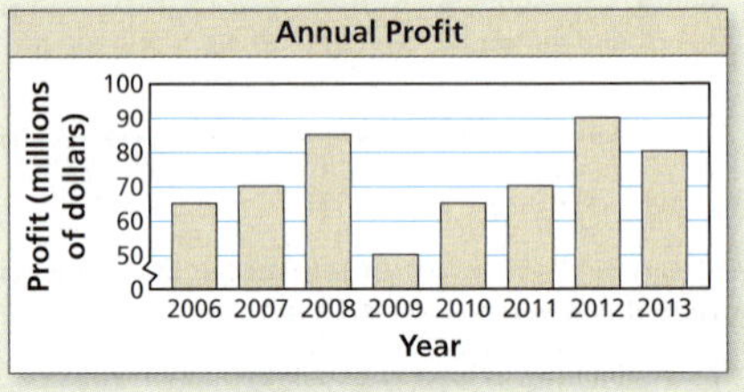

a. Write a compound inequality that represents the annual profits from 2006 to 2013.

b. You can use the formula $P = R - C$ to find the profit P, where R is the revenue and C is the cost. From 2006 to 2013, the company's annual cost was about \$125 million. Is it possible the company had an annual revenue of \$160 million from 2006 to 2013? Explain.

Maintaining Mathematical Proficiency

Reviewing what you learned in previous grades and lessons

Solve the equation. Graph the solutions, if possible. *(Section 1.4)*

35. $\left|\frac{d}{9}\right| = 6$ **36.** $7|5p - 7| = -21$ **37.** $|r + 2| = |3r - 4|$ **38.** $\left|\frac{1}{2}w - 6\right| = |w + 7|$

Find and interpret the mean absolute deviation of the data. *(Skills Review Handbook)*

39. 1, 1, 2, 5, 6, 8, 10, 12, 12, 13 **40.** 24, 26, 28, 28, 30, 30, 32, 32, 34, 36

Mini-Assessment

1. Write the sentence as an inequality. Graph the inequality. A number p is greater than −3 and less than 4.
$-3 < p < 4$

2. Solve $22 < 4x - 6 \le 38$. Graph the solution. $7 < x \le 11$

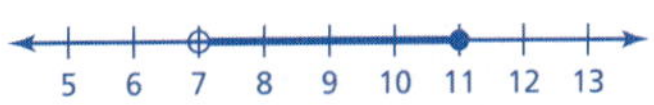

3. Solve $2c + 5 \le -11$ *or* $3c - 1 \ge -7$. Graph the solution.
$c \le -8$ *or* $c \ge -2$

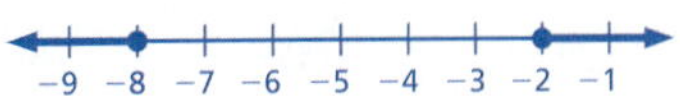

4. In a tropical rain forest, the temperature ranges from 22°C to 30°C. Write and solve a compound inequality that represents the possible temperatures (in degrees Fahrenheit) of a tropical rain forest.
$22 \le \frac{5}{9}(F - 32) \le 30$;
$71.6 \le F \le 86$; 71.6°F to 86°F

If students need help...	*If students got it...*
Resources by Chapter • Practice A and Practice B • Puzzle Time	Resources by Chapter • Enrichment and Extension • Cumulative Review
Student Journal • Practice	Start the *next* Section
Differentiating the Lesson Skills Review Handbook	

Laurie's Notes

Overview of Section 2.6

Introduction

- In Section 1.4, students learned to solve absolute value equations. The technique used for solving an equation such as $|x - 4| = 6$ was to consider when the expression inside the absolute value symbols was equal to 6 or -6.
- Now consider the inequality $|x - 4| \leq 6$. This question is asking when the expression inside the absolute value symbols is between -6 and 6, inclusive, or when is $-6 \leq x - 4 \leq 6$? This is an "and" compound inequality.
- For the inequality $|x - 4| \geq 6$, the question is asking when the expression inside the absolute value symbols is greater than or equal to 6 or less than or equal to -6. This translates into the "or" compound inequality $x - 4 \geq 6$ or $x - 4 \leq -6$.
- This lesson connects the lessons on solving absolute value equations (Section 1.4) and solving compound inequalities (Section 2.5).

Resources

- I have a long number line (8 feet) above one of my whiteboards. The number line is a helpful visual reference for students in this lesson.

Teaching Strategy

- **Human Graph:** Creating a human graph is an engaging visual way for students to display (plot) data. You need to have room to form a number line or an entire *xy*-coordinate grid in your classroom or adjacent space. If there are square tiles on the floor, the tiles can be used to help increment the axis (axes). You can also use masking tape and a dark pen to increment.
- In this lesson, a human graph is made to display the solution to an absolute value inequality. Students viewing the volunteers quickly see that the solution is either a segment (and) or two rays (or). A student volunteer who moves incorrectly sees the immediate feedback!

Another Way

- You can use the words "between" and "beyond" to describe absolute value inequalities. For example, $|x| < 2$ means that x is between -2 and 2; $|x| > 2$ means that x is beyond -2 or beyond 2.

Pacing Suggestion

- Students will work the same inequality three different ways in the explorations. Begin with the *Motivate* and discussion, followed by the explorations. When students have finished the explorations, begin the *Core Concept* in the formal lesson.

Dynamic Teaching Tools

- Dynamic Assessment & Progress Monitoring Tool
- Lesson Planning Tool
- Interactive Whiteboard Lesson Library
- Dynamic Classroom with Dynamic Investigations

What Your Students Will Learn

- Solve absolute value inequalities.

Laurie's Notes

Exploration

Motivate

- **Human Graph:** Have nine volunteers, each with a number from −4 to 4, form a number line at the front of the room facing their classmates.
- When you show an inequality, the volunteers holding a solution are to take a step forward. The others should remain in place. *Examples:*

  $|x| \leq 2$ The following step forward: −2, −1, 0, 1, 2

  $|x| > 1$ The following step forward: −4, −3, −2, 2, 3, 4

? "The less than inequalities remind you of what type of compound inequality?" and

? "The greater than inequalities remind you of what type of compound inequality?" or

Discuss

- Students solved absolute value equations in Section 1.4. Now they are combining that skill with their understanding of compound inequalities to solve absolute value inequalities.

? "What does $|x| = 2$ mean geometrically?" all values that are 2 units from 0

- Graph the solutions for $|x| = 2$.

? "What does $|x| < 2$ mean geometrically?" all values that are less than 2 units from 0

- Graph the solutions for $|x| < 2$.

? "What does $|x| > 2$ mean geometrically?" all values that are more than 2 units from 0

- Graph the solutions for $|x| > 2$.
- Connect these two types of absolute value inequalities to student understanding of "and" and "or" compound inequalities.

Exploration 1

- Students could use trial and error to see what values work. Connecting their knowledge of solving absolute value equations and compound inequalities, students should realize that they are looking for an interval bounded by $x = 1$ above and a negative number below.
- Students may be uncertain of how to describe the values of $x + 2$ that make the inequality true. Say, "The quantity inside the absolute value symbols is less than or equal to 3 or greater than or equal to −3." That leads to the inequalities $x + 2 \leq 3$ *and* $x + 2 \geq -3$.
- The solution is an "and" compound inequality.

Exploration 2

- **Connection:** When solving $|x| = 3$, you are finding the two values that are 3 units from 0, thus making 0 the midpoint. When solving $|x + 2| = 3$, you are still finding two values that are 3 units from a midpoint, but the midpoint has been translated 2 units to the left, where $x + 2 = 0$. Making the connection to the translation ($|x| = 3 \rightarrow |x + 2| = 3$) is a *Big Idea* that students have already seen and will continue to see in their study of mathematics.

? "Using the number line, what solutions did you have?" −5 and 1

- The solution to the absolute value inequality is all of the x-values between −5 and 1, inclusive.

Exploration 3

- Students can use the *table* feature on a graphing calculator. The solution, of course, will be the same as the two previous explorations.

Communicate Your Answer

- **Turn and Talk:** Have students share their thoughts about the three approaches with their partners.

Connecting to Next Step

- If you feel students are comfortable with solving simple absolute value inequalities, assign some of Exercises 3–8 on page 91.

2.6 Solving Absolute Value Inequalities

Essential Question How can you solve an absolute value inequality?

EXPLORATION 1 **Solving an Absolute Value Inequality Algebraically**

Work with a partner. Consider the absolute value inequality

$|x + 2| \leq 3.$

a. Describe the values of $x + 2$ that make the inequality true. Use your description to write two linear inequalities that represent the solutions of the absolute value inequality.

b. Use the linear inequalities you wrote in part (a) to find the solutions of the absolute value inequality.

c. How can you use linear inequalities to solve an absolute value inequality?

MAKING SENSE OF PROBLEMS

To be proficient in math, you need to explain to yourself the meaning of a problem and look for entry points to its solution.

EXPLORATION 2 **Solving an Absolute Value Inequality Graphically**

Work with a partner. Consider the absolute value inequality

$|x + 2| \leq 3.$

a. On a real number line, locate the point for which $x + 2 = 0$.

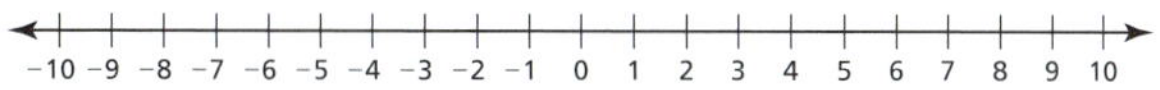

b. Locate the points that are within 3 units from the point you found in part (a). What do you notice about these points?

c. How can you use a number line to solve an absolute value inequality?

EXPLORATION 3 **Solving an Absolute Value Inequality Numerically**

Work with a partner. Consider the absolute value inequality

$|x + 2| \leq 3.$

a. Use a spreadsheet, as shown, to solve the absolute value inequality.

b. Compare the solutions you found using the spreadsheet with those you found in Explorations 1 and 2. What do you notice?

c. How can you use a spreadsheet to solve an absolute value inequality?

	A	B
1	x	|x + 2|
2	-6	4
3	-5	
4	-4	
5	-3	
6	-2	
7	-1	
8	0	
9	1	
10	2	
11		

abs(A2 + 2)

Communicate Your Answer

4. How can you solve an absolute value inequality?

5. What do you like or dislike about the algebraic, graphical, and numerical methods for solving an absolute value inequality? Give reasons for your answers.

Dynamic Teaching Tools

- Dynamic Assessment & Progress Monitoring Tool
- Lesson Planning Tool
- Interactive Whiteboard Lesson Library
- Dynamic Classroom with Dynamic Investigations

ANSWERS

1. **a.** The inequality is true if the distance between $x + 2$ and 0 is 3 or less; $x + 2 \leq 3$; $x + 2 \geq -3$
   b. $-5 \leq x \leq 1$
   c. Write a compound inequality representing the distance between the absolute value expression and 0.
2. See Additional Answers.
3. **a.** $-5 \leq x \leq 1$
   b. They are the same.
   c. Have the spreadsheet calculate the value of the absolute value expression for many values of x, and find the ones that give the expected solution.
4. solving algebraic equations, graphically on a number line, or by trial and error with a spreadsheet
5. *Sample answers:* The algebraic method is nice to use because it is the quickest method. The graphical method is nice to use because it helps to visualize absolute value. The numerical method is not favorable because setting up the spreadsheet is time-consuming.

Differentiated Instruction

Inclusion

When rewriting $|x + 7| \le 2$ as a compound inequality, students may have difficulty understanding why the direction of the inequality sign and the sign of the number 2 are changed to get $x + 7 \ge -2$. Remind them that the value of the expression inside the absolute value sign can be positive or negative. Write $-(x + 7) \le 2$ on the board and then multiply each side by -1 to get $x + 7 \ge -2$.

Extra Example 1

Solve each inequality. Graph each solution, if possible.

a. $|x + 3| \le 2$ $\quad -5 \le x \le -1$

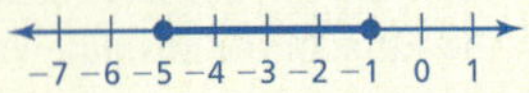

b. $|3c - 5| < -2$ The expression $|3c - 5|$ cannot be negative. So, the inequality has no solution.

2.6 Lesson

What You Will Learn

- Solve absolute value inequalities.
- Use absolute value inequalities to solve real-life problems.

Core Vocabulary

absolute value inequality, *p. 88*
absolute deviation, *p. 90*

Previous
compound inequality
mean

Solving Absolute Value Inequalities

An **absolute value inequality** is an inequality that contains an absolute value expression. For example, $|x| < 2$ and $|x| > 2$ are absolute value inequalities. Recall that $|x| = 2$ means the distance between x and 0 is 2.

The inequality $|x| < 2$ means the distance between x and 0 is *less than* 2.

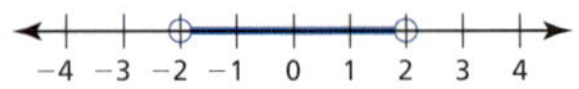

The graph of $|x| < 2$ is the graph of $x > -2$ *and* $x < 2$.

The inequality $|x| > 2$ means the distance between x and 0 is *greater than* 2.

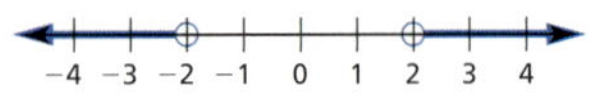

The graph of $|x| > 2$ is the graph of $x < -2$ *or* $x > 2$.

You can solve these types of inequalities by solving a compound inequality.

Core Concept

Solving Absolute Value Inequalities

To solve $|ax + b| < c$ for $c > 0$, solve the compound inequality

$$ax + b > -c \quad \textit{and} \quad ax + b < c.$$

To solve $|ax + b| > c$ for $c > 0$, solve the compound inequality

$$ax + b < -c \quad \textit{or} \quad ax + b > c.$$

In the inequalities above, you can replace < with ≤ and > with ≥.

EXAMPLE 1 Solving Absolute Value Inequalities

Solve each inequality. Graph each solution, if possible.

a. $|x + 7| \le 2$ **b.** $|8x - 11| < 0$

SOLUTION

a. Use $|x + 7| \le 2$ to write a compound inequality. Then solve.

$x + 7 \ge -2$	and	$x + 7 \le 2$	Write a compound inequality.
$\underline{-7} \quad \underline{-7}$		$\underline{-7} \quad \underline{-7}$	Subtract 7 from each side.
$x \ge -9$	and	$x \le -5$	Simplify.

REMEMBER
A compound inequality with "and" can be written as a single inequality.

▶ The solution is $-9 \le x \le -5$.

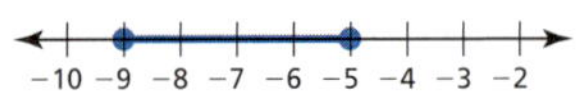

b. By definition, the absolute value of an expression must be greater than or equal to 0. The expression $|8x - 11|$ cannot be less than 0.

▶ So, the inequality has no solution.

Laurie's Notes Teacher Actions

- Write the *Core Concept* on the board.
- **Teaching Tip:** When solving an absolute value inequality such as $|2x + 1| < 4$, I place my fingers over the expression inside the absolute value bars and say, "We are looking for a quantity whose absolute value is less than 4 units from 0. What is under my fingers is between -4 *and* 4." That leads directly into writing the compound inequality $-4 < 2x + 1 < 4$. By covering up the expression, I find it helps focus student attention on the process, without getting overwhelmed by symbols. If the inequality had been $|2x + 1| > 4$, then I would have said, "We are looking for a quantity whose absolute value is more than 4 units from 0. What is under my fingers is less than -4 *or* greater than 4."
- Work through part (a) as shown.
- **?** Write the inequality in part (b) and ask, "What value(s) of x will make the expression on the left side of the inequality less than 0? Explain." none; Any value substituted for x and then multiplied by 8 and decreased by 11 will not be negative once you find the absolute value of the result.

Monitoring Progress Help in English and Spanish at *BigIdeasMath.com*

Solve the inequality. Graph the solution, if possible.

1. $|x| \leq 3.5$ **2.** $|k - 3| < -1$ **3.** $|2w - 1| < 11$

EXAMPLE 2 Solving Absolute Value Inequalities

Solve each inequality. Graph each solution.

a. $|c - 1| \geq 5$ **b.** $|10 - m| \geq -2$ **c.** $4|2x - 5| + 1 > 21$

SOLUTION

a. Use $|c - 1| \geq 5$ to write a compound inequality. Then solve.

$c - 1 \leq -5$	*or*	$c - 1 \geq 5$	Write a compound inequality.
$\underline{+1} \quad \underline{+1}$		$\underline{+1} \quad \underline{+1}$	Add 1 to each side.
$c \leq -4$	*or*	$c \geq 6$	Simplify.

▶ The solution is $c \leq -4$ *or* $c \geq 6$.

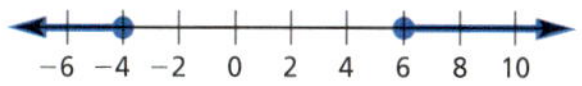

b. By definition, the absolute value of an expression must be greater than or equal to 0. The expression $|10 - m|$ will always be greater than -2.

▶ So, all real numbers are solutions.

c. First isolate the absolute value expression on one side of the inequality.

$4|2x - 5| + 1 > 21$	Write the inequality.
$\underline{-1} \quad \underline{-1}$	Subtract 1 from each side.
$4|2x - 5| > 20$	Simplify.
$\dfrac{4|2x - 5|}{4} > \dfrac{20}{4}$	Divide each side by 4.
$|2x - 5| > 5$	Simplify.

Then use $|2x - 5| > 5$ to write a compound inequality. Then solve.

$2x - 5 < -5$	*or*	$2x - 5 > 5$	Write a compound inequality.
$\underline{+5} \quad \underline{+5}$		$\underline{+5} \quad \underline{+5}$	Add 5 to each side.
$2x < 0$		$2x > 10$	Simplify.
$\dfrac{2x}{2} < \dfrac{0}{2}$		$\dfrac{2x}{2} > \dfrac{10}{2}$	Divide each side by 2.
$x < 0$	*or*	$x > 5$	Simplify.

▶ The solution is $x < 0$ *or* $x > 5$.

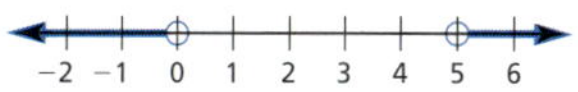

Monitoring Progress Help in English and Spanish at *BigIdeasMath.com*

Solve the inequality. Graph the solution.

4. $|x + 3| > 8$ **5.** $|n + 2| - 3 \geq -6$ **6.** $3|d + 1| - 7 \geq -1$

Extra Example 2

Solve each inequality. Graph each solution.

a. $|x + 1| \geq 3$ $x \leq -4$ *or* $x \geq 2$

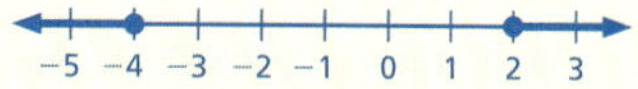

b. $|12 + w| \geq -5$ All real numbers are solutions.

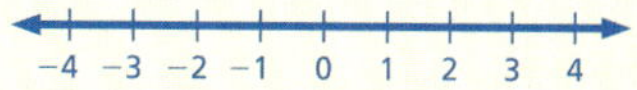

c. $3|2t - 3| - 5 > 16$ $t < -2$ *or* $t > 5$

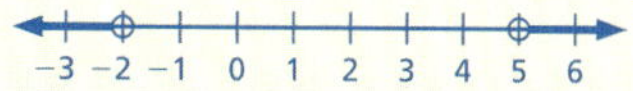

MONITORING PROGRESS ANSWERS

1. $-3.5 \leq x \leq 3.5$

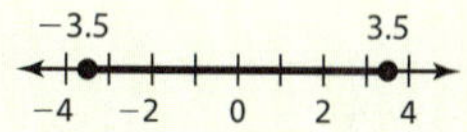

2. no solution
3. $-5 < w < 6$

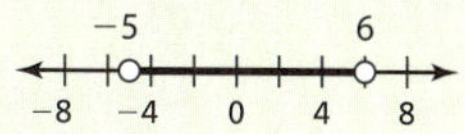

4. $x < -11$ *or* $x > 5$

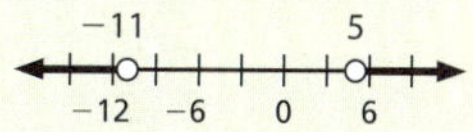

5. all real numbers

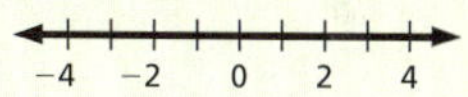

6. $d \leq -3$ *or* $d \geq 1$

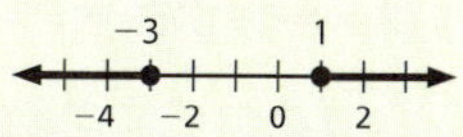

Laurie's Notes Teacher Actions

- **Think-Pair-Share:** Have partners work part (a) of Example 2.
- **? Wait Time:** "In part (b), is it possible for the absolute value of an expression, like $10 - m$, to be greater than or equal to -2?" Before soliciting an answer, ask students to give a *Thumbs Up* response. This can be a challenging inequality for students to make sense of. One approach might be to change the inequality to $|10 - m| < -2$. There are no values of m that make the inequality true. Now return to the original inequality with the $\geq$ symbol. yes; All values of m must be solutions.
- **Look For and Make Use of Structure:** When solving part (c), students need to think of $|2x - 5|$ as a term that needs to be isolated first. Students should think solving $4|2x - 5| + 1 > 21$ is similar to solving $4x + 1 > 21$. Additionally, solving $4x + 1 > 21$ is similar to solving $4x + 1 = 21$. To isolate the variable (or the absolute value expression), subtract 1 and then divide by 4.
- **Make Sense of Problems and Persevere in Solving Them:** By connecting the process back to prior skills, students generally can follow and make sense of the problems. Performing the process on their own is still a skill that needs to be practiced.

Extra Example 3

Use the table in Example 3. You are willing to pay the mean price with an absolute deviation of at most \$125. How many of the computers are priced to meet your condition? $539 \le x \le 789$; Six prices meet your condition: \$750, \$650, \$660, \$670, \$650, and \$725.

MONITORING PROGRESS ANSWER

7. 5

Solving Real-Life Problems

The **absolute deviation** of a number x from a given value is the absolute value of the difference of x and the given value.

absolute deviation $= |x - \text{given value}|$

EXAMPLE 3 Modeling with Mathematics

Computer prices	
\$890	\$750
\$650	\$370
\$660	\$670
\$450	\$650
\$725	\$825

You are buying a new computer. The table shows the prices of computers in a store advertisement. You are willing to pay the mean price with an absolute deviation of at most \$100. How many of the computer prices meet your condition?

SOLUTION

1. **Understand the Problem** You know the prices of 10 computers. You are asked to find how many computers are at most \$100 from the mean price.
2. **Make a Plan** Calculate the mean price by dividing the sum of the prices by the number of prices, 10. Use the absolute deviation and the mean price to write an absolute value inequality. Then solve the inequality and use it to answer the question.
3. **Solve the Problem**

The mean price is $\frac{6640}{10} = \$664$. Let x represent a price you are willing to pay.

$|x - 664| \le 100$ Write the absolute value inequality.

$-100 \le x - 664 \le 100$ Write a compound inequality.

$564 \le x \le 764$ Add 664 to each expression and simplify.

▶ The prices you will consider must be at least \$564 and at most \$764. Six prices meet your condition: \$750, \$650, \$660, \$670, \$650, and \$725.

4. **Look Back** You can check that your answer is correct by graphing the computer prices and the mean on a number line. Any point within 100 of 664 represents a price that you will consider.

STUDY TIP

The absolute deviation of at most \$100 from the mean, \$664, is given by the inequality $|x - 664| \le 100$.

Monitoring Progress

Help in English and Spanish at *BigIdeasMath.com*

7. **WHAT IF?** You are willing to pay the mean price with an absolute deviation of at most \$75. How many of the computer prices meet your condition?

Concept Summary

Solving Inequalities

One-Step and Multi-Step Inequalities

- Follow the steps for solving an equation. Reverse the inequality symbol when multiplying or dividing by a negative number.

Compound Inequalities

- If necessary, write the inequality as two separate inequalities. Then solve each inequality separately. Include *and* or *or* in the solution.

Absolute Value Inequalities

- If necessary, isolate the absolute value expression on one side of the inequality. Write the absolute value inequality as a compound inequality. Then solve the compound inequality.

Laurie's Notes Teacher Actions

- **Connection:** In middle school, students studied the *mean absolute deviation* of a data set. Define the *absolute deviation* of a number from a given value.
- **Teaching Tip:** Use the context of the speed of a car on the highway. There is a given value (speed limit) posted. If you drive too fast or too slow, you risk getting a ticket. The mean absolute deviation might be 10 miles per hour.
- **Make Sense of Problems and Persevere in Solving Them:** Discuss the context of the situation and why someone would be willing to pay the mean price with an absolute deviation of at most \$100. Students must first understand and make sense of the problem before they can make a plan to solve the problem.
- Work through the example as shown.
- **Whiteboarding:** Discuss each of the types of inequalities in the *Concept Summary* box. Have students solve 1 or 2 problems of each type with their partners.

Closure

- **Exit Ticket:** Solve and graph.

$|x - 7| < 13$ $\quad -6 < x < 20$

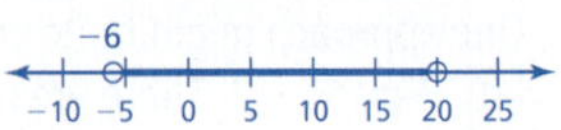

$|x + 4| \ge 8$ $\quad x \ge 4$ or $x \le -12$

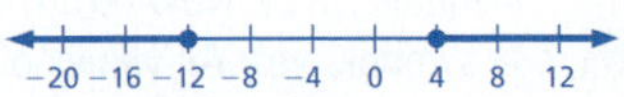

2.6 Exercises

Dynamic Solutions available at *BigIdeasMath.com*

Vocabulary and Core Concept Check

1. **REASONING** Can you determine the solution of $|4x - 2| \geq -6$ without solving? Explain.
2. **WRITING** Describe how solving $|w - 9| \leq 2$ is different from solving $|w - 9| \geq 2$.

Monitoring Progress and Modeling with Mathematics

In Exercises 3–18, solve the inequality. Graph the solution, if possible. *(See Examples 1 and 2.)*

3. $|x| < 3$
4. $|y| \geq 4.5$
5. $|d + 9| > 3$
6. $|h - 5| \leq 10$
7. $|2s - 7| \geq -1$
8. $|4c + 5| > 7$
9. $|5p + 2| < -4$
10. $|9 - 4n| < 5$
11. $|6t - 7| - 8 \geq 3$
12. $|3j - 1| + 6 > 0$
13. $3|14 - m| > 18$
14. $-4|6b - 8| \leq 12$
15. $2|3w + 8| - 13 \leq -5$
16. $-3|2 - 4u| + 5 < -13$
17. $6|-f + 3| + 7 > 7$
18. $\frac{2}{3}|4v + 6| - 2 \leq 10$

19. **MODELING WITH MATHEMATICS** The rules for an essay contest say that entries can have 500 words with an absolute deviation of at most 30 words. Write and solve an absolute value inequality that represents the acceptable numbers of words. *(See Example 3.)*

20. **MODELING WITH MATHEMATICS** The normal body temperature of a camel is 37°C. This temperature varies by up to 3°C throughout the day. Write and solve an absolute value inequality that represents the range of normal body temperatures (in degrees Celsius) of a camel throughout the day.

ERROR ANALYSIS In Exercises 21 and 22, describe and correct the error in solving the absolute value inequality.

21. ✗
$$|x - 5| < 20$$
$$x - 5 < 20$$
$$x < 25$$

22. ✗
$$|x + 4| > 13$$
$$x + 4 > -13 \quad \text{and} \quad x + 4 < 13$$
$$x > -17 \quad \text{and} \quad x < 9$$
$$-17 < x < 9$$

In Exercises 23–26, write the sentence as an absolute value inequality. Then solve the inequality.

23. A number is less than 6 units from 0.
24. A number is more than 9 units from 3.
25. Half of a number is at most 5 units from 14.
26. Twice a number is no less than 10 units from −1.

27. **PROBLEM SOLVING** An auto parts manufacturer throws out gaskets with weights that are not within 0.06 pound of the mean weight of the batch. The weights (in pounds) of the gaskets in a batch are 0.58, 0.63, 0.65, 0.53, and 0.61. Which gasket(s) should be thrown out?

28. **PROBLEM SOLVING** Six students measure the acceleration (in meters per second per second) of an object in free fall. The measured values are shown. The students want to state that the absolute deviation of each measured value x from the mean is at most d. Find the value of d.

10.56, 9.52, 9.73, 9.80, 9.78, 10.91

Assignment Guide and Homework Check

ASSIGNMENT

Basic: 1, 2, 3–15 odd, 19, 21, 27, 35, 38, 41–46

Average: 1, 2, 4–16 even, 20–28 even, 30, 35, 38, 41–46

Advanced: 1, 2, 10–34 even, 35–46

HOMEWORK CHECK

Basic: 5, 9, 11, 19

Average: 6, 10, 12, 20

Advanced: 10, 14, 16, 28

ANSWERS

1. yes; Because the absolute value must be positive, all values will be greater than −6.
2. Solving $|w - 9| \leq 2$ requires a compound inequality joined by "and." Solving $|w - 9| \geq 2$ requires a compound inequality joined by "or."
3. $-3 < x < 3$

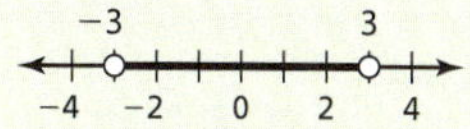

4. $y \leq -4.5$ *or* $y \geq 4.5$

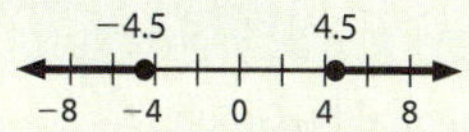

5. $d < -12$ *or* $d > -6$

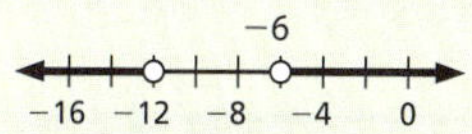

6. $-5 \leq h \leq 15$

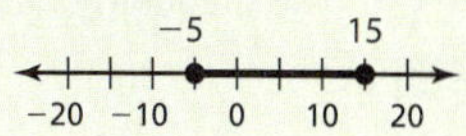

7. all real numbers

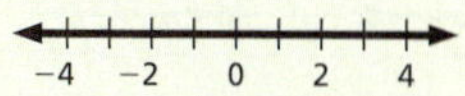

8. $c < -3$ *or* $c > \frac{1}{2}$

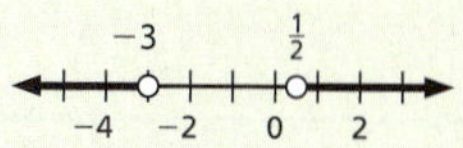

9. no solution
10. $1 < n < 3.5$

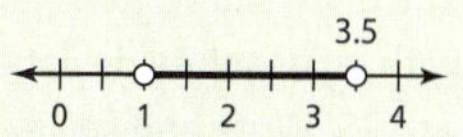

11. $t \leq -\frac{2}{3}$ *or* $t \geq 3$

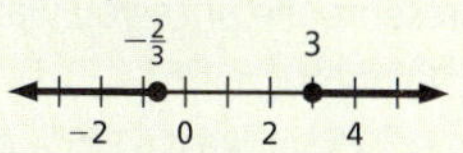

12. all real numbers

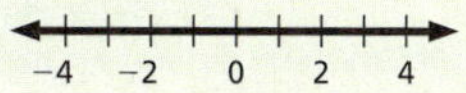

13. $m > 20$ *or* $m < 8$

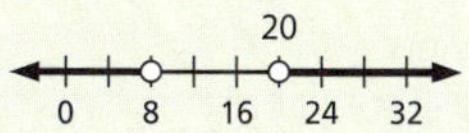

14. all real numbers

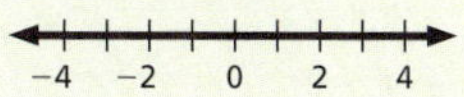

15. $-4 \leq w \leq -\frac{4}{3}$

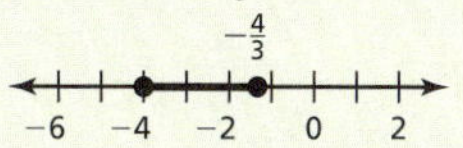

16. $u > 2$ *or* $u < -1$

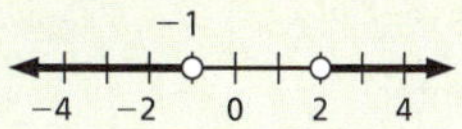

17. $f < 3$ *or* $f > 3$

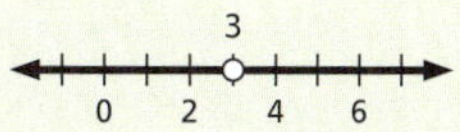

18. $-6 \leq v \leq 3$

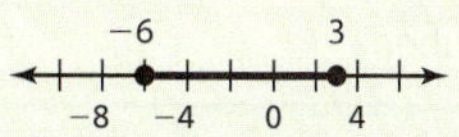

19–28. See Additional Answers.

Dynamic Teaching Tools

- Dynamic Assessment & Progress Monitoring Tool
- Interactive Whiteboard Lesson Library
- Dynamic Classroom with Dynamic Investigations

ANSWERS

29. $\left|\frac{1}{2}\cdot 4(x+6) - 2\cdot 6\right| < 2$; $-1 < x < 1$

30. $|2(x+1) + 2\cdot 3 - 4x| \le 3$; $\frac{5}{2} \le x \le \frac{11}{2}$

31. true

32. false; It could also be a solution of $x + 3 < -8$.

33. false; It has to be a solution of $x + 3 \le -8$ *or* $x + 3 \ge 8$.

34. true

35. no; If n is 0, the statement is false.

36. *Sample answer:*

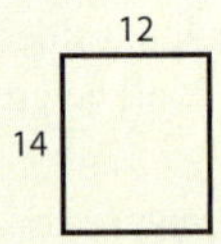

37–46. See Additional Answers.

Mini-Assessment

Solve the inequality. Graph the solution, if possible.

1. $|s - 2| < 3$ $-1 < s < 5$

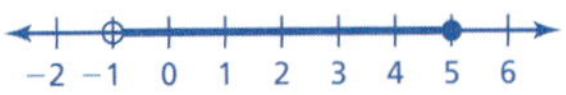

2. $|6x - 5| < 0$ The inequality has no solution.

3. $|n + 4| \ge 1$ $n \le -5$ *or* $n \ge -3$

4. $2|2y - 3| - 4 \ge 6$ $y \le -1$ *or* $y \ge 4$

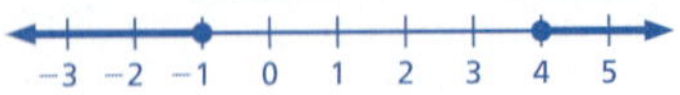

5. You are looking at ads for a new bicycle. The price you want to pay is \$200 with an absolute deviation of at most \$25. Write and solve an absolute value inequality that represents the price you want to pay. $|x - 200| \le 25$; $175 \le x \le 225$; at least \$175 and at most \$225

MATHEMATICAL CONNECTIONS **In Exercises 29 and 30, write an absolute value inequality that represents the situation. Then solve the inequality.**

29. The difference between the areas of the figures is less than 2.

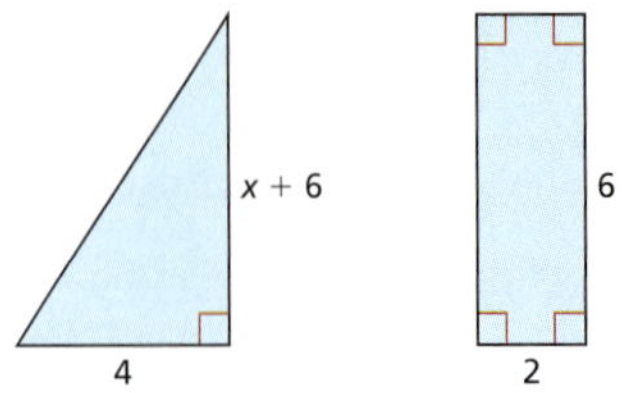

30. The difference between the perimeters of the figures is less than or equal to 3.

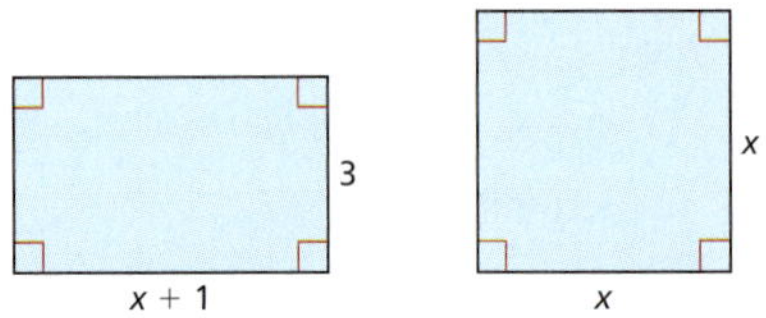

REASONING **In Exercises 31–34, tell whether the statement is true or false. If it is false, explain why.**

31. If a is a solution of $|x + 3| \le 8$, then a is also a solution of $x + 3 \ge -8$.

32. If a is a solution of $|x + 3| > 8$, then a is also a solution of $x + 3 > 8$.

33. If a is a solution of $|x + 3| \ge 8$, then a is also a solution of $x + 3 \ge -8$.

34. If a is a solution of $x + 3 \le -8$, then a is also a solution of $|x + 3| \ge 8$.

35. **MAKING AN ARGUMENT** One of your classmates claims that the solution of $|n| > 0$ is all real numbers. Is your classmate correct? Explain your reasoning.

36. **THOUGHT PROVOKING** Draw and label a geometric figure so that the perimeter P of the figure is a solution of the inequality $|P - 60| \le 12$.

37. **REASONING** What is the solution of the inequality $|ax + b| < c$, where $c < 0$? What is the solution of the inequality $|ax + b| > c$, where $c < 0$? Explain.

38. **HOW DO YOU SEE IT?** Write an absolute value inequality for each graph.

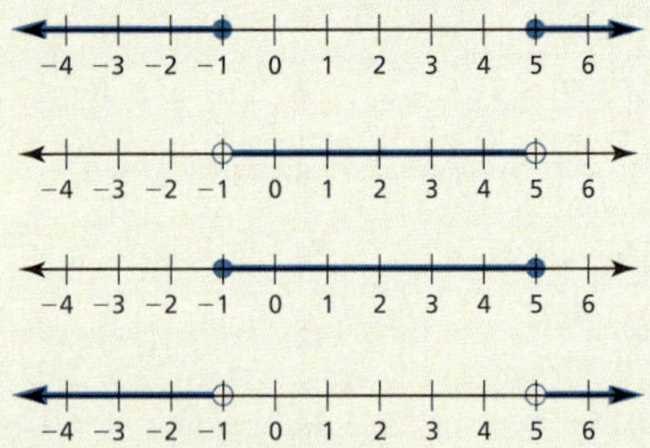

How did you decide which inequality symbol to use for each inequality?

39. **WRITING** Explain why the solution set of the inequality $|x| < 5$ is the *intersection* of two sets, while the solution set of the inequality $|x| > 5$ is the *union* of two sets.

40. **PROBLEM SOLVING** Solve the compound inequality below. Describe your steps.

$$|x - 3| < 4 \textit{ and } |x + 2| > 8$$

Maintaining Mathematical Proficiency

Reviewing what you learned in previous grades and lessons

Plot the ordered pair in a coordinate plane. Describe the location of the point. *(Skills Review Handbook)*

41. $A(1, 3)$ **42.** $B(0, -3)$ **43.** $C(-4, -2)$ **44.** $D(-1, 2)$

Copy and complete the table. *(Skills Review Handbook)*

45.

x	0	1	2	3	4
$5x + 1$					

46.

x	-2	-1	0	1	2
$-2x - 3$					

If students need help...	*If students got it...*
Resources by Chapter • Practice A and Practice B • Puzzle Time	Resources by Chapter • Enrichment and Extension • Cumulative Review
Student Journal • Practice	Start the *next* Section
Differentiating the Lesson Skills Review Handbook	

2.5–2.6 What Did You Learn?

Core Vocabulary

compound inequality, *p. 82*
absolute value inequality, *p. 88*
absolute deviation, *p. 90*

Core Concepts

Section 2.5

Writing and Graphing Compound Inequalities, *p. 82*
Solving Compound Inequalities, *p. 83*

Section 2.6

Solving Absolute Value Inequalities, *p. 88*

Mathematical Practices

1. How can you use a diagram to help you solve Exercise 12 on page 85?
2. In Exercises 13 and 14 on page 85, how can you use structure to break down the compound inequality into two inequalities?
3. Describe the given information and the overall goal of Exercise 27 on page 91.
4. For false statements in Exercises 31–34 on page 92, use examples to show the statements are false.

Performance Task

Grading Calculations

You are not doing as well as you had hoped in one of your classes. So, you want to figure out the minimum grade you need on the final exam to receive the semester grade that you want. Is it still possible to get an A? How would you explain your calculations to a classmate?

To explore the answers to this question and more, go to ***BigIdeasMath.com***.

Dynamic Teaching Tools

- Dynamic Assessment & Progress Monitoring Tool
- Interactive Whiteboard Lesson Library
- Dynamic Classroom with Dynamic Investigations

ANSWERS

1. A diagram can be used to show the different life zones, and the elevation ranges for each.
2. Take the left expression, first inequality symbol, and middle expression to form the first inequality. Take the middle expression, second inequality symbol, and last expression to form the second inequality.
3. The given information represents the weight in pounds of the gaskets in a batch, and the goal is to determine which gaskets should be thrown out because they are not within the tolerance range.
4. **32.** *Sample answer:* $a = -12$; $|-12 + 3| > 8$ is true, but $-12 + 3 > 8$ is false.

   33. *Sample answer:* $a = -14$; $|-14 + 3| \geq 8$ is true, but $-14 + 3 \geq -8$ is false.

ANSWERS

1. $d - 2 < -1$
2. $5h \leq 10$
3. 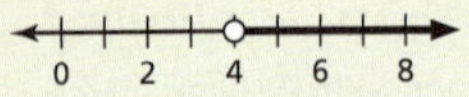
4. 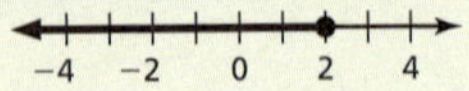
5. 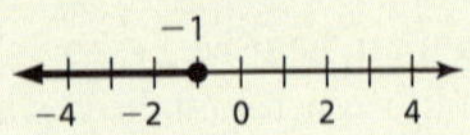
6. $p < 6$ 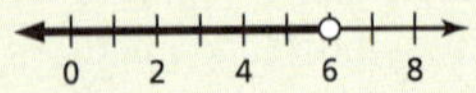
7. $r < -2$ 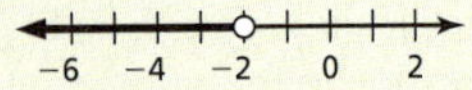
8. $m \leq 8.8$ 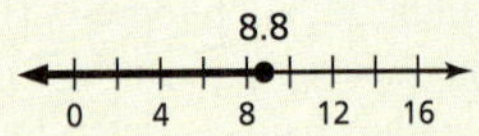

2.1 Writing and Graphing Inequalities *(pp. 53–60)*

a. A number x plus 36 is no more than 40. Write this sentence as an inequality.

A number x plus 36 | is no more than | 40.

$x + 36$ | $\leq$ | 40

▶ An inequality is $x + 36 \leq 40$.

b. Graph $w > -3$.

Test a number to the left of -3. $w = -4$ is *not* a solution.

Test a number to the right of -3. $w = 0$ is a solution.

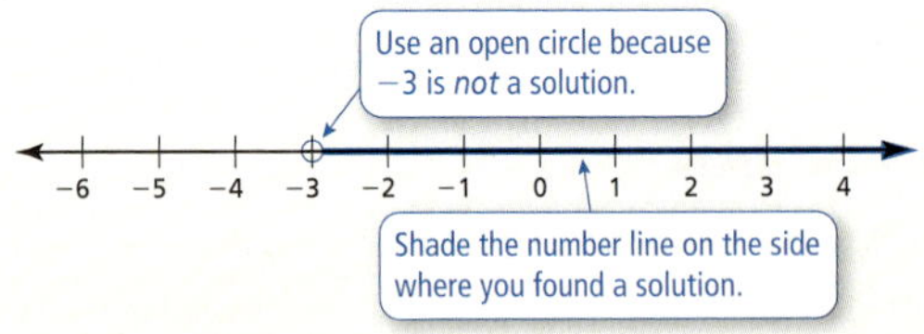

Write the sentence as an inequality.

1. A number d minus 2 is less than -1.
2. Ten is at least the product of a number h and 5.

Graph the inequality.

3. $x > 4$
4. $y \leq 2$
5. $-1 \geq z$

2.2 Solving Inequalities Using Addition or Subtraction *(pp. 61–66)*

Solve $x + 2.5 \leq -6$. Graph the solution.

$x + 2.5 \leq -6$ Write the inequality.

Subtraction Property of Inequality → $-\ 2.5 \quad -\ 2.5$ Subtract 2.5 from each side.

$x \leq -8.5$ Simplify.

▶ The solution is $x \leq -8.5$.

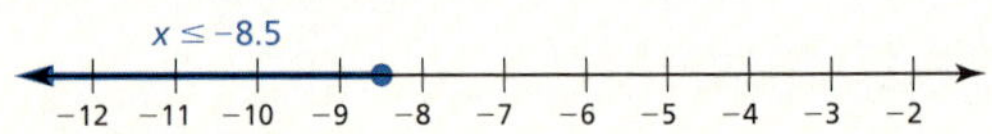

Solve the inequality. Graph the solution.

6. $p + 4 < 10$
7. $r - 4 < -6$
8. $2.1 \geq m - 6.7$

2.3 Solving Inequalities Using Multiplication or Division (pp. 67–72)

Solve $\frac{n}{-10} > 5$. Graph the solution.

$\frac{n}{-10} > 5$ — Write the inequality.

Multiplication Property of Inequality → $-10 \cdot \frac{n}{-10} < -10 \cdot 5$ — Multiply each side by −10. Reverse the inequality symbol.

$n < -50$ — Simplify.

▶ The solution is $n < -50$.

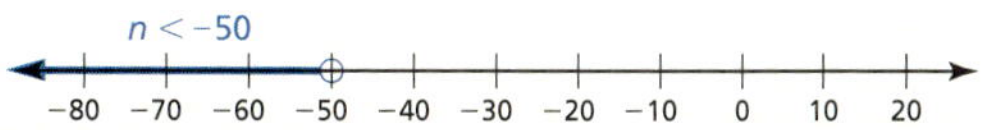

Solve the inequality. Graph the solution.

9. $3x > -21$

10. $-4 \le \frac{g}{5}$

11. $-\frac{3}{4}n \le 3$

12. $\frac{s}{-8} \ge 11$

13. $36 < 2q$

14. $-1.2k > 6$

2.4 Solving Multi-Step Inequalities (pp. 73–78)

Solve $22 + 3y \ge 4$. Graph the solution.

$22 + 3y \ge 4$ — Write the inequality.

$\underline{-22} \quad \underline{-22}$ — Subtract 22 from each side.

$3y \ge -18$ — Simplify.

$\frac{3y}{3} \ge \frac{-18}{3}$ — Divide each side by 3.

$y \ge -6$ — Simplify.

▶ The solution is $y \ge -6$.

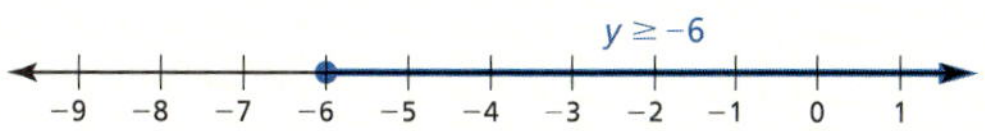

Solve the inequality. Graph the solution, if possible.

15. $3x - 4 > 11$

16. $-4 < \frac{b}{2} + 9$

17. $7 - 3n \le n + 3$

18. $2(-4s + 2) \ge -5s - 10$

19. $6(2t + 9) \le 12t - 1$

20. $3r - 8 > 3(r - 6)$

ANSWERS

9. $x > -7$

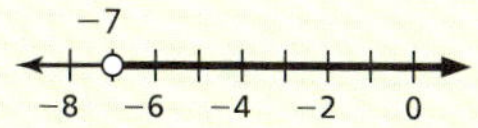

10. $g \ge -20$

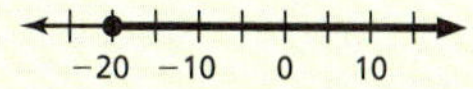

11. $n \ge -4$

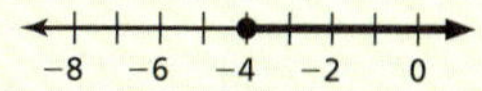

12. $s \le -88$

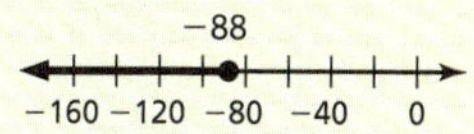

13. $q > 18$

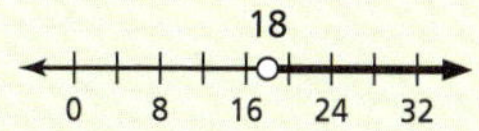

14. $k < -5$

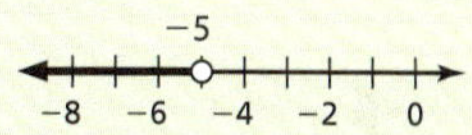

15. $x > 5$

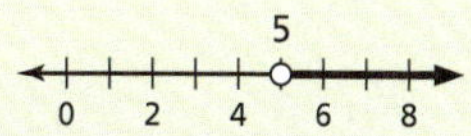

16. $b > -26$

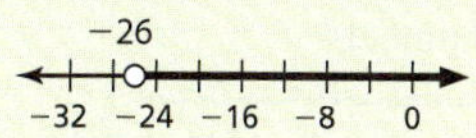

17. $n \ge 1$

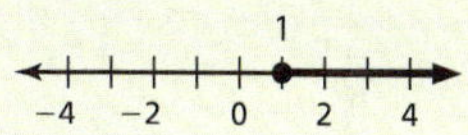

18. $s \le \frac{14}{3}$

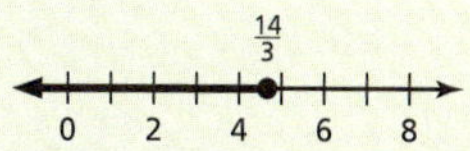

19. no solution

20. all real numbers

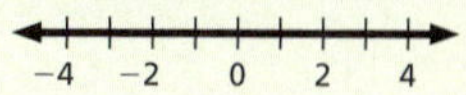

ANSWERS

21. $-6 < x \leq 8$

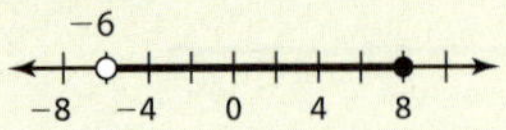

22. $-2 \leq z \leq 6$

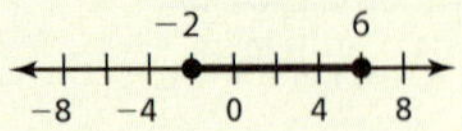

23. $r < -20$ *or* $r \geq -5$

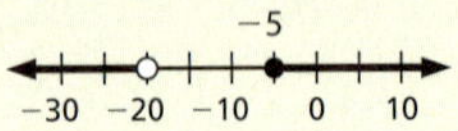

24. $m \geq 10$ *or* $m \leq -10$

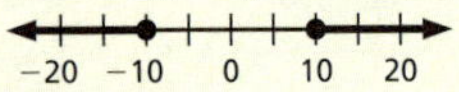

25. no solution

26. $3 \leq f \leq 9$

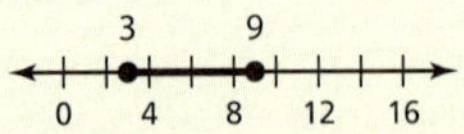

27. $b < -12$ *or* $b > -4$

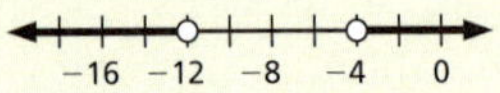

28. $-\frac{7}{3} < g < 1$

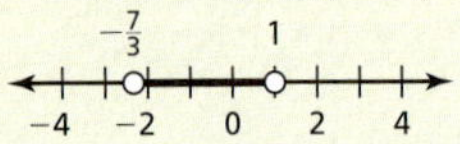

29. all real numbers

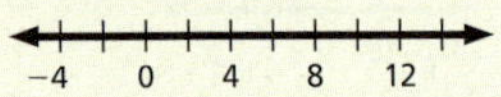

30. $|h - 106| \leq 7$; 99 cm to 113 cm

2.5 Solving Compound Inequalities (pp. 81–86)

Solve $-1 \leq -2d + 7 \leq 9$. Graph the solution.

$-1 \leq -2d + 7 \leq 9$	Write the inequality.
$\underline{-7} \quad \underline{-7} \quad \underline{-7}$	Subtract 7 from each expression.
$-8 \leq -2d \leq 2$	Simplify.
$\frac{-8}{-2} \geq \frac{-2d}{-2} \geq \frac{2}{-2}$	Divide each expression by -2. Reverse each inequality symbol.
$4 \geq d \geq -1$	Simplify.

▶ The solution is $-1 \leq d \leq 4$.

21. A number x is more than -6 and at most 8. Write this sentence as an inequality. Graph the inequality.

Solve the inequality. Graph the solution.

22. $19 \geq 3z + 1 \geq -5$

23. $\frac{r}{4} < -5$ *or* $-2r - 7 \leq 3$

2.6 Solving Absolute Value Inequalities (pp. 87–92)

Solve $|2x + 11| + 3 > 8$. Graph the solution.

$|2x + 11| + 3 > 8$	Write the inequality.
$\underline{-3} \quad \underline{-3}$	Subtract 3 from each side.
$|2x + 11| > 5$	Simplify.
$2x + 11 < -5$ *or* $2x + 11 > 5$	Write a compound inequality.
$\underline{-11} \quad \underline{-11} \qquad \underline{-11} \quad \underline{-11}$	Subtract 11 from each side.
$2x < -16 \qquad 2x > -6$	Simplify.
$\frac{2x}{2} < \frac{-16}{2} \qquad \frac{2x}{2} > \frac{-6}{2}$	Divide each side by 2.
$x < -8$ *or* $x > -3$	Simplify.

▶ The solution is $x < -8$ *or* $x > -3$.

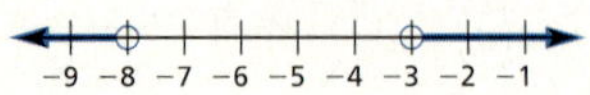

Solve the inequality. Graph the solution, if possible.

24. $|m| \geq 10$

25. $|k - 9| < -4$

26. $4|f - 6| \leq 12$

27. $5|b + 8| - 7 > 13$

28. $|-3g - 2| + 1 < 6$

29. $|9 - 2j| + 10 \geq 2$

30. A safety regulation states that the height of a guardrail should be 106 centimeters with an absolute deviation of no more than 7 centimeters. Write and solve an absolute value inequality that represents the acceptable heights of a guardrail.

2 Chapter Test

Write the sentence as an inequality.

1. The sum of a number y and 9 is at least -1.
2. A number r is more than 0 or less than or equal to -8.
3. A number k is less than 3 units from 10.

Solve the inequality. Graph the solution, if possible.

4. $\frac{x}{2} - 5 \geq -9$
5. $-4s < 6s + 1$
6. $4p + 3 \geq 2(2p + 1)$
7. $-7 < 2c - 1 < 10$
8. $-2 \leq 4 - 3a \leq 13$
9. $-5 < 2 - h$ *or* $6h + 5 > 71$
10. $|2q + 8| > 4$
11. $-2|y - 3| - 5 \geq -4$
12. $4|-3b + 5| - 9 < 7$

13. You start a small baking business, and you want to earn a profit of at least \$250 in the first month. The expenses in the first month are \$155. What are the possible revenues that you need to earn to meet the profit goal?

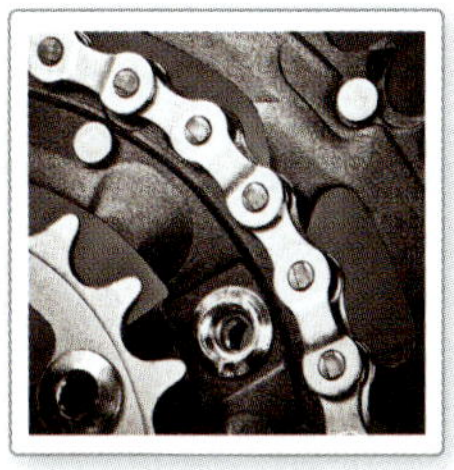

14. A manufacturer of bicycle parts requires that a bicycle chain have a width of 0.3 inch with an absolute deviation of at most 0.0003 inch. Write and solve an absolute value inequality that represents the acceptable widths.

15. Let a, b, c, and d be constants. Describe the possible solution sets of the inequality $ax + b < cx + d$.

Write and graph a compound inequality that represents the numbers that are *not* solutions of the inequality represented by the graph shown. Explain your reasoning.

16. 

17. 

18. A state imposes a sales tax on items of clothing that cost more than \$175. The tax applies only to the difference of the price of the item and \$175.
    a. Use the receipt shown to find the tax rate (as a percent).
    b. A shopper has \$430 to spend on a winter coat. Write and solve an inequality to find the prices p of coats that the shopper can afford. Assume that $p \geq 175$.
    c. Another state imposes a 5% sales tax on the entire price of an item of clothing. For which prices would paying the 5% tax be cheaper than paying the tax described above? Write and solve an inequality to find your answer and list three prices that are solutions.

If students need help...	*If students got it...*
Lesson Tutorials	Resources by Chapter • Enrichment and Extension • Cumulative Review
Skills Review Handbook	Performance Task
BigIdeasMath.com	Start the *next* Section

ANSWERS

1. $y + 9 \geq -1$
2. $r > 0$ *or* $r \leq -8$
3. $|k - 10| < 3$
4. $x \geq -8$
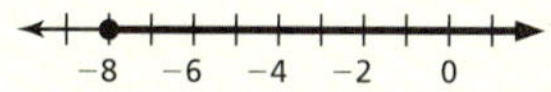
5. $s > -\frac{1}{10}$
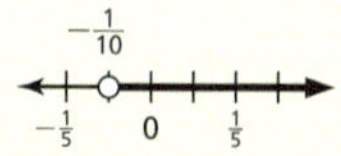
6. all real numbers
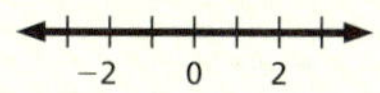
7. See Additional Answers.
8. $-3 \leq a \leq 2$
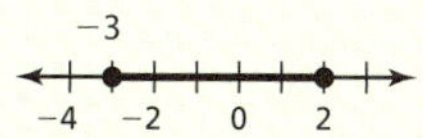
9. $h < 7$ *or* $h > 11$
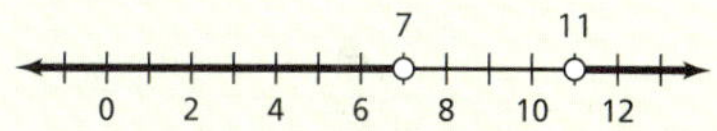
10. $q < -6$ *or* $q > -2$
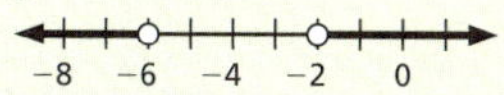
11. no solution
12. $\frac{1}{3} < b < 3$
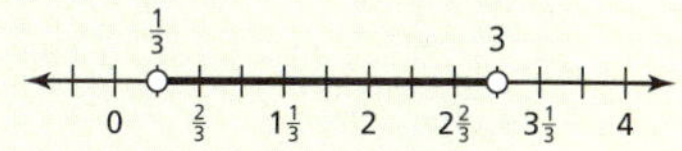
13. at least \$405
14. $|w - 0.3| \leq 0.0003$; between 0.2997 in. and 0.3003 in.
15. all real numbers; no solution; $x < \frac{d-b}{a-c}$; $x > \frac{d-b}{a-c}$
16. $-3 < x \leq 2$
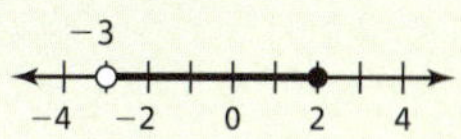
The values between -3 and 2, including 2, are not solutions.
17. $x < -4$ *or* $x > -1$
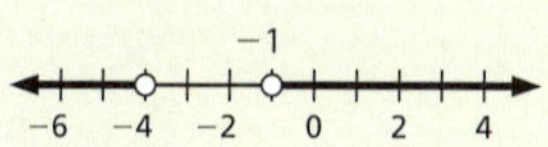
The values greater than -1 or less than -4 are not solutions.
18. a. 6.25%
    b. $175 + 1.0625(p - 175) \leq 430$; $p \leq 415$
    c. $1.05p < 175 + 1.0625(p - 175)$; $p > 875$; *Sample answers:* \$900, \$950, \$1000

ANSWERS

1. A
2. a. -2
   b. 3; *Sample answer:* 10
   c. 3; *Sample answer:* 0
3. At least one integer solution: $5x - 6 + x \geq 2x - 8$, $9x - 3 < 12$ *or* $6x + 2 > -10$, $5(x - 1) \leq 5x - 3$;
   No integer solutions: $2(3x + 8) > 3(2x + 6)$, $17 < 4x + 5 < 21$, $x - 8 + 4x \leq 3(x - 3) + 2x$
4. a. $180 < 25x$
   b. 0, 1, 2, 3, 4, 5, 6, 7

2 Cumulative Assessment

1. The expected attendance at a school event is 65 people. The actual attendance can vary by up to 30 people. Which equation can you use to find the minimum and maximum attendances?

Ⓐ $|x - 65| = 30$ Ⓑ $|x + 65| = 30$

Ⓒ $|x - 30| = 65$ Ⓓ $|x + 30| = 65$

2. Fill in values for a and b so that each statement is true for the inequality $ax + 4 \leq 3x + b$.

a. When $a = 5$ and $b =$ _____, $x \leq -3$.

b. When $a =$ _____ and $b =$ _____, the solution of the inequality is all real numbers.

c. When $a =$ _____ and $b =$ _____, the inequality has no solution.

3. Place each inequality into one of the two categories.

At least one integer solution	No integer solutions

$5x - 6 + x \geq 2x - 8$ $x - 8 + 4x \leq 3(x - 3) + 2x$

$2(3x + 8) > 3(2x + 6)$ $9x - 3 < 12$ *or* $6x + 2 > -10$

$17 < 4x + 5 < 21$ $5(x - 1) \leq 5x - 3$

4. Admission to a play costs \$25. A season pass costs \$180.

a. Write an inequality that represents the numbers x of plays you must attend for the season pass to be a better deal.

b. Select the numbers of plays for which the season pass is *not* a better deal.

0	1	2	3	4
5	6	7	8	9
10	11	12	13	14

5. Select the values of a that make the solution of the equation $3(2x - 4) = 4(ax - 2)$ positive.

−2 | −1 | 0 | 1 | 2 | 3 | 4 | 5

6. Fill in the compound inequality with <, ≤, =, ≥, or > so the solution is shown in the graph.

$$4x - 18 \;\square\; -x - 3 \text{ and } -3x - 9 \;\square\; -3$$

7. You have a $250 gift card to use at a sporting goods store.

a. Write an inequality that represents the possible numbers x of pairs of socks you can buy when you buy 2 pairs of sneakers. Can you buy 8 pairs of socks? Explain.

b. Describe what the inequality $60 + 80x \leq 250$ represents in this context.

8. Consider the equation shown, where a, b, c, and d are integers.

$$ax + b = cx + d$$

Student A claims the equation will always have one solution. Student B claims the equation will always have no solution. Use the numbers shown to answer parts (a)–(c).

−1 | 0 | 1 | 2 | 3 | 4 | 5 | 6

a. Select values for a, b, c, and d to create an equation that supports Student A's claim.

b. Select values for a, b, c, and d to create an equation that supports Student B's claim.

c. Select values for a, b, c, and d to create an equation that shows both Student A and Student B are incorrect.

ANSWERS

5. −2, −1, 0, 1

6. <; ≤

7. a. $160 + 12x \leq 250$; no; 8 is not a solution of the inequality.

   b. the possible number of pairs of sneakers you can buy when you buy 5 pairs of socks

8. a. *Sample answer:* 1; 2; 3; 4

   b. *Sample answer:* 2; 3; 2; 4

   c. *Sample answer:* 6; 5; 6; 5

Chapter 3 Pacing Guide

Chapter 3 Pacing Guide	
Chapter Opener/ Mathematical Practices	1 Day
Section 1	1 Day
Section 2	2 Days
Section 3	1 Day
Quiz	1 Day
Section 4	1 Day
Section 5	1 Day
Section 6	2 Days
Section 7	2 Days
Chapter Review/ Chapter Tests	2 Days
Total Chapter 3	14 Days
Year-to-Date	38 Days

3 Graphing Linear Functions

Submersible *(p. 140)*

Basketball *(p. 134)*

Speed of Light *(p. 125)*

Coins *(p. 116)*

Taxi Ride *(p. 109)*

Laurie's Notes

Chapter Summary

- Students should have a conceptual understanding of functions from a prior grade. Their understanding may be of a "function machine" where there is an input, a function is performed, and an output results. There is a pairing of the input and output, and each input is associated with exactly one output. Chapter 3 extends this introductory understanding of functions and presents the notation of functions. Consistent use of the notation and language of functions will help students become more confident.
- The early part of the chapter focuses on function notation, representing functions, discrete and continuous functions, and evaluating functions. Students may be resistant to using function notation, preferring the simpler "$y =$" notation. It is hard for students to appreciate what the broader notation enables us to do because they have not learned enough at this stage. When two equations are graphed on the same axes, we can clearly refer to f and g, versus saying "the first $y =$" and "the second $y =$." We compose functions and have functions with multiple inputs, two examples where function notation is useful.
- The middle portion of the chapter introduces two forms of linear equations—standard and slope-intercept. Students should be able to do quick sketches of each by inspecting information from the equation.
- The last two lessons of the chapter look at transformations of linear and absolute value functions. These same transformations will be applied to other types of functions, such as quadratic and trigonometric. Function notation is used to describe the transformations.

What Your Students Have Learned

Middle School

- Represent relations as ordered pairs and determine whether the relation is a function.
- Represent functions as equations, input-output tables, and graphs.
- Find the slope of a line.
- Graph linear equations written in standard form and slope-intercept form.
- Translate, reflect, and rotate figures in the coordinate plane.
- Understand the absolute value of a rational number as its distance from 0 on the number line.

What Your Students Will Learn

Algebra 1

- Find the domains and ranges of functions, and identify independent and dependent variables.
- Identify linear functions using graphs, tables, and equations in two variables.
- Use function notation to evaluate, interpret, solve, and graph functions.
- Graph linear equations written in standard form and slope-intercept form.
- Find slopes of nonvertical lines from graphs or tables.
- Translate, reflect, shrink, and stretch graphs of linear functions, and combine transformations of graphs of linear functions.
- Combine transformations of graphs of absolute value functions in the coordinate plane.

Dynamic Teaching Tools

- Dynamic Assessment & Progress Monitoring Tool
- Lesson Planning Tool
- Interactive Whiteboard Lesson Library
- Dynamic Classroom with Dynamic Investigations
- Real-Life STEM Videos

Scaffolding in the Classroom

Begin with a worked-out problem and work backwards.

Ask students to explain what happened in step 5 to get to step 6. Then ask what happened in step 4 to get to step 5, etc.

Questioning in the Classroom

How will this work?

Look for examples, exercises, problems, and questions that do not have a specific solution or answer. Ask students to explain how to solve the problem.

Laurie's Notes

Maintaining Mathematical Proficiency

Plotting Points

- Remind students of the signs of the coordinates in the four quadrants. In Quadrant I, both coordinates are positive. In Quadrant II, the first coordinate is negative and the second is positive. In Quadrant III, both coordinates are negative. In Quadrant IV, the first coordinate is positive and the second is negative.

COMMON ERROR Students may think that the first coordinate tells them how far to move up or down instead of right or left.

Evaluating Expressions

- Remind students that the first step in evaluating the expression is to replace the variable with the given value.
- Remind students that $4x$ means "4 times x," or the product of 4 and the value of x.

COMMON ERROR Students may not multiply the value of x and the coefficient. For example, in Exercise 7, they may replace $3x$ with 37 instead of 21.

Mathematical Practices (continued on page 102)

Using a Graphing Calculator

- The *Mathematical Practices* page focuses attention on how mathematics is learned—process versus content. This page demonstrates that the graphing calculator can be used as a tool to deepen student understanding of the slope of a line.
- Use the *Mathematical Practices* page to help students develop mathematical habits of mind—how mathematics can be explored and how mathematics is thought about.
- Discuss the *Core Concept* and have students practice setting the *standard viewing window* on their calculator. Change to a *square viewing window* and note the difference in the values for the maximum and minimum y-values.
- **Note:** The *square viewing window* can be set manually, as shown, or you can use the default feature of the calculator.

? "Why might you want a square viewing window when graphing an equation, say, $y = x$?" Listen for students to explain that a slope of 1 is a 45° angle.

- Make sure students understand and notice the spacing of the tick marks.
- Students will be graphing linear and absolute value functions as well as transformations of both these functions in this chapter. To see a nonrigid transformation (stretch or shrink), it is helpful to have a true perspective, meaning the square viewing window.

If students need help...	*If students got it...*
Student Journal • Maintaining Mathematical Proficiency	Game Closet at *BigIdeasMath.com*
Lesson Tutorials	Start the *next* Section
Skills Review Handbook	

Maintaining Mathematical Proficiency

Plotting Points

Example 1 **Plot the point $A(-3, 4)$ in a coordinate plane. Describe the location of the point.**

Start at the origin. Move 3 units left and 4 units up. Then plot the point. The point is in Quadrant II.

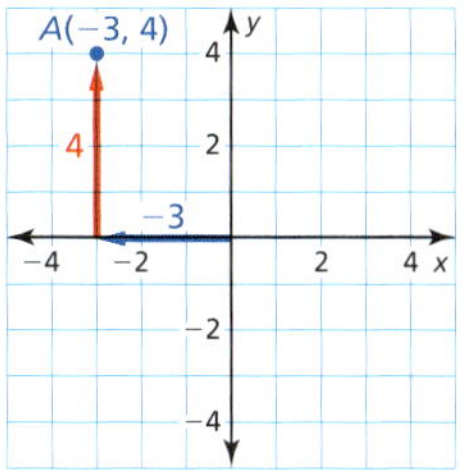

Plot the point in a coordinate plane. Describe the location of the point.

1. $A(3, 2)$ **2.** $B(-5, 1)$ **3.** $C(0, 3)$

4. $D(-1, -4)$ **5.** $E(-3, 0)$ **6.** $F(2, -1)$

Evaluating Expressions

Example 2 **Evaluate $4x - 5$ when $x = 3$.**

$4x - 5 = 4(3) - 5$ Substitute 3 for x.

$= 12 - 5$ Multiply.

$= 7$ Subtract.

Example 3 **Evaluate $-2x + 9$ when $x = -8$.**

$-2x + 9 = -2(-8) + 9$ Substitute −8 for x.

$= 16 + 9$ Multiply.

$= 25$ Add.

Evaluate the expression for the given value of *x*.

7. $3x - 4; x = 7$ **8.** $-5x + 8; x = 3$ **9.** $10x + 18; x = 5$

10. $-9x - 2; x = -4$ **11.** $24 - 8x; x = -2$ **12.** $15x + 9; x = -1$

13. **ABSTRACT REASONING** Let a and b be positive real numbers. Describe how to plot (a, b), $(-a, b)$, $(a, -b)$, and $(-a, -b)$.

Dynamic Solutions available at *BigIdeasMath.com*

What Your Students Have Learned

- Plot points in the coordinate plane.
- Evaluate algebraic expressions for a given value using Order of Operations.

ANSWERS

1–6.

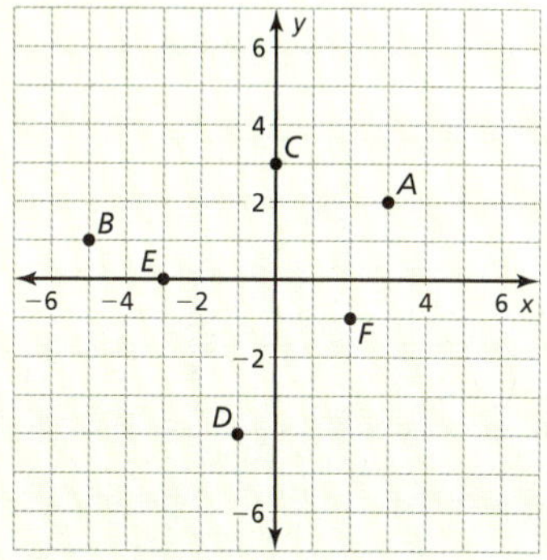

1. in Quadrant I
2. in Quadrant II
3. on positive y-axis
4. in Quadrant III
5. on negative x-axis
6. in Quadrant IV
7. 17
8. −7
9. 68
10. 34
11. 40
12. −6
13. Start at the origin. Move a units right and b units up. Plot the point; Start at the origin. Move a units left and b units up. Plot the point; Start at the origin. Move a units right and b units down. Plot the point; Start at the origin. Move a units left and b units down. Plot the point.

Vocabulary Review

Have students make a Four Square for each topic.

- Plotting points
- Evaluating expressions

MONITORING PROGRESS ANSWERS

1. yes; 10 : 15 is a 2 : 3 ratio.
2. no; 18 : 12 is not a 2 : 3 ratio.
3. yes; 24 : 36 is a 2 : 3 ratio.
4. 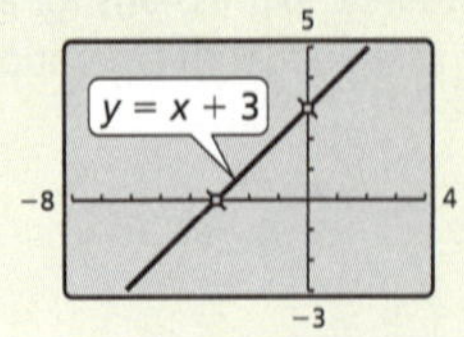

5. 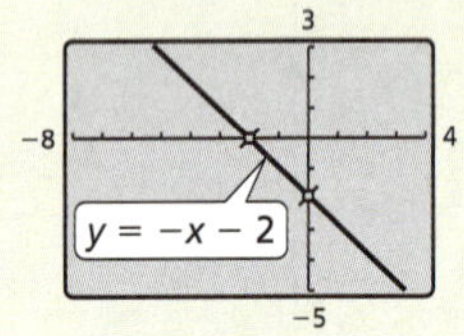

6. 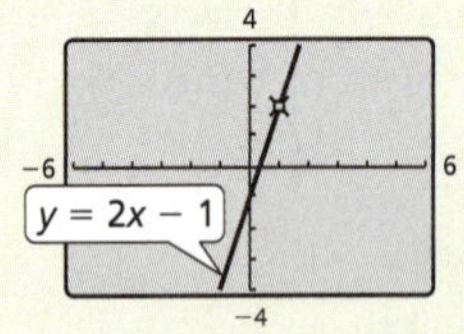

7. 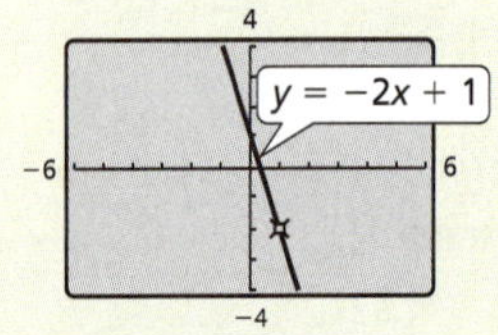

8. 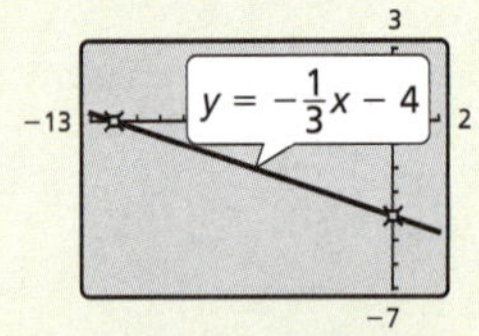

9. 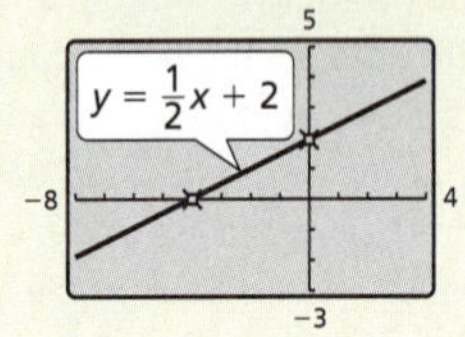

10. The slope of a line in a square viewing window appears to be steeper than in a standard viewing window.

Mathematical Practices

Mathematically proficient students use technological tools to explore concepts.

Using a Graphing Calculator

Core Concept

Standard and Square Viewing Windows

A typical graphing calculator screen has a height to width ratio of 2 to 3. This means that when you use the *standard viewing window* of −10 to 10 (on each axis), the graph will not be in its true perspective.

```
WINDOW
Xmin=-10
Xmax=10
Xscl=1
Ymin=-10
Ymax=10
Yscl=1
```

This is the standard viewing window.

To see a graph in its true perspective, you need to use a *square viewing window*, in which the tick marks on the x-axis are spaced the same as the tick marks on the y-axis.

```
WINDOW
Xmin=-9
Xmax=9
Xscl=1
Ymin=-6
Ymax=6
Yscl=1
```

This is a square viewing window.

EXAMPLE 1 **Using a Graphing Calculator**

Use a graphing calculator to graph $y = 2x + 5$.

SOLUTION

Enter the equation $y = 2x + 5$ into your calculator. Then graph the equation. The standard viewing window does not show the graph in its true perspective. Notice that the tick marks on the y-axis are closer together than the tick marks on the x-axis. To see the graph in its true perspective, use a square viewing window.

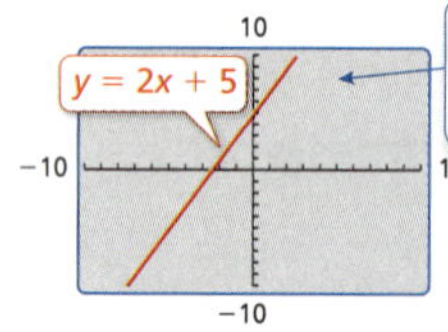

This is the graph in the standard viewing window.

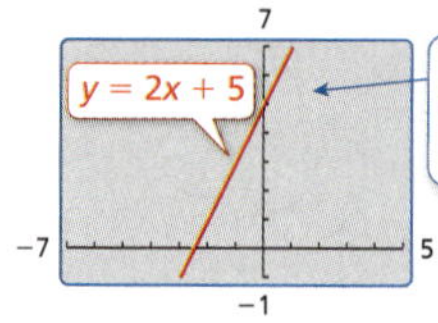

This is the graph in a square viewing window.

Monitoring Progress

Determine whether the viewing window is square. Explain.

1. $-8 \le x \le 7, -3 \le y \le 7$
2. $-6 \le x \le 6, -9 \le y \le 9$
3. $-18 \le x \le 18, -12 \le y \le 12$

Use a graphing calculator to graph the equation. Use a square viewing window.

4. $y = x + 3$
5. $y = -x - 2$
6. $y = 2x - 1$
7. $y = -2x + 1$
8. $y = -\frac{1}{3}x - 4$
9. $y = \frac{1}{2}x + 2$
10. How does the appearance of the slope of a line change between a standard viewing window and a square viewing window?

Laurie's Notes — Mathematical Practices (continued from page T-101)

- Figure 1 is a square viewing window for $f(x) = 2|x|$. Figure 2 is the same function with a window of [−4, 4] and [−8, 8]. Figure 3 is the same function with a window of [−8, 8] and [−4, 4].
- Throughout this course, you want students to develop the habit of reasoning about what effect the window has on how the function is viewed.
- Students could work with partners or in groups on *Monitoring Progress*. Allow private think time before dialogue begins.

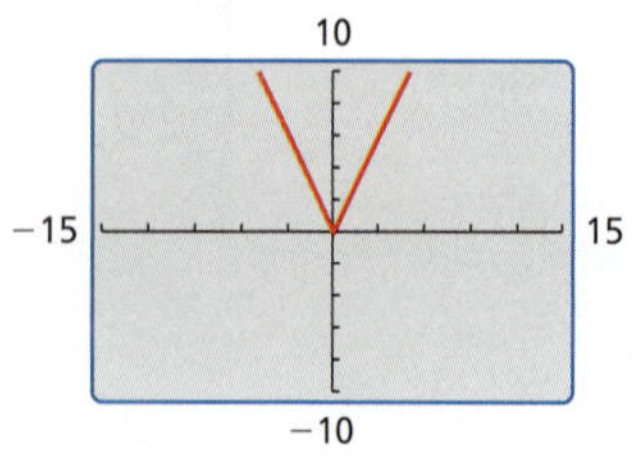

Figure 1

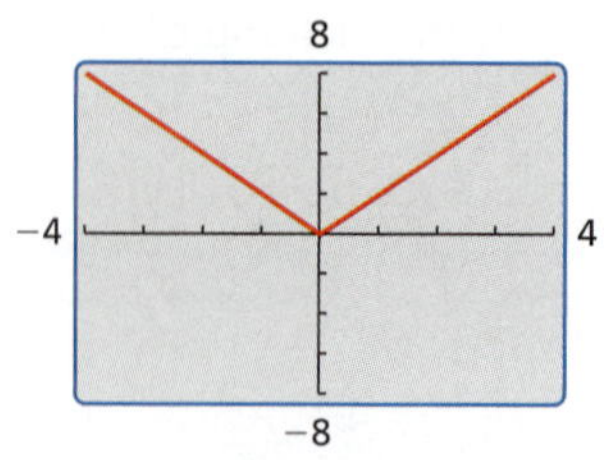

Figure 2

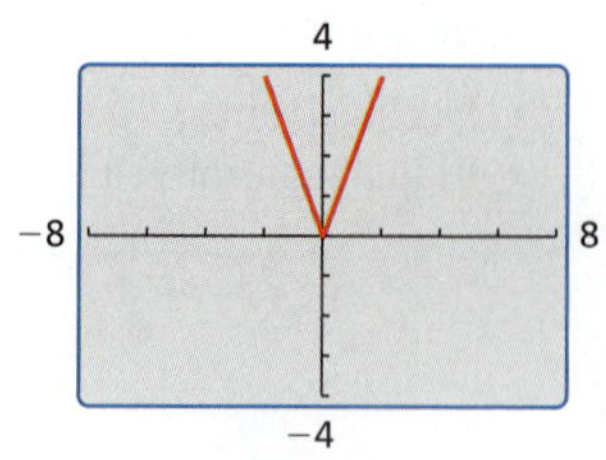

Figure 3

Laurie's Notes

Overview of Section 3.1

Introduction

- Students have prior knowledge of functions from middle school. Their understanding may be limited to discrete functions.

Words:
adding 2 to input to get output

Listing Ordered Pairs:
(1, 3), (2, 4), (3, 5)

Input-Output Table:

Input, *x*	1	2	3
Output, *y*	3	4	5

Mapping Diagram:

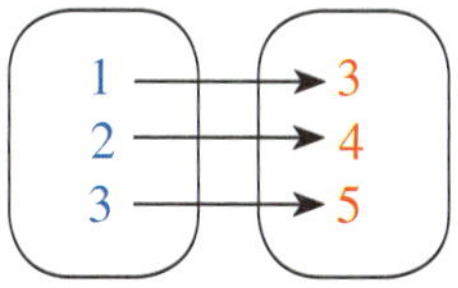

Graph:

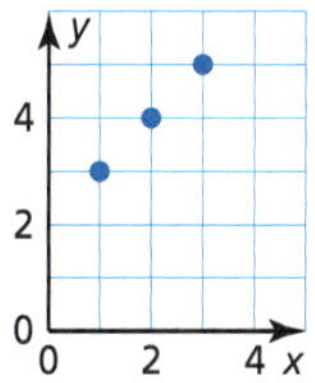

Equation:
$y = x + 2$

- In this lesson, students will be introduced to functions whose domain and/or range are infinite. There is a fair amount of vocabulary for students to learn and use in their discussions. Explain that just as smart phones and soccer have vocabulary associated with them, there is a vocabulary associated with functions.

Resources

- Student whiteboards can be used in a group of four to represent a function in different ways.

Teaching Strategy

- **Selective Responses:** When students are working alone or with partners on a problem, teachers have the opportunity to circulate about the room to see different approaches. Students' work varies, revealing their thinking and conceptions about the mathematics. Discourse about the mathematics and student comprehension can both be improved if we are thoughtful in how responses (students' work) are displayed.
- I carry a clipboard with me as I circulate so that I can make notes about the order in which I want to call on students. If I leave it to chance when I call on students to share their solutions, then I lose the ability to control the sequence of responses. I do not want the first response to be the most polished, efficient, or eloquent. Knowing the outcome(s) I want from the activity or problem, I look for particular work that clearly demonstrates the outcome. If there is a particular mathematical practice that is evident in the students' work, I also want to highlight that.

Pacing Suggestion

- The problems students will work in the explorations are similar to Example 1 in the formal lesson. If students recall prior work with functions, you might omit the explorations and start with Example 1. If student background is not adequate, spend sufficient time working through both explorations.

Dynamic Teaching Tools

- Dynamic Assessment & Progress Monitoring Tool
- Lesson Planning Tool
- Interactive Whiteboard Lesson Library
- Dynamic Classroom with Dynamic Investigations

What Your Students Will Learn

- Determine whether relations are functions.
- Find the domains and ranges of functions.
- Identify independent variables and dependent variables of functions.

Laurie's Notes

Exploration

Motivate

- Ask for a volunteer who will not mind you measuring his or her head for a hat. See the chart for sizes.

Hat size	$6\frac{1}{2}$	$6\frac{5}{8}$	$6\frac{3}{4}$	$6\frac{7}{8}$	7	$7\frac{1}{8}$	$7\frac{1}{4}$	$7\frac{3}{8}$	$7\frac{1}{2}$	$7\frac{5}{8}$	$7\frac{3}{4}$	$7\frac{7}{8}$	8
Inches	$20\frac{1}{2}$	$20\frac{7}{8}$	$21\frac{1}{4}$	$21\frac{5}{8}$	22	$22\frac{1}{2}$	$22\frac{7}{8}$	$23\frac{1}{4}$	$23\frac{5}{8}$	24	$24\frac{3}{8}$	$24\frac{3}{4}$	$25\frac{1}{4}$
Centimeters	52	53	54	55	56	57	58	59	60	61	62	63	64
	X-SMALL		SMALL		MEDIUM		LARGE		X-LARGE		XX-LARGE		

? "What is the input for determining your hat size?" size of your head in inches or centimeters

? "What are the outputs?" hat size as a number $\left(6\frac{1}{2}, 6\frac{5}{8}, \ldots\right)$ or a category (x-small, small, …)

- **Discuss:** A *relation* pairs inputs and outputs. The information in this example could be written as the ordered pair (measurement of head, hat size). Each head measurement is paired with exactly one hat size, and we call it a *function*.

Exploration 1

- Students should be familiar with different ways functions are represented.

? If students are stuck, ask, "Could this function be represented in an input-output table?" yes "Could this function be represented as a set of ordered pairs?" yes

- **Construct Viable Arguments and Critique the Reasoning of Others:** Circulate so you know what methods of representation students have used. Ask students to present based on the different methods they chose. Be selective in who you ask to present so that you have multiple methods covered. Representing the function as an equation is generally the most challenging, so hold this response for last.

Exploration 2

COMMON ERROR Students will often think that neither *a* nor *b* are functions because the number 8 is repeated. Have students go back to the definition of a function and determine when the conditions of the definition are met.

- **Construct Viable Arguments and Critique the Reasoning of Others:** When the relation is not a function, expect students to be explicit in their rationale of why the definition of a function is not met.
- **Extension:** If you are using the entire class period for the two explorations, you might ask students to look back at each of the relations in Exploration 2 and interchange the input and output. Is the new relation a function?

Communicate Your Answer

- Students should define a function versus how a function is represented.
- Challenge students to think of contextual examples of relations and functions.

Connecting to Next Step

- If you feel students are comfortable with determining when a relation is a function, assign items from Exercises 3–8 on page 108. In beginning the formal lesson, you might omit Example 1.

3.1 Functions

Essential Question What is a function?

A **relation** pairs inputs with outputs. When a relation is given as ordered pairs, the x-coordinates are inputs and the y-coordinates are outputs. A relation that pairs each input with *exactly one* output is a **function**.

EXPLORATION 1 Describing a Function

Work with a partner. Functions can be described in many ways.

- by an equation
- by an input-output table
- using words
- by a graph
- as a set of ordered pairs

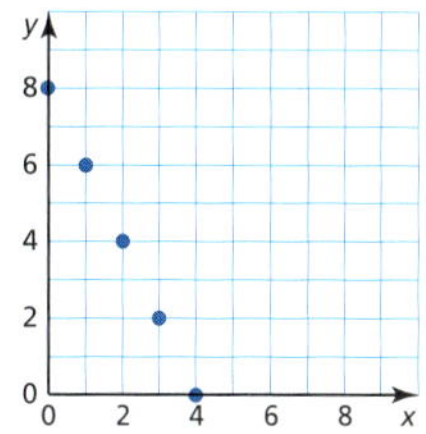

a. Explain why the graph shown represents a function.

b. Describe the function in two other ways.

ANALYZING RELATIONSHIPS

To be proficient in math, you need to analyze relationships mathematically to draw conclusions.

EXPLORATION 2 Identifying Functions

Work with a partner. Determine whether each relation represents a function. Explain your reasoning.

a.

Input, x	0	1	2	3	4
Output, y	8	8	8	8	8

b.

Input, x	8	8	8	8	8
Output, y	0	1	2	3	4

c. Input, x Output, y

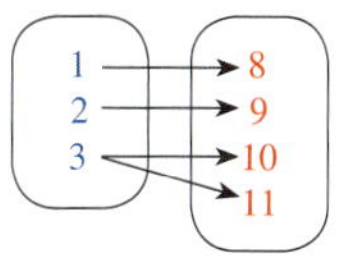

d.

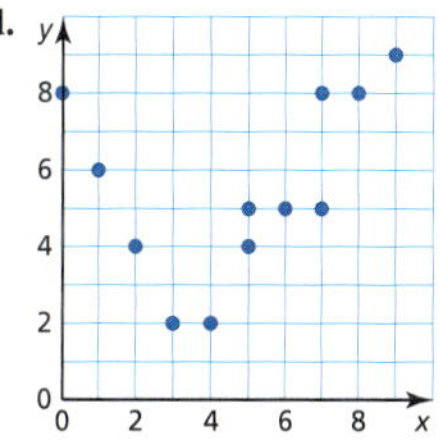

e. $(-2, 5), (-1, 8), (0, 6), (1, 6), (2, 7)$

f. $(-2, 0), (-1, 0), (-1, 1), (0, 1), (1, 2), (2, 2)$

g. Each radio frequency x in a listening area has exactly one radio station y.

h. The same television station x can be found on more than one channel y.

i. $x = 2$

j. $y = 2x + 3$

Communicate Your Answer

3. What is a function? Give examples of relations, other than those in Explorations 1 and 2, that (a) are functions and (b) are not functions.

Dynamic Teaching Tools

- Dynamic Assessment & Progress Monitoring Tool
- Lesson Planning Tool
- Interactive Whiteboard Lesson Library
- Dynamic Classroom with Dynamic Investigations

ANSWERS

1. **a.** Each x-coordinate is paired with exactly one y-coordinate.

   b. *Sample answer:* $(0, 8), (1, 6), (2, 4), (3, 2), (4, 0)$;

Input, x	0	1	2	3	4
Output, y	8	6	4	2	0

2. **a.** function; Each input value is paired with exactly one output value.

   b. not a function; The input value of 8 is paired with more than one output value.

   c. not a function; The input value of 3 is paired with more than one output value.

   d. not a function; The input values of 5 and 7 are each paired with more than one output value.

   e. function; Each input value is paired with exactly one output value.

   f. not a function; The input value of -1 is paired with more than one output value.

   g. function; Each input frequency is paired with exactly one output radio station.

   h. not a function; Each input television station is paired with more than one output channel.

   i. not a function; The input value of 2 is paired with every possible output value of y.

   j. function; Each input value of x is paired with exactly one output value of y.

3. a relation that pairs each input with exactly one output

   a. *Sample answer:* $(0, 10), (1, 20), (2, 30); y = -5x$

   b. *Sample answer:* $(0, 10), (0, 20), (0, 30); x = -5$

Differentiated Instruction

Organization

Have students choose a functional relationship and then make a chart to show the different ways to describe it.

Way to Describe	Example
Equation	
Input-output table	
Words	
Graph	
Set of ordered pairs	

Extra Example 1

Determine whether each relation is a function. Explain.

a. (−3, 1), (−2, 1), (−1, 1), (0, 1), (1, 1) The relation is a function. Every input has exactly one output.

b. (1, 3), (2, 4), (3, 5), (1, 5), (4, 6) The relation is not a function. The input 1 has two outputs, 3 and 5.

c.

Input, x	0	2	4	6	8	10
Output, y	0	1	2	3	4	5

The relation is a function. Every input has exactly one output.

d. Input, x Output, y

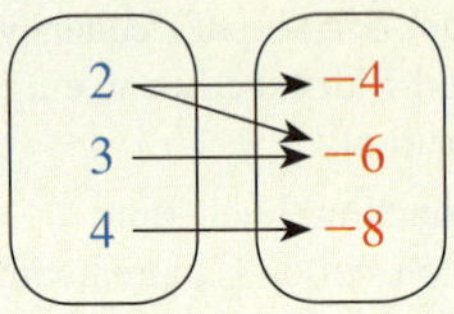

The relation is not a function. The input 2 has two outputs, −4 and −6.

MONITORING PROGRESS ANSWERS

1. not a function; The input 5 has two outputs, 0 and 10.
2. function; Every input has exactly one output.
3. function; Every input has exactly one output.
4. not a function; The input $\frac{1}{2}$ has three outputs, −2, 0, and 4.

3.1 Lesson

Core Vocabulary

relation, *p. 104*
function, *p. 104*
domain, *p. 106*
range, *p. 106*
independent variable, *p. 107*
dependent variable, *p. 107*

Previous
ordered pair
mapping diagram

What You Will Learn

- Determine whether relations are functions.
- Find the domain and range of a function.
- Identify the independent and dependent variables of functions.

Determining Whether Relations Are Functions

A **relation** pairs inputs with outputs. When a relation is given as ordered pairs, the x-coordinates are inputs and the y-coordinates are outputs. A relation that pairs each input with *exactly one* output is a **function**.

EXAMPLE 1 Determining Whether Relations Are Functions

Determine whether each relation is a function. Explain.

a. (−2, 2), (−1, 2), (0, 2), (1, 0), (2, 0)

b. (4, 0), (8, 7), (6, 4), (4, 3), (5, 2)

c.

Input, x	−2	−1	0	0	1	2
Output, y	3	4	5	6	7	8

REMEMBER

A relation can be represented by a mapping diagram.

d. Input, x Output, y

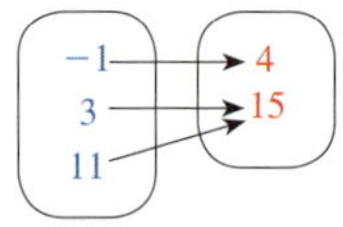

SOLUTION

a. Every input has exactly one output.

▶ So, the relation is a function.

b. The input 4 has two outputs, 0 and 3.

▶ So, the relation is *not* a function.

c. The input 0 has two outputs, 5 and 6.

▶ So, the relation is *not* a function.

d. Every input has exactly one output.

▶ So, the relation is a function.

Monitoring Progress

Help in English and Spanish at *BigIdeasMath.com*

Determine whether the relation is a function. Explain.

1. (−5, 0), (0, 0), (5, 0), (5, 10)
2. (−4, 8), (−1, 2), (2, −4), (5, −10)
3.

Input, x	Output, y
2	2.6
4	5.2
6	7.8

4. Input, x Output, y

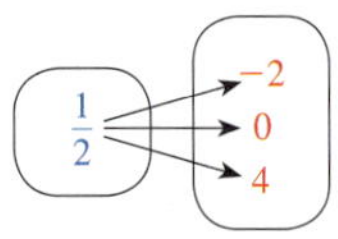

Laurie's Notes Teacher Actions

- The definition of *relation* and *function* were presented for the exploration. Check student understanding of the definitions.
- If the two explorations were completed, use *Monitoring Progress* as a closure for the previous class or a warm-up for the second day.

? Advancing Question: "Do the number of inputs and outputs in a relation have to be the same? Explain." no; The inputs just have to be paired with an output.

? Advancing Question: "Do the number of inputs and outputs in a function have to be the same? Explain." no; Each input just has to be paired with exactly one output. The output could be the same for two inputs.

Core Concept

Vertical Line Test

Words A graph represents a function when no vertical line passes through more than one point on the graph.

Examples Function

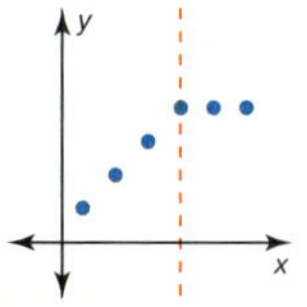

Not a function

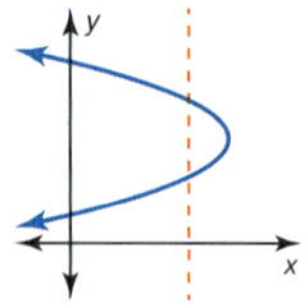

EXAMPLE 2 Using the Vertical Line Test

Determine whether each graph represents a function. Explain.

a. 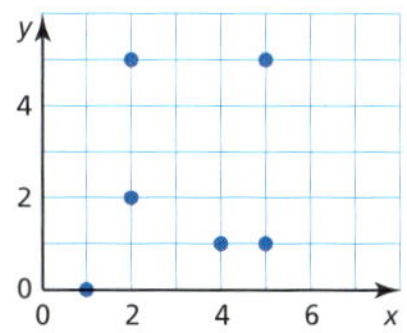

b. 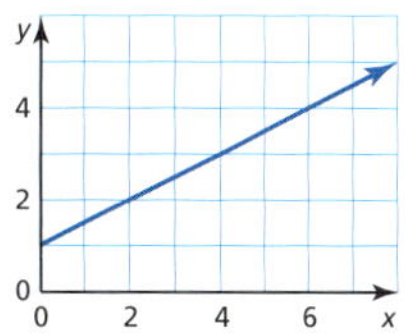

SOLUTION

a. You can draw a vertical line through (2, 2) and (2, 5).

▶ So, the graph does *not* represent a function.

b. No vertical line can be drawn through more than one point on the graph.

▶ So, the graph represents a function.

Monitoring Progress

Help in English and Spanish at *BigIdeasMath.com*

Determine whether the graph represents a function. Explain.

5. 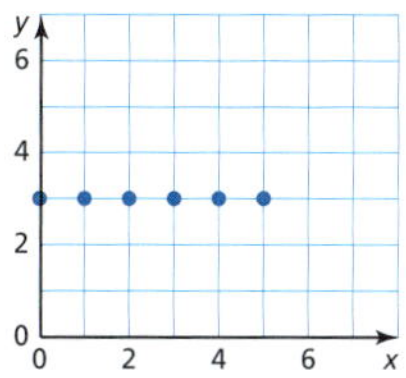

6. 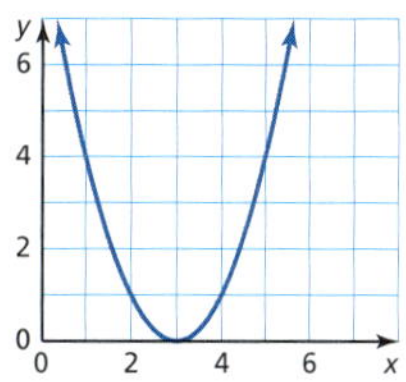

7. 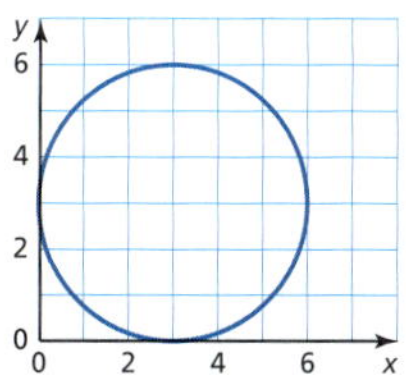

8. 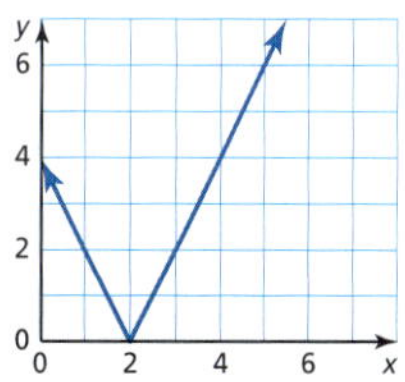

Extra Example 2

Determine whether each graph represents a function. Explain.

a. 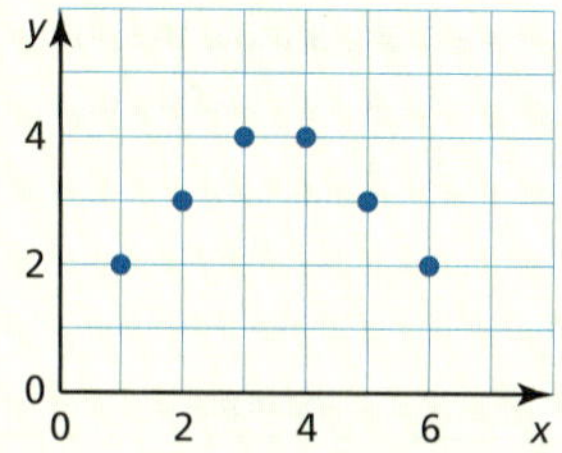

The graph represents a function. No vertical line can be drawn through more than one point on the graph.

b. 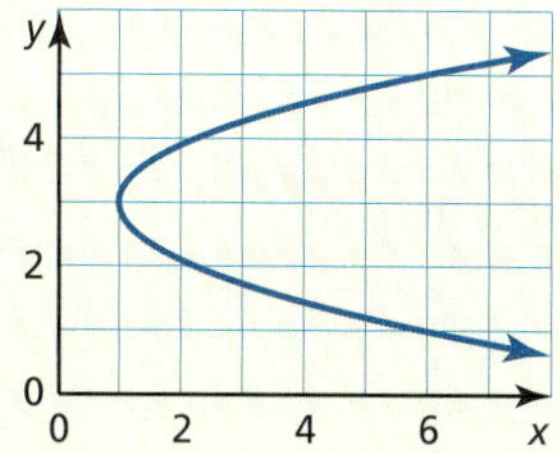

The graph does not represent a function. A vertical line can be drawn through more than one point on the graph, such as through points (6, 1) and (6, 5).

MONITORING PROGRESS ANSWERS

5. function; No vertical line can be drawn through more than one point on the graph.

6. function; No vertical line can be drawn through more than one point on the graph.

7. not a function; A vertical line can be drawn through more than one point on the graph at several places, such as (3, 0) and (3, 6).

8. function; No vertical line can be drawn through more than one point on the graph.

Laurie's Notes Teacher Actions

- **Teaching Tip:** Graph two different relations, $A = \{(1, 2), (1, 4), (2, 4)\}$ and $B = \{(1, 4), (2, 4), (3, 2)\}$. Ask whether they are both functions. Highlight (with a vertical line) the two ordered pairs that make A not a function.
- Write the *Core Concept* and refer to the relations A and B in the *Teaching Tip* or draw new examples.

? "What is true about ordered pairs whose orientation is vertical?" They have the same x-value (input) but different y-values (outputs).

English Language Learners

Vocabulary

Have students make a table to show how the terms used to describe a function are related.

Input	Output
x-values	y-values
Domain	Range
Independent variable	Dependent variable

Extra Example 3

Find the domain and range of the function represented by the graph.

a.

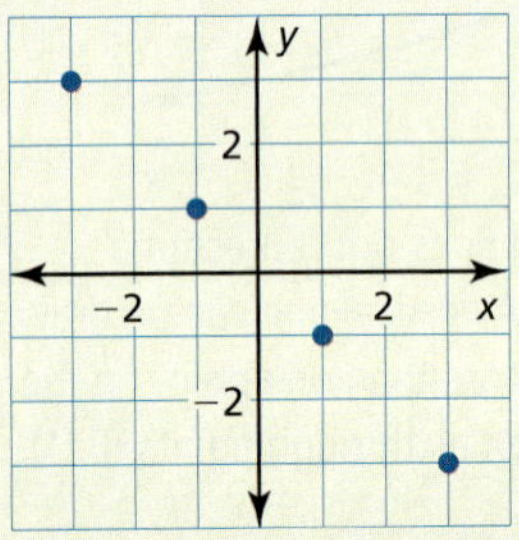

The domain is −3, −1, 1, and 3. The range is 3, 1, −1, and −3.

b.

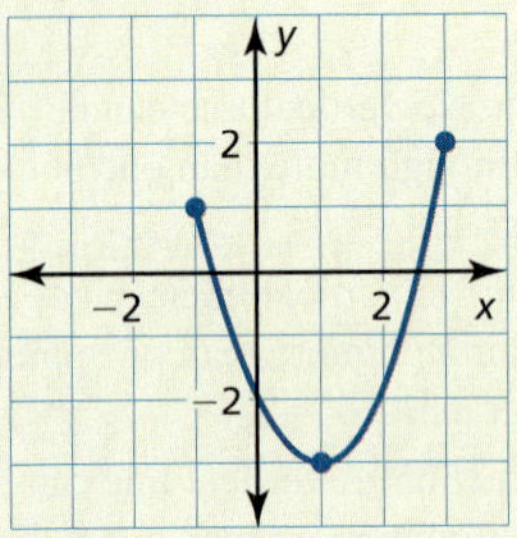

The domain is $-1 \le x \le 3$. The range is $-3 \le y \le 2$.

MONITORING PROGRESS ANSWERS

9. domain: −2, −1, 0, 1, 2; range: 0, 2, 4

10. domain: $1 \le x \le 5$; range: $0 \le y \le 4$

Finding the Domain and Range of a Function

Core Concept

The Domain and Range of a Function

The **domain** of a function is the set of all possible input values.

The **range** of a function is the set of all possible output values.

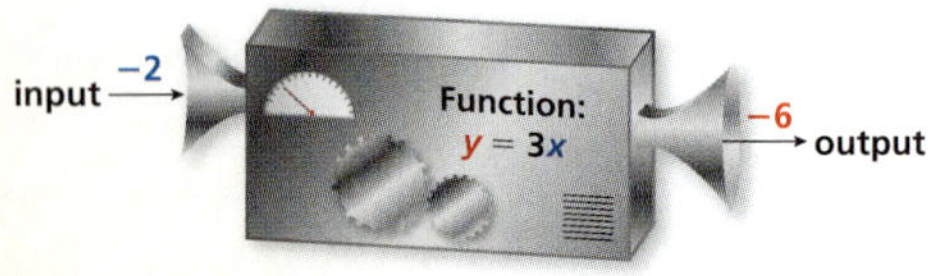

EXAMPLE 3 Finding the Domain and Range from a Graph

Find the domain and range of the function represented by the graph.

a.

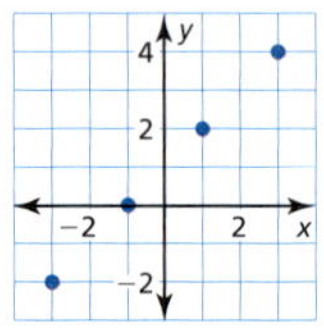

b.

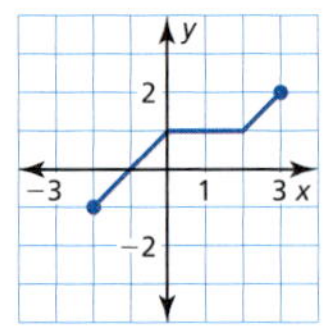

SOLUTION

a. Write the ordered pairs. Identify the inputs and outputs.

inputs

(−3, −2), (−1, 0), (1, 2), (3, 4)

outputs

▶ The domain is −3, −1, 1, and 3. The range is −2, 0, 2, and 4.

STUDY TIP

A relation also has a domain and a range.

b. Identify the x- and y-values represented by the graph.

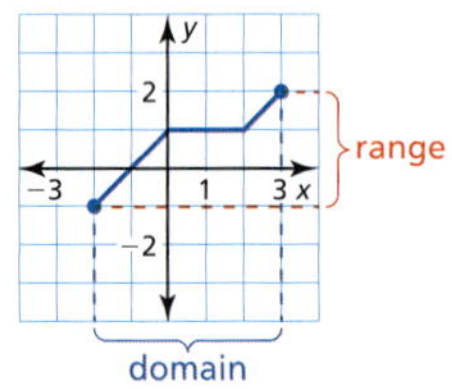

▶ The domain is $-2 \le x \le 3$. The range is $-1 \le y \le 2$.

Monitoring Progress

Help in English and Spanish at *BigIdeasMath.com*

Find the domain and range of the function represented by the graph.

9.

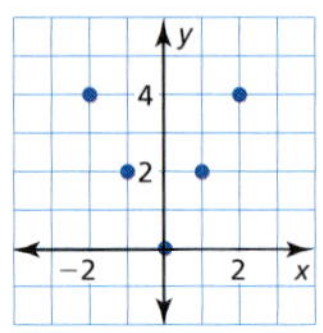

10.

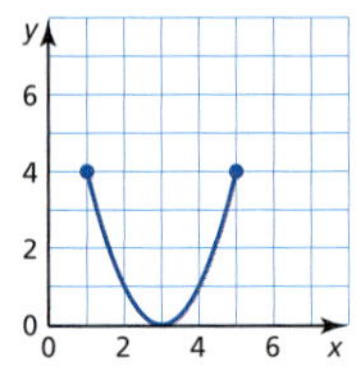

Laurie's Notes Teacher Actions

- Write the *Core Concept*. The graphic of a function machine should help students conceptualize the idea of entering an input value, applying a rule, and obtaining the output value. Relate this to a vending machine.
- **Common Misconception:** In finding the domain and range in part (b), students may only identify the integers associated with the endpoints of the segments.

? Turn and Talk: "In Example 3, how many different values are in the domain for each part?" (a) 4; (b) infinite number "How many different values are in the range?" (a) 4; (b) infinite number

Identifying Independent and Dependent Variables

The variable that represents the input values of a function is the **independent variable** because it can be *any* value in the domain. The variable that represents the output values of a function is the **dependent variable** because it *depends* on the value of the independent variable. When an equation represents a function, the dependent variable is defined in terms of the independent variable. The statement "y is a function of x" means that y varies depending on the value of x.

$$y = -x + 10$$

dependent variable, y — independent variable, x

EXAMPLE 4 Identifying Independent and Dependent Variables

The function $y = -3x + 12$ represents the amount y (in fluid ounces) of juice remaining in a bottle after you take x gulps.

a. Identify the independent and dependent variables.

b. The domain is 0, 1, 2, 3, and 4. What is the range?

SOLUTION

a. The amount y of juice remaining depends on the number x of gulps.

▶ So, y is the dependent variable, and x is the independent variable.

b. Make an input-output table to find the range.

Input, x	$-3x + 12$	Output, y
0	$-3(0) + 12$	12
1	$-3(1) + 12$	9
2	$-3(2) + 12$	6
3	$-3(3) + 12$	3
4	$-3(4) + 12$	0

▶ The range is 12, 9, 6, 3, and 0.

Monitoring Progress

Help in English and Spanish at *BigIdeasMath.com*

11. The function $a = -4b + 14$ represents the number a of avocados you have left after making b batches of guacamole.

a. Identify the independent and dependent variables.

b. The domain is 0, 1, 2, and 3. What is the range?

12. The function $t = 19m + 65$ represents the temperature t (in degrees Fahrenheit) of an oven after preheating for m minutes.

a. Identify the independent and dependent variables.

b. A recipe calls for an oven temperature of 350°F. Describe the domain and range of the function.

Extra Example 4

The function $c = 6t + 8$ represents the cost c for t T-shirts plus shipping.

a. Identify the independent and dependent variables. independent variable: t; dependent variable: c

b. The domain is 1, 2, 3, 4, and 5. What is the range? 14, 20, 26, 32, 38

MONITORING PROGRESS ANSWERS

11. **a.** a is the dependent variable and b is the independent variable.

b. 14, 10, 6, 2

12. **a.** t is the dependent variable and m is the independent variable.

b. domain: $0 \le m \le 15$; range: $65 \le t \le 350$

Laurie's Notes — Teacher Actions

- Discuss the meaning of *independent variables* and *dependent variables*. Use an example to help make the point of the dependent variable depending on the value of the independent variable.
- **Example:** In the equation $y = 0.99x$, the cost y in dollars of buying songs *depends* on the number of songs x. So, x is the independent variable (any number of songs may be purchased), and y is the dependent variable (the cost depends on the number of songs purchased).
- **Monitoring Progress:** Explain, if necessary, what preheating means in Question 12.

Closure

- **Exit Ticket:** Make an input-output table for the function $y = -2x + 3$ using the inputs $-2, 0, 2, 4$.

x	-2	0	2	4
y	7	3	-1	-5

Assignment Guide and Homework Check

ASSIGNMENT

Basic: 1, 2, 3–21 odd, 28, 29, 44–51

Average: 1, 2, 4–24 even, 28, 29, 36, 44–51

Advanced: 1, 2, 6–24 even, 25–30, 32–44 even, 44–51

HOMEWORK CHECK

Basic: 3, 7, 11, 13, 17

Average: 4, 8, 10, 14, 18

Advanced: 6, 12, 16, 18, 24

ANSWERS

1. The independent variable can be any value in the domain, but the dependent variable depends on the values of the independent variable.
2. Find the range of the function represented by the table; −1, 5, 7; −1, 0, 1
3. function; Every input has exactly one output.
4. not a function; The input 3 has two outputs, −8 and 6.
5. not a function; The input 2 has two outputs, 3 and 2.
6. function; Every input has exactly one output.
7. not a function; The input 16 has two outputs, −2 and 2, and the input 1 has two outputs, −1 and 1.
8. function; Every input has exactly one output.
9. function; No vertical line can be drawn through more than one point on the graph.
10. not a function; Three vertical lines can be drawn through more than one point on the graph, including one through (0, 2) and (0, 4).
11. not a function; A vertical line can be drawn through more than one point on the graph in many places, such as (4, 0) and (4, 6).
12. function; No vertical line can be drawn through more than one point on the graph.
13. domain: −2, −1, 0, 1, 2; range: −2, 0, 2
14. domain: −2, 0, 2, 4; range: 3
15. domain: $-4 \le x \le 2$; range: $2 \le y \le 6$
16. domain: $2 < x < 7$; range: $1 < y < 6$

3.1 Exercises

Dynamic Solutions available at *BigIdeasMath.com*

Vocabulary and Core Concept Check

1. **WRITING** How are independent variables and dependent variables different?

2. **DIFFERENT WORDS, SAME QUESTION** Which is different? Find "both" answers.

Find the range of the function represented by the table.

Find the inputs of the function represented by the table.

x	−1	0	1
y	7	5	−1

Find the x-values of the function represented by (−1, 7), (0, 5), and (1, −1).

Find the domain of the function represented by (−1, 7), (0, 5), and (1, −1).

Monitoring Progress and Modeling with Mathematics

In Exercises 3–8, determine whether the relation is a function. Explain. *(See Example 1.)*

3. (1, −2), (2, 1), (3, 6), (4, 13), (5, 22)

4. (7, 4), (5, −1), (3, −8), (1, −5), (3, 6)

5. Input, x Output, y

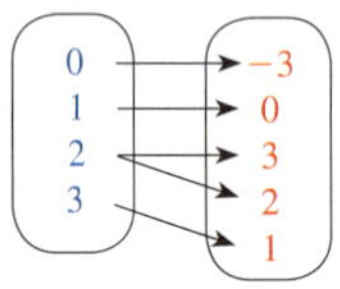

6. Input, x Output, y

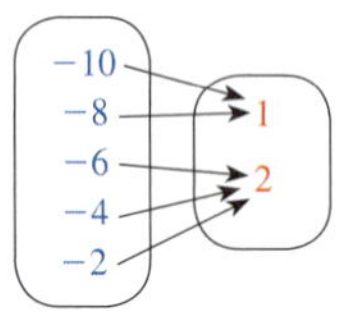

7.

Input, x	16	1	0	1	16
Output, y	−2	−1	0	1	2

8.

Input, x	−3	0	3	6	9
Output, y	11	5	−1	−7	−13

In Exercises 9–12, determine whether the graph represents a function. Explain. *(See Example 2.)*

9.

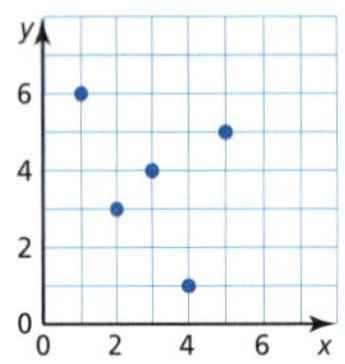

10.

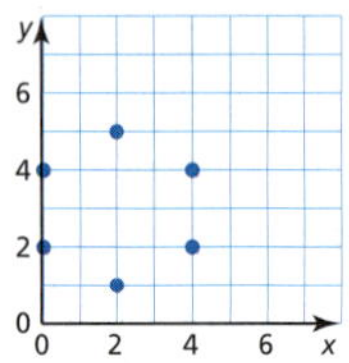

11.

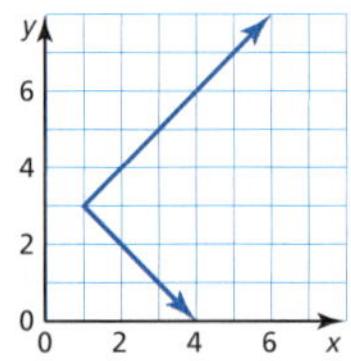

12.

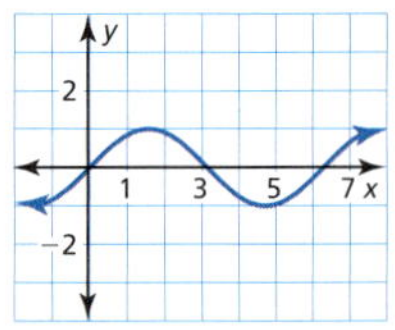

In Exercises 13–16, find the domain and range of the function represented by the graph. *(See Example 3.)*

13.

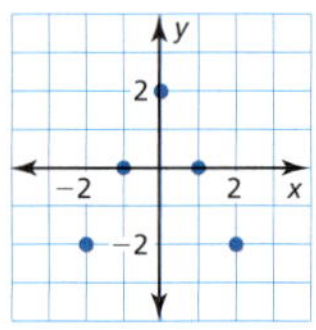

14.

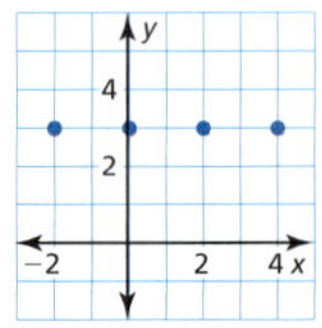

15.

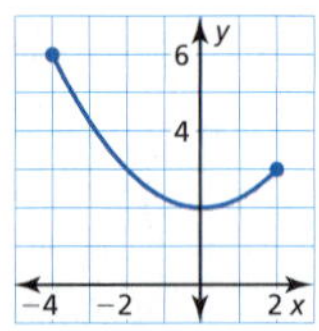

16.

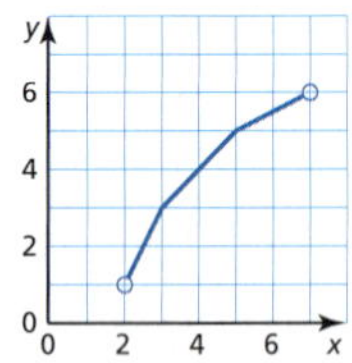

17. **MODELING WITH MATHEMATICS** The function $y = 25x + 500$ represents your monthly rent y (in dollars) when you pay x days late. *(See Example 4.)*

 a. Identify the independent and dependent variables.

 b. The domain is 0, 1, 2, 3, 4, and 5. What is the range?

17. a. y is the dependent variable and x is the independent variable.
 b. 500, 525, 550, 575, 600, 625

18. **MODELING WITH MATHEMATICS** The function $y = 3.5x + 2.8$ represents the cost y (in dollars) of a taxi ride of x miles.

a. Identify the independent and dependent variables.

b. You have enough money to travel at most 20 miles in the taxi. Find the domain and range of the function.

ERROR ANALYSIS In Exercises 19 and 20, describe and correct the error in the statement about the relation shown in the table.

Input, x	1	2	3	4	5
Output, y	6	7	8	6	9

19. ✗ The relation is *not* a function. One output is paired with two inputs.

20. ✗ The relation is a function. The range is 1, 2, 3, 4, and 5.

ANALYZING RELATIONSHIPS In Exercises 21 and 22, identify the independent and dependent variables.

21. The number of quarters you put into a parking meter affects the amount of time you have on the meter.

22. The battery power remaining on your MP3 player is based on the amount of time you listen to it.

23. **MULTIPLE REPRESENTATIONS** The balance y (in dollars) of your savings account is a function of the month x.

Month, x	0	1	2	3	4
Balance (dollars), y	100	125	150	175	200

a. Describe this situation in words.

b. Write the function as a set of ordered pairs.

c. Plot the ordered pairs in a coordinate plane.

24. **MULTIPLE REPRESENTATIONS** The function $1.5x + 0.5y = 12$ represents the number of hardcover books x and softcover books y you can buy at a used book sale.

a. Solve the equation for y.

b. Make an input-output table to find ordered pairs for the function.

c. Plot the ordered pairs in a coordinate plane.

25. **ATTENDING TO PRECISION** The graph represents a function. Find the input value corresponding to an output of 2.

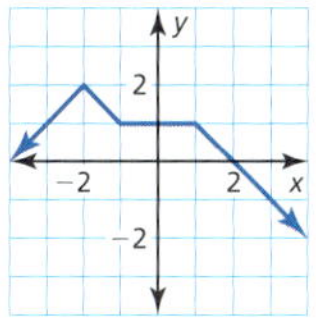

26. **OPEN-ENDED** Fill in the table so that when t is the independent variable, the relation is a function, and when t is the dependent variable, the relation is not a function.

t				
v				

27. **ANALYZING RELATIONSHIPS** You select items in a vending machine by pressing one letter and then one number.

a. Explain why the relation that pairs letter-number combinations with food or drink items is a function.

b. Identify the independent and dependent variables.

c. Find the domain and range of the function.

Dynamic Teaching Tools
- Dynamic Assessment & Progress Monitoring Tool
- Interactive Whiteboard Lesson Library
- Dynamic Classroom with Dynamic Investigations

ANSWERS

18. a. y is the dependent variable and x is the independent variable.
    b. domain: $0 \leq x \leq 20$; range: $2.8 \leq y \leq 72.8$
19. A function can have one output paired with two inputs, but cannot have one input paired with more than one output; The relation is a function. No input is paired with more than one output.
20. The given list is the domain of the function; The range is 6, 7, 8, 9.
21. The amount of time is the dependent variable and the number of quarters is the independent variable.
22. The battery power remaining is the dependent variable and the amount of time you listen is the independent variable.
23. a. *Sample answer:* The balance of the savings account is $100 in month 0 and increases by $25 per month through month 4.
    b. (0, 100), (1, 125), (2, 150), (3, 175), (4, 200)
    c.

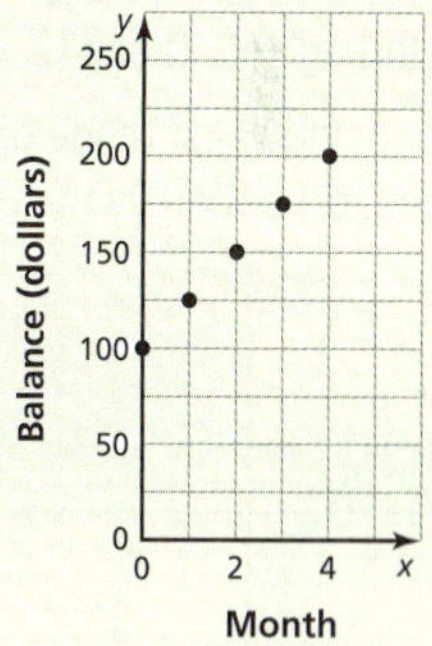

24. a. $y = 24 - 3x$
    b. (0, 24), (1, 21), (2, 18), (3, 15), (4, 12), (5, 9), (6, 6), (7, 3), (8, 0)
    c.

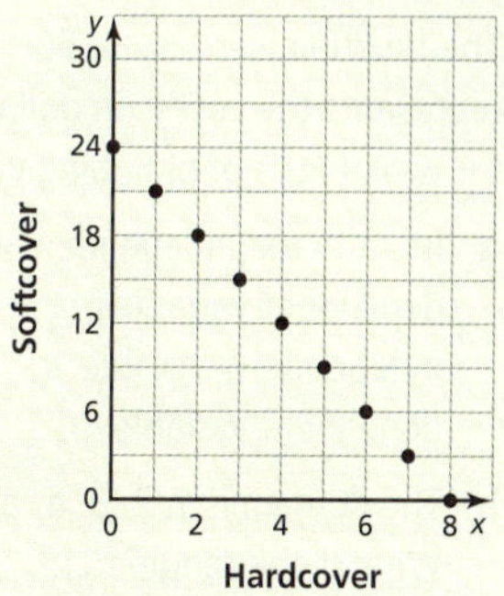

25. -2
26. *Sample answer:* 1, 2, 3, 4; 3, 0, 7, 0
27. a. Each letter-number combination is paired with exactly one food or drink item.
    b. The food item is the dependent variable and the letter-number combination is the independent variable.
    c. domain: A1, A2, A3, B1, B2, B3, B4, C1, C2, C3, C4; range: popcorn, nuts, pretzels, protein bar, granola bar, cereal, energy bar, orange juice, water, milk

ANSWERS

28. **a.** No vertical line can be drawn through more than one point on the graph.

b. *Sample answer:* 20 ft; 29 ft

c. *Sample answer:* $0 \le t \le 2.6$

d. no; There are many values of h that have two corresponding values of t.

29–51. See Additional Answers.

Mini-Assessment

1. Determine whether the relation is a function. Explain.
$(-5, -1), (-4, 1), (-3, 3), (-1, 5), (1, 7)$ The relation is a function. Every input has exactly one output.

2. Determine whether the graph represents a function. Explain.

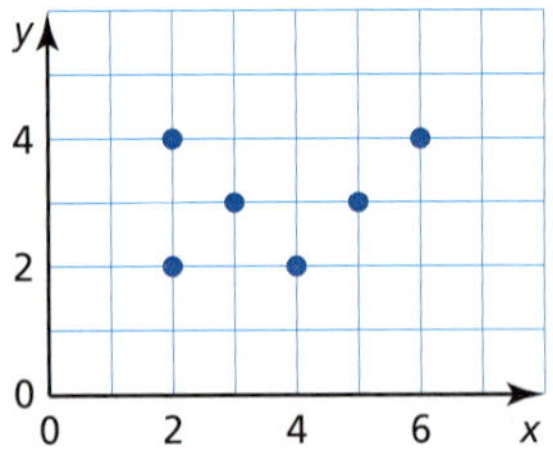

The graph does not represent a function. A vertical line can be drawn through (2, 2) and (2, 4).

3. Find the domain and range of the function represented by the graph.

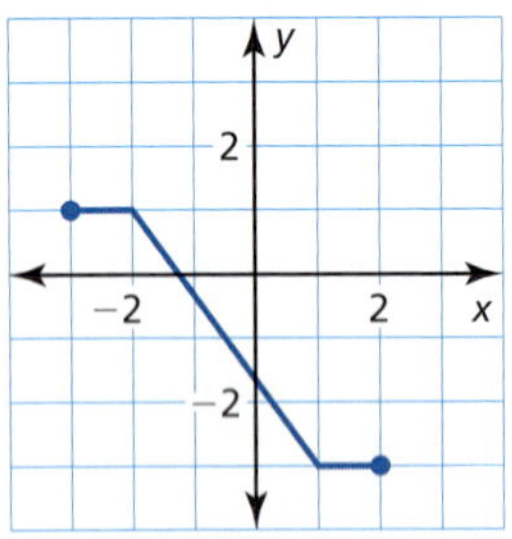

The domain is $-3 \le x \le 2$. The range is $-3 \le y \le 1$.

4. The function $y = 65x + 50$ represents the cost y (in dollars) for wwx hours of plumbing work.
   a. Identify the independent and dependent variables. independent variable: x; dependent variable: y
   b. The domain is 1, 2, 3, 4, and 5. What is the range? 115, 180, 245, 310, 375

28. HOW DO YOU SEE IT? The graph represents the height h of a projectile after t seconds.

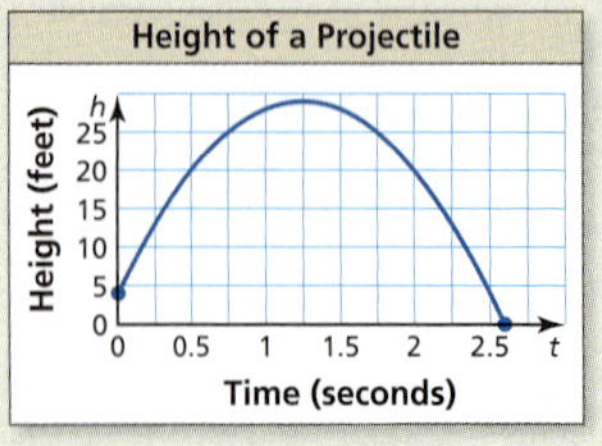

a. Explain why h is a function of t.

b. Approximate the height of the projectile after 0.5 second and after 1.25 seconds.

c. Approximate the domain of the function.

d. Is t a function of h? Explain.

29. MAKING AN ARGUMENT Your friend says that a line always represents a function. Is your friend correct? Explain.

30. THOUGHT PROVOKING Write a function in which the inputs and/or the outputs are not numbers. Identify the independent and dependent variables. Then find the domain and range of the function.

ATTENDING TO PRECISION In Exercises 31–34, determine whether the statement uses the word *function* in a way that is mathematically correct. Explain your reasoning.

31. The selling price of an item is a function of the cost of making the item.

32. The sales tax on a purchased item in a given state is a function of the selling price.

33. A function pairs each student in your school with a homeroom teacher.

34. A function pairs each chaperone on a school trip with 10 students.

REASONING In Exercises 35–38, tell whether the statement is true or false. If it is false, explain why.

35. Every function is a relation.

36. Every relation is a function.

37. When you switch the inputs and outputs of any function, the resulting relation is a function.

38. When the domain of a function has an infinite number of values, the range always has an infinite number of values.

39. MATHEMATICAL CONNECTIONS Consider the triangle shown.

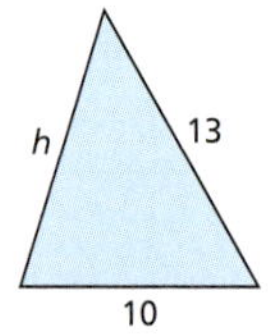

a. Write a function that represents the perimeter of the triangle.

b. Identify the independent and dependent variables.

c. Describe the domain and range of the function. (*Hint:* The sum of the lengths of any two sides of a triangle is greater than the length of the remaining side.)

REASONING In Exercises 40–43, find the domain and range of the function.

40. $y = |x|$ **41.** $y = -|x|$

42. $y = |x| - 6$ **43.** $y = 4 - |x|$

Maintaining Mathematical Proficiency

Reviewing what you learned in previous grades and lessons

Write the sentence as an inequality. *(Section 2.1)*

44. A number y is less than 16. **45.** Three is no less than a number x.

46. Seven is at most the quotient of a number d and -5.

47. The sum of a number w and 4 is more than -12.

Evaluate the expression. *(Skills Review Handbook)*

48. 11^2 **49.** $(-3)^4$ **50.** -5^2 **51.** 2^5

If students need help...	*If students got it...*
Resources by Chapter • Practice A and Practice B • Puzzle Time	Resources by Chapter • Enrichment and Extension • Cumulative Review
Student Journal • Practice	Start the *next* Section
Differentiating the Lesson Skills Review Handbook	

Laurie's Notes

Overview of Section 3.2

Introduction

- Students were introduced to linear functions in middle school. This lesson makes this content secure, along with introducing the concepts of discrete and continuous domains.
- The approach taken is to use multiple representations (table of values, graphs, and equations) in determining when the representation is a linear or nonlinear function. The additional concept of discrete versus continuous domains is then explored.
- The exploration makes use of two *Big Ideas* from middle school:
  - If two figures are similar, then the ratio of their perimeters is equal to the ratio of their corresponding side lengths. This will be a linear relationship.
  - If two figures are similar, then the ratio of their areas is equal to the square of the ratio of their corresponding side lengths. This will be a nonlinear relationship.

Resources

- Graphing calculators could be used in the exploration or formal lesson. Students can graph an equation and look at the associated table of values. For discrete data, enter ordered pairs in the STAT mode of the calculator and graph the associated scatter plot.

Formative Assessment Tips

- **Opposing Views:** This technique allows students to compare two or more solutions or ways of thinking about a problem. It can also be used to compare and contrast opposing views (i.e., yes, $y = \frac{4}{x}$ is a linear function, and no, $y = \frac{4}{x}$ is not a linear function).
- To be informative, students must present more than yes/no answers. Students must be able to explain their reasoning for classmates to hear. Once the different views are presented, all students should decide which view most closely reflects their own and give justification for their thinking. Students can respond in writing or you can solicit viewpoints as a discussion prompt.
- **Construct Viable Arguments and Critique the Reasoning of Others:** Using this formative assessment technique allows you to check students' conceptual knowledge and their ability to construct a viable argument.

Applications

- Students will start to see that many of the real-life applications in this book are defined for nonnegative domains, some discrete and others continuous.

Pacing Suggestion

- The formal lesson is quite long. You might have students complete Exploration 1 and then move to Example 3 in the formal lesson. Whatever examples are not finished could be done on the second day.

Dynamic Teaching Tools

- Dynamic Assessment & Progress Monitoring Tool
- Lesson Planning Tool
- Interactive Whiteboard Lesson Library
- Dynamic Classroom with Dynamic Investigations

What Your Students Will Learn

- Identify linear functions using graphs, tables, and equations in two variables.
- Determine whether domains of functions are discrete or continuous and graph functions in the coordinate plane.
- Describe situations represented by graphs of functions.

Laurie's Notes

Exploration

Motivate

- Hold up two objects that are similar but not congruent (two square pieces of paper, a baseball and a golf ball, a penny and a quarter).

? **Turn and Talk:** "Are these objects similar? Explain." Listen for responses beyond "same shape but different size."

- Hold up two pencils that are different lengths.

? **Turn and Talk:** "Are these objects similar? Explain." Listen for responses that specify that not all dimensions have been changed by the same factor.

- In this exploration, attributes of similar rectangles and circles are used as the context for exploring linear and nonlinear patterns. FYI: All circles are similar.

Exploration 1

- In parts (a) and (b), students are working with a rectangle whose length is twice its width. Students should be able to complete the table and graphs for the rectangles without difficulty.
- In parts (c) and (d), students may have forgotten or may switch the formulas for circumference and area of a circle.
- **Attend to Precision:** Using 3.14 for π and rounding results to the nearest tenth to graph will be precise enough to discern a visual pattern. If students leave their results in terms of π in the table, then numeric patterns should be discerned.

? **Turn and Talk:** "How do you decide whether the pattern is linear or nonlinear?" Listen for responses that go beyond looking at the graphs.

- **Whiteboarding:** Use whiteboards so that results can be quickly displayed by four students for parts (a) and (b), followed by four students for parts (c) and (d). Two students display the table of values and two students display the graphs.
- **Reason Abstractly and Quantitatively** and **Construct Viable Arguments and Critique the Reasoning of Others:** Ask a volunteer to explain how he or she knows the pattern is linear by looking at the table of values. Listen for language such as "when x increases by 1, y increases by the same amount each time." A student may even say that there is a constant rate of change!

Communicate Your Answer

- **Think-Pair-Share:** Circulate as students are sharing their answers with partners. Make note of insightful answers that would be beneficial for the whole class to hear.
- In Question 4, students may be challenged to think of a nonlinear example that is not geometric.

Connecting to Next Step

- If you feel students are comfortable with identifying linear and nonlinear patterns in tables of values or graphs, assign items from Exercises 5–14 on page 117. In beginning the formal lesson, you might omit Examples 1 and 2.

3.2 Linear Functions

Essential Question How can you determine whether a function is linear or nonlinear?

EXPLORATION 1 **Finding Patterns for Similar Figures**

Work with a partner. Copy and complete each table for the sequence of similar figures. (In parts (a) and (b), use the rectangle shown.) Graph the data in each table. Decide whether each pattern is linear or nonlinear. Justify your conclusion.

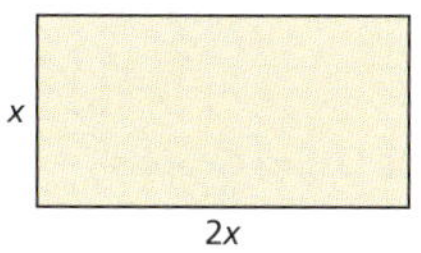

a. perimeters of similar rectangles

x	1	2	3	4	5
P					

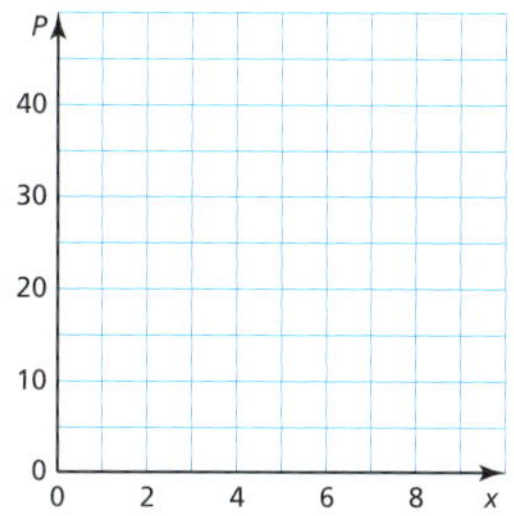

b. areas of similar rectangles

x	1	2	3	4	5
A					

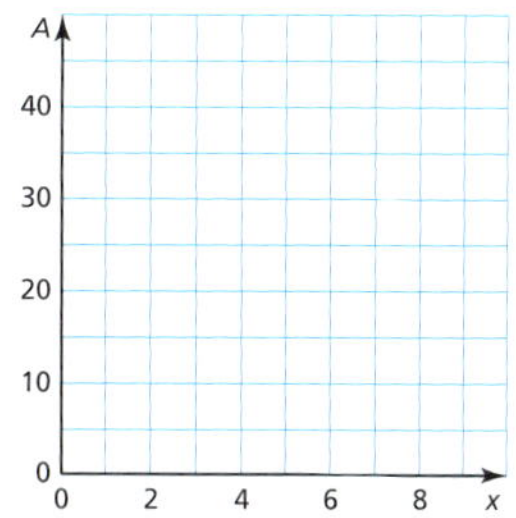

c. circumferences of circles of radius *r*

r	1	2	3	4	5
C					

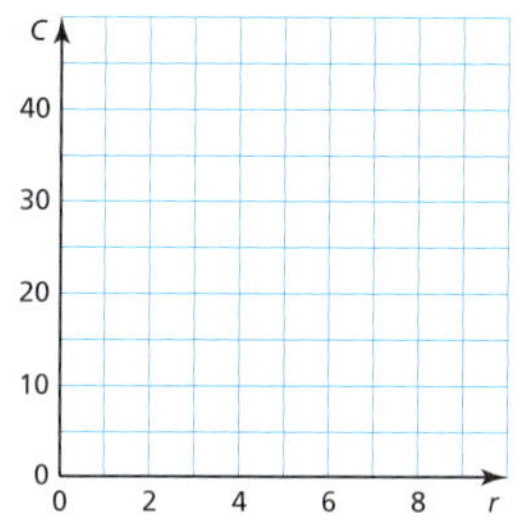

d. areas of circles of radius *r*

r	1	2	3	4	5
A					

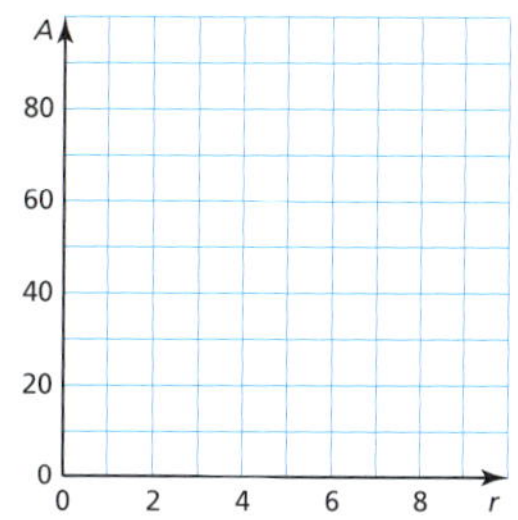

USING TOOLS STRATEGICALLY

To be proficient in math, you need to identify relationships using tools, such as tables and graphs.

Communicate Your Answer

2. How do you know that the patterns you found in Exploration 1 represent functions?
3. How can you determine whether a function is linear or nonlinear?
4. Describe two real-life patterns: one that is linear and one that is nonlinear. Use patterns that are different from those described in Exploration 1.

3. Sketch the graph of the function. When the graph is a line, the pattern is linear. When the graph is not a line, the pattern is not linear.
4. *Sample answer:* Distance traveled during a time interval at constant speed is linear. Distance traveled during a time interval while accelerating is nonlinear.

Dynamic Teaching Tools

- Dynamic Assessment & Progress Monitoring Tool
- Lesson Planning Tool
- Interactive Whiteboard Lesson Library
- Dynamic Classroom with Dynamic Investigations

ANSWERS

1. a. 6, 12, 18, 24, 30;

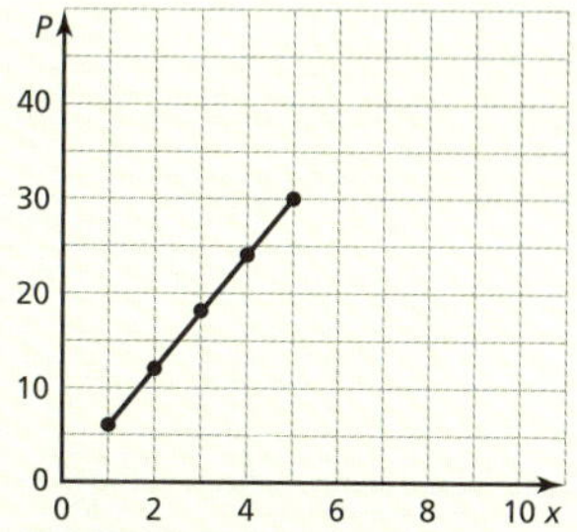

linear; The graph is a line.

b. 2, 8, 18, 32, 50;

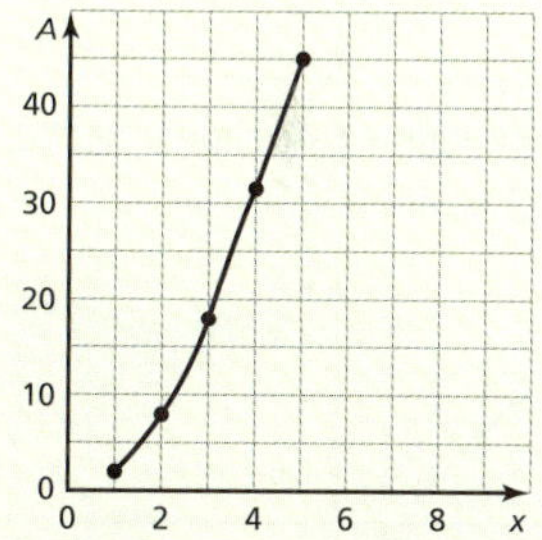

nonlinear; The graph is not a line.

c. 2π or about 6.3, 4π or about 12.6, 6π or about 18.8, 8π or about 25.1, 10π or about 31.4;

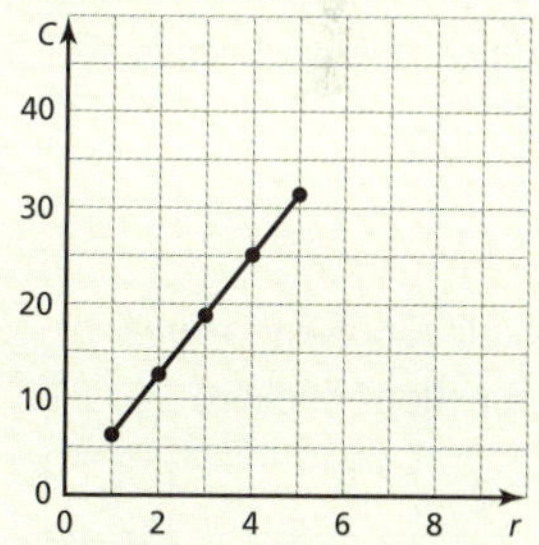

linear; The graph is a line.

d. π or about 3.1, 4π or about 12.6, 9π or about 28.3, 16π or about 50.3, 25π or about 78.5;

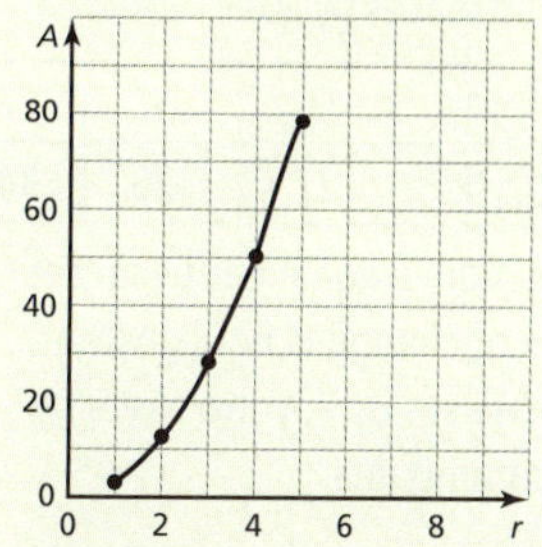

nonlinear; The graph is not a line.

2. No vertical line can be drawn through more than one point on any of the graphs.

Differentiated Instruction

Visual

In Example 2, have students graph the functions represented by tables of ordered pairs to reinforce their understanding of linear and nonlinear functions.

Extra Example 1

Does the graph represent a *linear* or *nonlinear* function? Explain.

a.

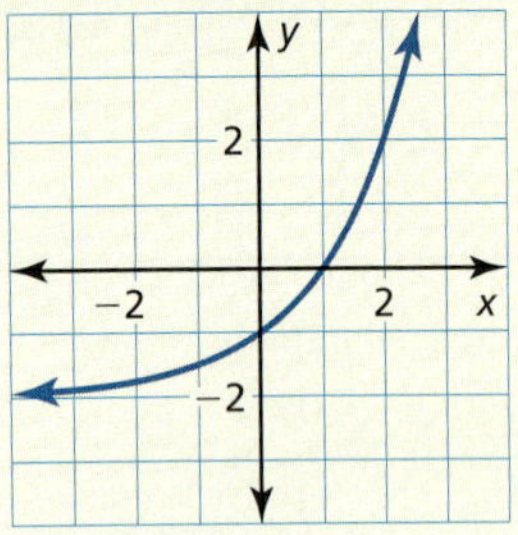

The function is nonlinear because the graph is not a line.

b.

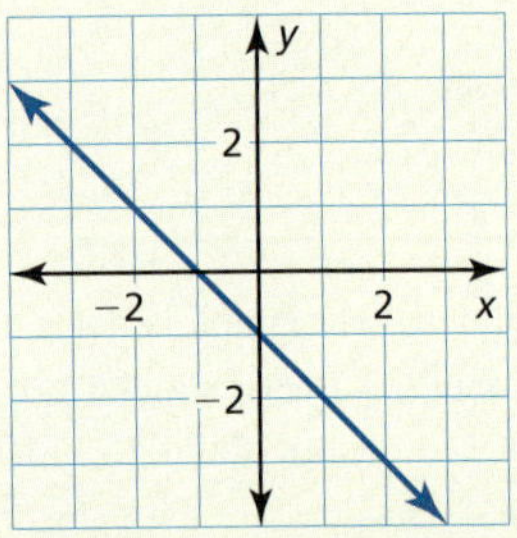

The function is linear because the graph is a line.

Extra Example 2

Does the table represent a *linear* or *nonlinear* function? Explain.

a.

x	2	4	6	8
y	9	18	27	36

The function is linear because the rate of change is constant. As x increases by 2, y increases by 9.

b.

x	1	3	5	7
y	1	27	125	343

The function is nonlinear because the rate of change is not constant. As x increases by 2, y increases by different amounts.

3.2 Lesson

Core Vocabulary

linear equation in two variables, *p. 112*
linear function, *p. 112*
nonlinear function, *p. 112*
solution of a linear equation in two variables, *p. 114*
discrete domain, *p. 114*
continuous domain, *p. 114*

Previous
whole number

What You Will Learn

- Identify linear functions using graphs, tables, and equations.
- Graph linear functions using discrete and continuous data.
- Write real-life problems to fit data.

Identifying Linear Functions

A **linear equation in two variables**, x and y, is an equation that can be written in the form $y = mx + b$, where m and b are constants. The graph of a linear equation is a line. Likewise, a **linear function** is a function whose graph is a nonvertical line. A linear function has a constant rate of change and can be represented by a linear equation in two variables. A **nonlinear function** does not have a constant rate of change. So, its graph is *not* a line.

EXAMPLE 1 Identifying Linear Functions Using Graphs

Does the graph represent a *linear* or *nonlinear* function? Explain.

a.

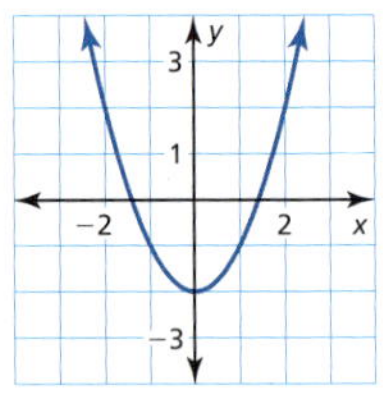

b.

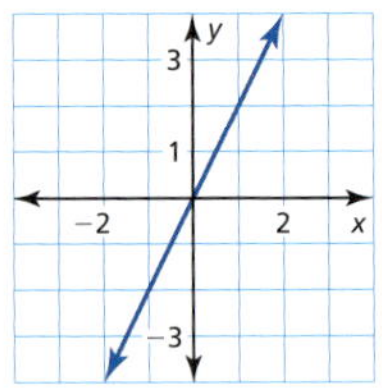

SOLUTION

a. The graph is *not* a line.

▶ So, the function is nonlinear.

b. The graph is a line.

▶ So, the function is linear.

EXAMPLE 2 Identifying Linear Functions Using Tables

Does the table represent a *linear* or *nonlinear* function? Explain.

a.

x	3	6	9	12
y	36	30	24	18

b.

x	1	3	5	7
y	2	9	20	35

REMEMBER

A constant rate of change describes a quantity that changes by equal amounts over equal intervals.

SOLUTION

a.

	+3	+3	+3	
x	3	6	9	12
y	36	30	24	18
	−6	−6	−6	

As x increases by 3, y decreases by 6. The rate of change is constant.

▶ So, the function is linear.

b.

	+2	+2	+2	
x	1	3	5	7
y	2	9	20	35
	+7	+11	+15	

As x increases by 2, y increases by different amounts. The rate of change is *not* constant.

▶ So, the function is nonlinear.

Laurie's Notes Teacher Actions

- Write and discuss the vocabulary, connecting it to the exploration with similar figures.

? Turn and Talk: "What does 'a constant rate of change' mean?" Listen for "as x increases by a constant rate, y is also increasing at a constant rate."

- Identifying linear equations from a graph is generally easier for students than identifying linear equations from a table of values.

COMMON ERROR In Example 2(b) there is a pattern by which y is increasing, but it is not a constant rate. Students may forget that the rate of change must be constant.

- **Teaching Tip:** Using the curved arrows to record the change between entries in the table of values helps students to conceptualize the change from one ordered pair to the next.

Monitoring Progress

Help in English and Spanish at *BigIdeasMath.com*

Does the graph or table represent a *linear* or *nonlinear* function? Explain.

1.

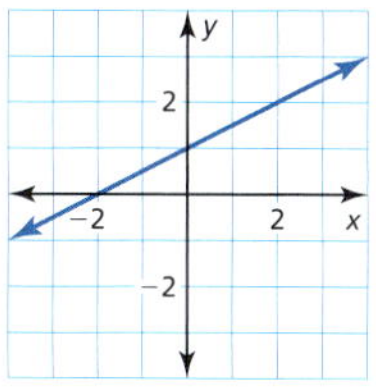

2.

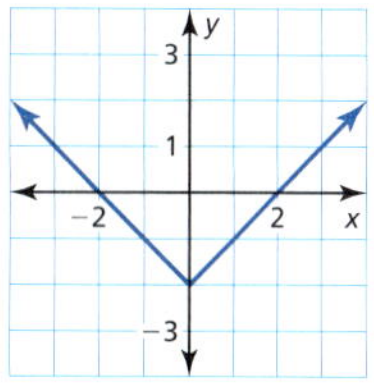

3.

x	0	1	2	3
y	3	5	7	9

4.

x	1	2	3	4
y	16	8	4	2

EXAMPLE 3 Identifying Linear Functions Using Equations

Which of the following equations represent linear functions? Explain.

$$y = 3.8,\ y = \sqrt{x},\ y = 3^x,\ y = \frac{2}{x},\ y = 6(x - 1), \text{ and } x^2 - y = 0$$

SOLUTION

You cannot rewrite the equations $y = \sqrt{x}$, $y = 3^x$, $y = \frac{2}{x}$, and $x^2 - y = 0$ in the form $y = mx + b$. So, these equations cannot represent linear functions.

▶ You can rewrite the equation $y = 3.8$ as $y = 0x + 3.8$ and the equation $y = 6(x - 1)$ as $y = 6x - 6$. So, they represent linear functions.

Monitoring Progress

Help in English and Spanish at *BigIdeasMath.com*

Does the equation represent a *linear* or *nonlinear* function? Explain.

5. $y = x + 9$

6. $y = \frac{3x}{5}$

7. $y = 5 - 2x^2$

Concept Summary

Representations of Functions

Words An output is 3 more than the input.

Equation $y = x + 3$

Input-Output Table

Input, *x*	Output, *y*
−1	2
0	3
1	4
2	5

Mapping Diagram

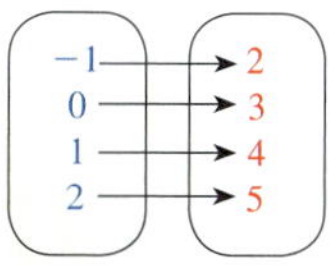

Graph

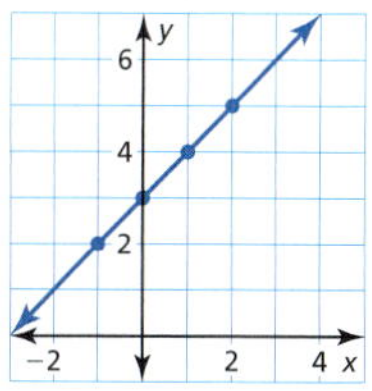

Extra Example 3

Which of the following equations represent linear functions? Explain.

$$y = 2^x,\ y = 5x,\ y = 3 - 2x^2,$$
$$x - y = 12,\ y = \sqrt{x} + 3$$

The equations $y = 5x$ and $x - y = 12$ represent linear functions because they can be written in the form $y = mx + b$. You can rewrite $y = 5x$ as $y = 5x + 0$ and you can rewrite $x - y = 12$ as $y = x - 12$. The other equations cannot be written in the form $y = mx + b$, so they cannot represent linear functions.

MONITORING PROGRESS ANSWERS

1. linear; The graph is a line.
2. nonlinear; The graph is not a line.
3. linear; As x increases by 1, y increases by 2. The rate of change is constant.
4. nonlinear; As x increases by 1, y decreases by different amounts. The rate of change is not constant.
5. linear; It can be written in the form $y = 1x + 9$.
6. linear; It can be written in the form $y = \frac{3}{5}x + 0$.
7. nonlinear; It cannot be written in the form $y = mx + b$.

Laurie's Notes Teacher Actions

- Have students check with their neighbors when they complete Questions 1–4.

? Pose Example 3 and give students time to discuss their thinking with their partners. You may have students who believe you need to have two variables, *x* and *y*, in the equation for it to be linear. The only input you should offer to student discussions is to ask students what a linear function is.

- **Opposing Views:** Take each function separately and ask two students with opposing views (is a linear function, is not a linear function) to explain the reasoning for their view.

COMMON ERROR In Question 6, students may not recognize that $y = \frac{3x}{5}$ is equivalent to $y = \frac{3}{5}x$.

Graphing Linear Functions

A **solution of a linear equation in two variables** is an ordered pair (x, y) that makes the equation true. The graph of a linear equation in two variables is the set of points (x, y) in a coordinate plane that represents all solutions of the equation. Sometimes the points are distinct, and other times the points are connected.

Core Concept

Discrete and Continuous Domains

A **discrete domain** is a set of input values that consists of only certain numbers in an interval.

Example: Integers from 1 to 5

A **continuous domain** is a set of input values that consists of all numbers in an interval.

Example: All numbers from 1 to 5

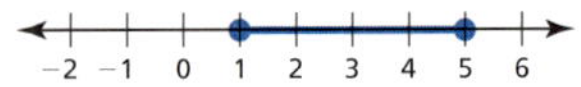

EXAMPLE 4 Graphing Discrete Data

The linear function $y = 15.95x$ represents the cost y (in dollars) of x tickets for a museum. Each customer can buy a maximum of four tickets.

a. Find the domain of the function. Is the domain discrete or continuous? Explain.

b. Graph the function using its domain.

STUDY TIP

The domain of a function depends on the real-life context of the function, not just the equation that represents the function.

SOLUTION

a. You cannot buy part of a ticket, only a certain number of tickets. Because x represents the number of tickets, it must be a whole number. The maximum number of tickets a customer can buy is four.

▶ So, the domain is 0, 1, 2, 3, and 4, and it is discrete.

b. Step 1 Make an input-output table to find the ordered pairs.

Input, x	$15.95x$	Output, y	(x, y)
0	15.95(0)	0	(0, 0)
1	15.95(1)	15.95	(1, 15.95)
2	15.95(2)	31.9	(2, 31.9)
3	15.95(3)	47.85	(3, 47.85)
4	15.95(4)	63.8	(4, 63.8)

Step 2 Plot the ordered pairs. The domain is discrete. So, the graph consists of individual points.

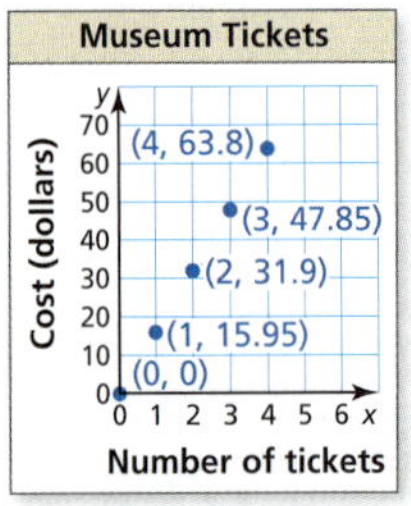

Monitoring Progress

Help in English and Spanish at *BigIdeasMath.com*

8. The linear function $m = 50 - 9d$ represents the amount m (in dollars) of money you have after buying d DVDs. (a) Find the domain of the function. Is the domain discrete or continuous? Explain. (b) Graph the function using its domain.

English Language Learners

Vocabulary

Discuss with students the everyday meanings of the terms *discrete* and *continuous*. Then relate those meanings to how the terms are used in mathematics.

Extra Example 4

The linear function $y = 1.25x$ represents the cost y (in dollars) of x bottles of orange juice. Each customer can buy a maximum of 5 bottles.

a. Find the domain of the function. Is the domain discrete or continuous? Explain. The maximum number of bottles a customer can buy is 5. The domain is 0, 1, 2, 3, 4, and 5. It is discrete because x represents the number of bottles, which must be a whole number.

b. Graph the function using its domain.

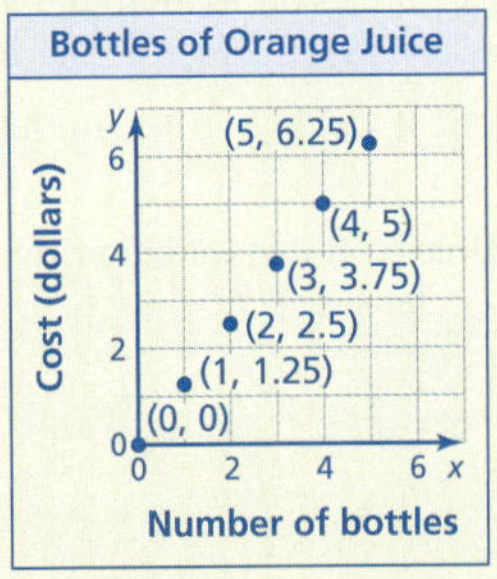

MONITORING PROGRESS ANSWER

8. a. 0, 1, 2, 3, 4, 5; discrete; *Sample answer:* The number of DVDs bought must be a whole number and m must be at least 0.

b.

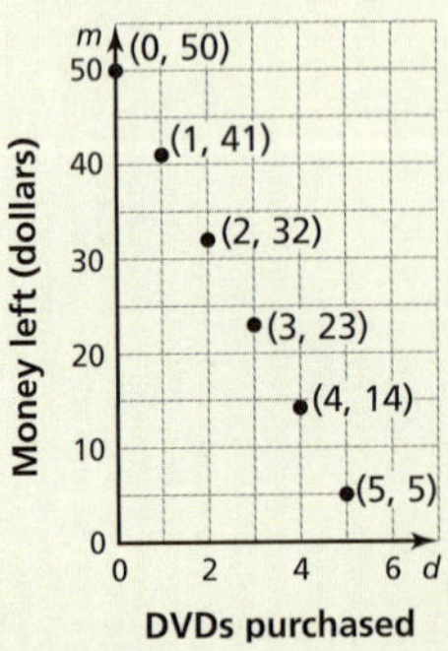

Laurie's Notes Teacher Actions

- Write and discuss what a *solution of a linear equation in two variables* is.
- Write the *Core Concept*.
- **Big Idea:** Continuous does not mean that the graph continues infinitely. Continuous is defined here for an interval.
- **FYI:** Students may be confused by the homophones *discreet* (judicious in one's conduct or speech; prudent) and *discrete* (apart or detached from others; distinct).
- ? Finish Example 4 and ask, "Can you think of other contexts where the domain would be discrete?" Give partners time to talk, and then use *Popsicle Sticks* to solicit answers. *Sample answer:* the number of seats on an airplane

EXAMPLE 5 **Graphing Continuous Data**

A cereal bar contains 130 calories. The number c of calories consumed is a function of the number b of bars eaten.

a. Does this situation represent a linear function? Explain.

b. Find the domain of the function. Is the domain discrete or continuous? Explain.

c. Graph the function using its domain.

STUDY TIP

When the domain of a linear function is not specified or cannot be obtained from a real-life context, it is understood to be all real numbers.

SOLUTION

a. As b increases by 1, c increases by 130. The rate of change is constant.

▶ So, this situation represents a linear function.

b. You can eat part of a cereal bar. The number b of bars eaten can be any value greater than or equal to 0.

▶ So, the domain is $b \geq 0$, and it is continuous.

c. Step 1 Make an input-output table to find ordered pairs.

Input, b	Output, c	(b, c)
0	0	(0, 0)
1	130	(1, 130)
2	260	(2, 260)
3	390	(3, 390)
4	520	(4, 520)

Step 2 Plot the ordered pairs.

Step 3 Draw a line through the points. The line should start at (0, 0) and continue to the right. Use an arrow to indicate that the line continues without end, as shown. The domain is continuous. So, the graph is a line with a domain of $b \geq 0$.

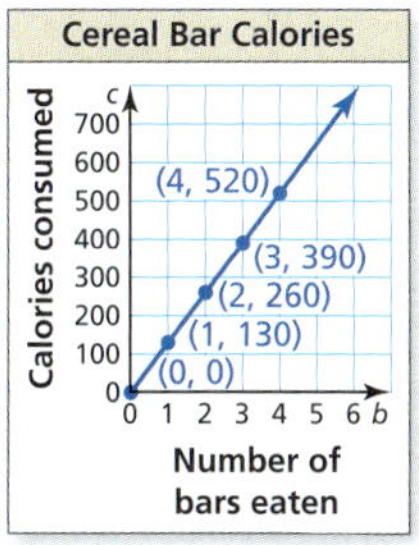

Monitoring Progress

Help in English and Spanish at BigIdeasMath.com

9. Is the domain discrete or continuous? Explain.

Input Number of stories, x	1	2	3
Output Height of building (feet), y	12	24	36

10. A 20-gallon bathtub is draining at a rate of 2.5 gallons per minute. The number g of gallons remaining is a function of the number m of minutes.

a. Does this situation represent a linear function? Explain.

b. Find the domain of the function. Is the domain discrete or continuous? Explain.

c. Graph the function using its domain.

Extra Example 5

A runner's speed is 5 miles per hour. The number m of miles run is a function of the number h of hours spent running.

a. Does this situation represent a linear function? Explain. This situation represents a linear function. As h increases by 1, m increases by 5. The rate of change is constant.

b. Find the domain of the function. Is the domain discrete or continuous? Explain. The domain is $h \geq 0$, and it is continuous. The number of hours spent running can be any value greater than or equal to 0.

c. Graph the function using its domain.

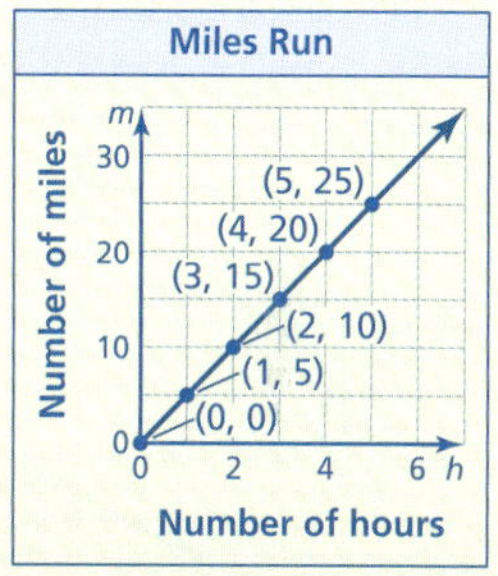

MONITORING PROGRESS ANSWERS

9. discrete; The number of stories must be a whole number.

10. a. yes; As m increases by 1, g decreases by 2.5. The rate of change is constant.

b. $0 \leq m \leq 8$; continuous; A tub can drain for part of a minute, so the time can be any value between 0 and 8.

c.

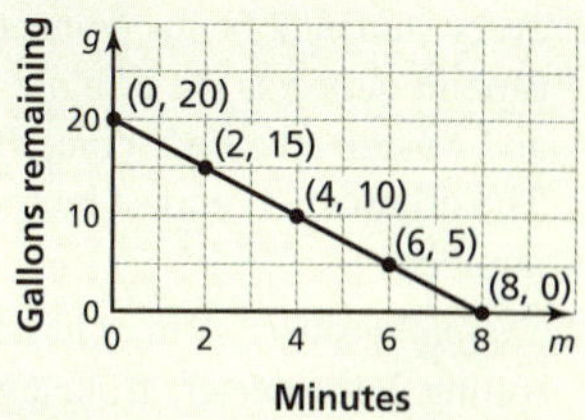

Laurie's Notes Teacher Actions

- Pose the scenario for Example 5, and give students time to consider the questions with their partners. The situation does not suggest that an entire cereal bar must be consumed, only that one bar contains 130 calories.
- **Construct Viable Arguments and Critique the Reasoning of Others:** Solicit students' responses for parts (a) and (b). Students should listen carefully and critique the reasoning of classmates.

? Assessing Question: "Is (0.5, 65) a solution to the equation?" yes "What does the ordered pair represent?" Eating half a cereal bar would be 65 calories.

- **Attend to Precision:** Note the clarity of reading the graph when the axes are labeled.

Writing Real-Life Problems

EXAMPLE 6 **Writing Real-Life Problems**

Write a real-life problem to fit the data shown in each graph. Is the domain of each function *discrete* or *continuous*? Explain.

a.

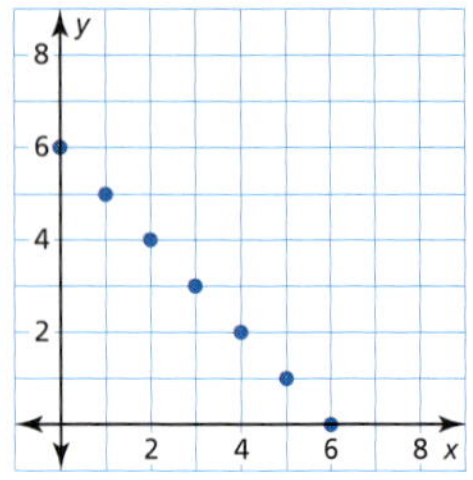

b.

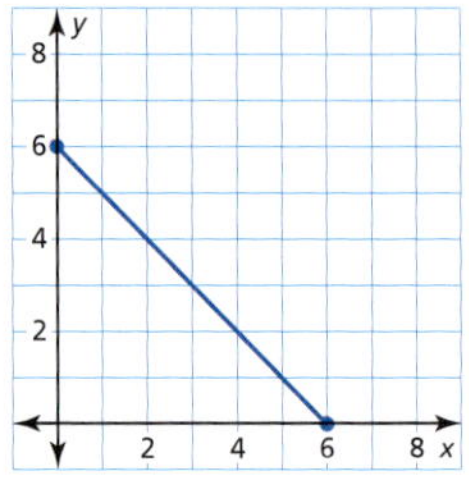

SOLUTION

a. You want to think of a real-life situation in which there are two variables, x and y. Using the graph, notice that the sum of the variables is always 6, and the value of each variable must be a whole number from 0 to 6.

x	0	1	2	3	4	5	6
y	6	5	4	3	2	1	0

Discrete domain

▶ One possibility is two people bidding against each other on six coins at an auction. Each coin will be purchased by one of the two people. Because it is not possible to purchase part of a coin, the domain is discrete.

b. You want to think of a real-life situation in which there are two variables, x and y. Using the graph, notice that the sum of the variables is always 6, and the value of each variable can be any real number from 0 to 6.

$$x + y = 6 \quad \text{or} \quad y = -x + 6$$

Continuous domain

▶ One possibility is two people bidding against each other on 6 ounces of gold dust at an auction. All the dust will be purchased by the two people. Because it is possible to purchase any portion of the dust, the domain is continuous.

Monitoring Progress

Help in English and Spanish at *BigIdeasMath.com*

Write a real-life problem to fit the data shown in the graph. Is the domain of the function *discrete* or *continuous*? Explain.

11.

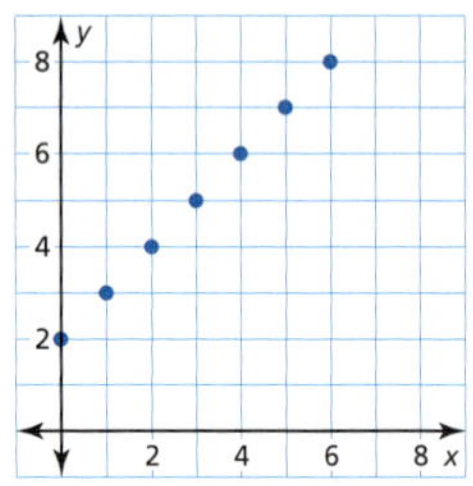

12.

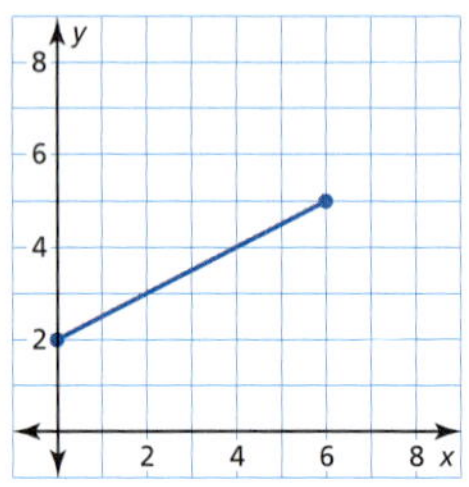

Laurie's Notes Teacher Actions

- **?** Display the two graphs and ask, "Can you describe the graph, explaining how x and y are changing?" x is increasing and y is decreasing. This question gets students used to reading a graph from left to right.
- The sample solutions make use of the pattern $x + y = 6$. Give students time to consider other scenarios. For example, in part (a) you could begin with \$6 and have it cost \$1 for a ticket. As the number of tickets x increases, the amount of money you have y decreases.

Closure

- **Exit Ticket:** Explain how to determine when a graph has a continuous or discrete domain. A continuous domain is a set of input values that consist of all numbers in an interval. A discrete domain is a set of input values that consist of only certain numbers.

Extra Example 6

Write a real-life problem to fit the data shown in each graph. Is the domain of each function *discrete* or *continuous*? Explain.

a.

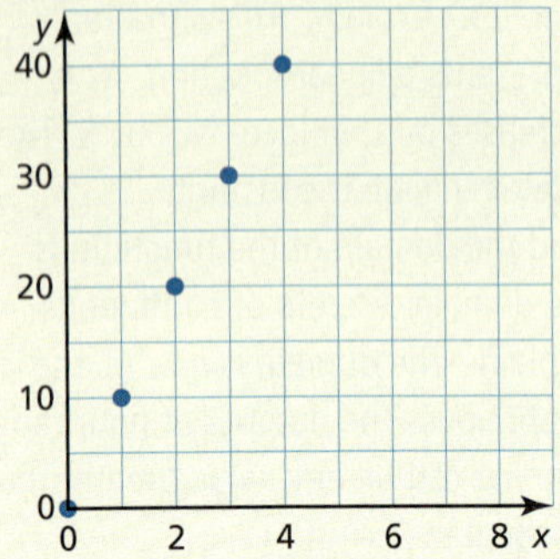

Sample answer: The books on one website each cost \$10. Because it is not possible to buy part of a book, the domain is discrete.

b.

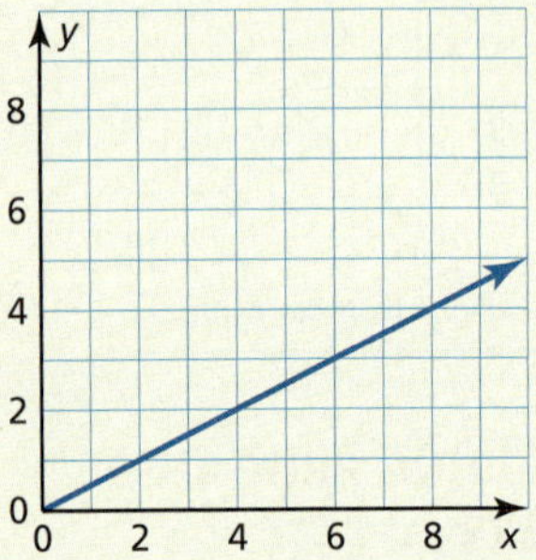

Sample answer: The frozen yogurt at one store costs \$0.50 for each ounce. Because it is possible to buy any portion of yogurt, the domain is continuous.

MONITORING PROGRESS ANSWERS

11. *Sample answer:* A teacher gives every student 2 bonus points on a 6-point quiz. x is the original score and y is the recorded score; discrete; The original score must be a whole number.

12. *Sample answer:* A marble starts rolling 2 centimeters from a wall and rolls at a constant rate of 0.5 centimeters per second for 6 seconds; continuous; The rolling time can be any value between 0 and 6.

3.2 Exercises

Dynamic Solutions available at *BigIdeasMath.com*

Vocabulary and Core Concept Check

1. **COMPLETE THE SENTENCE** A linear equation in two variables is an equation that can be written in the form ________, where m and b are constants.

2. **VOCABULARY** Compare linear functions and nonlinear functions.

3. **VOCABULARY** Compare discrete domains and continuous domains.

4. **WRITING** How can you tell whether a graph shows a discrete domain or a continuous domain?

Monitoring Progress and Modeling with Mathematics

In Exercises 5–10, determine whether the graph represents a *linear* or *nonlinear* function. Explain. *(See Example 1.)*

5. 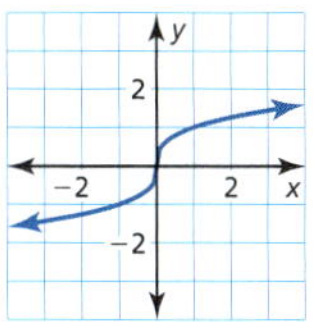

6. 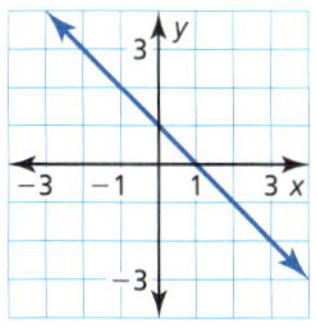

7. 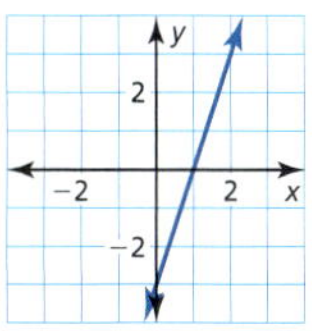

8. 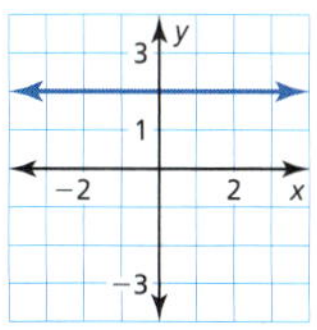

9. 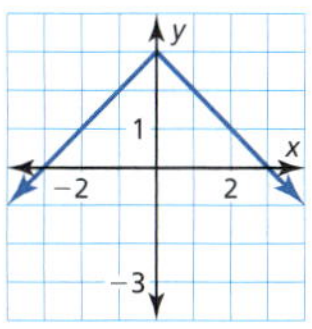

10. 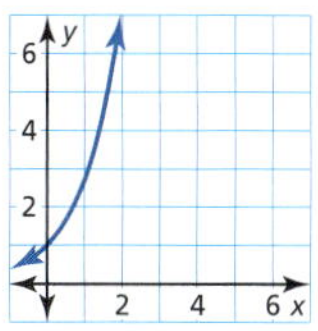

In Exercises 11–14, determine whether the table represents a *linear* or *nonlinear* function. Explain. *(See Example 2.)*

11.

x	1	2	3	4
y	5	10	15	20

12.

x	5	7	9	11
y	−9	−3	−1	3

13.

x	4	8	12	16
y	16	12	7	1

14.

x	−1	0	1	2
y	35	20	5	−10

ERROR ANALYSIS In Exercises 15 and 16, describe and correct the error in determining whether the table or graph represents a linear function.

15. ✗

+2 +2 +2

x	2	4	6	8
y	4	16	64	256

×4 ×4 ×4

As x increases by 2, y increases by a constant factor of 4. So, the function is linear.

16. ✗

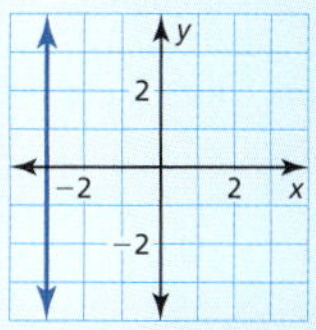

The graph is a line. So, the graph represents a linear function.

Assignment Guide and Homework Check

ASSIGNMENT

Basic: 1–4, 5–19 odd, 27–39 odd, 44, 52, 55–61

Average: 1–4, 6–34 even, 35–38, 40–46 even, 52, 55–61

Advanced: 1–4, 8–24 even, 25, 26–54 even, 55–61

HOMEWORK CHECK

Basic: 5, 11, 17, 35, 39

Average: 6, 12, 20, 36, 40

Advanced: 10, 14, 22, 36, 42

ANSWERS

1. $y = mx + b$
2. A linear function has a constant rate of change and a graph that is a nonvertical line. A nonlinear function does not have a constant rate of change and has a graph that is not a line.
3. Discrete domains consist of only certain numbers in an interval. Continuous domains consist of all numbers in an interval.
4. A graph with a discrete domain will consist of individual points, a graph with a continuous domain will show a line or curve.
5. nonlinear; The graph is not a line.
6. linear; The graph is a line.
7. linear; The graph is a line.
8. linear; The graph is a line.
9. nonlinear; The graph is not a line.
10. nonlinear; The graph is not a line.
11. linear; As x increases by 1, y increases by 5. The rate of change is constant.
12. nonlinear; As x increases by 2, y increases by different amounts. The rate of change is not constant.
13. nonlinear; As x increases by 4, y decreases by different amounts. The rate of change is not constant.
14. linear; As x increases by 1, y decreases by 15. The rate of change is constant.
15. The increase in y needs to be done by adding or subtracting the same amount to be linear, not multiplying; As x increases by 2, y increases by different amounts. The rate of change is not constant. So, the function is nonlinear.
16. A vertical line is not a function; The graph is a vertical line. So, the graph does not represent a function.

Dynamic Teaching Tools

- Dynamic Assessment & Progress Monitoring Tool
- Interactive Whiteboard Lesson Library
- Dynamic Classroom with Dynamic Investigations

ANSWERS

17. nonlinear; It cannot be rewritten in the form $y = mx + b$.

18. linear; It can be rewritten as $y = -3x + 7$.

19. linear; It can be rewritten as $y = -1x + 2$.

20. nonlinear; It cannot be rewritten in the form $y = mx + b$.

21. linear; It can be rewritten as $y = 18x + 12$.

22. linear; It can be rewritten as $y = \frac{9}{5}x + 0$.

23. linear; It can be rewritten as $y = 9x - 13$.

24. nonlinear; It cannot be rewritten in the form $y = mx + b$.

25. A, C, F; None of these can be rewritten in the form $y = mx + b$.

26. 2, 5, 8

27. 2, 4, 6; discrete; The graph consists of individual points.

28. $0 \le x \le 7$; continuous; The graph is a line segment.

29. discrete; The number of bags must be a whole number.

30. continuous; The number of years can be any value greater than or equal to 0.

31. continuous; The time can be any value greater than or equal to 0.

32. discrete; The number of relay teams must be a whole number.

33. There is no point with an x-value of 2.5; 2.5 is not in the domain.

34. The graph shows a continuous domain; The graph ends at $x = 6$, so the domain is $0 \le x \le 6$.

35. **a.** 0, 1, 2, 3, 4, 5, 6; discrete; The number of books must be a whole number.

b.

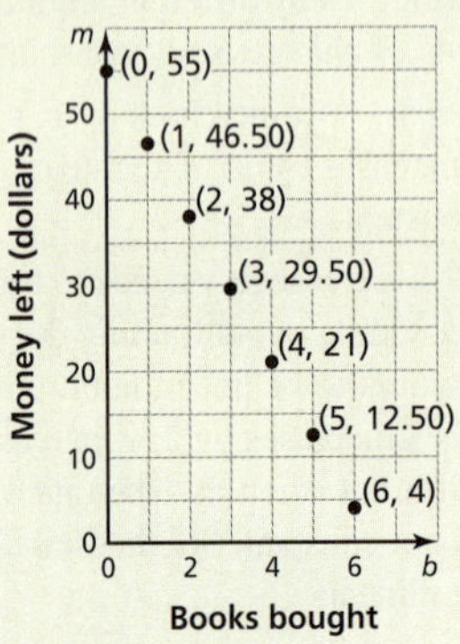

In Exercises 17–24, determine whether the equation represents a *linear* or *nonlinear* function. Explain. *(See Example 3.)*

17. $y = x^2 + 13$

18. $y = 7 - 3x$

19. $y = \sqrt[3]{8} - x$

20. $y = 4x(8 - x)$

21. $2 + \frac{1}{6}y = 3x + 4$

22. $y - x = 2x - \frac{2}{3}y$

23. $18x - 2y = 26$

24. $2x + 3y = 9xy$

25. CLASSIFYING FUNCTIONS Which of the following equations *do not* represent linear functions? Explain.

Ⓐ $12 = 2x^2 + 4y^2$ Ⓑ $y - x + 3 = x$

Ⓒ $x = 8$ Ⓓ $x = 9 - \frac{3}{4}y$

Ⓔ $y = \frac{5x}{11}$ Ⓕ $y = \sqrt{x} + 3$

26. USING STRUCTURE Fill in the table so it represents a linear function.

x	5	10	15	20	25
y	−1				11

In Exercises 27 and 28, find the domain of the function represented by the graph. Determine whether the domain is *discrete* or *continuous*. Explain.

27.

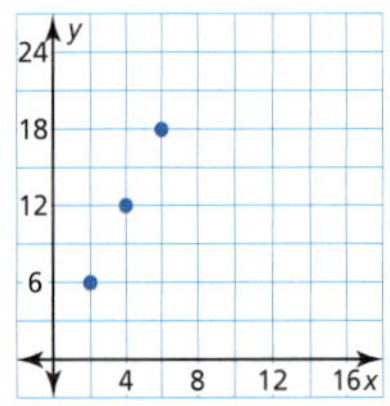

28.

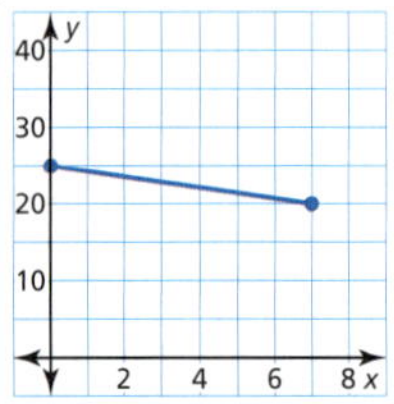

In Exercises 29–32, determine whether the domain is *discrete* or *continuous*. Explain.

29.

Input Bags, *x*	2	4	6
Output Marbles, *y*	20	40	60

30.

Input Years, *x*	1	2	3
Output Height of tree (feet), *y*	6	9	12

31.

Input Time (hours), *x*	3	6	9
Output Distance (miles), *y*	150	300	450

32.

Input Relay teams, *x*	0	1	2
Output Athletes, *y*	0	4	8

ERROR ANALYSIS In Exercises 33 and 34, describe and correct the error in the statement about the domain.

33.

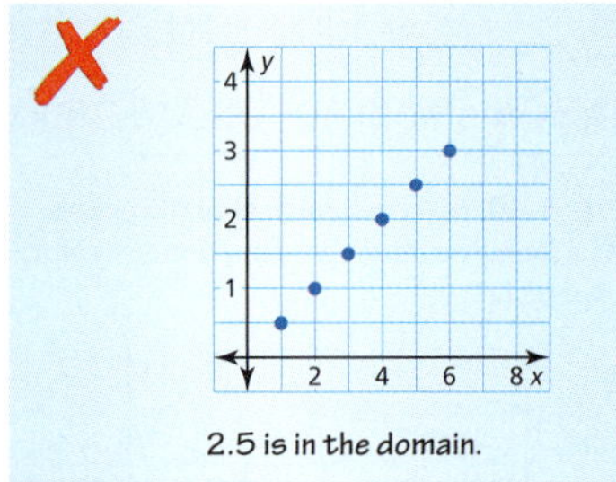

34.

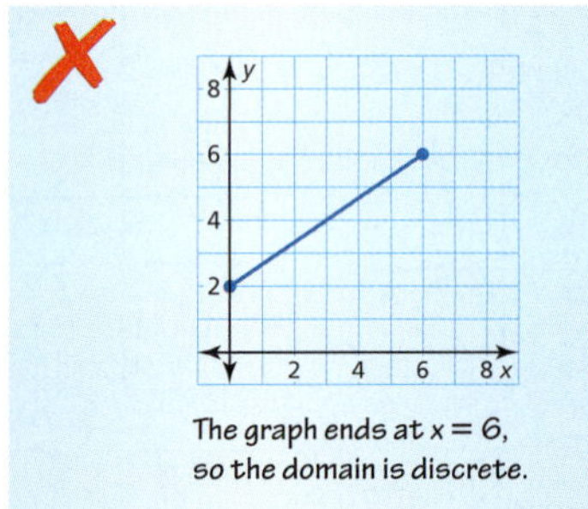

35. MODELING WITH MATHEMATICS The linear function $m = 55 - 8.5b$ represents the amount m (in dollars) of money that you have after buying b books. *(See Example 4.)*

a. Find the domain of the function. Is the domain discrete or continuous? Explain.

b. Graph the function using its domain.

36. **MODELING WITH MATHEMATICS** The number y of calories burned after x hours of rock climbing is represented by the linear function $y = 650x$.

a. Find the domain of the function. Is the domain discrete or continuous? Explain.

b. Graph the function using its domain.

37. **MODELING WITH MATHEMATICS** You are researching the speed of sound waves in dry air at 86°F. The table shows the distances d (in miles) sound waves travel in t seconds. *(See Example 5.)*

Time (seconds), t	Distance (miles), d
2	0.434
4	0.868
6	1.302
8	1.736
10	2.170

a. Does this situation represent a linear function? Explain.

b. Find the domain of the function. Is the domain discrete or continuous? Explain.

c. Graph the function using its domain.

38. **MODELING WITH MATHEMATICS** The function $y = 30 + 5x$ represents the cost y (in dollars) of having your dog groomed and buying x extra services.

a. Does this situation represent a linear function? Explain.

b. Find the domain of the function. Is the domain discrete or continuous? Explain.

c. Graph the function using its domain.

WRITING In Exercises 39–42, write a real-life problem to fit the data shown in the graph. Determine whether the domain of the function is *discrete* or *continuous*. Explain. *(See Example 6.)*

39.

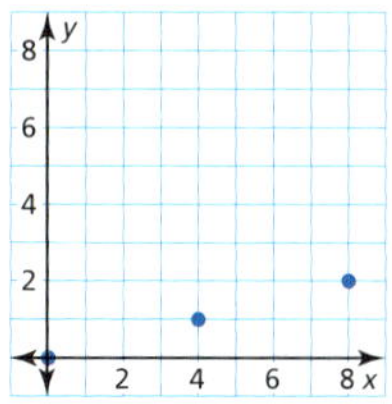

40.

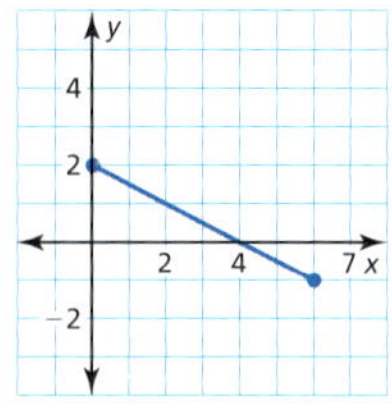

41.

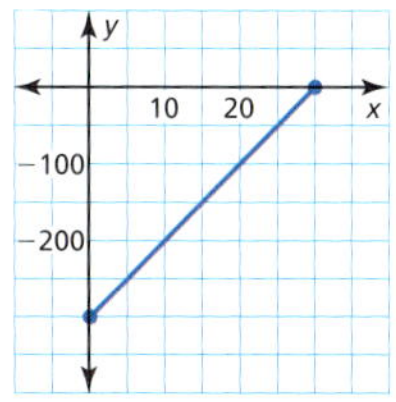

42.

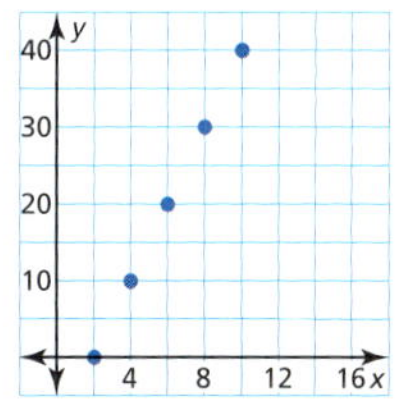

43. **USING STRUCTURE** The table shows your earnings y (in dollars) for working x hours.

a. What is the missing y-value that makes the table represent a linear function?

b. What is your hourly pay rate?

Time (hours), x	Earnings (dollars), y
4	40.80
5	
6	61.20
7	71.40

44. **MAKING AN ARGUMENT** The linear function $d = 50t$ represents the distance d (in miles) Car A is from a car rental store after t hours. The table shows the distances Car B is from the rental store.

Time (hours), t	Distance (miles), d
1	60
3	180
5	310

a. Does the table represent a linear or nonlinear function? Explain.

b. Your friend claims Car B is moving at a faster rate. Is your friend correct? Explain.

ANSWERS

36. a. $x \geq 0$; continuous; The time can be any number of hours greater than or equal to 0.

b. 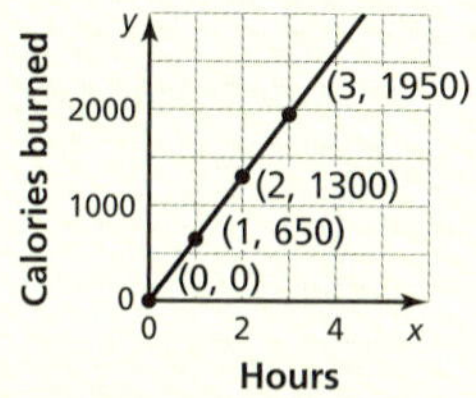

37. a. yes; As t increases by 2, d increases by 0.434. The rate of change is constant.

b. $t \geq 0$; continuous; The time can be any value greater than or equal to 0.

c. 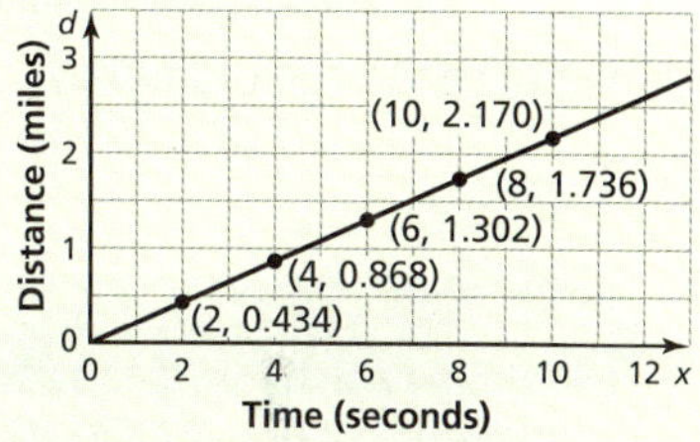

38. a. yes; The equation can be rewritten as $y = 5x + 30$.

b. 0, 1, 2, 3, 4, 5; discrete; The number of extra services must be a whole number.

c. 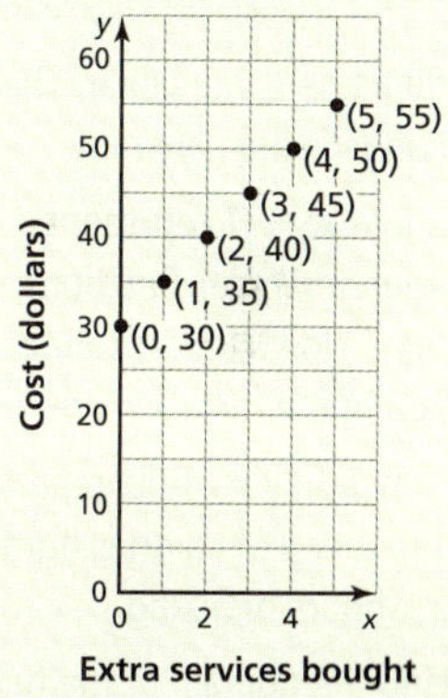

39. *Sample answer:* The number of hours on a parking meter is a function of the number of tokens used, 4 tokens for 1 hour and a maximum time of 2 hours; discrete; The number of tokens used must be 0, 4, or 8.

40. *Sample answer:* The temperature of a substance in degrees Celsius as a function of time in minutes; continuous; The time can be any value between 0 and 6 minutes.

41. *Sample answer:* The depth (in feet) of a scuba diver returning to the surface of an ocean as a function of the time; continuous; The time can be any value from 0 to 30.

42. *Sample answer:* The total discount on gloves as a function of the number of gloves purchased on sale, where a discount of \$10 per pair is applied after paying full price for the first pair, with a maximum of 5 pairs; discrete; The number of gloves bought must be an even number no more than 10.

43. a. 51.00

b. \$10.20

44. a. nonlinear; As t increases by 2, d increases by different amounts. The rate of change is not constant.

b. yes; Using $d = 50t$, Car A has traveled a shorter distance for each of the times given in the chart for Car B.

ANSWERS

45. nonlinear; $V = 9s^2$ cannot be written in linear form.

46. linear; The formula can be written in the form $V = 6b + 0$.

47. linear; The formula can be written in the form $V = (4\pi)h + 0$.

48. nonlinear; $V = 5\pi r^2$ cannot be written in linear form.

49. **a.** the total gallons of water in one jug of each type; continuous

b. the total number of jugs of both types; discrete

c. the total gallons of water in all the jugs of the first type; continuous

d. the total gallons of water in all the jugs of both types; continuous

50–61. See Additional Answers.

Mini-Assessment

1. Does the table represent a *linear* or *nonlinear* function? Explain.

x	3	5	7	9
y	−6	0	4	2

The function is nonlinear. The rate of change is not constant.

2. Does $y = x^2 + 1$ represent a *linear* or *nonlinear* function? Explain. Nonlinear; It cannot be written in the form $y = mx + b$.

3. The function $y = 408x$ represents the number y of calories burned after x hours of hiking.

   a. Find the domain of the function. Is the domain discrete or continuous? Explain. The domain is $x \geq 0$. It is continuous because the number of hours spent hiking can be any value greater than or equal to 0.

   b. Graph the function.

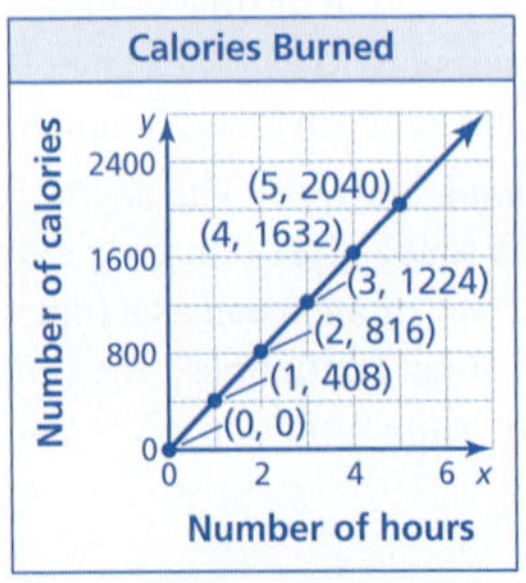

MATHEMATICAL CONNECTIONS **In Exercises 45–48, tell whether the volume of the solid is a linear or nonlinear function of the missing dimension(s). Explain.**

45.

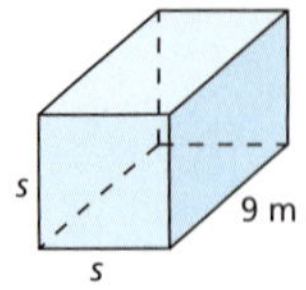

46.

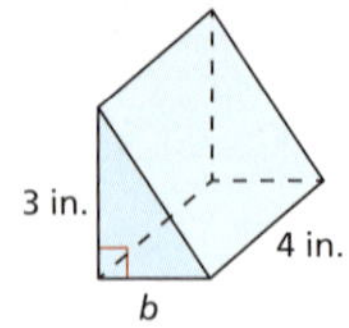

47.

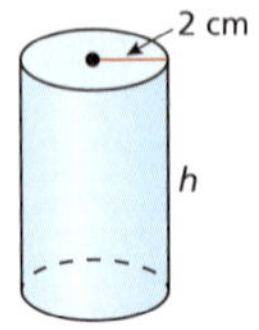

48.

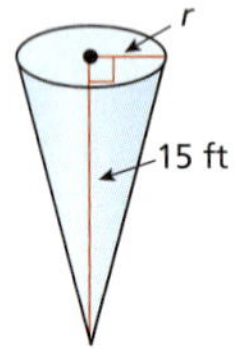

49. **REASONING** A water company fills two different-sized jugs. The first jug can hold x gallons of water. The second jug can hold y gallons of water. The company fills A jugs of the first size and B jugs of the second size. What does each expression represent? Does each expression represent a set of discrete or continuous values?

a. $x + y$

b. $A + B$

c. Ax

d. $Ax + By$

50. **THOUGHT PROVOKING** You go to a farmer's market to buy tomatoes. Graph a function that represents the cost of buying tomatoes. Explain your reasoning.

51. **CLASSIFYING A FUNCTION** Is the function represented by the ordered pairs linear or nonlinear? Explain your reasoning.

(0, 2), (3, 14), (5, 22), (9, 38), (11, 46)

52. **HOW DO YOU SEE IT?** You and your friend go running. The graph shows the distances you and your friend run.

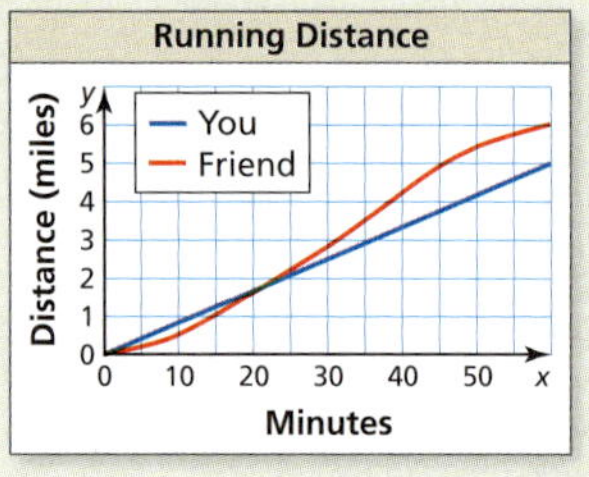

a. Describe your run and your friend's run. Who runs at a constant rate? How do you know? Why might a person not run at a constant rate?

b. Find the domain of each function. Describe the domains using the context of the problem.

WRITING **In Exercises 53 and 54, describe a real-life situation for the constraints.**

53. The function has at least one negative number in the domain. The domain is continuous.

54. The function gives at least one negative number as an output. The domain is discrete.

Maintaining Mathematical Proficiency

Reviewing what you learned in previous grades and lessons

Tell whether x and y show direct variation. Explain your reasoning. *(Skills Review Handbook)*

55.

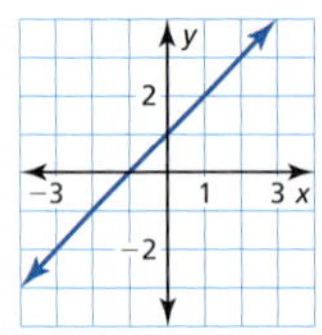

56.

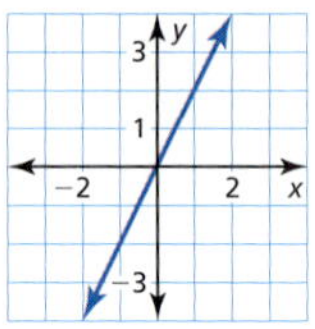

57.

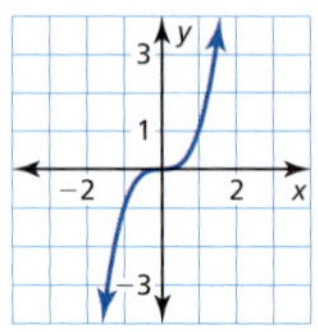

Evaluate the expression when $x = 2$. *(Skills Review Handbook)*

58. $6x + 8$ **59.** $10 - 2x + 8$ **60.** $4(x + 2 - 5x)$ **61.** $\frac{x}{2} + 5x - 7$

If students need help...	*If students got it...*
Resources by Chapter • Practice A and Practice B • Puzzle Time	Resources by Chapter • Enrichment and Extension • Cumulative Review
Student Journal • Practice	Start the *next* Section
Differentiating the Lesson Skills Review Handbook	

Laurie's Notes

Overview of Section 3.3

Introduction

- While students have worked with functions in the prior lesson and prior year, function notation will likely be new to them. Notation can be confusing to students, so it is important to spend time on the explanation. I like to do the *Motivate* on the next page as a way to get students comfortable with naming a function by a single letter.
- Exploration 1 allows students to work with two linear functions that should feel comfortable to them. Using the notation of $f(x) = 2x - 3$ to write $y = 2x - 3$ does not change the fact that it is a linear function with a slope of 2 and y-intercept of -3. That part should be secure with students.
- Exploration 2 is about evaluating a function, just as Example 1 in the formal lesson is about evaluating a function. The notation (x, y) is replaced with $(x, f(x))$. Again, this notation will need to be discussed and explained.
- This section also reviews the prior skills of solving an equation in two variables and graphing a linear function by plotting ordered pairs. The new content that frames this lesson is the use of function notation and the language associated with it.

Resources

- Graphing calculators could be used in Example 4 of the formal lesson.

Common Misconceptions

- Function notation, when first introduced, is confusing to students. Make clear that the notation $f(x)$ is read as "the value of f at x" or "f of x." It does not mean "f times x."
- It is not surprising that students believe that $f(x)$ is a multiplication problem when they see the parentheses. Ask them to think of other contexts in math where they have seen parentheses that do not mean multiplication.

  $8 + (-3) = 5$ $\quad\quad$ $\text{GCF}(8, 12) = 4$

  $\text{LCM}(8, 12) = 24$ $\quad\quad$ $P(\text{tossing heads}) = \frac{1}{2}$

Formative Assessment Tips

- **Writing Prompt:** This technique asks students to give feedback at the end of the lesson, activity, or learning experience. The *Writing Prompt* allows you to collect feedback in a short period of time (two minutes) on student learning.
- Used at the end of the class, a *Writing Prompt* is similar to an *Exit Ticket*. The written responses give students time to reflect on their learning, how well they believe they understand the concept or skill, and where they are still uncertain.
- It is important on the following day to share responses with students and let them know how the lesson has been adjusted to reflect their responses. Students need to know that you value their responses and that instruction is modified accordingly. Students will take the writing prompts more seriously when they are valued and used.

Pacing Suggestion

- Exploration 2 is similar to Example 1 in the formal lesson. If you use the *Motivate* and both explorations, you might omit Example 1 in the formal lesson.

Dynamic Teaching Tools

- Dynamic Assessment & Progress Monitoring Tool
- Lesson Planning Tool
- Interactive Whiteboard Lesson Library
- Dynamic Classroom with Dynamic Investigations

What Your Students Will Learn

- Use function notation to evaluate functions for given domain values.
- Interpret the meanings of statements written in function notation.
- Solve for the independent variable given the function value.
- Graph linear functions in function notation.

Laurie's Notes

Exploration

Motivate

- **Human Graph:** Make a coordinate grid on the classroom floor or foyer of the school using masking tape. Use a dark marker to scale the axes $[-6, 6]$ in each direction.
- Form two groups of about five students each. Call the groups f and g. Assign the x-values -1, 0, 1, 2, and 3 to each group, one x-value to each student. (Expand the set of x-values for a group of more than five.)
- Have the students in group f plot their respective points for the linear function $y = x - 1$, followed by group g plotting the points for the linear function $y = x - 3$.
- ? Probe students' recall. "What is the slope and y-intercept of f?" 1, -1 "What is the domain and range of f?" domain: $\{-1, 0, 1, 2, 3\}$, range: $\{-2, -1, 0, 1, 2\}$ "What is the slope and y-intercept of g?" 1, -3 "What is the domain and range of g?" domain: $\{-1, 0, 1, 2, 3\}$, range: $\{-4, -3, -2, -1, 0\}$

Discuss

- Introduce function notation. Connect the notation to the groups who plotted themselves as ordered pairs. In function notation, the original equations for each group are $f(x) = x - 1$ and $g(x) = x - 3$.
- Discuss how to read function notation.

Exploration 1

- Students should be able to match the linear functions using the slopes and y-intercepts as clues. To match the quadratic functions, students can consider the y-intercepts.
- **Popsicle Sticks:** Draw *Popsicle Sticks* and ask students how they made the matches.
- ? **Assessing Question:** "How did you know that graphs A and C did not match $f(x)$ or $g(x)$?" The graphs were not lines.
- If time permits, evaluate the function $j(x) = 2x^2 - 3$ for several values of x to confirm that the ordered pairs are on the graph and to give practice evaluating functions written in function notation.

Exploration 2

- Discuss how the ordered pairs are related to the form (x, y). The ordered pair $(2, f(2))$ is the same as $(2, y)$, where y is the value of the function when $x = 2$. So, the notation $f(2)$ represents the value of the function f when $x = 2$. You can read $f(2)$ as "f of 2."
- Students may be confused by the notation. Continue to remind them of what $f(-1)$ means, what $f(0)$ means, and so on.

Communicate Your Answer

- Student responses to Question 3 will inform your instruction during the formal lesson.

Connecting to Next Step

- If you feel students are comfortable with evaluating a function written in function notation, assign items from Exercises 3–10 on page 125. In beginning the formal lesson, you might omit Example 1.

3.3 Function Notation

Essential Question

How can you use function notation to represent a function?

The notation $f(x)$, called **function notation**, is another name for y. This notation is read as "the value of f at x" or "f of x." The parentheses do not imply multiplication. You can use letters other than f to name a function. The letters g, h, j, and k are often used to name functions.

EXPLORATION 1 Matching Functions with Their Graphs

Work with a partner. Match each function with its graph.

a. $f(x) = 2x - 3$

b. $g(x) = -x + 2$

c. $h(x) = x^2 - 1$

d. $j(x) = 2x^2 - 3$

ATTENDING TO PRECISION

To be proficient in math, you need to use clear definitions and state the meanings of the symbols you use.

A.
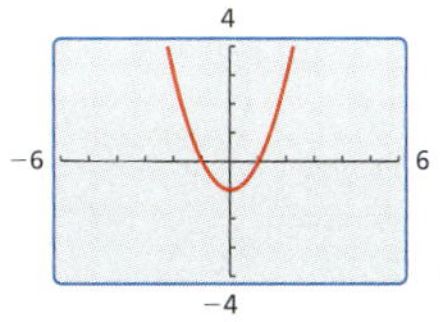

B.
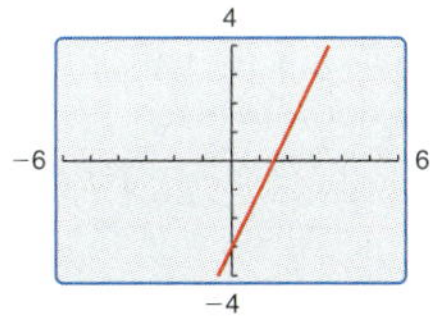

C.
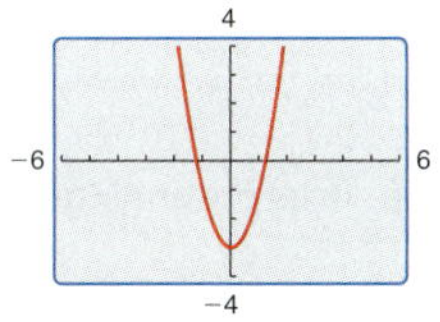

D.
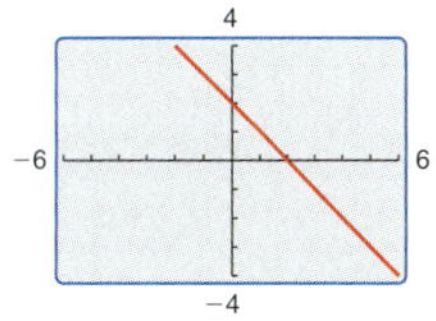

EXPLORATION 2 Evaluating a Function

Work with a partner. Consider the function

$f(x) = -x + 3.$

Locate the points $(x, f(x))$ on the graph. Explain how you found each point.

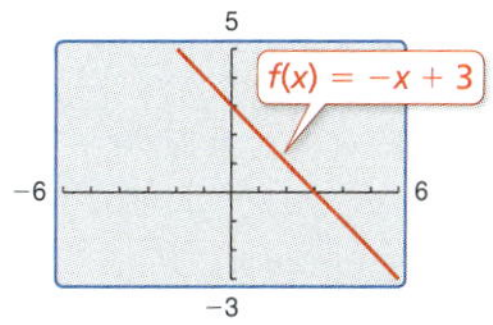

a. $(-1, f(-1))$

b. $(0, f(0))$

c. $(1, f(1))$

d. $(2, f(2))$

Communicate Your Answer

3. How can you use function notation to represent a function? How are standard notation and function notation similar? How are they different?

Standard Notation	*Function Notation*
$y = 2x + 5$	$f(x) = 2x + 5$

Dynamic Teaching Tools

- Dynamic Assessment & Progress Monitoring Tool
- Lesson Planning Tool
- Interactive Whiteboard Lesson Library
- Dynamic Classroom with Dynamic Investigations

ANSWERS

1. a. B

b. D

c. A

d. C

2. a. $(-1, 4)$

b. $(0, 3)$

c. $(1, 2)$

d. $(2, 1)$

Locate the x-coordinate, trace it up to the line, and then trace over to the y-axis to determine the y-coordinate.

3. Replace "y" with "$f(x)$" in the equation; They both give the rule for determining the output; Standard notation uses "y" to indicate the output, function notation uses "$f(x)$" to indicate the output.

3.3 Lesson

What You Will Learn

- Use function notation to evaluate and interpret functions.
- Use function notation to solve and graph functions.
- Solve real-life problems using function notation.

Core Vocabulary

function notation, *p. 122*

Previous
linear function
quadrant

Using Function Notation to Evaluate and Interpret

You know that a linear function can be written in the form $y = mx + b$. By naming a linear function f, you can also write the function using **function notation**.

$f(x) = mx + b$ — Function notation

The notation $f(x)$ is another name for y. If f is a function, and x is in its domain, then $f(x)$ represents the output of f corresponding to the input x. You can use letters other than f to name a function, such as g or h.

READING

The notation $f(x)$ is read as "the value of f at x" or "f of x." It does not mean "f times x."

EXAMPLE 1 Evaluating a Function

Evaluate $f(x) = -4x + 7$ when $x = 2$ and $x = -2$.

SOLUTION

$f(x) = -4x + 7$	Write the function.	$f(x) = -4x + 7$
$f(2) = -4(2) + 7$	Substitute for x.	$f(-2) = -4(-2) + 7$
$= -8 + 7$	Multiply.	$= 8 + 7$
$= -1$	Add.	$= 15$

▶ When $x = 2$, $f(x) = -1$, and when $x = -2$, $f(x) = 15$.

EXAMPLE 2 Interpreting Function Notation

Let $f(t)$ be the outside temperature (°F) t hours after 6 A.M. Explain the meaning of each statement.

a. $f(0) = 58$ **b.** $f(6) = n$ **c.** $f(3) < f(9)$

SOLUTION

a. The initial value of the function is 58. So, the temperature at 6 A.M. is 58°F.

b. The output of f when $t = 6$ is n. So, the temperature at noon (6 hours after 6 A.M.) is n°F.

c. The output of f when $t = 3$ is less than the output of f when $t = 9$. So, the temperature at 9 A.M. (3 hours after 6 A.M.) is less than the temperature at 3 P.M. (9 hours after 6 A.M.).

Monitoring Progress

Help in English and Spanish at *BigIdeasMath.com*

Evaluate the function when $x = -4$, 0, and 3.

1. $f(x) = 2x - 5$ **2.** $g(x) = -x - 1$

3. WHAT IF? In Example 2, let $f(t)$ be the outside temperature (°F) t hours after 9 A.M. Explain the meaning of each statement.

a. $f(4) = 75$ **b.** $f(m) = 70$ **c.** $f(2) = f(9)$ **d.** $f(6) > f(0)$

English Language Learners

Comprehension

Make sure that students understand that $f(x)$ is another name for y. Some students may confuse the function notation with the use of parentheses as a multiplication symbol because they know that the expression $2(x)$ means 2 times x.

Extra Example 1

Evaluate $f(x) = 2x - 7$ when $x = 4$ and $x = -4$. When $x = 4$, $f(x) = 1$ and when $x = -4$, $f(x) = -15$.

Extra Example 2

Let $f(t)$ be the outside temperature (°F) t hours after 10 A.M. Explain the meaning of each statement.

a. $f(0) = 75$
The temperature at 10 A.M. is 75°F.

b. $f(3) = n$
The temperature at 1 P.M. (3 hours after 10 A.M.) is n°F.

c. $f(1) > f(8)$
The temperature at 11 A.M. (1 hour after 10 A.M.) is greater than the temperature at 6 P.M. (8 hours after 10 A.M.).

MONITORING PROGRESS ANSWERS

1. -13; -5; 1
2. 3; -1; -4
3. a. The temperature is 75°F at 1 P.M.
   b. The temperature is 70°F m hours after 9 A.M.
   c. The temperature at 11 A.M. is the same as the temperature at 6 P.M.
   d. The temperature at 3 P.M. is greater than the temperature at 9 A.M.

Laurie's Notes — Teacher Actions

- If the explorations were not used, discuss *function notation*.
- ? "What does it mean to evaluate an expression?" Substitute a value for the variable and perform the computations to find the value of the expression.
- ? "What does it mean to evaluate an equation?" Substitute a value for the input variable and perform the computations to find the value of the output variable.
- **Teaching Tip:** Write $f(x) = -4x + 7$ and $y = -4x + 7$ separated by a vertical line. Evaluate each when $x = 2$. Students will feel comfortable with the "y" notation. Seeing the parallel work with function notation helps them understand that the output is the same. The language and notation have changed, the process has not.
- **Think-Pair-Share:** Use *Think-Pair-Share* to assess student understanding of function notation.

Using Function Notation to Solve and Graph

EXAMPLE 3 **Solving for the Independent Variable**

For $h(x) = \frac{2}{3}x - 5$, find the value of x for which $h(x) = -7$.

SOLUTION

$h(x) = \frac{2}{3}x - 5$	Write the function.
$-7 = \frac{2}{3}x - 5$	Substitute −7 for $h(x)$.
$+5 \qquad +5$	Add 5 to each side.
$-2 = \frac{2}{3}x$	Simplify.
$\frac{3}{2} \cdot (-2) = \frac{3}{2} \cdot \frac{2}{3}x$	Multiply each side by $\frac{3}{2}$.
$-3 = x$	Simplify.

▶ When $x = -3$, $h(x) = -7$.

EXAMPLE 4 **Graphing a Linear Function in Function Notation**

Graph $f(x) = 2x + 5$.

SOLUTION

Step 1 Make an input-output table to find ordered pairs.

x	−2	−1	0	1	2
$f(x)$	1	3	5	7	9

Step 2 Plot the ordered pairs.

Step 3 Draw a line through the points.

STUDY TIP
The graph of $y = f(x)$ consists of the points $(x, f(x))$.

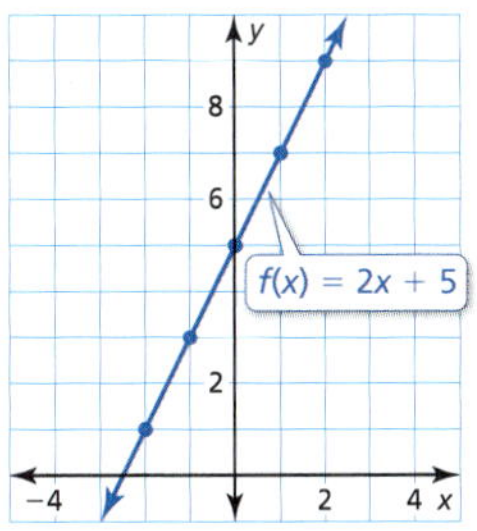

Monitoring Progress

Help in English and Spanish at *BigIdeasMath.com*

Find the value of x so that the function has the given value.

4. $f(x) = 6x + 9$; $f(x) = 21$ **5.** $g(x) = -\frac{1}{2}x + 3$; $g(x) = -1$

Graph the linear function.

6. $f(x) = 3x - 2$ **7.** $g(x) = -x + 4$ **8.** $h(x) = -\frac{3}{4}x - 1$

Extra Example 3

For $g(x) = \frac{1}{3}x - 2$, find the value of x for which $g(x) = -4$.
When $x = -6$, $g(x) = -4$.

Extra Example 4

Graph $f(x) = 3 - x$.

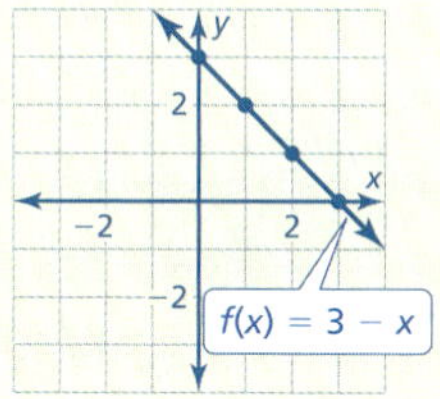

MONITORING PROGRESS ANSWERS

4. $x = 2$

5. $x = 8$

6.

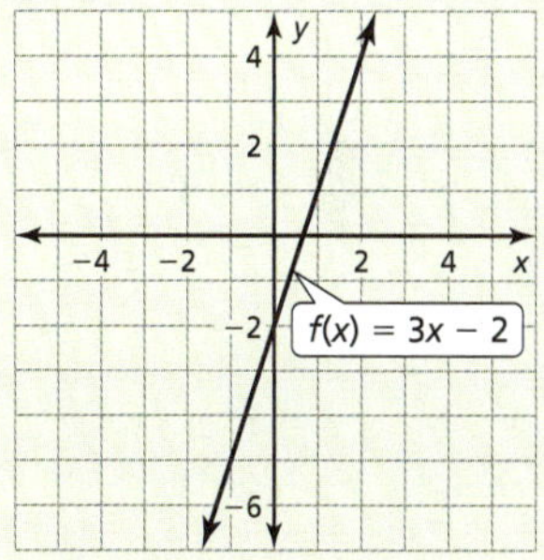

7.

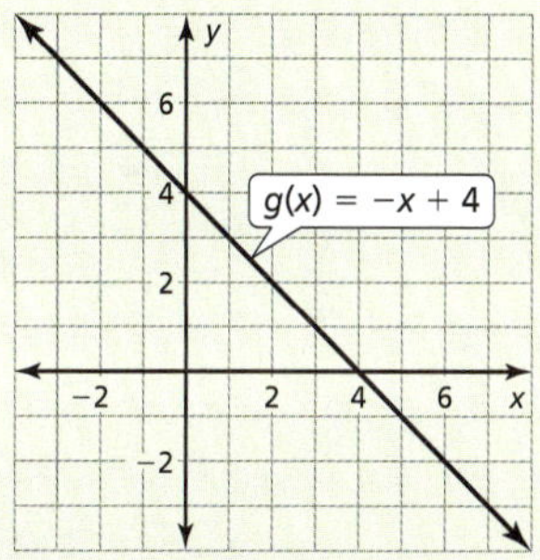

8. See Additional Answers.

Laurie's Notes Teacher Actions

- **Teaching Tip:** Before presenting Example 3, write $h(x) = \frac{2}{3}x - 5$ on the board. Ask students to evaluate the function when $x = 9$. This is asking what number in the range pairs with 9 in the domain. In Example 3, the question is asking what domain value pairs with −7 in the range.
- **Thumbs Up:** Write Example 3. Ask students to show a *Thumbs Up* indicating whether they understand what the question is asking. Assess to what extent the notation is a barrier to students' understanding of the question.
- **Look For and Express Regularity in Repeated Reasoning:** Make the table of values in Example 4. Ask students to describe the pattern in the data. As the input increases by 1, the output increases by 2.

? Attend to Precision: To practice the language of functions you can ask, "What is $f(1)$?" 7 "What is the value of the function at $x = -2$?" 1

Extra Example 5

Use the graph in Example 5 for the first flight of the helicopter. Let $f(x) = 300 - 75x$ represent the second flight, where $f(x)$ is the number of miles the helicopter is from its destination after x hours. Which flight takes less time? Explain. The first flight takes less time. For the first flight, when $f(x) = 0$, $x = 3$. For the second flight, when $f(x) = 0$, $x = 4$.

MONITORING PROGRESS ANSWER

9. the first flight; The graph of the second flight shows that when $f(x) = 0$, x is slightly more than 3.

Solving Real-Life Problems

EXAMPLE 5 **Modeling with Mathematics**

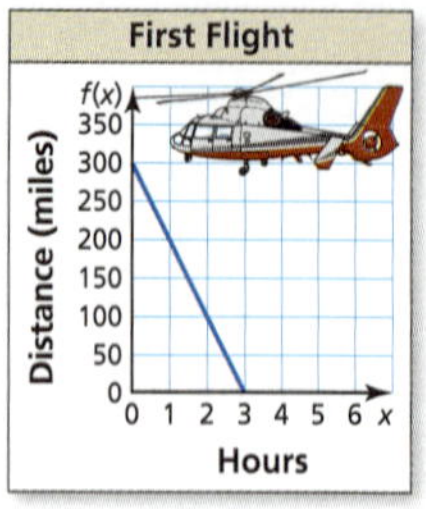

The graph shows the number of miles a helicopter is from its destination after x hours on its first flight. On its second flight, the helicopter travels 50 miles farther and increases its speed by 25 miles per hour. The function $f(x) = 350 - 125x$ represents the second flight, where $f(x)$ is the number of miles the helicopter is from its destination after x hours. Which flight takes less time? Explain.

SOLUTION

1. **Understand the Problem** You are given a graph of the first flight and an equation of the second flight. You are asked to compare the flight times to determine which flight takes less time.
2. **Make a Plan** Graph the function that represents the second flight. Compare the graph to the graph of the first flight. The x-value that corresponds to $f(x) = 0$ represents the flight time.
3. **Solve the Problem** Graph $f(x) = 350 - 125x$.

Step 1 Make an input-output table to find the ordered pairs.

x	0	1	2	3
$f(x)$	350	225	100	−25

Step 2 Plot the ordered pairs.

Step 3 Draw a line through the points. Note that the function only makes sense when x and $f(x)$ are positive. So, only draw the line in the first quadrant.

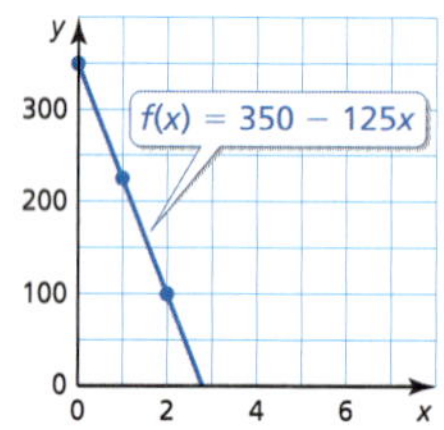

▶ From the graph of the first flight, you can see that when $f(x) = 0$, $x = 3$. From the graph of the second flight, you can see that when $f(x) = 0$, x is slightly less than 3. So, the second flight takes less time.

4. **Look Back** You can check that your answer is correct by finding the value of x for which $f(x) = 0$.

$f(x) = 350 - 125x$ Write the function.

$0 = 350 - 125x$ Substitute 0 for $f(x)$.

$-350 = -125x$ Subtract 350 from each side.

$2.8 = x$ Divide each side by −125.

So, the second flight takes 2.8 hours, which is less than 3.

Monitoring Progress

Help in English and Spanish at *BigIdeasMath.com*

9. **WHAT IF?** Let $f(x) = 250 - 75x$ represent the second flight, where $f(x)$ is the number of miles the helicopter is from its destination after x hours. Which flight takes less time? Explain.

Laurie's Notes Teacher Actions

- **Make Sense of Problems and Persevere in Solving Them:** Give partners A and B time to discuss the problem. Ask a partner A to describe what is known in the problem. Ask a partner B to describe what we are trying to find.
- **Attend to Precision:** Students should notice that the vertical axis is labeled $f(x)$ instead of y.

? Work through the example and ask, "What is the slope and what does it mean in the context of the situation?" slope = −125; Each time x increases by 1 hour, the miles from destination decrease by 125 miles, so the speed of the helicopter is 125 miles per hour.

? "On the second flight, how far is the helicopter from its destination to start?" 350 miles

? "Why are the graphs only drawn in Quadrant I?" Only values in Quadrant I make sense in the context of the situation.

Closure

- **Writing Prompt:** What do you understand when you see $f(x) = 4x - 3$? *Sample answer:* The equation is a function with the name of f. To find the value of f at x, substitute a value of x into the function and evaluate.

3.3 Exercises

Dynamic Solutions available at *BigIdeasMath.com*

Vocabulary and Core Concept Check

1. **COMPLETE THE SENTENCE** When you write the function $y = 2x + 10$ as $f(x) = 2x + 10$, you are using ____________.

2. **REASONING** Your height can be represented by a function h, where the input is your age. What does $h(14)$ represent?

Monitoring Progress and Modeling with Mathematics

In Exercises 3–10, evaluate the function when $x = -2$, 0, and 5. *(See Example 1.)*

3. $f(x) = x + 6$
4. $g(x) = 3x$
5. $h(x) = -2x + 9$
6. $r(x) = -x - 7$
7. $p(x) = -3 + 4x$
8. $b(x) = 18 - 0.5x$
9. $v(x) = 12 - 2x - 5$
10. $n(x) = -1 - x + 4$

11. **INTERPRETING FUNCTION NOTATION** Let $c(t)$ be the number of customers in a restaurant t hours after 8 A.M. Explain the meaning of each statement. *(See Example 2.)*
    a. $c(0) = 0$
    b. $c(3) = c(8)$
    c. $c(n) = 29$
    d. $c(13) < c(12)$

12. **INTERPRETING FUNCTION NOTATION** Let $H(x)$ be the percent of U.S. households with Internet use x years after 1980. Explain the meaning of each statement.
    a. $H(23) = 55$
    b. $H(4) = k$
    c. $H(27) \geq 61$
    d. $H(17) + H(21) \approx H(29)$

In Exercises 13–18, find the value of x so that the function has the given value. *(See Example 3.)*

13. $h(x) = -7x$; $h(x) = 63$
14. $t(x) = 3x$; $t(x) = 24$
15. $m(x) = 4x + 15$; $m(x) = 7$
16. $k(x) = 6x - 12$; $k(x) = 18$
17. $q(x) = \frac{1}{2}x - 3$; $q(x) = -4$
18. $j(x) = -\frac{4}{5}x + 7$; $j(x) = -5$

In Exercises 19 and 20, find the value of x so that $f(x) = 7$.

19. 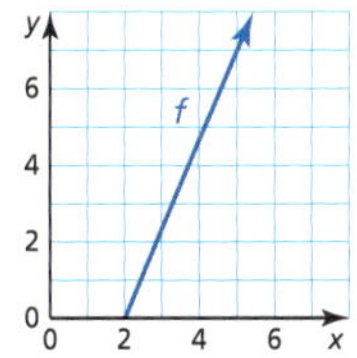

20. 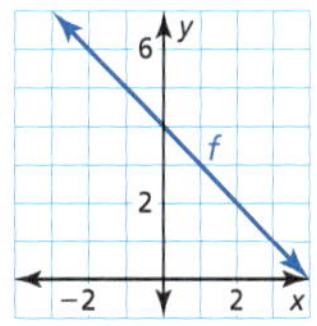

21. **MODELING WITH MATHEMATICS** The function $C(x) = 17.5x - 10$ represents the cost (in dollars) of buying x tickets to the orchestra with a \$10 coupon.
    a. How much does it cost to buy five tickets?
    b. How many tickets can you buy with \$130?

22. **MODELING WITH MATHEMATICS** The function $d(t) = 300{,}000t$ represents the distance (in kilometers) that light travels in t seconds.
    a. How far does light travel in 15 seconds?
    b. How long does it take light to travel 12 million kilometers?

In Exercises 23–28, graph the linear function. *(See Example 4.)*

23. $p(x) = 4x$
24. $h(x) = -5$
25. $d(x) = -\frac{1}{2}x - 3$
26. $w(x) = \frac{3}{5}x + 2$
27. $g(x) = -4 + 7x$
28. $f(x) = 3 - 6x$

Assignment Guide and Homework Check

ASSIGNMENT

Basic: 1, 2, 3–15 odd, 19–31 odd, 34, 37–42

Average: 1, 2–30 even, 31, 33, 34, 37–42

Advanced: 1, 2, 8–30 even, 31–42

HOMEWORK CHECK

Basic: 3, 11, 13, 23, 29

Average: 6, 12, 14, 24, 30

Advanced: 8, 12, 18, 26, 30

ANSWERS

1. function notation
2. your height at 14 years old
3. 4; 6; 11
4. −6; 0; 15
5. 13; 9; −1
6. −5; −7; −12
7. −11; −3; 17
8. 19; 18; 15.5
9. 11; 7; −3
10. 5; 3; −2
11. a. There are no customers in the restaurant at 8 A.M.
    b. There are the same number of customers in the restaurant at 11 A.M. as there are at 4 P.M.
    c. There are 29 customers in the restaurant n hours after 8 A.M.
    d. There are fewer customers in the restaurant at 9 P.M. than there are at 8 P.M.
12. a. 55% of U.S. households had Internet use in 2003.
    b. k% of U.S. households had Internet use in 1984.
    c. At least 61% of U.S. households had Internet use in 2007.
    d. The sum of the percents of U.S. households with Internet use in 1997 and 2001 is about the same as the percent of U.S. households with Internet use in 2009.
13. $x = -9$
14. $x = 8$
15. $x = -2$
16. $x = 5$
17. $x = -2$
18. $x = 15$
19. $x = 5$
20. $x = -3$
21. a. \$77.50
    b. 8 tickets
22. a. 4,500,000 km
    b. 40 sec
23. 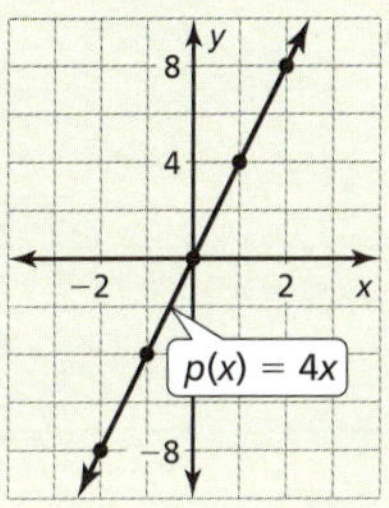

24. 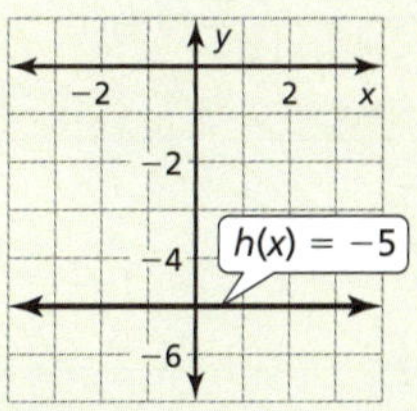

25–28. See Additional Answers.

Dynamic Teaching Tools

- Dynamic Assessment & Progress Monitoring Tool
- Interactive Whiteboard Lesson Library
- Dynamic Classroom with Dynamic Investigations

ANSWERS

29. the tablet computer; The graph of the tablet computer shows that when $p(t) = 0$, t is 6, which is greater than 5.

30–42. See Additional Answers.

Mini-Assessment

1. Evaluate $f(x) = 3x - 6$ when $x = 2$ and $x = -4$. When $x = 2$, $f(x) = 0$, and when $x = -4$, $f(x) = -18$.
2. Let $f(r)$ be the number of riders on a bus r hours after 8 A.M. Explain the meaning of the statement $f(5) = 23$. At 1 P.M. there are 23 riders on the bus.
3. For $g(x) = -5x + 1$, find the value of x for which $g(x) = -9$. When $x = 2$, $g(x) = -9$.
4. Graph $f(x) = -3x - 2$.

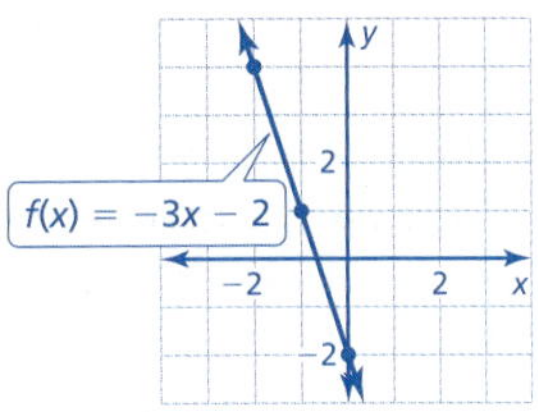

5. The table shows the number of miles a car is from its destination after x hours on its first trip.

Hours	Miles
1	150
2	100
3	50

On its second trip, the car travels 10 miles farther and increases its speed by 10 miles per hour. The function $f(x) = 210 - 60x$ represents the second trip. Which trip takes less time? Explain. The second trip takes less time. The first trip takes 4 hours, and the second trip takes 3.5 hours.

29. PROBLEM SOLVING The graph shows the percent p (in decimal form) of battery power remaining in a laptop computer after t hours of use. A tablet computer initially has 75% of its battery power remaining and loses 12.5% per hour. Which computer's battery will last longer? Explain. *(See Example 5.)*

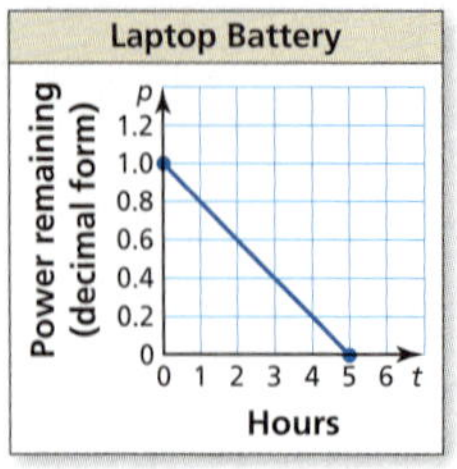

30. PROBLEM SOLVING The function $C(x) = 25x + 50$ represents the labor cost (in dollars) for Certified Remodeling to build a deck, where x is the number of hours of labor. The table shows sample labor costs from its main competitor, Master Remodeling. The deck is estimated to take 8 hours of labor. Which company would you hire? Explain.

Hours	Cost
2	\$130
4	\$160
6	\$190

31. MAKING AN ARGUMENT Let $P(x)$ be the number of people in the U.S. who own a cell phone x years after 1990. Your friend says that $P(x + 1) > P(x)$ for any x because $x + 1$ is always greater than x. Is your friend correct? Explain.

32. THOUGHT PROVOKING Let $B(t)$ be your bank account balance after t days. Describe a situation in which $B(0) < B(4) < B(2)$.

33. MATHEMATICAL CONNECTIONS Rewrite each geometry formula using function notation. Evaluate each function when $r = 5$ feet. Then explain the meaning of the result.

a. Diameter, $d = 2r$

b. Area, $A = \pi r^2$

c. Circumference, $C = 2\pi r$

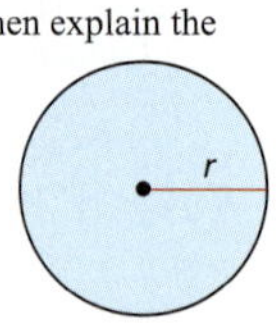

34. HOW DO YOU SEE IT? The function $y = A(x)$ represents the attendance at a high school x weeks after a flu outbreak. The graph of the function is shown.

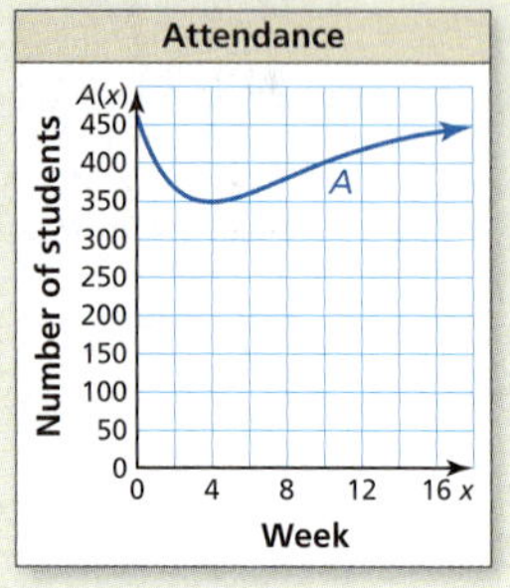

a. What happens to the school's attendance after the flu outbreak?

b. Estimate $A(13)$ and explain its meaning.

c. Use the graph to estimate the solution(s) of the equation $A(x) = 400$. Explain the meaning of the solution(s).

d. What was the least attendance? When did that occur?

e. How many students do you think are enrolled at this high school? Explain your reasoning.

35. INTERPRETING FUNCTION NOTATION Let f be a function. Use each statement to find the coordinates of a point on the graph of f.

a. $f(5)$ is equal to 9.

b. A solution of the equation $f(n) = -3$ is 5.

36. REASONING Given a function f, tell whether the statement

$$f(a + b) = f(a) + f(b)$$

is true or false for all inputs a and b. If it is false, explain why.

Maintaining Mathematical Proficiency

Reviewing what you learned in previous grades and lessons

Solve the inequality. Graph the solution. *(Section 2.5)*

37. $-2 \le x - 11 \le 6$

38. $5a < -35$ *or* $a - 14 > 1$

39. $-16 < 6k + 2 < 0$

40. $2d + 7 < -9$ *or* $4d - 1 > -3$

41. $5 \le 3y + 8 < 17$

42. $4v + 9 \le 5$ *or* $-3v \ge -6$

If students need help...	*If students got it...*
Resources by Chapter • Practice A and Practice B • Puzzle Time	Resources by Chapter • Enrichment and Extension • Cumulative Review
Student Journal • Practice	Start the *next* Section
Differentiating the Lesson Skills Review Handbook	

3.1–3.3 What Did You Learn?

Core Vocabulary

relation, *p. 104*
function, *p. 104*
domain, *p. 106*
range, *p. 106*
independent variable, *p. 107*
dependent variable, *p. 107*
linear equation in two variables, *p. 112*
linear function, *p. 112*
nonlinear function, *p. 112*
solution of a linear equation in two variables, *p. 114*
discrete domain, *p. 114*
continuous domain, *p. 114*
function notation, *p. 122*

Core Concepts

Section 3.1

Determining Whether Relations Are Functions, *p. 104*
Vertical Line Test, *p. 105*
The Domain and Range of a Function, *p. 106*
Independent and Dependent Variables, *p. 107*

Section 3.2

Linear and Nonlinear Functions, *p. 112*
Representations of Functions, *p. 113*
Discrete and Continuous Domains, *p. 114*

Section 3.3

Using Function Notation, *p. 122*

Mathematical Practices

1. How can you use technology to confirm your answers in Exercises 40–43 on page 110?
2. How can you use patterns to solve Exercise 43 on page 119?
3. How can you make sense of the quantities in the function in Exercise 21 on page 125?

Study Skills

Staying Focused during Class

As soon as class starts, quickly review your notes from the previous class and start thinking about math.

Repeat what you are writing in your head.

When a particular topic is difficult, ask for another example.

127

Dynamic Teaching Tools

Dynamic Assessment & Progress Monitoring Tool
Interactive Whiteboard Lesson Library
Dynamic Classroom with Dynamic Investigations

ANSWERS

1. Use a graphing calculator to graph the functions and check the domain and range of the graphs.
2. Look for a pattern in the change in earnings.
3. The function represents the cost in dollars for the tickets at $17.50 each minus the $10 discount. All quantities are in dollars.

3.1–3.3 Quiz

Determine whether the relation is a function. Explain. *(Section 3.1)*

1.

Input, *x*	−1	0	1	2	3
Output, *y*	0	1	4	4	8

2. $(-10, 2), (-8, 3), (-6, 5), (-8, 8), (-10, 6)$

Find the domain and range of the function represented by the graph. *(Section 3.1)*

3.

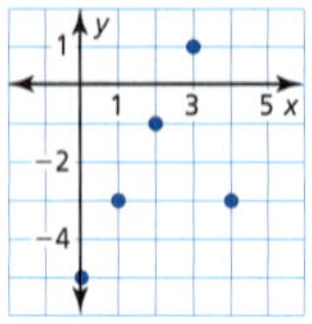

4.

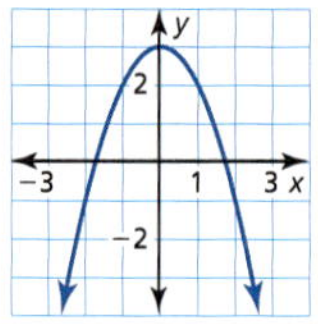

5.

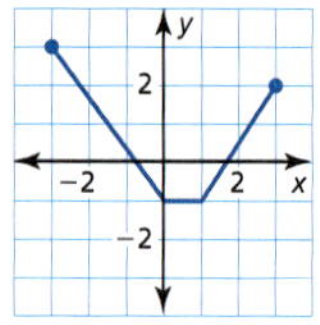

Determine whether the graph, table, or equation represents a *linear* or *nonlinear* function. Explain. *(Section 3.2)*

6.

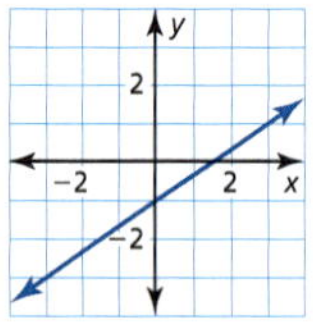

7.

x	*y*
−5	3
0	7
5	10

8. $y = x(2 - x)$

Determine whether the domain is *discrete* or *continuous*. Explain. *(Section 3.2)*

9.

Depth (feet), *x*	33	66	99
Pressure (ATM), *y*	2	3	4

10.

Hats, *x*	2	3	4
Cost (dollars), *y*	36	54	72

11. For $w(x) = -2x + 7$, find the value of x for which $w(x) = -3$. *(Section 3.3)*

Graph the linear function. *(Section 3.3)*

12. $g(x) = x + 3$ **13.** $p(x) = -3x - 1$ **14.** $m(x) = \frac{2}{3}x$

15. The function $m = 30 - 3r$ represents the amount m (in dollars) of money you have after renting r video games. *(Section 3.1 and Section 3.2)*

a. Identify the independent and dependent variables.

b. Find the domain and range of the function. Is the domain discrete or continuous? Explain.

c. Graph the function using its domain.

16. The function $d(x) = 1375 - 110x$ represents the distance (in miles) a high-speed train is from its destination after x hours. *(Section 3.3)*

a. How far is the train from its destination after 8 hours?

b. How long does the train travel before reaching its destination?

ANSWERS

1. function; Each input has exactly one output.
2. not a function; The input −10 has two outputs, 2 and 6, and the input −8 has two outputs, 3 and 8.
3. domain: 0, 1, 2, 3, 4; range: −5, −3, −1, 1
4. domain: all real numbers; range: $y \le 3$
5. domain: $-3 \le x \le 3$; range: $-1 \le y \le 3$
6. linear; The graph is a line.
7. nonlinear; As x increases by 5, y increases by different amounts. The rate of change is not constant.
8. nonlinear; It cannot be written in the form $y = mx + b$.
9. continuous; The depth can be any number of feet.
10. discrete; The number of hats must be a whole number.
11. $x = 5$
12. 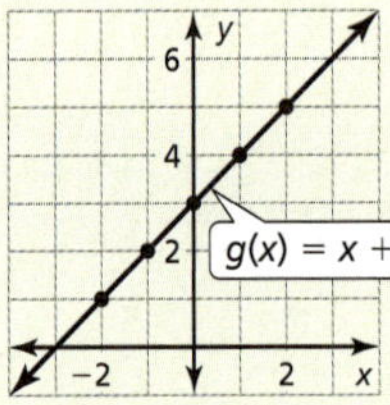

13. 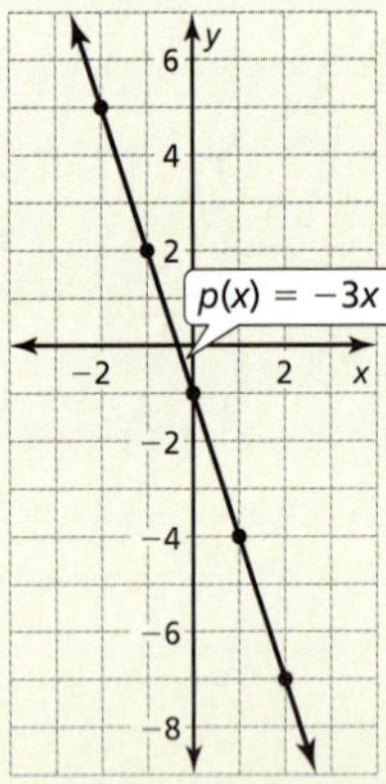

14. 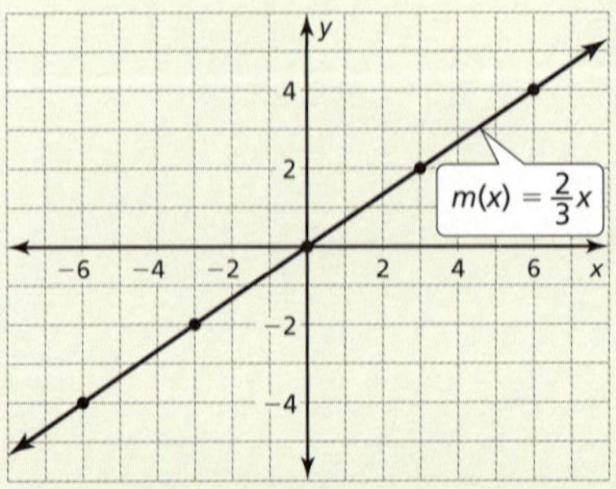

15. a. m is the dependent variable and r is the independent variable.

b. domain: 0, 1, 2, 3, 4, 5, 6, 7, 8, 9, 10; range: 0, 3, 6, 9, 12, 15, 18, 21, 24, 27, 30; discrete; The number of video games rented must be a whole number.

c.

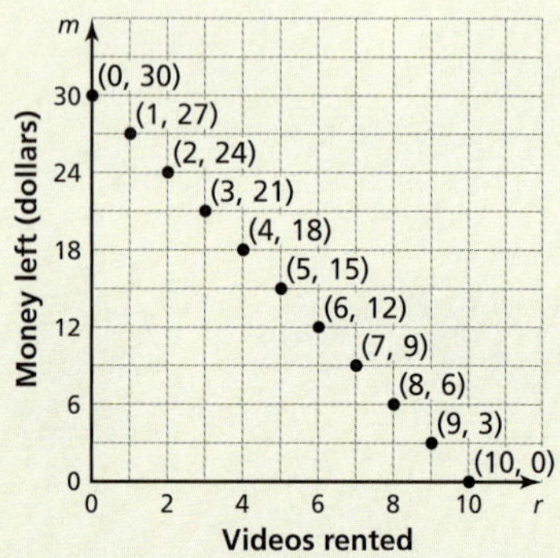

16. a. 495 mi

b. 12.5 h

Laurie's Notes

Overview of Section 3.4

Introduction

- Students were introduced to the standard form of a linear equation in middle school. They solved systems of equations, some of which were in standard form.
- The two explorations use verbal models written in standard form. Students plot points to graph the equation.
- In Exploration 2, students determine the x- and y-intercepts of the graphs. The connection is then made between the intercepts of the graph and the original equation. Plotting the two intercepts is an efficient way to graph an equation in standard form.
- In the formal lesson, students consider the special case of when $A = 0$ or $B = 0$. Students then practice graphing and using verbal models. As the coefficients and the constant term C increase in absolute value, the efficiency of doing a quick sketch using the intercepts becomes more evident to students.

Resources

- Students are asked to use a graphing calculator in Exploration 2. The skill of solving an equation for y becomes clear to students because equations must be in function form to enter them in most calculators. Setting the calculator for an appropriate viewing window will also be necessary when the intercepts are greater than 10, as some are in this lesson.

Teaching Strategy

- At this point in the year, most students are comfortable with evaluating expressions by saying the substitution aloud and performing the mental math. For instance, if $y = 3x - 7$ and we want to know y when $x = 3$, then we say aloud, "3 times 3 is 9 minus 7 is 2." We do not need to write $y = 3(3) - 7$.
- To find the intercepts, we are substituting $x = 0$ (for the y-intercept) and $y = 0$ (for the x-intercept). Substituting 0 and performing the remaining computation should be easy for most students.
- To find the intercepts for $6x + 10y = 180$, I first write the equation. Off to the side I write (0, ______) and (______, 0) to represent the y- and x-intercepts, respectively.
  - When $x = 0$, the term $6x$ becomes 0 and we are left solving $10y = 180$. At this point I place my hand over "$6x +$" so students only see $10y = 180$. They see that to solve for y they need to divide both sides by 10 and get $y = 18$.
  - When $y = 0$, the term $10y$ becomes 0 and we are left solving $6x = 180$. I then place my hand over "$+ 10y$" so students only see $6x = 180$. They see that to solve for x they need to divide both sides by 6 and get $x = 30$.
  - The visual of only seeing one term set equal to the constant seems to improve the ability of students to efficiently solve for the intercepts.
  - **Note:** When the coefficient of y is negative, be sure to leave the sign visible.

Pacing Suggestion

- If students work through both explorations, you might omit Example 2 in the formal lesson.

Dynamic Teaching Tools

- Dynamic Assessment & Progress Monitoring Tool
- Lesson Planning Tool
- Interactive Whiteboard Lesson Library
- Dynamic Classroom with Dynamic Investigations

What Your Students Will Learn

- Graph equations of horizontal and vertical lines in the coordinate plane.
- Graph linear equations in standard form by finding the *x*- and *y*-intercepts of the graphs.

Laurie's Notes

Exploration

Motivate

? "How many pairs of numbers can you think of that have a sum of 5?" Encourage students to write their numbers on paper as ordered pairs. Example: (2, 3)

? "Did any of you include numbers that are not whole numbers?" Check to see whether anyone had negative numbers or rational numbers.

- Plot the ordered pairs in a coordinate plane.

? "What do you think the equation of this line would be?" $x + y = 5$

Discuss

- $x + y = 5$ is another form of a linear equation. The *Essential Question* today asks how to describe the graph of the equation $Ax + By = C$.

? "Do you recall what this form is called?" standard form

- The goal is for students to recognize that the intercepts are an efficient way to graph an equation in standard form.

Exploration 1

- **Connection:** This exploration is an extension of the *Motivate*. Instead of just thinking of two numbers that sum to \$16, students must multiply each of the numbers by a factor (\$4 and \$2). Although the context may seem far-fetched for students, it is related to what recently happened to my daughter. She was asked to sell tickets to the Friday night football game. Red student tickets were \$2 and blue adult tickets were \$4. The principal forgot to mention that my daughter was to keep track of how many of each she sold! She called me Saturday to relay the story and said, "I know I could have estimated, but there are lots of combinations for \$964." Bravo Sarah, who took algebra more than 10 years ago.
- **Model with Mathematics:** The verbal model helps students to think about the computations needed and to remember the units associated with each.
- Circulate and read explanations in part (d). Make notes so that you may elicit *Selective Responses* from students.

Exploration 2

- The verbal model represents how the total cost of \$48 is found when Swiss cheese is \$8 per pound and cheddar cheese is \$6 per pound.
- **Turn and Talk:** Partners should work together in solving the equation for *y*.
- Circulate to determine the form in which students write the equation:

$$y = -\tfrac{4}{3}x + 8 \qquad y = \tfrac{1}{6}(-8x + 48) \qquad y = -\tfrac{8}{6}x + \tfrac{48}{6}$$

COMMON ERROR In solving for *y*, dividing the expression $-8x + 48$ by 6 is not always done correctly. If students are using graphing calculators, have they entered the equation correctly? Have they used parentheses as needed?

- **Big Idea:** Parts (c) and (d) help students focus on two key ordered pairs that can help in sketching the graph of a line when it is written in standard form. If the equation represents a contextual situation, then the intercepts generally have a meaning that should be interpreted. This will become evident in the formal lesson.

Communicate Your Answer

- Students should describe finding the two intercepts in order to graph the function.
- Equations written for Question 4 will involve purchasing two different items!

Connecting to Next Step

- If you feel students are comfortable working with the verbal models, assign Exercise 23 or 24 on page 133.

3.4 Graphing Linear Equations in Standard Form

Essential Question

How can you describe the graph of the equation $Ax + By = C$?

EXPLORATION 1 Using a Table to Plot Points

Work with a partner. You sold a total of \$16 worth of tickets to a fundraiser. You lost track of how many of each type of ticket you sold. Adult tickets are \$4 each. Child tickets are \$2 each.

$$\frac{\square}{\text{adult}} \cdot \text{Number of adult tickets} + \frac{\square}{\text{child}} \cdot \text{Number of child tickets} = \square$$

a. Let x represent the number of adult tickets. Let y represent the number of child tickets. Use the verbal model to write an equation that relates x and y.

b. Copy and complete the table to show the different combinations of tickets you might have sold.

x					
y					

c. Plot the points from the table. Describe the pattern formed by the points.

d. If you remember how many adult tickets you sold, can you determine how many child tickets you sold? Explain your reasoning.

EXPLORATION 2 Rewriting and Graphing an Equation

Work with a partner. You sold a total of \$48 worth of cheese. You forgot how many pounds of each type of cheese you sold. Swiss cheese costs \$8 per pound. Cheddar cheese costs \$6 per pound.

$$\frac{\square}{\text{pound}} \cdot \text{Pounds of Swiss} + \frac{\square}{\text{pound}} \cdot \text{Pounds of cheddar} = \square$$

a. Let x represent the number of pounds of Swiss cheese. Let y represent the number of pounds of cheddar cheese. Use the verbal model to write an equation that relates x and y.

b. Solve the equation for y. Then use a graphing calculator to graph the equation. Given the real-life context of the problem, find the domain and range of the function.

c. The **x-intercept** of a graph is the x-coordinate of a point where the graph crosses the x-axis. The **y-intercept** of a graph is the y-coordinate of a point where the graph crosses the y-axis. Use the graph to determine the x- and y-intercepts.

d. How could you use the equation you found in part (a) to determine the x- and y-intercepts? Explain your reasoning.

e. Explain the meaning of the intercepts in the context of the problem.

Communicate Your Answer

3. How can you describe the graph of the equation $Ax + By = C$?
4. Write a real-life problem that is similar to those shown in Explorations 1 and 2.

FINDING AN ENTRY POINT

To be proficient in math, you need to find an entry point into the solution of a problem. Determining what information you know, and what you can do with that information, can help you find an entry point.

Dynamic Teaching Tools

- Dynamic Assessment & Progress Monitoring Tool
- Lesson Planning Tool
- Interactive Whiteboard Lesson Library
- Dynamic Classroom with Dynamic Investigations

ANSWERS

1. a. $4x + 2y = 16$
   b. 0, 1, 2, 3, 4; 8, 6, 4, 2, 0
   c.
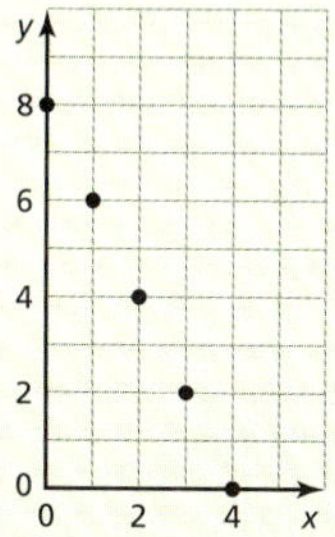

   The points create a line with a slope of -2. For each adult ticket sold, the number of child tickets sold decreases by 2.
   d. yes; The relationship can be expressed as an equation with 2 variables. If the value of one of the variables is known, the other can be found.
2. a. $8x + 6y = 48$
   b. $y = 8 - \frac{4}{3}x$;
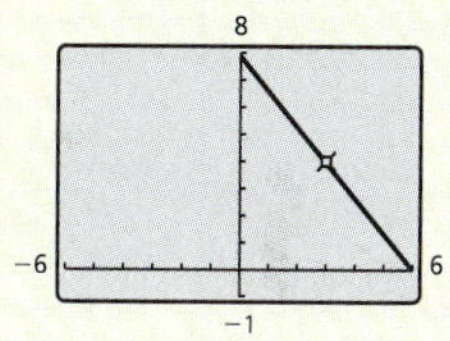

   domain: $0 \le x \le 6$; range: $0 \le x \le 8$
   c. x-intercept: 6; y-intercept: 8
   d. To find the x-intercept, set $y = 0$ and solve for x, to find the y-intercept, set $x = 0$ and solve for y; Because the intercepts are where the graph crosses the axes, the y-coordinate of the x-intercept must be 0, and the x-coordinate of the y-intercept must be 0.
   e. The intercepts represent the maximum number of pounds of each type of cheese you can buy if you only buy one type of cheese.
3. a straight line
4. *Sample answer:* You sold a total of \$375 worth of DVDs. You forgot to record how many of each type you sold. TV season DVDs sold for \$30 each and movie DVDs sold for \$15 each.

Differentiated Instruction

Auditory/Visual

Show students three different graphs—one with only an x-intercept, one with only a y-intercept, and one with both an x- and y-intercept. Ask one student to name each x-intercept and a second student to name each y-intercept. Emphasize that an intercept is not an ordered pair, but rather it is the x- or y-coordinate of a point where the line crosses an axis.

Extra Example 1

a. Graph $y = -1.5$.

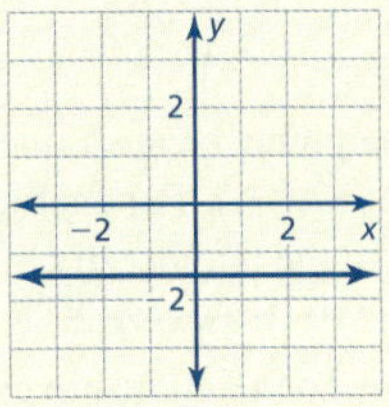

b. Graph $x = 1$.

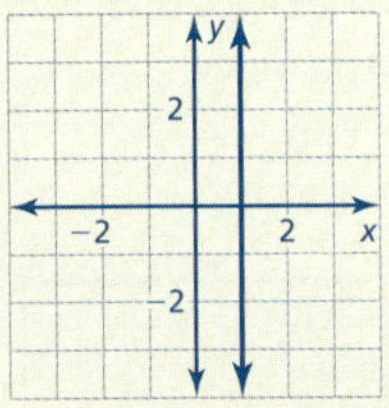

MONITORING PROGRESS ANSWERS

1–2. See Additional Answers.

3.4 Lesson

Core Vocabulary

standard form, *p. 130*
x-intercept, *p. 131*
y-intercept, *p. 131*

Previous
ordered pair
quadrant

What You Will Learn

- Graph equations of horizontal and vertical lines.
- Graph linear equations in standard form using intercepts.
- Use linear equations in standard form to solve real-life problems.

Horizontal and Vertical Lines

The **standard form** of a linear equation is $Ax + By = C$, where A, B, and C are real numbers and A and B are not both zero.

Consider what happens when $A = 0$ or when $B = 0$. When $A = 0$, the equation becomes $By = C$, or $y = \frac{C}{B}$. Because $\frac{C}{B}$ is a constant, you can write $y = b$. Similarly, when $B = 0$, the equation becomes $Ax = C$, or $x = \frac{C}{A}$, and you can write $x = a$.

Core Concept

Horizontal and Vertical Lines

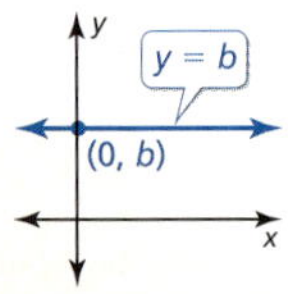

The graph of $y = b$ is a horizontal line. The line passes through the point $(0, b)$.

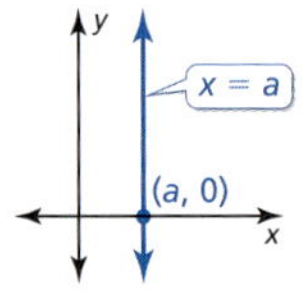

The graph of $x = a$ is a vertical line. The line passes through the point $(a, 0)$.

EXAMPLE 1 Horizontal and Vertical Lines

Graph (a) $y = 4$ and (b) $x = -2$.

SOLUTION

a. For every value of x, the value of y is 4. The graph of the equation $y = 4$ is a horizontal line 4 units above the x-axis.

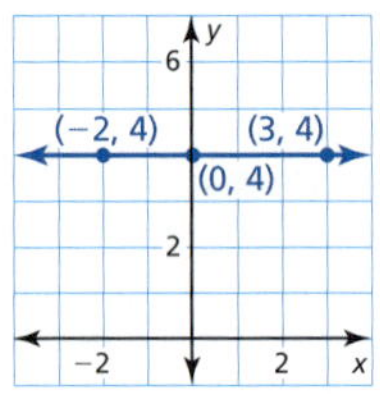

b. For every value of y, the value of x is -2. The graph of the equation $x = -2$ is a vertical line 2 units to the left of the y-axis.

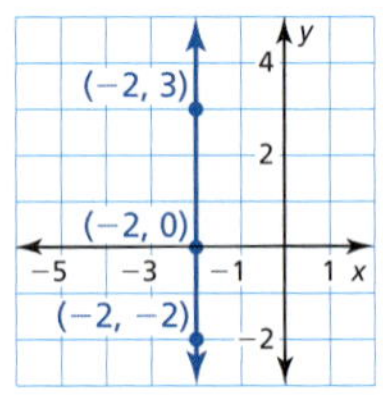

STUDY TIP

For every value of x, the ordered pair $(x, 4)$ is a solution of $y = 4$.

Monitoring Progress

Help in English and Spanish at *BigIdeasMath.com*

Graph the linear equation.

1. $y = -2.5$
2. $x = 5$

Laurie's Notes Teacher Actions

- Write the standard form of a linear equation and the conditions for A, B, and C.
- ? "Why can't A and B both equal 0?" You would have 0 equal to some number, which only makes sense when $C = 0$. It would not be a line.
- ? "What would the equation look like if $A = 0$ or $B = 0$?" If $A = 0$, then $By = C$. If $B = 0$, then $Ax = C$.
- Write the *Core Concept*.
- ? **Turn and Talk:** "Are horizontal and vertical lines examples of functions?" The horizontal line is a function. The vertical line is not a function.
- **Construct Viable Arguments and Critique the Reasoning of Others:** After partners have had a chance to discuss the questions, ask volunteers to share their thinking.
- **Common Misconception:** Students look at the equation $y = 4$ and want to substitute values of x in order to plot points. Help students understand that the equation means that for every value of x, the value of y is 4 by writing $0x + y = 4$ (standard form).

Using Intercepts to Graph Linear Equations

You can use the fact that two points determine a line to graph a linear equation. Two convenient points are the points where the graph crosses the axes.

Core Concept

Using Intercepts to Graph Equations

The **x-intercept** of a graph is the x-coordinate of a point where the graph crosses the x-axis. It occurs when $y = 0$.

The **y-intercept** of a graph is the y-coordinate of a point where the graph crosses the y-axis. It occurs when $x = 0$.

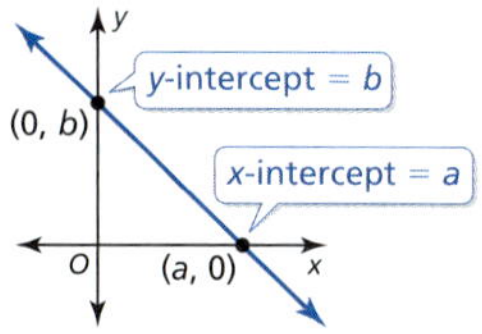

To graph the linear equation $Ax + By = C$, find the intercepts and draw the line that passes through the two intercepts.

- To find the x-intercept, let $y = 0$ and solve for x.
- To find the y-intercept, let $x = 0$ and solve for y.

EXAMPLE 2 Using Intercepts to Graph a Linear Equation

Use intercepts to graph the equation $3x + 4y = 12$.

SOLUTION

Step 1 Find the intercepts.

To find the x-intercept, substitute 0 for y and solve for x.

$3x + 4y = 12$ Write the original equation.

$3x + 4(0) = 12$ Substitute 0 for y.

$x = 4$ Solve for x.

To find the y-intercept, substitute 0 for x and solve for y.

$3x + 4y = 12$ Write the original equation.

$3(0) + 4y = 12$ Substitute 0 for x.

$y = 3$ Solve for y.

STUDY TIP

As a check, you can find a third solution of the equation and verify that the corresponding point is on the graph. To find a third solution, substitute any value for one of the variables and solve for the other variable.

Step 2 Plot the points and draw the line.

The x-intercept is 4, so plot the point (4, 0). The y-intercept is 3, so plot the point (0, 3). Draw a line through the points.

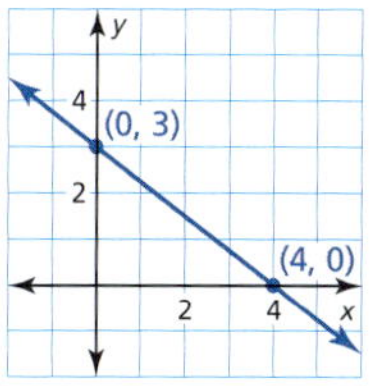

Monitoring Progress

Help in English and Spanish at *BigIdeasMath.com*

Use intercepts to graph the linear equation. Label the points corresponding to the intercepts.

3. $2x - y = 4$

4. $x + 3y = -9$

Extra Example 2

Use intercepts to graph the equation $x + 2y = 6$.

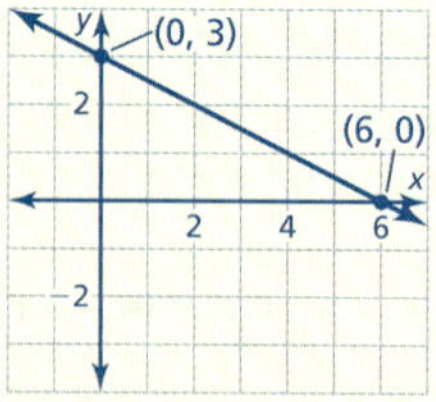

MONITORING PROGRESS ANSWERS

3.

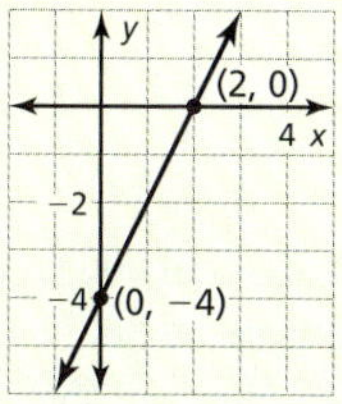

4.

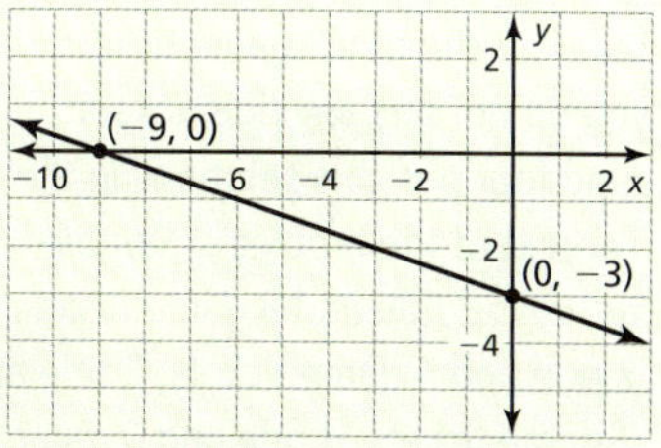

Laurie's Notes Teacher Actions

? "How many points does it take to determine a line?" two

- Write the *Core Concept* and discuss the difference between the y-intercept and the ordered pair where the graph crosses the y-axis.
- Note that all of the work is shown in the example. The *Teaching Strategy* for this lesson discusses a way to help students solve for the intercepts efficiently.
- **Monitoring Progress:** Use whiteboards to get a quick visual of student success with graphing equations in standard form.

COMMON ERROR In Question 3, students may forget to consider the sign of B. Before beginning the two examples, you might first ask students to identify A, B, and C for each example.

Extra Example 3

You are going to the fair and have \$16 to spend on rides and games. Ride tickets cost \$4 and game tickets cost \$2. The equation $4x + 2y = 16$ models this situation, where x is the number of ride tickets you buy and y is the number of game tickets you buy.

a. Graph the equation. Interpret the intercepts.

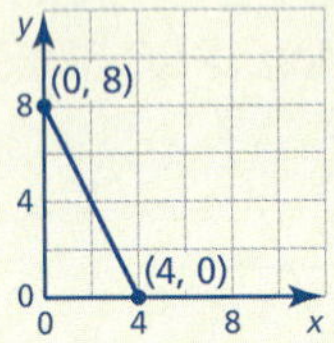

The x-intercept shows that you can buy 4 ride tickets when you do not buy any game tickets. The y-intercept shows that you can buy 8 game tickets when you do not buy any ride tickets.

b. Find four possible solutions in the context of the problem. *Sample answer:* 0 ride and 8 game tickets, 2 ride and 4 game tickets, 3 ride and 2 game tickets, 4 ride and 0 game tickets

MONITORING PROGRESS ANSWER

5.

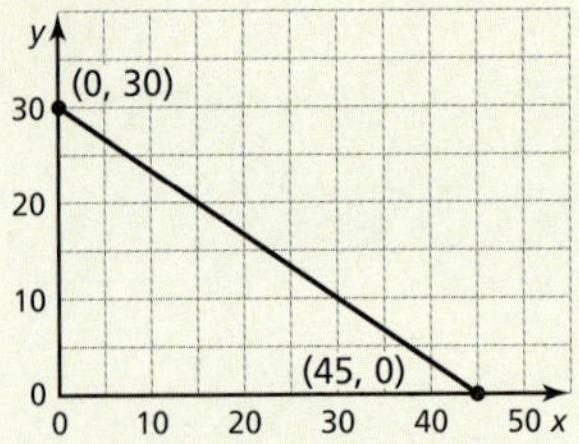

The x-intercept shows that you can rent 45 small tables and 0 large tables. The y-intercept shows that you can rent 30 large tables and 0 small tables.

Solving Real-Life Problems

EXAMPLE 3 **Modeling with Mathematics**

You are planning an awards banquet for your school. You need to rent tables to seat 180 people. Tables come in two sizes. Small tables seat 6 people, and large tables seat 10 people. The equation $6x + 10y = 180$ models this situation, where x is the number of small tables and y is the number of large tables.

a. Graph the equation. Interpret the intercepts.

b. Find four possible solutions in the context of the problem.

SOLUTION

1. **Understand the Problem** You know the equation that models the situation. You are asked to graph the equation, interpret the intercepts, and find four solutions.
2. **Make a Plan** Use intercepts to graph the equation. Then use the graph to interpret the intercepts and find other solutions.
3. **Solve the Problem**

a. Use intercepts to graph the equation. Neither x nor y can be negative, so only graph the equation in the first quadrant.

STUDY TIP

Although x and y represent whole numbers, it is convenient to draw a line segment that includes points whose coordinates are not whole numbers.

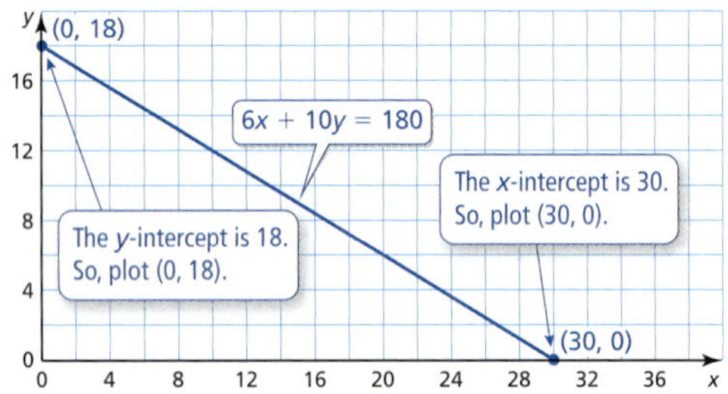

▶ The x-intercept shows that you can rent 30 small tables when you do not rent any large tables. The y-intercept shows that you can rent 18 large tables when you do not rent any small tables.

b. Only whole-number values of x and y make sense in the context of the problem. Besides the intercepts, it appears that the line passes through the points (10, 12) and (20, 6). To verify that these points are solutions, check them in the equation, as shown.

Check

$$6x + 10y = 180$$
$$6(10) + 10(12) \stackrel{?}{=} 180$$
$$180 = 180 \checkmark$$
$$6x + 10y = 180$$
$$6(20) + 10(6) \stackrel{?}{=} 180$$
$$180 = 180 \checkmark$$

▶ So, four possible combinations of tables that will seat 180 people are 0 small and 18 large, 10 small and 12 large, 20 small and 6 large, and 30 small and 0 large.

4. **Look Back** The graph shows that as the number x of small tables increases, the number y of large tables decreases. This makes sense in the context of the problem. So, the graph is reasonable.

Monitoring Progress

Help in English and Spanish at *BigIdeasMath.com*

5. **WHAT IF?** You decide to rent tables from a different company. The situation can be modeled by the equation $4x + 6y = 180$, where x is the number of small tables and y is the number of large tables. Graph the equation and interpret the intercepts.

Laurie's Notes Teacher Actions

- Read the example and write $6x + 10y = 180$ on the board. Students should work with their partners to solve the problem. To facilitate discussion when students have finished, students should do their work on whiteboards. If a document camera is available, work could be displayed from student notebooks.
- **Extension:** When students have finished, use a graphing calculator to graph the equation. Students need to consider what an appropriate viewing window will be for this situation.

? Extension: How would the equation change if there were 6-person and 8-person tables? The equation would be $6x + 8y = 180$. "How would the equation change if there were 200 people attending the banquet?" The equation would be $6x + 10y = 200$.

Closure

- **Writing Prompt:** To graph the equation $2x + y = 4$ … *Sample answer:* find and plot the points for the x- and y-intercepts, and then draw a line through these two points.

3.4 Exercises

Dynamic Solutions available at *BigIdeasMath.com*

Vocabulary and Core Concept Check

1. **WRITING** How are x-intercepts and y-intercepts alike? How are they different?

2. **WHICH ONE DOESN'T BELONG?** Which point does not belong with the other three? Explain your reasoning.

$(0, -3)$ $(0, 0)$ $(4, -3)$ $(4, 0)$

Monitoring Progress and Modeling with Mathematics

In Exercises 3–6, graph the linear equation. *(See Example 1.)*

3. $x = 4$
4. $y = 2$
5. $y = -3$
6. $x = -1$

In Exercises 7–12, find the x- and y-intercepts of the graph of the linear equation.

7. $2x + 3y = 12$
8. $3x + 6y = 24$
9. $-4x + 8y = -16$
10. $-6x + 9y = -18$
11. $3x - 6y = 2$
12. $-x + 8y = 4$

In Exercises 13–22, use intercepts to graph the linear equation. Label the points corresponding to the intercepts. *(See Example 2.)*

13. $5x + 3y = 30$
14. $4x + 6y = 12$
15. $-12x + 3y = 24$
16. $-2x + 6y = 18$
17. $-4x + 3y = -30$
18. $-2x + 7y = -21$
19. $-x + 2y = 7$
20. $3x - y = -5$
21. $-\frac{5}{2}x + y = 10$
22. $-\frac{1}{2}x + y = -4$

23. **MODELING WITH MATHEMATICS** A football team has an away game, and the bus breaks down. The coaches decide to drive the players to the game in cars and vans. Four players can ride in each car. Six players can ride in each van. There are 48 players on the team. The equation $4x + 6y = 48$ models this situation, where x is the number of cars and y is the number of vans. *(See Example 3.)*
   a. Graph the equation. Interpret the intercepts.
   b. Find four possible solutions in the context of the problem.

24. **MODELING WITH MATHEMATICS** You are ordering shirts for the math club at your school. Short-sleeved shirts cost \$10 each. Long-sleeved shirts cost \$12 each. You have a budget of \$300 for the shirts. The equation $10x + 12y = 300$ models the total cost, where x is the number of short-sleeved shirts and y is the number of long-sleeved shirts.
   a. Graph the equation. Interpret the intercepts.
   b. Twelve students decide they want short-sleeved shirts. How many long-sleeved shirts can you order?

ERROR ANALYSIS **In Exercises 25 and 26, describe and correct the error in finding the intercepts of the graph of the equation.**

25. ✗

$3x + 12y = 24$	$3x + 12y = 24$
$3x + 12(0) = 24$	$3(0) + 12y = 24$
$3x = 24$	$12y = 24$
$x = 8$	$y = 2$

The intercept is at (8, 2).

26. ✗

$4x + 10y = 20$	$4x + 10y = 20$
$4x + 10(0) = 20$	$4(0) + 10y = 20$
$4x = 20$	$10y = 20$
$x = 5$	$y = 2$

The x-intercept is at (0, 5), and the y-intercept is at (2, 0).

Assignment Guide and Homework Check

ASSIGNMENT

Basic: 1, 2, 3–19 odd, 23–27 odd, 34, 39–42

Average: 1, 2, 4–26 even, 27, 30–34 even, 39–42

Advanced: 1, 2, 10, 12, 20–27, 28–38 even, 39–42

HOMEWORK CHECK

Basic: 3, 7, 13, 23

Average: 4, 8, 14, 24

Advanced: 10, 20, 24, 30

ANSWERS

1. They are both places where a graph crosses an axis; The x-intercept is where a graph crosses the x-axis. The y-intercept is where a graph crosses the y-axis.
2. $(4, -3)$; It is the only one that is not on an axis.
3. 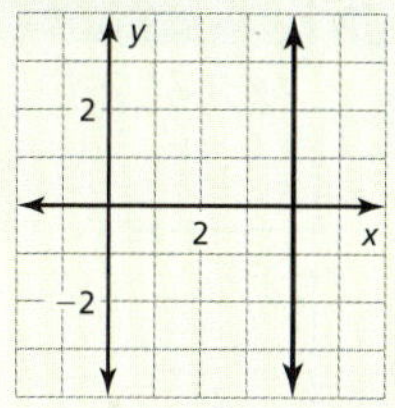

4. 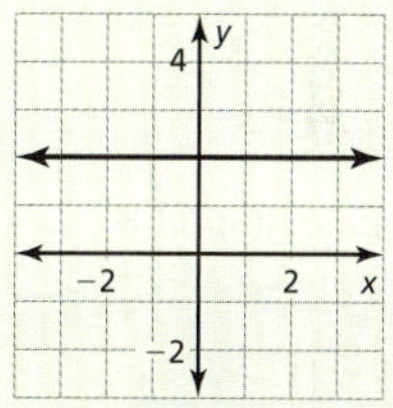

5. 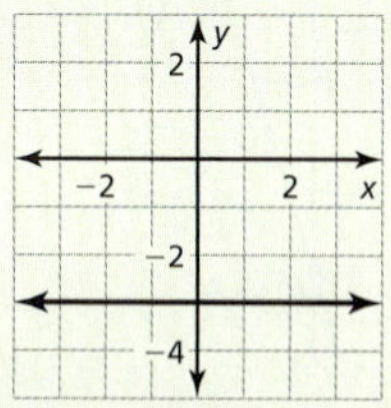

6. 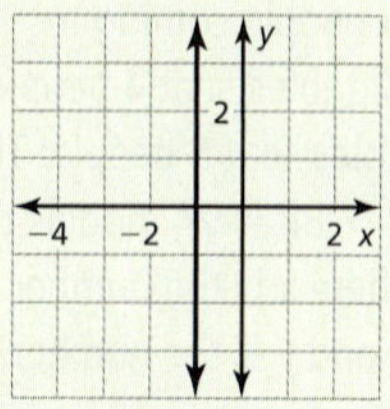

7. 6; 4
8. 8; 4
9. 4; −2
10. 3; −2
11. $\frac{2}{3}$; $-\frac{1}{3}$
12. −4; $\frac{1}{2}$
13. See Additional Answers.
14. 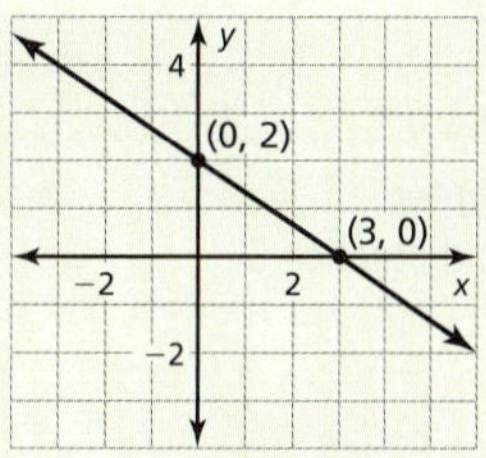

15–26. See Additional Answers.

Dynamic Teaching Tools

- Dynamic Assessment & Progress Monitoring Tool
- Interactive Whiteboard Lesson Library
- Dynamic Classroom with Dynamic Investigations

ANSWERS

27. no; You have to substitute 0 for y to find the x-intercept, not for x.

28. See Additional Answers.

29. A **30.** C

31. D **32.** B

33–34. See Additional Answers.

35. -3; 6

36. *Sample answer:* $2x + 3y = 18$; 18 is divisible by both 2 and 3.

37. yes; $x = a$ can be written as $1x + 0y = a$, and $y = b$ can be written as $0x + 1y = b$.

38. They are multiples of 15; For the intercepts to be integers, k must be divisible by both 3 and 5.

39–42. See Additional Answers.

Mini-Assessment

1. Graph $y = 3$.

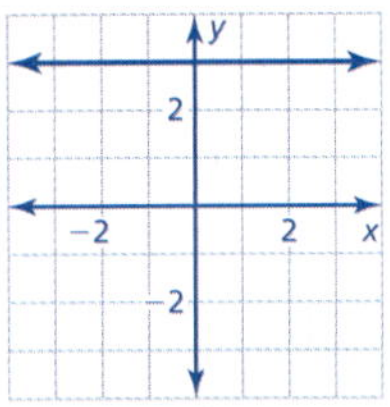

2. Use intercepts to graph the equation $3x - y = 6$.

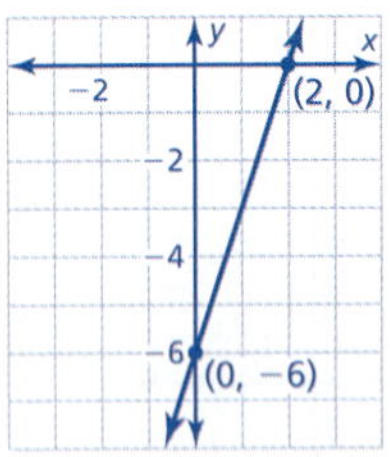

3. There are 24 people coming to dinner. Small tables seat 4 people and large tables seat 8 people. The equation $4x + 8y = 24$ models this situation, where x is the number of small tables and y is the number of large tables. Find three possible solutions in the context of the problem. *Sample answer:* 0 small and 3 large tables, 4 small and 1 large table, 6 small and 0 large tables

27. MAKING AN ARGUMENT You overhear your friend explaining how to find intercepts to a classmate. Your friend says, "When you want to find the x-intercept, just substitute 0 for x and continue to solve the equation." Is your friend's explanation correct? Explain.

28. ANALYZING RELATIONSHIPS You lose track of how many 2-point baskets and 3-point baskets a team makes in a basketball game. The team misses all the 1-point baskets and still scores 54 points. The equation $2x + 3y = 54$ models the total points scored, where x is the number of 2-point baskets made and y is the number of 3-point baskets made.

a. Find and interpret the intercepts.

b. Can the number of 3-point baskets made be odd? Explain your reasoning.

c. Graph the equation. Find two more possible solutions in the context of the problem.

MULTIPLE REPRESENTATIONS In Exercises 29–32, match the equation with its graph.

29. $5x + 3y = 30$ **30.** $5x + 3y = -30$

31. $5x - 3y = 30$ **32.** $5x - 3y = -30$

A.

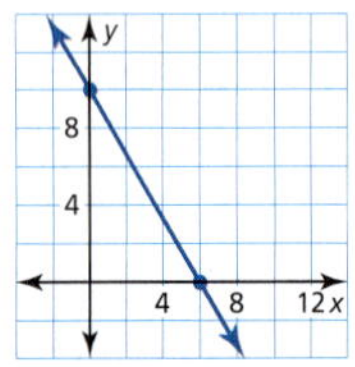

B.

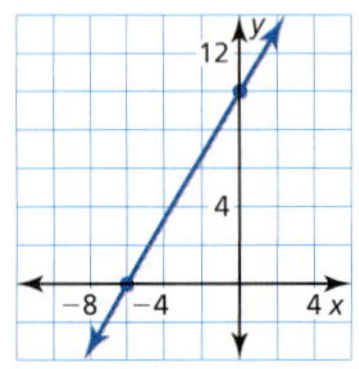

C.

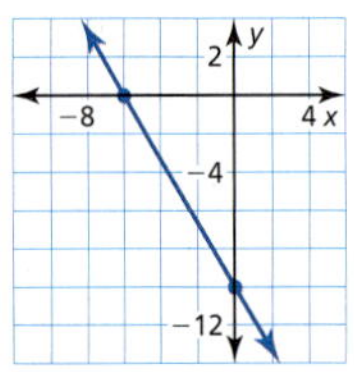

D.

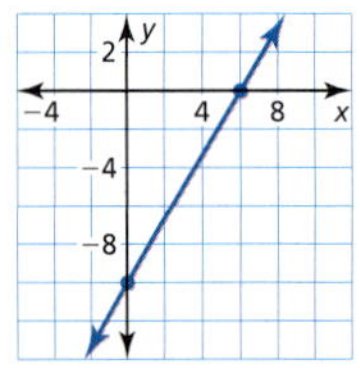

33. MATHEMATICAL CONNECTIONS Graph the equations $x = 5$, $x = 2$, $y = -2$, and $y = 1$. What enclosed shape do the lines form? Explain your reasoning.

34. HOW DO YOU SEE IT? You are organizing a class trip to an amusement park. The cost to enter the park is \$30. The cost to enter with a meal plan is \$45. You have a budget of \$2700 for the trip. The equation $30x + 45y = 2700$ models the total cost for the class to go on the trip, where x is the number of students who do not choose the meal plan and y is the number of students who do choose the meal plan.

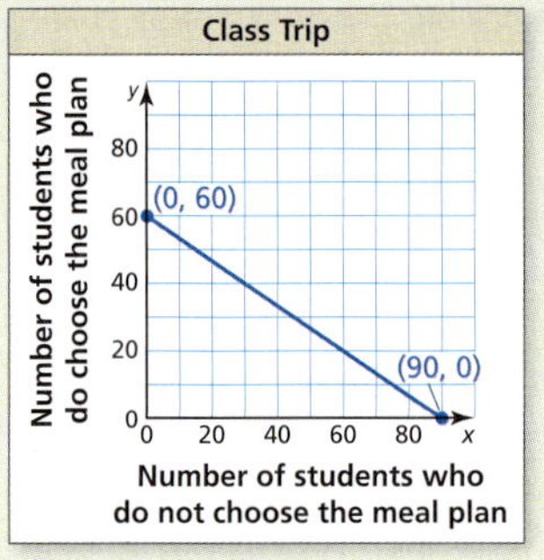

a. Interpret the intercepts of the graph.

b. Describe the domain and range in the context of the problem.

35. REASONING Use the values to fill in the equation $\square x + \square y = 30$ so that the x-intercept of the graph is -10 and the y-intercept of the graph is 5.

-10	-3	1	5	6

36. THOUGHT PROVOKING Write an equation in standard form of a line whose intercepts are integers. Explain how you know the intercepts are integers.

37. WRITING Are the equations of horizontal and vertical lines written in standard form? Explain your reasoning.

38. ABSTRACT REASONING The x- and y-intercepts of the graph of the equation $3x + 5y = k$ are integers. Describe the values of k. Explain your reasoning.

Maintaining Mathematical Proficiency

Reviewing what you learned in previous grades and lessons

Simplify the expression. *(Skills Review Handbook)*

39. $\dfrac{2-(-2)}{4-(-4)}$ **40.** $\dfrac{14-18}{0-2}$ **41.** $\dfrac{-3-9}{8-(-7)}$ **42.** $\dfrac{12-17}{-5-(-2)}$

If students need help...	*If students got it...*
Resources by Chapter • Practice A and Practice B • Puzzle Time	Resources by Chapter • Enrichment and Extension • Cumulative Review
Student Journal • Practice	Start the *next* Section
Differentiating the Lesson Skills Review Handbook	

Laurie's Notes

Overview of Section 3.5

Introduction

- Students were introduced to the slope-intercept form of a linear equation in middle school. They interpreted the equation $y = mx + b$ as defining a linear function whose graph is a straight line. They solved systems of equations, some of which were in slope-intercept form.
- Depending upon your students' background and familiarity with the content, this lesson may only need light coverage. Discussions earlier in this chapter should have given you enough evidence to know what depth this lesson requires. Core concepts to check include: computing the slope of a line from a graph or table of values; rewriting equations in slope-intercept form; interpreting slope and y-intercept in the context of a situation; interpreting a linear equation written in function notation.

Resources

- A graphing calculator could be used to assess concepts about slope.

Common Misconceptions

- Given the equation $2x + y = 5$, students may incorrectly say that the slope is 2. They remember that the slope is the coefficient of x, but they forget that the equation must be written in slope-intercept form first.

Formative Assessment Tips

- **Point of Most Significance:** This technique asks students to identify the most significant idea, learning, or concept they gained in the lesson today. Students reflect on the lesson and are asked to identify the key example, problem, or point that was significant in their learning today. If the goal or learning intention was identified at the beginning of the class, then students assess what contributed to their attainment of the goal.
- It is important for teachers to know whether the lesson as designed helped move learning forward for students or whether the lesson needs to be modified. Share with students what you learned from their reflections. Students will take the reflections more seriously when they are valued and used.

Pacing Suggestion

- Although the formal lesson is long, you may find that students move through the examples quickly given their background knowledge. Use class time for students to work applications in the exercise set with partners.

Dynamic Teaching Tools

- Dynamic Assessment & Progress Monitoring Tool
- Lesson Planning Tool
- Interactive Whiteboard Lesson Library
- Dynamic Classroom with Dynamic Investigations

What Your Students Will Learn

- Describe and find slopes of nonvertical lines from graphs or tables.
- Identify the slopes and y-intercepts of the graphs of equations written in slope-intercept form.
- Graph linear equations written in slope-intercept form or given a verbal description.

Laurie's Notes

Exploration

Motivate

- **Preparation:** Make three demonstration cards on 8.5 inch by 11 inch paper. The x-axis is labeled "time" and the y-axis is labeled "distance from home."
- Sample cards A, B, and C are shown.

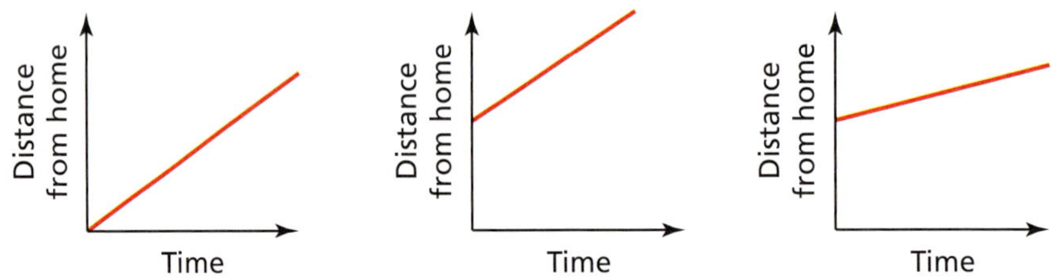

- Ask three students to hold the cards for the class to see.

? "Consider how the axes are labeled. What does the slope of the line represent?" $\frac{\text{distance}}{\text{time}} = \text{rate}$

? "What story does each card tell? How are the stories similar and different?" A: you begin at home; B: you travel at the same rate, but you start away from home; C: you start away from home, but you travel at a slower rate

- **Management Tip:** If you plan to use the demonstration cards again next year, laminate them.

Exploration 1

- The *Motivate* should act as a warm-up and reminder of slope. It also draws attention to the y-intercept of the graph.
- If your students have good recall of computing slope from last year, this exploration should not take long to complete.

? **Assessing Question:** "How are the two slopes different?" One line is increasing (positive slope) and the other is decreasing (negative slope).

- If your students do not recall how slope is computed, provide adequate time for the exploration.

COMMON ERROR Students may have the wrong sign for the slope.

Exploration 2

- Again, if your students are secure with interpreting equations written in slope-intercept form, you might consider omitting the graphing portion of the exploration and having partners record their results on whiteboards.
- **Differentiate:** You might assign one of the three remaining parts to different groups or pairs of students in the class. Work could be recorded on a whiteboard, enabling presentation of the work to the whole class.

Communicate Your Answer

- **Turn and Talk:** Circulate and listen as students share their understanding of the effect of m and b on the graph of $y = mx + b$.

Connecting to Next Step

- If you feel students are comfortable finding the slope from ordered pairs on a graph, assign Exercises 5–8 on page 141. You could also have students copy the *Core Concept, Slope,* into their notebooks.

3.5 Graphing Linear Equations in Slope-Intercept Form

Essential Question How can you describe the graph of the equation $y = mx + b$?

Slope is the rate of change between any two points on a line. It is the measure of the *steepness* of the line.

To find the slope of a line, find the ratio of the change in y (vertical change) to the change in x (horizontal change).

$$\text{slope} = \frac{\text{change in } y}{\text{change in } x}$$

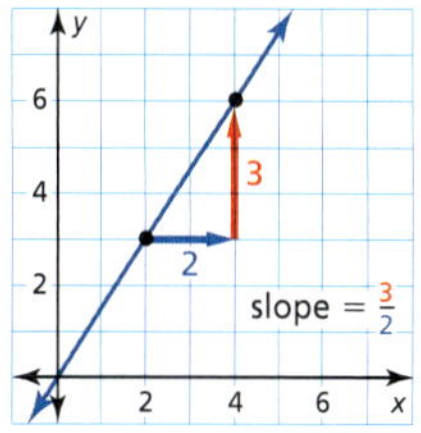

EXPLORATION 1 **Finding Slopes and y-Intercepts**

Work with a partner. Find the slope and y-intercept of each line.

a.

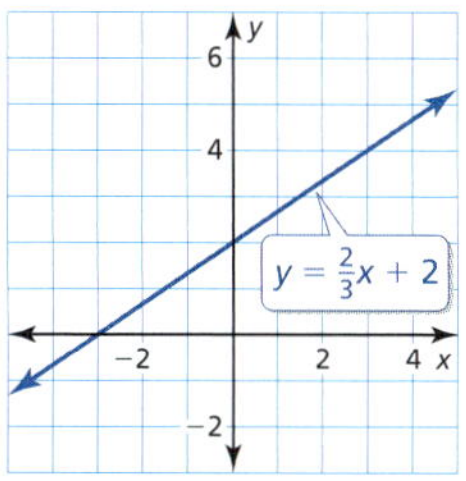

b.

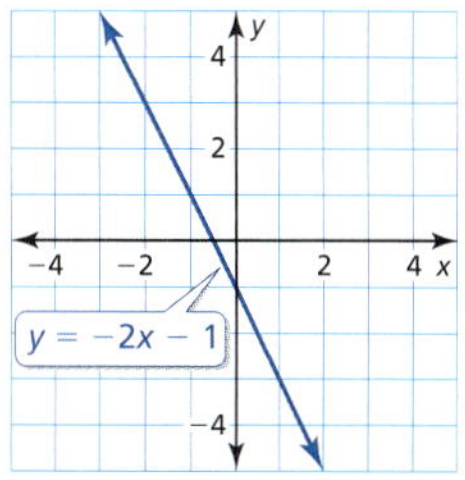

MAKING CONJECTURES

To be proficient in math, you first need to collect and organize data. Then make conjectures about the patterns you observe in the data.

EXPLORATION 2 **Writing a Conjecture**

Work with a partner. Graph each equation. Then copy and complete the table. Use the completed table to write a conjecture about the relationship between the graph of $y = mx + b$ and the values of m and b.

Equation	Description of graph	Slope of graph	y-Intercept
a. $y = -\frac{2}{3}x + 3$	Line	$-\frac{2}{3}$	3
b. $y = 2x - 2$			
c. $y = -x + 1$			
d. $y = x - 4$			

Communicate Your Answer

3. How can you describe the graph of the equation $y = mx + b$?
   - **a.** How does the value of m affect the graph of the equation?
   - **b.** How does the value of b affect the graph of the equation?
   - **c.** Check your answers to parts (a) and (b) by choosing one equation from Exploration 2 and (1) varying only m and (2) varying only b.

3. c. **(1)** *Sample answer:*

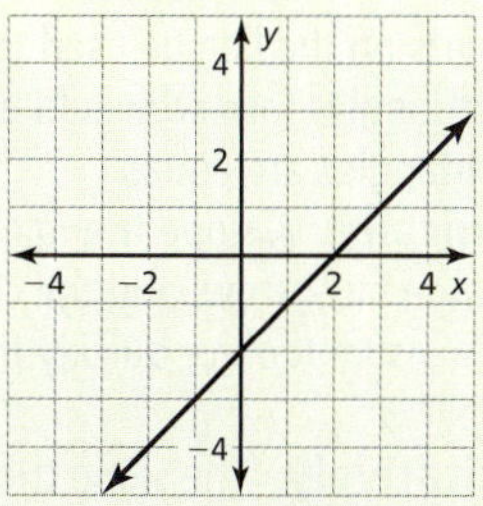

$y = x - 2$; slope: 1; y-intercept: -2

(2) *Sample answer:*

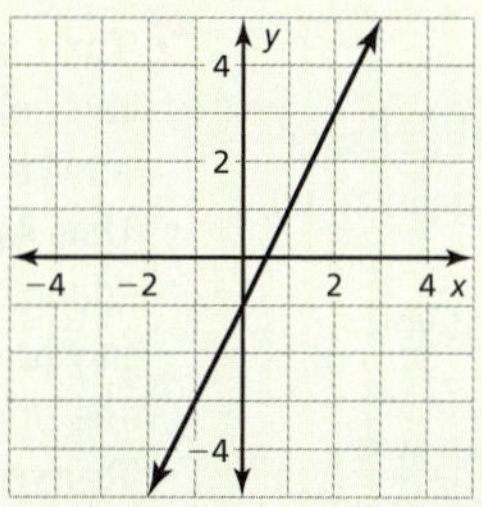

$y = 2x - 1$; slope: 2; y-intercept: -1

Dynamic Teaching Tools

- Dynamic Assessment & Progress Monitoring Tool
- Lesson Planning Tool
- Interactive Whiteboard Lesson Library
- Dynamic Classroom with Dynamic Investigations

ANSWERS

1. a. $\frac{2}{3}$; 2

b. -2; -1

2. a.

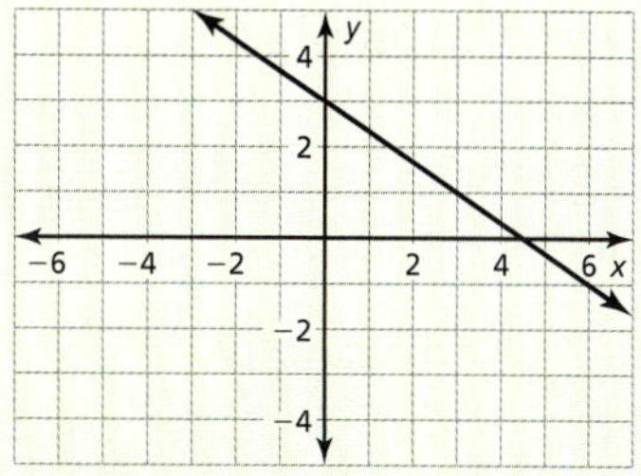

line; $-\frac{2}{3}$; 3

b.

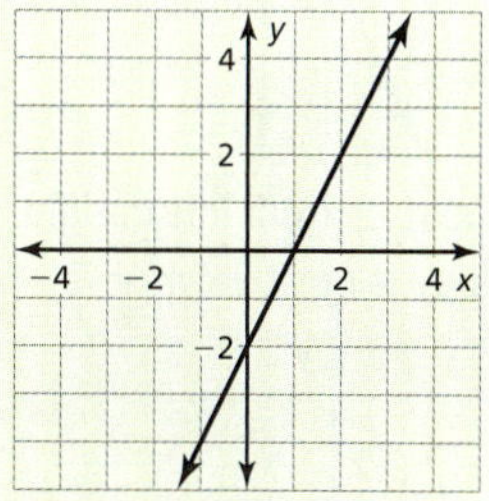

line; 2; -2

c.

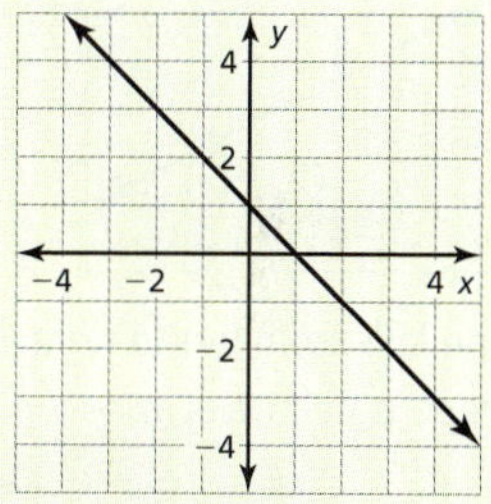

line; -1; 1

d.

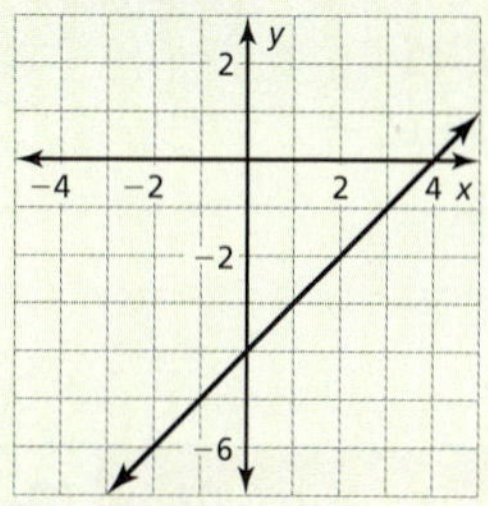

line; 1; -4

In the graph of $y = mx + b$, m is the slope of the line and b is the y-intercept of the line.

3. a line with slope m and y-intercept b

a. The value of m affects the steepness of the graph and whether the graph rises or falls from left to right.

b. The value of b affects the location of the y-intercept.

Differentiated Instruction

Visual/Kinesthetic

Have students graph the equations $y = 2x$, $y = x$, $y = -x$, and $y = -2x$ on the same coordinate plane, using a different color for each graph. Have students describe what happens to the line as the slope decreases.

Extra Example 1

Describe the slope of each line. Then find the slope.

a.

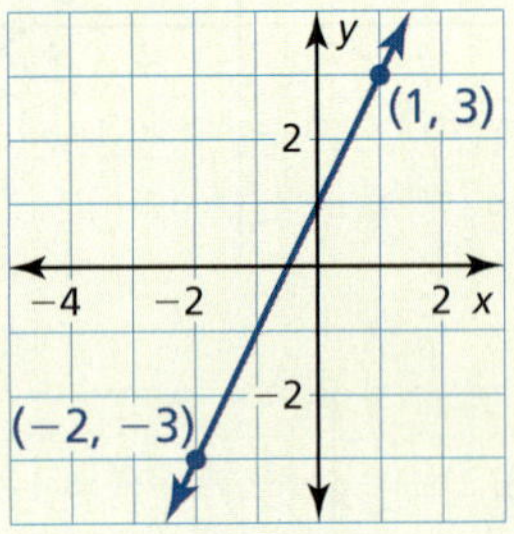

The line rises from left to right. The slope is positive; $m = 2$.

b.

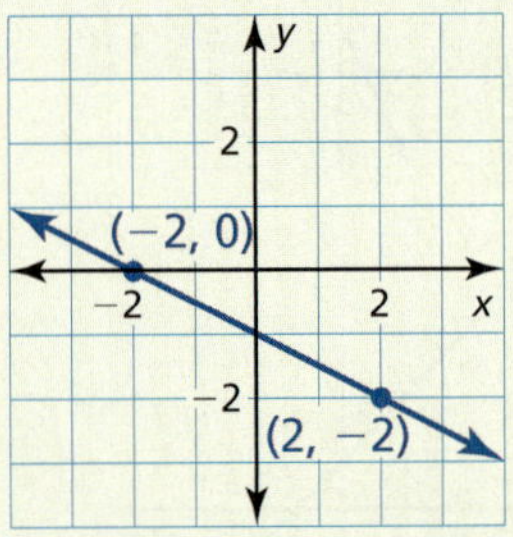

The line falls from left to right. The slope is negative; $m = -\frac{1}{2}$.

MONITORING PROGRESS ANSWERS

1. negative; $-\frac{2}{5}$
2. positive; $\frac{2}{3}$
3. positive; $\frac{7}{3}$

3.5 Lesson

Core Vocabulary

slope, *p. 136*
rise, *p. 136*
run, *p. 136*
slope-intercept form, *p. 138*
constant function, *p. 138*

Previous
dependent variable
independent variable

What You Will Learn

- Find the slope of a line.
- Use the slope-intercept form of a linear equation.
- Use slopes and y-intercepts to solve real-life problems.

The Slope of a Line

Core Concept

Slope

The **slope** m of a nonvertical line passing through two points (x_1, y_1) and (x_2, y_2) is the ratio of the **rise** (change in y) to the **run** (change in x).

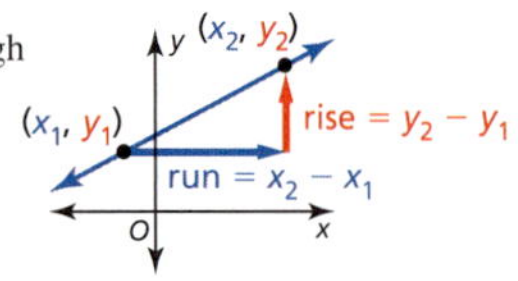

$$\text{slope} = m = \frac{\text{rise}}{\text{run}} = \frac{\text{change in } y}{\text{change in } x} = \frac{y_2 - y_1}{x_2 - x_1}$$

When the line rises from left to right, the slope is positive. When the line falls from left to right, the slope is negative.

EXAMPLE 1 Finding the Slope of a Line

Describe the slope of each line. Then find the slope.

a.

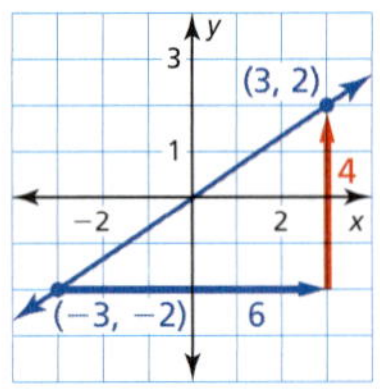

b.

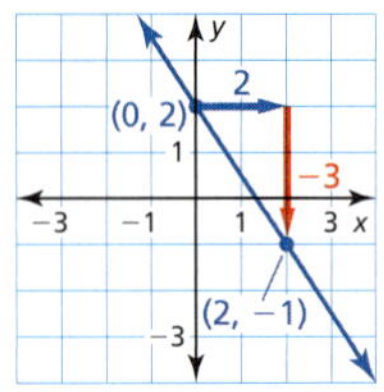

STUDY TIP

When finding slope, you can label either point as (x_1, y_1) and the other point as (x_2, y_2). The result is the same.

READING

In the slope formula, x_1 is read as "x sub one" and y_2 is read as "y sub two." The numbers 1 and 2 in x_1 and y_2 are called *subscripts*.

SOLUTION

a. The line rises from left to right. So, the slope is positive. Let $(x_1, y_1) = (-3, -2)$ and $(x_2, y_2) = (3, 2)$.

$$m = \frac{y_2 - y_1}{x_2 - x_1} = \frac{2 - (-2)}{3 - (-3)} = \frac{4}{6} = \frac{2}{3}$$

b. The line falls from left to right. So, the slope is negative. Let $(x_1, y_1) = (0, 2)$ and $(x_2, y_2) = (2, -1)$.

$$m = \frac{y_2 - y_1}{x_2 - x_1} = \frac{-1 - 2}{2 - 0} = \frac{-3}{2} = -\frac{3}{2}$$

Monitoring Progress

Help in English and Spanish at *BigIdeasMath.com*

Describe the slope of the line. Then find the slope.

1.

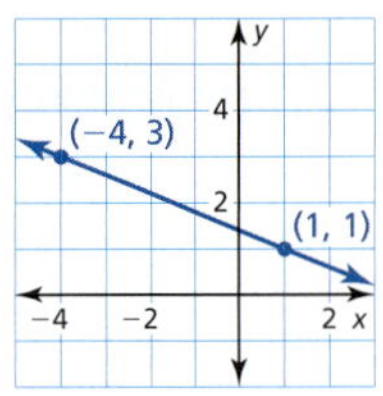

2.

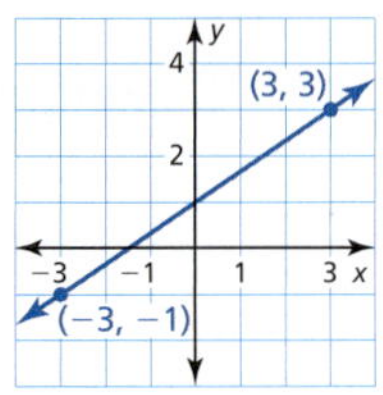

3.

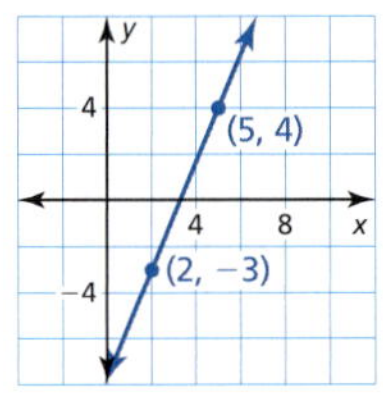

Laurie's Notes Teacher Actions

- Note the use of color in the definition and on the graph. The change in y and the vertical change arrow are both red. The change in x and the horizontal change arrow are both blue.
- Work through Example 1 as shown if your students are not yet secure with slope.
- **Common Misconception:** When finding the slope from two ordered pairs on the graph, the change arrows do not need to go from left to right. In part (a), if the change arrows began at (3, 2), then the change in y and the change in x would both be negative. The ratio (slope) would still be positive.

? **Turn and Talk:** "Can any two points on the line be used to find the slope? Explain." yes; The slope triangles formed would be similar and corresponding sides of similar triangles are equal.

- **Use Appropriate Tools Strategically:** A slope triangle is the triangle formed by the line, the change in x arrow, and the change in y arrow. For many students, it is a visual tool for quickly finding the slope.
- **Reason Abstractly and Quantitatively:** The slope triangle can be drawn above or below the line. The change arrows can be drawn starting with the left-most point or the right-most point. Can students explain why?

EXAMPLE 2 Finding Slope from a Table

The points represented by each table lie on a line. How can you find the slope of each line from the table? What is the slope of each line?

a.

x	y
4	20
7	14
10	8
13	2

b.

x	y
−1	2
1	2
3	2
5	2

c.

x	y
−3	−3
−3	0
−3	6
−3	9

STUDY TIP

As a check, you can plot the points represented by the table to verify that the line through them has a slope of −2.

SOLUTION

a. Choose any two points from the table and use the slope formula. Use the points $(x_1, y_1) = (4, 20)$ and $(x_2, y_2) = (7, 14)$.

$$m = \frac{y_2 - y_1}{x_2 - x_1} = \frac{14 - 20}{7 - 4} = \frac{-6}{3}, \text{ or } -2$$

▶ The slope is −2.

b. Note that there is no change in y. Choose any two points from the table and use the slope formula. Use the points $(x_1, y_1) = (-1, 2)$ and $(x_2, y_2) = (5, 2)$.

$$m = \frac{y_2 - y_1}{x_2 - x_1} = \frac{2 - 2}{5 - (-1)} = \frac{0}{6}, \text{ or } 0$$

The change in y is 0.

▶ The slope is 0.

c. Note that there is no change in x. Choose any two points from the table and use the slope formula. Use the points $(x_1, y_1) = (-3, 0)$ and $(x_2, y_2) = (-3, 6)$.

$$m = \frac{y_2 - y_1}{x_2 - x_1} = \frac{6 - 0}{-3 - (-3)} = \frac{6}{0} \quad ✗$$

The change in x is 0.

▶ Because division by zero is undefined, the slope of the line is undefined.

Monitoring Progress

Help in English and Spanish at *BigIdeasMath.com*

The points represented by the table lie on a line. How can you find the slope of the line from the table? What is the slope of the line?

4.

x	2	4	6	8
y	10	15	20	25

5.

x	5	5	5	5
y	−12	−9	−6	−3

Concept Summary

Slope

Positive slope

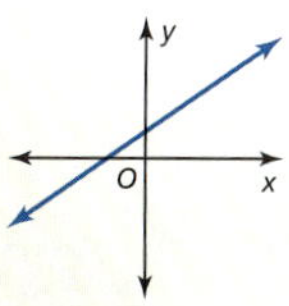

The line rises from left to right.

Negative slope

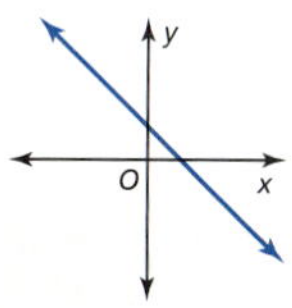

The line falls from left to right.

Slope of 0

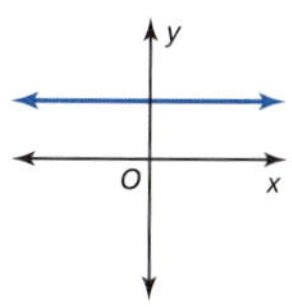

The line is horizontal.

Undefined slope

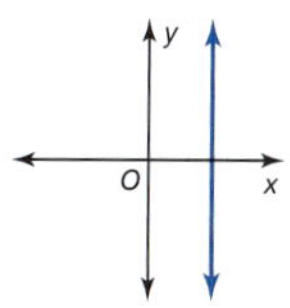

The line is vertical.

Extra Example 2

The points represented by each table lie on a line. How can you find the slope of each line from the table? What is the slope of each line?

a.

x	−3	−1	0	3
y	−1	3	5	11

Choose any two points from the table and use the slope formula. The slope is 2.

b.

x	−4	0	4	8
y	−2	−2	−2	−2

Note that there is no change in y. Choose any two points from the table and use the slope formula. The slope is 0.

c.

x	4	4	4	4
y	−5	−2	−1	4

Note that there is no change in x. Choose any two points from the table and use the slope formula. Because division by 0 is undefined, the slope is undefined.

MONITORING PROGRESS ANSWERS

4. Choose any 2 points from the table and use the slope formula; $\frac{5}{2}$

5. Choose any 2 points from the table and use the slope formula; undefined

Laurie's Notes Teacher Actions

? "The direction line states that the points represented in the table lie in a line. If this had not been stated, then how could you tell this?" Some students might suggest that the points be plotted. Other students might note that in part (a) the x-values are increasing by 3 and the y-values are decreasing by 6.

- **Teaching Tip:** In each part, you might have different groups or partners work with different pairs of ordered pairs. The resulting slopes should still be the same!
- Students can be confused by the results in parts (b) and (c). Discuss why $\frac{0}{6} \neq \frac{6}{0}$.
- Students should copy the *Concept Summary* into their notebooks.

English Language Learners

Labels

Have students use colors to identify and label the slope and y-intercept in equations written in slope-intercept form.

$y = 2x + 1$

slope — y-intercept

Extra Example 3

Find the slope and the y-intercept of the graph of each linear equation.

a. $y = 2x - 3$

The slope is 2, and the y-intercept is -3.

b. $y = -4.5$

The slope is 0, and the y-intercept is -4.5.

c. $6x - y = 8$

The slope is 6, and the y-intercept is -8.

MONITORING PROGRESS ANSWERS

6. slope: -6; y-intercept: 1

7. slope: 0; y-intercept: 8

8. slope: $-\frac{1}{4}$; y-intercept: $-\frac{5}{2}$

Using the Slope-Intercept Form of a Linear Equation

Core Concept

Slope-Intercept Form

Words A linear equation written in the form $y = mx + b$ is in **slope-intercept form**. The slope of the line is m, and the y-intercept of the line is b.

Algebra $y = mx + b$

slope (m) — y-intercept (b)

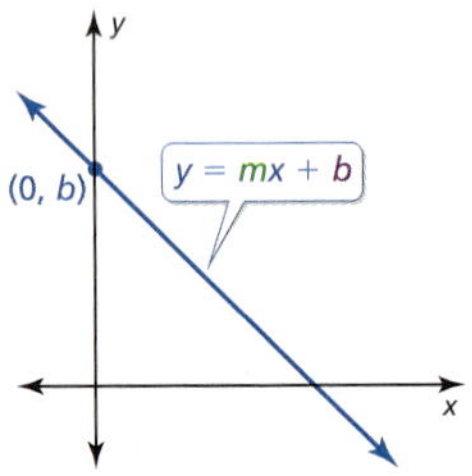

A linear equation written in the form $y = 0x + b$, or $y = b$, is a **constant function**. The graph of a constant function is a horizontal line.

EXAMPLE 3 Identifying Slopes and y-Intercepts

Find the slope and the y-intercept of the graph of each linear equation.

a. $y = 3x - 4$ **b.** $y = 6.5$ **c.** $-5x - y = -2$

SOLUTION

a. $y = mx + b$ — Write the slope-intercept form.

slope — y-intercept

$y = 3x + (-4)$ — Rewrite the original equation in slope-intercept form.

▶ The slope is 3, and the y-intercept is -4.

STUDY TIP

For a constant function, every input has the same output. For instance, in Example 3b, every input has an output of 6.5.

b. The equation represents a constant function. The equation can also be written as $y = 0x + 6.5$.

▶ The slope is 0, and the y-intercept is 6.5.

c. Rewrite the equation in slope-intercept form by solving for y.

$-5x - y = -2$	Write the original equation.
$+5x \qquad +5x$	Add $5x$ to each side.
$-y = 5x - 2$	Simplify.
$\frac{-y}{-1} = \frac{5x - 2}{-1}$	Divide each side by -1.
$y = -5x + 2$	Simplify.

▶ The slope is -5, and the y-intercept is 2.

STUDY TIP

When you rewrite a linear equation in slope-intercept form, you are expressing y as a function of x.

Monitoring Progress

Help in English and Spanish at *BigIdeasMath.com*

Find the slope and the y-intercept of the graph of the linear equation.

6. $y = -6x + 1$ **7.** $y = 8$ **8.** $x + 4y = -10$

Laurie's Notes — Teacher Actions

- Write the *Core Concept*. This should be familiar to students from middle school. If Exploration 2 was done, connect the *Core Concept* to the results of the exploration.
- Discuss the special case of a constant function, when the slope of the line is 0. The graph of a constant function is horizontal.
- Example 3 is an important skill that you want to make sure students are secure with. In Section 1.5, students practiced this skill of rewriting a linear equation, expressing y as a function of x.
- **Monitoring Progress:** In Question 8, check that students have the correct slope of $-\frac{1}{4}$.

EXAMPLE 4 Using Slope-Intercept Form to Graph

Graph $2x + y = 2$. Identify the x-intercept.

SOLUTION

Step 1 Rewrite the equation in slope-intercept form.

$$y = -2x + 2$$

Step 2 Find the slope and the y-intercept.

$$m = -2 \text{ and } b = 2$$

Step 3 The y-intercept is 2. So, plot (0, 2).

Step 4 Use the slope to find another point on the line.

$$\text{slope} = \frac{\text{rise}}{\text{run}} = \frac{-2}{1}$$

Plot the point that is 1 unit right and 2 units down from (0, 2). Draw a line through the two points.

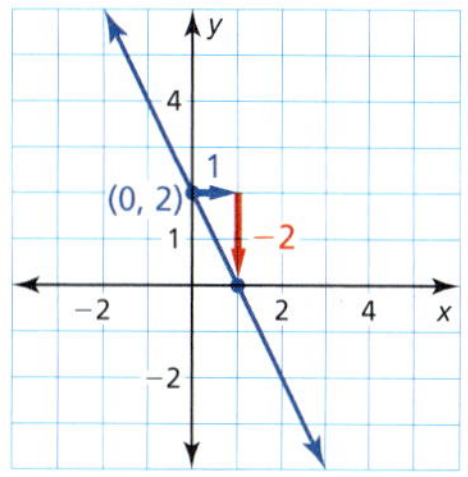

▶ The line crosses the x-axis at (1, 0). So, the x-intercept is 1.

STUDY TIP

You can use the slope to find points on a line in either direction. In Example 4, note that the slope can be written as $\frac{2}{-1}$. So, you could move 1 unit left and 2 units up from (0, 2) to find the point (−1, 4).

REMEMBER

You can also find the x-intercept by substituting 0 for y in the equation $2x + y = 2$ and solving for x.

EXAMPLE 5 Graphing from a Verbal Description

A linear function g models a relationship in which the dependent variable increases 3 units for every 1 unit the independent variable increases. Graph g when $g(0) = 3$. Identify the slope, y-intercept, and x-intercept of the graph.

SOLUTION

Because the function g is linear, it has a constant rate of change. Let x represent the independent variable and y represent the dependent variable.

Step 1 Find the slope. When the dependent variable increases by 3, the change in y is $+3$. When the independent variable increases by 1, the change in x is $+1$.

So, the slope is $\frac{3}{1}$, or 3.

Step 2 Find the y-intercept. The statement $g(0) = 3$ indicates that when $x = 0$, $y = 3$. So, the y-intercept is 3. Plot (0, 3).

Step 3 Use the slope to find another point on the line. A slope of 3 can be written as $\frac{-3}{-1}$. Plot the point that is 1 unit left and 3 units down from (0, 3). Draw a line through the two points. The line crosses the x-axis at $(-1, 0)$. So, the x-intercept is -1.

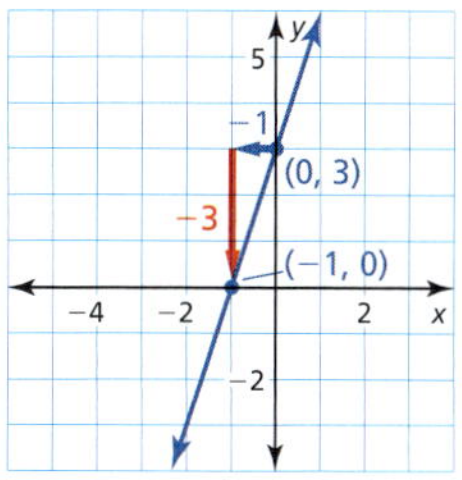

▶ The slope is 3, the y-intercept is 3, and the x-intercept is -1.

Monitoring Progress

Help in English and Spanish at *BigIdeasMath.com*

Graph the linear equation. Identify the x-intercept.

9. $y = 4x - 4$ **10.** $3x + y = -3$ **11.** $x + 2y = 6$

12. A linear function h models a relationship in which the dependent variable decreases 2 units for every 5 units the independent variable increases. Graph h when $h(0) = 4$. Identify the slope, y-intercept, and x-intercept of the graph.

Laurie's Notes Teacher Actions

- ? "What form is the equation $2x + y = 2$?" standard form "How would you graph this line?" Find the x- and y-intercepts and draw the line between them. "Can the equation be written in slope-intercept form?" yes
- **Whiteboarding:** Give partners time to work the example. Circulate and ask advancing questions to assist students in making progress.
- Discuss the *Study Tip*.
- After completing Example 5, ask students to give a *Thumbs Up* to assess their comfort with graphing a line in slope-intercept form.

Extra Example 4

Graph $y + x = -2$. Identify the x-intercept.

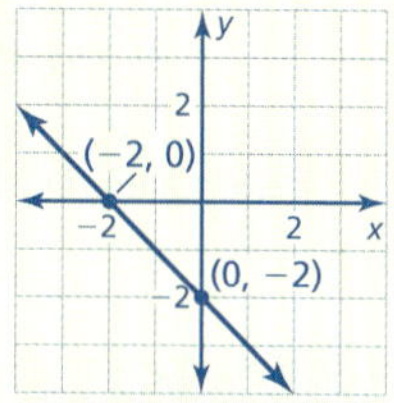

The line crosses the x-axis at $(-2, 0)$. So, the x-intercept is -2.

Extra Example 5

A linear function g models a relationship in which the dependent variable increases 2 units for every 3 units the independent variable increases. Graph g when $g(0) = 2$. Identify the slope, y-intercept, and x-intercept of the graph.

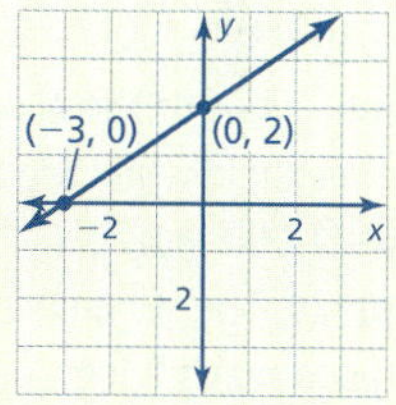

The slope is $\frac{2}{3}$, the y-intercept is 2, and the x-intercept is -3.

MONITORING PROGRESS ANSWERS

9.

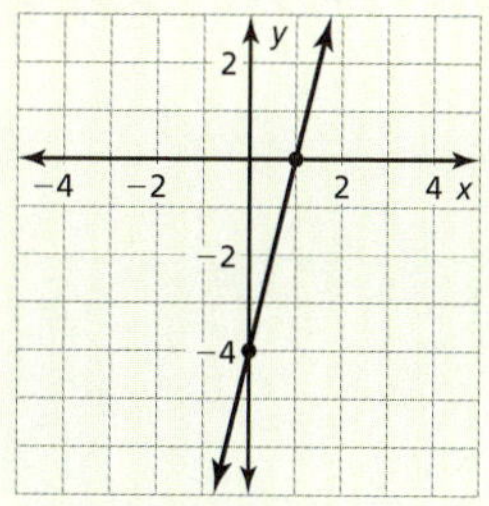

x-intercept: 1

10.

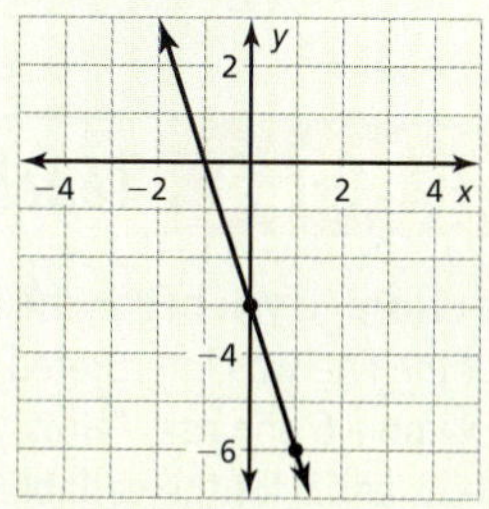

x-intercept: -1

11–12. See Additional Answers.

Extra Example 6

The elevation h (in feet) of a submersible is modeled by the function $h(t) = 550t - 11{,}000$, where t is the time (in minutes) since the submersible began to ascend.

a. Graph the function and identify its domain and range.

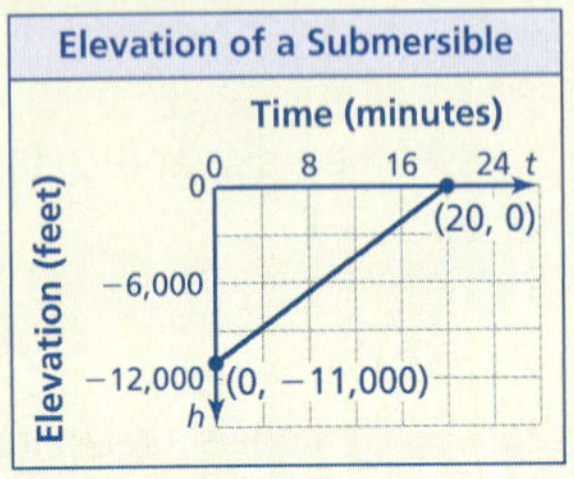

domain: $0 \le t \le 20$;
range: $-11{,}000 \le h \le 0$

b. Interpret the slope and the intercepts of the graph.

slope: 550; The submersible ascends at a rate of 550 ft/min.
h-intercept: −11,000; The elevation of the submersible when the ascent begins is −11,000 feet.
t-intercept: 20; The submersible takes 20 minutes to reach 0 feet, or sea level.

MONITORING PROGRESS ANSWER

13. See Additional Answers.

Solving Real-Life Problems

In most real-life problems, slope is interpreted as a rate, such as miles per hour, dollars per hour, or people per year.

EXAMPLE 6 **Modeling with Mathematics**

A submersible that is exploring the ocean floor begins to ascend to the surface. The elevation h (in feet) of the submersible is modeled by the function $h(t) = 650t - 13{,}000$, where t is the time (in minutes) since the submersible began to ascend.

a. Graph the function and identify its domain and range.

b. Interpret the slope and the intercepts of the graph.

SOLUTION

1. **Understand the Problem** You know the function that models the elevation. You are asked to graph the function and identify its domain and range. Then you are asked to interpret the slope and intercepts of the graph.

2. **Make a Plan** Use the slope-intercept form of a linear equation to graph the function. Only graph values that make sense in the context of the problem. Examine the graph to interpret the slope and the intercepts.

3. **Solve the Problem**

STUDY TIP
Because t is the independent variable, the horizontal axis is the t-axis and the graph will have a "t-intercept." Similarly, the vertical axis is the h-axis and the graph will have an "h-intercept."

a. The time t must be greater than or equal to 0. The elevation h is below sea level and must be less than or equal to 0. Use the slope of 650 and the h-intercept of −13,000 to graph the function in Quadrant IV.

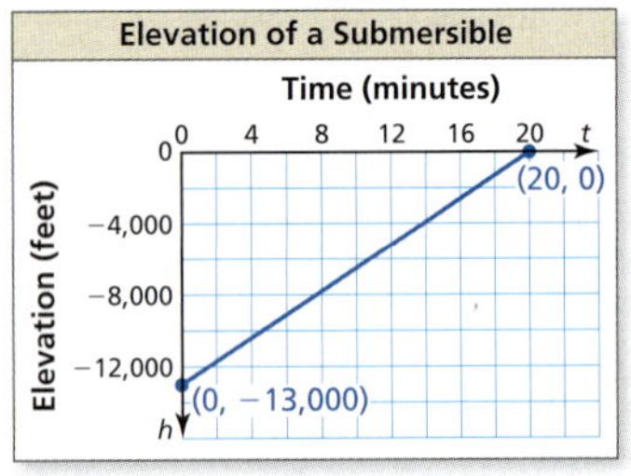

▶ The domain is $0 \le t \le 20$, and the range is $-13{,}000 \le h \le 0$.

b. The slope is 650. So, the submersible ascends at a rate of 650 feet per minute. The h-intercept is −13,000. So, the elevation of the submersible after 0 minutes, or when the ascent begins, is −13,000 feet. The t-intercept is 20. So, the submersible takes 20 minutes to reach an elevation of 0 feet, or sea level.

4. **Look Back** You can check that your graph is correct by substituting the t-intercept for t in the function. If $h = 0$ when $t = 20$, the graph is correct.

$h = 650(20) - 13{,}000$ Substitute 20 for t in the original equation.

$h = 0$ ✓ Simplify.

Monitoring Progress Help in English and Spanish at *BigIdeasMath.com*

13. **WHAT IF?** The elevation of the submersible is modeled by $h(t) = 500t - 10{,}000$. (a) Graph the function and identify its domain and range. (b) Interpret the slope and the intercepts of the graph.

Laurie's Notes **Teacher Actions**

- **Connection:** If you began the lesson with the *Motivate*, then the slope of the line was interpreted as a rate.
- ? Read Example 6 and ask, "How would the equation appear if we had used x and y for the variables?" $y = 650x - 13{,}000$ "What values will make sense for the domain and range in this situation?" The time (domain) should be positive and the elevation (range) will be negative.
- This is an example where a quick sketch can be done once students identify the slope (650) and y-intercept (−13,000).
- ? **Turn and Talk:** "How can you determine the t-intercept for this problem?" Substitute 0 for $h(t)$ and solve for t.

Closure

- **Point of Most Significance:** Ask students to identify, aloud or on a paper to be collected, the most significant point (or part) in the lesson that aided their learning.

3.5 Exercises

Dynamic Solutions available at *BigIdeasMath.com*

Vocabulary and Core Concept Check

1. **COMPLETE THE SENTENCE** The ________ of a nonvertical line passing through two points is the ratio of the rise to the run.
2. **VOCABULARY** What is a constant function? What is the slope of a constant function?
3. **WRITING** What is the slope-intercept form of a linear equation? Explain why this form is called the slope-intercept form.
4. **WHICH ONE DOESN'T BELONG?** Which equation does *not* belong with the other three? Explain your reasoning.

$y = -5x - 1$ $\quad$ $2x - y = 8$ $\quad$ $y = x + 4$ $\quad$ $y = -3x + 13$

Monitoring Progress and Modeling with Mathematics

In Exercises 5–8, describe the slope of the line. Then find the slope. *(See Example 1.)*

5. 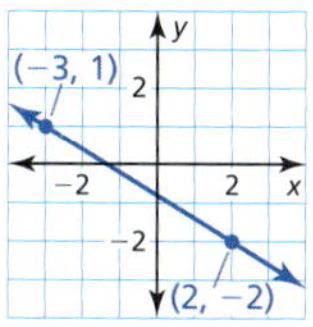

6. 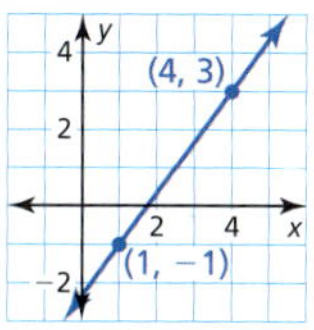

7. 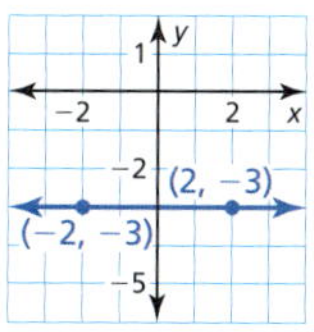

8. 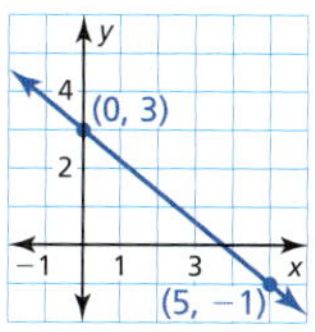

In Exercises 9–12, the points represented by the table lie on a line. Find the slope of the line. *(See Example 2.)*

9.

x	−9	−5	−1	3
y	−2	0	2	4

10.

x	−1	2	5	8
y	−6	−6	−6	−6

11.

x	0	0	0	0
y	−4	0	4	8

12.

x	−4	−3	−2	−1
y	2	−5	−12	−19

13. **ANALYZING A GRAPH** The graph shows the distance y (in miles) that a bus travels in x hours. Find and interpret the slope of the line.

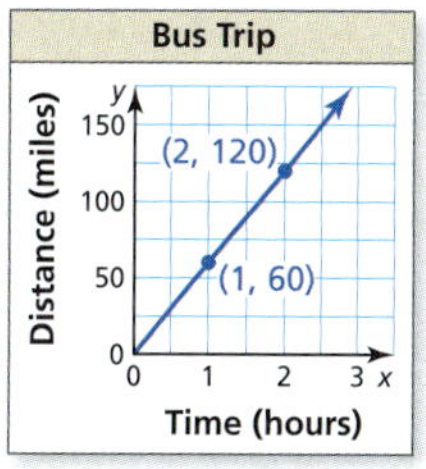

14. **ANALYZING A TABLE** The table shows the amount x (in hours) of time you spend at a theme park and the admission fee y (in dollars) to the park. The points represented by the table lie on a line. Find and interpret the slope of the line.

Time (hours), x	Admission (dollars), y
6	54.99
7	54.99
8	54.99

Assignment Guide and Homework Check

ASSIGNMENT

Basic: 1–4, 5–33 odd, 37, 41, 46, 50, 54–60

Average: 1–4, 6–42 even, 46, 50, 54–60

Advanced: 1–4, 8–14 even, 20–24 even, 30–52 even, 54–60

HOMEWORK CHECK

Basic: 5, 15, 25, 33, 37

Average: 6, 16, 26, 34, 38

Advanced: 8, 20, 32, 34, 38

ANSWERS

1. slope
2. a linear function of the form $y = b$; 0
3. $y = mx + b$; The equation gives the slope m and the y-intercept b.
4. $2x - y = 8$; It is in standard form, not slope-intercept form.
5. negative; $-\frac{3}{5}$
6. positive; $\frac{4}{3}$
7. zero; 0
8. negative; $-\frac{4}{5}$
9. $\frac{1}{2}$
10. 0
11. undefined
12. −7
13. $m = 60$; The bus is traveling at a speed of 60 miles per hour.
14. $m = 0$; The admission fee does not change based on the time spent in the park.

Dynamic Teaching Tools

- Dynamic Assessment & Progress Monitoring Tool
- Interactive Whiteboard Lesson Library
- Dynamic Classroom with Dynamic Investigations

ANSWERS

15. slope: -3; y-intercept: 2

16. slope: 4; y-intercept: -7

17. slope: 6; y-intercept: 0

18. slope: 0; y-intercept: -1

19. slope: 2; y-intercept: 4

20. slope: -1; y-intercept: -6

21. slope: 5; y-intercept: 8

22. slope: 7; y-intercept: $\frac{1}{2}$

23. To be in slope-intercept form the equation needs to be solved for y, not x; $y = -\frac{1}{4}x$; The slope is $-\frac{1}{4}$ and the y-intercept is 0.

24. To be in slope-intercept form, the equation needs to be rewritten as $y = 3x + (-6)$; The slope is 3 and the y-intercept is -6.

25.

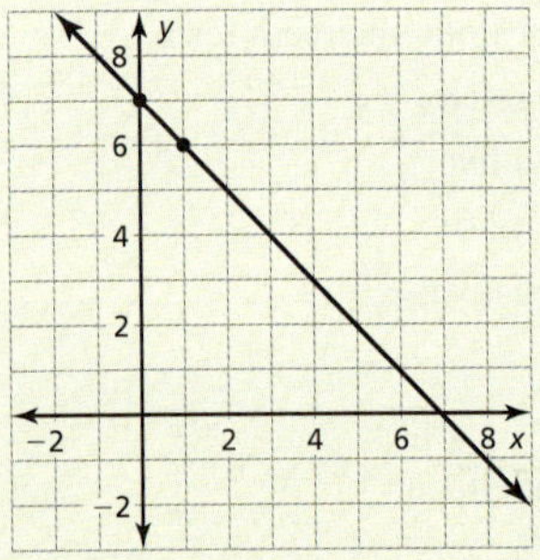

x-intercept: 7

26.

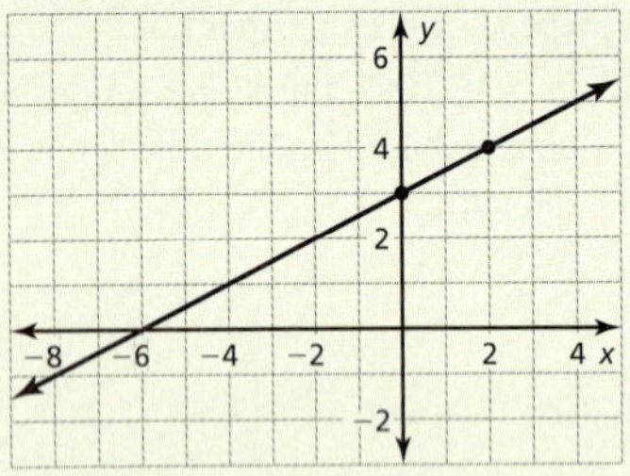

x-intercept: -6

In Exercises 15–22, find the slope and the y-intercept of the graph of the linear equation. *(See Example 3.)*

15. $y = -3x + 2$ **16.** $y = 4x - 7$

17. $y = 6x$ **18.** $y = -1$

19. $-2x + y = 4$ **20.** $x + y = -6$

21. $-5x = 8 - y$ **22.** $0 = 1 - 2y + 14x$

ERROR ANALYSIS In Exercises 23 and 24, describe and correct the error in finding the slope and the y-intercept of the graph of the equation.

23. ✗ $x = -4y$

The slope is -4, and the y-intercept is 0.

24. ✗ $y = 3x - 6$

The slope is 3, and the y-intercept is 6.

In Exercises 25–32, graph the linear equation. Identify the x-intercept. *(See Example 4.)*

25. $y = -x + 7$ **26.** $y = \frac{1}{2}x + 3$

27. $y = 2x$ **28.** $y = -x$

29. $3x + y = -1$ **30.** $x + 4y = 8$

31. $-y + 5x = 0$ **32.** $2x - y + 6 = 0$

In Exercises 33 and 34, graph the function with the given description. Identify the slope, y-intercept, and x-intercept of the graph. *(See Example 5.)*

33. A linear function f models a relationship in which the dependent variable decreases 4 units for every 2 units the independent variable increases. The value of the function at 0 is -2.

34. A linear function h models a relationship in which the dependent variable increases 1 unit for every 5 units the independent variable decreases. The value of the function at 0 is 3.

35. GRAPHING FROM A VERBAL DESCRIPTION A linear function r models the growth of your right index fingernail. The length of the fingernail increases 0.7 millimeter every week. Graph r when $r(0) = 12$. Identify the slope and interpret the y-intercept of the graph.

36. GRAPHING FROM A VERBAL DESCRIPTION A linear function m models the amount of milk sold by a farm per month. The amount decreases 500 gallons for every \$1 increase in price. Graph m when $m(0) = 3000$. Identify the slope and interpret the x- and y-intercepts of the graph.

37. MODELING WITH MATHEMATICS The function shown models the depth d (in inches) of snow on the ground during the first 9 hours of a snowstorm, where t is the time (in hours) after the snowstorm begins. *(See Example 6.)*

a. Graph the function and identify its domain and range.

b. Interpret the slope and the d-intercept of the graph.

38. MODELING WITH MATHEMATICS The function $c(x) = 0.5x + 70$ represents the cost c (in dollars) of renting a truck from a moving company, where x is the number of miles you drive the truck.

a. Graph the function and identify its domain and range.

b. Interpret the slope and the c-intercept of the graph.

39. COMPARING FUNCTIONS A linear function models the cost of renting a truck from a moving company. The table shows the cost y (in dollars) when you drive the truck x miles. Graph the function and compare the slope and the y-intercept of the graph with the slope and the c-intercept of the graph in Exercise 38.

Miles, x	Cost (dollars), y
0	40
50	80
100	120

27.

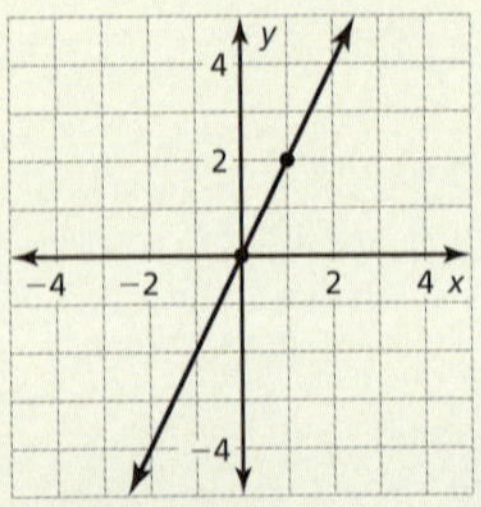

x-intercept: 0

28.

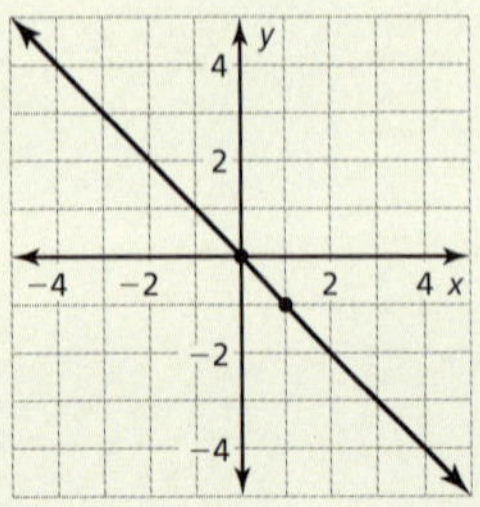

x-intercept: 0

29.

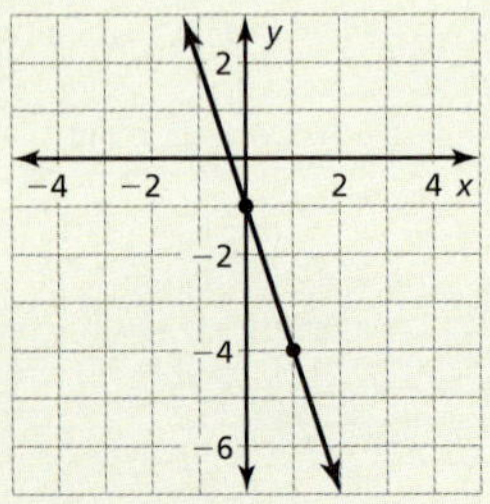

x-intercept: $-\frac{1}{3}$

30–39. See Additional Answers.

ERROR ANALYSIS In Exercises 40 and 41, describe and correct the error in graphing the function.

40.

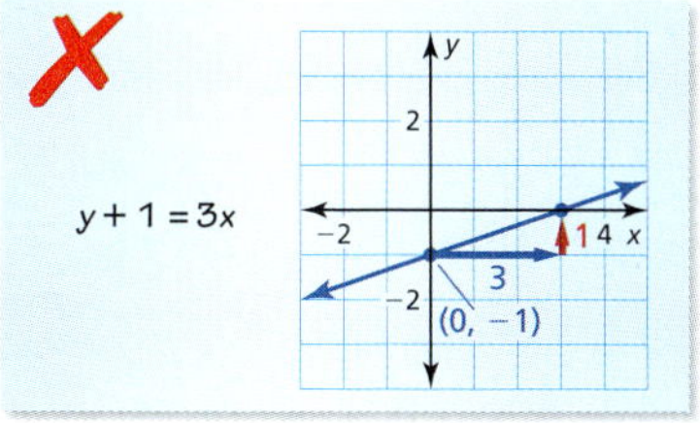

41.

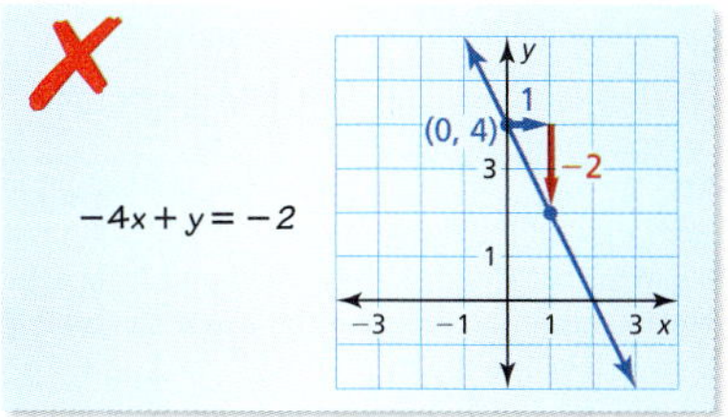

42. **MATHEMATICAL CONNECTIONS** Graph the four equations in the same coordinate plane.

$$3y = -x - 3$$
$$2y - 14 = 4x$$
$$4x - 3 - y = 0$$
$$x - 12 = -3y$$

a. What enclosed shape do you think the lines form? Explain.

b. Write a conjecture about the equations of parallel lines.

43. **MATHEMATICAL CONNECTIONS** The graph shows the relationship between the width y and the length x of a rectangle in inches. The perimeter of a second rectangle is 10 inches less than the perimeter of the first rectangle.

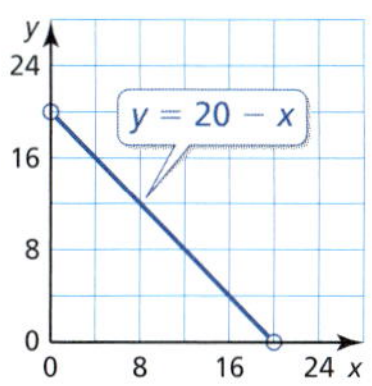

a. Graph the relationship between the width and length of the second rectangle.

b. How does the graph in part (a) compare to the the graph shown?

44. **MATHEMATICAL CONNECTIONS** The graph shows the relationship between the base length x and the side length (of the two equal sides) y of an isosceles triangle in meters. The perimeter of a second isosceles triangle is 8 meters more than the perimeter of the first triangle.

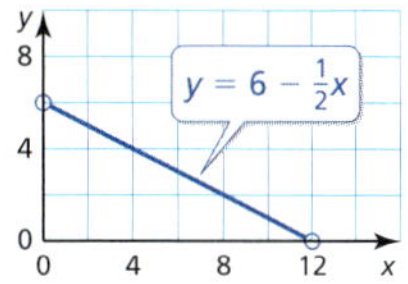

a. Graph the relationship between the base length and the side length of the second triangle.

b. How does the graph in part (a) compare to the graph shown?

45. **ANALYZING EQUATIONS** Determine which of the equations could be represented by each graph.

$y = -3x + 8$	$y = -x - \frac{4}{3}$
$y = -7x$	$y = 2x - 4$
$y = \frac{7}{4}x - \frac{1}{4}$	$y = \frac{1}{3}x + 5$
$y = -4x - 9$	$y = 6$

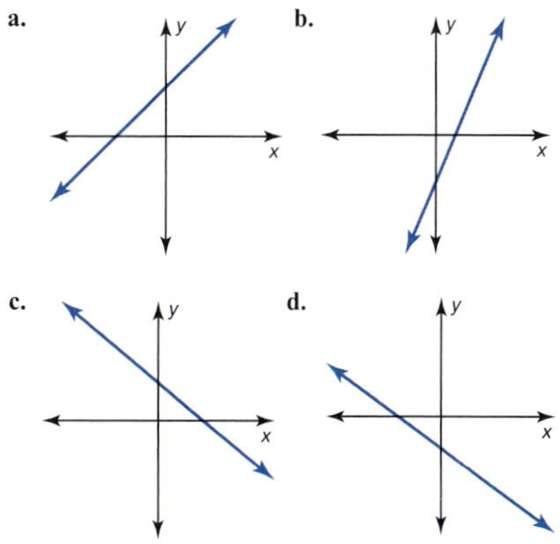

46. **MAKING AN ARGUMENT** Your friend says that you can write the equation of any line in slope-intercept form. Is your friend correct? Explain your reasoning.

ANSWERS

40. The slope is 3, but the graph shows a slope of $\frac{1}{3}$.

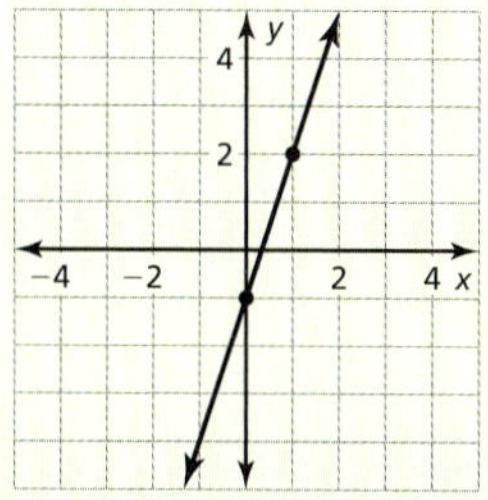

41. The slope was interpreted as the y-intercept, and the y-intercept was interpreted as the slope.

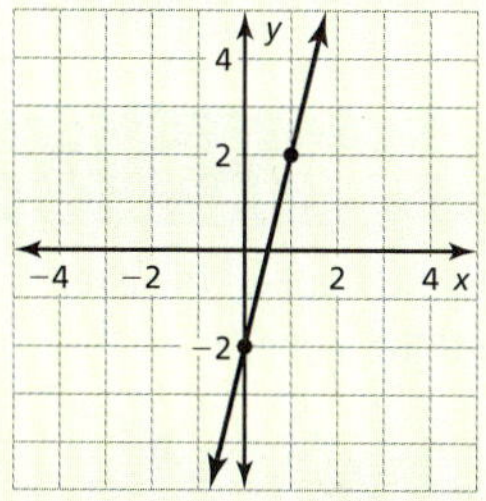

42. See Additional Answers.

43. a.

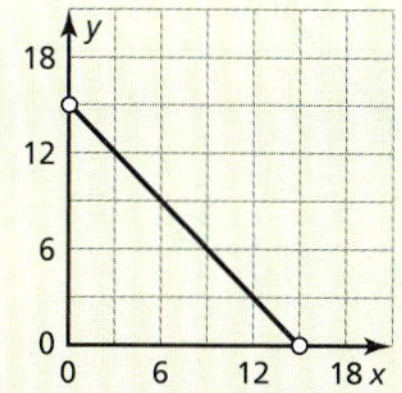

b. The slope of each graph is the same. The open circles in the graph in part (a) are closer to the origin.

44. a.

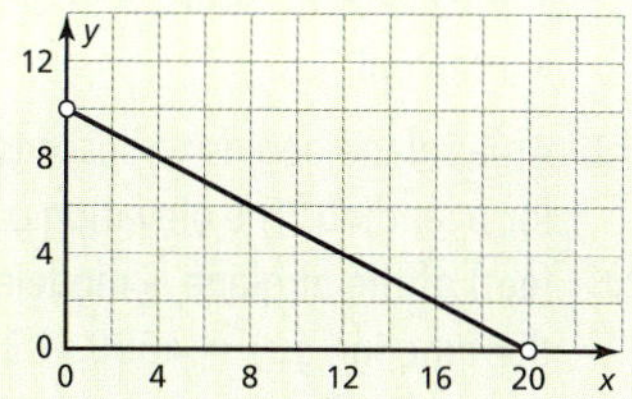

b. The slope of each graph is the same. The open circles in the graph in part (a) are farther from the origin.

45. a. $y = \frac{1}{3}x + 5$

b. $y = \frac{7}{4}x - \frac{1}{4}$; $y = 2x - 4$

c. $y = -3x + 8$

d. $y = -4x - 9$; $y = -x - \frac{4}{3}$

46. no; The equation of a vertical line cannot be written in slope-intercept form because its slope is undefined and it either has no y-intercept or infinite y-intercepts.

ANSWERS

47. *Sample answer:* the ratio of the rise to the run; the ratio of the change in y to the change in x

48–50. See Additional Answers.

51. $k = \frac{1}{8}$

52. $k = -12$

53–57. See Additional Answers.

58. linear; It can be rewritten as $y = \frac{1}{15}x - \frac{1}{5}$.

59. linear; It can be rewritten as $y = -3x + 12$.

60. nonlinear; It cannot be rewritten in the form $y = mx + b$.

Mini-Assessment

1. Describe the slope of the line. Then find the slope.

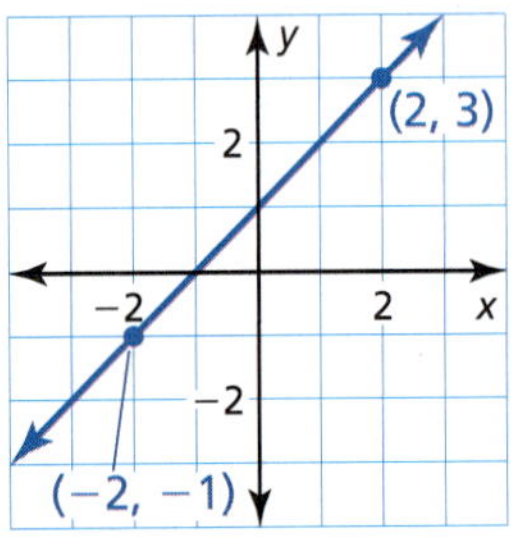

The line rises from left to right. The slope is positive; $m = 1$.

2. Find the slope and the y-intercept of the graph of the linear equation $y - 5x = -6$. slope: 5; y-intercept: -6

3. An airplane begins to descend for a landing. The elevation g (in feet) of the airplane is modeled by the function $g(t) = 4800 - 300t$, where t is the time (in minutes) since the airplane began to descend.
   a. Use the graph of the function to identify its domain and range. domain: $0 \le t \le 16$; range: $0 \le g \le 4800$
   b. Interpret the slope and the intercepts of the graph. slope: −300; The airplane descends at a rate of 300 ft/min. g-intercept: 4800; The elevation of the airplane when the descent begins is 4800 ft. t-intercept: 16; The airplane takes 16 min to land.

47. **WRITING** Write the definition of the slope of a line in two different ways.

48. **THOUGHT PROVOKING** Your family goes on vacation to a beach 300 miles from your house. You reach your destination 6 hours after departing. Draw a graph that describes your trip. Explain what each part of your graph represents.

49. **ANALYZING A GRAPH** The graphs of the functions $g(x) = 6x + a$ and $h(x) = 2x + b$, where a and b are constants, are shown. They intersect at the point (p, q).

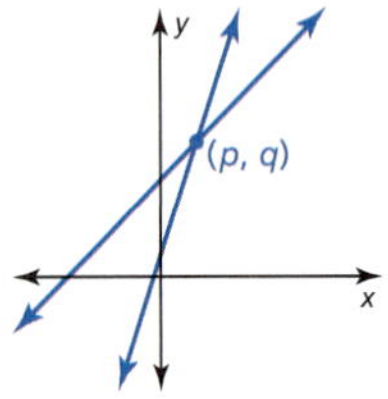

   a. Label the graphs of g and h.
   b. What do a and b represent?
   c. Starting at the point (p, q), trace the graph of g until you get to the point with the x-coordinate $p + 2$. Mark this point C. Do the same with the graph of h. Mark this point D. How much greater is the y-coordinate of point C than the y-coordinate of point D?

50. **HOW DO YOU SEE IT?** You commute to school by walking and by riding a bus. The graph represents your commute.

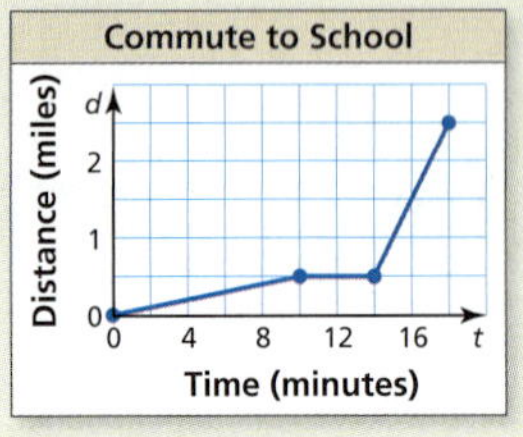

   a. Describe your commute in words.
   b. Calculate and interpret the slopes of the different parts of the graph.

PROBLEM SOLVING In Exercises 51 and 52, find the value of k so that the graph of the equation has the given slope or y-intercept.

51. $y = 4kx - 5$; $m = \frac{1}{2}$

52. $y = -\frac{1}{3}x + \frac{5}{6}k$; $b = -10$

53. **ABSTRACT REASONING** To show that the slope of a line is constant, let (x_1, y_1) and (x_2, y_2) be any two points on the line $y = mx + b$. Use the equation of the line to express y_1 in terms of x_1 and y_2 in terms of x_2. Then use the slope formula to show that the slope between the points is m.

Maintaining Mathematical Proficiency

Reviewing what you learned in previous grades and lessons

Find the coordinates of the figure after the transformation. *(Skills Review Handbook)*

54. Translate the rectangle 4 units left.

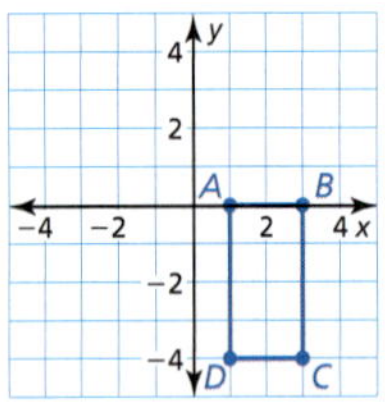

55. Dilate the triangle with respect to the origin using a scale factor of 2.

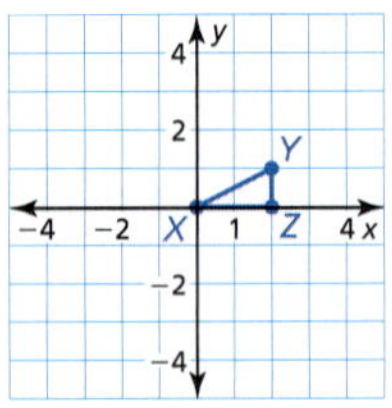

56. Reflect the trapezoid in the y-axis.

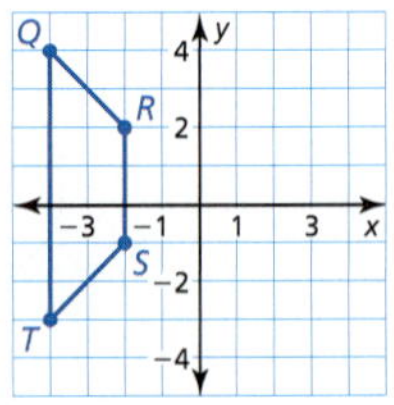

Determine whether the equation represents a *linear* or *nonlinear* function. Explain. *(Section 3.2)*

57. $y - 9 = \frac{2}{x}$

58. $x = 3 + 15y$

59. $\frac{x}{4} + \frac{y}{12} = 1$

60. $y = 3x^4 - 6$

If students need help...	*If students got it...*
Resources by Chapter • Practice A and Practice B • Puzzle Time	Resources by Chapter • Enrichment and Extension • Cumulative Review
Student Journal • Practice	Start the *next* Section
Differentiating the Lesson Skills Review Handbook	

Laurie's Notes

Overview of Section 3.6

Introduction

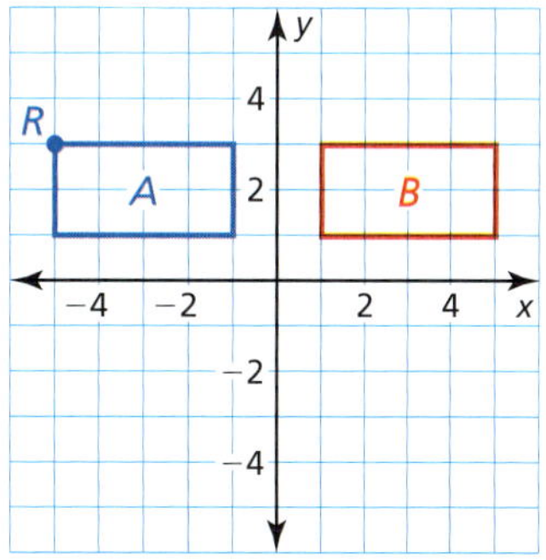

- Students worked with transformations of geometric figures in middle school, where they were expected to describe the effect of dilations, translations, rotations, and reflections on two-dimensional figures using coordinates.
- Knowing how the vertices are mapped from the preimage to the image allows a student to determine when figure B is the result of a reflection, rotation, or translation of figure A.
- If $R(-5, 3)$ is mapped to $R'(5, 3)$, then it is a reflection. If $R(-5, 3)$ is mapped to $R'(1, 3)$, then it is a translation. If $R(-5, 3)$ is mapped to $R'(5, 1)$, then it is a rotation.

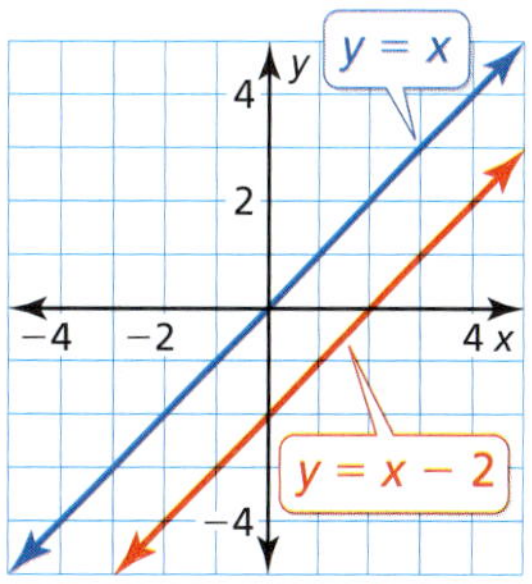

- In this lesson, students will perform transformations on lines. Without knowing the mapping rule, the line $y = x$ could be translated horizontally or vertically to the second line.
- Transformations are easier to visually identify for quadratics and absolute value functions than for linear functions.
- It is through the function notation that the transformation is identified, and that is more an abstract understanding than visual. Graphing $f(x) = 2x - 4$ is not difficult. Viewing it as a stretch by a factor of 2 of $f(x) = x$ followed by a translation 4 units down is what is challenging for students.
- **Connection:** The transformations defined in this lesson will be the same for absolute value functions (Section 3.7) and quadratic functions (Section 8.4). In later mathematics courses, students will study transformations on other functions, such as exponential, trigonometric, and rational.
- Transformations introduced in this lesson include translations, reflections, stretches, and shrinks.

Formative Assessment Tips

- **Muddiest Point:** This technique is the opposite of *Point of Most Significance.* Students are asked to reflect on the most difficult or confusing point in the lesson. Their written comments are collected at the end of the lesson so that the following day's instruction can address the confusion.
- This technique can be used at any time during instruction.
- It is important for teachers to know whether there was a point in the lesson that was confusing for students so that the lesson can be modified. Share with students what you learned from their reflections. Students will take reflections more seriously when they are valued and used.

Pacing Suggestion

- The formal lesson is quite long. You might have students complete only Exploration 3 as a warm-up and then move to the formal lesson.

Dynamic Teaching Tools

- Dynamic Assessment & Progress Monitoring Tool
- Lesson Planning Tool
- Interactive Whiteboard Lesson Library
- Dynamic Classroom with Dynamic Investigations

What Your Students Will Learn

- Translate graphs of linear functions horizontally and vertically.
- Reflect graphs of linear functions in the x-axis and in the y-axis.
- Transform graphs of linear functions using horizontal shrinks, horizontal stretches, vertical shrinks, and vertical stretches.
- Transform graphs of linear functions using combinations of translations, reflections, stretches, and shrinks.

Laurie's Notes

Exploration

Motivate

- If you have computer projection capability in your classroom, use a word processor or simple graphic program to draw a rectangle. Make a copy of the rectangle. Change the color of the first rectangle to red and the second rectangle to blue. Demonstrate (or ask about) the various transformations of the blue rectangle depending on how you manipulate it. You should be able to translate, flip, dilate, stretch, and shrink. Compare each version to the red (parent) rectangle.
- Explain that points and figures (rectangles) can be transformed. In algebra, we study transformation of functions and examine the effect on their graphs. The transformation is represented symbolically in the equation we write and graphically by the graphs of the functions.

Discuss

- Comfort with function notation is essential in these explorations and in this lesson. Review the language of the statement $f(x) = x$.
- Remind students that $f(x)$ is another name for y.

Exploration 1

- Students will not have difficulty graphing the function $f(x) = x$. In graphing each of the g functions, remind students to plot them on the same axes with $f(x) = x$. Students should use a graphing calculator, as needed, to check.
- Students will recognize that all five functions have a slope of 1. The y-intercepts are different.
- **Connection:** You want students to compare each of the g functions to the original function $f(x) = x$. Students will describe the graphs as being parallel. The challenge is that translating the line $f(x) = x$ left 4 units or up 4 units looks the same on the coordinate plane!
- Write and highlight: If $f(x) = x$ and $g(x) = x + 4$, then $g(x) = f(x) + 4$.

? "Why this is true?" The outputs for g are 4 more than the outputs for f.

- Repeat this strategy for each of the g functions.

Exploration 2

- Students should quickly sketch all five functions. Use a graphing calculator as needed to check. While the slopes are different, all five lines have the same y-intercept.
- Write and highlight: If $f(x) = x$ and $h(x) = \frac{1}{2}x$, then $h(x) = \frac{1}{2}f(x)$.

? "Why this is true?" The outputs for h are $\frac{1}{2}$ times the outputs for f.

- Repeat this strategy for each of the h functions.

Exploration 3

? "What strategies did you use to match the functions?" Listen for information about the slopes and/or y-intercepts.

- If time permits and function notation is feeling comfortable for students, ask how the k functions could be written in terms of the f function. For example, in part (a) $k(x) = 2f(x) - 4$.

Communicate Your Answer

- Students may recognize the vertical shift; however, the horizontal stretch or shrink is harder for students to see, especially for linear functions.

Connecting to Next Step

- If you have been able to discuss the types of translations, assign items from Exercises 5–8 on page 151.

3.6 Transformations of Graphs of Linear Functions

Essential Question How does the graph of the linear function $f(x) = x$ compare to the graphs of $g(x) = f(x) + c$ and $h(x) = f(cx)$?

EXPLORATION 1 **Comparing Graphs of Functions**

Work with a partner. The graph of $f(x) = x$ is shown. Sketch the graph of each function, along with f, on the same set of coordinate axes. Use a graphing calculator to check your results. What can you conclude?

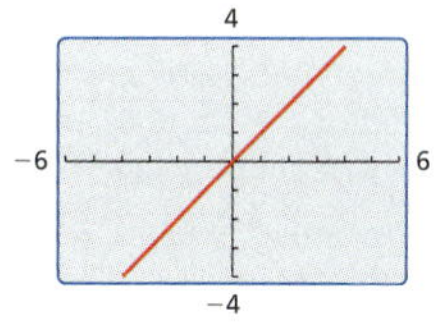

a. $g(x) = x + 4$ **b.** $g(x) = x + 2$

c. $g(x) = x - 2$ **d.** $g(x) = x - 4$

USING TOOLS STRATEGICALLY

To be proficient in math, you need to use the appropriate tools, including graphs, tables, and technology, to check your results.

EXPLORATION 2 **Comparing Graphs of Functions**

Work with a partner. Sketch the graph of each function, along with $f(x) = x$, on the same set of coordinate axes. Use a graphing calculator to check your results. What can you conclude?

a. $h(x) = \frac{1}{2}x$ **b.** $h(x) = 2x$ **c.** $h(x) = -\frac{1}{2}x$ **d.** $h(x) = -2x$

EXPLORATION 3 **Matching Functions with Their Graphs**

Work with a partner. Match each function with its graph. Use a graphing calculator to check your results. Then use the results of Explorations 1 and 2 to compare the graph of k to the graph of $f(x) = x$.

a. $k(x) = 2x - 4$ **b.** $k(x) = -2x + 2$

c. $k(x) = \frac{1}{2}x + 4$ **d.** $k(x) = -\frac{1}{2}x - 2$

A.

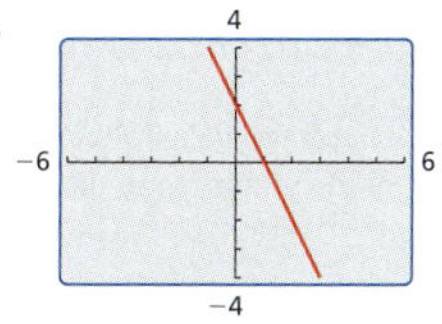

B.

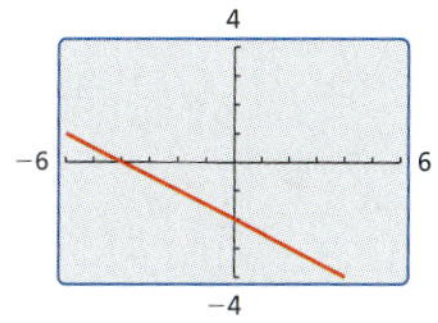

C.

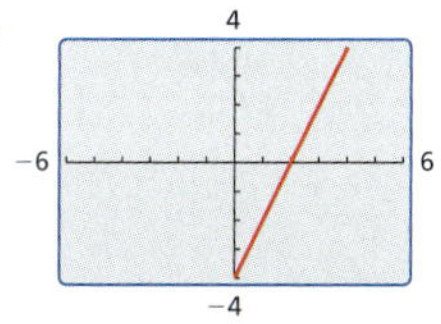

D.

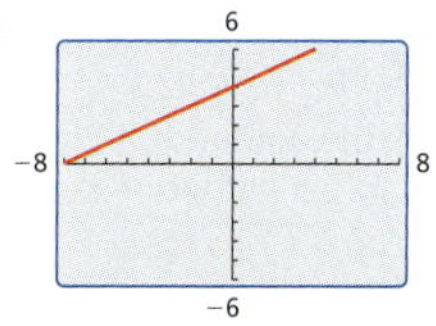

Communicate Your Answer

4. How does the graph of the linear function $f(x) = x$ compare to the graphs of $g(x) = f(x) + c$ and $h(x) = f(cx)$?

Dynamic Teaching Tools

- Dynamic Assessment & Progress Monitoring Tool
- Lesson Planning Tool
- Interactive Whiteboard Lesson Library
- Dynamic Classroom with Dynamic Investigations

ANSWERS

1. a. See Additional Answers.

b.

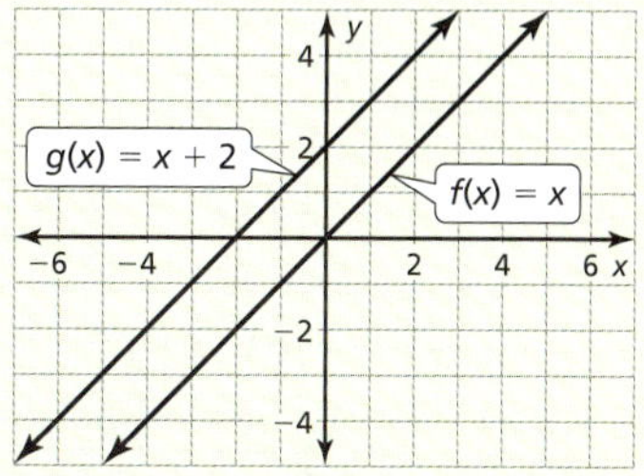

c.

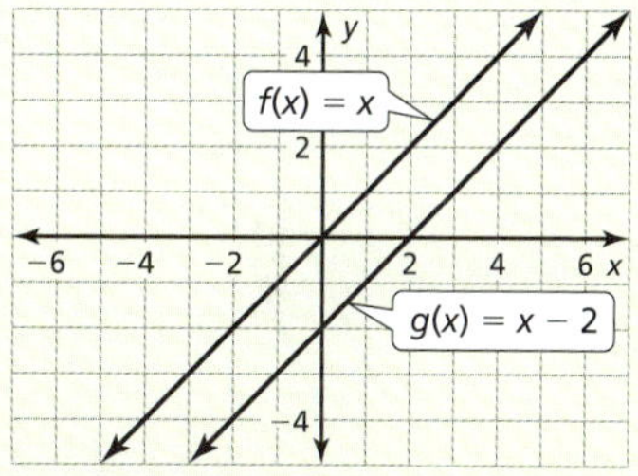

d.

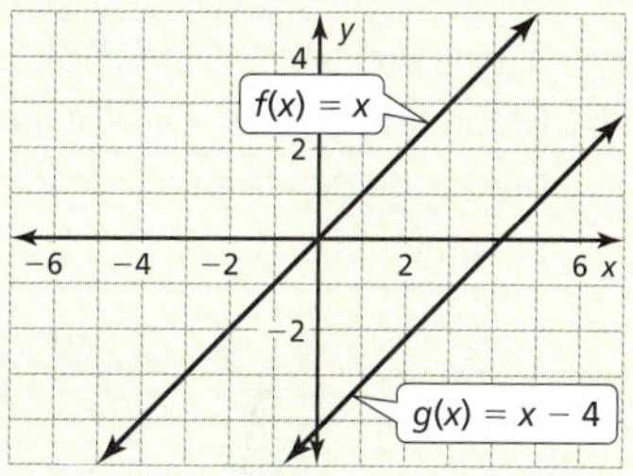

Adding or subtracting a constant to a function causes a vertical shift of the graph.

2. a. See Additional Answers.

b.

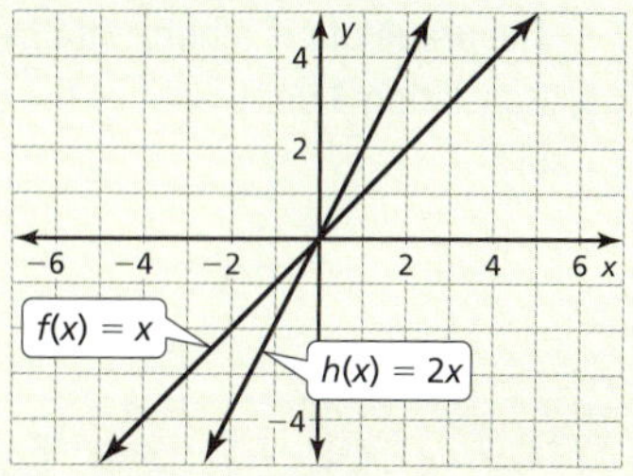

c.

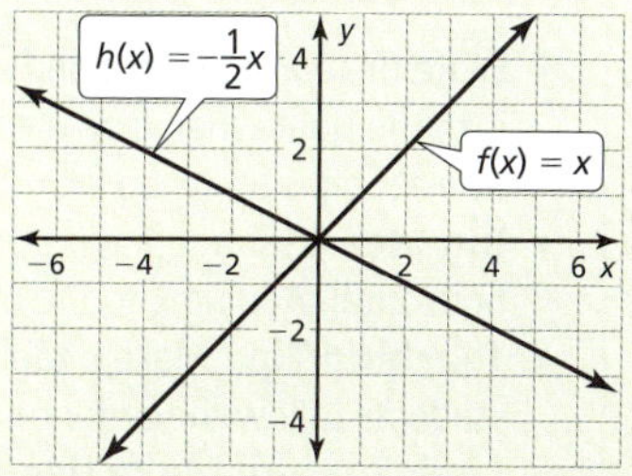

2. d.

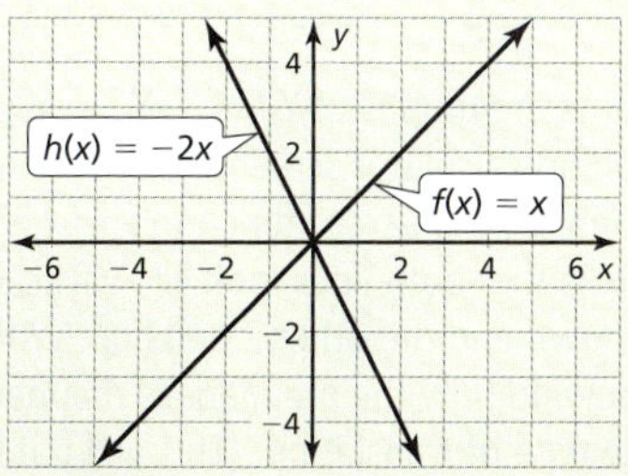

The graph of $h(x) = cx$ has a steeper slope than the graph of $f(x) = x$ when $c < -1$ or $c > 1$, is not as steep as the graph of $f(x) = x$ when $-1 < c < 1$, and slopes downward from left to right when c is negative.

3. a. C; The graph is shifted down 4 units and has a steeper slope.

b. A; The graph is shifted up 2 units, has a steeper slope, and slopes downward from left to right.

c. D; The graph is shifted up 4 units and is not as steep.

d. B; The graph is shifted down 2 units, is not as steep, and slopes downward from left to right.

4. The graph of $g(x) = f(x) + c$ is shifted vertically from $f(x) = x$; The graph of $h(x) = f(cx)$ has a different slope than $f(x) = x$.

Differentiated Instruction

Visual

Students may have difficulty remembering which graph is the parent function and which is the transformation. Choose one color for students to use exclusively for graphing the parent function. Have them use any other color to graph the transformation.

Extra Example 1

Let $f(x) = 2x + 1$.

a. Graph $g(x) = f(x) - 3$. Describe the transformation from the graph of f to the graph of g.

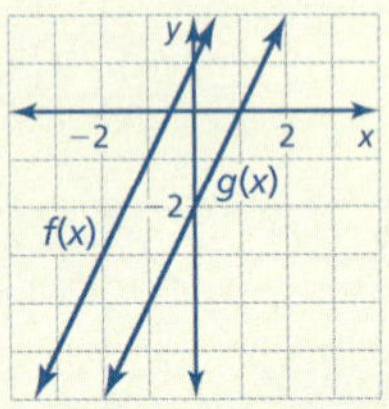

The graph of g is a vertical translation 3 units down of the graph of f.

b. Graph $t(x) = f(x - 3)$. Describe the transformation from the graph of f to the graph of t.

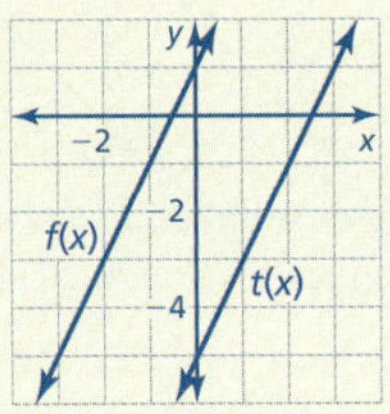

The graph of t is a horizontal translation 3 units right of the graph of f.

3.6 Lesson

Core Vocabulary

family of functions, *p. 146*
parent function, *p. 146*
transformation, *p. 146*
translation, *p. 146*
reflection, *p. 147*
horizontal shrink, *p. 148*
horizontal stretch, *p. 148*
vertical stretch, *p. 148*
vertical shrink, *p. 148*

Previous
linear function

What You Will Learn

- Translate and reflect graphs of linear functions.
- Stretch and shrink graphs of linear functions.
- Combine transformations of graphs of linear functions.

Translations and Reflections

A **family of functions** is a group of functions with similar characteristics. The most basic function in a family of functions is the **parent function**. For nonconstant linear functions, the parent function is $f(x) = x$. The graphs of all other nonconstant linear functions are *transformations* of the graph of the parent function. A **transformation** changes the size, shape, position, or orientation of a graph.

Core Concept

A **translation** is a transformation that shifts a graph horizontally or vertically but does not change the size, shape, or orientation of the graph.

Horizontal Translations

The graph of $y = f(x - h)$ is a horizontal translation of the graph of $y = f(x)$, where $h \neq 0$.

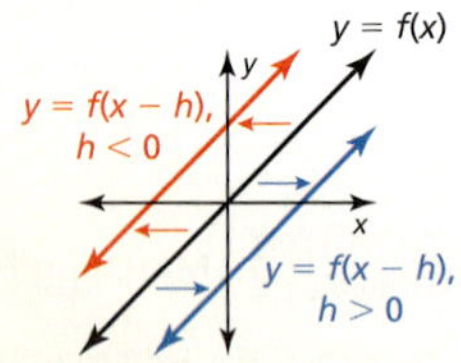

Subtracting h from the *inputs* before evaluating the function shifts the graph left when $h < 0$ and right when $h > 0$.

Vertical Translations

The graph of $y = f(x) + k$ is a vertical translation of the graph of $y = f(x)$, where $k \neq 0$.

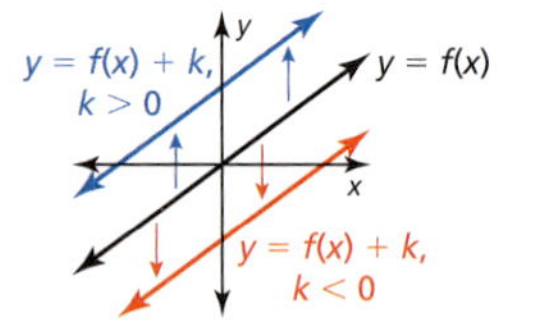

Adding k to the *outputs* shifts the graph down when $k < 0$ and up when $k > 0$.

EXAMPLE 1 **Horizontal and Vertical Translations**

Let $f(x) = 2x - 1$. Graph (a) $g(x) = f(x) + 3$ and (b) $t(x) = f(x + 3)$. Describe the transformations from the graph of f to the graphs of g and t.

SOLUTION

a. The function g is of the form $y = f(x) + k$, where $k = 3$. So, the graph of g is a vertical translation 3 units up of the graph of f.

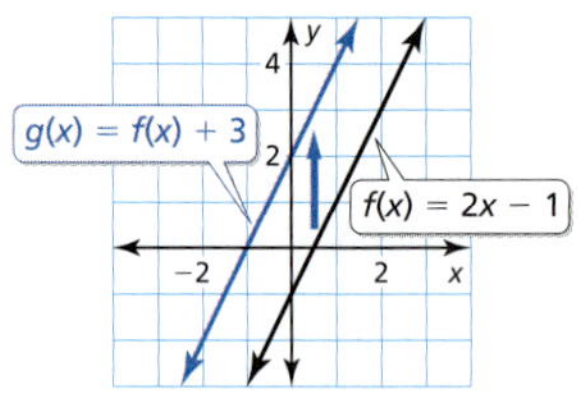

b. The function t is of the form $y = f(x - h)$, where $h = -3$. So, the graph of t is a horizontal translation 3 units left of the graph of f.

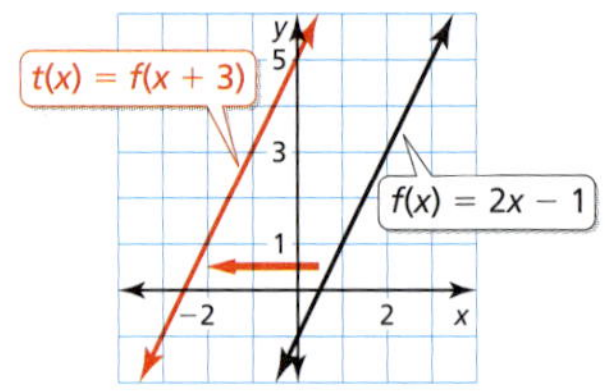

LOOKING FOR A PATTERN

In part (a), the output of g is equal to the output of f plus 3.

In part (b), the output of t is equal to the output of f when the input of f is 3 more than the input of t.

Laurie's Notes — Teacher Actions

- Write the *Core Concept* and summarize. "A horizontal translation is to the right when $h > 0$ and to the left when $h < 0$. A vertical translation is up when $k > 0$ and down when $k < 0$."
- **Big Idea:** Subtracting h from the input means the translation is horizontal. Adding k to the output means the translation is vertical.
- **Teaching Tip:** Make a table of values as done on page 147 to help students follow the order of evaluating $g(x)$ and $t(x)$.
- **Look For and Make Use of Structure:** The table enables students to follow the steps in performing transformations on a function.
- **Extension:** Show students the graphs of $Y_1 = X$, $Y_2 = Y_1(X-5)$, and $Y_3 = Y_1+2$ on a graphing calculator screen. Ask them to describe the two translations of the parent function $y = x$. What is challenging for students is that both lines look like vertical translations, one 5 units down and the other 2 units up. Without seeing the notation or knowing how the preimage is mapped to the image, you cannot tell the type of translation. (This will not be true for absolute value and quadratic functions.) Display the equation editor screen to show that one line is a horizontal translation 5 units right of the graph of $y = x$.

Core Concept

A **reflection** is a transformation that flips a graph over a line called the *line of reflection*.

Reflections in the x-axis

The graph of $y = -f(x)$ is a reflection in the x-axis of the graph of $y = f(x)$.

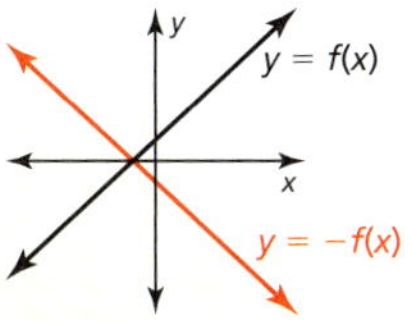

Multiplying the outputs by -1 changes their signs.

Reflections in the y-axis

The graph of $y = f(-x)$ is a reflection in the y-axis of the graph of $y = f(x)$.

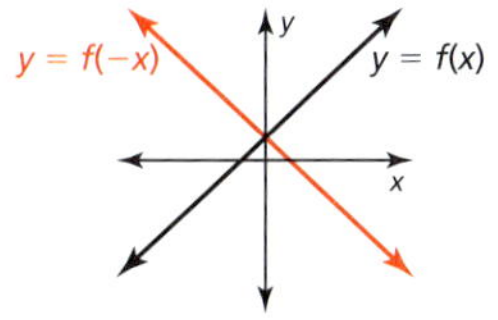

Multiplying the inputs by -1 changes their signs.

STUDY TIP

A reflected point is the same distance from the line of reflection as the original point but on the opposite side of the line.

EXAMPLE 2 Reflections in the x-axis and the y-axis

Let $f(x) = \frac{1}{2}x + 1$. Graph (a) $g(x) = -f(x)$ and (b) $t(x) = f(-x)$. Describe the transformations from the graph of f to the graphs of g and t.

SOLUTION

a. To find the outputs of g, multiply the outputs of f by -1. The graph of g consists of the points $(x, -f(x))$.

x	-4	-2	0
$f(x)$	-1	0	1
$-f(x)$	1	0	-1

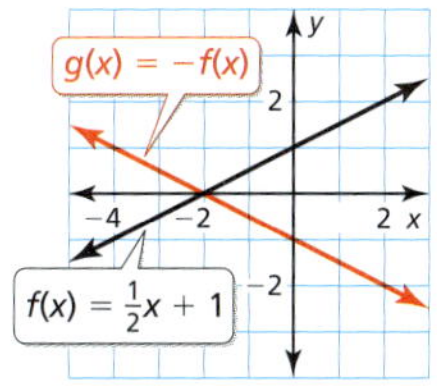

▶ The graph of g is a reflection in the x-axis of the graph of f.

b. To find the outputs of t, multiply the inputs by -1 and then evaluate f. The graph of t consists of the points $(x, f(-x))$.

x	-2	0	2
$-x$	2	0	-2
$f(-x)$	2	1	0

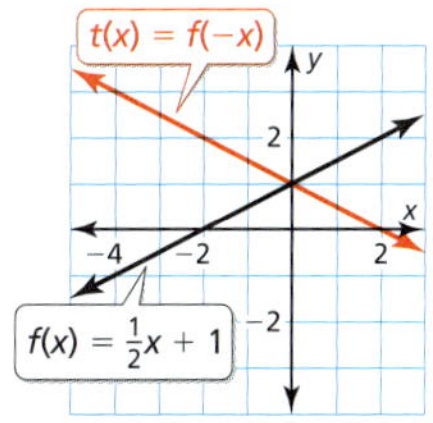

▶ The graph of t is a reflection in the y-axis of the graph of f.

Monitoring Progress

Help in English and Spanish at *BigIdeasMath.com*

Using f, graph (a) g and (b) h. Describe the transformations from the graph of f to the graphs of g and h.

1. $f(x) = 3x + 1$; $g(x) = f(x) - 2$; $h(x) = f(x - 2)$
2. $f(x) = -4x - 2$; $g(x) = -f(x)$; $h(x) = f(-x)$

English Language Learners

Notebook Development

Have students include a page on transformations in their notebooks. They should give a description and a sketch for each of the transformations in this section.

Extra Example 2

Let $f(x) = 3x + 2$.

a. Graph $g(x) = -f(x)$. Describe the transformation from the graph of f to the graph of g.

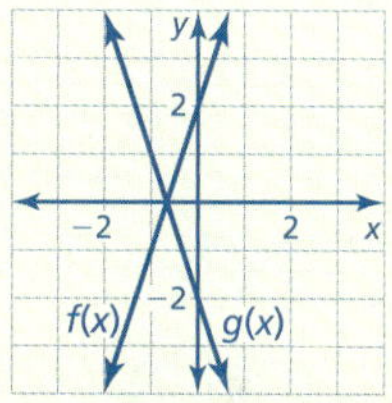

The graph of g is a reflection in the x-axis of the graph of f.

b. Graph $t(x) = f(-x)$. Describe the transformation from the graph of f to the graph of t.

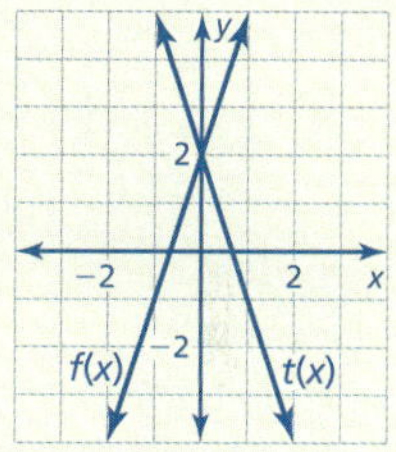

The graph of t is a reflection in the y-axis of the graph of f.

MONITORING PROGRESS ANSWERS

1–2. See Additional Answers.

Laurie's Notes Teacher Actions

- Write the *Core Concept* and summarize.
- **Teaching Tip:** Just like horizontal and vertical translations, students need to decide when they are performing the transformation to the input or the output. $y = -f(x)$ is affecting the output (sign change), so the reflection is over the x-axis. $y = f(-x)$ is affecting the input (sign change), so the reflection is over the y-axis.

? "How do you use the table of values to graph $g(x)$?" Plot the ordered pairs $(x, -f(x))$. "How do you use the table of values to graph $t(x)$?" Plot the ordered pairs $(x, f(-x))$.

Extra Example 3

Let $f(x) = -x + 2$.

a. Graph $g(x) = f\left(\frac{1}{2}x\right)$. Describe the transformation from the graph of f to the graph of g.

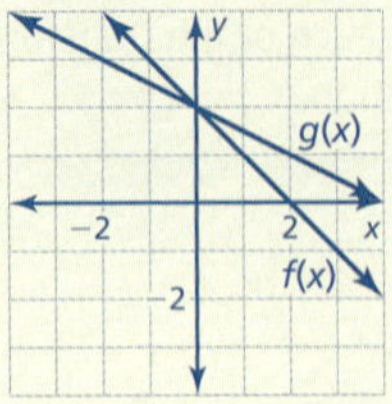

The graph of g is a horizontal stretch of the graph of f by a factor of $1 \div \frac{1}{2} = 2$.

b. Graph $h(x) = 2f(x)$. Describe the transformation from the graph of f to the graph of h.

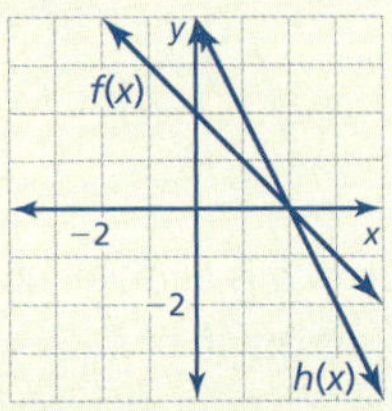

The graph of h is a vertical stretch of the graph of f by a factor of 2.

Stretches and Shrinks

You can transform a function by multiplying all the x-coordinates (inputs) by the same factor a. When $a > 1$, the transformation is a **horizontal shrink** because the graph shrinks toward the y-axis. When $0 < a < 1$, the transformation is a **horizontal stretch** because the graph stretches away from the y-axis. In each case, the y-intercept stays the same.

You can also transform a function by multiplying all the y-coordinates (outputs) by the same factor a. When $a > 1$, the transformation is a **vertical stretch** because the graph stretches away from the x-axis. When $0 < a < 1$, the transformation is a **vertical shrink** because the graph shrinks toward the x-axis. In each case, the x-intercept stays the same.

STUDY TIP

The graphs of $y = f(-ax)$ and $y = -a \cdot f(x)$ represent a stretch or shrink *and* a reflection in the x- or y-axis of the graph of $y = f(x)$.

Core Concept

Horizontal Stretches and Shrinks

The graph of $y = f(ax)$ is a horizontal stretch or shrink by a factor of $\frac{1}{a}$ of the graph of $y = f(x)$, where $a > 0$ and $a \neq 1$.

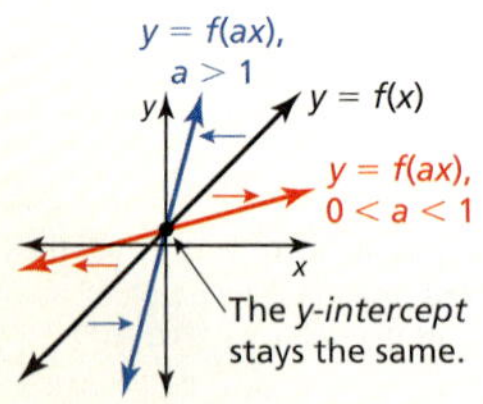

Vertical Stretches and Shrinks

The graph of $y = a \cdot f(x)$ is a vertical stretch or shrink by a factor of a of the graph of $y = f(x)$, where $a > 0$ and $a \neq 1$.

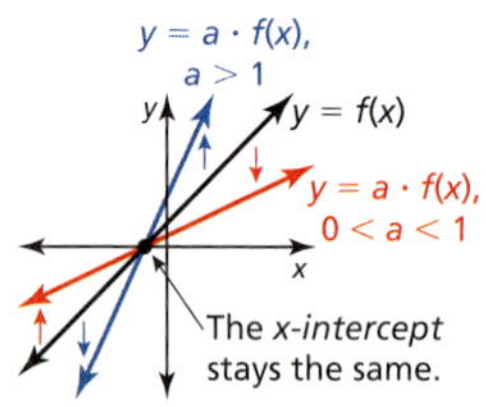

EXAMPLE 3 **Horizontal and Vertical Stretches**

Let $f(x) = x - 1$. Graph (a) $g(x) = f\left(\frac{1}{3}x\right)$ and (b) $h(x) = 3f(x)$. Describe the transformations from the graph of f to the graphs of g and h.

SOLUTION

a. To find the outputs of g, multiply the inputs by $\frac{1}{3}$. Then evaluate f. The graph of g consists of the points $\left(x, f\left(\frac{1}{3}x\right)\right)$.

x	−3	0	3
$\frac{1}{3}x$	−1	0	1
$f\left(\frac{1}{3}x\right)$	−2	−1	0

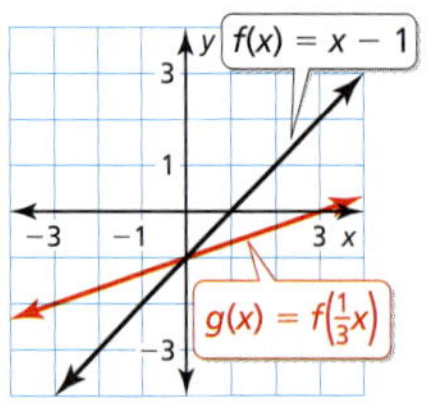

▶ The graph of g is a horizontal stretch of the graph of f by a factor of $1 \div \frac{1}{3} = 3$.

b. To find the outputs of h, multiply the outputs of f by 3. The graph of h consists of the points $(x, 3f(x))$.

x	0	1	2
$f(x)$	−1	0	1
$3f(x)$	−3	0	3

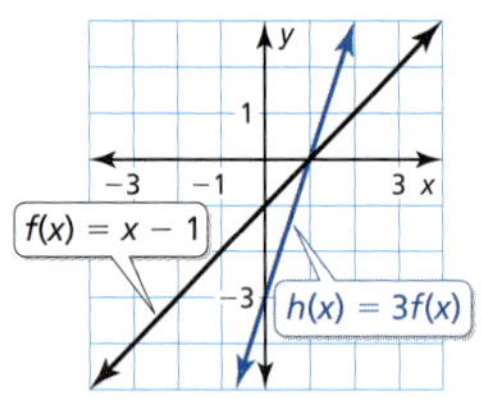

▶ The graph of h is a vertical stretch of the graph of f by a factor of 3.

Laurie's Notes Teacher Actions

- **Teaching Tip:** Before discussing stretches and shrinks, draw a square with vertices (2, 2), (2, −2), (−2, −2), and (−2, 2). Explore what happens when you multiply the x-coordinates by 3 and by $\frac{1}{2}$. Explore what happens when you multiply the y-coordinates by 3 and by $\frac{1}{2}$. You want students to see the stretches and shrinks as well as their direction.
- Have students read the introduction at the top of the page.
- Write the *Core Concept* and summarize.
- Work through the example as shown. The table of values will help students see the order when evaluating $g(x)$ and $h(x)$.

EXAMPLE 4 Horizontal and Vertical Shrinks

Let $f(x) = x + 2$. Graph (a) $g(x) = f(4x)$ and (b) $h(x) = \frac{1}{4}f(x)$. Describe the transformations from the graph of f to the graphs of g and h.

SOLUTION

a. To find the outputs of g, multiply the inputs by 4. Then evaluate f. The graph of g consists of the points $(x, f(4x))$.

x	−1	0	1
4x	−4	0	4
f(4x)	−2	2	6

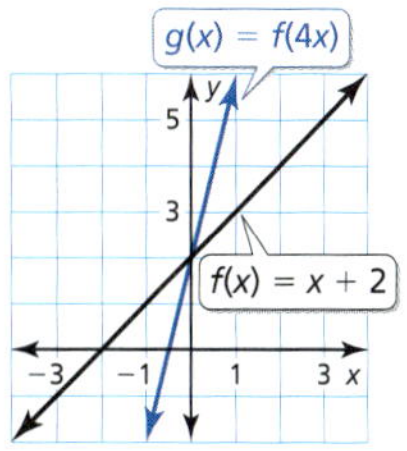

▶ The graph of g is a horizontal shrink of the graph of f by a factor of $\frac{1}{4}$.

b. To find the outputs of h, multiply the outputs of f by $\frac{1}{4}$. The graph of h consists of the points $\left(x, \frac{1}{4}f(x)\right)$.

x	−2	0	2
f(x)	0	2	4
$\frac{1}{4}$f(x)	0	$\frac{1}{2}$	1

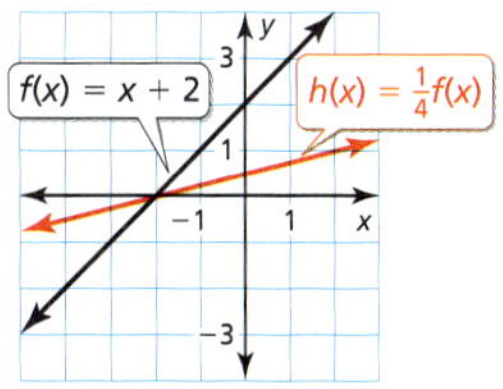

▶ The graph of h is a vertical shrink of the graph of f by a factor of $\frac{1}{4}$.

Monitoring Progress

Help in English and Spanish at *BigIdeasMath.com*

Using f, graph (a) g and (b) h. Describe the transformations from the graph of f to the graphs of g and h.

3. $f(x) = 4x - 2$; $g(x) = f\left(\frac{1}{2}x\right)$; $h(x) = 2f(x)$

4. $f(x) = -3x + 4$; $g(x) = f(2x)$; $h(x) = \frac{1}{2}f(x)$

Combining Transformations

STUDY TIP

You can perform transformations on the graph of *any* function f using these steps.

Core Concept

Transformations of Graphs

The graph of $y = a \cdot f(x - h) + k$ or the graph of $y = f(ax - h) + k$ can be obtained from the graph of $y = f(x)$ by performing these steps.

Step 1 Translate the graph of $y = f(x)$ horizontally h units.

Step 2 Use a to stretch or shrink the resulting graph from Step 1.

Step 3 Reflect the resulting graph from Step 2 when $a < 0$.

Step 4 Translate the resulting graph from Step 3 vertically k units.

Extra Example 4

Let $f(x) = 2x - 2$.

a. Graph $g(x) = f(2x)$. Describe the transformation from the graph of f to the graph of g.

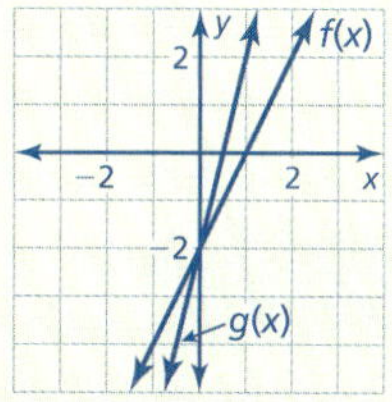

The graph of g is a horizontal shrink of the graph of f by a factor of $\frac{1}{2}$.

b. Graph $h(x) = \frac{1}{2}f(x)$. Describe the transformation from the graph of f to the graph of h.

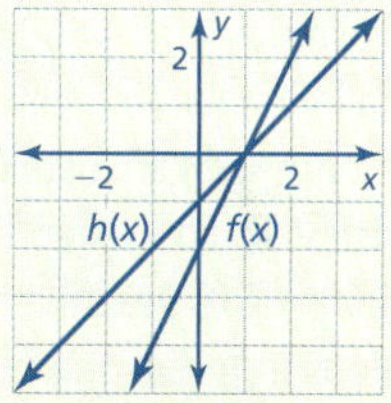

The graph of h is a vertical shrink of the graph of f by a factor of $\frac{1}{2}$.

MONITORING PROGRESS ANSWERS

3. See Additional Answers.

4. a.

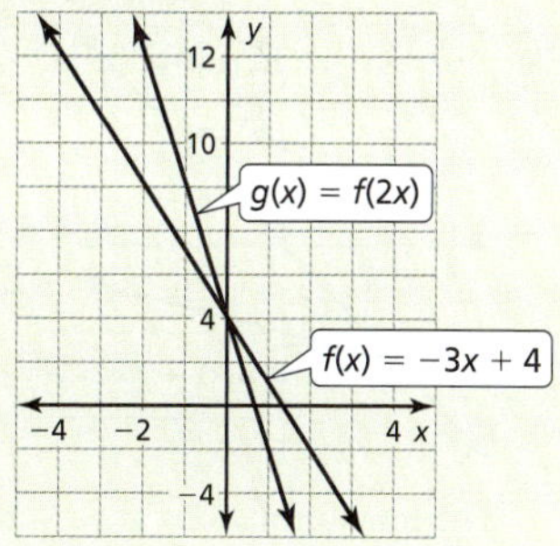

The graph of g is a horizontal shrink of the graph of f by a factor of $\frac{1}{2}$.

b.

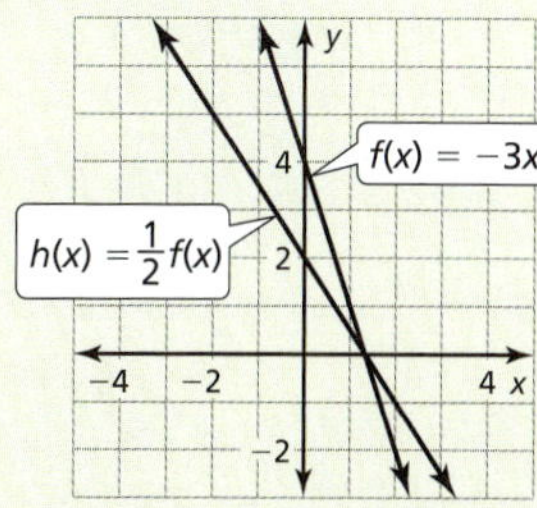

The graph of h is a vertical shrink of the graph of f by a factor of $\frac{1}{2}$.

Laurie's Notes Teacher Actions

- Work through the example as shown. The table of values will help students see the order when evaluating $g(x)$ and $h(x)$.

? "Do you think it is possible to perform more than one transformation on the graph of a function?" yes

- Write the *Core Concept*. Take time to connect the parameters to the impact each has on the graph of the function $y = f(x)$.

Extra Example 5

Graph $f(x) = x$ and $g(x) = 3x - 4$. Describe the transformations from the graph of f to the graph of g.

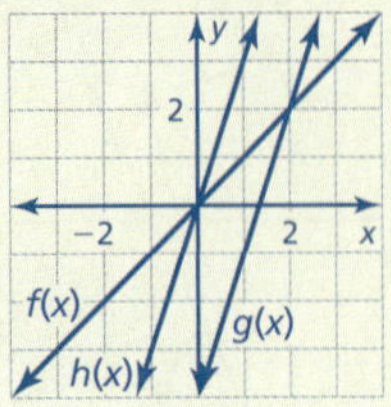

Stretch the graph of f vertically by a factor of 3 to get the graph of $h(x) = 3x$. Translate the graph of h vertically 4 units down to get the graph of $g(x) = 3x - 4$.

Extra Example 6

In Example 6, the cable company changes the cost to customers from \$60 per month with no installation fee to a new rate of \$45 per month with an installation fee of \$30. The new cost is represented by $s(m) = 45m + 30$. Describe the transformations from the graph of $c(m) = 60m$ to the graph of s. The transformations are a vertical shrink by a factor of $\frac{3}{4}$ and then a vertical translation 30 units up.

MONITORING PROGRESS ANSWER

5.

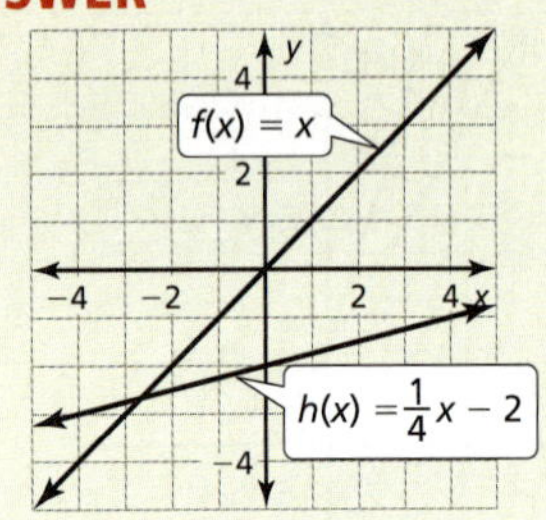

The transformations are a vertical shrink by a factor of $\frac{1}{4}$ and then a vertical translation 2 units down.

EXAMPLE 5 Combining Transformations

Graph $f(x) = x$ and $g(x) = -2x + 3$. Describe the transformations from the graph of f to the graph of g.

SOLUTION

Note that you can rewrite g as $g(x) = -2f(x) + 3$.

Step 1 There is no horizontal translation from the graph of f to the graph of g.

Step 2 Stretch the graph of f vertically by a factor of 2 to get the graph of $h(x) = 2x$.

Step 3 Reflect the graph of h in the x-axis to get the graph of $r(x) = -2x$.

Step 4 Translate the graph of r vertically 3 units up to get the graph of $g(x) = -2x + 3$.

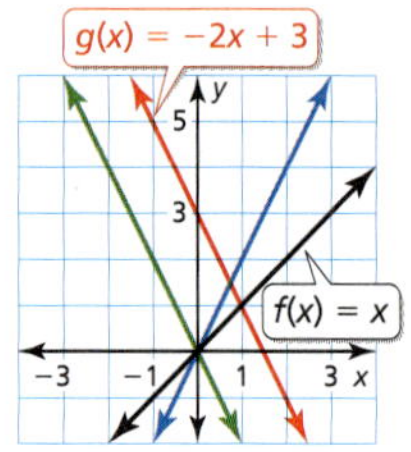

> **ANOTHER WAY**
>
> You could also rewrite g as $g(x) = f(-2x) + 3$. In this case, the transformations from the graph of f to the graph of g will be different from those in Example 5.

EXAMPLE 6 Solving a Real-Life Problem

A cable company charges customers \$60 per month for its service, with no installation fee. The cost to a customer is represented by $c(m) = 60m$, where m is the number of months of service. To attract new customers, the cable company reduces the monthly fee to \$30 but adds an installation fee of \$45. The cost to a new customer is represented by $r(m) = 30m + 45$, where m is the number of months of service. Describe the transformations from the graph of c to the graph of r.

SOLUTION

Note that you can rewrite r as $r(m) = \frac{1}{2}c(m) + 45$. In this form, you can use the order of operations to get the outputs of r from the outputs of c. First, multiply the outputs of c by $\frac{1}{2}$ to get $h(m) = 30m$. Then add 45 to the outputs of h to get $r(m) = 30m + 45$.

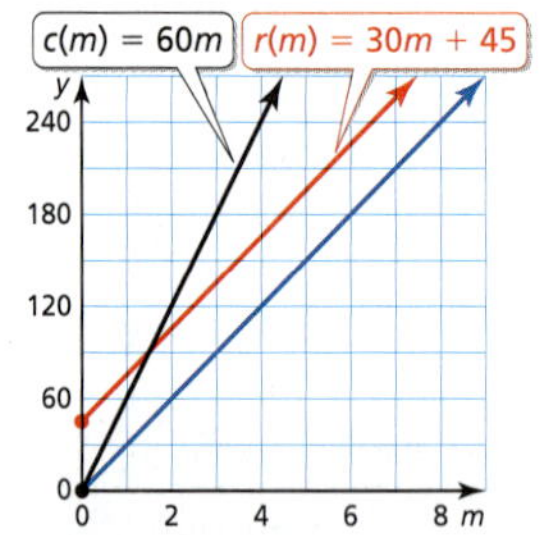

▶ The transformations are a vertical shrink by a factor of $\frac{1}{2}$ and then a vertical translation 45 units up.

Monitoring Progress

Help in English and Spanish at *BigIdeasMath.com*

5. Graph $f(x) = x$ and $h(x) = \frac{1}{4}x - 2$. Describe the transformations from the graph of f to the graph of h.

Laurie's Notes Teacher Actions

- **Use Appropriate Tools Strategically:** Note the color coding in Example 5 to connect the transformation and its graph. Color is a helpful tool for many students as they make connections in mathematics.

Closure

- **Muddiest Point:** Ask students to identify, aloud or on a paper to be collected, the muddiest point(s) about the lesson. What was difficult to understand?

3.6 Exercises

Dynamic Solutions available at *BigIdeasMath.com*

Vocabulary and Core Concept Check

1. **WRITING** Describe the relationship between $f(x) = x$ and all other nonconstant linear functions.

2. **VOCABULARY** Name four types of transformations. Give an example of each and describe how it affects the graph of a function.

3. **WRITING** How does the value of a in the equation $y = f(ax)$ affect the graph of $y = f(x)$? How does the value of a in the equation $y = af(x)$ affect the graph of $y = f(x)$?

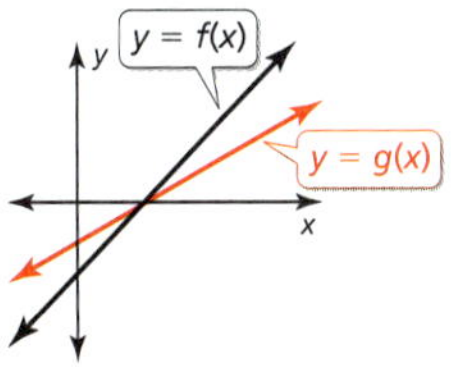

4. **REASONING** The functions f and g are linear functions. The graph of g is a vertical shrink of the graph of f. What can you say about the x-intercepts of the graphs of f and g? Is this always true? Explain.

Monitoring Progress and Modeling with Mathematics

In Exercises 5–10, use the graphs of f and g to describe the transformation from the graph of f to the graph of g. *(See Example 1.)*

5.

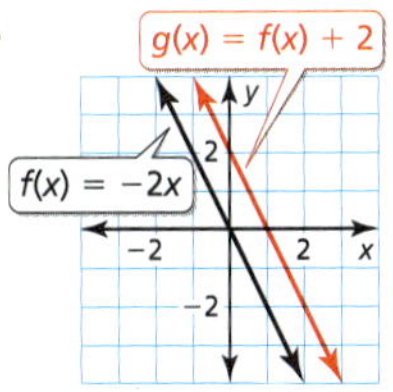

6.

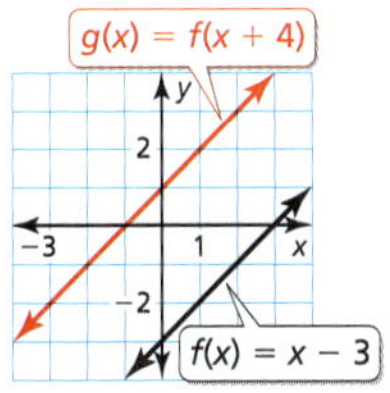

7. $f(x) = \frac{1}{3}x + 3$; $g(x) = f(x) - 3$

8. $f(x) = -3x + 4$; $g(x) = f(x) + 1$

9. $f(x) = -x - 2$; $g(x) = f(x + 5)$

10. $f(x) = \frac{1}{2}x - 5$; $g(x) = f(x - 3)$

11. **MODELING WITH MATHEMATICS** You and a friend start biking from the same location. Your distance d (in miles) after t minutes is given by the function $d(t) = \frac{1}{5}t$. Your friend starts biking 5 minutes after you. Your friend's distance f is given by the function $f(t) = d(t - 5)$. Describe the transformation from the graph of d to the graph of f.

12. **MODELING WITH MATHEMATICS** The total cost C (in dollars) to cater an event with p people is given by the function $C(p) = 18p + 50$. The set-up fee increases by \$25. The new total cost T is given by the function $T(p) = C(p) + 25$. Describe the transformation from the graph of C to the graph of T.

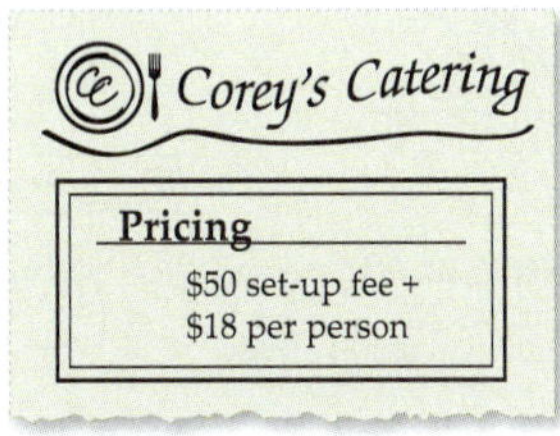

In Exercises 13–16, use the graphs of f and h to describe the transformation from the graph of f to the graph of h. *(See Example 2.)*

13.

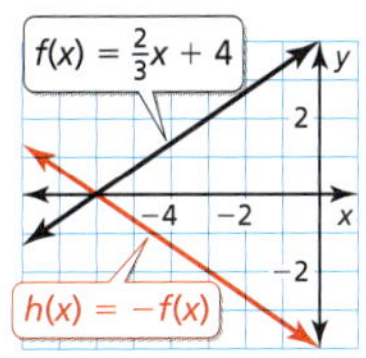

14.

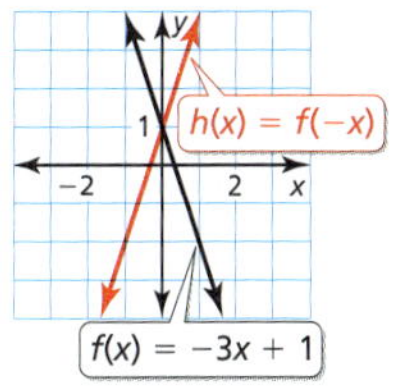

15. $f(x) = -5 - x$; $h(x) = f(-x)$

16. $f(x) = \frac{1}{4}x - 2$; $h(x) = -f(x)$

Assignment Guide and Homework Check

ASSIGNMENT

Basic: 1–4, 5–27 odd, 39–45 odd, 48, 62, 68–73

Average: 1–4, 6–32 even, 36–44 even, 48–52 even, 62, 68–73

Advanced: 1–4, 8–16 even, 20, 22, 26–66 even, 68–73

HOMEWORK CHECK

Basic: 5, 15, 17, 25, 41

Average: 6, 16, 18, 26, 42

Advanced: 8, 16, 20, 28, 42

ANSWERS

1. The graphs of all other nonconstant linear functions are transformations of the graph of $f(x) = x$.
2. translation, reflection, stretch, shrink; *Sample answer:* $y = f(x) + 2$ is a vertical translation 2 units up of the graph of f. $y = -f(x)$ is a reflection in the x-axis of the graph of f. $y = 2f(x)$ is a vertical stretch by a factor of 2 of the graph of f. $y = \frac{1}{2}f(x)$ is a vertical shrink by a factor of $\frac{1}{2}$ of the graph of f.
3. causes a horizontal stretch or shrink; causes a vertical stretch or shrink
4. They are the same; yes; *Sample answer:* The x-intercept is 0. Multiplying 0 by any factor, a, will also give the new function an x-intercept of 0.
5. The graph of g is a vertical translation 2 units up of the graph of f.
6. The graph of g is a horizontal translation 4 units left of the graph of f.
7. The graph of g is a vertical translation 3 units down of the graph of f.
8. The graph of g is a vertical translation 1 unit up of the graph of f.
9. The graph of g is a horizontal translation 5 units left of the graph of f.
10. The graph of g is a horizontal translation 3 units right of the graph of f.
11. The graph of f is a horizontal translation 5 units right of the graph of d.
12. The graph of T is a vertical translation 25 units up of the graph of C.
13. The graph of h is a reflection in the x-axis of the graph of f.
14. The graph of h is a reflection in the y-axis of the graph of f.
15. The graph of h is a reflection in the y-axis of the graph of f.
16. The graph of h is a reflection in the x-axis of the graph of f.

In Exercises 17–22, use the graphs of f and r to describe the transformation from the graph of f to the graph of r. *(See Example 3.)*

17. 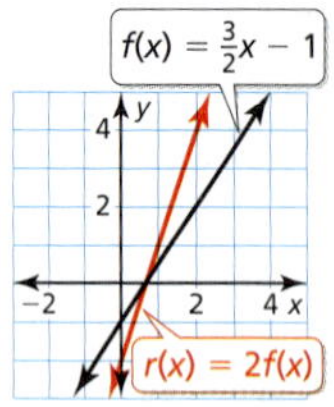

18. 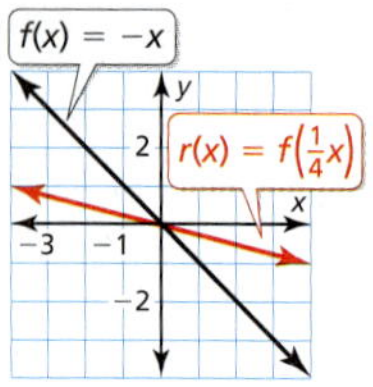

19. $f(x) = -2x - 4$; $r(x) = f\left(\frac{1}{2}x\right)$

20. $f(x) = 3x + 5$; $r(x) = f\left(\frac{1}{3}x\right)$

21. $f(x) = \frac{2}{3}x + 1$; $r(x) = 3f(x)$

22. $f(x) = -\frac{1}{4}x - 2$; $r(x) = 4f(x)$

In Exercises 23–28, use the graphs of f and h to describe the transformation from the graph of f to the graph of h. *(See Example 4.)*

23. 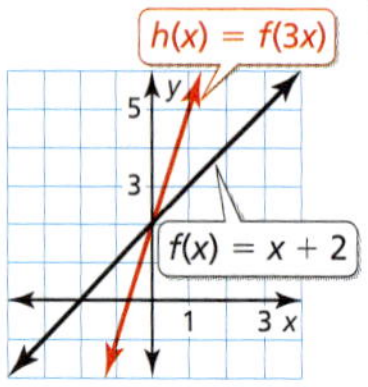

24. 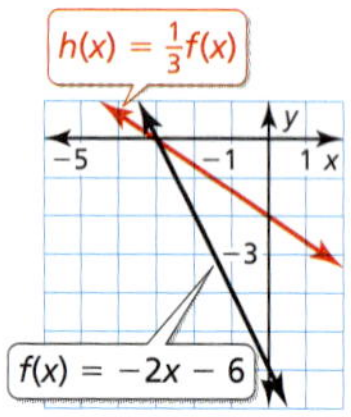

25. $f(x) = 3x - 12$; $h(x) = \frac{1}{6}f(x)$

26. $f(x) = -x + 1$; $h(x) = f(2x)$

27. $f(x) = -2x - 2$; $h(x) = f(5x)$

28. $f(x) = 4x + 8$; $h(x) = \frac{3}{4}f(x)$

In Exercises 29–34, use the graphs of f and g to describe the transformation from the graph of f to the graph of g.

29. $f(x) = x - 2$; $g(x) = \frac{1}{4}f(x)$

30. $f(x) = -4x + 8$; $g(x) = -f(x)$

31. $f(x) = -2x - 7$; $g(x) = f(x - 2)$

32. $f(x) = 3x + 8$; $g(x) = f\left(\frac{2}{3}x\right)$

33. $f(x) = x - 6$; $g(x) = 6f(x)$

34. $f(x) = -x$; $g(x) = f(x) - 3$

In Exercises 35–38, write a function g in terms of f so that the statement is true.

35. The graph of g is a horizontal translation 2 units right of the graph of f.

36. The graph of g is a reflection in the y-axis of the graph of f.

37. The graph of g is a vertical stretch by a factor of 4 of the graph of f.

38. The graph of g is a horizontal shrink by a factor of $\frac{1}{5}$ of the graph of f.

ERROR ANALYSIS In Exercises 39 and 40, describe and correct the error in graphing g.

39. 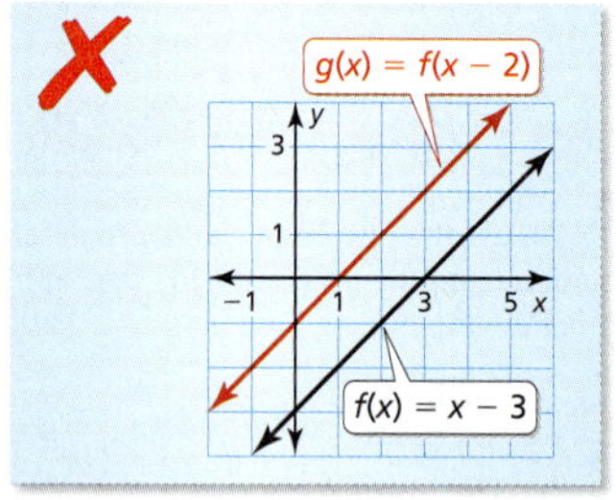

40. 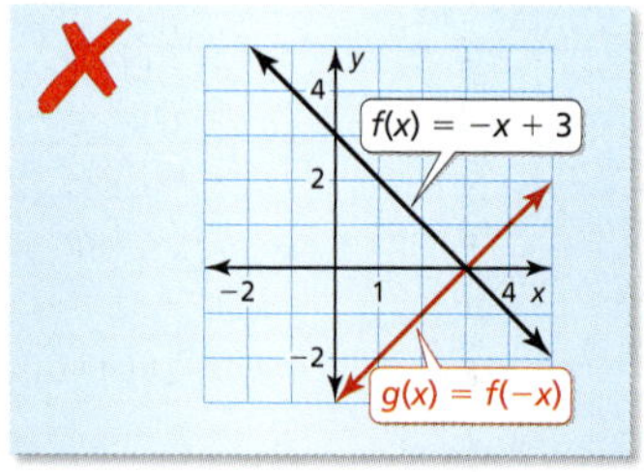

In Exercises 41–46, graph f and h. Describe the transformations from the graph of f to the graph of h. *(See Example 5.)*

41. $f(x) = x$; $h(x) = \frac{1}{3}x + 1$

42. $f(x) = x$; $h(x) = 4x - 2$

43. $f(x) = x$; $h(x) = -3x - 4$

44. $f(x) = x$; $h(x) = -\frac{1}{2}x + 3$

45. $f(x) = 2x$; $h(x) = 6x - 5$

46. $f(x) = 3x$; $h(x) = -3x - 7$

Dynamic Teaching Tools

- Dynamic Assessment & Progress Monitoring Tool
- Interactive Whiteboard Lesson Library
- Dynamic Classroom with Dynamic Investigations

ANSWERS

17. The graph of r is a vertical stretch of the graph of f by a factor of 2.

18. The graph of r is a horizontal stretch of the graph of f by a factor of 4.

19. The graph of r is a horizontal stretch of the graph of f by a factor of 2.

20. The graph of r is a horizontal stretch of the graph of f by a factor of 3.

21. The graph of r is a vertical stretch of the graph of f by a factor of 3.

22. The graph of r is a vertical stretch of the graph of f by a factor of 4.

23. The graph of h is a horizontal shrink of the graph of f by a factor of $\frac{1}{3}$.

24. The graph of h is a vertical shrink of the graph of f by a factor of $\frac{1}{3}$.

25. The graph of h is a vertical shrink of the graph of f by a factor of $\frac{1}{6}$.

26. The graph of h is a horizontal shrink of the graph of f by a factor of $\frac{1}{2}$.

27. The graph of h is a horizontal shrink of the graph of f by a factor of $\frac{1}{5}$.

28. The graph of h is a vertical shrink of the graph of f by a factor of $\frac{3}{4}$.

29. The graph of g is a vertical shrink of the graph of f by a factor of $\frac{1}{4}$.

30. The graph of g is a reflection in the x-axis of the graph of f.

31. The graph of g is a horizontal translation 2 units right of the graph of f.

32. The graph of g is a horizontal stretch of the graph of f by a factor of $\frac{3}{2}$.

33. The graph of g is a vertical stretch of the graph of f by a factor of 6.

34. The graph of g is a vertical translation 3 units down of the graph of f.

35. $g(x) = f(x - 2)$

36. $g(x) = f(-x)$

37. $g(x) = 4f(x)$

38. $g(x) = f(5x)$

39. $f(x - 2)$ is a translation to the right, not to the left.

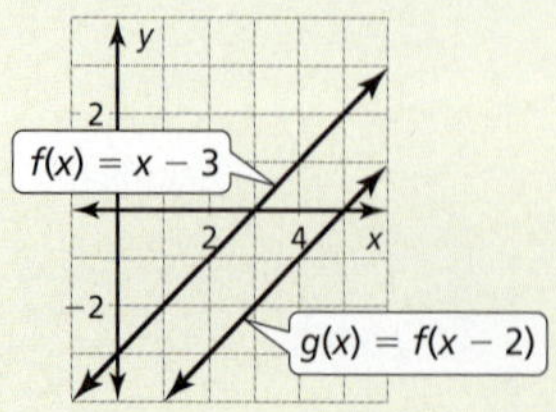

40. $f(-x)$ is a reflection in the y-axis, not the x-axis.

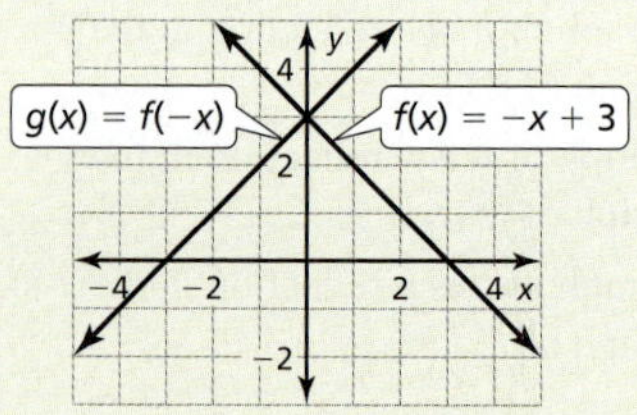

41–46. See Additional Answers.

47. **MODELING WITH MATHEMATICS** The function $t(x) = -4x + 72$ represents the temperature from 5 P.M. to 11 P.M., where x is the number of hours after 5 P.M. The function $d(x) = 4x + 72$ represents the temperature from 10 A.M. to 4 P.M., where x is the number of hours after 10 A.M. Describe the transformation from the graph of t to the graph of d.

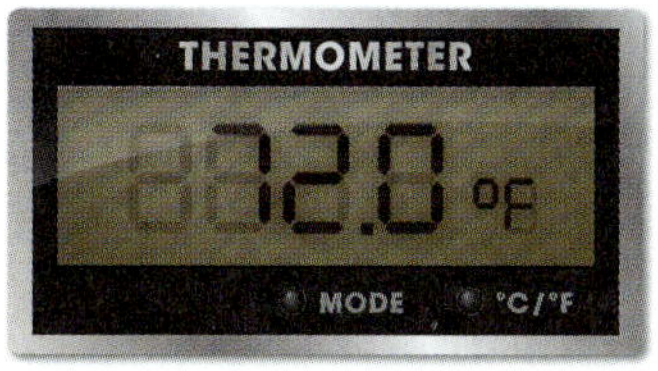

48. **MODELING WITH MATHEMATICS** A school sells T-shirts to promote school spirit. The school's profit is given by the function $P(x) = 8x - 150$, where x is the number of T-shirts sold. During the play-offs, the school increases the price of the T-shirts. The school's profit during the play-offs is given by the function $Q(x) = 16x - 200$, where x is the number of T-shirts sold. Describe the transformations from the graph of P to the graph of Q. *(See Example 6.)*

49. **USING STRUCTURE** The graph of $g(x) = a \cdot f(x - b) + c$ is a transformation of the graph of the linear function f. Select the word or value that makes each statement true.

reflection	translation	-1
stretch	shrink	0
left	right	1
y-axis	x-axis	

a. The graph of g is a vertical ______ of the graph of f when $a = 4$, $b = 0$, and $c = 0$.

b. The graph of g is a horizontal translation ______ of the graph of f when $a = 1$, $b = 2$, and $c = 0$.

c. The graph of g is a vertical translation 1 unit up of the graph of f when $a = 1$, $b = 0$, and $c =$ ____.

50. **USING STRUCTURE** The graph of $h(x) = a \cdot f(bx - c) + d$ is a transformation of the graph of the linear function f. Select the word or value that makes each statement true.

vertical	horizontal	0
stretch	shrink	$\frac{1}{5}$
y-axis	x-axis	5

a. The graph of h is a ______ shrink of the graph of f when $a = \frac{1}{3}$, $b = 1$, $c = 0$, and $d = 0$.

b. The graph of h is a reflection in the ______ of the graph of f when $a = 1$, $b = -1$, $c = 0$, and $d = 0$.

c. The graph of h is a horizontal stretch of the graph of f by a factor of 5 when $a = 1$, $b =$ _____, $c = 0$, and $d = 0$.

51. **ANALYZING GRAPHS** Which of the graphs are related by only a translation? Explain.

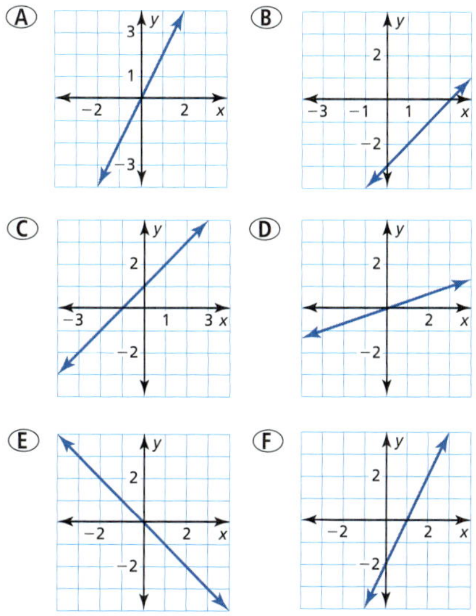

52. **ANALYZING RELATIONSHIPS** A swimming pool is filled with water by a hose at a rate of 1020 gallons per hour. The amount v (in gallons) of water in the pool after t hours is given by the function $v(t) = 1020t$. How does the graph of v change in each situation?

a. A larger hose is found. Then the pool is filled at a rate of 1360 gallons per hour.

b. Before filling up the pool with a hose, a water truck adds 2000 gallons of water to the pool.

ANSWERS

47. The graph of d is a reflection in the y-axis of the graph of t.

48. The transformations are a horizontal shrink by a factor of $\frac{1}{2}$ then a vertical translation 50 units down.

49. a. stretch
b. right
c. 1

50. a. vertical
b. y-axis
c. $\frac{1}{5}$

51. B and C, A and F; Adding 4 to the y-coordinate of each point on B gives the corresponding y-coordinate on C. Adding 2 to the y-coordinate of each point on F gives you the corresponding y-coordinate on A.

52. a. The new graph is a horizontal shrink of the old graph by a factor of $\frac{3}{4}$.
b. The new graph is a vertical translation of 2000 units up of the old graph.

ANSWERS

53. a. The new graph is a vertical translation 10 units up of the old graph.
    b. The new graph is a horizontal stretch of the old graph by a factor of 2.

54. *Sample answer:* $g(x) = \frac{1}{4}f(x) + 1$

55.

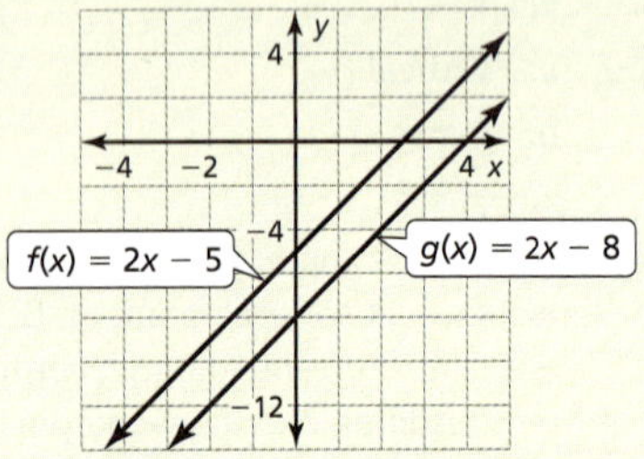

$g(x) = f(x) - 3$; The graph of g is a vertical translation 3 units down of the graph of f.

56–67. See Additional Answers.

68. $h = \dfrac{V}{\pi r^2}$

69. $w = \dfrac{P - 2\ell}{2}$ or $w = \frac{1}{2}P - \ell$

70. $-11 \leq x \leq 17$

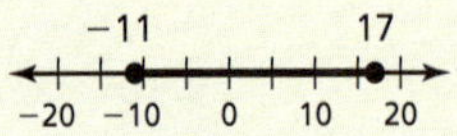

71–73. See Additional Answers.

Mini-Assessment

Use the graphs of f and g to describe the transformation from the graph of f to the graph of g.

1. $f(x) = 2x - 3$; $g(x) = f(x) - 4$
   The graph of g is a vertical translation 4 units down of the graph of f.
2. $f(x) = 3x$; $g(x) = -f(x)$
   The graph of g is a reflection in the x-axis of the graph of f.
3. $f(x) = -x + 4$; $g(x) = 2f(x)$
   The graph of g is a vertical stretch of the graph of f by a factor of 2.
4. $f(x) = \frac{1}{2}x - 2$; $g(x) = f(3x)$
   The graph of g is a horizontal shrink of the graph of f by a factor of $\frac{1}{3}$.
5. $f(x) = x$; $g(x) = 2x + 3$
   Stretch the graph of f vertically by a factor of 2 to get the graph of $h(x) = 2x$. Translate the graph of h vertically 3 units up to get the graph of $g(x) = 2x + 3$.

53. **ANALYZING RELATIONSHIPS** You have \$50 to spend on fabric for a blanket. The amount m (in dollars) of money you have after buying y yards of fabric is given by the function $m(y) = -9.98y + 50$. How does the graph of m change in each situation?

   a. You receive an additional \$10 to spend on the fabric.
   b. The fabric goes on sale, and each yard now costs \$4.99.

54. **THOUGHT PROVOKING** Write a function g whose graph passes through the point (4, 2) and is a transformation of the graph of $f(x) = x$.

In Exercises 55–60, graph f and g. Write g in terms of f. Describe the transformation from the graph of f to the graph of g.

55. $f(x) = 2x - 5$; $g(x) = 2x - 8$
56. $f(x) = 4x + 1$; $g(x) = -4x - 1$
57. $f(x) = 3x + 9$; $g(x) = 3x + 15$
58. $f(x) = -x - 4$; $g(x) = x - 4$
59. $f(x) = x + 2$; $g(x) = \frac{2}{3}x + 2$
60. $f(x) = x - 1$; $g(x) = 3x - 3$

61. **REASONING** The graph of $f(x) = x + 5$ is a vertical translation 5 units up of the graph of $f(x) = x$. How can you obtain the graph of $f(x) = x + 5$ from the graph of $f(x) = x$ using a horizontal translation?

62. **HOW DO YOU SEE IT?** Match each function with its graph. Explain your reasoning.

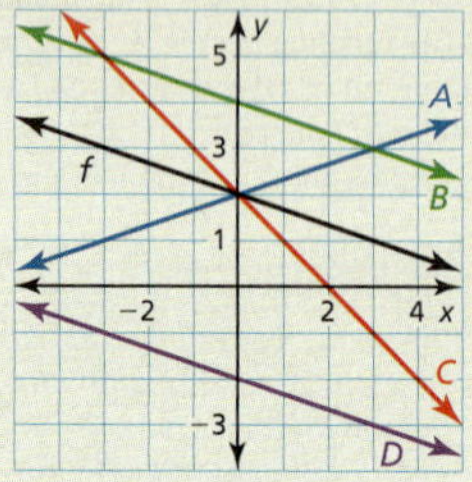

   a. $a(x) = f(-x)$ b. $g(x) = f(x) - 4$
   c. $h(x) = f(x) + 2$ d. $k(x) = f(3x)$

REASONING **In Exercises 63–66, find the value of r.**

63.

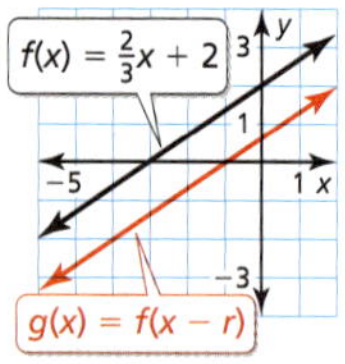

64.

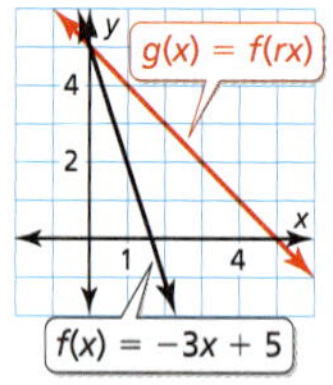

65.

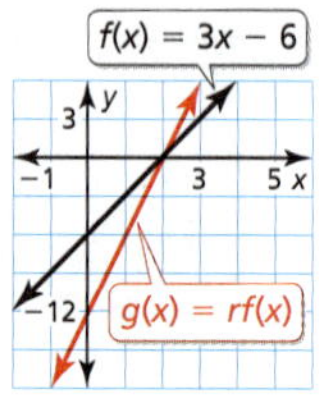

66.

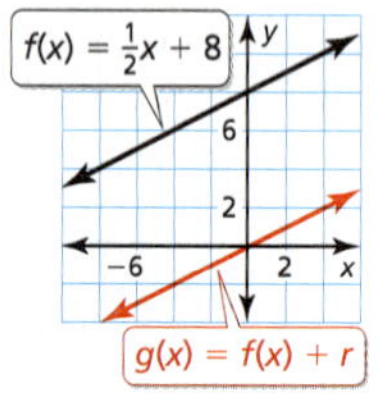

67. **CRITICAL THINKING** When is the graph of $y = f(x) + w$ the same as the graph of $y = f(x + w)$ for linear functions? Explain your reasoning.

Maintaining Mathematical Proficiency

Reviewing what you learned in previous grades and lessons

Solve the formula for the indicated variable. *(Section 1.5)*

68. Solve for h.

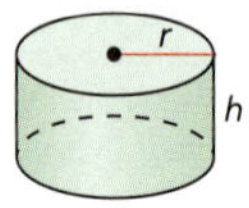

$V = \pi r^2 h$

69. Solve for w.

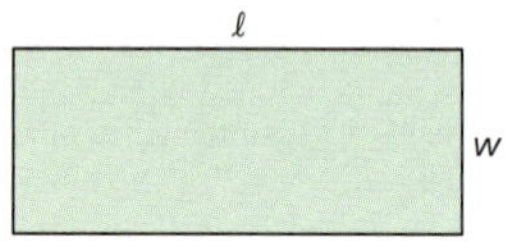

$P = 2\ell + 2w$

Solve the inequality. Graph the solution, if possible. *(Section 2.6)*

70. $|x - 3| \leq 14$
71. $|2x + 4| > 16$
72. $5|x + 7| < 25$
73. $-2|x + 1| \geq 18$

If students need help...	*If students got it...*
Resources by Chapter • Practice A and Practice B • Puzzle Time	Resources by Chapter • Enrichment and Extension • Cumulative Review
Student Journal • Practice	Start the *next* Section
Differentiating the Lesson Skills Review Handbook	

Laurie's Notes

Overview of Section 3.7

Introduction

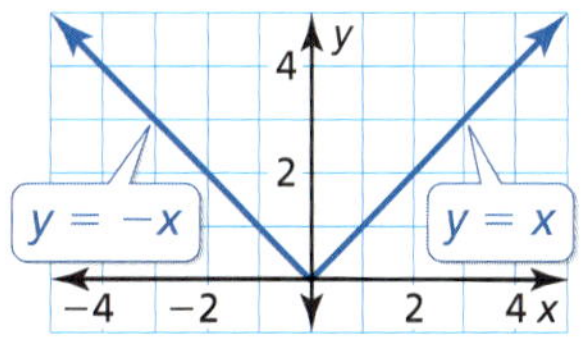

- The graph of the absolute value function will be new to students. I think it is important for them to discover the V-shape of the graph by making a simple table of values versus using the calculator. This can be done as the *Motivate* to begin the lesson before engaging in the explorations.
- The geometric definition that students have learned for absolute value is that the absolute value of a number is the distance the number is from 0. The algebraic definition is a split-rule function:

$$|x| = \begin{matrix} x, \text{ for } x \geq 0 \\ -x, \text{ for } x < 0 \end{matrix}.$$

- When students make a quick table of values, the domain and range of the parent function are obvious.
- The last lesson presented transformations to the graph of linear functions. All of the transformations of the last lesson apply to absolute value functions. Transformations of absolute value functions are much easier for students to see and understand. Whether you reflect $f(x) = x$ vertically or horizontally, it looks the same. That is not the case with $f(x) = |x|$.

Resources

- The graphing calculator is used in the exploration as a way to verify that the matches are correct. Ensure that students are making a reasoned attempt to do the matching before using the graphing calculator. The graphing calculator can also be a helpful teaching tool in the formal lesson.

Formative Assessment Tips

- **Look Back:** This technique is used when you want students to look back over a period of instruction and summarize their learning. What do they understand now that they did not before? It is helpful for students to articulate what helped them to learn a concept or skill.
- Sections 3.6 and 3.7 present the same transformations on the graphs of linear and absolute value functions. Having students summarize what they understand about transformations and what helped them learn it will enable you to make instruction more effective for your students in the future.
- If this is the first time you have used this technique, allow 6–8 minutes of class time to model what it looks like. Say, "Over the last two lessons we have been studying transformations to the graphs of $f(x) = x$ and $f(x) = |x|$. Take a piece of paper and draw a vertical line down the middle. I'm going to give you 5 minutes. On the left side, make a list of the things you learned that you did not know or understand before we began. On the right side, describe the how you learned these things. What technique(s) were effective in helping you learn?"
- Evaluate what worked for your students: particular examples, talking with partners, making tables, using the calculator, seeing the parent function and transformed function on the same graph, discussion of a particular problem, and so on.

Pacing Suggestion

- The formal lesson is long. Important discussion and understanding can occur during the exploration. Take time for students to make sense of each problem.

Dynamic Teaching Tools

- Dynamic Assessment & Progress Monitoring Tool
- Lesson Planning Tool
- Interactive Whiteboard Lesson Library
- Dynamic Classroom with Dynamic Investigations

What Your Students Will Learn

- Translate graphs of absolute value functions in the coordinate plane.
- Stretch, shrink, and reflect graphs of absolute value functions in the coordinate plane.
- Combine transformations of graphs of absolute value functions in the coordinate plane.

Laurie's Notes

Exploration

Motivate

- Distribute whiteboards to seven or nine volunteers. Tell the volunteers that they represent the domain values of a function. For seven volunteers they represent the integers -3 to 3. Orientation is important. They should stand at the front of the room facing their classmates with the left-most person representing $x = -3$ and the right-most person representing $x = 3$.
- Explain that you are going to say a function and the volunteers are to evaluate the function for the x-value they represent. They write the outputs on their whiteboards and hold the whiteboards in front of them. Students seated in the class will see the range values for the function. In other words, students will see the outputs for a table of values.
- Begin with a few linear functions: $y = x$; $y = x + 3$; $y = -x - 2$. Any miscalculations are obvious to spot because the slopes are 1 or -1. You could also ask about y-intercepts.
- Now ask about $y = |x|$.

? "What do you notice about the outputs?" Listen for students to observe that the outputs are all positive and that there is a symmetry in the outputs.

? **Turn and Talk:** "What will $y = |x|$ look like when graphed?" It is not uncommon for students to make a V-shape with their fingers, uncertain what to call it!

- Describe the parent function $f(x) = |x|$, and explain to students that in this lesson they will explore transformations of the parent function graph. Give students vocabulary to use in discussing the graph of the absolute value function. The *vertex* is the point where the graph changes direction, and the graph is symmetric about a vertical line through the vertex.

Exploration 1

- The goal is for students to use their understanding of transformations from the last lesson to match the equations and graphs.
- As you circulate, you should hear students make observations about three of the graphs being upward V's and three graphs being upside down V's. Students quickly observe that three of the functions have a negative sign outside the absolute value function notation.
- Students will need to make a table of values to help them do the matching.
- **Look For and Make Use of Structure:** Mathematically proficient students will pay attention to whether the transformation is done before or after the absolute value function is evaluated.
- **Selective Responses:** As you circulate, record the different methods and reasoning students use to match functions with graphs. When it is time for a whole class discussion, refer to your notes. Begin with a student who used little reasoning—he or she made a table of values. Work toward the student who used more reasoning, perhaps using insights from the previous lesson.
- The graphing calculator only verifies students' answers. It should not be used in advance of thorough discussions by partners about the process.

Communicate Your Answer

- Students should describe the effect of a, h, and k on the graph of the absolute value function.
- **Use Appropriate Tools Strategically:** The graphing calculator is a useful tool in checking students' equations for Question 3.

Connecting to Next Step

- Discussion of the reasoning used by students to do the matching should be followed by Example 1 in the formal lesson.

3.7 Graphing Absolute Value Functions

Essential Question

How do the values of a, h, and k affect the graph of the absolute value function $g(x) = a|x - h| + k$?

The parent absolute value function is

$f(x) = |x|$. Parent absolute value function

The graph of f is V-shaped.

EXPLORATION 1 Identifying Graphs of Absolute Value Functions

Work with a partner. Match each absolute value function with its graph. Then use a graphing calculator to verify your answers.

a. $g(x) = -|x - 2|$ **b.** $g(x) = |x - 2| + 2$ **c.** $g(x) = -|x + 2| - 2$

d. $g(x) = |x - 2| - 2$ **e.** $g(x) = 2|x - 2|$ **f.** $g(x) = -|x + 2| + 2$

A.

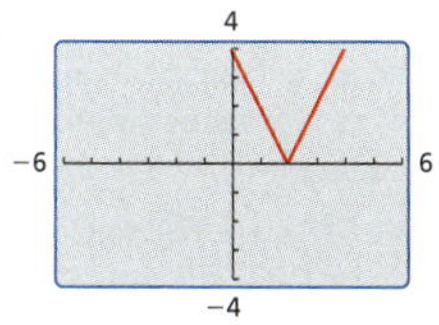

B.

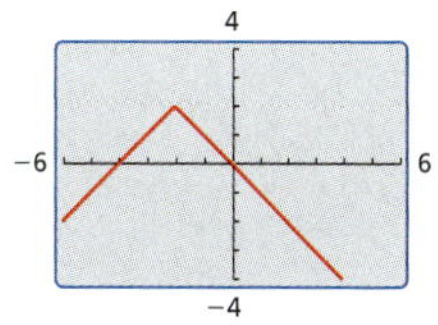

C.

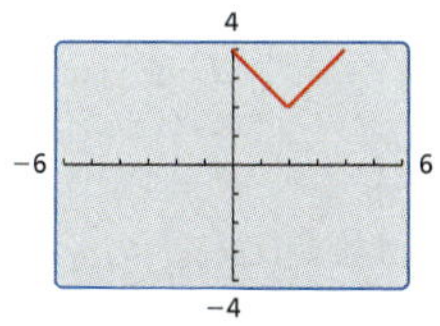

D.

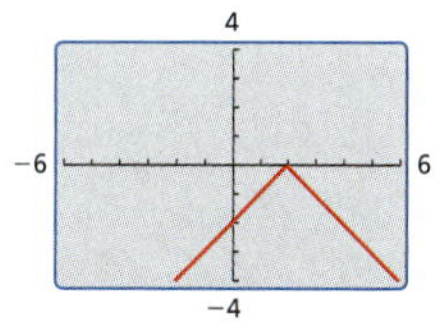

E.

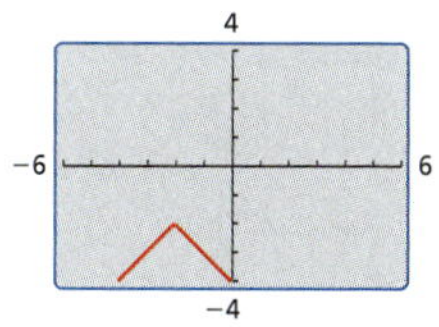

F.

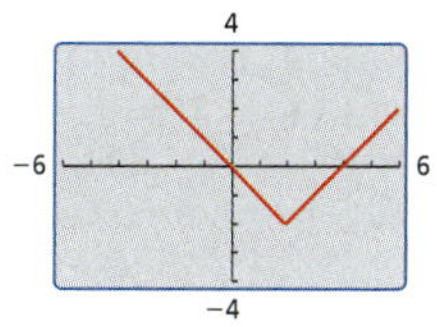

LOOKING FOR STRUCTURE

To be proficient in math, you need to look closely to discern a pattern or structure.

Communicate Your Answer

2. How do the values of a, h, and k affect the graph of the absolute value function

$g(x) = a|x - h| + k$?

3. Write the equation of the absolute value function whose graph is shown. Use a graphing calculator to verify your equation.

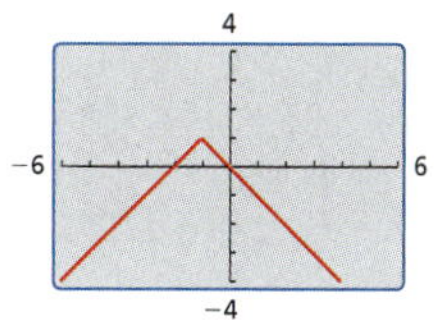

Dynamic Teaching Tools

- Dynamic Assessment & Progress Monitoring Tool
- Lesson Planning Tool
- Interactive Whiteboard Lesson Library
- Dynamic Classroom with Dynamic Investigations

ANSWERS

1. a. D

b. C

c. E

d. F

e. A

f. B

2. a stretches or shrinks the graph and determines whether the graph opens up or down; h translates the graph horizontally; k translates the graph vertically.

3. $g(x) = -|x + 1| + 1$

English Language Learners

Build on Past Knowledge

Some students may know the definition of a vertex as it applies to angles and other geometric figures. Connect their understanding of the vertex of an angle to the vertex of the graph of an absolute value equation.

Extra Example 1

Graph each function. Compare each graph to the graph of $f(x) = |x|$. Describe the domain and range.

a. $g(x) = |x| + 2$

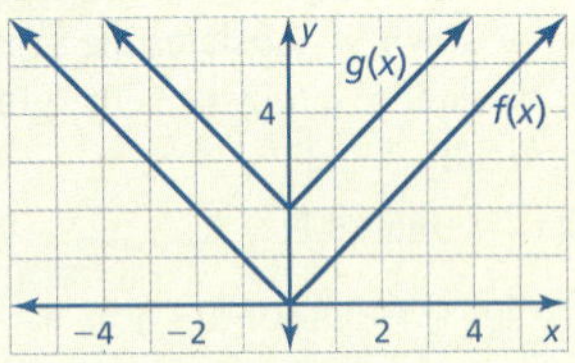

The graph of g is a vertical translation 2 units up of the graph of f. The domain is all real numbers. The range is $y \geq 2$.

b. $m(x) = |x + 3|$

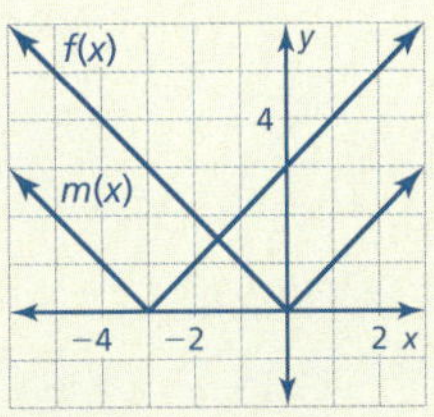

The graph of m is a horizontal translation 3 units left of the graph of f. The domain is all real numbers. The range is $y \geq 0$.

MONITORING PROGRESS ANSWERS

1–2. See Additional Answers.

3.7 Lesson

Core Vocabulary

absolute value function, *p. 156*
vertex, *p. 156*
vertex form, *p. 158*

Previous
domain
range

What You Will Learn

- Translate graphs of absolute value functions.
- Stretch, shrink, and reflect graphs of absolute value functions.
- Combine transformations of graphs of absolute value functions.

Translating Graphs of Absolute Value Functions

Core Concept

Absolute Value Function

An **absolute value function** is a function that contains an absolute value expression. The parent absolute value function is $f(x) = |x|$. The graph of $f(x) = |x|$ is V-shaped and symmetric about the y-axis. The **vertex** is the point where the graph changes direction. The vertex of the graph of $f(x) = |x|$ is (0, 0).

The domain of $f(x) = |x|$ is all real numbers. The range is $y \geq 0$.

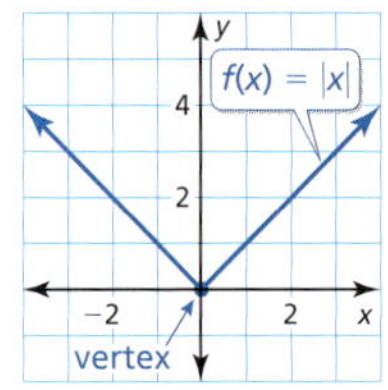

The graphs of all other absolute value functions are transformations of the graph of the parent function $f(x) = |x|$. The transformations presented in Section 3.6 also apply to absolute value functions.

EXAMPLE 1 **Graphing $g(x) = |x| + k$ and $g(x) = |x - h|$**

Graph each function. Compare each graph to the graph of $f(x) = |x|$. Describe the domain and range.

a. $g(x) = |x| + 3$ **b.** $m(x) = |x - 2|$

SOLUTION

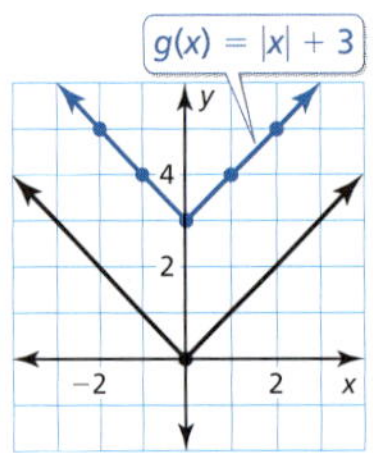

a. Step 1 Make a table of values.

x	−2	−1	0	1	2
$g(x)$	5	4	3	4	5

Step 2 Plot the ordered pairs.

Step 3 Draw the V-shaped graph.

The function g is of the form $y = f(x) + k$, where $k = 3$. So, the graph of g is a vertical translation 3 units up of the graph of f. The domain is all real numbers. The range is $y \geq 3$.

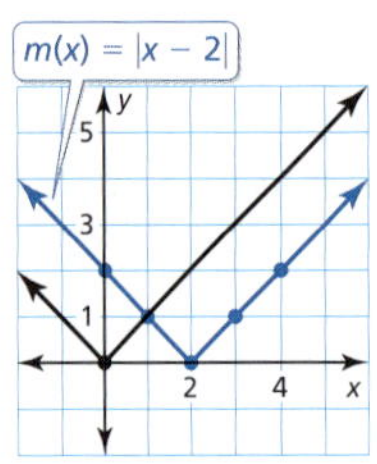

b. Step 1 Make a table of values.

x	0	1	2	3	4
$m(x)$	2	1	0	1	2

Step 2 Plot the ordered pairs.

Step 3 Draw the V-shaped graph.

The function m is of the form $y = f(x - h)$, where $h = 2$. So, the graph of m is a horizontal translation 2 units right of the graph of f. The domain is all real numbers. The range is $y \geq 0$.

Monitoring Progress

Help in English and Spanish at *BigIdeasMath.com*

Graph the function. Compare the graph to the graph of $f(x) = |x|$. Describe the domain and range.

1. $h(x) = |x| - 1$ **2.** $n(x) = |x + 4|$

Laurie's Notes Teacher Actions

- Students who have worked through the exploration should have the *Core Concept* in their notes and be ready for the first example.
- Making the table of values helps students focus on the vertex and where the symmetry occurs in the y-values.

? **Advancing Question:** "You can subtract (or add) a number before or after the function is evaluated. In which direction will the transformation occur?" *Sample answer:* before: $f(x) = |x - 2|$, right 2 units; after: $f(x) = |x| - 2$, down 2 units

- Subtracting 2 before evaluating the absolute value function affects the x-value. The misconception for students is that they think the graph should shift 2 units left. Looking at the table of values helps students see that when $x = 2$, you are evaluating $|2 - 2| = |0|$, which is the x-value of the vertex of the parent function.

? "How does the graph of $f(x) = |x| + k$ compare to $f(x) = |x|$?" It is a translation up ($k > 0$) or down ($k < 0$). "Does the range change?" yes; $y \geq k$

? "How does the graph of $f(x) = |x - h|$ compare to $f(x) = |x|$?" It is a translation right ($h > 0$) or left ($h < 0$). "Does the range change?" no

Stretching, Shrinking, and Reflecting

EXAMPLE 2 **Graphing $g(x) = a|x|$**

Graph each function. Compare each graph to the graph of $f(x) = |x|$. Describe the domain and range.

a. $q(x) = 2|x|$ **b.** $p(x) = -\frac{1}{2}|x|$

SOLUTION

a. Step 1 Make a table of values.

x	−2	−1	0	1	2
$q(x)$	4	2	0	2	4

Step 2 Plot the ordered pairs.

Step 3 Draw the V-shaped graph.

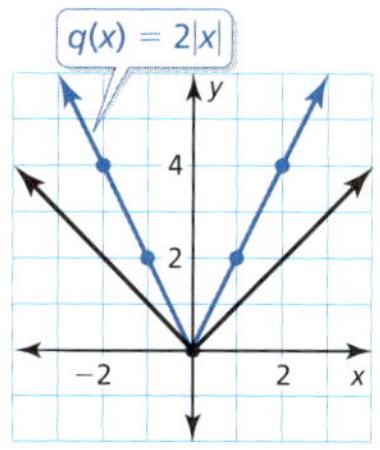

STUDY TIP

A vertical stretch of the graph of $f(x) = |x|$ is *narrower* than the graph of $f(x) = |x|$.

The function q is of the form $y = a \cdot f(x)$, where $a = 2$. So, the graph of q is a vertical stretch of the graph of f by a factor of 2. The domain is all real numbers. The range is $y \geq 0$.

b. Step 1 Make a table of values.

x	−2	−1	0	1	2
$p(x)$	−1	$-\frac{1}{2}$	0	$-\frac{1}{2}$	−1

Step 2 Plot the ordered pairs.

Step 3 Draw the V-shaped graph.

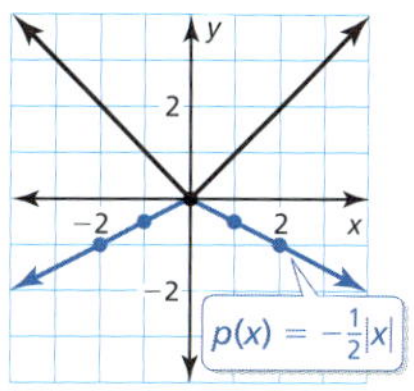

STUDY TIP

A vertical shrink of the graph of $f(x) = |x|$ is *wider* than the graph of $f(x) = |x|$.

The function p is of the form $y = -a \cdot f(x)$, where $a = \frac{1}{2}$. So, the graph of p is a vertical shrink of the graph of f by a factor of $\frac{1}{2}$ and a reflection in the x-axis. The domain is all real numbers. The range is $y \leq 0$.

Monitoring Progress

Help in English and Spanish at *BigIdeasMath.com*

Graph the function. Compare the graph to the graph of $f(x) = |x|$. Describe the domain and range.

3. $t(x) = -3|x|$ **4.** $v(x) = \frac{1}{4}|x|$

Differentiated Instruction

Kinesthetic/Visual

Some students will benefit by having access to graphing calculators or the graphing software programs available on many computer operating systems. Students can enter the absolute value functions, view and compare their graphs, and see the domain and range.

Extra Example 2

Graph each function. Compare each graph to the graph of $f(x) = |x|$. Describe the domain and range.

a. $s(x) = \frac{1}{2}|x|$

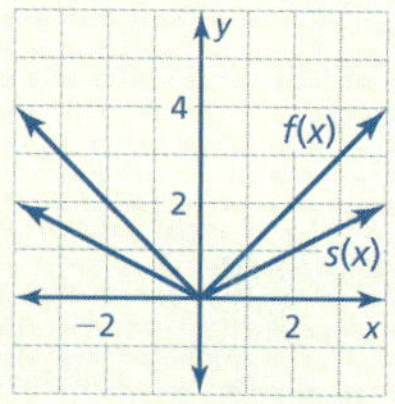

The graph of s is a vertical shrink of the graph of f by a factor of $\frac{1}{2}$. The domain is all real numbers. The range is $y \geq 0$.

b. $r(x) = -2|x|$

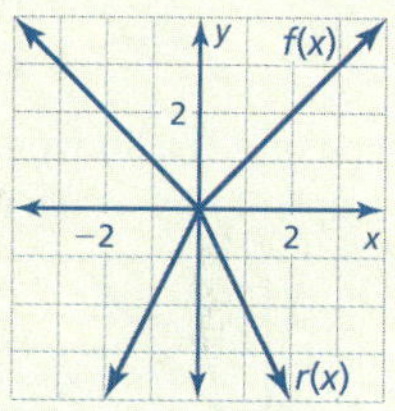

The graph of r is a vertical stretch of the graph of f by a factor of 2 and a reflection in the x-axis. The domain is all real numbers. The range is $y \leq 0$.

MONITORING PROGRESS ANSWERS

3–4. See Additional Answers.

Laurie's Notes — Teacher Actions

- **? Turn and Talk:** "Write the function $q(x) = 2|x|$. What effect do you think the 2 has on the graph of $f(x) = |x|$?" Draw *Popsicle Sticks* to solicit answers. *Sample answer:* The graph stretches upward.
- **Teaching Tip:** Graph the parent function and the transformed function so that students can compare and see the effect of the transformation.
- **?** "When you graph $y = x$ and $y = mx$ on the same graph, what does m do to the graph of $y = x$? It changes the steepness of the line.
- **Connection:** Discuss the transformations of the graph of $f(x) = a|x|$. In part (a), $a = 2$. Because $|a| > 1$, it makes the V-shape steeper than the parent function. This is a stretch. In part (b), $a = -\frac{1}{2}$. Because $|a| < 1$, it makes the V-shape wider than the parent function, and the graph opens down because $a < 0$. This is a shrink and a reflection over the x-axis.
- Discuss the range of each function.

Extra Example 3

Graph $f(x) = |x - 3| + 1$ and $g(x) = |3x - 3| + 1$. Compare the graph of g to the graph of f.

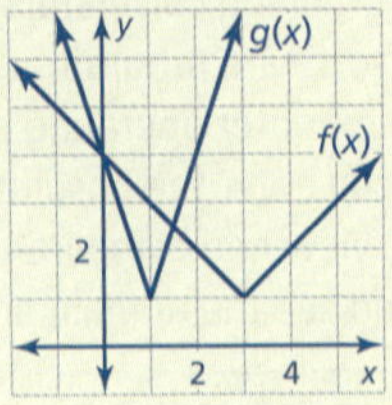

The graph of g is a horizontal shrink of the graph of f by a factor of $\frac{1}{3}$. The y-intercept is the same for both graphs and the points on the graph of f move $\frac{2}{3}$ closer to the y-axis, resulting in the graph of g. When the input values of f are 3 times the input values of g, the outputs are the same.

MONITORING PROGRESS ANSWERS

5.

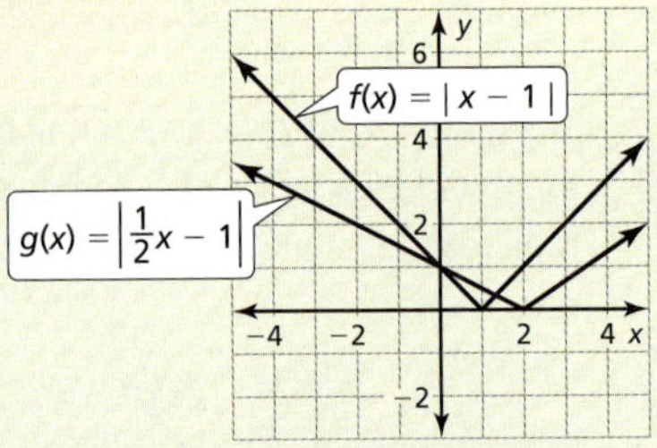

The graph of g is a horizontal stretch of the graph of f by a factor of 2.

6.

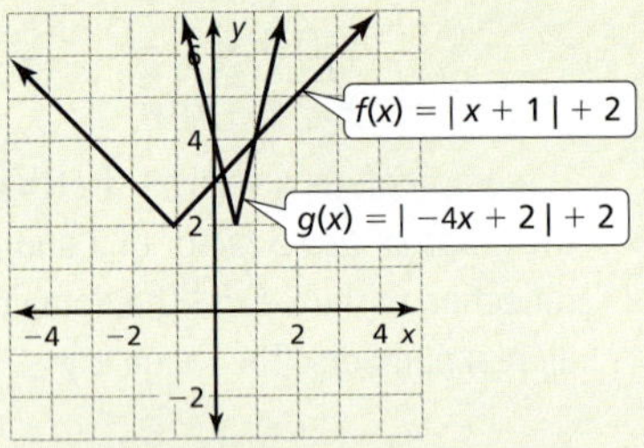

The graph of g is a horizontal shrink of the graph of f by a factor of $\frac{1}{4}$ and a reflection in the y-axis.

Core Concept

Vertex Form of an Absolute Value Function

An absolute value function written in the form $g(x) = a|x - h| + k$, where $a \neq 0$, is in **vertex form**. The vertex of the graph of g is (h, k).

Any absolute value function can be written in vertex form, and its graph is symmetric about the line $x = h$.

STUDY TIP

The function g is *not* in vertex form because the x variable does not have a coefficient of 1.

EXAMPLE 3 Graphing $f(x) = |x - h| + k$ and $g(x) = f(ax)$

Graph $f(x) = |x + 2| - 3$ and $g(x) = |2x + 2| - 3$. Compare the graph of g to the graph of f.

SOLUTION

Step 1 Make a table of values for each function.

x	−4	−3	−2	−1	0	1	2
$f(x)$	−1	−2	−3	−2	−1	0	1

x	−2	−1.5	−1	−0.5	0	0.5	1
$g(x)$	−1	−2	−3	−2	−1	0	1

Step 2 Plot the ordered pairs.

Step 3 Draw the V-shaped graph of each function. Notice that the vertex of the graph of f is $(-2, -3)$ and the graph is symmetric about $x = -2$.

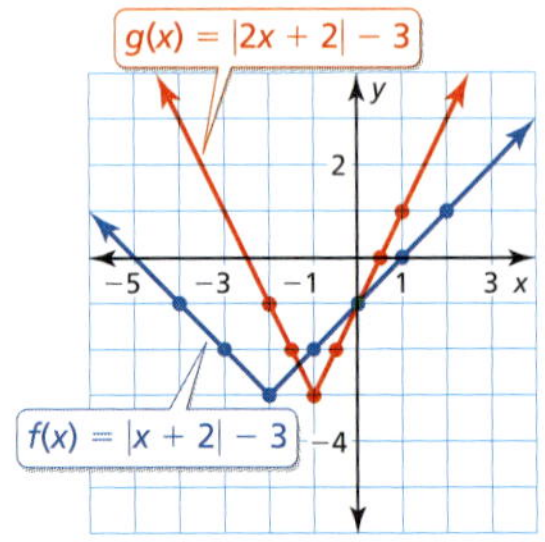

Note that you can rewrite g as $g(x) = f(2x)$, which is of the form $y = f(ax)$, where $a = 2$. So, the graph of g is a horizontal shrink of the graph of f by a factor of $\frac{1}{2}$. The y-intercept is the same for both graphs. The points on the graph of f move halfway closer to the y-axis, resulting in the graph of g. When the input values of f are 2 times the input values of g, the output values of f and g are the same.

Monitoring Progress

Help in English and Spanish at *BigIdeasMath.com*

5. Graph $f(x) = |x - 1|$ and $g(x) = \left|\frac{1}{2}x - 1\right|$. Compare the graph of g to the graph of f.

6. Graph $f(x) = |x + 2| + 2$ and $g(x) = |-4x + 2| + 2$. Compare the graph of g to the graph of f.

Laurie's Notes Teacher Actions

- **? Use Appropriate Tools Strategically:** Have students use a graphing calculator to graph $f(x) = 2|x - 3| + 1$. "What is the vertex?" (3, 1)
- **? Reason Abstractly and Quantitatively:** "Could you predict what the vertex of $f(x) = \frac{1}{2}|x + 5| - 4$ would be?" Give time for students to talk with neighbors. Ask volunteers to share their reasoning. (−5, −4)
- Write the *Core Concept*. Make connections to the transformation of lines in the previous lesson. Note that the coefficient of x is 1.
- Write the two functions in Example 3 and give students time to work with their partners.
- **Discuss:** $g(x)$ is a horizontal shrink of the graph of f by a factor of $\frac{1}{2}$.

Combining Transformations

EXAMPLE 4 **Graphing $g(x) = a|x - h| + k$**

Let $g(x) = -2|x - 1| + 3$. (a) Describe the transformations from the graph of $f(x) = |x|$ to the graph of g. (b) Graph g.

REMEMBER
You can obtain the graph of $y = a \cdot f(x - h) + k$ from the graph of $y = f(x)$ using the steps you learned in Section 3.6.

SOLUTION

a. Step 1 Translate the graph of f horizontally 1 unit right to get the graph of $t(x) = |x - 1|$.

Step 2 Stretch the graph of t vertically by a factor of 2 to get the graph of $h(x) = 2|x - 1|$.

Step 3 Reflect the graph of h in the x-axis to get the graph of $r(x) = -2|x - 1|$.

Step 4 Translate the graph of r vertically 3 units up to get the graph of $g(x) = -2|x - 1| + 3$.

b. Method 1

Step 1 Make a table of values.

x	−1	0	1	2	3
$g(x)$	−1	1	3	1	−1

Step 2 Plot the ordered pairs.

Step 3 Draw the V-shaped graph.

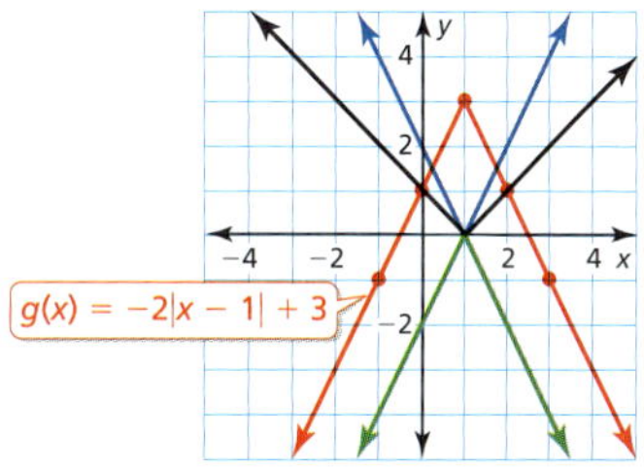

Method 2

Step 1 Identify and plot the vertex. $(h, k) = (1, 3)$

Step 2 Plot another point on the graph, such as (2, 1). Because the graph is symmetric about the line $x = 1$, you can use symmetry to plot a third point, (0, 1).

Step 3 Draw the V-shaped graph.

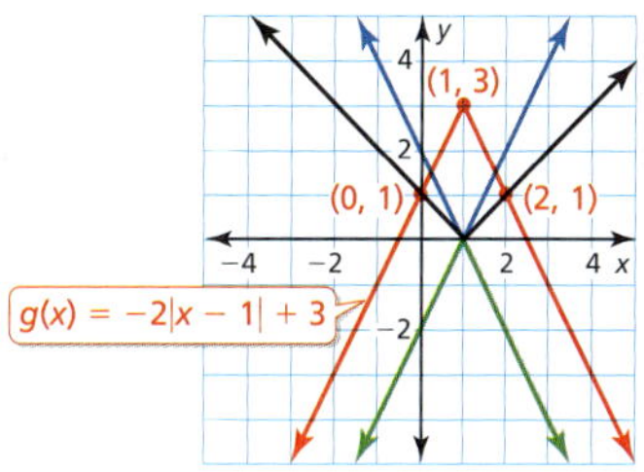

Monitoring Progress

Help in English and Spanish at BigIdeasMath.com

7. Let $g(x) = \left|-\frac{1}{2}x + 2\right| + 1$. (a) Describe the transformations from the graph of $f(x) = |x|$ to the graph of g. (b) Graph g.

Extra Example 4

Let $g(x) = -|x + 1| + 2$.

a. Describe the transformations from the graph of $f(x) = |x|$ to the graph of g. Translate the graph of f horizontally 1 unit left to get the graph of $h(x) = |x + 1|$. Reflect the graph of h in the x-axis to get the graph of $r(x) = -|x + 1|$. Translate the graph of r vertically 2 units up to get the graph of $g(x) = -|x + 1| + 2$.

b. Graph g.

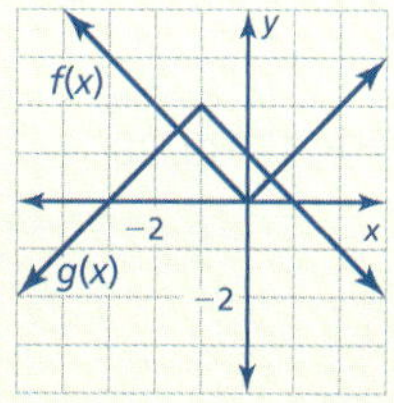

MONITORING PROGRESS ANSWER

7. **a.** The transformations are a horizontal translation 2 units left, then a horizontal stretch by a factor of 2, then a reflection in the y-axis, then a vertical translation 1 unit up.

b.

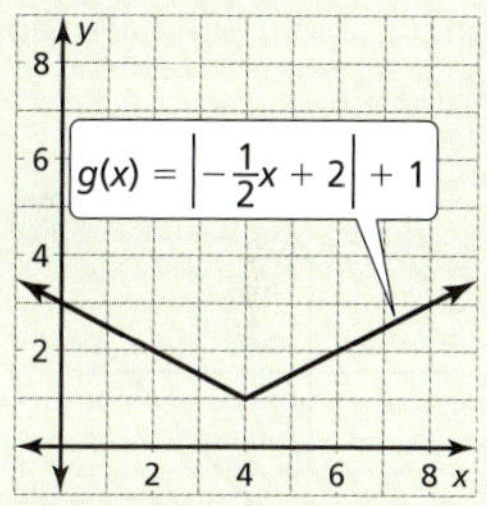

Laurie's Notes Teacher Actions

- Like linear functions, all of the transformations can be performed to the graph of an absolute value function. The vertex form showed this.
- **? Turn and Talk:** "Describe the effect of the −2, 1, and 3 on the parent function $f(x) = |x|$." The 1 translates the graph 1 unit right. The −2 reflects the graph in the x-axis and stretches the graph by a factor of 2. The 3 translates the graph 3 units up.
- **Whiteboarding:** Have partners reveal their descriptions of the transformations.
- **Big Idea:** Students often ask what values they should use in the table of values to be sure it includes the vertex and a few values on either side of the vertex. In the parent function, the vertex occurs at $x = 0$. In the transformed function, the vertex occurs when the expression inside the absolute value symbols equals 0. In other words, when does $x - h = 0$?

Closure

- **Look Back:** Have students *Look Back* over Sections 3.6 and 3.7. Have them summarize what they understand about transformations and what helped them learn it.

Assignment Guide and Homework Check

ASSIGNMENT

Basic: 1–4, 5–13 odd, 19, 23, 27–41 odd, 45, 50, 62, 64–70

Average: 1–4, 6–24 even, 28–46 even, 50–56 even, 62, 64–70

Advanced: 1–4, 8–26 even, 30, 32, 38–62 even, 64–70

HOMEWORK CHECK

Basic: 5, 9, 27, 35

Average: 6, 12, 28, 36

Advanced: 8, 12, 30, 38

ANSWERS

1. vertex
2. If $|a| > 1$, the graph is a vertical stretch. If $0 < |a| < 1$, the graph is a vertical shrink.
3. *Sample answer:* $g(x) = a|x|$ is a vertical stretch or shrink, $g(x) = |x - h|$ is a horizontal translation, and $g(x) = |x| + k$ is a vertical translation.
4. $g(x) = |3x - 2| + 5$; g is a horizontal shrink of f, so it will have the same y-intercept.
5.

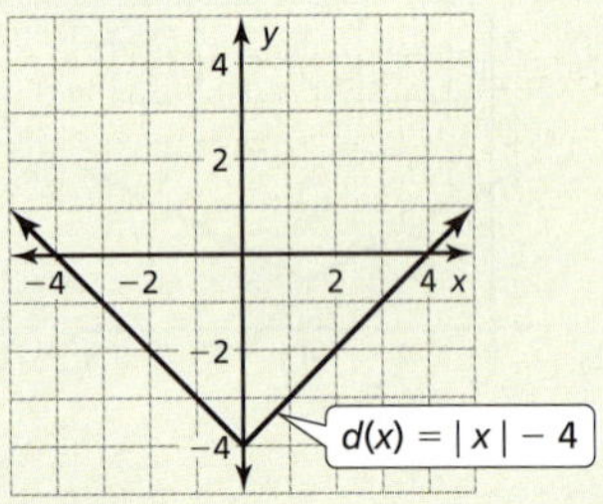

The graph of d is a vertical translation 4 units down of the graph of f; domain: all real numbers; range: $y \geq -4$

6.

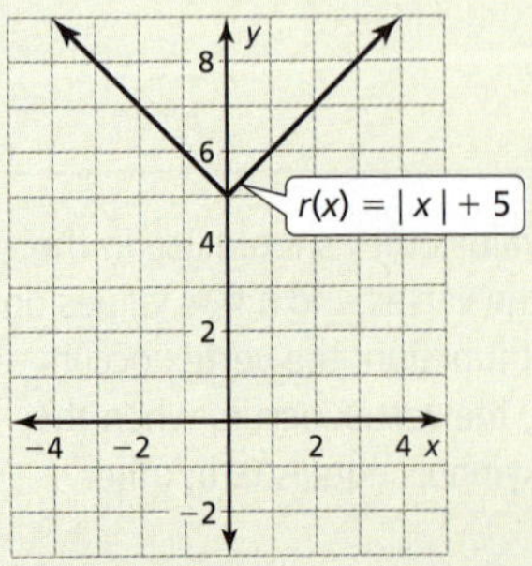

The graph of r is a vertical translation 5 units up of the graph of f; domain: all real numbers; range: $y \geq 5$

3.7 Exercises

Dynamic Solutions available at *BigIdeasMath.com*

Vocabulary and Core Concept Check

1. **COMPLETE THE SENTENCE** The point $(1, -4)$ is the _______ of the graph of $f(x) = -3|x - 1| - 4$.
2. **USING STRUCTURE** How do you know whether the graph of $f(x) = a|x - h| + k$ is a vertical stretch or a vertical shrink of the graph of $f(x) = |x|$?
3. **WRITING** Describe three different types of transformations of the graph of an absolute value function.
4. **REASONING** The graph of which function has the same y-intercept as the graph of $f(x) = |x - 2| + 5$? Explain.

   $g(x) = |3x - 2| + 5$ $\quad$ $h(x) = 3|x - 2| + 5$

Monitoring Progress and Modeling with Mathematics

In Exercises 5–12, graph the function. Compare the graph to the graph of $f(x) = |x|$. Describe the domain and range. *(See Examples 1 and 2.)*

5. $d(x) = |x| - 4$
6. $r(x) = |x| + 5$
7. $m(x) = |x + 1|$
8. $v(x) = |x - 3|$
9. $p(x) = \frac{1}{3}|x|$
10. $j(x) = 3|x|$
11. $a(x) = -5|x|$
12. $q(x) = -\frac{3}{2}|x|$

In Exercises 13–16, graph the function. Compare the graph to the graph of $f(x) = |x - 6|$.

13. $h(x) = |x - 6| + 2$
14. $n(x) = \frac{1}{2}|x - 6|$
15. $k(x) = -3|x - 6|$
16. $g(x) = |x - 1|$

In Exercises 17 and 18, graph the function. Compare the graph to the graph of $f(x) = |x + 3| - 2$.

17. $y(x) = |x + 4| - 2$
18. $b(x) = |x + 3| + 3$

In Exercises 19–22, compare the graphs. Find the value of h, k, or a.

19.

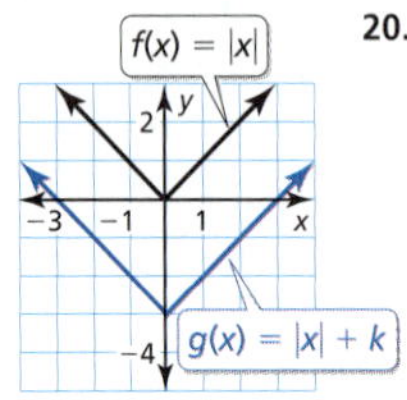

20.

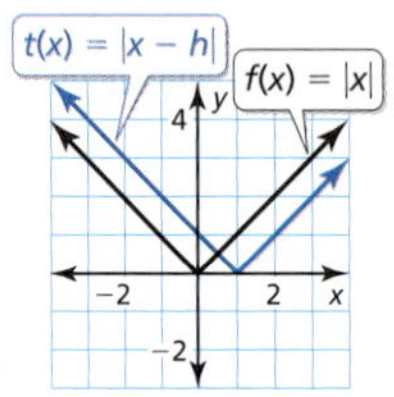

21.

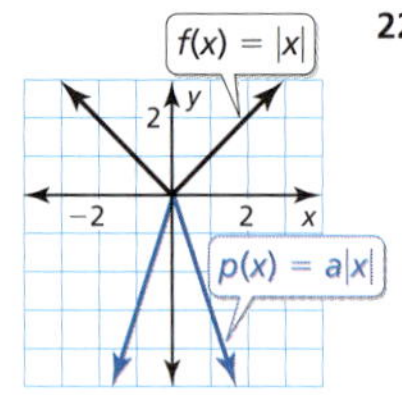

22.

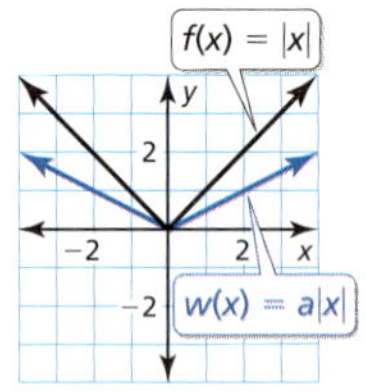

In Exercises 23–26, write an equation that represents the given transformation(s) of the graph of $g(x) = |x|$.

23. vertical translation 7 units down
24. horizontal translation 10 units left
25. vertical shrink by a factor of $\frac{1}{4}$
26. vertical stretch by a factor of 3 and a reflection in the x-axis

In Exercises 27–32, graph and compare the two functions. *(See Example 3.)*

27. $f(x) = |x - 4|$; $g(x) = |3x - 4|$
28. $h(x) = |x + 5|$; $t(x) = |2x + 5|$
29. $p(x) = |x + 1| - 2$; $q(x) = \left|\frac{1}{4}x + 1\right| - 2$
30. $w(x) = |x - 3| + 4$; $y(x) = |5x - 3| + 4$
31. $a(x) = |x + 2| + 3$; $b(x) = |-4x + 2| + 3$
32. $u(x) = |x - 1| + 2$; $v(x) = \left|-\frac{1}{2}x - 1\right| + 2$

7.

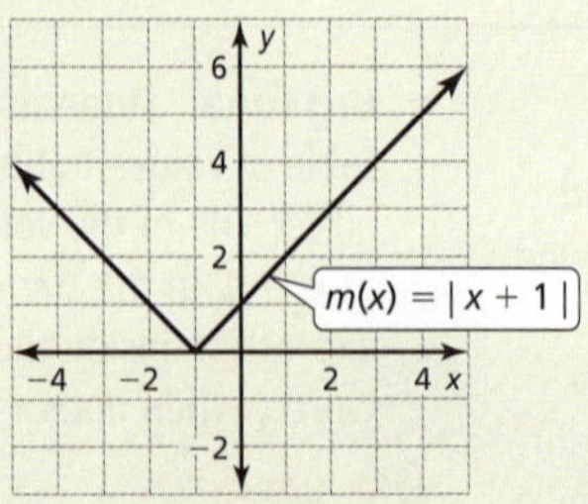

The graph of m is a horizontal translation 1 unit left of the graph of f; domain: all real numbers; range: $y \geq 0$

8–32. See Additional Answers.

In Exercises 33–40, describe the transformations from the graph of $f(x) = |x|$ to the graph of the given function. Then graph the given function. *(See Example 4.)*

33. $r(x) = |x + 2| - 6$ **34.** $c(x) = |x + 4| + 4$

35. $d(x) = -|x - 3| + 5$ **36.** $v(x) = -3|x + 1| + 4$

37. $m(x) = \frac{1}{2}|x + 4| - 1$ **38.** $s(x) = |2x - 2| - 3$

39. $j(x) = |-x + 1| - 5$ **40.** $n(x) = \left|-\frac{1}{3}x + 1\right| + 2$

41. MODELING WITH MATHEMATICS The number of pairs of shoes sold s (in thousands) increases and then decreases as described by the function $s(t) = -2|t - 15| + 50$, where t is the time (in weeks).

a. Graph the function.

b. What is the greatest number of pairs of shoes sold in 1 week?

42. MODELING WITH MATHEMATICS On the pool table shown, you bank the five ball off the side represented by the x-axis. The path of the ball is described by the function $p(x) = \frac{4}{3}\left|x - \frac{5}{4}\right|$.

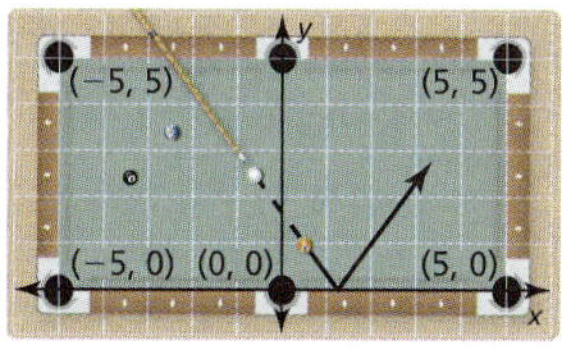

a. At what point does the five ball bank off the side?

b. Do you make the shot? Explain your reasoning.

43. USING TRANSFORMATIONS The points $A\left(-\frac{1}{2}, 3\right)$, $B(1, 0)$, and $C(-4, -2)$ lie on the graph of the absolute value function f. Find the coordinates of the points corresponding to A, B, and C on the graph of each function.

a. $g(x) = f(x) - 5$ **b.** $h(x) = f(x - 3)$

c. $j(x) = -f(x)$ **d.** $k(x) = 4f(x)$

44. USING STRUCTURE Explain how the graph of each function compares to the graph of $y = |x|$ for positive and negative values of k, h, and a.

a. $y = |x| + k$

b. $y = |x - h|$

c. $y = a|x|$

d. $y = |ax|$

ERROR ANALYSIS In Exercises 45 and 46, describe and correct the error in graphing the function.

45.

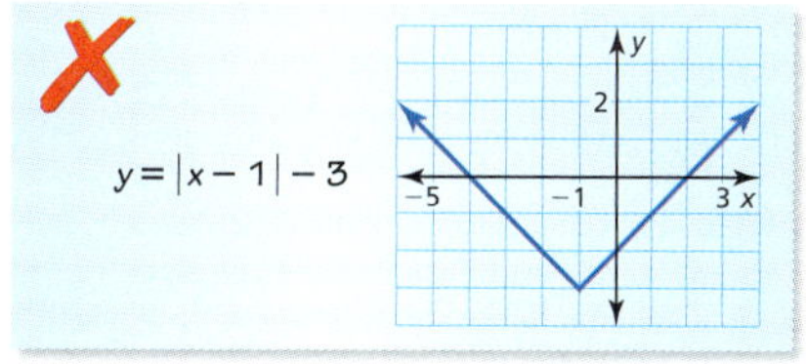

46.

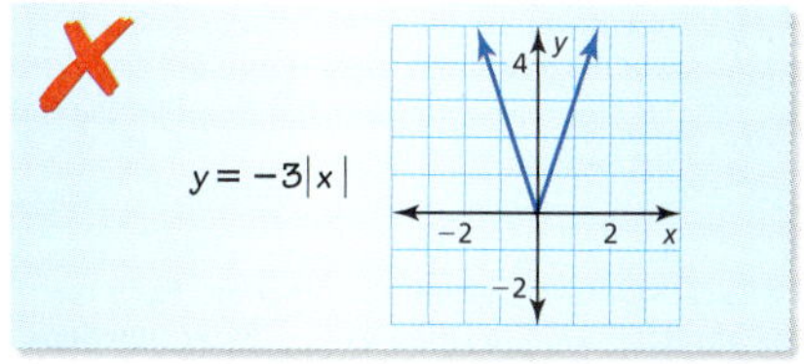

MATHEMATICAL CONNECTIONS In Exercises 47 and 48, write an absolute value function whose graph forms a square with the given graph.

47.

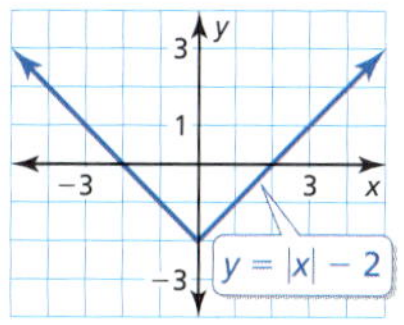

48.

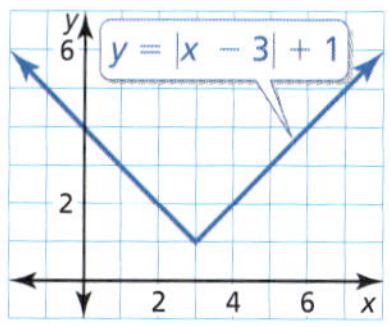

49. WRITING Compare the graphs of $p(x) = |x - 6|$ and $q(x) = |x| - 6$.

Dynamic Teaching Tools

- Dynamic Assessment & Progress Monitoring Tool
- Lesson Planning Tool
- Interactive Whiteboard Lesson Library
- Dynamic Classroom with Dynamic Investigations

ANSWERS

33. The transformations are a horizontal translation 2 units left then a vertical translation 6 units down.

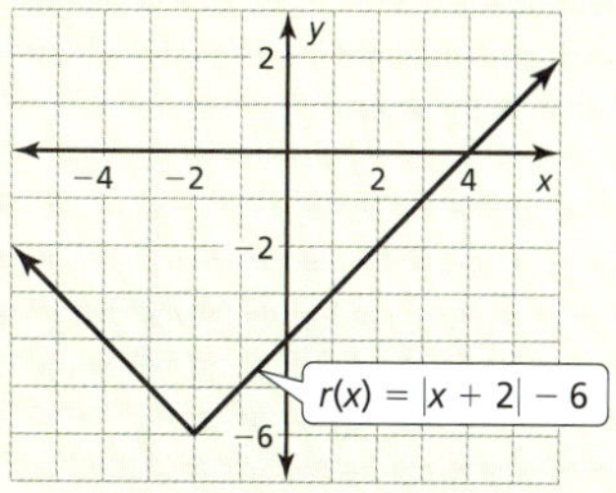

34. The transformations are a horizontal translation 4 units left then a vertical translation 4 units up.

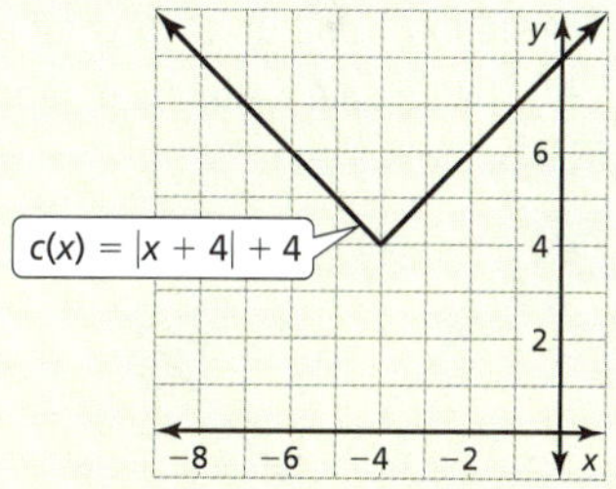

35. The transformations are a horizontal translation 3 units right, then a reflection in the x-axis, then a vertical translation 5 units up.

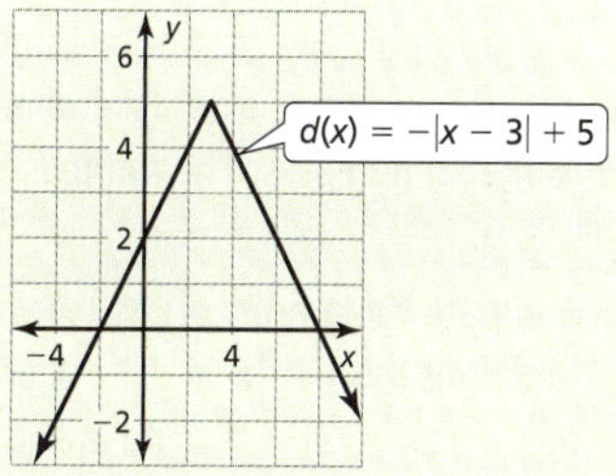

36. The transformations are a horizontal translation 1 unit left, then a vertical stretch by a factor of 3, then a reflection in the x-axis, then a vertical translation 4 units up.

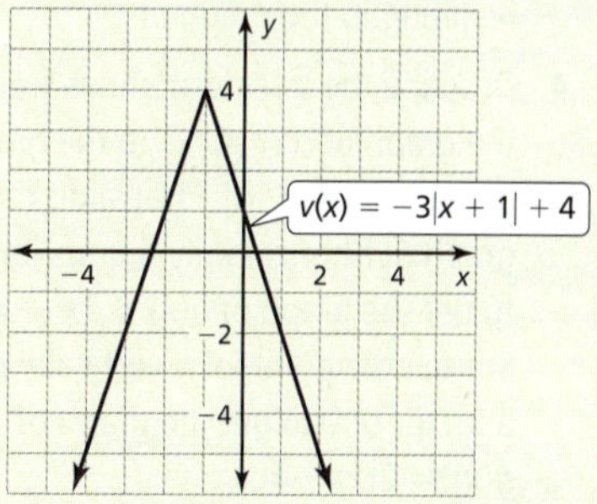

37. The transformations are a horizontal translation 4 units left, then a vertical shrink by a factor of $\frac{1}{2}$, then a vertical translation 1 unit down.

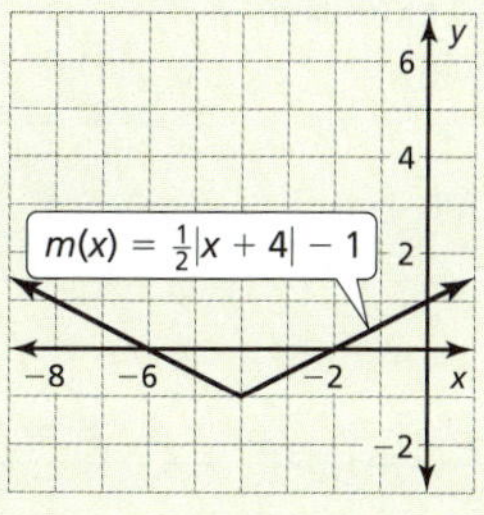

38–49. See Additional Answers.

ANSWERS

50. a. (7, 1)

b. Write an equation in vertex form with (7, 1) as the vertex; $f(x) = |x - 7| + 1$

51–70. See Additional Answers.

Mini-Assessment

Graph the function. Compare the graph to the graph of $f(x) = |x|$. Describe the domain and range.

1. $g(x) = |x| + 1$

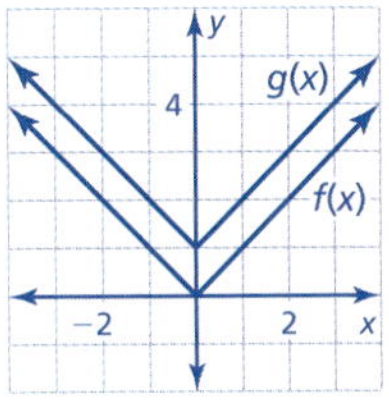

The graph of g is a vertical translation 1 unit up of the graph of f. The domain is all real numbers. The range is $y \geq 1$.

2. $d(x) = -\frac{1}{3}|x|$

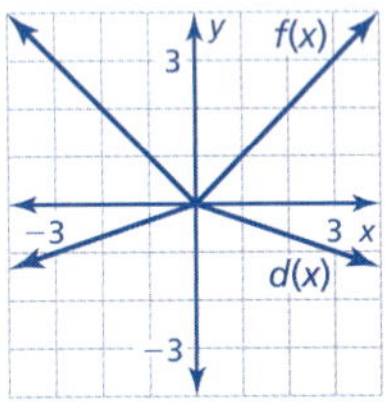

The graph of d is a vertical shrink of the graph of f by a factor of $\frac{1}{3}$ and a reflection in the x-axis. The domain is all real numbers. The range is $y \leq 0$.

3. Compare the graphs of the two functions $m(x) = |x + 2| - 1$ and $t(x) = |2x + 2| - 1$. The graph of t is a horizontal shrink of the graph of m by a factor of $\frac{1}{2}$. The y-intercept is the same for both graphs and the points on the graph of m move halfway closer to the y-axis, resulting in the graph of t.

4. Describe the transformations from the graph of $f(x) = |x|$ to the graph of $g(x) = 2|x + 4|$. Translate the graph of f horizontally 4 units left to get the graph of $h(x) = |x + 4|$. Stretch the graph of h vertically by a factor of 2 to get the graph of $g(x) = 2|x + 4|$.

50. HOW DO YOU SEE IT? The object of a computer game is to break bricks by deflecting a ball toward them using a paddle. The graph shows the current path of the ball and the location of the last brick.

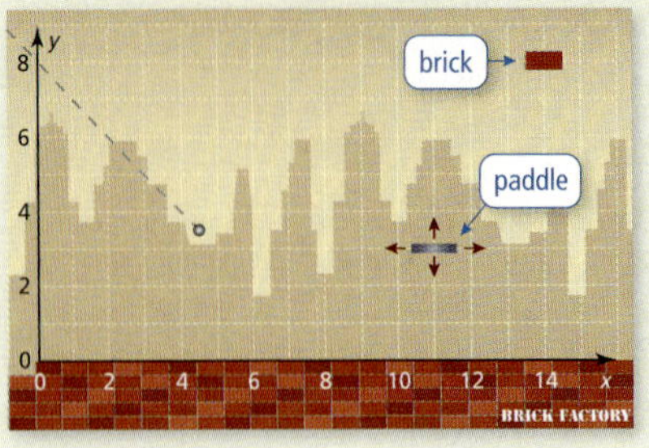

a. You can move the paddle up, down, left, and right. At what coordinates should you place the paddle to break the last brick? Assume the ball deflects at a right angle.

b. You move the paddle to the coordinates in part (a), and the ball is deflected. How can you write an absolute value function that describes the path of the ball?

In Exercises 51–54, graph the function. Then rewrite the absolute value function as two linear functions, one that has the domain $x < 0$ and one that has the domain $x \geq 0$.

51. $y = |x|$

52. $y = |x| - 3$

53. $y = -|x| + 9$

54. $y = -4|x|$

In Exercises 55–58, graph and compare the two functions.

55. $f(x) = |x - 1| + 2$; $g(x) = 4|x - 1| + 8$

56. $s(x) = |2x - 5| - 6$; $t(x) = \frac{1}{2}|2x - 5| - 3$

57. $v(x) = -2|3x + 1| + 4$; $w(x) = 3|3x + 1| - 6$

58. $c(x) = 4|x + 3| - 1$; $d(x) = -\frac{4}{3}|x + 3| + \frac{1}{3}$

59. REASONING Describe the transformations from the graph of $g(x) = -2|x + 1| + 4$ to the graph of $h(x) = |x|$. Explain your reasoning.

60. THOUGHT PROVOKING Graph an absolute value function f that represents the route a wide receiver runs in a football game. Let the x-axis represent distance (in yards) across the field horizontally. Let the y-axis represent distance (in yards) down the field. Be sure to limit the domain so the route is realistic.

61. SOLVING BY GRAPHING Graph $y = 2|x + 2| - 6$ and $y = -2$ in the same coordinate plane. Use the graph to solve the equation $2|x + 2| - 6 = -2$. Check your solutions.

62. MAKING AN ARGUMENT Let p be a positive constant. Your friend says that because the graph of $y = |x| + p$ is a *positive* vertical translation of the graph of $y = |x|$, the graph of $y = |x + p|$ is a *positive* horizontal translation of the graph of $y = |x|$. Is your friend correct? Explain.

63. ABSTRACT REASONING Write the vertex of the absolute value function $f(x) = |ax - h| + k$ in terms of a, h, and k.

Maintaining Mathematical Proficiency

Reviewing what you learned in previous grades and lessons

Solve the inequality. *(Section 2.4)*

64. $8a - 7 \leq 2(3a - 1)$

65. $-3(2p + 4) > -6p - 5$

66. $4(3h + 1.5) \geq 6(2h - 2)$

67. $-4(x + 6) < 2(2x - 9)$

Find the slope of the line. *(Section 3.5)*

68.

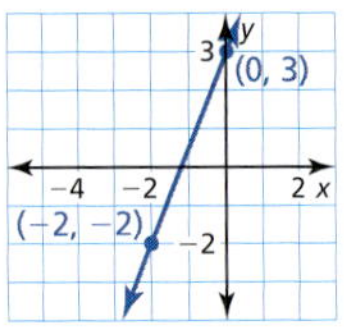

69.

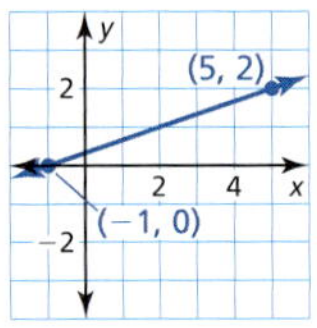

70.

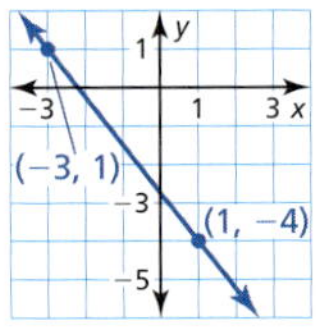

If students need help...	*If students got it...*
Resources by Chapter • Practice A and Practice B • Puzzle Time	Resources by Chapter • Enrichment and Extension • Cumulative Review
Student Journal • Practice	Start the *next* Section
Differentiating the Lesson Skills Review Handbook	

3.4–3.7 What Did You Learn?

Core Vocabulary

standard form, *p. 130*
x-intercept, *p. 131*
y-intercept, *p. 131*
slope, *p. 136*
rise, *p. 136*
run, *p. 136*
slope-intercept form, *p. 138*
constant function, *p. 138*
family of functions, *p. 146*
parent function, *p. 146*
transformation, *p. 146*
translation, *p. 146*
reflection, *p. 147*
horizontal shrink, *p. 148*
horizontal stretch, *p. 148*
vertical stretch, *p. 148*
vertical shrink, *p. 148*
absolute value function, *p. 156*
vertex, *p. 156*
vertex form, *p. 158*

Core Concepts

Section 3.4

Horizontal and Vertical Lines, *p. 130*
Using Intercepts to Graph Equations, *p. 131*

Section 3.5

Slope, *p. 136*
Slope-Intercept Form, *p. 138*

Section 3.6

Horizontal Translations, *p. 146*
Vertical Translations, *p. 146*
Reflections in the x-axis, *p. 147*
Reflections in the y-axis, *p. 147*
Horizontal Stretches and Shrinks, *p. 148*
Vertical Stretches and Shrinks, *p. 148*
Transformations of Graphs, *p. 149*

Section 3.7

Absolute Value Function, *p. 156*
Vertex Form of an Absolute Value Function, *p. 158*

Mathematical Practices

1. Explain how you determined what units of measure to use for the horizontal and vertical axes in Exercise 37 on page 142.
2. Explain your plan for solving Exercise 48 on page 153.

Performance Task

The Cost of a T-Shirt

You receive bids for making T-shirts for your class fundraiser from four companies. To present the pricing information, one company uses a table, one company uses a written description, one company uses an equation, and one company uses a graph. How will you compare the different representations and make the final choice?

To explore the answers to this question and more, go to ***BigIdeasMath.com***.

163

Dynamic Teaching Tools

- Dynamic Assessment & Progress Monitoring Tool
- Interactive Whiteboard Lesson Library
- Dynamic Classroom with Dynamic Investigations

ANSWERS

1. The units were given in the statement of the problem. Because t was the independent variable, the units for t went on the horizontal axis. Because d was the dependent variable, the units for d went on the vertical axis.
2. Rewrite Q as $2(8x) - 200$ to match the coefficient in P. Then rewrite the constant term as $-150 - 50$ to get an expression that is an exact match of P. Express Q in terms of P to describe the transformations.

ANSWERS

1. function; Every input has exactly one output.
2. not a function; A vertical line can be drawn through more than one point on the graph in many places.
3. function; Every input has exactly one output.
4. **a.** The amount of money in the bank account is the dependent variable and the hours you babysat is the independent variable.

   b. domain: $0 \le x \le 4$; range: $100 \le y \le 140$
5. linear; As x increases by 5, y decreases by 3. The rate of change is constant.
6. nonlinear; The graph is not a line.
7. **a.** 0, 1, 2, 3, 4, 5, 6, 7; discrete; The number of tickets bought must be a whole number.

   b.

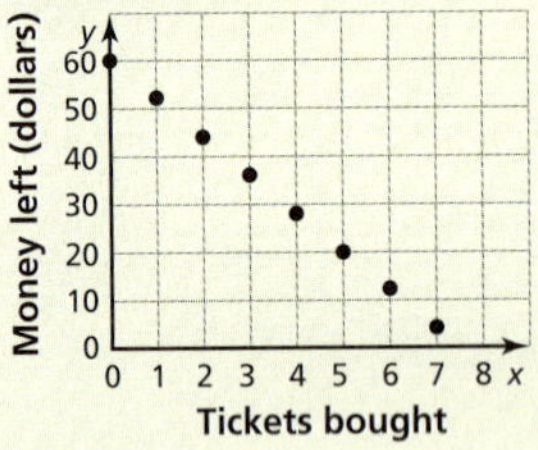

3 Chapter Review

Dynamic Solutions available at *BigIdeasMath.com*

3.1 Functions (pp. 103–110)

Determine whether the relation is a function. Explain.

Input, x	2	5	7	9	14
Output, y	5	11	19	12	3

Every input has exactly one output.

▶ So, the relation is a function.

Determine whether the relation is a function. Explain.

1. (0, 1), (5, 6), (7, 9)

2.

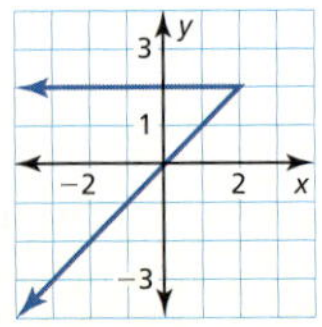

3.

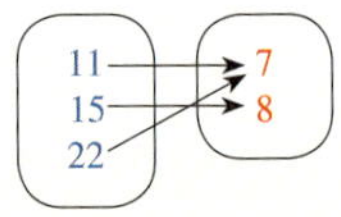

4. The function $y = 10x + 100$ represents the amount y (in dollars) of money in your bank account after you babysit for x hours.

a. Identify the independent and dependent variables.

b. You babysit for 4 hours. Find the domain and range of the function.

3.2 Linear Functions (pp. 111–120)

Does the table or equation represent a *linear* or *nonlinear* function? Explain.

a.

x	6	10	14	18
y	5	9	14	20

(x: +4, +4, +4; y: +4, +5, +6)

As x increases by 4, y increases by different amounts. The rate of change is *not* constant.

▶ So, the function is nonlinear.

b. $y = 3x - 4$

The equation is in the form $y = mx + b$.

▶ So, the equation represents a linear function.

Does the table or graph represent a *linear* or *nonlinear* function? Explain.

5.

x	2	7	12	17
y	2	−1	−4	−7

6.

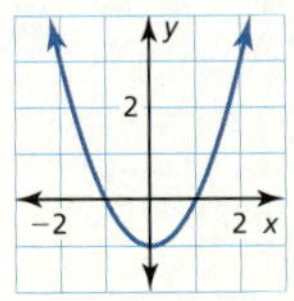

7. The function $y = 60 - 8x$ represents the amount y (in dollars) of money you have after buying x movie tickets. (a) Find the domain of the function. Is the domain discrete or continuous? Explain. (b) Graph the function using its domain.

3.3 Function Notation (pp. 121–126)

a. Evaluate $f(x) = 3x - 9$ when $x = 2$.

$f(x) = 3x - 9$ Write the function.

$f(2) = 3(2) - 9$ Substitute 2 for x.

$= 6 - 9$ Multiply.

$= -3$ Subtract.

▶ When $x = 2$, $f(x) = -3$.

b. For $f(x) = 4x$, find the value of x for which $f(x) = 12$.

$f(x) = 4x$ Write the function.

$12 = 4x$ Substitute 12 for $f(x)$.

$3 = x$ Divide each side by 4.

▶ When $x = 3$, $f(x) = 12$.

Evaluate the function when $x = -3$, 0, and 5.

8. $f(x) = x + 8$

9. $g(x) = 4 - 3x$

Find the value of x so that the function has the given value.

10. $k(x) = 7x$; $k(x) = 49$

11. $r(x) = -5x - 1$; $r(x) = 19$

Graph the linear function.

12. $g(x) = -2x - 3$

13. $h(x) = \frac{2}{3}x + 4$

3.4 Graphing Linear Equations in Standard Form (pp. 129–134)

Use intercepts to graph the equation $2x + 3y = 6$.

Step 1 Find the intercepts.

To find the x-intercept, substitute 0 for y and solve for x.

$2x + 3y = 6$

$2x + 3(0) = 6$

$x = 3$

To find the y-intercept, substitute 0 for x and solve for y.

$2x + 3y = 6$

$2(0) + 3y = 6$

$y = 2$

Step 2 Plot the points and draw the line.
The x-intercept is 3, so plot the point (3, 0).
The y-intercept is 2, so plot the point (0, 2).
Draw a line through the points.

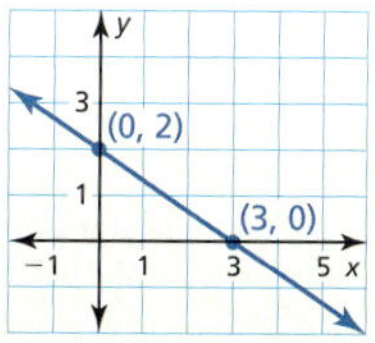

Graph the linear equation.

14. $8x - 4y = 16$

15. $-12x - 3y = 36$

16. $y = -5$

17. $x = 6$

ANSWERS

8. 5, 8, 13

9. 13, 4, −11

10. $x = 7$

11. $x = -4$

12.

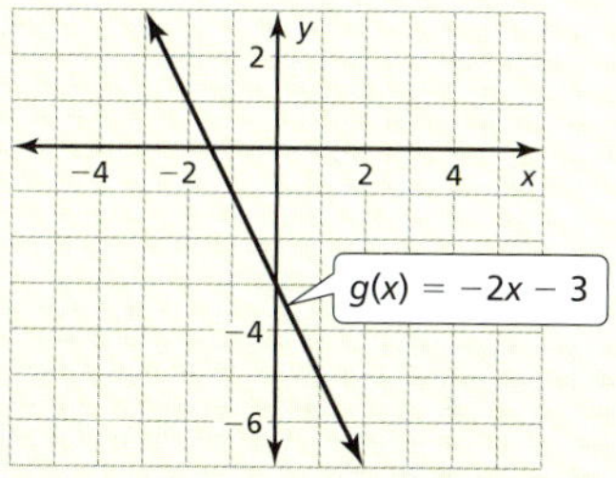

13.

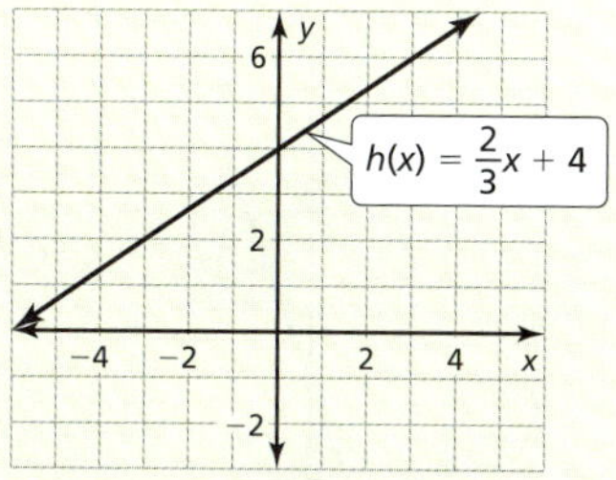

14.

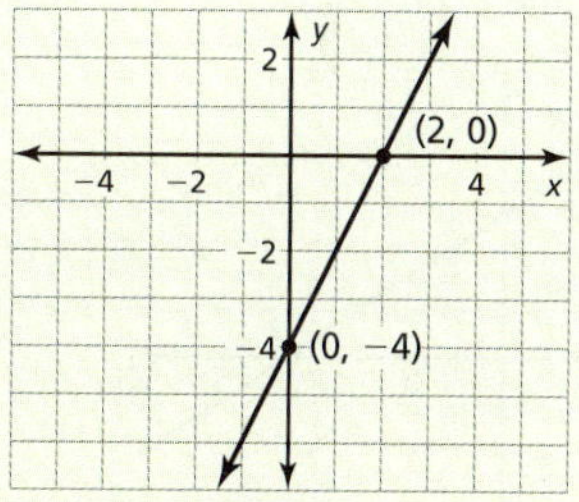

15.

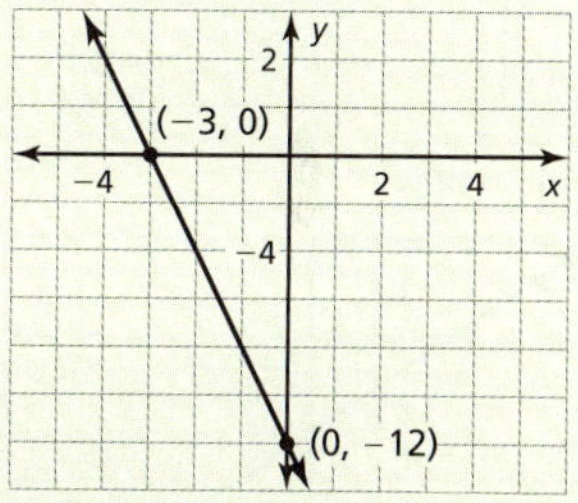

16.

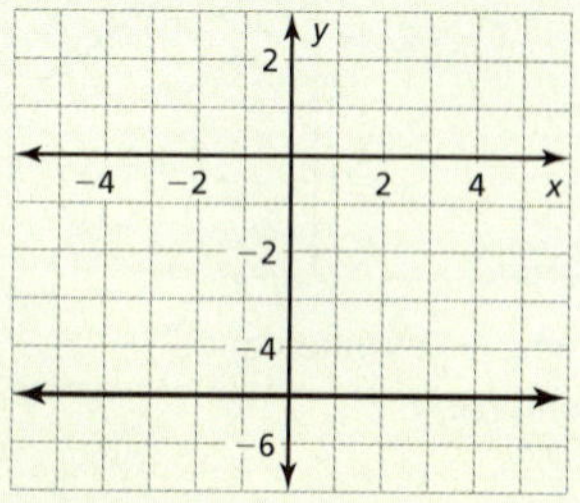

17.

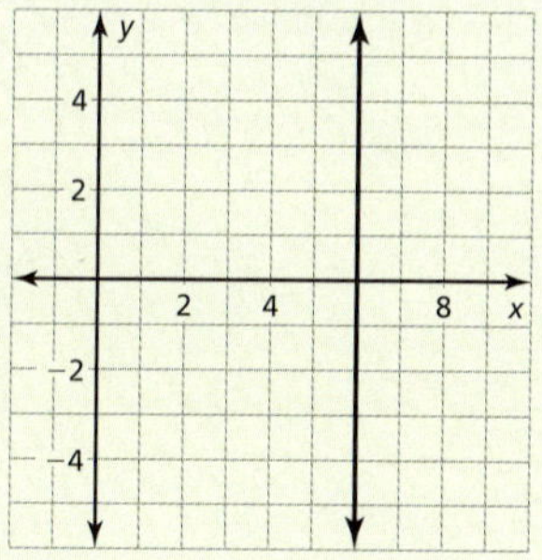

ANSWERS

18. $\frac{6}{5}$

19. undefined

20. 0

21.

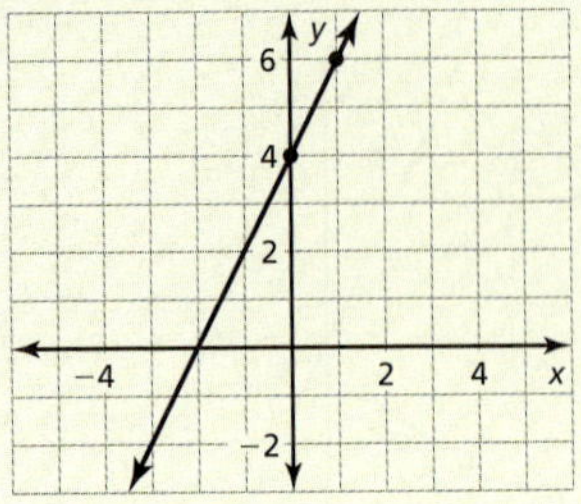

x-intercept: −2

22.

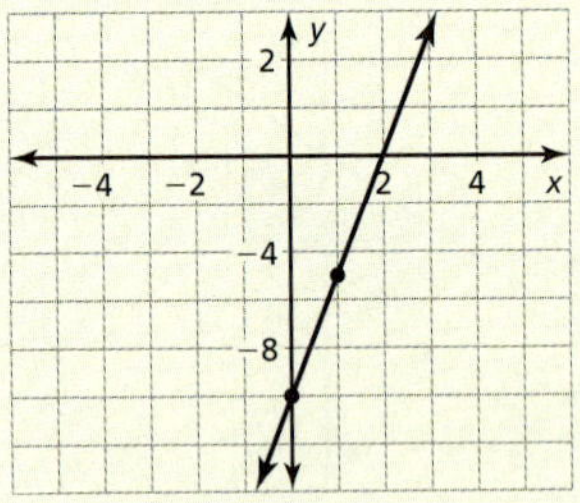

x-intercept: 2

23.

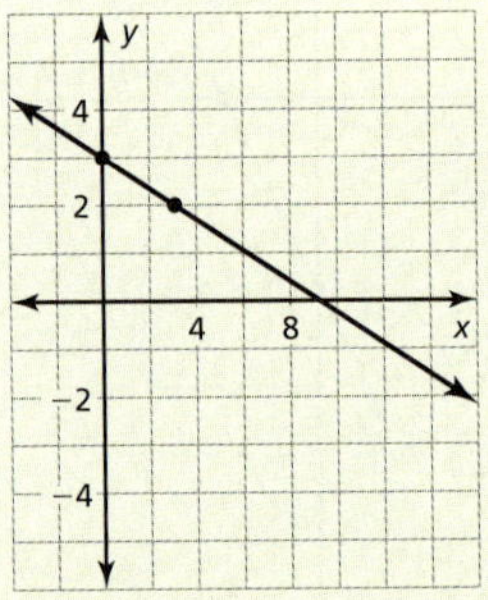

x-intercept: 9

24.

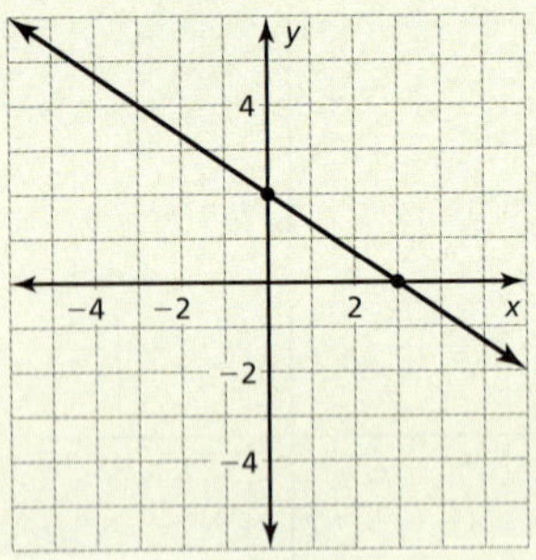

slope: $-\frac{2}{3}$; y-intercept: 2; x-intercept: 3

3.5 Graphing Linear Equations in Slope-Intercept Form (pp. 135–144)

a. The points represented by the table lie on a line. How can you find the slope of the line from the table? What is the slope of the line?

x	y
1	−7
4	2
7	11
10	20

Choose any two points from the table and use the slope formula. Use the points $(x_1, y_1) = (1, -7)$ and $(x_2, y_2) = (4, 2)$.

$$\text{slope} = \frac{y_2 - y_1}{x_2 - x_1} = \frac{2 - (-7)}{4 - 1} = \frac{9}{3}, \text{ or } 3$$

▶ The slope is 3.

b. Graph $-\frac{1}{2}x + y = 1$. Identify the x-intercept.

Step 1 Rewrite the equation in slope-intercept form.

$$y = \frac{1}{2}x + 1$$

Step 2 Find the slope and the y-intercept.

$$m = \frac{1}{2} \text{ and } b = 1$$

Step 3 The y-intercept is 1. So, plot (0, 1).

Step 4 Use the slope to find another point on the line.

$$\text{slope} = \frac{\text{rise}}{\text{run}} = \frac{1}{2}$$

Plot the point that is 2 units right and 1 unit up from (0, 1). Draw a line through the two points.

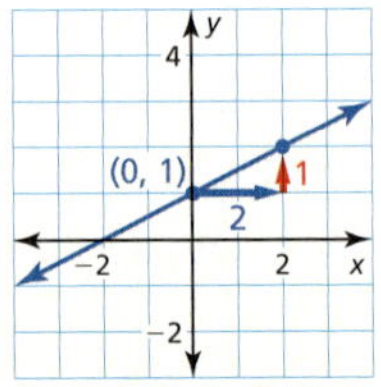

▶ The line crosses the x-axis at (−2, 0). So, the x-intercept is −2.

The points represented by the table lie on a line. Find the slope of the line.

18.

x	y
6	9
11	15
16	21
21	27

19.

x	y
3	−5
3	−2
3	5
3	8

20.

x	y
−4	−1
−3	−1
1	−1
9	−1

Graph the linear equation. Identify the x-intercept.

21. $y = 2x + 4$

22. $-5x + y = -10$

23. $x + 3y = 9$

24. A linear function h models a relationship in which the dependent variable decreases 2 units for every 3 units the independent variable increases. Graph h when $h(0) = 2$. Identify the slope, y-intercept, and x-intercept of the graph.

3.6 Transformations of Graphs of Linear Functions (pp. 145–154)

a. Let $f(x) = -\frac{1}{2}x + 2$. Graph $t(x) = f(x - 3)$. Describe the transformation from the graph of f to the graph of t.

The function t is of the form $y = f(x - h)$, where $h = 3$. So, the graph of t is a horizontal translation 3 units right of the graph of f.

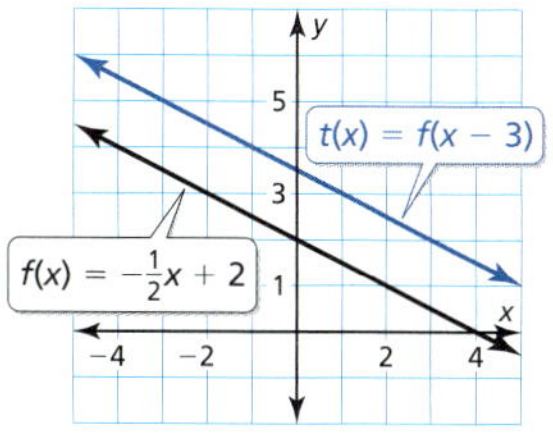

b. Graph $f(x) = x$ and $g(x) = -3x - 2$. Describe the transformations from the graph of f to the graph of g.

Note that you can rewrite g as $g(x) = -3f(x) - 2$.

Step 1 There is no horizontal translation from the graph of f to the graph of g.

Step 2 Stretch the graph of f vertically by a factor of 3 to get the graph of $h(x) = 3x$.

Step 3 Reflect the graph of h in the x-axis to get the graph of $r(x) = -3x$.

Step 4 Translate the graph of r vertically 2 units down to get the graph of $g(x) = -3x - 2$.

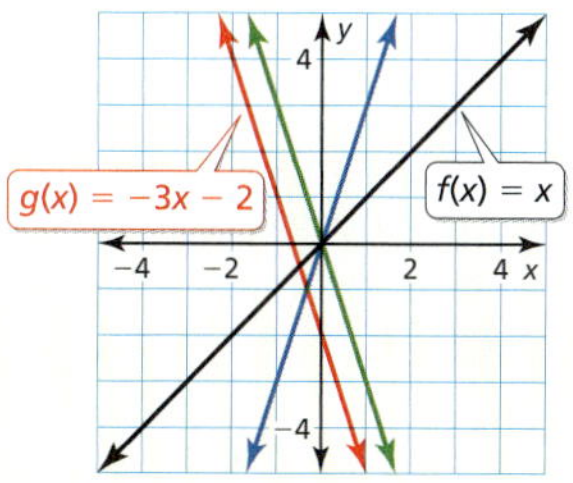

Let $f(x) = 3x + 4$. Graph f and h. Describe the transformation from the graph of f to the graph of h.

25. $h(x) = f(x + 3)$ | **26.** $h(x) = f(x) + 1$

27. $h(x) = f(-x)$ | **28.** $h(x) = -f(x)$

29. $h(x) = 3f(x)$ | **30.** $h(x) = f(6x)$

31. Graph $f(x) = x$ and $g(x) = 5x + 1$. Describe the transformations from the graph of f to the graph of g.

ANSWERS

25.

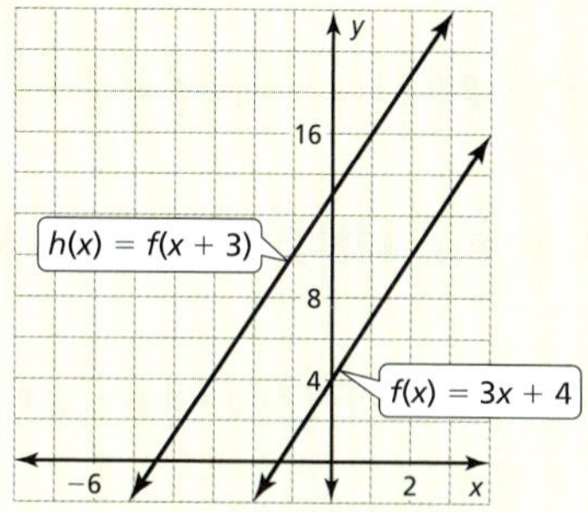

The graph of h is a horizontal translation 3 units left of the graph of f.

26.

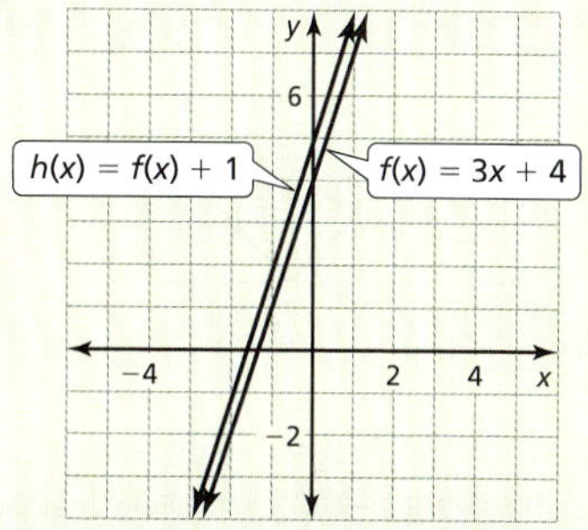

The graph of h is a vertical translation 1 unit up of the graph of f.

27.

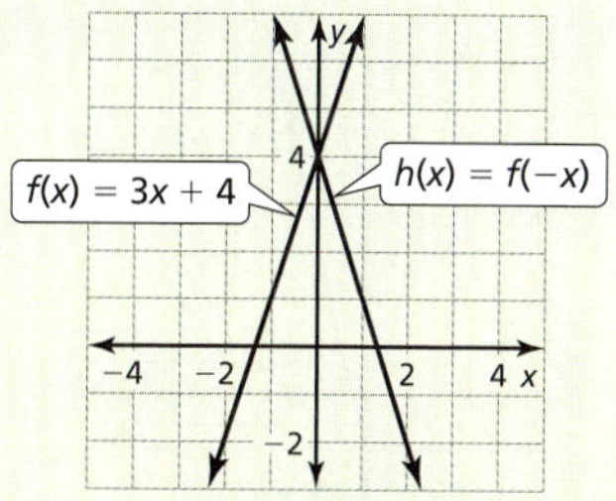

The graph of h is a reflection in the y-axis of the graph of f.

28.

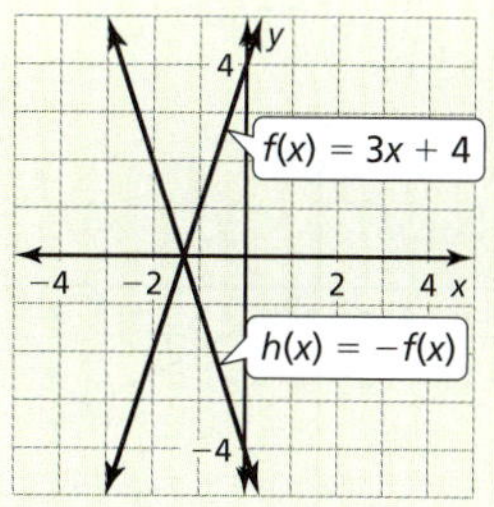

The graph of h is a reflection in the x-axis of the graph of f.

29.

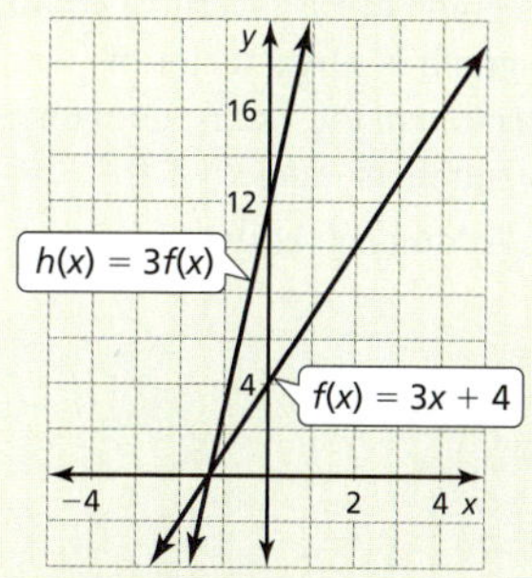

The graph of h is a vertical stretch of the graph of f by a factor of 3.

30.

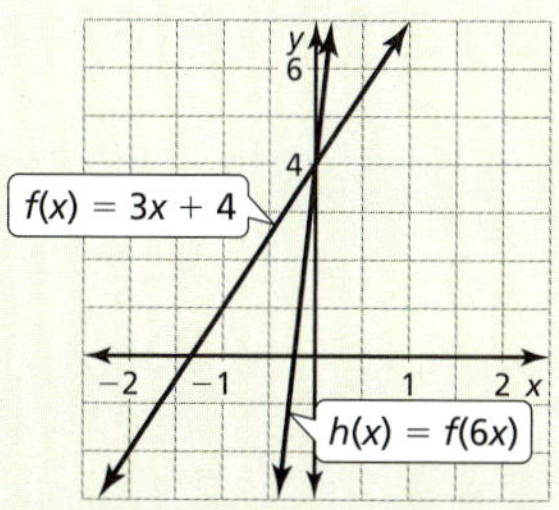

The graph of h is a horizontal shrink of the graph of f by a factor of $\frac{1}{6}$.

31.

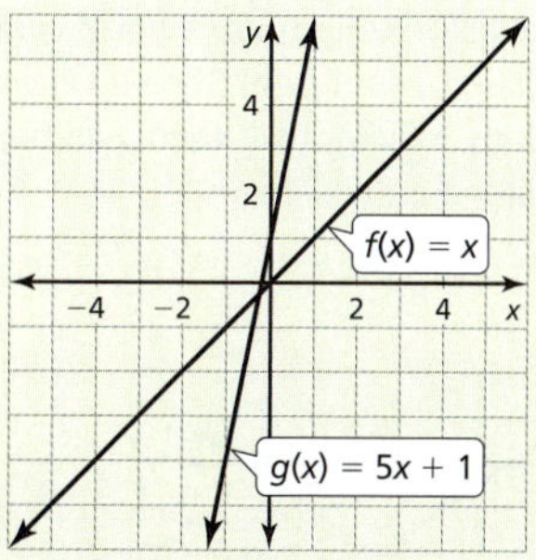

The transformations are a vertical stretch by a factor of 5, then a vertical translation 1 unit up.

ANSWERS

32.

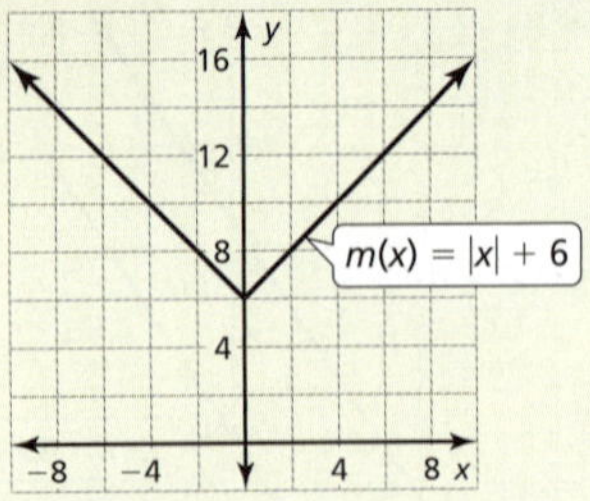

The graph of m is a vertical translation 6 units up of the graph of f; domain: all real numbers; range: $y \geq 6$

33.

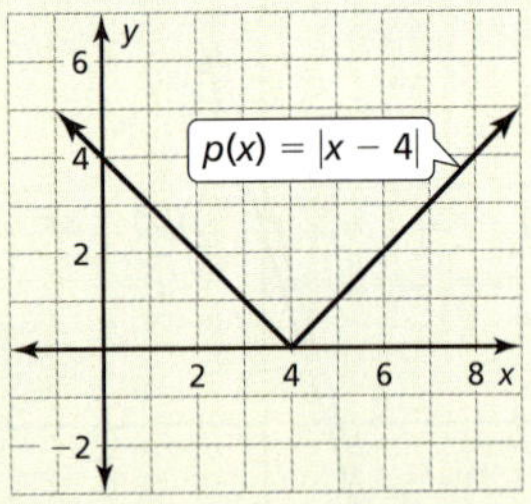

The graph of p is a horizontal translation 4 units right of the graph of f; domain: all real numbers; range: $y \geq 0$

34.

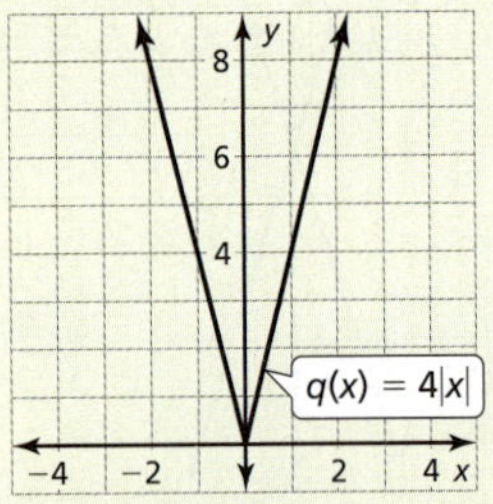

The graph of q is a vertical stretch of the graph of f by a factor of 4; domain: all real numbers; range: $y \geq 0$

35.

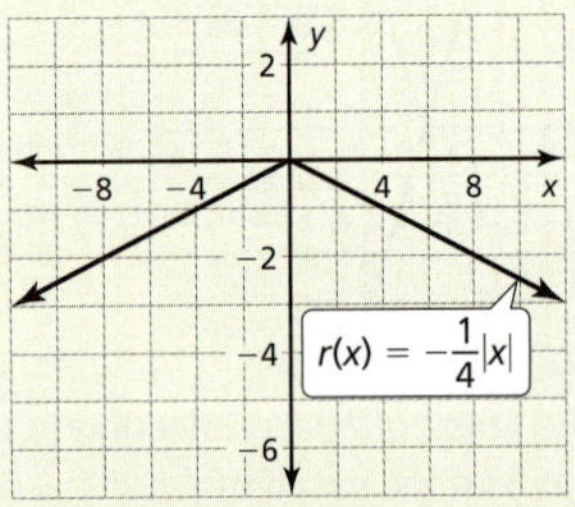

The graph of r is a vertical shrink of the graph of f by a factor of $\frac{1}{4}$ and a reflection in the x-axis; domain: all real numbers; range: $y \leq 0$

36. See Additional Answers.

3.7 Graphing Absolute Value Functions (pp. 155–162)

Let $g(x) = -3|x + 1| + 2$. (a) Describe the transformations from the graph of $f(x) = |x|$ to the graph of g. (b) Graph g.

a. Step 1 Translate the graph of f horizontally 1 unit left to get the graph of $t(x) = |x + 1|$.

Step 2 Stretch the graph of t vertically by a factor of 3 to get the graph of $h(x) = 3|x + 1|$.

Step 3 Reflect the graph of h in the x-axis to get the graph of $r(x) = -3|x + 1|$.

Step 4 Translate the graph of r vertically 2 units up to get the graph of $g(x) = -3|x + 1| + 2$.

b. Method 1

Step 1 Make a table of values.

x	−3	−2	−1	0	1
$g(x)$	−4	−1	2	−1	−4

Step 2 Plot the ordered pairs.

Step 3 Draw the V-shaped graph.

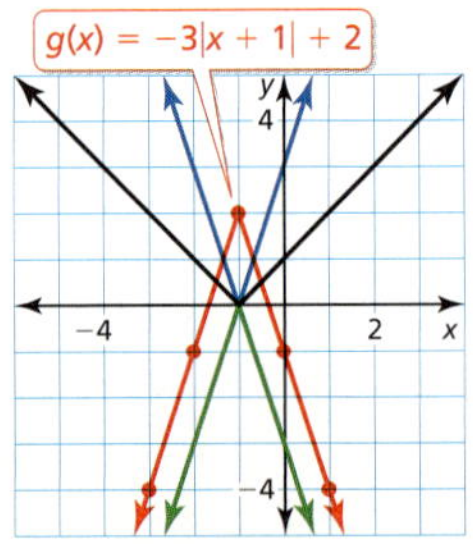

Method 2

Step 1 Identify and plot the vertex.

$(h, k) = (-1, 2)$

Step 2 Plot another point on the graph such as $(0, -1)$. Because the graph is symmetric about the line $x = -1$, you can use symmetry to plot a third point, $(-2, -1)$.

Step 3 Draw the V-shaped graph.

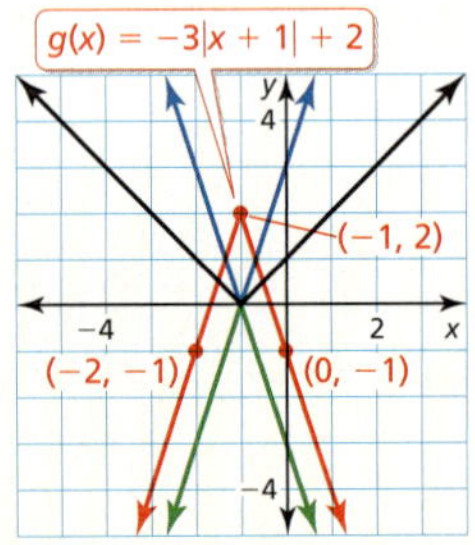

Graph the function. Compare the graph to the graph of $f(x) = |x|$. Describe the domain and range.

32. $m(x) = |x| + 6$ **33.** $p(x) = |x - 4|$ **34.** $q(x) = 4|x|$ **35.** $r(x) = -\frac{1}{4}|x|$

36. Graph $f(x) = |x - 2| + 4$ and $g(x) = |3x - 2| + 4$. Compare the graph of g to the graph of f.

37. Let $g(x) = \frac{1}{3}|x - 1| - 2$. (a) Describe the transformations from the graph of $f(x) = |x|$ to the graph of g. (b) Graph g.

37. a. The transformations are a horizontal translation 1 unit right, then a vertical shrink by a factor of $\frac{1}{3}$, then a vertical translation 2 units down.

b.

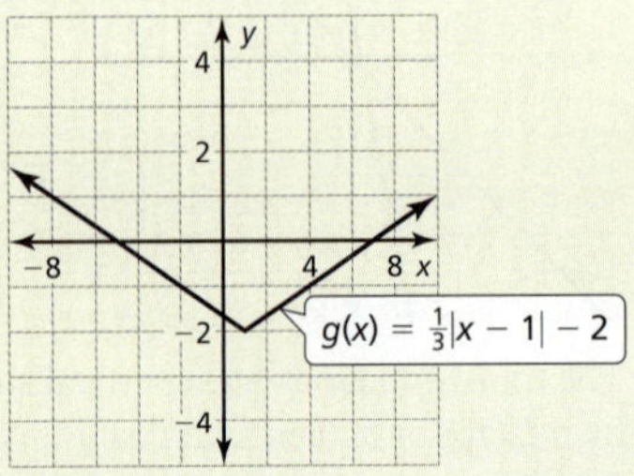

3 Chapter Test

Determine whether the relation is a function. If the relation is a function, determine whether the function is *linear* or *nonlinear*. Explain.

1.

x	-1	0	1	2
y	6	5	9	14

2. $y = -2x + 3$

3. $x = -2$

Graph the equation and identify the intercept(s). If the equation is linear, find the slope of the line.

4. $2x - 3y = 6$

5. $y = 4.5$

6. $y = |x - 1| - 2$

Find the domain and range of the function represented by the graph. Determine whether the domain is *discrete* or *continuous*. Explain.

7.

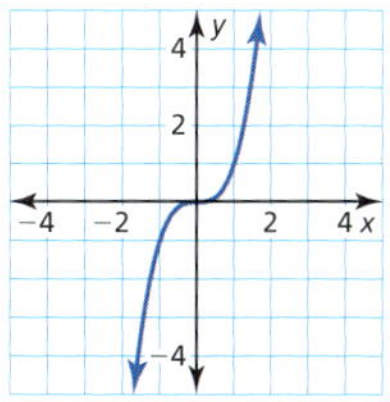

8.

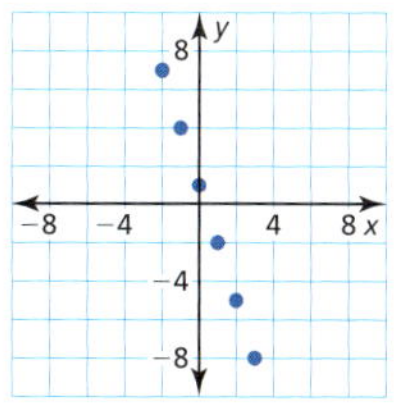

Graph f and g. Describe the transformations from the graph of f to the graph of g.

9. $f(x) = x$; $g(x) = -x + 3$

10. $f(x) = |x|$; $g(x) = |2x + 4|$

11. Function A represents the amount of money in a jar based on the number of quarters in the jar. Function B represents your distance from home over time. Compare the domains.

12. A mountain climber is scaling a 500-foot cliff. The graph shows the elevation of the climber over time.

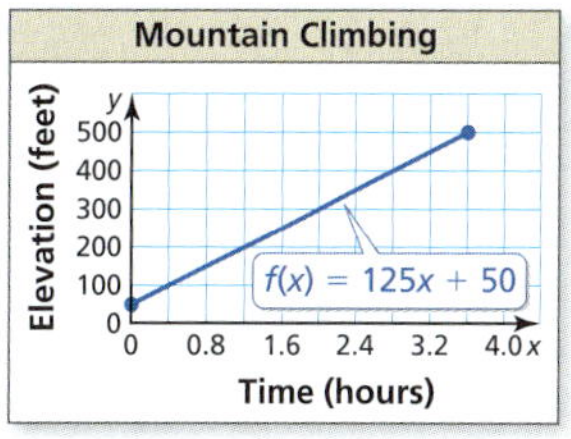

a. Find and interpret the slope and the y-intercept of the graph.

b. Explain two ways to find $f(3)$. Then find $f(3)$ and interpret its meaning.

c. How long does it take the climber to reach the top of the cliff? Justify your answer.

13. Without graphing, compare the slopes and the intercepts of the graphs of the functions $f(x) = x + 1$ and $g(x) = f(2x)$.

14. A rock band releases a new single. Weekly sales s (in thousands of dollars) increase and then decrease as described by the function $s(t) = -2|t - 20| + 40$, where t is the time (in weeks).

a. Identify the independent and dependent variables.

b. Graph s. Describe the transformations from the graph of $f(x) = |x|$ to the graph of s.

If students need help...	*If students got it...*
Lesson Tutorials	Resources by Chapter • Enrichment and Extension • Cumulative Review
Skills Review Handbook	Performance Task
BigIdeasMath.com	Start the *next* Section

ANSWERS

1. function; Every input has exactly one output; not linear; As x increases by 1, y changes by different amounts. The rate of change is not constant.

2. function; Every input has exactly one output; linear; It is in the form $y = mx + b$.

3. not a function; The input -2 has infinite output values.

4.

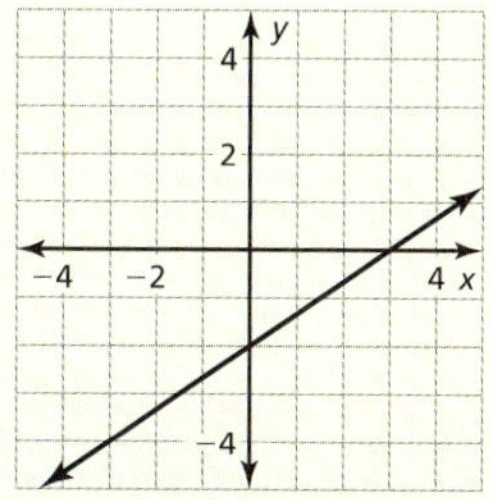

x-intercept: 3; y-intercept: -2; slope: $\frac{2}{3}$

5.

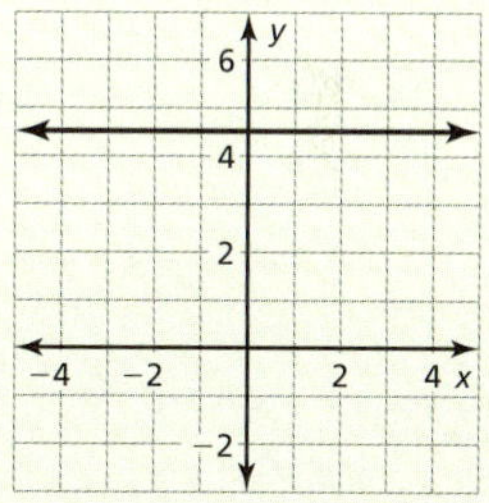

no x-intercept; y-intercept: 4.5; slope: 0

6.

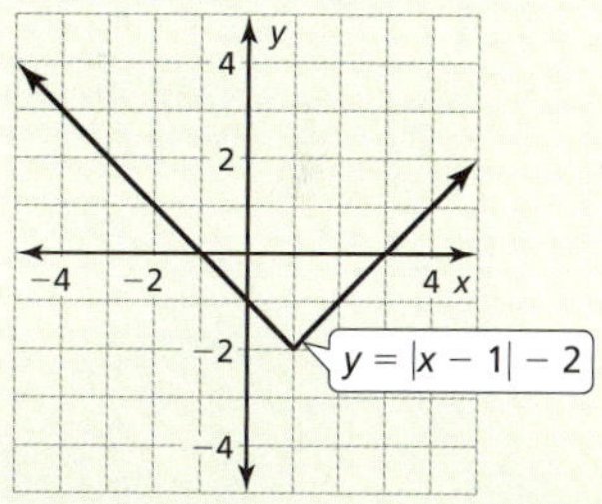

x-intercepts: -1, 3; y-intercept: -1

7. domain: all real numbers; range: all real numbers; continuous; The graph is a curve.

8. domain: $-2, -1, 0, 1, 2, 3$; range: $-8, -5, -2, 1, 4, 7$; discrete; The graph is individual points.

9.

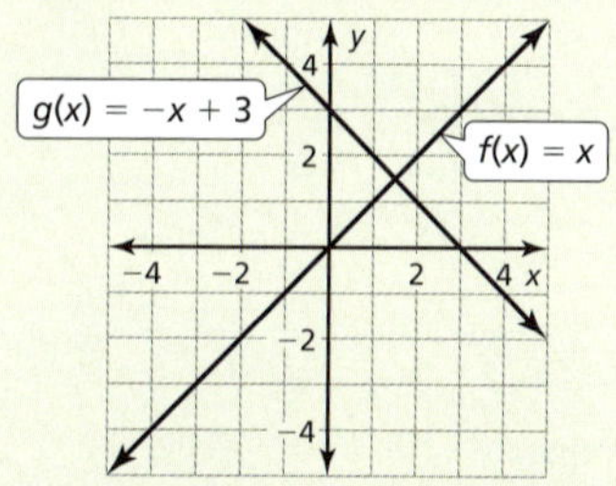

The transformations are a reflection in the x-axis then a vertical translation 3 units up.

10–14. See Additional Answers.

3 Cumulative Assessment

1. You claim you can create a table of values that represents a linear function. Your friend claims he can create a table of values that represents a nonlinear function. Using the given numbers, what values can you use for x (the input) and y (the output) to support your claim? What values can your friend use?

	Your claim			
x				
y				

	Friend's claim			
x				
y				

−4   −3   −2   −1   0

1   2   3   4   5

2. A car rental company charges an initial fee of \$42 and a daily fee of \$12.

   a. Use the numbers and symbols to write a function that represents this situation.

   x $f(x)$ 42   12   138

   $+$ $-$ $\times$ $\div$ $=$

   b. The bill is \$138. How many days did you rent the car?

3. Fill in values for a and b so that each statement is true for the inequality $ax - b > 0$.

   a. When $a =$ _____ and $b =$ _____, $x > \frac{b}{a}$.

   b. When $a =$ _____ and $b =$ _____, $x < \frac{b}{a}$.

4. Fill in the inequality with $<$, $\leq$, $>$, or $\geq$ so that the solution of the inequality is represented by the graph.

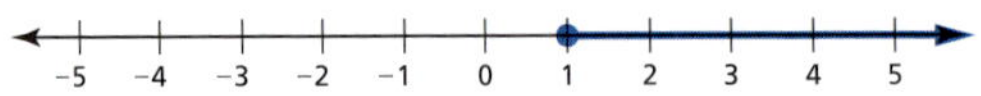

$-3(x + 7)$ ▭ -24

ANSWERS

1. *Sample answer:* x: −4, −3, −2, −1 and y: 1, 2, 3, 4; x: −4, −3, −2, −1 and y: 0, 1, 3, 5
2. **a.** $f(x) = 12x + 42$

   b. 8 days
3. **a.** *Sample answer:* 2, 10

   b. *Sample answer:* −2, 10
4. $\leq$

5. Use the numbers to fill in the coefficients of $ax + by = 40$ so that when you graph the function, the x-intercept is -10 and the y-intercept is 8.

−10	−8	−5	−4	4	5	8	10

6. Solve each equation. Then classify each equation based on the solution. Explain your reasoning.

   a. $2x - 9 = 5x - 33$
   b. $5x - 6 = 10x + 10$
   c. $2(8x - 3) = 4(4x + 7)$
   d. $-7x + 5 = 2(x - 10.1)$
   e. $6(2x + 4) = 4(4x + 10)$
   f. $8(3x + 4) = 2(12x + 16)$

7. The table shows the cost of bologna at a deli. Plot the points represented by the table in a coordinate plane. Decide whether you should connect the points with a line. Explain your reasoning.

Pounds, x	0.5	1	1.5	2
Cost, y	\$3	\$6	\$9	\$12

8. The graph of g is a horizontal translation right, then a vertical stretch, then a vertical translation down of the graph of $f(x) = x$. Use the numbers and symbols to create g.

-3	-1	$-\frac{1}{2}$	0	$\frac{1}{2}$	1	3
x	$g(x)$	$+$	$-$	$\times$	$\div$	$=$

9. What is the sum of the integer solutions of the compound inequality $2|x - 5| < 16$?

   Ⓐ 72 Ⓑ 75 Ⓒ 85 Ⓓ 88

10. Your bank offers a text alert service that notifies you when your checking account balance drops below a specific amount. You set it up so you are notified when your balance drops below \$700. The balance is currently \$3000. You only use your account for paying your rent (no other deposits or deductions occur). Your rent each month is \$625.

    a. Write an inequality that represents the number of months m you can pay your rent without receiving a text alert.

    b. What is the maximum number of months you can pay your rent without receiving a text alert?

    c. Suppose you start paying rent in June. Select all the months you can pay your rent without making a deposit.

June	July	August	September	October

ANSWERS

5. $-4x + 5y = 40$

6. a. $x = 8$; one solution; 8 is the only value that makes the equation true.

   b. $x = -\frac{16}{5}$; one solution; $-\frac{16}{5}$ is the only value that makes the equation true.

   c. no solution; no solution; $-6 = 28$ is never true.

   d. $x = 2.8$; one solution; 2.8 is the only value that makes the equation true.

   e. $x = -4$; one solution; -4 is the only value that makes the equation true.

   f. all real numbers; identity; $32 = 32$ is always true.

7. 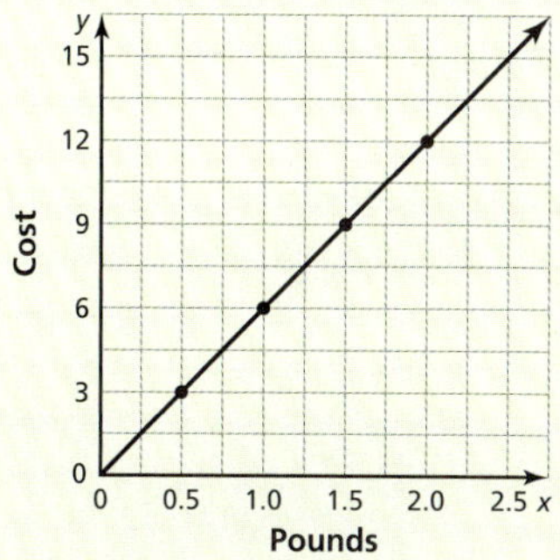

   yes; The pounds can be any value, so the domain is continuous.

8. *Sample answer:* $g(x) = 3(x - 1) - 3$

9. B

10. a. $3000 - 625m \geq 700$

    b. 3 months

    c. June, July, August, September

Chapter 4 Pacing Guide

Chapter Opener/ Mathematical Practices	1 Day
Section 1	1 Day
Section 2	1 Day
Section 3	1 Day
Quiz	1 Day
Section 4	1 Day
Section 5	2 Days
Section 6	2 Days
Section 7	2 Days
Chapter Review/ Chapter Tests	2 Days
Total Chapter 4	14 Days
Year-to-Date	52 Days

4 Writing Linear Functions

Karaoke Machine (p. 220)

Old Faithful Geyser (p. 204)

Helicopter Rescue (p. 190)

School Spirit (p. 184)

Renewable Energy (p. 178)

Laurie's Notes

Chapter Summary

- Students have just finished a chapter on graphing linear equations in various forms, including transformations of graphs of linear functions.
- This chapter begins with two lessons on writing linear equations in slope-intercept form and standard form. These forms are extended in the next lesson to include the cases of parallel and perpendicular lines.
- Many real-life applications involve data that can be modeled by a linear equation. The next two lessons focus on graphing scatter plots and writing a best-fit line for the data. Students first approximate a line of best fit and then use graphing calculators (or spreadsheets) to generate the regression equation of the line of best fit.
- The last two lessons of the chapter are connections to linear equations. Students should make the connection between the common difference in an arithmetic sequence and the slope of a linear equation. Further, a_0 is the y-intercept of a linear equation. In the last lesson, students review skills built in the early part of the chapter when they write the linear equations for each part of a piecewise function.
- New formative assessment tips are offered in many of the lessons, and assessment tips from previous chapters are referenced throughout the notes at point of use. Additional graphing calculator techniques are presented as helpful teaching strategies.

What Your Students Have Learned

Middle School

- Use slope to determine whether lines are parallel or perpendicular.
- Write equations of lines in slope-intercept form.
- Write equations of lines given a slope and a point or two points.
- Make scatter plots and describe the relationships between data.
- Identify lines of fit for scatter plots.

What Your Students Will Learn

Algebra 1

- Write equations of lines in slope-intercept form and in point-slope form.
- Use slope to identify and write equations of parallel lines and perpendicular lines.
- Use scatter plots to describe relationships and draw lines of fit to write equations to model data.
- Analyze lines of fit for sets of data using residuals, technology, and correlation coefficients.
- Write the terms of arithmetic sequences and graph arithmetic sequences.
- Write arithmetic sequences as equations and in function notation.
- Evaluate, graph, and write piecewise functions, including step and absolute value functions.

Dynamic Teaching Tools

- Dynamic Assessment & Progress Monitoring Tool
- Lesson Planning Tool
- Interactive Whiteboard Lesson Library
- Dynamic Classroom with Dynamic Investigations
- Real-Life STEM Videos

Scaffolding in the Classroom

Build on prior knowledge.

Instead of reteaching a topic, use review. Begin with the statement, "Remember when we did ___________?" This lets the students know that you know they have previously learned the concept or process. Use a problem to review. Then say, "Today we will use this idea to ___________."

Questioning in the Classroom

What was that answer?

Once a student has answered a question, ask another student to paraphrase the answer. Ask yet another student to illustrate or explain what was said.

Laurie's Notes

Maintaining Mathematical Proficiency

Using a Coordinate Plane

- Remind students that the first coordinate in an ordered pair is the x-coordinate. Its sign indicates whether the point is left or right of the y-axis. The second coordinate is the y-coordinate, and its sign indicates whether the point is above or below the x-axis.
- Remind students of the positions of the four quadrants. They are numbered in the counterclockwise direction, beginning with the upper right quadrant, Quadrant I.

COMMON ERROR Students may reverse the order of the coordinates in the ordered pairs.

Rewriting Equations

- Remind students that both sides of the equation must remain equal. So each operation must be performed on both sides of the equation. If students divide the left side by 2, they must also divide the right side by 2.

COMMON ERROR Students may use different operations on the left and right sides of equations. For example, they may rewrite $0 = 2y - 8x + 10$ as $10 = 2y - 8x$ by subtracting 10 from the right side and adding 10 to the left side.

Mathematical Practices (continued on page 174)

- The *Mathematical Practices* page focuses attention on how mathematics is learned—process versus content. This page demonstrates that solving a simpler problem and then applying that strategy to the original problem is a trait of a good problem solver. This example also demonstrates the use of unit analysis, which helps verify that the process was correct.
- Use the *Mathematical Practices* page to help students develop mathematical habits of mind—how mathematics can be explored and how mathematics is thought about.

If students need help...	*If students got it...*
Student Journal • Maintaining Mathematical Proficiency	Game Closet at *BigIdeasMath.com*
Lesson Tutorials	Start the *next* Section
Skills Review Handbook	

Maintaining Mathematical Proficiency

Using a Coordinate Plane

Example 1 What ordered pair corresponds to point A?

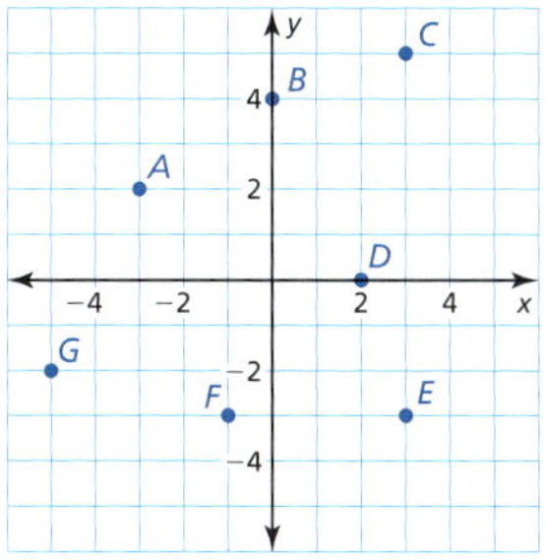

Point A is 3 units to the left of the origin and 2 units up. So, the x-coordinate is -3 and the y-coordinate is 2.

▶ The ordered pair $(-3, 2)$ corresponds to point A.

Use the graph to answer the question.

1. What ordered pair corresponds to point G?
2. What ordered pair corresponds to point D?
3. Which point is located in Quadrant I?
4. Which point is located in Quadrant IV?

Rewriting Equations

Example 2 Solve the equation $3x - 2y = 8$ for y.

$3x - 2y = 8$	Write the equation.
$3x - 2y - 3x = 8 - 3x$	Subtract $3x$ from each side.
$-2y = 8 - 3x$	Simplify.
$\frac{-2y}{-2} = \frac{8 - 3x}{-2}$	Divide each side by -2.
$y = -4 + \frac{3}{2}x$	Simplify.

Solve the equation for y.

5. $x - y = 5$
6. $6x + 3y = -1$
7. $0 = 2y - 8x + 10$
8. $-x + 4y - 28 = 0$
9. $2y + 1 - x = 7x$
10. $y - 4 = 3x + 5y$
11. **ABSTRACT REASONING** Both coordinates of the point (x, y) are multiplied by a negative number. How does this change the location of the point? Be sure to consider points originally located in all four quadrants.

Dynamic Solutions available at *BigIdeasMath.com*

What Your Students Have Learned

- Identify ordered pairs in the coordinate plane.
- Solve literal equations for one variable in terms of the other variables.

ANSWERS

1. $(-5, -2)$
2. $(2, 0)$
3. C
4. E
5. $y = -5 + x$
6. $y = -\frac{1}{3} - 2x$
7. $y = 4x - 5$
8. $y = \frac{1}{4}x + 7$
9. $y = 4x - \frac{1}{2}$
10. $y = -\frac{3}{4}x - 1$
11. Points in Quadrant I move to Quadrant III, points in Quadrant II move to Quadrant IV, points in Quadrant III move to Quadrant I, and points in Quadrant IV move to Quadrant II.

Vocabulary Review

Have students make Notetaking Organizers for the following topics.

- Identifying points in the coordinate plane
- Rewriting equations

MONITORING PROGRESS ANSWERS

1. $9.40 per hour
2. 26.2 mi/gal
3. about 8.8 h

Mathematical Practices

Mathematically proficient students try simpler forms of the original problem.

Problem-Solving Strategies

Core Concept

Solve a Simpler Problem

When solving a real-life problem, if the numbers in the problem seem complicated, then try solving a simpler form of the problem. After you have solved the simpler problem, look for a general strategy. Then apply that strategy to the original problem.

EXAMPLE 1 Using a Problem-Solving Strategy

In the deli section of a grocery store, a half pound of sliced roast beef costs $3.19. You buy 1.81 pounds. How much do you pay?

SOLUTION

Step 1 Solve a simpler problem.

Suppose the roast beef costs $3 per half pound, and you buy 2 pounds.

$$\text{Total cost} = \frac{\$3}{1/2\text{ lb}} \cdot 2\text{ lb} \qquad \text{Use unit analysis to write a verbal model.}$$

$$= \frac{\$6}{1\text{ lb}} \cdot 2\text{ lb} \qquad \text{Rewrite \$3 per } \tfrac{1}{2} \text{ pound as \$6 per pound.}$$

$$= \$12 \qquad \text{Simplify.}$$

▶ In the simpler problem, you pay $12.

Step 2 Apply the strategy to the original problem.

$$\text{Total cost} = \frac{\$3.19}{1/2\text{ lb}} \cdot 1.81\text{ lb} \qquad \text{Use unit analysis to write a verbal model.}$$

$$= \frac{\$6.38}{1\text{ lb}} \cdot 1.81\text{ lb} \qquad \text{Rewrite \$3.19 per } \tfrac{1}{2} \text{ pound as \$6.38 per pound.}$$

$$= \$11.55 \qquad \text{Simplify.}$$

▶ In the original problem, you pay $11.55.

Your answer is reasonable because you bought about 2 pounds.

Monitoring Progress

1. You work $37\frac{1}{2}$ hours and earn $352.50. What is your hourly wage?
2. You drive 1244.5 miles and use 47.5 gallons of gasoline. What is your car's gas mileage (in miles per gallon)?
3. You drive 236 miles in 4.6 hours. At the same rate, how long will it take you to drive 450 miles?

Laurie's Notes — Mathematical Practices (continued from page T-173)

- Allow students time to read through the *Core Concept* and Example 1.
- Ask probing questions to assess student understanding of the simpler problem that was solved first.
- Students could work with partners or in groups on *Monitoring Progress*. Allow private think time before dialogue begins.

Laurie's Notes

Overview of Section 4.1

Introduction

- Students graphed lines in slope-intercept form in an earlier chapter. In this lesson, they write the equation. The slope and y-intercept are given in different forms: stated directly, interpreted from information in a graph, determined from two ordered pairs, and stated in a contextual situation.
- Students should have little difficulty writing the equation in slope-intercept form when given the slope and the coordinates of the y-intercept. Students generally find the contextual word problems more challenging. Have students read and interpret one sentence at a time. When students see lots of text, they can become overwhelmed and shut down.
- The lesson also introduces the term *linear model* and refers back to function notation.

Formative Assessment Tips

- **Think-Alouds:** This technique is used when you want to hear how well partners comprehend a process involved with solving a problem. It is important to model the process first so that students have a sense of what is expected in *Think-Alouds*.
- *Think-Alouds* give students the opportunity to hear the metacognitive processes used by someone who is a proficient problem solver.
- **Make Sense of Problems and Persevere in Solving Them** and **Attend to Precision:** Hearing someone else describe a process using mathematical language will improve all students' problem-solving abilities; they can now apply the process with their partners to the problem being solved.
- Use this technique with a multistep problem. Model using a starter sentence such as: "The problem is asking …," "I can use the strategy of …," "The steps I will use in solving this problem are …," "This problem is similar to …," and "I can check my answer by …."
- Use this technique for a variety of problem types. Listen for comprehension of skills, concepts, and procedures as well as for precision of language.

Pacing Suggestion

- The problems students will work in the explorations are similar to Examples 1 and 2 in the formal lesson. If students have been successful, you may want to omit these examples and begin with Example 3.

Dynamic Teaching Tools

- Dynamic Assessment & Progress Monitoring Tool
- Lesson Planning Tool
- Interactive Whiteboard Lesson Library
- Dynamic Classroom with Dynamic Investigations

What Your Students Will Learn

- Use slopes and y-intercepts to write equations in slope-intercept form.
- Understand that when a quantity changes at a constant rate, the relationship can be modeled with an equation in slope-intercept form.

Laurie's Notes

Exploration

Motivate

- If there is sufficient space in your classroom, hallway, or school foyer, make coordinate axes using masking tape. Use a marker to scale each axis with integers –5 through 5.
- Take turns having two students be the rope anchors who will make a line on the coordinate axes while other students observe.
- Here are a series of directions you can give as well as some follow-up questions. Remind students that slope is rise over run and that the equation of a line in slope-intercept form is $y = mx + b$.
  - ? Make the line $y = x$. "What is the slope?" 1 "What is the y-intercept?" 0
  - ? Keep the same slope, but make the y-intercept 2. "What is the equation of this line?" $y = x + 2$
  - ? Use the y-intercept 2, but make the slope steeper. "What is the slope of this line?" Answers will vary.
  - ? Keep the same y-intercept, but make the slope $\frac{1}{2}$. "What is the equation of this line?" $y = \frac{1}{2}x + 2$
- **Management Tip:** This activity can also be done by drawing the axes on the board and having students hold the rope against the board.

Exploration 1

- For all four parts, students will need to determine the slope. Some students may use the slope formula and compute using paper and pencil. Others may determine the slope visually, saying "over 2, up 4, so the slope $= 4 \div 2 = 2$."
- The y-intercept is given for the first two parts. For the last two parts, some students may solve for the y-intercept using paper and pencil. Others may determine the y-intercept visually, "down 2, over 3, and you are on the ordered pair (0, 1), so the y-intercept is 1."
- **Make Sense of Problems and Persevere in Solving Them** and **Reason Abstractly and Quantitatively:** It is important that students share their strategies with one another. Hearing other strategies enhances each student's problem solving and reasoning.

Exploration 2

- The goal of this exploration is for students to read information from a graph and to interpret the results in the context of the situation.
- **Teaching Tip:** Remind students to look at the labels used to label the axes, including the units.

Communicate Your Answer

- Students should state that the slope and y-intercept are needed to write the equation.

Connecting to Next Step

- If students have been very successful with solving the explorations, be selective in examples and questions you work in the formal lesson. It may be sufficient to try a few questions from *Monitoring Progress* and the real-life example.

4.1 Writing Equations in Slope-Intercept Form

Essential Question Given the graph of a linear function, how can you write an equation of the line?

EXPLORATION 1 **Writing Equations in Slope-Intercept Form**

Work with a partner.

- Find the slope and y-intercept of each line.
- Write an equation of each line in slope-intercept form.
- Use a graphing calculator to verify your equation.

a.

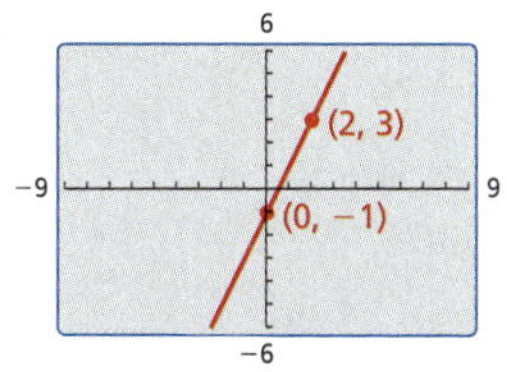

b.

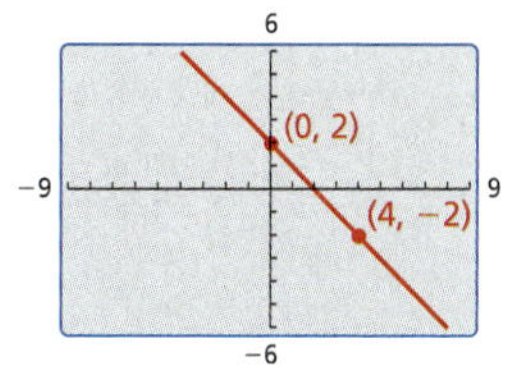

c.

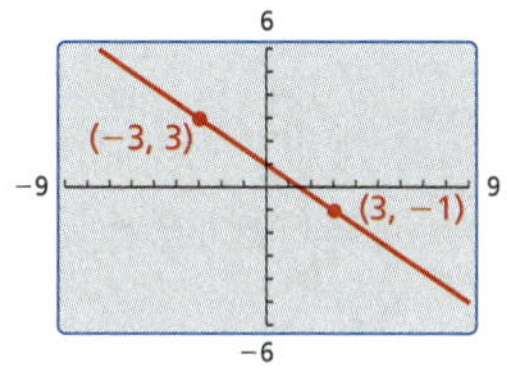

d.

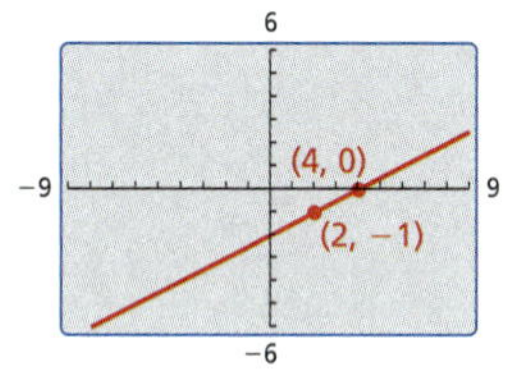

INTERPRETING MATHEMATICAL RESULTS

To be proficient in math, you need to routinely interpret your results in the context of the situation. The reason for studying mathematics is to enable you to model and solve real-life problems.

EXPLORATION 2 **Mathematical Modeling**

Work with a partner. The graph shows the cost of a smartphone plan.

a. What is the y-intercept of the line? Interpret the y-intercept in the context of the problem.

b. Approximate the slope of the line. Interpret the slope in the context of the problem.

c. Write an equation that represents the cost as a function of data usage.

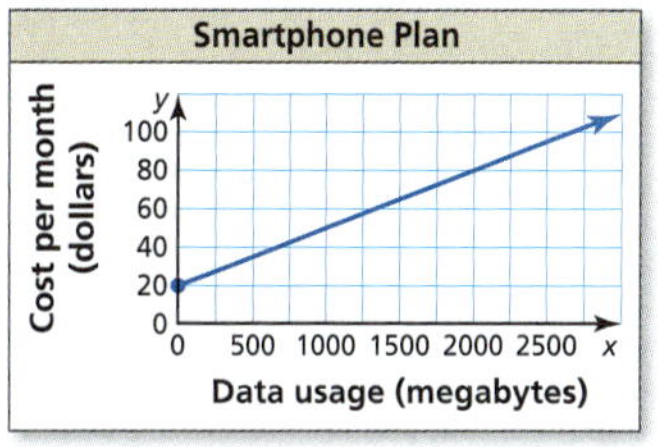

Communicate Your Answer

3. Given the graph of a linear function, how can you write an equation of the line?
4. Give an example of a graph of a linear function that is different from those above. Then use the graph to write an equation of the line.

Dynamic Teaching Tools

- Dynamic Assessment & Progress Monitoring Tool
- Lesson Planning Tool
- Interactive Whiteboard Lesson Library
- Dynamic Classroom with Dynamic Investigations

ANSWERS

1. **a.** slope is 2; y-intercept is -1; $y = 2x - 1$
   b. slope is -1; y-intercept is 2; $y = -x + 2$
   c. slope is $-\frac{2}{3}$; y-intercept is 1; $y = -\frac{2}{3}x + 1$
   d. slope is $\frac{1}{2}$; y-intercept is -2; $y = \frac{1}{2}x - 2$
2. **a.** 20; The base cost of the plan is \$20 per month.
   b. *Sample answer:* 0.03; Each megabyte of data costs \$0.03 to use.
   c. *Sample answer:* $y = 0.03x + 20$
3. Use the graph to find the slope and the y-intercept, and substitute these values into the slope-intercept form.
4. *Sample answer:*

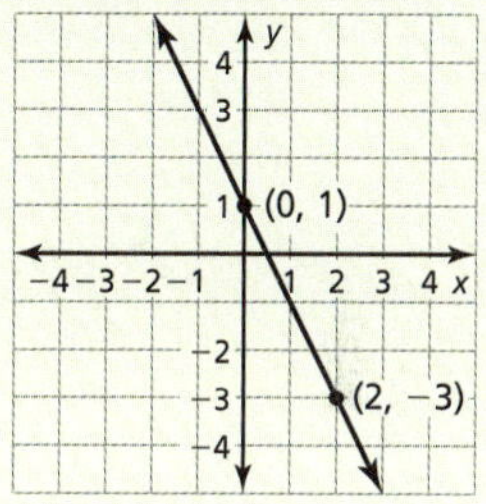

$y = -2x + 1$

Extra Example 1

Write an equation of each line with the given slope and y-intercept.

a. slope = 3; y-intercept = -1

$y = 3x - 1$

b. slope = $\frac{1}{2}$; y-intercept = 0

$y = \frac{1}{2}x$

Extra Example 2

Write an equation of each line in slope-intercept form.

a.

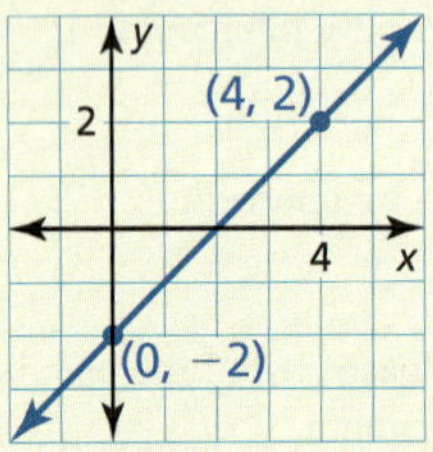

$y = x - 2$

b.

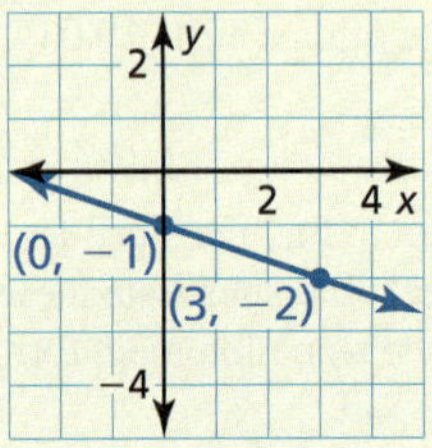

$y = -\frac{1}{3}x - 1$

4.1 Lesson

What You Will Learn

- Write equations in slope-intercept form.
- Use linear equations to solve real-life problems.

Core Vocabulary

linear model, *p. 178*

Previous
slope-intercept form
function
rate

Writing Equations in Slope-Intercept Form

EXAMPLE 1 Using Slopes and y-Intercepts to Write Equations

Write an equation of each line with the given slope and y-intercept.

a. slope = -3; y-intercept = $\frac{1}{2}$ **b.** slope = 0; y-intercept = -2

SOLUTION

a. $y = mx + b$ Write the slope-intercept form.

$y = -3x + \frac{1}{2}$ Substitute -3 for m and $\frac{1}{2}$ for b.

▶ An equation is $y = -3x + \frac{1}{2}$.

b. $y = mx + b$ Write the slope-intercept form.

$y = 0x + (-2)$ Substitute 0 for m and -2 for b.

$y = -2$ Simplify.

▶ An equation is $y = -2$.

EXAMPLE 2 Using Graphs to Write Equations

Write an equation of each line in slope-intercept form.

a.

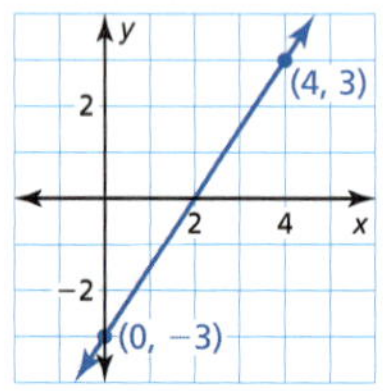

b.

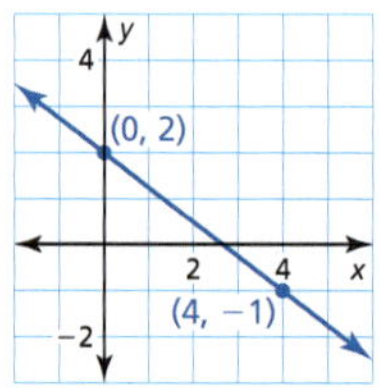

STUDY TIP

You can use any two points on a line to find the slope.

SOLUTION

a. Find the slope and y-intercept.

Let $(x_1, y_1) = (0, -3)$ and $(x_2, y_2) = (4, 3)$.

$$m = \frac{y_2 - y_1}{x_2 - x_1} = \frac{3 - (-3)}{4 - 0} = \frac{6}{4}, \text{ or } \frac{3}{2}$$

Because the line crosses the y-axis at $(0, -3)$, the y-intercept is -3.

▶ So, the equation is $y = \frac{3}{2}x - 3$.

STUDY TIP

After writing an equation, check that the given points are solutions of the equation.

b. Find the slope and y-intercept.

Let $(x_1, y_1) = (0, 2)$ and $(x_2, y_2) = (4, -1)$.

$$m = \frac{y_2 - y_1}{x_2 - x_1} = \frac{-1 - 2}{4 - 0} = \frac{-3}{4}, \text{ or } -\frac{3}{4}$$

Because the line crosses the y-axis at $(0, 2)$, the y-intercept is 2.

▶ So, the equation is $y = -\frac{3}{4}x + 2$.

Laurie's Notes Teacher Actions

- **Think-Alouds:** Pose Example 1 and say, "To solve this problem I need to …." Ask partner A to think aloud for partner B to hear the problem-solving process. When students have finished the example, use *Popsicle Sticks* to solicit responses.

? **Assessing Question:** "How is Example 2 different from Example 1?" The slope needs to be computed first. The y-intercept is given, so write the equation of the line in slope-intercept form.

EXAMPLE 3 Using Points to Write Equations

Write an equation of each line that passes through the given points.

a. $(-3, 5)$, $(0, -1)$ **b.** $(0, -5)$, $(8, -5)$

SOLUTION

a. Find the slope and y-intercept.

$$m = \frac{-1 - 5}{0 - (-3)} = -2$$

Because the line crosses the y-axis at $(0, -1)$, the y-intercept is -1.

▶ So, an equation is $y = -2x - 1$.

b. Find the slope and y-intercept.

$$m = \frac{-5 - (-5)}{8 - 0} = 0$$

Because the line crosses the y-axis at $(0, -5)$, the y-intercept is -5.

▶ So, an equation is $y = -5$.

REMEMBER

If f is a function and x is in its domain, then $f(x)$ represents the output of f corresponding to the input x.

EXAMPLE 4 Writing a Linear Function

Write a linear function f with the values $f(0) = 10$ and $f(6) = 34$.

SOLUTION

Step 1 Write $f(0) = 10$ as $(0, 10)$ and $f(6) = 34$ as $(6, 34)$.

Step 2 Find the slope of the line that passes through $(0, 10)$ and $(6, 34)$.

$$m = \frac{34 - 10}{6 - 0} = \frac{24}{6}, \text{ or } 4$$

Step 3 Write an equation of the line. Because the line crosses the y-axis at $(0, 10)$, the y-intercept is 10.

$y = mx + b$ Write the slope-intercept form.

$y = 4x + 10$ Substitute 4 for m and 10 for b.

▶ A function is $f(x) = 4x + 10$.

Monitoring Progress

Help in English and Spanish at *BigIdeasMath.com*

Write an equation of the line with the given slope and y-intercept.

1. slope $= 7$; y-intercept $= 2$
2. slope $= \frac{1}{3}$; y-intercept $= -1$

Write an equation of the line in slope-intercept form.

3. 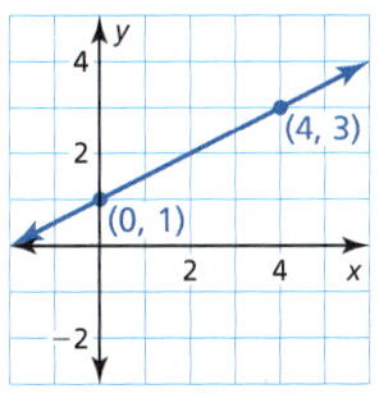

4. 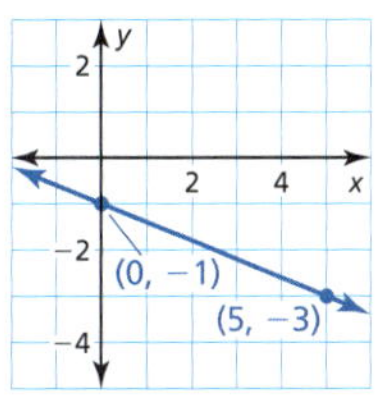

5. Write an equation of the line that passes through $(0, -2)$ and $(4, 10)$.
6. Write a linear function g with the values $g(0) = 9$ and $g(8) = 7$.

Differentiated Instruction

Organization

Have students make and complete a table to describe the methods shown in this section for writing equations in slope-intercept form.

What I Know	How to Find Slope	How to Find y-Intercept
slope and y-intercept		
graph		
two points		
function values		

Extra Example 3

Write an equation of each line that passes through the given points.

a. $(0, -6)$, $(3, 9)$ $y = 5x - 6$

b. $(4, -1)$, $(0, 2)$ $y = -\frac{3}{4}x + 2$

Extra Example 4

Write a linear function f with the values $f(0) = -4$ and $f(3) = 8$.

$f(x) = 4x - 4$

MONITORING PROGRESS ANSWERS

1. $y = 7x + 2$
2. $y = \frac{1}{3}x - 1$
3. $y = \frac{1}{2}x + 1$
4. $y = -\frac{2}{5}x - 1$
5. $y = 3x - 2$
6. $g(x) = -\frac{1}{4}x + 9$

Laurie's Notes Teacher Actions

- **Think-Alouds:** Pose Example 3 and say, "This example is similar to Example 2, so I need to …." Ask partner B to think aloud for partner A to hear the change in the problem-solving process. When students have finished the example, use *Popsicle Sticks* to solicit responses.

? "In Example 4, what does the function notation $f(0) = 10$ mean?" It means that $(0, 10)$ is an ordered pair that satisfies the function f.

- **Think-Pair-Share:** Have students answer *Monitoring Progress* Questions 5 and 6, and then share and discuss as a class.

Extra Example 5

In 2000, the average cost for college tuition in the United States was about \$14,000. In 2010, the cost was about \$18,000. Write a linear model that represents the cost of tuition as a function of the number of years since 2000. Use the model to predict the cost in 2020.

Let x represent the number of years since 2000 and let y represent the cost. The linear model is $y = 400x + 14,000$. The model predicts tuition will cost about \$22,000 in 2020.

MONITORING PROGRESS ANSWER

7. $y = 5.8x + 248$

Solving Real-Life Problems

A **linear model** is a linear function that models a real-life situation. When a quantity y changes at a constant rate with respect to a quantity x, you can use the equation $y = mx + b$ to model the relationship. The value of m is the constant rate of change, and the value of b is the initial, or starting, value of y.

EXAMPLE 5 Modeling with Mathematics

Excluding hydropower, U.S. power plants used renewable energy sources to generate 105 million megawatt hours of electricity in 2007. By 2012, the amount of electricity generated had increased to 219 million megawatt hours. Write a linear model that represents the number of megawatt hours generated by non-hydropower renewable energy sources as a function of the number of years since 2007. Use the model to predict the number of megawatt hours that will be generated in 2017.

SOLUTION

1. **Understand the Problem** You know the amounts of electricity generated in two distinct years. You are asked to write a linear model that represents the amount of electricity generated each year since 2007 and then predict a future amount.
2. **Make a Plan** Break the problem into parts and solve each part. Then combine the results to help you solve the original problem.

   Part 1 Define the variables. Find the initial value and the rate of change.

   Part 2 Write a linear model and predict the amount in 2017.
3. **Solve the Problem**

   Part 1 Let x represent the time (in years) since 2007 and let y represent the number of megawatt hours (in millions). Because time x is defined in years since 2007, 2007 corresponds to $x = 0$ and 2012 corresponds to $x = 5$. Let $(x_1, y_1) = (0, 105)$ and $(x_2, y_2) = (5, 219)$. The initial value is the y-intercept b, which is 105. The rate of change is the slope m.

$$m = \frac{y_2 - y_1}{x_2 - x_1} = \frac{219 - 105}{5 - 0} = \frac{114}{5} = 22.8$$

   Part 2

Megawatt hours (millions)	=	Initial value	+	Rate of change	•	Years since 2007
y	=	105	+	22.8	•	x

$y = 105 + 22.8x$ Write the equation.

2017 corresponds to $x = 10$. → $y = 105 + 22.8(10)$ Substitute 10 for x.

$y = 333$ Simplify.

▶ The linear model is $y = 22.8x + 105$. The model predicts non-hydropower renewable energy sources will generate 333 million megawatt hours in 2017.

4. **Look Back** To check that your model is correct, verify that (0, 105) and (5, 219) are solutions of the equation.

Monitoring Progress

Help in English and Spanish at *BigIdeasMath.com*

7. The corresponding data for electricity generated by hydropower are 248 million megawatt hours in 2007 and 277 million megawatt hours in 2012. Write a linear model that represents the number of megawatt hours generated by hydropower as a function of the number of years since 2007.

Laurie's Notes Teacher Actions

- Define *linear model*. Many of the real-life applications have been linear models, but they were not referred to as such.
- **Reason Abstractly and Quantitatively:** Defining x as the number of years since 2007 means input values will not need to be 4-digit numbers.
- **Turn and Talk:** Give partners time to discuss the example and make their attempt at the solution. Discuss as a class.

Closure

- **Writing Prompt:** For a line that has been graphed in a coordinate plane, you can write the equation by … finding the slope and y-intercept.

4.1 Exercises

Dynamic Solutions available at *BigIdeasMath.com*

Vocabulary and Core Concept Check

1. **COMPLETE THE SENTENCE** A linear function that models a real-life situation is called a __________.

2. **WRITING** Explain how you can use slope-intercept form to write an equation of a line given its slope and y-intercept.

Monitoring Progress and Modeling with Mathematics

In Exercises 3–8, write an equation of the line with the given slope and y-intercept. *(See Example 1.)*

3. slope: 2
   y-intercept: 9

4. slope: 0
   y-intercept: 5

5. slope: -3
   y-intercept: 0

6. slope: -7
   y-intercept: 1

7. slope: $\frac{2}{3}$
   y-intercept: -8

8. slope: $-\frac{3}{4}$
   y-intercept: -6

In Exercises 9–12, write an equation of the line in slope-intercept form. *(See Example 2.)*

9.

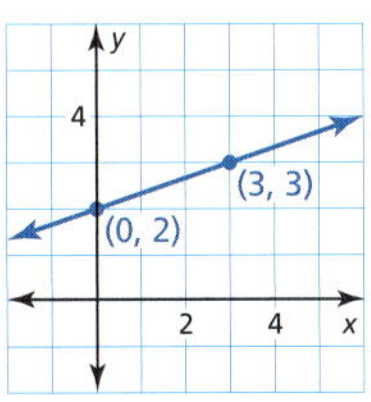

10.

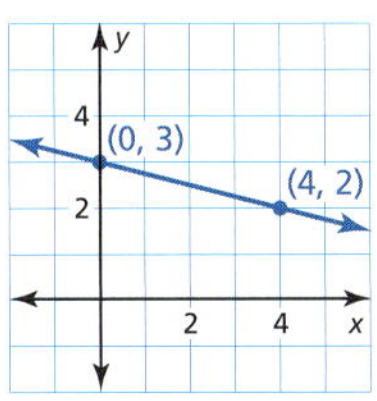

11.

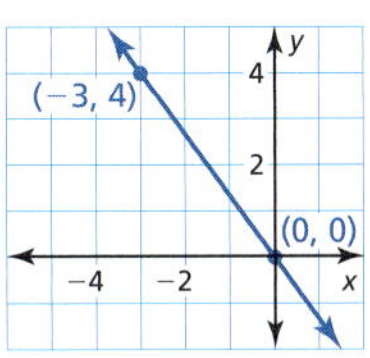

12.

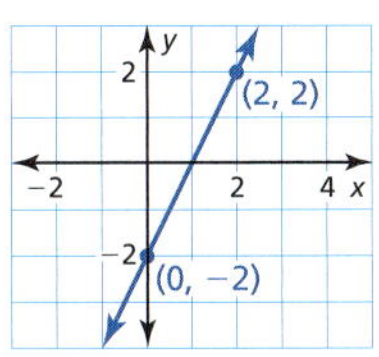

In Exercises 13–18, write an equation of the line that passes through the given points. *(See Example 3.)*

13. (3, 1), (0, 10)

14. (2, 7), (0, −5)

15. (2, −4), (0, −4)

16. (−6, 0), (0, −24)

17. (0, 5), (−1.5, 1)

18. (0, 3), (−5, 2.5)

In Exercises 19–24, write a linear function f with the given values. *(See Example 4.)*

19. $f(0) = 2, f(2) = 4$

20. $f(0) = 7, f(3) = 1$

21. $f(4) = -3, f(0) = -2$

22. $f(5) = -1, f(0) = -5$

23. $f(-2) = 6, f(0) = -4$

24. $f(0) = 3, f(-6) = 3$

In Exercises 25 and 26, write a linear function f with the given values.

25.

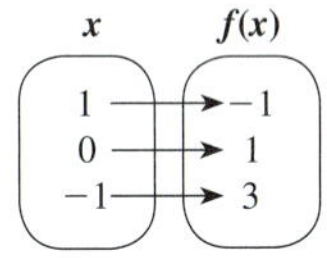

26.

x	$f(x)$
−4	−2
−2	−1
0	0

27. **ERROR ANALYSIS** Describe and correct the error in writing an equation of the line with a slope of 2 and a y-intercept of 7.

   ✗ $y = 7x + 2$

28. **ERROR ANALYSIS** Describe and correct the error in writing an equation of the line shown.

   ✗ $\text{slope} = \frac{1-4}{0-5}$
   $= \frac{-3}{-5} = \frac{3}{5}$
   $y = \frac{3}{5}x + 4$

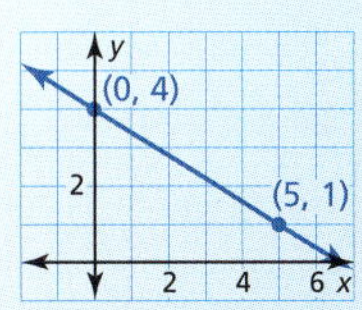

Assignment Guide and Homework Check

ASSIGNMENT

Basic: 1, 2, 3–23 odd, 27, 29, 34, 36, 38–45

Average: 1, 2–28 even, 29, 30, 34, 36, 38–45

Advanced: 1, 2, 6–12 even, 16, 18, 22–28 even, 29–45

HOMEWORK CHECK

Basic: 3, 11, 13, 19, 29

Average: 6, 12, 14, 20, 29

Advanced: 8, 12, 18, 24, 30

ANSWERS

1. linear model
2. Substitute the slope for m and the y-intercept for b in the slope-intercept form.
3. $y = 2x + 9$
4. $y = 5$
5. $y = -3x$
6. $y = -7x + 1$
7. $y = \frac{2}{3}x - 8$
8. $y = -\frac{3}{4}x - 6$
9. $y = \frac{1}{3}x + 2$
10. $y = -\frac{1}{4}x + 3$
11. $y = -\frac{4}{3}x$
12. $y = 2x - 2$
13. $y = -3x + 10$
14. $y = 6x - 5$
15. $y = -4$
16. $y = -4x - 24$
17. $y = \frac{8}{3}x + 5$
18. $y = \frac{1}{10}x + 3$
19. $f(x) = x + 2$
20. $f(x) = -2x + 7$
21. $f(x) = -\frac{1}{4}x - 2$
22. $f(x) = \frac{4}{5}x - 5$
23. $f(x) = -5x - 4$
24. $f(x) = 3$
25. $f(x) = -2x + 1$
26. $f(x) = \frac{1}{2}x$
27. The slope and the y-intercept were substituted incorrectly; $y = 2x + 7$
28. The x-coordinates and y-coordinates were not substituted in the same order in the slope formula;
    $\text{slope} = \frac{1-4}{5-0} = -\frac{3}{5}$;
    $y = -\frac{3}{5}x + 4$

Dynamic Teaching Tools

- Dynamic Assessment & Progress Monitoring Tool
- Interactive Whiteboard Lesson Library
- Dynamic Classroom with Dynamic Investigations

ANSWERS

29. **a.** $y = -0.005t + 3.91$

b. 3.71 min; 3.61 min

30. **a.** $C = 75t + 50$

b. software program; *Sample answer:* The total cost for 12 hours of recording time is \$950, which is more than the cost of the software program.

31. no; Because the x-coordinates are the same, the line is vertical and has an undefined slope.

32. *Sample answer:* Jet ski rental costs \$20 for one hour and \$15 for each additional hour.

33–45. See Additional Answers.

Mini-Assessment

1. Write an equation of the line with a slope of -2 and y-intercept of 3. $y = -2x + 3$
2. Write an equation of the line in slope-intercept form.

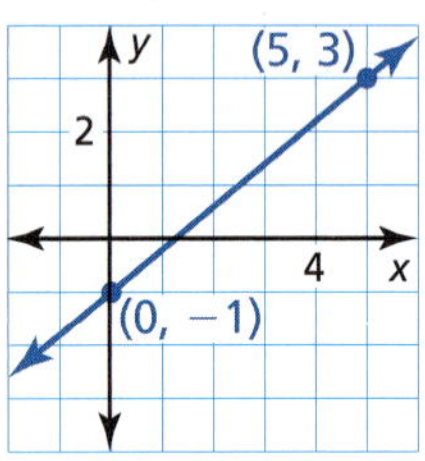

$y = \frac{4}{5}x - 1$

3. Write an equation of the line that passes through the points (0, −5) and (4, 7). $y = 3x - 5$
4. Write a linear function f with the values $f(0) = 6$ and $f(4) = -6$. $f(x) = -3x + 6$
5. In 2006, a company had sales of \$10 million. In 2011, sales were \$12.5 million. Write a linear model that represents the company's sales as a function of the number of years since 2006. Use the model to predict the sales in 2021. $y = 0.5x + 10$; \$17.5 million

29. MODELING WITH MATHEMATICS In 1960, the world record for the men's mile was 3.91 minutes. In 1980, the record time was 3.81 minutes. *(See Example 5.)*

a. Write a linear model that represents the world record (in minutes) for the men's mile as a function of the number of years since 1960.

b. Use the model to estimate the record time in 2000 and predict the record time in 2020.

30. MODELING WITH MATHEMATICS A recording studio charges musicians an initial fee of \$50 to record an album. Studio time costs an additional \$75 per hour.

a. Write a linear model that represents the total cost of recording an album as a function of studio time (in hours).

b. Is it less expensive to purchase 12 hours of recording time at the studio or a \$750 music software program that you can use to record on your own computer? Explain.

31. WRITING A line passes through the points (0, −2) and (0, 5). Is it possible to write an equation of the line in slope-intercept form? Justify your answer.

32. THOUGHT PROVOKING Describe a real-life situation involving a linear function whose graph passes through the points.

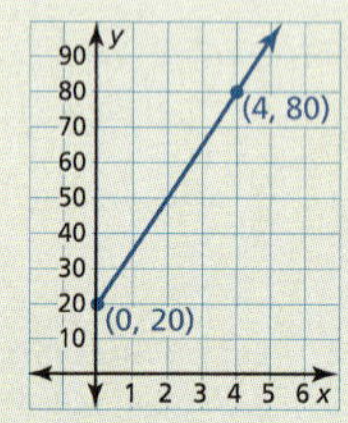

33. REASONING Recall that the standard form of a linear equation is $Ax + By = C$. Rewrite this equation in slope-intercept form. Use your answer to find the slope and y-intercept of the graph of the equation $-6x + 5y = 9$.

34. MAKING AN ARGUMENT Your friend claims that given $f(0)$ and any other value of a linear function f, you can write an equation in slope-intercept form that represents the function. Your cousin disagrees, claiming that the two points could lie on a vertical line. Who is correct? Explain.

35. ANALYZING A GRAPH Line ℓ is a reflection in the x-axis of line k. Write an equation that represents line k.

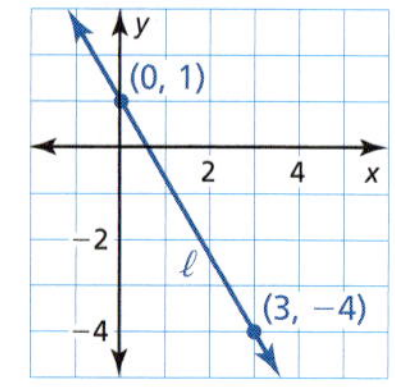

36. HOW DO YOU SEE IT? The graph shows the approximate U.S. box office revenues (in billions of dollars) from 2000 to 2012, where $x = 0$ represents the year 2000.

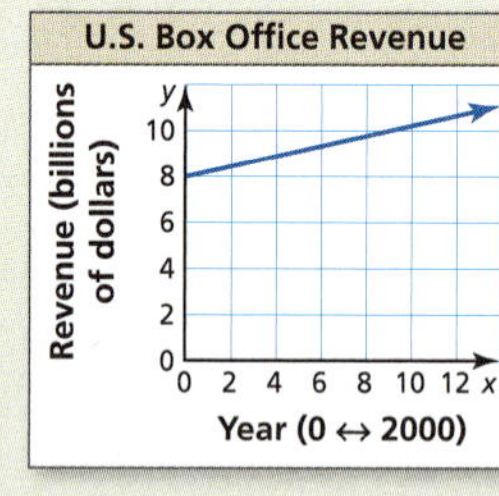

a. Estimate the slope and y-intercept of the graph.

b. Interpret your answers in part (a) in the context of the problem.

c. How can you use your answers in part (a) to predict the U.S. box office revenue in 2018?

37. ABSTRACT REASONING Show that the equation of the line that passes through the points $(0, b)$ and $(1, b + m)$ is $y = mx + b$. Explain how you can be sure that the point $(-1, b - m)$ also lies on the line.

Maintaining Mathematical Proficiency

Reviewing what you learned in previous grades and lessons

Solve the equation. *(Section 1.3)*

38. $3(x - 15) = x + 11$

39. $-4y - 10 = 4(y - 3)$

40. $2(3d + 3) = 7 + 6d$

41. $-5(4 - 3n) = 10(n - 2)$

Use intercepts to graph the linear equation. *(Section 3.4)*

42. $-4x + 2y = 16$

43. $3x + 5y = -15$

44. $x - 6y = 24$

45. $-7x - 2y = -21$

If students need help...	*If students got it...*
Resources by Chapter • Practice A and Practice B • Puzzle Time	Resources by Chapter • Enrichment and Extension • Cumulative Review
Student Journal • Practice	Start the *next* Section
Differentiating the Lesson Skills Review Handbook	

Laurie's Notes

Overview of Section 4.2

Introduction

- In this lesson, students will write the equation of a line in *point-slope form*. The point-slope form is first derived by writing the formula for slope between (x, y) and (x_1, y_1). To write the equation, students are given the slope and a point, two points, or a context from which the slope and an ordered pair can be deduced.
- The real-life example presents a table of data, and students must first determine whether the data are linear or not. There are different ways in which students might justify that the data are linear, which provides a good opportunity to both review and assess students' understanding.

Formative Assessment Tips

- **Fact-First Questioning:** This is a higher-order questioning technique that goes beyond asking straight recall questions. Instead, this strategy allows you to assess students' growing understanding of a concept or skill.
- First, state a fact. Then ask students why, or how, or to explain. Student thinking is activated, and you gain insight into the depth of students' conceptual understanding.
- *Example:* Make the statement that all linear functions have a y-intercept. Then ask, "Why is this true?"

Extensions

- **Connection to Transformations:** Help students make connections to the lesson on transformations in the last chapter. Vertical and horizontal shifts can be represented by subtraction. Consider $y = f(x - h) + k$. This could be written as $y - k = f(x - h)$. Written in this form, it can be seen that the vertical shift information is $y - k$ and the horizontal shift is $x - h$.
- Make a connection to the point-slope form $y - y_1 = m(x - x_1)$, which has vertical shift up y_1 units and horizontal shift right x_1 units. Initially this is confusing to students because it makes sense when positive means UP or RIGHT and negative means DOWN or LEFT. Explain to students that an equation such as $y + 3 = 4(x - 2)$ represents the shifting of the axes, not the graph, 3 units up and 2 units left, but the effect looks like the graph moved down 3 units and 2 units right.

Pacing Suggestion

- Complete at least the first two explorations and then begin the formal lesson with the *Core Concept*.

Dynamic Teaching Tools

- Dynamic Assessment & Progress Monitoring Tool
- Lesson Planning Tool
- Interactive Whiteboard Lesson Library
- Dynamic Classroom with Dynamic Investigations

What Your Students Will Learn

- Write equations of lines in point-slope form given the slope and a point on the line or given two points on the line.

Laurie's Notes

Exploration

Motivate

- Hold a piece of ribbon and a pair of scissors in your hands. Snip a one-foot piece of ribbon off. Repeat once or twice more.

? "Do you know how long my ribbon was when I first started?" no

- Your question should prompt students to ask two obvious questions:
- "How much are you cutting off each time?" and "How many times have you made a cut?" How much you cut off each time is the slope (−1). How many times you cut the ribbon helps students work backward to find the length before any cuts were made, which is the *y*-intercept.

Exploration 1

? Assessing Question: "What does it mean for a line to have a slope of $\frac{1}{2}$?" For every unit it rises, it runs 2 units.

- **Reason Abstractly and Quantitatively:** The *y*-intercept in part (a) is not an integer. Students will reason quantitatively and recognize that the coordinates are $(0, \frac{1}{2})$.

Exploration 2

- **Big Idea:** This exploration helps develop the formula for finding an equation of a line given its slope and a point on the line. Students should see the relationship between the slope formula and the point-slope form.
- **Make Sense of Problems and Persevere in Solving Them:** The steps of the derivation are provided. Encourage students to read carefully, discuss with their partners, and think about the process. Do not jump in too quickly to rescue students!
- **Think Alouds:** Allow sufficient time before having a class discussion of this exploration. Ask volunteers to share their work on the board. Sharing their process aloud helps students become more confident in their reasoning.

Exploration 3

? "What is the slope for this real-life situation?" Slope = \$25/month "What point is given?" (4, 175)

- **Teaching Tip:** After students substitute known information in the point-slope formula, have them transform it into slope-intercept form.

Communicate Your Answer

- **Think-Pair-Share:** Have students work independently and then share with partners. Discuss as a class.

Connecting to Next Step

- Students have now developed a sense of how to write a linear equation given the slope and a point. In the formal lesson, the point-slope formula will not need to be derived.

4.2 Writing Equations in Point-Slope Form

Essential Question How can you write an equation of a line when you are given the slope and a point on the line?

EXPLORATION 1 **Writing Equations of Lines**

Work with a partner.

- Sketch the line that has the given slope and passes through the given point.
- Find the y-intercept of the line.
- Write an equation of the line.

a. $m = \frac{1}{2}$

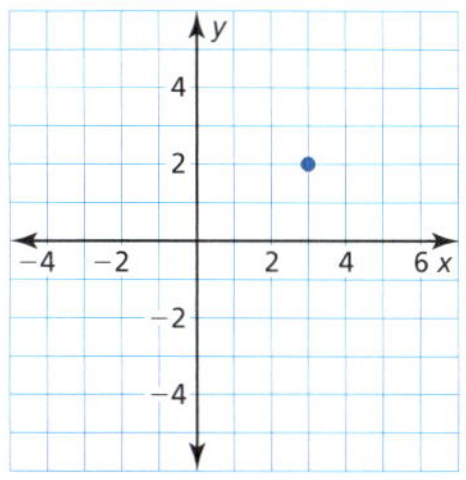

b. $m = -2$

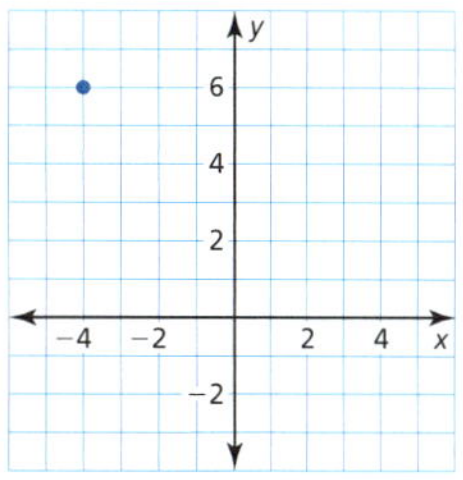

EXPLORATION 2 **Writing a Formula**

Work with a partner.

The point (x_1, y_1) is a given point on a nonvertical line. The point (x, y) is any other point on the line. Write an equation that represents the slope m of the line. Then rewrite this equation by multiplying each side by the difference of the x-coordinates to obtain the **point-slope form** of a linear equation.

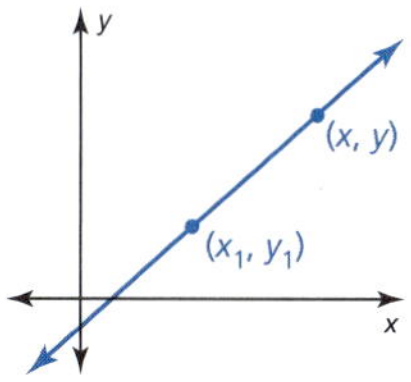

USING A GRAPHING CALCULATOR

To be proficient in math, you need to understand the feasibility, appropriateness, and limitations of the technological tools at your disposal. For instance, in real-life situations such as the one given in Exploration 3, it may not be feasible to use a square viewing window on a graphing calculator.

EXPLORATION 3 **Writing an Equation**

Work with a partner.

For four months, you have saved \$25 per month. You now have \$175 in your savings account.

a. Use your result from Exploration 2 to write an equation that represents the balance A after t months.

b. Use a graphing calculator to verify your equation.

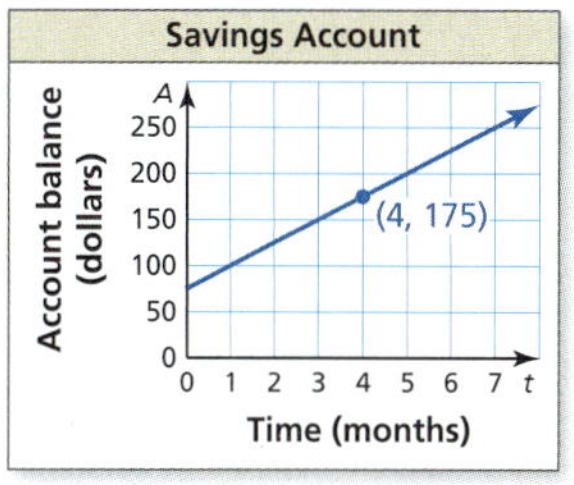

Communicate Your Answer

4. How can you write an equation of a line when you are given the slope and a point on the line?
5. Give an example of how to write an equation of a line when you are given the slope and a point on the line. Your example should be different from those above.

Dynamic Teaching Tools

- Dynamic Assessment & Progress Monitoring Tool
- Lesson Planning Tool
- Interactive Whiteboard Lesson Library
- Dynamic Classroom with Dynamic Investigations

ANSWERS

1. a.

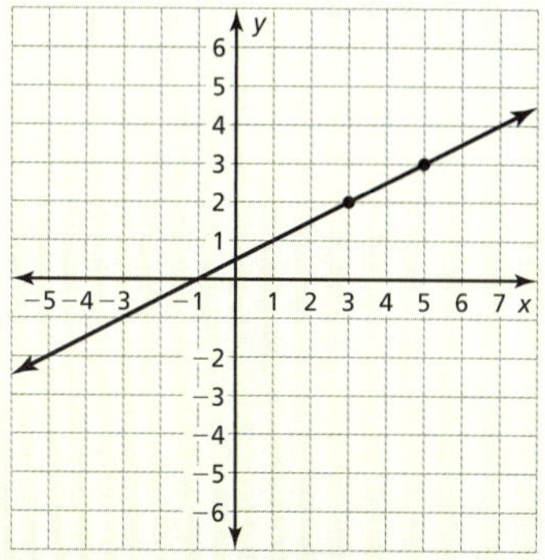

$\frac{1}{2}$; $y = \frac{1}{2}x + \frac{1}{2}$

b.

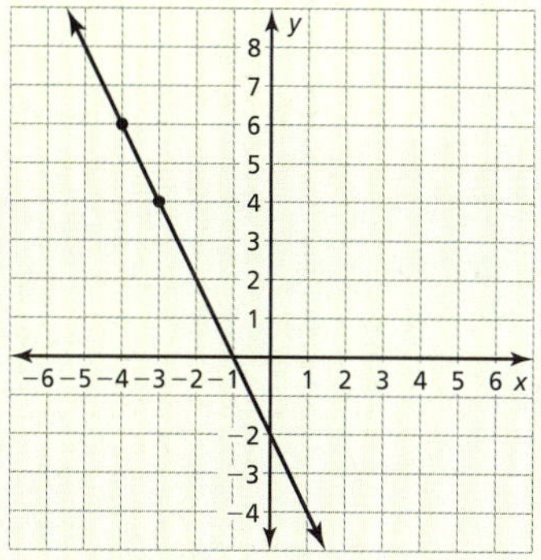

-2; $y = -2x - 2$

2. $m = \dfrac{y - y_1}{x - x_1}$; $y - y_1 = m(x - x_1)$

3. a. $A = 25t + 75$

b. Check students' work.

4. Substitute the slope for m and the coordinates of the point for x_1 and y_1 in $y - y_1 = m(x - x_1)$.

5. *Sample answer:* An equation of the line that passes through the point $(2, -9)$ and has a slope of $-\frac{1}{3}$ is $y + 9 = -\frac{1}{3}(x - 2)$.

English Language Learners

Labels

Explain to students that the name "point-slope form" relates to the information shown in the equation. Have students use colors to label and practice identifying the point and the slope when given equations in this format.

slope

$y - 5 = 2(x - 1)$

point (1, 5)

Extra Example 1

Write an equation in point-slope form of the line that passes through the point (1, 6) and has a slope of $-\frac{1}{2}$.

$y - 6 = -\frac{1}{2}(x - 1)$

MONITORING PROGRESS ANSWERS

1. $y + 1 = -2(x - 3)$
2. $y = -\frac{2}{3}(x - 4)$

4.2 Lesson

Core Vocabulary

point-slope form, *p. 182*

Previous

slope-intercept form
function
linear model
rate

What You Will Learn

- Write an equation of a line given its slope and a point on the line.
- Write an equation of a line given two points on the line.
- Use linear equations to solve real-life problems.

Writing Equations of Lines in Point-Slope Form

Given a point on a line and the slope of the line, you can write an equation of the line. Consider the line that passes through (2, 3) and has a slope of $\frac{1}{2}$. Let (x, y) be another point on the line where $x \neq 2$. You can write an equation relating x and y using the slope formula with $(x_1, y_1) = (2, 3)$ and $(x_2, y_2) = (x, y)$.

$$m = \frac{y_2 - y_1}{x_2 - x_1}$$ Write the slope formula.

$$\frac{1}{2} = \frac{y - 3}{x - 2}$$ Substitute values.

$$\frac{1}{2}(x - 2) = y - 3$$ Multiply each side by $(x - 2)$.

The equation in *point-slope form* is $y - 3 = \frac{1}{2}(x - 2)$.

Core Concept

Point-Slope Form

Words A linear equation written in the form $y - y_1 = m(x - x_1)$ is in **point-slope form**. The line passes through the point (x_1, y_1), and the slope of the line is m.

Algebra $y - y_1 = m(x - x_1)$

slope

passes through (x_1, y_1)

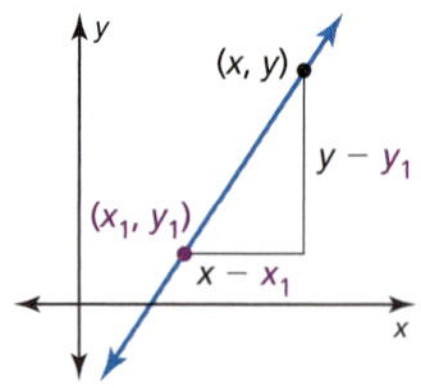

EXAMPLE 1 Using a Slope and a Point to Write an Equation

Write an equation in point-slope form of the line that passes through the point $(-8, 3)$ and has a slope of $\frac{1}{4}$.

SOLUTION

$$y - y_1 = m(x - x_1)$$ Write the point-slope form.

$$y - 3 = \frac{1}{4}[x - (-8)]$$ Substitute $\frac{1}{4}$ for m, -8 for x_1, and 3 for y_1.

$$y - 3 = \frac{1}{4}(x + 8)$$ Simplify.

▶ The equation is $y - 3 = \frac{1}{4}(x + 8)$.

Check

$y - 3 = \frac{1}{4}(x + 8)$

$3 - 3 \stackrel{?}{=} \frac{1}{4}(-8 + 8)$

$0 = 0$ ✓

Monitoring Progress

Help in English and Spanish at *BigIdeasMath.com*

Write an equation in point-slope form of the line that passes through the given point and has the given slope.

1. $(3, -1); m = -2$
2. $(4, 0); m = -\frac{2}{3}$

Laurie's Notes Teacher Actions

? Fact-First Questioning: "There is only one line through (1, 4) with a slope of 2. Explain why this is true." Listen for explanations that involve the fixed point and a constant rate of change or ratio of change of y to change in x. Students may even make a sketch.

- **Teaching Tip:** On a side board, write the formula for slope as $\frac{y - y_1}{x - x_1} = m$ so that students are reminded of how this form of the equation was derived. Again, subscripts are arbitrary; (x_2, y_2) has been replaced with (x, y).

Writing Equations of Lines Given Two Points

When you are given two points on a line, you can write an equation of the line using the following steps.

Step 1 Find the slope of the line.

Step 2 Use the slope and one of the points to write an equation of the line in point-slope form.

ANOTHER WAY

You can use either of the given points to write an equation of the line.

Use $m = -2$ and $(3, -2)$.

$$y - (-2) = -2(x - 3)$$
$$y + 2 = -2x + 6$$
$$y = -2x + 4$$

EXAMPLE 2 **Using Two Points to Write an Equation**

Write an equation in slope-intercept form of the line shown.

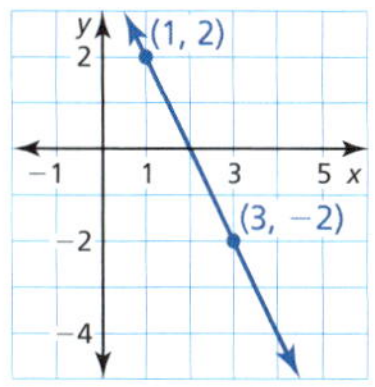

SOLUTION

Step 1 Find the slope of the line.

$$m = \frac{-2 - 2}{3 - 1} = \frac{-4}{2}, \text{ or } -2$$

Step 2 Use the slope $m = -2$ and the point $(1, 2)$ to write an equation of the line.

$y - y_1 = m(x - x_1)$ Write the point-slope form.

$y - 2 = -2(x - 1)$ Substitute -2 for m, 1 for x_1, and 2 for y_1.

$y - 2 = -2x + 2$ Distributive Property

$y = -2x + 4$ Write in slope-intercept form.

▶ The equation is $y = -2x + 4$.

EXAMPLE 3 **Writing a Linear Function**

Write a linear function f with the values $f(4) = -2$ and $f(8) = 4$.

SOLUTION

Note that you can rewrite $f(4) = -2$ as $(4, -2)$ and $f(8) = 4$ as $(8, 4)$.

Step 1 Find the slope of the line that passes through $(4, -2)$ and $(8, 4)$.

$$m = \frac{4 - (-2)}{8 - 4} = \frac{6}{4}, \text{ or } 1.5$$

Step 2 Use the slope $m = 1.5$ and the point $(8, 4)$ to write an equation of the line.

$y - y_1 = m(x - x_1)$ Write the point-slope form.

$y - 4 = 1.5(x - 8)$ Substitute 1.5 for m, 8 for x_1, and 4 for y_1.

$y - 4 = 1.5x - 12$ Distributive Property

$y = 1.5x - 8$ Write in slope-intercept form.

▶ A function is $f(x) = 1.5x - 8$.

Monitoring Progress

Help in English and Spanish at *BigIdeasMath.com*

Write an equation in slope-intercept form of the line that passes through the given points.

3. $(1, 4), (3, 10)$

4. $(-4, -1), (8, -4)$

5. Write a linear function g with the values $g(2) = 3$ and $g(6) = 5$.

Extra Example 2

Write an equation in slope-intercept form of the line shown.

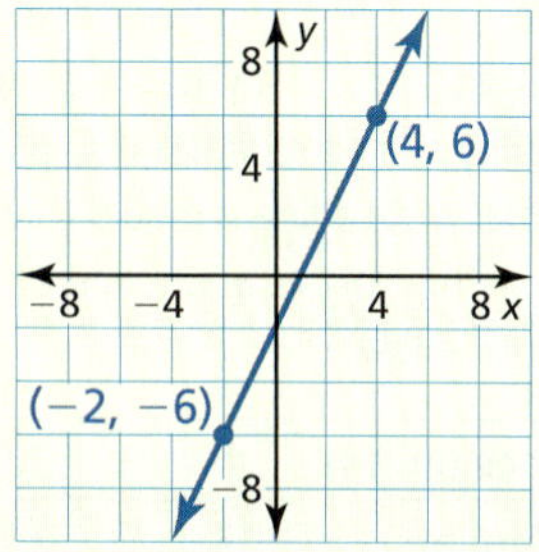

$y = 2x - 2$

Extra Example 3

Write a linear function f with the values $f(8) = -1$ and $f(6) = 0$.

$f(x) = -0.5x + 3$

MONITORING PROGRESS ANSWERS

3. $y = 3x + 1$

4. $y = -\frac{1}{4}x - 2$

5. $g(x) = \frac{1}{2}x + 2$

Laurie's Notes Teacher Actions

- **? Probing Question:** "Can you use the point-slope formula when given two points? Explain." yes; Find the slope first and substitute the answer along with one of the ordered pairs into the formula. "Which ordered pair should you use?" either; Have students verify this!
- **?** "In Example 3, what information is given by $f(4) = -2$?" You are given an ordered pair $(4, -2)$ for the function f.
- **Think-Pair-Share:** Have students work independently on Questions 3–5 and then share with partners. Discuss as a class.

Extra Example 4

To ride in a Best-in-Town taxi, you pay a start fee and a mileage fee. The table shows the total cost of a taxi ride for different distances. Can the situation be modeled by a linear equation? Explain. If possible, write a linear model that represents the cost as a function of the number of miles.

Distance (miles)	4	8	10	12
Cost (dollars)	13	23	28	33

Yes, the cost increases at a constant rate. The linear model is $C = 2.5d + 3$.

MONITORING PROGRESS ANSWER

6. yes; The total cost increases at a constant rate; $C = 42n + 50$

Solving Real-Life Problems

EXAMPLE 4 **Modeling with Mathematics**

The student council is ordering customized foam hands to promote school spirit. The table shows the cost of ordering different numbers of foam hands. Can the situation be modeled by a linear equation? Explain. If possible, write a linear model that represents the cost as a function of the number of foam hands.

Number of foam hands	4	6	8	10	12
Cost (dollars)	34	46	58	70	82

SOLUTION

1. **Understand the Problem** You know five data pairs from the table. You are asked whether the data are linear. If so, write a linear model that represents the cost.
2. **Make a Plan** Find the rate of change for consecutive data pairs in the table. If the rate of change is constant, use the point-slope form to write an equation. Rewrite the equation in slope-intercept form so that the cost is a function of the number of foam hands.
3. **Solve the Problem**

Step 1 Find the rate of change for consecutive data pairs in the table.

$$\frac{46-34}{6-4} = 6,\ \frac{58-46}{8-6} = 6,\ \frac{70-58}{10-8} = 6,\ \frac{82-70}{12-10} = 6$$

Because the rate of change is constant, the data are linear. So, use the point-slope form to write an equation that represents the data.

Step 2 Use the constant rate of change (slope) $m = 6$ and the data pair (4, 34) to write an equation. Let C be the cost (in dollars) and n be the number of foam hands.

$C - C_1 = m(n - n_1)$ — Write the point-slope form.

$C - 34 = 6(n - 4)$ — Substitute 6 for m, 4 for n_1, and 34 for C_1.

$C - 34 = 6n - 24$ — Distributive Property

$C = 6n + 10$ — Write in slope-intercept form.

▶ Because the cost increases at a constant rate, the situation can be modeled by a linear equation. The linear model is $C = 6n + 10$.

4. **Look Back** To check that your model is correct, verify that the other data pairs are solutions of the equation.

$46 = 6(6) + 10$ ✓ $58 = 6(8) + 10$ ✓

$70 = 6(10) + 10$ ✓ $82 = 6(12) + 10$ ✓

Monitoring Progress

Help in English and Spanish at *BigIdeasMath.com*

6. You pay an installation fee and a monthly fee for Internet service. The table shows the total cost for different numbers of months. Can the situation be modeled by a linear equation? Explain. If possible, write a linear model that represents the total cost as a function of the number of months.

Number of months	Total cost (dollars)
3	176
6	302
9	428
12	554

Laurie's Notes Teacher Actions

? "How will you know whether the table of data represents a linear equation?" The slope between consecutive ordered pairs will be the same.

- Work through the example as shown. Ask students to transform the equation to slope-intercept form and interpret the slope and y-intercept in the context of the situation.

Closure

- **Exit Ticket:** Write an equation of the line with a slope of 2 that passes through the point (−1, 4) in point-slope form and slope-intercept form. $y - 4 = 2(x + 1); y = 2x + 6$

4.2 Exercises

Dynamic Solutions available at *BigIdeasMath.com*

Vocabulary and Core Concept Check

1. **USING STRUCTURE** Without simplifying, identify the slope of the line given by the equation $y - 5 = -2(x + 5)$. Then identify one point on the line.

2. **WRITING** Explain how you can use the slope formula to write an equation of the line that passes through $(3, -2)$ and has a slope of 4.

Monitoring Progress and Modeling with Mathematics

In Exercises 3–10, write an equation in point-slope form of the line that passes through the given point and has the given slope. *(See Example 1.)*

3. $(2, 1)$; $m = 2$
4. $(3, 5)$; $m = -1$
5. $(7, -4)$; $m = -6$
6. $(-8, -2)$; $m = 5$
7. $(9, 0)$; $m = -3$
8. $(0, 2)$; $m = 4$
9. $(-6, 6)$; $m = \frac{3}{2}$
10. $(5, -12)$; $m = -\frac{2}{5}$

In Exercises 11–14, write an equation in slope-intercept form of the line shown. *(See Example 2.)*

11. 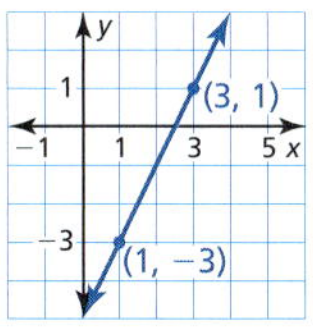

12. 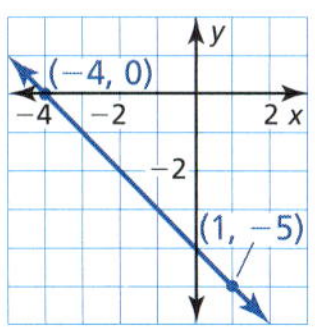

13. 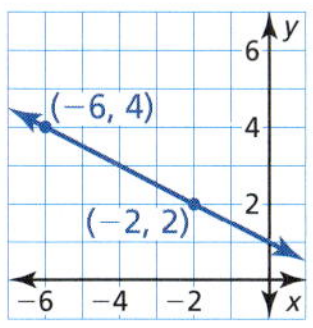

14. 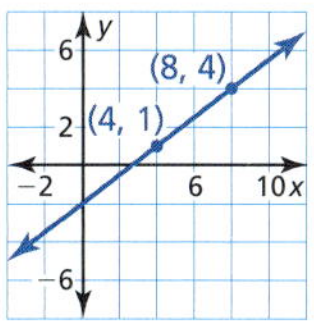

In Exercises 15–20, write an equation in slope-intercept form of the line that passes through the given points.

15. $(7, 2), (2, 12)$
16. $(6, -2), (12, 1)$
17. $(6, -1), (3, -7)$
18. $(-2, 5), (-4, -5)$
19. $(1, -9), (-3, -9)$
20. $(-5, 19), (5, 13)$

In Exercises 21–26, write a linear function f with the given values. *(See Example 3.)*

21. $f(2) = -2, f(1) = 1$
22. $f(5) = 7, f(-2) = 0$
23. $f(-4) = 2, f(6) = -3$
24. $f(-10) = 4, f(-2) = 4$
25. $f(-3) = 1, f(13) = 5$
26. $f(-9) = 10, f(-1) = -2$

In Exercises 27–30, tell whether the data in the table can be modeled by a linear equation. Explain. If possible, write a linear equation that represents y as a function of x. *(See Example 4.)*

27.

x	2	4	6	8	10
y	−1	5	15	29	47

28.

x	−3	−1	1	3	5
y	16	10	4	−2	−8

29.

x	y
0	1.2
1	1.4
2	1.6
4	2

30.

x	y
1	18
2	15
4	12
8	9

31. **ERROR ANALYSIS** Describe and correct the error in writing a linear function g with the values $g(5) = 4$ and $g(3) = 10$.

✗ $m = \frac{10 - 4}{3 - 5} = \frac{6}{-2} = -3$

$y - y_1 = mx - x_1$

$y - 4 = -3x - 5$

$y = -3x - 1$

A function is $g(x) = -3x - 1$.

31. In point-slope form, the slope is multiplied by the quantity, $x - x_1$; $y - y_1 = m(x - x_1)$; $y - 4 = -3(x - 5)$; $y - 4 = -3x + 15$; $y = -3x + 19$; A function is $g(x) = -3x + 19$.

Assignment Guide and Homework Check

ASSIGNMENT

Basic: 1, 2, 3–33 odd, 38, 41–44

Average: 1, 2–34 even, 38, 40–44

Advanced: 1, 2, 8–12 even, 18, 20, 26, 30–44

HOMEWORK CHECK

Basic: 3, 13, 21, 27

Average: 4, 14, 22, 28

Advanced: 8, 12, 18, 30

ANSWERS

1. -2; $(-5, 5)$
2. Substitute 4 for m, 3 for x_1, and -2 for y_1 in the slope formula, eliminate the subscripts for the second point, and solve for y.
3. $y - 1 = 2(x - 2)$
4. $y - 5 = -(x - 3)$
5. $y + 4 = -6(x - 7)$
6. $y + 2 = 5(x + 8)$
7. $y = -3(x - 9)$
8. $y - 2 = 4x$
9. $y - 6 = \frac{3}{2}(x + 6)$
10. $y + 12 = -\frac{2}{5}(x - 5)$
11. $y = 2x - 5$
12. $y = -x - 4$
13. $y = -\frac{1}{2}x + 1$
14. $y = \frac{3}{4}x - 2$
15. $y = -2x + 16$
16. $y = \frac{1}{2}x - 5$
17. $y = 2x - 13$
18. $y = 5x + 15$
19. $y = -9$
20. $y = -\frac{3}{5}x + 16$
21. $f(x) = -3x + 4$
22. $f(x) = x + 2$
23. $f(x) = -\frac{1}{2}x$
24. $f(x) = 4$
25. $f(x) = \frac{1}{4}x + \frac{7}{4}$
26. $f(x) = -\frac{3}{2}x - \frac{7}{2}$
27. no; y does not increase at a constant rate.
28. yes, y decreases at a constant rate; $y = -3x + 7$
29. yes; y increases at a constant rate; $y = 0.2x + 1.2$
30. no; y does not decrease at a constant rate.

Dynamic Teaching Tools

- Dynamic Assessment & Progress Monitoring Tool
- Interactive Whiteboard Lesson Library
- Dynamic Classroom with Dynamic Investigations

ANSWERS

32. The coordinates substituted into the point-slope form were not from the same point; $y - 2 = \frac{1}{3}(x - 1)$ or $y - 3 = \frac{1}{3}(x - 4)$

33. **a.** $C = 80n + 145$

b. \$865

34. **a.** yes; The total cost increases at a constant rate.

b. \$42; \$102

c. 11 days

35–44. See Additional Answers.

Mini-Assessment

1. Write an equation in point-slope form of the line that passes through the point $(3, -5)$ and has a slope of 2. $y + 5 = 2(x - 3)$

2. Write an equation in slope-intercept form of the line shown.

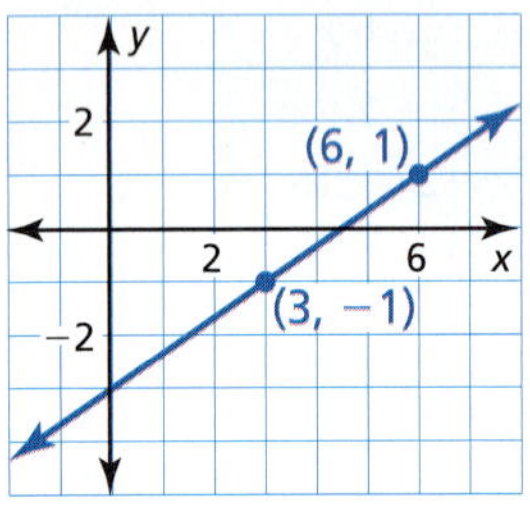

$y = \frac{2}{3}x - 3$

3. Write a linear function f with the values $f(1) = 1$ and $f(-3) = 17$. $f(x) = -4x + 5$

4. Tell whether the data in the table can be modeled by a linear equation. Explain. If possible, write a linear equation that represents y as a function of x.

x	3	5	12	20
y	7	10	16	26

The data cannot be modeled by a linear equation, because the rate of change is not constant.

32. ERROR ANALYSIS Describe and correct the error in writing an equation of the line that passes through the points (1, 2) and (4, 3).

✗ $m = \frac{3-2}{4-1} = \frac{1}{3}$ $\quad y - 2 = \frac{1}{3}(x - 4)$

33. MODELING WITH MATHEMATICS You are designing a sticker to advertise your band. A company charges \$225 for the first 1000 stickers and \$80 for each additional 1000 stickers.

a. Write an equation that represents the total cost (in dollars) of the stickers as a function of the number (in thousands) of stickers ordered.

b. Find the total cost of 9000 stickers.

34. MODELING WITH MATHEMATICS You pay a processing fee and a daily fee to rent a beach house. The table shows the total cost of renting the beach house for different numbers of days.

Days	2	4	6	8
Total cost (dollars)	246	450	654	858

a. Can the situation be modeled by a linear equation? Explain.

b. What is the processing fee? the daily fee?

c. You can spend no more than \$1200 on the beach house rental. What is the maximum number of days you can rent the beach house?

35. WRITING Describe two ways to graph the equation $y - 1 = \frac{3}{2}(x - 4)$.

36. THOUGHT PROVOKING The graph of a linear function passes through the point $(12, -5)$ and has a slope of $\frac{2}{5}$. Represent this function in two other ways.

37. REASONING You are writing an equation of the line that passes through two points that are not on the y-axis. Would you use slope-intercept form or point-slope form to write the equation? Explain.

38. HOW DO YOU SEE IT? The graph shows two points that lie on the graph of a linear function.

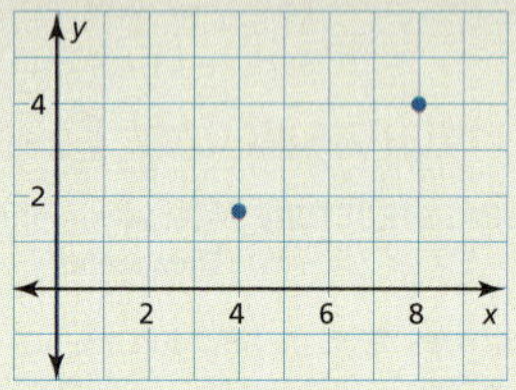

a. Does the y-intercept of the graph of the linear function appear to be positive or negative? Explain.

b. Estimate the coordinates of the two points. How can you use your estimates to confirm your answer in part (a)?

39. CONNECTION TO TRANSFORMATIONS Compare the graph of $y = 2x$ to the graph of $y - 1 = 2(x + 3)$. Make a conjecture about the graphs of $y = mx$ and $y - k = m(x - h)$.

40. COMPARING FUNCTIONS Three siblings each receive money for a holiday and then spend it at a constant weekly rate. The graph describes Sibling A's spending, the table describes Sibling B's spending, and the equation $y = -22.5x + 90$ describes Sibling C's spending. The variable y represents the amount of money left after x weeks.

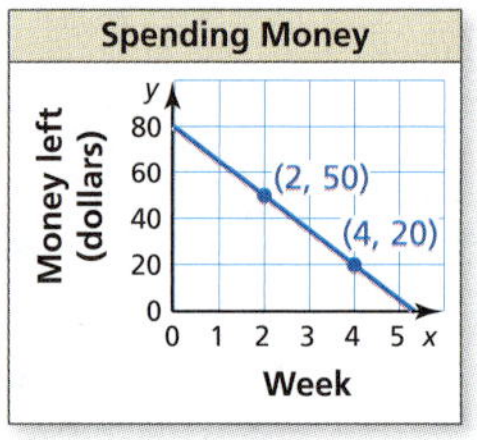

Week, x	Money left, y
1	\$100
2	\$75
3	\$50
4	\$25

a. Which sibling received the most money? the least money?

b. Which sibling spends money at the fastest rate? the slowest rate?

c. Which sibling runs out of money first? last?

Maintaining Mathematical Proficiency
Reviewing what you learned in previous grades and lessons

Write the reciprocal of the number. *(Skills Review Handbook)*

41. 5 **42.** −8 **43.** $-\frac{2}{7}$ **44.** $\frac{3}{2}$

If students need help...	*If students got it...*
Resources by Chapter • Practice A and Practice B • Puzzle Time	Resources by Chapter • Enrichment and Extension • Cumulative Review
Student Journal • Practice	Start the *next* Section
Differentiating the Lesson Skills Review Handbook	

Laurie's Notes

Overview of Section 4.3

Introduction

- In this lesson, students will write equations of lines that are parallel and perpendicular to a given line. The explorations are designed for students to discover that parallel lines have the same slope and perpendicular lines have slopes that are negative reciprocals (that is, their product is −1).
- Integrated in this lesson is the ability to find the slope of a line given two points, given a graph, or given an equation written in slope-intercept or standard form. Also reviewed is how to write the equation of a line using the point-slope form, or by solving for the slope and y-intercept.
- The real-life example is a lengthy problem involving a number of skills.

Common Misconceptions

- When lines are written in standard form, students may say that the coefficient of x is the slope. Remind students that they must first write the equation in slope-intercept form before stating that the coefficient of x is the slope.

Formative Assessment Tips

- **Always-Sometimes-Never True (AT-ST-NT):** This strategy is useful in assessing whether students overgeneralize or under generalize a particular concept.
- **Reason Abstractly and Quantitatively** and **Construct Viable Arguments and Critique the Reasoning of Others:** When answering, a student should be asked to justify his or her answer, and other students listening to the justification should critique the reasoning.
- *AT-ST-NT* statements help students practice the habit of checking validity when a statement (conjecture) is made. Are there different cases that need to be checked? Is there a counterexample that would show the conjecture to be false?
- Using *AT-ST-NT* statements encourages discourse. To develop these statements for a lesson, consider the common errors or misconceptions that students have relating to the goal(s) of the lesson. Allow private think time before students share their thinking with partners or the whole class.

Pacing Suggestion

- Complete the two explorations, and then start the formal lesson by stating the *Core Concept*.

Dynamic Teaching Tools

- Dynamic Assessment & Progress Monitoring Tool
- Lesson Planning Tool
- Interactive Whiteboard Lesson Library
- Dynamic Classroom with Dynamic Investigations

What Your Students Will Learn

- Use slope to identify and write equations of parallel lines.
- Use slope to identify and write equations of perpendicular lines.

Laurie's Notes

Exploration

Motivate

- Ask whether anyone has visited Salt Lake City, Utah. If they have, ask whether they know anything about the manner in which the streets are named.
- Salt Lake City's method of numbering streets is similar to the idea of latitude and longitude. The city is set up like a grid, with each address having coordinates that indicate the distance west or east and north or south from Temple Square.
- Share that today students will write equations of lines that are parallel and perpendicular, like a grid.

Exploration 1

- In this exploration, students need to rewrite equations from standard form to slope-intercept form. The goal is for students to recognize that lines with the same slope are parallel.
- **Use Appropriate Tools Strategically:** Figure 1 demonstrates that it is possible for students to enter the equations in the equation editor without solving. The equations do not need to be simplified. If students enter the equations in this form, then they will miss the goal of the exploration.
- **Attend to Precision:** Figures 2 and 3 show what will happen as students zoom out. The question asks which lines appear parallel. Students should question what happens beyond the viewing window they are seeing. If they continue to zoom out, then the first two lines appear to be collinear.

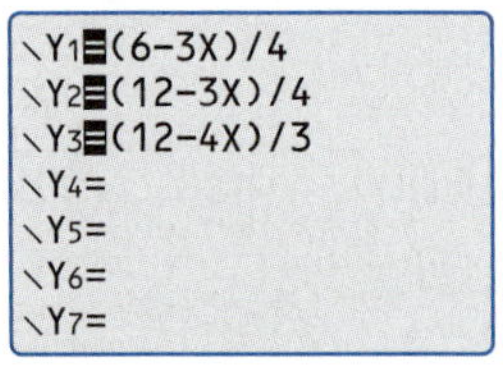

Figure 1

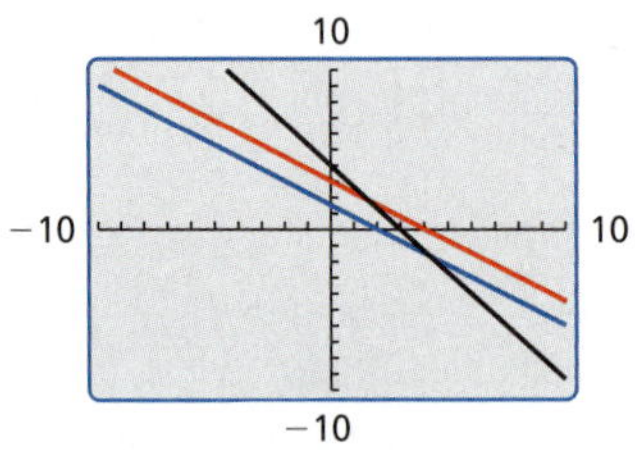

Figure 2

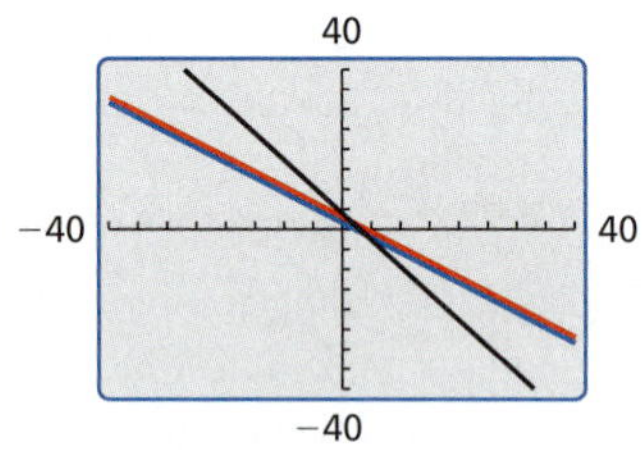

Figure 3

? Always-Sometimes-Never True: "If you graph two lines on your calculator and they do not intersect, then the lines are parallel." sometimes; If the viewing window is expanded, then the lines might intersect.

Exploration 2

- The goal of this exploration is for students to recognize that lines with slopes that are negative reciprocals appear to be perpendicular.
- Depending on how students write the slopes, they may or may not recognize a relationship. You may need to tell students to leave the slopes as simplified fractions.

Communicate Your Answer

- Students can describe parallel and perpendicular lines from their graphs. After completing each exploration, students should make a connection between slopes and whether lines are parallel (or perpendicular).

Connecting to Next Step

- Students should have discovered how to determine whether two lines are parallel or perpendicular by looking at the slopes of the lines. You should not need to do each example in the formal lesson.

4.3 Writing Equations of Parallel and Perpendicular Lines

Essential Question How can you recognize lines that are parallel or perpendicular?

EXPLORATION 1 Recognizing Parallel Lines

Work with a partner. Write each linear equation in slope-intercept form. Then use a graphing calculator to graph the three equations in the same square viewing window. (The graph of the first equation is shown.) Which two lines appear parallel? How can you tell?

a. $3x + 4y = 6$
$3x + 4y = 12$
$4x + 3y = 12$

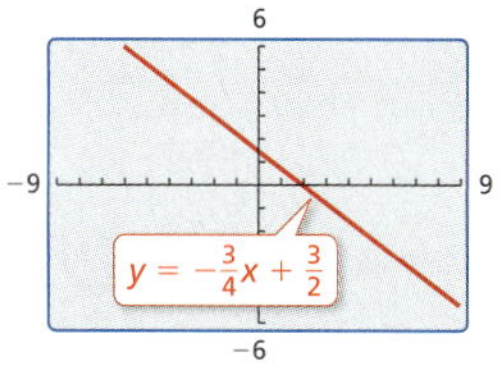

b. $5x + 2y = 6$
$2x + y = 3$
$2.5x + y = 5$

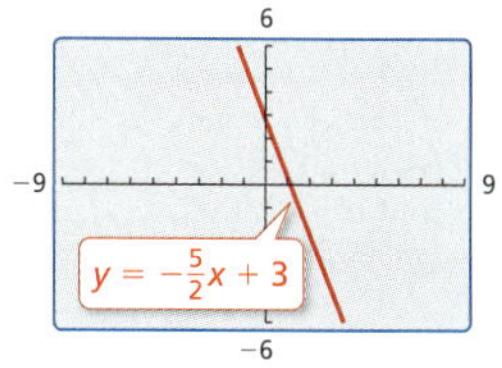

USING TOOLS STRATEGICALLY

To be proficient in math, you need to use a graphing calculator and other available technological tools, as appropriate, to help you explore relationships and deepen your understanding of concepts.

EXPLORATION 2 Recognizing Perpendicular Lines

Work with a partner. Write each linear equation in slope-intercept form. Then use a graphing calculator to graph the three equations in the same square viewing window. (The graph of the first equation is shown.) Which two lines appear perpendicular? How can you tell?

a. $3x + 4y = 6$
$3x - 4y = 12$
$4x - 3y = 12$

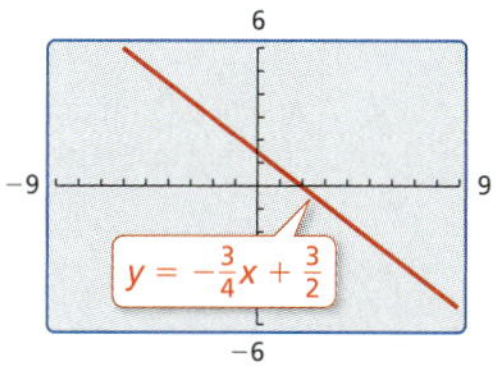

b. $2x + 5y = 10$
$-2x + y = 3$
$2.5x - y = 5$

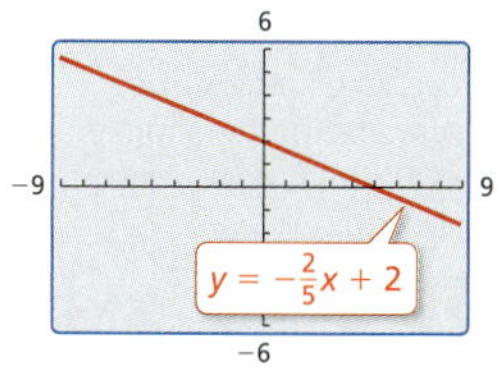

Communicate Your Answer

3. How can you recognize lines that are parallel or perpendicular?
4. Compare the slopes of the lines in Exploration 1. How can you use slope to determine whether two lines are parallel? Explain your reasoning.
5. Compare the slopes of the lines in Exploration 2. How can you use slope to determine whether two lines are perpendicular? Explain your reasoning.

Dynamic Teaching Tools

- Dynamic Assessment & Progress Monitoring Tool
- Lesson Planning Tool
- Interactive Whiteboard Lesson Library
- Dynamic Classroom with Dynamic Investigations

ANSWERS

1. a. $y = -\frac{3}{4}x + \frac{3}{2}$; $y = -\frac{3}{4}x + 3$; $y = -\frac{4}{3}x + 4$

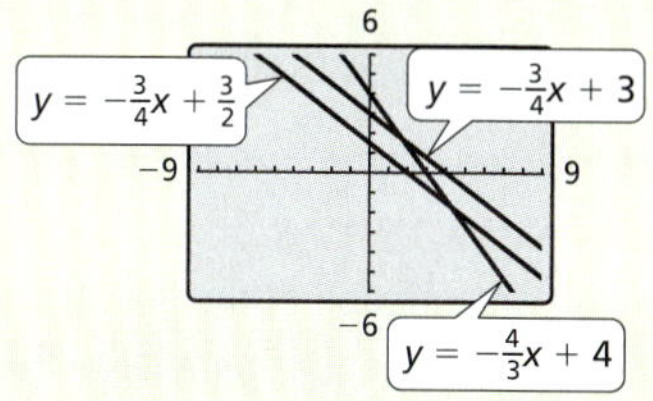

$3x + 4y = 6$ and $3x + 4y = 12$; *Sample answer:* The lines are always the same distance apart.

b. $y = -\frac{5}{2}x + 3$; $y = -2x + 3$; $y = -2.5x + 5$;

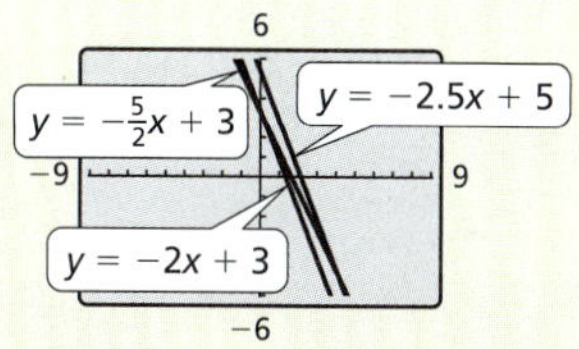

$5x + 2y = 6$ and $2.5x + y = 5$; *Sample answer:* The lines are always the same distance apart.

2. a. $y = -\frac{3}{4}x + \frac{3}{2}$; $y = \frac{3}{4}x - 3$; $y = \frac{4}{3}x - 4$;

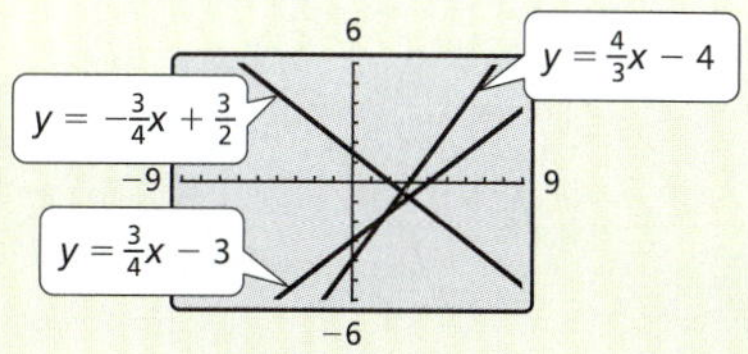

$3x + 4y = 6$ and $4x - 3y = 12$; *Sample answer:* The lines intersect at a right angle.

b. $y = -\frac{2}{5}x + 2$; $y = 2x + 3$; $y = 2.5x - 5$;

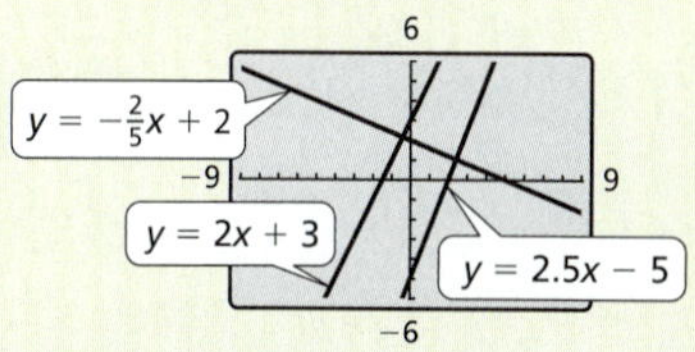

$2x + 5y = 10$ and $2.5x - y = 5$; *Sample answer:* The lines intersect at a right angle.

3. Parallel lines are always the same distance apart; Perpendicular lines intersect at right angles.

4. Parallel lines have the same slope; The lines with the same slope in Exploration 1 were parallel.

5. Perpendicular lines have slopes that are negative reciprocals of each other; The lines with negative reciprocal slopes in Exploration 2 were perpendicular.

Extra Example 1

Determine which of the lines are parallel.

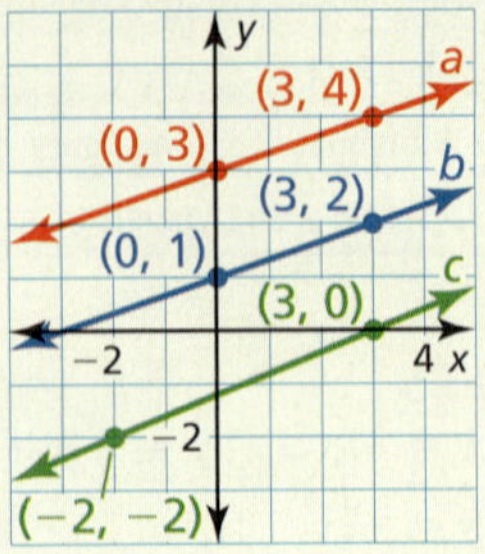

Lines *a* and *b* have the same slope, so they are parallel.

Extra Example 2

Write an equation of the line that passes through (3, 1) and is parallel to the line $y = 3x + 1$. $y = 3x - 8$

MONITORING PROGRESS ANSWERS

1. no; They do not have the same slope.
2. $y = \frac{1}{4}x + 3$

4.3 Lesson

What You Will Learn

- Identify and write equations of parallel lines.
- Identify and write equations of perpendicular lines.
- Use parallel and perpendicular lines in real-life problems.

Core Vocabulary

parallel lines, *p. 188*
perpendicular lines, *p. 189*

Previous

reciprocal

Identifying and Writing Equations of Parallel Lines

READING

The phrase "*A if and only if B*" is a way of writing two conditional statements at once. It means that if *A* is true, then *B* is true. It also means that if *B* is true, then *A* is true.

Core Concept

Parallel Lines and Slopes

Two lines in the same plane that never intersect are **parallel lines**. Two distinct nonvertical lines are parallel if and only if they have the same slope.

All vertical lines are parallel.

EXAMPLE 1 Identifying Parallel Lines

Determine which of the lines are parallel.

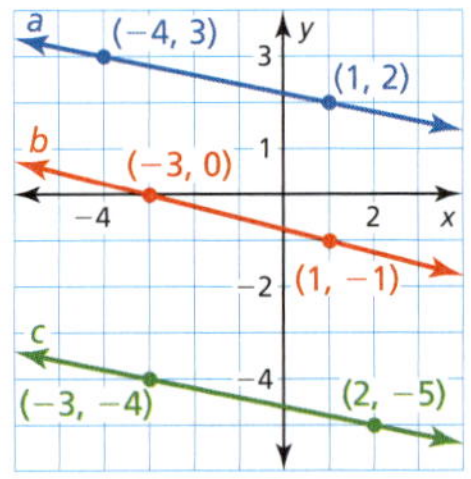

SOLUTION

Find the slope of each line.

Line *a*: $m = \frac{2 - 3}{1 - (-4)} = -\frac{1}{5}$

Line *b*: $m = \frac{-1 - 0}{1 - (-3)} = -\frac{1}{4}$

Line *c*: $m = \frac{-5 - (-4)}{2 - (-3)} = -\frac{1}{5}$

▶ Lines *a* and *c* have the same slope, so they are parallel.

EXAMPLE 2 Writing an Equation of a Parallel Line

Write an equation of the line that passes through (5, −4) and is parallel to the line $y = 2x + 3$.

SOLUTION

Step 1 Find the slope of the parallel line. The graph of the given equation has a slope of 2. So, the parallel line that passes through (5, −4) also has a slope of 2.

Step 2 Use the slope-intercept form to find the *y*-intercept of the parallel line.

$y = mx + b$ — Write the slope-intercept form.

$-4 = 2(5) + b$ — Substitute 2 for *m*, 5 for *x*, and −4 for *y*.

$-14 = b$ — Solve for *b*.

▶ Using $m = 2$ and $b = -14$, an equation of the parallel line is $y = 2x - 14$.

ANOTHER WAY

You can also use the slope $m = 2$ and the point-slope form to write an equation of the line that passes through (5, −4).

$$y - y_1 = m(x - x_1)$$
$$y - (-4) = 2(x - 5)$$
$$y = 2x - 14$$

Monitoring Progress

Help in English and Spanish at *BigIdeasMath.com*

1. Line *a* passes through (−5, 3) and (−6, −1). Line *b* passes through (3, −2) and (2, −7). Are the lines parallel? Explain.
2. Write an equation of the line that passes through (−4, 2) and is parallel to the line $y = \frac{1}{4}x + 1$.

Laurie's Notes Teacher Actions

? Write the *Core Concept*. "Why are vertical lines a special case?" Vertical lines have undefined slope.

- **Turn and Talk:** Have partners work Example 1. Use *Thumbs Up* to assess students' understanding and confidence.
- **No-Hands Questioning:** Pose Example 2 and ask students to write the equation. Some students will solve for the *y*-intercept as shown, while others may use the point-slope form. Give sufficient *Wait Time*, and then ask volunteers to share their solution method.

Identifying and Writing Equations of Perpendicular Lines

REMEMBER

The product of a nonzero number m and its negative reciprocal is -1:

$$m\left(-\frac{1}{m}\right) = -1.$$

Core Concept

Perpendicular Lines and Slopes

Two lines in the same plane that intersect to form right angles are **perpendicular lines**. Nonvertical lines are perpendicular if and only if their slopes are negative reciprocals.

Vertical lines are perpendicular to horizontal lines.

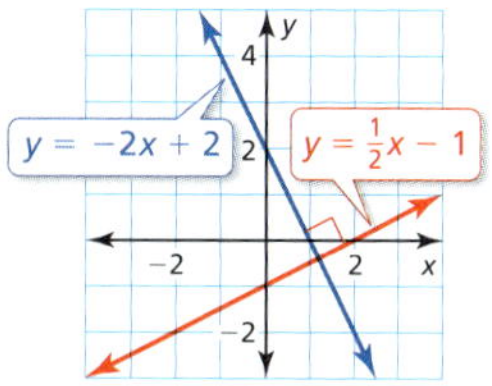

EXAMPLE 3 Identifying Parallel and Perpendicular Lines

Determine which of the lines, if any, are parallel or perpendicular.

Line a: $y = 4x + 2$ Line b: $x + 4y = 3$ Line c: $-8y - 2x = 16$

SOLUTION

Write the equations in slope-intercept form. Then compare the slopes.

Line a: $y = 4x + 2$ Line b: $y = -\frac{1}{4}x + \frac{3}{4}$ Line c: $y = -\frac{1}{4}x - 2$

▶ Lines b and c have slopes of $-\frac{1}{4}$, so they are parallel. Line a has a slope of 4, the negative reciprocal of $-\frac{1}{4}$, so it is perpendicular to lines b and c.

EXAMPLE 4 Writing an Equation of a Perpendicular Line

Write an equation of the line that passes through $(-3, 1)$ and is perpendicular to the line $y = \frac{1}{2}x + 3$.

SOLUTION

Step 1 Find the slope of the perpendicular line. The graph of the given equation has a slope of $\frac{1}{2}$. Because the slopes of perpendicular lines are negative reciprocals, the slope of the perpendicular line that passes through $(-3, 1)$ is -2.

Step 2 Use the slope $m = -2$ and the point-slope form to write an equation of the perpendicular line that passes through $(-3, 1)$.

$y - y_1 = m(x - x_1)$	Write the point-slope form.
$y - 1 = -2[x - (-3)]$	Substitute -2 for m, -3 for x_1, and 1 for y_1.
$y - 1 = -2x - 6$	Simplify.
$y = -2x - 5$	Write in slope-intercept form.

▶ An equation of the perpendicular line is $y = -2x - 5$.

ANOTHER WAY

You can also use the slope $m = -2$ and the slope-intercept form to write an equation of the line that passes through $(-3, 1)$.

$y = mx + b$

$1 = -2(-3) + b$

$-5 = b$

So, $y = -2x - 5$.

Monitoring Progress

Help in English and Spanish at *BigIdeasMath.com*

3. Determine which of the lines, if any, are parallel or perpendicular. Explain.
   Line a: $2x + 6y = -3$ Line b: $y = 3x - 8$ Line c: $-6y + 18x = 9$
4. Write an equation of the line that passes through $(-3, 5)$ and is perpendicular to the line $y = -3x - 1$.

English Language Learners

Build on Past Knowledge

Some students may have difficulty writing the negative reciprocal of the slope. Remind students that the product of a number and its reciprocal is 1. So the product of a number and its negative reciprocal is −1. Have students check each negative reciprocal by multiplying it by the original slope.

$$\frac{1}{4}(-4) = \frac{1}{4}\left(\frac{-4}{1}\right) = -1$$

Extra Example 3

Determine which of the lines, if any, are parallel or perpendicular.

Line a: $y = \frac{2}{3}x + 3$

Line b: $3x + 2y = 8$

Line c: $3y - 2x = -9$

Lines a and c have slopes of $\frac{2}{3}$, so they are parallel. Line b has a slope of $-\frac{3}{2}$, the negative reciprocal of $\frac{2}{3}$, so it is perpendicular to lines a and c.

Extra Example 4

Write an equation of the line that passes through $(4, -3)$ and is perpendicular to the line $y = 4x - 1$.

$y = -\frac{1}{4}x - 2$

MONITORING PROGRESS ANSWERS

3. Lines b and c are parallel; Line a is perpendicular to lines b and c; Lines b and c have the same slope and the slope of line a is the negative reciprocal of the slopes of lines b and c.
4. $y = \frac{1}{3}x + 6$

Laurie's Notes Teacher Actions

? Write the *Core Concept*. "Why are vertical and horizontal lines a special case?" Horizontal lines have a slope of 0. The product of 0 and any number is 0. So, vertical and horizontal lines are a special case.

- **Turn and Talk:** Have partners work Example 3. Use *Popsicle Sticks* to solicit answers.
- Pose Example 4. Ask what the slope of the perpendicular line will be. Have students finish working the example with their partners.

Extra Example 5

Use the graph in Example 5. Another helicopter is at (8, 5). The shortest flight path to the shoreline is one that is perpendicular to the shoreline. Write an equation that represents this path.

$y = \frac{3}{2}x - 7$

MONITORING PROGRESS ANSWER

5. $y = -\frac{2}{3}x + 9$

Writing Equations for Real-Life Problems

EXAMPLE 5 Writing an Equation of a Perpendicular Line

The position of a helicopter search and rescue crew is shown in the graph. The shortest flight path to the shoreline is one that is perpendicular to the shoreline. Write an equation that represents this path.

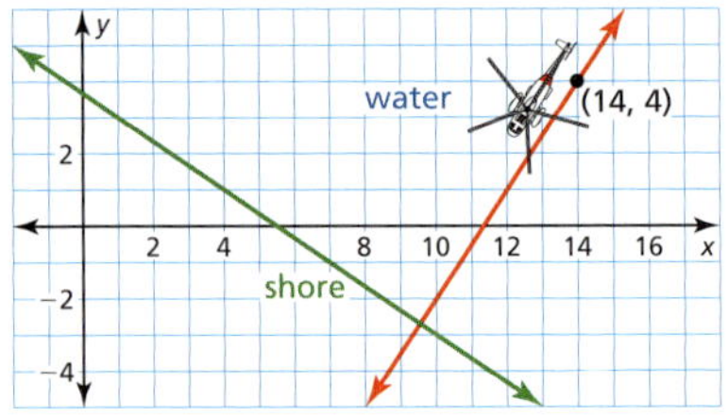

SOLUTION

1. **Understand the Problem** You can see the line that represents the shoreline. You know the coordinates of the helicopter. You are asked to write an equation that represents the shortest flight path to the shoreline.
2. **Make a Plan** Find the slope of the line that represents the shoreline. Use the negative reciprocal of this slope, the coordinates of the helicopter, and the point-slope form to write an equation.
3. **Solve the Problem**

Step 1 Find the slope of the line that represents the shoreline. The line passes through points (1, 3) and (4, 1). So, the slope is

$$m = \frac{1 - 3}{4 - 1} = -\frac{2}{3}.$$

Because the shoreline and shortest flight path are perpendicular, the slopes of their respective graphs are negative reciprocals. So, the slope of the graph of the shortest flight path is $\frac{3}{2}$.

Step 2 Use the slope $m = \frac{3}{2}$ and the point-slope form to write an equation of the shortest flight path that passes through (14, 4).

$y - y_1 = m(x - x_1)$ — Write the point-slope form.

$y - 4 = \frac{3}{2}(x - 14)$ — Substitute $\frac{3}{2}$ for m, 14 for x_1, and 4 for y_1.

$y - 4 = \frac{3}{2}x - 21$ — Distributive Property

$y = \frac{3}{2}x - 17$ — Write in slope-intercept form.

▶ An equation that represents the shortest flight path is $y = \frac{3}{2}x - 17$.

4. **Look Back** To check that your equation is correct, verify that (14, 4) is a solution of the equation.

$4 = \frac{3}{2}(14) - 17$ ✓

Monitoring Progress

Help in English and Spanish at *BigIdeasMath.com*

5. In Example 5, a boat is traveling parallel to the shoreline and passes through (9, 3). Write an equation that represents the path of the boat.

Laurie's Notes Teacher Actions

- ? **Always-Sometimes-Never True:** "The shortest distance from a point to a line is a perpendicular segment." always true
- **Make Sense of Problems and Persevere in Solving Them:** Step back and let students work through the example!
- ? **Assessing Question:** "What information will be needed to write the equation of the path of the helicopter? How can that information be found?" Listen and record the plan developed by students.

Closure

- **Writing Prompt:** To write an equation of the line that passes through (−2, 4) and is parallel to $y = -3x + 4$, ...

4.3 Exercises

Dynamic Solutions available at *BigIdeasMath.com*

Vocabulary and Core Concept Check

1. **COMPLETE THE SENTENCE** Two distinct nonvertical lines that have the same slope are ________.

2. **VOCABULARY** Two lines are perpendicular. The slope of one line is $-\frac{5}{7}$. What is the slope of the other line? Justify your answer.

Monitoring Progress and Modeling with Mathematics

In Exercises 3–8, determine which of the lines, if any, are parallel. Explain. *(See Example 1.)*

3.

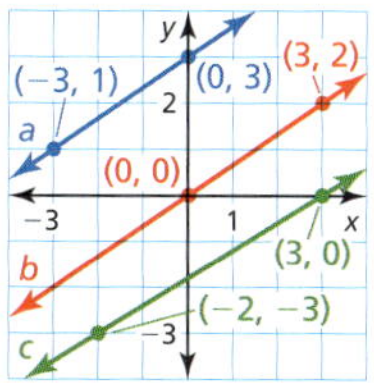

4.

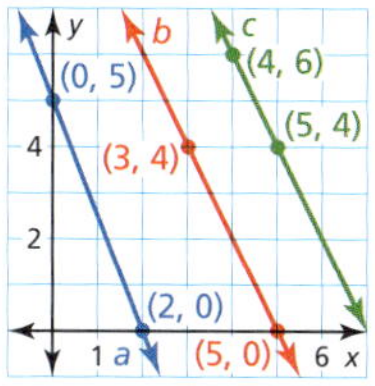

5. Line a passes through $(-1, -2)$ and $(1, 0)$.
Line b passes through $(4, 2)$ and $(2, -2)$.
Line c passes through $(0, 2)$ and $(-1, 1)$.

6. Line a passes through $(-1, 3)$ and $(1, 9)$.
Line b passes through $(-2, 12)$ and $(-1, 14)$.
Line c passes through $(3, 8)$ and $(6, 10)$.

7. Line a: $4y + x = 8$
Line b: $2y + x = 4$
Line c: $2y = -3x + 6$

8. Line a: $3y - x = 6$
Line b: $3y = x + 18$
Line c: $3y - 2x = 9$

In Exercises 9–12, write an equation of the line that passes through the given point and is parallel to the given line. *(See Example 2.)*

9. $(-1, 3)$; $y = 2x + 2$

10. $(1, 2)$; $y = -5x + 4$

11. $(18, 2)$; $3y - x = -12$

12. $(2, -5)$; $2y = 3x + 10$

In Exercises 13–18, determine which of the lines, if any, are parallel or perpendicular. Explain. *(See Example 3.)*

13.

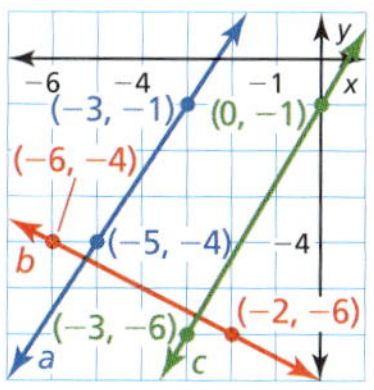

14.

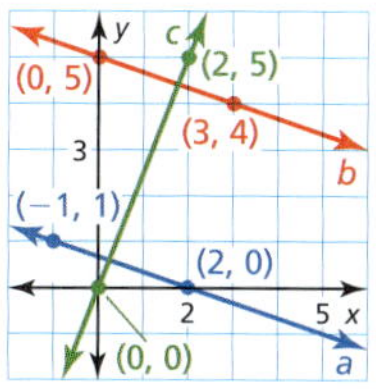

15. Line a passes through $(-2, 1)$ and $(0, 3)$.
Line b passes through $(4, 1)$ and $(6, 4)$.
Line c passes through $(1, 3)$ and $(4, 1)$.

16. Line a passes through $(2, 10)$ and $(4, 13)$.
Line b passes through $(4, 9)$ and $(6, 12)$.
Line c passes through $(2, 10)$ and $(4, 9)$.

17. Line a: $4x - 3y = 2$
Line b: $y = \frac{4}{3}x + 2$
Line c: $4y + 3x = 4$

18. Line a: $y = 6x - 2$
Line b: $6y = -x$
Line c: $y + 6x = 1$

In Exercises 19–22, write an equation of the line that passes through the given point and is perpendicular to the given line. *(See Example 4.)*

19. $(7, 10)$; $y = \frac{1}{2}x - 9$

20. $(-4, -1)$; $y = \frac{4}{3}x + 6$

21. $(-3, 3)$; $2y = 8x - 6$

22. $(8, 1)$; $2y + 4x = 12$

In Exercises 23 and 24, write an equation of the line that passes through the given point and is (a) parallel and (b) perpendicular to the given line.

23.

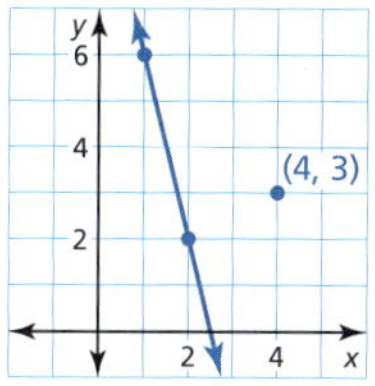

24.

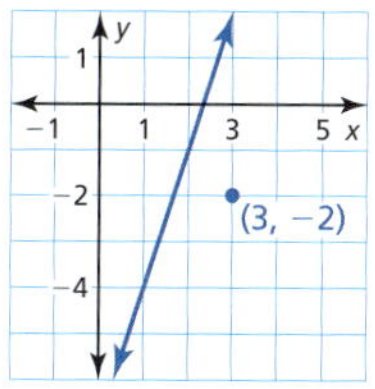

25. **ERROR ANALYSIS** Describe and correct the error in writing an equation of the line that passes through $(1, 3)$ and is parallel to the line $y = \frac{1}{4}x + 2$.

✗

$$y - y_1 = m(x - x_1)$$
$$y - 3 = -4(x - 1)$$
$$y - 3 = -4x + 4$$
$$y = -4x + 7$$

Assignment Guide and Homework Check

ASSIGNMENT

Basic: 1, 2, 3–27 odd, 31, 32, 37, 38

Average: 1, 2–26 even, 27, 28, 31, 32, 37, 38

Advanced: 1, 2, 6–12 even, 16–26 even, 27–38

HOMEWORK CHECK

Basic: 3, 9, 13, 19, 27

Average: 4, 10, 14, 20, 27

Advanced: 8, 12, 18, 22, 27

ANSWERS

1. parallel
2. $\frac{7}{5}$; $\frac{7}{5}$ is the negative reciprocal of $-\frac{5}{7}$.
3. lines a and b; They have the same slope.
4. lines b and c; They have the same slope.
5. lines a and c; They have the same slope.
6. none; None of the lines have the same slope.
7. none; None of the lines have the same slope.
8. lines a and b; They have the same slope.
9. $y = 2x + 5$
10. $y = -5x + 7$
11. $y = \frac{1}{3}x - 4$
12. $y = \frac{3}{2}x - 8$
13. None are parallel or perpendicular; None of the lines have the same slope or slopes that are negative reciprocals of each other.
14. Lines a and b are parallel; None are perpendicular; Lines a and b have the same slope and none of the lines have slopes that are negative reciprocals of each other.
15. None are parallel; Lines b and c are perpendicular; None of the lines have the same slope and the slope of line b is the negative reciprocal of the slope of line c.
16. Lines a and b are parallel; None are perpendicular; Lines a and b have the same slope and none of the lines have slopes that are negative reciprocals of each other.
17. Lines a and b are parallel; Line c is perpendicular to lines a and b; Lines a and b have the same slope and the slope of line c is the negative reciprocal of the slopes of lines a and b.
18. None are parallel; Lines a and b are perpendicular; None of the lines have the same slope and the slope of line a is the negative reciprocal of the slope of line b.
19. $y = -2x + 24$
20. $y = -\frac{3}{4}x - 4$
21. $y = -\frac{1}{4}x + \frac{9}{4}$
22. $y = \frac{1}{2}x - 3$
23. a. $y = -4x + 19$
    b. $y = \frac{1}{4}x + 2$
24. a. $y = 3x - 11$
    b. $y = -\frac{1}{3}x - 1$
25. Parallel lines have the same slope, not negative reciprocal slopes; $y - 3 = \frac{1}{4}(x - 1)$; $y - 3 = \frac{1}{4}x - \frac{1}{4}$; $y = \frac{1}{4}x + \frac{11}{4}$

Dynamic Teaching Tools

- Dynamic Assessment & Progress Monitoring Tool
- Interactive Whiteboard Lesson Library
- Dynamic Classroom with Dynamic Investigations

ANSWERS

26. Perpendicular lines have negative reciprocal slopes, not just reciprocal slopes; $y - (-5) = -3(x - 4)$; $y + 5 = -3x + 12$; $y = -3x + 7$

27. $y = -\frac{3}{4}x + \frac{3}{2}$

28. $y = \frac{4}{3}x - \frac{1}{3}$

29. **a.** yes; Opposite sides are parallel.

b. no; Adjacent sides are not perpendicular.

30. $a = -\frac{2}{3}$; $a = 6$

31. no; The lines that form the angle are not perpendicular.

32–38. See Additional Answers.

Mini-Assessment

1. Line a passes through $(-1, -5)$ and $(1, 3)$. Line b passes through $(-3, -7)$ and $(1, 9)$. Line c passes through $(0, -2)$ and $(4, -3)$. Which lines are parallel or perpendicular? Explain. Lines a and b (slopes: 4) are parallel. Line c (slope: $-\frac{1}{4}$) is perpendicular to lines a and b.
2. Write an equation of the line that passes through $(-2, 3)$ and is parallel to the line $y = 2x - 1$. $y = 2x + 7$
3. Write an equation of the line that passes through $(0, -2)$ and is perpendicular to the line $y = \frac{1}{5}x - 2$. $y = -5x - 2$
4. A road is constructed perpendicular to South Street. Write an equation that represents this new road.

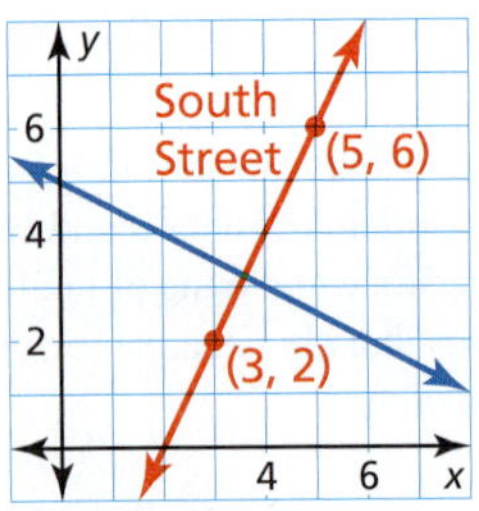

$y = -\frac{1}{2}x + 5$

26. ERROR ANALYSIS Describe and correct the error in writing an equation of the line that passes through $(4, -5)$ and is perpendicular to the line $y = \frac{1}{3}x + 5$.

✗
$$y - y_1 = m(x - x_1)$$
$$y - (-5) = 3(x - 4)$$
$$y + 5 = 3x - 12$$
$$y = 3x - 17$$

27. MODELING WITH MATHEMATICS A city water department is proposing the construction of a new water pipe, as shown. The new pipe will be perpendicular to the old pipe. Write an equation that represents the new pipe. *(See Example 5.)*

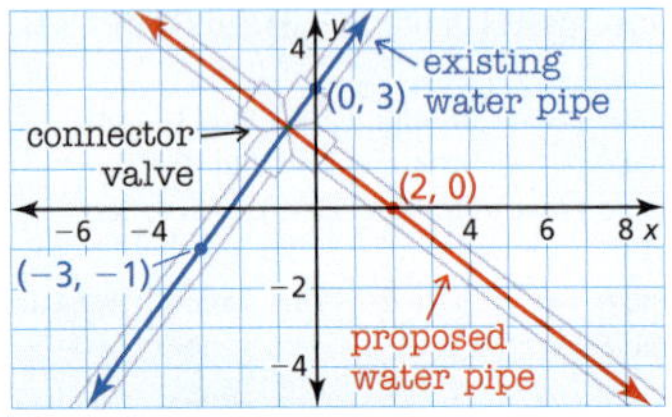

28. MODELING WITH MATHEMATICS A parks and recreation department is constructing a new bike path. The path will be parallel to the railroad tracks shown and pass through the parking area at the point (4, 5). Write an equation that represents the path.

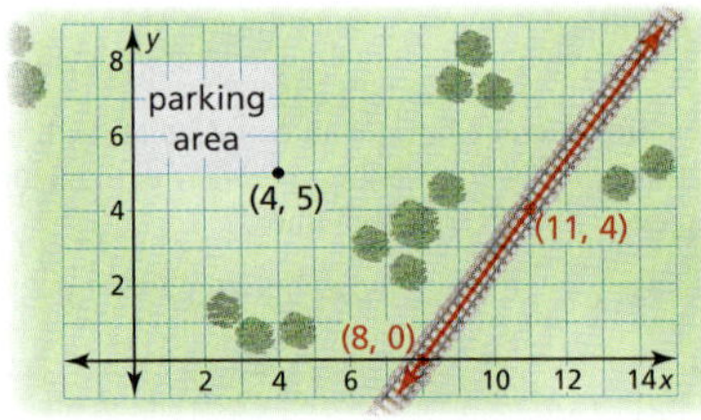

29. MATHEMATICAL CONNECTIONS The vertices of a quadrilateral are $A(2, 2)$, $B(6, 4)$, $C(8, 10)$, and $D(4, 8)$.

a. Is quadrilateral $ABCD$ a parallelogram? Explain.

b. Is quadrilateral $ABCD$ a rectangle? Explain.

30. USING STRUCTURE For what value of a are the graphs of $6y = -2x + 4$ and $2y = ax - 5$ parallel? perpendicular?

31. MAKING AN ARGUMENT A hockey puck leaves the blade of a hockey stick, bounces off a wall, and travels in a new direction, as shown. Your friend claims the path of the puck forms a right angle. Is your friend correct? Explain.

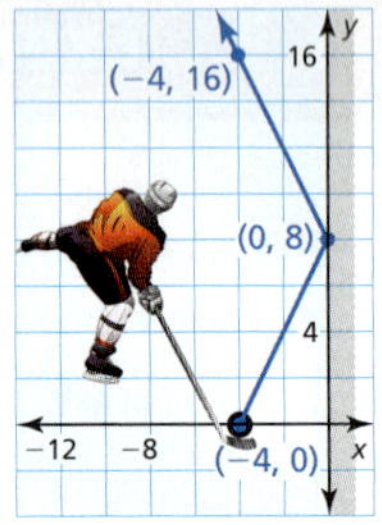

32. HOW DO YOU SEE IT? A softball academy charges students an initial registration fee plus a monthly fee. The graph shows the total amounts paid by two students over a 4-month period. The lines are parallel.

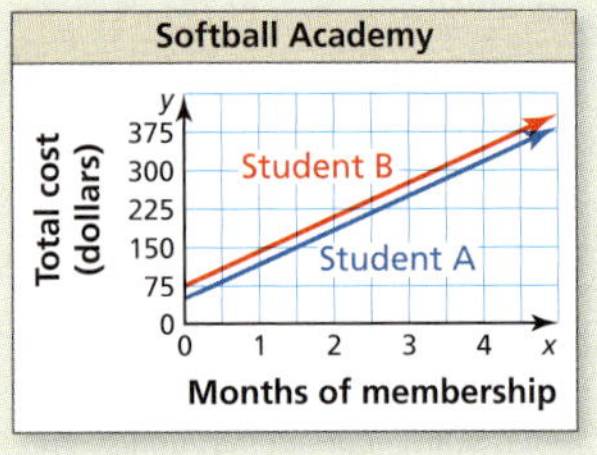

a. Did one of the students pay a greater registration fee? Explain.

b. Did one of the students pay a greater monthly fee? Explain.

REASONING **In Exercises 33–35, determine whether the statement is *always*, *sometimes*, or *never* true. Explain your reasoning.**

33. Two lines with positive slopes are perpendicular.

34. A vertical line is parallel to the y-axis.

35. Two lines with the same y-intercept are perpendicular.

36. THOUGHT PROVOKING You are designing a new logo for your math club. Your teacher asks you to include at least one pair of parallel lines and at least one pair of perpendicular lines. Sketch your logo in a coordinate plane. Write the equations of the parallel and perpendicular lines.

Maintaining Mathematical Proficiency

Reviewing what you learned in previous grades and lessons

Determine whether the relation is a function. Explain. *(Section 3.1)*

37. (3, 6), (4, 8), (5, 10), (6, 9), (7, 14)

38. (−1, 6), (1, 4), (−1, 2), (1, 6), (−1, 5)

If students need help...	*If students got it...*
Resources by Chapter • Practice A and Practice B • Puzzle Time	Resources by Chapter • Enrichment and Extension • Cumulative Review
Student Journal • Practice	Start the *next* Section
Differentiating the Lesson Skills Review Handbook	

4.1–4.3 What Did You Learn?

Core Vocabulary

linear model, *p. 178*
point-slope form, *p. 182*
parallel lines, *p. 188*
perpendicular lines, *p. 189*

Core Concepts

Section 4.1
Using Slope-Intercept Form, *p. 176*

Section 4.2
Using Point-Slope Form, *p. 182*

Section 4.3
Parallel Lines and Slopes, *p. 188*
Perpendicular Lines and Slopes, *p. 189*

Mathematical Practices

1. How can you explain to yourself the meaning of the graph in Exercise 36 on page 180?
2. How did you use the structure of the equations in Exercise 39 on page 186 to make a conjecture?
3. How did you use the diagram in Exercise 31 on page 192 to determine whether your friend was correct?

Study Skills

Getting Actively Involved in Class

If you do not understand something at all and do not even know how to phrase a question, just ask for clarification. You might say something like, "Could you please explain the steps in this problem one more time?"

If your teacher asks for someone to go up to the board, volunteer. The student at the board often receives additional attention and instruction to complete the problem.

Dynamic Teaching Tools
- Dynamic Assessment & Progress Monitoring Tool
- Interactive Whiteboard Lesson Library
- Dynamic Classroom with Dynamic Investigations

ANSWERS

1. The graph shows the increase in U.S. box office revenue over time.
2. The slopes were the same, so the lines were parallel and the constants in the point-slope form indicated similar translations.
3. The points on the diagram were used to determine the slopes of both parts of the path. The slopes were not negative reciprocals, so the lines were not perpendicular.

ANSWERS

1. $y = 5x - 2$
2. $y = -\frac{1}{3}x + 5$
3. $y = -2x$
4. $y - 5 = -2(x + 2)$ or $y + 1 = -2(x - 1)$
5. $y + 2 = \frac{1}{5}(x + 3)$ or $y + 1 = \frac{1}{5}(x - 2)$
6. $y = \frac{4}{3}(x - 1)$ or $y - 4 = \frac{4}{3}(x - 4)$
7. $f(x) = -x + 2$
8. $f(x) = -6$
9. $f(x) = 5x + 13$
10. None are parallel; Lines a and b are perpendicular; None of the lines have the same slope and the slope of line a is the negative reciprocal of the slope of line b.
11. Lines b and c are parallel; None are perpendicular; Lines b and c have the same slope and none of the lines have negative reciprocal slopes.
12. **a.** $y = 3x - 16$
    b. $y = -\frac{1}{3}x + 4$
13. **a.** $y = -\frac{1}{2}x - 4$
    b. $y = 2x + 1$
14. **a.** $y = -4x - 16$
    b. $y = \frac{1}{4}x + 1$
15. **a.** $C = 44m + 48$
    b. $312
    c. the original company; You can keep the site running for 13 months at the original company. You can keep the site running for 10 months at the new company.
16. yes; The amount of water decreases at a constant rate; $W = -\frac{5}{2}t + 175$

4.1–4.3 Quiz

Write an equation of the line in slope-intercept form. *(Section 4.1)*

1.

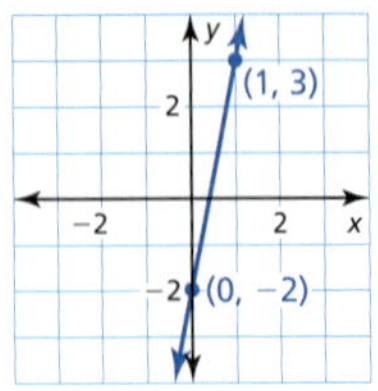

2.

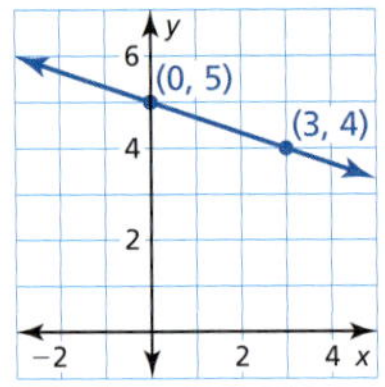

3.

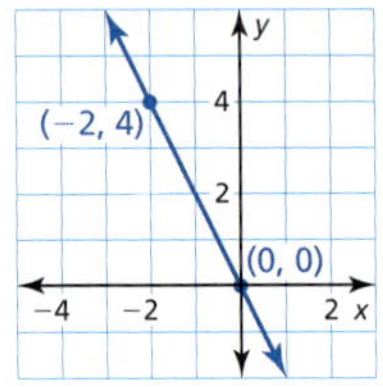

Write an equation in point-slope form of the line that passes through the given points. *(Section 4.2)*

4. $(-2, 5), (1, -1)$ **5.** $(-3, -2), (2, -1)$ **6.** $(1, 0), (4, 4)$

Write a linear function f with the given values. *(Section 4.1 and Section 4.2)*

7. $f(0) = 2, f(5) = -3$ **8.** $f(-1) = -6, f(4) = -6$ **9.** $f(-3) = -2, f(-2) = 3$

Determine which of the lines, if any, are parallel or perpendicular. Explain. *(Section 4.3)*

10. Line a passes through $(-2, 2)$ and $(2, 1)$.
Line b passes through $(1, -8)$ and $(3, 0)$.
Line c passes through $(-4, -3)$ and $(0, -2)$.

11. Line a: $2x + 6y = -12$
Line b: $y = \frac{3}{2}x - 5$
Line c: $3x - 2y = -4$

Write an equation of the line that passes through the given point and is (a) parallel and (b) perpendicular to the given line. *(Section 4.3)*

12.

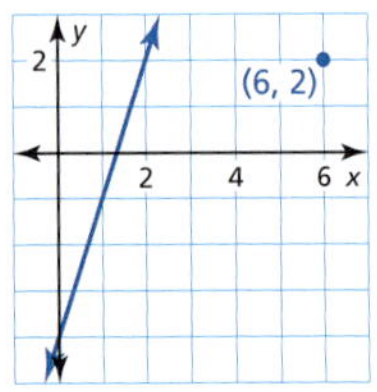

13.

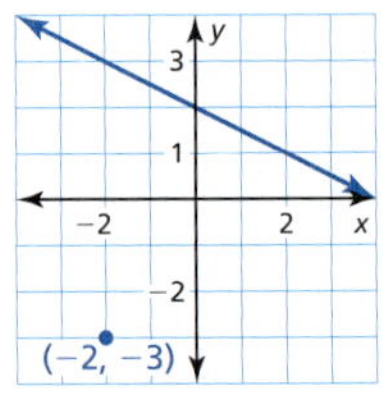

14.

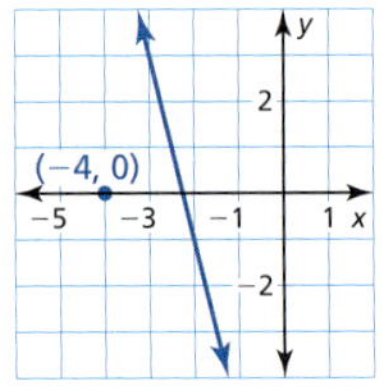

15. A website hosting company charges an initial fee of $48 to set up a website. The company charges $44 per month to maintain the website. *(Section 4.1)*

a. Write a linear model that represents the total cost of setting up and maintaining a website as a function of the number of months it is maintained.

b. Find the total cost of setting up a website and maintaining it for 6 months.

c. A different website hosting company charges $62 per month to maintain a website, but there is no initial set-up fee. You have $620. At which company can you set up and maintain a website for the greatest amount of time? Explain.

16. The table shows the amount of water remaining in a water tank as it drains. Can the situation be modeled by a linear equation? Explain. If possible, write a linear model that represents the amount of water remaining in the tank as a function of time. *(Section 4.2)*

Time (minutes)	8	10	12	14	16
Water (gallons)	155	150	145	140	135

Laurie's Notes

Overview of Section 4.4

Introduction

- **Note:** Use the *Motivate* before the formal lesson, rather than before the explorations.
- The goals of this lesson are for students to understand how to read and interpret a scatter plot and, when appropriate, fit a line to the data. The explorations present two plotted data sets for students to work with, one with many more data points than the other. In the formal lesson, the concept of correlation is introduced with a connection made to slope.
- The purpose of writing the equation of the line of fit is to make predictions. The equation becomes a model for the data set, describing its behavior. Additionally, students should always consider the slope and y-intercept and then interpret each in the context of the situation in which it makes sense.

Resources

- A rough scatter plot on grid paper or a whiteboard allows students to determine whether there is a linear correlation. To fit a line, students should not be working from a rough sketch. Students should have the experience of fitting a line manually (graph paper) and with technology. Fitting it manually can be modeled using spaghetti and then using a straightedge to draw the line. Regression is not introduced in this lesson.

Teaching Strategy

- When working with real-life data, it is very possible that no function will be a perfect fit. The strategy is to find a function that best fits the data, and the first step is to make a scatter plot using a good viewing window for the data. Students should learn to look at the tendency of the data as displayed in a scatter plot before deciding what type of model to fit to the data.

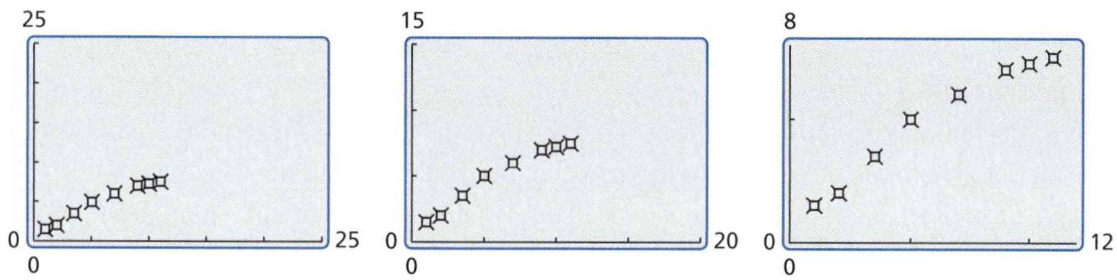

- When deciding how to set the window, it is wise to fill as much of the viewing window as possible. If there is a curvature, it will be more noticeable in the window that is the tightest.

Pacing Suggestion

- If students work through both explorations, you might elect to integrate correlation in the discussions and then check for understanding with a few of the *Monitoring Progress* questions. The goals of the formal lesson would be met.

Dynamic Teaching Tools

- Dynamic Assessment & Progress Monitoring Tool
- Lesson Planning Tool
- Interactive Whiteboard Lesson Library
- Dynamic Classroom with Dynamic Investigations

What Your Students Will Learn

- Interpret relationships between two data sets using scatter plots.
- Describe relationships as having positive, negative, or no correlation.
- Draw lines of fit on scatter plots and write equations to model data.

Laurie's Notes

Exploration

Motivate

- **Preparation:** Stop by any fast food restaurant to pick up a pamphlet, or go online to find nutritional information about the menu items.

? "Do you think there is a relationship between the grams of fat and the number of calories in a sandwich?" yes

- Share the information about a few of the sandwiches from your pamphlet or printout to confirm students' opinions.

Exploration 1

- In a data set of this type, it is easy for students to be confused about what each ordered pair represents. Each ordered pair is a married couple: (husband's age, wife's age). The two people may be the same age, the husband may be older, or the husband may be younger. The data are not about how old they were when they married, simply how old they were when surveyed. There is a spread of about 50 years from the youngest couple to the oldest couple.
- **Reason Abstractly and Quantitatively:** Ask questions about the data set to ensure that students understand what is being represented.
- **Attend to Precision:** Students will ask what a line "that approximates the data" means. You should explain that it is a line that approximates the trend of the data points. A line cannot pass through every single point, so there will be data points roughly spread on each side of the line. Use a straightedge to lightly draw the line.
- **Selective Responses:** As you circulate, make note of the different strategies employed for writing the equation of the line. Use your notes to select the order in which responses are shared and discussed.
- Discuss what slopes greater than 1 or slopes less than 1 mean.

Exploration 2

- The goal for this exploration is the same as the first, the difference being the size of the data set. Here, there can be a longer discussion about the placement of the line of fit.
- **Note:** The x-values in the equation represent the number of years since 1960. Using this technique means that instead of working with a 4-digit number, the year 1970 can be represented as 10.
- Have partners work together on the exploration. Record equations found by different pairs. Discuss similarities and any major differences.

? **Probing Questions:** "What does the slope mean in the context of the situation?" the average rate of change of the median age of the women each year "What has been the trend?" Women are waiting longer to get married. "Would you trust your equation to predict the median age of American women at their first marriage in 1890?" most likely, no

Communicate Your Answer

- Listen for an explanation that includes the importance of drawing the line to begin with, followed by calculating the equation of the line, which can be used to make predictions.

Connecting to Next Step

- Students have now developed a sense of how to fit a line to a scatter plot. In the formal lesson, they will also discuss correlation.

4.4 Scatter Plots and Lines of Fit

Essential Question How can you use a scatter plot and a line of fit to make conclusions about data?

A **scatter plot** is a graph that shows the relationship between two data sets. The two data sets are graphed as ordered pairs in a coordinate plane.

EXPLORATION 1 **Finding a Line of Fit**

Work with a partner. A survey was taken of 179 married couples. Each person was asked his or her age. The scatter plot shows the results.

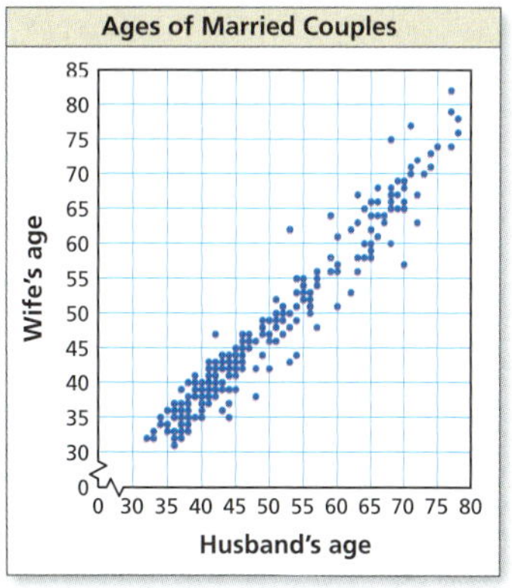

a. Draw a line that approximates the data. Write an equation of the line. Explain the method you used.

b. What conclusions can you make from the equation you wrote? Explain your reasoning.

> **REASONING QUANTITATIVELY**
> To be proficient in math, you need to make sense of quantities and their relationships in problem situations.

EXPLORATION 2 **Finding a Line of Fit**

Work with a partner. The scatter plot shows the median ages of American women at their first marriage for selected years from 1960 through 2010.

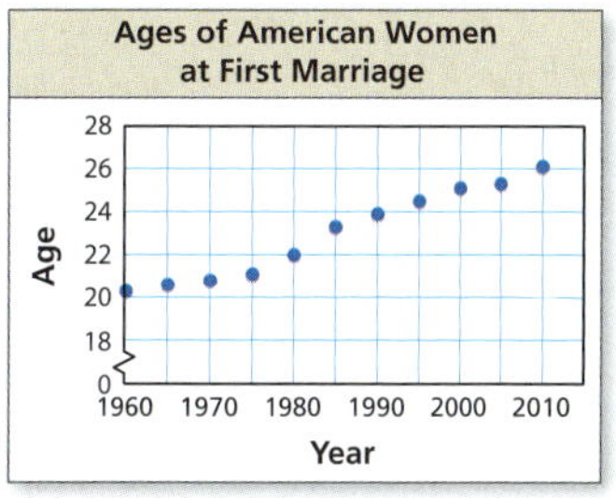

a. Draw a line that approximates the data. Write an equation of the line. Let x represent the number of years since 1960. Explain the method you used.

b. What conclusions can you make from the equation you wrote?

c. Use your equation to predict the median age of American women at their first marriage in the year 2020.

Communicate Your Answer

3. How can you use a scatter plot and a line of fit to make conclusions about data?

4. Use the Internet or some other reference to find a scatter plot of real-life data that is different from those given above. Then draw a line that approximates the data and write an equation of the line. Explain the method you used.

Dynamic Teaching Tools

- Dynamic Assessment & Progress Monitoring Tool
- Lesson Planning Tool
- Interactive Whiteboard Lesson Library
- Dynamic Classroom with Dynamic Investigations

ANSWERS

1. a. *Sample answer:*

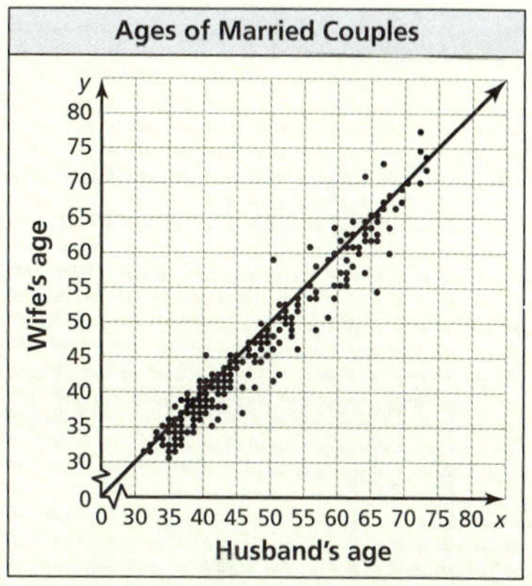

$y = x$; Estimate a line through the middle of the plotted data, then use two points on the line to find the equation.

b. *Sample answers:* The ages of the husband and wife in a married couple tend to be about the same; The equation shows they are equal.

2. a. *Sample answer:*

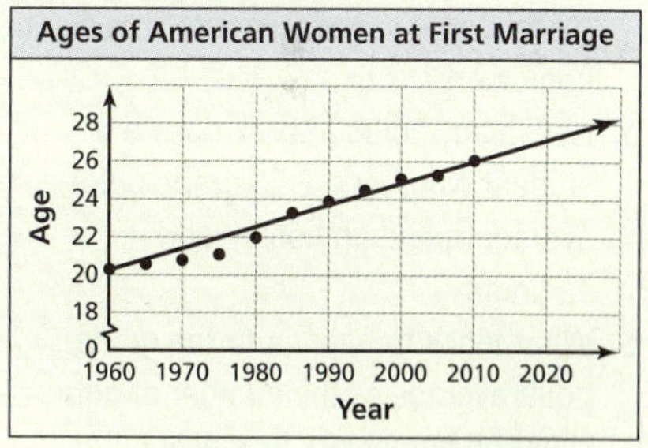

$y = 0.1x + 20.1$; Estimate a line through the middle of the plotted data, then use two points on that line to find the equation.

b. *Sample answer:* The median age at which American women first marry increases about 1 year every 10 years.

c. *Sample answer:* about 26.1 years old

3. Draw a line of fit on the scatter plot, find its equation, and interpret the slope and y-intercept.

4. See Additional Answers.

4.4 Lesson

Core Vocabulary

scatter plot, *p. 196*
correlation, *p. 197*
line of fit, *p. 198*

What You Will Learn

- Interpret scatter plots.
- Identify correlations between data sets.
- Use lines of fit to model data.

Interpreting Scatter Plots

Core Concept

Scatter Plot

A **scatter plot** is a graph that shows the relationship between two data sets. The two data sets are graphed as ordered pairs in a coordinate plane. Scatter plots can show trends in the data.

EXAMPLE 1 Interpreting a Scatter Plot

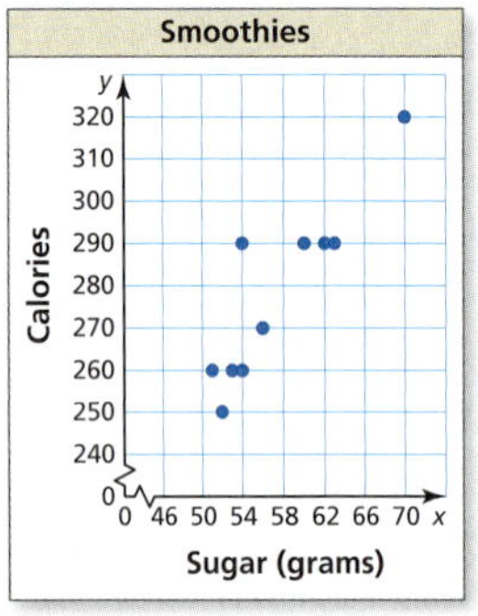

The scatter plot shows the amounts x (in grams) of sugar and the numbers y of calories in 10 smoothies.

a. How many calories are in the smoothie that contains 56 grams of sugar?

b. How many grams of sugar are in the smoothie that contains 320 calories?

c. What tends to happen to the number of calories as the number of grams of sugar increases?

SOLUTION

a. Draw a horizontal line from the point that has an x-value of 56. It crosses the y-axis at 270.

▶ So, the smoothie has 270 calories.

b. Draw a vertical line from the point that has a y-value of 320. It crosses the x-axis at 70.

▶ So, the smoothie has 70 grams of sugar.

c. Looking at the graph, the plotted points go up from left to right.

▶ So, as the number of grams of sugar increases, the number of calories increases.

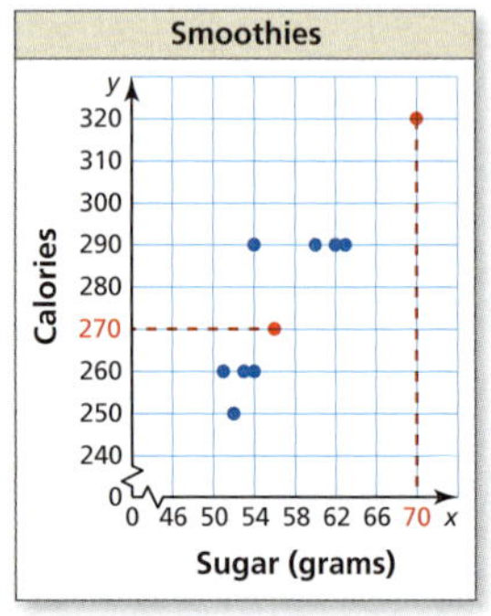

Monitoring Progress

Help in English and Spanish at *BigIdeasMath.com*

1. How many calories are in the smoothie that contains 51 grams of sugar?

2. How many grams of sugar are in the smoothie that contains 250 calories?

Differentiated Instruction

Visual

Some students may have difficulty naming the coordinates of points on a scatter plot that are not located on intersecting grid lines. Have these students work with a partner to identify coordinates of points and to plot additional data on a scatter plot.

Extra Example 1

The scatter plot shows the amounts of time x (in hours) that 8 students spend on homework each week and their grade point averages y.

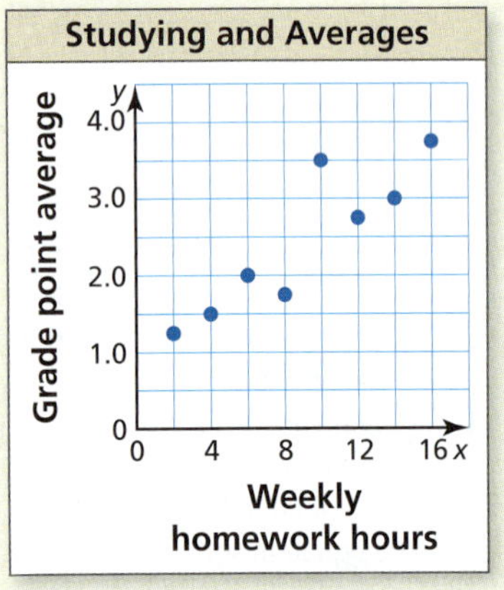

a. What is the grade point average of the student who spends 6 hours a week on homework? 2.0

b. How many hours a week does the student who has a 2.75 grade point average spend on homework? 12 hours

c. What tends to happen to the grade point average as the number of hours spent on homework increases? The grade point average increases.

MONITORING PROGRESS ANSWERS

1. 260 cal

2. about 52 g

Laurie's Notes Teacher Actions

- **Make Sense of Problems and Persevere in Solving Them:** This example helps students understand how a *scatter plot* is read and interpreted. Discuss the labels on the axes and what an ordered pair represents: (grams of sugar, number of calories) in a smoothie. There are 10 different smoothies represented.
- **Big Idea:** A scatter plot allows you to see trends in the data. You read a scatter plot from left to right. As the x-coordinate increases, is the y-coordinate increasing, decreasing, staying the same, or is there no pattern?

Identifying Correlations between Data Sets

STUDY TIP

You can think of a positive correlation as having a positive slope and a negative correlation as having a negative slope.

A **correlation** is a relationship between data sets. You can use a scatter plot to describe the correlation between data.

Positive Correlation

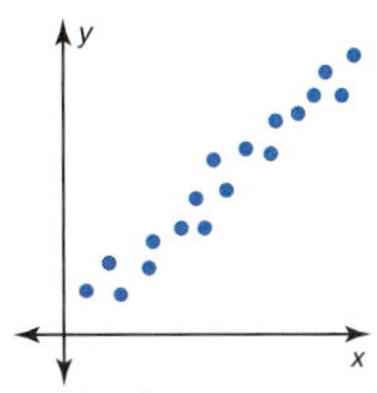

As *x* increases, *y* increases.

Negative Correlation

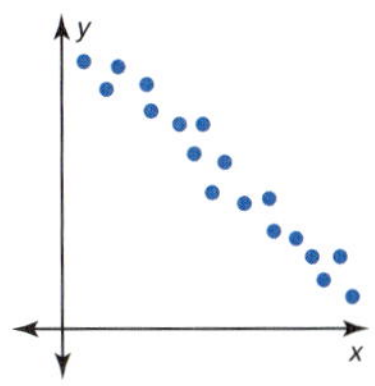

As *x* increases, *y* decreases.

No Correlation

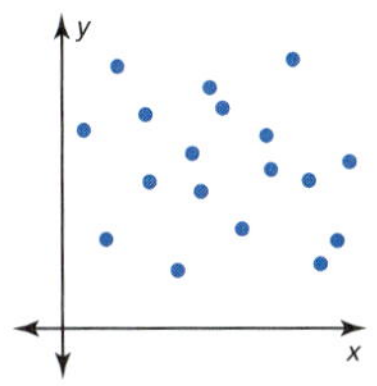

The points show no pattern.

EXAMPLE 2 Identifying Correlations

Tell whether the data show a *positive*, a *negative*, or *no* correlation.

a. age and vehicles owned

b. temperature and coat sales at a store

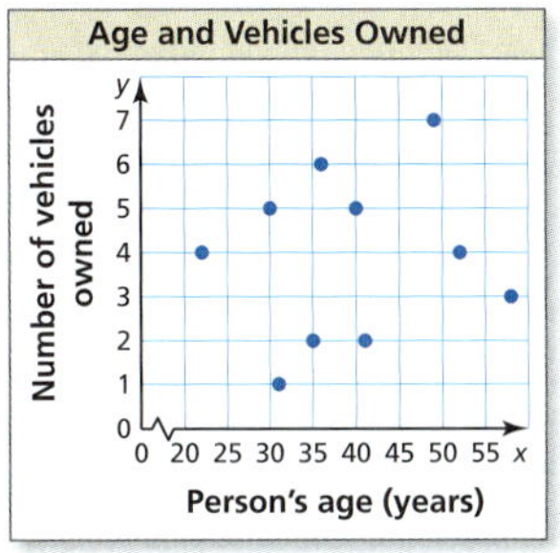

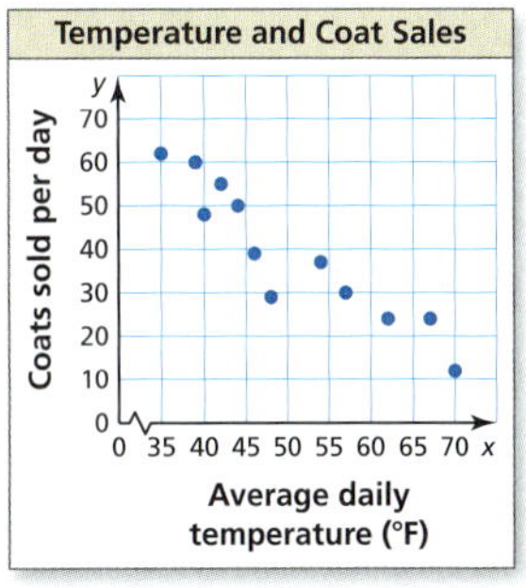

SOLUTION

a. The points show no pattern. The number of vehicles owned does not depend on a person's age.

▶ So, the scatter plot shows no correlation.

b. As the average temperature increases, the number of coats sold decreases.

▶ So, the scatter plot shows a negative correlation.

Monitoring Progress

Help in English and Spanish at *BigIdeasMath.com*

Make a scatter plot of the data. Tell whether the data show a *positive*, a *negative*, or *no* correlation.

3.

Temperature (°F), x	82	78	68	87	75	71	92	84
Attendees (thousands), y	4.5	4.0	1.7	5.5	3.8	2.9	4.7	5.3

4.

Age of a car (years), x	1	2	3	4	5	6	7	8
Value (thousands), y	\$24	\$21	\$19	\$18	\$15	\$12	\$8	\$7

Extra Example 2

Tell whether the data show a *positive*, a *negative*, or *no* correlation.

a. stopping distance of an automobile

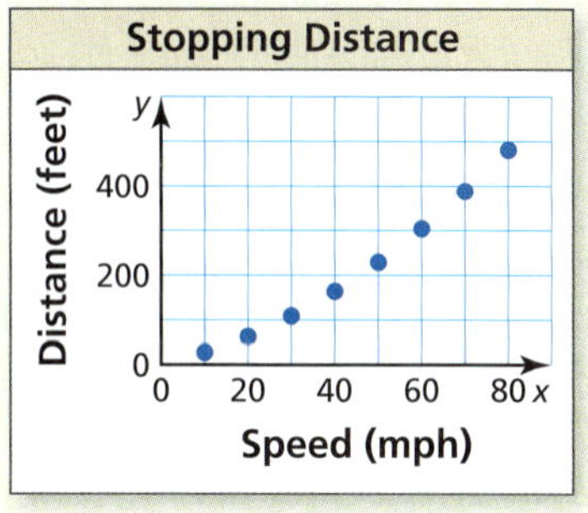

The scatter plot shows a positive correlation.

b. birth month and hours of sleep

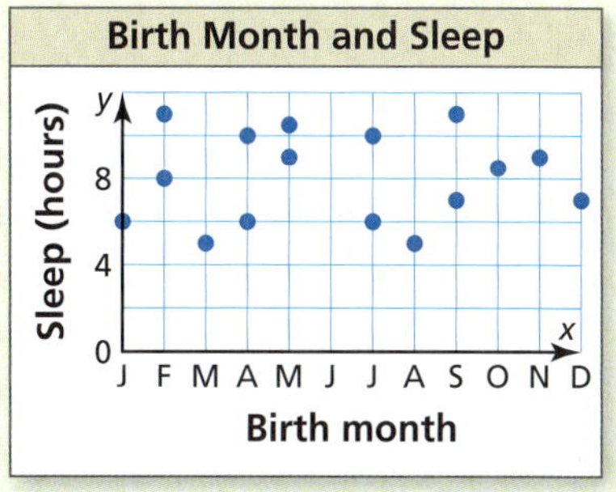

The scatter plot shows no correlation.

MONITORING PROGRESS ANSWERS

3. See Additional Answers.

4.

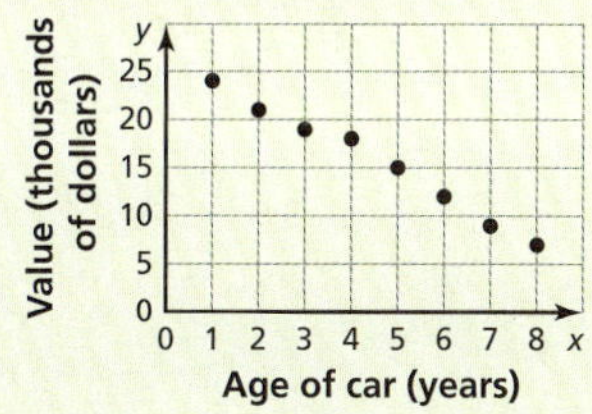

negative

Laurie's Notes Teacher Actions

- Introduce and discuss the concept of *correlation*. Note the *Study Tip* related to slope. Both explorations had positive correlations.

? "Should there be a correlation between a person's age and the number of vehicles owned?" no
"Should there be a correlation between the temperature and the number of coats sold?" yes

- **Use Appropriate Tools Strategically:** Students could use whiteboards or graphing calculators to make a scatter plot of the data in Questions 3 and 4.

Extra Example 3

The table shows the salaries y (in thousands of dollars) of 5 architects and the numbers x of years of experience. Write an equation that models the salaries as a function of the number of years of experience. Interpret the slope and y-intercept of the line of fit.

x	1	2	3	4	5
y	50	55	56	60	62

Sample answer: Using points (3, 56) and (5, 62), an equation of the line of fit is $y = 3x + 47$. The slope 3 means the salaries are increasing by about \$3000 each year. The y-intercept 47 means an architect's starting salary is about \$47,000.

MONITORING PROGRESS ANSWER

5. *Sample answer:* $y = 0.2x$; The slope of 0.2 means the monthly car payment increases by \$20 for every \$100 increase in monthly income. The y-intercept of 0 means a person with no monthly income has no car payment.

Using Lines of Fit to Model Data

When data show a positive or negative correlation, you can model the *trend* in the data using a line of fit. A **line of fit** is a line drawn on a scatter plot that is close to most of the data points.

STUDY TIP
A line of fit is also called a *trend line*.

Core Concept

Using a Line of Fit to Model Data

Step 1 Make a scatter plot of the data.

Step 2 Decide whether the data can be modeled by a line.

Step 3 Draw a line that appears to fit the data closely. There should be approximately as many points above the line as below it.

Step 4 Write an equation using two points on the line. The points do not have to represent actual data pairs, but they must lie on the line of fit.

EXAMPLE 3 Finding a Line of Fit

The table shows the weekly sales of a DVD and the number of weeks since its release. Write an equation that models the DVD sales as a function of the number of weeks since its release. Interpret the slope and y-intercept of the line of fit.

Week, x	1	2	3	4	5	6	7	8
Sales (millions), y	\$19	\$15	\$13	\$11	\$10	\$8	\$7	\$5

SOLUTION

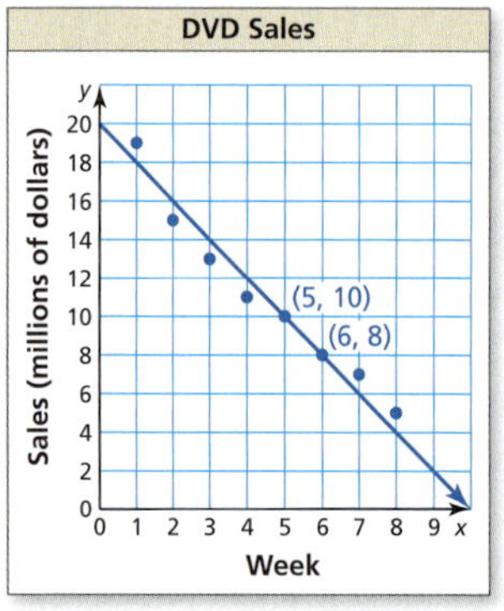

Step 1 Make a scatter plot of the data.

Step 2 Decide whether the data can be modeled by a line. Because the scatter plot shows a negative correlation, you can fit a line to the data.

Step 3 Draw a line that appears to fit the data closely.

Step 4 Write an equation using two points on the line. Use (5, 10) and (6, 8).

The slope of the line is $m = \frac{8 - 10}{6 - 5} = -2$.

Use the slope $m = -2$ and the point (6, 8) to write an equation of the line.

$y - y_1 = m(x - x_1)$ Write the point-slope form.

$y - 8 = -2(x - 6)$ Substitute -2 for m, 6 for x_1, and 8 for y_1.

$y = -2x + 20$ Solve for y.

▶ An equation of the line of fit is $y = -2x + 20$. The slope of the line is -2. This means the sales are decreasing by about \$2 million each week. The y-intercept is 20. The y-intercept has no meaning in this context because there are no sales in week 0.

Monitoring Progress

Help in English and Spanish at *BigIdeasMath.com*

5. The following data pairs show the monthly income x (in dollars) and the monthly car payment y (in dollars) of six people: (2100, 410), (1650, 315), (1950, 405), (1500, 295), (2250, 440), and (1800, 375). Write an equation that models the monthly car payment as a function of the monthly income. Interpret the slope and y-intercept of the line of fit.

Laurie's Notes Teacher Actions

- **Use Appropriate Tools Strategically:** Define and discuss a *line of fit*. You can model this with a piece of spaghetti that you overlay on a transparency. Some graphing technology has moveable lines that can also be used to model the process of eyeing a line of fit.
- Discuss the steps involved in fitting a line. See the *Teaching Strategy* on page T-194.
- **Turn and Talk:** Have partners work through the example.

Closure

- Have students fit a line for *Monitoring Progress* Questions 3 and/or 4. Check students' work.

4.4 Exercises

Dynamic Solutions available at *BigIdeasMath.com*

Vocabulary and Core Concept Check

1. **COMPLETE THE SENTENCE** When data show a positive correlation, the dependent variable tends to ___________ as the independent variable increases.

2. **VOCABULARY** What is a line of fit?

Monitoring Progress and Modeling with Mathematics

In Exercises 3–6, use the scatter plot to fill in the missing coordinate of the ordered pair.

3. (16, ▢)

4. (3, ▢)

5. (▢, 12)

6. (▢, 17)

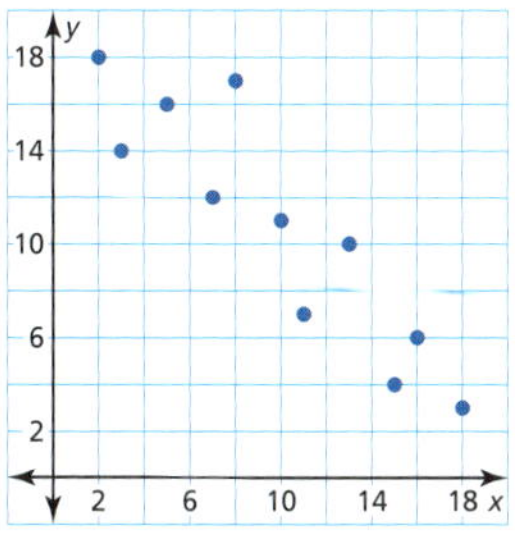

7. **INTERPRETING A SCATTER PLOT** The scatter plot shows the hard drive capacities (in gigabytes) and the prices (in dollars) of 10 laptops. *(See Example 1.)*

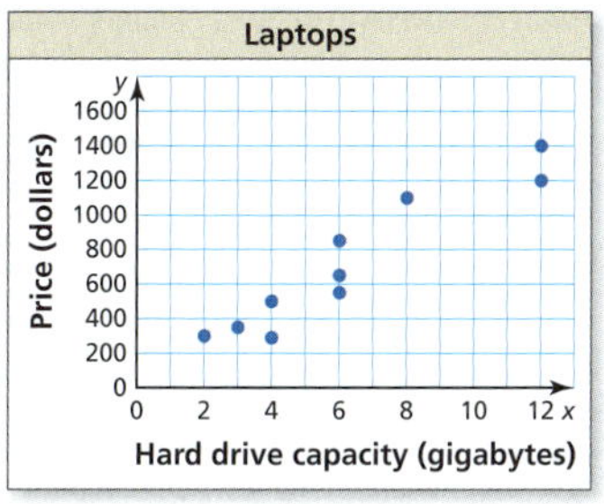

   a. What is the price of the laptop with a hard drive capacity of 8 gigabytes?

   b. What is the hard drive capacity of the $1200 laptop?

   c. What tends to happen to the price as the hard drive capacity increases?

8. **INTERPRETING A SCATTER PLOT** The scatter plot shows the earned run averages and the winning percentages of eight pitchers on a baseball team.

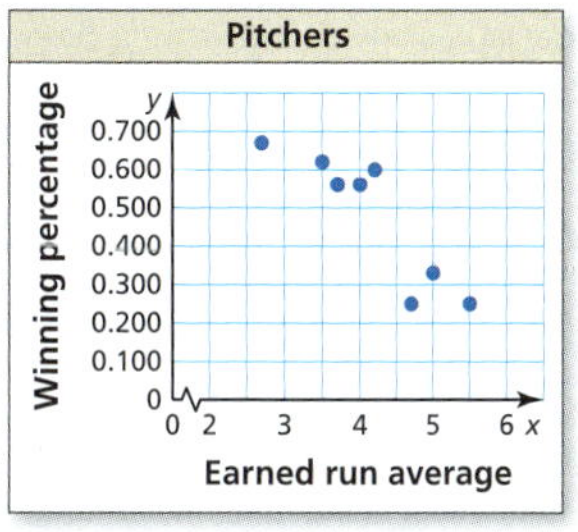

   a. What is the winning percentage of the pitcher with an earned run average of 4.2?

   b. What is the earned run average of the pitcher with a winning percentage of 0.33?

   c. What tends to happen to the winning percentage as the earned run average increases?

In Exercises 9–12, tell whether x and y show a *positive*, a *negative*, or *no* correlation. *(See Example 2.)*

9. 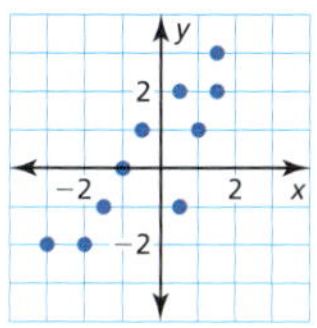

10. 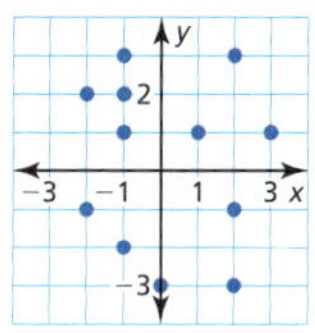

11. 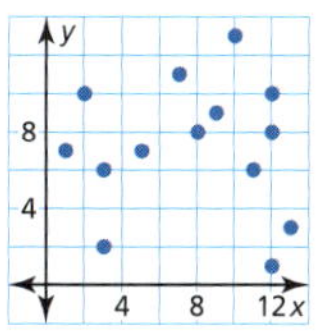

12. 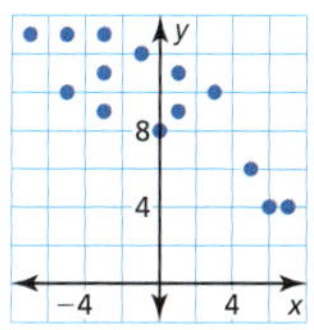

Assignment Guide and Homework Check

ASSIGNMENT

Basic: 1, 2, 3–15 odd, 18, 22, 25–28

Average: 1, 2–18 even, 22, 25–28

Advanced: 1, 2, 8–16 even, 17–28

HOMEWORK CHECK

Basic: 7, 9, 13, 15

Average: 8, 10, 14, 16

Advanced: 8, 12, 14, 16

ANSWERS

1. increase
2. a line drawn on a scatter plot that is close to most of the data points
3. 6
4. 14
5. 7
6. 8
7. a. $1100
   b. 12 GB
   c. increases
8. a. 0.600
   b. 5
   c. decreases
9. positive
10. no
11. no
12. negative

Dynamic Teaching Tools

- Dynamic Assessment & Progress Monitoring Tool
- Interactive Whiteboard Lesson Library
- Dynamic Classroom with Dynamic Investigations

ANSWERS

13.

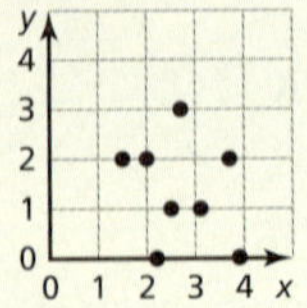

no

14–28. See Additional Answers.

Mini-Assessment

1. The scatter plot shows the days x of practice and the numbers y of free throws made during practice.

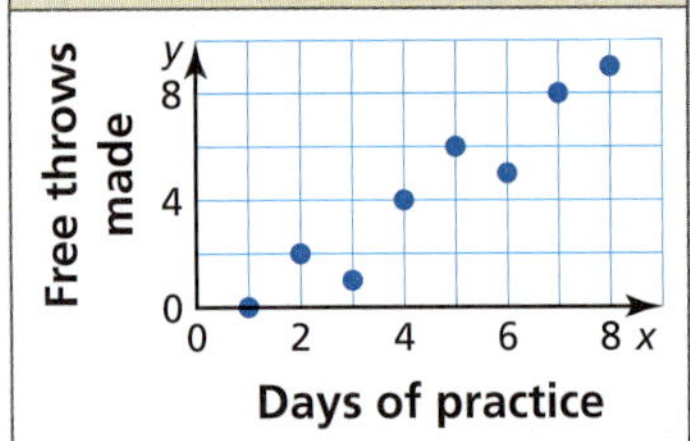

a. How many free throws were made on day 5 of practice? 6 free throws

b. On which day were 5 free throws made? Day 6

The table shows savings y (in dollars) over time x (in months).

x	1	2	3	4	5	6
y	20	18	14	16	12	10

2. Make a scatter plot of the data. Tell the type of correlation, if any, that the data show.

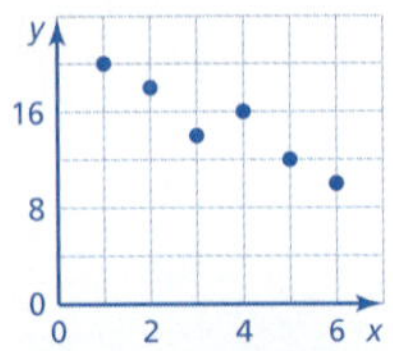

a negative correlation

3. Write an equation that models y as a function of x. *Sample answer:* $y = -2x + 22$, using points $(1, 20)$ and $(2, 18)$

In Exercises 13 and 14, make a scatter plot of the data. Tell whether x and y show a *positive*, a *negative*, or *no* correlation.

13.

x	3.1	2.2	2.5	3.7	3.9	1.5	2.7	2.0
y	1	0	1	2	0	2	3	2

14.

x	3	4	5	6	7	8	9	10
y	67	67	50	33	25	21	19	4

15. MODELING WITH MATHEMATICS The table shows the world birth rates y (number of births per 1000 people) x years since 1960. *(See Example 3.)*

x	0	10	20	30	40	50
y	35.4	33.6	28.3	27.0	22.4	20.0

a. Write an equation that models the birthrate as a function of the number of years since 1960.

b. Interpret the slope and y-intercept of the line of fit.

16. MODELING WITH MATHEMATICS The table shows the total earnings y (in dollars) of a food server who works x hours.

x	0	1	2	3	4	5	6
y	0	18	40	62	77	85	113

a. Write an equation that models the server's earnings as a function of the number of hours the server works.

b. Interpret the slope and y-intercept of the line of fit.

17. OPEN-ENDED Give an example of a real-life data set that shows a negative correlation.

18. MAKING AN ARGUMENT Your friend says that the data in the table show a negative correlation because the dependent variable y is decreasing. Is your friend correct? Explain.

x	14	12	10	8	6	4	2
y	4	1	0	−1	−2	−4	−5

19. USING TOOLS Use a ruler or a yardstick to find the heights and arm spans of five people.

a. Make a scatter plot using the data you collected. Then draw a line of fit for the data.

b. Interpret the slope and y-intercept of the line of fit.

20. THOUGHT PROVOKING A line of fit for a scatter plot is given by the equation $y = 5x + 20$. Describe a real-life data set that could be represented by the scatter plot.

21. WRITING When is data best displayed in a scatter plot, rather than another type of display, such as a bar graph or circle graph?

22. HOW DO YOU SEE IT? The scatter plot shows part of a data set and a line of fit for the data set. Four data points are missing. Choose possible coordinates for these data points.

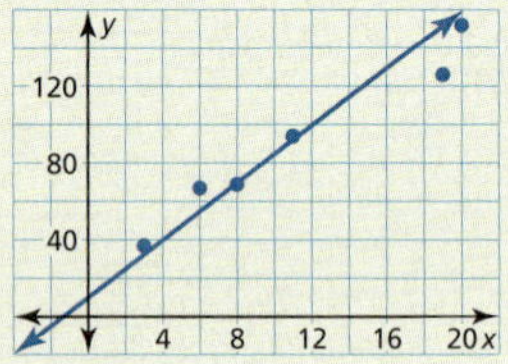

23. REASONING A data set has no correlation. Is it possible to find a line of fit for the data? Explain.

24. ANALYZING RELATIONSHIPS Make a scatter plot of the data in the tables. Describe the relationship between the variables. Is it possible to fit a line to the data? If so, write an equation of the line. If not, explain why.

x	−12	−9	−7	−4	−3	−1
y	150	76	50	15	10	1

x	2	5	6	7	9	15
y	5	22	37	52	90	226

Maintaining Mathematical Proficiency

Reviewing what you learned in previous grades and lessons

Evaluate the function when $x = -3, 0,$ and 4. *(Section 3.3)*

25. $g(x) = 6x$

26. $h(x) = -10x$

27. $f(x) = 5x - 8$

28. $v(x) = 14 - 3x$

If students need help...	If students got it...
Resources by Chapter • Practice A and Practice B • Puzzle Time	Resources by Chapter • Enrichment and Extension • Cumulative Review
Student Journal • Practice	Start the *next* Section
Differentiating the Lesson Skills Review Handbook	

Laurie's Notes

Overview of Section 4.5

Introduction

- The goal of this lesson is to introduce an analytic method for determining a *line of best fit*. In the last lesson, students used eyesight and quantitative reasoning to calculate a line of fit. Residuals, introduced in this lesson, are used in a more analytic approach to finding a best-fit line. Technology is used to find the regression equation. The lesson ends with a discussion of correlation and *causation*.

Resources

- Students should be familiar with techniques specific to their graphing calculator. Spreadsheets could also be used.

Common Misconceptions

- Students often confuse the residual plot with the graph of the actual data. While the residuals can give insights about the graph of the actual data, the plots are not the same. Take time to graph the model against the actual data and then look at the residual plot.

Teaching Strategy

- Residual plots can be calculated and plotted on a graphing calculator. One method is given below for Example 3.
  - After graphing the model and data, return to the Stat Editor screen. To compute the predicted y-values from the model, set the cursor on the L_3 heading and enter "$Y_1(L_1)$". (See Figure 1.) Press ENTER to populate the list with the predicted values.
  - To compute the residuals, set the cursor on the L_4 heading and enter "$L_2 - L_3$" and then press ENTER to populate the list with the residual values. (See Figure 2.)
  - Make a scatter plot of (actual x-value, residual), as shown in Figure 3.

L1	L2	L3
2	60	
3.7	83	
4.2	84	
1.9	58	
3.1	72	
2.5	62	
4.4	85	
3.9	85	

L3=Y1(L1)

Figure 1

L2	L3	L4
60	59	1
83	79.4	3.6
84	85.4	-1.4
58	57.8	.2
72	72.2	-.2
62	65	-3
85	87.8	-2.8
85	81.8	3.2

L4(1)=1

Figure 2

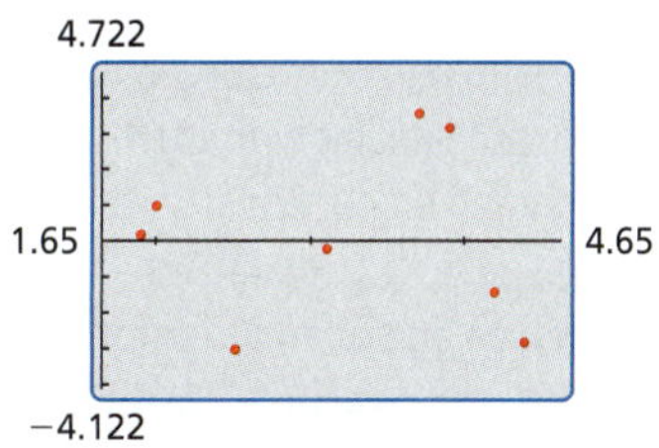

Figure 3

- Note that a negative residual occurs when the model predicts a y-value greater than the actual y-value, and conversely, a positive residual occurs when the model predicts a y-value less than the actual y-value.

Another Way

- Another way to analyze residuals is to find the sum of the squared residuals. The closer to 0 the sum is, the better the line fits the data. By using squares of residuals, positive and negative residuals do not "cancel out," and residuals with squares greater than 1 have a magnified effect on the sum. A regression line, or line of best fit, is the line for which the sum of the squares of the residuals is a minimum. The process of finding this line is linear regression. It can be calculated by hand, but that is very tedious. A graphing calculator can do it very easily, as shown in the exploration.

Pacing Suggestion

- The exploration introduces how to enter a bivariate data set in a graphing calculator and then perform linear regression on the data.

Dynamic Teaching Tools

- Dynamic Assessment & Progress Monitoring Tool
- Lesson Planning Tool
- Interactive Whiteboard Lesson Library
- Dynamic Classroom with Dynamic Investigations

What Your Students Will Learn

- Calculate and use residuals to determine whether lines are good models for sets of data.
- Use technology and interpret the correlation coefficient to find lines of best fit for sets of data.
- Identify correlations and causations for sets of data.

Laurie's Notes

Exploration

Motivate

- Write the following table on the board.

Number of weeks	1	2	3	4	6	7
Cumulative snowfall (in.)	11	18	32	52	73	82

- Use the data in the table to review the process of making a scatter plot using a graphing calculator. The important skills are entering the data, selecting the scatter plot option, setting an appropriate viewing window, and making the plot.

? "Do you think you could draw a line that models the data?" yes

? "Could your line help me estimate the cumulative snowfall in week 5? Explain." yes; Use the equation to find the y-value when $x = 5$.

For Your Information

- It is possible that a scatter plot shows there is a relationship between two variables. This does not imply that one variable causes the other, but there an association between the two variables. A scatter plot is helpful in determining the strength of the relationship.
- If there appears to be no association between the proposed explanatory variable and the dependent variable, then fitting a regression model to the data will have little meaning. Conversely, if there does appear to be an association (increasing or decreasing trend), then fitting a regression model to the data may well provide a useful model.
- A valuable numerical measure of association between two variables is the correlation coefficient, which is a value between −1 and 1 that indicates the strength of the association of the observed data for the two variables.

Exploration 1

- Keystrokes vary between models of calculators. Have students consult their manuals if you are not familiar with their calculators. Likewise, the syntax for the spreadsheet or computer is going to be particular to the system.

? **Assessing Question:** "What does the ordered pair (5, 20.6) represent?" In 1965, the median age of an American woman at her first marriage was 20.6 years.

- **Use Appropriate Tools Strategically:** While you can use eyesight to roughly approximate the line of fit, there is an analytic method called regression. A common regression method called the least-squares regression minimizes the sum of the squares of the residuals. Residuals are defined in the formal lesson.

COMMON ERROR Students may miss one of the ordered pairs or one of the coordinates of the ordered pair. Students should look for this error when they finish entering the data.

? "Did everyone get the same regression equation? Should we?" yes; If the technology is the same, then the results will be the same.

? Compare the regression equation to the line of fit found in Section 4.4. "Are they the same? Should they be?" They should be close, but they would not be exact. Fitting a line by eyesight involves finding the equation between two points. Regression takes all points into consideration.

Communicate Your Answer

- Listen for an explanation of *linear regression* to find the line of best fit analytically.

Connecting to Next Step

- Students have now been introduced to linear regression. In the formal lesson, students will use linear regression for a data set.

4.5 Analyzing Lines of Fit

Essential Question How can you *analytically* find a line of best fit for a scatter plot?

EXPLORATION 1 **Finding a Line of Best Fit**

Work with a partner.
The scatter plot shows the median ages of American women at their first marriage for selected years from 1960 through 2010. In Exploration 2 in Section 4.4, you approximated a line of fit graphically. To find the line of *best* fit, you can use a computer, spreadsheet, or graphing calculator that has a *linear regression* feature.

a. The data from the scatter plot is shown in the table. Note that 0, 5, 10, and so on represent the numbers of years since 1960. What does the ordered pair (25, 23.3) represent?

L1	L2	L3
0	20.3	
5	20.6	
10	20.8	
15	21.1	
20	22	
25	23.3	
30	23.9	
35	24.5	
40	25.1	
45	25.3	
50	26.1	

L1(55)=		

b. Use the *linear regression* feature to find an equation of the line of best fit. You should obtain results such as those shown below.

```
LinReg
 y=ax+b
 a=.1261818182
 b=19.84545455
 r²=.9738676804
 r=.986847344
```

c. Write an equation of the line of best fit. Compare your result with the equation you obtained in Exploration 2 in Section 4.4.

CONSTRUCTING VIABLE ARGUMENTS

To be proficient in math, you need to reason inductively about data.

Communicate Your Answer

2. How can you *analytically* find a line of best fit for a scatter plot?

3. The data set relates the number of chirps per second for striped ground crickets and the outside temperature in degrees Fahrenheit. Make a scatter plot of the data. Then find an equation of the line of best fit. Use your result to estimate the outside temperature when there are 19 chirps per second.

Chirps per second	20.0	16.0	19.8	18.4	17.1
Temperature (°F)	88.6	71.6	93.3	84.3	80.6

Chirps per second	14.7	15.4	16.2	15.0	14.4
Temperature (°F)	69.7	69.4	83.3	79.6	76.3

Dynamic Teaching Tools

- Dynamic Assessment & Progress Monitoring Tool
- Lesson Planning Tool
- Interactive Whiteboard Lesson Library
- Dynamic Classroom with Dynamic Investigations

ANSWERS

1. a. The median age of American women at their first marriage in 1985 was 23.3.

b. Check students' work.

c. $y = 0.13x + 19.8$; The equations are similar.

2. Use a computer, spreadsheet, or graphing calculator that has a linear regression feature.

3.

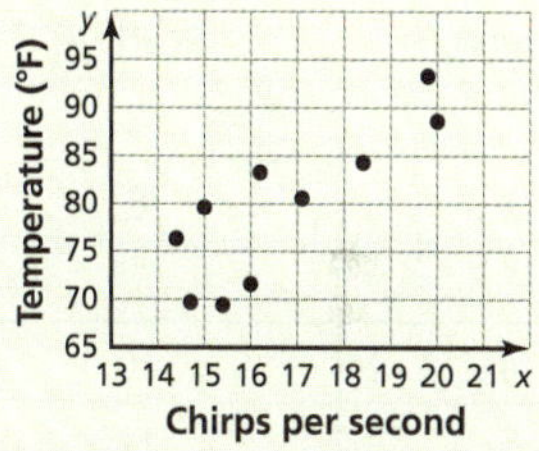

$y = 3.27x + 25.0$; about 87.1°F

Differentiated Instruction

Auditory

Some students may benefit from working with a partner to find the residuals. Students can take turns using the equation to find each residual, describing each step in the process to their partners.

Extra Example 1

The table shows the number y of gloves sold for each amount x (in inches) of snowfall. The equation $y = 1.5x + 3.5$ models the data. Is the model a good fit?

Snowfall (inches), x	Gloves Sold, y
1	5
2	8
3	7
4	10
5	11
6	12
7	14

The points (x, residual) are evenly dispersed about the horizontal axis. So, the equation $y = 1.5x + 3.5$ is a good fit.

4.5 Lesson

Core Vocabulary

What You Will Learn

- Use residuals to determine how well lines of fit model data.
- Use technology to find lines of best fit.
- Distinguish between correlation and causation.

Analyzing Residuals

One way to determine how well a line of fit models a data set is to analyze *residuals*.

Core Concept

Residuals

A **residual** is the difference of the y-value of a data point and the corresponding y-value found using the line of fit. A residual can be positive, negative, or zero.

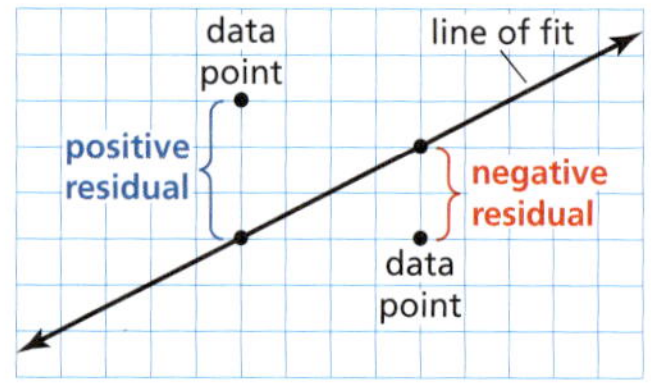

A scatter plot of the residuals shows how well a model fits a data set. If the model is a good fit, then the absolute values of the residuals are relatively small, and the residual points will be more or less evenly dispersed about the horizontal axis. If the model is not a good fit, then the residual points will form some type of pattern that suggests the data are not linear. Wildly scattered residual points suggest that the data might have no correlation.

EXAMPLE 1 Using Residuals

In Example 3 in Section 4.4, the equation $y = -2x + 20$ models the data in the table shown. Is the model a good fit?

Week, x	Sales (millions), y
1	$19
2	$15
3	$13
4	$11
5	$10
6	$8
7	$7
8	$5

SOLUTION

Step 1 Calculate the residuals. Organize your results in a table.

Step 2 Use the points (x, residual) to make a scatter plot.

x	y	y-Value from model	Residual
1	19	18	$19 - 18 = 1$
2	15	16	$15 - 16 = -1$
3	13	14	$13 - 14 = -1$
4	11	12	$11 - 12 = -1$
5	10	10	$10 - 10 = 0$
6	8	8	$8 - 8 = 0$
7	7	6	$7 - 6 = 1$
8	5	4	$5 - 4 = 1$

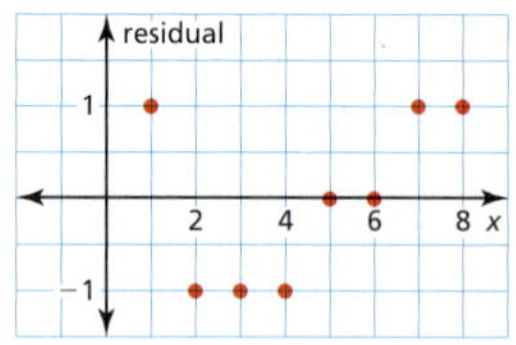

The points are evenly dispersed about the horizontal axis. So, the equation $y = -2x + 20$ is a good fit.

Laurie's Notes Teacher Actions

- Use a graph to help define a residual, the difference between the actual and predicted y-values. This can be done manually, or by using a spreadsheet, or by computing in the Stat Editor of a calculator, as shown in the *Teaching Strategy*.
- The *Core Concept* describes how the residuals give an indication about the goodness of fit of the line.

? Probing Question: "If the first half of the residuals was all positive and the second half was all negative, would this be a good fit? Explain." no; That would be a pattern and would suggest the data might not be a linear model.

EXAMPLE 2 Using Residuals

The table shows the ages x and salaries y (in thousands of dollars) of eight employees at a company. The equation $y = 0.2x + 38$ models the data. Is the model a good fit?

Age, x	35	37	41	43	45	47	53	55
Salary, y	42	44	47	50	52	51	49	45

SOLUTION

Step 1 Calculate the residuals. Organize your results in a table.

Step 2 Use the points (x, residual) to make a scatter plot.

x	y	y-Value from model	Residual
35	42	45.0	$42 - 45.0 = -3.0$
37	44	45.4	$44 - 45.4 = -1.4$
41	47	46.2	$47 - 46.2 = 0.8$
43	50	46.6	$50 - 46.6 = 3.4$
45	52	47.0	$52 - 47.0 = 5.0$
47	51	47.4	$51 - 47.4 = 3.6$
53	49	48.6	$49 - 48.6 = 0.4$
55	45	49.0	$45 - 49.0 = -4.0$

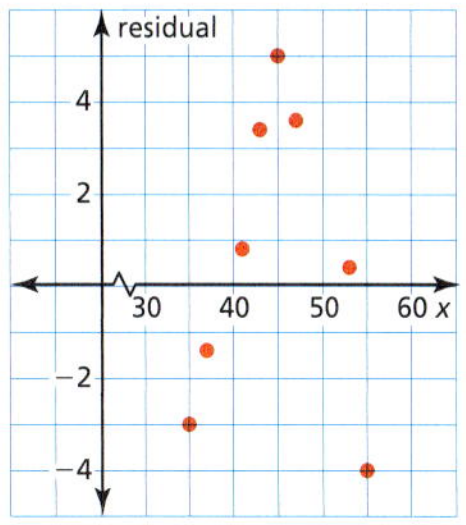

The residual points form a ∩-shaped pattern, which suggests the data are not linear. So, the equation $y = 0.2x + 38$ does not model the data well.

Monitoring Progress

Help in English and Spanish at *BigIdeasMath.com*

1. The table shows the attendances y (in thousands) at an amusement park from 2005 to 2014, where $x = 0$ represents the year 2005. The equation $y = -9.8x + 850$ models the data. Is the model a good fit?

Year, x	0	1	2	3	4	5	6	7	8	9
Attendance, y	850	845	828	798	800	792	785	781	775	760

STUDY TIP

You know how to use two points to find an equation of a line of fit. When finding an equation of the line of best fit, every point in the data set is used.

Finding Lines of Best Fit

Graphing calculators use a method called **linear regression** to find a precise line of fit called a **line of best fit**. This line best models a set of data. A calculator often gives a value r, called the **correlation coefficient**. This value tells whether the correlation is positive or negative and how closely the equation models the data. Values of r range from -1 to 1. When r is close to 1 or -1, there is a strong correlation between the variables. As r, gets closer to 0, the correlation becomes weaker.

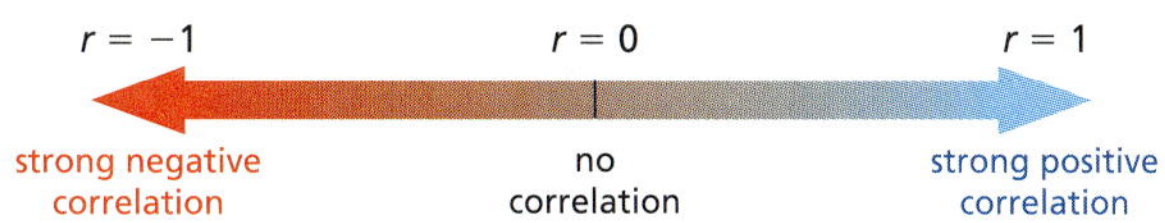

Extra Example 2

The table shows the high temperatures on Monday for eight weeks. The equation $y = 7x - 8$ models the data. Is the model a good fit?

Week	High Temperature (°F)
1	48
2	43
3	42
4	40
5	45
6	56
7	67
8	76

No, the residual points form a V-shaped pattern, which suggests the data are not linear. So, the equation $y = 7x - 8$ is not a good fit.

MONITORING PROGRESS ANSWER

1. no

Laurie's Notes — Teacher Actions

- Have students complete the table and the residual plot.
- ? "What does the pattern of residuals suggest about the ages and salary data?" not linear; Actual data is ∩-shaped.
- Compare the residual plot with the model graphed against the actual data.
- Remind students of the exploration. A method called *linear regression* was used to find a *line of best fit.* If the *correlation coefficient* does not appear when performing the regression, then select DiagnosticOn in the catalog.

English Language Learners

Vocabulary

Relate the word *interpolation* on page 205 with the word *intermission*. An *intermission* is a break *between* the parts of a performance. Similarly, an *interpolation* is a value *between* two known values. Relate the word *extrapolation* to *extraordinary*. An *extraordinary* achievement is one that is beyond or *outside* what is normally expected. An *extrapolation* is a value *outside* the range of known values.

Extra Example 3

The table shows the number x of teachers and the number y of students in each of several schools.

Teachers, x	Students, y
10	250
12	275
13	312
16	360
15	385
18	406

a. Use a graphing calculator to find an equation of the line of best fit. Then plot the data and graph the equation in the same viewing window. *Sample answer:* $y = 21x + 42$

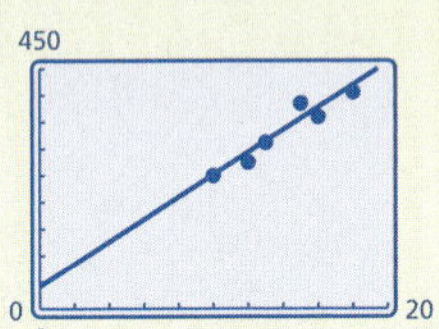

b. Identify and interpret the correlation coefficient. about 0.960; The relationship between the number of students and the number of teachers has a strong positive correlation and the equation closely models the data.

c. Interpret the slope and y-intercept of the line of best fit. The slope 21 means the number of students increases by about 21 for each additional teacher. The y-intercept 42 has no meaning in this context because a school would not have 0 teachers.

MONITORING PROGRESS ANSWER

2. See Additional Answers.

EXAMPLE 3 Finding a Line of Best Fit Using Technology

The table shows the durations x (in minutes) of several eruptions of the geyser Old Faithful and the times y (in minutes) until the next eruption. (a) Use a graphing calculator to find an equation of the line of best fit. Then plot the data and graph the equation in the same viewing window. (b) Identify and interpret the correlation coefficient. (c) Interpret the slope and y-intercept of the line of best fit.

Duration, x	2.0	3.7	4.2	1.9	3.1	2.5	4.4	3.9
Time, y	60	83	84	58	72	62	85	85

SOLUTION

a. Step 1 Enter the data from the table into two lists.

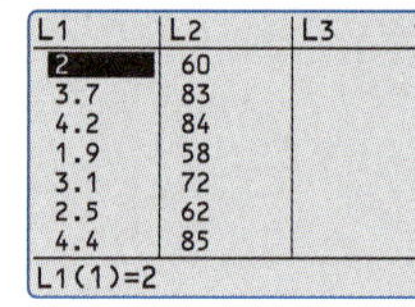

Step 2 Use the *linear regression* feature. The values in the equation can be rounded to obtain $y = 12.0x + 35$.

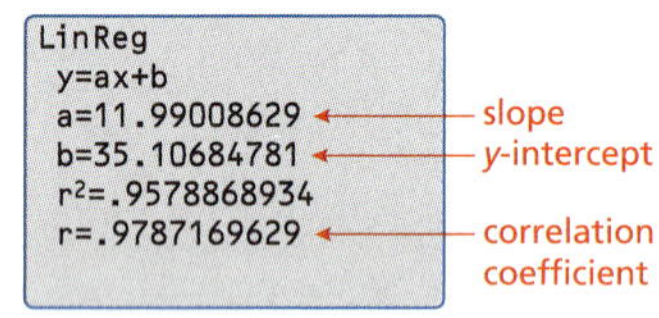

PRECISION
Be sure to analyze the data values to select an appropriate viewing window for your graph.

Step 3 Enter the equation $y = 12.0x + 35$ into the calculator. Then plot the data and graph the equation in the same viewing window.

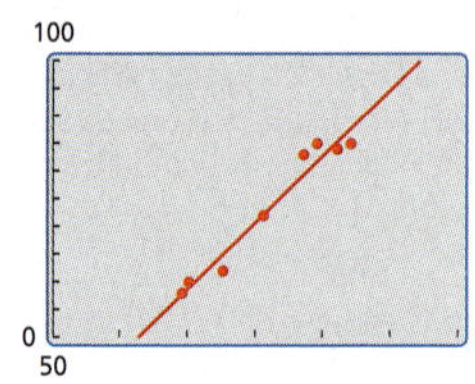

b. The correlation coefficient is about 0.979. This means that the relationship between the durations and the times until the next eruption has a strong positive correlation and the equation closely models the data, as shown in the graph.

c. The slope of the line is 12. This means the time until the next eruption increases by about 12 minutes for each minute the duration increases. The y-intercept is 35, but it has no meaning in this context because the duration cannot be 0 minutes.

Monitoring Progress

Help in English and Spanish at *BigIdeasMath.com*

2. Use the data in Monitoring Progress Question 1. (a) Use a graphing calculator to find an equation of the line of best fit. Then plot the data and graph the equation in the same viewing window. (b) Identify and interpret the correlation coefficient. (c) Interpret the slope and y-intercept of the line of best fit.

Laurie's Notes Teacher Actions

- **Make Sense of Problems and Persevere in Solving Them:** Have students interpret what an ordered pair (x, y) means in Example 3.
- **? Turn and Talk:** "Is there a discernable pattern in the data? Explain." Answers will vary. Students may note that as the eruption time increases, so too does the wait time for the next eruption.
- **? Probing Question:** "Would an eruption time of 5 minutes cause the time between eruptions to increase also?" Students may be unsure of this; however, they should recognize that other geologic factors are at work.

Using a graph or its equation to *approximate* a value between two known values is called **interpolation**. Using a graph or its equation to *predict* a value outside the range of known values is called **extrapolation**. In general, the farther removed a value is from the known values, the less confidence you can have in the accuracy of the prediction.

STUDY TIP

To approximate or predict an unknown value, you can evaluate the model algebraically or graph the model with a graphing calculator and use the *trace* feature.

EXAMPLE 4 Interpolating and Extrapolating Data

Refer to Example 3. Use the equation of the line of best fit.

a. Approximate the duration before a time of 77 minutes.

b. Predict the time after an eruption lasting 5.0 minutes.

SOLUTION

a. $y = 12.0x + 35$ — Write the equation.

$77 = 12.0x + 35$ — Substitute 77 for y.

$3.5 = x$ — Solve for x.

▶ An eruption lasts about 3.5 minutes before a time of 77 minutes.

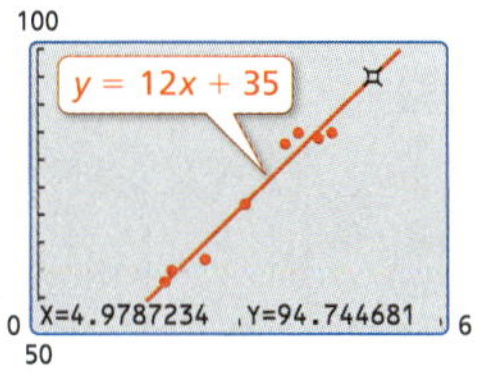

b. Use a graphing calculator to graph the equation. Use the *trace* feature to find the value of y when $x \approx 5.0$, as shown.

▶ A time of about 95 minutes will follow an eruption of 5.0 minutes.

Monitoring Progress — Help in English and Spanish at *BigIdeasMath.com*

3. Refer to Monitoring Progress Question 2. Use the equation of the line of best fit to predict the attendance at the amusement park in 2017.

Correlation and Causation

When a change in one variable causes a change in another variable, it is called **causation**. Causation produces a strong correlation between the two variables. The converse is *not* true. In other words, correlation does not imply causation.

READING

A causal relationship exists when one variable causes a change in another variable.

EXAMPLE 5 Identifying Correlation and Causation

Tell whether a correlation is likely in the situation. If so, tell whether there is a causal relationship. Explain your reasoning.

a. time spent exercising and the number of calories burned

b. the number of banks and the population of a city

SOLUTION

a. There is a positive correlation and a causal relationship because the more time you spend exercising, the more calories you burn.

b. There may be a positive correlation but no causal relationship. Building more banks will not cause the population to increase.

Monitoring Progress — Help in English and Spanish at *BigIdeasMath.com*

4. Is there a correlation between time spent playing video games and grade point average? If so, is there a causal relationship? Explain your reasoning.

Extra Example 4

Refer to Extra Example 3. Use the equation of the line of best fit $y = 21x + 42$.

a. Approximate the number of teachers when there are 336 students. about 14 teachers

b. Predict the number of students when there are 20 teachers. about 462 students

Extra Example 5

Tell whether a correlation is likely in the situation. If so, tell whether there is a causal relationship. Explain your reasoning.

a. time spent babysitting and the amount of money earned There is a positive correlation and a causal relationship because the more time you spend working, the more money you earn.

b. eating organic food and getting a good score on a test There may be a positive correlation but no causal relationship. Eating organic food will not cause you to get a good score on a test.

MONITORING PROGRESS ANSWERS

3. 729,300

4. There may be a negative correlation; A causal relationship is possible because the more time you spend playing video games, the less time you spend studying.

Laurie's Notes — Teacher Actions

- **Reason Abstractly and Quantitatively** and **Attend to Precision:** Define and discuss *interpolation* and *extrapolation*, using data from Example 3.
- Define and discuss *causation*.
- **Turn and Talk:** Have partners discuss Example 5.

Closure

- **Exit Ticket:** Find the line of best fit for the data in the *Motivate*. Use the line to estimate the cumulative snowfall in week 5. $y = 12.5x - 3$; 59.5 in.

Assignment Guide and Homework Check

ASSIGNMENT

Basic: 1–4, 5–21 odd, 26, 28, 32, 33

Average: 1–3, 4–26 even, 27, 28, 32, 33

Advanced: 1–4, 8–28 even, 29–33

HOMEWORK CHECK

Basic: 5, 9, 17, 19, 21

Average: 6, 10, 18, 20, 22

Advanced: 8, 10, 18, 20, 22

ANSWERS

1. when the actual y-value is greater than the y-value from the model; when the actual y-value is less than the y-value from the model.
2. If the model is a good fit, then the absolute values of the residuals are relatively small and the residual points will be more or less evenly dispersed about the horizontal axis. If the model is not a good fit, then the residual points will form some type of pattern that suggests the data are not linear.
3. Interpolation is using a graph or its equation to approximate a value between two known values, and extrapolation is using a graph or its equation to predict a value outside the range of known values.
4. $r = -0.09$; It is the only one that suggests no correlation.
5. no; The residual points are not evenly dispersed about the horizontal axis.
6. yes; The residual points are evenly dispersed about the horizontal axis.
7. yes; The residual points are evenly dispersed about the horizontal axis.
8. no; The residual points form a nonlinear pattern.
9. yes; The residual points are evenly dispersed about the horizontal axis.
10. no; The residual points are not evenly dispersed about the horizontal axis.
11. $y = 2.1x - 8$; $r = 0.980$; strong positive correlation
12. $y = -1.3x + 8$; $r = -0.886$; strong negative correlation
13. $y = 1.4x + 16$; $r = 0.999$; strong positive correlation
14. $y = -x + 11$; $r = -0.444$; weak negative correlation

4.5 Exercises

Dynamic Solutions available at *BigIdeasMath.com*

Vocabulary and Core Concept Check

1. **VOCABULARY** When is a residual positive? When is it negative?
2. **WRITING** Explain how you can use residuals to determine how well a line of fit models a data set.
3. **VOCABULARY** Compare interpolation and extrapolation.
4. **WHICH ONE DOESN'T BELONG?** Which correlation coefficient does *not* belong with the other three? Explain your reasoning.

$r = -0.98$ $r = 0.96$ $r = -0.09$ $r = 0.97$

Monitoring Progress and Modeling with Mathematics

In Exercises 5–8, use residuals to determine whether the model is a good fit for the data in the table. Explain. *(See Examples 1 and 2.)*

5. $y = 4x - 5$

x	−4	−3	−2	−1	0	1	2	3	4
y	−18	−13	−10	−7	−2	0	6	10	15

6. $y = 6x + 4$

x	1	2	3	4	5	6	7	8	9
y	13	14	23	26	31	42	45	52	62

7. $y = -1.3x + 1$

x	−8	−6	−4	−2	0	2	4	6	8
y	9	10	5	8	−1	1	−4	−12	−7

8. $y = -0.5x - 2$

x	4	6	8	10	12	14	16	18	20
y	−1	−3	−6	−8	−10	−10	−10	−9	−9

9. **ANALYZING RESIDUALS** The table shows the growth y (in inches) of an elk's antlers during week x. The equation $y = -0.7x + 6.8$ models the data. Is the model a good fit? Explain.

Week, x	1	2	3	4	5
Growth, y	6.0	5.5	4.7	3.9	3.3

10. **ANALYZING RESIDUALS** The table shows the approximate numbers y (in thousands) of movie tickets sold from January to June for a theater. In the table, $x = 1$ represents January. The equation $y = 1.3x + 27$ models the data. Is the model a good fit? Explain.

Month, x	Ticket sales, y
1	27
2	28
3	36
4	28
5	32
6	35

In Exercises 11–14, use a graphing calculator to find an equation of the line of best fit for the data. Identify and interpret the correlation coefficient.

11.

x	0	1	2	3	4	5	6	7
y	−8	−5	−2	−1	−1	2	5	8

12.

x	−4	−2	0	2	4	6	8	10
y	17	7	8	1	5	−2	2	−8

13.

x	−15	−10	−5	0	5	10	15	20
y	−4	2	7	16	22	30	37	43

14.

x	5	6	7	8	9	10	11	12
y	12	−2	8	3	−1	−4	6	0

ERROR ANALYSIS **In Exercises 15 and 16, describe and correct the error in interpreting the graphing calculator display.**

```
LinReg
 y=ax+b
 a=-4.47
 b=23.16
 r²=.9989451055
 r=-.9994724136
```

15. ✗ An equation of the line of best fit is $y = 23.16x - 4.47$.

16. ✗ The data have a strong positive correlation.

17. **MODELING WITH MATHEMATICS** The table shows the total numbers y of people who reported an earthquake x minutes after it ended. *(See Example 3.)*

Minutes, x	People, y
1	10
2	100
3	400
4	900
5	1400
6	1800
7	2100

a. Use a graphing calculator to find an equation of the line of best fit. Then plot the data and graph the equation in the same viewing window.

b. Identify and interpret the correlation coefficient.

c. Interpret the slope and y-intercept of the line of best fit.

18. **MODELING WITH MATHEMATICS** The table shows the numbers y of people who volunteer at an animal shelter on each day x.

Day, x	1	2	3	4	5	6	7	8
People, y	9	5	13	11	10	11	19	12

a. Use a graphing calculator to find an equation of the line of best fit. Then plot the data and graph the equation in the same viewing window.

b. Identify and interpret the correlation coefficient.

c. Interpret the slope and y-intercept of the line of best fit.

19. **MODELING WITH MATHEMATICS** The table shows the mileages x (in thousands of miles) and the selling prices y (in thousands of dollars) of several used automobiles of the same year and model. *(See Example 4.)*

Mileage, x	22	14	18	30	8	24
Price, y	16	17	17	14	18	15

a. Use a graphing calculator to find an equation of the line of best fit.

b. Identify and interpret the correlation coefficient.

c. Interpret the slope and y-intercept of the line of best fit.

d. Approximate the mileage of an automobile that costs $15,500.

e. Predict the price of an automobile with 6000 miles.

20. **MODELING WITH MATHEMATICS** The table shows the lengths x and costs y of several sailboats.

Length (feet), x	Cost (thousands of dollars), y
27	94
18	56
25	58
32	123
18	60
26	87
36	145

a. Use a graphing calculator to find an equation of the line of best fit.

b. Identify and interpret the correlation coefficient.

c. Interpret the slope and y-intercept of the line of best fit.

d. Approximate the cost of a sailboat that is 20 feet long.

e. Predict the length of a sailboat that costs $147,000.

In Exercises 21–24, tell whether a correlation is likely in the situation. If so, tell whether there is a causal relationship. Explain your reasoning. *(See Example 5.)*

21. the amount of time spent talking on a cell phone and the remaining battery life

22. the height of a toddler and the size of the toddler's vocabulary

23. the number of hats you own and the size of your head

24. the weight of a dog and the length of its tail

Dynamic Teaching Tools

- Dynamic Assessment & Progress Monitoring Tool
- Interactive Whiteboard Lesson Library
- Dynamic Classroom with Dynamic Investigations

ANSWERS

15. The slope and y-intercept were reversed; $y = -4.47x + 23.16$

16. The data do not have a strong positive correlation; The data have a strong negative correlation.

17. a. $y = 381x - 566$

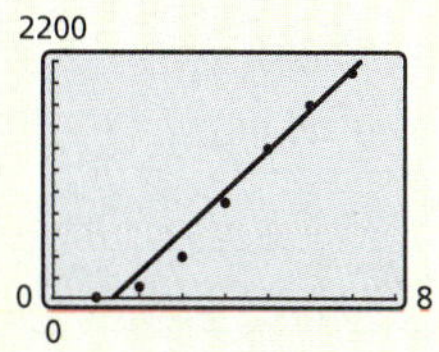

b. $r = 0.989$; strong positive correlation

c. The slope of 381 means the number of people who reported an earthquake increased by about 381 each minute after the earthquake ended. The y-intercept of -566 has no meaning in this context because the number of people cannot be negative.

18. a. $y = x + 7$

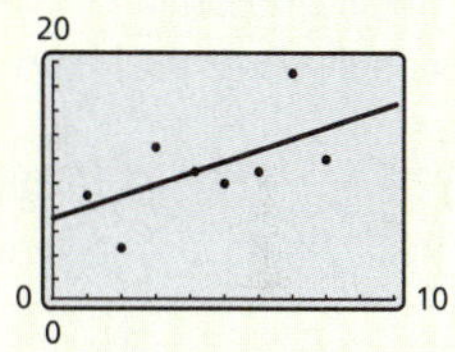

b. $r = 0.619$; weak positive correlation

c. The slope of 1 means the number of people who volunteer increases by about 1 each day. The y-intercept of 6.75 has no meaning in this context because there is no day 0.

19. a. $y = -0.2x + 20$

b. $r = -0.968$; strong negative correlation

c. The slope of -0.2 means the selling price decreases by about $200 for every increase in mileage of 1000 miles. The y-intercept of 19.7 has no meaning in this context because a used car cannot have 0 mileage.

d. 22,500 mi

e. $18,800

20. a. $y = 4.9x - 38$

b. $r = 0.936$; strong positive correlation

c. The slope of 4.9 means the cost of a sailboat increases by about $4900 for each foot the length of the sailboat increases. The y-intercept of -37.7 has no meaning in this context because the length of a sailboat cannot be negative.

d. $60,000

e. about 38 ft

21. There is a negative correlation and a causal relationship because the more time you spend talking on the phone, the less charge there is left in the battery.

22. There may be a positive correlation but no causal relationship. The height of a toddler does not determine the toddler's vocabulary.

23. A correlation is unlikely. The number of hats you own is not related to the size of your head.

24. There may be a positive correlation but no causal relationship. The weight of a dog does not determine the length of its tail.

ANSWERS

25. *Sample answer:* ACT math score and SAT math score

26–33. See Additional Answers.

Mini-Assessment

1. Is the model $y = 2x - 3$ a good fit for the data in the table? Explain.

x	1	3	5	7	9
y	0	3	8	11	14

Yes; The residual points are evenly dispersed about the horizontal axis.

2. Tell whether a correlation is likely in the situation: the number of cars in a store parking lot and the number of people in the store. If so, tell whether there is a causal relationship. Explain. There is a positive correlation and a causal relationship because the more cars there are in the parking lot, the more people there are in the store.

The table shows the distance x (in miles) Tom rode his bicycle and the time y (in minutes) of each ride.

Distance (miles), x	Time (minutes), y
3	15
5	27
9	42
10	50
12	58
15	80

3. Use a graphing calculator to find an equation of the line of best fit for the data. Identify and interpret the correlation coefficient. $y = 5x - 1$; about 0.993; a strong positive correlation between distance traveled and time riding

4. Use the equation of the line of best fit to approximate the distance for a 30-minute ride. about 6.2 miles

5. Predict the time for a 20-mile ride. about 99 minutes

25. OPEN-ENDED Describe a data set that has a strong correlation but does not have a causal relationship.

26. HOW DO YOU SEE IT? Match each graph with its correlation coefficient. Explain your reasoning.

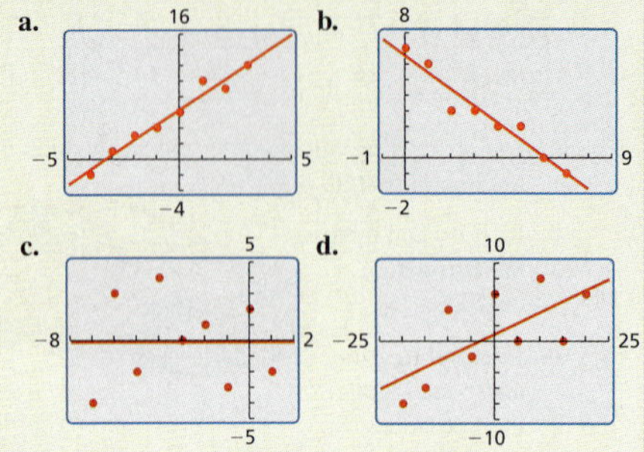

A. $r = 0$ **B.** $r = 0.98$

C. $r = -0.97$ **D.** $r = 0.69$

27. ANALYZING RELATIONSHIPS The table shows the grade point averages y of several students and the numbers x of hours they spend watching television each week.

Hours, x	Grade point average, y
10	3.0
5	3.4
3	3.5
12	2.7
20	2.1
15	2.8
8	3.0
4	3.7
16	2.5

a. Use a graphing calculator to find an equation of the line of best fit. Identify and interpret the correlation coefficient.

b. Interpret the slope and y-intercept of the line of best fit.

c. Another student watches about 14 hours of television each week. Approximate the student's grade point average.

d. Do you think there is a causal relationship between time spent watching television and grade point average? Explain.

28. MAKING AN ARGUMENT A student spends 2 hours watching television each week and has a grade point average of 2.4. Your friend says including this information in the data set in Exercise 27 will weaken the correlation. Is your friend correct? Explain.

29. USING MODELS Refer to Exercise 17.

a. Predict the total numbers of people who reported an earthquake 9 minutes and 15 minutes after it ended.

b. The table shows the actual data. Describe the accuracy of your extrapolations in part (a).

Minutes, x	9	15
People, y	2750	3200

30. THOUGHT PROVOKING A data set consists of the numbers x of people at Beach 1 and the numbers y of people at Beach 2 recorded daily for 1 week. Sketch a possible graph of the data set. Describe the situation shown in the graph and give a possible correlation coefficient. Determine whether there is a causal relationship. Explain.

31. COMPARING METHODS The table shows the numbers y (in billions) of text messages sent each year in a five-year period, where $x = 1$ represents the first year in the five-year period.

Year, x	1	2	3	4	5
Text messages (billions), y	241	601	1360	1806	2206

a. Use a graphing calculator to find an equation of the line of best fit. Identify and interpret the correlation coefficient.

b. Is there a causal relationship? Explain your reasoning.

c. Calculate the residuals. Then make a scatter plot of the residuals and interpret the results.

d. Compare the methods you used in parts (a) and (c) to determine whether the model is a good fit. Which method do you prefer? Explain.

Maintaining Mathematical Proficiency

Reviewing what you learned in previous grades and lessons

Determine whether the table represents a *linear* or *nonlinear* function. Explain. *(Section 3.2)*

32.

x	5	6	7	8
y	−4	4	−4	4

33.

x	2	4	6	8
y	13	8	3	−2

If students need help...	*If students got it...*
Resources by Chapter • Practice A and Practice B • Puzzle Time	Resources by Chapter • Enrichment and Extension • Cumulative Review
Student Journal • Practice	Start the *next* Section
Differentiating the Lesson Skills Review Handbook	

Laurie's Notes

Overview of Section 4.6

Introduction

- This lesson will make connections between the arithmetic sequences and linear equations. Sequence notation and vocabulary associated with arithmetic sequences are introduced.
- Students will write the terms of an arithmetic sequence and extend a given sequence. When the arithmetic sequence is graphed, the result is a line and the slope is the common difference.
- The last skill in the lesson is writing the arithmetic sequence as an equation and as a function.

Common Misconceptions

- Students may think that all sequences involving addition are arithmetic. For example, the sequence 1, 2, 4, 7, 11, ... has an observable addition pattern (add 1, add 2, add 3, add 4, ...), but it is not arithmetic. There must be a constant value being added for it to be arithmetic.

Formative Assessment Tips

- **3-2-1:** This formative assessment strategy is useful in giving students a structured way in which to reflect on their learning, particularly at the conclusion of a unit of study. Students are asked to respond to three writing prompts, giving three responses to the first prompt, two responses to the second prompt, and one response to the third prompt. All six responses relate to what students have learned during the unit of study—in this case, a chapter on linear functions.
- Distribute a printed *3-2-1* reflection sheet and give time for students to reflect on their learning and to write.
  - 3 new things (concepts, skills, procedures, ...) I learned in this chapter
  - 2 things (concepts, skills, procedures, ...) I am still struggling with
  - 1 thing that will help me tomorrow
- Collect and review student reflections, and plan instruction for tomorrow accordingly. The reflection allows students to see what learning has occurred and where they should focus their attention.

Pacing Suggestion

- Students should work through the exploration and then proceed with the formal lesson.

Dynamic Teaching Tools

- Dynamic Assessment & Progress Monitoring Tool
- Lesson Planning Tool
- Interactive Whiteboard Lesson Library
- Dynamic Classroom with Dynamic Investigations

What Your Students Will Learn

- Determine common differences between consecutive terms of arithmetic sequences, and then write the terms of the sequences.
- Graph arithmetic sequences and identify arithmetic sequences from graphs.
- Write arithmetic sequences as equations and in function notation.

Laurie's Notes

Exploration

Motivate

- Before students arrive, write five numbers of an arithmetic sequence on separate index cards. (*Example:* 13, 17, 21, 25, 29) Hand an index card to each of the first five students who enter the room.
- Tell the five students that they are holding numbers in an arithmetic sequence and to position themselves in order.
- Students may realize that the order could be increasing or decreasing.

  Version A: 13, 17, 21, 25, 29 **Version B:** 29, 25, 21, 17, 13

? "When 13 is in the first position, what number is in the fourth position?" 25 "What can you add to each number to get the next number?" 4

? "When 29 is in the first position, what number is in the fourth position?" 17 "What can you add to each number to get the next number?" −4

Discuss

- It is quite natural for teachers to see a sequence of numbers and to refer to each as a *term* in the sequence. Take time to describe what is meant by the terms in an arithmetic sequence.

Exploration 1

- Have students work with their partners to complete each part of the exploration.
- Students may describe the patterns in an additive manner or a multiplicative manner. For instance, in part (a), students may say that you are adding 10 each time to get the next y-value, or they may say you multiply the x-value by 10 to get the y-value.
- Students should recognize that each pattern can be described by a linear equation.

? **Probing Question:** "When you know the difference between terms in an arithmetic sequence, what do you know about the equation that describes the sequence?" You know the slope of the linear equation.

Communicate Your Answer

- Have students share their real-life examples.

Connecting to Next Step

- The exploration should be completed quickly and should help students recognize the connection between arithmetic sequences and linear equations.

4.6 Arithmetic Sequences

Essential Question How can you use an arithmetic sequence to describe a pattern?

An **arithmetic sequence** is an ordered list of numbers in which the difference between each pair of consecutive **terms**, or numbers in the list, is the same.

EXPLORATION 1 **Describing a Pattern**

Work with a partner. Use the figures to complete the table. Plot the points given by your completed table. Describe the pattern of the y-values.

a. $n = 1$ $n = 2$ $n = 3$ $n = 4$ $n = 5$

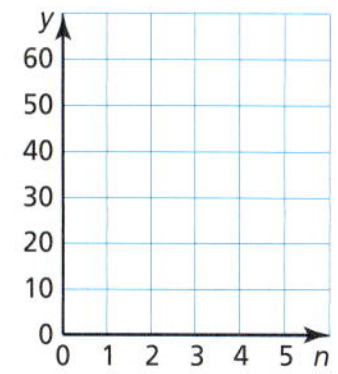

Number of stars, n	1	2	3	4	5
Number of sides, y					

b. $n = 1$ $n = 2$ $n = 3$ $n = 4$ $n = 5$

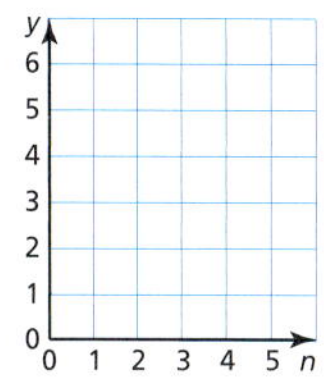

n	1	2	3	4	5
Number of circles, y					

c. $n = 1$ $n = 2$ $n = 3$ $n = 4$ $n = 5$

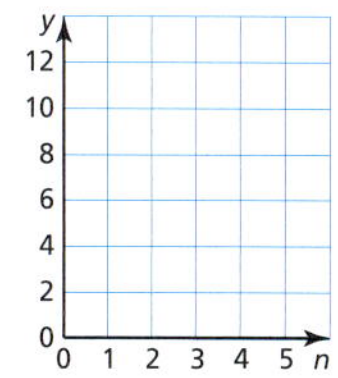

Number of rows, n	1	2	3	4	5
Number of dots, y					

LOOKING FOR A PATTERN

To be proficient in math, you need to look closely to discern patterns and structure.

Communicate Your Answer

2. How can you use an arithmetic sequence to describe a pattern? Give an example from real life.

3. In chemistry, water is called H_2O because each molecule of water has two hydrogen atoms and one oxygen atom. Describe the pattern shown below. Use the pattern to determine the number of atoms in 23 molecules.

$n = 1$ $n = 2$ $n = 3$ $n = 4$ $n = 5$

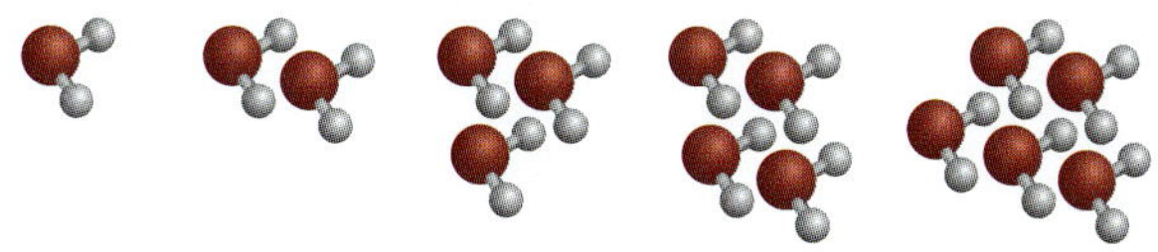

Dynamic Teaching Tools

- Dynamic Assessment & Progress Monitoring Tool
- Lesson Planning Tool
- Interactive Whiteboard Lesson Library
- Dynamic Classroom with Dynamic Investigations

ANSWERS

1. a. 10, 20, 30, 40, 50

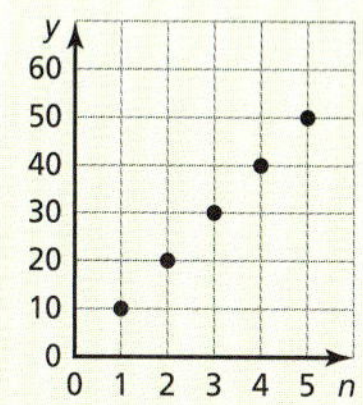

The y-value increases by 10 each time.

b. 2, 3, 4, 5, 6

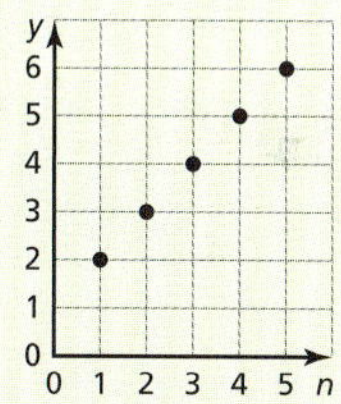

The y-value increases by 1 each time.

c. 2, 4, 6, 8, 10

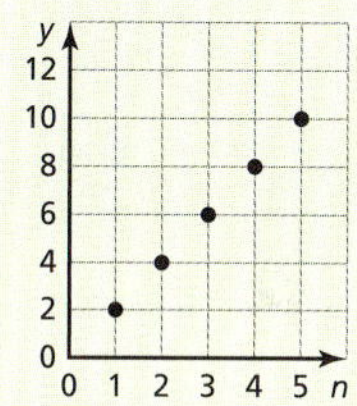

The y-value increases by 2 each time.

2. An arithmetic sequence can describe a pattern in which the difference between consecutive terms is the same; *Sample answer:* The amount of money earned for selling candy bars at \$2 each.

3. Each molecule added to the group increases the number of atoms by 3; 69 atoms

Differentiated Instruction

Auditory

Students may benefit from articulating the process for extending an arithmetic sequence by using the common difference. Have students take turns working at the board with one of several tables showing the terms of a sequence. Students should calculate the common difference and then use it to find the next 3 terms, explaining each step in the process.

Extra Example 1

Write the next three terms of the arithmetic sequence.

$-2, 6, 14, 22, \ldots$

The next three terms are 30, 38, and 46.

MONITORING PROGRESS ANSWERS

1. 36, 48, 60
2. 1.8, 2.2, 2.6
3. $3, 2\frac{3}{4}, 2\frac{1}{2}$

4.6 Lesson

What You Will Learn

- Write the terms of arithmetic sequences.
- Graph arithmetic sequences.
- Write arithmetic sequences as functions.

Core Vocabulary

sequence, *p. 210*
term, *p. 210*
arithmetic sequence, *p. 210*
common difference, *p. 210*

Previous
point-slope form
function notation

Writing the Terms of Arithmetic Sequences

A **sequence** is an ordered list of numbers. Each number in a sequence is called a **term**. Each term a_n has a specific position n in the sequence.

$5, \; 10, \; 15, \; 20, \; 25, \ldots, a_n, \ldots$

1st position (5)   3rd position (15)   *n*th position (a_n)

Core Concept

Arithmetic Sequence

In an **arithmetic sequence**, the difference between each pair of consecutive terms is the same. This difference is called the **common difference**. Each term is found by adding the common difference to the previous term.

$5, \; 10, \; 15, \; 20, \ldots$ Terms of an arithmetic sequence

+5   +5   +5   ← common difference

READING

An ellipsis (. . .) is a series of dots that indicates an intentional omission of information. In mathematics, the . . . notation means "and so forth." The ellipsis indicates that there are more terms in the sequence that are not shown.

EXAMPLE 1 Extending an Arithmetic Sequence

Write the next three terms of the arithmetic sequence.

$-7, -14, -21, -28, \ldots$

SOLUTION

Use a table to organize the terms and find the pattern.

Position	1	2	3	4
Term	−7	−14	−21	−28

+(−7)   +(−7)   +(−7)

Each term is 7 less than the previous term. So, the common difference is −7.

Add −7 to a term to find the next term.

Position	1	2	3	4	5	6	7
Term	−7	−14	−21	−28	−35	−42	−49

+(−7)   +(−7)   +(−7)

▶ The next three terms are −35, −42, and −49.

Monitoring Progress

Help in English and Spanish at *BigIdeasMath.com*

Write the next three terms of the arithmetic sequence.

1. $-12, 0, 12, 24, \ldots$
2. $0.2, 0.6, 1, 1.4, \ldots$
3. $4, 3\frac{3}{4}, 3\frac{1}{2}, 3\frac{1}{4}, \ldots$

Laurie's Notes Teacher Actions

- Write a *sequence*. Introduce the notation a_n and identify various positions in a sequence. Explain that when there is *common difference*, it is called an *arithmetic sequence*. The use of curved arrows helps to visually identify the common difference.
- **? Fact-First Questioning:** "Arithmetic sequences are not all increasing sequences. How can this be the case?" When the common difference is negative, it means a negative quantity is being added to each successive term.
- **Use Appropriate Tools Strategically:** The table is a helpful tool to organize the terms in the arithmetic sequence.

Graphing Arithmetic Sequences

To graph a sequence, let a term's position number n in the sequence be the x-value. The term a_n is the corresponding y-value. Plot the ordered pairs (n, a_n).

EXAMPLE 2 Graphing an Arithmetic Sequence

Graph the arithmetic sequence 4, 8, 12, 16, ... What do you notice?

SOLUTION

Make a table. Then plot the ordered pairs (n, a_n).

Position, n	Term, a_n
1	4
2	8
3	12
4	16

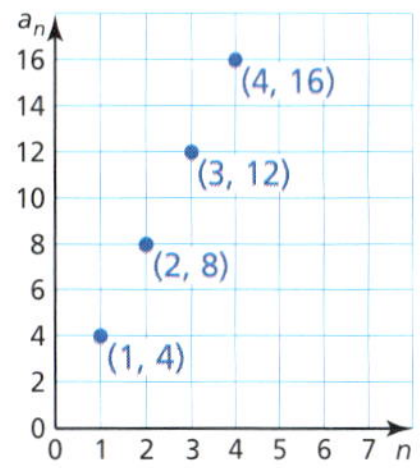

▶ The points lie on a line.

EXAMPLE 3 Identifying an Arithmetic Sequence from a Graph

Does the graph represent an arithmetic sequence? Explain.

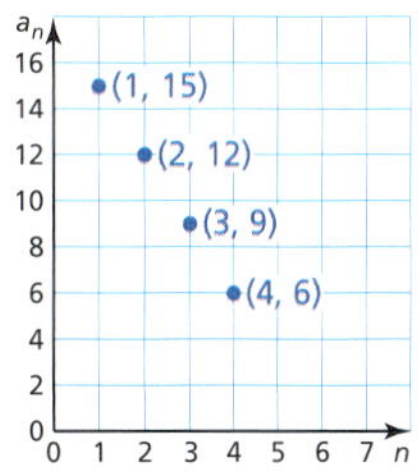

SOLUTION

Make a table to organize the ordered pairs. Then determine whether there is a common difference.

Position, n	1	2	3	4
Term, a_n	15	12	9	6

+(−3) +(−3) +(−3)

Each term is 3 less than the previous term. So, the common difference is −3.

▶ Consecutive terms have a common difference of −3. So, the graph represents the arithmetic sequence 15, 12, 9, 6, ...

Monitoring Progress

Help in English and Spanish at *BigIdeasMath.com*

Graph the arithmetic sequence. What do you notice?

4. 3, 6, 9, 12, . . . **5.** 4, 2, 0, −2, . . . **6.** 1, 0.8, 0.6, 0.4, . . .

7. Does the graph shown represent an arithmetic sequence? Explain.

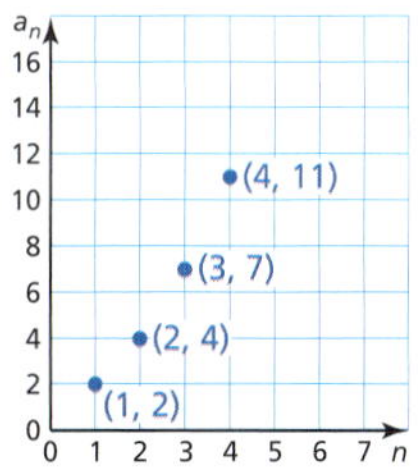

Laurie's Notes Teacher Actions

- **Attend to Precision:** Note that the axes are labeled n and a_n. These are the input and output variables. Point out that the scaling on the axes is different.
- **Big Idea:** The points on the graph of an arithmetic sequence lie on a line. The slope of the line is the common difference, or the constant rate of change. Use *Thumbs Up* to assess students' understanding of this *Big Idea*.
- **Think-Pair-Share:** Have students answer the *Monitoring Progress* questions and then share and discuss as a class.

Extra Example 2

Graph the arithmetic sequence 24, 18, 12, 6, ... What do you notice?

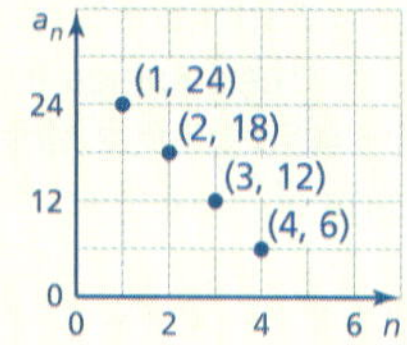

The points lie on a line.

Extra Example 3

Does the graph represent an arithmetic sequence? Explain.

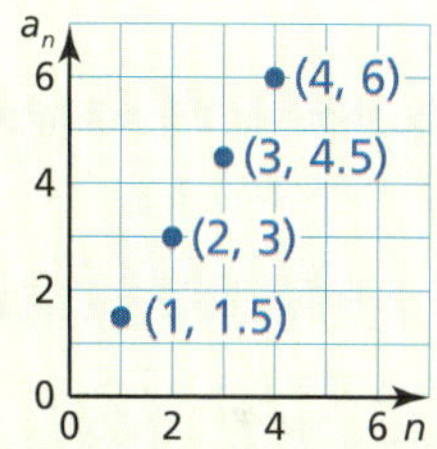

Consecutive terms have a common difference of 1.5. So, the graph represents the arithmetic sequence 1.5, 3, 4.5, 6, ...

MONITORING PROGRESS ANSWERS

4.

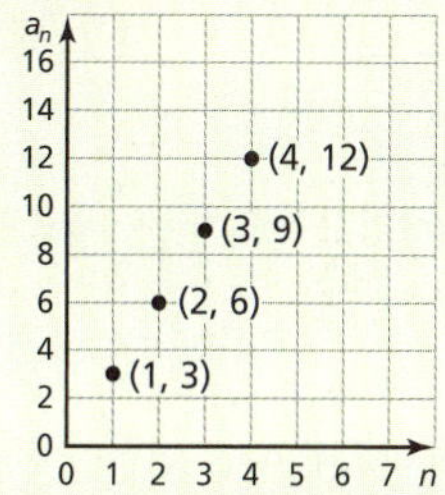

The points lie on a line.

5.

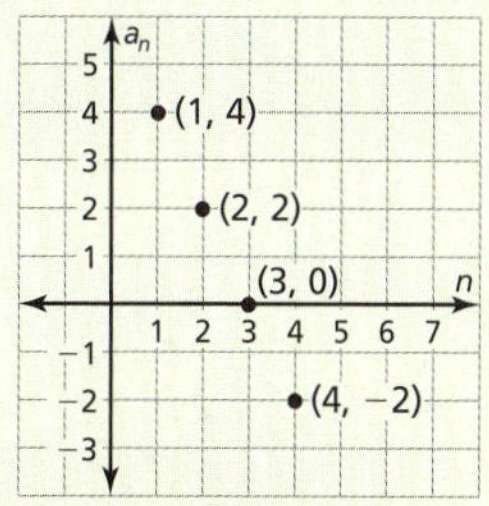

The points lie on a line.

6. See Additional Answers.

7. no; Consecutive terms do not have a common difference.

English Language Learners

Build on Past Knowledge

Focus students' attention on the Study Tip. Show how the equation from Example 4 is written in the slope-intercept form of a linear equation.

$$a_n = -3n + 17$$
$$y = mx + b$$

Emphasize that the equation they use to find the nth term of an arithmetic sequence can be written in this form.

Extra Example 4

Write an equation for the nth term of the arithmetic sequence 7, 3, −1, −5, ... Then find a_{30}.

$a_n = -4n + 11$; The 30th term of the arithmetic sequence is -109.

MONITORING PROGRESS ANSWERS

8. $a_n = n + 3$; 28

9. $a_n = 8n$; 200

10. $a_n = -n + 2$; -23

Writing Arithmetic Sequences as Functions

Because consecutive terms of an arithmetic sequence have a common difference, the sequence has a constant rate of change. So, the points represented by any arithmetic sequence lie on a line. You can use the first term and the common difference to write a linear function that describes an arithmetic sequence. Let $a_1 = 4$ and $d = 3$.

Position, n	Term, a_n	Written using a_1 and d	Numbers
1	first term, a_1	a_1	4
2	second term, a_2	$a_1 + d$	$4 + 3 = 7$
3	third term, a_3	$a_1 + 2d$	$4 + 2(3) = 10$
4	fourth term, a_4	$a_1 + 3d$	$4 + 3(3) = 13$
⋮	⋮	⋮	⋮
n	nth term, a_n	$a_1 + (n - 1)d$	$4 + (n - 1)(3)$

ANOTHER WAY

An *arithmetic sequence* is a linear function whose domain is the set of positive integers. You can think of d as the slope and $(1, a_1)$ as a point on the graph of the function. An equation in point-slope form for the function is

$$a_n - a_1 = d(n - 1).$$

This equation can be rewritten as

$$a_n = a_1 + (n - 1)d.$$

Core Concept

Equation for an Arithmetic Sequence

Let a_n be the nth term of an arithmetic sequence with first term a_1 and common difference d. The nth term is given by

$$a_n = a_1 + (n - 1)d.$$

EXAMPLE 4 Finding the nth Term of an Arithmetic Sequence

Write an equation for the nth term of the arithmetic sequence 14, 11, 8, 5, ... Then find a_{50}.

SOLUTION

The first term is 14, and the common difference is -3.

$a_n = a_1 + (n - 1)d$ — Equation for an arithmetic sequence

$a_n = 14 + (n - 1)(-3)$ — Substitute 14 for a_1 and -3 for d.

$a_n = -3n + 17$ — Simplify.

Use the equation to find the 50th term.

$a_n = -3n + 17$ — Write the equation.

$a_{50} = -3(50) + 17$ — Substitute 50 for n.

$= -133$ — Simplify.

▶ The 50th term of the arithmetic sequence is -133.

STUDY TIP

Notice that the equation in Example 4 is of the form $y = mx + b$, where y is replaced by a_n and x is replaced by n.

Monitoring Progress

Help in English and Spanish at *BigIdeasMath.com*

Write an equation for the nth term of the arithmetic sequence. Then find a_{25}.

8. 4, 5, 6, 7, . . .

9. 8, 16, 24, 32, . . .

10. 1, 0, −1, −2, . . .

Laurie's Notes Teacher Actions

- Generate consecutive terms in a sequence as shown. This helps students understand why $a_n = a_1 + (n - 1)d$ makes sense. You can also read the equation based on the meanings of the variables: "The nth term of the sequence is equal to the first term plus one less than the number of terms multiplied by the common difference."
- **Connection:** Show how the point-slope form could be used to write the equation, as shown in *Another Way*.

You can rewrite the equation for an arithmetic sequence with first term a_1 and common difference d in function notation by replacing a_n with $f(n)$.

$$f(n) = a_1 + (n - 1)d$$

The domain of the function is the set of positive integers.

EXAMPLE 5 Writing Real-Life Functions

Online bidding for a purse increases by \$5 for each bid after the \$60 initial bid.

Bid number	1	2	3	4
Bid amount	\$60	\$65	\$70	\$75

a. Write a function that represents the arithmetic sequence.

b. Graph the function.

c. The winning bid is \$105. How many bids were there?

SOLUTION

a. The first term is 60, and the common difference is 5.

$f(n) = a_1 + (n - 1)d$ — Function for an arithmetic sequence

$f(n) = 60 + (n - 1)5$ — Substitute 60 for a_1 and 5 for d.

$f(n) = 5n + 55$ — Simplify.

▶ The function $f(n) = 5n + 55$ represents the arithmetic sequence.

b. Make a table. Then plot the ordered pairs (n, a_n).

REMEMBER
The domain is the set of positive integers.

Bid number, n	Bid amount, a_n
1	60
2	65
3	70
4	75

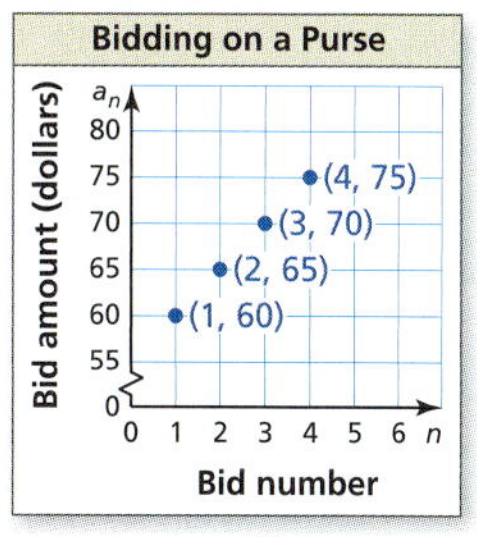

c. Use the function to find the value of n for which $f(n) = 105$.

$f(n) = 5n + 55$ — Write the function.

$105 = 5n + 55$ — Substitute 105 for $f(n)$.

$10 = n$ — Solve for n.

▶ There were 10 bids.

Monitoring Progress

Help in English and Spanish at *BigIdeasMath.com*

Games	Total cost
1	\$7
2	\$9
3	\$11
4	\$13

11. A carnival charges \$2 for each game after you pay a \$5 entry fee.

a. Write a function that represents the arithmetic sequence.

b. Graph the function.

c. How many games can you play when you take \$29 to the carnival?

Extra Example 5

Juan saves \$12 the first week of the year. He will increase the amount he saves each week by \$4.

Week	1	2	3	4
Amount	\$12	\$16	\$20	\$24

a. Write a function that represents the arithmetic sequence.
$f(n) = 4n + 8$

b. Graph the function.

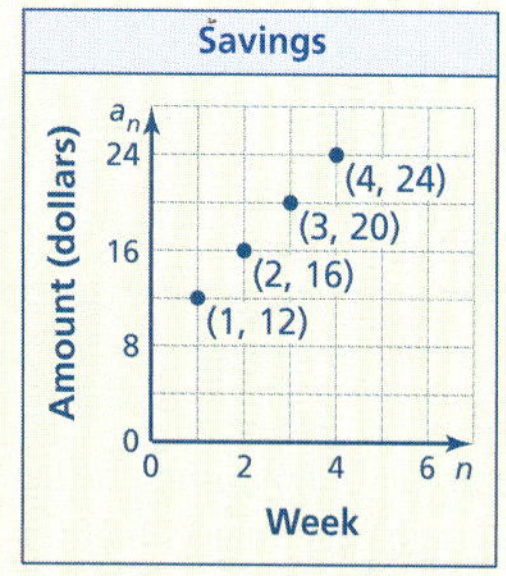

c. Juan's goal is to save \$68 in one week. In which week will he save that amount? Week 15

MONITORING PROGRESS ANSWER

11. a. $f(n) = 2n + 5$

b.

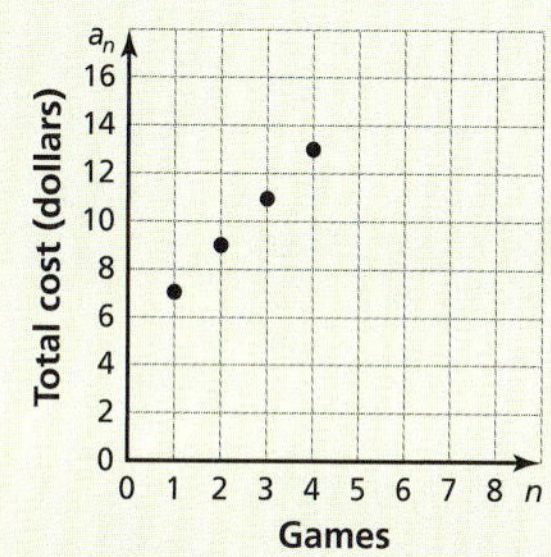

c. 12 games

Laurie's Notes Teacher Actions

? "Why is the domain of an arithmetic sequence the set of positive integers?" The input is the term number, which is 1, 2, 3, ...

- Writing the function generally makes sense to students when they do not get bogged down with notation. Work through the example.

Closure

- **3-2-1:** Hand out a reflection sheet as described in the *Formative Assessment Tips*.

Assignment Guide and Homework Check

ASSIGNMENT

Basic: 1, 2, 3–25 odd, 33–39 odd, 45, 50, 54, 58–65

Average: 1, 2–40 even, 46–50 even, 54, 58–65

Advanced: 1, 2, 16, 22, 26–48 even, 50–65

HOMEWORK CHECK

Basic: 11, 17, 23, 33, 45

Average: 12, 18, 24, 34, 46

Advanced: 16, 22, 26, 38, 46

4.6 Exercises

Dynamic Solutions available at *BigIdeasMath.com*

Vocabulary and Core Concept Check

1. **WRITING** Describe the graph of an arithmetic sequence.

2. **DIFFERENT WORDS, SAME QUESTION** Consider the arithmetic sequence represented by the graph. Which is different? Find "both" answers.

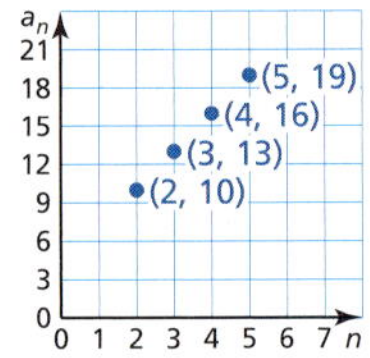

Find the slope of the linear function.	Find the difference between consecutive terms of the arithmetic sequence.
Find the difference between the terms a_2 and a_4.	Find the common difference of the arithmetic sequence.

Monitoring Progress and Modeling with Mathematics

In Exercises 3 and 4, write the next three terms of the arithmetic sequence.

3. First term: 2
   Common difference: 13

4. First term: 18
   Common difference: –6

In Exercises 5–10, find the common difference of the arithmetic sequence.

5. 13, 18, 23, 28, . . .
6. 175, 150, 125, 100, . . .
7. $-16, -12, -8, -4, \ldots$
8. $4, 3\frac{2}{3}, 3\frac{1}{3}, 3, \ldots$
9. 6.5, 5, 3.5, 2, . . .
10. $-16, -7, 2, 11, \ldots$

In Exercises 11–16, write the next three terms of the arithmetic sequence. *(See Example 1.)*

11. 19, 22, 25, 28, . . .
12. 1, 12, 23, 34, . . .
13. 16, 21, 26, 31, . . .
14. $60, 30, 0, -30, \ldots$
15. 1.3, 1, 0.7, 0.4, . . .
16. $\frac{5}{6}, \frac{2}{3}, \frac{1}{2}, \frac{1}{3}, \ldots$

In Exercises 17–22, graph the arithmetic sequence. *(See Example 2.)*

17. 4, 12, 20, 28, . . .
18. $-15, 0, 15, 30, \ldots$
19. –1, –3, –5, –7, . . .
20. 2, 19, 36, 53, . . .
21. $0, 4\frac{1}{2}, 9, 13\frac{1}{2}, \ldots$
22. 6, 5.25, 4.5, 3.75, . . .

In Exercises 23–26, determine whether the graph represents an arithmetic sequence. Explain. *(See Example 3.)*

23.
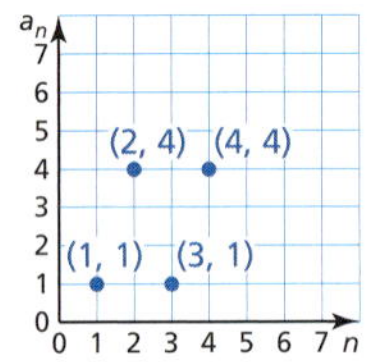

24.
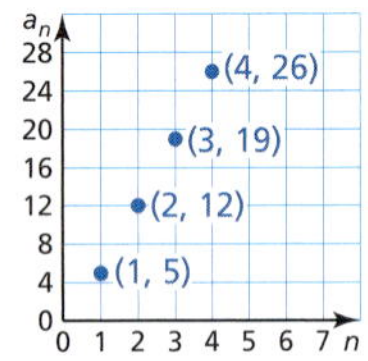

25.
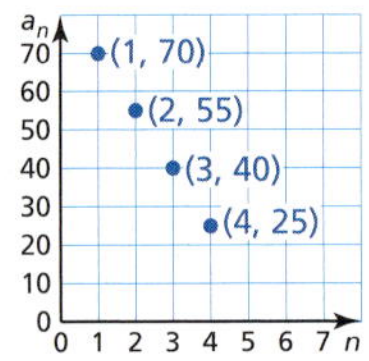

26.
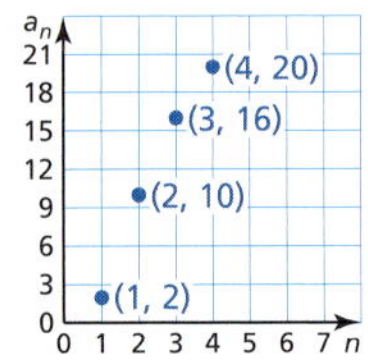

In Exercises 27–30, determine whether the sequence is arithmetic. If so, find the common difference.

27. 13, 26, 39, 52, . . .
28. 5, 9, 14, 20, . . .
29. 48, 24, 12, 6, . . .
30. 87, 81, 75, 69, . . .

31. **FINDING A PATTERN** Write a sequence that represents the number of smiley faces in each group. Is the sequence arithmetic? Explain.

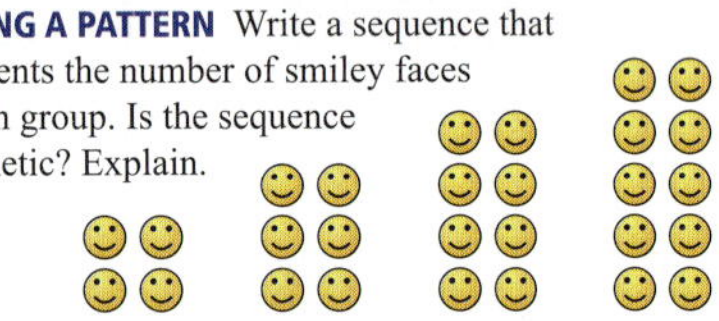

ANSWERS

1. The graph of an arithmetic sequence is a graph of a linear function whose domain is the set of positive integers.
2. Find the difference between the terms a_2 and a_4; 6; 3
3. 15, 28, 41
4. 12, 6, 0
5. 5
6. -25
7. 4
8. $-\frac{1}{3}$
9. -1.5
10. 9
11. 31, 34, 37
12. 45, 56, 67
13. 36, 41, 46
14. $-60, -90, -120$
15. $0.1, -0.2, -0.5$
16. $\frac{1}{6}, 0, -\frac{1}{6}$
17.
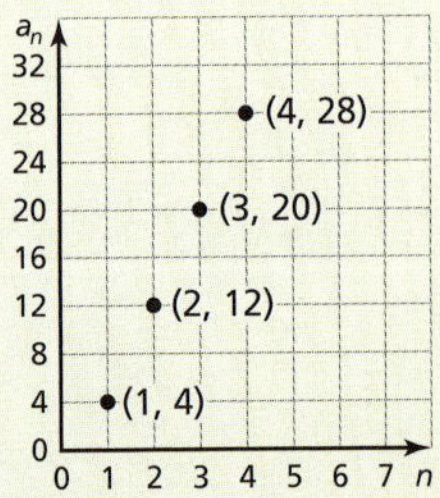

18.
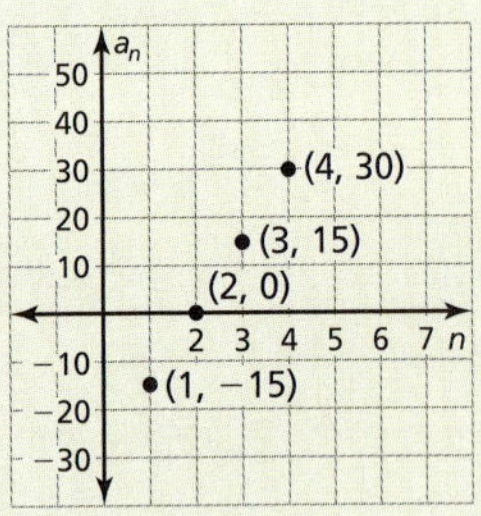

19.
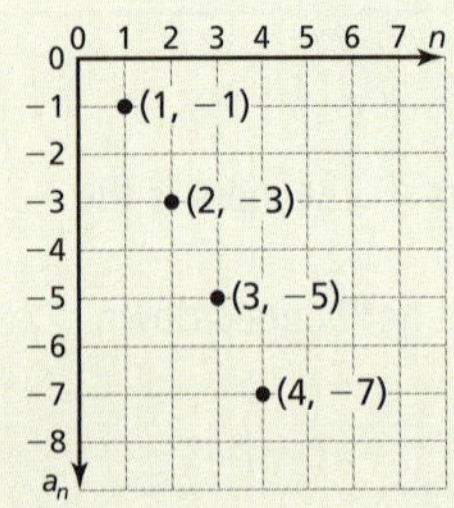

20.
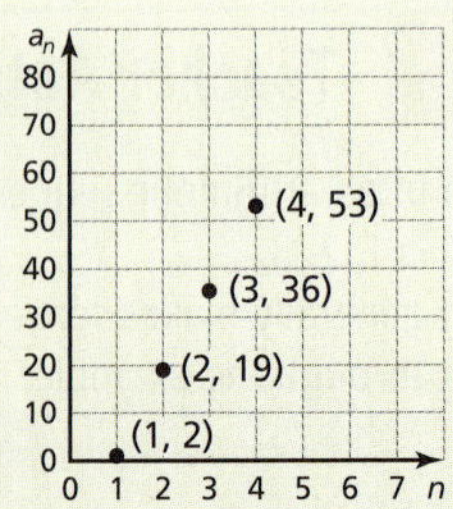

21–31. See Additional Answers.

32. **FINDING A PATTERN** Write a sequence that represents the sum of the numbers in each roll. Is the sequence arithmetic? Explain.

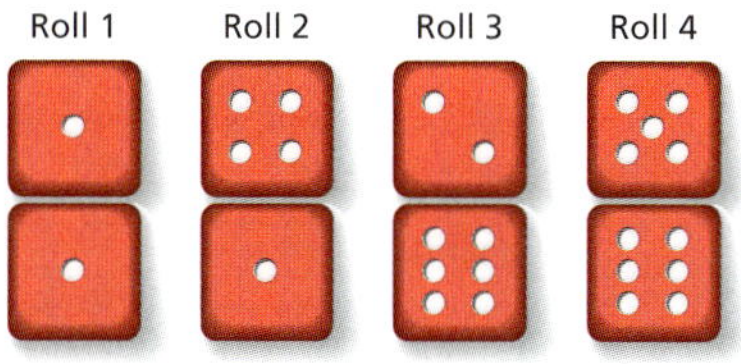

In Exercises 33–38, write an equation for the nth term of the arithmetic sequence. Then find a_{10}. *(See Example 4.)*

33. $-5, -4, -3, -2, \ldots$
34. $-6, -9, -12, -15, \ldots$
35. $\frac{1}{2}, 1, 1\frac{1}{2}, 2, \ldots$
36. $100, 110, 120, 130, \ldots$
37. $10, 0, -10, -20, \ldots$
38. $\frac{3}{7}, \frac{4}{7}, \frac{5}{7}, \frac{6}{7}, \ldots$

39. **ERROR ANALYSIS** Describe and correct the error in finding the common difference of the arithmetic sequence.

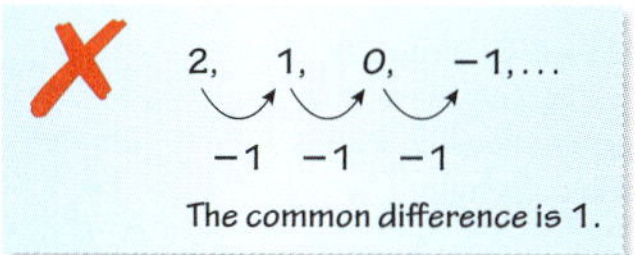

40. **ERROR ANALYSIS** Describe and correct the error in writing an equation for the nth term of the arithmetic sequence.

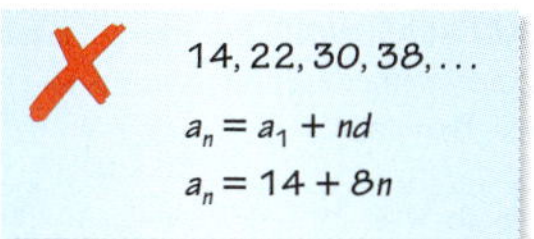

41. **NUMBER SENSE** The first term of an arithmetic sequence is 3. The common difference of the sequence is 1.5 times the first term. Write the next three terms of the sequence. Then graph the sequence.

42. **NUMBER SENSE** The first row of a dominoes display has 10 dominoes. Each row after the first has two more dominoes than the row before it. Write the first five terms of the sequence that represents the number of dominoes in each row. Then graph the sequence.

REPEATED REASONING In Exercises 43 and 44, (a) draw the next three figures in the sequence and (b) describe the 20th figure in the sequence.

43.

44.

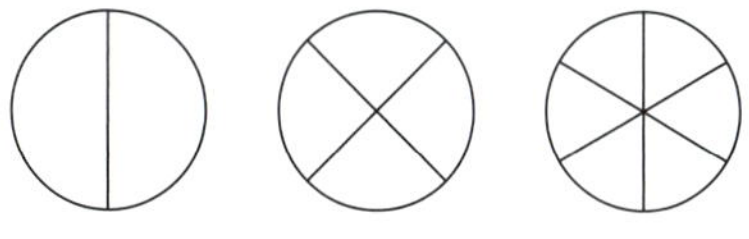

45. **MODELING WITH MATHEMATICS** The total number of babies born in a country each minute after midnight January 1st can be estimated by the sequence shown in the table. *(See Example 5.)*

Minutes after midnight January 1st	1	2	3	4
Total babies born	5	10	15	20

a. Write a function that represents the arithmetic sequence.

b. Graph the function.

c. Estimate how many minutes after midnight January 1st it takes for 100 babies to be born.

46. **MODELING WITH MATHEMATICS** The amount of money a movie earns each week after its release can be approximated by the sequence shown in the graph.

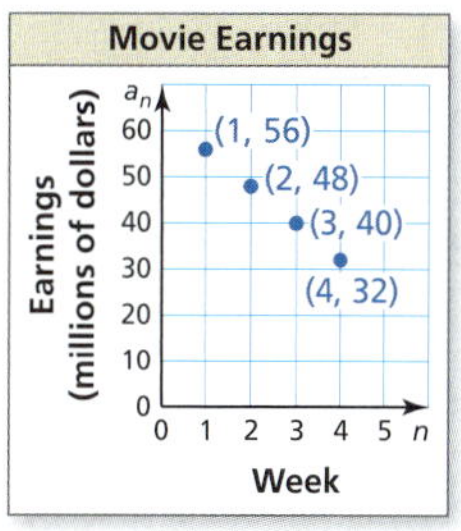

a. Write a function that represents the arithmetic sequence.

b. In what week does the movie earn \$16 million?

c. How much money does the movie earn overall?

Dynamic Teaching Tools

Dynamic Assessment & Progress Monitoring Tool
Interactive Whiteboard Lesson Library
Dynamic Classroom with Dynamic Investigations

ANSWERS

32. 2, 5, 8, 11; yes; Consecutive terms have a common difference of 3.

33. $a_n = n - 6; 4$

34. $a_n = -3n - 3; -33$

35. $a_n = \frac{1}{2}n; 5$

36. $a_n = 10n + 90; 190$

37. $a_n = -10n + 20; -80$

38. $a_n = \frac{1}{7}n + \frac{2}{7}; 1\frac{5}{7};$

39. -1 is added each time, not 1; The common difference is -1.

40. n should be $n - 1$;
$a_n = a_1 + (n - 1)d$;
$a_n = 14 + (n - 1)8; a_n = 8n + 6$

41. 7.5, 12, 16.5

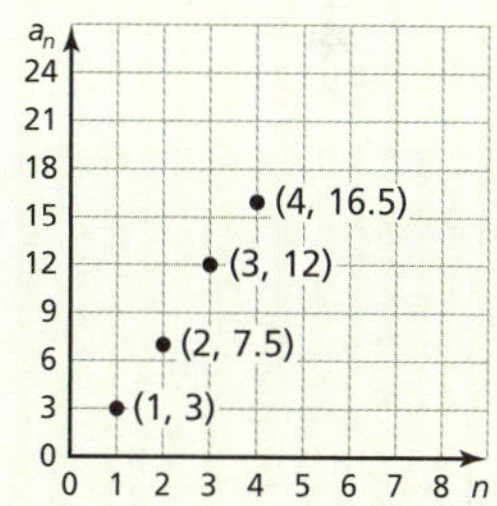

42. 10, 12, 14, 16, 18

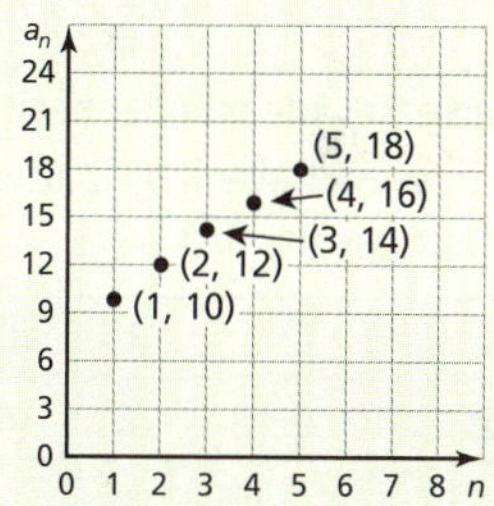

43–44. See Additional Answers.

45. a. $f(n) = 5n$

b.

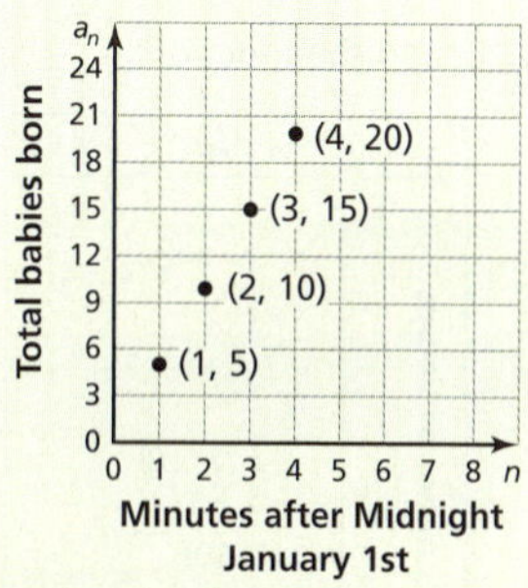

c. 20 min

46. a. $f(n) = -8n + 64$

b. week 6

c. \$224 million

ANSWERS

47. arithmetic sequence; $f(n) = 4n - 3$; 117

48. not an arithmetic sequence

49. discrete; discrete

50. no; If the first term of the sequence is negative and the output values constantly increase, or if the first term of the sequence is positive and the output values constantly decrease, the range contains positive and negative numbers.

51–65. See Additional Answers.

Mini-Assessment

1. Write the next three terms of the arithmetic sequence.
   $-4, -8, -12, -16, \ldots$
   -20, -24, and -28
2. Graph the arithmetic sequence.
   $5, 11, 17, 23, \ldots$

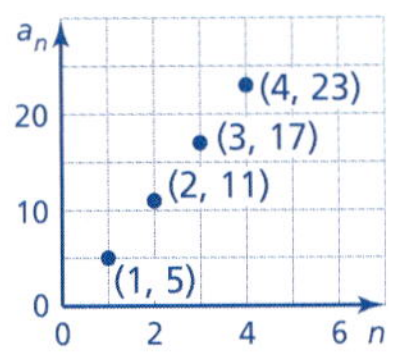

3. Dorrie increases the number of sit-ups she does each week by 8 after doing 10 sit-ups the first week.

Week	1	2	3	4
Sit-Ups	10	18	26	34

   a. Write a function that represents the arithmetic sequence.
   $f(n) = 8n + 2$

   b. Graph the function.

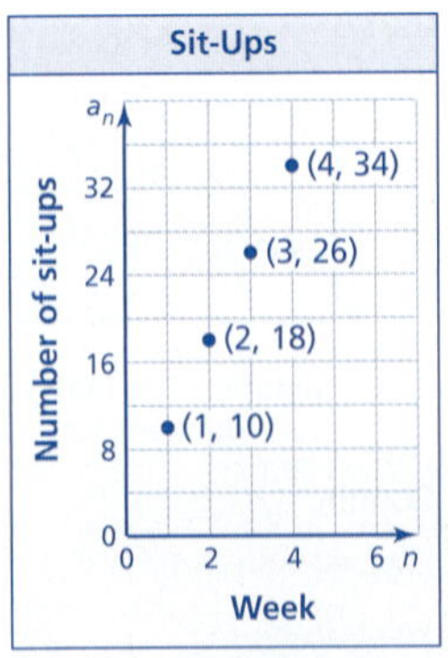

   c. Dorrie's goal is to do 74 sit-ups in one week. In which week will she meet that goal? Week 9

MATHEMATICAL CONNECTIONS **In Exercises 47 and 48, each small square represents 1 square inch. Determine whether the areas of the figures form an arithmetic sequence. If so, write a function f that represents the arithmetic sequence and find $f(30)$.**

47.

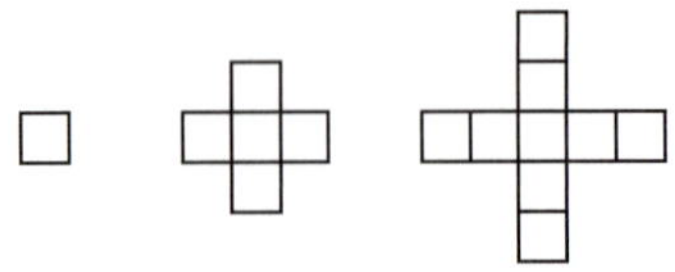

48.

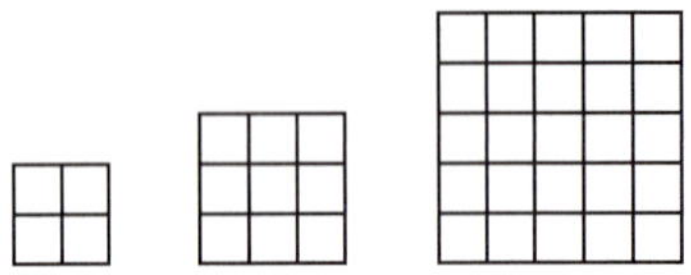

49. REASONING Is the domain of an arithmetic sequence discrete or continuous? Is the range of an arithmetic sequence discrete or continuous?

50. MAKING AN ARGUMENT Your friend says that the range of a function that represents an arithmetic sequence always contains only positive numbers or only negative numbers. Your friend claims this is true because the domain is the set of positive integers and the output values either constantly increase or constantly decrease. Is your friend correct? Explain.

51. OPEN-ENDED Write the first four terms of two different arithmetic sequences with a common difference of -3. Write an equation for the nth term of each sequence.

52. THOUGHT PROVOKING Describe an arithmetic sequence that models the numbers of people in a real-life situation.

53. REPEATED REASONING Firewood is stacked in a pile. The bottom row has 20 logs, and the top row has 14 logs. Each row has one more log than the row above it. How many logs are in the pile?

54. HOW DO YOU SEE IT? The bar graph shows the costs of advertising in a magazine.

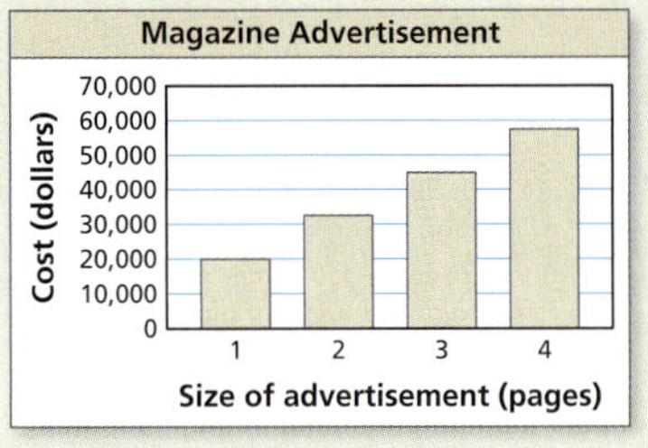

a. Does the graph represent an arithmetic sequence? Explain.

b. Explain how you would estimate the cost of a six-page advertisement in the magazine.

55. REASONING Write a function f that represents the arithmetic sequence shown in the mapping diagram.

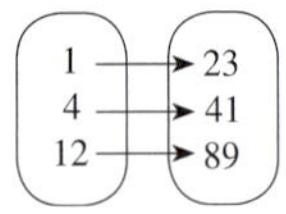

56. PROBLEM SOLVING A train stops at a station every 12 minutes starting at 6:00 A.M. You arrive at the station at 7:29 A.M. How long must you wait for the train?

57. ABSTRACT REASONING Let x be a constant. Determine whether each sequence is an arithmetic sequence. Explain.

a. $x + 6, 3x + 6, 5x + 6, 7x + 6, \ldots$

b. $x + 1, 3x + 1, 9x + 1, 27x + 1, \ldots$

Maintaining Mathematical Proficiency

Reviewing what you learned in previous grades and lessons

Solve the inequality. Graph the solution. *(Section 2.2)*

58. $x + 8 \geq -9$ **59.** $15 < b - 4$ **60.** $t - 21 < -12$ **61.** $7 + y \leq 3$

Graph the function. Compare the graph to the graph of $f(x) = |x|$. Describe the domain and range. *(Section 3.7)*

62. $h(x) = 3|x|$ **63.** $v(x) = |x - 5|$

64. $g(x) = |x| + 1$ **65.** $r(x) = -2|x|$

If students need help...	*If students got it...*
Resources by Chapter • Practice A and Practice B • Puzzle Time	Resources by Chapter • Enrichment and Extension • Cumulative Review
Student Journal • Practice	Start the *next* Section
Differentiating the Lesson Skills Review Handbook	

Laurie's Notes

Overview of Section 4.7

Introduction

- In this last section of the chapter, students will apply skills learned in the early part of the chapter when they evaluate, graph, and write *piecewise functions*. Another name for piecewise function is step function. Two special types of piecewise functions, the step function and absolute value functions, are presented at the end of the lesson.

Teaching Strategy

- Piecewise functions can be graphed on a calculator using Boolean logic.
- To graph the function

$$f(x) = \begin{cases} x - 4, & \text{if } x < 0 \\ -2, & \text{if } 0 < x < 2, \\ -x + 6, & \text{if } x > 2 \end{cases}$$

enter the following in the equation editor.

$(x - 4)(x < 0) + (-2)(x > 0 \text{ and } x < 2) + (-x + 6)(x > 2)$

- Each of the three expressions (or rules) is entered in parentheses and is followed by a specified domain. The domain is the Boolean operator. When the value of x is in that domain, it returns a Boolean value of 1 (true) and multiplies the expression by 1. When the value of x is *not* in that domain, it returns a Boolean value of 0 (false) and multiplies the expression by 0. By entering the three expressions with addition signs between them, only one of the three xpressions will be true at a time and the other two expressions will be 0.

```
\Y1=(X-4)(X<0)+(-2)
    (X>0andX<2)+(-X
    +6)(X>2)
\Y2=
\Y3=
\Y4=
\Y5=
```

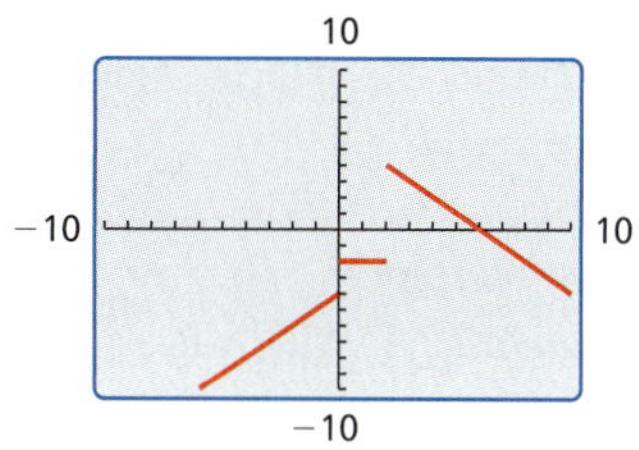

Another Way

- To sketch a piecewise function, lightly graph each line using the slope and y-intercept. Then use the domains to decide what portion of each line to keep and what to erase. This method is sometimes easier for students when the change between the expressions does not occur at $x = 0$ and when there are more than two expressions.

Pacing Suggestion

- Once students have worked the explorations, continue with the formal lesson.

Dynamic Teaching Tools

- Dynamic Assessment & Progress Monitoring Tool
- Lesson Planning Tool
- Interactive Whiteboard Lesson Library
- Dynamic Classroom with Dynamic Investigations

What Your Students Will Learn

- Evaluate piecewise functions.
- Graph and write piecewise functions and describe the domains and ranges.
- Graph and write step functions.
- Write absolute value functions as piecewise functions.

Laurie's Notes

Exploration

Motivate

- Ask students whether they have seen a utility bill where the cost per kilowatt-hour depended on the amount used. For instance, the first 5000 kilowatt-hours might be \$0.1782 per kilowatt-hour, then \$0.1219 per kilowatt-hour for the next 3000 kilowatt-hours, and so on.
- Explain to students that this type of billing scheme is an example of the type of function they will learn about today.

Discuss

- Review with students how to determine whether a graph is a function. If you have described the "vertical line test," then bring this into the discussion. Review the language associated with functions: domain, range, independent and dependent variables. Finally, mention the notation used for functions and how it is read. *Example:* $f(4) = -6$ is read, "f of 4 is equal to -6" or "the value of the function f at 4 is -6."
- You may also need to review compound inequalities, including how they are read and interpreted.

Exploration 1

- **Connection:** When solutions are plotted on a number line, the open and closed circles are used to represent when the value of x is or is not included in the solution set. When graphing piecewise-defined functions, also known as split-rule functions, the open and closed circles are used to represent when the value of x is included as a solution for each piece of the function.
- The first two questions help focus attention on where the graph changes from one rule to a new rule.

? "What information is needed to write the equation of either piece of the graph?" You need the slope and y-intercept.

- What can be confusing for students is the notation of specifying the domain for x (if $x < 0$). Explain that each rule applies to a different part of the domain.

Exploration 2

- In this exploration, students should recognize a series of constant functions. It is very common for students to describe the graph as looking like steps!

? Assessing Question: When students have finished, ask, "How do you know the graph represents a function?" Answers will vary. Listen for understanding that each value of the domain is associated with only one range value.

- **Extension:** Ask students whether this pattern, meaning the steps, could continue. You could also ask how the function would change if the segments were an open circle on the left endpoint and a closed circle on the right endpoint.

Communicate Your Answer

- In Question 4, check that students have only one function rule defined when $x = 0$.

Connecting to Next Step

- The short explorations provide an introduction to piecewise functions. You might consider omitting Example 3 in the formal lesson and checking for understanding with a few *Monitoring Progress* questions.

4.7 Piecewise Functions

Essential Question How can you describe a function that is represented by more than one equation?

EXPLORATION 1 **Writing Equations for a Function**

Work with a partner.

a. Does the graph represent y as a function of x? Justify your conclusion.

b. What is the value of the function when $x = 0$? How can you tell?

c. Write an equation that represents the values of the function when $x \le 0$.

$f(x) = $ ______, if $x \le 0$

d. Write an equation that represents the values of the function when $x > 0$.

$f(x) = $ ______, if $x > 0$

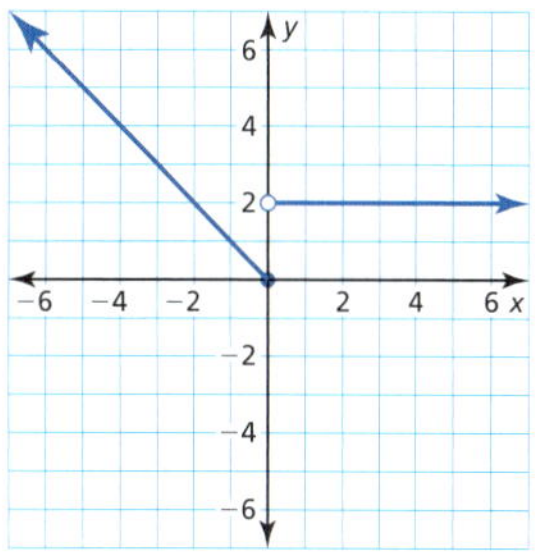

e. Combine the results of parts (c) and (d) to write a single description of the function.

$$f(x) = \begin{cases} \_\_\_\_\_\_, & \text{if } x \le 0 \\ \_\_\_\_\_\_, & \text{if } x > 0 \end{cases}$$

CONSTRUCTING VIABLE ARGUMENTS

To be proficient in math, you need to justify your conclusions and communicate them to others.

EXPLORATION 2 **Writing Equations for a Function**

Work with a partner.

a. Does the graph represent y as a function of x? Justify your conclusion.

b. Describe the values of the function for the following intervals.

$$f(x) = \begin{cases} \_\_\_\_\_\_, & \text{if } -6 \le x < -3 \\ \_\_\_\_\_\_, & \text{if } -3 \le x < 0 \\ \_\_\_\_\_\_, & \text{if } 0 \le x < 3 \\ \_\_\_\_\_\_, & \text{if } 3 \le x < 6 \end{cases}$$

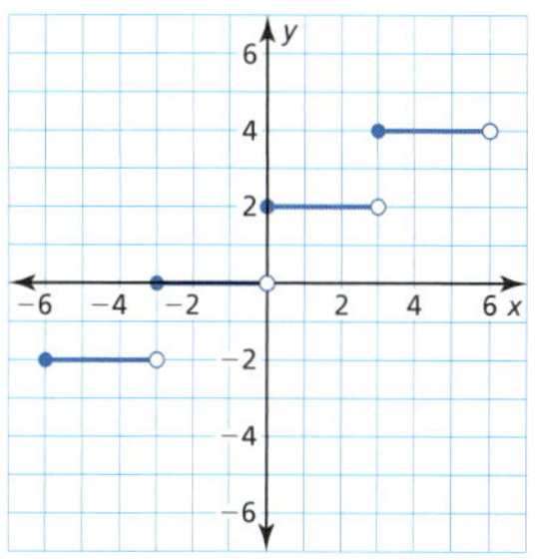

Communicate Your Answer

3. How can you describe a function that is represented by more than one equation?

4. Use two equations to describe the function represented by the graph.

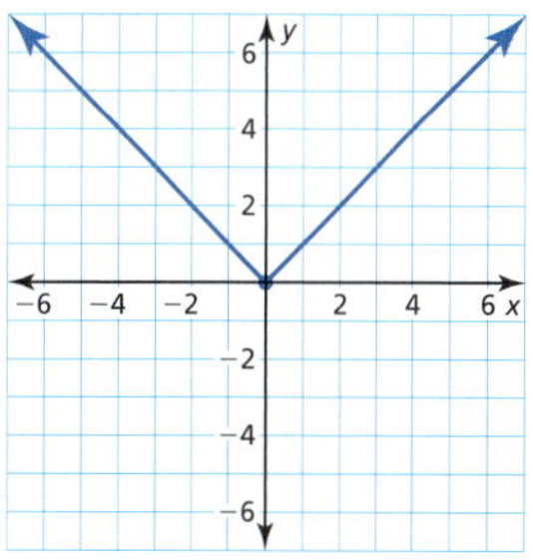

Dynamic Teaching Tools

- Dynamic Assessment & Progress Monitoring Tool
- Lesson Planning Tool
- Interactive Whiteboard Lesson Library
- Dynamic Classroom with Dynamic Investigations

ANSWERS

1. **a.** yes; No vertical line can be drawn through more than one point on the graph.
   b. 0; The point (0, 0) is plotted.
   c. $-x$
   d. 2
   e. $-x$; 2
2. **a.** yes; No vertical line can be drawn through more than one point on the graph.
   b. -2; 0; 2; 4
3. Write the expression for each part of the function along with the part of the domain to which the expression applies.
4. $f(x) = \begin{cases} -x, & \text{if } x \le 0 \\ x, & \text{if } x > 0 \end{cases}$ or $f(x) = \begin{cases} -x, & \text{if } x < 0 \\ x, & \text{if } x \ge 0 \end{cases}$

Differentiated Instruction

Inclusion

Some students may have difficulty associating piecewise functions with parts of a graph. Remind them that they can use two points from each piece of the graph to find the slope for each piece. They can also locate the y-intercept for each piece. Have them use the slope and y-intercept to verify the function given for each piece.

Extra Example 1

Evaluate the function.

$$f(x) = \begin{cases} x - 2, & \text{if } x \le 0 \\ 2x + 1, & \text{if } x > 0 \end{cases}$$

a. when $x = -1$

The value of f is -3 when $x = -1$.

b. when $x = 1$

The value of f is 3 when $x = 1$.

MONITORING PROGRESS ANSWERS

1. 3
2. 0
3. 2
4. 5
5. 7
6. 40

4.7 Lesson

Core Vocabulary

piecewise function, *p. 218*
step function, *p. 220*

Previous
absolute value function
vertex form
vertex

What You Will Learn

- Evaluate piecewise functions.
- Graph and write piecewise functions.
- Graph and write step functions.
- Write absolute value functions.

Evaluating Piecewise Functions

Core Concept

Piecewise Function

A **piecewise function** is a function defined by two or more equations. Each "piece" of the function applies to a different part of its domain. An example is shown below.

$$f(x) = \begin{cases} x - 2, & \text{if } x \le 0 \\ 2x + 1, & \text{if } x > 0 \end{cases}$$

- The expression $x - 2$ represents the value of f when x is less than or equal to 0.
- The expression $2x + 1$ represents the value of f when x is greater than 0.

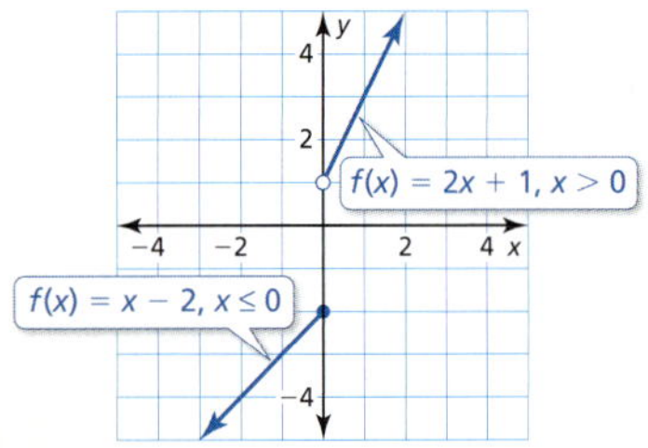

EXAMPLE 1 Evaluating a Piecewise Function

Evaluate the function f above when (a) $x = 0$ and (b) $x = 4$.

SOLUTION

a. $f(x) = x - 2$ — Because $0 \le 0$, use the first equation.

$f(0) = 0 - 2$ — Substitute 0 for x.

$f(0) = -2$ — Simplify.

▶ The value of f is -2 when $x = 0$.

b. $f(x) = 2x + 1$ — Because $4 > 0$, use the second equation.

$f(4) = 2(4) + 1$ — Substitute 4 for x.

$f(4) = 9$ — Simplify.

▶ The value of f is 9 when $x = 4$.

Monitoring Progress

Help in English and Spanish at *BigIdeasMath.com*

Evaluate the function.

$$f(x) = \begin{cases} 3, & \text{if } x < -2 \\ x + 2, & \text{if } -2 \le x \le 5 \\ 4x, & \text{if } x > 5 \end{cases}$$

1. $f(-8)$ **2.** $f(-2)$
3. $f(0)$ **4.** $f(3)$
5. $f(5)$ **6.** $f(10)$

Laurie's Notes — Teacher Actions

- Define a *piecewise function* and connect to the explorations.
- **Teaching Tip:** Graph or project a piecewise function. Cover the right portion of the function and ask students to describe the left portion. Switch and describe the right portion. Write the two descriptions together with the appropriate domains.
- **Monitoring Progress:** These questions are a quick check for understanding of how to evaluate a piecewise function.

Graphing and Writing Piecewise Functions

EXAMPLE 2 **Graphing a Piecewise Function**

Graph $y = \begin{cases} -x - 4, & \text{if } x < 0 \\ x, & \text{if } x \geq 0 \end{cases}$. Describe the domain and range.

SOLUTION

Step 1 Graph $y = -x - 4$ for $x < 0$. Because x is not equal to 0, use an open circle at $(0, -4)$.

Step 2 Graph $y = x$ for $x \geq 0$. Because x is greater than or equal to 0, use a closed circle at $(0, 0)$.

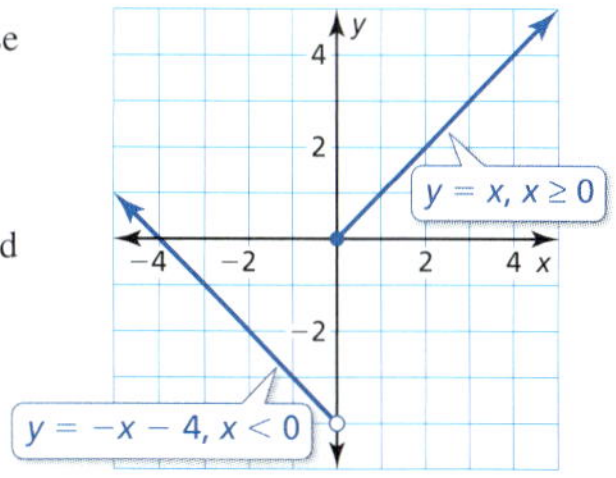

▶ The domain is all real numbers. The range is $y > -4$.

Monitoring Progress

Help in English and Spanish at *BigIdeasMath.com*

Graph the function. Describe the domain and range.

7. $y = \begin{cases} x + 1, & \text{if } x \leq 0 \\ -x, & \text{if } x > 0 \end{cases}$

8. $y = \begin{cases} x - 2, & \text{if } x < 0 \\ 4x, & \text{if } x \geq 0 \end{cases}$

EXAMPLE 3 **Writing a Piecewise Function**

Write a piecewise function for the graph.

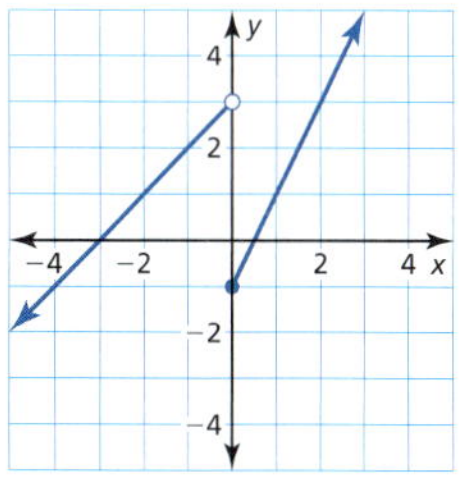

SOLUTION

Each "piece" of the function is linear.

Left Piece When $x < 0$, the graph is the line given by $y = x + 3$.

Right Piece When $x \geq 0$, the graph is the line given by $y = 2x - 1$.

▶ So, a piecewise function for the graph is

$$f(x) = \begin{cases} x + 3, & \text{if } x < 0 \\ 2x - 1, & \text{if } x \geq 0 \end{cases}.$$

Monitoring Progress

Help in English and Spanish at *BigIdeasMath.com*

Write a piecewise function for the graph.

9.

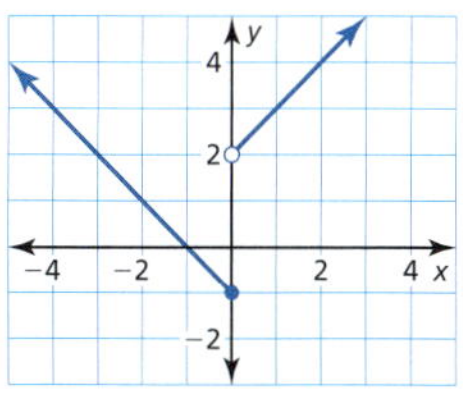

10.

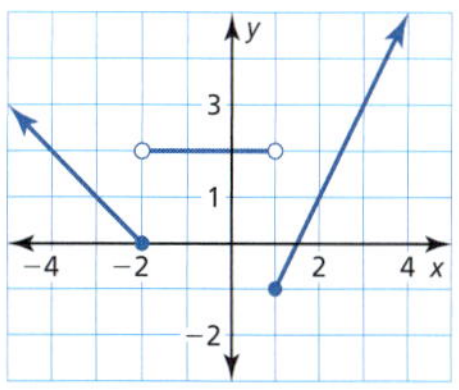

Laurie's Notes Teacher Actions

- **Use Appropriate Tools Strategically:** It is possible to graph piecewise functions with a graphing calculator. See the *Teaching Strategy*.
- In stating the range for Example 2, students should notice that all of the y-values are greater than -4. Sometimes students will find this difficult to answer until I move my hand in an upward motion starting at $y = -4$.
- **Whiteboarding:** Have students work with partners on Examples 2 and 3. Compare and discuss as a class.

Extra Example 2

Graph $y = \begin{cases} -x + 2, & \text{if } x \leq 0 \\ 2x, & \text{if } x > 0 \end{cases}$.

Describe the domain and range.

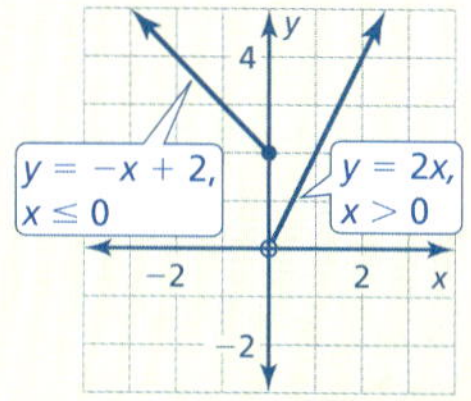

The domain is all real numbers. The range is $y > 0$.

Extra Example 3

Write a piecewise function for the graph.

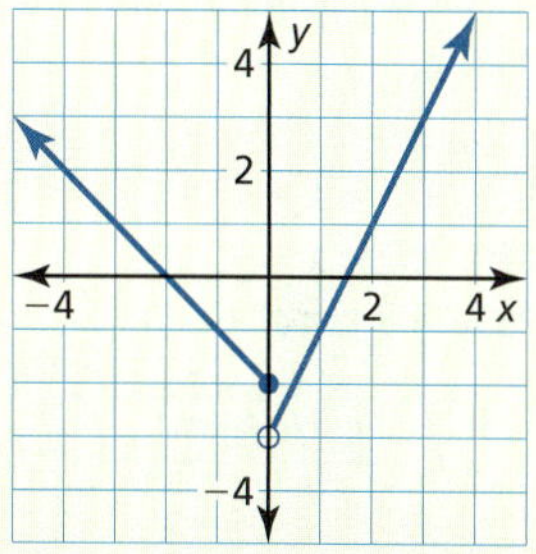

$y = \begin{cases} -x - 2, & \text{if } x \leq 0 \\ 2x - 3, & \text{if } x > 0 \end{cases}$

MONITORING PROGRESS ANSWERS

7.

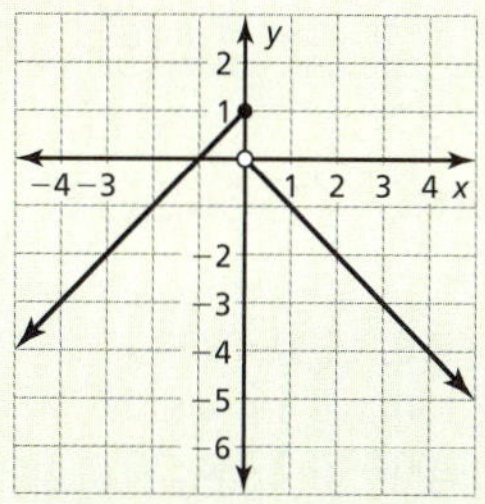

domain: all real numbers; range: $y \leq 1$

8.

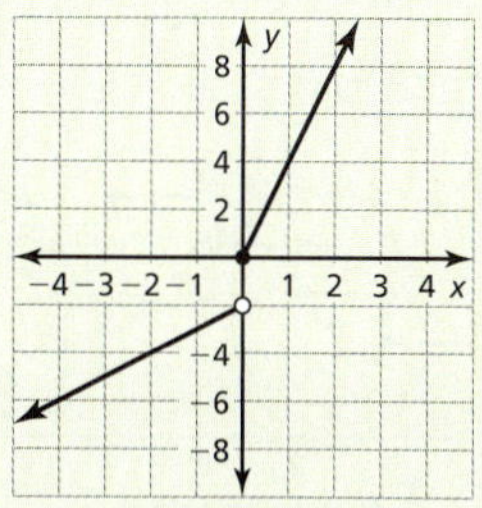

domain: all real numbers; range: $y < -2$ or $y \geq 0$

9. $f(x) = \begin{cases} -x - 1, & \text{if } x \leq 0 \\ x + 2, & \text{if } x > 0 \end{cases}$

10. $f(x) = \begin{cases} -x - 2, & \text{if } x \leq -2 \\ 2, & \text{if } -2 < x < 1 \\ 2x - 3, & \text{if } x \geq 1 \end{cases}$

English Language Learners

Culture

Explain to students that a karaoke machine is an entertainment device that plays the music of popular songs. People sing the words of the songs as the music plays.

Extra Example 4

You rent a bicycle for 4 days. The bike store charges \$20 for the first day and \$15 for each additional day. Write and graph a step function that represents the relationship between the number x of days and the total cost y (in dollars) of renting the bicycle.

$$f(x) = \begin{cases} 20, & \text{if } 0 < x \le 1 \\ 35, & \text{if } 1 < x \le 2 \\ 50, & \text{if } 2 < x \le 3 \\ 65, & \text{if } 3 < x \le 4 \end{cases}$$

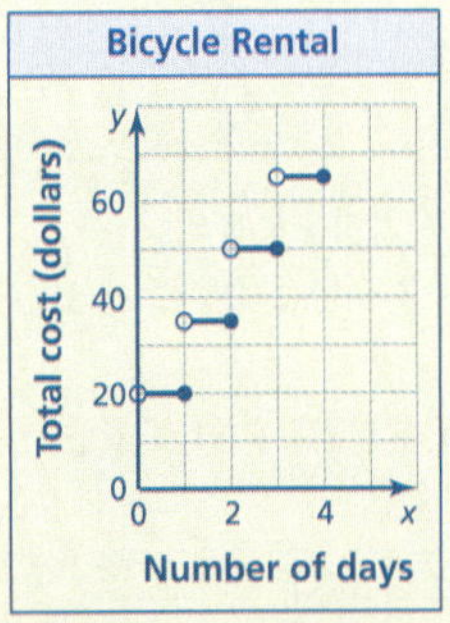

MONITORING PROGRESS ANSWER

11. $f(x) = \begin{cases} 100, & \text{if } 0 < x \le 1 \\ 150, & \text{if } 1 < x \le 2 \\ 200, & \text{if } 2 < x \le 3 \\ 250, & \text{if } 3 < x \le 4 \end{cases}$

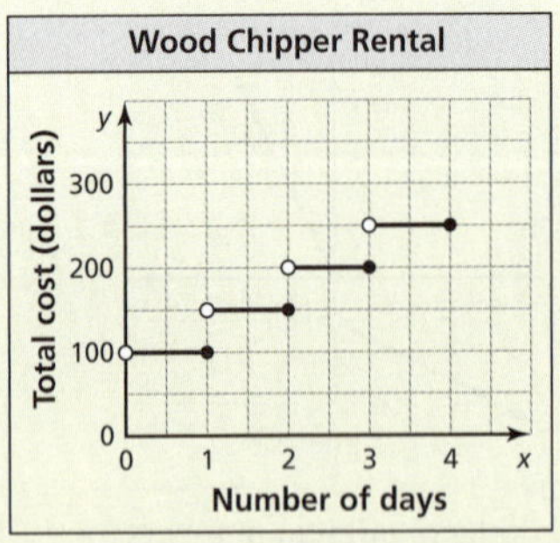

Graphing and Writing Step Functions

A **step function** is a piecewise function defined by a constant value over each part of its domain. The graph of a step function consists of a series of line segments.

STUDY TIP

The graph of a step function looks like a staircase.

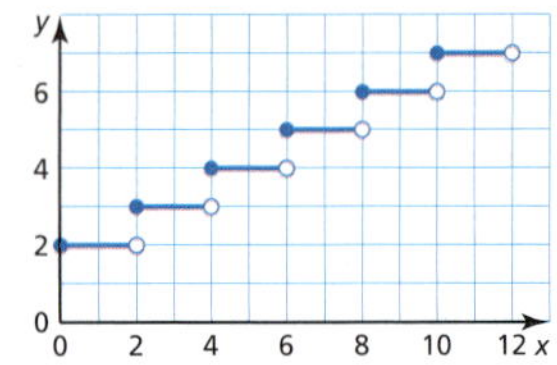

$$f(x) = \begin{cases} 2, & \text{if } 0 \le x < 2 \\ 3, & \text{if } 2 \le x < 4 \\ 4, & \text{if } 4 \le x < 6 \\ 5, & \text{if } 6 \le x < 8 \\ 6, & \text{if } 8 \le x < 10 \\ 7, & \text{if } 10 \le x < 12 \end{cases}$$

EXAMPLE 4 Graphing and Writing a Step Function

You rent a karaoke machine for 5 days. The rental company charges \$50 for the first day and \$25 for each additional day. Write and graph a step function that represents the relationship between the number x of days and the total cost y (in dollars) of renting the karaoke machine.

SOLUTION

Step 1 Use a table to organize the information.

Number of days	Total cost (dollars)
$0 < x \le 1$	50
$1 < x \le 2$	75
$2 < x \le 3$	100
$3 < x \le 4$	125
$4 < x \le 5$	150

Step 2 Write the step function.

$$f(x) = \begin{cases} 50, & \text{if } 0 < x \le 1 \\ 75, & \text{if } 1 < x \le 2 \\ 100, & \text{if } 2 < x \le 3 \\ 125, & \text{if } 3 < x \le 4 \\ 150, & \text{if } 4 < x \le 5 \end{cases}$$

Step 3 Graph the step function.

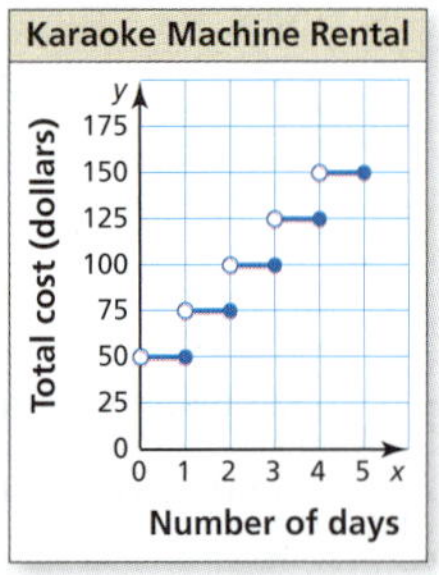

Monitoring Progress

Help in English and Spanish at *BigIdeasMath.com*

11. A landscaper rents a wood chipper for 4 days. The rental company charges \$100 for the first day and \$50 for each additional day. Write and graph a step function that represents the relationship between the number x of days and the total cost y (in dollars) of renting the chipper.

Laurie's Notes Teacher Actions

- **Make Sense of Problems and Persevere in Solving Them** and **Model with Mathematics:** The table helps organize the data. Compound inequalities are used to define the days. Because there are five days, the step function will have five expressions, each being a constant.

? "Why are the steps open at the left and closed at the right?" Once you go beyond a whole day, you pay for the next full day.

- **Think-Pair-Share:** Have students answer the *Monitoring Progress* question and then share and discuss as a class.

Writing Absolute Value Functions

The absolute value function $f(x) = |x|$ can be written as a piecewise function.

$$f(x) = \begin{cases} -x, & \text{if } x < 0 \\ x, & \text{if } x \geq 0 \end{cases}$$

REMEMBER

The vertex form of an absolute value function is $g(x) = a|x - h| + k$, where $a \neq 0$. The vertex of the graph of g is (h, k).

Similarly, the vertex form of an absolute value function $g(x) = a|x - h| + k$ can be written as a piecewise function.

$$g(x) = \begin{cases} a[-(x - h)] + k, & \text{if } x - h < 0 \\ a(x - h) + k, & \text{if } x - h \geq 0 \end{cases}$$

EXAMPLE 5 Writing an Absolute Value Function

In holography, light from a laser beam is split into two beams, a reference beam and an object beam. Light from the object beam reflects off an object and is recombined with the reference beam to form images on film that can be used to create three-dimensional images.

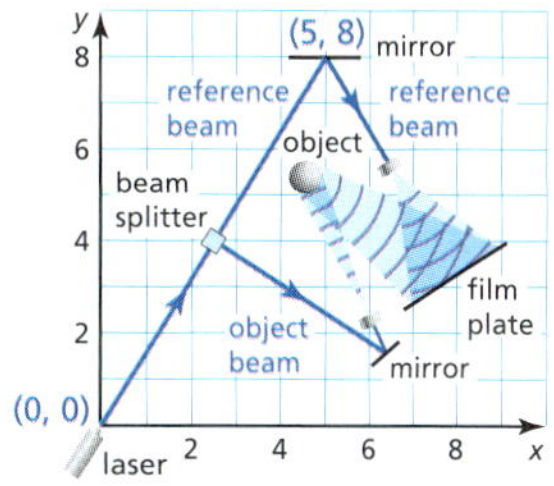

a. Write an absolute value function that represents the path of the reference beam.

b. Write the function in part (a) as a piecewise function.

SOLUTION

a. The vertex of the path of the reference beam is (5, 8). So, the function has the form $g(x) = a|x - 5| + 8$. Substitute the coordinates of the point (0, 0) into the equation and solve for a.

$g(x) = a|x - 5| + 8$ — Vertex form of the function

$0 = a|0 - 5| + 8$ — Substitute 0 for x and 0 for $g(x)$.

$-1.6 = a$ — Solve for a.

▶ So, the function $g(x) = -1.6|x - 5| + 8$ represents the path of the reference beam.

b. Write $g(x) = -1.6|x - 5| + 8$ as a piecewise function.

$$g(x) = \begin{cases} -1.6[-(x - 5)] + 8, & \text{if } x - 5 < 0 \\ -1.6(x - 5) + 8, & \text{if } x - 5 \geq 0 \end{cases}$$

Simplify each expression and solve the inequalities.

▶ So, a piecewise function for $g(x) = -1.6|x - 5| + 8$ is

$$g(x) = \begin{cases} 1.6x, & \text{if } x < 5 \\ -1.6x + 16, & \text{if } x \geq 5 \end{cases}$$

STUDY TIP

Recall that the graph of an absolute value function is symmetric about the line $x = h$. So, it makes sense that the piecewise definition "splits" the function at $x = 5$.

Monitoring Progress

Help in English and Spanish at *BigIdeasMath.com*

12. **WHAT IF?** The reference beam originates at (3, 0) and reflects off a mirror at (5, 4).

    a. Write an absolute value function that represents the path of the reference beam.

    b. Write the function in part (a) as a piecewise function.

Extra Example 5

The reference beam in Example 5 originates at (1, 0) and reflects off a mirror at (5, 6).

a. Write an absolute value function that represents the path of the reference beam.

$g(x) = -1.5|x - 5| + 6$

b. Write the function in part (a) as a piecewise function.

$$g(x) = \begin{cases} 1.5x - 1.5, & \text{if } x < 5 \\ -1.5x + 13.5, & \text{if } x \geq 5 \end{cases}$$

MONITORING PROGRESS ANSWER

12. **a.** $g(x) = -2|x - 5| + 4$

    b. $g(x) = \begin{cases} 2x - 6, & \text{if } x < 5 \\ -2x + 14, & \text{if } x \geq 5 \end{cases}$

Laurie's Notes Teacher Actions

- Discuss how to write the absolute value function as a piecewise function. Have students think about the equation for each part. Discuss the vertex form and how it can be written in two parts.
- **Turn and Talk:** Pose Example 5 and have partners discuss ideas for writing the functions for each part.

Closure

- **Point of Most Significance:** Ask students to identify, aloud or on a paper to be collected, the most significant point (or part) in the lesson that aided their learning.

Assignment Guide and Homework Check

ASSIGNMENT

Basic: 1, 2, 3–17 odd, 21–31 odd, 35–41 odd, 47, 52, 55–60

Average: 1, 2–18 even, 22–48 even, 49, 52, 55–60

Advanced: 1, 2, 10–14 even, 18–22 even, 28–36 even, 46, 48–60

HOMEWORK CHECK

Basic: 3, 15, 23, 35, 47

Average: 4, 16, 24, 36, 48

Advanced: 14, 18, 28, 36, 48

4.7 Exercises

Dynamic Solutions available at *BigIdeasMath.com*

Vocabulary and Core Concept Check

1. **VOCABULARY** Compare piecewise functions and step functions.
2. **WRITING** Use a graph to explain why you can write the absolute value function $y = |x|$ as a piecewise function.

Monitoring Progress and Modeling with Mathematics

In Exercises 3–12, evaluate the function. *(See Example 1.)*

$$f(x) = \begin{cases} 5x - 1, & \text{if } x < -2 \\ x + 3, & \text{if } x \geq -2 \end{cases}$$

$$g(x) = \begin{cases} -x + 4, & \text{if } x \leq -1 \\ 3, & \text{if } -1 < x < 2 \\ 2x - 5, & \text{if } x \geq 2 \end{cases}$$

3. $f(-3)$
4. $f(-2)$
5. $f(0)$
6. $f(5)$
7. $g(-4)$
8. $g(-1)$
9. $g(0)$
10. $g(1)$
11. $g(2)$
12. $g(5)$
13. **MODELING WITH MATHEMATICS** On a trip, the total distance (in miles) you travel in x hours is represented by the piecewise function

$$d(x) = \begin{cases} 55x, & \text{if } 0 \leq x \leq 2 \\ 65x - 20, & \text{if } 2 < x \leq 5 \end{cases}.$$

How far do you travel in 4 hours?

14. **MODELING WITH MATHEMATICS** The total cost (in dollars) of ordering x custom shirts is represented by the piecewise function

$$c(x) = \begin{cases} 17x + 20, & \text{if } 0 \leq x < 25 \\ 15.80x + 20, & \text{if } 25 \leq x < 50 \\ 14x + 20, & \text{if } x \geq 50 \end{cases}.$$

Determine the total cost of ordering 26 shirts.

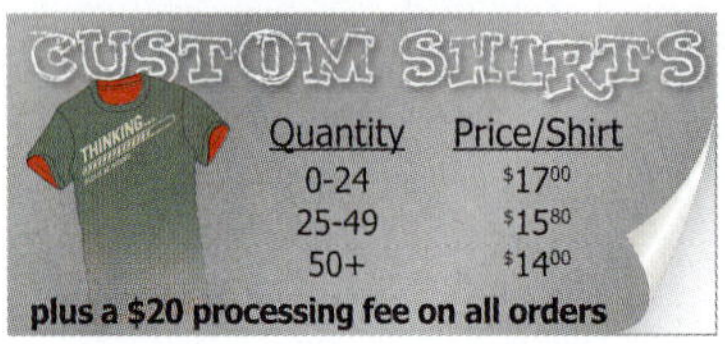

In Exercises 15–20, graph the function. Describe the domain and range. *(See Example 2.)*

15. $y = \begin{cases} -x, & \text{if } x < 2 \\ x - 6, & \text{if } x \geq 2 \end{cases}$
16. $y = \begin{cases} 2x, & \text{if } x \leq -3 \\ -2x, & \text{if } x > -3 \end{cases}$
17. $y = \begin{cases} -3x - 2, & \text{if } x \leq -1 \\ x + 2, & \text{if } x > -1 \end{cases}$
18. $y = \begin{cases} x + 8, & \text{if } x < 4 \\ 4x - 4, & \text{if } x \geq 4 \end{cases}$
19. $y = \begin{cases} 1, & \text{if } x < -3 \\ x - 1, & \text{if } -3 \leq x \leq 3 \\ -2x + 4, & \text{if } x > 3 \end{cases}$
20. $y = \begin{cases} 2x + 1, & \text{if } x \leq -1 \\ -x + 2, & \text{if } -1 < x < 2 \\ -3, & \text{if } x \geq 2 \end{cases}$
21. **ERROR ANALYSIS** Describe and correct the error in finding $f(5)$ when $f(x) = \begin{cases} 2x - 3, & \text{if } x < 5 \\ x + 8, & \text{if } x \geq 5 \end{cases}$.

✗ $f(5) = 2(5) - 3$
$= 7$

22. **ERROR ANALYSIS** Describe and correct the error in graphing $y = \begin{cases} x + 6, & \text{if } x \leq -2 \\ 1, & \text{if } x > -2 \end{cases}$.

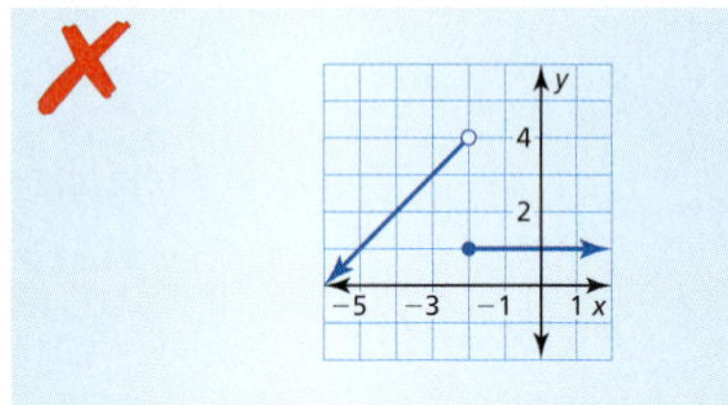

ANSWERS

1. A piecewise function is a function defined by two or more equations. A step function is a special piecewise function defined by a constant value over each part of its domain.
2. It can be written as a piecewise function by writing a linear function for each of the two lines that form the graph.
3. −16
4. 1
5. 3
6. 8
7. 8
8. 5
9. 3
10. 3
11. −1
12. 5
13. 240 mi
14. $430.80
15.

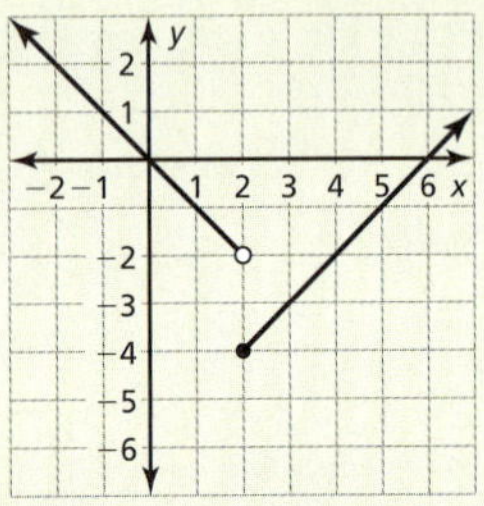

domain: all real numbers; range: $y \geq -4$

16.

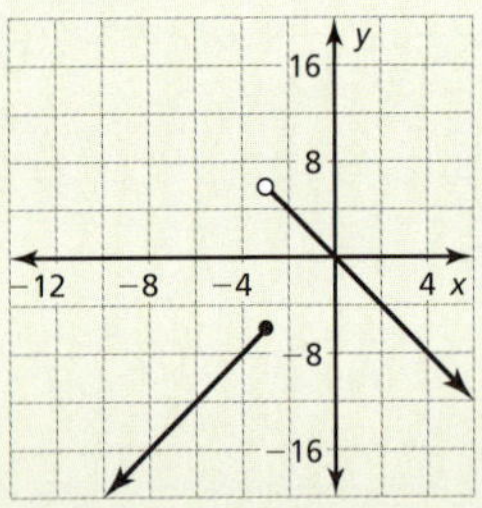

domain: all real numbers; range: $y < 6$

17.

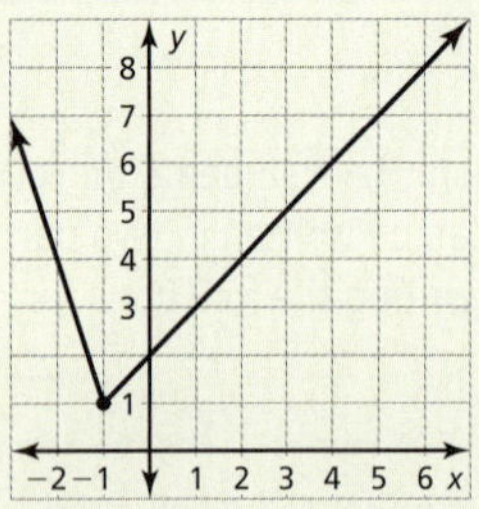

domain: all real numbers; range: $y \geq 1$

18.

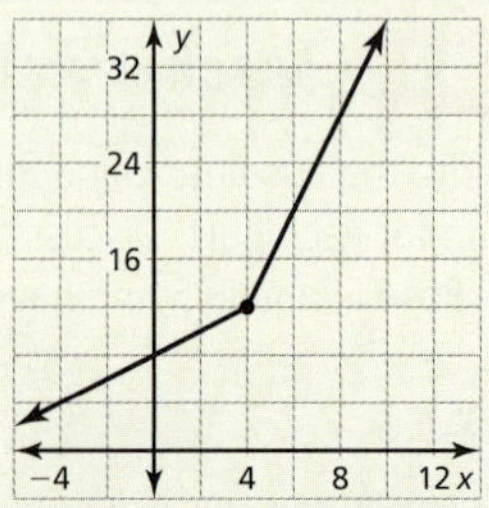

domain: all real numbers; range: all real numbers

19–22. See Additional Answers.

In Exercises 23–30, write a piecewise function for the graph. *(See Example 3.)*

23. 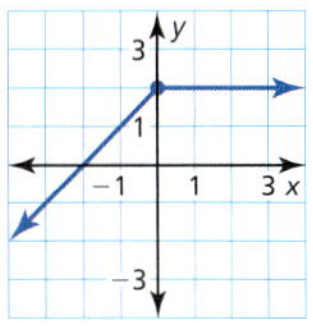

24. 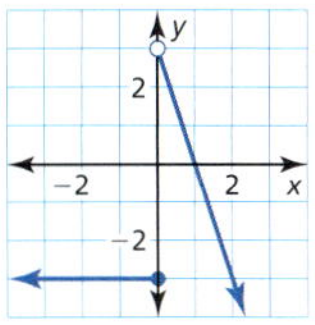

25. 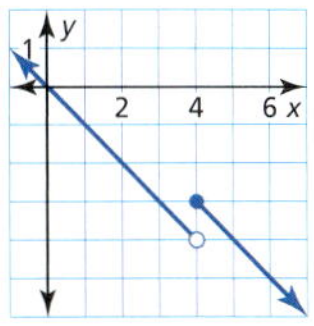

26. 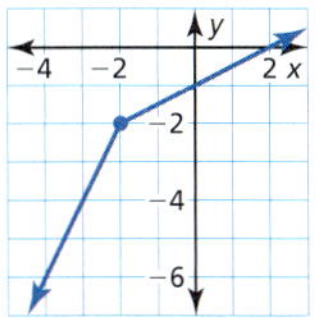

27. 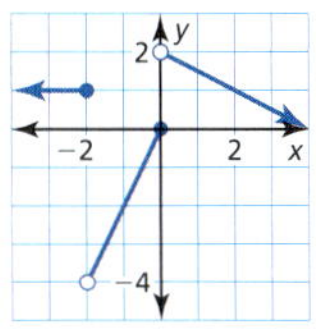

28. 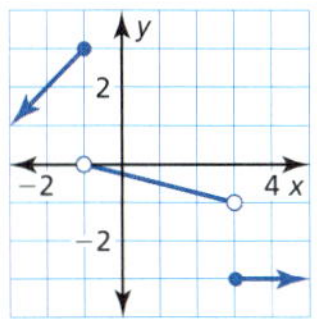

29. 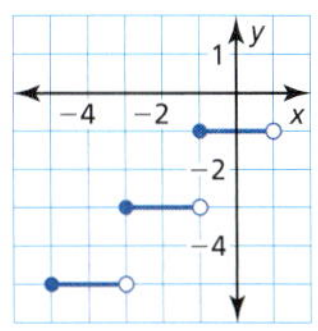

30. 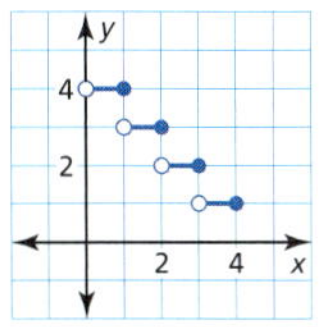

In Exercises 31–34, graph the step function. Describe the domain and range.

31. $f(x) = \begin{cases} 3, & \text{if } 0 \le x < 2 \\ 4, & \text{if } 2 \le x < 4 \\ 5, & \text{if } 4 \le x < 6 \\ 6, & \text{if } 6 \le x < 8 \end{cases}$

32. $f(x) = \begin{cases} -4, & \text{if } 1 < x \le 2 \\ -6, & \text{if } 2 < x \le 3 \\ -8, & \text{if } 3 < x \le 4 \\ -10, & \text{if } 4 < x \le 5 \end{cases}$

33. $f(x) = \begin{cases} 9, & \text{if } 1 < x \le 2 \\ 6, & \text{if } 2 < x \le 4 \\ 5, & \text{if } 4 < x \le 9 \\ 1, & \text{if } 9 < x \le 12 \end{cases}$

34. $f(x) = \begin{cases} -2, & \text{if } -6 \le x < -5 \\ -1, & \text{if } -5 \le x < -3 \\ 0, & \text{if } -3 \le x < -2 \\ 1, & \text{if } -2 \le x < 0 \end{cases}$

35. MODELING WITH MATHEMATICS The cost to join an intramural sports league is \$180 per team and includes the first five team members. For each additional team member, there is a \$30 fee. You plan to have nine people on your team. Write and graph a step function that represents the relationship between the number p of people on your team and the total cost of joining the league. *(See Example 4.)*

36. MODELING WITH MATHEMATICS The rates for a parking garage are shown. Write and graph a step function that represents the relationship between the number x of hours a car is parked in the garage and the total cost of parking in the garage for 1 day.

In Exercises 37–46, write the absolute value function as a piecewise function.

37. $y = |x| + 1$ **38.** $y = |x| - 3$

39. $y = |x - 2|$ **40.** $y = |x + 5|$

41. $y = 2|x + 3|$ **42.** $y = 4|x - 1|$

43. $y = -5|x - 8|$ **44.** $y = -3|x + 6|$

45. $y = -|x - 3| + 2$ **46.** $y = 7|x + 1| - 5$

47. MODELING WITH MATHEMATICS You are sitting on a boat on a lake. You can get a sunburn from the sunlight that hits you directly and also from the sunlight that reflects off the water. *(See Example 5.)*

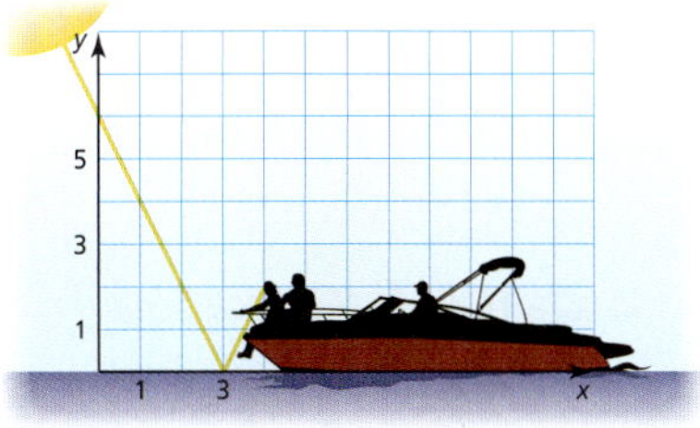

a. Write an absolute value function that represents the path of the sunlight that reflects off the water.

b. Write the function in part (a) as a piecewise function.

Dynamic Teaching Tools

Dynamic Assessment & Progress Monitoring Tool

Interactive Whiteboard Lesson Library

Dynamic Classroom with Dynamic Investigations

ANSWERS

23. $f(x) = \begin{cases} x + 2, & \text{if } x < 0 \\ 2, & \text{if } x \ge 0 \end{cases}$

24. $f(x) = \begin{cases} -3, & \text{if } x \le 0 \\ -3x + 3, & \text{if } x > 0 \end{cases}$

25. $f(x) = \begin{cases} -x, & \text{if } x < 4 \\ -x + 1, & \text{if } x \ge 4 \end{cases}$

26. $f(x) = \begin{cases} 2x + 2, & \text{if } x \le -2 \\ \frac{1}{2}x - 1, & \text{if } x > -2 \end{cases}$

27. $f(x) = \begin{cases} 1, & \text{if } x \le -2 \\ 2x, & \text{if } -2 < x \le 0 \\ -\frac{1}{2}x + 2, & \text{if } x > 0 \end{cases}$

28. $f(x) = \begin{cases} x + 4, & \text{if } x \le -1 \\ -\frac{1}{4}x - \frac{1}{4}, & \text{if } -1 < x < 3 \\ -3, & \text{if } x \ge 3 \end{cases}$

29. $f(x) = \begin{cases} -5, & \text{if } -5 \le x < -3 \\ -3, & \text{if } -3 \le x < -1 \\ -1, & \text{if } -1 \le x < 1 \end{cases}$

30. $f(x) = \begin{cases} 4, & \text{if } 0 < x \le 1 \\ 3, & \text{if } 1 < x \le 2 \\ 2, & \text{if } 2 < x \le 3 \\ 1, & \text{if } 3 < x \le 4 \end{cases}$

31. 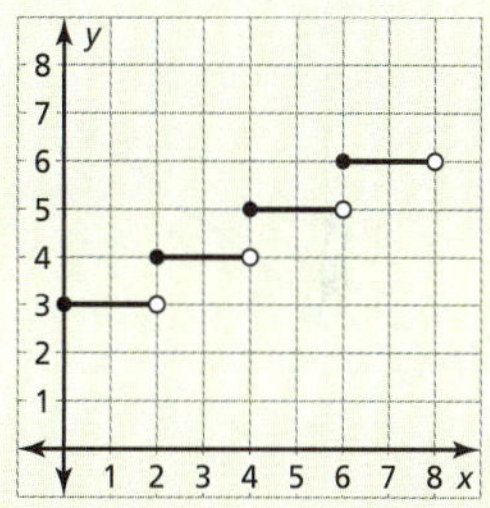

domain: $0 \le x < 8$; range: 3, 4, 5, 6

32. 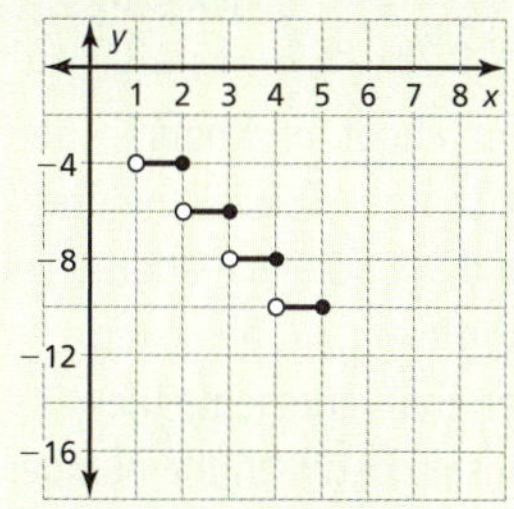

domain: $1 < x \le 5$; range: $-10, -8, -6, -4$

33. 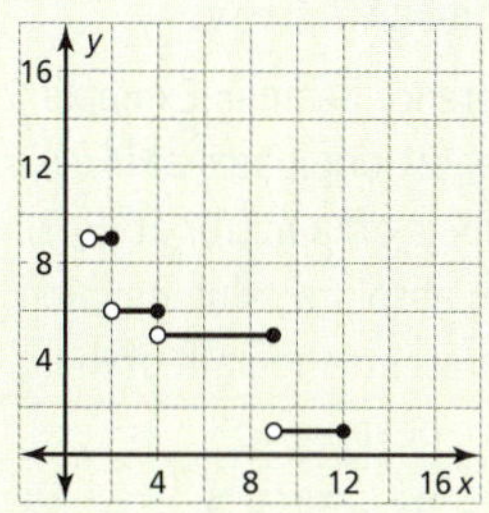

domain: $1 < x \le 12$; range: 1, 5, 6, 9

34–36. See Additional Answers.

37. $y = \begin{cases} -x + 1, & \text{if } x < 0 \\ x + 1, & \text{if } x \ge 0 \end{cases}$

38. $y = \begin{cases} -x - 3, & \text{if } x < 0 \\ x - 3, & \text{if } x \ge 0 \end{cases}$

39. $y = \begin{cases} -x + 2, & \text{if } x < 2 \\ x - 2, & \text{if } x \ge 2 \end{cases}$

40. $y = \begin{cases} -x - 5, & \text{if } x < -5 \\ x + 5, & \text{if } x \ge -5 \end{cases}$

41. $y = \begin{cases} -2x - 6, & \text{if } x < -3 \\ 2x + 6, & \text{if } x \ge -3 \end{cases}$

42. $y = \begin{cases} -4x + 4, & \text{if } x < 1 \\ 4x - 4, & \text{if } x \ge 1 \end{cases}$

43. $y = \begin{cases} 5x - 40, & \text{if } x < 8 \\ -5x + 40, & \text{if } x \ge 8 \end{cases}$

44. $y = \begin{cases} 3x + 18, & \text{if } x < -6 \\ -3x - 18, & \text{if } x \ge -6 \end{cases}$

45. $y = \begin{cases} x - 1, & \text{if } x < 3 \\ -x + 5, & \text{if } x \ge 3 \end{cases}$

46. $y = \begin{cases} -7x - 12, & \text{if } x < -1 \\ 7x + 2, & \text{if } x \ge -1 \end{cases}$

47. a. $f(x) = 2|x - 3|$

b. $f(x) = \begin{cases} -2x + 6, & \text{if } x < 3 \\ 2x - 6, & \text{if } x \ge 3 \end{cases}$

ANSWERS

48–60. See Additional Answers.

Mini-Assessment

1. Evaluate the function $f(x) = \begin{cases} -2x + 1, & \text{if } x \le 0 \\ x - 3, & \text{if } x > 0 \end{cases}$ when $x = 1$. $f(1) = -2$

2. Graph $y = \begin{cases} -x - 3, & \text{if } x \le 0 \\ 2x - 1 & \text{if } x > 0 \end{cases}$.

   Describe the domain and range.

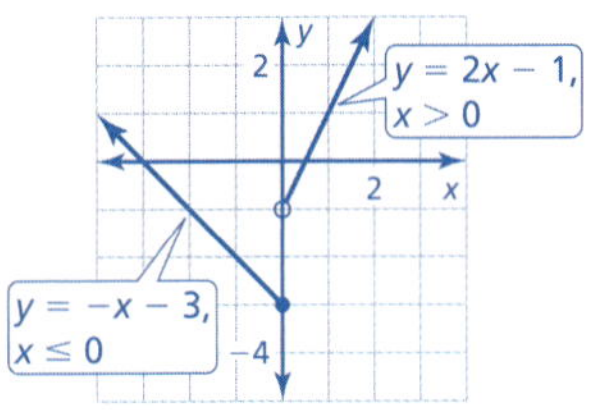

   The domain is all real numbers. The range is $y \ge -3$.

3. Write a piecewise function for the graph.

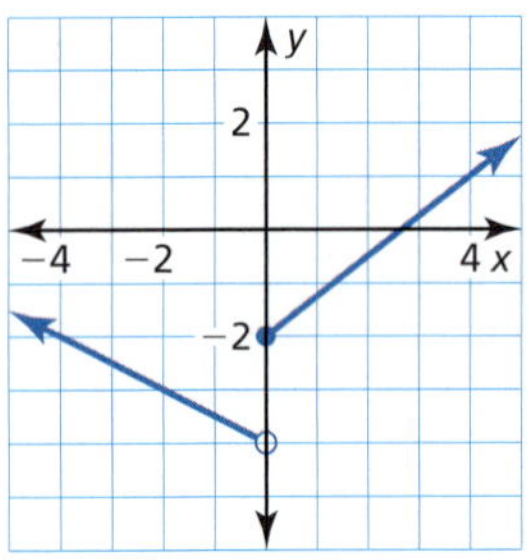

   $f(x) = \begin{cases} -0.5x - 4, & \text{if } x < 0 \\ 0.75x - 2, & \text{if } x \ge 0 \end{cases}$

4. You rent a boat for 4 hours. The marina charges \$60 for the first hour and \$40 for each additional hour. Write and graph a step function that represents the relationship between the number x of hours and the total cost y (in dollars) of renting the boat. See Additional Answers.

5. The reference beam in Example 5 on page 221 originates at (4, 0) and reflects off a mirror at (5, 2). Write an absolute value function that represents the path of the reference beam.

   $g(x) = -2|x - 5| + 2$

48. MODELING WITH MATHEMATICS You are trying to make a hole in one on the miniature golf green.

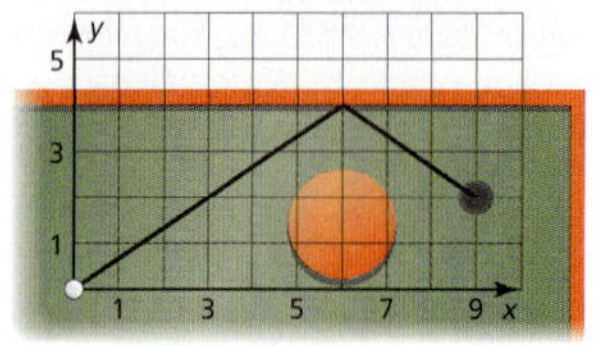

a. Write an absolute value function that represents the path of the golf ball.

b. Write the function in part (a) as a piecewise function.

49. REASONING The piecewise function f consists of two linear "pieces." The graph of f is shown.

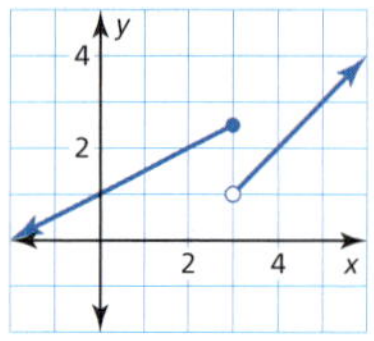

a. What is the value of $f(-10)$?

b. What is the value of $f(8)$?

50. CRITICAL THINKING Describe how the graph of each piecewise function changes when $<$ is replaced with $\le$ and $\ge$ is replaced with $>$. Do the domain and range change? Explain.

a. $f(x) = \begin{cases} x + 2, & \text{if } x < 2 \\ -x - 1, & \text{if } x \ge 2 \end{cases}$

b. $f(x) = \begin{cases} \frac{1}{2}x + \frac{3}{2}, & \text{if } x < 1 \\ -x + 3, & \text{if } x \ge 1 \end{cases}$

51. USING STRUCTURE Graph

$$y = \begin{cases} -x + 2, & \text{if } x \le -2 \\ |x|, & \text{if } x > -2 \end{cases}.$$

Describe the domain and range.

52. HOW DO YOU SEE IT? The graph shows the total cost C of making x photocopies at a copy shop.

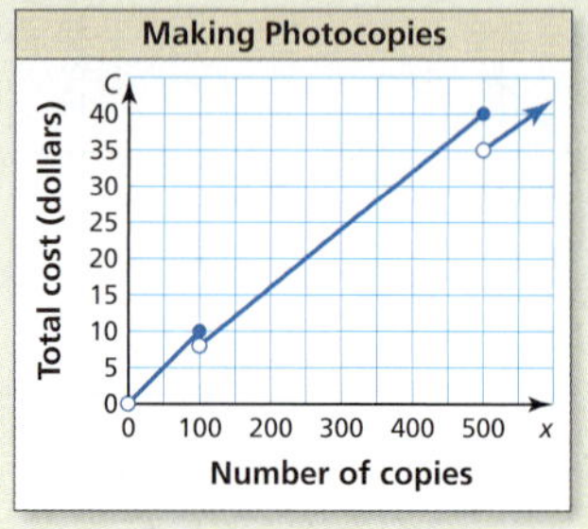

a. Does it cost more money to make 100 photocopies or 101 photocopies? Explain.

b. You have \$40 to make photocopies. Can you buy more than 500 photocopies? Explain.

53. USING STRUCTURE The output y of the *greatest integer function* is the greatest integer less than or equal to the input value x. This function is written as $f(x) = [\![x]\!]$. Graph the function for $-4 \le x < 4$. Is it a piecewise function? a step function? Explain.

54. THOUGHT PROVOKING Explain why

$$y = \begin{cases} 2x - 2, & \text{if } x \le 3 \\ -3, & \text{if } x \ge 3 \end{cases}$$

does not represent a function. How can you redefine y so that it does represent a function?

55. MAKING AN ARGUMENT During a 9-hour snowstorm, it snows at a rate of 1 inch per hour for the first 2 hours, 2 inches per hour for the next 6 hours, and 1 inch per hour for the final hour.

a. Write and graph a piecewise function that represents the depth of the snow during the snowstorm.

b. Your friend says 12 inches of snow accumulated during the storm. Is your friend correct? Explain.

Maintaining Mathematical Proficiency

Reviewing what you learned in previous grades and lessons

Write the sentence as an inequality. Graph the inequality. *(Section 2.5)*

56. A number r is greater than -12 and no more than 13.

57. A number t is less than or equal to 4 or no less than 18.

Graph f and h. Describe the transformations from the graph of f to the graph of h. *(Section 3.6)*

58. $f(x) = x$; $h(x) = 4x + 3$

59. $f(x) = x$; $h(x) = -x - 8$

60. $f(x) = x$; $h(x) = -\frac{1}{2}x + 5$

If students need help...	*If students got it...*
Resources by Chapter • Practice A and Practice B • Puzzle Time	Resources by Chapter • Enrichment and Extension • Cumulative Review
Student Journal • Practice	Start the *next* Section
Differentiating the Lesson Skills Review Handbook	

4.4–4.7 What Did You Learn?

Core Vocabulary

scatter plot, *p. 196*
correlation, *p. 197*
line of fit, *p. 198*
residual, *p. 202*
linear regression, *p. 203*
line of best fit, *p. 203*
correlation coefficient, *p. 203*
interpolation, *p. 205*
extrapolation, *p. 205*
causation, *p. 205*
sequence, *p. 210*
term, *p. 210*
arithmetic sequence, *p. 210*
common difference, *p. 210*
piecewise function, *p. 218*
step function, *p. 220*

Core Concepts

Section 4.4

Scatter Plot, *p. 196*
Identifying Correlations, *p. 197*
Using a Line of Fit to Model Data, *p. 198*

Section 4.5

Residuals, *p. 202*
Lines of Best Fit, *p. 203*
Correlation and Causation, *p. 205*

Section 4.6

Arithmetic Sequence, *p. 210*
Equation for an Arithmetic Sequence, *p. 212*

Section 4.7

Piecewise Function, *p. 218*
Step Function, *p. 220*
Writing Absolute Value Functions, *p. 221*

Mathematical Practices

1. What resources can you use to help you answer Exercise 17 on page 200?
2. What calculations are repeated in Exercises 11–16 on page 214? When finding a term such as a_{50}, is there a general method or shortcut you can use instead of repeating calculations?
3. Describe the definitions you used when you explained your answer in Exercise 53 on page 224.

Performance Task

Any Beginning

With so many ways to represent a linear relationship, where do you start? Use what you know to move between equations, graphs, tables, and contexts.

To explore the answers to this question and more, go to ***BigIdeasMath.com***.

Dynamic Teaching Tools

- Dynamic Assessment & Progress Monitoring Tool
- Interactive Whiteboard Lesson Library
- Dynamic Classroom with Dynamic Investigations

ANSWERS

1. *Sample answer:* the Internet, an almanac
2. addition; yes, substitute the values of a_1 and d in the formula $a_n = a_1 + (n - 1)d$, and then substitute 50 for n to calculate a_{50}.
3. The definition of piecewise function, which is a function defined by two or more equations, and the definition of step function, which is a special piecewise function defined by a constant value over each part of its domain.

ANSWERS

1. $y = -\frac{1}{2}x + 1$
2. $y - 7 = -(x - 4)$
3. $f(x) = x - 5$
4. $f(x) = -4$
5. $f(x) = -\frac{5}{3}x + 18$

4 Chapter Review

Dynamic Solutions available at *BigIdeasMath.com*

4.1 Writing Equations in Slope-Intercept Form *(pp. 175–180)*

Write an equation of the line in slope-intercept form.

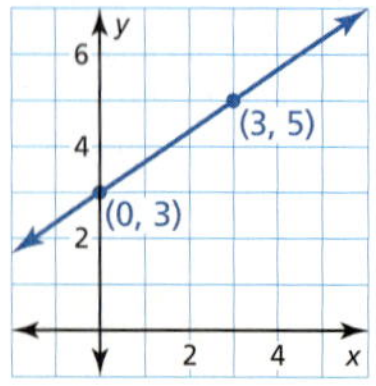

Find the slope and y-intercept.

Let $(x_1, y_1) = (0, 3)$ and $(x_2, y_2) = (3, 5)$.

$$m = \frac{y_2 - y_1}{x_2 - x_1} = \frac{5 - 3}{3 - 0} = \frac{2}{3}$$

Because the line crosses the y-axis at $(0, 3)$, the y-intercept is 3.

▶ So, the equation is $y = \frac{2}{3}x + 3$.

1. Write an equation of the line in slope-intercept form.

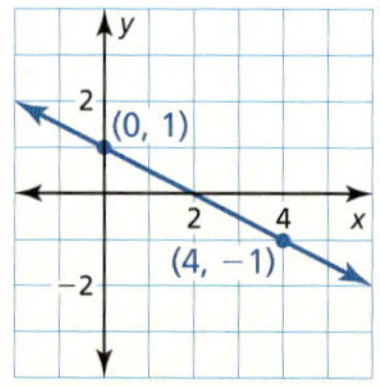

4.2 Writing Equations in Point-Slope Form *(pp. 181–186)*

Write an equation in point-slope form of the line that passes through the point $(-1, -8)$ and has a slope of 3.

$y - y_1 = m(x - x_1)$	Write the point-slope form.
$y - (-8) = 3[x - (-1)]$	Substitute 3 for m, -1 for x_1, and -8 for y_1.
$y + 8 = 3(x + 1)$	Simplify.

▶ The equation is $y + 8 = 3(x + 1)$.

2. Write an equation in point-slope form of the line that passes through the point $(4, 7)$ and has a slope of -1.

Write a linear function f with the given values.

3. $f(10) = 5, f(2) = -3$ **4.** $f(3) = -4, f(5) = -4$ **5.** $f(6) = 8, f(9) = 3$

4.3 Writing Equations of Parallel and Perpendicular Lines *(pp. 187–192)*

Determine which of the lines, if any, are parallel or perpendicular.

Line a: $y = 2x + 3$ Line b: $2y + x = 5$ Line c: $4y - 8x = -4$

Write the equations in slope-intercept form. Then compare the slopes.

Line a: $y = 2x + 3$ Line b: $y = -\frac{1}{2}x + \frac{5}{2}$ Line c: $y = 2x - 1$

▶ Lines a and c have slopes of 2, so they are parallel. Line b has a slope of $-\frac{1}{2}$, the negative reciprocal of 2, so it is perpendicular to lines a and c.

Determine which of the lines, if any, are parallel or perpendicular. Explain.

6. Line a passes through (0, 4) and (4, 3).
   Line b passes through (0, 1) and (4, 0).
   Line c passes through (2, 0) and (4, 4).

7. Line a: $2x - 7y = 14$
   Line b: $y = \frac{7}{2}x - 8$
   Line c: $2x + 7y = -21$

8. Write an equation of the line that passes through (1, 5) and is parallel to the line $y = -4x + 2$.

9. Write an equation of the line that passes through (2, −3) and is perpendicular to the line $y = -2x - 3$.

4.4 Scatter Plots and Lines of Fit (pp. 195–200)

The scatter plot shows the roasting times (in hours) and weights (in pounds) of seven turkeys. Tell whether the data show a *positive*, a *negative*, or *no* correlation.

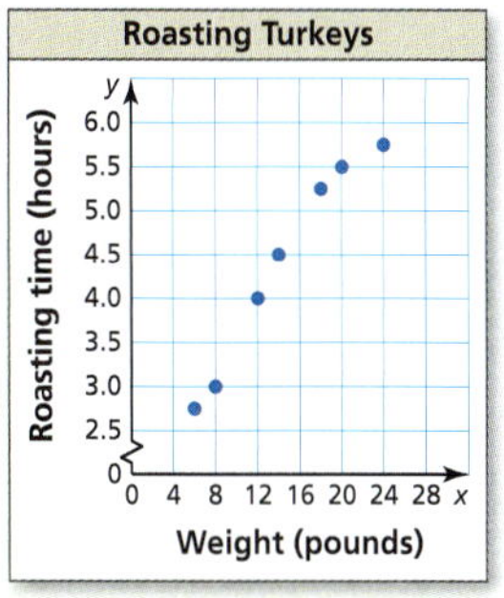

As the weight of a turkey increases, the roasting time increases.

▶ So, the scatter plot shows a positive correlation.

Use the scatter plot in the example.

10. What is the roasting time for a 12-pound turkey?

11. Write an equation that models the roasting time as a function of the weight of a turkey. Interpret the slope and y-intercept of the line of fit.

4.5 Analyzing Lines of Fit (pp. 201–208)

The table shows the heights x (in inches) and shoe sizes y of several students. Use a graphing calculator to find an equation of the line of best fit. Identify and interpret the correlation coefficient.

Height, x	Shoe size, y
64	9
62	7
70	12
63	8
72	13
68	9.5
66	9
74	13.5
68	10
59	6.5

Step 1 Enter the data from the table into two lists.

Step 2 Use the *linear regression* feature.

```
LinReg
 y=ax+b
 a=.4989919355
 b=-23.4828629
 r²=.9477256904
 r=.9735120392
```

▶ An equation of the line of best fit is $y = 0.50x - 23.5$. The correlation coefficient is about 0.974. This means that the relationship between the heights and the shoe sizes has a strong positive correlation and the equation closely models the data.

12. Make a scatter plot of the residuals to verify that the model in the example is a good fit.

13. Use the data in the example. (a) Approximate the height of a student whose shoe size is 9. (b) Predict the shoe size of a student whose height is 60 inches.

14. Is there a causal relationship in the data in the example? Explain.

ANSWERS

6. Lines a and b are parallel; None of the lines are perpendicular; Lines a and b have the same slope and none of them have negative reciprocal slopes.
7. None of the lines are parallel; Lines b and c are perpendicular; None of the lines have the same slope and the slope of line b is the negative reciprocal of the slope of line c.
8. $y = -4x + 9$
9. $y = \frac{1}{2}x - 4$
10. 4h
11. *Sample answer:* $y = \frac{1}{5}x + 2$; The slope of $\frac{1}{5}$ means the roasting time increases by about $\frac{1}{5}$ hour for each pound the weight of the turkey increases. The y-intercept of 2 has no meaning in this context because the weight of the turkey cannot be 0 pounds.
12.

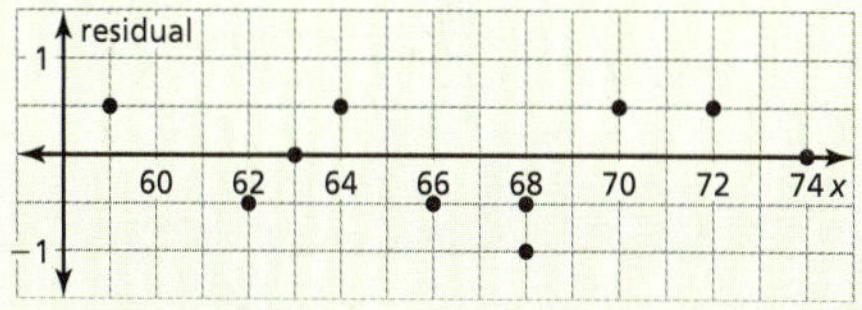

13. **a.** 65 in.
    b. 6.5
14. *Sample answer:* no; Height does not determine shoe size.

ANSWERS

15. $a_n = -n + 12$; -18

16. $a_n = 6n$; 180

17. $a_n = 3n - 12$; 78

18. **a.** 3

b. -10

19.

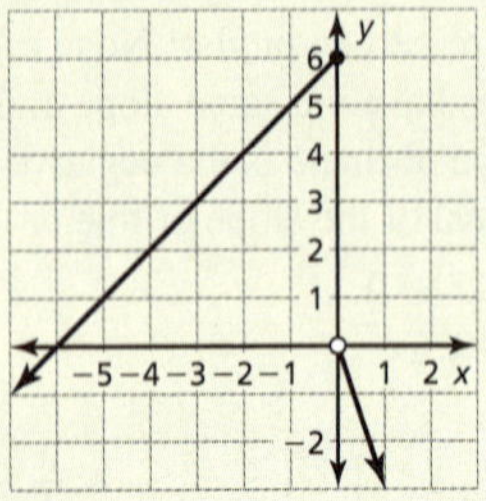

domain: all real numbers; range: $y \le 6$

20.

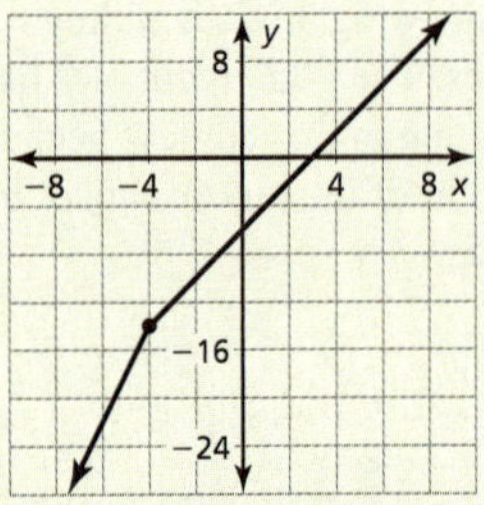

domain: all real numbers; range: all real numbers

21. $y = \begin{cases} -x + 15, & \text{if } x < 0 \\ x + 15, & \text{if } x \ge 0 \end{cases}$

22. $y = \begin{cases} -4x - 20, & \text{if } x < -5 \\ 4x + 20, & \text{if } x \ge -5 \end{cases}$

23. $y = \begin{cases} -2x - 7, & \text{if } x < -2 \\ 2x + 1, & \text{if } x \ge -2 \end{cases}$

24. $f(x) = \begin{cases} 65, & \text{if } 0 < x \le 1 \\ 100, & \text{if } 1 < x \le 2 \\ 135, & \text{if } 2 < x \le 3 \end{cases}$

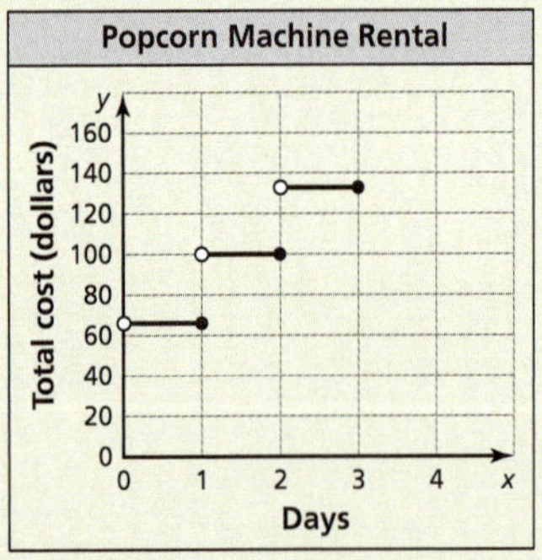

4.6 Arithmetic Sequences (pp. 209–216)

Write an equation for the *n*th term of the arithmetic sequence −3, −5, −7, −9, ... Then find a_{20}.

The first term is −3, and the common difference is −2.

$a_n = a_1 + (n - 1)d$ — Equation for an arithmetic sequence

$a_n = -3 + (n - 1)(-2)$ — Substitute −3 for a_1 and −2 for d.

$a_n = -2n - 1$ — Simplify.

Use the equation to find the 20th term.

$a_{20} = -2(20) - 1$ — Substitute 20 for n.

$= -41$ — Simplify.

▶ The 20th term of the arithmetic sequence is −41.

Write an equation for the *n*th term of the arithmetic sequence. Then find a_{30}.

15. 11, 10, 9, 8, . . . **16.** 6, 12, 18, 24, . . . **17.** −9, −6, −3, 0, . . .

4.7 Piecewise Functions (pp. 217–224)

Graph $y = \begin{cases} \frac{3}{2}x + 3, & \text{if } x \le 0 \\ -2x, & \text{if } x > 0 \end{cases}$. Describe the domain and range.

Step 1 Graph $y = \frac{3}{2}x + 3$ for $x \le 0$. Because x is less than or equal to 0, use a closed circle at (0, 3).

Step 2 Graph $y = -2x$ for $x > 0$. Because x is not equal to 0, use an open circle at (0, 0).

▶ The domain is all real numbers. The range is $y \le 3$.

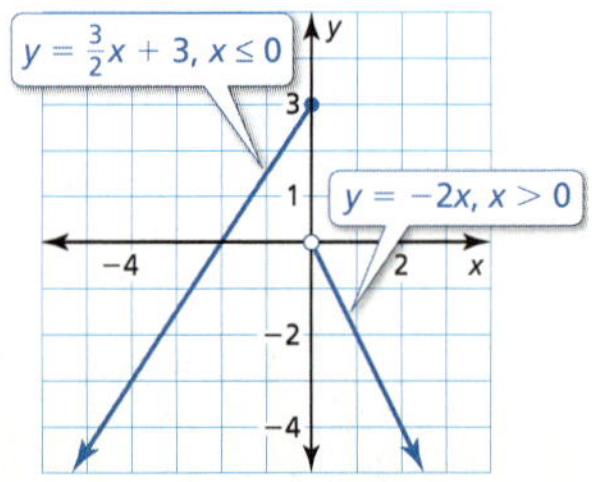

18. Evaluate the function in the example when (a) $x = 0$ and (b) $x = 5$.

Graph the function. Describe the domain and range.

19. $y = \begin{cases} x + 6, & \text{if } x \le 0 \\ -3x, & \text{if } x > 0 \end{cases}$ **20.** $y = \begin{cases} 4x + 2, & \text{if } x < -4 \\ 2x - 6, & \text{if } x \ge -4 \end{cases}$

Write the absolute value function as a piecewise function.

21. $y = |x| + 15$ **22.** $y = 4|x + 5|$ **23.** $y = 2|x + 2| - 3$

24. You are organizing a school fair and rent a popcorn machine for 3 days. The rental company charges $65 for the first day and $35 for each additional day. Write and graph a step function that represents the relationship between the number x of days and the total cost y (in dollars) of renting the popcorn machine.

4 Chapter Test

Graph the function. Describe the domain and range.

1. $y = \begin{cases} 2x + 4, & \text{if } x \le -1 \\ \frac{1}{3}x - 1, & \text{if } x > -1 \end{cases}$

2. $y = \begin{cases} 1, & \text{if } 0 \le x < 3 \\ 0, & \text{if } 3 \le x < 6 \\ -1, & \text{if } 6 \le x < 9 \\ -2, & \text{if } 9 \le x < 12 \end{cases}$

Write an equation in slope-intercept form of the line with the given characteristics.

3. slope $= \frac{2}{5}$; y-intercept $= -7$

4. passes through $(0, 6)$ and $(3, -3)$

5. parallel to the line $y = 3x - 1$; passes through $(-2, -8)$

6. perpendicular to the line $y = \frac{1}{4}x - 9$; passes through $(1, 1)$

Write an equation in point-slope form of the line with the given characteristics.

7. slope $= 10$; passes through $(6, 2)$

8. passes through $(-3, 2)$ and $(6, -1)$

9. The first row of an auditorium has 42 seats. Each row after the first has three more seats than the row before it.

a. Find the number of seats in Row 25.

b. Which row has 90 seats?

10. The table shows the amount x (in dollars) spent on advertising for a neighborhood festival and the attendance y of the festival for several years.

a. Make a scatter plot of the data. Describe the correlation.

b. Write an equation that models the attendance as a function of the amount spent on advertising.

c. Interpret the slope and y-intercept of the line of fit.

Advertising (dollars), x	Yearly attendance, y
500	400
1000	550
1500	550
2000	800
2500	650
3000	800
3500	1050
4000	1100

11. Consider the data in the table in Exercise 10.

a. Use a graphing calculator to find an equation of the line of best fit.

b. Identify and interpret the correlation coefficient.

c. What would you expect the scatter plot of the residuals to look like?

d. Is there a causal relationship in the data? Explain your reasoning.

e. Predict the amount that must be spent on advertising to get 2000 people to attend the festival.

12. Let a, b, c, and d be constants. Determine which of the lines, if any, are parallel or perpendicular. Explain.

Line 1: $y - c = ax$ Line 2: $ay = -x - b$ Line 3: $ax + y = d$

13. Write a piecewise function defined by three equations that has a domain of all real numbers and a range of $-3 < y \le 1$.

If students need help...	*If students got it...*
Lesson Tutorials	Resources by Chapter • Enrichment and Extension • Cumulative Review
Skills Review Handbook	Performance Task
BigIdeasMath.com	Start the *next* Section

ANSWERS

1.

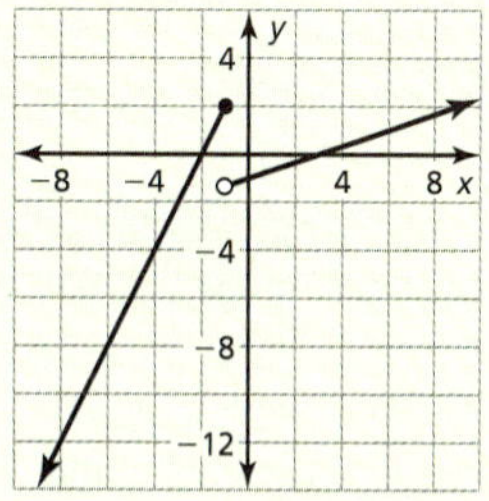

domain: all real numbers; range: all real numbers

2.

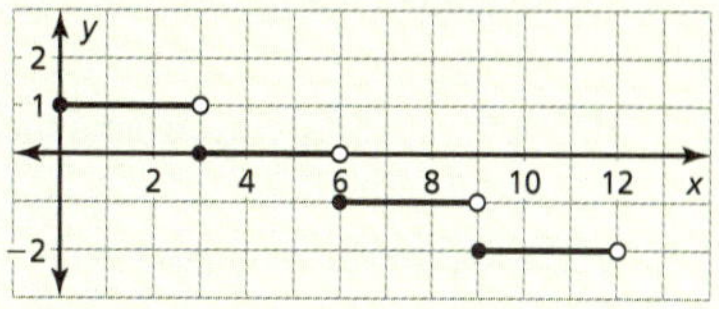

domain: $0 \le x < 12$; range: $-2, -1, 0, 1$

3. $y = \frac{2}{5}x - 7$

4. $y = -3x + 6$

5. $y = 3x - 2$

6. $y = -4x + 5$

7. $y - 2 = 10(x - 6)$

8. $y - 2 = -\frac{1}{3}(x + 3)$ or $y + 1 = -\frac{1}{3}(x - 6)$

9. a. 114

b. row 17

10. a.

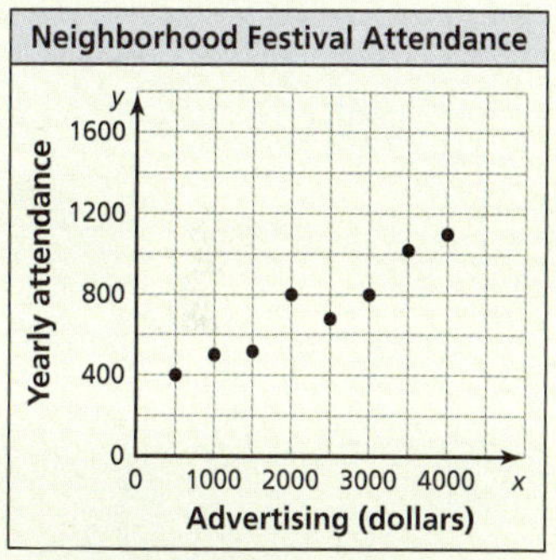

positive correlation

b. *Sample answer:* $y = \frac{1}{5}x + 300$

c. *Sample answer:* The slope of $\frac{1}{5}$ means the attendance increases by about 1 person for every additional \$5 spent on advertising. The y-intercept of 300 means that if no money is spent on advertising, about 300 people attend the festival.

11. a. $y = 0.19x + 309$

b. $r = 0.943$; strong positive correlation

c. more or less evenly dispersed about the horizontal axis

11d–13. See Additional Answers.

ANSWERS

1. C
2. $3x + 6 \le 8 + 2x$, $5x - 5 \ge 7x - 9$, and $-2 - \frac{3}{2}x \ge -3 - x$
3. a. yes; yes; The greater the price of the pants the fewer that will be sold.
   b. no; no; The number of cell phones is not related to the number of taxis in a city.
   c. no; no; A person's IQ is not related to the person's athletic ability.
   d. yes; yes; Studying affects the score earned.
4. $g(x) = f(x + 2)$, $h(x) = f(3x)$, $k(x) = f(x) + 4$, $q(x) = -f(x)$; The equation of the green line is $x + 1 = g(x)$. The equation of the red line is $3x - 1 = h(x)$. The equation of the purple line is $x + 3 = k(x)$. The equation of the blue line is $-x + 1 = q(x)$.

4 Cumulative Assessment

1. Which function represents the arithmetic sequence shown in the graph?

Ⓐ $f(n) = 15 + 3n$

Ⓑ $f(n) = 4 - 3n$

Ⓒ $f(n) = 27 - 3n$

Ⓓ $f(n) = 24 - 3n$

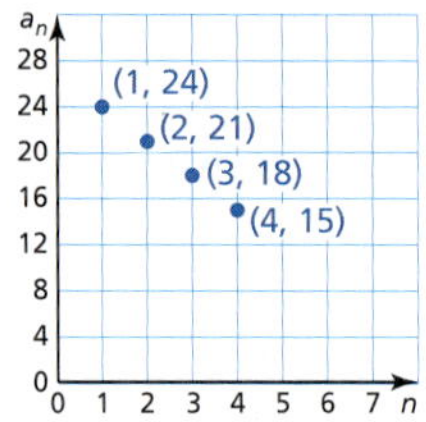

2. Which of the inequalities are equivalent?

$3x + 6 \le 8 + 2x$ | $5x - 5 \ge 7x - 9$ | $12 - 3x \le 18$ | $-2 - \frac{3}{2}x \ge -3 - x$

3. Complete the table for the four situations below. Explain your reasoning.

a. the price of a pair of pants and the number sold

b. the number of cell phones and the number of taxis in a city

c. a person's IQ and the time it takes the person to run 50 meters

d. the amount of time spent studying and the score earned

Situation	Correlation		Causation	
	Yes	No	Yes	No
a.				
b.				
c.				
d.				

4. Consider the function $f(x) = x - 1$. Select the functions that are shown in the graph. Explain your reasoning.

$g(x) = f(x + 2)$ | $h(x) = f(3x)$

$k(x) = f(x) + 4$ | $p(x) = f(-x)$

$r(x) = 3f(x)$ | $q(x) = -f(x)$

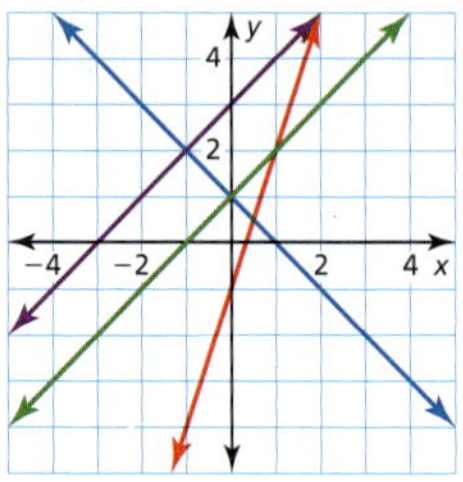

5. Use the numbers to fill in values for m and b in the equation $y = mx + b$ so that its graph passes through the points (6, 1) and (−2, −3).

−5	−2	−1	$-\frac{1}{2}$	0	$\frac{1}{2}$	1	2	5

6. Fill in the piecewise function with −, +, <, ≤, >, or ≥ so that the function is represented by the graph.

$$y = \begin{cases} 2x \; \square \; 3, & \text{if } x \; \square \; 0 \\ 2x \; \square \; 3, & \text{if } x \; \square \; 0 \end{cases}$$

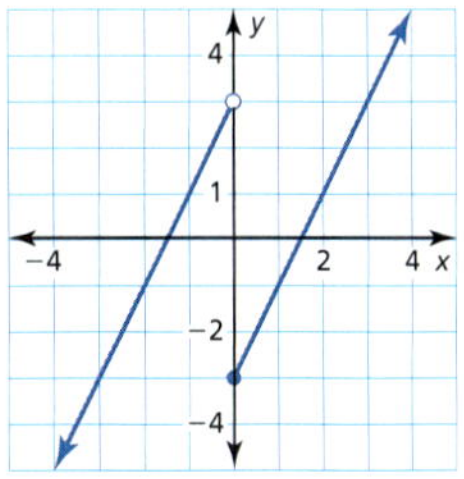

7. You claim that you can create a relation that is a function, and your friend claims that she can create a relation that is not a function. Using the given numbers, create a relation of five ordered pairs that supports your claim. What relation of five ordered pairs can your friend use to support her claim?

−4	−3	−2	−1	0	1	2	3	4

8. You have two coupons you can use at a restaurant. Write and solve an equation to determine how much your total bill must be for both coupons to save you the same amount of money.

9. The table shows the daily high temperatures x (in degrees Fahrenheit) and the numbers y of frozen fruit bars sold on eight randomly selected days. The equation $y = 3x - 50$ models the data.

Temperature (°F), x	54	60	68	72	78	84	92	98
Frozen fruit bars, y	40	120	180	260	280	260	220	180

a. Select the points that appear on a scatter plot of the residuals.

(92, −6)	(78, 96)	(60, −10)	(84, 58)	(98, −64)
(72, 94)	(54, −72)	(96, 78)	(60, 10)	(68, 26)

b. Determine whether the model is a good fit for the data. Explain your reasoning.

ANSWERS

5. $y = \frac{1}{2}x + (-2)$
6. +; <; −; ≥
7. *Sample answer*: function: (−4, 4), (−3, 3), (−2, 2), (−1, 1), (0, 0); not a function: (−4, 4), (−4, 3), (−4, 2), (−1, −4), (0, −4)
8. $0.2x = 5$; $25
9. a. (92, −6), (78, 96), (60, −10), (84, 58), (98, −64), (72, 94), (54, −72), (68, 26)

   b. no; The residuals form a ∩-shaped pattern.

Chapter 5 Pacing Guide

Chapter Opener/ Mathematical Practices	1 Day
Section 1	1 Day
Section 2	1 Day
Section 3	2 Days
Section 4	1 Day
Quiz	1 Day
Section 5	2 Days
Section 6	2 Days
Section 7	2 Days
Chapter Review/ Chapter Tests	2 Days
Total Chapter 5	15 Days
Year-to-Date	67 Days

5 Solving Systems of Linear Equations

5.1 Solving Systems of Linear Equations by Graphing
5.2 Solving Systems of Linear Equations by Substitution
5.3 Solving Systems of Linear Equations by Elimination
5.4 Solving Special Systems of Linear Equations
5.5 Solving Equations by Graphing
5.6 Graphing Linear Inequalities in Two Variables
5.7 Systems of Linear Inequalities

Fishing ***(p. 279)***

Pets ***(p. 266)***

Delivery Vans ***(p. 250)***

Drama Club ***(p. 244)***

Roofing Contractor ***(p. 238)***

Laurie's Notes

Chapter Summary

- There are three common techniques for solving a system of equations: graphing, substitution, and elimination. These techniques are presented in the first three sections of this chapter. Students are introduced to the definition of a linear system, and they learn to check their solutions.
- The fourth section looks at special linear systems, where there is no solution because the lines are parallel or there are infinitely many solutions because the lines coincide.
- In Section 5.5, students learn that solving a system by graphing can be used to solve an equation with variables on both sides. This is actually a technique that students will use extensively in future mathematics courses.
- The last lessons combine prior skills to allow students to graph systems of linear inequalities.

What Your Students Have Learned

Middle School
- Determine whether numbers are solutions of linear equations and linear inequalities.
- Solve systems of linear equations in two variables by graphing, substitution, and elimination.
- Solve systems of linear equations in two variables that have no solution or infinitely many solutions.
- Write, graph, and solve inequalities in one variable.
- Recognize that inequalities have infinitely many solutions.

What Your Students Will Learn

Algebra 1
- Solve systems of linear equations in two variables by graphing, substitution, and elimination.
- Determine whether systems of linear equations have one solution, no solution, or infinitely many solutions.
- Solve linear equations and absolute value equations by graphing.
- Check solutions of linear inequalities in two variables algebraically.
- Graph linear inequalities in one variable and in two variables in the coordinate plane.
- Graph and write systems of linear inequalities in two variables.

Dynamic Teaching Tools
- Dynamic Assessment & Progress Monitoring Tool
- Lesson Planning Tool
- Interactive Whiteboard Lesson Library
- Dynamic Classroom with Dynamic Investigations
- Real-Life STEM Videos

Scaffolding in the Classroom

Graphic Organizers: Example and Non-Example Chart

An Example and Non-Example Chart can be used to list examples and non-examples of a vocabulary word or term. Students write examples of the word or term in the left column and non-examples in the right column. This type of organizer serves as a good tool for assessing students' knowledge of pairs of topics that have subtle but important differences, such as complementary and supplementary angles. Blank Example and Non-Example Charts can be included on tests or quizzes for this purpose.

Questioning in the Classroom

How would you begin?

Before beginning the homework, select one challenging problem from the exercise set. Have students work in pairs or small groups to discuss how they would begin the problem. Students share their ideas before actually solving the problems, discussing the pros and cons of different methods.

Laurie's Notes

Maintaining Mathematical Proficiency

Graphing Linear Functions

- Remind students that slope-intercept form is $y = mx + b$, where m is the slope and b is the y-intercept.
- Remind students that the *rise* in the slope is the change in y (vertical change) and the *run* is the change in x (horizontal change).

COMMON ERROR Students use the value of the run for the change in y and the value of the rise as the change in x.

Solving and Graphing Linear Inequalities

- Remind students that when they multiply or divide each side of the inequality by a negative number, they must reverse the inequality symbol.

COMMON ERROR Students do not reverse the inequality symbol when they multiply or divide each side by a negative number. For example, they may rewrite $24 \leq -6t$ as $-4 \leq t$.

Mathematical Practices (continued on page 234)

- The *Mathematical Practices* page focuses attention on how mathematics is learned—process versus content. This page demonstrates that technological tools can be used to explore and deepen student understanding of a concept.
- Use the *Mathematical Practices* page to help students develop mathematical habits of mind—how mathematics can be explored and how mathematics is thought about.
- Example 1 shows a technique that will be used frequently by students in their study of mathematics. In this chapter, students will learn that solving a system of linear equations means you want to find the ordered pairs that are solutions to each equation. One way to do this is by graphing the two equations. When the two lines intersect at an ordered pair, it means that the ordered pair is a solution of each equation.

If students need help...	*If students got it...*
Student Journal • Maintaining Mathematical Proficiency	Game Closet at *BigIdeasMath.com*
Lesson Tutorials	Start the *next* Section
Skills Review Handbook	

Maintaining Mathematical Proficiency

Graphing Linear Functions

Example 1 Graph $3 + y = \frac{1}{2}x$.

Step 1 Rewrite the equation in slope-intercept form.

$$y = \frac{1}{2}x - 3$$

Step 2 Find the slope and the y-intercept.

$$m = \frac{1}{2} \text{ and } b = -3$$

Step 3 The y-intercept is -3. So, plot $(0, -3)$.

Step 4 Use the slope to find another point on the line.

$$\text{slope} = \frac{\text{rise}}{\text{run}} = \frac{1}{2}$$

Plot the point that is 2 units right and 1 unit up from $(0, -3)$. Draw a line through the two points.

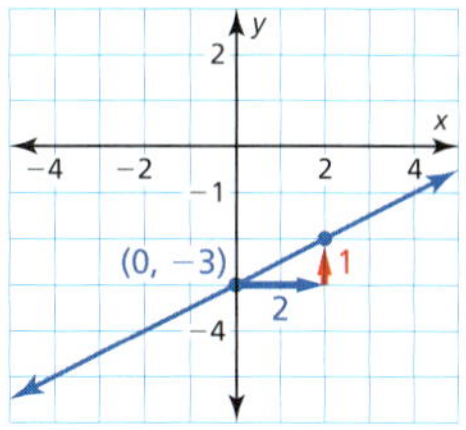

Graph the equation.

1. $y + 4 = x$ **2.** $6x - y = -1$ **3.** $4x + 5y = 20$ **4.** $-2y + 12 = -3x$

Solving and Graphing Linear Inequalities

Example 2 Solve $2x - 17 \leq 8x - 5$. Graph the solution.

$2x - 17 \leq 8x - 5$	Write the inequality.
$+5 \qquad +5$	Add 5 to each side.
$2x - 12 \leq 8x$	Simplify.
$-2x \qquad -2x$	Subtract $2x$ from each side.
$-12 \leq 6x$	Simplify.
$\frac{-12}{6} \leq \frac{6x}{6}$	Divide each side by 6.
$-2 \leq x$	Simplify.

▶ The solution is $x \geq -2$.

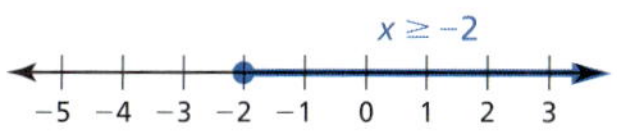

Solve the inequality. Graph the solution.

5. $m + 4 > 9$ **6.** $24 \leq -6t$ **7.** $2a - 5 \leq 13$

8. $-5z + 1 < -14$ **9.** $4k - 16 < k + 2$ **10.** $7w + 12 \geq 2w - 3$

11. ABSTRACT REASONING The graphs of the linear functions g and h have different slopes. The value of both functions at $x = a$ is b. When g and h are graphed in the same coordinate plane, what happens at the point (a, b)?

Dynamic Solutions available at *BigIdeasMath.com*

Vocabulary Review

Have students make Process Diagrams for the following topics.

- Graphing a linear function using slope-intercept form
- Solving and graphing a linear inequality

What Your Students Have Learned

- Graph linear equations written in slope-intercept form.
- Solve and graph multi-step linear inequalities in one variable using inverse operations, including multiplying and dividing by negative numbers.

ANSWERS

1.

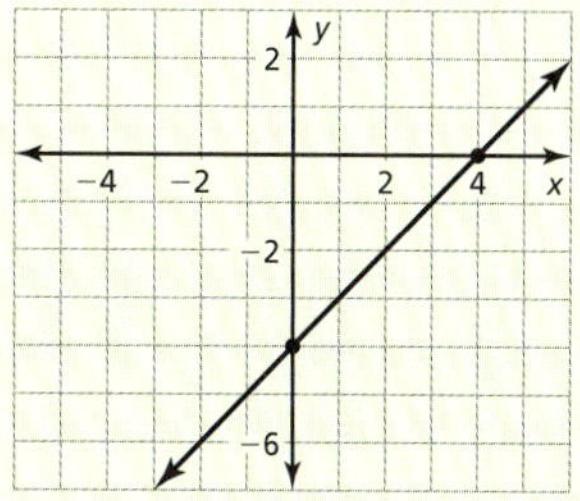

2.

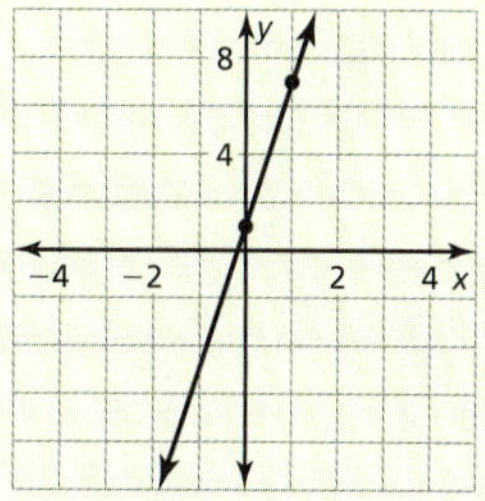

3.

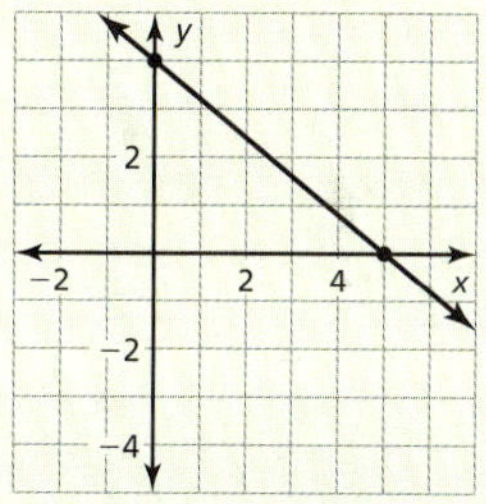

4.

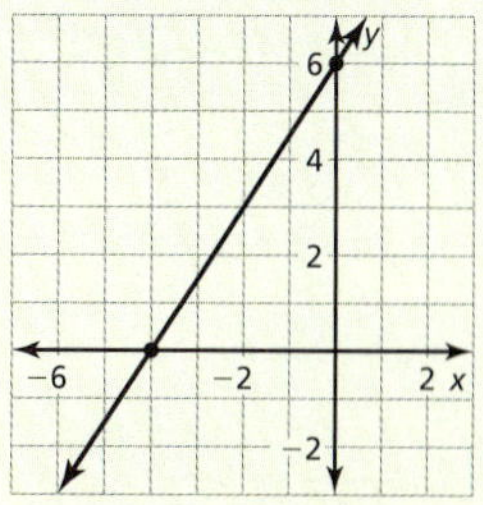

5. $m > 5$

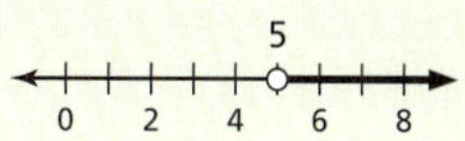

6. $t \leq -4$

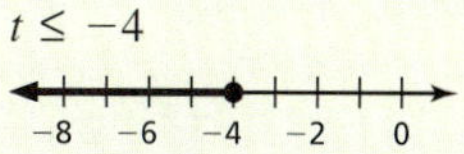

7. $a \leq 9$

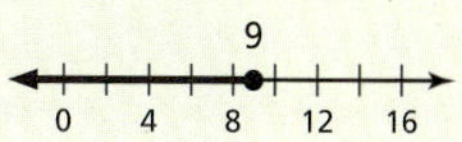

8–11. See Additional Answers.

MONITORING PROGRESS ANSWERS

1. $(0, -3)$
2. $\left(\frac{3}{2}, -\frac{1}{2}\right)$
3. $(2, 2)$

Mathematical Practices

Mathematically proficient students use technological tools to explore concepts.

Using a Graphing Calculator

Core Concept

Finding the Point of Intersection

You can use a graphing calculator to find the point of intersection, if it exists, of the graphs of two linear equations.

1. Enter the equations into a graphing calculator.
2. Graph the equations in an appropriate viewing window, so that the point of intersection is visible.
3. Use the *intersect* feature of the graphing calculator to find the point of intersection.

EXAMPLE 1 Using a Graphing Calculator

Use a graphing calculator to find the point of intersection, if it exists, of the graphs of the two linear equations.

$y = -\frac{1}{2}x + 2$ Equation 1

$y = 3x - 5$ Equation 2

SOLUTION

The slopes of the lines are not the same, so you know that the lines intersect. Enter the equations into a graphing calculator. Then graph the equations in an appropriate viewing window.

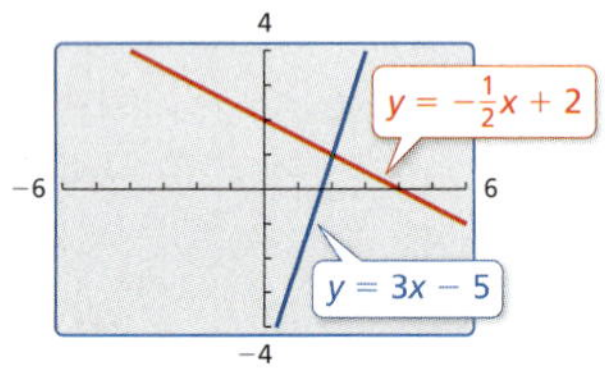

Use the *intersect* feature to find the point of intersection of the lines.

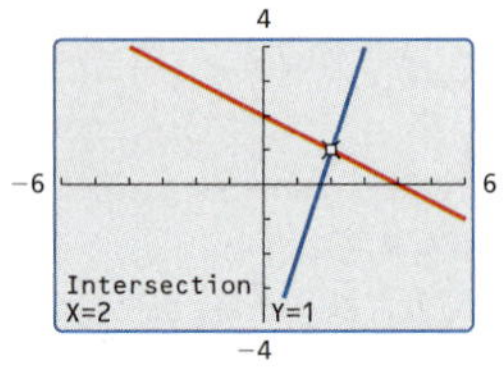

▶ The point of intersection is (2, 1).

Monitoring Progress

Use a graphing calculator to find the point of intersection of the graphs of the two linear equations.

1. $y = -2x - 3$
   $y = \frac{1}{2}x - 3$
2. $y = -x + 1$
   $y = x - 2$
3. $3x - 2y = 2$
   $2x - y = 2$

Laurie's Notes — Mathematical Practices (continued from page T-233)

- **Make Sense of Problems and Persevere in Solving Them** and **Use Appropriate Tools Strategically:** Although the example shown is for linear equations, this technique could also be used for more complex equations. Where other methods of solution such as an analytical approach might be more difficult, technology allows for an accurate solution to be found quickly.
- Give time for students to work through the questions in the *Monitoring Progress*, and then discuss as a class.

Laurie's Notes

Overview of Section 5.1

Introduction

- In this lesson, a system of linear equations, or simply a linear system, is defined. The solution of a linear system is an ordered pair that satisfies both equations.
- There are several strategies for solving a system, and this lesson presents the first strategy, solving by graphing.
- Students should be very familiar with graphing linear equations in slope-intercept form and in standard form. Students will estimate, by eyesight, the point of intersection. Alternately, you may wish to have students use graphing technology to solve some of the systems.

Teaching Strategy

- The equation editor of a graphing calculator is set up to accept an equation that has been solved for *y*. This does not mean that the equation must necessarily be in slope-intercept form.
- Students graphed equations in standard form in Section 3.4. Finding both intercepts is generally an efficient method when graphing with paper and pencil.
- To graph an equation in standard form using a calculator, it is not necessary to simplify the expression after solving for *y*.

$$Ax + By = C$$
$$By = C - Ax$$
$$y = \frac{C - Ax}{B}$$

- Enter $3.2x - 5y = -8$ as shown. Graph in a standard viewing window.

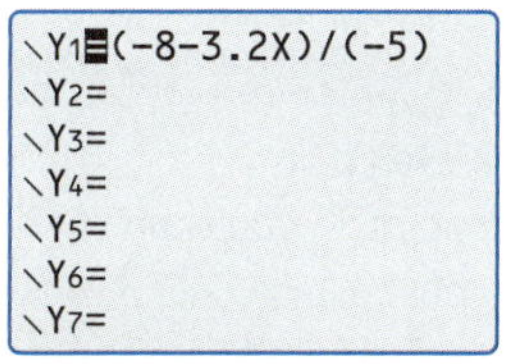

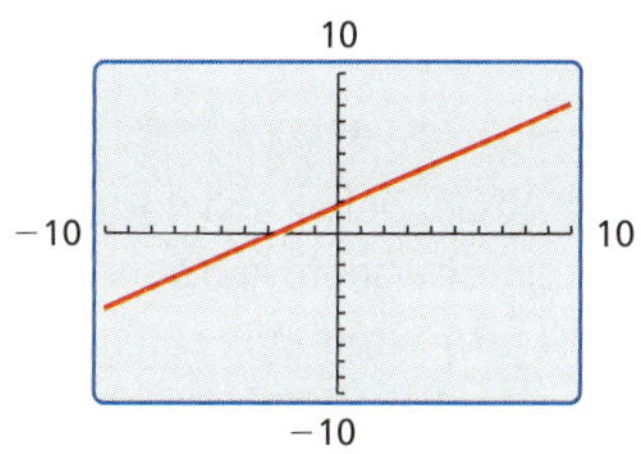

Pacing Suggestion

- The explorations should be review content for students, though seeing the connection between them is important. Students may be ready so that you can omit Example 1 in the formal lesson.

Dynamic Teaching Tools

- Dynamic Assessment & Progress Monitoring Tool
- Lesson Planning Tool
- Interactive Whiteboard Lesson Library
- Dynamic Classroom with Dynamic Investigations

What Your Students Will Learn

- Check solutions of systems of linear equations algebraically.
- Solve systems of linear equations in two variables by graphing.

Laurie's Notes

Exploration

Motivate

- Share a story about a trip to Indianapolis, where Market Street intersects Meridian Street at Monument Circle. Draw a sketch.

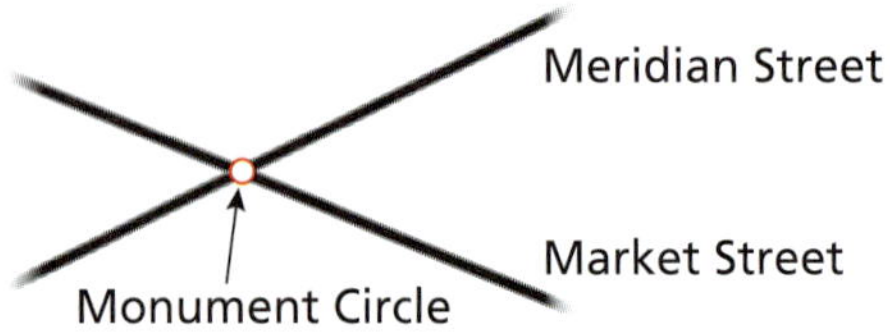

? Ask whether students have ever visited a town or city where a monument was located in the middle of two streets. Answers will vary.

- **Connection:** The monument is located on both streets. In other words, you will find the monument where the streets intersect.

Exploration 1

- Discuss what is known about your costs and your income.
- **Financial Literacy:** Do not assume that students are knowledgeable about concepts such as *costs* (fixed and variable) and *revenue* (income). Explain these words as you use them.
- Point out to students that the units in the verbal model agree. This means that "dollars per night × nights" is equal to dollars. So, the units in the equation are dollars = dollars + dollars.
- Ask a pair of students to share the equations they wrote.
- Discuss the definition of a system of linear equations.

Exploration 2

- Read the problem aloud. Define and discuss the break-even point.

? "Why would a business want to know the break-even point?" You want to know how many nights it will take before your revenue exceeds your cost.

- Provided that students have graphed the equations carefully, the lines should intersect at (10, 750).
- In answering part (d), you should expect students to use a complete sentence, meaning that they identify what the numbers $x = 10$ and $C = 750$ mean.

Communicate Your Answer

- **Attend to Precision:** In Question 3, students should realize the importance of checking solutions of systems of equations.
- **Use Appropriate Tools Strategically:** In Question 4, students are choosing between a graphing utility and pencil and paper.

Connecting to Next Step

- If you feel students are comfortable with checking whether an ordered pair is a solution of a system of equations, assign a few items from Exercises 3–8 on page 239 and begin with Example 2 in the formal lesson.

5.1 Solving Systems of Linear Equations by Graphing

Essential Question

How can you solve a system of linear equations?

EXPLORATION 1 Writing a System of Linear Equations

Work with a partner. Your family opens a bed-and-breakfast. They spend \$600 preparing a bedroom to rent. The cost to your family for food and utilities is \$15 per night. They charge \$75 per night to rent the bedroom.

a. Write an equation that represents the costs.

Cost, C (in dollars) = \$15 per night • Number of nights, x + \$600

b. Write an equation that represents the revenue (income).

Revenue, R (in dollars) = \$75 per night • Number of nights, x

c. A set of two (or more) linear equations is called a **system of linear equations**. Write the system of linear equations for this problem.

MODELING WITH MATHEMATICS

To be proficient in math, you need to identify important quantities in real-life problems and map their relationships using tools such as diagrams, tables, and graphs.

EXPLORATION 2 Using a Table or Graph to Solve a System

Work with a partner. Use the cost and revenue equations from Exploration 1 to determine how many nights your family needs to rent the bedroom before recovering the cost of preparing the bedroom. This is the *break-even point*.

a. Copy and complete the table.

x (nights)	0	1	2	3	4	5	6	7	8	9	10	11
C (dollars)												
R (dollars)												

b. How many nights does your family need to rent the bedroom before breaking even?

c. In the same coordinate plane, graph the cost equation and the revenue equation from Exploration 1.

d. Find the point of intersection of the two graphs. What does this point represent? How does this compare to the break-even point in part (b)? Explain.

Communicate Your Answer

3. How can you solve a system of linear equations? How can you check your solution?

4. Solve each system by using a table or sketching a graph. Explain why you chose each method. Use a graphing calculator to check each solution.

a. $y = -4.3x - 1.3$
$y = 1.7x + 4.7$

b. $y = x$
$y = -3x + 8$

c. $y = -x - 1$
$y = 3x + 5$

Dynamic Teaching Tools

- Dynamic Assessment & Progress Monitoring Tool
- Lesson Planning Tool
- Interactive Whiteboard Lesson Library
- Dynamic Classroom with Dynamic Investigations

ANSWERS

1. a. $C = 15x + 600$

b. $R = 75x$

c. $C = 15x + 600$
$R = 75x$

2. a. See Additional Answers.

b. 10 nights

c.

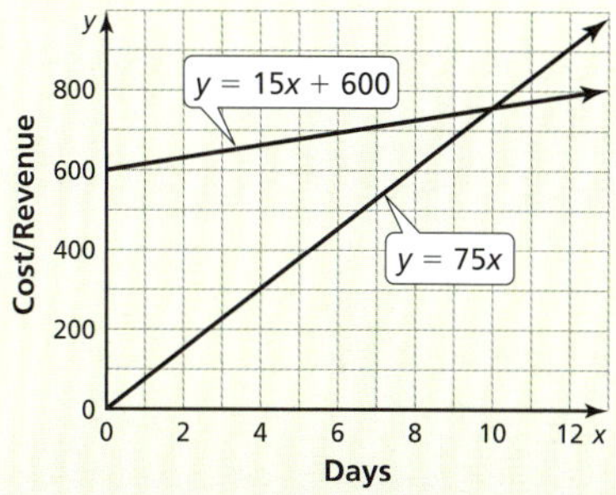

d. (10, 750); the point where both functions have the same x- and y-values; It is the same point; The break-even point is the point where both functions have the same x- and y-values.

3. Graph both equations and find the point of intersection; Substitute the x-coordinate for x in both of the equations and verify that both results are the y-coordinate of the point of intersection.

4. a. $(-1, 3)$; *Sample answer:* table because decimals are difficult to graph

b. (2, 2); *Sample answer:* graph because it is easier to draw

c. $\left(-\frac{3}{2}, \frac{1}{2}\right)$; *Sample answer:* graph because it is easier to draw

Extra Example 1

Tell whether the ordered pair is a solution of the system of linear equations.

a. (5, 1)

$x + y = 6$

$2x - y = 9$

Because the ordered pair (5, 1) is a solution of each equation, it is a solution of the linear system.

b. (−1, −5)

$y = -x + 2$

$y = 3x - 2$

The ordered pair (−1, −5) is *not* a solution of the first equation. So, it is *not* a solution of the linear system.

MONITORING PROGRESS ANSWERS

1. no
2. yes

5.1 Lesson

What You Will Learn

- Check solutions of systems of linear equations.
- Solve systems of linear equations by graphing.
- Use systems of linear equations to solve real-life problems.

Core Vocabulary

system of linear equations, *p. 236*
solution of a system of linear equations, *p. 236*

Previous
linear equation
ordered pair

Systems of Linear Equations

A **system of linear equations** is a set of two or more linear equations in the same variables. An example is shown below.

$x + y = 7$ Equation 1

$2x - 3y = -11$ Equation 2

A **solution of a system of linear equations** in two variables is an ordered pair that is a solution of each equation in the system.

EXAMPLE 1 Checking Solutions

Tell whether the ordered pair is a solution of the system of linear equations.

a. (2, 5); $x + y = 7$ Equation 1; $2x - 3y = -11$ Equation 2

b. (−2, 0); $y = -2x - 4$ Equation 1; $y = x + 4$ Equation 2

SOLUTION

a. Substitute 2 for *x* and 5 for *y* in each equation.

Equation 1	Equation 2
$x + y = 7$	$2x - 3y = -11$
$2 + 5 \stackrel{?}{=} 7$	$2(2) - 3(5) \stackrel{?}{=} -11$
$7 = 7$ ✓	$-11 = -11$ ✓

READING
A system of linear equations is also called a *linear system*.

▶ Because the ordered pair (2, 5) is a solution of each equation, it is a solution of the linear system.

b. Substitute −2 for *x* and 0 for *y* in each equation.

Equation 1	Equation 2
$y = -2x - 4$	$y = x + 4$
$0 \stackrel{?}{=} -2(-2) - 4$	$0 \stackrel{?}{=} -2 + 4$
$0 = 0$ ✓	$0 \neq 2$ ✗

▶ The ordered pair (−2, 0) is a solution of the first equation, but it is not a solution of the second equation. So, (−2, 0) is *not* a solution of the linear system.

Monitoring Progress

Help in English and Spanish at *BigIdeasMath.com*

Tell whether the ordered pair is a solution of the system of linear equations.

1. (1, −2); $2x + y = 0$, $-x + 2y = 5$

2. (1, 4); $y = 3x + 1$, $y = -x + 5$

Laurie's Notes Teacher Actions

COMMON ERROR When students hear the word *solution*, they often think of a single number, not an ordered pair. Explain that each line has an *x*-value of 0 associated with it, but the corresponding *y*-values do not have to be the same. They are looking for ordered pairs!

- An ordered pair can be a solution to one equation but not the other. An ordered pair could be a solution to both (or neither) equations.
- **Think-Pair-Share:** Have students answer Questions 1 and 2, and then share and discuss as a class.

Solving Systems of Linear Equations by Graphing

The solution of a system of linear equations is the point of intersection of the graphs of the equations.

Core Concept

Solving a System of Linear Equations by Graphing

Step 1 Graph each equation in the same coordinate plane.

Step 2 Estimate the point of intersection.

Step 3 Check the point from Step 2 by substituting for x and y in each equation of the original system.

REMEMBER

Note that the linear equations are in slope-intercept form. You can use the method presented in Section 3.5 to graph the equations.

EXAMPLE 2 Solving a System of Linear Equations by Graphing

Solve the system of linear equations by graphing.

$y = -2x + 5$ Equation 1

$y = 4x - 1$ Equation 2

SOLUTION

Step 1 Graph each equation.

Step 2 Estimate the point of intersection.
The graphs appear to intersect at (1, 3).

Step 3 Check your point from Step 2.

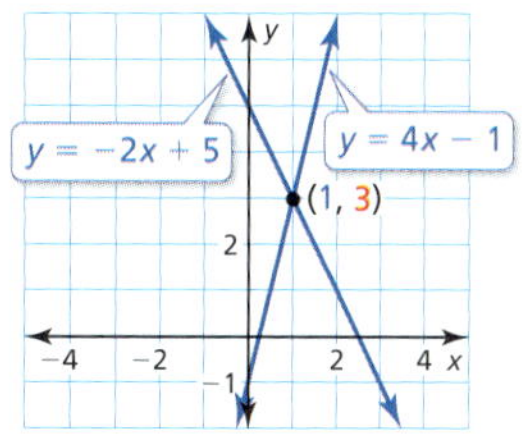

Equation 1	Equation 2
$y = -2x + 5$	$y = 4x - 1$
$3 \stackrel{?}{=} -2(1) + 5$	$3 \stackrel{?}{=} 4(1) - 1$
$3 = 3$ ✓	$3 = 3$ ✓

▶ The solution is (1, 3).

Check

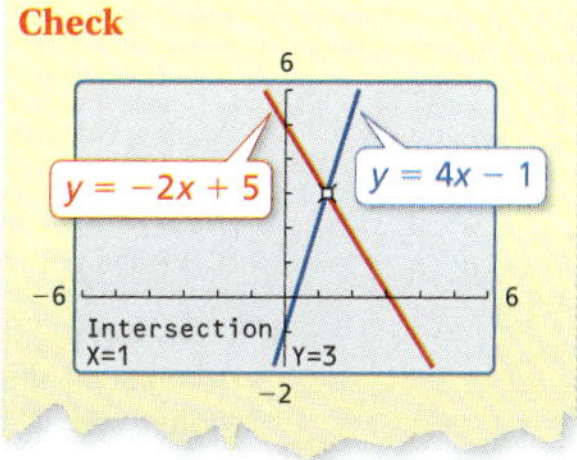

Monitoring Progress

Help in English and Spanish at *BigIdeasMath.com*

Solve the system of linear equations by graphing.

3. $y = x - 2$
$y = -x + 4$

4. $y = \frac{1}{2}x + 3$
$y = -\frac{3}{2}x - 5$

5. $2x + y = 5$
$3x - 2y = 4$

Differentiated Instruction

Visual

When graphing a system of equations, some students may benefit from using color to distinguish between the equations in the system. Have students use one color for the first equation and its graph and a different color for the second equation and its graph.

Extra Example 2

Solve the system of linear equations by graphing.

$y = -x + 5$

$y = 2x - 1$

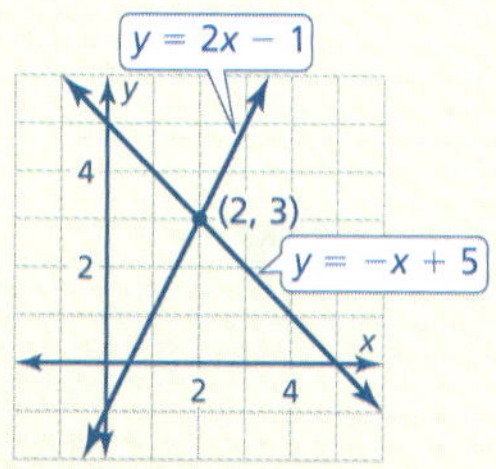

The solution is (2, 3).

MONITORING PROGRESS ANSWERS

3. (3, 1)

4. (−4, 1)

5. (2, 1)

Laurie's Notes Teacher Actions

- Discuss the steps for solving a system.

? "Why do you have to check your answer in both equations?" It may satisfy only one equation.

- Ask a volunteer to review the process for graphing a line in slope-intercept form.

? "Should we trust our eyes?" Answers will vary. "What if the solution is actually (1.1, 3.1)?" Checking the solution in the original equations will confirm our guess.

Extra Example 3

For the first tour on Monday, a museum sells 4 child tickets and 8 adult tickets for \$128. For the second tour, the museum sells 6 adult tickets for \$72. Find the price of one child ticket and the price of one adult ticket.

The price of one child ticket is \$8, and the price of one adult ticket is \$12.

MONITORING PROGRESS ANSWER

6. 12 math problems, 6 science problems

Solving Real-Life Problems

EXAMPLE 3 **Modeling with Mathematics**

A roofing contractor buys 30 bundles of shingles and 4 rolls of roofing paper for \$1040. In a second purchase (at the same prices), the contractor buys 8 bundles of shingles for \$256. Find the price per bundle of shingles and the price per roll of roofing paper.

SOLUTION

1. **Understand the Problem** You know the total price of each purchase and how many of each item were purchased. You are asked to find the price of each item.
2. **Make a Plan** Use a verbal model to write a system of linear equations that represents the problem. Then solve the system of linear equations.
3. **Solve the Problem**

Words 30 • Price per bundle + 4 • Price per roll = 1040

8 • Price per bundle + 0 • Price per roll = 256

Variables Let x be the price (in dollars) per bundle and let y be the price (in dollars) per roll.

System $30x + 4y = 1040$ Equation 1

$8x = 256$ Equation 2

Step 1 Graph each equation. Note that only the first quadrant is shown because x and y must be positive.

Step 2 Estimate the point of intersection. The graphs appear to intersect at (32, 20).

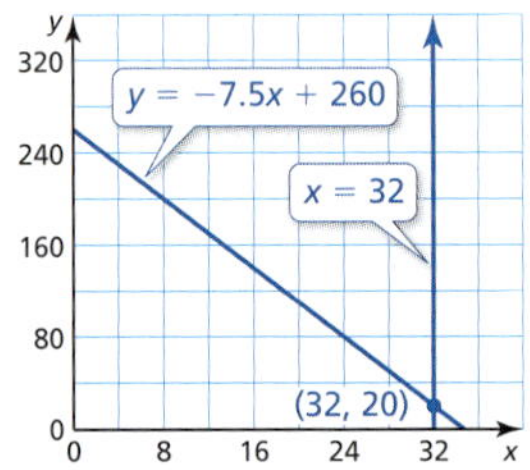

Step 3 Check your point from Step 2.

Equation 1	Equation 2
$30x + 4y = 1040$	$8x = 256$
$30(32) + 4(20) \stackrel{?}{=} 1040$	$8(32) \stackrel{?}{=} 256$
$1040 = 1040$ ✓	$256 = 256$ ✓

▶ The solution is (32, 20). So, the price per bundle of shingles is \$32, and the price per roll of roofing paper is \$20.

4. **Look Back** You can use estimation to check that your solution is reasonable. A bundle of shingles costs about \$30. So, 30 bundles of shingles and 4 rolls of roofing paper (at \$20 per roll) cost about 30(30) + 4(20) = \$980, and 8 bundles of shingles costs about 8(30) = \$240. These prices are close to the given values, so the solution seems reasonable.

Monitoring Progress

Help in English and Spanish at *BigIdeasMath.com*

6. You have a total of 18 math and science exercises for homework. You have six more math exercises than science exercises. How many exercises do you have in each subject?

Laurie's Notes Teacher Actions

- Ask a volunteer to describe what is known in this problem and what you are trying to solve for. Define the variables.
- **Construct Viable Arguments and Critique the Reasoning of Others:** Some students might suggest using the second purchase to solve for the cost of a bundle of shingles and then use this information to figure out the cost of a roll of roofing paper. Critique their reasoning. "What if the second purchase also included roofing paper?"

Closure

- **Phone Call:** Write a brief script for a phone conversation with a friend who was not in class today. Explain what a system of linear equations is and how you solve a system of linear equations by graphing.

5.1 Exercises

Dynamic Solutions available at *BigIdeasMath.com*

Vocabulary and Core Concept Check

1. **VOCABULARY** Do the equations $5y - 2x = 18$ and $6x = -4y - 10$ form a system of linear equations? Explain.

2. **DIFFERENT WORDS, SAME QUESTION** Consider the system of linear equations $-4x + 2y = 4$ and $4x - y = -6$. Which is different? Find "both" answers.

Solve the system of linear equations.	Solve each equation for y.
Find the point of intersection of the graphs of the equations.	Find an ordered pair that is a solution of each equation in the system.

Monitoring Progress and Modeling with Mathematics

In Exercises 3–8, tell whether the ordered pair is a solution of the system of linear equations. *(See Example 1.)*

3. $(2, 6)$; $x + y = 8$, $3x - y = 0$

4. $(8, 2)$; $x - y = 6$, $2x - 10y = 4$

5. $(-1, 3)$; $y = -7x - 4$, $y = 8x + 5$

6. $(-4, -2)$; $y = 2x + 6$, $y = -3x - 14$

7. $(-2, 1)$; $6x + 5y = -7$, $2x - 4y = -8$

8. $(5, -6)$; $6x + 3y = 12$, $4x + y = 14$

In Exercises 9–12, use the graph to solve the system of linear equations. Check your solution.

9. $x - y = 4$
   $4x + y = 1$

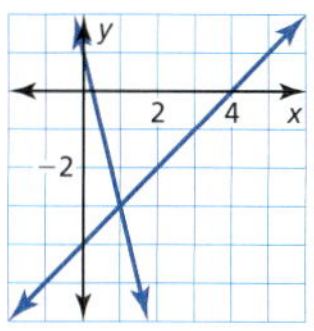

10. $x + y = 5$
    $y - 2x = -4$

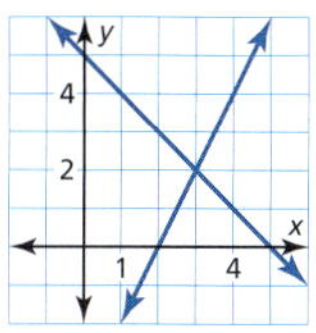

11. $6y + 3x = 18$
    $-x + 4y = 24$

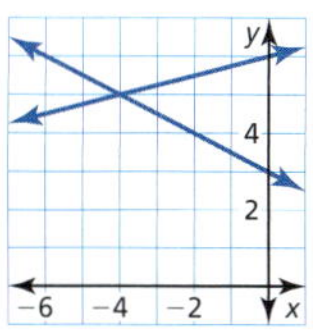

12. $2x - y = -2$
    $2x + 4y = 8$

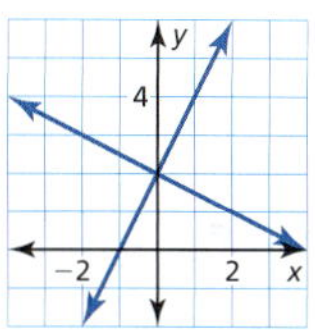

In Exercises 13–20, solve the system of linear equations by graphing. *(See Example 2.)*

13. $y = -x + 7$
    $y = x + 1$

14. $y = -x + 4$
    $y = 2x - 8$

15. $y = \frac{1}{3}x + 2$
    $y = \frac{2}{3}x + 5$

16. $y = \frac{3}{4}x - 4$
    $y = -\frac{1}{2}x + 11$

17. $9x + 3y = -3$
    $2x - y = -4$

18. $4x - 4y = 20$
    $y = -5$

19. $x - 4y = -4$
    $-3x - 4y = 12$

20. $3y + 4x = 3$
    $x + 3y = -6$

ERROR ANALYSIS In Exercises 21 and 22, describe and correct the error in solving the system of linear equations.

21.

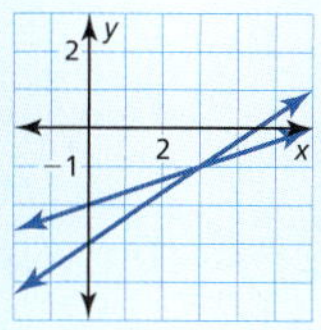

The solution of the linear system $x - 3y = 6$ and $2x - 3y = 3$ is $(3, -1)$.

22.

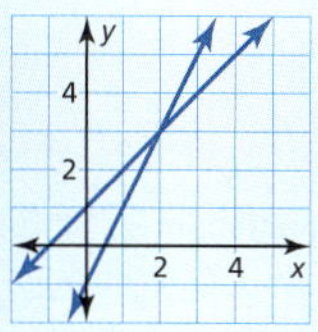

The solution of the linear system $y = 2x - 1$ and $y = x + 1$ is $x = 2$.

Assignment Guide and Homework Check

ASSIGNMENT

Basic: 1, 2, 3–21 odd, 27, 32–36

Average: 1, 2–28 even, 32–36

Advanced: 1, 2, 6–10 even, 18–28 even, 29–36

HOMEWORK CHECK

Basic: 3, 9, 13, 27

Average: 4, 10, 14, 28

Advanced: 6, 10, 18, 28

ANSWERS

1. yes; They are two equations in the same variables.
2. Solve each equation for y; $y = 2x + 2$, $y = 4x + 6$; $(-2, -2)$
3. yes
4. no
5. no
6. yes
7. yes
8. yes
9. $(1, -3)$
10. $(3, 2)$
11. $(-4, 5)$
12. $(0, 2)$
13. $(3, 4)$
14. $(4, 0)$
15. $(-9, -1)$
16. $(12, 5)$
17. $(-1, 2)$
18. $(0, -5)$
19. $(-4, 0)$
20. $(3, -3)$
21. The graph of the second equation should be $y = \frac{2}{3}x - 1$.

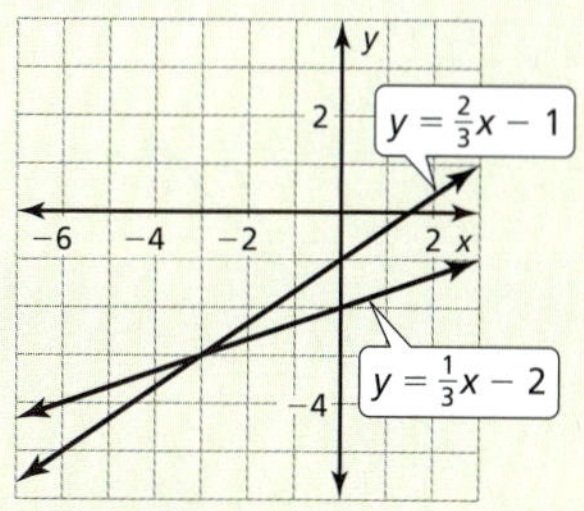

$(-3, -3)$

22. The solution needs to be an ordered pair; $(2, 3)$

Dynamic Teaching Tools

- Dynamic Assessment & Progress Monitoring Tool
- Interactive Whiteboard Lesson Library
- Dynamic Classroom with Dynamic Investigations

ANSWERS

23. (10, 5)

24. (5, 5)

25. (1.2, 1.4)

26. (0.5, 0.5)

27. 30 min on the elliptical trainer, 10 min on the stationary bike

28. 12 small candles, 16 large candles

29. $A = 18x - 18$; $P = 6x + 6$; (2, 18); The value of the perimeter and the area of the rectangle will both be 18 when $x = 2$.

30–36. See Additional Answers.

Mini-Assessment

1. Tell whether the ordered pair $(-2, -2)$ is a solution of the system of linear equations.

$2x - y = -2$

$x - 2y = 2$

$(-2, -2)$ is a solution of the system.

2. Solve the system of linear equations by graphing.

$y = 2x - 3$

$y = x + 6$

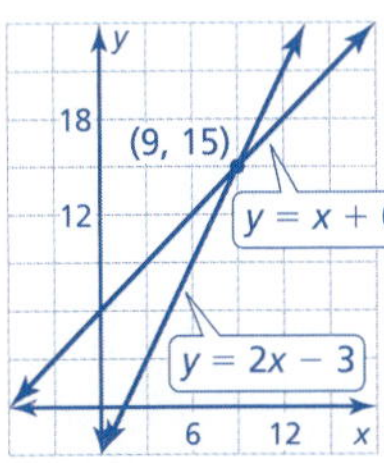

The solution is (9, 15).

3. Rob has 14 pens and markers. He has six more markers than pens. How many of each item does Rob have? 10 markers and 4 pens

USING TOOLS In Exercises 23–26, use a graphing calculator to solve the system of linear equations.

23. $0.2x + 0.4y = 4$
$-0.6x + 0.6y = -3$

24. $-1.6x - 3.2y = -24$
$2.6x + 2.6y = 26$

25. $-7x + 6y = 0$
$0.5x + y = 2$

26. $4x - y = 1.5$
$2x + y = 1.5$

27. MODELING WITH MATHEMATICS You have 40 minutes to exercise at the gym, and you want to burn 300 calories total using both machines. How much time should you spend on each machine? *(See Example 3.)*

Elliptical Trainer

8 calories per minute

Stationary Bike

6 calories per minute

28. MODELING WITH MATHEMATICS You sell small and large candles at a craft fair. You collect \$144 selling a total of 28 candles. How many of each type of candle did you sell?

29. MATHEMATICAL CONNECTIONS Write a linear equation that represents the area and a linear equation that represents the perimeter of the rectangle. Solve the system of linear equations by graphing. Interpret your solution.

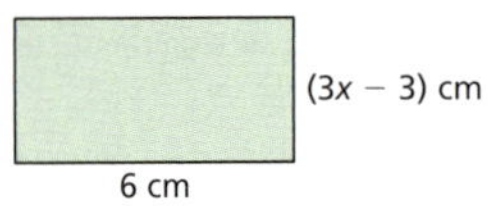

30. THOUGHT PROVOKING Your friend's bank account balance (in dollars) is represented by the equation $y = 25x + 250$, where x is the number of months. Graph this equation. After 6 months, you want to have the same account balance as your friend. Write a linear equation that represents your account balance. Interpret the slope and y-intercept of the line that represents your account balance.

31. COMPARING METHODS Consider the equation $x + 2 = 3x - 4$.

a. Solve the equation using algebra.

b. Solve the system of linear equations $y = x + 2$ and $y = 3x - 4$ by graphing.

c. How is the linear system and the solution in part (b) related to the original equation and the solution in part (a)?

32. HOW DO YOU SEE IT? A teacher is purchasing binders for students. The graph shows the total costs of ordering x binders from three different companies.

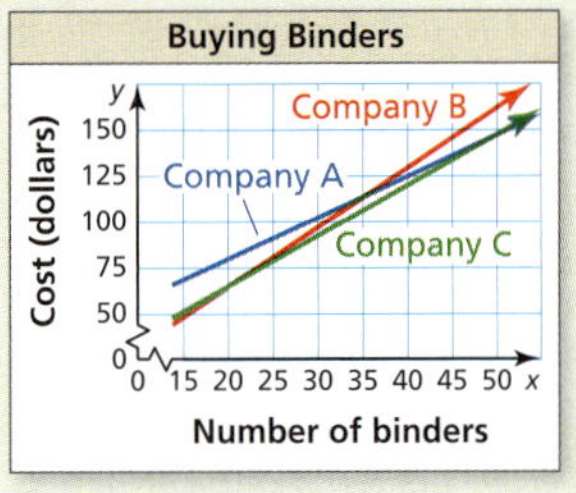

a. For what numbers of binders are the costs the same at two different companies? Explain.

b. How do your answers in part (a) relate to systems of linear equations?

33. MAKING AN ARGUMENT You and a friend are going hiking but start at different locations. You start at the trailhead and walk 5 miles per hour. Your friend starts 3 miles from the trailhead and walks 3 miles per hour.

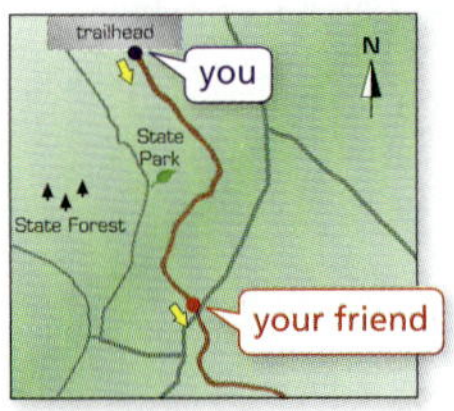

a. Write and graph a system of linear equations that represents this situation.

b. Your friend says that after an hour of hiking you will both be at the same location on the trail. Is your friend correct? Use the graph from part (a) to explain your answer.

Maintaining Mathematical Proficiency

Reviewing what you learned in previous grades and lessons

Solve the literal equation for *y*. *(Section 1.5)*

34. $10x + 5y = 5x + 20$

35. $9x + 18 = 6y - 3x$

36. $\frac{3}{4}x + \frac{1}{4}y = 5$

If students need help...	*If students got it...*
Resources by Chapter • Practice A and Practice B • Puzzle Time	Resources by Chapter • Enrichment and Extension • Cumulative Review
Student Journal • Practice	Start the *next* Section
Differentiating the Lesson Skills Review Handbook	

Laurie's Notes

Overview of Section 5.2

Introduction

- The second method for solving a linear system is substitution. Students generally have little trouble with this process. Their errors, if there are any, tend to be in making a mistake in their symbolic manipulations.
- Students are again encouraged to check their solution algebraically or graphically.

Common Misconceptions

- When asked to solve a linear system, the solution is an ordered pair. Students sometimes will solve for one variable and forget to solve for the second.

Teaching Strategy

- Students may not recognize that they have a linear system when both equations are in slope-intercept form to begin with. Substitution is a very logical process!
- There is no need to graph the system.
- Set the two expressions with x equal to one another and simply solve.

  $$\begin{aligned} y &= 2x + 5 \\ y &= 3x - 1 \end{aligned} \quad \rightarrow \quad \begin{aligned} 2x + 5 &= 3x - 1 \\ 6 &= x \end{aligned}$$

  Then substitute $x = 6$ into one of the original equations and solve for y.

  $$y = 2(6) + 5 = 17$$

  So the solution is (6, 17).

Pacing Suggestion

- Once students have worked the explorations, continue with the formal lesson.

Dynamic Teaching Tools

- Dynamic Assessment & Progress Monitoring Tool
- Lesson Planning Tool
- Interactive Whiteboard Lesson Library
- Dynamic Classroom with Dynamic Investigations

What Your Students Will Learn

- Solve systems of linear equations in two variables by substitution.

Laurie's Notes

Exploration

Motivate

- Play a game of "Zip, Zap, Zoop," which is a combination of several games.
- **Directions:** Stand in a circle. Count around the circle. When your number is a *multiple* of 4, say "**Zip**" instead of the number. When your number *contains* a 4, say "**Zap**." When your number is *both* a multiple of 4 *and* contains a 4, say "**Zoop**."
- The counting will go: 1, 2, 3, zoop, 5, 6, 7, zip, 9, 10, 11, zip, 13, zap, …
- Students are *substituting* an expression for a number.
- The faster students count, the funnier it becomes!

Discuss

- Students may not be familiar with the phrase "substitute the expression for x." It may be helpful to highlight "$-7 - y$" and identify it as "an expression for x" in Exploration 1(a).
- As a hint, tell students to think about the game they just played. They replaced certain numbers with words.

Exploration 1

- Introduce the language to students before they begin the exploration.
- Circulate to be sure that students are checking in with partners and the process is correct.
- To save time, have different groups do different pairs of problems.
- Discuss whether the results for each method are the same.

Exploration 2

- Students can struggle with trying to write any equation that passes through a given point. For the example $(-2, -3)$, ask students to think about the relationship between the two numbers: x is one more than y, so subtracting 1 from x is equal to y. This suggests the equation $y = x - 1$. A second equation can be generated by thinking, "You can double the x-value and add 1 to get the y-value." This suggests $y = 2x + 1$.
- Take time for students to share their strategies for generating their systems and to solve each other's system.

Communicate Your Answer

- **Look For and Make Use of Structure:** Students discovered in Exploration 1 that given a system of two equations in two variables, they can solve one equation for one of the variables and then substitute the expression for that variable in the other equation to find the value of the other variable.

Connecting to Next Step

- If you are confident that students are able to solve a system by substitution, practice a few problems in the formal lesson and spend time with applications.

5.2 Solving Systems of Linear Equations by Substitution

Essential Question How can you use substitution to solve a system of linear equations?

EXPLORATION 1 Using Substitution to Solve Systems

Work with a partner. Solve each system of linear equations using two methods.

Method 1 Solve for x first.

Solve for x in one of the equations. Substitute the expression for x into the other equation to find y. Then substitute the value of y into one of the original equations to find x.

Method 2 Solve for y first.

Solve for y in one of the equations. Substitute the expression for y into the other equation to find x. Then substitute the value of x into one of the original equations to find y.

Is the solution the same using both methods? Explain which method you would prefer to use for each system.

a. $x + y = -7$
$-5x + y = 5$

b. $x - 6y = -11$
$3x + 2y = 7$

c. $4x + y = -1$
$3x - 5y = -18$

EXPLORATION 2 Writing and Solving a System of Equations

Work with a partner.

a. Write a random ordered pair with integer coordinates. One way to do this is to use a graphing calculator. The ordered pair generated at the right is $(-2, -3)$.

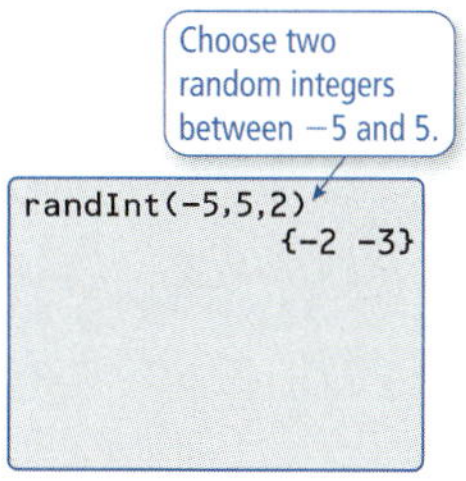

b. Write a system of linear equations that has your ordered pair as its solution.

c. Exchange systems with your partner and use one of the methods from Exploration 1 to solve the system. Explain your choice of method.

ATTENDING TO PRECISION

To be proficient in math, you need to communicate precisely with others.

Communicate Your Answer

3. How can you use substitution to solve a system of linear equations?

4. Use one of the methods from Exploration 1 to solve each system of linear equations. Explain your choice of method. Check your solutions.

a. $x + 2y = -7$
$2x - y = -9$

b. $x - 2y = -6$
$2x + y = -2$

c. $-3x + 2y = -10$
$-2x + y = -6$

d. $3x + 2y = 13$
$x - 3y = -3$

e. $3x - 2y = 9$
$-x - 3y = 8$

f. $3x - y = -6$
$4x + 5y = 11$

Dynamic Teaching Tools

- Dynamic Assessment & Progress Monitoring Tool
- Lesson Planning Tool
- Interactive Whiteboard Lesson Library
- Dynamic Classroom with Dynamic Investigations

ANSWERS

1. a. $(-2, -5)$; yes; *Sample answer:* Method 2 because both equations can be solved for y easily.

b. $(1, 2)$; yes; *Sample answer:* Method 1 because the first equation can be solved for x easily.

c. $(-1, 3)$; yes; *Sample answer:* Method 2 because the first equation can be solved for y easily.

2. a. *Sample answer:* $(2, 4)$

b. *Sample answer:* $y = x + 2$ and $3x - 4y = -10$

c. *Sample answer:* $y = x + 2$ and $3x - 4y = -10$; $(2, 4)$; Because the first equation is already solved for y, substitute that expression for y in the second equation.

3. Solve one of the equations for one of the variables. Substitute the expression for that variable into the other equation to find the value of the other variable. Substitute this value into one of the original equations to find the value of the remaining variable.

4. a. $(-5, -1)$; *Sample answer:* The first equation can be solved for x easily.

b. $(-2, 2)$; *Sample answer:* The first equation can be solved for x easily.

c. $(2, -2)$; *Sample answer:* The second equation can be solved for y easily.

d. $(3, 2)$; *Sample answer:* The second equation can be solved for x easily.

e. $(1, -3)$; *Sample answer:* The second equation can be solved for x easily.

f. $(-1, 3)$; *Sample answer:* The first equation can be solved for y easily.

Differentiated Instruction

Inclusion

Some students may have difficulty deciding which equation to use for the expression to be substituted. Tell them to look first for an equation that is already solved for x or y and use that equation. If neither equation is written like that, then they should look for an equation that has a variable with a coefficient of 1 or -1 and solve for that variable.

Extra Example 1

Solve the system of linear equations by substitution.

$y = -x + 3$

$3y + 5x = -1$

The solution is $(-5, 8)$.

MONITORING PROGRESS ANSWERS

1. $(-2, 8)$
2. $\left(\frac{1}{2}, -\frac{3}{4}\right)$
3. $(-1, 1)$

5.2 Lesson

Core Vocabulary

Previous
system of linear equations
solution of a system of linear equations

What You Will Learn

- Solve systems of linear equations by substitution.
- Use systems of linear equations to solve real-life problems.

Solving Linear Systems by Substitution

Another way to solve a system of linear equations is to use substitution.

Core Concept

Solving a System of Linear Equations by Substitution

Step 1 Solve one of the equations for one of the variables.

Step 2 Substitute the expression from Step 1 into the other equation and solve for the other variable.

Step 3 Substitute the value from Step 2 into one of the original equations and solve.

EXAMPLE 1 Solving a System of Linear Equations by Substitution

Solve the system of linear equations by substitution.

$y = -2x - 9$ Equation 1

$6x - 5y = -19$ Equation 2

SOLUTION

Step 1 Equation 1 is already solved for y.

Step 2 Substitute $-2x - 9$ for y in Equation 2 and solve for x.

$6x - 5y = -19$	Equation 2
$6x - 5(-2x - 9) = -19$	Substitute $-2x - 9$ for y.
$6x + 10x + 45 = -19$	Distributive Property
$16x + 45 = -19$	Combine like terms.
$16x = -64$	Subtract 45 from each side.
$x = -4$	Divide each side by 16.

Step 3 Substitute -4 for x in Equation 1 and solve for y.

$y = -2x - 9$	Equation 1
$= -2(-4) - 9$	Substitute -4 for x.
$= 8 - 9$	Multiply.
$= -1$	Subtract.

▶ The solution is $(-4, -1)$.

Check

Equation 1

$y = -2x - 9$

$-1 \stackrel{?}{=} -2(-4) - 9$

$-1 = -1$ ✓

Equation 2

$6x - 5y = -19$

$6(-4) - 5(-1) \stackrel{?}{=} -19$

$-19 = -19$ ✓

Monitoring Progress

Help in English and Spanish at *BigIdeasMath.com*

Solve the system of linear equations by substitution. Check your solution.

1. $y = 3x + 14$
   $y = -4x$
2. $3x + 2y = 0$
   $y = \frac{1}{2}x - 1$
3. $x = 6y - 7$
   $4x + y = -3$

Laurie's Notes Teacher Actions

- Write the *Core Concept*. At the end of Step 2, you can say, "You now know either the x- or y-coordinate of the ordered pair that satisfies both equations. Next, you need to find the other coordinate."
- **Attend to Precision:** Students may get sloppy and say they are "plugging in for y." "Plugging in" is not a mathematical operation or process. It is better to say that they are "substituting for y" so that they become familiar with the math terminology they are expected to know.

ANOTHER WAY

You could also begin by solving for x in Equation 1, solving for y in Equation 2, or solving for x in Equation 2.

EXAMPLE 2 **Solving a System of Linear Equations by Substitution**

Solve the system of linear equations by substitution.

$-x + y = 3$ — Equation 1

$3x + y = -1$ — Equation 2

SOLUTION

Step 1 Solve for y in Equation 1.

$y = x + 3$ — Revised Equation 1

Step 2 Substitute $x + 3$ for y in Equation 2 and solve for x.

$3x + y = -1$ — Equation 2

$3x + (x + 3) = -1$ — Substitute $x + 3$ for y.

$4x + 3 = -1$ — Combine like terms.

$4x = -4$ — Subtract 3 from each side.

$x = -1$ — Divide each side by 4.

Step 3 Substitute -1 for x in Equation 1 and solve for y.

$-x + y = 3$ — Equation 1

$-(-1) + y = 3$ — Substitute -1 for x.

$y = 2$ — Subtract 1 from each side.

The solution is $(-1, 2)$.

Algebraic Check

Equation 1

$-x + y = 3$

$-(-1) + 2 \stackrel{?}{=} 3$

$3 = 3$ ✓

Equation 2

$3x + y = -1$

$3(-1) + 2 \stackrel{?}{=} -1$

$-1 = -1$ ✓

Graphical Check

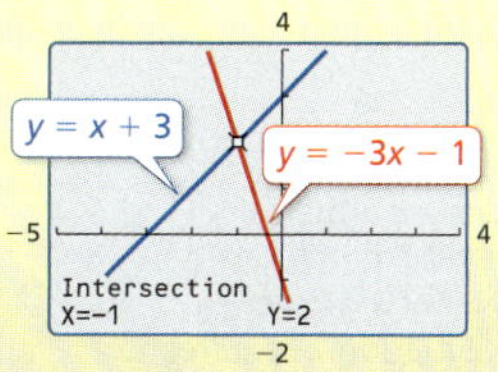

Monitoring Progress

Help in English and Spanish at *BigIdeasMath.com*

Solve the system of linear equations by substitution. Check your solution.

4. $x + y = -2$
$-3x + y = 6$

5. $-x + y = -4$
$4x - y = 10$

6. $2x - y = -5$
$3x - y = 1$

7. $x - 2y = 7$
$3x - 2y = 3$

Extra Example 2

Solve the system of linear equations by substitution.

$x - y = -4$

$2x + y = 4$

The solution is (0, 4).

MONITORING PROGRESS ANSWERS

4. $(-2, 0)$

5. $(2, -2)$

6. $(6, 17)$

7. $\left(-2, -\frac{9}{2}\right)$

Laurie's Notes — Teacher Actions

- **Turn and Talk:** Write the linear system in Example 2. "Which variable in which equation do you want to solve for and why?"
- **Construct Viable Arguments and Critique the Reasoning of Others:** There is no wrong answer provided students are able to offer a valid justification for their choice.

? "What does it means to check a solution algebraically?" Substitute the answer into both equations, simplify each side, and verify that you have true statements. "Graphically?" Graph both equations and verify that your solution is the point of intersection.

- **Think-Pair-Share:** Have students answer Questions 4–7, and then share and discuss as a class.

Extra Example 3

A theater earns \$3640 from a concert. An orchestra ticket costs three times as much as a balcony ticket.

Ticket Type	Number Sold
orchestra	148
balcony	76

Write a system of linear equations that represents this situation. What is the price of each type of ticket?

$148x + 76y = 3640$
$x = 3y$
The solution is (21, 7); orchestra ticket: \$21, balcony ticket: \$7.

MONITORING PROGRESS ANSWER

8. $x + y = 64$ and $x = y + 10$; 37 drama club students, 27 yearbook club students

Solving Real-Life Problems

EXAMPLE 3 **Modeling with Mathematics**

A drama club earns \$1040 from a production. A total of 64 adult tickets and 132 student tickets are sold. An adult ticket costs twice as much as a student ticket. Write a system of linear equations that represents this situation. What is the price of each type of ticket?

SOLUTION

1. **Understand the Problem** You know the amount earned, the total numbers of adult and student tickets sold, and the relationship between the price of an adult ticket and the price of a student ticket. You are asked to write a system of linear equations that represents the situation and find the price of each type of ticket.
2. **Make a Plan** Use a verbal model to write a system of linear equations that represents the problem. Then solve the system of linear equations.
3. **Solve the Problem**

Words 64 • Adult ticket price + 132 • Student ticket price = 1040

Adult ticket price = 2 • Student ticket price

Variables Let x be the price (in dollars) of an adult ticket and let y be the price (in dollars) of a student ticket.

System $64x + 132y = 1040$ Equation 1

$x = 2y$ Equation 2

Step 1 Equation 2 is already solved for x.

Step 2 Substitute $2y$ for x in Equation 1 and solve for y.

$64x + 132y = 1040$ Equation 1

$64(2y) + 132y = 1040$ Substitute $2y$ for x.

$260y = 1040$ Simplify.

$y = 4$ Simplify.

STUDY TIP
You can use either of the original equations to solve for x. However, using Equation 2 requires fewer calculations.

Step 3 Substitute 4 for y in Equation 2 and solve for x.

$x = 2y$ Equation 2

$x = 2(4)$ Substitute 4 for y.

$x = 8$ Simplify.

▶ The solution is (8, 4). So, an adult ticket costs \$8 and a student ticket costs \$4.

4. **Look Back** To check that your solution is correct, substitute the values of x and y into both of the original equations and simplify.

$64(8) + 132(4) = 1040$ $\quad$ $8 = 2(4)$

$1040 = 1040$ ✓ $\quad$ $8 = 8$ ✓

Monitoring Progress Help in English and Spanish at *BigIdeasMath.com*

8. There are a total of 64 students in a drama club and a yearbook club. The drama club has 10 more students than the yearbook club. Write a system of linear equations that represents this situation. How many students are in each club?

Laurie's Notes Teacher Actions

- **Common Misconception:** Students do not always interpret "an adult ticket costs twice as much as a student ticket" as a possible equation.
- **Attend to Precision:** Students can be sloppy in defining variables. They may say that x = adult tickets. Say, "What about the adult ticket? Number sold? Size of ticket? Cost of ticket? Color of ticket?" Students will get the idea when you are consistent in expecting precise definitions.

Closure

- **Exit Ticket:** Solve the system by substitution and by graphing.
  $y = 3x + 1$
  $y = x + 3$
  (1, 4); check students' graphs.

5.2 Exercises

Dynamic Solutions available at *BigIdeasMath.com*

Vocabulary and Core Concept Check

1. **WRITING** Describe how to solve a system of linear equations by substitution.

2. **NUMBER SENSE** When solving a system of linear equations by substitution, how do you decide which variable to solve for in Step 1?

Monitoring Progress and Modeling with Mathematics

In Exercises 3–8, tell which equation you would choose to solve for one of the variables. Explain.

3. $x + 4y = 30$
   $x - 2y = 0$

4. $3x - y = 0$
   $2x + y = -10$

5. $5x + 3y = 11$
   $5x - y = 5$

6. $3x - 2y = 19$
   $x + y = 8$

7. $x - y = -3$
   $4x + 3y = -5$

8. $3x + 5y = 25$
   $x - 2y = -6$

In Exercises 9–16, solve the sytem of linear equations by substitution. Check your solution. *(See Examples 1 and 2.)*

9. $x = 17 - 4y$
   $y = x - 2$

10. $6x - 9 = y$
    $y = -3x$

11. $x = 16 - 4y$
    $3x + 4y = 8$

12. $-5x + 3y = 51$
    $y = 10x - 8$

13. $2x = 12$
    $x - 5y = -29$

14. $2x - y = 23$
    $x - 9 = -1$

15. $5x + 2y = 9$
    $x + y = -3$

16. $11x - 7y = -14$
    $x - 2y = -4$

17. **ERROR ANALYSIS** Describe and correct the error in solving for one of the variables in the linear system $8x + 2y = -12$ and $5x - y = 4$.

✗

Step 1 $5x - y = 4$
$-y = -5x + 4$
$y = 5x - 4$

Step 2 $5x - (5x - 4) = 4$
$5x - 5x + 4 = 4$
$4 = 4$

18. **ERROR ANALYSIS** Describe and correct the error in solving for one of the variables in the linear system $4x + 2y = 6$ and $3x + y = 9$.

✗

Step 1 $3x + y = 9$
$y = 9 - 3x$

Step 2 $4x + 2(9 - 3x) = 6$
$4x + 18 - 6x = 6$
$-2x = -12$
$x = 6$

Step 3 $3x + y = 9$
$3x + 6 = 9$
$3x = 3$
$x = 1$

19. **MODELING WITH MATHEMATICS** A farmer plants corn and wheat on a 180-acre farm. The farmer wants to plant three times as many acres of corn as wheat. Write a system of linear equations that represents this situation. How many acres of each crop should the farmer plant? *(See Example 3.)*

20. **MODELING WITH MATHEMATICS** A company that offers tubing trips down a river rents tubes for a person to use and "cooler" tubes to carry food and water. A group spends \$270 to rent a total of 15 tubes. Write a system of linear equations that represents this situation. How many of each type of tube does the group rent?

Assignment Guide and Homework Check

ASSIGNMENT

Basic: 1, 2, 3–19 odd, 25, 30, 32, 36–41

Average: 1, 2–26 even, 30, 32, 36–41

Advanced: 1, 2, 8–28 even, 29–41

HOMEWORK CHECK

Basic: 3, 9, 13, 19

Average: 6, 10, 14, 20

Advanced: 8, 12, 14, 20

ANSWERS

1. Solve one of the equations for one of the variables. Substitute the expression for that variable into the other equation to find the value of the other variable. Substitute this value into one of the original equations to find the value of the remaining variable.
2. *Sample answer:* Look for the equation that has a variable with a coefficient of 1 or −1.
3. *Sample answer:* $x - 2y = 0$; This equation can be solved for x easily.
4. *Sample answer:* $2x + y = -10$; This equation can be solved for y easily.
5. *Sample answer:* $5x - y = 5$; This equation can be solved for y easily.
6. *Sample answer:* $x + y = 8$; This equation can be solved for x easily.
7. *Sample answer:* $x - y = -3$; This equation can be solved for x easily.
8. *Sample answer:* $x - 2y = -6$; This equation can be solved for x easily.
9. $(5, 3)$
10. $(1, -3)$
11. $(-4, 5)$
12. $(3, 22)$
13. $(6, 7)$
14. $(8, -7)$
15. $(5, -8)$
16. $(0, 2)$
17. In Step 2, the expression for y needs to be substituted in the other equation; $8x + 2(5x - 4) = -12$, $8x + 10x - 8 = -12$, $18x - 8 = -12$, $18x = -4$, $x = -\frac{2}{9}$
18. In Step 3, 6 should be substituted for x, not for y; $3(6) + y = 9$, $18 + y = 9$, $y = -9$
19. $x + y = 180$ and $x = 3y$; 135 acres of corn, 45 acres of wheat
20. $x + y = 15$ and $20x + 12.5y = 270$; 11 one-person tubes, 4 cooler tubes

Dynamic Teaching Tools

- Dynamic Assessment & Progress Monitoring Tool
- Interactive Whiteboard Lesson Library
- Dynamic Classroom with Dynamic Investigations

ANSWERS

21. *Sample answer:* $y = x + 2$ and $x + 2y = 13$

22. *Sample answer:* $y - 2x = 12$ and $x - 2y = -18$

23. *Sample answer:* $x = \frac{1}{3}y$ and $x + y = -16$

24. *Sample answer:* $2x + y = 5$ and $2x - y = 55$

25. 8 five-point problems, 30 two-point problems

26. 80 shares of Stock A, 120 shares of Stock B

27. **a.** $x + y + 90 = 180$
b. $x = 67, y = 23$

28. **a.** $x + 2y - 18 = 180$
b. $x = 86, y = 56$

29. $a = 4, b = 5$

30. yes; Both equations only have 1 variable and the variables are different.

31. *Sample answer:* $y = -3x + 4$ and $x + y = 6$

32–41. See Additional Answers.

In Exercises 21–24, write a system of linear equations that has the ordered pair as its solution.

21. (3, 5) **22.** (−2, 8)

23. (−4, −12) **24.** (15, −25)

25. **PROBLEM SOLVING** A math test is worth 100 points and has 38 problems. Each problem is worth either 5 points or 2 points. How many problems of each point value are on the test?

26. **PROBLEM SOLVING** An investor owns shares of Stock A and Stock B. The investor owns a total of 200 shares with a total value of \$4000. How many shares of each stock does the investor own?

Stock	Price
A	\$9.50
B	\$27.00

MATHEMATICAL CONNECTIONS **In Exercises 27 and 28, (a) write an equation that represents the sum of the angle measures of the triangle and (b) use your equation and the equation shown to find the values of x and y.**

27.

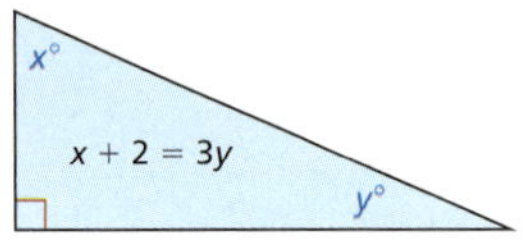

28.

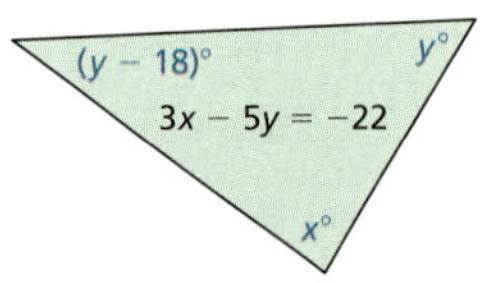

29. **REASONING** Find the values of a and b so that the solution of the linear system is (−9, 1).

$ax + by = -31$ Equation 1
$ax - by = -41$ Equation 2

30. **MAKING AN ARGUMENT** Your friend says that given a linear system with an equation of a horizontal line and an equation of a vertical line, you cannot solve the system by substitution. Is your friend correct? Explain.

31. **OPEN-ENDED** Write a system of linear equations in which (3, −5) is a solution of Equation 1 but not a solution of Equation 2, and (−1, 7) is a solution of the system.

32. **HOW DO YOU SEE IT?** The graphs of two linear equations are shown.

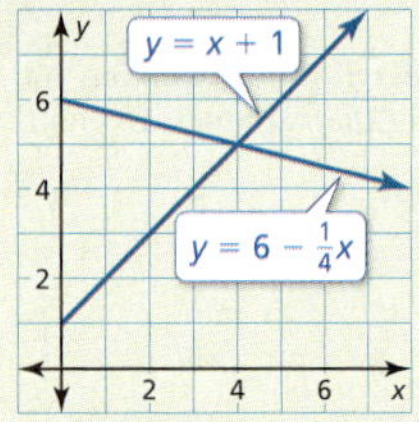

a. At what point do the lines appear to intersect?

b. Could you solve a system of linear equations by substitution to check your answer in part (a)? Explain.

33. **REPEATED REASONING** A radio station plays a total of 272 pop, rock, and hip-hop songs during a day. The number of pop songs is 3 times the number of rock songs. The number of hip-hop songs is 32 more than the number of rock songs. How many of each type of song does the radio station play?

34. **THOUGHT PROVOKING** You have \$2.65 in coins. Write a system of equations that represents this situation. Use variables to represent the number of each type of coin.

35. **NUMBER SENSE** The sum of the digits of a two-digit number is 11. When the digits are reversed, the number increases by 27. Find the original number.

Maintaining Mathematical Proficiency

Reviewing what you learned in previous grades and lessons

Find the sum or difference. (*Skills Review Handbook*)

36. $(x - 4) + (2x - 7)$ **37.** $(5y - 12) + (-5y - 1)$

38. $(t - 8) - (t + 15)$ **39.** $(6d + 2) - (3d - 3)$

40. $4(m + 2) + 3(6m - 4)$ **41.** $2(5v + 6) - 6(-9v + 2)$

Mini-Assessment

Solve the system of linear equations by substitution.

1. $x = 5 - y$
   $2x + 5y = 4$ (7, −2)
2. $x - y = -2$
   $3x - y = 2$ (2, 4)
3. There are a total of 124 students in the band. There are 12 more boys than girls in the band. Write a system of linear equations that represents this situation. How many boys and how many girls are in the band?
   $x + y = 124$
   $x = y + 12$
   The solution is (68, 56). So, there are 68 boys and 56 girls in the band.

If students need help...	*If students got it...*
Resources by Chapter • Practice A and Practice B • Puzzle Time	Resources by Chapter • Enrichment and Extension • Cumulative Review
Student Journal • Practice	Start the *next* Section
Differentiating the Lesson Skills Review Handbook	

Laurie's Notes

Overview of Section 5.3

Introduction

- The third method for solving a linear system is elimination. Once the justification for the method is understood, students generally have little trouble with this process.

COMMON ERROR There are two common errors that students make. The first is when multiplying through by a constant; students do not multiply every term by the constant. The second common error is when subtracting equations; students forget that every term in the second equation is being subtracted from a like term in the first equation.

- Students are again encouraged to check their solution algebraically or graphically.

Common Misconceptions

- Subtracting one equation from another, or adding one equation to another, feels unnatural to students. Some students will even say it is wrong and cannot be done. Students are used to adding or subtracting a single term, or constant, but not an equation.
- What feels wrong is that it does not appear that you are performing the same operation on both sides of the equation. Remind students of what they did in the *Motivate* to keep the scale balanced.

Formative Assessment Tips

- **Example/Non-Example (student generated):** This technique asks students to generate examples and non-examples to demonstrate their understanding of a concept. To generate the examples and non-examples, students must have a deeper understanding of the concept than just memorization would yield.
- In generating the examples, students can also be asked why—meaning they should provide justification. Listen carefully to students' language as they provide the justification.
- Note that this technique is an instructional technique used by teachers to introduce concepts, such as making a list of expressions that are monomials and those that are not monomials. The difference here is that students are using their cognitive skills to generate the examples and non-examples.
- This technique can be used by individuals, partners, or small groups.
- Be sure to allow sufficient time so that students receive feedback from others and/or you about their examples and non-examples.

Pacing Suggestion

- The explorations provide an opportunity for students to develop a conceptual understanding of what it means to add and subtract equations to eliminate a variable. You might have students do only one or two of the parts in Exploration 2. Transition to the formal lesson as soon as students have finished.

Dynamic Teaching Tools

- Dynamic Assessment & Progress Monitoring Tool
- Lesson Planning Tool
- Interactive Whiteboard Lesson Library
- Dynamic Classroom with Dynamic Investigations

What Your Students Will Learn

- Solve systems of linear equations in two variables by elimination.

Laurie's Notes

Exploration

Motivate

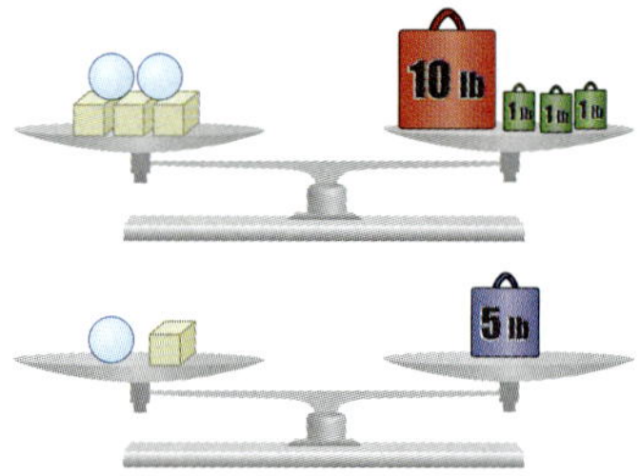

- Draw the sketch on the board. Make it clear to students that the right sides represent 13 pounds of weight and 5 pounds of weight.
- Ask students to draw a sketch of a balance scale with 4 cubes and 3 balls on the left and 18 pounds on the right. Relate this to adding equations.
- Ask students to draw a sketch of a balance scale with 2 cubes and 1 ball on the left and 8 pounds on the right. Relate this to subtracting equations.
- Explain to students that today they will perform operations with equations similar to those they represented with the balance scales.

Exploration 1

- A verbal model is given for this scenario, which will help students write the linear system.
- ? "What is the same in both orders?" You each purchase one drink. "What is the same in both equations?" There is a $1x$ term in each equation.
- Given the stepped-out instructions, students will be able to solve the system.
- When students interpret the solution, units and complete descriptions should be given: \$1.50 is the price of a drink, and the price of a sandwich is \$3.

Exploration 2

- Students are solving each system twice, once by adding and once by subtracting. Clearly this will not be possible in most linear systems. The goal is for students to see that because the coefficients of each variable are either the same or opposites, you can subtract or add to eliminate a variable.
- The obvious question that you want students to ask, and they will, is, "What if the coefficients are not the same or opposites?" They are ready for the next exploration!

Exploration 3

- As students consider part (a) and might be stuck, ask them to think about the balance scale and doubling the amount of items on each side.
- Give students time to consider their options to see whether ideas emerge. Relating this question to the balance scale should be enough to get students thinking about their options.

Communicate Your Answer

- **Look For and Make Use of Structure:** Students need to realize that one pair of like variable terms in a system must have the same or opposite coefficients before you can eliminate a variable by adding or subtracting the equations.

Connecting to Next Step

- The explorations help students make sense of adding or subtracting equations to eliminate a variable in solving a linear system. More complex linear systems will be found in the lesson.

5.3 Solving Systems of Linear Equations by Elimination

Essential Question How can you use elimination to solve a system of linear equations?

EXPLORATION 1 Writing and Solving a System of Equations

Work with a partner. You purchase a drink and a sandwich for $4.50. Your friend purchases a drink and five sandwiches for $16.50. You want to determine the price of a drink and the price of a sandwich.

a. Let x represent the price (in dollars) of one drink. Let y represent the price (in dollars) of one sandwich. Write a system of equations for the situation. Use the following verbal model.

Number of drinks • Price per drink + Number of sandwiches • Price per sandwich = Total price

Label one of the equations Equation 1 and the other equation Equation 2.

b. Subtract Equation 1 from Equation 2. Explain how you can use the result to solve the system of equations. Then find and interpret the solution.

CHANGING COURSE

To be proficient in math, you need to monitor and evaluate your progress and change course using a different solution method, if necessary.

EXPLORATION 2 Using Elimination to Solve Systems

Work with a partner. Solve each system of linear equations using two methods.

Method 1 Subtract. Subtract Equation 2 from Equation 1. Then use the result to solve the system.

Method 2 Add. Add the two equations. Then use the result to solve the system.

Is the solution the same using both methods? Which method do you prefer?

a. $3x - y = 6$
$3x + y = 0$

b. $2x + y = 6$
$2x - y = 2$

c. $x - 2y = -7$
$x + 2y = 5$

EXPLORATION 3 Using Elimination to Solve a System

Work with a partner.

$2x + y = 7$ Equation 1
$x + 5y = 17$ Equation 2

a. Can you eliminate a variable by adding or subtracting the equations as they are? If not, what do you need to do to one or both equations so that you can?

b. Solve the system individually. Then exchange solutions with your partner and compare and check the solutions.

Communicate Your Answer

4. How can you use elimination to solve a system of linear equations?
5. When can you add or subtract the equations in a system to solve the system? When do you have to multiply first? Justify your answers with examples.
6. In Exploration 3, why can you multiply an equation in the system by a constant and not change the solution of the system? Explain your reasoning.

Dynamic Teaching Tools

- Dynamic Assessment & Progress Monitoring Tool
- Lesson Planning Tool
- Interactive Whiteboard Lesson Library
- Dynamic Classroom with Dynamic Investigations

ANSWERS

1. **a.** Equation 1: $x + y = 4.5$; Equation 2: $x + 5y = 16.5$
   b. $4y = 12$; Solve the resulting equation for y. Then substitute the value of y into one of the original equations and solve for x; (1.5, 3); Drinks cost $1.50 and sandwiches cost $3.
2. **a.** $(1, -3)$; yes; *Sample answer:* method 2
   b. (2, 2); yes; *Sample answer:* method 2
   c. $(-1, 3)$; yes; *Sample answer:* method 1
3. **a.** no; *Sample answer:* Multiply each side of Equation 2 by –2.
   b. (2, 3)
4. Multiply, if necessary, one or both equations by a constant so one pair of like terms has the same or opposite coefficients. Add or subtract the equations to eliminate one of the variables. Solve the resulting equation. Then substitute the value of the variable found into one of the original equations and solve for the other variable.
5. when one of the variables has the same or opposite coefficients in both equations; *Sample answer:* $x + 3y = 2$ and $-x + 5y = 7$; when neither of the variables has the same nor opposite coefficients in both equations; *Sample answer*: $3x - 2y = 12$ and $2x - 3y = -15$
6. Multiplication Property of Equality; Multiplying each side of an equation by the same nonzero number produces an equivalent equation.

English Language Learners

Pair Activity

Pair each English learner with an English speaker. Ask one student to solve a system of equations by elimination and the other student to describe the steps in the process. Have students alternate solving and describing the process as they continue to solve several systems of equations.

Extra Example 1

Solve the system of linear equations by elimination.

$-2x + 3y = 4$

$2x - y = -8$

The solution is $(-5, -2)$.

5.3 Lesson

What You Will Learn

- Solve systems of linear equations by elimination.
- Use systems of linear equations to solve real-life problems.

Core Vocabulary

Previous
coefficient

Solving Linear Systems by Elimination

Core Concept

Solving a System of Linear Equations by Elimination

Step 1 Multiply, if necessary, one or both equations by a constant so at least one pair of like terms has the same or opposite coefficients.

Step 2 Add or subtract the equations to eliminate one of the variables.

Step 3 Solve the resulting equation.

Step 4 Substitute the value from Step 3 into one of the original equations and solve for the other variable.

You can use elimination to solve a system of equations because replacing one equation in the system with the sum of that equation and a multiple of the other produces a system that has the same solution. Here is why.

System 1

$a = b$ Equation 1
$c = d$ Equation 2

System 2

$a + kc = b + kd$ Equation 3
$c = d$ Equation 2

Consider System 1. In this system, a and c are algebraic expressions, and b and d are constants. Begin by multiplying each side of Equation 2 by a constant k. By the Multiplication Property of Equality, $kc = kd$. You can rewrite Equation 1 as Equation 3 by adding kc on the left and kd on the right. You can rewrite Equation 3 as Equation 1 by subtracting kc on the left and kd on the right. Because you can rewrite either system as the other, System 1 and System 2 have the same solution.

EXAMPLE 1 Solving a System of Linear Equations by Elimination

Solve the system of linear equations by elimination.

$3x + 2y = 4$ Equation 1
$3x - 2y = -4$ Equation 2

SOLUTION

Step 1 Because the coefficients of the y-terms are opposites, you do not need to multiply either equation by a constant.

Step 2 Add the equations.

$3x + 2y = 4$ Equation 1
$\underline{3x - 2y = -4}$ Equation 2
$6x \quad = 0$ Add the equations.

Step 3 Solve for x.

$6x = 0$ Resulting equation from Step 2
$x = 0$ Divide each side by 6.

Step 4 Substitute 0 for x in one of the original equations and solve for y.

$3x + 2y = 4$ Equation 1
$3(0) + 2y = 4$ Substitute 0 for x.
$y = 2$ Solve for y.

▶ The solution is (0, 2).

Check

Equation 1

$3x + 2y = 4$
$3(0) + 2(2) \stackrel{?}{=} 4$
$4 = 4$ ✓

Equation 2

$3x - 2y = -4$
$3(0) - 2(2) \stackrel{?}{=} -4$
$-4 = -4$ ✓

Laurie's Notes Teacher Actions

- The paragraph preceding Example 1 provides justification for the elimination method. Students can also make sense of it with an example like Exploration 1. The base price of the items (solution) remains the same regardless of what multiples and combinations are bought.
- **Teaching Tip:** Discuss the steps for solving a linear system by elimination. Students often find the steps wordy, but the process is fairly simple and straightforward. As you work through the first example, refer back to the steps in the *Core Concept* as you begin each step.

EXAMPLE 2 Solving a System of Linear Equations by Elimination

Solve the system of linear equations by elimination.

$-10x + 3y = 1$ Equation 1

$-5x - 6y = 23$ Equation 2

ANOTHER WAY

To use subtraction to eliminate one of the variables, multiply Equation 2 by 2 and then subtract the equations.

$$\begin{aligned} -10x + 3y &= 1 \\ -(-10x - 12y &= 46) \\ \hline 15y &= -45 \end{aligned}$$

SOLUTION

Step 1 Multiply Equation 2 by -2 so that the coefficients of the x-terms are opposites.

$-10x + 3y = 1$ → $-10x + 3y = 1$ Equation 1

$-5x - 6y = 23$ **Multiply by −2.** → $10x + 12y = -46$ Revised Equation 2

Step 2 Add the equations.

$-10x + 3y = 1$ Equation 1

$10x + 12y = -46$ Revised Equation 2

$15y = -45$ Add the equations.

Step 3 Solve for y.

$15y = -45$ Resulting equation from Step 2

$y = -3$ Divide each side by 15.

Step 4 Substitute -3 for y in one of the original equations and solve for x.

$-5x - 6y = 23$ Equation 2

$-5x - 6(-3) = 23$ Substitute -3 for y.

$-5x + 18 = 23$ Multiply.

$-5x = 5$ Subtract 18 from each side.

$x = -1$ Divide each side by -5.

▶ The solution is $(-1, -3)$.

Check

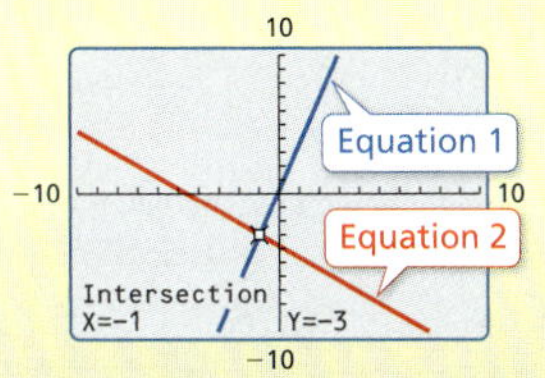

Monitoring Progress

Help in English and Spanish at *BigIdeasMath.com*

Solve the system of linear equations by elimination. Check your solution.

1. $3x + 2y = 7$
$-3x + 4y = 5$

2. $x - 3y = 24$
$3x + y = 12$

3. $x + 4y = 22$
$4x + y = 13$

Concept Summary

Methods for Solving Systems of Linear Equations

Method	When to Use
Graphing *(Lesson 5.1)*	To estimate solutions
Substitution *(Lesson 5.2)*	When one of the variables in one of the equations has a coefficient of 1 or −1
Elimination *(Lesson 5.3)*	When at least one pair of like terms has the same or opposite coefficients
Elimination (Multiply First) *(Lesson 5.3)*	When one of the variables cannot be eliminated by adding or subtracting the equations

Extra Example 2

Solve the system of linear equations by elimination.

$3x + 2y = -2$

$-6x - 5y = 2$

The solution is $(-2, 2)$.

MONITORING PROGRESS ANSWERS

1. (1, 2)
2. (6, −6)
3. (2, 5)

Laurie's Notes Teacher Actions

? Write the system. "Can you think of a way to rewrite the equations so that either the x-terms or the y-terms have coefficients that are the same or opposite?" Multiply through one equation by a constant.

COMMON ERROR When multiplying through by the constant, be sure that students multiply every term by the constant.

- Discuss the *Concept Summary* and have students write it in their notebooks.

Extra Example 3

A business with two locations buys seven large delivery trucks and five small delivery trucks. Location A receives three large trucks and two small trucks for a total cost of \$270,000. Location B receives four large trucks and three small trucks for a total cost of \$375,000. What is the cost of each type of truck?

The solution is (60,000, 45,000). So, a large truck costs \$60,000 and a small truck costs \$45,000.

MONITORING PROGRESS ANSWER

4. a large van costs \$35,000, a small van costs \$30,000

Solving Real-Life Problems

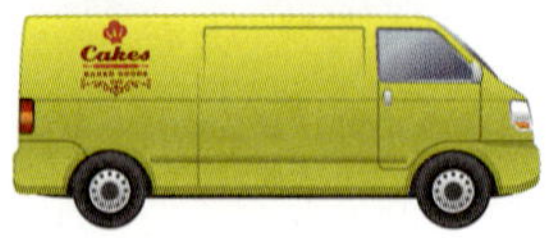

EXAMPLE 3 **Modeling with Mathematics**

A business with two locations buys seven large delivery vans and five small delivery vans. Location A receives five large vans and two small vans for a total cost of \$235,000. Location B receives two large vans and three small vans for a total cost of \$160,000. What is the cost of each type of van?

SOLUTION

1. **Understand the Problem** You know how many of each type of van each location receives. You also know the total cost of the vans for each location. You are asked to find the cost of each type of van.
2. **Make a Plan** Use a verbal model to write a system of linear equations that represents the problem. Then solve the system of linear equations.
3. **Solve the Problem**

Words 5 • Cost of large van + 2 • Cost of small van = 235,000

2 • Cost of large van + 3 • Cost of small van = 160,000

Variables Let x be the cost (in dollars) of a large van and let y be the cost (in dollars) of a small van.

System $5x + 2y = 235{,}000$ Equation 1

$2x + 3y = 160{,}000$ Equation 2

Step 1 Multiply Equation 1 by −3. Multiply Equation 2 by 2.

$5x + 2y = 235{,}000$ **Multiply by −3.** → $-15x - 6y = -705{,}000$ Revised Equation 1

$2x + 3y = 160{,}000$ **Multiply by 2.** → $4x + 6y = 320{,}000$ Revised Equation 2

Step 2 Add the equations.

$$\begin{aligned} -15x - 6y &= -705{,}000 && \text{Revised Equation 1} \\ 4x + 6y &= 320{,}000 && \text{Revised Equation 2} \\ \hline -11x &= -385{,}000 && \text{Add the equations.} \end{aligned}$$

Step 3 Solving the equation $-11x = -385{,}000$ gives $x = 35{,}000$.

Step 4 Substitute 35,000 for x in one of the original equations and solve for y.

$$\begin{aligned} 5x + 2y &= 235{,}000 && \text{Equation 1} \\ 5(35{,}000) + 2y &= 235{,}000 && \text{Substitute 35,000 for } x. \\ y &= 30{,}000 && \text{Solve for } y. \end{aligned}$$

▶ The solution is (35,000, 30,000). So, a large van costs \$35,000 and a small van costs \$30,000.

4. **Look Back** Check to make sure your solution makes sense with the given information. For Location A, the total cost is 5(35,000) + 2(30,000) = \$235,000. For Location B, the total cost is 2(35,000) + 3(30,000) = \$160,000. So, the solution makes sense.

STUDY TIP

In Example 3, both equations are multiplied by a constant so that the coefficients of the y-terms are opposites.

Monitoring Progress

Help in English and Spanish at *BigIdeasMath.com*

4. Solve the system in Example 3 by eliminating x.

Laurie's Notes Teacher Actions

? **Make Sense of Problems and Persevere in Solving Them:** Have students read through the example. "What is known?" number of each size van that went to each location "What are you solving for?" cost of each size van "What will the variables represent?" cost of each size van

- This is an example in which it is more efficient to check the solution analytically than by graphing.

Closure

- **Example/Non-Example:** Students have learned to solve systems of linear equations by graphing, substitution, and elimination. Although any linear system can be solved by any of the three methods, the approach selected usually depends on what the system looks like. For each of the three methods, have partners write a system that they would (example) and would not (non-example) solve using that method. When presented, listen for justification of why students would or would not use that method for the system written. Give feedback as needed.

5.3 Exercises

Dynamic Solutions available at *BigIdeasMath.com*

Vocabulary and Core Concept Check

1. **OPEN-ENDED** Give an example of a system of linear equations that can be solved by first adding the equations to eliminate one variable.

2. **WRITING** Explain how to solve the system of linear equations by elimination.

$$2x - 3y = -4 \quad \text{Equation 1}$$
$$-5x + 9y = 7 \quad \text{Equation 2}$$

Monitoring Progress and Modeling with Mathematics

In Exercises 3–10, solve the system of linear equations by elimination. Check your solution. *(See Example 1.)*

3. $x + 2y = 13$
   $-x + y = 5$

4. $9x + y = 2$
   $-4x - y = -17$

5. $5x + 6y = 50$
   $x - 6y = -26$

6. $-x + y = 4$
   $x + 3y = 4$

7. $-3x - 5y = -7$
   $-4x + 5y = 14$

8. $4x - 9y = -21$
   $-4x - 3y = 9$

9. $-y - 10 = 6x$
   $5x + y = -10$

10. $3x - 30 = y$
    $7y - 6 = 3x$

In Exercises 11–18, solve the system of linear equations by elimination. Check your solution. *(See Examples 2 and 3.)*

11. $x + y = 2$
    $2x + 7y = 9$

12. $8x - 5y = 11$
    $4x - 3y = 5$

13. $11x - 20y = 28$
    $3x + 4y = 36$

14. $10x - 9y = 46$
    $-2x + 3y = 10$

15. $4x - 3y = 8$
    $5x - 2y = -11$

16. $-2x - 5y = 9$
    $3x + 11y = 4$

17. $9x + 2y = 39$
    $6x + 13y = -9$

18. $12x - 7y = -2$
    $8x + 11y = 30$

19. **ERROR ANALYSIS** Describe and correct the error in solving for one of the variables in the linear system $5x - 7y = 16$ and $x + 7y = 8$.

✗

$$\begin{aligned} 5x - 7y &= 16 \\ x + 7y &= 8 \\ \hline 4x \quad &= 24 \\ x &= 6 \end{aligned}$$

20. **ERROR ANALYSIS** Describe and correct the error in solving for one of the variables in the linear system $4x + 3y = 8$ and $x - 2y = -13$.

✗

$4x + 3y = 8$ → $4x + 3y = 8$

$x - 2y = -13$ **Multiply by −4.** → $-4x + 8y = -13$

$$11y = -5$$
$$y = \frac{-5}{11}$$

21. **MODELING WITH MATHEMATICS** A service center charges a fee of x dollars for an oil change plus y dollars per quart of oil used. A sample of its sales record is shown. Write a system of linear equations that represents this situation. Find the fee and cost per quart of oil.

	A	B	C
1	Customer	Oil Tank Size (quarts)	Total Cost
2	A	5	$22.45
3	B	7	$25.45
4			

22. **MODELING WITH MATHEMATICS** A music website charges x dollars for individual songs and y dollars for entire albums. Person A pays $25.92 to download 6 individual songs and 2 albums. Person B pays $33.93 to download 4 individual songs and 3 albums. Write a system of linear equations that represents this situation. How much does the website charge to download a song? an entire album?

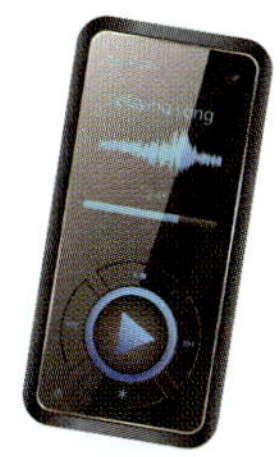

Assignment Guide and Homework Check

ASSIGNMENT

Basic: 1, 2, 3–21 odd, 22, 23, 28, 29, 36–42

Average: 1, 2–28 even, 29, 34, 36–42

Advanced: 1, 2, 8, 10, 16–26 even, 27–42

HOMEWORK CHECK

Basic: 3, 11, 15, 21

Average: 4, 12, 16, 22

Advanced: 8, 16, 18, 22

ANSWERS

1. *Sample answer:* $x + 3y = 5$ and $-x + 4y = 10$
2. *Sample answer:* Multiply each side of Equation 1 by 3. Add the equations to eliminate y. Solve the resulting equation for x. Substitute the value of x into one of the original equations and solve for y.
3. $(1, 6)$
4. $(-3, 29)$
5. $(4, 5)$
6. $(-2, 2)$
7. $(-1, 2)$
8. $(-3, 1)$
9. $(0, -10)$
10. $(12, 6)$
11. $(1, 1)$
12. $(2, 1)$
13. $(8, 3)$
14. $(19, 16)$
15. $(-7, -12)$
16. $(-17, 5)$
17. $(5, -3)$
18. $(1, 2)$
19. $5x + x \neq 4x$; $6x = 24$, $x = 4$
20. The right side of the second equation did not get multiplied by −4; $-4x + 8y = 52$, $11y = 60$, $y = \frac{60}{11}$
21. $x + 5y = 22.45$ and $x + 7y = 25.45$; $14.95 fee, $1.50 per quart of oil
22. $6x + 2y = 25.92$ and $4x + 3y = 33.93$; $0.99 per song; $9.99 per album

Dynamic Teaching Tools

- Dynamic Assessment & Progress Monitoring Tool
- Interactive Whiteboard Lesson Library
- Dynamic Classroom with Dynamic Investigations

ANSWERS

23. $(2, -1)$; *Sample answer:* elimination because y has the same coefficient in both equations

24. $(-5, -3)$; *Sample answer:* substitution because the second equation is already solved for x

25. $(4, 6)$; *Sample answer:* substitution because the second equation is already solved for y

26. $\left(\frac{1}{3}, -\frac{2}{3}\right)$; *Sample answer:* substitution because y has a coefficient of 1 in the first equation

27. 4 and –4; These values yield x terms with either the same or opposite coefficients.

28–42. See Additional Answers.

Mini-Assessment

Solve the system of linear equations by elimination.

1. $3x + y = 14$
   $4x - y = 0$
   $(2, 8)$
2. $2x + y = 7$
   $-3x - 2y = -9$
   $(5, -3)$
3. A restaurant with two locations buys twelve large bags of flour and eight large bags of sugar. Location A receives seven bags of flour and five bags of sugar for a total cost of \$584. Location B receives five bags of flour and three bags of sugar for a total cost of \$384. What is the cost of each bag of flour and each bag of sugar?
   The solution is $(42, 58)$. So, each bag of flour costs \$42 and each bag of sugar costs \$58.

In Exercises 23–26, solve the system of linear equations using any method. Explain why you chose the method.

23. $3x + 2y = 4$
$2y = 8 - 5x$

24. $-6y + 2 = -4x$
$y - 2 = x$

25. $y - x = 2$
$y = -\frac{1}{4}x + 7$

26. $3x + y = \frac{1}{3}$
$2x - 3y = \frac{8}{3}$

27. WRITING For what values of a can you solve the linear system $ax + 3y = 2$ and $4x + 5y = 6$ by elimination without multiplying first? Explain.

28. HOW DO YOU SEE IT? The circle graph shows the results of a survey in which 50 students were asked about their favorite meal.

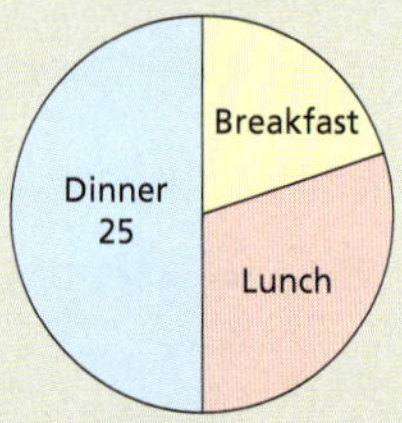

a. Estimate the numbers of students who chose breakfast and lunch.

b. The number of students who chose lunch was 5 more than the number of students who chose breakfast. Write a system of linear equations that represents the numbers of students who chose breakfast and lunch.

c. Explain how you can solve the linear system in part (b) to check your answers in part (a).

29. MAKING AN ARGUMENT Your friend says that any system of equations that can be solved by elimination can be solved by substitution in an equal or fewer number of steps. Is your friend correct? Explain.

30. THOUGHT PROVOKING Write a system of linear equations that can be added to eliminate a variable *or* subtracted to eliminate a variable.

31. MATHEMATICAL CONNECTIONS A rectangle has a perimeter of 18 inches. A new rectangle is formed by doubling the width w and tripling the length ℓ, as shown. The new rectangle has a perimeter P of 46 inches.

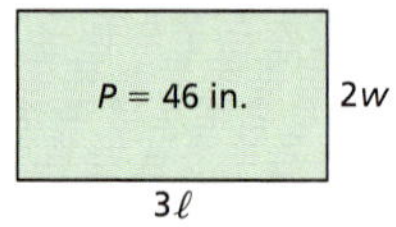

a. Write and solve a system of linear equations to find the length and width of the original rectangle.

b. Find the length and width of the new rectangle.

32. CRITICAL THINKING Refer to the discussion of System 1 and System 2 on page 248. Without solving, explain why the two systems shown have the same solution.

System 1		System 2	
$3x - 2y = 8$	Equation 1	$5x = 20$	Equation 3
$x + y = 6$	Equation 2	$x + y = 6$	Equation 2

33. PROBLEM SOLVING You are making 6 quarts of fruit punch for a party. You have bottles of 100% fruit juice and 20% fruit juice. How many quarts of each type of juice should you mix to make 6 quarts of 80% fruit juice?

34. PROBLEM SOLVING A motorboat takes 40 minutes to travel 20 miles downstream. The return trip takes 60 minutes. What is the speed of the current?

35. CRITICAL THINKING Solve for x, y, and z in the system of equations. Explain your steps.

$x + 7y + 3z = 29$ Equation 1
$3z + x - 2y = -7$ Equation 2
$5y = 10 - 2x$ Equation 3

Maintaining Mathematical Proficiency

Reviewing what you learned in previous grades and lessons

Solve the equation. Determine whether the equation has *one solution*, *no solution*, or *infinitely many solutions*. *(Section 1.3)*

36. $5d - 8 = 1 + 5d$

37. $9 + 4t = 12 - 4t$

38. $3n + 2 = 2(n - 3)$

39. $-3(4 - 2v) = 6v - 12$

Write an equation of the line that passes through the given point and is parallel to the given line. *(Section 4.3)*

40. $(4, -1)$; $y = -2x + 7$

41. $(0, 6)$; $y = 5x - 3$

42. $(-5, -2)$; $y = \frac{2}{3}x + 1$

If students need help...	*If students got it...*
Resources by Chapter • Practice A and Practice B • Puzzle Time	Resources by Chapter • Enrichment and Extension • Cumulative Review
Student Journal • Practice	Start the *next* Section
Differentiating the Lesson Skills Review Handbook	

Laurie's Notes

Overview of Section 5.4

Introduction

- Students are now familiar with three methods for solving a linear system. There are other methods that students may study in later mathematics courses.
- This lesson presents two special systems of linear equations. If the lines are parallel, then there is no solution. If the lines coincide, then there are infinitely many solutions—namely, all of the points on the line.

Teaching Strategy

- If you ask students in an algebra class versus those in a geometry class to think about what a line is, the two groups likely will have different images emerge. Algebra students likely think about a coordinate system and equations. Geometry students likely think about notations and transversals intersecting parallel lines.
- Spaghetti can be used often, in both classes! To introduce this lesson, I place two pieces of spaghetti on an overhead projector or under a document camera and say, "These represent two lines. Right now they are intersecting."

? "Is there any other relationship they could have?" Listen for parallel (nonintersecting) and a reasonable description of coinciding lines.

- Now place a transparency of a coordinate grid on top of the spaghetti. Suggest that lines, when graphed, do not always have to intersect, or they may indeed be the same line.
- Having a physical model of the lines seems to be helpful in reminding students that lines graphed on the coordinate plane are models, representing some quantities.

Extensions

- There is vocabulary used to describe the special systems in this lesson that is not included in the student edition. A linear system with no solution is called an *inconsistent system*. A linear system with infinitely many solutions is called a *consistent dependent system*.

Pacing Suggestion

- Complete the explorations and then start the formal lesson by stating the *Core Concept*.

Dynamic Teaching Tools

- Dynamic Assessment & Progress Monitoring Tool
- Lesson Planning Tool
- Interactive Whiteboard Lesson Library
- Dynamic Classroom with Dynamic Investigations

What Your Students Will Learn

- Analyze systems of equations in two variables to determine whether the systems have one solution, no solution, or infinitely many solutions.

Laurie's Notes

Exploration

Motivate

- The goal of this quick activity is to review the slope-intercept form of an equation.
- Make a set of cards in advance: equation card, slope card, and y-intercept card. Make enough cards so that everyone in the class has a card.

Examples:

$y = 3x - 4$	$m = 3$	$b = -4$
$3x + y = 4$	$m = -3$	$b = 4$
$y = 4x - 2$	$m = 4$	$b = -2$
$y = 2 - 4x$	$m = -4$	$b = 2$

? "What is the general form of an equation written in slope-intercept form?" $y = mx + b$

- Distribute all of the cards and have students form a matching set of three.

Exploration 1

- Take time to discuss questions students may have relating to the basics of financial literacy.

COMMON ERROR Students may forget that the cost starts at \$450, not \$0. The cost to make one skateboard is \$470, not \$20.

? "What do you notice about the numbers in each row of the table?" They are both going up by \$20. "When will your company break even? Explain." It won't. The revenue must exceed the cost.

- **Extension:** Write a linear equation that represents the cost and one that represents the revenue. Cost: $C = 450 + 20x$; Revenue: $R = 20x$. *Note*: The slope of the two equations is the same. The lines are parallel.

Exploration 2

? To get students thinking about this problem ask, "Can you think of something in this room that weighs about 1 gram?" Answers will vary. A reasonable answer would be a standard paper clip, which is about 1.4 grams.

- **FYI:** The point is that a gram is a very light: 1 ounce $\approx$ 28.35 grams.
- Students will have little difficulty writing the equations. Graphing will be more challenging. Students are used to the variables representing the number of an item you have or the cost of an item. The weight of an item feels different to students, especially in this context where the weight is a small number.
- Students should recall that when graphing an equation in standard form, finding the intercepts is an efficient method. If students graph each equation using the intercepts, they will immediately recognize that the two lines coincide, meaning they are the same line.

Communicate Your Answer

- Question 3 tries to help students summarize the two additional cases for the solution of a system of linear equations.

Connecting to Next Step

- The explorations help students recognize that there are other cases for the solution of a linear solution besides one ordered pair. Begin with the *Core Concept* in the formal lesson.

5.4 Solving Special Systems of Linear Equations

Essential Question

Can a system of linear equations have no solution or infinitely many solutions?

EXPLORATION 1 Using a Table to Solve a System

Work with a partner. You invest $450 for equipment to make skateboards. The materials for each skateboard cost $20. You sell each skateboard for $20.

a. Write the cost and revenue equations. Then copy and complete the table for your cost C and your revenue R.

x (skateboards)	0	1	2	3	4	5	6	7	8	9	10
C (dollars)											
R (dollars)											

b. When will your company break even? What is wrong?

MODELING WITH MATHEMATICS

To be proficient in math, you need to interpret mathematical results in real-life contexts.

EXPLORATION 2 Writing and Analyzing a System

Work with a partner. A necklace and matching bracelet have two types of beads. The necklace has 40 small beads and 6 large beads and weighs 10 grams. The bracelet has 20 small beads and 3 large beads and weighs 5 grams. The threads holding the beads have no significant weight.

a. Write a system of linear equations that represents the situation. Let x be the weight (in grams) of a small bead and let y be the weight (in grams) of a large bead.

b. Graph the system in the coordinate plane shown. What do you notice about the two lines?

c. Can you find the weight of each type of bead? Explain your reasoning.

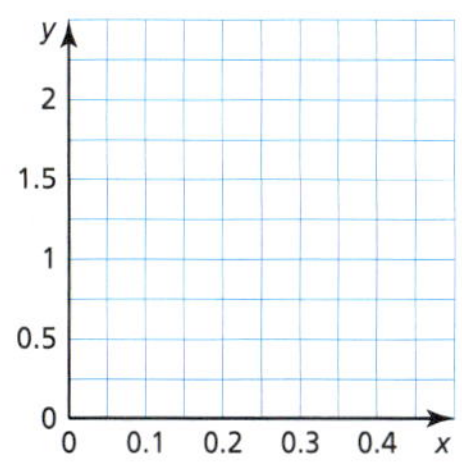

Communicate Your Answer

3. Can a system of linear equations have no solution or infinitely many solutions? Give examples to support your answers.

4. Does the system of linear equations represented by each graph have *no solution*, *one solution*, or *infinitely many solutions*? Explain.

a.

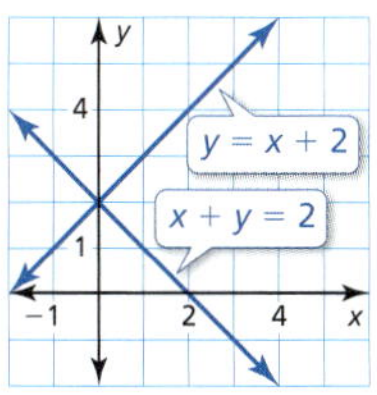

b.

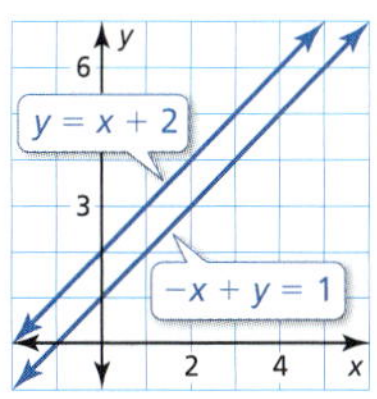

c.

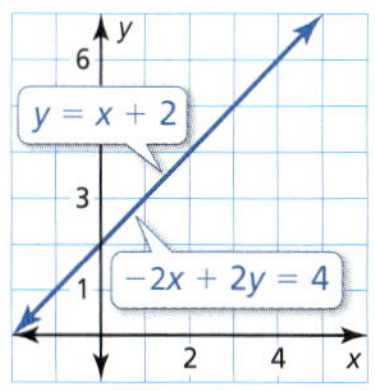

Dynamic Teaching Tools

- Dynamic Assessment & Progress Monitoring Tool
- Lesson Planning Tool
- Interactive Whiteboard Lesson Library
- Dynamic Classroom with Dynamic Investigations

ANSWERS

1. a. See Additional Answers.

b. never; Both cost and revenue increase at the same rate, but have different initial values.

2. a. $40x + 6y = 10$; $20x + 3y = 5$

b.

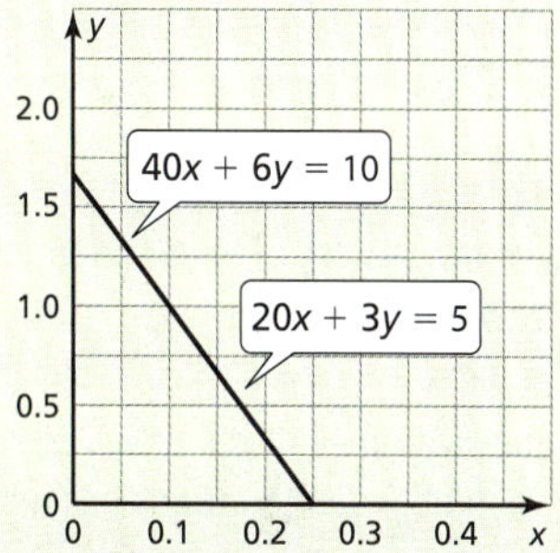

They are the same line.

c. no; Because the equations are equivalent, there are infinitely many solutions.

3. yes; *Sample answer:* The system $y = 3x$ and $y = 3x + 1$ has no solution. The system $y = 3x$ and $2y = 6x$ has infinitely many solutions.

4. a. one solution; The lines intersect at one point.

b. no solution; The lines do not intersect.

c. infinitely many solutions; The lines are the same line.

English Language Learners

Group Activity

Form groups of English learners and English speakers. Give each group three large index cards and have students write one of the following phrases on each card: *one solution*, *no solution*, and *infinitely many solutions*. One at a time, show students graphs of linear systems that illustrate all three types of solutions. Have each group discuss the graph and then hold up the card with the correct description for the solution of the system.

Extra Example 1

Solve the system of linear equations.

$y = 3x + 2$

$y = 3x - 1$

The equation $-1 = 2$ is never true. So, the system of linear equations has no solution.

5.4 Lesson

What You Will Learn

- Determine the numbers of solutions of linear systems.
- Use linear systems to solve real-life problems.

Core Vocabulary

Previous
parallel

The Numbers of Solutions of Linear Systems

Core Concept

Solutions of Systems of Linear Equations

A system of linear equations can have *one solution*, *no solution*, or *infinitely many solutions*.

One solution	No solution	Infinitely many solutions

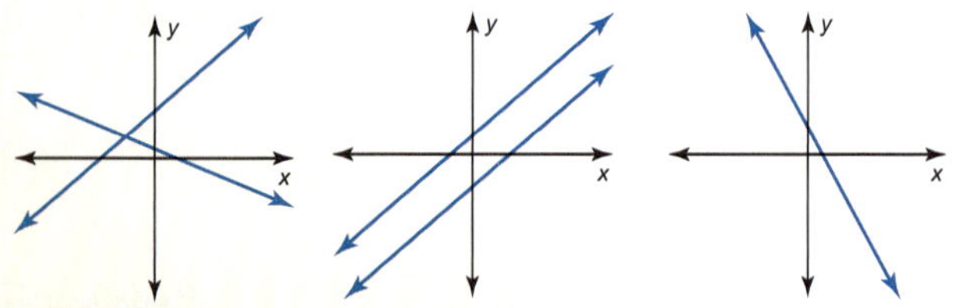

The lines intersect. The lines are parallel. The lines are the same.

ANOTHER WAY

You can solve some linear systems by inspection. In Example 1, notice you can rewrite the system as

$-2x + y = 1$

$-2x + y = -5.$

This system has no solution because $-2x + y$ cannot be equal to both 1 and -5.

EXAMPLE 1 **Solving a System: No Solution**

Solve the system of linear equations.

$y = 2x + 1$ Equation 1

$y = 2x - 5$ Equation 2

SOLUTION

Method 1 Solve by graphing.

Graph each equation.

The lines have the same slope and different y-intercepts. So, the lines are parallel.

Because parallel lines do not intersect, there is no point that is a solution of both equations.

- So, the system of linear equations has no solution.

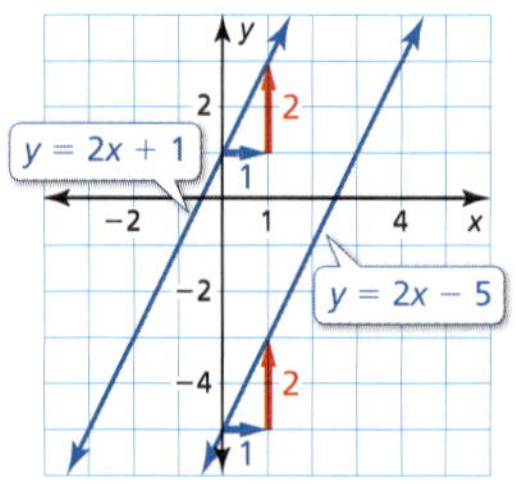

Method 2 Solve by substitution.

Substitute $2x - 5$ for y in Equation 1.

$y = 2x + 1$ Equation 1

$2x - 5 = 2x + 1$ Substitute $2x - 5$ for y.

$-5 = 1$ ✗ Subtract $2x$ from each side.

- The equation $-5 = 1$ is never true. So, the system of linear equations has no solution.

STUDY TIP

A linear system with no solution is called an *inconsistent system*.

Laurie's Notes Teacher Actions

- **? Always-Sometimes-Never True:** "If you graph two lines in the coordinate plane, then they will intersect or be parallel." sometimes true; They could also coincide.
- Because both equations are in slope-intercept form, students will not be surprised that the lines are parallel and therefore do not intersect.
- Method 2 is important so that students recognize what happens when there is no solution and they are trying to solve algebraically.
- **Look For and Make Use of Structure:** Because both equations are solved for y, substitution is an approach that makes sense.

EXAMPLE 2 Solving a System: Infinitely Many Solutions

Solve the system of linear equations.

$-2x + y = 3$ Equation 1

$-4x + 2y = 6$ Equation 2

SOLUTION

Method 1 Solve by graphing.

Graph each equation.

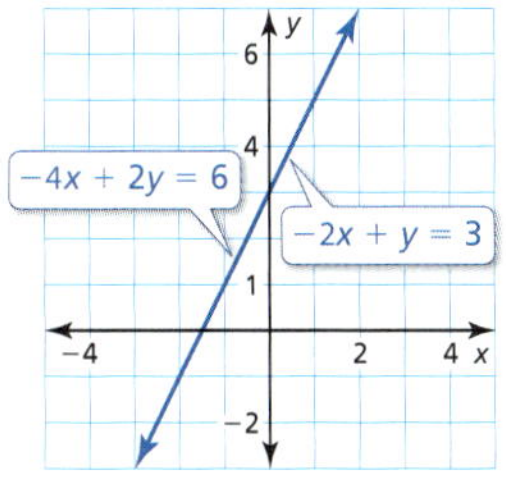

The lines have the same slope and the same y-intercept. So, the lines are the same. Because the lines are the same, all points on the line are solutions of both equations.

▶ So, the system of linear equations has infinitely many solutions.

Method 2 Solve by elimination.

Step 1 Multiply Equation 1 by −2.

$-2x + y = 3$	Multiply by −2.	$4x - 2y = -6$	Revised Equation 1
$-4x + 2y = 6$		$-4x + 2y = 6$	Equation 2

Step 2 Add the equations.

$4x - 2y = -6$ Revised Equation 1

$-4x + 2y = 6$ Equation 2

$0 = 0$ Add the equations.

▶ The equation $0 = 0$ is always true. So, the solutions are all the points on the line $-2x + y = 3$. The system of linear equations has infinitely many solutions.

STUDY TIP

A linear system with infinitely many solutions is called a *consistent dependent system.*

Monitoring Progress

Help in English and Spanish at *BigIdeasMath.com*

Solve the system of linear equations.

1. $x + y = 3$
   $2x + 2y = 6$
2. $y = -x + 3$
   $2x + 2y = 4$
3. $x + y = 3$
   $x + 2y = 4$
4. $y = -10x + 2$
   $10x + y = 10$

Extra Example 2

Solve the system of linear equations.

$x - 3y = 6$

$3x - 9y = 18$

The equation $0 = 0$ is always true. The solutions are all the points on the line $x - 3y = 6$. So, the system of linear equations has infinitely many solutions.

MONITORING PROGRESS ANSWERS

1. infinitely many solutions
2. no solution
3. (2, 1)
4. no solution

Laurie's Notes Teacher Actions

- Again, two approaches are used to solve the linear system, graphing and elimination. If students graph using intercepts, they will recognize that the lines coincide. If they graph using a calculator, a common comment is "My calculator only graphed one line!"

? **Always-Sometimes-Never True:** "In a linear system, if the coefficients of x are the same and the coefficients of y are opposites, then the system will have no solution. Explain." never true; The lines have different slopes, so they will intersect. There will be one solution.

Extra Example 3

Use the diagrams in Example 3. The perimeter of the trapezoidal piece of land is 72 kilometers. The perimeter of the rectangular piece of land is 216 kilometers. Write and solve a system of linear equations to find the values of x and y.

$6x + 12y = 72$

$18x + 36y = 216$

The system has infinitely many solutions.

MONITORING PROGRESS ANSWER

5. The system has no solution; The lines still have the same slope, but they now have different y-intercepts, so they are parallel.

Solving Real-Life Problems

EXAMPLE 3 **Modeling with Mathematics**

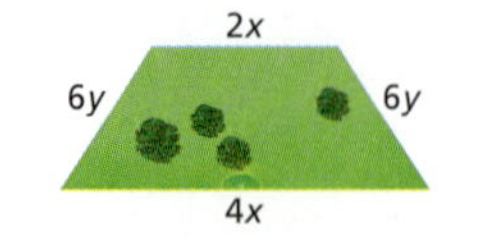

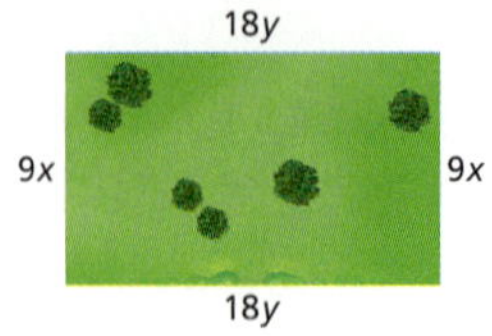

The perimeter of the trapezoidal piece of land is 48 kilometers. The perimeter of the rectangular piece of land is 144 kilometers. Write and solve a system of linear equations to find the values of x and y.

SOLUTION

1. **Understand the Problem** You know the perimeter of each piece of land and the side lengths in terms of x or y. You are asked to write and solve a system of linear equations to find the values of x and y.
2. **Make a Plan** Use the figures and the definition of perimeter to write a system of linear equations that represents the problem. Then solve the system of linear equations.
3. **Solve the Problem**

Perimeter of trapezoid

$2x + 4x + 6y + 6y = 48$

$6x + 12y = 48$ Equation 1

Perimeter of rectangle

$9x + 9x + 18y + 18y = 144$

$18x + 36y = 144$ Equation 2

System $6x + 12y = 48$ Equation 1

$18x + 36y = 144$ Equation 2

Method 1 Solve by graphing.

Graph each equation.

The lines have the same slope and the same y-intercept. So, the lines are the same.

In this context, x and y must be positive. Because the lines are the same, all the points on the line in Quadrant I are solutions of both equations.

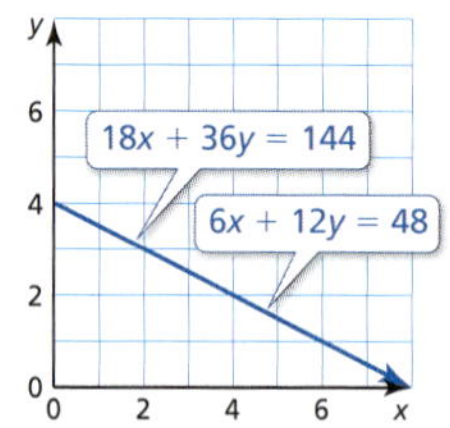

▶ So, the system of linear equations has infinitely many solutions.

Method 2 Solve by elimination.

Multiply Equation 1 by -3 and add the equations.

$6x + 12y = 48$ **Multiply by −3.** → $-18x - 36y = -144$ Revised Equation 1

$18x + 36y = 144$ → $18x + 36y = 144$ Equation 2

$0 = 0$ Add the equations.

▶ The equation $0 = 0$ is always true. In this context, x and y must be positive. So, the solutions are all the points on the line $6x + 12y = 48$ in Quadrant I. The system of linear equations has infinitely many solutions.

4. **Look Back** Choose a few of the ordered pairs (x, y) that are solutions of Equation 1. You should find that no matter which ordered pairs you choose, they will also be solutions of Equation 2. So, *infinitely many solutions* seems reasonable.

Monitoring Progress

Help in English and Spanish at *BigIdeasMath.com*

5. **WHAT IF?** What happens to the solution in Example 3 when the perimeter of the trapezoidal piece of land is 96 kilometers? Explain.

Laurie's Notes Teacher Actions

- **Look For and Express Regularity in Repeated Reasoning:** Work through Method 2. Point out that solving the system algebraically leads to a statement that is always true, $0 = 0$. You can interpret this identity to mean that the system has infinitely many solutions, because there are infinitely many ordered pairs that satisfy both equations and the lines coincide.

Closure

- **Exit Ticket:** Write a system of equations that has no solution. *Sample answer:* $2x + 3y = 4$; $4x + 6y = 12$ Write a system of equations that has infinitely many solutions. *Sample answer:* $x - 2y = 6$; $-2x + 4y = -12$

5.4 Exercises

Dynamic Solutions available at *BigIdeasMath.com*

Vocabulary and Core Concept Check

1. **REASONING** Is it possible for a system of linear equations to have exactly two solutions? Explain.

2. **WRITING** Compare the graph of a system of linear equations that has infinitely many solutions and the graph of a system of linear equations that has no solution.

Monitoring Progress and Modeling with Mathematics

In Exercises 3–8, match the system of linear equations with its graph. Then determine whether the system has *one solution*, *no solution*, or *infinitely many solutions*.

3. $-x + y = 1$
   $x - y = 1$

4. $2x - 2y = 4$
   $-x + y = -2$

5. $2x + y = 4$
   $-4x - 2y = -8$

6. $x - y = 0$
   $5x - 2y = 6$

7. $-2x + 4y = 1$
   $3x - 6y = 9$

8. $5x + 3y = 17$
   $x - 3y = -2$

A.
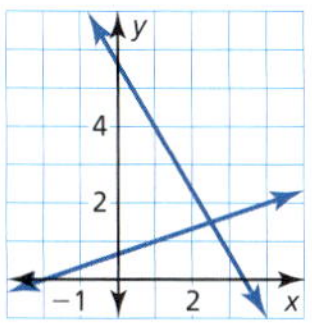

B.
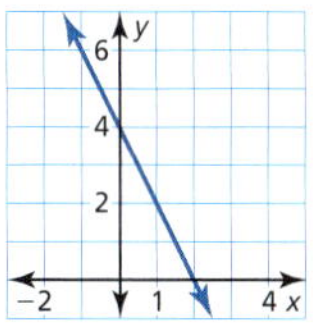

C.
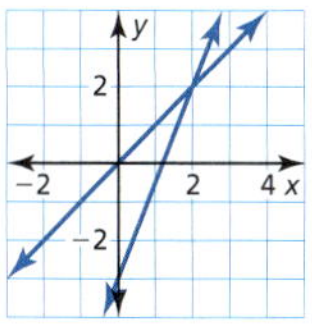

D.
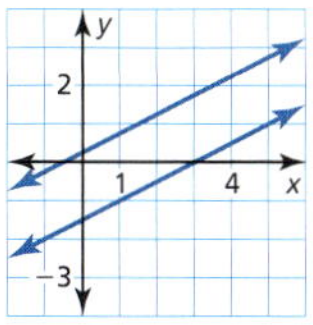

E.
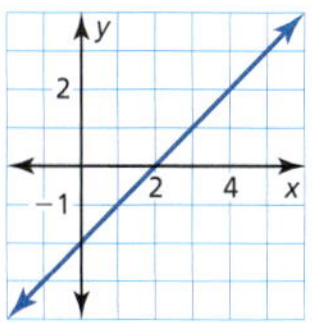

F.
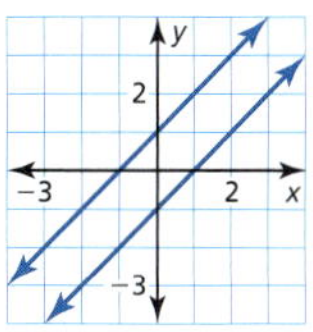

In Exercises 9–16, solve the system of linear equations. *(See Examples 1 and 2.)*

9. $y = -2x - 4$
   $y = 2x - 4$

10. $y = -6x - 8$
    $y = -6x + 8$

11. $3x - y = 6$
    $-3x + y = -6$

12. $-x + 2y = 7$
    $x - 2y = 7$

13. $4x + 4y = -8$
    $-2x - 2y = 4$

14. $15x - 5y = -20$
    $-3x + y = 4$

15. $9x - 15y = 24$
    $6x - 10y = -16$

16. $3x - 2y = -5$
    $4x + 5y = 47$

In Exercises 17–22, use only the slopes and *y*-intercepts of the graphs of the equations to determine whether the system of linear equations has *one solution*, *no solution*, or *infinitely many solutions*. Explain.

17. $y = 7x + 13$
    $-21x + 3y = 39$

18. $y = -6x - 2$
    $12x + 2y = -6$

19. $4x + 3y = 27$
    $4x - 3y = -27$

20. $-7x + 7y = 1$
    $2x - 2y = -18$

21. $-18x + 6y = 24$
    $3x - y = -2$

22. $2x - 2y = 16$
    $3x - 6y = 30$

ERROR ANALYSIS **In Exercises 23 and 24, describe and correct the error in solving the system of linear equations.**

23. ✗
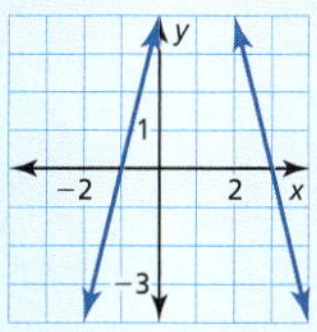

$-4x + y = 4$
$4x + y = 12$

The lines do not intersect. So, the system has no solution.

24. ✗ $y = 3x - 8$
$y = 3x - 12$

The lines have the same slope. So, the system has infinitely many solutions.

23. The lines are not parallel, so they must intersect.

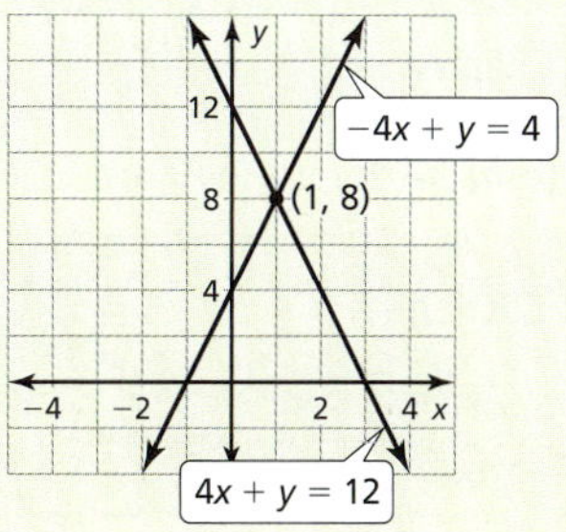

(1, 8)

24. The lines have the same slope but different *y*-intercepts, so the system has no solution.

Assignment Guide and Homework Check

ASSIGNMENT

Basic: 1, 2, 3–25 odd, 30, 32–36

Average: 1, 2–26 even, 27, 30, 32–36

Advanced: 1, 2, 8, 14–24 even, 25–36

HOMEWORK CHECK

Basic: 3, 5, 9, 17, 25

Average: 4, 10, 14, 18, 26

Advanced: 8, 14, 18, 25, 26

ANSWERS

1. no; Two lines cannot intersect at exactly two points.
2. The graph of a system of linear equations that has infinitely many solutions is a single line and the graph of a system of linear equations that has no solution is two parallel lines.
3. F; no solution
4. E; infinitely many solutions
5. B; infinitely many solutions
6. C; one solution
7. D; no solution
8. A; one solution
9. $(0, -4)$
10. no solution
11. infinitely many solutions
12. no solution
13. infinitely many solutions
14. infinitely many solutions
15. no solution
16. $(3, 7)$
17. infinitely many solutions; The lines have the same slope and the same *y*-intercept, so they are the same line.
18. no solution; The lines have the same slope but different *y*-intercepts, so they are parallel.
19. one solution; The lines have different slopes, so they will intersect.
20. no solution; The lines have the same slope but different *y*-intercepts, so they are parallel.
21. no solution; The lines have the same slope but different *y*-intercepts, so they are parallel.
22. one solution; The lines have different slopes, so they will intersect.

Dynamic Teaching Tools

- Dynamic Assessment & Progress Monitoring Tool
- Interactive Whiteboard Lesson Library
- Dynamic Classroom with Dynamic Investigations

ANSWERS

25. $3f + 4a = 6$ and $\frac{9}{2}f + 6a = 9$; infinitely many solutions

26. $d = 6t + 2$ and $d = 6t$; no; The system has no solution.

27. yes; The system of equations $150c + 80b = 22{,}860$ and $170c + 100b = 27{,}280$ has a solution.

28. *Sample answer:* $y = 2x$, $y = \frac{1}{2}x$, and $y = -x + 6$

29. one; The lines have different slopes, so they will intersect at exactly one point.

30. **a.** *Sample answer:* 40 m

b. yes; The lines that represent Team A and Team C's runners have different slopes, so they will intersect.

c. no; The lines that represent Team A and Team B's runners have the same slope, so they will never intersect.

31–36. See Additional Answers.

Mini-Assessment

Solve the system of linear equations.

1. $y = 4x - 2$
$y = 4x - 4$
no solution

2. $x - y = -2$
$-4x + 4y = 8$
infinitely many solutions

3. A small fruit basket costs $8 and a large fruit basket costs $20. The small basket contains 6 apples and 4 pears. The large basket contains 18 apples and 12 pears. Write a system of linear equations to represent this situation. Can you find the unit price of each fruit? Explain. $6x + 4y = 8$; $18x + 12y = 20$; The system has no solution, so you cannot find the unit price of each fruit.

25. MODELING WITH MATHEMATICS A small bag of trail mix contains 3 cups of dried fruit and 4 cups of almonds. A large bag contains $4\frac{1}{2}$ cups of dried fruit and 6 cups of almonds. Write and solve a system of linear equations to find the price of 1 cup of dried fruit and 1 cup of almonds. *(See Example 3.)*

$9 $6

26. MODELING WITH MATHEMATICS In a canoe race, Team A is traveling 6 miles per hour and is 2 miles ahead of Team B. Team B is also traveling 6 miles per hour. The teams continue traveling at their current rates for the remainder of the race. Write a system of linear equations that represents this situation. Will Team B catch up to Team A? Explain.

27. PROBLEM SOLVING A train travels from New York City to Washington, D.C., and then back to New York City. The table shows the number of tickets purchased for each leg of the trip. The cost per ticket is the same for each leg of the trip. Is there enough information to determine the cost of one coach ticket? Explain.

Destination	Coach tickets	Business class tickets	Money collected (dollars)
Washington, D.C.	150	80	22,860
New York City	170	100	27,280

28. THOUGHT PROVOKING Write a system of three linear equations in two variables so that any two of the equations have exactly one solution, but the entire system of equations has no solution.

29. REASONING In a system of linear equations, one equation has a slope of 2 and the other equation has a slope of $-\frac{1}{3}$. How many solutions does the system have? Explain.

30. HOW DO YOU SEE IT? The graph shows information about the last leg of a 4×200-meter relay for three relay teams. Team A's runner ran about 7.8 meters per second, Team B's runner ran about 7.8 meters per second, and Team C's runner ran about 8.8 meters per second.

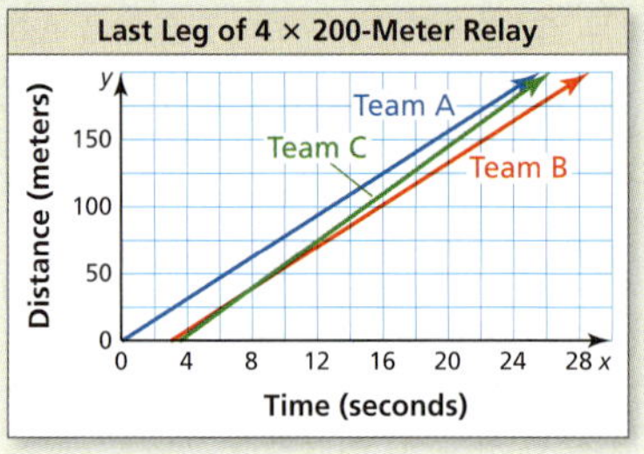

a. Estimate the distance at which Team C's runner passed Team B's runner.

b. If the race was longer, could Team C's runner have passed Team A's runner? Explain.

c. If the race was longer, could Team B's runner have passed Team A's runner? Explain.

31. ABSTRACT REASONING Consider the system of linear equations $y = ax + 4$ and $y = bx - 2$, where a and b are real numbers. Determine whether each statement is *always*, *sometimes*, or *never* true. Explain your reasoning.

a. The system has infinitely many solutions.

b. The system has no solution.

c. When $a < b$, the system has one solution.

32. MAKING AN ARGUMENT One admission to an ice skating rink costs x dollars, and renting a pair of ice skates costs y dollars. Your friend says she can determine the exact cost of one admission and one skate rental. Is your friend correct? Explain.

Maintaining Mathematical Proficiency

Reviewing what you learned in previous grades and lessons

Solve the equation. Check your solutions. *(Section 1.4)*

33. $|2x + 6| = |x|$

34. $|3x - 45| = |12x|$

35. $|x - 7| = |2x - 8|$

36. $|2x + 1| = |3x - 11|$

If students need help...	*If students got it...*
Resources by Chapter • Practice A and Practice B • Puzzle Time	Resources by Chapter • Enrichment and Extension • Cumulative Review
Student Journal • Practice	Start the *next* Section
Differentiating the Lesson Skills Review Handbook	

5.1–5.4 What Did You Learn?

Core Vocabulary

system of linear equations, *p. 236*

solution of a system of linear equations, *p. 236*

Core Concepts

Section 5.1

Solving a System of Linear Equations by Graphing, *p. 237*

Section 5.2

Solving a System of Linear Equations by Substitution, *p. 242*

Section 5.3

Solving a System of Linear Equations by Elimination, *p. 248*

Section 5.4

Solutions of Systems of Linear Equations, *p. 254*

Mathematical Practices

1. Describe the given information in Exercise 33 on page 246 and your plan for finding the solution.
2. Describe another real-life situation similar to Exercise 22 on page 251 and the mathematics that you can apply to solve the problem.
3. What question(s) can you ask your friend to help her understand the error in the statement she made in Exercise 32 on page 258?

Study Skills

Analyzing Your Errors

Study Errors

What Happens: You do not study the right material or you do not learn it well enough to remember it on a test without resources such as notes.

How to Avoid This Error: Take a practice test. Work with a study group. Discuss the topics on the test with your teacher. Do not try to learn a whole chapter's worth of material in one night.

Dynamic Teaching Tools

- Dynamic Assessment & Progress Monitoring Tool
- Interactive Whiteboard Lesson Library
- Dynamic Classroom with Dynamic Investigations

ANSWERS

1. The given information is the total number of songs played and the number of pop and hip-hop songs in relation to rock songs. The solution can be found by writing a system of linear equations and then solving the system using the substitution method.
2. *Sample answer:* A store sells candy bars for x dollars and eight-packs of candy bars for y dollars. Person A buys 4 candy bars and 2 eight-packs for $18. Person B buys 3 candy bars and 5 eight-packs for $38. To solve, write and solve a system of linear equations that represents this situation.
3. *Sample answer:* Are the equations derived from the two receipts equations for different lines?

ANSWERS

1. (3, 1)
2. (−2, −2)
3. (0, 1)
4. (−22, −26)
5. (2, −3)
6. (6, 1)
7. (2, 2)
8. (4, −1)
9. (2, 3)
10. no solution
11. (2, 2)
12. infinitely many solutions
13. **a.** $y = 4x + 14$ and $y = 6x + 8$
    b. (3, 26); Both trees will be 26 in. tall after 3 years.
14. **a.** 1 h at 55 mi/h, 2 h at 40 mi/h
    b. 55 mi; 80 mi
15. $7t + 3f = 26$ and $t + f = 6$; 2 touchdowns, 4 field goals

5.1–5.4 Quiz

Use the graph to solve the system of linear equations. Check your solution. *(Section 5.1)*

1. $y = -\frac{1}{3}x + 2$
$y = x - 2$

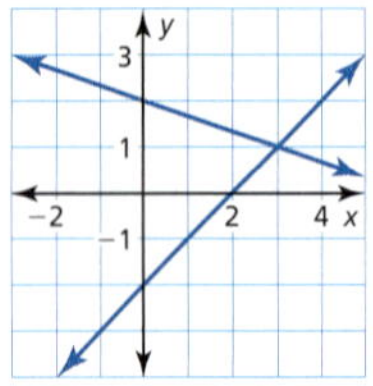

2. $y = \frac{1}{2}x - 1$
$y = 4x + 6$

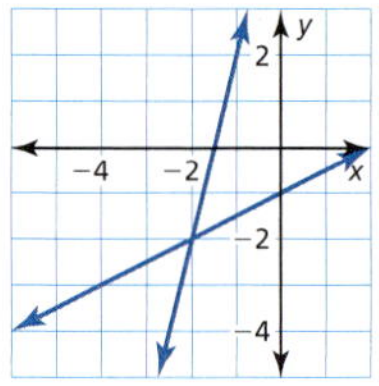

3. $y = 1$
$y = 2x + 1$

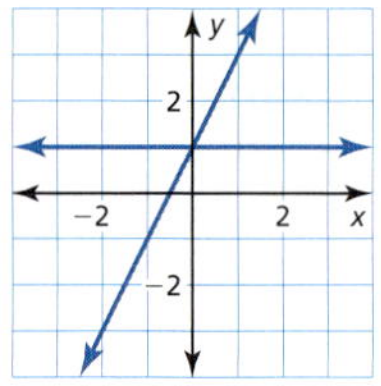

Solve the system of linear equations by substitution. Check your solution. *(Section 5.2)*

4. $y = x - 4$
$-2x + y = 18$

5. $2y + x = -4$
$y - x = -5$

6. $3x - 5y = 13$
$x + 4y = 10$

Solve the system of linear equations by elimination. Check your solution. *(Section 5.3)*

7. $x + y = 4$
$-3x - y = -8$

8. $x + 3y = 1$
$5x + 6y = 14$

9. $2x - 3y = -5$
$5x + 2y = 16$

Solve the system of linear equations. *(Section 5.4)*

10. $x - y = 1$
$x - y = 6$

11. $6x + 2y = 16$
$2x - y = 2$

12. $3x - 3y = -2$
$-6x + 6y = 4$

13. You plant a spruce tree that grows 4 inches per year and a hemlock tree that grows 6 inches per year. The initial heights are shown. *(Section 5.1)*

a. Write a system of linear equations that represents this situation.

b. Solve the system by graphing. Interpret your solution.

14. It takes you 3 hours to drive to a concert 135 miles away. You drive 55 miles per hour on highways and 40 miles per hour on the rest of the roads. *(Section 5.1, Section 5.2, and Section 5.3)*

a. How much time do you spend driving at each speed?

b. How many miles do you drive on highways? the rest of the roads?

15. In a football game, all of the home team's points are from 7-point touchdowns and 3-point field goals. The team scores six times. Write and solve a system of linear equations to find the numbers of touchdowns and field goals that the home team scores. *(Section 5.1, Section 5.2, and Section 5.3)*

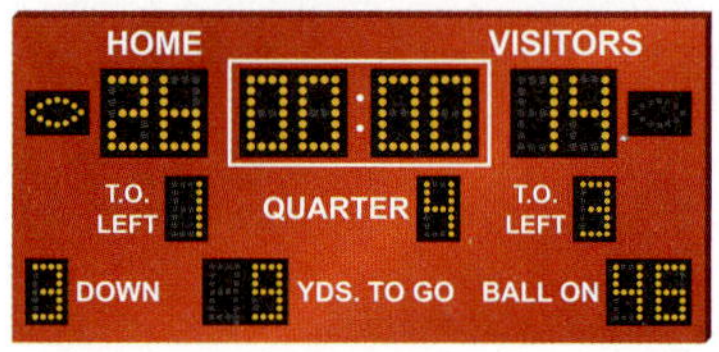

Laurie's Notes

Overview of Section 5.5

Introduction

- Solving a system by graphing can now be used to solve a problem from an earlier lesson—namely, solving an equation with variables on both sides. The approach in this lesson is to treat each side of the equation as a function, thus creating a system of two equations in two variables.

Teaching Strategy

- Example 2 in this lesson asks students to solve $|x + 1| = |2x - 4|$. In the lesson, this equation is rewritten as two related equations:

$$|x + 1| = |2x - 4| \quad \rightarrow \quad \begin{aligned} x + 1 &= 2x - 4 \\ x + 1 &= -(2x - 4) \end{aligned}$$

- Students may ask why you could not write the following related equations:

$$|2x - 4| = |x + 1| \quad \rightarrow \quad \begin{aligned} 2x - 4 &= x + 1 \\ 2x - 4 &= -(x + 1) \end{aligned}$$

- Take time to graph each pair to show that the solutions ($x = 1$ and $x = 5$) are the same.
- Graphs of absolute value equations were presented in Section 3.7 so students may also ask whether you can just graph each absolute value equation as they are. Yes, the solutions are still $x = 1$ and $x = 5$.

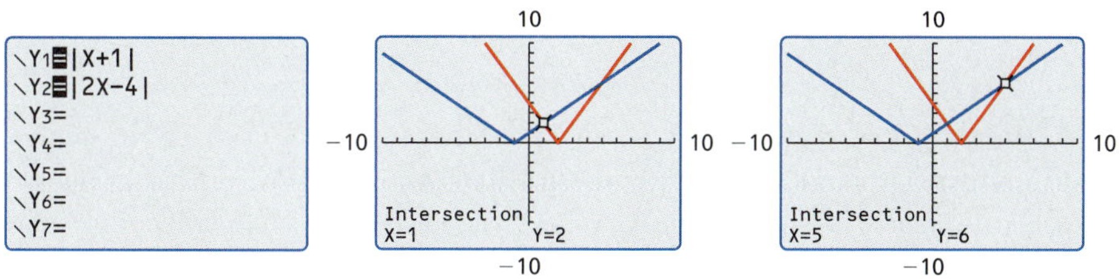

Pacing Suggestion

- Complete Exploration 1 and jigsaw the problems in Exploration 2. Transition to the formal lesson.

Dynamic Teaching Tools

- Dynamic Assessment & Progress Monitoring Tool
- Lesson Planning Tool
- Interactive Whiteboard Lesson Library
- Dynamic Classroom with Dynamic Investigations

What Your Students Will Learn

- Solve linear equations in one variable with variables on both sides by graphing.
- Solve absolute value equations in one variable with absolute values on both sides by graphing.

Laurie's Notes

Exploration

Motivate

? "How many of you have traveled with your family or friends and needed to rent a car?" Answers will vary.

? "What location claims to have the largest car rental market in the world?" Give time for students to offer their answers.

- According to the Orlando airport website, "Orlando is the largest rental car market in the world, with most of the major car rental companies located on-airport, without the need for a shuttle bus to pick up your rental car."
- Discuss why there might be so many cars available for rent in Orlando. Share with students that in today's lesson they will compare the cost of renting a car from two different companies.

Discuss

- **Connection:** Students previously solved equations with variables on both sides. In this lesson, they will use a system of linear equations as another way to solve an equation with variables on both sides.

Exploration 1

- Students may not understand the phrase "use the left side to write a linear equation." It may be helpful to write the equation and identify each side as y_1 and y_2 as shown.

$$\underbrace{2x - 1}_{y_1} = \underbrace{-\tfrac{1}{2}x + 4}_{y_2}$$

- **Reason Abstractly and Quantitatively:** In part (c), students should explain that the x-coordinate of the solution is the value of x that solves the original equation. They can verify this by substituting $x = 2$ into each side of the equation.

Exploration 2

- To save time, have different groups do different pairs of problems.
- Discuss whether the results for each method are the same.

Communicate Your Answer

- **Use Appropriate Tools Strategically:** Students should recognize that using a graphing calculator is an effective tool in solving equations with variables on both sides.

Connecting to Next Step

- The explorations help students see the connection between solving equations with variables on both sides and solving a linear system. This technique will be used with more complex equations in the formal lesson.

5.5 Solving Equations by Graphing

Essential Question How can you use a system of linear equations to solve an equation with variables on both sides?

Previously, you learned how to use algebra to solve equations with variables on both sides. Another way is to use a system of linear equations.

EXPLORATION 1 **Solving an Equation by Graphing**

Work with a partner. Solve $2x - 1 = -\frac{1}{2}x + 4$ by graphing.

a. Use the left side to write a linear equation. Then use the right side to write another linear equation.

b. Graph the two linear equations from part (a). Find the x-value of the point of intersection. Check that the x-value is the solution of

$$2x - 1 = -\frac{1}{2}x + 4.$$

c. Explain why this "graphical method" works.

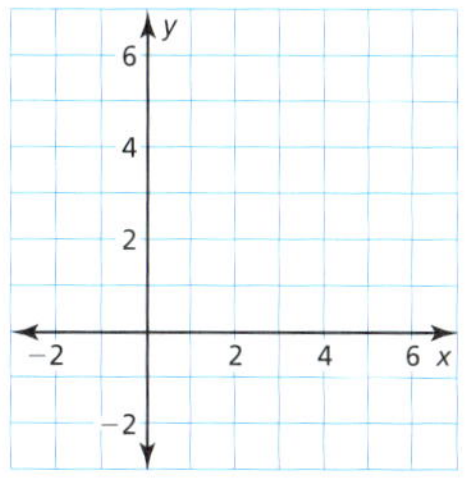

USING TOOLS STRATEGICALLY

To be proficient in math, you need to consider the available tools, which may include pencil and paper or a graphing calculator, when solving a mathematical problem.

EXPLORATION 2 **Solving Equations Algebraically and Graphically**

Work with a partner. Solve each equation using two methods.

Method 1 Use an algebraic method.

Method 2 Use a graphical method.

Is the solution the same using both methods?

a. $\frac{1}{2}x + 4 = -\frac{1}{4}x + 1$

b. $\frac{2}{3}x + 4 = \frac{1}{3}x + 3$

c. $-\frac{2}{3}x - 1 = \frac{1}{3}x - 4$

d. $\frac{4}{5}x + \frac{7}{5} = 3x - 3$

e. $-x + 2.5 = 2x - 0.5$

f. $-3x + 1.5 = x + 1.5$

Communicate Your Answer

3. How can you use a system of linear equations to solve an equation with variables on both sides?

4. Compare the algebraic method and the graphical method for solving a linear equation with variables on both sides. Describe the advantages and disadvantages of each method.

Dynamic Teaching Tools

- Dynamic Assessment & Progress Monitoring Tool
- Lesson Planning Tool
- Interactive Whiteboard Lesson Library
- Dynamic Classroom with Dynamic Investigations

ANSWERS

1. a. $y = 2x - 1$; $y = -\frac{1}{2}x + 4$

b.

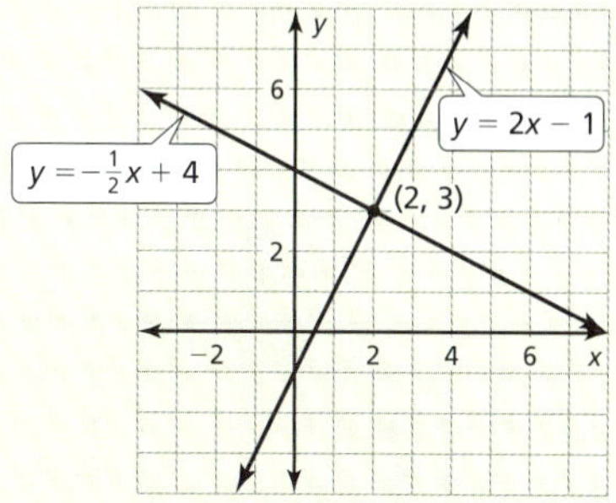

$x = 2$

c. The two sides of the equation are equal to each other. If you set one side of the equation equal to y, the transitive property allows you to set the other side of the equation equal to y.

2. a. $x = -4$; yes

b. $x = -3$; yes

c. $x = 3$; yes

d. $x = 2$; yes

e. $x = 1$; yes

f. $x = 0$; yes

3. Write two linear equations setting y equal to each side. Solve the system of linear equations. The x-value of the solution of the system of linear equations is the solution of the equation.

4. The algebraic method will always give the exact solution, but it is difficult to work with fractions as coefficients. The graphical method is easier with fractional coefficients of x because these are the slopes of the lines, but solutions may sometimes be estimates, especially when the solution does not fall on a grid line.

English Language Learners

Organization

Have students write in their notebooks the steps for solving an equation by graphing a system of related linear equations.

Step 1: Use each side of the original equation to write a system of linear equations.

Step 2: Graph the system.

Step 3: Find the x-value of the point of intersection. Use it to write the solution.

Students should include a note to check their results by substitution into the original equation.

Extra Example 1

Solve $2x + 3 = x + 2$ by graphing. Check your solution.

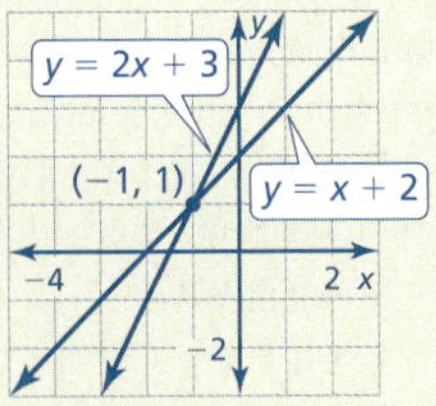

$x = -1$

MONITORING PROGRESS ANSWERS

1. $x = -2$
2. $x = \frac{1}{2}$

5.5 Lesson

What You Will Learn

- Solve linear equations by graphing.
- Solve absolute value equations by graphing.
- Use linear equations to solve real-life problems.

Core Vocabulary

Previous
absolute value equation

Solving Linear Equations by Graphing

You can use a system of linear equations to solve an equation with variables on both sides.

Core Concept

Solving Linear Equations by Graphing

Step 1 To solve the equation $ax + b = cx + d$, write two linear equations.

$$ax + b = cx + d$$

$y = ax + b$ and $y = cx + d$

Step 2 Graph the system of linear equations. The x-value of the solution of the system of linear equations is the solution of the equation $ax + b = cx + d$.

EXAMPLE 1 Solving an Equation by Graphing

Solve $-x + 1 = 2x - 5$ by graphing. Check your solution.

SOLUTION

Step 1 Write a system of linear equations using each side of the original equation.

$$-x + 1 = 2x - 5$$

$y = -x + 1$ $y = 2x - 5$

Step 2 Graph the system.

$y = -x + 1$ Equation 1

$y = 2x - 5$ Equation 2

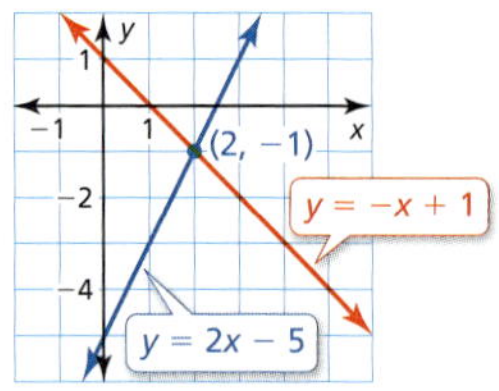

The graphs intersect at $(2, -1)$.

Check

$-x + 1 = 2x - 5$

$-(2) + 1 \stackrel{?}{=} 2(2) - 5$

$-1 = -1$ ✓

So, the solution of the equation is $x = 2$.

Monitoring Progress

Help in English and Spanish at *BigIdeasMath.com*

Solve the equation by graphing. Check your solution.

1. $\frac{1}{2}x - 3 = 2x$
2. $-4 + 9x = -3x + 2$

Laurie's Notes Teacher Actions

- **Use Appropriate Tools Strategically** and **Look For and Make Use of Structure:** Students have solved systems by graphing. Today, they will solve a linear equation with variables on both sides. The *Big Idea* is to think of each side of the equation as a linear equation by writing $y =$ left side of equation and $y =$ right side of equation.

? **Probing Question:** "How is the intersection of the lines related to the original question?" The graph of each linear equation shows the value of the expression for different values of x. So, the intersection shows the value of x for which the expressions are equal.

Solving Absolute Value Equations by Graphing

EXAMPLE 2 **Solving an Absolute Value Equation by Graphing**

Solve $|x + 1| = |2x - 4|$ by graphing. Check your solutions.

SOLUTION

Recall that an absolute value equation of the form $|ax + b| = |cx + d|$ has two related equations.

$ax + b = cx + d$ Equation 1

$ax + b = -(cx + d)$ Equation 2

So, the related equations of $|x + 1| = |2x - 4|$ are as follows.

$x + 1 = 2x - 4$ Equation 1

$x + 1 = -(2x - 4)$ Equation 2

Apply the steps for solving an equation by graphing to each of the related equations.

Step 1 Write a system of linear equations for each related equation.

Equation 1

$x + 1 = 2x - 4$

$y = x + 1$ $y = 2x - 4$

System 1

Equation 2

$x + 1 = -(2x - 4)$

$x + 1 = -2x + 4$

$y = x + 1$ $y = -2x + 4$

System 2

Step 2 Graph each system.

System 1

$y = x + 1$

$y = 2x - 4$

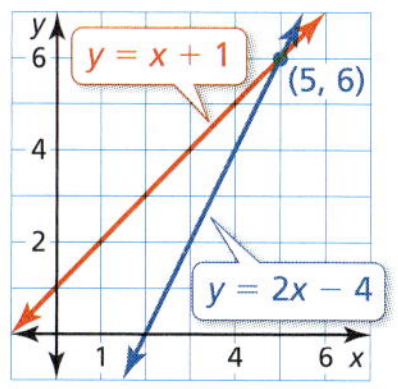

The graphs intersect at (5, 6).

System 2

$y = x + 1$

$y = -2x + 4$

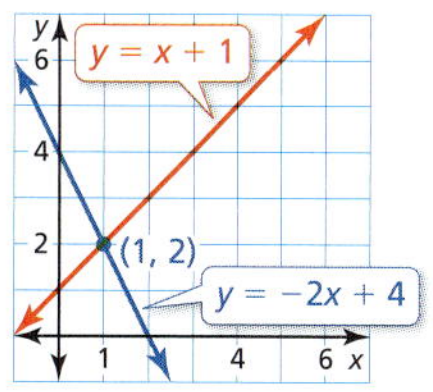

The graphs intersect at (1, 2).

▶ So, the solutions of the equation are $x = 5$ and $x = 1$.

Check

$|x + 1| = |2x - 4|$

$|5 + 1| \stackrel{?}{=} |2(5) - 4|$

$|6| \stackrel{?}{=} |6|$

$6 = 6$ ✓

$|x + 1| = |2x - 4|$

$|1 + 1| \stackrel{?}{=} |2(1) - 4|$

$|2| \stackrel{?}{=} |-2|$

$2 = 2$ ✓

Monitoring Progress

Help in English and Spanish at *BigIdeasMath.com*

Solve the equation by graphing. Check your solutions.

3. $|2x + 2| = |x - 2|$

4. $|x - 6| = |-x + 4|$

Extra Example 2

Solve $|x + 3| = |3x + 1|$ by graphing. Check your solutions.

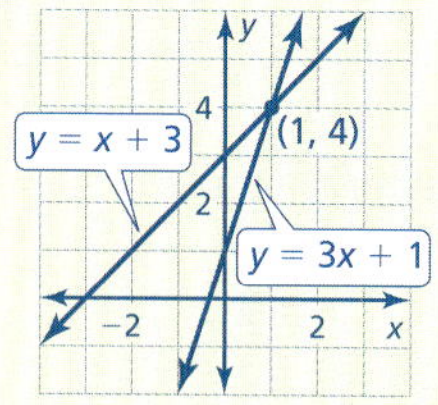

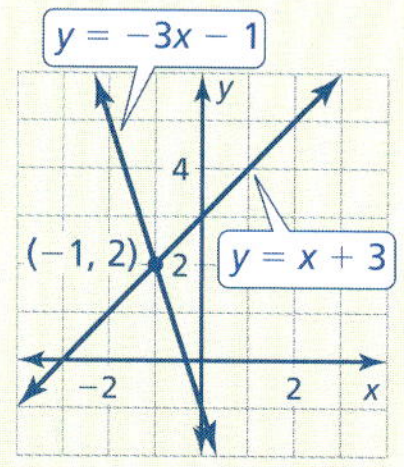

$x = 1$ and $x = -1$

MONITORING PROGRESS ANSWERS

3. $x = -4, x = 0$

4. $x = 5$

Laurie's Notes — Teacher Actions

- See Section 1.4 for absolute value equations like Example 2. Remind students of the two related equations that result.
- **Turn and Talk:** "What are the two systems that result from the absolute value equation?" Students will likely write the system shown. There is an alternate but equivalent system. See the *Teaching Strategy* on page T-260.
- Discuss with students the power of this graphing approach to solving equations.

Extra Example 3

Your family needs to rent a car for a week while on vacation. Company B charges \$3 per mile plus a flat fee of \$150 per week. Company D charges \$4 per mile plus a flat fee of \$75 per week. After how many miles are the total costs the same at Company B and Company D? The total costs are the same after 75 miles.

MONITORING PROGRESS ANSWER

5. 200 mi

Solving Real-Life Problems

EXAMPLE 3 **Modeling with Mathematics**

Your family needs to rent a car for a week while on vacation. Company A charges \$3.25 per mile plus a flat fee of \$125 per week. Company B charges \$3 per mile plus a flat fee of \$150 per week. After how many miles of travel are the total costs the same at both companies?

SOLUTION

1. **Understand the Problem** You know the costs of renting a car from two companies. You are asked to determine how many miles of travel will result in the same total costs at both companies.
2. **Make a Plan** Use a verbal model to write an equation that represents the problem. Then solve the equation by graphing.
3. **Solve the Problem**

Words Company A: Cost per mile • Miles + Flat fee = Company B: Cost per mile • Miles + Flat fee

Variable Let x be the number of miles traveled.

Equation $3.25x + 125 = 3x + 150$

Solve the equation by graphing.

Step 1 Write a system of linear equations using each side of the original equation.

$$3.25x + 125 = 3x + 150$$

$y = 3.25x + 125$ $\qquad$ $y = 3x + 150$

Step 2 Use a graphing calculator to graph the system.

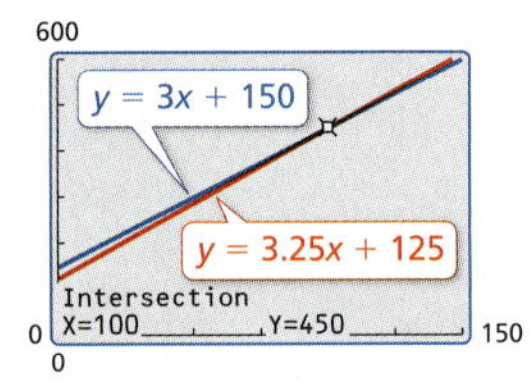

Check

$$3.25x + 125 = 3x + 150$$
$$0.25x + 125 = 150$$
$$0.25x = 25$$
$$x = 100$$

Because the graphs intersect at (100, 450), the solution of the equation is $x = 100$.

▶ So, the total costs are the same after 100 miles.

4. **Look Back** One way to check your solution is to solve the equation algebraically, as shown.

Monitoring Progress

Help in English and Spanish at *BigIdeasMath.com*

5. **WHAT IF?** Company C charges \$3.30 per mile plus a flat fee of \$115 per week. After how many miles are the total costs the same at Company A and Company C?

Laurie's Notes Teacher Actions

- Read the problem with students and probe to determine what is known and what they are trying to solve.

COMMON ERROR Students try to let x = the cost of renting at Company A and y = the cost of renting at Company B. Explain that a model is known for each company. What is not known is how many miles you drive, x. Now write the model for each company and solve.

- **Use Appropriate Tools Strategically:** The calculator is helpful in solving. Students will have to set an appropriate viewing window.

Closure

- **Exit Ticket:** Solve the equation $3x - 4 = \frac{1}{2}x + 1$ by graphing. Check your solution. $x = 2$

5.5 Exercises

Dynamic Solutions available at *BigIdeasMath.com*

Vocabulary and Core Concept Check

1. **REASONING** The graphs of the equations $y = 3x - 20$ and $y = -2x + 10$ intersect at the point $(6, -2)$. Without solving, find the solution of the equation $3x - 20 = -2x + 10$.

2. **WRITING** Explain how to rewrite the absolute value equation $|2x - 4| = |-5x + 1|$ as two systems of linear equations.

Monitoring Progress and Modeling with Mathematics

In Exercises 3–6, use the graph to solve the equation. Check your solution.

3. $-2x + 3 = x$

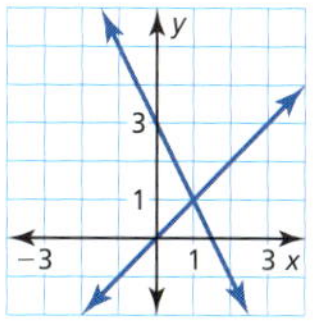

4. $-3 = 4x + 1$

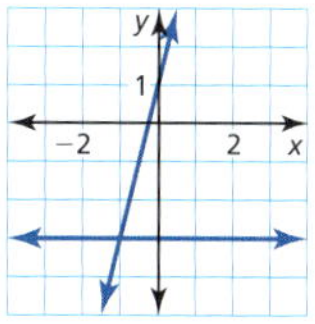

5. $-x - 1 = \frac{1}{3}x + 3$

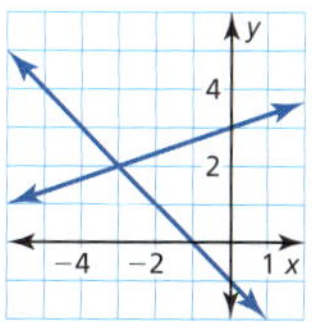

6. $-\frac{3}{2}x - 2 = -4x + 3$

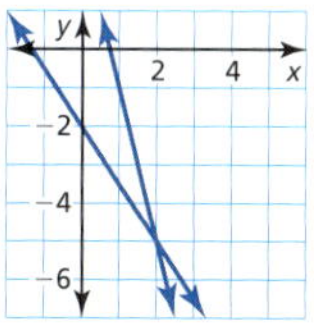

In Exercises 7–14, solve the equation by graphing. Check your solution. *(See Example 1.)*

7. $x + 4 = -x$
8. $4x = x + 3$
9. $x + 5 = -2x - 4$
10. $-2x + 6 = 5x - 1$
11. $\frac{1}{2}x - 2 = 9 - 5x$
12. $-5 + \frac{1}{4}x = 3x + 6$
13. $5x - 7 = 2(x + 1)$
14. $-6(x + 4) = -3x - 6$

In Exercises 15–20, solve the equation by graphing. Determine whether the equation has *one solution*, *no solution*, or *infinitely many solutions*.

15. $3x - 1 = -x + 7$
16. $5x - 4 = 5x + 1$
17. $-4(2 - x) = 4x - 8$
18. $-2x - 3 = 2(x - 2)$
19. $-x - 5 = -\frac{1}{3}(3x + 5)$
20. $\frac{1}{2}(8x + 3) = 4x + \frac{3}{2}$

In Exercises 21 and 22, use the graphs to solve the equation. Check your solutions.

21. $|x - 4| = |3x|$

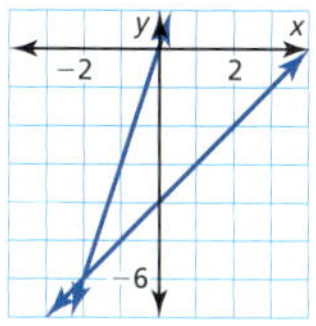

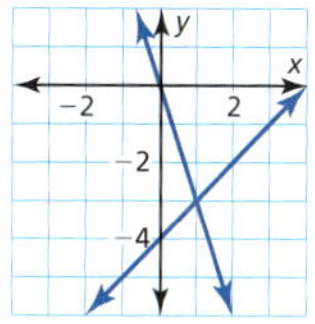

22. $|2x + 4| = |x - 1|$

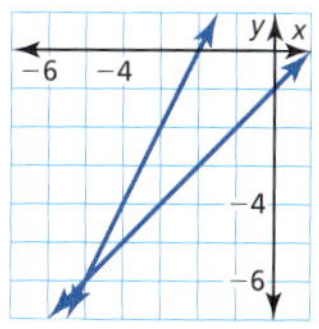

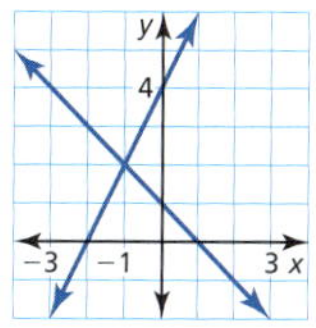

In Exercises 23–30, solve the equation by graphing. Check your solutions. *(See Example 2.)*

23. $|2x| = |x + 3|$
24. $|2x - 6| = |x|$
25. $|-x + 4| = |2x - 2|$
26. $|x + 2| = |-3x + 6|$
27. $|x + 1| = |x - 5|$
28. $|2x + 5| = |-2x + 1|$
29. $|x - 3| = 2|x|$
30. $4|x + 2| = |2x + 7|$

Assignment Guide and Homework Check

ASSIGNMENT

Basic: 1, 2, 3–33 odd, 36, 38, 42–49

Average: 1, 2–38 even, 39, 42–49

Advanced: 1, 2, 6, 12, 14, 20, 22, 28, 30–49

HOMEWORK CHECK

Basic: 3, 7, 23, 33

Average: 4, 8, 26, 34

Advanced: 6, 12, 28, 34

ANSWERS

1. $x = 6$
2. For the first system, set y equal to both expressions inside the absolute value symbols, $2x - 4$ and $-5x + 1$. For the second system, set y equal to the expression inside the first absolute value symbol, $2x - 4$, and the opposite of the expression inside the second absolute value symbol, $5x - 1$.
3. $x = 1$
4. $x = -1$
5. $x = -3$
6. $x = 2$
7. $x = -2$
8. $x = 1$
9. $x = -3$
10. $x = 1$
11. $x = 2$
12. $x = -4$
13. $x = 3$
14. $x = -6$
15. $x = 2$; one solution
16. no solution
17. all real numbers; infinitely many solutions
18. $x = \frac{1}{4}$; one solution
19. no solution
20. all real numbers; infinitely many solutions
21. $x = -2, x = 1$
22. $x = -5, x = -1$
23. $x = 3, x = -1$
24. $x = 6, x = 2$
25. $x = 2, x = -2$
26. $x = 1, x = 4$
27. $x = 2$
28. $x = -1$
29. $x = -3, x = 1$
30. $x = -\frac{1}{2}, x = -\frac{5}{2}$

Dynamic Teaching Tools

- Dynamic Assessment & Progress Monitoring Tool
- Interactive Whiteboard Lesson Library
- Dynamic Classroom with Dynamic Investigations

ANSWERS

31. $x = -2$

32. $x = 1.8$

33. 75 guests

34. 4 human years

35. 30 sec

36. no; Only the x-coordinate is the solution, so the solution is $x = 4$.

37. *Sample answer:* $m = 1, b = 8$

38. **a.** *Sample answer:* (7, 5.5)

b. *Sample answer:* After about 7 years, the company's revenue and expenses will each be about $5.5 million.

39. $x = 10$

40–49. See Additional Answers.

Mini-Assessment

1. Solve $x - 4 = 3x$ by graphing. Check your solution.

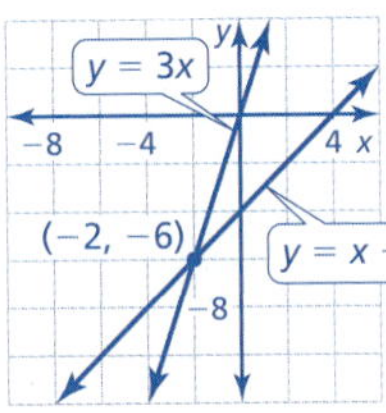

$x = -2$

2. Solve $|x + 5| = |-2x - 4|$ by graphing. Check your solutions.

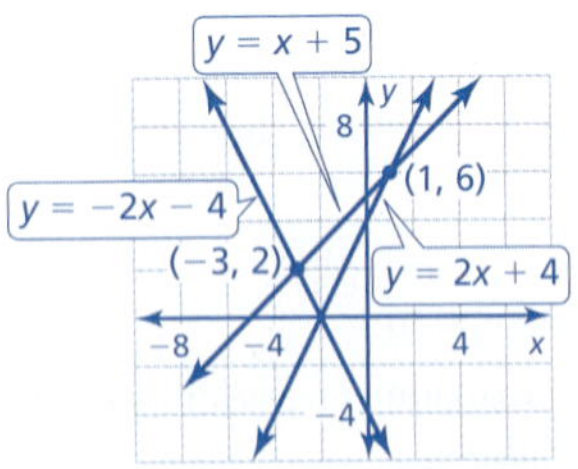

$x = -3$ and $x = 1$

3. You need to hire an electrician to install new wiring in your house. Electrician A charges $75 plus $45 per hour. Electrician B charges $100 plus $40 per hour. For how many hours of work are the total costs the same for both electricians?
5 hours

USING TOOLS In Exercises 31 and 32, use a graphing calculator to solve the equation.

31. $0.7x + 0.5 = -0.2x - 1.3$

32. $2.1x + 0.6 = -1.4x + 6.9$

33. **MODELING WITH MATHEMATICS** You need to hire a catering company to serve meals to guests at a wedding reception. Company A charges $500 plus $20 per guest. Company B charges $800 plus $16 per guest. For how many guests are the total costs the same at both companies? *(See Example 3.)*

34. **MODELING WITH MATHEMATICS** Your dog is 16 years old in dog years. Your cat is 28 years old in cat years. For every human year, your dog ages by 7 dog years and your cat ages by 4 cat years. In how many human years will both pets be the same age in their respective types of years?

35. **MODELING WITH MATHEMATICS** You and a friend race across a field to a fence and back. Your friend has a 50-meter head start. The equations shown represent you and your friend's distances d (in meters) from the fence t seconds after the race begins. Find the time at which you catch up to your friend.

You: $d = |-5t + 100|$

Your friend: $d = \left|-3\frac{1}{3}t + 50\right|$

36. **MAKING AN ARGUMENT** The graphs of $y = -x + 4$ and $y = 2x - 8$ intersect at the point (4, 0). So, your friend says the solution of the equation $-x + 4 = 2x - 8$ is (4, 0). Is your friend correct? Explain.

37. **OPEN-ENDED** Find values for m and b so that the solution of the equation $mx + b = -2x - 1$ is $x = -3$.

38. **HOW DO YOU SEE IT?** The graph shows the total revenue and expenses of a company x years after it opens for business.

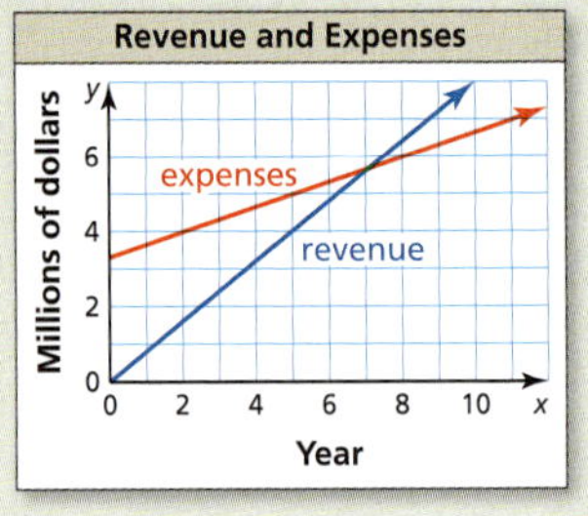

a. Estimate the point of intersection of the graphs.

b. Interpret your answer in part (a).

39. **MATHEMATICAL CONNECTIONS** The value of the perimeter of the triangle (in feet) is equal to the value of the area of the triangle (in square feet). Use a graph to find x.

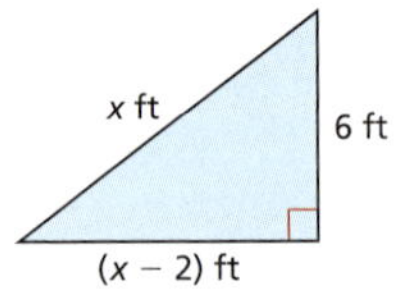

40. **THOUGHT PROVOKING** A car has an initial value of $20,000 and decreases in value at a rate of $1500 per year. Describe a different car that will be worth the same amount as this car in exactly 5 years. Specify the initial value and the rate at which the value decreases.

41. **ABSTRACT REASONING** Use a graph to determine the sign of the solution of the equation $ax + b = cx + d$ in each situation.

a. $0 < b < d$ and $a < c$ **b.** $d < b < 0$ and $a < c$

Maintaining Mathematical Proficiency

Reviewing what you learned in previous grades and lessons

Graph the inequality. *(Section 2.1)*

42. $y > 5$ **43.** $x \le -2$ **44.** $n \ge 9$ **45.** $c < -6$

Use the graphs of f and g to describe the transformation from the graph of f to the graph of g. *(Section 3.6)*

46. $f(x) = x - 5$; $g(x) = f(x + 2)$

47. $f(x) = 6x$; $g(x) = -f(x)$

48. $f(x) = -2x + 1$; $g(x) = f(4x)$

49. $f(x) = \frac{1}{2}x - 2$; $g(x) = f(x - 1)$

If students need help...	*If students got it...*
Resources by Chapter • Practice A and Practice B • Puzzle Time	Resources by Chapter • Enrichment and Extension • Cumulative Review
Student Journal • Practice	Start the *next* Section
Differentiating the Lesson Skills Review Handbook	

Laurie's Notes

Overview of Section 5.6

Introduction

- This lesson presents how to graph a linear inequality in the coordinate plane so that in the next lesson students can graph systems of linear inequalities.
- **Connection:** Students have graphed inequalities on a number line. Relate the solid and dashed lines in the coordinate plane to the closed and open circles used when inequalities in one variable are graphed on a number line. The closed and open circles are boundaries, creating two half-lines. The solid and dashed lines are boundaries, creating two half-planes.

Common Misconceptions

- When an inequality has only one variable, such as $x > -2$ or $y \leq 3$, students think the solution is graphed on a number line. That would be true, and that is why in this lesson students are directed to graph $x > -2$ in a coordinate plane.

Teaching Strategy

- Students are excited to graph their first linear inequality on a graphing calculator! It is cool to see the shading occur. Some calculators have apps that can be used to graph linear inequalities, while others have an option from the equation editor for shading.
- To use the calculator to graph an inequality, it is necessary to specify which side of the boundary line is to be shaded. In Figure 1, note that this has been selected to the left of Y_1. The graph for Example 4 is shown in Figure 2.

Shading selected →

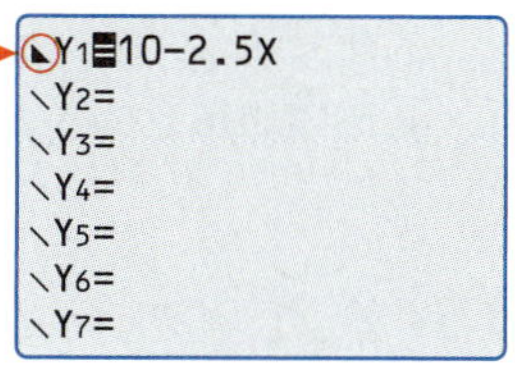

Figure 1

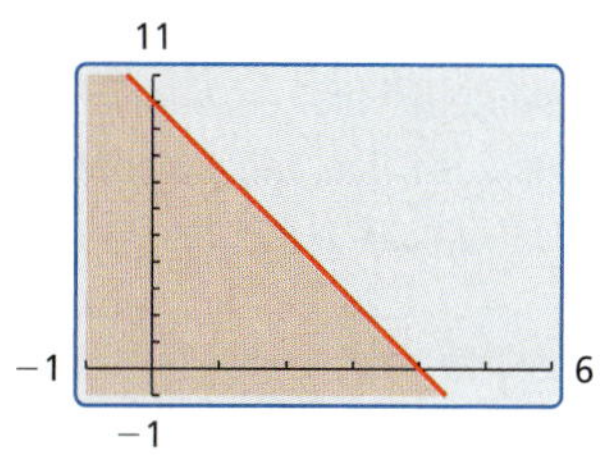

Figure 2

- Note that the calculator does not show whether the boundary line is a solution to the inequality.

Pacing Suggestion

- The first two explorations provide an introduction to graphing linear inequalities. Continue with Exploration 3 or transition to the formal lesson.

Dynamic Teaching Tools

- Dynamic Assessment & Progress Monitoring Tool
- Lesson Planning Tool
- Interactive Whiteboard Lesson Library
- Dynamic Classroom with Dynamic Investigations

What Your Students Will Learn

- Check solutions of linear inequalities in two variables algebraically.
- Graph linear inequalities in one variable and in two variables in the coordinate plane.

Laurie's Notes

Exploration

Motivate

- Before class begins, place a piece of rope, yarn, or tape on the floor so that the classroom is divided into two *halves*. The two halves do not need to be the same size, but you do want some student desks in each half.
- Draw a line on the board and plot a point.
- Explain that the point divides the line into two parts called rays (half-lines). The point is the "boundary" between the two rays.

? "What would divide a plane into two half-planes?" a line

- A line is the "boundary" that divides the plane into two half-planes.
- Refer to the boundary line in the classroom. Students are on one side of the line or the other, or perhaps they are "on" the line.
- Explain that today students will graph linear inequalities in two variables. A line will divide the coordinate plane into two parts. The solution will be on one side of the line or the other.

Exploration 1

- Students should be able to determine the slope and *y*-intercept from the graph. Because part (a) asks for an equation and not an inequality, students should be able to write $y = x - 3$.
- **Attend to Precision:** In describing the solution of the inequality in part (b), students often use language such as "shading on one side of a line." In other words, they describe the graphical representation versus what it means algebraically to be a solution.
- This is students' first encounter with a dashed line versus a solid line. My experience has been that students select the $<$ symbol for the dashed line, reasoning that it is analogous to the open circle on a number line.

Exploration 2

- **Teacher Tip:** It is possible that more than one type of graphing calculator is used in your classroom. You may want to group students that are using the same type of calculator together to act as a support group.

? "How do you enter the fraction $\frac{1}{4}$ on your calculator?" Answers depend on the calculator. Students should recognize that the decimal 0.25 could also be used.

- Explain the steps necessary to produce the shading for the inequality. Show students the icon for shading the region below the graph.
- Explain to students that some calculators always use the solid line. They will need to determine whether ordered pairs on the line are part of the solution set.

Exploration 3

- Students should have no difficulty graphing the related equations, but they may be unsure of when and why to use a dashed line versus a solid line and how to decide which side of the line to shade. If students are doing less exploring and you are doing more teaching, transition to the formal lesson.

Communicate Your Answer

- Have students share the real-life situations they wrote for Question 5.

Connecting to Next Step

- Students have now been introduced to graphing linear inequalities. In the formal lesson, students will graph additional examples and solve a real-life problem.

5.6 Graphing Linear Inequalities in Two Variables

Essential Question

How can you graph a linear inequality in two variables?

A **solution of a linear inequality in two variables** is an ordered pair (x, y) that makes the inequality true. The **graph of a linear inequality** in two variables shows all the solutions of the inequality in a coordinate plane.

EXPLORATION 1 Writing a Linear Inequality in Two Variables

Work with a partner.

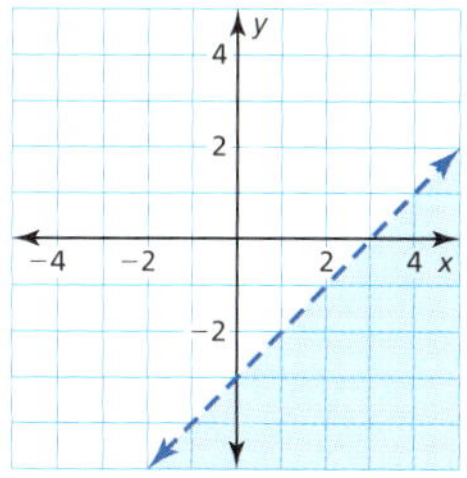

a. Write an equation represented by the dashed line.

b. The solutions of an inequality are represented by the shaded region. In words, describe the solutions of the inequality.

c. Write an inequality represented by the graph. Which inequality symbol did you use? Explain your reasoning.

EXPLORATION 2 Using a Graphing Calculator

USING TOOLS STRATEGICALLY

To be proficient in math, you need to use technological tools to explore and deepen your understanding of concepts.

Work with a partner. Use a graphing calculator to graph $y \geq \frac{1}{4}x - 3$.

a. Enter the equation $y = \frac{1}{4}x - 3$ into your calculator.

b. The inequality has the symbol ≥. So, the region to be shaded is above the graph of $y = \frac{1}{4}x - 3$, as shown. Verify this by testing a point in this region, such as (0, 0), to make sure it is a solution of the inequality.

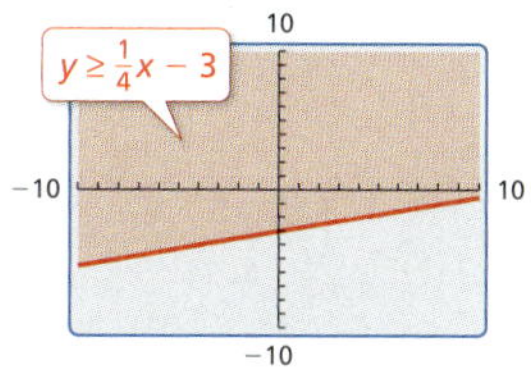

Because the inequality symbol is *greater than or equal to*, the line is solid and not dashed. Some graphing calculators always use a solid line when graphing inequalities. In this case, you have to determine whether the line should be solid or dashed, based on the inequality symbol used in the original inequality.

EXPLORATION 3 Graphing Linear Inequalities in Two Variables

Work with a partner. Graph each linear inequality in two variables. Explain your steps. Use a graphing calculator to check your graphs.

a. $y > x + 5$ **b.** $y \leq -\frac{1}{2}x + 1$ **c.** $y \geq -x - 5$

Communicate Your Answer

4. How can you graph a linear inequality in two variables?
5. Give an example of a real-life situation that can be modeled using a linear inequality in two variables.

Dynamic Teaching Tools

- Dynamic Assessment & Progress Monitoring Tool
- Lesson Planning Tool
- Interactive Whiteboard Lesson Library
- Dynamic Classroom with Dynamic Investigations

ANSWERS

1. a. $y = x - 3$

b. all ordered pairs below the graph of $y = x - 3$

c. $y < x - 3$; <; *Sample answer:* The point (4, 0) is in the shaded region, and to make the inequality true for that point the < symbol is needed.

2. $0 \geq -3$

3. a.

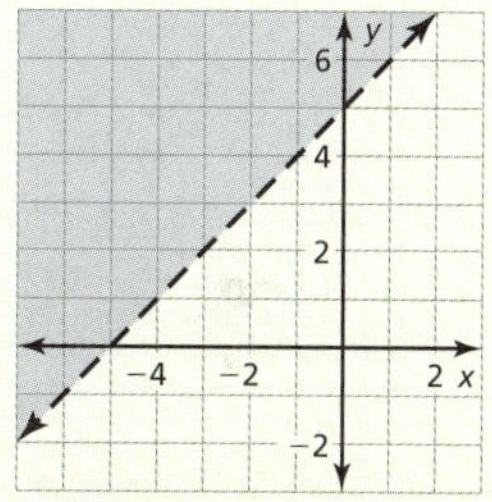

Sample answer: Graph $y = x + 5$ with a dashed line. Test the point (0, 0), which does not make the inequality true. Shade the half-plane that does not contain the point (0, 0).

b.

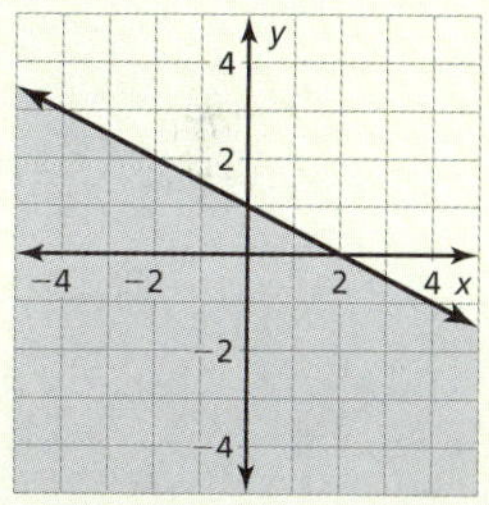

Sample answer: Graph $y = -\frac{1}{2}x + 1$ with a solid line. Test the point (0, 0), which does make the inequality true. Shade the half-plane that contains the point (0, 0).

3. c.

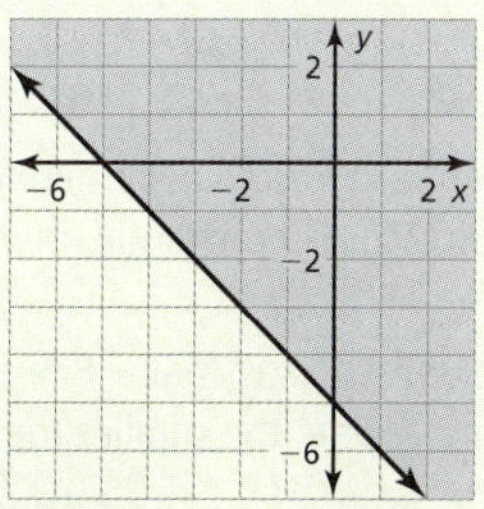

Sample answer: Graph $y = -x - 5$ with a solid line. Test the point (0, 0), which does make the inequality true. Shade the half-plane that contains the point (0, 0).

4. Graph the boundary line for the inequality. Use a dashed line for < or >. Use a solid line for ≤ or ≥. Test a point that is not on the boundary line to determine whether it is a solution of the inequality. When the test point is a solution, shade the half-plane that contains the point. When the test point is not a solution, shade the half-plane that does not contain the point.

5. *Sample answer:* You want to spend no more than \$15 at the deli for bologna at \$2.99 per pound and cheese at \$1.99 per pound. How many pounds of each can you purchase?

English Language Learners

Build on Past Knowledge

Remind students that when they graph inequalities on a number line, they use an open circle when the inequality symbol is $<$ or $>$. They use a closed circle when the inequality symbol is $\leq$ or $\geq$. Explain that when graphing inequalities in a coordinate plane, they will use a dashed line when the inequality symbol is $<$ or $>$, and a solid line when the inequality symbol is $\leq$ or $\geq$.

Extra Example 1

Tell whether the ordered pair is a solution of the inequality.

a. $3x - y < 2$; $(-2, 2)$

$(-2, 2)$ is a solution of the inequality.

b. $4x - y \geq 5$; $(1, 3)$

$(1, 3)$ is *not* a solution of the inequality.

MONITORING PROGRESS ANSWERS

1. no
2. no
3. yes
4. yes

5.6 Lesson

What You Will Learn

- Check solutions of linear inequalities.
- Graph linear inequalities in two variables.
- Use linear inequalities to solve real-life problems.

Core Vocabulary

linear inequality in two variables, *p. 268*
solution of a linear inequality in two variables, *p. 268*
graph of a linear inequality, *p. 268*
half-planes, *p. 268*

Previous
ordered pair

Linear Inequalities

A **linear inequality in two variables**, x and y, can be written as

$$ax + by < c \qquad ax + by \leq c \qquad ax + by > c \qquad ax + by \geq c$$

where a, b, and c are real numbers. A **solution of a linear inequality in two variables** is an ordered pair (x, y) that makes the inequality true.

EXAMPLE 1 Checking Solutions

Tell whether the ordered pair is a solution of the inequality.

a. $2x + y < -3$; $(-1, 9)$ **b.** $x - 3y \geq 8$; $(2, -2)$

SOLUTION

a. $2x + y < -3$ Write the inequality.

$2(-1) + 9 \overset{?}{<} -3$ Substitute -1 for x and 9 for y.

$7 \nless -3$ ✗ Simplify. 7 is *not* less than -3.

▶ So, $(-1, 9)$ is *not* a solution of the inequality.

b. $x - 3y \geq 8$ Write the inequality.

$2 - 3(-2) \overset{?}{\geq} 8$ Substitute 2 for x and -2 for y.

$8 \geq 8$ ✓ Simplify. 8 is equal to 8.

▶ So, $(2, -2)$ is a solution of the inequality.

Monitoring Progress

Help in English and Spanish at *BigIdeasMath.com*

Tell whether the ordered pair is a solution of the inequality.

1. $x + y > 0$; $(-2, 2)$ **2.** $4x - y \geq 5$; $(0, 0)$

3. $5x - 2y \leq -1$; $(-4, -1)$ **4.** $-2x - 3y < 15$; $(5, -7)$

Graphing Linear Inequalities in Two Variables

The **graph of a linear inequality** in two variables shows all the solutions of the inequality in a coordinate plane.

READING

A dashed boundary line means that points on the line are *not* solutions. A solid boundary line means that points on the line are solutions.

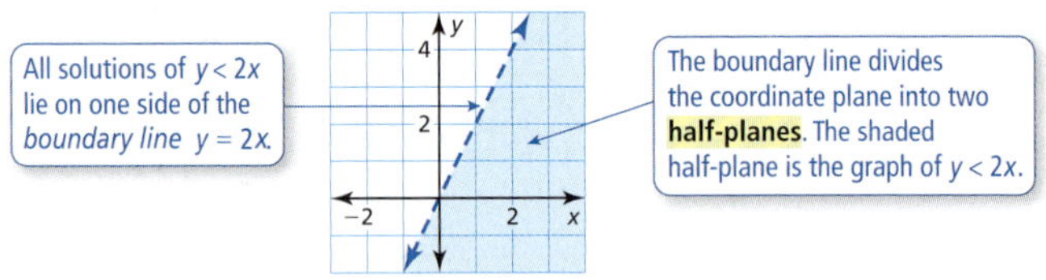

Laurie's Notes Teacher Actions

- Explain that while the examples are in a form that resembles standard form, linear inequalities can also be written in a form that resembles slope-intercept form. Most importantly, there are four inequality symbols: $<$, $\leq$, $>$, and $\geq$.

COMMON ERROR When checking solutions, it is common for students to say that $-8 > -6$. Remind students that the further to the left a number is on the number line, the smaller the value becomes. Another common error is $4 < 4$ (false) and $4 \leq 4$ (true).

Core Concept

Graphing a Linear Inequality in Two Variables

Step 1 Graph the boundary line for the inequality. Use a dashed line for $<$ or $>$. Use a solid line for $\leq$ or $\geq$.

Step 2 Test a point that is not on the boundary line to determine whether it is a solution of the inequality.

Step 3 When the test point is a solution, shade the half-plane that contains the point. When the test point is *not* a solution, shade the half-plane that does *not* contain the point.

STUDY TIP

It is often convenient to use the origin as a test point. However, you must choose a different test point when the origin is on the boundary line.

EXAMPLE 2 Graphing a Linear Inequality in One Variable

Graph $y \leq 2$ in a coordinate plane.

SOLUTION

Step 1 Graph $y = 2$. Use a solid line because the inequality symbol is $\leq$.

Step 2 Test $(0, 0)$.

$y \leq 2$ — Write the inequality.

$0 \leq 2$ ✓ — Substitute.

Step 3 Because $(0, 0)$ is a solution, shade the half-plane that contains $(0, 0)$.

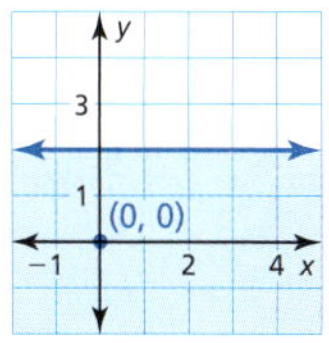

EXAMPLE 3 Graphing a Linear Inequality in Two Variables

Graph $-x + 2y > 2$ in a coordinate plane.

SOLUTION

Step 1 Graph $-x + 2y = 2$, or $y = \frac{1}{2}x + 1$. Use a dashed line because the inequality symbol is $>$.

Step 2 Test $(0, 0)$.

$-x + 2y > 2$ — Write the inequality.

$-(0) + 2(0) \overset{?}{>} 2$ — Substitute.

$0 \not> 2$ ✗ — Simplify.

Step 3 Because $(0, 0)$ is *not* a solution, shade the half-plane that does *not* contain $(0, 0)$.

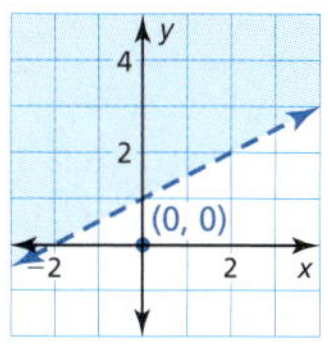

Check

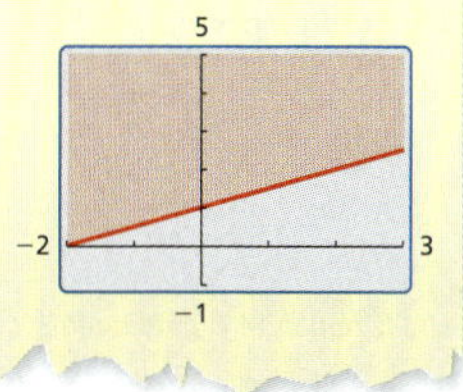

Monitoring Progress

Help in English and Spanish at *BigIdeasMath.com*

Graph the inequality in a coordinate plane.

5. $y > -1$

6. $x \leq -4$

7. $x + y \leq -4$

8. $x - 2y < 0$

Extra Example 2

Graph $y \geq -3$ in a coordinate plane.

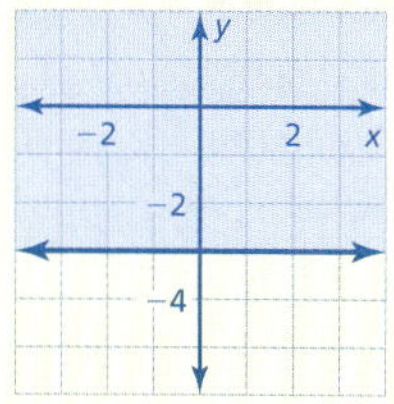

Extra Example 3

Graph $x + y < 2$ in a coordinate plane.

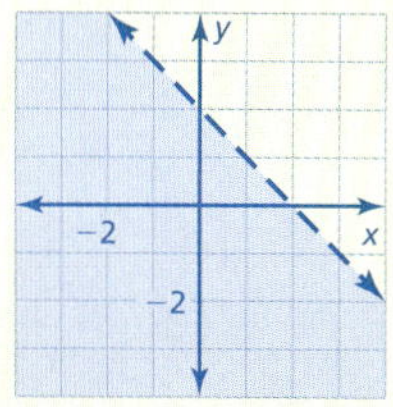

MONITORING PROGRESS ANSWERS

5.

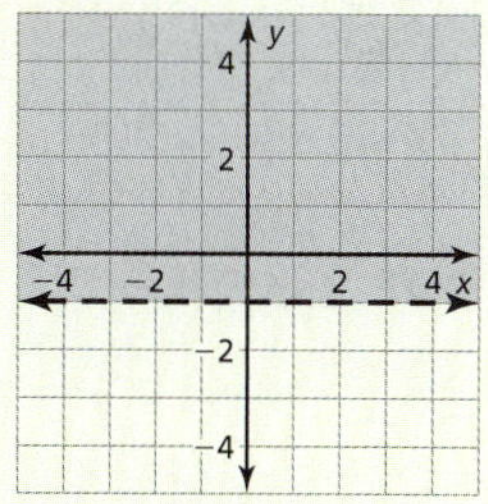

6.

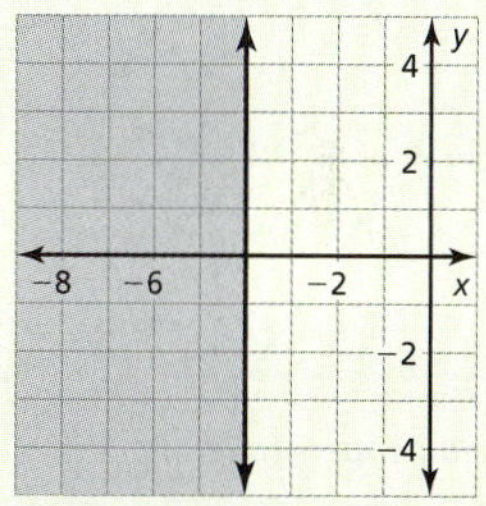

7.

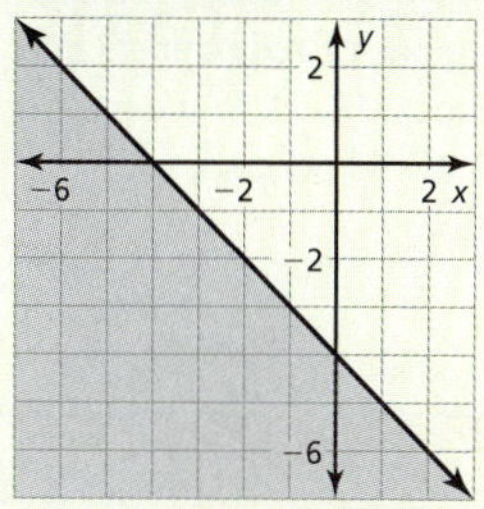

8.

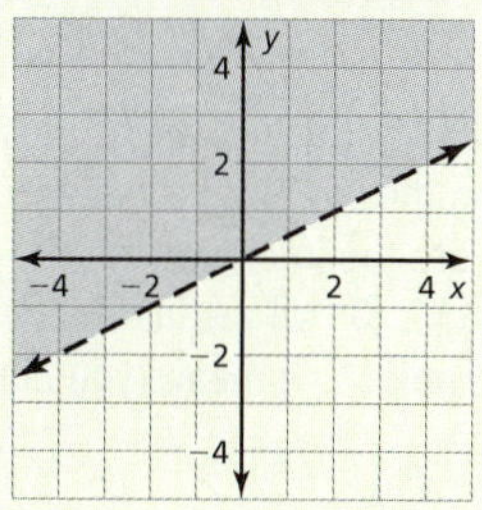

Laurie's Notes — Teacher Actions

- **?** Write the *Core Concept* and ask, "What test point would simplify the computation?" Listen for at least one student to suggest (0, 0).
- **Make Sense of Problems and Persevere in Solving Them:** After students complete Example 2, they should verbalize that the points in the solution set have *y*-coordinates less than or equal to 2.
- **? Attend to Precision:** Work through Example 3 as shown. "How would the process change if the inequality were $-x - 2y > 2$?" When solving for *y*, it would be necessary to change the direction of the inequality symbol.

Extra Example 4

You can spend at most \$9 on potatoes and carrots for stew. Potatoes cost \$3 per pound, and carrots cost \$1.50 per pound. Write and graph an inequality that represents the amounts of potatoes and carrots you can buy. Identify and interpret two solutions of the inequality.

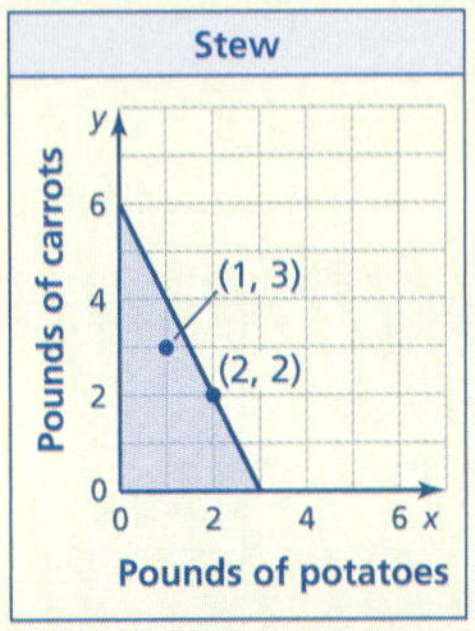

$3x + 1.50y \le 9$; *Sample answer:* (1, 3) because it lies in the shaded half-plane and (2, 2) because it lies on the solid line. So, you can buy 1 pound of potatoes and 3 pounds of carrots, or 2 pounds of potatoes and 2 pounds of carrots.

MONITORING PROGRESS ANSWER

9. $y \le -\frac{4}{3}x + 4$

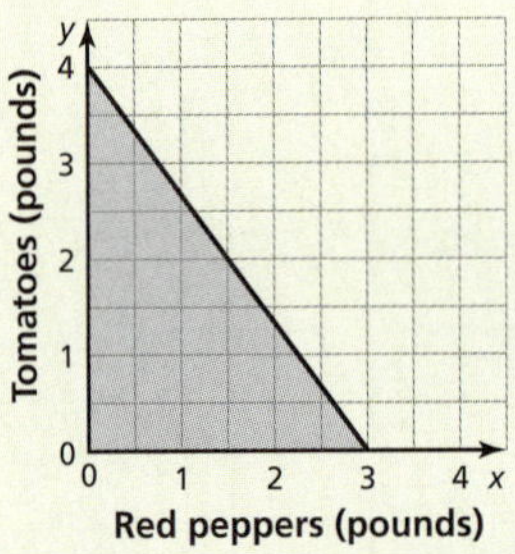

Sample answer: (3, 0), You can buy 3 pounds of red peppers and no tomatoes; (1, 1), You can buy 1 pound of red peppers and 1 pound of tomatoes.

Solving Real-Life Problems

EXAMPLE 4 **Modeling with Mathematics**

You can spend at most \$10 on grapes and apples for a fruit salad. Grapes cost \$2.50 per pound, and apples cost \$1 per pound. Write and graph an inequality that represents the amounts of grapes and apples you can buy. Identify and interpret two solutions of the inequality.

SOLUTION

1. **Understand the Problem** You know the most that you can spend and the prices per pound for grapes and apples. You are asked to write and graph an inequality and then identify and interpret two solutions.
2. **Make a Plan** Use a verbal model to write an inequality that represents the problem. Then graph the inequality. Use the graph to identify two solutions. Then interpret the solutions.
3. **Solve the Problem**

Words	Cost per pound of grapes	•	Pounds of grapes	+	Cost per pound of apples	•	Pounds of apples	≤	Amount you can spend

Variables Let x be pounds of grapes and y be pounds of apples.

Inequality $2.50 \cdot x + 1 \cdot y \le 10$

Step 1 Graph $2.5x + y = 10$, or $y = -2.5x + 10$. Use a solid line because the inequality symbol is ≤. Restrict the graph to positive values of x and y because negative values do not make sense in this real-life context.

Step 2 Test (0, 0).

$2.5x + y \le 10$ Write the inequality.

$2.5(0) + 0 \overset{?}{\le} 10$ Substitute.

$0 \le 10$ ✓ Simplify.

Step 3 Because (0, 0) is a solution, shade the half-plane that contains (0, 0).

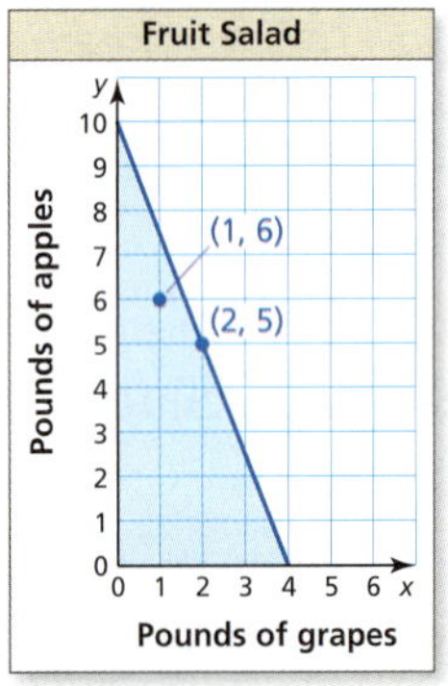

Check

$2.5x + y \le 10$

$2.5(1) + 6 \overset{?}{\le} 10$

$8.5 \le 10$ ✓

$2.5x + y \le 10$

$2.5(2) + 5 \overset{?}{\le} 10$

$10 \le 10$ ✓

▶ One possible solution is (1, 6) because it lies in the shaded half-plane. Another possible solution is (2, 5) because it lies on the solid line. So, you can buy 1 pound of grapes and 6 pounds of apples, or 2 pounds of grapes and 5 pounds of apples.

4. **Look Back** Check your solutions by substituting them into the original inequality, as shown.

Monitoring Progress

Help in English and Spanish at *BigIdeasMath.com*

9. You can spend at most \$12 on red peppers and tomatoes for salsa. Red peppers cost \$4 per pound, and tomatoes cost \$3 per pound. Write and graph an inequality that represents the amounts of red peppers and tomatoes you can buy. Identify and interpret two solutions of the inequality.

Laurie's Notes Teacher Actions

- Help students write the verbal model for this example. Students often think, "Cost of grapes plus cost of apples cannot exceed \$10." Defining the variables is a step students often omit.

? "What are the intercepts for the boundary line? Interpret their meaning." (0, 10) means you bought 10 pounds of apples and no grapes, and (4, 0) means you bought 4 pounds of grapes and no apples.

Closure

- **Phone Message:** Write a brief phone message that you would leave for a friend who missed today's class. Explain how to graph a linear inequality in two variables. *Sample answer:* First, graph the boundary line. If the inequality symbol is ≤ or ≥, then use a solid line. Otherwise, use a dashed line. Test a point on one side of the line by substituting the x-value and y-value into the inequality. If the point makes the inequality true, then shade the half of the plane that contains the point. Otherwise, shade the half of the plane on the other side of the line.

5.6 Exercises

Dynamic Solutions available at *BigIdeasMath.com*

Vocabulary and Core Concept Check

1. **VOCABULARY** How can you tell whether an ordered pair is a solution of a linear inequality?

2. **WRITING** Compare the graph of a linear inequality in two variables with the graph of a linear equation in two variables.

Monitoring Progress and Modeling with Mathematics

In Exercises 3–10, tell whether the ordered pair is a solution of the inequality. *(See Example 1.)*

3. $x + y < 7$; $(2, 3)$
4. $x - y \le 0$; $(5, 2)$
5. $x + 3y \ge -2$; $(-9, 2)$
6. $8x + y > -6$; $(-1, 2)$
7. $-6x + 4y \le 6$; $(-3, -3)$
8. $3x - 5y \ge 2$; $(-1, -1)$
9. $-x - 6y > 12$; $(-8, 2)$
10. $-4x - 8y < 15$; $(-6, 3)$

In Exercises 11–16, tell whether the ordered pair is a solution of the inequality whose graph is shown.

11. $(0, -1)$
12. $(-1, 3)$
13. $(1, 4)$
14. $(0, 0)$
15. $(3, 3)$
16. $(2, 1)$

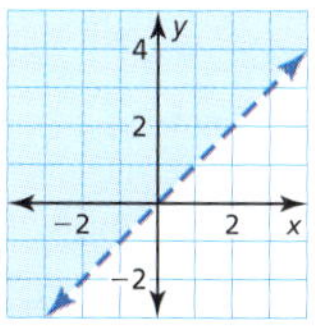

17. **MODELING WITH MATHEMATICS** A carpenter has at most \$250 to spend on lumber. The inequality $8x + 12y \le 250$ represents the numbers x of 2-by-8 boards and the numbers y of 4-by-4 boards the carpenter can buy. Can the carpenter buy twelve 2-by-8 boards and fourteen 4-by-4 boards? Explain.

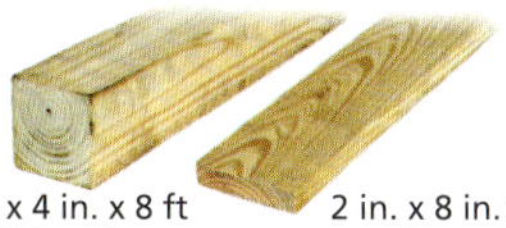

18. **MODELING WITH MATHEMATICS** The inequality $3x + 2y \ge 93$ represents the numbers x of multiple-choice questions and the numbers y of matching questions you can answer correctly to receive an A on a test. You answer 20 multiple-choice questions and 18 matching questions correctly. Do you receive an A on the test? Explain.

In Exercises 19–24, graph the inequality in a coordinate plane. *(See Example 2.)*

19. $y \le 5$
20. $y > 6$
21. $x < 2$
22. $x \ge -3$
23. $y > -7$
24. $x < 9$

In Exercises 25–30, graph the inequality in a coordinate plane. *(See Example 3.)*

25. $y > -2x - 4$
26. $y \le 3x - 1$
27. $-4x + y < -7$
28. $3x - y \ge 5$
29. $5x - 2y \le 6$
30. $-x + 4y > -12$

ERROR ANALYSIS **In Exercises 31 and 32, describe and correct the error in graphing the inequality.**

31. $y < -x + 1$

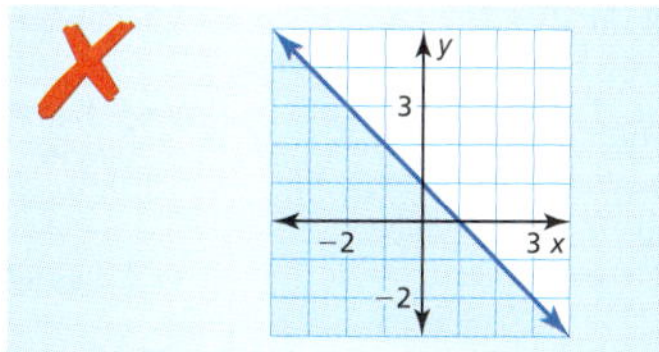

32. $y \le 3x - 2$

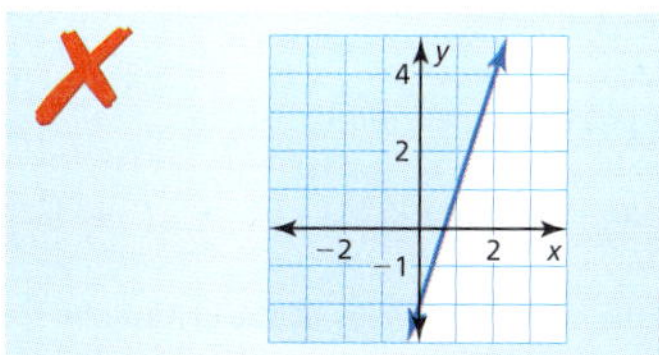

Assignment Guide and Homework Check

ASSIGNMENT

Basic: 1, 2, 3–33 odd, 40, 46–48

Average: 1, 2–36 even, 39, 40, 46–48

Advanced: 1, 2, 8, 10, 16–18, 24, 30–38 even, 39–48

HOMEWORK CHECK

Basic: 3, 19, 25, 33

Average: 4, 20, 26, 34

Advanced: 8, 24, 30, 34

ANSWERS

1. Substitute the values into the inequality and verify that the statement is true.
2. The graph of a linear equation in two variables is a line. The graph of a linear inequality in two variables has a line as a boundary and a shaded half-plane.
3. yes
4. no
5. no
6. no
7. yes
8. yes
9. no
10. yes
11. no
12. yes
13. yes
14. no
15. no
16. no
17. no; (12, 14) is not a solution of the inequality.
18. yes; (20, 18) is a solution of the inequality.
19.

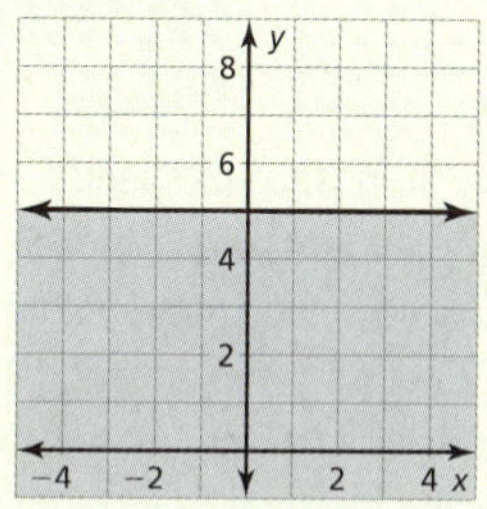

20.

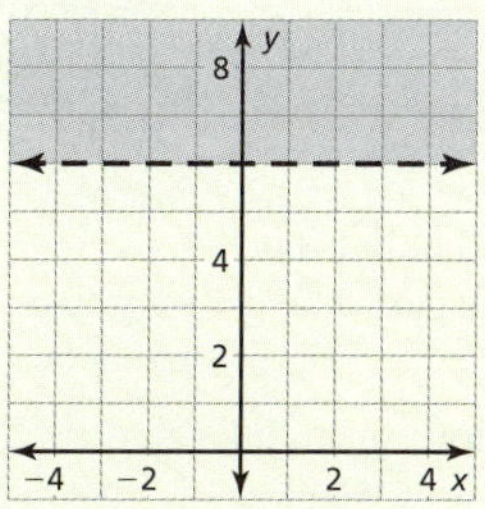

21.

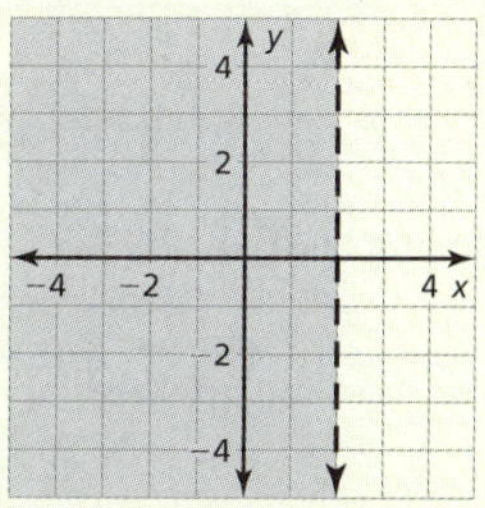

22–32. See Additional Answers.

Dynamic Teaching Tools

- Dynamic Assessment & Progress Monitoring Tool
- Interactive Whiteboard Lesson Library
- Dynamic Classroom with Dynamic Investigations

ANSWERS

33–34. See Additional Answers.

35. $y > 2x + 1$

36. $y > \frac{1}{2}x + 2$

37. $y \leq -\frac{1}{2}x - 2$

38. $y \leq 3x - 3$

39–45. See Additional Answers.

46. 40, 48, 56

47. −20, −23, −26

48. $\frac{7}{2}, \frac{9}{2}, \frac{11}{2}$

Mini-Assessment

1. Tell whether the ordered pair is a solution of the inequality.
   $4x + y > 3$; (2, 4)
   (2, 4) is a solution.
2. Graph $y < -1$ in a coordinate plane.
   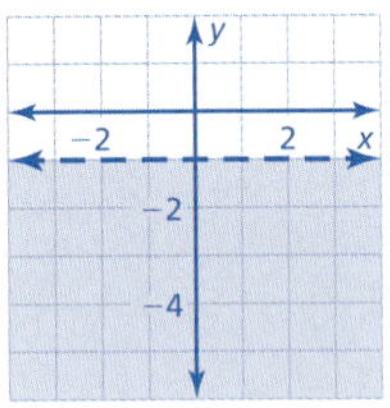
3. Graph $2x - y \geq -4$ in a coordinate plane.
   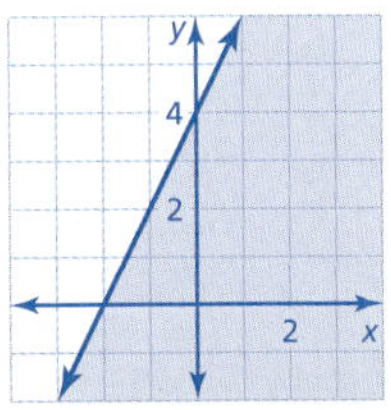
4. You have at most \$5 to spend on notebooks and markers. Markers cost \$1.00 each, and notebooks cost \$1.25 each. Write an inequality that represents this situation. Identify and interpret two solutions of the inequality.
   $x + 1.25y \leq 5$; *Sample answer:* (1, 3) and (5, 0). You can buy 1 marker and 3 notebooks, or 5 markers and 0 notebooks.

33. MODELING WITH MATHEMATICS You have at most \$20 to spend at an arcade. Arcade games cost \$0.75 each, and snacks cost \$2.25 each. Write and graph an inequality that represents the numbers of games you can play and snacks you can buy. Identify and interpret two solutions of the inequality. *(See Example 4.)*

34. MODELING WITH MATHEMATICS A drama club must sell at least \$1500 worth of tickets to cover the expenses of producing a play. Write and graph an inequality that represents how many adult and student tickets the club must sell. Identify and interpret two solutions of the inequality.

In Exercises 35–38, write an inequality that represents the graph.

35.
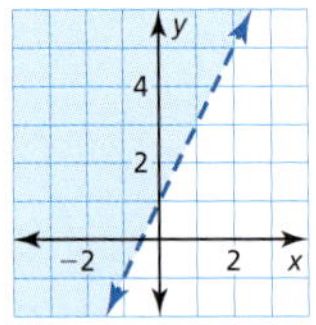

36.
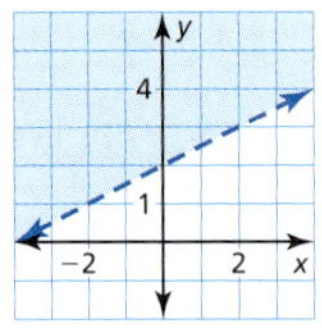

37.
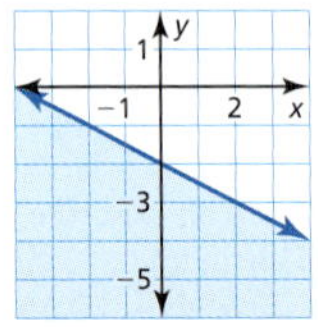

38.
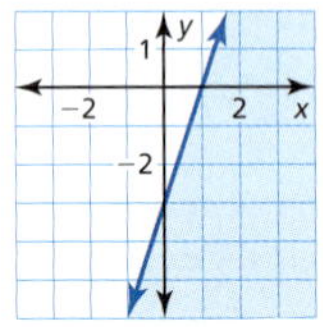

39. PROBLEM SOLVING Large boxes weigh 75 pounds, and small boxes weigh 40 pounds.

a. Write and graph an inequality that represents the numbers of large and small boxes a 200-pound delivery person can take on the elevator.

b. Explain why some solutions of the inequality might not be practical in real life.

40. HOW DO YOU SEE IT? Match each inequality with its graph.

a. $3x - 2y \leq 6$ b. $3x - 2y < 6$

c. $3x - 2y > 6$ d. $3x - 2y \geq 6$

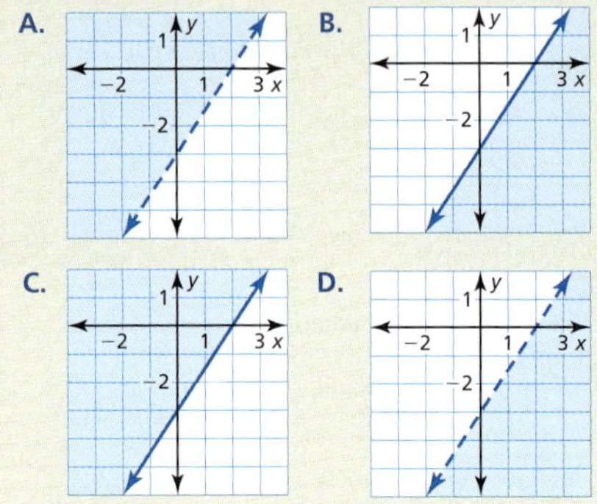

41. REASONING When graphing a linear inequality in two variables, why must you choose a test point that is *not* on the boundary line?

42. THOUGHT PROVOKING Write a linear inequality in two variables that has the following two properties.

- (0, 0), (0, −1), and (0, 1) are not solutions.
- (1, 1), (3, −1), and (−1, 3) are solutions.

43. WRITING Can you always use (0, 0) as a test point when graphing an inequality? Explain.

CRITICAL THINKING In Exercises 44 and 45, write and graph an inequality whose graph is described by the given information.

44. The points (2, 5) and (−3, −5) lie on the boundary line. The points (6, 5) and (−2, −3) are solutions of the inequality.

45. The points (−7, −16) and (1, 8) lie on the boundary line. The points (−7, 0) and (3, 14) are *not* solutions of the inequality.

Maintaining Mathematical Proficiency

Reviewing what you learned in previous grades and lessons

Write the next three terms of the arithmetic sequence. *(Section 4.6)*

46. 0, 8, 16, 24, 32, . . . **47.** −5, −8, −11, −14, −17, . . . **48.** $-\frac{3}{2}, -\frac{1}{2}, \frac{1}{2}, \frac{3}{2}, \frac{5}{2}, \ldots$

If students need help...	*If students got it...*
Resources by Chapter • Practice A and Practice B • Puzzle Time	Resources by Chapter • Enrichment and Extension • Cumulative Review
Student Journal • Practice	Start the *next* Section
Differentiating the Lesson Skills Review Handbook	

Laurie's Notes

Overview of Section 5.7

Introduction

- Students have graphed linear systems and linear inequalities in prior lessons. These two skills are combined in this lesson as students graph a system of linear inequalities.

Resources

- Colored pencils are very helpful as students are graphing systems of linear inequalities.

Teaching Strategy

- When graphing more than three inequalities, and sometimes when you are only graphing two inequalities, the shading can make it difficult for some students to see the solution. It becomes confusing for them, and for some students the graph can get pretty messy.
- Consider the graph of this system of inequalities:

  $y \le -2x + 8$

  $y \le 1.5x - 1$

  $y \ge -1$

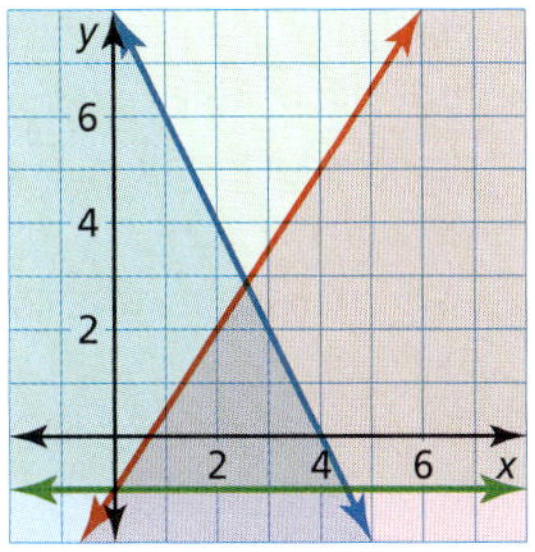

Figure 1

- The solution is the small triangular region where all of the solutions overlap. (See Figure 1.)
- Another strategy besides doing the shading is to put small arrows at either end of the line to indicate the direction in which the shading should be done. In other words, the arrows point to the half-plane where the solutions of the inequality are located. (See Figure 2.)
- When all three inequalities have been graphed, it is not difficult for students to look at the graph and say, "The solutions of the system are all the points below the first line, below the second line, and above the third line." They can then shade in the triangular region representing the solution. (See Figure 3.)

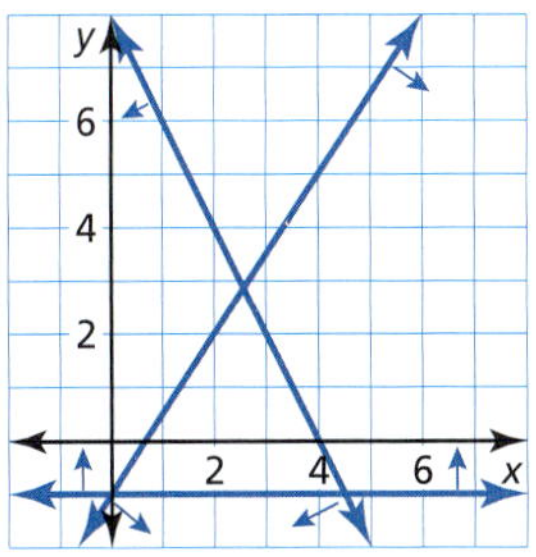

Figure 2

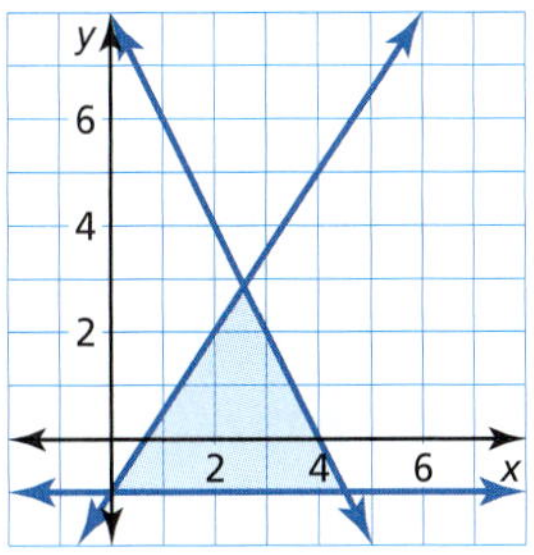

Figure 3

Applications

- Linear programming is an area of mathematics in which more complex systems of linear inequalities are studied. Real-life applications often have many more than two or three constraints. Linear programming is a topic students may indeed encounter in a future mathematics course.

Pacing Suggestion

- The explorations provide an independent introduction to graphing linear inequalities. When students have finished, transition to the formal lesson.

Dynamic Teaching Tools

- Dynamic Assessment & Progress Monitoring Tool
- Lesson Planning Tool
- Interactive Whiteboard Lesson Library
- Dynamic Classroom with Dynamic Investigations

What Your Students Will Learn

- Check solutions of systems of linear inequalities in two variables algebraically.
- Graph systems of linear inequalities in two variables in the coordinate plane.
- Write systems of linear inequalities in two variables represented by graphs.

Laurie's Notes

Exploration

Motivate

- In a coordinate plane, sketch a square region bounded by $x = -2$, $x = 2$, $y = -2$, and $y = 2$. Shade the interior.
- Ask students to write the equations for the lines forming the sides of the square.
- ? "The equations represent the sides of the square. How might you represent the interior of the square?" Listen for ideas about writing inequalities. "Name ordered pairs in the interior and in the exterior of the square." Answers will vary.
- Explain to students that in a previous lesson they graphed linear inequalities. In this lesson, they will graph two or more linear inequalities in the same coordinate plane.

Exploration 1

- Graphing a system of linear inequalities combines the skills of graphing linear inequalities and graphing linear systems.
- It should not take students long to match each linear inequality with its graph. Some may simply do a test point such as (0, 0). Others may solve the inequality for y and consider the slope of the line.
- ? "Which graph matches Inequality 1, and how did you decide?" graph A; Explanations will vary.
- ? "How do the graphs of $2x + y \leq 4$ and $2x + y < 4$ differ?" The boundary line is solid for $2x + y \leq 4$ and dashed for $2x + y < 4$.
- ? "What does the shaded portion of a linear inequality represent?" ordered pairs that are solutions of the inequality

Exploration 2

- Students use one color to graph the solution of the first inequality and a second color to graph the solution of the second inequality.
- **Teaching Tip:** If you do not have sufficient colored pencils, use a series of lines that are perpendicular to the boundary line. Where the two solutions overlap, the series of lines will have different slopes and hence intersect.
- Listen as students describe each of the four regions in the graph and interpret what each region means.

Communicate Your Answer

- **Look For and Make Use of Structure:** Students discern the relationships between the systems of linear inequalities they worked with and their graphs to develop a general strategy for graphing a system of linear inequalities.
- **Construct Viable Arguments and Critique the Reasoning of Others** and **Turn and Talk:** Have students discuss their thoughts about Question 5 with one another.

Connecting to Next Step

- The explorations help students make sense of what it means to solve a system of linear inequalities. More complex systems of linear inequalities will be found in the lesson.

5.7 Systems of Linear Inequalities

Essential Question How can you graph a system of linear inequalities?

EXPLORATION 1 **Graphing Linear Inequalities**

Work with a partner. Match each linear inequality with its graph. Explain your reasoning.

$2x + y \leq 4$ Inequality 1

$2x - y \leq 0$ Inequality 2

A.

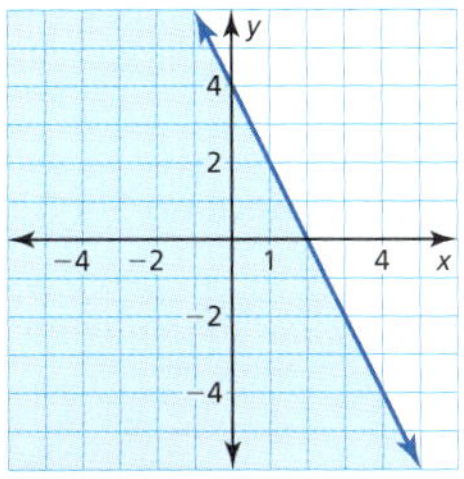

B.

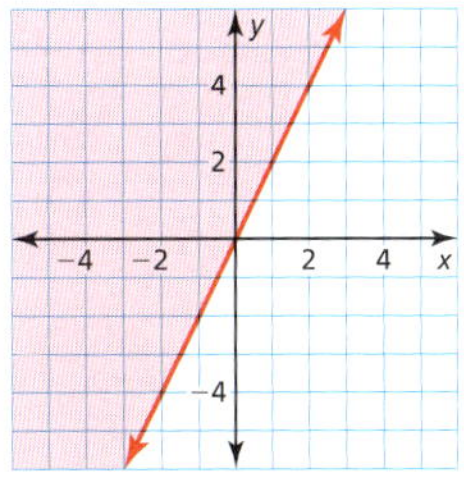

EXPLORATION 2 **Graphing a System of Linear Inequalities**

Work with a partner. Consider the linear inequalities given in Exploration 1.

$2x + y \leq 4$ Inequality 1

$2x - y \leq 0$ Inequality 2

a. Use two different colors to graph the inequalities in the same coordinate plane. What is the result?

b. Describe each of the shaded regions of the graph. What does the unshaded region represent?

MAKING SENSE OF PROBLEMS

To be proficient in math, you need to explain to yourself the meaning of a problem.

Communicate Your Answer

3. How can you graph a system of linear inequalities?

4. When graphing a system of linear inequalities, which region represents the solution of the system?

5. Do you think all systems of linear inequalities have a solution? Explain your reasoning.

6. Write a system of linear inequalities represented by the graph.

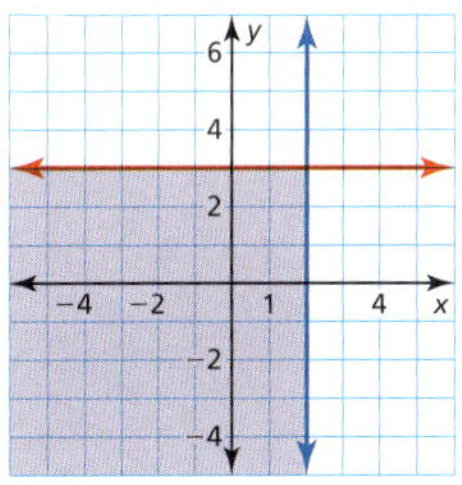

Dynamic Teaching Tools

- Dynamic Assessment & Progress Monitoring Tool
- Lesson Planning Tool
- Interactive Whiteboard Lesson Library
- Dynamic Classroom with Dynamic Investigations

ANSWERS

1. Inequality 1: A; The graph has a boundary line of $y = -2x + 4$; Inequality 2: B; The graph has a boundary line of $y = 2x$.

2. a.

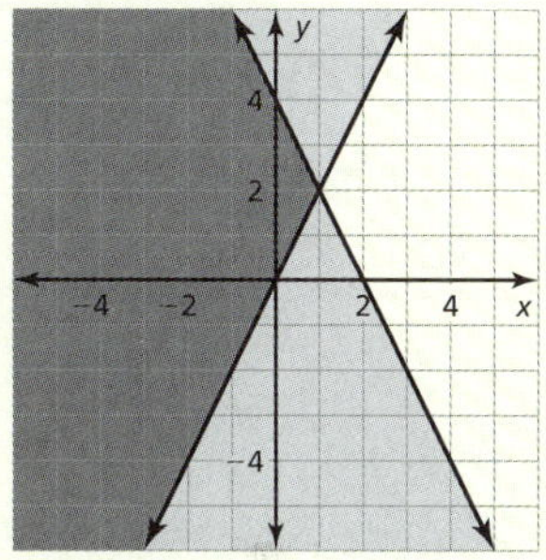

Parts of the two half-planes overlap.

b. One shaded region is only shaded the first color, one shaded region is only shaded the second color, and one shaded region is shaded both colors; the values where both inequalities are false

3. Graph each inequality in the same coordinate plane. Find the intersection of the half-planes that are solutions of the inequalities. This intersection is the graph of the system.

4. the region where the shaded half-planes of the inequalities overlap

5. no; When the boundary lines are parallel, it is possible the shaded regions will not overlap.

6. $x \leq 2$ and $y \leq 3$

Differentiated Instruction

Kinesthetic

Some students may benefit by using graphing calculators or a graphing software program to check solutions of systems of inequalities. Students can enter the inequalities, view their graphs, and determine whether or not the given points are in the intersection of the half-planes.

Extra Example 1

Tell whether each ordered pair is a solution of the system of linear inequalities.

$y > 3x$

$y \le -x - 2$

a. $(-2, -1)$

$(-2, -1)$ is a solution of the system.

b. $(0, 4)$

$(0, 4)$ is *not* a solution of the system.

MONITORING PROGRESS ANSWERS

1. no

2. yes

5.7 Lesson

Core Vocabulary

system of linear inequalities, *p. 274*
solution of a system of linear inequalities, *p. 274*
graph of a system of linear inequalities, *p. 275*

Previous
linear inequality in two variables

What You Will Learn

- Check solutions of systems of linear inequalities.
- Graph systems of linear inequalities.
- Write systems of linear inequalities.
- Use systems of linear inequalities to solve real-life problems.

Systems of Linear Inequalities

A **system of linear inequalities** is a set of two or more linear inequalities in the same variables. An example is shown below.

$y < x + 2$ Inequality 1

$y \ge 2x - 1$ Inequality 2

A **solution of a system of linear inequalities** in two variables is an ordered pair that is a solution of each inequality in the system.

EXAMPLE 1 Checking Solutions

Tell whether each ordered pair is a solution of the system of linear inequalities.

$y < 2x$ Inequality 1

$y \ge x + 1$ Inequality 2

a. $(3, 5)$ **b.** $(-2, 0)$

SOLUTION

a. Substitute 3 for x and 5 for y in each inequality.

Inequality 1	Inequality 2
$y < 2x$	$y \ge x + 1$
$5 \overset{?}{<} 2(3)$	$5 \overset{?}{\ge} 3 + 1$
$5 < 6$ ✓	$5 \ge 4$ ✓

▶ Because the ordered pair $(3, 5)$ is a solution of each inequality, it is a solution of the system.

b. Substitute -2 for x and 0 for y in each inequality.

Inequality 1	Inequality 2
$y < 2x$	$y \ge x + 1$
$0 \overset{?}{<} 2(-2)$	$0 \overset{?}{\ge} -2 + 1$
$0 \nless -4$ ✗	$0 \ge -1$ ✓

▶ Because $(-2, 0)$ is not a solution of each inequality, it is *not* a solution of the system.

Monitoring Progress

Help in English and Spanish at *BigIdeasMath.com*

Tell whether the ordered pair is a solution of the system of linear inequalities.

1. $(-1, 5)$; $y < 5$, $y > x - 4$

2. $(1, 4)$; $y \ge 3x + 1$, $y > x - 1$

Laurie's Notes Teacher Actions

- Students worked with a system of linear inequalities yesterday without the formal definition. Write and discuss the two definitions. Note that the definition alludes to the possibility of a system consisting of more than two inequalities.
- Remind students that $4 \le 4$ is a true statement, but $4 < 4$ is not a true statement.
- **Think-Pair-Share:** Have students answer Questions 1 and 2, and then share and discuss as a class.

Graphing Systems of Linear Inequalities

The **graph of a system of linear inequalities** is the graph of all the solutions of the system.

Core Concept

Graphing a System of Linear Inequalities

Step 1 Graph each inequality in the same coordinate plane.

Step 2 Find the intersection of the half-planes that are solutions of the inequalities. This intersection is the graph of the system.

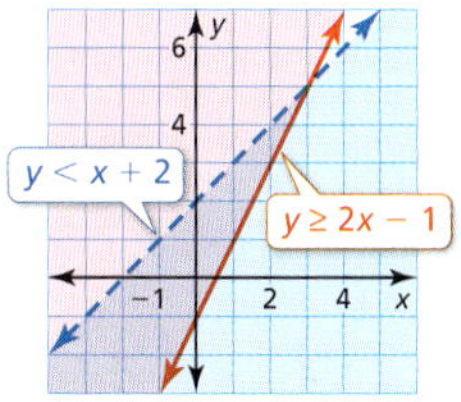

EXAMPLE 2 **Graphing a System of Linear Inequalities**

Graph the system of linear inequalities.

$y \le 3$ Inequality 1

$y > x + 2$ Inequality 2

SOLUTION

Step 1 Graph each inequality.

Step 2 Find the intersection of the half-planes. One solution is $(-3, 1)$.

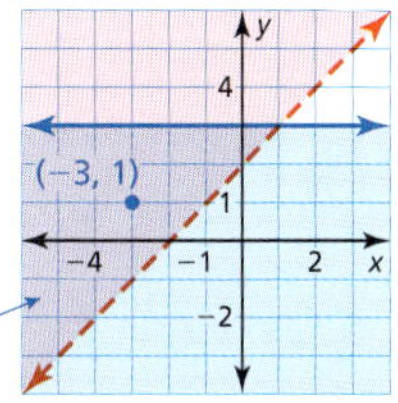

Check

Verify that $(-3, 1)$ is a solution of each inequality.

Inequality 1

$y \le 3$

$1 \le 3$ ✓

Inequality 2

$y > x + 2$

$1 \overset{?}{>} -3 + 2$

$1 > -1$ ✓

EXAMPLE 3 **Graphing a System of Linear Inequalities: No Solution**

Graph the system of linear inequalities.

$2x + y < -1$ Inequality 1

$2x + y > 3$ Inequality 2

SOLUTION

Step 1 Graph each inequality.

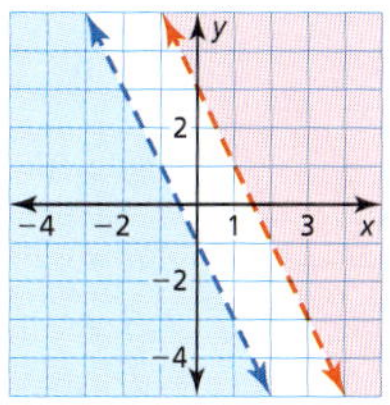

Step 2 Find the intersection of the half-planes. Notice that the lines are parallel, and the half-planes do not intersect.

▶ So, the system has no solution.

Monitoring Progress

Help in English and Spanish at *BigIdeasMath.com*

Graph the system of linear inequalities.

3. $y \ge -x + 4$
$x + y \le 0$

4. $y > 2x - 3$
$y \ge \frac{1}{2}x + 1$

5. $-2x + y < 4$
$2x + y > 4$

English Language Learners

Pair Activity

Pair each English learner with an English speaker. Have one student graph the first inequality of the system and the other student graph the second inequality on separate coordinate planes. Students should check each other's graphs and then work together to copy the graphs onto one coordinate plane and identify the solution of the system.

Extra Example 2

Graph the system of linear inequalities.

$y \ge -1$

$y < -x + 2$

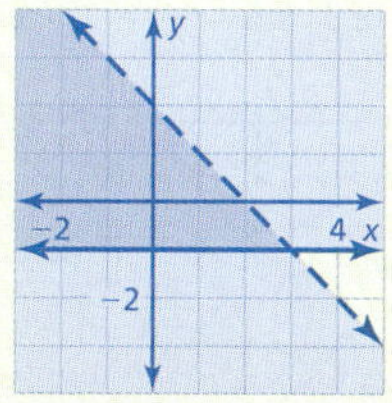

Extra Example 3

Graph the system of linear inequalities.

$x - y > 2$

$x - y \le -3$

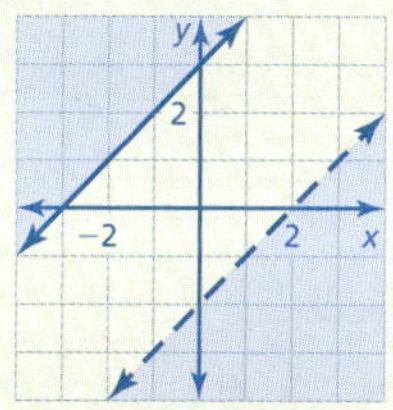

The system has no solution.

MONITORING PROGRESS ANSWERS

3.

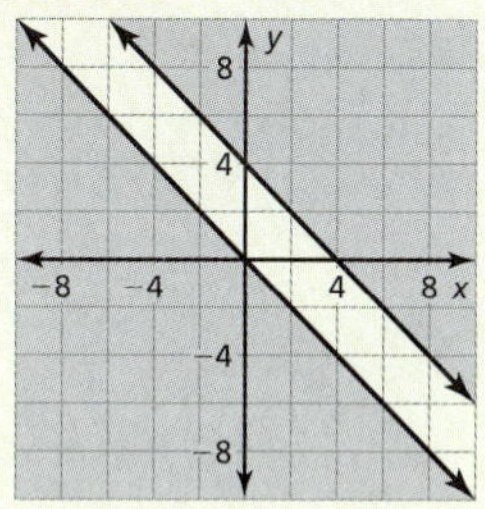

4–5. See Additional Answers.

Laurie's Notes — Teacher Actions

- Write the *Core Concept* and discuss. Students will find two colored pencils helpful.
- Work through Example 2 and share that you can visually select a solution point from where the shading overlaps.
- **Look For and Make Use of Structure:** In Example 3, the inequalities are written in standard form. Students may use intercepts to graph, or they may rewrite in slope-intercept form.
- **Look For and Express Regularity in Repeated Reasoning** and **Turn and Talk:** "How do you know which side of each boundary line to shade?"

Writing Systems of Linear Inequalities

EXAMPLE 4 Writing a System of Linear Inequalities

Write a system of linear inequalities represented by the graph.

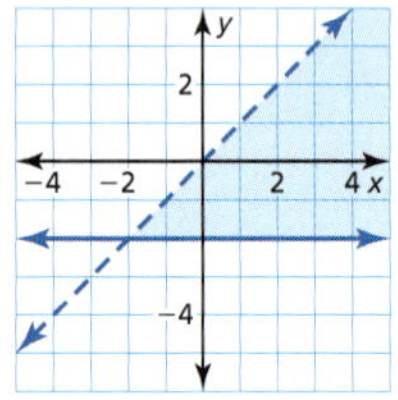

SOLUTION

Inequality 1 The horizontal boundary line passes through $(0, -2)$. So, an equation of the line is $y = -2$. The shaded region is *above* the *solid* boundary line, so the inequality is $y \geq -2$.

Inequality 2 The slope of the other boundary line is 1, and the y-intercept is 0. So, an equation of the line is $y = x$. The shaded region is *below* the *dashed* boundary line, so the inequality is $y < x$.

▶ The system of linear inequalities represented by the graph is

$y \geq -2$ Inequality 1

$y < x.$ Inequality 2

EXAMPLE 5 Writing a System of Linear Inequalities

Write a system of linear inequalities represented by the graph.

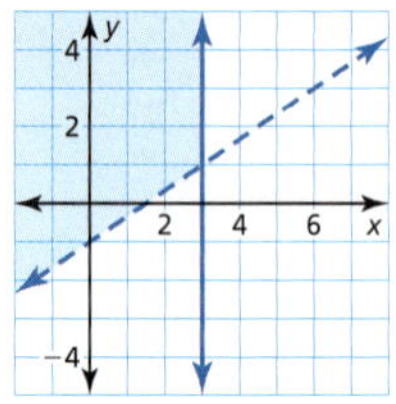

SOLUTION

Inequality 1 The vertical boundary line passes through $(3, 0)$. So, an equation of the line is $x = 3$. The shaded region is to the *left* of the *solid* boundary line, so the inequality is $x \leq 3$.

Inequality 2 The slope of the other boundary line is $\frac{2}{3}$, and the y-intercept is -1. So, an equation of the line is $y = \frac{2}{3}x - 1$. The shaded region is *above* the *dashed* boundary line, so the inequality is $y > \frac{2}{3}x - 1$.

▶ The system of linear inequalities represented by the graph is

$x \leq 3$ Inequality 1

$y > \frac{2}{3}x - 1.$ Inequality 2

Monitoring Progress

Help in English and Spanish at *BigIdeasMath.com*

Write a system of linear inequalities represented by the graph.

6.

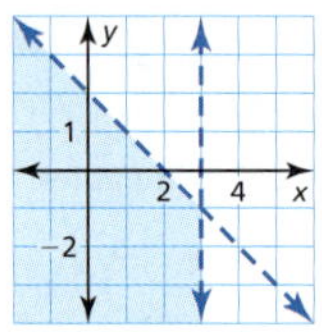

7.

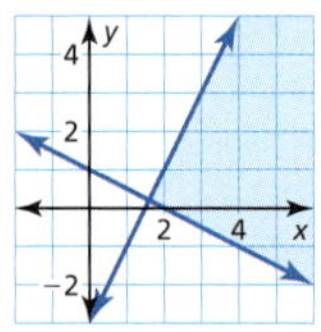

Extra Example 4

Write a system of linear inequalities represented by the graph.

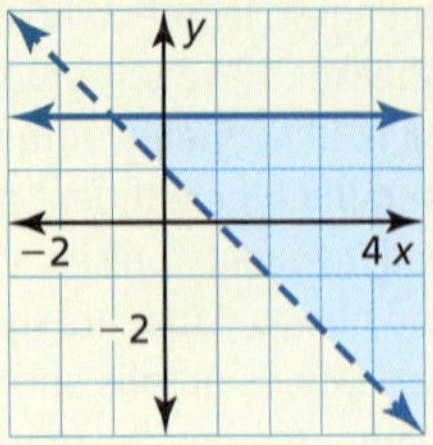

$y \leq 2$
$y > -x + 1$

Extra Example 5

Write a system of linear inequalities represented by the graph.

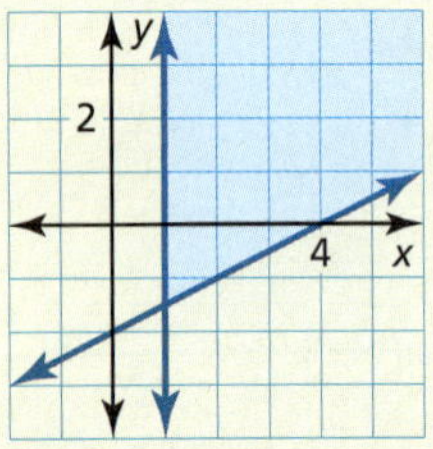

$x \geq 1$
$y \geq \frac{1}{2}x - 2$

MONITORING PROGRESS ANSWERS

6. $y < -x + 2$ and $x < 3$

7. $y \geq -\frac{1}{2}x + 1$ and $y \leq 2x - 3$

Laurie's Notes Teacher Actions

- **Think-Alouds** and **Popsicle Sticks:** Pose Example 4 and say, "To write this system I need to …." Ask partner A to think aloud for partner B to hear the problem-solving process. When students have finished writing the system, *Popsicle Stick* responses.
- **Think-Alouds** and **Popsicle Sticks:** Pose Example 5 and say, "To write this system I need to …." Ask partner B to think aloud for partner A to hear the problem-solving process. When students have finished writing the system, *Popsicle Stick* responses.

Solving Real-Life Problems

EXAMPLE 6 **Modeling with Mathematics**

You have at most 8 hours to spend at the mall and at the beach. You want to spend at least 2 hours at the mall and more than 4 hours at the beach. Write and graph a system that represents the situation. How much time can you spend at each location?

SOLUTION

1. **Understand the Problem** You know the total amount of time you can spend at the mall and at the beach. You also know how much time you want to spend at each location. You are asked to write and graph a system that represents the situation and determine how much time you can spend at each location.
2. **Make a Plan** Use the given information to write a system of linear inequalities. Then graph the system and identify an ordered pair in the solution region.
3. **Solve the Problem** Let x be the number of hours at the mall and let y be the number of hours at the beach.

$x + y \leq 8$ at most 8 hours at the mall and at the beach

$x \geq 2$ at least 2 hours at the mall

$y > 4$ more than 4 hours at the beach

Graph the system.

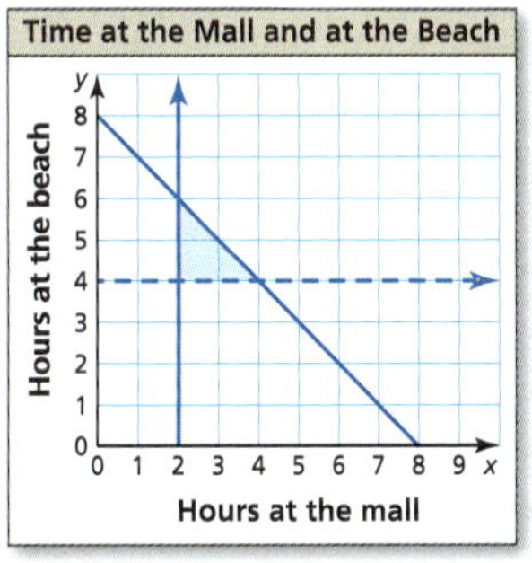

One ordered pair in the solution region is (2.5, 5).

▶ So, you can spend 2.5 hours at the mall and 5 hours at the beach.

4. **Look Back** Check your solution by substituting it into the inequalities in the system, as shown.

Check

$x + y \leq 8$

$2.5 + 5 \overset{?}{\leq} 8$

$7.5 \leq 8$ ✓

$x \geq 2$

$2.5 \geq 2$ ✓

$y > 4$

$5 > 4$ ✓

Monitoring Progress

Help in English and Spanish at *BigIdeasMath.com*

8. Name another solution of Example 6.
9. **WHAT IF?** You want to spend at least 3 hours at the mall. How does this change the system? Is (2.5, 5) still a solution? Explain.

Extra Example 6

You have at most 7 hours to spend playing soccer and swimming. You want to spend at least 2 hours playing soccer and more than 2 hours swimming. Write and graph a system that represents the situation. How much time can you spend on each activity? Let x be the number of hours playing soccer and y be the number of hours swimming.

$x + y \leq 7$

$x \geq 2$

$y > 2$

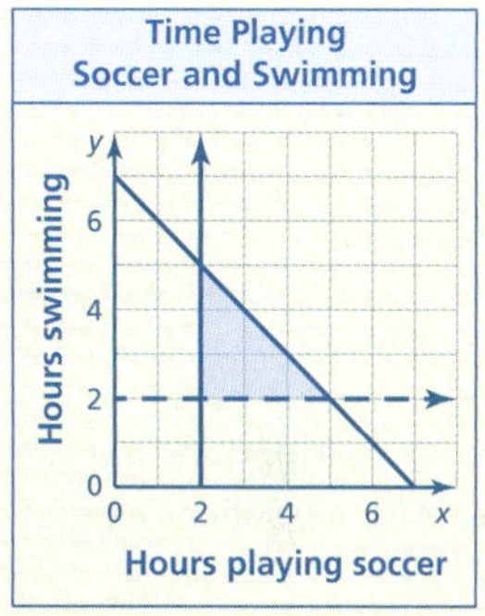

Sample answer: (3, 3); You can spend 3 hours playing soccer and 3 hours swimming.

MONITORING PROGRESS ANSWERS

8. *Sample answer:* 3 hours at the mall, 4.5 hours at the beach
9. $x \geq 2$ becomes $x \geq 3$; no; 2.5 is not greater than 3.

Laurie's Notes Teacher Actions

- Ask students what the word *constraint* means. Explain that in this example there are constraints on the variables. Each constraint can be represented by an inequality.
- **Attend to Precision:** Have students define the variables. Let "x = the number of hours at the mall" and "y = the number of hours at the beach."

Closure

- Name an ordered pair that is a solution of the system of linear inequalities. *Sample answer:* (0, 0)
- Name an ordered pair that is not a solution of the system of linear inequalities. *Sample answer:* (3, 3)

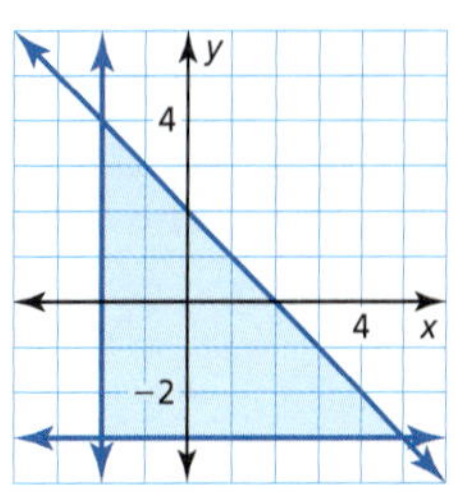

5.7 Exercises

Dynamic Solutions available at *BigIdeasMath.com*

Vocabulary and Core Concept Check

1. **VOCABULARY** How can you verify that an ordered pair is a solution of a system of linear inequalities?

2. **WHICH ONE DOESN'T BELONG?** Use the graph shown. Which of the ordered pairs does *not* belong with the other three? Explain your reasoning.

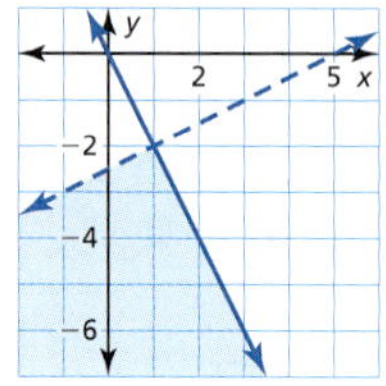

(1, −2) (0, −4) (−1, −6) (2, −4)

Monitoring Progress and Modeling with Mathematics

In Exercises 3–6, tell whether the ordered pair is a solution of the system of linear inequalities.

3. (−4, 3)
4. (−3, −1)
5. (−2, 5)
6. (1, 1)

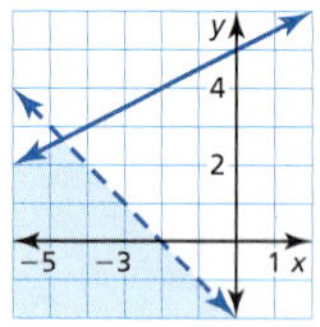

In Exercises 7–10, tell whether the ordered pair is a solution of the system of linear inequalities. *(See Example 1.)*

7. (−5, 2); $y < 4$, $y > x + 3$
8. (1, −1); $y > -2$, $y > x - 5$
9. (0, 0); $y \le x + 7$, $y \ge 2x + 3$
10. (4, −3); $y \le -x + 1$, $y \le 5x - 2$

In Exercises 11–20, graph the system of linear inequalities. *(See Examples 2 and 3.)*

11. $y > -3$; $y \ge 5x$
12. $y < -1$; $x > 4$
13. $y < -2$; $y > 2$
14. $y < x - 1$; $y \ge x + 1$
15. $y \ge -5$; $y - 1 < 3x$
16. $x + y > 4$; $y \ge \frac{3}{2}x - 9$
17. $x + y > 1$; $-x - y < -3$
18. $2x + y \le 5$; $y + 2 \ge -2x$
19. $x < 4$; $y > 1$; $y \ge -x + 1$
20. $x + y \le 10$; $x - y \ge 2$; $y > 2$

In Exercises 21–26, write a system of linear inequalities represented by the graph. *(See Examples 4 and 5.)*

21.

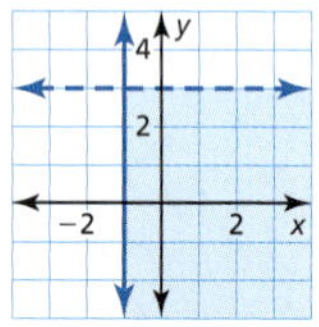

22.

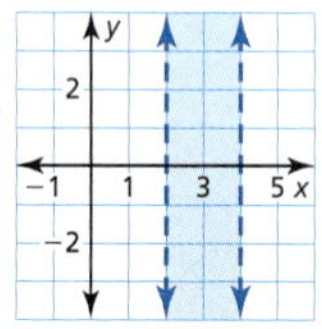

23.

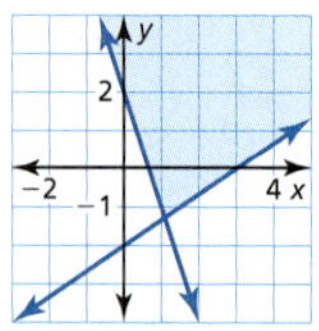

24.

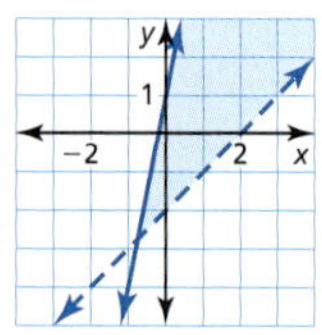

25.

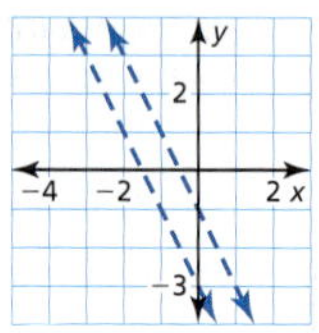

26.

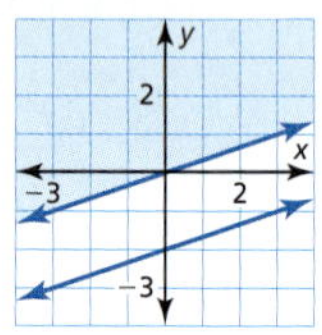

Assignment Guide and Homework Check

ASSIGNMENT

Basic: 1, 2, 3–29 odd, 35, 38, 40, 49–55

Average: 1, 2–40 even, 49–55

Advanced: 1, 2, 10, 18, 20, 24–40 even, 41–55

HOMEWORK CHECK

Basic: 7, 11, 21, 29

Average: 8, 14, 22, 30

Advanced: 10, 18, 24, 30

ANSWERS

1. Substitute the values into both inequalities and verify that both inequalities are true.
2. (1, −2); It is the only ordered pair that is not a solution of the system.
3. no
4. yes
5. no
6. no
7. yes
8. yes
9. no
10. yes
11.

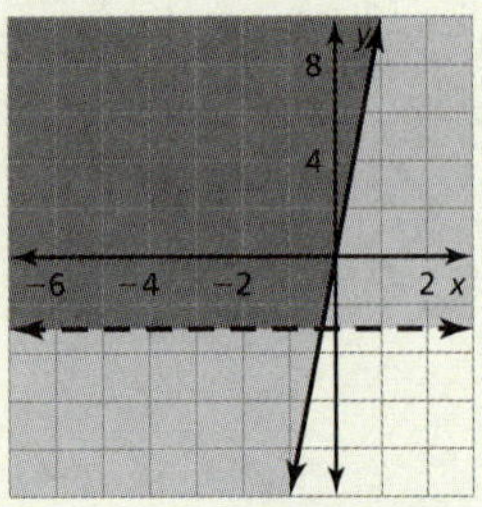

12.

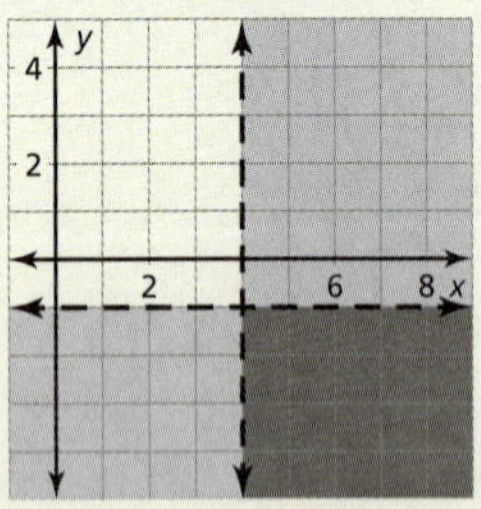

13.

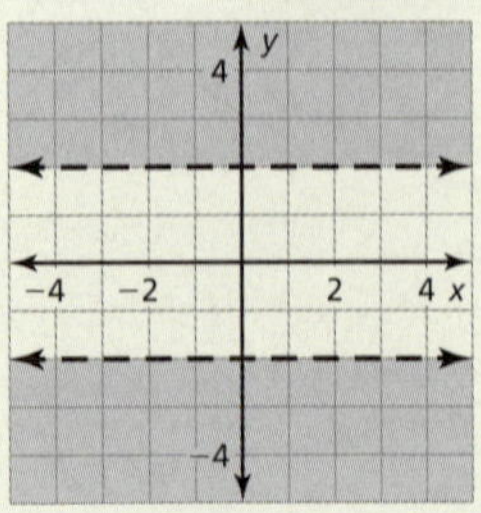

14.

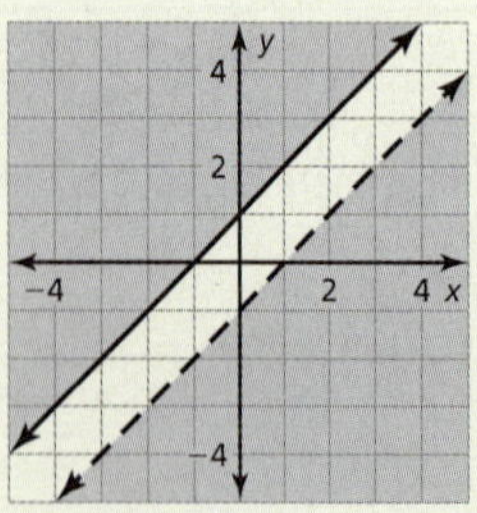

15.

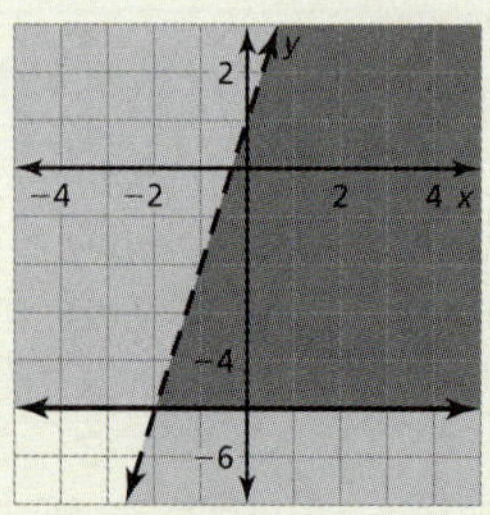

16–26. See Additional Answers.

ERROR ANALYSIS In Exercises 27 and 28, describe and correct the error in graphing the system of linear inequalities.

27.

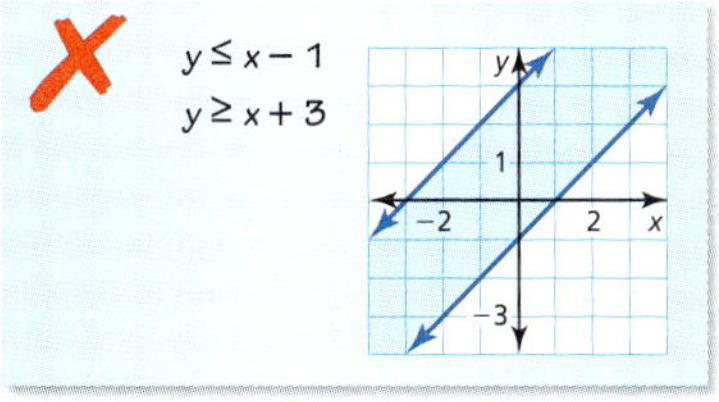

28.

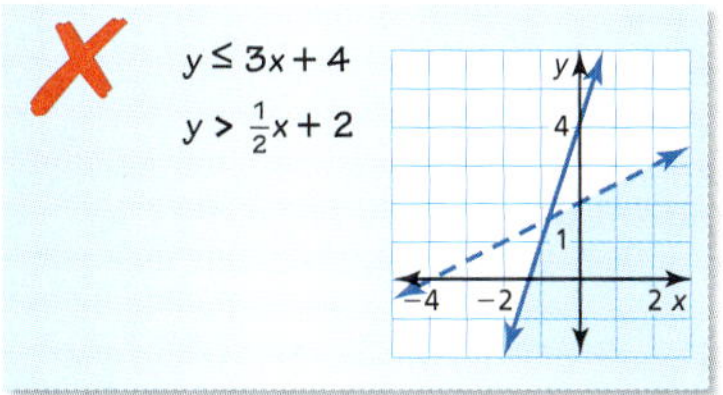

29. **MODELING WITH MATHEMATICS** You can spend at most \$21 on fruit. Blueberries cost \$4 per pound, and strawberries cost \$3 per pound. You need at least 3 pounds of fruit to make muffins. *(See Example 6.)*

   a. Write and graph a system of linear inequalities that represents the situation.

   b. Identify and interpret a solution of the system.

   c. Use the graph to determine whether you can buy 4 pounds of blueberries and 1 pound of strawberries.

30. **MODELING WITH MATHEMATICS** You earn \$10 per hour working as a manager at a grocery store. You are required to work at the grocery store at least 8 hours per week. You also teach music lessons for \$15 per hour. You need to earn at least \$120 per week, but you do not want to work more than 20 hours per week.

   a. Write and graph a system of linear inequalities that represents the situation.

   b. Identify and interpret a solution of the system.

   c. Use the graph to determine whether you can work 8 hours at the grocery store and teach 1 hour of music lessons.

31. **MODELING WITH MATHEMATICS** You are fishing for surfperch and rockfish, which are species of bottomfish. Gaming laws allow you to catch no more than 15 surfperch per day, no more than 10 rockfish per day, and no more than 20 total bottomfish per day.

   a. Write and graph a system of linear inequalities that represents the situation.

   b. Use the graph to determine whether you can catch 11 surfperch and 9 rockfish in 1 day.

surfperch

rockfish

32. **REASONING** Describe the intersection of the half-planes of the system shown.

$$x - y \le 4$$
$$x - y \ge 4$$

33. **MATHEMATICAL CONNECTIONS** The following points are the vertices of a shaded rectangle.

$$(-1, 1), (6, 1), (6, -3), (-1, -3)$$

   a. Write a system of linear inequalities represented by the shaded rectangle.

   b. Find the area of the rectangle.

34. **MATHEMATICAL CONNECTIONS** The following points are the vertices of a shaded triangle.

$$(2, 5), (6, -3), (-2, -3)$$

   a. Write a system of linear inequalities represented by the shaded triangle.

   b. Find the area of the triangle.

35. **PROBLEM SOLVING** You plan to spend less than half of your monthly \$2000 paycheck on housing and savings. You want to spend at least 10% of your paycheck on savings and at most 30% of it on housing. How much money can you spend on savings and housing?

36. **PROBLEM SOLVING** On a road trip with a friend, you drive about 70 miles per hour, and your friend drives about 60 miles per hour. The plan is to drive less than 15 hours and at least 600 miles each day. Your friend will drive more hours than you. How many hours can you and your friend each drive in 1 day?

Dynamic Teaching Tools

- Dynamic Assessment & Progress Monitoring Tool
- Interactive Whiteboard Lesson Library
- Dynamic Classroom with Dynamic Investigations

ANSWERS

27. The wrong regions are shaded for both inequalities.

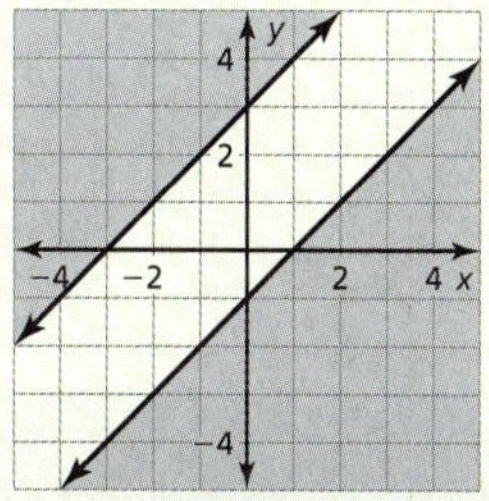

28. The wrong region is shaded for $y > \frac{1}{2}x + 2$.

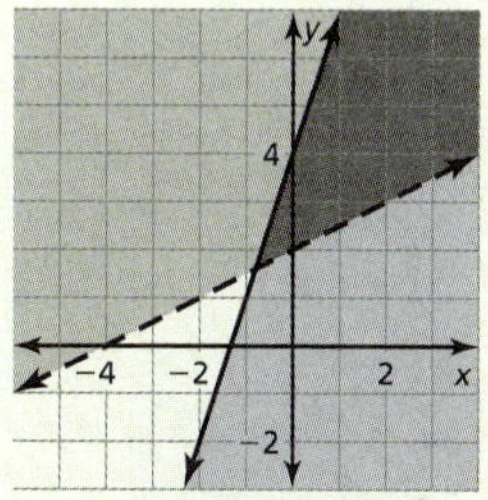

29. a. $4x + 3y \le 21$ and $x + y \ge 3$

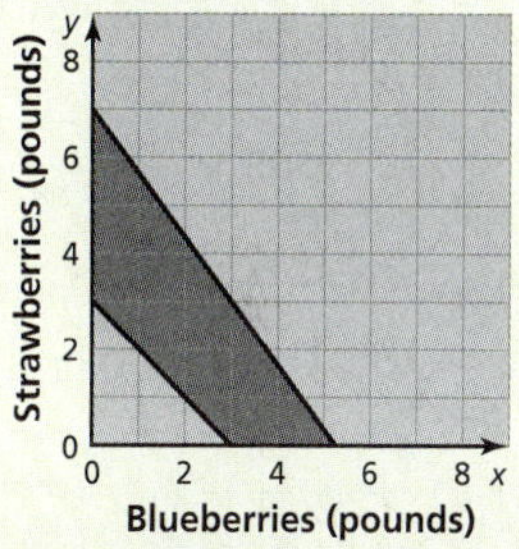

   b. *Sample answer:* (2, 4), You can buy 2 pounds of blueberries and 4 pounds of strawberries.

   c. yes

30. a. $x \ge 8$, $10x + 15y \ge 120$, and $x + y \le 20$

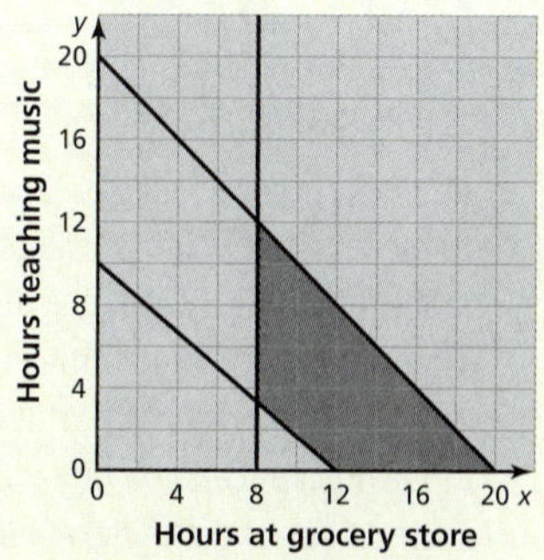

   b. *Sample answer:* (12, 2), You can work 12 hours at the grocery store and spend 2 hours teaching music lessons.

   c. no

31. a. $x \le 15$, $y \le 10$, and $x + y \le 20$

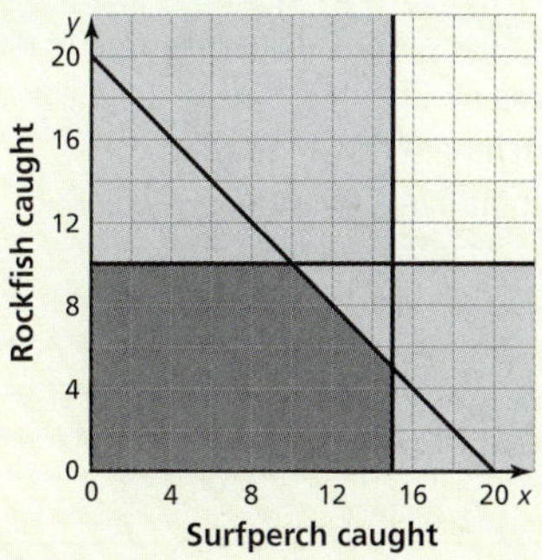

   b. yes

32. the line $x - y = 4$

33. a. $x \ge -1$, $x \le 6$, $y \ge -3$, and $y \le 1$

   b. 28 square units

34. a. $y \le 2x + 1$, $y \le -2x + 9$, and $y \ge -3$

   b. 32 square units

35. *Sample answer:* \$300 on savings, \$500 on housing

36. *Sample answer:* you 6 hours, your friend 8 hours

ANSWERS

37. Both can be done by graphing; Solving a system of linear inequalities requires finding overlapping half-planes, solving a system of linear equations does not.

38. >; <; This puts point C in the overlapping half-planes and leaves the other points out.

39–55. See Additional Answers.

Mini-Assessment

1. Tell whether the ordered pair $(-4, 2)$ is a solution of the system of linear inequalities.
   $y < -4x, y \geq x - 4$ yes
2. Graph the system of linear inequalities.
   $y \geq 0, y < 2x - 3$

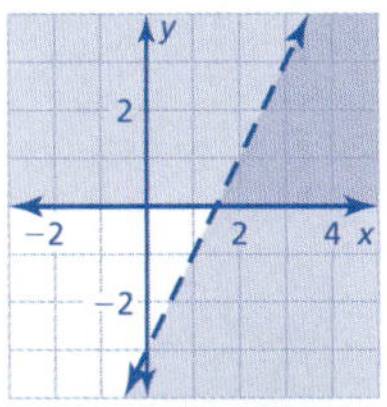

3. Write a system of linear inequalities represented by the graph.

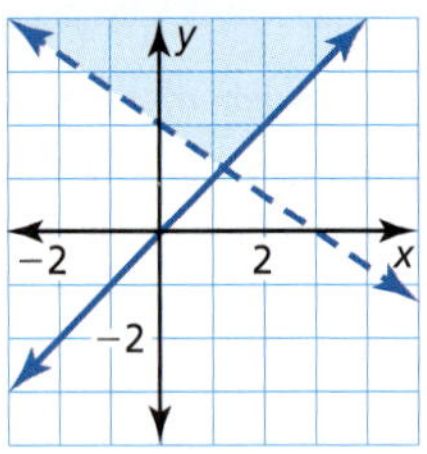

   $y \geq x, y > -\frac{2}{3}x + 2$

4. You can spend at most $24 at the fair. Game tickets cost $2 each, and ride tickets cost $4 each. You want to buy at least 3 game tickets and more than 2 ride tickets. Write a system of linear inequalities that represents the situation. How many of each ticket can you buy? Let x be the number of game tickets and y be the number of ride tickets: $2x + 4y \leq 24, x \geq 3, y > 2$. *Sample answer:* (4, 3); You can buy 4 game tickets and 3 ride tickets.

37. **WRITING** How are solving systems of linear inequalities and solving systems of linear equations similar? How are they different?

38. **HOW DO YOU SEE IT?** The graphs of two linear equations are shown.

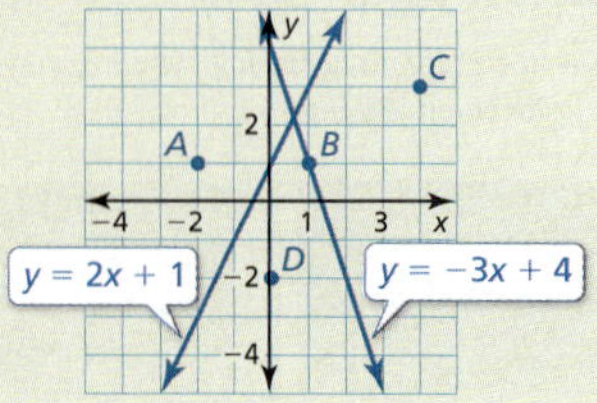

Replace the equal signs with inequality symbols to create a system of linear inequalities that has point C as a solution, but not points A, B, and D. Explain your reasoning.

$y \;\square\; -3x + 4$

$y \;\square\; 2x + 1$

39. **USING STRUCTURE** Write a system of linear inequalities that is equivalent to $|y| < x$, where $x > 0$. Graph the system.

40. **MAKING AN ARGUMENT** Your friend says that a system of linear inequalities in which the boundary lines are parallel must have no solution. Is your friend correct? Explain.

41. **CRITICAL THINKING** Is it possible for the solution set of a system of linear inequalities to be all real numbers? Explain your reasoning.

OPEN-ENDED **In Exercises 42–44, write a system of linear inequalities with the given characteristic.**

42. All solutions are in Quadrant I.

43. All solutions have one positive coordinate and one negative coordinate.

44. There are no solutions.

45. **OPEN-ENDED** One inequality in a system is $-4x + 2y > 6$. Write another inequality so the system has (a) no solution and (b) infinitely many solutions.

46. **THOUGHT PROVOKING** You receive a gift certificate for a clothing store and plan to use it to buy T-shirts and sweatshirts. Describe a situation in which you can buy 9 T-shirts and 1 sweatshirt, but you cannot buy 3 T-shirts and 8 sweatshirts. Write and graph a system of linear inequalities that represents the situation.

47. **CRITICAL THINKING** Write a system of linear inequalities that has exactly one solution.

48. **MODELING WITH MATHEMATICS** You make necklaces and key chains to sell at a craft fair. The table shows the amounts of time and money it takes to make a necklace and a key chain, and the amounts of time and money you have available for making them.

	Necklace	Key chain	Available
Time to make (hours)	0.5	0.25	20
Cost to make (dollars)	2	3	120

a. Write and graph a system of four linear inequalities that represents the number x of necklaces and the number y of key chains that you can make.

b. Find the vertices (corner points) of the graph of the system.

c. You sell each necklace for $10 and each key chain for $8. The revenue R is given by the equation $R = 10x + 8y$. Find the revenue corresponding to each ordered pair in part (b). Which vertex results in the maximum revenue?

Maintaining Mathematical Proficiency

Reviewing what you learned in previous grades and lessons

Write the product using exponents. *(Skills Review Handbook)*

49. $4 \cdot 4 \cdot 4 \cdot 4 \cdot 4$ **50.** $(-13) \cdot (-13) \cdot (-13)$ **51.** $x \cdot x \cdot x \cdot x \cdot x \cdot x$

Write an equation of the line with the given slope and *y*-intercept. *(Section 4.1)*

52. slope: 1; y-intercept: -6

53. slope: -3; y-intercept: 5

54. slope: $-\frac{1}{4}$; y-intercept: -1

55. slope: $\frac{4}{3}$; y-intercept: 0

If students need help...	*If students got it...*
Resources by Chapter • Practice A and Practice B • Puzzle Time	Resources by Chapter • Enrichment and Extension • Cumulative Review
Student Journal • Practice	Start the *next* Section
Differentiating the Lesson Skills Review Handbook	

5.5–5.7 What Did You Learn?

Core Vocabulary

linear inequality in two variables, *p. 268*
solution of a linear inequality in two variables, *p. 268*
graph of a linear inequality, *p. 268*
half-planes, *p. 268*
system of linear inequalities, *p. 274*
solution of a system of linear inequalities, *p. 274*
graph of a system of linear inequalities, *p. 275*

Core Concepts

Section 5.5

Solving Linear Equations by Graphing, *p. 262*
Solving Absolute Value Equations by Graphing, *p. 263*

Section 5.6

Graphing a Linear Inequality in Two Variables, *p. 269*

Section 5.7

Graphing a System of Linear Inequalities, *p. 275*
Writing a System of Linear Inequalities, *p. 276*

Mathematical Practices

1. Why do the equations in Exercise 35 on page 266 contain absolute value expressions?
2. Why is it important to be precise when answering part (a) of Exercise 39 on page 272?
3. Describe the overall step-by-step process you used to solve Exercise 35 on page 279.

Performance Task

Prize Patrol

You have been selected to drive a prize patrol cart and place prizes on the competing teams' predetermined paths. You know the teams' routes and you can only make one pass. Where will you place the prizes so that each team will have a chance to find a prize on their route?

To explore the answers to these questions and more, go to *BigIdeasMath.com*.

Dynamic Teaching Tools

- Dynamic Assessment & Progress Monitoring Tool
- Interactive Whiteboard Lesson Library
- Dynamic Classroom with Dynamic Investigations

ANSWERS

1. The expression inside the absolute value symbol is positive as we run towards the fence and negative as we run away from the fence, but the distance must always be positive.
2. Exceeding the maximum weight could be unsafe.
3. Write the inequalities $h + s < \frac{1}{2}(2000)$, $s \geq 10\%(2000)$, and $h \leq 30\%(2000)$. Simplify the expressions on the right side of each inequality. Choose values for s and h that make the last two inequalities true and verify that they also make the first inequality true.

ANSWERS

1. (2, −5)
2. (1, −1)
3. (4, −1)
4. (−2, −3)
5. (2, 1)
6. (2, 0)
7. 8 brushes, 4 tubes of paint

5 Chapter Review

Dynamic Solutions available at *BigIdeasMath.com*

5.1 Solving Systems of Linear Equations by Graphing *(pp. 235–240)*

Solve the system by graphing. $y = x - 2$ Equation 1
$y = -3x + 2$ Equation 2

Step 1 Graph each equation.

Step 2 Estimate the point of intersection.
The graphs appear to intersect at $(1, -1)$.

Step 3 Check your point from Step 2.

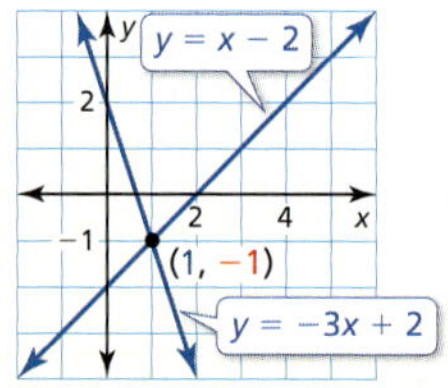

Equation 1	Equation 2
$y = x - 2$	$y = -3x + 2$
$-1 \stackrel{?}{=} 1 - 2$	$-1 \stackrel{?}{=} -3(1) + 2$
$-1 = -1$ ✓	$-1 = -1$ ✓

▶ The solution is $(1, -1)$.

Solve the system of linear equations by graphing.

1. $y = -3x + 1$
$y = x - 7$

2. $y = -4x + 3$
$4x - 2y = 6$

3. $5x + 5y = 15$
$2x - 2y = 10$

5.2 Solving Systems of Linear Equations by Substitution *(pp. 241–246)*

Solve the system by substitution. $-2x + y = -8$ Equation 1
$7x + y = 10$ Equation 2

Step 1 Solve for y in Equation 1.

$y = 2x - 8$ Revised Equation 1

Step 2 Substitute $2x - 8$ for y in Equation 2 and solve for x.

$7x + y = 10$ Equation 2
$7x + (2x - 8) = 10$ Substitute $2x - 8$ for y.
$9x - 8 = 10$ Combine like terms.
$9x = 18$ Add 8 to each side.
$x = 2$ Divide each side by 9.

Step 3 Substituting 2 for x in Equation 1 and solving for y gives $y = -4$.

▶ The solution is $(2, -4)$.

Solve the system of linear equations by substitution. Check your solution.

4. $3x + y = -9$
$y = 5x + 7$

5. $x + 4y = 6$
$x - y = 1$

6. $2x + 3y = 4$
$y + 3x = 6$

7. You spend \$20 total on tubes of paint and disposable brushes for an art project. Tubes of paint cost \$4.00 each and paintbrushes cost \$0.50 each. You purchase twice as many brushes as tubes of paint. How many brushes and tubes of paint do you purchase?

5.3 Solving Systems of Linear Equations by Elimination (pp. 247–252)

Solve the system by elimination. $4x + 6y = -8$ Equation 1
$x - 2y = -2$ Equation 2

Step 1 Multiply Equation 2 by 3 so that the coefficients of the y-terms are opposites.

$4x + 6y = -8$ → $4x + 6y = -8$ Equation 1

$x - 2y = -2$ **Multiply by 3.** → $3x - 6y = -6$ Revised Equation 2

Step 2 Add the equations.

$4x + 6y = -8$ Equation 1

$3x - 6y = -6$ Revised Equation 2

$7x = -14$ Add the equations.

Step 3 Solve for x.

$7x = -14$ Resulting equation from Step 2

$x = -2$ Divide each side by 7.

Step 4 Substitute -2 for x in one of the original equations and solve for y.

$4x + 6y = -8$ Equation 1

$4(-2) + 6y = -8$ Substitute -2 for x.

$-8 + 6y = -8$ Multiply.

$y = 0$ Solve for y.

Check

Equation 1

$4x + 6y = -8$

$4(-2) + 6(0) \stackrel{?}{=} -8$

$-8 = -8$ ✓

Equation 2

$x - 2y = -2$

$(-2) - 2(0) \stackrel{?}{=} -2$

$-2 = -2$ ✓

▶ The solution is $(-2, 0)$.

Solve the system of linear equations by elimination. Check your solution.

8. $9x - 2y = 34$
$5x + 2y = -6$

9. $x + 6y = 28$
$2x - 3y = -19$

10. $8x - 7y = -3$
$6x - 5y = -1$

5.4 Solving Special Systems of Linear Equations (pp. 253–258)

Solve the system. $4x + 2y = -14$ Equation 1
$y = -2x - 6$ Equation 2

Solve by substitution. Substitute $-2x - 6$ for y in Equation 1.

$4x + 2y = -14$ Equation 1

$4x + 2(-2x - 6) = -14$ Substitute $-2x - 6$ for y.

$4x - 4x - 12 = -14$ Distributive Property

$-12 = -14$ ✗ Combine like terms.

▶ The equation $-12 = -14$ is never true. So, the system has no solution.

Solve the system of linear equations.

11. $x = y + 2$
$-3x + 3y = 6$

12. $3x - 6y = -9$
$-5x + 10y = 10$

13. $-4x + 4y = 32$
$3x + 24 = 3y$

ANSWERS

8. $(2, -8)$
9. $(-2, 5)$
10. $(4, 5)$
11. no solution
12. no solution
13. infinitely many solutions

ANSWERS

14. $x = -3$

15. $x = -5$

16. $x = 13, x = 1$

17. 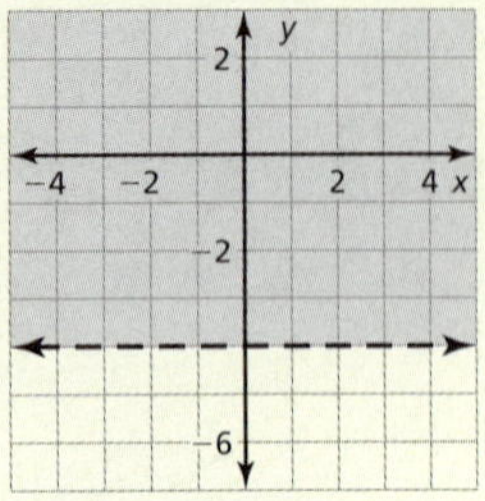

18. 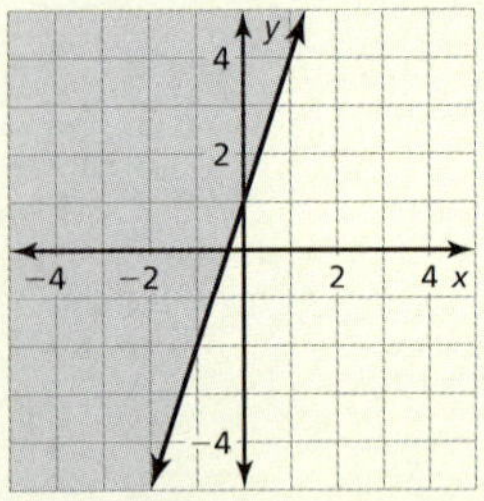

19. 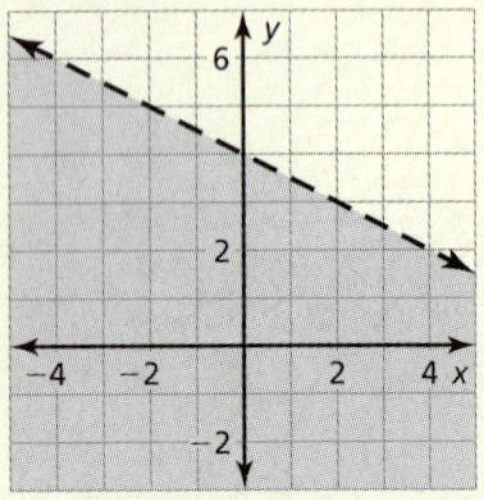

20. 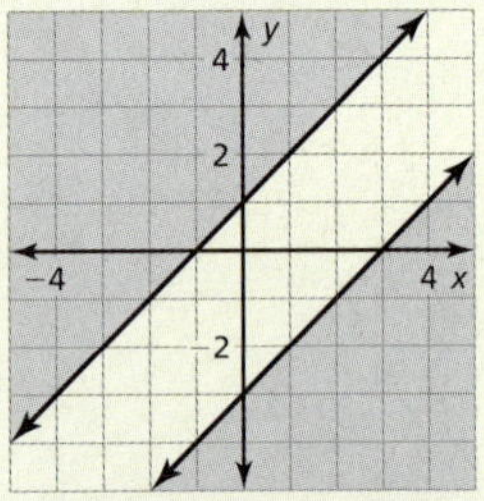

21. 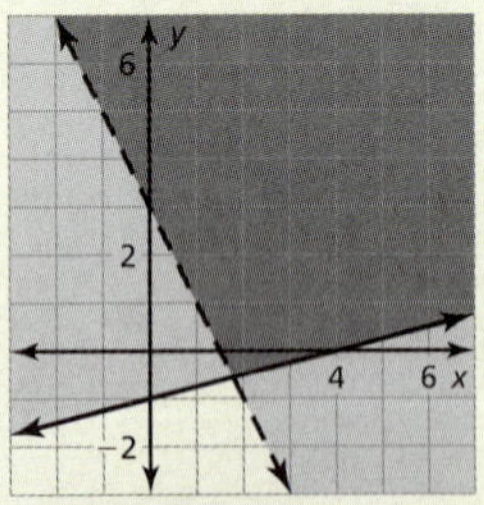

22. 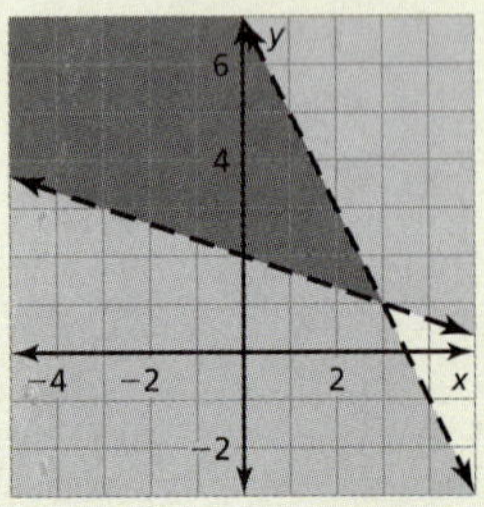

5.5 Solving Equations by Graphing (pp. 261–266)

Solve $3x - 1 = -2x + 4$ by graphing. Check your solution.

Step 1 Write a system of linear equations using each side of the original equation.

$y = 3x - 1$ → $3x - 1 = -2x + 4$ ← $y = -2x + 4$

Step 2 Graph the system.

$y = 3x - 1$ Equation 1

$y = -2x + 4$ Equation 2

The graphs intersect at (1, 2).

▶ So, the solution of the equation is $x = 1$.

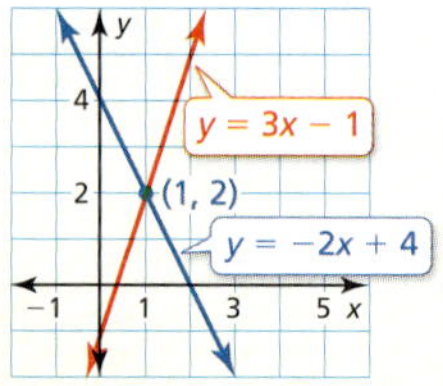

Check

$3x - 1 = -2x + 4$

$3(1) - 1 \stackrel{?}{=} -2(1) + 4$

$2 = 2$ ✓

Solve the equation by graphing. Check your solution(s).

14. $\frac{1}{3}x + 5 = -2x - 2$ **15.** $|x + 1| = |-x - 9|$ **16.** $|2x - 8| = |x + 5|$

5.6 Graphing Linear Inequalities in Two Variables (pp. 267–272)

Graph $4x + 2y \geq -6$ in a coordinate plane.

Step 1 Graph $4x + 2y = -6$, or $y = -2x - 3$. Use a solid line because the inequality symbol is $\geq$.

Step 2 Test (0, 0).

$4x + 2y \geq -6$ Write the inequality.

$4(0) + 2(0) \stackrel{?}{\geq} -6$ Substitute.

$0 \geq -6$ ✓ Simplify.

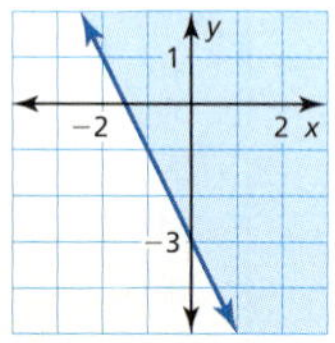

Step 3 Because (0, 0) is a solution, shade the half-plane that contains (0, 0).

Graph the inequality in a coordinate plane.

17. $y > -4$ **18.** $-9x + 3y \geq 3$ **19.** $5x + 10y < 40$

5.7 Systems of Linear Inequalities (pp. 273–280)

Graph the system. $y < x - 2$ Inequality 1

$y \geq 2x - 4$ Inequality 2

Step 1 Graph each inequality.

Step 2 Find the intersection of the half-planes. One solution is $(0, -3)$.

The solution is the purple-shaded region.

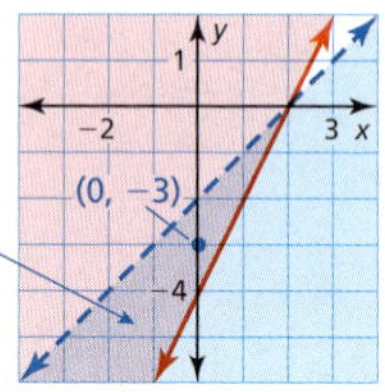

Graph the system of linear inequalities.

20. $y \leq x - 3$
$y \geq x + 1$

21. $y > -2x + 3$
$y \geq \frac{1}{4}x - 1$

22. $x + 3y > 6$
$2x + y < 7$

5 Chapter Test

Solve the system of linear equations using any method. Explain why you chose the method.

1. $8x + 3y = -9$
$-8x + y = 29$

2. $\frac{1}{2}x + y = -6$
$y = \frac{3}{5}x + 5$

3. $y = 4x + 4$
$-8x + 2y = 8$

4. $x = y - 11$
$x - 3y = 1$

5. $6x - 4y = 9$
$9x - 6y = 15$

6. $y = 5x - 7$
$-4x + y = -1$

7. Write a system of linear inequalities so the points (1, 2) and (4, −3) are solutions of the system, but the point (−2, 8) is not a solution of the system.

8. How is solving the equation $|2x + 1| = |x - 7|$ by graphing similar to solving the equation $4x + 3 = -2x + 9$ by graphing? How is it different?

Graph the system of linear inequalities.

9. $y > \frac{1}{2}x + 4$
$2y \le x + 4$

10. $x + y < 1$
$5x + y > 4$

11. $y \ge -\frac{2}{3}x + 1$
$-3x + y > -2$

12. You pay \$45.50 for 10 gallons of gasoline and 2 quarts of oil at a gas station. Your friend pays \$22.75 for 5 gallons of the same gasoline and 1 quart of the same oil.

a. Is there enough information to determine the cost of 1 gallon of gasoline and 1 quart of oil? Explain.

b. The receipt shown is for buying the same gasoline and same oil. Is there now enough information to determine the cost of 1 gallon of gasoline and 1 quart of oil? Explain.

c. Determine the cost of 1 gallon of gasoline and 1 quart of oil.

13. Describe the advantages and disadvantages of solving a system of linear equations by graphing.

14. You have at most \$60 to spend on trophies and medals to give as prizes for a contest.

a. Write and graph an inequality that represents the numbers of trophies and medals you can buy. Identify and interpret a solution of the inequality.

b. You want to purchase at least 6 items. Write and graph a system that represents the situation. How many of each item can you buy?

15. Compare the slopes and y-intercepts of the graphs of the equations in the linear system $8x + 4y = 12$ and $3y = -6x - 15$ to determine whether the system has one solution, no solution, or infinitely many solutions. Explain.

If students need help...	*If students got it...*
Lesson Tutorials	Resources by Chapter • Enrichment and Extension • Cumulative Review
Skills Review Handbook	Performance Task
BigIdeasMath.com	Start the *next* Section

ANSWERS

1. (−3, 5); *Sample answer:* elimination because one of the variables has opposite coefficients

2. (−10, −1); *Sample answer:* substitution because one of the equations was solved for y

3. infinitely many solutions; *Sample answer:* substitution because one of the equations was solved for y

4. (−17, −6); *Sample answer:* substitution because one of the equations was solved for x

5. no solution; *Sample answer:* elimination because no variable had a coefficient of 1 or −1

6. (6, 23); *Sample answer:* substitution because one of the equations was solved for y

7. *Sample answer:* $x > 0$ and $y < 6$

8. Both include graphing a system of equations formed by each side of the equation; Solving an absolute value equation requires graphing two systems, solving a linear equation requires graphing one system.

9.

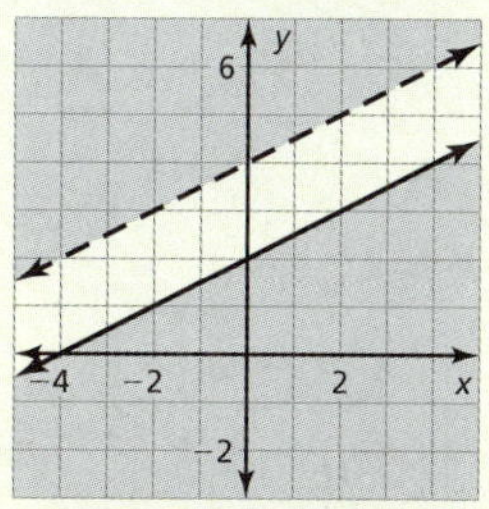

10.

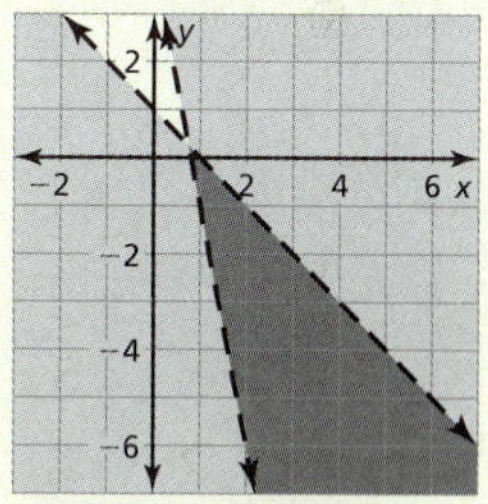

11.

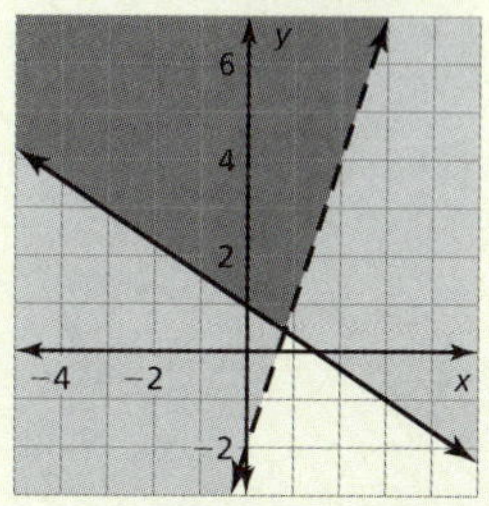

12. a. no; There are infinitely many solutions for the system of linear equations that represents this situation.

12b–15. See Additional Answers.

ANSWERS

1. D
2. 130; 140; 160; 180
3. ≥; >
4. no; The slopes of the line are the same, so either the lines are parallel or they are the same line.
5. reflection in the x-axis; horizontal translation; vertical translation

5 Cumulative Assessment

1. The graph of which equation is shown?

Ⓐ $9x - 2y = -18$

Ⓑ $-9x - 2y = 18$

Ⓒ $9x + 2y = 18$

Ⓓ $-9x + 2y = -18$

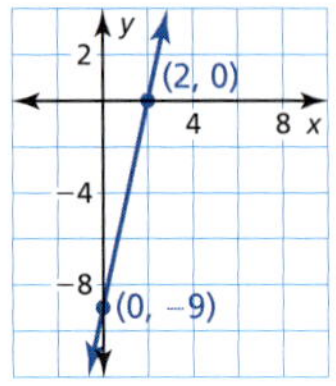

2. A van rental company rents out 6-, 8-, 12-, and 16-passenger vans. The function $C(x) = 100 + 5x$ represents the cost C (in dollars) of renting an x-passenger van for a day. Choose the numbers that are in the range of the function.

130 | 140 | 150 | 160 | 170 | 180 | 190 | 200

3. Fill in the system of linear inequalities with <, ≤, >, or ≥ so that the graph represents the system.

y ▭ $3x - 2$

y ▭ $-x + 5$

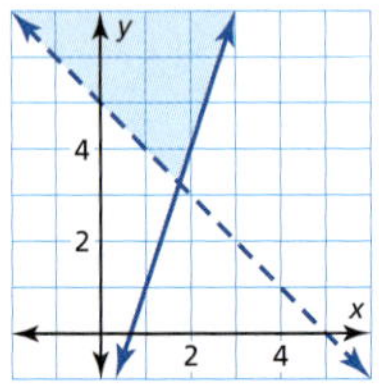

4. Your friend claims to be able to fill in each box with a constant so that when you set each side of the equation equal to y and graph the resulting equations, the lines will intersect exactly once. Do you support your friend's claim? Explain.

$4x +$ ▭ $= 4x +$ ▭

5. Select the phrases you should use when describing the transformations from the graph of f to the graph of g.

reflection in the x-axis	reflection in the y-axis
horizontal translation	vertical translation
horizontal stretch	vertical stretch
horizontal shrink	vertical shrink

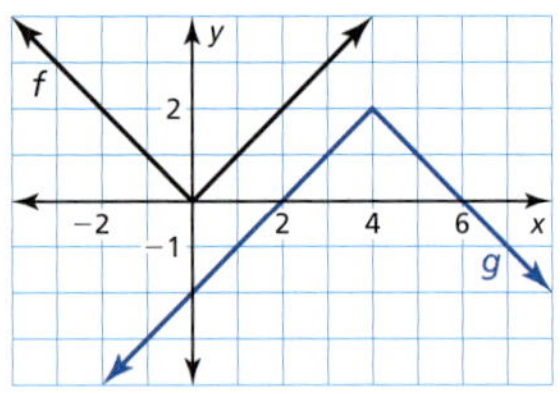

6. Which two equations form a system of linear equations that has no solution?

$y = 3x + 2$	$y = \frac{1}{3}x + 2$	$y = 2x + 3$	$y = 3x + \frac{1}{2}$

7. Fill in a value for a so that each statement is true for the equation $ax - 8 = 4 - x$.

   a. When $a =$ ▢, the solution is $x = -2$.

   b. When $a =$ ▢, the solution is $x = 12$.

   c. When $a =$ ▢, the solution is $x = 3$.

8. Which ordered pair is a solution of the linear inequality whose graph is shown?

   Ⓐ $(1, 1)$

   Ⓑ $(-1, 1)$

   Ⓒ $(-1, -1)$

   Ⓓ $(1, -1)$

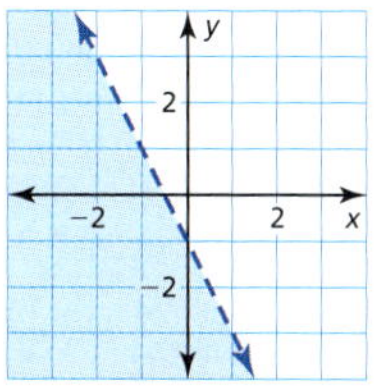

9. Which of the systems of linear equations are equivalent?

$4x - 5y = 3$ $2x + 15y = -1$	$4x - 5y = 3$ $-4x - 30y = 2$	$4x - 5y = 3$ $4x + 30y = -1$	$12x - 15y = 9$ $2x + 15y = -1$

10. The value of x is more than 9. Which of the inequalities correctly describe the triangle? The perimeter (in feet) is represented by P, and the area (in square feet) is represented by A.

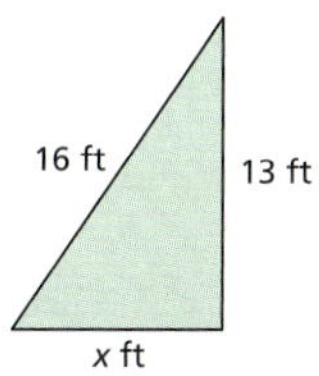

$P < 29$	$A > 117$	$P > 38$	$A > 58.5$	$A > 104$

ANSWERS

6. $y = 3x + 2$; $y = 3x + \frac{1}{2}$

7. **a.** -7

   b. 0

   c. 3

8. C

9. $\begin{array}{l} 4x - 5y = 3 \\ 2x + 15y = -1 \end{array}$; $\begin{array}{l} 4x - 5y = 3 \\ -4x - 30y = 2 \end{array}$; $\begin{array}{l} 12x - 15y = 9 \\ 2x + 15y = -1 \end{array}$

10. $P > 38$; $A > 58.5$

Chapter 6 Pacing Guide

Chapter Opener/ Mathematical Practices	1 Day
Section 1	1 Day
Section 2	2 Days
Section 3	2 Days
Section 4	2 Days
Quiz	1 Day
Section 5	2 Days
Section 6	2 Days
Section 7	2 Days
Chapter Review/ Chapter Tests	2 Days
Total Chapter 6	17 Days
Year-to-Date	84 Days

6 Exponential Functions and Sequences

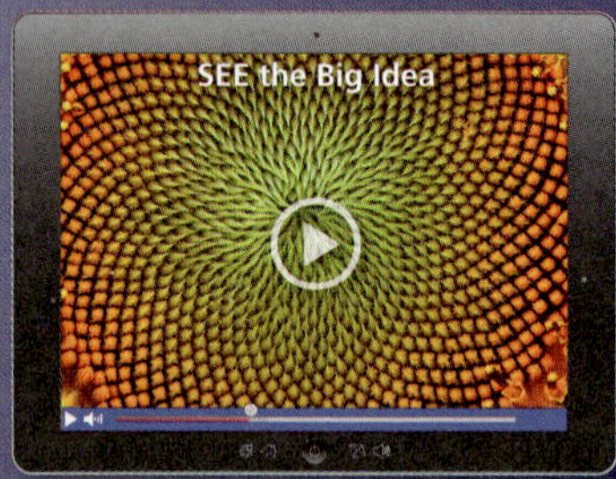

Fibonacci and Flowers *(p. 343)*

Soup Kitchen *(p. 338)*

Bacterial Culture *(p. 330)*

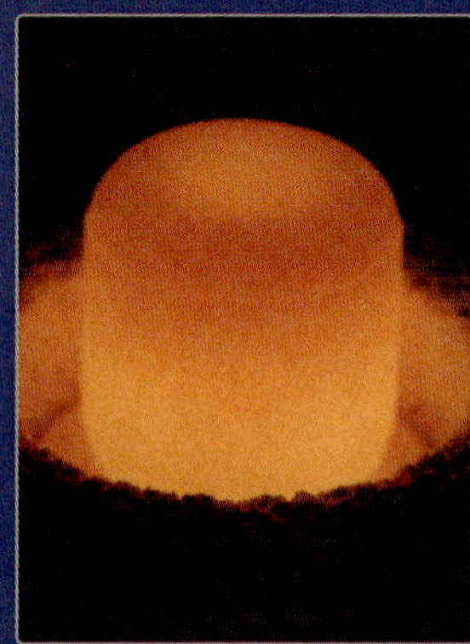

Plutonium Decay *(p. 321)*

Coyote Population *(p. 311)*

Laurie's Notes

Chapter Summary

- We are in the middle of the year, having finished work with linear equations. The work that follows this chapter involves polynomials and work with quadratics. This chapter introduces students to exponential functions and sequences. Students will revisit exponential functions and learn about logarithmic functions in Algebra 2.
- The properties of exponents presented in the first lesson should be a review for students. Many of the problems involve numeric expressions, although there are algebraic expressions as well. The same properties of exponents are applied to radicals and rational exponents in the next lesson.
- The next three lessons are about exponential functions and the attributes of exponential growth and decay functions. Exponential equations are solved using the properties of exponents from the beginning of the chapter and graphically with a graphing calculator.
- At the end of the chapter, geometric sequences are introduced with the connection made to exponential functions. This is similar to the lesson on arithmetic sequences and the connection to linear functions earlier in the book.
- The last lesson on recursively defined sequences looks at both arithmetic and geometric sequences, writing them as recursive rules.

What Your Students Have Learned

Middle School
- Write and evaluate numerical expressions involving integer exponents.
- Use properties of exponents and Order of Operations to evaluate expressions.
- Solve linear equations algebraically and graphically.
- Write functions to model linear relationships, and determine and interpret rates of change.
- Compare linear and nonlinear functions represented by tables, graphs, and equations.

What Your Students Will Learn

Algebra 1
- Evaluate and simplify expressions with exponents, including rational exponents.
- Identify, evaluate, and graph exponential functions and compare the graphs of exponential functions.
- Identify exponential growth and exponential decay functions and the percent rate of change.
- Solve exponential equations algebraically and graphically.
- Identify, extend, and graph geometric sequences.
- Write geometric sequences as equations and in function notation.
- Write terms and rules of recursively defined sequences and translate between recursive and explicit rules.

Dynamic Teaching Tools
- Dynamic Assessment & Progress Monitoring Tool
- Lesson Planning Tool
- Interactive Whiteboard Lesson Library
- Dynamic Classroom with Dynamic Investigations
- Real-Life STEM Videos

Scaffolding in the Classroom

Use vocabulary as a connector. When introducing vocabulary, ask students whether they are familiar with the word, and what they think it means. Have them explain how they think it relates to math. Students then find the definition and compare their original ideas to the actual meaning.

Questioning in the Classroom

One thing leads to another.

Try to build questions from the responses given. This requires students to listen to other students' responses.

Laurie's Notes

Maintaining Mathematical Proficiency

Using Order of Operations

- Remind students that sometimes evaluating a power is the first thing they need to do inside parentheses. For example, students should evaluate 4^2 in Exercise 3 before subtracting.

COMMON ERROR Students may ignore the order of operations and evaluate expressions from left to right.

Finding Square Roots

- Remind students that every positive number has a negative square root and a positive square root. If there is no sign in front of the radical sign, as in $\sqrt{64}$, students find the positive square root. If there is a negative sign in front, students have to find the negative square root. If there is a $\pm$ sign in front, students have to find both square roots.

COMMON ERROR Students may ignore the sign of the expressions and find only the positive square root of a number.

Writing Equations for Arithmetic Sequences

- Remind students that in an arithmetic sequence, the difference between each pair of consecutive terms is the same. This difference is called the common difference.

COMMON ERROR In the equation for an arithmetic sequence, students may not multiply both terms inside the parentheses by the common difference. For example, they may rewrite $a_n = 12 + (n - 1)(2)$ as $a_n = 12 + n - 2$ instead of $a_n = 12 + 2n - 2$.

Mathematical Practices (continued on page 290)

- The *Mathematical Practices* page focuses attention on how mathematics is learned—process versus content. This page demonstrates that when you work with a set of data, you should look to discern whether there is a pattern.
- Use the *Mathematical Practices* page to help students develop mathematical habits of mind—how mathematics can be explored and how mathematics is thought about.
- The volumes of successive chambers in a chambered nautilus shell are presented in Example 1. With the information presented in a table, students would at least observe that the volumes increase in each successive chamber. If asked how the volumes are increasing, students would likely look for an additive relationship. A multiplicative relationship, especially one increasing by a factor of ≈ 1.063, is not likely to be recognized by eyesight.

If students need help...	*If students got it...*
Student Journal • Maintaining Mathematical Proficiency	Game Closet at *BigIdeasMath.com*
Lesson Tutorials	Start the *next* Section
Skills Review Handbook	

Maintaining Mathematical Proficiency

Using Order of Operations

Example 1 Evaluate $10^2 \div (30 \div 3) - 4(3 - 9) + 5^1$.

First:	Parentheses	$10^2 \div (30 \div 3) - 4(3-9) + 5^1 = 10^2 \div 10 - 4(-6) + 5^1$
Second:	Exponents	$= 100 \div 10 - 4(-6) + 5$
Third:	Multiplication and Division (from left to right)	$= 10 + 24 + 5$
Fourth:	Addition and Subtraction (from left to right)	$= 39$

Evaluate the expression.

1. $12\left(\frac{14}{2}\right) - 3^3 + 15 - 9^2$ **2.** $5^2 \cdot 8 \div 2^2 + 20 \cdot 3 - 4$ **3.** $-7 + 16 \div 2^4 + (10 - 4^2)$

Finding Square Roots

Example 2 Find $-\sqrt{81}$.

▶ $-\sqrt{81}$ represents the negative square root. Because $9^2 = 81$, $-\sqrt{81} = -\sqrt{9^2} = -9$.

Find the square root(s).

4. $\sqrt{64}$ **5.** $-\sqrt{4}$ **6.** $-\sqrt{25}$ **7.** $\pm\sqrt{121}$

Writing Equations for Arithmetic Sequences

Example 3 Write an equation for the *n*th term of the arithmetic sequence 5, 15, 25, 35, ...

The first term is 5, and the common difference is 10.

$a_n = a_1 + (n-1)d$ Equation for an arithmetic sequence

$a_n = 5 + (n-1)(10)$ Substitute 5 for a_1 and 10 for d.

$a_n = 10n - 5$ Simplify.

Write an equation for the *n*th term of the arithmetic sequence.

8. 12, 14, 16, 18, . . . **9.** 6, 3, 0, −3, . . . **10.** 22, 15, 8, 1, . . .

11. ABSTRACT REASONING Recall that a perfect square is a number with integers as its square roots. Is the product of two perfect squares always a perfect square? Is the quotient of two perfect squares always a perfect square? Explain your reasoning.

Dynamic Solutions available at *BigIdeasMath.com*

What Your Students Have Learned

- Use Order of Operations to evaluate expressions, including those involving whole number exponents.
- Find square roots and evaluate expressions involving square roots.
- Write arithmetic sequences as equations.

ANSWERS

1. −9
2. 106
3. −12
4. 8
5. −2
6. −5
7. ±11
8. $a_n = 2n + 10$
9. $a_n = -3n + 9$
10. $a_n = -7n + 29$
11. yes; no; The product of two perfect squares can be represented by $m^2n^2 = (mm)(nn) = (mn)(mn) = (mn)^2$. If m and n are integers, their product is also an integer, so $(mn)^2$ is an integer. There are many counterexamples illustrating that the quotient of two perfect squares does not have to be a perfect square, such as $9 \div 4$.

Vocabulary Review

Have students make Information Wheels for the following topics.

- Order of operations
- Square roots
- Arithmetic sequences

MONITORING PROGRESS ANSWERS

1. 3435
2. 512

Mathematical Practices

Mathematically proficient students look closely to find a pattern.

Problem-Solving Strategies

Core Concept

Finding a Pattern

When solving a real-life problem, look for a pattern in the data. The pattern could include repeating items, numbers, or events. After you find the pattern, describe it and use it to solve the problem.

EXAMPLE 1 Using a Problem-Solving Strategy

The volumes of seven chambers of a chambered nautilus are given. Find the volume of Chamber 10.

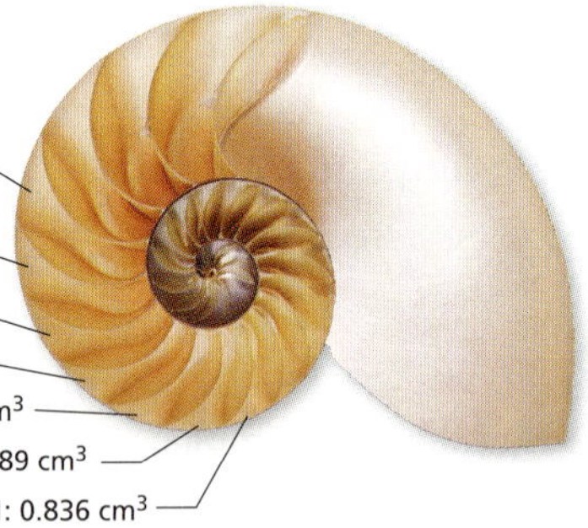

SOLUTION

To find a pattern, try dividing each volume by the volume of the previous chamber.

$$\frac{0.889}{0.836} \approx 1.063 \qquad \frac{0.945}{0.889} \approx 1.063 \qquad \frac{1.005}{0.945} \approx 1.063$$

$$\frac{1.068}{1.005} \approx 1.063 \qquad \frac{1.135}{1.068} \approx 1.063 \qquad \frac{1.207}{1.135} \approx 1.063$$

From this, you can see that the volume of each chamber is about 6.3% greater than the volume of the previous chamber. To find the volume of Chamber 10, multiply the volume of Chamber 7 by 1.063 three times.

$1.207(1.063) \approx \mathbf{1.283}$ (volume of Chamber 8) $\qquad 1.283(1.063) \approx \mathbf{1.364}$ (volume of Chamber 9) $\qquad 1.364(1.063) \approx \mathbf{1.450}$ (volume of Chamber 10)

▶ The volume of Chamber 10 is about 1.450 cubic centimeters.

Monitoring Progress

1. A rabbit population over 8 consecutive years is given by 50, 80, 128, 205, 328, 524, 839, 1342. Find the population in the tenth year.
2. The sums of the numbers in the first eight rows of Pascal's Triangle are 1, 2, 4, 8, 16, 32, 64, 128. Find the sum of the numbers in the tenth row.

Laurie's Notes Mathematical Practices (continued from page T-289)

- In this chapter, students will study exponential functions. The volume data is an example of an exponential function.
- **Look For and Make Use of Structure:** The example shown demonstrates how finding successive ratios helps you discern the pattern. Once the pattern is found, the problem can be solved.
- Give time for students to work through the *Monitoring Progress* questions, and then discuss as a class.

Laurie's Notes

Overview of Section 6.1

Introduction

- Students have worked with exponents previously, and the properties presented in this lesson should be familiar from middle school.
- Seven properties of exponents are presented in this lesson. Each of the properties is practiced independently to simplify expressions. Applying more than one property to the simplifying of an expression is where students can become challenged.

Formative Assessment Tips

- We are now at the middle portion of the book and the year. You should be comfortable with a variety of formative assessment techniques, and students should be comfortable with speaking aloud in class and with partners. The goal of your efforts is certainly to improve teaching and learning as result of the activities carried out during the instructional process. Although formative assessment is often described as a tool to identify certain student conceptions and misconceptions during the instructional phase, it is also a process by which your instruction is improved.
- Some formative assessment techniques are carried out over a period or unit of instruction. Throughout this chapter, you are encouraged to try a particular technique while still integrating and practicing previously learned techniques.
- **Response (Thinking) Logs:** This is a type of writing journal that will be used throughout this chapter after computations, during a conceptual activity, or as part of a problem-solving activity. Students will respond to a short writing prompt or sentence stem, reflecting on what they are learning. In the response log, students record the process they have gone through to learn something new, any questions or information they may need to have clarified, or what they find challenging. The response log allows students to make connections to what they have learned, set goals, and reflect on their learning process. The act of writing helps students become more aware of their own learning and what they can do to self-direct it.
- The *Response Log* is private. The purpose is to promote metacognition. If there are times when you want to collect the response logs, perhaps because you sense a need to redirect your instruction, students should be told this. If there is a possibility that the response log will not remain private, I let my students know this.
- Provide 4–6 pieces of lined paper stapled together or, alternatively, have students set aside a place in their spiral notebooks. On the left one-third of the paper, the writing stem or prompt is written. The right two-thirds of the paper is where the response is written. I have a large poster in my room with writing stems listed. Throughout the year, students suggest additions to the writing stems.
- I suggest writing stems throughout, though you should feel free to substitute or leave open for student selection any of the writing stems. For a beginning list of writing stems, see page T-298.

Pacing Suggestion

- The explorations provide an opportunity for students to develop a conceptual understanding of the properties of exponents. Complete all of the explorations and then transition to the formal lesson.

Dynamic Teaching Tools

- Dynamic Assessment & Progress Monitoring Tool
- Lesson Planning Tool
- Interactive Whiteboard Lesson Library
- Dynamic Classroom with Dynamic Investigations

What Your Students Will Learn

- Evaluate and simplify expressions involving zero and negative exponents.
- Use the properties of exponents to rewrite expressions using only positive exponents.
- Solve real-life problems involving exponents.

Laurie's Notes

Exploration

Motivate

- The $10,000 bill, which is no longer in circulation, would be much easier to carry than the same amount in pennies.

? Ask a few questions about money.
  - "How many pennies equal $10,000?" $100 \times 10{,}000 = 1{,}000{,}000$, or 10^6
  - "How many dimes equal $10,000?" $10 \times 10{,}000 = 100{,}000$, or 10^5
  - "How many $10 bills equal $10,000?" $\frac{1}{10} \times 10{,}000 = 1000$, or 10^3
  - "How many $100 bills equal $10,000?" $\frac{1}{100} \times 10{,}000 = 100$, or 10^2

Discuss

- Inductive reasoning is the use of specific examples to make a general statement. Encourage students to focus on the directions for each part and the language that is used. Each part begins with a question and ends with a statement about writing a general rule that answers the question that was posed. In writing the general rule, students may notice a pattern with the exponents, but they may need help in writing the rule using variables for the exponents. It may be necessary to model how to write the first rule after students have recognized the pattern.

Exploration 1

- For each of the problems in each part, students may need to expand the expression written and then simplify to see a pattern. For example:

  $(2^2)(2^3) = (2 \cdot 2)(2 \cdot 2 \cdot 2) = 2^5$

  $\frac{4^3}{4^2} = \frac{4 \cdot \cancel{4} \cdot \cancel{4}}{\cancel{4} \cdot \cancel{4}} = 4$

  $(2^2)^4 = (2^2)(2^2)(2^2)(2^2) = (2 \cdot 2)(2 \cdot 2)(2 \cdot 2)(2 \cdot 2) = 2^8$
- Some students may not need to write the expanded version, as they will recognize a pattern, or understand the structure, and will be able to write the general rule.
- **FYI:** All the quotients in the exploration result in nonnegative exponents. The formal lesson includes quotients that result in negative exponents.

Communicate Your Answer

- **Look For and Express Regularity in Repeated Reasoning:** Both questions involve repeated reasoning.

Connecting to Next Step

- Each part of the exploration is related to a property of exponents which will be stated in the formal lesson.

6.1 Properties of Exponents

Essential Question How can you write general rules involving properties of exponents?

EXPLORATION 1 **Writing Rules for Properties of Exponents**

Work with a partner.

WRITING GENERAL RULES

To be proficient in math, you need to understand and use stated assumptions, definitions, and previously established results in writing general rules.

a. What happens when you multiply two powers with the same base? Write the product of the two powers as a single power. Then write a *general rule* for finding the product of two powers with the same base.

i. $(2^2)(2^3) =$ ______ **ii.** $(4^1)(4^5) =$ ______

iii. $(5^3)(5^5) =$ ______ **iv.** $(x^2)(x^6) =$ ______

b. What happens when you divide two powers with the same base? Write the quotient of the two powers as a single power. Then write a *general rule* for finding the quotient of two powers with the same base.

i. $\frac{4^3}{4^2} =$ ______ **ii.** $\frac{2^5}{2^2} =$ ______

iii. $\frac{x^6}{x^3} =$ ______ **iv.** $\frac{3^4}{3^4} =$ ______

c. What happens when you find a power of a power? Write the expression as a single power. Then write a *general rule* for finding a power of a power.

i. $(2^2)^4 =$ ______ **ii.** $(7^3)^2 =$ ______

iii. $(y^3)^3 =$ ______ **iv.** $(x^4)^2 =$ ______

d. What happens when you find a power of a product? Write the expression as the product of two powers. Then write a *general rule* for finding a power of a product.

i. $(2 \cdot 5)^2 =$ ______ **ii.** $(5 \cdot 4)^3 =$ ______

iii. $(6a)^2 =$ ______ **iv.** $(3x)^2 =$ ______

e. What happens when you find a power of a quotient? Write the expression as the quotient of two powers. Then write a *general rule* for finding a power of a quotient.

i. $\left(\frac{2}{3}\right)^2 =$ ______ **ii.** $\left(\frac{4}{3}\right)^3 =$ ______

iii. $\left(\frac{x}{2}\right)^3 =$ ______ **iv.** $\left(\frac{a}{b}\right)^4 =$ ______

Communicate Your Answer

2. How can you write general rules involving properties of exponents?

3. There are 3^3 small cubes in the cube below. Write an expression for the number of small cubes in the large cube at the right.

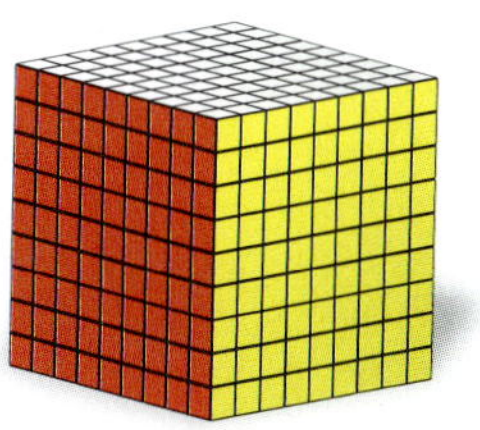

Dynamic Teaching Tools

- Dynamic Assessment & Progress Monitoring Tool
- Lesson Planning Tool
- Interactive Whiteboard Lesson Library
- Dynamic Classroom with Dynamic Investigations

ANSWERS

1. a. $a^m a^n = a^{m+n}$

i. 2^5
ii. 4^6
iii. 5^8
iv. x^8

b. $\frac{a^m}{a^n} = a^{m-n}$

i. 4^1
ii. 2^3
iii. x^3
iv. 3^0

c. $(a^m)^n = a^{mn}$

i. 2^8
ii. 7^6
iii. y^9
iv. x^8

d. $(ab)^m = a^m b^m$

i. $2^2 \cdot 5^2$
ii. $5^3 \cdot 4^3$
iii. $6^2 a^2$
iv. $3^2 x^2$

e. $\left(\frac{a}{b}\right)^m = \frac{a^m}{b^m}$

i. $\frac{2^2}{3^2}$
ii. $\frac{4^3}{3^3}$
iii. $\frac{x^3}{2^3}$
iv. $\frac{a^4}{b^4}$

2. Try several examples to find a pattern, then express the pattern using variables.

3. 9^3

Differentiated Instruction

Inclusion

Challenge students to write expressions that are equivalent to a given simplified expression. For example, give students the expression x^2y. Have them write two different expressions that can be simplified to x^2y.

Sample answer: $\frac{x^3}{xy^{-1}}$, $\frac{x^{-1}y^2}{x^{-3}y}$

Extra Example 1

Evaluate each expression.

a. 2.5^0 1

b. $(-3)^{-2}$ $\frac{1}{9}$

Extra Example 2

Simplify the expression $\frac{2^0x^2}{y^{-2}}$. Write your answer using only positive exponents. x^2y^2

MONITORING PROGRESS ANSWERS

1. 1
2. $\frac{1}{27}$
3. -4
4. $\frac{1}{9x^5}$

6.1 Lesson

Core Vocabulary

Previous
power
exponent
base
scientific notation

What You Will Learn

- Use zero and negative exponents.
- Use the properties of exponents.
- Solve real-life problems involving exponents.

Using Zero and Negative Exponents

Core Concept

Zero Exponent

Words For any nonzero number a, $a^0 = 1$. The power 0^0 is undefined.

Numbers $4^0 = 1$ **Algebra** $a^0 = 1$, where $a \neq 0$

Negative Exponents

Words For any integer n and any nonzero number a, a^{-n} is the reciprocal of a^n.

Numbers $4^{-2} = \frac{1}{4^2}$ **Algebra** $a^{-n} = \frac{1}{a^n}$, where $a \neq 0$

EXAMPLE 1 Using Zero and Negative Exponents

Evaluate each expression.

a. 6.7^0 **b.** $(-2)^{-4}$

SOLUTION

a. $6.7^0 = 1$ Definition of zero exponent

b. $(-2)^{-4} = \frac{1}{(-2)^4}$ Definition of negative exponent

$= \frac{1}{16}$ Simplify.

EXAMPLE 2 Simplifying an Expression

Simplify the expression $\frac{4x^0}{y^{-3}}$. Write your answer using only positive exponents.

SOLUTION

$\frac{4x^0}{y^{-3}} = 4x^0y^3$ Definition of negative exponent

$= 4y^3$ Definition of zero exponent

Monitoring Progress

Help in English and Spanish at *BigIdeasMath.com*

Evaluate the expression.

1. $(-9)^0$
2. 3^{-3}
3. $\frac{-5^0}{2^{-2}}$
4. Simplify the expression $\frac{3^{-2}x^{-5}}{y^0}$. Write your answer using only positive exponents.

Laurie's Notes Teacher Actions

- Write and discuss the two properties. Zero and negative exponents can be the most challenging for students to make sense of.
- Take time to discuss negative exponents in the denominator. A corollary would be $\frac{1}{a^{-n}} = a^n$. This situation shows up in Example 2.
- **Think-Pair-Share:** Have students answer Questions 1–4, and then share and discuss as a class.
- Students may ask what order to apply the properties. Different orders are possible. Model this with Question 4.

Using the Properties of Exponents

REMEMBER

The expression x^3 is called a *power*. The *base*, x, is used as a factor 3 times because the *exponent* is 3.

Core Concept

Product of Powers Property

Let a be a real number, and let m and n be integers.

Words To multiply powers with the same base, add their exponents.

Numbers $4^6 \cdot 4^3 = 4^{6+3} = 4^9$ **Algebra** $a^m \cdot a^n = a^{m+n}$

Quotient of Powers Property

Let a be a nonzero real number, and let m and n be integers.

Words To divide powers with the same base, subtract their exponents.

Numbers $\frac{4^6}{4^3} = 4^{6-3} = 4^3$ **Algebra** $\frac{a^m}{a^n} = a^{m-n}$, where $a \neq 0$

Power of a Power Property

Let a be a real number, and let m and n be integers.

Words To find a power of a power, multiply the exponents.

Numbers $(4^6)^3 = 4^{6 \cdot 3} = 4^{18}$ **Algebra** $(a^m)^n = a^{mn}$

EXAMPLE 3 Using Properties of Exponents

Simplify each expression. Write your answer using only positive exponents.

a. $3^2 \cdot 3^6$ **b.** $\frac{(-4)^2}{(-4)^7}$ **c.** $(z^4)^{-3}$

SOLUTION

a. $3^2 \cdot 3^6 = 3^{2+6}$ Product of Powers Property

$= 3^8 = 6561$ Simplify.

b. $\frac{(-4)^2}{(-4)^7} = (-4)^{2-7}$ Quotient of Powers Property

$= (-4)^{-5}$ Simplify.

$= \frac{1}{(-4)^5} = -\frac{1}{1024}$ Definition of negative exponent

c. $(z^4)^{-3} = z^{4 \cdot (-3)}$ Power of a Power Property

$= z^{-12}$ Simplify.

$= \frac{1}{z^{12}}$ Definition of negative exponent

Monitoring Progress

Help in English and Spanish at *BigIdeasMath.com*

Simplify the expression. Write your answer using only positive exponents.

5. $10^4 \cdot 10^{-6}$ **6.** $x^9 \cdot x^{-9}$ **7.** $\frac{-5^8}{-5^4}$

8. $\frac{y^6}{y^7}$ **9.** $(6^{-2})^{-1}$ **10.** $(w^{12})^5$

English Language Learners

Build on Past Knowledge

Remind students from their study of exponents that the expressions $(-4)^2$ and -4^2 are not equivalent. The expression $(-4)^2 = (-4) \cdot (-4) = 16$ while the expression $-4^2 = -(4 \cdot 4) = -16$. Remembering this expanded form will help students as they evaluate expressions similar to the ones in Example 3b and Exercise 7 in *Monitoring Progress*.

Extra Example 3

Simplify each expression. Write your answer using only positive exponents.

a. $2^3 \cdot 2^5$ $2^8 = 256$

b. $\frac{(-7)^2}{(-7)^4}$ $\frac{1}{(-7)^2} = \frac{1}{49}$

c. $(d^{-3})^5$ $d^{-15} = \frac{1}{d^{15}}$

MONITORING PROGRESS ANSWERS

5. $\frac{1}{100}$

6. 1

7. 625

8. $\frac{1}{y}$

9. 36

10. w^{60}

Laurie's Notes Teacher Actions

- These three properties are grouped together because they are about an operation being done to one or more powers.
- The expression in part (b) is different from the examples in the exploration because the exponent in the denominator is greater than the exponent in the numerator.

COMMON ERROR Students believe $(-4)^{-5}$ is the final answer. Remind them that answers must have positive exponents.

- **Response Logs:** "I figured out that ..."

Extra Example 4

Simplify each expression. Write your answer using only positive exponents.

a. $(-3x)^3$ $-27x^3$

b. $\left(\frac{b}{-3}\right)^4$ $\frac{b^4}{81}$

c. $\left(\frac{2x}{5}\right)^3$ $\frac{8x^3}{125}$

d. $\left(\frac{3c}{4}\right)^{-2}$ $\frac{16}{9c^2}$

MONITORING PROGRESS ANSWERS

11. $\frac{1}{1000y^3}$

12. $\frac{-1024}{n^5}$

13. $\frac{1}{32k^{10}}$

14. $\frac{49}{36c^2}$

Core Concept

Power of a Product Property

Let a and b be real numbers, and let m be an integer.

Words To find a power of a product, find the power of each factor and multiply.

Numbers $(3 \cdot 2)^5 = 3^5 \cdot 2^5$ **Algebra** $(ab)^m = a^m b^m$

Power of a Quotient Property

Let a and b be real numbers with $b \neq 0$, and let m be an integer.

Words To find the power of a quotient, find the power of the numerator and the power of the denominator and divide.

Numbers $\left(\frac{3}{2}\right)^5 = \frac{3^5}{2^5}$ **Algebra** $\left(\frac{a}{b}\right)^m = \frac{a^m}{b^m}$, where $b \neq 0$

EXAMPLE 4 Using Properties of Exponents

Simplify each expression. Write your answer using only positive exponents.

a. $(-1.5y)^2$ **b.** $\left(\frac{a}{-10}\right)^3$ **c.** $\left(\frac{3d}{2}\right)^4$ **d.** $\left(\frac{2x}{3}\right)^{-5}$

SOLUTION

a. $(-1.5y)^2 = (-1.5)^2 \cdot y^2$ Power of a Product Property

$= 2.25y^2$ Simplify.

b. $\left(\frac{a}{-10}\right)^3 = \frac{a^3}{(-10)^3}$ Power of a Quotient Property

$= -\frac{a^3}{1000}$ Simplify.

c. $\left(\frac{3d}{2}\right)^4 = \frac{(3d)^4}{2^4}$ Power of a Quotient Property

$= \frac{3^4 d^4}{2^4}$ Power of a Product Property

$= \frac{81d^4}{16}$ Simplify.

d. $\left(\frac{2x}{3}\right)^{-5} = \frac{(2x)^{-5}}{3^{-5}}$ Power of a Quotient Property

$= \frac{3^5}{(2x)^5}$ Definition of negative exponent

$= \frac{3^5}{2^5 x^5}$ Power of a Product Property

$= \frac{243}{32x^5}$ Simplify.

ANOTHER WAY

Because the exponent is negative, you could find the reciprocal of the base first. Then simplify.

$\left(\frac{2x}{3}\right)^{-5} = \left(\frac{3}{2x}\right)^5 = \frac{243}{32x^5}$

Monitoring Progress

Help in English and Spanish at *BigIdeasMath.com*

Simplify the expression. Write your answer using only positive exponents.

11. $(10y)^{-3}$ **12.** $\left(-\frac{4}{n}\right)^5$ **13.** $\left(\frac{1}{2k^2}\right)^5$ **14.** $\left(\frac{6c}{7}\right)^{-2}$

Laurie's Notes Teacher Actions

- **Common Misconception:** Students often refer to both of these properties as "distributing" the exponent. This is not!
- Work through each part as shown, pausing to ask questions. There are numerous places where common errors occur, such as when raising the product $(3d)^4$—students raise d to the fourth but do not raise 3 to the fourth.
- Students may suggest or ask about alternate solution methods. Try *Another Way* for the expression in part (d).
- **Response Logs:** "The hardest part right now is …"

Solving Real-Life Problems

EXAMPLE 5 Simplifying a Real-Life Expression

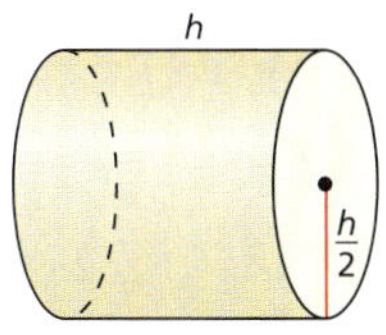

Volume = ?

$2\pi r^3$	$\pi h^3 2^{-2}$	$\pi h 4^{-1}$
$\frac{\pi h^2}{4}$	$\frac{\pi h^3}{4}$	$\frac{\pi h^3}{2}$

Which of the expressions shown represent the volume of the cylinder, where r is the radius and h is the height?

SOLUTION

$V = \pi r^2 h$ Formula for the volume of a cylinder

$= \pi\left(\frac{h}{2}\right)^2(h)$ Substitute $\frac{h}{2}$ for r.

$= \pi\left(\frac{h^2}{2^2}\right)(h)$ Power of a Quotient Property

$= \frac{\pi h^3}{4}$ Simplify.

Any expression equivalent to $\frac{\pi h^3}{4}$ represents the volume of the cylinder.

- You can use the properties of exponents to write $\pi h^3 2^{-2}$ as $\frac{\pi h^3}{4}$.
- Note $h = 2r$. When you substitute $2r$ for h in $\frac{\pi h^3}{4}$, you can write $\frac{\pi(2r)^3}{4}$ as $2\pi r^3$.
- None of the other expressions are equivalent to $\frac{\pi h^3}{4}$.

▶ The expressions $2\pi r^3$, $\pi h^3 2^{-2}$, and $\frac{\pi h^3}{4}$ represent the volume of the cylinder.

REMEMBER

A number is written in scientific notation when it is of the form $a \times 10^b$, where $1 \le a < 10$ and b is an integer.

EXAMPLE 6 Solving a Real-Life Problem

A jellyfish emits about 1.25×10^8 particles of light, or photons, in 6.25×10^{-4} second. How many photons does the jellyfish emit each second? Write your answer in scientific notation and in standard form.

SOLUTION

Divide to find the unit rate.

$\frac{1.25 \times 10^8}{6.25 \times 10^{-4}}$ ← photons / ← seconds Write the rate.

$= \frac{1.25}{6.25} \times \frac{10^8}{10^{-4}}$ Rewrite.

$= 0.2 \times 10^{12}$ Simplify.

$= 2 \times 10^{11}$ Write in scientific notation.

▶ The jellyfish emits 2×10^{11}, or 200,000,000,000 photons per second.

Monitoring Progress

Help in English and Spanish at *BigIdeasMath.com*

15. Write two expressions that represent the area of a base of the cylinder in Example 5.

16. It takes the Sun about 2.3×10^8 years to orbit the center of the Milky Way. It takes Pluto about 2.5×10^2 years to orbit the Sun. How many times does Pluto orbit the Sun while the Sun completes one orbit around the center of the Milky Way? Write your answer in scientific notation.

Extra Example 5

Which of the expressions shown represent the volume of the cone, where r is the radius and h is the height?

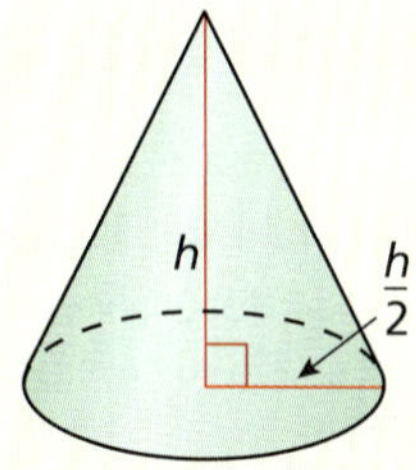

$\frac{\pi h^3}{12}$ $\frac{\pi r^3}{3}$ $\frac{\pi h^2}{6}$

$\frac{\pi h^2}{12}$ $\frac{2\pi r^3}{3}$ $8\pi r^3$ $\frac{\pi h^3}{12}, \frac{2\pi r^3}{3}$

Extra Example 6

A laptop computer uses approximately 8.4×10^{-3} kilowatt-hours of energy in 3.5×10^{-1} hours. How many kilowatt-hours of energy does the laptop use per hour? Write your answer in scientific notation and in standard form. The laptop uses 2.4×10^{-2}, or 0.024 kilowatt-hour of energy per hour.

MONITORING PROGRESS ANSWERS

15. πr^2; $\frac{\pi h^2}{4}$

16. 9.2×10^5

Laurie's Notes Teacher Actions

? "How do you find the volume of a cylinder?" area of the base times the height Have students work with their partners to simplify the six expressions to determine which are equivalent to $\frac{\pi h^3}{4}$.

- **Look For and Make Use of Structure:** In Example 6, help students see this problem as simplifying $\frac{a \cdot b^n}{c \cdot d^m}$, which is equivalent to $\frac{a}{c} \cdot \frac{b^n}{d^m}$.

Closure

- **Response Logs:** Select from "What is confusing me the most is …" or "I was successful in …."

Assignment Guide and Homework Check

ASSIGNMENT

Basic: 1–4, 5–45 odd, 55, 64, 67, 70–75

Average: 1–3, 4–44 even, 45, 46–58 even, 64, 67, 70–75

Advanced: 1–4, 10–24 even, 32–44 even, 45, 46–64 even, 65–75

HOMEWORK CHECK

Basic: 5, 13, 23, 37, 55

Average: 8, 16, 24, 38, 56

Advanced: 10, 18, 32, 40, 56

ANSWERS

1. Product of Powers Property, Power of a Power Property, definition of negative exponents; Use the Product of Powers Property to simplify the expression inside the parentheses to 4^4. Then, use the Power of a Power Property to simplify the entire expression to 4^{-8}. Then, use the definition of negative exponents to produce the final answer, $\frac{1}{4^8} = \frac{1}{65{,}536}$.
2. The Power of a Product Property is used when finding a power of a product by finding the power of each factor and multiplying.
3. The Quotient of Powers Property is used when dividing powers that have the same base. The answer is the common base raised to the difference of the exponents of the numerator and denominator.
4. Simplify $3^{6 \cdot 3}$; 3^{18}; 3^9
5. 1
6. 1
7. $\frac{1}{625}$
8. $-\frac{1}{32}$
9. $\frac{1}{16}$
10. $-\frac{1}{5}$
11. $-\frac{4}{3}$
12. $\frac{81}{64}$
13. $\frac{1}{x^7}$
14. 1
15. $\frac{9}{y^3}$
16. $\frac{15}{c^8}$

6.1 Exercises

Dynamic Solutions available at *BigIdeasMath.com*

Vocabulary and Core Concept Check

1. **VOCABULARY** Which definitions or properties would you use to simplify the expression $(4^8 \cdot 4^{-4})^{-2}$? Explain.
2. **WRITING** Explain when and how to use the Power of a Product Property.
3. **WRITING** Explain when and how to use the Quotient of Powers Property.
4. **DIFFERENT WORDS, SAME QUESTION** Which is different? Find "both" answers.

Simplify $3^3 \cdot 3^6$.	Simplify 3^{3+6}.	Simplify $3^{6 \cdot 3}$.	Simplify $3^6 \cdot 3^3$.

Monitoring Progress and Modeling with Mathematics

In Exercises 5–12, evaluate the expression. *(See Example 1.)*

5. $(-7)^0$
6. 4^0
7. 5^{-4}
8. $(-2)^{-5}$
9. $\frac{2^{-4}}{4^0}$
10. $\frac{5^{-1}}{-9^0}$
11. $\frac{-3^{-3}}{6^{-2}}$
12. $\frac{(-8)^{-2}}{3^{-4}}$

In Exercises 13–22, simplify the expression. Write your answer using only positive exponents. *(See Example 2.)*

13. x^{-7}
14. y^0
15. $9x^0y^{-3}$
16. $15c^{-8}d^0$
17. $\frac{2^{-2}m^{-3}}{n^0}$
18. $\frac{10^0r^{-11}s}{3^2}$
19. $\frac{4^{-3}a^0}{b^{-7}}$
20. $\frac{p^{-8}}{7^{-2}q^{-9}}$
21. $\frac{2^2y^{-6}}{8^{-1}z^0x^{-7}}$
22. $\frac{13x^{-5}y^0}{5^{-3}z^{-10}}$

In Exercises 23–32, simplify the expression. Write your answer using only positive exponents. *(See Example 3.)*

23. $\frac{5^6}{5^2}$
24. $\frac{(-6)^8}{(-6)^5}$
25. $(-9)^2 \cdot (-9)^2$
26. $4^{-5} \cdot 4^5$
27. $(p^6)^4$
28. $(s^{-5})^3$
29. $6^{-8} \cdot 6^5$
30. $-7 \cdot (-7)^{-4}$
31. $\frac{x^5}{x^4} \cdot x$
32. $\frac{z^8 \cdot z^2}{z^5}$

33. **USING PROPERTIES** A microscope magnifies an object 10^5 times. The length of an object is 10^{-7} meter. What is its magnified length?

34. **USING PROPERTIES** The area of the rectangular computer chip is $112a^3b^2$ square microns. What is the length?

width = $8ab$ microns

ERROR ANALYSIS In Exercises 35 and 36, describe and correct the error in simplifying the expression.

35. ✗
$$2^4 \cdot 2^5 = (2 \cdot 2)^{4+5} = 4^9$$

36. ✗
$$\frac{x^5 \cdot x^3}{x^4} = \frac{x^8}{x^4} = x^{8/4} = x^2$$

17. $\frac{1}{4m^3}$
18. $\frac{s}{9r^{11}}$
19. $\frac{b^7}{64}$
20. $\frac{49q^9}{p^8}$
21. $\frac{32x^7}{y^6}$
22. $\frac{1625z^{10}}{x^5}$
23. 625
24. -216
25. 6561
26. 1
27. p^{24}
28. $\frac{1}{s^{15}}$
29. $\frac{1}{216}$
30. $-\frac{1}{343}$
31. x^2
32. z^5
33. 10^{-2} m
34. $14a^2b$ microns
35. The product has a base of 2, not $2 \cdot 2$; $2^4 \cdot 2^5 = 2^9$
36. The exponent of the quotient should be the difference of the exponents, not the quotient of the exponents; $\frac{x^8}{x^4} = x^4$

In Exercises 37–44, simplify the expression. Write your answer using only positive exponents. *(See Example 4.)*

37. $(-5z)^3$

38. $(4x)^{-4}$

39. $\left(\frac{6}{n}\right)^{-2}$

40. $\left(\frac{-t}{3}\right)^2$

41. $(3s^8)^{-5}$

42. $(-5p^3)^3$

43. $\left(-\frac{w^3}{6}\right)^{-2}$

44. $\left(\frac{1}{2r^6}\right)^{-6}$

45. USING PROPERTIES Which of the expressions represent the volume of the sphere? Explain. *(See Example 5.)*

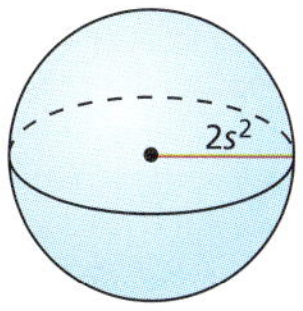

Ⓐ $\left(\frac{3s^2}{2^4\pi s^8}\right)^{-1}$

Ⓑ $(2^5\pi s^6)(3^{-1})$

Ⓒ $\frac{32\pi s^6}{3}$

Ⓓ $(2s)^5 \cdot \frac{\pi s}{3}$

Ⓔ $\left(\frac{3\pi s^6}{32}\right)^{-1}$

Ⓕ $\frac{32}{3}\pi s^5$

46. MODELING WITH MATHEMATICS Diffusion is the movement of molecules from one location to another. The time t (in seconds) it takes molecules to diffuse a distance of x centimeters is given by $t = \frac{x^2}{2D}$, where D is the diffusion coefficient. The diffusion coefficient for a drop of ink in water is about 10^{-5} square centimeters per second. How long will it take the ink to diffuse 1 micrometer (10^{-4} centimeter)?

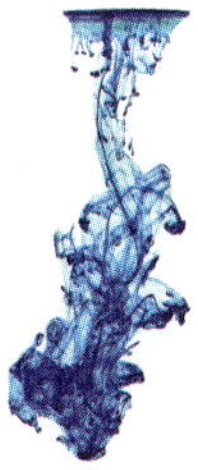

In Exercises 47–50, simplify the expression. Write your answer using only positive exponents.

47. $\left(\frac{2x^{-2}y^3}{3xy^{-4}}\right)^4$

48. $\left(\frac{4s^5t^{-7}}{-2s^{-2}t^4}\right)^3$

49. $\left(\frac{3m^{-5}n^2}{4m^{-2}n^0}\right)^2 \cdot \left(\frac{mn^4}{9n}\right)^2$

50. $\left(\frac{3x^3y^0}{x^{-2}}\right)^4 \cdot \left(\frac{y^2x^{-4}}{5xy^{-8}}\right)^3$

In Exercises 51–54, evaluate the expression. Write your answer in scientific notation and standard form.

51. $(3 \times 10^2)(1.5 \times 10^{-5})$

52. $(6.1 \times 10^{-3})(8 \times 10^9)$

53. $\frac{(6.4 \times 10^7)}{(1.6 \times 10^5)}$

54. $\frac{(3.9 \times 10^{-5})}{(7.8 \times 10^{-8})}$

55. PROBLEM SOLVING In 2012, on average, about 9.46×10^{-1} pound of potatoes was produced for every 2.3×10^{-5} acre harvested. How many pounds of potatoes on average were produced for each acre harvested? Write your answer in scientific notation and in standard form. *(See Example 6.)*

56. PROBLEM SOLVING The speed of light is approximately 3×10^5 kilometers per second. How long does it take sunlight to reach Jupiter? Write your answer in scientific notation and in standard form.

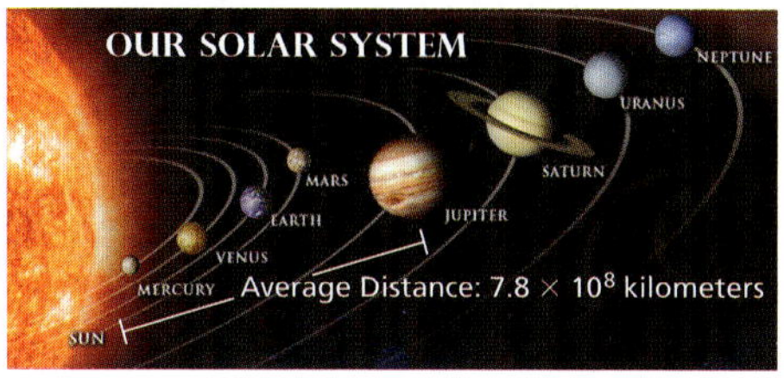

57. MATHEMATICAL CONNECTIONS Consider Cube A and Cube B.

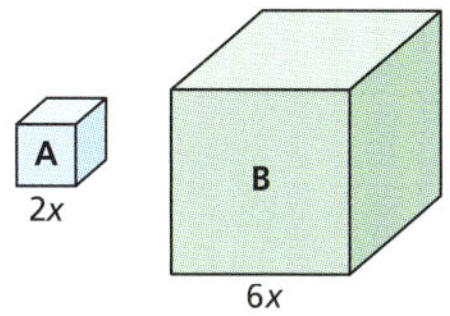

a. Which property of exponents should you use to simplify an expression for the volume of each cube?

b. How can you use the Power of a Quotient Property to find how many times greater the volume of Cube B is than the volume of Cube A?

58. PROBLEM SOLVING A byte is a unit used to measure a computer's memory. The table shows the numbers of bytes in several units of measure.

Unit	kilobyte	megabyte	gigabyte	terabyte
Number of bytes	2^{10}	2^{20}	2^{30}	2^{40}

a. How many kilobytes are in 1 terabyte?

b. How many megabytes are in 16 gigabytes?

c. Another unit used to measure a computer's memory is a bit. There are 8 bits in a byte. How can you convert the number of bytes in each unit of measure given in the table to bits? Can you still use a base of 2? Explain.

Dynamic Teaching Tools
- Dynamic Assessment & Progress Monitoring Tool
- Interactive Whiteboard Lesson Library
- Dynamic Classroom with Dynamic Investigations

ANSWERS

37. $-125z^3$

38. $\frac{1}{256x^4}$

39. $\frac{n^2}{36}$

40. $\frac{t^2}{9}$

41. $\frac{1}{243s^{40}}$

42. $-125p^9$

43. $\frac{36}{w^6}$

44. $64r^{36}$

45. B, C, D; These expressions simplify to be the volume of the sphere, which is $\frac{32\pi s^6}{3}$.

46. 5×10^{-4} sec, or 0.0005 sec

47. $\frac{16y^{28}}{81x^{12}}$

48. $\frac{-8s^{21}}{t^{33}}$

49. $\frac{n^{10}}{144m^4}$

50. $\frac{81x^5y^{30}}{125}$

51. 4.5×10^{-3}; 0.0045

52. 4.88×10^7; 48,800,000

53. 4×10^2; 400

54. 5×10^2; 500

55. about 4.113×10^4 lb/acre; about 41,130 lb/acre

56. 2.6×10^3 sec; 2600 sec

57. a. Power of a Product Property

b. Express $\frac{(6x)^3}{(2x)^3}$ as $\left(\frac{6x}{2x}\right)^3$. Simplify the expression inside the parentheses to produce $(3)^3$, so the volume is 27 times greater.

58. a. 2^{30} kilobytes

b. 16,384 megabytes

c. Multiply each number in the table by 8; yes; The number 8 can be expressed as 2^3, so multiply each number in the table by 2^3. Because the values have a common base of 2, they can be simplified using the Product of Powers Property.

ANSWERS

59. $(2ab)^3$

60. $(4rs)^2$

61. $(2w^3z^2)^6$

62. $(9x^2y^4)^2$ or $(3xy^2)^4$

63. **a.** $\left(\frac{1}{6}\right)^n$

b. $\frac{1}{1296}$

c. $\frac{1}{32}$; The probability of flipping heads once is $\frac{1}{2}$, and $\left(\frac{1}{2}\right)^5$ is $\frac{1}{32}$.

64. **a.** $\frac{1}{2}; \frac{1}{4}; \frac{1}{8}; \frac{1}{16}$

b. $2^{-1}; 2^{-2}; 2^{-3}; 2^{-4}$

65. $x = 8, y = -1$; Using the Quotient of Powers Property, you can conclude from the first equation that $x - y = 9$. Using the Product of Powers Property and the Quotient of Powers Property, you can conclude from the second equation that $x + 2 - 3y = 13$. Use these equations to solve a system of linear equations.

66. *Sample answer:* $r = x^2, h = 81x^4$

67. no; The mass of the seed from the double coconut palm is 10 kilograms.

68. **a.** 2^{13}

b. 2^{23}

c. They become 3^{13} and 3^{23}.

69–75. See Additional Answers.

REWRITING EXPRESSIONS In Exercises 59–62, rewrite the expression as a power of a product.

59. $8a^3b^3$ **60.** $16r^2s^2$

61. $64w^{18}z^{12}$ **62.** $81x^4y^8$

63. USING STRUCTURE The probability of rolling a 6 on a number cube is $\frac{1}{6}$. The probability of rolling a 6 twice in a row is $\left(\frac{1}{6}\right)^2 = \frac{1}{36}$.

a. Write an expression that represents the probability of rolling a 6 n times in a row.

b. What is the probability of rolling a 6 four times in a row?

c. What is the probability of flipping heads on a coin five times in a row? Explain.

64. HOW DO YOU SEE IT? The shaded part of Figure n represents the portion of a piece of paper visible after folding the paper in half n times.

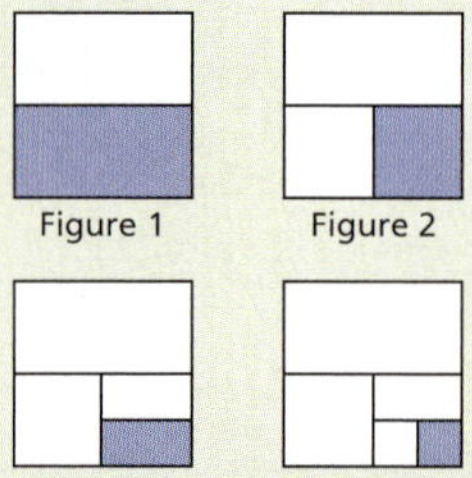

a. What fraction of the original piece of paper is each shaded part?

b. Rewrite each fraction from part (a) in the form 2^x.

65. REASONING Find x and y when $\frac{b^x}{b^y} = b^9$ and $\frac{b^x \cdot b^2}{b^{3y}} = b^{13}$. Explain how you found your answer.

66. THOUGHT PROVOKING Write expressions for r and h so that the volume of the cone can be represented by the expression $27\pi x^8$. Find r and h.

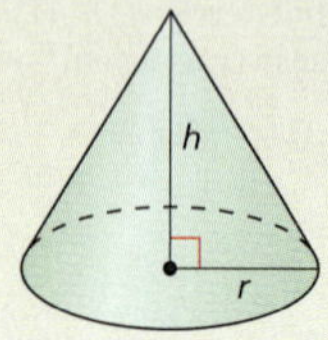

67. MAKING AN ARGUMENT One of the smallest plant seeds comes from an orchid, and one of the largest plant seeds comes from a double coconut palm. A seed from an orchid has a mass of 10^{-6} gram. The mass of a seed from a double coconut palm is 10^{10} times the mass of the seed from the orchid. Your friend says that the seed from the double coconut palm has a mass of about 1 kilogram. Is your friend correct? Explain.

68. CRITICAL THINKING Your school is conducting a survey. Students can answer the questions in either part with "agree" or "disagree."

Part 1: 13 questions		
Part 2: 10 questions		
Part 1: Classroom	Agree	Disagree
1. I come prepared for class.	O	O
2. I enjoy my assignments.	O	O

a. What power of 2 represents the number of different ways that a student can answer all the questions in Part 1?

b. What power of 2 represents the number of different ways that a student can answer all the questions on the entire survey?

c. The survey changes, and students can now answer "agree," "disagree," or "no opinion." How does this affect your answers in parts (a) and (b)?

69. ABSTRACT REASONING Compare the values of a^n and a^{-n} when $n < 0$, when $n = 0$, and when $n > 0$ for (a) $a > 1$ and (b) $0 < a < 1$. Explain your reasoning.

Maintaining Mathematical Proficiency

Reviewing what you learned in previous grades and lessons

Find the square root(s). *(Skills Review Handbook)*

70. $\sqrt{25}$ **71.** $-\sqrt{100}$ **72.** $\pm\sqrt{\frac{1}{64}}$

Classify the real number in as many ways as possible. *(Skills Review Handbook)*

73. 12 **74.** $\frac{65}{9}$ **75.** $\frac{\pi}{4}$

Mini-Assessment

1. Evaluate the expression $\frac{6^0}{7^{-2}}$. 49

Simplify the expression. Write your answer using only positive exponents.

2. $(x^{-4})^3$ $x^{-12} = \frac{1}{x^{12}}$

3. $\left(\frac{5n}{-3}\right)^{-3}$ $-\frac{27}{125n^3}$

4. A DVD player uses approximately 1.35×10^{-3} kilowatt-hours of energy in 4.5×10^{-2} hours. How many kilowatt-hours of energy does the DVD player use per hour? Write your answer in scientific notation and in standard form. 3.0×10^{-2}, or 0.03 kilowatt-hour of energy per hour

If students need help...	*If students got it...*
Resources by Chapter • Practice A and Practice B • Puzzle Time	Resources by Chapter • Enrichment and Extension • Cumulative Review
Student Journal • Practice	Start the *next* Section
Differentiating the Lesson Skills Review Handbook	

Laurie's Notes

Overview of Section 6.2

Introduction

- This lesson introduces the language and notation of *n*th roots and how to evaluate expressions involving *n*th roots. The examples presented are numeric so that it is possible to evaluate them. Simplifying algebraic expressions is left for another course.
- Real-life applications are presented at the end of the lesson.

Resources

- Calculators will be necessary to evaluate *n*th roots. Students may need instruction on how to evaluate expressions, both in radical form and rational exponent form.

Formative Assessment Tips

- Refer to page T-290 for a description of *Response Logs*. In this lesson and throughout this chapter, there will be sentence stems to prompt student writing. Use your judgment for placement in the lesson, amount of time to allow for writing, and the actual sentence stem given to students.
- Beginning list of writing stems:

  A sticky part for me is …
  I am feeling good about …
  I feel confident about …
  I figured out that …
  I got stuck …
  I made progress with …
  I was really surprised when …
  I will understand this better if I …
  I wish I had …
  I'm still not sure about …
  Right now I am thinking about …
  Right now I need to …
  Tomorrow I need to find out …
  What is confusing me the most is …
  At first I thought … but now I think …
  I didn't expect …
  I feel like I made progress today …
  I figured these out because …
  I learned to …
  I need to rethink …
  I was successful in …
  I wish I could …
  I'm not sure …
  It helped me today when …
  Right now I know …
  The hardest part right now is …
  What confused me the most was …
  What is puzzling me the most is …

Another Way

- Consider the cube root of 8, or $8^{1/3}$, rewritten as $(2 \cdot 2 \cdot 2)^{1/3}$. The index 1/3 indicates that the answer is 1 of 3 equal factors of 8, or 2.
- This helps students when they get to $8^{2/3}$, rewritten as $(2 \cdot 2 \cdot 2)^{2/3}$. Here the answer is 2 of the 3 equal factors of 8, or $2 \cdot 2 = 4$.
- This will also work with negatives, bases with variables, and the like.

Pacing Suggestion

- Complete the explorations and then start the formal lesson by defining the *n*th root of a number.

Dynamic Teaching Tools

- Dynamic Assessment & Progress Monitoring Tool
- Lesson Planning Tool
- Interactive Whiteboard Lesson Library
- Dynamic Classroom with Dynamic Investigations

What Your Students Will Learn

- Find nth roots and write the nth root of a as a power of a.
- Evaluate expressions with rational exponents.
- Solve real-life problems involving rational exponents.

Laurie's Notes

Exploration

Motivate

- "What do you think 8 yards of gravel looks like?" Students may not know that gravel is generally sold by the cubic yard, and the word *cubic* is rarely stated. So, 8 yards refers to 8 cubic yards.
- ? "If the gravel were dumped inside a cube-shaped container, what would the dimensions of the cube be?" 2 yards on each edge
- ? "What would the dimensions be in feet?" 6 feet on each edge
- ? "What would the volume be in cubic feet?" 216 ft^3
- Make the connection that 8 $\text{yd}^3 = 216\ \text{ft}^3$.

Discuss

- Introduce the cube root symbol.

Exploration 1

- Students will find a calculator helpful in testing their guesses for side length. For instance, they probably do not know that $15^3 = 3375$, although the 5 in the ones place should be an important clue!
- ? **Reason Abstractly and Quantitatively:** When students have finished, spend time discussing the two additional questions posed. "How did you determine which cube was the largest?" Listen for something about converting units and that 1 meter is longer than 1 yard.
- After finishing this exploration, students should be comfortable with the idea that cubing and finding a cube root are inverse operations. Have students write the results for each cube and then substitute to write an equivalent statement involving the cube. For example, in part (a), students should write $\sqrt[3]{27} = 3 \rightarrow \sqrt[3]{3^3} = 3$.

Exploration 2

- Write: "When you raise the nth root of a number to the nth power, you get the original number." Relate this to the solution statements in Exploration 1, and note the distinction. In Exploration 1, the cube root of a cubed number is the number. In the statement above, the cube of the cube root of a number is the number.

  Exploration 1: $\sqrt[3]{8^3} = 8$ Exploration 2: $\left(\sqrt[3]{8}\right)^3 = 8$

- Use this exploration to talk about nth roots for values of n other than 3.
- **Reason Abstractly and Quantitatively:** Discuss the justifications for part (a). Listen for student reasoning, such as $\sqrt[4]{25}$ is between 2 and 3 because $2^4 = 16$ and $3^4 = 81$. Also, $\sqrt[4]{25}$ is closer to 2 than 3.
- **Construct Viable Arguments and Critique the Reasoning of Others:** Ask volunteers to share their answers, explaining how they made their matches.

Communicate Your Answer

- ? "Could you lift 4000 kilograms? About how many pounds is 4000 kilograms?" no; 8800 lb
- ? "Where is the femur on a human? Do you think the size of your femur is greater than that of a dinosaur?" thighbone; yes, for some dinosaurs

Connecting to Next Step

- The explorations give students an opportunity to explore nth roots, particularly cube roots. In the formal lesson, the language and notation of nth roots are presented.

6.2 Radicals and Rational Exponents

Essential Question How can you write and evaluate an *n*th root of a number?

Recall that you cube a number as follows.

3rd power

$$2^3 = 2 \cdot 2 \cdot 2 = 8$$ 2 cubed is 8.

To "undo" cubing a number, take the cube root of the number.

Symbol for cube root is $\sqrt[3]{}$.

$$\sqrt[3]{8} = \sqrt[3]{2^3} = 2$$ The cube root of 8 is 2.

EXPLORATION 1 Finding Cube Roots

Work with a partner. Use a cube root symbol to write the side length of each cube. Then find the cube root. Check your answers by multiplying. Which cube is the largest? Which two cubes are the same size? Explain your reasoning.

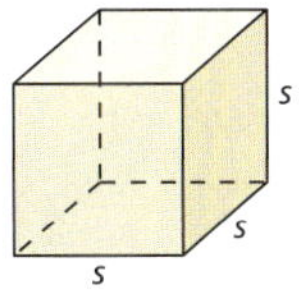

a. Volume = 27 ft^3 **b.** Volume = 125 cm^3 **c.** Volume = 3375 in.^{3}

d. Volume = 3.375 m^3 **e.** Volume = 1 yd^3 **f.** Volume = $\frac{125}{8}$ mm^3

JUSTIFYING CONCLUSIONS

To be proficient in math, you need to justify your conclusions and communicate them to others.

EXPLORATION 2 Estimating *n*th Roots

Work with a partner. Estimate each positive *n*th root. Then match each *n*th root with the point on the number line. Justify your answers.

a. $\sqrt[4]{25}$ **b.** $\sqrt{0.5}$ **c.** $\sqrt[5]{2.5}$

d. $\sqrt[3]{65}$ **e.** $\sqrt[3]{55}$ **f.** $\sqrt[6]{20{,}000}$

Communicate Your Answer

3. How can you write and evaluate an *n*th root of a number?

4. The body mass *m* (in kilograms) of a dinosaur that walked on two feet can be modeled by

$$m = (0.00016)C^{2.73}$$

where *C* is the circumference (in millimeters) of the dinosaur's femur. The mass of a *Tyrannosaurus rex* was 4000 kilograms. Use a calculator to approximate the circumference of its femur.

Dynamic Teaching Tools

- Dynamic Assessment & Progress Monitoring Tool
- Lesson Planning Tool
- Interactive Whiteboard Lesson Library
- Dynamic Classroom with Dynamic Investigations

ANSWERS

1. **a.** $\sqrt[3]{27}$ ft; 3 ft
   b. $\sqrt[3]{125}$ cm; 5 cm
   c. $\sqrt[3]{3375}$ in.; 15 in.
   d. $\sqrt[3]{3.375}$ m; 1.5 m
   e. $\sqrt[3]{1}$ yd; 1 yd
   f. $\sqrt[3]{\frac{125}{8}}$; 2.5 mm

   The cube in part (d) has the largest side length of 1.5 meters. The cubes in parts (a) and (e) have equal side lengths because 3 feet = 1 yard.

2. **a.** *Sample answer:* 2.2; C; 25 is between $2^4 = 16$ and $3^4 = 81$, and C is the only point on the graph between 2 and 3.
   b. *Sample answer:* 0.7; A; 0.5 is between $0^2 = 0$ and $1^2 = 1$, and A is the only point on the graph between 0 and 1.
   c. *Sample answer:* 1.2; B; 2.5 is between $1^5 = 1$ and $2^5 = 32$, and B is the only point on the graph between 1 and 2.
   d. *Sample answer:* 4.0; E; 65 is between $4^3 = 64$ and $5^3 = 125$, and E is the only point on the graph between 4 and 5.
   e. *Sample answer:* 3.8; D; 55 is between $3^3 = 27$ and $4^3 = 64$, and D is the only point on the graph between 3 and 4.
   f. *Sample answer:* 5.2; F; 20,000 is between $5^6 = 15{,}625$ and $6^6 = 46{,}656$, and F is the only point on the graph between 5 and 6.

3. Find what real number multiplied by itself *n* times gives you that number. If that is not possible, determine which *n*th powers the number is between and estimate the decimal part.

4. about 512.7 mm

Extra Example 1

Find the indicated real nth root(s) of a.

a. $n = 3, a = -64$

$\sqrt[3]{-64} = -4$, or $(-64)^{1/3} = -4$

b. $n = 2, a = 81$

$\pm\sqrt{81} = \pm 9$, or $\pm(81)^{1/2} = \pm 9$

MONITORING PROGRESS ANSWERS

1. -5
2. ± 2

6.2 Lesson

What You Will Learn

- Find nth roots.
- Evaluate expressions with rational exponents.
- Solve real-life problems involving rational exponents.

Core Vocabulary

*n*th root of *a*, *p. 300*
radical, *p. 300*
index of a radical, *p. 300*

Previous
square root

Finding *n*th Roots

You can extend the concept of a square root to other types of roots. For example, 2 is a cube root of 8 because $2^3 = 8$, and 3 is a fourth root of 81 because $3^4 = 81$. In general, for an integer n greater than 1, if $b^n = a$, then b is an ***n*th root of *a***. An nth root of a is written as $\sqrt[n]{a}$, where the expression $\sqrt[n]{a}$ is called a **radical** and n is the **index** of the radical.

You can also write an nth root of a as a power of a. If you assume the Power of a Power Property applies to rational exponents, then the following is true.

$$(a^{1/2})^2 = a^{(1/2) \cdot 2} = a^1 = a$$
$$(a^{1/3})^3 = a^{(1/3) \cdot 3} = a^1 = a$$
$$(a^{1/4})^4 = a^{(1/4) \cdot 4} = a^1 = a$$

Because $a^{1/2}$ is a number whose square is a, you can write $\sqrt{a} = a^{1/2}$. Similarly, $\sqrt[3]{a} = a^{1/3}$ and $\sqrt[4]{a} = a^{1/4}$. In general, $\sqrt[n]{a} = a^{1/n}$ for any integer n greater than 1.

READING

$\pm\sqrt[n]{a}$ represents both the positive and negative nth roots of a.

Core Concept

Real *n*th Roots of *a*

Let n be an integer greater than 1, and let a be a real number.

- If n is odd, then a has one real nth root: $\sqrt[n]{a} = a^{1/n}$
- If n is even and $a > 0$, then a has two real nth roots: $\pm\sqrt[n]{a} = \pm a^{1/n}$
- If n is even and $a = 0$, then a has one real nth root: $\sqrt[n]{0} = 0$
- If n is even and $a < 0$, then a has no real nth roots.

The nth roots of a number may be real numbers or *imaginary numbers*. You will study imaginary numbers in a future course.

EXAMPLE 1 Finding *n*th Roots

Find the indicated real nth root(s) of a.

a. $n = 3, a = -27$

b. $n = 4, a = 16$

SOLUTION

a. The index $n = 3$ is odd, so -27 has one real cube root. Because $(-3)^3 = -27$, the cube root of -27 is $\sqrt[3]{-27} = -3$, or $(-27)^{1/3} = -3$.

b. The index $n = 4$ is even, and $a > 0$. So, 16 has two real fourth roots. Because $2^4 = 16$ and $(-2)^4 = 16$, the fourth roots of 16 are $\pm\sqrt[4]{16} = \pm 2$, or $\pm 16^{1/4} = \pm 2$.

Monitoring Progress

Help in English and Spanish at *BigIdeasMath.com*

Find the indicated real *n*th root(s) of *a*.

1. $n = 3, a = -125$
2. $n = 6, a = 64$

Laurie's Notes Teacher Actions

- Introduce the vocabulary and notation associated with nth roots. Connect the radical form with the rational exponent form.
- **? Probing Question** and **Turn and Talk:** "Will the nth root of 64 always be a real number? yes Will the nth root of -64 always be a real number?" no You want students to activate their thinking about the various cases of $\sqrt[n]{a}$ (n is odd or even; sign of a) before writing the *Core Concept*.
- Students will say that $\sqrt{-64}$ is impossible or cannot be done. Explain that imaginary numbers will be introduced in a later course.

Evaluating Expressions with Rational Exponents

Recall that the radical $\sqrt{a}$ indicates the positive square root of a. Similarly, an nth root of a, $\sqrt[n]{a}$, with an *even* index indicates the positive nth root of a.

REMEMBER

The expression under the radical sign is the radicand.

EXAMPLE 2 Evaluating nth Root Expressions

Evaluate each expression.

a. $\sqrt[3]{-8}$ **b.** $-\sqrt[3]{8}$ **c.** $16^{1/4}$ **d.** $(-16)^{1/4}$

SOLUTION

a. $\sqrt[3]{-8} = \sqrt[3]{(-2)\cdot(-2)\cdot(-2)}$ Rewrite the expression showing factors.

$= -2$ Evaluate the cube root.

b. $-\sqrt[3]{8} = -\left(\sqrt[3]{2\cdot 2\cdot 2}\right)$ Rewrite the expression showing factors.

$= -(2)$ Evaluate the cube root.

$= -2$ Simplify.

c. $16^{1/4} = \sqrt[4]{16}$ Rewrite the expression in radical form.

$= \sqrt[4]{2\cdot 2\cdot 2\cdot 2}$ Rewrite the expression showing factors.

$= 2$ Evaluate the fourth root.

d. $(-16)^{1/4}$ is not a real number because there is no real number that can be multiplied by itself four times to produce -16.

A rational exponent does not have to be of the form $1/n$. Other rational numbers such as $3/2$ can also be used as exponents. You can use the properties of exponents to evaluate or simplify expressions involving rational exponents.

STUDY TIP

You can rewrite $27^{2/3}$ as $27^{(1/3)\cdot 2}$ and then use the Power of a Power Property to show that

$27^{(1/3)\cdot 2} = (27^{1/3})^2$.

Core Concept

Rational Exponents

Let $a^{1/n}$ be an nth root of a, and let m be a positive integer.

Algebra $a^{m/n} = (a^{1/n})^m = (\sqrt[n]{a})^m$

Numbers $27^{2/3} = (27^{1/3})^2 = \left(\sqrt[3]{27}\right)^2$

EXAMPLE 3 Evaluating Expressions with Rational Exponents

Evaluate (a) $16^{3/4}$ and (b) $27^{4/3}$.

SOLUTION

a. $16^{3/4} = (16^{1/4})^3$ Rational exponents

$= 2^3$ Evaluate the nth root.

$= 8$ Evaluate the power.

b. $27^{4/3} = (27^{1/3})^4$

$= 3^4$

$= 81$

Monitoring Progress Help in English and Spanish at *BigIdeasMath.com*

Evaluate the expression.

3. $\sqrt[3]{-125}$ **4.** $(-64)^{2/3}$ **5.** $9^{5/2}$ **6.** $256^{3/4}$

Differentiated Instruction

Kinesthetic

Some students may benefit by having access to a scientific or graphing calculator. After evaluating the expressions, students can enter the expressions in the calculator, find the values, and compare the calculated values to their answers.

Extra Example 2

Evaluate each expression.

a. $\sqrt[4]{81}$ 3

b. $-\sqrt[4]{81}$ -3

c. $25^{1/2}$ 5

d. $(-25)^{1/2}$ $(-25)^{1/2}$ is not a real number because there is no real number that can be multiplied by itself to produce -25.

Extra Example 3

a. Evaluate $243^{2/5}$. 9

b. Evaluate $16^{3/2}$. 64

MONITORING PROGRESS ANSWERS

3. -5

4. 16

5. 243

6. 64

Laurie's Notes Teacher Actions

- Work through each part of Example 2.

? Attend to Precision: Make sure students read $\sqrt[3]{-8}$ and $-\sqrt[3]{8}$ correctly and understand why the solutions are the same. Ask, "Are $\sqrt[4]{-16}$ and $-\sqrt[4]{16}$ also equivalent?" no; $\sqrt[4]{-16}$ has no real roots, and $-\sqrt[4]{16} = -2$.

- Write the expression in Example 3(a). Explain that you can rewrite the exponent as $3 \cdot \frac{1}{4}$, but it is usually easier to evaluate the root first.
- **Response Logs:** "Right now I am thinking about . . ."

Extra Example 4

The radius r of a sphere is given by the equation $r = \left(\frac{3V}{4\pi}\right)^{1/3}$, where V is the volume of the sphere.

Volume = 463 cubic inches

Find the radius of the basketball to the nearest inch. Use 3.14 for π. The radius of the basketball is about 5 inches.

Extra Example 5

To calculate the annual inflation rate r (in decimal form) of an item that increases in value from P to F over a period of n years, you can use the equation $r = \left(\frac{F}{P}\right)^{1/n} - 1$. Find the annual inflation rate to the nearest tenth of a percent of a house that increases in value from \$300,000 to \$360,000 over a period of 8 years. The annual inflation rate is about 2.3%.

MONITORING PROGRESS ANSWERS

7. about 16 in.
8. about 6.0%

Solving Real-Life Problems

EXAMPLE 4 **Solving a Real-Life Problem**

Volume = 113 cubic feet

The radius r of a sphere is given by the equation $r = \left(\frac{3V}{4\pi}\right)^{1/3}$, where V is the volume of the sphere. Find the radius of the beach ball to the nearest foot. Use 3.14 for π.

SOLUTION

1. **Understand the Problem** You know the equation that represents the radius of a sphere in terms of its volume. You are asked to find the radius for a given volume.
2. **Make a Plan** Substitute the given volume into the equation. Then evaluate to find the radius.
3. **Solve the Problem**

$$r = \left(\frac{3V}{4\pi}\right)^{1/3} \quad \text{Write the equation.}$$

$$= \left(\frac{3(113)}{4(3.14)}\right)^{1/3} \quad \text{Substitute 113 for } V \text{ and 3.14 for } \pi.$$

$$= \left(\frac{339}{12.56}\right)^{1/3} \quad \text{Multiply.}$$

$$\approx 3 \quad \text{Use a calculator.}$$

▶ The radius of the beach ball is about 3 feet.

4. **Look Back** To check that your answer is reasonable, compare the size of the ball to the size of the woman pushing the ball. The ball appears to be slightly taller than the woman. The average height of a woman is between 5 and 6 feet. So, a radius of 3 feet, or height of 6 feet, seems reasonable for the beach ball.

EXAMPLE 5 **Solving a Real-Life Problem**

To calculate the annual inflation rate r (in decimal form) of an item that increases in value from P to F over a period of n years, you can use the equation $r = \left(\frac{F}{P}\right)^{1/n} - 1$. Find the annual inflation rate to the nearest tenth of a percent of a house that increases in value from \$200,000 to \$235,000 over a period of 5 years.

SOLUTION

$$r = \left(\frac{F}{P}\right)^{1/n} - 1 \quad \text{Write the equation.}$$

$$= \left(\frac{235{,}000}{200{,}000}\right)^{1/5} - 1 \quad \text{Substitute 235,000 for } F\text{, 200,000 for } P\text{, and 5 for } n.$$

$$= 1.175^{1/5} - 1 \quad \text{Divide.}$$

$$\approx 0.03278 \quad \text{Use a calculator.}$$

▶ The annual inflation rate is about 3.3%.

REMEMBER
To write a decimal as a percent, move the decimal point two places to the right. Then add a percent symbol.

Monitoring Progress

Help in English and Spanish at *BigIdeasMath.com*

7. **WHAT IF?** In Example 4, the volume of the beach ball is 17,000 cubic inches. Find the radius to the nearest inch. Use 3.14 for π.
8. The average cost of college tuition increases from \$8500 to \$13,500 over a period of 8 years. Find the annual inflation rate to the nearest tenth of a percent.

Laurie's Notes Teacher Actions

? Make Sense of Problems and Persevere in Solving Them: "Can you use the photo in Example 4 to estimate the radius?" yes; about 2 to 3 feet Discuss and let students work the problem with partners.

- The solution calls for a calculator, but without a calculator you could simplify $339 \div 12.56 \approx 26.99$ and estimate that $\sqrt[3]{26.99} \approx 3.0$.
- **Thumbs Up:** Discuss inflation rate. Have students work Example 5 and share as a class. Ask for a *Thumbs Up* self-assessment at this point.

Closure

- **Response Logs:** Select from "I figured these out because …" or "I feel confident about …" or "At first I thought … but now I think …."

6.2 Exercises

Dynamic Solutions available at *BigIdeasMath.com*

Vocabulary and Core Concept Check

1. **WRITING** Explain how to evaluate $81^{1/4}$.

2. **WHICH ONE DOESN'T BELONG?** Which expression does *not* belong with the other three? Explain your reasoning.

$(\sqrt[3]{27})^2$	$27^{2/3}$	3^2	$(\sqrt[2]{27})^3$

Monitoring Progress and Modeling with Mathematics

In Exercises 3 and 4, rewrite the expression in rational exponent form.

3. $\sqrt{10}$
4. $\sqrt[5]{34}$

In Exercises 5 and 6, rewrite the expression in radical form.

5. $15^{1/3}$
6. $140^{1/8}$

In Exercises 7–10, find the indicated real *n*th root(s) of *a*. *(See Example 1.)*

7. $n = 2, a = 36$
8. $n = 4, a = 81$
9. $n = 3, a = 1000$
10. $n = 9, a = -512$

MATHEMATICAL CONNECTIONS **In Exercises 11 and 12, find the dimensions of the cube. Check your answer.**

11. Volume = 64 in.^{3}

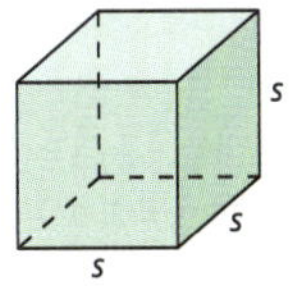

12. Volume = 216 cm^3

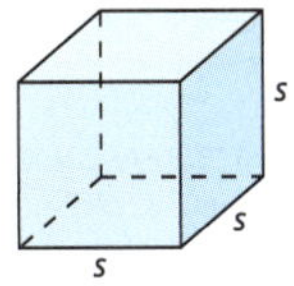

In Exercises 13–18, evaluate the expression. *(See Example 2.)*

13. $\sqrt[4]{256}$
14. $\sqrt[3]{-216}$
15. $\sqrt[3]{-343}$
16. $-\sqrt[5]{1024}$
17. $128^{1/7}$
18. $(-64)^{1/2}$

In Exercises 19 and 20, rewrite the expression in rational exponent form.

19. $(\sqrt[5]{8})^4$
20. $(\sqrt[5]{-21})^6$

In Exercises 21 and 22, rewrite the expression in radical form.

21. $(-4)^{2/7}$
22. $9^{5/2}$

In Exercises 23–28, evaluate the expression. *(See Example 3.)*

23. $32^{3/5}$
24. $125^{2/3}$
25. $(-36)^{3/2}$
26. $(-243)^{2/5}$
27. $(-128)^{5/7}$
28. $343^{4/3}$

29. **ERROR ANALYSIS** Describe and correct the error in rewriting the expression in rational exponent form.

✗ $(\sqrt[3]{2})^4 = 2^{3/4}$

30. **ERROR ANALYSIS** Describe and correct the error in evaluating the expression.

✗
$$(-81)^{3/4} = [(-81)^{1/4}]^3$$
$$= (-3)^3$$
$$= -27$$

In Exercises 31–34, evaluate the expression.

31. $\left(\frac{1}{1000}\right)^{1/3}$
32. $\left(\frac{1}{64}\right)^{1/6}$
33. $(27)^{-2/3}$
34. $(9)^{-5/2}$

Assignment Guide and Homework Check

ASSIGNMENT

Basic: 1, 2, 3–29 odd, 37, 40, 41, 44, 57–60

Average: 1, 2–44 even, 57–60

Advanced: 1, 2, 8–12 even, 16–44 even, 45–60

HOMEWORK CHECK

Basic: 7, 13, 23, 37, 41

Average: 8, 14, 26, 38, 42

Advanced: 10, 16, 26, 38, 42

ANSWERS

1. Find the fourth root of 81, or what real number multiplied by itself four times produces 81.
2. $(\sqrt[2]{27})^3$; It is the only one that is not equivalent to 9.
3. $10^{1/2}$
4. $34^{1/5}$
5. $\sqrt[3]{15}$
6. $\sqrt[8]{140}$
7. ± 6
8. ± 3
9. 10
10. -2
11. $s = 4$ in.
12. $s = 6$ cm
13. 4
14. -6
15. -7
16. -4
17. 2
18. not a real number
19. $8^{4/5}$
20. $-21^{6/5}$
21. $(\sqrt[7]{-4})^2$
22. $(\sqrt{9})^5$
23. 8
24. 25
25. not a real number
26. 9
27. -32
28. 2401
29. The numerator and denominator are reversed; $(\sqrt[3]{2})^4 = 2^{4/3}$
30. The number -81 does not have a real fourth root; not a real number
31. $\frac{1}{10}$
32. $\frac{1}{2}$
33. $\frac{1}{9}$
34. $\frac{1}{243}$

Dynamic Teaching Tools

- Dynamic Assessment & Progress Monitoring Tool
- Interactive Whiteboard Lesson Library
- Dynamic Classroom with Dynamic Investigations

ANSWERS

35. 6 ft^2

36. 243 mm

37. about 1 in.

38. about 44 m^3

39. Write the radicand, a, as the base and write the exponent as a fraction with the power, m, as the numerator and the index, n, as the denominator.

40. $x^{1/2}$ in.

41. about 5.5%

42. about 9.2%

43. -1, 0, and 1

44. no; If n is odd and a is negative, $\sqrt[n]{a}$ will be negative, and $-\sqrt[n]{a}$ will be positive.

45. $(xy)^{1/2}$

46. y^2

47. $2xy^2$

48. x^3y^5

49. about 1.38 ft

50. *Sample answer:* period of a pendulum: $T = 2\pi\left(\frac{\ell}{g}\right)^{1/2}$

51. always; Power of a Power Property

52. sometimes; true when $x = 1$, otherwise false

53–60. See Additional Answers.

Mini-Assessment

1. Find the indicated real nth roots of a.
   $n = 4, a = 625$ $\pm\sqrt[4]{625} = \pm5$, or $\pm(625)^{1/4} = \pm5$

Evaluate the expression.

2. $-\sqrt[4]{1296}$ -6
3. $64^{4/3}$ 256
4. Use the equation $r = \left(\frac{F}{P}\right)^{1/n} - 1$ to calculate the annual inflation rate r (in decimal form) of an item that increases in value from P to F over n years. The cost of a gallon of milk increased from \$2.81 to \$3.48 over 10 years. Find the annual inflation rate to the nearest tenth of a percent. about 2.2%

35. PROBLEM SOLVING A math club is having a bake sale. Find the area of the bake sale sign.

36. PROBLEM SOLVING The volume of a cube-shaped box is 27^5 cubic millimeters. Find the length of one side of the box.

37. MODELING WITH MATHEMATICS The radius r of the base of a cone is given by the equation

$$r = \left(\frac{3V}{\pi h}\right)^{1/2}$$

where V is the volume of the cone and h is the height of the cone. Find the radius of the paper cup to the nearest inch. Use 3.14 for π. *(See Example 4.)*

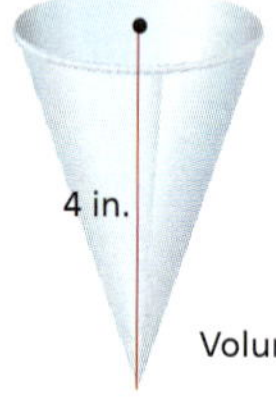

38. MODELING WITH MATHEMATICS The volume of a sphere is given by the equation $V = \frac{1}{6\sqrt{\pi}}S^{3/2}$, where S is the surface area of the sphere. Find the volume of a sphere, to the nearest cubic meter, that has a surface area of 60 square meters. Use 3.14 for π.

39. WRITING Explain how to write $(\sqrt[n]{a})^m$ in rational exponent form.

40. HOW DO YOU SEE IT? Write an expression in rational exponent form that represents the side length of the square.

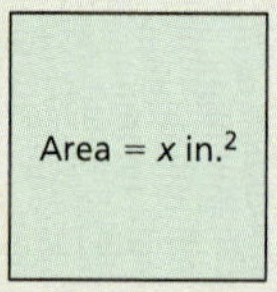

In Exercises 41 and 42, use the formula $r = \left(\frac{F}{P}\right)^{1/n} - 1$ to find the annual inflation rate to the nearest tenth of a percent. *(See Example 5.)*

41. A farm increases in value from \$800,000 to \$1,100,000 over a period of 6 years.

42. The cost of a gallon of gas increases from \$1.46 to \$3.53 over a period of 10 years.

43. REASONING For what values of x is $x = x^{1/5}$?

44. MAKING AN ARGUMENT Your friend says that for a real number a and a positive integer n, the value of $\sqrt[n]{a}$ is always positive and the value of $-\sqrt[n]{a}$ is always negative. Is your friend correct? Explain.

In Exercises 45–48, simplify the expression.

45. $(y^{1/6})^3 \cdot \sqrt{x}$

46. $(y \cdot y^{1/3})^{3/2}$

47. $x \cdot \sqrt[3]{y^6} + y^2 \cdot \sqrt[3]{x^3}$

48. $(x^{1/3} \cdot y^{1/2})^9 \cdot \sqrt{y}$

49. PROBLEM SOLVING The formula for the volume of a regular dodecahedron is $V \approx 7.66\ell^3$, where ℓ is the length of an edge. The volume of the dodecahedron is 20 cubic feet. Estimate the edge length.

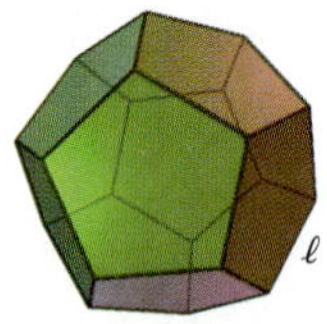

50. THOUGHT PROVOKING Find a formula (for instance, from geometry or physics) that contains a radical. Rewrite the formula using rational exponents.

ABSTRACT REASONING In Exercises 51–56, let x be a nonnegative real number. Determine whether the statement is *always*, *sometimes*, or *never* true. Justify your answer.

51. $(x^{1/3})^3 = x$

52. $x^{1/3} = x^{-3}$

53. $x^{1/3} = \sqrt[3]{x}$

54. $x^{1/3} = x^3$

55. $\frac{x^{2/3}}{x^{1/3}} = \sqrt[3]{x}$

56. $x = x^{1/3} \cdot x^3$

Maintaining Mathematical Proficiency

Reviewing what you learned in previous grades and lessons

Evaluate the function when $x = -3$, 0, and 8. *(Section 3.3)*

57. $f(x) = 2x - 10$

58. $w(x) = -5x - 1$

59. $h(x) = 13 - x$

60. $g(x) = 8x + 16$

If students need help...	*If students got it...*
Resources by Chapter • Practice A and Practice B • Puzzle Time	Resources by Chapter • Enrichment and Extension • Cumulative Review
Student Journal • Practice	Start the *next* Section
Differentiating the Lesson Skills Review Handbook	

Laurie's Notes

Overview of Section 6.3

Introduction

- In this lesson, exponential equations are defined. Students distinguish between linear and exponential equations, helping to focus on the definition of each. A linear function has a constant rate of change, and an exponential function has a constant percent rate of change. The constant percent rate of change is the constant factor that y is being multiplied by over equal intervals of the x-values.
- Once exponential equations are evaluated, they are graphed. The parent functions $y = 2^x$ and $y = \left(\frac{1}{2}\right)^x$ are used to explore transformations.

Resources

- The graphing calculator can be used to graph the exponential equations, giving students the ability to explore a series of graphs. Transformations of exponential equations can also be explored graphically, as well as numerically.

Formative Assessment Tips

- Refer to page T-290 for a description of *Response Logs*. In this lesson and throughout this chapter, there will be sentence stems to prompt student writing. Use your judgment for placement in the lesson, amount of time to allow for writing, and the actual sentence stem given to students.
- For a beginning list of writing stems, see page T-298.

Another Way

- **Look For and Express Regularity in Repeated Reasoning:** Show your students another way to rewrite the expression $(0.5)^{-2}$ in Example 2(b).

  $(0.5)^{-2} = (0.5)^{(2)(-1)} = \left[\left(\frac{1}{2}\right)^2\right]^{-1} = \left(\frac{1}{4}\right)^{-1} = 4$

  After a few examples, students learn the pattern and quit using all the steps.

Pacing Suggestion

- Complete at least the first two explorations and then start the formal lesson by defining exponential functions.

Dynamic Teaching Tools

- Dynamic Assessment & Progress Monitoring Tool
- Lesson Planning Tool
- Interactive Whiteboard Lesson Library
- Dynamic Classroom with Dynamic Investigations

What Your Students Will Learn

- Identify exponential functions represented by tables of values.
- Evaluate exponential functions for integer domain values.
- Graph exponential functions, including transformations, and compare the graphs of exponential functions.

Laurie's Notes

Exploration

Motivate

- Pose this scenario. You give a test and scores range from 40% to 100%. To help your students, you decide to increase each score by 10% of the score (10% scaling).

? "Does everyone's score increase?" yes

? "Does everyone's score increase the same amount?" no

? "Which students receive the least benefit? the greatest benefit? Explain." A 40% test score receives the least benefit (4 points). A 100% test score receives the greatest benefit (10 points).

- Without asking about fairness, discuss that the *percent of change* is the same (10%) for all tests, but the *amount of change* is not. In a linear function, the amount of change is the same, so this function is not linear. Explain that this section will explore this idea further.

Exploration 1

? **Fact-First Questioning:** "Only 2 is being raised to the x. Why is this true?" Listen for an understanding that this is not a power of a product. The constant 16 would need to be inside the parentheses in order to raise it to the x: $(16 \cdot 2)^x$.

- Students should quickly complete the tables of values.

? "What did you notice about the values of x?" x increases by 1 in the first table and increases by 2 in the second table. "What did you notice about the values of y?" y increases by a factor of 2 in the first table and increases by a factor of 4 in the second table.

Exploration 2

- Students should quickly complete the tables of values.

? "What did you notice about the values of x?" x increases by 1 in the first table and increases by 2 in the second table. "What did you notice about the values of y?" y decreases by a factor of 2 in the first table and decreases by a factor of 4 in the second table.

- **Construct Viable Arguments and Critique the Reasoning of Others:** Students' ability to offer justification to the question posed will vary. Improving the ability to justify their thinking is a goal that students should be working on.

Exploration 3

- Students could do a quick sketch on graph paper, or you might want them to enter the data points in their calculator to get a more accurate graph. The *Big Idea* is certainly that two of the graphs are increasing and two are decreasing.
- Students should mention that all of the graphs are located in the first and second quadrants, meaning that $y > 0$.

Communicate Your Answer

- You might consider a jigsaw approach to Question 5 to differentiate or when time is short.

Connecting to Next Step

- Students have now developed a sense of how exponential equations increase (or decrease) and what their graphs look like. The formal definition follows in the lesson, along with transformations of the parent function.

6.3 Exponential Functions

Essential Question What are some of the characteristics of the graph of an exponential function?

EXPLORATION 1 **Exploring an Exponential Function**

Work with a partner. Copy and complete each table for the *exponential function* $y = 16(2)^x$. In each table, what do you notice about the values of x? What do you notice about the values of y?

x	$y = 16(2)^x$
0	
1	
2	
3	
4	
5	

x	$y = 16(2)^x$
0	
2	
4	
6	
8	
10	

JUSTIFYING CONCLUSIONS

To be proficient in math, you need to justify your conclusions and communicate them to others.

EXPLORATION 2 **Exploring an Exponential Function**

Work with a partner. Repeat Exploration 1 for the exponential function $y = 16\left(\frac{1}{2}\right)^x$. Do you think the statement below is true for *any* exponential function? Justify your answer.

"As the independent variable x changes by a constant amount, the dependent variable y is multiplied by a constant factor."

EXPLORATION 3 **Graphing Exponential Functions**

Work with a partner. Sketch the graphs of the functions given in Explorations 1 and 2. How are the graphs similar? How are they different?

Communicate Your Answer

4. What are some of the characteristics of the graph of an exponential function?
5. Sketch the graph of each exponential function. Does each graph have the characteristics you described in Question 4? Explain your reasoning.

a. $y = 2^x$ **b.** $y = 2(3)^x$ **c.** $y = 3(1.5)^x$

d. $y = \left(\frac{1}{2}\right)^x$ **e.** $y = 3\left(\frac{1}{2}\right)^x$ **f.** $y = 2\left(\frac{3}{4}\right)^x$

Dynamic Teaching Tools

- Dynamic Assessment & Progress Monitoring Tool
- Lesson Planning Tool
- Interactive Whiteboard Lesson Library
- Dynamic Classroom with Dynamic Investigations

ANSWERS

1. 16, 32, 64, 128, 256, 512; 16, 64, 256, 1024, 4096, 16,384; Each value of x increases by the same amount; Each value of y is multiplied by the same factor.
2. 16, 8, 4, 2, 1, $\frac{1}{2}$; 16, 4, 1, $\frac{1}{4}$, $\frac{1}{16}$, $\frac{1}{64}$; yes; As the exponent increases by a constant amount, the base is multiplied by itself the same number of additional times.
3. 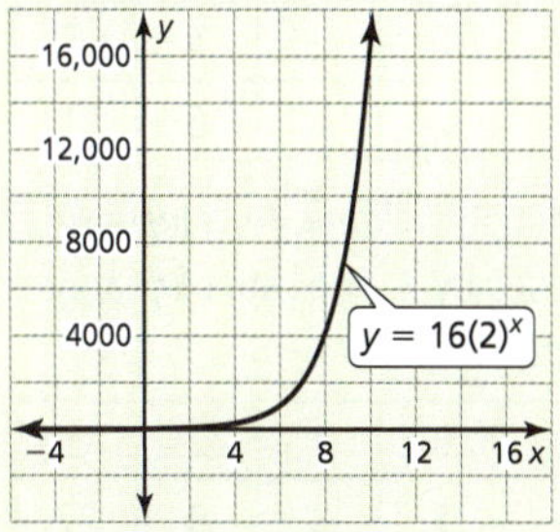

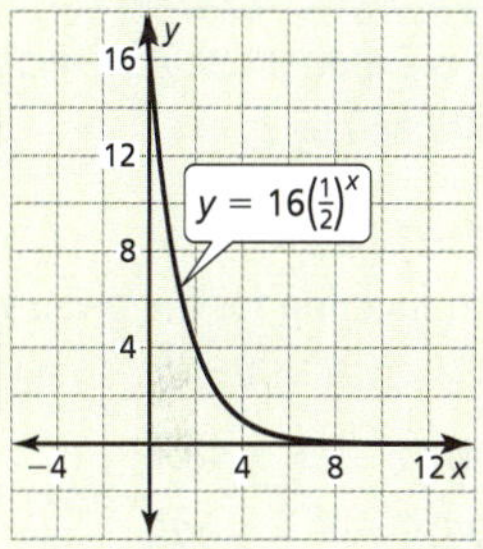

Both are curved and do not intersect the x-axis; The graph from Exploration 1 is increasing, the graph from Exploration 2 is decreasing.

4. *Sample answer:* curved shape, does not intersect the x-axis
5. a. 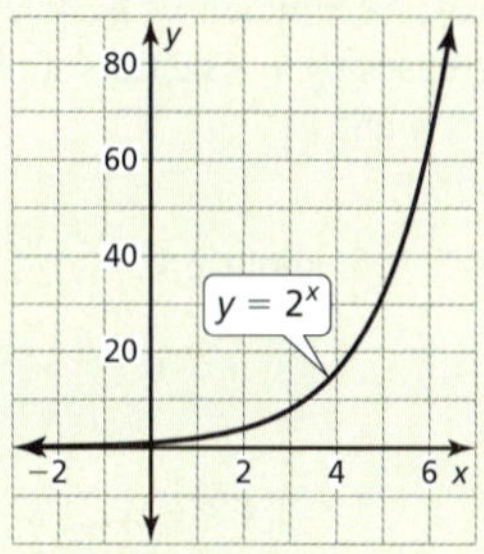

5. b. 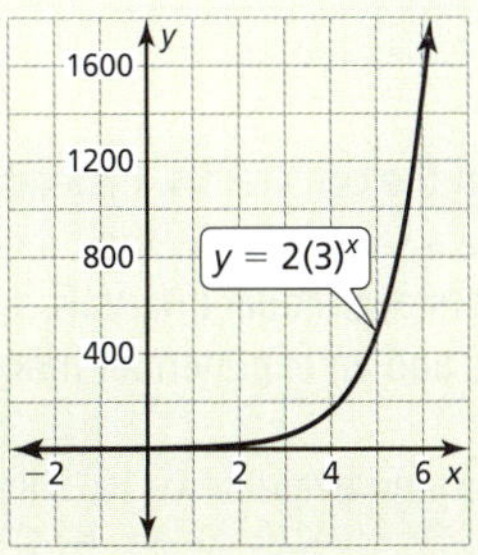

c. 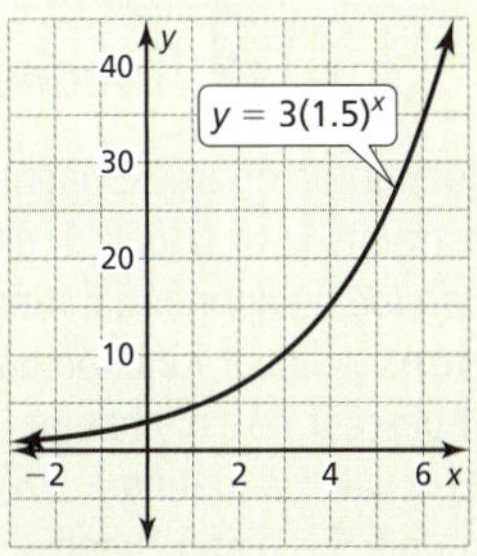

5d–f. See Additional Answers.

Differentiated Instruction

Kinesthetic/Visual

Some students may benefit from creating their own function tables. Have students work in pairs. Each student creates a table to represent a linear or exponential function and then exchanges tables with his or her partner. Students determine which type of function the table represents, and then explain their reasoning to their partners.

Extra Example 1

Does each table represent a *linear* or an *exponential* function? Explain.

a.

x	0	2	4	6
y	2	4	6	8

The function is linear. As x increases by 2, y increases by 2. The rate of change is constant.

b.

x	0	1	2	3
y	1	3	9	27

The function is exponential. As x increases by 1, y is multiplied by 3.

Extra Example 2

Evaluate each function for the given value of x.

a. $y = -3(4)^x$; $x = 2$ -48

b. $y = 2(0.25)^x$; $x = -3$ 128

MONITORING PROGRESS ANSWERS

1. exponential; As x increases by 1, y is multiplied by $\frac{1}{2}$.
2. linear; As x increases by 4, y decreases by 1. The rate of change is constant.
3. $\frac{2}{81}$; 2; 6
4. 0.375; 1.5; about 2.12

6.3 Lesson

Core Vocabulary

exponential function, *p. 306*

Previous

independent variable
dependent variable
parent function

What You Will Learn

- Identify and evaluate exponential functions.
- Graph exponential functions.
- Solve real-life problems involving exponential functions.

Identifying and Evaluating Exponential Functions

An **exponential function** is a nonlinear function of the form $y = ab^x$, where $a \neq 0$, $b \neq 1$, and $b > 0$. As the independent variable x changes by a constant amount, the dependent variable y is multiplied by a constant factor, which means consecutive y-values form a constant ratio.

EXAMPLE 1 **Identifying Functions**

Does each table represent a *linear* or an *exponential* function? Explain.

a.

x	0	1	2	3
y	2	4	6	8

b.

x	0	1	2	3
y	4	8	16	32

SOLUTION

a.

+1 +1 +1

x	0	1	2	3
y	2	4	6	8

+2 +2 +2

▶ As x increases by 1, y increases by 2. The rate of change is constant. So, the function is linear.

b.

+1 +1 +1

x	0	1	2	3
y	4	8	16	32

×2 ×2 ×2

▶ As x increases by 1, y is multiplied by 2. So, the function is exponential.

STUDY TIP

In Example 1b, consecutive y-values form a constant ratio.

$\frac{8}{4} = 2$, $\frac{16}{8} = 2$, $\frac{32}{16} = 2$

EXAMPLE 2 **Evaluating Exponential Functions**

Evaluate each function for the given value of x.

a. $y = -2(5)^x$; $x = 3$

b. $y = 3(0.5)^x$; $x = -2$

SOLUTION

a.

$y = -2(5)^x$ Write the function.

$= -2(5)^3$ Substitute for x.

$= -2(125)$ Evaluate the power.

$= -250$ Multiply.

b.

$y = 3(0.5)^x$

$= 3(0.5)^{-2}$

$= 3(4)$

$= 12$

Monitoring Progress

Help in English and Spanish at *BigIdeasMath.com*

Does the table represent a *linear* or an *exponential* function? Explain.

1.

x	0	1	2	3
y	8	4	2	1

2.

x	−4	0	4	8
y	1	0	−1	−2

Evaluate the function when $x = -2$, 0, and $\frac{1}{2}$.

3. $y = 2(9)^x$

4. $y = 1.5(2)^x$

Laurie's Notes Teacher Actions

- Write the definition of an exponential function. Do not dwell on the constraints. Instead, return to the constraints ($a \neq 0$, $b \neq 1$, and $b > 0$) after students have a sense of how the function behaves. *Big Idea*: Exponential functions change by equal factors over equal intervals.
- **Connection:** A linear function has a constant rate of change, and an exponential function has a constant percent rate of change.

COMMON ERROR Evaluating $-2(5)^3$, students may multiply first and evaluate $(-10)^3$. Remind students that only 5 is being cubed.

Graphing Exponential Functions

The graph of a function $y = ab^x$ is a vertical stretch or shrink by a factor of $|a|$ of the graph of the parent function $y = b^x$. When $a < 0$, the graph is also reflected in the x-axis. The y-intercept of the graph of $y = ab^x$ is a.

STUDY TIP

The graph of $y = ab^x$ approaches the x-axis but never intersects it.

Core Concept

Graphing $y = ab^x$ When $b > 1$

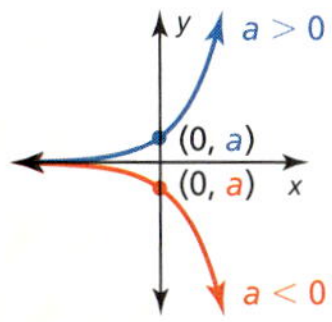

Graphing $y = ab^x$ When $0 < b < 1$

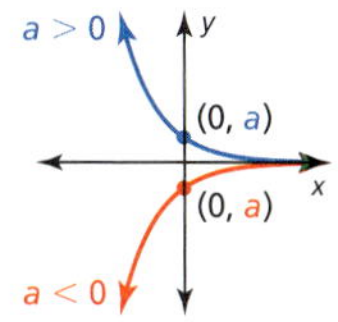

EXAMPLE 3 Graphing $y = ab^x$ When $b > 1$

Graph $f(x) = 4(2)^x$. Compare the graph to the graph of the parent function. Describe the domain and range of f.

SOLUTION

Step 1 Make a table of values.

Step 2 Plot the ordered pairs.

Step 3 Draw a smooth curve through the points.

x	−2	−1	0	1	2
$f(x)$	1	2	4	8	16

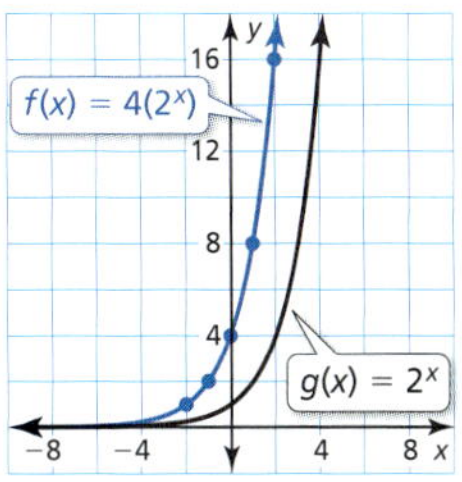

▶ The parent function is $g(x) = 2^x$. The graph of f is a vertical stretch by a factor of 4 of the graph of g. The y-intercept of the graph of f, 4, is above the y-intercept of the graph of g, 1. From the graph of f, you can see that the domain is all real numbers and the range is $y > 0$.

EXAMPLE 4 Graphing $y = ab^x$ When $0 < b < 1$

Graph $f(x) = -\left(\frac{1}{2}\right)^x$. Compare the graph to the graph of the parent function. Describe the domain and range of f.

SOLUTION

Step 1 Make a table of values.

Step 2 Plot the ordered pairs.

Step 3 Draw a smooth curve through the points.

x	−2	−1	0	1	2
$f(x)$	−4	−2	−1	$-\frac{1}{2}$	$-\frac{1}{4}$

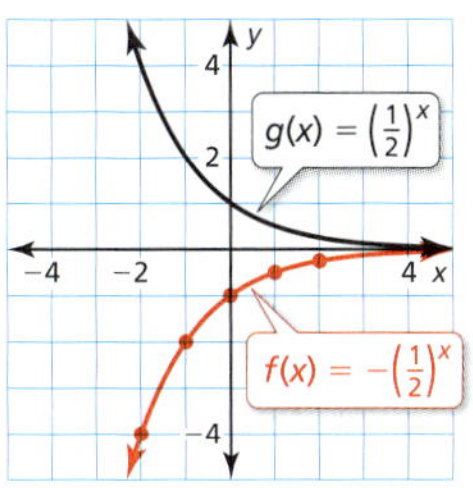

▶ The parent function is $g(x) = \left(\frac{1}{2}\right)^x$. The graph of f is a reflection in the x-axis of the graph of g. The y-intercept of the graph of f, −1, is below the y-intercept of the graph of g, 1. From the graph of f, you can see that the domain is all real numbers and the range is $y < 0$.

Monitoring Progress

Help in English and Spanish at *BigIdeasMath.com*

Graph the function. Compare the graph to the graph of the parent function. Describe the domain and range of *f*.

5. $f(x) = -2(4)^x$

6. $f(x) = 2\left(\frac{1}{4}\right)^x$

English Language Learners

Pair Activity

Pair each English learner with an English speaker to graph exponential functions. Ask one student to make the table of values and the other student to plot the ordered pairs and draw a smooth curve through the points. Have the pair compare the graph to the graph of the parent function and describe the domain and range.

Extra Example 3

Graph $f(x) = 5(2)^x$. Compare the graph to the graph of the parent function. Describe the domain and range of f.

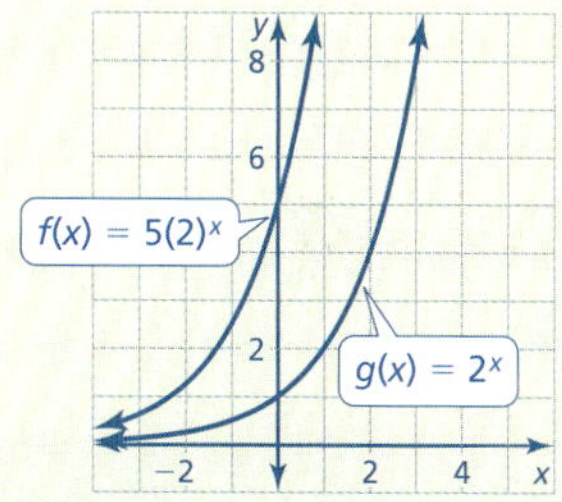

The parent function is $g(x) = 2^x$. The graph of f is a vertical stretch by a factor of 5 of the graph of g. The domain of f is all real numbers and the range is $y > 0$.

Extra Example 4

Graph $f(x) = -(0.25)^x$. Compare the graph to the graph of the parent function. Describe the domain and range of f.

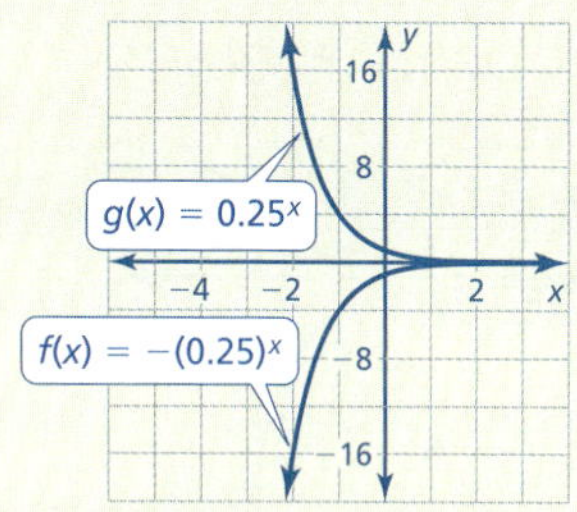

The parent function is $g(x) = (0.25)^x$. The graph of f is a reflection in the x-axis of the graph of g. The domain of f is all real numbers and the range is $y < 0$.

MONITORING PROGRESS ANSWERS

5–6. See Additional Answers.

Laurie's Notes — Teacher Actions

- Students should have an understanding of what the graphs of $f(x) = 2^x$ and $f(x) = \left(\frac{1}{2}\right)^x$ look like before performing any transformations.
- **Connection:** Transformations were presented in Section 3.6. It is hard to see a vertical stretch or shrink of an exponential equation.
- ? **Probing Question:** "Is there a way to determine the y-intercept from the equation as with linear equations?" Answers will vary.
- **Response Logs:** "I was really surprised when …"

Extra Example 5

Graph $y = 3(2)^{x-1} - 3$. Describe the domain and range.

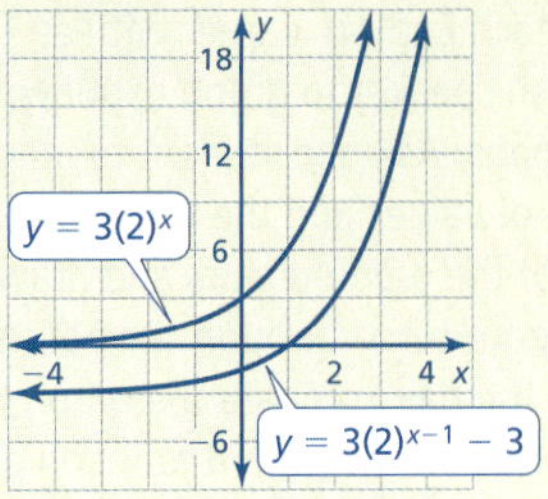

The domain is all real numbers and the range is $y > -3$.

Extra Example 6

An exponential function g models a relationship in which the dependent variable is multiplied by 2.5 for every 1 unit the independent variable x increases. Graph g when $g(0) = 4$. Compare g and the function f from Example 3 over the interval $x = 0$ to $x = 2$.

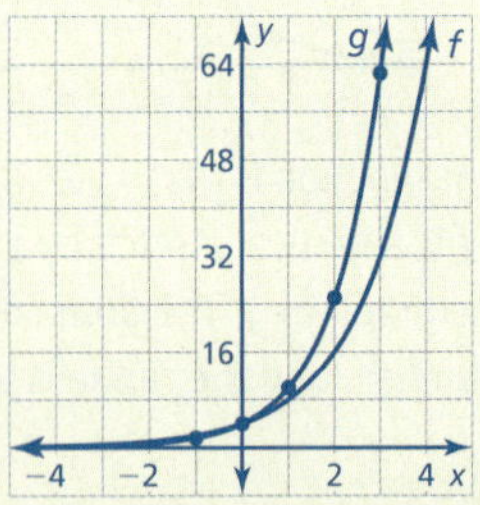

Both functions have the same value when $x = 0$, but the value of f is less than the value of g over the rest of the interval.

MONITORING PROGRESS ANSWERS

7.

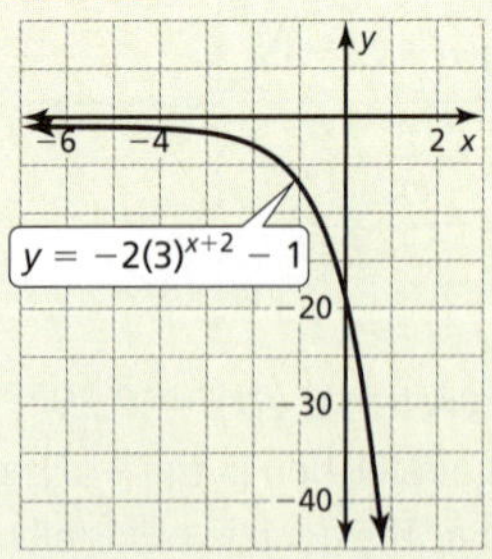

domain: all real numbers, range: $y < -1$

8–9. See Additional Answers.

To graph a function of the form $y = ab^{x-h} + k$, begin by graphing $y = ab^x$. Then translate the graph horizontally h units and vertically k units.

EXAMPLE 5 Graphing $y = ab^{x-h} + k$

Graph $y = 4(2)^{x-3} + 2$. Describe the domain and range.

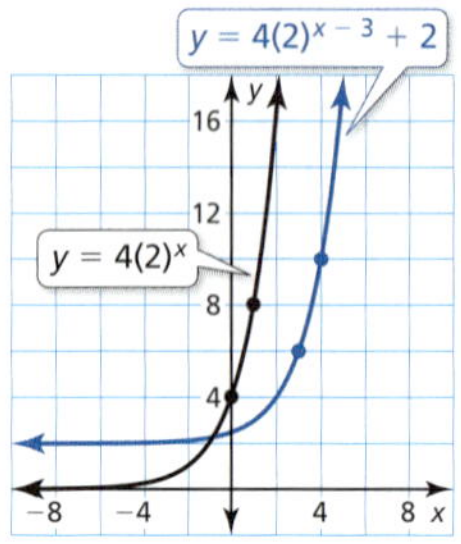

SOLUTION

Step 1 Graph $y = 4(2)^x$. This is the same function that is in Example 3, which passes through (0, 4) and (1, 8).

Step 2 Translate the graph 3 units right and 2 units up. The graph passes through (3, 6) and (4, 10).

Notice that the graph approaches the line $y = 2$ but does not intersect it.

▶ From the graph, you can see that the domain is all real numbers and the range is $y > 2$.

EXAMPLE 6 Comparing Exponential Functions

An exponential function g models a relationship in which the dependent variable is multiplied by 1.5 for every 1 unit the independent variable x increases. Graph g when $g(0) = 4$. Compare g and the function f from Example 3 over the interval $x = 0$ to $x = 2$.

SOLUTION

You know (0, 4) is on the graph of g. To find points to the right of (0, 4), multiply $g(x)$ by 1.5 for every 1 unit increase in x. To find points to the left of (0, 4), divide $g(x)$ by 1.5 for every 1 unit decrease in x.

Step 1 Make a table of values.

x	−1	0	1	2	3
$g(x)$	2.7	4	6	9	13.5

Step 2 Plot the ordered pairs.

Step 3 Draw a smooth curve through the points.

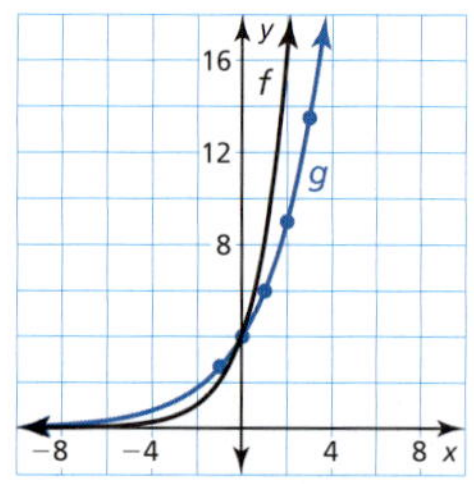

STUDY TIP

Note that f is increasing faster than g to the right of $x = 0$.

▶ Both functions have the same value when $x = 0$, but the value of f is greater than the value of g over the rest of the interval.

Monitoring Progress Help in English and Spanish at *BigIdeasMath.com*

Graph the function. Describe the domain and range.

7. $y = -2(3)^{x+2} - 1$

8. $f(x) = (0.25)^x + 3$

9. **WHAT IF?** In Example 6, the dependent variable of g is multiplied by 3 for every 1 unit the independent variable x increases. Graph g when $g(0) = 4$. Compare g and the function f from Example 3 over the interval $x = 0$ to $x = 2$.

Laurie's Notes Teacher Actions

- ? "In the equation $y = 4(2)^{x-3} + 2$, what is the parent function?" $y = 2^x$ "What do the 4, 3, and 2 do to the parent function?" The parent function is stretched vertically by a factor of 4; this function is shifted 3 units right and 2 units up.
- ? Pose Example 6. Ask, "What is known in this problem?" the y-intercept and the factor of increase Have students compare the bases of 2 and 1.5.
- **Think-Pair-Share:** Have students answer Questions 7–9, and then share and discuss as a class.

Solving Real-Life Problems

For an exponential function of the form $y = ab^x$, the y-values change by a factor of b as x increases by 1. You can use this fact to write an exponential function when you know the y-intercept, a. The table represents the exponential function $y = 2(5)^x$.

x	0	1	2	3	4
y	2	10	50	250	1250

(x: +1, +1, +1, +1; y: ×5, ×5, ×5, ×5)

EXAMPLE 7 Modeling with Mathematics

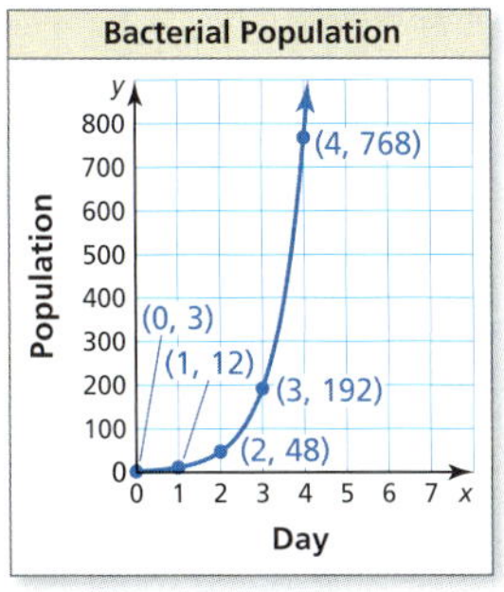

The graph represents a bacterial population y after x days.

a. Write an exponential function that represents the population.

b. Find the population after 12 hours and after 5 days.

SOLUTION

1. **Understand the Problem** You have a graph of the population that shows some data points. You are asked to write an exponential function that represents the population and find the population after different amounts of time.
2. **Make a Plan** Use the graph to make a table of values. Use the table and the y-intercept to write an exponential function. Then evaluate the function to find the populations.
3. **Solve the Problem**

   a. Use the graph to make a table of values.

x	0	1	2	3	4
y	3	12	48	192	768

(x: +1, +1, +1, +1; y: ×4, ×4, ×4, ×4)

The y-intercept is 3. The y-values increase by a factor of 4 as x increases by 1.

▶ So, the population can be modeled by $y = 3(4)^x$.

b. ***Population after 12 hours***

Population after 12 hours		Population after 5 days
$y = 3(4)^x$	Write the function.	$y = 3(4)^x$
$= 3(4)^{1/2}$ (12 hours $= \frac{1}{2}$ day)	Substitute for x.	$= 3(4)^5$
$= 3(2)$	Evaluate the power.	$= 3(1024)$
$= 6$	Multiply.	$= 3072$

▶ There are 6 bacteria after 12 hours and 3072 bacteria after 5 days.

4. **Look Back** The graph resembles an exponential function of the form $y = ab^x$, where $b > 1$ and $a > 0$. So, the exponential function $y = 3(4)^x$ is reasonable.

Monitoring Progress

Help in English and Spanish at *BigIdeasMath.com*

10. A bacterial population y after x days can be represented by an exponential function whose graph passes through (0, 100) and (1, 200). (a) Write a function that represents the population. (b) Find the population after 6 days. (c) Does this bacterial population grow faster than the bacterial population in Example 7? Explain.

Extra Example 7

The graph represents a bacterial population y after x days.

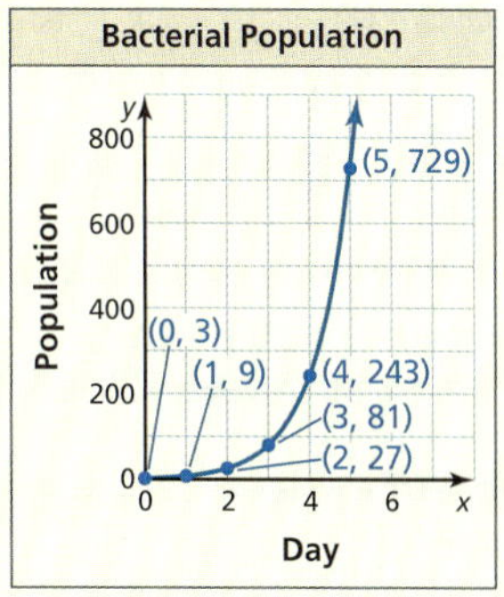

a. Write an exponential function that represents the population.
The population can be modeled by $y = 3(3)^x$.

b. Find the population after 12 hours and after 6 days.
There are about 5 bacteria after 12 hours and 2187 bacteria after 6 days.

MONITORING PROGRESS ANSWER

10. a. $y = 100(2)^x$

b. 6400 bacteria

c. no; The bacteria population in Example 7 is growing by a factor of 4, and the bacteria population in this problem is only growing by a factor of 2.

Laurie's Notes Teacher Actions

- **Teaching Tip:** Have students use their calculators to graph $y = 2^x$, $y = 3(2)^x$, and $y = 6(2)^x$. Have them use the *trace* or *table* feature to determine the y-intercepts of the graphs. This will help students in the next example where $a = 3$, so the equation is $y = 3(b)^x$.

? "What information can be told from the graph?" The y-intercept is 3, and y-values are increasing by a factor of 4.

Closure

- **Response Logs:** Select from "I made progress with …" or "Tomorrow I need to find out …" or "Right now I know …."

Assignment Guide and Homework Check

ASSIGNMENT

Basic: 1–4, 5–35 odd, 41–51 odd, 54, 58, 64–67

Average: 1–3, 4–50 even, 51, 52–58 even, 64–67

Advanced: 1–4, 10–14 even, 20–28 even, 36–48 even, 50–67

HOMEWORK CHECK

Basic: 11, 15, 25, 31, 51

Average: 12, 16, 26, 32, 52

Advanced: 14, 20, 28, 36, 52

6.3 Exercises

Dynamic Solutions available at *BigIdeasMath.com*

Vocabulary and Core Concept Check

1. **OPEN-ENDED** Sketch an increasing exponential function whose graph has a y-intercept of 2.
2. **REASONING** Why is a the y-intercept of the graph of the function $y = ab^x$?
3. **WRITING** Compare the graph of $y = 2(5)^x$ with the graph of $y = 5^x$.
4. **WHICH ONE DOESN'T BELONG?** Which equation does *not* belong with the other three? Explain your reasoning.

$y = 3^x$ $f(x) = 2(4)^x$ $f(x) = (-3)^x$ $y = 5(3)^x$

Monitoring Progress and Modeling with Mathematics

In Exercises 5–10, determine whether the equation represents an exponential function. Explain.

5. $y = 4(7)^x$
6. $y = -6x$
7. $y = 2x^3$
8. $y = -3^x$
9. $y = 9(-5)^x$
10. $y = \frac{1}{2}(1)^x$

In Exercises 11–14, determine whether the table represents a *linear* or an *exponential* function. Explain. *(See Example 1.)*

11.

x	y
1	−2
2	0
3	2
4	4

12.

x	y
1	6
2	12
3	24
4	48

13.

x	−1	0	1	2	3
y	0.25	1	4	16	64

14.

x	−3	0	3	6	9
y	10	1	−8	−17	−26

In Exercises 15–20, evaluate the function for the given value of x. *(See Example 2.)*

15. $y = 3^x$; $x = 2$
16. $f(x) = 3(2)^x$; $x = -1$
17. $y = -4(5)^x$; $x = 2$
18. $f(x) = 0.5^x$; $x = -3$
19. $f(x) = \frac{1}{3}(6)^x$; $x = 3$
20. $y = \frac{1}{4}(4)^x$; $x = \frac{3}{2}$

USING STRUCTURE In Exercises 21–24, match the function with its graph.

21. $f(x) = 2(0.5)^x$
22. $y = -2(0.5)^x$
23. $y = 2(2)^x$
24. $f(x) = -2(2)^x$

A.
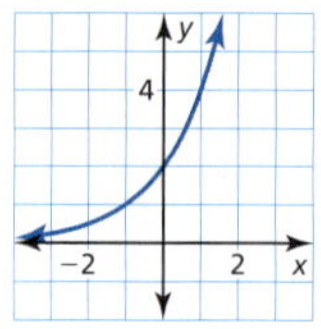

B.
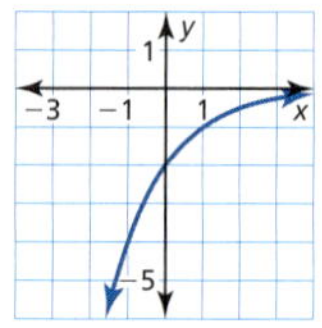

C.
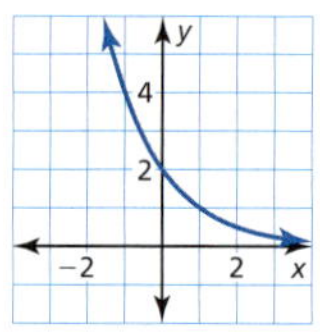

D.
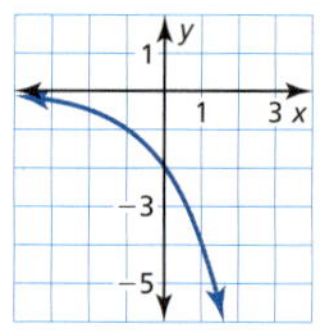

In Exercises 25–30, graph the function. Compare the graph to the graph of the parent function. Describe the domain and range of f. *(See Examples 3 and 4.)*

25. $f(x) = 3(0.5)^x$
26. $f(x) = -4^x$
27. $f(x) = -2(7)^x$
28. $f(x) = 6\left(\frac{1}{3}\right)^x$
29. $f(x) = \frac{1}{2}(8)^x$
30. $f(x) = \frac{3}{2}(0.25)^x$

In Exercises 31–36, graph the function. Describe the domain and range. *(See Example 5.)*

31. $f(x) = 3^x - 1$
32. $f(x) = 4^{x+3}$

ANSWERS

1. *Sample answer:*

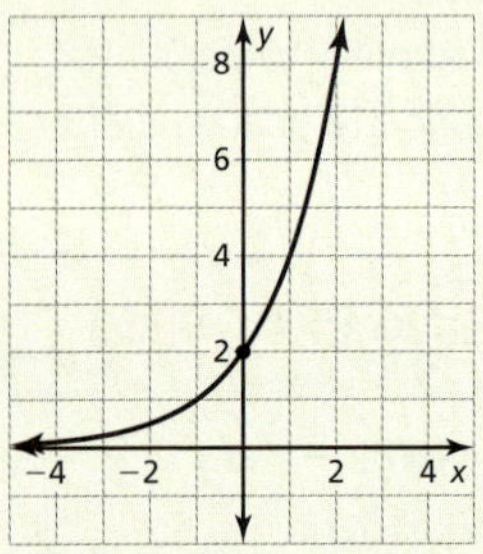

2. The y-intercept occurs when $x = 0$. So, $y = ab^0 = a \cdot 1 = a$.
3. The graph of $y = 2(5)^x$ is a vertical stretch by a factor of 2 of the graph of $y = 5^x$. The y-intercept of $y = 2(5)^x$, 2, is above the y-intercept of $y = 5^x$, 1.
4. $f(x) = (-3)^x$; It is the only one that is not an exponential function.
5. yes; It fits the pattern $y = ab^x$.
6. no; It is a linear function.
7. no; The exponent is a constant.
8. yes: It fits the pattern $y = ab^x$.
9. no; Although it fits the pattern $y = ab^x$, b cannot be negative.
10. no; Although it fits the pattern $y = ab^x$, b cannot be 1.
11. linear; As x increases by 1, y increases by 2. The rate of change is constant.
12. exponential; As x increases by 1, y is multiplied by 2.
13. exponential; As x increases by 1, y is multiplied by 4.
14. linear; As x increases by 3, y decreases by 9. The rate of change is constant.
15. 9
16. $\frac{3}{2}$
17. −100
18. 8
19. 72
20. 2
21. C
22. B
23. A
24. D
25.

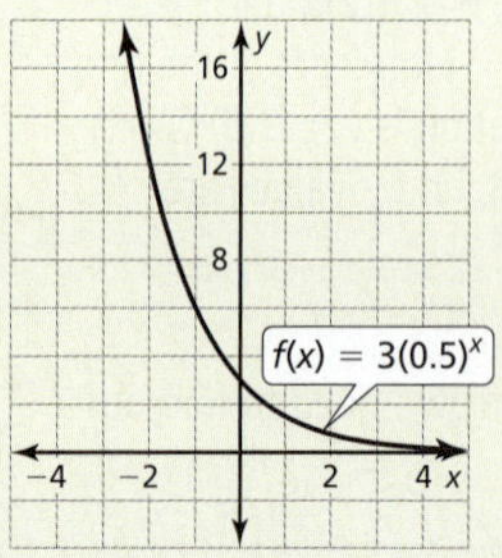

The graph of f is a vertical stretch by a factor of 3 of the graph of $g(x) = 0.5^x$. The y-intercept of the graph of f, 3, is above the y-intercept of the graph of g, 1; domain: all real numbers, range: $y > 0$

26–32. See Additional Answers.

33. $y = 5^{x-2} + 7$

34. $y = -\left(\frac{1}{2}\right)^{x+1} - 3$

35. $y = -8(0.75)^{x+2} - 2$

36. $f(x) = 3(6)^{x-1} - 5$

In Exercises 37–40, compare the graphs. Find the value of h, k, or a.

37.

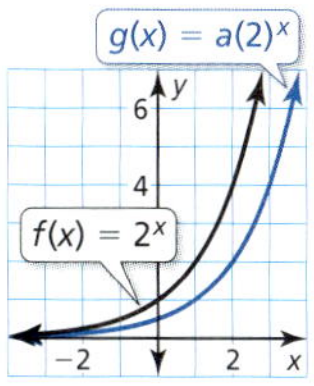

38.

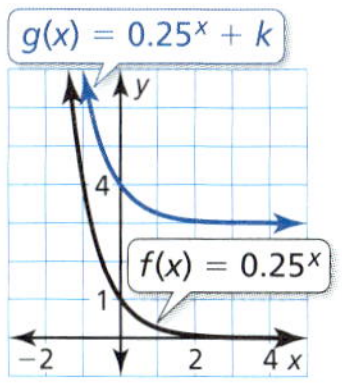

39.

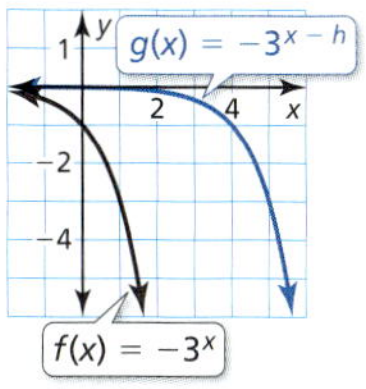

40.

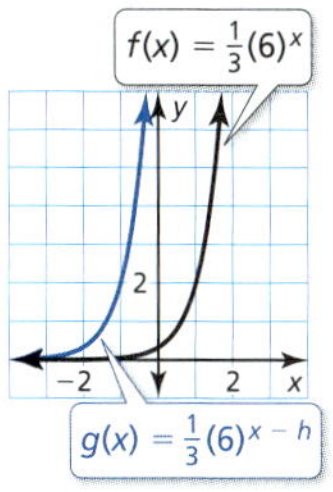

41. **ERROR ANALYSIS** Describe and correct the error in evaluating the function.

✗

$g(x) = 6(0.5)^x;\ x = -2$

$g(-2) = 6(0.5)^{-2}$

$= 3^{-2}$

$= \frac{1}{9}$

42. **ERROR ANALYSIS** Describe and correct the error in finding the domain and range of the function.

✗

The domain is all real numbers, and the range is $y < 0$.

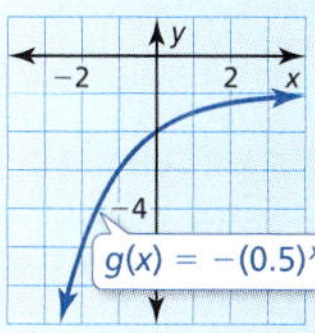

In Exercises 43 and 44, graph the function with the given description. Compare the function to $f(x) = 0.5(4)^x$ over the interval $x = 0$ to $x = 2$. *(See Example 6.)*

43. An exponential function g models a relationship in which the dependent variable is multiplied by 2.5 for every 1 unit the independent variable x increases. The value of the function at 0 is 8.

44. An exponential function h models a relationship in which the dependent variable is multiplied by $\frac{1}{2}$ for every 1 unit the independent variable x increases. The value of the function at 0 is 32.

45. **MODELING WITH MATHEMATICS** You graph an exponential function on a calculator. You zoom in repeatedly to 25% of the screen size. The function $y = 0.25^x$ represents the percent (in decimal form) of the original screen display that you see, where x is the number of times you zoom in.

a. Graph the function. Describe the domain and range.

b. Find and interpret the y-intercept.

c. You zoom in twice. What percent of the original screen do you see?

46. **MODELING WITH MATHEMATICS** A population y of coyotes in a national park triples every 20 years. The function $y = 15(3)^x$ represents the population, where x is the number of 20-year periods.

a. Graph the function. Describe the domain and range.

b. Find and interpret the y-intercept.

c. How many coyotes are in the national park in 40 years?

In Exercises 47–50, write an exponential function represented by the table or graph. *(See Example 7.)*

47.

x	0	1	2	3
y	2	14	98	686

48.

x	0	1	2	3
y	−50	−10	−2	−0.4

49.

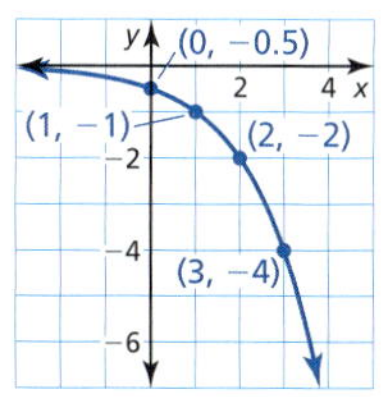

50.

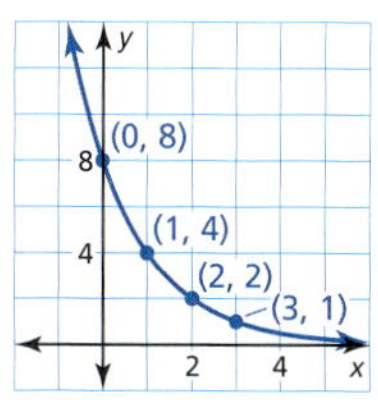

Dynamic Teaching Tools

Dynamic Assessment & Progress Monitoring Tool

Interactive Whiteboard Lesson Library

Dynamic Classroom with Dynamic Investigations

ANSWERS

33.

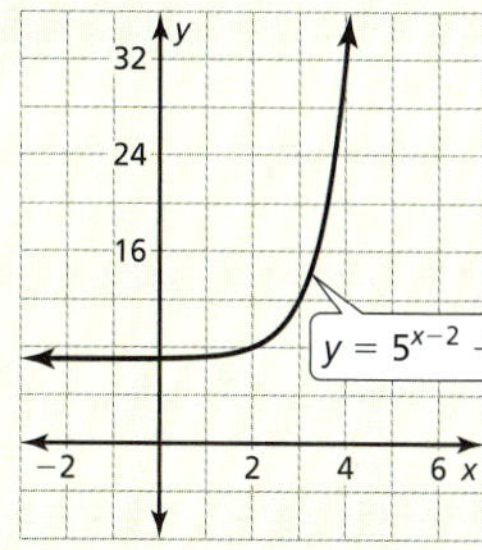

domain: all real numbers, range: $y > 7$

34.

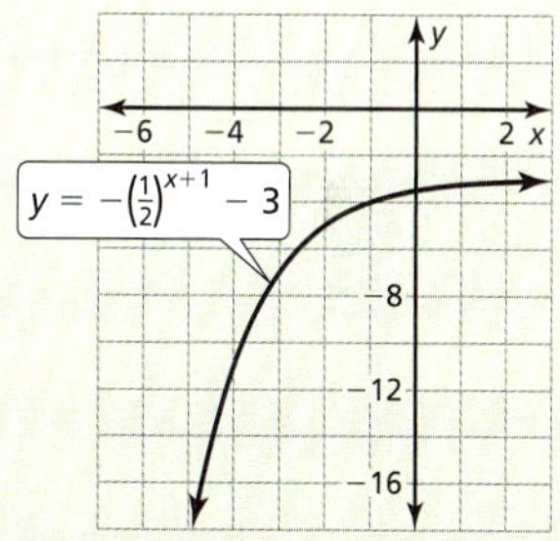

domain: all real numbers, range: $y < -3$

35.

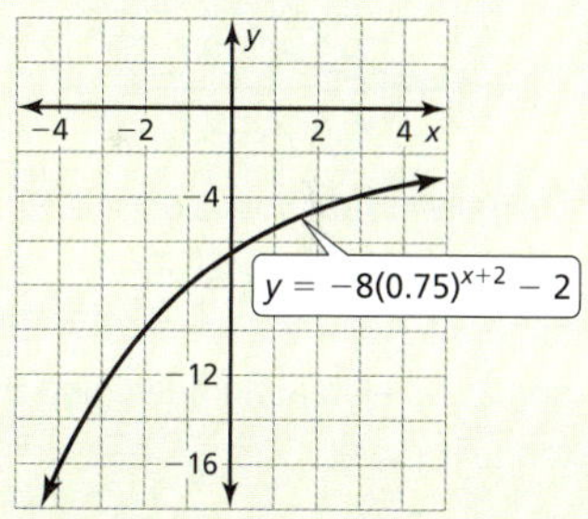

domain: all real numbers, range: $y < -2$

36.

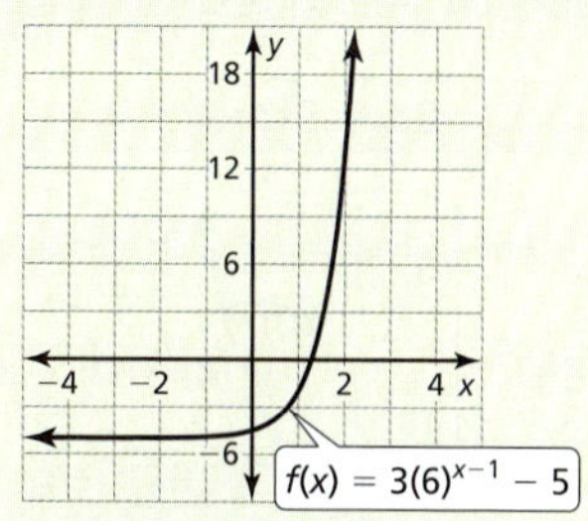

domain: all real numbers, range: $y > -5$

37. The graph of g is a vertical shrink by a factor of $\frac{1}{2}$ of the graph of f; $a = \frac{1}{2}$

38. The graph of g is a vertical translation 3 units up of the graph of f; $k = 3$

39. The graph of g is a horizontal translation 4 units right of the graph of f; $h = 4$

40. The graph of g is a horizontal translation 2 units left of the graph of f; $h = -2$

41. need to simplify the power before multiplying; $6(0.5)^{-2} = 6(4) = 24$

42. The graph approaches the line $y = -1$, not $y = 0$; The range is $y < -1$.

43.

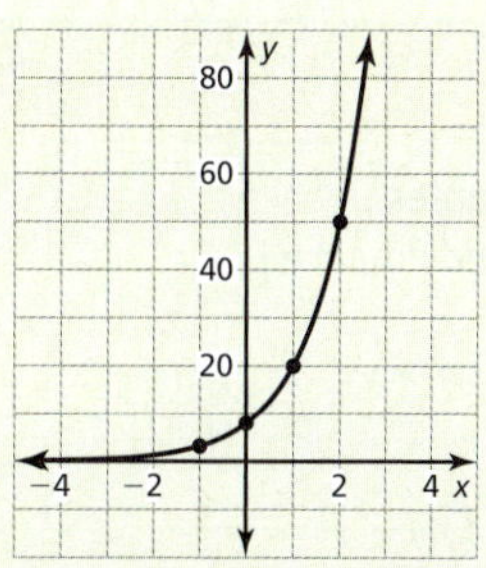

The value of f is less than the value of g over the entire interval.

44–50. See Additional Answers.

ANSWERS

51. **a.** $y = 40\left(\frac{3}{2}\right)^x$

b. about 304 visitors

52. about 4681 grills; $y = 3300(1.06)^x$

53–67. See Additional Answers.

Mini-Assessment

1. Does the table represent a *linear* or an *exponential* function? Explain.

x	0	1	2	3
y	−2	−6	−18	−54

The function is exponential. As x increases by 1, y is multiplied by 3.

2. Evaluate the function $y = 5(3)^x$ for $x = 2$. 45

3. Graph $f(x) = 2.5(2)^x$. Compare the graph to the graph of the parent function. Describe the domain and range of f.

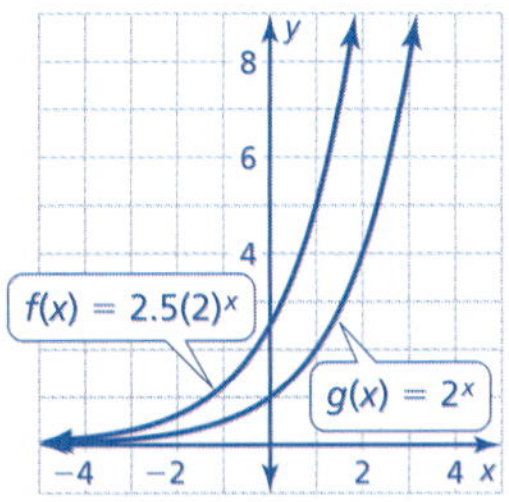

The parent function is $g(x) = 2^x$. The graph of f is a vertical stretch by a factor of 2.5 of the graph of g. The domain is all real numbers and the range is $y > 0$.

4. The graph represents the number y of customers at a new bookstore after x weeks.

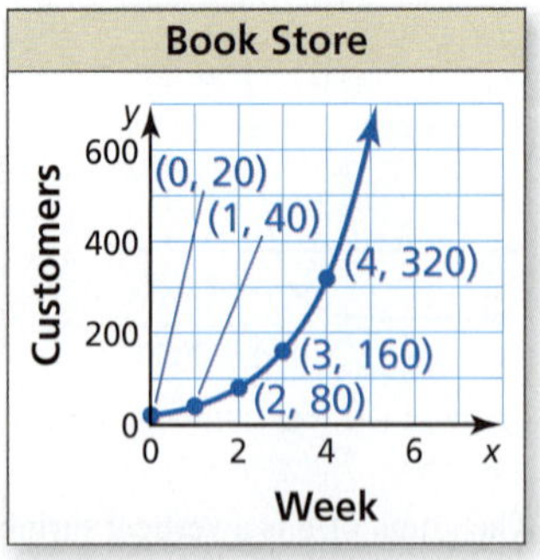

a. Write an exponential function that represents the number of customers. $y = 20(2)^x$

b. Find the number of customers after 5 weeks. 640 customers

51. **MODELING WITH MATHEMATICS** The graph represents the number y of visitors to a new art gallery after x months.

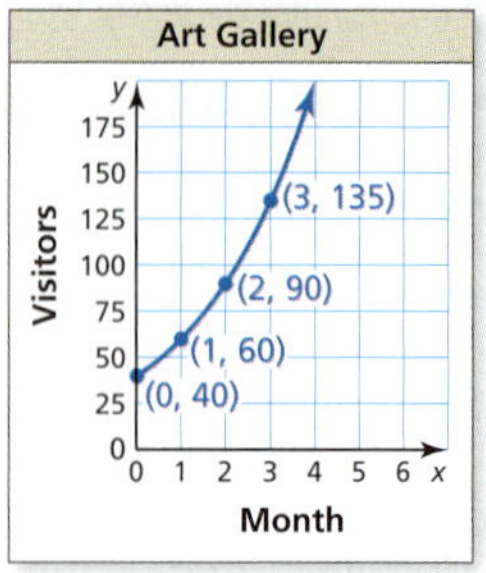

a. Write an exponential function that represents this situation.

b. Approximate the number of visitors after 5 months.

52. **PROBLEM SOLVING** A sales report shows that 3300 gas grills were purchased from a chain of hardware stores last year. The store expects grill sales to increase 6% each year. About how many grills does the store expect to sell in Year 6? Use an equation to justify your answer.

53. **WRITING** Graph the function $f(x) = -2^x$. Then graph $g(x) = -2^x - 3$. How are the y-intercept, domain, and range affected by the translation?

54. **MAKING AN ARGUMENT** Your friend says that the table represents an exponential function because y is multiplied by a constant factor. Is your friend correct? Explain.

x	0	1	3	6
y	2	10	50	250

55. **WRITING** Describe the effect of a on the graph of $y = a \cdot 2^x$ when a is positive and when a is negative.

56. **OPEN-ENDED** Write a function whose graph is a horizontal translation of the graph of $h(x) = 4^x$.

57. **USING STRUCTURE** The graph of g is a translation 4 units up and 3 units right of the graph of $f(x) = 5^x$. Write an equation for g.

58. **HOW DO YOU SEE IT?** The exponential function $y = V(x)$ represents the projected value of a stock x weeks after a corporation loses an important legal battle. The graph of the function is shown.

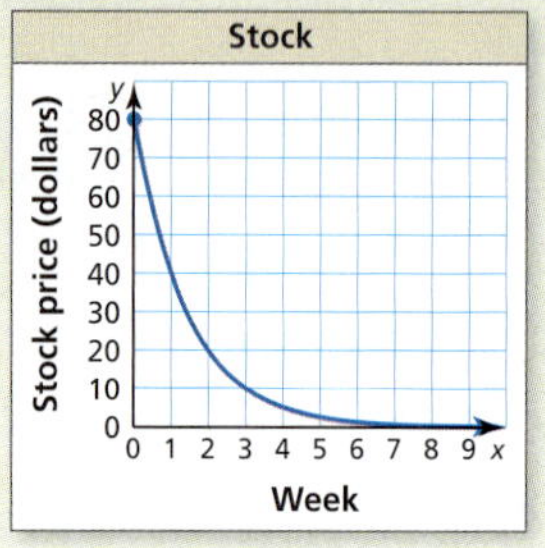

a. After how many weeks will the stock be worth $20?

b. Describe the change in the stock price from Week 1 to Week 3.

59. **USING GRAPHS** The graph represents the exponential function f. Find $f(7)$.

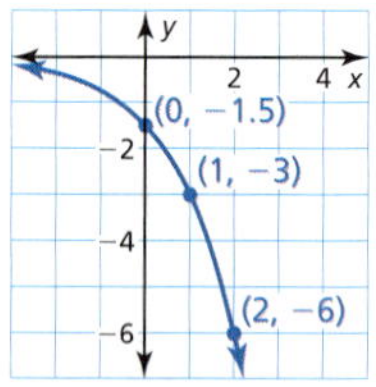

60. **THOUGHT PROVOKING** Write a function of the form $y = ab^x$ that represents a real-life population. Explain the meaning of each of the constants a and b in the real-life context.

61. **REASONING** Let $f(x) = ab^x$. Show that when x is increased by a constant k, the quotient $\dfrac{f(x+k)}{f(x)}$ is always the same regardless of the value of x.

62. **PROBLEM SOLVING** A function g models a relationship in which the dependent variable is multiplied by 4 for every 2 units the independent variable increases. The value of the function at 0 is 5. Write an equation that represents the function.

63. **PROBLEM SOLVING** Write an exponential function f so that the slope from the point $(0, f(0))$ to the point $(2, f(2))$ is equal to 12.

Maintaining Mathematical Proficiency

Reviewing what you learned in previous grades and lessons

Write the percent as a decimal. *(Skills Review Handbook)*

64. 4% **65.** 35% **66.** 128% **67.** 250%

If students need help...	*If students got it...*
Resources by Chapter • Practice A and Practice B • Puzzle Time	Resources by Chapter • Enrichment and Extension • Cumulative Review
Student Journal • Practice	Start the *next* Section
Differentiating the Lesson Skills Review Handbook	

Laurie's Notes

Overview of Section 6.4

Introduction

- This is a long lesson that introduces exponential growth and exponential decay. In each case, the connection needs to be made to the general exponential equation $y = ab^x$. Students should be able to identify and distinguish between exponential growth and decay from inspecting an equation, given a table of values, or by looking at a graph.
- There are many applications of exponential growth and decay. In this lesson, sufficient information is given so that exponential models can be written to solve a real-life problem.
- The compound interest formula is presented. The formula is used to compare balances of two different investment models.
- The last application presented is about the depreciation of an automobile.

Formative Assessment Tips

- Refer to page T-290 for a description of *Response Logs*. In this lesson and throughout this chapter, there will be sentence stems to prompt student writing. Use your judgment for placement in the lesson, amount of time to allow for writing, and the actual sentence stem given to students.
- For a beginning list of writing stems, see page T-298.

Applications

- **Effective Rate:** The annual interest rate for compound interest and the actual (effective) rate are not equal when compounded more than once a year. For example, a deposit of \$2000 earning interest at an annual percentage rate of 8% compounded monthly accrues \$166 in interest over a period of one year. The 8% is called the *stated rate* or *nominal rate*. However, 8% of \$2000 is \$160, which means that the compounding produced an *effective rate* that is greater than the *nominal rate*. To find the effective rate (for a time period of one year), you need to determine what percent \$166 is of \$2000. Because $\frac{166}{2000} = 0.083$, you conclude that the effective rate for the investment is 8.3%.

Pacing Suggestion

- The two explorations are similar in design, and each will take time for students to make sense of and to solve. Use one or both to pique student interest, and then transition to the formal lesson.

Dynamic Teaching Tools

- Dynamic Assessment & Progress Monitoring Tool
- Lesson Planning Tool
- Interactive Whiteboard Lesson Library
- Dynamic Classroom with Dynamic Investigations

What Your Students Will Learn

- Identify exponential growth and exponential decay functions as functions that increase or decrease by the same factor over equal intervals of time.
- Identify the percent rate of change for exponential growth and exponential decay functions.
- Use the compound interest formula to solve real-life problems.

Laurie's Notes

Exploration

Motivate

- Simulate the rumor game. You tell two students a rumor (no homework tonight). In 15 seconds, they each tell two students the rumor. This process continues.

? "How many students will hear the rumor after 1 minute?" 32

- In a small class, all the students will know the rumor in less than a minute.

Exploration 1

- **Construct Viable Arguments and Critique the Reasoning of Others:** Students construct an argument about whether the data are exponential. Although the x-values are not stated explicitly, students should be able to judge that they occur at equal intervals (about 5 years apart). The y-values are labeled on the graph.
- Give students time to discuss strategies and organize their work. Using a calculator will help students to focus on the process and concept rather than the computations.
- Expect students to find and compare the factors between consecutive terms.

$1875 \div 1188 \approx 1.58$ $\quad 3399 \div 1875 \approx 1.81$ $\quad 5094 \div 3399 \approx 1.50$

$6846 \div 5094 \approx 1.34$ $\quad 9789 \div 6846 \approx 1.43$

? "What is the approximate rate of growth of the population?" The population is growing by about 50% every 5 years.

- Students can solve the equation $100{,}000 = 1.5^x$ to find the number of 5-year time periods that will pass before the population catches up to that of the late 1700s.

Exploration 2

- **FYI:** This problem is an application of Newton's Law of Cooling, which students may have studied in science. Newton's Law of Cooling states that when an object at one temperature is exposed to air of another temperature, the difference in the two temperatures drops by the same percent each hour.
- Work through part (a) with students to ensure they get started correctly.

Time	Body Temp.	Room Temp.	Difference
Midnight	80.5°F	60°F	20.5°F
1:00 A.M.	78.5°F	60°F	18.5°F

- The percent decrease is $(20.5 - 18.5)/20.5 \approx 9.8\%$, or about 10%.
- The body temperature is dropping by 10% of the difference between the body temperature and 60°F each hour.
- Suggest to students that they make a table of values for time and body temperature. How many hours before the body temperature reaches 80.5°F?

Time (hours)	0				
Temperature (°F)	98.6				

- Students need to be organized with their data and careful with how they round their data.

Communicate Your Answer

- Students should make the connection with the graphs in the last lesson, when $b > 1$ and when $0 < b < 1$.

Connecting to Next Step

- The explorations should pique student interest in learning more about exponential growth and decay, which are presented in the formal lesson.

6.4 Exponential Growth and Decay

Essential Question What are some of the characteristics of exponential growth and exponential decay functions?

EXPLORATION 1 **Predicting a Future Event**

Work with a partner. It is estimated, that in 1782, there were about 100,000 nesting pairs of bald eagles in the United States. By the 1960s, this number had dropped to about 500 nesting pairs. In 1967, the bald eagle was declared an endangered species in the United States. With protection, the nesting pair population began to increase. Finally, in 2007, the bald eagle was removed from the list of endangered and threatened species.

MODELING WITH MATHEMATICS

To be proficient in math, you need to apply the mathematics you know to solve problems arising in everyday life.

Describe the pattern shown in the graph. Is it exponential growth? Assume the pattern continues. When will the population return to that of the late 1700s? Explain your reasoning.

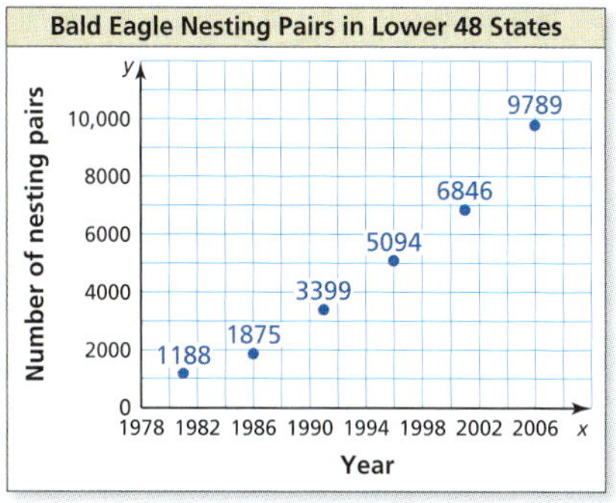

EXPLORATION 2 **Describing a Decay Pattern**

Work with a partner. A forensic pathologist was called to estimate the time of death of a person. At midnight, the body temperature was 80.5°F and the room temperature was a constant 60°F. One hour later, the body temperature was 78.5°F.

a. By what percent did the difference between the body temperature and the room temperature drop during the hour?

b. Assume that the original body temperature was 98.6°F. Use the percent decrease found in part (a) to make a table showing the decreases in body temperature. Use the table to estimate the time of death.

Communicate Your Answer

3. What are some of the characteristics of exponential growth and exponential decay functions?

4. Use the Internet or some other reference to find an example of each type of function. Your examples should be different than those given in Explorations 1 and 2.

a. exponential growth **b.** exponential decay

Dynamic Teaching Tools

- Dynamic Assessment & Progress Monitoring Tool
- Lesson Planning Tool
- Interactive Whiteboard Lesson Library
- Dynamic Classroom with Dynamic Investigations

ANSWERS

1. As x increases by 5, y is multiplied by about 1.5; yes; 2036; The situation can be approximately modeled by $y = 1188(1.5)^t$, where t represents the number of 5-year intervals since 1981. When $t = 11$, $y \approx 100{,}000$. Because $t = 11$ represents eleven 5-year intervals, or 55 years, the population will be about 100,000 in 2036.

2. a. $2/20.5 \approx 9.8\%$

b. See Additional Answers.

3. As the independent variable changes by a constant amount, the dependent variable is multiplied by a constant factor.

4. a. *Sample answer:* the value of a CD each year that earns 2.5% interest compounded annually

b. *Sample answer:* the worth of a farm tractor that depreciates at a rate of 10% per year

English Language Learners

Notebook Development

Have students create an exponential function page in their notebooks. Include the definition of exponential function from Section 6.3 along with the definitions of exponential growth and decay functions from this section. Have students include an example with each definition.

Extra Example 1

The inaugural attendance of an annual art festival is 80,000. The attendance y increases by 6% each year.

a. Write an exponential growth function that represents the attendance after t years.
   The festival attendance can be represented by $y = 80{,}000(1.06)^t$.

b. How many people will attend the festival in the fifth year? Round your answer to the nearest thousand.
   About 101,000 people will attend the festival in the fifth year.

MONITORING PROGRESS ANSWER

1. a. $y = 500{,}000(1.15)^t$
   b. about 1,160,000

6.4 Lesson

Core Vocabulary

exponential growth, *p. 314*
exponential growth function, *p. 314*
exponential decay, *p. 315*
exponential decay function, *p. 315*
compound interest, *p. 317*

What You Will Learn

- Use and identify exponential growth and decay functions.
- Interpret and rewrite exponential growth and decay functions.
- Solve real-life problems involving exponential growth and decay.

Exponential Growth and Decay Functions

Exponential growth occurs when a quantity increases by the same factor over equal intervals of time.

Core Concept

Exponential Growth Functions

A function of the form $y = a(1 + r)^t$, where $a > 0$ and $r > 0$, is an **exponential growth function**.

$$y = a(1 + r)^t$$

initial amount (a); rate of growth (in decimal form) (r); final amount (y); time (t); growth factor ($1 + r$)

STUDY TIP

Notice that an exponential growth function is of the form $y = ab^x$, where b is replaced by $1 + r$ and x is replaced by t.

EXAMPLE 1 Using an Exponential Growth Function

The inaugural attendance of an annual music festival is 150,000. The attendance y increases by 8% each year.

a. Write an exponential growth function that represents the attendance after t years.

b. How many people will attend the festival in the fifth year? Round your answer to the nearest thousand.

SOLUTION

a. The initial amount is 150,000, and the rate of growth is 8%, or 0.08.

$y = a(1 + r)^t$	Write the exponential growth function.
$= 150{,}000(1 + 0.08)^t$	Substitute 150,000 for a and 0.08 for r.
$= 150{,}000(1.08)^t$	Add.

▶ The festival attendance can be represented by $y = 150{,}000(1.08)^t$.

b. The value $t = 4$ represents the fifth year because $t = 0$ represents the first year.

$y = 150{,}000(1.08)^t$	Write the exponential growth function.
$= 150{,}000(1.08)^4$	Substitute 4 for t.
$\approx 204{,}073$	Use a calculator.

▶ About 204,000 people will attend the festival in the fifth year.

Monitoring Progress

Help in English and Spanish at *BigIdeasMath.com*

1. A website has 500,000 members in 2010. The number y of members increases by 15% each year. (a) Write an exponential growth function that represents the website membership t years after 2010. (b) How many members will there be in 2016? Round your answer to the nearest ten thousand.

Laurie's Notes Teacher Actions

- **Teaching Tip:** Write the definition of an exponential growth function. The number of variables in the general equation will create anxiety for some students. If possible, use different colors to highlight the different parts of the equation.
- **Connection:** Help students see that the base in the general exponential equation $y = ab^x$ is replaced by $1 + r$ and x is replaced by t.

? Read Example 1. "What is the initial amount? What does it mean?" 150,000; the number of people that attended the festival in the first year ($t = 0$)

Exponential decay occurs when a quantity decreases by the same factor over equal intervals of time.

STUDY TIP

Notice that an exponential decay function is of the form $y = ab^x$, where b is replaced by $1 - r$ and x is replaced by t.

Core Concept

Exponential Decay Functions

A function of the form $y = a(1 - r)^t$, where $a > 0$ and $0 < r < 1$, is an **exponential decay function**.

$$y = a(1 - r)^t$$

final amount → y; initial amount → a; rate of decay (in decimal form) → r; time → t; decay factor → $(1 - r)$

For exponential growth, the value inside the parentheses is greater than 1 because r is added to 1. For exponential decay, the value inside the parentheses is less than 1 because r is subtracted from 1.

EXAMPLE 2 Identifying Exponential Growth and Decay

Determine whether each table represents an *exponential growth function*, an *exponential decay function*, or *neither*.

a.

x	y
0	270
1	90
2	30
3	10

b.

x	0	1	2	3
y	5	10	20	40

SOLUTION

a.

x	y
0	270
1	90
2	30
3	10

(+1 for each step in x; $\times\frac{1}{3}$ for each step in y)

As x increases by 1, y is multiplied by $\frac{1}{3}$. So, the table represents an exponential decay function.

b.

x	0	1	2	3
y	5	10	20	40

(+1 for each step in x; $\times 2$ for each step in y)

As x increases by 1, y is multiplied by 2. So, the table represents an exponential growth function.

Monitoring Progress

Help in English and Spanish at *BigIdeasMath.com*

Determine whether the table represents an *exponential growth function*, an *exponential decay function*, or *neither*. Explain.

2.

x	0	1	2	3
y	64	16	4	1

3.

x	1	3	5	7
y	4	11	18	25

Differentiated Instruction

Auditory

Have students verbally describe the difference between exponential growth and exponential decay. Then have them verbally describe how they can determine whether a table of values represents an exponential growth function or an exponential decay function.

Extra Example 2

Determine whether each table represents an *exponential growth function*, an *exponential decay function*, or *neither*.

a.

x	y
0	24
1	12
2	6
3	3

The table represents an exponential decay function, because as x increases by 1, y is multiplied by $\frac{1}{2}$.

b.

x	0	1	2	3
y	4	40	400	4000

The table represents an exponential growth function, because as x increases by 1, y is multiplied by 10.

MONITORING PROGRESS ANSWERS

2. exponential decay; As x increases by 1, y is multiplied by $\frac{1}{4}$.
3. neither; As x increases by 2, y increases by 7.

Laurie's Notes Teacher Actions

- Introduce *exponential decay.* Students should think of the decay factor as what remains. If a quantity decreases 25%, then 75% remains.
- ? "Are the data in part (a) decreasing by the same factor over equal intervals? Explain." yes; The y-values are multiplied by the same factor, $\frac{1}{3}$.
- The decay factor is $\frac{1}{3}$. This is what remains of the previous amount. The rate of decay is $\frac{2}{3}$, meaning $\frac{2}{3}$ of the previous amount is lost.
- **Response Logs:** "Right now I need to …"

Extra Example 3

Determine whether each function represents *exponential growth* or *exponential decay*. Identify the percent rate of change.

a. $y = 4(1.09)^t$ The function represents exponential growth and the rate of growth is 9%.

b. $f(t) = 24(0.93)^t$ The function represents exponential decay and the rate of decay is 7%.

Extra Example 4

Rewrite each function to determine whether it represents *exponential growth* or *exponential decay*.

a. $y = 20(0.81)^{t/2}$ $y = 20(0.9)^t$; The function represents exponential decay.

b. $f(t) = (2.5)^{t-2}$ $f(t) = 0.16(2.5)^t$; The function represents exponential growth.

MONITORING PROGRESS ANSWERS

4. exponential decay; 8%
5. exponential growth; 20%
6. $f(t) \approx 3(1.22)^t$; exponential growth
7. $y \approx 0.9(0.95)^t$; exponential decay

Interpreting and Rewriting Exponential Functions

EXAMPLE 3 Interpreting Exponential Functions

Determine whether each function represents *exponential growth* or *exponential decay*. Identify the percent rate of change.

a. $y = 5(1.07)^t$ **b.** $f(t) = 0.2(0.98)^t$

SOLUTION

a. The function is of the form $y = a(1 + r)^t$, where $1 + r > 1$, so it represents exponential growth. Use the growth factor $1 + r$ to find the rate of growth.

$1 + r = 1.07$ Write an equation.

$r = 0.07$ Solve for r.

▶ So, the function represents exponential growth and the rate of growth is 7%.

b. The function is of the form $y = a(1 - r)^t$, where $1 - r < 1$, so it represents exponential decay. Use the decay factor $1 - r$ to find the rate of decay.

$1 - r = 0.98$ Write an equation.

$r = 0.02$ Solve for r.

▶ So, the function represents exponential decay and the rate of decay is 2%.

STUDY TIP

You can rewrite exponential expressions and functions using the properties of exponents. Changing the form of an exponential function can reveal important attributes of the function.

EXAMPLE 4 Rewriting Exponential Functions

Rewrite each function to determine whether it represents *exponential growth* or *exponential decay*.

a. $y = 100(0.96)^{t/4}$ **b.** $f(t) = (1.1)^{t-3}$

SOLUTION

a. $y = 100(0.96)^{t/4}$ Write the function.

$= 100(0.96^{1/4})^t$ Power of a Power Property

$\approx 100(0.99)^t$ Evaluate the power.

▶ So, the function represents exponential decay.

b. $f(t) = (1.1)^{t-3}$ Write the function.

$= \dfrac{(1.1)^t}{(1.1)^3}$ Quotient of Powers Property

$\approx 0.75(1.1)^t$ Evaluate the power and simplify.

▶ So, the function represents exponential growth.

Monitoring Progress

Help in English and Spanish at *BigIdeasMath.com*

Determine whether the function represents *exponential growth* or *exponential decay*. Identify the percent rate of change.

4. $y = 2(0.92)^t$ 5. $f(t) = (1.2)^t$

Rewrite the function to determine whether it represents *exponential growth* or *exponential decay*.

6. $f(t) = 3(1.02)^{10t}$ 7. $y = (0.95)^{t+2}$

Laurie's Notes Teacher Actions

? **Always-Sometimes-Never True:** "Given the equation $y = ab^x$, when $b > 1$, the equation represents exponential growth." always true; When the base of the exponential is greater than 1, the y-values will increase and you have exponential growth.

- **Popsicle Sticks:** Give students time to work on Example 3 with their partners. Use *Popsicle Sticks* to solicit responses.
- **Look For and Make Use of Structure:** Exponential equations are not always in the form $y = ab^x$. Rewriting equations requires applying properties of exponents.

Solving Real-Life Problems

Exponential growth functions are used in real-life situations involving *compound interest*. Although interest earned is expressed as an *annual* rate, the interest is usually compounded more frequently than once per year. So, the formula $y = a(1 + r)^t$ must be modified for compound interest problems.

Core Concept

Compound Interest

Compound interest is the interest earned on the principal *and* on previously earned interest. The balance y of an account earning compound interest is

$$y = P\left(1 + \frac{r}{n}\right)^{nt}.$$

P = principal (initial amount)
r = annual interest rate (in decimal form)
t = time (in years)
n = number of times interest is compounded per year

STUDY TIP

For interest compounded yearly, you can substitute 1 for n in the formula to get $y = P(1 + r)^t$.

EXAMPLE 5 Writing a Function

You deposit \$100 in a savings account that earns 6% annual interest compounded monthly. Write a function that represents the balance after t years.

SOLUTION

$y = P\left(1 + \frac{r}{n}\right)^{nt}$ — Write the compound interest formula.

$= 100\left(1 + \frac{0.06}{12}\right)^{12t}$ — Substitute 100 for P, 0.06 for r, and 12 for n.

$= 100(1.005)^{12t}$ — Simplify.

EXAMPLE 6 Solving a Real-Life Problem

The table shows the balance of a money market account over time.

a. Write a function that represents the balance after t years.

b. Graph the functions from part (a) and from Example 5 in the same coordinate plane. Compare the account balances.

Year, t	Balance
0	\$100
1	\$110
2	\$121
3	\$133.10
4	\$146.41
5	\$161.05

SOLUTION

a. From the table, you know the initial balance is \$100, and it increases 10% each year. So, $P = 100$ and $r = 0.1$.

$y = P(1 + r)^t$ — Write the compound interest formula when $n = 1$.

$= 100(1 + 0.1)^t$ — Substitute 100 for P and 0.1 for r.

$= 100(1.1)^t$ — Add.

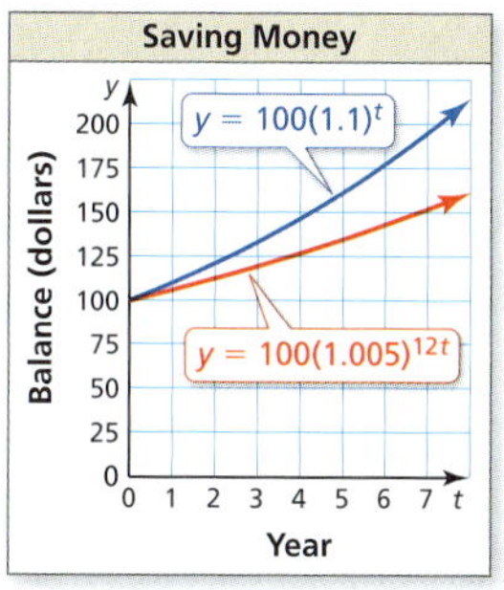

b. The money market account earns 10% interest each year, and the savings account earns 6% interest each year. So, the balance of the money market account increases faster.

Monitoring Progress

Help in English and Spanish at *BigIdeasMath.com*

8. You deposit \$500 in a savings account that earns 9% annual interest compounded monthly. Write and graph a function that represents the balance y (in dollars) after t years.

Extra Example 5

You deposit \$300 in a savings account that earns 3% annual interest compounded monthly. Write a function that represents the balance after t years.
$y = 300(1.0025)^{12t}$

Extra Example 6

The table shows the balance of a money market account over time.

Year, t	Balance
0	\$300
1	\$312
2	\$324.48
3	\$337.46
4	\$350.96
5	\$365

a. Write a function that represents the balance after t years.
$y = 300(1.04)^t$

b. Graph the functions from part (a) and from Extra Example 5 in the same coordinate plane. Compare the account balances.

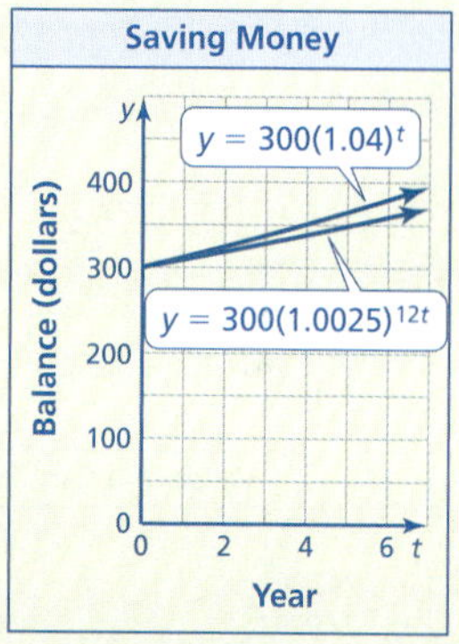

The money market account earns 4% interest each year, and the savings account earns 3% interest each year. So, the balance of the money market account increases faster.

Laurie's Notes Teacher Actions

- Write and discuss the compound interest formula. Explain what compounding means.
- **Extension:** In Example 5, after students have written the equation, have them use a calculator to find the monthly balance for 6 months.
- **? Turn and Talk:** "What information do you know from the table in Example 6?" The principal is \$100. The interest rate is 10% compounded once a year.
- **Use Appropriate Tools Strategically:** Graph the equations from Examples 5 and 6. A vertical line at any time t allows you to quickly compare balances in the accounts.

MONITORING PROGRESS ANSWER

8. $y = 500(1.0075)^{12t}$

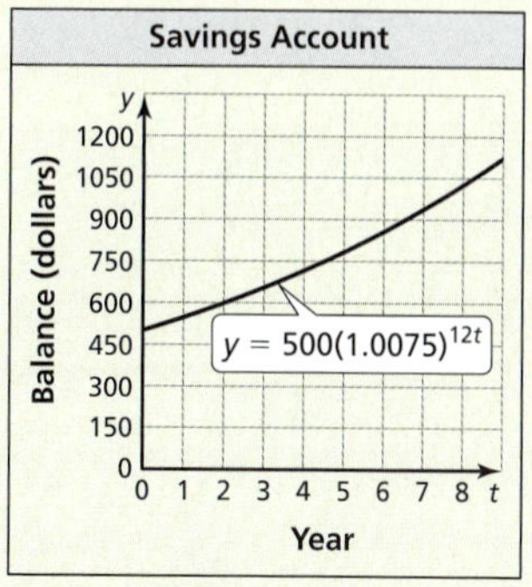

EXAMPLE 7 **Solving a Real-Life Problem**

The value of a car is \$21,500. It loses 12% of its value every year. (a) Write a function that represents the value y (in dollars) of the car after t years. (b) Find the approximate monthly percent decrease in value. (c) Graph the function from part (a). Use the graph to estimate the value of the car after 6 years.

SOLUTION

1. **Understand the Problem** You know the value of the car and its annual percent decrease in value. You are asked to write a function that represents the value of the car over time and approximate the monthly percent decrease in value. Then graph the function and use the graph to estimate the value of the car in the future.
2. **Make a Plan** Use the initial amount and the annual percent decrease in value to write an exponential decay function. Note that the annual percent decrease represents the rate of decay. Rewrite the function using the properties of exponents to approximate the monthly percent decrease (rate of decay). Then graph the original function and use the graph to estimate the y-value when the t-value is 6.
3. **Solve the Problem**

STUDY TIP
In real life, the percent decrease in value of an asset is called the *depreciation rate*.

a. The initial value is \$21,500, and the rate of decay is 12%, or 0.12.

$y = a(1 - r)^t$ — Write the exponential decay function.

$= 21{,}500(1 - 0.12)^t$ — Substitute 21,500 for a and 0.12 for r.

$= 21{,}500(0.88)^t$ — Subtract.

▶ The value of the car can be represented by $y = 21{,}500(0.88)^t$.

b. Use the fact that $t = \frac{1}{12}(12t)$ and the properties of exponents to rewrite the function in a form that reveals the monthly rate of decay.

$y = 21{,}500(0.88)^t$ — Write the original function.

$= 21{,}500(0.88)^{(1/12)(12t)}$ — Rewrite the exponent.

$= 21{,}500(0.88^{1/12})^{12t}$ — Power of a Power Property

$\approx 21{,}500(0.989)^{12t}$ — Evaluate the power.

Use the decay factor $1 - r \approx 0.989$ to find the rate of decay $r \approx 0.011$.

▶ So, the monthly percent decrease is about 1.1%.

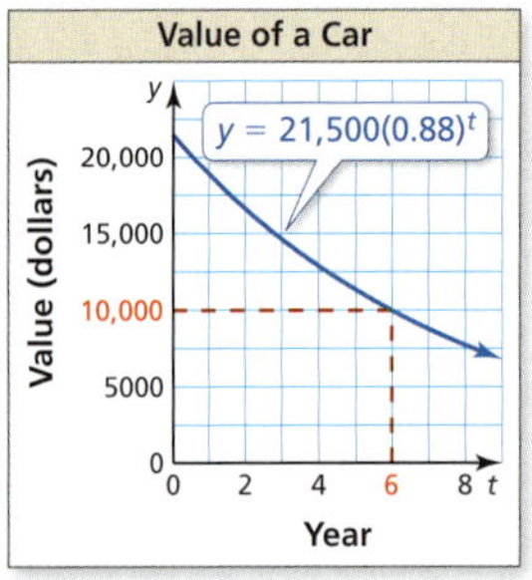

c. From the graph, you can see that the y-value is about 10,000 when $t = 6$.

▶ So, the value of the car is about \$10,000 after 6 years.

4. **Look Back** To check that the monthly percent decrease is reasonable, multiply it by 12 to see if it is close in value to the annual percent decrease of 12%.

$1.1\% \times 12 = 13.2\%$ — 13.2% is close to 12%, so 1.1% is reasonable.

When you evaluate $y = 21{,}500(0.88)^t$ for $t = 6$, you get about \$9985. So, \$10,000 is a reasonable estimation.

Monitoring Progress Help in English and Spanish at *BigIdeasMath.com*

9. **WHAT IF?** The car loses 9% of its value every year. (a) Write a function that represents the value y (in dollars) of the car after t years. (b) Find the approximate monthly percent decrease in value. (c) Graph the function from part (a). Use the graph to estimate the value of the car after 12 years. Round your answer to the nearest thousand.

Extra Example 7

The value of a car is \$15,000. It loses 10% of its value every year.

a. Write a function that represents the value y (in dollars) of the car after t years.
The value of the car can be represented by $y = 15{,}000(0.9)^t$.

b. Find the approximate monthly percent decrease in value.
The monthly percent decrease is about 0.9%.

c. Graph the function from part (a). Use the graph to estimate the value of the car after 6 years.

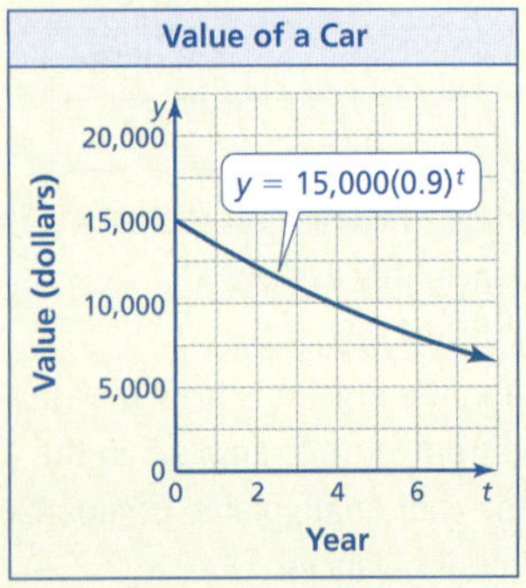

The value of the car is about \$8,000 after 6 years.

MONITORING PROGRESS ANSWER

9. **a.** $y = 21{,}500(0.91)^t$
   b. about 0.8%
   c.

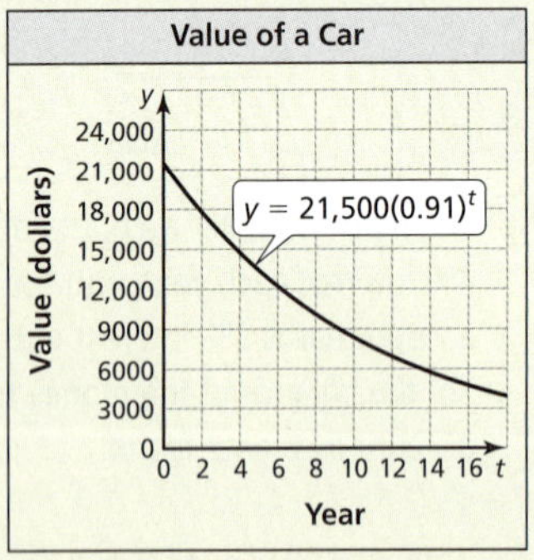

about \$7000

Laurie's Notes Teacher Actions

- Discuss depreciation of a car. Help students write the depreciation equation from the given information.
- ? In part (b), you are asked to find the monthly percent decrease in value. "What is $\frac{1}{12}(12t)$?" t Identify this substitution in the problem.
- ? "Is the amount the car loses in value the same each year? Explain." no; It loses more the first year than any other year. The percent of the value that it loses is the same.

Closure

- **Response Logs:** Select from "I got stuck …" or "What is puzzling me the most is …" or "I am feeling good about…."

6.4 Exercises

Dynamic Solutions available at *BigIdeasMath.com*

Vocabulary and Core Concept Check

1. **COMPLETE THE SENTENCE** In the exponential growth function $y = a(1 + r)^t$, the quantity r is called the ________.

2. **VOCABULARY** What is the decay factor in the exponential decay function $y = a(1 - r)^t$?

3. **VOCABULARY** Compare exponential growth and exponential decay.

4. **WRITING** When does the function $y = ab^x$ represent exponential growth? exponential decay?

Monitoring Progress and Modeling with Mathematics

In Exercises 5–12, identify the initial amount a and the rate of growth r (as a percent) of the exponential function. Evaluate the function when $t = 5$. Round your answer to the nearest tenth.

5. $y = 350(1 + 0.75)^t$
6. $y = 10(1 + 0.4)^t$
7. $y = 25(1.2)^t$
8. $y = 12(1.05)^t$
9. $f(t) = 1500(1.074)^t$
10. $h(t) = 175(1.028)^t$
11. $g(t) = 6.72(2)^t$
12. $p(t) = 1.8^t$

In Exercises 13–16, write a function that represents the situation.

13. Sales of \$10,000 increase by 65% each year.

14. Your starting annual salary of \$35,000 increases by 4% each year.

15. A population of 210,000 increases by 12.5% each year.

16. An item costs \$4.50, and its price increases by 3.5% each year.

17. **MODELING WITH MATHEMATICS** The population of a city has been increasing by 2% annually. The sign shown is from the year 2000. *(See Example 1.)*

    a. Write an exponential growth function that represents the population t years after 2000.

    b. What will the population be in 2020? Round your answer to the nearest thousand.

18. **MODELING WITH MATHEMATICS** A young channel catfish weighs about 0.1 pound. During the next 8 weeks, its weight increases by about 23% each week.

    a. Write an exponential growth function that represents the weight of the catfish after t weeks during the 8-week period.

    b. About how much will the catfish weigh after 4 weeks? Round your answer to the nearest thousandth.

In Exercises 19–26, identify the initial amount a and the rate of decay r (as a percent) of the exponential function. Evaluate the function when $t = 3$. Round your answer to the nearest tenth.

19. $y = 575(1 - 0.6)^t$
20. $y = 8(1 - 0.15)^t$
21. $g(t) = 240(0.75)^t$
22. $f(t) = 475(0.5)^t$
23. $w(t) = 700(0.995)^t$
24. $h(t) = 1250(0.865)^t$
25. $y = \left(\frac{7}{8}\right)^t$
26. $y = 0.5\left(\frac{3}{4}\right)^t$

In Exercises 27–30, write a function that represents the situation.

27. A population of 100,000 decreases by 2% each year.

28. A \$900 sound system decreases in value by 9% each year.

29. A stock valued at \$100 decreases in value by 9.5% each year.

Assignment Guide and Homework Check

ASSIGNMENT

Basic: 1–4, 5–21 odd, 31–61 odd, 63, 68, 71, 73–79

Average: 1–3, 4–68 even, 71, 73–79

Advanced: 1–4, 12, 16, 18, 24–44 even, 50–64 even, 65–79

HOMEWORK CHECK

Basic: 17, 33, 41, 53, 61

Average: 18, 34, 42, 50, 62

Advanced: 18, 38, 44, 54, 64

ANSWERS

1. rate of growth
2. $1 - r$
3. Exponential growth occurs when a quantity increases by the same factor over equal intervals of time. Exponential decay occurs when a quantity decreases by the same factor over equal intervals of time.
4. when $b > 1$ and x represents time; when $0 < b < 1$ and x represents time
5. $a = 350$, $r = 75\%$; about 5744.6
6. $a = 10$, $r = 40\%$; about 53.8
7. $a = 25$, $r = 20\%$; about 62.2
8. $a = 12$, $r = 5\%$; about 15.3
9. $a = 1500$, $r = 7.4\%$; about 2143.4
10. $a = 175$, $r = 2.8\%$; about 200.9
11. $a = 6.72$, $r = 100\%$; about 215.0
12. $a = 1$, $r = 80\%$; about 18.9
13. $y = 10{,}000(1.65)^t$
14. $y = 35{,}000(1.04)^t$
15. $y = 210{,}000(1.125)^t$
16. $y = 4.50(1.035)^t$
17. a. $y = 315{,}000(1.02)^t$
    b. about 468,000
18. a. $y = 0.1(1.23)^t$
    b. about 0.229 lb
19. $a = 575$, r = 60%; about 36.8
20. $a = 8$, $r = 15\%$; about 4.9
21. $a = 240$, $r = 25\%$; about 101.3
22. $a = 475$, $r = 50\%$; about 59.4
23. $a = 700$, $r = 0.5\%$; about 689.6
24. $a = 1250$, $r = 13.5\%$; about 809.0
25. $a = 1$, $r = 12.5\%$; about 0.7
26. $a = 0.5$, $r = 25\%$; about 0.2
27. $y = 100{,}000(0.98)^t$
28. $y = 900(0.91)^t$
29. $y = 100(0.905)^t$

Dynamic Teaching Tools

- Dynamic Assessment & Progress Monitoring Tool
- Interactive Whiteboard Lesson Library
- Dynamic Classroom with Dynamic Investigations

ANSWERS

30. $y = 20{,}000(0.866)^t$

31. The growth rate is $1 + 1.5$, not just 1.5; $b(t) = 10(2.5)^t$; $b(8) = 10(2.5)^8 \approx 15{,}258.8$; After 8 hours, there are about 15,259 bacteria in the culture.

32. The decay rate is $1 - 0.14$, not $1 + 0.14$; $v(t) = 25{,}000(0.86)^5 \approx 11{,}761$; The value of the car in 2015 is about $12,000.

33. exponential decay; As x increases by 1, y is multiplied by $\frac{1}{5}$.

34. neither; As x increases by 1, y decreases by 4.

35. neither; As x increases by 1, y decreases by 6.

36. exponential growth; As x increases by 1, y is multiplied by 3.

37. exponential growth; As x increases by 5, y is multiplied by 4.

38. exponential decay; As x increases by 2, y is multiplied by $\frac{1}{6}$.

39. a. exponential decay

b. about $15,155

40. a. exponential growth

b. about 17,716 people

41. exponential decay; 20%

42. exponential growth; 10%

43. exponential decay; 5%

44. exponential growth; 8%

45. exponential growth; 6%

46. exponential decay; 52%

47. exponential growth; 25%

48. exponential decay; 20%

49. $y \approx 1.52(0.9)^t$; exponential decay

50. $y \approx 14.8(1.4)^t$; exponential growth

51. $y \approx 2(1.69)^t$; exponential growth

52. $y \approx 5(0.96)^t$; exponential decay

53. $x(t) \approx (1.20)^t$; exponential growth

54. $f(t) \approx 0.34(1.16)^t$; exponential growth

55. $b(t) \approx 0.67(0.55)^t$; exponential decay

56. $r(t) \approx (0.60)^t$; exponential decay

30. A company profit of $20,000 decreases by 13.4% each year.

31. ERROR ANALYSIS The growth rate of a bacterial culture is 150% each hour. Initially, there are 10 bacteria. Describe and correct the error in finding the number of bacteria in the culture after 8 hours.

> ✗
> $b(t) = 10(1.5)^t$
> $b(8) = 10(1.5)^8 \approx 256.3$
> After 8 hours, there are about 256 bacteria in the culture.

32. ERROR ANALYSIS You purchase a car in 2010 for $25,000. The value of the car decreases by 14% annually. Describe and correct the error in finding the value of the car in 2015.

> ✗
> $v(t) = 25{,}000(1.14)^t$
> $v(5) = 25{,}000(1.14)^5 \approx 48{,}135$
> The value of the car in 2015 is about $48,000.

In Exercises 33–38, determine whether the table represents an *exponential growth function*, an *exponential decay function*, or *neither*. Explain. *(See Example 2.)*

33.

x	y
−1	50
0	10
1	2
2	0.4

34.

x	y
0	32
1	28
2	24
3	20

35.

x	y
0	35
1	29
2	23
3	17

36.

x	y
1	17
2	51
3	153
4	459

37.

x	y
5	2
10	8
15	32
20	128

38.

x	y
3	432
5	72
7	12
9	2

39. ANALYZING RELATIONSHIPS The table shows the value of a camper t years after it is purchased.

t	Value
1	$37,000
2	$29,600
3	$23,680
4	$18,944

a. Determine whether the table represents an exponential growth function, an exponential decay function, or neither.

b. What is the value of the camper after 5 years?

40. ANALYZING RELATIONSHIPS The table shows the total numbers of visitors to a website t days after it is online.

t	42	43	44	45
Visitors	11,000	12,100	13,310	14,641

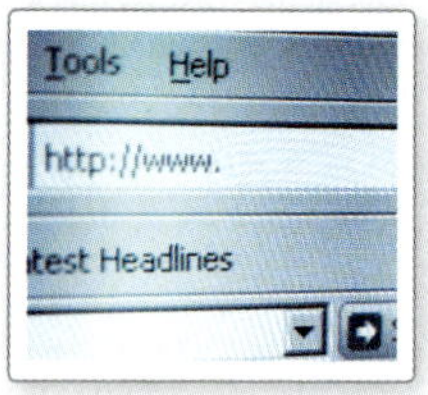

a. Determine whether the table represents an exponential growth function, an exponential decay function, or neither.

b. How many people will have visited the website after it is online 47 days?

In Exercises 41–48, determine whether each function represents *exponential growth* or *exponential decay*. Identify the percent rate of change. *(See Example 3.)*

41. $y = 4(0.8)^t$

42. $y = 15(1.1)^t$

43. $y = 30(0.95)^t$

44. $y = 5(1.08)^t$

45. $r(t) = 0.4(1.06)^t$

46. $s(t) = 0.65(0.48)^t$

47. $g(t) = 2\left(\frac{5}{4}\right)^t$

48. $m(t) = \left(\frac{4}{5}\right)^t$

In Exercises 49–56, rewrite the function to determine whether it represents *exponential growth* or *exponential decay*. *(See Example 4.)*

49. $y = (0.9)^{t-4}$

50. $y = (1.4)^{t+8}$

51. $y = 2(1.06)^{9t}$

52. $y = 5(0.82)^{t/5}$

53. $x(t) = (1.45)^{t/2}$

54. $f(t) = 0.4(1.16)^{t-1}$

55. $b(t) = 4(0.55)^{t+3}$

56. $r(t) = (0.88)^{4t}$

In Exercises 57–60, write a function that represents the balance after t years. *(See Example 5.)*

57. \$2000 deposit that earns 5% annual interest compounded quarterly

58. \$1400 deposit that earns 10% annual interest compounded semiannually

59. \$6200 deposit that earns 8.4% annual interest compounded monthly

60. \$3500 deposit that earns 9.2% annual interest compounded quarterly

61. PROBLEM SOLVING The cross-sectional area of a tree 4.5 feet from the ground is called its *basal area.* The table shows the basal areas (in square inches) of Tree A over time. *(See Example 6.)*

Year, t	0	1	2	3	4
Basal area, A	120	132	145.2	159.7	175.7

Growth rate: 6%

Initial basal area: 154 in.^{2}

a. Write functions that represent the basal areas of the trees after t years.

b. Graph the functions from part (a) in the same coordinate plane. Compare the basal areas.

62. PROBLEM SOLVING You deposit \$300 into an investment account that earns 12% annual interest compounded quarterly. The graph shows the balance of a savings account over time.

a. Write functions that represent the balances of the accounts after t years.

b. Graph the functions from part (a) in the same coordinate plane. Compare the account balances.

63. PROBLEM SOLVING A city has a population of 25,000. The population is expected to increase by 5.5% annually for the next decade. *(See Example 7.)*

a. Write a function that represents the population y after t years.

b. Find the approximate monthly percent increase in population.

c. Graph the function from part (a). Use the graph to estimate the population after 4 years.

64. PROBLEM SOLVING Plutonium-238 is a material that generates steady heat due to decay and is used in power systems for some spacecraft. The function $y = a(0.5)^{t/x}$ represents the amount y of a substance remaining after t years, where a is the initial amount and x is the length of the half-life (in years).

a. A scientist is studying a 3-gram sample. Write a function that represents the amount y of plutonium-238 after t years.

b. What is the yearly percent decrease of plutonium-238?

c. Graph the function from part (a). Use the graph to estimate the amount remaining after 12 years.

65. COMPARING FUNCTIONS The three given functions describe the amount y of ibuprofen (in milligrams) in a person's bloodstream t hours after taking the dosage.

$y \approx 800(0.71)^t$

$y \approx 800(0.9943)^{60t}$

$y \approx 800(0.843)^{2t}$

a. Show that these expressions are approximately equivalent.

b. Describe the information given by each of the functions.

ANSWERS

57. $y = 2000(1.0125)^{4t}$

58. $y = 1400(1.05)^{2t}$

59. $y = 6200(1.007)^{12t}$

60. $y = 3500(1.023)^{4t}$

61. a. $A_A = 120(1.1)^t$; $A_B = 154(1.06)^t$

b.

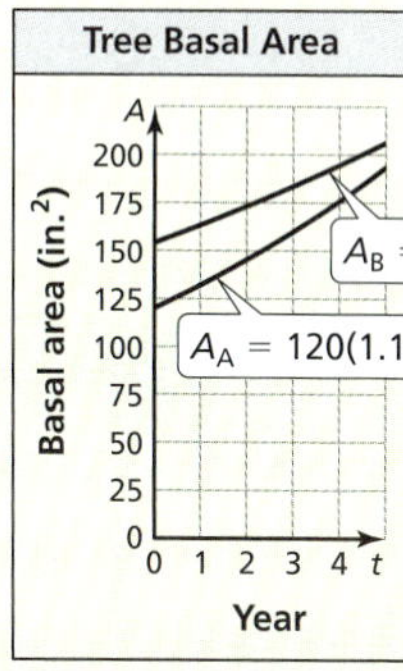

The basal area of Tree B is larger than the basal area of Tree A, but the difference between the basal areas is decreasing.

62. a. investment account: $y = 300(1.03)^{4t}$; savings account: $y = 300(1.08)^t$

b.

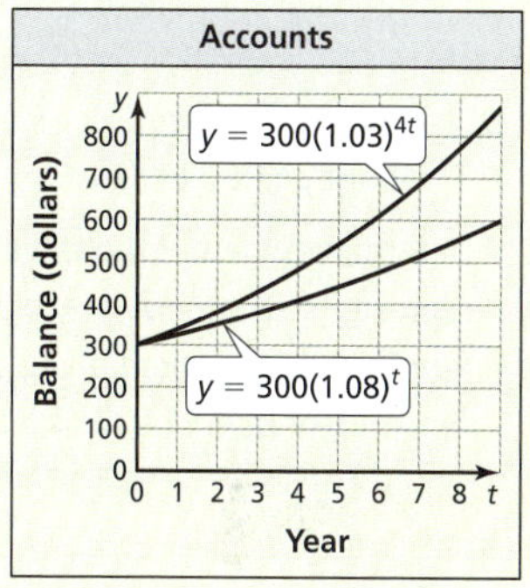

Both accounts start with the same balance. The investment account balance is increasing at a faster rate, so it is greater than the savings account balance after the start.

63. a. $y = 25{,}000(1.055)^t$

b. about 0.45%

c.

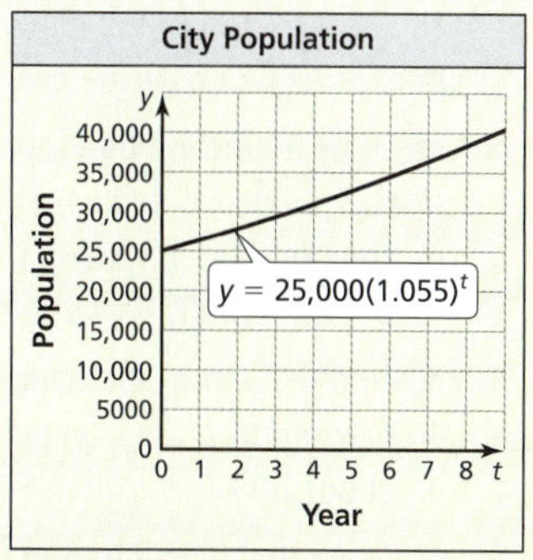

about 30,971 people

64. a. $y = 3(0.5)^{t/88}$

b. about 0.8%

64. c.

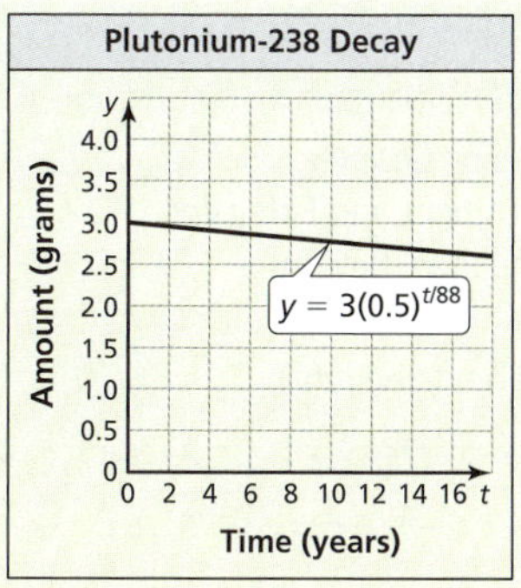

about 2.7 g

65. a. $y \approx 800(0.9943)^{60t} \approx 800(0.9943^{60})^t \approx 800(0.7097)^t$; $y \approx 800(0.843)^{2t} \approx 800(0.843^2)^t \approx 800(0.7106)^t$

b. All three functions indicate the initial amount of ibuprofen in a person's bloodstream is 800 mg. The first function indicates the amount of ibuprofen in a person's bloodstream decreases by about 29% each hour. The second function indicates the amount of ibuprofen in a person's bloodstream decreases by about 0.57% each minute. The third function indicates the amount of ibuprofen in a person's bloodstream decreases by about 15.7% each half-hour.

ANSWERS

66. $C(t) = 9000(1.003)^{12t} + 480t$; The total amount of money saved both in the savings account and at home.

67. 200%; The growth factor is 3, which is also $r + 1$, so r, the growth rate, is 2, or 200%.

68–79. See Additional Answers.

Mini-Assessment

1. The inaugural attendance of an annual music festival is 20,000. The attendance y increases by 5% each year.
   a. Write an exponential growth function that represents the attendance after t years.
   $y = 20{,}000(1.05)^t$
   b. How many people will attend the festival in the fifth year? Round your answer to the nearest thousand.
   about 24,000 people
2. Determine whether the table represents an *exponential growth function*, an *exponential decay function*, or *neither*.

x	0	1	2	3
y	7	35	175	875

   an exponential growth function

3. Rewrite the function $f(t) = 300(0.7)^{t/4}$ to determine whether it represents *exponential growth* or *exponential decay*.
   $f(t) = 300(0.91)^t$; exponential decay
4. The value of a computer is \$1200. It loses 20% of its value every year.
   a. Write a function that represents the value y (in dollars) of the computer after t years.
   $y = 1200(0.8)^t$
   b. Find the approximate monthly percent decrease in value.
   about 1.8%
   c. Graph the function from part (a). Use the graph to estimate the value of the computer after 6 years.
   See Additional Answers.

66. COMBINING FUNCTIONS You deposit \$9000 in a savings account that earns 3.6% annual interest compounded monthly. You also save \$40 per month in a safe at home. Write a function $C(t) = b(t) + h(t)$, where $b(t)$ represents the balance of your savings account and $h(t)$ represents the amount in your safe after t years. What does $C(t)$ represent?

67. NUMBER SENSE During a flu epidemic, the number of sick people triples every week. What is the growth rate as a percent? Explain your reasoning.

68. HOW DO YOU SEE IT? Match each situation with its graph. Explain your reasoning.

a. A bacterial population doubles each hour.

b. The value of a computer decreases by 18% each year.

c. A deposit earns 11% annual interest compounded yearly.

d. A radioactive element decays 5.5% each year.

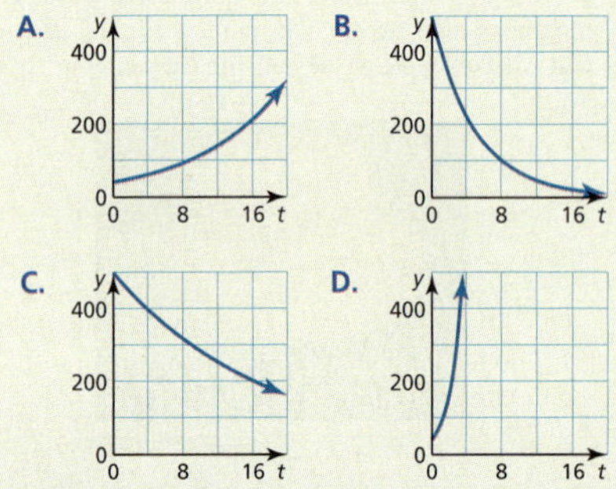

69. WRITING Give an example of an equation in the form $y = ab^x$ that does not represent an exponential growth function or an exponential decay function. Explain your reasoning.

70. THOUGHT PROVOKING Describe two account options into which you can deposit \$1000 and earn compound interest. Write a function that represents the balance of each account after t years. Which account would you rather use? Explain your reasoning.

71. MAKING AN ARGUMENT A store is having a sale on sweaters. On the first day, the prices of the sweaters are reduced by 20%. The prices will be reduced another 20% each day until the sweaters are sold. Your friend says the sweaters will be free on the fifth day. Is your friend correct? Explain.

72. COMPARING FUNCTIONS The graphs of f and g are shown.

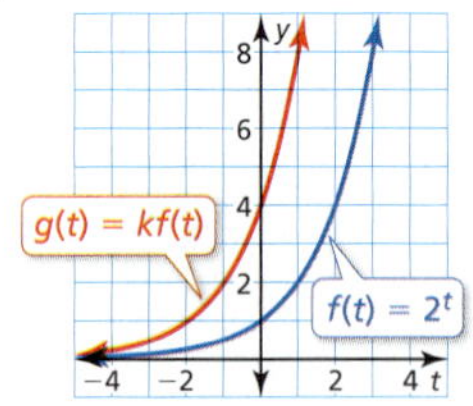

a. Explain why f is an exponential growth function. Identify the rate of growth.

b. Describe the transformation from the graph of f to the graph of g. Determine the value of k.

c. The graph of g is the same as the graph of $h(t) = f(t + r)$. Use properties of exponents to find the value of r.

Maintaining Mathematical Proficiency

Reviewing what you learned in previous grades and lessons

Solve the equation. Check your solution. *(Section 1.3)*

73. $8x + 12 = 4x$ **74.** $5 - t = 7t + 21$ **75.** $6(r - 2) = 2r + 8$

Find the slope and the y-intercept of the graph of the linear equation. *(Section 3.5)*

76. $y = -6x + 7$ **77.** $y = \frac{1}{4}x + 7$

78. $3y = 6x - 12$ **79.** $2y + x = 8$

If students need help...	*If students got it...*
Resources by Chapter • Practice A and Practice B • Puzzle Time	Resources by Chapter • Enrichment and Extension • Cumulative Review
Student Journal • Practice	Start the *next* Section
Differentiating the Lesson Skills Review Handbook	

6.1–6.4 What Did You Learn?

Core Vocabulary

*n*th root of *a*, *p. 300*
radical, *p. 300*
index of a radical, *p. 300*
exponential function, *p. 306*
exponential growth, *p. 314*
exponential growth function, *p. 314*
exponential decay, *p. 315*
exponential decay function, *p. 315*
compound interest, *p. 317*

Core Concepts

Section 6.1

Zero Exponent, *p. 292*
Negative Exponents, *p. 292*
Product of Powers Property, *p. 293*
Quotient of Powers Property, *p. 293*
Power of a Power Property, *p. 293*
Power of a Product Property, *p. 294*
Power of a Quotient Property, *p. 294*

Section 6.2

Real *n*th Roots of *a*, *p. 300*
Rational Exponents, *p. 301*

Section 6.3

Graphing $y = ab^x$ When $b > 1$, *p. 307*
Graphing $y = ab^x$ When $0 < b < 1$, *p. 307*

Section 6.4

Exponential Growth Functions, *p. 314*
Exponential Decay Functions, *p. 315*
Compound Interest, *p. 317*

Mathematical Practices

1. How did you apply what you know to simplify the complicated situation in Exercise 56 on page 297?
2. How can you use previously established results to construct an argument in Exercise 44 on page 304?
3. How is the form of the function you wrote in Exercise 66 on page 322 related to the forms of other types of functions you have learned about in this course?

Study Skills

Analyzing Your Errors

Misreading Directions

- **What Happens:** You incorrectly read or do not understand directions.
- **How to Avoid This Error:** Read the instructions for exercises at least twice and make sure you understand what they mean. Make this a habit and use it when taking tests.

323

Dynamic Teaching Tools

Dynamic Assessment & Progress Monitoring Tool
Interactive Whiteboard Lesson Library
Dynamic Classroom with Dynamic Investigations

ANSWERS

1. *Sample answer:* Apply the distance formula ($d = rt$) to find an equation solved for time, and apply the laws of exponents to simplify the powers of 10.
2. *Sample answer:* Apply what you previously established about the sign of a radical with an odd index, which is the same as the sign of the radicand.
3. *Sample answer:* The first term, $b(t)$, is in the form of an exponential function, and the second term, $h(t)$, is in the form of a linear function.

6.1–6.4 Quiz

Simplify the expression. Write your answer using only positive exponents. *(Section 6.1)*

1. $3^2 \cdot 3^4$ **2.** $(k^4)^{-3}$ **3.** $\left(\frac{4r^2}{3s^5}\right)^3$ **4.** $\left(\frac{2x^0}{4x^{-2}y^4}\right)^2$

Evaluate the expression. *(Section 6.2)*

5. $\sqrt[3]{27}$ **6.** $\left(\frac{1}{16}\right)^{1/4}$ **7.** $512^{2/3}$ **8.** $(\sqrt{4})^5$

Graph the function. Describe the domain and range. *(Section 6.3)*

9. $y = 5^x$ **10.** $y = -2\left(\frac{1}{6}\right)^x$ **11.** $y = 6(2)^{x-4} - 1$

Determine whether the table represents an *exponential growth function*, an *exponential decay function*, or *neither*. Explain. *(Section 6.4)*

12.

x	0	1	2	3
y	7	21	63	189

13.

x	1	2	3	4
y	14,641	1331	121	11

Determine whether the function represents *exponential growth* or *exponential decay*. Identify the percent rate of change. *(Section 6.4)*

14. $y = 3(1.88)^t$ **15.** $f(t) = \frac{1}{3}(1.26)^t$ **16.** $f(t) = 80\left(\frac{3}{5}\right)^t$

17. The table shows several units of mass. *(Section 6.1)*

Unit of mass	kilogram	hectogram	dekagram	decigram	centigram	milligram	microgram	nanogram
Mass (in grams)	10^3	10^2	10^1	10^{-1}	10^{-2}	10^{-3}	10^{-6}	10^{-9}

a. How many times larger is a kilogram than a nanogram? Write your answer using only positive exponents.

b. How many times smaller is a milligram than a hectogram? Write your answer using only positive exponents.

c. Which is greater, 10,000 milligrams or 1000 decigrams? Explain your reasoning.

18. You store blankets in a cedar chest. What is the volume of the cedar chest? *(Section 6.2)*

19. The function $f(t) = 5(4)^t$ represents the number of frogs in a pond after t years. *(Section 6.3 and Section 6.4)*

a. Does the function represent *exponential growth* or *exponential decay*? Explain.

b. Graph the function. Describe the domain and range.

c. What is the yearly percent change? the approximate monthly percent change?

d. How many frogs are in the pond after 4 years?

ANSWERS

1. 3^6

2. $\frac{1}{k^{12}}$

3. $\frac{64r^6}{27s^{15}}$

4. $\frac{x^4}{4y^8}$

5. 3

6. $\frac{1}{2}$

7. 64

8. 32

9.

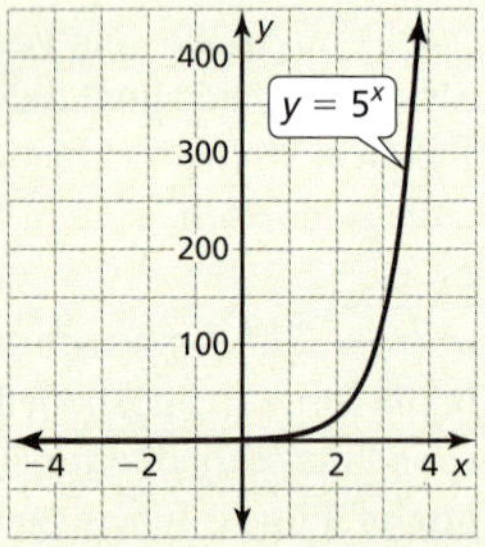

domain: all real numbers, range: $y > 0$

10.

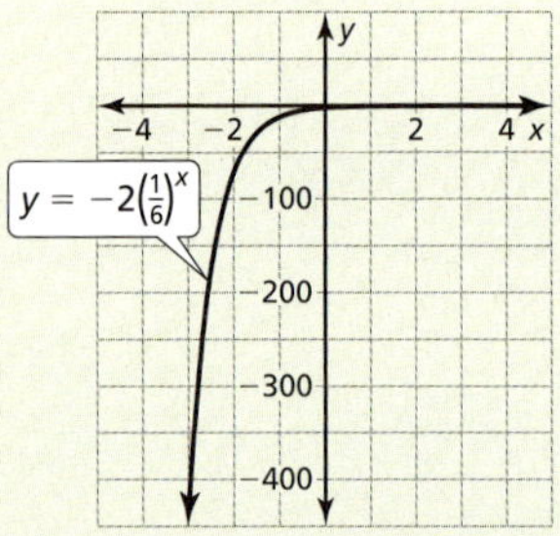

domain: all real numbers, range: $y < 0$

11.

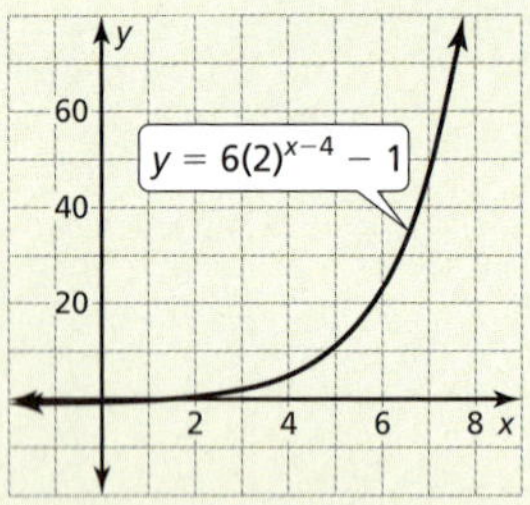

domain: all real numbers, range: $y > -1$

12. exponential growth; As x increases by 1, y is multiplied by 3.

13. exponential decay; As x increases by 1, y is multiplied by $\frac{1}{11}$.

14. exponential growth; 88%

15. exponential growth; 26%

16. exponential decay; 40%

17. a. 10^{12}

b. 10^5

c. 1000 dg; 10,000 mg = 10 g, 1000 dg = 100 g

18. 48 ft^3

19. a. exponential growth; As t increases by 1, y is multiplied by 4.

b.

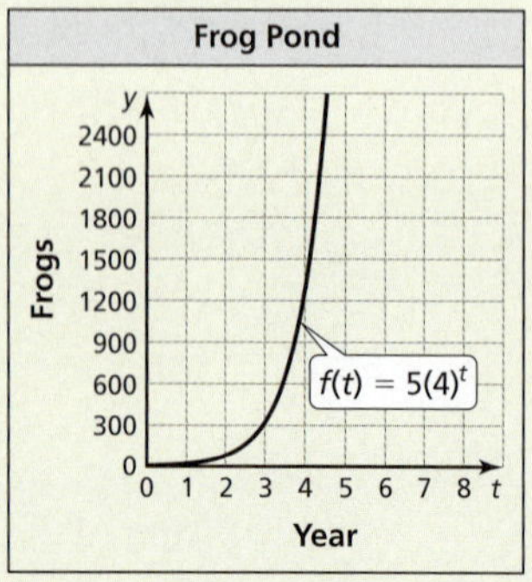

domain: $t \geq 0$, range: $y \geq 5$

c. 300%; about 12%

d. 1280 frogs

Laurie's Notes

Overview of Section 6.5

Introduction

- This is a short lesson on solving exponential equations. The first part of the lesson uses algebraic technique and looks at examples that already have the same base, as well as examples in which the exponential expressions must be rewritten to have a common base.
- In the last part of the lesson, exponential equations are solved graphically.

Resources

- The graphing calculator can be used to graph the exponential equations, giving students the ability to solve the equations graphically.

Formative Assessment Tips

- Refer to page T-290 for a description of *Response Logs*. In this lesson and throughout this chapter, there will be sentence stems to prompt student writing. Use your judgment for placement in the lesson, amount of time to allow for writing, and the actual sentence stem given to students.
- For a beginning list of writing stems, see page T-298.

Pacing Suggestion

- The explorations provide an opportunity for students to think about what solving exponential equations means. When students finish the explorations, transition to the formal lesson.

Dynamic Teaching Tools

- Dynamic Assessment & Progress Monitoring Tool
- Lesson Planning Tool
- Interactive Whiteboard Lesson Library
- Dynamic Classroom with Dynamic Investigations

What Your Students Will Learn

- Solve exponential equations algebraically with the same base and with different bases.
- Solve exponential equations using a graphing calculator.

Laurie's Notes

Exploration

Motivate

- Before class, write the following values on index cards.

2^4	3^4	2^6	$9^{3/2}$	2^5
4^2	9^2	8^2	3^3	$4^{5/2}$

- Play a quick matching game with 10 students. Hand each student a card and ask him or her to find the student whose card has an equivalent value.
- When the students have found their matches, have each pair of students explain why the values are equivalent.

Exploration 1

- Students may not remember to graph each side of the equation and look for the point of intersection. Instead they may use the *trace* key or the table of values.
- **Use Appropriate Tools Strategically:** Circulate as students are exploring this problem so that you will be able to solicit answers from students who used different methods.

Exploration 2

- There are many possible solutions for parts (b) and (c). Solicit a variety of answers.
- Students may not make the connection between Exploration 1 and the two questions they have just answered.

? "What does it mean for an equation to have a solution?" Listen for language such as, "There is a value of *x* that makes the function equal a certain value."

- **Construct Viable Arguments and Critique the Reasoning of Others:** Have students share their thinking for the questions posed in part (d).

Exploration 3

- You may choose a jigsaw approach with these problems, with each group of students solving about three problems.
- Circulate to observe the techniques students are using when solving.

Communicate Your Answer

- Question 5 connects to the prior lesson. Solicit answers from students who used different methods. They should describe their method, not just give an answer.

Connecting to Next Step

- In the first half of the formal lesson, students will solve exponential equations algebraically. In the last part of the lesson, they will solve exponential equations graphically.

6.5 Solving Exponential Equations

Essential Question How can you solve an exponential equation graphically?

EXPLORATION 1 Solving an Exponential Equation Graphically

Work with a partner. Use a graphing calculator to solve the exponential equation $2.5^{x-3} = 6.25$ graphically. Describe your process and explain how you determined the solution.

EXPLORATION 2 The Number of Solutions of an Exponential Equation

Work with a partner.

a. Use a graphing calculator to graph the equation $y = 2^x$.

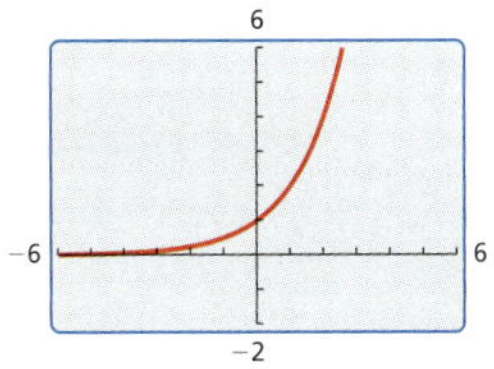

b. In the same viewing window, graph a linear equation (if possible) that does not intersect the graph of $y = 2^x$.

c. In the same viewing window, graph a linear equation (if possible) that intersects the graph of $y = 2^x$ in more than one point.

d. Is it possible for an exponential equation to have no solution? more than one solution? Explain your reasoning.

USING APPROPRIATE TOOLS

To be proficient in math, you need to use technological tools to explore and deepen your understanding of concepts.

EXPLORATION 3 Solving Exponential Equations Graphically

Work with a partner. Use a graphing calculator to solve each equation.

a. $2^x = \frac{1}{2}$ **b.** $2^{x+1} = 0$ **c.** $2^x = \sqrt{2}$

d. $3^x = 9$ **e.** $3^{x-1} = 0$ **f.** $4^{2x} = 2$

g. $2^{x/2} = \frac{1}{4}$ **h.** $3^{x+2} = \frac{1}{9}$ **i.** $2^{x-2} = \frac{3}{2}x - 2$

Communicate Your Answer

4. How can you solve an exponential equation graphically?

5. A population of 30 mice is expected to double each year. The number p of mice in the population each year is given by $p = 30(2^n)$. In how many years will there be 960 mice in the population?

Dynamic Teaching Tools
- Dynamic Assessment & Progress Monitoring Tool
- Lesson Planning Tool
- Interactive Whiteboard Lesson Library
- Dynamic Classroom with Dynamic Investigations

ANSWERS

1. $x = 5$; Graph $y = 2.5^{x-3}$ and $y = 6.25$. The x-coordinate of the point of intersection, 5, is the solution of the equation.

2. a–c. *Sample answer:*

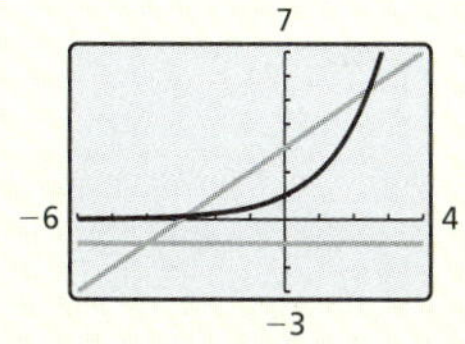

d. yes; yes; *Sample answer:* The graphs from parts (a)–(c) show that $2^x = -1$ has no solution and $2^x = x + 3$ has two solutions.

3. a. $x = -1$

b. no solution

c. $x = \frac{1}{2}$

d. $x = 2$

e. no solution

f. $x = \frac{1}{4}$

g. $x = -4$

h. $x = -4$

i. $x = 2, x = 4$

4. Form two new equations by setting each side of the original equation equal to y. Graph the new equations. The x-coordinate(s) of the point(s) of intersection is/are the solution(s).

5. 5 years

Extra Example 1

Solve each equation.

a. $4^{x+3} = 4^9$ $x = 6$

b. $2 = 2^{3x-2}$ $x = 1$

c. $8^{3x-4} = 8^{2x+1}$ $x = 5$

MONITORING PROGRESS ANSWERS

1. $x = 3$
2. $x = 1$
3. $x = -2$

6.5 Lesson

Core Vocabulary

exponential equation, *p. 326*

What You Will Learn

- Solve exponential equations with the same base.
- Solve exponential equations with unlike bases.
- Solve exponential equations by graphing.

Solving Exponential Equations with the Same Base

Exponential equations are equations in which variable expressions occur as exponents.

Core Concept

Property of Equality for Exponential Equations

Words Two powers with the *same positive base b*, where $b \neq 1$, are equal if and only if their exponents are equal.

Numbers If $2^x = 2^5$, then $x = 5$. If $x = 5$, then $2^x = 2^5$.

Algebra If $b > 0$ and $b \neq 1$, then $b^x = b^y$ if and only if $x = y$.

EXAMPLE 1 Solving Exponential Equations with the Same Base

Solve each equation.

a. $3^{x+1} = 3^5$ b. $6 = 6^{2x-3}$ c. $10^{3x} = 10^{2x+3}$

SOLUTION

a. $3^{x+1} = 3^5$ Write the equation.

$x + 1 = 5$ Equate the exponents.

$-1 \quad -1$ Subtract 1 from each side.

$x = 4$ Simplify.

b. $6 = 6^{2x-3}$ Write the equation.

$1 = 2x - 3$ Equate the exponents.

$+3 \quad +3$ Add 3 to each side.

$4 = 2x$ Simplify.

$\frac{4}{2} = \frac{2x}{2}$ Divide each side by 2.

$2 = x$ Simplify.

Check

$6 = 6^{2x-3}$

$6 \stackrel{?}{=} 6^{2(2)-3}$

$6 = 6$ ✓

c. $10^{3x} = 10^{2x+3}$ Write the equation.

$3x = 2x + 3$ Equate the exponents.

$-2x \quad -2x$ Subtract $2x$ from each side.

$x = 3$ Simplify.

Monitoring Progress

Help in English and Spanish at *BigIdeasMath.com*

Solve the equation. Check your solution.

1. $2^{2x} = 2^6$ 2. $5^{2x} = 5^{x+1}$ 3. $7^{3x+5} = 7^{x+1}$

Laurie's Notes Teacher Actions

- **Use Appropriate Tools Strategically:** Today, students will solve exponential equations by using properties of exponents and by using a graphing calculator.
- **Look For and Make Use of Structure:** To solve an exponential equation of the form $b^x = b^y$, set the exponents equal and solve. Future courses present other strategies.
- **Popsicle Sticks:** All the equations in Example 1 have the same base. Have students work with their partners to solve. Use *Popsicle Sticks* to solicit responses.
- **Think-Pair-Share:** Have students answer Questions 1–3, and then share and discuss as a class.

Solving Exponential Equations with Unlike Bases

To solve some exponential equations, you must first rewrite each side of the equation using the same base.

EXAMPLE 2 **Solving Exponential Equations with Unlike Bases**

Solve (a) $5^x = 125$, (b) $4^x = 2^{x-3}$, and (c) $9^{x+2} = 27^x$.

SOLUTION

a. $5^x = 125$ — Write the equation.
$5^x = 5^3$ — Rewrite 125 as 5^3.
$x = 3$ — Equate the exponents.

b. $4^x = 2^{x-3}$ — Write the equation.
$(2^2)^x = 2^{x-3}$ — Rewrite 4 as 2^2.
$2^{2x} = 2^{x-3}$ — Power of a Power Property
$2x = x - 3$ — Equate the exponents.
$x = -3$ — Solve for x.

c. $9^{x+2} = 27^x$ — Write the equation.
$(3^2)^{x+2} = (3^3)^x$ — Rewrite 9 as 3^2 and 27 as 3^3.
$3^{2x+4} = 3^{3x}$ — Power of a Power Property
$2x + 4 = 3x$ — Equate the exponents.
$4 = x$ — Solve for x.

Check
$4^x = 2^{x-3}$
$4^{-3} \stackrel{?}{=} 2^{-3-3}$
$\frac{1}{64} = \frac{1}{64}$ ✓

Check
$9^{x+2} = 27^x$
$9^{4+2} \stackrel{?}{=} 27^4$
$531{,}441 = 531{,}441$ ✓

EXAMPLE 3 **Solving Exponential Equations When $0 < b < 1$**

Solve (a) $\left(\frac{1}{2}\right)^x = 4$ and (b) $4^{x+1} = \frac{1}{64}$.

SOLUTION

a. $\left(\frac{1}{2}\right)^x = 4$ — Write the equation.
$(2^{-1})^x = 2^2$ — Rewrite $\frac{1}{2}$ as 2^{-1} and 4 as 2^2.
$2^{-x} = 2^2$ — Power of a Power Property
$-x = 2$ — Equate the exponents.
$x = -2$ — Solve for x.

b. $4^{x+1} = \frac{1}{64}$ — Write the equation.
$4^{x+1} = \frac{1}{4^3}$ — Rewrite 64 as 4^3.
$4^{x+1} = 4^{-3}$ — Definition of negative exponent
$x + 1 = -3$ — Equate the exponents.
$x = -4$ — Solve for x.

Check
$4^{x+1} = \frac{1}{64}$
$4^{-4+1} \stackrel{?}{=} \frac{1}{64}$
$\frac{1}{64} = \frac{1}{64}$ ✓

Monitoring Progress

Help in English and Spanish at *BigIdeasMath.com*

Solve the equation. Check your solution.

4. $4^x = 256$ **5.** $9^{2x} = 3^{x-6}$ **6.** $4^{3x} = 8^{x+1}$ **7.** $\left(\frac{1}{3}\right)^{x-1} = 27$

Differentiated Instruction

Inclusion

Some students may have difficulty solving exponential equations when the base of a term is a fraction between 0 and 1. Remind students from their study of exponents in Section 6.1 that $\frac{1}{a^n} = a^{-n}$. So, students can rewrite the fraction base as a base with a negative exponent. For example, the term $\left(\frac{1}{4}\right)^x$ can be rewritten as $(4^{-1})^x$ or $(2^{-2})^x$.

Extra Example 2

a. Solve $6^x = 1296$. $x = 4$
b. Solve $9^x = 3^{x+2}$. $x = 2$
c. Solve $4^{2x} = 8^{x-2}$. $x = -6$

Extra Example 3

a. Solve $\left(\frac{1}{6}\right)^x = 216$. $x = -3$
b. Solve $5^{x+3} = \frac{1}{25}$. $x = -5$

MONITORING PROGRESS ANSWERS

4. $x = 4$
5. $x = -2$
6. $x = 1$
7. $x = -2$

Laurie's Notes Teacher Actions

- In Example 2(a) ask, "What power of 5 is 125?" Students will quickly say 3, but work through the example anyway to show the process.

? "What common base can you use to rewrite the equation in Example 2(b)?" Use a base of 2. "What will you use in part (c)?" Use a base of 3.

COMMON ERROR In part (c), when the Power of a Power Property is used, the Distributive Property is also used to multiply 2 and $x + 2$.

- **Think-Alouds:** Use *Think-Alouds* for each part of Example 3.

Extra Example 4

a. Use a graphing calculator to solve $\left(\frac{1}{4}\right)^{x+1} = 5$.

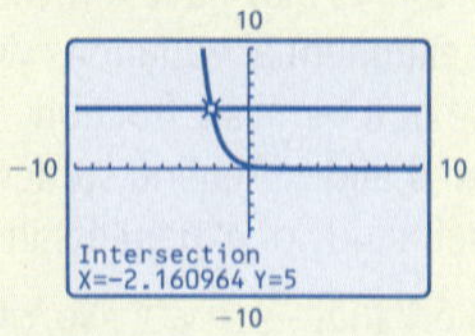

The solution is $x \approx -2.16$.

b. Use a graphing calculator to solve $2^{x+3} = x - 3$.

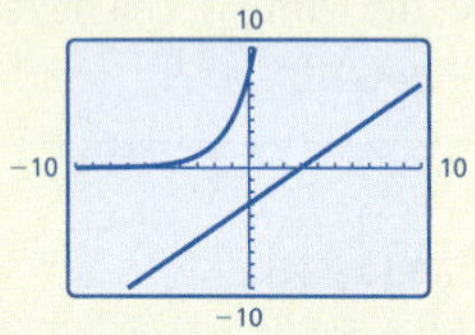

The equation has no solution, because the graphs do not intersect.

MONITORING PROGRESS ANSWERS

8. $x \approx 0.85$
9. $x \approx -2, x \approx 4.33$
10. no solution

Solving Exponential Equations by Graphing

Sometimes, it is impossible to rewrite each side of an exponential equation using the same base. You can solve these types of equations by graphing each side and finding the point(s) of intersection. Exponential equations can have no solution, one solution, or more than one solution depending on the number of points of intersection.

EXAMPLE 4 Solving Exponential Equations by Graphing

Use a graphing calculator to solve (a) $\left(\frac{1}{2}\right)^{x-1} = 7$ and (b) $3^{x+2} = x + 1$.

SOLUTION

a. Step 1 Write a system of equations using each side of the equation.

$y = \left(\frac{1}{2}\right)^{x-1}$ Equation 1

$y = 7$ Equation 2

Step 2 Enter the equations into a calculator. Then graph the equations in a viewing window that shows where the graphs could intersect.

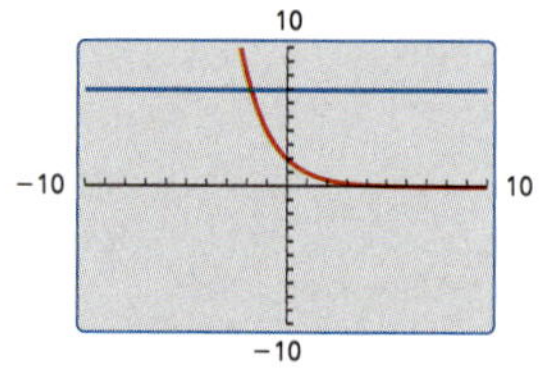

Step 3 Use the *intersect* feature to find the point of intersection. The graphs intersect at about $(-1.81, 7)$.

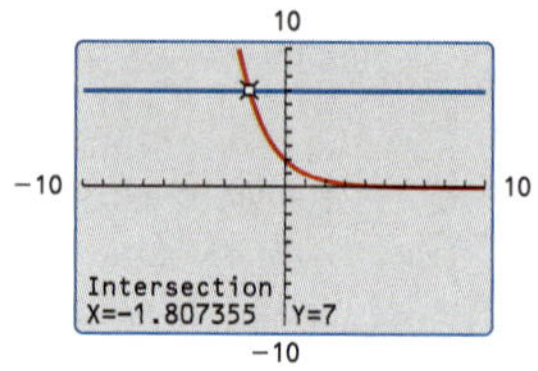

Check

$\left(\frac{1}{2}\right)^{x-1} = 7$

$\left(\frac{1}{2}\right)^{-1.81-1} \stackrel{?}{=} 7$

$7.01 \approx 7$ ✓

▶ So, the solution is $x \approx -1.81$.

b. Step 1 Write a system of equations using each side of the equation.

$y = 3^{x+2}$ Equation 1

$y = x + 1$ Equation 2

Step 2 Enter the equations into a calculator. Then graph the equations in a viewing window that shows where the graphs could intersect.

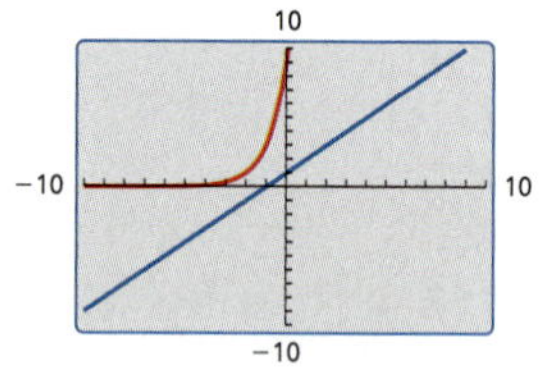

▶ The graphs do not intersect. So, the equation has no solution.

Monitoring Progress

Help in English and Spanish at *BigIdeasMath.com*

Use a graphing calculator to solve the equation.

8. $2^x = 1.8$ **9.** $4^{x-3} = x + 2$ **10.** $\left(\frac{1}{4}\right)^x = -2x - 3$

Laurie's Notes Teacher Actions

- **Use Appropriate Tools Strategically:** When an exponential equation cannot easily be rewritten in the form $b^x = b^y$, a graphing calculator is an appropriate tool.
- Before graphing, ask students whether they have a sense as to what the graphs will look like. There should be a familiarity with the graphs.
- Students should recall how to use the *intersect* feature on their calculators. When solutions are not rational, an approximation is used.

Closure

- **Response Logs:** Select from "What is confusing me the most is …" or "It helped me today when …" or "I learned to …."

6.5 Exercises

Dynamic Solutions available at *BigIdeasMath.com*

Vocabulary and Core Concept Check

1. **WRITING** Describe how to solve an exponential equation with unlike bases.

2. **WHICH ONE DOESN'T BELONG?** Which equation does *not* belong with the other three? Explain your reasoning.

$2^x = 4^{x+6}$	$5^{3x+8} = 5^{2x}$	$3^4 = x + 4^2$	$2^{x-7} = 2^7$

Monitoring Progress and Modeling with Mathematics

In Exercises 3–12, solve the equation. Check your solution. *(See Examples 1 and 2.)*

3. $4^{5x} = 4^{10}$
4. $7^{x-4} = 7^8$
5. $3^{9x} = 3^{7x+8}$
6. $2^{4x} = 2^{x+9}$
7. $2^x = 64$
8. $3^x = 243$
9. $7^{x-5} = 49^x$
10. $216^x = 6^{x+10}$
11. $64^{2x+4} = 16^{5x}$
12. $27^x = 9^{x-2}$

In Exercises 13–18, solve the equation. Check your solution. *(See Example 3.)*

13. $\left(\frac{1}{5}\right)^x = 125$
14. $\left(\frac{1}{4}\right)^x = 256$
15. $\frac{1}{128} = 2^{5x+3}$
16. $3^{4x-9} = \frac{1}{243}$
17. $36^{-3x+3} = \left(\frac{1}{216}\right)^{x+1}$
18. $\left(\frac{1}{27}\right)^{4-x} = 9^{2x-1}$

ERROR ANALYSIS In Exercises 19 and 20, describe and correct the error in solving the exponential equation.

19. ✗

$$5^{3x+2} = 25^{x-8}$$
$$3x + 2 = x - 8$$
$$x = -5$$

20. ✗

$$\left(\tfrac{1}{8}\right)^{5x} = 32^{x+8}$$
$$(2^3)^{5x} = (2^5)^{x+8}$$
$$2^{15x} = 2^{5x+40}$$
$$15x = 5x + 40$$
$$x = 4$$

In Exercises 21–24, match the equation with the graph that can be used to solve it. Then solve the equation.

21. $2^x = 6$
22. $4^{2x-5} = 6$
23. $5^{x+2} = 6$
24. $3^{-x-1} = 6$

A.
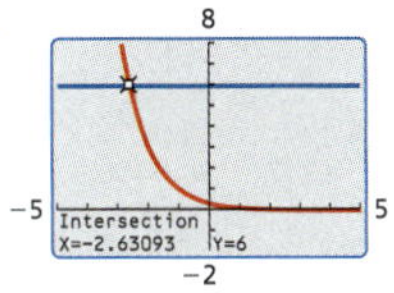

B.
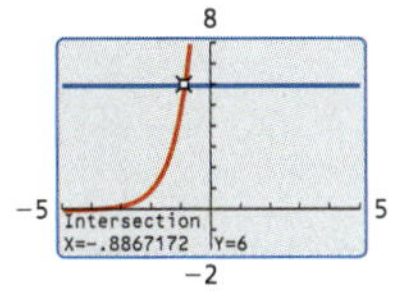

C.
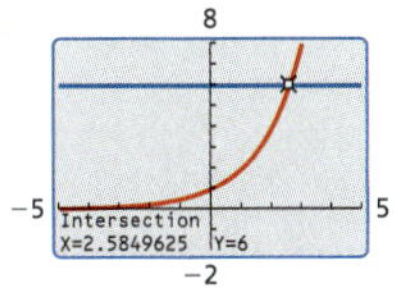

D.
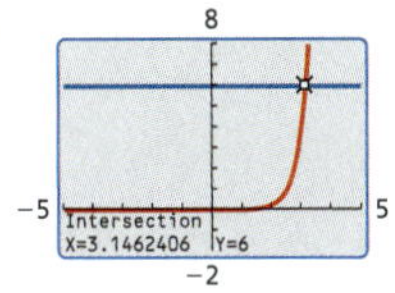

In Exercises 25–36, use a graphing calculator to solve the equation. *(See Example 4.)*

25. $6^{x+2} = 12$
26. $5^{x-4} = 8$
27. $\left(\frac{1}{2}\right)^{7x+1} = -9$
28. $\left(\frac{1}{3}\right)^{x+3} = 10$
29. $2^{x+6} = 2x + 15$
30. $3x - 2 = 5^{x-1}$
31. $\frac{1}{2}x - 1 = \left(\frac{1}{3}\right)^{2x-1}$
32. $2^{-x+1} = -\frac{3}{4}x + 3$
33. $5^x = -4^{-x+4}$
34. $7^{x-2} = 2^{-x}$
35. $2^{-x-3} = 3^{x+1}$
36. $5^{-2x+3} = -6^{x+5}$

Assignment Guide and Homework Check

ASSIGNMENT

Basic: 1, 2, 3–31 odd, 43, 50, 59–63

Average: 1, 2–50 even, 59–63

Advanced: 1, 2, 6, 10–22 even, 30–46 even, 47–52, 56–63

HOMEWORK CHECK

Basic: 3, 7, 13, 25

Average: 4, 8, 14, 26

Advanced: 6, 10, 16, 30

ANSWERS

1. Rewrite each side of the equation using the same base, then set the exponents equal to each other and solve. If it is impossible to rewrite each side of an exponential equation using the same base, you can solve the equation by graphing each side and finding the point(s) of intersection.
2. $3^4 = x + 4^2$; It is the only one that is not an exponential equation.
3. $x = 2$
4. $x = 12$
5. $x = 4$
6. $x = 3$
7. $x = 6$
8. $x = 5$
9. $x = -5$
10. $x = 5$
11. $x = 3$
12. $x = -4$
13. $x = -3$
14. $x = -4$
15. $x = -2$
16. $x = 1$
17. $x = 3$
18. $x = -10$
19. The exponents are not equal because there is not a common base; $5^{3x+2} = (5^2)^{x-8}$; $5^{3x+2} = 5^{2x-16}$; $3x + 2 = 2x - 16$; $x = -18$
20. $\frac{1}{8}$ is 2^{-3}, not 2^3; $(2^{-3})^{5x} = (2^5)^{x+8}$; $2^{-15x} = 2^{5x+40}$; $-15x = 5x + 40$; $x = -2$
21. C; $x \approx 2.58$
22. D; $x \approx 3.15$
23. B; $x \approx -0.89$
24. A; $x \approx -2.63$
25. $x \approx -0.61$
26. $x \approx 5.29$
27. no solution
28. $x \approx -5.10$
29. $x \approx -7.30$ and $x \approx -2.75$
30. $x = 1$; $x \approx 1.71$
31. $x \approx 2.06$
32. $x \approx -0.87$ and $x \approx 3.81$
33. no solution
34. $x \approx 1.47$
35. $x \approx -1.77$
36. no solution

Dynamic Teaching Tools

- Dynamic Assessment & Progress Monitoring Tool
- Interactive Whiteboard Lesson Library
- Dynamic Classroom with Dynamic Investigations

ANSWERS

37. $x = -2$

38. $x = 8$

39. $x = -3$

40. $x = 1$

41. 3 times

42. about 8.01 h

43. all real numbers

44. no solution

45. no solution

46. all real numbers

47. Any number to the zero power is 1, so $x - 4 = 0$, and $x = 4$.

48. $16 = 128\left(\frac{1}{2}\right)^x$; round 3

49. $800 = 500(1.06)^x$; about 8.07 years

50. **a.** 2007

b. Graph the two equations on a graphing calculator and use the *intersect* feature to find the point of intersection.

51. One to any power is 1, even if the powers are not equal; *Sample answer:* $1^3 = 1^5$, but $3 \neq 5$.

52. yes; *Sample answer:* $2^x = x + 3$

53. $x = \frac{5}{2}$

54. $x = -\frac{7}{2}$

55. $x = -\frac{5}{3}$

56. $x = \frac{4}{7}$

57. $x = 18$

58. $x = 4$

59–63. See Additional Answers.

Mini-Assessment

Solve the equation.

1. $6^{x+1} = 6^5$ $x = 4$
2. $64^x = 8^{3x-6}$ $x = 6$
3. $2^{x+2} = \frac{1}{32}$ $x = -7$
4. Use a graphing calculator to solve $\left(\frac{1}{2}\right)^{2x+3} = 4$.

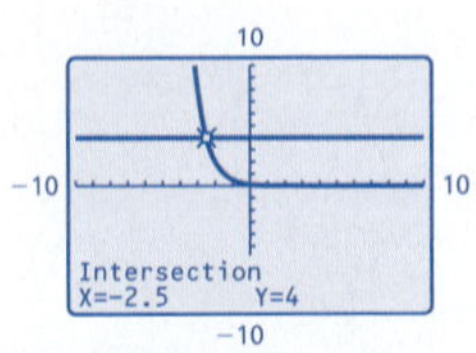

The solution is $x = -2.5$.

In Exercises 37–40, solve the equation by using the Property of Equality for Exponential Equations.

37. $30 \cdot 5^{x+3} = 150$

38. $12 \cdot 2^{x-7} = 24$

39. $4(3^{-2x-4}) = 36$

40. $2(4^{2x+1}) = 128$

41. MODELING WITH MATHEMATICS You scan a photo into a computer at four times its original size. You continue to increase its size repeatedly by 100% using the computer. The new size of the photo y in comparison to its original size after x enlargements on the computer is represented by $y = 2^{x+2}$. How many times must the photo be enlarged on the computer so the new photo is 32 times the original size?

42. MODELING WITH MATHEMATICS A bacterial culture quadruples in size every hour. You begin observing the number of bacteria 3 hours after the culture is prepared. The amount y of bacteria x hours after the culture is prepared is represented by $y = 192(4^{x-3})$. When will there be 200,000 bacteria?

In Exercises 43–46, solve the equation.

43. $3^{3x+6} = 27^{x+2}$

44. $3^{4x+3} = 81^x$

45. $4^{x+3} = 2^{2(x+1)}$

46. $5^{8(x-1)} = 625^{2x-2}$

47. NUMBER SENSE Explain how you can use mental math to solve the equation $8^{x-4} = 1$.

48. PROBLEM SOLVING There are a total of 128 teams at the start of a citywide 3-on-3 basketball tournament. Half the teams are eliminated after each round. Write and solve an exponential equation to determine after which round there are 16 teams left.

49. PROBLEM SOLVING You deposit \$500 in a savings account that earns 6% annual interest compounded yearly. Write and solve an exponential equation to determine when the balance of the account will be \$800.

50. HOW DO YOU SEE IT? The graph shows the annual attendance at two different events. Each event began in 2004.

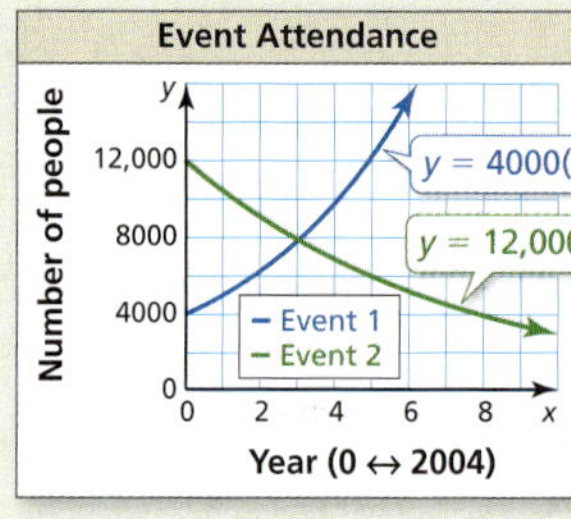

a. Estimate when the events will have about the same attendance.

b. Explain how you can verify your answer in part (a).

51. REASONING Explain why the Property of Equality for Exponential Equations does not work when $b = 1$. Give an example to justify your answer.

52. THOUGHT PROVOKING Is it possible for an exponential equation to have two different solutions? If not, explain your reasoning. If so, give an example.

USING STRUCTURE **In Exercises 53–58, solve the equation.**

53. $8^{x-2} = \sqrt{8}$

54. $\sqrt{5} = 5^{x+4}$

55. $\left(\sqrt[5]{7}\right)^x = 7^{2x+3}$

56. $12^{2x-1} = \left(\sqrt[4]{12}\right)^x$

57. $\left(\sqrt[3]{6}\right)^{2x} = \left(\sqrt{6}\right)^{x+6}$

58. $\left(\sqrt[5]{3}\right)^{5x-10} = \left(\sqrt[8]{3}\right)^{4x}$

59. MAKING AN ARGUMENT Consider the equation $\left(\frac{1}{a}\right)^x = b$, where $a > 1$ and $b > 1$. Your friend says the value of x will always be negative. Is your friend correct? Explain.

Maintaining Mathematical Proficiency

Reviewing what you learned in previous grades and lessons

Determine whether the sequence is arithmetic. If so, find the common difference. *(Section 4.6)*

60. $-20, -26, -32, -38, \ldots$

61. $9, 18, 36, 72, \ldots$

62. $-5, -8, -12, -17, \ldots$

63. $10, 20, 30, 40, \ldots$

If students need help...	*If students got it...*
Resources by Chapter • Practice A and Practice B • Puzzle Time	Resources by Chapter • Enrichment and Extension • Cumulative Review
Student Journal • Practice	Start the *next* Section
Differentiating the Lesson Skills Review Handbook	

Laurie's Notes

Overview of Section 6.6

Introduction
- In this lesson, students will make connections between geometric sequences and exponential equations. Sequence notation and vocabulary associated with geometric sequences are introduced.
- Students will write the terms of a geometric sequence and extend a given sequence.
- The last part of the lesson presents a real-life application in which students write a geometric sequence as an equation and also represent it as a function.

Common Misconceptions
- Discuss that geometric sequences can be increasing, decreasing, or neither. It would be neither when the common ratio is negative. The terms alternate between positive and negative.

Formative Assessment Tips
- Refer to page T-290 for a description of *Response Logs*. In this lesson and throughout this chapter, there will be sentence stems to prompt student writing. Use your judgment for placement in the lesson, amount of time to allow for writing, and the actual sentence stem given to students.
- For a beginning list of writing stems, see page T-298.

Extensions
- A connection can be made between the graph of a geometric sequence and the graph of an exponential equation.
- Take time to compare and contrast arithmetic sequences (Section 4.6) and geometric sequences.

Pacing Suggestion
- Complete both explorations, and then start the formal lesson by stating the *Core Concept* that defines a geometric sequence.

Dynamic Teaching Tools
- Dynamic Assessment & Progress Monitoring Tool
- Lesson Planning Tool
- Interactive Whiteboard Lesson Library
- Dynamic Classroom with Dynamic Investigations

What Your Students Will Learn

- Identify geometric sequences by finding the ratio between each pair of consecutive terms.
- Extend geometric sequences and graph geometric sequences in the coordinate plane.
- Write geometric sequences as equations and in function notation.

Laurie's Notes

Exploration

Motivate

- Before class, write the numbers 6, 12, 24, 48, and 96 on separate index cards. Hand an index card to the first five students who enter the room. Tell them that they are holding numbers in a geometric sequence and to position themselves in order.
- Students may realize that the order could be increasing or decreasing.

Version A: 6, 12, 24, 48, 96 **Version B:** 96, 48, 24, 12, 6

? "If 6 is in the first position, how can you generate the terms that follow?" multiply by 2

? "If 96 is in the first position, how can you generate the terms that follow?" multiply by $\frac{1}{2}$
"Is this the same number you would multiply 48 by to get 24?" yes

Exploration Note

- **Model with Mathematics** and **Use Appropriate Tools Strategically:** Students have worked with exponential functions for three sections and should have a good sense now of how exponential functions model real-life applications. The calculator is a useful tool in performing the repeated computations so that the focus is on the process (repeated common factor) versus the computation skill.

Exploration 1

- Students should be able to follow the keystrokes shown. Individual calculators may have a feature (such as ANS) that will appear on the screens.
- Part (a) multiplies by 2, and therefore it is generating powers of 2 ($2^1 = 2$, $2^2 = 4$, $2^3 = 8$, $2^4 = 16$, $2^5 = 32$). Part (b) multiplies by 0.5, or $\frac{1}{2}$, starting with 64.
- Ask students to describe the connection between parts (a) and (b).
- ? **Extension:** "What happens if you continue multiplying by 0.5 in part (b)?" 64, 32, 16, 8, 4, 2, 1, 0.5, 0.25, 0.125, ... Students may predict that the numbers will become negative. The numbers are decreasing, but they will always be positive.
- **Teaching Tip:** Students love to explore how many times they can repeat a particular task on a calculator, such as multiplying by 3. Give time for students to explore this question if it comes up. It does not take much time to reach an overflow error on the calculator, which is a good experience.

Exploration 2

- Provide paper for students to explore this activity. They may be surprised at the results. Search the Internet (folding paper in half) for websites about folding paper.
- **Extension:** "How many times would you need to fold the paper so that the stack would reach the moon (assuming that you could fold the paper that many times)?" Students will need to find the distance to the moon and then convert to millimeters. The answer of 42 folds will seem surprisingly low to students.

Communicate Your Answer

- Although no formal definition of geometric sequences is given in the exploration, students should realize that they have been multiplying by a constant rather than adding a constant as with arithmetic sequences.

Connecting to Next Step

- The explorations help students develop a conceptual understanding of what a geometric sequence is and how to generate successive terms in the sequence. Begin the formal lesson with the definition of a geometric sequence.

6.6 Geometric Sequences

Essential Question

How can you use a geometric sequence to describe a pattern?

In a **geometric sequence**, the ratio between each pair of consecutive terms is the same. This ratio is called the **common ratio**.

EXPLORATION 1 Describing Calculator Patterns

Work with a partner. Enter the keystrokes on a calculator and record the results in the table. Describe the pattern.

a. Step 1 [2] [=]
Step 2 [×] [2] [=]
Step 3 [×] [2] [=]
Step 4 [×] [2] [=]
Step 5 [×] [2] [=]

Step	1	2	3	4	5
Calculator display					

b. Step 1 [6] [4] [=]
Step 2 [×] [.] [5] [=]
Step 3 [×] [.] [5] [=]
Step 4 [×] [.] [5] [=]
Step 5 [×] [.] [5] [=]

Step	1	2	3	4	5
Calculator display					

c. Use a calculator to make your own sequence. Start with any number and multiply by 3 each time. Record your results in the table.

Step	1	2	3	4	5
Calculator display					

d. Part (a) involves a geometric sequence with a common ratio of 2. What is the common ratio in part (b)? part (c)?

LOOKING FOR REGULARITY IN REPEATED REASONING

To be proficient in math, you need to notice when calculations are repeated and look both for general methods and for shortcuts.

EXPLORATION 2 Folding a Sheet of Paper

Work with a partner. A sheet of paper is about 0.1 millimeter thick.

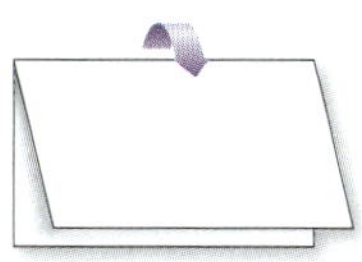

a. How thick will it be when you fold it in half once? twice? three times?

b. What is the greatest number of times you can fold a piece of paper in half? How thick is the result?

c. Do you agree with the statement below? Explain your reasoning.

"If it were possible to fold the paper in half 15 times, it would be taller than you."

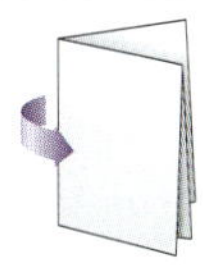

Communicate Your Answer

3. How can you use a geometric sequence to describe a pattern?

4. Give an example of a geometric sequence from real life other than paper folding.

Dynamic Teaching Tools

- Dynamic Assessment & Progress Monitoring Tool
- Lesson Planning Tool
- Interactive Whiteboard Lesson Library
- Dynamic Classroom with Dynamic Investigations

ANSWERS

1. a. 2, 4, 8, 16, 32; Each number is twice the preceding number.

b. 64; 32; 16; 8; 4; Each number is half of the preceding number.

c. *Sample answer:* 10, 30, 90, 270, 810

d. $\frac{1}{2}$; 3

2. a. 0.2 mm; 0.4 mm; 0.8 mm

b. *Sample answer:* 6 times; 6.4 mm

c. yes; The thickness is a geometric sequence that doubles with each fold. After 15 folds, the thickness would be 3276.8 millimeters, or 3.2768 meters, which is taller than a human being.

3. Indicate the common ratio between each consecutive pair of terms.

4. *Sample answer:* If you drop a tennis ball, the height after each bounce will be a common fraction of the height of the preceding bounce.

6.6 Lesson

What You Will Learn

- Identify geometric sequences.
- Extend and graph geometric sequences.
- Write geometric sequences as functions.

Core Vocabulary

geometric sequence, *p. 332*
common ratio, *p. 332*

Previous
arithmetic sequence
common difference

Identifying Geometric Sequences

Core Concept

Geometric Sequence

In a **geometric sequence**, the ratio between each pair of consecutive terms is the same. This ratio is called the **common ratio**. Each term is found by multiplying the previous term by the common ratio.

1, 5, 25, 125, . . . Terms of a geometric sequence

×5 ×5 ×5 ← common ratio

EXAMPLE 1 Identifying Geometric Sequences

Decide whether each sequence is *arithmetic*, *geometric*, or *neither*. Explain your reasoning.

a. 120, 60, 30, 15, . . .   **b.** 2, 6, 11, 17, . . .

SOLUTION

a. Find the ratio between each pair of consecutive terms.

120   60   30   15

$\frac{60}{120} = \frac{1}{2}$ $\frac{30}{60} = \frac{1}{2}$ $\frac{15}{30} = \frac{1}{2}$ The ratios are the same. The common ratio is $\frac{1}{2}$.

- So, the sequence is geometric.

b. Find the ratio between each pair of consecutive terms.

2   6   11   17

$\frac{6}{2} = 3$ $\frac{11}{6} = 1\frac{5}{6}$ $\frac{17}{11} = 1\frac{6}{11}$ There is no common ratio, so the sequence is *not* geometric.

Find the difference between each pair of consecutive terms.

2   6   11   17

$6 - 2 = 4$ $11 - 6 = 5$ $17 - 11 = 6$ There is no common difference, so the sequence is *not* arithmetic.

- So, the sequence is *neither* geometric nor arithmetic.

Monitoring Progress

Help in English and Spanish at *BigIdeasMath.com*

Decide whether the sequence is *arithmetic*, *geometric*, or *neither*. Explain your reasoning.

1. 5, 1, −3, −7, . . .   **2.** 1024, 128, 16, 2, . . .   **3.** 2, 6, 10, 16, . . .

Extra Example 1

Decide whether each sequence is *arithmetic*, *geometric*, or *neither*. Explain your reasoning.

a. 8, 24, 72, 216, ...
The sequence is geometric. The ratios are the same. The common ratio is $\frac{3}{1}$, or 3.

b. 8, 3, −2, −7, ...
The sequence is arithmetic. The differences are the same. The common difference is −5.

MONITORING PROGRESS ANSWERS

1. arithmetic; There is a common difference of −4.
2. geometric; There is a common ratio of $\frac{1}{8}$.
3. neither; There is no common difference or common ratio.

Laurie's Notes Teacher Actions

- Write and discuss the *Core Concept*. In the example shown, the common ratio is 5. *Note:* (any term) ÷ (previous term) = 5.
- Example 1 models finding successive ratios, though students may perform mental math to decide.

? Fact-First Questioning: "A geometric sequence might not be increasing or decreasing. Explain why." If the common ratio is negative, then the terms in the sequence will alternate between positive and negative.

Extending and Graphing Geometric Sequences

EXAMPLE 2 Extending Geometric Sequences

Write the next three terms of each geometric sequence.

a. 3, 6, 12, 24, . . . **b.** 64, −16, 4, −1, . . .

SOLUTION

Use tables to organize the terms and extend each sequence.

a.

Position	1	2	3	4	5	6	7
Term	3	6	12	24	48	96	192

×2 ×2 ×2 ×2 ×2 ×2

Each term is twice the previous term. So, the common ratio is 2.

Multiply a term by 2 to find the next term.

▶ The next three terms are 48, 96, and 192.

b.

Position	1	2	3	4	5	6	7
Term	64	−16	4	−1	$\frac{1}{4}$	$-\frac{1}{16}$	$\frac{1}{64}$

$\times\left(-\frac{1}{4}\right)$ $\times\left(-\frac{1}{4}\right)$ $\times\left(-\frac{1}{4}\right)$ $\times\left(-\frac{1}{4}\right)$ $\times\left(-\frac{1}{4}\right)$ $\times\left(-\frac{1}{4}\right)$

Multiply a term by $-\frac{1}{4}$ to find the next term.

▶ The next three terms are $\frac{1}{4}$, $-\frac{1}{16}$, and $\frac{1}{64}$.

LOOKING FOR STRUCTURE

When the terms of a geometric sequence alternate between positive and negative terms, or vice versa, the common ratio is negative.

EXAMPLE 3 Graphing a Geometric Sequence

Graph the geometric sequence 32, 16, 8, 4, 2, ... What do you notice?

SOLUTION

Make a table. Then plot the ordered pairs (n, a_n).

Position, n	1	2	3	4	5
Term, a_n	32	16	8	4	2

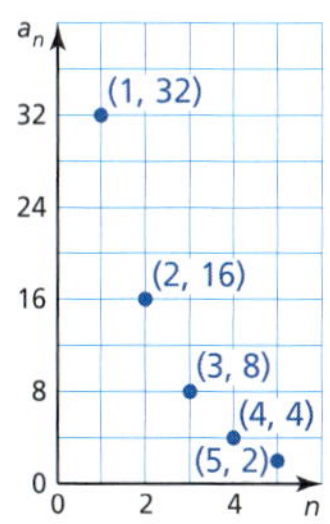

▶ The points appear to lie on an exponential curve.

STUDY TIP

The points of any geometric sequence with a *positive* common ratio lie on an exponential curve.

Monitoring Progress

Help in English and Spanish at *BigIdeasMath.com*

Write the next three terms of the geometric sequence. Then graph the sequence.

4. 1, 3, 9, 27, . . . **5.** 2500, 500, 100, 20, . . .

6. 80, −40, 20, −10, . . . **7.** −2, 4, −8, 16, . . .

Differentiated Instruction

Auditory

Students may benefit from explaining the process for extending a geometric sequence and how to use the common ratio in that process. Have students take turns working at the board with one of several geometric sequences. Students should calculate the common ratio and then use it to find the next three terms, explaining each step in the process.

Extra Example 2

Write the next three terms of each geometric sequence.

a. 972, 324, 108, 36, …
12, 4, $\frac{4}{3}$

b. 7, −14, 28, −56, …
112, −224, 448

Extra Example 3

Graph the geometric sequence 6, 12, 24, 48, 96, …. What do you notice?

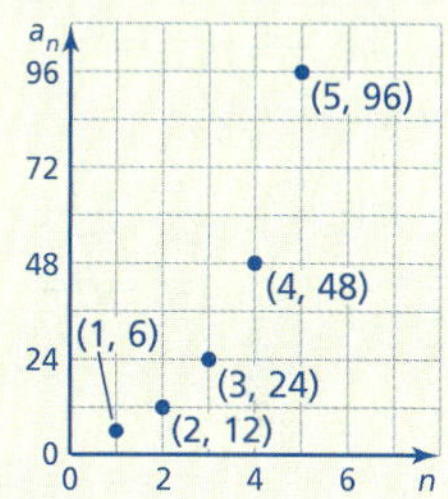

The points appear to lie on an exponential curve.

MONITORING PROGRESS ANSWERS

4. 81, 243, 729

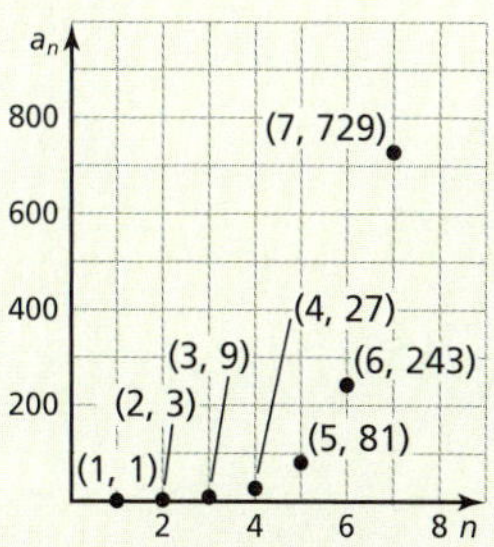

5–7. See Additional Answers.

Laurie's Notes Teacher Actions

- Explain that a table is used to organize the terms in a geometric sequence because it helps in recognizing patterns.
- Remind students that they can find the common ratio by finding the ratio of a term to the previous term.
- In Example 3, discuss using the notation n and a_n instead of x and y. The position is n and the value of the term at the nth position is a_n.
- **Response Logs:** "I didn't expect …"

English Language Learners

Build on Past Knowledge

Remind students of the exponential form of an equation: $y = ab^x$. Point out the *Study Tip* on page 334. Show how the equation $a_n = 2(6)^{n-1}$ in Example 4 is in the exponential function form.

$$a_n = 2(6)^{n-1}$$
$$y = ab^x$$

Emphasize that equations students write for the nth term of a geometric sequence should have this form.

Extra Example 4

Write an equation for the nth term of the geometric sequence 3, 12, 48, 192, ... Then find a_{10}.

$a_n = 3(4)^{n-1}$; The 10th term of the geometric sequence is 786,432.

MONITORING PROGRESS ANSWERS

8. $a_n = (-5)^{n-1}$; 15,625

9. $a_n = 13(2)^{n-1}$; 832

10. $a_n = 432\left(\frac{1}{6}\right)^{n-1}$; $\frac{1}{108}$

11. $a_n = 4(2.5)^{n-1}$; 976.5625

Writing Geometric Sequences as Functions

Because consecutive terms of a geometric sequence have a common ratio, you can use the first term a_1 and the common ratio r to write an exponential function that describes a geometric sequence. Let $a_1 = 1$ and $r = 5$.

Position, n	Term, a_n	Written using a_1 and r	Numbers
1	first term, a_1	a_1	1
2	second term, a_2	a_1r	$1 \cdot 5 = 5$
3	third term, a_3	a_1r^2	$1 \cdot 5^2 = 25$
4	fourth term, a_4	a_1r^3	$1 \cdot 5^3 = 125$
⋮	⋮	⋮	⋮
n	nth term, a_n	a_1r^{n-1}	$1 \cdot 5^{n-1}$

STUDY TIP

Notice that the equation $a_n = a_1r^{n-1}$ is of the form $y = ab^x$.

Core Concept

Equation for a Geometric Sequence

Let a_n be the nth term of a geometric sequence with first term a_1 and common ratio r. The nth term is given by

$$a_n = a_1r^{n-1}.$$

EXAMPLE 4 Finding the nth Term of a Geometric Sequence

Write an equation for the nth term of the geometric sequence 2, 12, 72, 432, ... Then find a_{10}.

SOLUTION

The first term is 2, and the common ratio is 6.

$a_n = a_1r^{n-1}$ Equation for a geometric sequence

$a_n = 2(6)^{n-1}$ Substitute 2 for a_1 and 6 for r.

Use the equation to find the 10th term.

$a_n = 2(6)^{n-1}$ Write the equation.

$a_{10} = 2(6)^{10-1}$ Substitute 10 for n.

$= 20,155,392$ Simplify.

▶ The 10th term of the geometric sequence is 20,155,392.

Monitoring Progress

Help in English and Spanish at *BigIdeasMath.com*

Write an equation for the nth term of the geometric sequence. Then find a_7.

8. 1, −5, 25, −125, . . .

9. 13, 26, 52, 104, . . .

10. 432, 72, 12, 2, . . .

11. 4, 10, 25, 62.5, . . .

Laurie's Notes Teacher Actions

- Spend time talking about the nth term and why the number of times the common factor has been used is $n - 1$, not n.
- **Connection:** Help students see the connection between the equation $a_n = a_1r^{n-1}$ and the general form of an exponential equation $y = ab^x$.

? "What information is needed to write the nth term of the geometric sequence?" the first term and the common ratio

- **Think-Pair-Share:** Have students answer Questions 8–11, and then share and discuss as a class.

You can rewrite the equation for a geometric sequence with first term a_1 and common ratio r in function notation by replacing a_n with $f(n)$.

$$f(n) = a_1 r^{n-1}$$

The domain of the function is the set of positive integers.

EXAMPLE 5 Modeling with Mathematics

Clicking the *zoom-out* button on a mapping website doubles the side length of the square map. After how many clicks on the *zoom-out* button is the side length of the map 640 miles?

Zoom-out clicks	1	2	3
Map side length (miles)	5	10	20

SOLUTION

1. **Understand the Problem** You know that the side length of the square map doubles after each click on the *zoom-out* button. So, the side lengths of the map represent the terms of a geometric sequence. You need to find the number of clicks it takes for the side length of the map to be 640 miles.

2. **Make a Plan** Begin by writing a function f for the nth term of the geometric sequence. Then find the value of n for which $f(n) = 640$.

3. **Solve the Problem** The first term is 5, and the common ratio is 2.

$f(n) = a_1 r^{n-1}$ — Function for a geometric sequence

$f(n) = 5(2)^{n-1}$ — Substitute 5 for a_1 and 2 for r.

The function $f(n) = 5(2)^{n-1}$ represents the geometric sequence. Use this function to find the value of n for which $f(n) = 640$. So, use the equation $640 = 5(2)^{n-1}$ to write a system of equations.

$y = 5(2)^{n-1}$ — Equation 1

$y = 640$ — Equation 2

Then use a graphing calculator to graph the equations and find the point of intersection. The point of intersection is (8, 640).

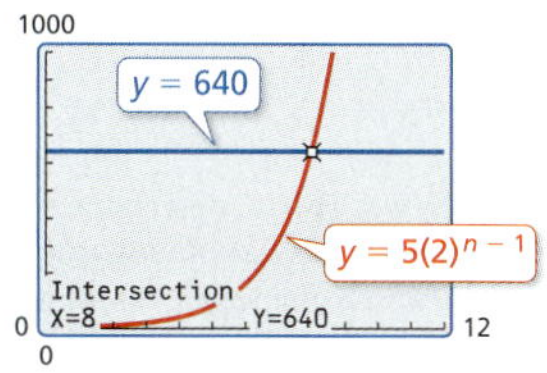

► So, after eight clicks, the side length of the map is 640 miles.

4. **Look Back** Find the value of n for which $f(n) = 640$ algebraically.

$640 = 5(2)^{n-1}$ — Write the equation.

$128 = (2)^{n-1}$ — Divide each side by 5.

$2^7 = (2)^{n-1}$ — Rewrite 128 as 2^7.

$7 = n - 1$ — Equate the exponents.

$8 = n$ ✓ — Add 1 to each side.

USING APPROPRIATE TOOLS STRATEGICALLY

You can also use the *table* feature of a graphing calculator to find the value of n for which $f(n) = 640$.

X	Y1	Y2
3	20	640
4	40	640
5	80	640
6	160	640
7	320	640
8	640	640
9	1280	640

X=8

Monitoring Progress

Help in English and Spanish at *BigIdeasMath.com*

12. **WHAT IF?** After how many clicks on the *zoom-out* button is the side length of the map 2560 miles?

Extra Example 5

Use the information in Example 5. After how many clicks on the *zoom-out* button is the side length of the map 5120 miles? After 11 clicks, the side length of the map is 5120 miles.

MONITORING PROGRESS ANSWER

12. 10 clicks

Laurie's Notes Teacher Actions

- Introduce the function notation.
- Ask students whether they have used a mapping application on a computer or smartphone. Pose Example 5.

? "What is the first term, meaning what is the side length of the map after one click?" 5 "What is the common ratio?" 2

- **Whiteboarding:** Have students work with their partners to solve the problem. Solicit volunteers to share their work.

Closure

- **Response Logs:** Select from "I will understand this better if I . . ." or "I feel like I made progress today . . ." or "I didn't expect . . . ."

Assignment Guide and Homework Check

ASSIGNMENT

Basic: 1, 2, 3–37 odd, 44, 46, 55, 56

Average: 1, 2–50 even, 55, 56

Advanced: 1, 2, 8, 12–24 even, 30–44 even, 45–56

HOMEWORK CHECK

Basic: 9, 19, 25, 37

Average: 12, 22, 26, 38

Advanced: 14, 24, 30, 40

ANSWERS

1. The first sequence is an arithmetic sequence with a common difference of 2. The second sequence is a geometric sequence with a common ratio of 2.
2. A negative number raised to an odd exponent is negative, and raised to an even exponent is positive. So, if the common ratio is negative, the terms will alternate signs.
3. 3
4. $\frac{1}{6}$
5. -8
6. 10
7. $\frac{3}{4}$
8. $-\frac{1}{3}$
9. arithmetic; There is a common difference of 8.
10. neither; There is no common difference or common ratio.
11. neither; There is no common difference or common ratio.
12. geometric; There is a common ratio of 7.
13. geometric; There is a common ratio of $\frac{1}{8}$.
14. arithmetic; There is a common difference of 7.
15. geometric; There is a common ratio of 5.
16. neither; There is no common difference or common ratio.
17. neither; There is no common difference or common ratio.
18. arithmetic; There is a common difference of 9.

6.6 Exercises

Dynamic Solutions available at *BigIdeasMath.com*

Vocabulary and Core Concept Check

1. **WRITING** Compare the two sequences.

   2, 4, 6, 8, 10, . . . 2, 4, 8, 16, 32, . . .

2. **CRITICAL THINKING** Why do the points of a geometric sequence lie on an exponential curve only when the common ratio is positive?

Monitoring Progress and Modeling with Mathematics

In Exercises 3–8, find the common ratio of the geometric sequence.

3. 4, 12, 36, 108, . . .
4. 36, 6, 1, $\frac{1}{6}$, . . .
5. $\frac{3}{8}$, -3, 24, -192, . . .
6. 0.1, 1, 10, 100, . . .
7. 128, 96, 72, 54, . . .
8. -162, 54, -18, 6, . . .

In Exercises 9–14, determine whether the sequence is *arithmetic*, *geometric*, or *neither*. Explain your reasoning. *(See Example 1.)*

9. -8, 0, 8, 16, . . .
10. -1, 4, -7, 10, . . .
11. 9, 14, 20, 27, . . .
12. $\frac{3}{49}$, $\frac{3}{7}$, 3, 21, . . .
13. 192, 24, 3, $\frac{3}{8}$, . . .
14. -25, -18, -11, -4, . . .

In Exercises 15–18, determine whether the graph represents an *arithmetic sequence*, a *geometric sequence*, or *neither*. Explain your reasoning.

15. 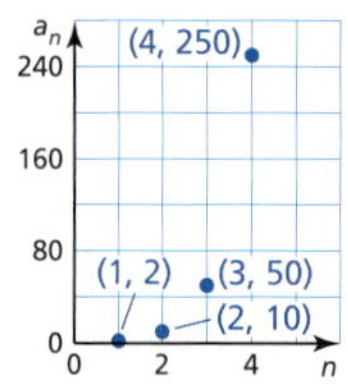

16. 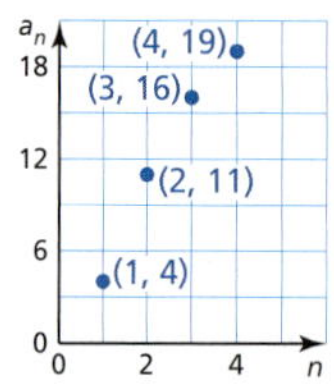

17. 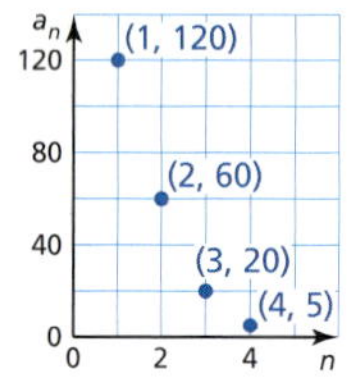

18. 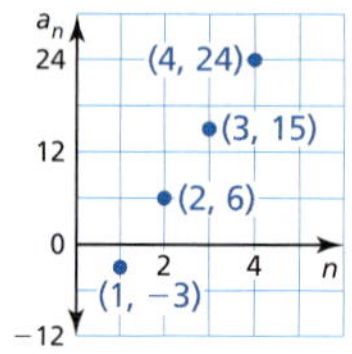

In Exercises 19–24, write the next three terms of the geometric sequence. Then graph the sequence. *(See Examples 2 and 3.)*

19. 5, 20, 80, 320, . . .
20. -3, 12, -48, 192, . . .
21. 81, -27, 9, -3, . . .
22. -375, -75, -15, -3, . . .
23. 32, 8, 2, $\frac{1}{2}$, . . .
24. $\frac{16}{9}$, $\frac{8}{3}$, 4, 6, . . .

In Exercises 25–32, write an equation for the *n*th term of the geometric sequence. Then find a_6. *(See Example 4.)*

25. 2, 8, 32, 128, . . .
26. 0.6, -3, 15, -75, . . .
27. $-\frac{1}{8}$, $-\frac{1}{4}$, $-\frac{1}{2}$, -1, . . .
28. 0.1, 0.9, 8.1, 72.9, . . .

29.

n	1	2	3	4
a_n	7640	764	76.4	7.64

30.

n	1	2	3	4
a_n	-192	48	-12	3

31. 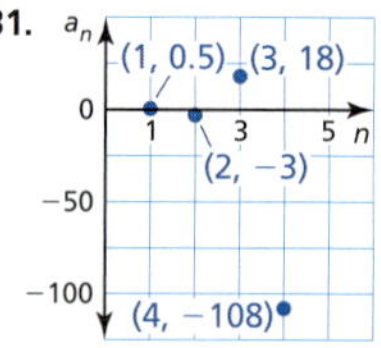

32. 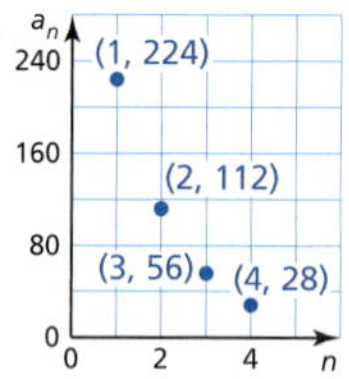

33. **PROBLEM SOLVING** A badminton tournament begins with 128 teams. After the first round, 64 teams remain. After the second round, 32 teams remain. How many teams remain after the third, fourth, and fifth rounds?

19. 1280; 5120; 20,480

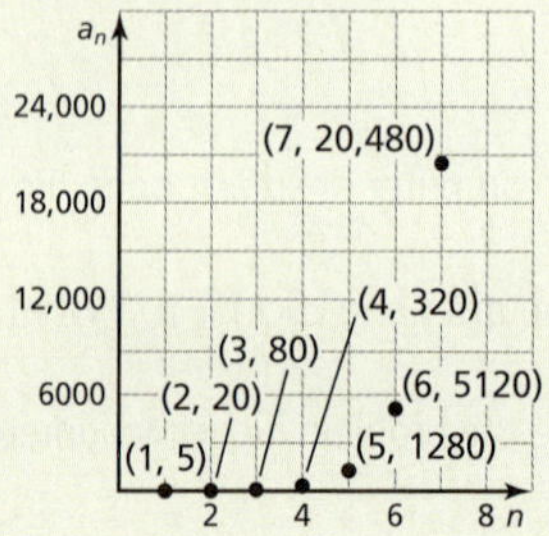

20. -768; 3072; $-12,288$

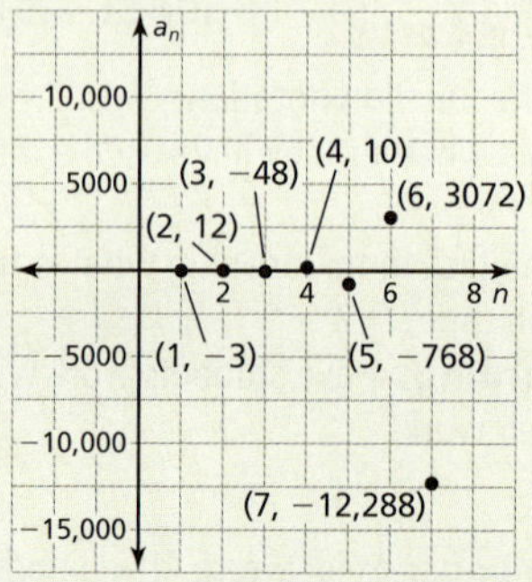

21–33. See Additional Answers.

34. **PROBLEM SOLVING** The graphing calculator screen displays an area of 96 square units. After you zoom out once, the area is 384 square units. After you zoom out a second time, the area is 1536 square units. What is the screen area after you zoom out four times?

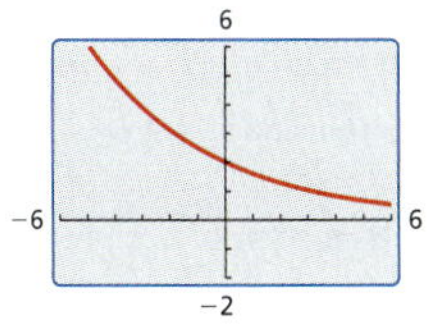

35. **ERROR ANALYSIS** Describe and correct the error in writing the next three terms of the geometric sequence.

> ✗ $-8, \; 4, \; -2, \; 1, \ldots$ with $\times(-2)$ $\times(-2)$ $\times(-2)$
>
> The next three terms are -2, 4, and -8.

36. **ERROR ANALYSIS** Describe and correct the error in writing an equation for the nth term of the geometric sequence.

> ✗ $-2, -12, -72, -432, \ldots$
>
> The first term is -2, and the common ratio is -6.
>
> $a_n = a_1 r^{n-1}$
>
> $a_n = -2(-6)^{n-1}$

37. **MODELING WITH MATHEMATICS** The distance (in millimeters) traveled by a swinging pendulum decreases after each swing, as shown in the table. *(See Example 5.)*

Swing	1	2	3
Distance (in millimeters)	625	500	400

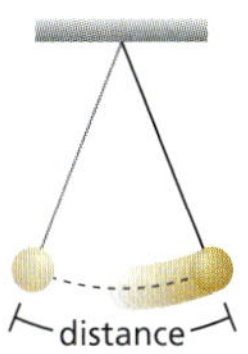

a. Write a function that represents the distance the pendulum swings on its nth swing.

b. After how many swings is the distance 256 millimeters?

38. **MODELING WITH MATHEMATICS** You start a chain email and send it to six friends. The next day, each of your friends forwards the email to six people. The process continues for a few days.

a. Write a function that represents the number of people who have received the email after n days.

b. After how many days will 1296 people have received the email?

MATHEMATICAL CONNECTIONS **In Exercises 39 and 40, (a) write a function that represents the sequence of figures and (b) describe the 10th figure in the sequence.**

39.

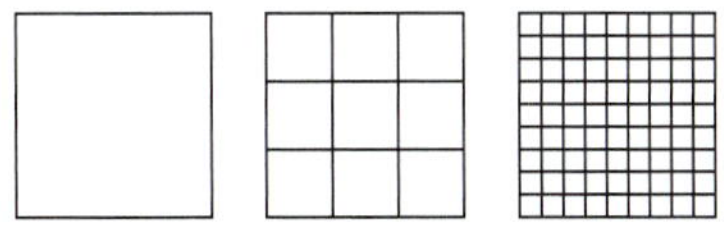

40.

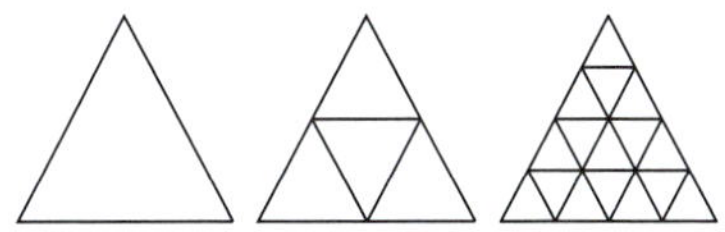

41. **REASONING** Write a sequence that represents the number of teams that have been eliminated after n rounds of the badminton tournament in Exercise 33. Determine whether the sequence is *arithmetic*, *geometric*, or *neither*. Explain your reasoning.

42. **REASONING** Write a sequence that represents the perimeter of the graphing calculator screen in Exercise 34 after you zoom out n times. Determine whether the sequence is *arithmetic*, *geometric*, or *neither*. Explain your reasoning.

43. **WRITING** Compare the graphs of arithmetic sequences to the graphs of geometric sequences.

44. **MAKING AN ARGUMENT** You are given two consecutive terms of a sequence.

$\ldots, -8, 0, \ldots$

Your friend says that the sequence is not geometric. A classmate says that is impossible to know given only two terms. Who is correct? Explain.

Dynamic Teaching Tools

- Dynamic Assessment & Progress Monitoring Tool
- Interactive Whiteboard Lesson Library
- Dynamic Classroom with Dynamic Investigations

ANSWERS

34. 24,576 square units

35. The common factor is $-\frac{1}{2}$, not -2;

$-8, \; 4, \; -2, \; 1, \ldots$ with $\times\left(-\frac{1}{2}\right)$ $\times\left(-\frac{1}{2}\right)$ $\times\left(-\frac{1}{2}\right)$

The next three terms are $-\frac{1}{2}$, $\frac{1}{4}$, and $-\frac{1}{8}$.

36. The common ratio is 6, not -6; The first term is -2 and the common ratio is 6; $a_n = a_1 r^{n-1}$; $a_n = -2(6)^{n-1}$

37. a. $a_n = 625\left(\frac{4}{5}\right)^{n-1}$
    b. 5 swings

38. a. $a_n = 6^n$
    b. 4 days

39. a. $a_n = 9^{n-1}$
    b. a large square containing 387,420,489 small squares

40. a. $a_n = 4^{n-1}$
    b. a large triangle containing 262,144 small triangles

41. $a_n = 64\left(\frac{1}{2}\right)^{n-1}$; geometric; There is a common ratio of $\frac{1}{2}$.

42. $a_n = 80(2)^{n-1}$; geometric; There is a common ratio of 2.

43. Graphs of arithmetic sequences form a linear pattern. Graphs of geometric sequences form an exponential pattern when the common ratio is positive, and a pattern of points alternating between Quadrants I and IV when the common ratio is negative.

44. your friend; The only way to get 0 as a term would be with a common ratio of 0. If the common ratio is 0, then there is no possible previous term that would result in -8 as the next term.

ANSWERS

45. yes; yes; It is an arithmetic sequence with a common difference of 0, and a geometric sequence with a common ratio of 1.

46. **a.** B

b. A

When the common ratio is greater than 1, the values in the sequence increase. When the common ratio is between 0 and 1, the values in the sequence decrease.

47. 59,049

48. *Sample answer:* 1, 1, 2, 3, 5, 8, 13, . . .; After the first two terms, each term is the sum of the preceding two terms.

49. dependent; Each term is calculated from the preceding term.

50–56. See Additional Answers.

Mini-Assessment

1. Decide whether the sequence 13, 23, 33, 43, ... is *arithmetic*, *geometric*, or *neither*. Explain your reasoning. arithmetic; The differences are the same. The common difference is 10.
2. Write the next three terms of the geometric sequence $\frac{1}{3}$, 1, 3, 9, ... Then graph the sequence. 27, 81, 243

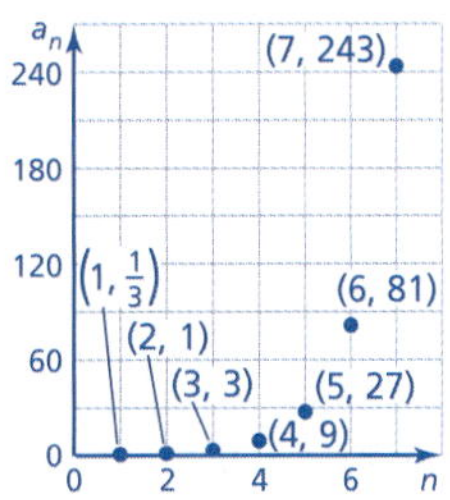

3. Write an equation for the nth term of the geometric sequence 11, −33, 99, −297, ... Then find a_8. $a_n = 11(-3)^{n-1}$; $a_8 = -24{,}057$
4. The number of bacteria in a culture doubles each hour. After how many hours is the number of bacteria 7680?

Hours	1	2	3
Number of bacteria	120	240	480

after 7 hours

45. **CRITICAL THINKING** Is the sequence shown an *arithmetic* sequence? a *geometric* sequence? Explain your reasoning.

3, 3, 3, 3, . . .

46. **HOW DO YOU SEE IT?** Without performing any calculations, match each equation with its graph. Explain your reasoning.

a. $a_n = 20\left(\frac{4}{3}\right)^{n-1}$

b. $a_n = 20\left(\frac{3}{4}\right)^{n-1}$

A.

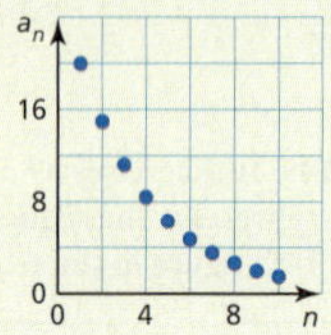

B.

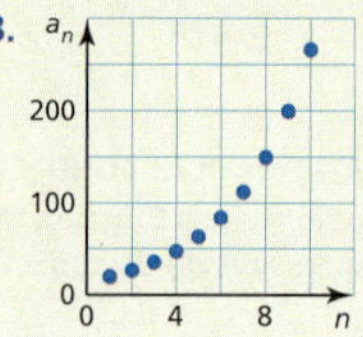

47. **REASONING** What is the 9th term of the geometric sequence where $a_3 = 81$ and $r = 3$?

48. **OPEN-ENDED** Write a sequence that has a pattern but is not arithmetic or geometric. Describe the pattern.

49. **ATTENDING TO PRECISION** Are the terms of a geometric sequence independent or dependent? Explain your reasoning.

50. **DRAWING CONCLUSIONS** A college student makes a deal with her parents to live at home instead of living on campus. She will pay her parents \$0.01 for the first day of the month, \$0.02 for the second day, \$0.04 for the third day, and so on.

a. Write an equation that represents the nth term of the geometric sequence.

b. What will she pay on the 25th day?

c. Did the student make a good choice or should she have chosen to live on campus? Explain.

51. **REPEATED REASONING** A soup kitchen makes 16 gallons of soup. Each day, a quarter of the soup is served and the rest is saved for the next day.

a. Write the first five terms of the sequence of the number of fluid ounces of soup left each day.

b. Write an equation that represents the nth term of the sequence.

c. When is all the soup gone? Explain.

52. **THOUGHT PROVOKING** Find the sum of the terms of the geometric sequence.

$$1, \frac{1}{2}, \frac{1}{4}, \frac{1}{8}, \ldots, \frac{1}{2^{n-1}}, \ldots$$

Explain your reasoning. Write a different infinite geometric sequence that has the same sum.

53. **OPEN-ENDED** Write a geometric sequence in which $a_2 < a_1 < a_3$.

54. **NUMBER SENSE** Write an equation that represents the nth term of each geometric sequence shown.

n	1	2	3	4
a_n	2	6	18	54

n	1	2	3	4
b_n	1	5	25	125

a. Do the terms $a_1 - b_1, a_2 - b_2, a_3 - b_3, \ldots$ form a geometric sequence? If so, how does the common ratio relate to the common ratios of the sequences above?

b. Do the terms $\frac{a_1}{b_1}, \frac{a_2}{b_2}, \frac{a_3}{b_3}, \ldots$ form a geometric sequence? If so, how does the common ratio relate to the common ratios of the sequences above?

Maintaining Mathematical Proficiency

Reviewing what you learned in previous grades and lessons

Use residuals to determine whether the model is a good fit for the data in the table. Explain. *(Section 4.5)*

55. $y = 3x - 8$

x	0	1	2	3	4	5	6
y	−10	−2	−1	2	1	7	10

56. $y = -5x + 1$

x	−3	−2	−1	0	1	2	3
y	6	4	6	1	2	−4	−3

If students need help...	*If students got it...*
Resources by Chapter • Practice A and Practice B • Puzzle Time	Resources by Chapter • Enrichment and Extension • Cumulative Review
Student Journal • Practice	Start the *next* Section
Differentiating the Lesson Skills Review Handbook	

Laurie's Notes

Overview of Section 6.7

Introduction

- **Make Sense of Problems and Persevere in Solving Them:** Recursive sequences are a new type of sequence for students to think about and work with. Conceptually, recursive rules can be simple to describe; however, the notation can be challenging for some students. The idea that you cannot determine the 12th term in the sequence without knowing the 11th term is also a new experience for students. The lesson is designed so that students can make connections to arithmetic and geometric sequences.
- The first skill presented is: given a recursive rule, generate the first few terms of the sequence. Then, given the terms, write the recursive rule. These two skills together should reinforce the need for a recursive rule and the first term in the sequence. If you have the rule and you do not know the first term, then you cannot write any of the terms.
- **Connection:** If the recursive rule describes an arithmetic or geometric sequence, then the explicit form would be a linear or exponential function with a discrete domain of $x = 1, 2, 3, \ldots$. This connection is used to help translate between explicit equations and recursive rules.
- The lesson ends with writing a recursive rule for a special sequence that is neither arithmetic nor geometric.

Common Misconceptions

- Explain the subscript notation. Students sometimes think that with a_{n-1} they are supposed to distribute, and they write $a_n - a$. Make sure students understand that a_{n-1} is the term before a_n and a_{n-3} would be the third term before a_n.

Formative Assessment Tips

- Refer to page T-290 for a description of *Response Logs*. In this lesson and throughout this chapter, there will be sentence stems to prompt student writing. Use your judgment for placement in the lesson, amount of time to allow for writing, and the actual sentence stem given to students.
- For a beginning list of writing stems, see page T-298.

Another Way

- In discussing recursive functions, the language "seed" is used to refer to the first term. The first term is the seed, and it gets the sequence started. This may be helpful for some students.

Pacing Suggestion

- The explorations provide an introduction to recursive functions. Transition to the formal lesson when students have discussed the explorations.

Dynamic Teaching Tools

- Dynamic Assessment & Progress Monitoring Tool
- Lesson Planning Tool
- Interactive Whiteboard Lesson Library
- Dynamic Classroom with Dynamic Investigations

What Your Students Will Learn

- Write terms of recursively defined arithmetic sequences and recursively defined geometric sequences.
- Organize the terms of sequences in a table and write recursive rules for the sequences.
- Write explicit rules for sequences represented by recursive rules.
- Write recursive rules for sequences represented by explicit rules.
- Write recursive rules for sequences that are neither arithmetic sequences nor geometric sequences.

Laurie's Notes

Exploration

Motivate

- **Story Time:** Tell students about your friend who was offered a job with the following salary plan:

  Year 1 = \$42,000

  Year 2 = Year 1 + 2 × \$500

  Year 3 = Year 2 + 3 × \$500

  Year 4 = Year 3 + 4 × \$500

? Your friend is curious what she will earn 20 years from now when she is ready to retire. "What do you need to know to find the salary in Year 20?" You need to know the salary in Year 19 to find the salary in Year 20.

Exploration Note

- Italian mathematician Leonardo Fibonacci first posed this classic problem in 1202 as an exercise in population biology. It concerns how a population of rabbits would grow from generation to generation when the rabbits reproduced according to three rules:
  - Each pair of fertile rabbits produces a new pair of offspring each month.
  - Rabbits become fertile in their second month of life.
  - Old rabbits never die.
- The rules are simple to state. Keeping track of rabbits after birth is a bit more challenging. The table shown gives a color to rabbit pairs to help keep track of offspring.

Exploration 1

- Have students read the introduction. Clarify the rules for how rabbit pairs reproduce if students have a question. You may need to discuss months 3 and 4 to ensure that students understand the problem.
- Students will come up with their own notation to keep track of rabbits and when they will become fertile.
- The big "aha" comes when students realize that at any month you always have the rabbits from the previous month (old rabbits never die) plus the new ones that have been born to fertile pairs.
- **Teaching Tip:** When students do their work on whiteboards, it allows them to more easily share their work with the whole class.

Exploration 2

- Answer questions about the use of subscripts to keep track of term numbers.

? "If $n = 4$, how would you read the recursive function?" $a_4 = a_3 + a_2$

- Students should recognize that the table has the same results as the first exploration. Students may also recognize this sequence of numbers as the Fibonacci sequence.

? "Could you find a_{20}?" yes, if you knew a_{19} and a_{18}

Communicate Your Answer

- Students' language may not be precise at this stage. Listen for the concept that a recursive sequence relates to knowing preceding terms in the sequence.

Connecting to Next Step

- The explorations give students the opportunity to explore a classic problem in mathematics related to recursive functions, which are formally defined in the lesson.

6.7 Recursively Defined Sequences

Essential Question

How can you define a sequence recursively?

A **recursive rule** gives the beginning term(s) of a sequence and a *recursive equation* that tells how a_n is related to one or more preceding terms.

EXPLORATION 1 Describing a Pattern

Work with a partner. Consider a hypothetical population of rabbits. Start with one breeding pair. After each month, each breeding pair produces another breeding pair. The total number of rabbits each month follows the exponential pattern 2, 4, 8, 16, 32, ... Now suppose that in the first month after each pair is born, the pair is too young to reproduce. Each pair produces another pair after it is 2 months old. Find the total number of pairs in months 6, 7, and 8.

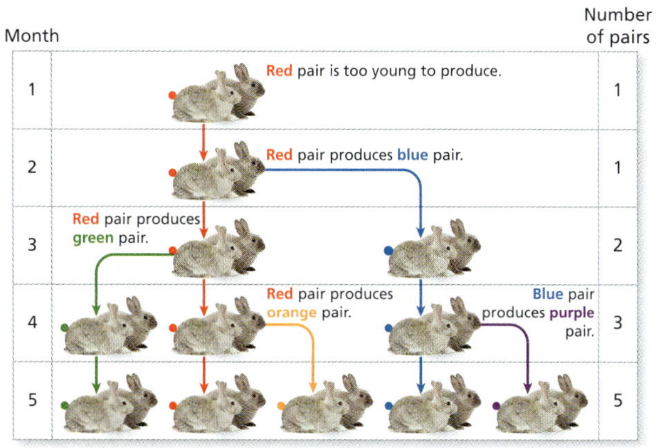

RECOGNIZING PATTERNS

To be proficient in math, you need to look closely to discern a pattern or structure.

EXPLORATION 2 Using a Recursive Equation

Work with a partner. Consider the following recursive equation.

$a_n = a_{n-1} + a_{n-2}$

Each term in the sequence is the sum of the two preceding terms.

Copy and complete the table. Compare the results with the sequence of the number of pairs in Exploration 1.

a_1	a_2	a_3	a_4	a_5	a_6	a_7	a_8
1	1						

Communicate Your Answer

3. How can you define a sequence recursively?
4. Use the Internet or some other reference to determine the mathematician who first described the sequences in Explorations 1 and 2.

Dynamic Teaching Tools

- Dynamic Assessment & Progress Monitoring Tool
- Lesson Planning Tool
- Interactive Whiteboard Lesson Library
- Dynamic Classroom with Dynamic Investigations

ANSWERS

1. 8 pairs, 13 pairs, 21 pairs
2. 2, 3, 5, 8, 13, 21; They are the same.
3. Give the beginning term(s) of a sequence and a recursive equation that tells how a_n is related to one or more preceding terms.
4. Leonardo Fibonacci

Differentiated Instruction

Inclusion

Some students may have difficulty understanding the difference between explicit and recursive equations for a sequence. Point out that an explicit equation allows you to find *any* term in a sequence. For example, given the first term and the explicit equation, you can find the second term, the 8th term, the 15th term, or any other term. A recursive equation allows you to find only the *next* term in a sequence. For example, given the first term and the recursive equation, you can find only the second term. Then you must use the second term and the recursive equation to find the third term, and so on.

Extra Example 1

Write the first six terms of each sequence. Then graph each sequence.

a. $a_1 = 3$, $a_n = a_{n-1} + 5$

3, 8, 13, 18, 23, 28

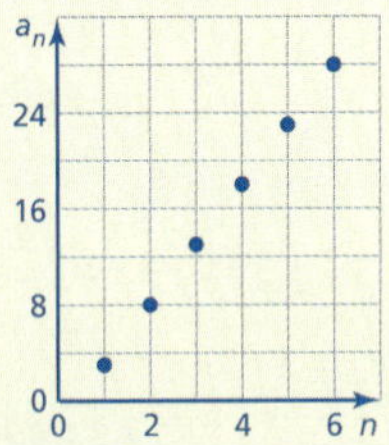

b. $a_1 = 0.5$, $a_n = 2a_{n-1}$

0.5, 1, 2, 4, 8, 16

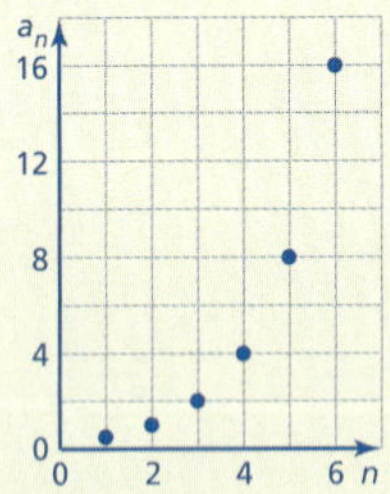

6.7 Lesson

Core Vocabulary

explicit rule, *p. 340*
recursive rule, *p. 340*

Previous
arithmetic sequence
geometric sequence

What You Will Learn

- Write terms of recursively defined sequences.
- Write recursive rules for sequences.
- Translate between recursive rules and explicit rules.
- Write recursive rules for special sequences.

Writing Terms of Recursively Defined Sequences

So far in this book, you have defined arithmetic and geometric sequences *explicitly*. An **explicit rule** gives a_n as a function of the term's position number n in the sequence. For example, an explicit rule for the arithmetic sequence 3, 5, 7, 9, . . . is $a_n = 3 + 2(n - 1)$, or $a_n = 2n + 1$.

Now, you will define arithmetic and geometric sequences *recursively*. A **recursive rule** gives the beginning term(s) of a sequence and a *recursive equation* that tells how a_n is related to one or more preceding terms.

Core Concept

Recursive Equation for an Arithmetic Sequence

$a_n = a_{n-1} + d$, where d is the common difference

Recursive Equation for a Geometric Sequence

$a_n = r \cdot a_{n-1}$, where r is the common ratio

EXAMPLE 1 Writing Terms of Recursively Defined Sequences

Write the first six terms of each sequence. Then graph each sequence.

a. $a_1 = 2$, $a_n = a_{n-1} + 3$ **b.** $a_1 = 1$, $a_n = 3a_{n-1}$

SOLUTION

You are given the first term. Use the recursive equation to find the next five terms.

a. $a_1 = 2$

$a_2 = a_1 + 3 = 2 + 3 = 5$

$a_3 = a_2 + 3 = 5 + 3 = 8$

$a_4 = a_3 + 3 = 8 + 3 = 11$

$a_5 = a_4 + 3 = 11 + 3 = 14$

$a_6 = a_5 + 3 = 14 + 3 = 17$

b. $a_1 = 1$

$a_2 = 3a_1 = 3(1) = 3$

$a_3 = 3a_2 = 3(3) = 9$

$a_4 = 3a_3 = 3(9) = 27$

$a_5 = 3a_4 = 3(27) = 81$

$a_6 = 3a_5 = 3(81) = 243$

STUDY TIP

A sequence is a discrete function. So, the points on the graph are not connected.

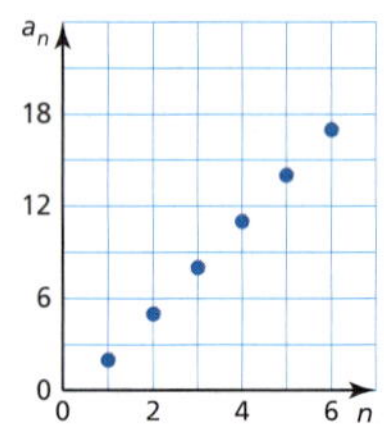

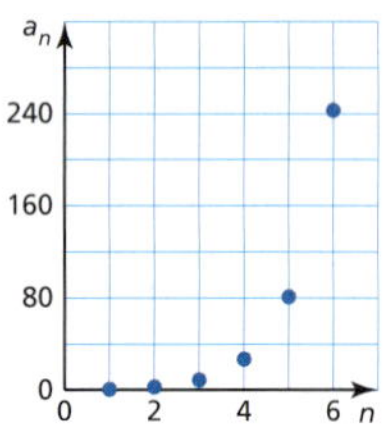

Laurie's Notes Teacher Actions

- Write the *Core Concept* and use additional words. For the arithmetic sequence, the *n*th term is the previous term plus some common difference. For the geometric sequence, the *n*th term is the previous term times some common ratio.
- After generating the sequence in Example 1(a) and plotting the points, anticipate that students will be confused about why this linear graph is not just written as $y = 3x - 1$ for $x = 1, 2, 3 \ldots$. It could be, if we wanted to write an explicit equation, which we will do in Example 3.

Monitoring Progress

Help in English and Spanish at *BigIdeasMath.com*

Write the first six terms of the sequence. Then graph the sequence.

1. $a_1 = 0, a_n = a_{n-1} - 8$

2. $a_1 = -7.5, a_n = a_{n-1} + 2.5$

3. $a_1 = -36, a_n = \frac{1}{2}a_{n-1}$

4. $a_1 = 0.7, a_n = 10a_{n-1}$

Writing Recursive Rules

EXAMPLE 2 Writing Recursive Rules

Write a recursive rule for each sequence.

a. $-30, -18, -6, 6, 18, \ldots$

b. $500, 100, 20, 4, 0.8, \ldots$

SOLUTION

Use a table to organize the terms and find the pattern.

a.

Position, n	1	2	3	4	5
Term, a_n	-30	-18	-6	6	18

+12 +12 +12 +12

The sequence is arithmetic, with first term $a_1 = -30$ and common difference $d = 12$.

$a_n = a_{n-1} + d$ Recursive equation for an arithmetic sequence

$a_n = a_{n-1} + 12$ Substitute 12 for d.

▶ So, a recursive rule for the sequence is $a_1 = -30, a_n = a_{n-1} + 12$.

COMMON ERROR

When writing a recursive rule for a sequence, you need to write both the beginning term(s) and the recursive equation.

b.

Position, n	1	2	3	4	5
Term, a_n	500	100	20	4	0.8

$\times\frac{1}{5}$ $\times\frac{1}{5}$ $\times\frac{1}{5}$ $\times\frac{1}{5}$

The sequence is geometric, with first term $a_1 = 500$ and common ratio $r = \frac{1}{5}$.

$a_n = r \cdot a_{n-1}$ Recursive equation for a geometric sequence

$a_n = \frac{1}{5}a_{n-1}$ Substitute $\frac{1}{5}$ for r.

▶ So, a recursive rule for the sequence is $a_1 = 500, a_n = \frac{1}{5}a_{n-1}$.

Monitoring Progress

Help in English and Spanish at *BigIdeasMath.com*

Write a recursive rule for the sequence.

5. $8, 3, -2, -7, -12, \ldots$

6. $1.3, 2.6, 3.9, 5.2, 6.5, \ldots$

7. $4, 20, 100, 500, 2500, \ldots$

8. $128, -32, 8, -2, 0.5, \ldots$

9. Write a recursive rule for the height of the sunflower over time.

English Language Learners

Comprehension

Make sure that students understand how a *recursive rule* is different from a *recursive equation*. Emphasize that the recursive rule for a sequence gives the first term(s) of the sequence followed by the recursive equation of the sequence.

Recursive rule:

$\underbrace{a_1 = -30,}_{\text{first term}}\ \underbrace{a_n = a_{n-1} + 12}_{\text{recursive equation}}$

Extra Example 2

Write a recursive rule for each sequence.

a. $-8, -1, 6, 13, 20, \ldots$

$a_1 = -8, a_n = a_{n-1} + 7$

b. $405, 135, 45, 15, 5, \ldots$

$a_1 = 405, a_n = \frac{1}{3}a_{n-1}$

MONITORING PROGRESS ANSWERS

1. $0, -8, -16, -24, -32, -40$

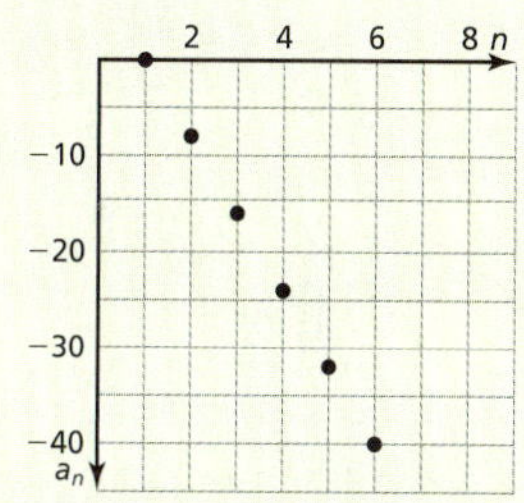

2. $-7.5, -5, -2.5, 0, 2.5, 5$

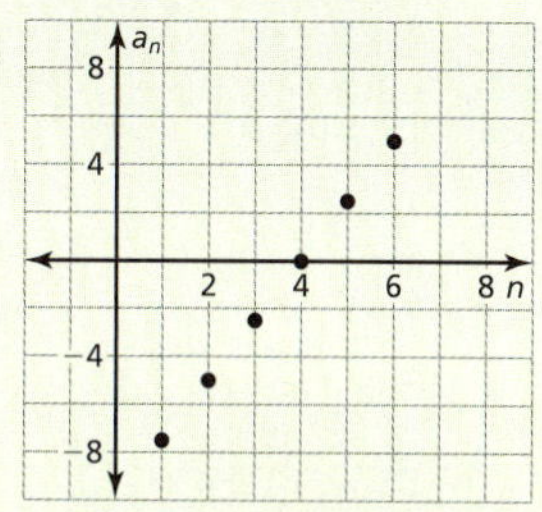

3. $-36, -18, -9, -\frac{9}{2}, -\frac{9}{4}, -\frac{9}{8}$

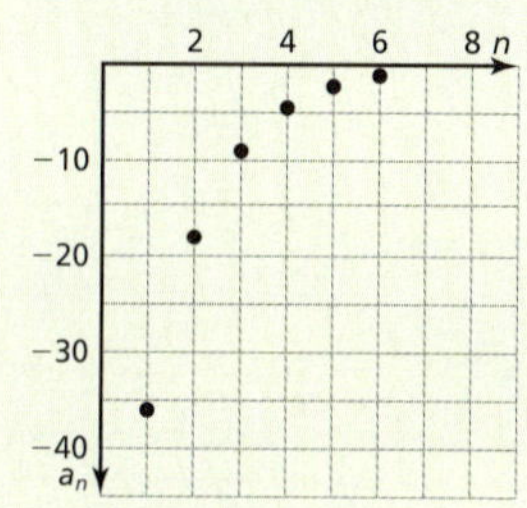

4–9. See Additional Answers.

Laurie's Notes Teacher Actions

- When terms of a sequence are given, it is often helpful to organize them in a table, labeling the position and related term.
- **Turn and Talk:** "For each sequence, is it arithmetic or geometric? How do you know?" List the common differences or common ratios.
- Discuss the *Common Error*. If this does not make sense, refer to $a_n = a_{n-1} + 12$. You cannot write the sequence terms without knowing a_1.
- **Response Logs:** Select from "I'm still not sure about …" or "I'm not sure about …."

Extra Example 3

Write an explicit rule for each recursive rule.

a. $a_1 = 9, a_n = a_{n-1} - 4$
   $a_n = -4n + 13$

b. $a_1 = 4.5, a_n = 3a_{n-1}$
   $a_n = 4.5(3)^{n-1}$

Extra Example 4

Write a recursive rule for each explicit rule.

a. $a_n = 6n - 5$
   $a_1 = 1, a_n = a_{n-1} + 6$

b. $a_n = 4(3)^{n-1}$
   $a_1 = 4, a_n = 3a_{n-1}$

MONITORING PROGRESS ANSWERS

10. $a_n = 20n - 65$

11. $a_n = 13(-3)^{n-1}$

12. $a_1 = 0, a_n = a_{n-1} - 1$

13. $a_1 = -2.5, a_n = 4a_{n-1}$

Translating between Recursive and Explicit Rules

EXAMPLE 3 Translating from Recursive Rules to Explicit Rules

Write an explicit rule for each recursive rule.

a. $a_1 = 25, a_n = a_{n-1} - 10$ **b.** $a_1 = 19.6, a_n = -0.5a_{n-1}$

SOLUTION

a. The recursive rule represents an arithmetic sequence, with first term $a_1 = 25$ and common difference $d = -10$.

$a_n = a_1 + (n - 1)d$ Explicit rule for an arithmetic sequence

$a_n = 25 + (n - 1)(-10)$ Substitute 25 for a_1 and -10 for d.

$a_n = -10n + 35$ Simplify.

► An explicit rule for the sequence is $a_n = -10n + 35$.

b. The recursive rule represents a geometric sequence, with first term $a_1 = 19.6$ and common ratio $r = -0.5$.

$a_n = a_1r^{n-1}$ Explicit rule for a geometric sequence

$a_n = 19.6(-0.5)^{n-1}$ Substitute 19.6 for a_1 and -0.5 for r.

► An explicit rule for the sequence is $a_n = 19.6(-0.5)^{n-1}$.

EXAMPLE 4 Translating from Explicit Rules to Recursive Rules

Write a recursive rule for each explicit rule.

a. $a_n = -2n + 3$ **b.** $a_n = -3(2)^{n-1}$

SOLUTION

a. The explicit rule represents an arithmetic sequence, with first term $a_1 = -2(1) + 3 = 1$ and common difference $d = -2$.

$a_n = a_{n-1} + d$ Recursive equation for an arithmetic sequence

$a_n = a_{n-1} + (-2)$ Substitute -2 for d.

► So, a recursive rule for the sequence is $a_1 = 1, a_n = a_{n-1} - 2$.

b. The explicit rule represents a geometric sequence, with first term $a_1 = -3$ and common ratio $r = 2$.

$a_n = r \cdot a_{n-1}$ Recursive equation for a geometric sequence

$a_n = 2a_{n-1}$ Substitute 2 for r.

► So, a recursive rule for the sequence is $a_1 = -3, a_n = 2a_{n-1}$.

Monitoring Progress

Help in English and Spanish at *BigIdeasMath.com*

Write an explicit rule for the recursive rule.

10. $a_1 = -45, a_n = a_{n-1} + 20$ **11.** $a_1 = 13, a_n = -3a_{n-1}$

Write a recursive rule for the explicit rule.

12. $a_n = -n + 1$ **13.** $a_n = -2.5(4)^{n-1}$

Laurie's Notes Teacher Actions

- **Turn and Talk:** "In Example 3, what are the first few terms of each sequence?" Listen for 25, 15, 5, … and 19.6, –9.8, 4.9, ….
- Write the explicit rule, substitute the known information, and simplify.
- Check results. Note that the sign of the common ratio (negative) causes the terms to alternate signs.
- In Example 4, remind students that the domain is 1, 2, 3, …; so when $n = 1$, students can find a_1 for the recursive rule.

Writing Recursive Rules for Special Sequences

You can write recursive rules for sequences that are neither arithmetic nor geometric. One way is to look for patterns in the sums of consecutive terms.

The sequence in Example 5 is called the *Fibonacci sequence*. This pattern is naturally occurring in many objects, such as flowers.

EXAMPLE 5 Writing Recursive Rules for Other Sequences

Use the sequence shown.

$$1, 1, 2, 3, 5, 8, \ldots$$

a. Write a recursive rule for the sequence.

b. Write the next three terms of the sequence.

SOLUTION

a. Find the difference and ratio between each pair of consecutive terms.

1, 1, 2, 3

$1 - 1 = 0$ $2 - 1 = 1$ $3 - 2 = 1$ — There is no common difference, so the sequence is *not* arithmetic.

1, 1, 2, 3

$\frac{1}{1} = 1$ $\frac{2}{1} = 2$ $\frac{3}{2} = 1\frac{1}{2}$ — There is no common ratio, so the sequence is *not* geometric.

Find the sum of each pair of consecutive terms.

$a_1 + a_2 = 1 + 1 = 2$ — 2 is the third term.

$a_2 + a_3 = 1 + 2 = 3$ — 3 is the fourth term.

$a_3 + a_4 = 2 + 3 = 5$ — 5 is the fifth term.

$a_4 + a_5 = 3 + 5 = 8$ — 8 is the sixth term.

Beginning with the third term, each term is the sum of the two previous terms. A recursive equation for the sequence is $a_n = a_{n-2} + a_{n-1}$.

▶ So, a recursive rule for the sequence is $a_1 = 1$, $a_2 = 1$, $a_n = a_{n-2} + a_{n-1}$.

b. Use the recursive equation $a_n = a_{n-2} + a_{n-1}$ to find the next three terms.

$a_7 = a_5 + a_6$	$a_8 = a_6 + a_7$	$a_9 = a_7 + a_8$
$= 5 + 8$	$= 8 + 13$	$= 13 + 21$
$= 13$	$= 21$	$= 34$

▶ The next three terms are 13, 21, and 34.

Monitoring Progress

Help in English and Spanish at *BigIdeasMath.com*

Write a recursive rule for the sequence. Then write the next three terms of the sequence.

14. $5, 6, 11, 17, 28, \ldots$

15. $-3, -4, -7, -11, -18, \ldots$

16. $1, 1, 0, -1, -1, 0, 1, 1, \ldots$

17. $4, 3, 1, 2, -1, 3, -4, \ldots$

Extra Example 5

Use the sequence shown.

$-4, -5, -9, -14, -23, -37, \ldots$

a. Write a recursive rule for the sequence.
$a_1 = -4, a_2 = -5, a_n = a_{n-2} + a_{n-1}$

b. Write the next three terms of the sequence. $-60, -97, -157$

MONITORING PROGRESS ANSWERS

14. $a_1 = 5, a_2 = 6, a_n = a_{n-2} + a_{n-1}$; 45, 73, 118

15. $a_1 = -3, a_2 = -4, a_n = a_{n-2} + a_{n-1}$; $-29, -47, -76$

16. $a_1 = 1, a_2 = 1, a_n = a_{n-1} - a_{n-2}$; $0, -1, -1$

17. $a_1 = 4, a_2 = 3, a_n = a_{n-2} - a_{n-1}$; $7, -11, 18$

Laurie's Notes Teacher Actions

- Example 5 is the Fibonacci sequence, which students worked with in the exploration. It is neither arithmetic nor geometric.
- Students may recall the pattern from the exploration, so writing the recursive rule should not be difficult at this point in the lesson.
- **Extension:** There are many different opportunities for students to explore Fibonacci numbers further. Much has been written.

Closure

- **Response Logs:** Select from "A sticky part for me is …" or "I need to rethink …" or "I wish I could …."

Assignment Guide and Homework Check

ASSIGNMENT

Basic: 1, 2, 3–45 odd, 54, 58, 60–67

Average: 1, 2–54 even, 58, 60–67

Advanced: 1, 2, 6, 8, 12, 16–22 even, 26–34 even, 38–48 even, 52–67

HOMEWORK CHECK

Basic: 7, 15, 23, 29, 39

Average: 8, 14, 26, 30, 40

Advanced: 12, 16, 28, 34, 44

ANSWERS

1. recursive equation
2. $a_n = 6n - 2$; It is the only explicit rule.
3. geometric
4. arithmetic
5. arithmetic
6. geometric
7. 0, 2, 4, 6, 8, 10

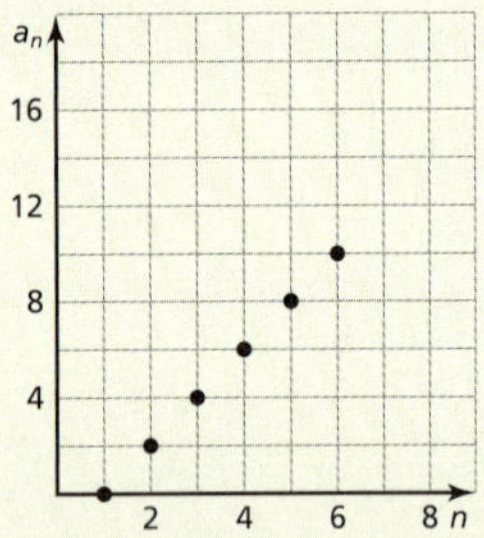

8. 10, 5, 0, −5, −10, −15

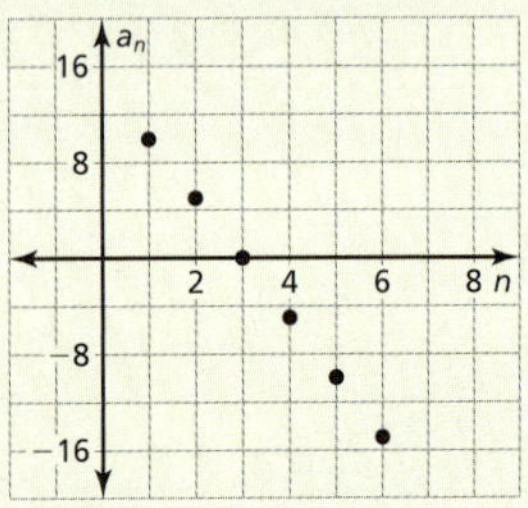

9. 2, 6, 18, 54, 162, 486

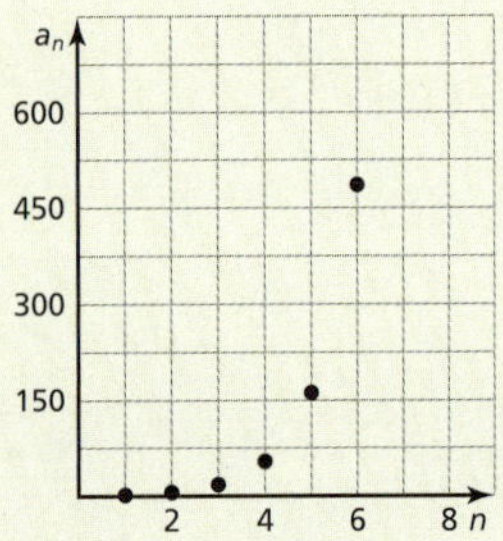

10. 8, 12, 18, 27, 40.5, 60.75

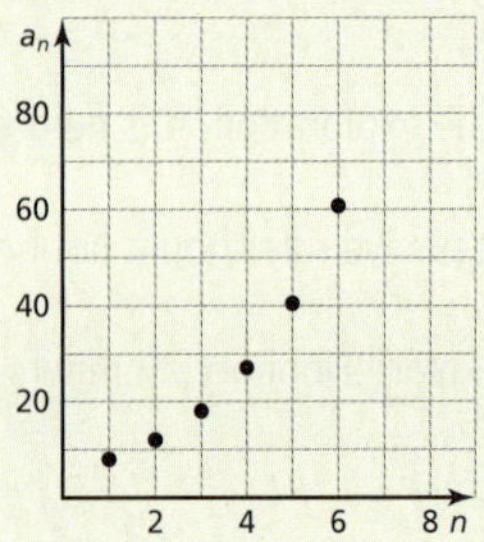

11. 80, −40, 20, −10, 5, $-\frac{5}{2}$

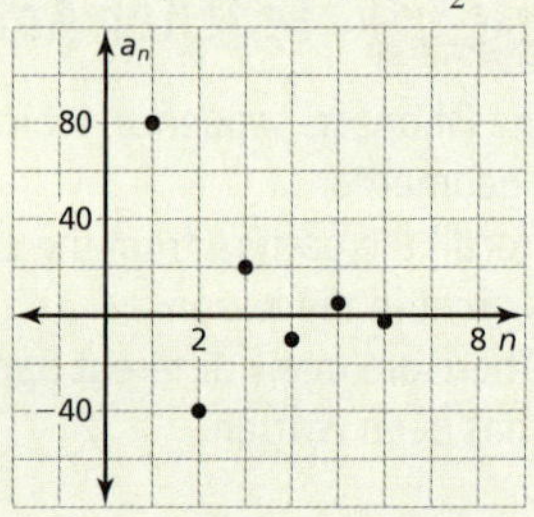

12–22. See Additional Answers.

6.7 Exercises

Dynamic Solutions available at *BigIdeasMath.com*

Vocabulary and Core Concept Check

1. **COMPLETE THE SENTENCE** A recursive rule gives the beginning term(s) of a sequence and a(n) ____________ that tells how a_n is related to one or more preceding terms.

2. **WHICH ONE DOESN'T BELONG?** Which rule does *not* belong with the other three? Explain your reasoning.

$a_1 = -1, a_n = 5a_{n-1}$	$a_n = 6n - 2$	$a_1 = -3, a_n = a_{n-1} + 1$	$a_1 = 9, a_n = 4a_{n-1}$

Monitoring Progress and Modeling with Mathematics

In Exercises 3–6, determine whether the recursive rule represents an *arithmetic sequence* or a *geometric sequence*.

3. $a_1 = 2, a_n = 7a_{n-1}$
4. $a_1 = 18, a_n = a_{n-1} + 1$
5. $a_1 = 5, a_n = a_{n-1} - 4$
6. $a_1 = 3, a_n = -6a_{n-1}$

In Exercises 7–12, write the first six terms of the sequence. Then graph the sequence. *(See Example 1.)*

7. $a_1 = 0, a_n = a_{n-1} + 2$
8. $a_1 = 10, a_n = a_{n-1} - 5$
9. $a_1 = 2, a_n = 3a_{n-1}$
10. $a_1 = 8, a_n = 1.5a_{n-1}$
11. $a_1 = 80, a_n = -\frac{1}{2}a_{n-1}$
12. $a_1 = -7, a_n = -4a_{n-1}$

In Exercises 13–20, write a recursive rule for the sequence. *(See Example 2.)*

13.

n	1	2	3	4
a_n	7	16	25	34

14.

n	1	2	3	4
a_n	8	24	72	216

15. 243, 81, 27, 9, 3, . . .
16. 3, 11, 19, 27, 35, . . .
17. 0, −3, −6, −9, −12, . . .
18. 5, −20, 80, −320, 1280, . . .

19.

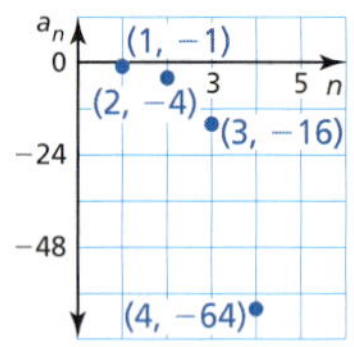

20.

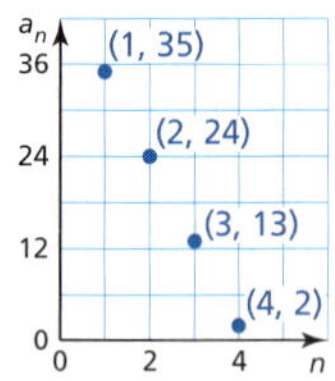

21. **MODELING WITH MATHEMATICS** Write a recursive rule for the number of bacterial cells over time.

22. **MODELING WITH MATHEMATICS** Write a recursive rule for the length of the deer antler over time.

In Exercises 23–28, write an explicit rule for the recursive rule. *(See Example 3.)*

23. $a_1 = -3, a_n = a_{n-1} + 3$

24. $a_1 = 8, a_n = a_{n-1} - 12$

25. $a_1 = 16, a_n = 0.5a_{n-1}$

26. $a_1 = -2, a_n = 9a_{n-1}$

27. $a_1 = 4, a_n = a_{n-1} + 17$

28. $a_1 = 5, a_n = -5a_{n-1}$

In Exercises 29–34, write a recursive rule for the explicit rule. *(See Example 4.)*

29. $a_n = 7(3)^{n-1}$ **30.** $a_n = -4n + 2$

31. $a_n = 1.5n + 3$ **32.** $a_n = 6n - 20$

33. $a_n = (-5)^{n-1}$ **34.** $a_n = -81\left(\frac{2}{3}\right)^{n-1}$

In Exercises 35–38, graph the first four terms of the sequence with the given description. Write a recursive rule and an explicit rule for the sequence.

35. The first term of a sequence is 5. Each term of the sequence is 15 more than the preceding term.

36. The first term of a sequence is 16. Each term of the sequence is half the preceding term.

37. The first term of a sequence is -1. Each term of the sequence is -3 times the preceding term.

38. The first term of a sequence is 19. Each term of the sequence is 13 less than the preceding term.

In Exercises 39–44, write a recursive rule for the sequence. Then write the next two terms of the sequence. *(See Example 5.)*

39. 1, 3, 4, 7, 11, . . . **40.** 10, 9, 1, 8, −7, 15, . . .

41. 2, 4, 2, −2, −4, −2, . . .

42. 6, 1, 7, 8, 15, 23, . . .

43.

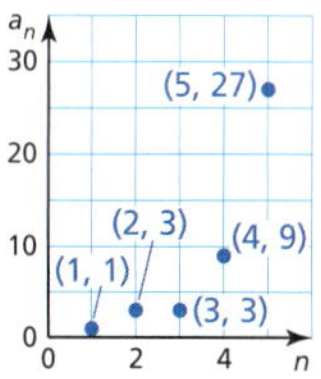

44.

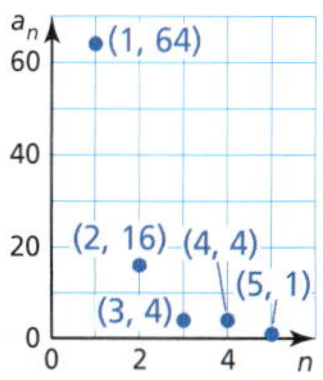

45. ERROR ANALYSIS Describe and correct the error in writing an explicit rule for the recursive rule $a_1 = 6$, $a_n = a_{n-1} - 12$.

✗
$a_n = a_1 + (n-1)d$
$a_n = 6 + (n-1)(12)$
$a_n = 6 + 12n - 12$
$a_n = -6 + 12n$

46. ERROR ANALYSIS Describe and correct the error in writing a recursive rule for the sequence 2, 4, 6, 10, 16, ...

✗
2, 4, 6, . . .
+2 +2

The sequence is arithmetic, with first term $a_1 = 2$ and common difference $d = 2$.

$a_n = a_{n-1} + d$
$a_1 = 2, a_n = a_{n-1} + 2$

In Exercises 47–51, the function *f* represents a sequence. Find the 2nd, 5th, and 10th terms of the sequence.

47. $f(1) = 3, f(n) = f(n-1) + 7$

48. $f(1) = -1, f(n) = 6f(n-1)$

49. $f(1) = 8, f(n) = -f(n-1)$

50. $f(1) = 4, f(2) = 5, f(n) = f(n-2) + f(n-1)$

51. $f(1) = 10, f(2) = 15, f(n) = f(n-1) - f(n-2)$

52. MODELING WITH MATHEMATICS The X-ray shows the lengths (in centimeters) of bones in a human hand.

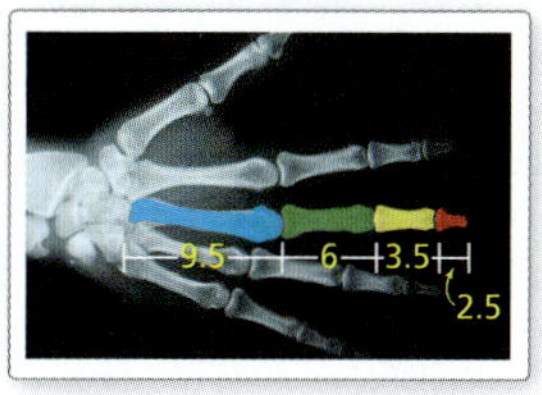

a. Write a recursive rule for the lengths of the bones.

b. Measure the lengths of different sections of your hand. Can the lengths be represented by a recursively defined sequence? Explain.

Dynamic Teaching Tools

Dynamic Assessment & Progress Monitoring Tool

Interactive Whiteboard Lesson Library

Dynamic Classroom with Dynamic Investigations

ANSWERS

23. $a_n = 3n - 6$

24. $a_n = -12n + 20$

25. $a_n = 16(0.5)^{n-1}$

26. $a_n = -2(9)^{n-1}$

27. $a_n = 17n - 13$

28. $a_n = 5(-5)^{n-1}$

29. $a_1 = 7, a_n = 3a_{n-1}$

30. $a_1 = -2, a_n = a_{n-1} - 4$

31. $a_1 = 4.5, a_n = a_{n-1} + 1.5$

32. $a_1 = -14, a_n = a_{n-1} + 6$

33. $a_1 = 1, a_n = -5a_{n-1}$

34. $a_1 = -81, a_n = \frac{2}{3}a_{n-1}$

35.

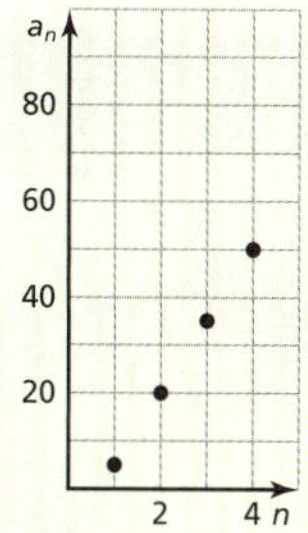

$a_1 = 5, a_n = a_{n-1} + 15;$
$a_n = 15n - 10$

36.

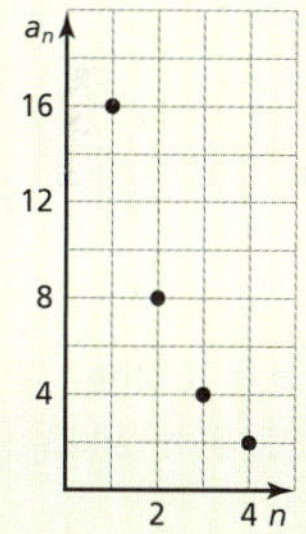

$a_1 = 16, a_n = \frac{1}{2}a_{n-1}; a_n = 16\left(\frac{1}{2}\right)^{n-1}$

37.

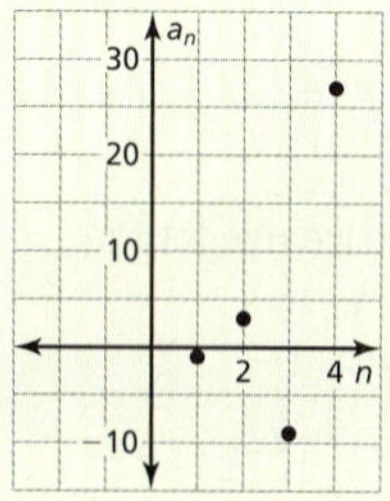

$a_1 = -1, a_n = -3a_{n-1};$
$a_n = -(-3)^{n-1}$

38.

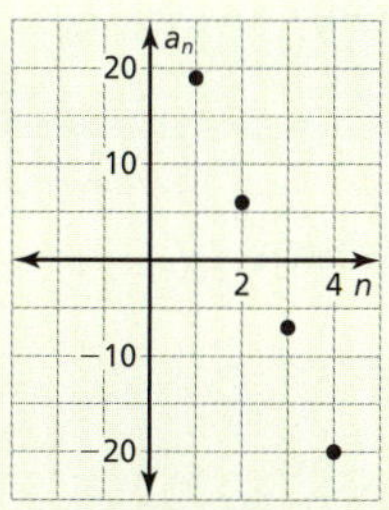

$a_1 = 19, a_n = a_{n-1} - 13; a_n = -13n + 32$

39. $a_1 = 1, a_2 = 3, a_n = a_{n-2} + a_{n-1}$; 18, 29

40. $a_1 = 10, a_2 = 9, a_n = a_{n-2} - a_{n-1}$; -22, 37

41. $a_1 = 2, a_2 = 4, a_n = a_{n-1} - a_{n-2}$; 2, 4

42. $a_1 = 6, a_2 = 1, a_n = a_{n-2} + a_{n-1}$; 38, 61

43. $a_1 = 1, a_2 = 3, a_n = (a_n - 1)(a_{n-2})$; 243, 6561

44. $a_1 = 64, a_2 = 16, a_n = \frac{a_{n-2}}{a_{n-1}}$; 4, $\frac{1}{4}$

45. The common difference is -12, not 12; $a_n = 6 + (n-1)(-12)$; $a_n = 6 - 12n + 12$; $a_n = 18 - 12n$

46. The sequence is not arithmetic; Each term in the sequence is the sum of the preceding two terms; $a_1 = 2, a_2 = 4, a_n = a_{n-2} + a_{n-1}$

47. 10; 31; 66

48–52. See Additional Answers.

ANSWERS

53. **a.** 3, 5, 7, 9, 11, 13, 15, 17, 19, 21

b. 3; 12; 48; 192; 768; 3072; 12,288; 49,152; 196,608; 786,432

c. 4, 7, 3, −4, −7, −3, 4, 7, 3, −4

54. **a.** 1, 1, 2, 3, 5, 8, or $a_{mn} = a_{(m-1)(n-1)} + a_{m(n-1)}$

b. Fibonacci sequence; 13

c. Check students' work. The diagram from the book should be drawn, along with an extended 13 by 13 grid at the bottom. The curve should continue into this grid and end in the bottom right corner. The grid should be labeled with a "7."

55. 5, 19, 61, 187, 565; neither; There is no common difference or common ratio.

56–67. See Additional Answers.

Mini-Assessment

1. Write the first six terms of the sequence. Then graph the sequence.
   $a_1 = 0$, $a_n = a_{n-1} + 6$
   0, 6, 12, 18, 24, 30

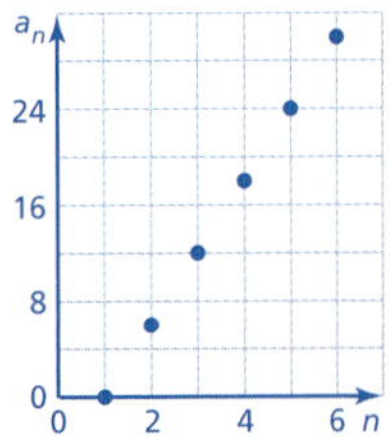

2. Write a recursive rule for the sequence 0.5, 4, 32, 256, 2048, ….
   $a_1 = 0.5$, $a_n = 8a_{n-1}$
3. Write an explicit rule for the recursive rule.
   $a_1 = -3$, $a_n = a_{n-1} + 5$
   $a_n = 5n - 8$
4. Write a recursive rule for the explicit rule.
   $a_n = -1.5(2)^{n-1}$
   $a_1 = -1.5$, $a_n = 2a_{n-1}$
5. Use the sequence shown.
   7, 9, 16, 25, 41, 66, …
   a. Write a recursive rule for the sequence.
   $a_1 = 7$, $a_2 = 9$,
   $a_n = a_{n-2} + a_{n-1}$
   b. Write the next three terms of the sequence. 107, 173, 280

53. **USING TOOLS** You can use a spreadsheet to generate the terms of a sequence.

A2	=A1+2		
	A	B	C
1	3		
2	5		
3			
4			

a. To generate the terms of the sequence $a_1 = 3$, $a_n = a_{n-1} + 2$, enter the value of a_1, 3, into cell A1. Then enter "=A1+2" into cell A2, as shown. Use the *fill down* feature to generate the first 10 terms of the sequence.

b. Use a spreadsheet to generate the first 10 terms of the sequence $a_1 = 3$, $a_n = 4a_{n-1}$. (*Hint:* Enter "=4*A1" into cell A2.)

c. Use a spreadsheet to generate the first 10 terms of the sequence $a_1 = 4$, $a_2 = 7$, $a_n = a_{n-1} - a_{n-2}$. (*Hint:* Enter "=A2-A1" into cell A3.)

54. **HOW DO YOU SEE IT?** Consider Squares 1–6 in the diagram.

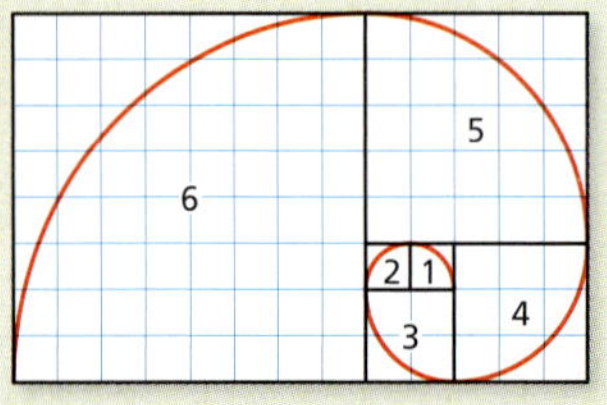

a. Write a sequence in which each term a_n is the side length of square n.

b. What is the name of this sequence? What is the next term of this sequence?

c. Use the term in part (b) to add another square to the diagram and extend the spiral.

55. **REASONING** Write the first 5 terms of the sequence $a_1 = 5$, $a_n = 3a_{n-1} + 4$. Determine whether the sequence is *arithmetic*, *geometric*, or *neither*. Explain your reasoning.

56. **THOUGHT PROVOKING** Describe the pattern for the numbers in Pascal's Triangle, shown below. Write a recursive rule that gives the mth number in the nth row.

1
1 1
1 2 1
1 3 3 1
1 4 6 4 1
1 5 10 10 5 1

57. **REASONING** The explicit rule $a_n = a_1 + (n-1)d$ defines an arithmetic sequence.

a. Explain why $a_{n-1} = a_1 + [(n-1) - 1]d$.

b. Justify each step in showing that a recursive equation for the sequence is $a_n = a_{n-1} + d$.

$a_n = a_1 + (n-1)d$
$= a_1 + [(n-1) + 0]d$
$= a_1 + [(n-1) - 1 + 1]d$
$= a_1 + [((n-1) - 1) + 1]d$
$= a_1 + [(n-1) - 1]d + d$
$= a_{n-1} + d$

58. **MAKING AN ARGUMENT** Your friend claims that the sequence

−5, 5, −5, 5, −5, . . .

cannot be represented by a recursive rule. Is your friend correct? Explain.

59. **PROBLEM SOLVING** Write a recursive rule for the sequence.

3, 7, 15, 31, 63, . . .

Maintaining Mathematical Proficiency

Reviewing what you learned in previous grades and lessons

Simplify the expression. *(Skills Review Handbook)*

60. $5x + 12x$ **61.** $9 - 6y - 14$ **62.** $2d - 7 - 8d$ **63.** $3 - 3m + 11m$

Write a linear function f with the given values. *(Section 4.2)*

64. $f(2) = 6, f(-1) = -3$ **65.** $f(-2) = 0, f(6) = -4$

66. $f(-3) = 5, f(-1) = 5$ **67.** $f(3) = -1, f(-4) = -15$

If students need help...	*If students got it...*
Resources by Chapter • Practice A and Practice B • Puzzle Time	Resources by Chapter • Enrichment and Extension • Cumulative Review
Student Journal • Practice	Start the *next* Section
Differentiating the Lesson Skills Review Handbook	

6.5–6.7 What Did You Learn?

Core Vocabulary

exponential equation, *p. 326*
geometric sequence, *p. 332*
common ratio, *p. 332*
explicit rule, *p. 340*
recursive rule, *p. 340*

Core Concepts

Section 6.5

Property of Equality for Exponential Equations, *p. 326*
Solving Exponential Equations by Graphing, *p. 328*

Section 6.6

Geometric Sequence, *p. 332*
Equation for a Geometric Sequence, *p. 334*

Section 6.7

Recursive Equation for an Arithmetic Sequence, *p. 340*
Recursive Equation for a Geometric Sequence, *p. 340*

Mathematical Practices

1. How did you decide on an appropriate level of precision for your answer in Exercise 49 on page 330?
2. Explain how writing a function in Exercise 39 part (a) on page 337 created a shortcut for answering part (b).
3. How did you choose an appropriate tool in Exercise 52 part (b) on page 345?

Performance Task

The New Car

There is so much more to buying a new car than the purchase price. Interest rates, depreciation, and inflation are all factors. So, what is the real cost of your new car?

To explore the answers to this question and more, go to ***BigIdeasMath.com***.

Dynamic Teaching Tools

- Dynamic Assessment & Progress Monitoring Tool
- Interactive Whiteboard Lesson Library
- Dynamic Classroom with Dynamic Investigations

ANSWERS

1. *Sample answer:* Because the growth factor ended in the hundredths place, the answer should be rounded to the hundredths place.
2. *Sample answer:* Use the function from part (a) to calculate the number of squares for the tenth figure in part (b) without determining the number of squares in the fourth through ninth figures.
3. *Sample answer:* Parts of a hand can be conveniently measured in centimeters, so a 30-centimeter ruler is an appropriate tool to use.

ANSWERS

1. $\frac{1}{y^2}$
2. $\frac{1}{x^3}$
3. y^6
4. $\frac{25y^8}{4x^4}$
5. 2
6. -3
7. 125
8. not a real number
9. 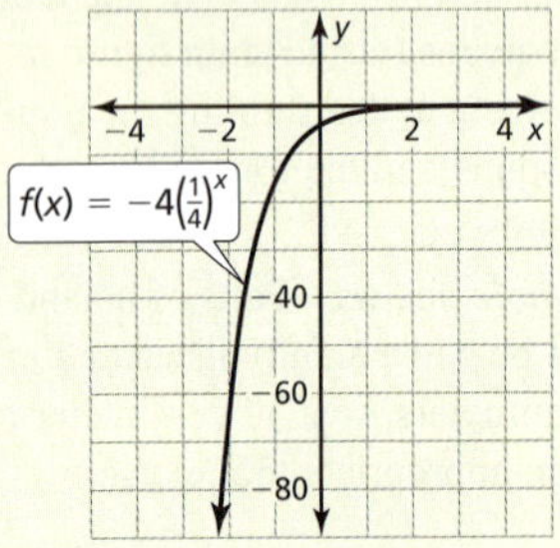

   domain: all real numbers, range: $y < 0$

10. 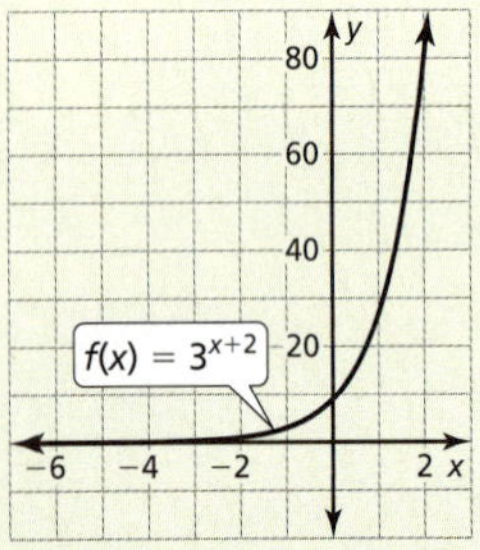

   domain: all real numbers, range: $y > 0$

11. 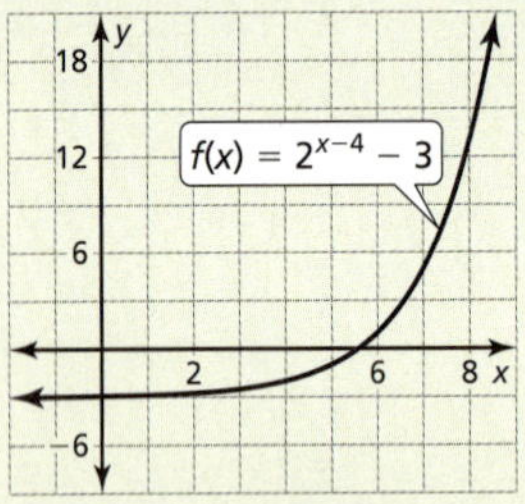

   domain: all real numbers, range: $y > -3$

12. $f(x) = 2\left(\frac{1}{2}\right)^x$

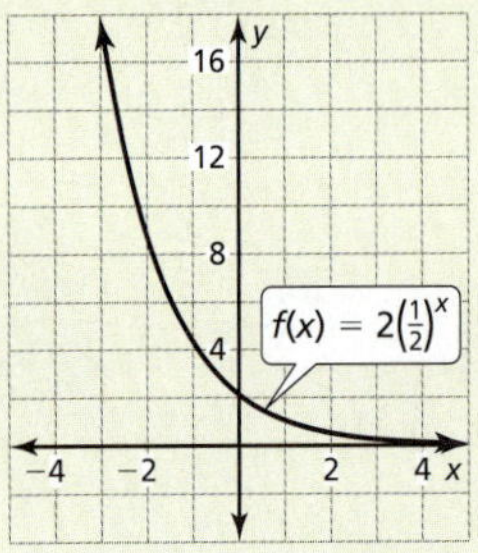

   The graph of f is a vertical stretch by a factor of 2 of the graph of g.

6 Chapter Review

Dynamic Solutions available at *BigIdeasMath.com*

6.1 Properties of Exponents *(pp. 291–298)*

Simplify $\left(\frac{x}{4}\right)^{-4}$. Write your answer using only positive exponents.

$$\left(\frac{x}{4}\right)^{-4} = \frac{x^{-4}}{4^{-4}} \quad \text{Power of a Quotient Property}$$
$$= \frac{4^4}{x^4} \quad \text{Definition of negative exponent}$$
$$= \frac{256}{x^4} \quad \text{Simplify.}$$

Simplify the expression. Write your answer using only positive exponents.

1. $y^3 \cdot y^{-5}$ **2.** $\frac{x^4}{x^7}$ **3.** $(x^0y^2)^3$ **4.** $\left(\frac{2x^2}{5y^4}\right)^{-2}$

6.2 Radicals and Rational Exponents *(pp. 299–304)*

Evaluate $512^{1/3}$.

$$512^{1/3} = \sqrt[3]{512} \quad \text{Rewrite the expression in radical form.}$$
$$= \sqrt[3]{8 \cdot 8 \cdot 8} \quad \text{Rewrite the expression showing factors.}$$
$$= 8 \quad \text{Evaluate the cube root.}$$

Evaluate the expression.

5. $\sqrt[3]{8}$ **6.** $\sqrt[5]{-243}$ **7.** $625^{3/4}$ **8.** $(-25)^{1/2}$

6.3 Exponential Functions *(pp. 305–312)*

Graph $f(x) = 9(3)^x$.

Step 1 Make a table of values.

x	-2	-1	0	1	2
$f(x)$	1	3	9	27	81

Step 2 Plot the ordered pairs.

Step 3 Draw a smooth curve through the points.

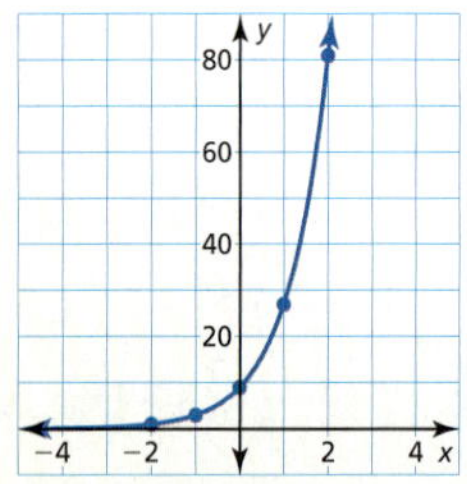

Graph the function. Describe the domain and range.

9. $f(x) = -4\left(\frac{1}{4}\right)^x$ **10.** $f(x) = 3^{x+2}$ **11.** $f(x) = 2^{x-4} - 3$

12. Write and graph an exponential function f represented by the table. Then compare the graph to the graph of $g(x) = \left(\frac{1}{2}\right)^x$.

x	0	1	2	3
y	2	1	0.5	0.25

6.4 Exponential Growth and Decay (pp. 313–322)

Rewrite the function $y = 10(0.65)^{t/8}$ to determine whether it represents *exponential growth* or *exponential decay*. Identify the percent rate of change.

$y = 10(0.65)^{t/8}$ — Write the function.

$= 10(0.65^{1/8})^t$ — Power of a Power Property

$\approx 10(0.95)^t$ — Evaluate the power.

The function is of the form $y = a(1 - r)^t$, where $1 - r < 1$, so it represents exponential decay. Use the decay factor $1 - r$ to find the rate of decay.

$1 - r = 0.95$ — Write an equation.

$r = 0.05$ — Solve for r.

▶ So, the function represents exponential decay, and the rate of decay is 5%.

Determine whether the table represents an *exponential growth function,* an *exponential decay function,* or *neither.* Explain.

13.

x	0	1	2	3
y	3	6	12	24

14.

x	1	2	3	4
y	162	108	72	48

Rewrite the function to determine whether it represents *exponential growth* or *exponential decay.* Identify the percent rate of change.

15. $f(t) = 4(1.25)^{t+3}$

16. $y = (1.06)^{8t}$

17. $f(t) = 6(0.84)^{t-4}$

18. You deposit \$750 in a savings account that earns 5% annual interest compounded quarterly. (a) Write a function that represents the balance after t years. (b) What is the balance of the account after 4 years?

19. The value of a TV is \$1500. Its value decreases by 14% each year. (a) Write a function that represents the value y (in dollars) of the TV after t years. (b) Find the approximate monthly percent decrease in value. (c) Graph the function from part (a). Use the graph to estimate the value of the TV after 3 years.

6.5 Solving Exponential Equations (pp. 325–330)

Solve $\frac{1}{9} = 3^{x+6}$.

$\frac{1}{9} = 3^{x+6}$ — Write the equation.

$3^{-2} = 3^{x+6}$ — Rewrite $\frac{1}{9}$ as 3^{-2}.

$-2 = x + 6$ — Equate the exponents.

$x = -8$ — Solve for x.

Solve the equation.

20. $5^x = 5^{3x-2}$

21. $3^{x-2} = 1$

22. $-4 = 6^{4x-3}$

23. $\left(\frac{1}{3}\right)^{2x+3} = 5$

24. $\left(\frac{1}{16}\right)^{3x} = 64^{2(x+8)}$

25. $27^{2x+2} = 81^{x+4}$

ANSWERS

13. exponential growth; As x increases by 1, y is multiplied by 2.

14. exponential decay; As x increases by 1, y is multiplied by $\frac{2}{3}$.

15. $f(t) \approx 7.81(1.25)^t$, exponential growth; 25%

16. $y \approx 1.59^t$, exponential growth; about 59%

17. $f(t) \approx 12.05(0.84)^t$, exponential decay; 16%

18. **a.** $y = 750(1.0125)^{4t}$

b. \$914.92

19. **a.** $y = 1500(0.86)^t$

b. about 1.2%

c.

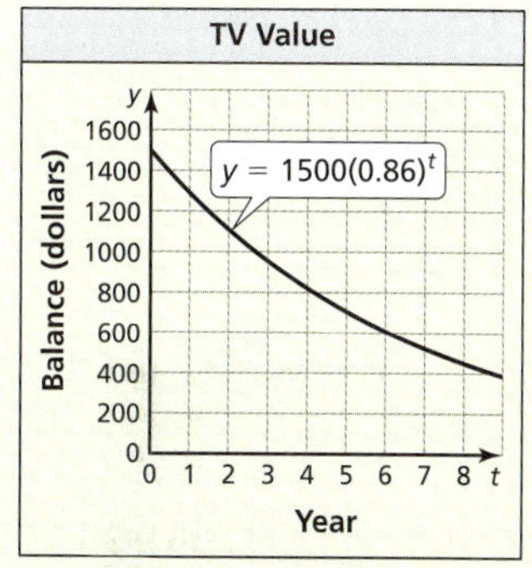

about \$950

20. $x = 1$

21. $x = 2$

22. no solution

23. $x \approx -2.23$

24. $x = -4$

25. $x = 5$

ANSWERS

26. geometric; There is a common ratio of 4; 768, 3072, 12,288

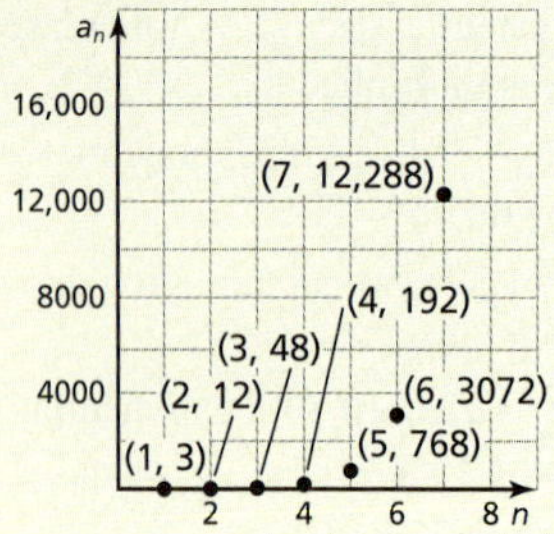

27. neither; There is no common ratio or common difference.

28. geometric; There is a common ratio of $-\frac{1}{5}$; $\frac{3}{5}$, $-\frac{3}{25}$, $\frac{3}{125}$

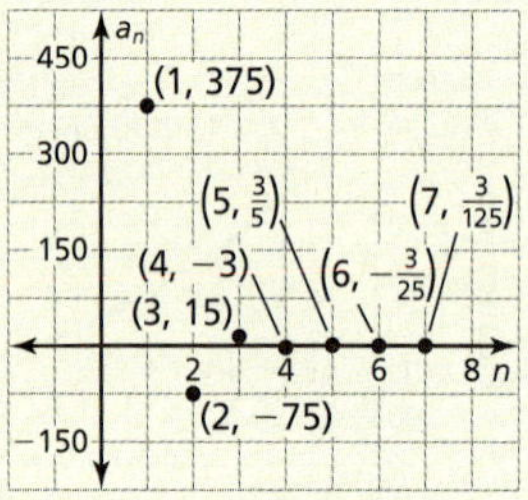

29. $a_n = 4^{n-1}$; 65,536

30. $a_n = 5(-2)^{n-1}$; 1280

31. $a_n = 486\left(\frac{1}{3}\right)^{n-1}$; $\frac{2}{27}$

32. 4, 9, 14, 19, 24, 29

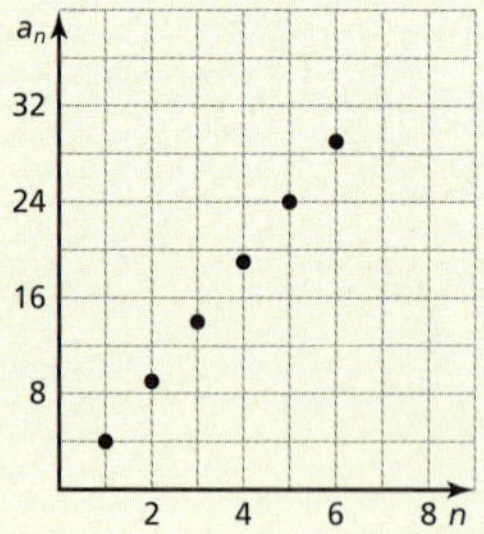

33. −4, 12, −36, 108, −324, 972

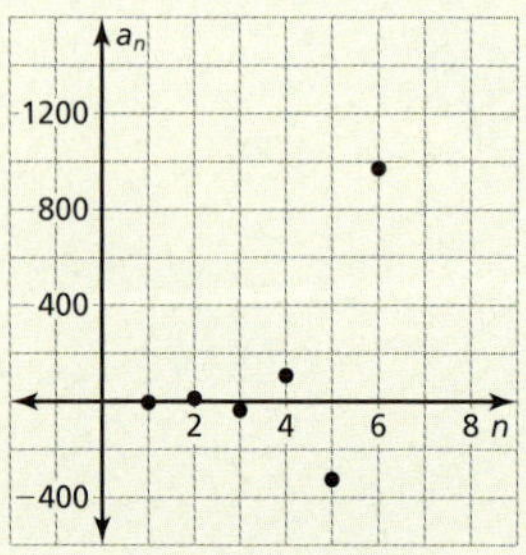

6.6 Geometric Sequences (pp. 331–338)

Write the next three terms of the geometric sequence 2, 6, 18, 54, ...

Use a table to organize the terms and extend the sequence.

Position	1	2	3	4	5	6	7
Term	2	6	18	54	162	486	1458

× 3 × 3 × 3 × 3 × 3 × 3

Each term is 3 times the previous term. So, the common ratio is 3.

Multiply a term by 3 to find the next term.

▶ The next three terms are 162, 486, and 1458.

Decide whether the sequence is *arithmetic*, *geometric*, or *neither*. Explain your reasoning. If the sequence is geometric, write the next three terms and graph the sequence.

26. 3, 12, 48, 192, . . . **27.** 9, −18, 27, −36, . . . **28.** 375, −75, 15, −3, . . .

Write an equation for the *n*th term of the geometric sequence. Then find a_9.

29. 1, 4, 16, 64, . . . **30.** 5, −10, 20, −40, . . . **31.** 486, 162, 54, 18, . . .

6.7 Recursively Defined Sequences (pp. 339–346)

Write a recursive rule for the sequence 5, 12, 19, 26, 33, ...

Use a table to organize the terms and find the pattern.

Position, n	1	2	3	4	5
Term, a_n	5	12	19	26	33

+ 7 + 7 + 7 + 7

The sequence is arithmetic, with first term $a_1 = 5$ and common difference $d = 7$.

$a_n = a_{n-1} + d$ Recursive equation for an arithmetic sequence

$a_n = a_{n-1} + 7$ Substitute 7 for d.

▶ So, a recursive rule for the sequence is $a_1 = 5$, $a_n = a_{n-1} + 7$.

Write the first six terms of the sequence. Then graph the sequence.

32. $a_1 = 4, a_n = a_{n-1} + 5$ **33.** $a_1 = -4, a_n = -3a_{n-1}$ **34.** $a_1 = 32, a_n = \frac{1}{4}a_{n-1}$

Write a recursive rule for the sequence.

35. 3, 8, 13, 18, 23, . . . **36.** 3, 6, 12, 24, 48, . . . **37.** 7, 6, 13, 19, 32, . . .

38. The first term of a sequence is 8. Each term of the sequence is 5 times the preceding term. Graph the first four terms of the sequence. Write a recursive rule and an explicit rule for the sequence.

34. 32, 8, 2, $\frac{1}{2}$, $\frac{1}{8}$, $\frac{1}{32}$

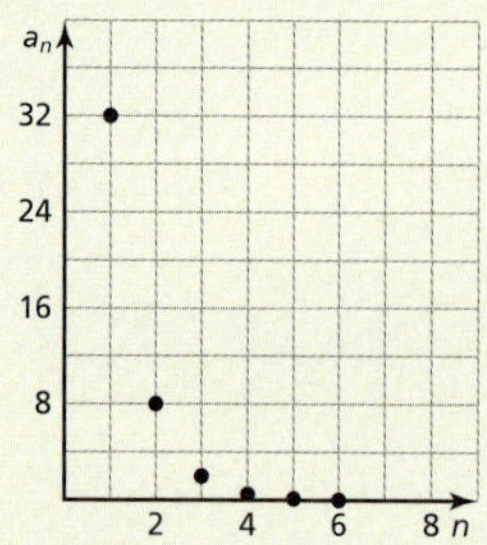

35. $a_1 = 3, a_n = a_{n-1} + 5$

36. $a_1 = 3, a_n = 2a_{n-1}$

37. $a_1 = 7, a_2 = 6, a_n = a_{n-2} + a_{n-1}$

38.

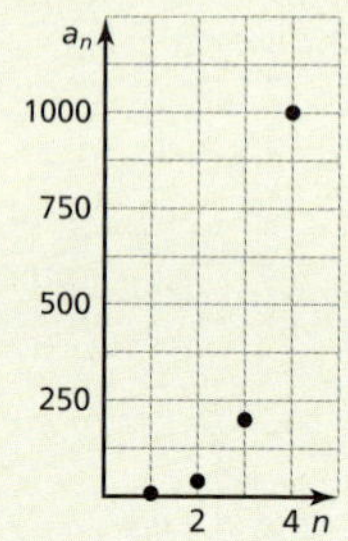

$a_1 = 8, a_n = 5a_{n-1}$; $a_n = 8(5)^{n-1}$

6 Chapter Test

Evaluate the expression.

1. $-\sqrt[4]{16}$
2. $729^{1/6}$
3. $(-32)^{7/5}$

Simplify the expression. Write your answer using only positive exponents.

4. $z^{-2} \cdot z^4$
5. $\dfrac{b^{-5}}{a^0 b^{-8}}$
6. $\left(\dfrac{2c^4}{5}\right)^{-3}$

Write and graph a function that represents the situation.

7. Your starting annual salary of \$42,500 increases by 3% each year.
8. You deposit \$500 in an account that earns 6.5% annual interest compounded yearly.

Write an explicit rule and a recursive rule for the sequence.

9.

n	1	2	3	4
a_n	−6	8	22	36

10.

n	1	2	3	4
a_n	400	100	25	6.25

Solve the equation. Check your solution.

11. $2^x = \dfrac{1}{128}$
12. $256^{x+2} = 16^{3x-1}$
13. Graph $f(x) = 2(6)^x$. Compare the graph to the graph of $g(x) = 6^x$. Describe the domain and range of f.

Use the equation to complete the statement "a ▒ b" with the symbol <, >, or =. Do not attempt to solve the equation.

14. $\dfrac{5^a}{5^b} = 5^{-3}$
15. $9^a \cdot 9^{-b} = 1$
16. The first two terms of a sequence are $a_1 = 3$ and $a_2 = -12$. Let a_3 be the third term when the sequence is arithmetic and let b_3 be the third term when the sequence is geometric. Find $a_3 - b_3$.
17. At sea level, Earth's atmosphere exerts a pressure of 1 atmosphere. Atmospheric pressure P (in atmospheres) decreases with altitude. It can be modeled by $P = (0.99988)^a$, where a is the altitude (in meters).
    a. Identify the initial amount, decay factor, and decay rate.
    b. Use a graphing calculator to graph the function. Use the graph to estimate the atmospheric pressure at an altitude of 5000 feet.
18. You follow the training schedule from your coach.
    a. Write an explicit rule and a recursive rule for the geometric sequence.
    b. On what day do you run approximately 3 kilometers?

Training On Your Own

Day 1: Run 1 km.

Each day after Day 1: Run 20% farther than the previous day.

If students need help...	*If students got it...*
Lesson Tutorials	Resources by Chapter • Enrichment and Extension • Cumulative Review
Skills Review Handbook	Performance Task
BigIdeasMath.com	Start the *next* Section

ANSWERS

1. −2
2. 3
3. −128
4. z^2
5. b^3
6. $\dfrac{125}{8c^{12}}$
7. $y = 42{,}500(1.03)^x$

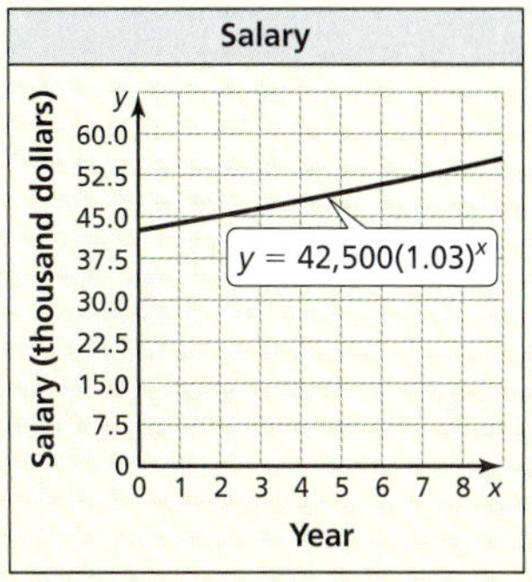

8. $y = 500(1.065)^x$

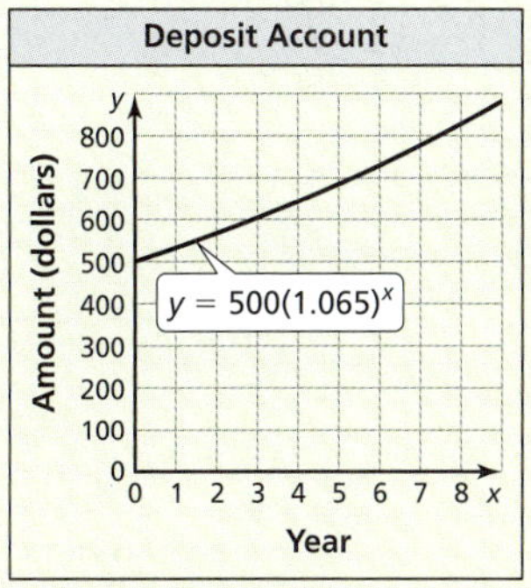

9. $a_n = 14n - 20$; $a_1 = -6$, $a_n = a_{n-1} + 14$
10. $a_n = 400\left(\frac{1}{4}\right)^{n-1}$; $a_1 = 400$, $a_n = \frac{1}{4}a_{n-1}$
11. $x = -7$
12. $x = 5$
13.

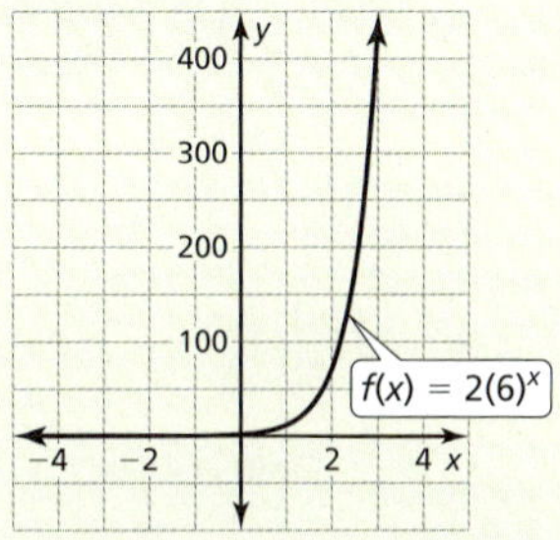

The graph of f is a vertical stretch by a factor of 2 of the graph of g; domain: all real numbers, range: $y > 0$

14. <
15. =
16. −75

17–18. See Additional Answers.

ANSWERS

1. 3
2. 128
3. a. *Sample answer:* $3x + y = 12$ and $3y + 9x = 36$
   b. *Sample answer:* $3x + y = 12$ and $3x + 2y = 12$
   c. *Sample answer:* $3x + y = 12$ and $6x + 2y = 6$
4. <
5. −3, 10; 10, 13

6 Cumulative Assessment

1. Fill in the exponent of x with a number to simplify the expression.

$$\frac{x^{5/3} \cdot x^{-1} \cdot \sqrt[3]{x}}{x^{-2} \cdot x^{0}} = x^{\square}$$

2. The graph of the exponential function f is shown. Find $f(-7)$.

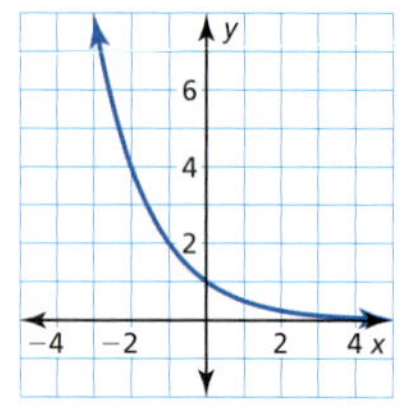

3. Student A claims he can form a linear system from the equations shown that has infinitely many solutions. Student B claims she can form a linear system from the equations shown that has one solution. Student C claims he can form a linear system from the equations shown that has no solution.

$3x + y = 12$	$3x + 2y = 12$	$6x + 2y = 6$
$3y + 9x = 36$	$2y - 6x = 12$	$9x - 3y = -18$

a. Select two equations to support Student A's claim.

b. Select two equations to support Student B's claim.

c. Select two equations to support Student C's claim.

4. Fill in the inequality with $<$, $\leq$, $>$, or $\geq$ so that the system of linear inequalities has no solution.

Inequality 1 $y - 2x \leq 4$

Inequality 2 $6x - 3y \ \square \ -12$

5. The second term of a sequence is 7. Each term of the sequence is 10 more than the preceding term. Fill in values to write a recursive rule and an explicit rule for the sequence.

$a_1 = \square$, $a_n = a_{n-1} + \square$

$a_n = \square\, n - \square$

6. A data set consists of the heights y (in feet) of a hot-air balloon t minutes after it begins its descent. An equation of the line of best fit is $y = 870 - 14.8t$. Which of the following is a correct interpretation of the line of best fit?

Ⓐ The initial height of the hot-air balloon is 870 feet. The slope has no meaning in this context.

Ⓑ The initial height of the hot-air balloon is 870 feet, and it descends 14.8 feet per minute.

Ⓒ The initial height of the hot-air balloon is 870 feet, and it ascends 14.8 feet per minute.

Ⓓ The hot-air balloon descends 14.8 feet per minute. The y-intercept has no meaning in this context.

7. Select all the functions whose x-value is an integer when $f(x) = 10$.

$f(x) = 3x - 2$	$f(x) = -2x + 4$	$f(x) = \frac{3}{2}x + 4$
$f(x) = -3x + 5$	$f(x) = \frac{1}{2}x - 6$	$f(x) = 4x + 14$

8. Place each function into one of the three categories. For exponential functions, state whether the function represents *exponential growth*, *exponential decay*, or *neither.*

Exponential	Linear	Neither
$f(x) = -2(8)^x$	$f(x) = 15 - x$	$f(x) = \frac{1}{2}(3)^x$
$f(x) = 6x^2 + 9$	$f(x) = 4(1.6)^{x/10}$	$f(x) = x(18 - x)$
$f(x) = 3\left(\frac{1}{6}\right)^x$	$f(x) = -3(4x + 1 - x)$	$f(x) = \sqrt[4]{16} + 2x$

9. How does the graph shown compare to the graph of $f(x) = 2^x$?

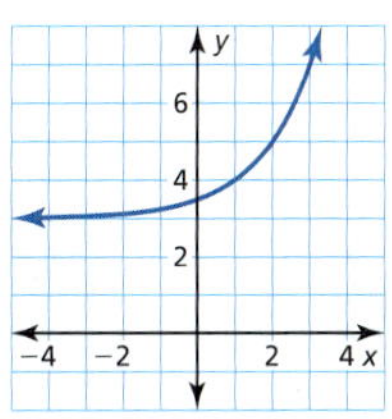

ANSWERS

6. B

7. $f(x) = 3x - 2, f(x) = -2x + 4,$
$f(x) = \frac{3}{2}x + 4, f(x) = \frac{1}{2}x - 6,$
$f(x) = 4x + 14$

8. exponential: $f(x) = -2(8)^x$ neither,
$f(x) = 3\left(\frac{1}{6}\right)^x$ exponential decay,
$f(x) = 4(1.6)^{x/10}$ exponential growth,
$f(x) = \frac{1}{2}(3)^x$ exponential growth;
linear: $f(x) = 15 - x,$
$f(x) = -3(4x + 1 - x),$
$f(x) = \sqrt[4]{16} + 2x;$
neither: $f(x) = 6x^2 + 9,$
$f(x) = x(18 - x)$

9. The graph shown is a horizontal translation 1 unit right and a vertical translation 3 units up of the graph of f.

Chapter 7 Pacing Guide

Chapter Opener/ Mathematical Practices	1 Day
Section 1	1 Day
Section 2	2 Days
Section 3	2 Days
Section 4	2 Days
Quiz	1 Day
Section 5	2 Days
Section 6	2 Days
Section 7	2 Days
Section 8	2 Days
Chapter Review/ Chapter Tests	2 Days
Total Chapter 7	19 Days
Year-to-Date	103 Days

7 Polynomial Equations and Factoring

Height of a Falling Object *(p. 400)*

Game Reserve *(p. 394)*

Photo Cropping *(p. 390)*

Gateway Arch *(p. 382)*

Framing a Photo *(p. 370)*

Laurie's Notes

Chapter Summary

- This is a long chapter about polynomial equations and factoring. It is positioned here in the book in preparation for upcoming work with quadratics.
- In the first few lessons, the vocabulary and representation of polynomials is introduced, along with operations with polynomials. Operations of addition, subtraction, and multiplication are presented.
- The remainder of the chapter is on solving polynomial equations, which can be done when the polynomial is written in factored form. Students will use the Zero-Product Property to solve polynomial equations in factored form. Students will learn a series of techniques for factoring polynomials, aided by visual explorations using algebra tiles.

What Your Students Have Learned

Middle School

- Solve linear equations using the Distributive Property and combining like terms.
- Find the greatest common factor (GCF) of two whole numbers.
- Add, subtract, factor, and expand linear expressions with rational coefficients.
- Know and apply the properties of integer exponents to generate equivalent numerical expressions.

What Your Students Will Learn

Algebra 1

- Classify, add, and subtract polynomials.
- Multiply two binomials using the FOIL Method, the square of a binomial pattern, and the sum and difference pattern.
- Multiply binomials and trinomials.
- Solve polynomial equations by using the Zero-Product Property and by factoring out the greatest common factor (GCF).
- Factor $x^2 + bx + c$, $ax^2 + bx + c$, and perfect square trinomials.
- Factor polynomials as the difference of two squares and by grouping.
- Use factoring to solve real-life problems involving polynomial equations.

Dynamic Teaching Tools

- Dynamic Assessment & Progress Monitoring Tool
- Lesson Planning Tool
- Interactive Whiteboard Lesson Library
- Dynamic Classroom with Dynamic Investigations
- Real-Life STEM Videos

Scaffolding in the Classroom

Graphic Organizers: Concept Circle

A Concept Circle can be used to organize information about a concept. Students write the concept above the circle. Then students write associated information in the sectors of the circle. Associated information can include (an explanation of) *Concept*, *Apply*, *Solve*, *Check*, *Example*, and *Justify*. Concept Circles can have any number of sectors. Students can place their concept circles on note cards to use as a quick study reference.

Questioning in the Classroom

Share ideas.

Have students discuss closure questions or start thinking questions in small groups and then share their answers using a document camera, chart paper, etc. Have different students from each group support their answers.

Laurie's Notes

Maintaining Mathematical Proficiency

Simplifying Algebraic Expressions

- Remind students that when they use the Distributive Property to write a product such as $-8(y - 3)$ as a sum or difference, they must multiply each number inside the parentheses by the factor outside the parentheses.

COMMON ERROR Students may not change the sign of the values inside the parentheses when multiplying by a negative factor. For example, students may rewrite $2m - 7(3 - m)$ in Exercise 5 as $2m - 21 - 7m$ instead of $2m - 21 + 7m$.

Finding the Greatest Common Factor

- Remind students to find the prime factorization of each number to determine the GCF. If the numbers have two or more common prime factors, the GCF is the product of those common prime factors.

COMMON ERROR Students may think that they need to identify only the greatest common prime factor and not all the common prime factors to find the GCF.

Mathematical Practices (continued on page 356)

- The *Mathematical Practices* page focuses attention on how mathematics is learned—process versus content. This page demonstrates that algebra tiles are an appropriate tool for modeling algebraic expressions. Algebra tiles will be used extensively in the lesson explorations in this chapter.
- Use the *Mathematical Practices* page to help students develop mathematical habits of mind—how mathematics can be explored and how mathematics is thought about.

If students need help...	*If students got it...*
Student Journal • Maintaining Mathematical Proficiency	Game Closet at *BigIdeasMath.com*
Lesson Tutorials	Start the *next* Section
Skills Review Handbook	

Maintaining Mathematical Proficiency

Simplifying Algebraic Expressions

Example 1 Simplify $6x + 5 - 3x - 4$.

$6x + 5 - 3x - 4 = 6x - 3x + 5 - 4$	Commutative Property of Addition
$= (6 - 3)x + 5 - 4$	Distributive Property
$= 3x + 1$	Simplify.

Example 2 Simplify $-8(y - 3) + 2y$.

$-8(y - 3) + 2y = -8(y) - (-8)(3) + 2y$	Distributive Property
$= -8y + 24 + 2y$	Multiply.
$= -8y + 2y + 24$	Commutative Property of Addition
$= (-8 + 2)y + 24$	Distributive Property
$= -6y + 24$	Simplify.

Simplify the expression.

1. $3x - 7 + 2x$
2. $4r + 6 - 9r - 1$
3. $-5t + 3 - t - 4 + 8t$
4. $3(s - 1) + 5$
5. $2m - 7(3 - m)$
6. $4(h + 6) - (h - 2)$

Finding the Greatest Common Factor

Example 3 Find the greatest common factor (GCF) of 42 and 70.

To find the GCF of two numbers, first write the prime factorization of each number. Then find the product of the common prime factors.

$42 = 2 \cdot 3 \cdot 7$

$70 = 2 \cdot 5 \cdot 7$

▶ The GCF of 42 and 70 is $2 \cdot 7 = 14$.

Find the greatest common factor.

7. 20, 36
8. 42, 63
9. 54, 81
10. 72, 84
11. 28, 64
12. 30, 77
13. **ABSTRACT REASONING** Is it possible for two integers to have no common factors? Explain your reasoning.

Dynamic Solutions available at *BigIdeasMath.com*

355

What Your Students Have Learned

- Simplify algebraic expressions using the Distributive Property and combining like terms.
- Find the greatest common factor (GCF) of two or more numbers by first writing the prime factorizations of the numbers.

ANSWERS

1. $5x - 7$
2. $-5r + 5$
3. $2t - 1$
4. $3s + 2$
5. $9m - 21$
6. $3h + 26$
7. 4
8. 21
9. 27
10. 12
11. 4
12. 1
13. no; The number 1 is a factor of every integer.

Vocabulary Review

Have students make Idea and Examples Charts for the following terms.

- Commutative Property of Addition
- Distributive Property
- Greatest common factor

MONITORING PROGRESS ANSWERS

1. $3x^2 - x + 1$
2. $-x^2 + x$
3. $-x^2 - 2x$
4. $x^2 + x - 1$
5. $x^2 + 2$
6. $x - 6$
7. $-x^2 + 2x$
8. 9
9. $2x^2$

Mathematical Practices

Mathematically proficient students consider concrete models when solving a mathematics problem.

Using Models

Core Concept

Using Algebra Tiles

When solving a problem, it can be helpful to use a model. For instance, you can use algebra tiles to model algebraic expressions and operations with algebraic expressions.

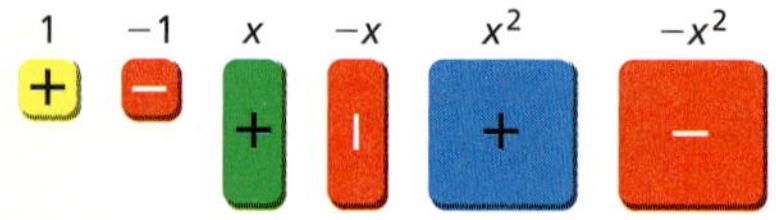

EXAMPLE 1 **Writing Expressions Modeled by Algebra Tiles**

Write the algebraic expression modeled by the algebra tiles.

a. 

b. 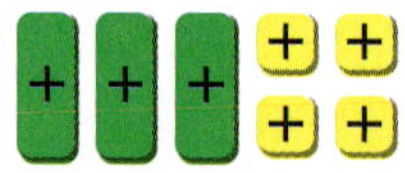

c. 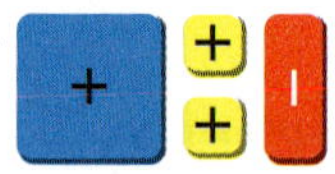

SOLUTION

a. The algebraic expression is x^2.

b. The algebraic expression is $3x + 4$.

c. The algebraic expression is $x^2 - x + 2$.

Monitoring Progress

Write the algebraic expression modeled by the algebra tiles.

1. 

2. 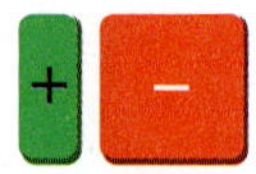

3. 

4. 

5. 

6. 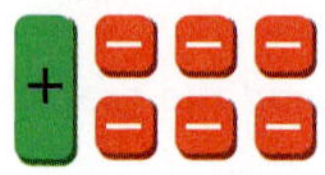

7. 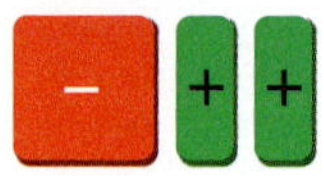

8. 

9. 

Laurie's Notes — Mathematical Practices (continued from page T-355)

- Throughout this chapter, algebra tiles will be used to model operations with algebraic expressions. The tiles can be used to help students understand why only like terms can be added or subtracted. The tiles also help students see a visual representation of two binomials being multiplied. Factoring trinomials can also be modeled with algebra tiles.
- It is assumed that students have worked with algebra tiles before, when learning about integer operations and when solving equations in one variable, as students did earlier in the year.
- Give time for students to work through the questions in the *Monitoring Progress,* and then discuss as a class.

Laurie's Notes

Overview of Section 7.1

Introduction

- This is the first chapter of three about polynomials and quadratics. In this first lesson, the language of polynomials is introduced. This language will help facilitate conversation for the remainder of the year and beyond.
- The second part of the lesson is on addition and subtraction of polynomials. The explorations help students recognize that only like terms can be combined by addition and subtraction.
- **Connection:** Students should recall that when you add or subtract fractions you need a common denominator (like terms). When you add or subtract whole numbers, you add or subtract like place values (like terms).

Formative Assessment Tips

- **Example/Non-Example (teacher generated):** This technique enables students to develop their own definition of a concept by looking at examples and non-examples of a concept. When students compare and contrast the attributes, characteristics, and features of the examples and non-examples, they develop a deeper understanding of what the defining characteristics of the concepts are.
- After examining the examples and non-examples, students are asked to write a definition of the concept presented. This requires very different cognitive skills than simply memorizing a definition given to them. Once the definition is written, students can be asked to generate additional examples supported by their definition.
- In generating the examples, students can also be asked why their example is correct—meaning they provide justification. Listen carefully to students' language as they provide the justification.
- Note that this is an instructional technique used by teachers to introduce concepts, such as making a list of expressions that are monomials and those that are not monomials. It can also be used after instruction to assess or gauge how well students have learned a particular concept.
- This technique can be used by individuals, partners, or small groups.
- Be sure to allow sufficient time so that students receive feedback from others and/or you about their definition(s) and examples.

Applications

- Example 6 uses the vertical motion model (the penny) and the falling object model (the paintbrush). These two models are commonly used in application questions in algebra.

Pacing Suggestion

- The explorations will not take long for students to complete. Transition to the formal lesson when they have finished.

Dynamic Teaching Tools

- Dynamic Assessment & Progress Monitoring Tool
- Lesson Planning Tool
- Interactive Whiteboard Lesson Library
- Dynamic Classroom with Dynamic Investigations

What Your Students Will Learn

- Find the degrees of monomials.
- Classify a polynomial by the number of terms, and identify the degree and leading coefficients.
- Add and subtract polynomials by combining like terms.
- Solve real-life problems involving the addition and subtraction of polynomials.

Laurie's Notes

Exploration

Motivate

- Share a bit of "Did You Know?" with students today.
- The modern addition and subtraction symbols + and − were first used in Germany during the fifteenth century. They have been in use for less than 600 years.
- Ancient Greeks, such as Aristotle, used letters to represent numbers. So, variables were being used long before addition and subtraction symbols.

Discuss

- **Look For and Make Use of Structure:** The shapes of algebra tiles are a visual clue that when you add $2x^2$ and $3x$, you do not get $5x^3$. The shapes are different, so you cannot combine them by addition or subtraction. Mathematically proficient students see that $2x^2 + 3x$ means you have 2 of something (x^2-term) being added to 3 of something different (x-term).

Exploration 1

- Students should be comfortable with integer operations and with using algebra tiles to represent polynomials. You may want to review the addition and subtraction of integers with algebra tiles.
- Once the expressions are added (pushed together), like terms (like tiles) are grouped and zero pairs are removed.
- **Attend to Precision:** Do not allow students to use the phrase "The positive and negative cancel each other, or cancel out." Encourage them to refer to this as the Additive Inverse Property, $a + (-a) = 0$.
- Write the problem and solution: $(3x + 2) + (x - 5) = 4x - 3$.

? "Do you think the sum of two binomials is always a binomial?" Students should be able to give examples for which the sum is not a binomial.

Exploration 2

? "By using algebra tiles to subtract integers, we discovered that subtraction was the same as what?" adding the opposite

- Explain that Step 2 shows how to add the opposite. Change the subtraction symbol to an addition symbol and flip over each of the tiles being subtracted (or swap them out if the tiles are not two-sided).
- Write the problem and solution: $(x^2 + 2x + 2) - (x - 1) = x^2 + x + 3$.

Communicate Your Answer

- You are listening for students to say that you can only add or subtract tiles that are the same size, meaning like terms.

Connecting to Next Step

- These brief explorations will prepare students for the second part of the formal lesson in which students will be adding and subtracting polynomials.

7.1 Adding and Subtracting Polynomials

Essential Question How can you add and subtract polynomials?

EXPLORATION 1 **Adding Polynomials**

Work with a partner. Write the expression modeled by the algebra tiles in each step.

Step 1 $(3x + 2) + (x - 5)$

Step 2

Step 3

Step 4

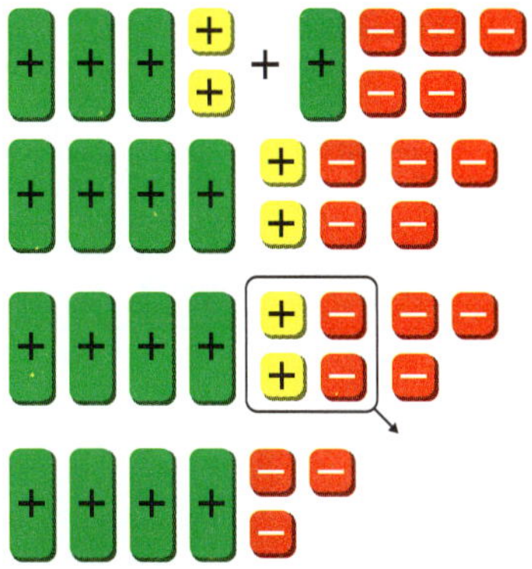

EXPLORATION 2 **Subtracting Polynomials**

Work with a partner. Write the expression modeled by the algebra tiles in each step.

Step 1 $(x^2 + 2x + 2) - (x - 1)$

Step 2

Step 3

Step 4

Step 5

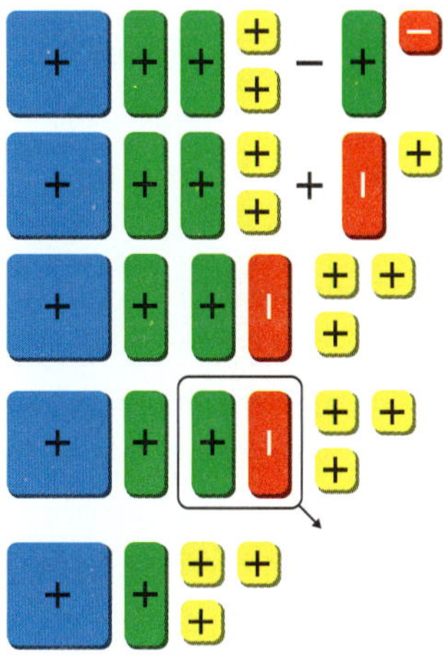

REASONING ABSTRACTLY

To be proficient in math, you need to represent a given situation using symbols.

Communicate Your Answer

3. How can you add and subtract polynomials?

4. Use your methods in Question 3 to find each sum or difference.

a. $(x^2 + 2x - 1) + (2x^2 - 2x + 1)$ **b.** $(4x + 3) + (x - 2)$

c. $(x^2 + 2) - (3x^2 + 2x + 5)$ **d.** $(2x - 3x) - (x^2 - 2x + 4)$

Dynamic Teaching Tools

- Dynamic Assessment & Progress Monitoring Tool
- Lesson Planning Tool
- Interactive Whiteboard Lesson Library
- Dynamic Classroom with Dynamic Investigations

ANSWERS

1. $4x + 2 - 5$; $4x + (2 - 2) - 3$; $4x - 3$

2. $(x^2 + 2x + 2) + (-x + 1)$; $x^2 + 2x - x + 3$; $x^2 + x + (x - x) + 3$; $x^2 + x + 3$

3. Add or subtract like terms.

4. a. $3x^2$

b. $5x + 1$

c. $-2x^2 - 2x - 3$

d. $-x^2 + x - 4$

Extra Example 1

Find the degree of each monomial.

a. $2x$

The degree of the monomial is 1.

b. $\frac{3}{4}x^2y^2$

The degree of the monomial is 4.

c. $5x^2y$

The degree of the monomial is 3.

d. -15

The degree of the monomial is 0.

MONITORING PROGRESS ANSWERS

1. 4
2. 5
3. 1
4. 0

7.1 Lesson

Core Vocabulary

monomial, *p. 358*
degree of a monomial, *p. 358*
polynomial, *p. 359*
binomial, *p. 359*
trinomial, *p. 359*
degree of a polynomial, *p. 359*
standard form, *p. 359*
leading coefficient, *p. 359*
closed, *p. 360*

What You Will Learn

- Find the degrees of monomials.
- Classify polynomials.
- Add and subtract polynomials.
- Solve real-life problems.

Finding the Degrees of Monomials

A **monomial** is a number, a variable, or the product of a number and one or more variables with whole number exponents.

The **degree of a monomial** is the sum of the exponents of the variables in the monomial. The degree of a nonzero constant term is 0. The constant 0 does not have a degree.

Monomial	Degree
10	0
$3x$	1
$\frac{1}{2}ab^2$	$1 + 2 = 3$
$-1.8m^5$	5

Not a monomial	Reason
$5 + x$	A sum is not a monomial.
$\frac{2}{n}$	A monomial cannot have a variable in the denominator.
4^a	A monomial cannot have a variable exponent.
x^{-1}	The variable must have a whole number exponent.

EXAMPLE 1 Finding the Degrees of Monomials

Find the degree of each monomial.

a. $5x^2$ **b.** $-\frac{1}{2}xy^3$ **c.** $8x^3y^3$ **d.** -3

SOLUTION

a. The exponent of x is 2.

▶ So, the degree of the monomial is 2.

b. The exponent of x is 1, and the exponent of y is 3.

▶ So, the degree of the monomial is $1 + 3$, or 4.

c. The exponent of x is 3, and the exponent of y is 3.

▶ So, the degree of the monomial is $3 + 3$, or 6.

d. You can rewrite -3 as $-3x^0$.

▶ So, the degree of the monomial is 0.

Monitoring Progress

Help in English and Spanish at *BigIdeasMath.com*

Find the degree of the monomial.

1. $-3x^4$ **2.** $7c^3d^2$ **3.** $\frac{5}{3}y$ **4.** -20.5

Laurie's Notes Teacher Actions

- **Example/Non-Example:** Make two lists: *Monomials* and *Not Monomials*. Give partners time to write a definition for monomials and degree of a monomial. Ask partners to write other examples of monomials. Use whiteboards to share and discuss results.
- **Turn and Talk** and **Popsicle Sticks:** Have partners discuss the degree of each monomial in Example 1. Use *Popsicle Sticks* to solicit answers.
- **Think-Pair-Share:** Have students answer Questions 1–4, and then share and discuss as a class.

Classifying Polynomials

Core Concept

Polynomials

A **polynomial** is a monomial or a sum of monomials. Each monomial is called a *term* of the polynomial. A polynomial with two terms is a **binomial**. A polynomial with three terms is a **trinomial**.

Binomial	Trinomial
$5x + 2$	$x^2 + 5x + 2$

The **degree of a polynomial** is the greatest degree of its terms. A polynomial in one variable is in **standard form** when the exponents of the terms decrease from left to right. When you write a polynomial in standard form, the coefficient of the first term is the **leading coefficient**.

leading coefficient → 2; degree → 3; constant term → 12

$$2x^3 + x^2 - 5x + 12$$

EXAMPLE 2 Writing a Polynomial in Standard Form

Write $15x - x^3 + 3$ in standard form. Identify the degree and leading coefficient of the polynomial.

SOLUTION

Consider the degree of each term of the polynomial.

$$15x - x^3 + 3$$

Degree is 1. ($15x$) Degree is 3. ($-x^3$) Degree is 0. (3)

▶ You can write the polynomial in standard form as $-x^3 + 15x + 3$. The greatest degree is 3, so the degree of the polynomial is 3, and the leading coefficient is -1.

EXAMPLE 3 Classifying Polynomials

Write each polynomial in standard form. Identify the degree and classify each polynomial by the number of terms.

a. $-3z^4$ **b.** $4 + 5x^2 - x$ **c.** $8q + q^5$

SOLUTION

Polynomial	Standard Form	Degree	Type of Polynomial
a. $-3z^4$	$-3z^4$	4	monomial
b. $4 + 5x^2 - x$	$5x^2 - x + 4$	2	trinomial
c. $8q + q^5$	$q^5 + 8q$	5	binomial

Monitoring Progress

Help in English and Spanish at *BigIdeasMath.com*

Write the polynomial in standard form. Identify the degree and leading coefficient of the polynomial. Then classify the polynomial by the number of terms.

5. $4 - 9z$ **6.** $t^2 - t^3 - 10t$ **7.** $2.8x + x^3$

English Language Learners

Notebook Development

Have students create in their notebooks a polynomial page that includes all the vocabulary words on pages 358 and 359. For each word, have students include the definition and an example.

Extra Example 2

Write $-8 + 9x^2 - 2x$ in standard form. Identify the degree and leading coefficient of the polynomial.

standard form: $9x^2 - 2x - 8$; degree of polynomial: 2; leading coefficient: 9

Extra Example 3

Write each polynomial in standard form. Identify the degree and classify each polynomial by the number of terms.

a. $3 + 7a^2$
standard form: $7a^2 + 3$; degree: 2; type: binomial

b. $-b^2 + 2b^4 + 6b$
standard form: $2b^4 - b^2 + 6b$; degree: 4; type: trinomial

c. $12c^5$
standard form: $12c^5$; degree: 5; type: monomial

MONITORING PROGRESS ANSWERS

5. $-9z + 4$; 1; -9; binomial

6. $-t^3 + t^2 - 10t$; 3; -1; trinomial

7. $x^3 + 2.8x$; 3; 1; binomial

Laurie's Notes Teacher Actions

- Have students read and write key information from the *Core Concept*. You might have students write several examples of each.
- Write Example 2 and have students share all that they know: not in standard form; degree = 3; trinomial; leading coefficient = -1; constant term = 3. As a follow-up, give students information about a polynomial and have them write the polynomial.
- Students can be distracted by the use of different variables. Remind students that polynomials can be written with any variable.

Differentiated Instruction

Auditory

Have students work in pairs. Ask one student to find a sum or difference of polynomials using the vertical format and the other student to find the same sum or difference using the horizontal format. Have students compare their answers. Partners should alternate methods as they continue to solve problems. After they solve several problems, have students tell which method they prefer and why.

Extra Example 4

Find the sum.

a. $(3x^2 + 4x + 3) + (2x + 5x^2 + 1)$
   $8x^2 + 6x + 4$

b. $(-5a^2 + a + 2) + (2a^2 - a - 9)$
   $-3a^2 - 7$

Extra Example 5

Find the difference.

a. $(3y^2 + 8) - (6y^2 + 4y - 2)$
   $-3y^2 - 4y + 10$

b. $(12t^2 + 5t - 7) - (-4t^2 + 3t - 1)$
   $16t^2 + 2t - 6$

Adding and Subtracting Polynomials

A set of numbers is **closed** under an operation when the operation performed on any two numbers in the set results in a number that is also in the set. For example, the set of integers is closed under addition, subtraction, and multiplication. This means that if a and b are two integers, then $a + b$, $a - b$, and ab are also integers.

The set of polynomials is closed under addition and subtraction. So, the sum or difference of any two polynomials is also a polynomial.

To add polynomials, add like terms. You can use a vertical or a horizontal format.

STUDY TIP
When a power of the variable appears in one polynomial but not the other, leave a space in that column, or write the term with a coefficient of 0.

EXAMPLE 4 Adding Polynomials

Find the sum.

a. $(2x^3 - 5x^2 + x) + (2x^2 + x^3 - 1)$ **b.** $(3x^2 + x - 6) + (x^2 + 4x + 10)$

SOLUTION

a. Vertical format: Align like terms vertically and add.

$$\begin{array}{r} 2x^3 - 5x^2 + x \phantom{{}-1} \\ +\ \ x^3 + 2x^2 \phantom{{}+x} - 1 \\ \hline 3x^3 - 3x^2 + x - 1 \end{array}$$

▶ The sum is $3x^3 - 3x^2 + x - 1$.

b. Horizontal format: Group like terms and simplify.

$$(3x^2 + x - 6) + (x^2 + 4x + 10) = (3x^2 + x^2) + (x + 4x) + (-6 + 10)$$
$$= 4x^2 + 5x + 4$$

▶ The sum is $4x^2 + 5x + 4$.

To subtract a polynomial, add its opposite. To find the opposite of a polynomial, multiply each of its terms by -1.

COMMON ERROR
Remember to multiply *each* term of the polynomial by -1 when you write the subtraction as addition.

EXAMPLE 5 Subtracting Polynomials

Find the difference.

a. $(4n^2 + 5) - (-2n^2 + 2n - 4)$ **b.** $(4x^2 - 3x + 5) - (3x^2 - x - 8)$

SOLUTION

a. Vertical format: Align like terms vertically and subtract.

$$\begin{array}{r} 4n^2 \phantom{{}+2n} + 5 \\ -\ (-2n^2 + 2n - 4) \\ \hline \end{array} \quad \Rightarrow \quad \begin{array}{r} 4n^2 \phantom{{}-2n} + 5 \\ +\ \ 2n^2 - 2n + 4 \\ \hline 6n^2 - 2n + 9 \end{array}$$

▶ The difference is $6n^2 - 2n + 9$.

b. Horizontal format: Group like terms and simplify.

$$(4x^2 - 3x + 5) - (3x^2 - x - 8) = 4x^2 - 3x + 5 - 3x^2 + x + 8$$
$$= (4x^2 - 3x^2) + (-3x + x) + (5 + 8)$$
$$= x^2 - 2x + 13$$

▶ The difference is $x^2 - 2x + 13$.

Laurie's Notes Teacher Actions

- Explain that when adding and subtracting polynomials, a vertical or horizontal format can be used, similar to whole-number operations.
- ? In Example 4(a) ask, "Why is a space left when writing $x^3 + 2x^2 - 1$?" so that like terms are lined up
- Subtracting polynomials is generally more challenging for students because they need to be careful with the signs of the terms.
- **Teaching Tip:** Before beginning Example 5(b), ask students to identify like terms.

Monitoring Progress

Help in English and Spanish at *BigIdeasMath.com*

Find the sum or difference.

8. $(b - 10) + (4b - 3)$

9. $(x^2 - x - 2) + (7x^2 - x)$

10. $(p^2 + p + 3) - (-4p^2 - p + 3)$

11. $(-k + 5) - (3k^2 - 6)$

Solving Real-Life Problems

EXAMPLE 6 Solving a Real-Life Problem

A penny is thrown straight down from a height of 200 feet. At the same time, a paintbrush is dropped from a height of 100 feet. The polynomials represent the heights (in feet) of the objects after t seconds.

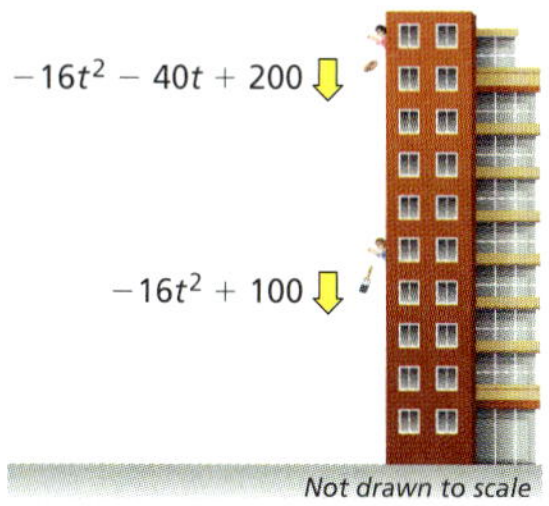

a. Write a polynomial that represents the distance between the penny and the paintbrush after t seconds.

b. Interpret the coefficients of the polynomial in part (a).

SOLUTION

a. To find the distance between the objects after t seconds, subtract the polynomials.

Penny	$-16t^2 - 40t + 200$		$-16t^2 - 40t + 200$
Paintbrush	$-\ (-16t^2 \qquad + 100)$	➡	$+\ \ 16t^2 \qquad - 100$
			$-40t + 100$

▶ The polynomial $-40t + 100$ represents the distance between the objects after t seconds.

b. When $t = 0$, the distance between the objects is $-40(0) + 100 = 100$ feet. So, the constant term 100 represents the distance between the penny and the paintbrush when both objects begin to fall.

As the value of t increases by 1, the value of $-40t + 100$ decreases by 40. This means that the objects become 40 feet closer to each other each second. So, -40 represents the amount that the distance between the objects changes each second.

Monitoring Progress

Help in English and Spanish at *BigIdeasMath.com*

12. WHAT IF? The polynomial $-16t^2 - 25t + 200$ represents the height of the penny after t seconds.

a. Write a polynomial that represents the distance between the penny and the paintbrush after t seconds.

b. Interpret the coefficients of the polynomial in part (a).

Extra Example 6

Use the information in Example 6. Suppose the polynomial $-16t^2 + 75$ represents the height (in feet) of the paintbrush after t seconds.

a. Write a polynomial that represents the distance between the penny and the paintbrush after t seconds.
$-40t + 125$

b. Interpret the coefficients of the polynomial in part (a). The constant term 125 represents the distance between the penny and the paintbrush when both objects begin to fall. The coefficient -40 represents the amount that the distance between the objects changes each second.

MONITORING PROGRESS ANSWERS

8. $5b - 13$

9. $8x^2 - 2x - 2$

10. $5p^2 + 2p$

11. $-3k^2 - k + 11$

12. a. $-25t + 100$

b. The constant term 100 indicates the distance between the penny and the paintbrush is 100 feet when they begin to fall. The coefficient of the linear term -25 indicates that the penny and the paintbrush become 25 feet closer to each other each second.

Laurie's Notes Teacher Actions

- Interpret the meaning of the answer for part (a). The model for the distance between the two objects is $-40t + 100$. This is a linear model. The initial difference is 100 feet, and the difference decreases at a rate of 40 feet per second.
- **Extension:** Make a table of values to evaluate the difference in the distances at 0, 0.5, 1, 1.5, and 2 seconds. Plot the points. The graph is linear with a slope of -40 and a y-intercept of 100.

Closure

- **Exit Ticket:** Write an example of the sum of a trinomial and a binomial that is a binomial.
Sample answer: $(3x^2 - 2x - 4) + (2x - 8) = 3x^2 - 12$

Assignment Guide and Homework Check

ASSIGNMENT

Basic: 1–4, 5–41 odd, 53, 56, 57, 62–64

Average: 1–4, 6–52 even, 53–57, 62–64

Advanced: 1–4, 8, 12, 18–22 even, 26–52 even, 53–64

HOMEWORK CHECK

Basic: 5, 13, 23, 33, 53

Average: 10, 16, 24, 34, 53

Advanced: 12, 18, 28, 36, 54

ANSWERS

1. when the exponents of the terms decrease from left to right
2. *Sample answer:* $2x^5 - 3x + 7$
3. Determine if performing the operation on any two numbers in the set always results in a number that is also in the set.
4. $x^2 - 8^x$; It is the only one that is not a polynomial.
5. 1
6. 4
7. 2
8. 0
9. 9
10. 6
11. 11
12. 11
13. $2c^4 + 6c^2 - c$; 4; 2; trinomial
14. $-w^{12} + 4w^{11}$; 12; -1; binomial
15. $3p^2 + 7$; 2; 3; binomial
16. $-4d^3 + 8d - 2$; 3; -4; trinomial
17. $3t^8$; 8; 3; monomial
18. $3z^4 + 2z^3 + 5z$; 4; 3; trinomial
19. $-\frac{5}{7}r^8 + 2r^5 + \pi r^2$; 8; $-\frac{5}{7}$; trinomial
20. $\sqrt{7}n^4$; 4; $\sqrt{7}$; monomial
21. It is the product of a number, $\frac{4}{3}\pi$, and a variable with a whole number exponent, r^3; 3
22. binomial; 8
23. $3y + 10$
24. $x - 8$
25. $n^2 - 8n + 5$
26. $-4p^3 - 3p^2 - 17p$
27. $6g^2 - 9g + 4$
28. $12r^2 + r - 7$
29. $a^3 - 5a^2 + 4a + 5$
30. $-5s^3 + 2s^2 - s - 9$

7.1 Exercises

Dynamic Solutions available at *BigIdeasMath.com*

Vocabulary and Core Concept Check

1. **VOCABULARY** When is a polynomial in one variable in standard form?
2. **OPEN-ENDED** Write a trinomial in one variable of degree 5 in standard form.
3. **VOCABULARY** How can you determine whether a set of numbers is closed under an operation?
4. **WHICH ONE DOESN'T BELONG?** Which expression does *not* belong with the other three? Explain your reasoning.

$a^3 + 4a$	$x^2 - 8^x$	$b - 2^{-1}$	$-\frac{\pi}{3} + 6y^8z$

Monitoring Progress and Modeling with Mathematics

In Exercises 5–12, find the degree of the monomial. *(See Example 1.)*

5. $4g$
6. $23x^4$
7. $-1.75k^2$
8. $-\frac{4}{9}$
9. s^8t
10. $8m^2n^4$
11. $9xy^3z^7$
12. $-3q^4rs^6$

In Exercises 13–20, write the polynomial in standard form. Identify the degree and leading coefficient of the polynomial. Then classify the polynomial by the number of terms. *(See Examples 2 and 3.)*

13. $6c^2 + 2c^4 - c$
14. $4w^{11} - w^{12}$
15. $7 + 3p^2$
16. $8d - 2 - 4d^3$
17. $3t^8$
18. $5z + 2z^3 + 3z^4$
19. $\pi r^2 - \frac{5}{7}r^8 + 2r^5$
20. $\sqrt{7}n^4$

21. **MODELING WITH MATHEMATICS** The expression $\frac{4}{3}\pi r^3$ represents the volume of a sphere with radius r. Why is this expression a monomial? What is its degree?

22. **MODELING WITH MATHEMATICS** The amount of money you have after investing \$400 for 8 years and \$600 for 6 years at the same interest rate is represented by $400x^8 + 600x^6$, where x is the growth factor. Classify the polynomial by the number of terms. What is its degree?

In Exercises 23–30, find the sum. *(See Example 4.)*

23. $(5y + 4) + (-2y + 6)$
24. $(-8x - 12) + (9x + 4)$
25. $(2n^2 - 5n - 6) + (-n^2 - 3n + 11)$
26. $(-3p^3 + 5p^2 - 2p) + (-p^3 - 8p^2 - 15p)$
27. $(3g^2 - g) + (3g^2 - 8g + 4)$
28. $(9r^2 + 4r - 7) + (3r^2 - 3r)$
29. $(4a - a^3 - 3) + (2a^3 - 5a^2 + 8)$
30. $(s^3 - 2s - 9) + (2s^2 - 6s^3 + s)$

In Exercises 31–38, find the difference. *(See Example 5.)*

31. $(d - 9) - (3d - 1)$
32. $(6x + 9) - (7x + 1)$
33. $(y^2 - 4y + 9) - (3y^2 - 6y - 9)$
34. $(4m^2 - m + 2) - (-3m^2 + 10m + 4)$
35. $(k^3 - 7k + 2) - (k^2 - 12)$
36. $(-r - 10) - (-4r^3 + r^2 + 7r)$

31. $-2d - 8$
32. $-x + 8$
33. $-2y^2 + 2y + 18$
34. $7m^2 - 11m - 2$
35. $k^3 - k^2 - 7k + 14$
36. $4r^3 - r^2 - 8r - 10$

37. $(t^4 - t^2 + t) - (12 - 9t^2 - 7t)$

38. $(4d - 6d^3 + 3d^2) - (10d^3 + 7d - 2)$

ERROR ANALYSIS In Exercises 39 and 40, describe and correct the error in finding the sum or difference.

39.

✗ $(x^2 + x) - (2x^2 - 3x) = x^2 + x - 2x^2 - 3x$
$= (x^2 - 2x^2) + (x - 3x)$
$= -x^2 - 2x$

40.

✗
$$\begin{array}{r} x^3 - 4x^2 + 3 \\ + \; -3x^3 + 8x \; - 2 \\ \hline -2x^3 + 4x^2 + 1 \end{array}$$

41. **MODELING WITH MATHEMATICS** The cost (in dollars) of making b bracelets is represented by $4 + 5b$. The cost (in dollars) of making b necklaces is represented by $8b + 6$. Write a polynomial that represents how much more it costs to make b necklaces than b bracelets.

42. **MODELING WITH MATHEMATICS** The number of individual memberships at a fitness center in m months is represented by $142 + 12m$. The number of family memberships at the fitness center in m months is represented by $52 + 6m$. Write a polynomial that represents the total number of memberships at the fitness center.

In Exercises 43–46, find the sum or difference.

43. $(2s^2 - 5st - t^2) - (s^2 + 7st - t^2)$

44. $(a^2 - 3ab + 2b^2) + (-4a^2 + 5ab - b^2)$

45. $(c^2 - 6d^2) + (c^2 - 2cd + 2d^2)$

46. $(-x^2 + 9xy) - (x^2 + 6xy - 8y^2)$

REASONING In Exercises 47–50, complete the statement with *always*, *sometimes*, or *never*. Explain your reasoning.

47. The terms of a polynomial are ________ monomials.

48. The difference of two trinomials is ________ a trinomial.

49. A binomial is ________ a polynomial of degree 2.

50. The sum of two polynomials is ________ a polynomial.

MODELING WITH MATHEMATICS The polynomial $-16t^2 + v_0t + s_0$ represents the height (in feet) of an object, where v_0 is the initial vertical velocity (in feet per second), s_0 is the initial height of the object (in feet), and t is the time (in seconds). In Exercises 51 and 52, write a polynomial that represents the height of the object. Then find the height of the object after 1 second.

51. You throw a water balloon from a building.

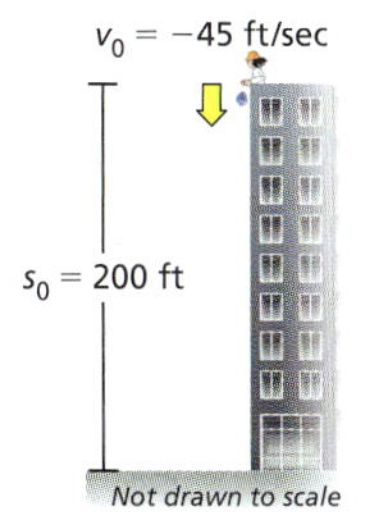

52. You bounce a tennis ball on a racket.

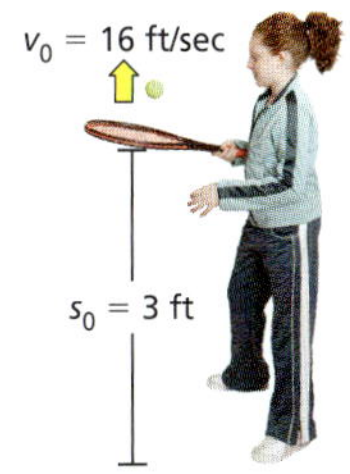

53. **MODELING WITH MATHEMATICS** You drop a ball from a height of 98 feet. At the same time, your friend throws a ball upward. The polynomials represent the heights (in feet) of the balls after t seconds. *(See Example 6.)*

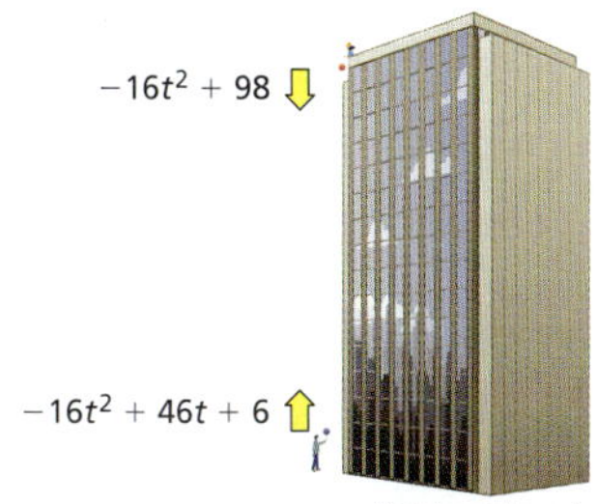

a. Write a polynomial that represents the distance between your ball and your friend's ball after t seconds.

b. Interpret the coefficients of the polynomial in part (a).

Dynamic Teaching Tools

- Dynamic Assessment & Progress Monitoring Tool
- Interactive Whiteboard Lesson Library
- Dynamic Classroom with Dynamic Investigations

ANSWERS

37. $t^4 + 8t^2 + 8t - 12$

38. $-16d^3 + 3d^2 - 3d + 2$

39. When writing the subtraction as addition, the last term of the polynomial was not multiplied by -1; $= (x^2 + x) + (-2x^2 + 3x) = (x^2 - 2x^2) + (x + 3x) = -x^2 + 4x$

40. $-4x^2$ and $8x$ are not like terms, so they cannot be added;
$$\begin{array}{r} x^3 - 4x^2 \qquad + 3 \\ + \; -3x^3 + \qquad 8x - 2 \\ \hline -2x^3 - 4x^2 + 8x + 1 \end{array}$$

41. $3b + 2$

42. $194 + 18m$

43. $s^2 - 12st$

44. $-3a^2 + 2ab + b^2$

45. $2c^2 - 2cd - 4d^2$

46. $-2x^2 + 3xy + 8y^2$

47. always; A polynomial is a monomial or a sum of monomials, and each monomial is a term of the polynomial.

48. sometimes; If like terms have the same coefficient, they will cancel when subtracted so the difference will have fewer than 3 terms. Or, if the terms in the trinomial are not all of the same degree, then the difference could have more than 3 terms.

49. sometimes; The two terms in the binomial can be of any degree.

50. always; Polynomials are closed under addition.

51. $-16t^2 - 45t + 200$; 139 ft

52. $-16t^2 + 16t + 3$; 3 ft

53. a. $-46t + 92$

b. The constant term 92 indicates that the distance between the two balls is 92 feet when they begin. The coefficient of the linear term -46 indicates that the two balls become 46 feet closer to each other each second.

ANSWERS

54. **a.** $-0.028t^3 - 0.32t^2 + 1.6t + 59$

b. \$55.5 million

55. $12x - 3$

56. **a.** $-3x + 3$ or $3x - 3$

b. when $x = 1$; This is the x-coordinate of the point of intersection of the two lines.

57. yes; Addition is commutative and associative, so you can add in any order.

58. $\frac{1}{2}x^2 + \frac{1}{2}, \frac{1}{2}x^2 - \frac{1}{2}$

59–64. See Additional Answers.

Mini-Assessment

1. Find the degree of the monomial $15d^3$. 3
2. Write the polynomial $-8 - x^2 + 2x$ in standard form. Identify the degree and classify the polynomial by the number of terms. standard form: $-x^2 + 2x - 8$; degree: 2; type: trinomial

Find the sum or difference.

3. $(6n^2 + 2n + 3) + (-5n^2 - 4n - 7)$ $n^2 - 2n - 4$
4. $(5x^2 - 4x + 7) - (4x - 7)$ $5x^2 - 8x + 14$
5. A ball is thrown straight down from a height of 300 feet. At the same time, a quarter is dropped from a height of 150 feet. The polynomials represent the heights (in feet) of the objects after t seconds.

   Ball: $-16t^2 - 50t + 300$

   Quarter: $-16t^2 + 150$

   a. Write a polynomial that represents the distance between the ball and the quarter after t seconds. $-50t + 150$

   b. Interpret the coefficients of the polynomial in part (a). The constant term 150 represents the distance between the ball and the quarter when both objects begin to fall. The coefficient -50 represents the amount that the distance between the objects changes each second.

54. MODELING WITH MATHEMATICS During a 7-year period, the amounts (in millions of dollars) spent each year on buying new vehicles N and used vehicles U by United States residents are modeled by the equations

$$N = -0.028t^3 + 0.06t^2 + 0.1t + 17$$
$$U = -0.38t^2 + 1.5t + 42$$

where $t = 1$ represents the first year in the 7-year period.

a. Write a polynomial that represents the total amount spent each year on buying new and used vehicles in the 7-year period.

b. How much is spent on buying new and used vehicles in the fifth year?

55. MATHEMATICAL CONNECTIONS Write the polynomial in standard form that represents the perimeter of the quadrilateral.

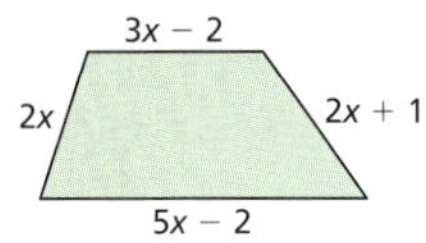

56. HOW DO YOU SEE IT? The right side of the equation of each line is a polynomial.

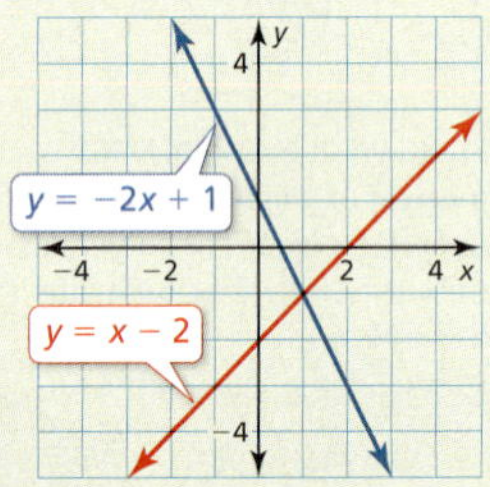

a. The absolute value of the difference of the two polynomials represents the vertical distance between points on the lines with the same x-value. Write this expression.

b. When does the expression in part (a) equal 0? How does this value relate to the graph?

57. MAKING AN ARGUMENT Your friend says that when adding polynomials, the order in which you add does not matter. Is your friend correct? Explain.

58. THOUGHT PROVOKING Write two polynomials whose sum is x^2 and whose difference is 1.

59. REASONING Determine whether the set is closed under the given operation. Explain.

a. the set of negative integers; multiplication

b. the set of whole numbers; addition

60. PROBLEM SOLVING You are building a multi-level deck.

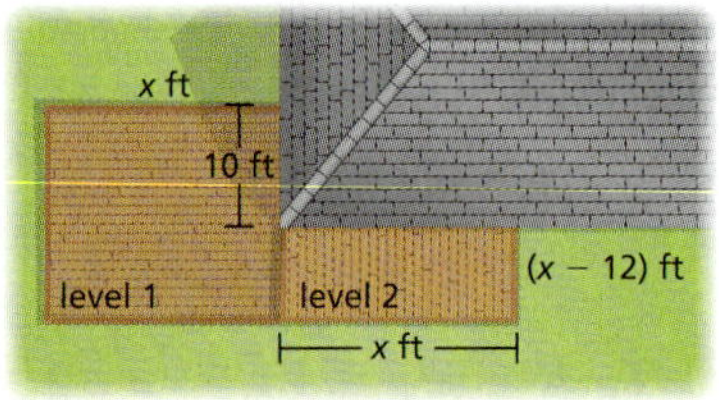

a. For each level, write a polynomial in standard form that represents the area of that level. Then write the polynomial in standard form that represents the total area of the deck.

b. What is the total area of the deck when $x = 20$?

c. A gallon of deck sealant covers 400 square feet. How many gallons of sealant do you need to cover the deck in part (b) once? Explain.

61. PROBLEM SOLVING A hotel installs a new swimming pool and a new hot tub.

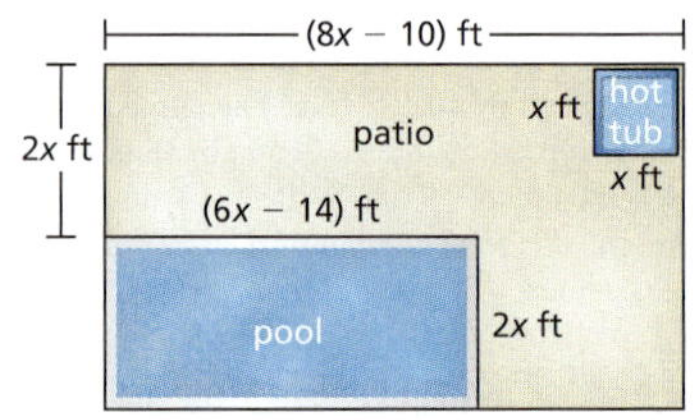

a. Write the polynomial in standard form that represents the area of the patio.

b. The patio will cost \$10 per square foot. Determine the cost of the patio when $x = 9$.

Maintaining Mathematical Proficiency

Reviewing what you learned in previous grades and lessons

Simplify the expression. *(Skills Review Handbook)*

62. $2(x - 1) + 3(x + 2)$ **63.** $8(4y - 3) + 2(y - 5)$ **64.** $5(2r + 1) - 3(-4r + 2)$

If students need help...	*If students got it...*
Resources by Chapter • Practice A and Practice B • Puzzle Time	Resources by Chapter • Enrichment and Extension • Cumulative Review
Student Journal • Practice	Start the *next* Section
Differentiating the Lesson Skills Review Handbook	

Laurie's Notes

Overview of Section 7.2

Introduction

- When students learned to multiply whole numbers, they likely began with an area model that was eventually represented using a box (table) model. This eventually led to an efficient algorithm. There is a connection between how whole numbers are multiplied and how binomials are multiplied.
- This lesson introduces multiplication of binomials, using three different approaches: the Distributive Property, a table, and FOIL (First-Outside-Inside-Last).
- The product of a binomial and a trinomial is also found in this lesson.

Teaching Strategy

- **Connection:** You want to be sure that students see the connection between multiplying two binomials and multiplying two 2-digit whole numbers. There are four multiplications that are performed.
- One early approach to multiplying whole numbers was to use an area model (base-ten blocks), which was later replaced with a simple table.
- A table model is shown in Example 2 for multiplying $(2x - 3)$ by $(x + 5)$.
- The table model can also be used to multiply a binomial by a trinomial. Example 4 is shown below.

	x^2	$-3x$	-2
x	x^3	$-3x^2$	$-2x$
5	$5x^2$	$-15x$	-10

Pacing Suggestion

- The explorations provide an opportunity for students to develop a conceptual understanding of what it means to square a binomial. Transition to the formal lesson as soon as students have discussed each exploration.

Dynamic Teaching Tools

- Dynamic Assessment & Progress Monitoring Tool
- Lesson Planning Tool
- Interactive Whiteboard Lesson Library
- Dynamic Classroom with Dynamic Investigations

What Your Students Will Learn

- Multiply binomials using the Distributive Property and using tables.
- Use the FOIL Method as a way to multiply two binomials.
- Multiply binomials and trinomials using the Distributive Property.

Laurie's Notes

Exploration

Motivate

- **Mental Math Challenge:** Ask students to find the following products using mental math. They should record only the answer.

  8×32 9×48 12×43

- Have students reveal each answer, and ask how they performed the computation. Listen for strategies that likely make use of the Distributive Property.

  $8 \times 32 = 8(30 + 2) = 240 + 16 = 256$

  $9 \times 48 = 9(50 - 2) = 450 - 18 = 432$

  $12 \times 43 = 12(40 + 3) = 480 + 36 = 516$

  OR $12 \times 43 = (10 + 2)(40 + 3)$

  $= 10 \times 40 + 10 \times 3 + 2 \times 40 + 2 \times 3 = 516$

- Make sure students recognize the use of the Distributive Property and that 12×43 can be done two different ways using the Distributive Property.

Exploration Note

- **Make Sense of Problems and Persevere in Solving Them** and **Look For and Make Use of Structure:** Working with algebra tiles in this exploration should be reminiscent of using base-ten blocks when performing multi-digit multiplication. The connection to whole-number multiplication, and how it is represented, should make sense to students. The underlying structure of the problem is the same.

Exploration 1

- In explaining their reasoning, you should expect students to write the problem using symbols, and then explain their answer. For instance, part (e) would be $(x)(-1) = -x$. The explanation might be that multiplying by -1 is the same as taking the opposite of the number.

Exploration 2

- In this exploration, the models for each factor and the product for parts (a) and (b) are shown. It is important to have a conversation about the product in part (a) before students begin. If students understand the area model, they should be able to fill in the product for parts (c) and (d).
- Make the connection to two-digit multiplication. Each binomial has two terms. There are four multiplications to perform, and the products are represented by the tiles in the lower right quadrant of the *t*-diagram.

? "In part (a), what are the factors in this product?" $(x + 3)$ and $(x - 2)$

? "What is the product of *x* and *x*?" x^2 "What is the product of *x* and –2?" $-2x$ "What is the product of 3 and *x*?" $3x$ "What is the product of 3 and -2?" -6

- Remind students to simplify expressions when writing a product.

Communicate Your Answer

- **Neighbor Check:** For Question 4, have students work independently and then have their neighbors check their work. Have students discuss any discrepancies.

Connecting to Next Step

- The explorations help students develop an understanding of the steps in squaring a binomial. When students see the symbolic approach in the formal lesson, they will have a visual model to connect it to.

7.2 Multiplying Polynomials

Essential Question How can you multiply two polynomials?

EXPLORATION 1 **Multiplying Monomials Using Algebra Tiles**

Work with a partner. Write each product. Explain your reasoning.

a. b. c. d. e. f. g. h. i. j.

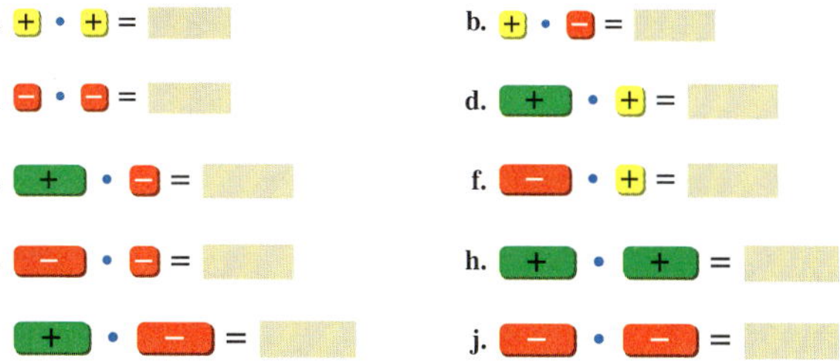

REASONING ABSTRACTLY

To be proficient in math, you need to reason abstractly and quantitatively. You need to pause as needed to recall the meanings of the symbols, operations, and quantities involved.

EXPLORATION 2 **Multiplying Binomials Using Algebra Tiles**

Work with a partner. Write the product of two binomials modeled by each rectangular array of algebra tiles. In parts (c) and (d), first draw the rectangular array of algebra tiles that models each product.

a. $(x + 3)(x - 2) =$

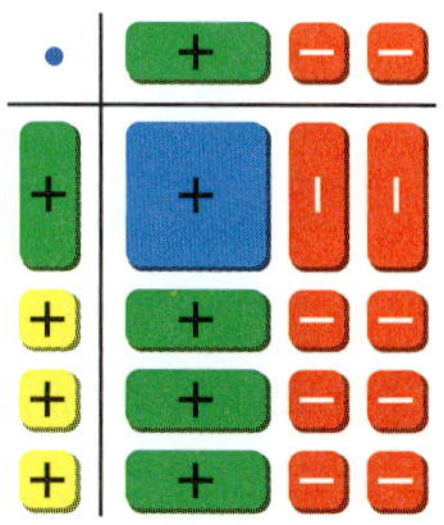

b. $(2x - 1)(2x + 1) =$

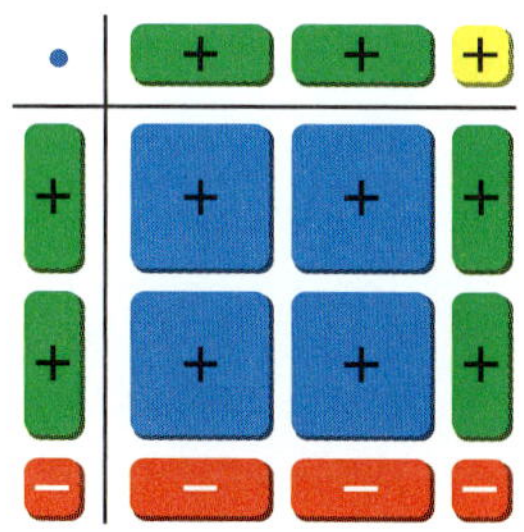

c. $(x + 2)(2x - 1) =$

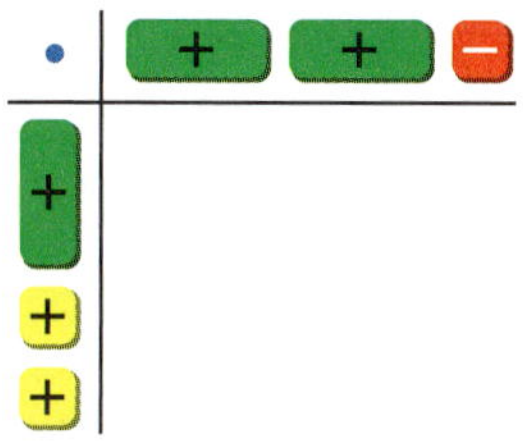

d. $(-x - 2)(x - 3) =$

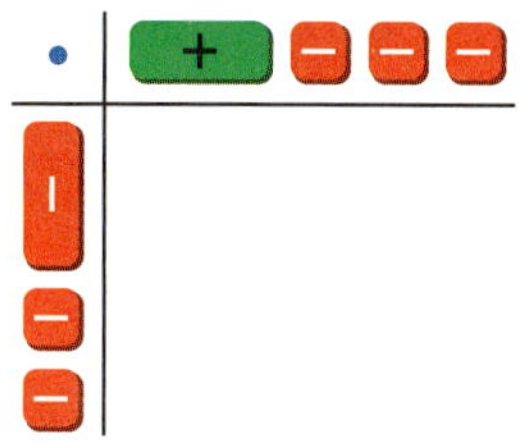

Communicate Your Answer

3. How can you multiply two polynomials?
4. Give another example of multiplying two binomials using algebra tiles that is similar to those in Exploration 2.

Dynamic Teaching Tools

- Dynamic Assessment & Progress Monitoring Tool
- Lesson Planning Tool
- Interactive Whiteboard Lesson Library
- Dynamic Classroom with Dynamic Investigations

ANSWERS

1. a. 1; The product of 1 and 1 is 1.
   b. -1; The product of 1 and -1 is -1.
   c. 1; The product of -1 and -1 is 1.
   d. x; The product of any number and 1 is that number.
   e. $-x$; The product of any number and -1 is the opposite of that number.
   f. $-x$; The product of any number and 1 is that number.
   g. x; The product of the opposite of a number and -1 is the number.
   h. x^2; The product of any number multiplied by itself is the number squared.
   i. $-x^2$; The product of any number and its opposite is the opposite of the number squared.
   j. x^2; The product of any number multiplied by itself is the number squared.
2. a. $x^2 + x - 6$
   b. $4x^2 - 1$
   c.

$2x^2 + 3x - 2$

   d.

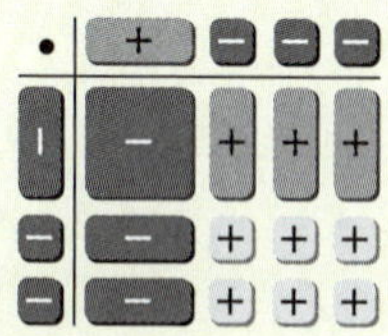

$-x^2 + x + 6$

3. Multiply each term in one polynomial by each term in the other polynomial, then combine like terms.
4. *Sample answer:* $(x + 1)(x - 2) = x^2 - x - 2$

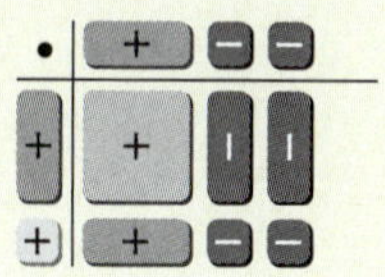

Extra Example 1

a. Find $(x + 8)(x + 2)$.
$x^2 + 10x + 16$

b. Find $(x - 5)(x + 1)$.
$x^2 - 4x - 5$

Extra Example 2

Find $(2x + 3)(x - 3)$.
$2x^2 - 3x - 9$

MONITORING PROGRESS ANSWERS

1. $y^2 + 5y + 4$
2. $z^2 + 4z - 12$
3. $p^2 - 5p - 24$
4. $2r^2 - 11r + 5$

7.2 Lesson

Core Vocabulary

FOIL Method, *p. 367*

Previous
polynomial
closed
binomial
trinomial

What You Will Learn

- Multiply binomials.
- Use the FOIL Method.
- Multiply binomials and trinomials.

Multiplying Binomials

The product of two polynomials is always a polynomial. So, like the set of integers, the set of polynomials is closed under multiplication. You can use the Distributive Property to multiply two binomials.

EXAMPLE 1 Multiplying Binomials Using the Distributive Property

Find (a) $(x + 2)(x + 5)$ and (b) $(x + 3)(x - 4)$.

SOLUTION

a. Use the horizontal method.

$(x + 2)(x + 5) = x(x + 5) + 2(x + 5)$	Distribute $(x + 5)$ to each term of $(x + 2)$.
$= x(x) + x(5) + 2(x) + 2(5)$	Distributive Property
$= x^2 + 5x + 2x + 10$	Multiply.
$= x^2 + 7x + 10$	Combine like terms.

▶ The product is $x^2 + 7x + 10$.

b. Use the vertical method.

	$x + 3$	
	$\times \quad x - 4$	Align like terms vertically.
Multiply $-4(x + 3)$.	$-4x - 12$	Distributive Property
Multiply $x(x + 3)$.	$x^2 + 3x$	Distributive Property
	$x^2 - x - 12$	Combine like terms.

▶ The product is $x^2 - x - 12$.

EXAMPLE 2 Multiplying Binomials Using a Table

Find $(2x - 3)(x + 5)$.

SOLUTION

Step 1 Write each binomial as a sum of terms.

$$(2x - 3)(x + 5) = [2x + (-3)](x + 5)$$

Step 2 Make a table of products.

	$2x$	-3
x	$2x^2$	$-3x$
5	$10x$	-15

▶ The product is $2x^2 - 3x + 10x - 15$, or $2x^2 + 7x - 15$.

Monitoring Progress

Help in English and Spanish at *BigIdeasMath.com*

Use the Distributive Property to find the product.

1. $(y + 4)(y + 1)$
2. $(z - 2)(z + 6)$

Use a table to find the product.

3. $(p + 3)(p - 8)$
4. $(r - 5)(2r - 1)$

Laurie's Notes Teacher Actions

- Point out that multiplying $(x + 2)$ by $(x + 5)$ means that both terms of $(x + 2)$ must be multiplied by $(x + 5)$.
- The order in which the multiplications are performed is not important because of the Commutative Property of Addition. However, in order to make the connection to the FOIL Method, perform the operations in the order shown.
- The box or table model can be used to multiply binomials. See the *Teaching Strategy* on page T-364.

Using the FOIL Method

The **FOIL Method** is a shortcut for multiplying two binomials.

Core Concept

FOIL Method

To multiply two binomials using the FOIL Method, find the sum of the products of the

First terms,	$(x + 1)(x + 2)$	$x(x) = x^2$
Outer terms,	$(x + 1)(x + 2)$	$x(2) = 2x$
Inner terms, and	$(x + 1)(x + 2)$	$1(x) = x$
Last terms.	$(x + 1)(x + 2)$	$1(2) = 2$

$$(x + 1)(x + 2) = x^2 + 2x + x + 2 = x^2 + 3x + 2$$

EXAMPLE 3 Multiplying Binomials Using the FOIL Method

Find each product.

a. $(x - 3)(x - 6)$ **b.** $(2x + 1)(3x - 5)$

SOLUTION

a. Use the FOIL Method.

First Outer Inner Last

$$(x - 3)(x - 6) = x(x) + x(-6) + (-3)(x) + (-3)(-6)$$ FOIL Method

$$= x^2 + (-6x) + (-3x) + 18$$ Multiply.

$$= x^2 - 9x + 18$$ Combine like terms.

▶ The product is $x^2 - 9x + 18$.

b. Use the FOIL Method.

First Outer Inner Last

$$(2x + 1)(3x - 5) = 2x(3x) + 2x(-5) + 1(3x) + 1(-5)$$ FOIL Method

$$= 6x^2 + (-10x) + 3x + (-5)$$ Multiply.

$$= 6x^2 - 7x - 5$$ Combine like terms.

▶ The product is $6x^2 - 7x - 5$.

Monitoring Progress

Help in English and Spanish at *BigIdeasMath.com*

Use the FOIL Method to find the product.

5. $(m - 3)(m - 7)$ **6.** $(x - 4)(x + 2)$

7. $\left(2u + \frac{1}{2}\right)\left(u - \frac{3}{2}\right)$ **8.** $(n + 2)(n^2 + 3)$

Differentiated Instruction

Auditory/Visual

Assign each group a different method for multiplying binominals, using the Distributive Property, a table, or FOIL. Have each student solve a different problem using the assigned method and share the solution within the group. Then have the groups solve the same problems again using the other two methods, and compare solutions.

Extra Example 3

Find each product.

a. $(x + 3)(x - 7)$
$x^2 - 4x - 21$

b. $(2x - 1)(5x - 4)$
$10x^2 - 13x + 4$

MONITORING PROGRESS ANSWERS

5. $m^2 - 10m + 21$

6. $x^2 - 2x - 8$

7. $2u^2 - \frac{5}{2}u - \frac{3}{4}$

8. $n^3 + 2n^2 + 3n + 6$

Laurie's Notes Teacher Actions

- Explain to students that the FOIL Method is simply an acronym to help them remember the four multiplications that need to be performed.
- Some students may be comfortable with saying the process aloud and simply writing the product, skipping the first step.
- **Teaching Tip:** Use your index fingers to point to the two terms being multiplied as you say, "first terms, outer terms, inner terms, and last terms."
- In part (b), students need to pay attention to the coefficients of the x-terms.

Extra Example 4

Find $(x - 3)(x^2 - 4x - 4)$.

$x^3 - 7x^2 + 8x + 12$

Extra Example 5

The purple piece of stained glass has a trapezoidal shape.

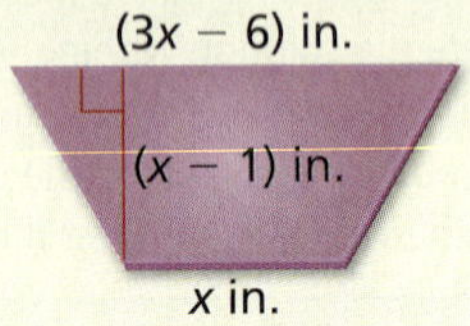

a. Write a polynomial that represents the area of the piece of glass.
$2x^2 - 5x + 3$

b. Find the area of the piece of glass when the shorter base is 7 inches.
66 square inches

MONITORING PROGRESS ANSWERS

9. $x^3 + 6x^2 + 13x + 8$
10. $n^3 - 5n^2 + 10n - 12$
11. It becomes $x^2 - \frac{3}{2}x - \frac{77}{2}$; The longer base becomes $x + 11$. Substituting this value in the formula for the area of a trapezoid along with the other unchanged values changes the linear and constant terms in the polynomial.

Multiplying Binomials and Trinomials

EXAMPLE 4 Multiplying a Binomial and a Trinomial

Find $(x + 5)(x^2 - 3x - 2)$.

SOLUTION

$$\begin{array}{rrrrl} & & x^2 & -3x & -2 \\ \times & & & x & +5 \\ \hline & & 5x^2 & -15x & -10 \\ & x^3 & -3x^2 & -2x & \\ \hline & x^3 & +2x^2 & -17x & -10 \end{array}$$

Multiply $5(x^2 - 3x - 2)$. → $5x^2 - 15x - 10$ — Distributive Property

Multiply $x(x^2 - 3x - 2)$. → $x^3 - 3x^2 - 2x$ — Distributive Property

Align like terms vertically. Combine like terms.

▶ The product is $x^3 + 2x^2 - 17x - 10$.

EXAMPLE 5 Solving a Real-Life Problem

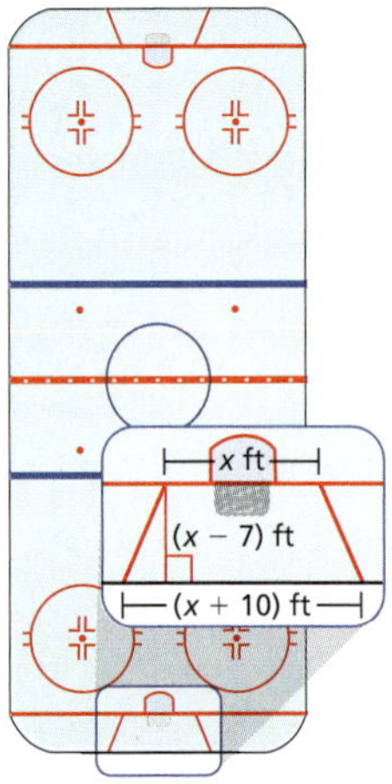

In hockey, a goalie behind the goal line can only play a puck in the trapezoidal region.

a. Write a polynomial that represents the area of the trapezoidal region.

b. Find the area of the trapezoidal region when the shorter base is 18 feet.

SOLUTION

a. $\frac{1}{2}h(b_1 + b_2) = \frac{1}{2}(x - 7)[x + (x + 10)]$ — Substitute.

$= \frac{1}{2}(x - 7)(2x + 10)$ — Combine like terms.

F O I L

$= \frac{1}{2}[2x^2 + 10x + (-14x) + (-70)]$ — FOIL Method

$= \frac{1}{2}(2x^2 - 4x - 70)$ — Combine like terms.

$= x^2 - 2x - 35$ — Distributive Property

▶ A polynomial that represents the area of the trapezoidal region is $x^2 - 2x - 35$.

b. Find the value of $x^2 - 2x - 35$ when $x = 18$.

$x^2 - 2x - 35 = 18^2 - 2(18) - 35$ — Substitute 18 for x.

$= 324 - 36 - 35$ — Simplify.

$= 253$ — Subtract.

▶ The area of the trapezoidal region is 253 square feet.

Monitoring Progress

Help in English and Spanish at *BigIdeasMath.com*

Find the product.

9. $(x + 1)(x^2 + 5x + 8)$
10. $(n - 3)(n^2 - 2n + 4)$
11. **WHAT IF?** In Example 5(a), how does the polynomial change when the longer base is extended by 1 foot? Explain.

Laurie's Notes Teacher Actions

- Solve Example 4 as shown, paying attention to like terms. See the *Teaching Strategy* on page T-364 for an alternate method.
- **Make Sense of Problems and Persevere in Solving Them** and **Thumbs Up:** Discuss the problem in Example 5. Do a *Thumbs Up* self-assessment to see whether students are secure in starting the problem independently.
- Have partners check their work on part (a) before beginning part (b).

Closure

- **Point of Most Significance:** Ask students to identify, aloud or on a paper to be collected, the most significant point (or part) in the lesson that aided their learning.

7.2 Exercises

Dynamic Solutions available at *BigIdeasMath.com*

Vocabulary and Core Concept Check

1. **VOCABULARY** Describe two ways to find the product of two binomials.

2. **WRITING** Explain how the letters of the word FOIL can help you to remember how to multiply two binomials.

Monitoring Progress and Modeling with Mathematics

In Exercises 3–10, use the Distributive Property to find the product. *(See Example 1.)*

3. $(x + 1)(x + 3)$
4. $(y + 6)(y + 4)$
5. $(z - 5)(z + 3)$
6. $(a + 8)(a - 3)$
7. $(g - 7)(g - 2)$
8. $(n - 6)(n - 4)$
9. $(3m + 1)(m + 9)$
10. $(5s + 6)(s - 2)$

In Exercises 11–18, use a table to find the product. *(See Example 2.)*

11. $(x + 3)(x + 2)$
12. $(y + 10)(y - 5)$
13. $(h - 8)(h - 9)$
14. $(c - 6)(c - 5)$
15. $(3k - 1)(4k + 9)$
16. $(5g + 3)(g + 8)$
17. $(-3 + 2j)(4j - 7)$
18. $(5d - 12)(-7 + 3d)$

ERROR ANALYSIS **In Exercises 19 and 20, describe and correct the error in finding the product of the binomials.**

19. ✗

$$(t - 2)(t + 5) = t - 2(t + 5)$$
$$= t - 2t - 10$$
$$= -t - 10$$

20. ✗

$(x - 5)(3x + 1)$

	$3x$	1
x	$3x^2$	x
5	$15x$	5

$(x - 5)(3x + 1) = 3x^2 + 16x + 5$

In Exercises 21–30, use the FOIL Method to find the product. *(See Example 3.)*

21. $(b + 3)(b + 7)$
22. $(w + 9)(w + 6)$
23. $(k + 5)(k - 1)$
24. $(x - 4)(x + 8)$
25. $\left(q - \frac{3}{4}\right)\left(q + \frac{1}{4}\right)$
26. $\left(z - \frac{5}{3}\right)\left(z - \frac{2}{3}\right)$
27. $(9 - r)(2 - 3r)$
28. $(8 - 4x)(2x + 6)$
29. $(w + 5)(w^2 + 3w)$
30. $(v - 3)(v^2 + 8v)$

MATHEMATICAL CONNECTIONS **In Exercises 31–34, write a polynomial that represents the area of the shaded region.**

31.

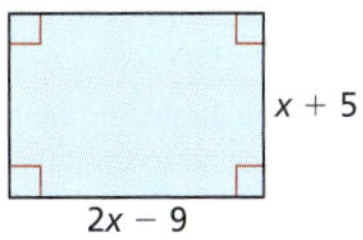

32.

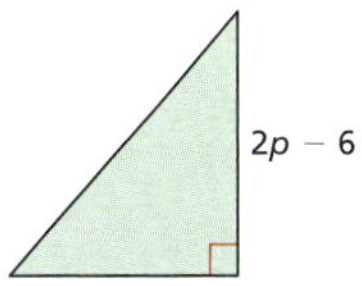

33.

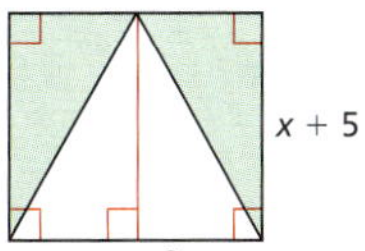

34.

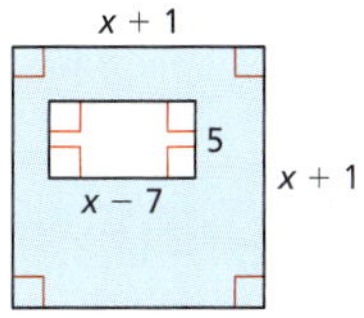

In Exercises 35–42, find the product. *(See Example 4.)*

35. $(x + 4)(x^2 + 3x + 2)$
36. $(f + 1)(f^2 + 4f + 8)$
37. $(y + 3)(y^2 + 8y - 2)$
38. $(t - 2)(t^2 - 5t + 1)$
39. $(4 - b)(5b^2 + 5b - 4)$
40. $(d + 6)(2d^2 - d + 7)$
41. $(3e^2 - 5e + 7)(6e + 1)$
42. $(6v^2 + 2v - 9)(4 - 5v)$

Assignment Guide and Homework Check

ASSIGNMENT

Basic: 1, 2, 3–43 odd, 47, 48, 52–58

Average: 1, 2–44 even, 47, 48, 52–58

Advanced: 1, 2, 8, 10, 16–20 even, 28–34 even, 40–44 even, 45–58

HOMEWORK CHECK

Basic: 3, 13, 21, 35, 43

Average: 4, 14, 22, 38, 44

Advanced: 8, 16, 28, 40, 44

ANSWERS

1. *Sample answer:* Distribute one of the binomials over each term in the other binomial and simplify; Write each binomial as a sum of terms and make a table of products.
2. The letters stand for the sets of terms to multiply: first, outer, inner, and last.
3. $x^2 + 4x + 3$
4. $y^2 + 10y + 24$
5. $z^2 - 2z - 15$
6. $a^2 + 5a - 24$
7. $g^2 - 9g + 14$
8. $n^2 - 10n + 24$
9. $3m^2 + 28m + 9$
10. $5s^2 - 4s - 12$
11. $x^2 + 5x + 6$
12. $y^2 + 5y - 50$
13. $h^2 - 17h + 72$
14. $c^2 - 11c + 30$
15. $12k^2 + 23k - 9$
16. $5g^2 + 43g + 24$
17. $8j^2 - 26j + 21$
18. $15d^2 - 71d + 84$
19. t also should be multiplied by $t + 5$; $= t(t + 5) - 2(t + 5) = t^2 + 5t - 2t - 10 = t^2 + 3t - 10$
20. The 5 in the left column should be -5;

	$3x$	1
x	$3x^2$	x
-5	$-15x$	-5

$(x - 5)(3x + 1) = 3x^2 - 14x - 5$

21. $b^2 + 10b + 21$
22. $w^2 + 15w + 54$
23. $k^2 + 4k - 5$
24. $x^2 + 4x - 32$
25. $q^2 - \frac{1}{2}q - \frac{3}{16}$
26. $z^2 - \frac{7}{3}z + \frac{10}{9}$
27. $3r^2 - 29r + 18$
28. $-8x^2 - 8x + 48$
29. $w^3 + 8w^2 + 15w$
30. $v^3 + 5v^2 - 24v$
31. $2x^2 + x - 45$
32. $p^2 - 2p - 3$
33. $\frac{1}{2}x^2 + \frac{11}{2}x + 15$
34. $x^2 - 3x + 36$
35. $x^3 + 7x^2 + 14x + 8$
36. $f^3 + 5f^2 + 12f + 8$
37. $y^3 + 11y^2 + 22y - 6$
38. $t^3 - 7t^2 + 11t - 2$
39. $-5b^3 + 15b^2 + 24b - 16$
40. $2d^3 + 11d^2 + d + 42$
41. $18e^3 - 27e^2 + 37e + 7$
42. $-30v^3 + 14v^2 + 53v - 36$

Dynamic Teaching Tools

- Dynamic Assessment & Progress Monitoring Tool
- Interactive Whiteboard Lesson Library
- Dynamic Classroom with Dynamic Investigations

ANSWERS

43. a. $(40x^2 + 240x + 200)$ ft^2
    b. $57{,}600$ ft^2
44. a. $(4x^2 + 84x + 440)$ in.^{2}
    b. 840 in.^{2}
45. The degree of the product is the sum of the degrees of each binomial.
46. *Sample answer:* $(2x - 6)(x^2 + 3x + 4)$
47. no; FOIL would leave out the products that include the middle terms of the two trinomials.
48. a. $(-8x - 9)(-4x + 3)$
    b. a and c are positive, b and d are negative.
49. yes; You are both multiplying the same binomials, and neither the order in which you multiply nor the method used will make a difference.
50. $(4x^3 + 9x^2 - x - 6)$ ft^3
51. a. They have the same signs.
    b. They have opposite signs.
52. $y = \begin{cases} -x + 4, & \text{if } x < 0 \\ x + 4, & \text{if } x \geq 0 \end{cases}$
53. $y = \begin{cases} -6x + 18, & \text{if } x < 3 \\ 6x - 18, & \text{if } x \geq 3 \end{cases}$
54. $y = \begin{cases} 4x + 8, & \text{if } x < -2 \\ -4x - 8, & \text{if } x \geq -2 \end{cases}$
55. 10^{11}
56. $\dfrac{1}{x^2}$
57. $\dfrac{1}{27z^{18}}$
58. $\dfrac{1}{4y^2}$

43. MODELING WITH MATHEMATICS The football field is rectangular. *(See Example 5.)*

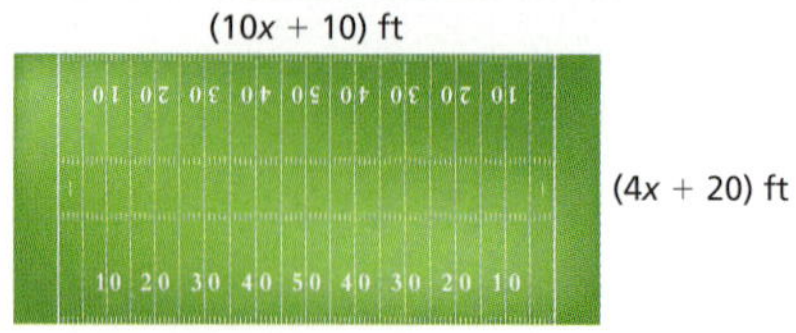

a. Write a polynomial that represents the area of the football field.

b. Find the area of the football field when the width is 160 feet.

44. MODELING WITH MATHEMATICS You design a frame to surround a rectangular photo. The width of the frame is the same on every side, as shown.

a. Write a polynomial that represents the combined area of the photo and the frame.

b. Find the combined area of the photo and the frame when the width of the frame is 4 inches.

45. WRITING When multiplying two binomials, explain how the degree of the product is related to the degree of each binomial.

46. THOUGHT PROVOKING Write two polynomials that are not monomials whose product is a trinomial of degree 3.

47. MAKING AN ARGUMENT Your friend says the FOIL Method can be used to multiply two trinomials. Is your friend correct? Explain your reasoning.

48. HOW DO YOU SEE IT? The table shows one method of finding the product of two binomials.

	$-4x$	3
$-8x$	a	b
-9	c	d

a. Write the two binomials being multiplied.

b. Determine whether a, b, c, and d will be positive or negative when $x > 0$.

49. COMPARING METHODS You use the Distributive Property to multiply $(x + 3)(x - 5)$. Your friend uses the FOIL Method to multiply $(x - 5)(x + 3)$. Should your answers be equivalent? Justify your answer.

50. USING STRUCTURE The shipping container is a rectangular prism. Write a polynomial that represents the volume of the container.

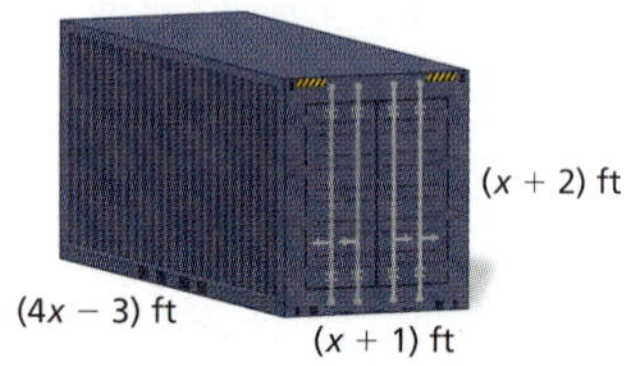

51. ABSTRACT REASONING The product of $(x + m)(x + n)$ is $x^2 + bx + c$.

a. What do you know about m and n when $c > 0$?

b. What do you know about m and n when $c < 0$?

Maintaining Mathematical Proficiency

Reviewing what you learned in previous grades and lessons

Write the absolute value function as a piecewise function. *(Section 4.7)*

52. $y = |x| + 4$
53. $y = 6|x - 3|$
54. $y = -4|x + 2|$

Simplify the expression. Write your answer using only positive exponents. *(Section 6.1)*

55. $10^2 \cdot 10^9$
56. $\dfrac{x^5 \cdot x}{x^8}$
57. $(3z^6)^{-3}$
58. $\left(\dfrac{2y^4}{y^3}\right)^{-2}$

Mini-Assessment

1. Use the Distributive Property to find $(x - 7)(x - 5)$.
   $x^2 - 12x + 35$
2. Use a table to find $(2a + 1)(a + 4)$.
   $2a^2 + 9a + 4$
3. Use the FOIL method to find $(3x - 1)(x + 5)$.
   $3x^2 + 14x - 5$
4. Find $(x - 2)(x^2 - 3x + 1)$.
   $x^3 - 5x^2 + 7x - 2$

If students need help...	*If students got it...*
Resources by Chapter • Practice A and Practice B • Puzzle Time	Resources by Chapter • Enrichment and Extension • Cumulative Review
Student Journal • Practice	Start the *next* Section
Differentiating the Lesson Skills Review Handbook	

Laurie's Notes

Overview of Section 7.3

Introduction

- Multiplication of binomials was presented in the last lesson. In this lesson, special products of binomials are computed: the square of a binomial pattern, and the sum and difference pattern.
- The lesson ends with a real-life application involving genetics.

Formative Assessment Tips

- **Give Me Five:** In the last chapter, students wrote a great deal of private reflections. This technique is public. Students are given a prompt and asked to think and write privately for a couple of minutes. You then ask five (or fewer) students to share aloud their reflection. In doing so, you are (a) continuing to encourage students to be reflective about their learning and (b) letting students know that you value their thoughts and insights.
- The feedback you receive will help guide subsequent instruction. If used at the end of the period, you will have insightful knowledge of students' learning that day and what subsequent instruction may be needed tomorrow. If used during the middle of the class after a critical point, be prepared to make course corrections to address what students have shared.
- As students share, you might have other students with similar thoughts indicate so by a show of hands. If many of the reflections are quite similar, it will not be necessary to have five students share.
- Sample prompts:
  - How did today's lesson help you better understand multiplication of binomials?
  - How well do you feel you understand how to add and subtract polynomial expressions?
  - As we have worked on multiplying polynomials this week, what has helped you the most in making sense of this topic?
  - How did today's explorations aid your understanding of multiplying expressions?

Pacing Suggestion

- The explorations provide an opportunity for students to recognize the patterns when special binomials are squared. Transition to the formal lesson as soon as students have discussed each exploration.

Dynamic Teaching Tools

- Dynamic Assessment & Progress Monitoring Tool
- Lesson Planning Tool
- Interactive Whiteboard Lesson Library
- Dynamic Classroom with Dynamic Investigations

What Your Students Will Learn

- Use the square of a binomial pattern to multiply expressions having the form $(a + b)^2$ or $(a - b)^2$.
- Use the sum and difference pattern to multiply expressions having the form $(a + b)(a - b)$.
- Use special product patterns to solve real-life problems.

Laurie's Notes

Exploration

Motivate

- Challenge your students to a multiplication race! Have the students choose a two-digit number. Then you choose a "compatible" two-digit number so that you can use special product patterns to multiply.
- *Example:* If students choose 42, you choose 38 or 58.

  $42 \times 38 = (40 + 2)(40 - 2) = 40^2 - 2^2 = 1600 - 4 = 1596$

  $42 \times 58 = (50 - 8)(50 + 8) = 50^2 - 8^2 = 2500 - 64 = 2436$
- **FYI:** Note that the number of hundreds is 1 less than a perfect square, and the last two digits are the difference between 100 and a perfect square.
- After several problems, help students to see the patterns that you are using.

Exploration Note

- **Make Sense of Problems and Persevere in Solving Them** and **Use Appropriate Tools Strategically:** Working with algebra tiles in this exploration is a visual and tactile experience that should help students make sense of, and recognize, the patterns that appear when finding special products.

Exploration 1

- In this exploration, the models for each factor and the product are shown.

? Ask the following questions.
  - "What are the factors in part (a)?" $(x + 2)$ and $(x - 2)$
  - "What is the product?" $x^2 - 4$
  - "What happened to the x-terms?" They added to zero.
  - "Did this also happen in part (b)?" yes
  - "How are the binomial factors different in each part?" The operation is different.

Exploration 2

- Students should be secure enough with algebra tiles at this point to be able to model the factors $(x + 2)$ in part (a) and $(2x - 1)$ in part (b).
- **Attend to Precision:** Part (a) is read, "x plus 2 quantity squared" not "x plus 2 squared."
- Students should recognize that there are two factors that are the same.

? Ask the following questions.
  - "What are the factors in part (a)?" $(x + 2)$ and $(x + 2)$
  - "Are the factors the same in each part?" yes
  - "What is the product in part (b)?" $4x^2 - 4x + 1$
  - "Do you see a pattern in the x-terms of the products?" Students may not notice a pattern at this stage, although they may comment on the symmetry of the product along the diagonal.

Communicate Your Answer

- **Neighbor Check:** Have students work independently and then have their neighbors check their work. Have students discuss any discrepancies.

Connecting to Next Step

- The explorations help students recognize patterns in special products of binomials. When students see the symbolic approach in the formal lesson, they will have a visual model to connect it to.

7.3 Special Products of Polynomials

Essential Question What are the patterns in the special products $(a + b)(a - b)$, $(a + b)^2$, and $(a - b)^2$?

EXPLORATION 1 Finding a Sum and Difference Pattern

Work with a partner. Write the product of two binomials modeled by each rectangular array of algebra tiles.

a. $(x + 2)(x - 2) =$ ______ **b.** $(2x - 1)(2x + 1) =$ ______

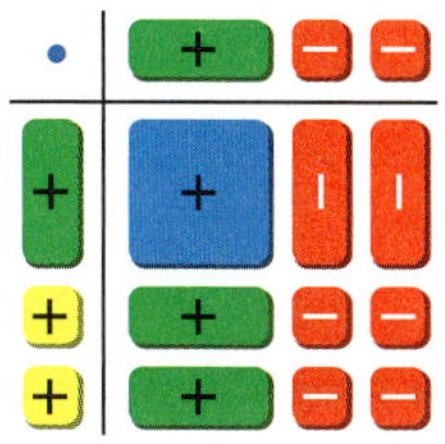

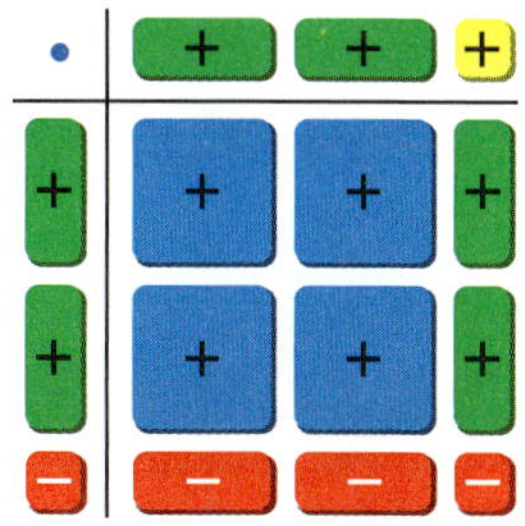

EXPLORATION 2 Finding the Square of a Binomial Pattern

Work with a partner. Draw the rectangular array of algebra tiles that models each product of two binomials. Write the product.

a. $(x + 2)^2 =$ ______ **b.** $(2x - 1)^2 =$ ______

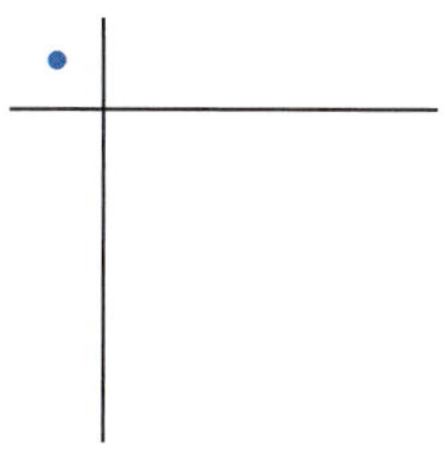

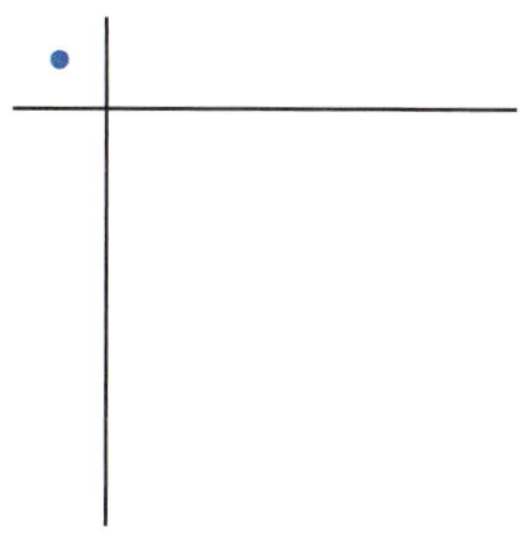

LOOKING FOR STRUCTURE

To be proficient in math, you need to look closely to discern a pattern or structure.

Communicate Your Answer

3. What are the patterns in the special products $(a + b)(a - b)$, $(a + b)^2$, and $(a - b)^2$?

4. Use the appropriate special product pattern to find each product. Check your answers using algebra tiles.

a. $(x + 3)(x - 3)$ **b.** $(x - 4)(x + 4)$ **c.** $(3x + 1)(3x - 1)$

d. $(x + 3)^2$ **e.** $(x - 2)^2$ **f.** $(3x + 1)^2$

Dynamic Teaching Tools

- Dynamic Assessment & Progress Monitoring Tool
- Lesson Planning Tool
- Interactive Whiteboard Lesson Library
- Dynamic Classroom with Dynamic Investigations

ANSWERS

1. a. $x^2 - 4$

b. $4x^2 - 1$

2. a.

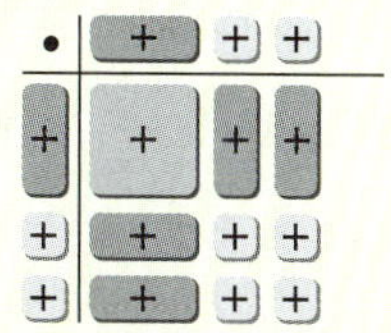

$x^2 + 4x + 4$

b.

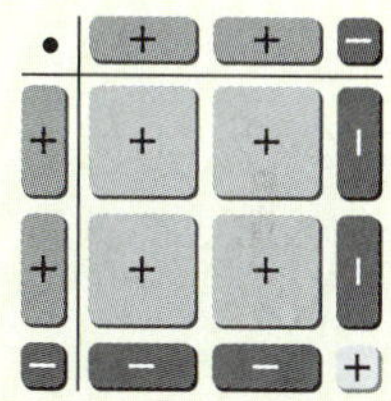

$4x^2 - 4x + 1$

3. $(a + b)(a - b) = a^2 - b^2$; $(a + b)^2 = a^2 + 2ab + b^2$; $(a - b)^2 = a^2 - 2ab + b^2$

4. a. $x^2 - 9$

b. $x^2 - 16$

c. $9x^2 - 1$

d. $x^2 + 6x + 9$

e. $x^2 - 4x + 4$

f. $9x^2 + 6x + 1$

Extra Example 1

Find each product.

a. $(2x + 6)^2$
$4x^2 + 24x + 36$

b. $(3x - y)^2$
$9x^2 - 6xy + y^2$

MONITORING PROGRESS ANSWERS

1. $x^2 + 14x + 49$
2. $49x^2 - 42x + 9$
3. $16x^2 - 8xy + y^2$
4. $9m^2 + 6mn + n^2$

7.3 Lesson

Core Vocabulary

Previous
binomial

What You Will Learn

- Use the square of a binomial pattern.
- Use the sum and difference pattern.
- Use special product patterns to solve real-life problems.

Using the Square of a Binomial Pattern

The diagram shows a square with a side length of $(a + b)$ units. You can see that the area of the square is

$$(a + b)^2 = a^2 + 2ab + b^2.$$

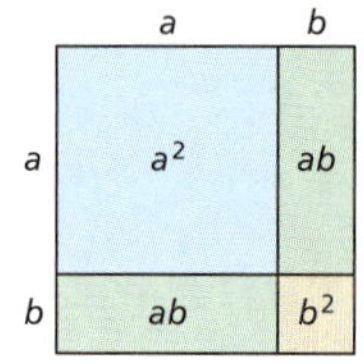

This is one version of a pattern called the square of a binomial. To find another version of this pattern, use algebra: replace b with $-b$.

$(a + (-b))^2 = a^2 + 2a(-b) + (-b)^2$ Replace b with $-b$ in the pattern above.

$(a - b)^2 = a^2 - 2ab + b^2$ Simplify.

Core Concept

Square of a Binomial Pattern

Algebra	Example
$(a + b)^2 = a^2 + 2ab + b^2$	$(x + 5)^2 = (x)^2 + 2(x)(5) + (5)^2$ $= x^2 + 10x + 25$
$(a - b)^2 = a^2 - 2ab + b^2$	$(2x - 3)^2 = (2x)^2 - 2(2x)(3) + (3)^2$ $= 4x^2 - 12x + 9$

LOOKING FOR STRUCTURE

When you use special product patterns, remember that a and b can be numbers, variables, or variable expressions.

EXAMPLE 1 Using the Square of a Binomial Pattern

Find each product.

a. $(3x + 4)^2$ **b.** $(5x - 2y)^2$

SOLUTION

a. $(3x + 4)^2 = (3x)^2 + 2(3x)(4) + 4^2$ Square of a binomial pattern

$= 9x^2 + 24x + 16$ Simplify.

▶ The product is $9x^2 + 24x + 16$.

b. $(5x - 2y)^2 = (5x)^2 - 2(5x)(2y) + (2y)^2$ Square of a binomial pattern

$= 25x^2 - 20xy + 4y^2$ Simplify.

▶ The product is $25x^2 - 20xy + 4y^2$.

Monitoring Progress

Help in English and Spanish at *BigIdeasMath.com*

Find the product.

1. $(x + 7)^2$ **2.** $(7x - 3)^2$ **3.** $(4x - y)^2$ **4.** $(3m + n)^2$

Laurie's Notes Teacher Actions

- Work through the examples as shown. You may want to use color coding to focus attention on the fact that the middle term of the product is twice the product of *ab*.
- **Make Sense of Problems and Persevere in Solving Them:** Some students will expand the square of a binomial and continue to use the FOIL Method. While this is not as efficient, it yields a correct solution and makes sense to the student.

Using the Sum and Difference Pattern

To find the product $(x + 2)(x - 2)$, you can multiply the two binomials using the FOIL Method.

$$(x + 2)(x - 2) = x^2 - 2x + 2x - 4 \quad \text{FOIL Method}$$
$$= x^2 - 4 \quad \text{Combine like terms.}$$

This suggests a pattern for the product of the sum and difference of two terms.

Core Concept

Sum and Difference Pattern

Algebra	Example
$(a + b)(a - b) = a^2 - b^2$	$(x + 3)(x - 3) = x^2 - 9$

EXAMPLE 2 Using the Sum and Difference Pattern

Find each product.

a. $(t + 5)(t - 5)$ **b.** $(3x + y)(3x - y)$

SOLUTION

a. $(t + 5)(t - 5) = t^2 - 5^2$ Sum and difference pattern
$= t^2 - 25$ Simplify.

▶ The product is $t^2 - 25$.

b. $(3x + y)(3x - y) = (3x)^2 - y^2$ Sum and difference pattern
$= 9x^2 - y^2$ Simplify.

▶ The product is $9x^2 - y^2$.

The special product patterns can help you use mental math to find certain products of numbers.

EXAMPLE 3 Using Special Product Patterns and Mental Math

Use special product patterns to find the product $26 \cdot 34$.

SOLUTION

Notice that 26 is 4 less than 30, while 34 is 4 more than 30.

$26 \cdot 34 = (30 - 4)(30 + 4)$ Write as product of difference and sum.
$= 30^2 - 4^2$ Sum and difference pattern
$= 900 - 16$ Evaluate powers.
$= 884$ Simplify.

▶ The product is 884.

Monitoring Progress

Help in English and Spanish at *BigIdeasMath.com*

Find the product.

5. $(x + 10)(x - 10)$ 6. $(2x + 1)(2x - 1)$ 7. $(x + 3y)(x - 3y)$
8. Describe how to use special product patterns to find 21^2.

Differentiated Instruction

Organization

Have students make a chart to show the patterns of special products. Have them include their own examples in the chart.

Patterns of Special Products
Square of a binomial (sum) $(a + b)^2 = a^2 + 2ab + b^2$ Example:
Square of a binomial (difference) $(a - b)^2 = a^2 - 2ab + b^2$ Example:
Sum and difference $(a + b)(a - b) = a^2 - b^2$ Example:

Extra Example 2

Find each product.

a. $(a + 8)(a - 8)$
$a^2 - 64$

b. $(4x + y)(4x - y)$
$16x^2 - y^2$

Extra Example 3

Use special product patterns to find the product $48 \cdot 52$.
$(50 - 2)(50 + 2) = 2500 - 4 = 2496$

MONITORING PROGRESS ANSWERS

5. $x^2 - 100$
6. $4x^2 - 1$
7. $x^2 - 9y^2$
8. Rewrite 21^2 as $(20 + 1)^2$, then use the square of a binomial pattern $(20 + 1)^2 = 20^2 + 2 \cdot 20 \cdot 1 + 1^2 = 400 + 40 + 1 = 441$.

Laurie's Notes Teacher Actions

? **Fact-First Questioning:** "The product of two binomials is not always a trinomial. Why?" Recalling the explorations, students should describe the sum and difference pattern. Write the *Core Concept.*

COMMON ERROR When substituting for *a* in Example 2(b), students may write $3x^2$ instead of $(3x)^2$. Remind them to use parentheses when the coefficient is not 1.

- Example 3 is related to the *Motivate*. Have students practice their mental math skills!

Extra Example 4

Each of two cats has one black patch gene (B) and one tan patch gene (t). Any gene combination with a B results in black patches in an offspring's fur. The Punnett square shows the possible gene combinations of the offspring and the resulting patch colors.

	B	t
B	BB	Bt
t	Bt	tt

a. What percent of the possible gene combinations result in tan patches? 25%

b. Show how you could use a polynomial to model the possible gene combinations of the offspring. $0.25B^2 + 0.5Bt + 0.25t^2$; The coefficient of t^2 shows that 25% of the possible gene combinations result in tan patches.

MONITORING PROGRESS ANSWER

9. a. 25%

b. $(0.5B + 0.5W)^2 = 0.25B^2 + 0.5BW + 0.25W^2$; The coefficients show 25% black, 50% gray, and 25% white.

Solving Real-Life Problems

EXAMPLE 4 Modeling with Mathematics

A combination of two genes determines the color of the dark patches of a border collie's coat. An offspring inherits one patch color gene from each parent. Each parent has two color genes, and the offspring has an equal chance of inheriting either one.

The gene B is for black patches, and the gene r is for red patches. Any gene combination with a B results in black patches. Suppose each parent has the same gene combination Br. The Punnett square shows the possible gene combinations of the offspring and the resulting patch colors.

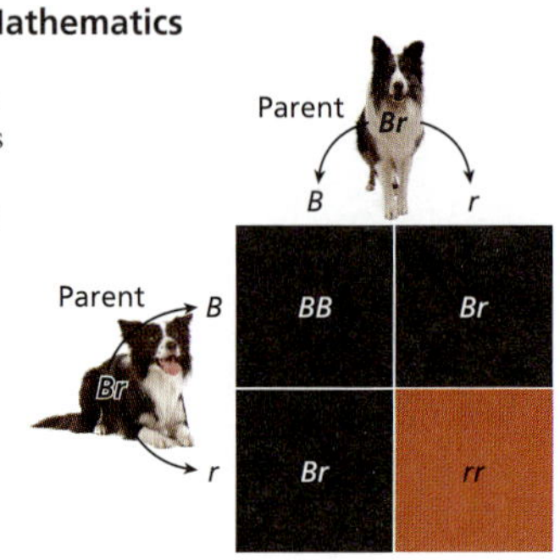

a. What percent of the possible gene combinations result in black patches?

b. Show how you could use a polynomial to model the possible gene combinations.

SOLUTION

a. Notice that the Punnett square shows four possible gene combinations of the offspring. Of these combinations, three result in black patches.

▶ So, 75% of the possible gene combinations result in black patches.

b. Model the gene from each parent with $0.5B + 0.5r$. There is an equal chance that the offspring inherits a black or a red gene from each parent.

You can model the possible gene combinations of the offspring with $(0.5B + 0.5r)^2$. Notice that this product also represents the area of the Punnett square.

Expand the product to find the possible patch colors of the offspring.

$$(0.5B + 0.5r)^2 = (0.5B)^2 + 2(0.5B)(0.5r) + (0.5r)^2$$
$$= 0.25B^2 + 0.5Br + 0.25r^2$$

Consider the coefficients in the polynomial.

$$0.25B^2 + 0.5Br + 0.25r^2$$

25% BB, black patches | 50% Br, black patches | 25% rr, red patches

The coefficients show that 25% + 50% = 75% of the possible gene combinations result in black patches.

Monitoring Progress

Help in English and Spanish at *BigIdeasMath.com*

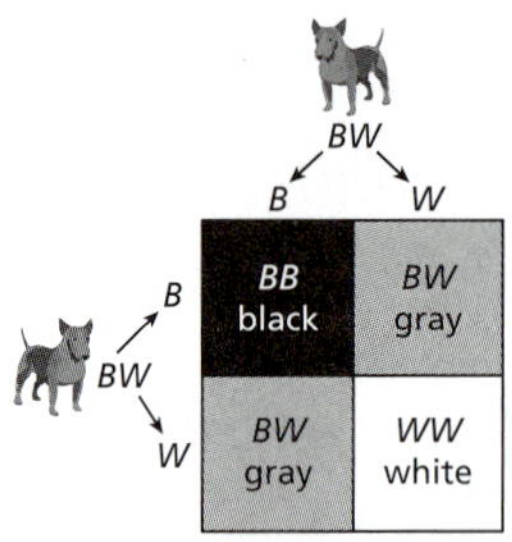

9. Each of two dogs has one black gene (B) and one white gene (W). The Punnett square shows the possible gene combinations of an offspring and the resulting colors.

a. What percent of the possible gene combinations result in black?

b. Show how you could use a polynomial to model the possible gene combinations of the offspring.

Laurie's Notes Teacher Actions

- Ask students whether they have studied genetics in science class. If they have, they will be familiar with the Punnett square, although they may not know it by that name.

COMMON ERROR Students may forget to square the decimal coefficients.

- **Think-Pair-Share:** Have students answer Question 9, and then share and discuss as a class.

Closure

- **Give Me Five:** "How did today's lesson help you better understand multiplication of binomials?"

7.3 Exercises

Dynamic Solutions available at *BigIdeasMath.com*

Vocabulary and Core Concept Check

1. **WRITING** Explain how to use the square of a binomial pattern.

2. **WHICH ONE DOESN'T BELONG?** Which expression does *not* belong with the other three? Explain your reasoning.

$(x + 1)(x - 1)$ $(3x + 2)(3x - 2)$ $(x + 2)(x - 3)$ $(2x + 5)(2x - 5)$

Monitoring Progress and Modeling with Mathematics

In Exercises 3–10, find the product. *(See Example 1.)*

3. $(x + 8)^2$
4. $(a - 6)^2$
5. $(2f - 1)^2$
6. $(5p + 2)^2$
7. $(-7t + 4)^2$
8. $(-12 - n)^2$
9. $(2a + b)^2$
10. $(6x - 3y)^2$

MATHEMATICAL CONNECTIONS **In Exercises 11–14, write a polynomial that represents the area of the square.**

11. 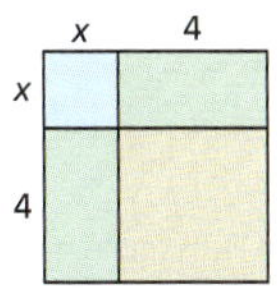

12. 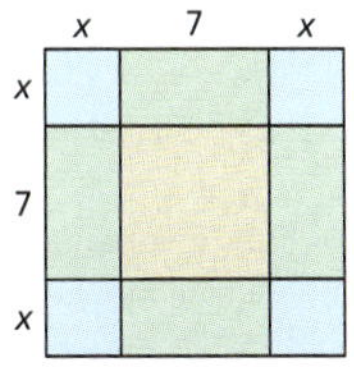

13. 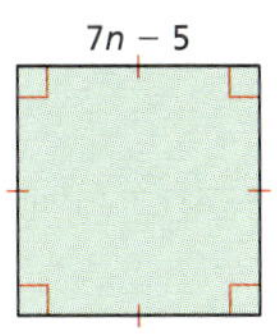

14. 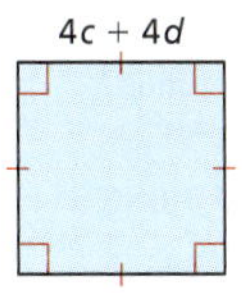

In Exercises 15–24, find the product. *(See Example 2.)*

15. $(t - 7)(t + 7)$
16. $(m + 6)(m - 6)$
17. $(4x + 1)(4x - 1)$
18. $(2k - 4)(2k + 4)$
19. $(8 + 3a)(8 - 3a)$
20. $\left(\frac{1}{2} - c\right)\left(\frac{1}{2} + c\right)$
21. $(p - 10q)(p + 10q)$
22. $(7m + 8n)(7m - 8n)$
23. $(-y + 4)(-y - 4)$
24. $(-5g - 2h)(-5g + 2h)$

In Exercises 25–30, use special product patterns to find the product. *(See Example 3.)*

25. $16 \cdot 24$
26. $33 \cdot 27$
27. 42^2
28. 29^2
29. 30.5^2
30. $10\frac{1}{3} \cdot 9\frac{2}{3}$

ERROR ANALYSIS **In Exercises 31 and 32, describe and correct the error in finding the product.**

31. ✗
$$(k + 4)^2 = k^2 + 4^2$$
$$= k^2 + 16$$

32. ✗
$$(s + 5)(s - 5) = s^2 + 2(s)(5) - 5^2$$
$$= s^2 + 10s - 25$$

33. **MODELING WITH MATHEMATICS** A contractor extends a house on two sides.

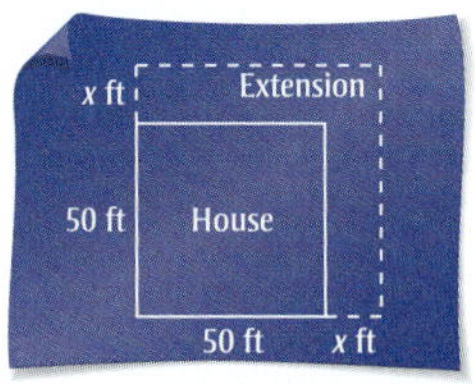

a. The area of the house after the renovation is represented by $(x + 50)^2$. Find this product.

b. Use the polynomial in part (a) to find the area when $x = 15$. What is the area of the extension?

31. The middle term in the square of a binomial pattern was not included; $= k^2 + 2(k)(4) + 4^2 = k^2 + 8k + 16$
32. There is no middle term in the sum and difference pattern; $= s^2 - 5^2 = s^2 - 25$
33. a. $(x^2 + 100x + 2500)$ ft^2
    b. 4225 ft^2; 1725 ft^2

Assignment Guide and Homework Check

ASSIGNMENT

Basic: 1, 2, 3–35 odd, 38, 43, 48–51

Average: 1, 2–34 even, 35–39, 43, 48–51

Advanced: 1, 2, 8–14 even, 22–34 even, 35–37, 38–42 even, 43–51

HOMEWORK CHECK

Basic: 3, 15, 17, 25, 35

Average: 4, 16, 18, 26, 35

Advanced: 10, 22, 24, 28, 35

ANSWERS

1. Substitute the first term of the binomial for a and the second term of the binomial for b in the square of a binomial pattern, then simplify.
2. $(x + 2)(x - 3)$; It is the only one that cannot be simplified using the sum and difference pattern.
3. $x^2 + 16x + 64$
4. $a^2 - 12a + 36$
5. $4f^2 - 4f + 1$
6. $25p^2 + 20p + 4$
7. $49t^2 - 56t + 16$
8. $n^2 + 24n + 144$
9. $4a^2 + 4ab + b^2$
10. $36x^2 - 36xy + 9y^2$
11. $x^2 + 8x + 16$
12. $4x^2 + 28x + 49$
13. $49n^2 - 70n + 25$
14. $16c^2 + 32cd + 16d^2$
15. $t^2 - 49$
16. $m^2 - 36$
17. $16x^2 - 1$
18. $4k^2 - 16$
19. $64 - 9a^2$
20. $\frac{1}{4} - c^2$
21. $p^2 - 100q^2$
22. $49m^2 - 64n^2$
23. $y^2 - 16$
24. $25g^2 - 4h^2$
25. 384
26. 891
27. 1764
28. 841
29. 930.25
30. $99\frac{8}{9}$

Dynamic Teaching Tools

- Dynamic Assessment & Progress Monitoring Tool
- Interactive Whiteboard Lesson Library
- Dynamic Classroom with Dynamic Investigations

ANSWERS

34. **a.** $(10{,}000 - x^2)$ ft^2

b. decrease; The original area is 10,000 ft^2, so the new area is the original area decreased by x^2.

c. 9559 ft^2

35. **a.** 25%

b. $(0.5N + 0.5a)^2 = 0.25N^2 + 0.5Na + 0.25a^2$

36–51. See Additional Answers.

Mini-Assessment

Find the product.

1. $(4x + 1)^2$ $16x^2 + 8x + 1$
2. $(3a - 5b)^2$ $9a^2 - 30ab + 25b^2$
3. $(2d + 3)(2d - 3)$ $4d^2 - 9$
4. Use special product patterns to find the product 43 • 37. $(40 + 3)(40 - 3) = 1600 - 9 = 1591$
5. Each of two pea plants has one tall gene (T) and one short gene (s). Any gene combination with a T results in a tall plant. The Punnett square shows the possible gene combinations of the offspring and the resulting plant heights.

	T	*s*
T	*TT*	*Ts*
s	*Ts*	*ss*

a. What percent of the possible gene combinations result in tall plants? 75%

b. Show how you could use a polynomial to model the possible gene combinations of the offspring. $0.25T^2 + 0.5Ts + 0.25s^2$; The coefficients show that 25% + 50% = 75% of the possible gene combinations result in tall plants.

34. MODELING WITH MATHEMATICS A square-shaped parking lot with 100-foot sides is reduced by x feet on one side and extended by x feet on an adjacent side.

a. The area of the new parking lot is represented by $(100 - x)(100 + x)$. Find this product.

b. Does the area of the parking lot increase, decrease, or stay the same? Explain.

c. Use the polynomial in part (a) to find the area of the new parking lot when $x = 21$.

35. MODELING WITH MATHEMATICS In deer, the gene N is for normal coloring and the gene a is for no coloring, or albino. Any gene combination with an N results in normal coloring. The Punnett square shows the possible gene combinations of an offspring and the resulting colors from parents that both have the gene combination Na. *(See Example 4.)*

a. What percent of the possible gene combinations result in albino coloring?

b. Show how you could use a polynomial to model the possible gene combinations of the offspring.

Parent A: *Na* → *N*, *a*; Parent B: *Na* → *N*, *a*

	N	*a*
N	*NN* normal	*Na* normal
a	*Na* normal	*aa* albino

36. MODELING WITH MATHEMATICS Your iris controls the amount of light that enters your eye by changing the size of your pupil.

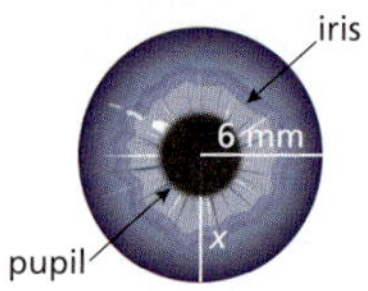

a. Write a polynomial that represents the area of your pupil. Write your answer in terms of π.

b. The width x of your iris decreases from 4 millimeters to 2 millimeters when you enter a dark room. How many times greater is the area of your pupil after entering the room than before entering the room? Explain.

37. CRITICAL THINKING Write two binomials that have the product $x^2 - 121$. Explain.

38. HOW DO YOU SEE IT? In pea plants, any gene combination with a green gene (G) results in a green pod. The Punnett square shows the possible gene combinations of the offspring of two Gy pea plants and the resulting pod colors.

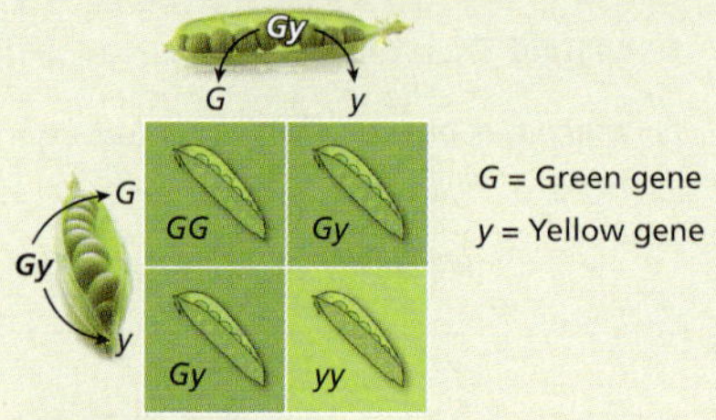

A polynomial that models the possible gene combinations of the offspring is

$$(0.5G + 0.5y)^2 = 0.25G^2 + 0.5Gy + 0.25y^2.$$

Describe two ways to determine the percent of possible gene combinations that result in green pods.

In Exercises 39–42, find the product.

39. $(x^2 + 1)(x^2 - 1)$ **40.** $(y^3 + 4)^2$

41. $(2m^2 - 5n^2)^2$ **42.** $(r^3 - 6t^4)(r^3 + 6t^4)$

43. MAKING AN ARGUMENT Your friend claims to be able to use a special product pattern to determine that $\left(4\frac{1}{3}\right)^2$ is equal to $16\frac{1}{9}$. Is your friend correct? Explain.

44. THOUGHT PROVOKING The area (in square meters) of the surface of an artificial lake is represented by x^2. Describe three ways to modify the dimensions of the lake so that the new area can be represented by the three types of special product patterns discussed in this section.

45. REASONING Find k so that $9x^2 - 48x + k$ is the square of a binomial.

46. REPEATED REASONING Find $(x + 1)^3$ and $(x + 2)^3$. Find a pattern in the terms and use it to write a pattern for the cube of a binomial $(a + b)^3$.

47. PROBLEM SOLVING Find two numbers a and b such that $(a + b)(a - b) < (a - b)^2 < (a + b)^2$.

Maintaining Mathematical Proficiency

Reviewing what you learned in previous grades and lessons

Factor the expression using the GCF. *(Skills Review Handbook)*

48. $12y - 18$ **49.** $9r + 27$ **50.** $49s + 35t$ **51.** $15x - 10y$

If students need help...	*If students got it...*
Resources by Chapter • Practice A and Practice B • Puzzle Time	Resources by Chapter • Enrichment and Extension • Cumulative Review
Student Journal • Practice	Start the *next* Section
Differentiating the Lesson Skills Review Handbook	

Laurie's Notes

Overview of Section 7.4

Introduction

- In this lesson, students will solve polynomial equations by using the Zero-Product Property. If the polynomial is already factored and set equal to zero, the equation can be solved by setting each of the factors equal to 0 and solving. If the equation is not in factored form, it can be rewritten by factoring out a common factor.
- Additional factoring techniques are presented in the next lesson.

Teaching Strategy

- In this lesson and in the remaining lessons in this chapter, students will be solving polynomials by factoring an expression and setting it equal to 0. Note that other techniques besides factoring will be presented in later chapters.
- When factors are set equal to 0, the Zero-Product Property is used to solve the polynomial. Many of the factors will be of the form $(ax + b)$, so students need to be proficient with solving $ax + b = 0$, where a and b are real numbers.
- Spend time solving a variety of equivalent forms so that students are efficient and accurate with their solutions, eventually using eyesight (inspection) to solve.

Examples: $-5x + 7 = 0$ $\quad 8 - 3x = 0$ $\quad 6 + \frac{1}{2}x = 0$

$-\frac{2}{3}x - 8 = 0$ $\quad 0.1x + 9 = 0$ $\quad -5 - 4x = 0$

Pacing Suggestion

- Complete at least the first two explorations before beginning the formal lesson. The third exploration probes prior learning and connects to the Zero-Product Property.

Dynamic Teaching Tools

- Dynamic Assessment & Progress Monitoring Tool
- Lesson Planning Tool
- Interactive Whiteboard Lesson Library
- Dynamic Classroom with Dynamic Investigations

What Your Students Will Learn

- Use the Zero-Product Property to solve polynomial equations in factored form.
- Factor polynomials by finding the greatest common factor (GCF).
- Use the Zero-Product Property to solve real-life problems involving polynomial equations.

Laurie's Notes

Exploration

Motivate

- Write the following equations on the board and ask which one does not belong. Have students explain their reasoning.

  A. $3(x-4)=12$ **B.** $\frac{3}{4}x=6$ **C.** $4-8x=3$ **D.** $6x-7=9+4x$
- Students should solve the equations and reason that C does not belong. The other three equations have the same solution, $x=8$. These equations are equivalent.

Discuss

- Explain what equivalent polynomial equations are, connecting to the *Motivate*.
- Review the language associated with polynomials: factored form, standard form, and nonstandard form. Check the solutions in each form.

Exploration 1

- Students may approach the exploration in one of two ways. If they recognize the solutions from the factored form, then they may look at the other two forms and determine which forms are equivalent by substituting the solutions from the factored form. The alternate method would be to expand the factored form to see which expressions are equivalent in standard form.
- Give students time to work through the problems. Then ask for volunteers to share their findings and their method of solving.

? "How did you determine which equations are equivalent?" Expect different methods.

Exploration 2

- For each part, students will substitute the given values and evaluate the polynomial.
- **Use Appropriate Tools Strategically:** A table is an appropriate tool to help organize what is known in this problem. The table might have seven columns. The first column lists the equation, and the next six columns are for the values of x being substituted in the equation. The value being substituted is either a solution or not a solution.
- **Big Idea:** Each polynomial has two solutions. When either or both of the solutions are substituted in the factored expression, the product is 0. If two numbers are substituted and neither is a solution, then the product is not equal to 0.

Exploration 3

- The numbers 0 and 1 have special properties. Students may not recall the formal names of the properties, but they should recall the essence of the properties.
- Remind students that they should check both 0 and 1 in each statement to determine whether the statement is true.
- When students are finished, ask for volunteers to share their answers and their reasoning.
- **Extension:** Ask whether any of the statements are true when $n=-1$.

Communicate Your Answer

- For Question 5, have students discuss which property they chose and why they chose it.

Connecting to Next Step

- Students have connected the solution of a quadratic in factored form with the Zero-Product Property, without necessarily knowing the name of the property. In the formal lesson, the approach is given a specific name.

7.4 Solving Polynomial Equations in Factored Form

Essential Question How can you solve a polynomial equation?

EXPLORATION 1 **Matching Equivalent Forms of an Equation**

Work with a partner. An equation is considered to be in *factored form* when the product of the factors is equal to 0. Match each factored form of the equation with its equivalent standard form and nonstandard form.

	Factored Form		Standard Form		Nonstandard Form
a.	$(x-1)(x-3)=0$	**A.**	$x^2-x-2=0$	**1.**	$x^2-5x=-6$
b.	$(x-2)(x-3)=0$	**B.**	$x^2+x-2=0$	**2.**	$(x-1)^2=4$
c.	$(x+1)(x-2)=0$	**C.**	$x^2-4x+3=0$	**3.**	$x^2-x=2$
d.	$(x-1)(x+2)=0$	**D.**	$x^2-5x+6=0$	**4.**	$x(x+1)=2$
e.	$(x+1)(x-3)=0$	**E.**	$x^2-2x-3=0$	**5.**	$x^2-4x=-3$

USING TOOLS STRATEGICALLY

To be proficient in math, you need to consider using tools such as a table or a spreadsheet to organize your results.

EXPLORATION 2 **Writing a Conjecture**

Work with a partner. Substitute 1, 2, 3, 4, 5, and 6 for x in each equation and determine whether the equation is true. Organize your results in a table. Write a conjecture describing what you discovered.

a. $(x-1)(x-2)=0$ **b.** $(x-2)(x-3)=0$

c. $(x-3)(x-4)=0$ **d.** $(x-4)(x-5)=0$

e. $(x-5)(x-6)=0$ **f.** $(x-6)(x-1)=0$

EXPLORATION 3 **Special Properties of 0 and 1**

Work with a partner. The numbers 0 and 1 have special properties that are shared by no other numbers. For each of the following, decide whether the property is true for 0, 1, both, or neither. Explain your reasoning.

a. When you add ____ to a number n, you get n.

b. If the product of two numbers is ____, then at least one of the numbers is 0.

c. The square of ____ is equal to itself.

d. When you multiply a number n by ____, you get n.

e. When you multiply a number n by ____, you get 0.

f. The opposite of ____ is equal to itself.

Communicate Your Answer

4. How can you solve a polynomial equation?

5. One of the properties in Exploration 3 is called the Zero-Product Property. It is one of the most important properties in all of algebra. Which property is it? Why do you think it is called the Zero-Product Property? Explain how it is used in algebra and why it is so important.

Dynamic Teaching Tools

- Dynamic Assessment & Progress Monitoring Tool
- Lesson Planning Tool
- Interactive Whiteboard Lesson Library
- Dynamic Classroom with Dynamic Investigations

ANSWERS

1. a. C; 5
b. D; 1
c. A; 3
d. B; 4
e. E; 2

2. See Additional Answers.

3. a. 0; Adding 0 does not change a value, but adding 1 increases the value.
b. 0; The product of 0 and any number is always 0, never 1.
c. both; $0^2=0$ and $1^2=1$
d. 1; The product of any number and 0 is 0, the product of any number and 1 is the number.
e. 0; The product of any number and 0 is 0, the product of any number and 1 is the number.
f. 0; Zero is neither positive nor negative, so it is its own opposite.

4. Set each polynomial factor equal to 0 and solve.

5. b; It describes what happens when you have a product that is equal to zero; It is used to solve polynomial equations by setting each polynomial factor equal to 0. It is important because it provides an easy way to solve polynomial equations.

Extra Example 1

Solve each equation.

a. $3x(x - 6) = 0$

The roots are $x = 0$ and $x = 6$.

b. $(x + 5)(x - 4) = 0$

The roots are $x = -5$ and $x = 4$.

MONITORING PROGRESS ANSWERS

1. $x = 0, x = 1$
2. $t = 0, t = -2$
3. $z = 4, z = 6$

7.4 Lesson

Core Vocabulary

factored form, *p. 378*
Zero-Product Property, *p. 378*
roots, *p. 378*
repeated roots, *p. 379*

Previous
polynomial
standard form
greatest common factor (GCF)
monomial

What You Will Learn

- Use the Zero-Product Property.
- Factor polynomials using the GCF.
- Use the Zero-Product Property to solve real-life problems.

Using the Zero-Product Property

A polynomial is in **factored form** when it is written as a product of factors.

Standard form	Factored form
$x^2 + 2x$	$x(x + 2)$
$x^2 + 5x - 24$	$(x - 3)(x + 8)$

When one side of an equation is a polynomial in factored form and the other side is 0, use the **Zero-Product Property** to solve the polynomial equation. The solutions of a polynomial equation are also called **roots**.

Core Concept

Zero-Product Property

Words If the product of two real numbers is 0, then at least one of the numbers is 0.

Algebra If a and b are real numbers and $ab = 0$, then $a = 0$ or $b = 0$.

EXAMPLE 1 Solving Polynomial Equations

Solve each equation.

a. $2x(x - 4) = 0$ **b.** $(x - 3)(x - 9) = 0$

SOLUTION

a. $2x(x - 4) = 0$ Write equation.

$2x = 0$ *or* $x - 4 = 0$ Zero-Product Property

$x = 0$ *or* $x = 4$ Solve for x.

▶ The roots are $x = 0$ and $x = 4$.

b. $(x - 3)(x - 9) = 0$ Write equation.

$x - 3 = 0$ *or* $x - 9 = 0$ Zero-Product Property

$x = 3$ *or* $x = 9$ Solve for x.

▶ The roots are $x = 3$ and $x = 9$.

Check

To check the solutions of Example 1(a), substitute each solution in the original equation.

$2(0)(0 - 4) \stackrel{?}{=} 0$

$0(-4) \stackrel{?}{=} 0$

$0 = 0$ ✓

$2(4)(4 - 4) \stackrel{?}{=} 0$

$8(0) \stackrel{?}{=} 0$

$0 = 0$ ✓

Monitoring Progress

Help in English and Spanish at *BigIdeasMath.com*

Solve the equation. Check your solutions.

1. $x(x - 1) = 0$
2. $3t(t + 2) = 0$
3. $(z - 4)(z - 6) = 0$

Laurie's Notes Teacher Actions

- Explain that solutions of a polynomial equation are also called *roots*. Use "roots" and "solutions" interchangeably so students are comfortable with the vocabulary. Likewise, continue to use "factor" and "product" to focus attention on the operation involved.

? **Teaching Tip:** In part (a), underline each of the factors and ask, "What are the factors that have a product of 0?" $2x$ and $(x - 4)$

? **Probing Question:** "If the product of two factors is 0, will there always be two roots? Explain." no; The roots could be the same.

When two or more roots of an equation are the same number, the equation has **repeated roots**.

EXAMPLE 2 **Solving Polynomial Equations**

Solve each equation.

a. $(2x + 7)(2x - 7) = 0$ **b.** $(x - 1)^2 = 0$ **c.** $(x + 1)(x - 3)(x - 2) = 0$

SOLUTION

a. $(2x + 7)(2x - 7) = 0$ Write equation.

$2x + 7 = 0$ *or* $2x - 7 = 0$ Zero-Product Property

$x = -\frac{7}{2}$ *or* $x = \frac{7}{2}$ Solve for x.

▶ The roots are $x = -\frac{7}{2}$ and $x = \frac{7}{2}$.

b. $(x - 1)^2 = 0$ Write equation.

$(x - 1)(x - 1) = 0$ Expand equation.

$x - 1 = 0$ *or* $x - 1 = 0$ Zero-Product Property

$x = 1$ *or* $x = 1$ Solve for x.

▶ The equation has repeated roots of $x = 1$.

c. $(x + 1)(x - 3)(x - 2) = 0$ Write equation.

$x + 1 = 0$ *or* $x - 3 = 0$ *or* $x - 2 = 0$ Zero-Product Property

$x = -1$ *or* $x = 3$ *or* $x = 2$ Solve for x.

▶ The roots are $x = -1$, $x = 3$, and $x = 2$.

STUDY TIP

You can extend the Zero-Product Property to products of more than two real numbers.

Monitoring Progress Help in English and Spanish at *BigIdeasMath.com*

Solve the equation. Check your solutions.

4. $(3s + 5)(5s + 8) = 0$ **5.** $(b + 7)^2 = 0$ **6.** $(d - 2)(d + 6)(d + 8) = 0$

Factoring Polynomials Using the GCF

To solve a polynomial equation using the Zero-Product Property, you may need to *factor* the polynomial, or write it as a product of other polynomials. Look for the *greatest common factor* (GCF) of the terms of the polynomial. This is a monomial that divides evenly into each term.

EXAMPLE 3 **Finding the Greatest Common Monomial Factor**

Factor out the greatest common monomial factor from $4x^4 + 24x^3$.

SOLUTION

The GCF of 4 and 24 is 4. The GCF of x^4 and x^3 is x^3. So, the greatest common monomial factor of the terms is $4x^3$.

▶ So, $4x^4 + 24x^3 = 4x^3(x + 6)$.

Monitoring Progress Help in English and Spanish at *BigIdeasMath.com*

7. Factor out the greatest common monomial factor from $8y^2 - 24y$.

Differentiated Instruction

Visual

When using the GCF to factor, some students may benefit from seeing the GCF in each term of the polynomial. For example, write $4x^4 + 24x^3$ in Example 3 as $4x^3(x) + 4x^3(6)$. This may help students see that $4x^3$ is a factor of each term and that $4x^4 + 24x^3 = 4x^3(x + 6)$.

Extra Example 2

Solve each equation.

a. $(4x + 5)(4x - 5) = 0$ The roots are $x = -\frac{5}{4}$ and $x = \frac{5}{4}$.

b. $(c + 6)^2 = 0$ The equation has repeated roots of $c = -6$.

c. $(a + 5)(a - 2)(a - 7) = 0$ The roots are $a = -5$, $a = 2$, and $a = 7$.

Extra Example 3

Factor out the greatest common monomial factor from $12x^3 + 3x^2$. $3x^2(4x + 1)$

MONITORING PROGRESS ANSWERS

4. $s = -\frac{5}{3}, s = -\frac{8}{5}$

5. $b = -7$

6. $d = 2, d = -6, d = -8$

7. $8y(y - 3)$

Laurie's Notes Teacher Actions

- In part (a), you want students to be fluent in solving $2x + 7 = 0$ and $2x - 7 = 0$ by inspection. See the *Teaching Strategy* on page T-376.
- Students benefit from initially seeing each factor set equal to 0. Remind students that this is the application of the *Zero-Product Property*.
- **Connection:** Work through Example 3 as shown. Students often fail to see that this is the Distributive Property. Instead of multiplying a factor times a binomial, they are factoring out a common factor from the binomial.

Extra Example 4

a. Solve $4x^2 + 12x = 0$. The roots are $x = 0$ and $x = -3$.

b. Solve $-10a^2 = 8a$. The roots are $a = 0$ and $a = -\frac{4}{5}$.

Extra Example 5

You can model the arch of an entrance to a train tunnel by using the equation $y = -\frac{5}{16}(x + 8)(x - 8)$, where x and y are measured in feet. The x-axis represents the ground. Find the width of the entrance at ground level. 16 feet

MONITORING PROGRESS ANSWERS

8. $a = 0, a = -5$
9. $s = 0, s = 3$
10. $x = 0, x = \frac{1}{2}$
11. 8 ft

EXAMPLE 4 Solving Equations by Factoring

Solve (a) $2x^2 + 8x = 0$ and (b) $6n^2 = 15n$.

SOLUTION

a.

$2x^2 + 8x = 0$	Write equation.
$2x(x + 4) = 0$	Factor left side.
$2x = 0$ *or* $x + 4 = 0$	Zero-Product Property
$x = 0$ *or* $x = -4$	Solve for x.

▶ The roots are $x = 0$ and $x = -4$.

b.

$6n^2 = 15n$	Write equation.
$6n^2 - 15n = 0$	Subtract $15n$ from each side.
$3n(2n - 5) = 0$	Factor left side.
$3n = 0$ *or* $2n - 5 = 0$	Zero-Product Property
$n = 0$ *or* $n = \frac{5}{2}$	Solve for n.

▶ The roots are $n = 0$ and $n = \frac{5}{2}$.

Monitoring Progress

Help in English and Spanish at *BigIdeasMath.com*

Solve the equation. Check your solutions.

8. $a^2 + 5a = 0$
9. $3s^2 - 9s = 0$
10. $4x^2 = 2x$

Solving Real-Life Problems

EXAMPLE 5 Modeling with Mathematics

You can model the arch of a fireplace using the equation $y = -\frac{1}{9}(x + 18)(x - 18)$, where x and y are measured in inches. The x-axis represents the floor. Find the width of the arch at floor level.

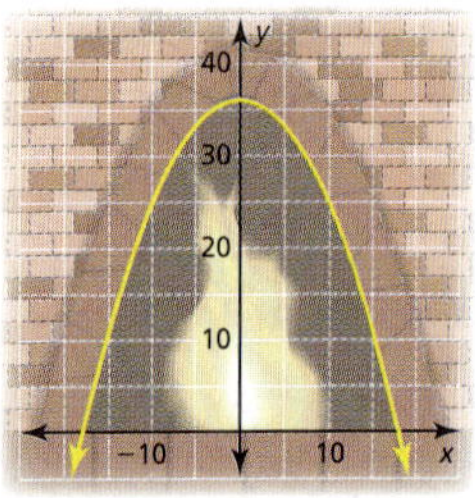

SOLUTION

Use the x-coordinates of the points where the arch meets the floor to find the width. At floor level, $y = 0$. So, substitute 0 for y and solve for x.

$y = -\frac{1}{9}(x + 18)(x - 18)$	Write equation.
$0 = -\frac{1}{9}(x + 18)(x - 18)$	Substitute 0 for y.
$0 = (x + 18)(x - 18)$	Multiply each side by -9.
$x + 18 = 0$ *or* $x - 18 = 0$	Zero-Product Property
$x = -18$ *or* $x = 18$	Solve for x.

The width is the distance between the x-coordinates, -18 and 18.

▶ So, the width of the arch at floor level is $|-18 - 18| = 36$ inches.

Monitoring Progress

Help in English and Spanish at *BigIdeasMath.com*

11. You can model the entrance to a mine shaft using the equation $y = -\frac{1}{2}(x + 4)(x - 4)$, where x and y are measured in feet. The x-axis represents the ground. Find the width of the entrance at ground level.

Laurie's Notes Teacher Actions

? "In part (a), how can you write $2x^2 + 8x$ as a product instead of a sum?" Factor out $2x$ so you have $2x(x + 4)$.

COMMON ERROR In Example 5, when multiplying each side by -9, students may think that all three factors on the right side must be multiplied by -9. To help students with this, ask them to find the product 3(4)(5). 60 Then ask them to find the product 2(3)(4)(5). 120

Closure

- **Exit Ticket:** Write a polynomial equation that has -6 and 4 as solutions. *Sample answer:* $(x + 6)(x - 4) = 0$

7.4 Exercises

Dynamic Solutions available at *BigIdeasMath.com*

Vocabulary and Core Concept Check

1. **WRITING** Explain how to use the Zero-Product Property to find the solutions of the equation $3x(x - 6) = 0$.

2. **DIFFERENT WORDS, SAME QUESTION** Which is different? Find *both* answers.

Solve the equation $(2k + 4)(k - 3) = 0$.	Find the values of k for which $2k + 4 = 0$ *or* $k - 3 = 0$.
Find the value of k for which $(2k + 4) + (k - 3) = 0$.	Find the roots of the equation $(2k + 4)(k - 3) = 0$.

Monitoring Progress and Modeling with Mathematics

In Exercises 3–8, solve the equation. *(See Example 1.)*

3. $x(x + 7) = 0$
4. $r(r - 10) = 0$
5. $12t(t - 5) = 0$
6. $-2v(v + 1) = 0$
7. $(s - 9)(s - 1) = 0$
8. $(y + 2)(y - 6) = 0$

In Exercises 9–20, solve the equation. *(See Example 2.)*

9. $(2a - 6)(3a + 15) = 0$
10. $(4q + 3)(q + 2) = 0$
11. $(5m + 4)^2 = 0$
12. $(h - 8)^2 = 0$
13. $(3 - 2g)(7 - g) = 0$
14. $(2 - 4d)(2 + 4d) = 0$
15. $z(z + 2)(z - 1) = 0$
16. $5p(2p - 3)(p + 7) = 0$
17. $(r - 4)^2(r + 8) = 0$
18. $w(w - 6)^2 = 0$
19. $(15 - 5c)(5c + 5)(-c + 6) = 0$
20. $(2 - n)\left(6 + \frac{2}{3}n\right)(n - 2) = 0$

In Exercises 21–24, find the x-coordinates of the points where the graph crosses the x-axis.

21.

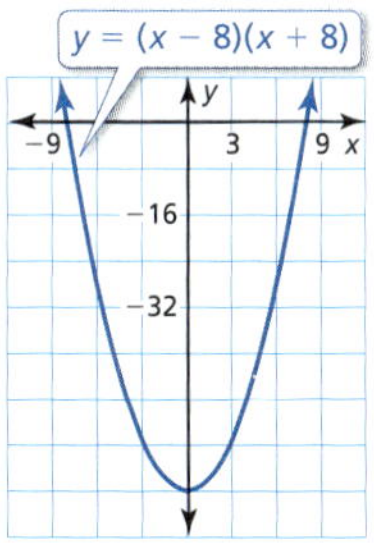

22.

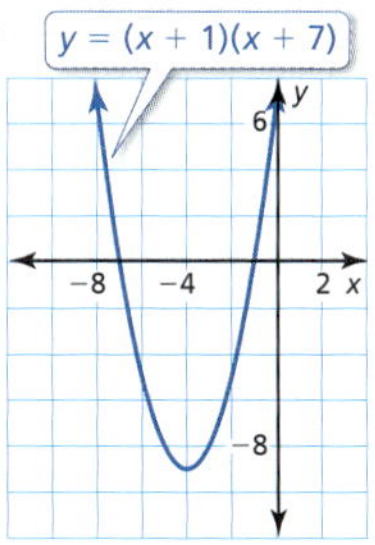

23.

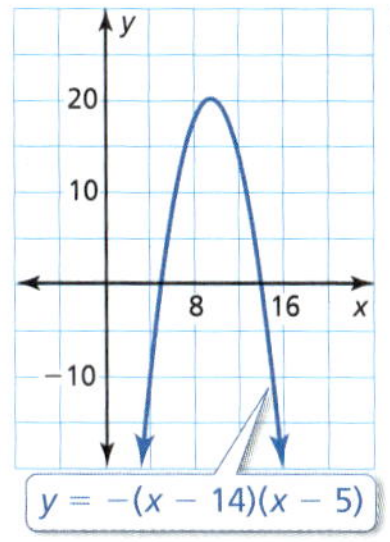

24.

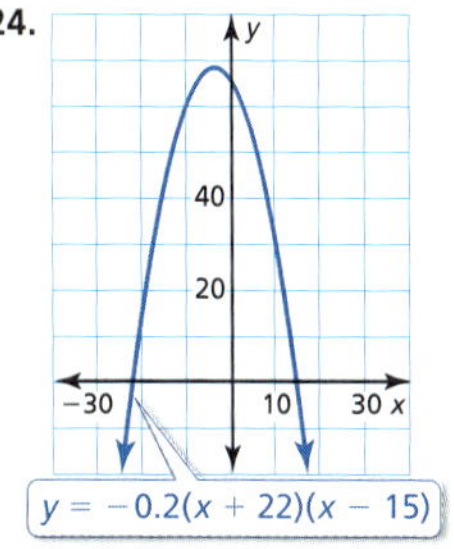

In Exercises 25–30, factor the polynomial. *(See Example 3.)*

25. $5z^2 + 45z$
26. $6d^2 - 21d$
27. $3y^3 - 9y^2$
28. $20x^3 + 30x^2$
29. $5n^6 + 2n^5$
30. $12a^4 + 8a$

In Exercises 31–36, solve the equation. *(See Example 4.)*

31. $4p^2 - p = 0$
32. $6m^2 + 12m = 0$
33. $25c + 10c^2 = 0$
34. $18q - 2q^2 = 0$
35. $3n^2 = 9n$
36. $-28r = 4r^2$

37. **ERROR ANALYSIS** Describe and correct the error in solving the equation.

✗
$$6x(x + 5) = 0$$
$$x + 5 = 0$$
$$x = -5$$
The root is $x = -5$.

34. $q = 0, q = 9$
35. $n = 0, n = 3$
36. $r = 0, r = -7$
37. also need to set $6x = 0$ and solve; $6x = 0$ or $x + 5 = 0$; $x = 0$ or $x = -5$; The roots are $x = 0$ and $x = -5$.

Assignment Guide and Homework Check

ASSIGNMENT

Basic: 1, 2, 3–39 odd, 42, 44, 49–52

Average: 1, 2–40 even, 41, 42, 44, 49–52

Advanced: 1, 2, 6, 8, 16–24 even, 28–40 even, 41–52

HOMEWORK CHECK

Basic: 3, 9, 25, 33, 39

Average: 4, 10, 26, 32, 40

Advanced: 8, 16, 28, 32, 40

ANSWERS

1. Set $3x = 0$ and $x - 6 = 0$, then solve both of the equations to get the solutions $x = 0$ and $x = 6$.
2. Find the value of k for which $(2k + 4) + (k - 3) = 0$; $k = -\frac{1}{3}$, $k = -2, k = 3$
3. $x = 0, x = -7$
4. $r = 0, r = 10$
5. $t = 0, t = 5$
6. $v = 0, v = -1$
7. $s = 9, s = 1$
8. $y = -2, y = 6$
9. $a = 3, a = -5$
10. $q = -\frac{3}{4}, q = -2$
11. $m = -\frac{4}{5}$
12. $h = 8$
13. $g = \frac{3}{2}, g = 7$
14. $d = \frac{1}{2}, d = -\frac{1}{2}$
15. $z = 0, z = -2, z = 1$
16. $p = 0, p = \frac{3}{2}, p = -7$
17. $r = 4, r = -8$
18. $w = 0, w = 6$
19. $c = 3, c = -1, c = 6$
20. $n = 2, n = -9$
21. $x = 8, x = -8$
22. $x = -1, x = -7$
23. $x = 14, x = 5$
24. $x = -22, x = 15$
25. $5z(z + 9)$
26. $3d(2d - 7)$
27. $3y^2(y - 3)$
28. $10x^2(2x + 3)$
29. $n^5(5n + 2)$
30. $4a(3a^3 + 2)$
31. $p = 0, p = \frac{1}{4}$
32. $m = 0, m = -2$
33. $c = 0, c = -\frac{5}{2}$

Dynamic Teaching Tools

- Dynamic Assessment & Progress Monitoring Tool
- Interactive Whiteboard Lesson Library
- Dynamic Classroom with Dynamic Investigations

ANSWERS

38. cannot divide both sides by y, because y could be 0 and division by 0 is undefined; $3y^2 - 21y = 0$; $3y(y - 7) = 0$; $3y = 0$ or $y - 7 = 0$; $y = 0$ or $y = 7$; The roots are $y = 0$ and $y = 7$.

39. 20 ft

40. **a.** 630 ft

b. 630 ft

41. $x = 0$, $x = 0.3$ sec; The roots represent the times when the penguin is at water level. $x = 0$ is when it leaves the water, and $x = 0.3$ sec is when it returns to the water after the leap.

42. $-$; $+$; The point of intersection on the positive x-axis is farther from the origin than the point of intersection on the negative x-axis, so the positive solution to the equation will be the one with the greater absolute value.

43. 2; x-intercepts occur when $y = 0$ and the equation has 2 roots when $y = 0$.

44. no; If $a = b$, then there is a repeated root and the graph will only have one x-intercept.

45–52. See Additional Answers.

Mini-Assessment

Solve the equation.

1. $5x(x - 4) = 0$
   The roots are $x = 0$ and $x = 4$.
2. $(3x + 8)(3x - 8) = 0$
   The roots are $x = -\frac{8}{3}$ and $x = \frac{8}{3}$.
3. $3x^5 + 21x^4 = 0$ The roots are $x = 0$ and $x = -7$.
4. Factor out the greatest common monomial factor from $6x^4 + 8x^3$.
   $2x^3(3x + 4)$
5. You can model the arch of a building entrance by using the equation $y = -\frac{1}{3}(x + 6)(x - 6)$, where x and y are measured in feet. The x-axis represents the ground. Find the width of the arch at the ground. 12 feet

38. ERROR ANALYSIS Describe and correct the error in solving the equation.

> ✗
> $3y^2 = 21y$
> $3y = 21$
> $y = 7$
> The root is $y = 7$.

39. MODELING WITH MATHEMATICS The entrance of a tunnel can be modeled by $y = -\frac{11}{50}(x - 4)(x - 24)$, where x and y are measured in feet. The x-axis represents the ground. Find the width of the tunnel at ground level. *(See Example 5.)*

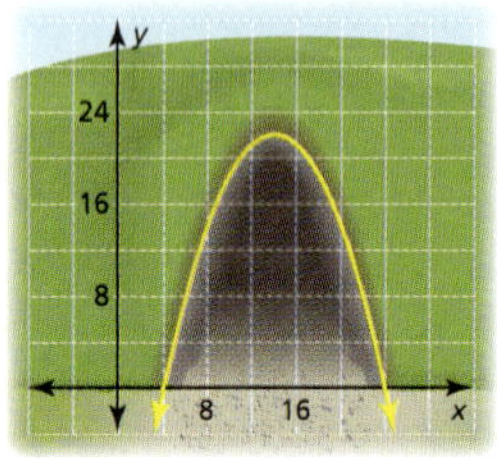

40. MODELING WITH MATHEMATICS The Gateway Arch in St. Louis can be modeled by $y = -\frac{2}{315}(x + 315)(x - 315)$, where x and y are measured in feet. The x-axis represents the ground.

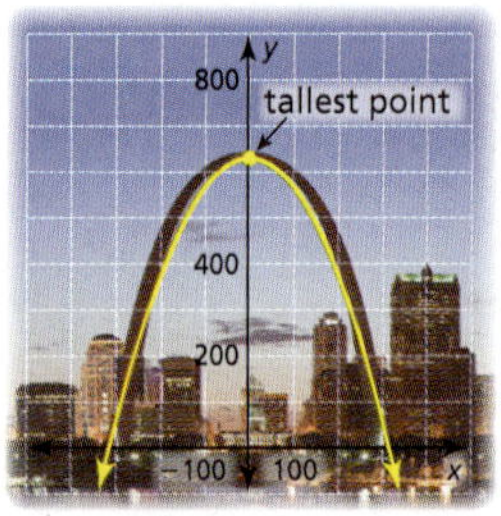

a. Find the width of the arch at ground level.

b. How tall is the arch?

41. MODELING WITH MATHEMATICS A penguin leaps out of the water while swimming. This action is called porpoising. The height y (in feet) of a porpoising penguin can be modeled by $y = -16x^2 + 4.8x$, where x is the time (in seconds) since the penguin leaped out of the water. Find the roots of the equation when $y = 0$. Explain what the roots mean in this situation.

42. HOW DO YOU SEE IT? Use the graph to fill in each blank in the equation with the symbol $+$ or $-$. Explain your reasoning.

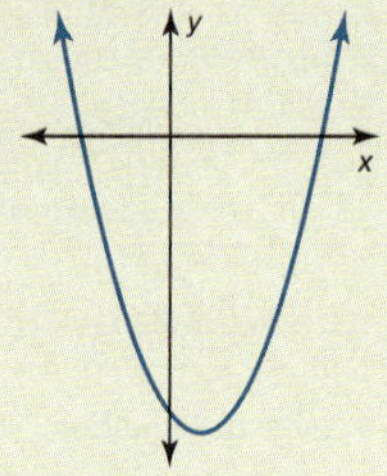

$y = (x \;\square\; 5)(x \;\square\; 3)$

43. CRITICAL THINKING How many x-intercepts does the graph of $y = (2x + 5)(x - 9)^2$ have? Explain.

44. MAKING AN ARGUMENT Your friend says that the graph of the equation $y = (x - a)(x - b)$ always has two x-intercepts for any values of a and b. Is your friend correct? Explain.

45. CRITICAL THINKING Does the equation $(x^2 + 3)(x^4 + 1) = 0$ have any real roots? Explain.

46. THOUGHT PROVOKING Write a polynomial equation of degree 4 whose only roots are $x = 1$, $x = 2$, and $x = 3$.

47. REASONING Find the values of x in terms of y that are solutions of each equation.

a. $(x + y)(2x - y) = 0$

b. $(x^2 - y^2)(4x + 16y) = 0$

48. PROBLEM SOLVING Solve the equation $(4^{x-5} - 16)(3^x - 81) = 0$.

Maintaining Mathematical Proficiency

Reviewing what you learned in previous grades and lessons

List the factor pairs of the number. *(Skills Review Handbook)*

49. 10 **50.** 18

51. 30 **52.** 48

If students need help...	*If students got it...*
Resources by Chapter • Practice A and Practice B • Puzzle Time	Resources by Chapter • Enrichment and Extension • Cumulative Review
Student Journal • Practice	Start the *next* Section
Differentiating the Lesson Skills Review Handbook	

7.1–7.4 What Did You Learn?

Core Vocabulary

monomial, *p. 358*
degree of a monomial, *p. 358*
polynomial, *p. 359*
binomial, *p. 359*
trinomial, *p. 359*
degree of a polynomial, *p. 359*
standard form, *p. 359*
leading coefficient, *p. 359*
closed, *p. 360*
FOIL Method, *p. 367*
factored form, *p. 378*
Zero-Product Property, *p. 378*
roots, *p. 378*
repeated roots, *p. 379*

Core Concepts

Section 7.1

Polynomials, *p. 359*
Adding Polynomials, *p. 360*
Subtracting Polynomials, *p. 360*

Section 7.2

Multiplying Binomials, *p. 366*
FOIL Method, *p. 367*
Multiplying Binomials and Trinomials, *p. 368*

Section 7.3

Square of a Binomial Pattern, *p. 372*
Sum and Difference Pattern, *p. 373*

Section 7.4

Zero-Product Property, *p. 378*
Factoring Polynomials Using the GCF, *p. 379*

Mathematical Practices

1. Explain how you wrote the polynomial in Exercise 11 on page 375. Is there another method you can use to write the same polynomial?
2. Find a shortcut for exercises like Exercise 7 on page 381 when the variable has a coefficient of 1. Does your shortcut work when the coefficient is *not* 1?

Study Skills

Preparing for a Test

- Review examples of each type of problem that could appear on the test.
- Review the homework problems your teacher assigned.
- Take a practice test.

Dynamic Teaching Tools

- Dynamic Assessment & Progress Monitoring Tool
- Interactive Whiteboard Lesson Library
- Dynamic Classroom with Dynamic Investigations

ANSWERS

1. *Sample answer:* Represent the sides as $x + 4$ and use the square of a binomial pattern; Yes, you could calculate the area of each shaded region and add those areas together.
2. The solutions are the constant terms with the opposite sign; no

ANSWERS

1. $-8q^3$; 3; -8; monomial
2. $d^2 - 3d + 9$; 2; 1; trinomial
3. $-\frac{5}{6}m^6 + \frac{2}{3}m^4$; 6; $-\frac{5}{6}$; binomial
4. $3z^4 + 7.4z^2 - 1.3z$; 4; 3; trinomial
5. $x^2 + 9$
6. $-5n^2 + n + 7$
7. $-2p^2 + 7p - 15$
8. $-2ab + 2b^2$
9. $w^2 + 13w + 42$
10. $-8d^2 + 26d - 15$
11. $y^3 + 11y^2 + 15y - 27$
12. $9z^2 - 25$
13. $t^2 + 10t + 25$
14. $4q^2 - 24q + 36$
15. $x = 0, x = 3$
16. $g = 8$
17. $p = -\frac{7}{3}, p = \frac{7}{3}, p = -8$
18. $y = 0, y = 8, y = -\frac{1}{2}$
19. a. $(8x + 240)$ in.
    b. $(4x^2 + 240x + 3456)$ in.^{2}
    c. 272 in.; 4480 in.^{2}
20. a. $1000r^2 + 2000r + 1000$
    b. \$1060.90
    c. no; The balance after 3 years is represented by the expression $1000(1 + 0.03)^3$, which is only about \$1092.73.
21. 144 in.

7.1–7.4 Quiz

Write the polynomial in standard form. Identify the degree and leading coefficient of the polynomial. Then classify the polynomial by the number of terms. *(Section 7.1)*

1. $-8q^3$

2. $9 + d^2 - 3d$

3. $\frac{2}{3}m^4 - \frac{5}{6}m^6$

4. $-1.3z + 3z^4 + 7.4z^2$

Find the sum or difference. *(Section 7.1)*

5. $(2x^2 + 5) + (-x^2 + 4)$

6. $(-3n^2 + n) - (2n^2 - 7)$

7. $(-p^2 + 4p) - (p^2 - 3p + 15)$

8. $(a^2 - 3ab + b^2) + (-a^2 + ab + b^2)$

Find the product. *(Section 7.2 and Section 7.3)*

9. $(w + 6)(w + 7)$

10. $(3 - 4d)(2d - 5)$

11. $(y + 9)(y^2 + 2y - 3)$

12. $(3z - 5)(3z + 5)$

13. $(t + 5)^2$

14. $(2q - 6)^2$

Solve the equation. *(Section 7.4)*

15. $5x^2 - 15x = 0$

16. $(8 - g)(8 - g) = 0$

17. $(3p + 7)(3p - 7)(p + 8) = 0$

18. $-3y(y - 8)(2y + 1) = 0$

19. You are making a blanket with a fringe border of equal width on each side. *(Section 7.1 and Section 7.2)*

a. Write a polynomial that represents the perimeter of the blanket including the fringe.

b. Write a polynomial that represents the area of the blanket including the fringe.

c. Find the perimeter and the area of the blanket including the fringe when the width of the fringe is 4 inches.

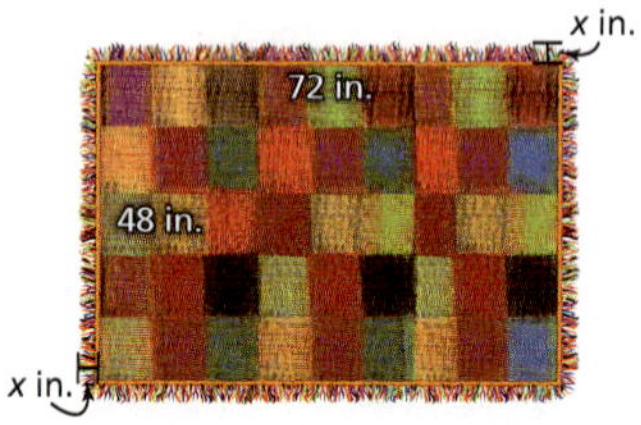

20. You are saving money to buy an electric guitar. You deposit \$1000 in an account that earns interest compounded annually. The expression $1000(1 + r)^2$ represents the balance after 2 years, where r is the annual interest rate in decimal form. *(Section 7.3)*

a. Write the polynomial in standard form that represents the balance of your account after 2 years.

b. The interest rate is 3%. What is the balance of your account after 2 years?

c. The guitar costs \$1100. Do you have enough money in your account *after 3 years*? Explain.

21. The front of a storage bunker can be modeled by $y = -\frac{5}{216}(x - 72)(x + 72)$, where x and y are measured in inches. The x-axis represents the ground. Find the width of the bunker at ground level. *(Section 7.4)*

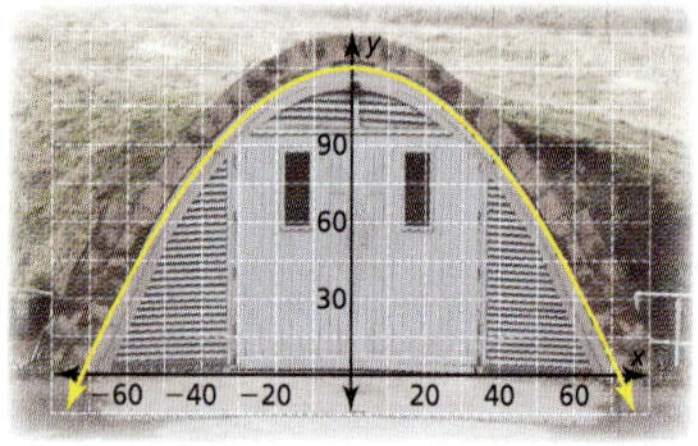

Laurie's Notes

Overview of Section 7.5

Introduction

- In this lesson, students factor trinomials of the form $x^2 + bx + c$. The different cases of $c > 0$ and $c < 0$ are presented. The lesson ends with a real-life problem.
- Although factoring of trinomials is the focus of the lesson, the ultimate goal is for students to be able to solve polynomial equations and simplify expressions with common factors.

Teaching Strategy

- **Use Appropriate Tools Strategically:** Algebra tiles can be used to help develop a greater understanding of what it means to factor a trinomial, in the same way base-ten blocks were used when developing an understanding of writing a whole number as the product of two factors. *Example:* Given 12 unit squares, they can be arranged into a rectangle of dimensions 1×12, 2×6, or 3×4.
- The polynomial $x^2 + 8x + 12$ can be modeled using 1 x^2-tile, 8 x-tiles, and 12 unit tiles.
- To arrange the tiles into a rectangle means that you need to have the 12 unit pieces arranged as an array with dimensions 1×12, 2×6, or 3×4. If factorable, the possible rectangular array must be one of the following:

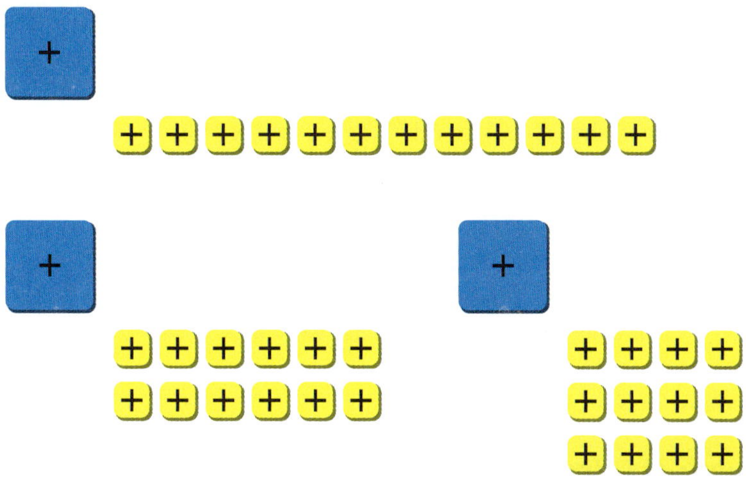

- The x-tiles now have to fill in the opposite regions of the array. The first array requires 13 x-tiles $(1 + 12)$, the second array requires 8 x-tiles $(2 + 6)$, and the third array requires 7 x-tiles $(3 + 4)$. Because there are 8 x-tiles, the second array is the correct arrangement. The trinomial $x^2 + 8x + 12$ equals $(x + 2)(x + 6)$.
- Consider the additional possibilities when the constant term is negative! Although this type of trinomial is not shown in the explorations, it can certainly be done.
- To factor $x^2 + 4x - 12$, make the same three arrangements as with the previous problem. You have 4 x-tiles. Two zero pairs of x-tiles are added to fill the rectangular array.

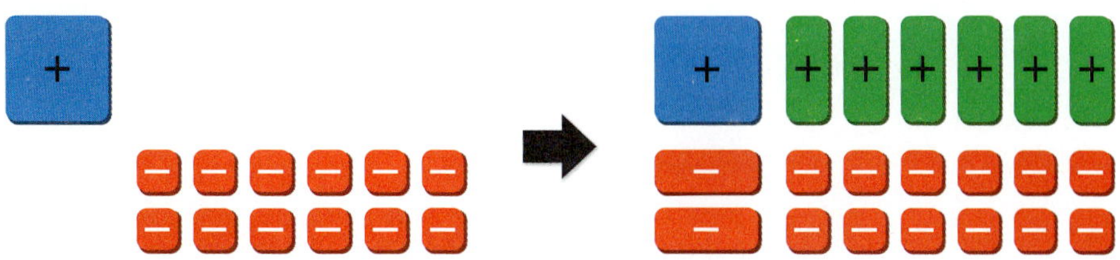

- So the trinomial $x^2 + 4x - 12$ equals $(x - 2)(x + 6)$.

Pacing Suggestion

- Complete the exploration and discuss students' observations. Transition to the formal lesson after the *Motivate*.

Dynamic Teaching Tools

- Dynamic Assessment & Progress Monitoring Tool
- Lesson Planning Tool
- Interactive Whiteboard Lesson Library
- Dynamic Classroom with Dynamic Investigations

What Your Students Will Learn

- Factor $x^2 + bx + c$ when b and c are positive, when b is negative and c is positive, and when c is negative.
- Use factoring to solve real-life problems involving polynomial equations.

Laurie's Notes

Exploration

Motivate

- Have students match the factored form with the correct standard form.

1. $(x + 2)(x + 6)$ B	**A.** $x^2 + 7x + 12$
2. $(x + 12)(x + 1)$ C	**B.** $x^2 + 8x + 12$
3. $(x + 3)(x + 4)$ A	**C.** $x^2 + 13x + 12$

? "How are the problems alike? How are they different?" Answers will vary.

- Circle the constant terms in the binomials (2, 6, 12, 1, 3, and 4), the coefficients of the linear terms (7, 8, and 13), and the constant terms (12, 12, and 12).

? "How are these numbers related?" The sum of each pair is the coefficient of the x-term, and the product of each pair is the constant term.

For Your Information

- Although students have used algebra tiles to multiply binomials, using tiles to factor trinomials can be challenging for them. See the *Teaching Strategy* on page T-384.

Exploration 1

- Model the sample problem with students. Ask students to model $x^2 + 5x + 6$ so that they form a rectangular array with the algebra tiles. There are two options for the six unit tiles: 1×6 or 2×3. If students try the 1×6, they will realize that they do not have enough x-tiles.

? "What are the dimensions of the rectangle?" $(x + 2)$ and $(x + 3)$

- Write and say, "So $x^2 + 5x + 6 = (x + 2)(x + 3)$." Discuss that $x^2 + 5x + 6$ is the product, or area of the rectangle, and $(x + 2)$ and $(x + 3)$ are the factors, or dimensions of the rectangle.
- As students work the four parts with their partners, encourage them to think about the dimensions of the rectangles. Students also need to pay attention to the signs of their tiles.
- In parts (c) and (d), students should recognize that when the constant term is 12, this product could be the result of (3)(4) or $(-3)(-4)$.
- Have students share their work at a document camera or overhead projector if possible.

Communicate Your Answer

- Students should recognize that to factor $x^2 + bx + c$, they need to find two numbers whose product is c and whose sum is b. Their language may not be precise at this point.

Connecting to Next Step

- The exploration helps students visualize and make sense of factoring a trinomial of the form $x^2 + bx + c$. Once explored, students will be better able to perform the algebraic manipulations.

7.5 Factoring $x^2 + bx + c$

Essential Question

How can you use algebra tiles to factor the trinomial $x^2 + bx + c$ into the product of two binomials?

EXPLORATION 1 **Finding Binomial Factors**

Work with a partner. Use algebra tiles to write each polynomial as the product of two binomials. Check your answer by multiplying.

Sample $x^2 + 5x + 6$

Step 1 Arrange algebra tiles that model $x^2 + 5x + 6$ into a rectangular array.

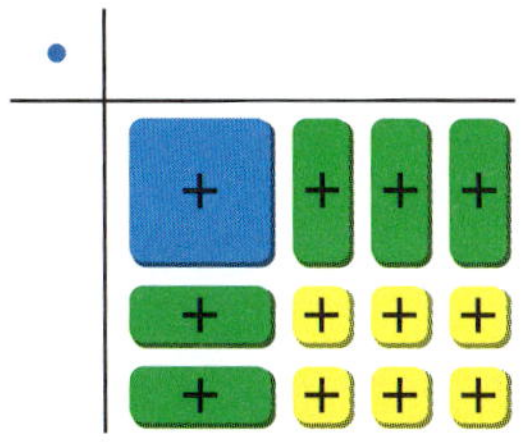

Step 2 Use additional algebra tiles to model the dimensions of the rectangle.

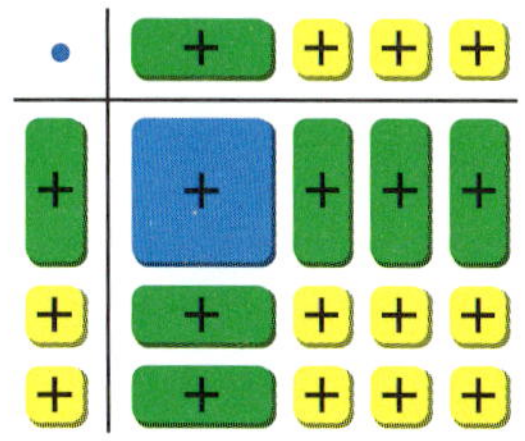

Step 3 Write the polynomial in factored form using the dimensions of the rectangle.

width, length

Area $= x^2 + 5x + 6 = (x + 2)(x + 3)$

a. $x^2 - 3x + 2 =$ ______

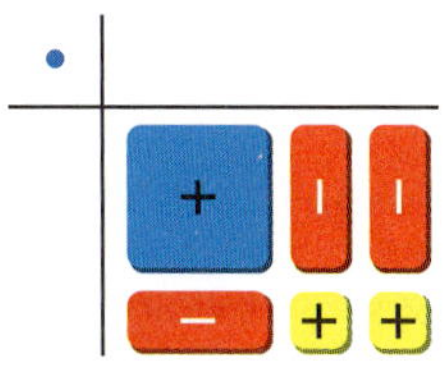

b. $x^2 + 5x + 4 =$ ______

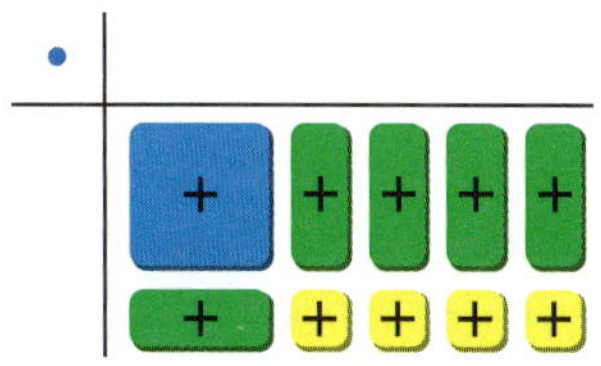

c. $x^2 - 7x + 12 =$ ______

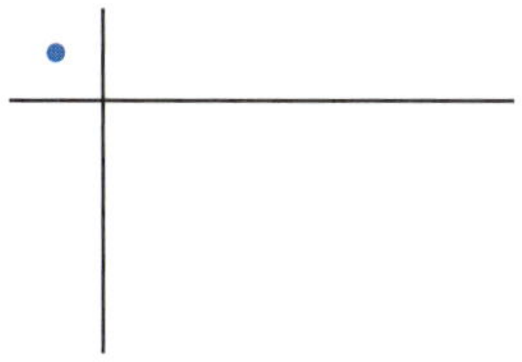

d. $x^2 + 7x + 12 =$ ______

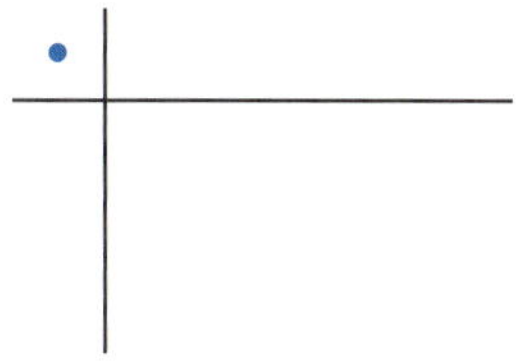

REASONING ABSTRACTLY

To be proficient in math, you need to understand a situation abstractly and represent it symbolically.

Communicate Your Answer

2. How can you use algebra tiles to factor the trinomial $x^2 + bx + c$ into the product of two binomials?

3. Describe a strategy for factoring the trinomial $x^2 + bx + c$ that does not use algebra tiles.

Dynamic Teaching Tools

- Dynamic Assessment & Progress Monitoring Tool
- Lesson Planning Tool
- Interactive Whiteboard Lesson Library
- Dynamic Classroom with Dynamic Investigations

ANSWERS

1. a. $(x - 1)(x - 2)$

b. $(x + 1)(x + 4)$

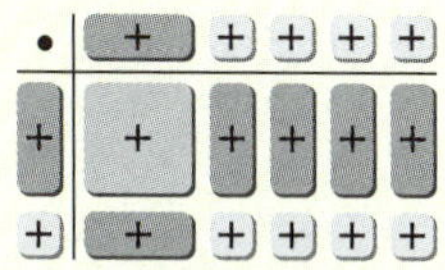

c. $(x - 4)(x - 3)$

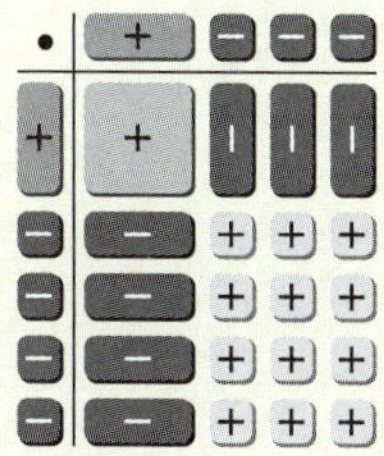

d. $(x + 4)(x + 3)$

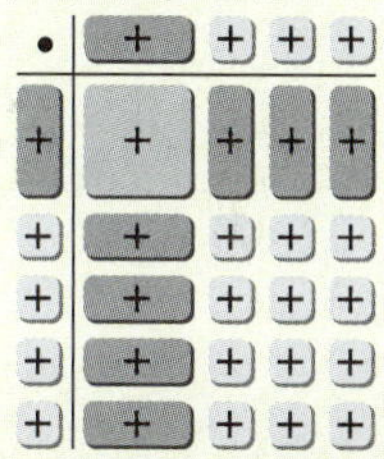

2. Arrange algebra tiles that model the trinomial into a rectangular array, use additional algebra tiles to model the dimensions of the rectangle, then write the polynomial in factored form using the dimensions of the rectangle.

3. Find two integer factors of c that have a sum of b, then write the binomial factors by adding each integer factor to x.

English Language Learners

Pair Activity

Have students work in pairs to factor polynomials. Each student factors a different polynomial. Then students take turns explaining their solutions to their partners.

Extra Example 1

Factor $x^2 + 9x + 14$.

$(x + 2)(x + 7)$

MONITORING PROGRESS ANSWERS

1. $(x + 1)(x + 6)$
2. $(x + 1)(x + 8)$

7.5 Lesson

Core Vocabulary

Previous
polynomial
FOIL Method
Zero-Product Property

What You Will Learn

- Factor $x^2 + bx + c$.
- Use factoring to solve real-life problems.

Factoring $x^2 + bx + c$

Writing a polynomial as a product of factors is called *factoring*. To factor $x^2 + bx + c$ as $(x + p)(x + q)$, you need to find p and q such that $p + q = b$ and $pq = c$.

$$(x + p)(x + q) = x^2 + px + qx + pq$$
$$= x^2 + (p + q)x + pq$$

Core Concept

Factoring $x^2 + bx + c$ When c Is Positive

Algebra $x^2 + bx + c = (x + p)(x + q)$ when $p + q = b$ and $pq = c$.
When c is positive, p and q have the same sign as b.

Examples $x^2 + 6x + 5 = (x + 1)(x + 5)$
$x^2 - 6x + 5 = (x - 1)(x - 5)$

EXAMPLE 1 Factoring $x^2 + bx + c$ When b and c Are Positive

Factor $x^2 + 10x + 16$.

SOLUTION

Notice that $b = 10$ and $c = 16$.

- Because c is positive, the factors p and q must have the same sign so that pq is positive.
- Because b is also positive, p and q must each be positive so that $p + q$ is positive.

Find two positive integer factors of 16 whose sum is 10.

Factors of 16	Sum of factors
1, 16	17
2, 8	10
4, 4	8

The values of p and q are 2 and 8.

So, $x^2 + 10x + 16 = (x + 2)(x + 8)$.

Check

Use the FOIL Method.

$(x + 2)(x + 8)$
$= x^2 + 8x + 2x + 16$
$= x^2 + 10x + 16$ ✓

Monitoring Progress

Help in English and Spanish at *BigIdeasMath.com*

Factor the polynomial.

1. $x^2 + 7x + 6$
2. $x^2 + 9x + 8$

Laurie's Notes Teacher Actions

- **Teaching Tip:** Write the example and say, "You need two numbers whose product is 16 and whose sum is 10."
- Some students may find it tedious to list the factor pairs. Share with them that in future problems the signs may not both be positive and the leading coefficient may be a number other than 1.
- Students should set up each problem before they factor. For example, $x^2 + 7x + 6 = (x + \_\_)(x + \_\_)$. "I need two numbers whose product is 6 and whose sum is 7."

EXAMPLE 2 Factoring $x^2 + bx + c$ When b Is Negative and c Is Positive

Factor $x^2 - 8x + 12$.

SOLUTION

Notice that $b = -8$ and $c = 12$.

- Because c is positive, the factors p and q must have the same sign so that pq is positive.
- Because b is negative, p and q must each be negative so that $p + q$ is negative.

Find two negative integer factors of 12 whose sum is -8.

Factors of 12	$-1, -12$	$-2, -6$	$-3, -4$
Sum of factors	-13	-8	-7

The values of p and q are -2 and -6.

▶ So, $x^2 - 8x + 12 = (x - 2)(x - 6)$.

Check

Use the FOIL Method.

$(x - 2)(x - 6)$

$= x^2 - 6x - 2x + 12$

$= x^2 - 8x + 12$ ✓

Core Concept

Factoring $x^2 + bx + c$ When c Is Negative

Algebra $x^2 + bx + c = (x + p)(x + q)$ when $p + q = b$ and $pq = c$.

When c is negative, p and q have different signs.

Example $x^2 - 4x - 5 = (x + 1)(x - 5)$

EXAMPLE 3 Factoring $x^2 + bx + c$ When c Is Negative

Factor $x^2 + 4x - 21$.

SOLUTION

Notice that $b = 4$ and $c = -21$. Because c is negative, the factors p and q must have different signs so that pq is negative.

Find two integer factors of -21 whose sum is 4.

Factors of -21	$-21, 1$	$-1, 21$	$-7, 3$	$-3, 7$
Sum of factors	-20	20	-4	4

The values of p and q are -3 and 7.

▶ So, $x^2 + 4x - 21 = (x - 3)(x + 7)$.

Check

Use the FOIL Method.

$(x - 3)(x + 7)$

$= x^2 + 7x - 3x - 21$

$= x^2 + 4x - 21$ ✓

Monitoring Progress

Help in English and Spanish at *BigIdeasMath.com*

Factor the polynomial.

3. $w^2 - 4w + 3$

4. $n^2 - 12n + 35$

5. $x^2 - 14x + 24$

6. $x^2 + 2x - 15$

7. $y^2 + 13y - 30$

8. $v^2 - v - 42$

Extra Example 2

Factor $x^2 - 12x + 27$.
$(x - 3)(x - 9)$

Extra Example 3

Factor $x^2 + 7x - 8$.
$(x - 1)(x + 8)$

MONITORING PROGRESS ANSWERS

3. $(w - 1)(w - 3)$

4. $(n - 5)(n - 7)$

5. $(x - 2)(x - 12)$

6. $(x - 3)(x + 5)$

7. $(y - 2)(y + 15)$

8. $(v + 6)(v - 7)$

Laurie's Notes Teacher Actions

- ? Write Example 2 and say, "You need two numbers whose product is 12 and whose sum is -8. What do you know about the two numbers?" The numbers must both be negative. Remind students to check their answers.
- **Turn and Talk:** In Example 3 say, "You need two numbers whose product is -21 and whose sum is 4. What are the possibilities?"
- **Think-Pair-Share:** Have students answer Questions 3–8, and then share and discuss as a class.

Extra Example 4

A gardener plants a rectangular lettuce patch in the southwest corner of a square garden. The area of the lettuce patch is 120 square feet. What is the area of the square garden?

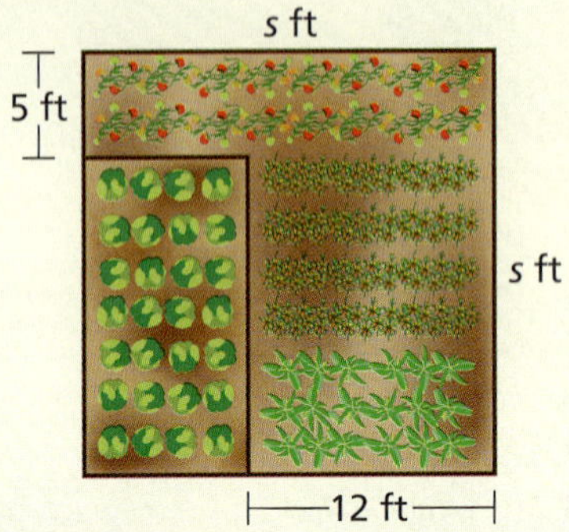

400 square feet

MONITORING PROGRESS ANSWER

9. 2500 m^2

Solving Real-Life Problems

EXAMPLE 4 **Solving a Real-Life Problem**

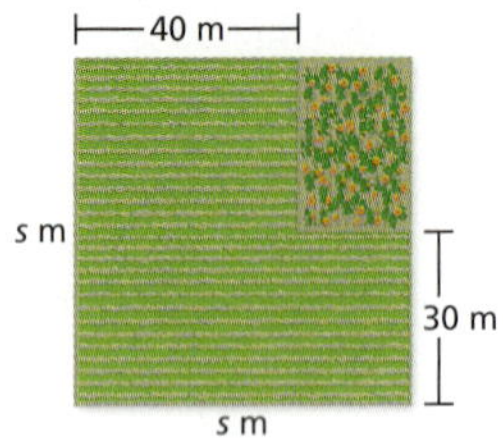

A farmer plants a rectangular pumpkin patch in the northeast corner of a square plot of land. The area of the pumpkin patch is 600 square meters. What is the area of the square plot of land?

SOLUTION

1. **Understand the Problem** You are given the area of the pumpkin patch, the difference of the side length of the square plot and the length of the pumpkin patch, and the difference of the side length of the square plot and the width of the pumpkin patch.
2. **Make a Plan** The length of the pumpkin patch is $(s - 30)$ meters and the width is $(s - 40)$ meters. Write and solve an equation to find the side length s. Then use the solution to find the area of the square plot of land.
3. **Solve the Problem** Use the equation for the area of a rectangle to write and solve an equation to find the side length s of the square plot of land.

$$600 = (s - 30)(s - 40) \quad \text{Write an equation.}$$
$$600 = s^2 - 70s + 1200 \quad \text{Multiply.}$$
$$0 = s^2 - 70s + 600 \quad \text{Subtract 600 from each side.}$$
$$0 = (s - 10)(s - 60) \quad \text{Factor the polynomial.}$$
$$s - 10 = 0 \quad or \quad s - 60 = 0 \quad \text{Zero-Product Property}$$
$$s = 10 \quad or \quad s = 60 \quad \text{Solve for } s.$$

STUDY TIP

The diagram shows that the side length is more than 40 meters, so a side length of 10 meters does not make sense in this situation. The side length is 60 meters.

▶ So, the area of the square plot of land is 60(60) = 3600 square meters.

4. **Look Back** Use the diagram to check that you found the correct side length. Using $s = 60$, the length of the pumpkin patch is $60 - 30 = 30$ meters and the width is $60 - 40 = 20$ meters. So, the area of the pumpkin patch is 600 square meters. This matches the given information and confirms the side length is 60 meters, which gives an area of 3600 square meters.

Monitoring Progress

Help in English and Spanish at *BigIdeasMath.com*

9. WHAT IF? The area of the pumpkin patch is 200 square meters. What is the area of the square plot of land?

Concept Summary

Factoring $x^2 + bx + c$ as $(x + p)(x + q)$

The diagram shows the relationships between the signs of b and c and the signs of p and q.

$x^2 + bx + c = (x + p)(x + q)$

- c is positive. b is positive. → p and q are positive.
- c is positive. b is negative. → p and q are negative.
- c is negative. → p and q have different signs.

Laurie's Notes Teacher Actions

- **Model with Mathematics:** In this problem, the diagram is a helpful model. When the problem has been solved, the results can be interpreted in the context of the problem, and again the diagram is helpful. It is clear that a side length of 10 is not possible.
- Students should recognize that in this problem they needed to subtract 600 from each side. The trinomial is set equal to 0 before factoring.

Closure

- Explain your strategy for factoring the following polynomials.
  a. $x^2 - 2x - 15$ $(x - 5)(x + 3)$
  b. $x^2 + 2x - 15$ $(x + 5)(x - 3)$
  c. $x^2 + 8x + 15$ $(x + 5)(x + 3)$
  d. $x^2 - 8x + 15$ $(x - 5)(x - 3)$

7.5 Exercises

Dynamic Solutions available at *BigIdeasMath.com*

Vocabulary and Core Concept Check

1. **WRITING** You are factoring $x^2 + 11x - 26$. What do the signs of the terms tell you about the factors? Explain.

2. **OPEN-ENDED** Write a trinomial that can be factored as $(x + p)(x + q)$, where p and q are positive.

Monitoring Progress and Modeling with Mathematics

In Exercises 3–8, factor the polynomial. *(See Example 1.)*

3. $x^2 + 8x + 7$
4. $z^2 + 10z + 21$
5. $n^2 + 9n + 20$
6. $s^2 + 11s + 30$
7. $h^2 + 11h + 18$
8. $y^2 + 13y + 40$

In Exercises 9–14, factor the polynomial. *(See Example 2.)*

9. $v^2 - 5v + 4$
10. $x^2 - 13x + 22$
11. $d^2 - 5d + 6$
12. $k^2 - 10k + 24$
13. $w^2 - 17w + 72$
14. $j^2 - 13j + 42$

In Exercises 15–24, factor the polynomial. *(See Example 3.)*

15. $x^2 + 3x - 4$
16. $z^2 + 7z - 18$
17. $n^2 + 4n - 12$
18. $s^2 + 3s - 40$
19. $y^2 + 2y - 48$
20. $h^2 + 6h - 27$
21. $x^2 - x - 20$
22. $m^2 - 6m - 7$
23. $-6t - 16 + t^2$
24. $-7y + y^2 - 30$

25. **MODELING WITH MATHEMATICS** A projector displays an image on a wall. The area (in square feet) of the projection is represented by $x^2 - 8x + 15$.

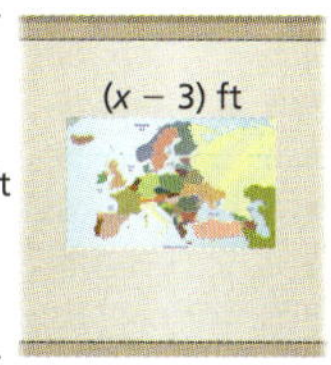

a. Write a binomial that represents the height of the projection.

b. Find the perimeter of the projection when the height of the wall is 8 feet.

26. **MODELING WITH MATHEMATICS** A dentist's office and parking lot are on a rectangular piece of land. The area (in square meters) of the land is represented by $x^2 + x - 30$.

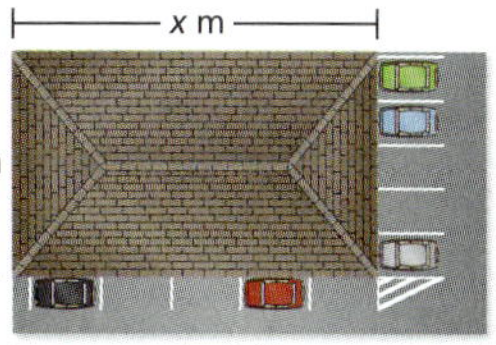

a. Write a binomial that represents the width of the land.

b. Find the area of the land when the length of the dentist's office is 20 meters.

ERROR ANALYSIS In Exercises 27 and 28, describe and correct the error in factoring the polynomial.

27. ✗ $x^2 + 14x + 48 = (x + 4)(x + 12)$

28. ✗ $s^2 - 17s - 60 = (s - 5)(s - 12)$

In Exercises 29–38, solve the equation.

29. $m^2 + 3m + 2 = 0$
30. $n^2 - 9n + 18 = 0$
31. $x^2 + 5x - 14 = 0$
32. $v^2 + 11v - 26 = 0$
33. $t^2 + 15t = -36$
34. $n^2 - 5n = 24$
35. $a^2 + 5a - 20 = 30$
36. $y^2 - 2y - 8 = 7$
37. $m^2 + 10 = 15m - 34$
38. $b^2 + 5 = 8b - 10$

Assignment Guide and Homework Check

ASSIGNMENT

Basic: 1, 2, 3–29 odd, 39, 41, 46, 52–55

Average: 1, 2–40 even, 41, 46, 47, 52–55

Advanced: 1, 2, 8, 14, 22–28 even, 36–40 even, 41–55

HOMEWORK CHECK

Basic: 3, 9, 15, 39

Average: 4, 10, 16, 40

Advanced: 8, 14, 22, 40

ANSWERS

1. They have opposite signs; When factoring $x^2 + bx + c = (x + p)(x + q)$, if c is negative, p and q must have opposite signs.
2. *Sample answer:* $x^2 + 5x + 6$
3. $(x + 1)(x + 7)$
4. $(z + 3)(z + 7)$
5. $(n + 4)(n + 5)$
6. $(s + 5)(s + 6)$
7. $(h + 2)(h + 9)$
8. $(y + 5)(y + 8)$
9. $(v - 1)(v - 4)$
10. $(x - 2)(x - 11)$
11. $(d - 2)(d - 3)$
12. $(k - 4)(k - 6)$
13. $(w - 8)(w - 9)$
14. $(j - 6)(j - 7)$
15. $(x - 1)(x + 4)$
16. $(z - 2)(z + 9)$
17. $(n - 2)(n + 6)$
18. $(s - 5)(s + 8)$
19. $(y - 6)(y + 8)$
20. $(h - 3)(h + 9)$
21. $(x + 4)(x - 5)$
22. $(m + 1)(m - 7)$
23. $(t + 2)(t - 8)$
24. $(y + 3)(y - 10)$
25. **a.** $(x - 5)$ ft
    b. 16 ft
26. **a.** $(x - 5)$ m
    b. 390 m^2
27. $4 + 12$ is not 14; $= (x + 6)(x + 8)$
28. c is -60, so the signs of the factors must be opposite; $= (s + 3)(s - 20)$
29. $m = -1, m = -2$
30. $n = 3, n = 6$
31. $x = 2, x = -7$
32. $v = 2, v = -13$
33. $t = -3, t = -12$
34. $n = -3, n = 8$
35. $a = 5, a = -10$
36. $y = -3, y = 5$
37. $m = 4, m = 11$
38. $b = 3, b = 5$

Dynamic Teaching Tools

- Dynamic Assessment & Progress Monitoring Tool
- Interactive Whiteboard Lesson Library
- Dynamic Classroom with Dynamic Investigations

ANSWERS

39. 100 in.^{2}

40. **a.** 6 in.

b. length: 13 in., width: 8 in.

41. yes; p and q must be factors of -12 that have a sum of b, and -12 has 6 sets of integer factors, -1 and 12, -2 and 6, -3 and 4, -4 and 3, -6 and 2, and -12 and 1.

42. **a.**

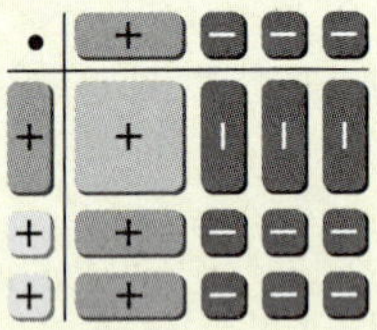

$x^2 - x - 6 = (x + 2)(x - 3)$

b.

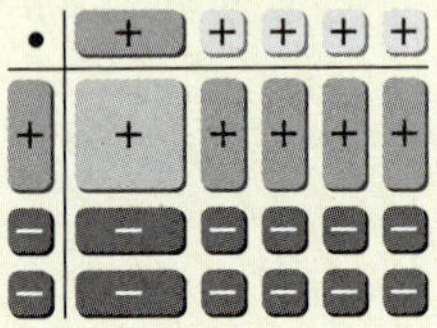

$x^2 + 2x - 8 = (x - 2)(x + 4)$

43. length: 11 ft, width: 4 ft

44. base: 10 m, height: 7 m

45–55. See Additional Answers.

Mini-Assessment

Factor the polynomial.

1. $x^2 + 13x + 12$ $(x + 1)(x + 12)$
2. $x^2 - 14x + 40$ $(x - 4)(x - 10)$
3. $x^2 + 3x - 54$ $(x - 6)(x + 9)$
4. A homeowner puts a patio in the northwest corner of a square back yard. The area of the patio is 750 square feet. What is the area of the backyard?

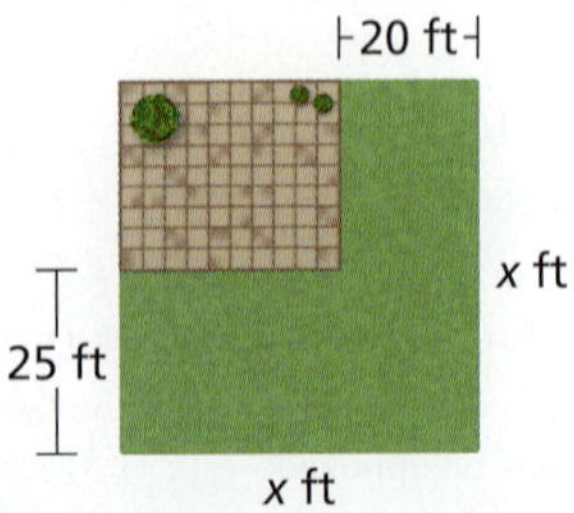

2500 square feet

39. MODELING WITH MATHEMATICS You trimmed a large square picture so that you could fit it into a frame. The area of the cut picture is 20 square inches. What is the area of the original picture? *(See Example 4.)*

40. MODELING WITH MATHEMATICS A web browser is open on your computer screen.

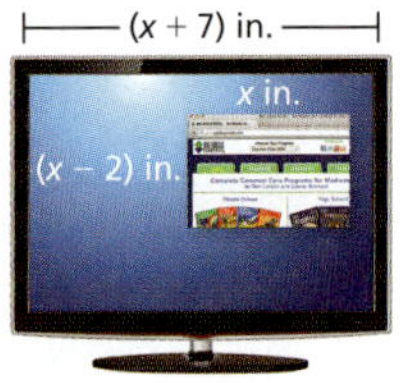

a. The area of the browser window is 24 square inches. Find the length of the browser window x.

b. The browser covers $\frac{3}{13}$ of the screen. What are the dimensions of the screen?

41. MAKING AN ARGUMENT Your friend says there are six integer values of b for which the trinomial $x^2 + bx - 12$ has two binomial factors of the form $(x + p)$ and $(x + q)$. Is your friend correct? Explain.

42. THOUGHT PROVOKING Use algebra tiles to factor each polynomial modeled by the tiles. Show your work.

a.

b.

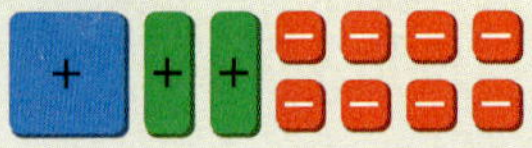

MATHEMATICAL CONNECTIONS In Exercises 43 and 44, find the dimensions of the polygon with the given area.

43. Area = 44 ft^2

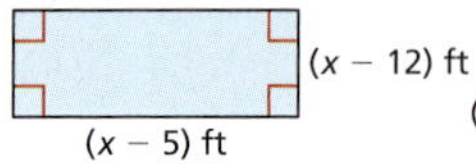

44. Area = 35 m^2

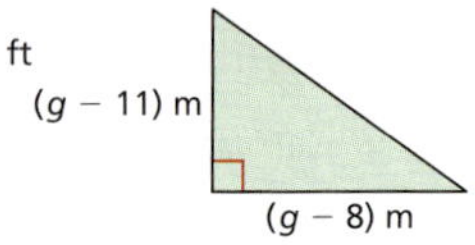

45. REASONING Write an equation of the form $x^2 + bx + c = 0$ that has the solutions $x = -4$ and $x = 6$. Explain how you found your answer.

46. HOW DO YOU SEE IT? The graph of $y = x^2 + x - 6$ is shown.

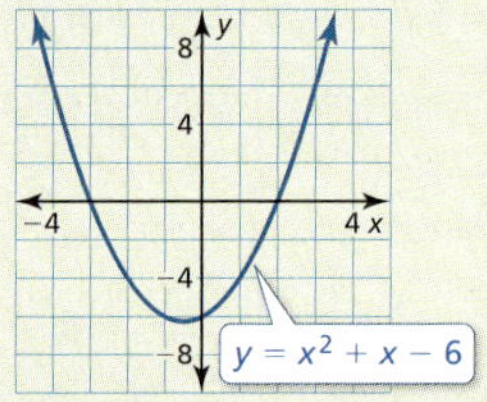

a. Explain how you can use the graph to factor the polynomial $x^2 + x - 6$.

b. Factor the polynomial.

47. PROBLEM SOLVING Road construction workers are paving the area shown.

a. Write an expression that represents the area being paved.

b. The area being paved is 280 square meters. Write and solve an equation to find the width of the road x.

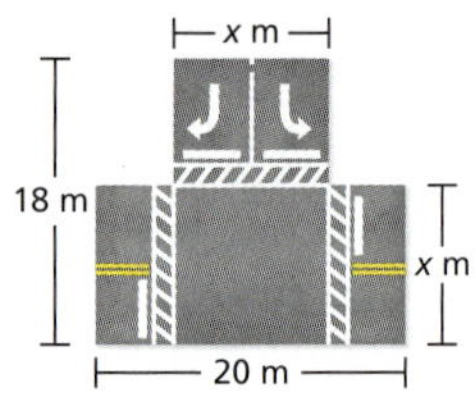

USING STRUCTURE In Exercises 48–51, factor the polynomial.

48. $x^2 + 6xy + 8y^2$

49. $r^2 + 7rs + 12s^2$

50. $a^2 + 11ab - 26b^2$

51. $x^2 - 2xy - 35y^2$

Maintaining Mathematical Proficiency

Reviewing what you learned in previous grades and lessons

Solve the equation. Check your solution. *(Section 1.1)*

52. $p - 9 = 0$

53. $z + 12 = -5$

54. $6 = \frac{c}{-7}$

55. $4k = 0$

If students need help...	*If students got it...*
Resources by Chapter • Practice A and Practice B • Puzzle Time	Resources by Chapter • Enrichment and Extension • Cumulative Review
Student Journal • Practice	Start the *next* Section
Differentiating the Lesson Skills Review Handbook	

Laurie's Notes

Overview of Section 7.6

Introduction

- In this lesson, students factor trinomials of the form $ax^2 + bx + c$. In factoring trinomials, any common factor should be factored out first, leaving the resulting coefficients with fewer factors to work with.
- Students will recognize that the fewest combinations occur when both a and c are prime. When either a or c is composite, there are more combinations to check. When both a and c are composite, there are even more combinations to check.
- The lesson ends with a real-life problem where once the polynomial has been factored and set equal to zero, the Zero-Product Property is used to solve the equation.

Teaching Strategy

- Factoring trinomials in which the leading coefficient is not equal to 1 can be challenging for some students. They struggle to hold several computations in their heads. It is helpful to have a structure that allows students to test out different combinations quickly.
- *Example:* If $2x^2 + 13x + 6$ factors over the integers, then the binomial factors must be $(x + \_\_)(2x + \_\_)$. We also know that the factors of the constant term c are {1, 2, 3, 6}. The goal is to find the combination that gives a linear term of $13x$.
- We could write out the four possible combinations:

  $(x + 1)(2x + 6)$ ➡ $1x \cdot 6 + 1 \cdot 2x = 8x$
  $(x + 6)(2x + 1)$ ➡ $1x \cdot 1 + 6 \cdot 2x = 13x$
  $(x + 2)(2x + 3)$ ➡ $1x \cdot 3 + 2 \cdot 2x = 7x$
  $(x + 3)(2x + 2)$ ➡ $1x \cdot 2 + 3 \cdot 2x = 8x$

- There is a symbolic way of representing all four cases. For each set of brackets, sum the products along both solid lines; sum the products along the dotted lines. The sum that is 13 is highlighted. Write the binomials that give the same calculations when multiplied.

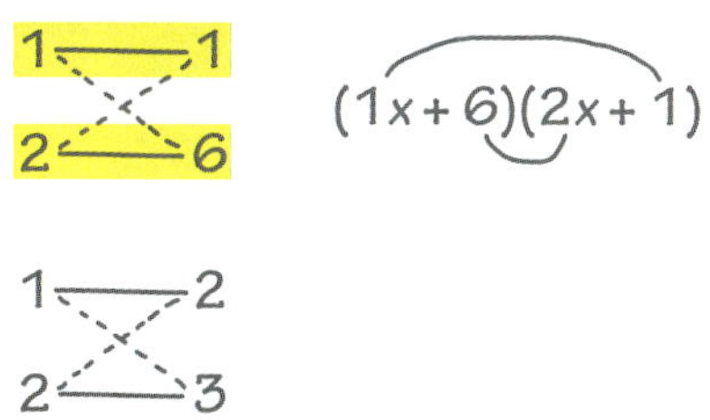

- The list of possible factor combinations can be written vertically for students to check. This helps students who need a structure for listing the possible combinations.

Pacing Suggestion

- Complete the exploration and discuss students' observations. Transition to the formal lesson.

Dynamic Teaching Tools

- Dynamic Assessment & Progress Monitoring Tool
- Lesson Planning Tool
- Interactive Whiteboard Lesson Library
- Dynamic Classroom with Dynamic Investigations

What Your Students Will Learn

- Factor $ax^2 + bx + c$ when ac is positive, when ac is negative, and when a is negative.
- Use factoring to solve real-life problems involving polynomial equations.

Laurie's Notes

Exploration

Motivate

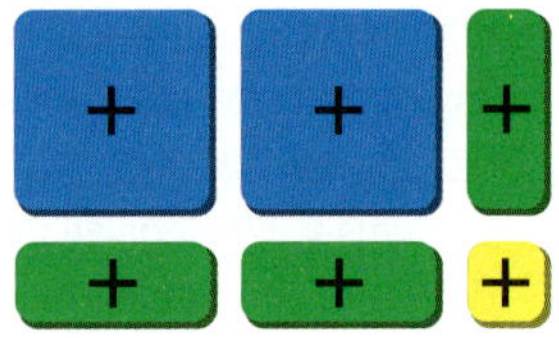

- Model $2x^2 + 3x + 1$ with tiles on the overhead. Have students do the same at their desks.

? "Can you arrange the algebra tiles so they form a rectangle with no holes or overlaps?" yes

- When students try to place all three x-tiles vertically, there is no place for the unit tile to fit. Model this when students do not make the attempt.

? "What are the dimensions of the rectangle?" $(x + 1)$ and $(2x + 1)$

- Label the dimensions. Explain that $2x^2 + 3x + 1 = (x + 1)(2x + 1)$.
- **Use Appropriate Tools Strategically:** This is similar to the visual model used when whole numbers are multiplied using base-ten blocks.

Exploration 1

- Ask students to form a rectangular array to model $2x^2 + 5x + 2$. The two x^2-tiles need to be placed side by side, either vertically or horizontally.
- Show the following three arrangements and explain that we use the one on the left because there is space for the two remaining unit tiles.

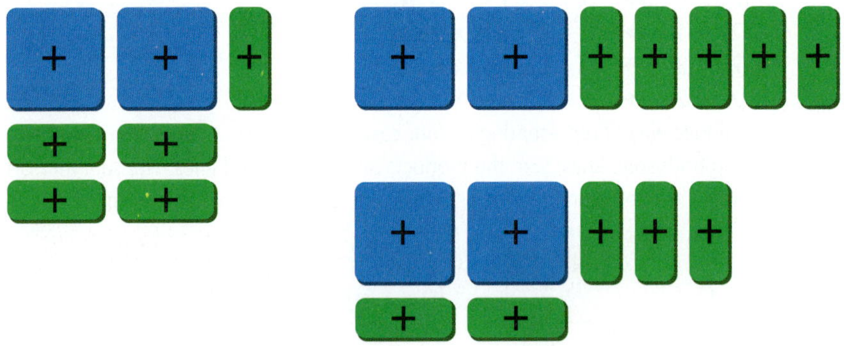

? "What are the dimensions of the rectangle?" $(x + 2)$ and $(2x + 1)$

- Label the dimensions. Explain that $2x^2 + 5x + 2 = (x + 2)(2x + 1)$.
- As students work part (a) with their partners, encourage them to think about the dimensions of the rectangle.
- Part (b) has been started so it eliminates the possibility of $(4x + \_\_)(x + \_\_)$.
- Part (c) has only one possibility for the x^2-tiles, though students need to pay attention to signs as they work the problem.

Communicate Your Answer

- **Use Appropriate Tools Strategically:** The algebra tiles provide a visual confirmation that the trinomial in Question 3 is not factorable.

Connecting to Next Step

- The exploration helps students visualize and make sense of factoring a trinomial of the form $ax^2 + bx + c$. Once explored, students will be better able to perform the algebraic manipulations in the formal lesson.

7.6 Factoring $ax^2 + bx + c$

Essential Question

How can you use algebra tiles to factor the trinomial $ax^2 + bx + c$ into the product of two binomials?

EXPLORATION 1 **Finding Binomial Factors**

Work with a partner. Use algebra tiles to write each polynomial as the product of two binomials. Check your answer by multiplying.

Sample $2x^2 + 5x + 2$

Step 1 Arrange algebra tiles that model $2x^2 + 5x + 2$ into a rectangular array.

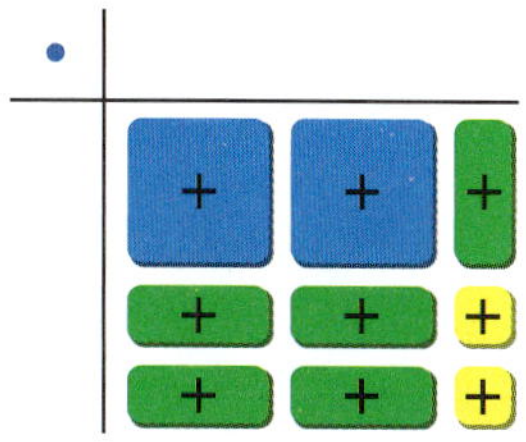

Step 2 Use additional algebra tiles to model the dimensions of the rectangle.

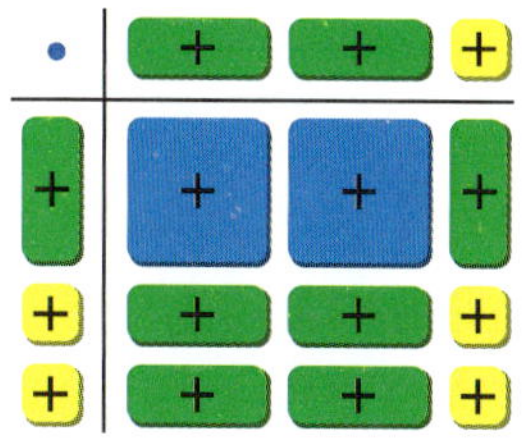

Step 3 Write the polynomial in factored form using the dimensions of the rectangle.

width, length

Area $= 2x^2 + 5x + 2 = (x + 2)(2x + 1)$

a. $3x^2 + 5x + 2 =$ ______

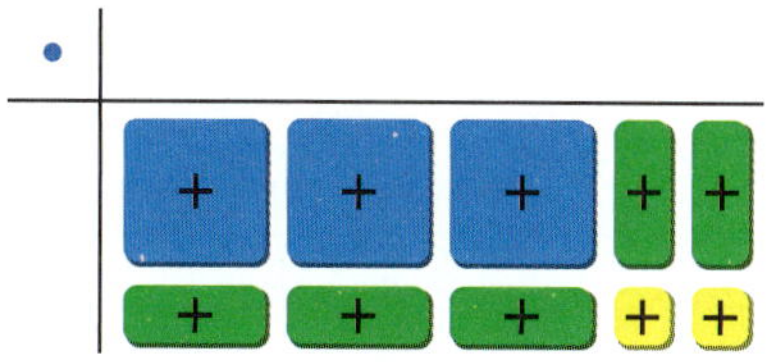

b. $4x^2 + 4x - 3 =$ ______

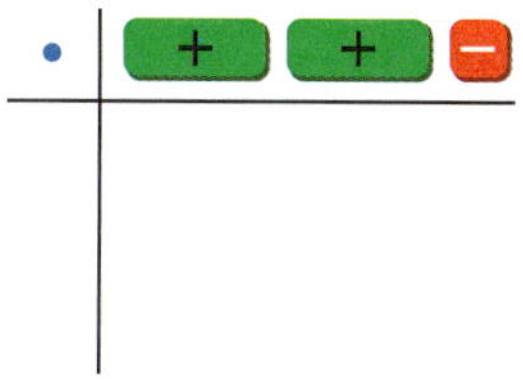

c. $2x^2 - 11x + 5 =$ ______

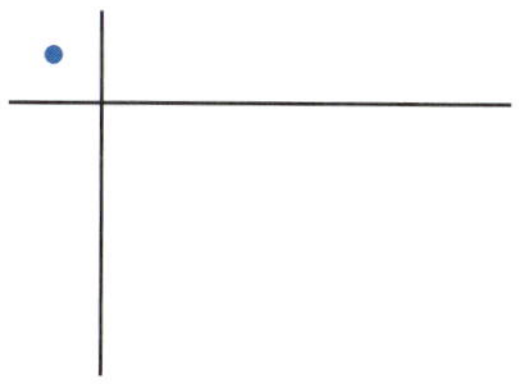

USING TOOLS STRATEGICALLY

To be proficient in math, you need to consider the available tools, including concrete models, when solving a mathematical problem.

Communicate Your Answer

2. How can you use algebra tiles to factor the trinomial $ax^2 + bx + c$ into the product of two binomials?

3. Is it possible to factor the trinomial $2x^2 + 2x + 1$? Explain your reasoning.

ANSWERS

1. a. $(x + 1)(3x + 2)$

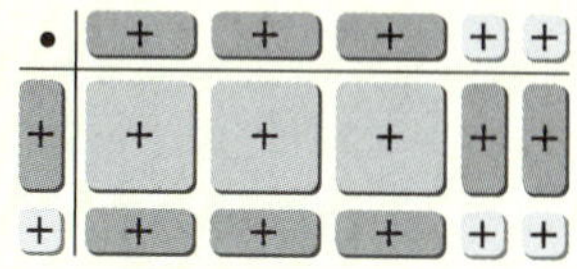

b. $(2x + 3)(2x - 1)$

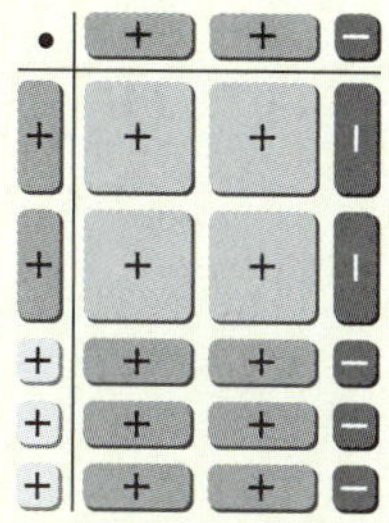

c. $(x - 5)(2x - 1)$

2. Arrange algebra tiles that model the trinomial into a rectangular array, use additional algebra tiles to model the dimensions of the rectangle, then write the polynomial in factored form using the dimensions of the rectangle.

3. no; There is no way to model this expression as a rectangular array.

Extra Example 1

Factor $4x^2 + 32x + 60$.

$4(x + 3)(x + 5)$

Extra Example 2

Factor each polynomial.

a. $2x^2 + 7x + 6$ $(2x + 3)(x + 2)$

b. $4x^2 - 7x + 3$ $(4x - 3)(x - 1)$

7.6 Lesson

Core Vocabulary

Previous
polynomial
greatest common factor (GCF)
Zero-Product Property

What You Will Learn

- Factor $ax^2 + bx + c$.
- Use factoring to solve real-life problems.

Factoring $ax^2 + bx + c$

In Section 7.5, you factored polynomials of the form $ax^2 + bx + c$, where $a = 1$. To factor polynomials of the form $ax^2 + bx + c$, where $a \neq 1$, first look for the GCF of the terms of the polynomial and then factor further if possible.

EXAMPLE 1 Factoring Out the GCF

Factor $5x^2 + 15x + 10$.

SOLUTION

Notice that the GCF of the terms $5x^2$, $15x$, and 10 is 5.

$5x^2 + 15x + 10 = 5(x^2 + 3x + 2)$ Factor out GCF.

$= 5(x + 1)(x + 2)$ Factor $x^2 + 3x + 2$.

▶ So, $5x^2 + 15x + 10 = 5(x + 1)(x + 2)$.

When there is no GCF, consider the possible factors of a and c.

EXAMPLE 2 Factoring $ax^2 + bx + c$ When ac Is Positive

Factor each polynomial.

a. $4x^2 + 13x + 3$ **b.** $3x^2 - 7x + 2$

SOLUTION

a. There is no GCF, so you need to consider the possible factors of a and c. Because b and c are both positive, the factors of c must be positive. Use a table to organize information about the factors of a and c.

STUDY TIP

You must consider the order of the factors of 3, because the middle terms formed by the possible factorizations are different.

Factors of 4	Factors of 3	Possible factorization	Middle term	
1, 4	1, 3	$(x + 1)(4x + 3)$	$3x + 4x = 7x$	✗
1, 4	3, 1	$(x + 3)(4x + 1)$	$x + 12x = 13x$	✓
2, 2	1, 3	$(2x + 1)(2x + 3)$	$6x + 2x = 8x$	✗

▶ So, $4x^2 + 13x + 3 = (x + 3)(4x + 1)$.

b. There is no GCF, so you need to consider the possible factors of a and c. Because b is negative and c is positive, both factors of c must be negative. Use a table to organize information about the factors of a and c.

Factors of 3	Factors of 2	Possible factorization	Middle term	
1, 3	−1, −2	$(x - 1)(3x - 2)$	$-2x - 3x = -5x$	✗
1, 3	−2, −1	$(x - 2)(3x - 1)$	$-x - 6x = -7x$	✓

▶ So, $3x^2 - 7x + 2 = (x - 2)(3x - 1)$.

Laurie's Notes Teacher Actions

- Students now understand that the ability to factor a trinomial is related to the values of a, b, and c. Factoring out the GCF will make sense!
- **Turn and Talk** and **Popsicle Sticks:** "When a and c are positive, what does the sign of b tell you?" Use *Popsicle Sticks* to solicit an explanation.
- The more factors a and c have, the more combinations must be checked. See the *Teaching Strategy* on page T-390 for a structure to list and check the various combinations.

EXAMPLE 3 Factoring $ax^2 + bx + c$ When ac Is Negative

Factor $2x^2 - 5x - 7$.

SOLUTION

There is no GCF, so you need to consider the possible factors of a and c. Because c is negative, the factors of c must have different signs. Use a table to organize information about the factors of a and c.

Factors of 2	Factors of −7	Possible factorization	Middle term	
1, 2	1, −7	$(x + 1)(2x - 7)$	$-7x + 2x = -5x$	✓
1, 2	7, −1	$(x + 7)(2x - 1)$	$-x + 14x = 13x$	✗
1, 2	−1, 7	$(x - 1)(2x + 7)$	$7x - 2x = 5x$	✗
1, 2	−7, 1	$(x - 7)(2x + 1)$	$x - 14x = -13x$	✗

▶ So, $2x^2 - 5x - 7 = (x + 1)(2x - 7)$.

STUDY TIP

When a is negative, factor −1 from each term of $ax^2 + bx + c$. Then factor the resulting trinomial as in the previous examples.

EXAMPLE 4 Factoring $ax^2 + bx + c$ When a Is Negative

Factor $-4x^2 - 8x + 5$.

SOLUTION

Step 1 Factor −1 from each term of the trinomial.

$$-4x^2 - 8x + 5 = -(4x^2 + 8x - 5)$$

Step 2 Factor the trinomial $4x^2 + 8x - 5$. Because c is negative, the factors of c must have different signs. Use a table to organize information about the factors of a and c.

Factors of 4	Factors of −5	Possible factorization	Middle term	
1, 4	1, −5	$(x + 1)(4x - 5)$	$-5x + 4x = -x$	✗
1, 4	5, −1	$(x + 5)(4x - 1)$	$-x + 20x = 19x$	✗
1, 4	−1, 5	$(x - 1)(4x + 5)$	$5x - 4x = x$	✗
1, 4	−5, 1	$(x - 5)(4x + 1)$	$x - 20x = -19x$	✗
2, 2	1, −5	$(2x + 1)(2x - 5)$	$-10x + 2x = -8x$	✗
2, 2	−1, 5	$(2x - 1)(2x + 5)$	$10x - 2x = 8x$	✓

▶ So, $-4x^2 - 8x + 5 = -(2x - 1)(2x + 5)$.

Monitoring Progress

Help in English and Spanish at *BigIdeasMath.com*

Factor the polynomial.

1. $8x^2 - 56x + 48$
2. $14x^2 + 31x + 15$
3. $2x^2 - 7x + 5$
4. $3x^2 - 14x + 8$
5. $4x^2 - 19x - 5$
6. $6x^2 + x - 12$
7. $-2y^2 - 5y - 3$
8. $-5m^2 + 6m - 1$
9. $-3x^2 - x + 2$

Differentiated Instruction

Organization

Some students may benefit from completing a table to show the signs of the factors of c. Students can use their tables to check the signs of the factors of c in their factorizations.

$ax^2 + bx + c$

b	c	Factors of c
positive	positive	both positive
positive	negative	different signs
negative	positive	both negative
negative	negative	different signs

Extra Example 3

Factor $3x^2 - 7x - 6$.

$(x - 3)(3x + 2)$

Extra Example 4

Factor $-9x^2 - 3x + 2$.

$-(3x - 1)(3x + 2)$

MONITORING PROGRESS ANSWERS

1. $8(x - 1)(x - 6)$
2. $(7x + 5)(2x + 3)$
3. $(x - 1)(2x - 5)$
4. $(3x - 2)(x - 4)$
5. $(4x + 1)(x - 5)$
6. $(2x + 3)(3x - 4)$
7. $-(y + 1)(2y + 3)$
8. $-(m - 1)(5m - 1)$
9. $-(x + 1)(3x - 2)$

Laurie's Notes Teacher Actions

? Write Example 3. "Because the constant term is −7, what do you know about the linear factors?" They will have different signs.

- Some students will have little difficulty using mental math to check the combinations. After modeling a problem, students must try problems independently. Watching someone factor trinomials cannot replicate the independent thought process that students need to experience.
- **Look For and Make Use of Structure:** In Example 4, factoring out a (−1) is not an intuitive step for students. Make the connection to the Distributive Property again.

Extra Example 5

The length of a rectangular state park is 2 miles longer than twice the width. The area of the park is 84 square miles. What is the width of the park? The width of the park is 6 miles.

MONITORING PROGRESS ANSWER

10. 8 mi

Solving Real-Life Problems

EXAMPLE 5 Solving a Real-Life Problem

The length of a rectangular game reserve is 1 mile longer than twice the width. The area of the reserve is 55 square miles. What is the width of the reserve?

SOLUTION

Use the formula for the area of a rectangle to write an equation for the area of the reserve. Let w represent the width. Then $2w + 1$ represents the length. Solve for w.

$w(2w + 1) = 55$ Area of the reserve

$2w^2 + w = 55$ Distributive Property

$2w^2 + w - 55 = 0$ Subtract 55 from each side.

Factor the left side of the equation. There is no GCF, so you need to consider the possible factors of a and c. Because c is negative, the factors of c must have different signs. Use a table to organize information about the factors of a and c.

Factors of 2	Factors of −55	Possible factorization	Middle term	
1, 2	1, −55	$(w + 1)(2w - 55)$	$-55w + 2w = -53w$	✗
1, 2	55, −1	$(w + 55)(2w - 1)$	$-w + 110w = 109w$	✗
1, 2	−1, 55	$(w - 1)(2w + 55)$	$55w - 2w = 53w$	✗
1, 2	−55, 1	$(w - 55)(2w + 1)$	$w - 110w = -109w$	✗
1, 2	5, −11	$(w + 5)(2w - 11)$	$-11w + 10w = -w$	✗
1, 2	11, −5	$(w + 11)(2w - 5)$	$-5w + 22w = 17w$	✗
1, 2	−5, 11	$(w - 5)(2w + 11)$	$11w - 10w = w$	✓
1, 2	−11, 5	$(w - 11)(2w + 5)$	$5w - 22w = -17w$	✗

So, you can rewrite $2w^2 + w - 55$ as $(w - 5)(2w + 11)$. Write the equation with the left side factored and continue solving for w.

$(w - 5)(2w + 11) = 0$ Rewrite equation with left side factored.

$w - 5 = 0$ *or* $2w + 11 = 0$ Zero-Product Property

$w = 5$ *or* $w = -\frac{11}{2}$ Solve for w.

Check

Use mental math.

The width is 5 miles, so the length is $5(2) + 1 = 11$ miles and the area is $5(11) = 55$ square miles. ✓

A negative width does not make sense, so you should use the positive solution.

▶ So, the width of the reserve is 5 miles.

Monitoring Progress

Help in English and Spanish at *BigIdeasMath.com*

10. WHAT IF? The area of the reserve is 136 square miles. How wide is the reserve?

Laurie's Notes Teacher Actions

- **Make Sense of Problems and Persevere in Solving Them** and **Use Appropriate Tools Strategically:** Ask a student to read the problem, make a sketch, and label the dimensions of the rectangle.

? Once the trinomial is set equal to 0 ask, "To get a sum of $1w$ for the middle term, what factors of -55 might be reasonable to try first? Why?" 5 and 11, because the 5 can be multiplied by $2w$ first.

Closure

- **Exit Ticket:** Factor $2x^2 - 7x + 3$. $(2x - 1)(x - 3)$

7.6 Exercises

Dynamic Solutions available at *BigIdeasMath.com*

Vocabulary and Core Concept Check

1. **REASONING** What is the greatest common factor of the terms of $3y^2 - 21y + 36$?
2. **WRITING** Compare factoring $6x^2 - x - 2$ with factoring $x^2 - x - 2$.

Monitoring Progress and Modeling with Mathematics

In Exercises 3–8, factor the polynomial. *(See Example 1.)*

3. $3x^2 + 3x - 6$
4. $8v^2 + 8v - 48$
5. $4k^2 + 28k + 48$
6. $6y^2 - 24y + 18$
7. $7b^2 - 63b + 140$
8. $9r^2 - 36r - 45$

In Exercises 9–16, factor the polynomial. *(See Examples 2 and 3.)*

9. $3h^2 + 11h + 6$
10. $8m^2 + 30m + 7$
11. $6x^2 - 5x + 1$
12. $10w^2 - 31w + 15$
13. $3n^2 + 5n - 2$
14. $4z^2 + 4z - 3$
15. $8g^2 - 10g - 12$
16. $18v^2 - 15v - 18$

In Exercises 17–22, factor the polynomial. *(See Example 4.)*

17. $-3t^2 + 11t - 6$
18. $-7v^2 - 25v - 12$
19. $-4c^2 + 19c + 5$
20. $-8h^2 - 13h + 6$
21. $-15w^2 - w + 28$
22. $-22d^2 + 29d - 9$

ERROR ANALYSIS **In Exercises 23 and 24, describe and correct the error in factoring the polynomial.**

23. ✗
$$2x^2 - 2x - 24 = 2(x^2 - 2x - 24) = 2(x - 6)(x + 4)$$

24. ✗
$$6x^2 - 7x - 3 = (3x - 3)(2x + 1)$$

In Exercises 25–28, solve the equation.

25. $5x^2 - 5x - 30 = 0$
26. $2k^2 - 5k - 18 = 0$
27. $-12n^2 - 11n = -15$
28. $14b^2 - 2 = -3b$

In Exercises 29–32, find the x-coordinates of the points where the graph crosses the x-axis.

29. 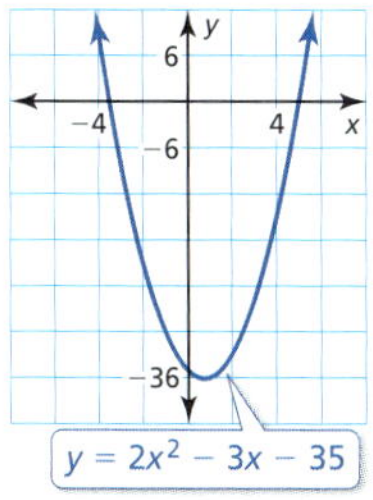

30. 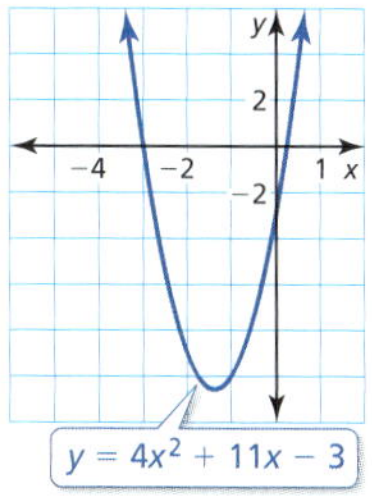

31. 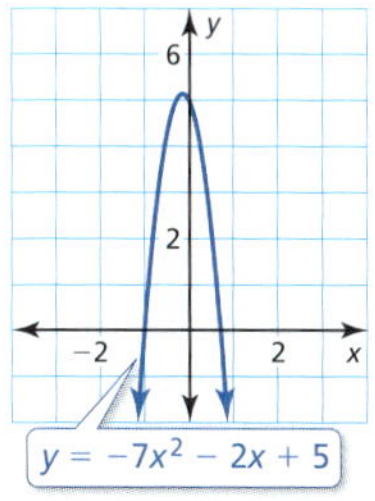

32. 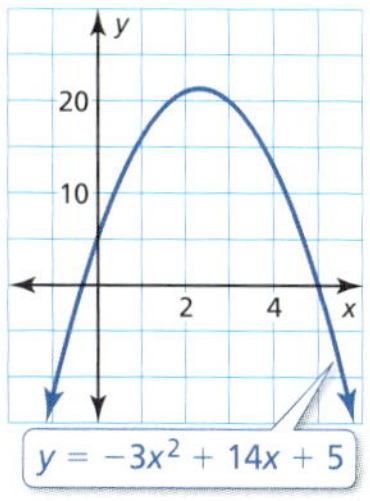

33. **MODELING WITH MATHEMATICS** The area (in square feet) of the school sign can be represented by $15x^2 - x - 2$.
    a. Write an expression that represents the length of the sign.
    b. Describe two ways to find the area of the sign when $x = 3$.

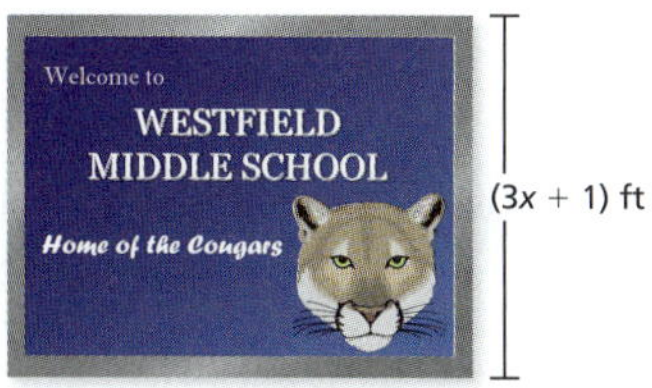

Assignment Guide and Homework Check

ASSIGNMENT

Basic: 1, 2, 3–25 odd, 33, 35, 38, 40, 49–56

Average: 1, 2–40 even, 49–56

Advanced: 1, 2, 8, 12–36 even, 37–56

HOMEWORK CHECK

Basic: 3, 9, 13, 17, 35

Average: 4, 10, 14, 18, 34

Advanced: 8, 12, 16, 20, 36

ANSWERS

1. 3
2. Factoring $6x^2 - x - 2$ requires considering factors of 6 and -2 in different combinations until the combination is found that produces the correct middle term. Factoring $x^2 - x - 2$ only requires finding the factors of -2 that add up to -1.
3. $3(x - 1)(x + 2)$
4. $8(v - 2)(v + 3)$
5. $4(k + 3)(k + 4)$
6. $6(y - 1)(y - 3)$
7. $7(b - 4)(b - 5)$
8. $9(r + 1)(r - 5)$
9. $(3h + 2)(h + 3)$
10. $(2m + 7)(4m + 1)$
11. $(2x - 1)(3x - 1)$
12. $(2w - 5)(5w - 3)$
13. $(n + 2)(3n - 1)$
14. $(2z - 1)(2z + 3)$
15. $2(g - 2)(4g + 3)$
16. $3(2v - 3)(3v + 2)$
17. $-(t - 3)(3t - 2)$
18. $-(v + 3)(7v + 4)$
19. $-(c - 5)(4c + 1)$
20. $-(h + 2)(8h - 3)$
21. $-(3w - 4)(5w + 7)$
22. $-(2d - 1)(11d - 9)$
23. need to factor 2 out of every term; $= 2(x^2 - x - 12) = 2(x + 3)(x - 4)$
24. These factors do not give the correct middle term; $= (2x - 3)(3x + 1)$
25. $x = -2, x = 3$
26. $k = -2, k = \frac{9}{2}$
27. $n = -\frac{5}{3}, n = \frac{3}{4}$
28. $b = -\frac{1}{2}, b = \frac{2}{7}$
29. $x = -\frac{7}{2}, x = 5$
30. $x = -3, x = \frac{1}{4}$
31. $x = -1, x = \frac{5}{7}$
32. $x = -\frac{1}{3}, x = 5$
33. a. $(5x - 2)$ ft
    b. Substitute 3 for x into the expression for the area $15x^2 - x - 2$, then simplify; Substitute 3 for x into the expressions for the length $(5x - 2)$ and width $(3x + 1)$, simplify each, then multiply these two numbers.

Dynamic Teaching Tools

- Dynamic Assessment & Progress Monitoring Tool
- Interactive Whiteboard Lesson Library
- Dynamic Classroom with Dynamic Investigations

ANSWERS

34. 2.5 sec

35. length: 70 m, width: 31 m

36. yes; The length of the invitation is 5 inches, which is less than $5\frac{1}{8}$ inches. The width of the invitation is 3 inches, which is less than $3\frac{5}{8}$ inches.

37. *Sample answer:* $6x^2 + 3x$

38. The graph of k represents function g, and the graph of ℓ represents function h; Because c is positive, the constant terms in the factors must have the same sign. Because g has a positive value of b, the constant terms of the factors will both be positive, which results in negative roots, and k has two negative x-intercepts. Because h has a negative value of b, the constant terms of the factors will both be negative, which results in positive roots, and ℓ has two positive x-intercepts.

39. when no combination of factors of a and c produce the correct middle term; *Sample answer:* $2x^2 + x + 1$

40. no; To use the Zero-Product Property, one side of the equation needs to be 0. So, you must first subtract 2 from each side of the equation, then factor.

41. $\pm 9, \pm 12, \pm 21$

42–56. See Additional Answers.

Mini-Assessment

Factor each polynomial.

1. $2x^2 + 14x + 24$ $2(x + 3)(x + 4)$
2. $3x^2 + 11x + 10$ $(3x + 5)(x + 2)$
3. $2x^2 - 7x - 4$ $(2x + 1)(x - 4)$
4. $-3x^2 - 4x + 7$ $-(3x + 7)(x - 1)$
5. The length of a rectangular garden is 4 yards less than twice the width. The area of the garden is 96 square yards. What is the width of the garden? The width of the garden is 8 yards.

34. MODELING WITH MATHEMATICS The height h (in feet) above the water of a cliff diver is modeled by $h = -16t^2 + 8t + 80$, where t is the time (in seconds). How long is the diver in the air?

35. MODELING WITH MATHEMATICS The Parthenon in Athens, Greece, is an ancient structure that has a rectangular base. The length of the base of the Parthenon is 8 meters more than twice its width. The area of the base is about 2170 square meters. Find the length and width of the base. *(See Example 5.)*

36. MODELING WITH MATHEMATICS The length of a rectangular birthday party invitation is 1 inch less than twice its width. The area of the invitation is 15 square inches. Will the invitation fit in the envelope shown without being folded? Explain.

37. OPEN-ENDED Write a binomial whose terms have a GCF of $3x$.

38. HOW DO YOU SEE IT? Without factoring, determine which of the graphs represents the function $g(x) = 21x^2 + 37x + 12$ and which represents the function $h(x) = 21x^2 - 37x + 12$. Explain your reasoning.

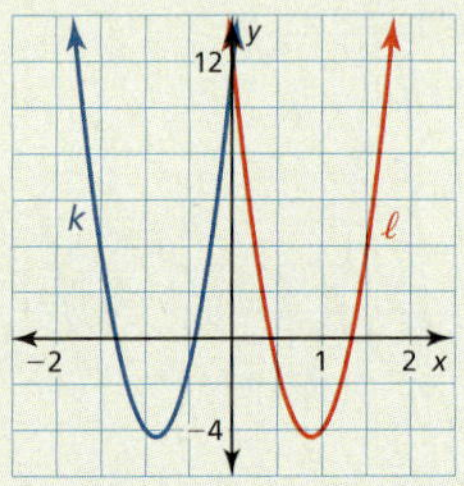

39. REASONING When is it not possible to factor $ax^2 + bx + c$, where $a \neq 1$? Give an example.

40. MAKING AN ARGUMENT Your friend says that to solve the equation $5x^2 + x - 4 = 2$, you should start by factoring the left side as $(5x - 4)(x + 1)$. Is your friend correct? Explain.

41. REASONING For what values of t can $2x^2 + tx + 10$ be written as the product of two binomials?

42. THOUGHT PROVOKING Use algebra tiles to factor each polynomial modeled by the tiles. Show your work.

43. MATHEMATICAL CONNECTIONS The length of a rectangle is 1 inch more than twice its width. The value of the area of the rectangle (in square inches) is 5 more than the value of the perimeter of the rectangle (in inches). Find the width.

44. PROBLEM SOLVING A rectangular swimming pool is bordered by a concrete patio. The width of the patio is the same on every side. The area of the surface of the pool is equal to the area of the patio. What is the width of the patio?

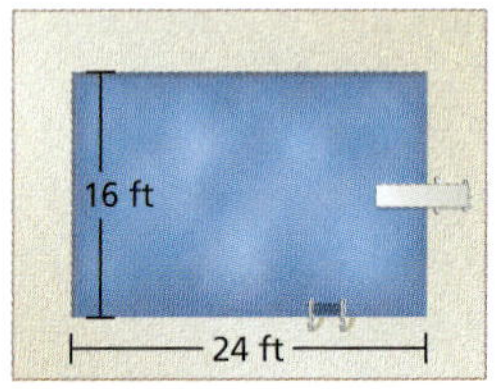

In Exercises 45–48, factor the polynomial.

45. $4k^2 + 7jk - 2j^2$

46. $6x^2 + 5xy - 4y^2$

47. $-6a^2 + 19ab - 14b^2$

48. $18m^3 + 39m^2n - 15mn^2$

Maintaining Mathematical Proficiency

Reviewing what you learned in previous grades and lessons

Find the square root(s). *(Skills Review Handbook)*

49. $\pm\sqrt{64}$ **50.** $\sqrt{4}$ **51.** $-\sqrt{225}$ **52.** $\pm\sqrt{81}$

Solve the system of linear equations by substitution. Check your solution. *(Section 5.2)*

53. $y = 3 + 7x$, $y - x = -3$

54. $2x = y + 2$, $-x + 3y = 14$

55. $5x - 2y = 14$, $-7 = -2x + y$

56. $-x - 8 = -y$, $9y - 12 + 3x = 0$

If students need help...	*If students got it...*
Resources by Chapter • Practice A and Practice B • Puzzle Time	Resources by Chapter • Enrichment and Extension • Cumulative Review
Student Journal • Practice	Start the *next* Section
Differentiating the Lesson Skills Review Handbook	

Laurie's Notes

Overview of Section 7.7

Introduction

- This lesson on factoring special products is the inverse of finding special products earlier in the chapter. Instead of multiplying to find the product, $(x - a)(x + a) = x^2 - a^2$, in this lesson students will factor and write $x^2 - a^2 = (x - a)(x + a)$.
- In addition to factoring the difference of two squares and factoring a perfect square trinomial, factoring is used to solve a real-life problem, again using the Zero-Product Property.

Teaching Strategy

- This is frequently a challenging lesson for students, mainly because some students will say that they do not see the need for memorizing more "rules," the name they give for these patterns.
- This is particularly true for the perfect square trinomial. Given the trinomial $n^2 + 8n + 16$, students say, "I need two numbers whose product is 16 and whose sum is 8." They write $(n + 4)(n + 4)$. Certainly the result is the same.
- I do not require my students to memorize the factor pattern for the perfect square trinomials. I do suggest that they at least examine each trinomial and ask whether there are any perfect squares in the expression.

Another Way

- To solve the quadratic $x^2 - 25 = 0$, students write the quadratic in factored form: $(x + 5)(x - 5) = 0$.
- In Chapter 9, after square roots are introduced, $x^2 - 25 = 0$ will be solved another way, by square rooting.

$$x^2 - 25 = 0$$
$$x^2 = 25$$
$$x = \pm 5$$

Pacing Suggestion

- Complete the exploration and discuss students' observations. Transition to the formal lesson.

Dynamic Teaching Tools

- Dynamic Assessment & Progress Monitoring Tool
- Lesson Planning Tool
- Interactive Whiteboard Lesson Library
- Dynamic Classroom with Dynamic Investigations

What Your Students Will Learn

- Use the special product pattern to factor the difference of two squares.
- Use special product patterns to factor perfect square trinomials.
- Use factoring to solve real-life problems involving polynomial equations.

Laurie's Notes

Exploration

Motivate

- Copy the four equations for students. If you have sets of numeral tiles (0–9), they can be used for this activity, or students can write the digits.
- **Directions:** Use the digits 0–9 to make the following statements true. Use each digit only once.

  $(x + \_\_)^2 = x^2 + 6x + \_\_$ 3; 9

  $(\_\_x - \_\_)(2x + 1) = 4x^2 - 1$ 2; 1

  $(x + \_\_)^2 = x^2 + 16x + \_\_\,\_\_$ 8; 64

  $(\_\_x - \_\_)^2 = 25x^2 - 7\_\_x + 49$ 5; 7; 0
- Completing these problems will help students review the special products from Section 7.3.

Exploration 1

- Say, "In this exploration, you are to determine the factors of the polynomial shown in the rectangular array. Pay attention to the signs."
- The standard form of the polynomial is already given. Students only need write the dimensions (factors).

? "What do you recognize about arrays (a), (b), and (c)?" Listen for students to recognize that each array is a square, not just a rectangle. Each is a special product from Section 7.3.

Exploration 2

- Before students begin, you may wish to write the three special products from Section 7.3. Ask students to help you.
- Note that all the tiles will be used. Students need to determine the signs of the missing pieces.

? "What special shape is each of the rectangular arrays?" square

Communicate Your Answer

- **Neighbor Check:** For Question 4, have students work independently, and then have their neighbors check their work. Have students discuss any discrepancies.

Connecting to Next Step

- The explorations help students visualize and make sense of factoring a trinomial that is one of the special products from an earlier lesson. Once explored, students will be better able to perform the algebraic manipulations in the formal lesson.

7.7 Factoring Special Products

Essential Question How can you recognize and factor special products?

EXPLORATION 1 Factoring Special Products

Work with a partner. Use algebra tiles to write each polynomial as the product of two binomials. Check your answer by multiplying. State whether the product is a "special product" that you studied in Section 7.3.

LOOKING FOR STRUCTURE

To be proficient in math, you need to see complicated things as single objects or as being composed of several objects.

a. $4x^2 - 1 =$ ______

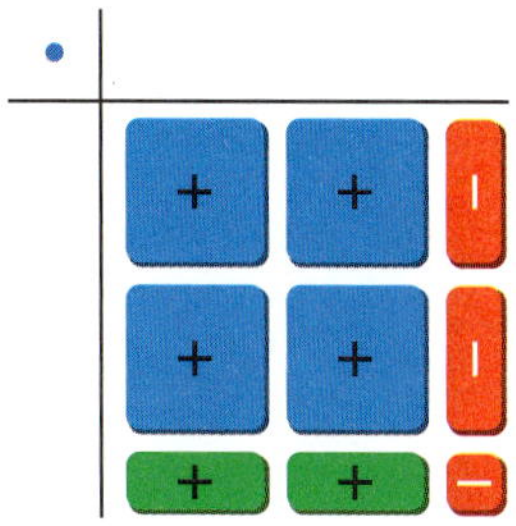

b. $4x^2 - 4x + 1 =$ ______

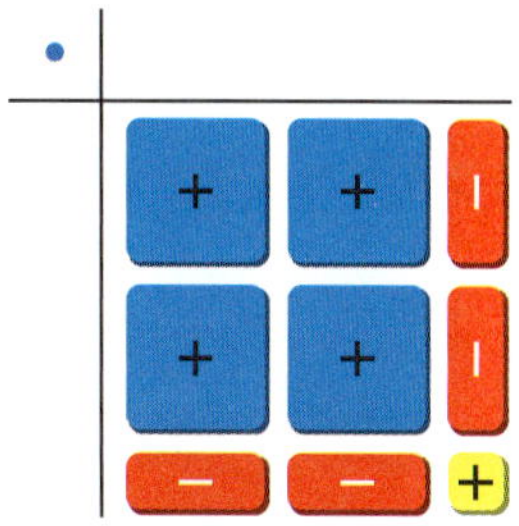

c. $4x^2 + 4x + 1 =$ ______

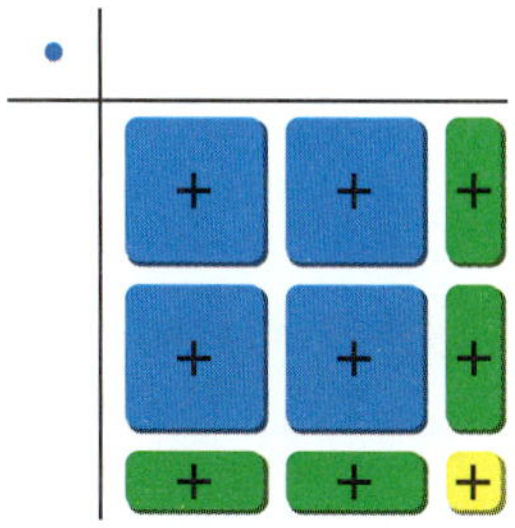

d. $4x^2 - 6x + 2 =$ ______

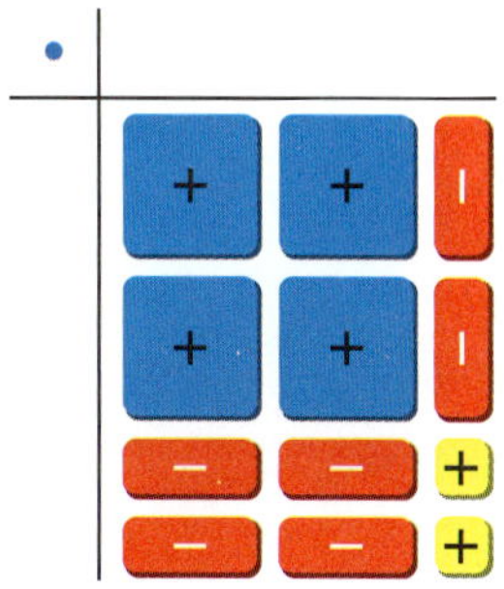

EXPLORATION 2 Factoring Special Products

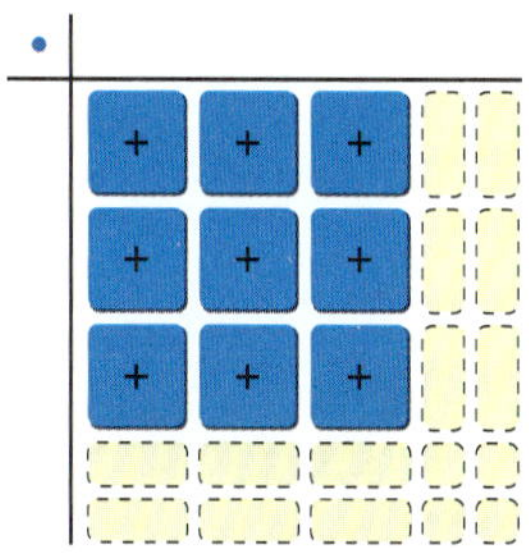

Work with a partner. Use algebra tiles to complete the rectangular array at the left in three different ways, so that each way represents a different special product. Write each special product in standard form and in factored form.

Communicate Your Answer

3. How can you recognize and factor special products? Describe a strategy for recognizing which polynomials can be factored as special products.
4. Use the strategy you described in Question 3 to factor each polynomial.
   a. $25x^2 + 10x + 1$ **b.** $25x^2 - 10x + 1$ **c.** $25x^2 - 1$

Dynamic Teaching Tools

Dynamic Assessment & Progress Monitoring Tool

Lesson Planning Tool

Interactive Whiteboard Lesson Library

Dynamic Classroom with Dynamic Investigations

ANSWERS

1. **a.** $(2x + 1)(2x - 1)$; yes, sum and difference pattern

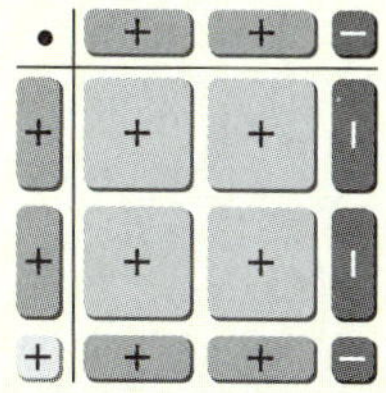

   b. $(2x - 1)^2$; yes, square of a binomial pattern

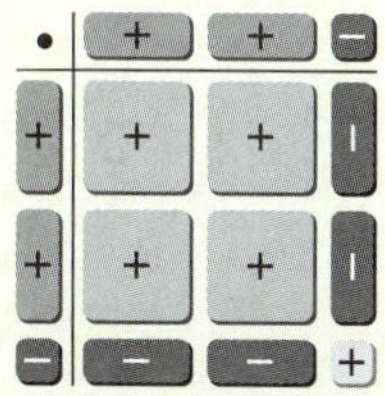

   c. $(2x + 1)^2$; yes, square of a binomial pattern

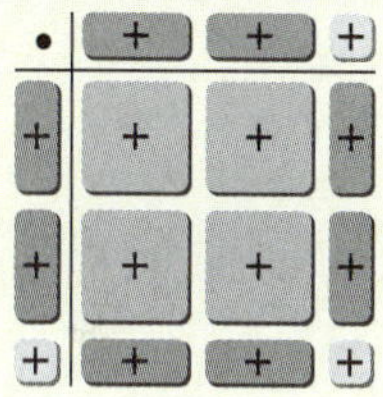

   d. $(2x - 2)(2x - 1)$; no, not a special pattern

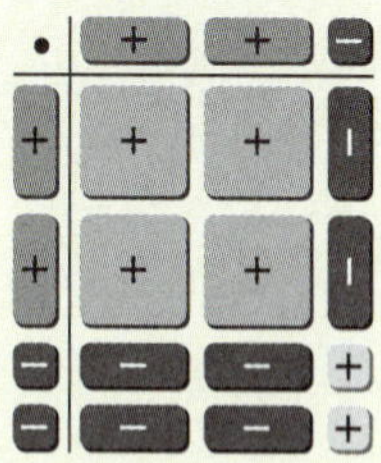

2. See Additional Answers.

3. *Sample answer:* The algebra tiles for special patterns will always form a square array, and the x^2 tiles will also form a square array. Factor special products by using the special patterns studied in Lesson 7.3, but in reversed order; For the sum and difference pattern, the number of x and $-x$ tiles will be the same, and there will only be -1 tiles to complete the array. For the square of a binomial pattern, the x tiles will either be all positive or all negative and there will only be $+1$ tiles to complete the array.

4. **a.** $(5x + 1)^2$
   b. $(5x - 1)^2$
   c. $(5x + 1)(5x - 1)$

Extra Example 1

a. Factor $x^2 - 64$. $(x + 8)(x - 8)$

b. Factor $25b^2 - 36$. $(5b + 6)(5b - 6)$

Extra Example 2

Use a special product pattern to evaluate the expression $68^2 - 62^2$.
$(68 + 62)(68 - 62) = 130(6) = 780$

MONITORING PROGRESS ANSWERS

1. $(x + 6)(x - 6)$
2. $(10 + m)(10 - m)$
3. $(3n + 4)(3n - 4)$
4. $(4h + 7)(4h - 7)$
5. 140
6. 273
7. 525
8. 208

7.7 Lesson

Core Vocabulary

Previous
polynomial
trinomial

What You Will Learn

- Factor the difference of two squares.
- Factor perfect square trinomials.
- Use factoring to solve real-life problems.

Factoring the Difference of Two Squares

You can use special product patterns to factor polynomials.

Core Concept

Difference of Two Squares Pattern

Algebra	**Example**
$a^2 - b^2 = (a + b)(a - b)$	$x^2 - 9 = x^2 - 3^2 = (x + 3)(x - 3)$

EXAMPLE 1 Factoring the Difference of Two Squares

Factor (a) $x^2 - 25$ and (b) $4z^2 - 1$.

SOLUTION

a. $x^2 - 25 = x^2 - 5^2$ — Write as $a^2 - b^2$.

$= (x + 5)(x - 5)$ — Difference of two squares pattern

▶ So, $x^2 - 25 = (x + 5)(x - 5)$.

b. $4z^2 - 1 = (2z)^2 - 1^2$ — Write as $a^2 - b^2$.

$= (2z + 1)(2z - 1)$ — Difference of two squares pattern

▶ So, $4z^2 - 1 = (2z + 1)(2z - 1)$.

EXAMPLE 2 Evaluating a Numerical Expression

Use a special product pattern to evaluate the expression $54^2 - 48^2$.

SOLUTION

Notice that $54^2 - 48^2$ is a difference of two squares. So, you can rewrite the expression in a form that it is easier to evaluate using the difference of two squares pattern.

$54^2 - 48^2 = (54 + 48)(54 - 48)$ — Difference of two squares pattern

$= 102(6)$ — Simplify.

$= 612$ — Multiply.

▶ So, $54^2 - 48^2 = 612$.

Monitoring Progress

Help in English and Spanish at *BigIdeasMath.com*

Factor the polynomial.

1. $x^2 - 36$
2. $100 - m^2$
3. $9n^2 - 16$
4. $16h^2 - 49$

Use a special product pattern to evaluate the expression.

5. $36^2 - 34^2$
6. $47^2 - 44^2$
7. $55^2 - 50^2$
8. $28^2 - 24^2$

Laurie's Notes Teacher Actions

- Write the *Core Concept*. Point out to students that they used the reverse of this pattern to multiply binomials earlier in this chapter.
- Write the expression in part (a). Point out that x^2 and 25 are both perfect squares. Some students find it helpful to write $x^2 + 0x - 25$.
- ? In part (b) ask, "What do you notice about the numbers 4 and 1, and the expression z^2?" They are all perfect squares.
- **Think-Pair-Share:** Have students answer Questions 1–8, and then share and discuss as a class.

Factoring Perfect Square Trinomials

Core Concept

Perfect Square Trinomial Pattern

Algebra	Example
$a^2 + 2ab + b^2 = (a + b)^2$	$x^2 + 6x + 9 = x^2 + 2(x)(3) + 3^2$ $= (x + 3)^2$
$a^2 - 2ab + b^2 = (a - b)^2$	$x^2 - 6x + 9 = x^2 - 2(x)(3) + 3^2$ $= (x - 3)^2$

EXAMPLE 3 **Factoring Perfect Square Trinomials**

Factor each polynomial.

a. $n^2 + 8n + 16$ **b.** $4x^2 - 12x + 9$

SOLUTION

a. $n^2 + 8n + 16 = n^2 + 2(n)(4) + 4^2$ Write as $a^2 + 2ab + b^2$.

$= (n + 4)^2$ Perfect square trinomial pattern

▶ So, $n^2 + 8n + 16 = (n + 4)^2$.

b. $4x^2 - 12x + 9 = (2x)^2 - 2(2x)(3) + 3^2$ Write as $a^2 - 2ab + b^2$.

$= (2x - 3)^2$ Perfect square trinomial pattern

▶ So, $4x^2 - 12x + 9 = (2x - 3)^2$.

EXAMPLE 4 **Solving a Polynomial Equation**

Solve $x^2 + \frac{2}{3}x + \frac{1}{9} = 0$.

SOLUTION

$x^2 + \frac{2}{3}x + \frac{1}{9} = 0$ Write equation.

$9x^2 + 6x + 1 = 0$ Multiply each side by 9.

$(3x)^2 + 2(3x)(1) + 1^2 = 0$ Write left side as $a^2 + 2ab + b^2$.

$(3x + 1)^2 = 0$ Perfect square trinomial pattern

$3x + 1 = 0$ Zero-Product Property

$x = -\frac{1}{3}$ Solve for x.

▶ The solution is $x = -\frac{1}{3}$.

LOOKING FOR STRUCTURE

Equations of the form $(x + a)^2 = 0$ always have repeated roots of $x = -a$.

Monitoring Progress

Help in English and Spanish at *BigIdeasMath.com*

Factor the polynomial.

9. $m^2 - 2m + 1$ **10.** $d^2 - 10d + 25$ **11.** $9z^2 + 36z + 36$

Solve the equation.

12. $a^2 + 6a + 9 = 0$ **13.** $w^2 - \frac{7}{3}w + \frac{49}{36} = 0$ **14.** $n^2 - 81 = 0$

Differentiated Instruction

Inclusion

Some students may have difficulty writing a trinomial such as $4x^2 - 12x + 9$ in the form $a^2 - 2ab + b^2$ or $a^2 + 2ab + b^2$. Pair each of these students with one who understands the process. Have the students rewrite a trinomial together while discussing the process. Then provide a second trinomial for each student to rewrite, and have partners compare and justify their answers.

Extra Example 3

Factor each polynomial.

a. $x^2 + 26x + 169$ $(x + 13)^2$

b. $9x^2 - 24x + 16$ $(3x - 4)^2$

Extra Example 4

Solve $x^2 + 3x + \frac{9}{4} = 0$. $x = -\frac{3}{2}$

MONITORING PROGRESS ANSWERS

9. $(m - 1)^2$

10. $(d - 5)^2$

11. $9(z + 2)^2$

12. $a = -3$

13. $w = \frac{7}{6}$

14. $n = -9, n = 9$

Laurie's Notes Teacher Actions

- Write the *Core Concept*. Point out again that students used the reverse of this pattern to square binomials earlier in this chapter.

? Always-Sometimes-Never True: "If a and c are perfect squares in the trinomial $ax^2 + bx + c$, then you have a perfect square trinomial." sometimes true; You need to check the value of b.

- In Example 4, students may factor the trinomial as $\left(x + \frac{1}{3}\right)^2$ and the solution, when set equal to 0, is the same as the method shown.

Extra Example 5

Use the information in Example 5. Suppose the golf ball hits the top of a 17-foot oak tree. After how many seconds does the ball hit the tree? The golf ball hits the tree after 2 seconds.

MONITORING PROGRESS ANSWER

15. 2.25 sec

Solving Real-Life Problems

EXAMPLE 5 **Modeling with Mathematics**

A bird picks up a golf ball and drops it while flying. The function represents the height y (in feet) of the golf ball t seconds after it is dropped. The ball hits the top of a 32-foot-tall pine tree. After how many seconds does the ball hit the tree?

SOLUTION

1. **Understand the Problem** You are given the height of the golf ball as a function of the amount of time after it is dropped and the height of the tree that the golf ball hits. You are asked to determine how many seconds it takes for the ball to hit the tree.
2. **Make a Plan** Use the function for the height of the golf ball. Substitute the height of the tree for y and solve for the time t.
3. **Solve the Problem** Substitute 32 for y and solve for t.

$y = 81 - 16t^2$	Write equation.
$32 = 81 - 16t^2$	Substitute 32 for y.
$0 = 49 - 16t^2$	Subtract 32 from each side.
$0 = 7^2 - (4t)^2$	Write as $a^2 - b^2$.
$0 = (7 + 4t)(7 - 4t)$	Difference of two squares pattern
$7 + 4t = 0$ *or* $7 - 4t = 0$	Zero-Product Property
$t = -\frac{7}{4}$ *or* $t = \frac{7}{4}$	Solve for t.

A negative time does not make sense in this situation.

▶ So, the golf ball hits the tree after $\frac{7}{4}$, or 1.75 seconds.

4. **Look Back** Check your solution, as shown, by substituting $t = \frac{7}{4}$ into the equation $32 = 81 - 16t^2$. Then verify that a time of $\frac{7}{4}$ seconds gives a height of 32 feet.

Check

$32 = 81 - 16t^2$

$32 \stackrel{?}{=} 81 - 16\left(\frac{7}{4}\right)^2$

$32 \stackrel{?}{=} 81 - 16\left(\frac{49}{16}\right)$

$32 \stackrel{?}{=} 81 - 49$

$32 = 32$ ✓

Monitoring Progress

Help in English and Spanish at *BigIdeasMath.com*

15. **WHAT IF?** The golf ball does not hit the pine tree. After how many seconds does the ball hit the ground?

Laurie's Notes Teacher Actions

- **FYI:** This is a falling object model where the polynomial is not written in standard form. The equation could have been written as $y = -16t^2 + 81$, but that does not fit the difference of two squares pattern.
- **Attend to Precision:** Because the coefficient of the linear term is 0, you could solve by square rooting. See *Another Way* on page T-396.

Closure

- **Exit Ticket:** Factor each polynomial.

  a. $81 - x^2$ $(9 - x)(9 + x)$

  b. $x^2 - 24x + 144$ $(x - 12)^2$

7.7 Exercises

Dynamic Solutions available at *BigIdeasMath.com*

Vocabulary and Core Concept Check

1. **REASONING** Can you use the perfect square trinomial pattern to factor $y^2 + 16y + 64$? Explain.

2. **WHICH ONE DOESN'T BELONG?** Which polynomial does *not* belong with the other three? Explain your reasoning.

$n^2 - 4$	$g^2 - 6g + 9$	$r^2 + 12r + 36$	$k^2 + 25$

Monitoring Progress and Modeling with Mathematics

In Exercises 3–8, factor the polynomial. *(See Example 1.)*

3. $m^2 - 49$
4. $z^2 - 81$
5. $64 - 81d^2$
6. $25 - 4x^2$
7. $225a^2 - 36b^2$
8. $16x^2 - 169y^2$

In Exercises 9–14, use a special product pattern to evaluate the expression. *(See Example 2.)*

9. $12^2 - 9^2$
10. $19^2 - 11^2$
11. $78^2 - 72^2$
12. $54^2 - 52^2$
13. $53^2 - 47^2$
14. $39^2 - 36^2$

In Exercises 15–22, factor the polynomial. *(See Example 3.)*

15. $h^2 + 12h + 36$
16. $p^2 + 30p + 225$
17. $y^2 - 22y + 121$
18. $x^2 - 4x + 4$
19. $a^2 - 28a + 196$
20. $m^2 + 24m + 144$
21. $25n^2 + 20n + 4$
22. $49a^2 - 14a + 1$

ERROR ANALYSIS In Exercises 23 and 24, describe and correct the error in factoring the polynomial.

23. ✗
$$n^2 - 64 = n^2 - 8^2 = (n - 8)^2$$

24. ✗
$$y^2 - 6y + 9 = y^2 - 2(y)(3) + 3^2 = (y - 3)(y + 3)$$

25. **MODELING WITH MATHEMATICS** The area (in square centimeters) of a square coaster can be represented by $d^2 + 8d + 16$.

   a. Write an expression that represents the side length of the coaster.

   b. Write an expression for the perimeter of the coaster.

26. **MODELING WITH MATHEMATICS** The polynomial represents the area (in square feet) of the square playground.

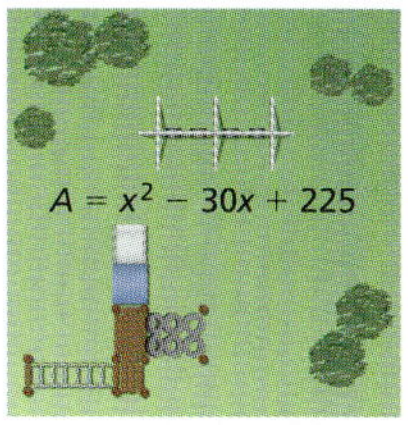

   a. Write a polynomial that represents the side length of the playground.

   b. Write an expression for the perimeter of the playground.

In Exercises 27–34, solve the equation. *(See Example 4.)*

27. $z^2 - 4 = 0$
28. $4x^2 = 49$
29. $k^2 - 16k + 64 = 0$
30. $s^2 + 20s + 100 = 0$
31. $n^2 + 9 = 6n$
32. $y^2 = 12y - 36$
33. $y^2 + \frac{1}{2}y = -\frac{1}{16}$
34. $-\frac{4}{3}x + \frac{4}{9} = -x^2$

In Exercises 35–40, factor the polynomial.

35. $3z^2 - 27$
36. $2m^2 - 50$
37. $4y^2 - 16y + 16$
38. $8k^2 + 80k + 200$
39. $50y^2 + 120y + 72$
40. $27m^2 - 36m + 12$

Assignment Guide and Homework Check

ASSIGNMENT

Basic: 1, 2, 3–35 odd, 41, 46, 49–56

Average: 1, 2–42 even, 46, 47, 49–56

Advanced: 1, 2, 8, 14, 20–26 even, 32–42 even, 43–56

HOMEWORK CHECK

Basic: 3, 9, 15, 29, 41

Average: 4, 12, 16, 30, 42

Advanced: 8, 14, 22, 32, 42

ANSWERS

1. yes; The square roots of the first and last terms are y and 8, and the middle term is $2 \cdot y \cdot 8$, so it fits the pattern.
2. $k^2 + 25$; It is the only one that cannot be factored using a special pattern.
3. $(m + 7)(m - 7)$
4. $(z + 9)(z - 9)$
5. $(8 + 9d)(8 - 9d)$
6. $(5 + 2x)(5 - 2x)$
7. $9(5a + 2b)(5a - 2b)$
8. $(4x + 13y)(4x - 13y)$
9. 63
10. 240
11. 900
12. 212
13. 600
14. 225
15. $(h + 6)^2$
16. $(p + 15)^2$
17. $(y - 11)^2$
18. $(x - 2)^2$
19. $(a - 14)^2$
20. $(m + 12)^2$
21. $(5n + 2)^2$
22. $(7a - 1)^2$
23. should follow the difference of two squares pattern; $= (n + 8)(n - 8)$
24. should follow the perfect square trinomial pattern; $= (y - 3)^2$
25. a. $(d + 4)$ cm
    b. $(4d + 16)$ cm
26. a. $(x - 15)$ ft
    b. $(4x - 60)$ ft
27. $z = -2, z = 2$
28. $x = -\frac{7}{2}, x = \frac{7}{2}$
29. $k = 8$
30. $s = -10$
31. $n = 3$
32. $y = 6$
33. $y = -\frac{1}{4}$
34. $x = \frac{2}{3}$
35. $3(z + 3)(z - 3)$
36. $2(m + 5)(m - 5)$
37. $4(y - 2)^2$
38. $8(k + 5)^2$
39. $2(5y + 6)^2$
40. $3(3m - 2)^2$

Dynamic Teaching Tools

- Dynamic Assessment & Progress Monitoring Tool
- Interactive Whiteboard Lesson Library
- Dynamic Classroom with Dynamic Investigations

ANSWERS

41. 1.25 sec

42. 0.25 sec

43. **a.** no; $w^2 + 18w + 81$

b. no; $y^2 - 10y + 25$

44. **a.**

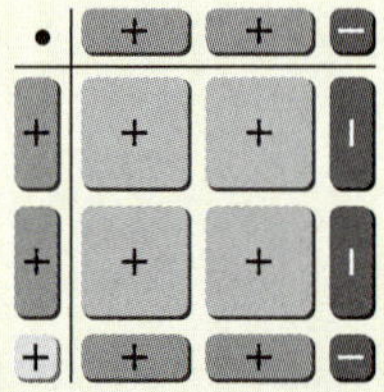

$4x^2 - 1 = (2x + 1)(2x - 1)$

b.

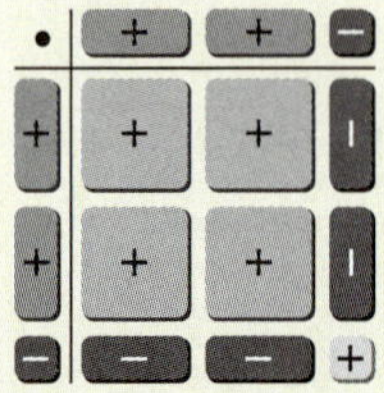

$4x^2 - 4x + 1 = (2x - 1)^2$

45. Square each binomial, then combine like terms; Use the difference of two squares pattern with each binomial as one of the terms, then simplify; *Sample answer:* The difference of two squares pattern; You do not need to square any binomials.

46–56. See Additional Answers.

Mini-Assessment

Factor each polynomial.

1. $9x^2 - 25$ $(3x + 5)(3x - 5)$
2. $x^2 - 40x + 400$ $(x - 20)^2$
3. Use a special product pattern to evaluate the expression $26^2 - 22^2$. $(26 + 22)(26 - 22) = 48(4) = 192$
4. Solve $16x^2 + 8x + 1 = 0$. $x = -\frac{1}{4}$
5. A bird picks up a stone and drops it while flying. The function $y = 49 - 16t^2$ represents the height y (in feet) of the stone t seconds after it is dropped. The stone hits the top of a 13-foot-tall building. After how many seconds does the stone hit the building? The stone hits the building after $\frac{3}{2}$, or 1.5, seconds.

41. MODELING WITH MATHEMATICS While standing on a ladder, you drop a paintbrush. The function represents the height y (in feet) of the paintbrush t seconds after it is dropped. After how many seconds does the paintbrush land on the ground? *(See Example 5.)*

42. MODELING WITH MATHEMATICS The function represents the height y (in feet) of a grasshopper jumping straight up from the ground t seconds after the start of the jump. After how many seconds is the grasshopper 1 foot off the ground?

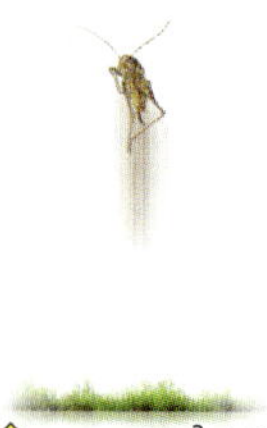

43. REASONING Tell whether the polynomial can be factored. If not, change the constant term so that the polynomial is a perfect square trinomial.

a. $w^2 + 18w + 84$ **b.** $y^2 - 10y + 23$

44. THOUGHT PROVOKING Use algebra tiles to factor each polynomial modeled by the tiles. Show your work.

a.

b.

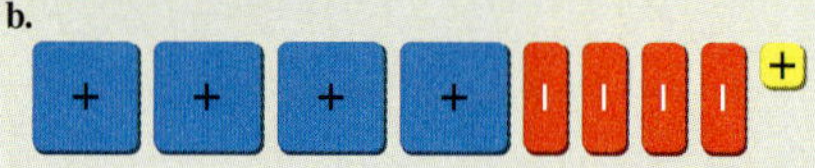

45. COMPARING METHODS Describe two methods you can use to simplify $(2x - 5)^2 - (x - 4)^2$. Which one would you use? Explain.

46. HOW DO YOU SEE IT? The figure shows a large square with an area of a^2 that contains a smaller square with an area of b^2.

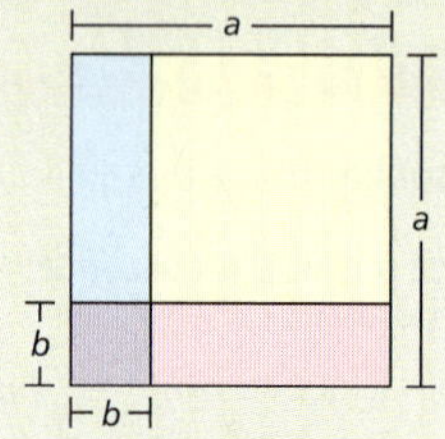

a. Describe the regions that represent $a^2 - b^2$. How can you rearrange these regions to show that $a^2 - b^2 = (a + b)(a - b)$?

b. How can you use the figure to show that $(a - b)^2 = a^2 - 2ab + b^2$?

47. PROBLEM SOLVING You hang nine identical square picture frames on a wall.

a. Write a polynomial that represents the area of the picture frames, not including the pictures.

b. The area in part (a) is 81 square inches. What is the side length of one of the picture frames? Explain your reasoning.

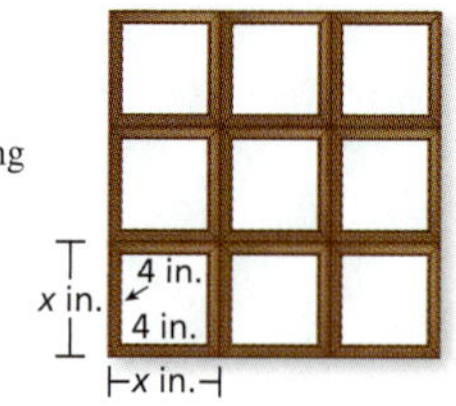

48. MATHEMATICAL CONNECTIONS The composite solid is made up of a cube and a rectangular prism.

a. Write a polynomial that represents the volume of the composite solid.

b. The volume of the composite solid is equal to $25x$. What is the value of x? Explain your reasoning.

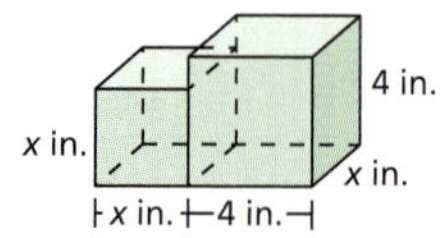

Maintaining Mathematical Proficiency

Reviewing what you learned in previous grades and lessons

Write the prime factorization of the number. *(Skills Review Handbook)*

49. 50 **50.** 44 **51.** 85 **52.** 96

Graph the inequality in a coordinate plane. *(Section 5.6)*

53. $y \leq 4x - 1$ **54.** $y > -\frac{1}{2}x + 3$ **55.** $4y - 12 \geq 8x$ **56.** $3y + 3 < x$

If students need help...	*If students got it...*
Resources by Chapter • Practice A and Practice B • Puzzle Time	Resources by Chapter • Enrichment and Extension • Cumulative Review
Student Journal • Practice	Start the *next* Section
Differentiating the Lesson Skills Review Handbook	

Laurie's Notes

Overview of Section 7.8

Introduction

- This last lesson, on factoring polynomials completely, requires many prior skills. Students will learn the technique of factoring by grouping, which involves the Distributive Property. Then they will factor polynomials completely by first looking for the greatest common factor in each term.
- Although factoring of polynomials is the focus of the lesson, the ultimate goal is for students to be able to solve polynomial equations and simplify expressions with common factors.

Teaching Strategy

- **Look For and Make Use of Structure:** All the following expressions have the same structure. Each expression has two terms and can be simplified using the Distributive Property.

  $ac + bc$ $\quad$ $a(x + 2) + b(x + 2)$ $\quad$ $a(x - 4) - b(x - 4)$
- While factoring out a common binomial factor is no different from factoring out a common monomial factor, students do not always recognize the underlying structure. To help, use colors and underlining.
- Notice how the use of color and underlining helps in recognizing common factors:

  $a\underline{c} + b\underline{c} = (a + b)\underline{c}$

  $a(\underline{x + 2}) + b(\underline{x + 2}) = (a + b)(\underline{x + 2})$

  $a(\underline{x - 4}) - b(\underline{x - 4}) = (a - b)(\underline{x - 4})$
- The Distributive Property is often viewed as a right-pointing arrow versus a statement of equality. In each of the examples above, a common factor has been factored out using the Distributive Property.

Pacing Suggestion

- Complete the explorations, perhaps using a jigsaw approach in each exploration, and then transition to the formal lesson.

Dynamic Teaching Tools

- Dynamic Assessment & Progress Monitoring Tool
- Lesson Planning Tool
- Interactive Whiteboard Lesson Library
- Dynamic Classroom with Dynamic Investigations

What Your Students Will Learn

- Factor polynomials by grouping by finding the GCF of pairs of terms.
- Factor polynomials with integer coefficients completely.
- Use factoring to solve real-life problems involving polynomial equations.

Laurie's Notes

Exploration

Motivate

- Today's lesson involves factoring polynomials. A quick warm-up with factoring out greatest common factors (GCFs) would be helpful.
- Ask students to factor the GCF out of each of the following polynomials:

  $3x^2 - 15x = 3x(x - 5)$

  $-16x^3 - 8x^2 - 24x = -8x(2x^2 + x + 3)$

  $30x^2 + 5x = 5x(6x + 1)$

Exploration Note

- Most of the examples students have worked with in this chapter involve two factors and second-degree polynomials. In both explorations, students are working with three factors and mostly third-degree polynomials in preparation for this last lesson in the chapter on factoring polynomials completely.

Exploration 1

- Explain that this first exploration is a good review. Students will write the expression modeled by the algebra tiles and then find the product and express it in standard form. This is a skill taught in the early part of the chapter.

COMMON ERROR When multiplying $(x + 1)(x + 1)(-2)$, students will sometimes multiply (-2) times each of the binomials and get $(-2x - 2)(-2x - 2)$ before multiplying the binomials. Remind them that when they were multiplying (3)(4)(2), they would not multiply the factor of 2 by each of the other factors and get (6)(8).

- If time is a factor, you might have half the students do parts (a), (c), and (e) and the rest of the students do parts (b), (d), and (f).

Exploration 2

- It is likely that most students will work with the factored form, multiply the expression, and look for their answer in the list of answers in standard form.
- Again, if time is a factor, you could split the class so they are doing every other problem. Half the class begins with A and half begins with B.
- **Popsicle Sticks:** Solicit responses when students have finished.

? Look back at the standard form expressions that are trinomials. "What do you notice about the terms in the trinomial?" Each of the terms has an x in it. "Could that x be factored out as a common factor?" yes

? Select one of the trinomial expressions, say part (d), and factor out an x. Write $x^3 - 4x^2 + 4x = x(x^2 - 4x + 4)$. "Can you factor the resulting trinomial?" yes; $(x^2 - 4x + 4) = (x - 2)^2$

- You could use some of the responses in the left-hand column (standard form) to review vocabulary associated with this chapter.
- **Extension:** Select a few of the factored form expressions and set them equal to 0 and ask what the solutions would be.

Communicate Your Answer

- **Neighbor Check:** For Question 4, have students work independently and then have their neighbors check their work. Have students discuss any discrepancies.

Connecting to Next Step

- The explorations help students visualize and make sense of factoring third-degree polynomials. In the formal lesson, students will work with factoring a polynomial completely.

7.8 Factoring Polynomials Completely

Essential Question

How can you factor a polynomial completely?

EXPLORATION 1 Writing a Product of Linear Factors

Work with a partner. Write the product represented by the algebra tiles. Then multiply to write the polynomial in standard form.

a. 

b. 

c. 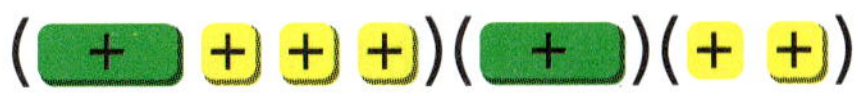

d. 

e. 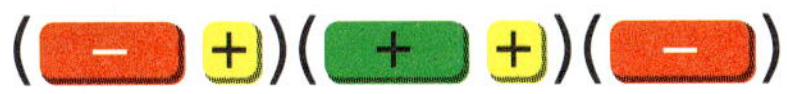

f. 

REASONING ABSTRACTLY

To be proficient in math, you need to know and flexibly use different properties of operations and objects.

EXPLORATION 2 Matching Standard and Factored Forms

Work with a partner. Match the standard form of the polynomial with the equivalent factored form. Explain your strategy.

a. $x^3 + x^2$	**A.** $x(x + 1)(x - 1)$
b. $x^3 - x$	**B.** $x(x - 1)^2$
c. $x^3 + x^2 - 2x$	**C.** $x(x + 1)^2$
d. $x^3 - 4x^2 + 4x$	**D.** $x(x + 2)(x - 1)$
e. $x^3 - 2x^2 - 3x$	**E.** $x(x - 1)(x - 2)$
f. $x^3 - 2x^2 + x$	**F.** $x(x + 2)(x - 2)$
g. $x^3 - 4x$	**G.** $x(x - 2)^2$
h. $x^3 + 2x^2$	**H.** $x(x + 2)^2$
i. $x^3 - x^2$	**I.** $x^2(x - 1)$
j. $x^3 - 3x^2 + 2x$	**J.** $x^2(x + 1)$
k. $x^3 + 2x^2 - 3x$	**K.** $x^2(x - 2)$
l. $x^3 - 4x^2 + 3x$	**L.** $x^2(x + 2)$
m. $x^3 - 2x^2$	**M.** $x(x + 3)(x - 1)$
n. $x^3 + 4x^2 + 4x$	**N.** $x(x + 1)(x - 3)$
o. $x^3 + 2x^2 + x$	**O.** $x(x - 1)(x - 3)$

Communicate Your Answer

3. How can you factor a polynomial completely?
4. Use your answer to Question 3 to factor each polynomial completely.
   a. $x^3 + 4x^2 + 3x$ **b.** $x^3 - 6x^2 + 9x$ **c.** $x^3 + 6x^2 + 9x$

Dynamic Teaching Tools

- Dynamic Assessment & Progress Monitoring Tool
- Lesson Planning Tool
- Interactive Whiteboard Lesson Library
- Dynamic Classroom with Dynamic Investigations

ANSWERS

1. a. $(x + 1)(x + 1)(-2)$; $-2x^2 - 4x - 2$
   b. $(x + 2)(x + 1)(-x)$; $-x^3 - 3x^2 - 2x$
   c. $(x + 3)(x)(2)$; $2x^2 + 6x$
   d. $(x + 1)(x - 1)(x)$; $x^3 - x$
   e. $(-x + 1)(x + 1)(-x)$; $x^3 - x$
   f. $(-x - 1)(x + 1)(-2)$; $2x^2 + 4x + 2$
2. a. J
   b. A
   c. D
   d. G
   e. N
   f. B
   g. F
   h. L
   i. I
   j. E
   k. M
   l. O
   m. K
   n. H
   o. C

   Factor out the greatest common monomial factor first, then factor the remaining expression if possible.
3. Factor out the greatest common monomial factor first, then factor the remaining expression if possible.
4. a. $x(x + 1)(x + 3)$
   b. $x(x - 3)^2$
   c. $x(x + 3)^2$

Extra Example 1

Factor each polynomial by grouping.

a. $x^3 + 4x^2 + 2x + 8$ $(x^2 + 2)(x + 4)$

b. $x^2 + 4y + 2x + 2xy$ $(x + 2)(x + 2y)$

MONITORING PROGRESS ANSWERS

1. $(a^2 + 1)(a + 3)$
2. $(y + 2)(y + x)$

7.8 Lesson

Core Vocabulary

factoring by grouping, *p. 404*
factored completely, *p. 404*

Previous
polynomial
binomial

What You Will Learn

- Factor polynomials by grouping.
- Factor polynomials completely.
- Use factoring to solve real-life problems.

Factoring Polynomials by Grouping

You have used the Distributive Property to factor out a greatest common monomial from a polynomial. Sometimes, you can factor out a common binomial. You may be able to use the Distributive Property to factor polynomials with four terms, as described below.

Core Concept

Factoring by Grouping

To factor a polynomial with four terms, group the terms into pairs. Factor the GCF out of each pair of terms. Look for and factor out the common binomial factor. This process is called **factoring by grouping**.

EXAMPLE 1 **Factoring by Grouping**

Factor each polynomial by grouping.

a. $x^3 + 3x^2 + 2x + 6$ **b.** $x^2 + y + x + xy$

SOLUTION

a. $x^3 + 3x^2 + 2x + 6 = (x^3 + 3x^2) + (2x + 6)$ Group terms with common factors.

Common binomial factor is $x + 3$. → $= x^2(x + 3) + 2(x + 3)$ Factor out GCF of each pair of terms.

$= (x + 3)(x^2 + 2)$ Factor out $(x + 3)$.

So, $x^3 + 3x^2 + 2x + 6 = (x + 3)(x^2 + 2)$.

b. $x^2 + y + x + xy = x^2 + x + xy + y$ Rewrite polynomial.

$= (x^2 + x) + (xy + y)$ Group terms with common factors.

Common binomial factor is $x + 1$. → $= x(x + 1) + y(x + 1)$ Factor out GCF of each pair of terms.

$= (x + 1)(x + y)$ Factor out $(x + 1)$.

So, $x^2 + y + x + xy = (x + 1)(x + y)$.

Monitoring Progress Help in English and Spanish at *BigIdeasMath.com*

Factor the polynomial by grouping.

1. $a^3 + 3a^2 + a + 3$
2. $y^2 + 2x + yx + 2y$

Factoring Polynomials Completely

You have seen that the polynomial $x^2 - 1$ can be factored as $(x + 1)(x - 1)$. This polynomial is factorable. Notice that the polynomial $x^2 + 1$ cannot be written as the product of polynomials with integer coefficients. This polynomial is unfactorable. A factorable polynomial with integer coefficients is **factored completely** when it is written as a product of unfactorable polynomials with integer coefficients.

Laurie's Notes Teacher Actions

- **Teaching Tip:** Use of color and underlining is helpful: $x^2(\underline{x + 3}) + 2(\underline{x + 3})$. This is the point where it is important for students to view this expression as the sum of two terms that have a common factor. The common factor $(\underline{x + 3})$ can be factored out just like a monomial.
- The polynomial in part (b) appears different to students because of the two variables and because terms with common factors are not adjacent. Group (put parentheses around) terms with common factors. Use the Distributive Property to factor.

Concept Summary

Guidelines for Factoring Polynomials Completely

To factor a polynomial completely, you should try each of these steps.

1.	Factor out the greatest common monomial factor.	$3x^2 + 6x = 3x(x + 2)$
2.	Look for a difference of two squares or a perfect square trinomial.	$x^2 + 4x + 4 = (x + 2)^2$
3.	Factor a trinomial of the form $ax^2 + bx + c$ into a product of binomial factors.	$3x^2 - 5x - 2 = (3x + 1)(x - 2)$
4.	Factor a polynomial with four terms by grouping.	$x^3 + x - 4x^2 - 4 = (x^2 + 1)(x - 4)$

EXAMPLE 2 Factoring Completely

Factor (a) $3x^3 + 6x^2 - 18x$ and (b) $7x^4 - 28x^2$.

SOLUTION

a. $3x^3 + 6x^2 - 18x = 3x(x^2 + 2x - 6)$ — Factor out $3x$.

$x^2 + 2x - 6$ is unfactorable, so the polynomial is factored completely.

▶ So, $3x^3 + 6x^2 - 18x = 3x(x^2 + 2x - 6)$.

b. $7x^4 - 28x^2 = 7x^2(x^2 - 4)$ — Factor out $7x^2$.

$= 7x^2(x^2 - 2^2)$ — Write as $a^2 - b^2$.

$= 7x^2(x + 2)(x - 2)$ — Difference of two squares pattern

▶ So, $7x^4 - 28x^2 = 7x^2(x + 2)(x - 2)$.

EXAMPLE 3 Solving an Equation by Factoring Completely

Solve $2x^3 + 8x^2 = 10x$.

SOLUTION

$2x^3 + 8x^2 = 10x$ — Original equation

$2x^3 + 8x^2 - 10x = 0$ — Subtract $10x$ from each side.

$2x(x^2 + 4x - 5) = 0$ — Factor out $2x$.

$2x(x + 5)(x - 1) = 0$ — Factor $x^2 + 4x - 5$.

$2x = 0$ *or* $x + 5 = 0$ *or* $x - 1 = 0$ — Zero-Product Property

$x = 0$ *or* $x = -5$ *or* $x = 1$ — Solve for x.

▶ The roots are $x = -5$, $x = 0$, and $x = 1$.

Monitoring Progress

Help in English and Spanish at *BigIdeasMath.com*

Factor the polynomial completely.

3. $3x^3 - 12x$ **4.** $2y^3 - 12y^2 + 18y$ **5.** $m^3 - 2m^2 - 8m$

Solve the equation.

6. $w^3 - 8w^2 + 16w = 0$ **7.** $x^3 - 25x = 0$ **8.** $c^3 - 7c^2 + 12c = 0$

English Language Learners

Organization

Students will benefit by writing down the steps for factoring polynomials completely, as shown in the *Concept Summary*. Have students write the steps in their notebooks. Consider also displaying a poster with the same information in the classroom.

Extra Example 2

a. Factor $2x^3 + 6x^2 - 2x$. $2x(x^2 + 3x - 1)$

b. Factor $5x^4 - 45x^2$. $5x^2(x + 3)(x - 3)$

Extra Example 3

Solve $3x^3 + 6x^2 = 24x$.

The roots are $x = -4$, $x = 0$, and $x = 2$.

MONITORING PROGRESS ANSWERS

3. $3x(x + 2)(x - 2)$

4. $2y(y - 3)^2$

5. $m(m + 2)(m - 4)$

6. $w = 0, w = 4$

7. $x = 0, x = -5, x = 5$

8. $c = 0, c = 3, c = 4$

Laurie's Notes — Teacher Actions

- Explain what it means to factor completely. Discuss the guidelines and review the sample problems.

COMMON ERROR Students may see a factor, but it may not be the greatest common factor. Always ask, "Are there any other factors?"

? Fact-First Questioning: Write $2x(x + 5)(x - 1) = 0$; $x(2x + 10)(x - 1) = 0$; $x(x + 5)(2x - 2) = 0$. Say, "All of these are equivalent. Explain why." They all have the same solutions, $x = -5$, $x = 0$, and $x = 1$.

Extra Example 4

A box in the shape of a rectangular prism has a volume of 96 cubic inches. The dimensions of the box in terms of its width are shown. Find the length, width, and height of the box.

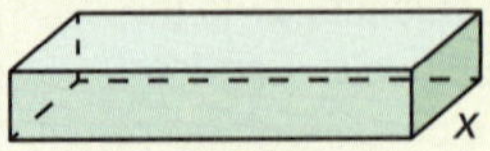

The width is 4 inches, the length is 12 inches, and the height is 2 inches.

MONITORING PROGRESS ANSWER

9. length: 3 ft, width: 2 ft, height: 12 ft

Solving Real-Life Problems

EXAMPLE 4 **Modeling with Mathematics**

A terrarium in the shape of a rectangular prism has a volume of 4608 cubic inches. Its length is more than 10 inches. The dimensions of the terrarium in terms of its width are shown. Find the length, width, and height of the terrarium.

SOLUTION

1. **Understand the Problem** You are given the volume of a terrarium in the shape of a rectangular prism and a description of the length. The dimensions are written in terms of its width. You are asked to find the length, width, and height of the terrarium.

2. **Make a Plan** Use the formula for the volume of a rectangular prism to write and solve an equation for the width of the terrarium. Then substitute that value in the expressions for the length and height of the terrarium.

3. **Solve the Problem**

Volume = length • width • height	Volume of a rectangular prism
$4608 = (36 - w)(w)(w + 4)$	Write equation.
$4608 = 32w^2 + 144w - w^3$	Multiply.
$0 = 32w^2 + 144w - w^3 - 4608$	Subtract 4608 from each side.
$0 = (-w^3 + 32w^2) + (144w - 4608)$	Group terms with common factors.
$0 = -w^2(w - 32) + 144(w - 32)$	Factor out GCF of each pair of terms.
$0 = (w - 32)(-w^2 + 144)$	Factor out $(w - 32)$.
$0 = -1(w - 32)(w^2 - 144)$	Factor -1 from $-w^2 + 144$.
$0 = -1(w - 32)(w - 12)(w + 12)$	Difference of two squares pattern
$w - 32 = 0$ *or* $w - 12 = 0$ *or* $w + 12 = 0$	Zero-Product Property
$w = 32$ *or* $w = 12$ *or* $w = -12$	Solve for w.

Disregard $w = -12$ because a negative width does not make sense. You know that the length is more than 10 inches. Test the solutions of the equation, 12 and 32, in the expression for the length.

length $= 36 - w = 36 - 12 = 24$ ✓ *or* length $= 36 - w = 36 - 32 = 4$ ✗

The solution 12 gives a length of 24 inches, so 12 is the correct value of w.

Use $w = 12$ to find the height, as shown.

height $= w + 4 = 12 + 4 = 16$

▶ The width is 12 inches, the length is 24 inches, and the height is 16 inches.

4. **Look Back** Check your solution. Substitute the values for the length, width, and height when the width is 12 inches into the formula for volume. The volume of the terrarium should be 4608 cubic inches.

Check

$V = \ell wh$

$4608 \stackrel{?}{=} 24(12)(16)$

$4608 = 4608$ ✓

Monitoring Progress

Help in English and Spanish at *BigIdeasMath.com*

9. A box in the shape of a rectangular prism has a volume of 72 cubic feet. The box has a length of x feet, a width of $(x - 1)$ feet, and a height of $(x + 9)$ feet. Find the dimensions of the box.

Laurie's Notes Teacher Actions

- **Make Sense of Problems and Persevere in Solving Them** and **Model with Mathematics:** Draw and label the diagram and ask, "How do you find the volume of this rectangular prism?" $V = \ell wh$, so by substituting, $4608 = (36 - w)(w)(w + 4)$.
- **Think-Alouds:** "To solve this problem I need to ...." Ask partner A to think aloud for partner B to hear the problem-solving process. Partner B should only speak when partner A is stuck, but he or she should not rush in too quickly.

Closure

- **Exit Ticket:** Solve $3x^5 - 6x^4 - 45x^3 = 0$. $x = -3, x = 0, x = 5$

7.8 Exercises

Dynamic Solutions available at *BigIdeasMath.com*

Vocabulary and Core Concept Check

1. **VOCABULARY** What does it mean for a polynomial to be factored completely?

2. **WRITING** Explain how to choose which terms to group together when factoring by grouping.

Monitoring Progress and Modeling with Mathematics

In Exercises 3–10, factor the polynomial by grouping. *(See Example 1.)*

3. $x^3 + x^2 + 2x + 2$
4. $y^3 - 9y^2 + y - 9$
5. $3z^3 + 2z - 12z^2 - 8$
6. $2s^3 - 27 - 18s + 3s^2$
7. $x^2 + xy + 8x + 8y$
8. $q^2 + q + 5pq + 5p$
9. $m^2 - 3m + mn - 3n$
10. $2a^2 + 8ab - 3a - 12b$

In Exercises 11–22, factor the polynomial completely. *(See Example 2.)*

11. $2x^3 - 2x$
12. $36a^4 - 4a^2$
13. $2c^2 - 7c + 19$
14. $m^2 - 5m - 35$
15. $6g^3 - 24g^2 + 24g$
16. $-15d^3 + 21d^2 - 6d$
17. $3r^5 + 3r^4 - 90r^3$
18. $5w^4 - 40w^3 + 80w^2$
19. $-4c^4 + 8c^3 - 28c^2$
20. $8t^2 + 8t - 72$
21. $b^3 - 5b^2 - 4b + 20$
22. $h^3 + 4h^2 - 25h - 100$

In Exercises 23–28, solve the equation. *(See Example 3.)*

23. $5n^3 - 30n^2 + 40n = 0$
24. $k^4 - 100k^2 = 0$
25. $x^3 + x^2 = 4x + 4$
26. $2t^5 + 2t^4 - 144t^3 = 0$
27. $12s - 3s^3 = 0$
28. $4y^3 - 7y^2 + 28 = 16y$

In Exercises 29–32, find the x-coordinates of the points where the graph crosses the x-axis.

29.
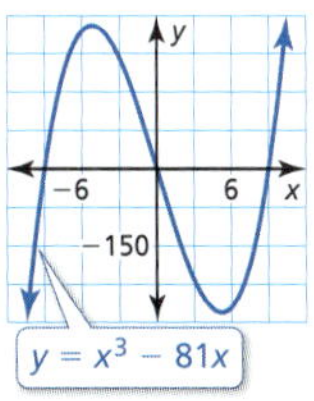

30.
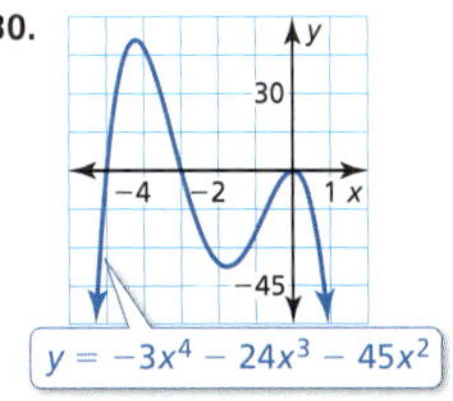

31.
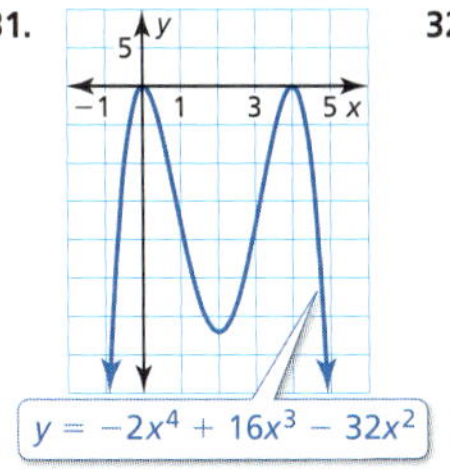

32.
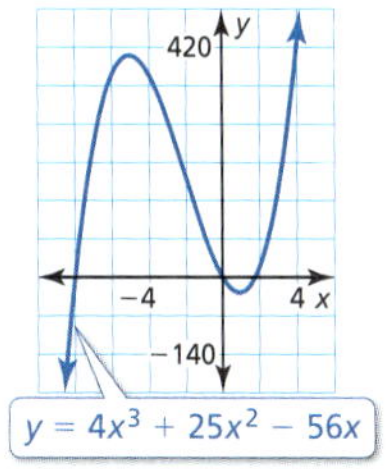

ERROR ANALYSIS **In Exercises 33 and 34, describe and correct the error in factoring the polynomial completely.**

33. ✗ $a^3 + 8a^2 - 6a - 48 = a^2(a + 8) + 6(a + 8)$
$= (a + 8)(a^2 + 6)$

34. ✗ $x^3 - 6x^2 - 9x + 54 = x^2(x - 6) - 9(x - 6)$
$= (x - 6)(x^2 - 9)$

35. **MODELING WITH MATHEMATICS** You are building a birdhouse in the shape of a rectangular prism that has a volume of 128 cubic inches. The dimensions of the birdhouse in terms of its width are shown. *(See Example 4.)*

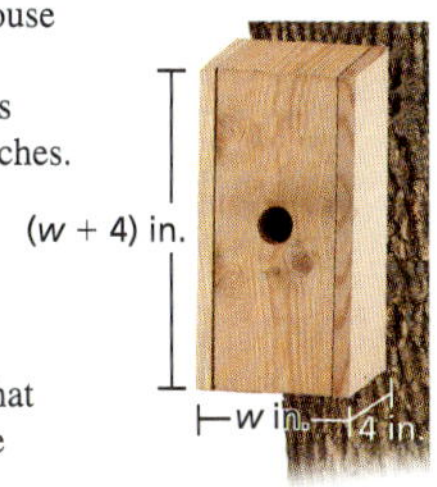

a. Write a polynomial that represents the volume of the birdhouse.

b. What are the dimensions of the birdhouse?

34. It is not factored completely because $x^2 - 9$ can be factored; $= (x - 6)(x + 3)(x - 3)$

35. a. $(4w^2 + 16w)$ in.^{3}

b. length: 4 in., width: 4 in., height: 8 in.

Assignment Guide and Homework Check

ASSIGNMENT

Basic: 1, 2, 3–35 odd, 42, 44, 50–57

Average: 1, 2–46 even, 50–57

Advanced: 1, 2, 8, 10, 20–40 even, 41–57

HOMEWORK CHECK

Basic: 3, 11, 23, 35

Average: 4, 12, 26, 36

Advanced: 8, 20, 28, 36

ANSWERS

1. It is written as a product of unfactorable polynomials with integer coefficients.
2. Look for terms with common factors.
3. $(x + 1)(x^2 + 2)$
4. $(y - 9)(y^2 + 1)$
5. $(z - 4)(3z^2 + 2)$
6. $(2s + 3)(s + 3)(s - 3)$
7. $(x + y)(x + 8)$
8. $(q + 1)(q + 5p)$
9. $(m - 3)(m + n)$
10. $(2a - 3)(a + 4b)$
11. $2x(x + 1)(x - 1)$
12. $4a^2(3a + 1)(3a - 1)$
13. unfactorable
14. unfactorable
15. $6g(g - 2)^2$
16. $-3d(d - 1)(5d - 2)$
17. $3r^3(r + 6)(r - 5)$
18. $5w^2(w - 4)^2$
19. $-4c^2(c^2 - 2c + 7)$
20. $8(t^2 + t - 9)$
21. $(b - 5)(b + 2)(b - 2)$
22. $(h + 4)(h + 5)(h - 5)$
23. $n = 0, n = 2, n = 4$
24. $k = 0, k = -10, k = 10$
25. $x = -1, x = -2, x = 2$
26. $t = 0, t = -9, t = 8$
27. $s = 0, s = 2, s = -2$
28. $y = \frac{7}{4}, y = -2, y = 2$
29. $x = 0, x = -9, x = 9$
30. $x = 0, x = -5, x = -3$
31. $x = 0, x = 4$
32. $x = 0, x = -8, x = \frac{7}{4}$
33. In the second group, factor out -6 instead of 6; $= a^2(a + 8) - 6(a + 8)$ $= (a + 8)(a^2 - 6)$

Dynamic Teaching Tools

- Dynamic Assessment & Progress Monitoring Tool
- Interactive Whiteboard Lesson Library
- Dynamic Classroom with Dynamic Investigations

ANSWERS

36. length: 16 in., width: 6 in., height: 12 in.

37. $(x + 2y)(x + 1)(x - 1)$

38. $(2b - a)(2b + 3)(2b - 3)$

39. $(4s - 1)(s + 3t)$

40. $(m^2 - 2n)(6m + n)$

41. no; The factors of the polynomial are $x^2 + 3$ and $x + 2$. Using the Zero-Product Property, $x + 2 = 0$ will give 1 real solution, but $x^2 + 3 = 0$ has no real solutions.

42. The x-intercepts occur when $y = 0$, so set each factor equal to 0 and solve for x to get the x-coordinates of the x-intercepts.

43. **a.** *Sample answer:* $x^3 + x^2 + x + 2$
b. *Sample answer:* $x^3 + x^2 + x + 1$

44. no; It is possible that the terms will have a common monomial factor.

45. $3z, 2z + 3, 2z - 3$

46–57. See Additional Answers.

Mini-Assessment

1. Factor $x^2 + 15y + 5x + 3xy$ by grouping. $(x + 5)(x + 3y)$
2. Factor $3x^5 - 48x^3$ completely. $3x^3(x + 4)(x - 4)$
3. Solve $6x^3 - 30x^2 = 36x$. The roots are $x = -1$, $x = 0$, and $x = 6$.
4. A box in the shape of a rectangular prism has a volume of 180 cubic inches. The dimensions of the box in terms of its height are shown. Find the length, width, and height of the box.

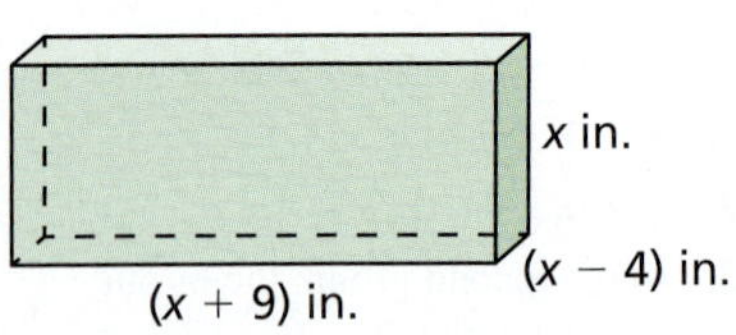

The length is 15 inches, the width is 2 inches, and the height is 6 inches.

36. **MODELING WITH MATHEMATICS** A gift bag shaped like a rectangular prism has a volume of 1152 cubic inches. The dimensions of the gift bag in terms of its width are shown. The height is greater than the width. What are the dimensions of the gift bag?

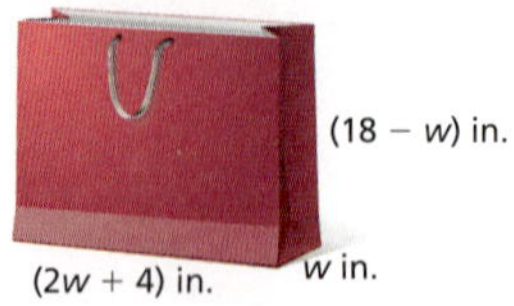

In Exercises 37–40, factor the polynomial completely.

37. $x^3 + 2x^2y - x - 2y$ **38.** $8b^3 - 4b^2a - 18b + 9a$

39. $4s^2 - s + 12st - 3t$

40. $6m^3 - 12mn + m^2n - 2n^2$

41. **WRITING** Is it possible to find three real solutions of the equation $x^3 + 2x^2 + 3x + 6 = 0$? Explain your reasoning.

42. **HOW DO YOU SEE IT?** How can you use the factored form of the polynomial $x^4 - 2x^3 - 9x^2 + 18x = x(x - 3)(x + 3)(x - 2)$ to find the x-intercepts of the graph of the function?

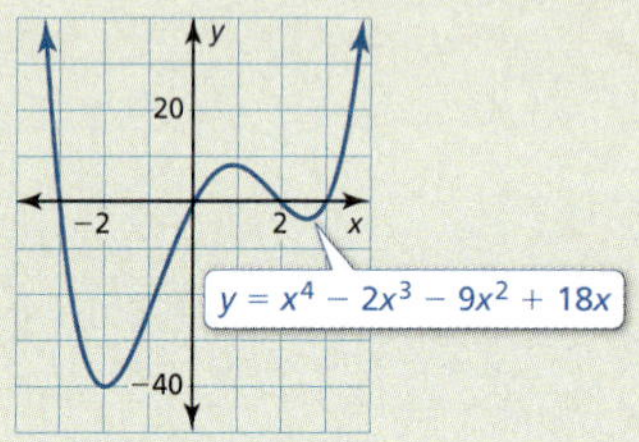

43. **OPEN-ENDED** Write a polynomial of degree 3 that satisfies each of the given conditions.

a. is not factorable **b.** can be factored by grouping

44. **MAKING AN ARGUMENT** Your friend says that if a trinomial cannot be factored as the product of two binomials, then the trinomial is factored completely. Is your friend correct? Explain.

45. **PROBLEM SOLVING** The volume (in cubic feet) of a room in the shape of a rectangular prism is represented by $12z^3 - 27z$. Find expressions that could represent the dimensions of the room.

46. **MATHEMATICAL CONNECTIONS** The width of a box in the shape of a rectangular prism is 4 inches more than the height h. The length is the difference of 9 inches and the height.

a. Write a polynomial that represents the volume of the box in terms of its height (in inches).

b. The volume of the box is 180 cubic inches. What are the possible dimensions of the box?

c. Which dimensions result in a box with the least possible surface area? Explain your reasoning.

47. **MATHEMATICAL CONNECTIONS** The volume of a cylinder is given by $V = \pi r^2 h$, where r is the radius of the base of the cylinder and h is the height of the cylinder. Find the dimensions of the cylinder.

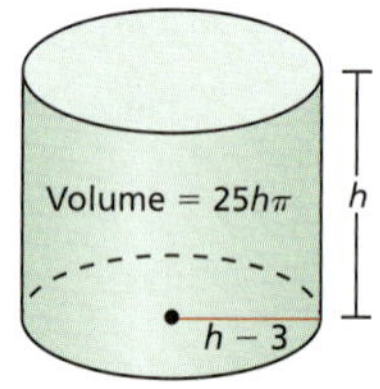

48. **THOUGHT PROVOKING** Factor the polynomial $x^5 - x^4 - 5x^3 + 5x^2 + 4x - 4$ completely.

49. **REASONING** Find a value for w so that the equation has (a) two solutions and (b) three solutions. Explain your reasoning.

$$5x^3 + wx^2 + 80x = 0$$

Maintaining Mathematical Proficiency

Reviewing what you learned in previous grades and lessons

Solve the system of linear equations by graphing. *(Section 5.1)*

50. $y = x - 4$; $y = -2x + 2$

51. $y = \frac{1}{2}x + 2$; $y = 3x - 3$

52. $5x - y = 12$; $\frac{1}{4}x + y = 9$

53. $x = 3y$; $y - 10 = 2x$

Graph the function. Describe the domain and range. *(Section 6.3)*

54. $f(x) = 5^x$ **55.** $y = 9\left(\frac{1}{3}\right)^x$ **56.** $y = -3(0.5)^x$ **57.** $f(x) = -3(4)^x$

If students need help...	*If students got it...*
Resources by Chapter • Practice A and Practice B • Puzzle Time	Resources by Chapter • Enrichment and Extension • Cumulative Review
Student Journal • Practice	Start the *next* Section
Differentiating the Lesson Skills Review Handbook	

7.5–7.8 What Did You Learn?

Core Vocabulary

factoring by grouping, *p. 404*
factored completely, *p. 404*

Core Concepts

Section 7.5

Factoring $x^2 + bx + c$ When c Is Positive, *p. 386*
Factoring $x^2 + bx + c$ When c Is Negative, *p. 387*

Section 7.6

Factoring $ax^2 + bx + c$ When ac Is Positive, *p. 392*
Factoring $ax^2 + bx + c$ When ac Is Negative, *p. 393*

Section 7.7

Difference of Two Squares Pattern, *p. 398*
Perfect Square Trinomial Pattern, *p. 399*

Section 7.8

Factoring by Grouping, *p. 404*
Factoring Polynomials Completely, *p. 404*

Mathematical Practices

1. How are the solutions of Exercise 29 on page 389 related to the graph of $y = m^2 + 3m + 2$?
2. The equation in part (b) of Exercise 47 on page 390 has two solutions. Are both solutions of the equation reasonable in the context of the problem? Explain your reasoning.

Performance Task

The View Matters

The way an equation or expression is written can help you interpret and solve problems. Which representation would you rather have when trying to solve for specific information? Why?

To explore the answers to these questions and more, go to ***BigIdeasMath.com***.

Dynamic Teaching Tools

- Dynamic Assessment & Progress Monitoring Tool
- Interactive Whiteboard Lesson Library
- Dynamic Classroom with Dynamic Investigations

ANSWERS

1. The solutions are the m-coordinates of the m-intercepts of the graph.
2. no; The solution $x = 28$ is not reasonable in the context of the problem because the diagram shows that $x < 18$.

ANSWERS

1. $2x^2 + 6$; 2; 2; binomial
2. $5p^6 - 3p^3 - 4$; 6; 5; trinomial
3. $9x^7 + 13x^5 - 6x^2$; 7; 9; trinomial
4. $8y^3 - 12y$; 3; 8; binomial
5. $4a + 6$
6. $3x^2 + 6x + 10$
7. $-2y^2 + 6y + 4$
8. $-6p^2 - 12p + 7$
9. $x^2 + 2x - 24$
10. $3y^2 - 7y - 40$
11. $x^3 + 11x^2 + 28x$
12. $-12y^3 + 7y^2 + 20y - 7$
13. $x^2 - 81$
14. $4y^2 - 16$
15. $p^2 + 8p + 16$
16. $4d^2 - 4d + 1$

7 Chapter Review

Dynamic Solutions available at *BigIdeasMath.com*

7.1 Adding and Subtracting Polynomials (pp. 357–364)

Find $(2x^3 + 6x^2 - x) - (-3x^3 - 2x^2 - 9x)$.

$$(2x^3 + 6x^2 - x) - (-3x^3 - 2x^2 - 9x) = (2x^3 + 6x^2 - x) + (3x^3 + 2x^2 + 9x)$$
$$= (2x^3 + 3x^3) + (6x^2 + 2x^2) + (-x + 9x)$$
$$= 5x^3 + 8x^2 + 8x$$

Write the polynomial in standard form. Identify the degree and leading coefficient of the polynomial. Then classify the polynomial by the number of terms.

1. $6 + 2x^2$ **2.** $-3p^3 + 5p^6 - 4$ **3.** $9x^7 - 6x^2 + 13x^5$ **4.** $-12y + 8y^3$

Find the sum or difference.

5. $(3a + 7) + (a - 1)$
6. $(x^2 + 6x - 5) + (2x^2 + 15)$
7. $(-y^2 + y + 2) - (y^2 - 5y - 2)$
8. $(p + 7) - (6p^2 + 13p)$

7.2 Multiplying Polynomials (pp. 365–370)

Find $(x + 7)(x - 9)$.

$(x + 7)(x - 9) = x(x - 9) + 7(x - 9)$ — Distribute $(x - 9)$ to each term of $(x + 7)$.

$= x(x) + x(-9) + 7(x) + 7(-9)$ — Distributive Property

$= x^2 + (-9x) + 7x + (-63)$ — Multiply.

$= x^2 - 2x - 63$ — Combine like terms.

Find the product.

9. $(x + 6)(x - 4)$ **10.** $(y - 5)(3y + 8)$ **11.** $(x + 4)(x^2 + 7x)$ **12.** $(-3y + 1)(4y^2 - y - 7)$

7.3 Special Products of Polynomials (pp. 371–376)

Find each product.

a. $(6x + 4y)^2$

$(6x + 4y)^2 = (6x)^2 + 2(6x)(4y) + (4y)^2$ — Square of a binomial pattern

$= 36x^2 + 48xy + 16y^2$ — Simplify.

b. $(2x + 3y)(2x - 3y)$

$(2x + 3y)(2x - 3y) = (2x)^2 - (3y)^2$ — Sum and difference pattern

$= 4x^2 - 9y^2$ — Simplify.

Find the product.

13. $(x + 9)(x - 9)$ **14.** $(2y + 4)(2y - 4)$ **15.** $(p + 4)^2$ **16.** $(-1 + 2d)^2$

7.4 Solving Polynomial Equations in Factored Form (pp. 377–382)

Solve $(x + 6)(x - 8) = 0$.

$(x + 6)(x - 8) = 0$ Write equation.

$x + 6 = 0$ *or* $x - 8 = 0$ Zero-Product Property

$x = -6$ *or* $x = 8$ Solve for x.

Solve the equation.

17. $x^2 + 5x = 0$ **18.** $(z + 3)(z - 7) = 0$ **19.** $(b + 13)^2 = 0$ **20.** $2y(y - 9)(y + 4) = 0$

7.5 Factoring $x^2 + bx + c$ (pp. 385–390)

Factor $x^2 + 6x - 27$.

Notice that $b = 6$ and $c = -27$. Because c is negative, the factors p and q must have different signs so that pq is negative.

Find two integer factors of -27 whose sum is 6.

Factors of −27	−27, 1	−1, 27	−9, 3	−3, 9
Sum of factors	−26	26	−6	6

The values of p and q are -3 and 9.

▶ So, $x^2 + 6x - 27 = (x - 3)(x + 9)$.

Factor the polynomial.

21. $p^2 + 2p - 35$ **22.** $b^2 + 18b + 80$ **23.** $z^2 - 4z - 21$ **24.** $x^2 - 11x + 28$

7.6 Factoring $ax^2 + bx + c$ (pp. 391–396)

Factor $5x^2 + 36x + 7$.

There is no GCF, so you need to consider the possible factors of a and c. Because b and c are both positive, the factors of c must be positive. Use a table to organize information about the factors of a and c.

Factors of 5	Factors of 7	Possible factorization	Middle term	
1, 5	1, 7	$(x + 1)(5x + 7)$	$7x + 5x = 12x$	✗
1, 5	7, 1	$(x + 7)(5x + 1)$	$x + 35x = 36x$	✓

▶ So, $5x^2 + 36x + 7 = (x + 7)(5x + 1)$.

Factor the polynomial.

25. $3t^2 + 16t - 12$ **26.** $-5y^2 - 22y - 8$ **27.** $6x^2 + 17x + 7$

28. $-2y^2 + 7y - 6$ **29.** $3z^2 + 26z - 9$ **30.** $10a^2 - 13a - 3$

ANSWERS

17. $x = 0, x = -5$

18. $z = -3, z = 7$

19. $b = -13$

20. $y = 0, y = 9, y = -4$

21. $(p + 7)(p - 5)$

22. $(b + 8)(b + 10)$

23. $(z + 3)(z - 7)$

24. $(x - 7)(x - 4)$

25. $(t + 6)(3t - 2)$

26. $-(y + 4)(5y + 2)$

27. $(2x + 1)(3x + 7)$

28. $-(y - 2)(2y - 3)$

29. $(z + 9)(3z - 1)$

30. $(2a - 3)(5a + 1)$

ANSWERS

31. $(x + 3)(x - 3)$
32. $(y + 10)(y - 10)$
33. $(z - 3)^2$
34. $(m + 8)^2$
35. $n(n + 3)(n - 3)$
36. $(x - 3)(x + 4a)$
37. $2x^2(x^2 + x - 10)$
38. $x = 0, x = 6, x = -3$
39. $x = -\frac{3}{2}, x = \frac{3}{2}$
40. $z = -3, z = 5, z = -5$
41. length: 12 ft, width: 4 ft, height: 2 ft

7.7 Factoring Special Products (pp. 397–402)

Factor each polynomial.

a. $x^2 - 16$

$x^2 - 16 = x^2 - 4^2$ — Write as $a^2 - b^2$.

$= (x + 4)(x - 4)$ — Difference of two squares pattern

b. $25x^2 - 30x + 9$

$25x^2 - 30x + 9 = (5x)^2 - 2(5x)(3) + 3^2$ — Write as $a^2 - 2ab + b^2$.

$= (5x - 3)^2$ — Perfect square trinomial pattern

Factor the polynomial.

31. $x^2 - 9$ **32.** $y^2 - 100$ **33.** $z^2 - 6z + 9$ **34.** $m^2 + 16m + 64$

7.8 Factoring Polynomials Completely (pp. 403–408)

Factor each polynomial completely.

a. $x^3 + 4x^2 - 3x - 12$

$x^3 + 4x^2 - 3x - 12 = (x^3 + 4x^2) + (-3x - 12)$ — Group terms with common factors.

$= x^2(x + 4) + (-3)(x + 4)$ — Factor out GCF of each pair of terms.

$= (x + 4)(x^2 - 3)$ — Factor out $(x + 4)$.

b. $2x^4 - 8x^2$

$2x^4 - 8x^2 = 2x^2(x^2 - 4)$ — Factor out $2x^2$.

$= 2x^2(x^2 - 2^2)$ — Write as $a^2 - b^2$.

$= 2x^2(x + 2)(x - 2)$ — Difference of two squares pattern

c. $2x^3 + 18x^2 - 72x$

$2x^3 + 18x^2 - 72x = 2x(x^2 + 9x - 36)$ — Factor out $2x$.

$= 2x(x + 12)(x - 3)$ — Factor $x^2 + 9x - 36$.

Factor the polynomial completely.

35. $n^3 - 9n$ **36.** $x^2 - 3x + 4ax - 12a$ **37.** $2x^4 + 2x^3 - 20x^2$

Solve the equation.

38. $3x^3 - 9x^2 - 54x = 0$ **39.** $16x^2 - 36 = 0$ **40.** $z^3 + 3z^2 - 25z - 75 = 0$

41. A box in the shape of a rectangular prism has a volume of 96 cubic feet. The box has a length of $(x + 8)$ feet, a width of x feet, and a height of $(x - 2)$ feet. Find the dimensions of the box.

7 Chapter Test

Find the sum or difference. Then identify the degree of the sum or difference and classify it by the number of terms.

1. $(-2p + 4) - (p^2 - 6p + 8)$

2. $(9c^6 - 5b^4) - (4c^6 - 5b^4)$

3. $(4s^4 + 2st + t) + (2s^4 - 2st - 4t)$

Find the product.

4. $(h - 5)(h - 8)$

5. $(2w - 3)(3w + 5)$

6. $(z + 11)(z - 11)$

7. Explain how you can determine whether a polynomial is a perfect square trinomial.

8. Is 18 a polynomial? Explain your reasoning.

Factor the polynomial completely.

9. $s^2 - 15s + 50$

10. $h^3 + 2h^2 - 9h - 18$

11. $-5k^2 - 22k + 15$

Solve the equation.

12. $(n - 1)(n + 6)(n + 5) = 0$

13. $d^2 + 14d + 49 = 0$

14. $6x^4 + 8x^2 = 26x^3$

15. The expression $\pi(r - 3)^2$ represents the area covered by the hour hand on a clock in one rotation, where r is the radius of the entire clock. Write a polynomial in standard form that represents the area covered by the hour hand in one rotation.

16. A magician's stage has a trapdoor.

a. The total area (in square feet) of the stage can be represented by $x^2 + 27x + 176$. Write an expression for the width of the stage.

b. Write an expression for the perimeter of the stage.

c. The area of the trapdoor is 10 square feet. Find the value of x.

d. The magician wishes to have the area of the stage be at least 20 times the area of the trapdoor. Does this stage satisfy his requirement? Explain.

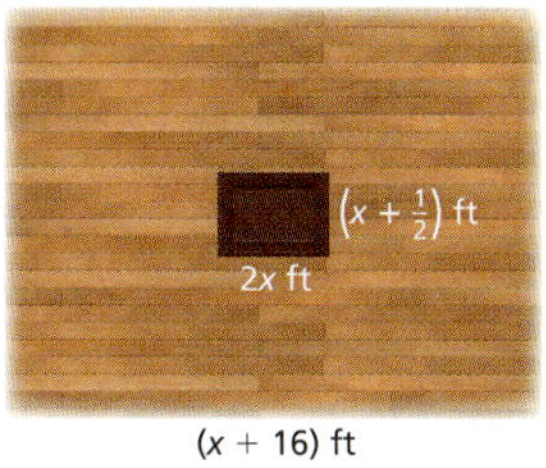

17. Write a polynomial equation in factored form that has three positive roots.

18. You are jumping on a trampoline. For one jump, your height y (in feet) above the trampoline after t seconds can be represented by $y = -16t^2 + 24t$. How many seconds are you in the air?

19. A cardboard box in the shape of a rectangular prism has the dimensions shown.

a. Write a polynomial that represents the volume of the box.

b. The volume of the box is 60 cubic inches. What are the length, width, and height of the box?

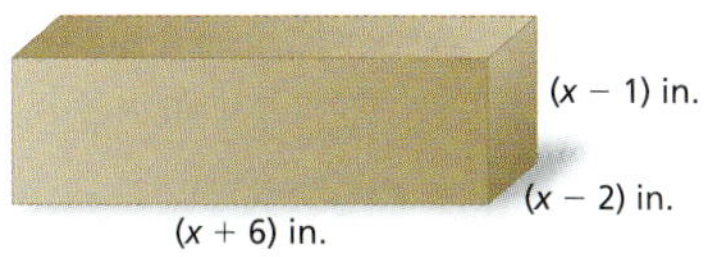

ANSWERS

1. $-p^2 + 4p - 4$; 2; trinomial

2. $5c^6$; 6; monomial

3. $6s^4 - 3t$; 4; binomial

4. $h^2 - 13h + 40$

5. $6w^2 + w - 15$

6. $z^2 - 121$

7. Check the first and last terms to verify they are perfect squares. If they are, find 2 times the product of their square roots. If this matches the middle term, or the opposite of the middle term, it is a perfect square trinomial.

8. yes; A monomial is a type of polynomial, and a number is a type of monomial.

9. $(s - 5)(s - 10)$

10. $(h + 2)(h + 3)(h - 3)$

11. $-(5k - 3)(k + 5)$

12. $n = 1, n = -6, n = -5$

13. $d = -7$

14. $x = 0, x = \frac{1}{3}, x = 4$

15. $\pi r^2 - 6\pi r + 9\pi$

16. a. $(x + 11)$ ft

b. $(4x + 54)$ ft

c. $x = 2$

d. yes; The area of the stage is 234 square feet, and twenty times the area of the trap door is 200 square feet, so the area of the stage is greater than twenty times the area of the trap door.

17. *Sample answer:* $(x - 1)(x - 2)(x - 3) = 0$

18. 1.5 sec

19. a. $(x^3 + 3x^2 - 16x + 12)$ in.^{3}

b. length: 10 in., width: 2 in., height: 3 in.

If students need help...	*If students got it...*
Lesson Tutorials	Resources by Chapter • Enrichment and Extension • Cumulative Review
Skills Review Handbook	Performance Task
BigIdeasMath.com	Start the *next* Section

ANSWERS

1. **a.** monomial
   b. binomial
   c. trinomial
   d. binomial
   e. binomial
   f. trinomial
   c, a, d, b, f, e
2. D
3. $-6, -2, 2$
4. $y = 61x + 1$
5. **a.**

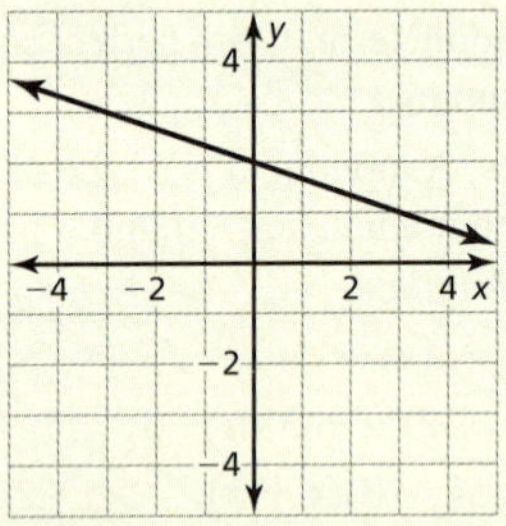

   b. linear
   c. continuous

7 Cumulative Assessment

1. Classify each polynomial by the number of terms. Then order the polynomials by degree from least to greatest.

a. $-4x^3$ **b.** $6y - 3y^5$

c. $c^2 + 2 + c$ **d.** $-10d^4 + 7d^2$

e. $-5z^{11} + 8z^{12}$ **f.** $3b^6 - 12b^8 + 4b^4$

2. Which exponential function is increasing the fastest over the interval $x = 0$ to $x = 2$?

Ⓐ $f(x) = 4(2.5)^x$

Ⓑ

x	0	1	2	3	4
$g(x)$	8	12	18	27	40.5

Ⓒ

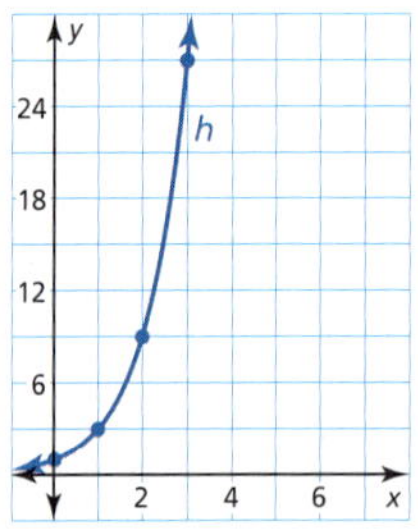

Ⓓ An exponential function j models a relationship in which the dependent variable is multiplied by 6 for every 1 unit the independent variable increases. The value of the function at 0 is 2.

3. Find all solutions of the equation $x^3 + 6x^2 - 4x = 24$.

-6 -4 -2 -1 0 1 2 4 6 24

4. The table shows the distances you travel over a 6-hour period. Create an equation that models the distance traveled as a function of the number of hours.

Hours, x	Distance (miles), y
1	62
2	123
3	184
4	245
5	306
6	367

5. Consider the equation $y = -\frac{1}{3}x + 2$.

a. Graph the equation in a coordinate plane.

b. Does the equation represent a linear or nonlinear function?

c. Is the domain discrete or continuous?

6. Which expressions are equivalent to $-2x + 15x^2 - 8$?

$15x^2 - 2x - 8$	$(5x + 4)(3x + 2)$
$(5x - 4)(3x + 2)$	$15x^2 + 2x - 8$
$(3x - 2)(5x - 4)$	$(3x + 2)(5x - 4)$

7. The graph shows the function $f(x) = 2(3)^x$.

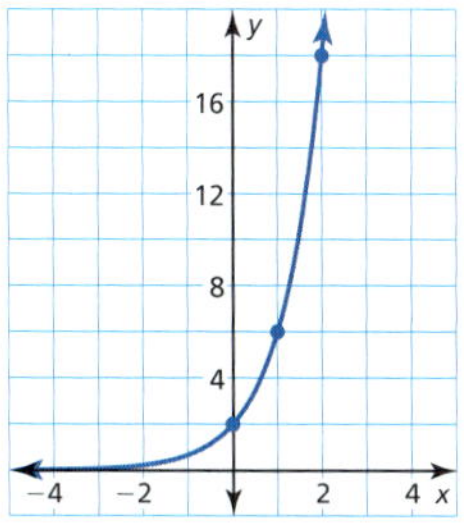

a. Is the function increasing or decreasing for increasing values of x?

b. Identify any x- and y-intercepts.

8. Which polynomial represents the product of $2x - 4$ and $x^2 + 6x - 2$?

Ⓐ $2x^3 + 8x^2 - 4x + 8$

Ⓑ $2x^3 + 8x^2 - 28x + 8$

Ⓒ $2x^3 + 8$

Ⓓ $2x^3 - 24x - 2$

9. You are playing miniature golf on the hole shown.

a. Write a polynomial that represents the area of the golf hole.

b. Write a polynomial that represents the perimeter of the golf hole.

c. Find the perimeter of the golf hole when the area is 216 square feet.

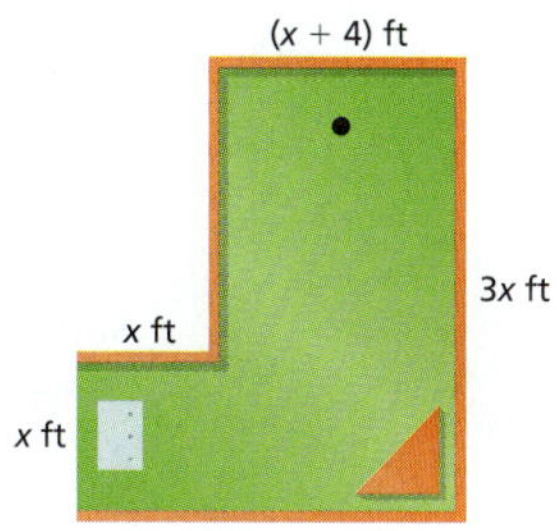

ANSWERS

6. $15x^2 - 2x - 8$, $(5x - 4)(3x + 2)$, $(3x + 2)(5x - 4)$

7. a. increasing

b. no x-intercept; y-intercept is 2

8. B

9. a. $(4x^2 + 12x)$ ft^2

b. $(10x + 8)$ ft

c. 68 ft

Chapter 8 Pacing Guide	
Chapter Opener/ Mathematical Practices	1 Day
Section 1	2 Days
Section 2	2 Days
Section 3	2 Days
Quiz	1 Day
Section 4	2 Days
Section 5	2 Days
Section 6	2 Days
Chapter Review/ Chapter Tests	2 Days
Total Chapter 8	16 Days
Year-to-Date	119 Days

8 Graphing Quadratic Functions

8.1 Graphing $f(x) = ax^2$
8.2 Graphing $f(x) = ax^2 + c$
8.3 Graphing $f(x) = ax^2 + bx + c$
8.4 Graphing $f(x) = a(x - h)^2 + k$
8.5 Using Intercept Form
8.6 Comparing Linear, Exponential, and Quadratic Functions

Town Population *(p. 464)*

Satellite Dish *(p. 457)*

Roller Coaster *(p. 448)*

Firework Explosion *(p. 437)*

Garden Waterfalls *(p. 430)*

Laurie's Notes

Chapter Summary

- This is the second of three chapters generally about polynomials and more specifically about quadratics. In the last chapter, students factored quadratics and used the Zero-Product Property to solve quadratics.
- The first four lessons of the chapter are about graphing quadratic functions. Connections to transformations of functions are made, just as students had seen with linear functions earlier. In addition to graphing, the concept of zeros of functions is introduced and connected to the *x*-intercept of a graph.
- There are three forms of quadratic equations: standard, vertex, and intercept. Students should become familiar with what the parameters of each equation tell about the graph of the function.
- The final lesson in the chapter compares the behavior of linear, exponential, and quadratic functions. This is done by looking at the data numerically and graphically, a theme emphasized throughout the chapter.

What Your Students Have Learned

Middle School
- Graph linear equations written in standard form and slope-intercept form.
- Write functions to model linear relationships, and determine and interpret rates of change.
- Compare linear and nonlinear functions represented by tables, graphs, and equations.

What Your Students Will Learn

Algebra 1
- Graph, write, and use quadratic functions in standard form, vertex form, and intercept form.
- Find the minimum and maximum values of quadratic functions.
- Identify even and odd functions algebraically and graphically.
- Use the intercept form of quadratic and cubic functions to find the zeros of the functions.
- Write quadratic functions to model data and write a recursive rule for a quadratic function.
- Compare linear, exponential, or quadratic functions using average rates of change.

Dynamic Teaching Tools
- Dynamic Assessment & Progress Monitoring Tool
- Lesson Planning Tool
- Interactive Whiteboard Lesson Library
- Dynamic Classroom with Dynamic Investigations
- Real-Life STEM Videos

Scaffolding in the Classroom
How do you solve that?
Verbally state the thought processes you use to solve a problem. What questions do you ask yourself? Help the students develop their own set of questions.

Questioning in the Classroom

All students need to participate. When a small group of students continually answers the questions in your classroom, no one else has to think. Try asking a different student whether they agree with the answer given, and then follow up with why or why not.

Laurie's Notes

Maintaining Mathematical Proficiency

Graphing Linear Equations

- Remind students that the graph of a linear equation shows the relationship between two variables in a coordinate plane.
- Review how to use the given equation to make a table of values, then plot the points and draw a line through them.

COMMON ERROR Students may find the coordinates of only two points on the line, and if one contains an error, will graph the line incorrectly. Remind students to use at least three points.

Evaluating Expressions

- Remind students to substitute the given value of x in the expression first, and then simplify the expression.
- Review how to use the order of operations to simplify expressions involving exponents.

COMMON ERROR Students may use the order of operations incorrectly.

Mathematical Practices (continued on page 418)

- The *Mathematical Practices* page focuses attention on how mathematics is learned—process versus content. This page demonstrates the need for students to try simpler cases of a problem—in this case, try graphing the parent function first. Just as with linear and exponential functions, student will see that when graphing quadratics, they should consider how the parent function has been transformed.

If students need help...	*If students got it...*
Student Journal • Maintaining Mathematical Proficiency	Game Closet at *BigIdeasMath.com*
Lesson Tutorials	Start the *next* Section
Skills Review Handbook	

Maintaining Mathematical Proficiency

Graphing Linear Equations

Example 1 Graph $y = -x - 1$.

Step 1 Make a table of values.

x	$y = -x - 1$	y	(x, y)
-1	$y = -(-1) - 1$	0	$(-1, 0)$
0	$y = -(0) - 1$	-1	$(0, -1)$
1	$y = -(1) - 1$	-2	$(1, -2)$
2	$y = -(2) - 1$	-3	$(2, -3)$

Step 2 Plot the ordered pairs.

Step 3 Draw a line through the points.

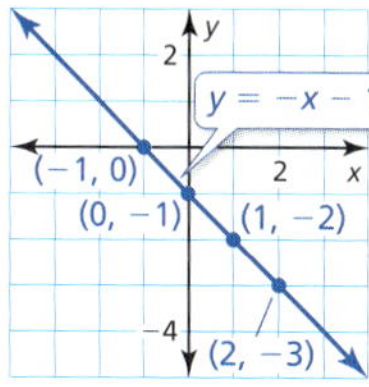

Graph the linear equation.

1. $y = 2x - 3$
2. $y = -3x + 4$
3. $y = -\frac{1}{2}x - 2$
4. $y = x + 5$

Evaluating Expressions

Example 2 Evaluate $2x^2 + 3x - 5$ when $x = -1$.

$$
\begin{aligned}
2x^2 + 3x - 5 &= 2(-1)^2 + 3(-1) - 5 && \text{Substitute } -1 \text{ for } x.\\
&= 2(1) + 3(-1) - 5 && \text{Evaluate the power.}\\
&= 2 - 3 - 5 && \text{Multiply.}\\
&= -6 && \text{Subtract.}
\end{aligned}
$$

Evaluate the expression when $x = -2$.

5. $5x^2 - 9$
6. $3x^2 + x - 2$
7. $-x^2 + 4x + 1$
8. $x^2 + 8x + 5$
9. $-2x^2 - 4x + 3$
10. $-4x^2 + 2x - 6$
11. **ABSTRACT REASONING** Complete the table. Find a pattern in the differences of consecutive y-values. Use the pattern to write an expression for y when $x = 6$.

x	1	2	3	4	5
$y = ax^2$					

What Your Students Have Learned

- Graph linear equations written in slope-intercept form.
- Use Order of Operations to evaluate expressions, including those involving whole number exponents.

ANSWERS

1.

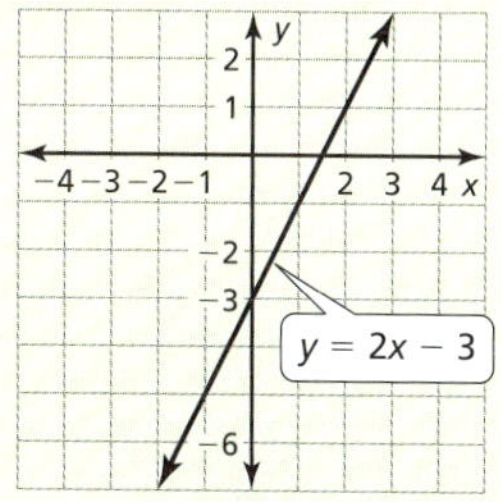

2.

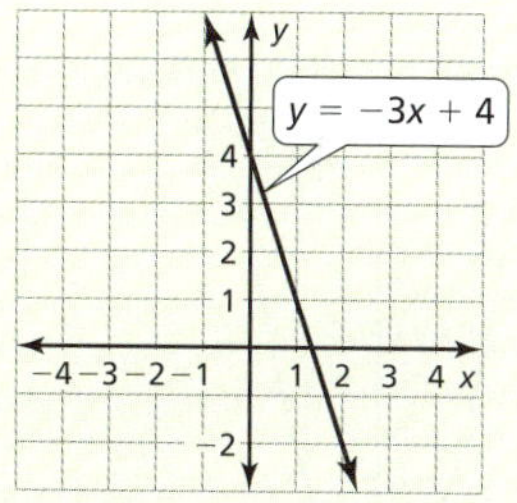

3.

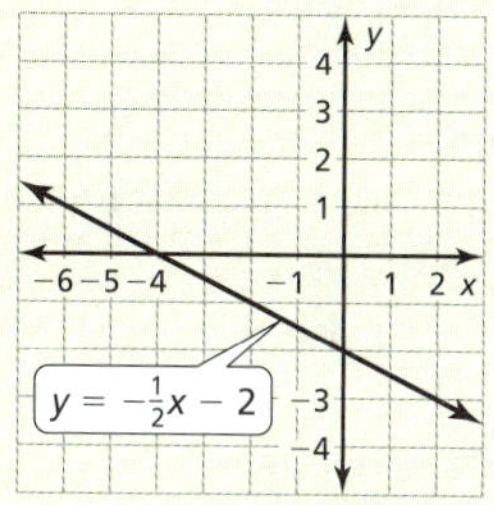

4.

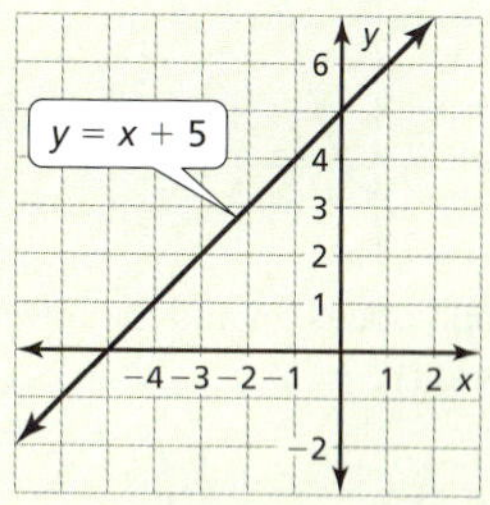

5. 11
6. 8
7. −11
8. −7
9. 3
10. −26
11. a, $4a$, $9a$, $16a$, $25a$; *Sample answer:* The coefficient of each difference is the next consecutive odd integer; $36a$

Vocabulary Review

Have students make Process Diagrams for the following topics.

- Graphing linear equations
- Evaluating algebraic expressions

MONITORING PROGRESS ANSWERS

1.

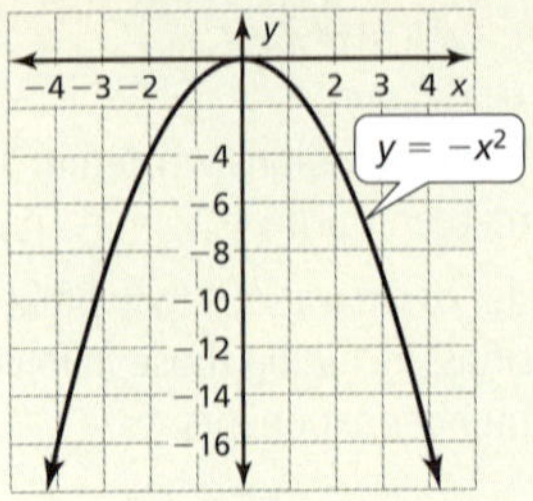

The graph opens down and the highest point is at the origin.

2.

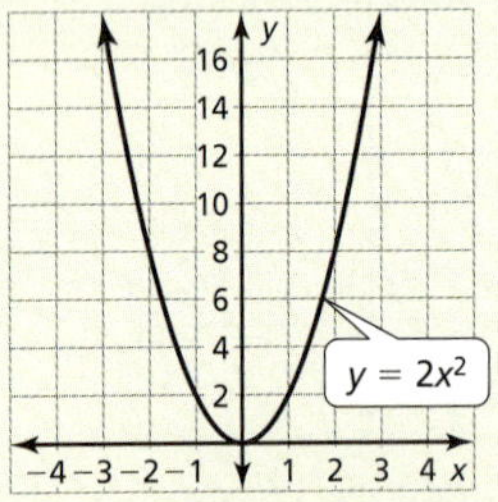

The graph opens up and the lowest point is at the origin.

3.

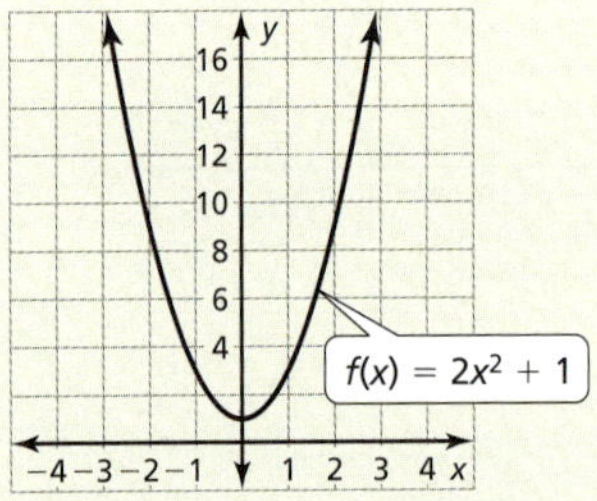

The graph opens up and the lowest point is at (0, 1).

4.

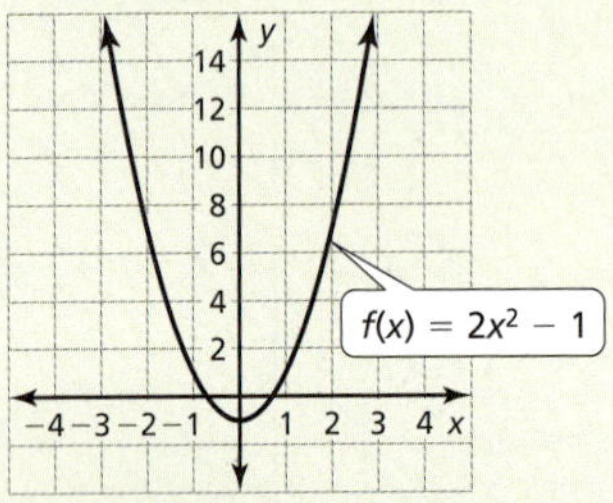

The graph opens up and the lowest point is at (0, −1).

5.

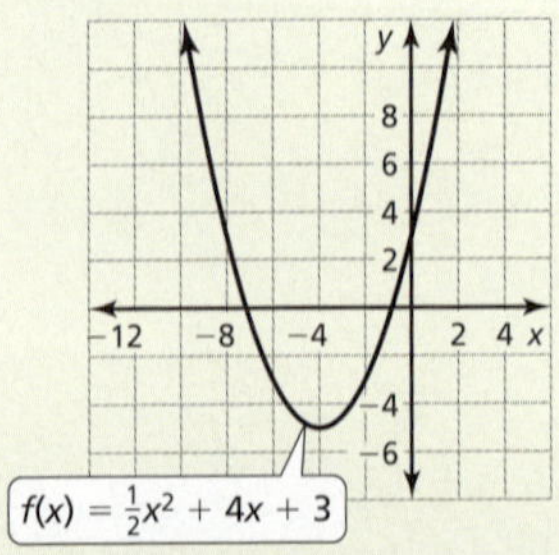

The graph opens up and the lowest point is at (−4, −5).

6–9. See Additional Answers.

Mathematical Practices

Mathematically proficient students try special cases of the original problem to gain insight into its solution.

Problem-Solving Strategies

Core Concept

Trying Special Cases

When solving a problem in mathematics, it can be helpful to try special cases of the original problem. For instance, in this chapter, you will learn to graph a quadratic function of the form $f(x) = ax^2 + bx + c$. The problem-solving strategy used is to first graph quadratic functions of the form $f(x) = ax^2$. From there, you progress to other forms of quadratic functions.

$f(x) = ax^2$	Section 8.1
$f(x) = ax^2 + c$	Section 8.2
$f(x) = ax^2 + bx + c$	Section 8.3
$f(x) = a(x - h)^2 + k$	Section 8.4

EXAMPLE 1 Graphing the Parent Quadratic Function

Graph the parent quadratic function $y = x^2$. Then describe its graph.

SOLUTION

The function is of the form $y = ax^2$, where $a = 1$. By plotting several points, you can see that the graph is U-shaped, as shown.

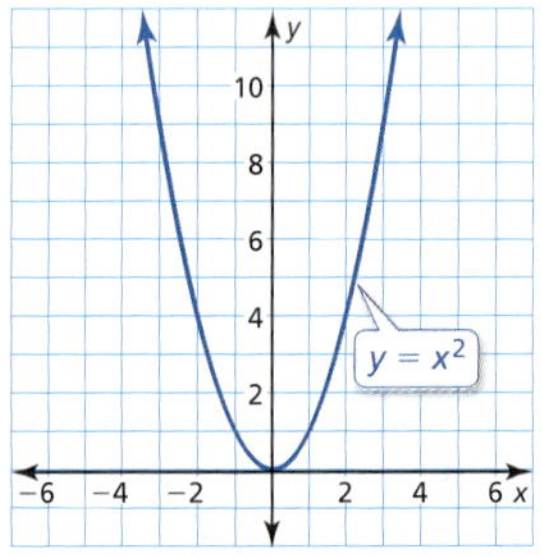

▶ The graph opens up, and the lowest point is at the origin.

Monitoring Progress

Graph the quadratic function. Then describe its graph.

1. $y = -x^2$ **2.** $y = 2x^2$ **3.** $f(x) = 2x^2 + 1$ **4.** $f(x) = 2x^2 - 1$

5. $f(x) = \frac{1}{2}x^2 + 4x + 3$ **6.** $f(x) = \frac{1}{2}x^2 - 4x + 3$ **7.** $y = -2(x + 1)^2 + 1$ **8.** $y = -2(x - 1)^2 + 1$

9. How are the graphs in Monitoring Progress Questions 1−8 similar? How are they different?

Laurie's Notes Mathematical Practices (continued from page T-417)

- Use the *Mathematical Practices* page to help students develop mathematical habits of mind—how mathematics can be explored and how mathematics is thought about.
- Transformations of functions will be a focus in this chapter. The basic quadratic $y = x^2$ will be shifted vertically and horizontally. It will be stretched and compressed vertically and horizontally.
- Give time for students to work through the questions in the *Monitoring Progress*, and then discuss as a class. Students should observe that the parent function has been shifted or reflected, but essentially students will describe seeing a U-shaped graph.

Laurie's Notes

Overview of Section 8.1

Introduction

- The lesson begins by describing the features of a *quadratic function*. The graph is called a *parabola*, and it has a *vertex* and an *axis of symmetry*. Unlike a linear function, which is increasing, decreasing, or constant, a quadratic function changes from a decreasing function to an increasing function or vice versa.
- Students are familiar with transformations from their work with linear functions. In this lesson, vertical stretches and shrinks of the parent function $y = x^2$ are performed, along with reflections in the x-axis. The connection for students is the effect of m on the graph of $y = mx$.

Resources

- Use dynamic geometry software to investigate the effect of a on the graph of $y = ax^2$. A slider can be used to generate values of a, $-5 < a < 5$.
- In lieu of creating your own, you can find online versions to download.

Teaching Strategy

- **Rule of 3:** Throughout this chapter, students will be graphing quadratic functions. Calculators can help students develop a stronger understanding of the behavior of quadratic functions, particularly understanding of transformations of these functions. This occurs by seeing the three representations of the function—symbolic, graphical, and numerical. The table of values can be particularly helpful in making sense of increasing and decreasing functions.
- In this example, the value of the coefficient is 2, meaning the parent function is stretched vertically. Graphically, you can see that the y-value of one is twice the y-value of the other. In the table, the values in the Y_2 column are twice the values in the Y_1 column for each value of x.

\Y1=X²
\Y2=2X²
\Y3=
\Y4=
\Y5=
\Y6=
\Y7=

Symbolic

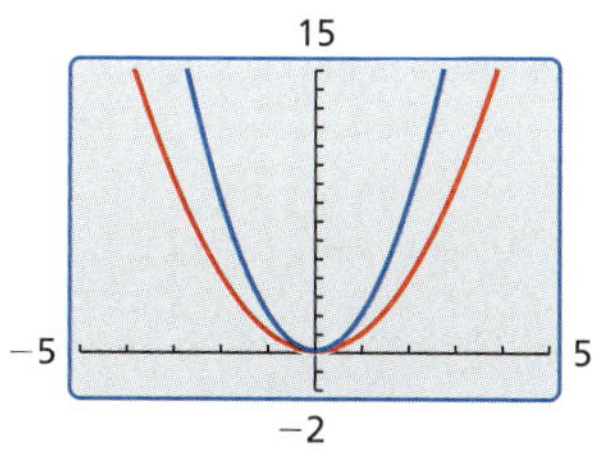

Graphical

X	Y1	Y2
-3	9	18
-2	4	8
-1	1	2
0	0	0
1	1	2
2	4	8
3	9	18

X=-3

Numerical

- Beginning in this lesson and throughout this chapter, look for opportunities to integrate this *Rule of 3* in your teaching.

Pacing Suggestion

- The exploration helps students discover the effect of the value of a on the graph of $y = ax^2$. You may choose to omit a few examples and check for understanding with the *Monitoring Progress* questions.

Dynamic Teaching Tools

- Dynamic Assessment & Progress Monitoring Tool
- Lesson Planning Tool
- Interactive Whiteboard Lesson Library
- Dynamic Classroom with Dynamic Investigations

What Your Students Will Learn

- Identify characteristics of the graphs of quadratic functions such as the vertex, axis of symmetry, and behavior of the graph.
- Graph and use quadratic functions of the form $f(x) = ax^2$ when $a > 0$ and when $a < 0$.

Laurie's Notes

Exploration

Motivate

- Use a flashlight to cast a shadow that students can describe. If you do not have a flashlight, use something that has a parabolic cross section, such as a bulb in an overhead projector. Tell students that the graphs they will be sketching represent a common shape.

Exploration 1

- Having the graph of $y = x^2$ already on the grid may help students compare the graph of $y = ax^2$ to the graph of $y = x^2$.
- There are two features that the value of a affects, and you may need to prod students' thinking as you circulate. The obvious effect is whether the graph opens up or down. The feature that is more difficult for students to explain is the vertical shrink or stretch. Informally, this can be referred to as "more bowl-shaped" or "narrower."
- Students often say that when $a < 1$, the graphs become more bowl-shaped. Remind students that $-5 < 1$. What they really are trying to describe are rational numbers between -1 and 1, or $|a| < 1$.
- If a student fails to notice the shape of the graph, give the student additional functions to graph.

? "Is the graph nonlinear when $a = 0$? Explain." no; When $a = 0$, the function becomes the constant function $y = 0$.

- Discuss as a class when all have finished.

? "Describe the graph when $a < 0$." opens downward "Describe the graph when $|a| > 1$." narrower "Describe the graph when $|a| < 1$." more bowl-shaped

- If time allows, use a graphing calculator to check conjectures. You can also use the calculator to graph several functions and ask for reasonable equations for each graph.

Communicate Your Answer

- Have students summarize their conjectures in a table.

Connecting to Next Step

- The exploration allows students to discover the effect of the value of a on the graph of $f(x) = ax^2$. In the formal lesson, students will graph additional functions of this type.

8.1 Graphing $f(x) = ax^2$

Essential Question What are some of the characteristics of the graph of a quadratic function of the form $f(x) = ax^2$?

EXPLORATION 1 **Graphing Quadratic Functions**

Work with a partner. Graph each quadratic function. Compare each graph to the graph of $f(x) = x^2$.

a. $g(x) = 3x^2$

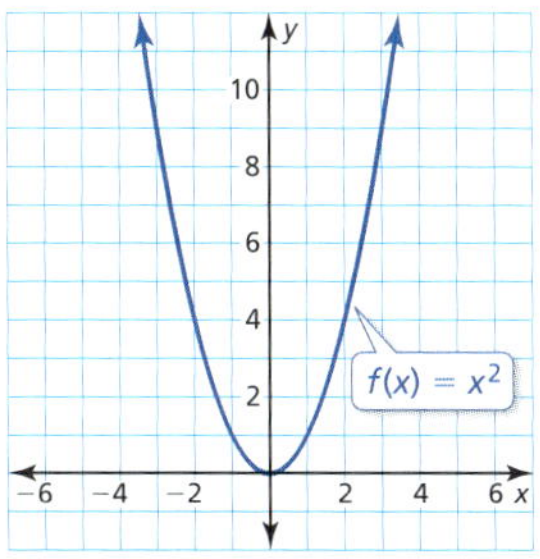

b. $g(x) = -5x^2$

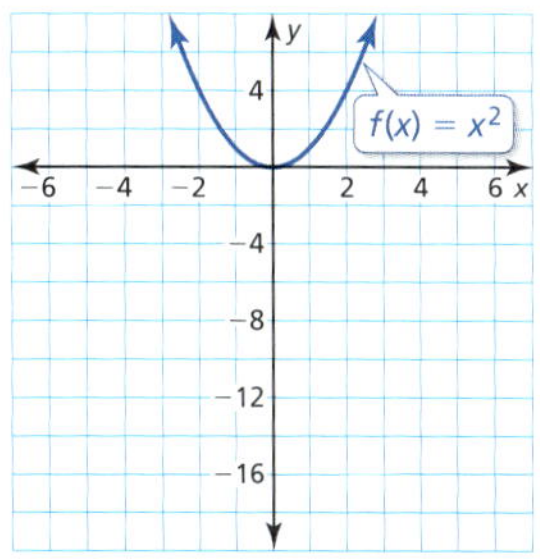

c. $g(x) = -0.2x^2$

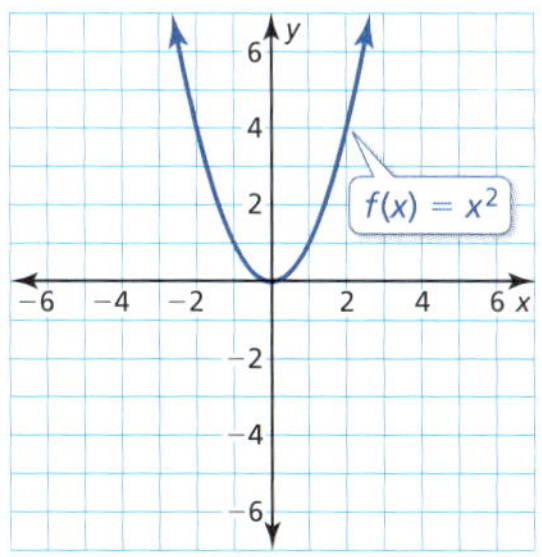

d. $g(x) = \frac{1}{10}x^2$

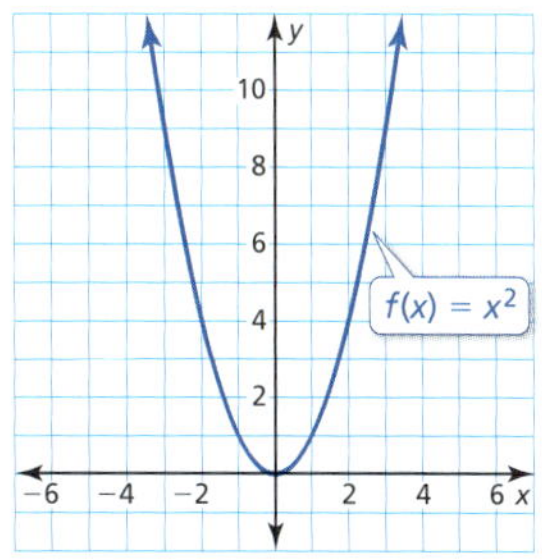

REASONING QUANTITATIVELY

To be proficient in math, you need to make sense of quantities and their relationships in problem situations.

Communicate Your Answer

2. What are some of the characteristics of the graph of a quadratic function of the form $f(x) = ax^2$?

3. How does the value of a affect the graph of $f(x) = ax^2$? Consider $0 < a < 1$, $a > 1$, $-1 < a < 0$, and $a < -1$. Use a graphing calculator to verify your answers.

4. The figure shows the graph of a quadratic function of the form $y = ax^2$. Which of the intervals in Question 3 describes the value of a? Explain your reasoning.

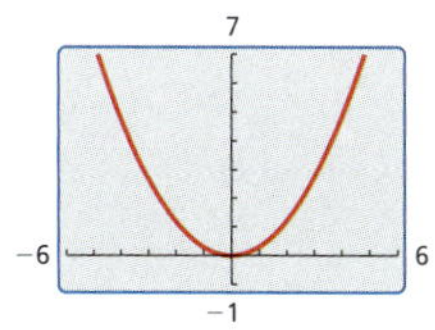

Dynamic Teaching Tools

- Dynamic Assessment & Progress Monitoring Tool
- Lesson Planning Tool
- Interactive Whiteboard Lesson Library
- Dynamic Classroom with Dynamic Investigations

ANSWERS

1. a.

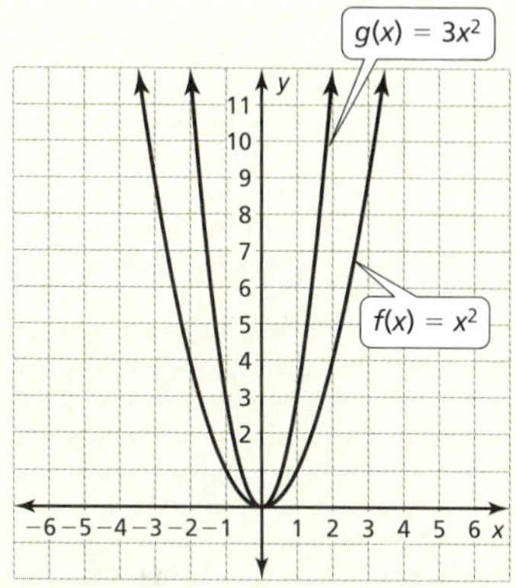

The graph of g is a vertical stretch by a factor of 3 of the graph of f.

b.

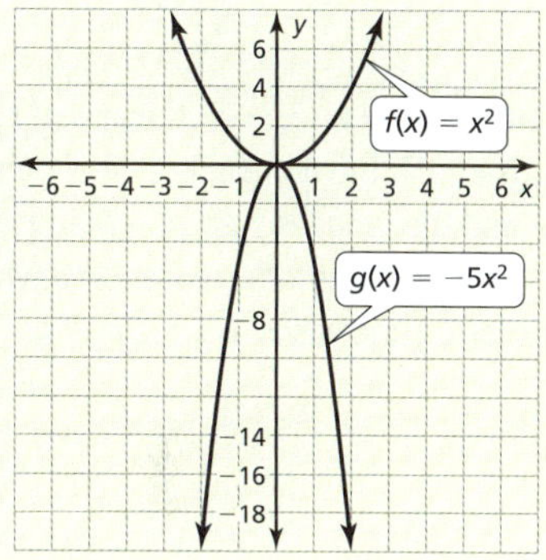

The graph of g is a vertical stretch by a factor of 5 and a reflection in the x-axis of the graph of f.

c.

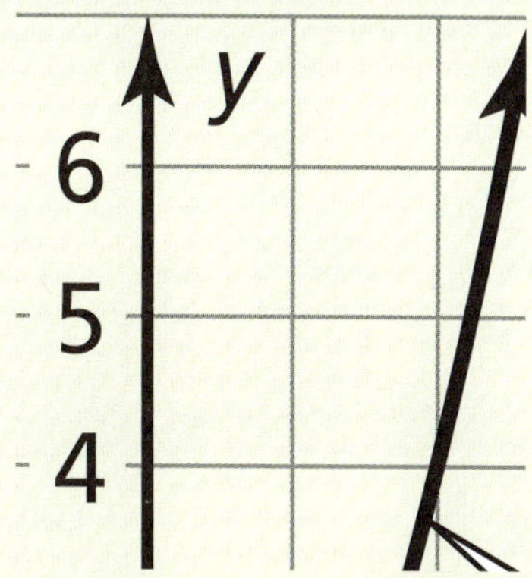

The graph of g is a vertical shrink by a factor of 0.2 and a reflection in the x-axis of the graph of f.

d. See Additional Answers.

2. *Sample answer:* They are U-shaped and symmetric. They either open up with the lowest point at the origin, or they open down with the highest point at the origin.

3. *Sample answer:* When $0 < a < 1$ the graph of $f(x) = ax^2$ is a vertical shrink of the graph of $f(x) = x^2$, when $a > 1$ the graph of $f(x) = ax^2$ is a vertical stretch of the graph of $f(x) = x^2$, when $-1 < a < 0$ the graph of $f(x) = ax^2$ is a vertical shrink and a reflection in the x-axis of the graph of $f(x) = x^2$, and when $a < -1$ the graph of $f(x) = ax^2$ is a vertical stretch and a reflection in the x-axis of the graph of $f(x) = x^2$.

4. $0 < a < 1$; *Sample answer:* The graph is a vertical shrink of the graph of $f(x) = x^2$.

Differentiated Instruction

Kinesthetic/Visual

Some students may benefit by having access to graphing calculators or graphing software programs available on many computers. Students can enter the quadratic function $f(x) = x^2$ and the given quadratic equation, view the graphs, and then compare the graphs.

Extra Example 1

Identify characteristics of the quadratic function and its graph.

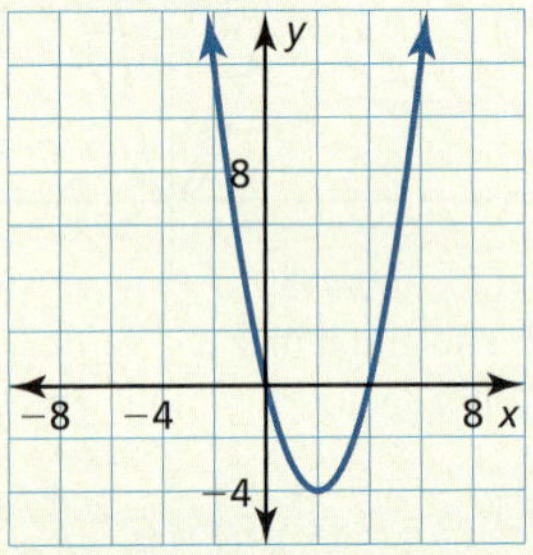

The vertex is (2, −4), the axis of symmetry is $x = 2$, the domain is all real numbers, and the range is $y \geq -4$. When $x < 2$, y increases as x decreases, and when $x > 2$, y increases as x increases.

8.1 Lesson

Core Vocabulary

quadratic function, *p. 420*
parabola, *p. 420*
vertex, *p. 420*
axis of symmetry, *p. 420*

Previous
domain
range
vertical shrink
vertical stretch
reflection

What You Will Learn

- Identify characteristics of quadratic functions.
- Graph and use quadratic functions of the form $f(x) = ax^2$.

Identifying Characteristics of Quadratic Functions

A **quadratic function** is a nonlinear function that can be written in the standard form $y = ax^2 + bx + c$, where $a \neq 0$. The U-shaped graph of a quadratic function is called a **parabola**. In this lesson, you will graph quadratic functions, where b and c equal 0.

REMEMBER
The notation $f(x)$ is another name for y.

Core Concept

Characteristics of Quadratic Functions

The *parent quadratic function* is $f(x) = x^2$. The graphs of all other quadratic functions are *transformations* of the graph of the parent quadratic function.

The lowest point on a parabola that opens up or the highest point on a parabola that opens down is the **vertex**. The vertex of the graph of $f(x) = x^2$ is (0, 0).

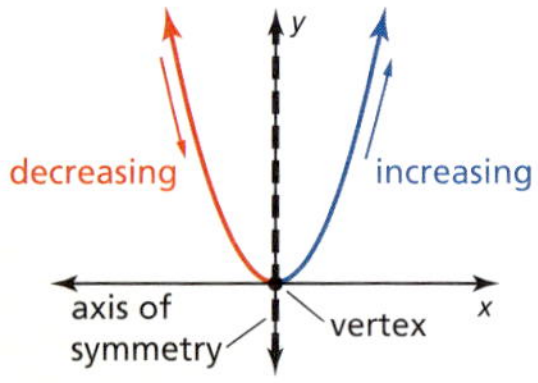

The vertical line that divides the parabola into two symmetric parts is the **axis of symmetry**. The axis of symmetry passes through the vertex. For the graph of $f(x) = x^2$, the axis of symmetry is the y-axis, or $x = 0$.

EXAMPLE 1 Identifying Characteristics of a Quadratic Function

Consider the graph of the quadratic function.

Using the graph, you can identify characteristics such as the vertex, axis of symmetry, and the behavior of the graph, as shown.

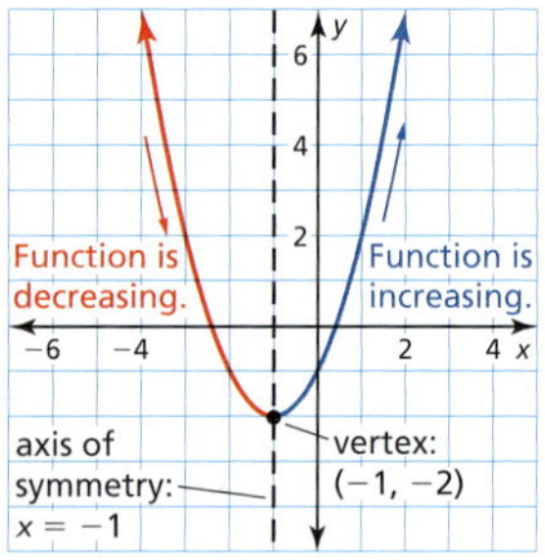

You can also determine the following:

- The domain is all real numbers.
- The range is all real numbers greater than or equal to −2.
- When $x < -1$, y increases as x decreases.
- When $x > -1$, y increases as x increases.

Laurie's Notes Teacher Actions

- **Look For and Make Use of Structure:** Define *quadratic function* in standard form. Explain that a, b, and c are parameters just as m and b are parameters in $y = mx + b$.
- Discuss the red and blue portions of the graph. Graphs are read left to right. So when the phrase "increasing function" is used, the values of the function (y-values) are increasing. When the phrase "decreasing function" is used, the values of the function are decreasing.
- In Example 1, help students understand that the range is $y \geq -2$ by moving your hand on the graph from (0, −2) in an upward motion.

Monitoring Progress

Help in English and Spanish at *BigIdeasMath.com*

Identify characteristics of the quadratic function and its graph.

1.

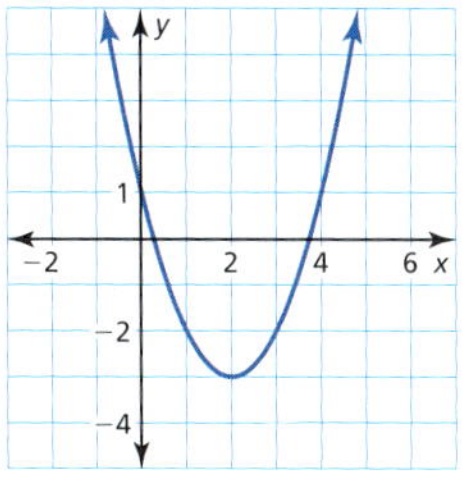

2.

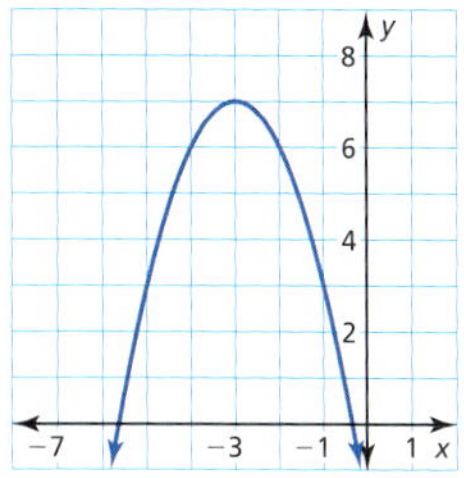

> **REMEMBER**
>
> The graph of $y = a \cdot f(x)$ is a vertical stretch or shrink by a factor of a of the graph of $y = f(x)$.
>
> The graph of $y = -f(x)$ is a reflection in the x-axis of the graph of $y = f(x)$.

Graphing and Using $f(x) = ax^2$

Core Concept

Graphing $f(x) = ax^2$ When $a > 0$

- When $0 < a < 1$, the graph of $f(x) = ax^2$ is a vertical shrink of the graph of $f(x) = x^2$.
- When $a > 1$, the graph of $f(x) = ax^2$ is a vertical stretch of the graph of $f(x) = x^2$.

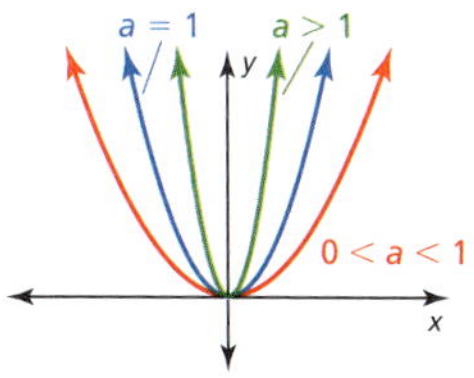

Graphing $f(x) = ax^2$ When $a < 0$

- When $-1 < a < 0$, the graph of $f(x) = ax^2$ is a vertical shrink with a reflection in the x-axis of the graph of $f(x) = x^2$.
- When $a < -1$, the graph of $f(x) = ax^2$ is a vertical stretch with a reflection in the x-axis of the graph of $f(x) = x^2$.

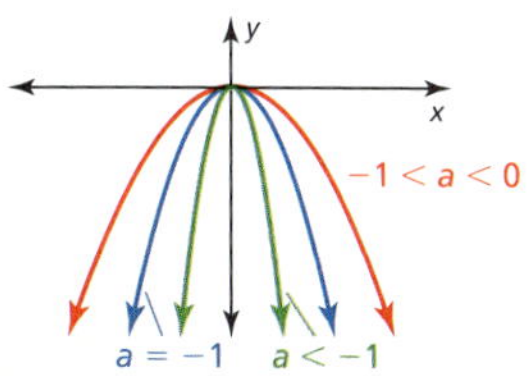

EXAMPLE 2 Graphing $y = ax^2$ When $a > 0$

Graph $g(x) = 2x^2$. Compare the graph to the graph of $f(x) = x^2$.

SOLUTION

Step 1 Make a table of values.

x	−2	−1	0	1	2
$g(x)$	8	2	0	2	8

Step 2 Plot the ordered pairs.

Step 3 Draw a smooth curve through the points.

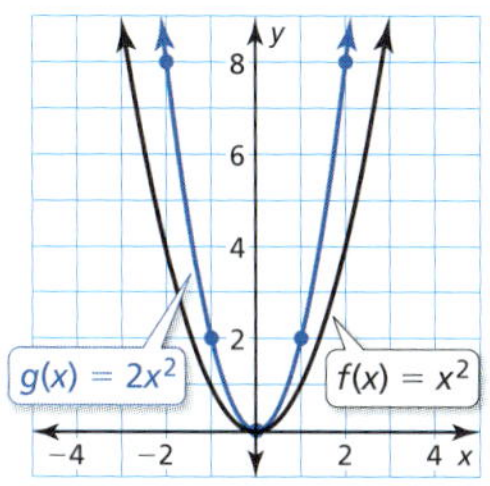

▶ Both graphs open up and have the same vertex, (0, 0), and the same axis of symmetry, $x = 0$. The graph of g is narrower than the graph of f because the graph of g is a vertical stretch by a factor of 2 of the graph of f.

Extra Example 2

Graph $g(x) = 4x^2$. Compare the graph to the graph of $f(x) = x^2$.

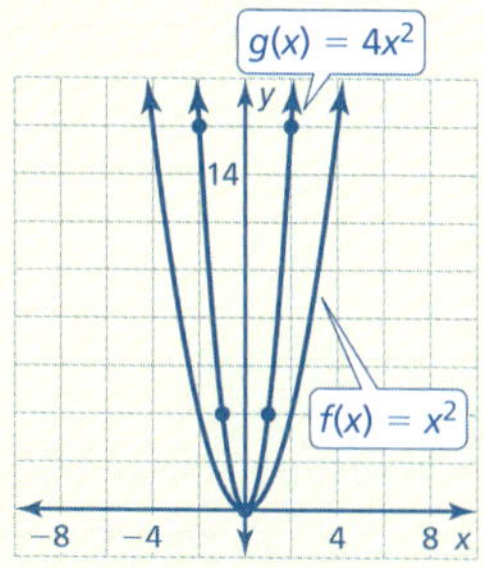

Both graphs open up and have the same vertex, (0, 0), and the same axis of symmetry, $x = 0$. The graph of g is a vertical stretch by a factor of 4 of the graph of f.

MONITORING PROGRESS ANSWERS

1. The vertex is $(2, -3)$. The axis of symmetry is $x = 2$. The domain is all real numbers. The range is $y \geq -3$. When $x < 2$, y increases as x decreases. When $x > 2$, y increases as x increases.
2. The vertex is $(-3, 7)$. The axis of symmetry is $x = -3$. The domain is all real numbers. The range is $y \leq 7$. When $x < -3$, y increases as x increases. When $x > -3$, y increases as x decreases.

Laurie's Notes Teacher Actions

- Relate the *Core Concept* to the exploration. Use colors to distinguish $0 < |a| < 1$ and $|a| > 1$ from $|a| = 1$. Relate to transformations: the vertical stretch ($|a| > 1$) that students may call "narrower" and vertical shrink $0 < |a| < 1$ that they may call "more bowl-shaped."

COMMON ERROR Students multiply x by 2 and then square it instead of squaring x and then multiplying by 2. Recall order of operations!

- **Connection:** The y-values in the table of values decrease and then increase. Refer to the graph and point out the same feature.

Extra Example 3

Graph $h(x) = -\frac{2}{9}x^2$. Compare the graph to the graph of $f(x) = x^2$.

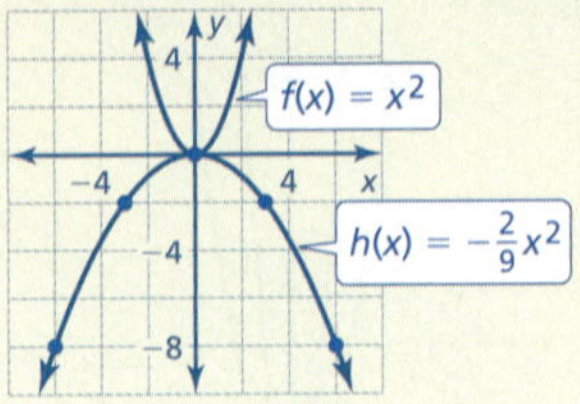

The graphs have the same vertex, (0, 0), and the same axis of symmetry, $x = 0$. The graph of h opens down and is wider than the graph of f. So, the graph of h is a vertical shrink by a factor of $\frac{2}{9}$ and a reflection in the x-axis of the graph of f.

Extra Example 4

The diagram shows the cross section of a parabolic mirror, where x and y are measured in feet. Find the width and the height of the mirror.

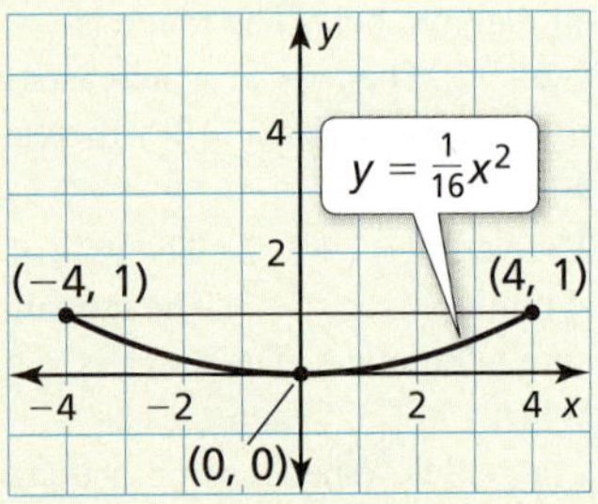

The mirror is 8 feet wide and 1 foot tall.

MONITORING PROGRESS ANSWERS

3–8. See Additional Answers.

9. 4 inches wide, 2 inches deep

STUDY TIP

To make the calculations easier, choose x-values that are multiples of 3.

EXAMPLE 3 Graphing $y = ax^2$ When $a < 0$

Graph $h(x) = -\frac{1}{3}x^2$. Compare the graph to the graph of $f(x) = x^2$.

SOLUTION

Step 1 Make a table of values.

x	−6	−3	0	3	6
$h(x)$	−12	−3	0	−3	−12

Step 2 Plot the ordered pairs.

Step 3 Draw a smooth curve through the points.

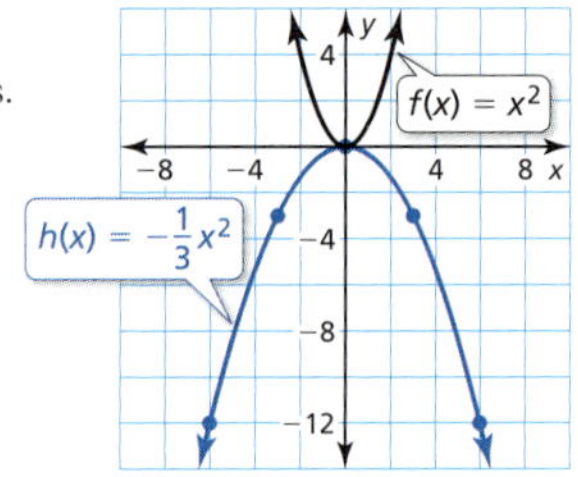

▶ The graphs have the same vertex, (0, 0), and the same axis of symmetry, $x = 0$, but the graph of h opens down and is wider than the graph of f. So, the graph of h is a vertical shrink by a factor of $\frac{1}{3}$ and a reflection in the x-axis of the graph of f.

Monitoring Progress

Help in English and Spanish at *BigIdeasMath.com*

Graph the function. Compare the graph to the graph of $f(x) = x^2$.

3. $g(x) = 5x^2$ **4.** $h(x) = \frac{1}{3}x^2$ **5.** $n(x) = \frac{3}{2}x^2$

6. $p(x) = -3x^2$ **7.** $q(x) = -0.1x^2$ **8.** $g(x) = -\frac{1}{4}x^2$

EXAMPLE 4 Solving a Real-Life Problem

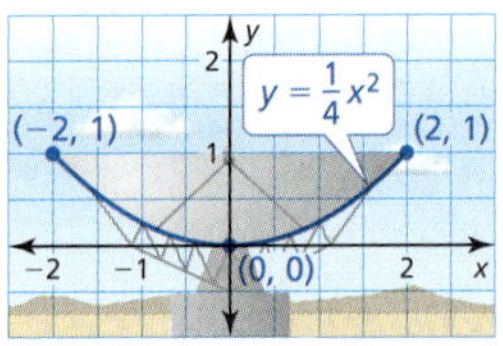

The diagram at the left shows the cross section of a satellite dish, where x and y are measured in meters. Find the width and depth of the dish.

SOLUTION

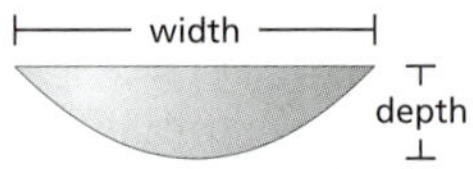

Use the domain of the function to find the width of the dish. Use the range to find the depth.

The leftmost point on the graph is (−2, 1), and the rightmost point is (2, 1). So, the domain is $-2 \le x \le 2$, which represents 4 meters.

The lowest point on the graph is (0, 0), and the highest points on the graph are (−2, 1) and (2, 1). So, the range is $0 \le y \le 1$, which represents 1 meter.

▶ So, the satellite dish is 4 meters wide and 1 meter deep.

Monitoring Progress

Help in English and Spanish at *BigIdeasMath.com*

9. The cross section of a spotlight can be modeled by the graph of $y = 0.5x^2$, where x and y are measured in inches and $-2 \le x \le 2$. Find the width and depth of the spotlight.

Laurie's Notes Teacher Actions

? "When $a = -\frac{1}{3}$, what do you know?" The graph opens down because $a < 0$. It is a vertical shrink (more bowl-shaped) because $0 < |a| < 1$.

- The parabolic shape is one that students have likely seen. The cross section of a satellite dish is parabolic.
- In Example 4, connect the domain and range to vertical and horizontal distances to finish the problem.

Closure

- **Exit Ticket:** Describe the differences between the graphs of $y = -3x^2$ and $y = \frac{1}{3}x^2$. The graph of $y = -3x^2$ opens down and the graph of $y = \frac{1}{3}x^2$ opens up. The graph of $y = -3x^2$ is narrower than the graph of $y = \frac{1}{3}x^2$. The range of $y = -3x^2$ is $y \le 0$. The range of $y = \frac{1}{3}x^2$ is $y \ge 0$.

8.1 Exercises

Dynamic Solutions available at *BigIdeasMath.com*

Vocabulary and Core Concept Check

1. **VOCABULARY** What is the U-shaped graph of a quadratic function called?

2. **WRITING** When does the graph of a quadratic function open up? open down?

Monitoring Progress and Modeling with Mathematics

In Exercises 3 and 4, identify characteristics of the quadratic function and its graph. *(See Example 1.)*

3. 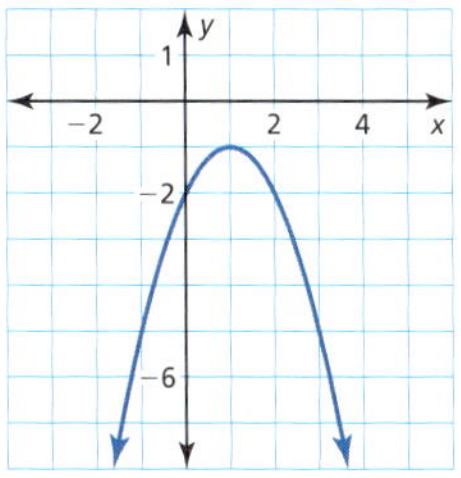

4. 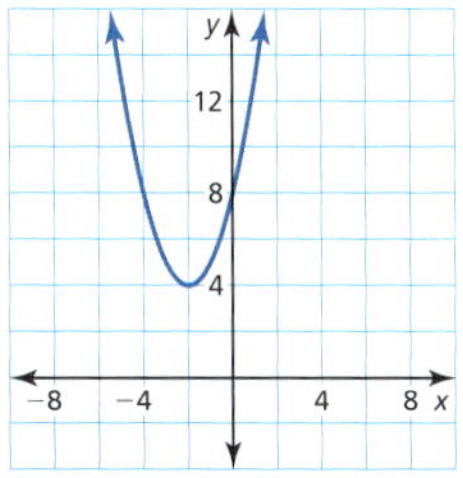

In Exercises 5–12, graph the function. Compare the graph to the graph of $f(x) = x^2$. *(See Examples 2 and 3.)*

5. $g(x) = 6x^2$
6. $b(x) = 2.5x^2$
7. $h(x) = \frac{1}{4}x^2$
8. $j(x) = 0.75x^2$
9. $m(x) = -2x^2$
10. $q(x) = -\frac{9}{2}x^2$
11. $k(x) = -0.2x^2$
12. $p(x) = -\frac{2}{3}x^2$

In Exercises 13–16, use a graphing calculator to graph the function. Compare the graph to the graph of $y = -4x^2$.

13. $y = 4x^2$
14. $y = -0.4x^2$
15. $y = -0.04x^2$
16. $y = -0.004x^2$

17. **ERROR ANALYSIS** Describe and correct the error in graphing and comparing $y = x^2$ and $y = 0.5x^2$.

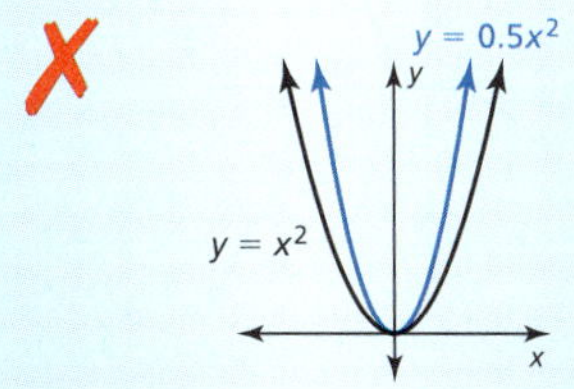

The graphs have the same vertex and the same axis of symmetry. The graph of $y = 0.5x^2$ is narrower than the graph of $y = x^2$.

18. **MODELING WITH MATHEMATICS** The arch support of a bridge can be modeled by $y = -0.0012x^2$, where x and y are measured in feet. Find the height and width of the arch. *(See Example 4.)*

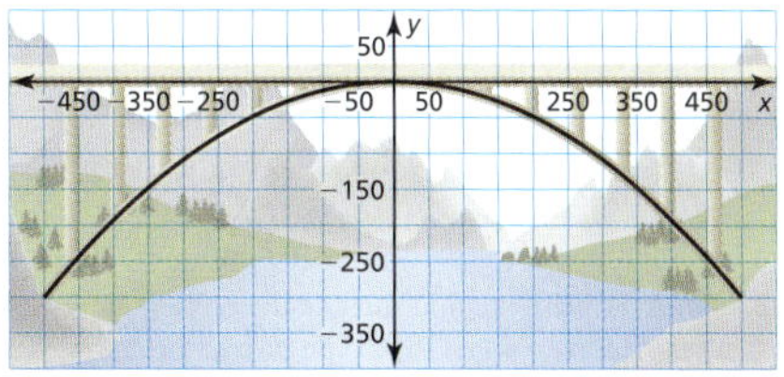

19. **PROBLEM SOLVING** The breaking strength z (in pounds) of a manila rope can be modeled by $z = 8900d^2$, where d is the diameter (in inches) of the rope.

a. Describe the domain and range of the function.

b. Graph the function using the domain in part (a).

c. A manila rope has four times the breaking strength of another manila rope. Does the stronger rope have four times the diameter? Explain.

Assignment Guide and Homework Check

ASSIGNMENT

Basic: 1, 2, 3–17 odd, 18, 20–22, 31–35

Average: 1, 2–16 even, 17–23, 31–35

Advanced: 1, 2, 4, 8, 12–16 even, 17–23, 24–30 even, 31–35

HOMEWORK CHECK

Basic: 3, 5, 9, 18

Average: 4, 6, 12, 18

Advanced: 4, 8, 12, 18

ANSWERS

1. parabola
2. when $a > 0$; when $a < 0$
3. The vertex is $(1, -1)$. The axis of symmetry is $x = 1$. The domain is all real numbers. The range is $y \le -1$. When $x < 1$, y increases as x increases. When $x > 1$, y increases as x decreases.
4. The vertex is $(-2, 4)$. The axis of symmetry is $x = -2$. The domain is all real numbers. The range is $y \ge 4$. When $x < -2$, y increases as x decreases. When $x > -2$, y increases as x increases.
5. 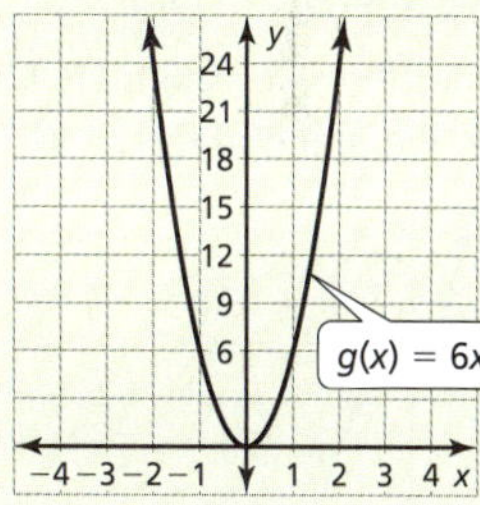

The graph of g is a vertical stretch by a factor of 6 of the graph of f.

6. 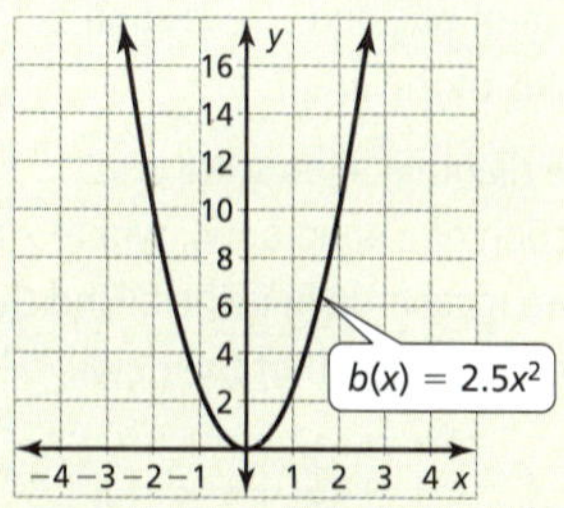

The graph of b is a vertical stretch by a factor of 2.5 of the graph of f.

7. 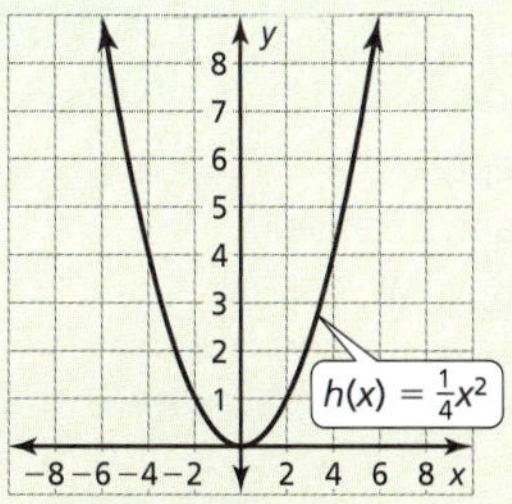

The graph of h is a vertical shrink by a factor of $\frac{1}{4}$ of the graph of f.

8. 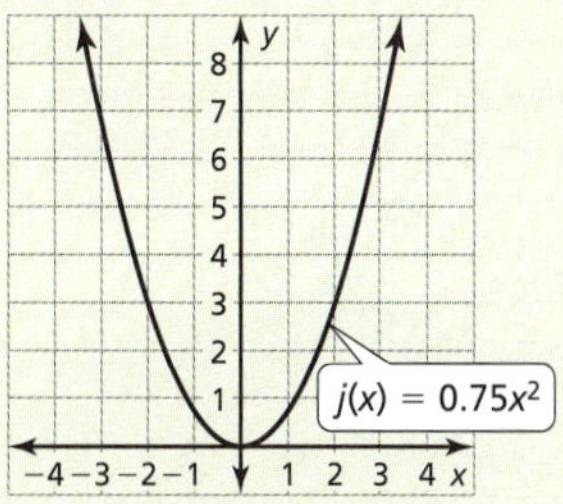

The graph of j is a vertical shrink by a factor of 0.75 of the graph of f.

9–19. See Additional Answers.

Dynamic Teaching Tools

- Dynamic Assessment & Progress Monitoring Tool
- Interactive Whiteboard Lesson Library
- Dynamic Classroom with Dynamic Investigations

ANSWERS

20. **a.** $a > 1$

b. $-1 < a < 0$

21. f is increasing when $x > 0$. g is increasing when $x < 0$.

22–35. See Additional Answers.

Mini-Assessment

1. Identify characteristics of the quadratic function and its graph.

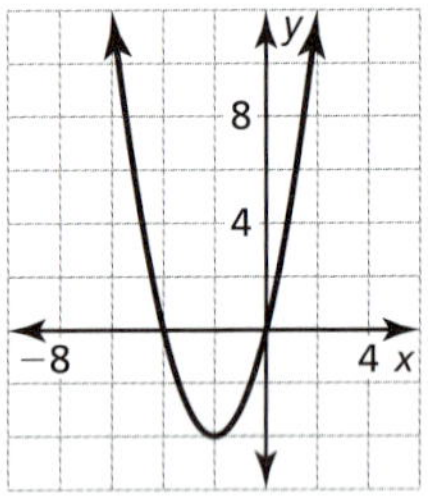

See Additional Answers.

2. Graph $g(x) = 3.5x^2$. Compare the graph to the graph of $f(x) = x^2$.
See Additional Answers.

3. Graph $n(x) = -\frac{3}{8}x^2$. Compare the graph to the graph of $f(x) = x^2$.

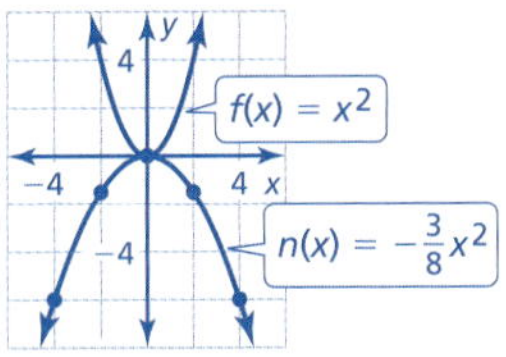

The graph of n is a vertical shrink by a factor of $\frac{3}{8}$ and a reflection in the x-axis of the graph of f.

4. The diagram shows the cross section of a soup bowl, where x and y are measured in inches. Find the width and depth of the bowl.

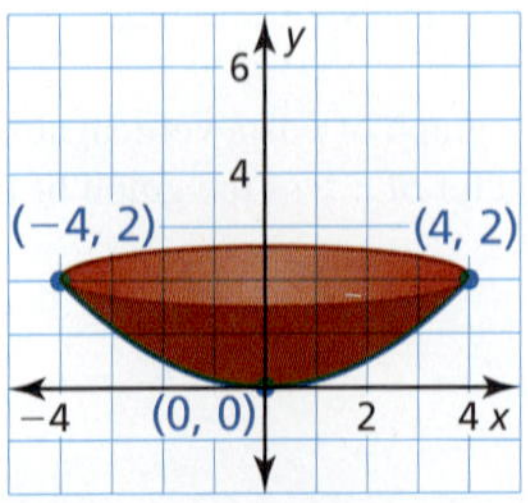

8 inches wide, 2 inches deep

20. **HOW DO YOU SEE IT?** Describe the possible values of a.

a.

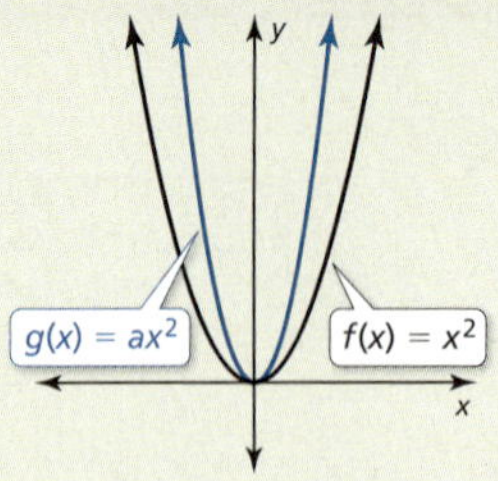

b.

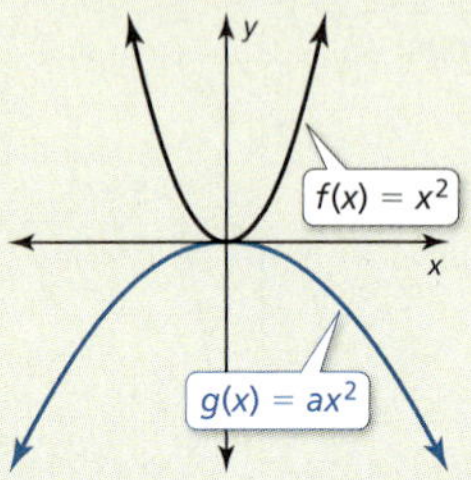

ANALYZING GRAPHS **In Exercises 21–23, use the graph.**

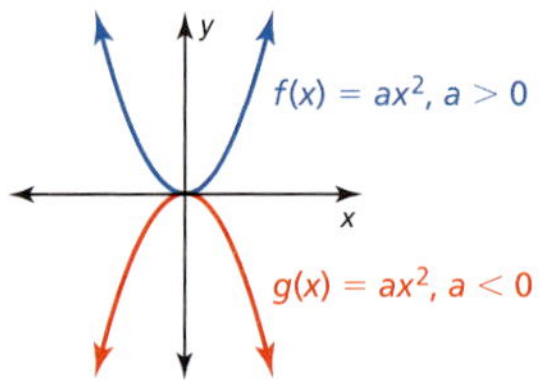

21. When is each function increasing?

22. When is each function decreasing?

23. Which function could include the point $(-2, 3)$? Find the value of a when the graph passes through $(-2, 3)$.

24. **REASONING** Is the x-intercept of the graph of $y = ax^2$ always 0? Justify your answer.

25. **REASONING** A parabola opens up and passes through $(-4, 2)$ and $(6, -3)$. How do you know that $(-4, 2)$ is not the vertex?

ABSTRACT REASONING **In Exercises 26–29, determine whether the statement is *always*, *sometimes*, or *never* true. Explain your reasoning.**

26. The graph of $f(x) = ax^2$ is narrower than the graph of $g(x) = x^2$ when $a > 0$.

27. The graph of $f(x) = ax^2$ is narrower than the graph of $g(x) = x^2$ when $|a| > 1$.

28. The graph of $f(x) = ax^2$ is wider than the graph of $g(x) = x^2$ when $0 < |a| < 1$.

29. The graph of $f(x) = ax^2$ is wider than the graph of $g(x) = dx^2$ when $|a| > |d|$.

30. **THOUGHT PROVOKING** Draw the isosceles triangle shown. Divide each leg into eight congruent segments. Connect the highest point of one leg with the lowest point of the other leg. Then connect the second highest point of one leg to the second lowest point of the other leg. Continue this process. Write a quadratic equation whose graph models the shape that appears.

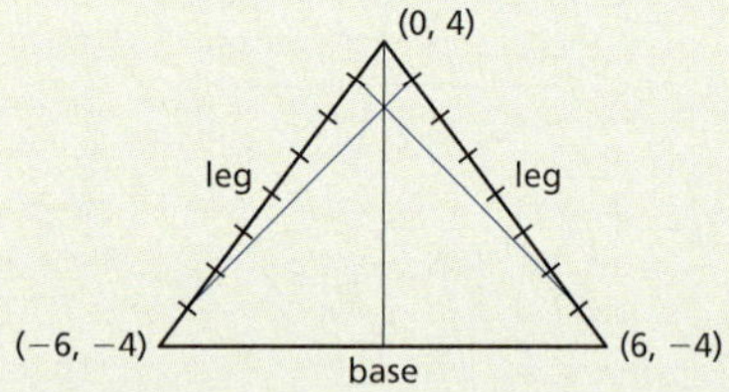

31. **MAKING AN ARGUMENT** The diagram shows the parabolic cross section of a swirling glass of water, where x and y are measured in centimeters.

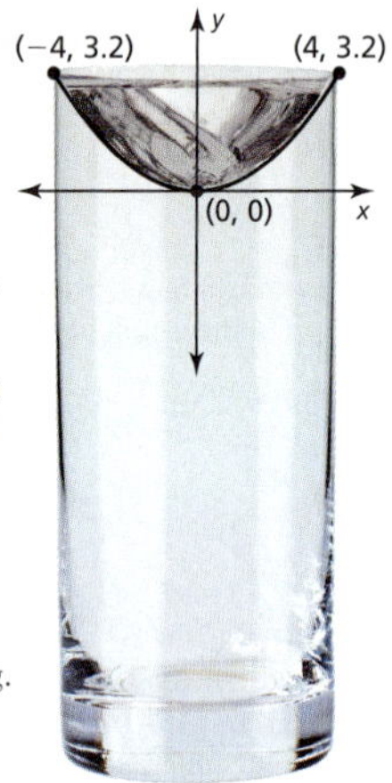

a. About how wide is the mouth of the glass?

b. Your friend claims that the rotational speed of the water would have to increase for the cross section to be modeled by $y = 0.1x^2$. Is your friend correct? Explain your reasoning.

Maintaining Mathematical Proficiency

Reviewing what you learned in previous grades and lessons

Evaluate the expression when $n = 3$ and $x = -2$. *(Skills Review Handbook)*

32. $n^2 + 5$ **33.** $3x^2 - 9$ **34.** $-4n^2 + 11$ **35.** $n + 2x^2$

If students need help...	*If students got it...*
Resources by Chapter • Practice A and Practice B • Puzzle Time	Resources by Chapter • Enrichment and Extension • Cumulative Review
Student Journal • Practice	Start the *next* Section
Differentiating the Lesson Skills Review Handbook	

Laurie's Notes

Overview of Section 8.2

Introduction

- The lesson begins by describing the effect of c on the graph of $y = ax^2 + c$. The connection for students is the effect of b on the graph of $y = mx + b$.
- Students should recognize that vertical translations do not change the axis of symmetry, meaning $y = 3x^2$ and $y = 3x^2 - 4$ have the same axis of symmetry. Also, the vertex of the parabola remains on the y-axis.
- Also introduced in this lesson is the definition of a zero of a function. The zero of a function is where there is an x-intercept on the graph. The zeros of a function are found in a real-life application using the falling-objects model.

Resources

- Use dynamic geometry software to investigate the effect of c on the graph of $y = ax^2 + c$. A slider can be used to generate values of c, $-5 < c < 5$.
- In lieu of creating your own, you can find online versions to download.

Teaching Strategy

- **Rule of 3:** In this lesson, the *Rule of 3* strategy, connecting the symbolic, graphical, and numerical representation of functions, is still very important.
- In Example 3 in the lesson, students are given $f(x) = -0.5x^2 + 2$ and $g(x) = f(x) - 7$ and are asked to describe the transformation from the graph of f to the graph of g. In the equation editor of the calculator, it is possible to enter the functions as given. $Y_1 = f(x)$ and $Y_2 = g(x)$. Note in Figure 1 how the functions are entered.
- In Figure 2, the graph of Y_2 is a vertical shift of Y_1 down 7 units. In Figure 3, the table of values is shown, and students should recognize that subtracting 7 from each value in Y_1 gives the associated value in Y_2.

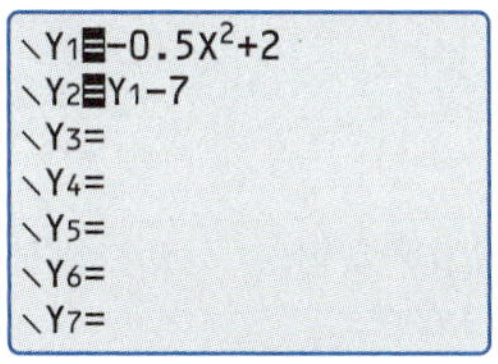

Figure 1

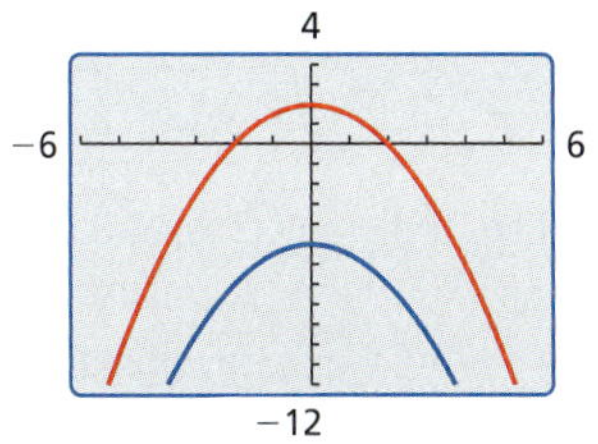

Figure 2

X	Y1	Y2
-3	-2.5	-9.5
-2	0	-7
-1	1.5	-5.5
0	2	-5
1	1.5	-5.5
2	0	-7
3	-2.5	-9.5

X=-3

Figure 3

Pacing Suggestion

- The explorations help students discover the effect of the value of c on the graph of $y = ax^2 + c$. You may choose to omit a few examples and check for understanding with the *Monitoring Progress* questions.

Dynamic Teaching Tools

- Dynamic Assessment & Progress Monitoring Tool
- Lesson Planning Tool
- Interactive Whiteboard Lesson Library
- Dynamic Classroom with Dynamic Investigations

What Your Students Will Learn

- Graph quadratic functions of the form $f(x) = ax^2 + c$ by identifying the vertical translations of $|c|$ units up or down.
- Solve real-life problems involving the falling-objects model of the form $f(x) = ax^2 + c$.

Laurie's Notes

Exploration

Motivate

? "How many y-intercepts does a parabola have?" Students may say 1 or 0. A parabola always has one y-intercept, even if it cannot be seen in the standard viewing window of a graphing calculator.

? "How many x-intercepts does a parabola have?" Students will likely say 0, 1, or 2. Have them quickly sketch a parabola for each case.

For Your Information

- **Make Sense of Problems and Persevere in Solving Them:** Students should be comfortable with graphing $y = x^2$ and $y = ax^2$ and explaining the effect of a on the graph of $y = ax^2$. Today, they will make sense of another parameter when they graph $y = ax^2 + c$. Using tabular and graphical representations, they make sense of the effect of c.

Exploration 1

- Students should be comfortable graphing functions of the form $y = ax^2$ by making a table of values, plotting the ordered pairs, and drawing a smooth curve through the points.
- Students are graphing a pair of functions in the same coordinate plane.
- Discuss the characteristics of each graph, such as opens up or opens down and wider or narrower than $y = x^2$.
- Students may only notice that the vertex is no longer at (0, 0). They may not think of "c" as a vertical shift.
- **Connection:** Discuss the idea of the vertex being the y-intercept of the graph.

? "Could there be more than one y-intercept? Explain." no; It would not be a function.

COMMON ERROR Students do not always see that the pairs of graphs are similar, with one graph being a translation of the other. This is sometimes the case because different parts of each graph may be plotted. This may happen when using a graphing calculator with an inappropriate viewing window.

Exploration 2

? "What is an x-intercept?" It is where the graph crosses the x-axis.

? "Can a parabola have more than one x-intercept? Explain." yes; *Sample answer*: A parabola can open up and have a vertex below the x-axis, giving it two x-intercepts.

- Explain to students that in this exploration they will be looking for the x-intercept(s) of the graph.
- Again, students are graphing by first making a table of values. When making the table, students may identify an x-intercept.
- When students have finished both parts, ask for volunteers to share their graphs and explain how they found the x-intercepts. From the graph in part (a), they know that the roots occur between -3 and -2, and between 2 and 3. Students are familiar with the definition of square roots, so it should seem reasonable to them that solving $x^2 = 7$ will result in $x = \pm\sqrt{7}$.

Communicate Your Answer

- Make sure students are using appropriate viewing windows when verifying their conclusions.

Connecting to Next Step

- The exploration allows students to discover the effect of the value of c on the graph of $f(x) = ax^2 + c$. In the formal lesson, students will graph additional functions of this type.

8.2 Graphing $f(x) = ax^2 + c$

Essential Question How does the value of c affect the graph of $f(x) = ax^2 + c$?

EXPLORATION 1 Graphing $y = ax^2 + c$

Work with a partner. Sketch the graphs of the functions in the same coordinate plane. What do you notice?

a. $f(x) = x^2$ and $g(x) = x^2 + 2$

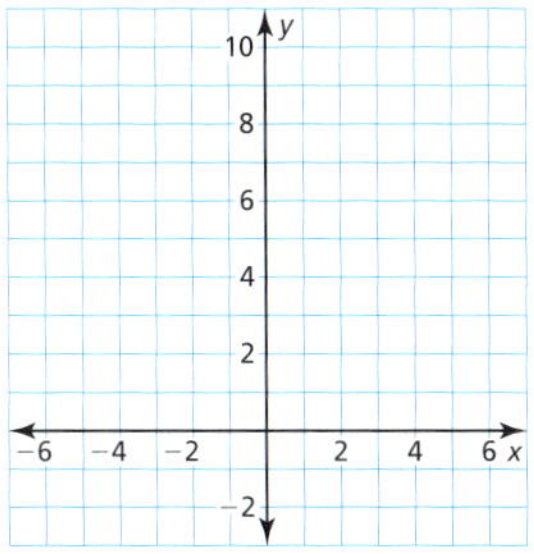

b. $f(x) = 2x^2$ and $g(x) = 2x^2 - 2$

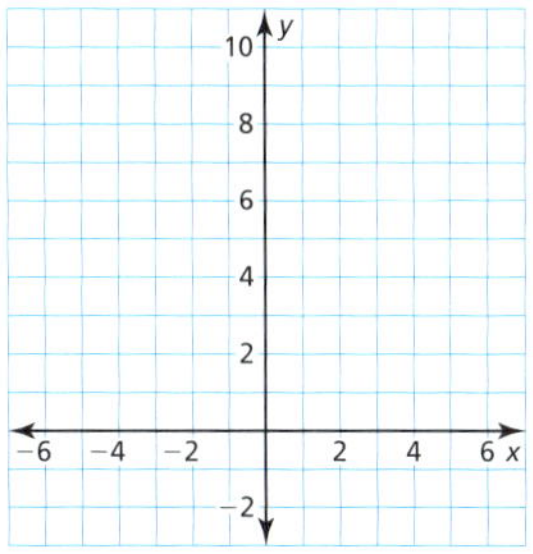

EXPLORATION 2 Finding x-Intercepts of Graphs

Work with a partner. Graph each function. Find the x-intercepts of the graph. Explain how you found the x-intercepts.

a. $y = x^2 - 7$

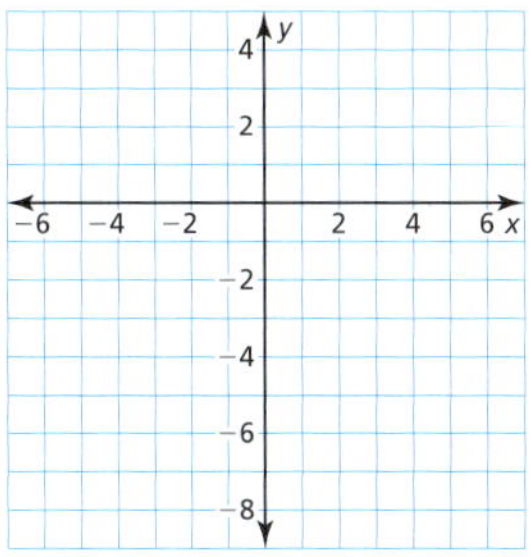

b. $y = -x^2 + 1$

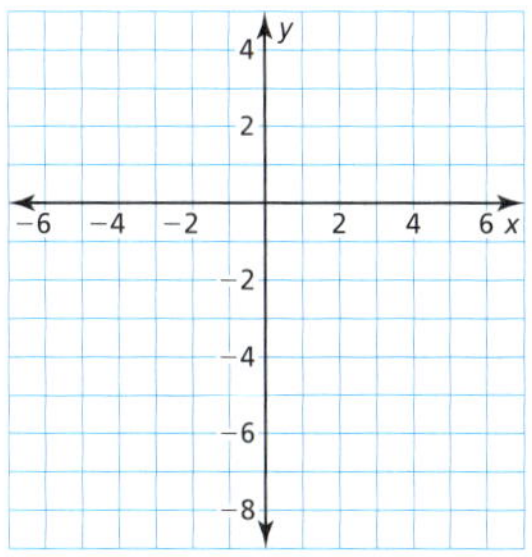

USING TOOLS STRATEGICALLY

To be proficient in math, you need to consider the available tools, such as a graphing calculator, when solving a mathematical problem.

Communicate Your Answer

3. How does the value of c affect the graph of $f(x) = ax^2 + c$?
4. Use a graphing calculator to verify your answers to Question 3.
5. The figure shows the graph of a quadratic function of the form $y = ax^2 + c$. Describe possible values of a and c. Explain your reasoning.

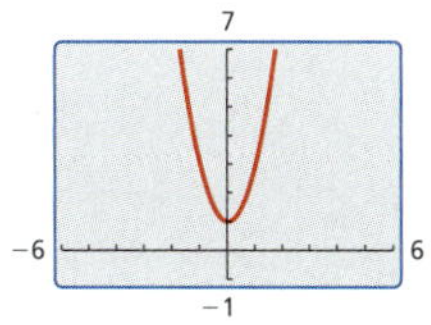

Dynamic Teaching Tools
- Dynamic Assessment & Progress Monitoring Tool
- Lesson Planning Tool
- Interactive Whiteboard Lesson Library
- Dynamic Classroom with Dynamic Investigations

ANSWERS

1. a.

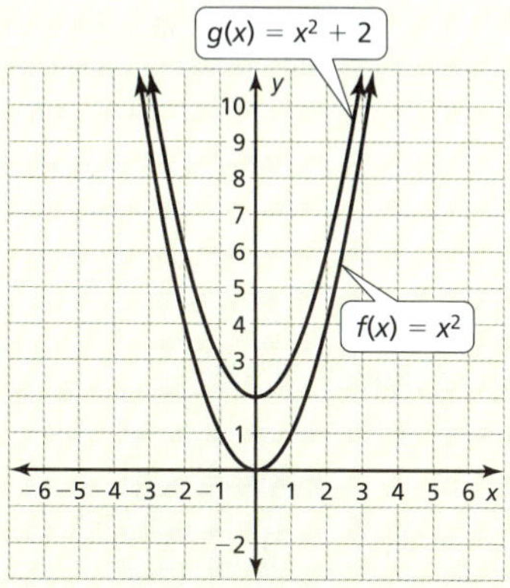

b.

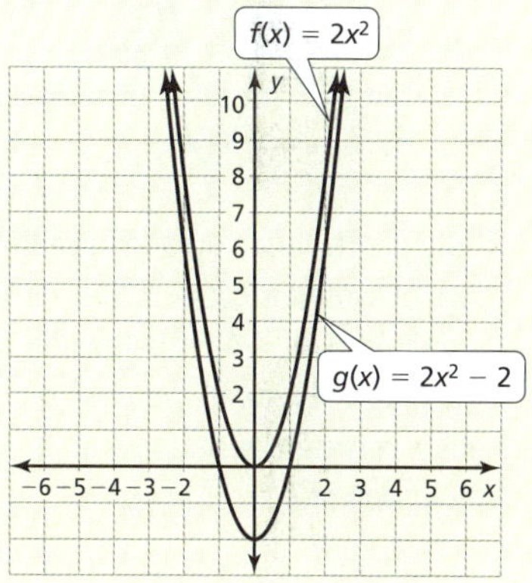

Sample answer: In $g(x) = ax^2 + c$, c causes a vertical shift in the graph of $f(x) = ax^2$.

2. a.

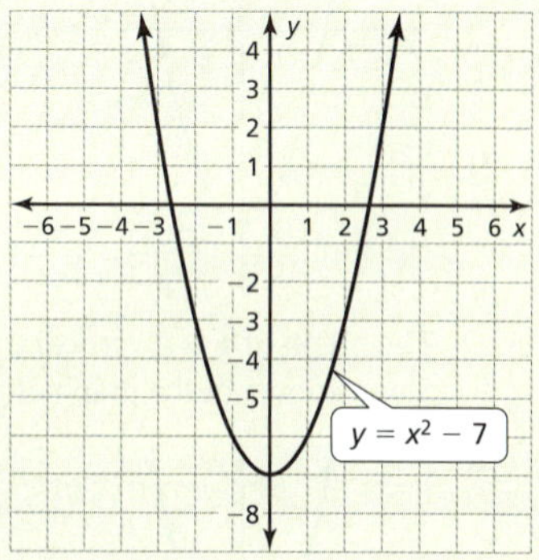

$x = \sqrt{7} \approx 2.6$, $x = -\sqrt{7} \approx -2.6$

2. b.

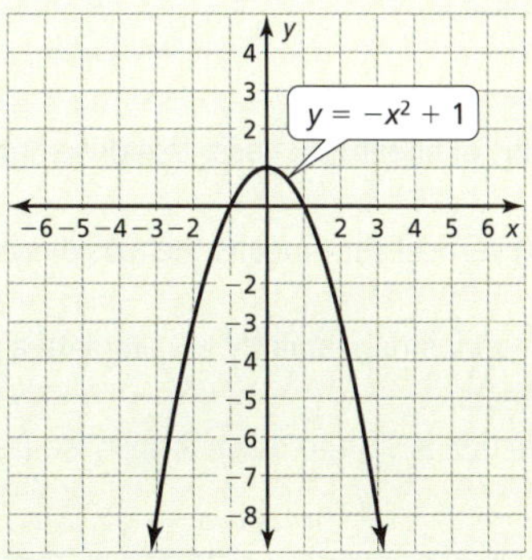

$x = 1, x = -1$

Sample answer: Estimate the points where the graph intersects the x-axis.

3. It causes a vertical shift of c units. When $c > 0$, the graph is translated up. When $c < 0$, the graph is translated down.
4. Check students' work.
5. *Sample answer:* $a > 1$, $c = 1$; The graph is narrower than $y = x^2$, so it is a vertical stretch. The vertex is 1 unit above the origin, so the graph is a vertical translation 1 unit up of the graph of $y = x^2$.

English Language Learners

Pair Activity

Pair each English learner with an English speaker. Have the pairs graph quadratic functions and compare their graphs to the graph of the function $f(x) = x^2$. Have one student make a table of values and the other student use the ordered pairs to draw the graph. Have partners work together to compare the graph of each function to the graph of $f(x) = x^2$.

Extra Example 1

Graph $g(x) = x^2 - 4$. Compare the graph to the graph of $f(x) = x^2$.

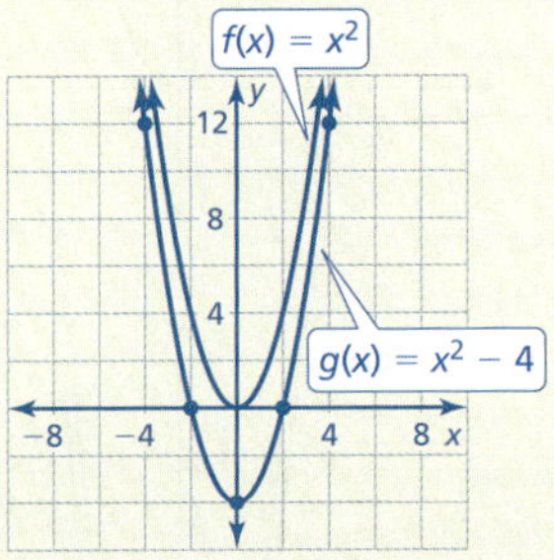

Both graphs open up and have the same axis of symmetry, $x = 0$. The vertex of the graph of g, $(0, -4)$, is below the vertex of the graph of f, $(0, 0)$, because the graph of g is a vertical translation 4 units down of the graph of f.

MONITORING PROGRESS ANSWERS

1.

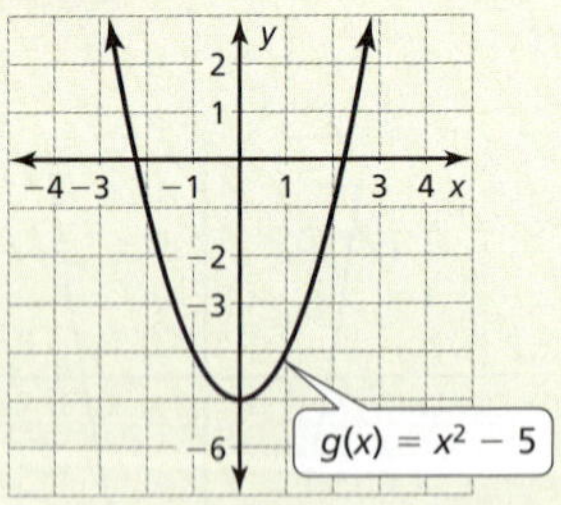

The graph of g is a vertical translation 5 units down of the graph of f.

2.

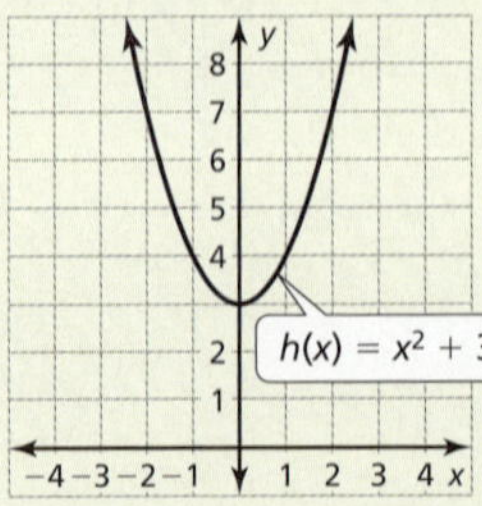

The graph of h is a vertical translation 3 units up of the graph of f.

8.2 Lesson

What You Will Learn

- Graph quadratic functions of the form $f(x) = ax^2 + c$.
- Solve real-life problems involving functions of the form $f(x) = ax^2 + c$.

Core Vocabulary

zero of a function, *p. 428*

Previous
translation
vertex of a parabola
axis of symmetry
vertical stretch
vertical shrink

Graphing $f(x) = ax^2 + c$

Core Concept

Graphing $f(x) = ax^2 + c$

- When $c > 0$, the graph of $f(x) = ax^2 + c$ is a vertical translation c units up of the graph of $f(x) = ax^2$.
- When $c < 0$, the graph of $f(x) = ax^2 + c$ is a vertical translation $|c|$ units down of the graph of $f(x) = ax^2$.

The vertex of the graph of $f(x) = ax^2 + c$ is $(0, c)$, and the axis of symmetry is $x = 0$.

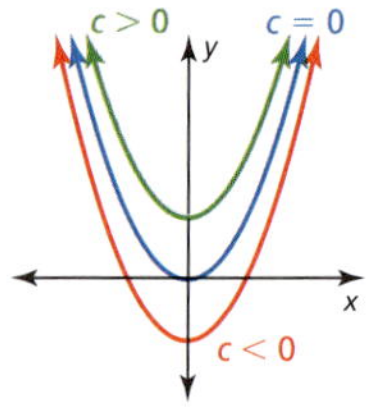

EXAMPLE 1 **Graphing $y = x^2 + c$**

Graph $g(x) = x^2 - 2$. Compare the graph to the graph of $f(x) = x^2$.

SOLUTION

Step 1 Make a table of values.

x	−2	−1	0	1	2
$g(x)$	2	−1	−2	−1	2

Step 2 Plot the ordered pairs.

Step 3 Draw a smooth curve through the points.

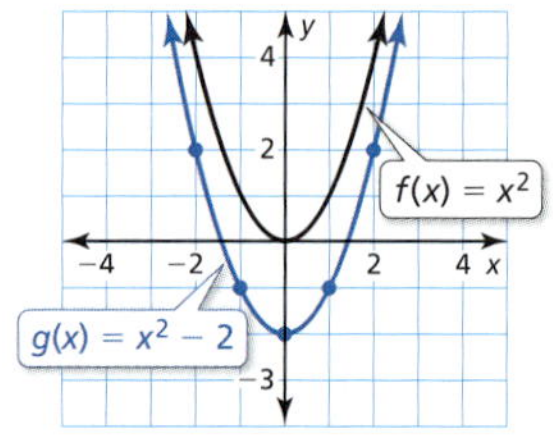

REMEMBER

The graph of $y = f(x) + k$ is a vertical translation, and the graph of $y = f(x - h)$ is a horizontal translation of the graph of f.

Both graphs open up and have the same axis of symmetry, $x = 0$. The vertex of the graph of g, $(0, -2)$, is below the vertex of the graph of f, $(0, 0)$, because the graph of g is a vertical translation 2 units down of the graph of f.

Monitoring Progress

Help in English and Spanish at *BigIdeasMath.com*

Graph the function. Compare the graph to the graph of $f(x) = x^2$.

1. $g(x) = x^2 - 5$

2. $h(x) = x^2 + 3$

Laurie's Notes Teacher Actions

- Write the *Core Concept*. Relate it to vertical translations of linear functions. Use different colors to connect the equations and the graphs.
- Explain that the shape of the graph is not changing. A vertical shift means the axis of symmetry is still $y = 0$ and the vertex is $(0, c)$.
- **Think-Pair-Share:** Pose Example 1. Have students work independently to complete a table of values and graph. Have them compare with their partners.

? "If $f(a) = 8$, what is $g(a)$? Explain." $g(a) = 6$; The function is shifted down 2 units, so $(a, 8)$ shifts down to $(a, 6)$.

EXAMPLE 2 Graphing $y = ax^2 + c$

Graph $g(x) = 4x^2 + 1$. Compare the graph to the graph of $f(x) = x^2$.

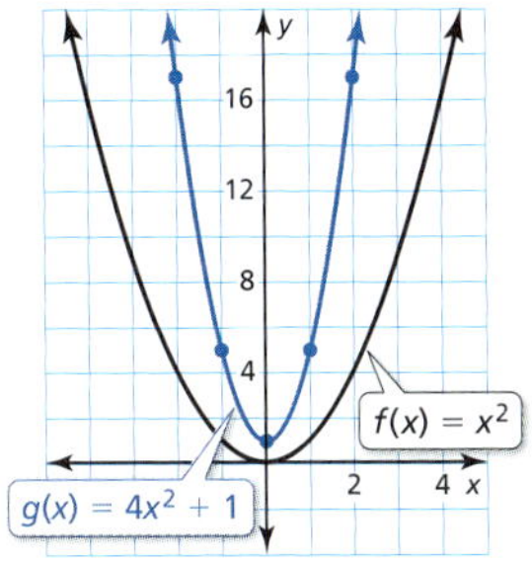

SOLUTION

Step 1 Make a table of values.

x	−2	−1	0	1	2
$g(x)$	17	5	1	5	17

Step 2 Plot the ordered pairs.

Step 3 Draw a smooth curve through the points.

▶ Both graphs open up and have the same axis of symmetry, $x = 0$. The graph of g is narrower, and its vertex, (0, 1), is above the vertex of the graph of f, (0, 0). So, the graph of g is a vertical stretch by a factor of 4 and a vertical translation 1 unit up of the graph of f.

EXAMPLE 3 Translating the Graph of $y = ax^2 + c$

Let $f(x) = -0.5x^2 + 2$ and $g(x) = f(x) - 7$.

a. Describe the transformation from the graph of f to the graph of g. Then graph f and g in the same coordinate plane.

b. Write an equation that represents g in terms of x.

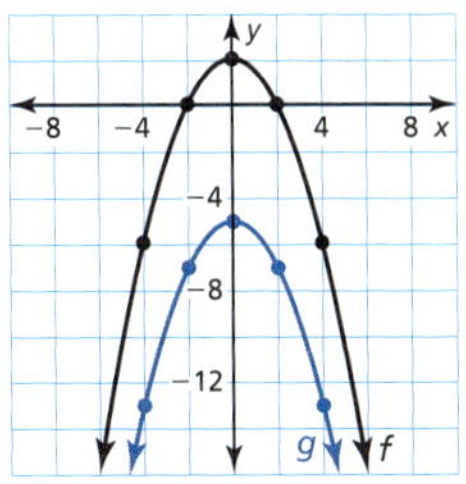

SOLUTION

a. The function g is of the form $y = f(x) + k$, where $k = -7$. So, the graph of g is a vertical translation 7 units down of the graph of f.

x	−4	−2	0	2	4
$f(x)$	−6	0	2	0	−6
$g(x)$	−13	−7	−5	−7	−13

($-0.5x^2 + 2$; $f(x) - 7$)

b. $g(x) = f(x) - 7$ — Write the function g.

$= -0.5x^2 + 2 - 7$ — Substitute for $f(x)$.

$= -0.5x^2 - 5$ — Subtract.

▶ So, the equation $g(x) = -0.5x^2 - 5$ represents g in terms of x.

Monitoring Progress

Help in English and Spanish at *BigIdeasMath.com*

Graph the function. Compare the graph to the graph of $f(x) = x^2$.

3. $g(x) = 2x^2 - 5$

4. $h(x) = -\frac{1}{4}x^2 + 4$

5. Let $f(x) = 3x^2 - 1$ and $g(x) = f(x) + 3$.

a. Describe the transformation from the graph of f to the graph of g. Then graph f and g in the same coordinate plane.

b. Write an equation that represents g in terms of x.

Laurie's Notes Teacher Actions

- Example 2 is similar to Example 1, except now there is a leading coefficient of $a = 4$. The vertical shift of 1 unit is to the function $y = 4x^2$.
- **Connection:** Students should see the symmetry numerically in the table and graphically in the graph.
- **Attend to Precision:** In Example 3, help students make sense of the notation. One function is called $f(x)$ and is defined: $f(x) = -0.5x^2 + 2$. The second function is $g(x)$ and is defined: $g(x) = f(x) - 7$. The $g(x)$ function is the $f(x)$ function shifted down 7 units. See the *Teaching Strategy* on page T-424.

Extra Example 2

Graph $g(x) = 2x^2 + 4$. Compare the graph to the graph of $f(x) = x^2$.

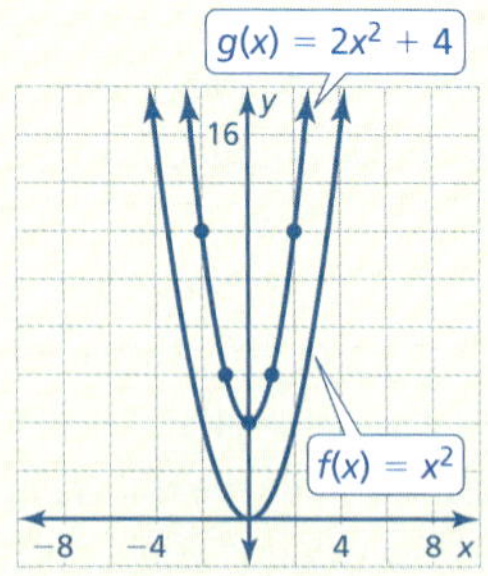

Both graphs open up and have the same axis of symmetry, $x = 0$. The graph of g is narrower, and its vertex, (0, 4), is above the vertex of the graph of f, (0, 0). So, the graph of g is a vertical stretch by a factor of 2 and a vertical translation 4 units up of the graph of f.

Extra Example 3

Let $f(x) = -0.75x^2 + 1$ and $g(x) = f(x) - 3$.

a. Describe the transformation from the graph of f to the graph of g. Then graph f and g in the same coordinate plane.

The graph of g is a vertical translation 3 units down of the graph of f.

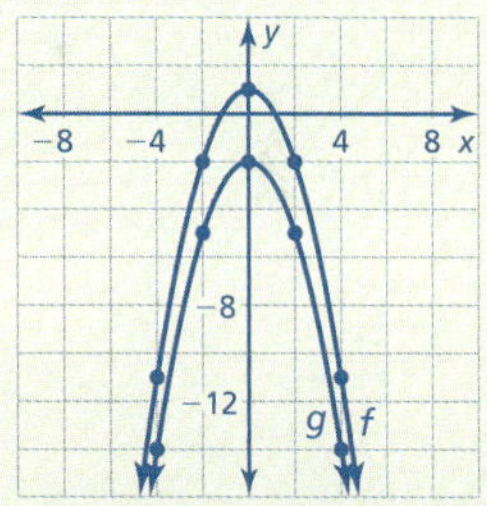

b. Write an equation that represents g in terms of x.

$g(x) = -0.75x^2 - 2$

MONITORING PROGRESS ANSWERS

3–5. See Additional Answers.

Extra Example 4

The function $h(t) = -4.9t^2 + s_0$ represents the approximate height (in meters) of a falling object t seconds after it is dropped from an initial height s_0 (in meters). A coin is dropped from an initial height of 44.1 meters.

a. After how many seconds does the coin hit the ground? The coin hits the ground 3 seconds after it is dropped.

b. Suppose the initial height is m meters lower. How will this affect part (a)? The coin will take less than 3 seconds to hit the ground.

MONITORING PROGRESS ANSWERS

6. *Sample answer:* Negative values of t would represent times before the egg was dropped, which have no meaning in the context of this problem.

7. 2.5 sec

Solving Real-Life Problems

A **zero of a function** f is an x-value for which $f(x) = 0$. A zero of a function is an x-intercept of the graph of the function.

EXAMPLE 4 Solving a Real-Life Problem

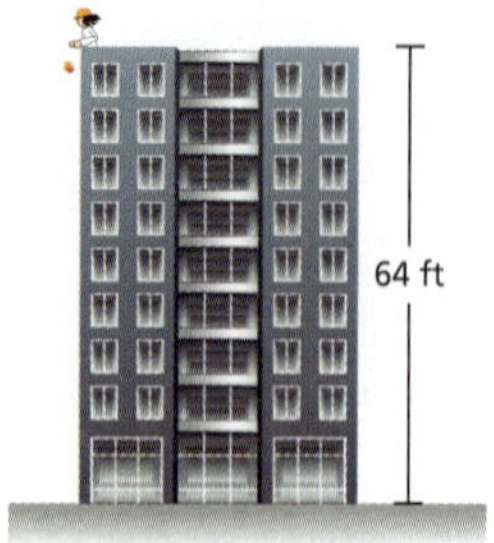

The function $f(t) = -16t^2 + s_0$ represents the approximate height (in feet) of a falling object t seconds after it is dropped from an initial height s_0 (in feet). An egg is dropped from a height of 64 feet.

a. After how many seconds does the egg hit the ground?

b. Suppose the initial height is adjusted by k feet. How will this affect part (a)?

SOLUTION

1. **Understand the Problem** You know the function that models the height of a falling object and the initial height of an egg. You are asked to find how many seconds it takes the egg to hit the ground when dropped from the initial height. Then you need to describe how a change in the initial height affects how long it takes the egg to hit the ground.

2. **Make a Plan** Use the initial height to write a function that models the height of the egg. Use a table to graph the function. Find the zero(s) of the function to answer the question. Then explain how vertical translations of the graph affect the zero(s) of the function.

3. **Solve the Problem**

COMMON ERROR
The graph in Step 1 shows the height of the object over time, not the path of the object.

a. The initial height is 64 feet. So, the function $f(t) = -16t^2 + 64$ represents the height of the egg t seconds after it is dropped. The egg hits the ground when $f(t) = 0$.

Step 1 Make a table of values and sketch the graph.

t	0	1	2
$f(t)$	64	48	0

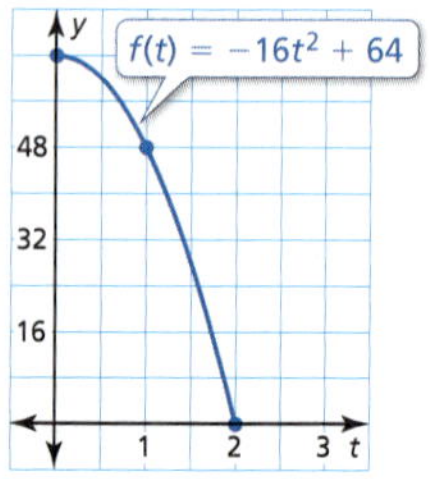

Step 2 Find the positive zero of the function. When $t = 2$, $f(t) = 0$. So, the zero is 2.

▶ The egg hits the ground 2 seconds after it is dropped.

b. When the initial height is adjusted by k feet, the graph of f is translated up k units when $k > 0$ or down $|k|$ units when $k < 0$. So, the x-intercept of the graph of f will move right when $k > 0$ or left when $k < 0$.

▶ When $k > 0$, the egg will take more than 2 seconds to hit the ground. When $k < 0$, the egg will take less than 2 seconds to hit the ground.

4. **Look Back** To check that the egg hits the ground 2 seconds after it is dropped, you can solve $0 = -16t^2 + 64$ by factoring.

Monitoring Progress Help in English and Spanish at BigIdeasMath.com

6. Explain why only nonnegative values of t are used in Example 4.

7. **WHAT IF?** The egg is dropped from a height of 100 feet. After how many seconds does the egg hit the ground?

Laurie's Notes Teacher Actions

- Define a *zero of a function.* Pose the egg drop problem. Introduce the falling-objects model.
- **Wait Time:** Ask questions to assess students' understanding of the problem, the model, and the known information. Give *Wait Time* for students to work with partners to graph the function and find the x-intercept.

? **Probing Question:** "If the graph shifts up (down), what happens to the zero?" Listen for correct reasoning.

Closure

- **Writing Prompt:** The graph of $y = -4x^2 + 12$ is …

8.2 Exercises

Dynamic Solutions available at *BigIdeasMath.com*

Vocabulary and Core Concept Check

1. **VOCABULARY** State the vertex and axis of symmetry of the graph of $y = ax^2 + c$.

2. **WRITING** How does the graph of $y = ax^2 + c$ compare to the graph of $y = ax^2$?

Monitoring Progress and Modeling with Mathematics

In Exercises 3–6, graph the function. Compare the graph to the graph of $f(x) = x^2$. *(See Example 1.)*

3. $g(x) = x^2 + 6$
4. $h(x) = x^2 + 8$
5. $p(x) = x^2 - 3$
6. $q(x) = x^2 - 1$

In Exercises 7–12, graph the function. Compare the graph to the graph of $f(x) = x^2$. *(See Example 2.)*

7. $g(x) = -x^2 + 3$
8. $h(x) = -x^2 - 7$
9. $s(x) = 2x^2 - 4$
10. $t(x) = -3x^2 + 1$
11. $p(x) = -\frac{1}{3}x^2 - 2$
12. $q(x) = \frac{1}{2}x^2 + 6$

In Exercises 13–16, describe the transformation from the graph of f to the graph of g. Then graph f and g in the same coordinate plane. Write an equation that represents g in terms of x. *(See Example 3.)*

13. $f(x) = 3x^2 + 4$
    $g(x) = f(x) + 2$
14. $f(x) = \frac{1}{2}x^2 + 1$
    $g(x) = f(x) - 4$
15. $f(x) = -\frac{1}{4}x^2 - 6$
    $g(x) = f(x) - 3$
16. $f(x) = 4x^2 - 5$
    $g(x) = f(x) + 7$

17. **ERROR ANALYSIS** Describe and correct the error in comparing the graphs.

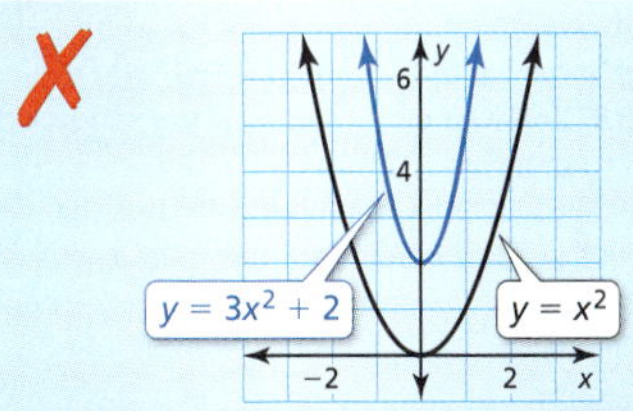

The graph of $y = 3x^2 + 2$ is a vertical shrink by a factor of 3 and a translation 2 units up of the graph of $y = x^2$.

18. **ERROR ANALYSIS** Describe and correct the error in graphing and comparing $f(x) = x^2$ and $g(x) = x^2 - 10$.

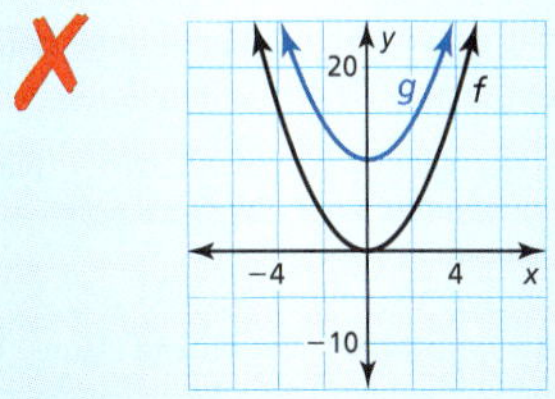

Both graphs open up and have the same axis of symmetry. However, the vertex of the graph of g, (0, 10), is 10 units above the vertex of the graph of f, (0, 0).

In Exercises 19–26, find the zeros of the function.

19. $y = x^2 - 1$
20. $y = x^2 - 36$
21. $f(x) = -x^2 + 25$
22. $f(x) = -x^2 + 49$
23. $f(x) = 4x^2 - 16$
24. $f(x) = 3x^2 - 27$
25. $f(x) = -12x^2 + 3$
26. $f(x) = -8x^2 + 98$

27. **MODELING WITH MATHEMATICS** A water balloon is dropped from a height of 144 feet. *(See Example 4.)*
    a. After how many seconds does the water balloon hit the ground?
    b. Suppose the initial height is adjusted by k feet. How does this affect part (a)?

28. **MODELING WITH MATHEMATICS** The function $y = -16x^2 + 36$ represents the height y (in feet) of an apple x seconds after falling from a tree. Find and interpret the x- and y-intercepts.

Assignment Guide and Homework Check

ASSIGNMENT

Basic: 1, 2, 3–19 odd, 27, 29, 34, 36, 42–45

Average: 1, 2–22 even, 27–30, 34–38 even, 42–45

Advanced: 1, 2, 6, 10–26 even, 27, 28–32 even, 33–45

HOMEWORK CHECK

Basic: 3, 7, 13, 27

Average: 4, 10, 14, 27

Advanced: 6, 10, 16, 28

ANSWERS

1. $(0, c)$, $x = 0$

2. When $c > 0$, the graph of $f(x) = ax^2 + c$ is a vertical translation c units up of the graph of $f(x) = ax^2$. When $c < 0$, the graph of $f(x) = ax^2 + c$ is a vertical translation $|c|$ units down of the graph of $f(x) = ax^2$.

3. 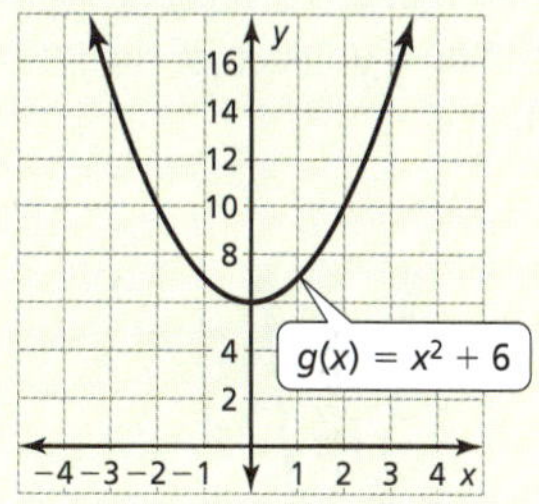

The graph of g is a vertical translation 6 units up of the graph of f.

4. 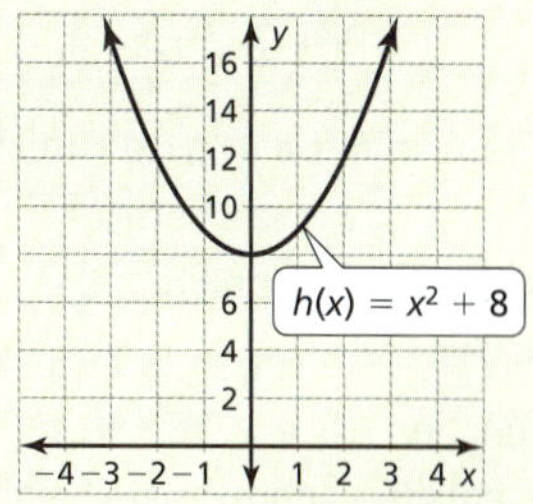

The graph of h is a vertical translation 8 units up of the graph of f.

5. 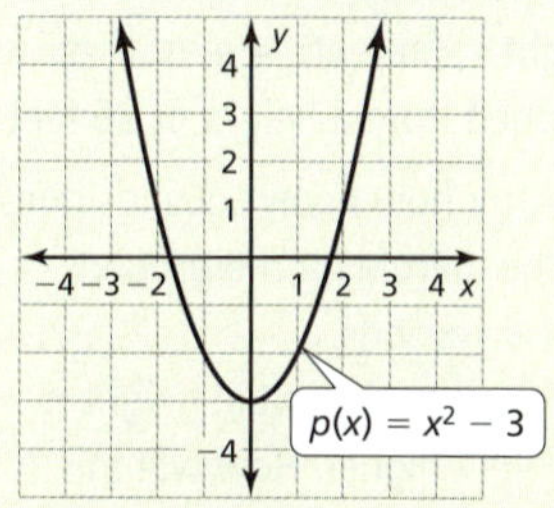

The graph of p is a vertical translation 3 units down of the graph of f.

6. 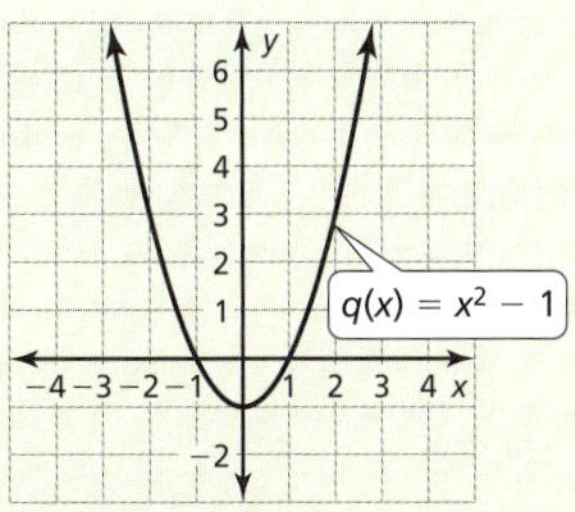

The graph of q is a vertical translation 1 unit down of the graph of f.

7. 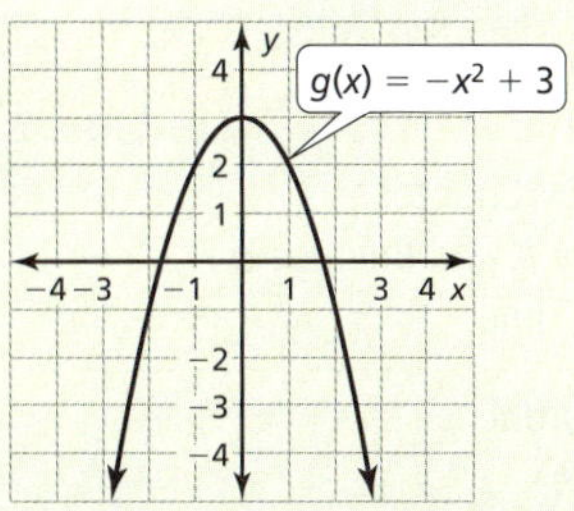

The graph of g is a reflection in the x-axis, and a vertical translation 3 units up of the graph of f.

8–28. See Additional Answers.

Dynamic Teaching Tools

- Dynamic Assessment & Progress Monitoring Tool
- Interactive Whiteboard Lesson Library
- Dynamic Classroom with Dynamic Investigations

ANSWERS

29–45. See Additional Answers.

Mini-Assessment

Graph the function. Compare the graph to the graph of $f(x) = x^2$.

1. $g(x) = x^2 + 4$

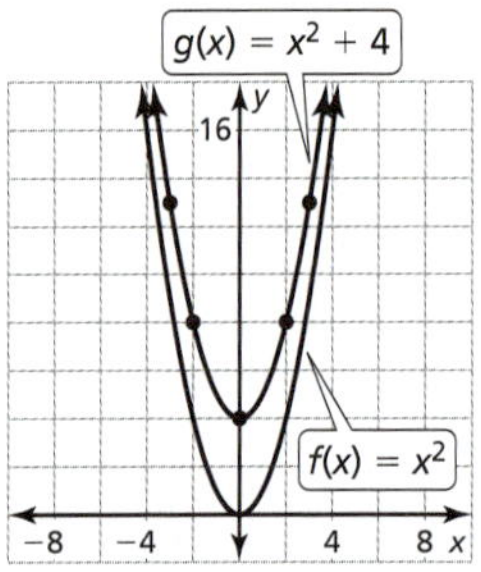

The graph of g is a vertical translation 4 units up of the graph of f.

2. $h(x) = 2x^2 - 6$

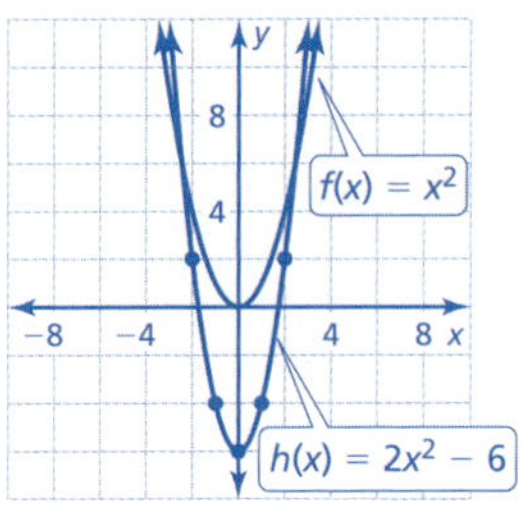

The graph of h is a vertical stretch by a factor of 2 and a vertical translation 6 units down of the graph of f.

3. The function $f(t) = -16t^2 + s_0$ represents the approximate height (in feet) of a falling object t seconds after it is dropped from an initial height s_0 (in feet). A marble is dropped from a height of 36 feet.
   a. After how many seconds does the marble hit the ground? 1.5 seconds
   b. Suppose the initial height is k feet higher. How will this affect part (a)? The marble will take more than 1.5 seconds to hit the ground.

In Exercises 29–32, sketch a parabola with the given characteristics.

29. The parabola opens up, and the vertex is (0, 3).

30. The vertex is (0, 4), and one of the x-intercepts is 2.

31. The related function is increasing when $x < 0$, and the zeros are −1 and 1.

32. The highest point on the parabola is (0, −5).

33. DRAWING CONCLUSIONS You and your friend both drop a ball at the same time. The function $h(x) = -16x^2 + 256$ represents the height (in feet) of your ball after x seconds. The function $g(x) = -16x^2 + 300$ represents the height (in feet) of your friend's ball after x seconds.

a. Write the function $T(x) = h(x) - g(x)$. What does $T(x)$ represent?

b. When your ball hits the ground, what is the height of your friend's ball? Use a graph to justify your answer.

34. MAKING AN ARGUMENT Your friend claims that in the equation $y = ax^2 + c$, the vertex changes when the value of a changes. Is your friend correct? Explain your reasoning.

35. MATHEMATICAL CONNECTIONS The area A (in square feet) of a square patio is represented by $A = x^2$, where x is the length of one side of the patio. You add 48 square feet to the patio, resulting in a total area of 192 square feet. What are the dimensions of the original patio? Use a graph to justify your answer.

36. HOW DO YOU SEE IT? The graph of $f(x) = ax^2 + c$ is shown. Points A and B are the same distance from the vertex of the graph of f. Which point is closer to the vertex of the graph of f as c increases?

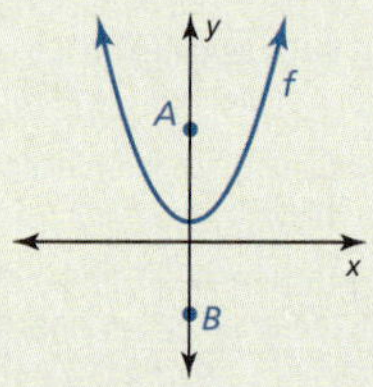

37. REASONING Describe two algebraic methods you can use to find the zeros of the function $f(t) = -16t^2 + 400$. Check your answer by graphing.

38. PROBLEM SOLVING The paths of water from three different garden waterfalls are given below. Each function gives the height h (in feet) and the horizontal distance d (in feet) of the water.

Waterfall 1 $h = -3.1d^2 + 4.8$

Waterfall 2 $h = -3.5d^2 + 1.9$

Waterfall 3 $h = -1.1d^2 + 1.6$

a. Which waterfall drops water from the highest point?

b. Which waterfall follows the narrowest path?

c. Which waterfall sends water the farthest?

39. WRITING EQUATIONS Two acorns fall to the ground from an oak tree. One falls 45 feet, while the other falls 32 feet.

a. For each acorn, write an equation that represents the height h (in feet) as a function of the time t (in seconds).

b. Describe how the graphs of the two equations are related.

40. THOUGHT PROVOKING One of two classic problems in calculus is to find the area under a curve. Approximate the area of the region bounded by the parabola and the x-axis. Show your work.

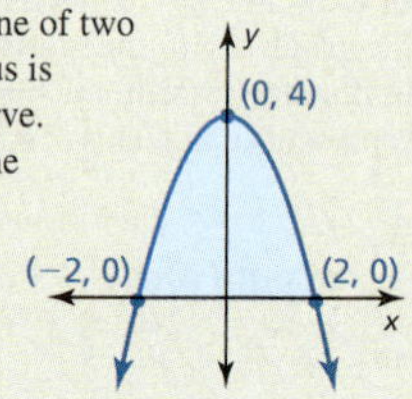

41. CRITICAL THINKING A cross section of the parabolic surface of the antenna shown can be modeled by $y = 0.012x^2$, where x and y are measured in feet. The antenna is moved up so that the outer edges of the dish are 25 feet above the x-axis. Where is the vertex of the cross section located? Explain.

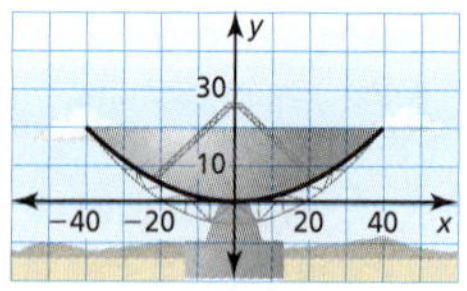

Maintaining Mathematical Proficiency

Reviewing what you learned in previous grades and lessons

Evaluate the expression when $a = 4$ and $b = -3$. *(Skills Review Handbook)*

42. $\dfrac{a}{4b}$ **43.** $-\dfrac{b}{2a}$ **44.** $\dfrac{a-b}{3a+b}$ **45.** $-\dfrac{b+2a}{ab}$

If students need help...	*If students got it...*
Resources by Chapter • Practice A and Practice B • Puzzle Time	Resources by Chapter • Enrichment and Extension • Cumulative Review
Student Journal • Practice	Start the *next* Section
Differentiating the Lesson Skills Review Handbook	

Laurie's Notes

Overview of Section 8.3

Introduction

- The first two lessons in the chapter looked at transformations of the parent function. In this lesson, the linear term, bx, is added to the equation. The lesson begins with the formula for the x-coordinate of the vertex of $y = ax^2 + bx + c$, which is also related to the equation of the axis of symmetry. This information is used to graph the quadratic function versus making a table of values that may or may not include the vertex.
- The second part of the lesson focuses on finding the maximum or minimum value of the function. This is done by evaluating the function using the x-coordinate of the vertex. The y-coordinate is the maximum or minimum value of the function.

Teaching Strategy

- In this lesson, students will use parameters from the equation to find the vertex of the parabola and also the maximum or minimum value of the function. Graphing technology could also be used.
- Example 5 in the lesson is about launching a water balloon. Enter the function $f(t) = -16t^2 + 80t + 5$, where f represents the height (in feet) of the water balloon t seconds after it is launched, in Y_1 of the equation editor, as shown in Figure 1.
- Go to the Calculate screen and choose "maximum," as shown in Figure 2. Follow the steps in the user's guide for your calculator to find and display the vertex and its value.
- Figure 3 shows the calculated value for the vertex.
- **Attend to Precision:** Note that the calculator generated a vertex of (2.4999994, 105). Calculated by hand, the vertex is (2.5, 105). Students need to know how to interpret results such as these and know that the vertex is indeed (2.5, 105).

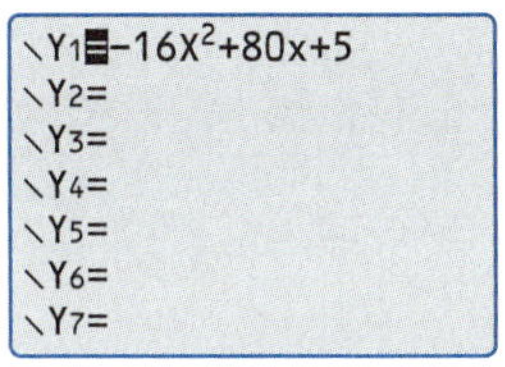

Figure 1

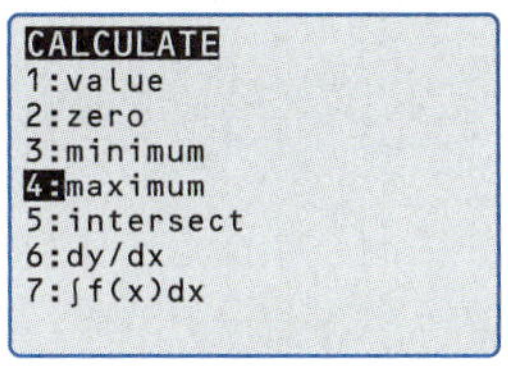

Figure 2

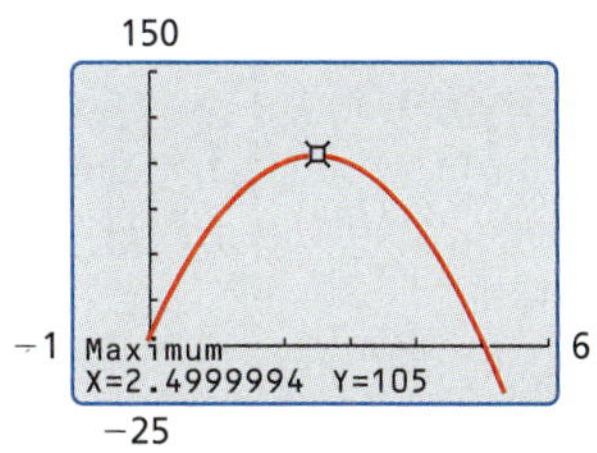

Figure 3

Pacing Suggestion

- Once students have worked the explorations, continue with the formal lesson.

Dynamic Teaching Tools

- Dynamic Assessment & Progress Monitoring Tool
- Lesson Planning Tool
- Interactive Whiteboard Lesson Library
- Dynamic Classroom with Dynamic Investigations

What Your Students Will Learn

- Graph quadratic functions of the form $f(x) = ax^2 + bx + c$ by finding the axis of symmetry, plotting the vertex, the y-intercept, and additional points, and drawing a smooth curve through the points.
- Find the minimum and maximum values of quadratic functions by finding the y-coordinate of the vertex.

Laurie's Notes

Exploration

Motivate

- Discuss with students that, in a future course, they will study conic sections: ellipses, parabolas, and hyperbolas. A form of each of those words is a part of spoken English.
- Have students find the words *ellipsis*, *hyperbole*, and *parable* in a dictionary. Discuss the definitions.
- In rhetoric, "hyperbolic" speech goes beyond the facts, "elliptic" speech falls short of the facts, and a "parable" is a story that fits the facts exactly.

For Your Information

- **Look For and Make Use of Structure:** All of the quadratic functions that students have graphed have been symmetric about the y-axis. In the equation $y = ax^2 + c$, the values of a and c do not affect the graph in a horizontal direction. In this section, quadratic functions contain an x-term, which shifts the axis of symmetry away from the y-axis.

Exploration 1

- Students should use a table of values to graph the functions. In doing so, they gain a sense of the differences, numerically, for the two functions. Evaluating the functions when $x = 2$, the arithmetic is the same except for the second function where 6 is added.
- In both functions, when students use domain values of -2, -1, 0, 1, and 2, the y-values are decreasing. When they evaluate either function for $x = 3$, the y-values start to increase. You want students to interpret this change. The vertex occurs when $x = 2$.
- The x-intercepts may have been found when generating the table of values. If not, students should substitute 0 for y and solve.
- Discuss the connections between the axis of symmetry, vertex, and x-intercepts.

Exploration 2

- Students may need guidance to factor $y = ax^2 + bx$. Suggest that they treat a and b as constants and focus on solving for x.
- When students complete the table, they are substituting the x-intercepts, 0 and $-\frac{b}{a}$, into the equation and solving. Both results should be $y = 0$.

Exploration 3

- Students need to reflect back on the two explorations they completed today.
- Students may describe the answer in words if they have difficulty finding an expression for x. For instance, in the first answer blank they may say that the vertex occurs halfway between 0 and $-\frac{b}{a}$, not realizing that they can express this as $x = -\frac{b}{2a}$. You may need to give examples and pose questions such as "What is halfway between 4 and 6? Halfway between 0 and a?"

Communicate Your Answer

- Listen for understanding of what students discovered in the explorations.

Connecting to Next Step

- The explorations help students make sense of why the x-coordinate of the vertex is $-\frac{b}{2a}$. In the formal lesson, this is stated at the beginning of the lesson.

8.3 Graphing $f(x) = ax^2 + bx + c$

Essential Question

How can you find the vertex of the graph of $f(x) = ax^2 + bx + c$?

EXPLORATION 1 Comparing x-Intercepts with the Vertex

Work with a partner.

a. Sketch the graphs of $y = 2x^2 - 8x$ and $y = 2x^2 - 8x + 6$.

b. What do you notice about the x-coordinate of the vertex of each graph?

c. Use the graph of $y = 2x^2 - 8x$ to find its x-intercepts. Verify your answer by solving $0 = 2x^2 - 8x$.

d. Compare the value of the x-coordinate of the vertex with the values of the x-intercepts.

EXPLORATION 2 Finding x-Intercepts

Work with a partner.

a. Solve $0 = ax^2 + bx$ for x by factoring.

b. What are the x-intercepts of the graph of $y = ax^2 + bx$?

c. Copy and complete the table to verify your answer.

x	$y = ax^2 + bx$
0	
$-\frac{b}{a}$	

CONSTRUCTING VIABLE ARGUMENTS

To be proficient in math, you need to make conjectures and build a logical progression of statements.

EXPLORATION 3 Deductive Reasoning

Work with a partner. Complete the following logical argument.

The x-intercepts of the graph of $y = ax^2 + bx$ are 0 and $-\frac{b}{a}$.

The vertex of the graph of $y = ax^2 + bx$ occurs when $x =$ ______.

The vertices of the graphs of $y = ax^2 + bx$ and $y = ax^2 + bx + c$ have the same x-coordinate.

The vertex of the graph of $y = ax^2 + bx + c$ occurs when $x =$ ______.

Communicate Your Answer

4. How can you find the vertex of the graph of $f(x) = ax^2 + bx + c$?

5. Without graphing, find the vertex of the graph of $f(x) = x^2 - 4x + 3$. Check your result by graphing.

Dynamic Teaching Tools

- Dynamic Assessment & Progress Monitoring Tool
- Lesson Planning Tool
- Interactive Whiteboard Lesson Library
- Dynamic Classroom with Dynamic Investigations

ANSWERS

1. a.

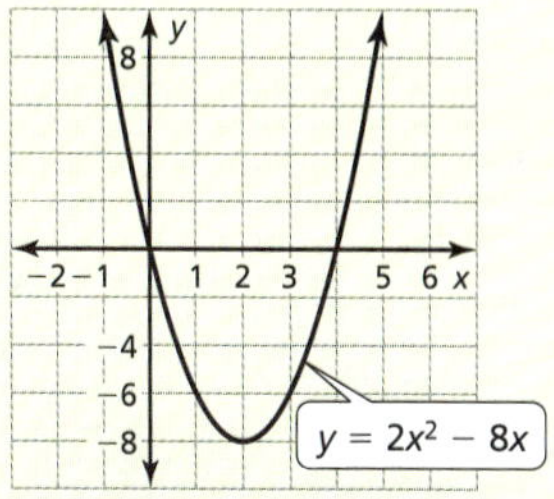

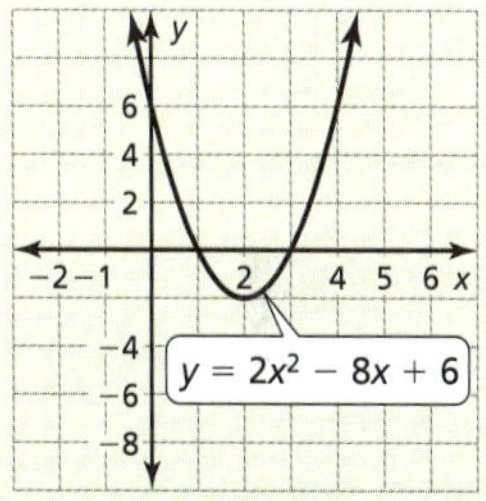

b. They are the same.

c. $x = 0, x = 4$

d. *Sample answer:* The value of the x-coordinate of the vertex is the average of the values of the x-intercepts.

2. a. $x = 0, x = -\frac{b}{a}$

b. $x = 0, x = -\frac{b}{a}$

c. 0, 0

3. $-\frac{b}{2a}; -\frac{b}{2a}$

4. *Sample answer:* Find the x-coordinate using $x = -\frac{b}{2a}$, then use the function to find the y-coordinate.

5. $(2, -1)$

Differentiated Instruction

Organization

Have students copy and complete the table below in their notebooks. Have students refer to the table to help them determine whether a function of the form $f(x) = ax^2 + bx + c$ has a maximum value or a minimum value.

$f(x) = ax^2 + bx + c$		
Value of a	$a > 0$	$a < 0$
Graph opens …	up	down
Maximum or minimum?		
Sketch		

Extra Example 1

a. Find the axis of symmetry of the graph of $f(x) = 2x^2 - 4x + 5$.
The axis of symmetry is $x = 1$.

b. Find the vertex of the graph of $f(x) = 2x^2 - 4x + 5$.
The vertex is $(1, 3)$.

MONITORING PROGRESS ANSWERS

1. a. $x = \frac{1}{3}$
b. $\left(\frac{1}{3}, -\frac{1}{3}\right)$

2. a. $x = -3$
b. $(-3, -4)$

3. a. $x = 7$
b. $\left(7, \frac{41}{2}\right)$

8.3 Lesson

Core Vocabulary

maximum value, *p. 433*
minimum value, *p. 433*

Previous
independent variable
dependent variable

What You Will Learn

- Graph quadratic functions of the form $f(x) = ax^2 + bx + c$.
- Find maximum and minimum values of quadratic functions.

Graphing $f(x) = ax^2 + bx + c$

Core Concept

Graphing $f(x) = ax^2 + bx + c$

- The graph opens up when $a > 0$, and the graph opens down when $a < 0$.
- The y-intercept is c.
- The x-coordinate of the vertex is $-\frac{b}{2a}$.
- The axis of symmetry is $x = -\frac{b}{2a}$.

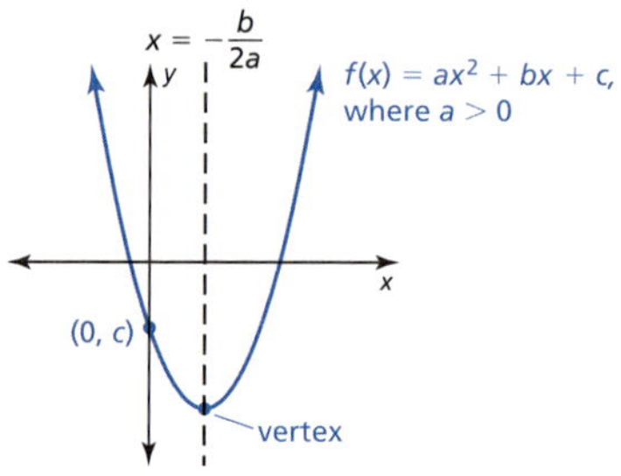

EXAMPLE 1 Finding the Axis of Symmetry and the Vertex

Find (a) the axis of symmetry and (b) the vertex of the graph of $f(x) = 2x^2 + 8x - 1$.

SOLUTION

a. Find the axis of symmetry when $a = 2$ and $b = 8$.

$x = -\frac{b}{2a}$	Write the equation for the axis of symmetry.
$x = -\frac{8}{2(2)}$	Substitute 2 for a and 8 for b.
$x = -2$	Simplify.

▶ The axis of symmetry is $x = -2$.

b. The axis of symmetry is $x = -2$, so the x-coordinate of the vertex is -2. Use the function to find the y-coordinate of the vertex.

$f(x) = 2x^2 + 8x - 1$	Write the function.
$f(-2) = 2(-2)^2 + 8(-2) - 1$	Substitute -2 for x.
$= -9$	Simplify.

▶ The vertex is $(-2, -9)$.

Check

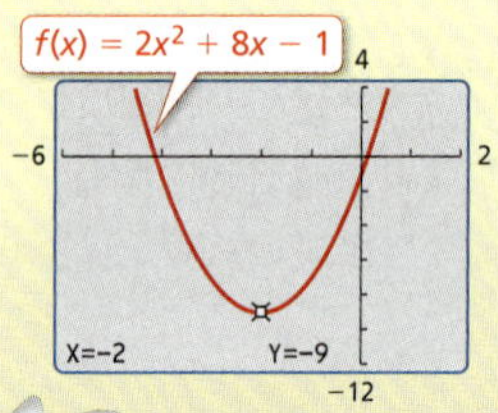

Monitoring Progress

Help in English and Spanish at *BigIdeasMath.com*

Find (a) the axis of symmetry and (b) the vertex of the graph of the function.

1. $f(x) = 3x^2 - 2x$ **2.** $g(x) = x^2 + 6x + 5$ **3.** $h(x) = -\frac{1}{2}x^2 + 7x - 4$

Laurie's Notes Teacher Actions

- **Connection:** Write the *Core Concept*. The constant term in linear and quadratic functions is the y-intercept.
- **Thumbs Up:** Write the remaining information in the *Core Concept* about the vertex and line of symmetry. Have students self-assess with a *Thumbs Up* sign.

? Pose Example 1. "How can you find the axis of symmetry?" Listen as students describe using the formula $x = -\frac{b}{2a}$.

? "Now that you know the axis of symmetry, how do you find the vertex?" Substitute -2 for x in the equation and simplify.

EXAMPLE 2 Graphing $f(x) = ax^2 + bx + c$

Graph $f(x) = 3x^2 - 6x + 5$. Describe the domain and range.

SOLUTION

Step 1 Find and graph the axis of symmetry.

$$x = -\frac{b}{2a} = -\frac{(-6)}{2(3)} = 1 \qquad \text{Substitute and simplify.}$$

Step 2 Find and plot the vertex.

The axis of symmetry is $x = 1$, so the x-coordinate of the vertex is 1. Use the function to find the y-coordinate of the vertex.

$f(x) = 3x^2 - 6x + 5$ — Write the function.

$f(1) = 3(1)^2 - 6(1) + 5$ — Substitute 1 for x.

$= 2$ — Simplify.

So, the vertex is $(1, 2)$.

Step 3 Use the y-intercept to find two more points on the graph.

Because $c = 5$, the y-intercept is 5. So, $(0, 5)$ lies on the graph. Because the axis of symmetry is $x = 1$, the point $(2, 5)$ also lies on the graph.

Step 4 Draw a smooth curve through the points.

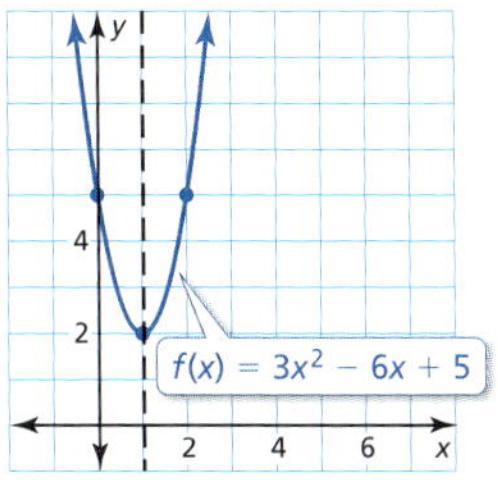

▶ The domain is all real numbers. The range is $y \geq 2$.

COMMON ERROR

Be sure to include the negative sign before the fraction when finding the axis of symmetry.

REMEMBER

The domain is the set of all possible input values of the independent variable x. The range is the set of all possible output values of the dependent variable y.

Monitoring Progress

Help in English and Spanish at *BigIdeasMath.com*

Graph the function. Describe the domain and range.

4. $h(x) = 2x^2 + 4x + 1$ **5.** $k(x) = x^2 - 8x + 7$ **6.** $p(x) = -5x^2 - 10x - 2$

Finding Maximum and Minimum Values

Core Concept

Maximum and Minimum Values

The y-coordinate of the vertex of the graph of $f(x) = ax^2 + bx + c$ is the **maximum value** of the function when $a < 0$ or the **minimum value** of the function when $a > 0$.

$f(x) = ax^2 + bx + c, a < 0$

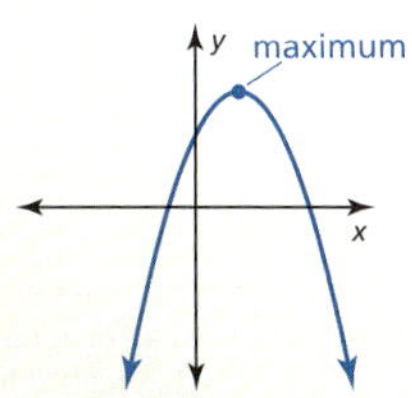

$f(x) = ax^2 + bx + c, a > 0$

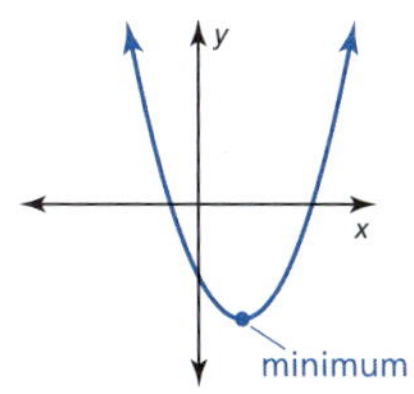

Laurie's Notes — Teacher Actions

- **Look For and Make Use of Structure:** Explain that besides a table of values and a calculator, information from the equation can be used to graph a quadratic. First, find the axis of symmetry, and then use the x-coordinate to find the vertex. To increase accuracy, use symmetry to find two additional points.
- **Think-Pair-Share:** Have students answer Questions 4–6, and then share and discuss as a class.
- **Misconception:** The vertex is not the maximum or minimum value. The y-coordinate of the vertex is the maximum or minimum value.

English Language Learners

Vocabulary

Use sketches of $f(x) = ax^2 + bx + c$ with positive and negative values of a to illustrate the maximum or minimum value of a quadratic function. Guide students to verbalize that "maximum" means "highest point" and "minimum" means "lowest point." If students are familiar with phrases such as "maximum speed" or "minimum wage," discuss the relationship between these phrases and the points on the graphs.

Extra Example 2

Graph $f(x) = x^2 - 2x - 3$. Describe the domain and range.

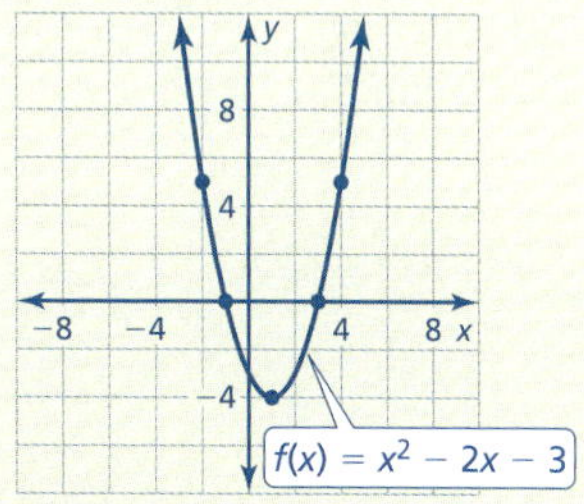

The domain is all real numbers. The range is $y \geq -4$.

MONITORING PROGRESS ANSWERS

4.

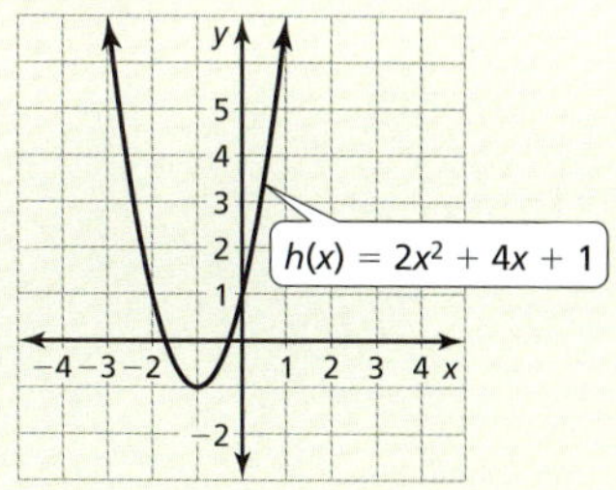

domain: all real numbers, range: $y \geq -1$

5.

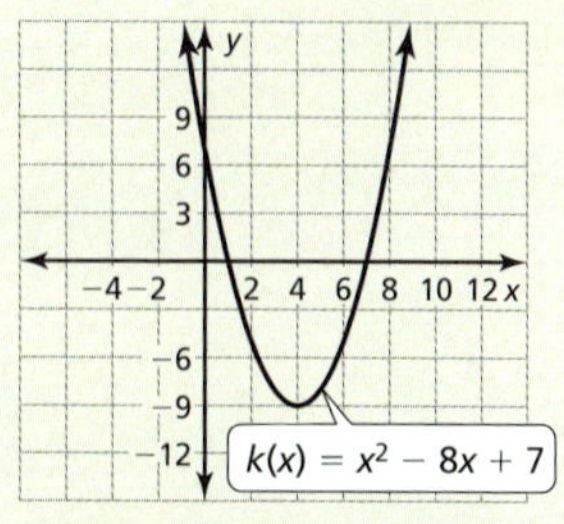

domain: all real numbers, range: $y \geq -9$

6. See Additional Answers.

Extra Example 3

Tell whether the function $f(x) = 5x^2 + 15x - 10$ has a minimum value or a maximum value. Then find the value. The minimum value is -21.25.

Extra Example 4

A company plans to have coffee mugs produced with the company logo on them. The cost of each mug depends on the number of mugs that are ordered, and is modeled by $y = 0.0001x^2 - 0.1x + 30$, where x is the number of mugs ordered and y is the cost (in dollars). What is the minimum cost of each mug? The minimum cost of each mug is $5.00.

MONITORING PROGRESS ANSWERS

7. minimum value; 4
8. maximum value; 10
9. about 176 ft

EXAMPLE 3 Finding a Maximum or Minimum Value

Tell whether the function $f(x) = -4x^2 - 24x - 19$ has a minimum value or a maximum value. Then find the value.

SOLUTION

For $f(x) = -4x^2 - 24x - 19$, $a = -4$ and $-4 < 0$. So, the parabola opens down and the function has a maximum value. To find the maximum value, find the y-coordinate of the vertex.

First, find the x-coordinate of the vertex. Use $a = -4$ and $b = -24$.

$$x = -\frac{b}{2a} = -\frac{(-24)}{2(-4)} = -3$$ Substitute and simplify.

Then evaluate the function when $x = -3$ to find the y-coordinate of the vertex.

$$f(-3) = -4(-3)^2 - 24(-3) - 19$$ Substitute -3 for x.

$$= 17$$ Simplify.

▶ The maximum value is 17.

EXAMPLE 4 Finding a Minimum Value

The suspension cables between the two towers of the Mackinac Bridge in Michigan form a parabola that can be modeled by $y = 0.000098x^2 - 0.37x + 552$, where x and y are measured in feet. What is the height of the cable above the water at its lowest point?

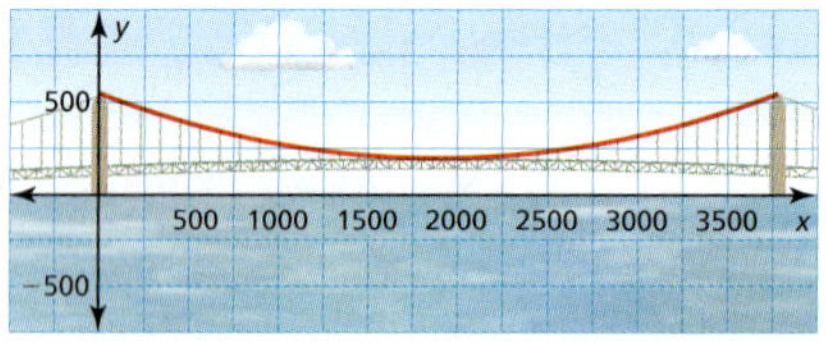

SOLUTION

The lowest point of the cable is at the vertex of the parabola. Find the x-coordinate of the vertex. Use $a = 0.000098$ and $b = -0.37$.

$$x = -\frac{b}{2a} = -\frac{(-0.37)}{2(0.000098)} \approx 1888$$ Substitute and use a calculator.

Substitute 1888 for x in the equation to find the y-coordinate of the vertex.

$$y = 0.000098(1888)^2 - 0.37(1888) + 552 \approx 203$$

▶ The cable is about 203 feet above the water at its lowest point.

Monitoring Progress

Help in English and Spanish at *BigIdeasMath.com*

Tell whether the function has a minimum value or a maximum value. Then find the value.

7. $g(x) = 8x^2 - 8x + 6$

8. $h(x) = -\frac{1}{4}x^2 + 3x + 1$

9. The cables between the two towers of the Tacoma Narrows Bridge in Washington form a parabola that can be modeled by $y = 0.00016x^2 - 0.46x + 507$, where x and y are measured in feet. What is the height of the cable above the water at its lowest point?

Laurie's Notes Teacher Actions

- **? Probing Question:** "Do you need to graph the quadratic to know whether it has a maximum or minimum value? Explain." no; When $a > 0$, the quadratic opens up and has a minimum value. When $a < 0$, the quadratic opens down and has a maximum value.
- **Make Sense of Problems and Persevere in Solving Them** and **Wait Time:** Pose Example 4, and then give partners time to work on the problem. Do not rush in to solve this for them. Give sufficient *Wait Time*.
- Students can use calculators to find the x-coordinate of the vertex and use this x-value to evaluate the function to find the minimum value.

EXAMPLE 5 Modeling with Mathematics

A group of friends is launching water balloons. The function $f(t) = -16t^2 + 80t + 5$ represents the height (in feet) of the first water balloon t seconds after it is launched. The height of the second water balloon t seconds after it is launched is shown in the graph. Which water balloon went higher?

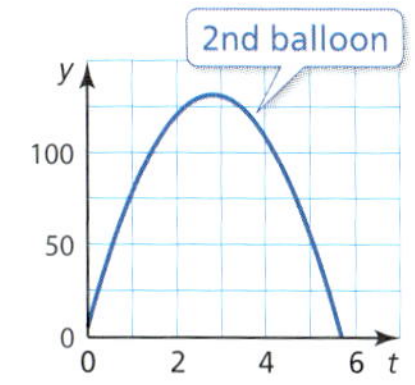

MODELING WITH MATHEMATICS

Because time cannot be negative, use only nonnegative values of t.

SOLUTION

1. **Understand the Problem** You are given a function that represents the height of the first water balloon. The height of the second water balloon is represented graphically. You need to find and compare the maximum heights of the water balloons.
2. **Make a Plan** To compare the maximum heights, represent both functions graphically. Use a graphing calculator to graph $f(t) = -16t^2 + 80t + 5$ in an appropriate viewing window. Then visually compare the heights of the water balloons.
3. **Solve the Problem** Enter the function $f(t) = -16t^2 + 80t + 5$ into your calculator and graph it. Compare the graphs to determine which function has a greater maximum value.

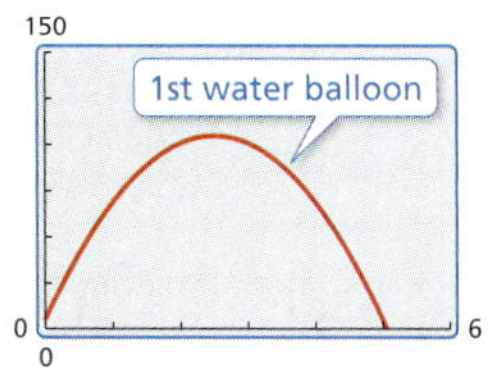

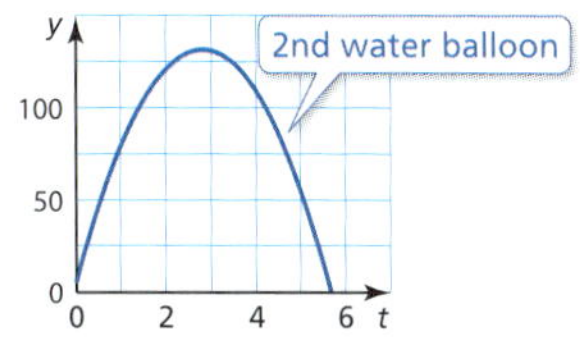

You can see that the second water balloon reaches a height of about 125 feet, while the first water balloon reaches a height of only about 100 feet.

▶ So, the second water balloon went higher.

4. **Look Back** Use the *maximum* feature to determine that the maximum value of $f(t) = -16t^2 + 80t + 5$ is 105. Use a straightedge to represent a height of 105 feet on the graph that represents the second water balloon to clearly see that the second water balloon went higher.

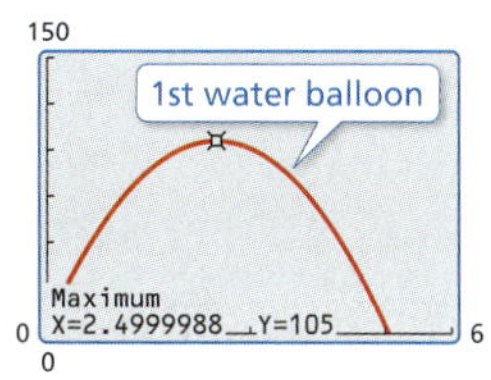

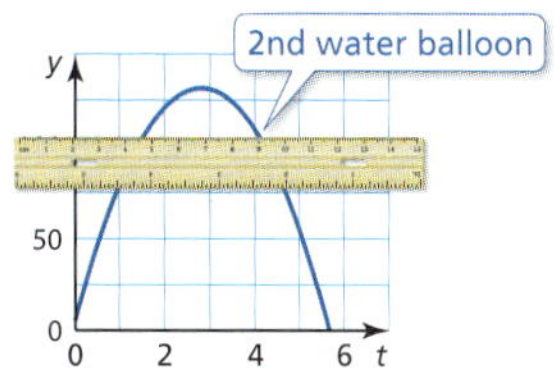

Monitoring Progress

Help in English and Spanish at *BigIdeasMath.com*

10. Which balloon is in the air longer? Explain your reasoning.
11. Which balloon reaches its maximum height faster? Explain your reasoning.

Extra Example 5

Two jugglers toss beanbags in the air. The function $h(t) = -5t^2 + 5t + 1$ represents the height (in meters) of a red beanbag after t seconds. The height of a blue beanbag t seconds after it is launched is shown in the graph. Which beanbag went higher?

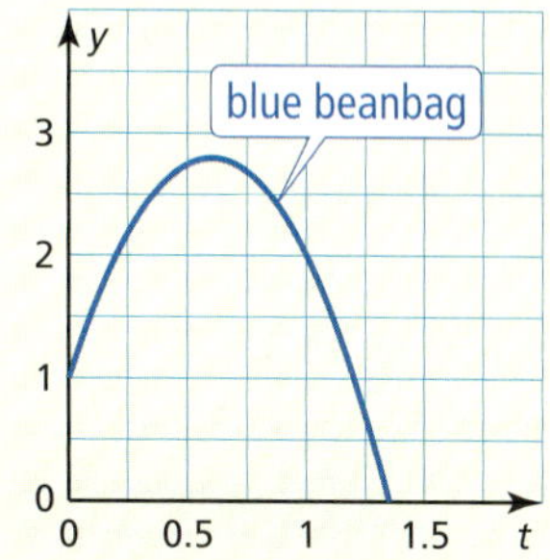

The blue beanbag went higher.

MONITORING PROGRESS ANSWERS

10. second water balloon; *Sample answer:* Comparing the graphs, the first water balloon has a height of 0 after about 5 seconds, and the second water balloon has a height of 0 after about 5.7 seconds.
11. first water balloon; *Sample answer:* Comparing the graphs, the first water balloon is at its maximum height after about 2.5 seconds, and the second water balloon is at its maximum height after about 2.8 seconds.

Laurie's Notes Teacher Actions

- **Model with Mathematics** and **Use Appropriate Tools Strategically:** Students are given two models represented differently—an equation and a graph. Have students work with partners to solve the problem. Students will likely graph both functions and use the maximum feature on their calculators. See the *Teaching Strategy* on page T-430.

? Extension: "When will the balloons hit the ground?" first water balloon: about 5 seconds, second water balloon: about 5.5 seconds

Closure

- Write an equation of a quadratic function that opens up, has a negative y-intercept, and is wider than the graph of $y = x^2$. *Sample answer:* $y = 0.25x^2 - 2$

Assignment Guide and Homework Check

ASSIGNMENT

Basic: 1, 2, 3–29 odd, 32, 37, 42, 50–53

Average: 1, 2–36 even, 37, 38, 42, 45, 50–53

Advanced: 1, 2, 6, 12, 16–20 even, 24–38 even, 39–53

HOMEWORK CHECK

Basic: 7, 13, 23, 27, 37

Average: 8, 16, 24, 28, 37

Advanced: 12, 16, 24, 32, 38

8.3 Exercises

Dynamic Solutions available at *BigIdeasMath.com*

Vocabulary and Core Concept Check

1. **VOCABULARY** Explain how you can tell whether a quadratic function has a maximum value or a minimum value without graphing the function.

2. **DIFFERENT WORDS, SAME QUESTION** Consider the quadratic function $f(x) = -2x^2 + 8x + 24$. Which is different? Find "both" answers.

   What is the maximum value of the function?

   What is the greatest number in the range of the function?

   What is the y-coordinate of the vertex of the graph of the function?

   What is the axis of symmetry of the graph of the function?

Monitoring Progress and Modeling with Mathematics

In Exercises 3–6, find the vertex, the axis of symmetry, and the y-intercept of the graph.

3. 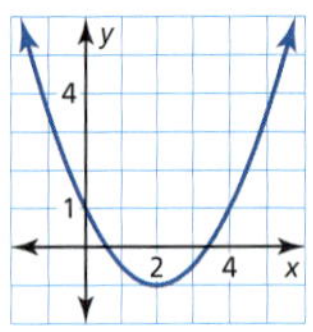

4. 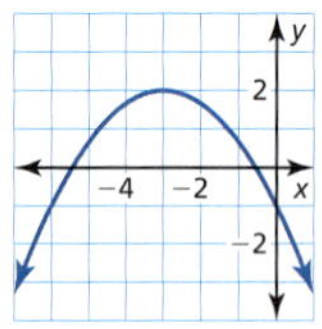

5. 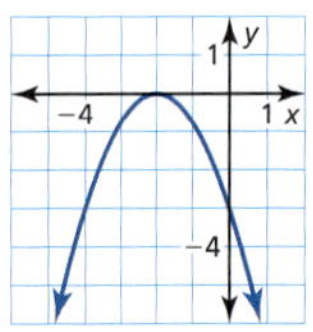

6. 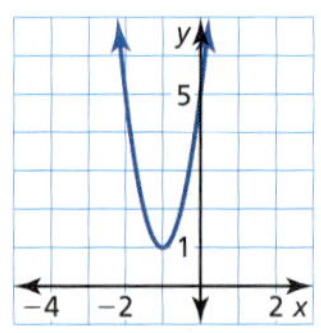

In Exercises 7–12, find (a) the axis of symmetry and (b) the vertex of the graph of the function. *(See Example 1.)*

7. $f(x) = 2x^2 - 4x$
8. $y = 3x^2 + 2x$
9. $y = -9x^2 - 18x - 1$
10. $f(x) = -6x^2 + 24x - 20$
11. $f(x) = \frac{2}{5}x^2 - 4x + 14$
12. $y = -\frac{3}{4}x^2 + 9x - 18$

In Exercises 13–18, graph the function. Describe the domain and range. *(See Example 2.)*

13. $f(x) = 2x^2 + 12x + 4$
14. $y = 4x^2 + 24x + 13$
15. $y = -8x^2 - 16x - 9$
16. $f(x) = -5x^2 + 20x - 7$
17. $y = \frac{2}{3}x^2 - 6x + 5$
18. $f(x) = -\frac{1}{2}x^2 - 3x - 4$

19. **ERROR ANALYSIS** Describe and correct the error in finding the axis of symmetry of the graph of $y = 3x^2 - 12x + 11$.

   ✗ $x = -\frac{b}{2a} = \frac{-12}{2(3)} = -2$

   The axis of symmetry is $x = -2$.

20. **ERROR ANALYSIS** Describe and correct the error in graphing the function $f(x) = x^2 + 4x + 3$.

   ✗ The axis of symmetry is $x = \frac{b}{2a} = \frac{4}{2(1)} = 2$.

   $f(2) = 2^2 + 4(2) + 3 = 15$

   So, the vertex is (2, 15).

   The y-intercept is 3. So, the points (0, 3) and (4, 3) lie on the graph.

   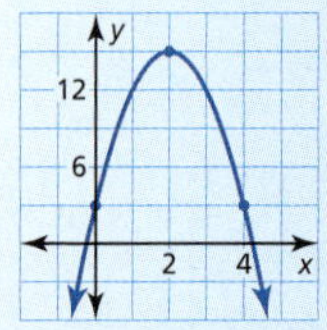

In Exercises 21–26, tell whether the function has a minimum value or a maximum value. Then find the value. *(See Example 3.)*

21. $y = 3x^2 - 18x + 15$
22. $f(x) = -5x^2 + 10x + 7$
23. $f(x) = -4x^2 + 4x - 2$
24. $y = 2x^2 - 10x + 13$
25. $y = -\frac{1}{2}x^2 - 11x + 6$
26. $f(x) = \frac{1}{5}x^2 - 5x + 27$

ANSWERS

1. *Sample answer:* If the leading coefficient is positive, the graph has a minimum value. If the leading coefficient is negative, the graph has a maximum value.
2. What is the axis of symmetry of the graph of the function?; $x = 2$; 32
3. $(2, -1)$; $x = 2$; 1
4. $(-3, 2)$; $x = -3$; -1
5. $(-2, 0)$; $x = -2$; -3
6. $(-1, 1)$; $x = -1$; 5
7. a. $x = 1$
   b. $(1, -2)$
8. a. $x = -\frac{1}{3}$
   b. $\left(-\frac{1}{3}, -\frac{1}{3}\right)$
9. a. $x = -1$
   b. $(-1, 8)$
10. a. $x = 2$
    b. $(2, 4)$
11. a. $x = 5$
    b. $(5, 4)$
12. a. $x = 6$
    b. $(6, 9)$
13. 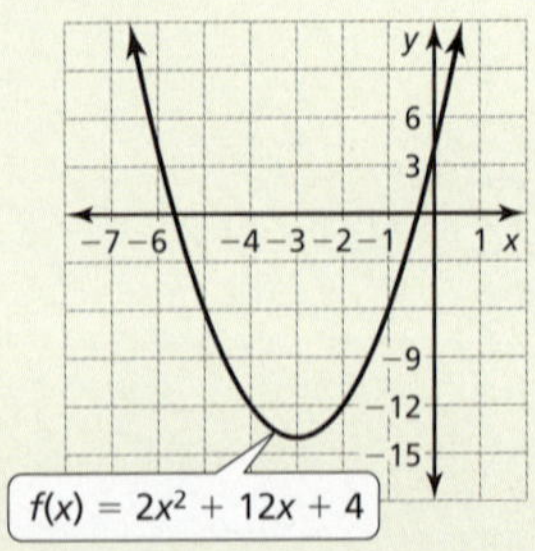

    domain: all real numbers, range: $y \geq -14$

14. 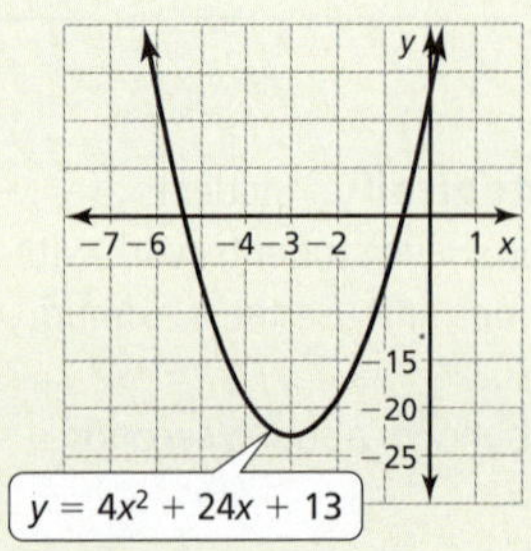

    domain: all real numbers, range: $y \geq -23$

15. 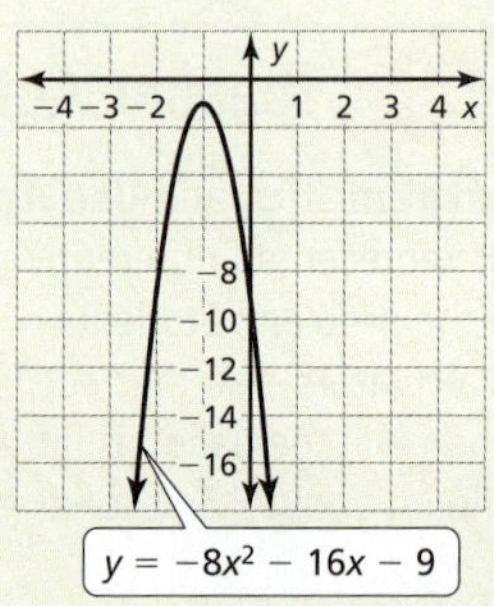

    domain: all real numbers, range: $y \leq -1$

16–26. See Additional Answers.

27. **MODELING WITH MATHEMATICS** The function shown represents the height h (in feet) of a firework t seconds after it is launched. The firework explodes at its highest point. *(See Example 4.)*

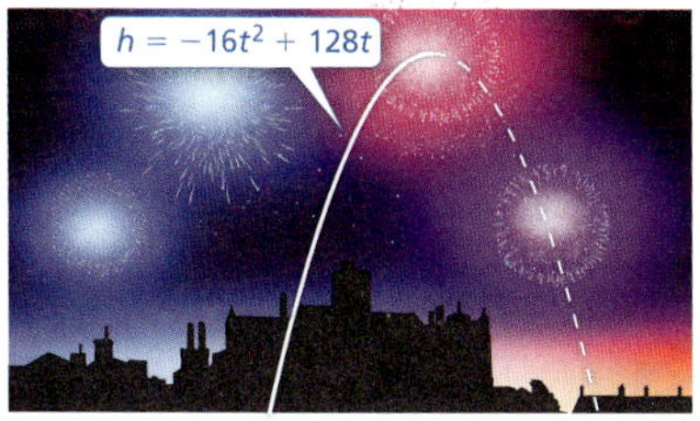

a. When does the firework explode?

b. At what height does the firework explode?

28. **MODELING WITH MATHEMATICS** The function $h(t) = -16t^2 + 16t$ represents the height (in feet) of a horse t seconds after it jumps during a steeplechase.

a. When does the horse reach its maximum height?

b. Can the horse clear a fence that is 3.5 feet tall? If so, by how much?

c. How long is the horse in the air?

29. **MODELING WITH MATHEMATICS** The cable between two towers of a suspension bridge can be modeled by the function shown, where x and y are measured in feet. The cable is at road level midway between the towers.

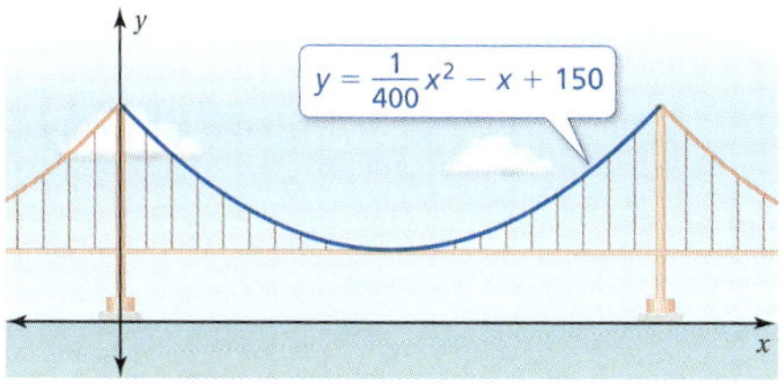

a. How far from each tower shown is the lowest point of the cable?

b. How high is the road above the water?

c. Describe the domain and range of the function shown.

30. **REASONING** Find the axis of symmetry of the graph of the equation $y = ax^2 + bx + c$ when $b = 0$. Can you find the axis of symmetry when $a = 0$? Explain.

31. **ATTENDING TO PRECISION** The vertex of a parabola is $(3, -1)$. One point on the parabola is $(6, 8)$. Find another point on the parabola. Justify your answer.

32. **MAKING AN ARGUMENT** Your friend claims that it is possible to draw a parabola through any two points with different x-coordinates. Is your friend correct? Explain.

USING TOOLS In Exercises 33–36, use the *minimum* or *maximum* feature of a graphing calculator to approximate the vertex of the graph of the function.

33. $y = 0.5x^2 + \sqrt{2}x - 3$

34. $y = -6.2x^2 + 4.8x - 1$

35. $y = -\pi x^2 + 3x$

36. $y = 0.25x^2 - 5^{2/3}x + 2$

37. **MODELING WITH MATHEMATICS** The opening of one aircraft hangar is a parabolic arch that can be modeled by the equation $y = -0.006x^2 + 1.5x$, where x and y are measured in feet. The opening of a second aircraft hangar is shown in the graph. *(See Example 5.)*

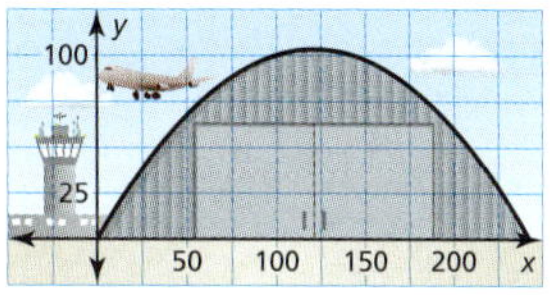

a. Which aircraft hangar is taller?

b. Which aircraft hangar is wider?

38. **MODELING WITH MATHEMATICS** An office supply store sells about 80 graphing calculators per month for \$120 each. For each \$6 decrease in price, the store expects to sell eight more calculators. The revenue from calculator sales is given by the function $R(n) =$ (unit price)(units sold), or $R(n) = (120 - 6n)(80 + 8n)$, where n is the number of \$6 price decreases.

a. How much should the store charge to maximize monthly revenue?

b. Using a different revenue model, the store expects to sell five more calculators for each \$4 decrease in price. Which revenue model results in a greater maximum monthly revenue? Explain.

Dynamic Teaching Tools

Dynamic Assessment & Progress Monitoring Tool

Interactive Whiteboard Lesson Library

Dynamic Classroom with Dynamic Investigations

ANSWERS

27. a. 4 sec
b. 256 ft

28. a. 0.5 sec
b. yes; 0.5 ft
c. 1 sec

29. a. 200 ft
b. 50 ft
c. domain: $0 \le x \le 400$, range: $50 \le y \le 150$

30. $x = 0$; no; *Sample answer:* If $a = 0$, the equation is linear, not quadratic, so it has no axis of symmetry.

31. (0, 8); *Sample answer:* Because the axis of symmetry is $x = 3$, the point (0, 8) would also lie on the graph.

32. yes; *Sample answer:* If the two points have the same y-coordinate, then a parabola can be drawn by selecting a vertex with its x-coordinate halfway between the x-coordinates of the two points, and with a different y-coordinate. If the two points have different y-coordinates, a parabola can be drawn by making one of the points the vertex and drawing the parabola through the other point.

33. $(-1.41, -4)$

34. $(0.39, -0.07)$

35. $(0.48, 0.72)$

36. $(5.85, -6.55)$

37. a. second aircraft hangar
b. first aircraft hangar

38. a. \$90
b. the first revenue model; *Sample answer:* The first revenue model has a maximum value of \$10,800 and the second revenue model has a maximum value of \$10,580, so the first has a greater maximum value.

ANSWERS

39. **a.** $x = 4.5$
b. 20.25 in.^{2}

40. **a.** $x = 2$
b. 32 ft^2

41. The graph of g is a reflection in the y-axis of the graph of h.

42. **a.** 1.5 m
b. about 1.6 m
c. about 90 m

43–53. See Additional Answers.

Mini-Assessment

1. Graph $f(x) = 3x^2 + 6x + 2$. Find the axis of symmetry and the vertex of the graph. Describe the domain and range.

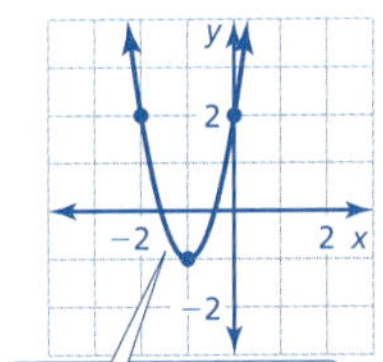

axis of symmetry: $x = -1$; vertex: $(-1, -1)$; domain: all real numbers; range: $y \geq -1$

2. Tell whether the function $f(x) = -6x^2 + 24x - 8$ has a minimum value or a maximum value. Then find the value. maximum value: 16

3. A group of friends is launching water balloons from the roof of a building. The function $f(t) = -16t^2 + 96t + 20$ represents the height (in feet) of the first balloon t seconds after it is launched. The height of the second balloon t seconds after it is launched is shown in the graph. Which balloon went higher?

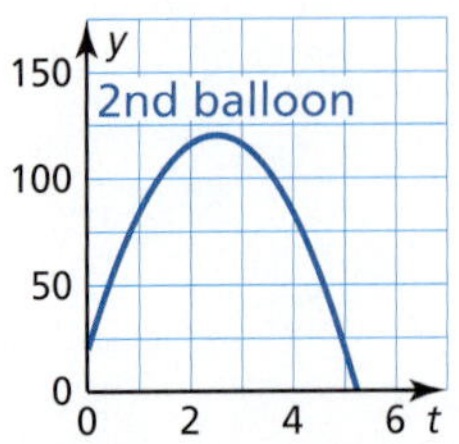

The first water balloon went higher.

MATHEMATICAL CONNECTIONS **In Exercises 39 and 40, (a) find the value of x that maximizes the area of the figure and (b) find the maximum area.**

39.

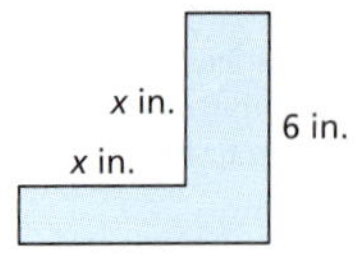

40.

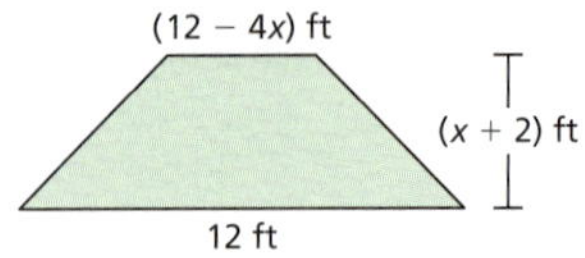

41. **WRITING** Compare the graph of $g(x) = x^2 + 4x + 1$ with the graph of $h(x) = x^2 - 4x + 1$.

42. **HOW DO YOU SEE IT?** During an archery competition, an archer shoots an arrow. The arrow follows the parabolic path shown, where x and y are measured in meters.

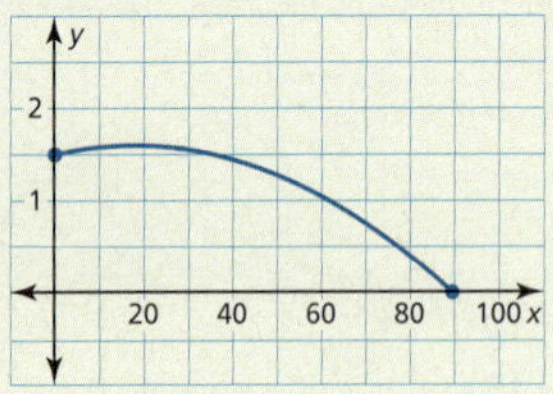

a. What is the initial height of the arrow?

b. Estimate the maximum height of the arrow.

c. How far does the arrow travel?

43. **USING TOOLS** The graph of a quadratic function passes through (3, 2), (4, 7), and (9, 2). Does the graph open up or down? Explain your reasoning.

44. **REASONING** For a quadratic function f, what does $f\left(-\frac{b}{2a}\right)$ represent? Explain your reasoning.

45. **PROBLEM SOLVING** Write a function of the form $y = ax^2 + bx$ whose graph contains the points (1, 6) and (3, 6).

46. **CRITICAL THINKING** Parabolas A and B contain the points shown. Identify characteristics of each parabola, if possible. Explain your reasoning.

Parabola A

x	y
2	3
6	4

Parabola B

x	y
1	4
3	−4
5	4

47. **MODELING WITH MATHEMATICS** At a basketball game, an air cannon launches T-shirts into the crowd. The function $y = -\frac{1}{8}x^2 + 4x$ represents the path of a T-shirt. The function $3y = 2x - 14$ represents the height of the bleachers. In both functions, y represents vertical height (in feet) and x represents horizontal distance (in feet). At what height does the T-shirt land in the bleachers?

48. **THOUGHT PROVOKING** One of two classic problems in calculus is finding the slope of a *tangent line* to a curve. An example of a tangent line, which just touches the parabola at one point, is shown.

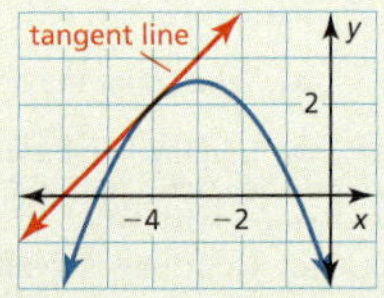

Approximate the slope of the tangent line to the graph of $y = x^2$ at the point (1, 1). Explain your reasoning.

49. **PROBLEM SOLVING** The owners of a dog shelter want to enclose a rectangular play area on the side of their building. They have k feet of fencing. What is the maximum area of the outside enclosure in terms of k? (*Hint:* Find the y-coordinate of the vertex of the graph of the area function.)

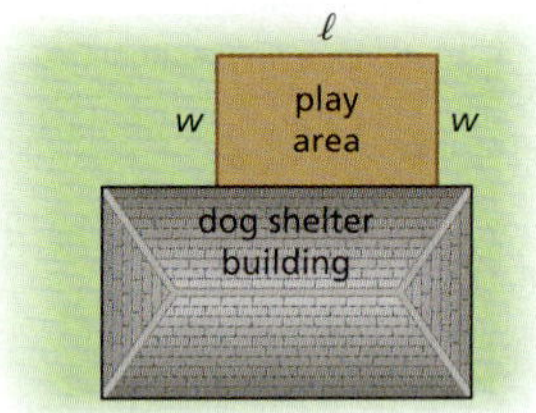

Maintaining Mathematical Proficiency

Reviewing what you learned in previous grades and lessons

Describe the transformation(s) from the graph of $f(x) = |x|$ to the graph of the given function. *(Section 3.7)*

50. $q(x) = |x + 6|$ **51.** $h(x) = -0.5|x|$ **52.** $g(x) = |x - 2| + 5$ **53.** $p(x) = 3|x + 1|$

If students need help...	*If students got it...*
Resources by Chapter • Practice A and Practice B • Puzzle Time	Resources by Chapter • Enrichment and Extension • Cumulative Review
Student Journal • Practice	Start the *next* Section
Differentiating the Lesson Skills Review Handbook	

8.1–8.3 What Did You Learn?

Core Vocabulary

quadratic function, *p. 420*
parabola, *p. 420*
vertex, *p. 420*
axis of symmetry, *p. 420*
zero of a function, *p. 428*
maximum value, *p. 433*
minimum value, *p. 433*

Core Concepts

Section 8.1

Characteristics of Quadratic Functions, *p. 420*
Graphing $f(x) = ax^2$ When $a > 0$, *p. 421*
Graphing $f(x) = ax^2$ When $a < 0$, *p. 421*

Section 8.2

Graphing $f(x) = ax^2 + c$, *p. 426*

Section 8.3

Graphing $f(x) = ax^2 + bx + c$, *p. 432*
Maximum and Minimum Values, *p. 433*

Mathematical Practices

1. Explain your plan for solving Exercise 18 on page 423.
2. How does graphing a function in Exercise 27 on page 429 help you answer the questions?
3. What definition and characteristics of the graph of a quadratic function did you use to answer Exercise 44 on page 438?

Study Skills

Learning Visually

- Draw a picture of a word problem before writing a verbal model. You do not have to be an artist.
- When making a review card for a word problem, include a picture. This will help you recall the information while taking a test.
- Make sure your notes are visually neat for easy recall.

Dynamic Teaching Tools

Dynamic Assessment & Progress Monitoring Tool
Interactive Whiteboard Lesson Library
Dynamic Classroom with Dynamic Investigations

ANSWERS

1. *Sample answer:* Because the highest point has a y-coordinate of 0, the height is the opposite of the y-coordinate of the lowest points. The width is the absolute value of the difference of the x-coordinates of the endpoints.
2. *Sample answer:* The t-intercept of the graph is the total time before the water balloon hits the ground.
3. *Sample answer:* the definition of maximum value/minimum value and the vertex

ANSWERS

1. The vertex is (1, 4). The axis of symmetry is $x = 1$. The domain is all real numbers. The range is $y \le 4$. When $x < 1$, y increases as x increases. When $x > 1$, y increases as x decreases.
2. The vertex is $(-2, 5)$. The axis of symmetry is $x = -2$. The domain is all real numbers. The range is $y \ge 5$. When $x < -2$, y increases as x decreases. When $x > -2$, y increases as x increases.
3. 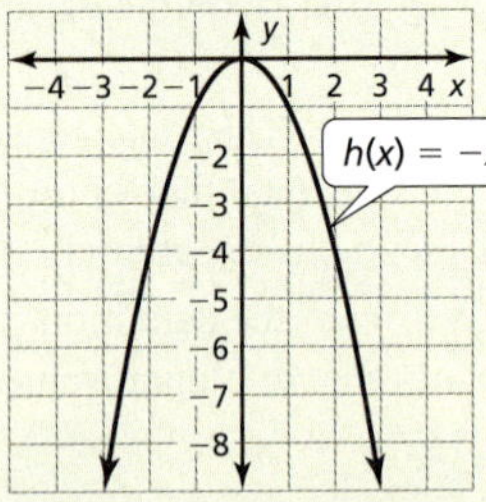

The graph of h is a reflection in the x-axis of the graph of f.
4. 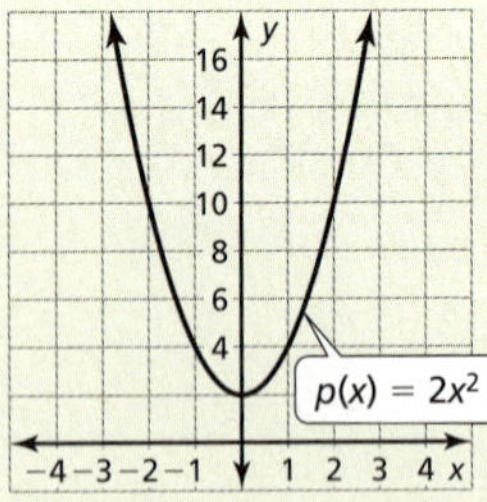

The graph of p is a vertical stretch by a factor of 2 and a vertical translation 2 units up of the graph of f.
5. 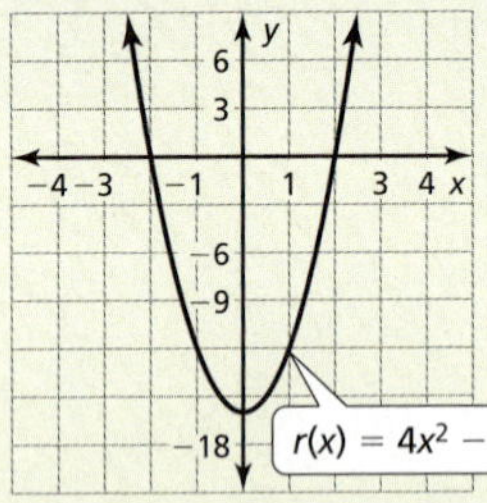

The graph of r is a vertical stretch by a factor of 4 and a vertical translation 16 units down of the graph of f.

8.1–8.3 Quiz

Identify characteristics of the quadratic function and its graph. *(Section 8.1)*

1. 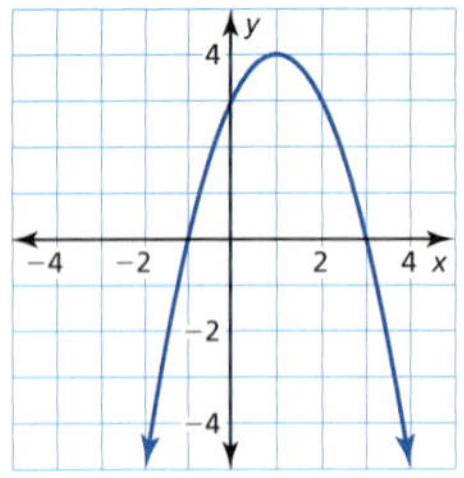
2. 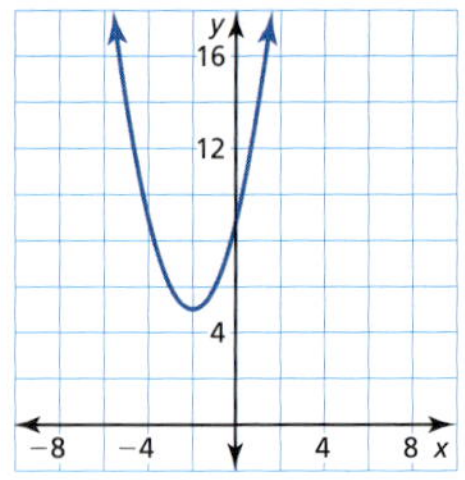

Graph the function. Compare the graph to the graph of $f(x) = x^2$. *(Section 8.1 and Section 8.2)*

3. $h(x) = -x^2$
4. $p(x) = 2x^2 + 2$
5. $r(x) = 4x^2 - 16$
6. $b(x) = 8x^2$
7. $g(x) = \frac{2}{5}x^2$
8. $m(x) = -\frac{1}{2}x^2 - 4$

Describe the transformation from the graph of f to the graph of g. Then graph f and g in the same coordinate plane. Write an equation that represents g in terms of x. *(Section 8.2)*

9. $f(x) = 2x^2 + 1$; $g(x) = f(x) + 2$
10. $f(x) = -3x^2 + 12$; $g(x) = f(x) - 9$
11. $f(x) = \frac{1}{2}x^2 - 2$; $g(x) = f(x) - 6$
12. $f(x) = 5x^2 - 3$; $g(x) = f(x) + 1$

Graph the function. Describe the domain and range. *(Section 8.3)*

13. $f(x) = -4x^2 - 4x + 7$
14. $f(x) = 2x^2 + 12x + 5$
15. $y = x^2 + 4x - 5$
16. $y = -3x^2 + 6x + 9$

Tell whether the function has a minimum value or a maximum value. Then find the value. *(Section 8.3)*

17. $f(x) = 5x^2 + 10x - 3$
18. $f(x) = -\frac{1}{2}x^2 + 2x + 16$
19. $y = -x^2 + 4x + 12$
20. $y = 2x^2 + 8x + 3$

21. The distance y (in feet) that a coconut falls after t seconds is given by the function $y = 16t^2$. Use a graph to determine how many seconds it takes for the coconut to fall 64 feet. *(Section 8.1)*

22. The function $y = -16t^2 + 25$ represents the height y (in feet) of a pinecone t seconds after falling from a tree. *(Section 8.2)*
    a. After how many seconds does the pinecone hit the ground?
    b. A second pinecone falls from a height of 36 feet. Which pinecone hits the ground in the least amount of time? Explain.

23. The function shown models the height (in feet) of a softball t seconds after it is pitched in an underhand motion. Describe the domain and range. Find the maximum height of the softball. *(Section 8.3)*

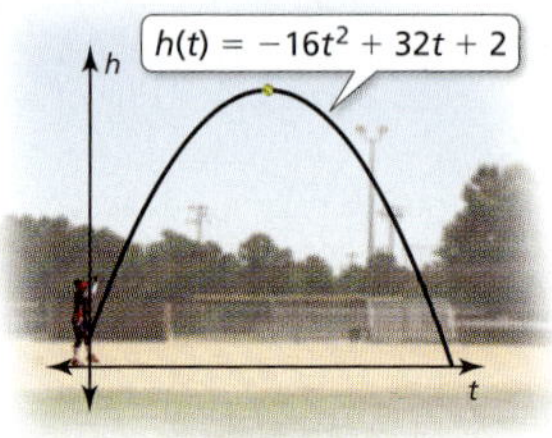

6. 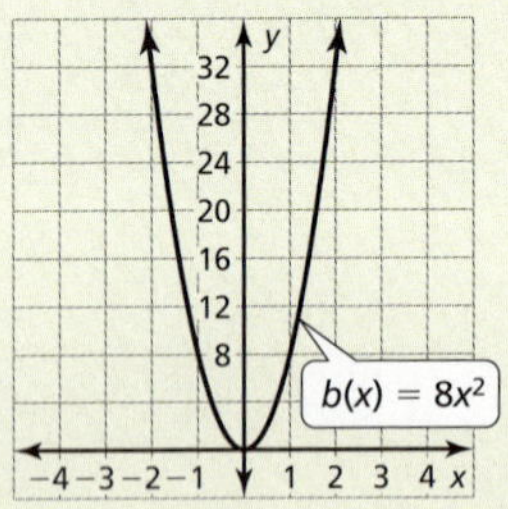

The graph of b is a vertical stretch by a factor of 8 of the graph of f.
7. 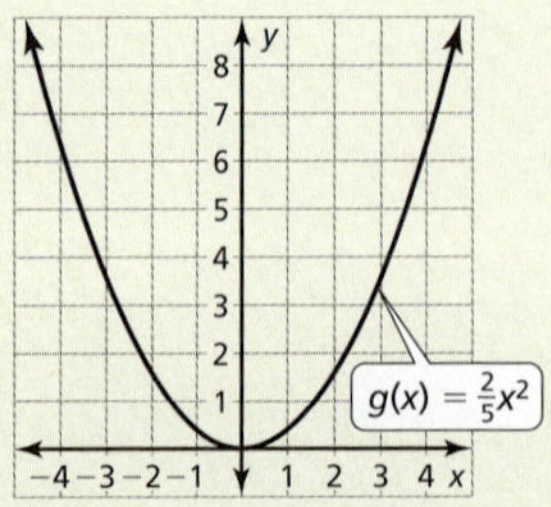

The graph of g is a vertical shrink by a factor of $\frac{2}{5}$ of the graph of f.
8. 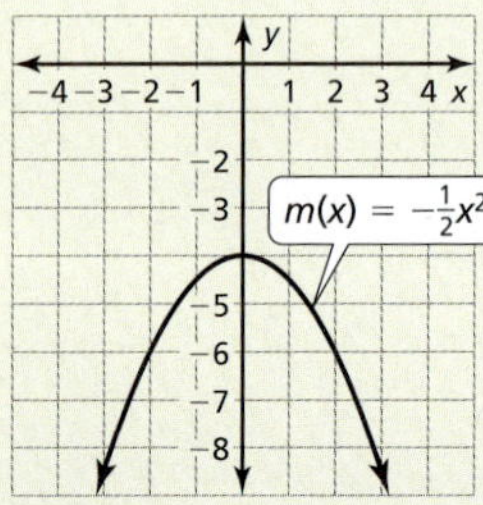

The graph of m is a vertical shrink by a factor of $\frac{1}{2}$, a reflection in the x-axis, and a vertical translation 4 units down of the graph of f.

9–23. See Additional Answers.

Laurie's Notes

Overview of Section 8.4

Introduction

- In Chapter 3, students were introduced to transformations of linear functions. In this chapter, students have performed several transformations on a quadratic function: translating vertically, stretching or shrinking vertically, and reflecting the function in the *x*-axis.
- In this lesson, the quadratic is translated horizontally, after the definitions of even and odd functions are given.
- The lesson develops the vertex form of a quadratic, $y = a(x - h)^2 + k$, from prior work with transformations. In the next chapter, when students learn the technique of completing the square, the vertex form could be derived from the standard form by completing the square.
- The lesson ends with a real-life example.

Resources

- Use dynamic geometry software to investigate the effect of h and k on the graph of $y = a(x - h)^2 + k$. Sliders can be used to generate values of h and k, $-5 < h < 5$ and $-5 < k < 5$.
- In lieu of creating your own, you can find online versions to download.

Formative Assessment Tips

- **Three-Minute Pause:** This technique is used when a lesson involves a particularly large amount of information or a lengthy process. The teacher takes a three-minute break after a period of instruction, and students work with partners or small groups to process the information they have just learned. They might ask one another questions or clarify what their understanding is, relative to the activity or instruction they have just experienced.
- The break in instruction allows students to resolve questions and ask for feedback from peers. Once the three minutes are up, the instruction continues until the next *Three-Minute Pause*.
- After the final *Three-Minute Pause*, any lingering questions can be written down and used by the teacher to clarify concepts taught in the lesson.

Pacing Suggestion

- The explorations prepare students to be thinking about horizontal translations before the vertex form of a quadratic is presented in the formal lesson. Have students work through the explorations, and then transition to the formal lesson.

Dynamic Teaching Tools

- Dynamic Assessment & Progress Monitoring Tool
- Lesson Planning Tool
- Interactive Whiteboard Lesson Library
- Dynamic Classroom with Dynamic Investigations

What Your Students Will Learn

- Identify even and odd functions algebraically and graphically.
- Graph quadratic functions of the form $f(x) = a(x - h)^2$ by identifying the horizontal translations of $|h|$ units right or left.
- Graph quadratic functions of the form $f(x) = a(x - h)^2 + k$ by identifying the vertex (h, k) and the axis of symmetry $x = h$.
- Model real-life problems using the vertex form of a quadratic function, $f(x) = a(x - h)^2 + k$.

Laurie's Notes

Exploration

Motivate

- Write two trinomials on the board for students to factor. Ask volunteers to share their answers.
  (a) $x^2 - 4x + 4$ $(x - 2)^2$ (b) $2x^2 + 4x + 2$ $2(x + 1)^2$
- Because $x^2 - 4x + 4 = (x - 2)^2$, the graphs of $y = x^2 - 4x + 4$ and $y = (x - 2)^2$ are the same.
- Today, students will graph quadratic functions in vertex form.

Discuss

- **Look For and Make Use of Structure:** All of the quadratic functions that students have graphed have been in standard form $y = ax^2 + bx + c$ with b and c sometimes equal to 0. In this lesson, students will investigate the vertex form of a quadratic function. Make the connection to linear functions and the different forms in which they are written. The structure of the equation gives information about the graph of the function.

Exploration 1

- There are several ways in which students will approach graphing $g(x) = (x - 2)^2$. They could make a table of values, expand the factored form to $x^2 - 4x + 4$ and find the vertex, and so on, or they may think about transformations and recall that there is a horizontal shift of 2 units right.
- **Make Sense of Problems and Persevere in Solving Them:** Take time to solicit different methods from students. In future problems, it will be helpful for students to recall a repertoire of approaches.
- In each part, students are graphing two functions. The goal is for students to recognize the translation.

? "How did the value of 2 affect the graph of $y = (x - 2)^2$?" The graph of $y = x^2$ was shifted 2 units right.

Exploration 2

- The approach in this exploration is the same as Exploration 1. There are different methods students may use, although they should be influenced by the results of the previous exploration!
- In each part, students are graphing two functions. The goal is for students to recognize the translation of the quadratic function that has been reflected in the x-axis.
- Be sure that students note that the general equation is written with the quantity $x - h$ in parentheses. Therefore, in the expression $(x + 2)$, $h = -2$.

? "How did the value of -2 affect the graph of $y = -(x + 2)^2$?" The graph of $y = -x^2$ was shifted 2 units left.

Communicate Your Answer

- **Turn and Talk:** Have partners discuss the problems together before sharing as a class.

Connecting to Next Step

- The explorations help students make sense of the horizontal translation for the graph of $y = (x - h)^2$. In the formal lesson, this is extended to the vertex form of a quadratic.

8.4 Graphing $f(x) = a(x - h)^2 + k$

Essential Question How can you describe the graph of $f(x) = a(x - h)^2$?

EXPLORATION 1 **Graphing $y = a(x - h)^2$ When $h > 0$**

Work with a partner. Sketch the graphs of the functions in the same coordinate plane. How does the value of h affect the graph of $y = a(x - h)^2$?

a. $f(x) = x^2$ and $g(x) = (x - 2)^2$ **b.** $f(x) = 2x^2$ and $g(x) = 2(x - 2)^2$

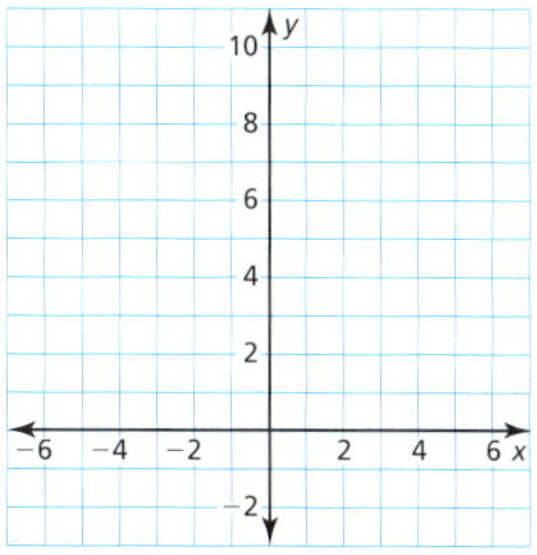
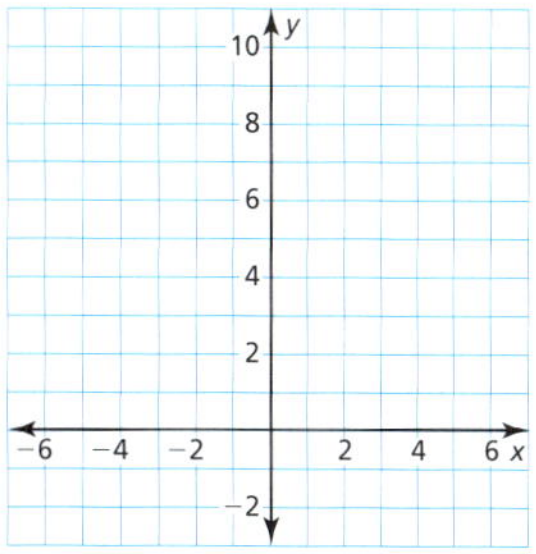

EXPLORATION 2 **Graphing $y = a(x - h)^2$ When $h < 0$**

Work with a partner. Sketch the graphs of the functions in the same coordinate plane. How does the value of h affect the graph of $y = a(x - h)^2$?

a. $f(x) = -x^2$ and $g(x) = -(x + 2)^2$ **b.** $f(x) = -2x^2$ and $g(x) = -2(x + 2)^2$

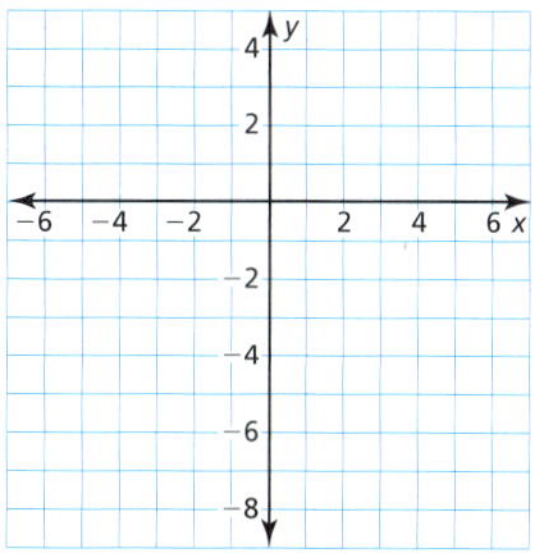
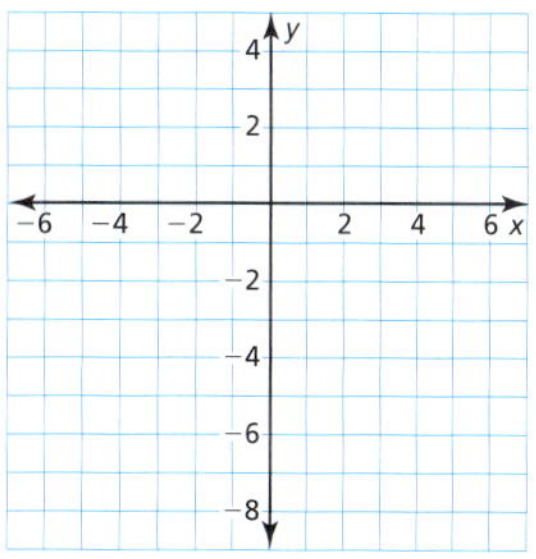

USING TOOLS STRATEGICALLY

To be proficient in math, you need to consider the available tools, such as a graphing calculator, when solving a mathematical problem.

Communicate Your Answer

3. How can you describe the graph of $f(x) = a(x - h)^2$?

4. Without graphing, describe the graph of each function. Use a graphing calculator to check your answer.
   a. $y = (x - 3)^2$
   b. $y = (x + 3)^2$
   c. $y = -(x - 3)^2$

Dynamic Teaching Tools
- Dynamic Assessment & Progress Monitoring Tool
- Lesson Planning Tool
- Interactive Whiteboard Lesson Library
- Dynamic Classroom with Dynamic Investigations

ANSWERS

1. a.

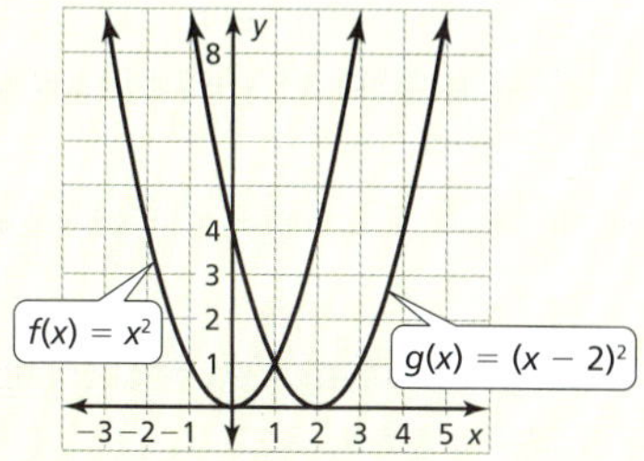

b.

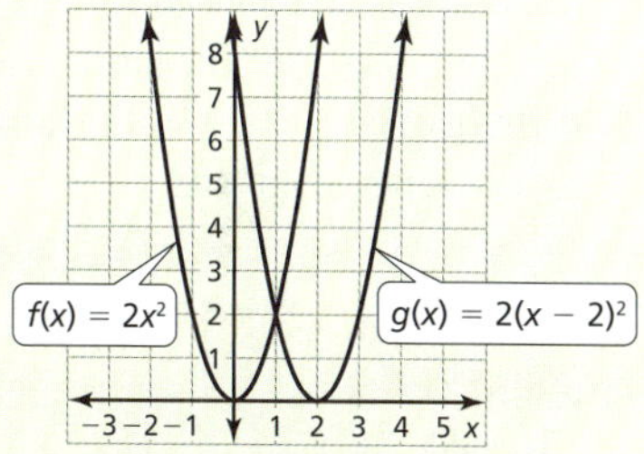

Sample answer: The value of h causes a horizontal translation of the graph of $y = ax^2$.

2. a.

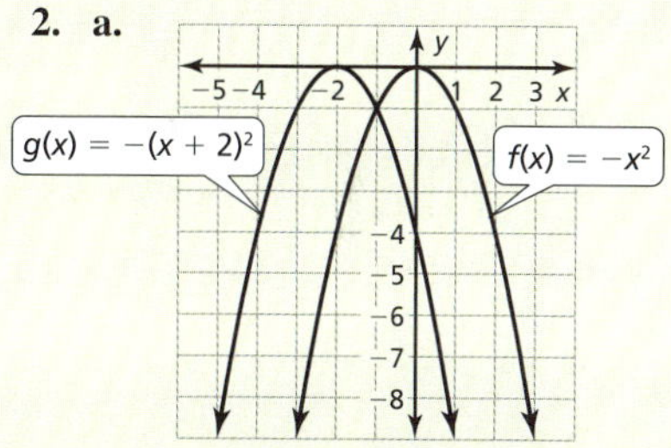

b.

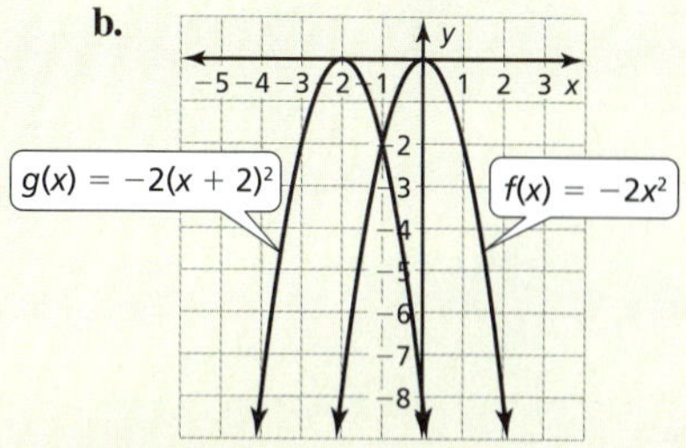

Sample answer: The value of h causes a horizontal translation of the graph of $y = ax^2$.

3. When $h > 0$, the graph of $f(x) = a(x - h)^2$ is a horizontal translation h units to the right of the graph of $f(x) = ax^2$. When $h < 0$, the graph of $f(x) = a(x - h)^2$ is a horizontal translation $|h|$ units to the left of the graph of $f(x) = ax^2$.

4. a. The graph of $y = (x - 3)^2$ is a horizontal translation 3 units right of the graph of $y = x^2$.
   b. The graph of $y = (x + 3)^2$ is a horizontal translation 3 units left of the graph of $y = x^2$.
   c. The graph of $y = -(x - 3)^2$ is a horizontal translation 3 units right and a reflection in the x-axis of the graph of $y = x^2$.

English Language Learners

Vocabulary

Students may be familiar with using the terms *odd* and *even* to describe numbers. Emphasize that, in this lesson, *odd* and *even* are used to describe functions. Sketch several functions and help students label each function as *odd*, *even*, or *neither*. Encourage them to refer to the sketches when needed.

Extra Example 1

Determine whether each function is *even*, *odd*, or *neither*.

a. $f(x) = 3x$ The function is odd.

b. $g(x) = 2x^2 - 6$ The function is even.

c. $h(x) = 3x^2 - 2x + 4$ The function is neither even nor odd.

MONITORING PROGRESS ANSWERS

1. odd
2. neither
3. even

8.4 Lesson

Core Vocabulary

even function, *p. 442*
odd function, *p. 442*
vertex form (of a quadratic function), *p. 444*

Previous
reflection

What You Will Learn

- Identify even and odd functions.
- Graph quadratic functions of the form $f(x) = a(x - h)^2$.
- Graph quadratic functions of the form $f(x) = a(x - h)^2 + k$.
- Model real-life problems using $f(x) = a(x - h)^2 + k$.

Identifying Even and Odd Functions

Core Concept

Even and Odd Functions

A function $y = f(x)$ is **even** when $f(-x) = f(x)$ for each x in the domain of f. The graph of an even function is symmetric about the y-axis.

A function $y = f(x)$ is **odd** when $f(-x) = -f(x)$ for each x in the domain of f. The graph of an odd function is symmetric about the origin. A graph is *symmetric about the origin* when it looks the same after reflections in the x-axis and then in the y-axis.

STUDY TIP

The graph of an odd function looks the same after a 180° rotation about the origin.

EXAMPLE 1 Identifying Even and Odd Functions

Determine whether each function is *even*, *odd*, or *neither*.

a. $f(x) = 2x$ **b.** $g(x) = x^2 - 2$ **c.** $h(x) = 2x^2 + x - 2$

SOLUTION

a.
$f(x) = 2x$ — Write the original function.
$f(-x) = 2(-x)$ — Substitute $-x$ for x.
$= -2x$ — Simplify.
$= -f(x)$ — Substitute $f(x)$ for $2x$.

Because $f(-x) = -f(x)$, the function is odd.

b.
$g(x) = x^2 - 2$ — Write the original function.
$g(-x) = (-x)^2 - 2$ — Substitute $-x$ for x.
$= x^2 - 2$ — Simplify.
$= g(x)$ — Substitute $g(x)$ for $x^2 - 2$.

Because $g(-x) = g(x)$, the function is even.

c.
$h(x) = 2x^2 + x - 2$ — Write the original function.
$h(-x) = 2(-x)^2 + (-x) - 2$ — Substitute $-x$ for x.
$= 2x^2 - x - 2$ — Simplify.

Because $h(x) = 2x^2 + x - 2$ and $-h(x) = -2x^2 - x + 2$, you can conclude that $h(-x) \neq h(x)$ and $h(-x) \neq -h(x)$. So, the function is neither even nor odd.

STUDY TIP

Most functions are neither even nor odd.

Monitoring Progress

Help in English and Spanish at *BigIdeasMath.com*

Determine whether the function is *even*, *odd*, or *neither*.

1. $f(x) = 5x$ **2.** $g(x) = 2^x$ **3.** $h(x) = 2x^2 + 3$

Laurie's Notes Teacher Actions

- Define *even functions* and *odd functions*.
- The notation can be confusing, so be sure to use additional words. Say, "A function is even when you evaluate the function at x and at $-x$ and the answers (y-values) are the same."
- Similarly, "A function is odd when you evaluate the function at x and at $-x$ and the answers (y-values) are opposites."
- The *Study Tip* is helpful in recognizing an *odd function* from its graph.
- Work through each part of the example as shown. Discuss the *Study Tip*. Most functions are neither even nor odd.

Graphing $f(x) = a(x - h)^2$

Core Concept

Graphing $f(x) = a(x - h)^2$

- When $h > 0$, the graph of $f(x) = a(x - h)^2$ is a horizontal translation h units right of the graph of $f(x) = ax^2$.
- When $h < 0$, the graph of $f(x) = a(x - h)^2$ is a horizontal translation $|h|$ units left of the graph of $f(x) = ax^2$.

The vertex of the graph of $f(x) = a(x - h)^2$ is $(h, 0)$, and the axis of symmetry is $x = h$.

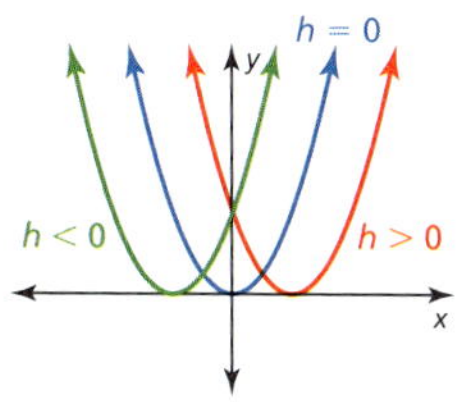

ANOTHER WAY

In Step 3, you could instead choose two x-values greater than the x-coordinate of the vertex.

EXAMPLE 2 Graphing $y = a(x - h)^2$

Graph $g(x) = \frac{1}{2}(x - 4)^2$. Compare the graph to the graph of $f(x) = x^2$.

SOLUTION

Step 1 Graph the axis of symmetry. Because $h = 4$, graph $x = 4$.

Step 2 Plot the vertex. Because $h = 4$, plot $(4, 0)$.

Step 3 Find and plot two more points on the graph. Choose two x-values less than the x-coordinate of the vertex. Then find $g(x)$ for each x-value.

When $x = 0$:

$g(0) = \frac{1}{2}(0 - 4)^2$

$= 8$

When $x = 2$:

$g(2) = \frac{1}{2}(2 - 4)^2$

$= 2$

So, plot $(0, 8)$ and $(2, 2)$.

Step 4 Reflect the points plotted in Step 3 in the axis of symmetry. So, plot $(8, 8)$ and $(6, 2)$.

Step 5 Draw a smooth curve through the points.

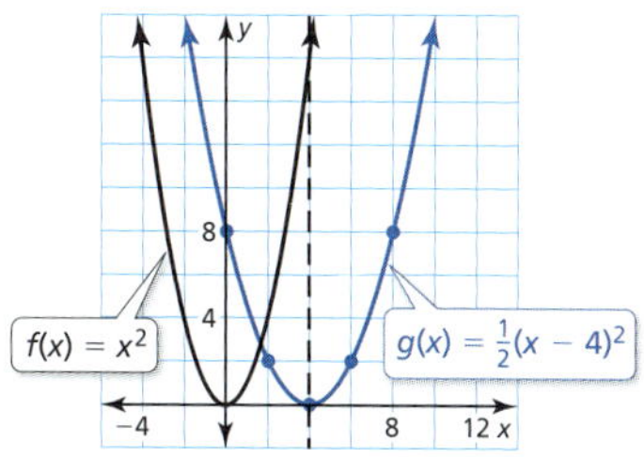

STUDY TIP

From the graph, you can see that $f(x) = x^2$ is an even function. However, $g(x) = \frac{1}{2}(x - 4)^2$ is neither even nor odd.

▶ Both graphs open up. The graph of g is wider than the graph of f. The axis of symmetry $x = 4$ and the vertex $(4, 0)$ of the graph of g are 4 units right of the axis of symmetry $x = 0$ and the vertex $(0, 0)$ of the graph of f. So, the graph of g is a translation 4 units right and a vertical shrink by a factor of $\frac{1}{2}$ of the graph of f.

Monitoring Progress

Help in English and Spanish at *BigIdeasMath.com*

Graph the function. Compare the graph to the graph of $f(x) = x^2$.

4. $g(x) = 2(x + 5)^2$

5. $h(x) = -(x - 2)^2$

Laurie's Notes Teacher Actions

- Write the *Core Concept*, and connect it to the explorations.
- What is confusing for students is that when $h > 0$, the expression is $(x - h)$ and the shift is h units right. In Example 2, $h = 4$ and the transformation is 4 units right. In *Monitoring Progress* Question 4, where $h = -5$, the expression becomes $(x + 5)$ and the shift is 5 units left.
- Work through Example 2 as shown. Take time to compare graphs of $f(x)$ and $g(x)$, discussing the transformations.

Differentiated Instruction

Graphic Organizer

Have students make a table like this one for Example 2 to organize the attributes of a function they are comparing to $f(x) = x^2$. Encourage students to complete a table for each function and then use the information to write their comparison.

Function	$f(x) = x^2$	$f(x) = \frac{1}{2}(x - 4)^2$
Axis of symmetry	$x = 0$	$x = 4$
Vertex	(0, 0)	(4, 0)
Type of shrink or stretch (factor)	—	vertical shrink $\left(\frac{1}{2}\right)$
Translation	—	4 units right
Reflection	—	—

Extra Example 2

Graph $g(x) = \frac{1}{2}(x + 2)^2$. Compare the graph to the graph of $f(x) = x^2$.

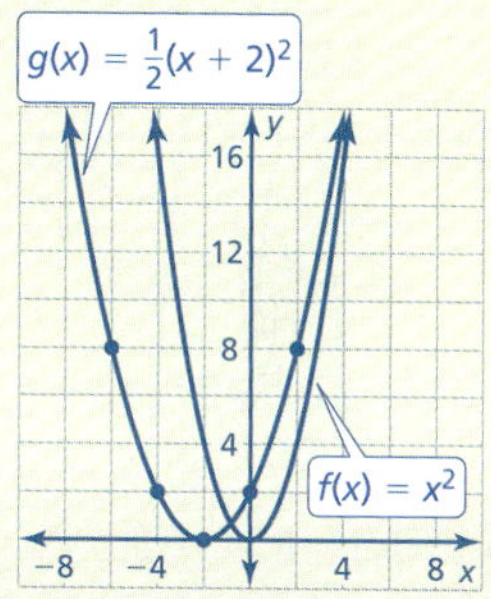

Both graphs open up. The graph of g is wider than the graph of f. The axis of symmetry, $x = -2$, and the vertex, $(-2, 0)$, of the graph of g are 2 units left of the axis of symmetry, $x = 0$, and the vertex, $(0, 0)$, of the graph of f. So, the graph of g is a translation 2 units left and a vertical shrink by a factor of $\frac{1}{2}$ of the graph of f.

MONITORING PROGRESS ANSWERS

4–5. See Additional Answers.

Extra Example 3

Graph $g(x) = -3(x + 2)^2 + 2$. Compare the graph to the graph of $f(x) = x^2$.

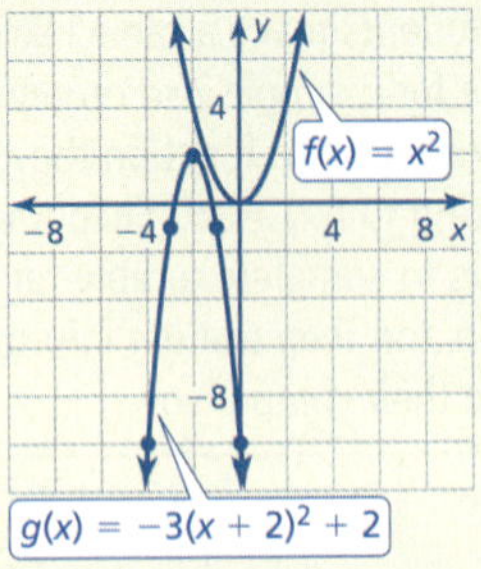

The graph of g opens down and is narrower than the graph of f. The vertex of the graph of g, $(-2, 2)$, is 2 units to the left and 2 units up from the vertex of the graph of f, $(0, 0)$. So, the graph of g is a vertical stretch by a factor of 3, a reflection in the x-axis, and a translation 2 units left and 2 units up of the graph of f.

Extra Example 4

Consider function g in Extra Example 3. Graph $f(x) = g(x - 4)$.

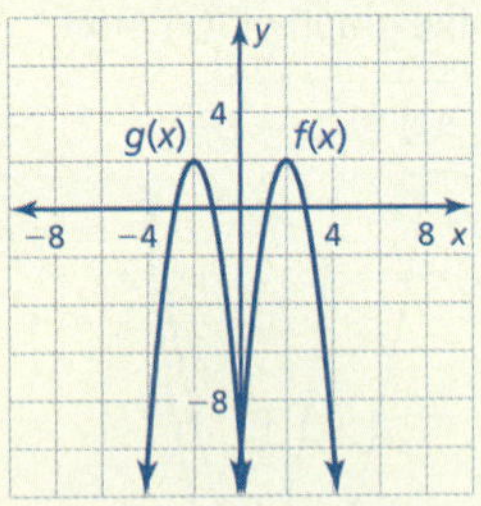

MONITORING PROGRESS ANSWERS

6\.

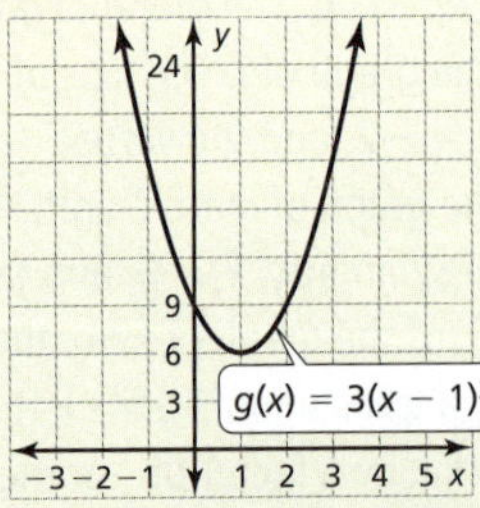

The graph of g is a vertical stretch by a factor of 3, and a translation 1 unit right and 6 units up of the graph of f.

7–8. See Additional Answers.

Graphing $f(x) = a(x - h)^2 + k$

Core Concept

Graphing $f(x) = a(x - h)^2 + k$

The **vertex form** of a quadratic function is $f(x) = a(x - h)^2 + k$, where $a \neq 0$. The graph of $f(x) = a(x - h)^2 + k$ is a translation h units horizontally and k units vertically of the graph of $f(x) = ax^2$.

The vertex of the graph of $f(x) = a(x - h)^2 + k$ is (h, k), and the axis of symmetry is $x = h$.

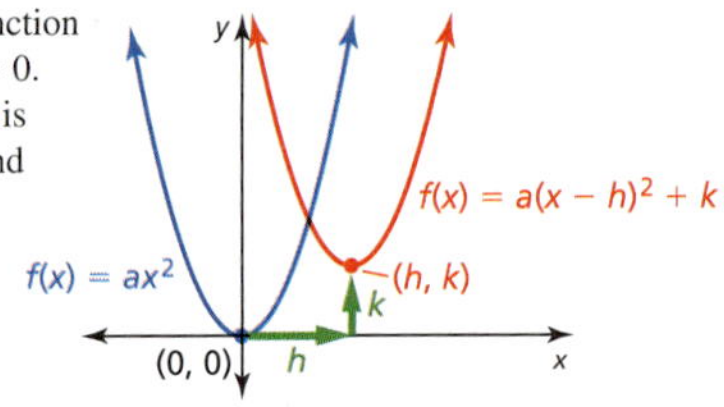

EXAMPLE 3 Graphing $y = a(x - h)^2 + k$

Graph $g(x) = -2(x + 2)^2 + 3$. Compare the graph to the graph of $f(x) = x^2$.

SOLUTION

Step 1 Graph the axis of symmetry. Because $h = -2$, graph $x = -2$.

Step 2 Plot the vertex. Because $h = -2$ and $k = 3$, plot $(-2, 3)$.

Step 3 Find and plot two more points on the graph. Choose two x-values less than the x-coordinate of the vertex. Then find $g(x)$ for each x-value. So, plot $(-4, -5)$ and $(-3, 1)$.

x	−4	−3
$g(x)$	−5	1

Step 4 Reflect the points plotted in Step 3 in the axis of symmetry. So, plot $(-1, 1)$ and $(0, -5)$.

Step 5 Draw a smooth curve through the points.

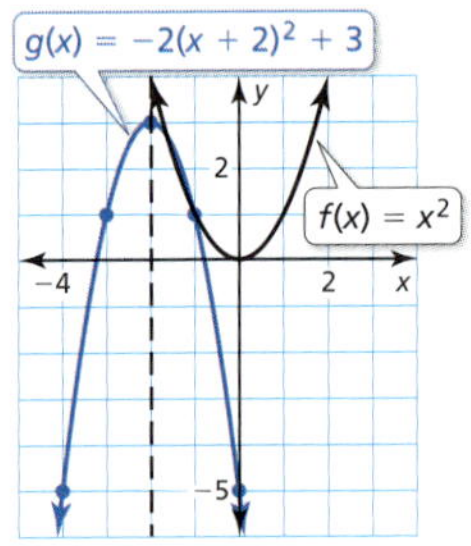

▶ The graph of g opens down and is narrower than the graph of f. The vertex of the graph of g, $(-2, 3)$, is 2 units left and 3 units up of the vertex of the graph of f, $(0, 0)$. So, the graph of g is a vertical stretch by a factor of 2, a reflection in the x-axis, and a translation 2 units left and 3 units up of the graph of f.

EXAMPLE 4 Transforming the Graph of $y = a(x - h)^2 + k$

Consider function g in Example 3. Graph $f(x) = g(x + 5)$.

SOLUTION

The function f is of the form $y = g(x - h)$, where $h = -5$. So, the graph of f is a horizontal translation 5 units left of the graph of g. To graph f, subtract 5 from the x-coordinates of the points on the graph of g.

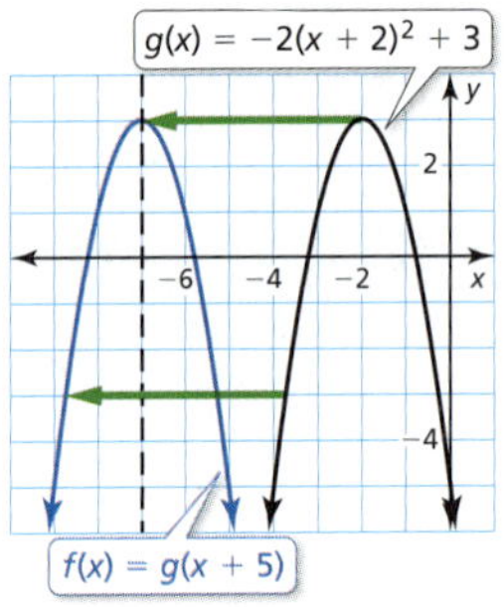

Monitoring Progress

Help in English and Spanish at *BigIdeasMath.com*

Graph the function. Compare the graph to the graph of $f(x) = x^2$.

6. $g(x) = 3(x - 1)^2 + 6$

7. $h(x) = \frac{1}{2}(x + 4)^2 - 2$

8. Consider function g in Example 3. Graph $f(x) = g(x) - 3$.

Laurie's Notes Teacher Actions

- **Three-Minute Pause:** For Example 3, pause three minutes for students to confer with partners or in small groups. Have them discuss what they know about transformations and the vertex form. There are four transformations (reflection, vertical stretch, horizontal shift, vertical shift). Does the order of the transformations matter? What is the vertex? Can they visualize the graph?
- Display the graph of Example 3 as you discuss Example 4. See the *Teaching Strategy* on page T-424 for graphing with technology.

Modeling Real-Life Problems

EXAMPLE 5 Modeling with Mathematics

Water fountains are usually designed to give a specific visual effect. For example, the water fountain shown consists of streams of water that are shaped like parabolas. Notice how the streams are designed to land on the underwater spotlights. Write and graph a quadratic function that models the path of a stream of water with a maximum height of 5 feet, represented by a vertex of (3, 5), landing on a spotlight 6 feet from the water jet, represented by (6, 0).

SOLUTION

1. **Understand the Problem** You know the vertex and another point on the graph that represents the parabolic path. You are asked to write and graph a quadratic function that models the path.
2. **Make a Plan** Use the given points and the vertex form to write a quadratic function. Then graph the function.
3. **Solve the Problem**
   Use the vertex form, vertex (3, 5), and point (6, 0) to find the value of a.

$f(x) = a(x - h)^2 + k$	Write the vertex form of a quadratic function.
$f(x) = a(x - 3)^2 + 5$	Substitute 3 for h and 5 for k.
$0 = a(6 - 3)^2 + 5$	Substitute 6 for x and 0 for $f(x)$.
$0 = 9a + 5$	Simplify.
$-\frac{5}{9} = a$	Solve for a.

So, $f(x) = -\frac{5}{9}(x - 3)^2 + 5$ models the path of a stream of water. Now graph the function.

Step 1 Graph the axis of symmetry. Because $h = 3$, graph $x = 3$.

Step 2 Plot the vertex, (3, 5).

Step 3 Find and plot two more points on the graph. Because the x-axis represents the water surface, the graph should only contain points with nonnegative values of $f(x)$. You know that (6, 0) is on the graph. To find another point, choose an x-value between $x = 3$ and $x = 6$. Then find the corresponding value of $f(x)$.

$f(4.5) = -\frac{5}{9}(4.5 - 3)^2 + 5 = 3.75$

So, plot (6, 0) and (4.5, 3.75).

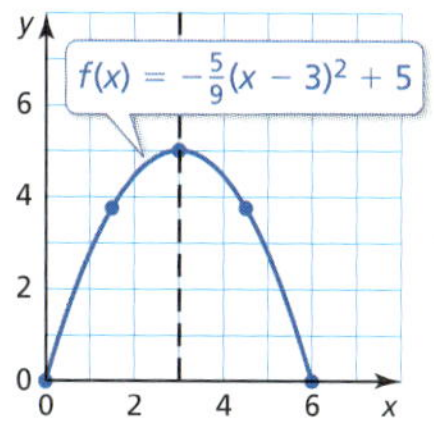

Step 4 Reflect the points plotted in Step 3 in the axis of symmetry. So, plot (0, 0) and (1.5, 3.75).

Step 5 Draw a smooth curve through the points.

4. **Look Back** Use a graphing calculator to graph $f(x) = -\frac{5}{9}(x - 3)^2 + 5$. Use the *maximum* feature to verify that the maximum value is 5. Then use the *zero* feature to verify that $x = 6$ is a zero of the function.

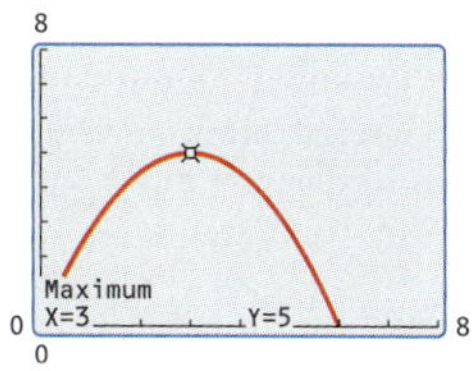

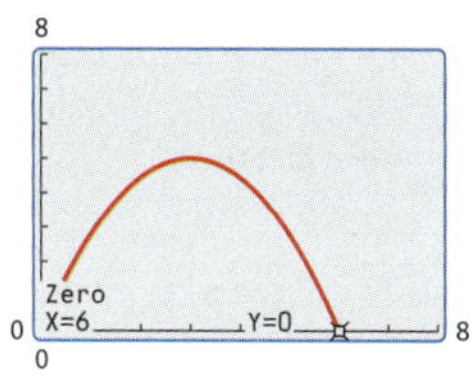

Monitoring Progress

Help in English and Spanish at *BigIdeasMath.com*

9. **WHAT IF?** The vertex is (3, 6). Write and graph a quadratic function that models the path.

Extra Example 5

Use the information in Example 5. Write and graph a quadratic function that models the path of a stream of water with a maximum height of 4 feet, represented by a vertex of (3, 4), landing on a spotlight 6 feet from the water jet, represented by (6, 0).

$f(x) = -\frac{4}{9}(x - 3)^2 + 4$

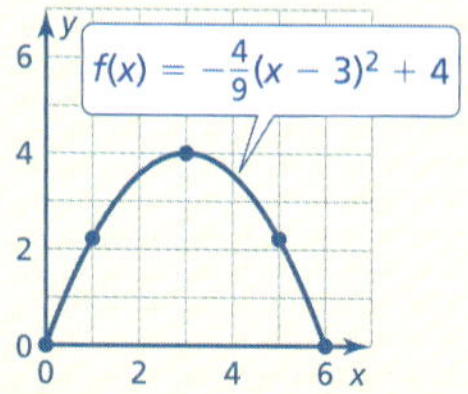

MONITORING PROGRESS ANSWER

9. $f(x) = -\frac{2}{3}(x - 3)^2 + 6$

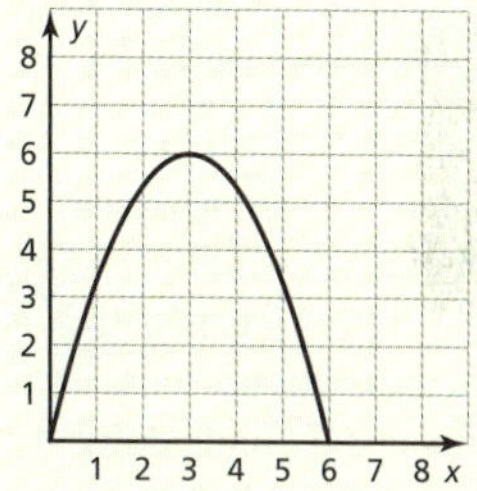

Laurie's Notes Teacher Actions

- **Make Sense of Problems and Persevere in Solving Them** and **Model with Mathematics:** A quick search will result in many online videos to motivate this example.
- **Three-Minute Pause:** Read the problem. Pause three minutes for students to confer with partners or in small groups. Discuss what is known and what they are trying to find. Ask probing questions without giving answers. Ask volunteers to share their progress, and then give additional work time. Trust that students can solve this problem.

Closure

- **Exit Ticket:** Given $f(x) = \frac{1}{2}(x + 8)^2 + 4$, tell what you know about the function and sketch its graph. The graph opens up and is wider than the graph of $y = x^2$. The axis of symmetry is $x = -8$. The vertex is $(-8, 4)$. The domain is all real numbers. The range is $y \geq 4$. So, the graph of f is a vertical shrink by a factor of $\frac{1}{2}$ and a translation 8 units left and 4 units up of the graph of $y = x^2$. Check students' graphs.

Assignment Guide and Homework Check

ASSIGNMENT

Basic: 1–4, 5–55 odd, 63, 70, 79–82

Average: 1–4, 6–66 even, 70, 79–82

Advanced: 1–4, 12–16 even, 22, 26–38 even, 42–48 even, 52–68 even, 69–82

HOMEWORK CHECK

Basic: 5, 23, 39, 49, 63

Average: 8, 24, 40, 50, 64

Advanced: 12, 26, 42, 52, 64

ANSWERS

1. *Sample answer:* The graph of an even function is symmetric about the y-axis. The graph of an odd function is symmetric about the origin.
2. *Sample answer:* $y = 5(x - 1)^2 + 2$
3. The graph of g is a horizontal translation h units right if h is positive or $|h|$ units left if h is negative, and a vertical translation k units up if k is positive or $|k|$ units down if k is negative of the graph of f.
4. $f(x) = 2(x + 0)^2$; It is the only function that is not a horizontal translation of the parent function $g(x) = x^2$.
5. neither
6. even
7. neither
8. neither
9. even
10. odd
11. neither
12. even
13. even
14. neither
15. neither
16. even
17. odd
18. neither
19. $(-1, 0)$; $x = -1$
20. $(6, 0)$; $x = 6$
21. $(4, 0)$; $x = 4$
22. $(-9, 0)$; $x = -9$

8.4 Exercises

Dynamic Solutions available at *BigIdeasMath.com*

Vocabulary and Core Concept Check

1. **VOCABULARY** Compare the graph of an even function with the graph of an odd function.
2. **OPEN-ENDED** Write a quadratic function whose graph has a vertex of (1, 2).
3. **WRITING** Describe the transformation from the graph of $f(x) = ax^2$ to the graph of $g(x) = a(x - h)^2 + k$.
4. **WHICH ONE DOESN'T BELONG?** Which function does *not* belong with the other three? Explain your reasoning.

$f(x) = 8(x + 4)^2$	$f(x) = (x - 2)^2 + 4$	$f(x) = 2(x + 0)^2$	$f(x) = 3(x + 1)^2 + 1$

Monitoring Progress and Modeling with Mathematics

In Exercises 5–12, determine whether the function is *even*, *odd*, or *neither*. *(See Example 1.)*

5. $f(x) = 4x + 3$
6. $g(x) = 3x^2$
7. $h(x) = 5^x + 2$
8. $m(x) = 2x^2 - 7x$
9. $p(x) = -x^2 + 8$
10. $f(x) = -\frac{1}{2}x$
11. $n(x) = 2x^2 - 7x + 3$
12. $r(x) = -6x^2 + 5$

In Exercises 13–18, determine whether the function represented by the graph is *even*, *odd*, or *neither*.

13. 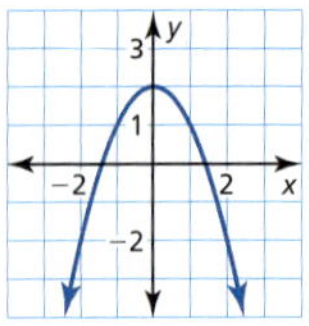

14. 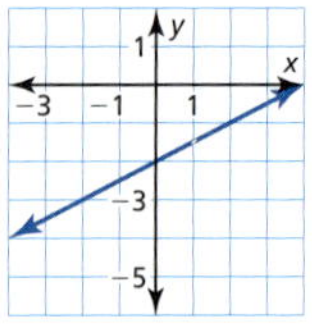

15. 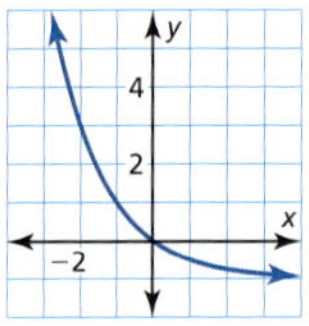

16. 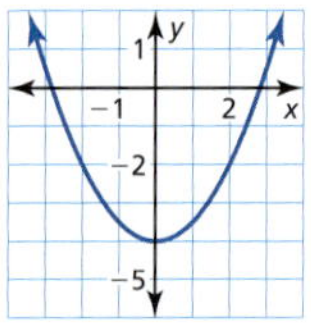

17. 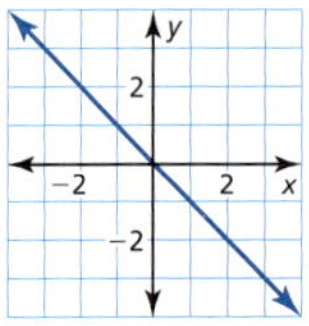

18. 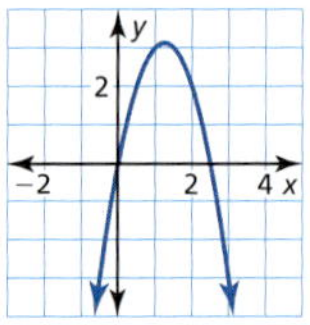

In Exercises 19–22, find the vertex and the axis of symmetry of the graph of the function.

19. $f(x) = 3(x + 1)^2$
20. $f(x) = \frac{1}{4}(x - 6)^2$
21. $y = -\frac{1}{8}(x - 4)^2$
22. $y = -5(x + 9)^2$

In Exercises 23–28, graph the function. Compare the graph to the graph of $f(x) = x^2$. *(See Example 2.)*

23. $g(x) = 2(x + 3)^2$
24. $p(x) = 3(x - 1)^2$
25. $r(x) = \frac{1}{4}(x + 10)^2$
26. $n(x) = \frac{1}{3}(x - 6)^2$
27. $d(x) = \frac{1}{5}(x - 5)^2$
28. $q(x) = 6(x + 2)^2$

29. **ERROR ANALYSIS** Describe and correct the error in determining whether the function $f(x) = x^2 + 3$ is even, odd, or neither.

✗
$f(x) = x^2 + 3$
$f(-x) = (-x)^2 + 3$
$= x^2 + 3$
$= f(x)$
So, $f(x)$ is an odd function.

30. **ERROR ANALYSIS** Describe and correct the error in finding the vertex of the graph of the function.

✗
$y = -(x + 8)^2$
Because $h = -8$, the vertex is $(0, -8)$.

23. 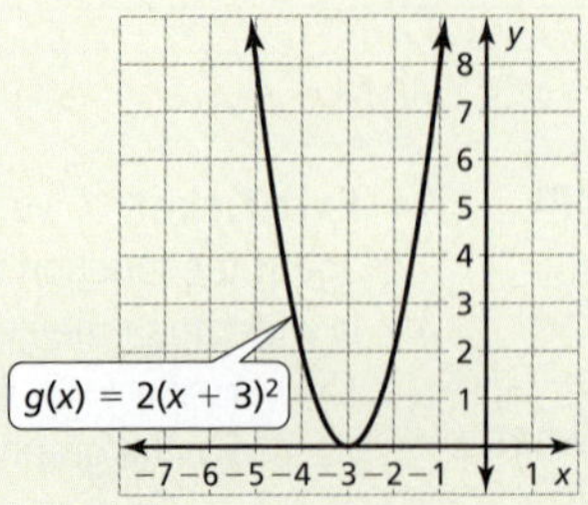

The graph of g is a horizontal translation 3 units left and a vertical stretch by a factor of 2 of the graph of f.

24. 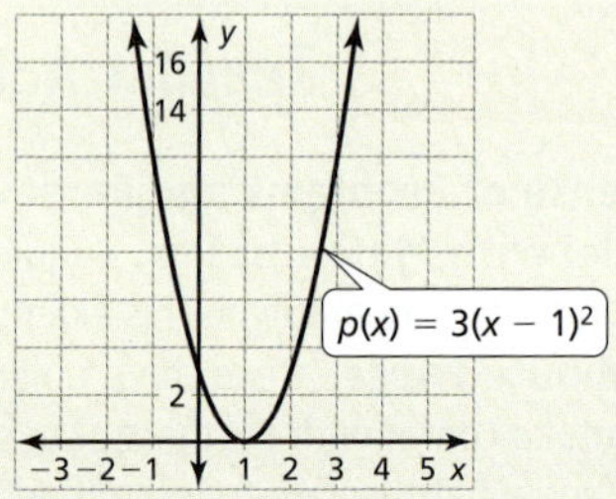

The graph of p is a horizontal translation 1 unit right and a vertical stretch by a factor of 3 of the graph of f.

25–30. See Additional Answers.

In Exercises 31–34, find the vertex and the axis of symmetry of the graph of the function.

31. $y = -6(x + 4)^2 - 3$ **32.** $f(x) = 3(x - 3)^2 + 6$

33. $f(x) = -4(x + 3)^2 + 1$ **34.** $y = -(x - 6)^2 - 5$

In Exercises 35–38, match the function with its graph.

35. $y = -(x + 1)^2 - 3$ **36.** $y = -\frac{1}{2}(x - 1)^2 + 3$

37. $y = \frac{1}{3}(x - 1)^2 + 3$ **38.** $y = 2(x + 1)^2 - 3$

A.

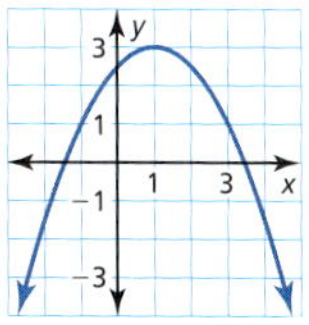

B.

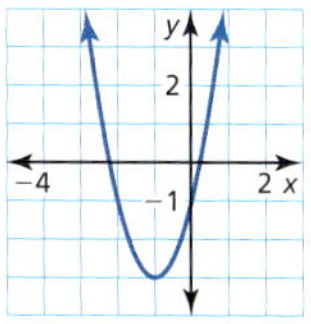

C.

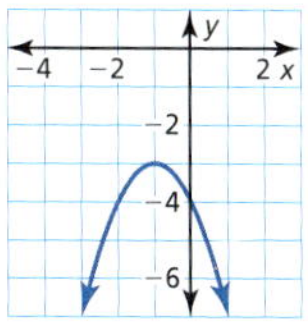

D.

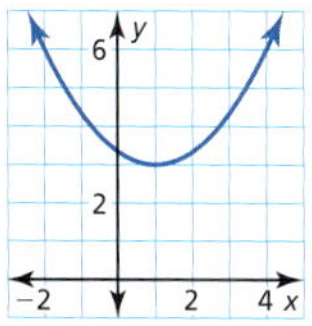

In Exercises 39–44, graph the function. Compare the graph to the graph of $f(x) = x^2$. *(See Example 3.)*

39. $h(x) = (x - 2)^2 + 4$ **40.** $g(x) = (x + 1)^2 - 7$

41. $r(x) = 4(x - 1)^2 - 5$ **42.** $n(x) = -(x + 4)^2 + 2$

43. $g(x) = -\frac{1}{3}(x + 3)^2 - 2$ **44.** $r(x) = \frac{1}{2}(x - 2)^2 - 4$

In Exercises 45–48, let $f(x) = (x - 2)^2 + 1$. Match the function with its graph.

45. $g(x) = f(x - 1)$ **46.** $r(x) = f(x + 2)$

47. $h(x) = f(x) + 2$ **48.** $p(x) = f(x) - 3$

A.

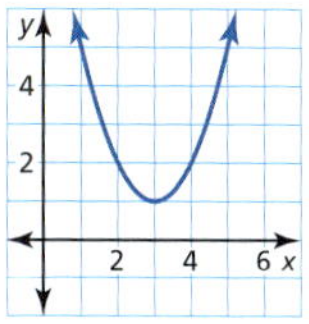

B.

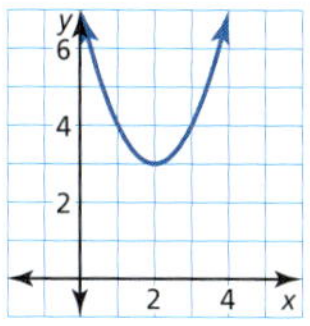

C.

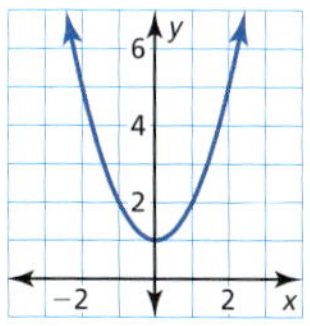

D.

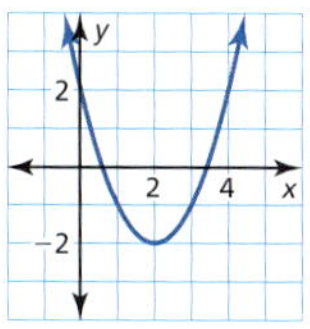

In Exercises 49–54, graph g. *(See Example 4.)*

49. $f(x) = 2(x - 1)^2 + 1$; $g(x) = f(x + 3)$

50. $f(x) = -(x + 1)^2 + 2$; $g(x) = \frac{1}{2}f(x)$

51. $f(x) = -3(x + 5)^2 - 6$; $g(x) = 2f(x)$

52. $f(x) = 5(x - 3)^2 - 1$; $g(x) = f(x) - 6$

53. $f(x) = (x + 3)^2 + 5$; $g(x) = f(x - 4)$

54. $f(x) = -2(x - 4)^2 - 8$; $g(x) = -f(x)$

55. MODELING WITH MATHEMATICS The height (in meters) of a bird diving to catch a fish is represented by $h(t) = 5(t - 2.5)^2$, where t is the number of seconds after beginning the dive.

a. Graph h.

b. Another bird's dive is represented by $r(t) = 2h(t)$. Graph r.

c. Compare the graphs. Which bird starts its dive from a greater height? Explain.

56. MODELING WITH MATHEMATICS A kicker punts a football. The height (in yards) of the football is represented by $f(x) = -\frac{1}{9}(x - 30)^2 + 25$, where x is the horizontal distance (in yards) from the kicker's goal line.

a. Graph f. Describe the domain and range.

b. On the next possession, the kicker punts the football. The height of the football is represented by $g(x) = f(x + 5)$. Graph g. Describe the domain and range.

c. Compare the graphs. On which possession does the kicker punt closer to his goal line? Explain.

In Exercises 57–62, write a quadratic function in vertex form whose graph has the given vertex and passes through the given point.

57. vertex: (1, 2); passes through (3, 10)

58. vertex: (−3, 5); passes through (0, −14)

59. vertex: (−2, −4); passes through (−1, −6)

60. vertex: (1, 8); passes through (3, 12)

61. vertex: (5, −2); passes through (7, 0)

62. vertex: (−5, −1); passes through (−2, 2)

Dynamic Teaching Tools

- Dynamic Assessment & Progress Monitoring Tool
- Interactive Whiteboard Lesson Library
- Dynamic Classroom with Dynamic Investigations

ANSWERS

31. $(-4, -3)$; $x = -4$

32. $(3, 6)$; $x = 3$

33. $(-3, 1)$; $x = -3$

34. $(6, -5)$; $x = 6$

35. C **36.** A

37. D **38.** B

39.

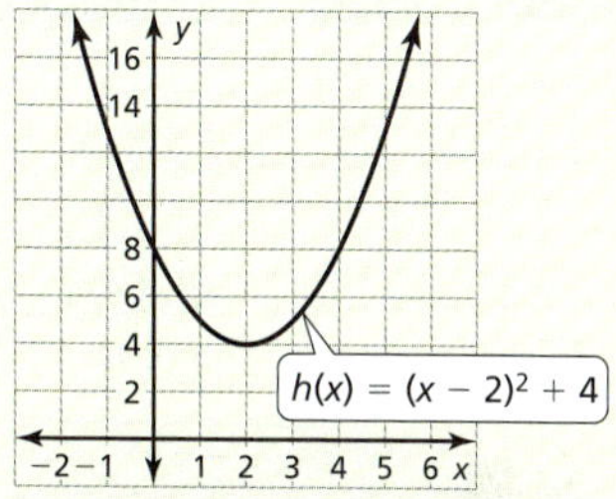

The graph of h is a translation 2 units right and 4 units up of the graph of f.

40.

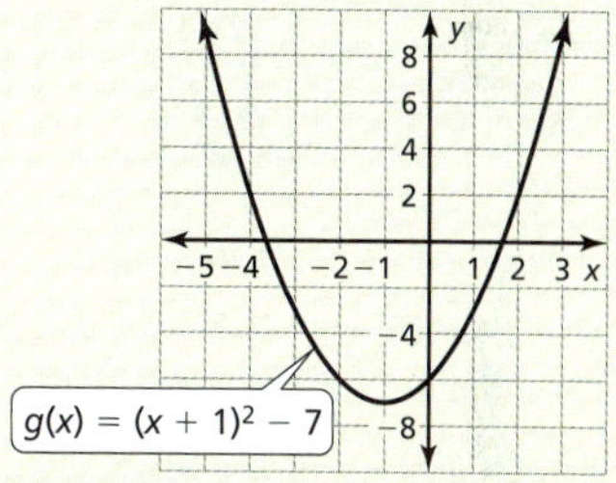

The graph of g is a translation 1 unit left and 7 units down of the graph of f.

41.

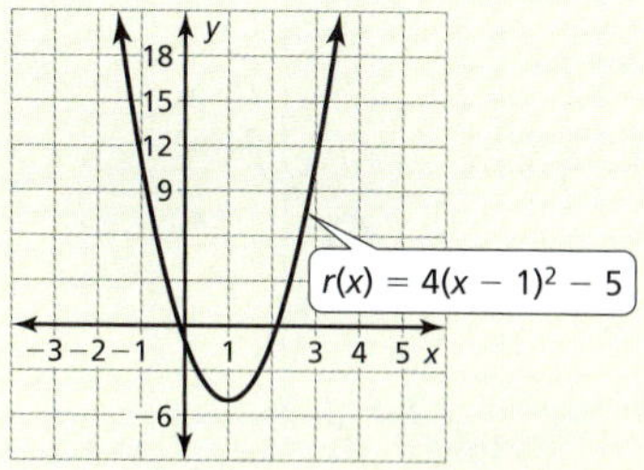

The graph of r is a vertical stretch by a factor of 4, and a translation 1 unit right and 5 units down of the graph of f.

42.

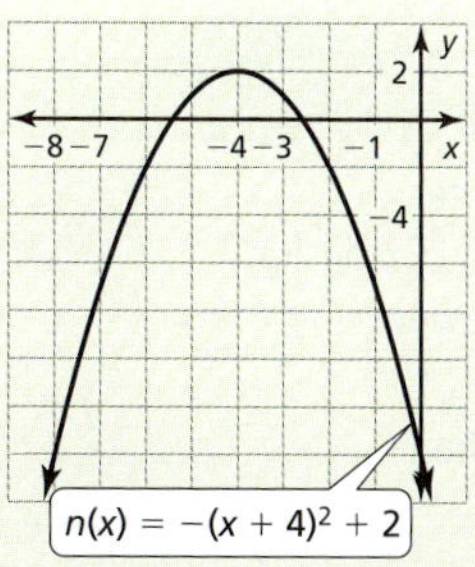

The graph of n is a reflection in the x-axis, and a translation 4 units left and 2 units up of the graph of f.

43–44. See Additional Answers.

45. A **46.** C

47. B **48.** D

49–56. See Additional Answers.

57. $f(x) = 2(x - 1)^2 + 2$

58. $f(x) = -\frac{19}{9}(x + 3)^2 + 5$

59. $f(x) = -2(x + 2)^2 - 4$

60. $f(x) = (x - 1)^2 + 8$

61. $f(x) = \frac{1}{2}(x - 5)^2 - 2$

62. $f(x) = \frac{1}{3}(x + 5)^2 - 1$

ANSWERS

63–78. See Additional Answers.

79. no; *Sample answer:* If the absolute value function includes a horizontal translation, it will not be symmetric about the y-axis.

80. $x = 0, x = 1$

81. $x = -3, x = 8$

82. $x = 3, x = -3$

Mini-Assessment

1. Determine whether the function $f(x) = x^2 + 5$ is *even*, *odd*, or *neither*. even

2. Graph $g(x) = 2(x - 3)^2$. Compare the graph to the graph of $f(x) = x^2$.

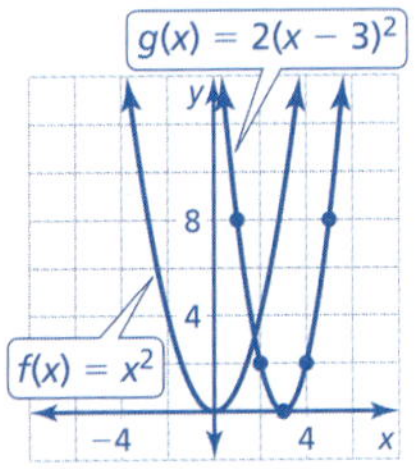

The graph of g is a translation 3 units right and a vertical stretch by a factor of 2 of the graph of f.

3. Graph $g(x) = -(x + 3)^2 - 1$. Compare the graph to the graph of $f(x) = x^2$.

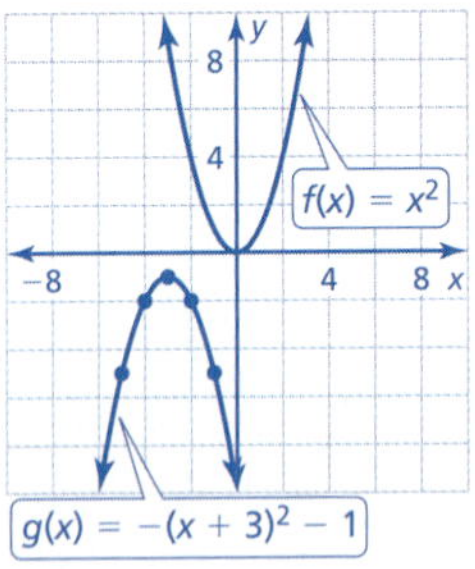

The graph of g is a reflection in the x-axis, and a translation 3 units left and 1 unit down of the graph of f.

4. A toy rocket is launched from the ground and travels in a path shaped like a parabola. Write and graph a quadratic equation that models the path of a rocket that lands 20 feet from where it was launched and reaches a maximum height of 36 feet, represented by the vertex (10, 36). See Additional Answers.

63. MODELING WITH MATHEMATICS A portion of a roller coaster track is in the shape of a parabola. Write and graph a quadratic function that models this portion of the roller coaster with a maximum height of 90 feet, represented by a vertex of (25, 90), passing through the point (50, 0). *(See Example 5.)*

64. MODELING WITH MATHEMATICS A flare is launched from a boat and travels in a parabolic path until reaching the water. Write and graph a quadratic function that models the path of the flare with a maximum height of 300 meters, represented by a vertex of (59, 300), landing in the water at the point (119, 0).

In Exercises 65–68, rewrite the quadratic function in vertex form.

65. $y = 2x^2 - 8x + 4$ **66.** $y = 3x^2 + 6x - 1$

67. $f(x) = -5x^2 + 10x + 3$

68. $f(x) = -x^2 - 4x + 2$

69. REASONING Can a function be symmetric about the x-axis? Explain.

70. HOW DO YOU SEE IT? The graph of a quadratic function is shown. Determine which symbols to use to complete the vertex form of the quadratic function. Explain your reasoning.

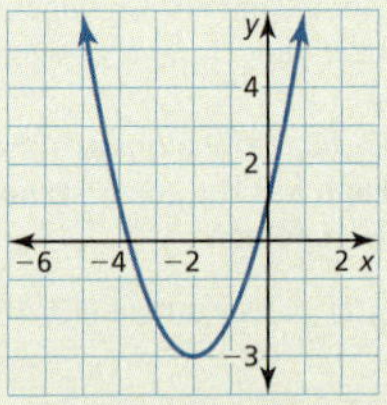

$y = a(x \;\square\; 2)^2 \;\square\; 3$

In Exercises 71–74, describe the transformation from the graph of f to the graph of h. Write an equation that represents h in terms of x.

71. $f(x) = -(x + 1)^2 - 2$
$h(x) = f(x) + 4$

72. $f(x) = 2(x - 1)^2 + 1$
$h(x) = f(x - 5)$

73. $f(x) = 4(x - 2)^2 + 3$
$h(x) = 2f(x)$

74. $f(x) = -(x + 5)^2 - 6$
$h(x) = \frac{1}{3}f(x)$

75. REASONING The graph of $y = x^2$ is translated 2 units right and 5 units down. Write an equation for the function in vertex form and in standard form. Describe advantages of writing the function in each form.

76. THOUGHT PROVOKING Which of the following are true? Justify your answers.

a. Any constant multiple of an even function is even.

b. Any constant multiple of an odd function is odd.

c. The sum or difference of two even functions is even.

d. The sum or difference of two odd functions is odd.

e. The sum or difference of an even function and an odd function is odd.

77. COMPARING FUNCTIONS A cross section of a birdbath can be modeled by $y = \frac{1}{81}(x - 18)^2 - 4$, where x and y are measured in inches. The graph shows the cross section of another birdbath.

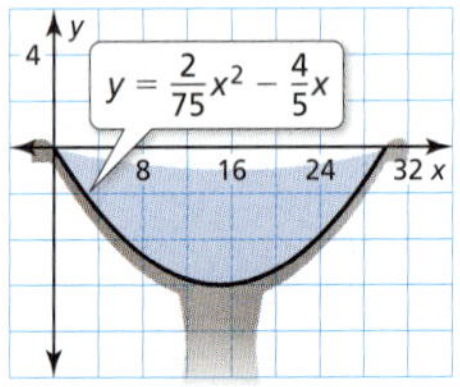

a. Which birdbath is deeper? Explain.

b. Which birdbath is wider? Explain.

78. REASONING Compare the graphs of $y = 2x^2 + 8x + 8$ and $y = x^2$ without graphing the functions. How can factoring help you compare the parabolas? Explain.

79. MAKING AN ARGUMENT Your friend says all absolute value functions are even because of their symmetry. Is your friend correct? Explain.

Maintaining Mathematical Proficiency

Reviewing what you learned in previous grades and lessons

Solve the equation. *(Section 7.4)*

80. $x(x - 1) = 0$ **81.** $(x + 3)(x - 8) = 0$ **82.** $(3x - 9)(4x + 12) = 0$

If students need help...	*If students got it...*
Resources by Chapter • Practice A and Practice B • Puzzle Time	Resources by Chapter • Enrichment and Extension • Cumulative Review
Student Journal • Practice	Start the *next* Section
Differentiating the Lesson Skills Review Handbook	

Laurie's Notes

Dynamic Teaching Tools

- Dynamic Assessment & Progress Monitoring Tool
- Lesson Planning Tool
- Interactive Whiteboard Lesson Library
- Dynamic Classroom with Dynamic Investigations

Overview of Section 8.5

Introduction

- Earlier in the chapter, the standard form and vertex form of a quadratic were studied. In this lesson, the intercept form is presented and connections are made to the zeros of a function.
- The lesson begins with students graphing a quadratic written in intercept form. Then students are asked to use the intercepts to find the zeros of the function.
- The second half of the lesson focuses on the characteristics of quadratic and cubic functions so that the equations can be written and the functions graphed.

Formative Assessment Tips

- **Agree-Disagree Statement:** This technique has two parts. First, students are given a statement with which they may agree or disagree, or they could indicate that they need additional information in order to decide. They are then asked to explain why they agree or disagree, or why they might need additional information. The second part of the technique involves students describing how they could investigate, figure out, or test their thinking.
- This technique gives students the opportunity to think about their own understanding of a concept or process. Further, it helps students practice the skill of actively investigating their own thinking. In listening to their own words, and in listening to the thinking of others, students can solidify or modify their own beliefs.
- This technique can be used at the beginning of a class or unit of study to assess students' understanding of a concept. Student responses will help inform your own instruction, knowing what experiences students need to help guide their learning.
- *Note:* The statements made are not low cognitive demand questions such as true-false or yes-no questions. The goal is to uncover student conception about bigger ideas, such as whether all functions have an inverse or whether multiplying positive numbers gives a product greater than either factor.

Pacing Suggestion

- This is a long lesson with the explorations helping students to make the connection between the zeros of a function and the *x*-intercepts of its graph. Refer to the explorations as you work through the formal lesson.

What Your Students Will Learn

- Graph quadratic functions of the form $f(x) = a(x - p)(x - q)$ by identifying the x-intercepts and axis of symmetry.
- Use the intercept form of a quadratic function to find the zeros of the function.
- Use characteristics to graph quadratic functions and to write quadratic functions in standard form or intercept form.
- Use characteristics to graph cubic functions and to write cubic functions in standard form or intercept form.

Laurie's Notes

Exploration

Motivate

- To motivate the explorations, share the graphs of two quadratic functions: $f(x) = (x - 3)^2$ and $g(x) = -(x + 5)^2$.
- Ask students to identify the vertex of each: (3, 0) and (−5, 0).

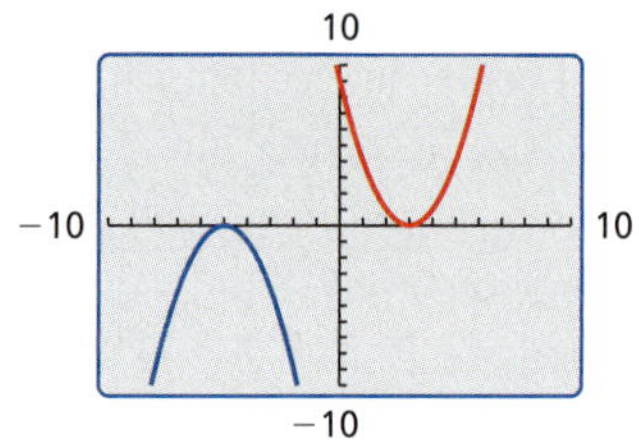

? "What are the zeros of each function?" 3 and −5, respectively

- Probe students to think of other ways in which the functions could be written, besides standard form.
- Write: $f(x) = (x - 3)^2 =$ ________ • ________. In other words, $f(x) = (x - 3)^2 = (x - 3)(x - 3)$. Say, "When $x = 3$, the function equals 0. So the zero of the function is the x-coordinate of the vertex."
- Ask students to represent $g(x)$ as the product of two factors. Interpret the vertex and zero of the function.

Exploration Note

- Before having students begin the exploration, be sure to complete the *Motivate* activity. This should be enough of an introduction for students to successfully explore the graphs shown.

Exploration 1

- Students can determine the x- and y-intercepts of each graph. Assure students that this is sufficient information in order to write the equations in the form $f(x) = (x - p)(x - q)$ or $f(x) = -(x - p)(x - q)$.
- Students should check their equations using a graphing calculator.
- **Reason Abstractly and Quantitatively:** Listen for correct reasoning as to why their equations make sense.

Communicate Your Answer

- **Use Appropriate Tools Strategically:** Have students explore the questions with their partners using a graphing calculator.

Connecting to Next Step

- The explorations help students make sense of the intercept form of a quadratic. In the formal lesson, this is stated along with how to find the axis of symmetry when the quadratic is written in the intercept form.

8.5 Using Intercept Form

Essential Question What are some of the characteristics of the graph of $f(x) = a(x - p)(x - q)$?

EXPLORATION 1 Using Zeros to Write Functions

Work with a partner. Each graph represents a function of the form $f(x) = (x - p)(x - q)$ or $f(x) = -(x - p)(x - q)$. Write the function represented by each graph. Explain your reasoning.

a.
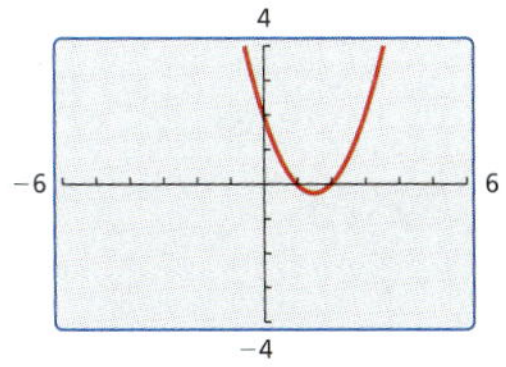

b.
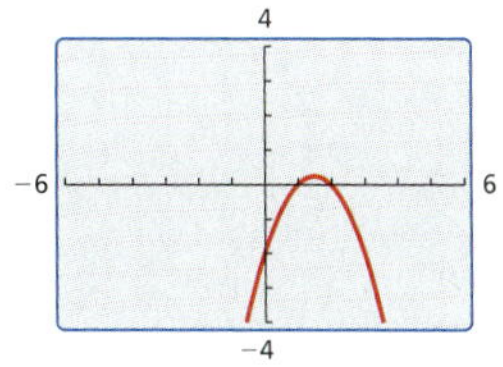

c.
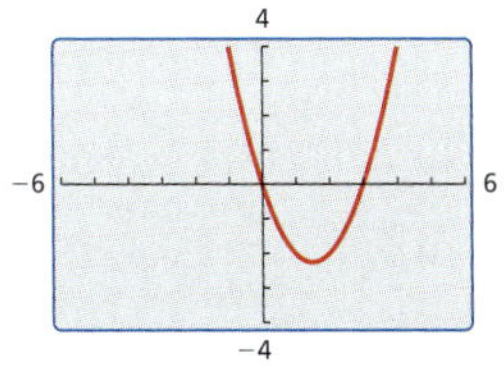

d.
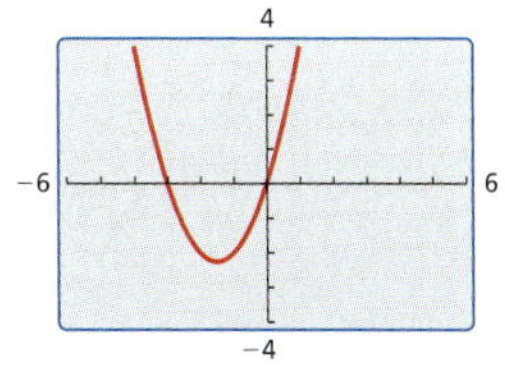

e.
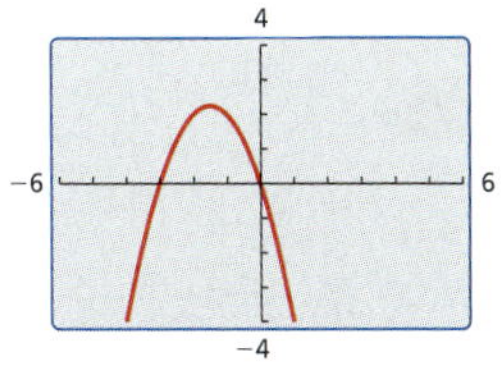

f.
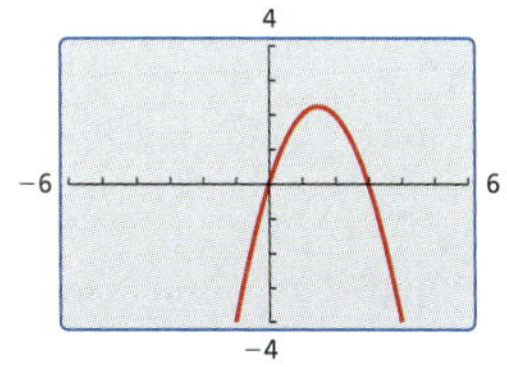

g.
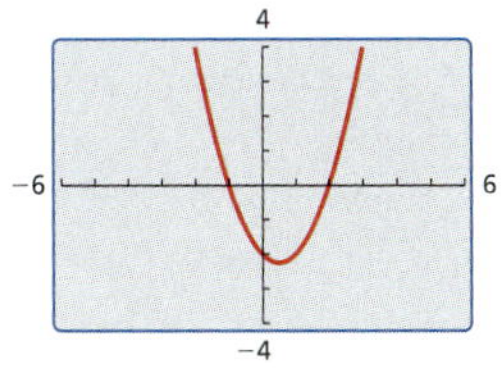

h.
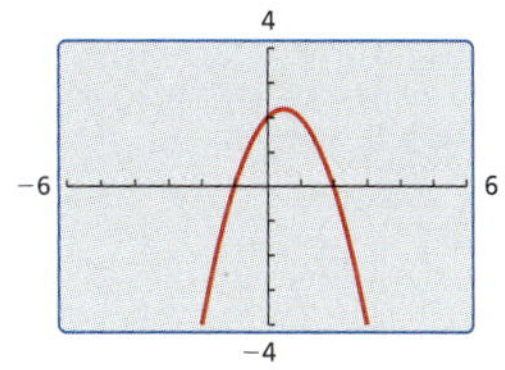

CONSTRUCTING VIABLE ARGUMENTS

To be proficient in math, you need to justify your conclusions and communicate them to others.

Communicate Your Answer

2. What are some of the characteristics of the graph of $f(x) = a(x - p)(x - q)$?
3. Consider the graph of $f(x) = a(x - p)(x - q)$.
   a. Does changing the sign of a change the x-intercepts? Does changing the sign of a change the y-intercept? Explain your reasoning.
   b. Does changing the value of p change the x-intercepts? Does changing the value of p change the y-intercept? Explain your reasoning.

Dynamic Teaching Tools

- Dynamic Assessment & Progress Monitoring Tool
- Lesson Planning Tool
- Interactive Whiteboard Lesson Library
- Dynamic Classroom with Dynamic Investigations

ANSWERS

1. a. $y = (x - 1)(x - 2)$; *Sample answer:* The graph opens up and the x-intercepts are 1 and 2.
   b. $y = -(x - 1)(x - 2)$; *Sample answer:* The graph opens down and the x-intercepts are 1 and 2.
   c. $y = x(x - 3)$; *Sample answer:* The graph opens up and the x-intercepts are 0 and 3.
   d. $y = x(x + 3)$; *Sample answer:* The graph opens up and the x-intercepts are 0 and -3.
   e. $y = -x(x + 3)$; *Sample answer:* The graph opens down and the x-intercepts are 0 and -3.
   f. $y = -x(x - 3)$; *Sample answer:* The graph opens down and the x-intercepts are 0 and 3.
   g. $y = (x + 1)(x - 2)$; *Sample answer:* The graph opens up and the x-intercepts are -1 and 2.
   h. $y = -(x + 1)(x - 2)$; *Sample answer:* The graph opens down and the x-intercepts are -1 and 2.
2. *Sample answer:* It is a parabola with x-intercepts p and q.
3. a. no; yes; *Sample answer:* The x-intercepts are still p and q; Changing the sign of a will also change the sign of the y-intercept when the y-intercept is not 0.
   b. yes; yes; *Sample answer:* One of the x-intercepts will change from the old value of p to the new value of p. The y-intercept is apq, so changing the value of p will change the value of the y-intercept when the y-intercept is not 0.

8.5 Lesson

Core Vocabulary

intercept form, *p. 450*

What You Will Learn

- Graph quadratic functions of the form $f(x) = a(x - p)(x - q)$.
- Use intercept form to find zeros of functions.
- Use characteristics to graph and write quadratic functions.
- Use characteristics to graph and write cubic functions.

Graphing $f(x) = a(x - p)(x - q)$

You have already graphed quadratic functions written in several different forms, such as $f(x) = ax^2 + bx + c$ (standard form) and $g(x) = a(x - h)^2 + k$ (vertex form). Quadratic functions can also be written in **intercept form**, $f(x) = a(x - p)(x - q)$, where $a \neq 0$. In this form, the polynomial that defines a function is in factored form and the x-intercepts of the graph can be easily determined.

Core Concept

Graphing $f(x) = a(x - p)(x - q)$

- The x-intercepts are p and q.
- The axis of symmetry is halfway between $(p, 0)$ and $(q, 0)$. So, the axis of symmetry is $x = \frac{p + q}{2}$.
- The graph opens up when $a > 0$, and the graph opens down when $a < 0$.

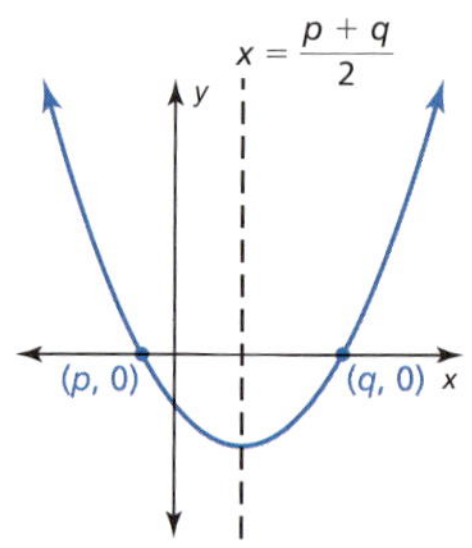

EXAMPLE 1 Graphing $f(x) = a(x - p)(x - q)$

Graph $f(x) = -(x + 1)(x - 5)$. Describe the domain and range.

SOLUTION

Step 1 Identify the x-intercepts. Because the x-intercepts are $p = -1$ and $q = 5$, plot $(-1, 0)$ and $(5, 0)$.

Step 2 Find and graph the axis of symmetry.

$$x = \frac{p + q}{2} = \frac{-1 + 5}{2} = 2$$

Step 3 Find and plot the vertex.

The x-coordinate of the vertex is 2. To find the y-coordinate of the vertex, substitute 2 for x and simplify.

$$f(2) = -(2 + 1)(2 - 5) = 9$$

So, the vertex is $(2, 9)$.

Step 4 Draw a parabola through the vertex and the points where the x-intercepts occur.

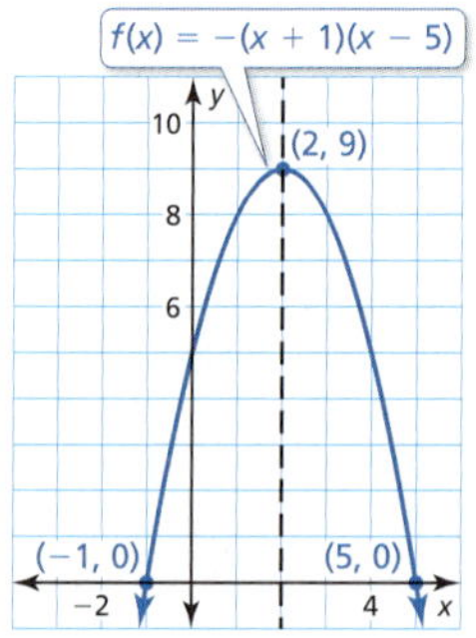

▶ The domain is all real numbers. The range is $y \leq 9$.

Differentiated Instruction

Kinesthetic

Ask two students to assist you at the board to solve a problem similar to Example 1 or Example 2. Assign finding the x-intercepts, the axis of symmetry, and the vertex of a function to one student. Assign graphing the x-intercepts, axis of symmetry, the vertex, and the parabola to the second student. Have students discuss how the intercepts, axis, and vertex relate to the function and its graph.

Extra Example 1

Graph $f(x) = (x - 1)(x + 3)$. Describe the domain and range.

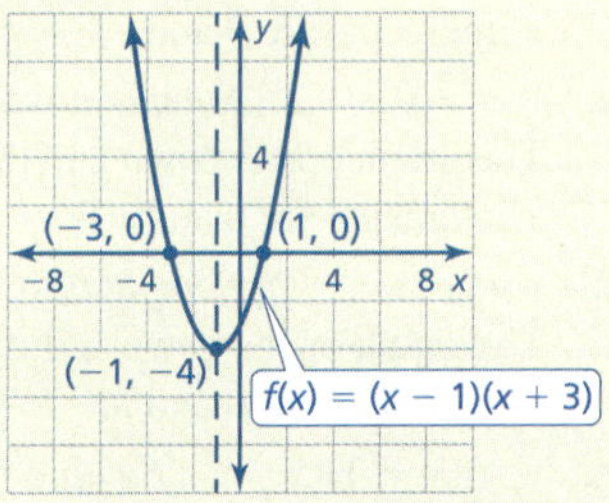

The domain is all real numbers. The range is $y \geq -4$.

Laurie's Notes — Teacher Actions

? Agree-Disagree Statement: "All quadratic functions can be written $f(x) = (x - p)(x - q)$, where p and q are any real numbers. Agree/Disagree. Explain." Students should consider work from Section 7.3 and the explorations on page 449. What may concern students is the parent function and writing $f(x) = (x - 0)(x - 0)$. They may also wonder "What if the quadratic doesn't have x-intercepts?"

- Write the *Core Concept*. For quadratics that do have x-intercepts, discuss the equation, the axis of symmetry, and the effect of the constant a.

EXAMPLE 2 Graphing a Quadratic Function

Graph $f(x) = 2x^2 - 8$. Describe the domain and range.

SOLUTION

Step 1 Rewrite the quadratic function in intercept form.

$f(x) = 2x^2 - 8$ — Write the function.

$= 2(x^2 - 4)$ — Factor out common factor.

$= 2(x + 2)(x - 2)$ — Difference of two squares pattern

Step 2 Identify the x-intercepts. Because the x-intercepts are $p = -2$ and $q = 2$, plot $(-2, 0)$ and $(2, 0)$.

Step 3 Find and graph the axis of symmetry.

$$x = \frac{p + q}{2} = \frac{-2 + 2}{2} = 0$$

Step 4 Find and plot the vertex.

The x-coordinate of the vertex is 0.
The y-coordinate of the vertex is

$$f(0) = 2(0)^2 - 8 = -8.$$

So, the vertex is $(0, -8)$.

Step 5 Draw a parabola through the vertex and the points where the x-intercepts occur.

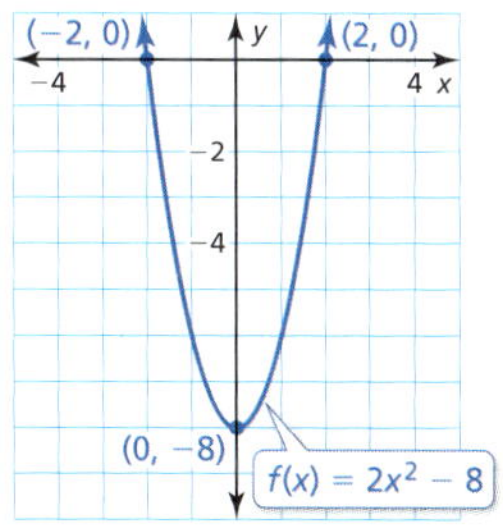

▶ The domain is all real numbers. The range is $y \geq -8$.

Monitoring Progress

Help in English and Spanish at *BigIdeasMath.com*

Graph the quadratic function. Label the vertex, axis of symmetry, and x-intercepts. Describe the domain and range of the function.

1. $f(x) = (x + 2)(x - 3)$ **2.** $g(x) = -2(x - 4)(x + 1)$ **3.** $h(x) = 4x^2 - 36$

REMEMBER

Functions have zeros, and *graphs* have x-intercepts.

Using Intercept Form to Find Zeros of Functions

In Section 8.2, you learned that a zero of a function is an x-value for which $f(x) = 0$. You can use the intercept form of a function to find the zeros of the function.

EXAMPLE 3 Finding Zeros of a Function

Find the zeros of $f(x) = (x - 1)(x + 2)$.

SOLUTION

To find the zeros, determine the x-values for which $f(x)$ is 0.

$f(x) = (x - 1)(x + 2)$ — Write the function.

$0 = (x - 1)(x + 2)$ — Substitute 0 for $f(x)$.

$x - 1 = 0$ *or* $x + 2 = 0$ — Zero-Product Property

$x = 1$ *or* $x = -2$ — Solve for x.

▶ So, the zeros of the function are -2 and 1.

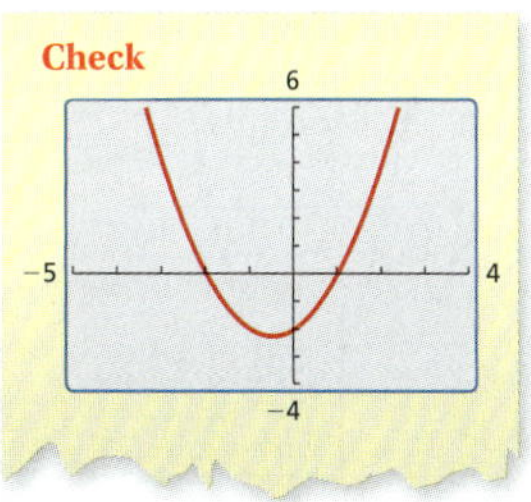

English Language Learners

Build on Past Knowledge

Remind students that a zero of a function $f(x)$ is a value of x that makes $f(x) = 0$, not the value $f(0)$.

Extra Example 2

Graph $f(x) = 3x^2 - 3$. Describe the domain and range.

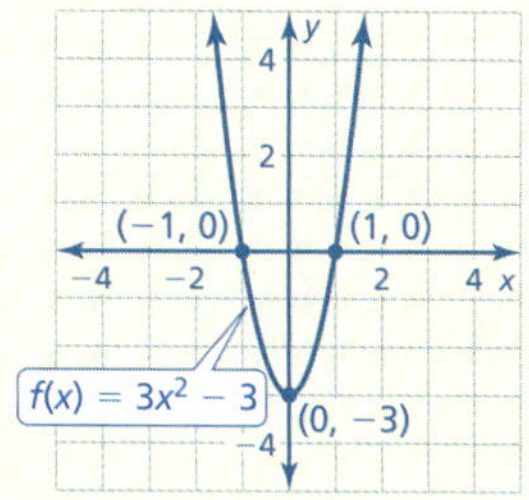

The domain is all real numbers. The range is $y \geq -3$.

Extra Example 3

Find the zeros of $f(x) = (x - 5)(x + 7)$.
The zeros of the function are -7 and 5.

MONITORING PROGRESS ANSWERS

1.

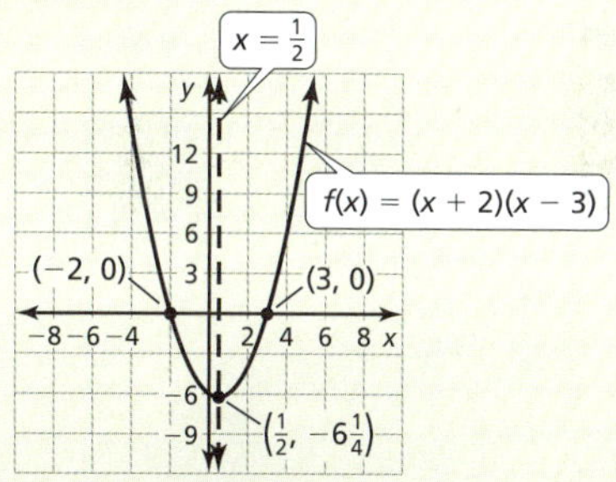

domain: all real numbers, range: $y \geq -6\frac{1}{4}$

2.

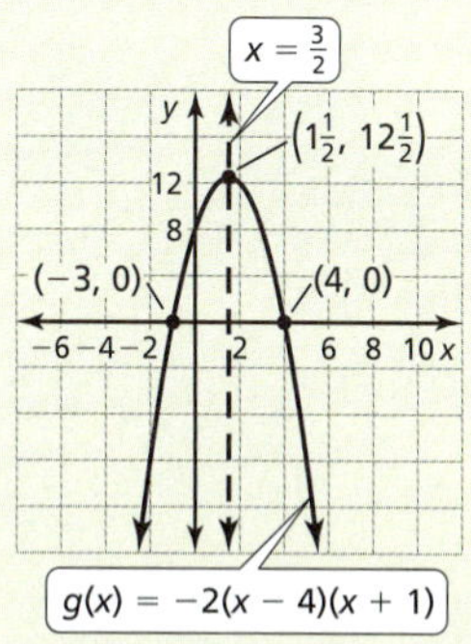

domain: all real numbers, range: $y \leq 12\frac{1}{2}$

3. See Additional Answers.

Laurie's Notes Teacher Actions

- ? "What do you know about the graph of $f(x) = 2x^2 - 8$?" It opens upward, is stretched vertically, and is shifted down 8 units. The vertex is $(0, -8)$.
- ? "Do you know the x-intercepts or could you determine them?" Factor $2x^2 - 8$ into $2(x - 2)(x + 2)$ and you have x-intercepts of $x = 2$ and $x = -2$.
- **Think-Pair-Share:** Have students answer Questions 1–3, and then share and discuss as a class.
- Say, "Recall the connection between zeros of a function and x-intercepts of a graph." Have students work independently on Example 3.

Extra Example 4

Find the zeros of each function.

a. $f(x) = 2x^2 - 8x - 24$

The zeros of the function are -2 and 6.

b. $h(x) = (x - 2)(x^2 - 25)$

The zeros of the function are -5, 2, and 5.

Extra Example 5

Use zeros to graph $h(x) = x^2 + 3x - 4$.

The zeros of the function are -4 and 1.

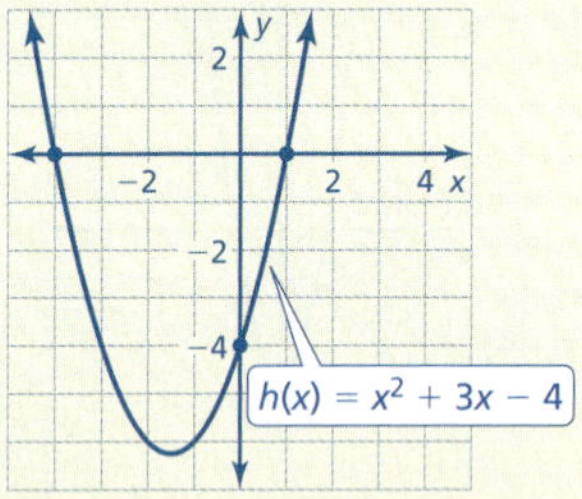

MONITORING PROGRESS ANSWERS

4. 1, 6

5. 2

6. $-1, 0, 1$

Core Concept

Factors and Zeros

For any factor $x - n$ of a polynomial, n is a zero of the function defined by the polynomial.

EXAMPLE 4 Finding Zeros of Functions

Find the zeros of each function.

a. $f(x) = -2x^2 - 10x - 12$ **b.** $h(x) = (x - 1)(x^2 - 16)$

SOLUTION

Write each function in intercept form to identify the zeros.

a. $f(x) = -2x^2 - 10x - 12$ Write the function.

$= -2(x^2 + 5x + 6)$ Factor out common factor.

$= -2(x + 3)(x + 2)$ Factor the trinomial.

▶ So, the zeros of the function are -3 and -2.

b. $h(x) = (x - 1)(x^2 - 16)$ Write the function.

$= (x - 1)(x + 4)(x - 4)$ Difference of two squares pattern

▶ So, the zeros of the function are -4, 1, and 4.

LOOKING FOR STRUCTURE

The function in Example 4(b) is called a *cubic function*. You can extend the concept of intercept form to cubic functions. You will graph a cubic function in Example 7.

Monitoring Progress

Help in English and Spanish at *BigIdeasMath.com*

Find the zero(s) of the function.

4. $f(x) = (x - 6)(x - 1)$ **5.** $g(x) = 3x^2 - 12x + 12$ **6.** $h(x) = x(x^2 - 1)$

Using Characteristics to Graph and Write Quadratic Functions

EXAMPLE 5 Graphing a Quadratic Function Using Zeros

Use zeros to graph $h(x) = x^2 - 2x - 3$.

SOLUTION

The function is in standard form. You know that the parabola opens up $(a > 0)$ and the y-intercept is -3. So, begin by plotting $(0, -3)$.

Notice that the polynomial that defines the function is factorable. So, write the function in intercept form and identify the zeros.

$h(x) = x^2 - 2x - 3$ Write the function.

$= (x + 1)(x - 3)$ Factor the trinomial.

The zeros of the function are -1 and 3. So, plot $(-1, 0)$ and $(3, 0)$. Draw a parabola through the points.

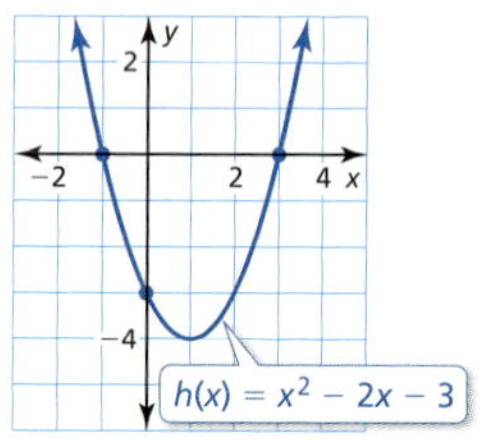

ATTENDING TO PRECISION

To sketch a more precise graph, make a table of values and plot other points on the graph.

Laurie's Notes Teacher Actions

- **Connection:** Students need to recall factoring polynomials in Chapter 7 to find the zeros of a function.
- **Wait Time** and **Whiteboarding:** Pose Example 4. Give *Wait Time* while students work with partners to solve. Use whiteboards for partners to record work. In part (a), there are several equivalent forms.
- ? **Agree-Disagree Statement:** "If you know the zeros of a quadratic, you can graph the function. Agree/Disagree. Explain." no; There are infinitely many quadratics that have the same zeros.
- In Example 5, the equation is known, so finding the zeros aids in graphing.

EXAMPLE 6 Writing Quadratic Functions

STUDY TIP

In part (a), many possible functions satisfy the given condition. The value a can be *any* nonzero number. To allow easier calculations, let $a = 1$. By letting $a = 2$, the resulting function would be $f(x) = 2x^2 + 12x + 22$.

Write a quadratic function in standard form whose graph satisfies the given condition(s).

a. vertex: $(-3, 4)$

b. passes through $(-9, 0)$, $(-2, 0)$, and $(-4, 20)$

SOLUTION

a. Because you know the vertex, use vertex form to write a function.

$f(x) = a(x - h)^2 + k$	Vertex form
$= 1(x + 3)^2 + 4$	Substitute for *a*, *h*, and *k*.
$= x^2 + 6x + 9 + 4$	Find the product $(x + 3)^2$.
$= x^2 + 6x + 13$	Combine like terms.

b. The given points indicate that the x-intercepts are -9 and -2. So, use intercept form to write a function.

$f(x) = a(x - p)(x - q)$	Intercept form
$= a(x + 9)(x + 2)$	Substitute for *p* and *q*.

Use the other given point, $(-4, 20)$, to find the value of a.

$20 = a(-4 + 9)(-4 + 2)$	Substitute -4 for *x* and 20 for *f*(*x*).
$20 = a(5)(-2)$	Simplify.
$-2 = a$	Solve for *a*.

Use the value of a to write the function.

$f(x) = -2(x + 9)(x + 2)$	Substitute -2 for *a*.
$= -2x^2 - 22x - 36$	Simplify.

Monitoring Progress

Help in English and Spanish at *BigIdeasMath.com*

Use zeros to graph the function.

7. $f(x) = (x - 1)(x - 4)$ **8.** $g(x) = x^2 + x - 12$

Write a quadratic function in standard form whose graph satisfies the given condition(s).

9. x-intercepts: -1 and 1 **10.** vertex: $(8, 8)$

11. passes through $(0, 0)$, $(10, 0)$, and $(4, 12)$

12. passes through $(-5, 0)$, $(4, 0)$, and $(3, -16)$

Using Characteristics to Graph and Write Cubic Functions

In Example 4, you extended the concept of intercept form to cubic functions.

$f(x) = a(x - p)(x - q)(x - r),\ a \neq 0$ Intercept form of a cubic function

The x-intercepts of the graph of f are p, q, and r.

Extra Example 6

Write a quadratic function in standard form whose graph satisfies the given condition(s).

a. x-intercepts: -1 and 5

Sample answer: $f(x) = x^2 - 4x - 5$

b. vertex: $(-3, -2)$

Sample answer: $f(x) = x^2 + 6x + 7$

MONITORING PROGRESS ANSWERS

7.

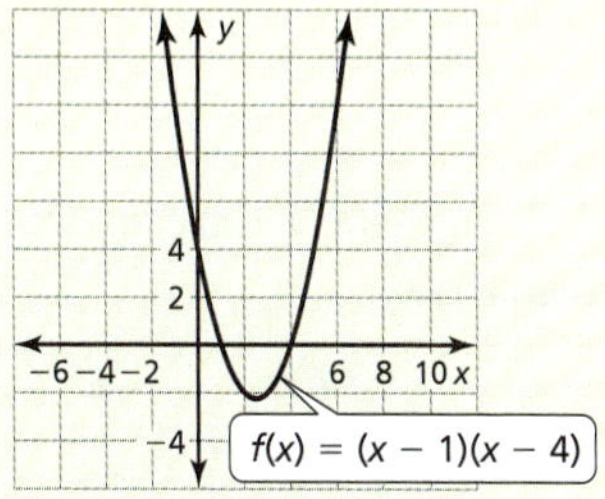

8.

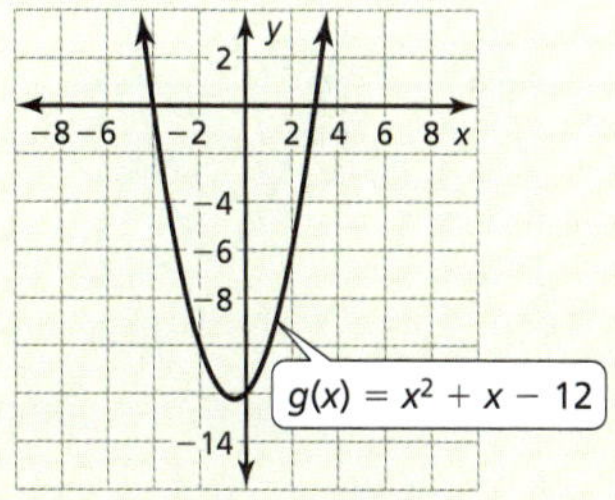

9. *Sample answer:* $f(x) = x^2 - 1$

10. *Sample answer:* $f(x) = x^2 - 16x + 72$

11. $f(x) = -\frac{1}{2}x^2 + 5x$

12. $f(x) = 2x^2 + 2x - 40$

Laurie's Notes Teacher Actions

- Example 6 helps students focus on the three forms of a quadratic function and what information is known about the graph from each equation.
- **Three-Minute Pause:** Work through both parts of the example as shown. Give students time to discuss the example with partners and reflect on their understanding. Then have students self-assess and determine what equations are needed to answer Questions 9–12.
- **Look For and Make Use of Structure:** Help students understand that they have now written a cubic polynomial in intercept form.

Extra Example 7

Use zeros to graph $f(x) = -x^3 + 4x^2 - 3x$.

The zeros of the function are 0, 1, and 3.

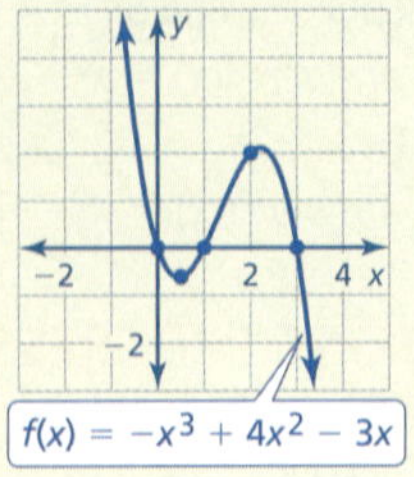

Extra Example 8

The graph represents a cubic function. Write the function.

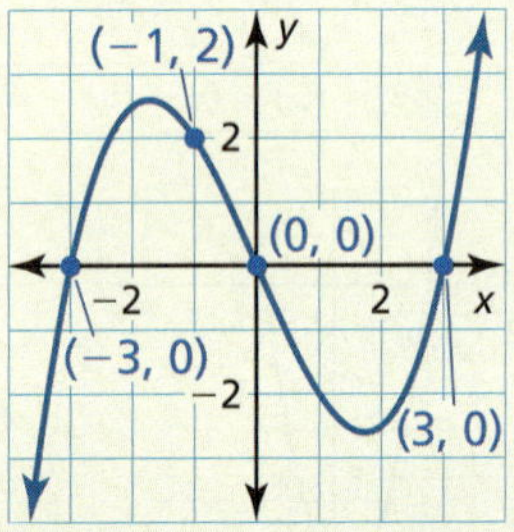

The function represented by the graph is $f(x) = \frac{1}{4}(x^3 - 9x)$.

MONITORING PROGRESS ANSWERS

13.

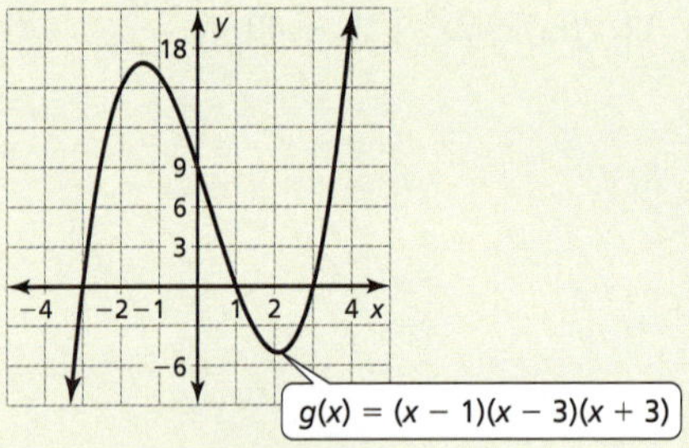

14.

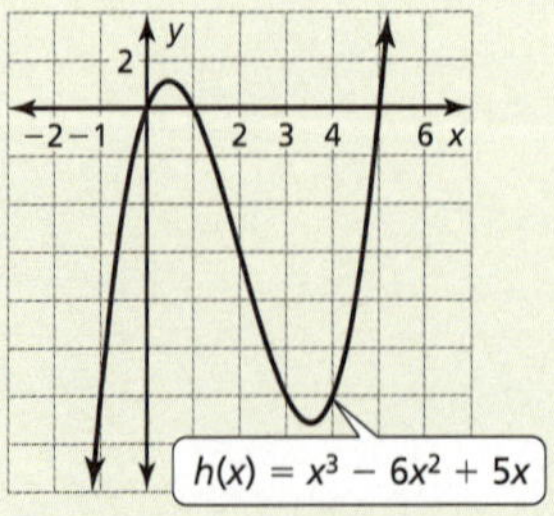

15. $f(x) = x^3 + 3x^2 - x - 3$

EXAMPLE 7 Graphing a Cubic Function Using Zeros

Use zeros to graph $f(x) = x^3 - 4x$.

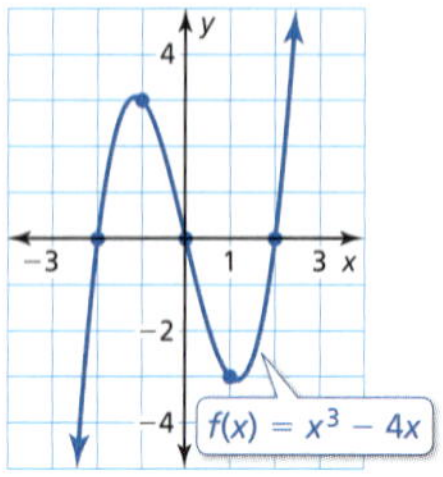

SOLUTION

Notice that the polynomial that defines the function is factorable. So, write the function in intercept form and identify the zeros.

$f(x) = x^3 - 4x$	Write the function.
$= x(x^2 - 4)$	Factor out x.
$= x(x + 2)(x - 2)$	Difference of two squares pattern

The zeros of the function are −2, 0, and 2. So, plot (−2, 0), (0, 0), and (2, 0).

To help determine the shape of the graph, find points between the zeros.

x	−1	1
$f(x)$	3	−3

Plot (−1, 3) and (1, −3). Draw a smooth curve through the points.

EXAMPLE 8 Writing a Cubic Function

The graph represents a cubic function. Write the function.

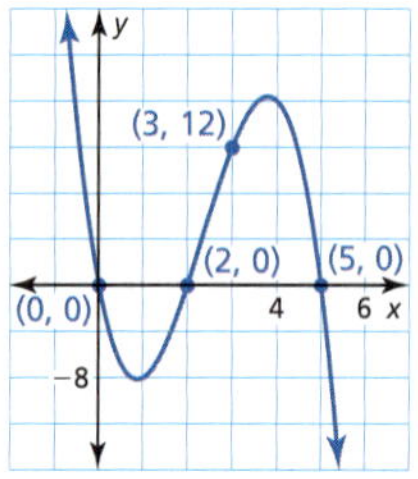

SOLUTION

From the graph, you can see that the x-intercepts are 0, 2, and 5. Use intercept form to write a function.

$f(x) = a(x - p)(x - q)(x - r)$	Intercept form
$= a(x - 0)(x - 2)(x - 5)$	Substitute for p, q, and r.
$= a(x)(x - 2)(x - 5)$	Simplify.

Use the other given point, (3, 12), to find the value of a.

$12 = a(3)(3 - 2)(3 - 5)$	Substitute 3 for x and 12 for $f(x)$.
$-2 = a$	Solve for a.

Use the value of a to write the function.

$f(x) = -2(x)(x - 2)(x - 5)$	Substitute −2 for a.
$= -2x^3 + 14x^2 - 20x$	Simplify.

▶ The function represented by the graph is $f(x) = -2x^3 + 14x^2 - 20x$.

Monitoring Progress

Help in English and Spanish at *BigIdeasMath.com*

Use zeros to graph the function.

13. $g(x) = (x - 1)(x - 3)(x + 3)$

14. $h(x) = x^3 - 6x^2 + 5x$

15. The zeros of a cubic function are −3, −1, and 1. The graph of the function passes through the point (0, −3). Write the function.

Laurie's Notes Teacher Actions

- **? Turn and Talk:** "How can you use zeros to graph $f(x) = x^3 - 4x$?" Listen for students deciding how to factor $x^3 - 4x$. Knowing the zeros is not enough. "What additional information is needed?" points between the zeros
- In Example 8, students will need to read the ordered pair (3, 12) from the graph to finish writing the equation.

Closure

- **Exit Ticket:** Use zeros to sketch the graphs of $f(x) = -(x + 2)(x - 3)$ and $g(x) = x^2 + 7x + 10$. Check students' work.

8.5 Exercises

Dynamic Solutions available at *BigIdeasMath.com*

Vocabulary and Core Concept Check

1. **COMPLETE THE SENTENCE** The values p and q are __________ of the graph of the function $f(x) = a(x - p)(x - q)$.

2. **WRITING** Explain how to find the maximum value or minimum value of a quadratic function when the function is given in intercept form.

Monitoring Progress and Modeling with Mathematics

In Exercises 3–6, find the x-intercepts and axis of symmetry of the graph of the function.

3. 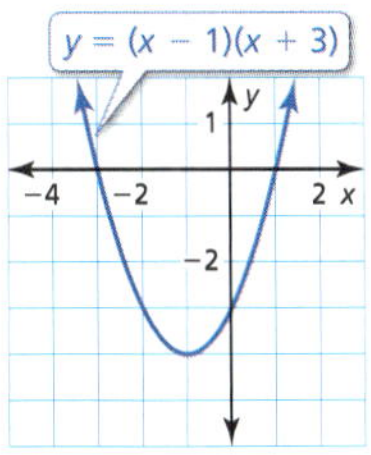

4. 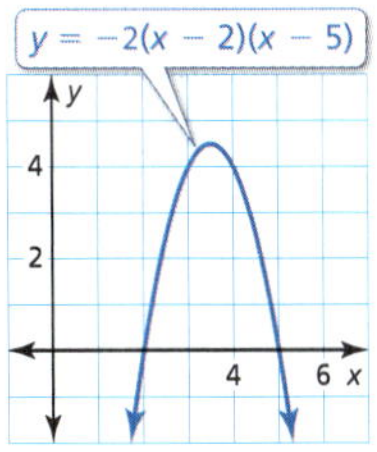

5. $f(x) = -5(x + 7)(x - 5)$
6. $g(x) = \frac{2}{3}x(x + 8)$

In Exercises 7–12, graph the quadratic function. Label the vertex, axis of symmetry, and x-intercepts. Describe the domain and range of the function. *(See Example 1.)*

7. $f(x) = (x + 4)(x + 1)$
8. $y = (x - 2)(x + 2)$
9. $y = -(x + 6)(x - 4)$
10. $h(x) = -4(x - 7)(x - 3)$
11. $g(x) = 5(x + 1)(x + 2)$
12. $y = -2(x - 3)(x + 4)$

In Exercises 13–20, graph the quadratic function. Label the vertex, axis of symmetry, and x-intercepts. Describe the domain and range of the function. *(See Example 2.)*

13. $y = x^2 - 9$
14. $f(x) = x^2 - 8x$
15. $h(x) = -5x^2 + 5x$
16. $y = 3x^2 - 48$
17. $q(x) = x^2 + 9x + 14$
18. $p(x) = x^2 + 6x - 27$
19. $y = 4x^2 - 36x + 32$
20. $y = -2x^2 - 4x + 30$

In Exercises 21–30, find the zero(s) of the function. *(See Examples 3 and 4.)*

21. $y = -2(x - 2)(x - 10)$
22. $f(x) = \frac{1}{3}(x + 5)(x - 1)$
23. $g(x) = x^2 + 5x - 24$
24. $y = x^2 - 17x + 52$
25. $y = 3x^2 - 15x - 42$
26. $g(x) = -4x^2 - 8x - 4$
27. $f(x) = (x + 5)(x^2 - 4)$
28. $h(x) = (x^2 - 36)(x - 11)$
29. $y = x^3 - 49x$
30. $y = x^3 - x^2 - 9x + 9$

In Exercises 31–36, match the function with its graph.

31. $y = (x + 5)(x + 3)$
32. $y = (x + 5)(x - 3)$
33. $y = (x - 5)(x + 3)$
34. $y = (x - 5)(x - 3)$
35. $y = (x + 5)(x - 5)$
36. $y = (x + 3)(x - 3)$

A. 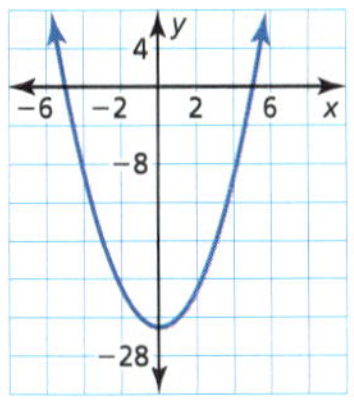

B. 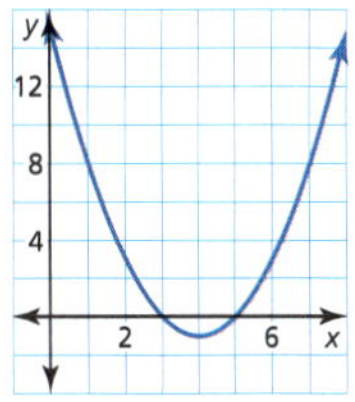

C. 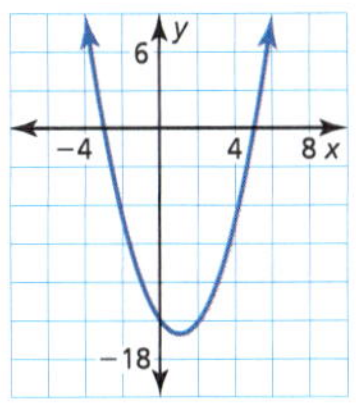

D. 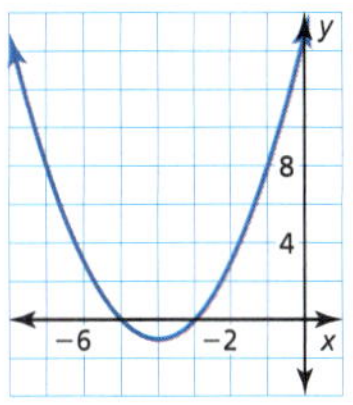

E. 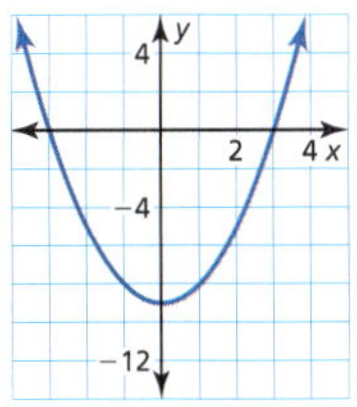

F. 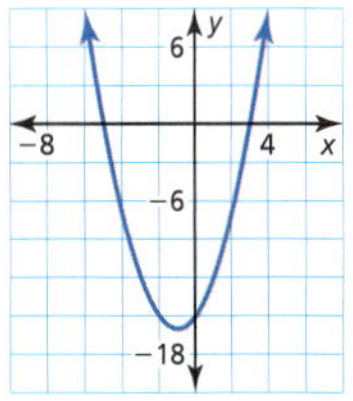

Assignment Guide and Homework Check

ASSIGNMENT

Basic: 1, 2, 3–71 odd, 85, 92, 96, 102–108

Average: 1, 2–92 even, 96, 102–108

Advanced: 1, 2, 10–44 even, 52–60 even, 66–90 even, 91–108

HOMEWORK CHECK

Basic: 7, 21, 45, 57, 61

Average: 8, 24, 46, 58, 62

Advanced: 10, 24, 52, 58, 66

ANSWERS

1. x-intercepts
2. *Sample answer:* Find the x-coordinate of the vertex using $x = \frac{p + q}{2}$, then evaluate the function to determine the maximum value or minimum value.
3. $-3, 1; x = -1$
4. $2, 5; x = \frac{7}{2}$
5. $-7, 5; x = -1$
6. $-8, 0; x = -4$
7. 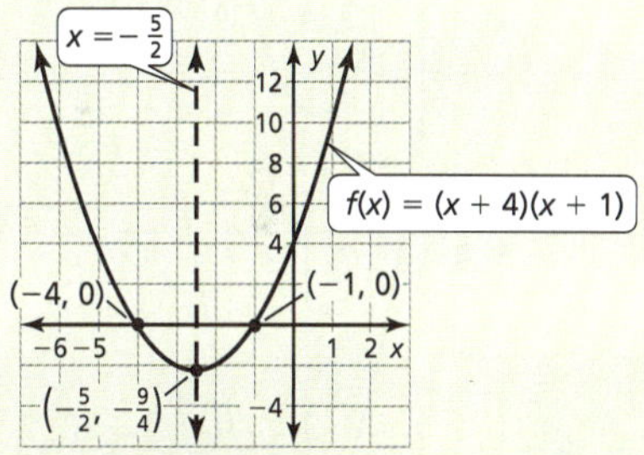

domain: all real numbers, range: $y \geq -\frac{9}{4}$

8. 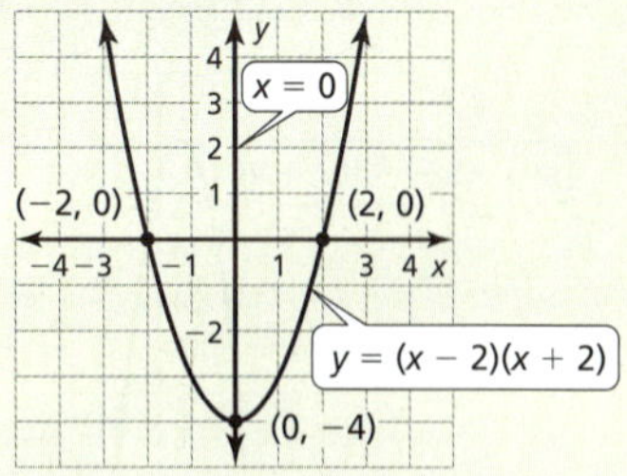

domain: all real numbers, range: $y \geq -4$

9. 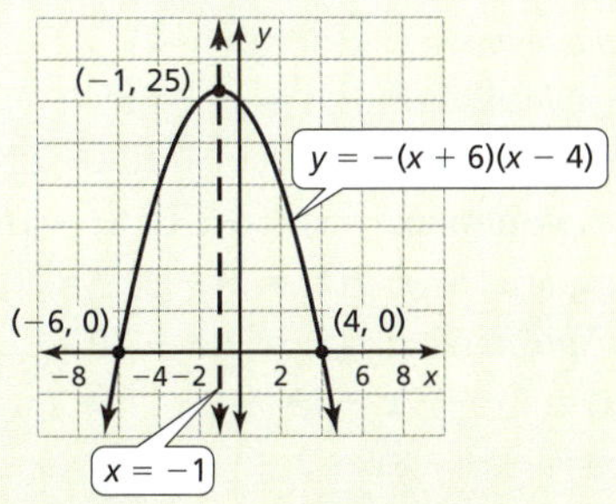

domain: all real numbers, range: $y \leq 25$

10. 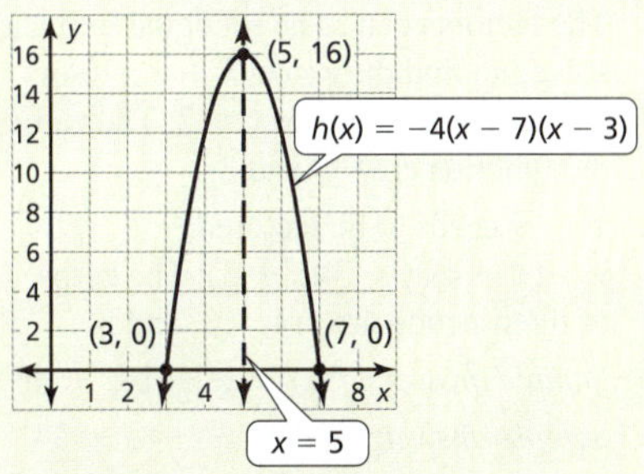

domain: all real numbers, range: $y \leq 16$

11–36. See Additional Answers.

Dynamic Teaching Tools

- Dynamic Assessment & Progress Monitoring Tool
- Interactive Whiteboard Lesson Library
- Dynamic Classroom with Dynamic Investigations

ANSWERS

37.

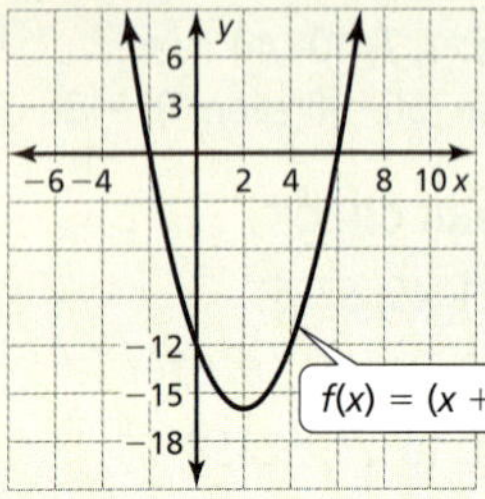

38.

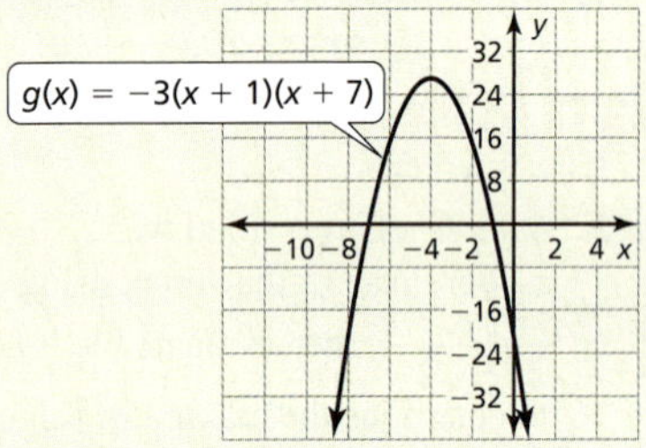

39.

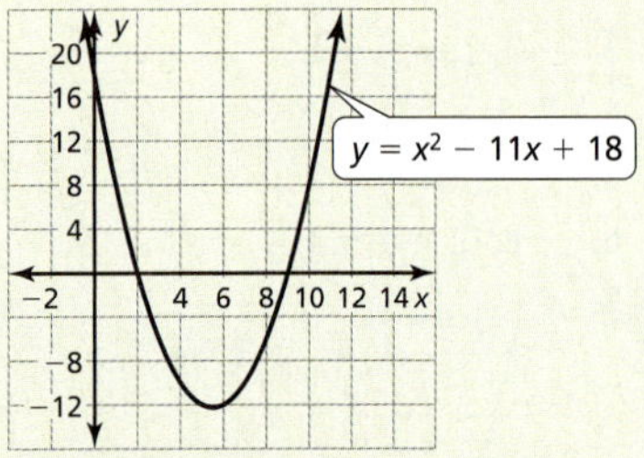

40.

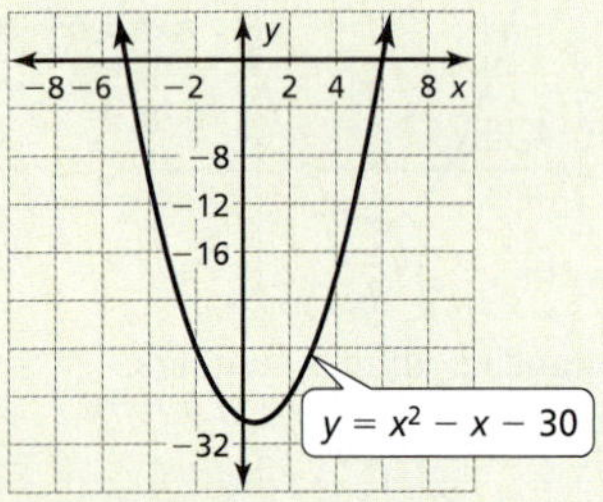

41.

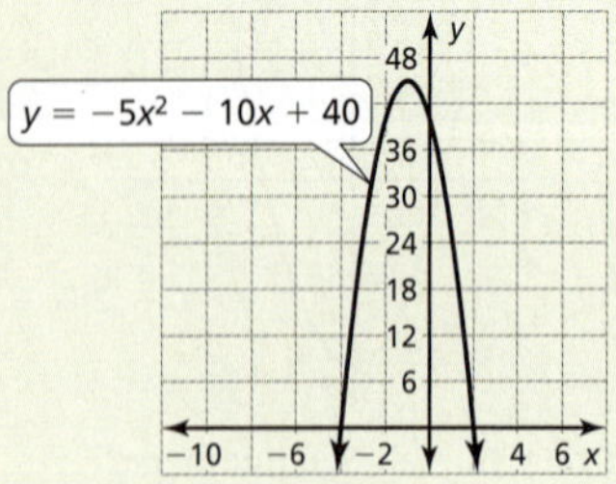

42.

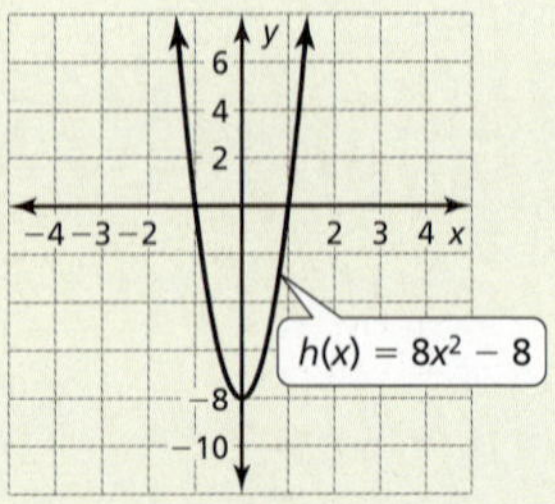

In Exercises 37–42, use zeros to graph the function. *(See Example 5.)*

37. $f(x) = (x + 2)(x - 6)$ **38.** $g(x) = -3(x + 1)(x + 7)$

39. $y = x^2 - 11x + 18$ **40.** $y = x^2 - x - 30$

41. $y = -5x^2 - 10x + 40$ **42.** $h(x) = 8x^2 - 8$

ERROR ANALYSIS In Exercises 43 and 44, describe and correct the error in finding the zeros of the function.

43.

✗ $y = 5(x + 3)(x - 2)$

The zeros of the function are 3 and −2.

44.

✗ $y = (x + 4)(x^2 - 9)$

The zeros of the function are −4 and 9.

In Exercises 45–56, write a quadratic function in standard form whose graph satisfies the given condition(s). *(See Example 6.)*

45. vertex: $(7, -3)$ **46.** vertex: $(4, 8)$

47. x-intercepts: 1 and 9 **48.** x-intercepts: -2 and -5

49. passes through $(-4, 0)$, $(3, 0)$, and $(2, -18)$

50. passes through $(-5, 0)$, $(-1, 0)$, and $(-4, 3)$

51. passes through $(7, 0)$

52. passes through $(0, 0)$ and $(6, 0)$

53. axis of symmetry: $x = -5$

54. y increases as x increases when $x < 4$; y decreases as x increases when $x > 4$.

55. range: $y \geq -3$ **56.** range: $y \leq 10$

In Exercises 57–60, write the quadratic function represented by the graph.

57.

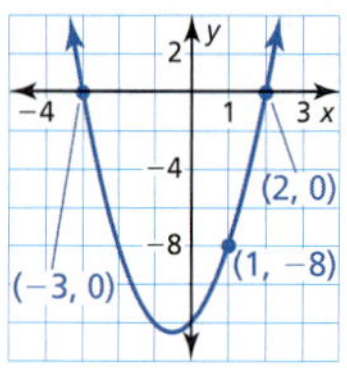

58.

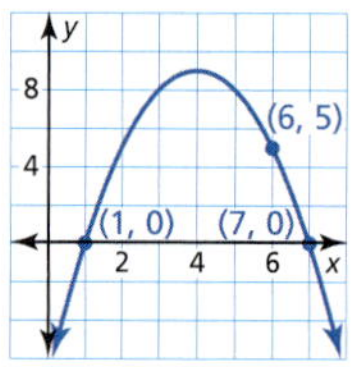

59.

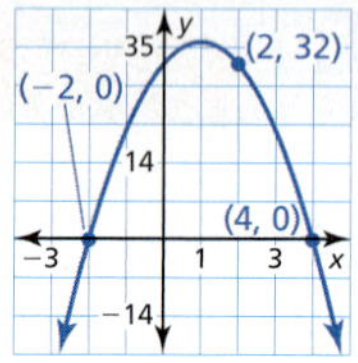

60.

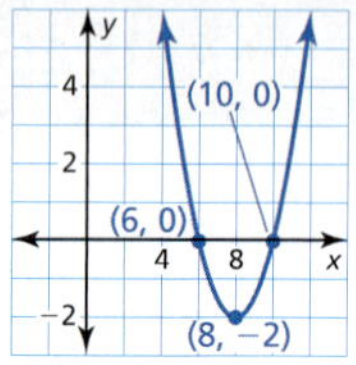

In Exercises 61–68, use zeros to graph the function. *(See Example 7.)*

61. $y = 5x(x + 2)(x - 6)$ **62.** $f(x) = -x(x + 9)(x + 3)$

63. $h(x) = (x - 2)(x + 2)(x + 7)$

64. $y = (x + 1)(x - 5)(x - 4)$

65. $f(x) = 3x^3 - 48x$ **66.** $y = -2x^3 + 20x^2 - 50x$

67. $y = -x^3 - 16x^2 - 28x$

68. $g(x) = 6x^3 + 30x^2 - 36x$

In Exercises 69–72, write the cubic function represented by the graph. *(See Example 8.)*

69.

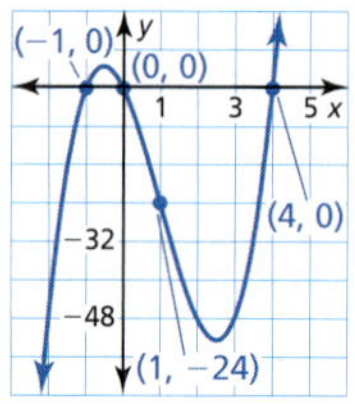

70.

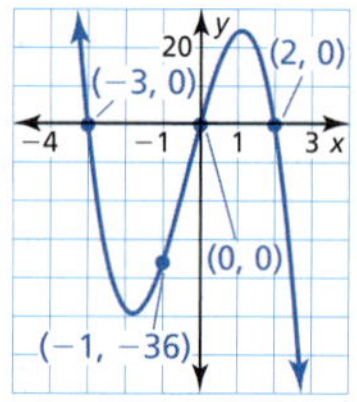

71.

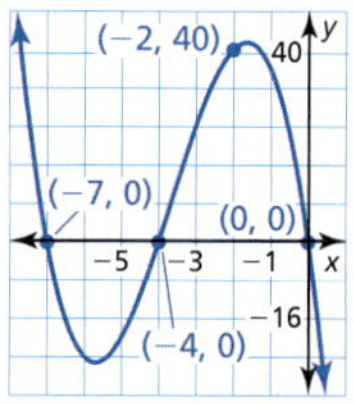

72.

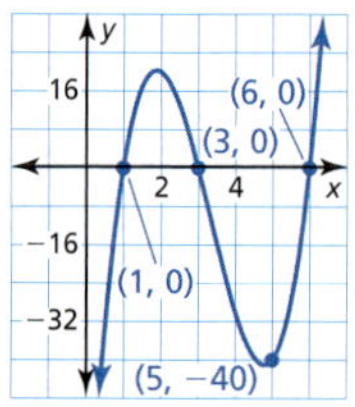

In Exercises 73–76, write a cubic function whose graph satisfies the given condition(s).

73. x-intercepts: -2, 3, and 8

74. x-intercepts: -7, -5, and 0

75. passes through $(1, 0)$ and $(7, 0)$

76. passes through $(0, 6)$

43. The factors need to be set equal to 0 and solved to find the zeros; $x + 3 = 0$ or $x - 2 = 0$; $x = -3$ or $x = 2$; The zeros of the function are -3 and 2.

44. $x^2 - 9$ needs to be factored; $y = (x + 4)(x + 3)(x - 3)$; The zeros of the function are -4, -3, and 3.

45. *Sample answer:* $f(x) = x^2 - 14x + 46$

46. *Sample answer:* $f(x) = x^2 - 8x + 24$

47. *Sample answer:* $f(x) = x^2 - 10x + 9$

48. *Sample answer:* $f(x) = x^2 + 7x + 10$

49. $f(x) = 3x^2 + 3x - 36$

50. $f(x) = -x^2 - 6x - 5$

51. *Sample answer:* $f(x) = x^2 - 7x$

52. *Sample answer:* $f(x) = x^2 - 6x$

53. *Sample answer:* $f(x) = x^2 + 10x + 25$

54. *Sample answer:* $f(x) = -x^2 + 8x - 16$

55. *Sample answer:* $f(x) = x^2 - 3$

56. *Sample answer:* $f(x) = -x^2 + 10$

57. $f(x) = 2x^2 + 2x - 12$

58. $f(x) = -x^2 + 8x - 7$

59. $f(x) = -4x^2 + 8x + 32$

60. $f(x) = \frac{1}{2}x^2 - 8x + 30$

61–76. See Additional Answers.

In Exercises 77–80, all the zeros of a function are given. Use the zeros and the other point given to write a quadratic or cubic function represented by the table.

77.

x	y
0	0
2	30
7	0

78.

x	y
−3	0
1	−72
4	0

79.

x	y
−4	0
−3	0
0	−180
3	0

80.

x	y
−8	0
−6	−36
−3	0
0	0

In Exercises 81–84, sketch a parabola that satisfies the given conditions.

81. x-intercepts: −4 and 2; range: $y \geq -3$

82. axis of symmetry: $x = 6$; passes through (4, 15)

83. range: $y \leq 5$; passes through (0, 2)

84. x-intercept: 6; y-intercept: 1; range: $y \geq -4$

85. MODELING WITH MATHEMATICS Satellite dishes are shaped like parabolas to optimally receive signals. The cross section of a satellite dish can be modeled by the function shown, where x and y are measured in feet. The x-axis represents the top of the opening of the dish.

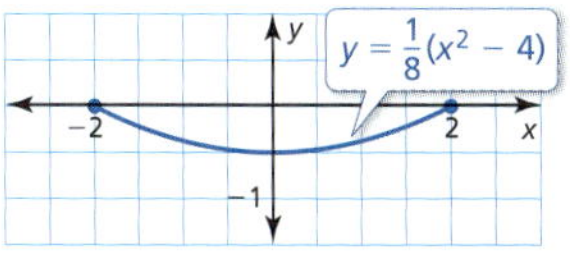

a. How wide is the satellite dish?

b. How deep is the satellite dish?

c. Write a quadratic function in standard form that models the cross section of a satellite dish that is 6 feet wide and 1.5 feet deep.

86. MODELING WITH MATHEMATICS A professional basketball player's shot is modeled by the function shown, where x and y are measured in feet.

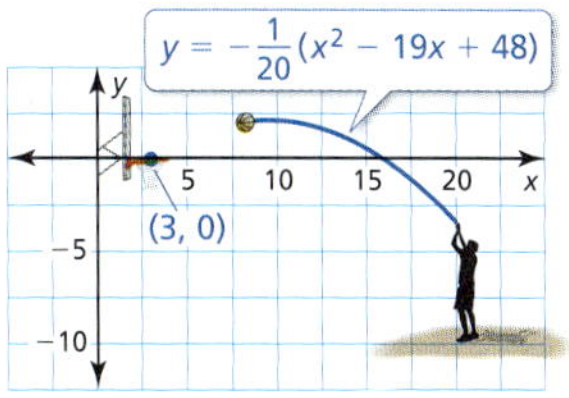

a. Does the player make the shot? Explain.

b. The basketball player releases another shot from the point (13, 0) and makes the shot. The shot also passes through the point (10, 1.4). Write a quadratic function in standard form that models the path of the shot.

USING STRUCTURE In Exercises 87–90, match the function with its graph.

87. $y = -x^2 + 5x$

88. $y = x^2 - x - 12$

89. $y = x^3 - 2x^2 - 8x$

90. $y = x^3 - 4x^2 - 11x + 30$

A.

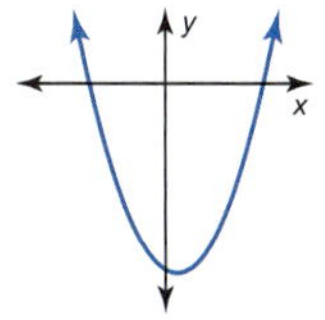

B.

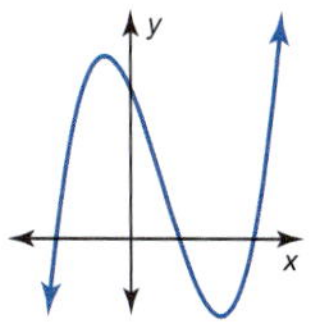

C.

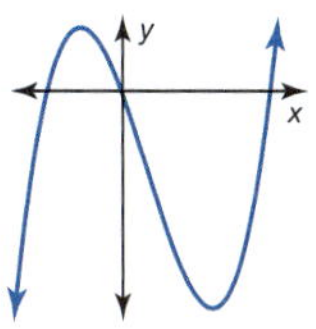

D.

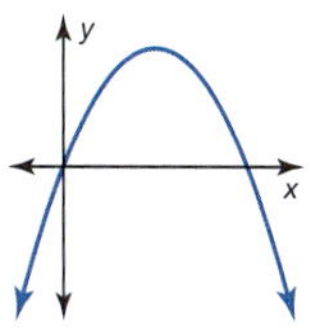

91. CRITICAL THINKING Write a quadratic function represented by the table, if possible. If not, explain why.

x	−5	−3	−1	1
y	0	12	4	0

ANSWERS

77. $f(x) = -3x^2 + 21x$

78. $f(x) = 6x^2 - 6x - 72$

79. $f(x) = 5x^3 + 20x^2 - 45x - 180$

80. $f(x) = -x^3 - 11x^2 - 24x$

81.

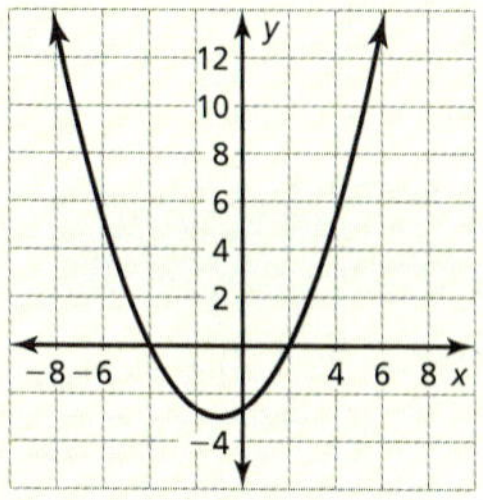

82. *Sample answer:*

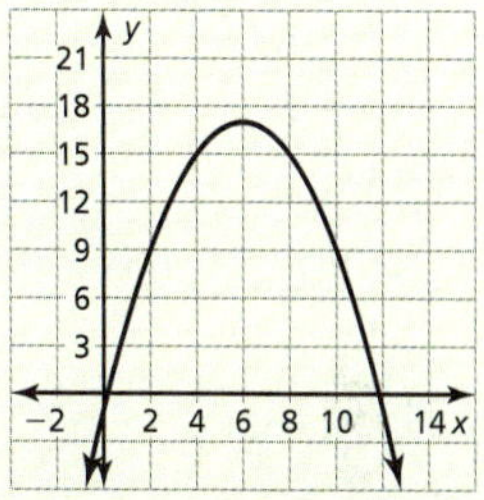

83. *Sample answer:*

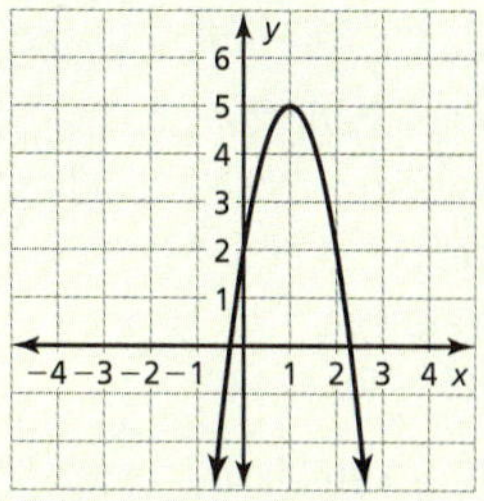

84. *Sample answer:*

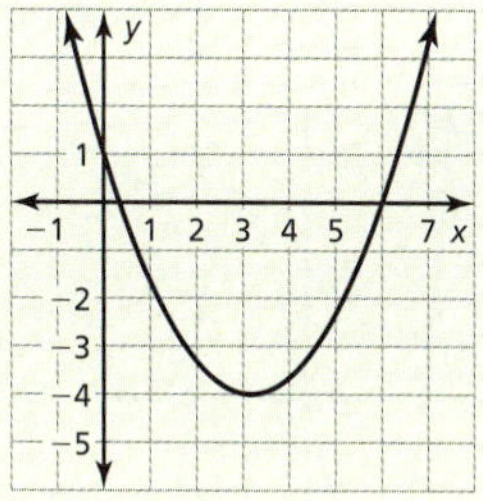

85. a. 4 ft

b. $\frac{1}{2}$ ft

c. $y = \frac{1}{6}(x^2 - 9)$

86. a. yes; *Sample answer:* Using the function, when the horizontal distance of the ball is 3 feet, the vertical distance is 0, which is the location of the basket.

b. $y = -\frac{1}{15}x^2 + \frac{16}{15}x - \frac{13}{5}$

87. D

88. A

89. C

90. B

91. not possible; *Sample answer:* Because −5 and 1 are the x-intercepts, the axis of symmetry is $x = -2$. The points (−3, 12) and (−1, 4) are the same horizontal distance from the axis of symmetry, so for both of them to lie on the parabola they would have to have the same y-coordinate.

ANSWERS

92–108. See Additional Answers.

Mini-Assessment

1. Graph $f(x) = (x - 2)(x + 4)$. Describe the domain and range.

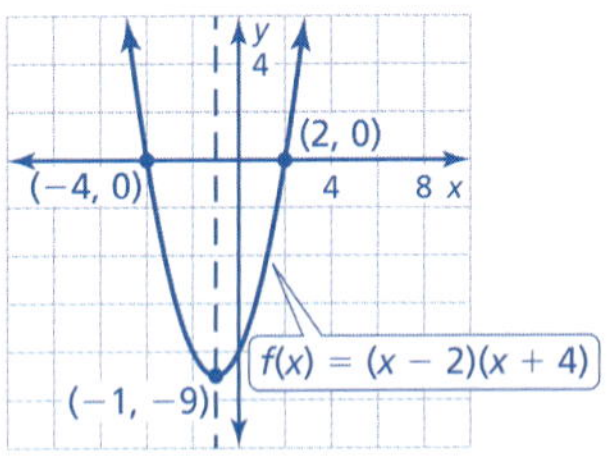

domain: all real numbers; range: $y \geq -9$

2. Find the zeros of the function $f(x) = 2x^2 + 2x - 60$. -6 and 5

3. Write a quadratic function in standard form whose graph has x-intercepts -4 and -3.
Sample answer: $f(x) = x^2 + 7x + 12$

4. Use zeros to graph $f(x) = -x^3 + 16x$.

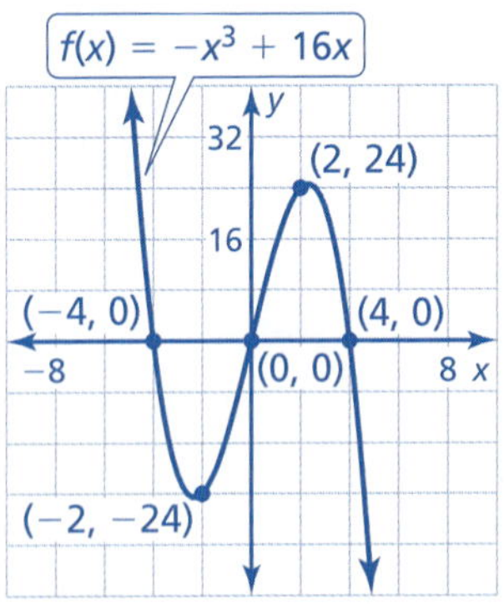

5. The graph represents a cubic function. Write the function.

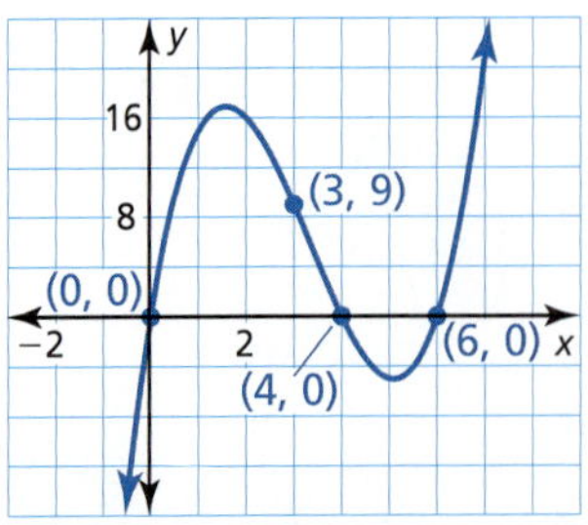

$f(x) = x^3 - 10x^2 + 24x$

---

92. HOW DO YOU SEE IT? The graph shows the parabolic arch that supports the roof of a convention center, where x and y are measured in feet.

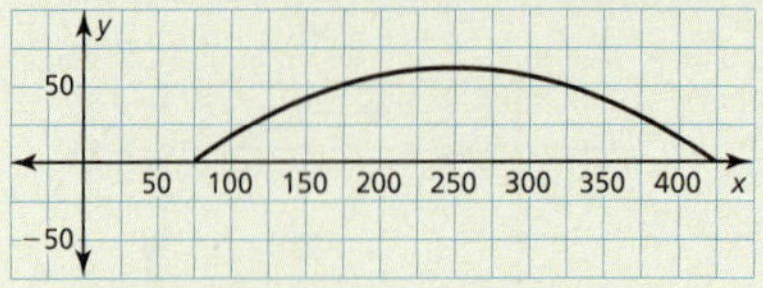

a. The arch can be represented by a function of the form $f(x) = a(x - p)(x - q)$. Estimate the values of p and q.

b. Estimate the width and height of the arch. Explain how you can use your height estimate to calculate a.

ANALYZING EQUATIONS **In Exercises 93 and 94, (a) rewrite the quadratic function in intercept form and (b) graph the function using any method. Explain the method you used.**

93. $f(x) = -3(x + 1)^2 + 27$

94. $g(x) = 2(x - 1)^2 - 2$

95. WRITING Can a quadratic function with exactly one real zero be written in intercept form? Explain.

96. MAKING AN ARGUMENT Your friend claims that any quadratic function can be written in standard form and in vertex form. Is your friend correct? Explain.

97. PROBLEM SOLVING Write the function represented by the graph in intercept form.

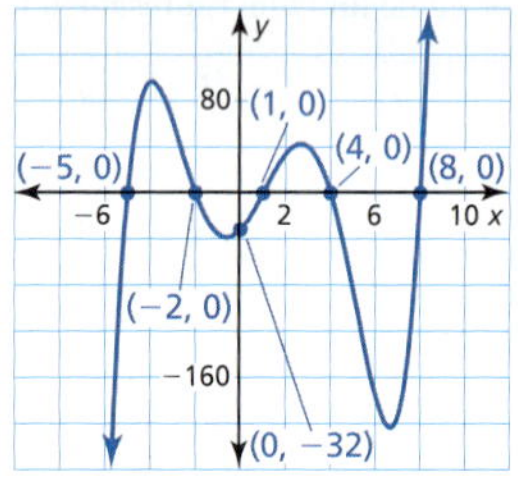

98. THOUGHT PROVOKING Sketch the graph of each function. Explain your procedure.

a. $f(x) = (x^2 - 1)(x^2 - 4)$

b. $g(x) = x(x^2 - 1)(x^2 - 4)$

99. REASONING Let k be a constant. Find the zeros of the function $f(x) = kx^2 - k^2x - 2k^3$ in terms of k.

PROBLEM SOLVING **In Exercises 100 and 101, write a system of two quadratic equations whose graphs intersect at the given points. Explain your reasoning.**

100. $(-4, 0)$ and $(2, 0)$

101. $(3, 6)$ and $(7, 6)$

Maintaining Mathematical Proficiency

Reviewing what you learned in previous grades and lessons

The scatter plot shows the amounts x (in grams) of fat and the numbers y of calories in 12 burgers at a fast-food restaurant. *(Section 4.4)*

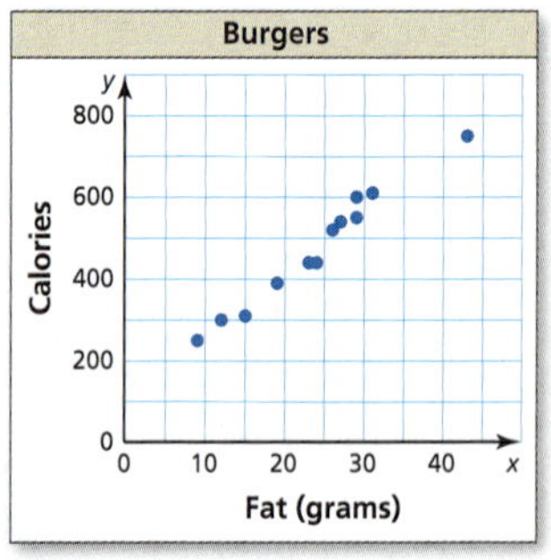

102. How many calories are in the burger that contains 12 grams of fat?

103. How many grams of fat are in the burger that contains 600 calories?

104. What tends to happen to the number of calories as the number of grams of fat increases?

Determine whether the sequence is *arithmetic*, *geometric*, or *neither*. Explain your reasoning. *(Section 6.6)*

105. 3, 11, 21, 33, 47, . . .

106. −2, −6, −18, −54, . . .

107. 26, 18, 10, 2, −6, . . .

108. 4, 5, 9, 14, 23, . . .

If students need help...	*If students got it...*
Resources by Chapter • Practice A and Practice B • Puzzle Time	Resources by Chapter • Enrichment and Extension • Cumulative Review
Student Journal • Practice	Start the *next* Section
Differentiating the Lesson Skills Review Handbook	

Laurie's Notes

Dynamic Teaching Tools

- Dynamic Assessment & Progress Monitoring Tool
- Lesson Planning Tool
- Interactive Whiteboard Lesson Library
- Dynamic Classroom with Dynamic Investigations

Overview of Section 8.6

Introduction

- The last lesson in this chapter is designed to help students take a look graphically and numerically at the three main functions they have studied this year.
- Given a set of data, students can graph the data to determine the type of function represented, provided sufficient behavior is evident from the graph. If the graph is incomplete—for instance, if only the increasing portion of a quadratic is displayed—it would be difficult to determine that the function is a quadratic.
- Examining the consecutive y-values in a data set is another technique for determining the type of function a data set represents. This is done by looking for patterns in the first and second differences of y-values, or by looking for a common ratio between consecutive y-values. This technique is used to help write a function for a data set.
- The lesson ends with a look at using average rates of change to compare different function types. The *Teaching Strategy* below helps explain this technique.

Teaching Strategy

- In the exploration, students compare the behavior of three functions over the domain $0 \leq x \leq 1$ and then for an expanded domain of $1 \leq x \leq 5$. The first function is linear ($y = x$), the second function is exponential ($y = 2^x - 1$), and the third is quadratic ($y = x^2$). These are represented in Y_1, Y_2, and Y_3, respectively, in Figure 1.
- Notice that over the domain $0 \leq x \leq 1$, for each value of x, the linear function has the greatest y-values, followed by the exponential, and then the quadratic. (See Figure 2 and Figure 3.)

```
\Y1=X
\Y2=2^X-1
\Y3=X²
\Y4=
\Y5=
\Y6=
\Y7=
```

Figure 1

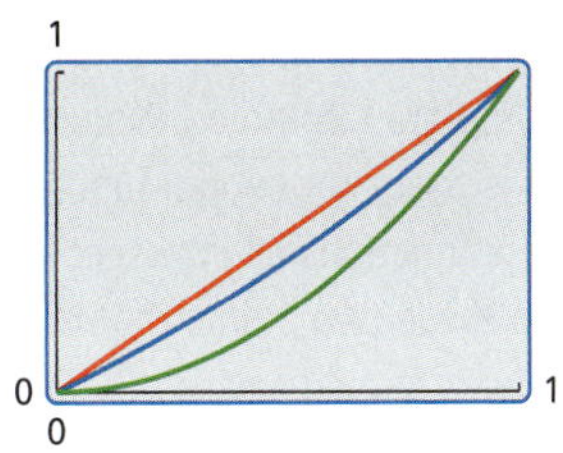

Figure 2

X	Y1	Y2	Y3
0	0	0	0
.1	.1	.07177	.01
.2	.2	.1487	.04
.3	.3	.23114	.09
.4	.4	.31951	.16
.5	.5	.41421	.25
.6	.6	.51572	.36
X=0			

Figure 3

- Now change the domain to $1 \leq x \leq 4$. For each value of x, the quadratic function has the greatest y-values, followed by the exponential, and then the linear. (See Figure 4.) The final views (Figures 5 and 6) are for a domain $4 \leq x \leq 9$, which shows that the exponential function has y-values that will continue to grow at a much faster rate than the quadratic or linear functions. *Note:* In Exploration 2 on page 459, students will use a domain of $1 \leq x \leq 5$.

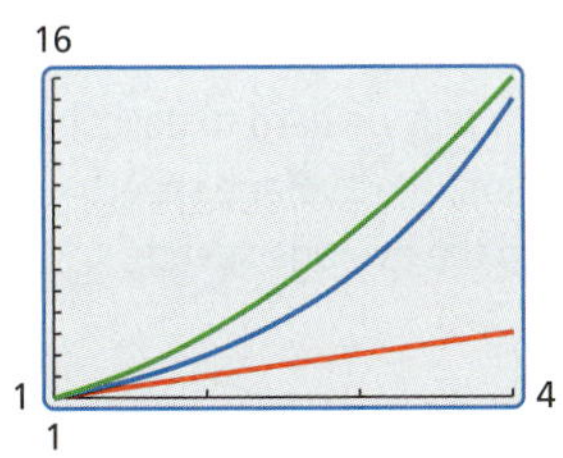

Figure 4

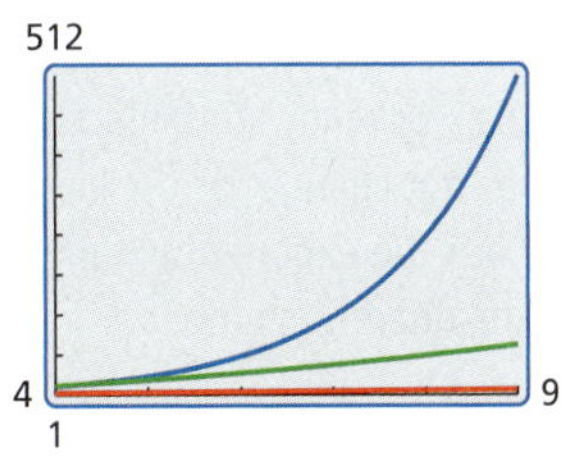

Figure 5

X	Y1	Y2	Y3
4	4	15	16
4.5	4.5	21.627	20.25
5	5	31	25
5.5	5.5	44.255	30.25
6	6	63	36
6.5	6.5	89.51	42.25
7	7	127	49
X=4			

Figure 6

Pacing Suggestion

- This is a long lesson with the explorations helping students to sense the behavior of linear, exponential, and quadratic functions. Refer to the explorations as you work through the formal lesson.

What Your Students Will Learn

- Plot sets of data and determine whether the data can be modeled with linear, exponential, or quadratic functions.
- Write functions to model data and write a recursive rule for a quadratic function.
- Compare functions using average rates of change over equal periods of time and interpret the average rates of change.

Laurie's Notes

Exploration

Motivate

- **Story Time:** Tell students that you test drove a sports car recently—a Lamborghini Reventón. Any student who follows cars will know that you are fibbing. The base price of a Reventón is approximately $1.6 million, a bit much for a teaching salary. So, instead of test driving, tell students you read about the car and the fact that it can go from 0 to 60 miles per hour in 3.3 seconds with a top speed of 211 miles per hour!
- A car that is a bit more affordable would be a Chevrolet Camaro ZL1, with a base price of about $53,000. It can go from 0 to 60 miles per hour in 3.9 seconds with a top speed of 184 miles per hour.
- Today, students will investigate some fast cars, but no speeding!

Exploration Note

- **Model with Mathematics:** Linear, exponential, and quadratic functions are common functions that can be used to model real-life applications. Mathematically proficient students can identify each of these functions from an equation, a complete graph, or a table of values. When examining a table of values, the behavior of the function becomes evident.
- In this exploration and lesson, students will compare linear, exponential, and quadratic functions.

Exploration 1

- This first exploration reacquaints students with the three types of functions they have studied so far this year: linear, exponential, and quadratic. Students will explore the growth rates of the three functions over a small, finite domain.
- Students work with partners to complete the table of values for each function.
- A common belief among students is that exponential functions grow most rapidly. What students discover quickly is that the functions all intersect at (0, 0) and (1, 1). It is the behavior of the functions between these two points that is of interest.

? "Were there any surprises?" Students should say yes.

- The car with a constant speed should be relatively easy to determine.
- Discuss the explanations offered for which car is accelerating the most.

Exploration 2

- Ask students to describe the three different speeds. Students can visually interpret the speeds in a graph, or they can calculate the speed of each car. Remember, speed is distance divided by time.
- If time permits, set up a table to show the change in average speed for each car. This connects to the answers in Questions 3 and 4.
- Ask students to interpret the table. For many students, the pattern may be hard to describe. The average speed of the car does not change for the linear function; the average speed "increases at an increasing rate" for the exponential function; and the average speed increases at a constant rate for the quadratic function.

Communicate Your Answer

- Students should discuss the questions with partners and then have a class discussion.

Connecting to Next Step

- These explorations help focus attention on the differences, numerically and graphically, between the three types of functions. In the formal lesson, students will do additional work with these three functions.

8.6 Comparing Linear, Exponential, and Quadratic Functions

Essential Question How can you compare the growth rates of linear, exponential, and quadratic functions?

EXPLORATION 1 **Comparing Speeds**

Work with a partner. Three cars start traveling at the same time. The distance traveled in t minutes is y miles. Complete each table and sketch all three graphs in the same coordinate plane. Compare the speeds of the three cars. Which car has a constant speed? Which car is accelerating the most? Explain your reasoning.

t	$y = t$
0	
0.2	
0.4	
0.6	
0.8	
1.0	

t	$y = 2^t - 1$
0	
0.2	
0.4	
0.6	
0.8	
1.0	

t	$y = t^2$
0	
0.2	
0.4	
0.6	
0.8	
1.0	

COMPARING PREDICTIONS

To be proficient in math, you need to visualize the results of varying assumptions, explore consequences, and compare predictions with data.

EXPLORATION 2 **Comparing Speeds**

Work with a partner. Analyze the speeds of the three cars over the given time periods. The distance traveled in t minutes is y miles. Which car eventually overtakes the others?

t	$y = t$
1.0	
1.5	
2.0	
2.5	
3.0	
3.5	
4.0	
4.5	
5.0	

t	$y = 2^t - 1$
1.0	
1.5	
2.0	
2.5	
3.0	
3.5	
4.0	
4.5	
5.0	

t	$y = t^2$
1.0	
1.5	
2.0	
2.5	
3.0	
3.5	
4.0	
4.5	
5.0	

Communicate Your Answer

3. How can you compare the growth rates of linear, exponential, and quadratic functions?
4. Which function has a growth rate that is eventually much greater than the growth rates of the other two functions? Explain your reasoning.

Dynamic Teaching Tools

- Dynamic Assessment & Progress Monitoring Tool
- Lesson Planning Tool
- Interactive Whiteboard Lesson Library
- Dynamic Classroom with Dynamic Investigations

ANSWERS

1. 0, 0.2, 0.4, 0.6, 0.8, 1.0;
0, 0.15, 0.32, 0.52, 0.74, 1.00;
0, 0.04, 0.16, 0.36, 0.64, 1.00

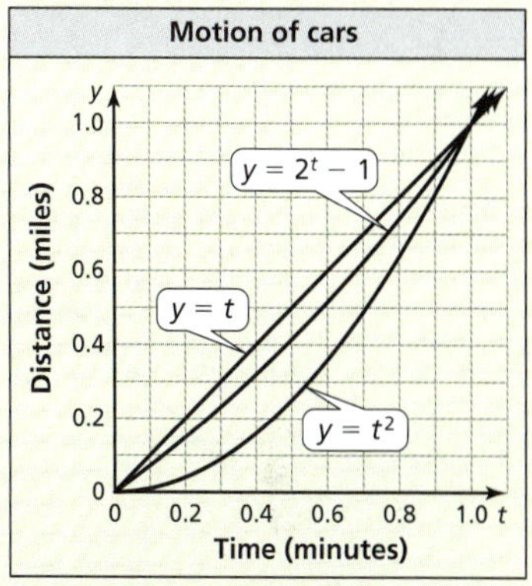

Sample answer: All three cars have the same average speed over the interval; first car; third car; The car modeled by a linear function is traveling at the same speed because it is traveling the same distance in each time interval. The car modeled by the quadratic function is accelerating the most because it has the greatest change in speed over each time interval.

2. 1, 1.5, 2, 2.5, 3, 3.5, 4, 4.5, 5;
1, 1.83, 3, 4.66, 7, 10.31, 15, 21.63, 31; 1, 2.25, 4, 6.25, 9, 12.25, 16, 20.25, 25;
The first car has a constant speed. The second and third car are accelerating; The car modeled by the exponential function eventually overtakes the others.
3. *Sample answer:* Graph functions of each type in the same coordinate plane and compare the shapes of the graphs.
4. exponential; *Sample answer:* An exponential graph begins with a small change in each interval, but the change increases more rapidly as the value of the exponent increases.

English Language Learners

Build on Past Knowledge

Remind students that they can identify the type of function by using both its graph and the form of its equation. The graph of a *linear* function is a *line*. The variable x in an *exponential* function equation is an *exponent*.

Extra Example 1

Plot the points. Tell whether the points appear to represent a *linear*, an *exponential*, or a *quadratic* function.

a. $(8, 4), (4, 0), (0, -4), (-2, -6), (-4, -8)$

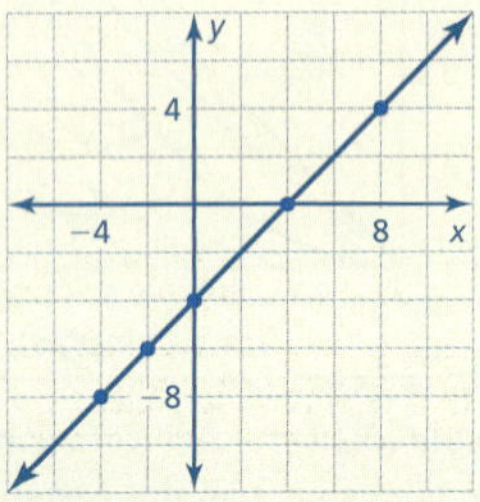

linear function

b. $(-3, 6), (-2, 1), (0, -3), (2, 1), (3, 6)$

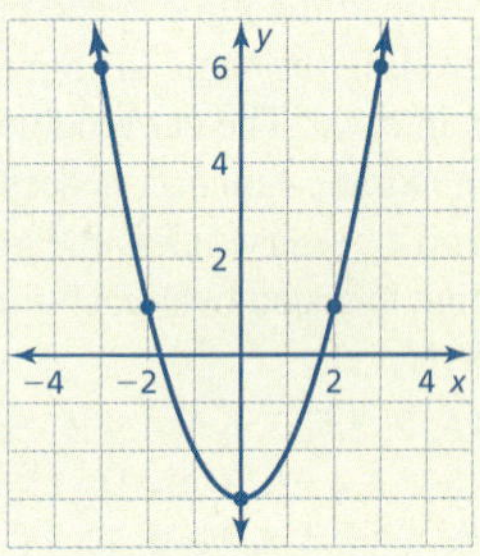

quadratic function

c. $(-3, 9), (-2, 5), (-1, 3), (0, 2), (1, 1.5)$

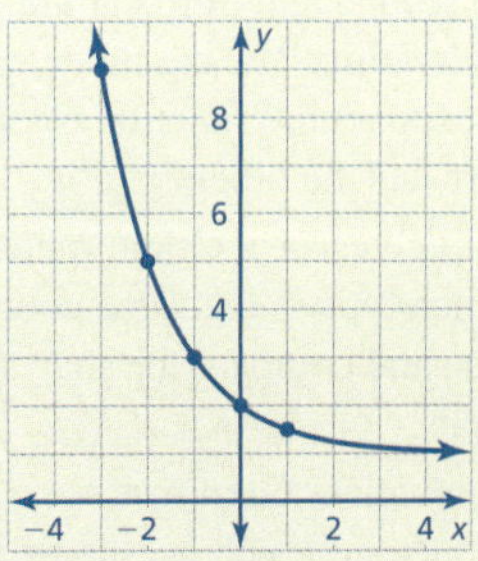

exponential function

MONITORING PROGRESS ANSWERS

1–3. See Additional Answers.

8.6 Lesson

Core Vocabulary

average rate of change, *p. 462*

Previous

slope

What You Will Learn

- Choose functions to model data.
- Write functions to model data.
- Compare functions using average rates of change.
- Solve real-life problems involving different function types.

Choosing Functions to Model Data

So far, you have studied linear functions, exponential functions, and quadratic functions. You can use these functions to model data.

Core Concept

Linear, Exponential, and Quadratic Functions

Linear Function

$y = mx + b$

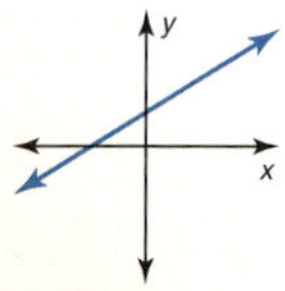

Exponential Function

$y = ab^x$

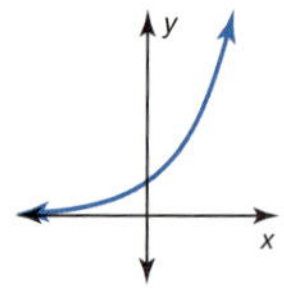

Quadratic Function

$y = ax^2 + bx + c$

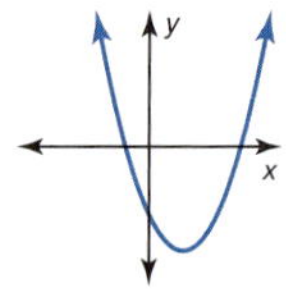

EXAMPLE 1 Using Graphs to Identify Functions

Plot the points. Tell whether the points appear to represent a *linear*, an *exponential*, or a *quadratic* function.

a. $(4, 4), (2, 0), (0, 0), \left(1, -\frac{1}{2}\right), (-2, 4)$

b. $(0, 1), (2, 4), (4, 7), (-2, -2), (-4, -5)$

c. $(0, 2), (2, 8), (1, 4), (-1, 1), \left(-2, \frac{1}{2}\right)$

SOLUTION

a.

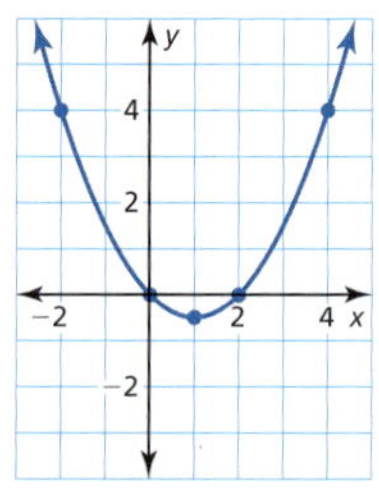

quadratic

b.

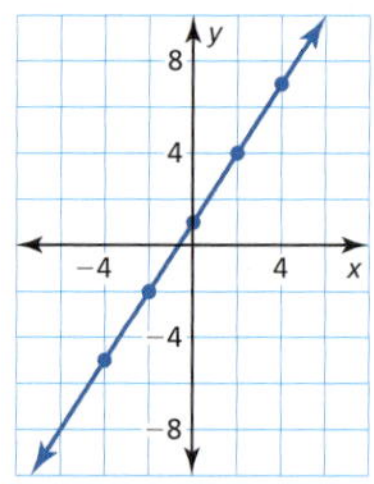

linear

c.

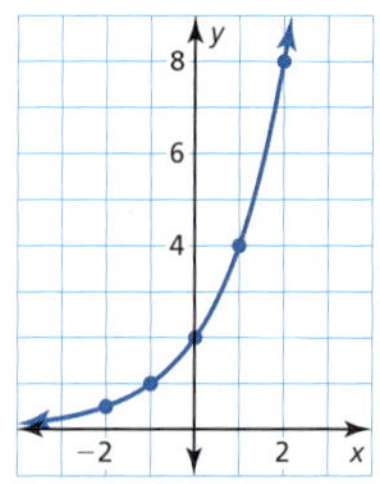

exponential

Monitoring Progress

Help in English and Spanish at *BigIdeasMath.com*

Plot the points. Tell whether the points appear to represent a *linear*, an *exponential*, or a *quadratic* function.

1. $(-1, 5), (2, -1), (0, -1), (3, 5), (1, -3)$
2. $(-1, 2), (-2, 8), (-3, 32), \left(0, \frac{1}{2}\right), \left(1, \frac{1}{8}\right)$
3. $(-3, 5), (0, -1), (2, -5), (-4, 7), (1, -3)$

Laurie's Notes Teacher Actions

- The *Core Concept* presents a summary of the three types of functions. It would be beneficial to briefly review what the parameters tell you about the graph of each function.
- **Thumbs Up:** Have students work independently on Example 1. Have students self-assess with a *Thumbs Up* signal.
- **Use Appropriate Tools Strategically:** If time permits, have students use a graphing calculator to enter the data and plot. If they know the related functions, enter them in the equation editor.

Core Concept

Differences and Ratios of Functions

You can use patterns between consecutive data pairs to determine which type of function models the data. The differences of consecutive y-values are called *first differences*. The differences of consecutive first differences are called *second differences*.

- **Linear Function** The first differences are constant.
- **Exponential Function** Consecutive y-values have a common *ratio*.
- **Quadratic Function** The second differences are constant.

In all cases, the differences of consecutive x-values need to be constant.

STUDY TIP

The first differences for exponential and quadratic functions are *not* constant.

EXAMPLE 2 Using Differences or Ratios to Identify Functions

STUDY TIP

First determine that the differences of consecutive x-values are constant. Then check whether the first differences are constant or consecutive y-values have a common ratio. If neither of these is true, check whether the second differences are constant.

Tell whether each table of values represents a *linear*, an *exponential*, or a *quadratic* function.

a.

x	−3	−2	−1	0	1
y	11	8	5	2	−1

b.

x	−2	−1	0	1	2
y	1	2	4	8	16

c.

x	−2	−1	0	1	2
y	−1	−2	−1	2	7

SOLUTION

a.

+1 +1 +1 +1

x	−3	−2	−1	0	1
y	11	8	5	2	−1

−3 −3 −3 −3

▶ The first differences are constant. So, the table represents a linear function.

b.

+1 +1 +1 +1

x	−2	−1	0	1	2
y	1	2	4	8	16

×2 ×2 ×2 ×2

▶ Consecutive y-values have a common ratio. So, the table represents an exponential function.

c.

+1 +1 +1 +1

x	−2	−1	0	1	2
y	−1	−2	−1	2	7

first differences → −1 +1 +3 +5

second differences → +2 +2 +2

▶ The second differences are constant. So, the table represents a quadratic function.

Monitoring Progress

Help in English and Spanish at *BigIdeasMath.com*

4. Tell whether the table of values represents a *linear*, an *exponential*, or a *quadratic* function.

x	−1	0	1	2	3
y	1	3	9	27	81

Differentiated Instruction

Inclusion

Some students may not understand the concept of "second difference." Give those students several tables of values with constant consecutive x-values. Have them find the y-value differences. Then have them find the differences of the y-value differences, or the "second differences."

Extra Example 2

Tell whether each table of values represents a *linear*, an *exponential*, or a *quadratic* function.

a.

x	−2	−1	0	1	2
y	11	5	3	5	11

quadratic function

b.

x	−2	−1	0	1	2
y	6	8	10	12	14

linear function

c.

x	0	1	2	3	4
y	0.5	3	18	108	648

exponential function

MONITORING PROGRESS ANSWER

4. exponential

Laurie's Notes Teacher Actions

- Equations and complete graphs can help determine what type of function you have. Differences and ratios can as well.
- Write the *Core Concept*. First differences and the ratio of consecutive y-values should make sense from the definitions of linear and exponential functions.
- In Example 2, note that consecutive x-values are used.
- Explain that often you can tell by inspection whether the first differences of the y-values are constant. Likewise, you might suspect a possible common ratio by inspection. When neither is true, try second differences.

Extra Example 3

Tell whether the table of values represents a *linear*, an *exponential*, or a *quadratic* function. Then write the function.

x	−2	−1	0	1	2
y	−3	−1	1	3	5

linear function; $y = 2x + 1$

MONITORING PROGRESS ANSWER

5. exponential; $y = 8\left(\frac{1}{2}\right)^x$

Writing Functions to Model Data

EXAMPLE 3 Writing a Function to Model Data

x	2	4	6	8	10
y	12	0	−4	0	12

Tell whether the table of values represents a *linear*, an *exponential*, or a *quadratic* function. Then write the function.

SOLUTION

Step 1 Determine which type of function the table of values represents.

	+2	+2	+2	+2	
x	2	4	6	8	10
y	12	0	−4	0	12

first differences → −12, −4, +4, +12

second differences → +8, +8, +8

The second differences are constant. So, the table represents a quadratic function.

Step 2 Write an equation of the quadratic function. Using the table, notice that the x-intercepts are 4 and 8. So, use intercept form to write a function.

$y = a(x - 4)(x - 8)$ — Substitute for p and q in intercept form.

Use another point from the table, such as (2, 12), to find a.

$12 = a(2 - 4)(2 - 8)$ — Substitute 2 for x and 12 for y.

$1 = a$ — Solve for a.

Use the value of a to write the function.

$y = (x - 4)(x - 8)$ — Substitute 1 for a.

$= x^2 - 12x + 32$ — Use the FOIL Method and combine like terms.

▶ So, the quadratic function is $y = x^2 - 12x + 32$.

STUDY TIP

To check your function in Example 3, substitute the other points from the table to verify that they satisfy the function.

Monitoring Progress

Help in English and Spanish at *BigIdeasMath.com*

5. Tell whether the table of values represents a *linear*, an *exponential*, or a *quadratic* function. Then write the function.

x	−1	0	1	2	3
y	16	8	4	2	1

Comparing Functions Using Average Rates of Change

For nonlinear functions, the rate of change is not constant. You can compare two nonlinear functions over the same interval using their *average rates of change*. The **average rate of change** of a function $y = f(x)$ between $x = a$ and $x = b$ is the slope of the line through $(a, f(a))$ and $(b, f(b))$.

$$\text{average rate of change} = \frac{\text{change in } y}{\text{change in } x} = \frac{f(b) - f(a)}{b - a}$$

Exponential Function

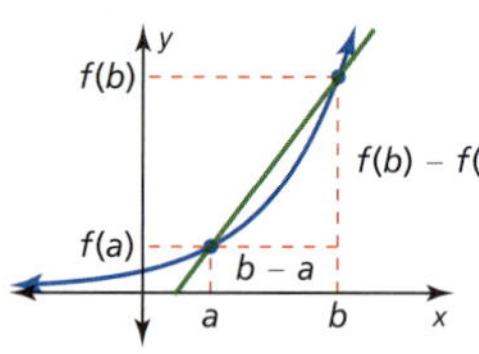

Quadratic Function

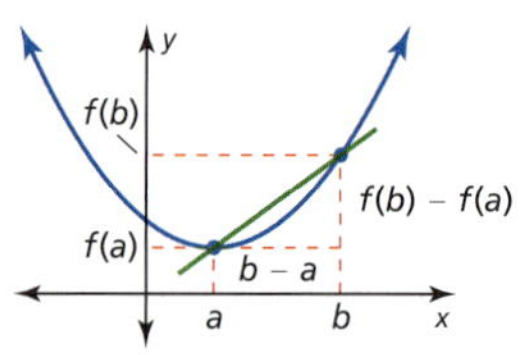

Laurie's Notes Teacher Actions

- **Misconception:** The x-values in a data set do not have to have a common difference of 1. In this example, x-values differ by 2.
- Present the data set in Example 3.
- **? Turn and Talk:** "What type of function do the data represent?" Give time for partners to discuss. Once they decide it is quadratic, ask, "What can you do with data points (4, 0) and (8, 0)?" Identify 4 and 8 as the x-intercepts and substitute in the intercept form.
- Explain how an *average rate of change* is found for nonlinear functions when comparing them over the same interval.

Core Concept

Comparing Functions Using Average Rates of Change

- As a and b increase, the average rate of change between $x = a$ and $x = b$ of an increasing exponential function $y = f(x)$ will eventually exceed the average rate of change between $x = a$ and $x = b$ of an increasing quadratic function $y = g(x)$ or an increasing linear function $y = h(x)$. So, as x increases, $f(x)$ will eventually exceed $g(x)$ or $h(x)$.
- As a and b increase, the average rate of change between $x = a$ and $x = b$ of an increasing quadratic function $y = g(x)$ will eventually exceed the average rate of change between $x = a$ and $x = b$ of an increasing linear function $y = h(x)$. So, as x increases, $g(x)$ will eventually exceed $h(x)$.

STUDY TIP

You can explore these concepts using a graphing calculator.

EXAMPLE 4 Using and Interpreting Average Rates of Change

Website A	
Day, *x*	**Members, *y***
0	650
5	1025
10	1400
15	1775
20	2150
25	2525

Two social media websites open their memberships to the public. (a) Compare the websites by calculating and interpreting the average rates of change from Day 10 to Day 20. (b) Predict which website will have more members after 50 days. Explain.

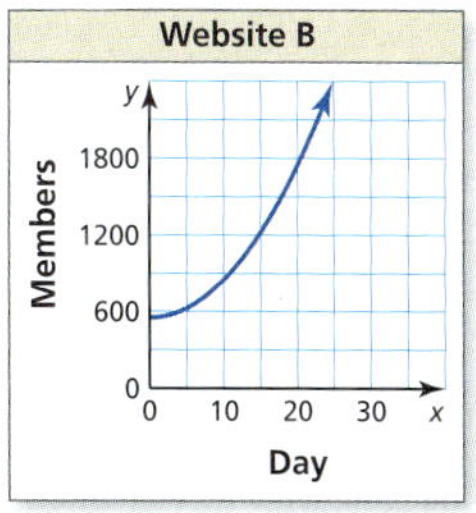

SOLUTION

a. Calculate the average rates of change by using the points whose x-coordinates are 10 and 20.

Website A: Use (10, 1400) and (20, 2150).

$$\text{average rate of change} = \frac{f(b) - f(a)}{b - a} = \frac{2150 - 1400}{20 - 10} = \frac{750}{10} = 75$$

Website B: Use the graph to estimate the points when $x = 10$ and $x = 20$. Use (10, 850) and (20, 1800).

$$\text{average rate of change} = \frac{f(b) - f(a)}{b - a} \approx \frac{1800 - 850}{20 - 10} = \frac{950}{10} = 95$$

▶ From Day 10 to Day 20, Website A membership increases at an average rate of 75 people per day, and Website B membership increases at an average rate of about 95 people per day. So, Website B membership is growing faster.

b. Using the table, membership increases and the average rates of change are constant. So, Website A membership can be represented by an increasing linear function. Using the graph, membership increases and the average rates of change are increasing. It appears that Website B membership can be represented by an increasing exponential or quadratic function.

After 25 days, the memberships of both websites are about equal and the average rate of change of Website B exceeds the average rate of change of Website A. So, Website B will have more members after 50 days.

Monitoring Progress

Help in English and Spanish at *BigIdeasMath.com*

6. Compare the websites in Example 4 by calculating and interpreting the average rates of change from Day 0 to Day 10.

Laurie's Notes Teacher Actions

- To help students understand the *Core Concept*, you might use a graphing calculator. See the *Teaching Strategy* on page T-458.
- **Wait Time:** Pose Example 4, and give *Wait Time* while partners discuss their understanding of the problem.
- Part (a) should make sense to students. When interpreting the results, make sure students understand that an average rate of increase of 75 members per day means there were days when there were fewer or more than 75 new members. Connect these rates to the table of values and the graph.

Extra Example 4

Two news websites open their memberships to the public.

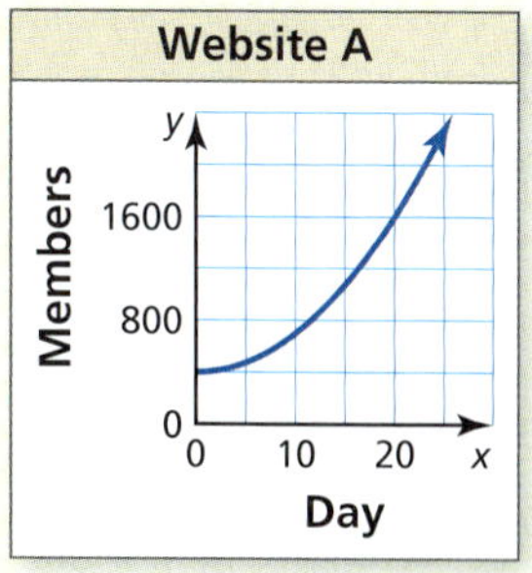

Website B	
Day, *x*	**Members, *y***
0	500
5	850
10	1200
15	1550
20	1900
25	2250

a. Compare the websites by calculating and interpreting the average rates of change from Day 10 to Day 20. From Day 10 to Day 20, Website A membership increases at an average rate of about 90 people per day, and Website B membership increases at an average rate of 70 people per day. So, Website A membership is growing faster.

b. Predict which website will have more members after 50 days. Explain. It appears that Website A membership can be represented by an increasing exponential or quadratic function. Website B membership can be represented by an increasing linear function. After 25 days, the memberships of both websites are about equal and the average rate of change of Website A exceeds the average rate of change of Website B. So, Website A will have more members after 50 days.

MONITORING PROGRESS ANSWER

6. From Day 0 to Day 10, Website A membership increases at an average rate of 75 people per day and Website B membership increases at an average rate of about 30 people per day. So, Website A membership is growing faster.

Extra Example 5

In 1900, Greenville had a population of 4000 people. Greenville's population increased by 100 people each year. In 1900, Springfield had a population of 1000 people. Springfield's population increased by 4% each year.

a. In what year were the populations about equal? *The populations were about equal in 1958.*

b. Suppose Greenville's initial population doubled to 8000 and maintained a constant rate of increase of 100 people each year. Did Springfield's population still catch up to Greenville's population? If so, in which year? *Yes, Springfield's population caught up to Greenville's population in 1968.*

c. Suppose Greenville's rate of increase doubled to 200 people each year, in addition to doubling the initial population. Did Springfield's population still catch up to Greenville's population? Explain. *Yes, Springfield's population caught up to Greenville's population in 1981. Because Greenville's population shows linear growth and Springfield's population shows exponential growth, Springfield's population eventually exceeded Greenville's population regardless of Greenville's constant rate or initial value.*

MONITORING PROGRESS ANSWER

7. 1917

Solving Real-Life Problems

EXAMPLE 5 Comparing Different Function Types

In 1900, Littleton had a population of 1000 people. Littleton's population increased by 50 people each year. In 1900, Tinyville had a population of 500 people. Tinyville's population increased by 5% each year.

a. In what year were the populations about equal?

b. Suppose Littleton's initial population doubled to 2000 and maintained a constant rate of increase of 50 people each year. Did Tinyville's population still catch up to Littleton's population? If so, in which year?

c. Suppose Littleton's rate of increase doubled to 100 people each year, in addition to doubling the initial population. Did Tinyville's population still catch up to Littleton's population? Explain.

SOLUTION

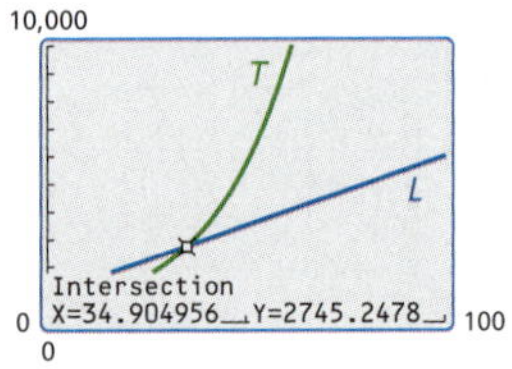

a. Let x represent the number of years since 1900. Write a function to model the population of each town.

Littleton: $L(x) = 50x + 1000$ — Linear function

Tinyville: $T(x) = 500(1.05)^x$ — Exponential function

Use a graphing calculator to graph each function in the same viewing window. Use the *intersect* feature to find the value of x for which $L(x) \approx T(x)$. The graphs intersect when $x \approx 34.9$.

▶ So, the populations were about equal in 1934.

b. Littleton's new population function is $f(x) = 50x + 2000$. Use a graphing calculator to graph f and T in the same viewing window. Use the *intersect* feature to find the value of x for which $f(x) \approx T(x)$. The graphs intersect when $x \approx 43.5$.

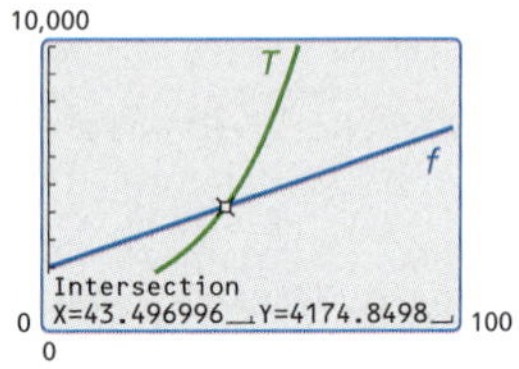

▶ So, Tinyville's population caught Littleton's population in 1943.

c. Littleton's new population function is $g(x) = 100x + 2000$. Use a graphing calculator to graph g and T in the same viewing window. Use the *intersect* feature to find the value of x for which $g(x) \approx T(x)$. The graphs intersect when $x \approx 55.7$.

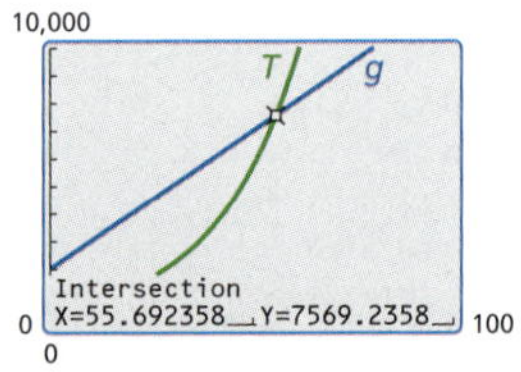

▶ So, Tinyville's population caught Littleton's population in 1955. Because Littleton's population shows linear growth and Tinyville's population shows exponential growth, Tinyville's population eventually exceeded Littleton's regardless of Littleton's constant rate or initial value.

Monitoring Progress

Help in English and Spanish at *BigIdeasMath.com*

7. **WHAT IF?** Tinyville's population increased by 8% each year. In what year were the populations about equal?

Laurie's Notes — Teacher Actions

- **Make Sense of Problems and Persevere in Solving Them:** Read through the problem. Answer questions about language.
- **Popsicle Sticks:** Have partners work on part (a). Use *Popsicle Sticks* to solicit a solution.
- **Use Appropriate Tools Strategically:** Use technology to work through the last two parts of the problem, as shown. Technology can help students explore the behavior of the two functions as x continues to increase. Use the table of values and graphs.

Closure

- Write a linear, an exponential, and a quadratic function that each has a y-intercept of 2.

Sample answer: $y = \frac{1}{2}x + 2$, $y = 2(3)^x$, $y = x^2 - 4x + 2$

8.6 Exercises

Dynamic Solutions available at *BigIdeasMath.com*

Vocabulary and Core Concept Check

1. **WRITING** Name three types of functions that you can use to model data. Describe the equation and graph of each type of function.

2. **WRITING** How can you decide whether to use a linear, an exponential, or a quadratic function to model a data set?

3. **VOCABULARY** Describe how to find the average rate of change of a function $y = f(x)$ between $x = a$ and $x = b$.

4. **WHICH ONE DOESN'T BELONG?** Which graph does *not* belong with the other three? Explain your reasoning.

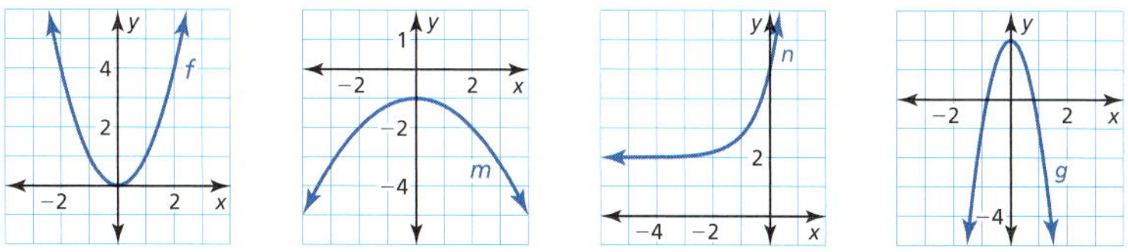

Monitoring Progress and Modeling with Mathematics

In Exercises 5–8, tell whether the points appear to represent a *linear*, an *exponential*, or a *quadratic* function.

5. 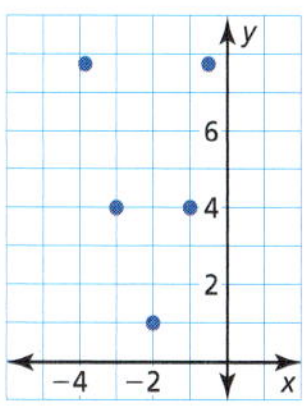

6. 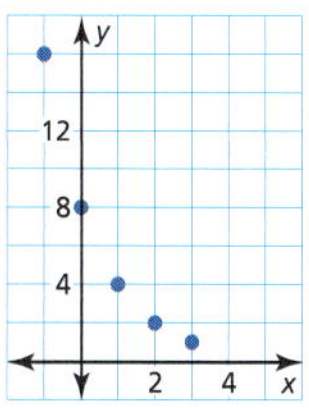

7. 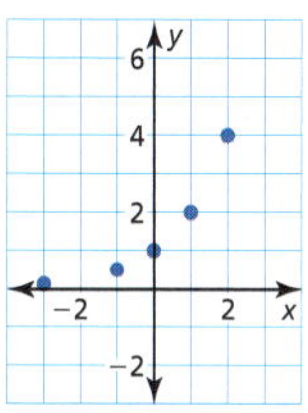

8. 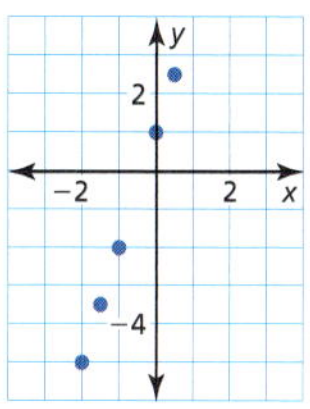

In Exercises 9–14, plot the points. Tell whether the points appear to represent a *linear*, an *exponential*, or a *quadratic* function. *(See Example 1.)*

9. $(-2, -1), (-1, 0), (1, 2), (2, 3), (0, 1)$

10. $\left(0, \frac{1}{4}\right), (1, 1), (2, 4), (3, 16), \left(-1, \frac{1}{16}\right)$

11. $(0, -3), (1, 0), (2, 9), (-2, 9), (-1, 0)$

12. $(-1, -3), (-3, 5), (0, -1), (1, 5), (2, 15)$

13. $(-4, -4), (-2, -3.4), (0, -3), (2, -2.6), (4, -2)$

14. $(0, 8), (-4, 0.25), (-3, 0.4), (-2, 1), (-1, 3)$

In Exercises 15–18, tell whether the table of values represents a *linear*, an *exponential*, or a *quadratic* function. *(See Example 2.)*

15.

x	−2	−1	0	1	2
y	0	0.5	1	1.5	2

16.

x	−1	0	1	2	3
y	0.2	1	5	25	125

17.

x	2	3	4	5	6
y	2	6	18	54	162

18.

x	−3	−2	−1	0	1
y	2	4.5	8	12.5	18

Assignment Guide and Homework Check

ASSIGNMENT

Basic: 1–4, 5–27 odd, 32–36 even, 41, 43–50

Average: 1–4, 6–38 even, 41, 43–50

Advanced: 1–4, 12–36 even, 37–50

HOMEWORK CHECK

Basic: 9, 15, 21, 32, 34

Average: 12, 16, 22, 32, 34

Advanced: 14, 18, 24, 32, 38

ANSWERS

1. linear, exponential, quadratic; $y = mx + b$, $y = ab^x$, $y = ax^2 + bx + c$; a straight line, a continuously increasing or decreasing curve, a parabola
2. *Sample answer:* Plot the points from the data set and determine what shape they appear to take.
3. Find the slope of the line through $(a, f(a))$ and $(b, f(b))$.
4. n; n is an exponential function, but the rest are quadratic functions.
5. quadratic
6. exponential
7. exponential
8. linear
9. 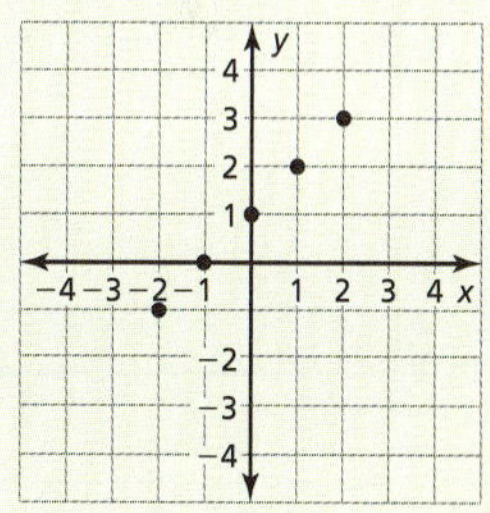

   linear
10. 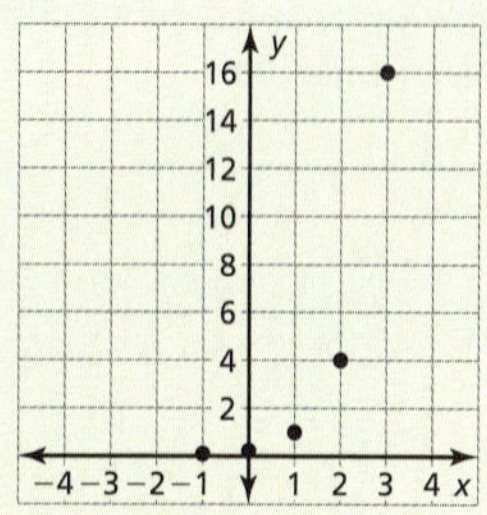

   exponential
11. 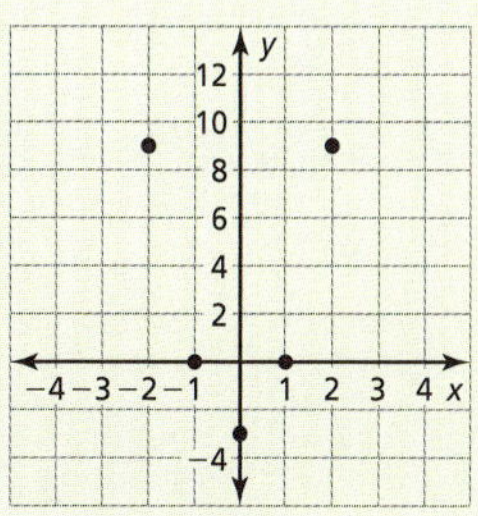

   quadratic
12. 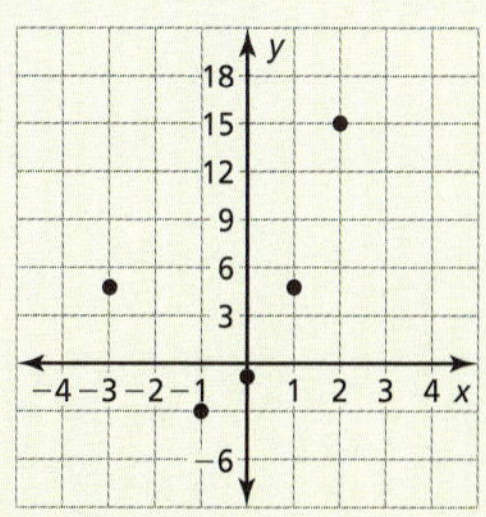

   quadratic

13–18. See Additional Answers.

Dynamic Teaching Tools

- Dynamic Assessment & Progress Monitoring Tool
- Interactive Whiteboard Lesson Library
- Dynamic Classroom with Dynamic Investigations

ANSWERS

19. linear; The first differences are constant.

20. quadratic; The second differences are constant.

21. quadratic; $y = 2x^2 - 2x - 4$

22. exponential; $y = \left(\frac{1}{2}\right)^x$

23. linear; $y = -3x - 2$

24. exponential; $y = 5(2)^x$

25. quadratic; $y = -3x^2$

26. linear; $y = -2x - 2$

27. Consecutive y-values have a constant ratio. They do not change by a constant amount; Consecutive y-values change by a constant ratio. So, the table represents an exponential function.

28. The factor for the first x-intercept should be $(x + 2)$, not $(x - 2)$; $f(x) = a(x + 2)(x - 1)$; $4 = a(-3 + 2)(-3 - 1)$; $1 = a$; $f(x) = 1(x + 2)(x - 1)$; $= x^2 + x - 2$; So, the function is $f(x) = x^2 + x - 2$.

29. a.

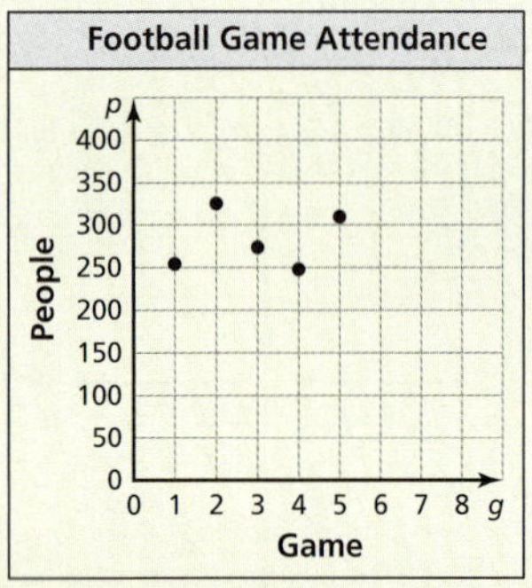

b. no; The points do not appear to follow the shape of any of these types of functions.

19. MODELING WITH MATHEMATICS A student takes a subway to a public library. The table shows the distances d (in miles) the student travels in t minutes. Let the time t represent the independent variable. Tell whether the data can be modeled by a *linear*, an *exponential*, or a *quadratic* function. Explain.

Time, t	0.5	1	3	5
Distance, d	0.335	0.67	2.01	3.35

20. MODELING WITH MATHEMATICS A store sells custom circular rugs. The table shows the costs c (in dollars) of rugs that have diameters of d feet. Let the diameter d represent the independent variable. Tell whether the data can be modeled by a *linear*, an *exponential*, or a *quadratic* function. Explain.

Diameter, d	3	4	5	6
Cost, c	63.90	113.60	177.50	255.60

In Exercises 21–26, tell whether the data represent a *linear*, an *exponential*, or a *quadratic* function. Then write the function. *(See Example 3.)*

21. $(-2, 8), (-1, 0), (0, -4), (1, -4), (2, 0), (3, 8)$

22. $(-3, 8), (-2, 4), (-1, 2), (0, 1), (1, 0.5)$

23.

x	−2	−1	0	1	2
y	4	1	−2	−5	−8

24.

x	−1	0	1	2	3
y	2.5	5	10	20	40

25.

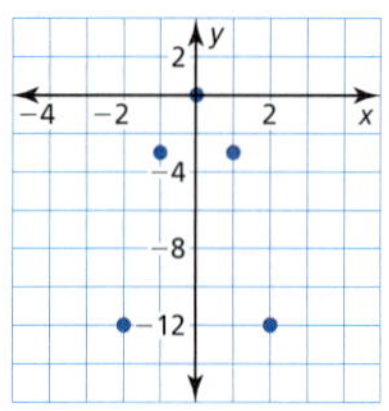

26.

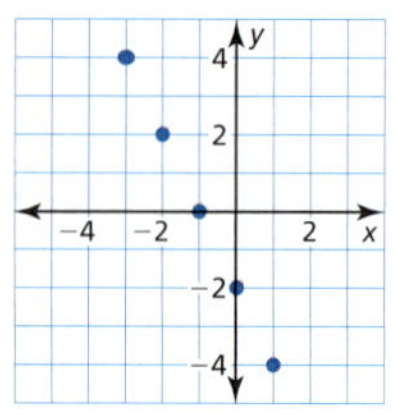

27. ERROR ANALYSIS Describe and correct the error in determining whether the table represents a linear, an exponential, or a quadratic function.

✗

x	1	2	3	4	5
y	3	9	27	81	243

x: +1, +1, +1, +1

y: ×3, ×3, ×3, ×3

Consecutive y-values change by a constant amount. So, the table represents a linear function.

28. ERROR ANALYSIS Describe and correct the error in writing the function represented by the table.

✗

x	−3	−2	−1	0	1
y	4	0	−2	−2	0

x: +1, +1, +1, +1

first differences → −4, −2, +0, +2

second differences → +2, +2, +2

The table represents a quadratic function.

$f(x) = a(x - 2)(x - 1)$

$4 = a(-3 - 2)(-3 - 1)$

$\frac{1}{5} = a$

$f(x) = \frac{1}{5}(x - 2)(x - 1)$

$= \frac{1}{5}x^2 - \frac{3}{5}x + \frac{2}{5}$

So, the function is $f(x) = \frac{1}{5}x^2 - \frac{3}{5}x + \frac{2}{5}$.

29. REASONING The table shows the numbers of people attending the first five football games at a high school.

Game, g	1	2	3	4	5
People, p	252	325	270	249	310

a. Plot the points. Let the game g represent the independent variable.

b. Can a linear, an exponential, or a quadratic function represent this situation? Explain.

30. **MODELING WITH MATHEMATICS** The table shows the breathing rates y (in liters of air per minute) of a cyclist traveling at different speeds x (in miles per hour).

Speed, x	20	21	22	23	24
Breathing rate, y	51.4	57.1	63.3	70.3	78.0

a. Plot the points. Let the speed x represent the independent variable. Then determine the type of function that best represents this situation.

b. Write a function that models the data.

c. Find the breathing rate of a cyclist traveling 18 miles per hour. Round your answer to the nearest tenth.

31. **ANALYZING RATES OF CHANGE** The function $f(t) = -16t^2 + 48t + 3$ represents the height (in feet) of a volleyball t seconds after it is hit into the air.

a. Copy and complete the table.

t	0	0.5	1	1.5	2	2.5	3
$f(t)$							

b. Plot the ordered pairs and draw a smooth curve through the points.

c. Describe where the function is increasing and decreasing.

d. Find the average rate of change for each 0.5-second interval in the table. What do you notice about the average rates of change when the function is increasing? decreasing?

32. **ANALYZING RELATIONSHIPS** The population of Town A in 1970 was 3000. The population of Town A increased by 20% every decade. Let x represent the number of decades since 1970. The graph shows the population of Town B. *(See Example 4.)*

a. Compare the populations of the towns by calculating and interpreting the average rates of change from 1990 to 2010.

b. Predict which town will have a greater population after 2030. Explain.

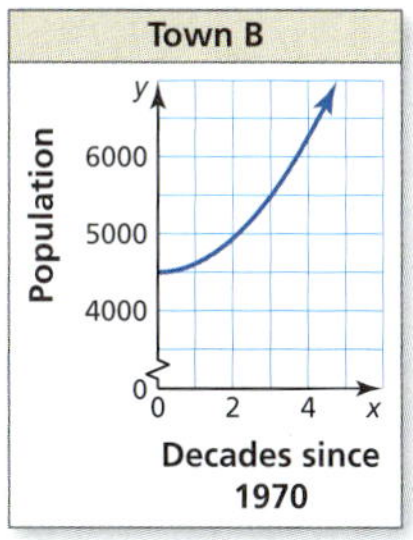

33. **ANALYZING RELATIONSHIPS** Three organizations are collecting donations for a cause. Organization A begins with one donation, and the number of donations quadruples each hour. The table shows the numbers of donations collected by Organization B. The graph shows the numbers of donations collected by Organization C.

Time (hours), t	**Number of donations, y**
0	0
1	4
2	8
3	12
4	16
5	20
6	24

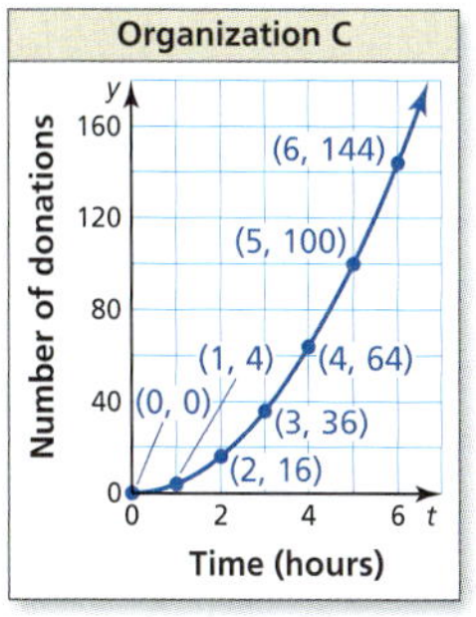

a. What type of function represents the numbers of donations collected by Organization A? B? C?

b. Find the average rates of change of each function for each 1-hour interval from $t = 0$ to $t = 6$.

c. For which function does the average rate of change increase most quickly? What does this tell you about the numbers of donations collected by the three organizations?

34. **COMPARING FUNCTIONS** The room expenses for two different resorts are shown. *(See Example 5.)*

a. For what length of vacation does each resort cost about the same?

b. Suppose Blue Water Resort charges $1450 for the first three nights and $105 for each additional night. Would Sea Breeze Resort ever be more expensive than Blue Water Resort? Explain.

c. Suppose Sea Breeze Resort charges $1200 for the first three nights. The charge increases 10% for each additional night. Would Blue Water Resort ever be more expensive than Sea Breeze Resort? Explain.

ANSWERS

30. a.

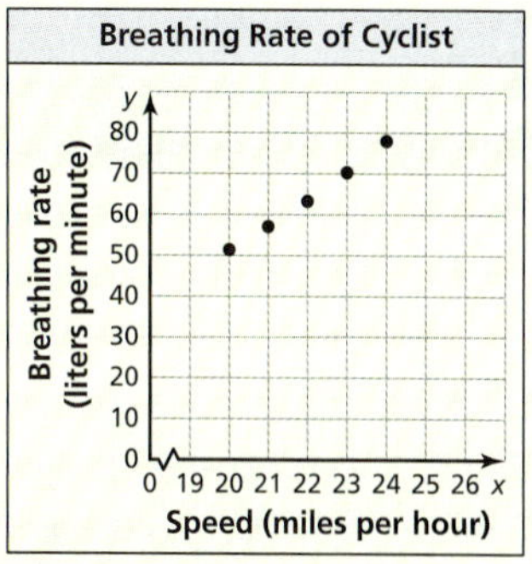

exponential function

b. $y = 51.4(1.11)^{x-20}$

c. about 41.7 L/min

31. a. 3, 23, 35, 39, 35, 23, 3

b.

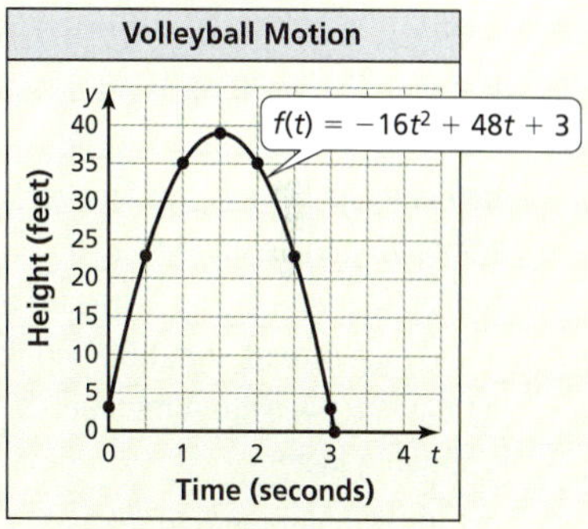

c. The function is increasing between 0 and 1.5 seconds, and the function is decreasing between 1.5 seconds and about 3 seconds.

d. 40, 24, 8, −8, −24, −40; The average rate of change decreases when the function is increasing, and the average rate of change increases in the negative direction when the function is decreasing.

32. a. From 1990 to 2010, the population of Town A increased at an average rate of about 950 people per decade and the population of Town B increased at an average rate of 660 people per decade. So, the population of Town A is growing faster.

b. Town A; In 2010, the populations of both towns are about equal and the average rate of change of Town A, which shows exponential growth, exceeds the average rate of change of Town B, which appears to show exponential or quadratic growth. So, Town A will have a greater population after 2030.

33. a. exponential; linear; quadratic

b. Organization A: 3, 12, 48, 192, 768, 3072; Organization B: 4, 4, 4, 4, 4, 4; Organization C: 4, 12, 20, 28, 36, 44

c. Organization A; Organization A will have the most donations, followed by Organization C, then Organization B.

34. a. 9 nights

b. yes; Because the cost at Sea Breeze Resort shows exponential growth, it will eventually be greater than the cost at Blue Water Resort, which shows linear growth.

c. no; The cost of the first three nights is greater at Sea Breeze Resort than at Blue Water Resort. Also, the cost of each additional night at Sea Breeze Resort is greater than the cost of each additional night at Blue Water Resort. So, the total cost is always greater at Sea Breeze Resort.

ANSWERS

35–50. See Additional Answers.

Mini-Assessment

1. Tell whether the table of values represents a *linear*, an *exponential*, or a *quadratic* function. Then write the function.

x	0	1	2	3	4
y	8	3	0	−1	0

quadratic function; $y = x^2 - 6x + 8$

2. Use the table and graph in Example 4. Compare the social media websites by calculating and interpreting the average rates of change from Day 0 to Day 20. From Day 0 to Day 20, Website A membership increases at an average rate of 75 people per day, and Website B membership increases at an average rate of about 60 people per day. So, Website A membership is growing faster.

3. In 1900, Smalltown had a population of 2000 people. Smalltown's population increased by 100 people each year. In 1900, Growton had a population of 1000 people. Growton's population increased by 6% each year.
   a. In what year were the populations about equal? 1926
   b. Suppose Smalltown's initial population doubled to 4000 and maintained a constant rate of increase of 100 people each year. Did Growton's population still catch up to Smalltown's population? If so, in which year? Yes; 1934
   c. Suppose Smalltown's rate of increase doubled to 200 people each year, in addition to doubling the initial population. Did Growton's population still catch up to Smalltown's population? Explain. Yes, in 1943; Growton's population shows exponential growth, so it will eventually exceed Smalltown's population, which shows linear growth.

35. REASONING Explain why the average rate of change of a linear function is constant and the average rate of change of a quadratic or exponential function is not constant.

36. HOW DO YOU SEE IT? Match each graph with its function. Explain your reasoning.

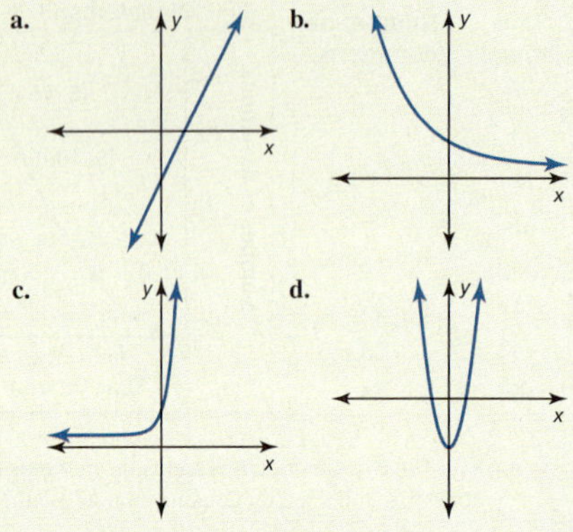

A. $y = 2x^2 - 4$ **B.** $y = 2(4)^x + 1$

C. $y = 2\left(\frac{3}{4}\right)^x + 1$ **D.** $y = 2x - 4$

37. CRITICAL THINKING In the ordered pairs below, the y-values are given in terms of n. Tell whether the ordered pairs represent a *linear*, an *exponential*, or a *quadratic* function. Explain.

$(1, 3n - 1), (2, 10n + 2), (3, 26n),$
$(4, 51n - 7), (5, 85n - 19)$

38. USING STRUCTURE Write a function that has constant second differences of 3.

39. CRITICAL THINKING Is the graph of a set of points enough to determine whether the points represent a linear, an exponential, or a quadratic function? Justify your answer.

40. THOUGHT PROVOKING Find four different patterns in the figure. Determine whether each pattern represents a *linear*, an *exponential*, or a *quadratic* function. Write a model for each pattern.

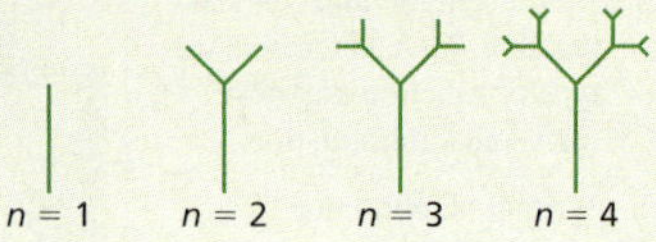

41. MAKING AN ARGUMENT Function p is an exponential function and function q is a quadratic function. Your friend says that after about $x = 3$, function q will always have a greater y-value than function p. Is your friend correct? Explain.

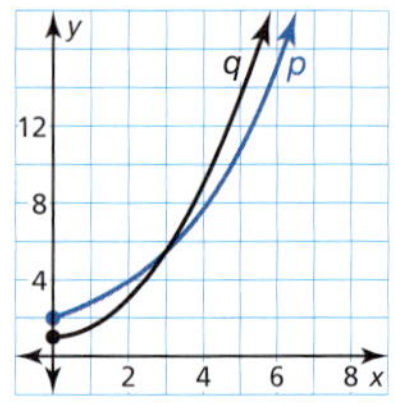

42. USING TOOLS The table shows the amount a (in billions of dollars) United States residents spent on pets or pet-related products and services each year for a 5-year period. Let the year x represent the independent variable. Using technology, find a function that models the data. How did you choose the model? Predict how much residents will spend on pets or pet-related products and services in Year 7.

Year, x	1	2	3	4	5
Amount, a	53.1	56.9	61.8	65.7	67.1

Maintaining Mathematical Proficiency

Reviewing what you learned in previous grades and lessons

Evaluate the expression. *(Section 6.2)*

43. $\sqrt{121}$ **44.** $\sqrt[3]{125}$

45. $\sqrt[3]{512}$ **46.** $\sqrt[5]{243}$

Find the product. *(Section 7.3)*

47. $(x + 8)(x - 8)$ **48.** $(4y + 2)(4y - 2)$

49. $(3a - 5b)(3a + 5b)$ **50.** $(-2r + 6s)(-2r - 6s)$

If students need help...	*If students got it...*
Resources by Chapter • Practice A and Practice B • Puzzle Time	Resources by Chapter • Enrichment and Extension • Cumulative Review
Student Journal • Practice	Start the *next* Section
Differentiating the Lesson Skills Review Handbook	

8.4–8.6 What Did You Learn?

Core Vocabulary

even function, *p. 442*
odd function, *p. 442*
vertex form (of a quadratic function), *p. 444*
intercept form, *p. 450*
average rate of change, *p. 462*

Core Concepts

Section 8.4

Even and Odd Functions, *p. 442*
Graphing $f(x) = a(x - h)^2$, *p. 443*
Graphing $f(x) = a(x - h)^2 + k$, *p. 444*
Writing Quadratic Functions of the Form $f(x) = a(x - h)^2 + k$, *p. 445*

Section 8.5

Graphing $f(x) = a(x - p)(x - q)$, *p. 450*
Factors and Zeros, *p. 452*
Using Characteristics to Graph and Write Quadratic Functions, *p. 452*
Using Characteristics to Graph and Write Cubic Functions, *p. 453*

Section 8.6

Linear, Exponential, and Quadratic Functions, *p. 460*
Differences and Ratios of Functions, *p. 461*
Writing Functions to Model Data, *p. 462*
Comparing Functions Using Average Rates of Change, *p. 463*

Mathematical Practices

1. How can you use technology to confirm your answer in Exercise 64 on page 448?
2. How did you use the structure of the equation in Exercise 85 on page 457 to solve the problem?
3. Describe why your answer makes sense considering the context of the data in Exercise 20 on page 466.

Performance Task

Asteroid Aim

Apps take a long time to design and program. One app in development is a game in which players shoot lasers at asteroids. They score points based on the number of hits per shot. The designer wants your feedback. Do you think students will like the game and want to play it? What changes would improve it?

To explore the answers to this question and more, go to ***BigIdeasMath.com***.

469

Dynamic Teaching Tools

- Dynamic Assessment & Progress Monitoring Tool
- Interactive Whiteboard Lesson Library
- Dynamic Classroom with Dynamic Investigations

ANSWERS

1. *Sample answer:* Graph the function on a graphing calculator and compare it to your graph.
2. *Sample answer:* The equation can be factored into intercept form, and the width can be determined from the intercepts.
3. *Sample answer:* The area of a circular rug, πr^2, is a quadratic function, so it makes sense that the cost, based on the amount of material needed to make the rug, would also be a quadratic function.

8 Chapter Review

Dynamic Solutions available at *BigIdeasMath.com*

8.1 Graphing $f(x) = ax^2$ (pp. 419–424)

Graph $g(x) = -4x^2$. Compare the graph to the graph of $f(x) = x^2$.

Step 1 Make a table of values.

x	−2	−1	0	1	2
$g(x)$	−16	−4	0	−4	−16

Step 2 Plot the ordered pairs.

Step 3 Draw a smooth curve through the points.

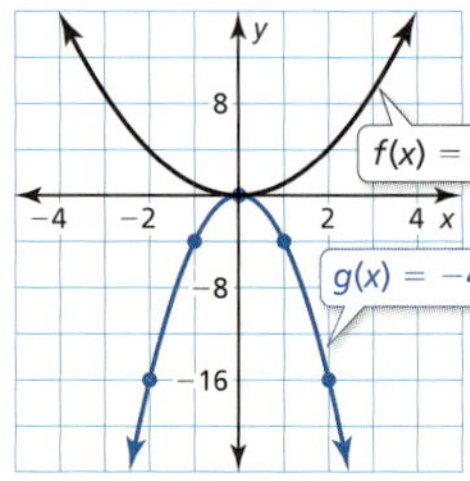

▶ The graphs have the same vertex, (0, 0), and the same axis of symmetry, $x = 0$, but the graph of g opens down and is narrower than the graph of f. So, the graph of g is a vertical stretch by a factor of 4 and a reflection in the x-axis of the graph of f.

Graph the function. Compare the graph to the graph of $f(x) = x^2$.

1. $p(x) = 7x^2$ **2.** $q(x) = \frac{1}{2}x^2$ **3.** $g(x) = -\frac{3}{4}x^2$ **4.** $h(x) = -6x^2$

5. Identify characteristics of the quadratic function and its graph.

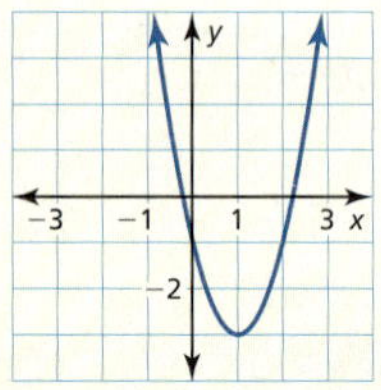

8.2 Graphing $f(x) = ax^2 + c$ (pp. 425–430)

Graph $g(x) = 2x^2 + 3$. Compare the graph to the graph of $f(x) = x^2$.

Step 1 Make a table of values.

x	−2	−1	0	1	2
$g(x)$	11	5	3	5	11

Step 2 Plot the ordered pairs.

Step 3 Draw a smooth curve through the points.

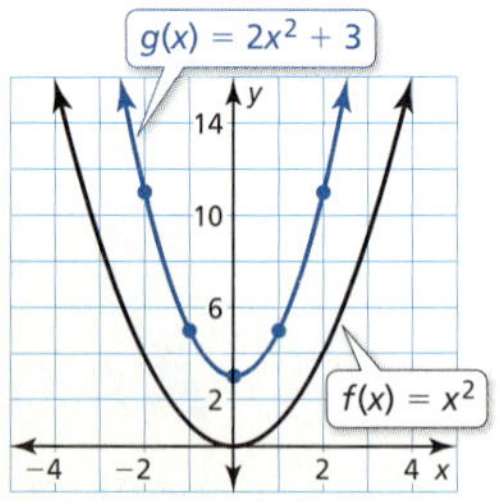

▶ Both graphs open up and have the same axis of symmetry, $x = 0$. The graph of g is narrower, and its vertex, (0, 3), is above the vertex of the graph of f, (0, 0). So, the graph of g is a vertical stretch by a factor of 2 and a vertical translation 3 units up of the graph of f.

Graph the function. Compare the graph to the graph of $f(x) = x^2$.

6. $g(x) = x^2 + 5$ **7.** $h(x) = -x^2 - 4$ **8.** $m(x) = -2x^2 + 6$ **9.** $n(x) = \frac{1}{3}x^2 - 5$

ANSWERS

1.

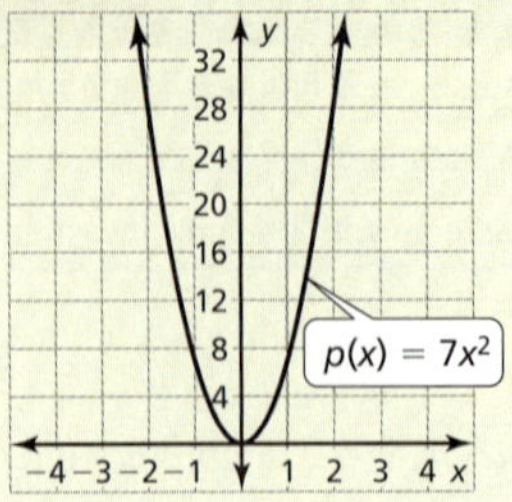

The graph of p is a vertical stretch by a factor of 7 of the graph of f.

2.

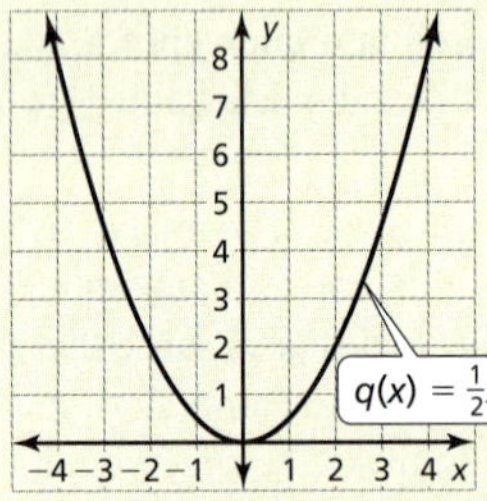

The graph of q is a vertical shrink by a factor of $\frac{1}{2}$ of the graph of f.

3.

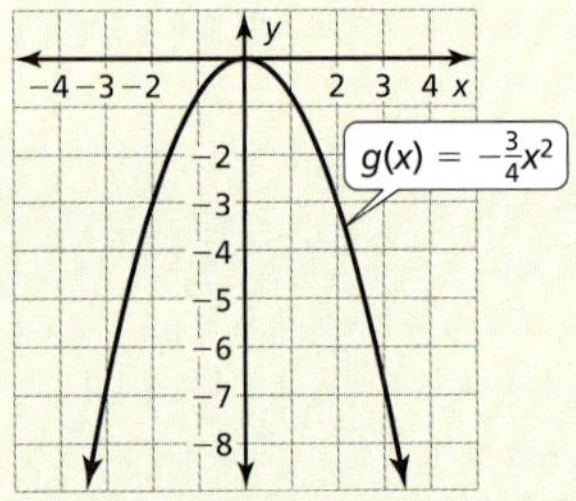

The graph of g is a vertical shrink by a factor of $\frac{3}{4}$ and a reflection in the x-axis of the graph of f.

4.

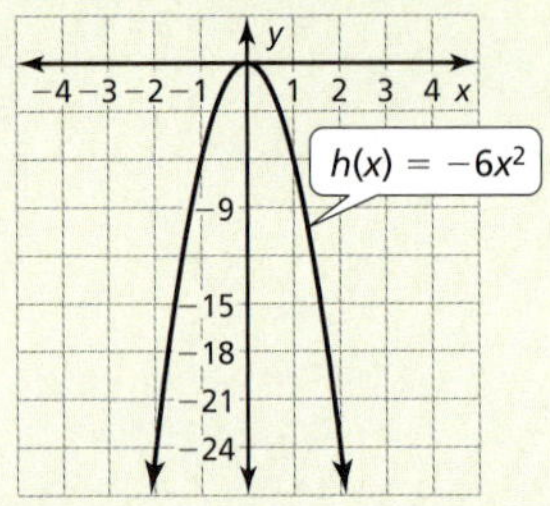

The graph of h is a vertical stretch by a factor of 6 and a reflection in the x-axis of the graph of f.

5. The vertex is (1, −3). The axis of symmetry is $x = 1$. The domain is all real numbers. The range is $y \geq -3$. When $x < 1$, y increases as x decreases. When $x > 1$, y increases as x increases.

6.

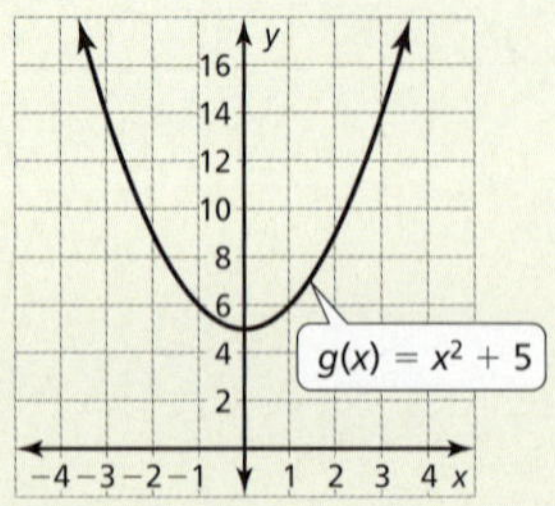

The graph of g is a vertical translation 5 units up of the graph of f.

7.

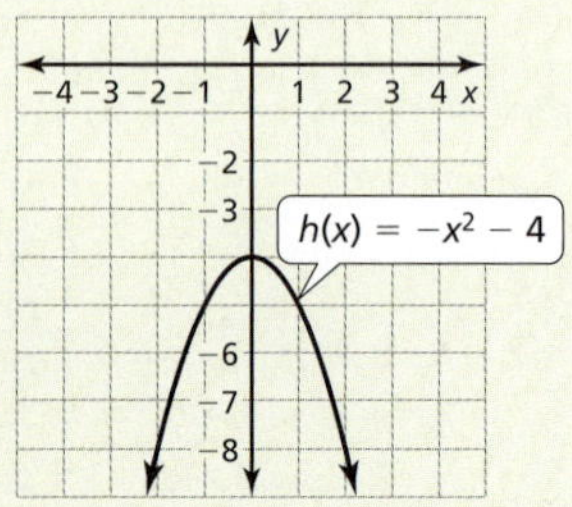

The graph of h is a reflection in the x-axis, and a vertical translation 4 units down of the graph of f.

8–9. See Additional Answers.

8.3 Graphing $f(x) = ax^2 + bx + c$ (pp. 431–438)

Graph $f(x) = 4x^2 + 8x - 1$. Describe the domain and range.

Step 1 Find and graph the axis of symmetry: $x = -\frac{b}{2a} = -\frac{8}{2(4)} = -1$.

Step 2 Find and plot the vertex. The axis of symmetry is $x = -1$. So, the x-coordinate of the vertex is -1. The y-coordinate of the vertex is $f(-1) = 4(-1)^2 + 8(-1) - 1 = -5$. So, the vertex is $(-1, -5)$.

Step 3 Use the y-intercept to find two more points on the graph. Because $c = -1$, the y-intercept is -1. So, $(0, -1)$ lies on the graph. Because the axis of symmetry is $x = -1$, the point $(-2, -1)$ also lies on the graph.

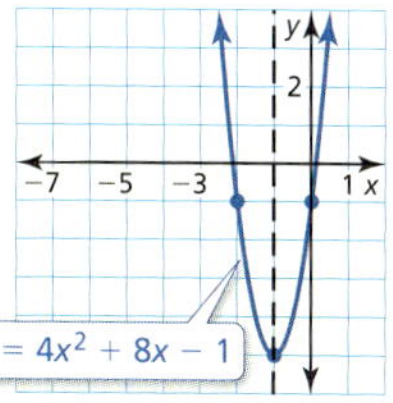

Step 4 Draw a smooth curve through the points.

▶ The domain is all real numbers. The range is $y \geq -5$.

Graph the function. Describe the domain and range.

10. $y = x^2 - 2x + 7$ **11.** $f(x) = -3x^2 + 3x - 4$ **12.** $y = \frac{1}{2}x^2 - 6x + 10$

13. The function $f(t) = -16t^2 + 88t + 12$ represents the height (in feet) of a pumpkin t seconds after it is launched from a catapult. When does the pumpkin reach its maximum height? What is the maximum height of the pumpkin?

8.4 Graphing $f(x) = a(x - h)^2 + k$ (pp. 441–448)

Determine whether $f(x) = 2x^2 + 4$ is *even*, *odd*, or *neither*.

$f(x) = 2x^2 + 4$ Write the original function.

$f(-x) = 2(-x)^2 + 4$ Substitute $-x$ for x.

$= 2x^2 + 4$ Simplify.

$= f(x)$ Substitute $f(x)$ for $2x^2 + 4$.

▶ Because $f(-x) = f(x)$, the function is even.

Determine whether the function is *even*, *odd*, or *neither*.

14. $w(x) = 5^x$ **15.** $r(x) = -8x$ **16.** $h(x) = 3x^2 - 2x$

Graph the function. Compare the graph to the graph of $f(x) = x^2$.

17. $h(x) = 2(x - 4)^2$ **18.** $g(x) = \frac{1}{2}(x - 1)^2 + 1$ **19.** $q(x) = -(x + 4)^2 + 7$

20. Consider the function $g(x) = -3(x + 2)^2 - 4$. Graph $h(x) = g(x - 1)$.

21. Write a quadratic function whose graph has a vertex of (3, 2) and passes through the point (4, 7).

ANSWERS

10.

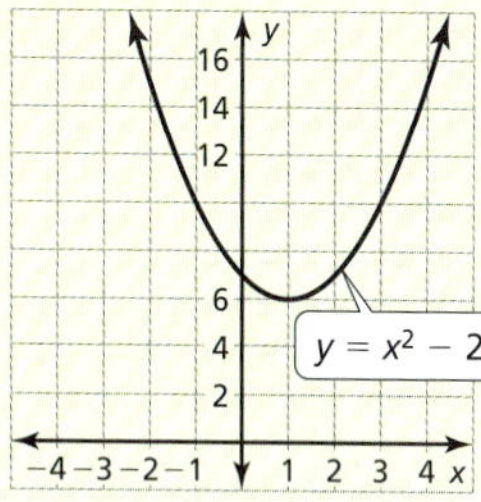

domain: all real numbers, range: $y \geq 6$

11.

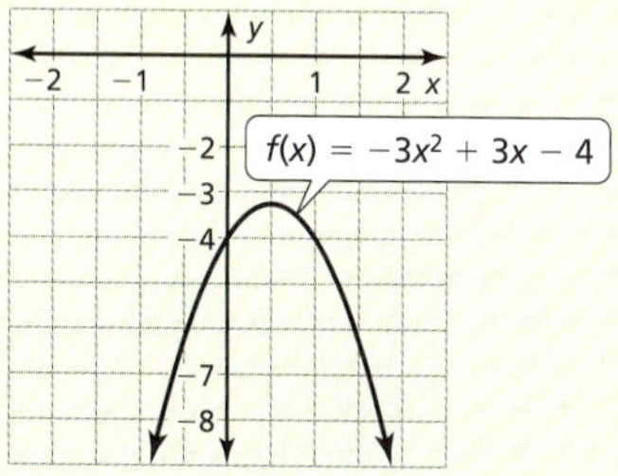

domain: all real numbers, range: $y \leq -\frac{13}{4}$

12.

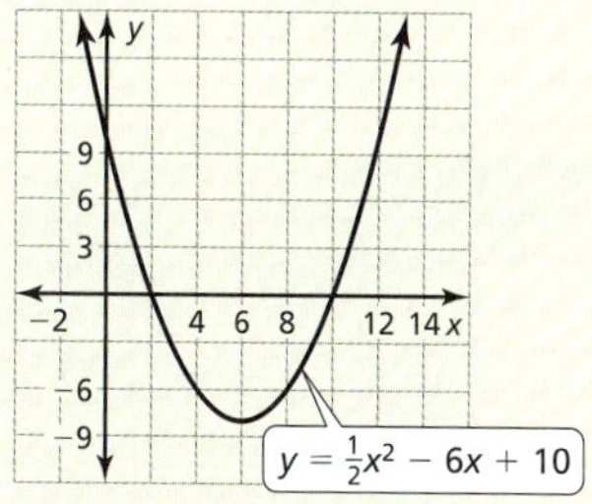

domain: all real numbers, range: $y \geq -8$

13. 2.75 sec; 133 ft

14. neither

15. odd

16. neither

17.

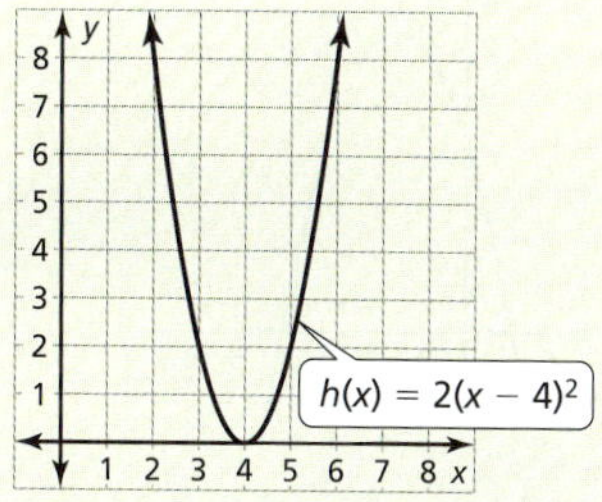

The graph of h is a vertical stretch by a factor of 2 and a horizontal translation 4 units right of the graph of f.

18.

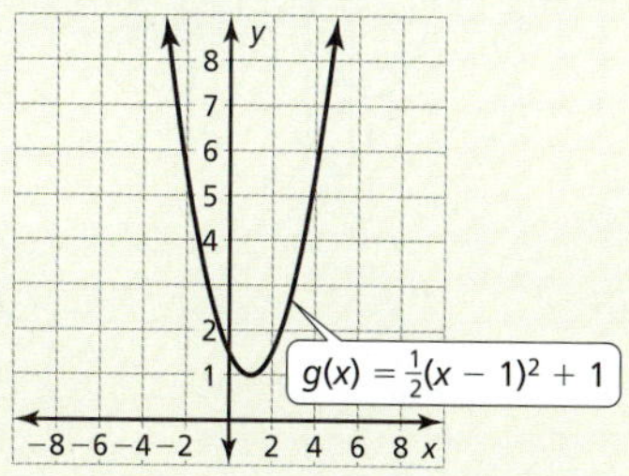

The graph of g is a vertical shrink by a factor of $\frac{1}{2}$, and a translation 1 unit right and 1 unit up of the graph of f.

19.

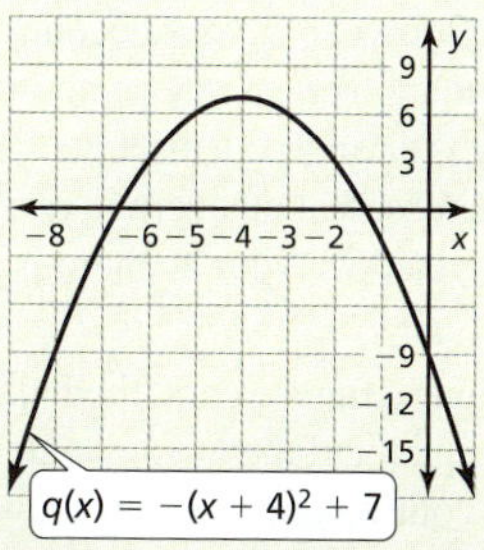

The graph of q is a reflection in the x-axis, and a translation 4 units left and 7 units up of the graph of f.

20.

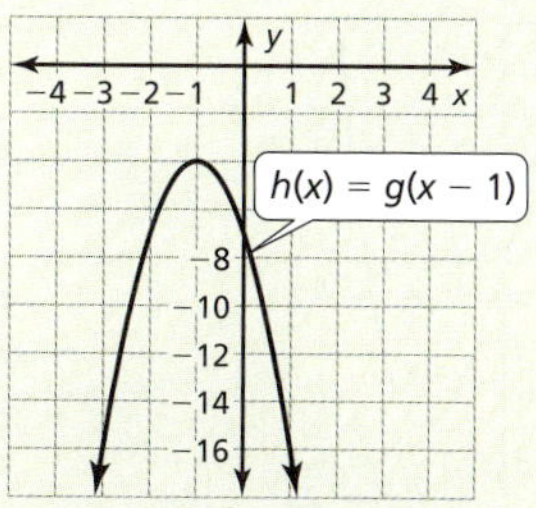

21. $f(x) = 5(x - 3)^2 + 2$

ANSWERS

22.

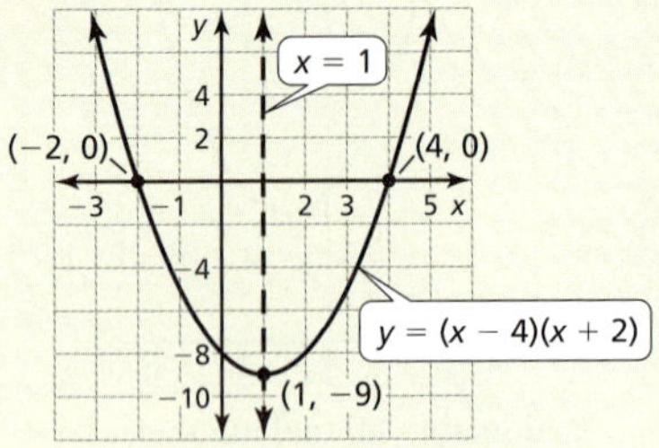

domain: all real numbers,
range: $y \geq -9$

23.

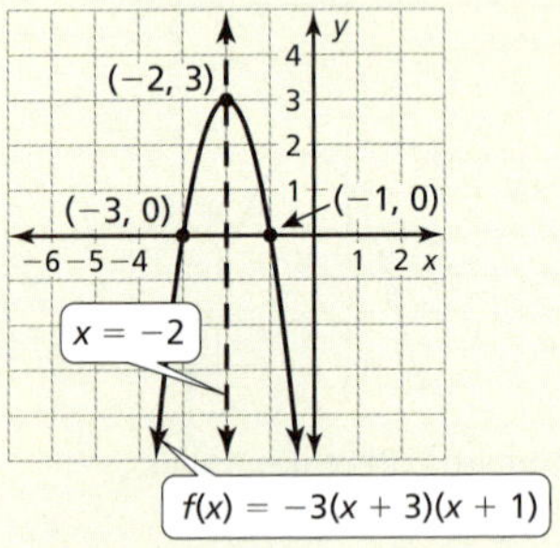

domain: all real numbers,
range: $y \leq 3$

24.

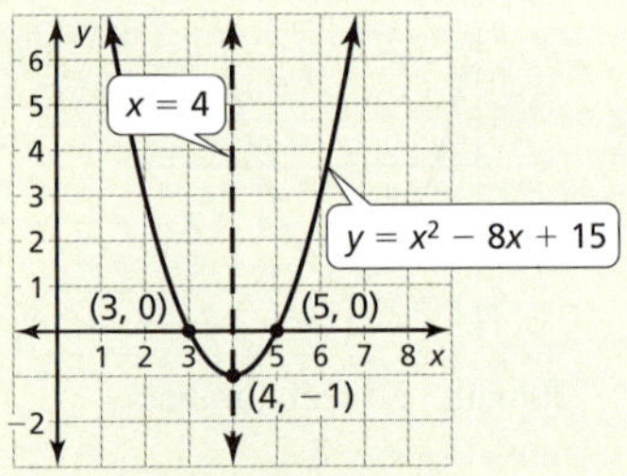

domain: all real numbers,
range: $y \geq -1$

25.

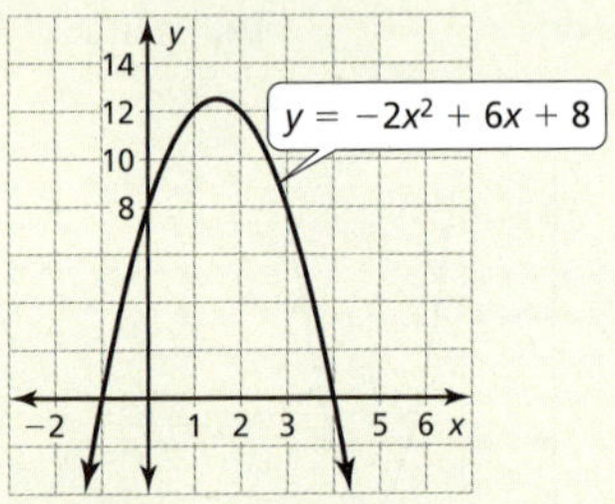

26.

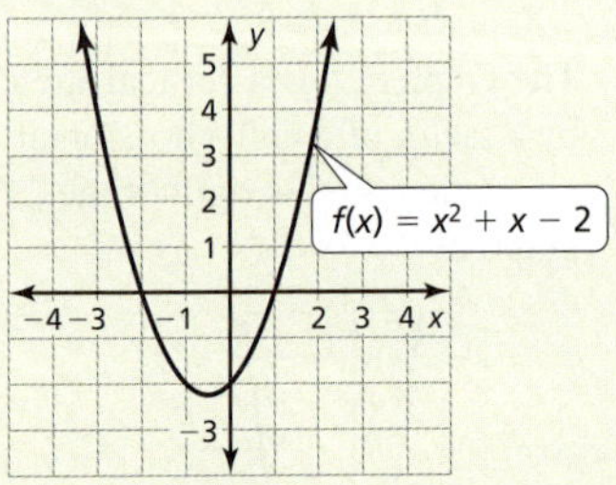

8.5 Using Intercept Form (pp. 449–458)

Use zeros to graph $h(x) = x^2 - 7x + 6$.

The function is in standard form. The parabola opens up ($a > 0$), and the y-intercept is 6. So, plot (0, 6).

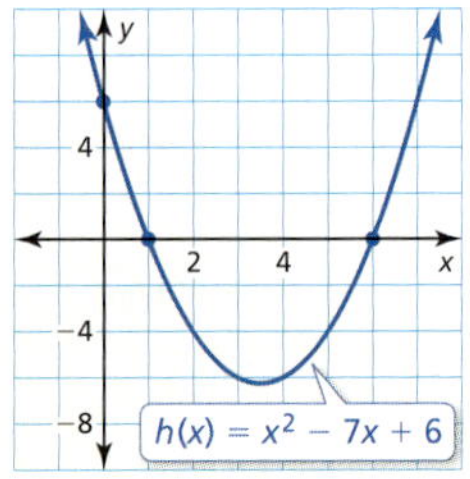

The polynomial that defines the function is factorable. So, write the function in intercept form and identify the zeros.

$h(x) = x^2 - 7x + 6$ Write the function.

$= (x - 6)(x - 1)$ Factor the trinomial.

The zeros of the function are 1 and 6. So, plot (1, 0) and (6, 0). Draw a parabola through the points.

Graph the quadratic function. Label the vertex, axis of symmetry, and x-intercepts. Describe the domain and range of the function.

22. $y = (x - 4)(x + 2)$ **23.** $f(x) = -3(x + 3)(x + 1)$ **24.** $y = x^2 - 8x + 15$

Use zeros to graph the function.

25. $y = -2x^2 + 6x + 8$ **26.** $f(x) = x^2 + x - 2$ **27.** $f(x) = 2x^3 - 18x$

28. Write a quadratic function in standard form whose graph passes through (4, 0) and (6, 0).

8.6 Comparing Linear, Exponential, and Quadratic Functions (pp. 459–468)

Tell whether the data represent a *linear*, an *exponential*, or a *quadratic* function.

a. (−4, 1), (−3, −2), (−2, −3) (−1, −2), (0, 1)

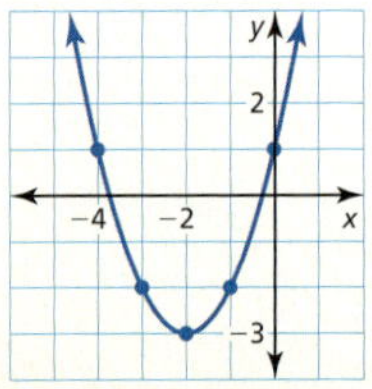

▶ The points appear to represent a quadratic function.

b.

x	−1	0	1	2	3
y	15	8	1	−6	−13

+1 +1 +1 +1

x	−1	0	1	2	3
y	15	8	1	−6	−13

−7 −7 −7 −7

▶ The first differences are constant. So, the table represents a linear function.

29. Tell whether the table of values represents a *linear*, an *exponential*, or *a quadratic* function. Then write the function.

x	−1	0	1	2	3
y	512	128	32	8	2

30. The balance y (in dollars) of your savings account after t years is represented by $y = 200(1.1)^t$. The beginning balance of your friend's account is $250, and the balance increases by $20 each year. (a) Compare the account balances by calculating and interpreting the average rates of change from $t = 2$ to $t = 7$. (b) Predict which account will have a greater balance after 10 years. Explain.

27.

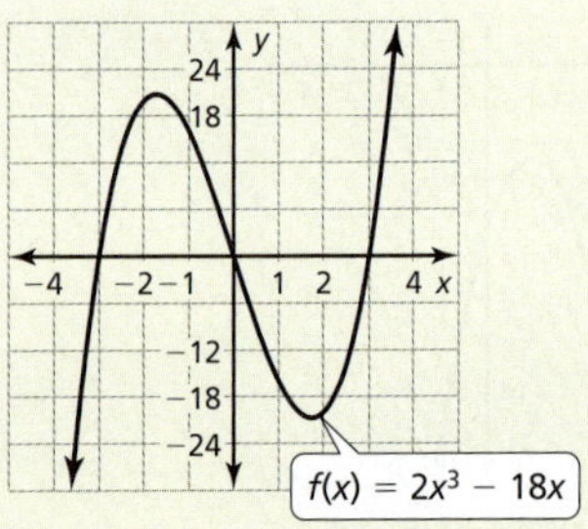

28. *Sample answer:* $x^2 - 10x + 24$

29. exponential; $y = 128\left(\frac{1}{4}\right)^x$

30. a. From year 2 to year 7, your account increased at an average rate of about $29.55 per year and your friend's account increased at an average rate of $20 per year. So, your account is growing faster.

b. your account; After 7 years, the balances are about equal and the average rate of change of your balance, which shows exponential growth, exceeds the average rate of change of your friend's balance, which shows linear growth. So, your account will have a greater balance after 10 years.

8 Chapter Test

Graph the function. Compare the graph to the graph of $f(x) = x^2$.

1. $h(x) = 2x^2 - 3$ **2.** $g(x) = -\frac{1}{2}x^2$ **3.** $p(x) = \frac{1}{2}(x + 1)^2 - 1$

4. Consider the graph of the function f.

a. Find the domain, range, and zeros of the function.

b. Write the function f in standard form.

c. Compare the graph of f to the graph of $g(x) = x^2$.

d. Graph $h(x) = f(x - 6)$.

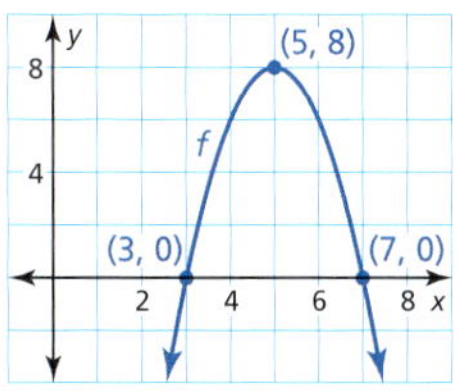

Use zeros to graph the function. Describe the domain and range of the function.

5. $f(x) = 2x^2 - 8x + 8$ **6.** $y = -(x + 5)(x - 1)$ **7.** $h(x) = 16x^2 - 4$

Tell whether the table of values represents a *linear*, an *exponential*, or a *quadratic* function. Explain your reasoning. Then write the function.

8.

x	−1	0	1	2	3
y	4	8	16	32	64

9.

x	−2	−1	0	1	2
y	−8	−2	0	−2	−8

Write a quadratic function in standard form whose graph satisfies the given conditions. Explain the process you used.

10. passes through $(-8, 0)$, $(-2, 0)$, and $(-6, 4)$

11. passes through $(0, 0)$, $(10, 0)$, and $(9, -27)$

12. is even and has a range of $y \geq 3$ **13.** passes through $(4, 0)$ and $(1, 9)$

14. The table shows the distances d (in miles) that Earth moves in its orbit around the Sun after t seconds. Let the time t be the independent variable. Tell whether the data can be modeled by a *linear*, an *exponential*, or a *quadratic* function. Explain. Then write a function that models the data.

Time, *t*	1	2	3	4	5
Distance, *d*	19	38	57	76	95

15. You are playing tennis with a friend. The path of the tennis ball after you return a serve can be modeled by the function $y = -0.005x^2 + 0.17x + 3$, where x is the horizontal distance (in feet) from where you hit the ball and y is the height (in feet) of the ball.

a. What is the maximum height of the tennis ball?

b. You are standing 30 feet from the net, which is 3 feet high. Will the ball clear the net? Explain your reasoning.

16. Find values of a, b, and c so that the function $f(x) = ax^2 + bx + c$ is (a) even, (b) odd, and (c) neither even nor odd.

17. Consider the function $f(x) = x^2 + 4$. Find the average rate of change from $x = 0$ to $x = 1$, from $x = 1$ to $x = 2$, and from $x = 2$ to $x = 3$. What do you notice about the average rates of change when the function is increasing?

If students need help...	*If students got it...*
Lesson Tutorials	Resources by Chapter • Enrichment and Extension • Cumulative Review
Skills Review Handbook	Performance Task
BigIdeasMath.com	Start the *next* Section

ANSWERS

1.

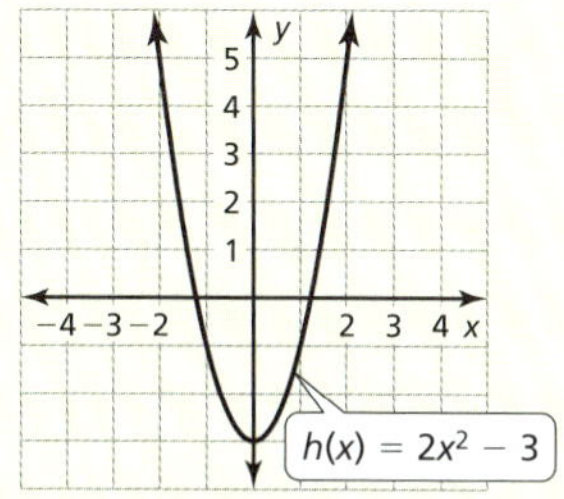

The graph of h is a vertical stretch by a factor of 2 and a vertical translation 3 units down of the graph of f.

2.

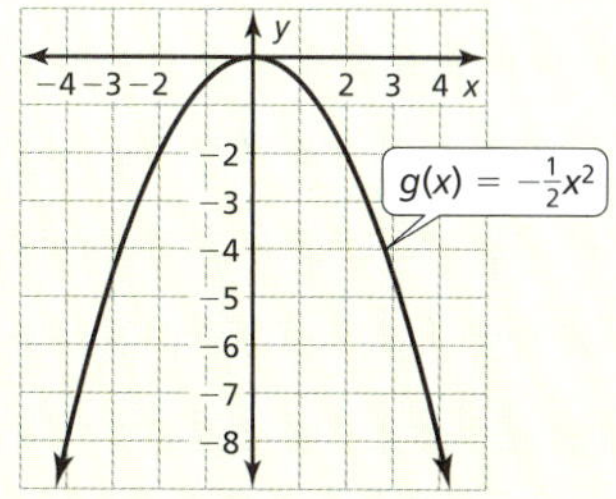

The graph of g is a vertical shrink by a factor of $\frac{1}{2}$ and a reflection in the x-axis of the graph of f.

3.

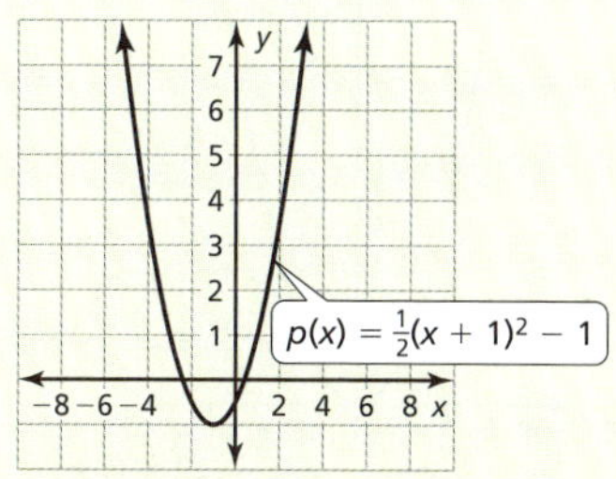

The graph of p is a vertical shrink by a factor of $\frac{1}{2}$, and a translation 1 unit left and 1 unit down of the graph of f.

4. a. domain: all real numbers, range: $y \leq 8$; 3, 7

b. $f(x) = -2x^2 + 20x - 42$

c. The graph of f is a vertical stretch by a factor of 2, a reflection in the x-axis, and a translation 5 units right and 8 units up of the graph of g.

d.

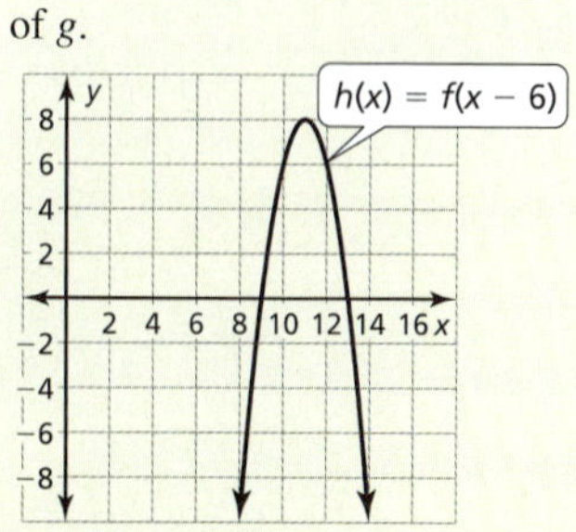

5–17. See Additional Answers.

ANSWERS

1. C
2. 1, 2, 3, 5, 8, 13, 21, 34, 55, 89
3. 28
4. one solution: $y = 6x + 9$, $y = -\frac{1}{6}x + 9$; $2x + 4y = -2$, $10x + 4y = -2$; no solution: $7x + 4y = 12$, $8y - 12 = -14x$; $y = -3x + 5$, $y = -3x + 9$; infinite solutions: $3x + y = 5$, $-15 + 3y + 9x = 0$; $y - 2x = \frac{3}{2}$, $-3 + 2y = 4x$
5. yes; *Sample answer:*

two real zeros

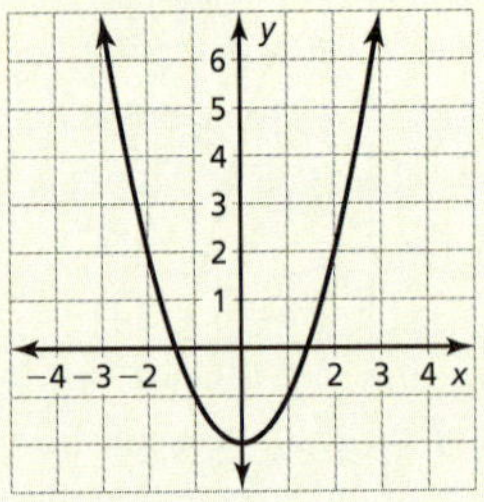

one real zero

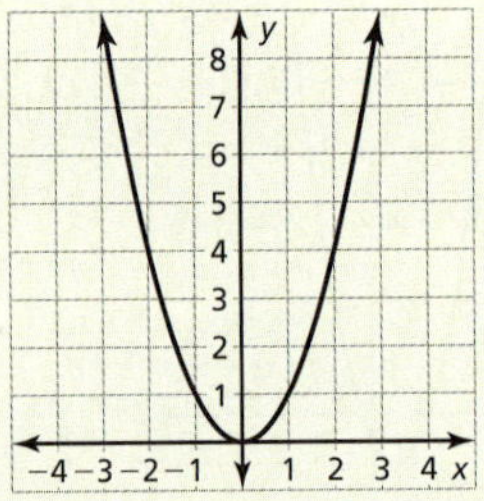

no real zeros

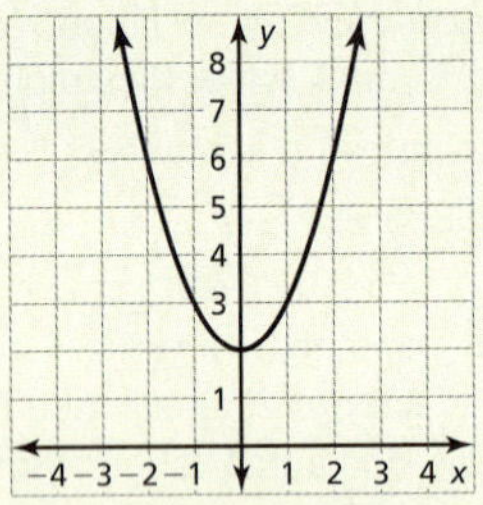

6. C

8 Cumulative Assessment

1. Which function is represented by the graph?

   Ⓐ $y = \frac{1}{2}x^2$

   Ⓑ $y = 2x^2$

   Ⓒ $y = -\frac{1}{2}x^2$

   Ⓓ $y = -2x^2$

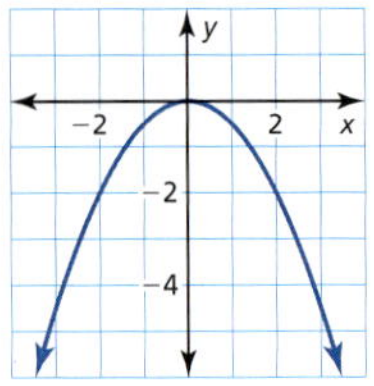

2. Find all numbers between 0 and 100 that are in the range of the function defined below.

$$f(1) = 1,\ f(2) = 1,\ f(n) = f(n-1) + f(n-2)$$

3. The function $f(t) = -16t^2 + v_0t + s_0$ represents the height (in feet) of a ball t seconds after it is thrown from an initial height s_0 (in feet) with an initial vertical velocity v_0 (in feet per second). The ball reaches its maximum height after $\frac{7}{8}$ second when it is thrown with an initial vertical velocity of ______ feet per second.

4. Classify each system of equations by the number of solutions.

$y = 6x + 9$ $y = -\frac{1}{6}x + 9$	$7x + 4y = 12$ $8y - 12 = -14x$	$2x + 4y = -2$ $10x + 4y = -2$
$3x + y = 5$ $-15 + 3y + 9x = 0$	$y - 2x = \frac{3}{2}$ $-3 + 2y = 4x$	$y = -3x + 5$ $y = -3x + 9$

5. Your friend claims that quadratic functions can have two, one, or no real zeros. Do you support your friend's claim? Use graphs to justify your answer.

6. Which polynomial represents the area (in square feet) of the shaded region of the figure?

   Ⓐ $a^2 - x^2$

   Ⓑ $x^2 - a^2$

   Ⓒ $x^2 - 2ax + a^2$

   Ⓓ $x^2 + 2ax + a^2$

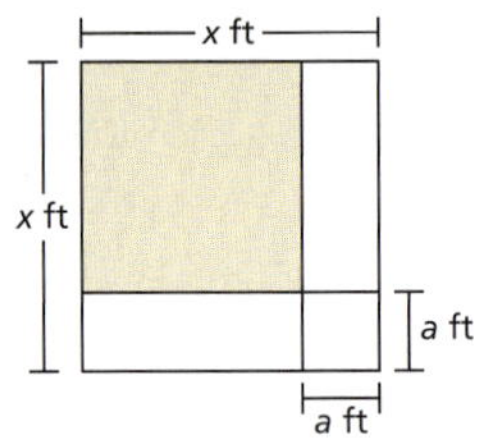

7. Consider the functions represented by the tables.

x	0	1	2	3
p(x)	−4	−16	−28	−40

x	1	2	3	4
r(x)	0	15	40	75

x	1	2	3	4
s(x)	72	36	18	9

x	1	3	5	7
t(x)	3	−5	−21	−45

a. Classify each function as *linear*, *exponential*, or *quadratic*.

b. Order the functions from least to greatest according to the average rates of change between $x = 1$ and $x = 3$.

8. Complete each function using the symbols + or −, so that the graph of the quadratic function satisfies the given conditions.

a. $f(x) = 5(x \;\square\; 3)^2 \;\square\; 4$; vertex: $(-3, 4)$

b. $g(x) = -(x \;\square\; 2)(x \;\square\; 8)$; x-intercepts: -8 and 2

c. $h(x) = \;\square\; 3x^2 \;\square\; 6$; range: $y \geq -6$

d. $j(x) = \;\square\; 4(x \;\square\; 1)(x \;\square\; 1)$; range: $y \leq 4$

9. The graph shows the amounts y (in dollars) that a referee earns for refereeing x high school volleyball games.

a. Does the graph represent a linear or nonlinear function? Explain.

b. Describe the domain of the function. Is the domain discrete or continuous?

c. Write a function that models the data.

d. Can the referee earn exactly \$500? Explain.

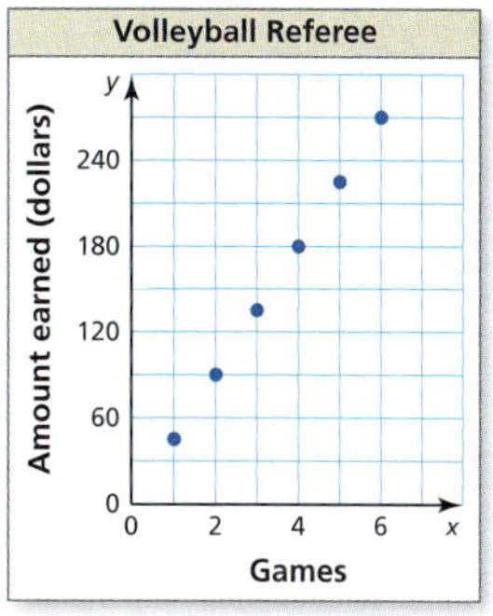

10. Which expressions are equivalent to $(b^{-5})^{-4}$?

b^{-20}	$b^{-6}b^{-14}$	$(b^{-4})^{-5}$	b^{-9}	$(b^{-2})^{-7}$
$b^{-5}b^{-4}$	$(b^{10})^2$	b^{20}	$(b^{-12})^3$	$b^{12}b^8$

ANSWERS

7. a. p is linear, r is quadratic, s is exponential, t is quadratic

b. s, p, t, r

8. a. +; +

b. −; +

c. +; −

d. −; +; − or −; −; +

9. a. linear; *Sample answer:* The first difference is constant.

b. domain: whole numbers; discrete

c. $y = 45x$

d. no; *Sample answer:* If y is 500, x is $11\frac{1}{9}$, which is not in the domain.

10. $(b^{-4})^{-5}$, $(b^{10})^2$, b^{20}, $b^{12}b^8$

Chapter 9 Pacing Guide

Chapter Opener/ Mathematical Practices	1 Day
Section 1	2 Days
Section 2	2 Days
Section 3	2 Days
Quiz	1 Day
Section 4	2 Days
Section 5	2 Days
Section 6	2 Days
Chapter Review/ Chapter Tests	2 Days
Total Chapter 9	16 Days
Year-to-Date	135 Days

9 Solving Quadratic Equations

Dolphin *(p. 521)*

Half-pipe *(p. 513)*

Pond *(p. 501)*

Kicker *(p. 493)*

Parthenon *(p. 483)*

Laurie's Notes

Chapter Summary

- This is a third chapter generally about polynomials and more specifically, about quadratics. In the previous chapters, students factored quadratics and graphed quadratic equations in different forms: standard, vertex, and intercept.
- This chapter is about solving quadratic equations, and depending upon the form in which the equations are written, different techniques are used.
- The chapter begins with a lesson on square roots and how to simplify them. These skills are needed for work later in the chapter.
- The next four lessons present different ways in which a quadratic can be solved: graphing, using square roots, completing the square, and using the Quadratic Formula. Students learn to distinguish between finding the zero of a function and finding the x-intercept of a graph.
- The final lesson in the chapter involves solving systems of nonlinear equations. This reviews techniques for solving systems from Chapter 5.

What Your Students Have Learned

Middle School

- Find square roots and evaluate expressions involving square roots.
- Solve linear systems algebraically and graphically.

What Your Students Will Learn

Algebra 1

- Simplify radical expressions and perform addition, subtraction, and multiplication with radicals.
- Solve quadratic equations with real solutions by graphing, using square roots, by completing the square, and using the Quadratic Formula.
- Interpret the discriminant to determine the number of real solutions of quadratic equations.
- Choose efficient methods for solving quadratic equations.
- Solve systems of nonlinear equations in two variables by graphing, substitution, and elimination.

Dynamic Teaching Tools

- Dynamic Assessment & Progress Monitoring Tool
- Lesson Planning Tool
- Interactive Whiteboard Lesson Library
- Dynamic Classroom with Dynamic Investigations
- Real-Life STEM Videos

Scaffolding in the Classroom

Graphic Organizers: Y-Chart

A Y-Chart can be used to compare two topics. Students list differences between the two topics in the branches of the Y and similarities in the base of the Y. A Y-Chart serves as a good tool for assessing students' knowledge of a pair of topics that have subtle but important differences. You can include blank Y-Charts on tests or quizzes for this purpose.

Questioning in the Classroom

What I really want to know is ...

Have each student write a "pertinent" question on a 3×5 card. Collect the cards (check the cards before using them) and randomly select a question to ask the class. After class discussion, the last person to participate randomly selects another question to ask the class.

Laurie's Notes

Maintaining Mathematical Proficiency

Factoring Perfect Square Trinomials

- Remind students that the sign of the middle term of a perfect square trinomial determines whether they should use addition or subtraction in the factors. For example, the second term, $14x$, in $x^2 + 14x + 49$ is positive, so addition is used in the factors: $(x + 7)(x + 7) = (x + 7)^2$.

COMMON ERROR Students confuse perfect square trinomials with the difference of two squares and write two different factors. For example, students may write $(x + 10)(x - 10)$ instead of $(x - 10)^2$ as the factors of $x^2 - 20x + 100$ in Exercise 2.

Solving Systems of Linear Equations by Graphing

- Point out that the equations are in slope-intercept form. Remind students that they can plot the y-intercept of each equation and then use the slope to find a second point on each line.

COMMON ERROR Students graph one or both of the equations incorrectly and do not check their solution in both equations.

Mathematical Practices (continued on page 478)

- The *Mathematical Practices* page focuses attention on how mathematics is learned—process versus content. This page demonstrates how a mathematically proficient student goes through the process of making adjustments in his or her thinking. Reasoning plays a role in these adjustments.
- Use the *Mathematical Practices* page to help students develop mathematical habits of mind—how mathematics can be explored and how mathematics is thought about.
- Students will be working with irrational numbers in this chapter. In the *Core Concept*, a decimal approximation of $\sqrt{2}$ is being found. Consider how students think about and understand irrational numbers. For instance, ask students the side length of a square with an area of 4 square units and they quickly reply 2 units. Repeat the same question for an area of 2 square units and common answers are 1 or 1 by 2. (They have forgotten it is a square.) By checking their answers students start to understand that they are looking for a number that when multiplied by itself has a product of 2. Their guess-and-check process will look much like the *Core Concept*.

If students need help...	*If students got it...*
Student Journal • Maintaining Mathematical Proficiency	Game Closet at *BigIdeasMath.com*
Lesson Tutorials	Start the *next* Section
Skills Review Handbook	

Maintaining Mathematical Proficiency

Factoring Perfect Square Trinomials

Example 1 Factor $x^2 + 14x + 49$.

$x^2 + 14x + 49 = x^2 + 2(x)(7) + 7^2$ — Write as $a^2 + 2ab + b^2$.

$= (x + 7)^2$ — Perfect square trinomial pattern

Factor the trinomial.

1. $x^2 + 10x + 25$
2. $x^2 - 20x + 100$
3. $x^2 + 12x + 36$
4. $x^2 - 18x + 81$
5. $x^2 + 16x + 64$
6. $x^2 - 30x + 225$

Solving Systems of Linear Equations by Graphing

Example 2 Solve the system of linear equations by graphing.

$y = 2x + 1$ — Equation 1

$y = -\frac{1}{3}x + 8$ — Equation 2

Step 1 Graph each equation.

Step 2 Estimate the point of intersection.
The graphs appear to intersect at (3, 7).

Step 3 Check your point from Step 2.

Equation 1	Equation 2
$y = 2x + 1$	$y = -\frac{1}{3}x + 8$
$7 \stackrel{?}{=} 2(3) + 1$	$7 \stackrel{?}{=} -\frac{1}{3}(3) + 8$
$7 = 7$ ✓	$7 = 7$ ✓

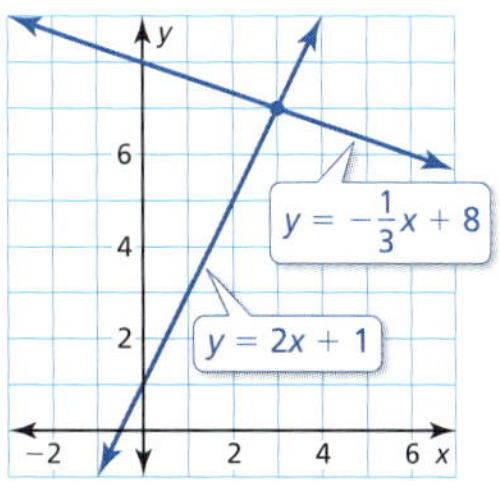

▶ The solution is (3, 7).

Solve the system of linear equations by graphing.

7. $y = -5x + 3$; $y = 2x - 4$
8. $y = \frac{3}{2}x - 2$; $y = -\frac{1}{4}x + 5$
9. $y = \frac{1}{2}x + 4$; $y = -3x - 3$

10. **ABSTRACT REASONING** What value of c makes $x^2 + bx + c$ a perfect square trinomial?

Dynamic Solutions available at *BigIdeasMath.com*

What Your Students Have Learned

- Use special product patterns to factor perfect square trinomials.
- Solve systems of linear equations in two variables by graphing.

ANSWERS

1. $(x + 5)^2$
2. $(x - 10)^2$
3. $(x + 6)^2$
4. $(x - 9)^2$
5. $(x + 8)^2$
6. $(x - 15)^2$
7. $(1, -2)$
8. $(4, 4)$
9. $(-2, 3)$
10. $\left(\frac{b}{2}\right)^2$

Vocabulary Review

Have students make Four Squares for the following topics.

- Factoring perfect square trinomials
- Solving systems of linear equations by graphing

MONITORING PROGRESS ANSWERS

1. $x \approx -1.618$
2. $x \approx 1.303, x \approx -2.303$

Mathematical Practices

Mathematically proficient students monitor their work and change course as needed.

Problem-Solving Strategies

Core Concept

Guess, Check, and Revise

When solving a problem in mathematics, it is often helpful to estimate a solution and then observe how close that solution is to being correct. For instance, you can use the guess, check, and revise strategy to find a decimal approximation of the square root of 2.

	Guess	Check	How to revise
1.	1.4	$1.4^2 = 1.96$	Increase guess.
2.	1.41	$1.41^2 = 1.9881$	Increase guess.
3.	1.415	$1.415^2 = 2.002225$	Decrease guess.

By continuing this process, you can determine that the square root of 2 is approximately 1.4142.

EXAMPLE 1 Approximating a Solution of an Equation

The graph of $y = x^2 + x - 1$ is shown. Approximate the positive solution of the equation $x^2 + x - 1 = 0$ to the nearest thousandth.

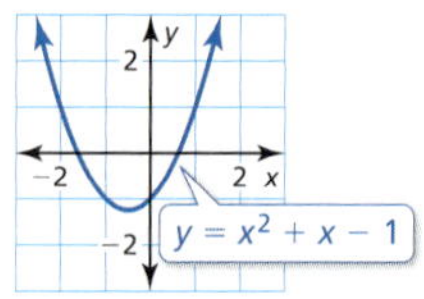

SOLUTION

Using the graph, you can make an initial estimate of the positive solution to be $x = 0.65$.

	Guess	Check	How to revise
1.	0.65	$0.65^2 + 0.65 - 1 = 0.0725$	Decrease guess.
2.	0.62	$0.62^2 + 0.62 - 1 = 0.0044$	Decrease guess.
3.	0.618	$0.618^2 + 0.618 - 1 = -0.000076$	Increase guess.
4.	0.6181	$0.6181^2 + 0.6181 - 1 \approx 0.00015$	The solution is between 0.618 and 0.6181.

▶ So, to the nearest thousandth, the positive solution of the equation is $x = 0.618$.

Monitoring Progress

1. Use the graph in Example 1 to approximate the negative solution of the equation $x^2 + x - 1 = 0$ to the nearest thousandth.

2. The graph of $y = x^2 + x - 3$ is shown. Approximate both solutions of the equation $x^2 + x - 3 = 0$ to the nearest thousandth.

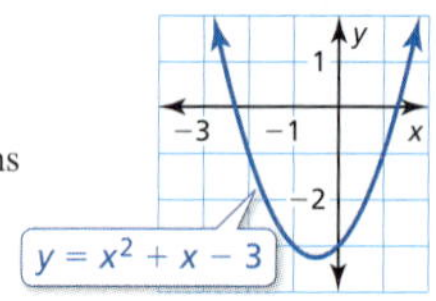

Laurie's Notes Mathematical Practices (continued from page T-477)

- Pose Example 1, and ask students to solve this without using the zero function on their calculators. Do they use correct reasoning in the process?
- Give time for students to work through the questions in the *Monitoring Progress*, and then discuss as a class. Have students make a list of their trials so that their reasoning and process is clear. It is not about random guessing!

Laurie's Notes

Overview of Section 9.1

Introduction

- Students have worked with radicals in an earlier lesson, Section 6.2. This included work with cube roots.
- This is a very long lesson that presents many skills and techniques. There are many connections to properties of exponents and real number properties.
- In the first part of the lesson, students will use the properties of radicals to simplify expressions. This includes work with both square roots and cube roots. This is followed by rationalizing the denominator by multiplying by the conjugate of the denominator. The last part of the lesson is on performing operations with radicals.
- The work in this lesson will help students when working with the Quadratic Formula later in the chapter.

Resources

- Example 7 uses the *golden rectangle*. In geometry, a golden rectangle is a rectangle whose side lengths are in the *golden ratio*, $1 : \frac{1 + \sqrt{5}}{2}$, which is 1: φ (the Greek letter phi), where φ is approximately 1.618. Artists and architects have been fascinated by the idea that the golden rectangle is considered aesthetically pleasing. The proportions of the golden rectangle have been observed in many famous buildings and works of art, such as the Parthenon, Egyptian pyramids, the Taj Mahal, Michelangelo's *David*, and Da Vinci's *Mona Lisa*.
- You might consider having students do a report or construction project related to the golden rectangle or golden ratio.

Formative Assessment Tips

- **Which One Doesn't Belong?:** This technique is one that you should be quite familiar with because it is often used in the *Vocabulary and Core Concept Check* at the beginning of exercise sets! Students are presented with four expressions, quantities, images, or words and asked which one does not belong with the other three. They are also expected to give a reason for their choice.
- This technique gives students the opportunity to analyze and compare items in a set and determine what is alike in three of the four cases. This technique challenges students' reasoning and understanding of some aspect of the lesson they have just learned or what knowledge and conceptions they have about content to be learned.
- Used at the end of instruction, this technique informs you as to how students have conceptualized and made connections in their learning. The reasoning or justification for their choice can be quite informative. Used at the beginning of a lesson, this technique can inform you about what knowledge students already have about the topic.
- Select four items for which it is not immediately obvious which one does not belong. You want to encourage deeper thinking.

Pacing Suggestion

- This is a very long lesson with many skills presented. Judge pacing and need for all of the *Monitoring Progress* questions based on frequent student self-assessment.

Dynamic Teaching Tools

- Dynamic Assessment & Progress Monitoring Tool
- Lesson Planning Tool
- Interactive Whiteboard Lesson Library
- Dynamic Classroom with Dynamic Investigations

What Your Students Will Learn

- Use properties of square roots and cube roots to simplify radical expressions.
- Simplify radical expressions by rationalizing the denominator and using conjugates when the denominator contains a sum or difference involving a square root.
- Perform addition, subtraction, and multiplication with like radicals.

Laurie's Notes

Exploration

Motivate

- As a warm-up, write several numeric sequences and ask students to *Think-Pair-Share* a description of the patterns. Examples:
  2, 5, 11, 23, … double the previous term and add 1
  2, 3, 5, 8, 13, … add the two previous terms
  2, 3, 5, 7, … prime numbers; Students may ask for more information.
  1, 4, 9, 16, … perfect squares
- Ask partners to share their descriptions.
- Be sure to include the last example. Perfect squares will be part of the investigation today.

Exploration 1

- Calculators are not necessary to perform these operations. The numbers involved are all perfect squares.
- Introduce the exploration by posing a question about a numeric expression that involves two operations, such as addition and multiplication.

? "Does the order in which the operations are performed matter?" yes

- Explain that in each part of this exploration, you will determine whether order matters when combining one of the four basic operations with taking square roots.
- **Construct Viable Arguments and Critique the Reasoning of Others:** Encourage students to build strong arguments to support their conclusions. They should understand that one counterexample is sufficient to say that order does matter. However, they should try other examples to support their reasoning when they feel that order does not matter. Suggest that working with perfect squares is helpful. For instance, they may compare $\sqrt{16} \cdot \sqrt{25}$ to $\sqrt{16 \cdot 25}$, and so on.
- Ask students to summarize their findings and discuss their arguments. They are often surprised by the results. They try to make the connection to the Commutative and Associative Properties, where addition and multiplication work but subtraction and division do not.

Exploration 2

- Students should recognize that using perfect squares in their counterexamples will help facilitate the computation.
- **Whiteboarding:** Have students record their counterexamples on a whiteboard so that they can be shared quickly and efficiently.

Communicate Your Answer

- Students may not be able to phrase the statement efficiently, but listen for "The product of the square roots of two numbers is equal to the square root of the product of the two numbers." A similar statement can be made for quotients.

Connecting to Next Step

- Exploration 1 allows students to make sense of the first two properties stated in the formal lesson, providing a great introduction to their learning.

9.1 Properties of Radicals

Essential Question How can you multiply and divide square roots?

EXPLORATION 1 **Operations with Square Roots**

Work with a partner. For each operation with square roots, compare the results obtained using the two indicated orders of operations. What can you conclude?

a. Square Roots and Addition

Is $\sqrt{36} + \sqrt{64}$ equal to $\sqrt{36 + 64}$?

In general, is $\sqrt{a} + \sqrt{b}$ equal to $\sqrt{a + b}$? Explain your reasoning.

b. Square Roots and Multiplication

Is $\sqrt{4} \cdot \sqrt{9}$ equal to $\sqrt{4 \cdot 9}$?

In general, is $\sqrt{a} \cdot \sqrt{b}$ equal to $\sqrt{a \cdot b}$? Explain your reasoning.

c. Square Roots and Subtraction

Is $\sqrt{64} - \sqrt{36}$ equal to $\sqrt{64 - 36}$?

In general, is $\sqrt{a} - \sqrt{b}$ equal to $\sqrt{a - b}$? Explain your reasoning.

d. Square Roots and Division

Is $\dfrac{\sqrt{100}}{\sqrt{4}}$ equal to $\sqrt{\dfrac{100}{4}}$?

In general, is $\dfrac{\sqrt{a}}{\sqrt{b}}$ equal to $\sqrt{\dfrac{a}{b}}$? Explain your reasoning.

REASONING ABSTRACTLY

To be proficient in math, you need to recognize and use counterexamples.

EXPLORATION 2 **Writing Counterexamples**

Work with a partner. A **counterexample** is an example that proves that a general statement is *not* true. For each general statement in Exploration 1 that is not true, write a counterexample different from the example given.

Communicate Your Answer

3. How can you multiply and divide square roots?
4. Give an example of multiplying square roots and an example of dividing square roots that are different from the examples in Exploration 1.
5. Write an algebraic rule for each operation.
   a. the product of square roots
   b. the quotient of square roots

Dynamic Teaching Tools

- Dynamic Assessment & Progress Monitoring Tool
- Lesson Planning Tool
- Interactive Whiteboard Lesson Library
- Dynamic Classroom with Dynamic Investigations

ANSWERS

1. a. no; no; Because $\sqrt{36} + \sqrt{64} \neq \sqrt{36 + 64}$, the general statement cannot be true.
   b. yes; yes; By the Power of a Product Property, $a^{1/2} \cdot b^{1/2} = (a \cdot b)^{1/2}$.
   c. no; no; Because $\sqrt{64} - \sqrt{36} \neq \sqrt{64 - 36}$, the general statement cannot be true.
   d. yes; yes; By the Power of a Quotient Property, $\dfrac{a^{1/2}}{b^{1/2}} = \left(\dfrac{a}{b}\right)^{1/2}$.
2. *Sample answer:* $\sqrt{9} + \sqrt{16} \neq \sqrt{25}$, $\sqrt{16} - \sqrt{9} \neq \sqrt{7}$
3. Multiply or divide the numbers inside the square root symbols and take the square root of the product or quotient.
4. *Sample answer:* $\sqrt{9} \cdot \sqrt{16} = \sqrt{9 \cdot 16} = \sqrt{144} = 12$, $\dfrac{\sqrt{16}}{\sqrt{4}} = \sqrt{\dfrac{16}{4}} = \sqrt{4} = 2$
5. a. $\sqrt{a} \cdot \sqrt{b} = \sqrt{a \cdot b}$
   b. $\dfrac{\sqrt{a}}{\sqrt{b}} = \sqrt{\dfrac{a}{b}}$

Differentiated Instruction

Inclusion

Some students may have difficulty remembering frequently used square roots and cube roots such as $\sqrt{144} = 12$ and $\sqrt[3]{64} = 4$. Have these students make a list in their notebooks of the first eight perfect squares and their square roots and the first eight perfect cubes and their cube roots. Students can refer to the list as they simplify radicals.

Extra Example 1

Simplify each expression.

a. $\sqrt{147}$ $7\sqrt{3}$

b. $\sqrt{25x^5}$ $5x^2\sqrt{x}$

MONITORING PROGRESS ANSWERS

1. $2\sqrt{6}$
2. $-4\sqrt{5}$
3. $7x\sqrt{x}$
4. $5n^2\sqrt{3n}$

9.1 Lesson

Core Vocabulary

counterexample, *p. 479*
radical expression, *p. 480*
simplest form of a radical, *p. 480*
rationalizing the denominator, *p. 482*
conjugates, *p. 482*
like radicals, *p. 484*

Previous
radicand
perfect cube

What You Will Learn

- Use properties of radicals to simplify expressions.
- Simplify expressions by rationalizing the denominator.
- Perform operations with radicals.

Using Properties of Radicals

A **radical expression** is an expression that contains a radical. An expression involving a radical with index n is in **simplest form** when these three conditions are met.

- No radicands have perfect nth powers as factors other than 1.
- No radicands contain fractions.
- No radicals appear in the denominator of a fraction.

You can use the property below to simplify radical expressions involving square roots.

Core Concept

Product Property of Square Roots

Words The square root of a product equals the product of the square roots of the factors.

Numbers $\sqrt{9 \cdot 5} = \sqrt{9} \cdot \sqrt{5} = 3\sqrt{5}$

Algebra $\sqrt{ab} = \sqrt{a} \cdot \sqrt{b}$, where $a, b \geq 0$

STUDY TIP

There can be more than one way to factor a radicand. An efficient method is to find the greatest perfect square factor.

STUDY TIP

In this course, whenever a variable appears in the radicand, assume that it has only *nonnegative* values.

EXAMPLE 1 Using the Product Property of Square Roots

a. $\sqrt{108} = \sqrt{36 \cdot 3}$ Factor using the greatest perfect square factor.

$= \sqrt{36} \cdot \sqrt{3}$ Product Property of Square Roots

$= 6\sqrt{3}$ Simplify.

b. $\sqrt{9x^3} = \sqrt{9 \cdot x^2 \cdot x}$ Factor using the greatest perfect square factor.

$= \sqrt{9} \cdot \sqrt{x^2} \cdot \sqrt{x}$ Product Property of Square Roots

$= 3x\sqrt{x}$ Simplify.

Monitoring Progress

Help in English and Spanish at *BigIdeasMath.com*

Simplify the expression.

1. $\sqrt{24}$
2. $-\sqrt{80}$
3. $\sqrt{49x^3}$
4. $\sqrt{75n^5}$

Core Concept

Quotient Property of Square Roots

Words The square root of a quotient equals the quotient of the square roots of the numerator and denominator.

Numbers $\sqrt{\frac{3}{4}} = \frac{\sqrt{3}}{\sqrt{4}} = \frac{\sqrt{3}}{2}$

Algebra $\sqrt{\frac{a}{b}} = \frac{\sqrt{a}}{\sqrt{b}}$, where $a \geq 0$ and $b > 0$

Laurie's Notes Teacher Actions

- Define *radical expression* and what it means for a radical to be in simplest form.
- Write both *Core Concepts* (product and quotient) because students have just completed the exploration.
- Work through Example 1, and make note of the *Study Tips*.

? **Which One Doesn't Belong?:** In lieu of the *Monitoring Progress*, write $4\sqrt{8x^2}$, $2\sqrt{32x^2}$, $x\sqrt{128}$, and $2\sqrt{16x^2}$. Ask, "Which one doesn't belong? Explain." The expression $2\sqrt{16x^2}$ simplifies to $8x$ and doesn't belong. The other three expressions simplify to $8x\sqrt{2}$.

EXAMPLE 2 Using the Quotient Property of Square Roots

a. $\sqrt{\dfrac{15}{64}} = \dfrac{\sqrt{15}}{\sqrt{64}}$ — Quotient Property of Square Roots

$= \dfrac{\sqrt{15}}{8}$ — Simplify.

b. $\sqrt{\dfrac{81}{x^2}} = \dfrac{\sqrt{81}}{\sqrt{x^2}}$ — Quotient Property of Square Roots

$= \dfrac{9}{x}$ — Simplify.

You can extend the Product and Quotient Properties of Square Roots to other radicals, such as cube roots. When using these *properties of cube roots,* the radicands may contain negative numbers.

EXAMPLE 3 Using Properties of Cube Roots

STUDY TIP
To write a cube root in simplest form, find factors of the radicand that are perfect cubes.

a. $\sqrt[3]{-128} = \sqrt[3]{-64 \cdot 2}$ — Factor using the greatest perfect cube factor.

$= \sqrt[3]{-64} \cdot \sqrt[3]{2}$ — Product Property of Cube Roots

$= -4\sqrt[3]{2}$ — Simplify.

b. $\sqrt[3]{125x^7} = \sqrt[3]{125 \cdot x^6 \cdot x}$ — Factor using the greatest perfect cube factors.

$= \sqrt[3]{125} \cdot \sqrt[3]{x^6} \cdot \sqrt[3]{x}$ — Product Property of Cube Roots

$= 5x^2\sqrt[3]{x}$ — Simplify.

c. $\sqrt[3]{\dfrac{y}{216}} = \dfrac{\sqrt[3]{y}}{\sqrt[3]{216}}$ — Quotient Property of Cube Roots

$= \dfrac{\sqrt[3]{y}}{6}$ — Simplify.

d. $\sqrt[3]{\dfrac{8x^4}{27y^3}} = \dfrac{\sqrt[3]{8x^4}}{\sqrt[3]{27y^3}}$ — Quotient Property of Cube Roots

$= \dfrac{\sqrt[3]{8 \cdot x^3 \cdot x}}{\sqrt[3]{27 \cdot y^3}}$ — Factor using the greatest perfect cube factors.

$= \dfrac{\sqrt[3]{8} \cdot \sqrt[3]{x^3} \cdot \sqrt[3]{x}}{\sqrt[3]{27} \cdot \sqrt[3]{y^3}}$ — Product Property of Cube Roots

$= \dfrac{2x\sqrt[3]{x}}{3y}$ — Simplify.

Monitoring Progress

Help in English and Spanish at *BigIdeasMath.com*

Simplify the expression.

5. $\sqrt{\dfrac{23}{9}}$
6. $-\sqrt{\dfrac{17}{100}}$
7. $\sqrt{\dfrac{36}{z^2}}$
8. $\sqrt{\dfrac{4x^2}{64}}$
9. $\sqrt[3]{54}$
10. $\sqrt[3]{16x^4}$
11. $\sqrt[3]{\dfrac{a}{-27}}$
12. $\sqrt[3]{\dfrac{25c^7d^3}{64}}$

English Language Learners

Notebook Development

Have students create a radicals page in their notebooks. They should include the vocabulary and properties used in this section. For each property, have students include an example. Be sure they include, when appropriate, a cube root example, as well as a square root example.

Extra Example 2

Simplify each expression.

a. $\sqrt{\dfrac{11}{36}}$ $\dfrac{\sqrt{11}}{6}$

b. $\sqrt{\dfrac{144}{n^2}}$ $\dfrac{12}{n}$

Extra Example 3

Simplify each expression.

a. $\sqrt[3]{135}$ $3\sqrt[3]{5}$

b. $\sqrt[3]{8x^4}$ $2x\sqrt[3]{x}$

c. $\sqrt[3]{\dfrac{n^2}{-1000}}$ $-\dfrac{\sqrt[3]{n^2}}{10}$

d. $\sqrt[3]{\dfrac{343a^8}{b^3}}$ $\dfrac{7a^2\sqrt[3]{a^2}}{b}$

MONITORING PROGRESS ANSWERS

5. $\dfrac{\sqrt{23}}{3}$
6. $-\dfrac{\sqrt{17}}{10}$
7. $\dfrac{6}{z}$
8. $\dfrac{x}{4}$
9. $3\sqrt[3]{2}$
10. $2x\sqrt[3]{2x}$
11. $-\dfrac{\sqrt[3]{a}}{3}$
12. $\dfrac{c^2d}{4}\sqrt[3]{25c}$

Laurie's Notes Teacher Actions

- **? Which One Doesn't Belong?:** Write $\sqrt{64}$, $\sqrt{-64}$, $\sqrt[3]{64}$, and $\sqrt[3]{-64}$. Ask, "Which one doesn't belong? Explain." All have real solutions except $\sqrt{-64}$. This question prompts memory of cube roots.
- Explain that the Product and Quotient Properties extend to cube roots. Students should be thinking perfect cubes: 1, 8, 27, 64, ….
- These expressions in Example 3 should be worked through slowly, and in some cases, showing alternate approaches. Explain why $\sqrt[3]{x^6} = x^2$.
- **Think-Pair-Share:** Give a *Three-Minute Pause* and question time before having students work independently on Questions 5–12.

Extra Example 4

Simplify each expression.

a. $\dfrac{\sqrt{7}}{\sqrt{2x}}$ $\dfrac{\sqrt{14x}}{2x}$

b. $\dfrac{5}{\sqrt[3]{4}}$ $\dfrac{5\sqrt[3]{2}}{2}$

Extra Example 5

Simplify $\dfrac{4}{\sqrt{3}+1}$. $2\sqrt{3} - 2$

MONITORING PROGRESS ANSWERS

13. $\dfrac{\sqrt{5}}{5}$
14. $\dfrac{\sqrt{30}}{3}$
15. $\dfrac{7\sqrt{2x}}{2x}$
16. $\dfrac{y\sqrt{6}}{3}$
17. $\dfrac{5\sqrt[3]{2}}{4}$
18. $-4 + 4\sqrt{3}$
19. $\sqrt{65} + 2\sqrt{13}$
20. $-\dfrac{12\sqrt{2} - 12\sqrt{7}}{5}$

Rationalizing the Denominator

When a radical is in the denominator of a fraction, you can multiply the fraction by an appropriate form of 1 to eliminate the radical from the denominator. This process is called **rationalizing the denominator**.

STUDY TIP

Rationalizing the denominator works because you multiply the numerator and denominator by the same nonzero number *a*, which is the same as multiplying by $\frac{a}{a}$, or 1.

EXAMPLE 4 Rationalizing the Denominator

a. $\dfrac{\sqrt{5}}{\sqrt{3n}} = \dfrac{\sqrt{5}}{\sqrt{3n}} \cdot \dfrac{\sqrt{3n}}{\sqrt{3n}}$ Multiply by $\dfrac{\sqrt{3n}}{\sqrt{3n}}$.

$= \dfrac{\sqrt{15n}}{\sqrt{9n^2}}$ Product Property of Square Roots

$= \dfrac{\sqrt{15n}}{\sqrt{9} \cdot \sqrt{n^2}}$ Product Property of Square Roots

$= \dfrac{\sqrt{15n}}{3n}$ Simplify.

b. $\dfrac{2}{\sqrt[3]{9}} = \dfrac{2}{\sqrt[3]{9}} \cdot \dfrac{\sqrt[3]{3}}{\sqrt[3]{3}}$ Multiply by $\dfrac{\sqrt[3]{3}}{\sqrt[3]{3}}$.

$= \dfrac{2\sqrt[3]{3}}{\sqrt[3]{27}}$ Product Property of Cube Roots

$= \dfrac{2\sqrt[3]{3}}{3}$ Simplify.

The binomials $a\sqrt{b} + c\sqrt{d}$ and $a\sqrt{b} - c\sqrt{d}$, where *a*, *b*, *c*, and *d* are rational numbers, are called **conjugates**. You can use conjugates to simplify radical expressions that contain a sum or difference involving square roots in the denominator.

LOOKING FOR STRUCTURE

Notice that the product of two conjugates $a\sqrt{b} + c\sqrt{d}$ and $a\sqrt{b} - c\sqrt{d}$ does not contain a radical and is a *rational* number.

$(a\sqrt{b} + c\sqrt{d})(a\sqrt{b} - c\sqrt{d})$
$= (a\sqrt{b})^2 - (c\sqrt{d})^2$
$= a^2b - c^2d$

EXAMPLE 5 Rationalizing the Denominator Using Conjugates

Simplify $\dfrac{7}{2-\sqrt{3}}$.

SOLUTION

$\dfrac{7}{2-\sqrt{3}} = \dfrac{7}{2-\sqrt{3}} \cdot \dfrac{2+\sqrt{3}}{2+\sqrt{3}}$ The conjugate of $2 - \sqrt{3}$ is $2 + \sqrt{3}$.

$= \dfrac{7(2+\sqrt{3})}{2^2 - (\sqrt{3})^2}$ Sum and difference pattern

$= \dfrac{14 + 7\sqrt{3}}{1}$ Simplify.

$= 14 + 7\sqrt{3}$ Simplify.

Monitoring Progress

Help in English and Spanish at *BigIdeasMath.com*

Simplify the expression.

13. $\dfrac{1}{\sqrt{5}}$ 14. $\dfrac{\sqrt{10}}{\sqrt{3}}$ 15. $\dfrac{7}{\sqrt{2x}}$ 16. $\sqrt{\dfrac{2y^2}{3}}$

17. $\dfrac{5}{\sqrt[3]{32}}$ 18. $\dfrac{8}{1+\sqrt{3}}$ 19. $\dfrac{\sqrt{13}}{\sqrt{5}-2}$ 20. $\dfrac{12}{\sqrt{2}+\sqrt{7}}$

Laurie's Notes Teacher Actions

- **Look For and Make Use of Structure:** Be sure students see the connection between rationalizing the denominator and writing equivalent fractions. See the *Study Tip*.
- **Think-Alouds:** Model Example 4(a) and then do a *Think-Aloud* as partner A solves part (b) aloud for partner B. "To simplify $\frac{2}{\sqrt[3]{9}}$, I first need to …"
- **Look For and Make Use of Structure** and **Connection:** Multiplying by the conjugate is connected to the Sum and Difference Pattern: $(a + b)(a - b) = a^2 - b^2$.
- **Thumbs Up:** Work through Example 5 as shown. Ask students to self-assess with a *Thumbs Up*. Address questions and concerns.

EXAMPLE 6 Solving a Real-Life Problem

The distance d (in miles) that you can see to the horizon with your eye level h feet above the water is given by $d = \sqrt{\frac{3h}{2}}$. How far can you see when your eye level is 5 feet above the water?

SOLUTION

$$d = \sqrt{\frac{3(5)}{2}} \qquad \text{Substitute 5 for } h.$$

$$= \frac{\sqrt{15}}{\sqrt{2}} \qquad \text{Quotient Property of Square Roots}$$

$$= \frac{\sqrt{15}}{\sqrt{2}} \cdot \frac{\sqrt{2}}{\sqrt{2}} \qquad \text{Multiply by } \frac{\sqrt{2}}{\sqrt{2}}.$$

$$= \frac{\sqrt{30}}{2} \qquad \text{Simplify.}$$

▶ You can see $\frac{\sqrt{30}}{2}$, or about 2.74 miles.

EXAMPLE 7 Modeling with Mathematics

The ratio of the length to the width of a *golden rectangle* is $(1 + \sqrt{5}) : 2$. The dimensions of the face of the Parthenon in Greece form a golden rectangle. What is the height h of the Parthenon?

SOLUTION

1. **Understand the Problem** Think of the length and height of the Parthenon as the length and width of a golden rectangle. The length of the rectangular face is 31 meters. You know the ratio of the length to the height. Find the height h.
2. **Make a Plan** Use the ratio $(1 + \sqrt{5}) : 2$ to write a proportion and solve for h.
3. **Solve the Problem**

$$\frac{1 + \sqrt{5}}{2} = \frac{31}{h} \qquad \text{Write a proportion.}$$

$$h(1 + \sqrt{5}) = 62 \qquad \text{Cross Products Property}$$

$$h = \frac{62}{1 + \sqrt{5}} \qquad \text{Divide each side by } 1 + \sqrt{5}.$$

$$h = \frac{62}{1 + \sqrt{5}} \cdot \frac{1 - \sqrt{5}}{1 - \sqrt{5}} \qquad \text{Multiply the numerator and denominator by the conjugate.}$$

$$h = \frac{62 - 62\sqrt{5}}{-4} \qquad \text{Simplify.}$$

$$h \approx 19.16 \qquad \text{Use a calculator.}$$

▶ The height is about 19 meters.

4. **Look Back** $\frac{1 + \sqrt{5}}{2} \approx 1.62$ and $\frac{31}{19.16} \approx 1.62$. So, your answer is reasonable.

Monitoring Progress

Help in English and Spanish at *BigIdeasMath.com*

21. **WHAT IF?** In Example 6, how far can you see when your eye level is 35 feet above the water?
22. The dimensions of a dance floor form a golden rectangle. The shorter side of the dance floor is 50 feet. What is the length of the longer side of the dance floor?

Extra Example 6

The distance d (in miles) that you can see to the horizon with your eye level h feet above the water is given by $d = \sqrt{\frac{3h}{2}}$. How far can you see when your eye level is 15 feet above the water?

You can see $\frac{3\sqrt{10}}{2}$, or about 4.74 miles.

Extra Example 7

The ratio of the length to the width of a *golden rectangle* is $(1 + \sqrt{5})$: 2. The dimensions of the picture frame form a golden rectangle. What is the height h of the frame?

The height is about 15 inches.

MONITORING PROGRESS ANSWERS

21. about 7.25 mi
22. about 81 ft

Laurie's Notes Teacher Actions

- **Attend to Precision:** In Example 6, an exact answer such as $\frac{\sqrt{30}}{2}$ is not meaningful in answering the question. Use the approximate answer of 2.74 miles.
- Give a bit of background on the *golden rectangle*. (See the *Resources* on page T-478.)
- Although the terms *length* and *width* are interchangeable, in this example the length is the longer of the two sides. Students should be comfortable setting up a proportion.

COMMON ERROR When finding the cross product, h is multiplied by the quantity $(1 + \sqrt{5})$, resulting in $h(1 + \sqrt{5})$, not $h + \sqrt{5}$.

Extra Example 8

Simplify each expression.

a. $3\sqrt{15} + 2\sqrt{13} - 7\sqrt{15}$ $-4\sqrt{15} + 2\sqrt{13}$

b. $5\sqrt{7} + \sqrt{63}$ $8\sqrt{7}$

c. $4\sqrt[3]{2x} + 5\sqrt[3]{2x}$ $9\sqrt[3]{2x}$

Extra Example 9

Simplify $\sqrt{7}\left(\sqrt{3} + \sqrt{48}\right)$. $5\sqrt{21}$

MONITORING PROGRESS ANSWERS

23. $13\sqrt{2} - \sqrt{6}$

24. $-14\sqrt{7}$

25. $-7\sqrt[3]{5x}$

26. $36\sqrt{6}$

27. $36 - 16\sqrt{5}$

28. 2

Performing Operations with Radicals

Radicals with the same index and radicand are called **like radicals**. You can add and subtract like radicals the same way you combine like terms by using the Distributive Property.

STUDY TIP

Do not assume that radicals with different radicands cannot be added or subtracted. Always check to see whether you can simplify the radicals. In some cases, the radicals will become like radicals.

EXAMPLE 8 Adding and Subtracting Radicals

a. $5\sqrt{7} + \sqrt{11} - 8\sqrt{7} = 5\sqrt{7} - 8\sqrt{7} + \sqrt{11}$ Commutative Property of Addition

$= (5 - 8)\sqrt{7} + \sqrt{11}$ Distributive Property

$= -3\sqrt{7} + \sqrt{11}$ Subtract.

b. $10\sqrt{5} + \sqrt{20} = 10\sqrt{5} + \sqrt{4 \cdot 5}$ Factor using the greatest perfect square factor.

$= 10\sqrt{5} + \sqrt{4} \cdot \sqrt{5}$ Product Property of Square Roots

$= 10\sqrt{5} + 2\sqrt{5}$ Simplify.

$= (10 + 2)\sqrt{5}$ Distributive Property

$= 12\sqrt{5}$ Add.

c. $6\sqrt[3]{x} + 2\sqrt[3]{x} = (6 + 2)\sqrt[3]{x}$ Distributive Property

$= 8\sqrt[3]{x}$ Add.

EXAMPLE 9 Multiplying Radicals

Simplify $\sqrt{5}\left(\sqrt{3} - \sqrt{75}\right)$.

SOLUTION

Method 1 $\sqrt{5}\left(\sqrt{3} - \sqrt{75}\right) = \sqrt{5} \cdot \sqrt{3} - \sqrt{5} \cdot \sqrt{75}$ Distributive Property

$= \sqrt{15} - \sqrt{375}$ Product Property of Square Roots

$= \sqrt{15} - 5\sqrt{15}$ Simplify.

$= (1 - 5)\sqrt{15}$ Distributive Property

$= -4\sqrt{15}$ Subtract.

Method 2 $\sqrt{5}\left(\sqrt{3} - \sqrt{75}\right) = \sqrt{5}\left(\sqrt{3} - 5\sqrt{3}\right)$ Simplify $\sqrt{75}$.

$= \sqrt{5}\left[(1 - 5)\sqrt{3}\right]$ Distributive Property

$= \sqrt{5}\left(-4\sqrt{3}\right)$ Subtract.

$= -4\sqrt{15}$ Product Property of Square Roots

Monitoring Progress

Help in English and Spanish at *BigIdeasMath.com*

Simplify the expression.

23. $3\sqrt{2} - \sqrt{6} + 10\sqrt{2}$

24. $4\sqrt{7} - 6\sqrt{63}$

25. $4\sqrt[3]{5x} - 11\sqrt[3]{5x}$

26. $\sqrt{3}\left(8\sqrt{2} + 7\sqrt{32}\right)$

27. $\left(2\sqrt{5} - 4\right)^2$

28. $\sqrt[3]{-4}\left(\sqrt[3]{2} - \sqrt[3]{16}\right)$

Laurie's Notes Teacher Actions

- **Look For and Make Use of Structure:** *Like radicals* and like terms can be combined using the Distributive Property.
- **Think-Alouds:** Model Example 8(a) and (b), and then do a *Think-Aloud* as partner B solves part (c) aloud for partner A. "To simplify $6\sqrt[3]{x} + 2\sqrt[3]{x}$, I need to …"
- Pose Example 9, and give independent time for students to work. If both approaches do not surface, share the alternate method.

Closure

- **Writing Prompt:** This was a very long lesson. Right now I am feeling …

9.1 Exercises

Dynamic Solutions available at *BigIdeasMath.com*

Vocabulary and Core Concept Check

1. **COMPLETE THE SENTENCE** The process of eliminating a radical from the denominator of a radical expression is called ______________.

2. **VOCABULARY** What is the conjugate of the binomial $\sqrt{6} + 4$?

3. **WRITING** Are the expressions $\frac{1}{3}\sqrt{2x}$ and $\sqrt{\frac{2x}{9}}$ equivalent? Explain your reasoning.

4. **WHICH ONE DOESN'T BELONG?** Which expression does *not* belong with the other three? Explain your reasoning.

$-\frac{1}{3}\sqrt{6}$	$6\sqrt{3}$	$\frac{1}{6}\sqrt{3}$	$-3\sqrt{3}$

Monitoring Progress and Modeling with Mathematics

In Exercises 5–12, determine whether the expression is in simplest form. If the expression is not in simplest form, explain why.

5. $\sqrt{19}$
6. $\sqrt{\frac{1}{7}}$
7. $\sqrt{48}$
8. $\sqrt{34}$
9. $\frac{5}{\sqrt{2}}$
10. $\frac{3\sqrt{10}}{4}$
11. $\frac{1}{2 + \sqrt[3]{2}}$
12. $6 - \sqrt[3]{54}$

In Exercises 13–20, simplify the expression. *(See Example 1.)*

13. $\sqrt{20}$
14. $\sqrt{32}$
15. $\sqrt{128}$
16. $-\sqrt{72}$
17. $\sqrt{125b}$
18. $\sqrt{4x^2}$
19. $-\sqrt{81m^3}$
20. $\sqrt{48n^5}$

In Exercises 21–28, simplify the expression. *(See Example 2.)*

21. $\sqrt{\frac{4}{49}}$
22. $-\sqrt{\frac{7}{81}}$
23. $-\sqrt{\frac{23}{64}}$
24. $\sqrt{\frac{65}{121}}$
25. $\sqrt{\frac{a^3}{49}}$
26. $\sqrt{\frac{144}{k^2}}$
27. $\sqrt{\frac{100}{4x^2}}$
28. $\sqrt{\frac{25v^2}{36}}$

In Exercises 29–36, simplify the expression. *(See Example 3.)*

29. $\sqrt[3]{16}$
30. $\sqrt[3]{-108}$
31. $\sqrt[3]{-64x^5}$
32. $-\sqrt[3]{343n^2}$
33. $\sqrt[3]{\frac{6c}{-125}}$
34. $\sqrt[3]{\frac{8h^4}{27}}$
35. $-\sqrt[3]{\frac{81y^2}{1000x^3}}$
36. $\sqrt[3]{\frac{21}{-64a^3b^6}}$

ERROR ANALYSIS **In Exercises 37 and 38, describe and correct the error in simplifying the expression.**

37. ✗
$$\sqrt{72} = \sqrt{4 \cdot 18} = \sqrt{4} \cdot \sqrt{18} = 2\sqrt{18}$$

38. ✗
$$\sqrt[3]{\frac{128y^3}{125}} = \frac{\sqrt[3]{128y^3}}{125} = \frac{\sqrt[3]{64 \cdot 2 \cdot y^3}}{125} = \frac{\sqrt[3]{64} \cdot \sqrt[3]{2} \cdot \sqrt[3]{y^3}}{125} = \frac{4y\sqrt[3]{2}}{125}$$

Assignment Guide and Homework Check

ASSIGNMENT

Basic: 1–4, 5–61 odd, 73–89 odd, 102, 104, 108–115

Average: 1–4, 6–90 even, 91, 94, 102, 104, 108–115

Advanced: 1–4, 20, 28, 34–40 even, 52–64 even, 68–74 even, 80–90 even, 91–115

HOMEWORK CHECK

Basic: 13, 29, 55, 75, 83

Average: 14, 32, 56, 76, 84

Advanced: 20, 34, 56, 80, 84

ANSWERS

1. rationalizing the denominator
2. $\sqrt{6} - 4$
3. yes; $\sqrt{\frac{2x}{9}} = \sqrt{\frac{1}{9} \cdot 2x} = \sqrt{\frac{1}{9}} \cdot \sqrt{2x} = \frac{1}{3}\sqrt{2x}$
4. $-\frac{1}{3}\sqrt{6}$; The other three expressions have like radicals of $\sqrt{3}$.
5. yes
6. no; The radicand is a fraction.
7. no; The radicand has a perfect square factor of 16.
8. yes
9. no; A radical appears in the denominator of a fraction.
10. yes
11. no; A radical appears in the denominator of a fraction.
12. no; The radicand has a perfect cube factor of 27.
13. $2\sqrt{5}$
14. $4\sqrt{2}$
15. $8\sqrt{2}$
16. $-6\sqrt{2}$
17. $5\sqrt{5b}$
18. $2x$
19. $-9m\sqrt{m}$
20. $4n^2\sqrt{3n}$
21. $\frac{2}{7}$
22. $-\frac{\sqrt{7}}{9}$
23. $-\frac{\sqrt{23}}{8}$
24. $\frac{\sqrt{65}}{11}$
25. $\frac{a\sqrt{a}}{7}$
26. $\frac{12}{k}$
27. $\frac{5}{x}$
28. $\frac{5v}{6}$
29. $2\sqrt[3]{2}$
30. $-3\sqrt[3]{4}$
31. $-4x\sqrt[3]{x^2}$
32. $-7\sqrt[3]{n^2}$
33. $-\frac{\sqrt[3]{6c}}{5}$
34. $\frac{2h\sqrt[3]{h}}{3}$
35. $-\frac{3\sqrt[3]{3y^2}}{10x}$
36. $-\frac{\sqrt[3]{21}}{4ab^2}$
37. The radicand 18 has a perfect square factor of 9; $\sqrt{72} = \sqrt{36 \cdot 2} = 6\sqrt{2}$
38. The denominator should be $\sqrt[3]{125}$; $\sqrt[3]{\frac{128y^3}{125}} = \frac{\sqrt[3]{128y^3}}{\sqrt[3]{125}} = \frac{4y\sqrt[3]{2}}{5}$

Dynamic Teaching Tools

- Dynamic Assessment & Progress Monitoring Tool
- Interactive Whiteboard Lesson Library
- Dynamic Classroom with Dynamic Investigations

ANSWERS

39. $\frac{\sqrt{6}}{\sqrt{6}}$
40. $\frac{\sqrt{13z}}{\sqrt{13z}}$
41. $\frac{\sqrt[3]{x}}{\sqrt[3]{x}}$
42. $\frac{\sqrt[3]{2}}{\sqrt[3]{2}}$
43. $\frac{\sqrt{5}+8}{\sqrt{5}+8}$
44. $\frac{\sqrt{3}-\sqrt{7}}{\sqrt{3}-\sqrt{7}}$
45. $\sqrt{2}$
46. $\frac{4\sqrt{3}}{3}$
47. $\frac{\sqrt{15}}{12}$
48. $\frac{\sqrt{13}}{13}$
49. $\frac{3\sqrt{a}}{a}$
50. $\frac{\sqrt{2x}}{2x}$
51. $\frac{d\sqrt{15}}{5}$
52. $\frac{2\sqrt{6n}}{3n^2}$
53. $\frac{4\sqrt[3]{5}}{5}$
54. $\frac{\sqrt[3]{2y}}{6y}$
55. $\frac{\sqrt{7}-1}{6}$
56. $\frac{5+\sqrt{3}}{11}$
57. $\frac{7\sqrt{10}+2\sqrt{5}}{47}$
58. $\frac{6\sqrt{5}-5}{31}$
59. $\sqrt{5}+\sqrt{2}$
60. $\frac{\sqrt{21}-3}{4}$
61. a. about 1.85 sec
    b. about 0.41 sec
62. a. $P = d\sqrt{d}$
    b. about 11.86 Earth years

In Exercises 39–44, write a factor that you can use to rationalize the denominator of the expression.

39. $\frac{4}{\sqrt{6}}$ 40. $\frac{1}{\sqrt{13z}}$

41. $\frac{2}{\sqrt[3]{x^2}}$ 42. $\frac{3m}{\sqrt[3]{4}}$

43. $\frac{\sqrt{2}}{\sqrt{5}-8}$ 44. $\frac{5}{\sqrt{3}+\sqrt{7}}$

In Exercises 45–54, simplify the expression. *(See Example 4.)*

45. $\frac{2}{\sqrt{2}}$ 46. $\frac{4}{\sqrt{3}}$

47. $\frac{\sqrt{5}}{\sqrt{48}}$ 48. $\sqrt{\frac{4}{52}}$

49. $\frac{3}{\sqrt{a}}$ 50. $\frac{1}{\sqrt{2x}}$

51. $\sqrt{\frac{3d^2}{5}}$ 52. $\frac{\sqrt{8}}{\sqrt{3n^3}}$

53. $\frac{4}{\sqrt[3]{25}}$ 54. $\sqrt[3]{\frac{1}{108y^2}}$

In Exercises 55–60, simplify the expression. *(See Example 5.)*

55. $\frac{1}{\sqrt{7}+1}$ 56. $\frac{2}{5-\sqrt{3}}$

57. $\frac{\sqrt{10}}{7-\sqrt{2}}$ 58. $\frac{\sqrt{5}}{6+\sqrt{5}}$

59. $\frac{3}{\sqrt{5}-\sqrt{2}}$ 60. $\frac{\sqrt{3}}{\sqrt{7}+\sqrt{3}}$

61. **MODELING WITH MATHEMATICS** The time t (in seconds) it takes an object to hit the ground is given by $t = \sqrt{\frac{h}{16}}$, where h is the height (in feet) from which the object was dropped. *(See Example 6.)*

    a. How long does it take an earring to hit the ground when it falls from the roof of the building?

    b. How much sooner does the earring hit the ground when it is dropped from two stories (22 feet) below the roof?

62. **MODELING WITH MATHEMATICS** The orbital period of a planet is the time it takes the planet to travel around the Sun. You can find the orbital period P (in Earth years) using the formula $P = \sqrt{d^3}$, where d is the average distance (in astronomical units, abbreviated AU) of the planet from the Sun.

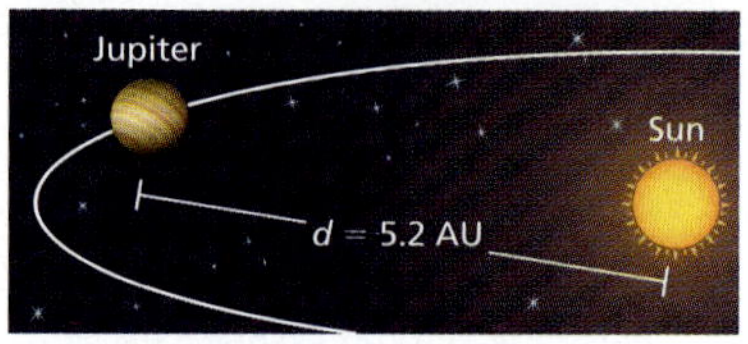

    a. Simplify the formula.

    b. What is Jupiter's orbital period?

63. **MODELING WITH MATHEMATICS** The electric current I (in amperes) an appliance uses is given by the formula $I = \sqrt{\frac{P}{R}}$, where P is the power (in watts) and R is the resistance (in ohms). Find the current an appliance uses when the power is 147 watts and the resistance is 5 ohms.

64. **MODELING WITH MATHEMATICS** You can find the average annual interest rate r (in decimal form) of a savings account using the formula $r = \sqrt{\frac{V_2}{V_0}} - 1$, where V_0 is the initial investment and V_2 is the balance of the account after 2 years. Use the formula to compare the savings accounts. In which account would you invest money? Explain.

Account	Initial investment	Balance after 2 years
1	\$275	\$293
2	\$361	\$382
3	\$199	\$214
4	\$254	\$272
5	\$386	\$406

63. about 5.42 amperes
64. Account 3; It has the greatest interest rate.

In Exercises 65–68, evaluate the function for the given value of x. Write your answer in simplest form and in decimal form rounded to the nearest hundredth.

65. $h(x) = \sqrt{5x};\ x = 10$

66. $g(x) = \sqrt{3x};\ x = 60$

67. $r(x) = \sqrt{\dfrac{3x}{3x^2 + 6}};\ x = 4$

68. $p(x) = \sqrt{\dfrac{x - 1}{5x}};\ x = 8$

In Exercises 69–72, evaluate the expression when $a = -2$, $b = 8$, and $c = \frac{1}{2}$. Write your answer in simplest form and in decimal form rounded to the nearest hundredth.

69. $\sqrt{a^2 + bc}$

70. $-\sqrt{4c - 6ab}$

71. $-\sqrt{2a^2 + b^2}$

72. $\sqrt{b^2 - 4ac}$

73. **MODELING WITH MATHEMATICS** The text in the book shown forms a golden rectangle. What is the width w of the text? *(See Example 7.)*

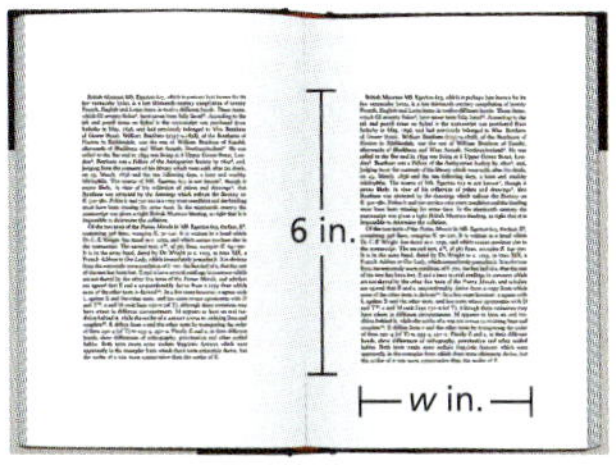

74. **MODELING WITH MATHEMATICS** The flag of Togo is approximately the shape of a golden rectangle. What is the width w of the flag?

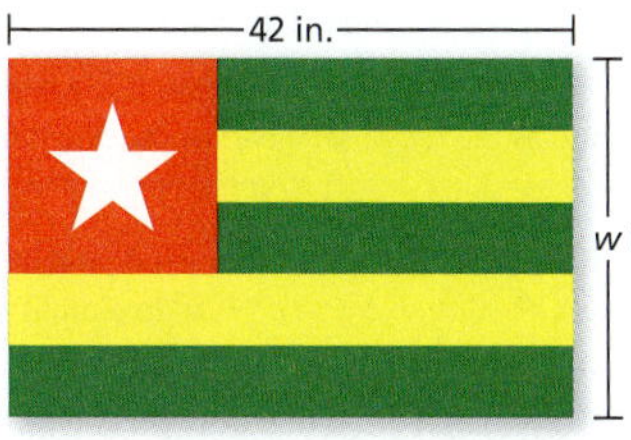

In Exercises 75–82, simplify the expression. *(See Example 8.)*

75. $\sqrt{3} - 2\sqrt{2} + 6\sqrt{2}$

76. $\sqrt{5} - 5\sqrt{13} - 8\sqrt{5}$

77. $2\sqrt{6} - 5\sqrt{54}$

78. $9\sqrt{32} + \sqrt{2}$

79. $\sqrt{12} + 6\sqrt{3} + 2\sqrt{6}$

80. $3\sqrt{7} - 5\sqrt{14} + 2\sqrt{28}$

81. $\sqrt[3]{-81} + 4\sqrt[3]{3}$

82. $6\sqrt[3]{128t} - 2\sqrt[3]{2t}$

In Exercises 83–90, simplify the expression. *(See Example 9.)*

83. $\sqrt{2}(\sqrt{45} + \sqrt{5})$

84. $\sqrt{3}(\sqrt{72} - 3\sqrt{2})$

85. $\sqrt{5}(2\sqrt{6x} - \sqrt{96x})$

86. $\sqrt{7y}(\sqrt{27y} + 5\sqrt{12y})$

87. $(4\sqrt{2} - \sqrt{98})^2$

88. $(\sqrt{3} + \sqrt{48})(\sqrt{20} - \sqrt{5})$

89. $\sqrt[3]{3}(\sqrt[3]{4} + \sqrt[3]{32})$

90. $\sqrt[3]{2}(\sqrt[3]{135} - 4\sqrt[3]{5})$

91. **MODELING WITH MATHEMATICS** The circumference C of the art room in a mansion is approximated by the formula $C \approx 2\pi\sqrt{\dfrac{a^2 + b^2}{2}}$. Approximate the circumference of the room.

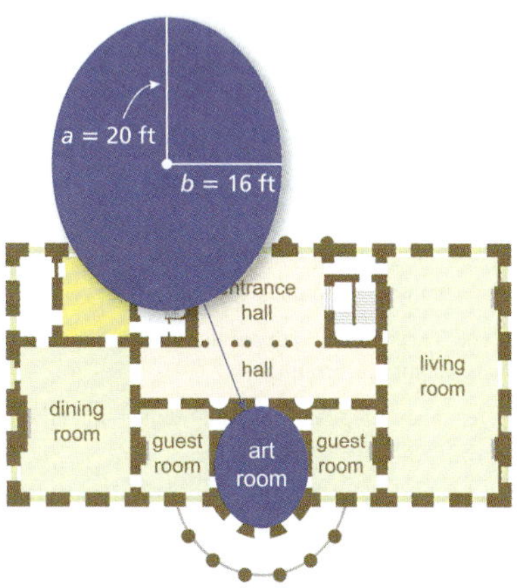

92. **CRITICAL THINKING** Determine whether each expression represents a *rational* or an *irrational* number. Justify your answer.

a. $4 + \sqrt{6}$

b. $\dfrac{\sqrt{48}}{\sqrt{3}}$

c. $\dfrac{8}{\sqrt{12}}$

d. $\sqrt{3} + \sqrt{7}$

e. $\dfrac{a}{\sqrt{10} - \sqrt{2}}$, where a is a positive integer

f. $\dfrac{2 + \sqrt{5}}{2b + \sqrt{5b^2}}$, where b is a positive integer

In Exercises 93–98, simplify the expression.

93. $\sqrt[5]{\dfrac{13}{5x^5}}$

94. $\sqrt[4]{\dfrac{10}{81}}$

95. $\sqrt[4]{256y}$

96. $\sqrt[5]{160x^6}$

97. $6\sqrt[4]{9} - \sqrt[5]{9} + 3\sqrt[4]{9}$

98. $\sqrt[5]{2}(\sqrt[4]{7} + \sqrt[5]{16})$

ANSWERS

65. $5\sqrt{2}$, about 7.07

66. $6\sqrt{5}$, about 13.42

67. $\dfrac{\sqrt{2}}{3}$, about 0.47

68. $\dfrac{\sqrt{70}}{20}$, about 0.42

69. $2\sqrt{2}$, about 2.83

70. $-7\sqrt{2}$, about -9.90

71. $-6\sqrt{2}$, about -8.49

72. $2\sqrt{17}$, about 8.25

73. about 3.71 in.

74. about 25.96 in.

75. $\sqrt{3} + 4\sqrt{2}$

76. $-7\sqrt{5} - 5\sqrt{13}$

77. $-13\sqrt{6}$

78. $37\sqrt{2}$

79. $8\sqrt{3} + 2\sqrt{6}$

80. $7\sqrt{7} - 5\sqrt{14}$

81. $\sqrt[3]{3}$

82. $22\sqrt[3]{2t}$

83. $4\sqrt{10}$

84. $3\sqrt{6}$

85. $-2\sqrt{30x}$

86. $13y\sqrt{21}$

87. 18

88. $5\sqrt{15}$

89. $3\sqrt[3]{12}$

90. $-\sqrt[3]{10}$

91. about 114 ft

92. a. irrational; Six is not a perfect square.

b. rational; Four is a rational number.

c. irrational; Twelve is not a perfect square.

d. irrational; Three and seven are not perfect squares.

e. irrational; Two and ten are not perfect squares.

f. rational; $\dfrac{1}{b}$ is rational when b is a positive integer.

93. $\dfrac{\sqrt[5]{8125}}{5x}$

94. $\dfrac{\sqrt[4]{10}}{3}$

95. $4\sqrt[4]{y}$

96. $2x\sqrt[5]{5x}$

97. $9\sqrt[4]{9} - \sqrt[5]{9}$

98. $\sqrt[5]{2}\sqrt[4]{7} + 2$

ANSWERS

99. **a.** $4, 2\frac{1}{4}, 2, 2+\sqrt{3}, 2-\sqrt{3}, 2+\pi;$
$2\frac{1}{4}, \frac{1}{2}, \frac{1}{4}, \frac{1}{4}+\sqrt{3}, \frac{1}{4}-\sqrt{3}, \frac{1}{4}+\pi;$
$2, \frac{1}{4}, 0, \sqrt{3}, -\sqrt{3}, \pi;$
$2+\sqrt{3}, \frac{1}{4}+\sqrt{3}, \sqrt{3}, 2\sqrt{3}, 0, \pi+\sqrt{3};$
$2-\sqrt{3}, \frac{1}{4}-\sqrt{3}, -\sqrt{3}, 0, -2\sqrt{3}, \pi-\sqrt{3};$
$2+\pi, \frac{1}{4}+\pi, \pi, \pi+\sqrt{3}, \pi-\sqrt{3}, 2\pi$

b. $4, \frac{1}{2}, 0, 2\sqrt{3}, -2\sqrt{3}, 2\pi;$
$\frac{1}{2}, \frac{1}{16}, 0, \frac{\sqrt{3}}{4}, -\frac{\sqrt{3}}{4}, \frac{\pi}{4};$
$0, 0, 0, 0, 0, 0;$
$2\sqrt{3}, \frac{\sqrt{3}}{4}, 0, 3, -3, \pi\sqrt{3};$
$-2\sqrt{3}, -\frac{\sqrt{3}}{4}, 0, -3, 3, -\pi\sqrt{3};$
$2\pi, \frac{\pi}{4}, 0, \pi\sqrt{3}, -\pi\sqrt{3}, \pi^2$

100. **a.** always; The sum of two fractions can be written as a fraction.
b. always; The sum cannot be written as the ratio of two integers.
c. sometimes; $\sqrt{3}+\sqrt{3}=2\sqrt{3}$, but $\sqrt{3}+(-\sqrt{3})=0$.
d. always; The product of two fractions can be written as a fraction.
e. always; The product cannot be written as the ratio of two integers.
f. sometimes; $\sqrt{3}\cdot\pi=\pi\sqrt{3}$, but $\sqrt{3}\cdot\sqrt{3}=3$.

101. odd; even; When m is even, 2^m is a perfect square.

102. *Sample answer:* $\sqrt[3]{2}$

103. $a^2 < ab < b^2$ when $a < b$.

104–115. See Additional Answers.

REASONING **In Exercises 99 and 100, use the table shown.**

	2	$\frac{1}{4}$	0	$\sqrt{3}$	$-\sqrt{3}$	π
2						
$\frac{1}{4}$						
0						
$\sqrt{3}$						
$-\sqrt{3}$						
π						

99. Copy and complete the table by (a) finding each sum $\left(2+2, 2+\frac{1}{4}, \text{etc.}\right)$ and (b) finding each product $\left(2\cdot 2, 2\cdot\frac{1}{4}, \text{etc.}\right)$.

100. Use your answers in Exercise 99 to determine whether each statement is *always*, *sometimes*, or *never* true. Justify your answer.

- **a.** The sum of a rational number and a rational number is rational.
- **b.** The sum of a rational number and an irrational number is irrational.
- **c.** The sum of an irrational number and an irrational number is irrational.
- **d.** The product of a rational number and a rational number is rational.
- **e.** The product of a nonzero rational number and an irrational number is irrational.
- **f.** The product of an irrational number and an irrational number is irrational.

101. REASONING Let m be a positive integer. For what values of m will the simplified form of the expression $\sqrt{2^m}$ contain a radical? For what values will it *not* contain a radical? Explain.

102. HOW DO YOU SEE IT? The edge length s of a cube is an irrational number, the surface area is an irrational number, and the volume is a rational number. Give a possible value of s.

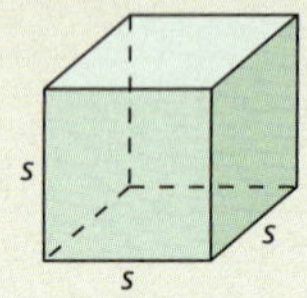

103. REASONING Let a and b be positive numbers. Explain why $\sqrt{ab}$ lies between a and b on a number line. (*Hint:* Let $a < b$ and multiply each side of $a < b$ by a. Then let $a < b$ and multiply each side by b.)

104. MAKING AN ARGUMENT Your friend says that you can rationalize the denominator of the expression $\frac{2}{4+\sqrt[3]{5}}$ by multiplying the numerator and denominator by $4-\sqrt[3]{5}$. Is your friend correct? Explain.

105. PROBLEM SOLVING The ratio of consecutive terms $\frac{a_n}{a_{n-1}}$ in the Fibonacci sequence gets closer and closer to the golden ratio $\frac{1+\sqrt{5}}{2}$ as n increases. Find the term that precedes 610 in the sequence.

106. THOUGHT PROVOKING Use the golden ratio $\frac{1+\sqrt{5}}{2}$ and the golden ratio conjugate $\frac{1-\sqrt{5}}{2}$ for each of the following.

- **a.** Show that the golden ratio and golden ratio conjugate are both solutions of $x^2 - x - 1 = 0$.
- **b.** Construct a geometric diagram that has the golden ratio as the length of a part of the diagram.

107. CRITICAL THINKING Use the special product pattern $(a+b)(a^2-ab+b^2)=a^3+b^3$ to simplify the expression $\frac{2}{\sqrt[3]{x}+1}$. Explain your reasoning.

Maintaining Mathematical Proficiency
Reviewing what you learned in previous grades and lessons

Graph the linear equation. Identify the x-intercept. *(Section 3.5)*

108. $y = x - 4$ **109.** $y = -2x + 6$ **110.** $y = -\frac{1}{3}x - 1$ **111.** $y = \frac{3}{2}x + 6$

Solve the equation. Check your solution. *(Section 6.5)*

112. $32 = 2^x$ **113.** $27^x = 3^{x-6}$ **114.** $\left(\frac{1}{6}\right)^{2x} = 216^{1-x}$ **115.** $625^x = \left(\frac{1}{25}\right)^{x+2}$

If students need help...	*If students got it...*
Resources by Chapter • Practice A and Practice B • Puzzle Time	Resources by Chapter • Enrichment and Extension • Cumulative Review
Student Journal • Practice	Start the *next* Section
Differentiating the Lesson Skills Review Handbook	

Mini-Assessment

Simplify the expression.

1. $\sqrt{144x^3}$ $12x\sqrt{x}$
2. $\sqrt[3]{-54x^3}$ $-3x\sqrt[3]{2}$
3. $\frac{7}{\sqrt{2}+3}$ $-\sqrt{2}+3$
4. $4\sqrt{5}-\sqrt{3}+3\sqrt{5}$ $7\sqrt{5}-\sqrt{3}$
5. $\sqrt{3}(\sqrt{2}-\sqrt{8})$ $-\sqrt{6}$

Laurie's Notes

Overview of Section 9.2

Introduction

- In this lesson, students will use their prior knowledge of factoring quadratics and graphing quadratic equations in order to solve a quadratic equation in one variable and to find the zeros of a quadratic function.
- Equations and functions presented have 0, 1, or 2 solutions or zeros. One problem involving a cubic has 3 zeros.
- Real-life problems do not always have whole number or rational solutions, so students will judge what degree of precision is required for the problem.

Formative Assessment Tips

- **Response Cards:** This technique is used as a quick check to see whether students' knowledge of a skill, technique, precept, or procedure is correct. Students are given cards at the beginning of class that are held up in front of them in response to a question. The cards are prepared in advance with particular responses (A, B, C, D, or 1, 2, 3, or True, False) or may be left blank for students to write their response on.
- This is the low-tech version of a clicker response system. Whiteboards could be used for students to write responses on. If cards are going to be used multiple times, and for several years, consider laminating the cards.
- This technique gives all students the opportunity to participate in the lesson. As a teacher, you are soliciting information from all students and not just those who raise their hands. Because the cards are held facing the teacher, when all students are also facing the teacher, it is a private way to gather quick information about students' understanding.
- The information learned from the responses should guide your next instructional move. If the responses are evenly distributed, you could gather like responses together and ask each group to present to the others the reason(s) for their response. If most of the responses are incorrect, there is reteaching to do. If only a few students have a wrong response, you might probe for the reason or be mindful of their response as the next question is posed that might clarify earlier confusion.
- In this lesson, response cards with 0, 1, 2, and 3 are distributed, which will refer to the number of solutions for the problems presented.

Pacing Suggestion

- If you are confident in students' ability to graph quadratics, you might have students use a graphing calculator for the explorations. These explorations allow students to discover that quadratic equations in one variable can have 0, 1, or 2 solutions. Transition to the formal lesson after the explorations have been discussed.

Dynamic Teaching Tools

- Dynamic Assessment & Progress Monitoring Tool
- Lesson Planning Tool
- Interactive Whiteboard Lesson Library
- Dynamic Classroom with Dynamic Investigations

What Your Students Will Learn

- Solve quadratic equations by graphing and finding the x-intercepts.
- Use graphs to find and approximate the zeros of functions.
- Solve real-life problems using tables and graphs of quadratic functions.

Laurie's Notes

Exploration

Motivate

- Show a picture of a projectile being launched from a trebuchet or a catapult.
- Ask students what they would like to know about the projectile. Hopefully, a student will ask how long it takes for the projectile to land.
- Explain to students that in this section they will answer that type of question.

Discuss

- Connect the projectile motion to the x-intercept dialogue.
- Also, connect the previous two chapters to this section by reminding students that they solved quadratic equations by factoring and by graphing quadratic functions.

Exploration 1

? "What do you know about the graph of $y = x^2 - 2x$?" opens up; y-intercept is 0; x-coordinate of the vertex is 1; axis of symmetry is $x = 1$

- Students should make an input-output table that includes x-values that show the key features of the graph.
- While students work, question different pairs of students about the vertex and the minimum value.
- **Big Idea:** In part (d), students should discuss evaluating the left side of the equation for the x-values found in part (c), and they should also discuss the x-intercept. It is not enough to solve the equation graphically because computational errors can influence the graph. An algebraic check is also important.
- **Make Sense of Problems and Persevere in Solving Them:** This approach to solving a quadratic equation should make sense to students if they think back to solving systems of equations. They can think of solving $x^2 - 2x = 0$ as solving the system $y = x^2 - 2x$ and $y = 0$. The graph of $y = 0$ is the x-axis, which intersects the graph of $y = x^2 - 2x$ at its x-intercepts.

Exploration 2

- As they are solving each equation by graphing, students should be checking the reasonableness of their graphs.
- While students work, probe different pairs of students about the vertex, the minimum or maximum value, the y-intercept, and the general shape of the graph.
- The equations in parts (e) and (f) are different from the first four equations. It should occur to students to try factoring, which does not yield a solution. Next, graphing by finding the vertex and a few additional points should be possible. Students will recognize that the parabolas do not intersect the x-axis.

? "In Exploration 1, the equation had two solutions. Does each equation in Exploration 2 have two solutions?" no; The equation in part (d) has only one solution, and the equations in parts (e) and (f) have no solution.

? "Do you think it is possible to predict how many solutions a quadratic equation in one variable will have?" Answers will vary.

Communicate Your Answer

- **Make Sense of Problems and Persevere in Solving Them:** In Question 4, students are making sense of a problem in more than one way. It is important for students to realize that the x-intercepts correspond to the solutions of the equation, so substituting these values for x should result in a true equation.

Connecting to Next Step

- Exploration 1 is important in establishing what it means to solve a quadratic by graphing. This will be stated at the beginning of the formal lesson.

9.2 Solving Quadratic Equations by Graphing

Essential Question How can you use a graph to solve a quadratic equation in one variable?

Based on what you learned about the x-intercepts of a graph in Section 3.4, it follows that the x-intercept of the graph of the linear equation

$y = ax + b$ 2 variables

is the same value as the solution of

$ax + b = 0.$ 1 variable

You can use similar reasoning to solve *quadratic equations*.

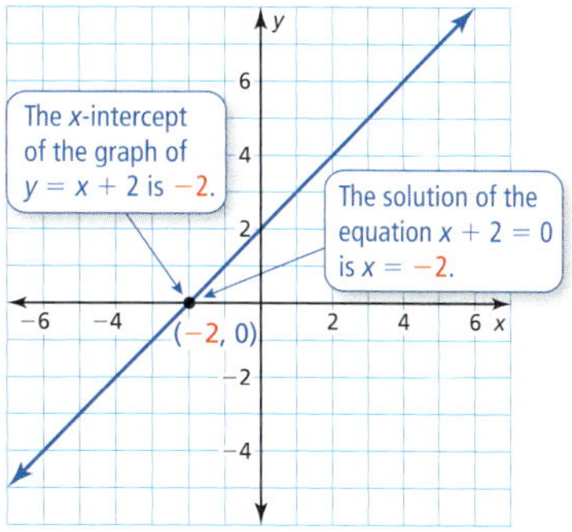

EXPLORATION 1 Solving a Quadratic Equation by Graphing

Work with a partner.

a. Sketch the graph of $y = x^2 - 2x$.

b. What is the definition of an x-intercept of a graph? How many x-intercepts does this graph have? What are they?

c. What is the definition of a solution of an equation in x? How many solutions does the equation $x^2 - 2x = 0$ have? What are they?

d. Explain how you can verify the solutions you found in part (c).

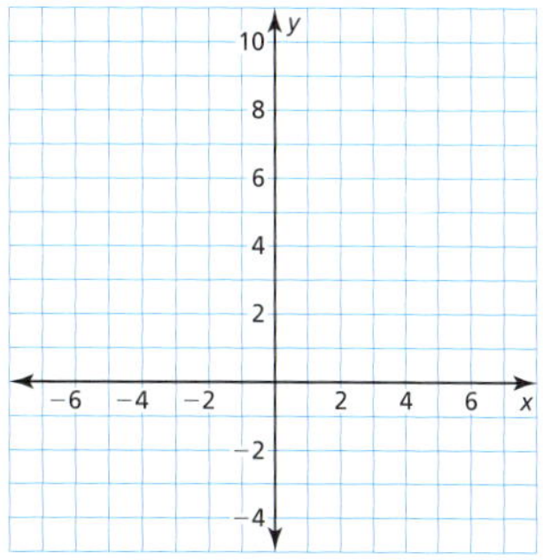

EXPLORATION 2 Solving Quadratic Equations by Graphing

Work with a partner. Solve each equation by graphing.

a. $x^2 - 4 = 0$ **b.** $x^2 + 3x = 0$

c. $-x^2 + 2x = 0$ **d.** $x^2 - 2x + 1 = 0$

e. $x^2 - 3x + 5 = 0$ **f.** $-x^2 + 3x - 6 = 0$

MAKING SENSE OF PROBLEMS

To be proficient in math, you need to check your answers to problems using a different method and continually ask yourself, "Does this make sense?"

Communicate Your Answer

3. How can you use a graph to solve a quadratic equation in one variable?

4. After you find a solution graphically, how can you check your result algebraically? Check your solutions for parts (a)–(d) in Exploration 2 algebraically.

5. How can you determine graphically that a quadratic equation has no solution?

Dynamic Teaching Tools

- Dynamic Assessment & Progress Monitoring Tool
- Lesson Planning Tool
- Interactive Whiteboard Lesson Library
- Dynamic Classroom with Dynamic Investigations

ANSWERS

1. a.

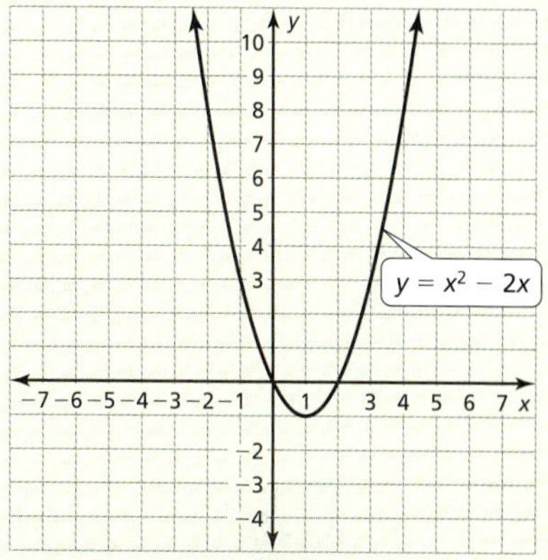

b. the x-coordinate of a point where the graph crosses the x-axis; 2; 0, 2

c. a value of x that makes the equation true; 2; 0, 2

d. Substitute the solutions into the equation.

2. a. $x = 2, x = -2$

b. $x = 0, x = -3$

c. $x = 0, x = 2$

d. $x = 1$

e. no solution

f. no solution

3. Write the equation in standard form $ax^2 + bx + c = 0$, graph the related function $y = ax^2 + bx + c$, and find the x-intercepts.

4. Substitute the solutions into the equation.

5. The related graph will have no x-intercepts.

Differentiated Instruction

Auditory

Have students verbally describe the steps they would use to graph a quadratic equation such as $y = x^2 - 3$. Ask them to explain how they can predict whether the graph will open up or down. Be sure students understand that this graph will open up because the coefficient of x^2 is greater than 0.

Extra Example 1

Solve $x^2 + 2x = 8$ by graphing.

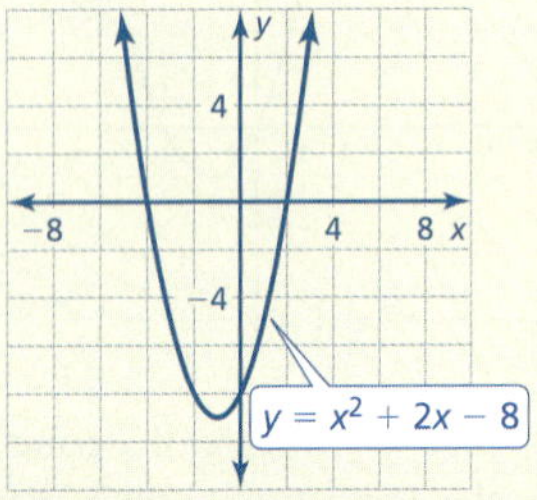

The solutions are $x = -4$ and $x = 2$.

MONITORING PROGRESS ANSWERS

1. $x = 2, x = -1$
2. $x = -5, x = -2$
3. $x = -4, x = 3$

9.2 Lesson

Core Vocabulary

quadratic equation, *p. 490*

Previous

x-intercept

root

zero of a function

What You Will Learn

- Solve quadratic equations by graphing.
- Use graphs to find and approximate the zeros of functions.
- Solve real-life problems using graphs of quadratic functions.

Solving Quadratic Equations by Graphing

A **quadratic equation** is a nonlinear equation that can be written in the standard form $ax^2 + bx + c = 0$, where $a \neq 0$.

In Chapter 7, you solved quadratic equations by factoring. You can also solve quadratic equations by graphing.

Core Concept

Solving Quadratic Equations by Graphing

Step 1 Write the equation in standard form, $ax^2 + bx + c = 0$.

Step 2 Graph the related function $y = ax^2 + bx + c$.

Step 3 Find the x-intercepts, if any.

The solutions, or *roots*, of $ax^2 + bx + c = 0$ are the x-intercepts of the graph.

EXAMPLE 1 Solving a Quadratic Equation: Two Real Solutions

Solve $x^2 + 2x = 3$ by graphing.

SOLUTION

Step 1 Write the equation in standard form.

$$x^2 + 2x = 3 \quad \text{Write original equation.}$$
$$x^2 + 2x - 3 = 0 \quad \text{Subtract 3 from each side.}$$

Step 2 Graph the related function $y = x^2 + 2x - 3$.

Step 3 Find the x-intercepts. The x-intercepts are -3 and 1.

▶ So, the solutions are $x = -3$ and $x = 1$.

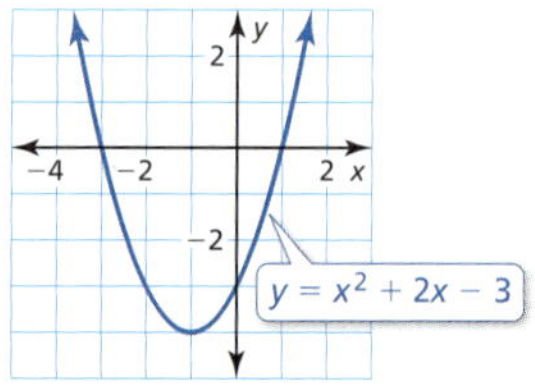

Check

$x^2 + 2x = 3$	Original equation	$x^2 + 2x = 3$
$(-3)^2 + 2(-3) \stackrel{?}{=} 3$	Substitute.	$1^2 + 2(1) \stackrel{?}{=} 3$
$3 = 3$ ✓	Simplify.	$3 = 3$ ✓

Monitoring Progress

Help in English and Spanish at *BigIdeasMath.com*

Solve the equation by graphing. Check your solutions.

1. $x^2 - x - 2 = 0$
2. $x^2 + 7x = -10$
3. $x^2 + x = 12$

Laurie's Notes Teacher Actions

- **Make Sense of Problems and Persevere in Solving Them:** Throughout this lesson, discuss with students the different graphing methods they can use to solve these types of equations, as well as the different checks they can use: factoring, using a graphing utility, or substituting solutions back into the original equation.

? "What is the first step in solving $x^2 + 2x = 3$ by graphing?" Write the equation in standard form. "What technique can be used to graph $y = x^2 + 2x - 3$?" Listen for valid methods, including factoring into $y = (x + 3)(x - 1)$. Discuss methods for checking the solutions.

EXAMPLE 2 Solving a Quadratic Equation: One Real Solution

Solve $x^2 - 8x = -16$ by graphing.

SOLUTION

Step 1 Write the equation in standard form.

$$x^2 - 8x = -16 \quad \text{Write original equation.}$$
$$x^2 - 8x + 16 = 0 \quad \text{Add 16 to each side.}$$

Step 2 Graph the related function $y = x^2 - 8x + 16$.

Step 3 Find the x-intercept. The only x-intercept is at the vertex, (4, 0).

▶ So, the solution is $x = 4$.

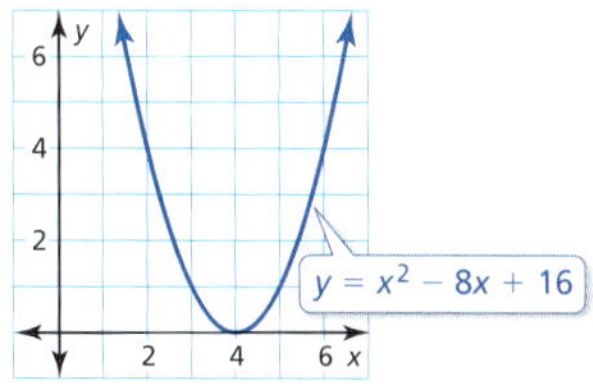

ANOTHER WAY

You can also solve the equation in Example 2 by factoring.

$x^2 - 8x + 16 = 0$

$(x - 4)(x - 4) = 0$

So, $x = 4$.

EXAMPLE 3 Solving a Quadratic Equation: No Real Solutions

Solve $-x^2 = 2x + 4$ by graphing.

SOLUTION

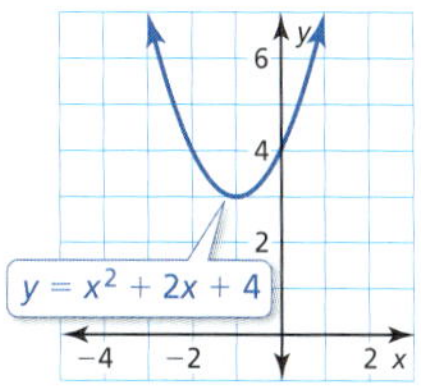

Method 1 Write the equation in standard form, $x^2 + 2x + 4 = 0$. Then graph the related function $y = x^2 + 2x + 4$, as shown at the left.

▶ There are no x-intercepts. So, $-x^2 = 2x + 4$ has no real solutions.

Method 2 Graph each side of the equation.

$$y = -x^2 \quad \text{Left side}$$
$$y = 2x + 4 \quad \text{Right side}$$

▶ The graphs do not intersect. So, $-x^2 = 2x + 4$ has no real solutions.

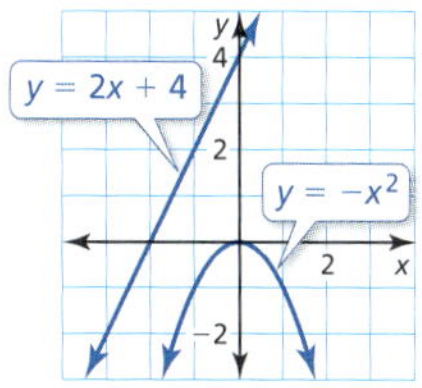

Monitoring Progress

Help in English and Spanish at *BigIdeasMath.com*

Solve the equation by graphing.

4. $x^2 + 36 = 12x$ **5.** $x^2 + 4x = 0$ **6.** $x^2 + 10x = -25$

7. $x^2 = 3x - 3$ **8.** $x^2 + 7x = -6$ **9.** $2x + 5 = -x^2$

Concept Summary

Number of Solutions of a Quadratic Equation

A quadratic equation has:

- two real solutions when the graph of its related function has two x-intercepts.
- one real solution when the graph of its related function has one x-intercept.
- no real solutions when the graph of its related function has no x-intercepts.

English Language Learners

Pair Activity

Have students work in pairs to solve quadratic equations by graphing. Each student solves a different equation. When both students are done, they take turns explaining their solution while the other person follows along.

Extra Example 2

Solve $x^2 + 4x = -4$ by graphing.

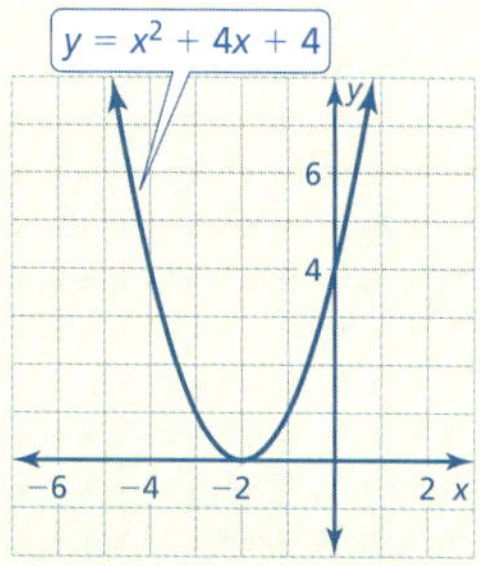

The solution is $x = -2$.

Extra Example 3

Solve $-x^2 = 4x + 8$ by graphing.

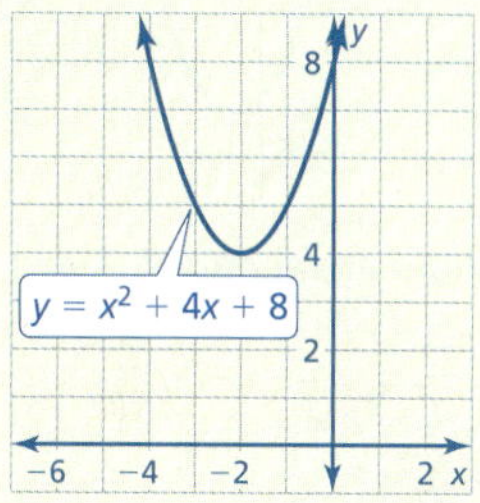

$-x^2 = 4x + 8$ has no real solutions.

MONITORING PROGRESS ANSWERS

4. $x = 6$

5. $x = 0, x = -4$

6. $x = -5$

7. no solution

8. $x = -6, x = -1$

9. no solution

Laurie's Notes Teacher Actions

- Pose Example 2, and have students work independently to solve.
- **Response Cards:** Have students hold up the *Response Card* showing the number of solutions found. Solicit the solution, $x = 4$.
- Pose Example 3, and have students work independently to solve.
- **Response Cards:** Again use the *Response Cards*.
- Explain Method 2, as it is likely that students would not have used this approach. It is a technique they will use frequently in future mathematics classes.

Extra Example 4

The graph of $f(x) = (x - 1)(x^2 - x - 6)$ is shown. Find the zeros of f.

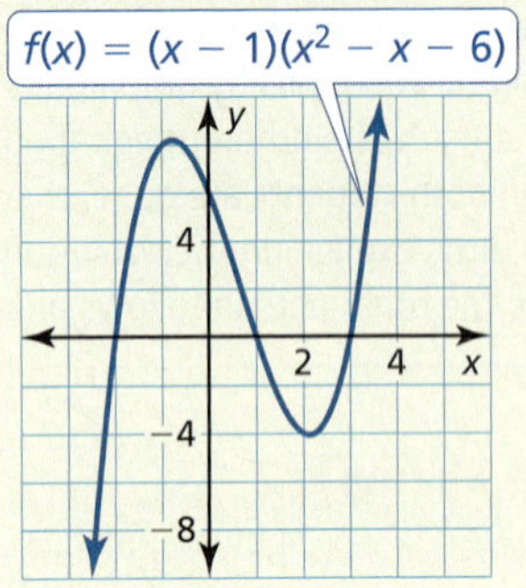

The zeros of f are -2, 1, and 3.

Extra Example 5

The graph of $f(x) = x^2 - x - 5$ is shown. Approximate the zeros of f to the nearest tenth.

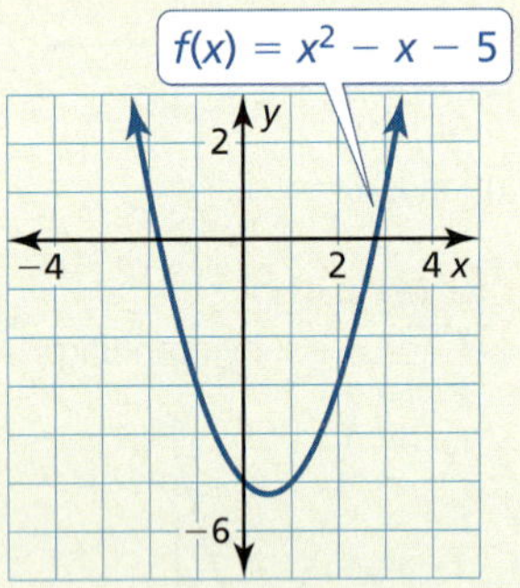

The zeros of f are about -1.8 and 2.8.

MONITORING PROGRESS ANSWERS

10.

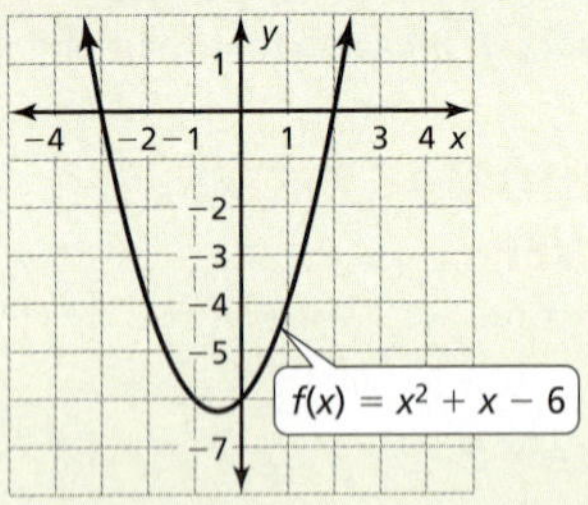

-3, 2

11.

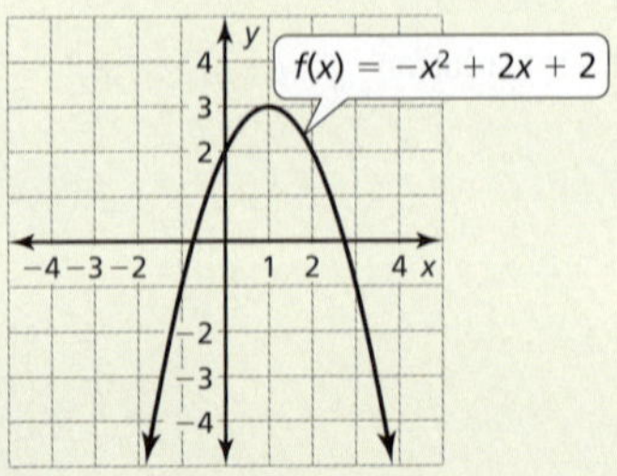

about -0.7, about 2.7

Finding Zeros of Functions

Recall that a zero of a function is an x-intercept of the graph of the function.

EXAMPLE 4 Finding the Zeros of a Function

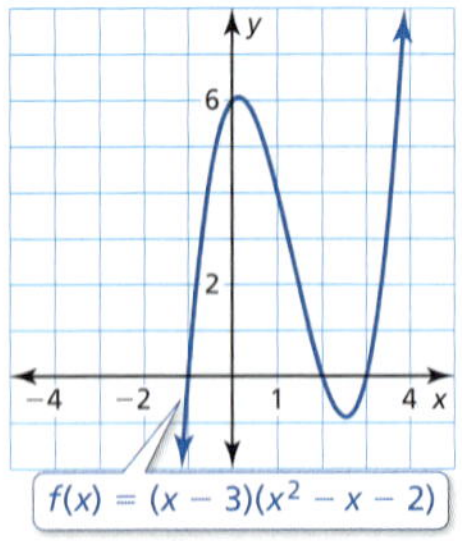

The graph of $f(x) = (x - 3)(x^2 - x - 2)$ is shown. Find the zeros of f.

SOLUTION

The x-intercepts are -1, 2, and 3.

▶ So, the zeros of f are -1, 2, and 3.

Check

$f(-1) = (-1 - 3)[(-1)^2 - (-1) - 2] = 0$ ✓

$f(2) = (2 - 3)(2^2 - 2 - 2) = 0$ ✓

$f(3) = (3 - 3)(3^2 - 3 - 2) = 0$ ✓

The zeros of a function are not necessarily integers. To approximate zeros, analyze the signs of function values. When two function values have different signs, a zero lies between the x-values that correspond to the function values.

EXAMPLE 5 Approximating the Zeros of a Function

The graph of $f(x) = x^2 + 4x + 1$ is shown. Approximate the zeros of f to the nearest tenth.

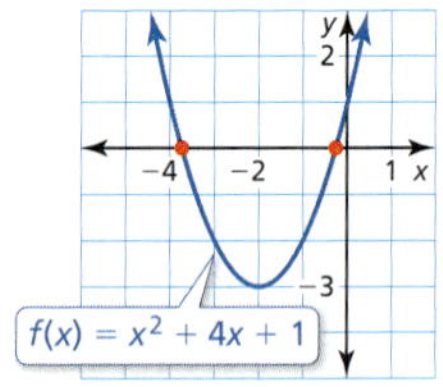

SOLUTION

There are two x-intercepts: one between -4 and -3, and another between -1 and 0.

Make tables using x-values between -4 and -3, and between -1 and 0. Use an increment of 0.1. Look for a change in the signs of the function values.

x	−3.9	−3.8	(−3.7)	−3.6	−3.5	−3.4	−3.3	−3.2	−3.1
f(x)	0.61	0.24	−0.11	−0.44	−0.75	−1.04	−1.31	−1.56	−1.79

change in signs

x	−0.9	−0.8	−0.7	−0.6	−0.5	−0.4	(−0.3)	−0.2	−0.1
f(x)	−1.79	−1.56	−1.31	−1.04	−0.75	−0.44	−0.11	0.24	0.61

change in signs

ANOTHER WAY

You could approximate one zero using a table and then use the axis of symmetry to find the other zero.

The function values that are closest to 0 correspond to x-values that best approximate the zeros of the function.

▶ In each table, the function value closest to 0 is -0.11. So, the zeros of f are about -3.7 and -0.3.

Monitoring Progress

Help in English and Spanish at BigIdeasMath.com

10. Graph $f(x) = x^2 + x - 6$. Find the zeros of f.

11. Graph $f(x) = -x^2 + 2x + 2$. Approximate the zeros of f to the nearest tenth.

Laurie's Notes Teacher Actions

- ? "What is another way to write $f(x) = (x - 3)(x^2 - x - 2)$?" Answers will vary. One approach is to factor completely: $f(x) = (x - 3)(x - 2)(x + 1)$.
- ? **Assessing Question:** "Can you factor every quadratic expression?" Students will assume rational solutions and say no.
- **Use Appropriate Tools Strategically:** In approximating the zeros of the function, use the table feature on the graphing calculator.
- **Big Idea:** As students adjust the increment, they are focusing on the entries where the function value changes sign.

Solving Real-Life Problems

EXAMPLE 6 Real-Life Application

A football player kicks a football 2 feet above the ground with an initial vertical velocity of 75 feet per second. The function $h = -16t^2 + 75t + 2$ represents the height h (in feet) of the football after t seconds. (a) Find the height of the football each second after it is kicked. (b) Use the results of part (a) to estimate when the height of the football is 50 feet. (c) Using a graph, after how many seconds is the football 50 feet above the ground?

SOLUTION

a. Make a table of values starting with $t = 0$ seconds using an increment of 1. Continue the table until a function value is negative.

Seconds, t	Height, h
0	2
1	61
2	88
3	83
4	46
5	−23

▶ The height of the football is 61 feet after 1 second, 88 feet after 2 seconds, 83 feet after 3 seconds, and 46 feet after 4 seconds.

b. From part (a), you can estimate that the height of the football is 50 feet between 0 and 1 second and between 3 and 4 seconds.

▶ Based on the function values, it is reasonable to estimate that the height of the football is 50 feet slightly less than 1 second and slightly less than 4 seconds after it is kicked.

c. To determine when the football is 50 feet above the ground, find the t-values for which $h = 50$. So, solve the equation $-16t^2 + 75t + 2 = 50$ by graphing.

Step 1 Write the equation in standard form.

$$-16t^2 + 75t + 2 = 50 \quad \text{Write the equation.}$$

$$-16t^2 + 75t - 48 = 0 \quad \text{Subtract 50 from each side.}$$

Step 2 Use a graphing calculator to graph the related function $h = -16t^2 + 75t - 48$.

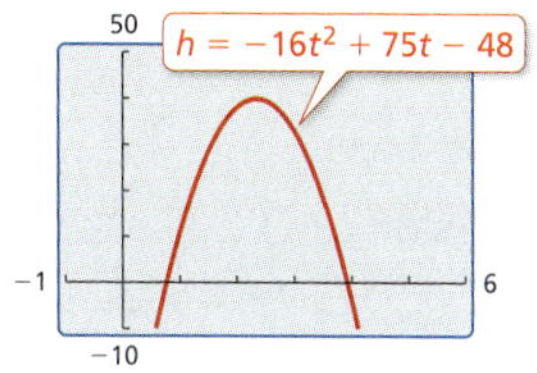

REMEMBER

Equations have *solutions*, or *roots*. Graphs have *x-intercepts*. Functions have *zeros*.

Step 3 Use the *zero* feature to find the zeros of the function.

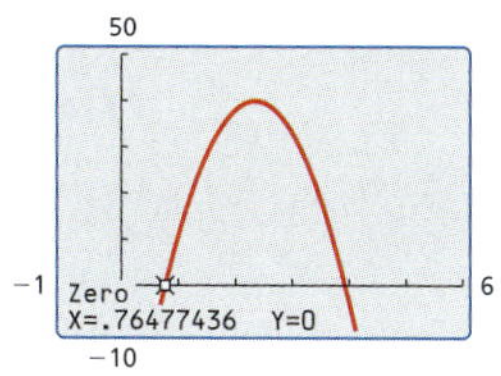

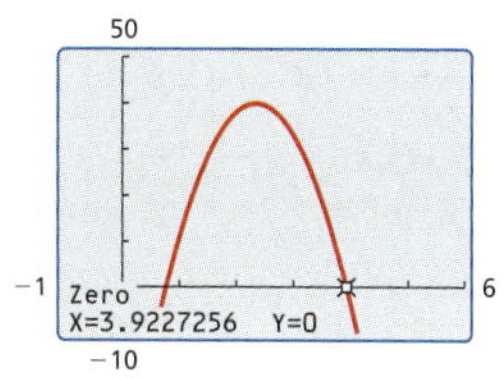

▶ The football is 50 feet above the ground after about 0.8 second and about 3.9 seconds, which supports the estimates in part (b).

Monitoring Progress

Help in English and Spanish at *BigIdeasMath.com*

12. **WHAT IF?** After how many seconds is the football 65 feet above the ground?

Extra Example 6

A soccer player kicks a soccer ball 2 feet above the ground with an initial vertical velocity of 60 feet per second. The function $h = -16t^2 + 60t + 2$ represents the height h (in feet) of the soccer ball after t seconds.

a. Find the height of the soccer ball each second after it is kicked. The height of the soccer ball is 46 feet after 1 second, 58 feet after 2 seconds, and 38 feet after 3 seconds.

b. Use the results of part (a) to estimate when the height of the soccer ball is 40 feet. The height of the soccer ball is 40 feet slightly less than 1 second and slightly less than 3 seconds after it is kicked.

c. Using a graph, after how many seconds is the soccer ball 40 feet above the ground? The soccer ball is 40 feet above the ground after about 0.8 second and about 2.9 seconds.

MONITORING PROGRESS ANSWER

12. about 1.1 sec, about 3.6 sec

Laurie's Notes Teacher Actions

- Remind students that the ordered pair (x, y) is replaced by (t, h) representing (time, height), with units associated with each variable.
- **Misconception:** The graph shows the height of the object over time, not the path of the object.
- **Connection:** This example could also be solved like Example 3, Method 2: Graph $y = -16t^2 + 75t + 2$ and $y = 50$.

Closure

- When a quadratic equation has two solutions, what do you know about the graph of its related function? It has two x-intercepts.

Assignment Guide and Homework Check

ASSIGNMENT

Basic: 1–4, 5–29 odd, 37–49 odd, 53, 58, 62, 66, 67

Average: 1–4, 6–58 even, 62, 66, 67

Advanced: 1–4, 8, 12, 20–32 even, 40–58 even, 59–67

HOMEWORK CHECK

Basic: 13, 19, 37, 43, 53

Average: 14, 20, 38, 44, 54

Advanced: 20, 24, 40, 44, 54

ANSWERS

1. an equation that can be written in the standard form $ax^2 + bx + c = 0$, where $a \neq 0$
2. $x^2 + x - 4 = 0$; It is the only equation written in standard form.
3. The number of x-intercepts is the number of solutions.
4. They are the same.
5. $x = 3, x = -1$
6. $x = 4, x = 2$
7. $x = -4$
8. no solution
9. $4x^2 - 12 = 0$ or $-4x^2 + 12 = 0$
10. $x^2 + 15 = 0$ or $-x^2 - 15 = 0$
11. $x^2 - 2x + 1 = 0$ or $-x^2 + 2x - 1 = 0$
12. $3x^2 - x - 5 = 0$ or $-3x^2 + x + 5 = 0$
13. $x = 0, x = 5$
14. $x = 2$
15. no solution
16. $x = 7, x = -1$
17. $x = 3$
18. no solution
19. $x = -1$
20. no solution
21. $x = -6, x = 2$
22. $x = 3, x = 2$
23. $x = -2, x = 1$
24. $x = -4$

9.2 Exercises

Dynamic Solutions available at *BigIdeasMath.com*

Vocabulary and Core Concept Check

1. **VOCABULARY** What is a quadratic equation?
2. **WHICH ONE DOESN'T BELONG?** Which equation does *not* belong with the other three? Explain your reasoning.

$x^2 + 5x = 20$	$x^2 + x - 4 = 0$	$x^2 - 6 = 4x$	$7x + 12 = x^2$

3. **WRITING** How can you use a graph to find the number of solutions of a quadratic equation?
4. **WRITING** How are solutions, roots, x-intercepts, and zeros related?

Monitoring Progress and Modeling with Mathematics

In Exercises 5–8, use the graph to solve the equation.

5. $-x^2 + 2x + 3 = 0$

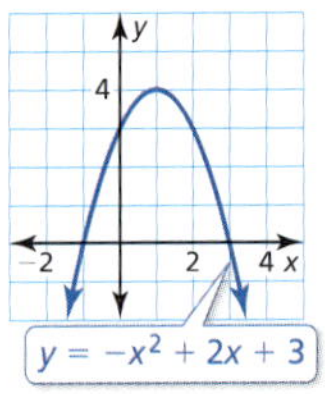

6. $x^2 - 6x + 8 = 0$

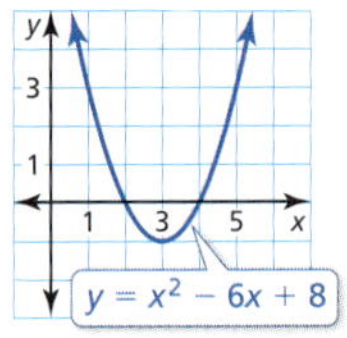

7. $x^2 + 8x + 16 = 0$

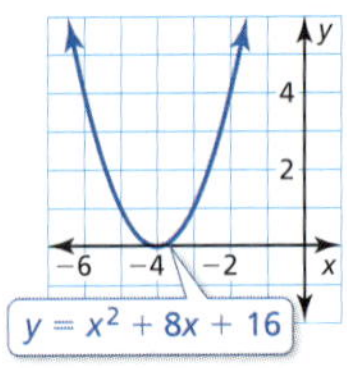

8. $-x^2 - 4x - 6 = 0$

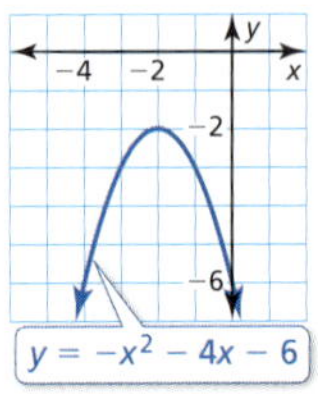

In Exercises 9–12, write the equation in standard form.

9. $4x^2 = 12$
10. $-x^2 = 15$
11. $2x - x^2 = 1$
12. $5 + x = 3x^2$

In Exercises 13–24, solve the equation by graphing. *(See Examples 1, 2, and 3.)*

13. $x^2 - 5x = 0$
14. $x^2 - 4x + 4 = 0$
15. $x^2 - 2x + 5 = 0$
16. $x^2 - 6x - 7 = 0$
17. $x^2 = 6x - 9$
18. $-x^2 = 8x + 20$
19. $x^2 = -1 - 2x$
20. $x^2 = -x - 3$
21. $4x - 12 = -x^2$
22. $5x - 6 = x^2$
23. $x^2 - 2 = -x$
24. $16 + x^2 = -8x$

25. **ERROR ANALYSIS** Describe and correct the error in solving $x^2 + 3x = 18$ by graphing.

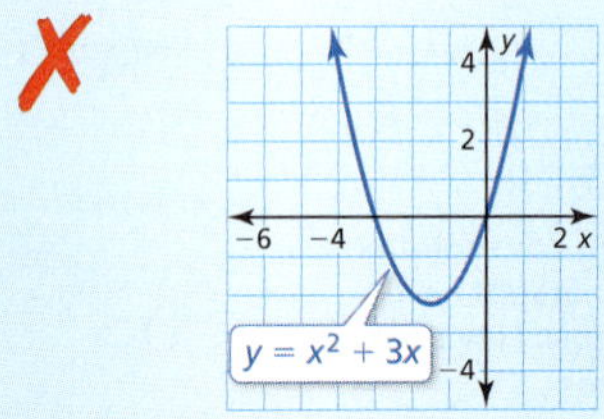

The solutions of the equation $x^2 + 3x = 18$ are $x = -3$ and $x = 0$.

26. **ERROR ANALYSIS** Describe and correct the error in solving $x^2 + 6x + 9 = 0$ by graphing.

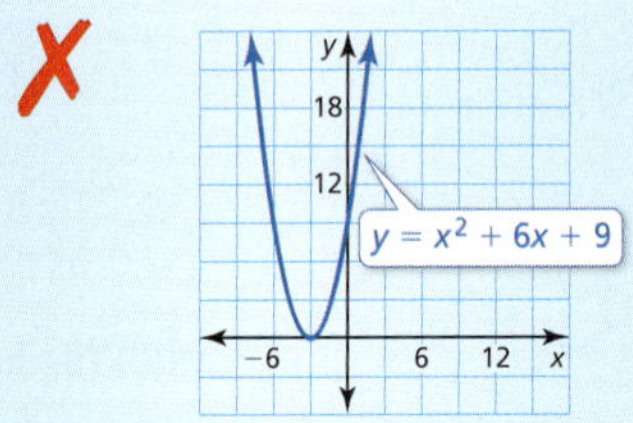

The solution of the equation $x^2 + 6x + 9 = 0$ is $x = 9$.

25. The equation needs to be in standard form; $x^2 + 3x - 18 = 0$

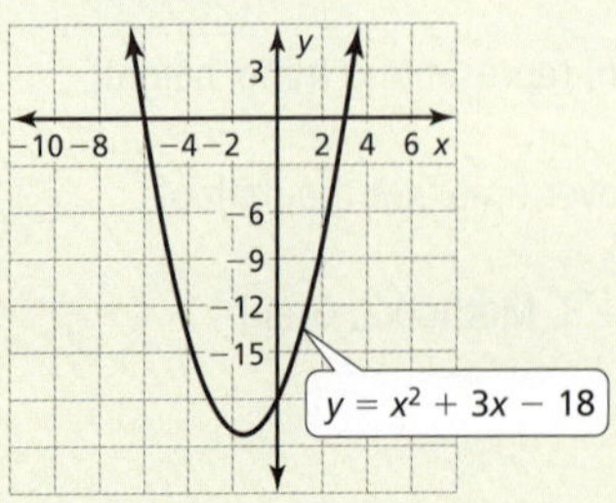

The solutions are $x = -6$ and $x = 3$.

26. The solution is the x-intercept, not the y-intercept; The solution is $x = -3$.

27. MODELING WITH MATHEMATICS The height y (in yards) of a flop shot in golf can be modeled by $y = -x^2 + 5x$, where x is the horizontal distance (in yards).

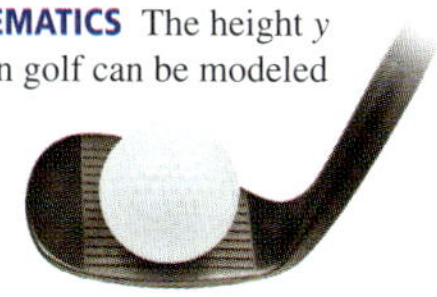

a. Interpret the x-intercepts of the graph of the equation.

b. How far away does the golf ball land?

28. MODELING WITH MATHEMATICS The height h (in feet) of an underhand volleyball serve can be modeled by $h = -16t^2 + 30t + 4$, where t is the time (in seconds).

a. Do both t-intercepts of the graph of the function have meaning in this situation? Explain.

b. No one receives the serve. After how many seconds does the volleyball hit the ground?

In Exercises 29–36, solve the equation by using Method 2 from Example 3.

29. $x^2 = 10 - 3x$

30. $2x - 3 = x^2$

31. $5x - 7 = x^2$

32. $x^2 = 6x - 5$

33. $x^2 + 12x = -20$

34. $x^2 + 8x = 9$

35. $-x^2 - 5 = -2x$

36. $-x^2 - 4 = -4x$

In Exercises 37–42, find the zero(s) of f. *(See Example 4.)*

37.

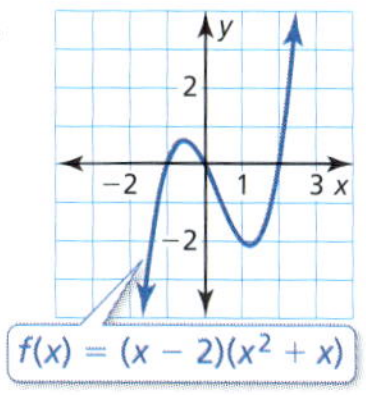

38.

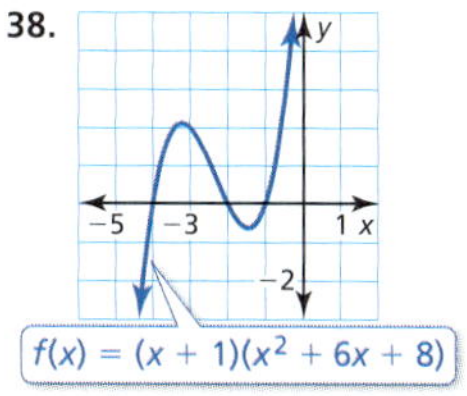

39.

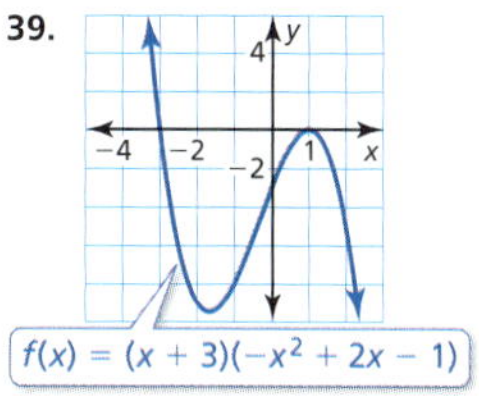

40.

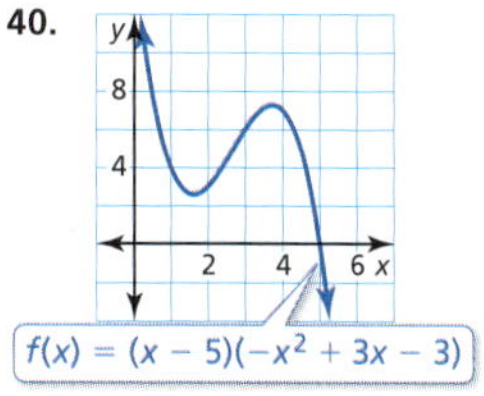

41.

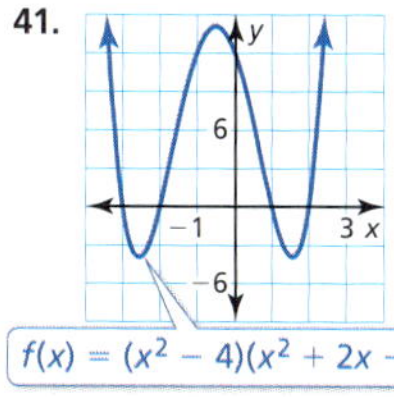

42.

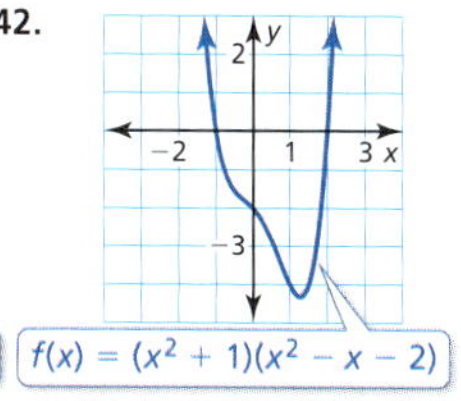

In Exercises 43–46, approximate the zeros of f to the nearest tenth. *(See Example 5.)*

43.

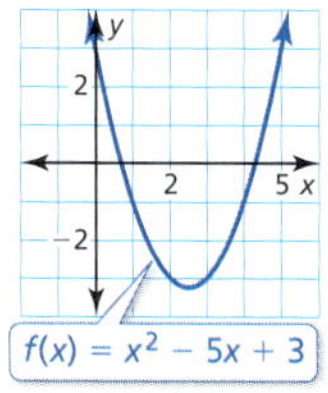

44.

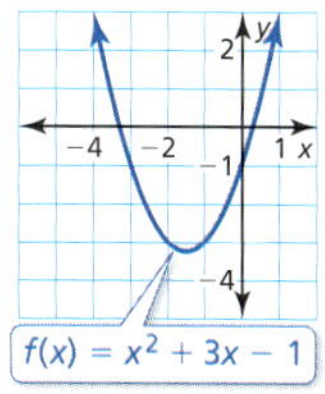

45.

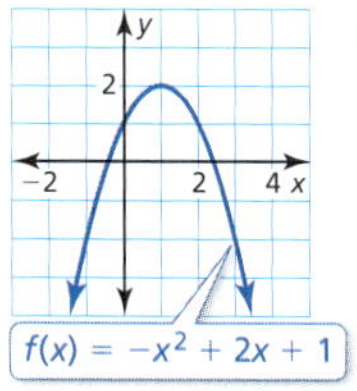

46.

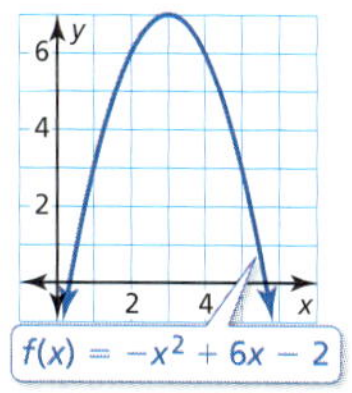

In Exercises 47–52, graph the function. Approximate the zeros of the function to the nearest tenth, if necessary.

47. $f(x) = x^2 + 6x + 1$

48. $f(x) = x^2 - 3x + 2$

49. $y = -x^2 + 4x - 2$

50. $y = -x^2 + 9x - 6$

51. $f(x) = \frac{1}{2}x^2 + 2x - 5$

52. $f(x) = -3x^2 + 4x + 3$

53. MODELING WITH MATHEMATICS At a Civil War reenactment, a cannonball is fired into the air with an initial vertical velocity of 128 feet per second. The release point is 6 feet above the ground. The function $h = -16t^2 + 128t + 6$ represents the height h (in feet) of the cannonball after t seconds. *(See Example 6.)*

a. Find the height of the cannonball each second after it is fired.

b. Use the results of part (a) to estimate when the height of the cannonball is 150 feet.

c. Using a graph, after how many seconds is the cannonball 150 feet above the ground?

Dynamic Teaching Tools

- Dynamic Assessment & Progress Monitoring Tool
- Interactive Whiteboard Lesson Library
- Dynamic Classroom with Dynamic Investigations

ANSWERS

27. a. The x-intercepts are the distances at which the height is 0 yards.

b. 5 yd

28. a. no; One of the t-intercepts is negative.

b. 2 sec

29. $x = -5, x = 2$

30. no solution

31. no solution

32. $x = 5, x = 1$

33. $x = -10, x = -2$

34. $x = -9, x = 1$

35. no solution

36. $x = 2$

37. 2, 0, −1

38. −1, −4, −2

39. −3, 1

40. 5

41. −2, 2, −3, 1

42. −1, 2

43. about 4.3, about 0.7

44. about 0.3, about −3.3

45. about 2.4, about −0.4

46. about 5.6, about 0.4

47.

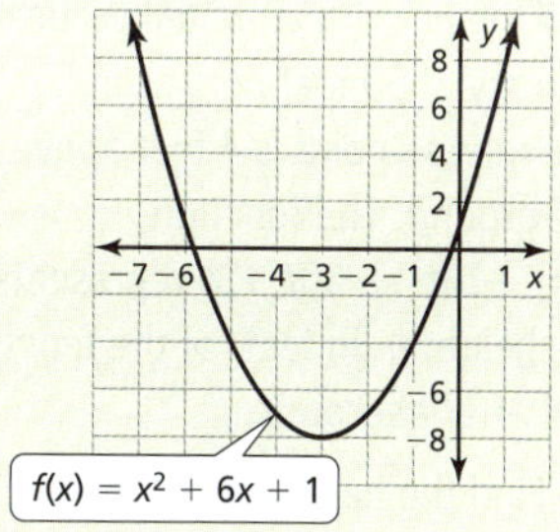

about −0.2, about −5.8

48.

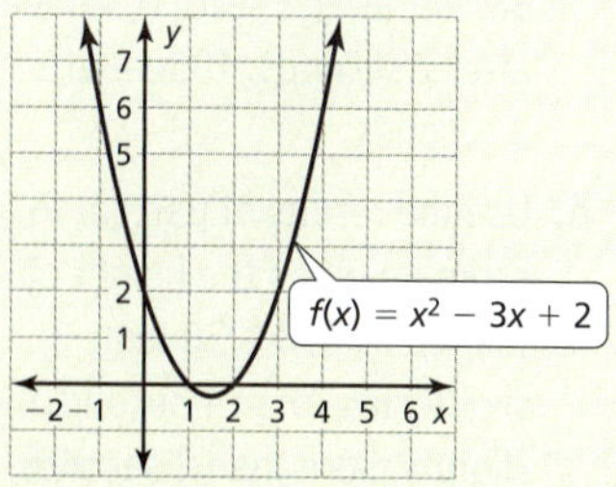

1, 2

49.

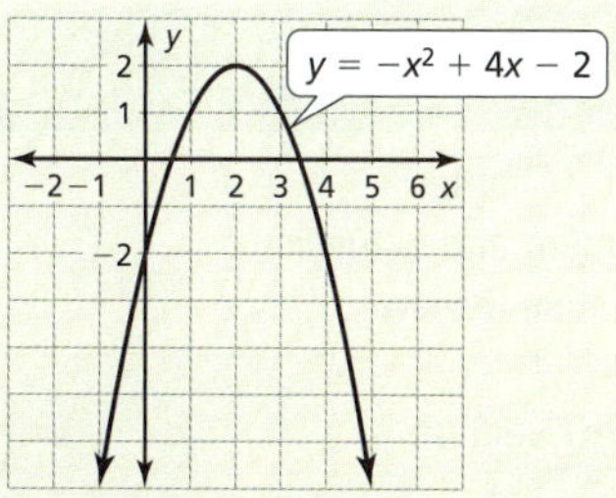

about 3.4, about 0.6

50.

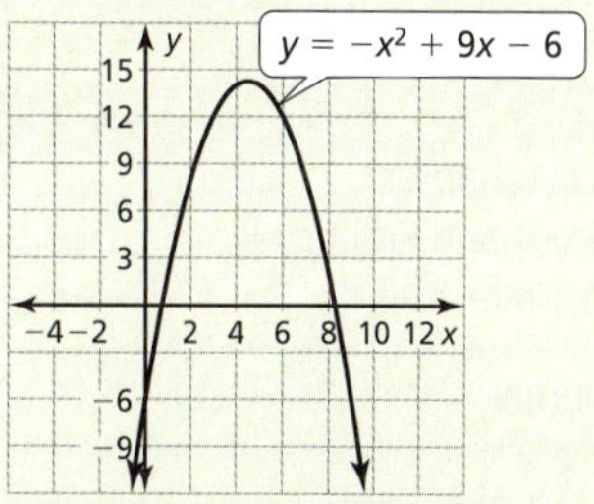

about 8.3, about 0.7

51–53. See Additional Answers.

ANSWERS

54. **a.** 29 ft, 21 ft

b. about 0.5 sec, about 2.3 sec

c. about 0.3 sec, about 2.2 sec

55–67. See Additional Answers.

Mini-Assessment

1. Solve $x^2 + 4x = 5$ by graphing.

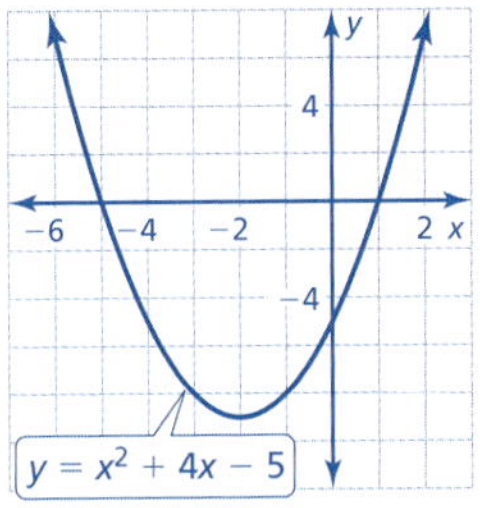

$x = -5$ and $x = 1$

2. Use the graph to find the zeros of $f(x) = (x - 2)(x^2 + 3x + 2)$.

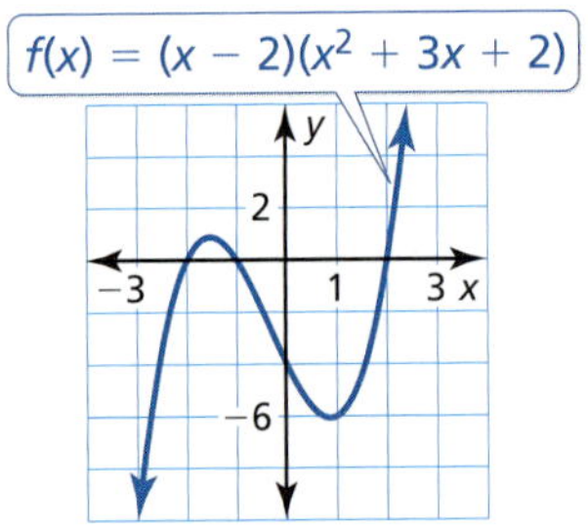

The zeros are −2, −1, and 2.

3. A tennis ball is thrown straight up into the air with an initial vertical velocity of 50 feet per second. The release point is 4 feet above the ground. The function $h = -16t^2 + 50t + 4$ represents the height h (in feet) of the tennis ball after t seconds.

   a. Find the height of the tennis ball each second after it is thrown.
   38 feet after 1 second, 40 feet after 2 seconds, 10 feet after 3 seconds

   b. Use the results of part (a) to estimate when the height of the tennis ball is 30 feet.
   slightly less than 1 second and slightly more than 2 seconds after it is thrown

   c. Using a graph, after how many seconds is the tennis ball 30 feet above the ground?
   after about 0.7 second and about 2.5 seconds

54. MODELING WITH MATHEMATICS You throw a softball straight up into the air with an initial vertical velocity of 40 feet per second. The release point is 5 feet above the ground. The function $h = -16t^2 + 40t + 5$ represents the height h (in feet) of the softball after t seconds.

a. Find the height of the softball each second after it is released.

b. Use the results of part (a) to estimate when the height of the softball is 15 feet.

c. Using a graph, after how many seconds is the softball 15 feet above the ground?

MATHEMATICAL CONNECTIONS In Exercises 55 and 56, use the given surface area S of the cylinder to find the radius r to the nearest tenth.

55. $S = 225\text{ ft}^2$

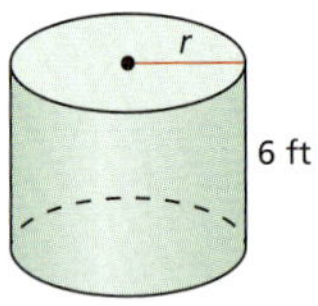

56. $S = 750\text{ m}^2$

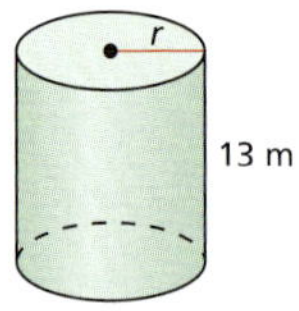

57. WRITING Explain how to approximate zeros of a function when the zeros are not integers.

58. HOW DO YOU SEE IT? Consider the graph shown.

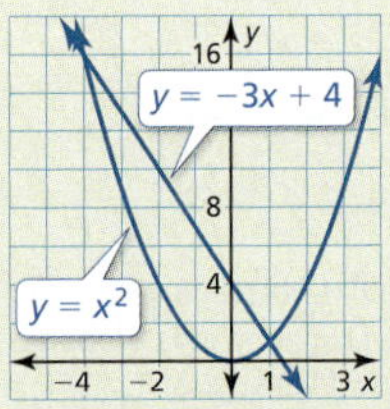

a. How many solutions does the quadratic equation $x^2 = -3x + 4$ have? Explain.

b. Without graphing, describe what you know about the graph of $y = x^2 + 3x - 4$.

59. COMPARING METHODS Example 3 shows two methods for solving a quadratic equation. Which method do you prefer? Explain your reasoning.

60. THOUGHT PROVOKING How many different parabolas have −2 and 2 as x-intercepts? Sketch examples of parabolas that have these two x-intercepts.

61. MODELING WITH MATHEMATICS To keep water off a road, the surface of the road is shaped like a parabola. A cross section of the road is shown in the diagram. The surface of the road can be modeled by $y = -0.0017x^2 + 0.041x$, where x and y are measured in feet. Find the width of the road to the nearest tenth of a foot.

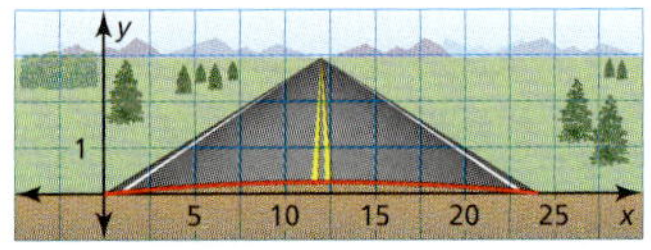

62. MAKING AN ARGUMENT A stream of water from a fire hose can be modeled by $y = -0.003x^2 + 0.58x + 3$, where x and y are measured in feet. A firefighter is standing 57 feet from a building and is holding the hose 3 feet above the ground. The bottom of a window of the building is 26 feet above the ground. Your friend claims the stream of water will pass through the window. Is your friend correct? Explain.

REASONING In Exercises 63–65, determine whether the statement is *always*, *sometimes*, or *never* true. Justify your answer.

63. The graph of $y = ax^2 + c$ has two x-intercepts when a is negative.

64. The graph of $y = ax^2 + c$ has no x-intercepts when a and c have the same sign.

65. The graph of $y = ax^2 + bx + c$ has more than two x-intercepts when $a \neq 0$.

Maintaining Mathematical Proficiency

Reviewing what you learned in previous grades and lessons

Determine whether the table represents an *exponential growth function*, an *exponential decay function*, or *neither*. Explain. *(Section 6.4)*

66.

x	−1	0	1	2
y	18	3	$\frac{1}{2}$	$\frac{1}{12}$

67.

x	0	1	2	3
y	2	8	32	128

If students need help...	*If students got it...*
Resources by Chapter • Practice A and Practice B • Puzzle Time	Resources by Chapter • Enrichment and Extension • Cumulative Review
Student Journal • Practice	Start the *next* Section
Differentiating the Lesson Skills Review Handbook	

Laurie's Notes

Overview of Section 9.3

Introduction

- In this lesson, students will solve square root equations using the same techniques they have used in solving linear equations. The one difference is that there is one additional step of taking the square root of each side of the equation.
- Not all square root equations have solutions that are rational numbers, and therefore in this lesson, students will approximate solutions.

Teaching Strategy

- In Example 1, students are asked to solve $3x^2 - 27 = 0$. After adding 27 to each side of the equation, the equation becomes $3x^2 = 27$. This problem is like many that students will encounter in a mathematics class.
- A technique that can be used is to treat each side of the equation as a function. Graph each function and look for solutions to the system of equations, meaning where the two graphs intersect.
- In the graph shown, the original equation is shown, a parabola with a y-intercept at -27. The other two graphs represent the system with equations $y = 3x^2$ and $y = 27$.

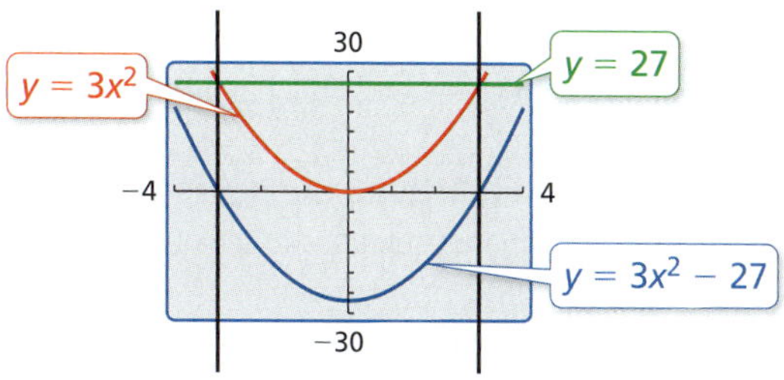

- The connection you want students to make is that the intersection points of the graphs have the same x-values as the x-intercepts of the original equation.

Pacing Suggestion

- Once students have worked the explorations, continue with the formal lesson.

Dynamic Teaching Tools

- Dynamic Assessment & Progress Monitoring Tool
- Lesson Planning Tool
- Interactive Whiteboard Lesson Library
- Dynamic Classroom with Dynamic Investigations

What Your Students Will Learn

- Solve quadratic equations of the form $ax^2 + c = 0$ using square roots.
- Approximate the solutions of quadratic equations.

Laurie's Notes

Exploration

Motivate

- All of today's graphs will be symmetric about the y-axis.
- Draw two collections of shapes. In the first, each shape has one line of symmetry. In the second, each shape has no symmetry or more than one line of symmetry.

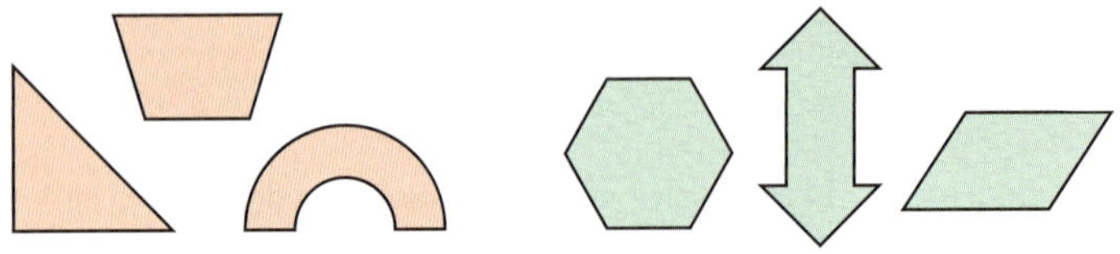

- Ask students to figure out how the groups have been sorted.

? "In which group would you place a parabola?" group on left

Discuss

- Today, you want students to think about the function $y = x^2$ (one x-intercept) and the effect of adding a constant c that shifts the graph up (no x-intercepts) or down (two x-intercepts).

Exploration 1

- Refer to the related function in part (c), $y = x^2$, as the parent function or basic quadratic function.
- **Construct Viable Arguments and Critique the Reasoning of Others:** Conjectures will likely relate the number of solutions to the number of x-intercepts of the graph.
- Students might go further and say that when c is positive, there are no solutions, and when c is negative, there are two solutions. Note, however, that this is not true when $a < 0$.

Exploration 2

- A calculator is a helpful tool for this exploration.

? "In part (a), what is happening to the x-values?" increasing by 0.01

? "How do the tables help you estimate the solutions of $x^2 - 5 = 0$?" The consecutive x-values at which the expression values change in sign from negative to positive indicate an approximate solution.

- **Reason Abstractly and Quantitatively:** In exploring the table of values, you are asking your students to reason quantitatively.

Exploration 3

- Students are using a calculator to confirm the estimates in Exploration 2.
- **Attend to Precision:** In part (b), students should have obtained approximate solutions such as $x \approx \pm 2.236067977$. In part (c), the exact solutions are $x = \pm\sqrt{5}$. Discuss the use of exact and approximate solutions. This draws attention to the concept of precision.

Communicate Your Answer

- **Reason Abstractly and Quantitatively:** Listen for correct reasoning as students share their answers.

Connecting to Next Step

- Students have now made the connection between the graph of the quadratic equation and the number of solutions the related quadratic equation has. This connection is stated in the formal lesson.

9.3 Solving Quadratic Equations Using Square Roots

Essential Question How can you determine the number of solutions of a quadratic equation of the form $ax^2 + c = 0$?

EXPLORATION 1 The Number of Solutions of $ax^2 + c = 0$

Work with a partner. Solve each equation by graphing. Explain how the number of solutions of $ax^2 + c = 0$ relates to the graph of $y = ax^2 + c$.

a. $x^2 - 4 = 0$

b. $2x^2 + 5 = 0$

c. $x^2 = 0$

d. $x^2 - 5 = 0$

EXPLORATION 2 Estimating Solutions

Work with a partner. Complete each table. Use the completed tables to estimate the solutions of $x^2 - 5 = 0$. Explain your reasoning.

a.

x	$x^2 - 5$
2.21	
2.22	
2.23	
2.24	
2.25	
2.26	

b.

x	$x^2 - 5$
−2.21	
−2.22	
−2.23	
−2.24	
−2.25	
−2.26	

ATTENDING TO PRECISION

To be proficient in math, you need to calculate accurately and express numerical answers with a level of precision appropriate for the problem's context.

EXPLORATION 3 Using Technology to Estimate Solutions

Work with a partner. Two equations are equivalent when they have the same solutions.

a. Are the equations $x^2 - 5 = 0$ and $x^2 = 5$ equivalent? Explain your reasoning.

b. Use the square root key on a calculator to estimate the solutions of $x^2 - 5 = 0$. Describe the accuracy of your estimates in Exploration 2.

c. Write the exact solutions of $x^2 - 5 = 0$.

Communicate Your Answer

4. How can you determine the number of solutions of a quadratic equation of the form $ax^2 + c = 0$?

5. Write the exact solutions of each equation. Then use a calculator to estimate the solutions.

   a. $x^2 - 2 = 0$

   b. $3x^2 - 18 = 0$

   c. $x^2 = 8$

Dynamic Teaching Tools

- Dynamic Assessment & Progress Monitoring Tool
- Lesson Planning Tool
- Interactive Whiteboard Lesson Library
- Dynamic Classroom with Dynamic Investigations

ANSWERS

1. **a.** $x = 2, x = -2$

   b. no solution

   c. $x = 0$

   d. $x \approx 2.2, x \approx -2.2$; The number of solutions is equal to the number of x-intercepts in the related graph.

2. **a.** −0.1159, −0.0716, −0.0271, 0.0176, 0.0625, 0.1076

   b. −0.1159, −0.0716, −0.0271, 0.0176, 0.0625, 0.1076 $x \approx 2.24$, $x \approx -2.24$; The value 0.0176 is closest to 0.

3. **a.** yes; Adding 5 to each side of $x^2 - 5 = 0$ gives $x^2 = 5$.

   b. $x \approx 2.236, x \approx -2.236$; The estimates in Exploration 2 were accurate to the hundredths place.

   c. $x = \sqrt{5}, x = -\sqrt{5}$

4. Graph the related equation $y = ax^2 + c$. The number of solutions will be the same as the number of x-intercepts.

5. **a.** $x = \sqrt{2}, x = -\sqrt{2}; x \approx 1.41, x \approx -1.41$

   b. $x = \sqrt{6}, x = -\sqrt{6}; x \approx 2.45, x \approx -2.45$

   c. $x = \sqrt{8}, x = -\sqrt{8}; x \approx 2.83, x \approx -2.83$

Extra Example 1

a. Solve $2x^2 - 32 = 0$ using square roots. The solutions are $x = 4$ and $x = -4$.

b. Solve $x^2 - 8 = -8$ using square roots. The only solution is $x = 0$.

c. Solve $-2x^2 + 3 = 27$ using square roots. The equation has no real solutions.

9.3 Lesson

Core Vocabulary

Previous
square root
zero of a function

What You Will Learn

- Solve quadratic equations using square roots.
- Approximate the solutions of quadratic equations.

Solving Quadratic Equations Using Square Roots

Earlier in this chapter, you studied properties of square roots. Now you will use square roots to solve quadratic equations of the form $ax^2 + c = 0$. First isolate x^2 on one side of the equation to obtain $x^2 = d$. Then solve by taking the square root of each side.

Core Concept

Solutions of $x^2 = d$

- When $d > 0$, $x^2 = d$ has two real solutions, $x = \pm\sqrt{d}$.
- When $d = 0$, $x^2 = d$ has one real solution, $x = 0$.
- When $d < 0$, $x^2 = d$ has no real solutions.

ANOTHER WAY

You can also solve $3x^2 - 27 = 0$ by factoring.

$$3(x^2 - 9) = 0$$
$$3(x - 3)(x + 3) = 0$$
$$x = 3 \text{ or } x = -3$$

EXAMPLE 1 **Solving Quadratic Equations Using Square Roots**

a. Solve $3x^2 - 27 = 0$ using square roots.

$3x^2 - 27 = 0$	Write the equation.
$3x^2 = 27$	Add 27 to each side.
$x^2 = 9$	Divide each side by 3.
$x = \pm\sqrt{9}$	Take the square root of each side.
$x = \pm 3$	Simplify.

► The solutions are $x = 3$ and $x = -3$.

b. Solve $x^2 - 10 = -10$ using square roots.

$x^2 - 10 = -10$	Write the equation.
$x^2 = 0$	Add 10 to each side.
$x = 0$	Take the square root of each side.

► The only solution is $x = 0$.

c. Solve $-5x^2 + 11 = 16$ using square roots.

$-5x^2 + 11 = 16$	Write the equation.
$-5x^2 = 5$	Subtract 11 from each side.
$x^2 = -1$	Divide each side by -5.

► The square of a real number cannot be negative. So, the equation has no real solutions.

Laurie's Notes Teacher Actions

- **Look For and Make Use of Structure:** Discuss similarities in solving $3x - 27 = 0$ and $3x^2 - 27 = 0$. The steps are the same, with one extra for $3x^2 - 27 = 0$.
- Note the side column suggestion for *Another Way* to solve. See the *Teaching Strategy* on page T-496 for a calculator technique.
- **Think-Alouds:** Model part (a), and then do a *Think-Aloud* as partner B solves part (b) aloud for partner A. "To solve $x^2 - 10 = -10$, I need to …." Now reverse roles and have partner A solve part (c) aloud for partner B.

STUDY TIP

Each side of the equation $(x-1)^2 = 25$ is a square. So, you can still solve by taking the square root of each side.

EXAMPLE 2 Solving a Quadratic Equation Using Square Roots

Solve $(x-1)^2 = 25$ using square roots.

SOLUTION

$(x-1)^2 = 25$	Write the equation.
$x - 1 = \pm 5$	Take the square root of each side.
$x = 1 \pm 5$	Add 1 to each side.

▶ So, the solutions are $x = 1 + 5 = 6$ and $x = 1 - 5 = -4$.

Check

Use a graphing calculator to check your answer. Rewrite the equation as $(x-1)^2 - 25 = 0$. Graph the related function $f(x) = (x-1)^2 - 25$ and find the zeros of the function. The zeros are -4 and 6.

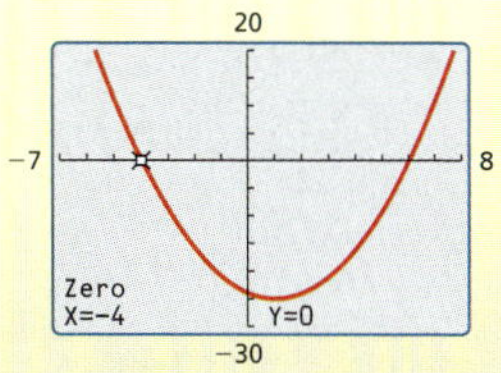

Monitoring Progress

Help in English and Spanish at *BigIdeasMath.com*

Solve the equation using square roots.

1. $-3x^2 = -75$
2. $x^2 + 12 = 10$
3. $4x^2 - 15 = -15$
4. $(x+7)^2 = 0$
5. $4(x-3)^2 = 9$
6. $(2x+1)^2 = 36$

Approximating Solutions of Quadratic Equations

EXAMPLE 3 Approximating Solutions of a Quadratic Equation

Solve $4x^2 - 13 = 15$ using square roots. Round the solutions to the nearest hundredth.

SOLUTION

$4x^2 - 13 = 15$	Write the equation.
$4x^2 = 28$	Add 13 to each side.
$x^2 = 7$	Divide each side by 4.
$x = \pm\sqrt{7}$	Take the square root of each side.
$x \approx \pm 2.65$	Use a calculator.

▶ The solutions are $x \approx -2.65$ and $x \approx 2.65$.

Check

Graph each side of the equation and find the points of intersection. The x-values of the points of intersection are about -2.65 and 2.65.

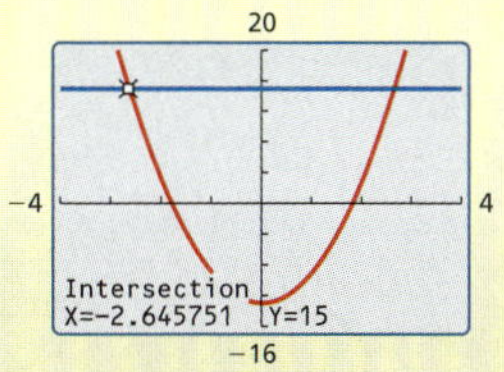

Monitoring Progress

Help in English and Spanish at *BigIdeasMath.com*

Solve the equation using square roots. Round your solutions to the nearest hundredth.

7. $x^2 + 8 = 19$
8. $5x^2 - 2 = 0$
9. $3x^2 - 30 = 4$

English Language Learners

Pair Activity

Pair an English learner with an English speaker. One student solves a quadratic equation using square roots. The other student solves the same equation using a graphing calculator. Partners compare solutions and then take turns explaining their solutions to one another. Students should repeat the activity with a different quadratic equation and should swap solution methods.

Extra Example 2

Solve $(x-3)^2 = 16$ using square roots. The solutions are $x = 7$ and $x = -1$.

Extra Example 3

Solve $2x^2 + 10 = 32$ using square roots. Round the solutions to the nearest hundredth. The solutions are $x \approx -3.32$ and $x \approx 3.32$.

MONITORING PROGRESS ANSWERS

1. $x = 5, x = -5$
2. no real solutions
3. $x = 0$
4. $x = -7$
5. $x = \frac{9}{2}, x = \frac{3}{2}$
6. $x = \frac{5}{2}, x = -\frac{7}{2}$
7. $x \approx 3.32, x \approx -3.32$
8. $x \approx 0.63, x \approx -0.63$
9. $x \approx 3.37, x \approx -3.37$

Laurie's Notes *Teacher Actions*

? **Attend to Precision:** "How do you read $(x-1)^2 = 25$?" Listen for something like "the quantity x minus 1 squared equals 25."

- **Connection:** Checking with a graphing calculator, the x-intercepts at -4 and 6 occur equidistant from the x-value of the vertex.
- **Think-Pair-Share:** Have students answer Questions 1–6, and then share and discuss as a class.
- **Turn and Talk:** Pose Example 3. "How do you solve $4x^2 - 13 = 15$, and how is this like solving $4x - 13 = 15$?"

Extra Example 4

A storage container has the shape of a rectangular prism. Its height is 6 feet. Its length is two times its width. The volume is 288 cubic feet. Find the length and width of the container.

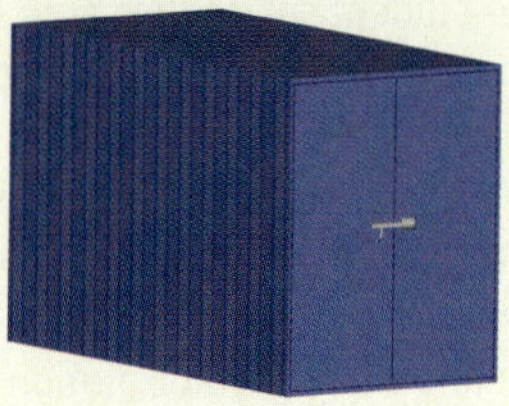

width: $\sqrt{24} \approx 4.9$ feet;
length: $2\sqrt{24} \approx 9.8$ feet

Extra Example 5

The surface area S of a cube with side length a is given by the formula $S = 6a^2$. Solve the formula for a. Then approximate the side length of a cube with a surface area of 724 square inches.

$a = \sqrt{\frac{S}{6}}$; about 11 inches

MONITORING PROGRESS ANSWERS

10. width: about 5.9 ft, length: about 17.7 ft

11. $r = \sqrt{\frac{S}{4\pi}}$; about 8 in.

EXAMPLE 4 Solving a Real-Life Problem

A touch tank has a height of 3 feet. Its length is three times its width. The volume of the tank is 270 cubic feet. Find the length and width of the tank.

SOLUTION

The length ℓ is three times the width w, so $\ell = 3w$. Write an equation using the formula for the volume of a rectangular prism.

$V = \ell wh$	Write the formula.
$270 = 3w(w)(3)$	Substitute 270 for V, $3w$ for ℓ, and 3 for h.
$270 = 9w^2$	Multiply.
$30 = w^2$	Divide each side by 9.
$\pm\sqrt{30} = w$	Take the square root of each side.

The solutions are $\sqrt{30}$ and $-\sqrt{30}$. Use the positive solution.

▶ So, the width is $\sqrt{30} \approx 5.5$ feet and the length is $3\sqrt{30} \approx 16.4$ feet.

INTERPRETING MATHEMATICAL RESULTS

Use the positive square root because negative solutions do not make sense in this context. Length and width cannot be negative.

EXAMPLE 5 Rearranging and Evaluating a Formula

The area A of an equilateral triangle with side length s is given by the formula $A = \frac{\sqrt{3}}{4}s^2$. Solve the formula for s. Then approximate the side length of the traffic sign that has an area of 390 square inches.

SOLUTION

Step 1 Solve the formula for s.

$A = \frac{\sqrt{3}}{4}s^2$	Write the formula.
$\frac{4A}{\sqrt{3}} = s^2$	Multiply each side by $\frac{4}{\sqrt{3}}$.
$\sqrt{\frac{4A}{\sqrt{3}}} = s$	Take the positive square root of each side.

Step 2 Substitute 390 for A in the new formula and evaluate.

$$s = \sqrt{\frac{4A}{\sqrt{3}}} = \sqrt{\frac{4(390)}{\sqrt{3}}} = \sqrt{\frac{1560}{\sqrt{3}}} \approx 30$$ Use a calculator.

▶ The side length of the traffic sign is about 30 inches.

ANOTHER WAY

Notice that you can rewrite the formula as $s = \frac{2}{3^{1/4}}\sqrt{A}$, or $s \approx 1.52\sqrt{A}$. This can help you efficiently find the value of s for various values of A.

Monitoring Progress

Help in English and Spanish at *BigIdeasMath.com*

10. **WHAT IF?** In Example 4, the volume of the tank is 315 cubic feet. Find the length and width of the tank.

11. The surface area S of a sphere with radius r is given by the formula $S = 4\pi r^2$. Solve the formula for r. Then find the radius of a globe with a surface area of 804 square inches.

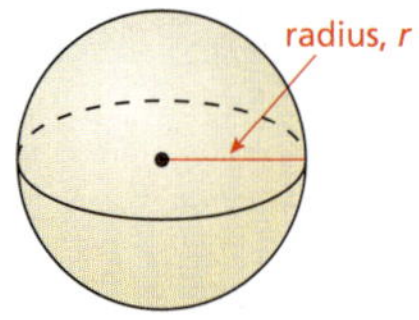

Laurie's Notes Teacher Actions

- ? "What is a reasonable estimate for $\sqrt{30}$? Explain." About 5.5, because $\sqrt{25} = 5$ and $\sqrt{36} = 6$, and 30 is about halfway between 25 and 36.
- Solving for s in Example 5 can be challenging for students because of the fraction and the radical. Explain that you still want to multiply by the reciprocal of the coefficient. Having an expression with a radical as a radicand can confuse students.
- Discuss *Another Way.*

Closure

- **Exit Ticket:** State the number of solutions for each equation.
  a. $2x^2 + 8 = 40$ 2 **b.** $2x^2 - 8 = -40$ 0 **c.** $2x^2 = 0$ 1

9.3 Exercises

Dynamic Solutions available at *BigIdeasMath.com*

Vocabulary and Core Concept Check

1. **COMPLETE THE SENTENCE** The equation $x^2 = d$ has ____ real solutions when $d > 0$.

2. **DIFFERENT WORDS, SAME QUESTION** Which is different? Find "both" answers.

Solve $x^2 = 144$ using square roots.	Solve $x^2 - 144 = 0$ using square roots.
Solve $x^2 + 146 = 2$ using square roots.	Solve $x^2 + 2 = 146$ using square roots.

Monitoring Progress and Modeling with Mathematics

In Exercises 3–8, determine the number of real solutions of the equation. Then solve the equation using square roots.

3. $x^2 = 25$
4. $x^2 = -36$
5. $x^2 = -21$
6. $x^2 = 400$
7. $x^2 = 0$
8. $x^2 = 169$

In Exercises 9–18, solve the equation using square roots. *(See Example 1.)*

9. $x^2 - 16 = 0$
10. $x^2 + 6 = 0$
11. $3x^2 + 12 = 0$
12. $x^2 - 55 = 26$
13. $2x^2 - 98 = 0$
14. $-x^2 + 9 = 9$
15. $-3x^2 - 5 = -5$
16. $4x^2 - 371 = 29$
17. $4x^2 + 10 = 11$
18. $9x^2 - 35 = 14$

In Exercises 19–24, solve the equation using square roots. *(See Example 2.)*

19. $(x + 3)^2 = 0$
20. $(x - 1)^2 = 4$
21. $(2x - 1)^2 = 81$
22. $(4x + 5)^2 = 9$
23. $9(x + 1)^2 = 16$
24. $4(x - 2)^2 = 25$

In Exercises 25–30, solve the equation using square roots. Round your solutions to the nearest hundredth. *(See Example 3.)*

25. $x^2 + 6 = 13$
26. $x^2 + 11 = 24$
27. $2x^2 - 9 = 11$
28. $5x^2 + 2 = 6$
29. $-21 = 15 - 2x^2$
30. $2 = 4x^2 - 5$

31. **ERROR ANALYSIS** Describe and correct the error in solving the equation $2x^2 - 33 = 39$ using square roots.

✗
$$2x^2 - 33 = 39$$
$$2x^2 = 72$$
$$x^2 = 36$$
$$x = 6$$
▶ The solution is $x = 6$.

32. **MODELING WITH MATHEMATICS** An in-ground pond has the shape of a rectangular prism. The pond has a depth of 24 inches and a volume of 72,000 cubic inches. The length of the pond is two times its width. Find the length and width of the pond. *(See Example 4.)*

33. **MODELING WITH MATHEMATICS** A person sitting in the top row of the bleachers at a sporting event drops a pair of sunglasses from a height of 24 feet. The function $h = -16x^2 + 24$ represents the height h (in feet) of the sunglasses after x seconds. How long does it take the sunglasses to hit the ground?

Assignment Guide and Homework Check

ASSIGNMENT

Basic: 1, 2, 3–33 odd, 34, 36, 40, 45–50

Average: 1, 2–30 even, 31–36, 40, 45–50

Advanced: 1, 2, 6, 16–30 even, 31, 32–36 even, 37–50

HOMEWORK CHECK

Basic: 9, 19, 25, 33, 36

Average: 10, 20, 26, 32, 36

Advanced: 16, 22, 28, 32, 36

ANSWERS

1. two
2. Solve $x^2 + 146 = 2$ using square roots; no real solutions; $x = 12, x = -12$
3. $2; x = 5, x = -5$
4. 0; no real solutions
5. 0; no real solutions
6. $2; x = 20, x = -20$
7. $1; x = 0$
8. $2; x = 13, x = -13$
9. $x = 4, x = -4$
10. no real solutions
11. no real solutions
12. $x = 9, x = -9$
13. $x = 7, x = -7$
14. $x = 0$
15. $x = 0$
16. $x = 10, x = -10$
17. $x = \frac{1}{2}, x = -\frac{1}{2}$
18. $x = \frac{7}{3}, x = -\frac{7}{3}$
19. $x = -3$
20. $x = 3, x = -1$
21. $x = 5, x = -4$
22. $x = -\frac{1}{2}, x = -2$
23. $x = \frac{1}{3}, x = -\frac{7}{3}$
24. $x = \frac{9}{2}, x = -\frac{1}{2}$
25. $x \approx 2.65, x \approx -2.65$
26. $x \approx 3.61, x \approx -3.61$
27. $x \approx 3.16, x \approx -3.16$
28. $x \approx 0.89, x \approx -0.89$
29. $x \approx 4.24, x \approx -4.24$
30. $x \approx 1.32, x \approx -1.32$
31. The number 36 has both a positive and negative square root; $x = \pm 6$
32. length: about 77.5 in., width: about 38.7 in.
33. about 1.2 sec

Dynamic Teaching Tools

- Dynamic Assessment & Progress Monitoring Tool
- Interactive Whiteboard Lesson Library
- Dynamic Classroom with Dynamic Investigations

ANSWERS

34. your cousin; Subtracting 4 from each side of the equation gives $x^2 = -4$, and a negative number has no real square roots.

35. 3 ft

36. **a.** $r = \sqrt{\frac{A}{\pi}}$

b. about 6 ft; about 24 in., about 13 m

c. The steps for solving only need to be completed once.

37. *Sample answer:* Use a calculator.

38. **a.** a and c have opposite signs.

b. a is not 0, and c is 0.

c. a and c have the same sign.

39. (3, 9), (−3, 9); When $y = 9$, $x = \pm 3$.

40. 1; The graph has one x-intercept.

41. $x = 1.2$, $x = -1.2$; $1.2^2 = 1.44$

42–50. See Additional Answers.

Mini-Assessment

Solve the equation using square roots.

1. $3x^2 - 108 = 0$ $x = 6$ and $x = -6$
2. $(x - 5)^2 = 81$ $x = 14$ and $x = -4$
3. Solve $4x^2 - 3 = 17$ using square roots. Round the solutions to the nearest hundredth.
   $x \approx -2.24$ and $x \approx 2.24$
4. A cardboard box has the shape of a rectangular prism. Its height is 10 inches. Its length is three times its width. The volume is 540 cubic inches. Find the length and width of the box.
   width: $\sqrt{18} \approx 4.2$ inches; length: $3\sqrt{18} \approx 12.7$ inches
5. The volume V of a cylinder with radius r and height h is given by the formula $V = \pi r^2 h$. Solve the formula for r. Then find the radius of a cylinder with a height of 15 centimeters and a volume of 750 cubic centimeters.
   $r = \sqrt{\frac{V}{\pi h}}$; about 4 centimeters

34. MAKING AN ARGUMENT Your friend says that the solution of the equation $x^2 + 4 = 0$ is $x = 0$. Your cousin says that the equation has no real solutions. Who is correct? Explain your reasoning.

35. MODELING WITH MATHEMATICS The design of a square rug for your living room is shown. You want the area of the inner square to be 25% of the total area of the rug. Find the side length x of the inner square.

36. MATHEMATICAL CONNECTIONS The area A of a circle with radius r is given by the formula $A = \pi r^2$. *(See Example 5.)*

a. Solve the formula for r.

b. Use the formula from part (a) to find the radius of each circle.

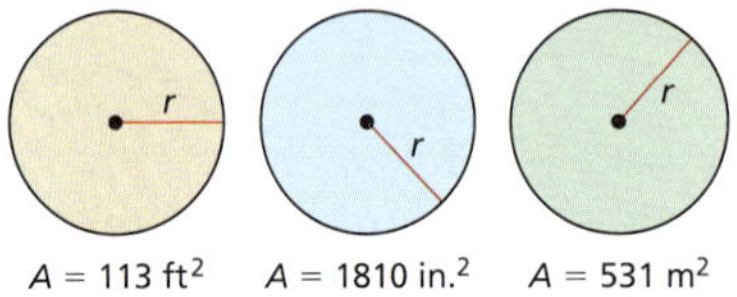

c. Explain why it is beneficial to solve the formula for r before finding the radius.

37. WRITING How can you approximate the roots of a quadratic equation when the roots are not integers?

38. WRITING Given the equation $ax^2 + c = 0$, describe the values of a and c so the equation has the following number of solutions.

a. two real solutions

b. one real solution

c. no real solutions

39. REASONING Without graphing, where do the graphs of $y = x^2$ and $y = 9$ intersect? Explain.

40. HOW DO YOU SEE IT? The graph represents the function $f(x) = (x - 1)^2$. How many solutions does the equation $(x - 1)^2 = 0$ have? Explain.

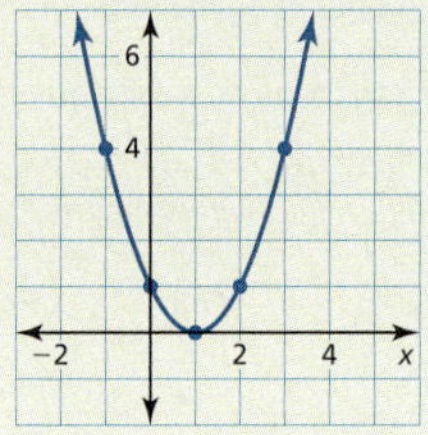

41. REASONING Solve $x^2 = 1.44$ without using a calculator. Explain your reasoning.

42. THOUGHT PROVOKING The quadratic equation

$$ax^2 + bx + c = 0$$

can be rewritten in the following form.

$$\left(x + \frac{b}{2a}\right)^2 = \frac{b^2 - 4ac}{4a^2}$$

Use this form to write the solutions of the equation.

43. REASONING An equation of the graph shown is $y = \frac{1}{2}(x - 2)^2 + 1$. Two points on the parabola have y-coordinates of 9. Find the x-coordinates of these points.

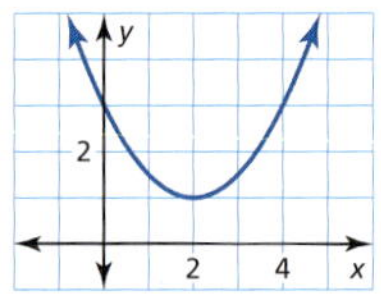

44. CRITICAL THINKING Solve each equation without graphing.

a. $x^2 - 12x + 36 = 64$

b. $x^2 + 14x + 49 = 16$

Maintaining Mathematical Proficiency

Reviewing what you learned in previous grades and lessons

Factor the polynomial. *(Section 7.7)*

45. $x^2 + 8x + 16$ **46.** $x^2 - 4x + 4$ **47.** $x^2 - 14x + 49$

48. $x^2 + 18x + 81$ **49.** $x^2 + 12x + 36$ **50.** $x^2 - 22x + 121$

If students need help...	*If students got it...*
Resources by Chapter • Practice A and Practice B • Puzzle Time	Resources by Chapter • Enrichment and Extension • Cumulative Review
Student Journal • Practice	Start the *next* Section
Differentiating the Lesson Skills Review Handbook	

9.1–9.3 What Did You Learn?

Core Vocabulary

counterexample, *p. 479*
radical expression, *p. 480*
simplest form of a radical, *p. 480*
rationalizing the denominator, *p. 482*
conjugates, *p. 482*
like radicals, *p. 484*
quadratic equation, *p. 490*

Core Concepts

Section 9.1

Product Property of Square Roots, *p. 480*
Quotient Property of Square Roots, *p. 480*
Rationalizing the Denominator, *p. 482*
Performing Operations with Radicals, *p. 484*

Section 9.2

Solving Quadratic Equations by Graphing, *p. 490*
Number of Solutions of a Quadratic Equation, *p. 491*
Finding Zeros of Functions, *p. 492*

Section 9.3

Solutions of $x^2 = d$, *p. 498*
Approximating Solutions of Quadratic Equations, *p. 499*

Mathematical Practices

1. For each part of Exercise 100 on page 488 that is *sometimes* true, list all examples and counterexamples from the table that represent the sum or product being described.
2. Which Examples can you use to help you solve Exercise 54 on page 496?
3. Describe how solving a simpler equation can help you solve the equation in Exercise 41 on page 502.

Study Skills

Keeping a Positive Attitude

Do you ever feel frustrated or overwhelmed by math? You're not alone. Just take a deep breath and assess the situation. Try to find a productive study environment, review your notes and the examples in the textbook, and ask your teacher or friends for help.

Dynamic Teaching Tools

- Dynamic Assessment & Progress Monitoring Tool
- Interactive Whiteboard Lesson Library
- Dynamic Classroom with Dynamic Investigations

ANSWERS

1. part (c) examples: $\sqrt{3} + \sqrt{3} = 2\sqrt{3}$, $\sqrt{3} + \pi = \pi + \sqrt{3}$, $-\sqrt{3} + (-\sqrt{3}) = -2\sqrt{3}$, $-\sqrt{3} + \pi = \pi - \sqrt{3}$, $\pi + \pi = 2\pi$; part (c) counterexample: $\sqrt{3} + (-\sqrt{3}) = 0$, part (f) examples: $\sqrt{3} \cdot \pi = \pi\sqrt{3}$, $-\sqrt{3} \cdot \pi = -\pi\sqrt{3}$, $\pi \cdot \pi = \pi^2$; part (f) counterexamples: $\sqrt{3} \cdot \sqrt{3} = 3$, $\sqrt{3} \cdot (-\sqrt{3}) = -3$, $-\sqrt{3} \cdot (-\sqrt{3}) = 3$
2. Example 6 on page 493 or Example 5 on page 492
3. Solving the simpler equation $x^2 = 144$ helps because the solution can be found by moving the decimal point in the solution to the simpler equation.

ANSWERS

1. $4x\sqrt{7x}$
2. $\dfrac{\sqrt{2}}{3}$
3. $-5\sqrt[3]{5}$
4. $\dfrac{3\sqrt{2}}{2}$
5. $\dfrac{4\sqrt{11}}{11}$
6. $\dfrac{12\sqrt{13}}{13}$
7. $\dfrac{3x\sqrt[3]{2x}}{7y^2}$
8. $\dfrac{x\sqrt{7z}}{7y^2z^3}$
9. $\dfrac{15 - 3\sqrt{3}}{11}$
10. $-4\sqrt{5} + 7\sqrt{10}$
11. $-10\sqrt{2} - 5\sqrt{10}$
12. $30\sqrt{2}$
13. $x = -1, x = 3$
14. no real solutions
15. $x = -5$
16. $x = -7, x = -2$
17. $x = 8, x = -1$
18. no real solutions
19. $x = 4, x = -4$
20. no real solutions
21. $x = 9, x = 7$
22. Because $100 > 0$, the equation has two real solutions.
23. length: about 17.4 m, width: about 4.4 m
24. **a.** 0.5 sec, 1 sec
    b. about 1.65 sec

9.1–9.3 Quiz

Simplify the expression. *(Section 9.1)*

1. $\sqrt{112x^3}$
2. $\sqrt{\dfrac{18}{81}}$
3. $\sqrt[3]{-625}$
4. $\dfrac{12}{\sqrt{32}}$
5. $\dfrac{4}{\sqrt{11}}$
6. $\sqrt{\dfrac{144}{13}}$
7. $\sqrt[3]{\dfrac{54x^4}{343y^6}}$
8. $\sqrt{\dfrac{4x^2}{28y^4z^5}}$
9. $\dfrac{6}{5 + \sqrt{3}}$
10. $2\sqrt{5} + 7\sqrt{10} - 3\sqrt{20}$
11. $\dfrac{10}{\sqrt{8} - \sqrt{10}}$
12. $\sqrt{6}(7\sqrt{12} - 4\sqrt{3})$

Use the graph to solve the equation. *(Section 9.2)*

13. $x^2 - 2x - 3 = 0$

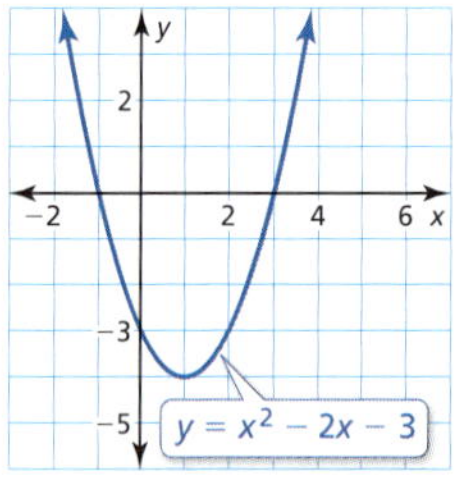

14. $x^2 - 2x + 3 = 0$

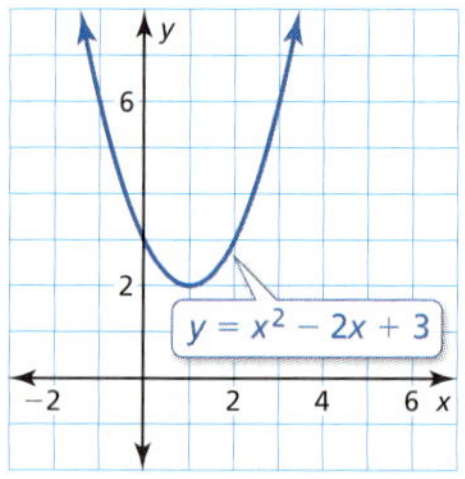

15. $x^2 + 10x + 25 = 0$

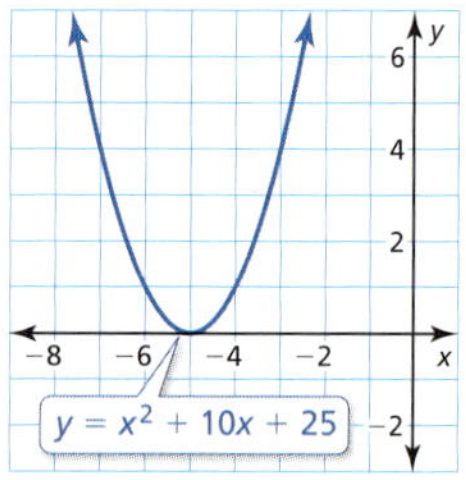

Solve the equation by graphing. *(Section 9.2)*

16. $x^2 + 9x + 14 = 0$
17. $x^2 - 7x = 8$
18. $x + 4 = -x^2$

Solve the equation using square roots. *(Section 9.3)*

19. $4x^2 = 64$
20. $-3x^2 + 6 = 10$
21. $(x - 8)^2 = 1$

22. Explain how to determine the number of real solutions of $x^2 = 100$ without solving. *(Section 9.3)*

23. The length of a rectangular prism is four times its width. The volume of the prism is 380 cubic meters. Find the length and width of the prism. *(Section 9.3)*

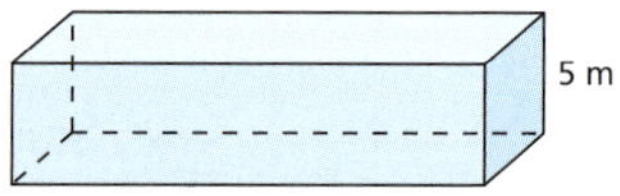

24. You cast a fishing lure into the water from a height of 4 feet above the water. The height h (in feet) of the fishing lure after t seconds can be modeled by the equation $h = -16t^2 + 24t + 4$. *(Section 9.2)*
    a. After how many seconds does the fishing lure reach a height of 12 feet?
    b. After how many seconds does the fishing lure hit the water?

Laurie's Notes

Overview of Section 9.4

Introduction

- In this lesson, the *completing the square* technique is used to solve quadratic equations and to rewrite quadratic equations from standard form to vertex form. When the equation is in vertex form, the maximum or minimum value of the function can be found.
- Real-life problems are solved by using the *completing the square* technique.
- In this lesson, students are expected to interpret information about a quadratic function from parameters in the equation.

Formative Assessment Tips

- **Pass the Problem:** This technique provides students the opportunity to work with others on the solution of a problem or a proof that requires more than a few steps. Seeing how others approach and work the problem helps students to reflect on their own process.
- There are different ways, or configurations, to using this technique. Begin by posing a problem that individuals or pairs begin to work on. After a fixed amount of time, the problem is swapped with partners or another pair of students. The recipients finish solving the problem or make modifications or corrections to the problem. If changes are made, they must explain why there was an error or why the strategy that was begun is not going to be followed.
- This technique gives all students the opportunity to participate in the lesson. All students receive feedback on their work, not just one student that the teacher might have called on. If the work passed to another is not clear because steps are missing, students receive this feedback from peers, not just the teacher. If students or partners were unable to get started on a problem, they will hopefully be exchanging with others who can help them make sense of the problem.
- When students have finished the problem, the two or four people involved in swapping problems will confer with one another to discuss the problem and offer additional feedback. One thing you hope to hear is positive feedback on the clarity of thinking that was recorded, allowing the recipients to make sense of the work.

Pacing Suggestion

- The explorations help students visualize what completing the square means. The explorations should not take long. Transition to the formal lesson.

Dynamic Teaching Tools

- Dynamic Assessment & Progress Monitoring Tool
- Lesson Planning Tool
- Interactive Whiteboard Lesson Library
- Dynamic Classroom with Dynamic Investigations

What Your Students Will Learn

- Complete the square for expressions of the form $ax^2 + bx$ by adding a constant c to make a perfect square trinomial.
- Solve quadratic equations of the form $ax^2 + bx + c = 0$ by completing the square and then using square roots.
- Find and use maximum and minimum values of quadratic functions by writing the functions in vertex form.
- Solve real-life problems by modeling the problems with a quadratic equation and completing the square.

Laurie's Notes

Exploration

Motivate

- Write the sequence on the board: 1, 4, ____, 16, ____, 36, ...
- ? "What numbers are missing and what is the pattern? Explain." 9 and 25; The terms of the sequence are perfect squares.
- ? "What is a perfect square trinomial?" a trinomial that can be factored as $(a \pm b)^2$

Exploration 1

- Review the names and dimensions of the algebra tiles.
- Students should model the equation with the goal of adding 1-tiles to form a square array on the left side. To make this possible, start by arranging half of the x-tiles vertically and half horizontally.
- Add the number of 1-tiles needed to form a square on the left side. Add the same number of 1-tiles to the right side.
- Remind students how multiplication of binomials was modeled (rectangular array).
- Explain how to use the dimensions of the square formed on the left side of the equation to write it as the square of a binomial. The vertical and horizontal dimensions are each $x + 2$, so the left side of the equation represents $(x + 2)^2$.
- **Reason Abstractly and Quantitatively:** Students have obtained a concrete model of $(x + 2)^2 = 2$. They will now reason abstractly, using algebra to manipulate the equation.

Exploration 2

- Having been guided through the first exploration, students should be able to work with partners on this exploration.
- ? "What equation is modeled by the algebra tiles?" $x^2 + 6x = -5$
- ? "What equation is represented by the tiles after completing the square?" $x^2 + 6x + 9 = 4$
- **Construct Viable Arguments and Critique the Reasoning of Others:** Asking students to explain the thinking behind their solutions is helpful for them and for their peers.
- It is important to work through the checking process.

Communicate Your Answer

- **Construct Viable Arguments and Critique the Reasoning of Others:** In Question 3, answers should include an understanding of writing the quadratic equation in a form that allows you to solve it using square roots.

Connecting to Next Step

- Students have now seen a visual model of what it means to "complete the square," and they should be ready to work through the same process symbolically in the formal lesson.

9.4 Solving Quadratic Equations by Completing the Square

Essential Question How can you use "completing the square" to solve a quadratic equation?

EXPLORATION 1 **Solving by Completing the Square**

Work with a partner.

a. Write the equation modeled by the algebra tiles. This is the equation to be solved.

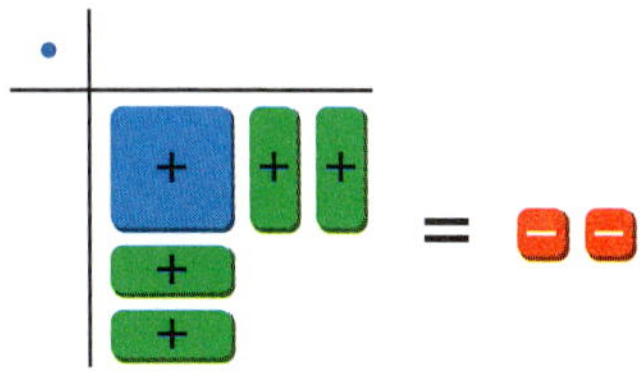

b. Four algebra tiles are added to the left side to "complete the square." Why are four algebra tiles also added to the right side?

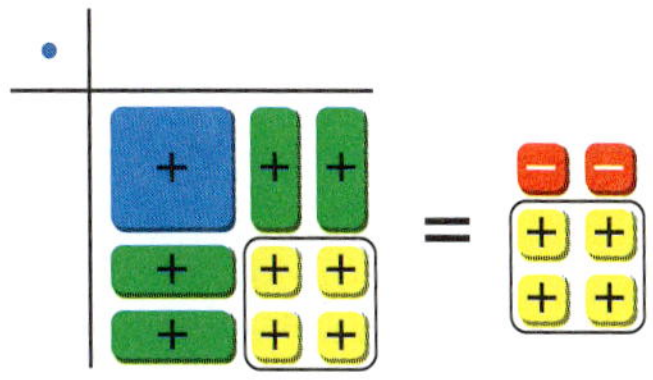

c. Use algebra tiles to label the dimensions of the square on the left side and simplify on the right side.

d. Write the equation modeled by the algebra tiles so that the left side is the square of a binomial. Solve the equation using square roots.

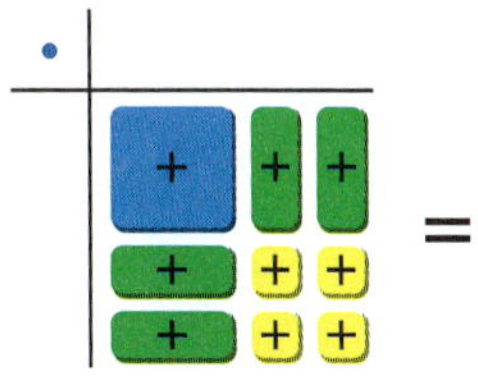

MAKING SENSE OF PROBLEMS

To be proficient in math, you need to explain to yourself the meaning of a problem. After that, you need to look for entry points to its solution.

EXPLORATION 2 **Solving by Completing the Square**

Work with a partner.

a. Write the equation modeled by the algebra tiles.

b. Use algebra tiles to "complete the square."

c. Write the solutions of the equation.

d. Check each solution in the original equation.

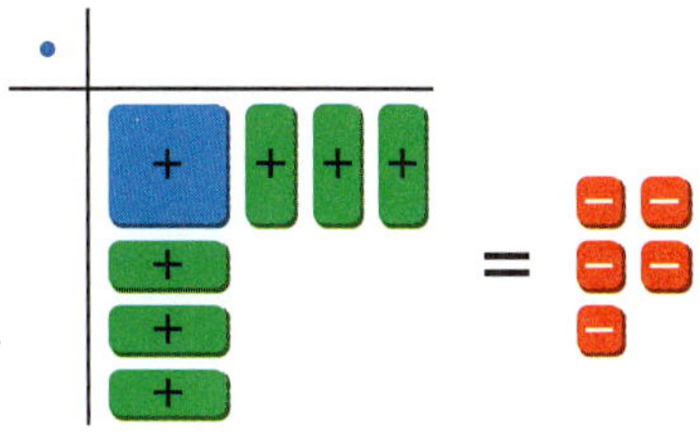

Communicate Your Answer

3. How can you use "completing the square" to solve a quadratic equation?

4. Solve each quadratic equation by completing the square.

a. $x^2 - 2x = 1$ **b.** $x^2 - 4x = -1$ **c.** $x^2 + 4x = -3$

Dynamic Teaching Tools

- Dynamic Assessment & Progress Monitoring Tool
- Lesson Planning Tool
- Interactive Whiteboard Lesson Library
- Dynamic Classroom with Dynamic Investigations

ANSWERS

1. a. $x^2 + 4x = -2$

b. To be equivalent equations, the same number must be added to each side.

c.

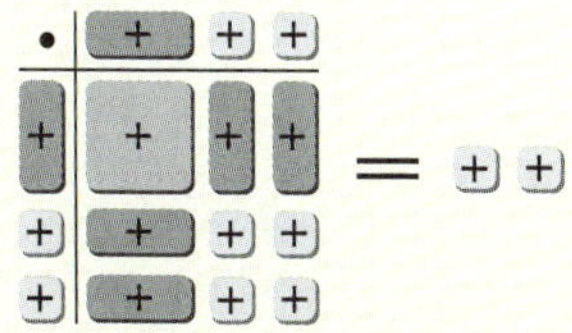

d. $(x + 2)^2 = 2$; $x \approx -0.59$, $x \approx -3.41$

2. a. $x^2 + 6x = -5$

b.

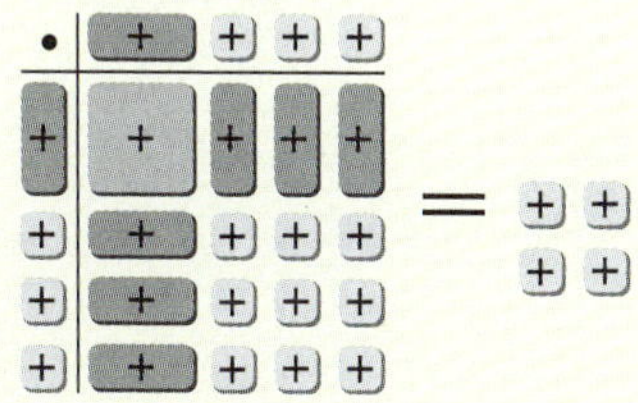

c. $x = -1, x = -5$

d. $(-1)^2 + 6(-1) = -5$; $(-5)^2 + 6(-5) = -5$

3. Write the equation in the form $x^2 + bx = d$. Add $\left(\frac{b}{2}\right)^2$ to each side of the equation. Factor the resulting expression on the left side as the square of a binomial. Solve the resulting equation using square roots.

4. a. $x \approx 2.41, x \approx -0.41$

b. $x \approx 3.73, x \approx 0.27$

c. $x = -1, x = -3$

English Language Learners

Organization

Students will benefit by writing down the steps for completing the square shown in the *Core Concept* box. Have students write the steps in their notebooks. A poster with the steps could be posted in the classroom.

Extra Example 1

Complete the square for each expression. Then factor the trinomial.

a. $x^2 + 18x$ $x^2 + 18x + 81$; $(x + 9)^2$

b. $x^2 - 17x$ $x^2 - 17x + \frac{289}{4}$; $\left(x - \frac{17}{2}\right)^2$

MONITORING PROGRESS ANSWERS

1. $x^2 + 10x + 25$; $(x + 5)^2$
2. $x^2 - 4x + 4$; $(x - 2)^2$
3. $x^2 + 7x + \frac{49}{4}$; $\left(x + \frac{7}{2}\right)^2$

9.4 Lesson

Core Vocabulary

completing the square, *p. 506*

Previous
perfect square trinomial
coefficient
maximum value
minimum value
vertex form of a quadratic function

What You Will Learn

- Complete the square for expressions of the form $x^2 + bx$.
- Solve quadratic equations by completing the square.
- Find and use maximum and minimum values.
- Solve real-life problems by completing the square.

Completing the Square

For an expression of the form $x^2 + bx$, you can add a constant c to the expression so that $x^2 + bx + c$ is a perfect square trinomial. This process is called **completing the square**.

Core Concept

Completing the Square

Words To complete the square for an expression of the form $x^2 + bx$, follow these steps.

Step 1 Find one-half of b, the coefficient of x.

Step 2 Square the result from Step 1.

Step 3 Add the result from Step 2 to $x^2 + bx$.

Factor the resulting expression as the square of a binomial.

Algebra $x^2 + bx + \left(\frac{b}{2}\right)^2 = \left(x + \frac{b}{2}\right)^2$

JUSTIFYING STEPS

In each diagram below, the combined area of the shaded regions is $x^2 + bx$. Adding $\left(\frac{b}{2}\right)^2$ completes the square in the second diagram.

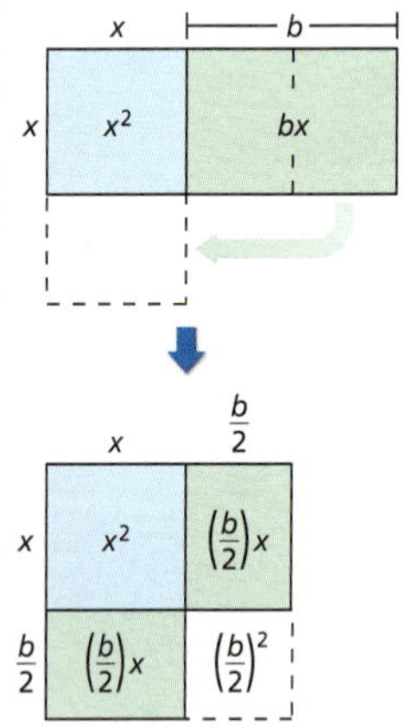

EXAMPLE 1 Completing the Square

Complete the square for each expression. Then factor the trinomial.

a. $x^2 + 6x$ **b.** $x^2 - 9x$

SOLUTION

a. Step 1 Find one-half of b. $\frac{b}{2} = \frac{6}{2} = 3$

Step 2 Square the result from Step 1. $3^2 = 9$

Step 3 Add the result from Step 2 to $x^2 + bx$. $x^2 + 6x + 9$

▶ $x^2 + 6x + 9 = (x + 3)^2$

b. Step 1 Find one-half of b. $\frac{b}{2} = \frac{-9}{2}$

Step 2 Square the result from Step 1. $\left(\frac{-9}{2}\right)^2 = \frac{81}{4}$

Step 3 Add the result from Step 2 to $x^2 + bx$. $x^2 - 9x + \frac{81}{4}$

▶ $x^2 - 9x + \frac{81}{4} = \left(x - \frac{9}{2}\right)^2$

Monitoring Progress

Help in English and Spanish at *BigIdeasMath.com*

Complete the square for the expression. Then factor the trinomial.

1. $x^2 + 10x$ **2.** $x^2 - 4x$ **3.** $x^2 + 7x$

Laurie's Notes Teacher Actions

- Write the *Core Concept*. The diagrams help to explain why $\left(\frac{b}{2}\right)^2$ is the amount added to complete the square.
- The examples help students become familiar with the technique of completing the square before they use it to solve an equation.
- It may be helpful to ask students what is missing that would make the expression a perfect square trinomial: $x^2 + 6x + \underline{?} = (x + \underline{?})^2$.
- **Think-Pair-Share:** Have students answer Questions 1–3, and then share and discuss as a class.

Solving Quadratic Equations by Completing the Square

The method of completing the square can be used to solve any quadratic equation. To solve a quadratic equation by completing the square, you must write the equation in the form $x^2 + bx = d$.

COMMON ERROR

When completing the square to solve an equation, be sure to add $\left(\frac{b}{2}\right)^2$ to each side of the equation.

EXAMPLE 2 **Solving a Quadratic Equation: $x^2 + bx = d$**

Solve $x^2 - 16x = -15$ by completing the square.

SOLUTION

$x^2 - 16x = -15$ — Write the equation.

$x^2 - 16x + (-8)^2 = -15 + (-8)^2$ — Complete the square by adding $\left(\frac{-16}{2}\right)^2$, or $(-8)^2$, to each side.

$(x - 8)^2 = 49$ — Write the left side as the square of a binomial.

$x - 8 = \pm 7$ — Take the square root of each side.

$x = 8 \pm 7$ — Add 8 to each side.

▶ The solutions are $x = 8 + 7 = 15$ and $x = 8 - 7 = 1$.

Check

$x^2 - 16x = -15$	Original equation	$x^2 - 16x = -15$
$15^2 - 16(15) \stackrel{?}{=} -15$	Substitute.	$1^2 - 16(1) \stackrel{?}{=} -15$
$-15 = -15$ ✓	Simplify.	$-15 = -15$ ✓

COMMON ERROR

Before you complete the square, be sure that the coefficient of the x^2-term is 1.

EXAMPLE 3 **Solving a Quadratic Equation: $ax^2 + bx + c = 0$**

Solve $2x^2 + 20x - 8 = 0$ by completing the square.

SOLUTION

$2x^2 + 20x - 8 = 0$ — Write the equation.

$2x^2 + 20x = 8$ — Add 8 to each side.

$x^2 + 10x = 4$ — Divide each side by 2.

$x^2 + 10x + 5^2 = 4 + 5^2$ — Complete the square by adding $\left(\frac{10}{2}\right)^2$, or 5^2, to each side.

$(x + 5)^2 = 29$ — Write the left side as the square of a binomial.

$x + 5 = \pm\sqrt{29}$ — Take the square root of each side.

$x = -5 \pm \sqrt{29}$ — Subtract 5 from each side.

▶ The solutions are $x = -5 + \sqrt{29} \approx 0.39$ and $x = -5 - \sqrt{29} \approx -10.39$.

Monitoring Progress

Help in English and Spanish at *BigIdeasMath.com*

Solve the equation by completing the square. Round your solutions to the nearest hundredth, if necessary.

4. $x^2 - 2x = 3$ **5.** $m^2 + 12m = -8$ **6.** $3g^2 - 24g + 27 = 0$

Differentiated Instruction

Inclusion

Have advanced students work in pairs to write and solve equations like those in Examples 2 and 3. Each student should write an equation, exchange equations with his or her partner, and solve the equation. After solving, students should exchange equations again and check the solutions.

Extra Example 2

Solve $x^2 - 18x = -17$ by completing the square. The solutions are $x = 9 + 8 = 17$ and $x = 9 - 8 = 1$.

Extra Example 3

Solve $2x^2 + 12x - 10 = 0$ by completing the square. The solutions are $x = -3 + \sqrt{14} \approx 0.74$ and $x = -3 - \sqrt{14} \approx -6.74$.

MONITORING PROGRESS ANSWERS

4. $x = 3, x = -1$

5. $m \approx -0.71, m \approx -11.29$

6. $g \approx 6.65, g \approx 1.35$

Laurie's Notes Teacher Actions

- Write $x^2 - 16x + \underline{?} = -15 + \underline{?}$ on the board. This helps students remember to add $\left(\frac{b}{2}\right)^2$ to both sides of the equation.
- Continue to solve the equation as shown. Remind students that when square rooting each side of the equation, there are + and − cases.
- **Pass the Problem:** Explain this technique and then pose Example 3. Give partners 1 to 2 minutes to discuss and begin a solution. Call "swap" and have partners continue to solve. Both groups debrief. Deciding how to work with the coefficient of x^2 should be a rich conversation.

Extra Example 4

Find the minimum value of $y = x^2 + 8x + 5$.
The function has a minimum value of -11.

Extra Example 5

Find the maximum value of $y = -x^2 + 4x + 2$. The function has a maximum value of 6.

MONITORING PROGRESS ANSWERS

7. maximum value; 8
8. minimum value; 4
9. minimum value; -3

Finding and Using Maximum and Minimum Values

One way to find the maximum or minimum value of a quadratic function is to write the function in vertex form by completing the square. Recall that the vertex form of a quadratic function is $y = a(x - h)^2 + k$, where $a \neq 0$. The vertex of the graph is (h, k).

EXAMPLE 4 **Finding a Minimum Value**

Find the minimum value of $y = x^2 + 4x - 1$.

SOLUTION

Write the function in vertex form.

$y = x^2 + 4x - 1$	Write the function.
$y + 1 = x^2 + 4x$	Add 1 to each side.
$y + 1 + 4 = x^2 + 4x + 4$	Complete the square for $x^2 + 4x$.
$y + 5 = x^2 + 4x + 4$	Simplify the left side.
$y + 5 = (x + 2)^2$	Write the right side as the square of a binomial.
$y = (x + 2)^2 - 5$	Write in vertex form.

The vertex is $(-2, -5)$. Because a is positive ($a = 1$), the parabola opens up and the y-coordinate of the vertex is the minimum value.

▶ So, the function has a minimum value of -5.

Check

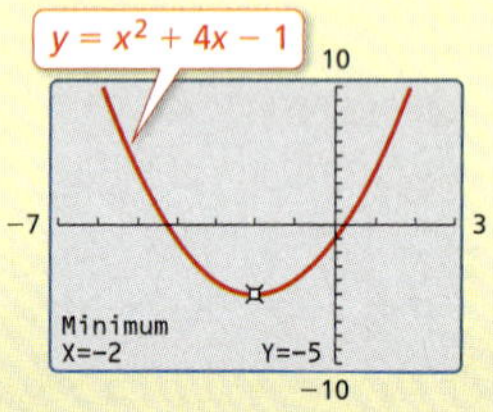

EXAMPLE 5 **Finding a Maximum Value**

Find the maximum value of $y = -x^2 + 2x + 7$.

SOLUTION

Write the function in vertex form.

$y = -x^2 + 2x + 7$	Write the function.
$y - 7 = -x^2 + 2x$	Subtract 7 from each side.
$y - 7 = -(x^2 - 2x)$	Factor out -1.
$y - 7 - 1 = -(x^2 - 2x + 1)$	Complete the square for $x^2 - 2x$.
$y - 8 = -(x^2 - 2x + 1)$	Simplify the left side.
$y - 8 = -(x - 1)^2$	Write $x^2 - 2x + 1$ as the square of a binomial.
$y = -(x - 1)^2 + 8$	Write in vertex form.

STUDY TIP

Adding 1 inside the parentheses results in subtracting 1 from the right side of the equation.

The vertex is $(1, 8)$. Because a is negative ($a = -1$), the parabola opens down and the y-coordinate of the vertex is the maximum value.

▶ So, the function has a maximum value of 8.

Monitoring Progress

Help in English and Spanish at *BigIdeasMath.com*

Determine whether the quadratic function has a maximum or minimum value. Then find the value.

7. $y = -x^2 - 4x + 4$ **8.** $y = x^2 + 12x + 40$ **9.** $y = x^2 - 2x - 2$

Laurie's Notes Teacher Actions

- **Connection:** Completing the square is a technique that can be used to rewrite a quadratic from standard form to vertex form.
- ? Pose Example 4. "Does this quadratic have a maximum or minimum value? Explain." It has a minimum value because it opens upward.
- **Make Sense of Problems and Persevere in Solving Them:** Have partners work together on Example 5. Circulate, but do not rush in to solve the example for them.
- **Probing Question:** Ask probing questions (see the *Study Tip*) and expect students to persevere.
- Graphing the original equation and the vertex form will tell students whether they are correct.

EXAMPLE 6 Interpreting Forms of Quadratic Functions

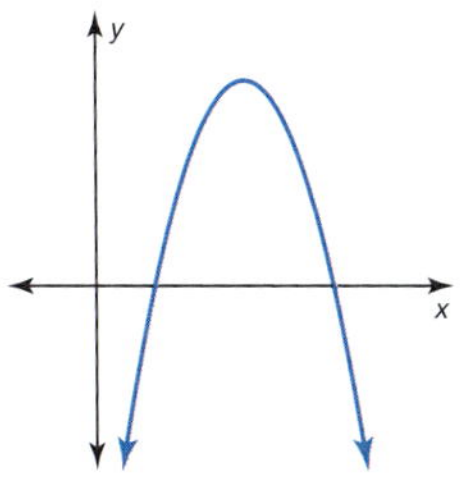

$f(x) = -\frac{1}{2}(x + 4)^2 + 8$

$g(x) = -(x - 5)^2 + 9$

$m(x) = (x - 3)(x - 12)$

$p(x) = -(x - 2)(x - 8)$

Which of the functions could be represented by the graph? Explain.

SOLUTION

You do not know the scale of either axis. To eliminate functions, consider the characteristics of the graph and information provided by the form of each function. The graph appears to be a parabola that opens down, which means the function has a maximum value. The vertex of the graph is in the first quadrant. Both x-intercepts are positive.

- The graph of f opens down because $a < 0$, which means f has a maximum value. However, the vertex $(-4, 8)$ of the graph of f is in the second quadrant. So, the graph does not represent f.
- The graph of g opens down because $a < 0$, which means g has a maximum value. The vertex $(5, 9)$ of the graph of g is in the first quadrant. By solving $0 = -(x - 5)^2 + 9$, you see that the x-intercepts of the graph of g are 2 and 8. So, the graph could represent g.
- The graph of m has two positive x-intercepts. However, its graph opens up because $a > 0$, which means m has a minimum value. So, the graph does not represent m.
- The graph of p has two positive x-intercepts, and its graph opens down because $a < 0$. This means that p has a maximum value and the vertex must be in the first quadrant. So, the graph could represent p.

▶ The graph could represent function g or function p.

EXAMPLE 7 Real-Life Application

The function $y = -16x^2 + 96x$ represents the height y (in feet) of a model rocket x seconds after it is launched. (a) Find the maximum height of the rocket. (b) Find and interpret the axis of symmetry.

SOLUTION

a. To find the maximum height, identify the maximum value of the function.

$y = -16x^2 + 96x$ — Write the function.

$y = -16(x^2 - 6x)$ — Factor out -16.

$y - 144 = -16(x^2 - 6x + 9)$ — Complete the square for $x^2 - 6x$.

$y = -16(x - 3)^2 + 144$ — Write in vertex form.

STUDY TIP
Adding 9 inside the parentheses results in subtracting 144 from the right side of the equation.

▶ Because the maximum value is 144, the model rocket reaches a maximum height of 144 feet.

b. The vertex is $(3, 144)$. So, the axis of symmetry is $x = 3$. On the left side of $x = 3$, the height increases as time increases. On the right side of $x = 3$, the height decreases as time increases.

Monitoring Progress

Help in English and Spanish at BigIdeasMath.com

Determine whether the function could be represented by the graph in Example 6. Explain.

10. $h(x) = (x - 8)^2 + 10$

11. $n(x) = -2(x - 5)(x - 20)$

12. WHAT IF? Repeat Example 7 when the function is $y = -16x^2 + 128x$.

Extra Example 6

Which of the functions could be represented by the graph? Explain.

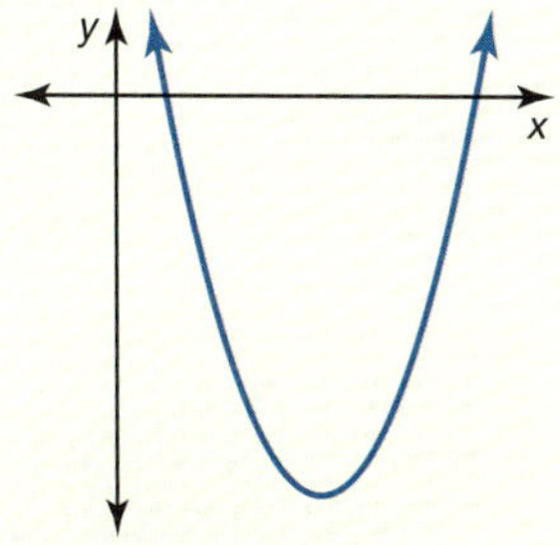

$f(x) = (x - 4)^2 + 9$
$g(x) = -(x - 3)^2 + 5$
$m(x) = -\frac{1}{4}(x + 2)(x + 6)$
$p(x) = (x - 1)(x - 7)$

The graph could represent function p.

The graph does not represent function f. The graph of f opens up because $a > 0$. However, the vertex $(4, 9)$ is in the first quadrant, so the graph of f does not cross the x-axis.

The graph does not represent functions g or m because $a < 0$ in both functions and these graphs open down.

The graph of p opens up because $a > 0$. It has two positive x-intercepts. So, p has a minimum value and could represent the graph shown.

Extra Example 7

The function $y = -16x^2 + 160x$ represents the height y (in feet) of a model rocket x seconds after it is launched.

a. Find the maximum height of the rocket. 400 feet

b. Find and interpret the axis of symmetry. The axis of symmetry is $x = 5$. On the left side of $x = 5$, the height increases as time increases. On the right side of $x = 5$, the height decreases as time increases.

MONITORING PROGRESS ANSWERS

10. no; The graph opens up.

11. yes; The graph has two positive x-intercepts and opens down.

12. a. 256 ft

b. $x = 4$; On the left side of $x = 4$, the height increases as time increases. On the right side of $x = 4$, the height decreases as time increases.

Laurie's Notes Teacher Actions

- **Reason Abstractly and Quantitatively** and **Construct Viable Arguments and Critique the Reasoning of Others:** Example 6 is done without calculators. Partners should discuss each function. In addition to selecting which function(s) could be represented by the graph, students should also explain why the other function(s) could not be represented by the graph.

? Popsicle Sticks: "Could $f(x)$ be the function for this graph? Explain." Listen for reasoning associated with information deduced from the function.

- **Look For and Make Use of Structure:** In Example 7, students should recognize that the Distributive Property is used in completing the square. Discuss the *Study Tip*.

Extra Example 8

You decide to put a rug on a floor. You want the rug to cover 100 square feet and to have a uniform border of floor visible, as shown. Find the width of the border to the nearest inch.

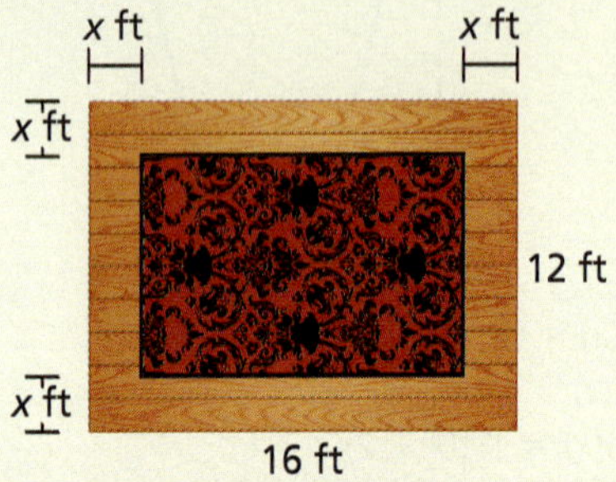

The width of the border should be about 23 inches.

MONITORING PROGRESS ANSWER

13. about 13 in.

Solving Real-Life Problems

EXAMPLE 8 Modeling with Mathematics

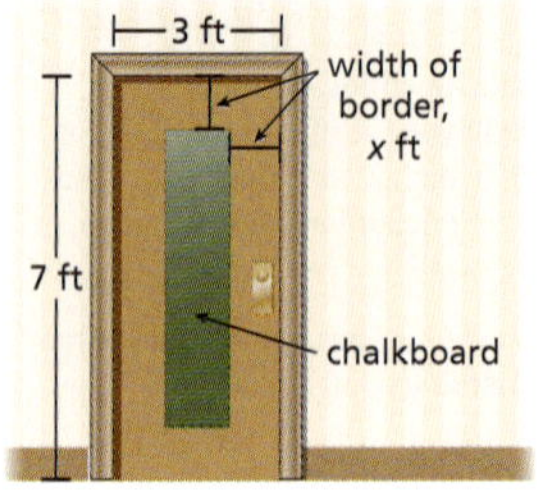

You decide to use chalkboard paint to create a chalkboard on a door. You want the chalkboard to cover 6 square feet and to have a uniform border, as shown. Find the width of the border to the nearest inch.

SOLUTION

1. **Understand the Problem** You know the dimensions (in feet) of the door from the diagram. You also know the area (in square feet) of the chalkboard and that it will have a uniform border. You are asked to find the width of the border to the nearest inch.
2. **Make a Plan** Use a verbal model to write an equation that represents the area of the chalkboard. Then solve the equation.
3. **Solve the Problem**

Let x be the width (in feet) of the border, as shown in the diagram.

Area of chalkboard (square feet)	=	Length of chalkboard (feet)	•	Width of chalkboard (feet)
6	=	$(7 - 2x)$	•	$(3 - 2x)$

$6 = (7 - 2x)(3 - 2x)$ Write the equation.

$6 = 21 - 20x + 4x^2$ Multiply the binomials.

$-15 = 4x^2 - 20x$ Subtract 21 from each side.

$-\frac{15}{4} = x^2 - 5x$ Divide each side by 4.

$-\frac{15}{4} + \frac{25}{4} = x^2 - 5x + \frac{25}{4}$ Complete the square for $x^2 - 5x$.

$\frac{5}{2} = x^2 - 5x + \frac{25}{4}$ Simplify the left side.

$\frac{5}{2} = \left(x - \frac{5}{2}\right)^2$ Write the right side as the square of a binomial.

$\pm\sqrt{\frac{5}{2}} = x - \frac{5}{2}$ Take the square root of each side.

$\frac{5}{2} \pm \sqrt{\frac{5}{2}} = x$ Add $\frac{5}{2}$ to each side.

The solutions of the equation are $x = \frac{5}{2} + \sqrt{\frac{5}{2}} \approx 4.08$ and $x = \frac{5}{2} - \sqrt{\frac{5}{2}} \approx 0.92$.

It is not possible for the width of the border to be 4.08 feet because the width of the door is 3 feet. So, the width of the border is about 0.92 foot.

$0.92 \text{ ft} \cdot \frac{12 \text{ in.}}{1 \text{ ft}} = 11.04 \text{ in.}$ Convert 0.92 foot to inches.

▶ The width of the border should be about 11 inches.

4. **Look Back** When the width of the border is slightly less than 1 foot, the length of the chalkboard is slightly more than 5 feet and the width of the chalkboard is slightly more than 1 foot. Multiplying these dimensions gives an area close to 6 square feet. So, an 11-inch border is reasonable.

Monitoring Progress

Help in English and Spanish at *BigIdeasMath.com*

13. **WHAT IF?** You want the chalkboard to cover 4 square feet. Find the width of the border to the nearest inch.

Laurie's Notes Teacher Actions

- **Model with Mathematics** and **Thumbs Up:** Read through the problem. Students should self-assess with *Thumbs Up* to indicate that they understand the problem. Discuss a plan for solving the problem.
- **Pass the Problem:** Follow the protocol for this formative assessment technique. Circulate and listen to student dialogue. When students have finished, the groups should debrief one another's work.

Closure

- **Point of Most Significance:** Ask students to identify, aloud or written, the most significant point in the lesson that aided their learning.

9.4 Exercises

Dynamic Solutions available at *BigIdeasMath.com*

Vocabulary and Core Concept Check

1. **COMPLETE THE SENTENCE** The process of adding a constant c to the expression $x^2 + bx$ so that $x^2 + bx + c$ is a perfect square trinomial is called ______________.

2. **VOCABULARY** Explain how to complete the square for an expression of the form $x^2 + bx$.

3. **WRITING** Is it more convenient to complete the square for $x^2 + bx$ when b is odd or when b is even? Explain.

4. **WRITING** Describe how you can use the process of completing the square to find the maximum or minimum value of a quadratic function.

Monitoring Progress and Modeling with Mathematics

In Exercises 5–10, find the value of c that completes the square.

5. $x^2 - 8x + c$
6. $x^2 - 2x + c$
7. $x^2 + 4x + c$
8. $x^2 + 12x + c$
9. $x^2 - 15x + c$
10. $x^2 + 9x + c$

In Exercises 11–16, complete the square for the expression. Then factor the trinomial. *(See Example 1.)*

11. $x^2 - 10x$
12. $x^2 - 40x$
13. $x^2 + 16x$
14. $x^2 + 22x$
15. $x^2 + 5x$
16. $x^2 - 3x$

In Exercises 17–22, solve the equation by completing the square. Round your solutions to the nearest hundredth, if necessary. *(See Example 2.)*

17. $x^2 + 14x = 15$
18. $x^2 - 6x = 16$
19. $x^2 - 4x = -2$
20. $x^2 + 2x = 5$
21. $x^2 - 5x = 8$
22. $x^2 + 11x = -10$

23. **MODELING WITH MATHEMATICS** The area of the patio is 216 square feet.

    a. Write an equation that represents the area of the patio.

    b. Find the dimensions of the patio by completing the square.

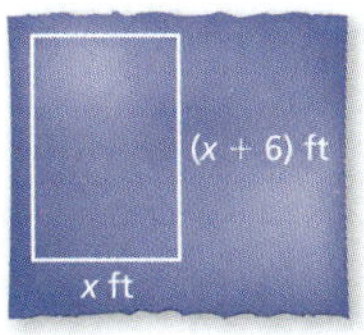

24. **MODELING WITH MATHEMATICS** Some sand art contains sand and water sealed in a glass case, similar to the one shown. When the art is turned upside down, the sand and water fall to create a new picture. The glass case has a depth of 1 centimeter and a volume of 768 cubic centimeters.

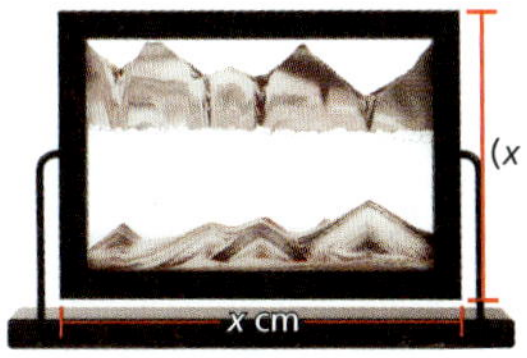

    a. Write an equation that represents the volume of the glass case.

    b. Find the dimensions of the glass case by completing the square.

In Exercises 25–32, solve the equation by completing the square. Round your solutions to the nearest hundredth, if necessary. *(See Example 3.)*

25. $x^2 - 8x + 15 = 0$
26. $x^2 + 4x - 21 = 0$
27. $2x^2 + 20x + 44 = 0$
28. $3x^2 - 18x + 12 = 0$
29. $-3x^2 - 24x + 17 = -40$
30. $-5x^2 - 20x + 35 = 30$
31. $2x^2 - 14x + 10 = 26$
32. $4x^2 + 12x - 15 = 5$

Assignment Guide and Homework Check

ASSIGNMENT

Basic: 1–4, 5–33 odd, 39–47 odd, 51–55 odd, 63, 66, 71, 75–80

Average: 1–4, 6–66 even, 71, 75–80

Advanced: 1–4, 10, 16, 22, 24, 30–38 even, 44–66 even, 67–80

HOMEWORK CHECK

Basic: 11, 17, 25, 41, 53

Average: 12, 20, 26, 42, 54

Advanced: 16, 22, 30, 44, 54

ANSWERS

1. completing the square
2. Add $\left(\frac{b}{2}\right)^2$ to $x^2 + bx$.
3. even; When b is even, $\frac{b}{2}$ is an integer.
4. Use completing the square to write the function in vertex form. The maximum or minimum is the y-coordinate of the vertex.
5. 16
6. 1
7. 4
8. 36
9. $\frac{225}{4}$
10. $\frac{81}{4}$
11. $x^2 - 10x + 25$; $(x - 5)^2$
12. $x^2 - 40x + 400$; $(x - 20)^2$
13. $x^2 + 16x + 64$; $(x + 8)^2$
14. $x^2 + 22x + 121$; $(x + 11)^2$
15. $x^2 + 5x + \frac{25}{4}$; $\left(x + \frac{5}{2}\right)^2$
16. $x^2 - 3x + \frac{9}{4}$; $\left(x - \frac{3}{2}\right)^2$
17. $x = 1, x = -15$
18. $x = 8, x = -2$
19. $x \approx 3.41, x \approx 0.59$
20. $x \approx 1.45, x \approx -3.45$
21. $x \approx 6.27, x \approx -1.27$
22. $x = -1, x = -10$
23. a. $x^2 + 6x = 216$

    b. width: 12 ft, length: 18 ft
24. a. $x^2 - 8x = 768$

    b. length: 32 cm, height: 24 cm
25. $x = 5, x = 3$
26. $x = 3, x = -7$
27. $x \approx -3.27, x \approx -6.73$
28. $x \approx 5.24, x \approx 0.76$
29. $x \approx 1.92, x \approx -9.92$
30. $x \approx 0.24, x \approx -4.24$
31. $x = 8, x = -1$
32. $x \approx 1.19, x \approx -4.19$

Dynamic Teaching Tools

- Dynamic Assessment & Progress Monitoring Tool
- Interactive Whiteboard Lesson Library
- Dynamic Classroom with Dynamic Investigations

ANSWERS

33. The number 16 should be added to each side of the equation; $x^2 + 8x + 16 = 10 + 16$; $x = -4 \pm \sqrt{26}$

34. The leading coefficient should be 1 before completing the square; $x^2 - x = 2$; $x^2 - x + \frac{1}{4} = 2 + \frac{1}{4}$

35. $b = 10$, $b = -10$; In a perfect square trinomial $c = \left(\frac{b}{2}\right)^2$, so $b = \pm 2\sqrt{c}$.

36. Divide each side of the equation by 3.

37. $y = (x + 3)^2 - 6$; D

38. $y = -(x - 4)^2 + 4$; A

39. $y = -(x + 2)^2 + 2$; B

40. $y = (x - 1)^2 + 3$; C

41. minimum value; -6

42. minimum value; 1

43. maximum value; -5

44. maximum value; 15

45. maximum value; -6

46. minimum value; -17

47. yes; The graph has two negative x-intercepts and it opens down.

48. no; The x-intercept is negative.

49. no; The x-intercepts are both positive.

50. yes; The graph has one positive x-intercept and one negative x-intercept, and it opens down.

51. f, m; The graph has two negative x-intercepts and it opens up.

52. r, n; The graph has one positive x-intercept and one negative x-intercept, and it opens down.

53. **a.** 36 ft

b. $x = \frac{3}{2}$; On the left side of $x = \frac{3}{2}$, the height increases as time increases. On the right side of $x = \frac{3}{2}$, the height decreases as time increases.

---

33. **ERROR ANALYSIS** Describe and correct the error in solving $x^2 + 8x = 10$ by completing the square.

✗
$$x^2 + 8x = 10$$
$$x^2 + 8x + 16 = 10$$
$$(x + 4)^2 = 10$$
$$x + 4 = \pm\sqrt{10}$$
$$x = -4 \pm \sqrt{10}$$

34. **ERROR ANALYSIS** Describe and correct the error in the first two steps of solving $2x^2 - 2x - 4 = 0$ by completing the square.

✗
$$2x^2 - 2x - 4 = 0$$
$$2x^2 - 2x = 4$$
$$2x^2 - 2x + 1 = 4 + 1$$

35. **NUMBER SENSE** Find all values of b for which $x^2 + bx + 25$ is a perfect square trinomial. Explain how you found your answer.

36. **REASONING** You are completing the square to solve $3x^2 + 6x = 12$. What is the first step?

In Exercises 37–40, write the function in vertex form by completing the square. Then match the function with its graph.

37. $y = x^2 + 6x + 3$ **38.** $y = -x^2 + 8x - 12$

39. $y = -x^2 - 4x - 2$ **40.** $y = x^2 - 2x + 4$

A.
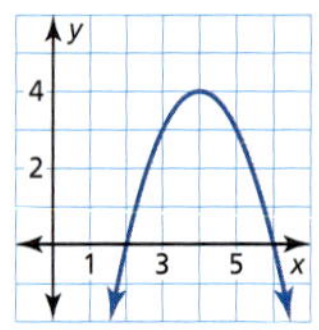

B.
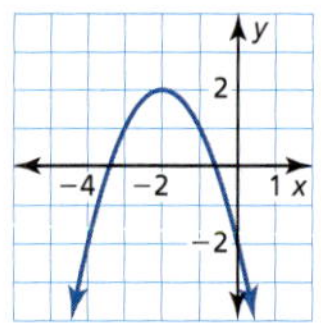

C.
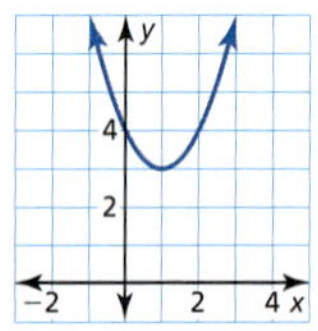

D.
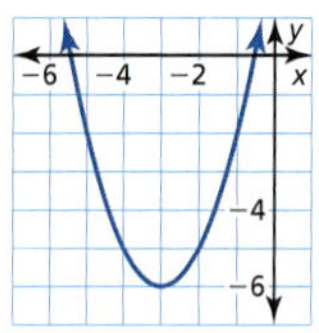

In Exercises 41–46, determine whether the quadratic function has a maximum or minimum value. Then find the value. *(See Examples 4 and 5.)*

41. $y = x^2 - 4x - 2$ **42.** $y = x^2 + 6x + 10$

43. $y = -x^2 - 10x - 30$ **44.** $y = -x^2 + 14x - 34$

45. $f(x) = -3x^2 - 6x - 9$ **46.** $f(x) = 4x^2 - 28x + 32$

In Exercises 47–50, determine whether the graph could represent the function. Explain.

47. $y = -(x + 8)(x + 3)$

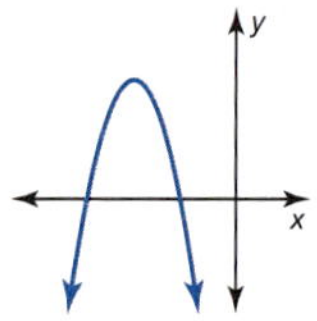

48. $y = (x - 5)^2$

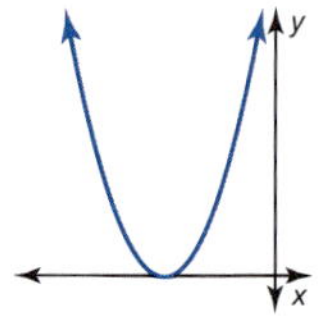

49. $y = \frac{1}{4}(x + 2)^2 - 4$

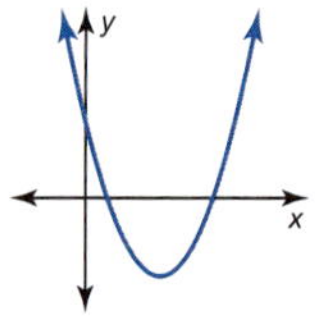

50. $y = -2(x - 1)(x + 2)$

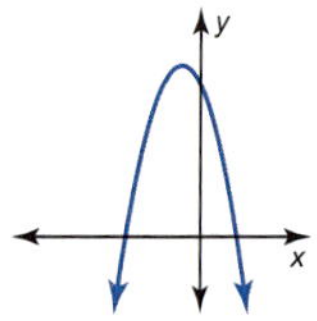

In Exercises 51 and 52, determine which of the functions could be represented by the graph. Explain. *(See Example 6.)*

51.

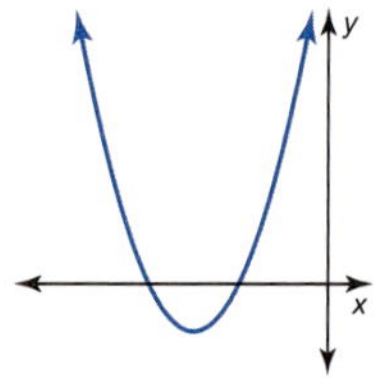

$h(x) = (x + 2)^2 + 3$

$f(x) = 2(x + 3)^2 - 2$

$g(x) = -\frac{1}{2}(x - 8)(x - 4)$

$m(x) = (x + 2)(x + 4)$

52.

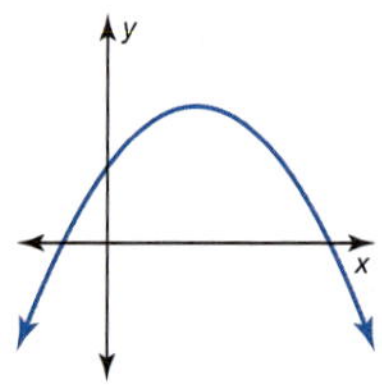

$r(x) = -\frac{1}{3}(x - 5)(x + 1)$

$p(x) = -2(x - 2)(x - 6)$

$q(x) = (x + 1)^2 + 4$

$n(x) = -(x - 2)^2 + 9$

53. **MODELING WITH MATHEMATICS** The function $h = -16t^2 + 48t$ represents the height h (in feet) of a kickball t seconds after it is kicked from the ground. *(See Example 7.)*

a. Find the maximum height of the kickball.

b. Find and interpret the axis of symmetry.

54. **MODELING WITH MATHEMATICS** You throw a stone from a height of 16 feet with an initial vertical velocity of 32 feet per second. The function $h = -16t^2 + 32t + 16$ represents the height h (in feet) of the stone after t seconds.

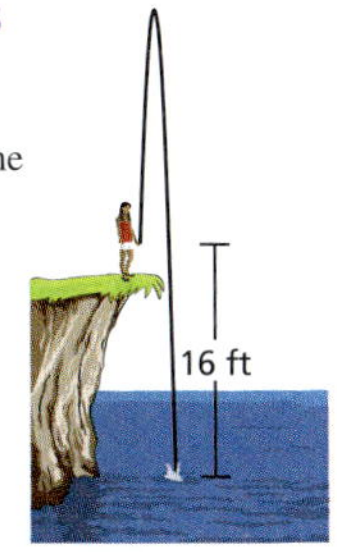

a. Find the maximum height of the stone.

b. Find and interpret the axis of symmetry.

55. **MODELING WITH MATHEMATICS** You are building a rectangular brick patio surrounded by a crushed stone border with a uniform width, as shown. You purchase patio bricks to cover 140 square feet. Find the width of the border. *(See Example 8.)*

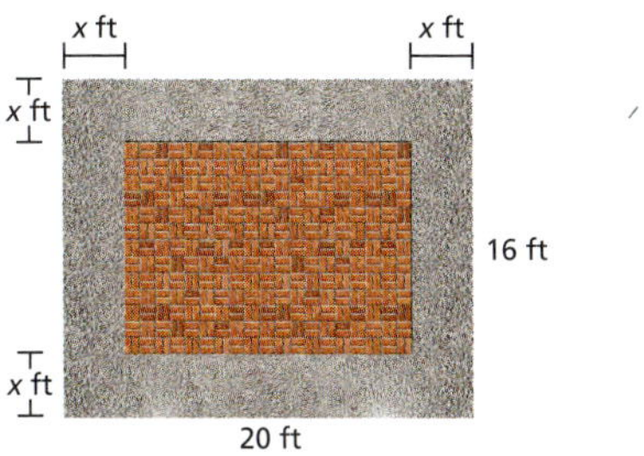

56. **MODELING WITH MATHEMATICS** You are making a poster that will have a uniform border, as shown. The total area of the poster is 722 square inches. Find the width of the border to the nearest inch.

MATHEMATICAL CONNECTIONS **In Exercises 57 and 58, find the value of x. Round your answer to the nearest hundredth, if necessary.**

57. $A = 108\text{ m}^2$

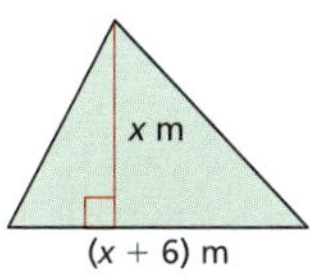

58. $A = 288\text{ in.}^2$

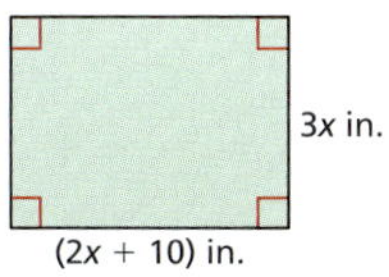

In Exercises 59–62, solve the equation by completing the square. Round your solutions to the nearest hundredth, if necessary.

59. $0.5x^2 + x - 2 = 0$

60. $0.75x^2 + 1.5x = 4$

61. $\frac{8}{3}x - \frac{2}{3}x^2 = -\frac{5}{6}$

62. $\frac{1}{4}x^2 + \frac{1}{2}x - \frac{5}{4} = 0$

63. **PROBLEM SOLVING** The distance d (in feet) that it takes a car to come to a complete stop can be modeled by $d = 0.05s^2 + 2.2s$, where s is the speed of the car (in miles per hour). A car has 168 feet to come to a complete stop. Find the maximum speed at which the car can travel.

64. **PROBLEM SOLVING** During a "big air" competition, snowboarders launch themselves from a half-pipe, perform tricks in the air, and land back in the half-pipe. The height h (in feet) of a snowboarder above the bottom of the half-pipe can be modeled by $h = -16t^2 + 24t + 16.4$, where t is the time (in seconds) after the snowboarder launches into the air. The snowboarder lands 3.2 feet lower than the height of the launch. How long is the snowboarder in the air? Round your answer to the nearest tenth of a second.

Cross section of a half-pipe

65. **PROBLEM SOLVING** You have 80 feet of fencing to make a rectangular horse pasture that covers 750 square feet. A barn will be used as one side of the pasture, as shown.

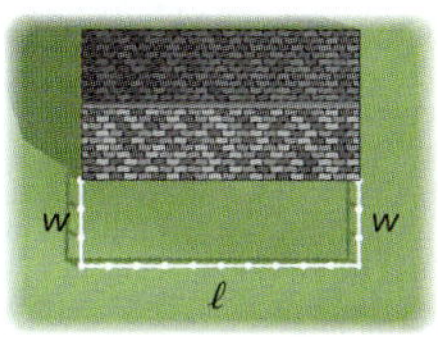

a. Write equations for the amount of fencing to be used and the area enclosed by the fencing.

b. Use substitution to solve the system of equations from part (a). What are the possible dimensions of the pasture?

ANSWERS

54. a. 32 ft

b. $x = 1$; On the left side of $x = 1$, the height increases as time increases. On the right side of $x = 1$, the height decreases as time increases.

55. 3 ft

56. about 1 in.

57. 12

58. about 4.87

59. $x \approx 1.24, x \approx -3.24$

60. $x \approx 1.52, x \approx -3.52$

61. $x \approx 4.29, x \approx -0.29$

62. $x \approx 1.45, x \approx -3.45$

63. 40 mi/h

64. about 1.6 sec

65. a. $\ell + 2w = 80, \ell w = 750$

b. length: 30 ft, width: 25 ft; length: 50 ft, width: 15 ft

ANSWERS

66. **a.** 3, 1

b. Substitute 3 for y in the equation, then solve the equation by completing the square.

67. **a.** $x \approx 0.8, x \approx -12.8$

b. $x \approx 0.78, x \approx -12.78$

c. *Sample answer:* completing the square; The result is more accurate.

68. See Additional Answers.

69. $x(x + 2) = 48$, 6 and 8

70. $x(x + 2) = 195$; $x = -15$ and $x = -13$

71–80. See Additional Answers.

Mini-Assessment

1. Complete the square for the expression $x^2 + 24x$. Then factor the trinomial. $x^2 + 24x + 144$; $(x + 12)^2$
2. Solve $x^2 - 12x = 45$ by completing the square. $x = 15$ and $x = -3$
3. Find the minimum value of $y = x^2 - 20x + 75$. -25
4. The function $y = -16x^2 + 32x$ represents the height y (in feet) of a football x seconds after it is kicked off the ground.
   a. Find the maximum height of the football. 16 feet
   b. Find and interpret the axis of symmetry. $x = 1$; The height increases when $x < 1$ and decreases when $x > 1$.
5. A rectangular city garden is surrounded by a sidewalk with a uniform width, as shown. The garden covers 340 square yards. Find the width of the sidewalk to the nearest foot.

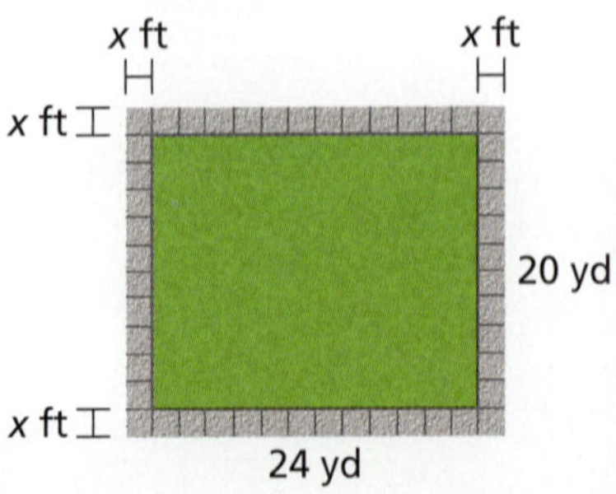

about 5 feet

66. HOW DO YOU SEE IT? The graph represents the quadratic function $y = x^2 - 4x + 6$.

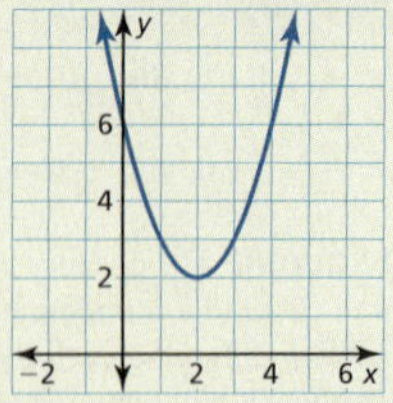

a. Use the graph to estimate the x-values for which $y = 3$.

b. Explain how you can use the method of completing the square to check your estimates in part (a).

67. COMPARING METHODS Consider the quadratic equation $x^2 + 12x + 2 = 12$.

a. Solve the equation by graphing.

b. Solve the equation by completing the square.

c. Compare the two methods. Which do you prefer? Explain.

68. THOUGHT PROVOKING Sketch the graph of the equation $x^2 - 2xy + y^2 - x - y = 0$. Identify the graph.

69. REASONING The product of two consecutive even integers that are positive is 48. Write and solve an equation to find the integers.

70. REASONING The product of two consecutive odd integers that are negative is 195. Write and solve an equation to find the integers.

71. MAKING AN ARGUMENT You purchase stock for \$16 per share. You sell the stock 30 days later for \$23.50 per share. The price y (in dollars) of a share during the 30-day period can be modeled by $y = -0.025x^2 + x + 16$, where x is the number of days after the stock is purchased. Your friend says you could have sold the stock earlier for \$23.50 per share. Is your friend correct? Explain.

72. REASONING You are solving the equation $x^2 + 9x = 18$. What are the advantages of solving the equation by completing the square instead of using other methods you have learned?

73. PROBLEM SOLVING You are knitting a rectangular scarf. The pattern results in a scarf that is 60 inches long and 4 inches wide. However, you have enough yarn to knit 396 square inches. You decide to increase the dimensions of the scarf so that you will use all your yarn. The increase in the length is three times the increase in the width. What are the dimensions of your scarf?

74. WRITING How many solutions does $x^2 + bx = c$ have when $c < -\left(\frac{b}{2}\right)^2$? Explain.

Maintaining Mathematical Proficiency

Reviewing what you learned in previous grades and lessons

Write a recursive rule for the sequence. *(Section 6.7)*

75.

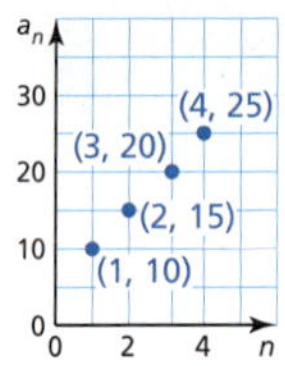

76.

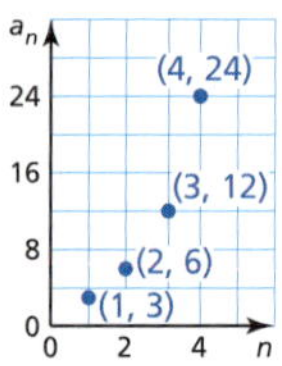

77.

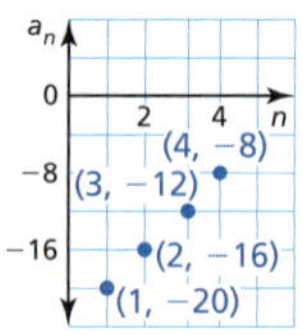

Simplify the expression $\sqrt{b^2 - 4ac}$ for the given values. *(Section 9.1)*

78. $a = 3, b = -6, c = 2$ **79.** $a = -2, b = 4, c = 7$ **80.** $a = 1, b = 6, c = 4$

If students need help...	*If students got it...*
Resources by Chapter • Practice A and Practice B • Puzzle Time	Resources by Chapter • Enrichment and Extension • Cumulative Review
Student Journal • Practice	Start the *next* Section
Differentiating the Lesson Skills Review Handbook	

Laurie's Notes

Overview of Section 9.5

Introduction

- The Quadratic Formula is presented as the fifth method for solving quadratic equations. The formula is derived in the explorations and stated formally in the lesson. The formula is then used to solve quadratic equations.
- The discriminant is discussed and then interpreted to determine the number of zeros of an equation or the number of x-intercepts of a graph.
- The lesson ends with a summary of the five methods used to solve quadratics, and then students are asked to choose efficient methods for solving various quadratic equations.

Resources

- There are many online videos of students singing the Quadratic Formula to the tune of "Pop Goes the Weasel." Share one with students. You could return to the video at the end of the period when students may be ready to sing along.

Formative Assessment Tips

- **Paired Verbal Fluency (PVF):** This technique is used between two partners where each person takes a turn speaking, uninterrupted, for a specified period of time. The roles reverse and the listener then speaks, uninterrupted, for the same amount of time.
- The technique can be used at the beginning, middle, or end of instruction. Used at the beginning of instruction, students share their prior knowledge about a particular topic, skill, or concept. Used in the middle of instruction, students can share their understanding of a particular problem. Used at the end of instruction, students reflect on learning that occurred during the lesson or at the end of a connected group of lessons.
- Verbalizing their understanding and being an attentive listener will activate student thinking and should help identify areas of difficulty or uncertainty.
- In this lesson, students will be asked to reflect on their understanding of solving quadratics using a variety of techniques.

Pacing Suggestion

- The explorations provide an opportunity for students to derive the Quadratic Formula and then practice using the formula to solve a few simple quadratic equations. You may want to omit Example 1 and begin with the application problem in Example 2.

Dynamic Teaching Tools

- Dynamic Assessment & Progress Monitoring Tool
- Lesson Planning Tool
- Interactive Whiteboard Lesson Library
- Dynamic Classroom with Dynamic Investigations

What Your Students Will Learn

- Solve quadratic equations of the form $ax^2 + bx + c = 0$ using the Quadratic Formula.
- Interpret the discriminant to determine the number of real solutions of quadratic equations.
- Choose efficient methods for solving quadratic equations and explain the advantages of the methods.

Laurie's Notes

Exploration

Motivate

- Share with students some history about the Quadratic Formula.
  - Around 400 B.C., the Babylonians and Chinese use a method called "completing the square" to solve problems involving areas.
  - Around 300 B.C., the Greek mathematicians Pythagoras and Euclid use geometry to find a general procedure for solving a quadratic equation, but their methods are not considered useful.
  - Around A.D. 700, a Hindu mathematician named Brahmagupta finds the general solution for the quadratic equation. He uses irrational numbers and also recognizes the two roots in the solution.
  - Around A.D. 1100, another Hindu mathematician named Baskhara finds the complete solution. He recognizes that any positive number has two square roots.
  - In 1637, the French mathematician René Descartes publishes *La Géométrie*, which presents the Quadratic Formula in its present form.

Discuss

- Tell students that today they will see one way of deriving the Quadratic Formula algebraically. Then they will derive the Quadratic Formula on their own by completing the square. They will also learn about the discriminant and discover how it relates to the number of solutions.

Exploration 1

- A derivation of the Quadratic Formula is shown in this exploration. Students justify each step.
- If students get stuck on a step, ask them what is different from the last step. Mathematically, what has changed?
- When students are finished, discuss the justifications.
- Students sometimes lose sight of what they have accomplished. Starting with the quadratic equation in general form, they solve for x algebraically to produce a formula for finding the solutions of any quadratic equation.

Exploration 2

- In the last lesson, students learned how to solve a quadratic equation by completing the square. They are using that method to derive the Quadratic Formula and compare it to the method used in Exploration 1.
- **Make Sense of Problems and Persevere in Solving Them:** Do not be too quick to rescue your students. Give them time to wrestle with the derivation. If they get stuck, have them refer back to their notes from the last section. Believe that your students can persevere in deriving the formula. Unlike many problems, they know what the formula should look like when they finish because of Exploration 1.

Communicate Your Answer

- To solve each quadratic equation in Question 4, tell students they need to evaluate the Quadratic Formula for the values of a, b, and c from the equation.

Connecting to Next Step

- Students have now derived the Quadratic Formula. The formula will be stated at the beginning of the formal lesson.

9.5 Solving Quadratic Equations Using the Quadratic Formula

Essential Question How can you derive a formula that can be used to write the solutions of any quadratic equation in standard form?

EXPLORATION 1 Deriving the Quadratic Formula

Work with a partner. The following steps show a method of solving $ax^2 + bx + c = 0$. Explain what was done in each step.

$ax^2 + bx + c = 0$ ← 1. Write the equation.

$4a^2x^2 + 4abx + 4ac = 0$ ← 2. What was done?

$4a^2x^2 + 4abx + 4ac + b^2 = b^2$ ← 3. What was done?

$4a^2x^2 + 4abx + b^2 = b^2 - 4ac$ ← 4. What was done?

$(2ax + b)^2 = b^2 - 4ac$ ← 5. What was done?

$2ax + b = \pm\sqrt{b^2 - 4ac}$ ← 6. What was done?

$2ax = -b \pm \sqrt{b^2 - 4ac}$ ← 7. What was done?

Quadratic Formula: $x = \dfrac{-b \pm \sqrt{b^2 - 4ac}}{2a}$ ← 8. What was done?

EXPLORATION 2 Deriving the Quadratic Formula by Completing the Square

Work with a partner.

a. Solve $ax^2 + bx + c = 0$ by completing the square. (*Hint:* Subtract c from each side, divide each side by a, and then proceed by completing the square.)

b. Compare this method with the method in Exploration 1. Explain why you think $4a$ and b^2 were chosen in Steps 2 and 3 of Exploration 1.

Communicate Your Answer

3. How can you derive a formula that can be used to write the solutions of any quadratic equation in standard form?

4. Use the Quadratic Formula to solve each quadratic equation.

a. $x^2 + 2x - 3 = 0$ **b.** $x^2 - 4x + 4 = 0$ **c.** $x^2 + 4x + 5 = 0$

5. Use the Internet to research *imaginary numbers.* How are they related to quadratic equations?

USING TOOLS STRATEGICALLY

To be proficient in math, you need to identify relevant external mathematical resources.

Dynamic Teaching Tools

- Dynamic Assessment & Progress Monitoring Tool
- Lesson Planning Tool
- Interactive Whiteboard Lesson Library
- Dynamic Classroom with Dynamic Investigations

ANSWERS

1. **2.** Multiply each side by $4a$.

3. Add b^2 to each side.

4. Subtract $4ac$ from each side.

5. Write the left side in factored form.

6. Take the square root of each side.

7. Subtract b from each side.

8. Divide each side by $2a$.

2. **a.** $x = \dfrac{-b \pm \sqrt{b^2 - 4ac}}{2a}$

b. to obtain a perfect square trinomial on the left side of the equation

3. Complete the square of the general form of the quadratic equation, $ax^2 + bx + c = 0$.

4. **a.** $x = 1, x = -3$

b. $x = 2$

c. no real solution

5. The imaginary number, i, is $\sqrt{-1}$. Quadratic equations with no real solution have complex solutions, which include an imaginary part.

Differentiated Instruction

Organization/Inclusion

Some students may view the Quadratic Formula as complicated and may not know what to do first when trying to use it. Suggest that they complete this table before they begin solving a quadratic equation in standard form.

Coefficient	Value
a	
b	
c	
Term	**Value**
$-b$	
b^2	
$4ac$	
$2a$	

After completing the table, students can substitute the appropriate values into the formula and simplify.

Extra Example 1

Solve $12x^2 - 4x - 5 = 0$ using the Quadratic Formula.

The solutions are $x = \frac{4+16}{24} = \frac{5}{6}$ and $x = \frac{4-16}{24} = -\frac{1}{2}$.

MONITORING PROGRESS ANSWERS

1. $x = 5, x = 1$
2. $x \approx 3.6, x \approx -5.6$
3. $x \approx -1.2, x \approx 1.9$
4. $x = \frac{1}{2}$

9.5 Lesson

Core Vocabulary

Quadratic Formula, *p. 516*
discriminant, *p. 518*

What You Will Learn

- Solve quadratic equations using the Quadratic Formula.
- Interpret the discriminant.
- Choose efficient methods for solving quadratic equations.

Using the Quadratic Formula

By completing the square for the quadratic equation $ax^2 + bx + c = 0$, you can develop a formula that gives the solutions of any quadratic equation in standard form. This formula is called the **Quadratic Formula**.

Core Concept

Quadratic Formula

The real solutions of the quadratic equation $ax^2 + bx + c = 0$ are

$$x = \frac{-b \pm \sqrt{b^2 - 4ac}}{2a} \qquad \text{Quadratic Formula}$$

where $a \neq 0$ and $b^2 - 4ac \geq 0$.

EXAMPLE 1 Using the Quadratic Formula

Solve $2x^2 - 5x + 3 = 0$ using the Quadratic Formula.

SOLUTION

$x = \frac{-b \pm \sqrt{b^2 - 4ac}}{2a}$ — Quadratic Formula

$= \frac{-(-5) \pm \sqrt{(-5)^2 - 4(2)(3)}}{2(2)}$ — Substitute 2 for *a*, −5 for *b*, and 3 for *c*.

$= \frac{5 \pm \sqrt{1}}{4}$ — Simplify.

$= \frac{5 \pm 1}{4}$ — Evaluate the square root.

So, the solutions are $x = \frac{5+1}{4} = \frac{3}{2}$ and $x = \frac{5-1}{4} = 1$.

STUDY TIP

You can use the roots of a quadratic equation to factor the related expression. In Example 1, you can use 1 and $\frac{3}{2}$ to factor $2x^2 - 5x + 3$ as $(x - 1)(2x - 3)$.

Check

$2x^2 - 5x + 3 = 0$	Original equation	$2x^2 - 5x + 3 = 0$
$2\left(\frac{3}{2}\right)^2 - 5\left(\frac{3}{2}\right) + 3 \stackrel{?}{=} 0$	Substitute.	$2(1)^2 - 5(1) + 3 \stackrel{?}{=} 0$
$\frac{9}{2} - \frac{15}{2} + 3 \stackrel{?}{=} 0$	Simplify.	$2 - 5 + 3 \stackrel{?}{=} 0$
$0 = 0$ ✓	Simplify.	$0 = 0$ ✓

Monitoring Progress

Help in English and Spanish at *BigIdeasMath.com*

Solve the equation using the Quadratic Formula. Round your solutions to the nearest tenth, if necessary.

1. $x^2 - 6x + 5 = 0$
2. $\frac{1}{2}x^2 + x - 10 = 0$
3. $-3x^2 + 2x + 7 = 0$
4. $4x^2 - 4x = -1$

Laurie's Notes Teacher Actions

- ? Write the quadratic equation. "Why must one side of the equation be equal to 0?" So you can determine the values of *a*, *b*, and *c*.
- ? "Why is there a restriction that $a \neq 0$?" When $a = 0$, you do not have a quadratic equation. Also, the denominator would be 0 in the Quadratic Formula and you cannot divide by 0.
- ? "Why is there a restriction that $b^2 - 4ac \geq 0$?" You cannot take the square root of a negative number.
- Work through Example 1. Review order of operations as you simplify. Explain how to interpret the ± symbol.

EXAMPLE 2 Modeling With Mathematics

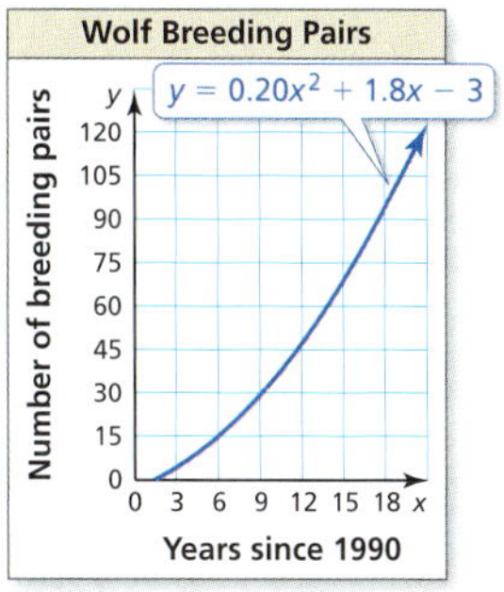

The number y of Northern Rocky Mountain wolf breeding pairs x years since 1990 can be modeled by the function $y = 0.20x^2 + 1.8x - 3$. When were there about 35 breeding pairs?

SOLUTION

1. **Understand the Problem** You are given a quadratic function that represents the number of wolf breeding pairs for years after 1990. You need to use the model to determine when there were 35 wolf breeding pairs.

2. **Make a Plan** To determine when there were 35 wolf breeding pairs, find the x-values for which $y = 35$. So, solve the equation $35 = 0.20x^2 + 1.8x - 3$.

3. **Solve the Problem**

$35 = 0.20x^2 + 1.8x - 3$ Write the equation.

$0 = 0.20x^2 + 1.8x - 38$ Write in standard form.

$x = \dfrac{-b \pm \sqrt{b^2 - 4ac}}{2a}$ Quadratic Formula

$= \dfrac{-1.8 \pm \sqrt{1.8^2 - 4(0.2)(-38)}}{2(0.2)}$ Substitute 0.2 for a, 1.8 for b, and -38 for c.

$= \dfrac{-1.8 \pm \sqrt{33.64}}{0.4}$ Simplify.

$= \dfrac{-1.8 \pm 5.8}{0.4}$ Simplify.

The solutions are $x = \dfrac{-1.8 + 5.8}{0.4} = 10$ and $x = \dfrac{-1.8 - 5.8}{0.4} = -19$.

INTERPRETING MATHEMATICAL RESULTS

You can ignore the solution $x = -19$ because -19 represents the year 1971, which is not in the given time period.

▶ Because x represents the number of years since 1990, x is greater than or equal to zero. So, there were about 35 breeding pairs 10 years after 1990, in 2000.

4. **Look Back** Use a graphing calculator to graph the equations $y = 0.20x^2 + 1.8x - 3$ and $y = 35$. Then use the *intersect* feature to find the point of intersection. The graphs intersect at (10, 35).

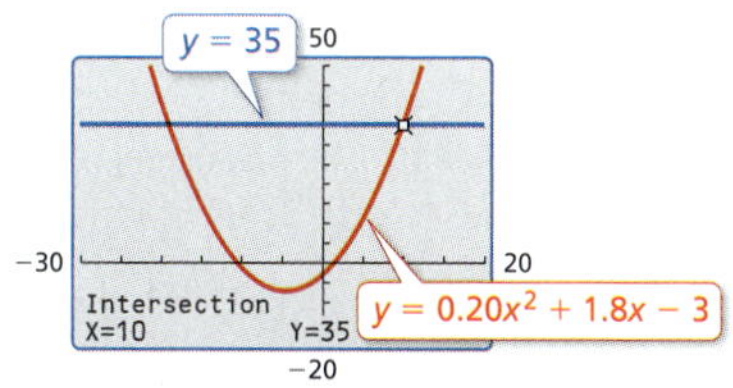

Monitoring Progress

Help in English and Spanish at *BigIdeasMath.com*

5. **WHAT IF?** When were there about 60 wolf breeding pairs?

6. The number y of bald eagle nesting pairs in a state x years since 2000 can be modeled by the function $y = 0.34x^2 + 13.1x + 51$.
   - **a.** When were there about 160 bald eagle nesting pairs?
   - **b.** How many bald eagle nesting pairs were there in 2000?

English Language Learners

Forming Answers

Encourage English learners to form complete sentences in their responses to real-life problems. Demonstrate how to use the question to help students form the answer.

Question: When were there about 35 breeding pairs?

Answer: There were about 35 breeding pairs 10 years after 1990, in the year 2000.

Extra Example 2

The number y of Northern Rocky Mountain wolf breeding pairs x years since 1990 can be modeled by the function $y = 0.20x^2 + 1.8x - 3$. When were there about 77 breeding pairs of wolves? There were about 77 breeding pairs 16 years after 1990, in 2006.

MONITORING PROGRESS ANSWERS

5. 2004
6. **a.** 2007
   b. 51 bald eagle nesting pairs

Laurie's Notes Teacher Actions

- **Reason Abstractly and Quantitatively:** Students decontextualize the model to solve algebraically. Once the solutions are found, the context is considered again.
- Consider using technology to graph the function $y = 0.20x^2 + 1.8x - 38$. Use the graph to approximate the zeros. Then solve the corresponding equation by using the Quadratic Formula.

? "Do both solutions make sense in the context of the problem? Explain." no; Only the positive solution makes sense.

? "Can the related equation be factored using only integers? Explain." yes; Dividing the equation by 0.2 results in the equation $x^2 + 9x - 190 = 0$, which can be factored as $(x + 19)(x - 10) = 0$.

Extra Example 3

a. Determine the number of real solutions of $2x^2 - 3x - 1 = 0$.
The discriminant is greater than 0, so the equation has two real solutions.

b. Determine the number of real solutions of $x^2 + 4 = x$.
The discriminant is less than 0, so the equation has no real solutions.

MONITORING PROGRESS ANSWERS

7. one real solution
8. no real solutions
9. two real solutions

Interpreting the Discriminant

The expression $b^2 - 4ac$ in the Quadratic Formula is called the **discriminant**.

$$x = \frac{-b \pm \sqrt{b^2 - 4ac}}{2a} \quad \leftarrow \text{discriminant}$$

Because the discriminant is under the radical symbol, you can use the value of the discriminant to determine the number of real solutions of a quadratic equation and the number of x-intercepts of the graph of the related function.

STUDY TIP

The solutions of a quadratic equation may be real numbers or *imaginary numbers*. You will study imaginary numbers in a future course.

Core Concept

Interpreting the Discriminant

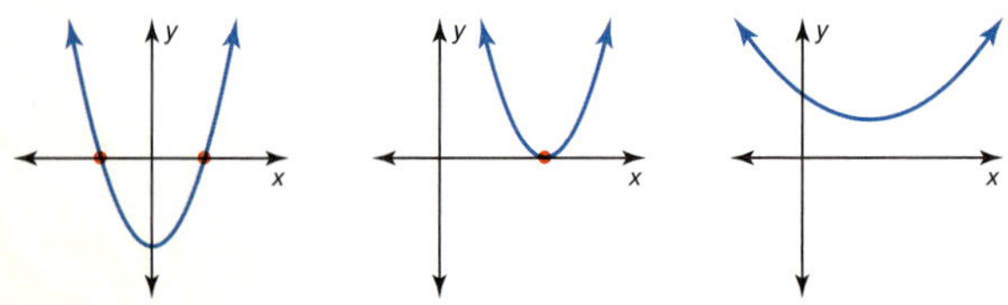

$b^2 - 4ac > 0$	$b^2 - 4ac = 0$	$b^2 - 4ac < 0$
• two real solutions • two x-intercepts	• one real solution • one x-intercept	• no real solutions • no x-intercepts

EXAMPLE 3 Determining the Number of Real Solutions

a. Determine the number of real solutions of $x^2 + 8x - 3 = 0$.

$b^2 - 4ac = 8^2 - 4(1)(-3)$ Substitute 1 for *a*, 8 for *b*, and −3 for *c*.

$= 64 + 12$ Simplify.

$= 76$ Add.

▶ The discriminant is greater than 0. So, the equation has two real solutions.

b. Determine the number of real solutions of $9x^2 + 1 = 6x$.

Write the equation in standard form: $9x^2 - 6x + 1 = 0$.

$b^2 - 4ac = (-6)^2 - 4(9)(1)$ Substitute 9 for *a*, −6 for *b*, and 1 for *c*.

$= 36 - 36$ Simplify.

$= 0$ Subtract.

▶ The discriminant is 0. So, the equation has one real solution.

Monitoring Progress

Help in English and Spanish at *BigIdeasMath.com*

Determine the number of real solutions of the equation.

7. $-x^2 + 4x - 4 = 0$
8. $6x^2 + 2x = -1$
9. $\frac{1}{2}x^2 = 7x - 1$

Laurie's Notes Teacher Actions

- **Connection:** Write the *Core Concept*. Discuss connections to previous lessons in the chapter and ways in which quadratics have been solved. Ask a volunteer to discuss the ways to recognize each number of solutions when solving by graphing, factoring, and the Quadratic Formula.

? After writing the equation in part (b) ask, "What are the values of *a*, *b*, and *c*?" Check that students subtract the 6*x* term first! $a = 9, b = -6, c = 1$

- **Think-Pair-Share:** Have students answer Questions 7–9, and then share and discuss as a class.

EXAMPLE 4 Finding the Number of x-Intercepts of a Parabola

Find the number of x-intercepts of the graph of $y = 2x^2 + 3x + 9$.

SOLUTION

Determine the number of real solutions of $0 = 2x^2 + 3x + 9$.

$$\begin{aligned} b^2 - 4ac &= 3^2 - 4(2)(9) && \text{Substitute 2 for } a\text{, 3 for } b\text{, and 9 for } c. \\ &= 9 - 72 && \text{Simplify.} \\ &= -63 && \text{Subtract.} \end{aligned}$$

Because the discriminant is less than 0, the equation has no real solutions.

▶ So, the graph of $y = 2x^2 + 3x + 9$ has no x-intercepts.

Check

Use a graphing calculator to check your answer. Notice that the graph of $y = 2x^2 + 3x + 9$ has no x-intercepts.

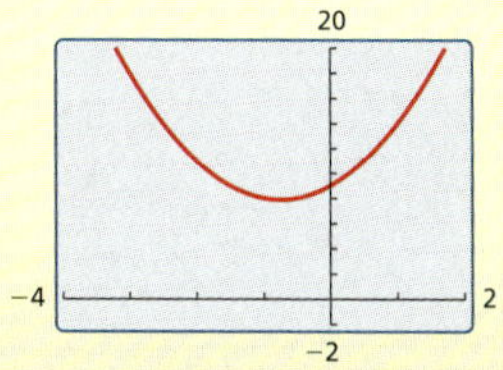

Monitoring Progress

Help in English and Spanish at *BigIdeasMath.com*

Find the number of x-intercepts of the graph of the function.

10. $y = -x^2 + x - 6$ **11.** $y = x^2 - x$ **12.** $f(x) = x^2 + 12x + 36$

Choosing an Efficient Method

The table shows five methods for solving quadratic equations. For a given equation, it may be more efficient to use one method instead of another. Some advantages and disadvantages of each method are shown.

Core Concept

Methods for Solving Quadratic Equations

Method	Advantages	Disadvantages
Factoring *(Lessons 7.5–7.8)*	• Straightforward when the equation can be factored easily	• Some equations are not factorable.
Graphing *(Lesson 9.2)*	• Can easily see the number of solutions • Use when approximate solutions are sufficient. • Can use a graphing calculator	• May not give exact solutions
Using Square Roots *(Lesson 9.3)*	• Use to solve equations of the form $x^2 = d$.	• Can only be used for certain equations
Completing the Square *(Lesson 9.4)*	• Best used when $a = 1$ and b is even	• May involve difficult calculations
Quadratic Formula *(Lesson 9.5)*	• Can be used for any quadratic equation • Gives exact solutions	• Takes time to do calculations

Extra Example 4

Find the number of x-intercepts of the graph of $y = 2x^2 - 4x - 3$.

The graph of $y = 2x^2 - 4x - 3$ has two x-intercepts.

MONITORING PROGRESS ANSWERS

10. no x-intercepts

11. two x-intercepts

12. one x-intercept

Laurie's Notes Teacher Actions

- **Connection:** In Example 3, students are finding the number of real solutions of an equation, and in Example 4, they are finding the number of x-intercepts of a graph. Discuss how the Quadratic Formula is used to solve both problems and how the problems are related.
- **Paired Verbal Fluency:** Have students pair up and follow the protocol described on page T-514. Ask students to describe the various methods they have learned to solve quadratic equations, noting the advantages and disadvantages of each method. Listen to students' dialogue.

Extra Example 5

Solve the equation using any method. Explain your choice of method.

a. $x^2 - 4x - 7 = 0$

$x = 2 + \sqrt{11} \approx 5.3$ and $x = 2 - \sqrt{11} \approx -1.3$; The coefficient of the x^2-term is 1, and the coefficient of the x-term is an even number. So, solve by completing the square.

b. $(x - 2)^2 = 64$

$x = 10$ and $x = -6$; The equation has the form $x^2 = d$. So, solve by using square roots.

c. $10x^2 + 7x - 12 = 0$

$x = \frac{-7 + 23}{20} = \frac{4}{5}$ and $x = \frac{-7 - 23}{20} = -\frac{3}{2}$; The equation is not easily factorable, and the numbers are somewhat large. So, solve using the Quadratic Formula.

MONITORING PROGRESS ANSWERS

13. $x = -12, x = 1$; *Sample answer:* The equation is easily factorable, so solve by factoring.
14. $x = 1, x = -1$; *Sample answer:* The equation can be written in the form $x^2 = d$, so solve using square roots.
15. $x \approx 0.56, x \approx -0.36$; *Sample answer:* The equation is not factorable and the coefficient of the x^2-term is not 1, so solve using the quadratic formula.
16. no real solutions; *Sample answer:* The coefficient of the x^2-term is 1 and b is even, so solve by completing the square.

EXAMPLE 5 Choosing a Method

Solve the equation using any method. Explain your choice of method.

a. $x^2 - 10x = 1$ **b.** $2x^2 - 13x - 24 = 0$ **c.** $x^2 + 8x + 12 = 0$

SOLUTION

a. The coefficient of the x^2-term is 1, and the coefficient of the x-term is an even number. So, solve by completing the square.

$x^2 - 10x = 1$	Write the equation.
$x^2 - 10x + 25 = 1 + 25$	Complete the square for $x^2 - 10x$.
$(x - 5)^2 = 26$	Write the left side as the square of a binomial.
$x - 5 = \pm\sqrt{26}$	Take the square root of each side.
$x = 5 \pm \sqrt{26}$	Add 5 to each side.

▶ So, the solutions are $x = 5 + \sqrt{26} \approx 10.1$ and $x = 5 - \sqrt{26} \approx -0.1$.

b. The equation is not easily factorable, and the numbers are somewhat large. So, solve using the Quadratic Formula.

$x = \frac{-b \pm \sqrt{b^2 - 4ac}}{2a}$	Quadratic Formula
$= \frac{-(-13) \pm \sqrt{(-13)^2 - 4(2)(-24)}}{2(2)}$	Substitute 2 for a, -13 for b, and -24 for c.
$= \frac{13 \pm \sqrt{361}}{4}$	Simplify.
$= \frac{13 \pm 19}{4}$	Evaluate the square root.

▶ So, the solutions are $x = \frac{13 + 19}{4} = 8$ and $x = \frac{13 - 19}{4} = -\frac{3}{2}$.

Check

Graph the related function $f(x) = x^2 + 8x + 12$ and find the zeros. The zeros are -6 and -2.

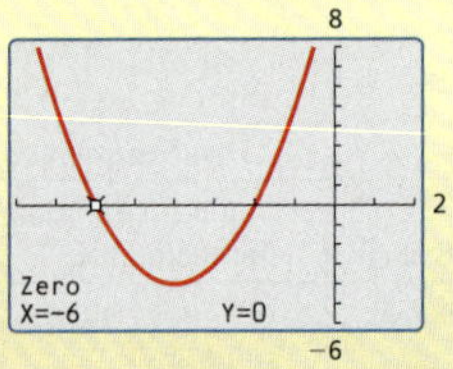

c. The equation is easily factorable. So, solve by factoring.

$x^2 + 8x + 12 = 0$	Write the equation.
$(x + 2)(x + 6) = 0$	Factor the polynomial.
$x + 2 = 0$ or $x + 6 = 0$	Zero-Product Property
$x = -2$ or $x = -6$	Solve for x.

▶ The solutions are $x = -2$ and $x = -6$.

Monitoring Progress

Help in English and Spanish at *BigIdeasMath.com*

Solve the equation using any method. Explain your choice of method.

13. $x^2 + 11x - 12 = 0$
14. $9x^2 - 5 = 4$
15. $5x^2 - x - 1 = 0$
16. $x^2 = 2x - 5$

Laurie's Notes Teacher Actions

- There is more than one method that could be used to solve the three quadratic equations. As students examine the equations, they should be considering why one method would be a wise choice versus another method.
- **Construct Viable Arguments and Critique the Reasoning of Others:** Have students critique each of the choices made by classmates. Allow personal preference. Listen for comments that are thoughtful and reasonable.

Closure

- Write a quadratic equation that you would *not* solve using (a) square roots, (b) factoring, and (c) completing the square. Check students' work. *Sample answers:* (a) $x^2 + 12x + 8 = 0$; (b) $x^2 = 211$; (c) $x^2 - 7x + 12 = 0$

9.5 Exercises

Dynamic Solutions available at *BigIdeasMath.com*

Vocabulary and Core Concept Check

1. **VOCABULARY** What formula can you use to solve any quadratic equation? Write the formula.

2. **VOCABULARY** In the Quadratic Formula, what is the discriminant? What does the value of the discriminant determine?

Monitoring Progress and Modeling with Mathematics

In Exercises 3–8, write the equation in standard form. Then identify the values of *a*, *b*, and *c* that you would use to solve the equation using the Quadratic Formula.

3. $x^2 = 7x$
4. $x^2 - 4x = -12$
5. $-2x^2 + 1 = 5x$
6. $3x + 2 = 4x^2$
7. $4 - 3x = -x^2 + 3x$
8. $-8x - 1 = 3x^2 + 2$

In Exercises 9–22, solve the equation using the Quadratic Formula. Round your solutions to the nearest tenth, if necessary. *(See Example 1.)*

9. $x^2 - 12x + 36 = 0$
10. $x^2 + 7x + 16 = 0$
11. $x^2 - 10x - 11 = 0$
12. $2x^2 - x - 1 = 0$
13. $2x^2 - 6x + 5 = 0$
14. $9x^2 - 6x + 1 = 0$
15. $6x^2 - 13x = -6$
16. $-3x^2 + 6x = 4$
17. $1 - 8x = -16x^2$
18. $x^2 - 5x + 3 = 0$
19. $x^2 + 2x = 9$
20. $5x^2 - 2 = 4x$
21. $2x^2 + 9x + 7 = 3$
22. $8x^2 + 8 = 6 - 9x$

23. **MODELING WITH MATHEMATICS** A dolphin jumps out of the water, as shown in the diagram. The function $h = -16t^2 + 26t$ models the height h (in feet) of the dolphin after t seconds. After how many seconds is the dolphin at a height of 5 feet? *(See Example 2.)*

24. **MODELING WITH MATHEMATICS** The amount of trout y (in tons) caught in a lake from 1995 to 2014 can be modeled by the equation $y = -0.08x^2 + 1.6x + 10$, where x is the number of years since 1995.

    a. When were about 15 tons of trout caught in the lake?

    b. Do you think this model can be used to determine the amounts of trout caught in future years? Explain your reasoning.

In Exercises 25–30, determine the number of real solutions of the equation. *(See Example 3.)*

25. $x^2 - 6x + 10 = 0$
26. $x^2 - 5x - 3 = 0$
27. $2x^2 - 12x = -18$
28. $4x^2 = 4x - 1$
29. $-\frac{1}{4}x^2 + 4x = -2$
30. $-5x^2 + 8x = 9$

In Exercises 31–36, find the number of *x*-intercepts of the graph of the function. *(See Example 4.)*

31. $y = x^2 + 5x - 1$
32. $y = 4x^2 + 4x + 1$
33. $y = -6x^2 + 3x - 4$
34. $y = -x^2 + 5x + 13$
35. $f(x) = 4x^2 + 3x - 6$
36. $f(x) = 2x^2 + 8x + 8$

In Exercises 37–44, solve the equation using any method. Explain your choice of method. *(See Example 5.)*

37. $-10x^2 + 13x = 4$
38. $x^2 - 3x - 40 = 0$
39. $x^2 + 6x = 5$
40. $-5x^2 = -25$
41. $x^2 + x - 12 = 0$
42. $x^2 - 4x + 1 = 0$
43. $4x^2 - x = 17$
44. $x^2 + 6x + 9 = 16$

Assignment Guide and Homework Check

ASSIGNMENT

Basic: 1, 2, 3–49 odd, 71, 74, 83–86

Average: 1, 2–48 even, 49, 50–62 even, 72, 74, 83–86

Advanced: 1, 2, 8, 20–30 even, 34–48 even, 49, 50–76 even, 77–86

HOMEWORK CHECK

Basic: 9, 23, 25, 31, 37

Average: 10, 24, 26, 32, 38

Advanced: 20, 24, 28, 34, 38

ANSWERS

1. the quadratic formula; $x = \dfrac{-b \pm \sqrt{b^2 - 4ac}}{2a}$
2. $b^2 - 4ac$; the number of real solutions of a quadratic equation
3. $x^2 - 7x = 0; a = 1, b = -7, c = 0$ or $-x^2 + 7x = 0; a = -1, b = 7, c = 0$
4. $x^2 - 4x + 12 = 0; a = 1, b = -4, c = 12$ or $-x^2 + 4x - 12 = 0; a = -1, b = 4, c = -12$
5. $-2x^2 - 5x + 1 = 0; a = -2, b = -5, c = 1$ or $2x^2 + 5x - 1 = 0; a = 2, b = 5, c = -1$
6. $-4x^2 + 3x + 2 = 0; a = -4, b = 3, c = 2$ or $4x^2 - 3x - 2 = 0; a = 4, b = -3, c = -2$
7. $x^2 - 6x + 4 = 0; a = 1, b = -6, c = 4$ or $-x^2 + 6x - 4 = 0; a = -1, b = 6, c = -4$
8. $-3x^2 - 8x - 3 = 0; a = -3, b = -8, c = -3$ or $3x^2 + 8x + 3 = 0; a = 3, b = 8, c = 3$
9. $x = 6$
10. no real solutions
11. $x = 11, x = -1$
12. $x = 1, x = -\frac{1}{2}$
13. no real solutions
14. $x = \frac{1}{3}$
15. $x = \frac{3}{2}, x = \frac{2}{3}$
16. no real solutions
17. $x = \frac{1}{4}$
18. $x \approx 4.3, x \approx 0.7$
19. $x \approx 2.2, x \approx -4.2$
20. $x \approx 1.1, x \approx -0.3$
21. $x = -\frac{1}{2}, x = -4$
22. $x \approx -0.3, x \approx -0.8$
23. about 0.2 sec, about 1.4 sec
24. a. 1999, 2011

    b. no; According to the model, the amount of trout caught after 2020 is negative.
25. no real solutions
26. two real solutions
27. one real solution
28. one real solution
29. two real solutions
30. no real solutions
31. two x-intercepts
32. one x-intercept
33. no x-intercepts
34. two x-intercepts
35. two x-intercepts
36. one x-intercept
37. $x = \frac{1}{2}, x = \frac{4}{5}$; *Sample answer:* The equation is not easily factorable and $a \neq 1$, so solve using the quadratic formula.
38. $x = -5, x = 8$; *Sample answer:* The equation is easily factorable, so solve by factoring.

39–44. See Additional Answers.

Dynamic Teaching Tools

- Dynamic Assessment & Progress Monitoring Tool
- Interactive Whiteboard Lesson Library
- Dynamic Classroom with Dynamic Investigations

ANSWERS

45. $-b$ should be $-(-7)$, not -7;

$x = \dfrac{-(-7) \pm \sqrt{(-7)^2 - 4(3)(-6)}}{2(3)}$;

$x = 3$ and $x = -\dfrac{2}{3}$

46. The equation needs to be in the form $ax^2 + bx + c = 0$, so $c = -4$ not 4;

$x = \dfrac{-9 \pm \sqrt{9^2 - 4(-2)(-4)}}{2(-2)}$;

$x = \dfrac{1}{2}$ and $x = 4$

47. yes; about 42 ft, about 158 ft

48. **a.** no

b. yes; 2 days

49. no; The discriminant is -47, so the equation has no real solutions.

50. no; Substituting 4 for y in the model results in no real solutions.

51. 5; length: 13 m, width: 7 m

52. 4; length: 19 ft, width: 11 ft

53. **a–c.** $x = -2$

Sample answer: factoring; The equation is easily factorable.

54. **a–c.** $x = -\frac{2}{3}$, $x = -3$

Sample answer: quadratic formula; The equation is not easily factorable and one x-intercept is not an integer.

55. 2; When a and c have different signs, ac is negative, so the discriminant is positive.

56. rational; When the discriminant is a perfect square, the square root of the discriminant is an integer.

57. **a.** *Sample answer:* $\frac{1}{2}$

b. 1

c. *Sample answer:* 2

58. **a.** *Sample answer:* 1

b. 16

c. *Sample answer:* 17

59. **a.** *Sample answer:* 1

b. 9

c. *Sample answer:* 10

45. ERROR ANALYSIS Describe and correct the error in solving the equation $3x^2 - 7x - 6 = 0$ using the Quadratic Formula.

✗

$$x = \frac{-7 \pm \sqrt{(-7)^2 - 4(3)(-6)}}{2(3)}$$

$$= \frac{-7 \pm \sqrt{121}}{6}$$

$$x = \frac{2}{3} \text{ and } x = -3$$

46. ERROR ANALYSIS Describe and correct the error in solving the equation $-2x^2 + 9x = 4$ using the Quadratic Formula.

✗

$$x = \frac{-9 \pm \sqrt{9^2 - 4(-2)(4)}}{2(-2)}$$

$$= \frac{-9 \pm \sqrt{113}}{-4}$$

$$x \approx -0.41 \text{ and } x \approx 4.91$$

47. MODELING WITH MATHEMATICS A fountain shoots a water arc that can be modeled by the graph of the equation $y = -0.006x^2 + 1.2x + 10$, where x is the horizontal distance (in feet) from the river's north shore and y is the height (in feet) above the river. Does the water arc reach a height of 50 feet? If so, about how far from the north shore is the water arc 50 feet above the water?

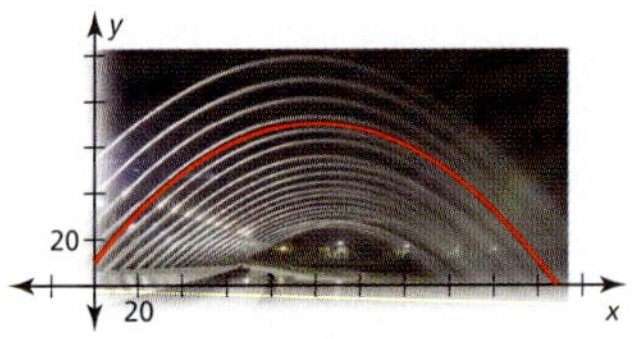

48. MODELING WITH MATHEMATICS Between the months of April and September, the number y of hours of daylight per day in Seattle, Washington, can be modeled by $y = -0.00046x^2 + 0.076x + 13$, where x is the number of days since April 1.

a. Do any of the days between April and September in Seattle have 17 hours of daylight? If so, how many?

b. Do any of the days between April and September in Seattle have 14 hours of daylight? If so, how many?

49. MAKING AN ARGUMENT Your friend uses the discriminant of the equation $2x^2 - 5x - 2 = -11$ and determines that the equation has two real solutions. Is your friend correct? Explain your reasoning.

50. MODELING WITH MATHEMATICS The frame of the tent shown is defined by a rectangular base and two parabolic arches that connect the opposite corners of the base. The graph of $y = -0.18x^2 + 1.6x$ models the height y (in feet) of one of the arches x feet along the diagonal of the base. Can a child who is 4 feet tall walk under one of the arches without having to bend over? Explain.

MATHEMATICAL CONNECTIONS In Exercises 51 and 52, use the given area A of the rectangle to find the value of x. Then give the dimensions of the rectangle.

51. $A = 91\text{ m}^2$

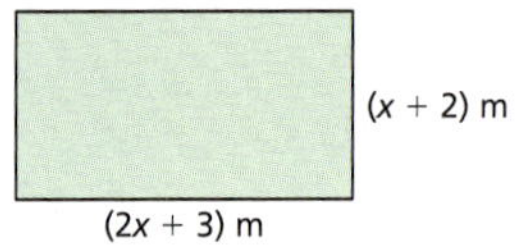

52. $A = 209\text{ ft}^2$

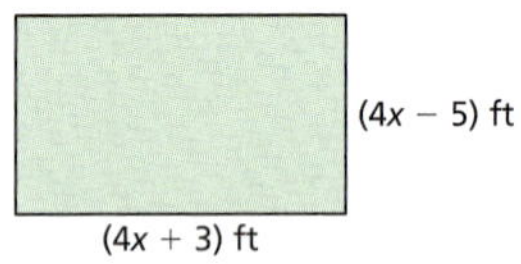

COMPARING METHODS In Exercises 53 and 54, solve the equation by (a) graphing, (b) factoring, and (c) using the Quadratic Formula. Which method do you prefer? Explain your reasoning.

53. $x^2 + 4x + 4 = 0$

54. $3x^2 + 11x + 6 = 0$

55. REASONING How many solutions does the equation $ax^2 + bx + c = 0$ have when a and c have different signs? Explain your reasoning.

56. REASONING When the discriminant is a perfect square, are the solutions of $ax^2 + bx + c = 0$ rational or irrational? (Assume a, b, and c are integers.) Explain your reasoning.

REASONING In Exercises 57–59, give a value of c for which the equation has (a) two solutions, (b) one solution, and (c) no solutions.

57. $x^2 - 2x + c = 0$

58. $x^2 - 8x + c = 0$

59. $4x^2 + 12x + c = 0$

60. **REPEATED REASONING** You use the Quadratic Formula to solve an equation.

a. You obtain solutions that are integers. Could you have used factoring to solve the equation? Explain your reasoning.

b. You obtain solutions that are fractions. Could you have used factoring to solve the equation? Explain your reasoning.

c. Make a generalization about quadratic equations with rational solutions.

61. **MODELING WITH MATHEMATICS** The fuel economy y (in miles per gallon) of a car can be modeled by the equation $y = -0.013x^2 + 1.25x + 5.6$, where $5 \le x \le 75$ and x is the speed (in miles per hour) of the car. Find the speed(s) at which you can travel and have a fuel economy of 32 miles per gallon.

62. **MODELING WITH MATHEMATICS** The depth d (in feet) of a river can be modeled by the equation $d = -0.25t^2 + 1.7t + 3.5$, where $0 \le t \le 7$ and t is the time (in hours) after a heavy rain begins. When is the river 6 feet deep?

ANALYZING EQUATIONS In Exercises 63–68, tell whether the vertex of the graph of the function lies above, below, or on the x-axis. Explain your reasoning without using a graph.

63. $y = x^2 - 3x + 2$

64. $y = 3x^2 - 6x + 3$

65. $y = 6x^2 - 2x + 4$

66. $y = -15x^2 + 10x - 25$

67. $f(x) = -3x^2 - 4x + 8$

68. $f(x) = 9x^2 - 24x + 16$

69. **REASONING** NASA creates a weightless environment by flying a plane in a series of parabolic paths. The height h (in feet) of a plane after t seconds in a parabolic flight path can be modeled by $h = -11t^2 + 700t + 21{,}000$. The passengers experience a weightless environment when the height of the plane is greater than or equal to 30,800 feet. For approximately how many seconds do passengers experience weightlessness on such a flight? Explain.

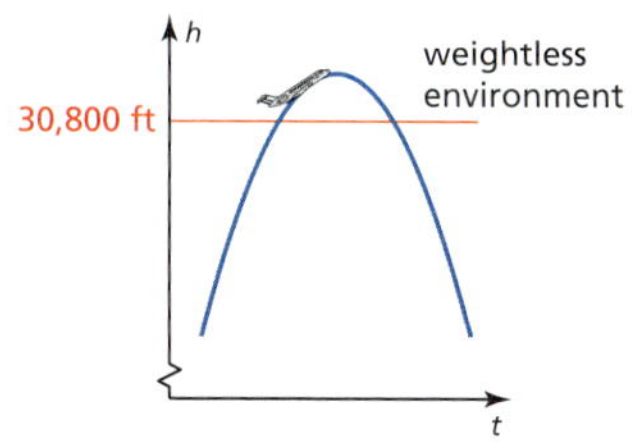

70. **WRITING EQUATIONS** Use the numbers to create a quadratic equation with the solutions $x = -1$ and $x = -\frac{1}{4}$.

$\_\_x^2 + \_\_x + \_\_ = 0$

−5	−4	−3	−2	−1
1	2	3	4	5

71. **PROBLEM SOLVING** A rancher constructs two rectangular horse pastures that share a side, as shown. The pastures are enclosed by 1050 feet of fencing. Each pasture has an area of 15,000 square feet.

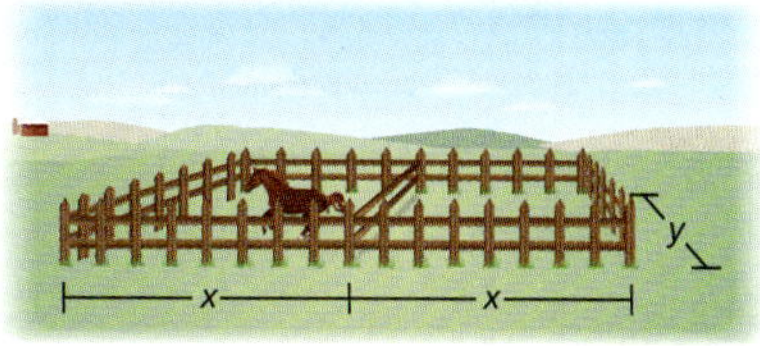

a. Show that $y = 350 - \frac{4}{3}x$.

b. Find the possible lengths and widths of each pasture.

72. **PROBLEM SOLVING** A kicker punts a football from a height of 2.5 feet above the ground with an initial vertical velocity of 45 feet per second.

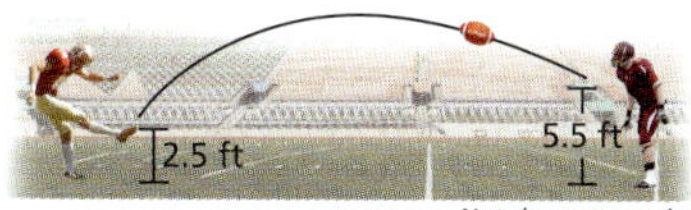

Not drawn to scale

a. Write an equation that models this situation using the function $h = -16t^2 + v_0t + s_0$, where h is the height (in feet) of the football, t is the time (in seconds) after the football is punted, v_0 is the initial vertical velocity (in feet per second), and s_0 is the initial height (in feet).

b. The football is caught 5.5 feet above the ground, as shown in the diagram. Find the amount of time that the football is in the air.

73. **CRITICAL THINKING** The solutions of the quadratic equation $ax^2 + bx + c = 0$ are $x = \frac{-b + \sqrt{b^2 - 4ac}}{2a}$ and $x = \frac{-b - \sqrt{b^2 - 4ac}}{2a}$. Find the mean of the solutions. How is the mean of the solutions related to the graph of $y = ax^2 + bx + c$? Explain.

ANSWERS

60. a. yes; The equation can be written in intercept form.

b. yes; The equation can be written in intercept form.

c. A quadratic equation with rational solutions can be solved by factoring.

61. about 31 mi/h, about 65 mi/h

62. after about 2.2 h, after about 4.6 h

63. below the x-axis; The discriminant is positive and $a > 0$.

64. on the x-axis; The discriminant is zero.

65. above the x-axis; The discriminant is negative and $a > 0$.

66. below the x-axis; The discriminant is negative and $a < 0$.

67. above the x-axis; The discriminant is positive and $a < 0$.

68. on the x-axis; The discriminant is zero.

69. about 22 sec; The height is 30,800 feet after about 20.8 seconds and after about 42.8 seconds.

70. 4; 5; 1 or −4; −5; −1

71. a. $4x + 3y = 1050$; $3y = 1050 - 4x$; $y = 350 - \frac{4}{3}x$

b. length: about 54 ft, width: about 278 ft; length: about 209 ft, width: about 72 ft

72. a. $h = -16t^2 + 45t + 2.5$

b. about 2.74 sec

73. $-\frac{b}{2a}$; The mean of the solutions is the x-coordinate of the vertex; The mean of the solutions is equal to the graph's axis of symmetry, which is where the vertex lies.

ANSWERS

74. **a.** C; The graph has two x-intercepts.

b. A; The graph has one x-intercept.

c. B; The graph has no x-intercepts.

75. about 24.7 ft/sec

76. $x = -\dfrac{b}{2a}$ is the axis of symmetry, and $\dfrac{\sqrt{b^2 - 4ac}}{2a}$ is the horizontal distance from the axis of symmetry to each x-intercept.

77. $\dfrac{-b}{a}$; $\dfrac{c}{a}$; *Sample answer:* $2x^2 - 4x + 1 = 0$

78. $x = \dfrac{-1 \pm \sqrt{1 - 4ac}}{2a}$; $x \approx -1.77$, $x \approx 2.27$

79. See Additional Answers.

80. **a.** $k < \frac{1}{24}$

b. $k = \frac{1}{24}$

c. $k > \frac{1}{24}$

81. **a.** $k < -3$ or $k > 3$

b. $k = -3$ or $k = 3$

c. $-3 < k < 3$

82–86. See Additional Answers.

Mini-Assessment

1. Solve $2x^2 - 5x + 3 = 0$ using the Quadratic Formula.
   $x = \frac{3}{2}$ and $x = 1$
2. Determine the number of real solutions of $-x^2 - 25 = -10x$.
   one real solution
3. Find the number of x-intercepts of the graph of $y = x^2 - 7x + 14$.
   no x-intercepts

Solve the equation using any method. Explain your choice of method.

4. $x^2 - 6x = 4$
   $x = 3 + \sqrt{13} \approx 6.6$ and $x = 3 - \sqrt{13} \approx -0.6$; The coefficient of the x^2-term is 1, and the coefficient of the x-term is an even number. So, solve by completing the square.
5. $x^2 - 2x = 15$
   $x = -3$ and $x = 5$; The equation is easily factorable to $(x - 5)(x + 3) = 0$. So, solve by factoring.

74. HOW DO YOU SEE IT? Match each graph with its discriminant. Explain your reasoning.

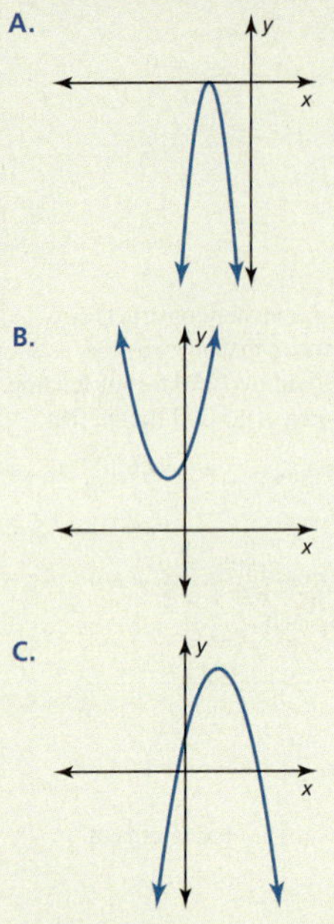

a. $b^2 - 4ac > 0$

b. $b^2 - 4ac = 0$

c. $b^2 - 4ac < 0$

75. CRITICAL THINKING You are trying to hang a tire swing. To get the rope over a tree branch that is 15 feet high, you tie the rope to a weight and throw it over the branch. You release the weight at a height s_0 of 5.5 feet. What is the minimum initial vertical velocity v_0 needed to reach the branch? (*Hint*: Use the equation $h = -16t^2 + v_0t + s_0$.)

76. THOUGHT PROVOKING Consider the graph of the standard form of a quadratic function $y = ax^2 + bx + c$. Then consider the Quadratic Formula as given by

$$x = -\frac{b}{2a} \pm \frac{\sqrt{b^2 - 4ac}}{2a}.$$

Write a graphical interpretation of the two parts of this formula.

77. ANALYZING RELATIONSHIPS Find the sum and product of $\dfrac{-b + \sqrt{b^2 - 4ac}}{2a}$ and $\dfrac{-b - \sqrt{b^2 - 4ac}}{2a}$. Then write a quadratic equation whose solutions have a sum of 2 and a product of $\frac{1}{2}$.

78. WRITING A FORMULA Derive a formula that can be used to find solutions of equations that have the form $ax^2 + x + c = 0$. Use your formula to solve $-2x^2 + x + 8 = 0$.

79. MULTIPLE REPRESENTATIONS If p is a solution of a quadratic equation $ax^2 + bx + c = 0$, then $(x - p)$ is a factor of $ax^2 + bx + c$.

a. Copy and complete the table for each pair of solutions.

Solutions	Factors	Quadratic equation
3, 4	$(x - 3), (x - 4)$	$x^2 - 7x + 12 = 0$
$-1, 6$		
0, 2		
$-\frac{1}{2}, 5$		

b. Graph the related function for each equation. Identify the zeros of the function.

CRITICAL THINKING **In Exercises 80–82, find all values of k for which the equation has (a) two solutions, (b) one solution, and (c) no solutions.**

80. $2x^2 + x + 3k = 0$

81. $x^2 - 4kx + 36 = 0$

82. $kx^2 + 5x - 16 = 0$

Maintaining Mathematical Proficiency

Reviewing what you learned in previous grades and lessons

Solve the system of linear equations using any method. Explain why you chose the method. *(Section 5.1, Section 5.2, and Section 5.3)*

83. $y = -x + 4$
$y = 2x - 8$

84. $x = 16 - 4y$
$3x + 4y = 8$

85. $2x - y = 7$
$2x + 7y = 31$

86. $3x - 2y = -20$
$x + 1.2y = 6.4$

If students need help...	*If students got it...*
Resources by Chapter • Practice A and Practice B • Puzzle Time	Resources by Chapter • Enrichment and Extension • Cumulative Review
Student Journal • Practice	Start the *next* Section
Differentiating the Lesson Skills Review Handbook	

Laurie's Notes

Overview of Section 9.6

Introduction

- This lesson revisits techniques studied in Chapter 5 for solving systems of linear equations: graphing, substitution, and elimination.
- The systems in this lesson contain at least one nonlinear equation. In some cases, the solutions are not rational and students will need to approximate the solution.

Formative Assessment Tips

- **Partner Speaks:** This strategy provides a student the opportunity to share his or her thinking about a problem or concept with a partner. The partner then shares this thinking with the whole class.
- Thinking through a problem with a partner is less intimidating for many students than sharing it with an entire class. For the listener, he or she needs to pay attention to the thoughts of the speaker and set aside his or her own thinking about the problem. As students are engaged in this dialogue, circulate to hear the discussion and to gain an understanding of how students are thinking about the problem or concept. You will be able to judge appropriate next steps.
- The opportunity to speak to a partner and then have him or her share the student's thinking should not be used for simple, less challenging problems. It should be used as a way to pair students differently than always having a conversation with the person to your right.

Pacing Suggestion

- The explorations provide an opportunity for students to recall methods for solving a system of equations. Transition to the formal lesson as soon as students have discussed each exploration.

Dynamic Teaching Tools

- Dynamic Assessment & Progress Monitoring Tool
- Lesson Planning Tool
- Interactive Whiteboard Lesson Library
- Dynamic Classroom with Dynamic Investigations

What Your Students Will Learn

- Solve systems of nonlinear equations in two variables by graphing.
- Solve systems of nonlinear equations in two variables algebraically by substitution and by elimination.
- Approximate solutions of nonlinear systems and equations.

Laurie's Notes

Exploration

Motivate

- **Story Time:** Share a story about shooting clay pigeons launched into the air by a machine. In the last round, you hit the target on the way down. (Sketch the diagram shown below.)

- Connect this to today's exploration. Explain that you prefer to hit the target on the way up, because it is closer to you.

Discuss

? "What is a system of linear equations?" a set of two or more linear equations in the same variables

? "How do you solve a system of linear equations?" graphing, substitution, or elimination

- Explain that today's exploration is about systems of equations that include quadratic equations.

Exploration 1

- As students graph the two equations, you should hear comments about slope, y-intercepts, the parabola opening up, and so on.
- A careful graph of the two equations should allow the solutions to be read.
- **Attend to Precision:** Students should confirm the solutions by substituting them into each equation to verify that they are indeed solutions.

Exploration 2

- In Exploration 1, the focus was on the different solution methods. In Exploration 2, the focus is on the number of solutions.
- The graphs visually suggest three different cases for the number of solutions. These are similar to the three cases for the number of x-intercepts of the graph of a quadratic function.
- **Attend to Precision:** Students could approximate the solution(s) for each system. Instead, make sure students find the exact solution(s).

? "What happened when you solved the system in part (c) algebraically?" found that the system has no real solutions

Communicate Your Answer

- Question 4 takes time, and students are likely to begin by using trial and error.

? "In Question 4, which of the three cases was the most challenging and why?" Answers will vary. Many students will say that finding a system with one solution was the most challenging.

Connecting to Next Step

- The explorations should remind students of the various ways in which systems of equations are solved. In the formal lesson, all the methods will be reviewed.

9.6 Solving Nonlinear Systems of Equations

Essential Question How can you solve a system of two equations when one is linear and the other is quadratic?

EXPLORATION 1 **Solving a System of Equations**

Work with a partner. Solve the system of equations by graphing each equation and finding the points of intersection.

System of Equations

$y = x + 2$ Linear

$y = x^2 + 2x$ Quadratic

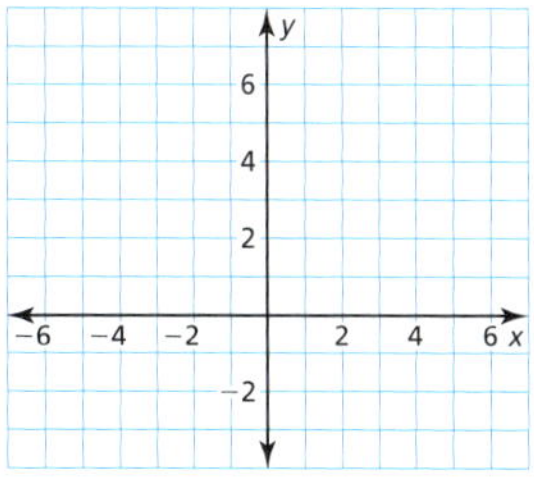

EXPLORATION 2 **Analyzing Systems of Equations**

Work with a partner. Match each system of equations with its graph. Then solve the system of equations.

a. $y = x^2 - 4$
$y = -x - 2$

b. $y = x^2 - 2x + 2$
$y = 2x - 2$

c. $y = x^2 + 1$
$y = x - 1$

d. $y = x^2 - x - 6$
$y = 2x - 2$

A.

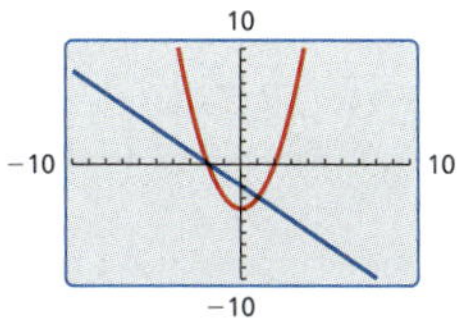

B.

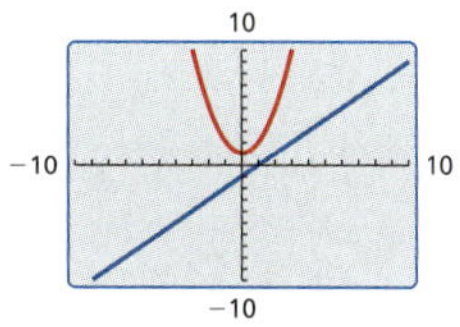

C.

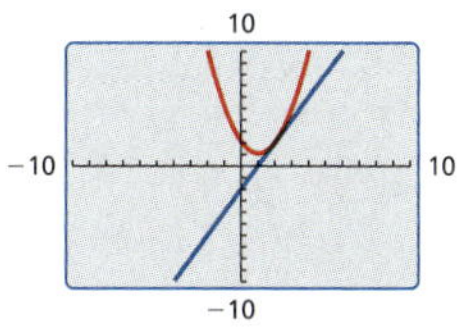

D.

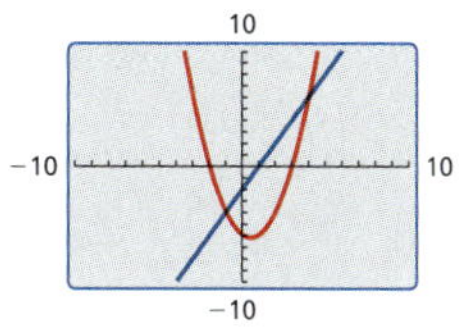

MAKING SENSE OF PROBLEMS

To be proficient in math, you need to analyze givens, relationships, and goals.

Communicate Your Answer

3. How can you solve a system of two equations when one is linear and the other is quadratic?

4. Write a system of equations (one linear and one quadratic) that has (a) no solutions, (b) one solution, and (c) two solutions. Your systems should be different from those in Explorations 1 and 2.

Dynamic Teaching Tools

- Dynamic Assessment & Progress Monitoring Tool
- Lesson Planning Tool
- Interactive Whiteboard Lesson Library
- Dynamic Classroom with Dynamic Investigations

ANSWERS

1. $(-2, 0)$, $(1, 3)$

2. a. A; $(-2, 0)$, $(1, -3)$

b. C; $(2, 2)$

c. B; no solutions

d. D; $(4, 6)$, $(-1, -4)$

3. Graph the equations in the same coordinate plane and find the point(s) of intersection.

4. a. *Sample answer:* $y = x + 1$, $y = x^2 + 5$

b. *Sample answer:* $y = 5$, $y = x^2 + 5$

c. *Sample answer:* $y = x + 8$, $y = x^2 + 5$

English Language Learners

Build on Past Knowledge

Remind students that a solution for a system of linear equations is an ordered pair that is a solution of each equation in the system. A linear system can have no solution, one solution, or infinitely many solutions. Explain that a nonlinear system that involves a linear equation and a quadratic equation may have no solution, one solution, or two solutions.

Extra Example 1

Solve the system by graphing.

$y = x^2 - 6x + 5$

$y = x - 1$

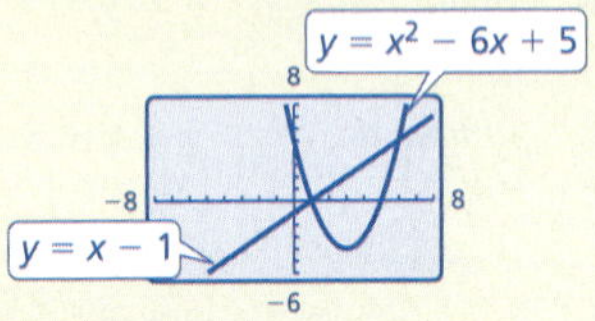

The solutions are (1, 0) and (6, 5).

MONITORING PROGRESS ANSWERS

1. $(-1, -7)$
2. no solutions
3. $(2, -9), (8, 9)$

9.6 Lesson

Core Vocabulary

system of nonlinear equations, *p. 526*

Previous

system of linear equations

What You Will Learn

- Solve systems of nonlinear equations by graphing.
- Solve systems of nonlinear equations algebraically.
- Approximate solutions of nonlinear systems and equations.

Solving Nonlinear Systems by Graphing

The methods for solving systems of linear equations can also be used to solve *systems of nonlinear equations*. A **system of nonlinear equations** is a system in which at least one of the equations is nonlinear.

When a nonlinear system consists of a linear equation and a quadratic equation, the graphs can intersect in zero, one, or two points. So, the system can have zero, one, or two solutions, as shown.

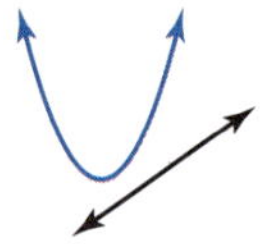

No solutions

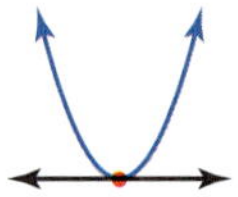

One solution

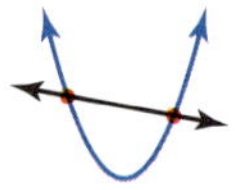

Two solutions

EXAMPLE 1 Solving a Nonlinear System by Graphing

Solve the system by graphing.

$y = 2x^2 + 5x - 1$ Equation 1

$y = x - 3$ Equation 2

SOLUTION

Step 1 Graph each equation.

Step 2 Estimate the point of intersection. The graphs appear to intersect at $(-1, -4)$.

Step 3 Check the point from Step 2 by substituting the coordinates into each of the original equations.

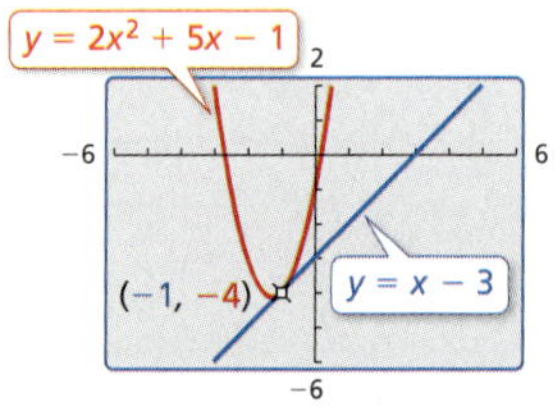

Equation 1	Equation 2
$y = 2x^2 + 5x - 1$	$y = x - 3$
$-4 \stackrel{?}{=} 2(-1)^2 + 5(-1) - 1$	$-4 \stackrel{?}{=} -1 - 3$
$-4 = -4$ ✓	$-4 = -4$ ✓

▶ The solution is $(-1, -4)$.

Monitoring Progress

Help in English and Spanish at *BigIdeasMath.com*

Solve the system by graphing.

1. $y = x^2 + 4x - 4$
$y = 2x - 5$

2. $y = -x + 6$
$y = -2x^2 - x + 3$

3. $y = 3x - 15$
$y = \frac{1}{2}x^2 - 2x - 7$

Laurie's Notes Teacher Actions

- From the explorations, students know what a *system of nonlinear equations* is and strategies for solving the system.
- **?** Pose Example 1. "Can you determine the solution(s) in a standard viewing window?" Answers will vary. Students may say no and that they need to zoom in.
- Students can zoom in near the point of intersection and check the table of values. Students should be able to reason why there is only one point of intersection and not two—two that would be very close together.

REMEMBER

The algebraic procedures that you use to solve nonlinear systems are similar to the procedures that you used to solve linear systems in Sections 5.2 and 5.3.

Solving Nonlinear Systems Algebraically

EXAMPLE 2 Solving a Nonlinear System by Substitution

Solve the system by substitution.

$y = x^2 + x - 1$ Equation 1

$y = -2x + 3$ Equation 2

SOLUTION

Step 1 The equations are already solved for y.

Step 2 Substitute $-2x + 3$ for y in Equation 1 and solve for x.

$-2x + 3 = x^2 + x - 1$ Substitute $-2x + 3$ for y in Equation 1.

$3 = x^2 + 3x - 1$ Add $2x$ to each side.

$0 = x^2 + 3x - 4$ Subtract 3 from each side.

$0 = (x + 4)(x - 1)$ Factor the polynomial.

$x + 4 = 0$ or $x - 1 = 0$ Zero-Product Property

$x = -4$ or $x = 1$ Solve for x.

Step 3 Substitute -4 and 1 for x in Equation 2 and solve for y.

$y = -2(-4) + 3$ Substitute for x in Equation 2. $y = -2(1) + 3$

$= 11$ Simplify. $= 1$

▶ So, the solutions are $(-4, 11)$ and $(1, 1)$.

Check

Use a graphing calculator to check your answer. Notice that the graphs have two points of intersection at $(-4, 11)$ and $(1, 1)$.

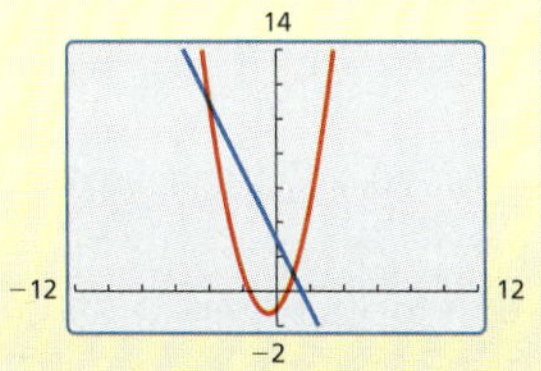

EXAMPLE 3 Solving a Nonlinear System by Elimination

Solve the system by elimination.

$y = x^2 - 3x - 2$ Equation 1

$y = -3x - 8$ Equation 2

SOLUTION

Step 1 Because the coefficients of the y-terms are the same, you do not need to multiply either equation by a constant.

Step 2 Subtract Equation 2 from Equation 1.

$y = x^2 - 3x - 2$ Equation 1

$y = \quad -3x - 8$ Equation 2

$0 = x^2 \quad + 6$ Subtract the equations.

Step 3 Solve for x.

$0 = x^2 + 6$ Resulting equation from Step 2

$-6 = x^2$ Subtract 6 from each side.

▶ The square of a real number cannot be negative. So, the system has no real solutions.

Check

Use a graphing calculator to check your answer. The graphs do not intersect.

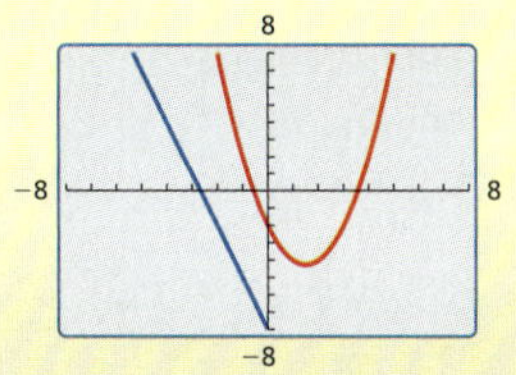

Differentiated Instruction

Auditory

Have students work in pairs to solve nonlinear systems algebraically. Ask one student to solve a system by substitution or by elimination and the other student to describe the steps in the process. Have students alternate solving and describing the process used as they continue to solve several nonlinear systems of equations.

Extra Example 2

Solve the system by substitution.

$y = x^2 + 4x$

$y = -4$

The solution is $(-2, -4)$.

Extra Example 3

Solve the system by elimination.

$y = x^2 + 2x - 5$

$y = 2x - 1$

The solutions are $(-2, -5)$ and $(2, 3)$.

Laurie's Notes Teacher Actions

- In Example 2, the equations are already solved for y. Solving by substitution should make sense.
- **Turn and Talk:** "Solve the system to determine whether there are two, one, or no solutions." Circulate, giving *Wait Time* for students to work through the problem. Solicit a solution.

? "How is Example 3 different from Example 2?" Answers will vary. Students should recognize that the equations have the same linear terms.

- Solving by the elimination method should make sense.

? "How do you interpret the result $x^2 = -6$?" There are no real solutions.

Extra Example 4

Approximate the solution(s) of the system to the nearest thousandth.

$y = x^2 - 2x - 5$

$y = 2^x + 1$

The solution of the system is about (–1.703, 1.307).

MONITORING PROGRESS ANSWERS

4. (0, 9)
5. (−3, 15), (1, −5)
6. no solutions
7. about (2.24, 7.24), about (−2.24, 2.76)
8. $\left(\frac{1}{3}, -\frac{7}{3}\right), \left(-\frac{2}{3}, -\frac{22}{3}\right)$
9. no solutions

Monitoring Progress

Help in English and Spanish at *BigIdeasMath.com*

Solve the system by substitution.

4. $y = x^2 + 9$, $y = 9$

5. $y = -5x$, $y = x^2 - 3x - 3$

6. $y = -3x^2 + 2x + 1$, $y = 5 - 3x$

Solve the system by elimination.

7. $y = x^2 + x$, $y = x + 5$

8. $y = 9x^2 + 8x - 6$, $y = 5x - 4$

9. $y = 2x + 5$, $y = -3x^2 + x - 4$

Approximating Solutions

When you cannot find the exact solution(s) of a system of equations, you can analyze output values to approximate the solution(s).

EXAMPLE 4 Approximating Solutions of a Nonlinear System

Approximate the solution(s) of the system to the nearest thousandth.

$y = \frac{1}{2}x^2 + 3$ — Equation 1

$y = 3^x$ — Equation 2

SOLUTION

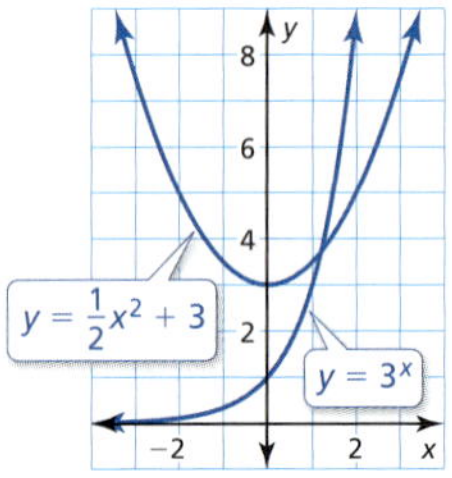

Sketch a graph of the system. You can see that the system has one solution between $x = 1$ and $x = 2$.

Substitute 3^x for y in Equation 1 and rewrite the equation.

$3^x = \frac{1}{2}x^2 + 3$ — Substitute 3^x for y in Equation 1.

$3^x - \frac{1}{2}x^2 - 3 = 0$ — Rewrite the equation.

Because you do not know how to solve this equation algebraically, let $f(x) = 3^x - \frac{1}{2}x^2 - 3$. Then evaluate the function for x-values between 1 and 2.

$f(1.1) \approx -0.26$

$f(1.2) \approx 0.02$

Because $f(1.1) < 0$ and $f(1.2) > 0$, the zero is between 1.1 and 1.2.

REMEMBER
The function values that are closest to 0 correspond to x-values that best approximate the zeros of the function.

$f(1.2)$ is closer to 0 than $f(1.1)$, so decrease your guess and evaluate $f(1.19)$.

$f(1.19) \approx -0.012$ — Because $f(1.19) < 0$ and $f(1.2) > 0$, the zero is between 1.19 and 1.2. So, increase guess.

$f(1.191) \approx -0.009$ — Result is negative. Increase guess.

$f(1.192) \approx -0.006$ — Result is negative. Increase guess.

$f(1.193) \approx -0.003$ — Result is negative. Increase guess.

$f(1.194) \approx -0.0002$ — Result is negative. Increase guess.

$f(1.195) \approx 0.003$ — Result is positive.

Because $f(1.194)$ is closest to 0, $x \approx 1.194$.

Substitute $x = 1.194$ into one of the original equations and solve for y.

$y = \frac{1}{2}x^2 + 3 = \frac{1}{2}(1.194)^2 + 3 \approx 3.713$

▶ So, the solution of the system is about (1.194, 3.713).

Laurie's Notes Teacher Actions

- **Think-Pair-Share:** Have students answer Questions 4–9, as needed, and then share and discuss as a class.
- ? **Reason Abstractly and Quantitatively** and **Attend to Precision:** Pose Example 4. "Describe this system. What are the intersection possibilities?" The system has a quadratic equation and exponential equation. There may be 0, 1, 2, or 3 points of intersection.
- Have students use the table feature of their calculators to find an approximate solution. They should be able to explain their reasoning as they get closer and closer to an approximation of the solution.

REMEMBER

When entering the equations, be sure to use an appropriate viewing window that shows all the points of intersection. For this system, an appropriate viewing window is $-4 \le x \le 4$ and $-4 \le y \le 4$.

Recall from Section 5.5 that you can use systems of equations to solve equations with variables on both sides.

EXAMPLE 5 Approximating Solutions of an Equation

Solve $-2(4)^x + 3 = 0.5x^2 - 2x$.

SOLUTION

You do not know how to solve this equation algebraically. So, use each side of the equation to write the system $y = -2(4)^x + 3$ and $y = 0.5x^2 - 2x$.

Method 1 Use a graphing calculator to graph the system. Then use the *intersect* feature to find the coordinates of each point of intersection.

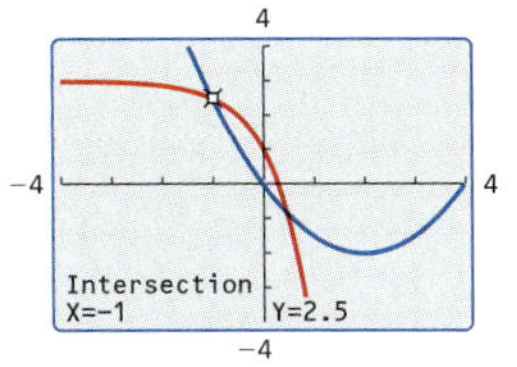

One point of intersection is $(-1, 2.5)$.

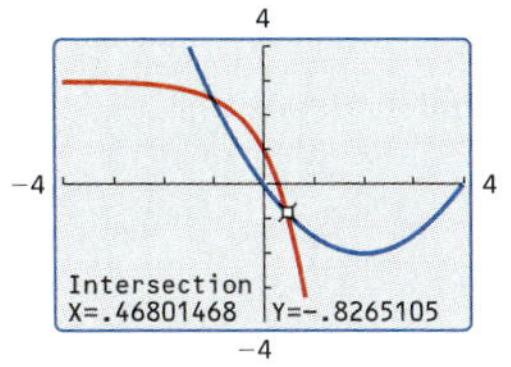

The other point of intersection is about $(0.47, -0.83)$.

▶ So, the solutions of the equation are $x = -1$ and $x \approx 0.47$.

STUDY TIP

You can use the differences between the corresponding *y*-values to determine the best approximation of a solution.

Method 2 Use the *table* feature to create a table of values for the equations. Find the *x*-values for which the corresponding *y*-values are approximately equal.

X	Y1	Y2
-1.03	2.5204	2.5905
-1.02	2.5137	2.5602
-1.01	2.5069	2.5301
-1	2.5	2.5
-.99	2.493	2.4701
-.98	2.4859	2.4402
-.97	2.4788	2.4105
X=-1		

When $x = -1$, the corresponding *y*-values are 2.5.

X	Y1	Y2
.44	-.6808	-.7832
.45	-.7321	-.7988
.46	-.7842	-.8142
.47	-.8371	-.8296
.48	-.8906	-.8448
.49	-.9449	-.86
.50	-1	-.875
X=.47		

When $x = 0.47$, the corresponding *y*-values are approximately -0.83.

▶ So, the solutions of the equation are $x = -1$ and $x \approx 0.47$.

Monitoring Progress

Help in English and Spanish at *BigIdeasMath.com*

Use the method in Example 4 to approximate the solution(s) of the system to the nearest thousandth.

10. $y = 4^x$
$y = x^2 + x + 3$

11. $y = 4x^2 - 1$
$y = -2(3)^x + 4$

12. $y = x^2 + 3x$
$y = -x^2 + x + 10$

Solve the equation. Round your solution(s) to the nearest hundredth.

13. $3^x - 1 = x^2 - 2x + 5$

14. $4x^2 + x = -2\left(\frac{1}{2}\right)^x + 5$

Extra Example 5

Solve $-(3^x) + 3 = x^2 - x$.
The solutions of the equation are $x \approx -1.23$ and $x = 1$.

MONITORING PROGRESS ANSWERS

10. about (1.285, 5.938)

11. about (−1.045, 3.368), about (0.565, 0.278)

12. about (1.791, 8.581), about (−2.791, −0.583)

13. $x \approx 1.51$

14. $x \approx 0.87, x \approx -0.77$

Laurie's Notes Teacher Actions

- Example 5 should remind students of an earlier lesson. Pose the problem and say, "Work independently to solve this equation. Then you will have time to share your thinking with partners."
- **Partner Speaks:** After students have had sufficient time, pair students. Only one person speaks. The listener asks for clarification or gives feedback. Ask the listener to describe the solution method used by his or her partner.

Closure

- **Exit Ticket:** Solve the system using any method.
$y = 2x + 2$
$y = x^2 - x - 2$ $(-1, 0)$ and $(4, 10)$

Assignment Guide and Homework Check

ASSIGNMENT

Basic: 1, 2, 3–45 odd, 52, 56, 63–70

Average: 1, 2–56 even, 63–70

Advanced: 1, 2, 12, 18, 24–30 even, 38, 44–54 even, 55–70

HOMEWORK CHECK

Basic: 7, 13, 19, 33, 39

Average: 8, 14, 20, 34, 40

Advanced: 12, 18, 24, 38, 44

ANSWERS

1. Solve one of the equations for one of the variables and substitute into the other equation and solve.
2. *Sample answer:* Both can be solved by graphing; Some nonlinear systems cannot be solved algebraically.
3. B; (0, 1), (3, 4)
4. D; no solutions
5. A; (0, −1)
6. C; (−2, 5), (1, 2)
7. (2, 9), (−1, 6)
8. no solutions
9. (−1, 2)
10. (6, 4), (2, 0)
11. (3, 6), (−3, −6)
12. (−3, 14), (1, 2)
13. (0, −5), (−3, −8)
14. (−1, −3)
15. no solutions
16. (−3, −2), (1, 6)
17. (0, 5)
18. (2, 10), $\left(-\frac{3}{2}, -4\right)$
19. (4, −11), (−4, 29)
20. no solutions
21. (0, 2), (−6, −22)
22. no solutions
23. (1, 1)
24. no solutions
25. about (0.87, −1.74), about (−6.87, 13.74)
26. about (6.19, 5.38), about (0.81, −5.38)

9.6 Exercises

Dynamic Solutions available at *BigIdeasMath.com*

Vocabulary and Core Concept Check

1. **VOCABULARY** Describe how to use substitution to solve a system of nonlinear equations.

2. **WRITING** How is solving a system of nonlinear equations similar to solving a system of linear equations? How is it different?

Monitoring Progress and Modeling with Mathematics

In Exercises 3–6, match the system of equations with its graph. Then solve the system.

3. $y = x^2 - 2x + 1$, $y = x + 1$
4. $y = x^2 + 3x + 2$, $y = -x - 3$
5. $y = x - 1$, $y = -x^2 + x - 1$
6. $y = -x + 3$, $y = -x^2 - 2x + 5$

A.
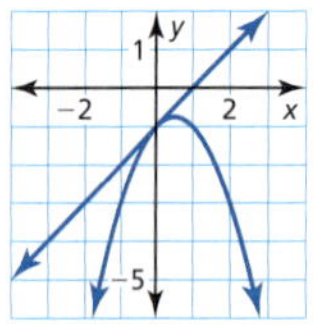

B.
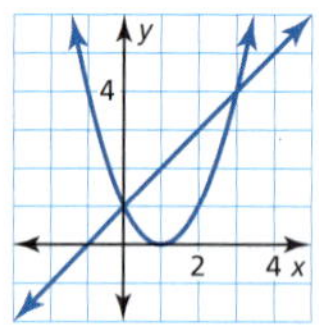

C.
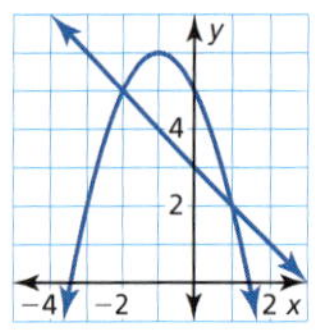

D.
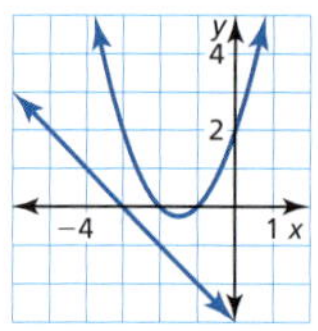

In Exercises 7–12, solve the system by graphing. *(See Example 1.)*

7. $y = 3x^2 - 2x + 1$, $y = x + 7$
8. $y = x^2 + 2x + 5$, $y = -2x - 5$
9. $y = -2x^2 - 4x$, $y = 2$
10. $y = \frac{1}{2}x^2 - 3x + 4$, $y = x - 2$
11. $y = \frac{1}{3}x^2 + 2x - 3$, $y = 2x$
12. $y = 4x^2 + 5x - 7$, $y = -3x + 5$

In Exercises 13–18, solve the system by substitution. *(See Example 2.)*

13. $y = x - 5$, $y = x^2 + 4x - 5$
14. $y = -3x^2$, $y = 6x + 3$
15. $y = -x + 7$, $y = -x^2 - 2x - 1$
16. $y = -x^2 + 7$, $y = 2x + 4$
17. $y - 5 = -x^2$, $y = 5$
18. $y = 2x^2 + 3x - 4$, $y - 4x = 2$

In Exercises 19–26, solve the system by elimination. *(See Example 3.)*

19. $y = x^2 - 5x - 7$, $y = -5x + 9$
20. $y = -3x^2 + x + 2$, $y = x + 4$
21. $y = -x^2 - 2x + 2$, $y = 4x + 2$
22. $y = -2x^2 + x - 3$, $y = 2x - 2$
23. $y = 2x - 1$, $y = x^2$
24. $y = x^2 + x + 1$, $y = -x - 2$
25. $y + 2x = 0$, $y = x^2 + 4x - 6$
26. $y = 2x - 7$, $y + 5x = x^2 - 2$

27. **ERROR ANALYSIS** Describe and correct the error in solving the system of equations by graphing.

✗ $y = x^2 - 3x + 4$
$y = 2x + 4$
The only solution of the system is (0, 4).

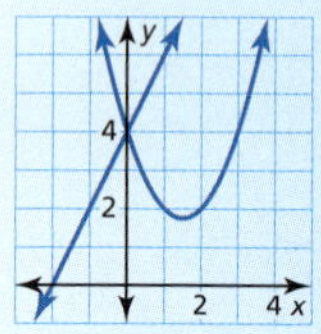

27. The graph does not show both solutions.

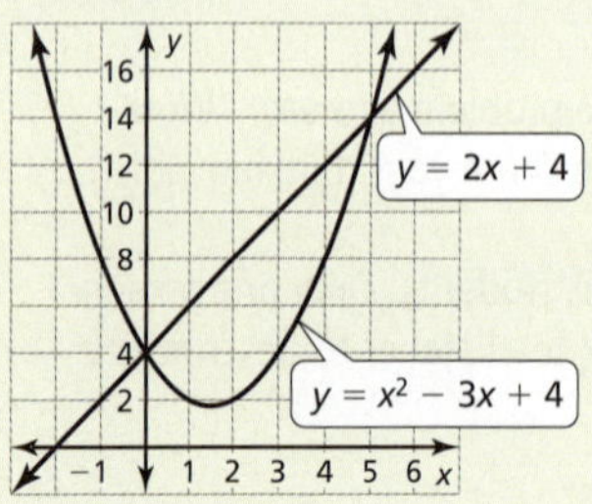

The solutions are (0, 4) and (5, 14).

28. ERROR ANALYSIS Describe and correct the error in solving for one of the variables in the system.

> ✗
> $y = 3x^2 - 6x + 4$
> $y = 4$
>
> $y = 3(4)^2 - 6(4) + 4$ Substitute.
> $y = 28$ Simplify.

In Exercises 29–32, use the table to describe the locations of the zeros of the quadratic function *f*.

29.

x	−4	−3	−2	−1	0	1
$f(x)$	−2	2	4	4	2	−2

30.

x	−1	0	1	2	3	4
$f(x)$	11	5	1	−1	−1	1

31.

x	−4	−3	−2	−1	0	1
$f(x)$	3	−1	−1	3	11	23

32.

x	1	2	3	4	5	6
$f(x)$	−25	−9	1	5	3	−5

In Exercises 33–38, use the method in Example 4 to approximate the solution(s) of the system to the nearest thousandth. *(See Example 4.)*

33. $y = x^2 + 2x + 3$
$y = 3^x$

34. $y = 2^x + 5$
$y = x^2 - 3x + 1$

35. $y = 2(4)^x - 1$
$y = 3x^2 + 8x$

36. $y = -x^2 - 4x - 4$
$y = -5^x - 2$

37. $y = -x^2 - x + 5$
$y = 2x^2 + 6x - 3$

38. $y = 2x^2 + x - 8$
$y = x^2 - 5$

In Exercises 39–46, solve the equation. Round your solution(s) to the nearest hundredth. *(See Example 5.)*

39. $3x + 1 = x^2 + 7x - 1$

40. $-x^2 + 2x = -2x + 5$

41. $x^2 - 6x + 4 = -x^2 - 2x$

42. $2x^2 + 8x + 10 = -x^2 - 2x + 5$

43. $-4\left(\frac{1}{2}\right)^x = -x^2 - 5$

44. $1.5(2)^x - 3 = -x^2 + 4x$

45. $8^{x-2} + 3 = 2\left(\frac{3}{2}\right)^x$

46. $-0.5(4)^x = 5^x - 6$

47. COMPARING METHODS Solve the system in Exercise 37 using substitution. Compare the exact solutions to the approximated solutions.

48. COMPARING METHODS Solve the system in Exercise 38 using elimination. Compare the exact solutions to the approximated solutions.

49. MODELING WITH MATHEMATICS The attendances y for two movies can be modeled by the following equations, where x is the number of days since the movies opened.

$y = -x^2 + 35x + 100$ Movie A

$y = -5x + 275$ Movie B

When is the attendance for each movie the same?

50. MODELING WITH MATHEMATICS You and a friend are driving boats on the same lake. Your path can be modeled by the equation $y = -x^2 - 4x - 1$, and your friend's path can be modeled by the equation $y = 2x + 8$. Do your paths cross each other? If so, what are the coordinates of the point(s) where the paths meet?

51. MODELING WITH MATHEMATICS The arch of a bridge can be modeled by $y = -0.002x^2 + 1.06x$, where x is the distance (in meters) from the left pylons and y is the height (in meters) of the arch above the water. The road can be modeled by the equation $y = 52$. To the nearest meter, how far from the left pylons are the two points where the road intersects the arch of the bridge?

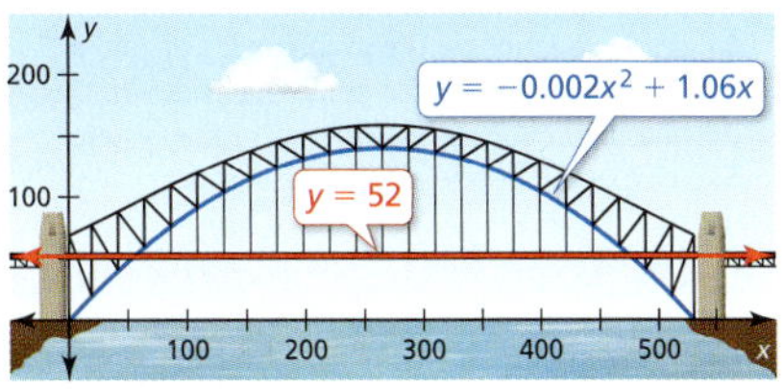

52. MAKING AN ARGUMENT Your friend says that a system of equations consisting of a linear equation and a quadratic equation can have zero, one, two, or infinitely many solutions. Is your friend correct? Explain.

Dynamic Teaching Tools
- Dynamic Assessment & Progress Monitoring Tool
- Interactive Whiteboard Lesson Library
- Dynamic Classroom with Dynamic Investigations

ANSWERS

28. The number 4 should be substituted for y, not for x; $4 = 3x^2 - 6x + 4$; $x = 0$ or $x = 2$

29. between −4 and −3, between 0 and 1

30. between 1 and 2, between 3 and 4

31. between −4 and −3, between −2 and −1

32. between 2 and 3, between 5 and 6

33. about (2.350, 13.225)

34. about (−1.092, 5.469)

35. about (−2.543, −0.941), about (0.185, 1.586), about (1.854, 25.152)

36. about (−3.416, −2.004), about (−0.415, −2.513), about (1.405, −11.593)

37. about (0.840, 3.454), about (−3.174, −1.898)

38. about (1.303, −3.303), about (−2.303, 0.303)

39. $x \approx 0.45, x \approx -4.45$

40. no solutions

41. no solutions

42. $x \approx -0.61, x \approx -2.72$

43. $x \approx -0.36$

44. $x \approx -0.43, x \approx 2.21$

45. $x \approx 1.13, x \approx 2.40$

46. $x \approx 0.90$

47. $\left(\frac{-7 + \sqrt{145}}{6}, \frac{7 + 2\sqrt{145}}{9}\right)$, $\left(\frac{-7 - \sqrt{145}}{6}, \frac{7 - 5\sqrt{145}}{9}\right)$; They are about the same.

48. $\left(\frac{-1 + \sqrt{13}}{2}, \frac{-3 - \sqrt{13}}{2}\right)$, $\left(\frac{-1 - \sqrt{13}}{2}, \frac{-3 + \sqrt{13}}{2}\right)$; They are about the same.

49. 5 days and 35 days after the movie opened

50. yes; (−3, 2)

51. about 55 m, about 475 m

52. no; A line and a parabola cannot be the same.

ANSWERS

53. **a–c.** (2, 11), (−2, −5)

Sample answer: elimination; The resulting equation can be written in the form $x^2 = d$.

54. **a–c.** (−4, 11), (3, 4)

Sample answer: substitution; The resulting equation is easy to factor.

55. **a.** $y = 30x + 290$

b. (1, 320), (34, 1310)

56. **a.** two solutions

b. no solutions

57. See Additional Answers.

58. **a.** year 84

b. yes; year 115

59. **a.** (0, 4)

b. (5, 9)

60. yes; yes; *Sample answer:* When one equation is quadratic in *x* and the other is quadratic in *y*, there can be three or four solutions.

61. (1, −6)

62. (4, −5), (−5, 4)

63–70. See Additional Answers.

Mini-Assessment

1. Solve the system by graphing.

$y = 2x^2 - 3x - 3$
$y = x + 3$

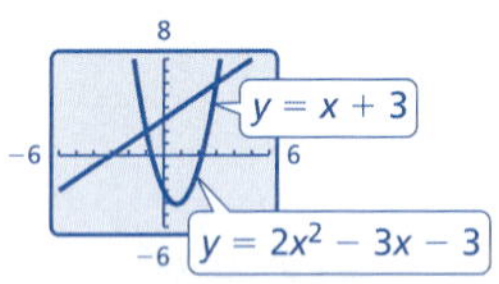

(−1, 2) and (3, 6)

2. Solve the system by substitution.

$y = x^2 + 5x - 1$
$y = x - 4$

(−3, −7) and (−1, −5)

3. Solve the system by elimination.

$y = x^2 + 3x + 2$
$y = 3x + 6$

(−2, 0) and (2, 12)

4. Approximate the solution(s) of the system to the nearest thousandth.

$y = x^2 + x + 2$
$y = 3^x - 3$

about (2.309, 9.642)

5. Solve $2^x + 1 = x^2 + 3x - 1$.

$x \approx -3.58$ and $x = 1$

COMPARING METHODS **In Exercises 53 and 54, solve the system of equations by (a) graphing, (b) substitution, and (c) elimination. Which method do you prefer? Explain your reasoning.**

53. $y = 4x + 3$
$y = x^2 + 4x - 1$

54. $y = x^2 - 5$
$y = -x + 7$

55. **MODELING WITH MATHEMATICS** The function $y = -x^2 + 65x + 256$ models the number y of subscribers to a website, where x is the number of days since the website launched. The number of subscribers to a competitor's website can be modeled by a linear function. The websites have the same number of subscribers on Days 1 and 34.

a. Write a linear function that models the number of subscribers to the competitor's website.

b. Solve the system to verify the function from part (a).

56. **HOW DO YOU SEE IT?** The diagram shows the graphs of two equations in a system that has one solution.

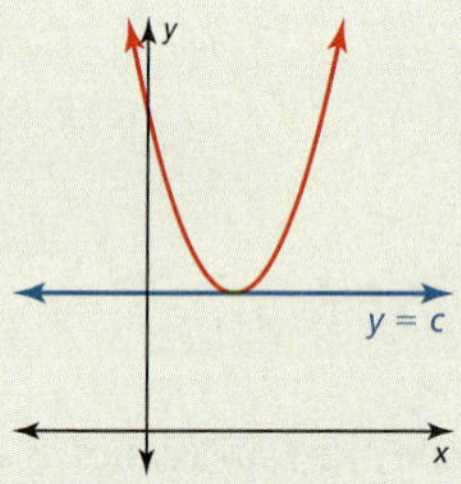

a. How many solutions will the system have when you change the linear equation to $y = c + 2$?

b. How many solutions will the system have when you change the linear equation to $y = c - 2$?

57. **WRITING** A system of equations consists of a quadratic equation whose graph opens up and a quadratic equation whose graph opens down. Describe the possible numbers of solutions of the system. Sketch examples to justify your answer.

58. **PROBLEM SOLVING** The population of a country is 2 million people and increases by 3% each year. The country's food supply is sufficient to feed 3 million people and increases at a constant rate that feeds 0.25 million additional people each year.

a. When will the country first experience a food shortage?

b. The country doubles the rate at which its food supply increases. Will food shortages still occur? If so, in what year?

59. **ANALYZING GRAPHS** Use the graphs of the linear and quadratic functions.

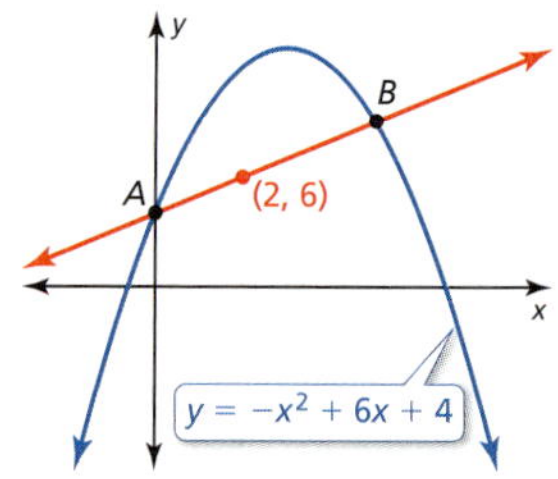

a. Find the coordinates of point *A*.

b. Find the coordinates of point *B*.

60. **THOUGHT PROVOKING** Is it possible for a system of two quadratic equations to have exactly three solutions? exactly four solutions? Explain your reasoning. (*Hint*: Rotations of the graphs of quadratic equations still represent quadratic equations.)

61. **PROBLEM SOLVING** Solve the system of three equations shown.

$y = 2x - 8$
$y = x^2 - 4x - 3$
$y = -3(2)^x$

62. **PROBLEM SOLVING** Find the point(s) of intersection, if any, of the line $y = -x - 1$ and the circle $x^2 + y^2 = 41$.

Maintaining Mathematical Proficiency

Reviewing what you learned in previous grades and lessons

Graph the system of linear inequalities. *(Section 5.7)*

63. $y > 2x$
$y > -x + 4$

64. $y \geq 4x + 1$
$y \leq 7$

65. $y - 3 \leq -2x$
$y + 5 < 3x$

66. $x + y > -6$
$2y \leq 3x + 4$

Graph the function. Describe the domain and range. *(Section 8.3)*

67. $y = 3x^2 + 2$

68. $y = -x^2 - 6x$

69. $y = -2x^2 + 12x - 7$

70. $y = 5x^2 + 10x - 3$

If students need help...	*If students got it...*
Resources by Chapter • Practice A and Practice B • Puzzle Time	Resources by Chapter • Enrichment and Extension • Cumulative Review
Student Journal • Practice	Start the *next* Section
Differentiating the Lesson Skills Review Handbook	

9.4–9.6 What Did You Learn?

Core Vocabulary

completing the square, *p. 506*
Quadratic Formula, *p. 516*
discriminant, *p. 518*
system of nonlinear equations, *p. 526*

Core Concepts

Section 9.4
Completing the Square, *p. 506*

Section 9.5
Quadratic Formula, *p. 516*
Interpreting the Discriminant, *p. 518*

Section 9.6
Solving Systems of Nonlinear Equations, *p. 526*

Mathematical Practices

1. How does your answer to Exercise 74 on page 514 help create a shortcut when solving some quadratic equations by completing the square?
2. What logical progression led you to your answer in Exercise 55 on page 522?
3. Compare the methods used to solve Exercise 53 on page 532. Discuss the similarities and differences among the methods.

Performance Task

Form Matters

Each form of a quadratic function has its pros and cons. Using one form, you can easily find the vertex, but the zeros are more difficult to find. Using another form, you can easily find the y-intercept, but the vertex is more difficult to find. Which form would you use in different situations? How can you convert one form into another?

To explore the answers to these questions and more, go to ***BigIdeasMath.com***.

533

Dynamic Teaching Tools

Dynamic Assessment & Progress Monitoring Tool
Interactive Whiteboard Lesson Library
Dynamic Classroom with Dynamic Investigations

ANSWERS

1. It provides a quick way to check for equations with no solutions.
2. When a and c have different signs, ac is negative, so subtracting $-4ac$ is subtracting a negative, which is the same as adding a positive. So, $b^2 - 4ac$ is positive.
3. *Sample answer:* All the methods had the same results. Substitution and elimination both resulted in equations that could easily be put in the form $x^2 = d$. Graphing did not require algebraic manipulation.

ANSWERS

1. $6p^3\sqrt{2p}$
2. $\dfrac{3\sqrt{35y}}{7y}$
3. $\dfrac{5x^3\sqrt[3]{2x^2}}{2}$
4. $4\sqrt{6} - 8$
5. $14\sqrt{3}$
6. $9\sqrt[3]{2}$
7. $88 + 30\sqrt{7}$
8. $10\sqrt{3}$
9. $x = 6, x = 3$
10. no solutions
11. $x = -4$
12. $-3, -1, 1$
13. 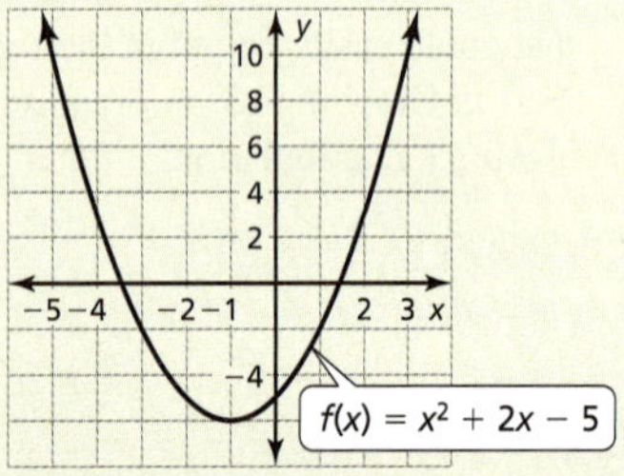

$-3.4, 1.4$

9 Chapter Review

Dynamic Solutions available at *BigIdeasMath.com*

9.1 Properties of Radicals *(pp. 479–488)*

a. Simplify $\sqrt[3]{27x^{10}}$.

$$\sqrt[3]{27x^{10}} = \sqrt[3]{27 \cdot x^9 \cdot x}$$ Factor using the greatest perfect cube factors.

$$= \sqrt[3]{27} \cdot \sqrt[3]{x^9} \cdot \sqrt[3]{x}$$ Product Property of Cube Roots

$$= 3x^3\sqrt[3]{x}$$ Simplify.

b. Simplify $\dfrac{12}{3 + \sqrt{5}}$.

$$\frac{12}{3 + \sqrt{5}} = \frac{12}{3 + \sqrt{5}} \cdot \frac{3 - \sqrt{5}}{3 - \sqrt{5}}$$ The conjugate of $3 + \sqrt{5}$ is $3 - \sqrt{5}$.

$$= \frac{12(3 - \sqrt{5})}{3^2 - (\sqrt{5})^2}$$ Sum and difference pattern

$$= \frac{36 - 12\sqrt{5}}{4}$$ Simplify.

$$= 9 - 3\sqrt{5}$$ Simplify.

Simplify the expression.

1. $\sqrt{72p^7}$
2. $\sqrt{\dfrac{45}{7y}}$
3. $\sqrt[3]{\dfrac{125x^{11}}{4}}$
4. $\dfrac{8}{\sqrt{6} + 2}$
5. $4\sqrt{3} + 5\sqrt{12}$
6. $15\sqrt[3]{2} - 2\sqrt[3]{54}$
7. $(3\sqrt{7} + 5)^2$
8. $\sqrt{6}(\sqrt{18} + \sqrt{8})$

9.2 Solving Quadratic Equations by Graphing *(pp. 489–496)*

Solve $x^2 + 3x = 4$ by graphing.

Step 1 Write the equation in standard form.

$x^2 + 3x = 4$ Write original equation.

$x^2 + 3x - 4 = 0$ Subtract 4 from each side.

Step 2 Graph the related function $y = x^2 + 3x - 4$.

Step 3 Find the x-intercepts. The x-intercepts are -4 and 1.

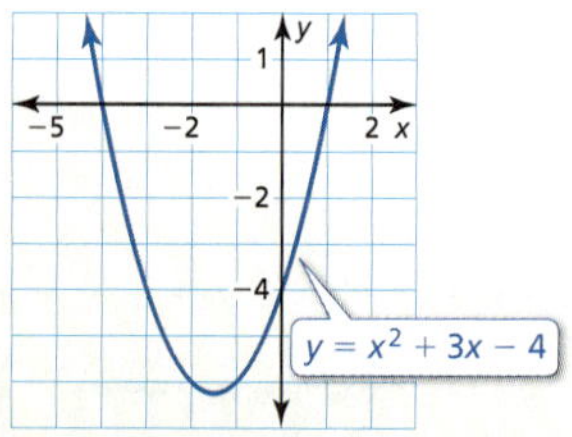

▶ So, the solutions are $x = -4$ and $x = 1$.

Solve the equation by graphing.

9. $x^2 - 9x + 18 = 0$
10. $x^2 - 2x = -4$
11. $-8x - 16 = x^2$
12. The graph of $f(x) = (x + 1)(x^2 + 2x - 3)$ is shown. Find the zeros of f.

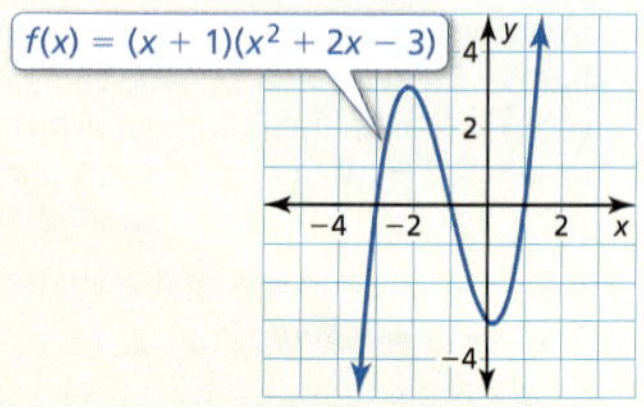

13. Graph $f(x) = x^2 + 2x - 5$. Approximate the zeros of f to the nearest tenth.

9.3 Solving Quadratic Equations Using Square Roots (pp. 497–502)

A sprinkler sprays water that covers a circular region of 90π square feet. Find the diameter of the circle.

Write an equation using the formula for the area of a circle.

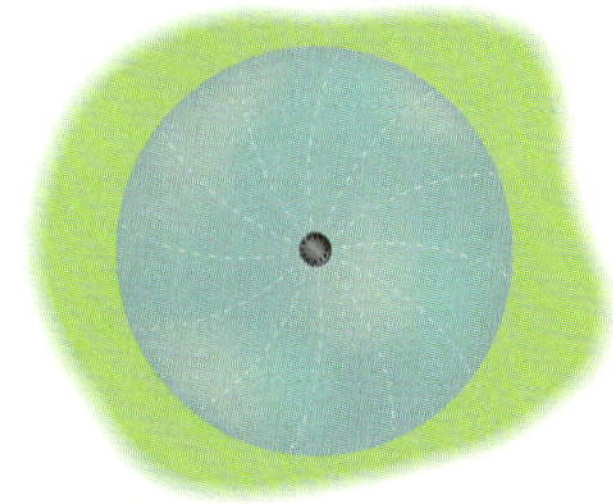

$A = \pi r^2$	Write the formula.
$90\pi = \pi r^2$	Substitute 90π for A.
$90 = r^2$	Divide each side by π.
$\pm\sqrt{90} = r$	Take the square root of each side.
$\pm 3\sqrt{10} = r$	Simplify.

A diameter cannot be negative, so use the positive square root. The diameter is twice the radius. So, the diameter is $6\sqrt{10}$.

▶ The diameter of the circle is $6\sqrt{10} \approx 19$ feet.

Solve the equation using square roots. Round your solutions to the nearest hundredth, if necessary.

14. $x^2 + 5 = 17$ **15.** $x^2 - 14 = -14$ **16.** $(x + 2)^2 = 64$

17. $4x^2 + 25 = -75$ **18.** $(x - 1)^2 = 0$ **19.** $19 = 30 - 5x^2$

9.4 Solving Quadratic Equations by Completing the Square (pp. 505–514)

Solve $x^2 - 6x + 4 = 11$ by completing the square.

$x^2 - 6x + 4 = 11$	Write the equation.
$x^2 - 6x = 7$	Subtract 4 from each side.
$x^2 - 6x + (-3)^2 = 7 + (-3)^2$	Complete the square by adding $\left(\frac{-6}{2}\right)^2$, or $(-3)^2$, to each side.
$(x - 3)^2 = 16$	Write the left side as the square of a binomial.
$x - 3 = \pm 4$	Take the square root of each side.
$x = 3 \pm 4$	Add 3 to each side.

▶ The solutions are $x = 3 + 4 = 7$ and $x = 3 - 4 = -1$.

Solve the equation by completing the square. Round your solutions to the nearest hundredth, if necessary.

20. $x^2 + 6x - 40 = 0$ **21.** $x^2 + 2x + 5 = 4$ **22.** $2x^2 - 4x = 10$

Determine whether the quadratic function has a maximum or minimum value. Then find the value.

23. $y = -x^2 + 6x - 1$ **24.** $f(x) = x^2 + 4x + 11$ **25.** $y = 3x^2 - 24x + 15$

26. The width w of a credit card is 3 centimeters shorter than the length ℓ. The area is 46.75 square centimeters. Find the perimeter.

ANSWERS

14. $x \approx 3.46, x \approx -3.46$
15. $x = 0$
16. $x = 6, x = -10$
17. no solutions
18. $x = 1$
19. $x \approx 1.48, x \approx -1.48$
20. $x = 4, x = -10$
21. $x = -1$
22. $x \approx 3.45, x \approx -1.45$
23. maximum value; 8
24. minimum value; 7
25. minimum value; -33
26. 28 cm

ANSWERS

27. $x = 3, x = -5$
28. $x \approx 2.3, x \approx -1.8$
29. $x = 1$
30. one x-intercept
31. no x-intercepts
32. two x-intercepts
33. $(1, -5)$
34. about $(4.87, 14.75)$, about $(-2.87, -0.75)$
35. about $(-1.88, 2.35)$, about $(2.48, -4.64)$

9.5 Solving Quadratic Equations Using the Quadratic Formula (pp. 515–524)

Solve $-3x^2 + x = -8$ using the Quadratic Formula.

$-3x^2 + x = -8$	Write the equation.
$-3x^2 + x + 8 = 0$	Write in standard form.
$x = \dfrac{-b \pm \sqrt{b^2 - 4ac}}{2a}$	Quadratic Formula
$x = \dfrac{-1 \pm \sqrt{1^2 - 4(-3)(8)}}{2(-3)}$	Substitute -3 for a, 1 for b, and 8 for c.
$x = \dfrac{-1 \pm \sqrt{97}}{-6}$	Simplify.

▶ So, the solutions are $x = \dfrac{-1 + \sqrt{97}}{-6} \approx -1.5$ and $x = \dfrac{-1 - \sqrt{97}}{-6} \approx 1.8$.

Solve the equation using the Quadratic Formula. Round your solutions to the nearest tenth, if necessary.

27. $x^2 + 2x - 15 = 0$ **28.** $2x^2 - x + 8 = 16$ **29.** $-5x^2 + 10x = 5$

Find the number of x-intercepts of the graph of the function.

30. $y = -x^2 + 6x - 9$ **31.** $y = 2x^2 + 4x + 8$ **32.** $y = -\frac{1}{2}x^2 + 2x$

9.6 Solving Nonlinear Systems of Equations (pp. 525–532)

Solve the system by substitution. $y = x^2 - 5$ Equation 1; $y = -x + 1$ Equation 2

Step 1 The equations are already solved for y.

Step 2 Substitute $-x + 1$ for y in Equation 1 and solve for x.

$-x + 1 = x^2 - 5$	Substitute $-x + 1$ for y in Equation 1.
$1 = x^2 + x - 5$	Add x to each side.
$0 = x^2 + x - 6$	Subtract 1 from each side.
$0 = (x + 3)(x - 2)$	Factor the polynomial.
$x + 3 = 0$ or $x - 2 = 0$	Zero-Product Property
$x = -3$ or $x = 2$	Solve for x.

Step 3 Substitute -3 and 2 for x in Equation 2 and solve for y.

$y = -(-3) + 1$	Substitute for x in Equation 2.	$y = -2 + 1$
$= 4$	Simplify.	$= -1$

▶ So, the solutions are $(-3, 4)$ and $(2, -1)$.

Solve the system using any method.

33. $y = x^2 - 2x - 4$; $y = -5$

34. $y = x^2 - 9$; $y = 2x + 5$

35. $y = 2\left(\frac{1}{2}\right)^x - 5$; $y = -x^2 - x + 4$

9 Chapter Test

Solve the equation using any method. Explain your choice of method.

1. $x^2 - 121 = 0$ **2.** $x^2 - 6x = 10$ **3.** $-2x^2 + 3x + 7 = 0$

4. $x^2 - 7x + 12 = 0$ **5.** $5x^2 + x - 4 = 0$ **6.** $(4x + 3)^2 = 16$

7. Describe how you can use the method of completing the square to determine whether the function $f(x) = 2x^2 + 4x - 6$ can be represented by the graph shown.

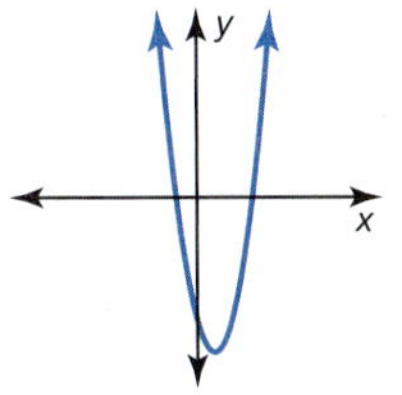

8. Write an expression involving radicals in which a conjugate can be used to simplify the expression.

Solve the system using any method.

9. $y = x^2 - 4x - 2$
$y = -4x + 2$

10. $y = -5x^2 + x - 1$
$y = -7$

11. $y = \frac{1}{2}(4)^x + 1$
$y = x^2 - 2x + 4$

12. A skier leaves an 8-foot-tall ramp with an initial vertical velocity of 28 feet per second. The function $h = -16t^2 + 28t + 8$ represents the height h (in feet) of the skier after t seconds. The skier has a perfect landing. How many points does the skier earn?

Criteria	Scoring
Maximum height	1 point per foot
Time in air	5 points per second
Perfect landing	25 points

13. An amusement park ride lifts seated riders 265 feet above the ground. The riders are then dropped and experience free fall until the brakes are activated 105 feet above the ground. The function $h = -16t^2 + 265$ represents the height h (in feet) of the riders t seconds after they are dropped. How long do the riders experience free fall? Round your solution to the nearest hundredth.

14. Write an expression in simplest form that represents the area of the painting shown.

$\frac{36}{\sqrt{3}}$ in.

$\sqrt{30x^7}$ in.

15. Explain how you can determine the number of times the graph of $y = 5x^2 - 10x + 5$ intersects the x-axis without graphing or solving an equation.

16. Consider the quadratic equation $ax^2 + bx + c = 0$. Find values of a, b, and c so that the graph of its related function has (a) two x-intercepts, (b) one x-intercept, and (c) no x-intercepts.

17. The numbers y of two types of bacteria after x hours are represented by the models below.

$y = 3x^2 + 8x + 20$ Type A
$y = 27x + 60$ Type B

a. When are there 400 Type A bacteria?

b. When are the number of Type A and Type B bacteria the same?

c. When are there more Type A bacteria than Type B? When are there more Type B bacteria than Type A? Use a graph to support your answer.

If students need help...	*If students got it...*
Lesson Tutorials	Resources by Chapter • Enrichment and Extension • Cumulative Review
Skills Review Handbook	Performance Task
BigIdeasMath.com	Start the *next* Section

ANSWERS

1. $x = 11, x = -11$; *Sample answer:* The equation can be written in the form $x^2 = d$, so solve using square roots.

2. $x \approx 7.36, x \approx -1.36$; *Sample answer:* $a = 1$ and b is even, so solve by completing the square.

3. $x \approx -1.27, x \approx 2.77$; *Sample answer:* The equation is not factorable and $a \neq 1$, so solve using the quadratic formula.

4. $x = 4, x = 3$; *Sample answer:* The equation is easily factorable, so solve by factoring.

5. $x = \frac{4}{5}, x = -1$; *Sample answer:* The equation is not easily factorable and $a \neq 1$, so solve using the quadratic formula.

6. $x = \frac{1}{4}, x = -\frac{7}{4}$; *Sample answer:* The equation is in the form $x^2 = d$ where x is a binomial, so solve using square roots.

7. The equation can be put in vertex form $y = 2(x + 1)^2 - 8$. The vertex $(-1, -8)$, is in the third quadrant, so the function cannot be represented by the graph shown.

8. *Sample answer:* $\frac{1}{2 + \sqrt{3}}$

9. $(2, -6), (-2, 10)$

10. $(-1, -7), \left(\frac{6}{5}, -7\right)$

11. $(1, 3)$

12. 55 points

13. about 3.16 sec

14. $36x^3\sqrt{10x}$ in.^{2}

15. Use the discriminant to determine the number of solutions.

16. a. *Sample answer:* $a = 3, b = 2, c = -1$

b. *Sample answer:* $a = 3, b = 6, c = 3$

c. *Sample answer:* $a = 2, b = 3, c = 4$

17. a. after 10 h

b. after 8 h

c. after 8 h; before 8 h

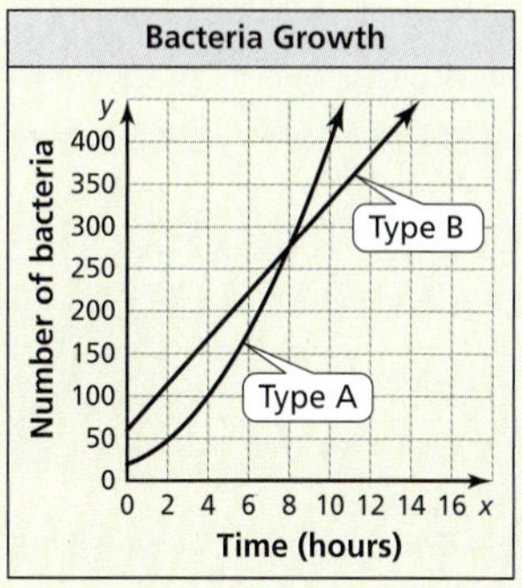

ANSWERS

1. f: negative; g: positive; h and j: zero
2. a. *Sample answer:* 400, 1.15
   b. *Sample answer:* 800, 1.15
   c. *Sample answer:* 600, 1.05
3. a. no; The height is unknown.
   b. yes; $r = \sqrt{\frac{S}{4\pi}}$
4. negative; positive

9 Cumulative Assessment

1. The graphs of four quadratic functions are shown. Determine whether the discriminants of the equations $f(x) = 0$, $g(x) = 0$, $h(x) = 0$, and $j(x) = 0$ are positive, negative, or zero.

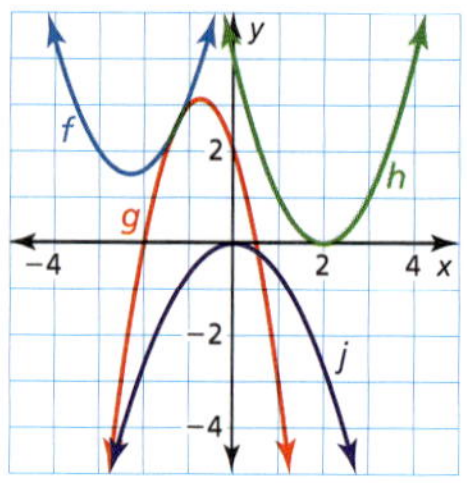

2. The function $f(x) = a(1.08)^x$ represents the total amount of money (in dollars) in Account A after x years. The function $g(x) = 600(b)^x$ represents the total amount of money (in dollars) in Account B after x years. Fill in values for a and b so that each statement is true.

   a. When $a =$ ____ and $b =$ ____, Account B has a greater initial amount and increases at a faster rate than Account A.

   b. When $a =$ ____ and $b =$ ____, Account B has a lesser initial amount than Account A but increases at a faster rate than Account A.

   c. When $a =$ ____ and $b =$ ____, Account B and Account A have the same initial amount, and Account B increases at a slower rate than Account A.

3. Your friend claims to be able to find the radius r of each figure, given the surface area S. Do you support your friend's claim? Justify your answer.

   a. $S = 2\pi r^2 + 2\pi rh$

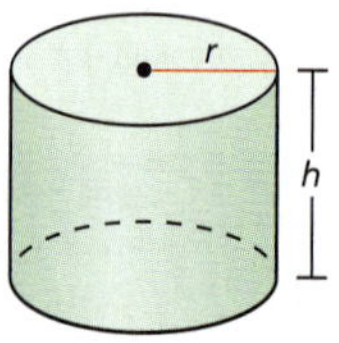

   b. $S = 4\pi r^2$

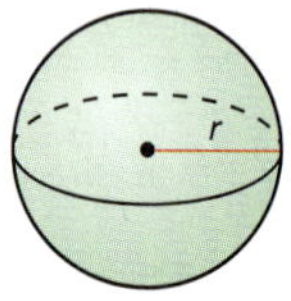

4. The tables represent the numbers of items sold at a concession stand on days with different average temperatures. Determine whether the data represented by each table show a *positive*, a *negative*, or *no* correlation.

Temperature (°F), x	14	27	32	41	48	62	73
Cups of hot chocolate, y	35	28	22	9	4	2	1

Temperature (°F), x	14	27	32	41	48	62	73
Bottles of sports drink, y	8	12	13	16	19	27	29

5. Which graph shows exponential growth?

Ⓐ
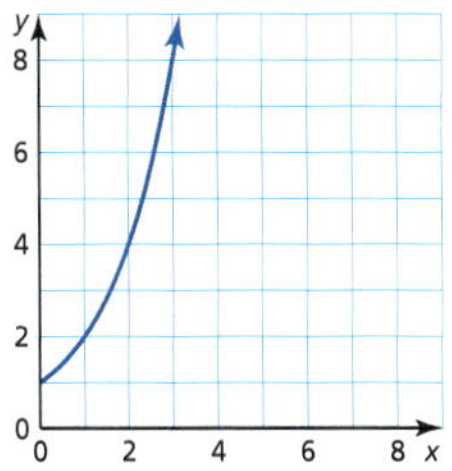

Ⓑ
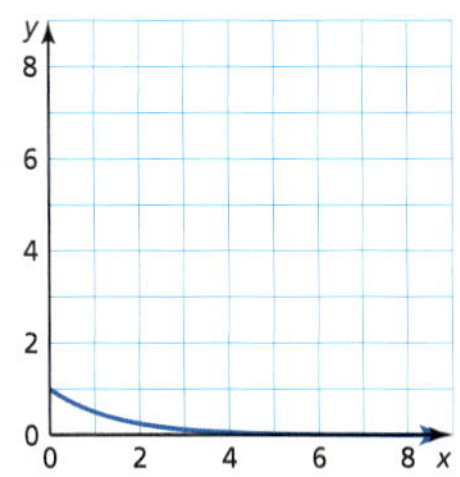

Ⓒ
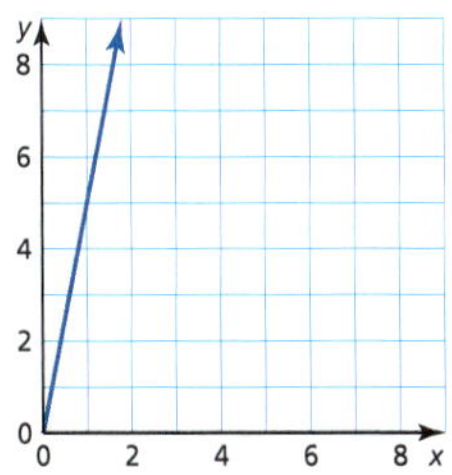

Ⓓ
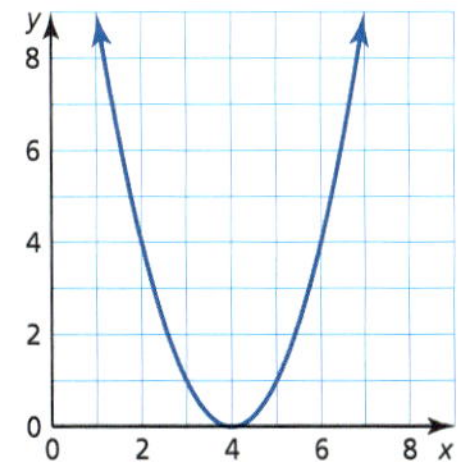

6. Which statement best describes the solution(s) of the system of equations?

$$y = x^2 + 2x - 8$$
$$y = 5x + 2$$

Ⓐ The graphs intersect at one point, $(-2, -8)$. So, there is one solution.

Ⓑ The graphs intersect at two points, $(-2, -8)$ and $(5, 27)$. So, there are two solutions.

Ⓒ The graphs do not intersect. So, there is no solution.

Ⓓ The graph of $y = x^2 + 2x - 8$ has two x-intercepts. So, there are two solutions.

7. Which expressions are in simplest form?

$x\sqrt{45x}$	$\dfrac{16}{\sqrt{5}}$	$\sqrt[3]{\dfrac{4}{9}}$	$16\sqrt{5}$	$3x\sqrt{5x}$
$\dfrac{\sqrt[3]{x^4}}{2}$	$\dfrac{4\sqrt{7}}{3}$	$\dfrac{\sqrt{16}}{5}$	$2\sqrt[3]{x^2}$	$3\dfrac{\sqrt{7}}{\sqrt{x}}$

8. The domain of the function shown is all integers in the interval $-3 < x \leq 3$. Find all the ordered pairs that are solutions of the equation $y = f(x)$.

$$f(x) = 4x - 5$$

ANSWERS

5. A
6. B
7. $16\sqrt{5}, 3x\sqrt{5x}, \dfrac{4\sqrt{7}}{3}, 2\sqrt[3]{x^2}$
8. $(-2, -13), (-1, -9), (0, -5), (1, -1), (2, 3), (3, 7)$

Chapter 10 Pacing Guide

Chapter 10 Pacing Guide	
Chapter Opener/ Mathematical Practices	1 Day
Section 1	2 Days
Section 2	2 Days
Section 3	2 Days
Section 4	2 Days
Chapter Review/ Chapter Tests	2 Days
Total Chapter 10	11 Days
Year-to-Date	146 Days

10 Radical Functions and Equations

Crow Feeding Habits *(p. 573)*

Trapeze Artist *(p. 565)*

Asian Elephant *(p. 554)*

Firefighting *(p. 549)*

Tsunami *(p. 547)*

Laurie's Notes

Chapter Summary

- This is a short chapter that introduces radical functions, both square root and cubic functions. The transformations of the parent functions give students another opportunity to examine horizontal and vertical translations, horizontal reflections in the *x*-axis, and vertical stretches and shrinks.
- A lesson on solving radical equations is a nice connection to solving equations in general. Students have used the properties of equality and have taken the square root of each side of an equation. Now they learn that squaring both sides of an equation is another technique for solving equations, although extraneous roots may be introduced.
- The last lesson presents inverse relations and inverse functions. It connects the functions studied in this chapter with functions studied earlier.

What Your Students Have Learned

Middle School
- Find square roots and evaluate expressions involving square roots.
- Write functions to model linear relationships, and determine and interpret rates of change.
- Solve linear equations in one variable using the Distributive Property and the properties of equality.
- Represent relations as ordered pairs and determine whether the relation is a function.
- Represent functions as equations, input-output tables, and graphs.

What Your Students Will Learn

Algebra 1
- Graph and compare the transformations of square root functions and cube root functions to the graphs of the parent functions.
- Compare square root functions and cube root functions using average rates of change.
- Solve radical equations and identify extraneous solutions.
- Explore inverses of functions algebraically, numerically, and graphically.
- Find inverses of linear, quadratic, and radical functions algebraically.
- Determine whether the inverse of a function is a function by applying the Horizontal Line Test.

Dynamic Teaching Tools
- Dynamic Assessment & Progress Monitoring Tool
- Lesson Planning Tool
- Interactive Whiteboard Lesson Library
- Dynamic Classroom with Dynamic Investigations
- Real-Life STEM Videos

Scaffolding in the Classroom

Graphic Organizers: Summary Triangle
A Summary Triangle can be used to explain a concept. Typically, the Summary Triangle is divided into 3 or 4 parts. In the top part, students write the concept being explained. In the middle part(s), students write any procedure, explanation, description, definition, theorem, and/or formula(s). In the bottom part, students write an example to illustrate the concept. A Summary Triangle can be used as an assessment tool, in which blanks are left for students to complete. Also, students can place their Summary Triangles on note cards to use as a quick study reference.

Questioning in the Classroom

Wait Time

Students need the opportunity to process the question before formulating an answer. Although the silence is difficult, make certain to give sufficient time before calling on someone or answering yourself.

Laurie's Notes

Maintaining Mathematical Proficiency

Evaluating Expressions Involving Square Roots

- Remind students to follow the order of operations when evaluating the expressions. For example, in Exercise 3, they should first evaluate the square root, next simplify the fraction, then subtract inside the parentheses, and finally multiply.

COMMON ERROR Students do not follow the order of operations. For example, in Exercise 1, students may first add 7 and 10 to get $17\sqrt{25}$ and then evaluate the square root.

Transforming Linear Functions

- Remind students of the types of transformations and the $f(x)$ notation for each: horizontal translation $f(x - h)$, vertical translation $f(x) + k$, reflection $f(-x)$ or $-f(x)$, horizontal stretch or shrink $f(ax)$, and vertical stretch or shrink $a \cdot f(x)$.

COMMON ERROR Students confuse horizontal shrinks and vertical stretches. For example, in Exercise 5, students may describe the transformation to get the graph of $h(x) = 2x$ as a horizontal shrink by a factor of $\frac{1}{2}$, instead of a vertical stretch by a factor of 2.

Mathematical Practices (continued on page 542)

- The *Mathematical Practices* page focuses attention on how mathematics is learned—process versus content. This page demonstrates how a student would reason logically about a statement he or she is trying to prove. Proof by contradiction is a style of proof that is sometimes referred to as an indirect proof.
- Use the *Mathematical Practices* page to help students develop mathematical habits of mind—how mathematics can be explored and how mathematics is thought about.
- Students were introduced to irrational numbers in the last chapter. Rational numbers were defined in middle school as any number that can be written as the ratio of two numbers, $\frac{a}{b}$, where $b \neq 0$.

If students need help...	*If students got it...*
Student Journal • Maintaining Mathematical Proficiency	Game Closet at *BigIdeasMath.com*
Lesson Tutorials	Start the *next* Section
Skills Review Handbook	

Maintaining Mathematical Proficiency

Evaluating Expressions Involving Square Roots

Example 1 Evaluate $-4(\sqrt{121} - 16)$.

$$-4(\sqrt{121} - 16) = -4(11 - 16) \quad \text{Evaluate the square root.}$$
$$= -4(-5) \quad \text{Subtract.}$$
$$= 20 \quad \text{Multiply.}$$

Evaluate the expression.

1. $7\sqrt{25} + 10$
2. $-8 - \sqrt{\frac{64}{16}}$
3. $5\left(\frac{\sqrt{81}}{3} - 7\right)$
4. $-2(3\sqrt{9} + 13)$

Transforming Linear Functions

Example 2 Graph $f(x) = x$ and $g(x) = -3x - 4$. Describe the transformations from the graph of f to the graph of g.

Note that you can rewrite g as $g(x) = -3f(x) - 4$.

Step 1 There is no horizontal translation from the graph of f to the graph of g.

Step 2 Stretch the graph of f vertically by a factor of 3 to get the graph of $h(x) = 3x$.

Step 3 Reflect the graph of h in the x-axis to get the graph of $r(x) = -3x$.

Step 4 Translate the graph of r vertically 4 units down to get the graph of $g(x) = -3x - 4$.

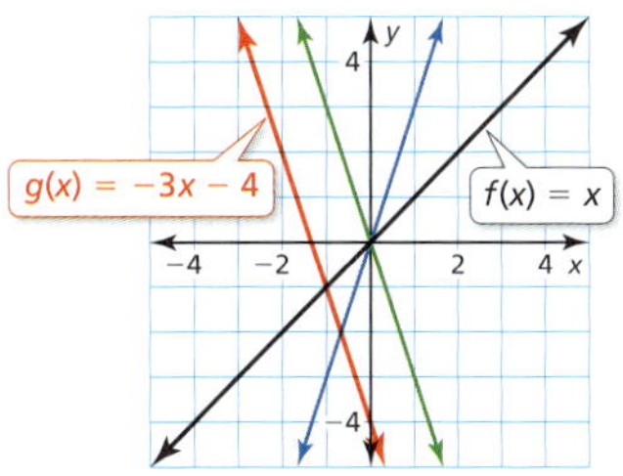

Graph f and g. Describe the transformations from the graph of f to the graph of g.

5. $f(x) = x$; $g(x) = 2x - 2$
6. $f(x) = x$; $g(x) = \frac{1}{3}x + 5$
7. $f(x) = x$; $g(x) = -x + 3$
8. **ABSTRACT REASONING** Let a and b represent constants, where $b \geq 0$. Describe the transformations from the graph of $m(x) = ax + b$ to the graph of $n(x) = -2ax - 4b$.

Vocabulary Review

Have students make Process Diagrams for the following topics.

- Evaluating expressions involving square roots
- Transforming linear functions

What Your Students Have Learned

- Find square roots and evaluate expressions involving square roots.
- Translate, reflect, shrink, and stretch graphs of linear functions, and combine transformations of graphs of linear functions.

ANSWERS

1. 45
2. −10
3. −20
4. −44
5. 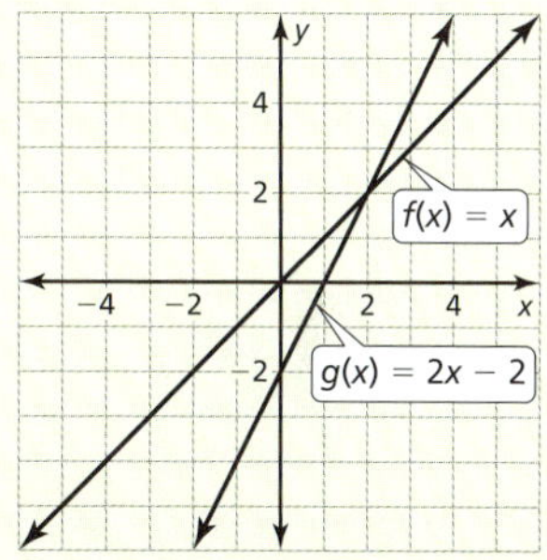

The graph of g is a vertical stretch by a factor of 2 then a vertical translation 2 units down of the graph of f.

6. 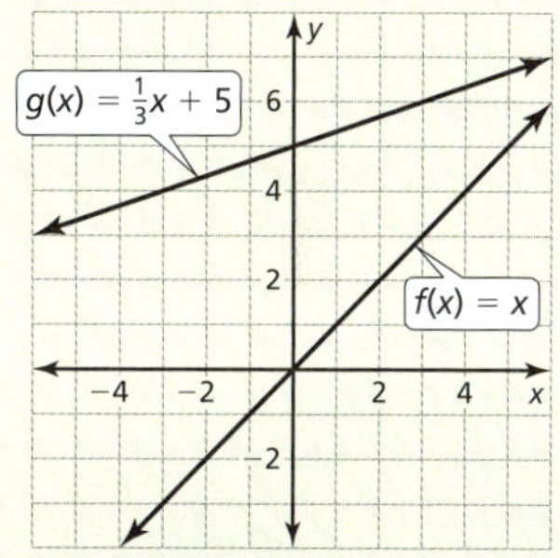

The graph of g is a vertical shrink by a factor of $\frac{1}{3}$ then a vertical translation 5 units up of the graph of f.

7. 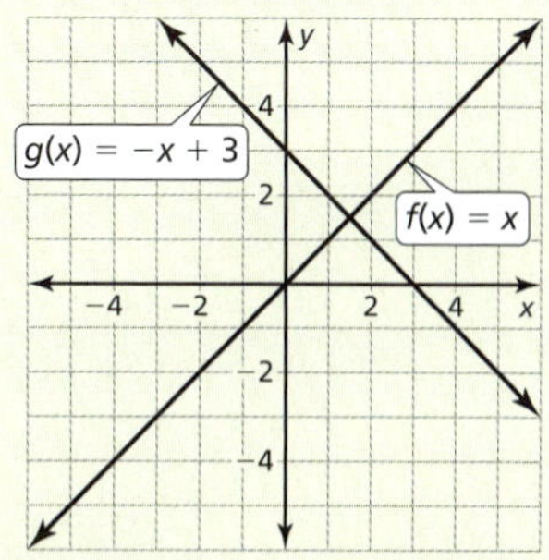

The graph of g is a reflection in the x-axis then a vertical translation 3 units up of the graph of f.

8. The graph of n is a vertical stretch by a factor of 2, a reflection in the x-axis, then a vertical translation $2b$ units down of the graph of m.

MONITORING PROGRESS ANSWERS

1. $\sqrt{0}, \sqrt{1}, \sqrt{4}, \sqrt{9}$; $\sqrt{0} = \frac{0}{2}$, $\sqrt{1} = \frac{1}{1}$, $\sqrt{4} = \frac{2}{1}$, $\sqrt{9} = \frac{3}{1}$
2. When $x = 1$, $1 - 1 = 0$, so the division by $x - 1$ is undefined.

Mathematical Practices

Mathematically proficient students distinguish correct reasoning from flawed reasoning.

Logical Reasoning

Core Concept

Logical Reasoning and Proof by Contradiction

Mathematics is a logical system that is built from only a few assumptions and undefined terms. The assumptions are called *axioms* or *postulates*. After starting with a collection of axioms and undefined terms, the remainder of mathematics is logically built using careful definitions and theorems (or rules), which are based on the axioms or on previously proven theorems.

To write an indirect proof, or *proof by contradiction*, identify the statement you want to prove. Assume temporarily that this statement is false by assuming that its opposite is true. Then reason logically until you reach a contradiction. Point out that the original statement must be true because the contradiction proves the temporary assumption false.

EXAMPLE 1 Understanding a Proof

A number is *rational* when it can be written as the ratio a/b of two integers, where $b \neq 0$. Use proof by contradiction to prove that $\sqrt{2}$ is not a rational number.

SOLUTION

Assume that $\sqrt{2}$ can be written as the ratio of two integers (in simplest form) and show that this assumption leads to a contradiction.

$\sqrt{2} = \frac{a}{b}$ — Assume $\sqrt{2}$ is rational.

$2 = \frac{a^2}{b^2}$ — Square each side.

$a^2 = 2b^2$ — Multiply each side by b^2 and interchange left and right sides.

This implies that a^2 is even, which is only true when a is even (divisible by 2). This implies that a^2 is divisible by 2^2, or 4. So, $2b^2$ is also divisible by 4, meaning that b^2 is divisible by 2, b^2 is even, and b is even. Because both a and b are even, they have a common factor (of at least 2).

This contradicts the assumption that the ratio a/b is written in simplest form. So, the initial assumption that $\sqrt{2}$ is rational must be false. Therefore, $\sqrt{2}$ is not a rational number.

Monitoring Progress

1. Which of the following square roots are rational numbers? Explain your reasoning.

   $\sqrt{0}, \sqrt{1}, \sqrt{3}, \sqrt{4}, \sqrt{5}, \sqrt{6}, \sqrt{7}, \sqrt{8}, \sqrt{9}$

2. The sequence of steps shown appears to prove that $1 = 0$. What is wrong with this argument?

   $x = 1$ — Let $x = 1$.

   $x - 1 = 0$ — Subtract 1 from each side.

   $x(x - 1) = 0$ — Multiply each side by x.

   $x = 0$ — Divide each side by $(x - 1)$.

Laurie's Notes — Mathematical Practices (continued from page T-541)

- In the *Core Concept*, the strategy and logic for writing a proof by contradiction is explained. Use a simple example, such as trying to prove that some nonzero real number x is positive. Instead of trying to prove that x is positive, you assume that x is negative. In doing so, some contradiction results, and therefore x cannot be negative. You then conclude that x must have been positive.
- Work through Example 1 as shown, interpreting the steps and results.
- Give time for students to work through the questions in the *Monitoring Progress*. Have students record their thinking about each problem, and then discuss as a class.

Laurie's Notes

Overview of Section 10.1

Introduction

- In this lesson, the graph of the square root function is introduced, along with transformations to the parent function.
- The square root function is nonlinear; therefore there is not a constant rate of change. Instead, the average rate of change is found along different intervals of the graph of a square root function.

Teaching Strategy

- Students need to pay attention to the order in which transformations are performed. For the equation $g(x) = -2\sqrt{x-3} - 2$ in Example 4, you want students to recognize that there are four different transformations of the parent function that are taking place in this equation.
- Begin with the parent function $y = \sqrt{x}$. The first transformation is $t(x) = \sqrt{x-3}$, which is a horizontal shift of 3 units (see Figure 1).
- The next transformation is $h(x) = 2t(x)$, which stretches it vertically by a factor of 2.
- Then the transformation is $r(x) = -h(x)$, which reflects $h(x)$ in the x-axis (see Figure 2).
- The last transformation is $g(x) = r(x) - 2$, which shifts $r(x)$ down 2 units (see Figure 3). Note that the final graph, $g(x) = -2\sqrt{x-3} - 2$, has a critical value at $(3, -2)$.

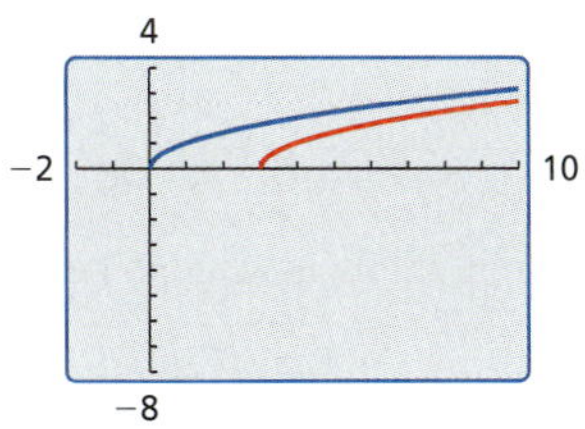

Figure 1

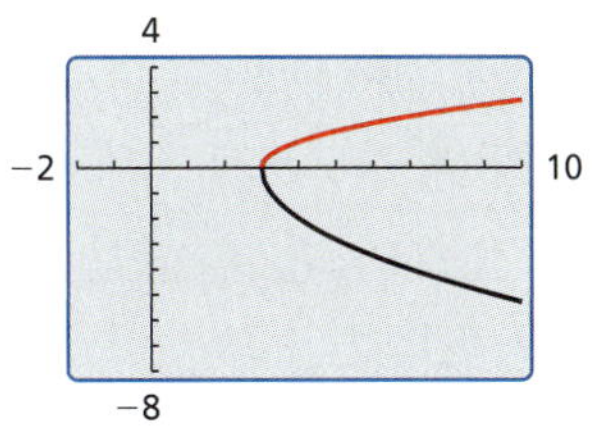

Figure 2

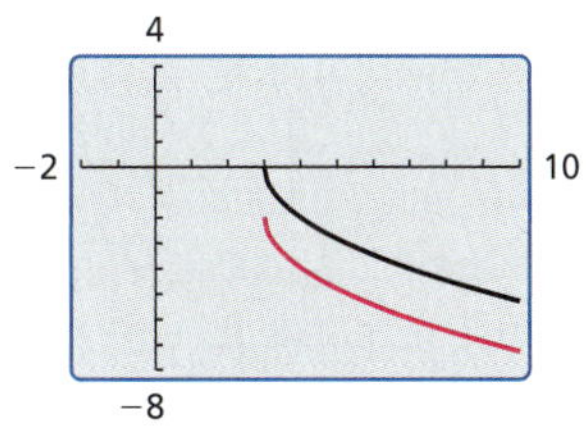

Figure 3

- Students should recognize that when the transformations are performed in a different order, the graph is not the same. For instance, when the parent function $y = \sqrt{x}$ is shifted down 2 units first (see Figure 4), shifted right 3 units (see Figure 5), and finally reflected in the x-axis while being stretched vertically by a factor of 2 (see Figure 6), the graph would be very different.

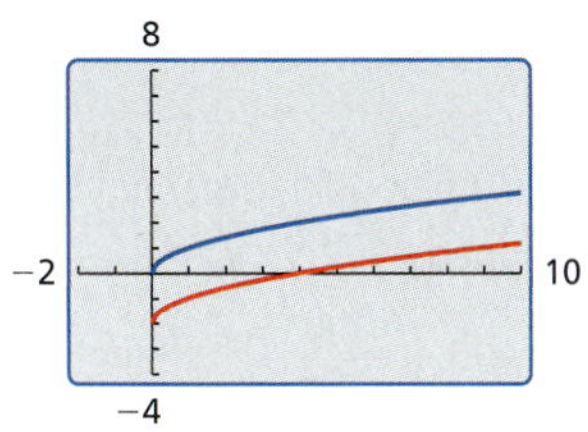

Figure 4

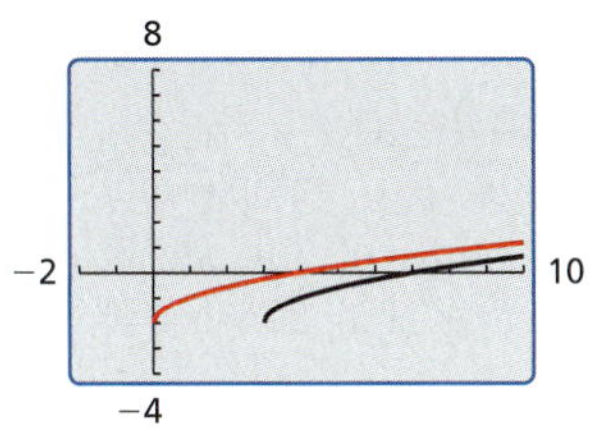

Figure 5

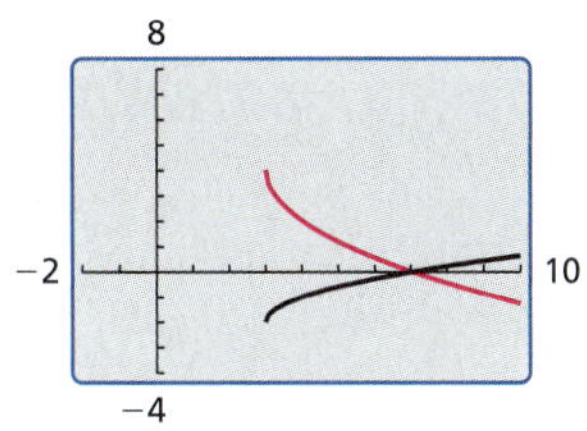

Figure 6

Pacing Suggestion

- The problems students will work in the explorations are similar to Examples 1 and 2 in the formal lesson. If students show good comprehension in the explorations, begin the formal lesson by stating both *Core Concepts*.

Dynamic Teaching Tools

- Dynamic Assessment & Progress Monitoring Tool
- Lesson Planning Tool
- Interactive Whiteboard Lesson Library
- Dynamic Classroom with Dynamic Investigations

What Your Students Will Learn

- Graph square root functions and describe the domain and range.
- Graph and compare the transformations of square root functions to the graph of the parent function.
- Compare square root functions using average rates of change.
- Solve real-life problems involving square root functions.

Laurie's Notes

Exploration

Motivate

- Play a quick round of "30-second Un-Do." In this game, students list as many actions and undoing actions as they can in 30 seconds. For example: filling a sink and draining a sink.
- Have students share a few examples after their time is up.
- Play another round, but this time the actions must involve math operations.
- Have students share a few examples. Hopefully at least one student will have an example that does not involve addition, subtraction, multiplication, or division.
- Explain that today's explorations use more precise language. They involve taking square roots, the inverse operation of squaring.

Exploration 1

- Students have graphed a variety of functions, so the directions should need no additional explanation.
- **Attend to Precision:** When using a calculator, students need to be aware of the syntax for square roots. Some calculators require the radicand to be enclosed in parentheses. Students need to recognize the difference between $\sqrt{(x)} + 2$ and $\sqrt{(x + 2)}$.

? "What is the domain of each function?" The domain of the first is $x \geq 0$, and the domain of the second is $x \geq -2$.

? "What is the range of each function?" The range of both is $y \geq 0$.

- After students have finished both graphs, you could have them check their work using graphing calculators.
- You could also ask volunteers to describe the differences in the graphs. Students should be able to describe the transformation in part (b).

Exploration 2

- Ordered pairs have been given so that students can determine the equation of the square root function.
- Students may need to use trial and error.
- Probe students as you walk around to see whether they recognize any patterns between the two tables. They will likely notice that for $x = -4$, -1, and 0, the y-values differ by 1.
- Have students identify the domain and range for each function.

Communicate Your Answer

- Students may comment that the square root graphs look like half of a parabola turned 90°.
- **Look For and Make Use of Structure:** Students have not graphed $y = a\sqrt{x}$ or $y = -a\sqrt{x}$. Ask them how a affects the graph of $y = \sqrt{x}$.

Connecting to Next Step

- If you feel students are comfortable with identifying the domain and range of a square root function from the equation or table of values, begin the formal lesson with Example 3.

10.1 Graphing Square Root Functions

Essential Question What are some of the characteristics of the graph of a square root function?

EXPLORATION 1 **Graphing Square Root Functions**

Work with a partner.

- Make a table of values for each function.
- Use the table to sketch the graph of each function.
- Describe the domain of each function.
- Describe the range of each function.

a. $y = \sqrt{x}$

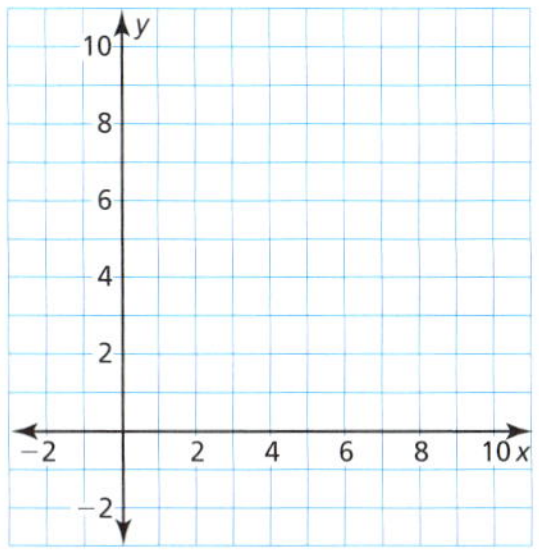

b. $y = \sqrt{x + 2}$

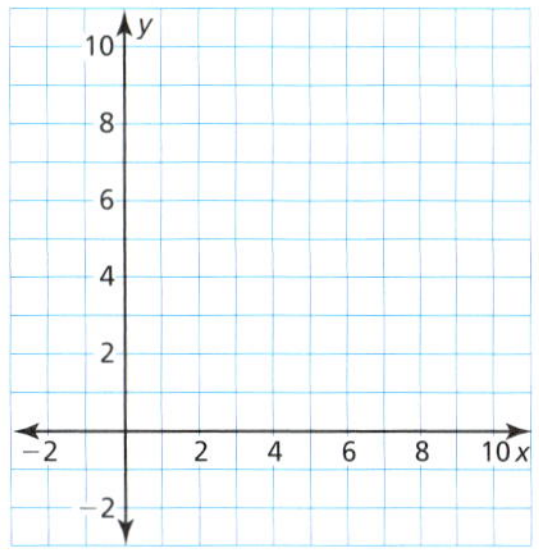

LOOKING FOR A PATTERN

To be proficient in math, you need to look closely to discern a pattern or structure.

EXPLORATION 2 **Writing Square Root Functions**

Work with a partner. Write a square root function, $y = f(x)$, that has the given values. Then use the function to complete the table.

a.

x	$f(x)$
-4	0
-3	
-2	
-1	$\sqrt{3}$
0	2
1	

b.

x	$f(x)$
-4	1
-3	
-2	
-1	$1 + \sqrt{3}$
0	3
1	

Communicate Your Answer

3. What are some of the characteristics of the graph of a square root function?

4. Graph each function. Then compare the graph to the graph of $f(x) = \sqrt{x}$.

a. $g(x) = \sqrt{x - 1}$ **b.** $g(x) = \sqrt{x} - 1$ **c.** $g(x) = 2\sqrt{x}$ **d.** $g(x) = -2\sqrt{x}$

Dynamic Teaching Tools

- Dynamic Assessment & Progress Monitoring Tool
- Lesson Planning Tool
- Interactive Whiteboard Lesson Library
- Dynamic Classroom with Dynamic Investigations

ANSWERS

1. See Additional Answers.

2. a. $y = \sqrt{x + 4}$; $1, \sqrt{2}, \sqrt{5}$

b. $y = \sqrt{x + 4} + 1$; $2, 1 + \sqrt{2}, 1 + \sqrt{5}$

3. The graph of a square root function is half of a parabola opening to one side.

4. a.

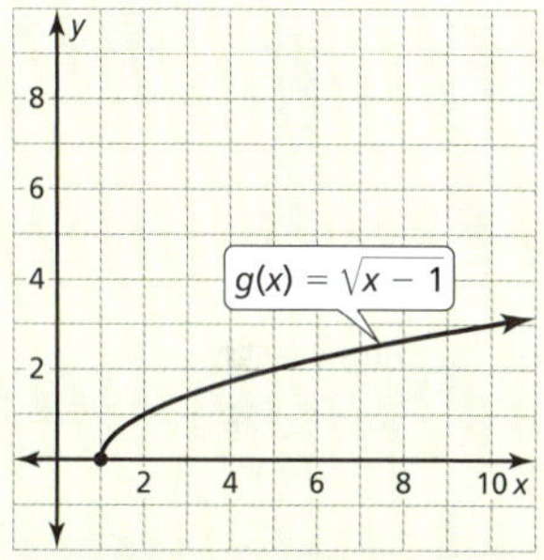

The graph of g is a translation 1 unit right of the graph of f.

b.

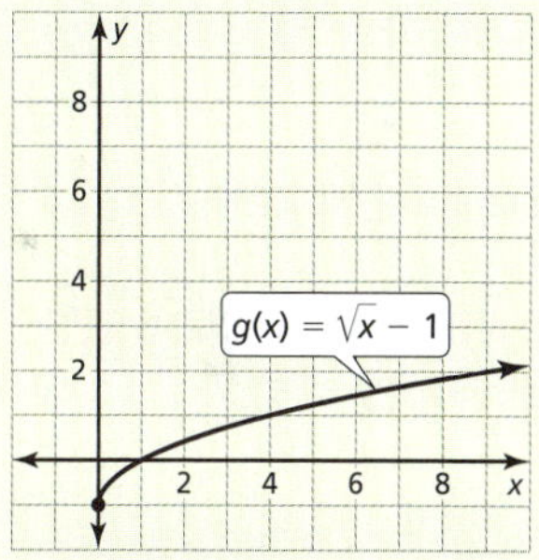

The graph of g is a translation 1 unit down of the graph of f.

c.

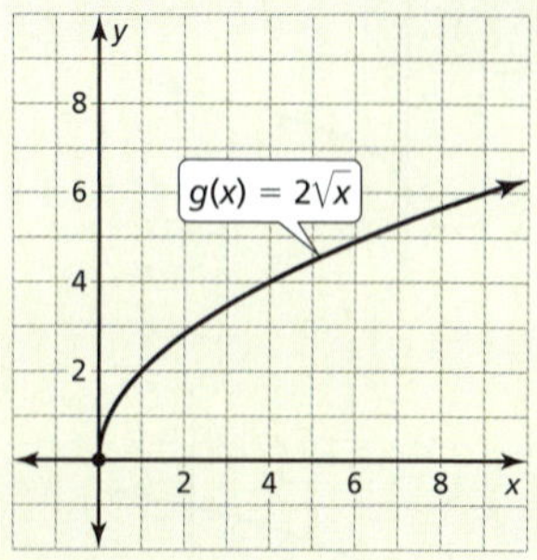

The graph of g is a vertical stretch by a factor of 2 of the graph of f.

4. d.

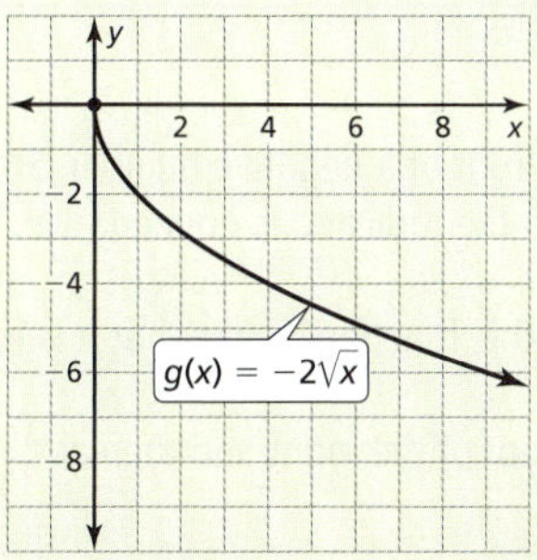

The graph of g is a vertical stretch by a factor of 2 and a reflection of the x-axis of the graph of f.

Differentiated Instruction

Kinesthetic/Auditory

Have students work in pairs to graph square root functions and describe the range. Have one student make the table of values and the other student graph the function. Then have partners work together to describe the range. Partners should alternate making the table and graphing the function as they solve several problems.

Extra Example 1

Describe the domain of $f(x) = 2\sqrt{x + 3}$.

The domain is the set of real numbers greater than or equal to -3.

Extra Example 2

Graph $f(x) = \sqrt{x} + 4$. Describe the range of the function.

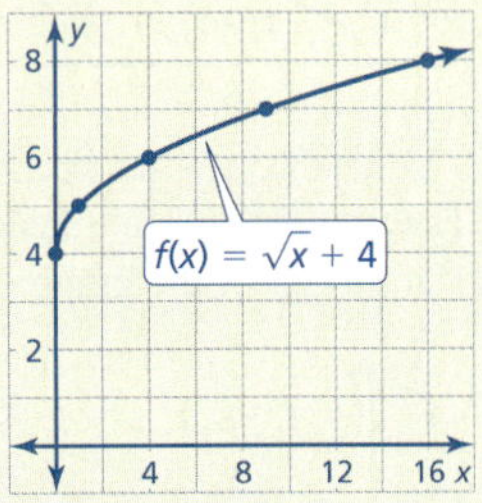

The range of f is $y \geq 4$.

MONITORING PROGRESS ANSWERS

1. $x \geq 0$
2. $x \geq 0$
3. $x \leq 1$
4. 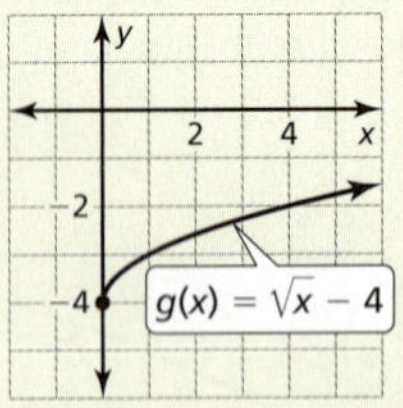

$y \geq -4$
5. 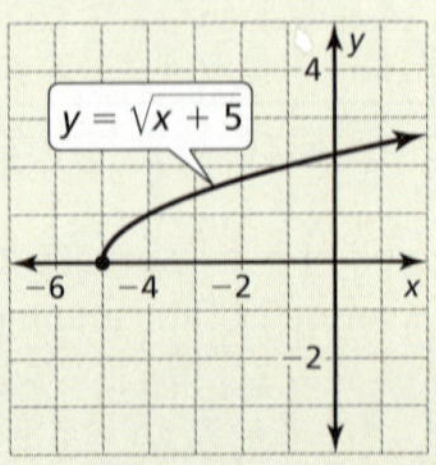

$y \geq 0$
6. See Additional Answers.

10.1 Lesson

Core Vocabulary

square root function, *p. 544*
radical function, *p. 545*

Previous
radicand
transformation
average rate of change

What You Will Learn

- Graph square root functions.
- Compare square root functions using average rates of change.
- Solve real-life problems involving square root functions.

Graphing Square Root Functions

Core Concept

Square Root Functions

A **square root function** is a function that contains a square root with the independent variable in the radicand. The parent function for the family of square root functions is $f(x) = \sqrt{x}$. The domain of f is $x \geq 0$, and the range of f is $y \geq 0$.

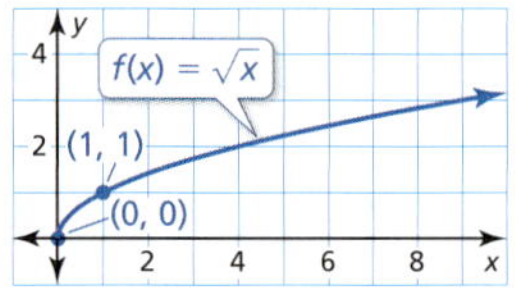

STUDY TIP

The graph of $f(x) = \sqrt{x}$ starts at (0, 0) and increases on the entire domain. So, the minimum value of f is 0.

The value of the radicand in a square root function cannot be negative. So, the domain of a square root function includes x-values for which the radicand is greater than or equal to 0.

EXAMPLE 1 Describing the Domain of a Square Root Function

Describe the domain of $f(x) = 3\sqrt{x - 5}$.

SOLUTION

The radicand cannot be negative. So, $x - 5$ is greater than or equal to 0.

$x - 5 \geq 0$ Write an inequality for the domain.

$x \geq 5$ Add 5 to each side.

▶ The domain is the set of real numbers greater than or equal to 5.

EXAMPLE 2 Graphing a Square Root Function

Graph $f(x) = \sqrt{x} + 3$. Describe the range of the function.

SOLUTION

Step 1 Use the domain of f, $x \geq 0$, to make a table of values.

x	0	1	4	9	16
$f(x)$	3	4	5	6	7

Step 2 Plot the ordered pairs.

Step 3 Draw a smooth curve through the points, starting at (0, 3).

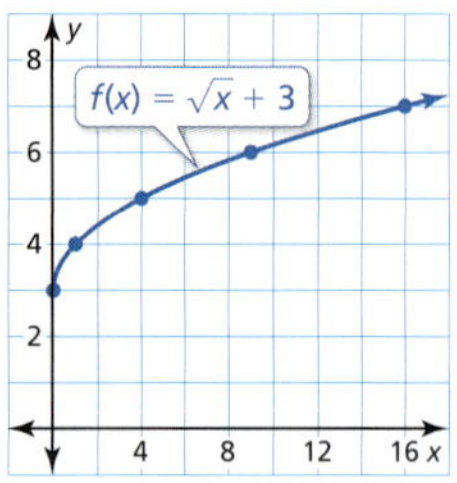

▶ From the graph, you can see that the range of f is $y \geq 3$.

Monitoring Progress

Help in English and Spanish at *BigIdeasMath.com*

Describe the domain of the function.

1. $f(x) = 10\sqrt{x}$
2. $y = \sqrt{2x} + 7$
3. $h(x) = \sqrt{-x + 1}$

Graph the function. Describe the range.

4. $g(x) = \sqrt{x} - 4$
5. $y = \sqrt{x + 5}$
6. $n(x) = 5\sqrt{x}$

Laurie's Notes Teacher Actions

- Write the *Core Concept*. Say, "You cannot take the square root of a negative number. So, the domain of a square root function includes inputs for which the radicand is nonnegative." Give examples to support this statement.
- ? "In Example 2, can you use $x = 1, 2, 3, 4$, and 5 to graph this function?" Yes, but it is easier to use perfect squares.
- **Attend to Precision:** Stress that you take the square root of the input and then add three, not add three and then take the square root.

STUDY TIP

You will study another type of radical function in the next section.

A **radical function** is a function that contains a radical expression with the independent variable in the radicand. A square root function is a type of radical function.

You can transform graphs of radical functions in the same way you transformed graphs of functions previously. In Example 2, notice that the graph of f is a vertical translation of the graph of the parent square root function.

Core Concept

Transformation	$f(x)$ Notation	Examples	
Horizontal Translation Graph shifts left or right.	$f(x - h)$	$g(x) = \sqrt{x - 2}$ $g(x) = \sqrt{x + 3}$	2 units right 3 units left
Vertical Translation Graph shifts up or down.	$f(x) + k$	$g(x) = \sqrt{x} + 7$ $g(x) = \sqrt{x} - 1$	7 units up 1 unit down
Reflection Graph flips over x- or y-axis.	$f(-x)$ $-f(x)$	$g(x) = \sqrt{-x}$ $g(x) = -\sqrt{x}$	in the y-axis in the x-axis
Horizontal Stretch or Shrink Graph stretches away from or shrinks toward y-axis.	$f(ax)$	$g(x) = \sqrt{3x}$ $g(x) = \sqrt{\frac{1}{2}x}$	shrink by a factor of $\frac{1}{3}$ stretch by a factor of 2
Vertical Stretch or Shrink Graph stretches away from or shrinks toward x-axis.	$a \cdot f(x)$	$g(x) = 4\sqrt{x}$ $g(x) = \frac{1}{5}\sqrt{x}$	stretch by a factor of 4 shrink by a factor of $\frac{1}{5}$

EXAMPLE 3 Comparing Graphs of Square Root Functions

Graph $g(x) = -\sqrt{x - 2}$. Compare the graph to the graph of $f(x) = \sqrt{x}$.

SOLUTION

Step 1 Use the domain of g, $x \geq 2$, to make a table of values.

x	2	3	4	5	6
$g(x)$	0	−1	−1.4	−1.7	−2

Step 2 Plot the ordered pairs.

Step 3 Draw a smooth curve through the points, starting at (2, 0).

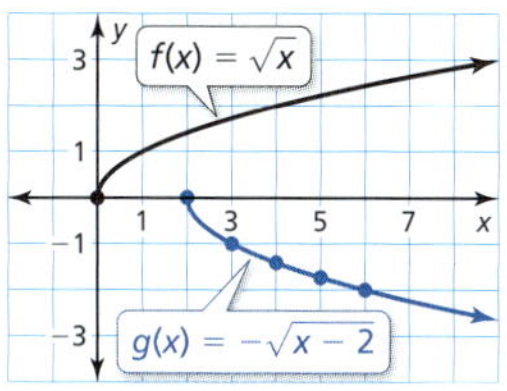

▶ The graph of g is a translation 2 units right and a reflection in the x-axis of the graph of f.

Monitoring Progress

Help in English and Spanish at *BigIdeasMath.com*

Graph the function. Compare the graph to the graph of $f(x) = \sqrt{x}$.

7. $h(x) = \sqrt{\frac{1}{4}x}$ **8.** $g(x) = \sqrt{x} - 6$ **9.** $m(x) = -3\sqrt{x}$

Laurie's Notes Teacher Actions

- **FYI:** The square root function is an example of a radical function. There are other radical functions, such as $y = \sqrt[3]{x}$. This should make sense to students from when they learned about rational exponents and nth roots.
- Discuss transformations of the parent function.
- **Turn and Talk:** "What will the graph of $g(x) = -\sqrt{x - 2}$ look like and why?" Listen for a shift 2 units right and a horizontal reflection.
- **Think-Pair-Share:** Have students answer Questions 7–9, and then share and discuss as a class.

English Language Learners

Notebook Development

Have students copy the *Core Concept* transformation table into their notebooks. Students can refer to the table when they compare graphs of square root functions like those in Example 3 and describe transformations of graphs like those in Example 4.

Extra Example 3

Graph $g(x) = -\sqrt{x} + 2$. Compare the graph to the graph of $f(x) = \sqrt{x}$.

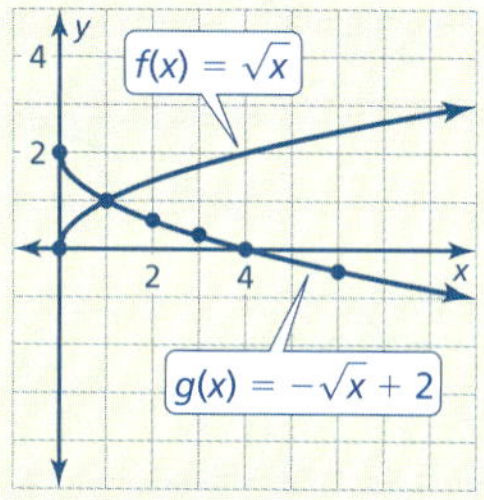

The graph of g is a reflection in the x-axis and a translation 2 units up of the graph of f.

MONITORING PROGRESS ANSWERS

7.

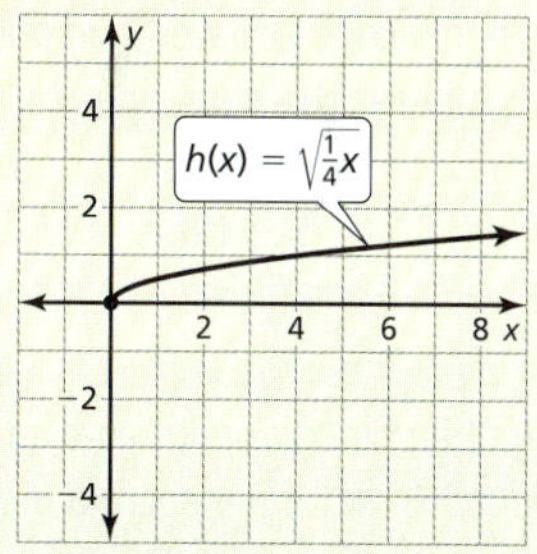

The graph of h is a horizontal stretch by a factor of 4 of the graph of f.

8.

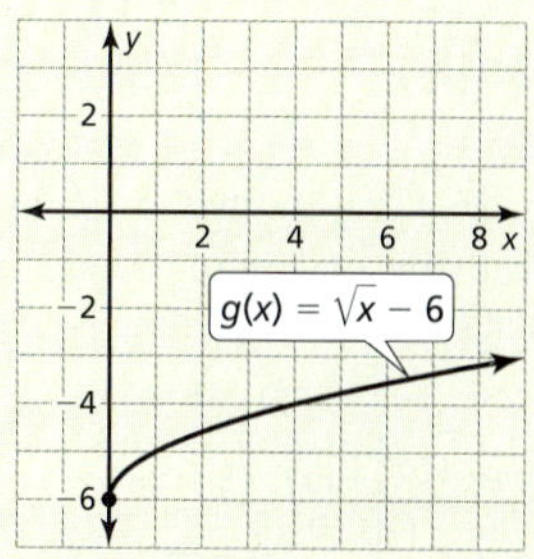

The graph of g is a translation 6 units down of the graph of f.

9. See Additional Answers.

Extra Example 4

Let $g(x) = 3\sqrt{x+1} - 3$.

a. Describe the transformations from the graph of $f(x) = \sqrt{x}$ to the graph of g.

Translate the graph of f horizontally 1 unit left to get the graph of $t(x) = \sqrt{x+1}$.

Stretch the graph of t vertically by a factor of 3 to get the graph of $h(x) = 3\sqrt{x+1}$.

Translate the graph of h vertically 3 units down to get the graph of $g(x) = 3\sqrt{x+1} - 3$.

b. Graph g.

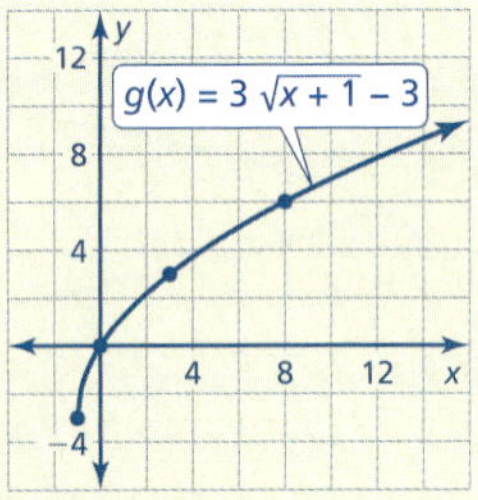

Extra Example 5

Use the graph in Example 5. The model $v(d) = \sqrt{2gd}$ represents the velocity v (in meters per second) of a free-falling object on the moon, where g is the constant 1.6 meters per second squared and d is the distance (in meters) the object has fallen. Compare the velocities by finding and interpreting their average rates of change over the interval $d = 10$ to $d = 20$.

From 10 to 20 meters, the velocity of a free-falling object increases at an average rate of about 0.6 meter per second per meter on Earth and about 0.23 meter per second per meter on the moon.

MONITORING PROGRESS ANSWER

10. Translate 4 units to the left, shrink vertically by a factor of $\frac{1}{2}$, and translate 1 unit up to obtain the graph of g.

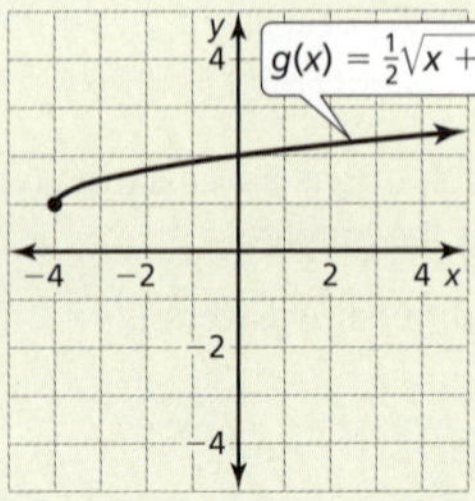

To graph a square root function of the form $y = a\sqrt{x-h} + k$, where $a \neq 0$, start at (h, k).

EXAMPLE 4 Graphing $y = a\sqrt{x-h} + k$

Let $g(x) = -2\sqrt{x-3} - 2$. (a) Describe the transformations from the graph of $f(x) = \sqrt{x}$ to the graph of g. (b) Graph g.

REMEMBER

The graph of $y = a \cdot f(x - h) + k$ can be obtained from the graph of $y = f(x)$ using the steps you learned in Section 3.6.

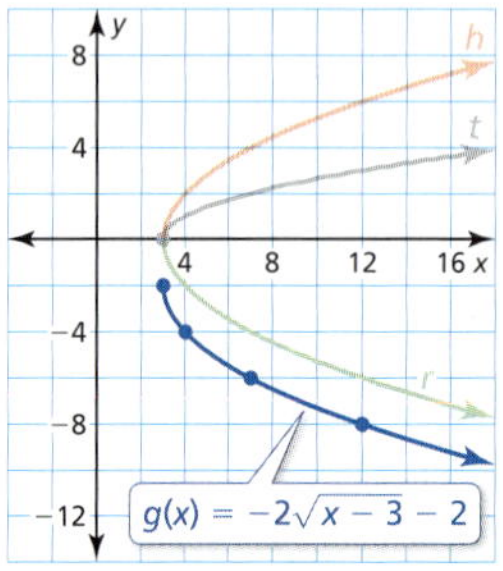

SOLUTION

a. **Step 1** Translate the graph of f horizontally 3 units right to get the graph of $t(x) = \sqrt{x-3}$.

Step 2 Stretch the graph of t vertically by a factor of 2 to get the graph of $h(x) = 2\sqrt{x-3}$.

Step 3 Reflect the graph of h in the x-axis to get the graph of $r(x) = -2\sqrt{x-3}$.

Step 4 Translate the graph of r vertically 2 units down to get the graph of $g(x) = -2\sqrt{x-3} - 2$.

b. **Step 1** Use the domain, $x \geq 3$, to make a table of values.

x	3	4	7	12
$g(x)$	−2	−4	−6	−8

Step 2 Plot the ordered pairs.

Step 3 Start at $(h, k) = (3, -2)$ and draw a smooth curve through the points.

Monitoring Progress

Help in English and Spanish at *BigIdeasMath.com*

10. Let $g(x) = \frac{1}{2}\sqrt{x+4} + 1$. Describe the transformations from the graph of $f(x) = \sqrt{x}$ to the graph of g. Then graph g.

Comparing Average Rates of Change

EXAMPLE 5 Comparing Square Root Functions

The model $v(d) = \sqrt{2gd}$ represents the velocity v (in meters per second) of a free-falling object on the moon, where g is the constant 1.6 meters per second squared and d is the distance (in meters) the object has fallen. The velocity of a free-falling object on Earth is shown in the graph. Compare the velocities by finding and interpreting their average rates of change over the interval $d = 0$ to $d = 10$.

Free-Falling Object on Earth

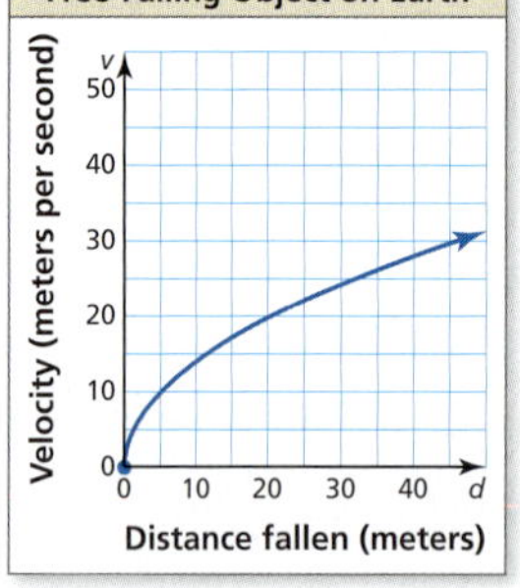

SOLUTION

To calculate the average rates of change, use points whose d-coordinates are 0 and 10.

Earth: Use the graph to estimate. Use (0, 0) and (10, 14).

$$\frac{v(10) - v(0)}{10 - 0} \approx \frac{14 - 0}{10} = 1.4$$ Average rate of change on Earth

Moon: Evaluate v when $d = 0$ and $d = 10$.

$v(0) = \sqrt{2(1.6)(0)} = 0$ and $v(10) = \sqrt{2(1.6)(10)} = \sqrt{32} \approx 5.7$

Use (0, 0) and $(10, \sqrt{32})$.

$$\frac{v(10) - v(0)}{10 - 0} = \frac{\sqrt{32} - 0}{10} \approx 0.57$$ Average rate of change on the moon

▶ From 0 to 10 meters, the velocity of a free-falling object increases at an average rate of about 1.4 meters per second per meter on Earth and about 0.57 meter per second per meter on the moon.

Laurie's Notes Teacher Actions

? "What do you recall about the graph of $y = a(x - h)^2 + k$?" The graph is a parabola with vertex at (h, k).

? "What do you know about the graph of $g(x) = -2\sqrt{x-3} - 2$?" Answers will vary. Listen for (3, −2) as a key ordered pair. See the *Teaching Strategy* on page T-542.

- Give time for students to read Example 5 and study the graph.

? "What does it mean to find an average rate of change?" to calculate the slope between two points on a curve

- **Thumbs Up:** Have students work with their partners to perform both computations. Use the *Thumbs Up* self-assessment.

Monitoring Progress

Help in English and Spanish at *BigIdeasMath.com*

11. In Example 5, compare the velocities by finding and interpreting their average rates of change over the interval $d = 30$ to $d = 40$.

Solving Real-Life Problems

EXAMPLE 6 Real-Life Application

The velocity v (in meters per second) of a tsunami can be modeled by the function $v(x) = \sqrt{9.8x}$, where x is the water depth (in meters). (a) Use a graphing calculator to graph the function. At what depth does the velocity of the tsunami exceed 200 meters per second? (b) What happens to the average rate of change of the velocity as the water depth increases?

SOLUTION

1. **Understand the Problem** You know the function that models the velocity of a tsunami based on water depth. You are asked to graph the function using a calculator and find the water depth where the velocity exceeds 200 meters per second. Then you are asked to describe the average rate of change of the velocity as the water depth increases.

2. **Make a Plan** Graph the function using a calculator. Use the *trace* feature to find the value of x when $v(x)$ is about 200. Then calculate and compare average rates of change of the velocity over different intervals.

3. **Solve the Problem**

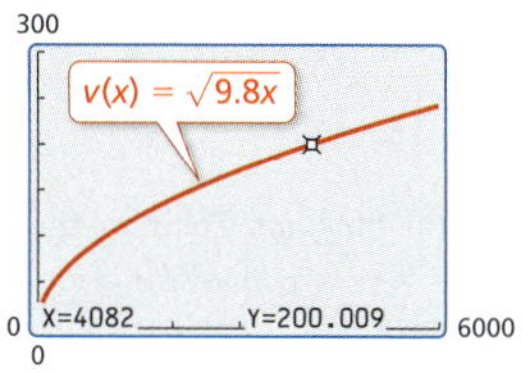

a. Step 1 Enter the function into your calculator and graph it.

Step 2 Use the *trace* feature to find the value of x when $v(x) \approx 200$.

▶ The velocity exceeds 200 meters per second at a depth of about 4082 meters.

b. Calculate the average rates of change over the intervals $x = 0$ to $x = 1000$, $x = 1000$ to $x = 2000$, and $x = 2000$ to $x = 3000$.

$$\frac{v(1000) - v(0)}{1000 - 0} = \frac{\sqrt{9800} - 0}{1000} \approx 0.099 \qquad \text{0 to 1000 meters}$$

$$\frac{v(2000) - v(1000)}{2000 - 1000} = \frac{\sqrt{19{,}600} - \sqrt{9800}}{1000} \approx 0.041 \qquad \text{1000 to 2000 meters}$$

$$\frac{v(3000) - v(2000)}{3000 - 2000} = \frac{\sqrt{29{,}400} - \sqrt{19{,}600}}{1000} \approx 0.031 \qquad \text{2000 to 3000 meters}$$

▶ The average rate of change of the velocity decreases as the water depth increases.

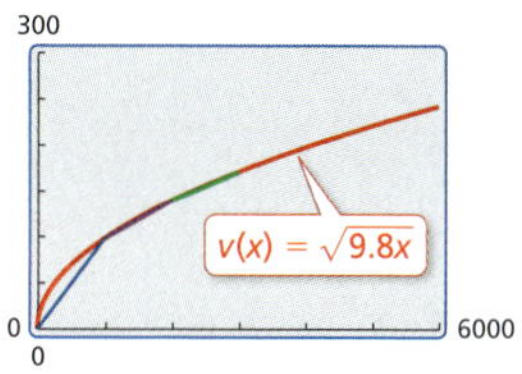

4. **Look Back** To check the answer in part (a), find $v(x)$ when $x = 4082$.

$$v(4082) = \sqrt{9.8(4082)} \approx 200 \checkmark$$

In part (b), the slopes of the line segments (shown at the left) that represent the average rates of change over the intervals are decreasing. So, the answer to part (b) is reasonable.

Monitoring Progress

Help in English and Spanish at *BigIdeasMath.com*

12. WHAT IF? At what depth does the velocity of the tsunami exceed 100 meters per second?

Extra Example 6

The distance d (in miles) that you can see to the horizon can be modeled by $d(x) = \sqrt{\frac{3x}{2}}$, where x is the height of your eye level (in feet) above the ground.

a. Use a graphing calculator to graph the function. At what eye-level height does the distance you can see to the horizon exceed 6 miles?

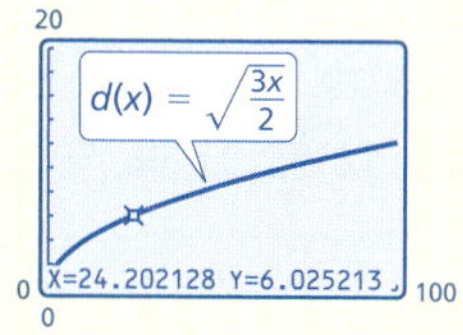

The distance exceeds 6 miles at an eye-level height of about 24 feet.

b. What happens to the average rate of change of the distance as your eye-level height increases?
The average rate of change of the distance decreases as your eye-level height increases.

MONITORING PROGRESS ANSWERS

11. From 30 to 40 meters, the velocity of a falling object increases at an average rate of about 0.40 meter per second per meter on Earth and about 0.15 meter per second per meter on the Moon.

12. about 1021 m

Laurie's Notes Teacher Actions

- Pose Example 6. Say, "Graph the function $v(x) = \sqrt{9.8x}$ in a good viewing window." Circulate as students work.

? "At what depth does the velocity exceed 200 meters?" at about 4082 meters

- Increments of 1000 meters are selected to explore the change in the average rate of change. Explain that 1000 meters is a reasonable increment versus using only one interval, 0 to 3000 meters.

Closure

- **Exit Ticket:** Compare the graph of $y = \sqrt{x + 1} - 3$ to the graph of $y = \sqrt{x}$. translation 1 unit left and 3 units down

10.1 Exercises

Dynamic Solutions available at *BigIdeasMath.com*

Vocabulary and Core Concept Check

1. **COMPLETE THE SENTENCE** A ________ is a function that contains a radical expression with the independent variable in the radicand.

2. **VOCABULARY** Is $y = 2x\sqrt{5}$ a square root function? Explain.

3. **WRITING** How do you describe the domain of a square root function?

4. **REASONING** Is the graph of $g(x) = 1.25\sqrt{x}$ a vertical stretch or a vertical shrink of the graph of $f(x) = \sqrt{x}$? Explain.

Monitoring Progress and Modeling with Mathematics

In Exercises 5–14, describe the domain of the function. *(See Example 1.)*

5. $y = 8\sqrt{x}$
6. $y = \sqrt{4x}$
7. $y = 4 + \sqrt{-x}$
8. $y = \sqrt{-\frac{1}{2}x} + 1$
9. $h(x) = \sqrt{x - 4}$
10. $p(x) = \sqrt{x + 7}$
11. $f(x) = \sqrt{-x + 8}$
12. $g(x) = \sqrt{-x - 1}$
13. $m(x) = 2\sqrt{x + 4}$
14. $n(x) = \frac{1}{2}\sqrt{-x - 2}$

In Exercises 15–18, match the function with its graph. Describe the range.

15. $y = \sqrt{x - 3}$
16. $y = 3\sqrt{x}$
17. $y = \sqrt{x} - 3$
18. $y = \sqrt{-x + 3}$

A.
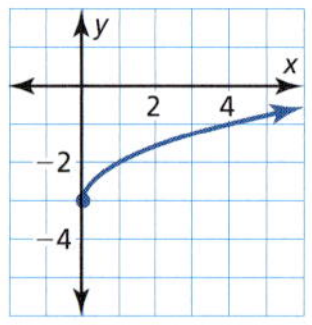

B.
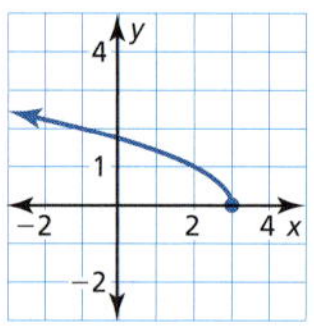

C.
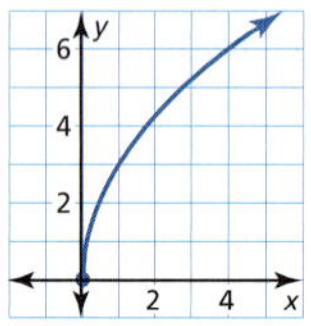

D.
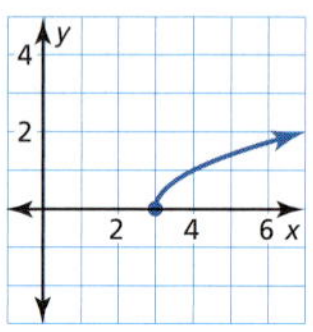

In Exercises 19–26, graph the function. Describe the range. *(See Example 2.)*

19. $y = \sqrt{3x}$
20. $y = 4\sqrt{-x}$
21. $y = \sqrt{x} + 5$
22. $y = -2 + \sqrt{x}$
23. $f(x) = -\sqrt{x - 3}$
24. $g(x) = \sqrt{x + 4}$
25. $h(x) = \sqrt{x + 2} - 2$
26. $f(x) = -\sqrt{x - 1} + 3$

In Exercises 27–34, graph the function. Compare the graph to the graph of $f(x) = \sqrt{x}$. *(See Example 3.)*

27. $g(x) = \frac{1}{4}\sqrt{x}$
28. $r(x) = \sqrt{2x}$
29. $h(x) = \sqrt{x + 3}$
30. $q(x) = \sqrt{x} + 8$
31. $p(x) = \sqrt{-\frac{1}{3}x}$
32. $g(x) = -5\sqrt{x}$
33. $m(x) = -\sqrt{x} - 6$
34. $n(x) = -\sqrt{x - 4}$

35. **ERROR ANALYSIS** Describe and correct the error in graphing the function $y = \sqrt{x + 1}$.

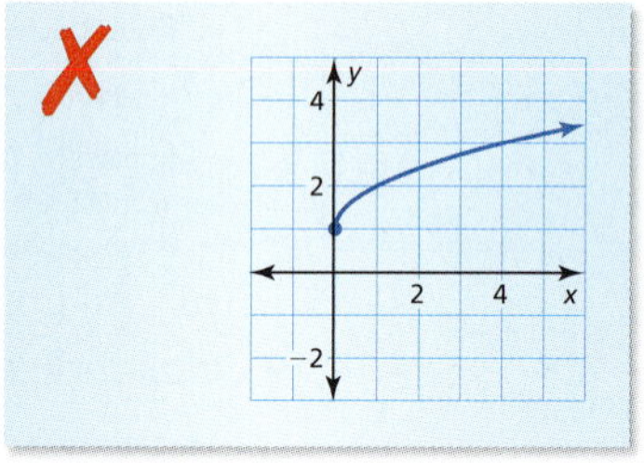

Assignment Guide and Homework Check

ASSIGNMENT

Basic: 1–4, 5–45 odd, 49, 54, 58–63

Average: 1–4, 6–50 even, 54, 58–63

Advanced: 1–4, 12–18 even, 24, 26, 32–36 even, 42–50 even, 51–63

HOMEWORK CHECK

Basic: 5, 19, 27, 45, 49

Average: 6, 20, 28, 46, 50

Advanced: 12, 24, 32, 46, 50

ANSWERS

1. radical function
2. no; The independent variable is not in the radicand.
3. Describe the x-values for which the radicand is greater than or equal to 0.
4. vertical stretch; The y-coordinates increase by a factor of 1.25.
5. $x \ge 0$
6. $x \ge 0$
7. $x \le 0$
8. $x \le 0$
9. $x \ge 4$
10. $x \ge -7$
11. $x \le 8$
12. $x \le -1$
13. $x \ge -4$
14. $x \le -2$
15. D; $y \ge 0$
16. C; $y \ge 0$
17. A; $y \ge -3$
18. B; $y \ge 0$
19. 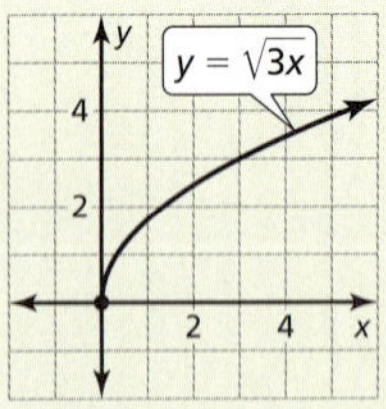

$y \ge 0$

20. 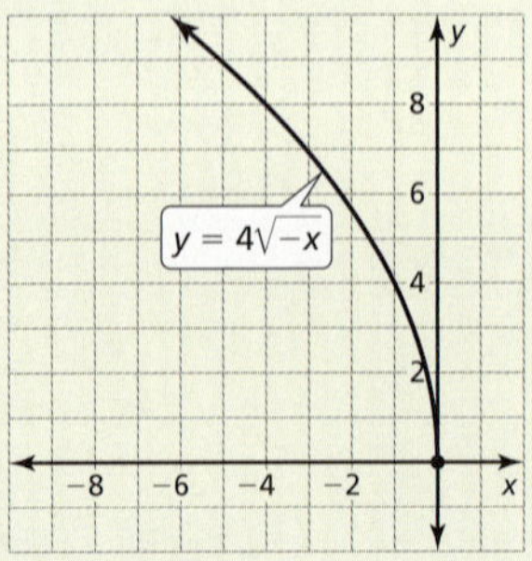

$y \ge 0$

21. 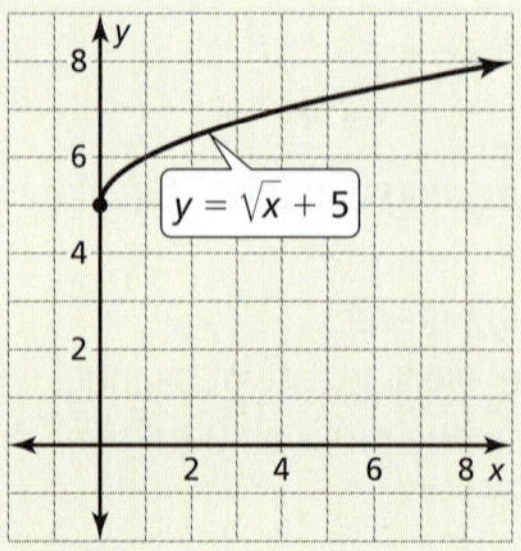

$y \ge 5$

22. 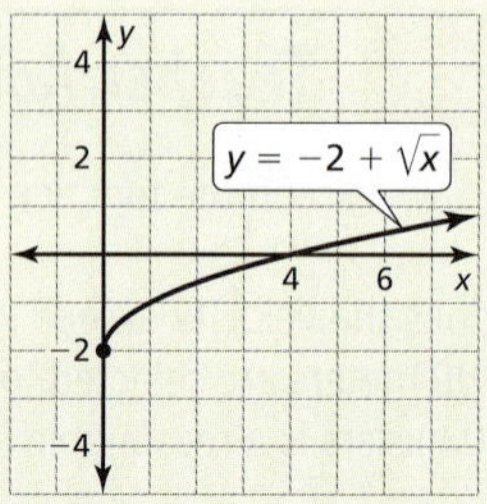

$y \ge -2$

23–35. See Additional Answers.

36. ERROR ANALYSIS Describe and correct the error in comparing the graph of $g(x) = -\frac{1}{4}\sqrt{x}$ to the graph of $f(x) = \sqrt{x}$.

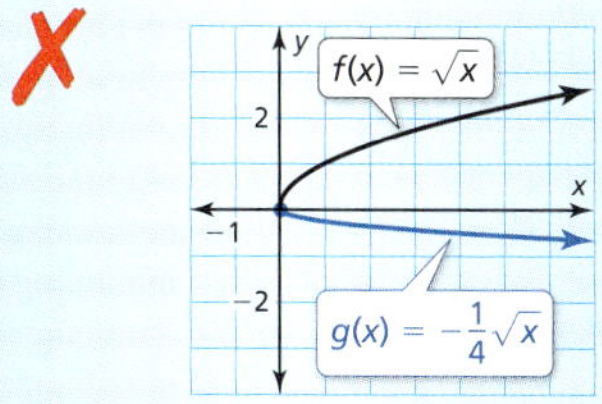

The graph of g is a horizontal stretch by a factor of 4 and a reflection in the x-axis of the graph of f.

In Exercises 37–44, describe the transformations from the graph of $f(x) = \sqrt{x}$ to the graph of h. Then graph h. *(See Example 4.)*

37. $h(x) = 4\sqrt{x+2} - 1$

38. $h(x) = \frac{1}{2}\sqrt{x-6} + 3$

39. $h(x) = 2\sqrt{-x} - 6$

40. $h(x) = -\sqrt{x-3} - 2$

41. $h(x) = \frac{1}{3}\sqrt{x+3} - 3$

42. $h(x) = 2\sqrt{x-1} + 4$

43. $h(x) = -2\sqrt{x-1} + 5$

44. $h(x) = -5\sqrt{x+2} - 1$

45. COMPARING FUNCTIONS The model $S(d) = \sqrt{30df}$ represents the speed S (in miles per hour) of a van before it skids to a stop, where f is the drag factor of the road surface and d is the length (in feet) of the skid marks. The drag factor of Road Surface A is 0.75. The graph shows the speed of the van on Road Surface B. Compare the speeds by finding and interpreting their average rates of change over the interval $d = 0$ to $d = 15$. *(See Example 5.)*

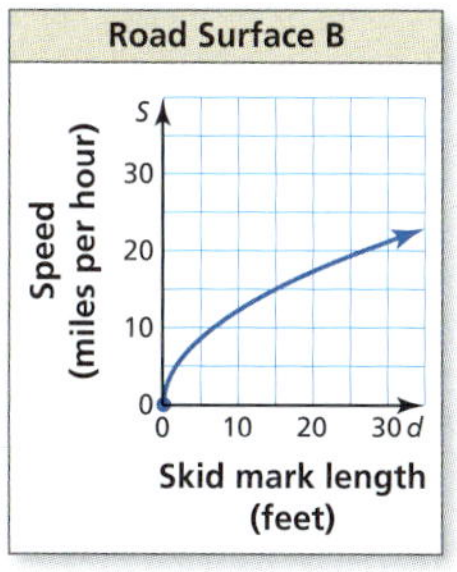

46. COMPARING FUNCTIONS The velocity v (in meters per second) of an object in motion is given by $v(E) = \sqrt{\frac{2E}{m}}$, where E is the kinetic energy of the object (in joules) and m is the mass of the object (in kilograms). The mass of Object A is 4 kilograms. The graph shows the velocity of Object B. Compare the velocities of the objects by finding and interpreting the average rates of change over the interval $E = 0$ to $E = 6$.

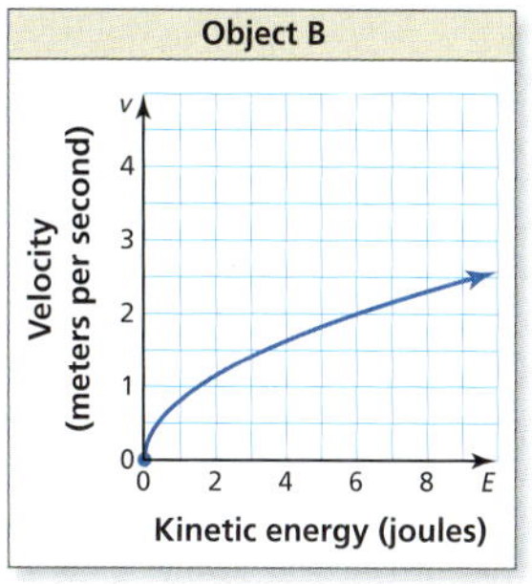

47. OPEN-ENDED Consider the graph of $y = \sqrt{x}$.

a. Write a function that is a vertical translation of the graph of $y = \sqrt{x}$.

b. Write a function that is a reflection of the graph of $y = \sqrt{x}$.

48. REASONING Can the domain of a square root function include negative numbers? Can the range include negative numbers? Explain your reasoning.

49. PROBLEM SOLVING The nozzle pressure of a fire hose allows firefighters to control the amount of water they spray on a fire. The flow rate f (in gallons per minute) can be modeled by the function $f = 120\sqrt{p}$, where p is the nozzle pressure (in pounds per square inch). *(See Example 6.)*

a. Use a graphing calculator to graph the function. At what pressure does the flow rate exceed 300 gallons per minute?

b. What happens to the average rate of change of the flow rate as the pressure increases?

Dynamic Teaching Tools

- Dynamic Assessment & Progress Monitoring Tool
- Interactive Whiteboard Lesson Library
- Dynamic Classroom with Dynamic Investigations

ANSWERS

36. A vertical shrink was confused with a horizontal stretch. The graph of g is a vertical shrink by a factor of $\frac{1}{4}$ and a reflection in the x-axis of the graph of f.

37. Translate 2 units left, stretch vertically by a factor of 4, and translate 1 unit down to obtain the graph of h.

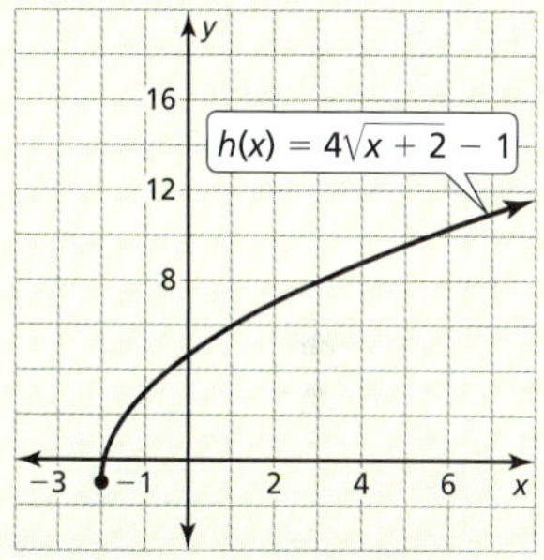

38. Translate 6 units right, shrink vertically by a factor of $\frac{1}{2}$, and translate 3 units up to obtain the graph of h.

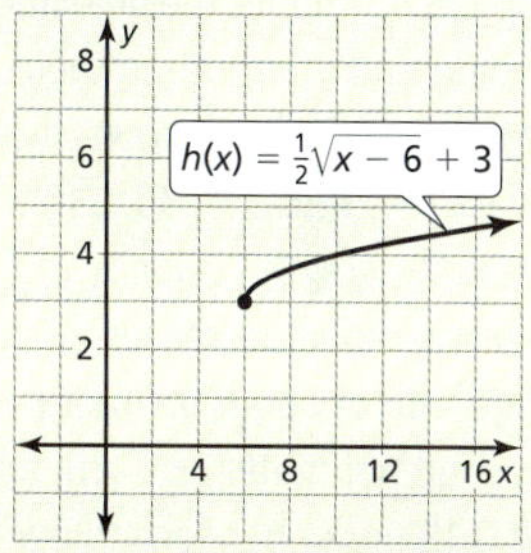

39. Reflect in the y-axis, stretch vertically by a factor of 2, and translate 6 units down to obtain the graph of h.

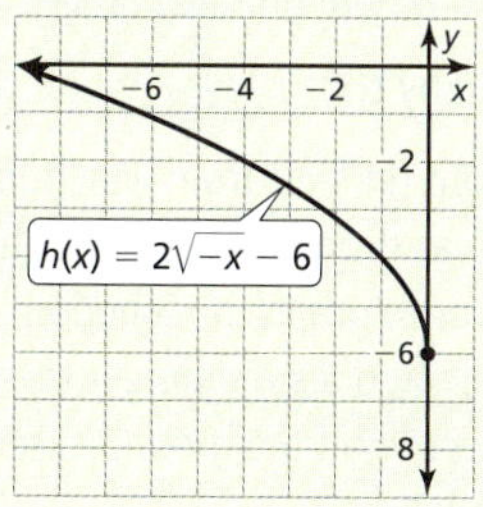

40. Translate 3 units right, reflect in the x-axis, and translate 2 units down to obtain the graph of h.

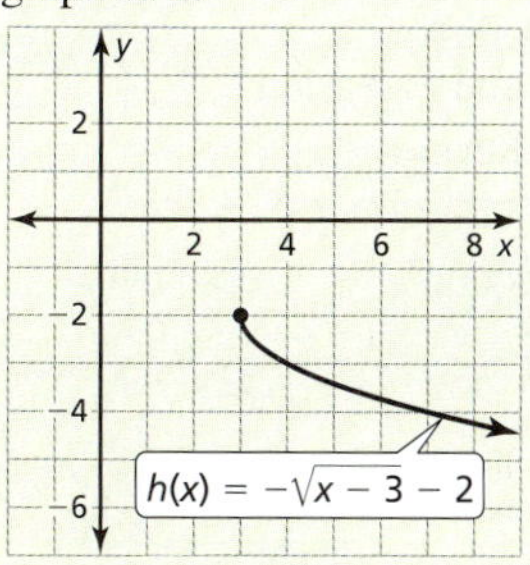

41. Translate 3 units left, shrink vertically by a factor of $\frac{1}{3}$, and translate 3 units down to obtain the graph of h.

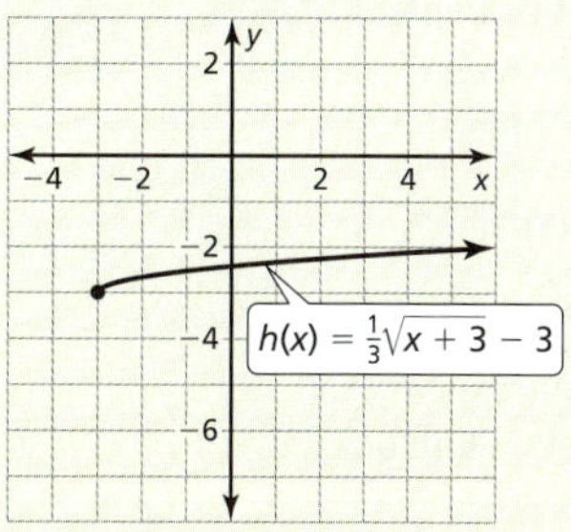

42–49. See Additional Answers.

ANSWERS

50. a.

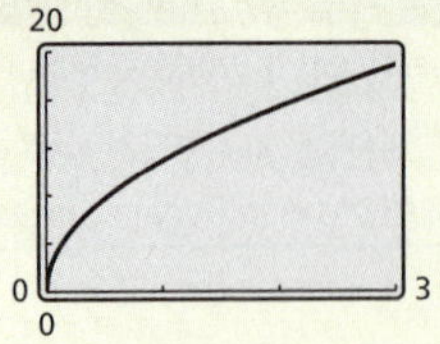

about 0.71 m

b. The graph of s would be translated up.

51–63. See Additional Answers.

Mini-Assessment

1. Graph $f(x) = \sqrt{2x} - 1$. Describe the range of the function.

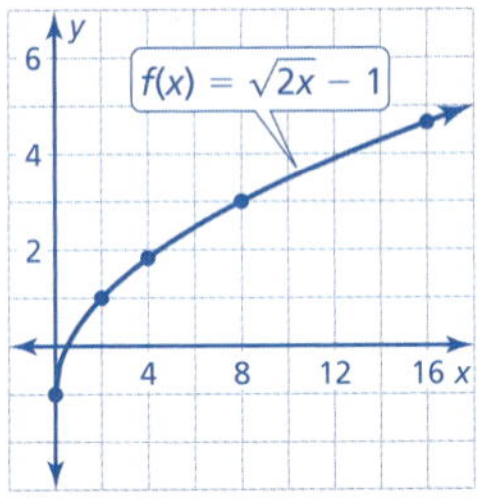

The range of f is $y \geq -1$.

2. Use the graph and function in Example 5. Compare the velocities by finding and interpreting their average rates of change over the interval $d = 20$ to $d = 30$. From 20 to 30 meters, the velocity of a free-falling object increases at an average rate of about 0.4 meter per second per meter on Earth and about 0.18 meter per second per meter on the moon.

3. The time t (in seconds) it takes an object to hit the ground can be modeled by $t(x) = \sqrt{\frac{x}{16}}$, where x is the height (in feet) from which the object was dropped. Use a graphing calculator to graph the function. At what height does the time it takes the object to hit the ground exceed 4 seconds?

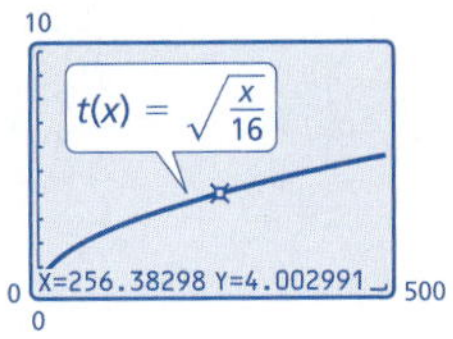

The time exceeds 4 seconds at a height of about 256 feet.

50. PROBLEM SOLVING The speed s (in meters per second) of a long jumper before jumping can be modeled by the function $s = 10.9\sqrt{h}$, where h is the maximum height (in meters from the ground) of the jumper.

a. Use a graphing calculator to graph the function. A jumper is running 9.2 meters per second. Estimate the maximum height of the jumper.

b. Suppose the runway and pit are raised on a platform slightly higher than the ground. How would the graph of the function be transformed?

51. MATHEMATICAL CONNECTIONS The radius r of a circle is given by $r = \sqrt{\frac{A}{\pi}}$, where A is the area of the circle.

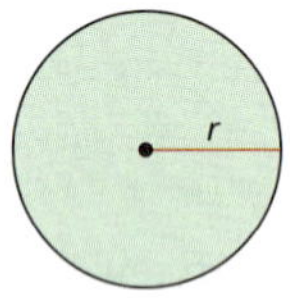

a. Describe the domain of the function. Use a graphing calculator to graph the function.

b. Use the *trace* feature to approximate the area of a circle with a radius of 5.4 inches.

52. REASONING Consider the function $f(x) = 8a\sqrt{x}$.

a. For what value of a will the graph of f be identical to the graph of the parent square root function?

b. For what values of a will the graph of f be a vertical stretch of the graph of the parent square root function?

c. For what values of a will the graph of f be a vertical shrink and a reflection of the graph of the parent square root function?

53. REASONING The graph represents the function $f(x) = \sqrt{x}$.

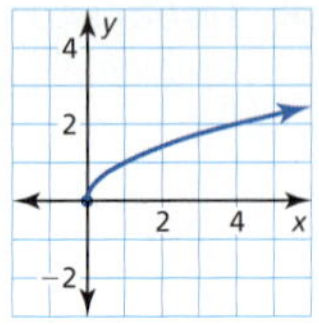

a. What is the minimum value of the function?

b. Does the function have a maximum value? Explain.

c. Write a square root function that has a maximum value. Does the function have a minimum value? Explain.

d. Write a square root function that has a minimum value of −4.

54. HOW DO YOU SEE IT? Match each function with its graph. Explain your reasoning.

A. $f(x) = \sqrt{x} + 2$ **B.** $m(x) = f(x) - 4$

C. $n(x) = f(-x)$ **D.** $p(x) = f(3x)$

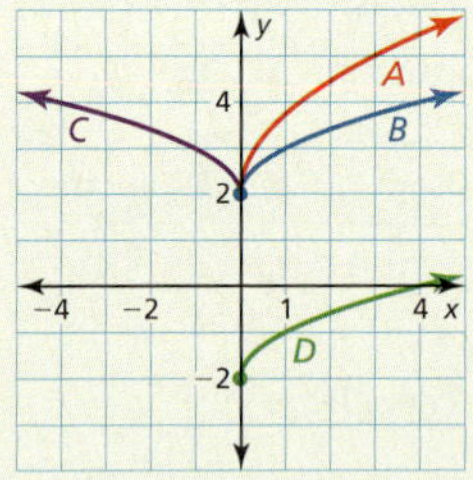

55. REASONING Without graphing, determine which function's graph rises more steeply, $f(x) = 5\sqrt{x}$ or $g(x) = \sqrt{5x}$. Explain your reasoning.

56. THOUGHT PROVOKING Use a graphical approach to find the solutions of $x - 1 = \sqrt{5x - 9}$. Show your work. Verify your solutions algebraically.

57. OPEN-ENDED Write a radical function that has a domain of all real numbers greater than or equal to −5 and a range of all real numbers less than or equal to 3.

Maintaining Mathematical Proficiency

Reviewing what you learned in previous grades and lessons

Evaluate the expression. *(Section 6.2)*

58. $\sqrt[3]{343}$ **59.** $\sqrt[3]{-64}$ **60.** $-\sqrt[3]{-\frac{1}{27}}$

Factor the polynomial. *(Section 7.5)*

61. $x^2 + 7x + 6$ **62.** $d^2 - 11d + 28$ **63.** $y^2 - 3y - 40$

If students need help...	*If students got it...*
Resources by Chapter • Practice A and Practice B • Puzzle Time	Resources by Chapter • Enrichment and Extension • Cumulative Review
Student Journal • Practice	Start the *next* Section
Differentiating the Lesson Skills Review Handbook	

Laurie's Notes

Overview of Section 10.2

Introduction
- In this lesson, the graph of the cube root function is introduced, along with transformations to the parent function.
- Cube root functions are compared using average rate of change.
- The lesson ends with a real-life application.

Teaching Strategy
- In the last lesson, students graphed transformations of the square root function. Transformations of the cube root function are the same.
- Unlike the square root function, the cube root function is defined over all real numbers.
- The cube root function is an odd function with symmetry about the origin. There are three key points that help in graphing the parent function and transformations of the parent function: (0, 0), (1, 1) and (−1, −1).

? "What happens to (0, 0) under the transformation $g(x) = 2\sqrt[3]{x-3} + 4$?" Ordered pairs are translated 3 units right and up 4 units, so (0, 0) maps to (3, 4).

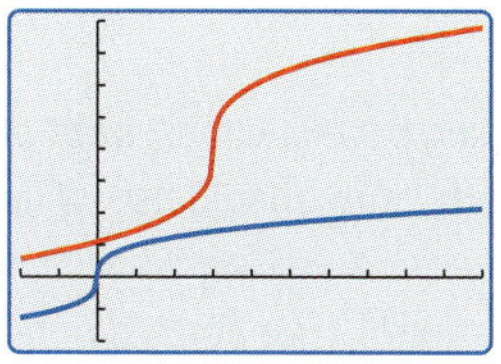

- In the table of values, you can see this point of inflection and also the images of (1, 1) and (−1, −1) of the parent function.

X	Y1	Y2
-2	-1.26	.58005
-1	-1	.8252
0	0	1.1155
1	1	1.4802
2	1.2599	2
3	1.4422	4
4	1.5874	6

X=-2

Pacing Suggestion
- The problems students will work in the explorations are similar to Example 1 in the formal lesson. If students show good comprehension in the explorations, you may begin the formal lesson with *Monitoring Progress* Questions 1–3.

Dynamic Teaching Tools
- Dynamic Assessment & Progress Monitoring Tool
- Lesson Planning Tool
- Interactive Whiteboard Lesson Library
- Dynamic Classroom with Dynamic Investigations

What Your Students Will Learn

- Graph cube root functions and describe the domain and range.
- Graph and compare the transformations of cube root functions to the graph of the parent function.
- Compare cube root functions using average rates of change.
- Solve real-life problems involving cube root functions.

Laurie's Notes

Exploration

Motivate

- Ask students what they know about Asian elephants.
- Elephants are the largest of Earth's land mammals, and Asian elephants are slightly smaller than their African cousins. Their ears help them to keep cool in the hot climate, and they also enjoy spraying themselves with water to cool off.
- Their average lifespan in the wild is up to 60 years. Their shoulder height ranges from 6.6 to 9.8 feet, and their weight ranges from 4500 to 11,000 pounds.
- Elephants eat roots, grasses, fruit, and bark, consuming up to 300 pounds of food in one day. These hungry animals do not sleep much, and they roam over great distances while foraging for food.

Exploration Note

- Students were introduced to radical expressions in Lesson 6.2, and therefore they should be comfortable with the notation $\sqrt[3]{x}$.

Exploration 1

- Students have graphed a variety of functions, so the directions should need no additional explanation.
- Check to see that students are using perfect cubes, both positive and negative, as their domain values in part (a). In part (b), they should use values that result in the radicand being a perfect cube.

? "What is the domain of each function?" The domain of both functions is all real numbers.

? "What is the range of each function?" The range of both functions is all real numbers.

- After students have finished both graphs, you could have them check their work using a graphing calculator.
- You could also ask volunteers to describe the differences in the graphs. Students should be able to describe the transformation in part (b).

Exploration 2

- Ordered pairs have been given so that students can determine the equation of the cube root function. In part (a), there are three ordered pairs, $(-4, 0)$, $(-1, \sqrt[3]{3})$, and $(4, 2)$. In part (b), there are three ordered pairs where each y-value is 1 more than the y-value of the related ordered pair in part (a).
- Students may need to use trial and error.
- Probe students as you walk around to see whether they recognize any patterns between the two tables.
- Have students identify the domain and range for each function.

Communicate Your Answer

- Students may comment that the cube root graphs look like a stretched out letter S.
- **Look For and Make Use of Structure:** Students have only graphed $y = \sqrt[3]{x - h}$. Ask them how a and k affect the graph of $y = a\sqrt[3]{x - h} + k$.

Connecting to Next Step

- If you feel students are comfortable with identifying the domain and range of a cube root function and graphing a vertical shift of the parent function, begin the formal lesson with Example 2.

10.2 Graphing Cube Root Functions

Essential Question What are some of the characteristics of the graph of a cube root function?

EXPLORATION 1 **Graphing Cube Root Functions**

Work with a partner.

- Make a table of values for each function. Use positive and negative values of x.
- Use the table to sketch the graph of each function.
- Describe the domain of each function.
- Describe the range of each function.

a. $y = \sqrt[3]{x}$

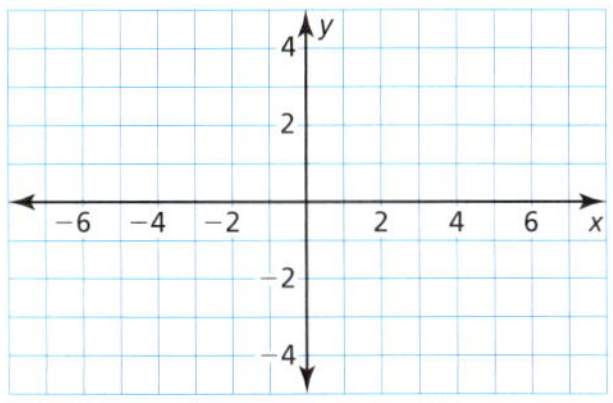

b. $y = \sqrt[3]{x + 3}$

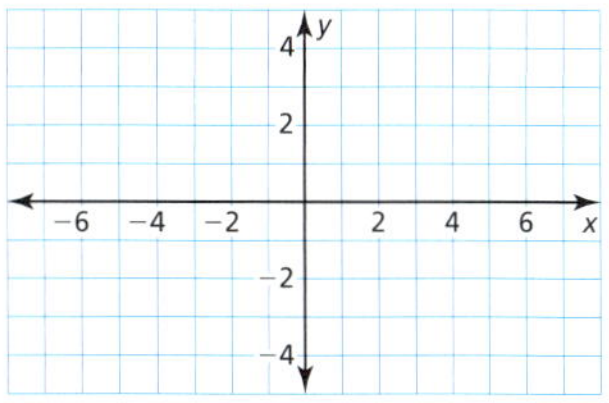

LOOKING FOR REGULARITY IN REPEATED REASONING

To be proficient in math, you need to notice whether calculations are repeated and look for both general methods and shortcuts.

EXPLORATION 2 **Writing Cube Root Functions**

Work with a partner. Write a cube root function, $y = f(x)$, that has the given values. Then use the function to complete the table.

a.

x	$f(x)$
−4	0
−3	
−2	
−1	$\sqrt[3]{3}$
0	

x	$f(x)$
1	
2	
3	
4	2
5	

b.

x	$f(x)$
−4	1
−3	
−2	
−1	$1 + \sqrt[3]{3}$
0	

x	$f(x)$
1	
2	
3	
4	3
5	

Communicate Your Answer

3. What are some of the characteristics of the graph of a cube root function?
4. Graph each function. Then compare the graph to the graph of $f(x) = \sqrt[3]{x}$.
   a. $g(x) = \sqrt[3]{x - 1}$
   b. $g(x) = \sqrt[3]{x} - 1$
   c. $g(x) = 2\sqrt[3]{x}$
   d. $g(x) = -2\sqrt[3]{x}$

Dynamic Teaching Tools

- Dynamic Assessment & Progress Monitoring Tool
- Lesson Planning Tool
- Interactive Whiteboard Lesson Library
- Dynamic Classroom with Dynamic Investigations

ANSWERS

1. a.

x	y
−8	−2
−1	−1
0	0
1	1
8	2

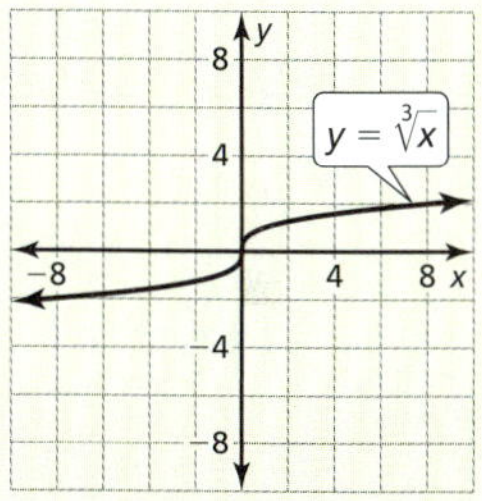

domain: all real numbers;
range: all real numbers

b.

x	y
−11	−2
−4	−1
−3	0
−2	1
5	2

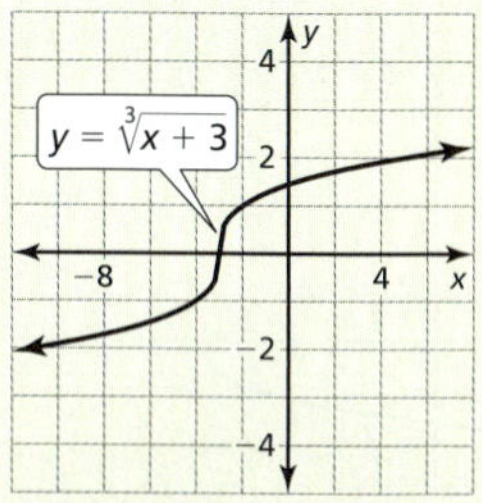

domain: all real numbers;
range: all real numbers

2. See Additional Answers.

3. *Sample answer:* The graph has a center of rotation and looks somewhat like a parabola that opens to one side with the top half reflected over the y-axis.

4. a.

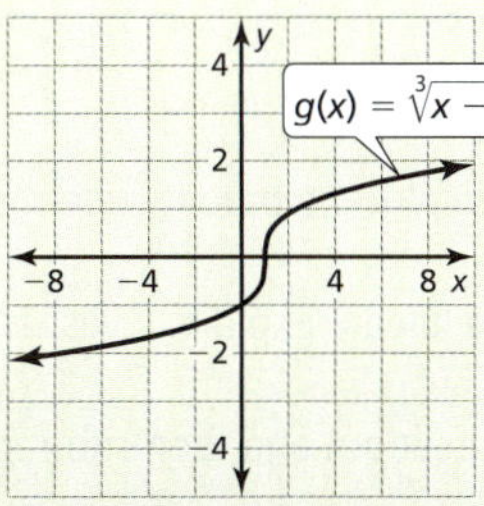

The graph of g is a translation 1 unit right of the graph of f.

b.

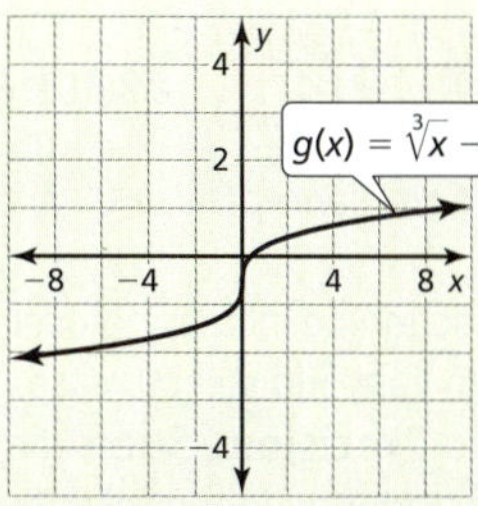

The graph of g is a translation 1 unit down of the graph of f.

4c–d. See Additional Answers.

Differentiated Instruction

Kinesthetic

Some students may benefit by having access to graphing calculators or a graphing software program. Students can enter the given function and the function $f(x) = \sqrt[3]{x}$, view the graphs, and compare the functions.

Extra Example 1

Graph $h(x) = \sqrt[3]{x} + 6$. Compare the graph to the graph of $f(x) = \sqrt[3]{x}$.

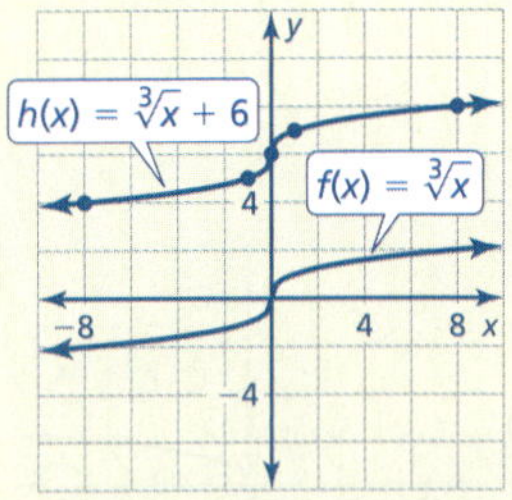

The graph of h is a translation 6 units up of the graph of f.

MONITORING PROGRESS ANSWERS

1.

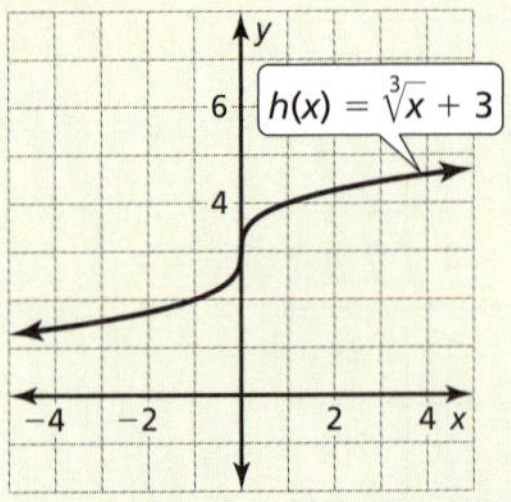

The graph of h is a translation 3 units up of the graph of f.

2.

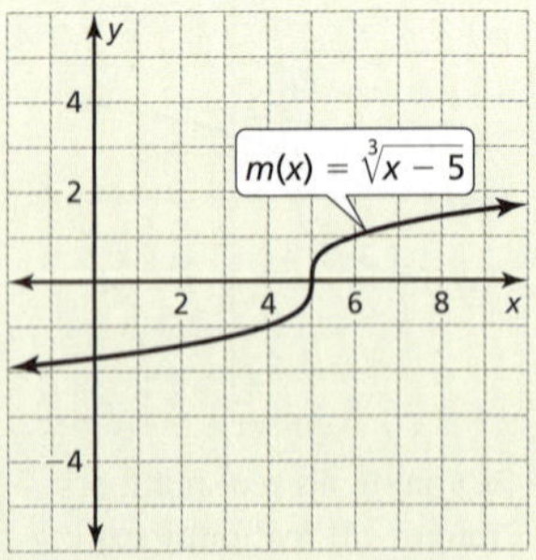

The graph of m is a translation 5 units right of the graph of f.

3. See Additional Answers.

10.2 Lesson

What You Will Learn

- Graph cube root functions.
- Compare cube root functions using average rates of change.
- Solve real-life problems involving cube root functions.

Core Vocabulary

cube root function, *p. 552*

Previous

radical function

index

Graphing Cube Root Functions

Core Concept

Cube Root Functions

A **cube root function** is a radical function with an index of 3. The parent function for the family of cube root functions is $f(x) = \sqrt[3]{x}$. The domain and range of f are all real numbers.

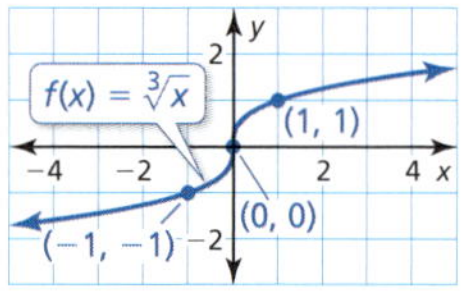

The graph of $f(x) = \sqrt[3]{x}$ increases on the entire domain.

You can transform graphs of cube root functions in the same way you transformed graphs of square root functions.

LOOKING FOR STRUCTURE

Use x-values so that the cube root of the radicand is an integer. This makes it easier to perform the calculations and plot the points.

EXAMPLE 1 Comparing Graphs of Cube Root Functions

Graph $h(x) = \sqrt[3]{x} - 4$. Compare the graph to the graph of $f(x) = \sqrt[3]{x}$.

SOLUTION

Step 1 Make a table of values.

x	−8	−1	0	1	8
$h(x)$	−6	−5	−4	−3	−2

Step 2 Plot the ordered pairs.

Step 3 Draw a smooth curve through the points.

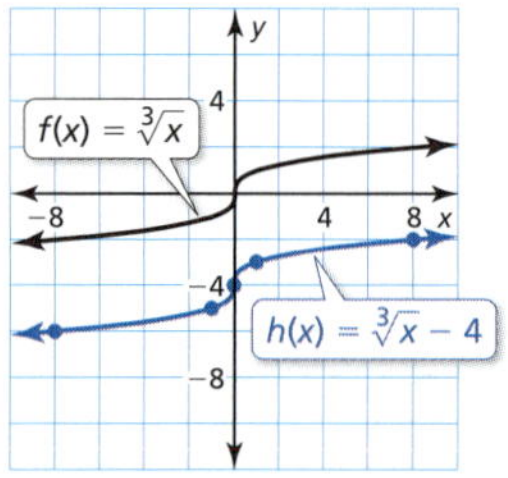

▶ The graph of h is a translation 4 units down of the graph of f.

Monitoring Progress

Help in English and Spanish at *BigIdeasMath.com*

Graph the function. Compare the graph to the graph of $f(x) = \sqrt[3]{x}$.

1. $h(x) = \sqrt[3]{x} + 3$
2. $m(x) = \sqrt[3]{x - 5}$
3. $g(x) = 4\sqrt[3]{x}$

Laurie's Notes Teacher Actions

- Write the *Core Concept*. Use 0, ±1, and ±8 as domain values to plot the parent cube root function.
- ? "Can a function be increasing over its entire domain and not be linear? Explain." yes; The rate of change can vary and still increase.
- **Attend to Precision:** Stress that you take the cube root of the input and then subtract four, not subtract four and then take the cube root.
- ? "How do you describe the transformation?" It is a vertical translation 4 units down of the parent function.

EXAMPLE 2 Comparing Graphs of Cube Root Functions

Graph $g(x) = -\sqrt[3]{x+2}$. Compare the graph to the graph of $f(x) = \sqrt[3]{x}$.

SOLUTION

Step 1 Make a table of values.

x	−10	−3	−2	−1	6
$g(x)$	2	1	0	−1	−2

Step 2 Plot the ordered pairs.

Step 3 Draw a smooth curve through the points.

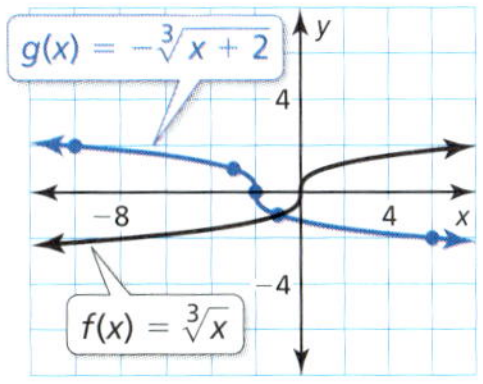

► The graph of g is a translation 2 units left and a reflection in the x-axis of the graph of f.

REMEMBER

The graph of $y = a \cdot f(x - h) + k$ can be obtained from the graph of $y = f(x)$ using the steps you learned in Section 3.6.

EXAMPLE 3 Graphing $y = a\sqrt[3]{x-h} + k$

Let $g(x) = 2\sqrt[3]{x-3} + 4$. (a) Describe the transformations from the graph of $f(x) = \sqrt[3]{x}$ to the graph of g. (b) Graph g.

SOLUTION

a. Step 1 Translate the graph of f horizontally 3 units right to get the graph of $t(x) = \sqrt[3]{x-3}$.

Step 2 Stretch the graph of t vertically by a factor of 2 to get the graph of $h(x) = 2\sqrt[3]{x-3}$.

Step 3 Because $a > 0$, there is no reflection.

Step 4 Translate the graph of h vertically 4 units up to get the graph of $g(x) = 2\sqrt[3]{x-3} + 4$.

b. Step 1 Make a table of values.

x	−5	2	3	4	11
$g(x)$	0	2	4	6	8

Step 2 Plot the ordered pairs.

Step 3 Draw a smooth curve through the points.

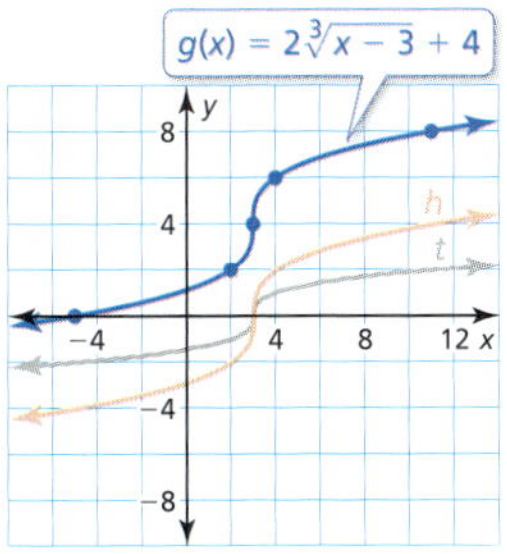

Monitoring Progress

Help in English and Spanish at BigIdeasMath.com

Graph the function. Compare the graph to the graph of $f(x) = \sqrt[3]{x}$.

4. $g(x) = \sqrt[3]{0.5x} + 5$ 5. $h(x) = 4\sqrt[3]{x-1}$ 6. $n(x) = \sqrt[3]{4-x}$

7. Let $g(x) = -\frac{1}{2}\sqrt[3]{x+2} - 4$. Describe the transformations from the graph of $f(x) = \sqrt[3]{x}$ to the graph of g. Then graph g.

Laurie's Notes Teacher Actions

? Reason Abstractly and Quantitatively: "In Example 2, what values should you use in evaluating $g(x) = -\sqrt[3]{x+2}$ and why?" Listen for x-values that result in cube rooting 0, ±1, or ±8.

? "In Example 3, what do you know about the graph of $g(x) = 2\sqrt[3]{x-3} + 4$?" Answers will vary. Listen for (3, 4) as a key ordered pair and a vertical stretch by a factor of 2. See the *Teaching Strategy* on page T-550. Note that (2, 2), (3, 4), and (4, 6) are the images of (−1, −1), (0, 0), and (1, 1) of the parent function.

- In graphing transformations of cube root functions, determine the ordered pairs related to (−1, −1), (0, 0), and (1, 1).

Extra Example 2

Graph $g(x) = -\sqrt[3]{x+4}$. Compare the graph to the graph of $f(x) = \sqrt[3]{x}$.

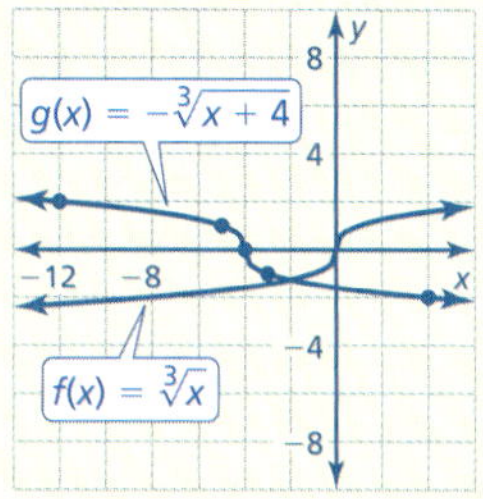

The graph of g is a translation 4 units left and a reflection in the x-axis of the graph of f.

Extra Example 3

Let $h(x) = 3\sqrt[3]{x+1} - 2$.

a. Describe the transformations from the graph of $f(x) = \sqrt[3]{x}$ to the graph of h.

Translate the graph of f horizontally 1 unit left to get the graph of $t(x) = \sqrt[3]{x+1}$.

Stretch the graph of t vertically by a factor of 3 to get the graph of $g(x) = 3\sqrt[3]{x+1}$.

Translate the graph of g vertically 2 units down to get the graph of $h(x) = 3\sqrt[3]{x+1} - 2$.

b. Graph h.

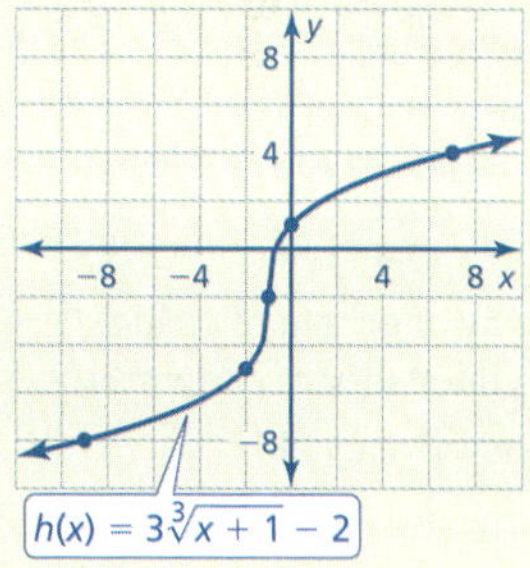

MONITORING PROGRESS ANSWERS

4.

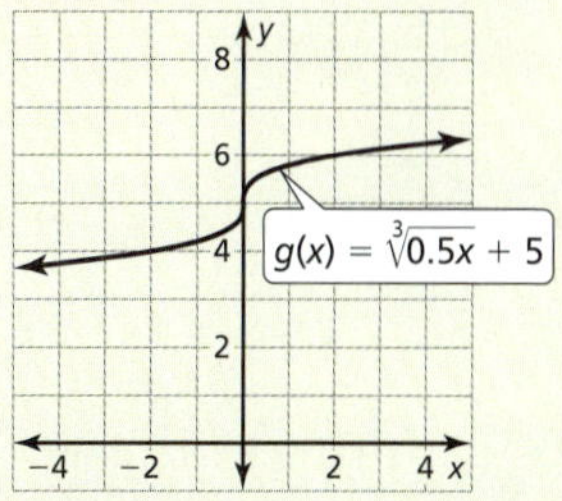

The graph of g is a horizontal stretch by a factor of 2 and a translation 5 units up of the graph of f.

5–7. See Additional Answers.

Extra Example 4

Use the graph in Example 4. Compare the average rate of change of m to the average rate of change of $h(x) = \sqrt[3]{2x}$ over the interval $x = 0$ to $x = 8$.

The average rate of change of m is $1 \div \frac{\sqrt[3]{2}}{4} \approx 3.2$ times greater than the average rate of change of h over the interval $x = 0$ to $x = 8$.

Extra Example 5

A regular dodecahedron is a solid with 12 regular pentagons as faces. The length s (in feet) of an edge in a regular solid can be modeled by the function $s = \sqrt[3]{\frac{V}{7.66}}$, where V is the volume (in cubic feet) of the solid. Use a graphing calculator to graph the function. Estimate the volume of a regular dodecahedron with an edge length of 2 feet.

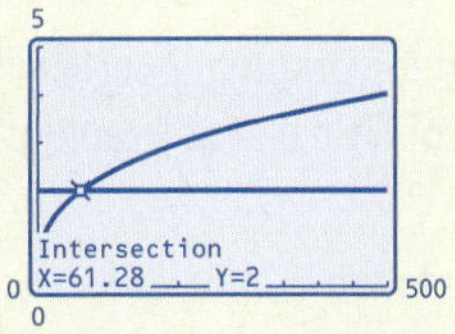

The volume is about 61 cubic feet.

MONITORING PROGRESS ANSWERS

8. The average rate of change of m is about $\frac{0.44}{0.07} \approx 6.3$ times greater than the average rate of change of h over the interval $x = 2$ to $x = 10$.
9. about 4 years old

Comparing Average Rates of Change

EXAMPLE 4 **Comparing Cube Root Functions**

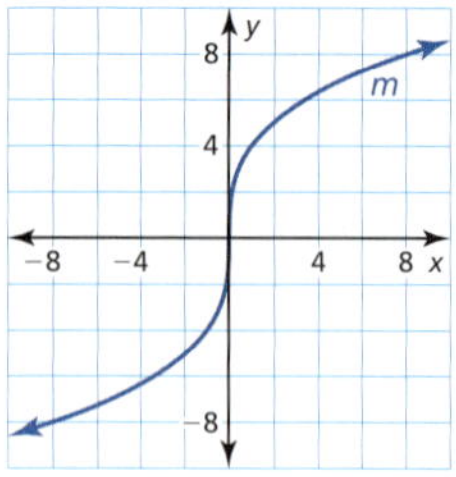

The graph of cube root function m is shown. Compare the average rate of change of m to the average rate of change of $h(x) = \sqrt[3]{\frac{1}{4}x}$ over the interval $x = 0$ to $x = 8$.

SOLUTION

To calculate the average rates of change, use points whose x-coordinates are 0 and 8.

Function m: Use the graph to estimate. Use (0, 0) and (8, 8).

$$\frac{m(8) - m(0)}{8 - 0} \approx \frac{8 - 0}{8} = 1 \qquad \text{Average rate of change of } m$$

Function h: Evaluate h when $x = 0$ and $x = 8$.

$$h(0) = \sqrt[3]{\frac{1}{4}(0)} = 0 \quad \text{and} \quad h(8) = \sqrt[3]{\frac{1}{4}(8)} = \sqrt[3]{2} \approx 1.3$$

Use (0, 0) and $(8, \sqrt[3]{2})$.

$$\frac{h(8) - h(0)}{8 - 0} = \frac{\sqrt[3]{2} - 0}{8} \approx 0.16 \qquad \text{Average rate of change of } h$$

▶ The average rate of change of m is $1 \div \frac{\sqrt[3]{2}}{8} \approx 6.3$ times greater than the average rate of change of h over the interval $x = 0$ to $x = 8$.

Monitoring Progress

Help in English and Spanish at *BigIdeasMath.com*

8. In Example 4, compare the average rates of change over the interval $x = 2$ to $x = 10$.

Solving Real-Life Problems

EXAMPLE 5 **Real-Life Application**

The shoulder height h (in centimeters) of a male Asian elephant can be modeled by the function $h = 62.5\sqrt[3]{t} + 75.8$, where t is the age (in years) of the elephant. Use a graphing calculator to graph the function. Estimate the age of an elephant whose shoulder height is 200 centimeters.

SOLUTION

Step 1 Enter $y_1 = 62.5\sqrt[3]{t} + 75.8$ and $y_2 = 200$ into your calculator and graph the equations. Choose a viewing window that shows the point where the graphs intersect.

Step 2 Use the *intersect* feature to find the x-coordinate of the intersection point.

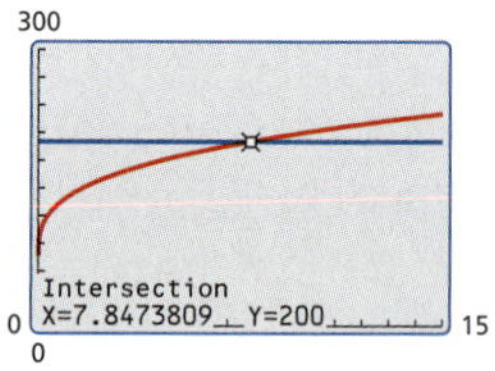

▶ The two graphs intersect at about (8, 200). So, the elephant is about 8 years old.

Monitoring Progress

Help in English and Spanish at *BigIdeasMath.com*

9. **WHAT IF?** Estimate the age of an elephant whose shoulder height is 175 centimeters.

Laurie's Notes Teacher Actions

- Example 4 compares the average rate of change for two cubic functions, one shown graphically and one given symbolically.
- **Reason Abstractly and Quantitatively** and **Use Appropriate Tools Strategically:** Students should be comfortable using a graphing calculator to solve Example 5. Understanding what the graph of the cubic function h looks like will enable students to determine a good viewing window for the graph.

Closure

- **Exit Ticket:** Graph $g(x) = 3\sqrt[3]{x - 4}$ and compare to the graph of $f(x) = \sqrt[3]{x}$. Check students' graphs. The graph of g is a translation 4 units right and a vertical stretch by a factor of 3 of the graph of f.

10.2 Exercises

Dynamic Solutions available at *BigIdeasMath.com*

Vocabulary and Core Concept Check

1. **COMPLETE THE SENTENCE** The __________ of the radical in a cube root function is 3.

2. **WRITING** Describe the domain and range of the function $f(x) = \sqrt[3]{x-4} + 1$.

Monitoring Progress and Modeling with Mathematics

In Exercises 3–6, match the function with its graph.

3. $y = \sqrt[3]{x+2}$

4. $y = \sqrt[3]{x-2}$

5. $y = \sqrt[3]{x} + 2$

6. $y = \sqrt[3]{x} - 2$

A. 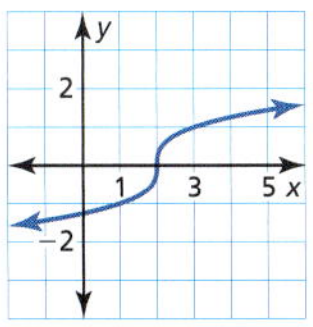

B. 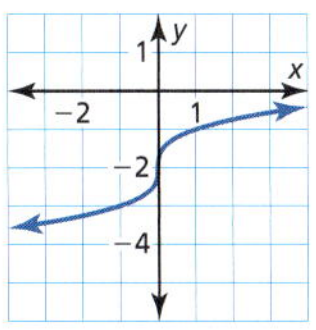

C. 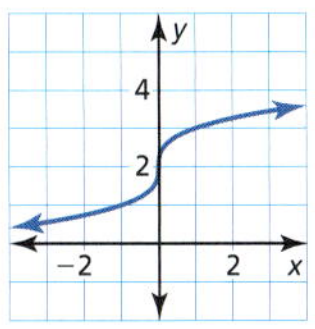

D. 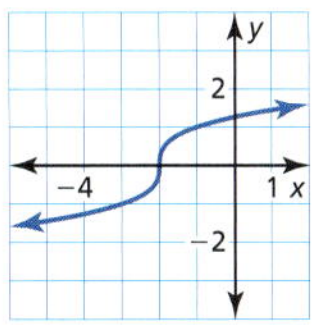

In Exercises 7–12, graph the function. Compare the graph to the graph of $f(x) = \sqrt[3]{x}$. *(See Example 1.)*

7. $h(x) = \sqrt[3]{x-4}$

8. $g(x) = \sqrt[3]{x+1}$

9. $m(x) = \sqrt[3]{x} + 5$

10. $q(x) = \sqrt[3]{x} - 3$

11. $p(x) = 6\sqrt[3]{x}$

12. $j(x) = \sqrt[3]{\frac{1}{2}x}$

In Exercises 13–16, compare the graphs. Find the value of *h*, *k*, or *a*.

13. 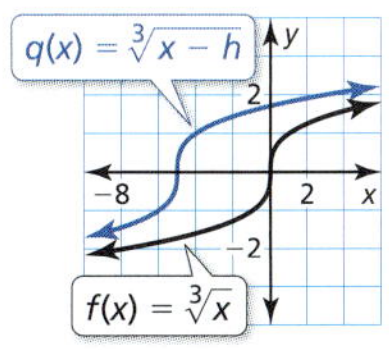

14. 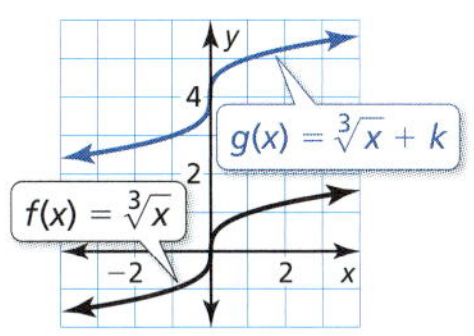

15. 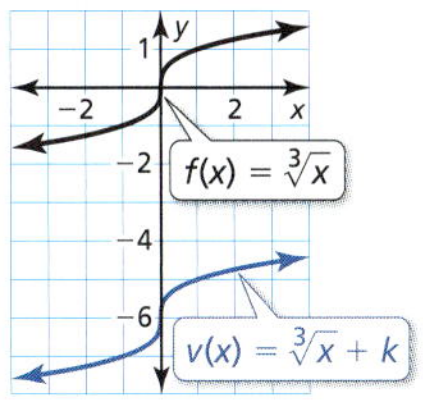

16. 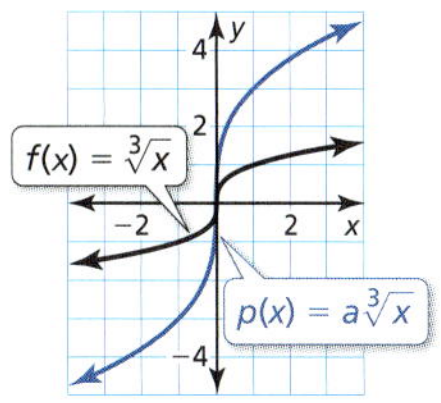

In Exercises 17–26, graph the function. Compare the graph to the graph of $f(x) = \sqrt[3]{x}$. *(See Example 2.)*

17. $r(x) = -\sqrt[3]{x-2}$

18. $h(x) = -\sqrt[3]{x} + 3$

19. $k(x) = 5\sqrt[3]{x+1}$

20. $j(x) = 0.5\sqrt[3]{x-4}$

21. $g(x) = 4\sqrt[3]{x} - 3$

22. $m(x) = 3\sqrt[3]{x} + 7$

23. $n(x) = \sqrt[3]{-8x} - 1$

24. $v(x) = \sqrt[3]{5x} + 2$

25. $q(x) = \sqrt[3]{2(x+3)}$

26. $p(x) = \sqrt[3]{3(1-x)}$

In Exercises 27–32, describe the transformations from the graph of $f(x) = \sqrt[3]{x}$ to the graph of the given function. Then graph the given function. *(See Example 3.)*

27. $g(x) = \sqrt[3]{x-4} + 2$

28. $n(x) = \sqrt[3]{x+1} - 3$

29. $j(x) = -5\sqrt[3]{x+3} + 2$

30. $k(x) = 6\sqrt[3]{x-9} - 5$

31. $v(x) = \frac{1}{3}\sqrt[3]{x-1} + 7$

32. $h(x) = -\frac{3}{2}\sqrt[3]{x+4} - 3$

33. **ERROR ANALYSIS** Describe and correct the error in graphing the function $f(x) = \sqrt[3]{x-3}$.

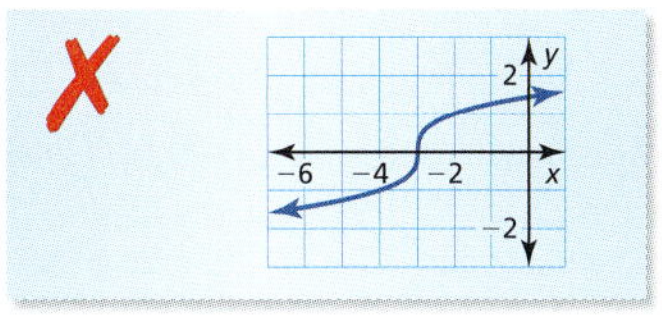

Assignment Guide and Homework Check

ASSIGNMENT

Basic: 1, 2, 3–21 odd, 27, 29, 33–39 odd, 40, 43–49

Average: 1, 2–38 even, 39, 40, 43–49

Advanced: 1, 2, 10–14 even, 24–38 even, 39–49

HOMEWORK CHECK

Basic: 7, 17, 27, 35, 37

Average: 8, 18, 28, 36, 38

Advanced: 10, 24, 30, 36, 38

ANSWERS

1. index
2. The domain and range of f are all real numbers.
3. D
4. A
5. C
6. B
7. 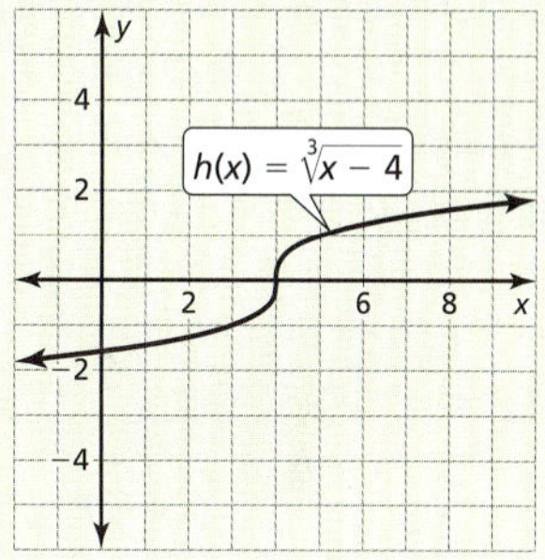

The graph of h is a translation 4 units right of the graph of f.
8. 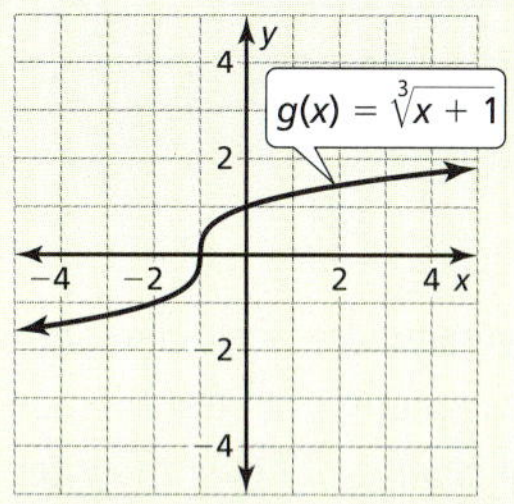

The graph of g is a translation 1 unit left of the graph of f.
9. 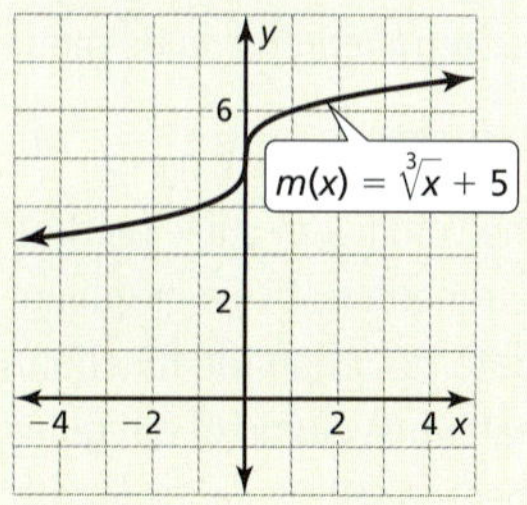

The graph of m is a translation 5 units up of the graph of f.
10. 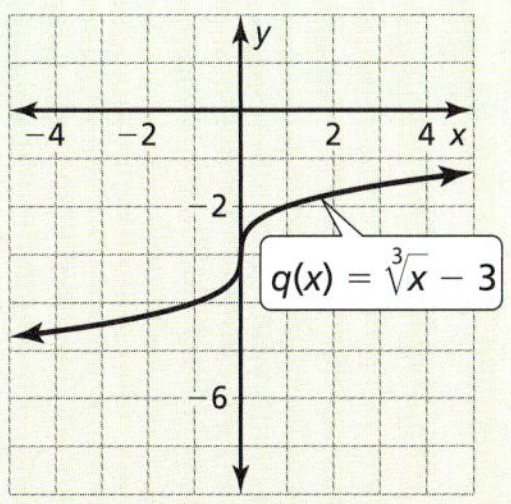

The graph of q is a translation 3 units down of the graph of f.
11. 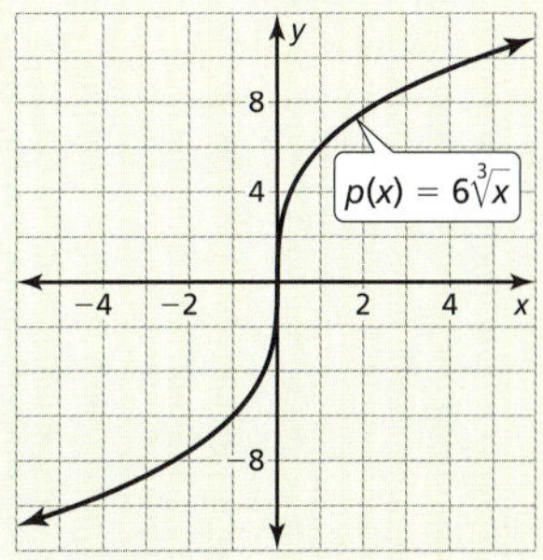

The graph of p is a vertical stretch by a factor of 6 of the graph of f.

12–33. See Additional Answers.

Dynamic Teaching Tools

- Dynamic Assessment & Progress Monitoring Tool
- Interactive Whiteboard Lesson Library
- Dynamic Classroom with Dynamic Investigations

ANSWERS

34–49. See Additional Answers.

Mini-Assessment

1. Graph $h(x) = 3\sqrt[3]{x}$. Compare the graph to the graph of $f(x) = \sqrt[3]{x}$.

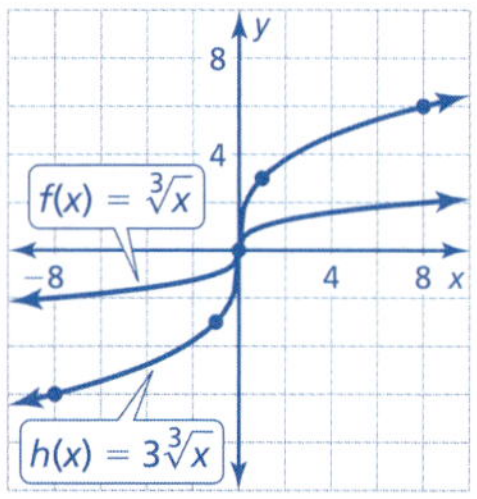

The graph of h is a vertical stretch by a factor of 3 of the graph of f.

2. The graphs of two cube root functions are shown. Compare the average rates of change of the two functions over the interval $x = 0$ to $x = 8$.

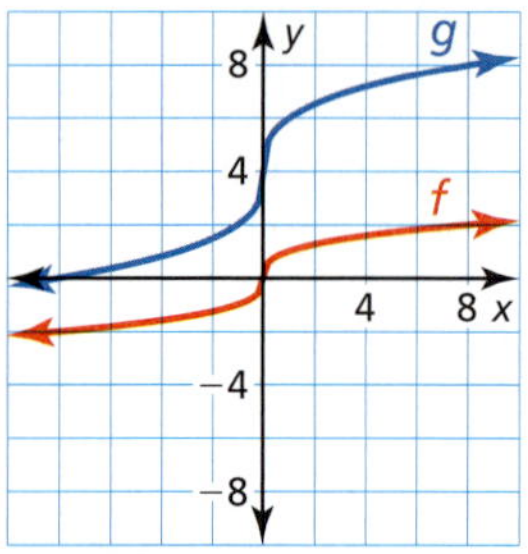

The average rate of change of g is 2 times greater than the average rate of change of f over the interval $x = 0$ to $x = 8$.

3. The length s (in feet) of an edge in a regular tetrahedron can be modeled by the function $s = \sqrt[3]{\frac{V}{0.118}}$, where V is the volume (in cubic feet) of the tetrahedron. Use a graphing calculator to graph the function. Estimate the volume of a regular tetrahedron with an edge length of 3 feet.

See Additional Answers.

34. ERROR ANALYSIS Describe and correct the error in graphing the function $h(x) = \sqrt[3]{x} + 1$.

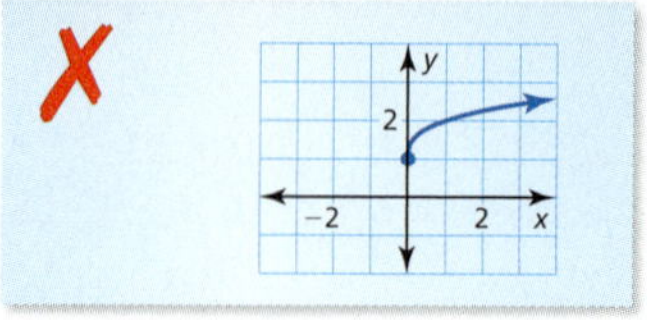

35. COMPARING FUNCTIONS The graph of cube root function q is shown. Compare the average rate of change of q to the average rate of change of $f(x) = 3\sqrt[3]{x}$ over the interval $x = 0$ to $x = 6$. *(See Example 4.)*

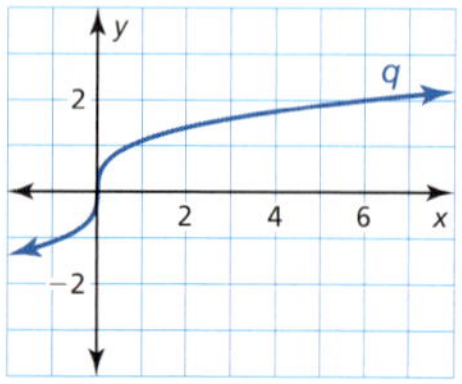

36. COMPARING FUNCTIONS The graphs of two cube root functions are shown. Compare the average rates of change of the two functions over the interval $x = -2$ to $x = 2$.

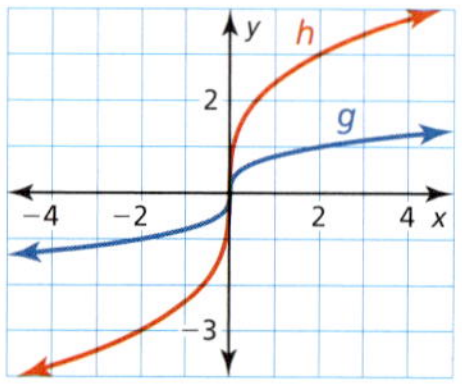

37. MODELING WITH MATHEMATICS For a drag race car that weighs 1600 kilograms, the velocity v (in kilometers per hour) reached by the end of a drag race can be modeled by the function $v = 23.8\sqrt[3]{p}$, where p is the car's power (in horsepower). Use a graphing calculator to graph the function. Estimate the power of a 1600-kilogram car that reaches a velocity of 220 kilometers per hour. *(See Example 5.)*

38. MODELING WITH MATHEMATICS The radius r of a sphere is given by the function $r = \sqrt[3]{\frac{3}{4\pi}V}$, where V is the volume of the sphere. Use a graphing calculator to graph the function. Estimate the volume of a spherical beach ball with a radius of 13 inches.

39. MAKING AN ARGUMENT Your friend says that all cube root functions are odd functions. Is your friend correct? Explain.

40. HOW DO YOU SEE IT? The graph represents the cube root function $f(x) = \sqrt[3]{x}$.

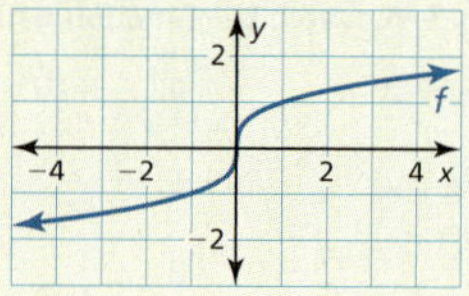

a. On what interval is f negative? positive?

b. On what interval, if any, is f decreasing? increasing?

c. Does f have a maximum or minimum value? Explain.

d. Find the average rate of change of f over the interval $x = -1$ to $x = 1$.

41. PROBLEM SOLVING Write a cube root function that passes through the point (3, 4) and has an average rate of change of -1 over the interval $x = -5$ to $x = 2$.

42. THOUGHT PROVOKING Write the cube root function represented by the graph. Use a graphing calculator to check your answer.

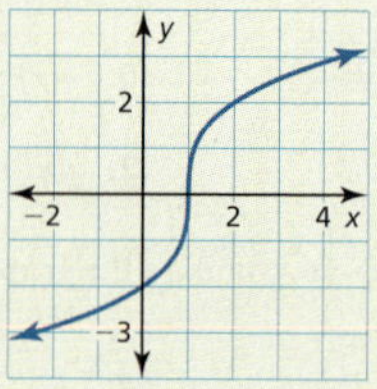

Maintaining Mathematical Proficiency

Reviewing what you learned in previous grades and lessons

Factor the polynomial. *(Section 7.6)*

43. $3x^2 + 12x - 36$ **44.** $2x^2 - 11x + 9$ **45.** $4x^2 + 7x - 15$

Solve the equation using square roots. *(Section 9.3)*

46. $x^2 - 36 = 0$ **47.** $5x^2 + 20 = 0$ **48.** $(x + 4)^2 = 81$ **49.** $25(x - 2)^2 = 9$

If students need help...	*If students got it...*
Resources by Chapter • Practice A and Practice B • Puzzle Time	Resources by Chapter • Enrichment and Extension • Cumulative Review
Student Journal • Practice	Start the *next* Section
Differentiating the Lesson Skills Review Handbook	

10.1–10.2 What Did You Learn?

Core Vocabulary

square root function, *p. 544*
radical function, *p. 545*
cube root function, *p. 552*

Core Concepts

Section 10.1

Square Root Functions, *p. 544*
Transformations of Square Root Functions, *p. 545*
Comparing Square Root Functions Using Average Rates of Change, *p. 546*

Section 10.2

Cube Root Functions, *p. 552*
Comparing Cube Root Functions Using Average Rates of Change, *p. 554*

Mathematical Practices

1. In Exercise 45 on page 549, what information are you given? What relationships are present? What is your goal?
2. What units of measure did you use in your answer to Exercise 38 on page 556? Explain your reasoning.

Study Skills

Making Note Cards

Invest in three different colors of note cards. Use one color for each of the following: vocabulary words, rules, and calculator keystrokes.

- Using the first color of note cards, write a vocabulary word on one side of a card. On the other side, write the definition and an example. If possible, put the definition in your own words.
- Using the second color of note cards, write a rule on one side of a card. On the other side, write an explanation and an example.
- Using the third color of note cards, write a calculation on one side of a card. On the other side, write the keystrokes required to perform the calculation.

Use the note cards as references while completing your homework. Quiz yourself once a day.

Dynamic Teaching Tools

Dynamic Assessment & Progress Monitoring Tool
Interactive Whiteboard Lesson Library
Dynamic Classroom with Dynamic Investigations

ANSWERS

1. You are given models that represent the speeds S of a van before it skids to a stop for two surfaces. One model is given algebraically and one model is given graphically; For each model, the speed is related to the drag factor f of the road surface and the length d of the skid marks by a square root function; Your goal is to compare the speeds by finding and interpreting average rates of change over an interval.
2. cubic inches; Because the radius of the beach ball is given in inches, the volume V as determined by the given formula is a measure of cubic inches.

ANSWERS

1. $x \ge 3$
2. $x \ge 0$
3. $x \le 3$
4. 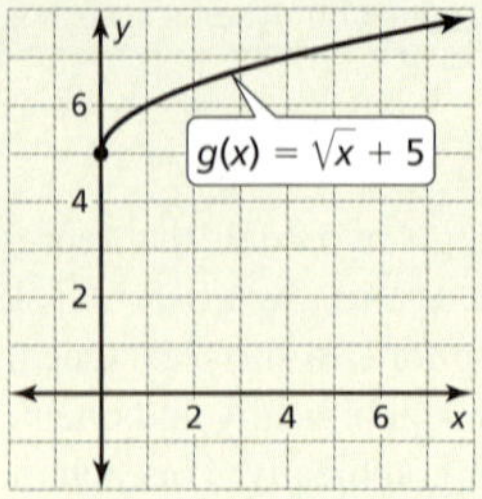

$y \ge 5$; The graph of g is a translation 5 units up of the graph of f.

5. 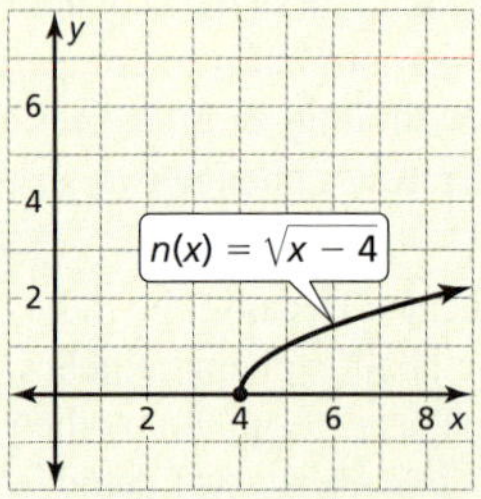

$y \ge 0$; The graph of n is a translation 4 units right of the graph of f.

6. 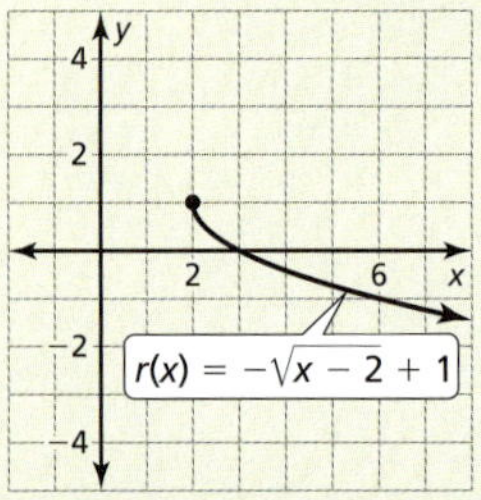

$y \le 1$; The graph of r is a translation 2 units right, a reflection in the x-axis, and a translation 1 unit up of the graph of f.

7. 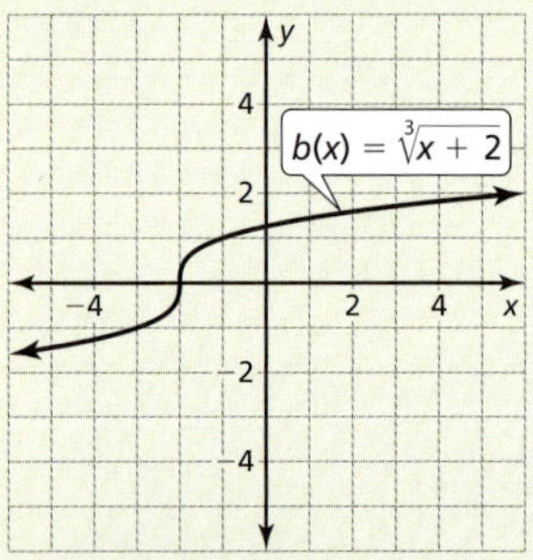

The graph of b is a translation 2 units left of the graph of f.

10.1–10.2 Quiz

Describe the domain of the function. *(Section 10.1)*

1. $y = \sqrt{x - 3}$
2. $f(x) = 15\sqrt{x}$
3. $y = \sqrt{3 - x}$

Graph the function. Describe the range. Compare the graph to the graph of $f(x) = \sqrt{x}$. *(Section 10.1)*

4. $g(x) = \sqrt{x} + 5$
5. $n(x) = \sqrt{x - 4}$
6. $r(x) = -\sqrt{x - 2} + 1$

Graph the function. Compare the graph to the graph of $f(x) = \sqrt[3]{x}$. *(Section 10.2)*

7. $b(x) = \sqrt[3]{x + 2}$
8. $h(x) = -3\sqrt[3]{x} - 6$
9. $q(x) = \sqrt[3]{-4 - x}$

Compare the graphs. Find the value of h, k, or a. *(Section 10.1 and Section 10.2)*

10. 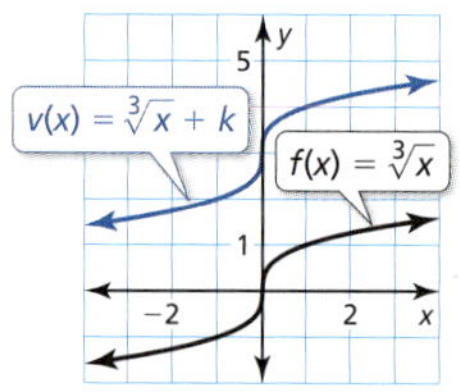

11. 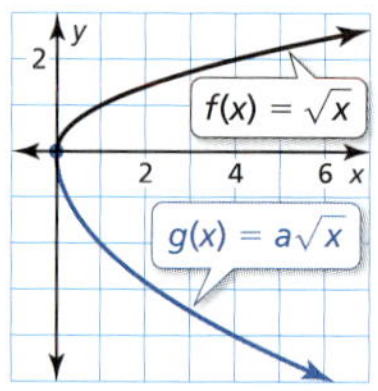

12. 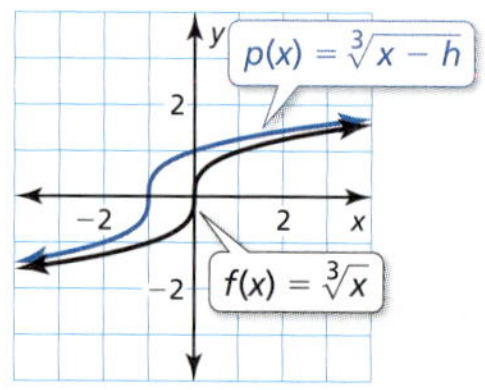

Describe the transformations from the graph of f to the graph of h. Then graph h. *(Section 10.1 and Section 10.2)*

13. $f(x) = \sqrt{x}$; $h(x) = -3\sqrt{x + 2} + 6$
14. $f(x) = \sqrt[3]{x}$; $h(x) = \frac{1}{2}\sqrt[3]{x} - 3$

15. The time t (in seconds) it takes a dropped object to fall h feet is given by $t = \frac{1}{4}\sqrt{h}$. *(Section 10.1)*
    a. Use a graphing calculator to graph the function. Describe the domain and range.
    b. It takes about 7.4 seconds for a stone dropped from the New River Gorge Bridge in West Virginia to reach the water below. About how high is the bridge above the New River?

16. The radius r of a sphere is given by the function $r = \sqrt[3]{\frac{3}{4\pi}V}$, where V is the volume of the sphere. Spaceship Earth is a spherical structure at Walt Disney World that has an inner radius of about 25 meters. Use a graphing calculator to graph the function. Estimate the volume of Spaceship Earth. *(Section 10.2)*

17. The graph of square root function g is shown. Compare the average rate of change of g to the average rate of change of $h(x) = \sqrt[3]{\frac{3}{2}x}$ over the interval $x = 0$ to $x = 3$. *(Section 10.1 and Section 10.2)*

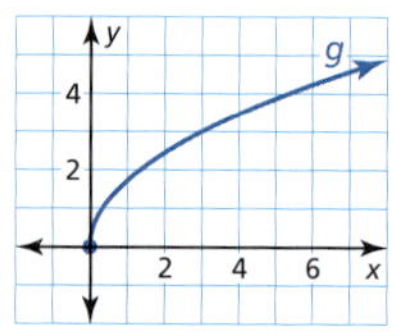

8. 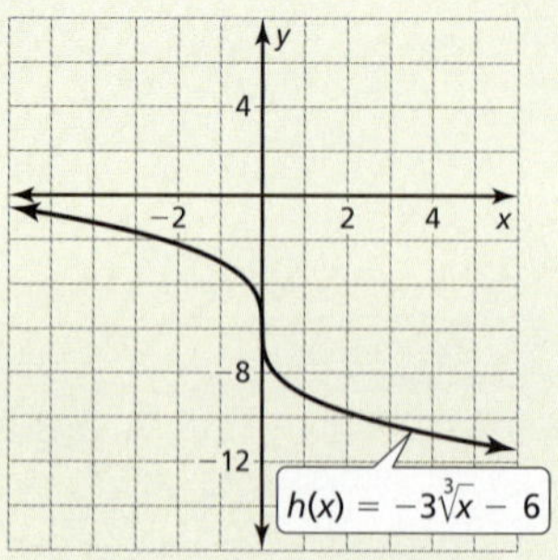

The graph of h is a vertical stretch by a factor of 3, a reflection in the x-axis, and a translation 6 units down of the graph of f.

9. 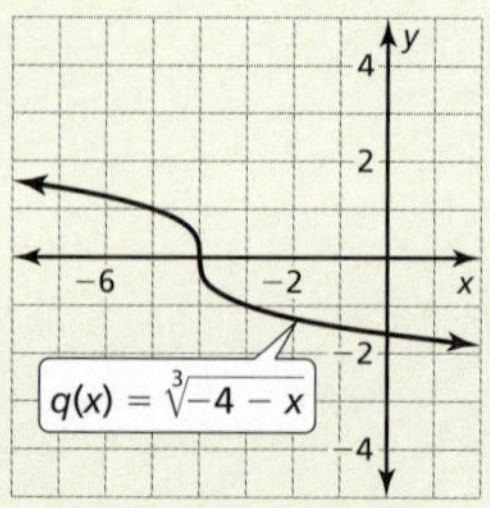

The graph of q is a translation 4 units right and a reflection in the y-axis of the graph of f.

10–17. See Additional Answers.

Laurie's Notes

Overview of Section 10.3

Introduction

- Students have solved many different types of equations. A connection is made to the structure of different types of equations and that the term involving the variable must be isolated using properties of equality.
- Radical equations, both square roots and cube roots, are solved by squaring (or cubing) both sides of the equation.
- The concept of an extraneous root is introduced.

Resources

- To add interest to Example 7, read an excerpt or find a brief online video from "The Pit and the Pendulum" by Edgar Allan Poe. Your students may have some familiarity with this classic literary work.

Teaching Strategy

- One way to help students think about extraneous roots is to look at the graphical representation of equations they are solving.
- In Example 2, students are working with an equation involving a square root function and a constant function. Students should recognize that a constant function and a square root function cannot have two points of intersection. Be sure that students recognize that even though the graphs do not intersect in the viewing window, it is possible that they may intersect for some value of x in the domain.

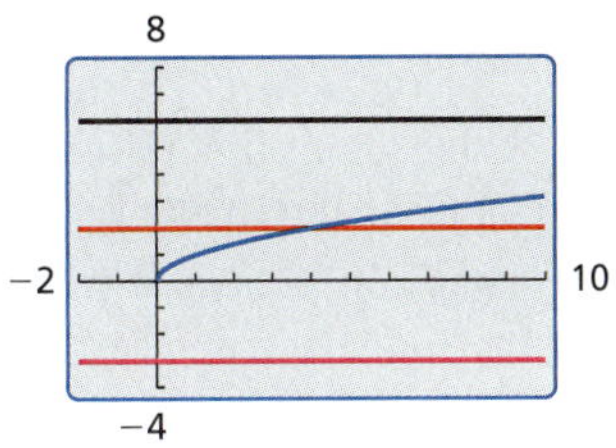

- Have students explore the intersection possibilities for a linear function and a square root function. Ask students to come up with examples where there are 0, 1, or 2 points of intersection.

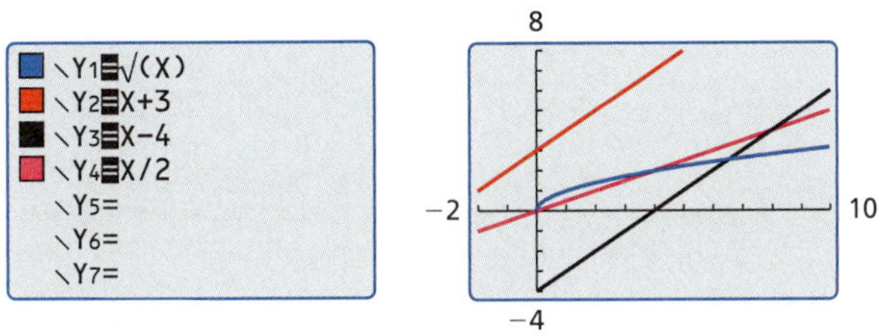

Pacing Suggestion

- As students work through the explorations, listen and probe recall of prior graphing and equation-solving skills. This recall helps students make connections in the formal lesson.

Dynamic Teaching Tools

- Dynamic Assessment & Progress Monitoring Tool
- Lesson Planning Tool
- Interactive Whiteboard Lesson Library
- Dynamic Classroom with Dynamic Investigations

What Your Students Will Learn

- Solve radical equations involving square roots by squaring each side.
- Solve radical equations involving cube roots by cubing each side.
- Identify extraneous solutions of radical equations by checking each solution in the original equation.
- Solve real-life problems involving radical equations.

Laurie's Notes

Exploration

Motivate

- Slide a math book off the edge of your desk or a table and let it hit the floor. This sound should get your students' attention.
- ? "How long do you think it took for the book to hit the floor from the moment it started falling?" Answers will vary.
- Drop the book again. This time, do it from your shoulder height.
- ? "How long do you think it took for the book to hit the floor from the moment it started falling?" Answers will vary. However, answers should be different from those for the first height.
- ? "Suppose I drop the book from the height of the ceiling. How long do you think it would take the book to hit the floor?" It would take longer than for the first two heights.
- Explain to students that today they will investigate the time it takes for an object to fall from different heights.

Exploration 1

- As students begin to work on this exploration, make sure that they are plotting (d, t) and not (t, d).
- ? "Should the plotted points be connected with a smooth curve?" yes
- ? "What is the independent variable?" d, the distance (in feet) that the free-falling object falls
- ? "What is the dependent variable?" t, the time (in seconds) that the free-falling object falls
- When answering part (b), students will most likely interpolate and guess a time about halfway between 3.74 and 4.00 seconds.
- Students use substitution to answer part (c).
- When answering part (d), students should substitute 5 for t and solve. To solve for d, students must square each side of the equation and then multiply each side by 16.
- If time permits, investigate the equation on a graphing calculator. Adjusting the viewing window to see the key features of the graph is a good exercise for students. In this graph, the scale for the x-axis is much different from the scale for the y-axis.

Exploration 2

- Have students read the introduction to this exploration.
- "Is the object falling at a constant rate? Explain." no; The rate at which the object is falling is modeled by a square root function. The rate at which the object is falling is increasing at a decreasing rate.
- Have partners work to solve for the distance d in each of the equations.
- **Construct Viable Arguments and Critique the Reasoning of Others:** Have students explain their strategies for solving the square root equations. Some may say it is trial and error. Some may enter the equation $y = \sqrt{64x}$ in their calculators and use the table or graph to solve. Some students may say that they figured out that they needed to set the radicand equal to the square of the speed, s.
- It is important to hear different strategies and for students to explain their reasoning.

Communicate Your Answer

- Students should have solved enough equations, including square root equations, so that they know to isolate the square root term.

Connecting to Next Step

- Students have now developed a sense of how to solve a square root equation. In the formal lesson, this procedure will be practiced.

10.3 Solving Radical Equations

Essential Question How can you solve an equation that contains square roots?

EXPLORATION 1 **Analyzing a Free-Falling Object**

Work with a partner. The table shows the time t (in seconds) that it takes a free-falling object (with no air resistance) to fall d feet.

d (feet)	t (seconds)
0	0.00
32	1.41
64	2.00
96	2.45
128	2.83
160	3.16
192	3.46
224	3.74
256	4.00
288	4.24
320	4.47

MODELING WITH MATHEMATICS

To be proficient in math, you need to routinely interpret your mathematical results in the context of the situation and reflect on whether the results make sense.

a. Use the data in the table to sketch the graph of t as a function of d. Use the coordinate plane below.

b. Use your graph to estimate the time it takes the object to fall 240 feet.

c. The relationship between d and t is given by the function

$$t = \sqrt{\frac{d}{16}}.$$

Use this function to check your estimate in part (b).

d. It takes 5 seconds for the object to hit the ground. How far did it fall? Explain your reasoning.

EXPLORATION 2 **Solving a Square Root Equation**

Work with a partner. The speed s (in feet per second) of the free-falling object in Exploration 1 is given by the function

$$s = \sqrt{64d}.$$

Find the distance the object has fallen when it reaches each speed.

a. $s = 8$ ft/sec **b.** $s = 16$ ft/sec **c.** $s = 24$ ft/sec

Communicate Your Answer

3. How can you solve an equation that contains square roots?
4. Use your answer to Question 3 to solve each equation.
   a. $5 = \sqrt{x + 20}$ **b.** $4 = \sqrt{x} - 18$
   c. $\sqrt{x} + 2 = 3$ **d.** $-3 = -2\sqrt{x}$

Dynamic Teaching Tools

- Dynamic Assessment & Progress Monitoring Tool
- Lesson Planning Tool
- Interactive Whiteboard Lesson Library
- Dynamic Classroom with Dynamic Investigations

ANSWERS

1. a.

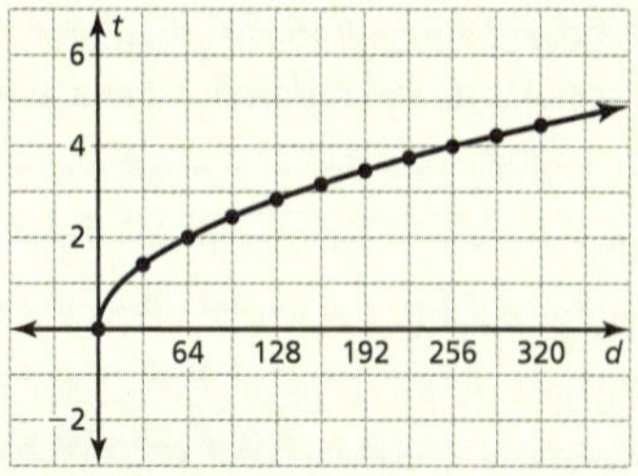

   b. about 3.9 sec

   c. $t = \sqrt{\frac{d}{16}} = \sqrt{\frac{240}{16}} \approx 3.87$

   d. 400 ft; Substitute 5 for t and solve for d.

2. a. 1 ft
   b. 4 ft
   c. 9 ft

3. Isolate the square root. Then square both sides of the equation to "undo" the square root and solve as usual.

4. a. $x = 5$
   b. $x = 34$
   c. $x = 1$
   d. $x = \frac{9}{4}$

Differentiated Instruction

Visual/Auditory

Ask one student to come to the board and solve the linear equation $x + 7 = 12$. Ask a second student to come to the board and solve the radical equation $\sqrt{x} + 7 = 12$. Then, ask a third student to use the two solutions to verbally describe how solving a radical equation is similar to and different from solving a linear equation.

Extra Example 1

Solve each equation.

a. $\sqrt{x} + 4 = 9$

The solution is $x = 25$.

b. $10 - \sqrt{x} = -5$

The solution is $x = 225$.

MONITORING PROGRESS ANSWERS

1. $x = 36$
2. $x = 100$
3. $y = 49$
4. $c = 9$

10.3 Lesson

Core Vocabulary

radical equation, *p. 560*

Previous

radical

radical expression

extraneous solution

What You Will Learn

- Solve radical equations.
- Identify extraneous solutions.
- Solve real-life problems involving radical equations.

Solving Radical Equations

A **radical equation** is an equation that contains a radical expression with a variable in the radicand. To solve a radical equation involving a square root, first use properties of equality to isolate the radical on one side of the equation. Then use the following property to eliminate the radical and solve for the variable.

Core Concept

Squaring Each Side of an Equation

Words If two expressions are equal, then their squares are also equal.

Algebra If $a = b$, then $a^2 = b^2$.

EXAMPLE 1 Solving Radical Equations

Solve each equation.

a. $\sqrt{x} + 5 = 13$ b. $3 - \sqrt{x} = 0$

SOLUTION

a. $\sqrt{x} + 5 = 13$ Write the equation.

$\sqrt{x} = 8$ Subtract 5 from each side.

$(\sqrt{x})^2 = 8^2$ Square each side of the equation.

$x = 64$ Simplify.

▶ The solution is $x = 64$.

Check

$\sqrt{x} + 5 = 13$

$\sqrt{64} + 5 \stackrel{?}{=} 13$

$8 + 5 \stackrel{?}{=} 13$

$13 = 13$ ✓

b. $3 - \sqrt{x} = 0$ Write the equation.

$3 = \sqrt{x}$ Add $\sqrt{x}$ to each side.

$3^2 = (\sqrt{x})^2$ Square each side of the equation.

$9 = x$ Simplify.

▶ The solution is $x = 9$.

Check

$3 - \sqrt{x} = 0$

$3 - \sqrt{9} \stackrel{?}{=} 0$

$3 - 3 \stackrel{?}{=} 0$

$0 = 0$ ✓

Monitoring Progress

Help in English and Spanish at *BigIdeasMath.com*

Solve the equation. Check your solution.

1. $\sqrt{x} = 6$
2. $\sqrt{x} - 7 = 3$
3. $\sqrt{y} + 15 = 22$
4. $1 - \sqrt{c} = -2$

Laurie's Notes Teacher Actions

- **Look For and Make Use of Structure:** Solving square root equations requires that students recognize the structure of the equation. These equations all involve subtracting 5 as the first step: $2x + 5 = 13$; $x^2 + 5 = 13$; $\sqrt{x} + 5 = 13$.

? "How do you undo the square root of x?" Square each side of the equation.

- In part (b), students may ask whether the result is the same when you subtract 3 from each side first and then square each side. Have students verify.

EXAMPLE 2 Solving a Radical Equation

Check

$4\sqrt{x+2}+3=19$

$4\sqrt{14+2}+3\stackrel{?}{=}19$

$4\sqrt{16}+3\stackrel{?}{=}19$

$4(4)+3\stackrel{?}{=}19$

$19=19$ ✓

$4\sqrt{x+2}+3=19$	Original equation
$4\sqrt{x+2}=16$	Subtract 3 from each side.
$\sqrt{x+2}=4$	Divide each side by 4.
$(\sqrt{x+2})^2=4^2$	Square each side of the equation.
$x+2=16$	Simplify.
$x=14$	Subtract 2 from each side.

▶ The solution is $x = 14$.

EXAMPLE 3 Solving an Equation with Radicals on Both Sides

Solve $\sqrt{2x-1}=\sqrt{x+4}$.

SOLUTION

Method 1

$\sqrt{2x-1}=\sqrt{x+4}$	Write the equation.
$(\sqrt{2x-1})^2=(\sqrt{x+4})^2$	Square each side of the equation.
$2x-1=x+4$	Simplify.
$x=5$	Solve for x.

▶ The solution is $x = 5$.

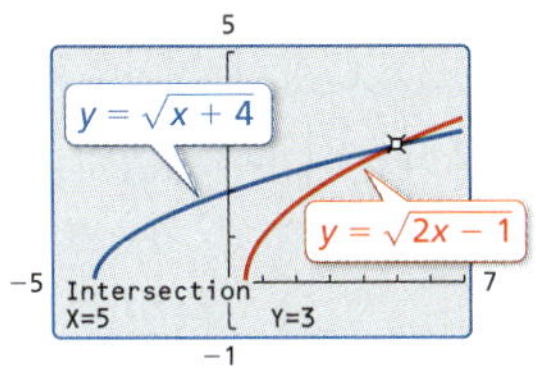

Method 2 Graph each side of the equation, as shown. Use the *intersect* feature to find the coordinates of the point of intersection. The x-value of the point of intersection is 5.

▶ So, the solution is $x = 5$.

EXAMPLE 4 Solving a Radical Equation Involving a Cube Root

Solve $\sqrt[3]{5x-2}=12$.

SOLUTION

$\sqrt[3]{5x-2}=12$	Write the equation.
$(\sqrt[3]{5x-2})^3=12^3$	Cube each side of the equation.
$5x-2=1728$	Simplify.
$x=346$	Solve for x.

▶ The solution is $x = 346$.

LOOKING FOR STRUCTURE

You can extend the concept taught in Examples 1–3 to solve a radical equation involving a cube root. Instead of squaring each side of the equation, you *cube* each side to eliminate the radical.

Monitoring Progress

Help in English and Spanish at *BigIdeasMath.com*

Solve the equation. Check your solution.

5. $\sqrt{x+4}+7=11$ **6.** $15=6+\sqrt{3w-9}$ **7.** $\sqrt{3x+1}=\sqrt{4x-7}$

8. $\sqrt{n}=\sqrt{5n-1}$ **9.** $\sqrt[3]{y}-4=1$ **10.** $\sqrt[3]{3c+7}=10$

Extra Example 2

Solve $3\sqrt{x+5}-1=11$.

The solution is $x = 11$.

Extra Example 3

Solve $\sqrt{4x-3}=\sqrt{x+9}$.

The solution is $x = 4$.

Extra Example 4

Solve $\sqrt[3]{2x-5}=3$.

The solution is $x = 16$.

MONITORING PROGRESS ANSWERS

5. $x = 12$

6. $w = 30$

7. $x = 8$

8. $n = \frac{1}{4}$

9. $y = 125$

10. $c = 331$

Laurie's Notes Teacher Actions

COMMON ERROR When squaring $(\sqrt{x+2})^2$ in Example 2, students may square the x and square the 2.

- Pose Example 3, and have students discuss what each side of the equation looks like graphically. Work through the example as shown.
- **Turn and Talk:** "Discuss a strategy for solving $\sqrt[3]{5x-2}=12$ in Example 4. Test out your strategy." Give partners time to talk and try out their strategy. Anticipate that most students will cube both sides of the equation and will successfully solve the equation independently.

English Language Learners

Vocabulary

For English learners, relate the word *extraneous* to the word *extra*. Something that is *extra* is something that is more than or in addition to what is usual.

Extra Example 5

Solve $x = \sqrt{20 - x}$.

Because $x = -5$ does not satisfy the original equation, it is an extraneous solution. The only solution is $x = 4$.

Extra Example 6

Solve $7 + \sqrt{3n} = 1$.

Because $n = 12$ does not satisfy the original equation, it is an extraneous solution. So, the equation has no solution.

MONITORING PROGRESS ANSWERS

11. $x = 1$

12. no solution

13. $p = 7$

Identifying Extraneous Solutions

Squaring each side of an equation can sometimes introduce an extraneous solution.

ATTEND TO PRECISION

To understand how extraneous solutions can be introduced, consider the equation $\sqrt{x} = -2$. This equation has no real solution, however, you obtain $x = 4$ after squaring each side.

EXAMPLE 5 Identifying an Extraneous Solution

Solve $x = \sqrt{x + 6}$.

SOLUTION

$x = \sqrt{x + 6}$	Write the equation.
$x^2 = (\sqrt{x + 6})^2$	Square each side of the equation.
$x^2 = x + 6$	Simplify.
$x^2 - x - 6 = 0$	Subtract x and 6 from each side.
$(x - 3)(x + 2) = 0$	Factor.
$x - 3 = 0$ or $x + 2 = 0$	Zero-Product Property
$x = 3$ or $x = -2$	Solve for x.

STUDY TIP

Be sure to always substitute your solutions into the original equation to check for extraneous solutions.

Check Check each solution in the original equation.

$3 \stackrel{?}{=} \sqrt{3 + 6}$	Substitute for x.	$-2 \stackrel{?}{=} \sqrt{-2 + 6}$
$3 \stackrel{?}{=} \sqrt{9}$	Simplify.	$-2 \stackrel{?}{=} \sqrt{4}$
$3 = 3$ ✓	Simplify.	$-2 \neq 2$ ✗

▶ Because $x = -2$ does not satisfy the original equation, it is an extraneous solution. The only solution is $x = 3$.

EXAMPLE 6 Identifying an Extraneous Solution

Solve $13 + \sqrt{5n} = 3$.

SOLUTION

$13 + \sqrt{5n} = 3$	Write the equation.
$\sqrt{5n} = -10$	Subtract 13 from each side.
$(\sqrt{5n})^2 = (-10)^2$	Square each side of the equation.
$5n = 100$	Simplify.
$n = 20$	Divide each side by 5.

Check

$13 + \sqrt{5n} = 3$

$13 + \sqrt{5(20)} \stackrel{?}{=} 3$

$13 + \sqrt{100} \stackrel{?}{=} 3$

$23 \neq 3$ ✗

▶ Because $n = 20$ does not satisfy the original equation, it is an extraneous solution. So, the equation has no solution.

Monitoring Progress

Help in English and Spanish at *BigIdeasMath.com*

Solve the equation. Check your solution(s).

11. $\sqrt{4 - 3x} = x$ **12.** $\sqrt{3m} + 10 = 1$ **13.** $p + 1 = \sqrt{7p + 15}$

Laurie's Notes Teacher Actions

? "What are the intersection possibilities for the graphs of a linear function and a square root function?" 0, 1, or 2 points of intersection

- Instead of telling students about extraneous solutions, work through Example 5 and when checking the solution, emphasize that something is wrong. Then explain the possibility of introducing extraneous roots when you square each side of an equation.
- **Extension:** For either example, graph each side of the equation. Use the intersect feature on the calculator to find the solution(s).

Solving Real-Life Problems

EXAMPLE 7 **Modeling with Mathematics**

> **STUDY TIP**
> The period of a pendulum is the amount of time it takes for the pendulum to swing back and forth.

The period P (in seconds) of a pendulum is given by the function $P = 2\pi\sqrt{\frac{L}{32}}$, where L is the pendulum length (in feet). A pendulum has a period of 4 seconds. Is this pendulum twice as long as a pendulum with a period of 2 seconds? Explain your reasoning.

SOLUTION

1. **Understand the Problem** You are given a function that represents the period P of a pendulum based on its length L. You need to find and compare the values of L for two values of P.
2. **Make a Plan** Substitute $P = 2$ and $P = 4$ into the function and solve for L. Then compare the values.
3. **Solve the Problem**

$P = 2\pi\sqrt{\frac{L}{32}}$	Write the function.	$P = 2\pi\sqrt{\frac{L}{32}}$
$2 = 2\pi\sqrt{\frac{L}{32}}$	Substitute for P.	$4 = 2\pi\sqrt{\frac{L}{32}}$
$\frac{2}{2\pi} = \sqrt{\frac{L}{32}}$	Divide each side by 2π.	$\frac{4}{2\pi} = \sqrt{\frac{L}{32}}$
$\frac{1}{\pi} = \sqrt{\frac{L}{32}}$	Simplify.	$\frac{2}{\pi} = \sqrt{\frac{L}{32}}$
$\frac{1}{\pi^2} = \frac{L}{32}$	Square each side and simplify.	$\frac{4}{\pi^2} = \frac{L}{32}$
$\frac{32}{\pi^2} = L$	Multiply each side by 32.	$\frac{128}{\pi^2} = L$
$3.24 \approx L$	Use a calculator.	$12.97 \approx L$

▶ No, the length of the pendulum with a period of 4 seconds is $\frac{128}{\pi^2} \div \frac{32}{\pi^2} = 4$ times longer than the length of a pendulum with a period of 2 seconds.

4. **Look Back** Use the *trace* feature of a graphing calculator to check your solutions.

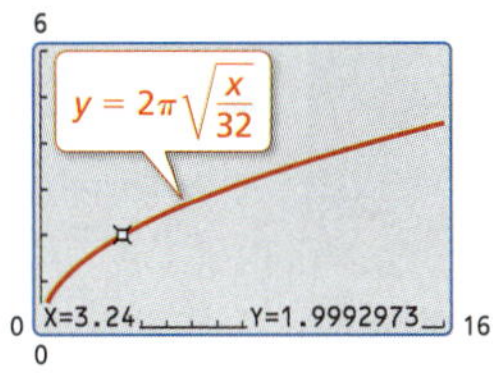

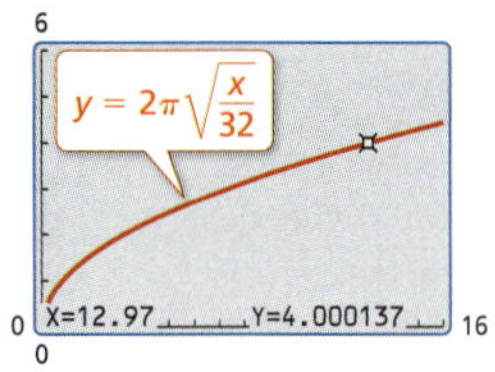

Monitoring Progress Help in English and Spanish at *BigIdeasMath.com*

14. What is the length of a pendulum that has a period of 2.5 seconds?

Extra Example 7

The period P (in seconds) of a pendulum is given by the function $P = 2\pi\sqrt{\frac{L}{32}}$, where L is the pendulum length (in feet). A pendulum has a period of 6 seconds. Is this pendulum three times as long as a pendulum with a period of 2 seconds? Explain your reasoning. No, the length of the pendulum with a period of 6 seconds is $\frac{288}{\pi^2} \div \frac{32}{\pi^2} = 9$ times longer than the length of a pendulum with a period of 2 seconds.

MONITORING PROGRESS ANSWER

14. about 5.07 ft

Laurie's Notes Teacher Actions

- **Teaching Tip:** Use a string with a weight on it to model the pendulum motion and to identify the period. Model several periods.

? "What is the independent variable?" L, the length of the pendulum in feet

? "What is the dependent variable?" P, the period in seconds

- Have students make a prediction about the different lengths L for periods of 2 seconds versus 4 seconds.

Closure

- Solve each equation.

a. $2x + 1 = 10$ $x = \frac{9}{2}$

b. $x^2 + 1 = 10$ $x = 3, x = -3$

c. $\sqrt{x} + 1 = 10$ $x = 81$

Assignment Guide and Homework Check

ASSIGNMENT

Basic: 1, 2, 3–59 odd, 63, 70, 81, 84–89

Average: 1, 2–70 even, 81, 84–89

Advanced: 1, 2, 10–24 even, 32–36 even, 42–48 even, 54–70 even, 71–89

HOMEWORK CHECK

Basic: 5, 13, 37, 49, 63

Average: 6, 14, 38, 50, 64

Advanced: 10, 16, 42, 54, 64

10.3 Exercises

Dynamic Solutions available at *BigIdeasMath.com*

Vocabulary and Core Concept Check

1. **VOCABULARY** Why should you check every solution of a radical equation?

2. **WHICH ONE DOESN'T BELONG?** Which equation does *not* belong with the other three? Explain your reasoning.

$\sqrt{x}+6=10$	$2\sqrt{x+3}=32$	$x\sqrt{3}-5=4$	$\sqrt{x-1}=16$

Monitoring Progress and Modeling with Mathematics

In Exercises 3–12, solve the equation. Check your solution. *(See Example 1.)*

3. $\sqrt{x}=9$
4. $\sqrt{y}=4$
5. $7=\sqrt{m}-5$
6. $\sqrt{p}-7=-1$
7. $\sqrt{c}+12=23$
8. $\sqrt{x}+6=8$
9. $4-\sqrt{a}=2$
10. $-8=7-\sqrt{r}$
11. $3\sqrt{y}-18=-3$
12. $2\sqrt{q}+5=11$

In Exercises 13–20, solve the equation. Check your solution. *(See Example 2.)*

13. $\sqrt{a-3}+5=9$
14. $\sqrt{b+7}-5=-2$
15. $2\sqrt{x+4}=16$
16. $5\sqrt{y-2}=10$
17. $-1=\sqrt{5r+1}-7$
18. $2=\sqrt{4s-4}-4$
19. $7+3\sqrt{3p-9}=25$
20. $19-4\sqrt{3c-11}=11$

21. **MODELING WITH MATHEMATICS** The Cave of Swallows is a natural open-air pit cave in the state of San Luis Potosí, Mexico. The 1220-foot-deep cave was a popular destination for BASE jumpers. The function $t=\frac{1}{4}\sqrt{d}$ represents the time t (in seconds) that it takes a BASE jumper to fall d feet. How far does a BASE jumper fall in 3 seconds?

22. **MODELING WITH MATHEMATICS** The edge length s of a cube with a surface area of A is given by $s=\sqrt{\frac{A}{6}}$. What is the surface area of a cube with an edge length of 4 inches?

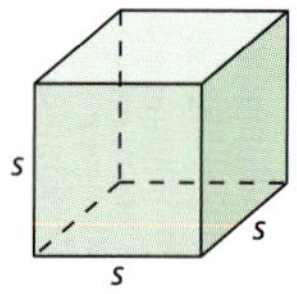

In Exercises 23–26, use the graph to solve the equation.

23. $\sqrt{2x+2}=\sqrt{x+3}$

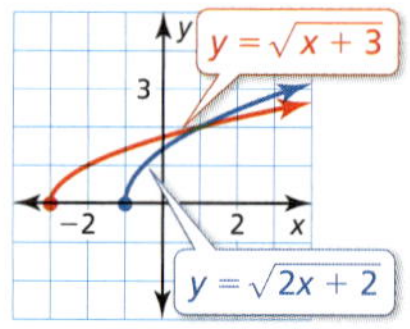

24. $\sqrt{3x+1}=\sqrt{4x-4}$

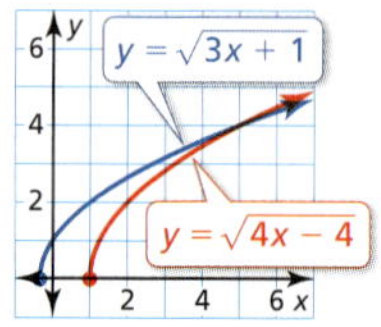

25. $\sqrt{x+2}-\sqrt{2x}=0$

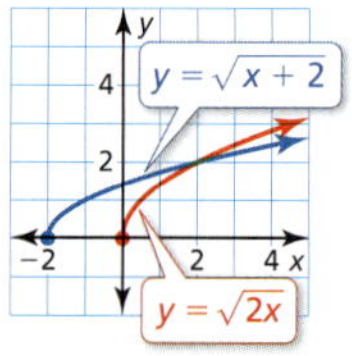

26. $\sqrt{x+5}-\sqrt{3x+7}=0$

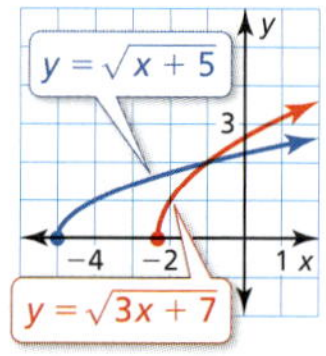

In Exercises 27–34, solve the equation. Check your solution. *(See Example 3.)*

27. $\sqrt{2x-9}=\sqrt{x}$
28. $\sqrt{y+1}=\sqrt{4y-8}$
29. $\sqrt{3g+1}=\sqrt{7g-19}$
30. $\sqrt{8h-7}=\sqrt{6h+7}$
31. $\sqrt{\frac{p}{2}-2}=\sqrt{p-8}$
32. $\sqrt{2v-5}=\sqrt{\frac{v}{3}+5}$
33. $\sqrt{2c+1}-\sqrt{4c}=0$
34. $\sqrt{5r}-\sqrt{8r-2}=0$

ANSWERS

1. Raising each side of a radical equation to an exponent can introduce extraneous solutions.
2. $x\sqrt{3}-5=4$; It is the only equation that is not a radical equation.
3. $x=81$
4. $y=16$
5. $m=144$
6. $p=36$
7. $c=121$
8. $x=4$
9. $a=4$
10. $r=225$
11. $y=25$
12. $q=9$
13. $a=19$
14. $b=2$
15. $x=60$
16. $y=6$
17. $r=7$
18. $s=10$
19. $p=15$
20. $c=5$
21. 144 ft
22. 96 in.^{2}
23. $x=1$
24. $x=5$
25. $x=2$
26. $x=-1$
27. $x=9$
28. $y=3$
29. $g=5$
30. $h=7$
31. $p=12$
32. $v=6$
33. $c=\frac{1}{2}$
34. $r=\frac{2}{3}$

MATHEMATICAL CONNECTIONS **In Exercises 35 and 36, find the value of x.**

35. Perimeter = 22 cm

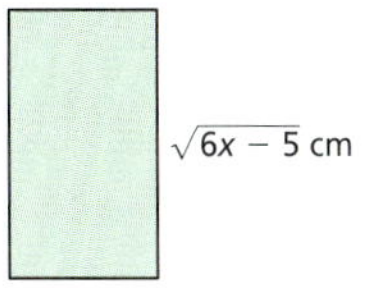

36. Area = $\sqrt{5x-4}$ ft^2

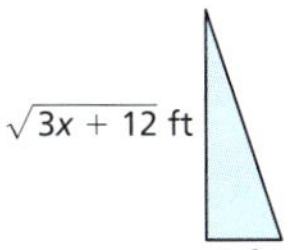

In Exercises 37–44, solve the equation. Check your solution. *(See Example 4.)*

37. $\sqrt[3]{x} = 4$

38. $\sqrt[3]{y} = 2$

39. $6 = \sqrt[3]{8g}$

40. $\sqrt[3]{r+19} = 3$

41. $\sqrt[3]{2s+9} = -3$

42. $-5 = \sqrt[3]{10x+15}$

43. $\sqrt[3]{y+6} = \sqrt[3]{5y-2}$

44. $\sqrt[3]{7j-2} = \sqrt[3]{j+4}$

In Exercises 45–48, determine which solution, if any, is an extraneous solution.

45. $\sqrt{6x-5} = x;\ x = 5,\ x = 1$

46. $\sqrt{2y+3} = y;\ y = -1,\ y = 3$

47. $\sqrt{12p+16} = -2p;\ p = -1,\ p = 4$

48. $-3g = \sqrt{-18-27g};\ g = -2,\ g = -1$

In Exercises 49–58, solve the equation. Check your solution(s). *(See Examples 5 and 6.)*

49. $y = \sqrt{5y-4}$

50. $\sqrt{-14-9x} = x$

51. $\sqrt{1-3a} = 2a$

52. $2q = \sqrt{10q-6}$

53. $9 + \sqrt{5p} = 4$

54. $\sqrt{3n} - 11 = -5$

55. $\sqrt{2m+2} - 3 = 1$

56. $15 + \sqrt{4b-8} = 13$

57. $r + 4 = \sqrt{-4r-19}$

58. $\sqrt{3-s} = s - 1$

ERROR ANALYSIS **In Exercises 59 and 60, describe and correct the error in solving the equation.**

59. ✗

$2 + 5\sqrt{x} = 12$
$5\sqrt{x} = 10$
$5x = 100$
$x = 20$

60. ✗

$x = \sqrt{12-4x}$
$x^2 = 12 - 4x$
$x^2 + 4x - 12 = 0$
$(x-2)(x+6) = 0$
$x = 2$ or $x = -6$
The solutions are $x = 2$ and $x = -6$.

61. **REASONING** Explain how to use mental math to solve $\sqrt{2x} + 5 = 1$.

62. **WRITING** Explain how you would solve $\sqrt[4]{m+4} - \sqrt[4]{3m} = 0$.

63. **MODELING WITH MATHEMATICS** The formula $V = \sqrt{PR}$ relates the voltage V (in volts), power P (in watts), and resistance R (in ohms) of an electrical circuit. The hair dryer shown is on a 120-volt circuit. Is the resistance of the hair dryer half as much as the resistance of the same hair dryer on a 240-volt circuit? Explain your reasoning. *(See Example 7.)*

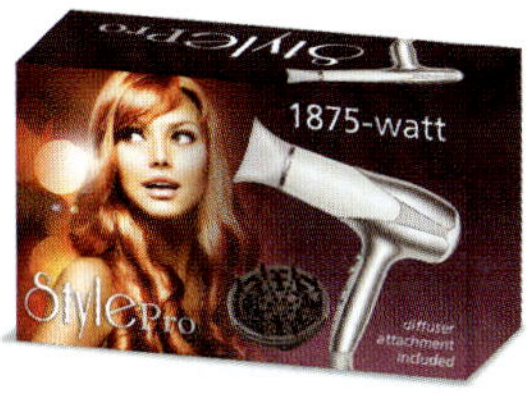

64. **MODELING WITH MATHEMATICS** The time t (in seconds) it takes a trapeze artist to swing back and forth is represented by the function $t = 2\pi\sqrt{\frac{r}{32}}$, where r is the rope length (in feet). It takes the trapeze artist 6 seconds to swing back and forth. Is this rope $\frac{3}{2}$ as long as the rope used when it takes the trapeze artist 4 seconds to swing back and forth? Explain your reasoning.

REASONING **In Exercises 65–68, determine whether the statement is *true* or *false*. If it is false, explain why.**

65. If $\sqrt{a} = b$, then $(\sqrt{a})^2 = b^2$.

66. If $\sqrt{a} = \sqrt{b}$, then $a = b$.

67. If $a^2 = b^2$, then $a = b$.

68. If $a^2 = \sqrt{b}$, then $a^4 = (\sqrt{b})^2$.

Dynamic Teaching Tools

- Dynamic Assessment & Progress Monitoring Tool
- Interactive Whiteboard Lesson Library
- Dynamic Classroom with Dynamic Investigations

ANSWERS

35. $x = 9$

36. $x = 8$

37. $x = 64$

38. $y = 8$

39. $g = 27$

40. $r = 8$

41. $s = -18$

42. $x = -14$

43. $y = 2$

44. $j = 1$

45. neither

46. $y = -1$

47. $p = 4$

48. neither

49. $y = 1,\ y = 4$

50. no solution

51. $a = \frac{1}{4}$

52. $q = 1,\ q = \frac{3}{2}$

53. no solution

54. $n = 12$

55. $m = 7$

56. no solution

57. no solution

58. $s = 2$

59. An error was made in squaring both sides because the radical was not isolated. The third equation should be $\sqrt{x} = 2$, and the fourth equation should be $x = 4$.

60. $x = -6$ should not be shown as a solution because it is extraneous.

61. This equation is equivalent to $\sqrt{2x} = -4$, which has no solution because the positive square root of a number cannot be negative.

62. Add $\sqrt[4]{3m}$ to each side to isolate a radical on each side. Then raise each side to the fourth power, solve the resulting linear equation, and check the solution.

63. no; The resistance is $\frac{7.68}{30.72} = \frac{1}{4}$ as much for a 120-volt circuit than for a 240-volt circuit.

64. no; The length of a rope with a period of 6 seconds is $\frac{288}{\pi^2} \div \frac{128}{\pi^2} = \frac{9}{4}$ longer than a rope with a period of 4 seconds.

The time is $\frac{6}{4} = \frac{3}{2}$ as long but because rope length r is under the radical sign in the formula, the rope length would need to be $\left(\frac{3}{2}\right)^2 = \frac{9}{4}$ as long for a time of 6 seconds than the length for a time of 4 seconds.

65. true

66. true

67. false; when $b = -a$, then $a^2 = (-a)^2$, but $a \neq -a$

68. true

ANSWERS

69. **a.** no solution; Graph $y = x + 2$ and $y = \sqrt{2x - 3}$ on the same set of coordinate axes. The graphs do not intersect, so there is no solution.

b. no solution; Square each side, simplify, and subtract $2x - 3$ from each side of the equation. Then, use the quadratic formula, which gives no real solutions.

c. *Sample answer:* the graphical approach, because it gives a visual picture of the situation

70. **a.** *Sample answer:* $\sqrt{2x + 3} = \sqrt{4x - 3}$

b. $x = 3$

71. first cone: $r = 2$ units; second cone: $2r = 4$ units

72. Squaring $\sqrt{x + 2}$ simply eliminates the radical sign and the result is $x + 2$. Squaring $\sqrt{x} + 2$ involves multiplying the terms of two sums, and the result is $x + 4\sqrt{x} + 4$.

73. $m = 5$

74. $x = -\frac{7}{16}$

75. $y = \frac{16}{5}$

76. no solution

77. $h = 0$

78. $z = -4$

79. *Sample answer:* $\sqrt{x - 4} = 1$

80. *Sample answer:* $x = \sqrt{7x - 12}$

81–89. See Additional Answers.

69. COMPARING METHODS Consider the equation $x + 2 = \sqrt{2x - 3}$.

a. Solve the equation by graphing. Describe the process.

b. Solve the equation algebraically. Describe the process.

c. Which method do you prefer? Explain your reasoning.

70. HOW DO YOU SEE IT? The graph shows two radical functions.

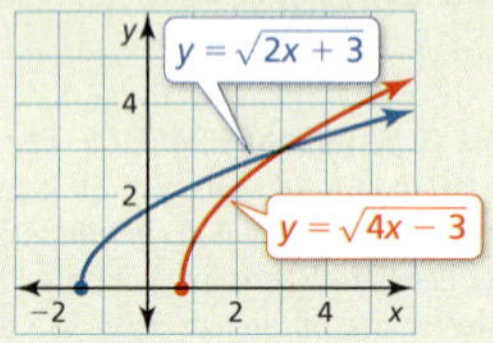

a. Write an equation whose solution is the x-coordinate of the point of intersection of the graphs.

b. Use the graph to solve the equation.

71. MATHEMATICAL CONNECTIONS The slant height s of a cone with a radius of r and a height of h is given by $s = \sqrt{r^2 + h^2}$. The slant heights of the two cones are equal. Find the radius of each cone.

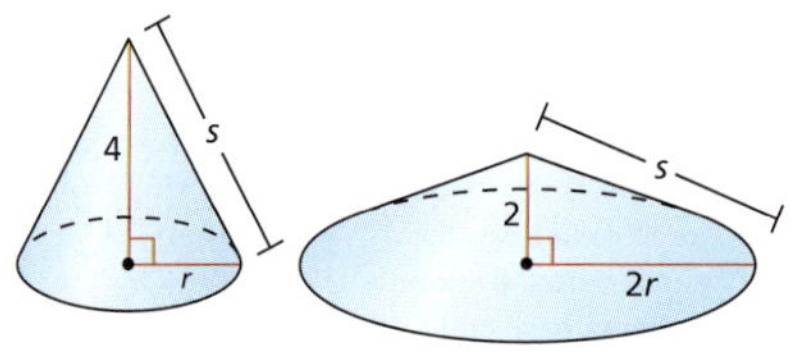

72. CRITICAL THINKING How is squaring $\sqrt{x + 2}$ different from squaring $\sqrt{x} + 2$?

USING STRUCTURE In Exercises 73–78, solve the equation. Check your solution.

73. $\sqrt{m + 15} = \sqrt{m} + \sqrt{5}$ **74.** $2 - \sqrt{x + 1} = \sqrt{x + 2}$

75. $\sqrt{5y + 9} + \sqrt{5y} = 9$

76. $\sqrt{2c - 8} - \sqrt{2c} - 4 = 0$

77. $2\sqrt{1 + 4h} - 4\sqrt{h} - 2 = 0$

78. $\sqrt{20 - 4z} + 2\sqrt{-z} = 10$

79. OPEN-ENDED Write a radical equation that has a solution of $x = 5$.

80. OPEN-ENDED Write a radical equation that has $x = 3$ and $x = 4$ as solutions.

81. MAKING AN ARGUMENT Your friend says the equation $\sqrt{(2x + 5)^2} = 2x + 5$ is always true, because after simplifying the left side of the equation, the result is an equation with infinitely many solutions. Is your friend correct? Explain.

82. THOUGHT PROVOKING Solve the equation $\sqrt[3]{x + 1} = \sqrt{x - 3}$. Show your work and explain your steps.

83. MODELING WITH MATHEMATICS The frequency f (in cycles per second) of a string of an electric guitar is given by the equation $f = \frac{1}{2\ell}\sqrt{\frac{T}{m}}$, where ℓ is the length of the string (in meters), T is the string's tension (in newtons), and m is the string's mass per unit length (in kilograms per meter). The high E string of an electric guitar is 0.64 meter long with a mass per unit length of 0.000401 kilogram per meter.

a. How much tension is required to produce a frequency of about 330 cycles per second?

b. Would you need more or less tension to create the same frequency on a string with greater mass per unit length? Explain.

Maintaining Mathematical Proficiency

Reviewing what you learned in previous grades and lessons

Find the product. *(Section 7.2)*

84. $(x + 8)(x - 2)$ **85.** $(3p - 1)(4p + 5)$ **86.** $(s + 2)(s^2 + 3s - 4)$

Graph the function. Compare the graph to the graph of $f(x) = x^2$. *(Section 8.1)*

87. $r(x) = 3x^2$ **88.** $g(x) = \frac{3}{4}x^2$ **89.** $h(x) = -5x^2$

Mini-Assessment

Solve the equation.

1. $\sqrt{x} - 5 = 12$ $x = 289$
2. $\sqrt{2x + 5} = \sqrt{4x + 1}$ $x = 2$
3. $\sqrt[3]{x + 3} = 4$ $x = 61$
4. $x = \sqrt{24 - 2x}$ $x = 4$
5. The radius r (in inches) of a circle is given by the function $r = \sqrt{\frac{A}{\pi}}$, where A is the area (in square inches) of the circle. A circle has a radius of 4 inches. Is the area of this circle twice the area of a circle with a radius of 2 inches? Explain your reasoning. No, the area of the circle with a radius of 4 inches is $(16\pi) \div (4\pi) = 4$ times the area of a circle with a radius of 2 inches.

If students need help...	*If students got it...*
Resources by Chapter • Practice A and Practice B • Puzzle Time	Resources by Chapter • Enrichment and Extension • Cumulative Review
Student Journal • Practice	Start the *next* Section
Differentiating the Lesson Skills Review Handbook	

Laurie's Notes

Overview of Section 10.4

Introduction

- This lesson presents inverse relations and inverse functions.
- The inverse of a linear equation is a linear equation. The inverse of a quadratic equation (with a restricted domain) is a square root function, the function studied earlier in this chapter.
- The inverse of a function can be found algebraically by solving $y = f(x)$ for x. When graphed, inverse functions are symmetric about the line $y = x$.

Formative Assessment Tips

- **Predict, Explain, Observe (P-E-O Probe):** This technique provides students the opportunity to make a prediction or to select a prediction from a set of options. Students explain or give reasons why their prediction makes sense, and then they test out their prediction. Finally, they analyze the results against their prediction and make adjustments to their thinking.
- This technique gives students the opportunity to reason about a problem, situation, or concept that is being probed, and students are asked to make a prediction related to the probe. Explaining the reason for their prediction is an important step. Curiosity piqued, students think of how to investigate their prediction. Students carry out their investigation and analyze the results. If their prediction is correct and it was based on valid reasoning (versus an educated guess), then their reasoning has been confirmed. If the results are not exactly what they had predicted, then the cognitive dissonance created should lead to more discussion, rethinking, and a new prediction.
- This technique is best used when you want to probe students' ideas. This technique does not need to be done only with individuals. Partners or small groups can make predictions, discussing their reasoning and rethinking their prediction after an investigation.

Pacing Suggestion

- The explorations are straightforward and should be relatively brief in terms of time. Transition to the formal lesson when students finish.

Dynamic Teaching Tools

- Dynamic Assessment & Progress Monitoring Tool
- Lesson Planning Tool
- Interactive Whiteboard Lesson Library
- Dynamic Classroom with Dynamic Investigations

What Your Students Will Learn

- Find inverses of relations by switching the coordinates of each ordered pair or by switching the inputs and outputs of a table.
- Explore inverses of functions algebraically, numerically, and graphically.
- Find inverses of linear functions algebraically.
- Find inverses of quadratic and radical functions.
- Determine whether the inverse of a function is a function by applying the Horizontal Line Test.

Laurie's Notes

Exploration

Motivate

- ? "What was the average daily temperature yesterday?" Answers will vary. Find this information in advance.
- Write a table on the board listing the date (input) and average daily temperature (output) for a specific location. Feel free to make up the data. Make sure that at least two of the temperatures are the same.
- ? "Does this table represent a function? Explain." yes; Every date is associated with exactly one average daily temperature.
- ? "If we switch the inputs and outputs, would it represent a function? Explain." no; An average daily temperature would be associated with more than one date.

Exploration 1

- The introduction mentions that f and g are inverses, but inverse functions are not defined. From looking at the two tables, students will quickly and imprecisely state, "The x's and y's are switched."
- Probe students to see whether there is alternate language they could use to describe the inverse functions. When the ordered pair (a, b) is a solution of one function, (b, a) is a solution of the inverse function.

Exploration 2

- Students should have little difficulty plotting the two functions.
- ? "What observation do you have about the two graphs?" Answers will vary. Some students may observe the symmetry about the line $y = x$, but if they don't, it will be discussed in the lesson.
- The ordered pairs that have integer coordinates should give students enough information to write the equation for each function.

Communicate Your Answer

- Listen for student understanding of how inverse functions are related.
- In Question 4, students should recognize that $f(x) = x + 1$, and they will quickly guess that the inverse of "adding 1" is "subtracting 1." So the inverse will be $g(x) = x - 1$.

Connecting to Next Step

- The explorations will help students develop a beginning understanding of inverse functions. Inverse relations and inverse functions are discussed in the lesson.

10.4 Inverse of a Function

Essential Question How are a function and its inverse related?

EXPLORATION 1 **Exploring Inverse Functions**

Work with a partner. The functions f and g are *inverses* of each other. Compare the tables of values of the two functions. How are the functions related?

x	0	0.5	1	1.5	2	2.5	3	3.5
$f(x)$	0	0.25	1	2.25	4	6.25	9	12.25

x	0	0.25	1	2.25	4	6.25	9	12.25
$g(x)$	0	0.5	1	1.5	2	2.5	3	3.5

EXPLORATION 2 **Exploring Inverse Functions**

Work with a partner.

a. Plot the two sets of points represented by the tables in Exploration 1. Use the coordinate plane below.

b. Connect each set of points with a smooth curve.

c. Describe the relationship between the two graphs.

d. Write an equation for each function.

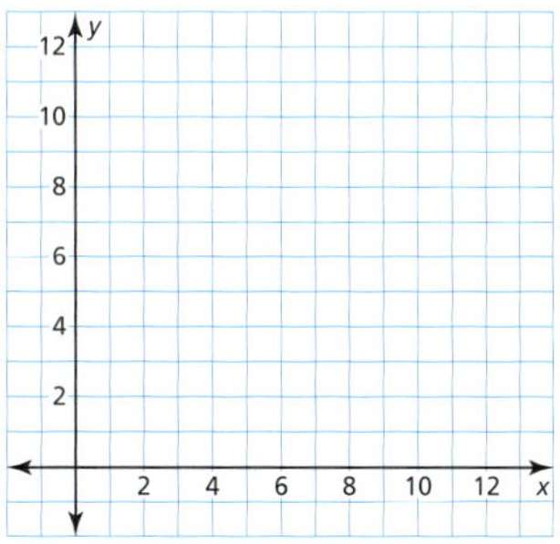

Communicate Your Answer

3. How are a function and its inverse related?

4. A table of values for a function f is given. Create a table of values for a function g, the inverse of f.

x	0	1	2	3	4	5	6	7
$f(x)$	1	2	3	4	5	6	7	8

5. Sketch the graphs of $f(x) = x + 4$ and its inverse in the same coordinate plane. Then write an equation of the inverse of f. Explain your reasoning.

ATTENDING TO PRECISION

To be proficient in math, you need to communicate precisely with others.

Dynamic Teaching Tools

- Dynamic Assessment & Progress Monitoring Tool
- Lesson Planning Tool
- Interactive Whiteboard Lesson Library
- Dynamic Classroom with Dynamic Investigations

ANSWERS

1. The x-values in each table are the same as the function values in the other table.

2. a–b.

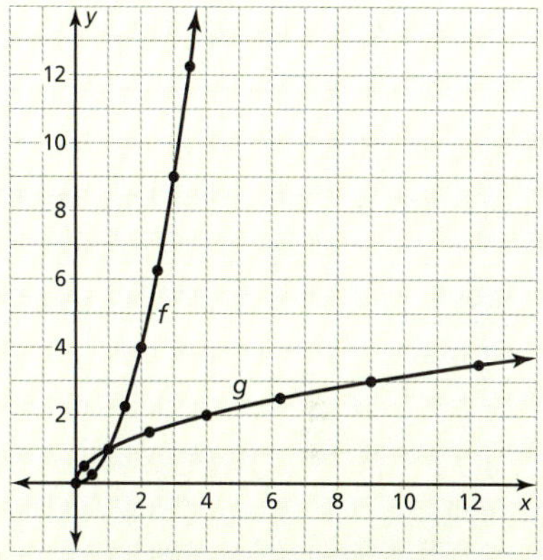

c. The graphs are reflections of each other in the line $y = x$.

d. $f(x) = x^2, x \geq 0$; $g(x) = \sqrt{x}$

3. The x-coordinates of a function are the y-coordinates of its inverse function, and the y-coordinates of the function are the x-coordinates of its inverse function.

4.

x	1	2	3	4	5	6	7	8
$g(x)$	0	1	2	3	4	5	6	7

5.

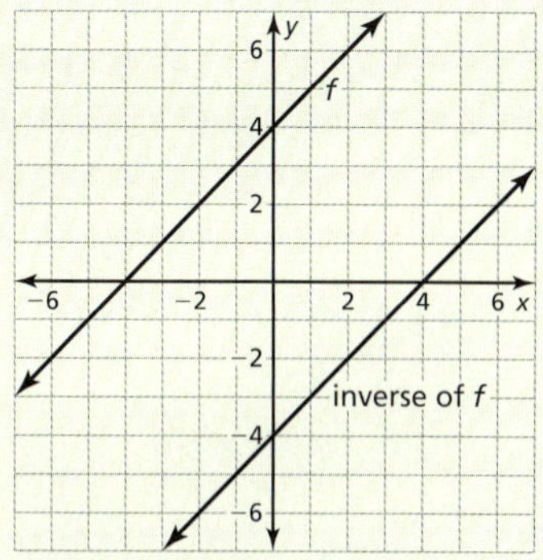

$y = x - 4$; This is an equation of the line containing the points obtained by switching the inputs and outputs of f to get the inverse of f.

English Language Learners

Notebook Development

Have students create an inverse relations and functions page in their notebooks. Students should write the definitions of *inverse relation* and *inverse function* and include examples with each definition.

Extra Example 1

Find the inverse of each relation.

a. (–2, 6), (1, 5), (4, 4), (7, 3), (10, 2)

(6, –2), (5, 1), (4, 4), (3, 7), (2, 10)

b.

Input	0	2	4	4	2	0
Output	2	4	6	8	10	12

Input	2	4	6	8	10	12
Output	0	2	4	4	2	0

MONITORING PROGRESS ANSWERS

1. $(-4, -3), (0, -2), (4, -1), (8, 0), (12, 1), (16, 2), (20, 3)$

2.

Input	4	1	0	1	4
Output	−2	−1	0	1	2

10.4 Lesson

Core Vocabulary

inverse relation, *p. 568*
inverse function, *p. 569*

Previous
input
output
inverse operations
reflection
line of reflection

What You Will Learn

- Find inverses of relations.
- Explore inverses of functions.
- Find inverses of functions algebraically.
- Find inverses of nonlinear functions.

Finding Inverses of Relations

Recall that a relation pairs inputs with outputs. An **inverse relation** switches the input and output values of the original relation.

Core Concept

Inverse Relation

When a relation contains (a, b), the inverse relation contains (b, a).

EXAMPLE 1 Finding Inverses of Relations

Find the inverse of each relation.

a. $(-4, 7), (-2, 4), (0, 1), (2, -2), (4, -5)$ — Switch the coordinates of each ordered pair.

$(7, -4), (4, -2), (1, 0), (-2, 2), (-5, 4)$ — Inverse relation

b.

Input	−1	0	1	2	3	4
Output	5	10	15	20	25	30

Switch the inputs and outputs.

Inverse relation:

Input	5	10	15	20	25	30
Output	−1	0	1	2	3	4

Monitoring Progress

Help in English and Spanish at *BigIdeasMath.com*

Find the inverse of the relation.

1. $(-3, -4), (-2, 0), (-1, 4), (0, 8), (1, 12), (2, 16), (3, 20)$

2.

Input	−2	−1	0	1	2
Output	4	1	0	1	4

Exploring Inverses of Functions

Throughout this book, you have used given inputs to find corresponding outputs of $y = f(x)$ for various types of functions. You have also used given outputs to find corresponding inputs. Now you will solve equations of the form $y = f(x)$ for x to obtain a formula for finding the input given a specific output of the function f.

Laurie's Notes Teacher Actions

- Students should recall relations, which were introduced earlier in the book. When the input and output of a relation are switched, an inverse relation is created. Color coding is used to draw attention to the switch of the input and output.
- **Think-Pair-Share:** Have students answer Questions 1 and 2, and then share and discuss as a class.
- **Turn and Talk:** "If you knew the equation (rule) for a relation, how could you find the equation (rule) for the inverse relation?"

EXAMPLE 2 Writing a Formula for the Input of a Function

Let $f(x) = 2x + 1$. Solve $y = f(x)$ for x. Then find the input when the output is -3.

SOLUTION

$y = 2x + 1$ — Set y equal to $f(x)$.

$y - 1 = 2x$ — Subtract 1 from each side.

$\frac{y-1}{2} = x$ — Divide each side by 2.

Find the input when $y = -3$.

$x = \frac{-3-1}{2}$ — Substitute -3 for y.

$= \frac{-4}{2}$ — Subtract.

$= -2$ — Divide.

Check

$f(-2) = 2(-2) + 1$

$= -4 + 1$

$= -3$ ✓

▶ So, the input is -2 when the output is -3.

Monitoring Progress

Help in English and Spanish at *BigIdeasMath.com*

Solve $y = f(x)$ for x. Then find the input when the output is 4.

3. $f(x) = x - 6$ **4.** $f(x) = \frac{1}{2}x + 3$ **5.** $f(x) = 4x^2$

UNDERSTANDING MATHEMATICAL TERMS

The term *inverse functions* does not refer to a new type of function. Rather, it describes any pair of functions that are inverses.

In Example 2, notice the steps involved after substituting for x in $y = 2x + 1$ and after substituting for y in $x = \frac{y-1}{2}$.

$y = 2x + 1$ | $x = \frac{y-1}{2}$

Step 1 Multiply by 2. → **Step 2** Divide by 2.

Step 2 Add 1. → **Step 1** Subtract 1.

inverse operations in the reverse order

Notice that these steps *undo* each other. **Inverse functions** are functions that undo each other. In Example 2, you can use the equation solved for x to write the inverse of f by switching the roles of x and y.

$f(x) = 2x + 1$ original function $\qquad g(x) = \frac{x-1}{2}$ inverse function

LOOKING FOR A PATTERN

Notice that the graph of the inverse function g is a reflection of the graph of the original function f. The line of reflection is $y = x$.

Because an inverse function interchanges the input and output values of the original function, the domain and range are also interchanged.

Original function: $f(x) = 2x + 1$

x	−2	−1	0	1	2
y	−3	−1	1	3	5

Inverse function: $g(x) = \frac{x-1}{2}$

x	−3	−1	1	3	5
y	−2	−1	0	1	2

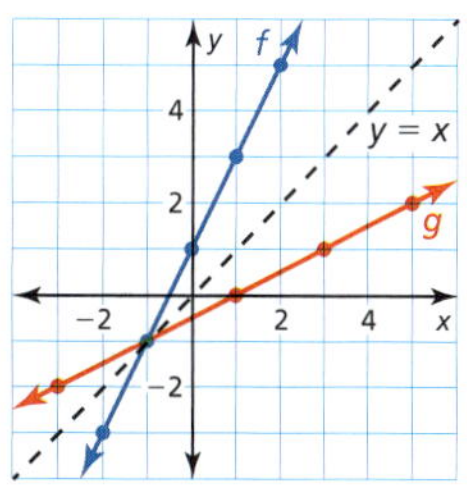

Extra Example 2

Let $f(x) = 3x - 5$. Solve $y = f(x)$ for x. Then find the input when the output is 4.

$x = \frac{y+5}{3}$; The input is 3 when the output is 4.

MONITORING PROGRESS ANSWERS

3. $x = y + 6$; 10

4. $x = 2y - 6$; 2

5. $x = \pm\frac{\sqrt{y}}{2}$; ± 1

Laurie's Notes Teacher Actions

- Pose Example 2. This question is asking students to solve the function for x, not simply find the input for a given output.
- Summarize by saying, "For the function $f(x) = 2x + 1$, the formula to find the input is $\frac{y-1}{2} = x$."
- Ask whether students observe a pattern. Most students make the connection that $f(x)$ multiplies the input by 2 and then adds 1. The inverse relation subtracts 1 and divides by 2.
- Discuss inverse operations in the reverse order.

? "If the ordered pair $(1, -3)$ satisfies the original function, what ordered pair satisfies the inverse function?" $(-3, 1)$

Finding Inverses of Functions Algebraically

Core Concept

Finding Inverses of Functions Algebraically

Step 1 Set y equal to $f(x)$.

Step 2 Switch x and y in the equation.

Step 3 Solve the equation for y.

STUDY TIP

On the previous page, you solved a function for x and switched the roles of x and y to find the inverse function. You can also find the inverse function by switching x and y first, and then solving for y.

EXAMPLE 3 Finding the Inverse of a Linear Function

Find the inverse of $f(x) = 4x - 9$.

SOLUTION

Method 1 Use the method above.

Step 1	$f(x) = 4x - 9$	Write the function.
	$y = 4x - 9$	Set y equal to $f(x)$.
Step 2	$x = 4y - 9$	Switch x and y in the equation.
Step 3	$x + 9 = 4y$	Add 9 to each side.
	$\frac{x+9}{4} = y$	Divide each side by 4.

▶ The inverse of f is $g(x) = \frac{x+9}{4}$, or $g(x) = \frac{1}{4}x + \frac{9}{4}$.

Method 2 Use inverse operations in the reverse order.

$f(x) = 4x - 9$ — Multiply the input x by 4 and then subtract 9.

To find the inverse, apply inverse operations in the reverse order.

$g(x) = \frac{x+9}{4}$ — Add 9 to the input x and then divide by 4.

▶ The inverse of f is $g(x) = \frac{x+9}{4}$, or $g(x) = \frac{1}{4}x + \frac{9}{4}$.

Check

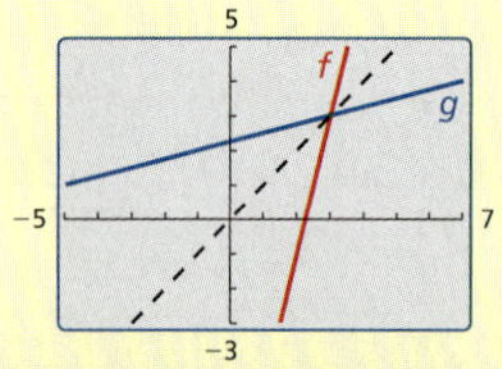

The graph of g appears to be a reflection of the graph of f in the line $y = x$. ✓

Monitoring Progress

Help in English and Spanish at *BigIdeasMath.com*

Find the inverse of the function. Then graph the function and its inverse.

6. $f(x) = 6x$ **7.** $f(x) = -x + 5$ **8.** $f(x) = \frac{1}{4}x - 1$

Finding Inverses of Nonlinear Functions

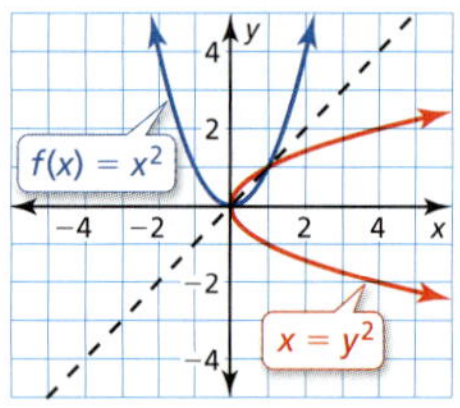

The inverse of the linear function in Example 3 is also a function. The inverse of a function, however, is *not* always a function. The graph of $f(x) = x^2$ is shown along with its reflection in the line $y = x$. Notice that the graph of the inverse of $f(x) = x^2$ does not pass the Vertical Line Test. So, the inverse is *not* a function.

When the domain of $f(x) = x^2$ is *restricted* to only nonnegative real numbers, the inverse of f *is* a function, as shown in the next example.

Differentiated Instruction

Organization

Some students may benefit by writing down the steps for finding the inverses of functions algebraically. Have them copy into their notebooks the steps shown in the *Core Concept*. Students can use the steps as a reference as they solve problems like Examples 3–5.

Extra Example 3

Find the inverse of $f(x) = 3x + 4$.

The inverse of f is $g(x) = \frac{x-4}{3}$, or $g(x) = \frac{1}{3}x - \frac{4}{3}$.

MONITORING PROGRESS ANSWERS

6. $g(x) = \frac{1}{6}x$

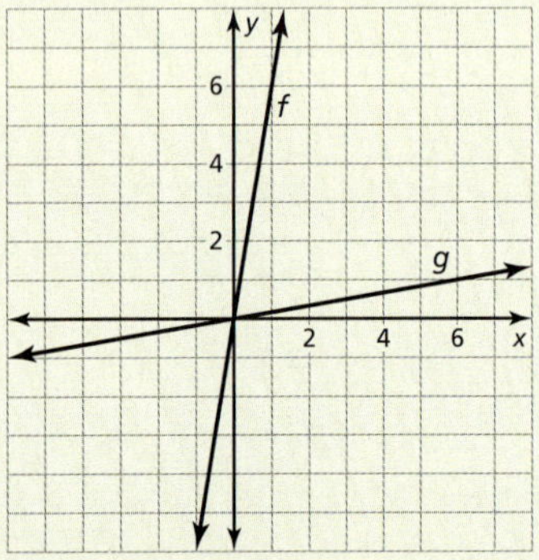

7. $g(x) = -x + 5$

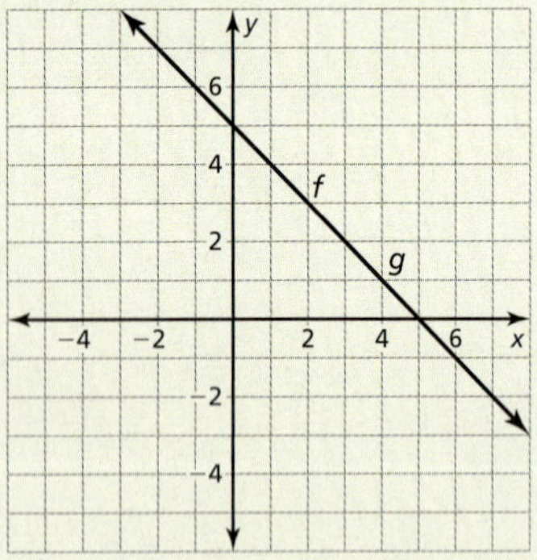

8. $g(x) = 4x + 4$

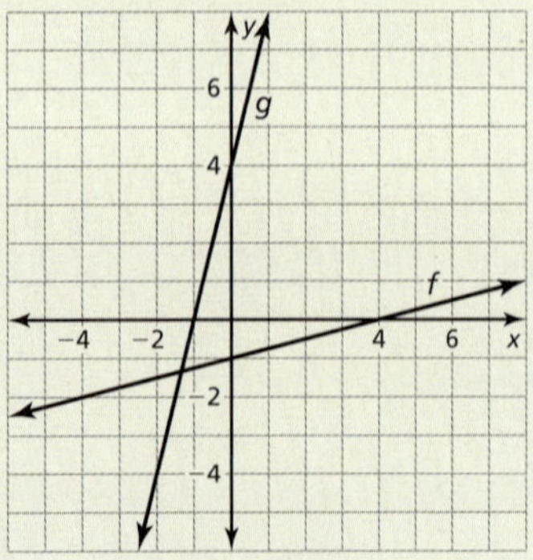

Laurie's Notes Teacher Actions

- Write the *Core Concept*. Work through Example 3, showing both methods.
- **Predict, Explain, Observe:** Group students (2–4 in a group) and ask, "Does every function have an inverse? Explain your reasoning." Give groups sufficient time to consider the question and to consider their reasoning. Probe with additional questions as students consider how to investigate their prediction. "Consider different functions you have studied." Students may need to revise their thinking!

EXAMPLE 4 Finding the Inverse of a Quadratic Function

Find the inverse of $f(x) = x^2$, $x \geq 0$. Then graph the function and its inverse.

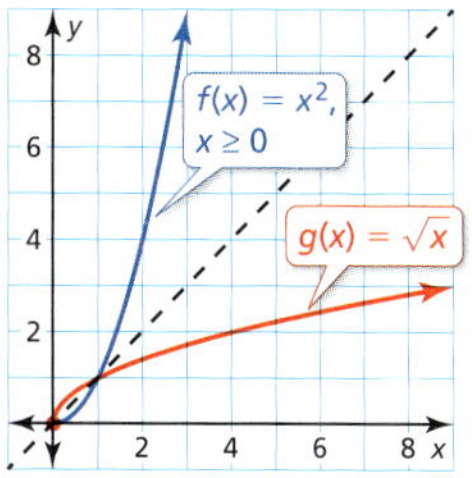

SOLUTION

$f(x) = x^2$	Write the function.
$y = x^2$	Set y equal to $f(x)$.
$x = y^2$	Switch x and y in the equation.
$\pm\sqrt{x} = y$	Take square root of each side.

▶ Because the domain of f is restricted to nonnegative values of x, the range of the inverse must also be restricted to nonnegative values. So, the inverse of f is $g(x) = \sqrt{x}$.

You can use the graph of a function f to determine whether the inverse of f is a function by applying the *Horizontal Line Test*.

Core Concept

Horizontal Line Test

The inverse of a function f is also a function if and only if no horizontal line intersects the graph of f more than once.

EXAMPLE 5 Finding the Inverse of a Radical Function

Consider the function $f(x) = \sqrt{x+2}$. Determine whether the inverse of f is a function. Then find the inverse.

SOLUTION

Graph the function f. Because no horizontal line intersects the graph more than once, the inverse of f is a function. Find the inverse.

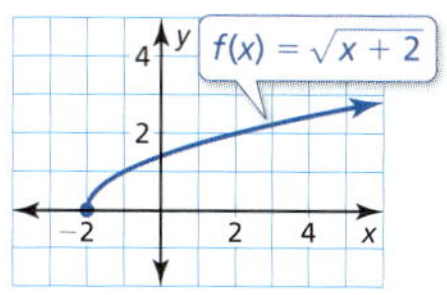

$y = \sqrt{x+2}$	Set y equal to $f(x)$.
$x = \sqrt{y+2}$	Switch x and y in the equation.
$x^2 = \left(\sqrt{y+2}\right)^2$	Square each side.
$x^2 = y + 2$	Simplify.
$x^2 - 2 = y$	Subtract 2 from each side.

▶ Because the range of f is $y \geq 0$, the domain of the inverse must be restricted to $x \geq 0$. So, the inverse of f is $g(x) = x^2 - 2$, where $x \geq 0$.

Check

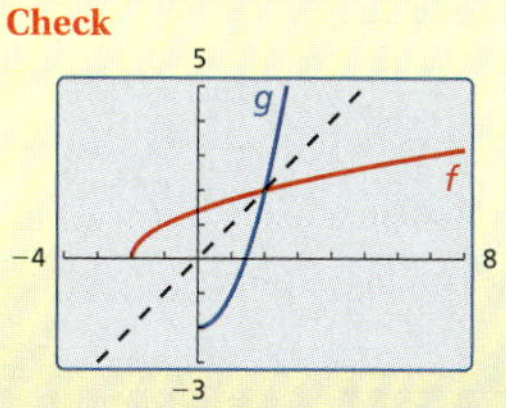

Monitoring Progress

Help in English and Spanish at *BigIdeasMath.com*

Find the inverse of the function. Then graph the function and its inverse.

9. $f(x) = -x^2$, $x \leq 0$

10. $f(x) = 4x^2 + 3$, $x \geq 0$

11. Is the inverse of $f(x) = \sqrt{2x - 1}$ a function? Find the inverse.

Extra Example 4

Find the inverse of $f(x) = \frac{1}{2}x^2$, $x \geq 0$. Then graph the function and its inverse.
Because the domain of f is restricted to nonnegative values of x, the range of the inverse must also be restricted to nonnegative values. So, the inverse of f is $g(x) = \sqrt{2x}$.

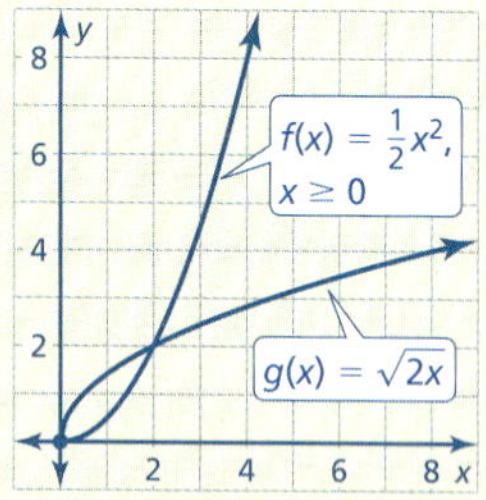

Extra Example 5

Consider the function $f(x) = \sqrt{x - 3}$. Determine whether the inverse of f is a function. Then find the inverse.
The inverse of f is a function and is $g(x) = x^2 + 3$, where $x \geq 0$.

MONITORING PROGRESS ANSWERS

9. $g(x) = -\sqrt{-x}$

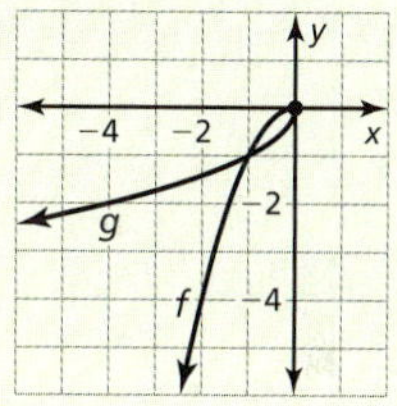

10. $g(x) = \dfrac{\sqrt{x-3}}{2}$

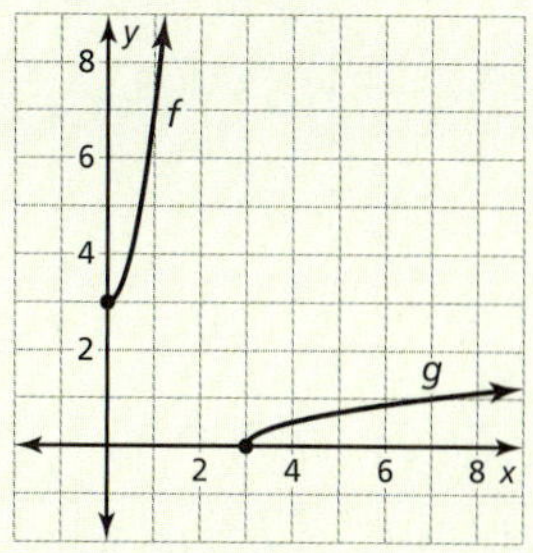

11. yes; $g(x) = \dfrac{x^2+1}{2}$, $x \geq 0$

Laurie's Notes Teacher Actions

- **Reason Abstractly and Quantitatively:** The domain in Example 4 is nonnegative, as shown in the graph. So, the range of the inverse must also be nonnegative.
- Return to the prediction made by students regarding all functions having inverses. Restrictions on the domain must sometimes be made. Introduce the Horizontal Line Test.

Closure

- **Exit Ticket:** Explain the steps you would take to find the inverse of $f(x) = 3x - 4$. Then find the inverse. Replace $f(x)$ with y, switch x and y, solve for y, and then replace y with $g(x)$ to obtain $g(x) = \frac{1}{3}x + \frac{4}{3}$.

Dynamic Solutions available at *BigIdeasMath.com*

Vocabulary and Core Concept Check

1. **COMPLETE THE SENTENCE** A relation contains the point $(-3, 10)$. The __________ contains the point $(10, -3)$.

2. **DIFFERENT WORDS, SAME QUESTION** Consider the function f represented by the graph. Which is different? Find "both" answers.

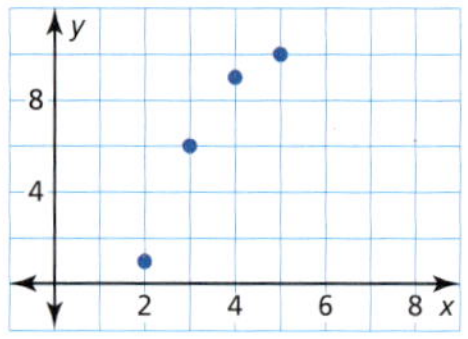

Graph the inverse of the function.

Reflect the graph of the function in the x-axis.

Reflect the graph of the function in the line $y = x$.

Switch the inputs and outputs of the function and graph the resulting function.

Monitoring Progress and Modeling with Mathematics

In Exercises 3–8, find the inverse of the relation. *(See Example 1.)*

3. $(1, 0), (3, -8), (4, -3), (7, -5), (9, -1)$

4. $(2, 1), (4, -3), (6, 7), (8, 1), (10, -4)$

5.

Input	-5	-5	0	5	10
Output	8	6	0	6	8

6.

Input	-12	-8	-5	-3	-2
Output	2	5	-1	10	-2

7. Input Output

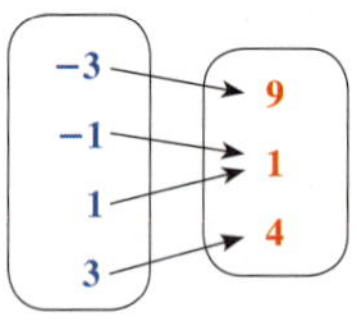

8. Input Output

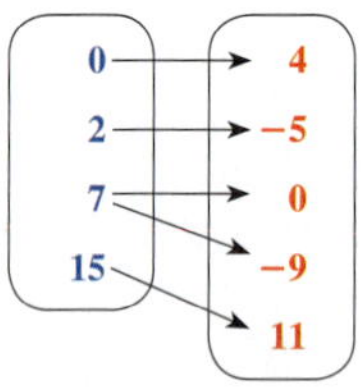

In Exercises 9–14, solve $y = f(x)$ for x. Then find the input when the output is 2. *(See Example 2.)*

9. $f(x) = x + 5$

10. $f(x) = 2x - 3$

11. $f(x) = \frac{1}{4}x - 1$

12. $f(x) = \frac{2}{3}x + 4$

13. $f(x) = 9x^2$

14. $f(x) = \frac{1}{2}x^2 - 7$

In Exercises 15 and 16, graph the inverse of the function by reflecting the graph in the line $y = x$. Describe the domain and range of the inverse.

15.

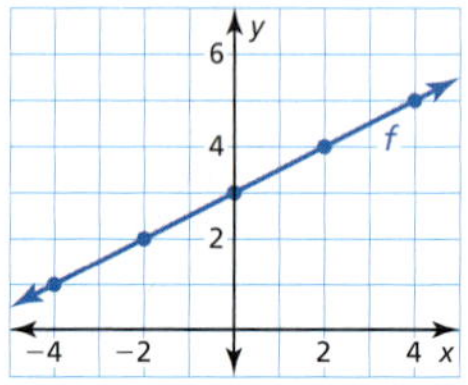

16.

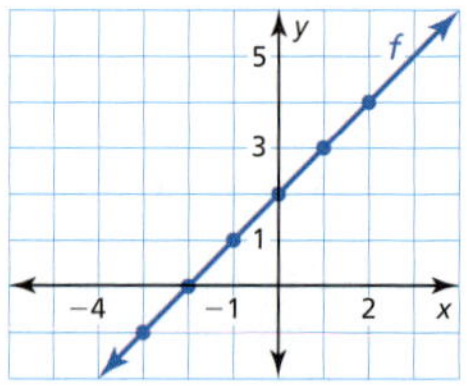

In Exercises 17–22, find the inverse of the function. Then graph the function and its inverse. *(See Example 3.)*

17. $f(x) = 4x - 1$

18. $f(x) = -2x + 5$

19. $f(x) = -3x - 2$

20. $f(x) = 2x + 3$

21. $f(x) = \frac{1}{3}x + 8$

22. $f(x) = -\frac{3}{2}x + \frac{7}{2}$

Assignment Guide and Homework Check

ASSIGNMENT

Basic: 1, 2, 3–45 odd, 55, 56, 64–71

Average: 1, 2–50 even, 55–57, 64–71

Advanced: 1, 2, 8, 14, 16, 22, 26–30 even, 38–54 even, 55–71

HOMEWORK CHECK

Basic: 3, 9, 17, 23, 33

Average: 6, 10, 18, 26, 34

Advanced: 8, 14, 22, 26, 38

ANSWERS

1. inverse relation

2. Reflect the graph of the function in the x-axis; This is the only statement that does not describe graphing the inverse of the function;

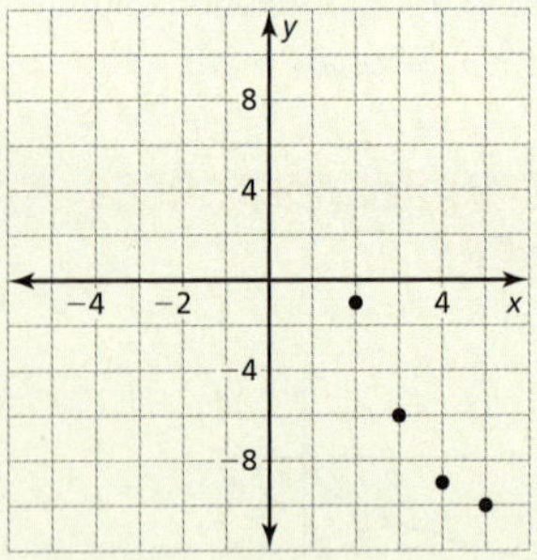

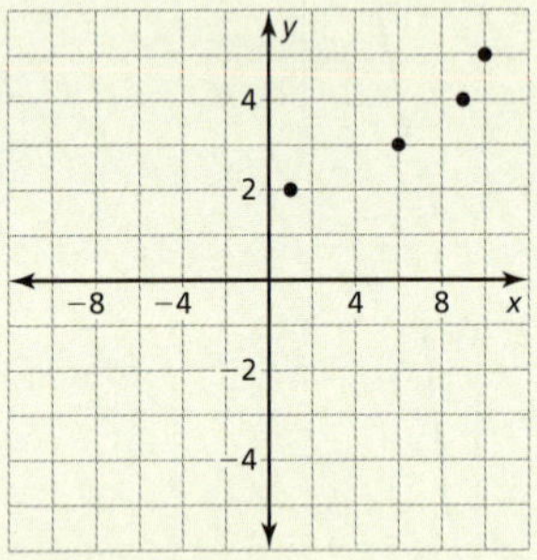

3. $(0, 1), (-8, 3), (-3, 4), (-5, 7), (-1, 9)$

4. $(1, 2), (-3, 4), (7, 6), (1, 8), (-4, 10)$

5.

Input	8	6	0	6	8
Output	-5	-5	0	5	10

6.

Input	2	5	-1	10	-2
Output	-12	-8	-5	-3	-2

7. Input Output

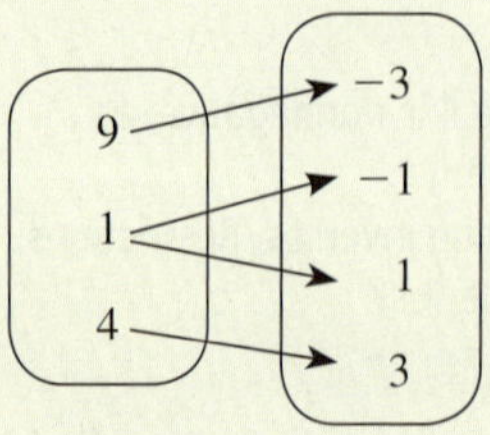

8. Input Output

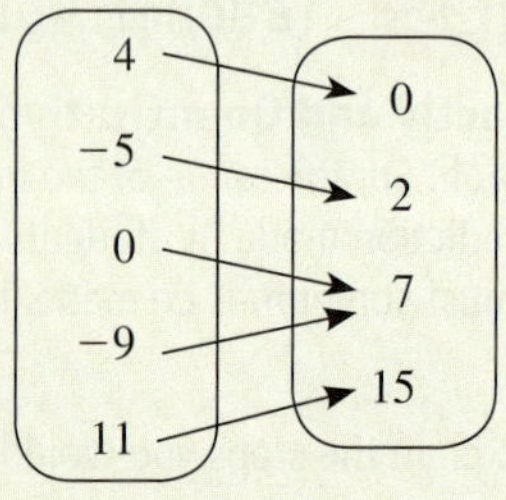

9. $x = y - 5; -3$

10. $x = \frac{y + 3}{2}; \frac{5}{2}$

11–22. See Additional Answers.

In Exercises 23–28, find the inverse of the function. Then graph the function and its inverse. *(See Example 4.)*

23. $f(x) = 4x^2, x \geq 0$

24. $f(x) = -\frac{1}{25}x^2, x \leq 0$

25. $f(x) = -x^2 + 10, x \leq 0$

26. $f(x) = 2x^2 + 6, x \geq 0$

27. $f(x) = \frac{1}{9}x^2 + 2, x \geq 0$

28. $f(x) = -4x^2 - 8, x \leq 0$

In Exercises 29–32, use the Horizontal Line Test to determine whether the inverse of f is a function.

29.

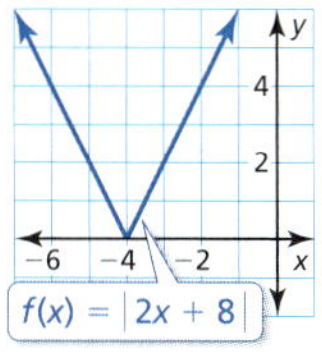

30.

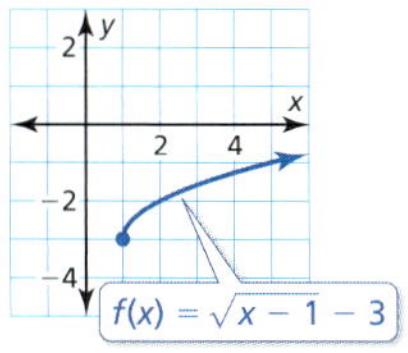

31.

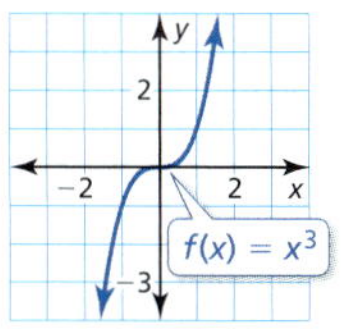

32.

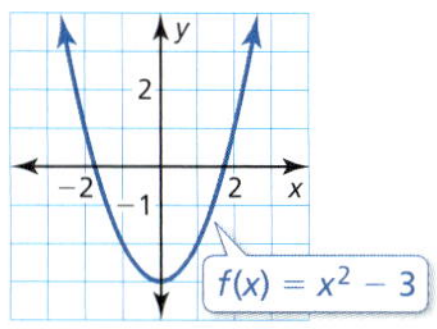

In Exercises 33–42, determine whether the inverse of f is a function. Then find the inverse. *(See Example 5.)*

33. $f(x) = \sqrt{x + 3}$

34. $f(x) = \sqrt{x - 5}$

35. $f(x) = \sqrt{2x - 6}$

36. $f(x) = \sqrt{4x + 1}$

37. $f(x) = 3\sqrt{x - 8}$

38. $f(x) = -\frac{1}{4}\sqrt{5x + 2}$

39. $f(x) = -\sqrt{3x + 5} - 2$

40. $f(x) = 2\sqrt{x - 7} + 6$

41. $f(x) = 2x^2$

42. $f(x) = |x|$

43. ERROR ANALYSIS Describe and correct the error in finding the inverse of the function $f(x) = 3x + 5$.

✗

$$y = 3x + 5$$
$$y - 5 = 3x$$
$$\frac{y - 5}{3} = x$$

The inverse of f is $g(x) = \frac{y - 5}{3}$, or $g(x) = \frac{y}{3} - \frac{5}{3}$.

44. ERROR ANALYSIS Describe and correct the error in finding and graphing the inverse of the function $f(x) = \sqrt{x - 3}$.

✗

$$y = \sqrt{x - 3}$$
$$x = \sqrt{y - 3}$$
$$x^2 = y - 3$$
$$x^2 + 3 = y$$

The inverse of f is $g(x) = x^2 + 3$.

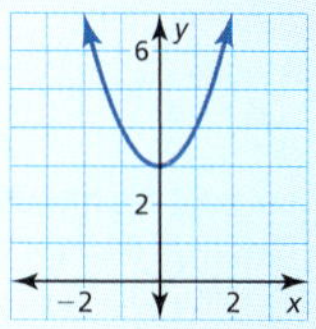

45. MODELING WITH MATHEMATICS The euro is the unit of currency for the European Union. On a certain day, the number E of euros that could be obtained for D U.S. dollars was represented by the formula shown.

$$E = 0.74683D$$

Solve the formula for D. Then find the number of U.S. dollars that could be obtained for 250 euros on that day.

46. MODELING WITH MATHEMATICS A crow is flying at a height of 50 feet when it drops a walnut to break it open. The height h (in feet) of the walnut above ground can be modeled by $h = -16t^2 + 50$, where t is the time (in seconds) since the crow dropped the walnut. Solve the equation for t. After how many seconds will the walnut be 15 feet above the ground?

MATHEMATICAL CONNECTIONS In Exercises 47 and 48, s is the side length of an equilateral triangle. Solve the formula for s. Then evaluate the new formula for the given value.

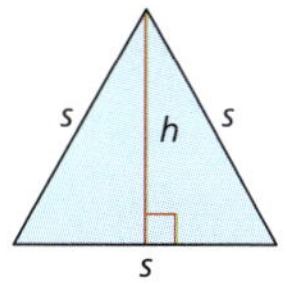

47. Height: $h = \frac{\sqrt{3}s}{2}$; $h = 16$ in.

48. Area: $A = \frac{\sqrt{3}s^2}{4}$; $A = 11$ ft^2

Dynamic Teaching Tools

- Dynamic Assessment & Progress Monitoring Tool
- Interactive Whiteboard Lesson Library
- Dynamic Classroom with Dynamic Investigations

ANSWERS

23. $g(x) = \frac{\sqrt{x}}{2}$

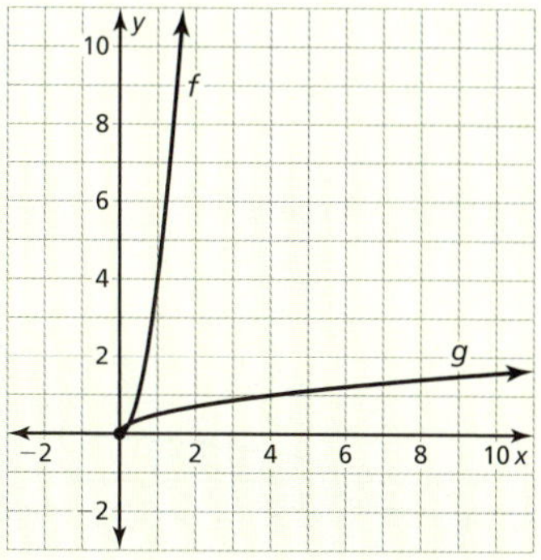

24. $g(x) = -5\sqrt{-x}$

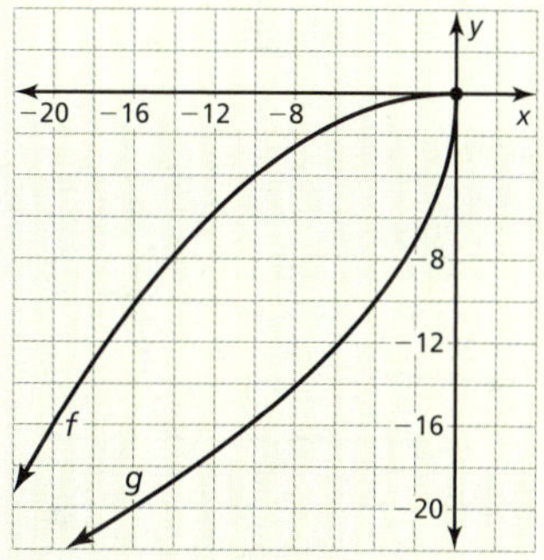

25. $g(x) = -\sqrt{10 - x}$

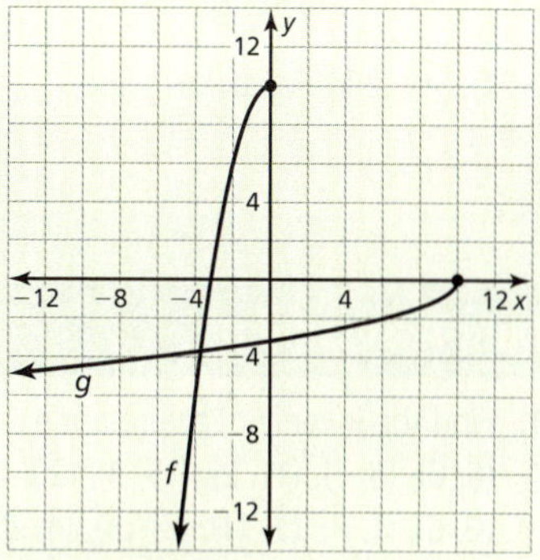

26. $g(x) = \sqrt{\frac{x - 6}{2}}$

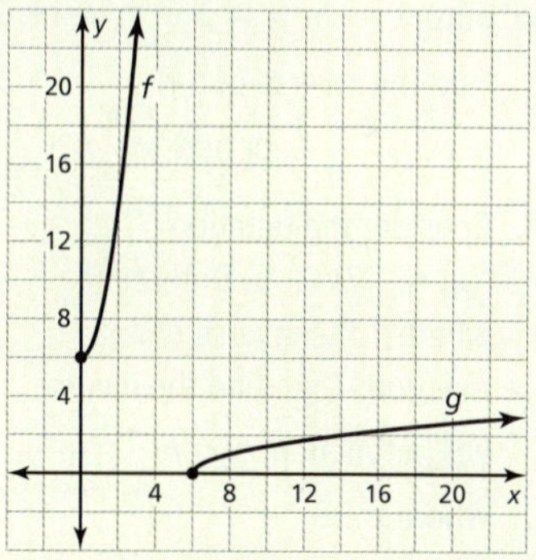

27. $g(x) = 3\sqrt{x - 2}$

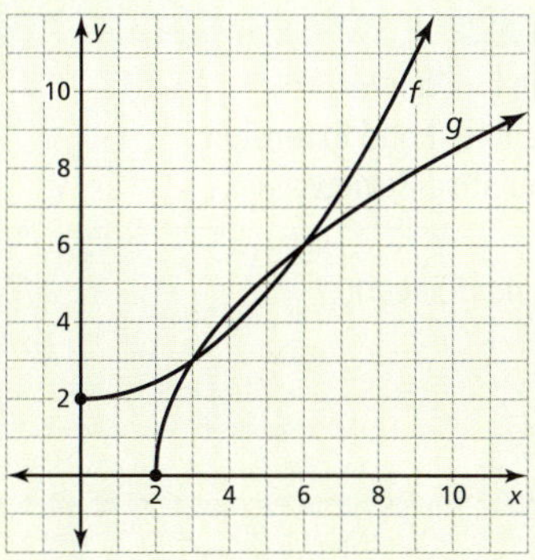

28. $g(x) = -\frac{1}{2}\sqrt{-x - 8}$

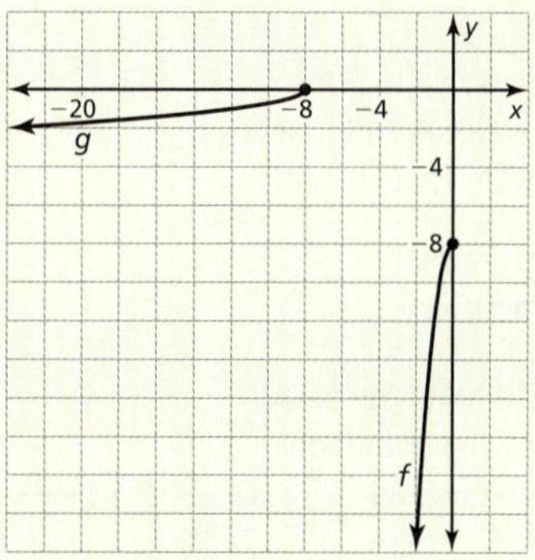

29–48. See Additional Answers.

ANSWERS

49. $g(x) = \sqrt[3]{\dfrac{x}{2}}$

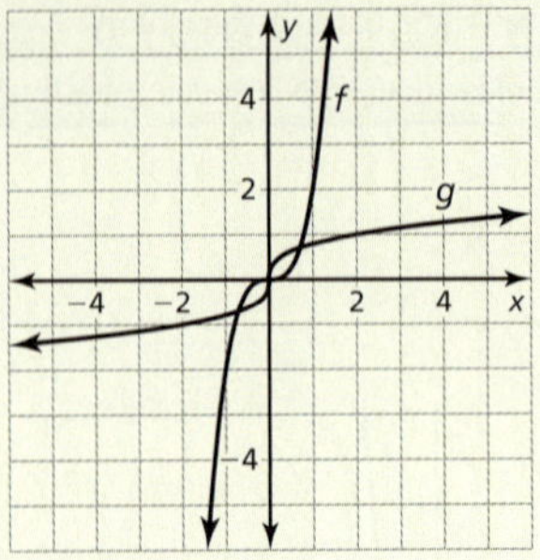

50. $g(x) = \sqrt[3]{x + 4}$

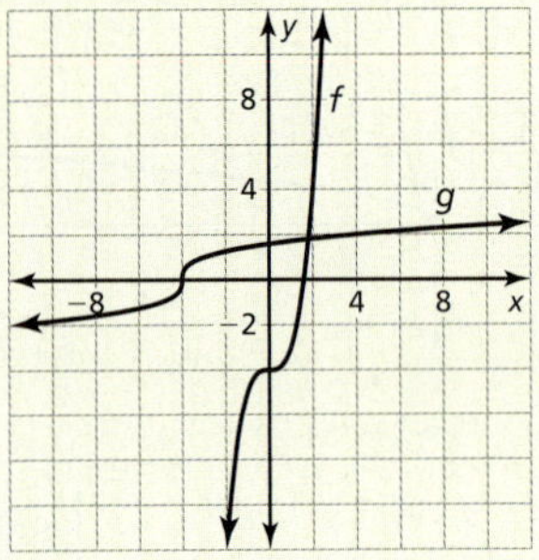

51. $g(x) = \sqrt[3]{x} + 5$

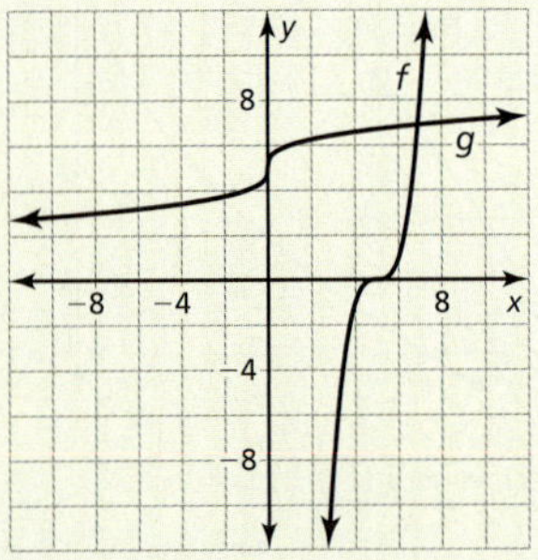

52–71. See Additional Answers.

Mini-Assessment

1. Find the inverse of the relation. (6, 0), (8, 1), (11, 2), (15, 3), (20, 4) (0, 6), (1, 8), (2, 11), (3, 15), (4, 20)
2. Let $f(x) = -x + 2$. Solve $y = f(x)$ for x. Then find the input when the output is –5. $x = 2 - y$; input: 7
3. Find the inverse of $f(x) = 4x - 3$. $g(x) = \dfrac{x + 3}{4}$, or $g(x) = \dfrac{1}{4}x + \dfrac{3}{4}$
4. Consider the function $f(x) = \sqrt{2x + 5}$. Determine whether the inverse of f is a function. Then find the inverse. yes, a function; $g(x) = \dfrac{x^2 - 5}{2}$, where $x \geq 0$

In Exercises 49–54, find the inverse of the function. Then graph the function and its inverse.

49. $f(x) = 2x^3$ **50.** $f(x) = x^3 - 4$

51. $f(x) = (x - 5)^3$ **52.** $f(x) = 8(x + 2)^3$

53. $f(x) = 4\sqrt[3]{x}$ **54.** $f(x) = -\sqrt[3]{x - 1}$

55. MAKING AN ARGUMENT Your friend says that the inverse of the function $f(x) = 3$ is a function because all linear functions pass the Horizontal Line Test. Is your friend correct? Explain.

56. HOW DO YOU SEE IT? Pair the graph of each function with the graph of its inverse.

A.
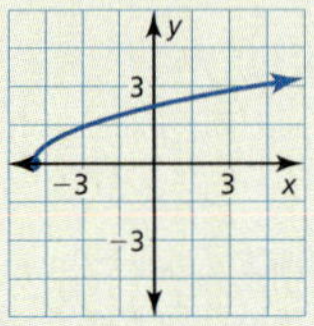

B.
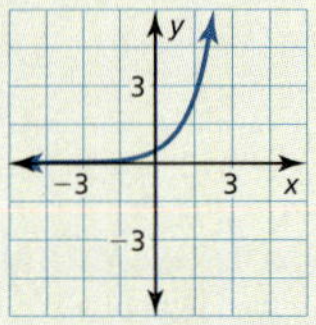

C.
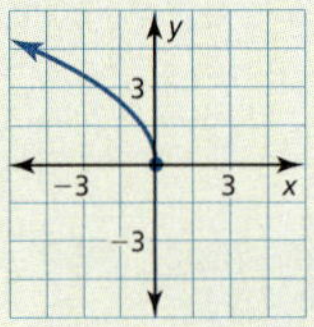

D.
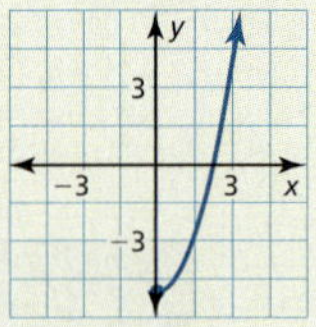

E.
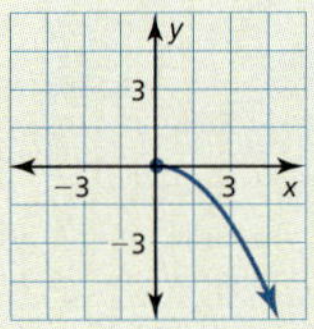

F.
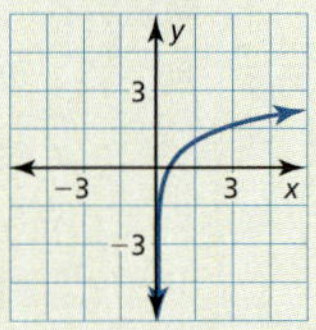

57. WRITING Describe changes you could make to the function $f(x) = x^2 - 5$ so that its inverse is a function. Describe the domain and range of the new function and its inverse.

58. CRITICAL THINKING Can an even function with at least two values in its domain have an inverse that is a function? Explain.

59. OPEN-ENDED Write a function such that the graph of its inverse is a line with a slope of 4.

60. CRITICAL THINKING Consider the function $g(x) = -x$.

a. Graph $g(x) = -x$ and explain why it is its own inverse.

b. Graph other linear functions that are their own inverses. Write equations of the lines you graph.

c. Use your results from part (b) to write a general equation that describes the family of linear functions that are their own inverses.

61. REASONING Show that the inverse of any linear function $f(x) = mx + b$, where $m \neq 0$, is also a linear function. Write the slope and y-intercept of the graph of the inverse in terms of m and b.

62. THOUGHT PROVOKING The graphs of $f(x) = x^3 - 3x$ and its inverse are shown. Find the greatest interval $-a \leq x \leq a$ for which the inverse of f is a function. Write an equation of the inverse function.

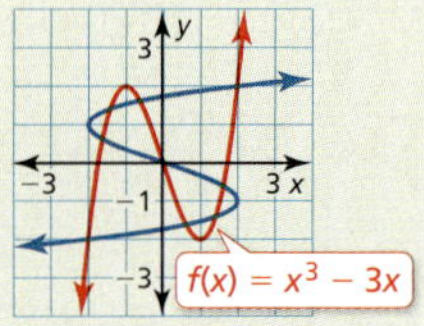

63. REASONING Is the inverse of $f(x) = 2|x + 1|$ a function? Are there any values of a, h, and k for which the inverse of $f(x) = a|x - h| + k$ is a function? Explain your reasoning.

Maintaining Mathematical Proficiency

Reviewing what you learned in previous grades and lessons

Find the sum or difference. *(Section 7.1)*

64. $(2x - 9) - (6x + 5)$ **65.** $(8y + 1) + (-y - 12)$

66. $(t^2 - 4t - 4) + (7t^2 + 12t + 3)$ **67.** $(-3d^2 + 10d - 8) - (7d^2 - d - 6)$

Graph the function. Compare the graph to the graph of $f(x) = x^2$. *(Section 8.2)*

68. $g(x) = x^2 + 6$ **69.** $h(x) = -x^2 - 2$ **70.** $p(x) = -4x^2 + 5$ **71.** $q(x) = \frac{1}{3}x^2 - 1$

If students need help...	*If students got it...*
Resources by Chapter • Practice A and Practice B • Puzzle Time	Resources by Chapter • Enrichment and Extension • Cumulative Review
Student Journal • Practice	Start the *next* Section
Differentiating the Lesson Skills Review Handbook	

10.3–10.4 What Did You Learn?

Core Vocabulary

radical equation, *p. 560*
inverse relation, *p. 568*
inverse function, *p. 569*

Core Concepts

Section 10.3

Squaring Each Side of an Equation, *p. 560*
Identifying Extraneous Solutions, *p. 562*

Section 10.4

Inverse Relation, *p. 568*
Finding Inverses of Functions Algebraically, *p. 570*
Finding Inverses of Nonlinear Functions, *p. 570*
Horizontal Line Test, *p. 571*

Mathematical Practices

1. Could you also solve Exercises 37–44 on page 565 by graphing? Explain.
2. What external resources could you use to check the reasonableness of your answer in Exercise 45 on page 573?

Performance Task

Medication and the Mosteller Formula

When taking medication, it is critical to take the correct dosage. For children in particular, body surface area (BSA) is a key component in calculating that dosage. The Mosteller Formula is commonly used to approximate body surface area. How will you use this formula to calculate BSA for the optimum dosage?

To explore the answers to this question and more, go to ***BigIdeasMath.com***.

Dynamic Teaching Tools

- Dynamic Assessment & Progress Monitoring Tool
- Interactive Whiteboard Lesson Library
- Dynamic Classroom with Dynamic Investigations

ANSWERS

1. yes; Set a dependent variable equal to each side of the equation and graph the two resulting equations. The solution of the original equation is the value of the independent variable at the point of intersection of the two graphs.
2. *Sample answer:* The Internet or a smart phone application that performs real-time currency conversions.

ANSWERS

1.

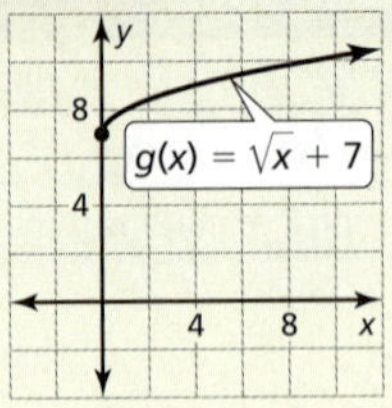

$x \geq 0$; $y \geq 7$; The graph of g is a translation 7 units up of the graph of f.

2.

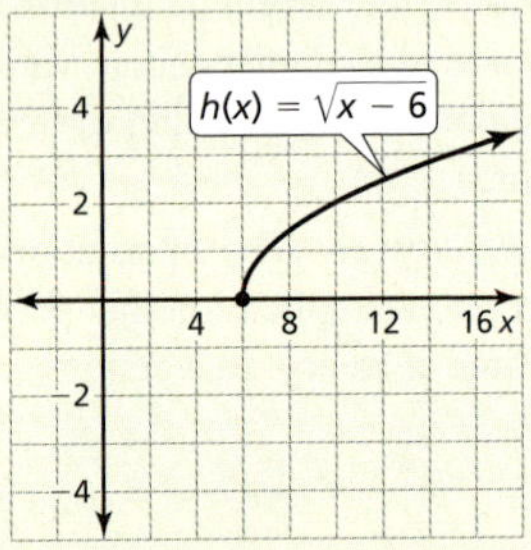

$x \geq 6$; $y \geq 0$; The graph of h is a translation 6 units right of the graph of f.

3.

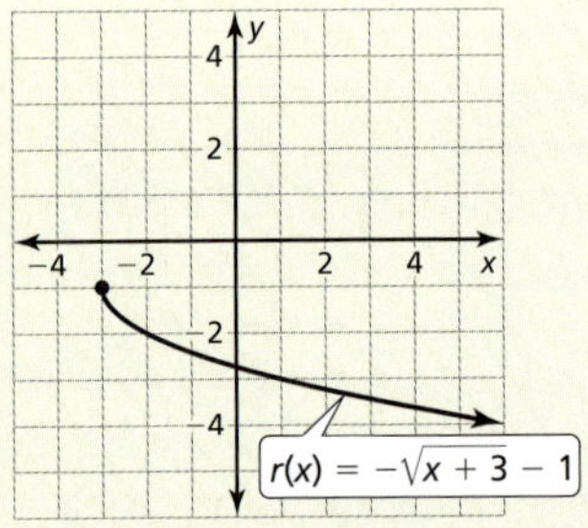

$x \geq -3$; $y \leq -1$; The graph of r is a translation 3 units left, a reflection in the x-axis, and a translation 1 unit down of the graph of f.

4. Translate 6 units right, shrink vertically by a factor of $\frac{1}{4}$, and translate 2 units up to obtain the graph of g.

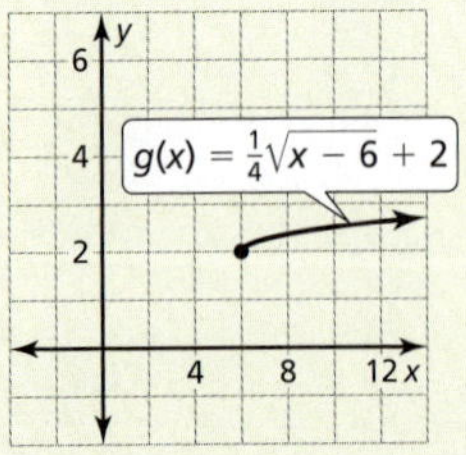

10.1 Graphing Square Root Functions (pp. 543–550)

a. Describe the domain of $f(x) = 4\sqrt{x + 2}$.

The radicand cannot be negative. So, $x + 2$ is greater than or equal to 0.

$x + 2 \geq 0$ Write an inequality for the domain.

$x \geq -2$ Subtract 2 from each side.

▶ The domain is the set of real numbers greater than or equal to -2.

b. Graph $g(x) = \sqrt{x} - 1$. Describe the range. Compare the graph to the graph of $f(x) = \sqrt{x}$.

Step 1 Use the domain of g, $x \geq 0$, to make a table of values.

x	0	1	4	9	16
$g(x)$	-1	0	1	2	3

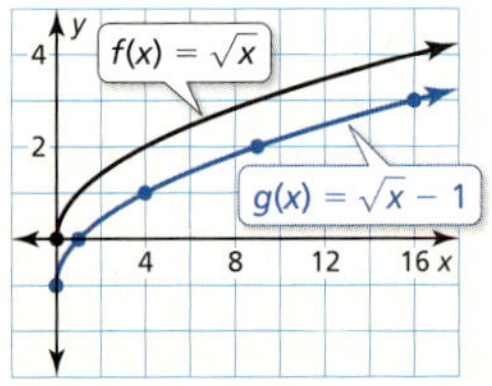

Step 2 Plot the ordered pairs.

Step 3 Draw a smooth curve through the points, starting at $(0, -1)$.

▶ The range of g is $y \geq -1$. The graph of g is a translation 1 unit down of the graph of f.

Graph the function. Describe the domain and range. Compare the graph to the graph of $f(x) = \sqrt{x}$.

1. $g(x) = \sqrt{x} + 7$
2. $h(x) = \sqrt{x - 6}$
3. $r(x) = -\sqrt{x + 3} - 1$
4. Let $g(x) = \frac{1}{4}\sqrt{x - 6} + 2$. Describe the transformations from the graph of $f(x) = \sqrt{x}$ to the graph of g. Then graph g.

10.2 Graphing Cube Root Functions (pp. 551–556)

Graph $g(x) = -\sqrt[3]{x - 2}$. Compare the graph to the graph of $f(x) = \sqrt[3]{x}$.

Step 1 Make a table of values.

x	-6	1	2	3	10
$g(x)$	2	1	0	-1	-2

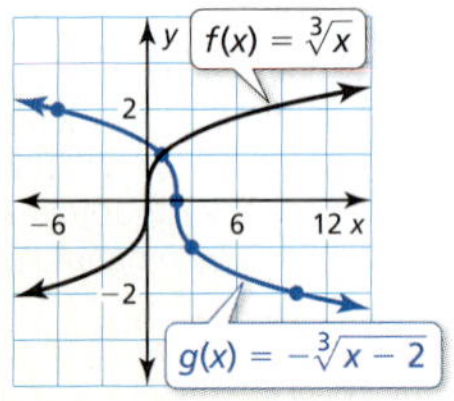

Step 2 Plot the ordered pairs.

Step 3 Draw a smooth curve through the points.

▶ The graph of g is a translation 2 units right and a reflection in the x-axis of the graph of f.

Graph the function. Compare the graph to the graph of $f(x) = \sqrt[3]{x}$.

5. $g(x) = \sqrt[3]{x} + 4$ **6.** $h(x) = -8\sqrt[3]{x}$ **7.** $s(x) = \sqrt[3]{-2(x-3)}$

8. Let $g(x) = -3\sqrt[3]{x+2} - 1$. Describe the transformations from the graph of $f(x) = \sqrt[3]{x}$ to the graph of g. Then graph g.

9. The graph of cube root function r is shown. Compare the average rate of change of r to the average rate of change of $p(x) = \sqrt[3]{\frac{1}{2}x}$ over the interval $x = 0$ to $x = 8$.

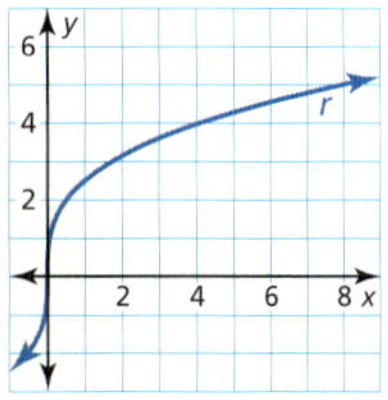

10.3 Solving Radical Equations (pp. 559–566)

Solve $\sqrt{10 - 3x} = x$.

$\sqrt{10-3x} = x$	Write the equation.
$(\sqrt{10-3x})^2 = x^2$	Square each side of the equation.
$10 - 3x = x^2$	Simplify.
$0 = x^2 + 3x - 10$	Subtract 10 and add $3x$ to each side.
$0 = (x-2)(x+5)$	Factor.
$x - 2 = 0$ or $x + 5 = 0$	Zero-Product Property
$x = 2$ or $x = -5$	Solve for x.

Check Check each solution in the original equation.

$\sqrt{10 - 3(2)} \stackrel{?}{=} 2$	Substitute for x.	$\sqrt{10 - 3(-5)} \stackrel{?}{=} -5$
$\sqrt{4} \stackrel{?}{=} 2$	Simplify.	$\sqrt{25} \stackrel{?}{=} -5$
$2 = 2$ ✓	Simplify.	$5 \neq -5$ ✗

▶ Because $x = -5$ does not satisfy the original equation, it is an extraneous solution. The only solution is $x = 2$.

Solve the equation. Check your solution(s).

10. $8 + \sqrt{x} = 18$ **11.** $\sqrt[3]{x-1} = 3$ **12.** $\sqrt{5x-9} = \sqrt{4x}$

13. $x = \sqrt{3x+4}$ **14.** $8\sqrt{x-5} + 34 = 58$ **15.** $\sqrt{5x} + 6 = 5$

16. The radius r of a cylinder is represented by the function $r = \sqrt{\frac{V}{\pi h}}$, where V is the volume and h is the height of the cylinder. What is the volume of the cylindrical can?

ANSWERS

5.

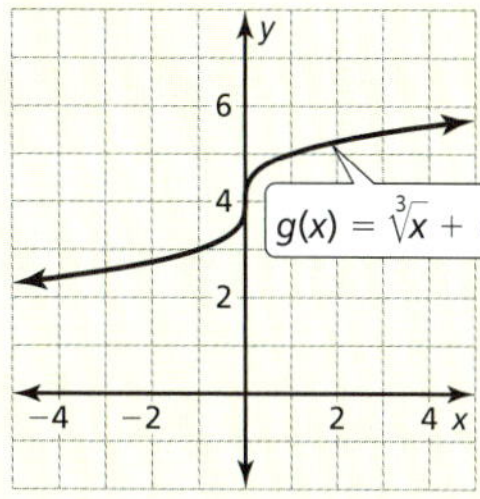

The graph of g is a translation 4 units up of the graph of f.

6.

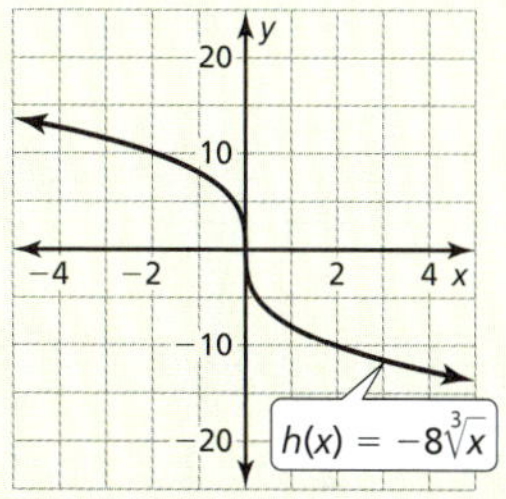

The graph of h is a vertical stretch by a factor of 8 and a reflection in the x-axis of the graph of f.

7.

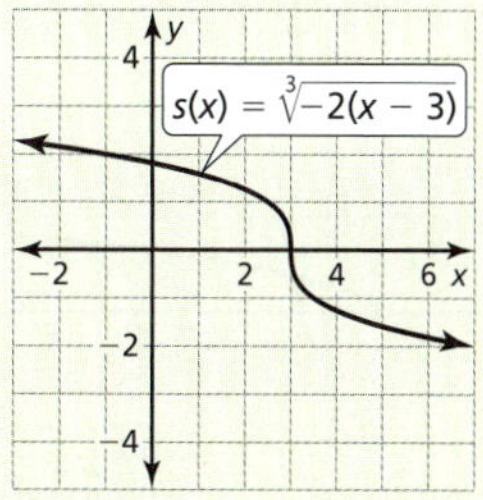

The graph of s is a reflection in the y-axis, a translation 3 units right, and a horizontal shrink by a factor of $\frac{1}{2}$ of the graph of f.

8. Stretch vertically by a factor of 3, reflect in the x-axis, translate 2 units left, and translate 1 unit down to obtain the graph of g.

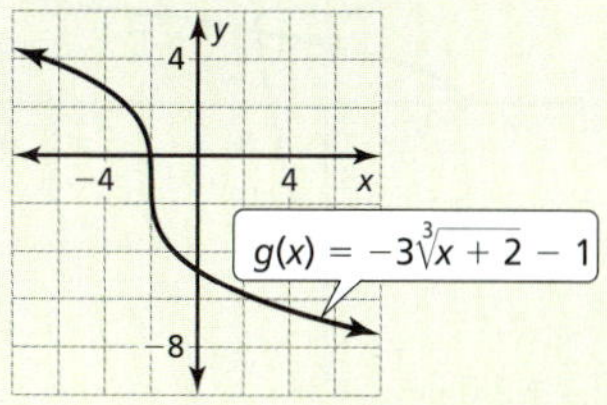

9. The average rate of change of r is about $0.625 \div \left(\frac{\sqrt[3]{4}}{8}\right) \approx 3.15$ times greater than the average rate of change of p over the interval $x = 0$ to $x = 8$.

10. $x = 100$

11. $x = 28$

12. $x = 9$

13. $x = 4$

14. $x = 14$

15. no solution

16. $16\pi \approx 50.3 \text{ in.}^3$

ANSWERS

17. $(-10, 1), (-4, 3), (4, 5), (14, 7), (26, 9)$

18.

Input	6	3	0	−3	−6
Output	−4	−2	0	2	4

19. $g(x) = \dfrac{10 - x}{5}$

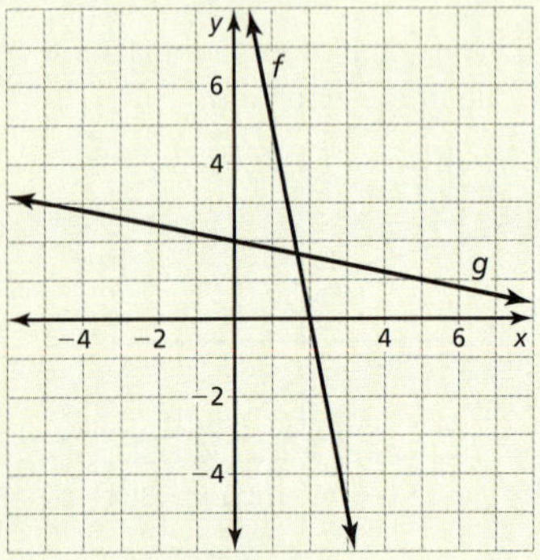

20. $g(x) = \frac{1}{3}\sqrt{3(x + 1)}$

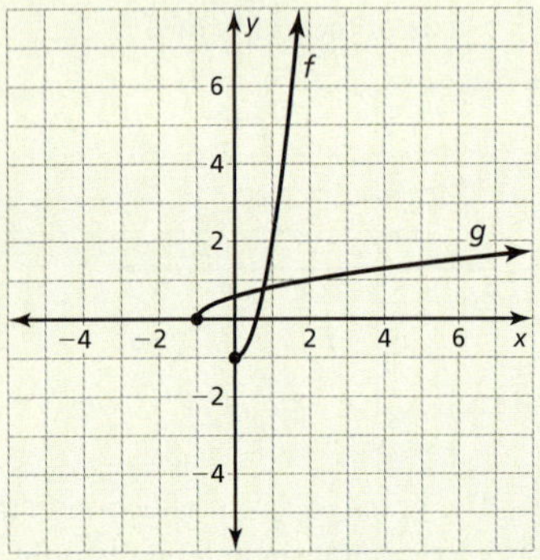

21. $g(x) = 2x^2 - 3, x \geq 0$

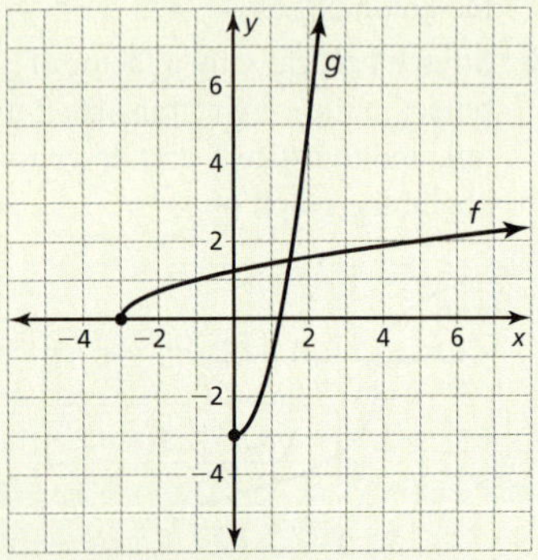

22. not a function

23. $a = 210 - 1.25h$; 175

10.4 Inverse of a Function (pp. 567–574)

a. Find the inverse of the relation.

Input	−4	−2	0	2	4	6
Output	−3	0	3	6	9	12

Switch the inputs and outputs.

Inverse relation:

Input	−3	0	3	6	9	12
Output	−4	−2	0	2	4	6

b. Find the inverse of $f(x) = \sqrt{x - 4}$. Then graph the function and its inverse.

$y = \sqrt{x - 4}$ Set y equal to $f(x)$.

$x = \sqrt{y - 4}$ Switch x and y in the equation.

$x^2 = \left(\sqrt{y - 4}\right)^2$ Square each side.

$x^2 = y - 4$ Simplify.

$x^2 + 4 = y$ Add 4 to each side.

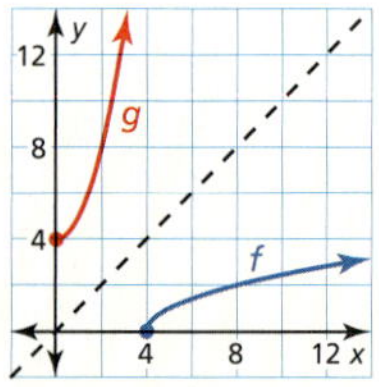

Because the range of f is $y \geq 0$, the domain of the inverse must be restricted to $x \geq 0$.

▶ So, the inverse of f is $g(x) = x^2 + 4$, where $x \geq 0$.

Find the inverse of the relation.

17. $(1, -10), (3, -4), (5, 4), (7, 14), (9, 26)$

18.

Input	−4	−2	0	2	4
Output	6	3	0	−3	−6

Find the inverse of the function. Then graph the function and its inverse.

19. $f(x) = -5x + 10$

20. $f(x) = 3x^2 - 1, x \geq 0$

21. $f(x) = \frac{1}{2}\sqrt{2x + 6}$

22. Consider the function $f(x) = x^2 + 4$. Use the Horizontal Line Test to determine whether the inverse of f is a function.

23. In bowling, a handicap is an adjustment to a bowler's score to even out differences in ability levels. In a particular league, you can find a bowler's handicap h by using the formula $h = 0.8(210 - a)$, where a is the bowler's average. Solve the formula for a. Then find a bowler's average when the bowler's handicap is 28.

10 Chapter Test

Find the inverse of the function.

1. $f(x) = 5x - 8$
2. $f(x) = 2\sqrt{x + 3} - 1$
3. $f(x) = -\frac{1}{3}x^2 + 4, x \geq 0$

Graph the function *f*. Describe the domain and range. Compare the graph of *f* to the graph of *g*.

4. $f(x) = -\sqrt{x + 6}$; $g(x) = \sqrt{x}$
5. $f(x) = \sqrt{x - 3} + 2$; $g(x) = \sqrt{x}$
6. $f(x) = \sqrt[3]{x} - 5$; $g(x) = \sqrt[3]{x}$
7. $f(x) = -2\sqrt[3]{x + 1}$; $g(x) = \sqrt[3]{x}$

Solve the equation. Check your solution(s).

8. $9 - \sqrt{x} = 3$
9. $\sqrt{2x - 7} - 3 = 6$
10. $\sqrt{8x - 21} = \sqrt{18 - 5x}$
11. $x + 5 = \sqrt{7x + 53}$

12. When solving the equation $x - 5 = \sqrt{ax + b}$, you obtain $x = 2$ and $x = 8$. Explain why at least one of these solutions must be extraneous.

Describe the transformations from the graph of $f(x) = \sqrt[3]{x}$ to the graph of the given function. Then graph the given function.

13. $h(x) = 4\sqrt[3]{x - 1} + 5$
14. $w(x) = -\sqrt[3]{x + 7} - 2$

15. The velocity v (in meters per second) of a roller coaster at the bottom of a hill is given by $v = \sqrt{19.6h}$, where h is the height (in meters) of the hill. (a) Use a graphing calculator to graph the function. Describe the domain and range. (b) How tall must the hill be for the velocity of the roller coaster at the bottom of the hill to be at least 28 meters per second? (c) What happens to the average rate of change of the velocity as the height of the hill increases?

16. The speed s (in meters per second) of sound through air is given by $s = 20\sqrt{T + 273}$, where T is the temperature (in degrees Celsius).
    a. What is the temperature when the speed of sound through air is 340 meters per second?
    b. How long does it take you to hear the wolf howl when the temperature is $-17°C$?

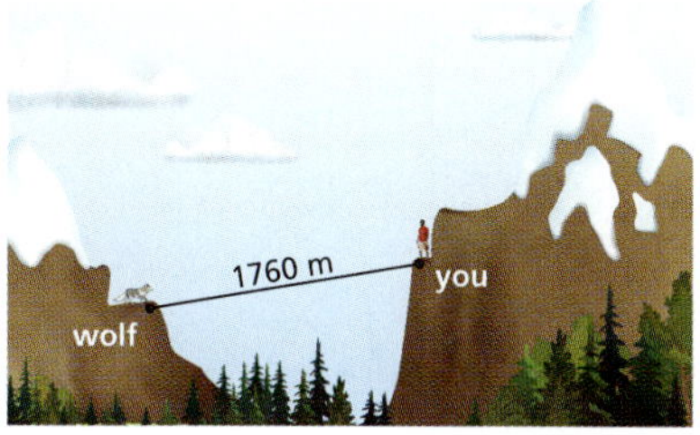

17. How can you restrict the domain of the function $f(x) = (x - 3)^2$ so that the inverse of f is a function?

18. Write a radical function that has a domain of all real numbers less than or equal to 0 and a range of all real numbers greater than or equal to 9.

If students need help...	*If students got it...*
Lesson Tutorials	Resources by Chapter • Enrichment and Extension • Cumulative Review
Skills Review Handbook	Performance Task
BigIdeasMath.com	Start the *next* Section

ANSWERS

1. $g(x) = \frac{x + 8}{5}$
2. $g(x) = \left(\frac{x + 1}{2}\right)^2 - 3, x \geq -1$
3. $g(x) = \sqrt{12 - 3x}$
4. 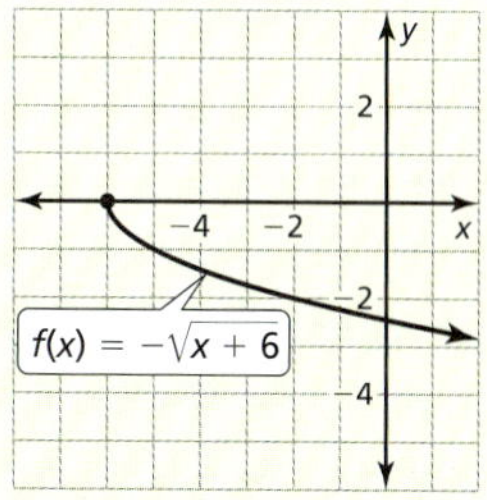

   $x \geq -6$; $y \leq 0$; The graph of f is a translation 6 units left and a reflection in the x-axis of the graph of g.
5. 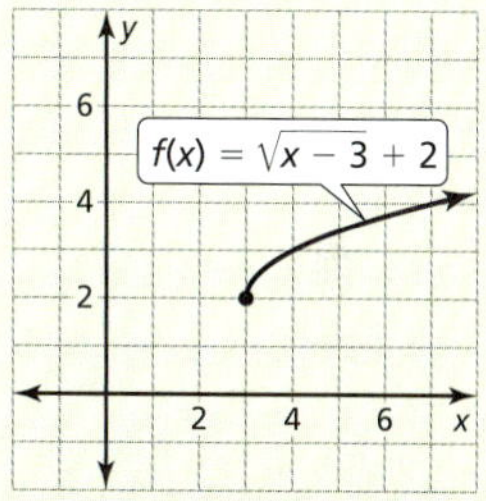

   $x \geq 3$; $y \geq 2$; The graph of f is a translation 3 units right and 2 units up of the graph of g.
6. 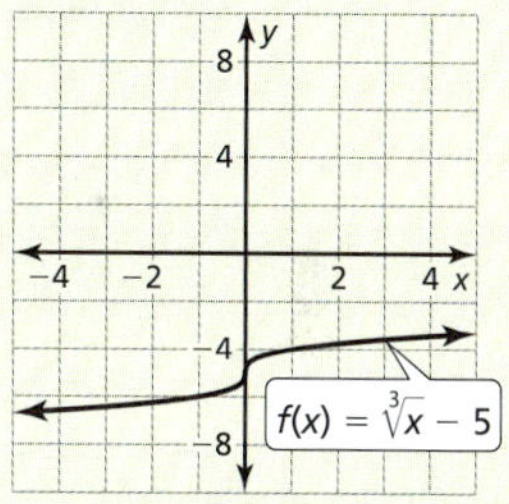

   all real numbers; all real numbers; The graph of f is a translation 5 units down of the graph of g.
7. 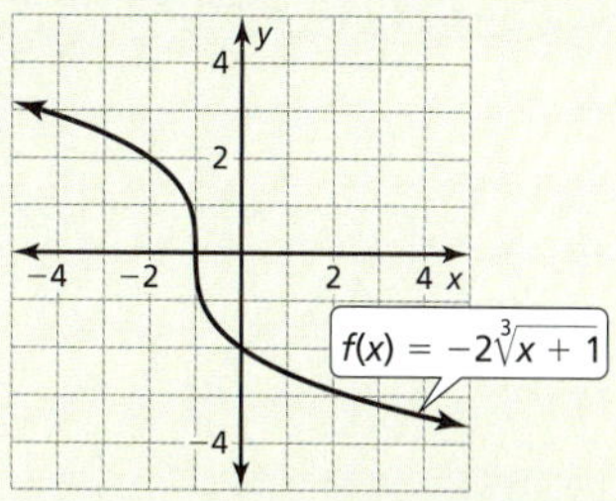

   all real numbers; all real numbers; The graph of f is a vertical stretch by a factor of 2, a reflection in the x-axis, and a translation 1 unit left of the graph of g.
8. $x = 36$
9. $x = 44$
10. $x = 3$
11. $x = 4$

12–18. See Additional Answers.

ANSWERS

1. −; 1; 2
2. a. *Sample answer:* 1; 5
   b. *Sample answer:* 1; −3
   c. $-\frac{1}{4}; \frac{25}{4}$
3. B
4. a. *Sample answer:* $a = 4, b = -4$
   b. *Sample answer:* $a = 3, b = -2$
   c. *Sample answer:* $a = -3,$ $b = -2$

10 Cumulative Assessment

1. Fill in the function so that it is represented by the graph.

$f(x) = \square\sqrt{x - \square} + \square$

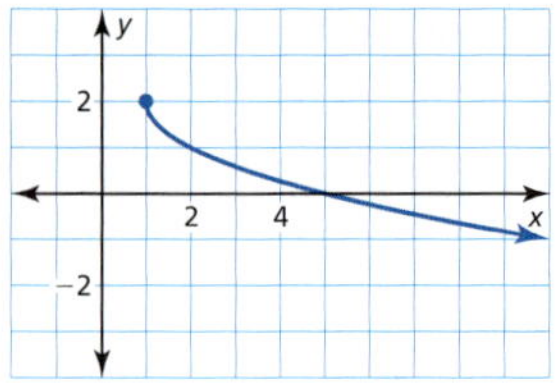

2. Consider the equation $y = mx + b$. Fill in values for m and b so that each statement is true.

a. When $m =$ ______ and $b =$ ______, the graph of the equation passes through the point $(-1, 4)$.

b. When $m =$ ______ and $b =$ ______, the graph of the equation has a positive slope and passes through the point $(-2, -5)$.

c. When $m =$ ______ and $b =$ ______, the graph of the equation is perpendicular to the graph of $y = 4x - 3$ and passes through the point $(1, 6)$.

3. Which graph represents the inverse of the function $f(x) = 2x + 4$?

Ⓐ
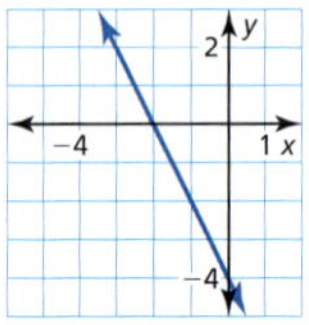

Ⓑ
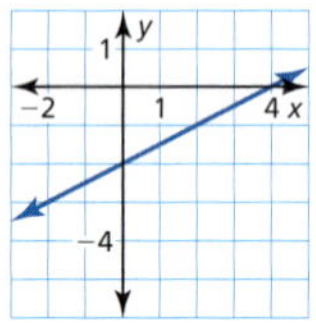

Ⓒ
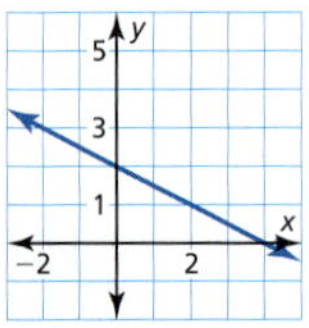

Ⓓ
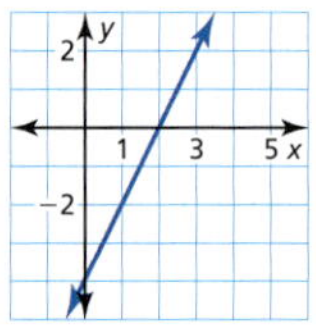

4. Consider the equation $x = \sqrt{ax + b}$. Student A claims this equation has one real solution. Student B claims this equation has two real solutions. Use the numbers to answer parts (a)–(c).

−4	−3	−2	−1	0	1	2	3	4

a. Choose values for a and b to create an equation that supports Student A's claim.

b. Choose values for a and b to create an equation that supports Student B's claim.

c. Choose values for a and b to create an equation that does not support either student's claim.

5. Which equation represents the nth term of the sequence 3, 12, 48, 192, . . .?

Ⓐ $a_n = 3(4)^{n-1}$ Ⓑ $a_n = 3(9)^{n-1}$

Ⓒ $a_n = 9n - 6$ Ⓓ $a_n = 9n + 3$

6. Consider the function $f(x) = \frac{1}{2}\sqrt[3]{x} + 3$. The graph represents function g. Select all the statements that are true.

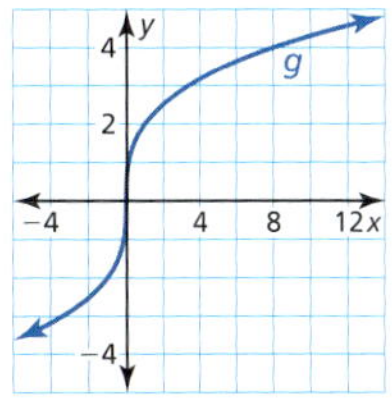

- The x-intercept of the graph of f is greater than the x-intercept of the graph of g.
- The graph of g is always increasing.
- The average rate of change of g decreases as x increases.
- The average rate of change of f increases as x increases.
- The average rate of change of g is greater than the average rate of change of f over the interval $x = 0$ to $x = 8$.

7. Place each function into one of the three categories.

No zeros	One zero	Two zeros

$f(x) = 3x^2 + 4x + 2$ $f(x) = 4x^2 - 8x + 4$ $f(x) = 7x^2$
$f(x) = -x^2 + 2x$ $f(x) = x^2 - 3x - 21$ $f(x) = -6x^2 - 5$

8. You are making a tabletop with a tiled center and a uniform mosaic border.

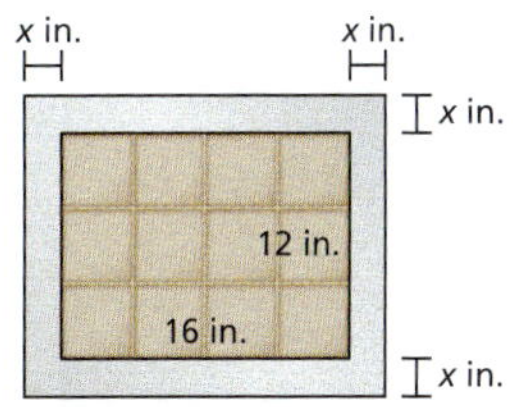

a. Write the polynomial in standard form that represents the perimeter of the tabletop.

b. Write the polynomial in standard form that represents the area of the tabletop.

c. The perimeter of the tabletop is less than 80 inches, and the area of tabletop is at least 252 square inches. Select all the possible values of x.

0.5 | 1 | 1.5 | 2 | 2.5 | 3 | 3.5 | 4

ANSWERS

5. A
6. The graph of g is always increasing; The average rate of change of g is greater than the average rate of change of f over the interval $x = 0$ to $x = 8$.
7. See Additional Answers.
8. a. $8x + 56$
   b. $4x^2 + 56x + 192$
   c. 1, 1.5, 2, 2.5

Chapter 11 Pacing Guide

Chapter 11 Pacing Guide	
Chapter Opener/ Mathematical Practices	1 Day
Section 1	2 Days
Section 2	2 Days
Section 3	2 Days
Quiz	1 Day
Section 4	2 Days
Section 5	2 Days
Chapter Review/ Chapter Tests	2 Days
Total Chapter 11	14 Days
Year-to-Date	160 Days

11 Data Analysis and Displays

Watching Sports on TV *(p. 616)*

Shoes *(p. 603)*

Backpacking *(p. 594)*

Bowling Scores *(p. 591)*

Altitudes of Airplanes *(p. 589)*

Laurie's Notes

Chapter Summary

- This last chapter on data analysis and displays contains and extends certain content studied in middle school. Students should be familiar with the different measures of center: mean, median, and mode, and the different measures of variation: range and mean absolute deviation.
- In this chapter, a new measure of variation, standard deviation, is introduced. Students can compute the standard deviation for a small data set or use technology to compute it for larger data sets.
- Box-and-whisker plots are analyzed, noting the variation or spread of the data. Other data displays are integrated throughout the chapter.
- The shapes of distributions, symmetric and skewed, are analyzed and compared.
- Two-way tables are used to organize categories of data from the same source.
- The final lesson looks at how to choose a data display for a given set of data.

What Your Students Have Learned

Middle School

- Find the means of data sets and discover the relationships between the means and outliers.
- Find the medians and modes of data sets, and identify the measures of center that best describes data sets.
- Find the ranges, quartiles, interquartile ranges, and mean absolute deviations of data sets.
- Construct and analyze two-way tables.
- Choose and construct appropriate data displays.

What Your Students Will Learn

Algebra 1

- Compare the means, medians, and modes of data sets, find the ranges and standard deviations of data sets, and identify the effects of transformations on data.
- Construct and interpret box-and-whisker plots, and use box-and-whisker plots to compare data sets.
- Describe the shapes of data distributions, choose appropriate measures to describe them, and compare data distributions.
- Make two-way tables, find and interpret marginal frequencies, and find relative and conditional relative frequencies.
- Recognize associations in data from reading two-way tables.
- Classify data as qualitative or quantitative, choose and create appropriate data displays, and analyze misleading graphs.

Dynamic Teaching Tools

- Dynamic Assessment & Progress Monitoring Tool
- Lesson Planning Tool
- Interactive Whiteboard Lesson Library
- Dynamic Classroom with Dynamic Investigations
- Real-Life STEM Videos

Scaffolding in the Classroom

Graphic Organizers: Four Square

A Four Square can be used to organize information about a topic. Students write the topic in the "bubble" in the middle of the Four Square. Then students write concepts related to the topic in the four squares surrounding the bubble. Any concept related to the topic can be used. Encourage students to include concepts that will help them learn the topic. Students can place their Four Squares on note cards to use as a quick study reference.

Questioning in the Classroom

What do you think?

Allow students time to write their answers on individual whiteboards. After the allotted time, they can hold up their boards. The teacher can see what students are thinking and follow up with discussion.

Laurie's Notes

Maintaining Mathematical Proficiency

Displaying Data

- Remind students that a histogram uses bars to show the frequency of data values in equal-size intervals. There should be no spaces between the bars.
- Point out that a circle graph shows data as parts of a whole. The sum of the degrees for the categories in the graph must equal 360°.

COMMON ERROR For the circle graph, students may not use the correct fraction (or decimal) to find the number of degrees for a category. Instead of using the total number as the denominator, students may use 100, thinking they are finding percents. Students may also use 360 as the denominator, thinking they need to divide the data value by 360 to find the degrees, instead of first dividing the data value by the total number.

Mathematical Practices (continued on page 584)

- The *Mathematical Practices* page focuses attention on how mathematics is learned—process versus content. This page reviews the various graphical displays a student should be familiar with and the type of analysis each is used for.
- Use the *Mathematical Practices* page to help students develop mathematical habits of mind—how mathematics can be explored and how mathematics is thought about.
- In the *Core Concept*, nine different graphical displays are shown with a brief description of what each does in terms of using and displaying data.
- Ask students to describe a set of data that could be displayed using each type of display.

If students need help...	*If students got it...*
Student Journal • Maintaining Mathematical Proficiency	Game Closet at *BigIdeasMath.com*
Lesson Tutorials	Start the *next* Section
Skills Review Handbook	

Maintaining Mathematical Proficiency

Displaying Data

Example 1 The frequency table shows the numbers of books that 12 people read last month. Display the data in a histogram.

Number of books	Frequency
0–1	6
2–3	4
4–5	0
6–7	2

Example 2 The table shows the results of a survey. Display the data in a circle graph.

Class trip location	Water park	Museum	Zoo	Other
Students	25	11	5	4

A total of 45 students took the survey.

Water park:

$\frac{25}{45} \cdot 360° = 200°$

Museum:

$\frac{11}{45} \cdot 360° = 88°$

Zoo:

$\frac{5}{45} \cdot 360° = 40°$

Other:

$\frac{4}{45} \cdot 360° = 32°$

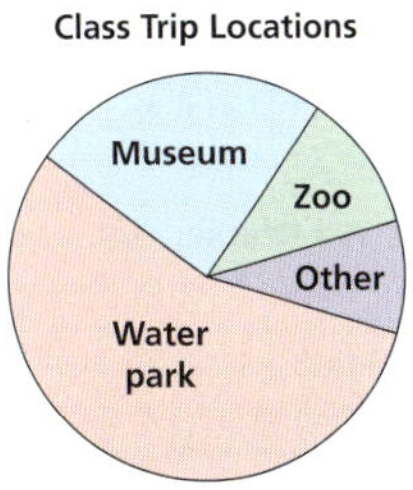

The table shows the results of a survey. Display the data in a histogram.

1.

After-school activities	Frequency
0–1	11
2–3	8
4–5	6
6–7	1

2.

Pets	Frequency
0–1	10
2–3	18
4–5	2

The table shows the results of a survey. Display the data in a circle graph.

3.

Favorite subject	Math	Science	English	History
Students	8	5	7	4

4. ABSTRACT REASONING Twenty people respond "yes" or "no" to a survey question. Let a and b represent the frequencies of the responses. What must be true about the sum of a and b? What must be true about the sum when "maybe" is an option for the response?

Dynamic Solutions available at *BigIdeasMath.com*

What Your Students Have Learned

- Construct histograms to represent data sets.
- Construct circle graphs to represent data sets.

ANSWERS

1.

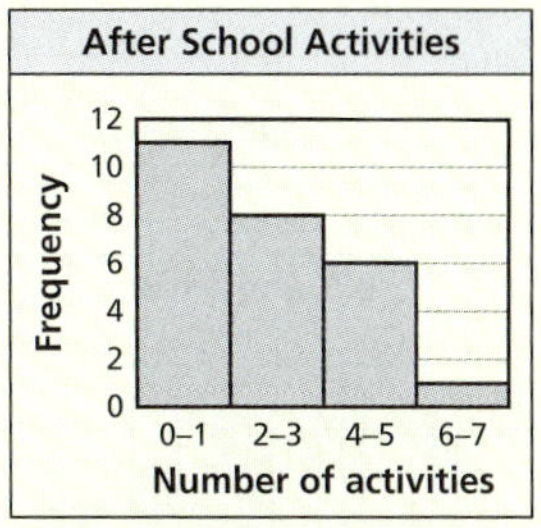

2.

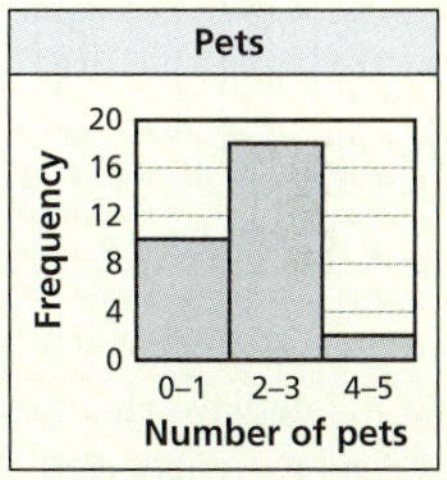

3. Students' Favorite Subjects

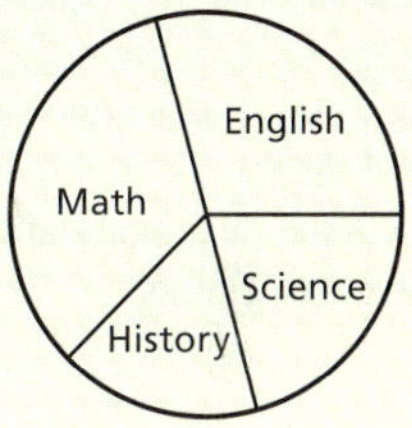

4. $a + b = 20$; $a + b \leq 20$

Vocabulary Review

Have students make Four Squares for the following terms.

- Histogram
- Circle graph

MONITORING PROGRESS ANSWERS

1. *Sample answer:*

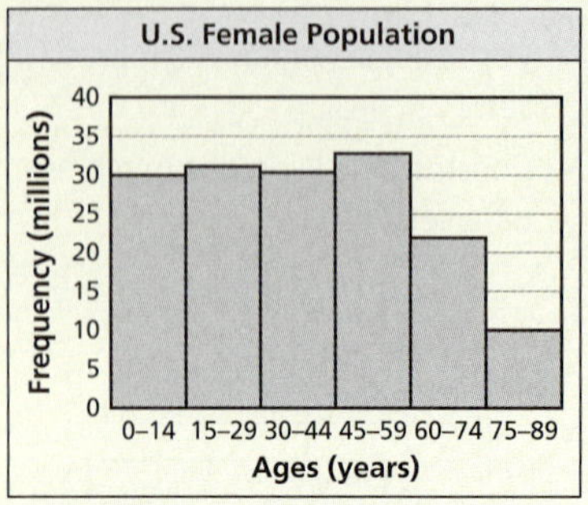

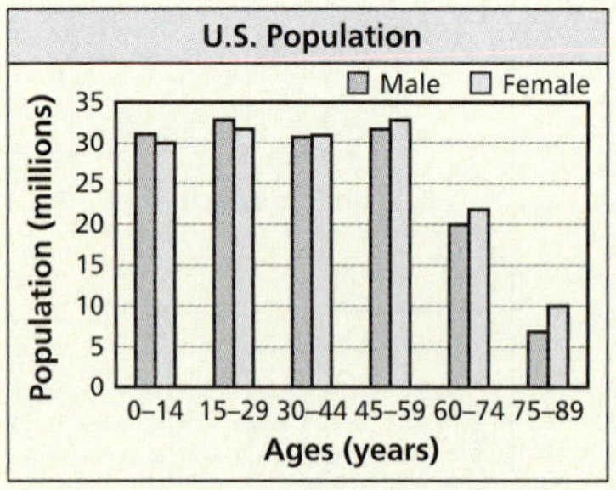

2. *Sample answer:* Histograms display frequencies in intervals, and double bar graphs display frequencies for each gender side-by-side; the populations are roughly equal for the first four age groups, but decrease significantly for the last two age groups and the female population is greater than the male population for the last two age groups.

Mathematical Practices

Mathematically proficient students use diagrams and graphs to show relationships between data. They also analyze data to draw conclusions.

Using Data Displays

Core Concept

Displaying Data Graphically

When solving a problem involving data, it is helpful to display the data graphically. This can be done in a variety of ways.

Data Display	What does it do?	
Pictograph	shows data using pictures	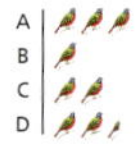
Bar Graph	shows data in specific categories	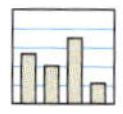
Circle Graph	shows data as parts of a whole	
Line Graph	shows how data change over time	
Histogram	shows frequencies of data values in intervals of the same size	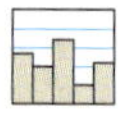
Stem-and-Leaf Plot	orders numerical data and shows how they are distributed	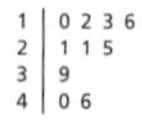
Box-and-Whisker Plot	shows the variability of a data set using quartiles	
Dot Plot	shows the number of times each value occurs in a data set	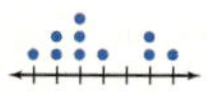
Scatter Plot	shows the relationship between two data sets using ordered pairs in a coordinate plane	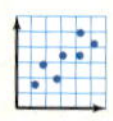

Monitoring Progress

1. The table shows the estimated populations of males and females by age in the United States in 2012. Use a spreadsheet, graphing calculator, or some other form of technology to make two different displays for the data.
2. Explain why you chose each type of data display in Monitoring Progress Question 1. What conclusions can you draw from your data displays?

U.S. Population by Age and Gender

Ages (years)	Males	Females
0–14	31,242,542	29,901,556
15–29	33,357,203	31,985,028
30–44	30,667,513	30,759,902
45–59	31,875,279	33,165,976
60–74	19,737,347	22,061,730
75–89	6,999,292	10,028,195

Laurie's Notes — Mathematical Practices (continued from page T-583)

- It is important for students to be able to select an appropriate data display when presented with a set of data. Students must consider what the raw data look like and the purpose of the display when determining which type of display to use.
- Give time for students to work through the questions in the *Monitoring Progress*. Have students record their thinking about each question, and then discuss as a class.

Laurie's Notes

Overview of Section 11.1

Introduction

- This lesson extends concepts studied in middle school. Students should be familiar with measures of center (mean, median, and mode) and measures of variation (range and mean absolute deviation).
- New in this lesson is a measure of variation called *standard deviation.*
- Two data sets can have the same mean but different standard deviations, and vice versa.
- Students should be able to explain how measures of center and measures of variation help to describe a data set.

Resources

- Students will need a calculator to find the standard deviation.

Teaching Strategy

- In order to examine measures of center and measures of variation effectively, it is necessary to have a sizable data set, perhaps at least 10 data values. It can be tedious to repeat computations when you want to examine transformations on a data set, or when an incorrect number is entered, or when you want to omit an outlier, and so on. Using the statistics feature of most scientific or graphing calculators is advised in this lesson.
- Use the *one-variable stats* feature on a graphing calculator to find the mean $\bar{x}$, median (Q2), and standard deviation (σ) of a data set.
- In Figure 1, the data for Example 1 are shown. Figure 2 shows the mean and standard deviation for this data set. Figure 3 shows the results when the outlier ($16.50) is removed.

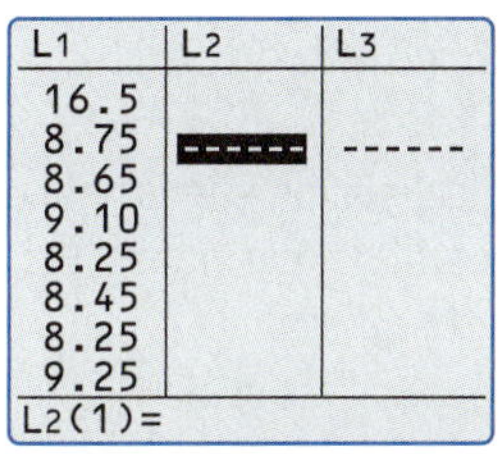

Figure 1

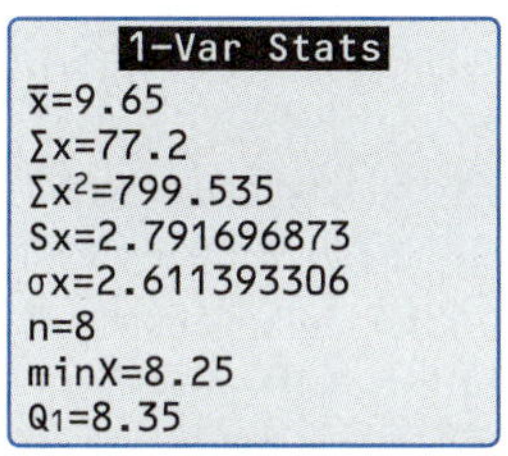

Figure 2

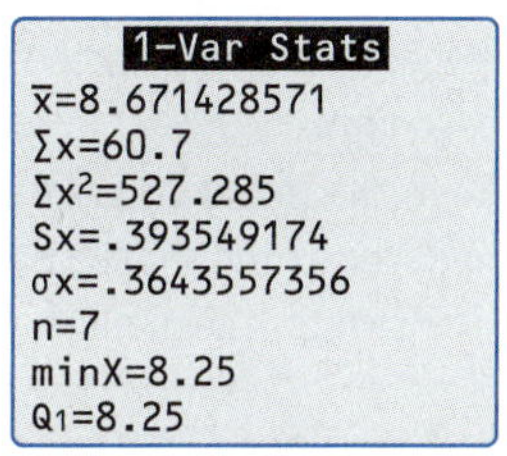

Figure 3

- Transformations to a list of data can be performed easily by adding a constant or multiplying by a factor.
- In Figures 4 and 5, the original data from Example 1 are transformed by adding $0.50 to each data value. The mean and standard deviation of the transformed data are shown in Figure 6.

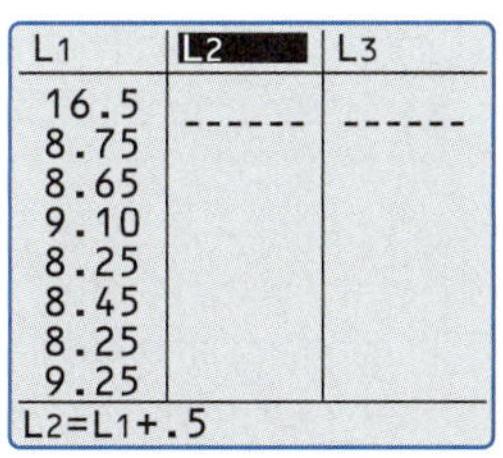

Figure 4

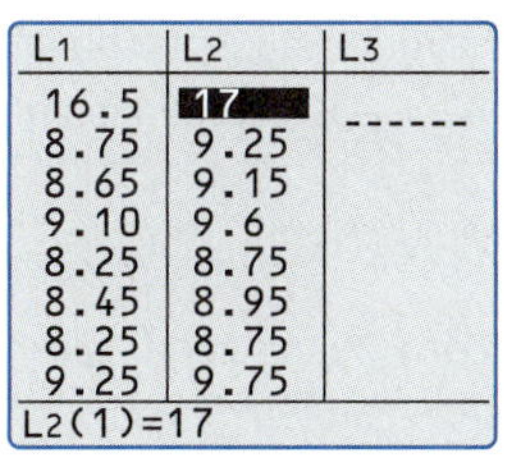

Figure 5

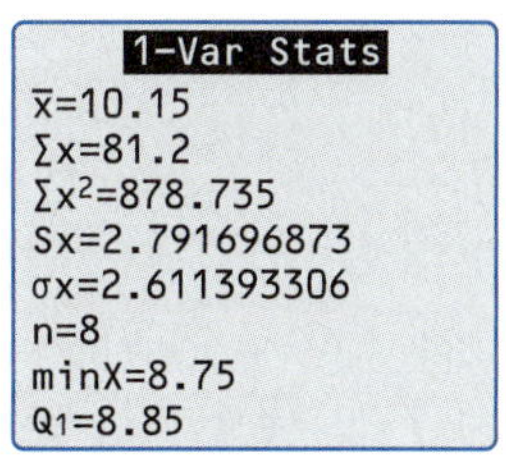

Figure 6

Pacing Suggestion

- As students work through the explorations, listen and probe for recall of prior skills, and then continue with the formal lesson.

Dynamic Teaching Tools

- Dynamic Assessment & Progress Monitoring Tool
- Lesson Planning Tool
- Interactive Whiteboard Lesson Library
- Dynamic Classroom with Dynamic Investigations

What Your Students Will Learn

- Compare the means, medians, and modes of data sets and describe the effects of outliers.
- Find the ranges and standard deviations of data sets.
- Identify the effects of adding a constant term to each value in a numerical data set.
- Identify the effects of multiplying each value in a numerical data set by a constant factor.

Laurie's Notes

Exploration

Motivate

- Play a game of "Did You Know?"
- Walter Camp is called the "Father of American Football."
- The game originated from the game of rugby.
- The American Professional Football Association began in 1920 at a car dealership in Canton, Ohio. This association would later become the National Football League (NFL).
- The Green Bay Packers won the first two Super Bowls in 1967 and 1968. It took them almost 30 years to win the Super Bowl again.
- A football is 11 inches long and weighs 14 to 15 ounces.

Exploration Note

- When data are presented in a histogram, the actual data are lost. In the first exploration, there is no vertical scale given for the histogram, though one might reason that each small rectangle represents one player. If we make the assumption that each rectangle is one player, we see that there are five football players in the range of 180 to 190 pounds. We do not know where in that range the actual weights fall. For that reason, we cannot find the actual mean.

Exploration 1

- Give students time to study the histograms so they can make sense of what the additional variable of color tells about the data.

? "How many football players on the team weigh 195? Explain how you know." You cannot determine how many weigh 195, only that four players are in the range of 190 to 200 pounds.

- **Reason Abstractly and Quantitatively:** Have students estimate a mean for each histogram and explain their reasoning. Compare across the class and listen for valid reasoning.
- When answering parts (a) and (b), students may describe only the variation (dispersion) of the outer extreme values. Students should mention the difference in the two graphs: the data vary from the mean much more for the football data than for the baseball data.
- **Construct Viable Arguments and Critique the Reasoning of Others:** Students should note that the different positions on the football team are more clustered than the different positions on the baseball team.

Exploration 2

? "What type of graph makes sense for displaying these data? Why?" The histogram shows the frequencies of the data values in certain intervals.

- If time is a concern, you could have a completed histogram for students to analyze.
- When comparing the dispersions of data, students should notice that the basketball data are similar to the football data in that there is a correlation between positions and weight.

Communicate Your Answer

- Displaying data in histograms is one way to see graphically the dispersion or spread of data. Other types of graphs, such as a box-and-whisker plot, stem-and-leaf plot, or dot plot, would also allow you to see how data are spread out or clustered.

Connecting to Next Step

- Students are likely to mention other measures of center while completing the explorations. Measures of center and measures of variation are formally defined in the lesson.

11.1 Measures of Center and Variation

Essential Question

How can you describe the variation of a data set?

EXPLORATION 1 Describing the Variation of Data

Work with a partner. The graphs show the weights of the players on a professional football team and a professional baseball team.

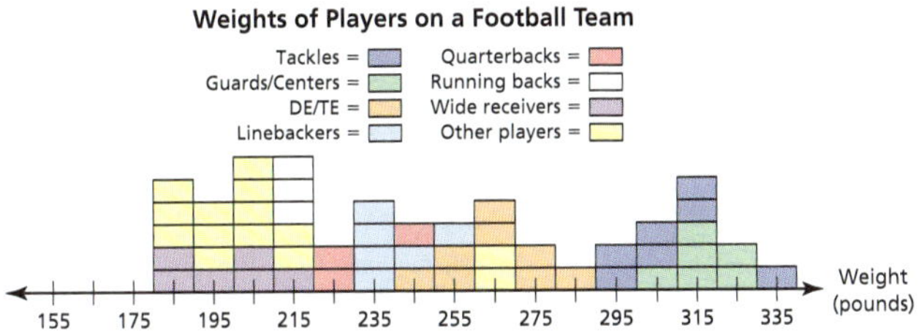

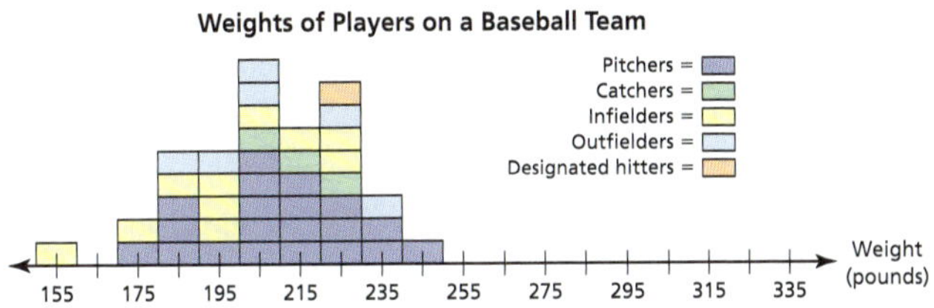

CONSTRUCTING VIABLE ARGUMENTS

To be proficient in math, you need to reason inductively about data, making plausible arguments that take into account the context from which the data arose.

a. Describe the data in each graph in terms of how much the weights vary from the mean. Explain your reasoning.

b. Compare how much the weights of the players on the football team vary from the mean to how much the weights of the players on the baseball team vary from the mean.

c. Does there appear to be a correlation between the body weights and the positions of players in professional football? in professional baseball? Explain.

EXPLORATION 2 Describing the Variation of Data

Work with a partner. The weights (in pounds) of the players on a professional basketball team by position are as follows.

Power forwards: 235, 255, 295, 245; small forwards: 235, 235; centers: 255, 245, 325; point guards: 205, 185, 205; shooting guards: 205, 215, 185

Make a graph that represents the weights and positions of the players. Does there appear to be a correlation between the body weights and the positions of players in professional basketball? Explain your reasoning.

Communicate Your Answer

3. How can you describe the variation of a data set?

Dynamic Teaching Tools

- Dynamic Assessment & Progress Monitoring Tool
- Lesson Planning Tool
- Interactive Whiteboard Lesson Library
- Dynamic Classroom with Dynamic Investigations

ANSWERS

1. a. The weights of the football players vary from the mean of about 245 pounds by up to 90 pounds. The weights of the baseball players vary from the mean of about 205 pounds by up to 40 pounds.

b. The weights of the football players vary from the mean much more than the weights of the baseball players.

c. yes; no; Each football position seems to be clustered around a similar weight; Each baseball position is more spread out.

2. See Additional Answers.

3. Describe how the data vary from the mean.

Differentiated Instruction

Auditory

Have students verbally describe what the mean, median, and mode of a data set are and how to find each of them. The description of how to find the median should include ordering the data as a first step.

Extra Example 1

A water park hires students for the summer. The students' hourly wages are shown in the table

Students' Hourly Wages	
\$8.15	\$8.70
\$8.50	\$8.50
\$9.00	\$8.25
\$14.50	\$9.20

a. Find the mean, median, and mode of the hourly wages. mean: \$9.35; median: \$8.60; mode: \$8.50

b. Which measure of center best represents the data? Explain. Both the median and the mode represent the data well. The mean is greater than most of the data.

MONITORING PROGRESS ANSWER

1. a. mean: about 9.52, decreases, median: 8.65, decreases, modes: 8.25 and 8.45; \$8.45 is less than the mean and median.

   b. median; The mean is greater than most of the data and the modes are less than most of the data.

11.1 Lesson

Core Vocabulary

measure of center, *p. 586*
mean, *p. 586*
median, *p. 586*
mode, *p. 586*
outlier, *p. 587*
measure of variation, *p. 587*
range, *p. 587*
standard deviation, *p. 588*
data transformation, *p. 589*

What You Will Learn

- Compare the mean, median, and mode of a data set.
- Find the range and standard deviation of a data set.
- Identify the effects of transformations on data.

Comparing the Mean, Median, and Mode

A **measure of center** is a measure that represents the center, or typical value, of a data set. The *mean*, *median*, and *mode* are measures of center.

Core Concept

Mean

The **mean** of a numerical data set is the sum of the data divided by the number of data values. The symbol $\overline{x}$ represents the mean. It is read as "x-bar."

Median

The **median** of a numerical data set is the middle number when the values are written in numerical order. When a data set has an even number of values, the median is the mean of the two middle values.

Mode

The **mode** of a data set is the value or values that occur most often. There may be one mode, no mode, or more than one mode.

EXAMPLE 1 Comparing Measures of Center

An amusement park hires students for the summer. The students' hourly wages are shown in the table.

Students' Hourly Wages	
\$16.50	\$8.25
\$8.75	\$8.45
\$8.65	\$8.25
\$9.10	\$9.25

a. Find the mean, median, and mode of the hourly wages.

b. Which measure of center best represents the data? Explain.

SOLUTION

a. Mean $\overline{x} = \frac{16.5 + 8.75 + 8.65 + 9.1 + 8.25 + 8.45 + 8.25 + 9.25}{8} = 9.65$

Median 8.25, 8.25, 8.45, 8.65, 8.75, 9.10, 9.25, 16.50 — Order the data.

$\frac{17.4}{2} = 8.7$ — Mean of two middle values

Mode 8.25, 8.25, 8.45, 8.65, 8.75, 9.10, 9.25, 16.50 — 8.25 occurs most often.

▶ The mean is \$9.65, the median is \$8.70, and the mode is \$8.25.

b. The median best represents the data. The mode is less than most of the data, and the mean is greater than most of the data.

STUDY TIP

Mode is the only measure of center that can represent a nonnumerical data set.

Monitoring Progress

Help in English and Spanish at *BigIdeasMath.com*

1. WHAT IF? The park hires another student at an hourly wage of \$8.45.
(a) How does this additional value affect the mean, median, and mode? Explain.
(b) Which measure of center best represents the data? Explain.

Laurie's Notes Teacher Actions

- **Turn and Talk:** "Give 5 numbers that have a mean of 80 and a median of 70." This will help you see how students understand these two measures. You can extend this by asking for 5 numbers that have a mean of 80, a median of 75, and a mode of 85.

COMMON ERROR When finding the median, students forget to sort the data. They also forget to account for frequency of repeated data.

- **Reason Abstractly and Quantitatively:** Computing the measures of center is not difficult. Deciding which one best describes the data requires reasoning.

An **outlier** is a data value that is much greater than or much less than the other values in a data set.

EXAMPLE 2 Removing an Outlier

Consider the data in Example 1. (a) Identify the outlier. How does the outlier affect the mean, median, and mode? (b) Describe one possible explanation for the outlier.

SOLUTION

a. The value $16.50 is much greater than the other wages. It is the outlier. Find the mean, median, and mode without the outlier.

Mean $\bar{x} = \dfrac{8.75 + 8.65 + 9.1 + 8.25 + 8.45 + 8.25 + 9.25}{7} \approx 8.67$

Median 8.25, 8.25, 8.45, 8.65, 8.75, 9.10, 9.25 — The middle value is 8.65.

Mode 8.25, 8.25, 8.45, 8.65, 8.75, 9.10, 9.25 — The mode is 8.25.

STUDY TIP
Outliers usually have the greatest effect on the mean.

▶ When you remove the outlier, the mean decreases $9.65 − $8.67 = $0.98, the median decreases $8.70 − $8.65 = $0.05, and the mode is the same.

b. The outlier could be a student who is hired to maintain the park's website, while the other students could be game attendants.

Monitoring Progress — Help in English and Spanish at *BigIdeasMath.com*

2. The table shows the annual salaries of the employees of an auto repair service. (a) Identify the outlier. How does the outlier affect the mean, median, and mode? (b) Describe one possible explanation for the outlier.

Annual Salaries	
$32,000	$42,000
$41,000	$38,000
$38,000	$45,000
$72,000	$35,000

Finding the Range and Standard Deviation

A **measure of variation** is a measure that describes the spread, or distribution, of a data set. One measure of variation is the *range*. The **range** of a data set is the difference of the greatest value and the least value.

EXAMPLE 3 Finding a Range

Two reality cooking shows select 12 contestants each. The ages of the contestants are shown in the tables. Find the range of the ages for each show. Compare your results.

Show A Ages	
20	29
19	22
25	27
27	29
30	20
21	31

Show B Ages	
25	19
20	27
22	25
27	22
48	21
32	24

SOLUTION

Show A 19, 20, 20, 21, 22, 25, 27, 27, 29, 29, 30, 31 — Order the data.
So, the range is 31 − 19, or 12 years.

Show B 19, 20, 21, 22, 22, 24, 25, 25, 27, 27, 32, 48 — Order the data.
So, the range is 48 − 19, or 29 years.

▶ The range of the ages for Show A is 12 years, and the range of the ages for Show B is 29 years. So, the ages for Show B are more spread out.

Monitoring Progress — Help in English and Spanish at *BigIdeasMath.com*

3. After the first week, the 25-year-old is voted off Show A and the 48-year-old is voted off Show B. How does this affect the range of the ages of the remaining contestants on each show in Example 3? Explain.

Extra Example 2

Consider the data in Extra Example 1.

Students' Hourly Wages	
$8.15	$8.70
$8.50	$8.50
$9.00	$8.25
$14.50	$9.20

a. Identify the outlier. How does the outlier affect the mean, median, and mode? The value $14.50 is the outlier. When you remove the outlier, the mean decreases $9.35 − $8.61 = $0.74, the median decreases $8.60 − $8.50 = $0.10, and the mode is the same.

b. Describe one possible explanation for the outlier. The outlier could be a more experienced student who was given greater responsibilities, while the other students could be new and inexperienced employees.

Extra Example 3

Two game shows select 12 contestants each. The ages of the contestants are shown in the tables. Find the range of the ages for each show. Compare your results.

Show A Ages	
35	52
32	35
25	29
19	20
21	43
40	33

Show B Ages	
28	38
32	24
35	34
26	30
40	27
20	35

The range of the ages for Show A is 33 years, and the range of the ages for Show B is 20 years. So, the ages for Show A are more spread out.

MONITORING PROGRESS ANSWERS

2. a. $72,000; It increases the mean and median and does not affect the mode.

b. *Sample answer:* The outlier may be the manager.

3. Show A: 12; Show B: 13; 31 and 19 are still values for Show A; The greatest value for Show B is now 32.

Laurie's Notes — Teacher Actions

? "Which measure of center is generally most affected by an outlier? Explain." the mean Listen for correct reasoning.

- See the *Teaching Strategy* on page T-584 for using technology to find the measures of center to demonstrate the effect of an outlier.

COMMON ERROR When finding the range, students often describe the interval (low-high) instead of subtracting to find a single number.

- **Think-Pair-Share:** Have students answer Question 3, and then share and discuss as a class.

English Language Learners

Notebook Development

Have students create a page in their notebooks for the measures of center and variation. Students should include all the core vocabulary in this section. For each vocabulary term, have them provide a definition and an example.

Extra Example 4

Find the standard deviation of the ages for Show A in Extra Example 3. Use a table to organize your work. Interpret your result.

Show A	
Ages	
35	52
32	35
25	29
19	20
21	43
40	33

The standard deviation is about 9.6. This means that the typical age of a contestant on Show A differs from the mean by about 9.6 years.

MONITORING PROGRESS ANSWERS

4. about 7.47 years; The typical age of a contestant differs from the mean by about 7.47 years.
5. The standard deviation for Show B is greater, so the ages are more spread out.

A disadvantage of using the range to describe the spread of a data set is that it uses only two data values. A measure of variation that uses all the values of a data set is the *standard deviation.*

Core Concept

Standard Deviation

The **standard deviation** of a numerical data set is a measure of how much a typical value in the data set differs from the mean. The symbol σ represents the standard deviation. It is read as "sigma." It is given by

$$\sigma = \sqrt{\frac{(x_1 - \overline{x})^2 + (x_2 - \overline{x})^2 + \cdots + (x_n - \overline{x})^2}{n}}$$

where n is the number of values in the data set. The deviation of a data value x is the difference of the data value and the mean of the data set, $x - \overline{x}$.

Step 1 Find the mean, $\overline{x}$.

Step 2 Find the deviation of each data value, $x - \overline{x}$.

Step 3 Square each deviation, $(x - \overline{x})^2$.

Step 4 Find the mean of the squared deviations. This is called the *variance.*

Step 5 Take the square root of the variance.

REMEMBER
An ellipsis " . . . " indicates that a pattern continues.

A small standard deviation means that the data are clustered around the mean. A large standard deviation means that the data are more spread out.

EXAMPLE 4 Finding a Standard Deviation

Find the standard deviation of the ages for Show A in Example 3. Use a table to organize your work. Interpret your result.

SOLUTION

x	$\overline{x}$	$x - \overline{x}$	$(x - \overline{x})^2$
20	25	−5	25
29	25	4	16
19	25	−6	36
22	25	−3	9
25	25	0	0
27	25	2	4
27	25	2	4
29	25	4	16
30	25	5	25
20	25	−5	25
21	25	−4	16
31	25	6	36

Step 1 Find the mean, $\overline{x}$.

$$\overline{x} = \frac{300}{12} = 25$$

Step 2 Find the deviation of each data value, $x - \overline{x}$, as shown.

Step 3 Square each deviation, $(x - \overline{x})^2$, as shown.

Step 4 Find the mean of the squared deviations, or variance.

$$\frac{(x_1 - \overline{x})^2 + (x_2 - \overline{x})^2 + \cdots + (x_n - \overline{x})^2}{n} = \frac{25 + 16 + \cdots + 36}{12} = \frac{212}{12} \approx 17.7$$

Step 5 Use a calculator to take the square root of the variance.

$$\sqrt{\frac{(x_1 - \overline{x})^2 + (x_2 - \overline{x})^2 + \cdots + (x_n - \overline{x})^2}{n}} = \sqrt{\frac{212}{12}} \approx 4.2$$

▶ The standard deviation is about 4.2. This means that the typical age of a contestant on Show A differs from the mean by about 4.2 years.

Monitoring Progress

Help in English and Spanish at *BigIdeasMath.com*

4. Find the standard deviation of the ages for Show B in Example 3. Interpret your result.
5. Compare the standard deviations for Show A and Show B. What can you conclude?

Laurie's Notes Teacher Actions

- **Connection:** Students learned the Mean Absolute Deviation (MAD) statistic in middle school.
- Write the *Core Concept.* You want students to understand the *Big Idea*—namely that the standard deviation is a measure of dispersion that takes into account all data values, not just the least and greatest.
- **Use Appropriate Tools Strategically:** Many computations are involved, so a small data set is used and a table is used to organize results. Emphasize that it measures how much a typical value in the data set differs from the mean.

Effects of Data Transformations

A **data transformation** is a procedure that uses a mathematical operation to change a data set into a different data set.

STUDY TIP

The standard deviation stays the same because the amount by which each data value deviates from the mean stays the same.

Core Concept

Data Transformations Using Addition

When a real number k is added to each value in a numerical data set

- the measures of center of the new data set can be found by adding k to the original measures of center.
- the measures of variation of the new data set are the *same* as the original measures of variation.

Data Transformations Using Multiplication

When each value in a numerical data set is multiplied by a real number k, where $k > 0$, the measures of center and variation can be found by multiplying the original measures by k.

EXAMPLE 5 Real-Life Application

Consider the data in Example 1. (a) Find the mean, median, mode, range, and standard deviation when each hourly wage increases by \$0.50. (b) Find the mean, median, mode, range, and standard deviation when each hourly wage increases by 10%.

Students' Hourly Wages	
\$17.00	\$8.75
\$9.25	\$8.95
\$9.15	\$8.75
\$9.60	\$9.75

SOLUTION

a. Method 1 Make a new table by adding \$0.50 to each hourly wage. Find the mean, median, mode, range, and standard deviation of the new data set.

▶ Mean: \$10.15 Median: \$9.20 Mode: \$8.75
Range: \$8.25 Standard deviation: \$2.61

Method 2 Find the mean, median, mode, range, and standard deviation of the original data set.

Mean: \$9.65 Median: \$8.70 Mode: \$8.25 *From Example 1*
Range: \$8.25 Standard deviation: \$2.61

Add \$0.50 to the mean, median, and mode. The range and standard deviation are the same as the original range and standard deviation.

▶ Mean: \$10.15 Median: \$9.20 Mode: \$8.75
Range: \$8.25 Standard deviation: \$2.61

b. Increasing by 10% means to multiply by 1.1. So, multiply the original mean, median, mode, range, and standard deviation from Method 2 of part (a) by 1.1.

▶ Mean: \$10.62 Median: \$9.57 Mode: \$9.08
Range: \$9.08 Standard deviation: \$2.87

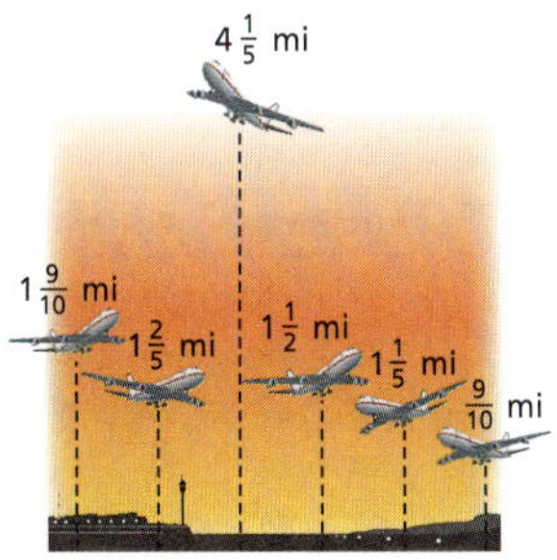

Monitoring Progress

Help in English and Spanish at *BigIdeasMath.com*

6. Find the mean, median, mode, range, and standard deviation of the altitudes of the airplanes when each altitude increases by $1\frac{1}{2}$ miles.

Extra Example 5

Consider the data in Extra Example 1.

Students' Hourly Wages	
\$8.15	\$8.70
\$8.50	\$8.50
\$9.00	\$8.25
\$14.50	\$9.20

a. Find the mean, median, mode, range, and standard deviation when each hourly wage increases by \$0.50.
mean: \$9.85; median: \$9.10; mode: \$9.00; range: \$6.35; standard deviation: about \$1.97

b. Find the mean, median, mode, range, and standard deviation when each hourly wage increases by 10%.
mean: \$10.29; median: \$9.46; mode: \$9.35; range: \$6.99; standard deviation: about \$2.17

MONITORING PROGRESS ANSWER

6. mean: 3.35 mi, median: 2.95 mi, mode: none, range: 3.3 mi, standard deviation: about 1.09 mi.

Laurie's Notes Teacher Actions

- **Predict, Explain, Observe:** Group students (2–4 in a group) and ask, "If each value in a data set is changed by a constant k, what happens to the measures of center? Measures of variation? Explain." Give groups sufficient time to consider the questions and to consider their reasoning. Repeat questions, but use a constant factor. After exploring, ask, "Were your predictions accurate?"

Closure

- **Writing Prompt:** Explain to a student missing from class today what measures of center and measures of variation are.

Assignment Guide and Homework Check

ASSIGNMENT

Basic: 1–4, 5–29 odd, 32, 34, 40–47

Average: 1–4, 6–30 even, 31, 32, 34, 39–47

Advanced: 1–4, 8–18 even, 22–34 even, 35–47

HOMEWORK CHECK

Basic: 5, 15, 17, 23, 27

Average: 6, 12, 18, 22, 28

Advanced: 8, 16, 18, 24, 28

11.1 Exercises

Dynamic Solutions available at *BigIdeasMath.com*

Vocabulary and Core Concept Check

1. **VOCABULARY** In a data set, what does a measure of center represent? What does a measure of variation describe?
2. **WRITING** Describe how removing an outlier from a data set affects the mean of the data set.
3. **OPEN-ENDED** Create a data set that has more than one mode.
4. **REASONING** What is an advantage of using the range to describe a data set? Why do you think the standard deviation is considered a more reliable measure of variation than the range?

Monitoring Progress and Modeling with Mathematics

In Exercises 5–8, (a) find the mean, median, and mode of the data set and (b) determine which measure of center best represents the data. Explain. *(See Example 1.)*

5. 3, 5, 1, 5, 1, 1, 2, 3, 15
6. 12, 9, 17, 15, 10
7. 13, 30, 16, 19, 20, 22, 25, 31
8. 14, 15, 3, 15, 14, 14, 18, 15, 8, 16

9. **ANALYZING DATA** The table shows the lengths of nine movies.

Movie Lengths (hours)		
$1\frac{1}{3}$	$1\frac{2}{3}$	2
3	$2\frac{1}{3}$	$1\frac{2}{3}$
2	2	$1\frac{2}{3}$

   a. Find the mean, median, and mode of the lengths.

   b. Which measure of center best represents the data? Explain.

10. **ANALYZING DATA** The table shows the daily changes in the value of a stock over 12 days.

Changes in Stock Value (dollars)			
1.05	2.03	−13.78	−2.41
2.64	0.67	4.02	1.39
0.66	−0.28	−3.01	2.20

   a. Find the mean, median, and mode of the changes in stock value.

   b. Which measure of center best represents the data? Explain.

   c. On the 13th day, the value of the stock increases by \$4.28. How does this additional value affect the mean, median, and mode? Explain.

In Exercises 11–14, find the value of x.

11. 2, 8, 9, 7, 6, x; The mean is 6.
12. 12.5, −10, −7.5, x; The mean is 11.5.
13. 9, 10, 12, x, 20, 25; The median is 14.
14. 30, 45, x, 100; The median is 51.

15. **ANALYZING DATA** The table shows the masses of eight polar bears. *(See Example 2.)*

Masses (kilograms)			
455	262	471	358
364	553	62	351

   a. Identify the outlier. How does the outlier affect the mean, median, and mode?

   b. Describe one possible explanation for the outlier.

16. **ANALYZING DATA** The sizes of emails (in kilobytes) in your inbox are 2, 3, 5, 2, 1, 46, 3, 7, 2, and 1.

   a. Identify the outlier. How does the outlier affect the mean, median, and mode?

   b. Describe one possible explanation for the outlier.

17. **ANALYZING DATA** The scores of two golfers are shown. Find the range of the scores for each golfer. Compare your results. *(See Example 3.)*

Golfer A	
83	88
84	95
91	89
90	87
98	95

Golfer B	
89	87
93	95
92	94
88	91
89	92

ANSWERS

1. the center, or typical value; the distribution of the data
2. The mean gets closer to the median.
3. *Sample answer:* 3, 4, 4, 7, 9, 9
4. It is easy to calculate; It uses all of the values of a data set.
5. a. mean: 4, median: 3, mode: 1
   b. median; The mean is greater than most of the data and the mode is less than most of the data.
6. a. mean: 12.6, median: 12, mode: none
   b. mean; There are no outliers.
7. a. mean: 22, median: 21, mode: none
   b. mean; There are no outliers.
8. a. mean: 13.2, median: 14.5, modes: 14 and 15
   b. median; The mean is less than most of the data and it is the mean of the two modes.
9. a. mean: about 1.96, median: 2, modes: $1\frac{2}{3}$ and 2
   b. median; The data are evenly distributed.
10. a. mean: about −0.402, median: 0.86, mode: none
    b. median; The mean is less than most of the data and there is no mode.
    c. mean: about −0.042, increases; median: 1.05, increases; mode: none; \$4.28 is greater than the mean and median.
11. 4
12. 51
13. 16
14. 57
15. a. 62; The outlier decreases the mean and median and does not affect the mode.
    b. *Sample answer:* The outlier could be the mass of a baby polar bear.
16. a. 46; The outlier increases the mean and median and does not affect the mode.
    b. *Sample answer:* The email could have contained a picture.
17. Golfer A: 15; Golfer B: 8; The range for Golfer A is greater.

18. **ANALYZING DATA** The graph shows a player's monthly home run totals in two seasons. Find the range of the number of home runs for each season. Compare your results.

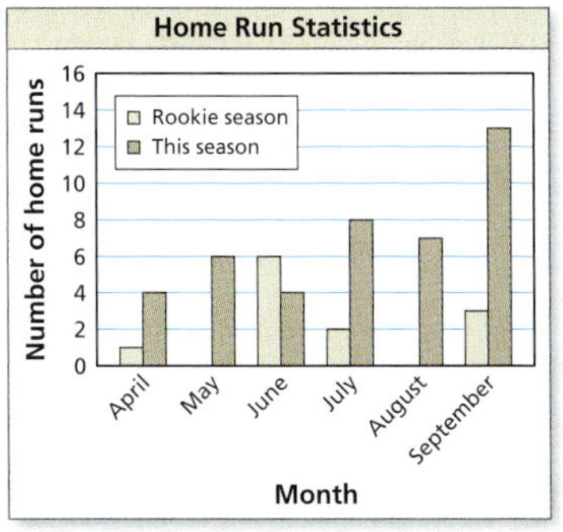

In Exercises 19–22, find (a) the range and (b) the standard deviation of the data set.

19. 40, 35, 45, 55, 60

20. 141, 116, 117, 135, 126, 121

21. 0.5, 2.0, 2.5, 1.5, 1.0, 1.5

22. 8.2, 10.1, 2.6, 4.8, 2.4, 5.6, 7.0, 3.3

23. **ANALYZING DATA** Consider the data in Exercise 17. *(See Example 4.)*
   a. Find the standard deviation of the scores of Golfer A. Interpret your result.
   b. Find the standard deviation of the scores of Golfer B. Interpret your result.
   c. Compare the standard deviations for Golfer A and Golfer B. What can you conclude?

24. **ANALYZING DATA** Consider the data in Exercise 18.
   a. Find the standard deviation of the monthly home run totals in the player's rookie season. Interpret your result.
   b. Find the standard deviation of the monthly home run totals in this season. Interpret your result.
   c. Compare the standard deviations for the rookie season and this season. What can you conclude?

In Exercises 25 and 26, find the mean, median, and mode of the data set after the given transformation.

25. In Exercise 5, each data value increases by 4.

26. In Exercise 6, each data value increases by 20%.

27. **TRANSFORMING DATA** Find the values of the measures shown when each value in the data set increases by 14. *(See Example 5.)*

Mean: 62 Median: 55 Mode: 49
Range: 46 Standard deviation: 15.5

28. **TRANSFORMING DATA** Find the values of the measures shown when each value in the data set is multiplied by 0.5.

Mean: 320 Median: 300 Mode: none
Range: 210 Standard deviation: 70.6

29. **ERROR ANALYSIS** Describe and correct the error in finding the median of the data set.

> ✗ 7, 4, 6, 2, 4, 6, 8, 8, 3
> The median is 4.

30. **ERROR ANALYSIS** Describe and correct the error in finding the range of the data set after the given transformation.

> ✗ −13, −12, −7, 2, 10, 13
> Add 10 to each value.
> The range is 26 + 10 = 36.

31. **PROBLEM SOLVING** In a bowling match, the team with the greater mean score wins. The scores of the members of two bowling teams are shown.

Team A: 172, 130, 173, 212
Team B: 136, 184, 168, 192

   a. Which team wins the match? If the team with the greater median score wins, is the result the same? Explain.
   b. Which team is more consistent? Explain.
   c. In another match between the two teams, all the members of Team A increase their scores by 15 and all the members of Team B increase their scores by 12.5%. Which team wins this match? Explain.

32. **MAKING AN ARGUMENT** Your friend says that when two data sets have the same range, you can assume the data sets have the same standard deviation, because both range and standard deviation are measures of variation. Is your friend correct? Explain.

Dynamic Teaching Tools

- Dynamic Assessment & Progress Monitoring Tool
- Interactive Whiteboard Lesson Library
- Dynamic Classroom with Dynamic Investigations

ANSWERS

18. rookie season: 6; this season: 9; The range for this season is greater.
19. a. 25
    b. about 9.27
20. a. 25
    b. about 9.24
21. a. 2
    b. about 0.65
22. a. 7.7
    b. about 2.60
23. a. about 4.6; The typical score differs from the mean by about 4.6 points.
    b. about 2.5; The typical score differs from the mean by about 2.5 points.
    c. The standard deviation for Golfer A is greater, so the scores are more spread out.
24. a. about 2.08; The typical number of home runs differs from the mean by about 2 home runs.
    b. 3.06; The typical number of home runs differs from the mean by about 3 home runs.
    c. The standard deviation for this season is greater, so the numbers of home runs are more spread out.
25. mean: 8, median: 7, mode: 5
26. mean: 15.12, median: 14.4, mode: none
27. mean: 76, median: 69, mode: 63, range: 46, standard deviation: 15.5
28. mean: 160, median: 150, mode: none, range: 105, standard deviation: 35.3
29. The numbers are not in numerical order; The median is 6.
30. When a number is added to each value in a data set, the range stays the same; The range is 26.
31. a. Team A; no; Team B has a greater median.
    b. Team B; The range and standard deviation are less than Team A's.
    c. Team B; The means are 186.75 and 191.25.
32. no; The values in between may be different.

ANSWERS

33. mode; The data are not numeric.

34. B; C; B has the greatest range, C has the least range.

35. mean: 89, median: 104, mode: 62, range: 123, standard deviation: 27

36. yes; no; All the data values could be the same; Standard deviation is a positive square root.

37. **a.** Answers will vary.

b. *Sample answer:* The mean, median, range, and standard deviation increase.

38. The geometric mean is always less than or equal to the arithmetic mean.

39. See Additional Answers.

40. $x \le -5$ **41.** no solution

42. no solution **43.** $w < \frac{3}{2}$

44. 64 **45.** $\frac{1}{49}$

46. 320 **47.** -162

Mini-Assessment

The table shows the prices of six cell phone cases.

Cell Phone Case Prices	
\$12	\$24
\$21	\$18
\$33	\$12

1. Find the mean, median, and mode of the prices. mean: \$20; median: \$19.50; mode: \$12
2. Which measure of center best represents the data? Explain. The mean and the median both represent the data well. The mode is less than most of the data.
3. Find the range of the prices. \$21
4. Find the standard deviation of the prices. \$7.28
5. Find the mean, median, mode, range, and standard deviation when each price increases by \$1. mean: \$21; median: \$20.50; mode: \$13; range: \$21; standard deviation: \$7.28

33. ANALYZING DATA The table shows the results of a survey that asked 12 students about their favorite meal. Which measure of center (mean, median, or mode) can be used to describe the data? Explain.

Favorite Meal			
spaghetti	pizza	steak	hamburger
steak	taco	pizza	chili
pizza	chicken	fish	spaghetti

34. HOW DO YOU SEE IT? The dot plots show the ages of the members of three different adventure clubs. Without performing calculations, which data set has the greatest standard deviation? Which has the least standard deviation? Explain your reasoning.

Ⓐ

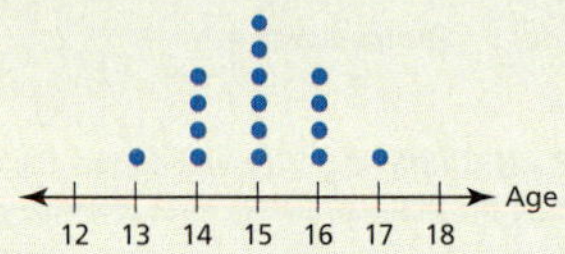

Ⓑ

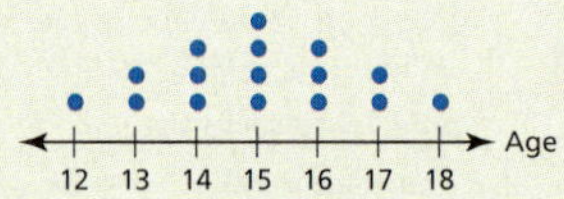

Ⓒ

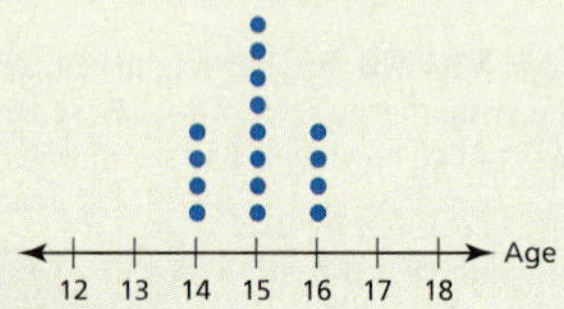

35. REASONING A data set is described by the measures shown.

Mean: 27 Median: 32 Mode: 18
Range: 41 Standard deviation: 9

Find the mean, median, mode, range, and standard deviation of the data set when each data value is multiplied by 3 and then increased by 8.

36. CRITICAL THINKING Can the standard deviation of a data set be 0? Can it be negative? Explain.

37. USING TOOLS Measure the heights (in inches) of the students in your class.

a. Find the mean, median, mode, range, and standard deviation of the heights.

b. A new student who is 7 feet tall joins your class. How would you expect this student's height to affect the measures in part (a)? Verify your answer.

38. THOUGHT PROVOKING To find the arithmetic mean of n numbers, divide the sum of the numbers by n. To find the geometric mean of n numbers $a_1, a_2, a_3, \ldots, a_n$, take the nth root of the product of the numbers.

$$\text{geometric mean} = \sqrt[n]{a_1 \cdot a_2 \cdot a_3 \cdot \ldots \cdot a_n}$$

Compare the arithmetic mean to the geometric mean of n numbers.

39. PROBLEM SOLVING The circle graph shows the distribution of the ages of 200 students in a college Psychology I class.

College Student Ages

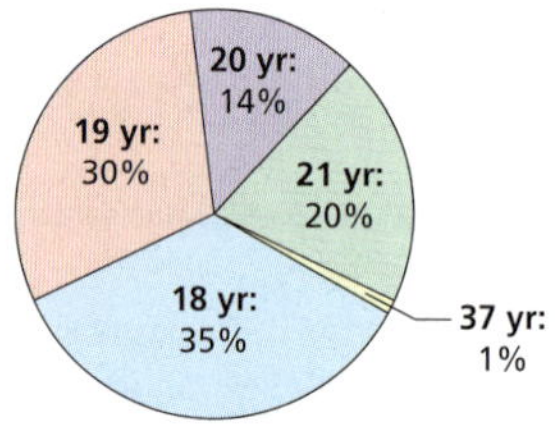

a. Find the mean, median, and mode of the students' ages.

b. Identify the outliers. How do the outliers affect the mean, median, and mode?

c. Suppose all 200 students take the same Psychology II class exactly 1 year later. Draw a new circle graph that shows the distribution of the ages of this class and find the mean, median, and mode of the students' ages.

Maintaining Mathematical Proficiency

Reviewing what you learned in previous grades and lessons

Solve the inequality. *(Section 2.4)*

40. $6x + 1 \le 4x - 9$ **41.** $-3(3y - 2) < 1 - 9y$ **42.** $2(5c - 4) \ge 5(2c + 8)$ **43.** $4(3 - w) > 3(4w - 4)$

Evaluate the function for the given value of x. *(Section 6.3)*

44. $f(x) = 4^x; x = 3$ **45.** $f(x) = 7^x; x = -2$ **46.** $f(x) = 5(2)^x; x = 6$ **47.** $f(x) = -2(3)^x; x = 4$

If students need help...	*If students got it...*
Resources by Chapter • Practice A and Practice B • Puzzle Time	Resources by Chapter • Enrichment and Extension • Cumulative Review
Student Journal • Practice	Start the *next* Section
Differentiating the Lesson Skills Review Handbook	

Laurie's Notes

Overview of Section 11.2

Introduction

- Students should be familiar with representing data using box-and-whisker plots from middle school.
- In this lesson, students will construct and interpret box-and-whisker plots.
- An important feature of box-and-whisker plots is that two or more data sets can be compared.

Teaching Strategy

- There are software programs, apps, and graphing calculators that can quickly display data as a box-and-whisker plot.
- Figures 1 and 2 show the data and the associated box-and-whisker plot from Example 1.
- Figure 3 shows the effect of adding a piece of data to the data set, namely a 40-year-old person joining the backpacking expedition. Ask students to describe the effect of the outlier on the plot.

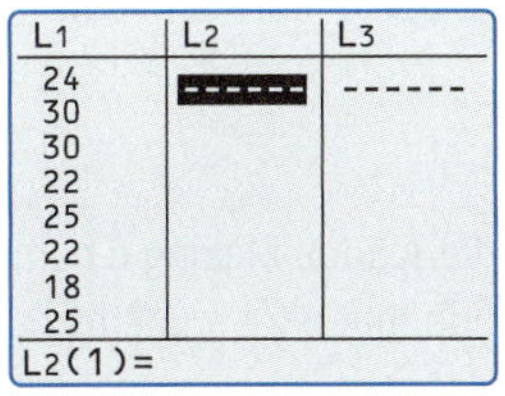

Figure 1

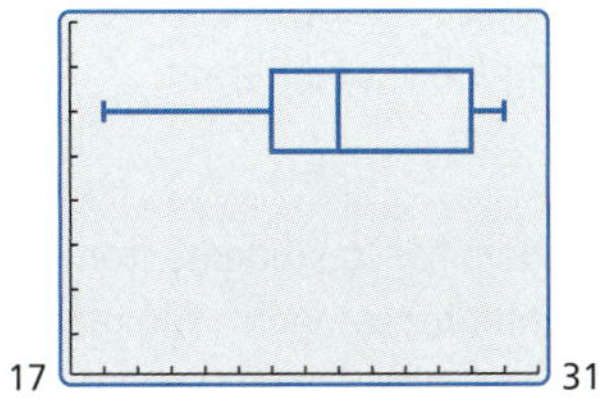

Figure 2

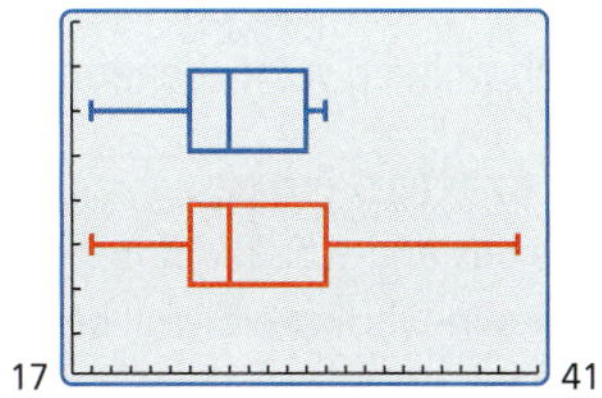

Figure 3

Extensions

- Explore transformations of data sets visually by making two box-and-whisker plots, one for the original data and one for the transformed data.

Pacing Suggestion

- Given the review nature of the content in this lesson, you should only spend one day. You could do the *Motivate* (or Exploration) and Examples 2 and 3, or you could simply present the formal lesson. Students would prefer the more active version!

Dynamic Teaching Tools

- Dynamic Assessment & Progress Monitoring Tool
- Lesson Planning Tool
- Interactive Whiteboard Lesson Library
- Dynamic Classroom with Dynamic Investigations

What Your Students Will Learn

- Construct and interpret box-and-whisker plots.
- Use box-and-whisker plots to compare two data sets.

Laurie's Notes

Exploration

Motivate

- The physical involvement of making a human box-and-whisker plot makes a lasting impression on students.
- **Preparation:** Give each student an index card with a number written on it. Include an outlier or two on one end of the data.

? "What is the first step in making a box-and-whisker plot?" sort the data

- Students should stand up and sort themselves. Have the median, the first and third quartiles, and the least and greatest data values take one step forward. If you have an even number of data values, the middle two students must figure out how to represent the mean and take a step forward.
- Make a number line on the floor or on the board. Position the five key values. If the plot is done on the floor, use string to form the whiskers. Students will have to visualize the box.
- **Reason Abstractly and Quantitatively:** Discuss features of the plot. If the plot includes an outlier, the length of the string becomes an instant topic of conversation. Students recognize that the same number of data values (25% of the class) is being represented by each whisker, yet the lengths of string are very different.

Exploration Note

- The box-and-whisker plot should be familiar to students from middle school. Making a human box-and-whisker plot will be interesting for students. You may decide that once students have done the *Motivate*, there is no need to do the *Exploration*.

Exploration 1

? "How many pieces of data are there?" 24

- Explain to students that today they are going to construct a box-and-whisker plot, a data display that is generally used for very large data sets. For instance, the results of a state test for all eighth-graders could be displayed using a box-and-whisker plot.
- The box-and-whisker plot uses a number line to visually represent the data. Specific data values are not graphed, but characteristics of the data are still conveyed.

? "How many numbers did you graph in making the box-and-whisker plot?" 5

- **Big Idea:** The five numbers graphed summarize the entire data set. The least and greatest values are the boundaries. The median separates the data into two parts. The first (or lower) quartile is the median of the lower half. The third (or upper) quartile is the median of the upper half. The box encloses the middle 50% of the data.

? "What percent of the data is represented by the upper whisker?" 25% "How many data values are in the upper whisker?" 6

? "What percent of the data is represented by the lower whisker?" 25% "How many data values are in the lower whisker?" 6

- **Reason Abstractly and Quantitatively:** These questions provide opportunities for students to reason quantitatively.

Communicate Your Answer

- **Think-Pair-Share:** Students should read each question independently and then work in pairs to answer the questions. When they have answered the questions, the pairs should compare their answers with another group and discuss any discrepancies.

Connecting to Next Step

- The exploration allows students to recall how to construct a box-and-whisker plot. In the formal lesson, they will make comparisons between two box-and-whisker plots.

11.2 Box-and-Whisker Plots

Essential Question How can you use a box-and-whisker plot to describe a data set?

EXPLORATION 1 **Drawing a Box-and-Whisker Plot**

Work with a partner. The numbers of first cousins of the students in a ninth-grade class are shown. A *box-and-whisker plot* is one way to represent the data visually.

Numbers of First Cousins

3	10	18	8
9	3	0	32
23	19	13	8
6	3	3	10
12	45	1	5
13	24	16	14

a. Order the data on a strip of grid paper with 24 equally spaced boxes.

b. Fold the paper in half again to divide the data into four groups. Because there are 24 numbers in the data set, each group should have 6 numbers. Find the least value, the greatest value, the first quartile, and the third quartile.

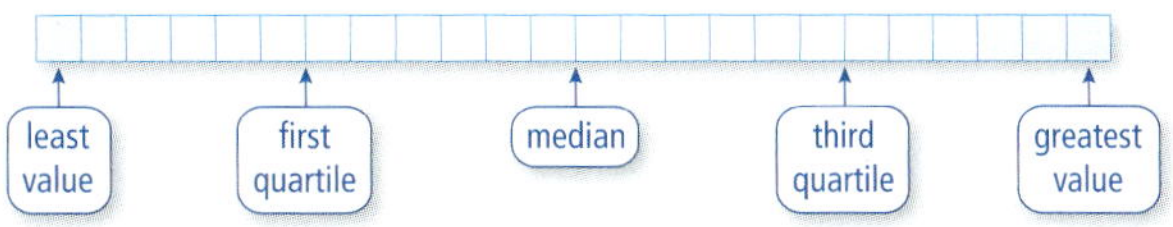

c. Explain how the box-and-whisker plot shown represents the data set.

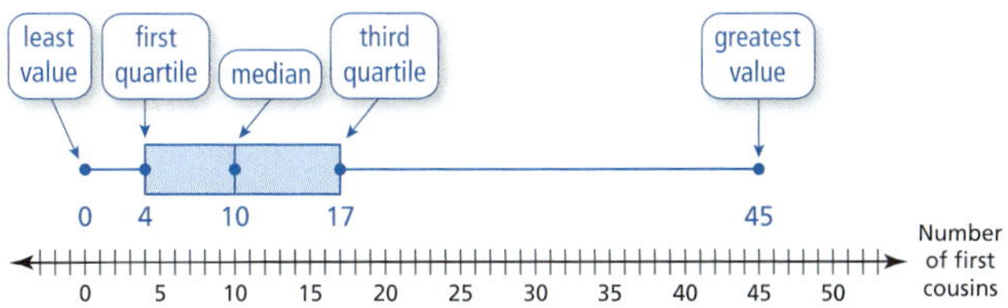

MODELING WITH MATHEMATICS

To be proficient in math, you need to identify important quantities in a practical situation.

Communicate Your Answer

2. How can you use a box-and-whisker plot to describe a data set?

3. Interpret each box-and-whisker plot.

a. body mass indices (BMI) of students in a ninth-grade class

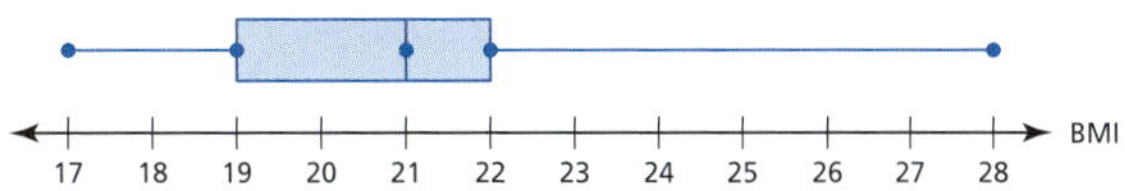

b. heights of roller coasters at an amusement park

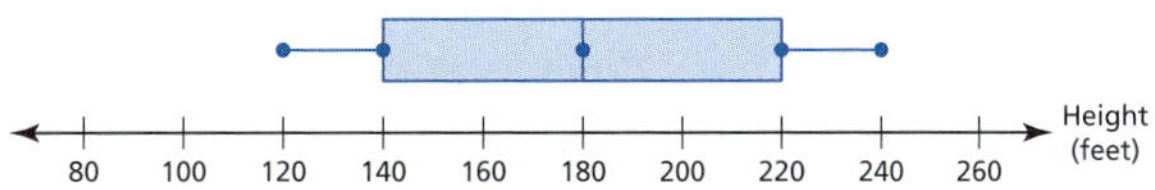

Dynamic Teaching Tools

- Dynamic Assessment & Progress Monitoring Tool
- Lesson Planning Tool
- Interactive Whiteboard Lesson Library
- Dynamic Classroom with Dynamic Investigations

ANSWERS

1. a. 0, 1, 3, 3, 3, 3, 5, 6, 8, 8, 9, 10, 10, 12, 13, 13, 14, 16, 18, 19, 23, 24, 32, 45; median: 10

b. least value: 0, first quartile: 4, third quartile: 17, greatest value: 45

c. The box represents the middle half of the data and the whiskers represent the bottom and top quarters of the data.

2. A box-and-whisker plot can be used to show the median, the first and third quartiles, and the least and greatest values of a data set.

3. a. The median is 21, the first quartile is 19, the third quartile is 22, the least value is 17, and the greatest value is 28.

b. The median is 180 feet, the first quartile is 140 feet, the third quartile is 220 feet, the least value is 120 feet, and the greatest value is 240 feet.

Extra Example 1

Make a box-and-whisker plot that represents the prices (in dollars) of DVDs in a store.

15, 18, 12, 18, 20, 24, 16, 13

MONITORING PROGRESS ANSWER

1.

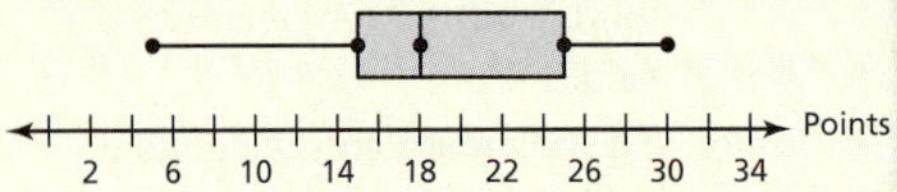

11.2 Lesson

Core Vocabulary

box-and-whisker plot, *p. 594*
quartile, *p. 594*
five-number summary, *p. 594*
interquartile range, *p. 595*

STUDY TIP

Sometimes, the first quartile is called the *lower quartile* and the third quartile is called the *upper quartile*.

What You Will Learn

- Use box-and-whisker plots to represent data sets.
- Interpret box-and-whisker plots.
- Use box-and-whisker plots to compare data sets.

Using Box-and-Whisker Plots to Represent Data Sets

Core Concept

Box-and-Whisker Plot

A **box-and-whisker plot** shows the variability of a data set along a number line using the least value, the greatest value, and the *quartiles* of the data. **Quartiles** divide the data set into four equal parts. The median (second quartile, Q2) divides the data set into two halves. The median of the lower half is the first quartile, Q1. The median of the upper half is the third quartile, Q3.

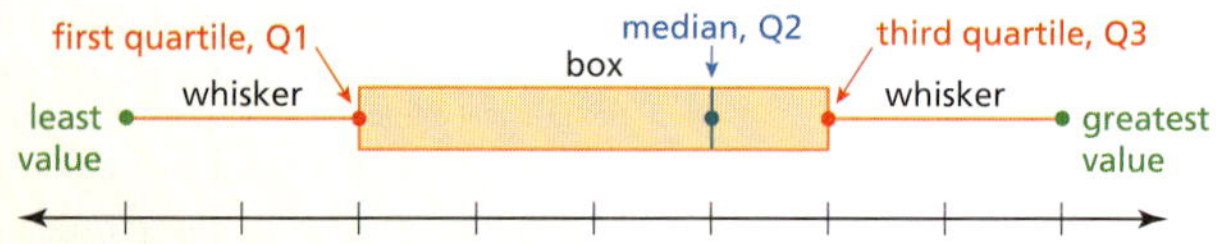

The five numbers that make up a box-and-whisker plot are called the **five-number summary** of the data set.

EXAMPLE 1 Making a Box-and-Whisker Plot

Make a box-and-whisker plot that represents the ages of the members of a backpacking expedition in the mountains.

24, 30, 30, 22, 25, 22, 18, 25, 28, 30, 25, 27

SOLUTION

Step 1 Order the data. Find the median and the quartiles.

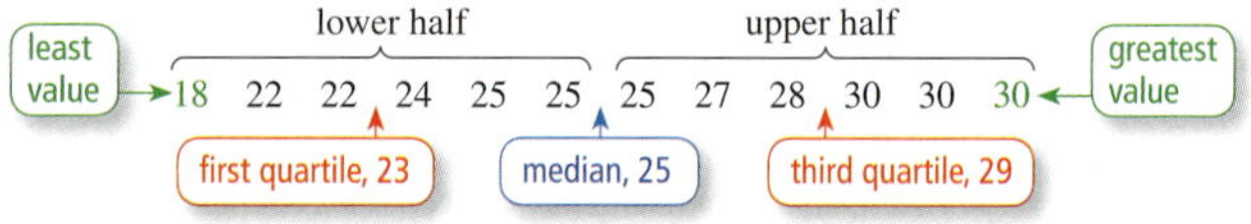

Step 2 Draw a number line that includes the least and greatest values. Graph points above the number line for the five-number summary.

Step 3 Draw a box using Q1 and Q3. Draw a line through the median. Draw whiskers from the box to the least and greatest values.

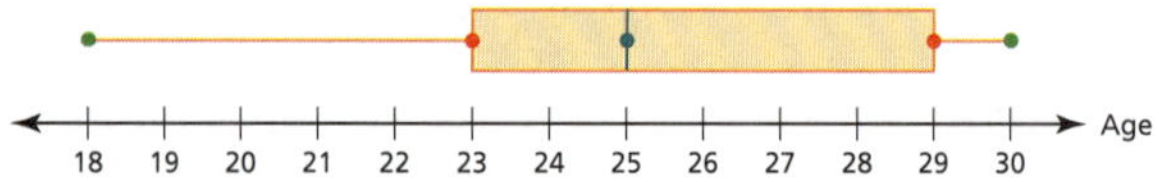

Monitoring Progress

Help in English and Spanish at *BigIdeasMath.com*

1. A basketball player scores 14, 16, 20, 5, 22, 30, 16, and 28 points during a tournament. Make a box-and-whisker plot that represents the data.

Laurie's Notes Teacher Actions

- Write the *Core Concept*. The box-and-whisker plot shows the variability of the data. Refer to this idea in each example today.
- In Example 1, notice that the first quartile (23) and the third quartile (29) are not data values from the set. This is fine. The five values are simply giving a marker for how the sorted data is spread out into four groups.

? "How many backpackers are represented in each quartile?" 3
? "What percent of the backpackers were older than 23?" 75%
? "What is the range?" 12

Interpreting Box-and-Whisker Plots

The figure shows how data are distributed in a box-and-whisker plot.

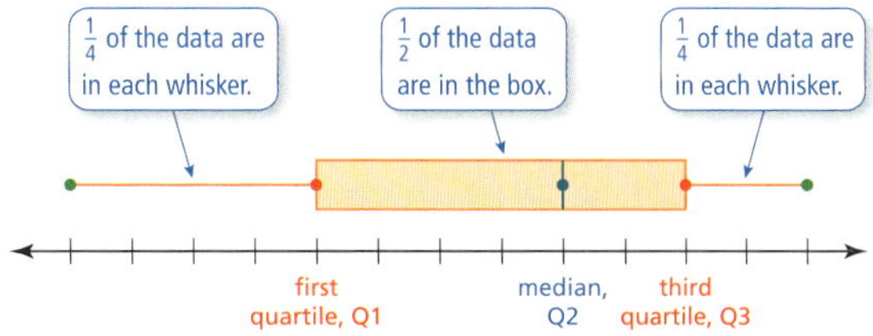

STUDY TIP

A long whisker or box indicates that the data are more spread out.

Another measure of variation for a data set is the **interquartile range** (IQR), which is the difference of the third quartile, Q3, and the first quartile, Q1. It represents the range of the middle half of the data.

EXAMPLE 2 Interpreting a Box-and-Whisker Plot

The box-and-whisker plot represents the lengths (in seconds) of the songs played by a rock band at a concert.

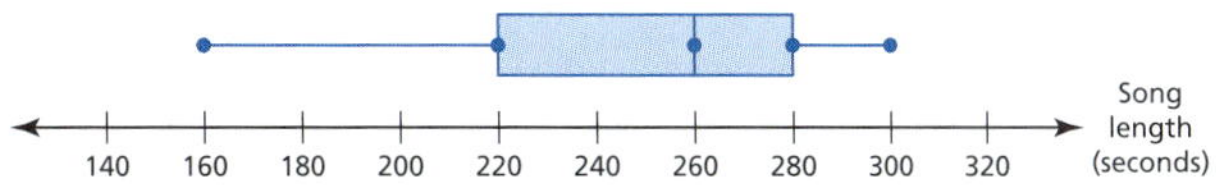

a. Find and interpret the range of the data.

b. Describe the distribution of the data.

c. Find and interpret the interquartile range of the data.

d. Are the data more spread out below Q1 or above Q3? Explain.

SOLUTION

a. The least value is 160. The greatest value is 300.

▶ So, the range is $300 - 160 = 140$ seconds. This means that the song lengths vary by no more than 140 seconds.

b. Each whisker represents 25% of the data. The box represents 50% of the data. So,

- 25% of the song lengths are between 160 and 220 seconds.
- 50% of the song lengths are between 220 and 280 seconds.
- 25% of the song lengths are between 280 and 300 seconds.

c. $\text{IQR} = \text{Q3} - \text{Q1} = 280 - 220 = 60$

▶ So, the interquartile range is 60 seconds. This means that the middle half of the song lengths vary by no more than 60 seconds.

d. The left whisker is longer than the right whisker.

▶ So, the data below Q1 are more spread out than data above Q3.

Monitoring Progress

Help in English and Spanish at *BigIdeasMath.com*

Use the box-and-whisker plot in Example 1.

2. Find and interpret the range and interquartile range of the data.

3. Describe the distribution of the data.

Laurie's Notes Teacher Actions

- Define *interquartile range*. In the sample shown, the lengths of the two parts of the box indicate that the data in the box are less spread out on the right side of the median than on the left side of the median. This interpretation is important for students to understand.

? "Do you know from the box-and-whisker plot how many songs the band played at the concert?" no

? "What does the length of each part of the box-and-whisker plot tell you about the distribution of the data?" *Sample answer:* 50% of the songs were between 260 and 300 seconds; 25% were 220 seconds or less.

English Language Learners

Graphic Organizer

Provide small groups with different data sets and a copy of the organizer below. Have each group find the measures for their data set and present their work to the class.

Data set:	
Least value	
Greatest value	
Median	
First quartile	
Third quartile	
Range	
Interquartile range	

Extra Example 2

The box-and-whisker plot represents the time (in hours) spent on a project by students in a history class.

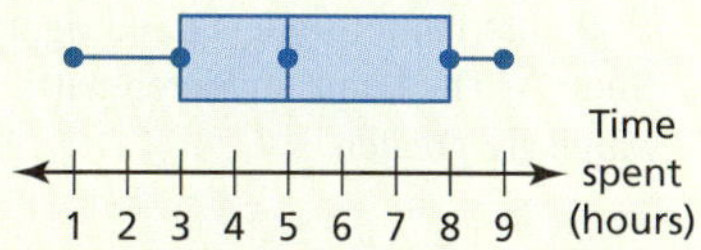

a. Find and interpret the range of the data. The range is $9 - 1 = 8$ hours. This means that the time spent varies by no more than 8 hours.

b. Describe the distribution of the data. 25% of the students spent between 1 and 3 hours on the project. 50% of the students spent between 3 and 8 hours on the project. 25% of the students spent between 8 and 9 hours on the project.

c. Find and interpret the interquartile range of the data. The interquartile range is 5 hours. This means that the middle half of the time spent on the project varies by no more than 5 hours.

d. Are the data more spread out below Q1 or above Q3? Explain. The data below Q1 are more spread out than the data above Q3. The left whisker is longer than the right whisker.

MONITORING PROGRESS ANSWERS

2. 12, 6; The ages vary by no more than 12 years and the middle half of the ages vary by no more than 6 years.

3. See Additional Answer.

Extra Example 3

The double box-and-whisker plot represents the lengths (in yards) of field goals made by two football kickers.

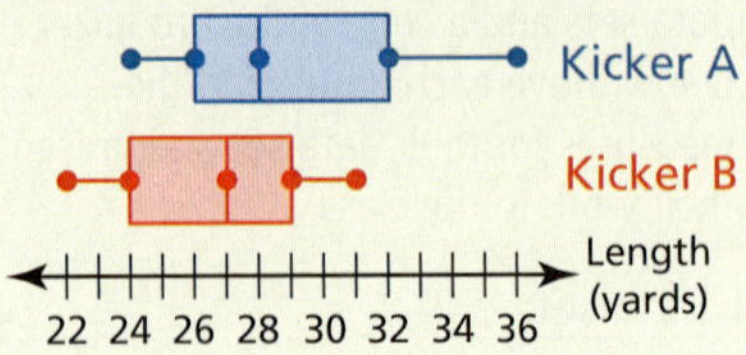

a. Identify the shape of each distribution. The distribution for Kicker A is skewed right, and the distribution for Kicker B is symmetric.

b. Which lengths are more spread out? Explain. The lengths for Kicker A are more spread out. The range and interquartile range of the lengths for Kicker A are greater than the range and interquartile range for Kicker B.

MONITORING PROGRESS ANSWER

4. Both distributions are skewed right; Shop A; The range and interquartile range are greater.

Using Box-and-Whisker Plots to Compare Data Sets

A box-and-whisker plot shows the shape of a distribution.

STUDY TIP

If you can draw a line through the median of a box-and-whisker plot, and each side is approximately a mirror image of the other, then the distribution is symmetric.

Core Concept

Shapes of Box-and-Whisker Plots

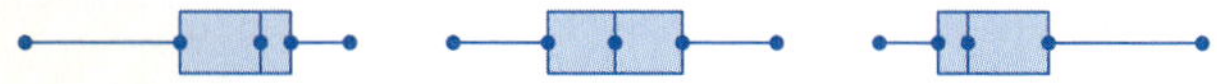

Skewed left	Symmetric	Skewed right
• The left whisker is longer than the right whisker. • Most of the data are on the right side of the plot.	• The whiskers are about the same length. • The median is in the middle of the plot.	• The right whisker is longer than the left whisker. • Most of the data are on the left side of the plot.

EXAMPLE 3 Comparing Box-and-Whisker Plots

The double box-and-whisker plot represents the test scores for your class and your friend's class.

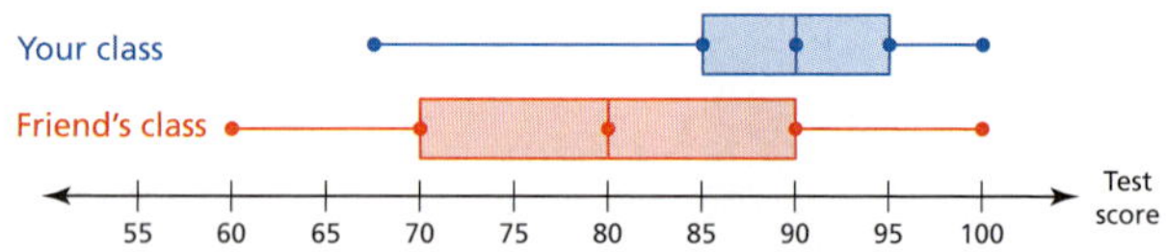

a. Identify the shape of each distribution.

b. Which test scores are more spread out? Explain.

SOLUTION

a. For your class, the left whisker is longer than the right whisker, and most of the data are on the right side of the plot. For your friend's class, the whisker lengths are equal, and the median is in the middle of the plot.

▶ So, the distribution for your class is skewed left, and the distribution for your friend's class is symmetric.

b. The range and interquartile range of the test scores in your friend's class are greater than the range and interquartile range in your class.

▶ So, the test scores in your friend's class are more spread out.

Monitoring Progress

Help in English and Spanish at *BigIdeasMath.com*

4. The double box-and-whisker plot represents the surfboard prices at Shop A and Shop B. Identify the shape of each distribution. Which shop's prices are more spread out? Explain.

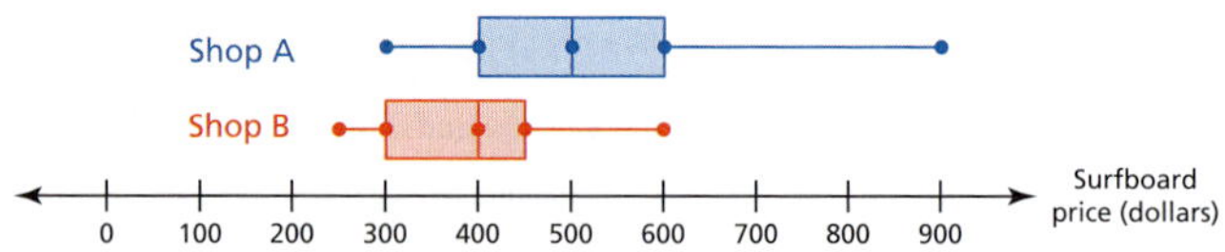

Laurie's Notes Teacher Actions

- In part (a), a student might comment that a low test score in "your class" could be the reason for it being skewed. Without the outlier, it could be a symmetric graph.
- Demonstrate using data in a graphing calculator. See the *Teaching Strategy* on page T-592.
- **Reason Abstractly and Quantitatively:** Supplement the discussion in part (b) with quantitative comparisons: 50% of your class scored as well as the top 25% of your friend's class.

Closure

- **Exit Ticket:** What are the five key values that are graphed in a box-and-whisker plot? least value, first quartile, median, third quartile, and greatest value How does an outlier affect the plot? One of the whiskers of the plot is longer.

11.2 Exercises

Dynamic Solutions available at *BigIdeasMath.com*

Vocabulary and Core Concept Check

1. **WRITING** Describe how to find the first quartile of a data set.

2. **DIFFERENT WORDS, SAME QUESTION** Consider the box-and-whisker plot shown. Which is different? Find "both" answers.

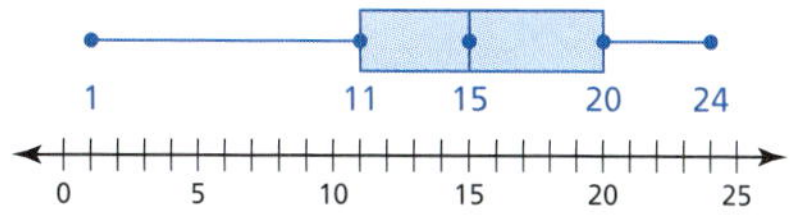

Find the interquartile range of the data.

Find the range of the middle half of the data.

Find the difference of the greatest value and the least value of the data set.

Find the difference of the third quartile and the first quartile.

Monitoring Progress and Modeling with Mathematics

In Exercises 3–8, use the box-and-whisker plot to find the given measure.

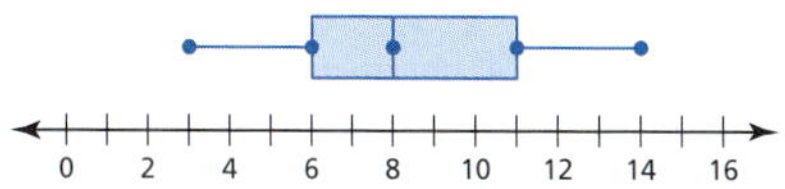

3. least value
4. greatest value
5. third quartile
6. first quartile
7. median
8. range

In Exercises 9–12, make a box-and-whisker plot that represents the data. *(See Example 1.)*

9. Hours of television watched: 0, 3, 4, 5, 2, 4, 6, 5

10. Cat lengths (in inches): 16, 18, 20, 25, 17, 22, 23, 21

11. Elevations (in feet): −2, 0, 5, −4, 1, −3, 2, 0, 2, −3, 6

12. MP3 player prices (in dollars): 124, 95, 105, 110, 95, 124, 300, 190, 114

13. **ANALYZING DATA** The dot plot represents the numbers of hours students spent studying for an exam. Make a box-and-whisker plot that represents the data.

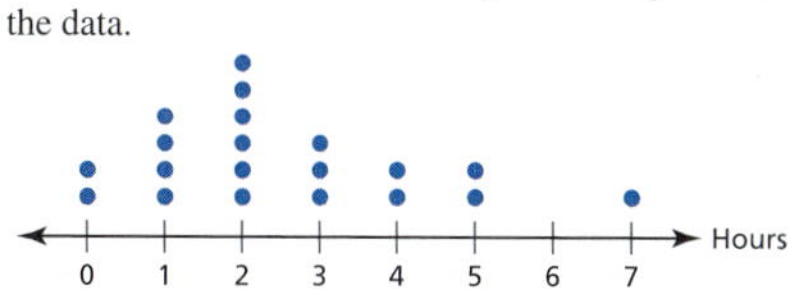

14. **ANALYZING DATA** The stem-and-leaf plot represents the lengths (in inches) of the fish caught on a fishing trip. Make a box-and-whisker plot that represents the data.

Stem	Leaf
0	6 7 8 8 9
1	0 0 2 2 3 4 4 7
2	1 2

Key: 1 | 0 = 10 inches

15. **ANALYZING DATA** The box-and-whisker plot represents the prices (in dollars) of the entrées at a restaurant. *(See Example 2.)*

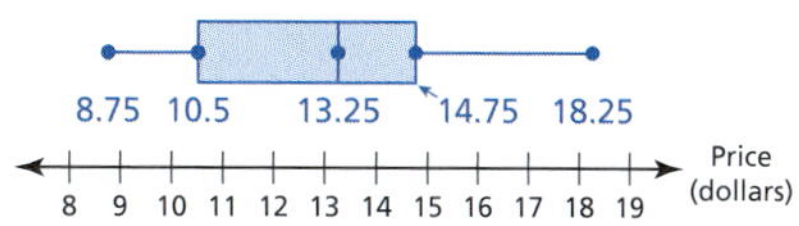

a. Find and interpret the range of the data.

b. Describe the distribution of the data.

c. Find and interpret the interquartile range of the data.

d. Are the data more spread out below Q1 or above Q3? Explain.

Assignment Guide and Homework Check

ASSIGNMENT

Basic: 1, 2, 3–17 odd, 18, 20, 24–27

Average: 1, 2–16 even, 17–21, 24–27

Advanced: 1, 2, 10–16 even, 17–27

HOMEWORK CHECK

Basic: 9, 13, 15, 17

Average: 10, 14, 16, 17

Advanced: 12, 14, 16, 17

ANSWERS

1. Order the data and find the median of the lower half.
2. Find the difference of the greatest value and the least value of the data set. 23; 9
3. 3
4. 14
5. 11
6. 6
7. 8
8. 11
9. 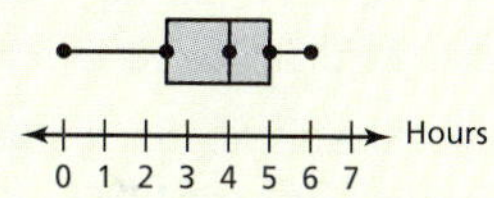

10. 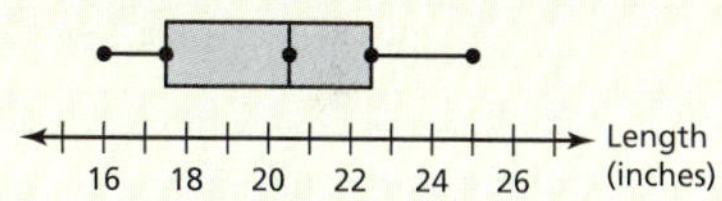

11. 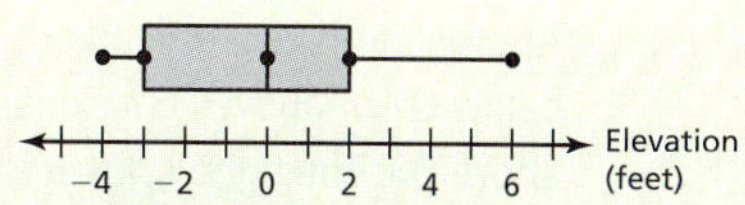

12. 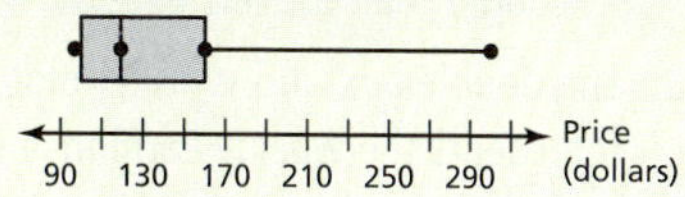

13. 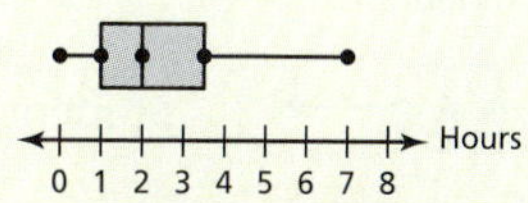

14. 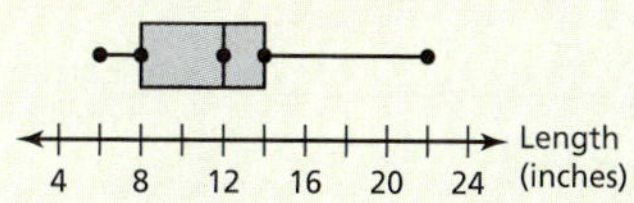

15. a. 9.5; The prices vary by no more than $9.50.

b. 25% of the prices are between $8.75 and $10.50, 50% of the prices are between $10.50 and $14.75, and 25% of the prices are between $14.75 and $18.25.

c. 4.25; The middle half of the prices vary by no more than $4.25.

d. above Q3; the whisker is longer

Dynamic Teaching Tools

- Dynamic Assessment & Progress Monitoring Tool
- Interactive Whiteboard Lesson Library
- Dynamic Classroom with Dynamic Investigations

ANSWERS

16–27. See Additional Answers.

Mini-Assessment

1. Make a box-and-whisker plot that represents the ages of the employees at a yogurt shop.
18, 23, 25, 34, 26, 22, 42, 18

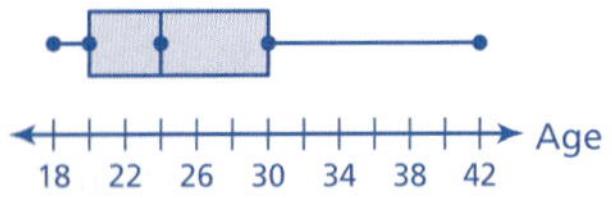

2. Use the box-and-whisker plot you made in Exercise 1.
   a. Find and interpret the range and interquartile range of the data. range: 24 years; The ages vary by no more than 24 years; interquartile range: 10 years; The middle half of the ages vary by no more than 10 years.
   b. Describe the distribution of the data. 25% between 18 and 20 years old; 50% between 20 and 30 years old; 25% between 30 and 42 years old
   c. Are the data more spread out below Q1 or above Q3? Explain. above Q3; The right whisker is longer than the left whisker.
3. The double box-and-whisker plot represents the weekly earnings of two students.

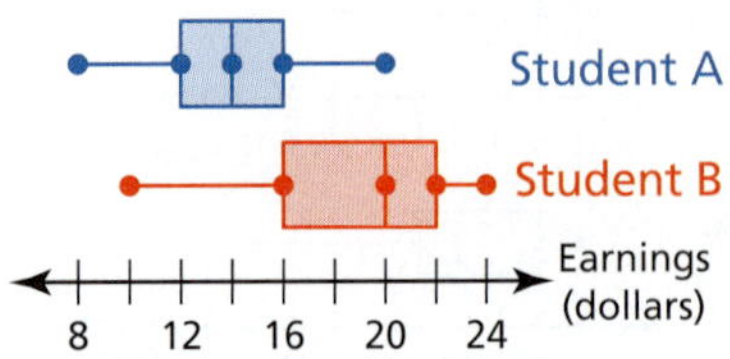

   a. Identify the shape of each distribution. Student A: symmetric; Student B: skewed left
   b. Whose earnings are more spread out? Explain. Student B; The range and interquartile range of the earnings of Student B are greater than those for Student A.

16. ANALYZING DATA A baseball player scores 101 runs in a season. The box-and-whisker plot represents the numbers of runs the player scores against different opposing teams.

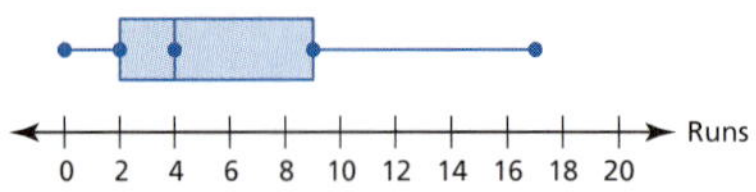

a. Find and interpret the range and interquartile range of the data.

b. Describe the distribution of the data.

c. Are the data more spread out between Q1 and Q2 or between Q2 and Q3? Explain.

17. ANALYZING DATA The double box-and-whisker plot represents the monthly car sales for a year for two sales representatives. *(See Example 3.)*

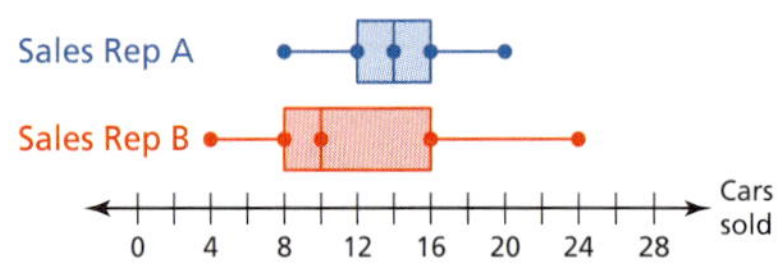

a. Identify the shape of each distribution.

b. Which representative's sales are more spread out? Explain.

c. Which representative had the single worst sales month during the year? Explain.

18. ERROR ANALYSIS Describe and correct the error in describing the box-and-whisker plot.

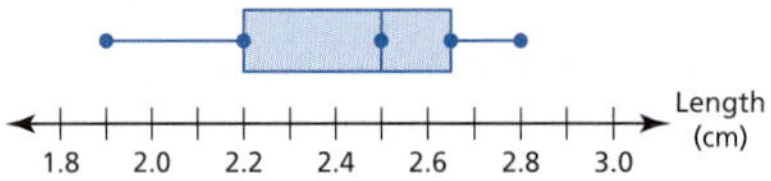

✗ The distribution is skewed left. So, most of the data are on the left side of the plot.

19. WRITING Given the numbers 36 and 12, identify which number is the range and which number is the interquartile range of a data set. Explain.

20. HOW DO YOU SEE IT? The box-and-whisker plot represents a data set. Determine whether each statement is always true. Explain your reasoning.

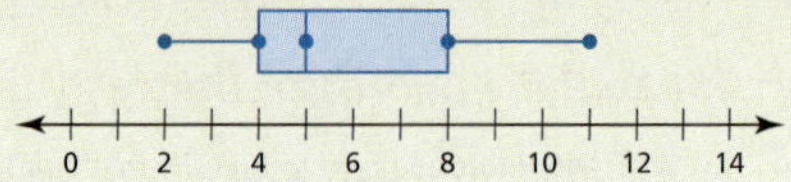

a. The data set contains the value 11.

b. The data set contains the value 6.

c. The distribution is skewed right.

d. The mean of the data is 5.

21. ANALYZING DATA The double box-and-whisker plot represents the battery lives (in hours) of two brands of cell phones.

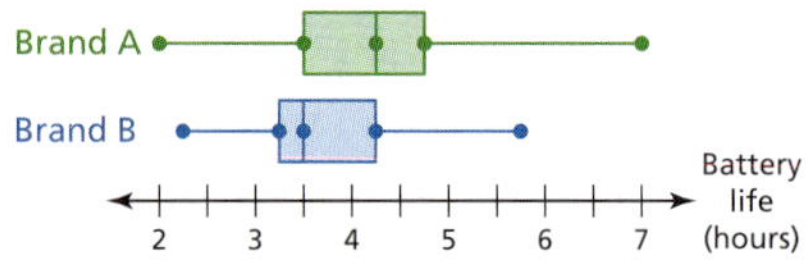

a. Identify the shape of each distribution.

b. What is the range of the upper 75% of each brand?

c. Compare the interquartile ranges of the two data sets.

d. Which brand do you think has a greater standard deviation? Explain.

e. You need a cell phone that has a battery life of more than 3.5 hours most of the time. Which brand should you buy? Explain.

22. THOUGHT PROVOKING Create a data set that can be represented by the box-and-whisker plot shown. Justify your answer.

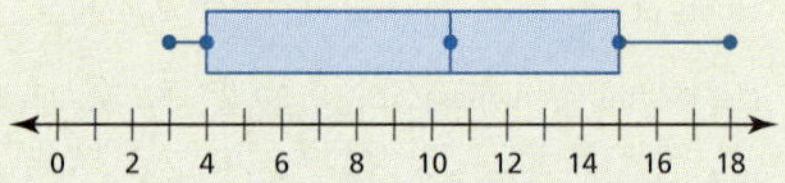

23. CRITICAL THINKING Two data sets have the same median, the same interquartile range, and the same range. Is it possible for the box-and-whisker plots of the data sets to be different? Justify your answer.

Maintaining Mathematical Proficiency

Reviewing what you learned in previous grades and lessons

Use zeros to graph the function. *(Section 8.5)*

24. $f(x) = -2(x + 9)(x - 3)$

25. $y = 3(x - 5)(x + 5)$

26. $y = 4x^2 - 16x - 48$

27. $h(x) = -x^2 + 5x + 14$

If students need help...	*If students got it...*
Resources by Chapter • Practice A and Practice B • Puzzle Time	Resources by Chapter • Enrichment and Extension • Cumulative Review
Student Journal • Practice	Start the *next* Section
Differentiating the Lesson Skills Review Handbook	

Laurie's Notes

Overview of Section 11.3

Introduction

- This lesson introduces language to describe data displays in terms of their shape. The shapes are then used to choose appropriate measures (center and variability) to describe them.
- Prior skills (constructing histograms and box-and-whisker plots, calculating measures of center and measures of variability) are then used to compare two data distributions.

Teaching Strategy

- Over the course of the year, there are many data sets we use that are fairly large. It can be challenging for students to enter data quickly and correctly in their calculators. For that reason, and because I want to use valuable class time wisely, I store data sets as programs.
- The sole result of executing (running) the program is that the data set is stored in a list in the calculator. Once in the list, the data can be plotted and/or analyzed.
- All the large data sets I use are named by page and example number. The speeding data on page 601 in Example 2 would be named P601EX2.
- Programs are sent to student calculators using a USB cable.
- Steps for writing the program are given. When the program is executed (i.e., in class three weeks later), these data are written to list L_1.
  1. Enter the data in list L_1 of your calculator. (See Figure 1.)
  2. Create a new program and name it P601EX2.
  3. To copy the data from the list to the program, press RCL , L1 and ENTER. (See Figure 2.)
  4. Press STO and L1. (See Figure 3.) Save and exit the program.

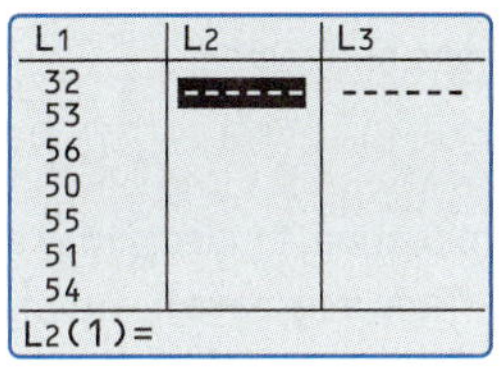

Figure 1

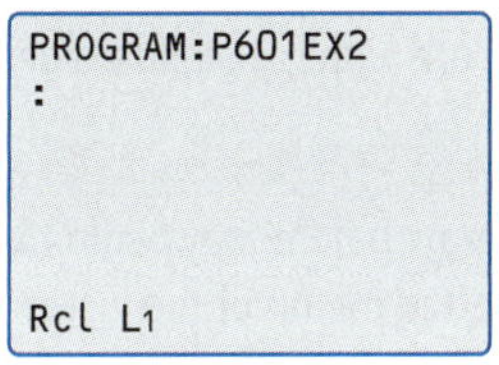

Figure 2

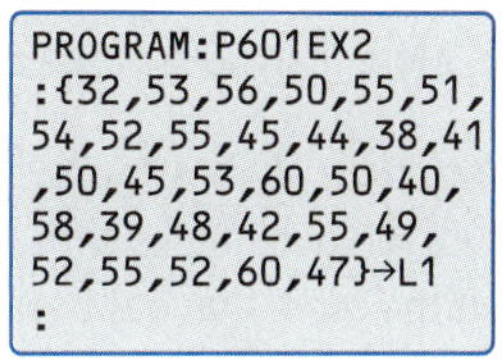

Figure 3

- There is much more that can be done with programming. This is a simple teaching strategy meant to save time over the course of the year.

Pacing Suggestion

- The explorations introduce students to the practice of analyzing a graphical display to see how the data are dispersed on either side of the mean. Have students complete the explorations, and transition to the formal lesson.

Dynamic Teaching Tools

- Dynamic Assessment & Progress Monitoring Tool
- Lesson Planning Tool
- Interactive Whiteboard Lesson Library
- Dynamic Classroom with Dynamic Investigations

What Your Students Will Learn

- Describe the shapes of data distributions and choose appropriate measures of center and variation to describe them.
- Compare two data distributions using measures of center and variation.

Laurie's Notes

Exploration

Motivate

- Ask students whether they know what a militia is. You could give them the hint that during the American Revolutionary War, the Minutemen were members of a militia. A militia refers to a military force made up of citizens that protect their community or town in times of emergency.
- Tell students about the Scottish Militia Bill, which was passed by the Houses of Parliament in Great Britain in the early 1700s. The purpose of this bill was to help create uniformity in governing the militias across Great Britain. This bill was not passed into law by Queen Anne, because she feared the militia in Scotland would not be loyal. This would mark the last time a British ruler would veto a bill passed by both Houses of Parliament.
- Data about the Scottish Militia is the subject of today's exploration.

Exploration 1

- The histogram in this exploration is shaded in a fashion that will help students answer the questions posed. Students may notice and comment on the symmetry of the graph.

? "In this exploration, what does it mean to lie within 1 standard deviation of the mean?" It means a chest size is within 2 inches on either side of the mean of 40 inches. So, it is between 38 inches and 42 inches.

- A calculator is helpful in calculating the percents. Students who read carefully estimate that there are about 5700 men represented in the histogram. The percents are cumulative. So, the men who are within 1 standard deviation are also within 2 standard deviations of the mean, and so on.
- **FYI:** More than 99% of the data are within 3 standard deviations of the mean.

Exploration 2

- Discuss the two graphs with students before they begin the exploration. Students may ask about the curve that has been plotted on top of the histogram. They may have heard of a "bell-shaped curve." It is fine to suggest that the curve is tracing the basic shape of the distribution of heights. It is smooth versus the line segment outline of the actual graph. The bell-shaped curve will be referred to in Question 4.
- The frequencies on the vertical axis will help students estimate the size of the populations plotted.
- When students have finished, have them share their reasoning for part (a). Students should discuss the symmetry of the distributions and what that means in terms of the distribution of the data.
- Discuss the shape of the distribution and what it means in the real-life context. The mean and standard deviation for males is greater than the mean and standard deviation for females.

Communicate Your Answer

- Make sure students understand the connection between a bell shape and the distribution of the data. In the lesson that follows, students will see that not all distributions are symmetric or bell-shaped.

? "Can a distribution be symmetric and not be bell-shaped?" yes; The distribution could be U-shaped or uniform.

Connecting to Next Step

- In the lesson, students will describe the shapes of distributions, some of which may be symmetric.

11.3 Shapes of Distributions

Essential Question

How can you use a histogram to characterize the basic shape of a distribution?

EXPLORATION 1 Analyzing a Famous Symmetric Distribution

Work with a partner. A famous data set was collected in Scotland in the mid-1800s. It contains the chest sizes, measured in inches, of 5738 men in the Scottish Militia. Estimate the percent of the chest sizes that lie within (a) 1 standard deviation of the mean, (b) 2 standard deviations of the mean, and (c) 3 standard deviations of the mean. Explain your reasoning.

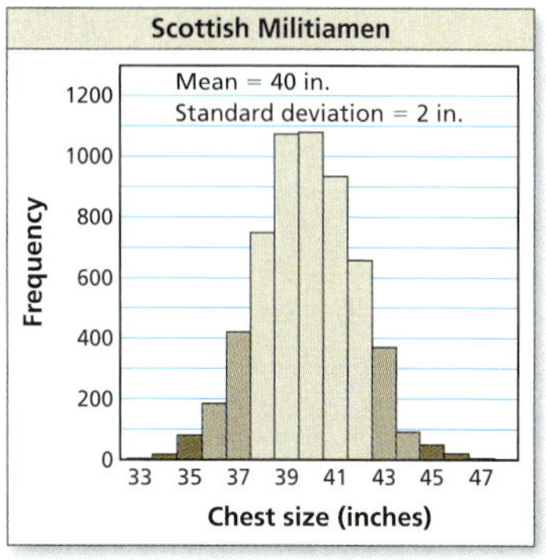

EXPLORATION 2 Comparing Two Symmetric Distributions

Work with a partner. The graphs show the distributions of the heights of 250 adult American males and 250 adult American females.

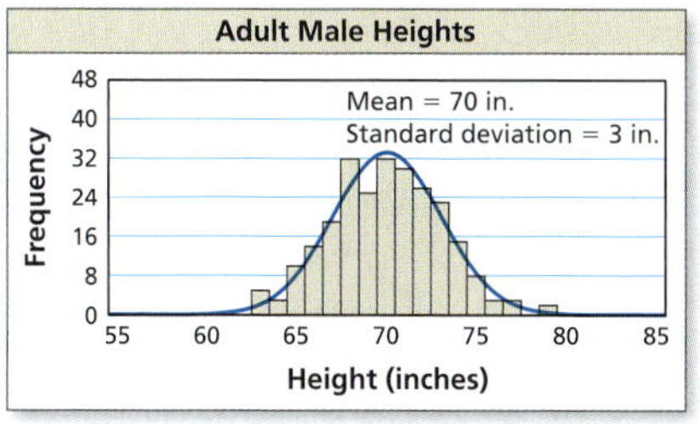

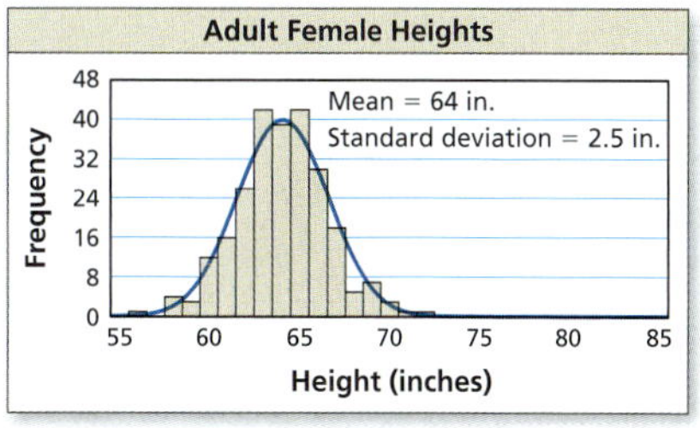

a. Which data set has a smaller standard deviation? Explain what this means in the context of the problem.

b. Estimate the percent of male heights between 67 inches and 73 inches.

ATTENDING TO PRECISION

To be proficient in math, you need to express numerical answers with a level of precision appropriate for the problem's context.

Communicate Your Answer

3. How can you use a histogram to characterize the basic shape of a distribution?

4. All three distributions in Explorations 1 and 2 are roughly symmetric. The histograms are called "bell-shaped."

a. What are the characteristics of a symmetric distribution?

b. Why is a symmetric distribution called "bell-shaped?"

c. Give two other real-life examples of symmetric distributions.

Dynamic Teaching Tools

- Dynamic Assessment & Progress Monitoring Tool
- Lesson Planning Tool
- Interactive Whiteboard Lesson Library
- Dynamic Classroom with Dynamic Investigations

ANSWERS

1. a. *Sample answer:* 77.5%; The green bars represent the data within 1 standard deviation of the mean.

b. *Sample answer:* 96.7%; The green and blue bars represent the data within 2 standard deviations of the mean.

c. *Sample answer:* 99.3%; The green, blue, and red bars represent the data within 3 standard deviations of the mean.

2. a. Adult female heights; The heights of adult females do not vary as far from the mean.

b. *Sample answer:* 75%

3. *Sample answer:* Draw a curve through the tops of the bars.

4. a. Most of the data is clustered around the mean in the center of the data.

b. The curve connecting the tops of the bars looks like a bell.

c. *Sample answer:* standardized test scores; length of daylight each day over a year

Differentiated Instruction

Visual/Kinesthetic

Form small groups of students. Give each group three large index cards and have students write one of the following phrases on each card: *skewed left*, *symmetric*, and *skewed right*. One at a time, show students histograms that illustrate each of the three distributions. Have each group discuss the histogram and then hold up the card with the correct description of the distribution.

Extra Example 1

The frequency table shows the number of hours that students in your grade spent on homework last week. Display the data in a histogram. Describe the shape of the distribution.

Number of hours	Frequency
0–1	5
2–3	25
4–5	20
6–7	18
8–9	12
10–11	6
12–13	2

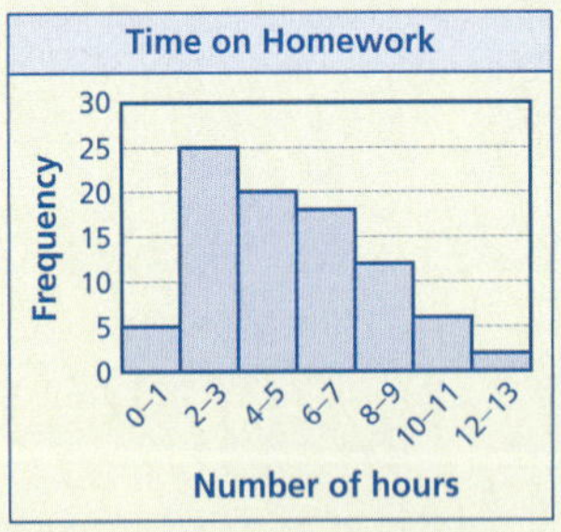

The distribution is skewed right.

MONITORING PROGRESS ANSWER

1. See Additional Answers.

11.3 Lesson

Core Vocabulary

Previous
histogram
frequency table

What You Will Learn

- Describe the shapes of data distributions.
- Use the shapes of data distributions to choose appropriate measures.
- Compare data distributions.

Describing the Shapes of Data Distributions

Recall that a histogram is a bar graph that shows the frequency of data values in intervals of the same size. A histogram is another useful data display that shows the shape of a distribution.

STUDY TIP

If all the bars of a histogram are about the same height, then the distribution is a *flat*, or *uniform*, distribution. A uniform distribution is also symmetric.

Core Concept

Symmetric and Skewed Distributions

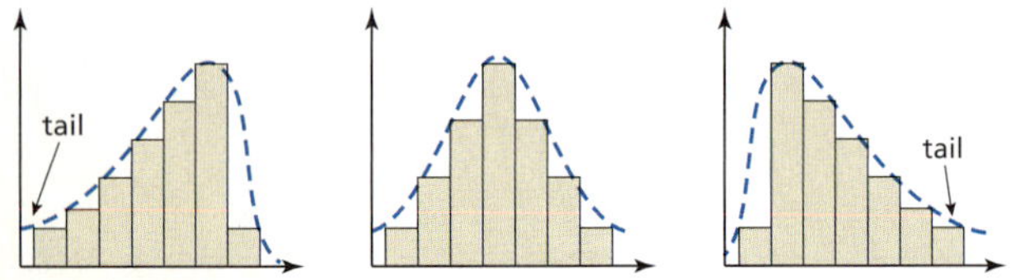

Skewed left
- The "tail" of the graph extends to the left.
- Most of the data are on the right.

Symmetric
- The data on the right of the distribution are approximately a mirror image of the data on the left of the distribution.

Skewed right
- The "tail" of the graph extends to the right.
- Most of the data are on the left.

EXAMPLE 1 Describing the Shape of a Distribution

The frequency table shows the numbers of raffle tickets sold by students in your grade. Display the data in a histogram. Describe the shape of the distribution.

Number of tickets sold	Frequency
1–8	5
9–16	9
17–24	16
25–32	25
33–40	20
41–48	8
49–56	7

SOLUTION

Step 1 Draw and label the axes.

Step 2 Draw a bar to represent the frequency of each interval.

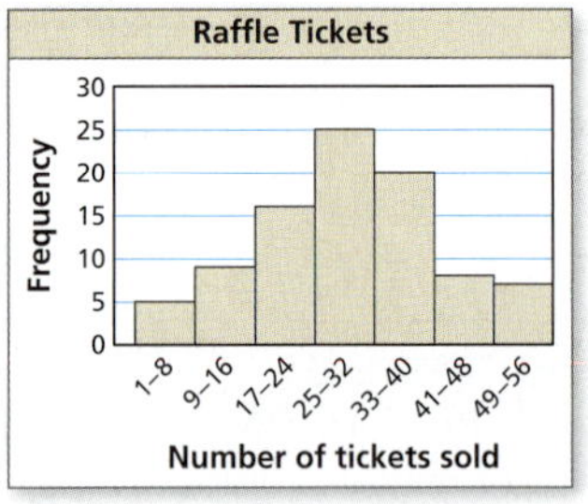

The data on the right of the distribution are approximately a mirror image of the data on the left of the distribution.

► So, the distribution is symmetric.

Monitoring Progress

Help in English and Spanish at *BigIdeasMath.com*

1. The frequency table shows the numbers of pounds of aluminum cans collected by classes for a fundraiser. Display the data in a histogram. Describe the shape of the distribution.

Number of pounds	Frequency
1–10	7
11–20	8
21–30	10
31–40	16
41–50	34
51–60	15

Laurie's Notes Teacher Actions

- Write the *Core Concept*. Explain that a dotted line can be drawn to help identify the greatest frequency and to help highlight the basic shape of a distribution.

? "Do all distributions fall into one of these types of distributions?" No. There are many different types of distributions.

- **Extension:** In Example 1, ask students to estimate the mean of the distribution even though the actual data values are not known. You could also ask about the maximum range (56 − 1 = 55).

Choosing Appropriate Measures

Use the shape of a distribution to choose the most appropriate measure of center and measure of variation to describe the data set.

STUDY TIP

When a distribution is symmetric, the mean and median are about the same. When a distribution is skewed, the mean will be in the direction in which the distribution is skewed while the median will be less affected.

Core Concept

Choosing Appropriate Measures

When a data distribution is symmetric,

- use the mean to describe the center and
- use the standard deviation to describe the variation.

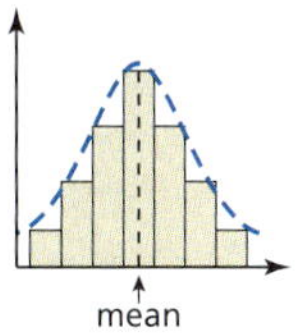

When a data distribution is skewed,

- use the median to describe the center and
- use the five-number summary to describe the variation.

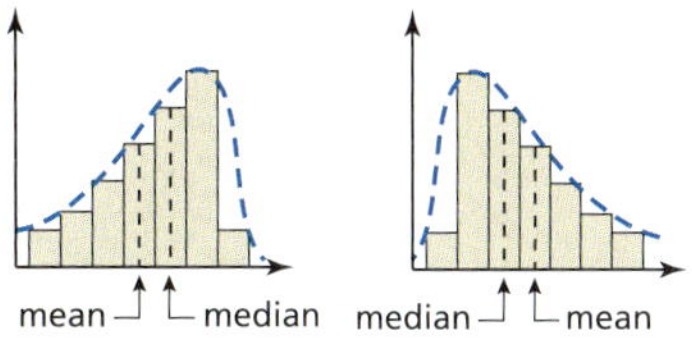

EXAMPLE 2 Choosing Appropriate Measures

Speeds (mi/h)		
32	44	39
53	38	48
56	41	42
50	50	55
55	45	49
51	53	52
54	60	55
52	50	52
55	40	60
45	58	47

A police officer measures the speeds (in miles per hour) of 30 motorists. The results are shown in the table at the left. (a) Display the data in a histogram using six intervals beginning with 31–35. (b) Which measures of center and variation best represent the data? (c) The speed limit is 45 miles per hour. How would you interpret these results?

SOLUTION

a. Make a frequency table using the described intervals. Then use the frequency table to make a histogram.

Speed (mi/h)	Frequency
31–35	1
36–40	3
41–45	5
46–50	6
51–55	11
56–60	4

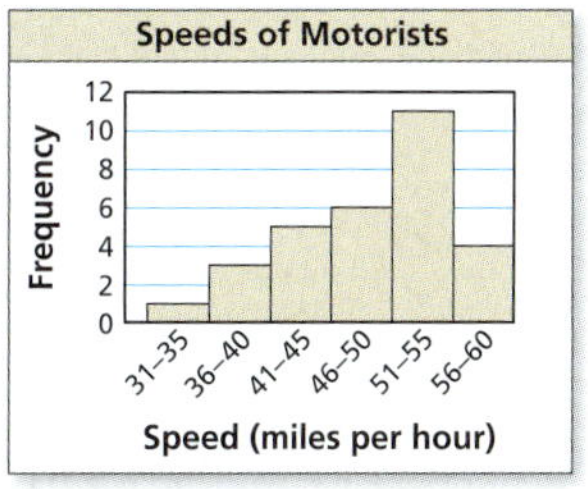

b. Because most of the data are on the right and the tail of the graph extends to the left, the distribution is skewed left. So, use the median to describe the center and the five-number summary to describe the variation.

c. Using the frequency table and the histogram, you can see that most of the speeds are more than 45 miles per hour. So, most of the motorists were speeding.

Monitoring Progress

Help in English and Spanish at *BigIdeasMath.com*

Email Attachments Sent				
74	105	98	68	64
85	75	60	48	51
65	55	58	45	38
64	52	65	30	70
72	5	45	77	83
42	25	95	16	120

2. You record the numbers of email attachments sent by 30 employees of a company in 1 week. Your results are shown in the table. (a) Display the data in a histogram using six intervals beginning with 1–20. (b) Which measures of center and variation best represent the data? Explain.

Laurie's Notes Teacher Actions

- Write the *Core Concept,* and explain that the shape of a distribution can help you decide the most appropriate measure of center when describing the data set. Discuss the two shapes (symmetric, skewed) and the appropriate measures.
- Example 2 reviews making a frequency table. Make the connection to the five-number summary for the box-and-whisker plot that gave a visual representation of the spread of the data.
- The histogram is skewed to the left, so motorists were speeding!

English Language Learners

Pair Activity

Have students work in pairs to make histograms. Have one student put in the title, labels, intervals, and scale. Have the other student draw the bars. Have partners discuss the shape of the distribution and determine the most appropriate measure of center and measure of variation to describe the data set.

Extra Example 2

A grocery store cashes checks for customers. The amounts cashed for one week are shown in the table.

Check Amounts (dollars)			
65	15	50	65
40	25	35	60
10	45	60	30
100	125	75	25
75	40	90	75
75	5	50	50
35	30	140	60

a. Display the data in a histogram using six intervals beginning with 1–25.

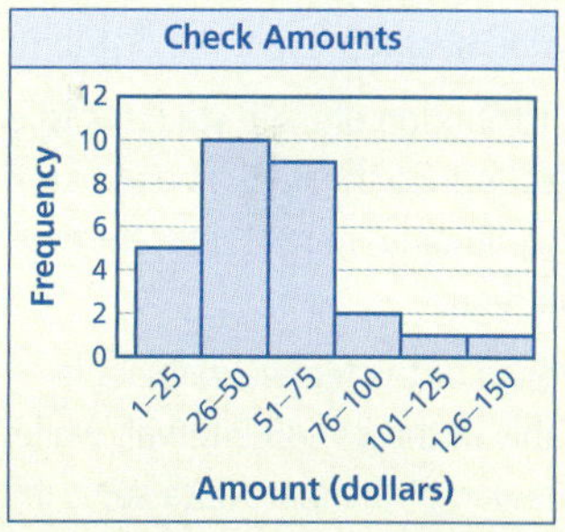

b. Which measures of center and variation best represent the data? *The distribution is skewed right, so use the median to describe the center and the five-number summary to describe the variation.*

c. The grocery store charges a fee for cashing checks that are greater than \$75. How would you interpret these results? *Most customers do not want to pay the fee.*

MONITORING PROGRESS ANSWER

2. See Additional Answers.

Extra Example 3

The double histogram shows the distributions of the numbers of minutes of exercise done by a group of female students and a group of male students during one day. Compare the distributions using their shapes and appropriate measures of center and variation.

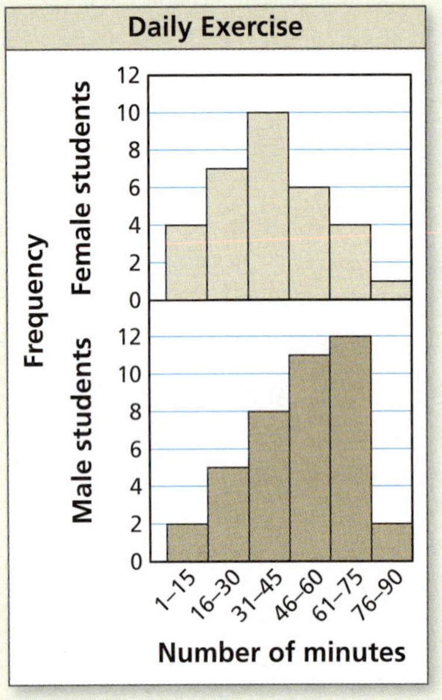

The distribution of the female students is symmetric, so the mean and standard deviation best represent the distribution.

The distribution of male students is skewed left, so the median and five-number summary best represent the distribution.

The mean of the female data set is probably in the 31–45 interval, while the median of the male data set is in the 46–60 interval. So, a male student is more likely to spend more time exercising than a female student.

The data for the female students is more variable than the data for the male students. This means that the time spent exercising tends to differ more from one female student to the next.

MONITORING PROGRESS ANSWER

3. Company employees are more likely to have more years of experience and the number of years of experience tends to differ more from one company employee to the next.

Comparing Data Distributions

EXAMPLE 3 **Comparing Data Distributions**

Emoticons are graphic symbols that represent facial expressions. They are used to convey a person's mood in a text message. The double histogram shows the distributions of emoticon messages sent by a group of female students and a group of male students during 1 week. Compare the distributions using their shapes and appropriate measures of center and variation.

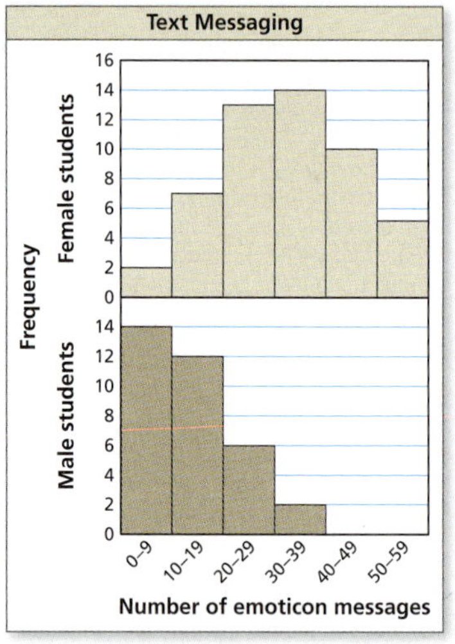

SOLUTION

Because the data on the right of the distribution for the female students are approximately a mirror image of the data on the left of the distribution, the distribution is symmetric. So, the mean and standard deviation best represent the distribution for female students.

Because most of the data are on the left of the distribution for the male students and the tail of the graph extends to the right, the distribution is skewed right. So, the median and five-number summary best represent the distribution for male students.

The mean of the female data set is probably in the 30–39 interval, while the median of the male data set is in the 10–19 interval. So, a typical female student is much more likely to use emoticons than a typical male student.

The data for the female students is more variable than the data for the male students. This means that the use of emoticons tends to differ more from one female student to the next.

Monitoring Progress

Help in English and Spanish at *BigIdeasMath.com*

3. Compare the distributions using their shapes and appropriate measures of center and variation.

Laurie's Notes — Teacher Actions

- **Turn and Talk:** "Describe the scales used for each histogram and what a piece of data represents." Check that students understand that the horizontal axis is the same for both histograms. A piece of data represents the number of emoticons sent by one student during one week.
- **Reason Abstractly and Quantitatively:** As students compare the histograms, be sure that they discuss the measures of center and the variation of each graph. Statements such as "Females send more emoticons than males" are too simplistic.

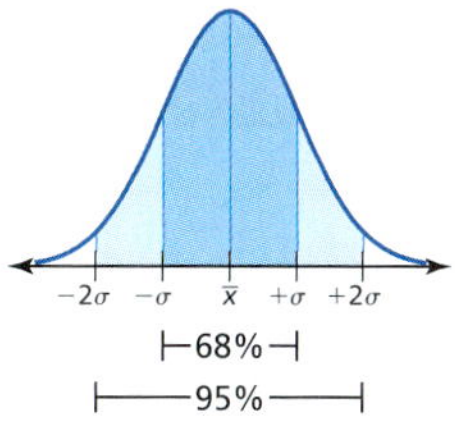

Many real-life data sets have distributions that are bell-shaped and approximately symmetric about the mean. In a future course, you will study this type of distribution in detail. For now, the following rules can help you see how valuable the standard deviation can be as a measure of variation.

- About 68% of the data lie within 1 standard deviation of the mean.
- About 95% of the data lie within 2 standard deviations of the mean.
- Data values that are more than 2 standard deviations from the mean are considered unusual.

Because the data are symmetric, you can deduce that 34% of the data lie within 1 standard deviation to the left of the mean, and 34% of the data lie within 1 standard deviation to the right of the mean.

EXAMPLE 4 **Comparing Data Distributions**

The table shows the results of a survey that asked men and women how many pairs of shoes they own.

a. Make a double box-and-whisker plot that represents the data. Describe the shape of each distribution.

b. Compare the number of pairs of shoes owned by men to the number of pairs of shoes owned by women.

c. About how many of the women surveyed would you expect to own between 10 and 18 pairs of shoes?

	Men	Women
Survey size	35	40
Minimum	2	5
Maximum	17	24
1st Quartile	5	12
Median	7	14
3rd Quartile	10	17
Mean	8	14
Standard deviation	3	4

SOLUTION

a.

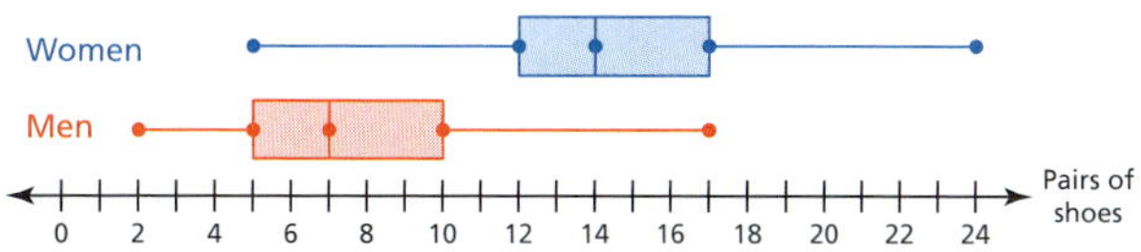

The distribution for men is skewed right, and the distribution for women is symmetric.

b. The centers and spreads of the two data sets are quite different from each other. The mean for women is twice the median for men, and there is more variability in the number of pairs of shoes owned by women.

c. Assuming the symmetric distribution is bell-shaped, you know about 68% of the data lie within 1 standard deviation of the mean. Because the mean is 14 and the standard deviation is 4, the interval from 10 to 18 represents about 68% of the data. So, you would expect about $0.68 \cdot 40 \approx 27$ of the women surveyed to own between 10 and 18 pairs of shoes.

Monitoring Progress

Help in English and Spanish at *BigIdeasMath.com*

4. Why is the mean greater than the median for the men?
5. If 50 more women are surveyed, about how many more would you expect to own between 10 and 18 pairs of shoes?

Extra Example 4

The table shows the results of a survey that asked men and women how many pairs of socks they own.

	Men	Women
Survey size	30	25
Minimum	4	8
Maximum	30	35
1st Quartile	8	12
Median	15	15
3rd Quartile	25	23
Mean	16	18
Standard deviation	10	8

a. Make a double box-and-whisker plot that represents the data. Describe the shape of each distribution.

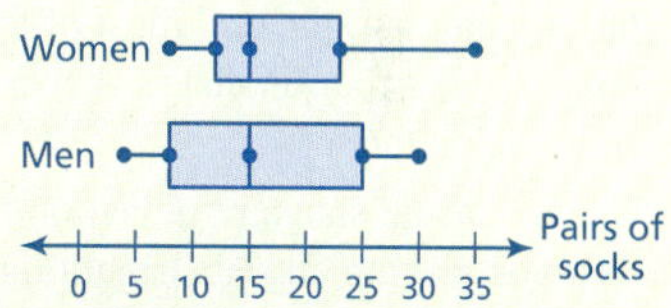

The distribution for men is symmetric, and the distribution for women is skewed right.

b. Compare the number of pairs of socks owned by men to the number of pairs of socks owned by women.
The centers and spreads of the two data sets are similar to each other. The median for women is just 1 less than the mean for men, and there is slightly more variability in the number of pairs of socks owned by men.

c. About how many of the men surveyed would you expect to own between 6 and 26 pairs of socks?
about 20 men; Assuming the symmetric distribution is bell-shaped, you know about 68% of the data lie within 1 standard deviation of the mean. Because the mean is 16 and the standard deviation is 10, the interval from 6 to 26 represents about 68% of the data. So, you would expect about $0.68 \cdot 30 \approx 20$ of the men surveyed to own between 6 and 26 pairs of socks.

Laurie's Notes Teacher Actions

- Discuss the introductory information about symmetric distributions and standard deviation. Connect this information to Exploration 1.
- **Wait Time:** Pose Example 4, and give sufficient *Wait Time* for partners to work all parts of the example.
- **Reason Abstractly and Quantitatively** and **Popsicle Sticks:** Circulate and listen for reasoning about measures of center and measures of variability. Use *Popsicle Sticks* to solicit responses.

Closure

- **Point of Most Significance:** Ask students to identify, aloud or on a paper, the most significant point (or part) in the lesson that aided their learning.

MONITORING PROGRESS ANSWERS

4. The distribution is skewed right.
5. about 34

Assignment Guide and Homework Check

ASSIGNMENT

Basic: 1, 2, 3–11 odd, 15, 19, 22, 25–27

Average: 1, 2–22 even, 25–27

Advanced: 1, 2–20 even, 21–27

HOMEWORK CHECK

Basic: 3, 9, 15, 19

Average: 4, 10, 16, 20

Advanced: 4, 10, 14, 20

ANSWERS

1. The data in a symmetric distribution are evenly distributed to the left and right of the highest bar, most of the data in a distribution that is skewed left are on the right, and most of the data in a distribution that is skewed right are on the left.
2. The mean and standard deviation best describe a symmetric distribution and the median and five-number summary best describe skewed distributions.

3–4. See Additional Answers.

5. skewed right; Most of the data are on the first two stems.
6. symmetric; Most of the data are in the center.
7. mean, standard deviation; The distribution is symmetric.
8. median, five-number summary; The distribution is skewed left.
9. **a.**

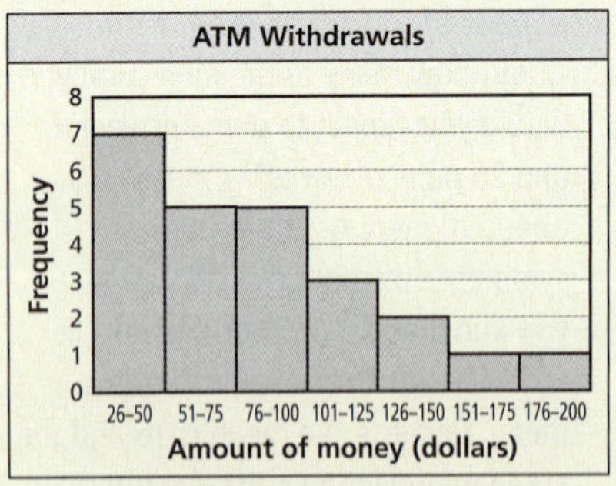

b. median, five-number summary; The distribution is skewed right.

c. Most people were charged a fee.

11.3 Exercises

Dynamic Solutions available at *BigIdeasMath.com*

Vocabulary and Core Concept Check

1. **VOCABULARY** Describe how data are distributed in a symmetric distribution, a distribution that is skewed left, and a distribution that is skewed right.
2. **WRITING** How does the shape of a distribution help you decide which measures of center and variation best describe the data?

Monitoring Progress and Modeling with Mathematics

3. **DESCRIBING DISTRIBUTIONS** The frequency table shows the numbers of hours that students volunteer per month. Display the data in a histogram. Describe the shape of the distribution. *(See Example 1.)*

Number of volunteer hours	1–2	3–4	5–6	7–8	9–10	11–12	13–14
Frequency	1	5	12	20	15	7	2

4. **DESCRIBING DISTRIBUTIONS** The frequency table shows the results of a survey that asked people how many hours they spend online per week. Display the data in a histogram. Describe the shape of the distribution.

Hours online	Frequency
0–3	5
4–7	7
8–11	12
12–15	14
16–19	26
20–23	45
24–27	33

In Exercises 5 and 6, describe the shape of the distribution of the data. Explain your reasoning.

5.

Stem	Leaf
1	1 1 3 4 8
2	2 3 4 7 8
3	1 2 4 9
4	0 3 2
5	7 9
6	6

Key: 3 | 1 = 31

6.

Stem	Leaf
5	0 0 1
6	3 6 7 9
7	1 4 5 8 9
8	2 4 5 5 7
9	4 6 8 9
10	1 3 4

Key: 6 | 3 = 63

In Exercises 7 and 8, determine which measures of center and variation best represent the data. Explain your reasoning.

7.

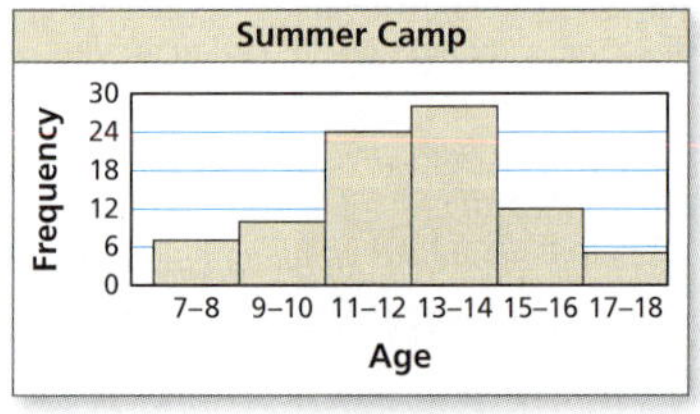

8.

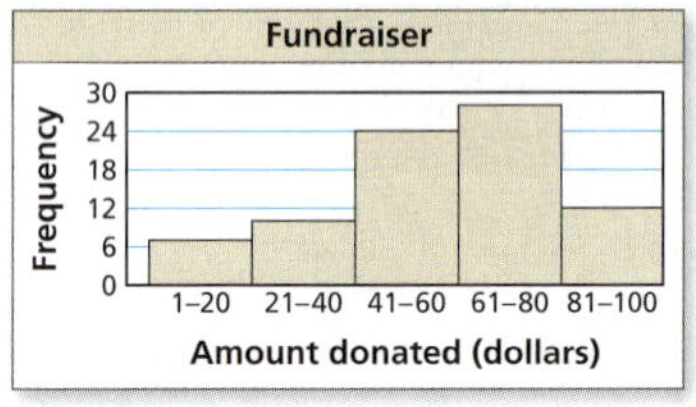

9. **ANALYZING DATA** The table shows the last 24 ATM withdrawals at a bank. *(See Example 2.)*
   a. Display the data in a histogram using seven intervals beginning with 26–50.
   b. Which measures of center and variation best represent the data? Explain.
   c. The bank charges a fee for any ATM withdrawal less than $150. How would you interpret the data?

ATM Withdrawals (dollars)		
120	100	70
60	40	80
150	80	50
120	60	175
30	50	50
60	200	30
100	150	110
70	40	100

10. **ANALYZING DATA** Measuring an IQ is an inexact science. However, IQ scores have been around for years in an attempt to measure human intelligence. The table shows some of the greatest known IQ scores.

IQ Scores		
170	190	180
160	180	210
154	170	180
195	230	160
170	186	180
225	190	170

   a. Display the data in a histogram using five intervals beginning with 151–166.

   b. Which measures of center and variation best represent the data? Explain.

   c. The distribution of IQ scores for the human population is symmetric. What happens to the shape of the distribution in part (a) as you include more and more IQ scores from the human population in the data set?

ERROR ANALYSIS In Exercises 11 and 12, describe and correct the error in the statements about the data displayed in the histogram.

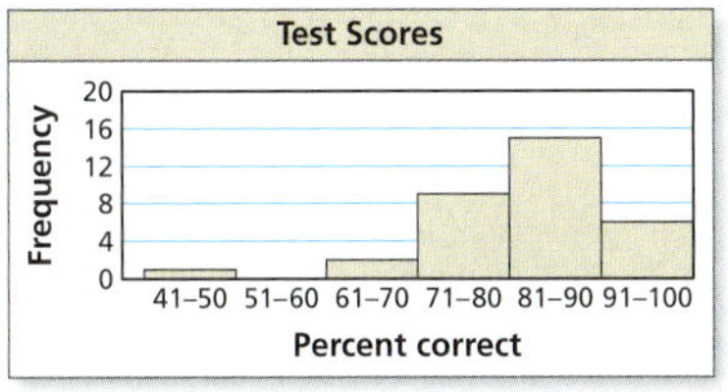

11. ✗ Most of the data are on the right. So, the distribution is skewed right.

12. ✗ Because the distribution is skewed, use the standard deviation to describe the variation of the data.

13. **USING TOOLS** For a large data set, would you use a stem-and-leaf plot or a histogram to show the distribution of the data? Explain.

14. **REASONING** For a symmetric distribution, why is the mean used to describe the center and the standard deviation used to describe the variation? For a skewed distribution, why is the median used to describe the center and the five-number summary used to describe the variation?

15. **COMPARING DATA SETS** The double histogram shows the distributions of daily high temperatures for two towns over a 50-day period. Compare the distributions using their shapes and appropriate measures of center and variation. *(See Example 3.)*

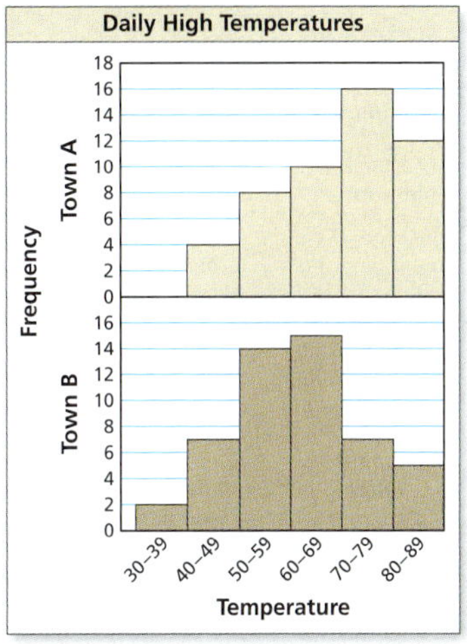

16. **COMPARING DATA SETS** The frequency tables show the numbers of entrées in certain price ranges (in dollars) at two different restaurants. Display the data in a double histogram. Compare the distributions using their shapes and appropriate measures of center and variation.

Restaurant A	
Price range	Frequency
8–10	5
11–13	9
14–16	12
17–19	4
20–22	3
23–25	0

Restaurant B	
Price range	Frequency
8–10	0
11–13	2
14–16	5
17–19	7
20–22	8
23–25	6

17. **OPEN-ENDED** Describe a real-life data set that has a distribution that is skewed right.

18. **OPEN-ENDED** Describe a real-life data set that has a distribution that is skewed left.

Dynamic Teaching Tools

Dynamic Assessment & Progress Monitoring Tool

Interactive Whiteboard Lesson Library

Dynamic Classroom with Dynamic Investigations

ANSWERS

10. a.

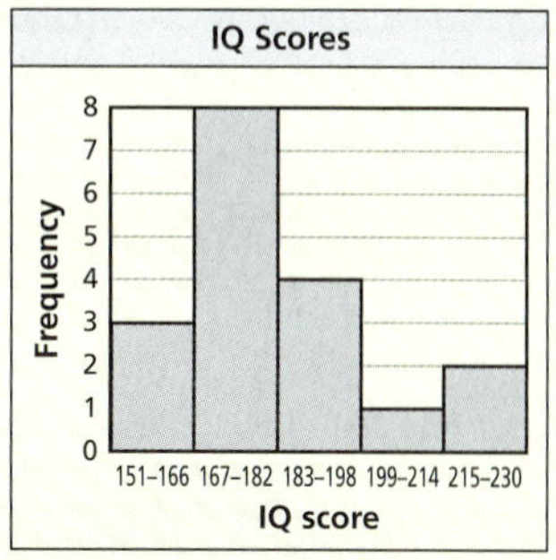

b. median, five-number summary; The distribution is skewed right.

c. It becomes more symmetric.

11. When most of the data are on the right, the distribution is skewed left, not right.

12. The standard deviation is used when the data are symmetric; Use the five-number summary to describe the variation of the data.

13. histogram; You do not need to list every value.

14. The mean is closest to the center and the variation is the same on each side; The median is closest to the center and the variation is different on each side.

15. Town A usually has a greater daily high temperature, and the daily high temperature tends to differ more for town B.

16. See Additional Answers.

17. *Sample answer:* salaries of employees at a large company

18. *Sample answer:* ages of residents and staff at a retirement home

ANSWERS

19–27. See Additional Answers.

Mini-Assessment

1. The table shows the sales yesterday at a gift shop.

Amount (dollars)			
8.25	15.80	5.85	17.90
12.50	22.30	16.75	17.25
15.22	14.50	14.50	16.20
19.20	19.99	18.20	18.90
16.10	18.75	10.50	9.10
18.65	23.50	20.85	21.35

   a. Display the data in a histogram using five intervals beginning with 4.00–7.99. See Additional Answers.

   b. Describe the shape of the distribution. skewed to the left

2. Which measures of center and variation best represent the data in Exercise 1? the median and the five-number summary

3. The double histogram shows the distributions of the numbers of CDs sold during each hour at two music stores. Compare the distributions using their shapes and appropriate measures of center and variation.

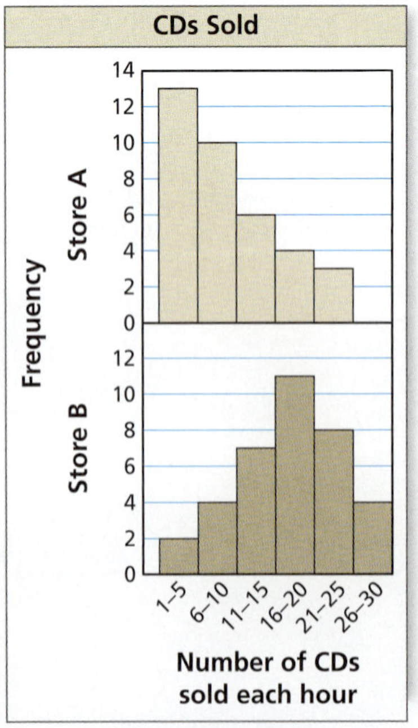

See Additional Answers.

19. COMPARING DATA SETS The table shows the results of a survey that asked freshmen and sophomores how many songs they have downloaded on their MP3 players. *(See Example 4.)*

	Freshmen	Sophomores
Survey size	45	54
Minimum	250	360
Maximum	2150	2400
1st Quartile	800	780
Median	1200	2000
3rd Quartile	1600	2200
Mean	1150	1650
Standard deviation	420	480

a. Make a double box-and-whisker plot that represents the data. Describe the shape of each distribution.

b. Compare the number of songs downloaded by freshmen to the number of songs downloaded by sophomores.

c. About how many of the freshmen surveyed would you expect to have between 730 and 1570 songs downloaded on their MP3 players?

d. If you survey 100 more freshmen, about how many would you expect to have downloaded between 310 and 1990 songs on their MP3 players?

20. COMPARING DATA SETS You conduct the same survey as in Exercise 19 but use a different group of freshmen. The results are as follows.
Survey size: 60; minimum: 200; maximum: 2400; 1st quartile: 640; median: 1670; 3rd quartile: 2150; mean: 1480; standard deviation: 500

a. Compare the number of songs downloaded by this group of freshmen to the number of songs downloaded by sophomores.

b. Why is the median greater than the mean for this group of freshmen?

21. REASONING A data set has a symmetric distribution. Every value in the data set is doubled. Describe the shape of the new distribution. Are the measures of center and variation affected? Explain.

22. HOW DO YOU SEE IT? Match the distribution with the corresponding box-and-whisker plot.

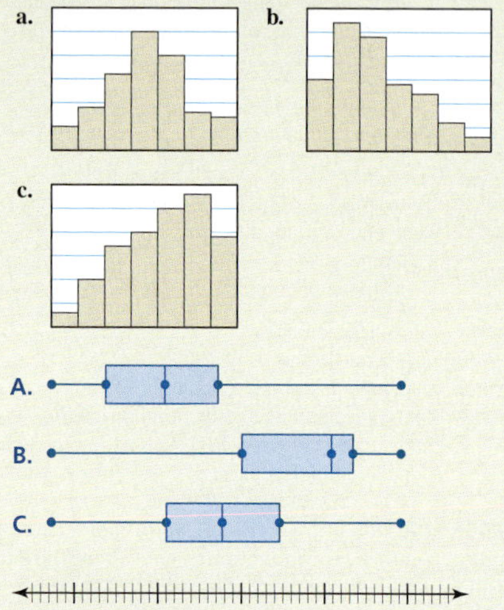

23. REASONING You record the following waiting times at a restaurant.

Waiting Times (minutes)									
26	38	15	8	22	42	25	20	17	18
40	35	24	31	42	29	25	0	30	13

a. Display the data in a histogram using five intervals beginning with 0–9. Describe the shape of the distribution.

b. Display the data in a histogram using 10 intervals beginning with 0–4. What happens when the number of intervals is increased?

c. Which histogram best represents the data? Explain your reasoning.

24. THOUGHT PROVOKING The shape of a *bimodal* distribution is shown. Describe a real-life example of a bimodal distribution.

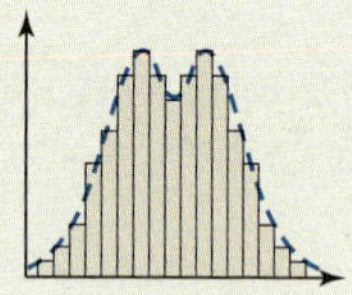

Maintaining Mathematical Proficiency

Reviewing what you learned in previous grades and lessons

Find the domain of the function. *(Section 10.1)*

25. $f(x) = \sqrt{x} + 6$ **26.** $f(x) = \sqrt{2x}$ **27.** $f(x) = \frac{1}{4}\sqrt{x} - 7$

If students need help...	*If students got it...*
Resources by Chapter • Practice A and Practice B • Puzzle Time	Resources by Chapter • Enrichment and Extension • Cumulative Review
Student Journal • Practice	Start the *next* Section
Differentiating the Lesson Skills Review Handbook	

11.1–11.3 What Did You Learn?

Core Vocabulary

measure of center, *p. 586*
mean, *p. 586*
median, *p. 586*
mode, *p. 586*
outlier, *p. 587*
measure of variation, *p. 587*
range, *p. 587*
standard deviation, *p. 588*
data transformation, *p. 589*
box-and-whisker plot, *p. 594*
quartile, *p. 594*
five-number summary, *p. 594*
interquartile range, *p. 595*

Core Concepts

Section 11.1

Measures of Center, *p. 586*
Measures of Variation, *p. 587*
Data Transformations Using Addition, *p. 589*
Data Transformations Using Multiplication, *p. 589*

Section 11.2

Box-and-Whisker Plot, *p. 594*
Shapes of Box-and-Whisker Plots, *p. 596*

Section 11.3

Symmetric and Skewed Distributions, *p. 600*
Choosing Appropriate Measures, *p. 601*

Mathematical Practices

1. Exercises 15 and 16 on page 590 are similar. For each data set, is the outlier much greater than or much less than the rest of the data values? Compare how the outliers affect the means. Explain why this makes sense.
2. In Exercise 18 on page 605, provide a possible reason for why the distribution is skewed left.

Study Skills

Studying for Finals

- Form a study group of three or four students several weeks before the final exam.
- Find out what material you must know for the final exam, even if your teacher has not yet covered it.
- Ask for a practice final exam or create one yourself and have your teacher look at it.
- Have each group member take the practice final exam.
- Decide when the group is going to meet and what you will cover during each session.
- During the sessions, make sure you stay on track.

Dynamic Teaching Tools

Dynamic Assessment & Progress Monitoring Tool
Interactive Whiteboard Lesson Library
Dynamic Classroom with Dynamic Investigations

ANSWERS

1. Exercise 15: less than, decreases the mean; Exercise 16: greater than, increases the mean; The mean depends on the sum of the data values.
2. *Sample answer:* The residents would be much older than the staff, so most data values would be on the right.

11.1–11.3 Quiz

Find the mean, median, and mode of the data set. Which measure of center best represents the data? Explain. *(Section 11.1)*

1.

Hours Spent on Project		
$3\frac{1}{2}$	5	$2\frac{1}{2}$
3	$3\frac{1}{2}$	$\frac{1}{2}$

2.

Waterfall Height (feet)		
1000	1267	1328
1200	1180	1000
2568	1191	1100

Find the range and standard deviation of each data set. Then compare your results. *(Section 11.1)*

3. Absent students during a week of school
Female: 6, 2, 4, 3, 4
Male: 5, 3, 6, 6, 9

4. Numbers of points scored
Juniors: 19, 15, 20, 10, 14, 21, 18, 15
Seniors: 22, 19, 29, 32, 15, 26, 30, 19

Make a box-and-whisker plot that represents the data. *(Section 11.2)*

5. Ages of family members:
60, 15, 25, 20, 55, 70, 40, 30

6. Minutes of violin practice:
20, 50, 60, 40, 40, 30, 60, 40, 50, 20, 20, 35

7. Display the data in a histogram. Describe the shape of the distribution. *(Section 11.3)*

Quiz score	0–2	3–5	6–8	9–11	12–14
Frequency	1	3	6	16	4

8. The table shows the prices of eight mountain bikes in a sporting goods store. *(Section 11.1 and Section 11.2)*

Price (dollars)	98	119	95	211	130	98	100	125

a. Find the mean, median, mode, range, and standard deviation of the prices.

b. Identify the outlier. How does the outlier affect the mean, median, and mode?

c. Make a box-and-whisker plot that represents the data. Find and interpret the interquartile range of the data. Identify the shape of the distribution.

d. Find the mean, median, mode, range, and standard deviation of the prices when the store offers a 5% discount on all mountain bikes.

9. The table shows the times of 20 presentations. *(Section 11.3)*

a. Display the data in a histogram using five intervals beginning with 3–5.

b. Which measures of center and variation best represent the data? Explain.

c. The presentations are supposed to be 10 minutes long. How would you interpret these results?

Time (minutes)			
9	7	10	12
10	11	8	10
10	17	11	5
9	10	4	12
6	14	8	10

ANSWERS

1. mean: 3; median: $3\frac{1}{4}$; mode: $3\frac{1}{2}$; median; The median is in the center of the data and it is the mean of the mean and the mode.

2. mean: $1314\frac{8}{9}$; median: 1191; mode: 1000; median; The mean is greater than most of the data, and the mode is less than most of the data.

3. female: range = 4, standard deviation ≈ 1.3; male: range = 6, standard deviation ≈ 1.9; Male absences are more spread out.

4. juniors: range = 11, standard deviation ≈ 3.4; seniors: range = 17, standard deviation ≈ 5.7; Senior points scored are more spread out.

5.

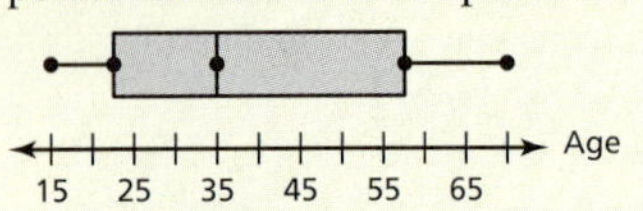

6.

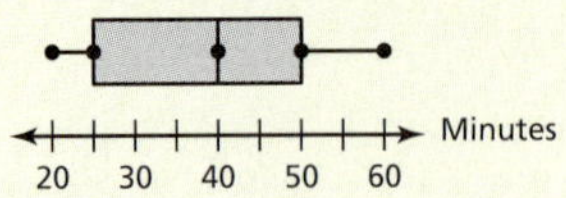

7.

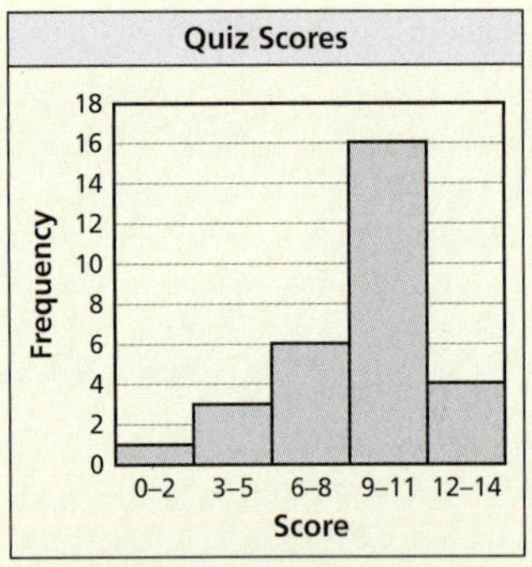

skewed left

8. a. mean: 122, median: 109.5, mode: 98, range: 116, standard deviation: 36

b. 211; The outlier increases the mean and median, and does not affect the mode.

c.

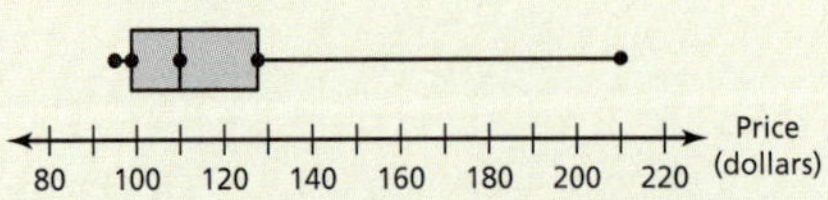

29.5; The middle half of the prices vary by no more than $29.50; skewed right

d. mean: 115.9, median: 104.025, mode: 93.1, range: 110.2, standard deviation: 34.2

9. a.

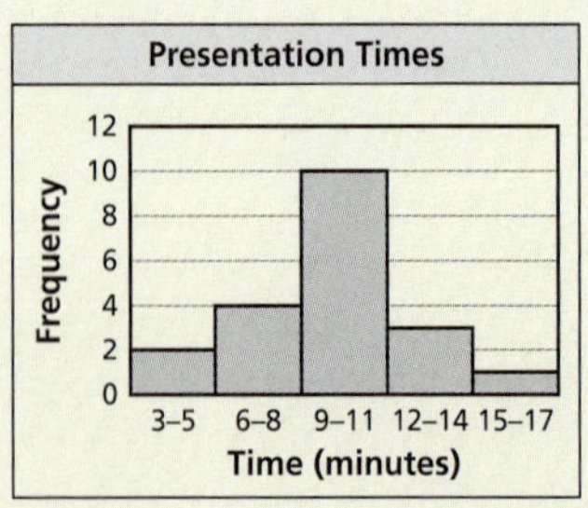

b. mean and standard deviation; The distribution is symmetric.

c. Most presentations are about the right length.

Laurie's Notes

Overview of Section 11.4

Introduction

- Two-way tables are introduced in this lesson, along with relative and conditional relative frequencies. Constructing the tables does not require a great deal of computational skills; instead it requires reasoning skills.
- The vocabulary used in the lesson can be confusing and intimidating. Having examples to refer to will be helpful to students.
- The lesson ends using two-way tables to recognize associations in data.

Formative Assessment Tips

- **Visitor Explanation:** This technique simulates what it would look like if a visitor were to enter your classroom during the middle of an exploration or activity. Could your students explain what they are doing and why they are doing it?
- This technique lets you know whether students understand the goal, or the essential question related to the exploration. Are students simply following directions or are they aware of the goal for today's exploration? Students are more engaged and their learning improves when the learning objective or purpose of the exploration is understood.
- This technique is best used when students are actively engaged in an exploration. This technique can be done as a *Think-Pair-Share* or as a *Writing Prompt*.
- The learning objective or essential question should have been made known to the students at the outset. This technique is not used to see whether students figured out the goal of the exploration or activity.

Pacing Suggestion

- The explorations allow students to figure out on their own what a two-way table is. You may consider omitting Example 1 in the formal lesson.

Dynamic Teaching Tools

- Dynamic Assessment & Progress Monitoring Tool
- Lesson Planning Tool
- Interactive Whiteboard Lesson Library
- Dynamic Classroom with Dynamic Investigations

What Your Students Will Learn

- Make two-way tables for data of two different categories collected from one source.
- Find and interpret marginal frequencies for two-way tables.
- Find joint relative, marginal relative, and conditional relative frequencies for two-way tables.
- Recognize associations in data from reading two-way tables.

Laurie's Notes

Exploration

Motivate

- **Story Time:** Tell students about a few of the sessions and workshops you attended at a 3-day math conference. When you returned, you submitted your expenses.

	Day 1	Day 2	Day 3	Totals
Meals				A
Lodging				
Taxi				
Totals		B		C

- The school district does not want to share your expenses publicly, so they are blacked out.

? "What do A, B, and C represent?" A is the total amount spent on meals for 3 days; B is the total expenses for day 2; C is the total expenses for all 3 days.

? "How do you find C?" Find the sum of the last column or the sum of the last row.

- If students do not know the vocabulary terms *column* and *row*, be sure to clarify.

Exploration 1

- Students should find that reading a two-way table is relatively easy. The term "two-way table" is new to students, yet they do not need a formal definition to make sense of the problem and what it is asking. In fact, some students will jump in and start adding the entries in the rows and columns without reading the introduction.
- When students have finished, discuss their responses to part (b).
- **Visitor Explanation:** Ask, "If a visitor entered the room right now, how would you explain what you are doing and why are you doing it?" Students should practice explanations with partners. Solicit oral responses or do a one-minute write.

Exploration 2

- Ask a few key questions to ensure that students know how to read a three-dimensional bar graph.
- **Reason Abstractly and Quantitatively:** In part (b), discuss students' observations and listen for evidence of their statements. For instance, instead of saying that more girls have part-time jobs than boys, it is a stronger comparison to say that about 40 more girls have part-time jobs than boys. Encourage students to give quantitative evidence in their reasoning.

Communicate Your Answer

- Listen for students' descriptions that accurately reference the rows and columns of the two-way table.

Connecting to Next Step

- Students should have figured out how to complete and interpret data in a two-way table. In the formal lesson, they will also work with relative frequencies in a two-way table.

11.4 Two-Way Tables

Essential Question How can you read and make a two-way table?

EXPLORATION 1 Reading a Two-Way Table

Work with a partner. You are the manager of a sports shop. The two-way tables show the numbers of soccer T-shirts in stock at your shop at the beginning and end of the selling season. (a) Complete the totals for the rows and columns in each table. (b) How would you alter the number of T-shirts you order for next season? Explain your reasoning.

Beginning of season		T-Shirt Size					
		S	M	L	XL	XXL	Total
Color	blue/white	5	6	7	6	5	
	blue/gold	5	6	7	6	5	
	red/white	5	6	7	6	5	
	black/white	5	6	7	6	5	
	black/gold	5	6	7	6	5	
	Total						145

End of season		T-Shirt Size					
		S	M	L	XL	XXL	Total
Color	blue/white	5	4	1	0	2	
	blue/gold	3	6	5	2	0	
	red/white	4	2	4	1	3	
	black/white	3	4	1	2	1	
	black/gold	5	2	3	0	2	
	Total						

MODELING WITH MATHEMATICS

To be proficient in math, you need to identify important quantities and map their relationships using tools such as graphs and two-way tables.

EXPLORATION 2 Making a Two-Way Table

Work with a partner. The three-dimensional bar graph shows the numbers of hours students work at part-time jobs.

a. Make a two-way table showing the data. Use estimation to find the entries in your table.

b. Write two observations that summarize the data in your table.

Part-Time Jobs of Students at a High School

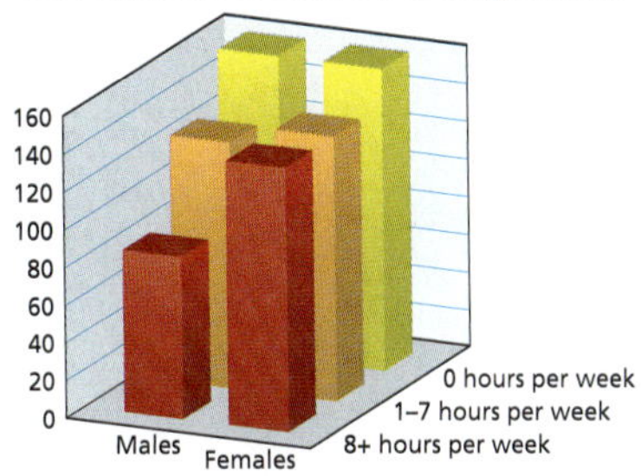

Communicate Your Answer

3. How can you read and make a two-way table?

Dynamic Teaching Tools

- Dynamic Assessment & Progress Monitoring Tool
- Lesson Planning Tool
- Interactive Whiteboard Lesson Library
- Dynamic Classroom with Dynamic Investigations

ANSWERS

1. a. Beginning: 29; 29; 29; 29; 29; 25; 30; 35; 30; 25; End: 12; 16; 14; 11; 12; 20; 18; 14; 5; 8; 65

b. For each shirt category, order the quantity equal to the difference between the number at the beginning of the season and the number at the end of the season.

2. a. *Sample answer:*

		Gender		
		Males	Females	Total
Hours	0 hours per week	145	145	290
	1–7 hours per week	110	115	225
	8+ hours per week	70	115	185
	Total	325	375	700

b. *Sample answer:* More females work 8+ hours per week than males. Approximately equal numbers of males and females work 1–7 hours per week and 0 hours per week.

3. *Sample answer:* Rows and columns represent different categories, each entry represents the number in both categories.

English Language Learners

Labels

Have students create their own two-way table that includes the Total column and Total row, like the one below. Have students use different colors to label and identify the joint frequencies and the marginal frequencies.

		Pet		
		Yes	No	Total
Gender	Female	12	10	22
	Male	15	9	24
	Total	27	19	46

joint frequencies (Yes/No entries); marginal frequencies (Total row and column)

Extra Example 1

The two-way table shows the results of a survey that asked students whether they ride the school bus. Find and interpret the marginal frequencies for the two-way table.

		Ride School Bus	
		Yes	No
Class	Junior	38	32
	Senior	21	44

		Ride School Bus		
		Yes	No	Total
Class	Junior	38	32	70
	Senior	21	44	65
	Total	59	76	135

70 juniors responded, 65 seniors responded, 59 students ride the school bus, 76 students do not ride the school bus, and 135 students were surveyed.

MONITORING PROGRESS ANSWER

1. 52 students own a tablet, 191 students do not own a tablet, 158 students own a cell phone, 85 students do not own a cell phone, 243 students were surveyed.

11.4 Lesson

What You Will Learn

- Find and interpret marginal frequencies.
- Make two-way tables.
- Find relative and conditional relative frequencies.
- Use two-way tables to recognize associations in data.

Core Vocabulary

two-way table, *p. 610*
joint frequency, *p. 610*
marginal frequency, *p. 610*
joint relative frequency, *p. 611*
marginal relative frequency, *p. 611*
conditional relative frequency, *p. 612*

Finding and Interpreting Marginal Frequencies

A **two-way table** is a frequency table that displays data collected from one source that belong to two different categories. One category of data is represented by rows, and the other is represented by columns. For instance, the two-way table below shows the results of a survey that asked freshmen and sophomores whether they access the Internet using a mobile device, such as a smartphone.

The two categories of data are *class* and *mobile access*. Class is further divided into *freshman* and *sophomore*, and mobile access is further divided into *yes* and *no*.

Each entry in the table is called a **joint frequency**. The sums of the rows and columns in a two-way table are called **marginal frequencies**.

		Mobile Access	
		Yes	No
Class	Freshman	55	22
	Sophomore	63	12

categories (Class, Mobile Access); joint frequency (55, 22, 63, 12)

REMEMBER
The frequency of an event is the number of times the event occurs.

EXAMPLE 1 Finding and Interpreting Marginal Frequencies

Find and interpret the marginal frequencies for the two-way table above.

SOLUTION

Create a new column and a new row for the marginal frequencies. Then add the entries in each row and column.

		Mobile Access		
		Yes	No	Total
Class	Freshman	55	22	77
	Sophomore	63	12	75
	Total	118	34	152

77 freshmen responded.
75 sophomores responded.
152 students were surveyed.
118 students access the Internet using a mobile device.
34 students do not access the Internet using a mobile device.

STUDY TIP
The sum of the "total" row should be equal to the sum of the "total" column. Place this sum of the marginal frequencies at the bottom right of your two-way table.

Monitoring Progress

Help in English and Spanish at *BigIdeasMath.com*

1. You conduct a technology survey to publish on your school's website. You survey students in the school cafeteria about the technological devices they own. The results are shown in the two-way table. Find and interpret the marginal frequencies.

		Tablet Computer	
		Yes	No
Cell Phone	Yes	34	124
	No	18	67

Laurie's Notes Teacher Actions

- Emphasize that information is known about two categories from the same source.
- Define *joint frequency*. Each entry in the two-way table is a frequency for two categories, hence the name joint frequency.
- *Marginal frequencies* are row and column sums, written in the margins.

? "How many freshmen have mobile access?" 55
? "How many students have no mobile access?" 34
? "What percent of sophomores have no mobile access?" $\frac{12}{75} = 0.16 = 16\%$

Making Two-Way Tables

EXAMPLE 2 **Making a Two-Way Table**

You conduct a survey that asks 286 students in your freshman class whether they play a sport or a musical instrument. One hundred eighteen of the students play a sport, and 64 of those students play an instrument. Ninety-three of the students do not play a sport or an instrument. Organize the results in a two-way table. Include the marginal frequencies.

SOLUTION

Step 1 Determine the two categories for the table: *sport* and *instrument*.

Step 2 Use the given joint and marginal frequencies to fill in parts of the table.

		Instrument		
		Yes	No	Total
Sport	Yes	64		118
	No		93	
	Total			286

Step 3 Use reasoning to find the missing joint and marginal frequencies. For instance, you can conclude that there are $286 - 118 = 168$ students who do not play a sport, and $118 - 64 = 54$ students who play a sport but do not play an instrument.

		Instrument		
		Yes	No	Total
Sport	Yes	64	54	118
	No	75	93	168
	Total	139	147	286

Monitoring Progress

Help in English and Spanish at *BigIdeasMath.com*

2. You survey students about whether they are getting a summer job. Seventy-five males respond, with 18 of them responding "no." Fifty-seven females respond, with 45 of them responding "yes." Organize the results in a two-way table. Include the marginal frequencies.

Finding Relative and Conditional Relative Frequencies

You can display entries of a two-way table as frequency counts (as in Examples 1 and 2) or as *relative frequencies*.

Core Concept

Relative Frequencies

A **joint relative frequency** is the ratio of a frequency that is not in the "total" row or the "total" column to the total number of values or observations.

A **marginal relative frequency** is the sum of the joint relative frequencies in a row or a column.

When finding relative frequencies in a two-way table, you can use the corresponding decimals or percents.

Extra Example 2

You conduct a survey that asks 248 students in your grade about whether they take a foreign language or art class. One hundred seventy-five of the students take a foreign language class, and 36 of those students take an art class. Twenty-two of the students do not take a foreign language or art class. Organize the results in a two-way table. Include the marginal frequencies.

		Art		
		Yes	No	Total
Foreign Language	Yes	36	139	175
	No	51	22	73
	Total	87	161	248

MONITORING PROGRESS ANSWER

2.

		Summer Job		
		Yes	No	Total
Gender	Male	57	18	75
	Female	45	12	57
	Total	102	30	132

Laurie's Notes Teacher Actions

? **Reason Abstractly and Quantitatively** and **Use Appropriate Tools Strategically:** Pose Example 2. "What are the two categories of data and what is the single source the data come from?" The two categories are sports and musical instruments. The data are from the 286 students in the freshman class.

- Have students work independently to construct the two-way table. Solicit a volunteer to construct the solution on the board.
- The language in the *Core Concept* can be confusing to students. Work through an example, and then refer back to this language.

Differentiated Instruction

Auditory

On the board, show a two-way table that includes joint frequencies. Ask one student to come to the board and explain how to find the marginal frequencies. Have a second student explain how to find the marginal relative frequencies, and have a third student explain how to find the conditional relative frequencies based on row totals and column totals.

Extra Example 3

The two-way table shows the results of a survey that asked college-bound high school students whether they plan to go to an in-state college. Make a two-way table that shows the joint and marginal relative frequencies.

Class		In-State College	
		Yes	No
	Junior	142	170
	Senior	204	124

Junior: $\frac{142}{640} \approx 0.22$, $\frac{170}{640} \approx 0.27$

Senior: $\frac{204}{640} \approx 0.32$, $\frac{124}{640} \approx 0.19$

Class		In-State College		
		Yes	No	Total
	Junior	0.22	0.27	0.49
	Senior	0.32	0.19	0.51
	Total	0.54	0.46	1

Extra Example 4

Use the survey results in Extra Example 3 to make a two-way table that shows the conditional relative frequencies based on the column totals.

Junior: $\frac{0.22}{0.54} \approx 0.41$, $\frac{0.27}{0.46} \approx 0.59$

Senior: $\frac{0.32}{0.54} \approx 0.59$, $\frac{0.19}{0.46} \approx 0.41$

Class		In-State College	
		Yes	No
	Junior	0.41	0.59
	Senior	0.59	0.41

MONITORING PROGRESS ANSWERS

3–4. See Additional Answers.

EXAMPLE 3 Finding Relative Frequencies

The two-way table shows the results of a survey that asked college-bound high school students whether they plan to major in a medical field. Make a two-way table that shows the joint and marginal relative frequencies.

Class		Major in Medical Field	
		Yes	No
	Junior	124	219
	Senior	101	236

SOLUTION

There are 124 + 219 + 101 + 236 = 680 students in the survey. To find the joint relative frequencies, divide each frequency by 680. Then find the sum of each row and each column to find the marginal relative frequencies.

Class		Major in Medical Field		
		Yes	No	Total
	Junior	$\frac{124}{680} \approx 0.18$	$\frac{219}{680} \approx 0.32$	0.50
	Senior	$\frac{101}{680} \approx 0.15$	$\frac{236}{680} \approx 0.35$	0.50
	Total	0.33	0.67	1

About 50% of the students are juniors.

About 35% of the students are seniors and are not planning to major in a medical field.

STUDY TIP

The sum of the marginal relative frequencies in the "total" row and the "total" column should each be equal to 1.

Core Concept

Conditional Relative Frequencies

A **conditional relative frequency** is the ratio of a joint relative frequency to the marginal relative frequency. You can find a conditional relative frequency using a row total or a column total of a two-way table.

EXAMPLE 4 Finding Conditional Relative Frequencies

Use the survey results in Example 3 to make a two-way table that shows the conditional relative frequencies based on the column totals.

SOLUTION

Use the marginal relative frequency of each *column* to calculate the conditional relative frequencies.

Class		Major in Medical Field	
		Yes	No
	Junior	$\frac{0.18}{0.33} \approx 0.55$	$\frac{0.32}{0.67} \approx 0.48$
	Senior	$\frac{0.15}{0.33} \approx 0.45$	$\frac{0.35}{0.67} \approx 0.52$

Given that a student is not planning to major in a medical field, the conditional relative frequency that he or she is a junior is about 48%.

STUDY TIP

When you use column totals, the sum of the conditional relative frequencies for each column should be equal to 1.

Monitoring Progress

Help in English and Spanish at *BigIdeasMath.com*

3. Use the survey results in Monitoring Progress Question 2 to make a two-way table that shows the joint and marginal relative frequencies. What percent of students are not getting a summer job?

4. Use the survey results in Example 3 to make a two-way table that shows the conditional relative frequencies based on the row totals. Given that a student is a senior, what is the conditional relative frequency that he or she is planning to major in a medical field?

Laurie's Notes Teacher Actions

- **Teaching Tip:** Display the two-way table in the side margin and ask, "How many students were surveyed?" 680 Say, "I want to replace each entry with the percent it is of the whole (680)." Have students work with their partners to compute the percentages. Use this information to complete the table of relative frequencies. Refer to the language *joint relative frequency* and *marginal relative frequency*.
- Discuss conditional relative frequency where the whole is either the row total or the column total.

Recognizing Associations in Data

EXAMPLE 5 Recognizing Associations in Data

You survey students and find that 40% exercise regularly, 35% eat fruits and vegetables each day, and 52% do not exercise and do not eat fruits and vegetables each day. Is there an association between exercising regularly and eating fruits and vegetables each day?

SOLUTION

Use the given information to make a two-way table. Use reasoning to find the missing joint and marginal relative frequencies.

		Exercise Regularly		
		Yes	No	Total
Eats Fruit/Vegetables	Yes	27%	8%	35%
	No	13%	52%	65%
	Total	40%	60%	100%

Use conditional relative frequencies based on the column totals to determine whether there is an association. Of the students who exercise regularly, 67.5% eat fruits and vegetables each day. Of the students who do not exercise regularly, only about 13% eat fruits and vegetables each day. It appears that students who exercise regularly are more likely to eat more fruits and vegetables than students who do not exercise regularly.

		Exercise Regularly	
		Yes	No
Eats Fruit/Vegetables	Yes	$\frac{0.27}{0.4} = 0.675$	$\frac{0.08}{0.6} \approx 0.133$
	No	$\frac{0.13}{0.4} = 0.325$	$\frac{0.52}{0.6} \approx 0.867$

▶ So, there is an association between exercising regularly and eating fruits and vegetables each day.

You can also find the conditional relative frequencies by dividing each joint frequency by its corresponding column total or row total.

EXAMPLE 6 Recognizing Associations in Data

The two-way table shows the results of a survey that asked students whether they share a computer at home with other family members. Is there an association between age and sharing a computer?

		Age			
		12–13	14–15	16–17	18–19
Share a Computer	Yes	40	47	42	22
	No	10	25	36	34

SOLUTION

Use conditional relative frequencies based on column totals to determine whether there is an association. Based on this sample, 80% of students ages 12–13 share a computer and only about 39% of students ages 18–19 share a computer.

▶ The table shows that as age increases, students are less likely to share a computer with other family members. So, there is an association.

		Age			
		12–13	14–15	16–17	18–19
Share a Computer	Yes	$\frac{40}{50} = 0.8$	$\frac{47}{72} \approx 0.65$	$\frac{42}{78} \approx 0.54$	$\frac{22}{56} \approx 0.39$
	No	$\frac{10}{50} \approx 0.2$	$\frac{25}{72} \approx 0.35$	$\frac{36}{78} \approx 0.46$	$\frac{34}{56} \approx 0.61$

Monitoring Progress Help in English and Spanish at *BigIdeasMath.com*

5. Using the results of the survey in Monitoring Progress Question 1, is there an association between owning a tablet computer and owning a cell phone? Explain your reasoning.

Extra Example 5

You survey students and find that 52% buy a snack at the movies, 55% buy a drink, and 22% do not buy a snack or a drink. Is there an association between buying a snack and buying a drink at the movies? Of the students who buy a snack, $\frac{0.29}{0.52} \approx 0.558$, or $\frac{0.26}{0.48} \approx 0.542$, or about 55.8% buy a drink. Of the students who do not buy a snack, about 54.2% buy a drink. It appears that students who buy a snack are not more likely to buy a drink than students who do not buy a snack. So, there is no association between buying a snack and buying a drink at the movies.

Extra Example 6

The two-way table shows the results of a survey that asked students whether they watch animated movies. Is there an association between age and watching animated movies?

		Age			
		12–13	14–15	16–17	18–19
Watch Animated Movies	Yes	52	31	37	15
	No	18	19	38	43

The conditional relative frequencies based on column totals for "Yes" are age 12–13: 74%, age 14–15: 62%, age 16–17: 49%, and age 18–19: 26%. The results show that as age increases, students are less likely to watch animated movies. So, there is an association.

MONITORING PROGRESS ANSWER

5. no; The conditional marginal frequencies based on column totals are about the same.

Laurie's Notes Teacher Actions

- **Reason Abstractly and Quantitatively:** Students must reason about the partial information given in a two-way table in order to find the missing entries.
- Discuss with students that there is no measure or formal rule they are using at this point to determine whether there is an association between different categories. In a statistics course, there would be. Reasonable evidence is being used to state that there is (is not) an association.

Closure

- **Writing Prompt:** To decide whether there is an association between data in a two-way table, you …

Assignment Guide and Homework Check

ASSIGNMENT

Basic: 1–4, 5–27 odd, 30, 33, 34

Average: 1–4, 6–30 even, 33, 34

Advanced: 1–4, 10–28 even, 29–34

HOMEWORK CHECK

Basic: 9, 15, 17, 23, 25

Average: 10, 16, 18, 24, 26

Advanced: 12, 16, 18, 24, 28

ANSWERS

1. joint frequency
2. when data is collected from one source and belongs to two different categories
3. A marginal relative frequency is the sum of the joint relative frequencies in a row or column. A conditional relative frequency is the ratio of a joint relative frequency to the marginal relative frequency.
4. Use the marginal relative frequency of each row or of each column; Find the ratio of a joint relative frequency to the corresponding marginal relative frequency.
5. 178
6. 168
7. 208
8. 138
9. 118 students have set academic goals, 310 students have not set academic goals, 232 males responded, 196 females responded, 428 students were surveyed.
10. 290 people own a cat, 306 people do not own a cat, 312 people own a dog, 284 people do not own a dog, 596 people were surveyed.
11. 204 students plan to participate, 124 students do not plan to participate, 86 students are undecided, 222 freshmen responded, 192 sophomores responded, 414 students were surveyed.
12. 120 seniors plan to receive an associate's degree, 244 seniors plan to receive a bachelor's degree, 90 seniors plan to receive a master's degree, 226 males responded, 228 females responded, 454 seniors were surveyed.
13. 90; 152; 16; 200; 78

11.4 Exercises

Dynamic Solutions available at *BigIdeasMath.com*

Vocabulary and Core Concept Check

1. **COMPLETE THE SENTENCE** Each entry in a two-way table is called a(n) __________.
2. **WRITING** When is it appropriate to use a two-way table to organize data?
3. **VOCABULARY** Explain the relationship between joint relative frequencies, marginal relative frequencies, and conditional relative frequencies.
4. **WRITING** Describe two ways you can find conditional relative frequencies.

Monitoring Progress and Modeling with Mathematics

You conduct a survey that asks 346 students whether they buy lunch at school. In Exercises 5–8, use the results of the survey shown in the two-way table.

		Buy Lunch at School	
		Yes	No
Class	Freshman	92	86
	Sophomore	116	52

5. How many freshmen were surveyed?
6. How many sophomores were surveyed?
7. How many students buy lunch at school?
8. How many students do not buy lunch at school?

In Exercises 9 and 10, find and interpret the marginal frequencies. *(See Example 1.)*

9.

		Set Academic Goals	
		Yes	No
Gender	Male	64	168
	Female	54	142

10.

		Cat	
		Yes	No
Dog	Yes	104	208
	No	186	98

11. **USING TWO-WAY TABLES** You conduct a survey that asks students whether they plan to participate in school spirit week. The results are shown in the two-way table. Find and interpret the marginal frequencies.

		Participate in Spirit Week		
		Yes	No	Undecided
Class	Freshman	112	56	54
	Sophomore	92	68	32

12. **USING TWO-WAY TABLES** You conduct a survey that asks college-bound high school seniors about the type of degree they plan to receive. The results are shown in the two-way table. Find and interpret the marginal frequencies.

		Type of Degree		
		Associate's	Bachelor's	Master's
Gender	Male	58	126	42
	Female	62	118	48

USING STRUCTURE In Exercises 13 and 14, complete the two-way table.

13.

		Traveled on an Airplane		
		Yes	No	Total
Class	Freshman		62	
	Sophomore	184		
	Total	274		352

14.

		Plan to Attend School Dance		
		Yes	No	Total
Gender	Male	38		
	Female		24	112
	Total			196

15. **MAKING TWO-WAY TABLES** You conduct a survey that asks 245 students in your school whether they have taken a Spanish or a French class. One hundred nine of the students have taken a Spanish class, and 45 of those students have taken a French class. Eighty-two of the students have not taken a Spanish or a French class. Organize the results in a two-way table. Include the marginal frequencies. *(See Example 2.)*

16. **MAKING TWO-WAY TABLES** A car dealership has 98 cars on its lot. Fifty-five of the cars are new. Of the new cars, 36 are domestic cars. There are 15 used foreign cars on the lot. Organize this information in a two-way table. Include the marginal frequencies.

In Exercises 17 and 18, make a two-way table that shows the joint and marginal relative frequencies. *(See Example 3.)*

17.

		Exercise Preference	
		Aerobic	Anaerobic
Gender	Male	88	104
	Female	96	62

18.

		Meat	
		Turkey	Ham
Bread	White	452	146
	Wheat	328	422

19. **USING TWO-WAY TABLES** Refer to Exercise 17. What percent of students prefer aerobic exercise? What percent of students are males who prefer anaerobic exercise?

20. **USING TWO-WAY TABLES** Refer to Exercise 18. What percent of the sandwiches are on wheat bread? What percent of the sandwiches are turkey on white bread?

ERROR ANALYSIS In Exercises 21 and 22, describe and correct the error in using the two-way table.

		Participate in Fundraiser	
		Yes	No
Class	Freshman	187	85
	Sophomore	123	93

21.

✗ One hundred eighty-seven freshmen responded to the survey.

22.

✗ The two-way table shows the joint relative frequencies.

		Participate in Fundraiser	
		Yes	No
Class	Freshman	$\frac{187}{272} \approx 0.69$	$\frac{85}{272} \approx 0.31$
	Sophomore	$\frac{123}{216} \approx 0.57$	$\frac{93}{216} \approx 0.43$

23. **USING TWO-WAY TABLES** A company is hosting an event for its employees to celebrate the end of the year. It asks the employees whether they prefer a lunch event or a dinner event. It also asks whether they prefer a catered event or a potluck. The results are shown in the two-way table. Make a two-way table that shows the conditional relative frequencies based on the row totals. Given that an employee prefers a lunch event, what is the conditional relative frequency that he or she prefers a catered event? *(See Example 4.)*

		Menu	
		Potluck	Catered
Meal	Lunch	36	48
	Dinner	44	72

Dynamic Teaching Tools

- Dynamic Assessment & Progress Monitoring Tool
- Interactive Whiteboard Lesson Library
- Dynamic Classroom with Dynamic Investigations

ANSWERS

14. 46; 84; 88; 126; 70

15–18. See Additional Answers.

19. 52%; 30%

20. 55%; 34%

21. This number is only the number of freshmen who participated in the fundraiser; 272 freshmen responded to the survey.

22. The two-way table does not show the joint relative frequencies; The two-way table shows conditional relative frequencies.

23.

		Menu	
		Potluck	Catered
Meal	Lunch	0.43	0.57
	Dinner	0.38	0.62

about 57%

ANSWERS

24–34. See Additional Answers.

Mini-Assessment

1. You conduct a survey that asks 250 students whether they have a sister or brother. One hundred ten students have a brother and 74 of those students have a sister. Thirty-two of the students do not have a brother or a sister. Organize the results in a two-way table. Find and interpret the marginal frequencies.

		Sister		
		Yes	No	Total
Brother	Yes	74	36	110
	No	108	32	140
	Total	182	68	250

See Additional Answers.

2. The two-way table shows the results of a survey about whether students skateboard or snowboard. Make a two-way table that shows the joint and marginal relative frequencies.

		Skateboard	
		Yes	No
Snowboard	Yes	82	38
	No	24	76

Divide each entry by 220, the number of students in the survey.

		Skateboard		
		Yes	No	Total
Snowboard	Yes	0.37	0.17	0.54
	No	0.11	0.35	0.46
	Total	0.48	0.52	1

3. Use the table you made for Exercise 2. Is there an association between skateboarding and snowboarding? Explain.
yes; Of the students who skateboard, about 77% snowboard. Of the students who do not skateboard, about 67% do not snowboard.

24. USING TWO-WAY TABLES The two-way table shows the results of a survey that asked students about their preference for a new school mascot. Make a two-way table that shows the conditional relative frequencies based on the column totals. Given that a student prefers a hawk as a mascot, what is the conditional relative frequency that he or she prefers a cartoon mascot?

		Type		
		Tiger	Hawk	Dragon
Style	Realistic	67	74	51
	Cartoon	58	18	24

25. ANALYZING TWO-WAY TABLES You survey college-bound seniors and find that 85% plan to live on campus, 35% plan to have a car while at college, and 5% plan to live off campus and not have a car. Is there an association between living on campus and having a car at college? Explain. *(See Example 5.)*

26. ANALYZING TWO-WAY TABLES You survey students and find that 70% watch sports on TV, 48% participate in a sport, and 16% do not watch sports on TV or participate in a sport. Is there an association between participating in a sport and watching sports on TV? Explain.

27. ANALYZING TWO-WAY TABLES The two-way table shows the results of a survey that asked adults whether they participate in recreational skiing. Is there an association between age and recreational skiing? *(See Example 6.)*

		Age				
		21–30	31–40	41–50	51–60	61–70
Ski	Yes	87	93	68	37	20
	No	165	195	148	117	125

28. ANALYZING TWO-WAY TABLES Refer to Exercise 12. Is there an association between gender and type of degree? Explain.

29. WRITING Compare Venn diagrams and two-way tables.

30. HOW DO YOU SEE IT? The graph shows the results of a survey that asked students about their favorite movie genre.

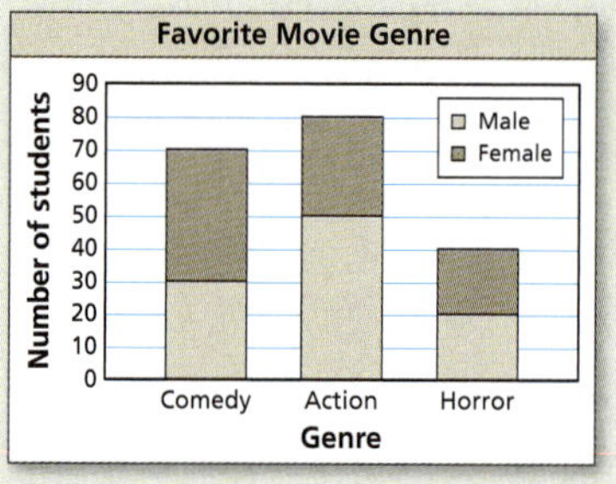

a. Display the given information in a two-way table.

b. Which of the data displays do you prefer? Explain.

31. PROBLEM SOLVING A box office sells 1809 tickets to a play, 800 of which are for the main floor. The tickets consist of $2x + y$ adult tickets on the main floor, $x - 40$ child tickets on the main floor, $x + 2y$ adult tickets in the balcony, and $3x - y - 80$ child tickets in the balcony.

a. Organize this information in a two-way table.

b. Find the values of x and y.

c. What percent of tickets are adult tickets?

d. What percent of child tickets are balcony tickets?

32. THOUGHT PROVOKING Compare "one-way tables" and "two-way tables." Is it possible to have a "three-way table?" If so, give an example of a three-way table.

Maintaining Mathematical Proficiency

Reviewing what you learned in previous grades and lessons

Tell whether the table of values represents a *linear*, an *exponential*, or a *quadratic* function. *(Section 8.6)*

33.

x	0	1	2	3	4
y	144	24	4	$\frac{2}{3}$	$\frac{1}{9}$

34.

x	−1	0	1	2	3
y	3	0	−1	0	3

If students need help...	*If students got it...*
Resources by Chapter • Practice A and Practice B • Puzzle Time	Resources by Chapter • Enrichment and Extension • Cumulative Review
Student Journal • Practice	Start the *next* Section
Differentiating the Lesson Skills Review Handbook	

Laurie's Notes

Overview of Section 11.5

Introduction

- Students have worked with different types of data displays since middle school. They made choices about what type of data display to use based on the type of data given.
- *Quantitative data* and *qualitative data* are defined in this lesson.
- Students will choose and construct appropriate data displays based on the given data.
- The lesson ends with analyzing misleading graphs.

Teaching Strategy

- It is relatively easy to demonstrate how to make a misleading graph. Bar graphs and line graphs are particularly susceptible to becoming misleading when the scale on the axes is changed or when the starting value of the axes is changed.
- Use a graphing calculator or graphing technology and change the viewing window so that the vertical scale is changed.

Pacing Suggestion

- You might jigsaw the explorations, having different groups of students working on just one of the data sets. Discuss the analysis for each part, and then transition to the formal lesson.

Dynamic Teaching Tools

- Dynamic Assessment & Progress Monitoring Tool
- Lesson Planning Tool
- Interactive Whiteboard Lesson Library
- Dynamic Classroom with Dynamic Investigations

What Your Students Will Learn

- Classify data as qualitative (or categorical) or quantitative.
- Choose and create appropriate data displays.
- Analyze statistical graphs and determine whether the graphs are misleading.

Laurie's Notes

Exploration

Motivate

- The theme for the exploration is road kill. While students may giggle at the thought, automobile accidents involving large animals can be serious.
- My only two accidents have involved hitting deer. In both instances, the deer were killed, my cars sustained substantial damage, and fortunately, neither I nor my children were injured.
- Allow time for students to share personal stories.
- Use the Internet to research and share vehicular data with students, such as the number of miles of roads in the United States, the number of registered vehicles, the number of accidents, and the number of animal-related accidents.

Discuss

- Discuss the data displays with which students are familiar: pictograph, bar graph, line graph, circle graph, stem-and-leaf plot, histogram, dot plot, box-and-whisker plot, and scatter plot. Have students describe the feature(s) of each display.
- Discuss the different numerical tools students have for describing data: mean, median, mode, range, standard deviation, quartile, and interquartile range.

Exploration 1

- Students need to decide which display makes sense for the type of data that they have. There may be more than one appropriate answer.
- **Construct Viable Arguments and Critique the Reasoning of Others:** Discuss students' choices and their explanations.
- Possible data displays:
  - Part (a): a circle graph (what part of the whole set is each animal) or a bar graph (compare the different categories, although there is a large difference in bar heights: 75 to 2746)
  - Part (b): a scatter plot and line of best fit (pair data, show trend over time, and make predictions for the future) or a line graph
  - Part (c): a stem-and-leaf plot (spread of data), along with calculating the mean (about 16.7) and median (17.1)
  - Part (d): a bar graph (compare the categories of data) or a circle graph (what part of the whole set is each animal)

Communicate Your Answer

- **Model with Mathematics:** Many students can make the displays when they are told which display to use. It is equally important that students be able to select the type of display to use based on the data and the question they hope to answer from making the display.
- The information gathered by students can be made into classroom posters.

Connecting to Next Step

- Various data displays have been reviewed in the explorations. In the formal lesson, qualitative and quantitative data are discussed, along with how to analyze misleading graphs.

11.5 Choosing a Data Display

Essential Question How can you display data in a way that helps you make decisions?

EXPLORATION 1 **Displaying Data**

Work with a partner. Analyze the data and then create a display that best represents the data. Explain your choice of data display.

a. A group of schools in New England participated in a 2-month study and reported 3962 animals found dead along roads.

birds: 307	mammals: 2746	amphibians: 145
reptiles: 75	unknown: 689	

b. The data below show the numbers of black bears killed on a state's roads from 1993 to 2012.

1993: 30	2000: 47	2007: 99
1994: 37	2001: 49	2008: 129
1995: 46	2002: 61	2009: 111
1996: 33	2003: 74	2010: 127
1997: 43	2004: 88	2011: 141
1998: 35	2005: 82	2012: 135
1999: 43	2006: 109	

c. A 1-week study along a 4-mile section of road found the following weights (in pounds) of raccoons that had been killed by vehicles.

13.4	14.8	17.0	12.9	21.3	21.5	16.8	14.8
15.2	18.7	18.6	17.2	18.5	9.4	19.4	15.7
14.5	9.5	25.4	21.5	17.3	19.1	11.0	12.4
20.4	13.6	17.5	18.5	21.5	14.0	13.9	19.0

d. A yearlong study by volunteers in California reported the following numbers of animals killed by motor vehicles.

raccoons: 1693	gray squirrels: 715
skunks: 1372	cottontail rabbits: 629
ground squirrels: 845	barn owls: 486
opossum: 763	jackrabbits: 466
deer: 761	gopher snakes: 363

USING TOOLS STRATEGICALLY

To be proficient in math, you need to identify relevant external mathematical resources.

Communicate Your Answer

2. How can you display data in a way that helps you make decisions?

3. Use the Internet or some other reference to find examples of the following types of data displays.

bar graph	circle graph	scatter plot
stem-and-leaf plot	pictograph	line graph
box-and-whisker plot	histogram	dot plot

Dynamic Teaching Tools

- Dynamic Assessment & Progress Monitoring Tool
- Lesson Planning Tool
- Interactive Whiteboard Lesson Library
- Dynamic Classroom with Dynamic Investigations

ANSWERS

1. See Additional Answers.

2. You can display data using different plots and graphs to make it easy to interpret and draw conclusions from the data.

3. Check students' work.

English Language Learners

Vocabulary

For English learners, relate the word *qualitative* with the word *quality* and the word *quantitative* with the word *quantity*. A quality of something is a characteristic that can be described but not measured, such as color or appearance. A quantity is an amount that can be counted or measured, such as length, speed, or time.

Extra Example 1

Tell whether the data are *qualitative* or *quantitative*.

a. heights of plants in an experiment
quantitative

b. license plates of cars on a road
qualitative

d. types of flowers in a bouquet
qualitative

c. number of hours worked each week
quantitative

MONITORING PROGRESS ANSWERS

1. qualitative; It does not make sense to compare or measure telephone numbers.
2. quantitative; Ages are numerical values.
3. quantitative; Lengths of videos are numerical values.
4. qualitative; Types of flowers are nonnumerical.

11.5 Lesson

Core Vocabulary

qualitative (categorical) data, *p. 618*
quantitative data, *p. 618*
misleading graph, *p. 620*

What You Will Learn

- Classify data as quantitative or qualitative.
- Choose and create appropriate data displays.
- Analyze misleading graphs.

Classifying Data

Data sets can consist of two types of data: *qualitative* or *quantitative.*

Core Concept

Types of Data

Qualitative data, or **categorical data**, consist of labels or nonnumerical entries that can be separated into different categories. When using qualitative data, operations such as adding or finding a mean do not make sense.

Quantitative data consist of numbers that represent counts or measurements.

STUDY TIP

Just because a frequency count can be shown for a data set does not make it quantitative. A frequency count can be shown for both qualitative and quantitative data.

EXAMPLE 1 Classifying Data

Tell whether the data are *qualitative* or *quantitative*.

a. prices of used cars at a dealership
b. jersey numbers on a basketball team
c. lengths of songs played at a concert
d. zodiac signs of students in your class

SOLUTION

a. Prices are numerical entries. So, the data are quantitative.

b. Jersey numbers are numerical, but they are labels. It does not make sense to compare them, and you cannot measure them. So, the data are qualitative.

c. Song lengths are numerical measurements. So, the data are quantitative.

d. Zodiac signs are nonnumerical entries that can be separated into different categories. So, the data are qualitative.

Monitoring Progress

Help in English and Spanish at *BigIdeasMath.com*

Tell whether the data are *qualitative* or *quantitative*. Explain your reasoning.

1. telephone numbers in a directory
2. ages of patients at a hospital
3. lengths of videos on a website
4. types of flowers at a florist

Qualitative and quantitative data can be collected from the same data source, as shown below. You can use these types of data together to obtain a more accurate description of a population.

Data Source	Quantitative Data	Qualitative Data
a student	How much do you earn per hour at your job? \$10.50	What is your occupation? painter
a house	How many square feet of living space is in the house? 2500 ft^2	In what city is the house located? Chicago

Laurie's Notes Teacher Actions

- Write the *Core Concept*, and discuss the differences between the types of data. Solicit examples of each.
- **Thumbs Up:** Pose Example 1. Anticipate that students can answer the questions independently. Have students do a *Thumbs Up* self-assessment.
- Check for understanding by having students quickly complete Questions 1–4.
- ? "Can qualitative and quantitative data be collected from the same source? Explain." Answers will vary. One answer will generate others.

Choosing and Creating Appropriate Data Displays

As shown on page 584, you have learned a variety of ways to display data sets graphically. Choosing an appropriate data display can depend on whether the data are qualitative or quantitative.

EXAMPLE 2 Choosing and Creating a Data Display

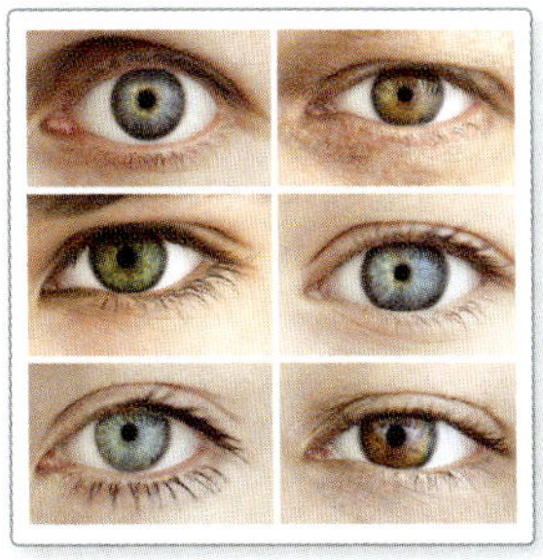

Analyze the data and then create a display that best represents the data. Explain your reasoning.

a.

Eye Color Survey	
Color	**Number of students**
brown	63
blue	37
hazel	25
green	10
gray	3
amber	2

b.

Speeds of Vehicles (mi/h)			
Interstate A		**Interstate B**	
65	67	67	72
68	71	70	78
72	70	65	71
68	65	71	80
65	68	84	81
75	82	77	79
68	59	69	70
62	68	66	69
75	80	73	75
77	75	84	79

SOLUTION

a. A circle graph is one appropriate way to display this qualitative data. It shows data as parts of a whole.

Step 1 Find the angle measure for each section of the circle graph by multiplying the fraction of students who have each eye color by 360°. Notice that there are 63 + 37 + 25 + 10 + 3 + 2 = 140 students in the survey.

Brown: $\frac{63}{140} \cdot 360° \approx 162°$ **Blue:** $\frac{37}{140} \cdot 360° \approx 95°$ **Hazel:** $\frac{25}{140} \cdot 360° \approx 64°$

Green: $\frac{10}{140} \cdot 360° \approx 26°$ **Gray:** $\frac{3}{140} \cdot 360° \approx 8°$ **Amber:** $\frac{2}{140} \cdot 360° \approx 5°$

Step 2 Use a protractor to draw the angle measures found in Step 1 on a circle. Then label each section and title the circle graph, as shown.

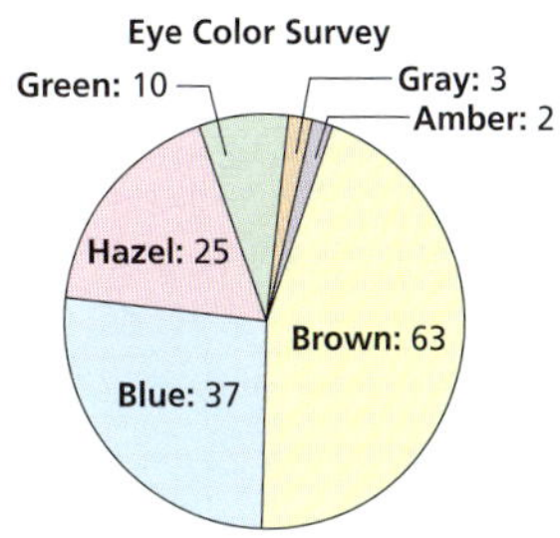

b. A double box-and-whisker plot is one appropriate way to display this quantitative data. Use the five-number summary of each data set to create a double box-and-whisker plot.

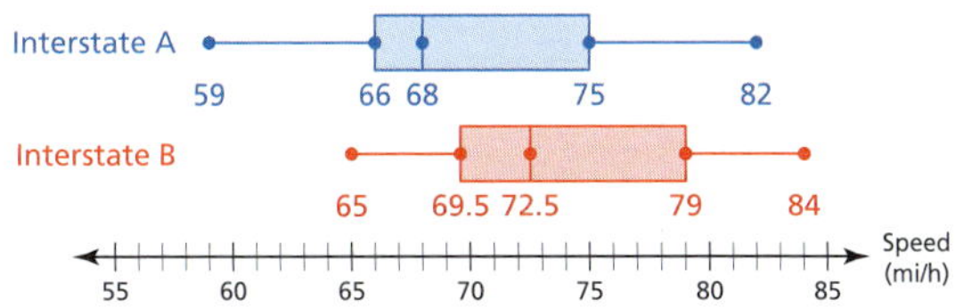

Extra Example 2

Analyze the data and then create a display that best represents the data. Explain your reasoning.

a.

Miles Walked	
Week	**Number of miles**
1	5
2	8
3	10
4	11

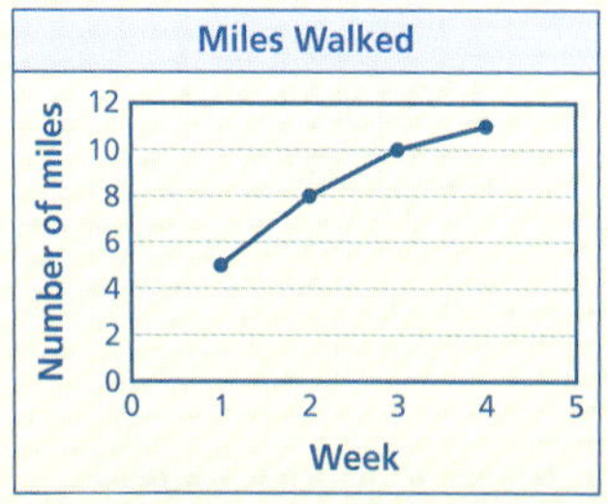

A line graph is one appropriate way to display this quantitative data. It shows how data change over time.

b.

Favorite Fruit Survey	
Fruit	**Number of students**
apple	35
banana	40
grape	15
orange	28
pear	12

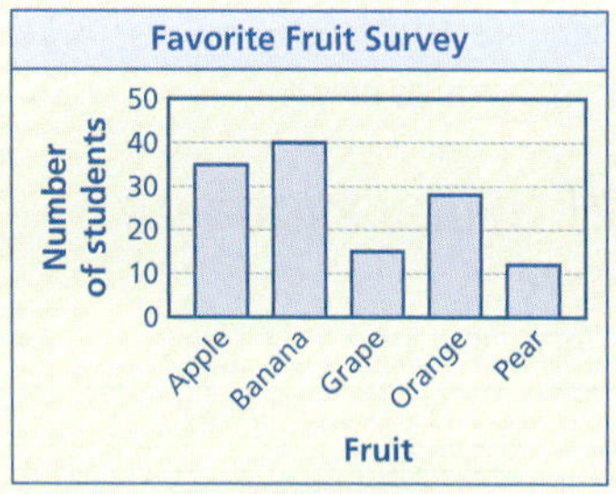

A bar graph is one appropriate way to display this qualitative data. It shows data in specific categories.

Laurie's Notes Teacher Actions

- **Turn and Talk:** Share the data in parts (a) and (b). "What type of display would best represent each set of data?" Answers will vary.
- Jigsaw the two parts of this example. In part (a), have some groups construct a circle graph and some a bar graph. In part (b), have some groups construct a double box-and-whisker and some a back-to-back stem-and-leaf plot. This approach replaces Questions 5 and 6 of the *Monitoring Progress.*
- Compare and contrast the various displays. What information can be seen from the displays?

Extra Example 3

Describe how each graph is misleading. Then explain how someone might misinterpret the graph.

a.

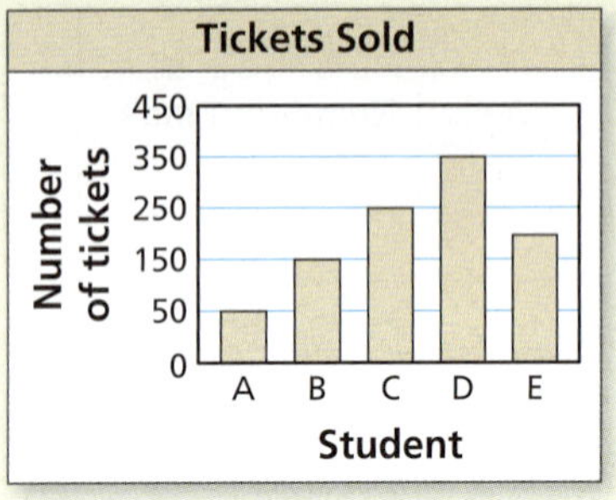

The vertical axis has increments that are not equal. Someone might believe that Student A sold half as many tickets as Student B, when Student A actually sold only one-third as many.

b.

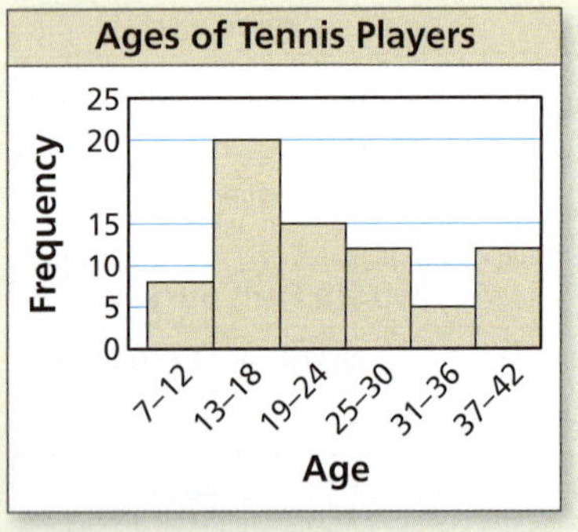

The horizontal bars are not equally spaced. Someone might believe that there are twice as many players aged 13–18 as there are aged 19–24, when there are actually only 5 more players, or $1\frac{1}{3}$ times as many.

MONITORING PROGRESS ANSWERS

5. *Sample answer:*

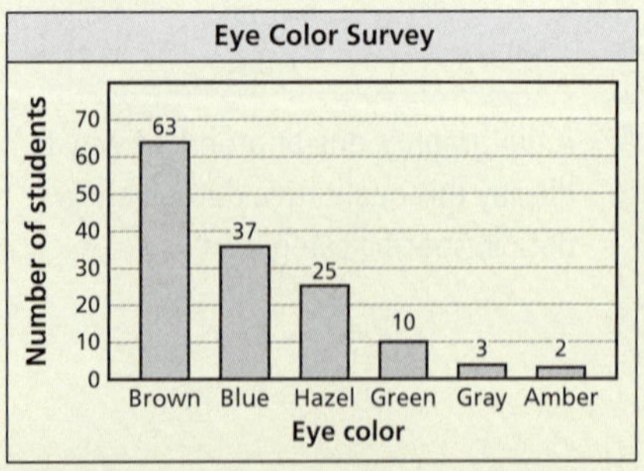

6–7. See Additional Answers.

Monitoring Progress

Help in English and Spanish at *BigIdeasMath.com*

5. Display the data in Example 2(a) in another way.

6. Display the data in Example 2(b) in another way.

Analyzing Misleading Graphs

Just as there are several ways to display data accurately using graphs, there are several ways to display data that are misleading. A **misleading graph** is a statistical graph that is not drawn appropriately. This may occur when the creator of a graph wants to give viewers the impression that results are better than they actually are. Below are some questions you can ask yourself when analyzing a statistical graph that will help you recognize when a graph is trying to deceive or mislead.

- Does the graph have a title?
- Are the numbers of the scale evenly spaced?
- Does the scale begin at zero? If not, is there a break?
- Does the graph need a key?
- Are all the axes or sections of the graph labeled?
- Are all the components of the graph, such as the bars, the same size?

EXAMPLE 3 Analyzing Misleading Graphs

Describe how each graph is misleading. Then explain how someone might misinterpret the graph.

a.

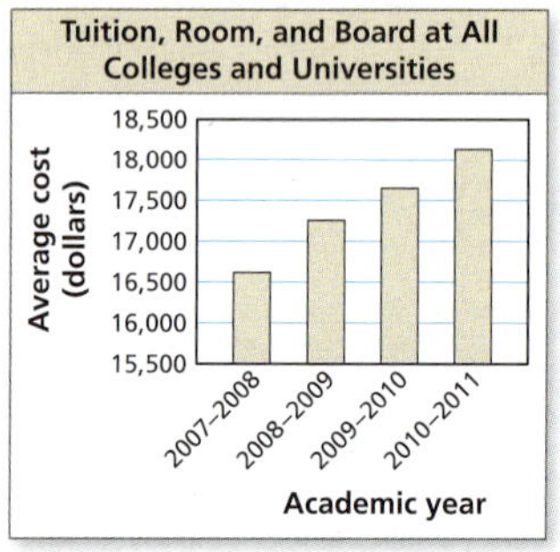

b.

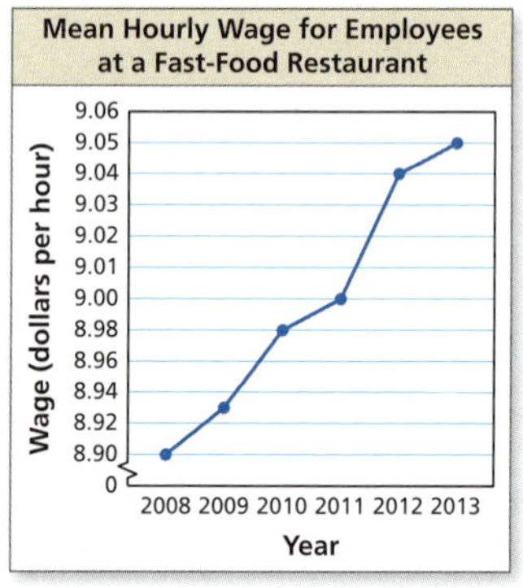

SOLUTION

a. The scale on the vertical axis of the graph starts at \$15,500 and does not have a break. This makes it appear that the average cost increased rapidly for the years given.

Someone might believe that the average cost more than doubled from 2007 to 2011, when actually, it increased by only about \$1500.

b. The scale on the vertical axis has very small increments that are not equal.

Someone might believe that the greatest increase in the mean hourly wage occurred from 2011 to 2012, when the greatest increase actually occurred from 2009 to 2010.

Monitoring Progress

Help in English and Spanish at *BigIdeasMath.com*

7. Redraw the graphs in Example 3 so they are not misleading.

Laurie's Notes Teacher Actions

- Have students "read" each display and summarize what information it describes.
- Some students may conclude from the bar graph that average costs for tuition, room, and board doubled from 2007 to 2011.

? "What makes the bar graph misleading?" The scale on the vertical axis does not begin at 0.

? "What makes the line graph misleading?" The scale on the vertical axis has increments that are not equal.

Closure

- **Exit Ticket:** Make a bar graph for Example 3(a) that would not be misleading. Check students' work.

11.5 Exercises

Dynamic Solutions available at *BigIdeasMath.com*

Vocabulary and Core Concept Check

1. **OPEN-ENDED** Describe two ways that a line graph can be misleading.

2. **WHICH ONE DOESN'T BELONG?** Which data set does *not* belong with the other three? Explain your reasoning.

ages of people attending a concert	heights of skyscrapers in a city
populations of counties in a state	breeds of dogs at a pet store

Monitoring Progress and Modeling with Mathematics

In Exercises 3–8, tell whether the data are *qualitative* or *quantitative*. Explain your reasoning. *(See Example 1.)*

3. brands of cars in a parking lot
4. weights of bears at a zoo
5. budgets of feature films
6. file formats of documents on a computer
7. shoe sizes of students in your class
8. street addresses in a phone book

In Exercises 9–12, choose an appropriate data display for the situation. Explain your reasoning.

9. the number of students in a marching band each year
10. a comparison of students' grades (out of 100) in two different classes
11. the favorite sports of students in your class
12. the distribution of teachers by age

In Exercises 13–16, analyze the data and then create a display that best represents the data. Explain your reasoning. *(See Example 2.)*

13.

Ages of World Cup Winners	
2010 Men's World Cup Winner (Spain)	2011 Women's World Cup Winner (Japan)
29 24 23 30 32 26 28 30 26 23 32 28 22 28 24 21 27 22 25 21 24 24 27	36 27 24 20 27 23 29 26 25 32 27 27 22 25 24 23 24 28 20 18 24

14.

Average Precipitation (inches)			
January	1.1	July	4.0
February	1.5	August	4.4
March	2.2	September	4.2
April	3.7	October	3.5
May	5.1	November	2.1
June	5.5	December	1.8

15.

Grades (out of 100) on a Test				
96	74	97	80	62
84	88	53	77	75
89	81	52	85	63
87	95	59	83	100

16.

Colors of Cars that Drive by Your House			
white	25	green	3
red	12	silver/gray	27
yellow	1	blue	6
black	21	brown/biege	5

17. **DISPLAYING DATA** Display the data in Exercise 13 in another way.

18. **DISPLAYING DATA** Display the data in Exercise 14 in another way.

19. **DISPLAYING DATA** Display the data in Exercise 15 in another way.

Assignment Guide and Homework Check

ASSIGNMENT

Basic: 1, 2, 3–17 odd, 21–27 odd, 28, 32, 33

Average: 1, 2–26 even, 27–29, 32, 33

Advanced: 1, 2–10 even, 14–18 even, 22–26 even, 27–33

HOMEWORK CHECK

Basic: 3, 5, 13, 21

Average: 4, 8, 14, 22

Advanced: 4, 8, 14, 24

ANSWERS

1. *Sample answer:* The numbers on the scale(s) could be unevenly spaced; The scale(s) may not begin at zero.
2. breeds of dogs at a pet store; The other three data sets are quantitative.
3. qualitative; Brands of cars are nonnumerical.
4. quantitative; Weights are numerical values.
5. quantitative; Budgets are numerical values.
6. qualitative; File formats are nonnumerical.
7. quantitative; Shoe sizes are numerical values.
8. qualitative; It does not make sense to compare or measure street addresses.
9. *Sample answer:* line graph; It shows data values over time.
10. *Sample answer:* back-to-back stem-and-leaf plot; It shows a comparison of data values from two different sources.
11. *Sample answer:* bar graph; It shows data in each specific category.
12. *Sample answer:* histogram; It shows frequencies of data values in equally-sized intervals.

13. *Sample answer:*

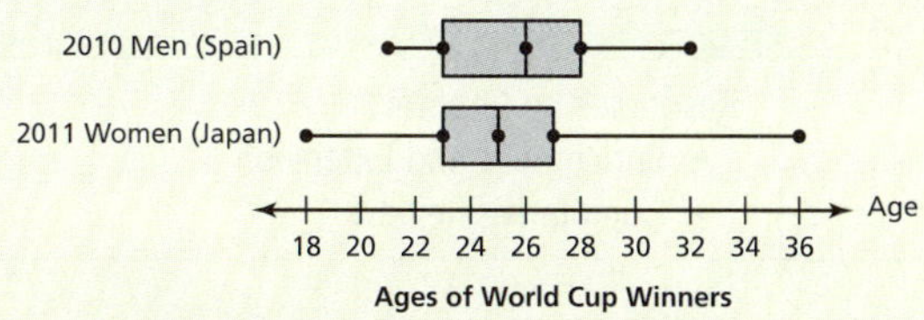

A double box-and-whisker plot shows the distributions of the data.

14–19. See Additional Answers.

Dynamic Teaching Tools

- Dynamic Assessment & Progress Monitoring Tool
- Interactive Whiteboard Lesson Library
- Dynamic Classroom with Dynamic Investigations

ANSWERS

20–33. See Additional Answers.

Mini-Assessment

1. Tell whether the data are *qualitative* or *quantitative*.
   a. prices of school supplies
   quantitative
   b. types of coins in a display
   qualitative
2. Analyze the data and then create a display that best represents the data. Explain your reasoning.

Favorite Color Survey	
Color	**Number of students**
blue	35
green	20
red	30
purple	15

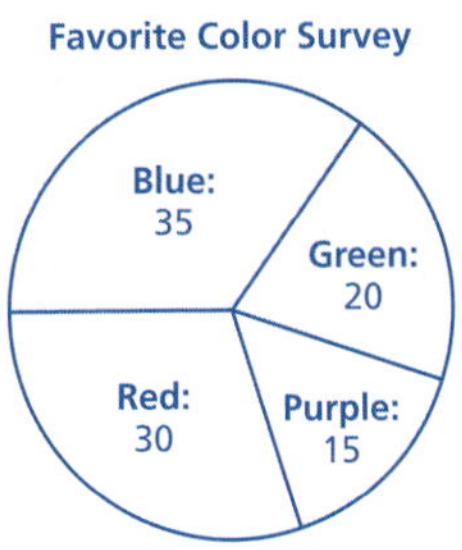

A circle graph shows data as parts of a whole.

3. Describe how the graph is misleading. Then explain how someone might misinterpret it.

See Additional Answers.

20. **DISPLAYING DATA** Display the data in Exercise 16 in another way.

In Exercises 21–24, describe how the graph is misleading. Then explain how someone might misinterpret the graph. *(See Example 3.)*

21.

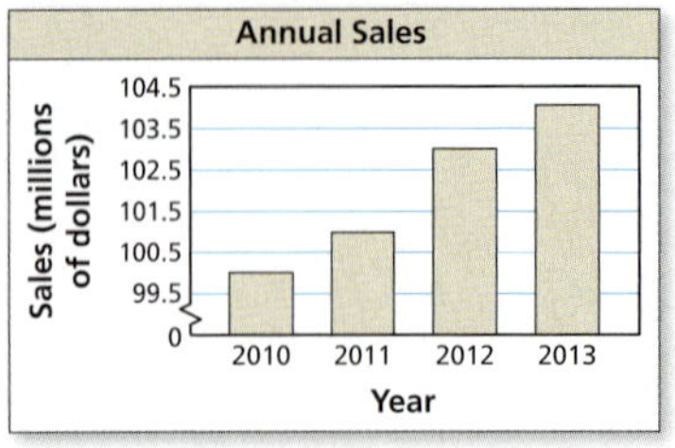

22.

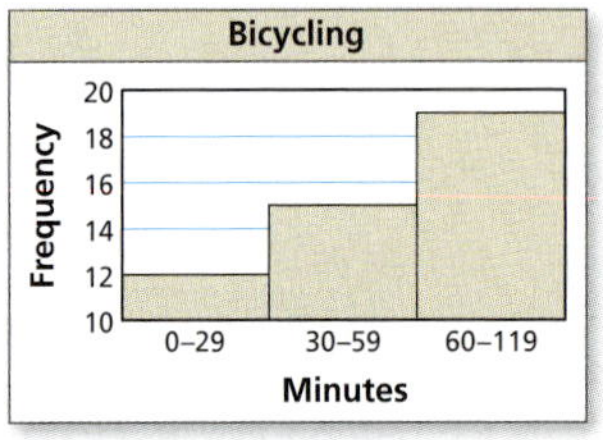

23.

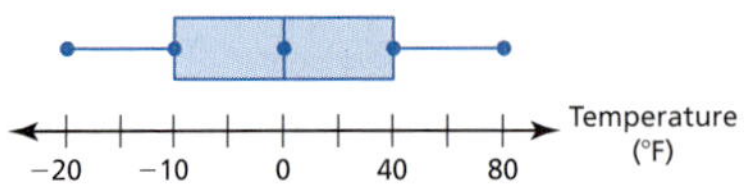

24.

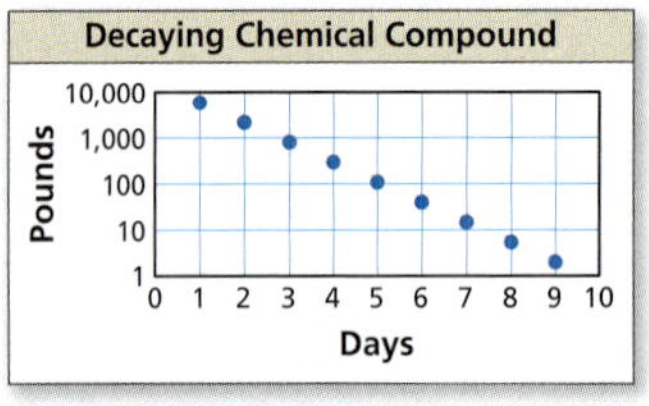

25. **DISPLAYING DATA** Redraw the graph in Exercise 21 so it is not misleading.

26. **DISPLAYING DATA** Redraw the graph in Exercise 22 so it is not misleading.

27. **MAKING AN ARGUMENT** A data set gives the ages of voters for a city election. Classmate A says the data should be displayed in a bar graph, while Classmate B says the data would be better displayed in a histogram. Who is correct? Explain.

28. **HOW DO YOU SEE IT?** The manager of a company sees the graph shown and concludes that the company is experiencing a decline. What is missing from the graph? Explain why the manager may be mistaken.

29. **REASONING** A survey asked 100 students about the sports they play. The results are shown in the circle graph.

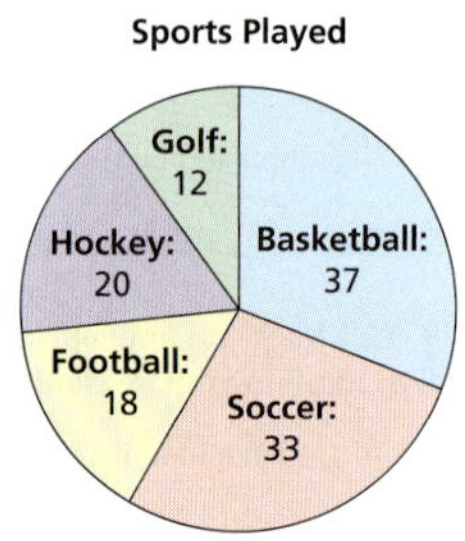

a. Explain why the graph is misleading.

b. What type of data display would be more appropriate for the data? Explain.

30. **THOUGHT PROVOKING** Use a spreadsheet program to create a type of data display that is not used in this section.

31. **REASONING** What type of data display shows the mode of a data set?

Maintaining Mathematical Proficiency

Reviewing what you learned in previous grades and lessons

Determine whether the relation is a function. Explain. *(Section 3.1)*

32. $(-5, -1), (-6, 0), (-5, 1), (-2, 2), (3, 3)$

33. $(0, 1), (4, 0), (8, 1), (12, 2), (16, 3)$

If students need help...	*If students got it...*
Resources by Chapter • Practice A and Practice B • Puzzle Time	Resources by Chapter • Enrichment and Extension • Cumulative Review
Student Journal • Practice	Start the *next* Section
Differentiating the Lesson Skills Review Handbook	

11.4–11.5 What Did You Learn?

Core Vocabulary

two-way table, *p. 610*
joint frequency, *p. 610*
marginal frequency, *p. 610*
joint relative frequency, *p. 611*
marginal relative frequency, *p. 611*
conditional relative frequency, *p. 612*
qualitative (categorical) data, *p. 618*
quantitative data, *p. 618*
misleading graph, *p. 620*

Core Concepts

Section 11.4

Joint and Marginal Frequencies, *p. 610*
Making Two-Way Tables, *p. 611*
Relative Frequencies, *p. 611*
Conditional Relative Frequencies, *p. 612*
Recognizing Associations in Data, *p. 613*

Section 11.5

Types of Data, *p. 618*
Choosing and Creating Appropriate Data Displays, *p. 619*
Analyzing Misleading Graphs, *p. 620*

Mathematical Practices

1. Consider the data given in the two-way table for Exercises 5–8 on page 614. Your sophomore friend responded to the survey. Is your friend more likely to have responded "yes" or "no" to buying a lunch? Explain.
2. Use your answer to Exercise 28 on page 622 to explain why it is important for a company manager to see accurate graphs.

Performance Task

College Student Study Time

Data from a small survey at a state university could provide insight into the amount of study time necessary to be successful in college. Based on the information you find when you organize the data, what advice should you give your peers? How will you support your conclusions?

To explore the answers to these questions and more, go to *BigIdeasMath.com*.

Dynamic Teaching Tools

- Dynamic Assessment & Progress Monitoring Tool
- Interactive Whiteboard Lesson Library
- Dynamic Classroom with Dynamic Investigations

ANSWERS

1. yes; More sophomores said yes than no.
2. *Sample answer:* Business decisions are based on data. Accurate graphs help to make correct decisions.

ANSWERS

1. mean: 4, same; median: 4.1, decreases; modes: 4.0 and 4.3; 4 is the mean and is less than the median.
2. mean: 4.5, increases; median: 4.3, increases; mode: 4.3; 10.0 is greater than the mean and median.
3. mean: 4, median: 3, mode: 10
4. mean: 1.7, median: 1, mode: 1
5. Player A: range = 56, standard deviation ≈ 15.56; Player B: range = 86, standard deviation ≈ 30.33; The scores for Player B are more spread out.
6. Store A: range = \$110, standard deviation ≈ \$39.37; Store B: range = \$150, standard deviation ≈ \$48.09; The prices at Store B are more spread out.
7. mean: 134, median: 129, mode: 121, range: 45, standard deviation: 3.6
8. mean: 65.4, median: 62.4, mode: 57.6, range: 27, standard deviation: 2.16

11 Chapter Review

Dynamic Solutions available at *BigIdeasMath.com*

11.1 Measures of Center and Variation *(pp. 585–592)*

The table shows the number of miles you ran each day for 10 days. Find the mean, median, and mode of the distances.

Miles Run	
3.5	4.1
4.0	4.3
4.4	4.5
3.9	2.0
4.3	5.0

Mean $\bar{x} = \frac{3.5 + 4.0 + 4.4 + 3.9 + 4.3 + 4.1 + 4.3 + 4.5 + 2.0 + 5.0}{10} = 4$

Median 2.0, 3.5, 3.9, 4.0, 4.1, 4.3, 4.3, 4.4, 4.5, 5.0 — Order the data.

$\frac{8.4}{2} = 4.2$ — Mean of two middle values

Mode 2.0, 3.5, 3.9, 4.0, 4.1, 4.3, 4.3, 4.4, 4.5, 5.0 — 4.3 occurs most often.

▶ The mean is 4 miles, the median is 4.2 miles, and the mode is 4.3 miles.

1. Use the data in the example above. You run 4.0 miles on Day 11. How does this additional value affect the mean, median, and mode? Explain.
2. Use the data in the example above. You run 10.0 miles on Day 11. How does this additional value affect the mean, median, and mode? Explain.

Find the mean, median, and mode of the data.

3.

Ski Resort Temperatures (°F)		
11	3	3
0	−9	−2
10	10	10

4. 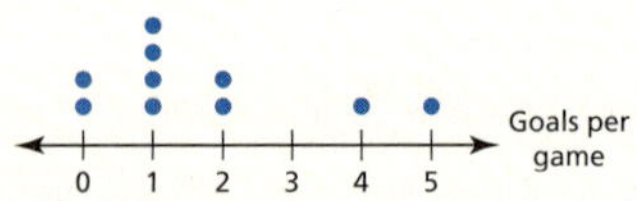

Find the range and standard deviation of each data set. Then compare your results.

5.

Bowling Scores			
Player A		Player B	
205	190	228	205
185	200	172	181
210	219	154	240
174	203	235	235
194	230	168	192

6.

Tablet Prices			
Store A		Store B	
\$140	\$180	\$225	\$310
\$200	\$250	\$260	\$190
\$150	\$190	\$190	\$285
\$250	\$160	\$160	\$240

Find the values of the measures shown after the given transformation.

Mean: 109 Median: 104 Mode: 96 Range: 45 Standard deviation: 3.6

7. Each value in the data set increases by 25.
8. Each value in the data set is multiplied by 0.6.

11.2 Box-and-Whisker Plots (pp. 593–598)

Make a box-and-whisker plot that represents the weights (in pounds) of pumpkins sold at a market.

16, 20, 11, 15, 10, 8, 8, 19, 11, 9, 16, 9

Step 1 Order the data. Find the median and the quartiles.

lower half — upper half

least value → 8 8 9 9 10 11 11 15 16 16 19 20 ← greatest value

first quartile, 9 median, 11 third quartile, 16

Step 2 Draw a number line that includes the least and greatest values. Graph points above the number line for the five-number summary.

Step 3 Draw a box using Q_1 and Q_3. Draw a line through the median. Draw whiskers from the box to the least and greatest values.

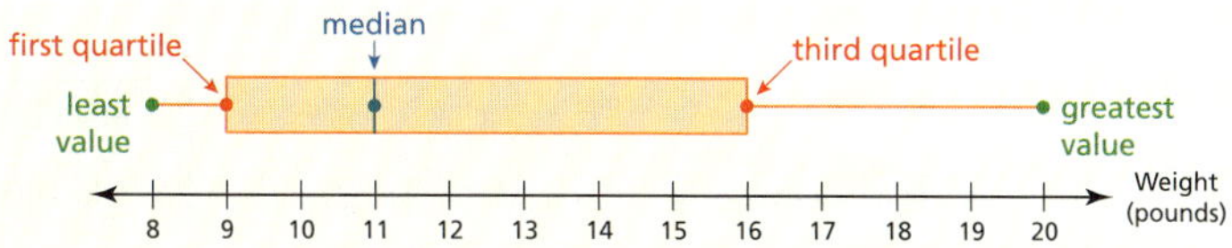

Make a box-and-whisker plot that represents the data. Identify the shape of the distribution.

9. Ages of volunteers at a hospital:
14, 17, 20, 16, 17, 14, 21, 18, 22

10. Masses (in kilograms) of lions:
120, 230, 180, 210, 200, 200, 230, 160

11.3 Shapes of Distributions (pp. 599–606)

The histogram shows the amounts of money a group of adults have in their pockets. Describe the shape of the distribution. Which measures of center and variation best represent the data?

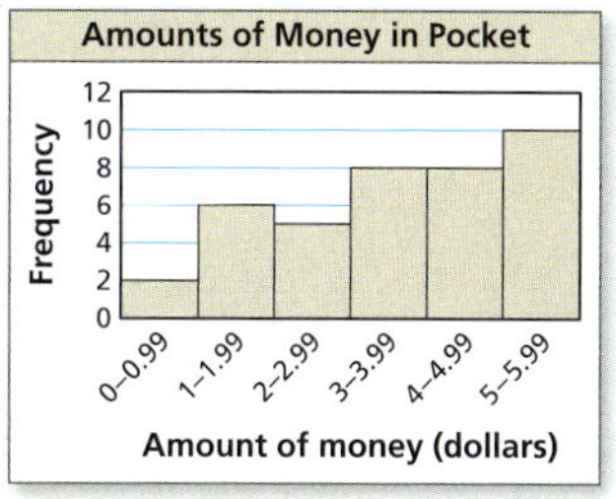

▶ The distribution is skewed left. So, use the median to describe the center and the five-number summary to describe the variation.

Amount	Frequency
0–0.99	9
1–1.99	10
2–2.99	9
3–3.99	7
4–4.99	4
5–5.99	1

11. The frequency table shows the amounts (in dollars) of money the students in a class have in their pockets.

a. Display the data in a histogram. Describe the shape of the distribution.

b. Which measures of center and variation best represent the data?

c. Compare this distribution with the distribution shown above using their shapes and appropriate measures of center and variation.

ANSWERS

9.

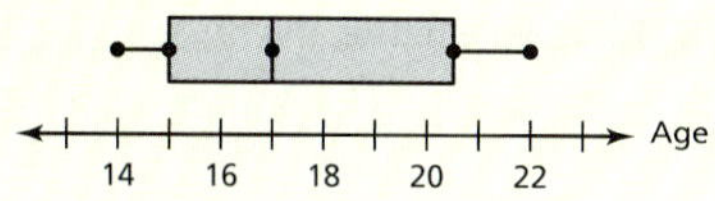

skewed right

10.

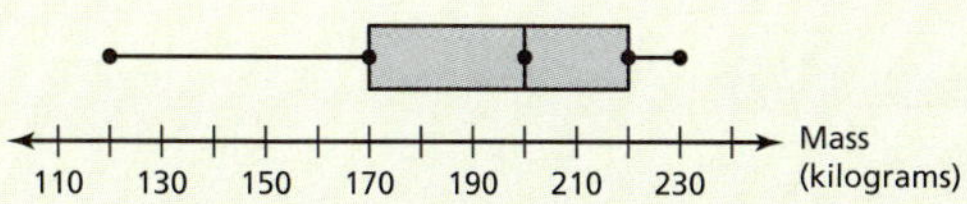

skewed left

11. a.

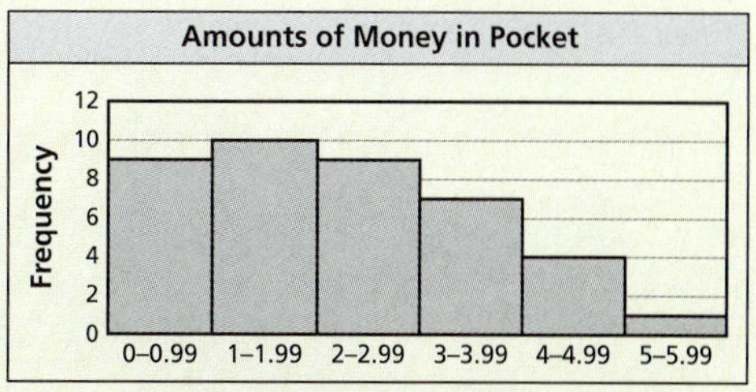

skewed right

b. median, five-number summary

c. The adults typically have more money in their pockets and the amounts vary more.

ANSWERS

12–13. See Additional Answers.

14. quantitative; Heights are numerical values.

15. qualitative; It does not make sense to measure grade levels.

11.4 Two-Way Tables (pp. 609–616)

You conduct a survey that asks 130 students about whether they have an after-school job. Sixty males respond, 38 of which have a job. Twenty-six females do not have a job. Organize the results in a two-way table. Find and interpret the marginal frequencies.

		After-School Job		
		Yes	No	Total
Gender	Males	38	22	60
	Females	44	26	70
	Total	82	48	130

60 males responded.
70 females responded.
130 students were surveyed.
82 students have a job.
48 students do not have a job.

12. The two-way table shows the results of a survey that asked shoppers at a mall about whether they like the new food court.

a. Make a two-way table that shows the joint and marginal relative frequencies.

b. Make a two-way table that shows the conditional relative frequencies based on the column totals.

		Food Court	
		Like	Dislike
Shoppers	Adults	21	79
	Teenagers	96	4

11.5 Choosing a Data Display (pp. 617–622)

Analyze the data and then create a display that best represents the data.

Ages of U.S. Presidents at Inauguration										
57	61	57	57	58	57	61	54	68	51	49
64	50	48	65	52	56	46	54	49	51	47
55	55	54	42	51	56	55	51	54	51	60
62	43	55	56	61	52	69	64	46	54	47

A stem-and-leaf plot is one appropriate way to display this quantitative data. It orders numerical data and shows how they are distributed.

Ages of U.S. Presidents at Inauguration

Stem	Leaf
4	2 3 6 6 7 7 8 9 9
5	0 1 1 1 1 1 2 2 4 4 4 4 4 5 5 5 5 6 6 6 7 7 7 7 8
6	0 1 1 1 2 4 4 5 8 9

Key: 5|0 = 50

13. Analyze the data in the table at the right and then create a display that best represents the data. Explain your reasoning.

Perfect Attendance	
Class	Number of students
freshman	84
sophomore	42
junior	67
senior	31

Tell whether the data are *qualitative* or *quantitative*. Explain.

14. heights of the members of a basketball team

15. grade level of students in an elementary school

11 Chapter Test

Describe the shape of the data distribution. Then determine which measures of center and variation best represent the data.

1. 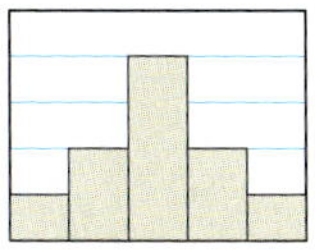

2. 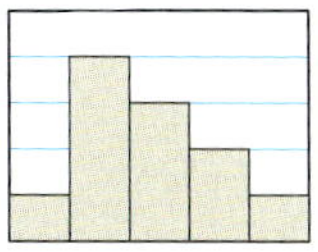

3. 

4. Determine whether each statement is *always*, *sometimes*, or *never* true. Explain your reasoning.

   a. The sum of the marginal relative frequencies in the "total" row and the "total" column of a two-way table should each be equal to 1.

   b. In a box-and-whisker plot, the length of the box to the left of the median and the length of the box to the right of the median are equal.

   c. Qualitative data are numerical.

5. Find the mean, median, mode, range, and standard deviation of the prices.

Prices of Shirts at a Clothing Store			
\$15.50	\$18.90	\$10.60	\$12.25
\$7.80	\$23.50	\$9.75	\$21.70

6. Repeat Exercise 5 when all the shirts in the clothing store are 20% off.

7. Which data display best represents the data, a histogram or a stem-and-leaf plot? Explain.

   15, 21, 18, 10, 12, 11, 17, 18, 16, 12, 20, 12, 17, 16

8. The tables show the battery lives (in hours) of two brands of laptops.

   a. Make a double box-and-whisker plot that represents the data.

   b. Identify the shape of each distribution.

   c. Which brand's battery lives are more spread out? Explain.

   d. Compare the distributions using their shapes and appropriate measures of center and variation.

Brand A	
20.75	18.5
13.5	16.25
8.5	13.5
14.5	15.5
11.5	16.75

Brand B	
10.5	12.5
9.5	10.25
9.0	9.75
8.5	8.5
9.0	7.0

Preferred method of exercise	Number of students
walking	20
jogging	28
biking	17
swimming	11
lifting weights	10
dancing	14

9. The table shows the results of a survey that asked students their preferred method of exercise. Analyze the data and then create a display that best represents the data. Explain your reasoning.

10. You conduct a survey that asks 271 students in your class whether they are attending the class field trip. One hundred twenty-one males respond, 92 of which are attending the field trip. Thirty-one females are not attending the field trip.

    a. Organize the results in a two-way table. Find and interpret the marginal frequencies.

    b. What percent of females are attending the class field trip?

ANSWERS

1. symmetric; mean, standard deviation
2. skewed right; median, five-number summary
3. skewed left; median, five-number summary
4. a. always; The marginal frequencies in the "total" row and the "total" column each represent the total population.

   b. sometimes; The distribution might not be symmetric.

   c. sometimes; Telephone numbers are qualitive data.
5. mean = \$15.00, median ≈ \$13.88, mode: none, range = \$15.70, standard deviation ≈ \$5.46
6. mean = \$12.00, median ≈ \$11.10, mode: none, range = \$12.56, standard deviation ≈ \$4.37
7. histogram; It shows the distribution better.

8–10. See Additional Answers.

If students need help...	*If students got it...*
Lesson Tutorials	Resources by Chapter • Enrichment and Extension • Cumulative Review
Skills Review Handbook	Performance Task
BigIdeasMath.com	Start the *next* Section

ANSWERS

1. yes; About 69% of males and about 65% of females have cell phones.
2. a. yes; $f(x)$ increases more rapidly than $h(x)$ when $x > 2$.
   b. yes; $g(x)$ increases more rapidly than $f(x)$ when $x > 5$.
   c. yes; $g(x)$ increases more rapidly than $h(x)$ when $x > 3$.
3. a. skewed right
   b. symmetric
   c. skewed left
   d. skewed right
4. -3; -2
5. $y = -3x^2, x \geq 0, y = \sqrt{-\frac{1}{3}x}$;
   $y = -x + 7, y = -x + 7$;
   $y = 2x - 4, y = \frac{1}{2}x + 2$;
   $y = x^2 - 5, x \geq 0, y = \sqrt{x + 5}$

11 Cumulative Assessment

1. You ask all the students in your grade whether they have a cell phone. The results are shown in the two-way table. Your friend claims that a greater percent of males in your grade have cell phones than females. Do you support your friend's claim? Justify your answer.

		Cell Phones	
		Yes	**No**
Gender	**Male**	27	12
	Female	31	17

2. Use the graphs of the functions to answer each question.

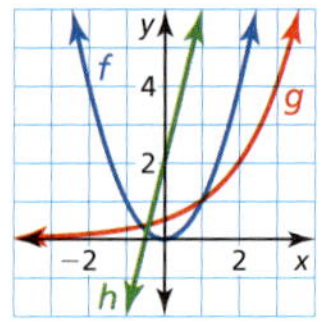

a. Are there any values of x greater than 0 where $f(x) > h(x)$? Explain.

b. Are there any values of x greater than 1 where $g(x) > f(x)$? Explain.

c. Are there any values of x greater than 0 where $g(x) > h(x)$? Explain.

3. Classify the shape of each distribution as *symmetric*, *skewed left*, or *skewed right*.

a. 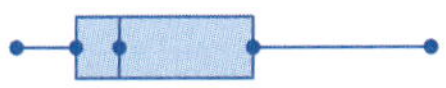

b. 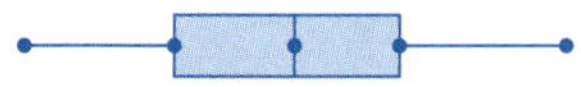

c. 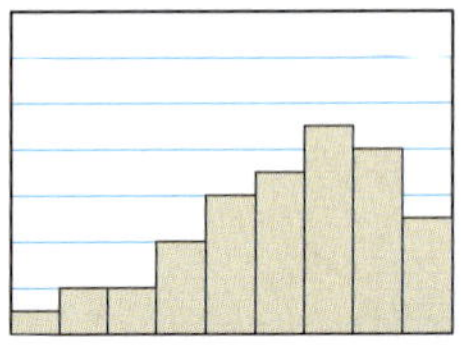

d. 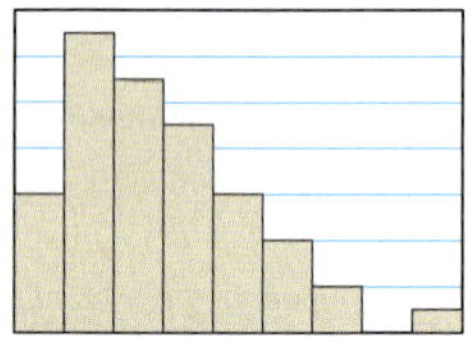

4. Complete the equation so that the solutions of the system of equations are $(-2, 4)$ and $(1, -5)$.

$$y = \square x + \square$$
$$y = 2x^2 - x - 6$$

5. Pair each function with its inverse.

$y = -3x^2, x \geq 0$	$y = -x + 7$	$y = 2x - 4$	$y = \sqrt{-\frac{1}{3}x}$
$y = \frac{1}{2}x + 2$	$y = x^2 - 5, x \geq 0$	$y = \sqrt{x + 5}$	$y = -x + 7$

6. The box-and-whisker plot represents the lengths (in minutes) of project presentations at a science fair. Find the interquartile range of the data. What does this represent in the context of the situation?

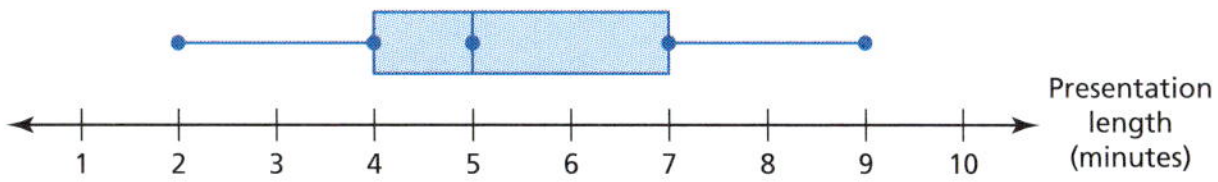

Ⓐ 7; The middle half of the presentation lengths vary by no more than 7 minutes.

Ⓑ 3; The presentation lengths vary by no more than 3 minutes.

Ⓒ 3; The middle half of the presentation lengths vary by no more than 3 minutes.

Ⓓ 7; The presentation lengths vary by no more than 7 minutes.

7. Scores in a video game can be between 0 and 100. Use the data set shown to fill in a value for x so that each statement is true.

a. When $x =$ ____, the mean of the scores is 45.5.

b. When $x =$ ____, the median of the scores is 47.

c. When $x =$ ____, the mode of the scores is 63.

d. When $x =$ ____, the range of the scores is 71.

Video Game Scores	
36	28
48	x
42	57
63	52

8. Select all the numbers that are in the range of the function shown.

$$y = \begin{cases} x^2 + 4x + 7, & \text{if } x \le -1 \\ \frac{1}{2}x + 2, & \text{if } x > -1 \end{cases}$$

0	$\frac{1}{2}$	1	$1\frac{1}{2}$	2	$2\frac{1}{2}$	3	$3\frac{1}{2}$	4

9. A traveler walks and takes a shuttle bus to get to a terminal of an airport. The function $y = D(x)$ represents the traveler's distance (in feet) after x minutes.

a. Estimate and interpret $D(2)$.

b. Use the graph to find the solution of the equation $D(x) = 3500$. Explain the meaning of the solution.

c. How long does the traveler wait for the shuttle bus?

d. How far does the traveler ride on the shuttle bus?

e. What is the total distance that the traveler walks before and after riding the shuttle bus?

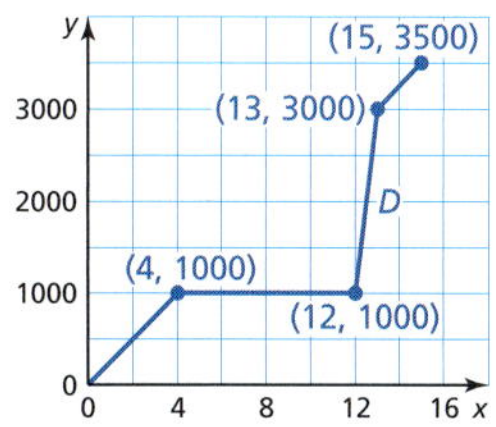

ANSWERS

6. C
7. a. 38
   b. 46
   c. 63
   d. 99
8. $2, 2\frac{1}{2}, 3, 3\frac{1}{2}, 4$
9. a. 500; The traveler is about 500 feet from the starting point after 2 minutes.
   b. $x = 15$; It takes the traveler 15 minutes to travel 3500 feet.
   c. 8 min
   d. 2000 ft
   e. 1500 ft

Additional Answers

Chapter 1

Maintaining Mathematical Proficiency

19. b. Add the opposite; *Sample answer:* $5 - (-3) = 5 + 3 = 8$

c. Multiply the absolute values.
If the signs are the same, then the product is positive.
If the signs are different, then the product is negative; *Sample answer:* $(-6)(-4) = 24$

d. Divide the absolute values. If the signs are the same, then the quotient is positive. If the signs are different, then the quotient is negative; *Sample answer:* $-15 \div 3 = -5$

1.2 Explorations

4. Connecting a vertex with each of the other vertices in a polygon creates $n - 2$ triangles, each of which has a total angle measure of 180°.

5. 8; Solve the equation $180(n - 2) = 1080$ for n.

1.2 Monitoring Progress and Modeling with Mathematics

48. *Sample answer:*

Component	Student's score	Weight	Score × weight
Class Participation	92%	0.2	18.4%
Homework	95%	0.3	28.5%
Midterm Exam	88%	0.2	17.6%
Final Exam	85%	0.3	25.5%
Total		1	90.0%

The student would need a score of 85% on the final exam to get a 90% because

$$92(0.2) + 95(0.3) + 88(0.2) + f(0.3) = 90$$
$$f = 85.$$

52. $x = \frac{3}{4} - a$

53. $x = \frac{12.5 + b}{a}$

54. $x = \frac{c - b}{a}$

55. $x = -\frac{8}{b}$

56. $x = \frac{9b}{c}$

1.2 Maintaining Mathematical Proficiency

57. $m + 5$

58. $9 - 2b$

59. -2

60. a

61. b

62. b

63. a

64. b

65. a

1.3 Monitoring Progress and Modeling with Mathematics

39. a. *Sample answer:* $3x + 12 = 2x + x$; simplifies to a statement that is never true

b. *Sample answer:* $5x + 3 = 2x + 3 + 3x$; simplifies to a statement that is always true

40. *Sample answer:*

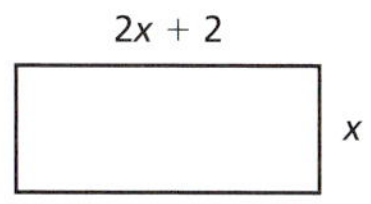

The perimeters are both $6x + 4$.

1.3 Maintaining Mathematical Proficiency

41. $-4, |2|, |-4|, 5, 9$

42. $-|21|, -16, |-10|, 22, |-32|$

43. $-19, -18, |-18|, |22|, |-24|$

44. $-|-3|, -2, -1, |0|, |2|$

1.4 Monitoring Progress and Modeling with Mathematics

61. 1; If $c = d$, then the absolute value expression will be equal to 0; 2; If $c > d$, then $d - c$ will be negative. When this value is divided by a negative value of a, the result will be positive.

1.4 Maintaining Mathematical Proficiency

62. Addition Property of Equality

63. Division Property of Equality

64. 9 m

65. 6 in.

66. 12 ft

67. 9 cm

1.5 Maintaining Mathematical Proficiency

47. 35

48. 34

49. 47

50. -29

51. $x = -2, x = 8$;

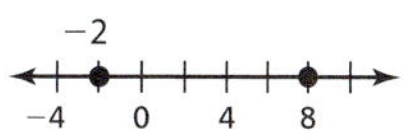

52. $y = 1, y = 7$;

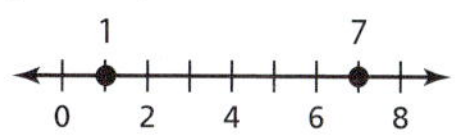

53. no solution

54. $s = -15, s = -3$;

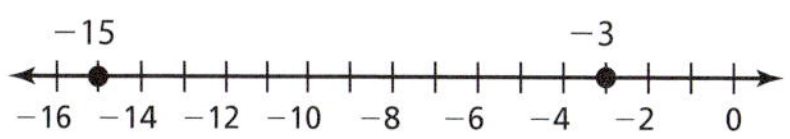

Chapter 2

2.1 Maintaining Mathematical Proficency

60. $x = 1$

61. $y = 14$

62. $y = 2$

63. $y = -1$

64. $x = \frac{v}{yz}$

65. $x = \frac{s - 2r}{3}$

66. $x = \frac{w - 2}{3}$

67. $x = n - \frac{1}{2}$

2.2 Monitoring Progress and Modeling with Mathematics

25. a. $38 + w \le 50$; $w \le 12$

b. no; The total being added is 14 pounds, which is not a solution of the inequality found in part (a).

26. $19.76 + x \ge 25$; $x \ge 5.24$

27. The graph is going in the wrong direction.

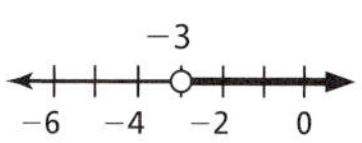

28. In the second step, 10 needs to be added to each side; $-10 + 10 + x \geq -9 + 10; x \geq 1$

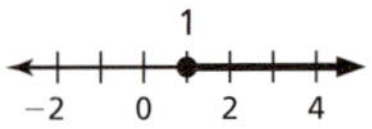

29. 33 or more goals

36. a. no; It is also possible that the number of students with brown eyes is greater than the number of students with brown hair.
   b. no; It is also possible that the number of students with brown eyes is greater than 10 more than the number of students with brown hair.
   c. yes; All students with both characteristics must have brown hair.
   d. yes; If H is greater than a number, then $H + 10$ must also be greater than the number.
   e. no; If all students with brown eyes also have brown hair, then the values would be equal.
   f. yes; If H is greater than a number, then $H + 10$ must also be greater than the number.

37. a. $x \geq 6$

   b. $x < 17.7$

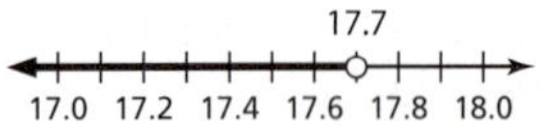

38. $b < a < d < c$

2.2 Maintaining Mathematical Proficiency

39. -63 40. 132
41. 9 42. -4
43. $x = 4$ 44. $y = 6$
45. $s = -104$ 46. $n = -29.2$

2.3 Monitoring Progress and Modeling with Mathematics

24. $n > -16$ 25. $x > \frac{3}{8}$
26. $y < 4$
27. The inequality should not be reversed when multiplying each side by $\frac{3}{2}$; $\frac{3}{2} \cdot (-6) > \frac{3}{2} \cdot \frac{2}{3}x$; $-\frac{18}{2} > x$; $-9 > x$; $x < -9$; The solution is $x < -9$.
28. The inequality should be reversed when dividing by -4; $\frac{-4y}{-4} \geq \frac{-32}{-4}$; $y \geq 8$; The solution is $y \geq 8$.
29. $(14 \cdot 14)c \leq 700$; ft, ft, dollars

   $196c \leq 700$; ft^2, dollars

   $c \leq 3.57$; dollars/ft^2

35. a. $A > B$ or $B < A$
   b. $-A < -B$ or $-B > -A$
   c. As numbers move farther away from zero, their absolute value becomes larger. $A > B$ and $|A| > |B|$. $-A < -B$ and $|A| > |B|$.
36. Because x may represent a positive or negative number, the inequality may need to be reversed when multiplying by x.
37. $\frac{C}{2\pi} > 5, C > 10\pi$ 38. $\frac{d}{0.75} < 18, d < 13.5$
39. $36p \geq 12, p \geq \frac{1}{3}$

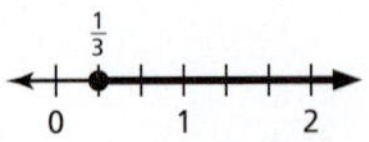

2.3 Maintaining Mathematical Proficiency

40. $x = 2$ 41. $y = -4$
42. $n = 1$ 43. $z = 6$
44. 85% 45. $\frac{16}{30}$
46. 120% 47. $\frac{2}{3}$

2.4 Monitoring Progress and Modeling with Mathematics

29. In the first step, you need to use the Distributive Property on the left side; $x + 24 \geq 12; x \geq -12$
30. Because the final statement is false, there is no solution.
31. $20n + 100 \leq 320, n \leq 11$
38. *Sample answer:* $\frac{25.5 + 24.3 + 24.8 + 23.5 + 3x}{7} < 24$; $x < 23.3$; run three more races in less than 23.3 minutes each
39. $a = 4$ 40. $a = 3$

2.4 Maintaining Mathematical Proficiency

41. $6y \leq 10$ 42. $p + 7 > 24$
43. $\frac{r}{7} \leq 18$

2.5 Monitoring Progress and Modeling with Mathematics

20. $b < -3$ *or* $b \geq 5$

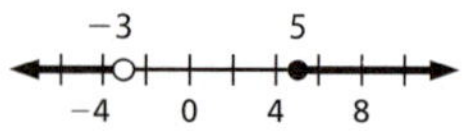

24. $1.16h < 150$ *or* $1.16h > 220$; heights less than about 129.3 cm or greater than about 189.7 cm
25. no solution
26. $m > -5$

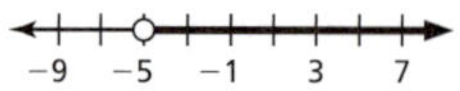

27. $y > 7$

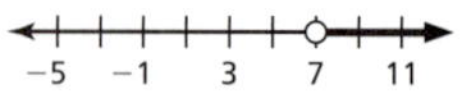

28. all real numbers

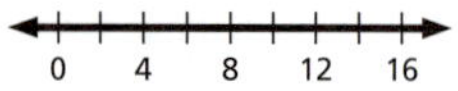

29. all real numbers

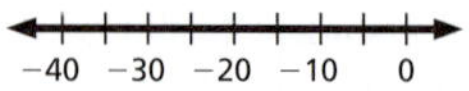

30. no solution 31. $\leq$
32. *Sample answer:* You are biking on a trail. The spot where you start on the trail is the 6-mile mark. You bike 4 miles on the trail, and then leave the trail for a different path. The graph represents where you bike on the trail.
33. $7 + 5 > x \rightarrow x < 12, 7 + x > 5 \rightarrow x > -2, 5 + x > 7 \rightarrow x > 2$; no; A value of 1 does not make the inequality $x > 2$ true.
34. a. $50 \leq P \leq 90$
   b. no; Solving the inequality $50 \leq R - 125 \leq 90$ indicates that the revenue must be between \$175 million and \$215 million.

2.5 Maintaining Mathematical Proficiency

35. $d = 54, d = -54$ 36. no solution

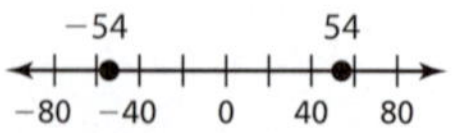

37. $r = 3, r = \frac{1}{2}$

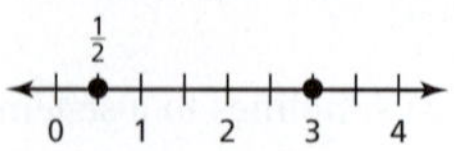

38. $w = -26, w = -\frac{2}{3}$

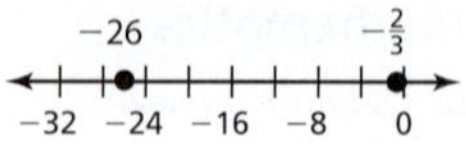

39. 4; The data values are clustered close together.
40. 2.8; The data values are clustered close together.

2.6 Explorations

2. **a.**

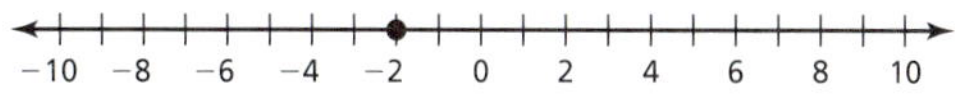

b.

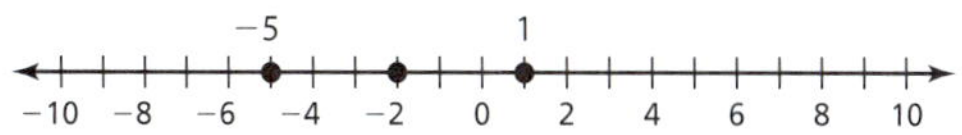

They are solutions of the absolute value inequality.

c. Plot the distances to determine the endpoints of the solution.

2.6 Monitoring Progress and Modeling with Mathematics

19. $|w - 500| \leq 30$; 470 to 530 words

20. $|T - 37| \leq 3$; 34°C to 40°C

21. did not rewrite the absolute value inequality as a compound inequality; $-20 < x - 5 < 20$; $-15 < x < 25$

22. The compound inequality should be joined by "or;" and both inequality symbols are reversed; $x + 4 < -13$ *or* $x + 4 > 13$; $x < -17$ *or* $x > 9$

23. $|n| < 6$; $-6 < n < 6$

24. $|n - 3| > 9$; $n < -6$ *or* $n > 12$

25. $|\frac{1}{2}n - 14| \leq 5$; $18 \leq n \leq 38$

26. $|2n + 1| \geq 10$; $n \leq -\frac{11}{2}$ *or* $n \geq \frac{9}{2}$

27. the gasket with a weight of 0.53 lb

28. $d = 0.86$

37. no solution; An absolute value cannot be negative; all real numbers; An absolute value must be positive or zero, so it will always be greater than any negative number.

38. *Sample answer:* $|x - 2| \geq 3$; $|x - 2| < 3$; $|x - 2| \leq 3$; $|x - 2| > 3$; A segment indicated <, two opposite rays indicated >, and an open or closed circle indicated whether or not to add "or equal to."

39. The solution of $|x| < 5$ is the set of values that make both of 2 inequalities true, the solution of $|x| > 5$ is the set of values that make either of 2 inequalities true.

40. $6 < x < 7$; *Sample answer:* Solve each inequality, then draw a sketch of the 2 solutions on the same number line to see where they intersect.

2.6 Maintaining Mathematical Proficiency

41–44.

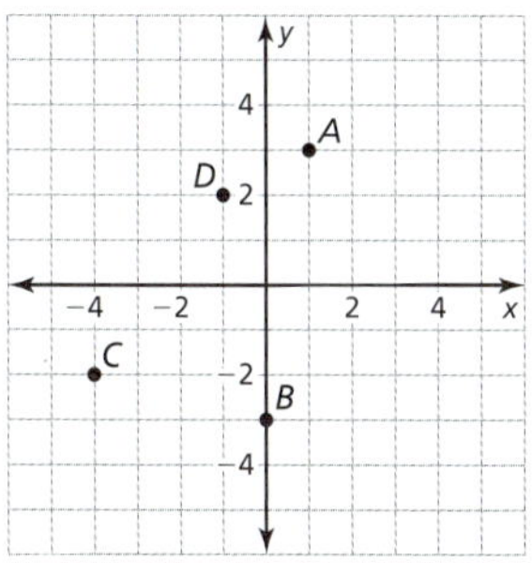

41. Quadrant I

42. y-axis

43. Quadrant III

44. Quadrant II

45. 1, 6, 11, 16, 21

46. 1, −1, −3, −5, −7

Chapter 2 Test

7. $-3 < c < 5.5$

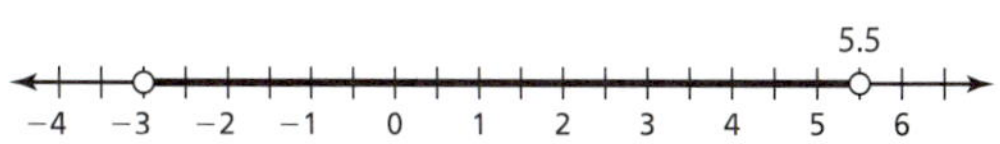

Chapter 3

3.1 Monitoring Progress and Modeling with Mathematics

29. no; A vertical line does not represent a function.

30. *Sample answers:* the score and letter grade on a 10-point quiz; The letter grade is the dependent variable and the score is the independent variable; domain: 0, 1, 2, 3, 4, 5, 6, 7, 8, 9, 10; range: A, B, C, D, F

31. no; Items that cost the same to make could be sold for different prices.

32. yes; Each possible selling price would have the same amount of tax charged.

33. yes; Each student has exactly one homeroom teacher.

34. no; Each chaperone has more than one student assigned.

35. true

36. false; A relation is not a function when an input value has more than one output value.

37. false; More than one input can have the same output in a function, so reversing the values may not produce a function.

38. false; Because more than one input can be paired with the same output, the outputs could be selected from a finite list.

39. **a.** $P = h + 23$

b. P is the dependent variable and h is the independent variable.

c. domain: $3 < h < 23$; range: $26 < P < 46$

40. domain: all real numbers; range: $y \geq 0$

41. domain: all real numbers; range: $y \leq 0$

42. domain: all real numbers; range: $y \geq -6$

43. domain: all real numbers; range: $y \leq 4$

3.1 Maintaining Mathematical Proficiency

44. $y < 16$

45. $3 \geq x$

46. $7 \leq \frac{d}{-5}$

47. $w + 4 > -12$

48. 121

49. 81

50. −25

51. 32

3.2 Monitoring Progress and Modeling with Mathematics

50. *Sample answer:*

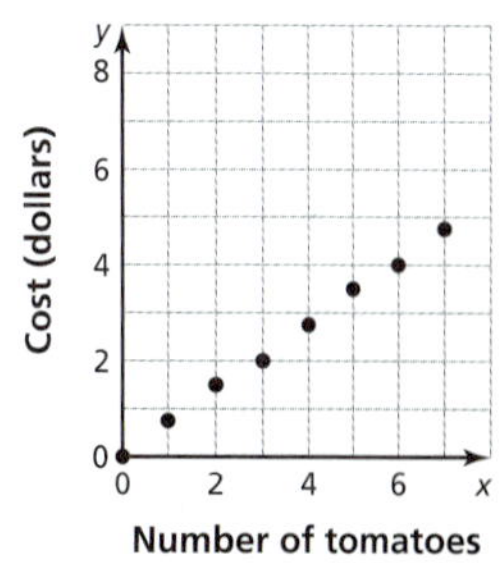

The tomatoes are sold for \$0.75 each or 3 for \$2.00.

51. linear; As x increases by 1, y increases by 4. The rate of change is constant.

52. **a.** Your friend ran farther in the same amount of time; You run at a constant rate and your friend does not; Your graph is a line, and your friend's graph is not a line; A person may not run at a constant rate because of fatigue.

b. The domain for both functions is $0 \leq x \leq 60$; You both run for a total of 60 minutes.

53. *Sample answer:* how long it takes an ice cube to melt as a function of its temperature in degrees Celsius

54. *Sample answer:* profit (or loss) on the sale of x units at \$1 each that cost \$0.75 to make, plus \$50 in fixed costs

3.2 Maintaining Mathematical Proficiency

55. no; The graph does not pass through the origin.

56. yes; The graph passes through the origin and shows a constant rate of change.

57. no; The graph does not show a constant rate of change.

58. 20

59. 14

60. -24

61. 4

3.3 Lesson Monitoring Progress

8. 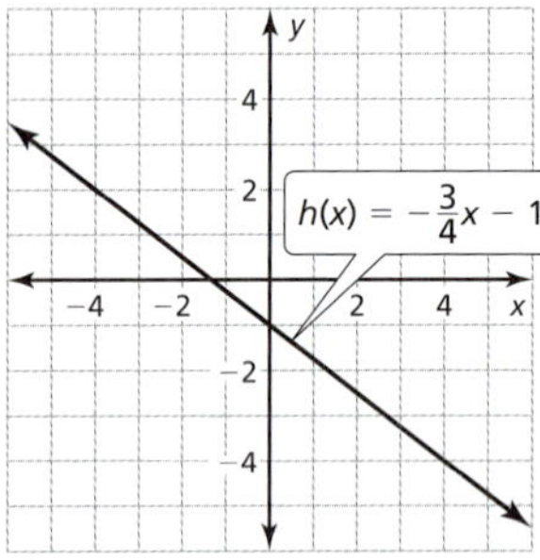

3.3 Monitoring Progress and Modeling with Mathematics

25. 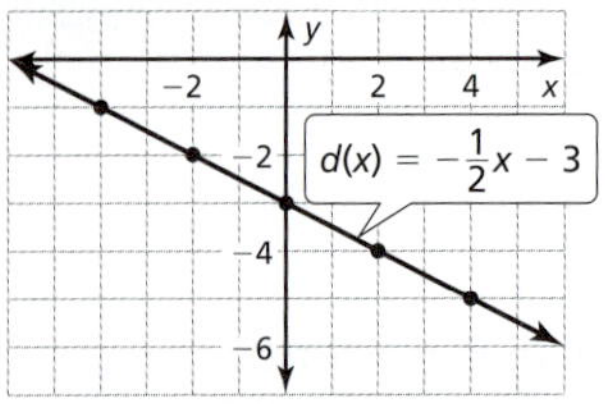

26. 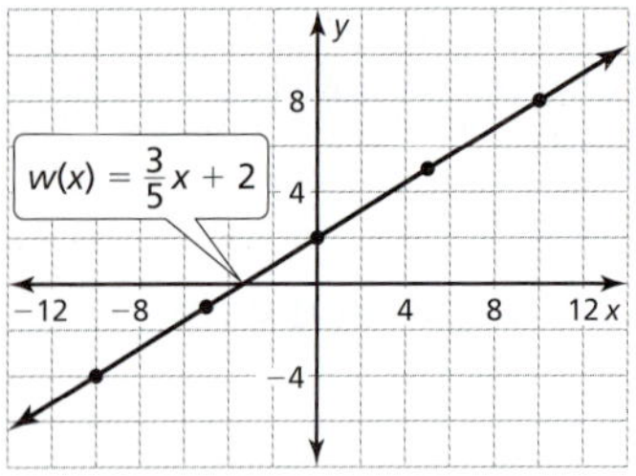

27. 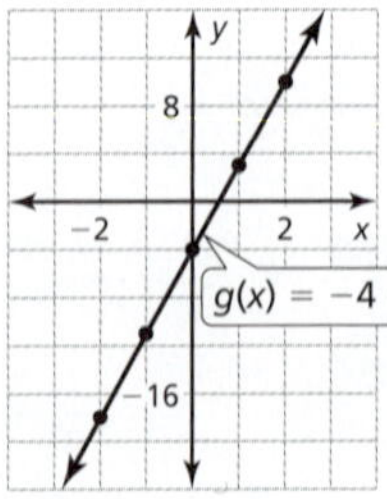

28. 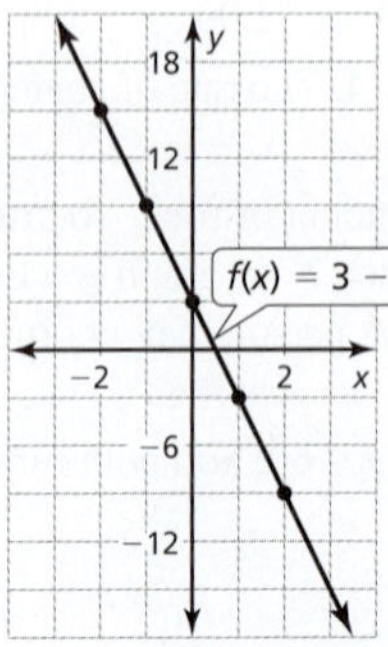

30. Master Remodeling; Using the function, the cost for Certified Remodeling is \$250. The table shows an increase of \$30 for every increase of 2 hours, so the cost for Master Remodeling is \$220.

31. no; Because the function rule is unknown, it is unknown whether an increase in x would result in an increase or decrease in the value of the function.

32. *Sample answer:* You receive a paycheck between days 0 and 2, and then spend some of it between days 2 and 4.

33. **a.** $d(r) = 2r$; 10; The diameter of the circle is 10 feet.
    b. $A(r) = \pi r^2$; 25π or about 78.5; The area of the circle is 25π square feet.
    c. $C(r) = 2\pi r$; 10π or about 31.4; The circumference of the circle is 10π feet.

34. **a.** It decreases rapidly, then slowly increases.
    b. *Sample answer:* 425; The school attendance was about 425 students 13 weeks after the flu outbreak.
    c. *Sample answer:* 1, 10; The school attendance was about 400 students 1 week and 10 weeks after the flu outbreak.
    d. *Sample answer:* about 350 students; about 4 weeks after the flu outbreak
    e. *Sample answer:* about 460 students; The graph does not show any attendance values greater than 460.

35. **a.** (5, 9)
    b. (5, −3)

36. false; The function could include operations that would not give the same result. For example if $f(x) = x^2$, $f(3 + 4) \neq f(3) + f(4)$.

3.3 Maintaining Mathematical Proficiency

37. $9 \le x \le 17$;

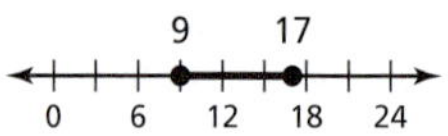

38. $a < -7$ *or* $a > 15$;

39. $-3 < k < -\frac{1}{3}$;

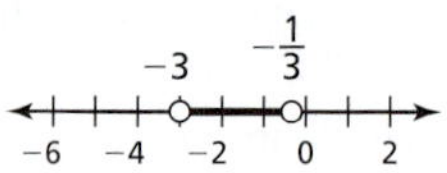

40. $d < -8$ *or* $d > -\frac{1}{2}$;

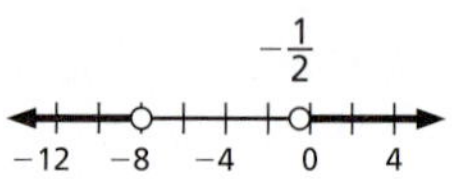

41. $-1 \le y < 3$;

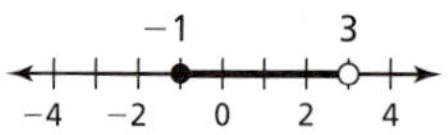

42. $v \le 2$;

3.4 Lesson Monitoring Progress

1. 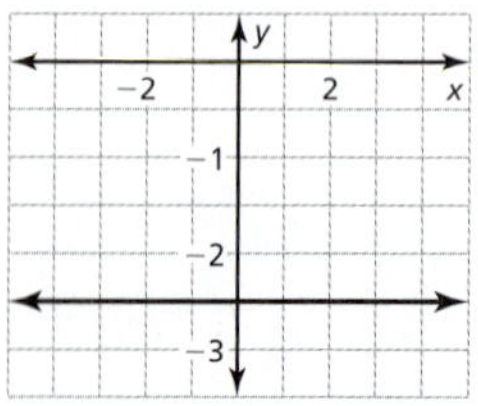

2. 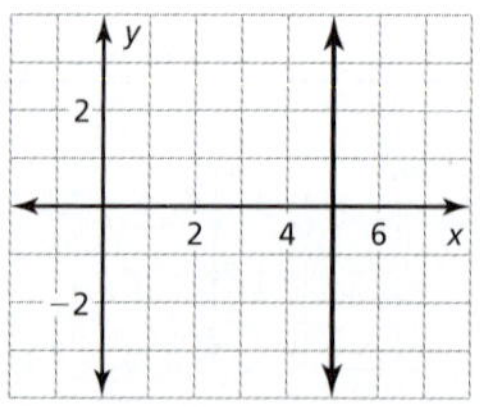

3.4 Monitoring Progress and Modeling with Mathematics

13.

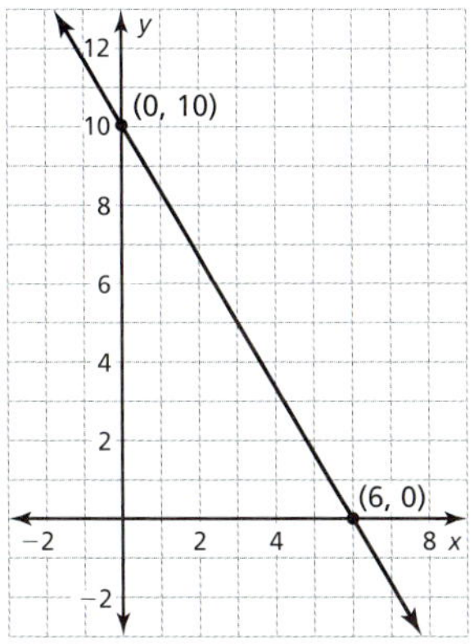

15.

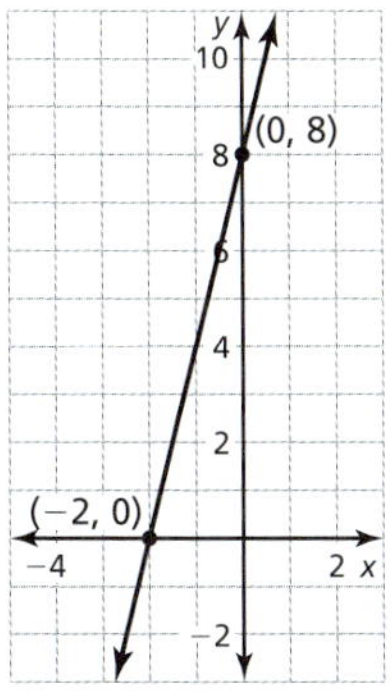

16.

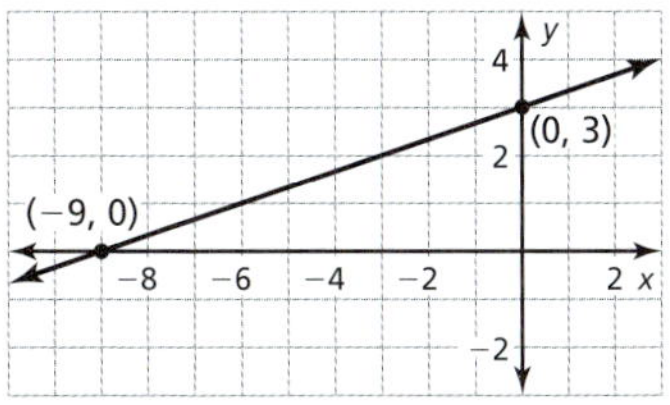

17.

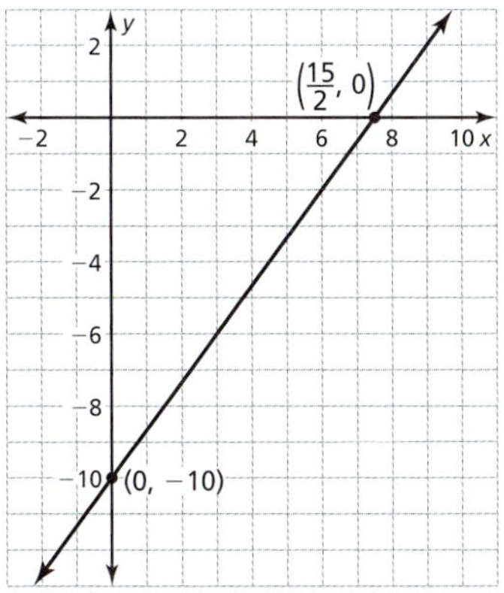

18.

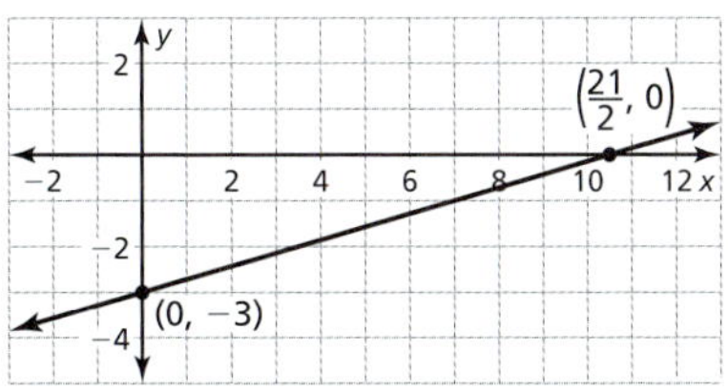

19.

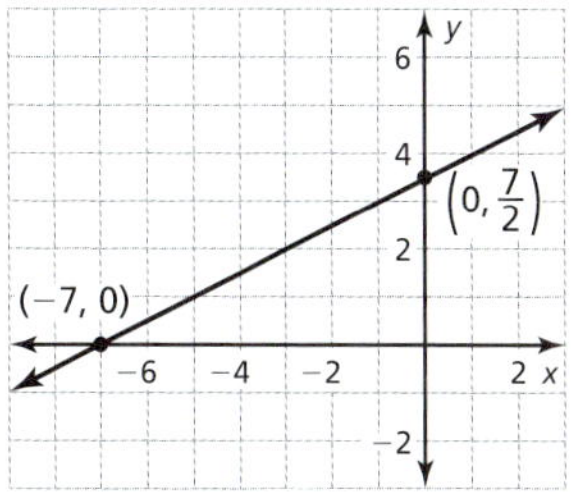

20.

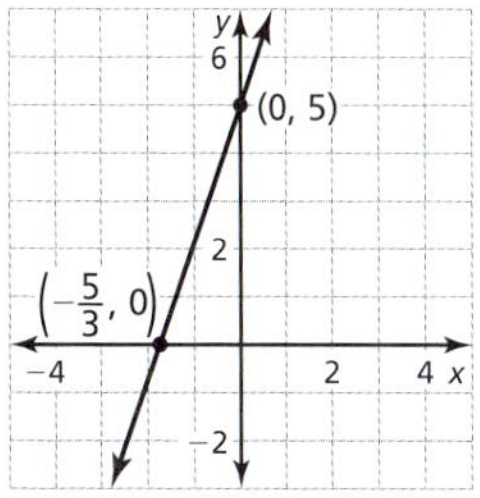

21.

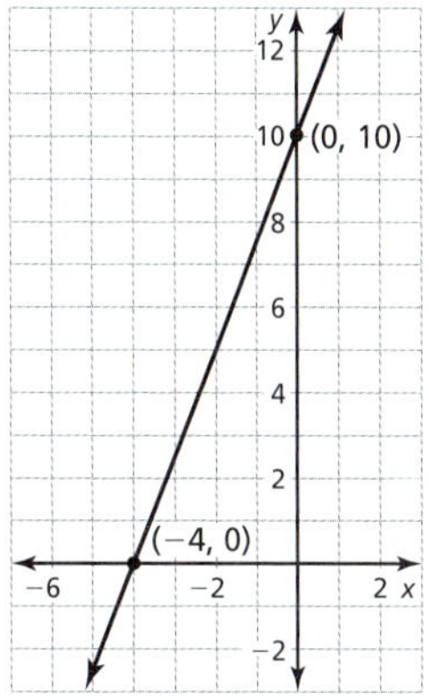

22.

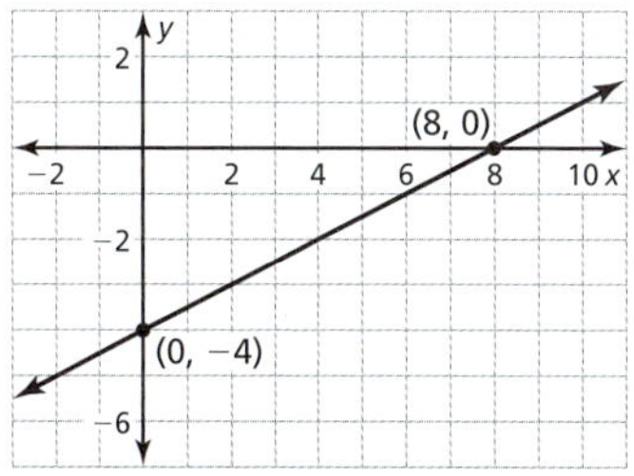

23. a.

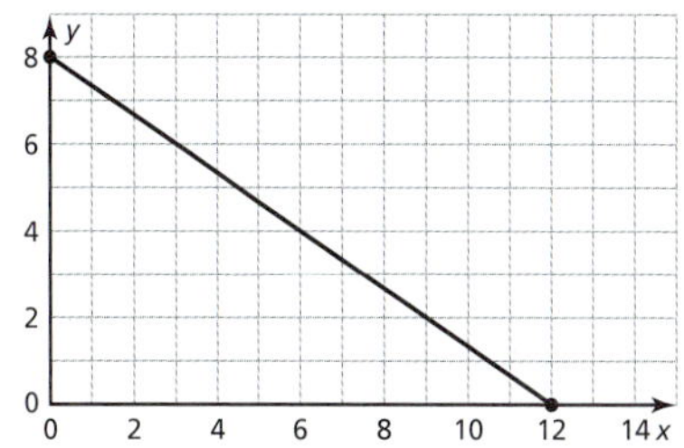

The x-intercept shows they can take 12 cars and 0 vans. The y-intercept shows they can take 8 vans and 0 cars.

b. *Sample answers:* 12 cars, 0 vans; 9 cars, 2 vans; 6 cars, 4 vans; 0 cars, 8 vans

24. a. 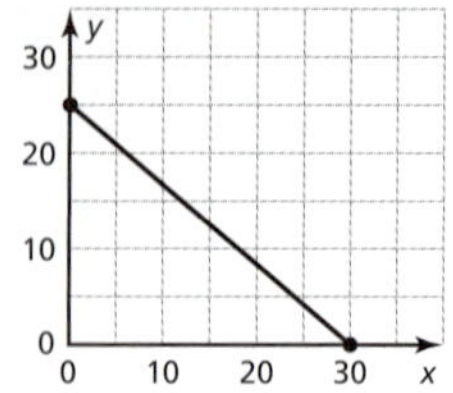

The x-intercept shows that you can buy 30 short-sleeve shirts and 0 long-sleeve shirts. The y-intercept shows that you can buy 25 long-sleeve shirts and 0 short-sleeve shirts.

b. 15

25. The equations locate 2 separate intercepts, not 1. The intercepts are 8 and 2.

26. The x- and y-coordinates of the intercepts are reversed. The x-intercept is at (5, 0) and the y-intercept is at (0, 2).

28. a. 27; 18; The x-intercept indicates 27 two-point baskets were made if there were no three-point baskets. The y-intercept indicates 18 three-point baskets were made if there were no two-point baskets.

b. no; If y is odd, then $3y$ will also be odd. Subtracting an odd number from 54 will also be odd, so dividing this result by 2 to find x will never give a whole number.

c. 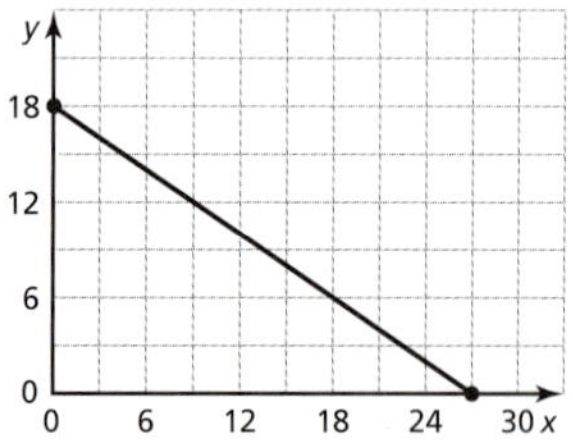

Sample answer: 24 two-point baskets and 2 three-point baskets; 9 two-point baskets and 12 three-point baskets

33. 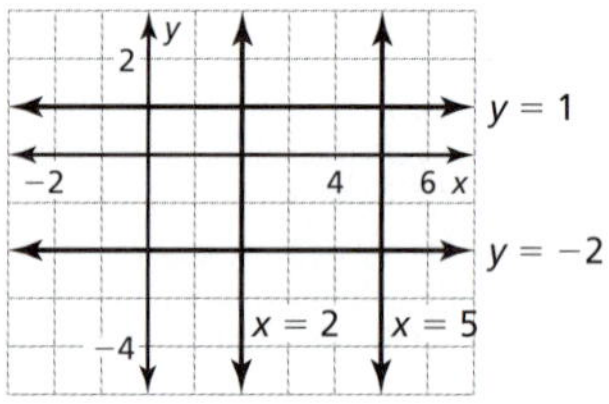

square; The graphs are 2 horizontal lines and 2 vertical lines, which will intersect at right angles. The length of each side is 3 units.

34. a. The x-intercept indicates 90 students can go if none of them choose the meal plan. The y-intercept indicates 60 students can go if all of them choose the meal plan.

b. domain: 0 to 90 students not choosing the meal plan; range: 0 to 60 students choosing the meal plan

3.4 Maintaining Mathematical Proficiency

39. $\frac{1}{2}$

40. 2

41. $-\frac{4}{5}$

42. $\frac{5}{3}$

3.5 Lesson Monitoring Progress

11. 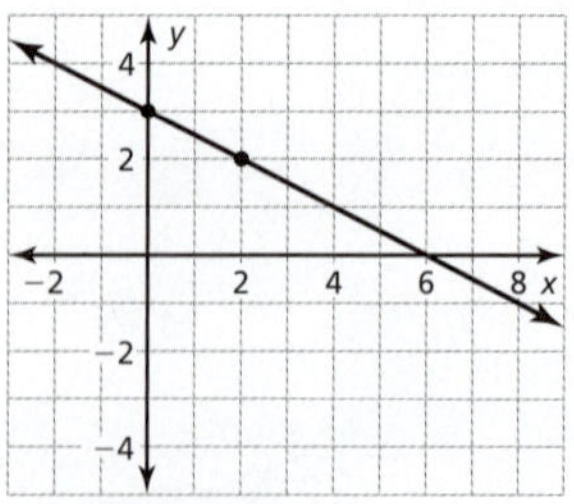

x-intercept: 6

12. 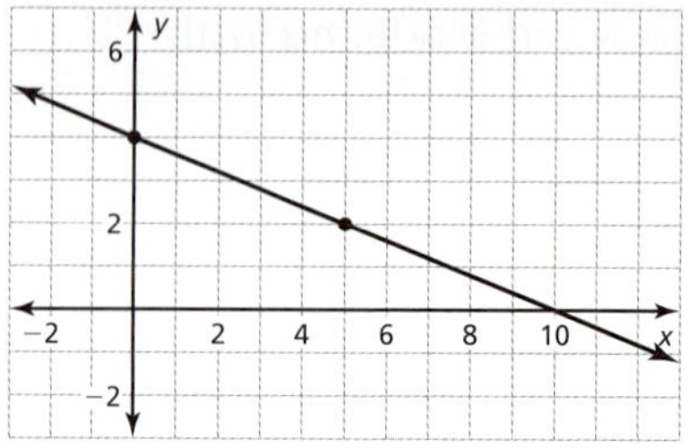

slope: $-\frac{2}{5}$; y-intercept: 4; x-intercept: 10

13. a. 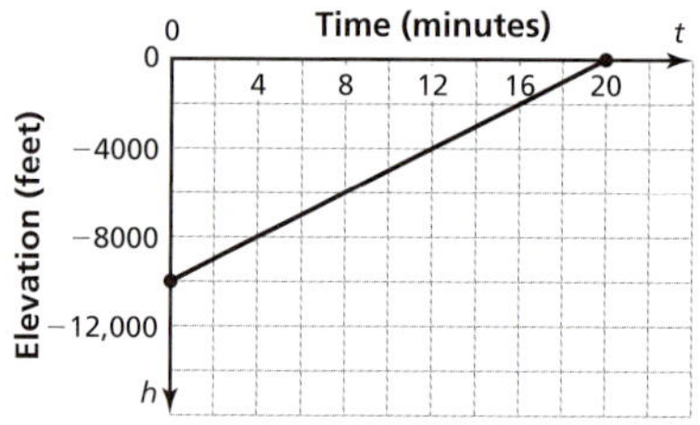

domain: $0 \le t \le 20$; range: $-10{,}000 \le h \le 0$

b. The slope is 500. So, the submersible ascends at a rate of 500 feet per minute; The h-intercept is $-10{,}000$. So, the elevation of the submersible when the ascent begins is $-10{,}000$ feet; The t-intercept is 20. So, the submersible takes 20 minutes to reach sea level.

3.5 Monitoring Progress and Modeling with Mathematics

30. 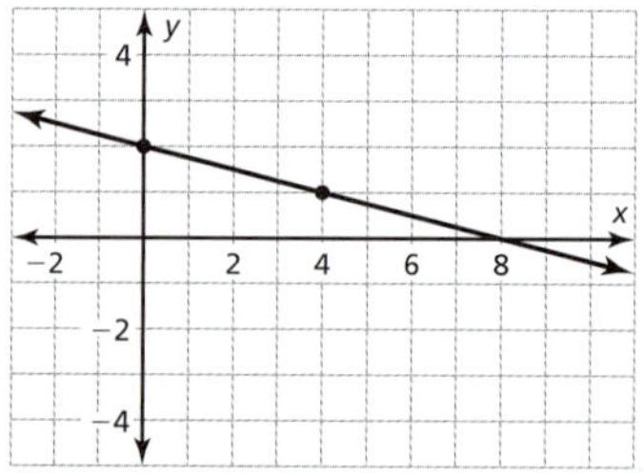

x-intercept: 8

31. 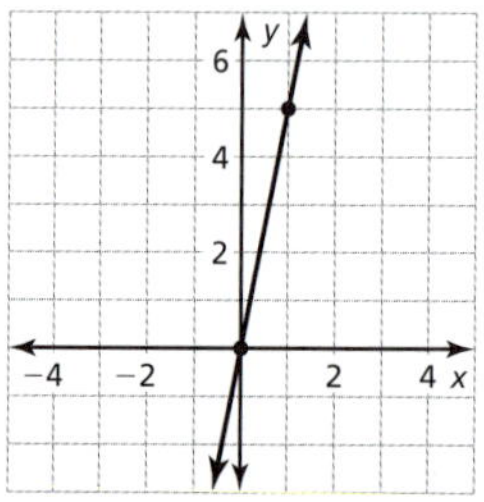

x-intercept: 0

32. 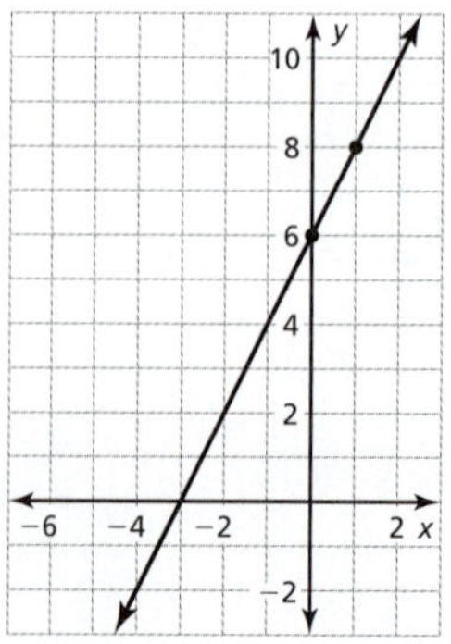

x-intercept: -3

33. 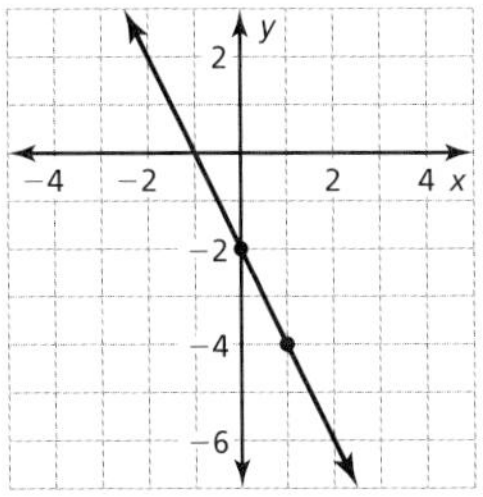

slope: -2; y-intercept: -2; x-intercept: -1

34. 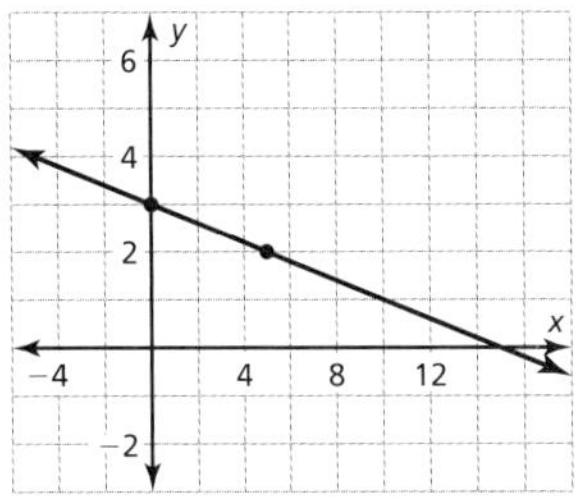

slope: $-\frac{1}{5}$; y-intercept: 3; x-intercept: 15

35. 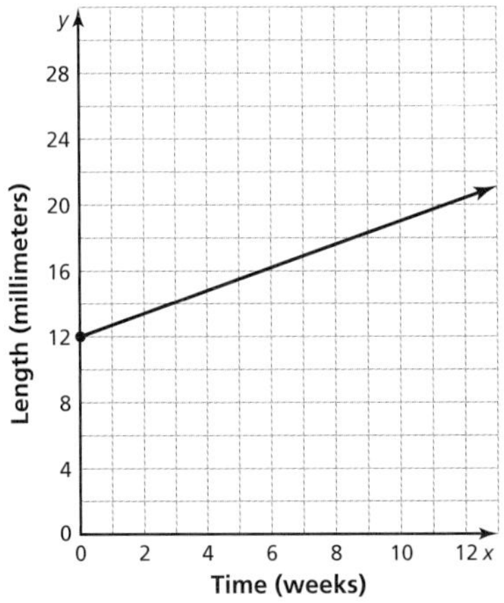

slope: $\frac{7}{10}$; y-intercept: 12; So, the right index fingernail is initially 12 millimeters long.

36. 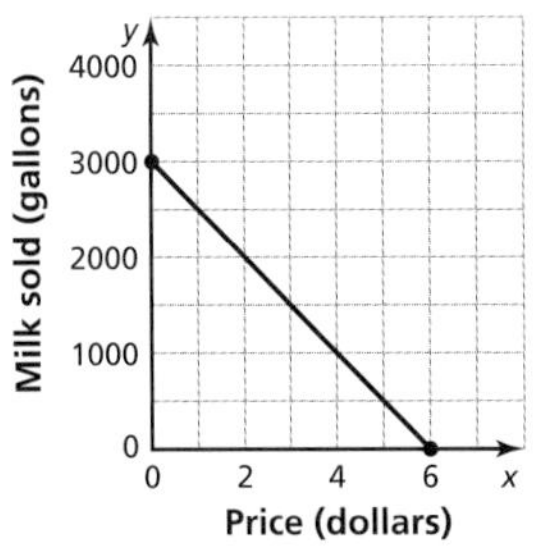

slope: -500; x-intercept: 6; So, no milk is sold when the price reaches \$6 per gallon; y-intercept: 3000; So, the maximum amount of milk sold is 3000 gallons.

37. a. 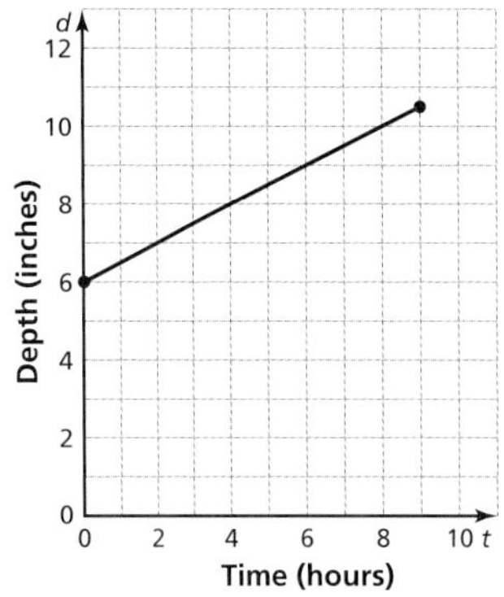

domain: $0 \le t \le 9$; range: $6 \le d \le 10\frac{1}{2}$

b. slope: $\frac{1}{2}$; So, $\frac{1}{2}$ inch of snow falls every hour during the storm; d-intercept: 6; So, there were 6 inches of snow already on the ground at the start of the storm.

38. a. 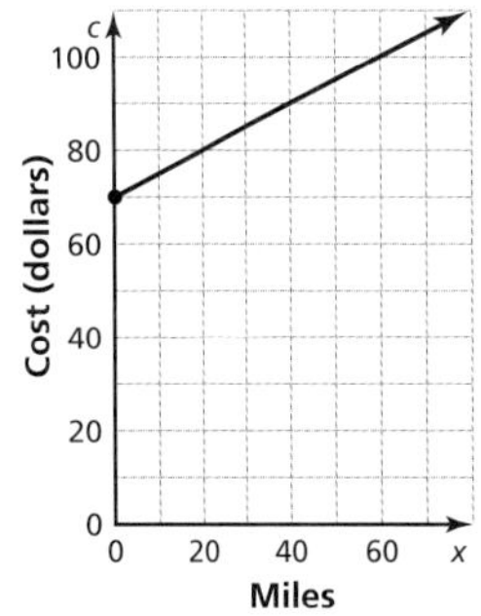

domain: $x \ge 0$; range: $c \ge 70$

b. slope: 0.5; So, the mileage rate is \$0.50 per mile; c-intercept: 70; So, the basic rental fee is \$70.

39. 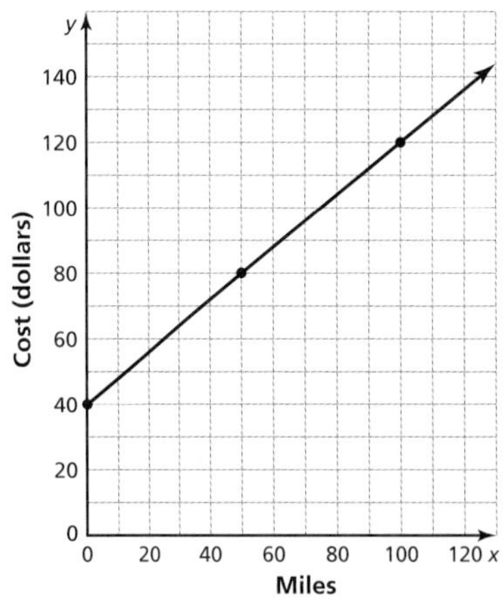

The slope of $\frac{4}{5}$ is greater than the slope in Exercise 38. The y-intercept of 40 is less than the c-intercept in Exercise 38.

42.

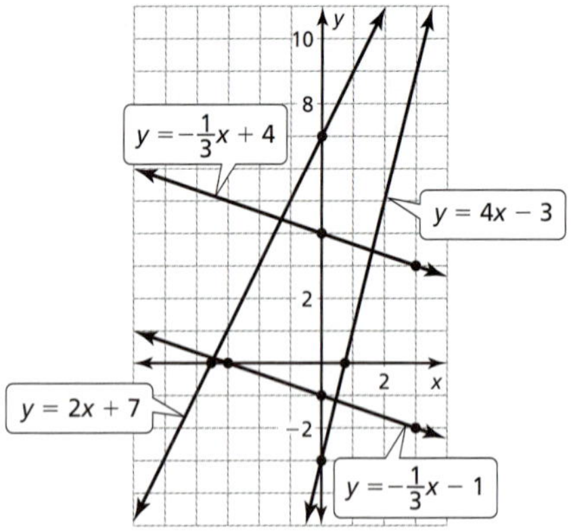

a. trapezoid; The 2 lines with a slope of $-\frac{1}{3}$ are parallel, and the other 2 lines are not parallel.

b. Parallel lines have the same slope.

48. *Sample answer:*

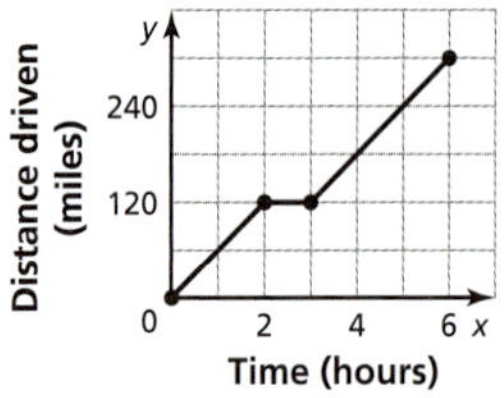

The first line on the graph represents your family driving 60 miles per hour for 2 hours. The second line represents your family not driving at all for 1 hour. The third line represents your family driving 60 miles per hour for 3 hours to complete your trip to the beach.

49. a.

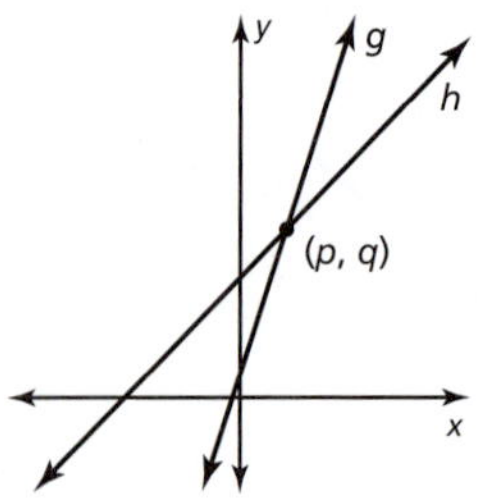

b. a is the y-intercept of the graph of g; b is the y-intercept of the graph of h

c. 8 units greater

50. a. You commute to school by walking $\frac{1}{2}$ mile to the bus stop in 10 minutes, then waiting 4 minutes for the bus, then riding the bus for 2 miles in 4 minutes.

b. The slope of the first part is 0.05. So, you walk at an average speed of 0.05 mile per minute; The slope of the second part is 0. So, you do not move while waiting for the bus; The slope of the last part is 0.5. So, the bus travels at an average speed of 0.5 mile per minute.

53.

$$\frac{y_2 - y_1}{x_2 - x_1} = \frac{(mx_2 + b) - (mx_1 + b)}{x_2 - x_1}$$
$$= \frac{mx_2 + b - mx_1 - b}{x_2 - x_1}$$
$$= \frac{mx_2 - mx_1}{x_2 - x_1}$$
$$= \frac{m(x_2 - x_1)}{x_2 - x_1}$$
$$= m$$

3.5 Maintaining Mathematical Proficiency

54. $A'(-3, 0)$, $B'(-1, 0)$, $C'(-1, -4)$, $D'(-3, -4)$

55. $X'(0, 0)$, $Y'(4, 2)$, $Z'(4, 0)$

56. $Q'(4, 4)$, $R'(2, 2)$, $S'(2, -1)$, $T'(4, -3)$

57. nonlinear; It cannot be rewritten in the form $y = mx + b$.

3.6 Explorations

1. a.

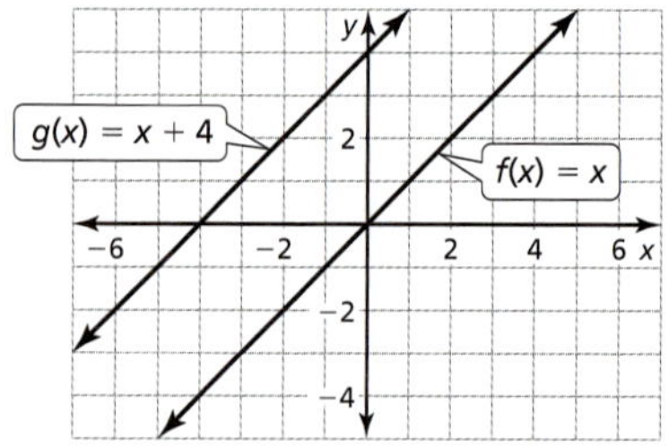

2. a.

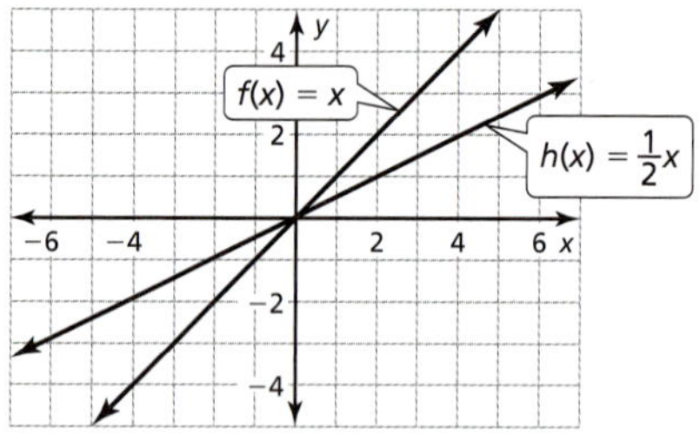

3.6 Lesson Monitoring Progress

1. a.

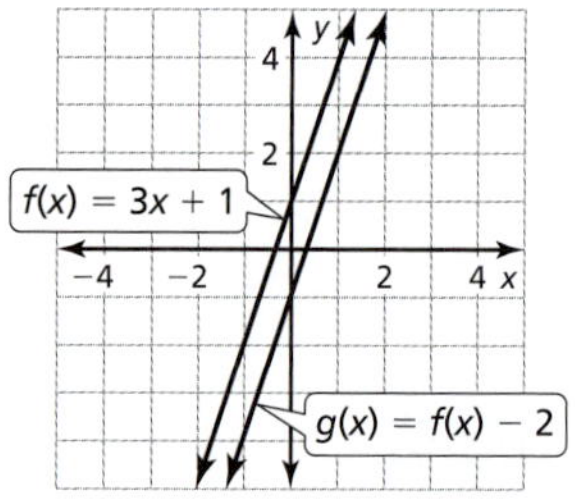

The graph of g is a vertical translation 2 units down of the graph of f.

b.

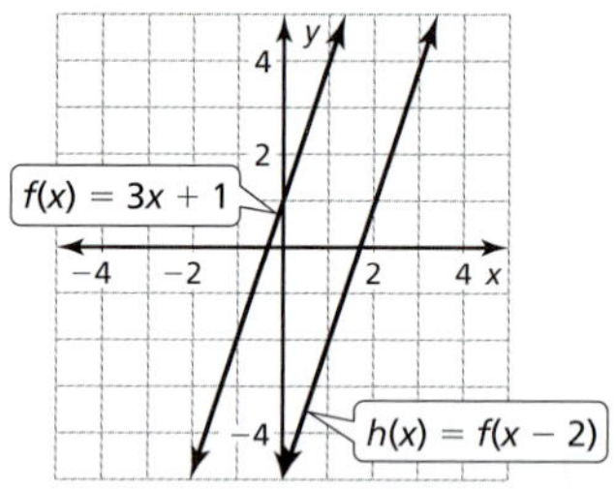

The graph of h is a horizontal translation 2 units right of the graph of f.

2. a.

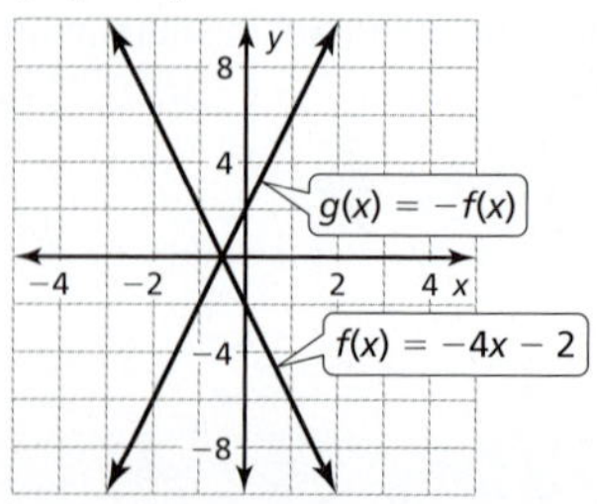

The graph of g is a reflection in the x-axis of the graph of f.

b.

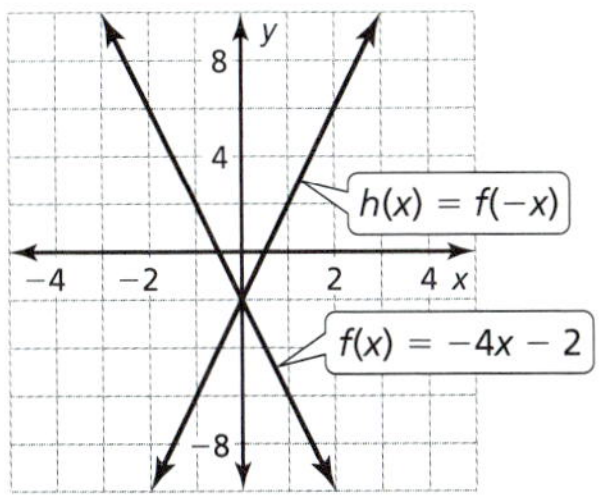

The graph of h is a reflection in the y-axis of the graph of f.

3. a.

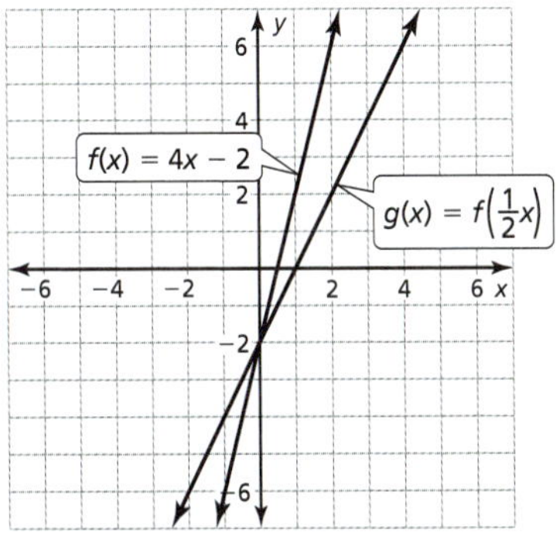

The graph of g is a horizontal stretch of the graph of f by a factor of 2.

b.

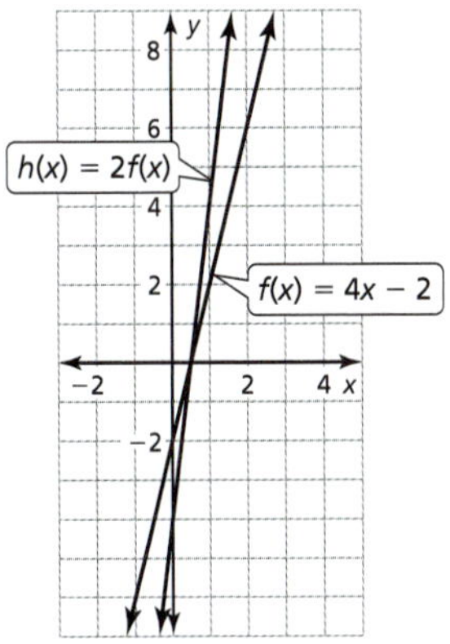

The graph of h is a vertical stretch of the graph of f by a factor of 2.

3.6 Monitoring Progress and Modeling with Mathematics

41.

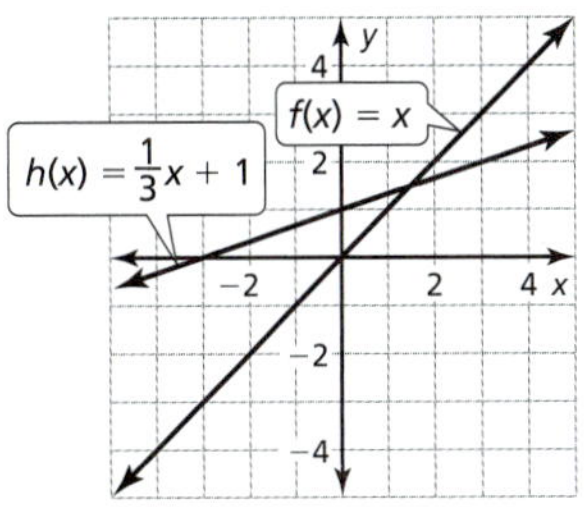

The transformations are a vertical shrink by a factor of $\frac{1}{3}$ then a vertical translation 1 unit up.

42.

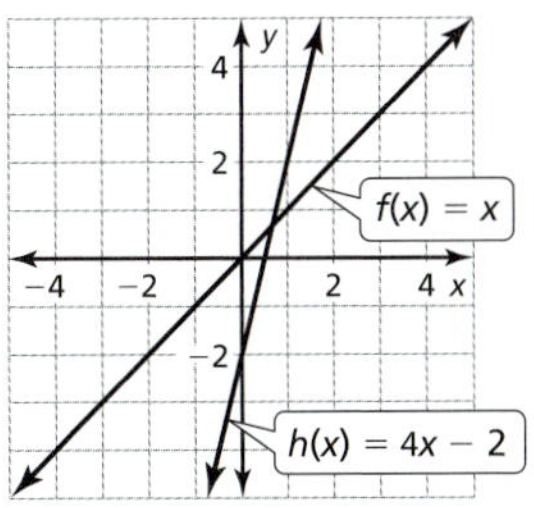

The transformations are a vertical stretch by a factor of 4 then a vertical translation 2 units down.

43.

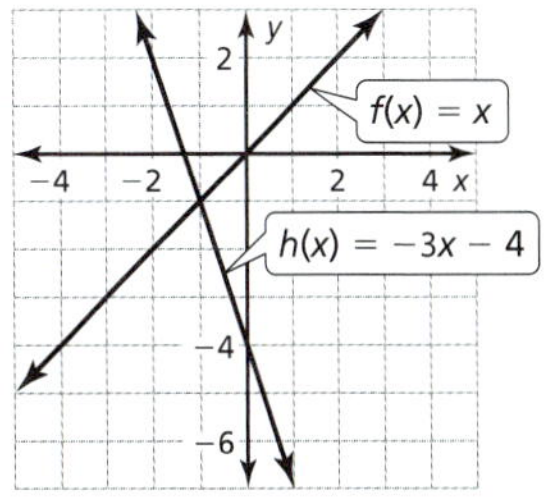

The transformations are a vertical stretch by a factor of 3, then a reflection in the x-axis, then a vertical translation 4 units down.

44.

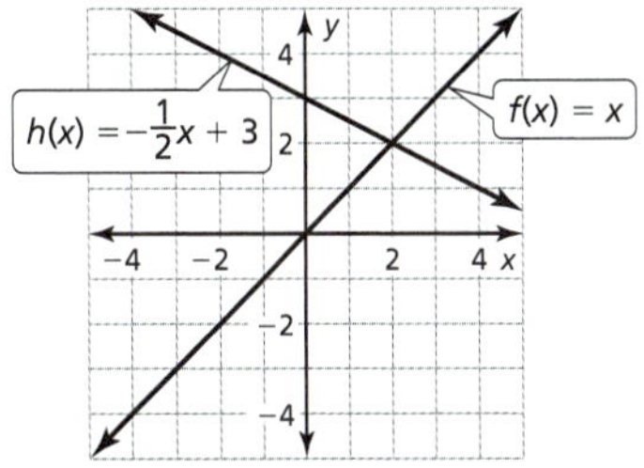

The transformations are a vertical shrink by a factor of $\frac{1}{2}$, then a reflection in the x-axis, then a vertical translation 3 units up.

45.

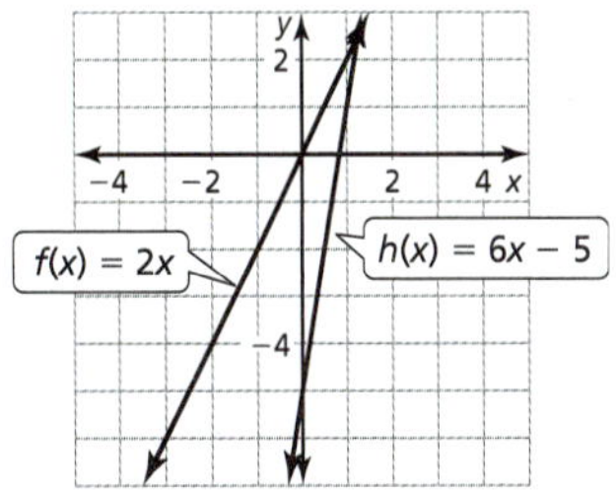

The transformations are a vertical stretch by a factor of 3 then a vertical translation 5 units down.

46.

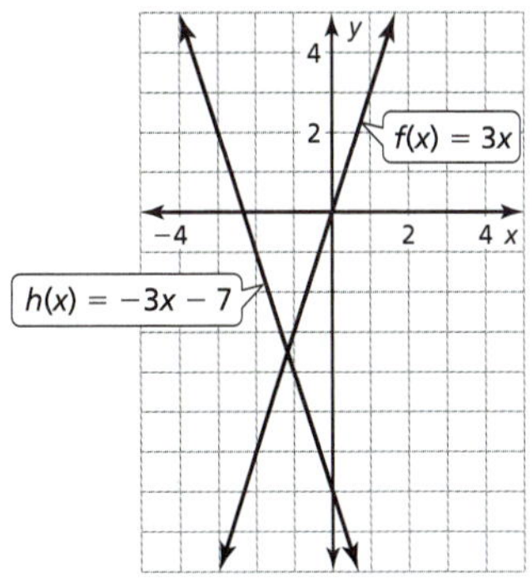

The transformations are a reflection in the x-axis then a vertical translation 7 units down.

56.

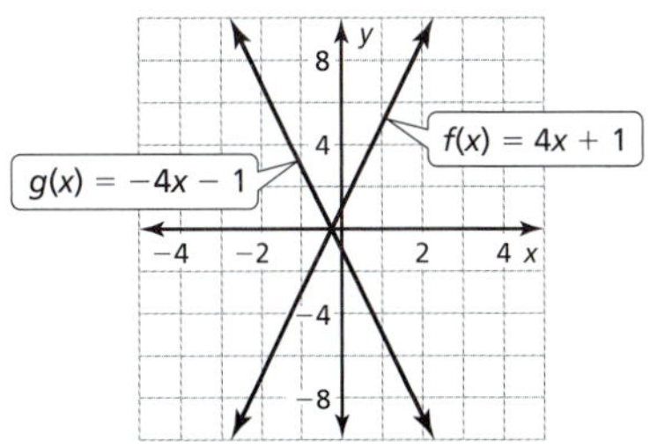

$g(x) = -f(x)$; The graph of g is a reflection in the x-axis of the graph of f.

57. 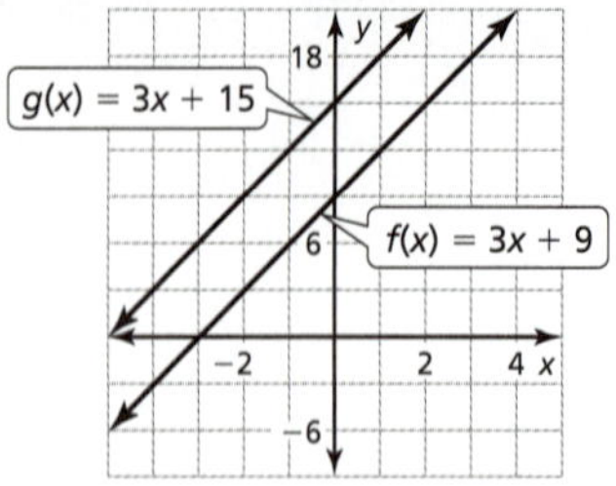

$g(x) = f(x) + 6$; The graph of g is a vertical translation 6 units up of the graph of f.

58. 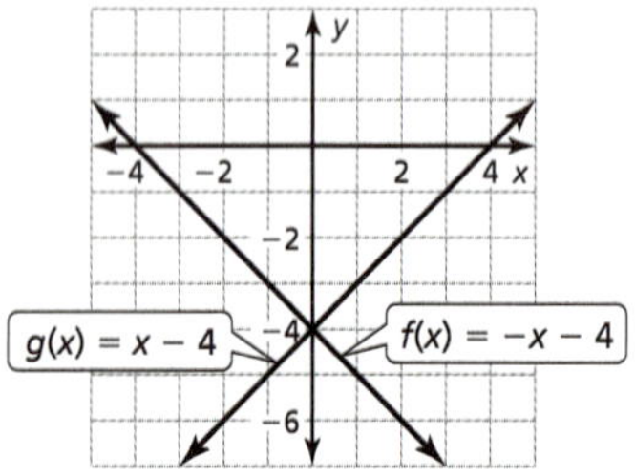

$g(x) = f(-x)$; The graph of g is a reflection in the y-axis of the graph of f.

59. 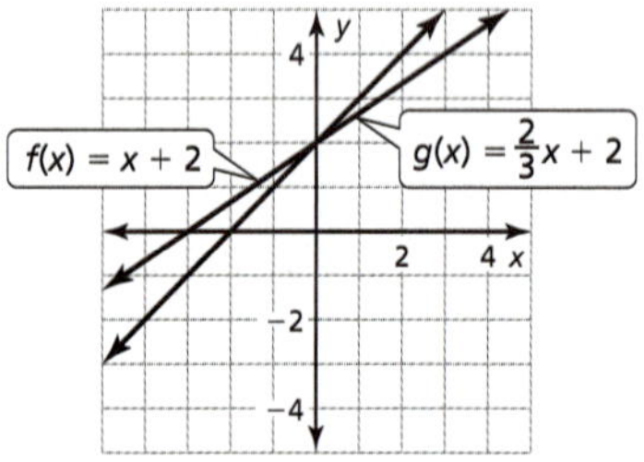

$g(x) = f\left(\frac{2}{3}x\right)$; The graph of g is a horizontal stretch of the graph of f by a factor of $\frac{3}{2}$.

60. 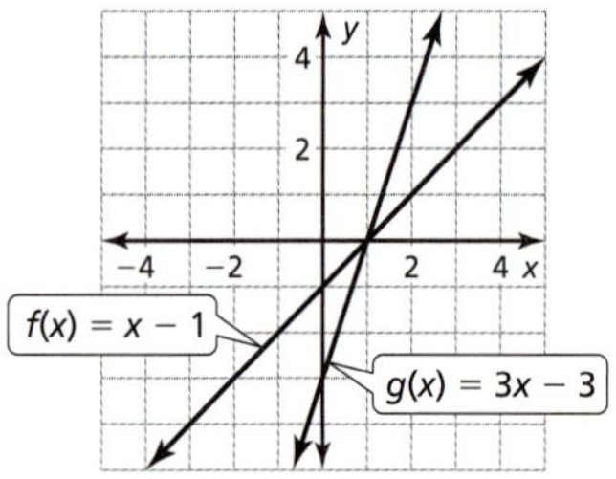

$g(x) = 3f(x)$; The graph of g is a vertical stretch of the graph of f by a factor of 3.

61. Translate the graph of $f(x) = x$ horizontally 5 units left.

62. **a.** A; reflection in the y-axis
 b. D: vertical translation down
 c. B; vertical translation up
 d. C; horizontal shrink

63. $r = 2$ 64. $r = \frac{1}{3}$

65. $r = 2$ 66. $r = -8$

67. when the slope is 1; A slope of 1 occurs when the ratio of the vertical change to the horizontal change is 1, meaning the vertical change and horizontal change are the same.

3.6 Maintaining Mathematical Proficiency

71. $x < -10$ or $x > 6$

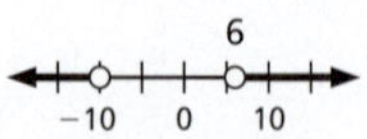

72. $-12 < x < -2$

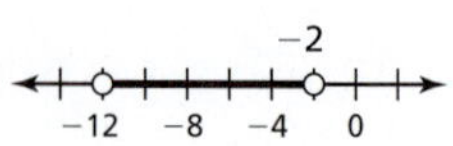

73. no solution

3.7 Lesson Monitoring Progress

1. 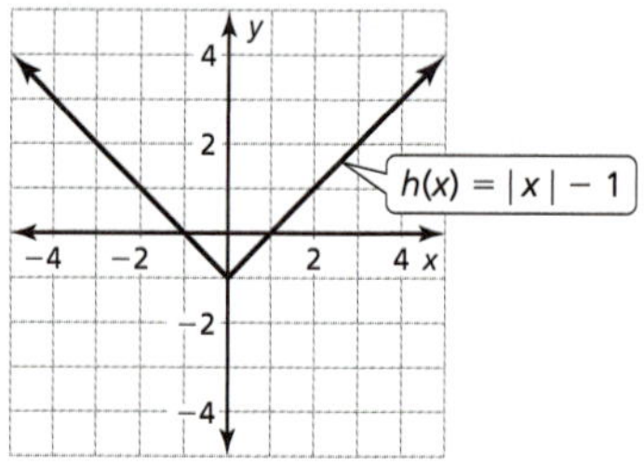

The graph of h is a vertical translation 1 unit down of the graph of f; domain: all real numbers; range: $y \geq -1$

2. 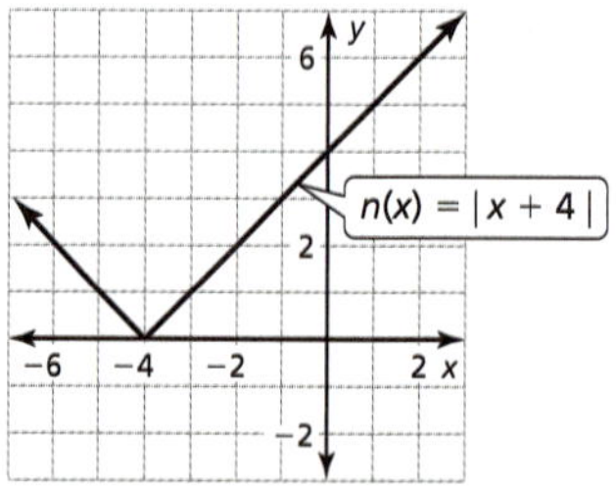

The graph of n is a horizontal translation 4 units left of the graph of f; domain: all real numbers; range: $y \geq 0$

3. 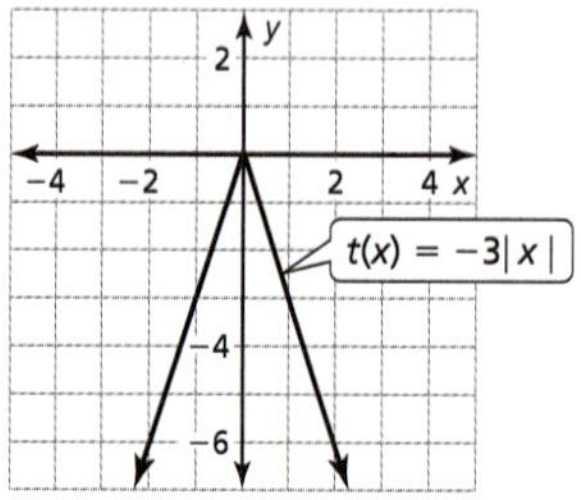

The graph of t is a vertical stretch of the graph of f by a factor of 3 and a reflection in the x-axis; domain: all real numbers; range: $y \leq 0$

4. 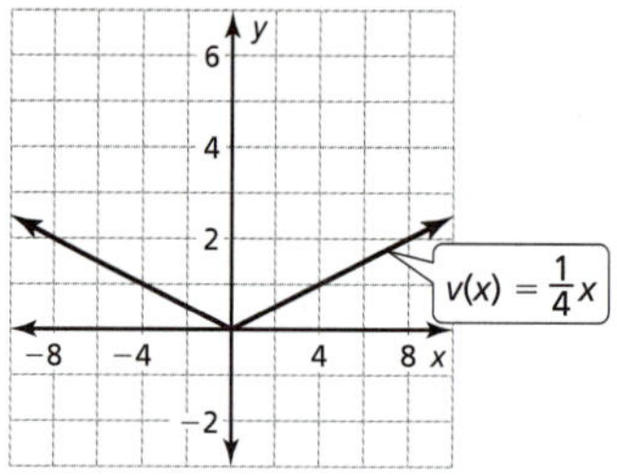

The graph of v is a vertical shrink of the graph of f by a factor of $\frac{1}{4}$; domain: all real numbers; range: $y \geq 0$

3.7 Monitoring Progress and Modeling with Mathematics

8. 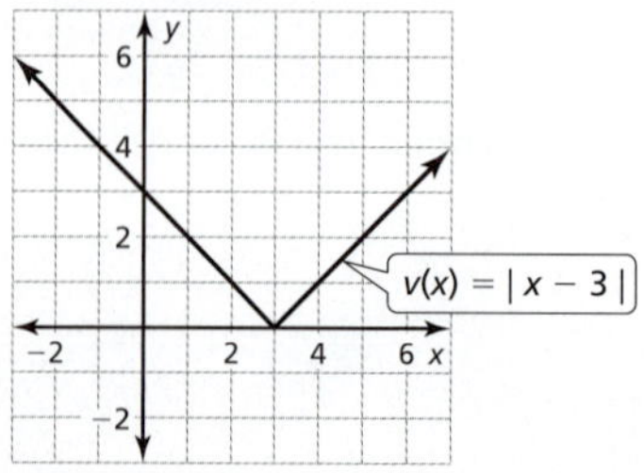

The graph of v is a horizontal translation 3 units right of the graph of f; domain: all real numbers; range: $y \geq 0$

9.

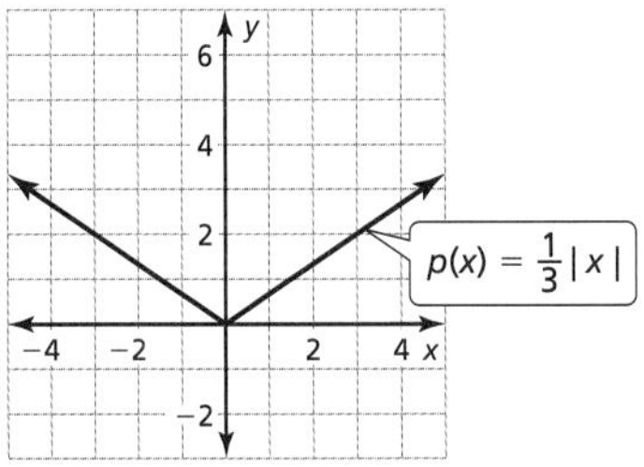

The graph of p is a vertical shrink of the graph of f by a factor of $\frac{1}{3}$; domain: all real numbers; range: $y \geq 0$

10.

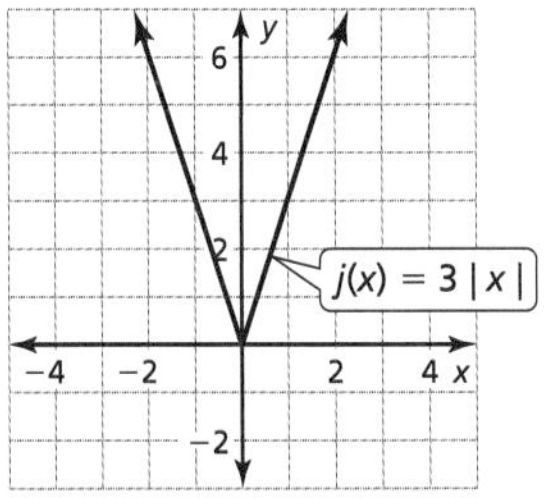

The graph of j is a vertical stretch of the graph of f by a factor of 3; domain: all real numbers; range: $y \geq 0$

11.

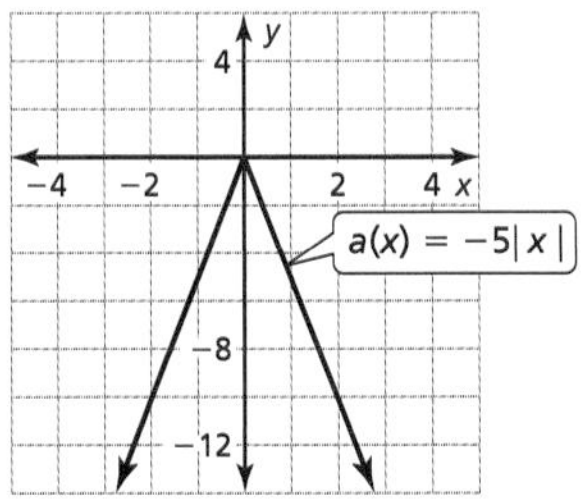

The graph of a is a vertical stretch of the graph of f by a factor of 5 and a reflection in the x-axis; domain: all real numbers; range: $y \leq 0$

12.

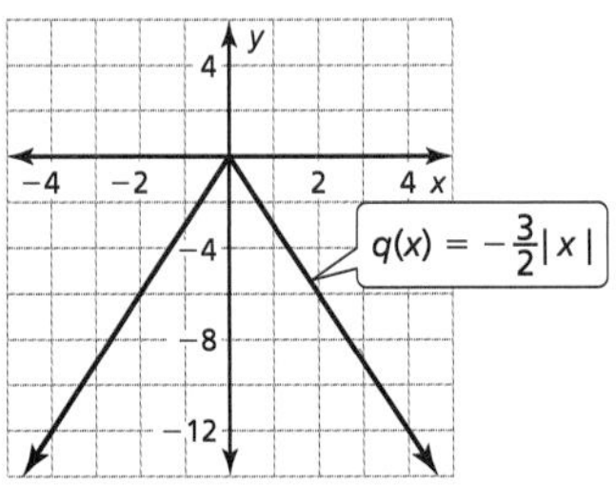

The graph of q is a vertical stretch of the graph of f by a factor of $\frac{3}{2}$ and a reflection in the x-axis; domain: all real numbers; range: $y \leq 0$

13.

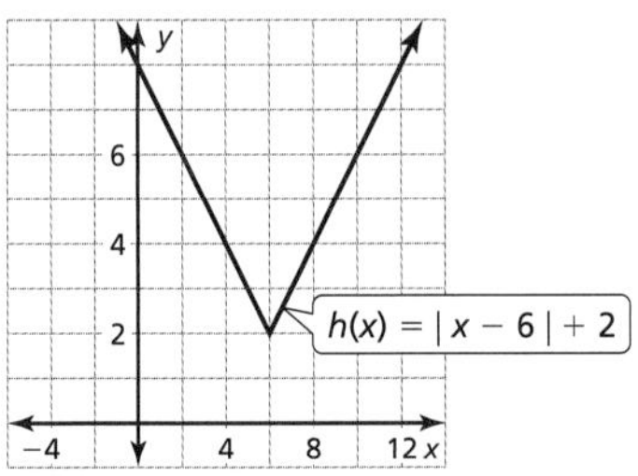

The graph of h is a vertical translation 2 units up of the graph of f.

14.

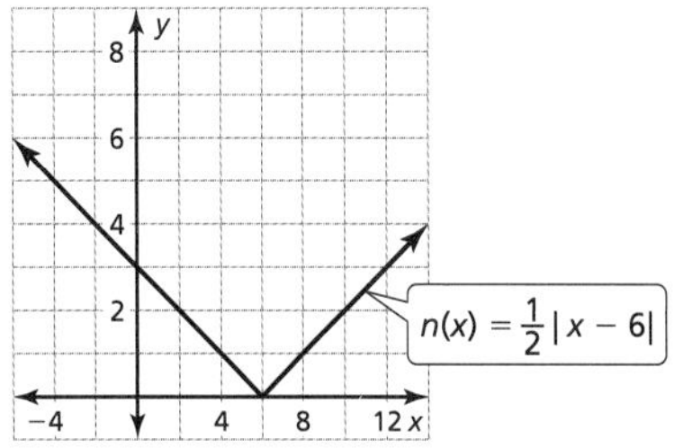

The graph of n is a vertical shrink of the graph of f by a factor of $\frac{1}{2}$.

15.

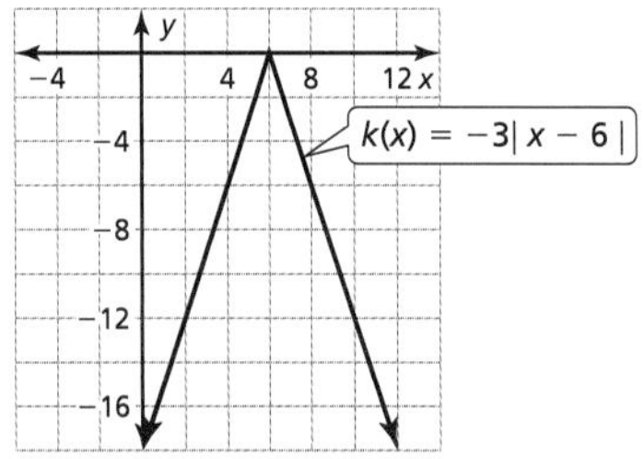

The graph of k is a vertical stretch of the graph of f by a factor of 3 and a reflection in the x-axis.

16.

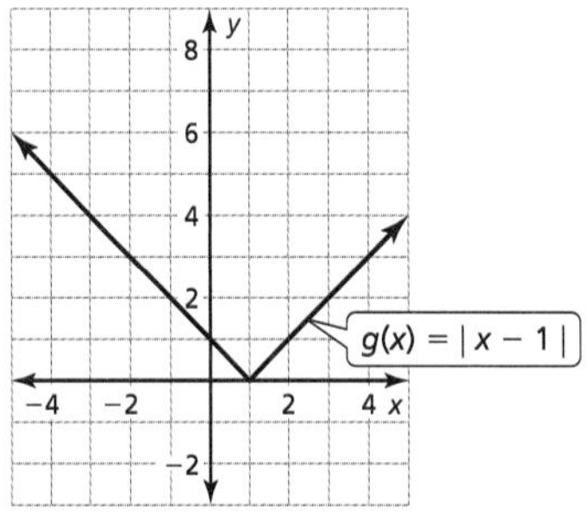

The graph of g is a horizontal translation 5 units left of the graph of f.

17.

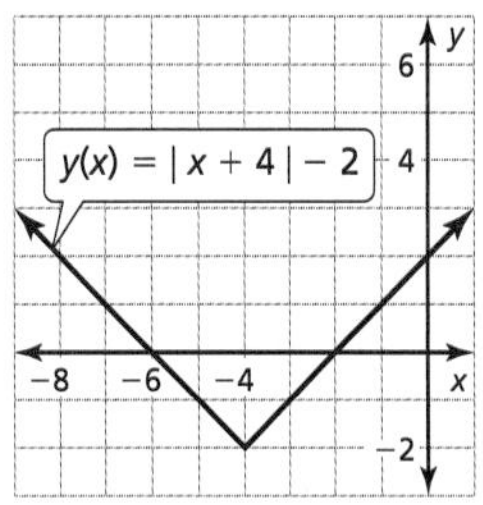

The graph of y is a horizontal translation 1 unit left of the graph of f.

18.

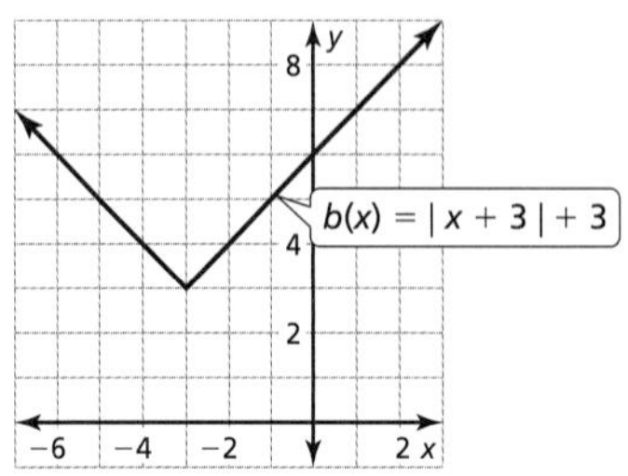

The graph of b is a vertical translation 5 units up of the graph of f.

19. The graph of g is a vertical translation 3 units down of the graph of f; $k = -3$
20. The graph of t is a horizontal translation 1 unit right of the graph of f; $h = 1$
21. The graph of p is a vertical stretch of the graph of f by a factor of 3 and a reflection in the x-axis; $a = -3$
22. The graph of w is a vertical shrink of the graph of f by a factor of 0.5; $a = 0.5$

23. $h(x) = |x| - 7$

24. $h(x) = |x + 10|$

25. $h(x) = \frac{1}{4}|x|$

26. $h(x) = -3|x|$

27.

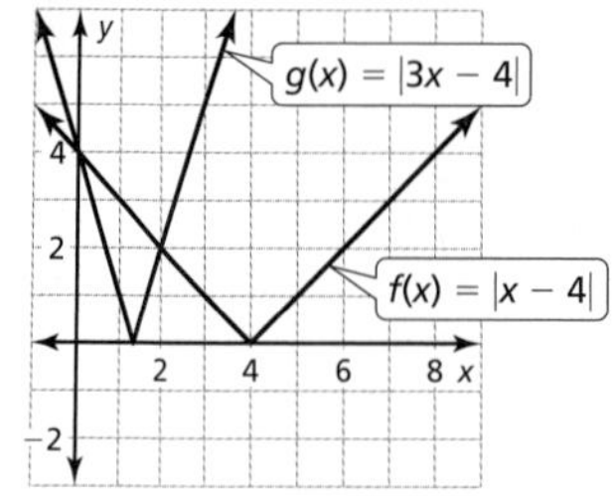

The graph of g is a horizontal shrink of the graph of f by a factor of $\frac{1}{3}$.

28.

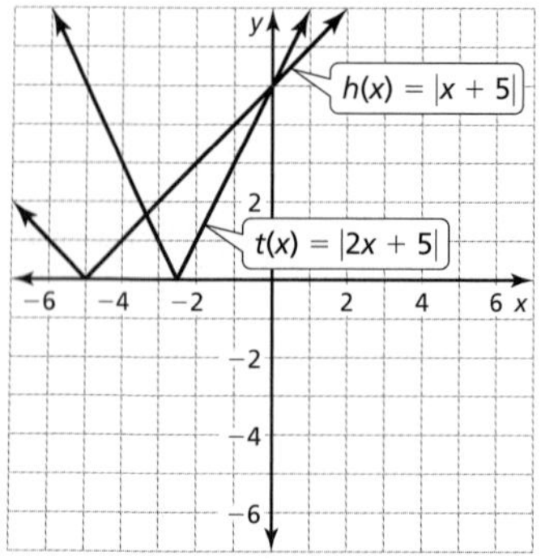

The graph of t is a horizontal shrink of the graph of h by a factor of $\frac{1}{2}$.

29.

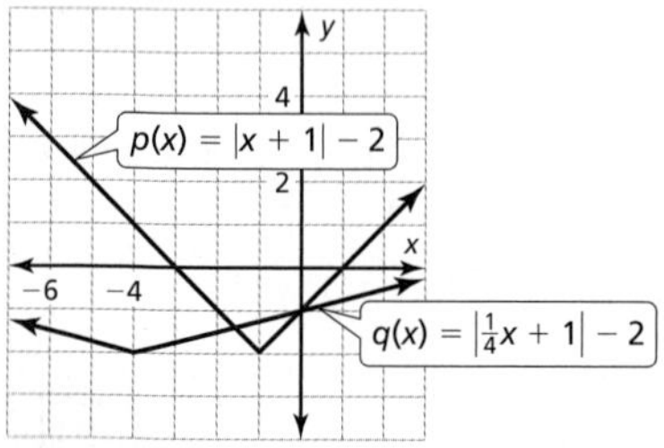

The graph of q is a horizontal stretch of the graph of p by a factor of 4.

30.

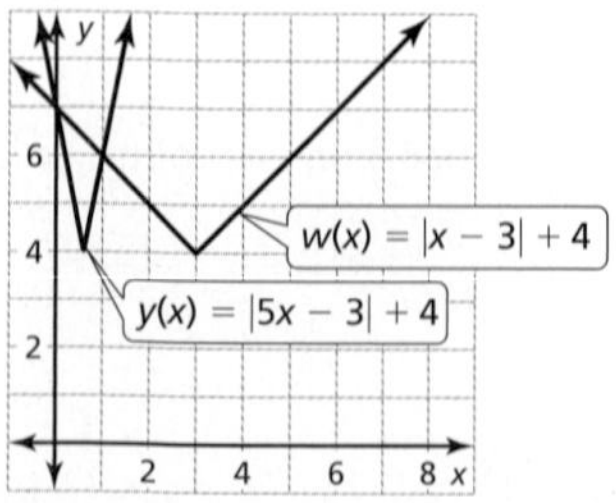

The graph of y is a horizontal shrink of the graph of w by a factor of $\frac{1}{5}$.

31.

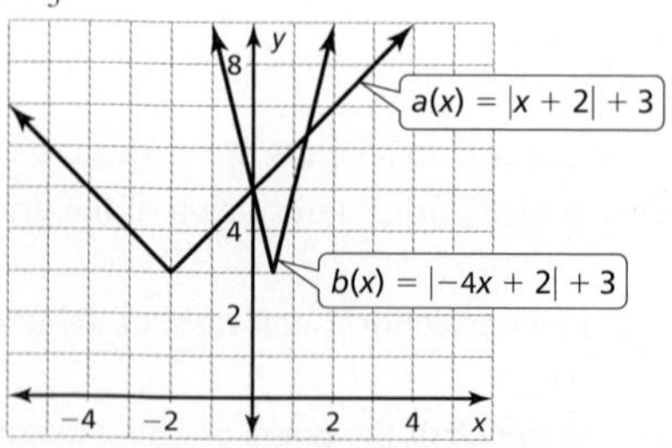

The graph of b is a horizontal shrink of the graph of a by a factor of $\frac{1}{4}$ and a reflection in the y-axis.

32.

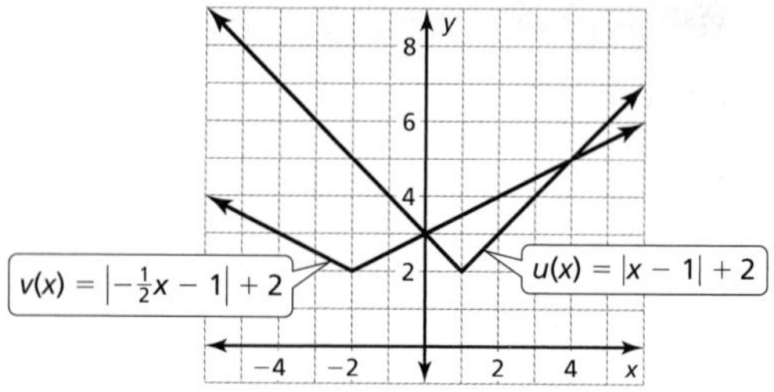

The graph of v is a horizontal stretch of the graph of u by a factor of 2 and a reflection in the y-axis.

38. The transformations are a horizontal translation 2 units right, then a horizontal shrink by a factor of $\frac{1}{2}$, then a vertical translation 3 units down.

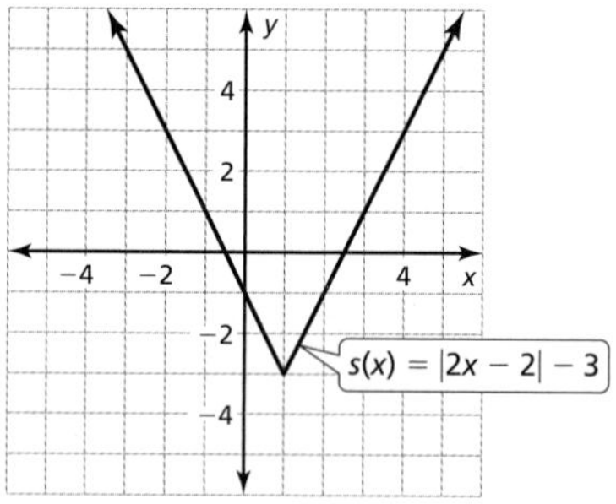

39. The transformations are a horizontal translation 1 unit left, then a reflection in the y-axis, then a vertical translation 5 units down.

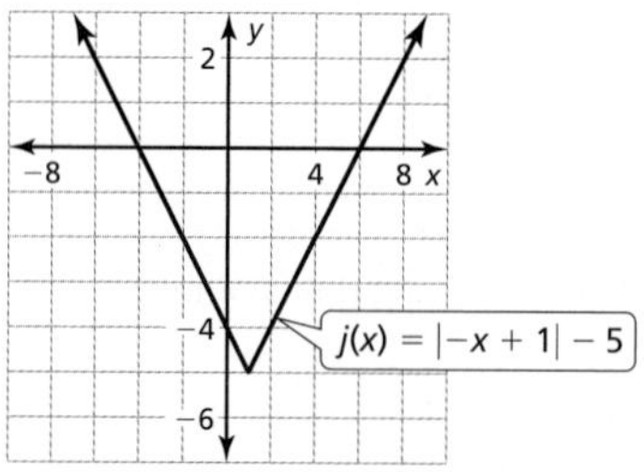

40. The transformations are a horizontal translation 1 unit left, then a horizontal stretch by a factor of 3, then a reflection in the y-axis, then a vertical translation 2 units up.

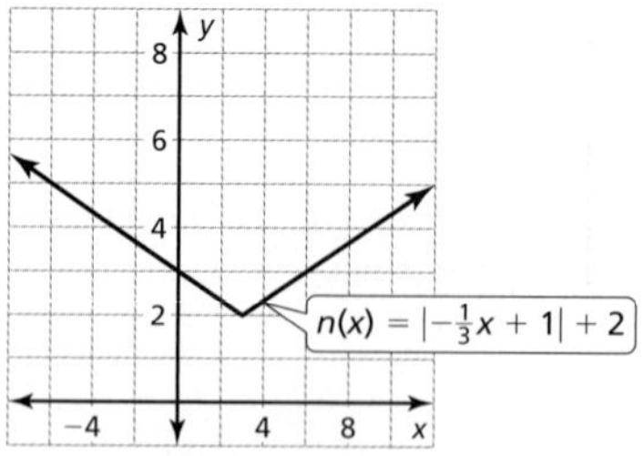

41. a.

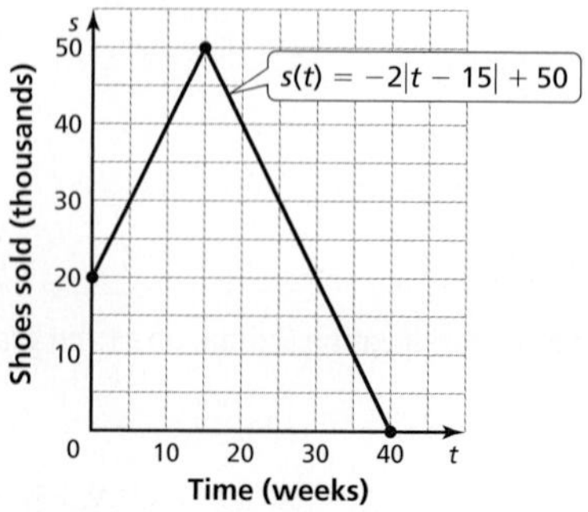

b. 50,000 pairs of shoes

42. a. $\left(\frac{5}{4}, 0\right)$

b. yes; The location of the pocket (5, 5) is a solution of p.

43. **a.** $A'\left(-\frac{1}{2}, -2\right)$; $B'(1, -5)$; $C'(-4, -7)$

b. $A'\left(\frac{5}{2}, 3\right)$; $B'(4, 0)$; $C'(-1, -2)$

c. $A'\left(-\frac{1}{2}, -3\right)$; $B'(1, 0)$; $C'(-4, 2)$

d. $A'\left(-\frac{1}{2}, 12\right)$; $B'(1, 0)$; $C'(-4, -8)$

44. **a.** $y = |x| + k$ is a vertical translation up of the graph of $y = |x|$ for positive values of k, and down for negative values of k.

b. $y = |x - h|$ is a horizontal translation right of the graph of $y = |x|$ for positive values of h, and left for negative values of h.

c. $y = a|x|$ is a vertical stretch or shrink by a factor of a of the graph of $y = |x|$ for positive values of a. It is both a vertical stretch or shrink by a factor of a and a reflection in the x-axis for negative values of a.

d. $y = |ax|$ is a reflection in the y-axis of the graph of $y = |x|$ for negative values of a. $y = |ax|$ is a horizontal stretch of the graph of $y = |x|$ by a factor of $\frac{1}{a}$ when $0 < |a| < 1$, or a horizontal shrink by a factor of $\frac{1}{a}$ when $|a| > 1$.

45. The graph should have a horizontal shift right, not left.

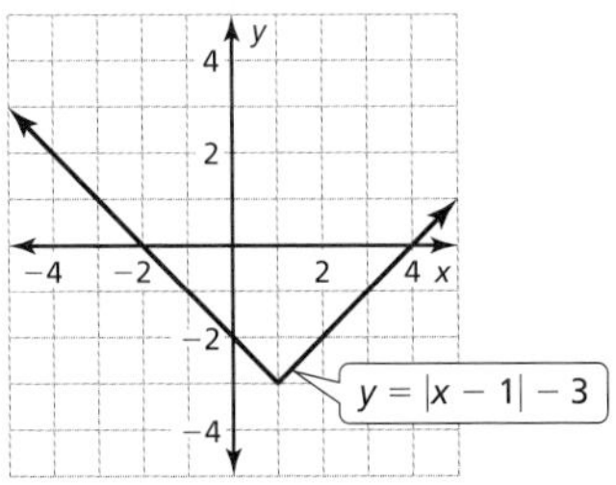

46. The graph should have a reflection in the x-axis.

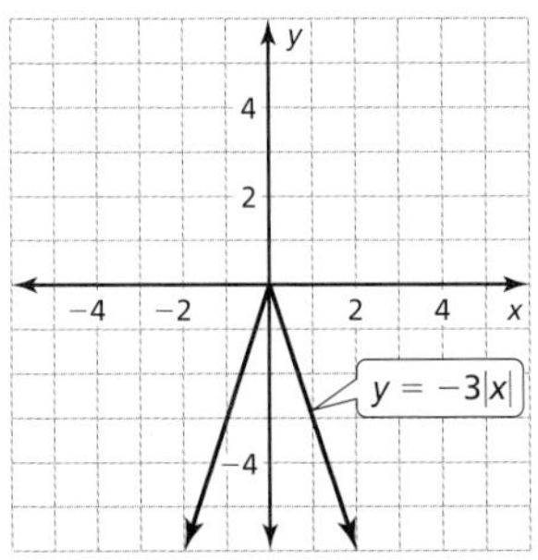

47. *Sample answer:* $f(x) = -|x| + 2$

48. *Sample answer:* $f(x) = -|x - 3| + 3$

49. The graph of p is a horizontal translation 6 units right of the graph of $f(x) = |x|$. The graph of q is a vertical translation 6 units down of the graph of $f(x) = |x|$.

51.

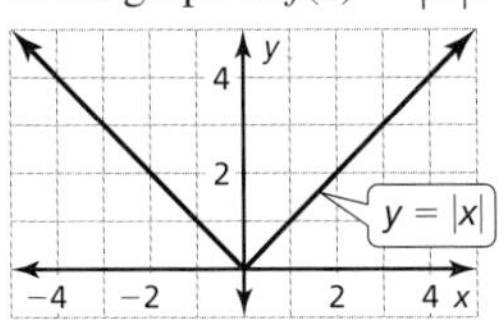

$y = -x$; $y = x$

52.

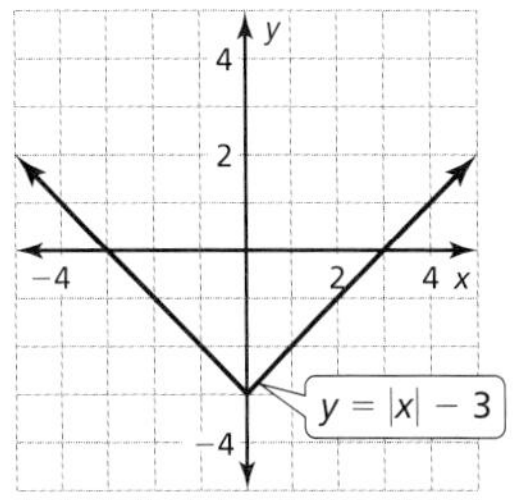

$y = x - 3$; $y = -x - 3$

53.

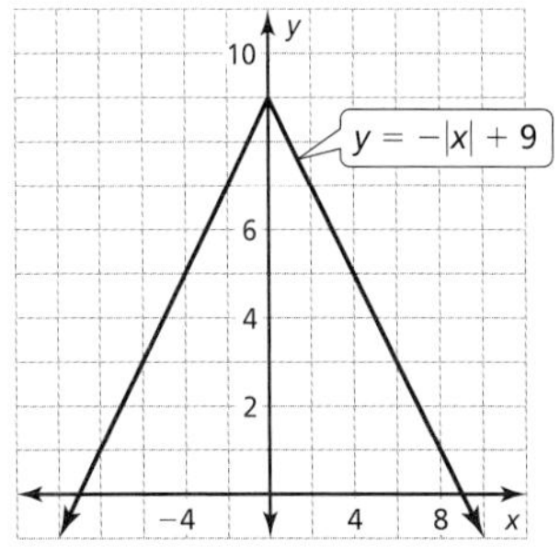

$y = x + 9$; $y = -x + 9$

54.

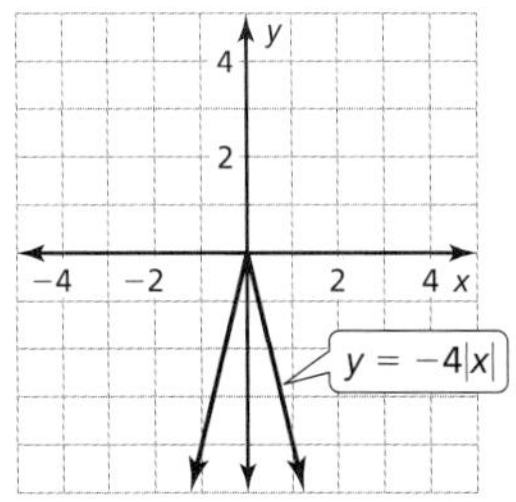

$y = 4x$; $y = -4x$

55.

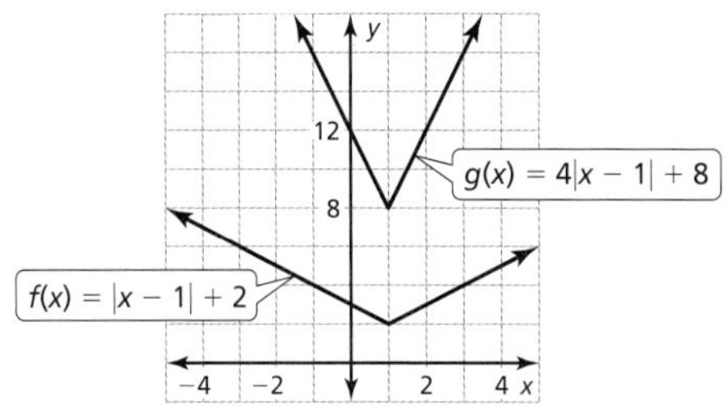

The graph of g is a vertical stretch by a factor of 4 of the graph of f.

56.

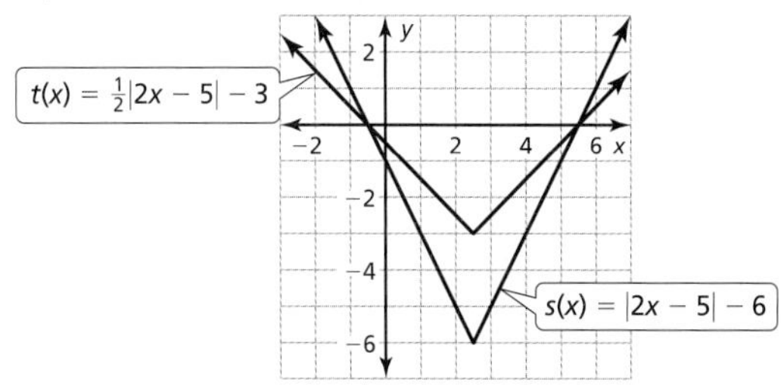

The graph of t is a vertical shrink by a factor of $\frac{1}{2}$ of the graph of s.

57. 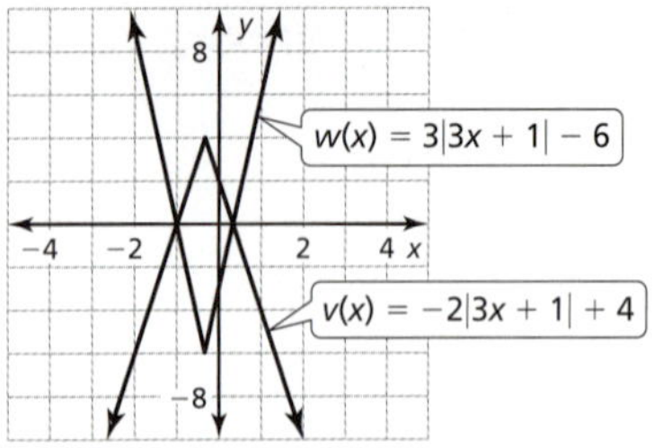

The graph of w is a vertical stretch of the graph of v by a factor of $\frac{3}{2}$ and a reflection in the x-axis.

58. 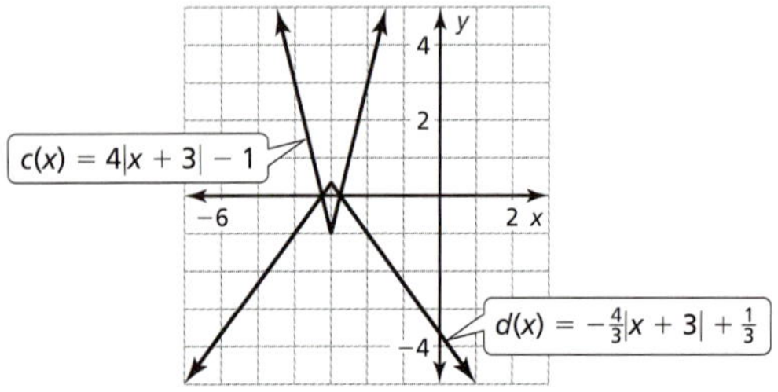

The graph of d is a vertical shrink of the graph of c by a factor of $\frac{1}{3}$ and a reflection in the x-axis.

59. The transformations are a vertical translation 4 units down, then a reflection in the x-axis, then a vertical shrink by a factor of $\frac{1}{2}$, then a horizontal translation 1 unit right; *Sample answer:* Determine the transformations from h to g, then perform the opposite transformations in reverse order.

60. *Sample answer:* $y = -|2x - 40| + 40$ on [0, 25]; The receiver runs a diagonal route that cuts back at an angle at the point (20, 40) and catches the ball at (25, 30).

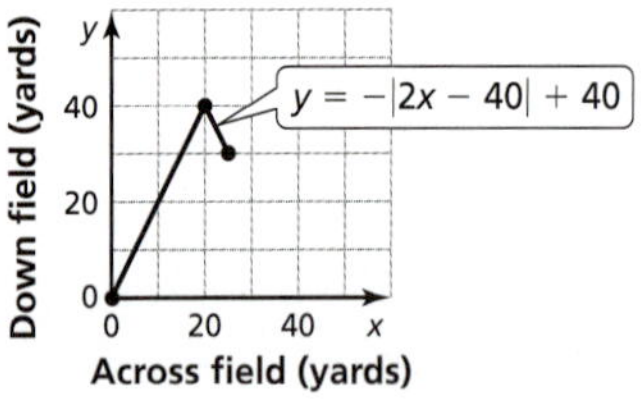

61. 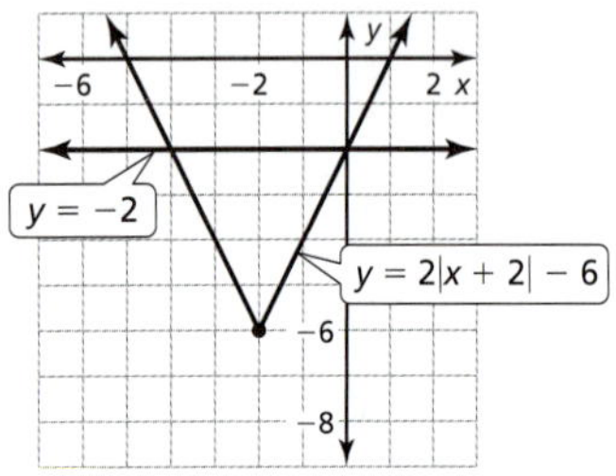

$x = 0, x = -4$

62. no; The form of a horizontal transformation is of $y = f(x)$ is $y = f(x - h)$. So, if p is positive, the graph of $y = |x + p|$ is a negative horizontal translation.

63. $\left(\frac{h}{a}, k\right)$

3.7 Maintaining Mathematical Proficiency

64. $a \le \frac{5}{2}$

65. no solution

66. all real numbers

67. $x > -\frac{3}{4}$

68. $\frac{5}{2}$

69. $\frac{1}{3}$

70. $-\frac{5}{4}$

Chapter 3 Review

36. 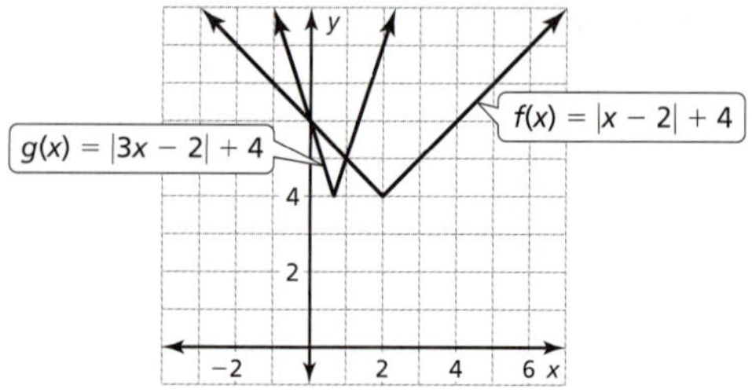

The graph of g is a horizontal shrink of the graph of f by a factor of $\frac{1}{3}$.

Chapter 3 Test

10. 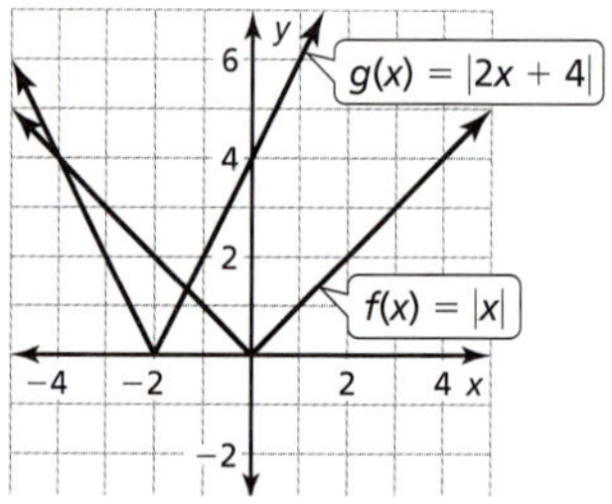

The transformations are a horizontal translation 4 units left then a horizontal shrink by a factor of $\frac{1}{2}$.

11. The domain of A is discrete because the number of quarters must be a whole number, and the domain of B is continuous because the time can be any value greater than or equal to 0.

12. **a.** slope: 125; So, the climber ascends at a rate of 125 feet per hour; y-intercept: 50; So, the climber began the ascent at 50 feet above sea level.

b. $f(3)$ can be found by reading the point off the graph or by substituting 3 into the function definition; $f(3) = 425$; So, the climber is 425 feet above sea level after climbing for 3 hours.

c. 3.6 h; At the top, $f(x) = 500$. Setting 500 equal to $125x + 50$ and solving for x gives the time it takes to reach the top.

13. The slope of g is twice the slope of f. Both functions have the same y-intercept; The x-intercept of g is half the x-intercept of f.

14. **a.** Weekly sales is the dependent variable and time is the independent variable.

b. 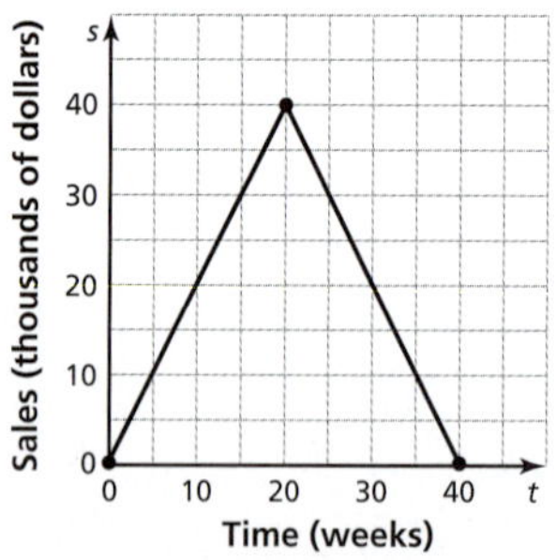

The transformations are a horizontal translation 20 units right, then a vertical stretch by a factor of 2, then a reflection in the x-axis, then a vertical translation 40 units up.

Chapter 4

4.1 Monitoring Progress and Modeling with Mathematics

33. $y = -\frac{A}{B}x + \frac{C}{B}$; slope is $\frac{6}{5}$; y-intercept is $\frac{9}{5}$

34. your friend; If f is a function, then the line is not vertical.

35. $y = \frac{5}{3}x - 1$

36. **a.** *Sample answer:* about 0.22; 8

b. *Sample answer:* The approximate U.S. box office revenue increased by about \$0.22 billion each year from 2000 to 2012 and was \$8 billion in 2000.

c. Write a linear model in slope-intercept form, then substitute 18 for x and solve for y.

37. slope $= \dfrac{(b + m) - b}{1 - 0} = m$; y-intercept $= b$; $y = mx + b$;

Substitute -1 for x in $y = mx + b$ and verify that $y = b - m$.

4.1 Maintaining Mathematical Proficiency

38. $x = 28$

39. $y = \frac{1}{4}$

40. no solution

41. $n = 0$

42.

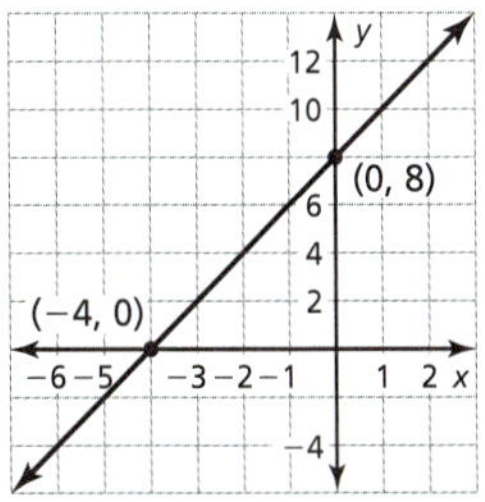

43.

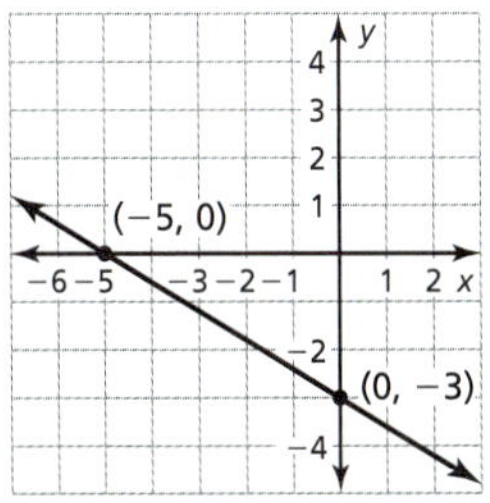

44.

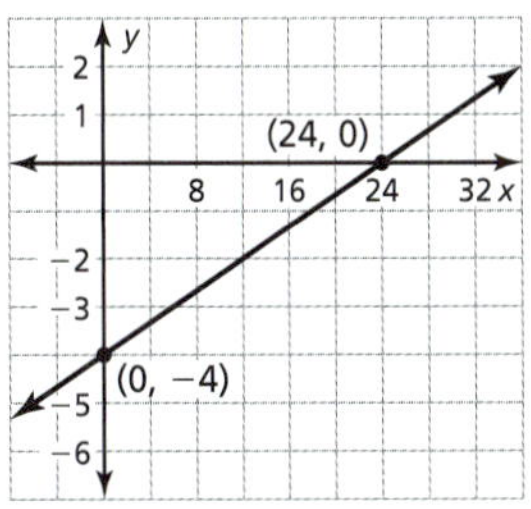

45.

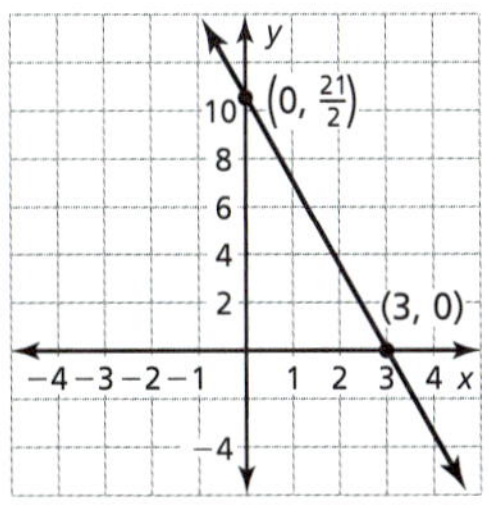

4.2 Monitoring Progress and Modeling with Mathematics

35. *Sample answer:* Plot the point (4, 1), then use the slope of $\frac{3}{2}$ to find a second point on the graph and draw a line through the points; Rewrite the equation in slope-intercept form, then use the y-intercept, -5, and the slope of $\frac{3}{2}$ to graph the equation.

36. *Sample answer:* $y + 5 = \frac{2}{5}(x - 12)$; $y = \frac{2}{5}x - \frac{49}{5}$

37. *Sample answer:* point-slope form; The value of the y-intercept is unknown.

38. **a.** *Sample answer:* negative; It appears the line through the two points will extend below the x-axis before crossing the y-axis.

b. *Sample answer:* (8, 4); $\left(4, \frac{5}{3}\right)$; Use the two points to find the slope of the line. Substitute the slope and one of the points into $y - y_1 = m(x - x_1)$. Rewrite the equation in slope-intercept form and determine the y-intercept from the rewritten equation.

39. *Sample answer:* The graph of $y - k = m(x - h)$ is a translation h units to the right and k units upward of the graph of $y = mx$.

40. **a.** B; A

b. B; A

c. C; A

4.2 Maintaining Mathematical Proficiency

41. $\frac{1}{5}$

42. $-\frac{1}{8}$

43. $-\frac{7}{2}$

44. $\frac{2}{3}$

4.3 Monitoring Progress and Modeling with Mathematics

32. **a.** yes; Student B paid a greater registration fee.

b. no; Both graphs have the same slope.

33. never; Perpendicular lines have opposite reciprocal slopes, so one must be positive and the other must be negative.

34. always; The y-axis is a vertical line, and all vertical lines are parallel.

35. sometimes; They are perpendicular when the slopes are negative reciprocals, otherwise they will not be perpendicular.

36. *Sample answer:*

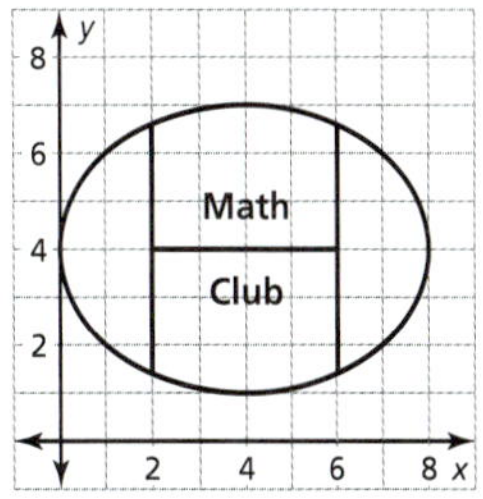

$y = 4$; $x = 2$; $x = 6$

4.3 Maintaining Mathematical Proficiency

37. function; Each input value is paired with exactly one output value.

38. not a function; The input values of -1 and 1 are each paired with more than one output value.

4.4 Explorations

4. *Sample answer:*

The data in the graph is from the Pennsylvania Department of Transportation.

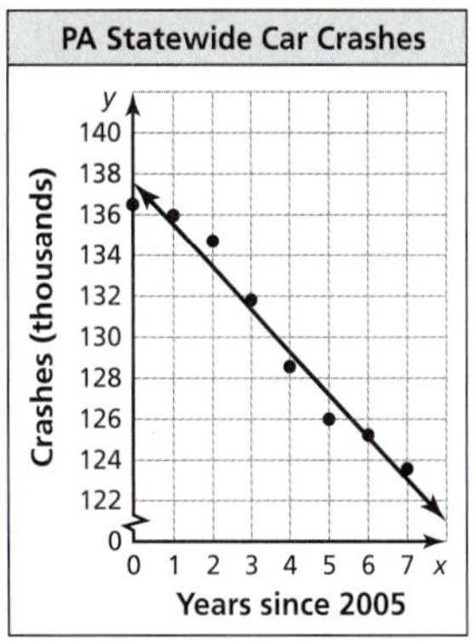

$y = -2.1x + 137.5$; Estimate a line through the middle of the plotted data, then use two points on that line to find the equation.

4.4 Lesson Monitoring Progress

3.

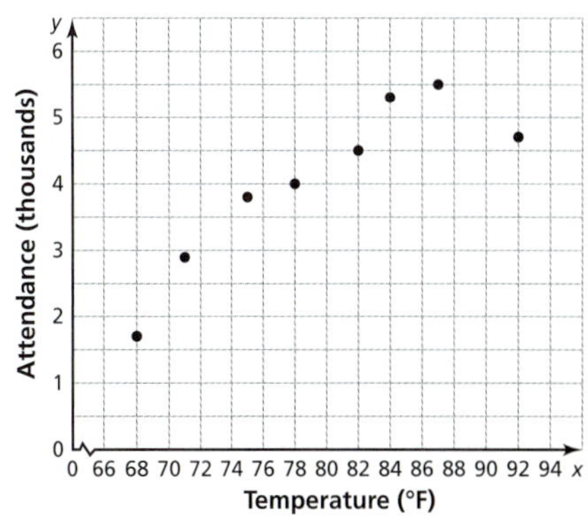

positive

4.4 Monitoring Progress and Modeling with Mathematics

14. 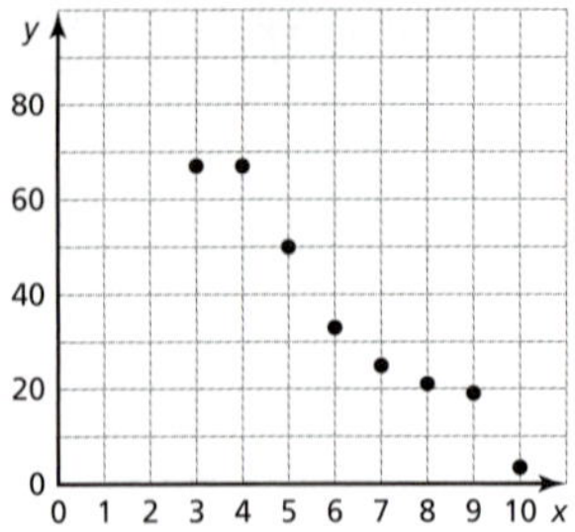

negative

15. **a.** *Sample answer:* $y = -0.3x + 35$

b. *Sample answer:* The slope of -0.3 means the birthrate is decreasing by about 3 births per 1000 people every 10 years. The y-intercept of 35 means in 1960 the birth rate was about 35 births per 1000 people.

16. **a.** *Sample answer:* $y = 20x$

b. *Sample answer:* The slope of 20 means the server's earnings increase by about \$20 per hour. The y-intercept of 0 means the server makes \$0 if he doesn't work.

17. *Sample answer:*

Weight of car (pounds), x	2400	2500	2900	3000
Gas mileage (mpg), y	39	38	25	32

Weight of car (pounds), x	3400	3500	3700	5100
Gas mileage (mpg), y	30	24	21	16

18. no; Because both variables are decreasing, the data show a positive correlation.

19. **a.** *Sample answer:*

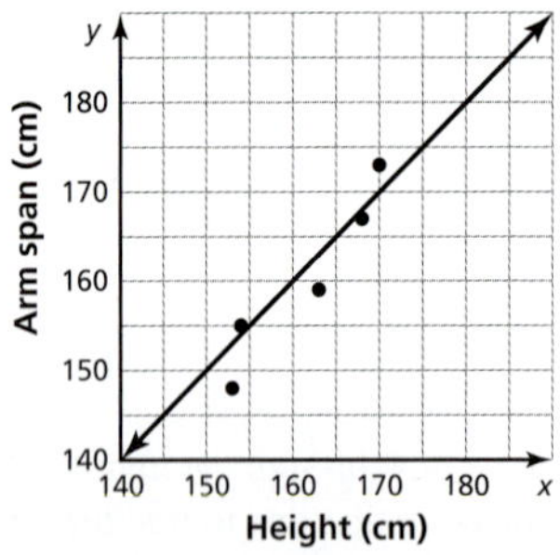

b. *Sample answer:* The slope of 1 means a person's arm span increases by about 1 centimeter for every 1 centimeter increase in height. The y-intercept of 0 has no meaning in this context because the height cannot be 0.

20. *Sample answer*: The number of students y in a school club x years after the club was founded.

21. *Sample answer:* When the data are from two sets such as age and time.

22. *Sample answer:* (2, 5), (9, 57), (12, 99), (15, 102)

23. no; The data points do not have a linear trend.

24. 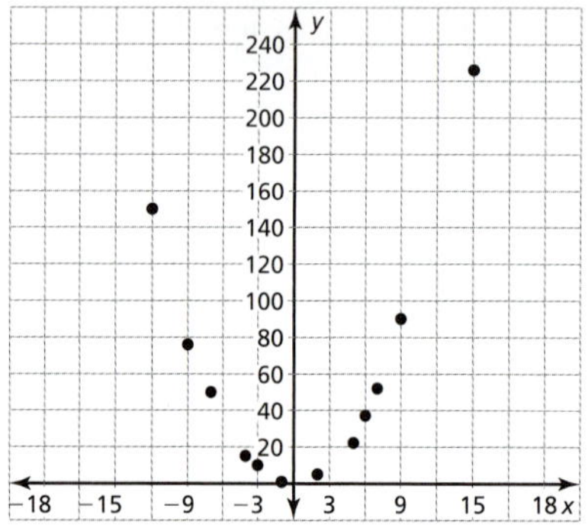

no correlation; no; The data points do not have a linear trend.

4.4 Maintaining Mathematical Proficiency

25. $-18, 0, 24$

26. $30, 0, -40$

27. $-23, -8, 12$

28. $23, 14, 2$

4.5 Lesson Monitoring Progress

2. **a.** $y = -9.6x + 844.5$

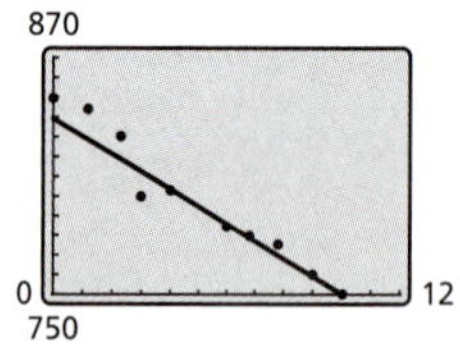

b. -0.964; strong negative correlation

c. The slope of -9.6 means the attendance decreases by about 9600 each year. The y-intercept of 844.5 means the attendance in 2005 was about 844,500.

4.5 Monitoring Progress and Modeling with Mathematics

26. **a.** B; The points appear linear with a positive slope.

b. C; The points appear linear with a negative slope.

c. A; The points are scattered.

d. D; The points appear somewhat linear with a positive slope.

27. **a.** $y = -0.08x + 3.8$; $r = -0.965$; strong negative correlation

b. The slope of -0.08 means the GPA decreases by about 0.08 for every hour spent watching television in a week. The y-intercept of 3.8 means that a student who watches no television has a GPA of about 3.8.

c. 2.7

d. *Sample answer:* no; The time spent watching television does not determine GPA.

28. yes; This point is an outlier.

29. **a.** 2863 people; 5149 people

b. relatively accurate at 9 min, but not accurate at 15 min.

30. *Sample answer:*

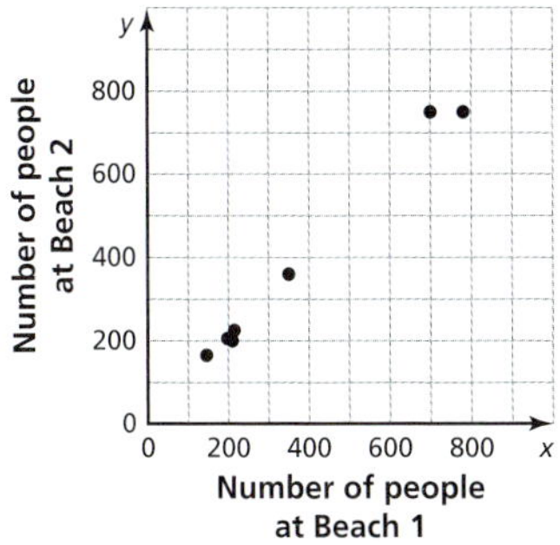

As the number of people at Beach 1 increases, the number of people at Beach 2 also increases; possible $r = 0.99$; There is not a causal relationship; Warm weather and holidays will increase the number of people at both beaches, but the number of people at Beach 1 does not determine the number of people at Beach 2.

31. a. $y = 513.5x - 298$; $r = 0.993$; strong positive correlation

b. no; The year does not determine the number of text messages sent.

c. 25.5; −128; 117.5; 50; −63.5

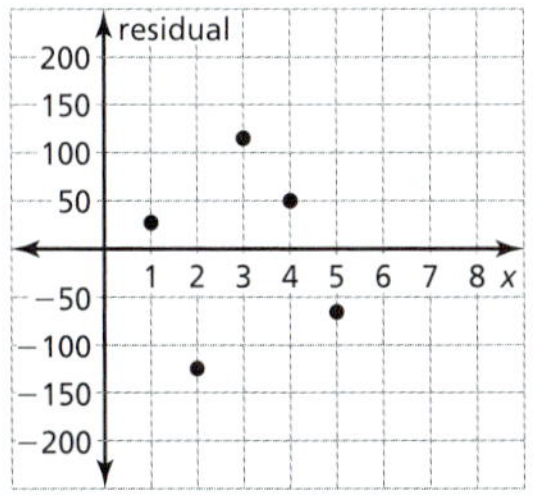

The equation $y = 513.5x - 298$ is a good fit.

d. *Sample answer:* part (a); The correlation coefficient is a single value, which is easily interpreted where as interpreting the scatter plot of the residuals is more subjective.

4.5 Maintaining Mathematical Proficiency

32. nonlinear; The rate of change is not constant.

33. linear; The rate of change is constant.

4.6 Lesson Monitoring Progress

6.

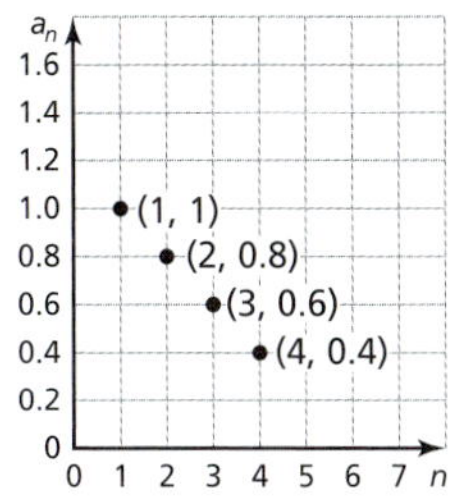

The points lie on a line.

4.6 Monitoring Progress and Modeling with Mathematics

21.

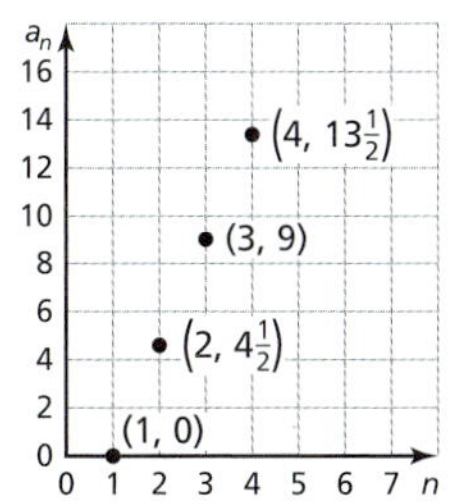

22.

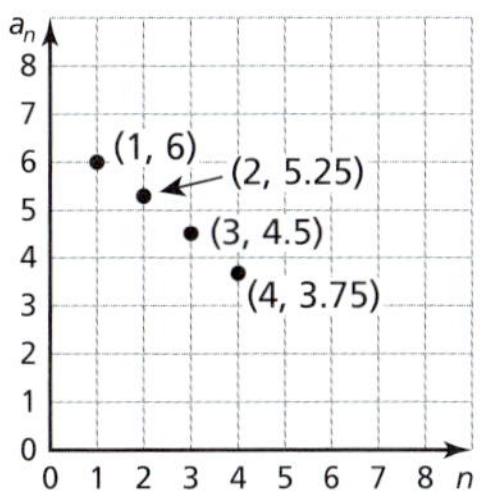

23. not an arithmetic sequence; Consecutive terms do not have a common difference.

24. arithmetic sequence; Consecutive terms have a common difference of 7.

25. arithmetic sequence; Consecutive terms have a common difference of −15.

26. not an arithmetic sequence; Consecutive terms do not have a common difference.

27. arithmetic sequence; 13

28. not an arithmetic sequence

29. not an arithmetic sequence

30. arithmetic sequence; −6

31. 4, 6, 8, 10; yes; Consecutive terms have a common difference of 2.

43. a.

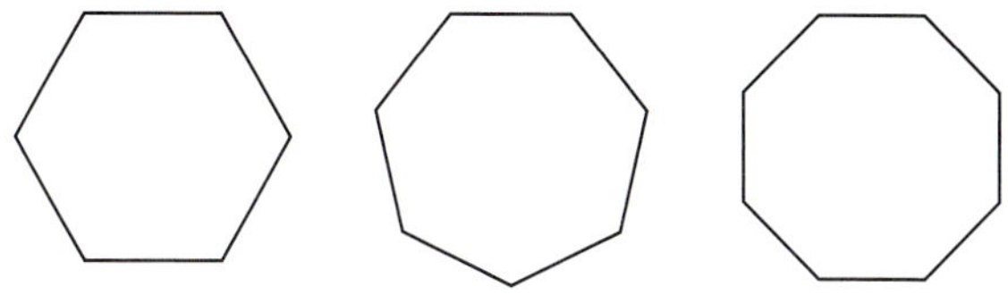

b. a regular 22-sided polygon

44. a.

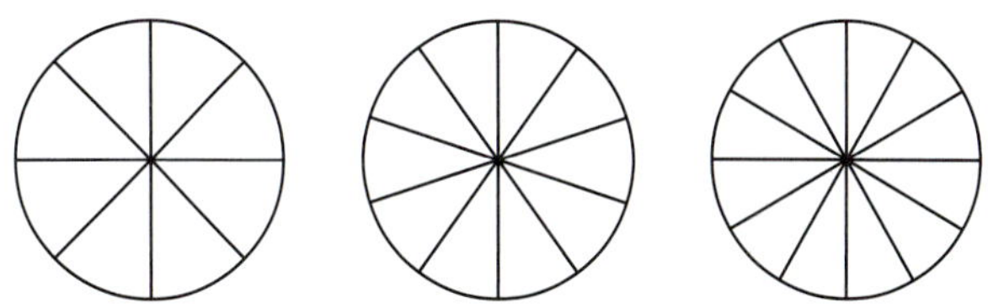

b. a circle divided into 40 equal sections

51. *Sample answer:* −12, −15, −18, −21; $a_n = -3n - 9$; 12, 9, 6, 3; $a_n = -3n + 15$

52. *Sample answer:* The number of people who have ridden a ride at an amusement park that seats 25 people at a time, when every seat is filled each time.

53. 119

54. a. yes; Consecutive terms have a common difference of 12,500.

b. Write a function that represents an arithmetic sequence with $a_1 = 20{,}000$ and $d = 12{,}500$, then find a_6.

55. $f(n) = 6n + 17$

56. 7 min

57. a. arithmetic sequence; Consecutive terms have a common difference of $2x$.

b. not an arithmetic sequence; Consecutive terms do not have a common difference.

4.6 Maintaining Mathematical Proficiency

58. $x \geq -17$

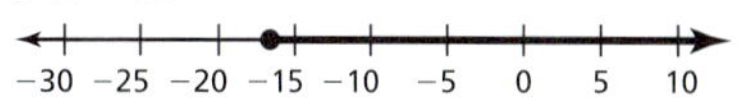

59. $b > 19$

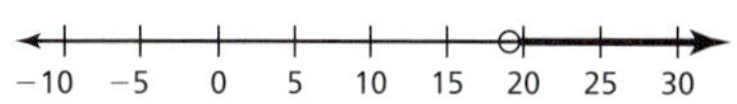

60. $t < 9$

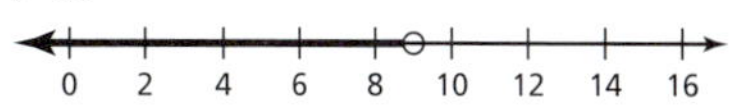

61. $y \leq -4$

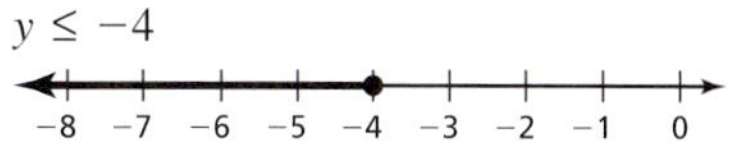

Additional Answers

62. 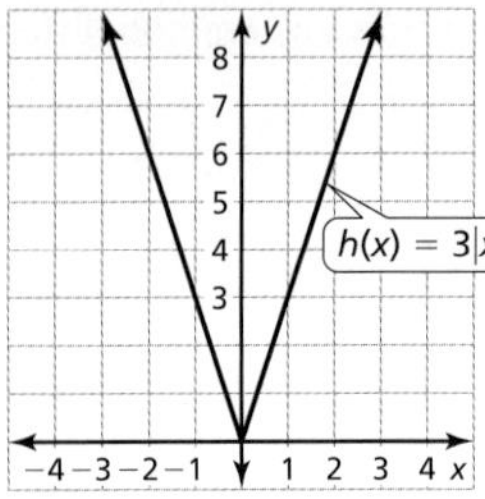

The graph of h is a vertical stretch of the graph of f by a factor of 3; domain: all real numbers; range: $y \geq 0$

63. 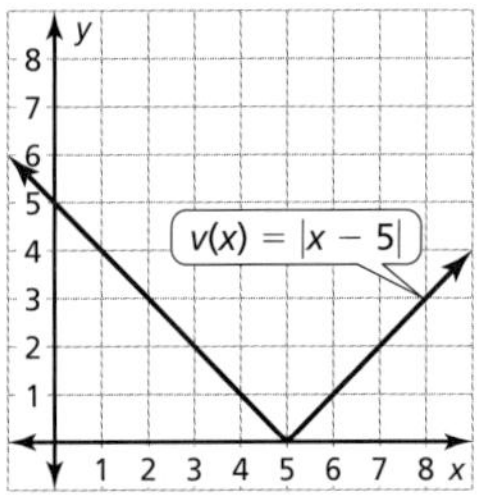

The graph of v is a horizontal translation 5 units right of the graph of f; domain: all real numbers; range: $y \geq 0$

64. 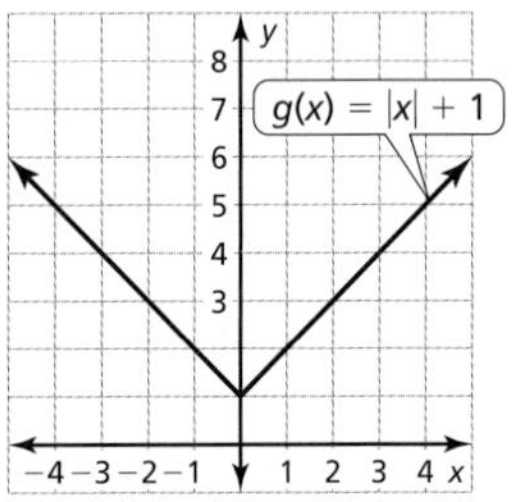

The graph of g is a vertical translation 1 unit up of the graph of f; domain: all real numbers; range: $y \geq 1$

65. 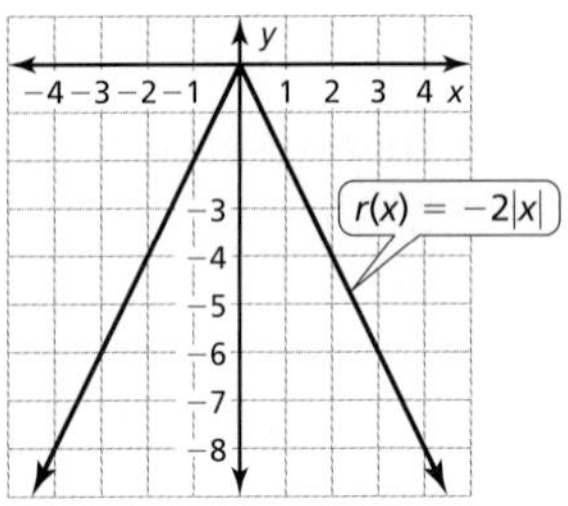

The graph of r is a vertical stretch of the graph of f by a factor of 2 and a reflection in the x-axis; domain: all real numbers; range: $y \leq 0$

4.7 Monitoring Progress and Modeling with Mathematics

19. 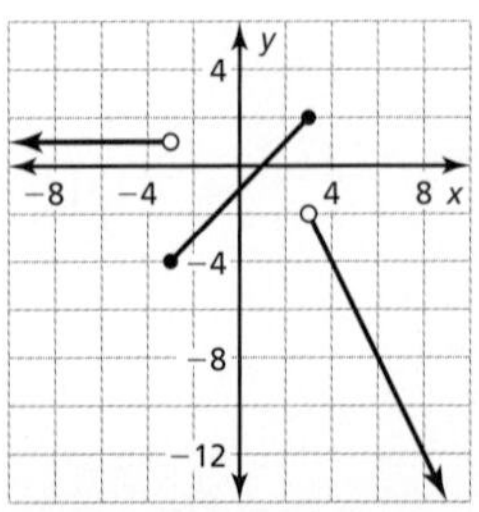

domain: all real numbers; range: $y \leq 2$

20. 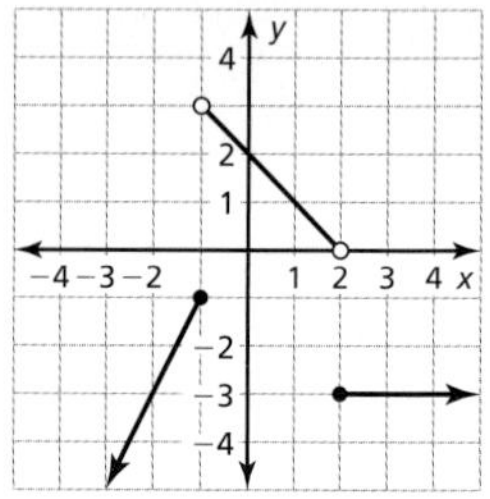

domain: all real numbers; range: $y \leq -1$ or $0 < y < 3$

21. The wrong expression was used to evaluate the function; $f(5) = 5 + 8 = 13$

22. The domain of $y = x + 6$ should be $x \leq 2$, and the domain of $y = 1$ should be $x > 2$.

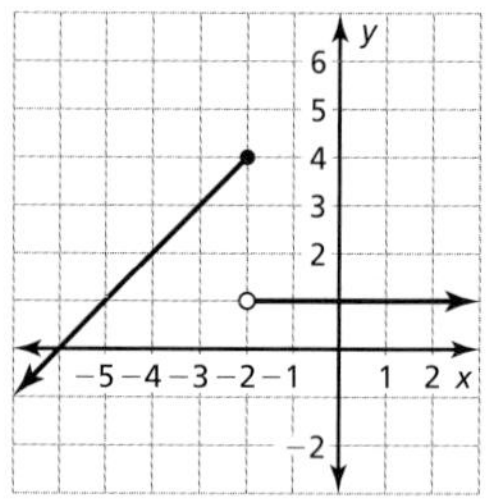

34. 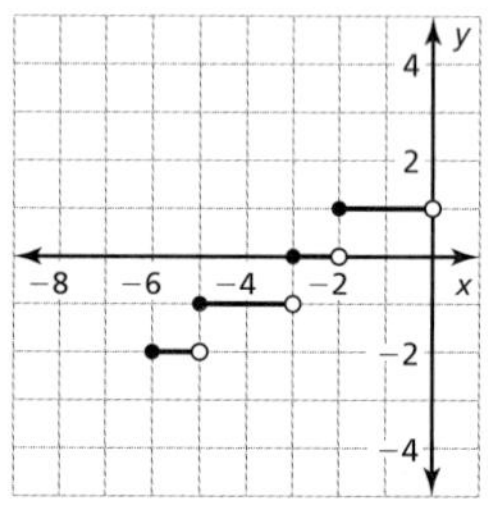

domain: $-6 \leq x < 0$; range: $-2, -1, 0, 1$

35. $$C(p) = \begin{cases} 180, & \text{if } 0 < p \leq 5 \\ 210, & \text{if } 5 < p \leq 6 \\ 240, & \text{if } 6 < p \leq 7 \\ 270, & \text{if } 7 < p \leq 8 \\ 300, & \text{if } 8 < p \leq 9 \end{cases}$$

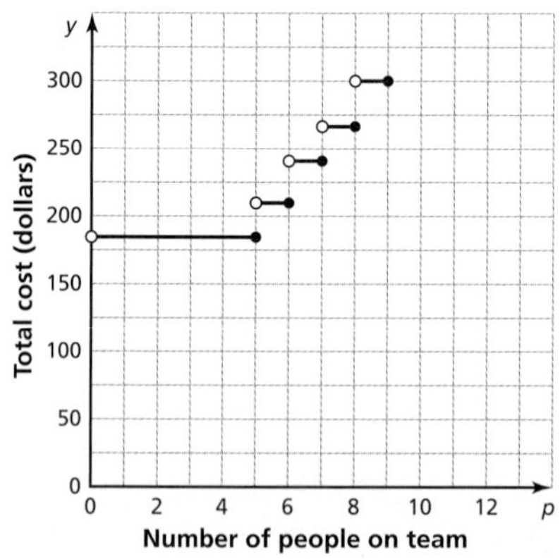

36. $f(x) = \begin{cases} 4, & \text{if } 0 < x \le 1 \\ 8, & \text{if } 1 < x \le 2 \\ 12, & \text{if } 2 < x \le 3 \\ 15, & \text{if } 3 < x \le 24 \end{cases}$

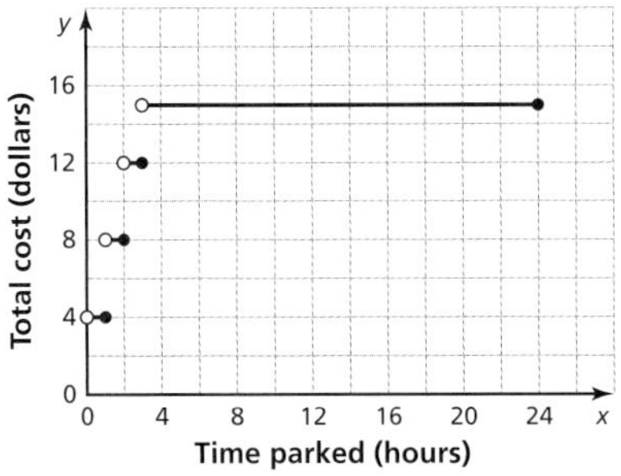

48. **a.** $f(x) = -\frac{2}{3}|x - 6| + 4$

b. $f(x) = \begin{cases} \frac{2}{3}x, & \text{if } x < 6 \\ -\frac{2}{3}x + 8, & \text{if } x \ge 6 \end{cases}$

49. **a.** -4

b. 6

50. **a.** The open circle at (2, 4) changes to a closed circle and the closed circle at (2, -3) changes to an open circle; The domain does not change, the range changes from $y < 4$ to $y \le 4$.

b. It does not change; The domain and range do not change; Because the value of each expression at $x = 1$ is 2, the two pieces of the function have the same endpoint. So, the graph does not change.

51.

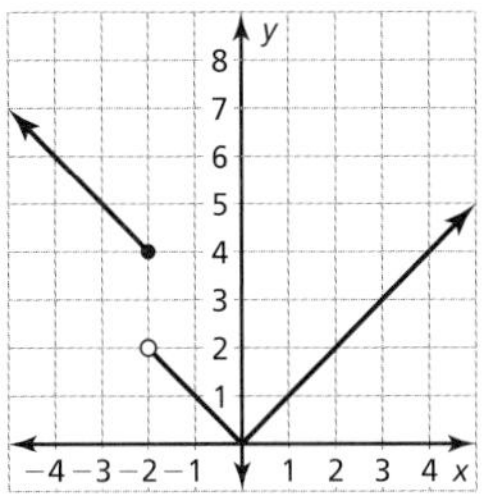

domain: all real numbers; range: $y \ge 0$

52. **a.** 100; At $x = 101$, $C < 10$.

b. yes; From the graph, you can make more than 550 copies with \$40.

53.

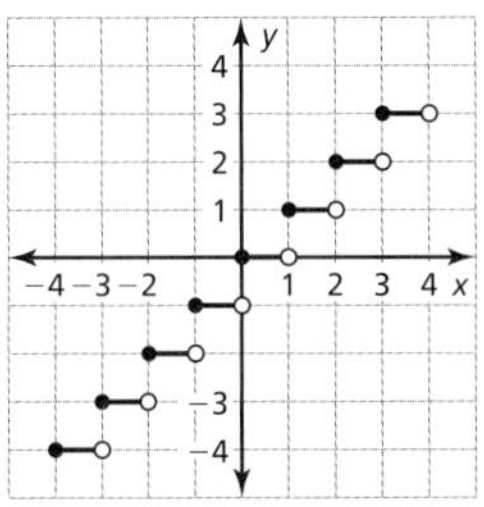

yes; yes; It is a function defined by two or more equations and it is defined by a constant value over each part of its domain.

54. The input value of 3 has two outputs; Either change ≤ 3 to < 3 in the first expression or change ≥ 3 to > 3 in the second expression.

55. **a.** $f(x) = \begin{cases} x, & \text{if } 0 \le x \le 2 \\ 2x - 2, & \text{if } 2 < x \le 8 \\ x + 6, & \text{if } 8 < x \le 9 \end{cases}$

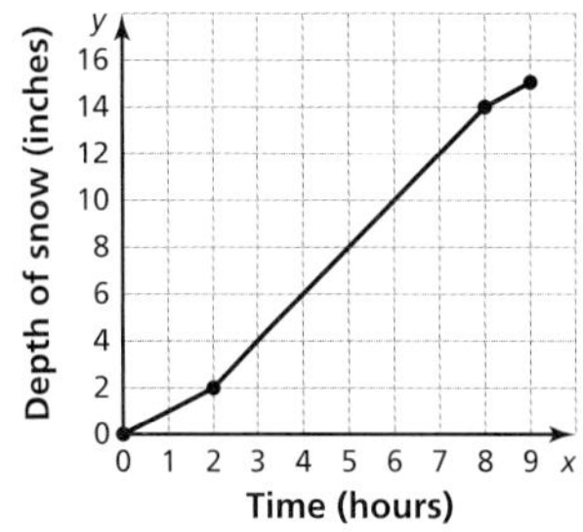

b. no; The total accumulation is 15 inches.

4.7 Maintaining Mathematical Proficiency

56. $-12 < r \le 13$

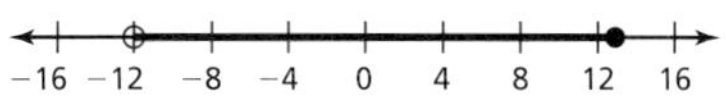

57. $t \le 4$ or $t \ge 18$

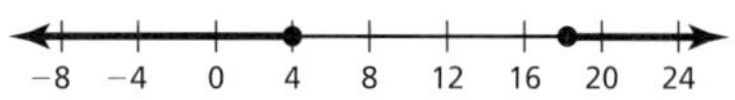

58.

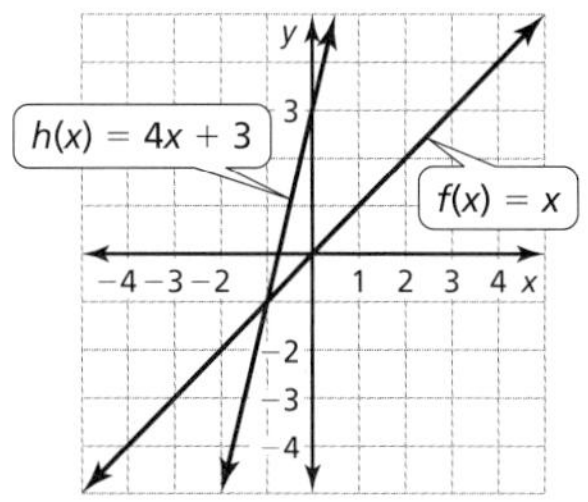

The graph of h is a vertical stretch by a factor of 4 and a vertical translation 3 units up of the graph of f.

59.

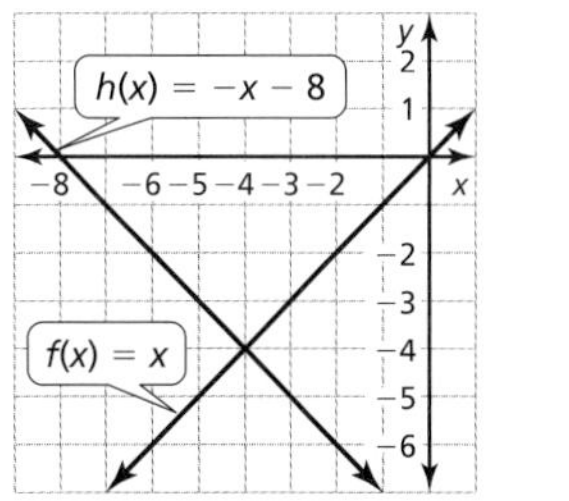

The graph of h is a reflection in the x-axis and a vertical translation 8 units down of the graph of f.

60.

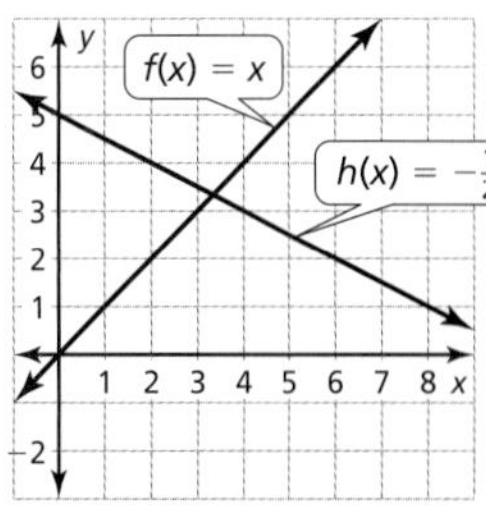

The graph of h is a vertical shrink by a factor of $\frac{1}{2}$, a reflection in the x-axis, and a vertical translation 5 units up of the graph of f.

Additional Answers

4.7 Mini-Assessment

4. $f(x) = \begin{cases} 60, & \text{if } 0 < x \leq 1 \\ 100, & \text{if } 1 < x \leq 2 \\ 140, & \text{if } 2 < x \leq 3 \\ 180, & \text{if } 3 < x \leq 4 \end{cases}$

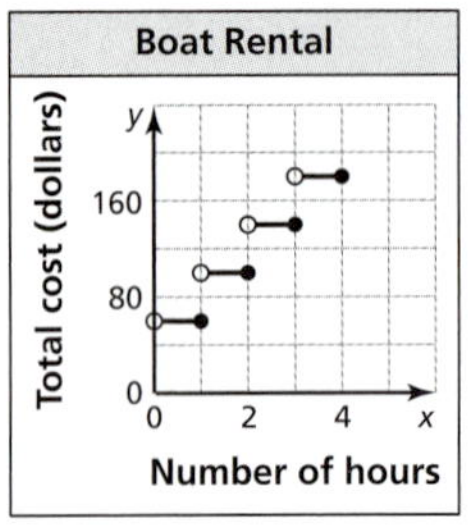

Chapter 4 Test

11. d. There may be a causal relationship in the data but the correlation may be caused by other factors, such as the quality of the attractions each year.

e. \$8900

12. None of the lines are parallel; Lines 1 and 2 are perpendicular; None of the lines have the same slope, but the slope of line 1 is the negative reciprocal of the slope of line 2.

13. *Sample answer:* $y = \begin{cases} 0, & \text{if } x \leq -3 \\ x, & \text{if } -3 < x \leq 1 \\ 1, & \text{if } x > 1 \end{cases}$

Chapter 5

Maintaining Mathematical Proficiency

8. $z > 3$

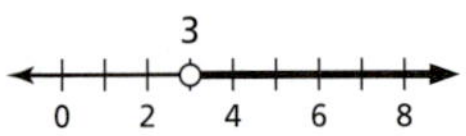

9. $k < 6$

10. $w \geq -3$

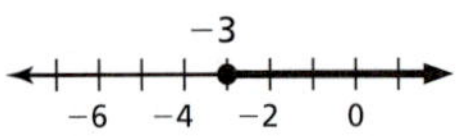

11. The two lines intersect.

5.1 Explorations

2. a.

x (nights)	0	1	2	3	4	5
C (dollars)	600	615	630	645	660	675
R (dollars)	0	75	150	225	300	375

x (nights)	6	7	8	9	10	11
C (dollars)	690	705	720	735	750	765
R (dollars)	450	525	600	675	750	825

5.1 Monitoring Progress and Modeling with Mathematics

30.

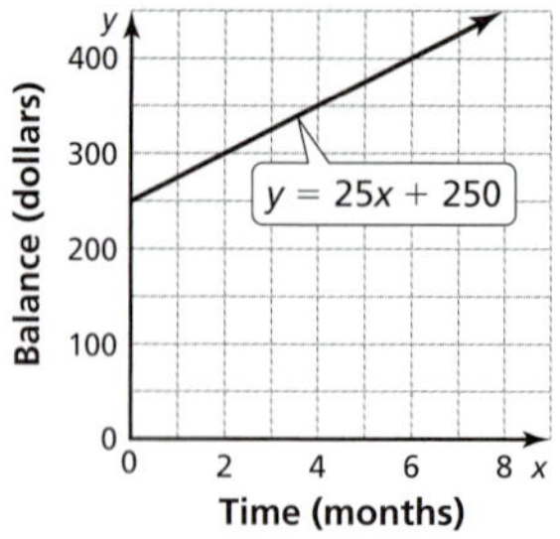

Sample answer: $y = 20x + 280$; The slope of the line is 20. This means you deposit \$20 each month. The y-intercept is 280. This means you opened the account with \$280.

31. a. $x = 3$ **b.** $(3, 5)$

c. The equation in part (a) shows that the 2 values of y in the system in part (b) are equal. The x-coordinate of the solution of the system in part (b) is the solution of the equation in part (a).

32. a. 20, 35, 50; Two of the lines intersect at the points where the number of binders is each of these values.

b. Each of these values is the x-coordinate of the solutions of the systems of linear equations formed by each pair of equations.

33. a. $y = 5x$ and $y = 3x + 3$

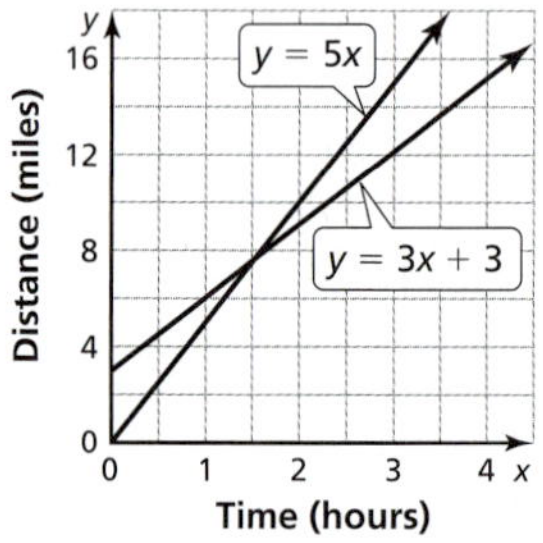

b. no; The point of intersection does not occur at $x = 1$.

5.1 Maintaining Mathematical Proficiency

34. $y = -x + 4$ **35.** $y = 2x + 3$

36. $y = -3x + 20$

5.2 Monitoring Progress and Modeling with Mathematics

32. a. $(4, 5)$

b. yes; Write an equation of each line. Then solve this system of linear equations using substitution.

33. 144 pop songs, 48 rock songs, 80 hip-hop songs

34. *Sample answer:* $x + y = 13$ and $25x + 10y = 265$, x = number of quarters and y = number of dimes

35. 47

5.2 Maintaining Mathematical Proficiency

36. $3x - 11$ **37.** -13

38. -23 **39.** $3d + 5$

40. $22m - 4$ **41.** $64v$

5.3 Monitoring Progress and Modeling with Mathematics

28. a. *Sample answer:* 10 breakfast, 15 lunch

b. $y = x + 5$ and $x + y = 25$

c. Solve the linear system in part (b) using substitution and compare the solution to the answers in part (a).

29. no; *Sample answer:* The system $8x - 5y = 11$ and $4x - 3y = 5$ can be solved by elimination in fewer steps than it can be solved by substitution.

30. *Sample answer:* $3x + 4y = 12$ and $3x - 4y = 6$

31. a. $2\ell + 2w = 18$ and $6\ell + 4w = 46$; length: 5 in.; width: 4 in.

b. length: 15 in.; width: 8 in.

32. Equation 1 can be rewritten as Equation 3 by adding $2(x + y)$ to the left and $2(6)$ to the right, and Equation 3 can be rewritten as Equation 1 by subtracting $2(x + y)$ from the left and $2(6)$ from the right. Because either system can be rewritten as the other, System 1 and System 2 have the same solution.

33. 4.5 qt of 100% fruit juice, 1.5 qt of 20% fruit juice

34. 5 mi/h

35. $(-5, 4, 2)$; *Sample answer:* Subtract Equation 2 from Equation 1. The resulting equation only has 1 variable, y, so use it to solve for y. Substitute this result in Equation 3 and solve for x. Substitute the values of x and y in Equation 2 and solve for z.

5.3 Maintaining Mathematical Proficiency

36. no solution

37. $t = \frac{3}{8}$; one solution

38. $n = -8$; one solution

39. all real numbers; infinitely many solutions

40. $y = -2x + 7$

41. $y = 5x + 6$

42. $y = \frac{2}{3}x + \frac{4}{3}$

5.4 Explorations

1. a. $C = 450 + 20x$; $R = 20x$

x (skateboards)	0	1	2	3	4
C (dollars)	450	470	490	510	530
R (dollars)	0	20	40	60	80

x (skateboards)	5	6	7	8	9	10
C (dollars)	550	570	590	610	630	650
R (dollars)	100	120	140	160	180	200

5.4 Monitoring Progress and Modeling with Mathematics

31. a. never; The y-intercepts are different, so they can never be equations for the same line.

b. sometimes; When $a = b$, the lines are parallel and there is no solution.

c. always; When $a < b$, the slopes are different and the lines intersect at one point.

32. no; The system $3x + 2y = 38$ and $15x + 10y = 190$ has infinitely many solutions.

5.4 Maintaining Mathematical Proficiency

33. $x = -6, x = -2$

34. $x = -5, x = 3$

35. $x = 1, x = 5$

36. $x = 12, x = 2$

5.5 Monitoring Progress and Modeling with Mathematics

40. *Sample answer:* initial value \$22,500, decrease in value at a rate of \$2000 per year

41. a. negative

b. positive

5.5 Maintaining Mathematical Proficiency

42.

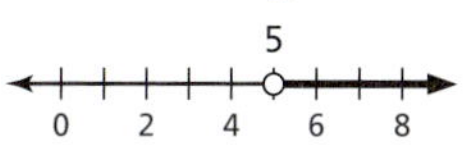

43.

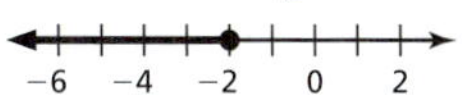

44.

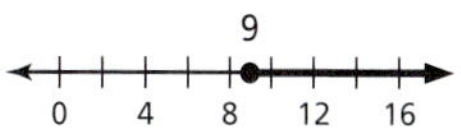

45.

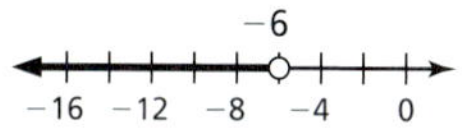

46. The graph of g is a horizontal translation 2 units left of the graph of f.

47. The graph of g is a reflection in the x-axis of the graph of f.

48. The graph of g is a horizontal shrink of the graph of f by a factor of $\frac{1}{4}$.

49. The graph of g is a horizontal translation 1 unit right of the graph of f.

5.6 Monitoring Progress and Modeling with Mathematics

22.

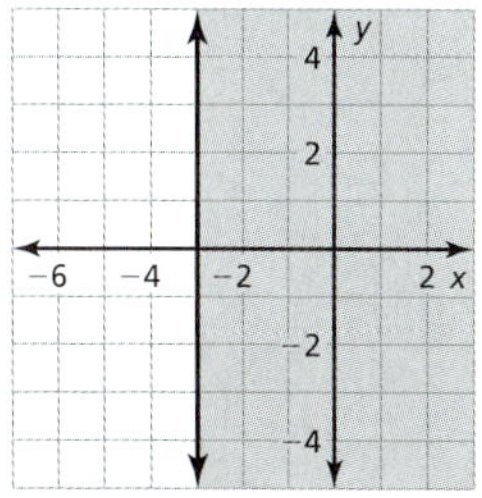

23.

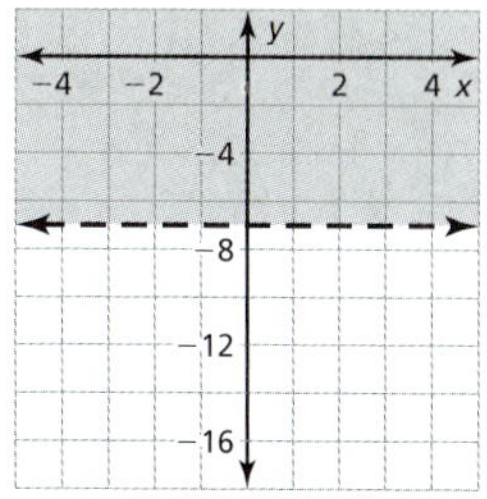

24.

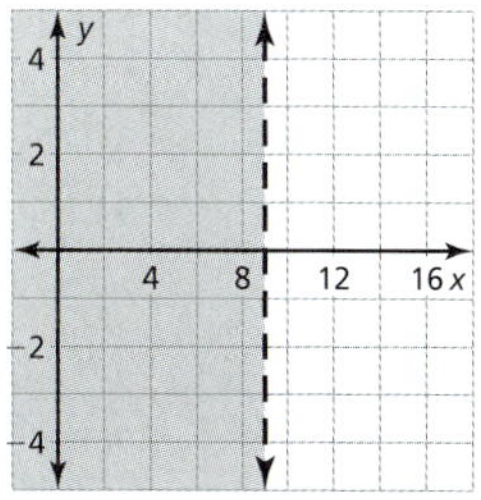

25.

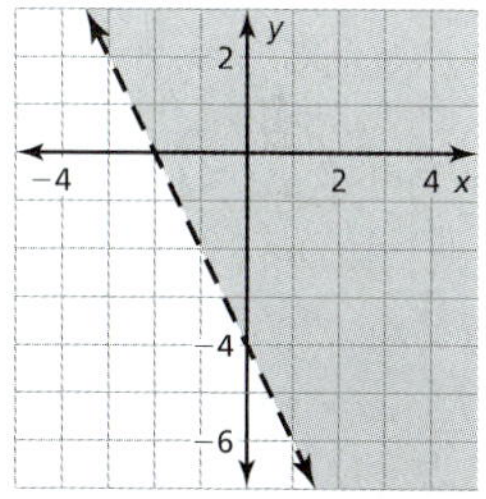

26.

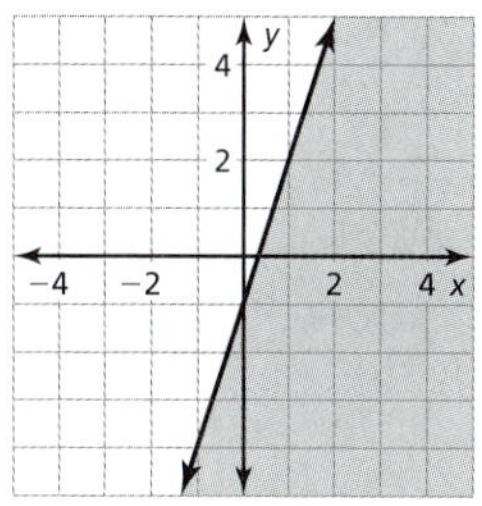

27.

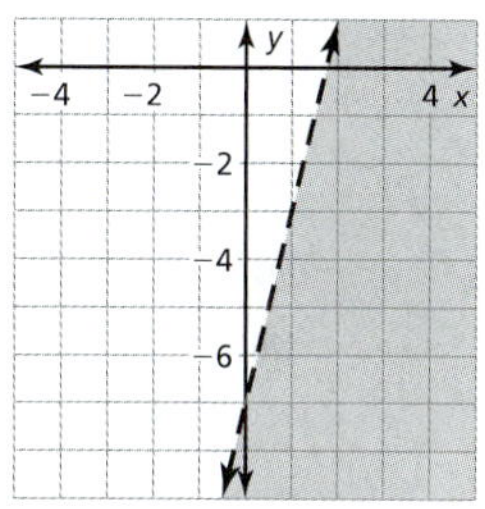

28.

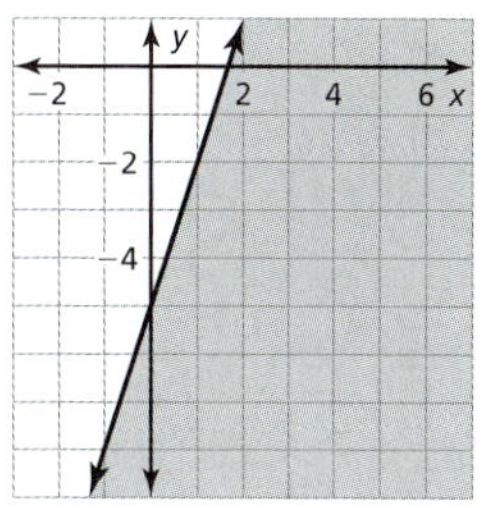

29.

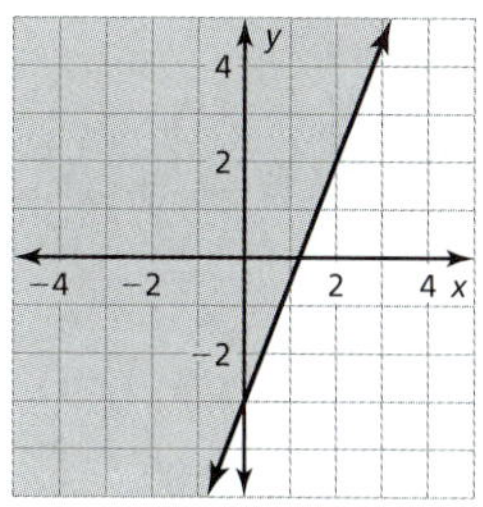

30.

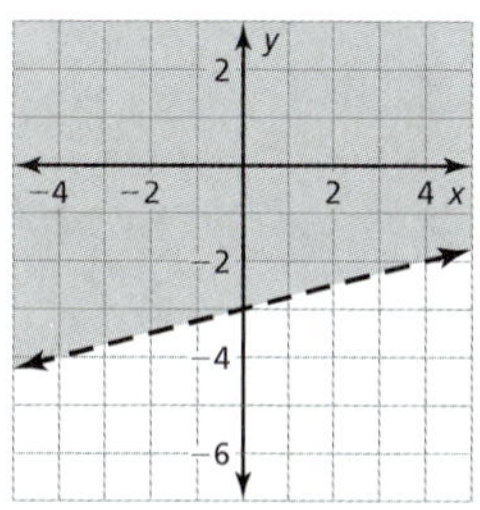

31. The line should be dashed.

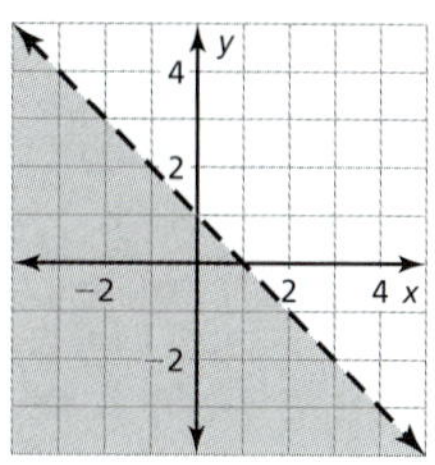

32. The wrong half-plane is shaded.

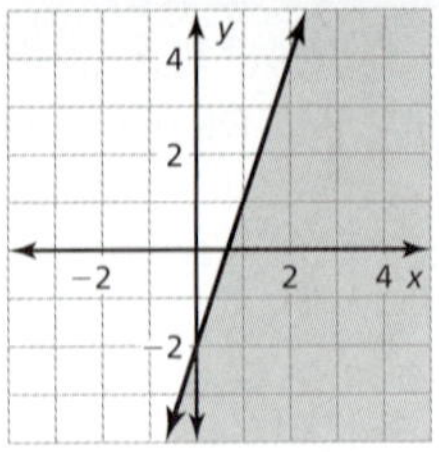

33. $0.75x + 2.25y \le 20$

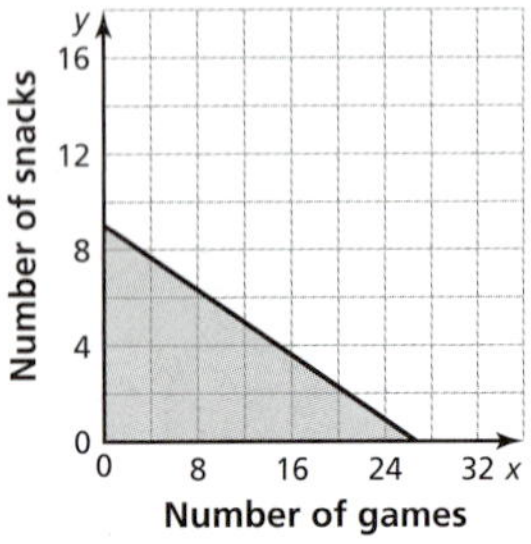

Sample answer: (12, 4), You can play 12 games and buy 4 snacks; (5, 2), You can play 5 games and buy 2 snacks.

34. $10x + 6y \ge 1500$

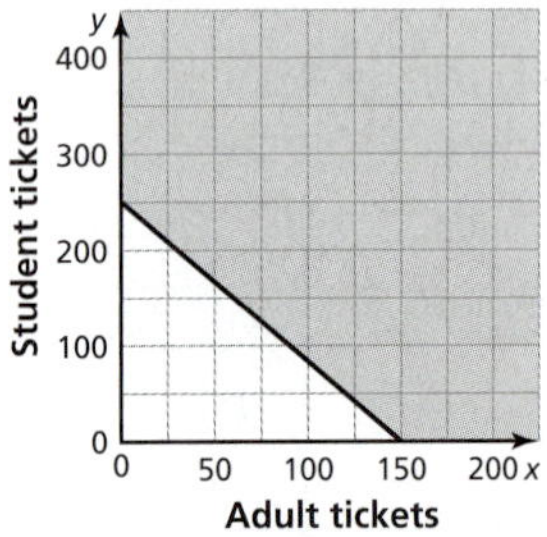

Sample answer: (150, 0), 150 adult tickets were sold and no student tickets were sold; (100, 300), 100 adult tickets were sold and 300 student tickets were sold.

39. **a.** $75x + 40y \le 1800$

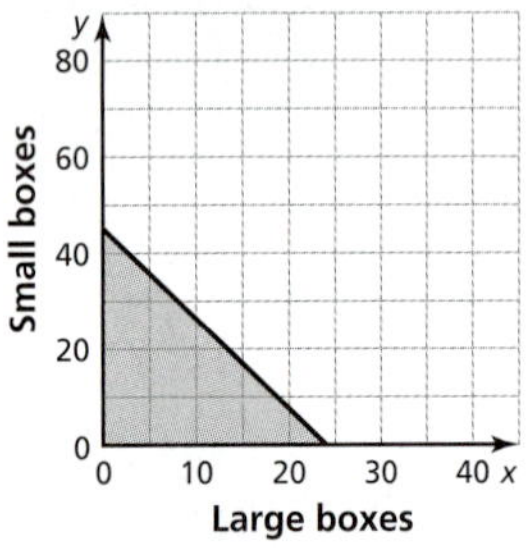

b. *Sample answer:* The number of boxes must be a positive whole number. It is also unlikely for 1 person to carry 45 boxes onto an elevator.

40. **a.** C **b.** A **c.** D **d.** B

41. A test point on the boundary line is not in either half-plane, so it will not indicate which half-plane to shade.

42. *Sample answer:* $y > -x + 1$

43. no; You cannot use (0, 0) as a test point when it is on the boundary line.

44. $y \le 2x + 1$

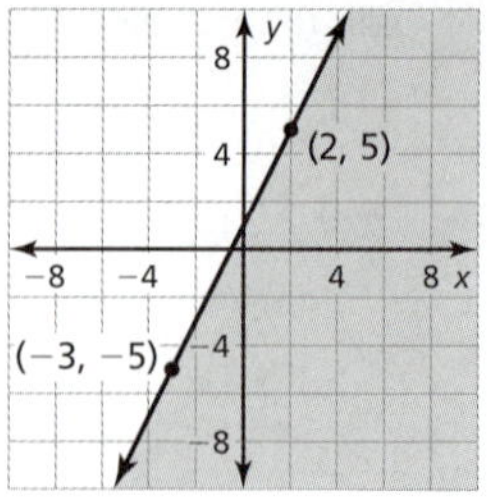

45. $y < 3x + 5$

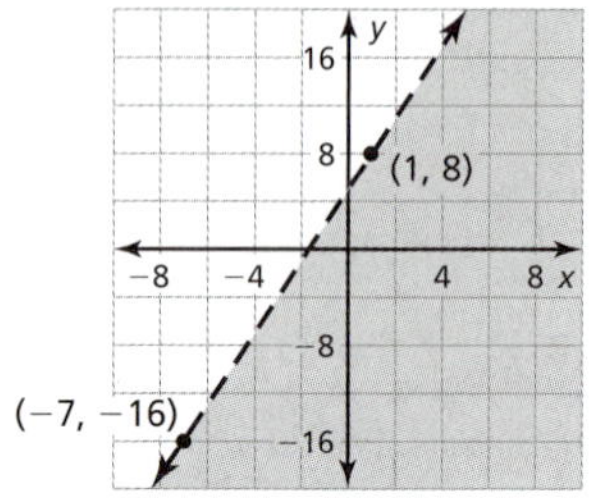

5.7 Lesson Monitoring Progress

4.

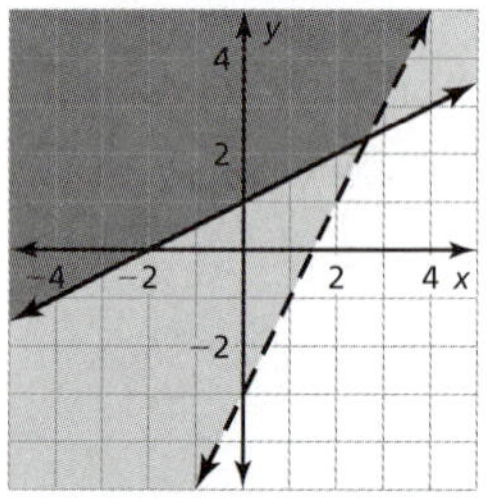

5.

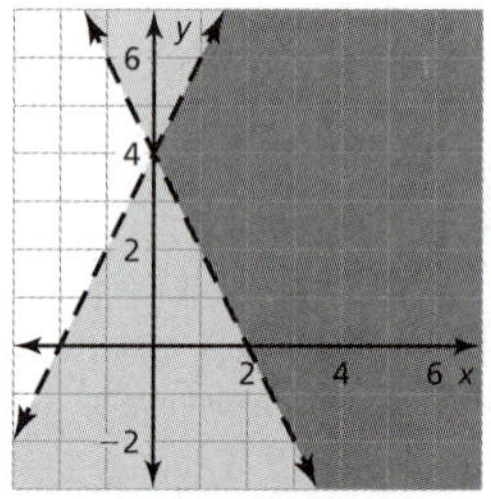

5.7 Monitoring Progress and Modeling with Mathematics

16.

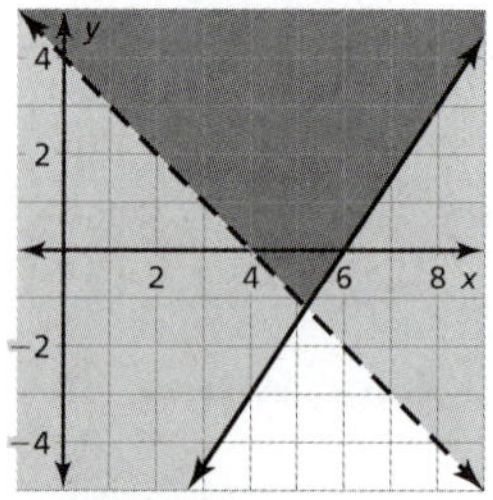

17.

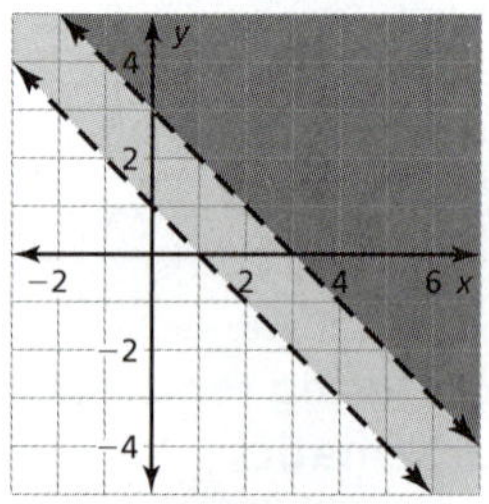

18.

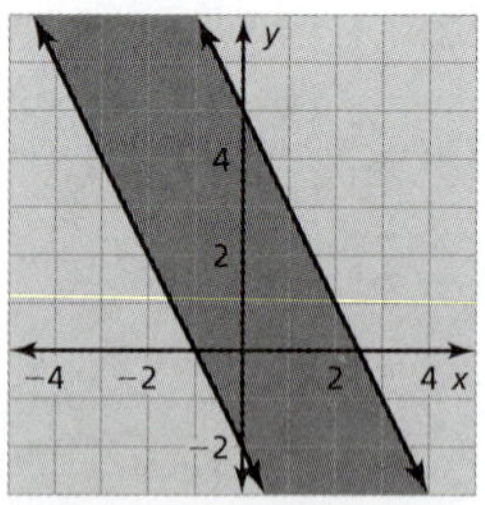

19.

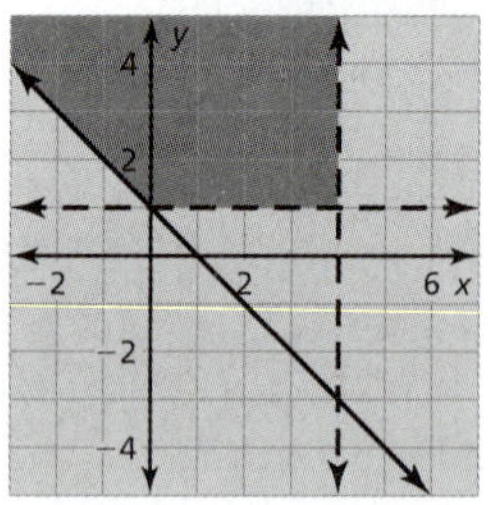

20.

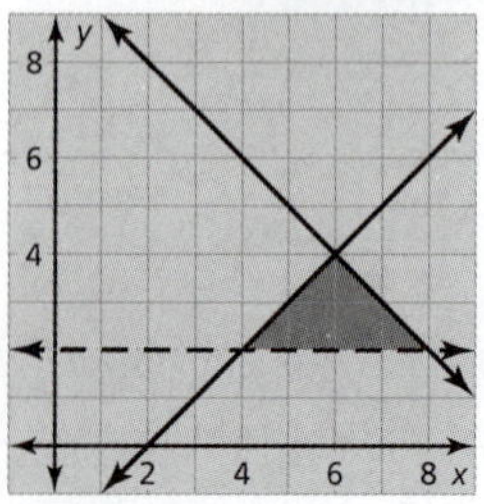

21. $x \ge -1$ and $y < 3$

22. $x > 2$ and $x < 4$

23. $y \ge \frac{2}{3}x - 2$ and $y \ge -3x + 2$

24. $y \leq 5x + 1$ and $y > x - 2$

25. $y > -2x - 1$ and $y < -2x - 3$

26. $y \geq \frac{1}{3}x$ and $y \geq \frac{1}{3}x - 2$

39. $y > -x$; $y < x$; and $x > 0$

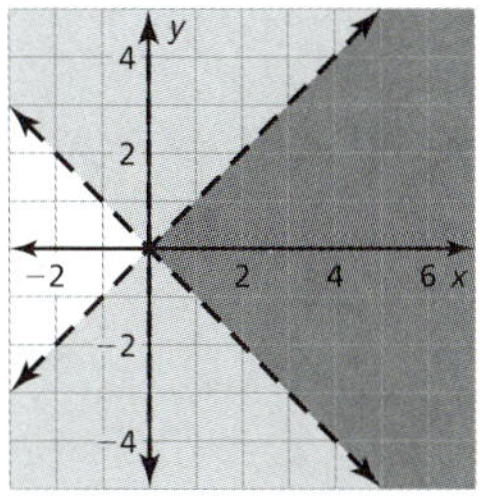

40. no; The half-planes could overlap depending on which side of the boundaries they are on.

41. no; The solution of each inequality is a half-plane, and so the intersection can be at most a half-plane.

42. *Sample answer:* $x > 2$ and $y > 3$

43. *Sample answer:* $x > 2$ and $y < -4$

44. *Sample answer:* $y > 2$ and $y < -2$

45. **a.** *Sample answer:* $-4x + 2y < 6$

    b. *Sample answer:* $-2x + y > 3$

46. *Sample answer:* T-shirts cost \$5, sweatshirts cost \$20, gift certificate is worth \$80, want to buy at least 6 shirts; $5x + 20y \leq 80$ and $x + y \geq 6$

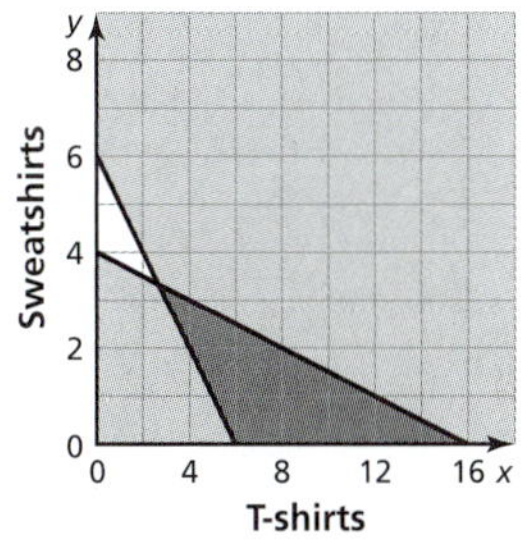

47. *Sample answer:* $x \geq 4$, $x \leq 4$, $y \geq 5$, and $y \leq 5$

48. **a.** $0.5x + 0.25y \leq 20$, $2x + 3y \leq 120$, $x \geq 0$, and $y \geq 0$

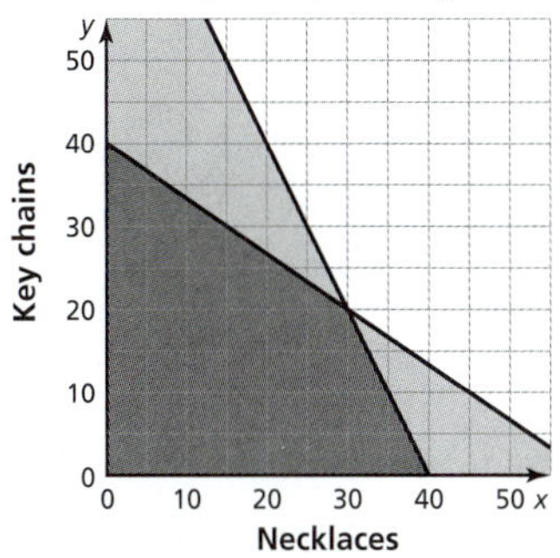

b. (0, 0); (0, 40); (30, 20); (40, 0)

c. \$0, \$320, \$460, \$400; (30, 20)

5.7 Maintaining Mathematical Proficiency

49. 4^5

50. $(-13)^3$

51. x^6

52. $y = x - 6$

53. $y = -3x + 5$

54. $y = -\frac{1}{4}x - 1$

55. $y = \frac{4}{3}x$

Chapter 5 Test

12. **b.** yes; The equation representing the receipt and either of the equations from part (a) form a linear system that has a single solution.

    c. \$3.55/gallon of gas, \$5/quart of oil

13. can be quick when an estimate is sufficient; does not always give exact answers

14. **a.** $12x + 3y \leq 60$

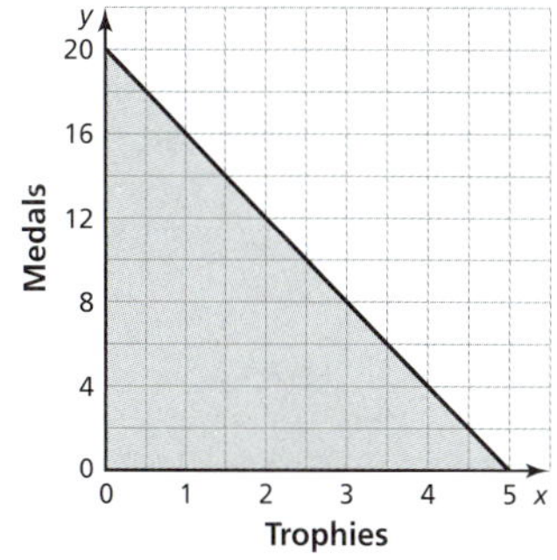

Sample answer: (2, 8); You can buy 2 trophies and 8 medals.

b. $12x + 3y \leq 60$ and $x + y \geq 6$

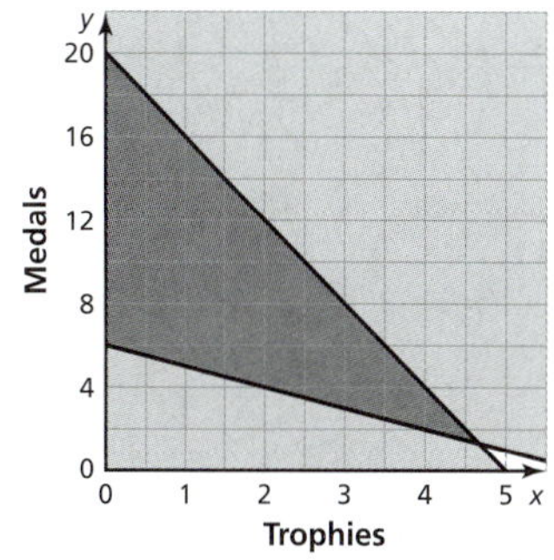

Sample answer: 2 trophies, 6 medals

15. no solution; The lines have the same slope but different y-intercepts, so they are parallel.

Chapter 6

6.1 Monitoring Progress and Modeling with Mathematics

69. **a.** When $a > 1$ and $n < 0$, $a^n < a^{-n}$ because a^n will be less than 1 and a^{-n} will be greater than 1. When $a > 1$ and $n = 0$, $a^n = a^{-n} = 1$, because any number to the zero power is 1. When $a > 1$ and $n > 0$, $a^n > a^{-n}$ because a^n will be greater than 1 and a^{-n} will be less than 1.

    b. When $0 < a < 1$ and $n < 0$, $a^n > a^{-n}$, because a^n will be greater than 1 and a^{-n} will be less than 1. When $0 < a < 1$ and $n = 0$, $a^n = a^{-n} = 1$, because any number to the zero power is 1. When $0 < a < 1$ and $n > 0$, $a^n < a^{-n}$ because a^n will be less than 1 and a^{-n} will be greater than 1.

6.1 Maintaining Mathematical Proficiency

70. 5

71. -10

72. $\pm\frac{1}{8}$

73. natural number, whole number, integer, rational number, real number

74. rational number, real number

75. irrational number, real number

6.2 Monitoring Progress and Modeling with Mathematics

53. always; definition of rational exponent

54. sometimes; true when $x = 0$, $x = -1$, or $x = 1$, otherwise false

55. sometimes; false when $x = 0$ because division by 0 is undefined

56. sometimes: true when $x = 0$ or $x = 1$, otherwise false

6.2 Maintaining Mathematical Proficiency

57. $f(-3) = -16; f(0) = -10; f(8) = 6$

58. $w(-3) = 14; w(0) = -1; w(8) = -41$

59. $h(-3) = 16; h(0) = 13; h(8) = 5$

60. $g(-3) = -8; g(0) = 16; g(8) = 80$

6.3 Explorations

5. d.

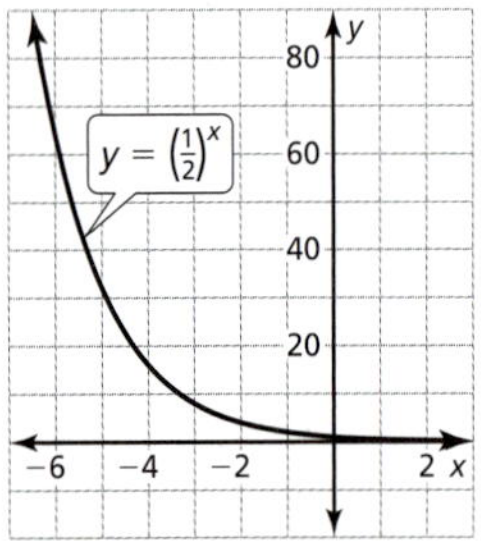

e.

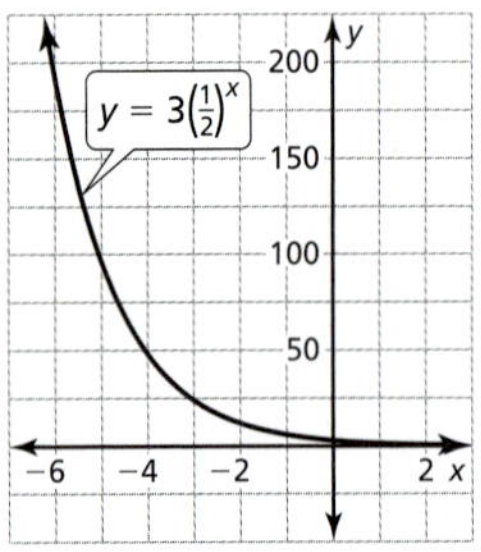

f.

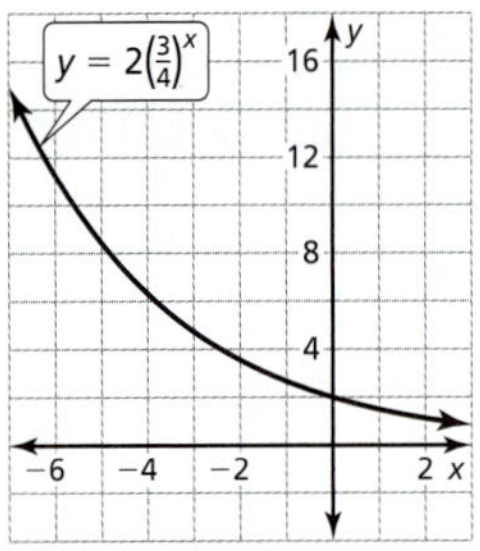

yes; They have the same general curved shape, and they do not intersect the x-axis.

6.3 Lesson Monitoring Progress

5.

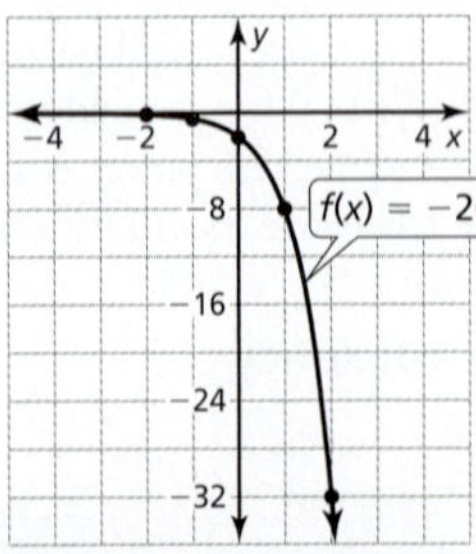

The graph of f is a vertical stretch by a factor of 2 and a reflection in the x-axis of the graph of $g(x) = 4^x$. The y-intercept of the graph of f, -2, is below the y-intercept of the graph of g, 1; domain: all real numbers, range: $y < 0$

6.

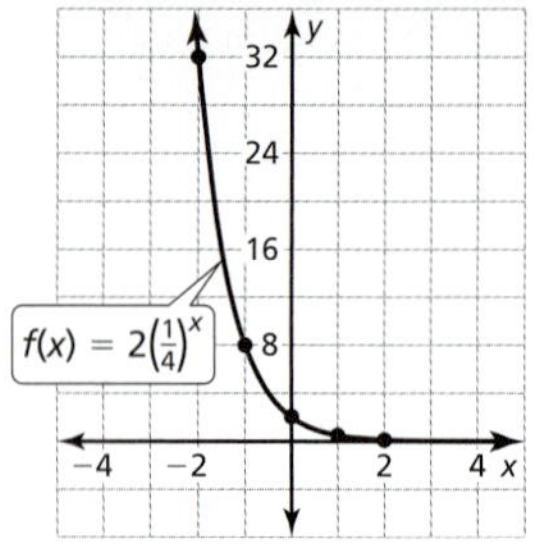

The graph of f is a vertical stretch by a factor of 2 of the graph of $g(x) = \left(\frac{1}{4}\right)^x$. The y-intercept of the graph of f, 2, is above the y-intercept of the graph of g, 1; domain: all real numbers, range: $y > 0$

8.

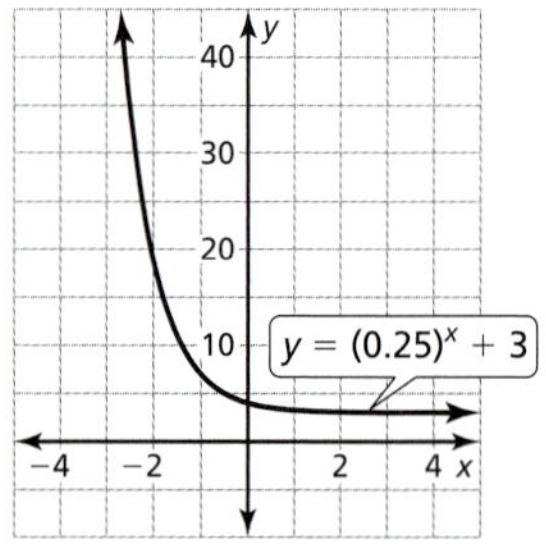

domain: all real numbers, range: $y > 3$

9.

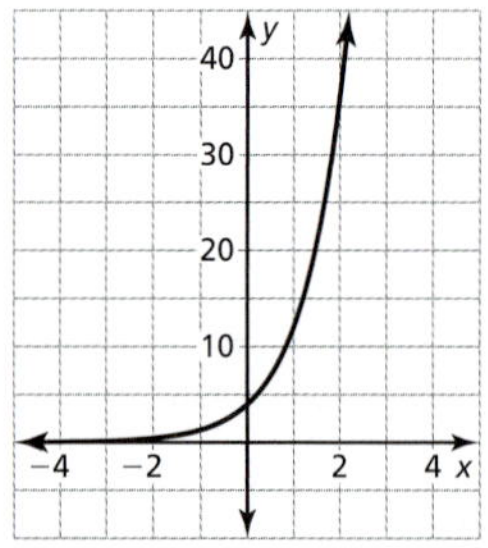

Both functions have the same value when $x = 0$, but the value of f is less than the value of g over the rest of the interval.

6.3 Monitoring Progress and Modeling with Mathematics

26.

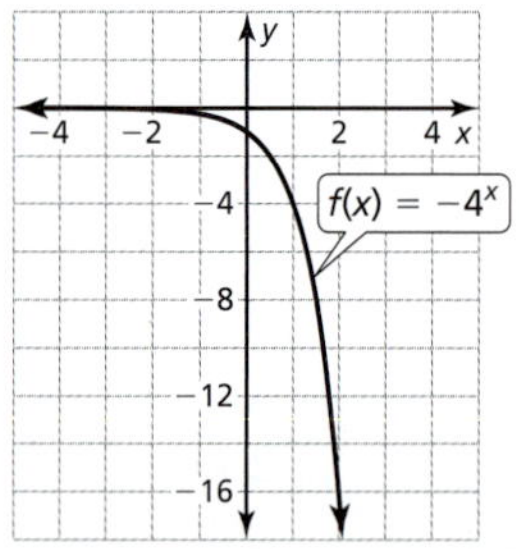

The graph of f is a reflection in the x-axis of the graph of $g(x) = 4^x$. The y-intercept of the graph of f, -1, is below the y-intercept of the graph of g, 1; domain: all real numbers, range: $y < 0$

27. 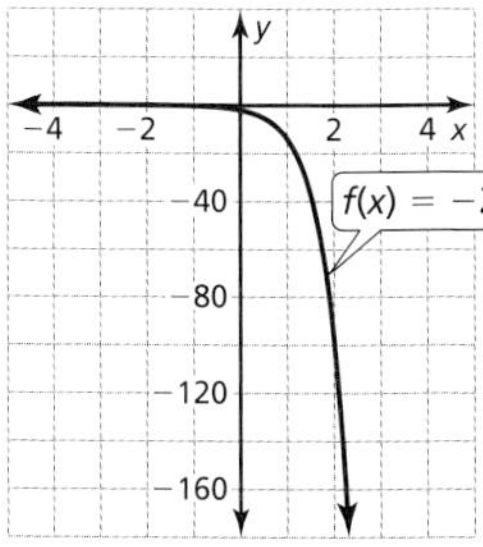

The graph of f is a vertical stretch by a factor of 2 and a reflection in the x-axis of the graph of $g(x) = 7^x$. The y-intercept of the graph of f, -2, is below the y-intercept of the graph of g, 1; domain: all real numbers, range: $y < 0$

28. 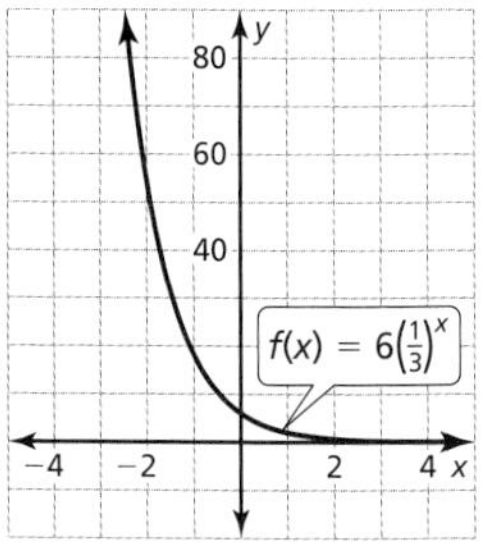

The graph of f is a vertical stretch by a factor of 6 of the graph of $g(x) = \left(\frac{1}{3}\right)^x$. The y-intercept of the graph of f, 6, is above the y-intercept of the graph of g, 1; domain: all real numbers, range: $y > 0$

29. 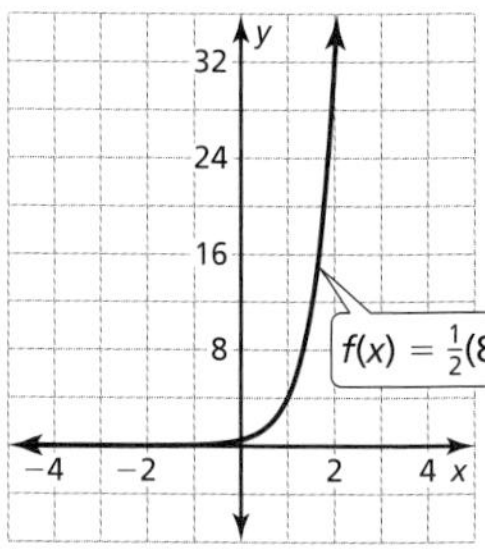

The graph of f is a vertical shrink by a factor of $\frac{1}{2}$ of the graph of $g(x) = 8^x$. The y-intercept of the graph of f, $\frac{1}{2}$, is below the y-intercept of the graph of g, 1; domain: all real numbers, range: $y > 0$

30. 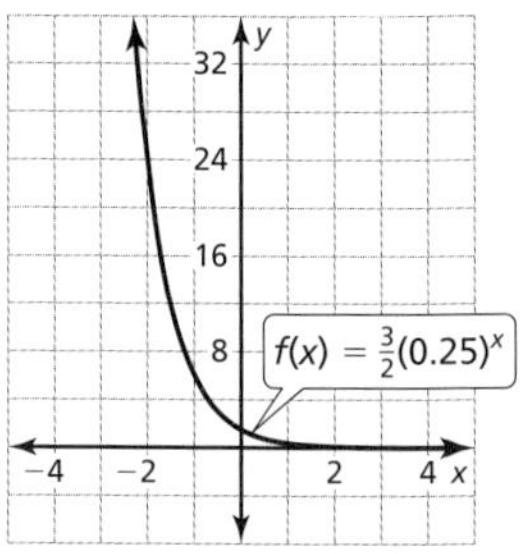

The graph of f is a vertical stretch by a factor of $\frac{3}{2}$ of the graph of $g(x) = 0.25^x$. The y-intercept of the graph of f, $\frac{3}{2}$, is above the y-intercept of the graph of g, 1; domain: all real numbers, range: $y > 0$

31. 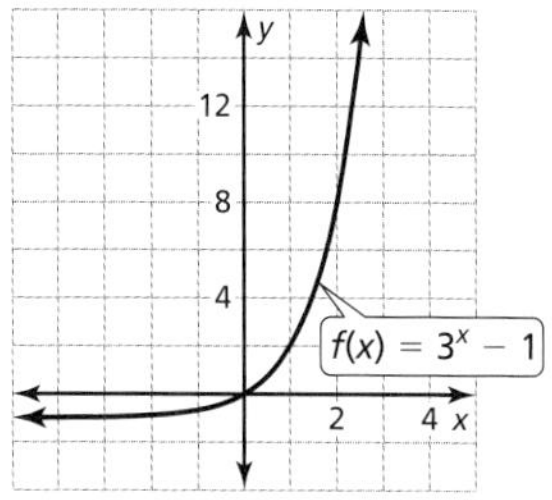

domain: all real numbers, range: $y > -1$

32. 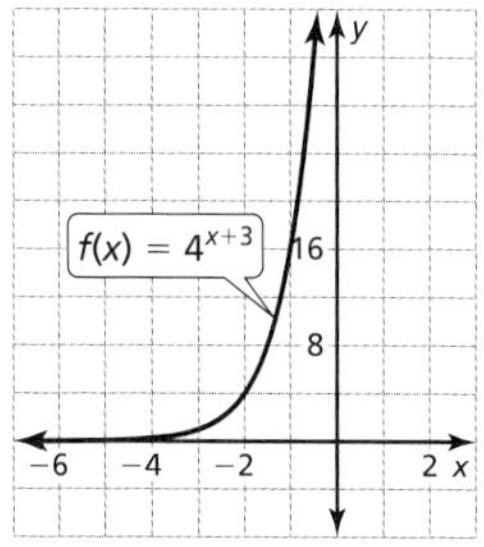

domain: all real numbers, range: $y > 0$

44. 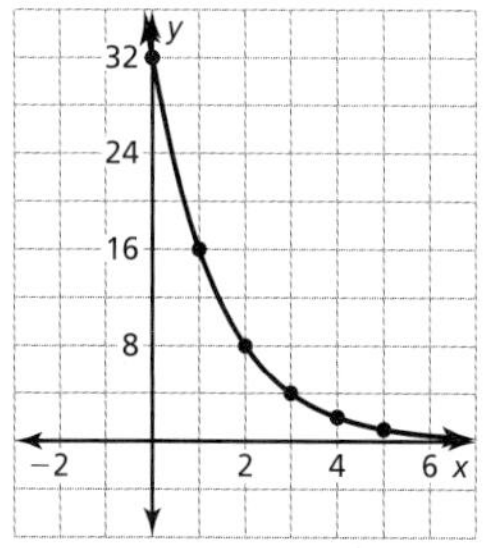

Both functions have the same value when $x = 2$, but the value of h is greater than the value of f over the rest of the interval.

45. a. 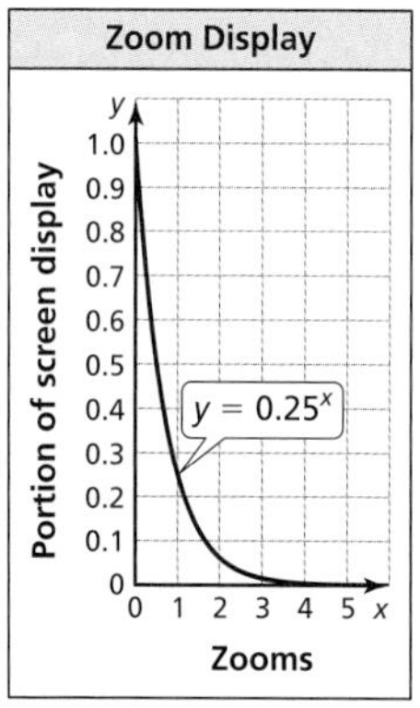

domain: $x \geq 0$, range: $0 < y \leq 1$

b. The y-intercept is 1. This means that when you do not zoom in, 100%, or all, of the original screen display is seen.

c. $6\frac{1}{4}\%$

46. a.

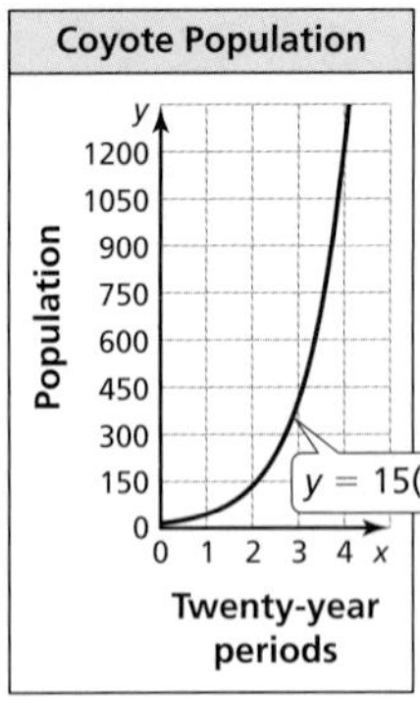

domain: $x \geq 0$, range: $y \geq 15$

b. The y-intercept is 15. This means that the coyote population was 15 at the beginning of the first interval.

c. 135 coyotes

47. $y = 2(7)^x$

48. $y = -50\left(\frac{1}{5}\right)^x$

49. $y = -\frac{1}{2}(2)^x$

50. $y = 8\left(\frac{1}{2}\right)^x$

53.

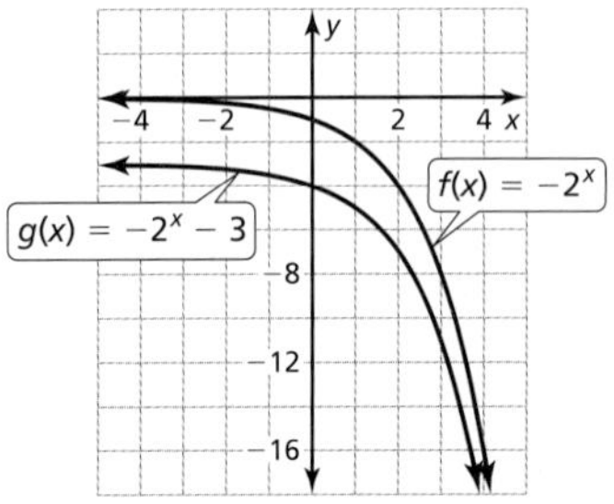

The y-intercept of g is 3 units below the y-intercept of f. The domain of both functions is all real numbers. The range of g is $y < -3$ and the range of f is $y < 0$.

54. no; The values of x are not increasing by a constant amount.

55. When a is positive, it causes a vertical stretch or shrink on the graph. When a is negative, it causes a vertical stretch or shrink and a reflection in the x-axis on the graph.

56. *Sample answer:* $f(x) = 4^{x-2}$

57. $g(x) = 5^{x-3} + 4$

58. a. 2 weeks

b. The stock price drops $30.

59. -192

60. *Sample answer:* The number of bacteria, y, in a Petri dish after x hours is modeled by $y = 100(16)^x$; The constant 100 indicates that there were 100 bacteria in the Petri dish at the beginning, and the constant 16 indicates the number of bacteria is 16 times greater every hour.

61. $\dfrac{f(x+k)}{f(x)} = \dfrac{ab^{x+k}}{ab^x} = \dfrac{b^{x+k}}{b^x} = b^{(x+k)-x} = b^k$

62. $g(x) = 5(2)^x$

63. *Sample answer:* $y = 8(2)^x$

6.3 Maintaining Mathematical Proficiency

64. 0.04

65. 0.35

66. 1.28

67. 2.5

6.4 Explorations

2. b.

Time (h)	0	1	2	3
Temperature difference (°F)	38.6	34.8	31.4	28.3
Body temperature (°F)	98.6	94.8	91.4	88.3

Time (h)	4	5	6
Temperature difference (°F)	25.5	23.0	20.7
Body temperature (°F)	85.5	83.0	80.7

about 6 P.M.

6.4 Monitoring Progress and Modeling with Mathematics

68. a. D; The graph is increasing, and the value from $x = 0$ to $x = 2$ increases from 40 to 160, which is the expected value.

b. B; The graph is decreasing, and the value from $x = 0$ to $x = 2$ decreases from 500 to about 350, which is the expected value.

c. A; The graph is increasing, and the value from $x = 0$ to $x = 2$ increases from 40 to about 50, which is the expected value.

d. C; The graph is decreasing, and the value from $x = 0$ to $x = 2$ decreases from 500 to about 450, which is the expected value.

69. *Sample answer:* $y = 5(1)^x$; Any power of one is 1, so this is a constant function equivalent to $y = 5$.

70. *Sample answer:* account 1: 4% compounded annually, account 2: 8% compounded quarterly; $y = 1000(1.04)^t$; $y = 1000(1.02)^{4t}$; account 2; Graphing both functions shows the balance is always greater in account 2.

71. no; The discount is 20% of the preceding day's price, not always the original price, so the amount of the discount is less each day.

72. a. As t increases by 1, y is multiplied by 2; 100%

b. The graph of g is a vertical stretch of the graph of f; $k = 4$

c. $r = 2$

6.4 Maintaining Mathematical Proficiency

73. $x = -3$

74. $t = -2$

75. $r = 5$

76. slope: -6; y-intercept: 7

77. slope: $\frac{1}{4}$; y-intercept: 7

78. slope: 2; y-intercept: -4

79. slope: $-\frac{1}{2}$; y-intercept: 4

6.4 Mini-Assessment

4. c.

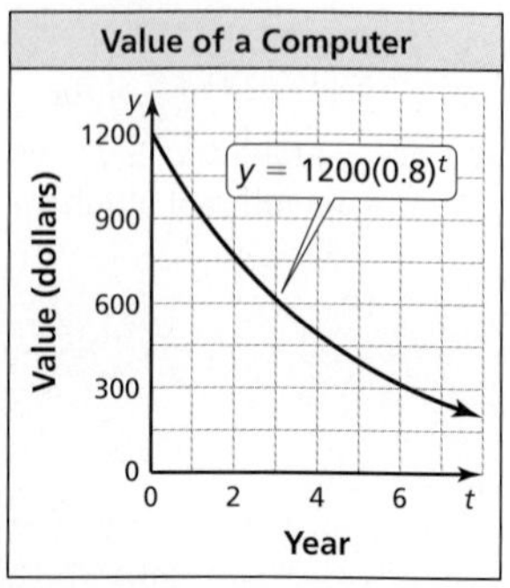

about $300

6.5 Monitoring Progress and Modeling with Mathematics

59. yes; If x was 0, b would equal 1. By the definition of negative exponents, $\left(\frac{1}{a}\right)^x = (a^{-1})^x$. By the Power of a Power Property, $(a^{-1})^x = a^{-x}$. a must be raised to a positive exponent to stay positive. So, x must be negative.

6.5 Maintaining Mathematical Proficiency

60. yes, -6

61. no

62. no

63. yes, 10

6.6 Lesson Monitoring Progress

5. 4, 0.8, 0.16

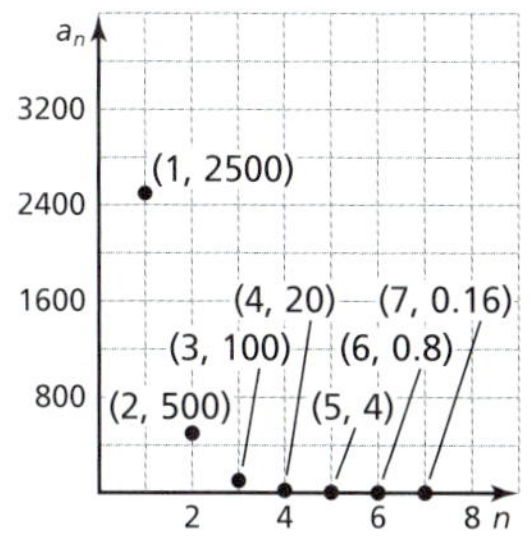

6. 5, -2.5, 1.25

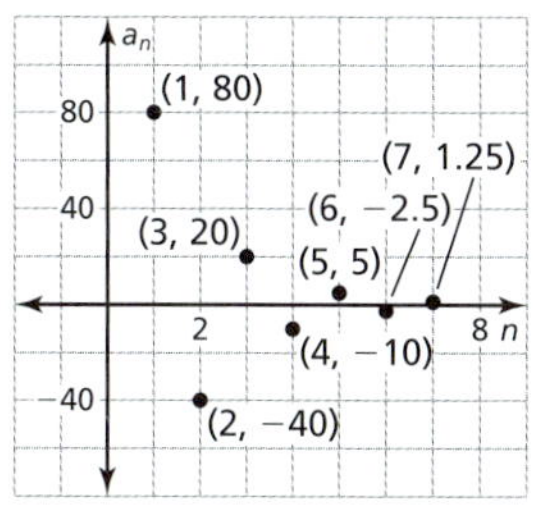

7. -32, 64, -128

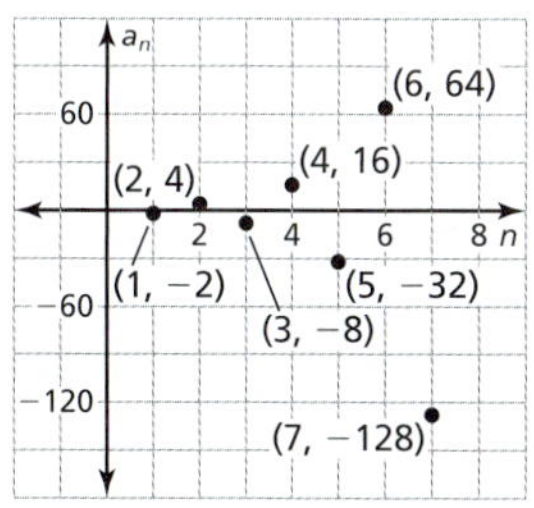

6.6 Monitoring Progress and Modeling with Mathematics

21. $1, -\frac{1}{3}, \frac{1}{9}$

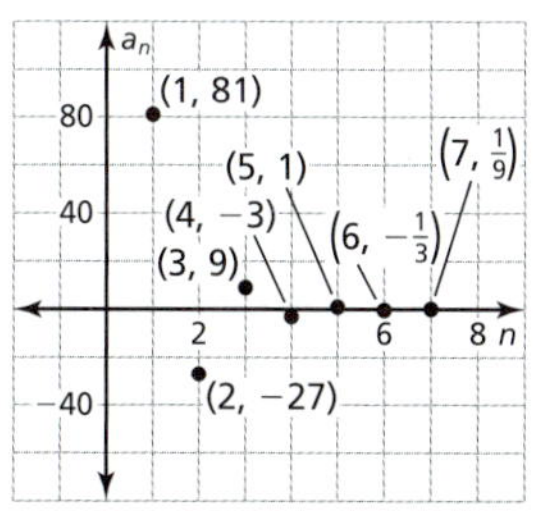

22. $-\frac{3}{5}, -\frac{3}{25}, -\frac{3}{125}$

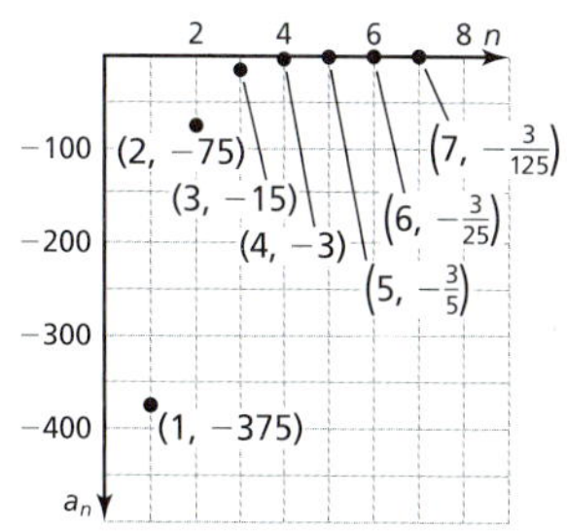

23. $\frac{1}{8}, \frac{1}{32}, \frac{1}{128}$

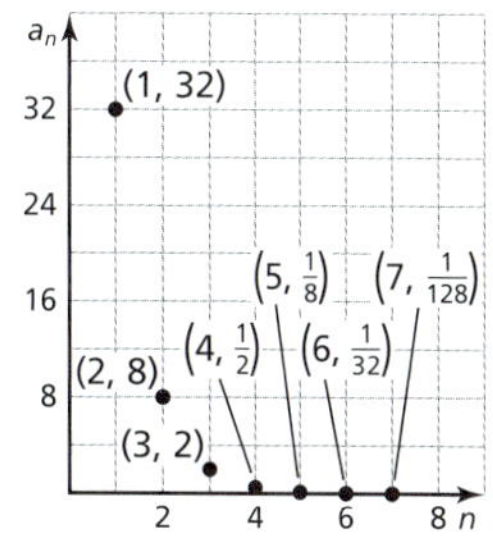

24. $9, \frac{27}{2}, \frac{81}{4}$

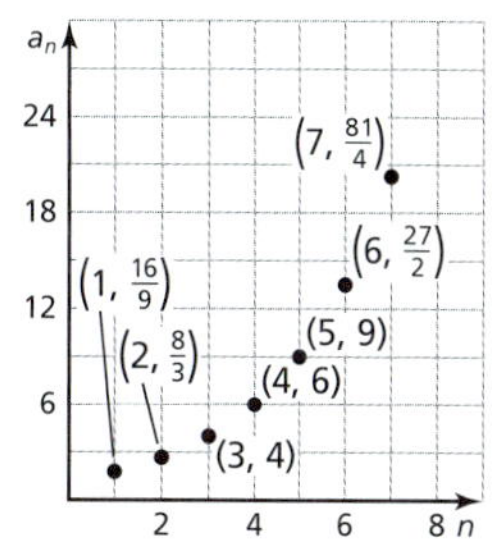

25. $a_n = 2(4)^{n-1}$; 2048

26. $a_n = 0.6(-5)^{n-1}$; -1875

27. $a_n = -\frac{1}{8}(2)^{n-1}$; -4

28. $a_n = 0.1(9)^{n-1}$; 5904.9

29. $a_n = 7640(0.1)^{n-1}$; 0.0764

30. $a_n = -192\left(-\frac{1}{4}\right)^{n-1}$; $\frac{3}{16}$

31. $a_n = 0.5(-6)^{n-1}$; -3888

32. $a_n = 224\left(\frac{1}{2}\right)^{n-1}$; 7

33. 16 teams; 8 teams; 4 teams

50. **a.** $a_n = 0.01(2)^{n-1}$

b. \$167,772.16

c. live on campus; By the end of the month, she will be paying over \$5 million.

51. **a.** 2048, 1536, 1152, 864, 648

b. $a_n = 2048\left(\frac{3}{4}\right)^{n-1}$

c. Theoretically, all of the soup is never gone because an exponential function approaches 0, but never actually equals 0. However, in real life, eventually there will not be enough soup to serve.

52. 2; Putting the values in a spreadsheet and calculating the sum after each term is added shows that the sum approaches 2;

Sample answer: $\frac{1}{2}, \frac{3}{8}, \frac{9}{32}, \frac{27}{128}, \ldots, \frac{1}{2}\left(\frac{3}{4}\right)^{n-1}, \ldots$

53. *Sample answer:* 1, -2, 4, -8, . . .

54. $a_n = 2(3)^{n-1}$; $b_n = 5^{n-1}$

a. no

b. yes; It is the quotient of the common ratios of each sequence.

6.6 Maintaining Mathematical Proficiency

55. yes; The residual points are evenly dispersed about the horizontal axis.

56. no; The residual points are not evenly dispersed about the horizontal axis.

6.7 Lesson Monitoring Progress

4. 0.7; 7; 70; 700; 7000; 70,000

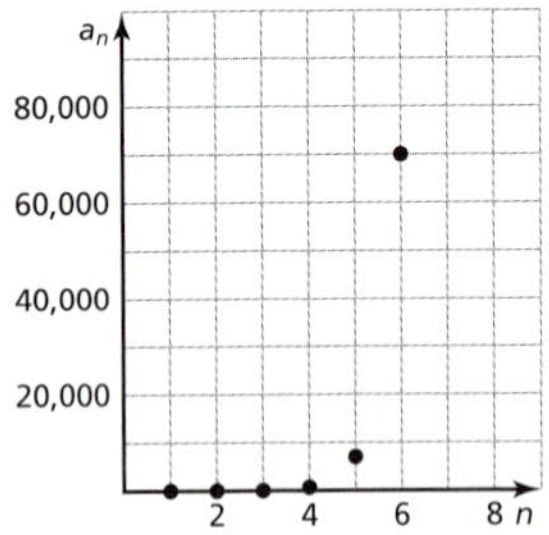

5. $a_1 = 8, a_n = a_{n-1} - 5$

6. $a_1 = 1.3, a_n = a_{n-1} + 1.3$

7. $a_1 = 4, a_n = 5a_{n-1}$

8. $a_1 = 128, a_n = -\frac{1}{4}a_{n-1}$

9. $a_1 = 2$ ft, $a_n = (a_{n-1} + 1.5)$ ft

6.7 Monitoring Progress and Modeling with Mathematics

12. −7, 28, −112, 448, −1792, 7168

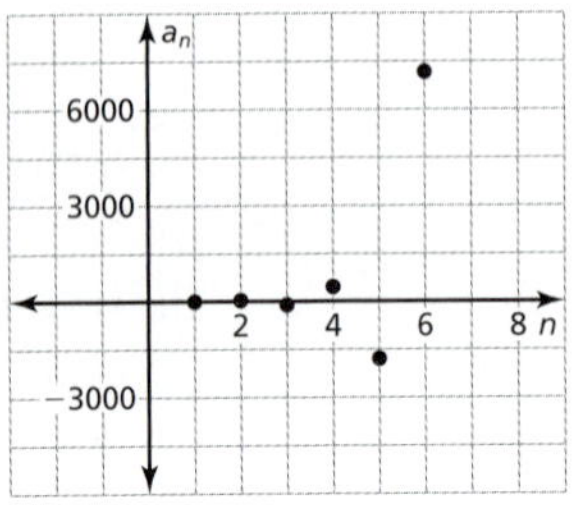

13. $a_1 = 7, a_n = a_{n-1} + 9$

14. $a_1 = 8, a_n = 3a_{n-1}$

15. $a_1 = 243, a_n = \frac{1}{3}a_{n-1}$

16. $a_1 = 3, a_n = a_{n-1} + 8$

17. $a_1 = 0, a_n = a_{n-1} - 3$

18. $a_1 = 5, a_n = -4a_{n-1}$

19. $a_1 = -1, a_n = 4a_{n-1}$

20. $a_1 = 35, a_n = a_{n-1} - 11$

21. $a_1 = 1$ cell, $a_n = (2a_{n-1})$ cells

22. $a_1 = 4\frac{1}{2}$ in., $a_n = \left(a_{n-1} + \frac{1}{4}\right)$ in.

48. −6; −1296; −10,077,696

49. −8; 8; −8

50. 5; 23; 254

51. 15; −15; −10

52. a. $a_1 = 2.5$ cm, $a_2 = 3.5$ cm, $a_n = (a_{n-2} + a_{n-1})$ cm

b. *Sample answer:* 2.7 cm, 2.4 cm, 2.5 cm, 8.5 cm; no; They do not follow a pattern.

56. Each row begins and ends with 1, and each term is the sum of the two terms above it in the previous row; $a_{1,n} = 1$, $a_{n,n} = 1$, $a_{m,n} = a_{m-1,n-1} + a_{m,n-1}$

57. a. Substituting $n - 1$ for n in the explicit rule that defines an arithmetic sequence results in this expression.

b. Write the equation; Identity Property of Addition; Additive Inverse Property; Associative Property of Addition; Distributive Property; Substitution Property of Equality

58. no; The sequence is a geometric sequence with a common factor of −1, so it can be represented by $a_1 = -5$, $a_n = -a_{n-1}$.

59. $a_1 = 3, a_n = a_{n-1} + 2^n$

6.7 Maintaining Mathematical Proficiency

60. $17x$

61. $-6y - 5$

62. $-6d - 7$

63. $8m + 3$

64. $f(x) = 3x$

65. $f(x) = -\frac{1}{2}x - 1$

66. $f(x) = 5$

67. $f(x) = 2x - 7$

Chapter 6 Test

17. a. 1; 0.99988; 0.012%

b.

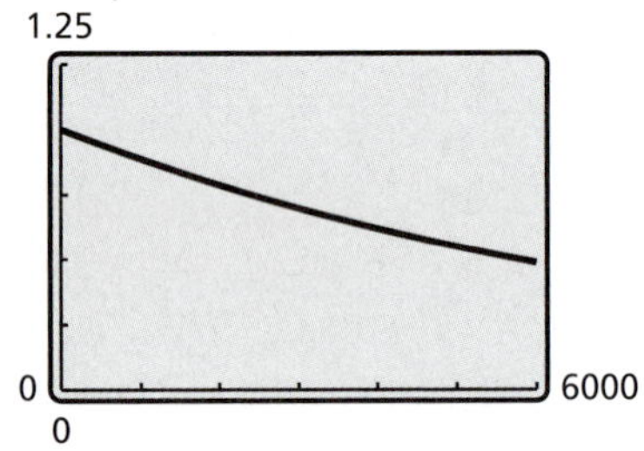

about 0.549 atmosphere

18. a. $a_n = 1.2^{n-1}$; $a_1 = 1, a_n = 1.2a_{n-1}$

b. Day 7

Chapter 7

7.1 Monitoring Progress and Modeling with Mathematics

59. a. no; The product of two negative integers is always a positive integer.

b. yes; The sum of two whole numbers is always a whole number.

60. a. $x^2 - 2x$; $x^2 - 12x$; $2x^2 - 14x$

b. 520 ft^2

c. 1.3 gal; Divide the total area of the deck by the area that can be covered by 1 gallon of sealant to get the amount of sealant needed.

61. a. $19x^2 - 12x$

b. $14,310

7.1 Maintaining Mathematical Proficiency

62. $5x + 4$

63. $34y - 34$

64. $22r - 1$

7.3 Monitoring Progress and Modeling with Mathematics

36. a. $(\pi x^2 - 12\pi x + 36\pi)$ mm^2

b. 4 times greater; The area increases from 4π square millimeters to 16π square millimeters, which is 4 times greater.

37. $(x + 11)(x - 11)$; $x^2 - 121$ fits the product side of the sum and difference pattern, so working backwards, a and b are the square roots of a^2 and b^2.

38. Use the Punnett Square which shows that $\frac{3}{4} = 75\%$ of the combinations result in green pods; Because both the G^2 and the Gy combinations result in green pods, add the coefficients of these terms in the polynomial, $0.25 + 0.5 = 0.75 = 75\%$.

39. $x^4 - 1$ **40.** $y^6 + 8y^3 + 16$

41. $4m^4 - 20m^2n^2 + 25n^4$ **42.** $r^6 - 36t^8$

43. no; $\left(4\frac{1}{3}\right)^2$ can be written as $\left(4 + \frac{1}{3}\right)^2$, however using the square of a binomial pattern results in $16 + \frac{8}{3} + \frac{1}{9}$, which is $18\frac{7}{9}$, not $16\frac{1}{9}$.

44. *Sample answer:* Increase both dimensions by 2, which can be modeled by $A = (x + 2)^2$; Decrease both dimensions by 3, which can be modeled by $A = (x - 3)^2$; Increase one dimension by 4, and decrease the other dimension by 4, which can be modeled by $A = (x + 4)(x - 4)$.

45. $k = 64$

46. $x^3 + 3x^2 + 3x + 1$; $x^3 + 6x^2 + 12x + 8$; $a^3 + 3a^2b + 3ab^2 + b^3$

47. *Sample answer:* $a = 3, b = 4$

7.3 Maintaining Mathematical Proficiency

48. $6(2y - 3)$ **49.** $9(r + 3)$

50. $7(7s + 5t)$ **51.** $5(3x - 2y)$

7.4 Explorations

2.

	x =	1	2	3	4	5	6
a.	(x − 1)(x − 2) = 0	T	T	F	F	F	F
b.	(x − 2)(x − 3) = 0	F	T	T	F	F	F
c.	(x − 3)(x − 4) = 0	F	F	T	T	F	F
d.	(x − 4)(x − 5) = 0	F	F	F	T	T	F
e.	(x − 5)(x − 6) = 0	F	F	F	F	T	T
f.	(x − 6)(x − 1) = 0	T	F	F	F	F	T

If the product of two values is 0, then at least one of the values must be 0. If $(x - a)$ is a factor of an equation, then $x = a$ is a solution.

7.4 Monitoring Progress and Modeling with Mathematics

45. no; Roots will occur if $x^2 + 3 = 0$ or $x^4 + 1 = 0$. However, solving these equations results in $x^2 = -3$ or $x^4 = -1$, and even powers of any number cannot be negative.

46. *Sample answer:* $(x - 1)^2(x - 2)(x - 3)$

47. **a.** $x = -y, x = \frac{1}{2}y$

b. $x = \pm y, x = -4y$

48. $x = 7, x = 4$

7.4 Maintaining Mathematical Proficiency

49. 1, 10; 2, 5 **50.** 1, 18; 2, 9; 3, 6

51. 1, 30; 2, 15; 3, 10; 5, 6 **52.** 1, 48; 2, 24; 3, 16; 4, 12; 6, 8

7.5 Monitoring Progress and Modeling with Mathematics

45. $x^2 - 2x - 24 = 0$; Multiply $(x + 4)(x - 6)$.

46. **a.** Look for the x-intercepts and substitute them for p and q in $(x - p)(x - q)$.

b. $(x - 2)(x + 3)$

47. **a.** $-x^2 + 38x$

b. $-x^2 + 38x = 280$; 10 m

48. $(x + 2y)(x + 4y)$ **49.** $(r + 3s)(r + 4s)$

50. $(a - 2b)(a + 13b)$ **51.** $(x + 5y)(x - 7y)$

7.5 Maintaining Mathematical Proficiency

52. $p = 9$ **53.** $z = -17$

54. $c = -42$ **55.** $k = 0$

7.6 Monitoring Progress and Modeling with Mathematics

42. **a.**

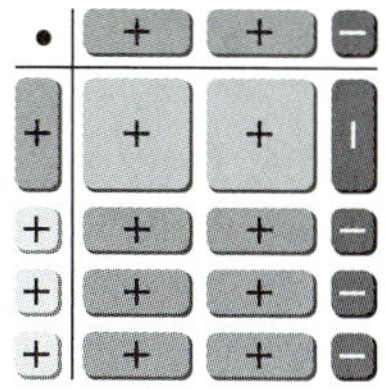

$2x^2 + 5x - 3 = (x + 3)(2x - 1)$

b.

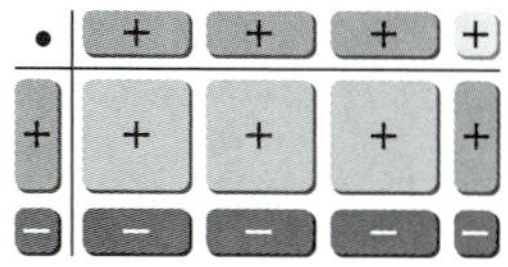

$3x^2 - 2x - 1 = (x - 1)(3x + 1)$

43. 3.5 in. **44.** 4 ft

45. $(k + 2j)(4k - j)$ **46.** $(2x - y)(3x + 4y)$

47. $-(a - 2b)(6a - 7b)$ **48.** $3m(2m + 5n)(3m - n)$

7.6 Maintaining Mathematical Proficiency

49. ± 8 **50.** 2

51. -15 **52.** ± 9

53. $(-1, -4)$ **54.** $(4, 6)$

55. $(0, -7)$ **56.** $(-5, 3)$

7.7 Explorations

2.

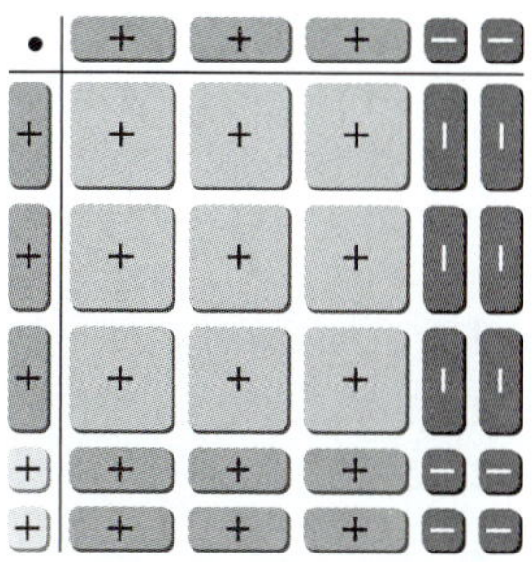

$9x^2 - 4$; $(3x + 2)(3x - 2)$

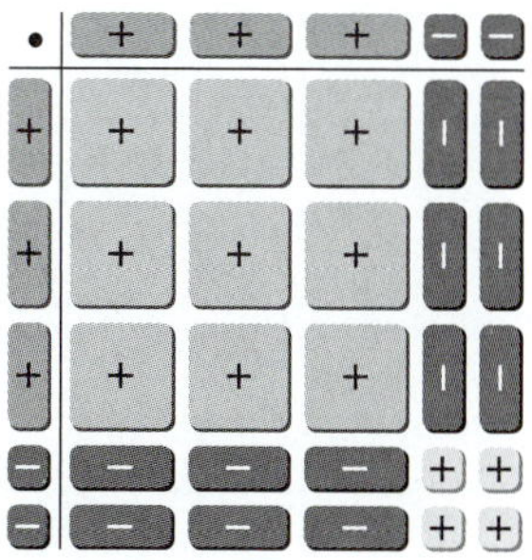

$9x^2 - 12x + 4$; $(3x - 2)^2$

(continued)

Additional Answers

2. (continued)

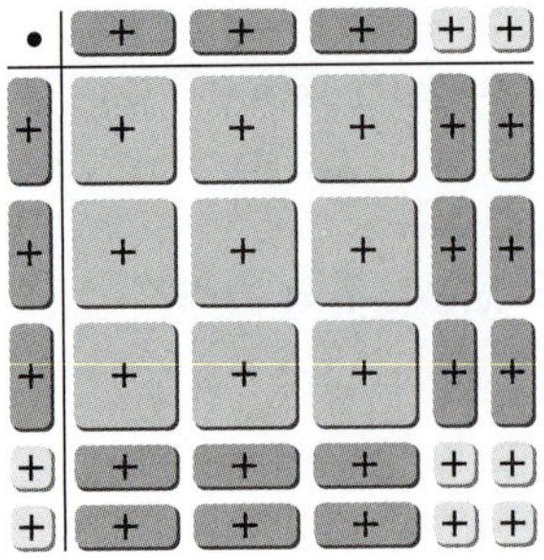

$9x^2 + 12x + 4$; $(3x + 2)^2$

7.7 Monitoring Progress and Modeling with Mathematics

46. a. the blue, yellow, and red regions; *Sample answer:*

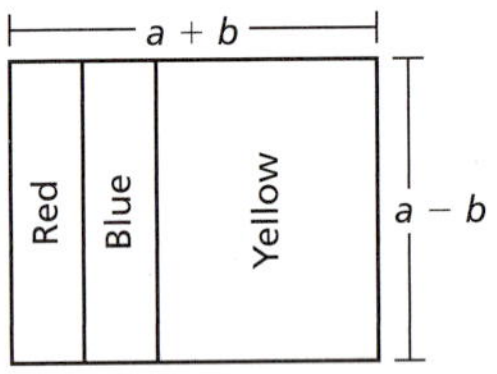

b. $(a - b)^2$ is represented by the yellow region. a^2 is the entire square. Subtracting $2ab$ requires removing 2 regions that each represent ab, which would be the blue and purple regions, and the red and purple regions. There is only one purple region, and subtracting $2ab$ required removing it twice, so one purple region needs to be replaced, which adds b^2.

47. a. $(9x^2 - 144)$ in.^{2}

b. 5 in.; Setting the polynomial in part (a) equal to 81 and solving results in $x = -5$ or $x = 5$. Length cannot be negative, so 5 is the solution.

48. a. $(x^3 + 16x)$ in.^{3}

b. $x = 3$; Setting the polynomial in part (a) equal to $25x$ and solving results in $x = 0$, $x = \;\; 3$, or $x = 3$. Because x is also a side length, and length cannot be negative or zero, 3 is the solution.

7.7 Maintaining Mathematical Proficiency

49. $2 \cdot 5^2$

50. $2^2 \cdot 11$

51. $5 \cdot 17$

52. $2^5 \cdot 3$

53.

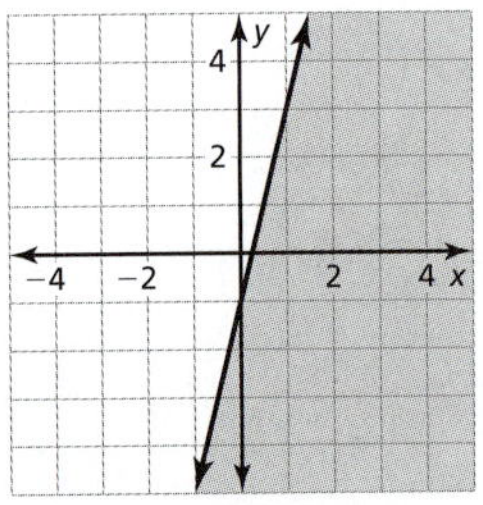

54.

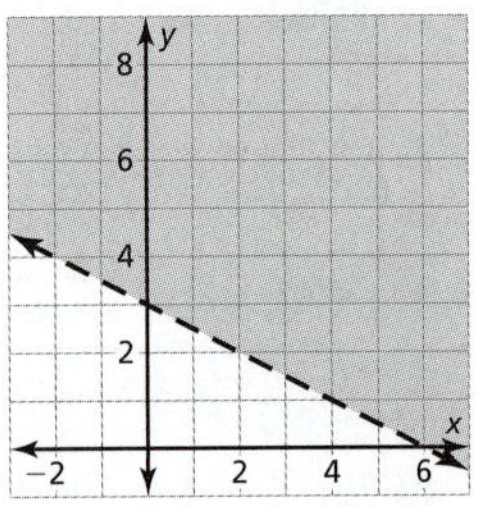

55.

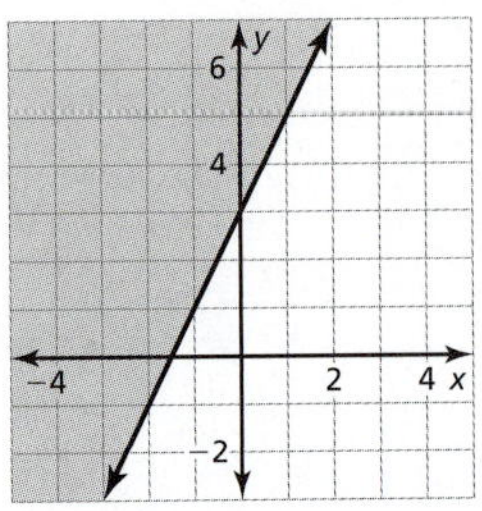

56.

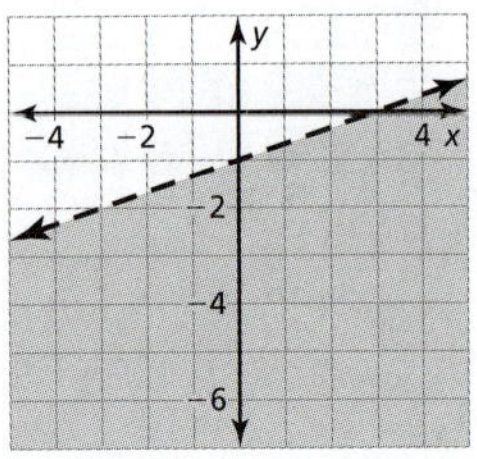

7.8 Monitoring Progress and Modeling with Mathematics

46. a. $(-h^3 + 5h^2 + 36h)$ in.^{3}

b. length: 3 in., width: 10 in., height: 6 in.; length: 4 in., width: 9 in., height: 5 in.

c. length: 4 in., width: 9 in., height: 5 in.; These dimensions result in a surface area of 202 square inches. The other possible dimensions result in a surface area of 216 square inches.

47. radius: 5, height: 8

48. $(x - 1)(x - 1)(x + 1)(x + 2)(x - 2)$

49. a. *Sample answer:* $w = 40$; When $w = 40$, factoring out $5x$ will leave a perfect square trinomial, so there will be 2 factors.

b. *Sample answer:* $w = 50$; When $w = 50$, factoring out $5x$ will leave a factorable trinomial that is not a perfect square, so there will be 3 factors.

7.8 Maintaining Mathematical Proficiency

50. $(2, -2)$

51. $(2, 3)$

52. $(4, 8)$

53. $(-6, -2)$

54.

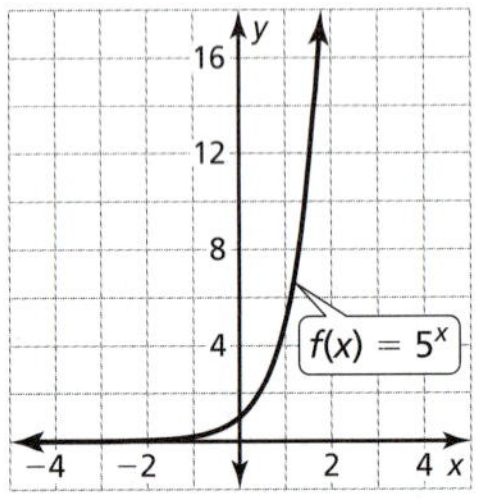

domain: all real numbers, range: $y > 0$

55.

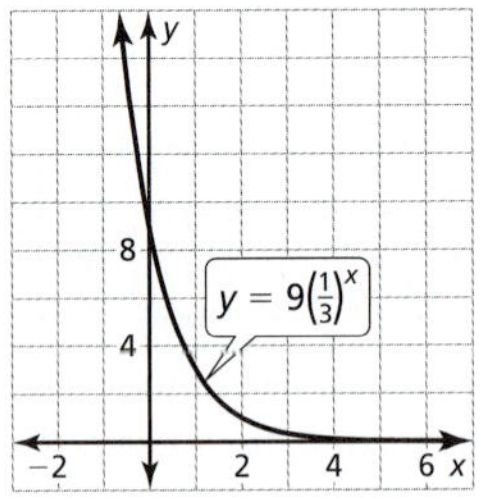

domain: all real numbers, range: $y > 0$

56.

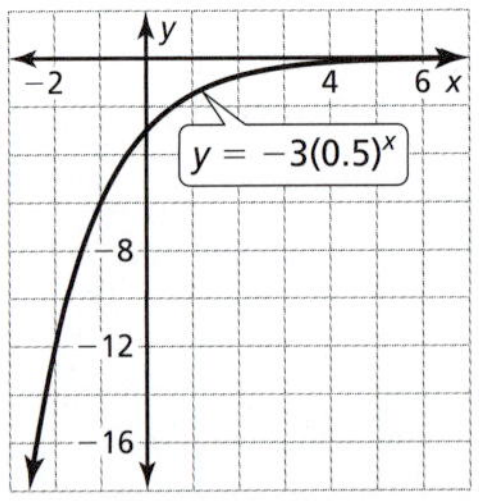

domain: all real numbers, range: $y < 0$

57.

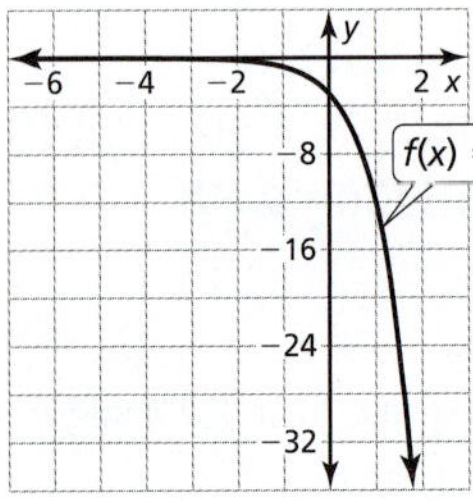

domain: all real numbers, range: $y < 0$

Chapter 8

Mathematical Practices

6. 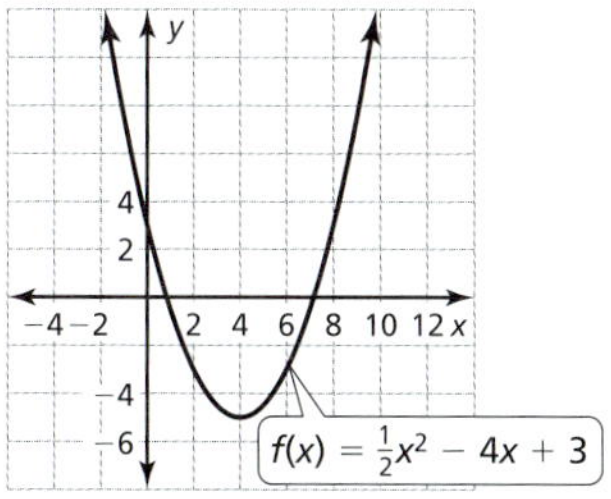

The graph opens up and the lowest point is at (4, −5).

7. 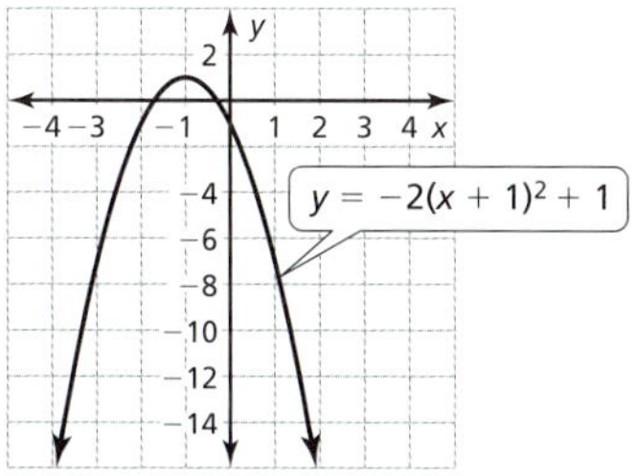

The graph opens down and the highest point is at (−1, 1).

8. 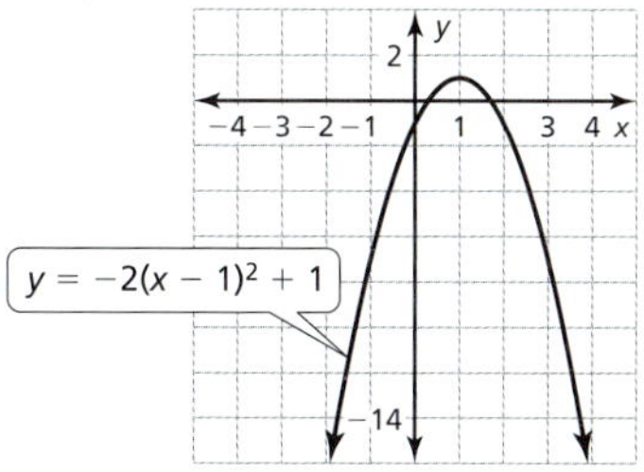

The graph opens down and the highest point is at (1, 1).

9. *Sample answer:* They are all U-shaped; Some open up and some open down.

8.1 Explorations

1. d. 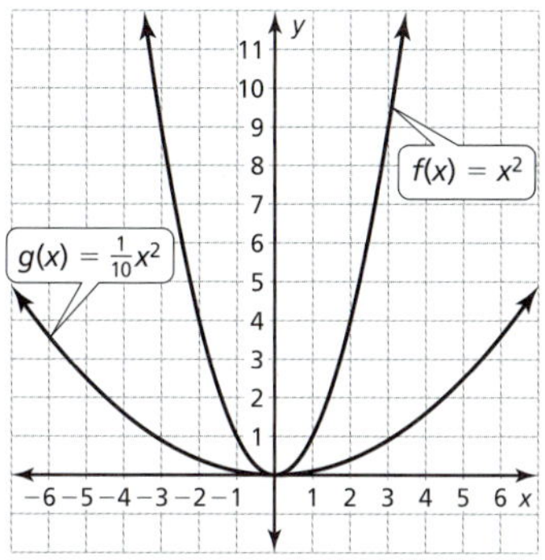

The graph of g is a vertical shrink by a factor of $\frac{1}{10}$ of the graph of f.

8.1 Lesson Monitoring Progress

3. 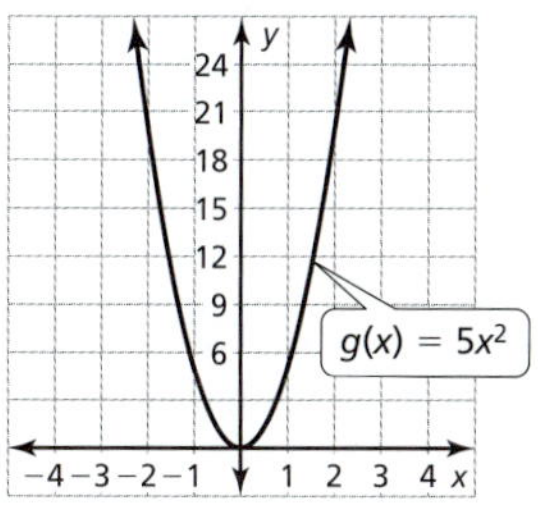

The graph of g is a vertical stretch by a factor of 5 of the graph of f.

4. 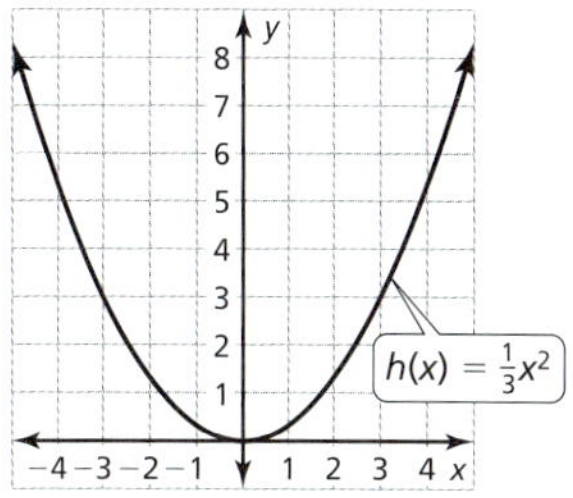

The graph of h is a vertical shrink by a factor of $\frac{1}{3}$ of the graph of f.

5. 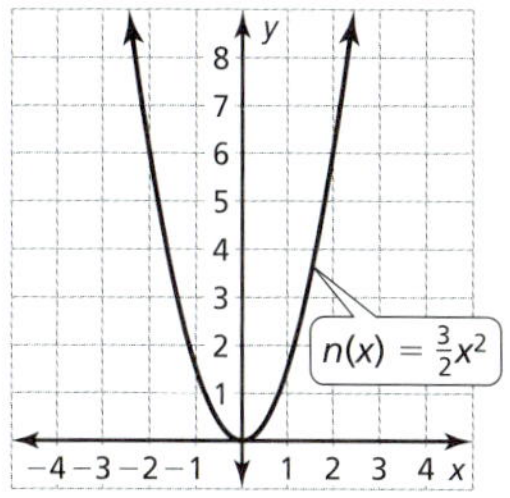

The graph of n is a vertical stretch by a factor of $\frac{3}{2}$ of the graph of f.

6. 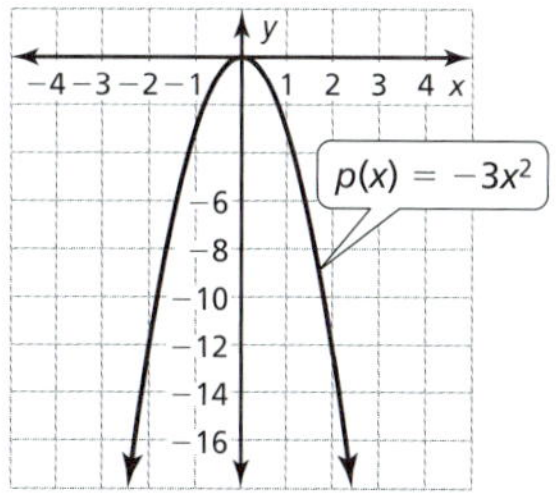

The graph of p is a vertical stretch by a factor of 3 and a reflection in the x-axis of the graph of f.

7. 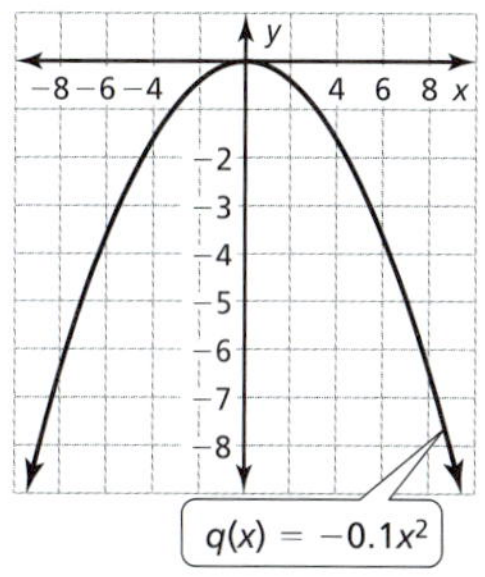

The graph of q is a vertical shrink by a factor of 0.1 and a reflection in the x-axis of the graph of f.

8. 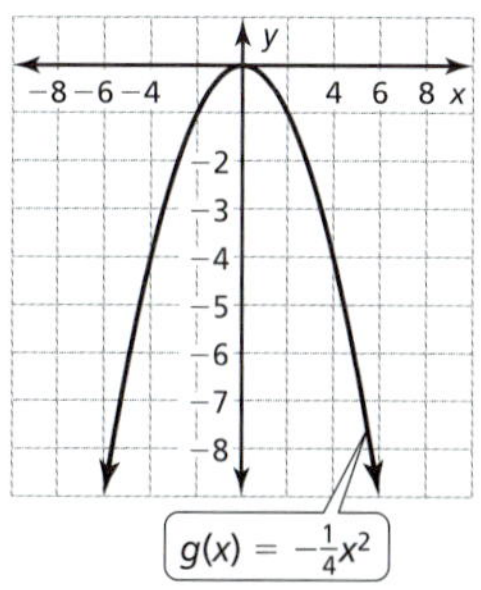

The graph of g is a vertical shrink by a factor of $\frac{1}{4}$ and a reflection in the x-axis of the graph of f.

8.1 Monitoring Progress and Modeling with Mathematics

9.

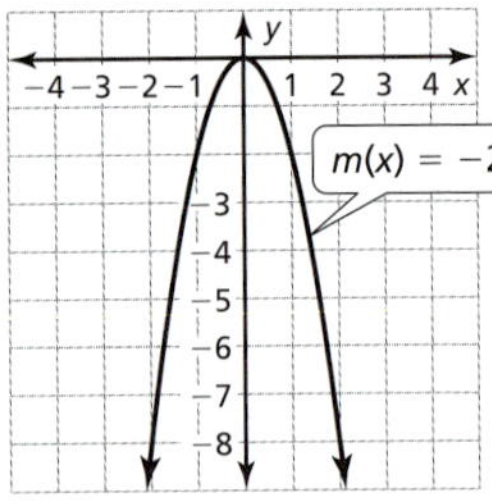

The graph of m is a vertical stretch by a factor of 2 and a reflection in the x-axis of the graph of f.

10.

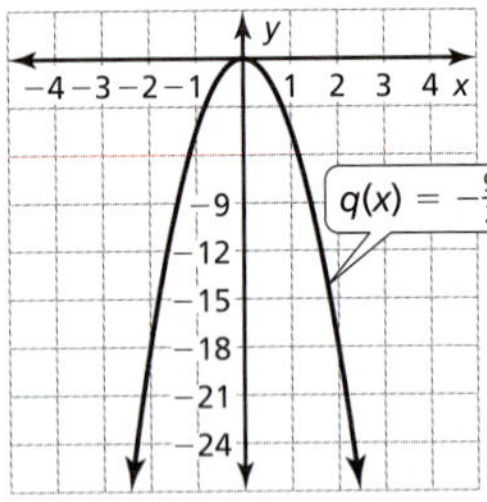

The graph of q is a vertical stretch by a factor of $\frac{9}{2}$ and a reflection in the x-axis of the graph of f.

11.

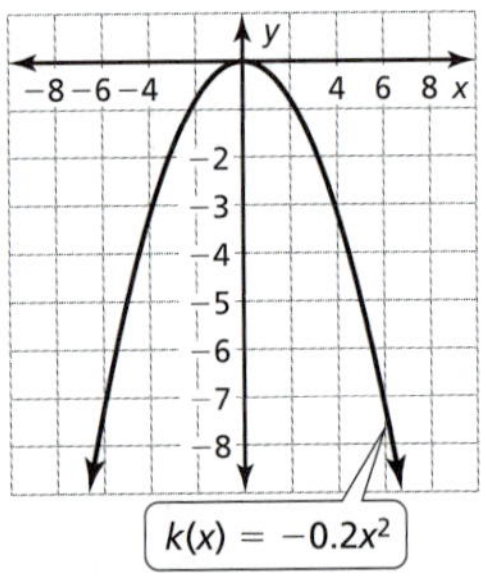

The graph of k is a vertical shrink by a factor of 0.2 and a reflection in the x-axis of the graph of f.

12.

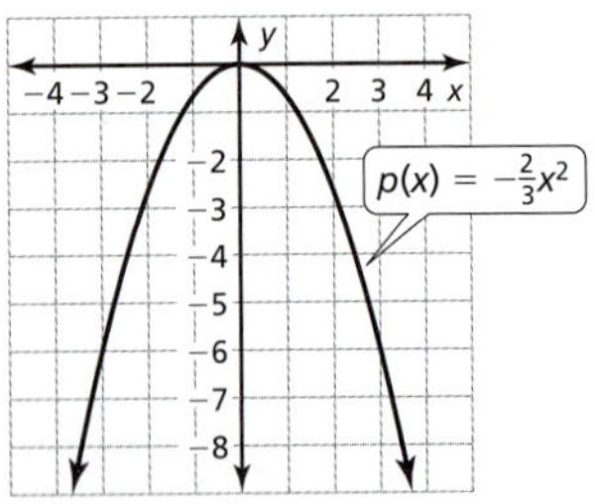

The graph of p is a vertical shrink by a factor of $\frac{2}{3}$ and a reflection in the x-axis of the graph of f.

13.

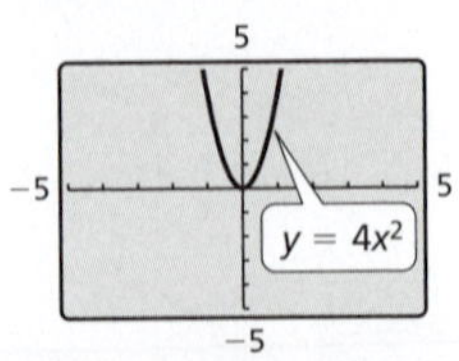

The graph of $y = 4x^2$ is a reflection in the x-axis of the graph of $y = -4x^2$.

14.

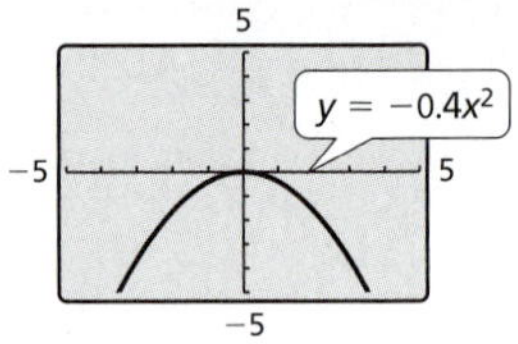

The graph of $y = -0.4x^2$ is a vertical shrink by a factor of $\frac{1}{10}$ of the graph of $y = -4x^2$.

15.

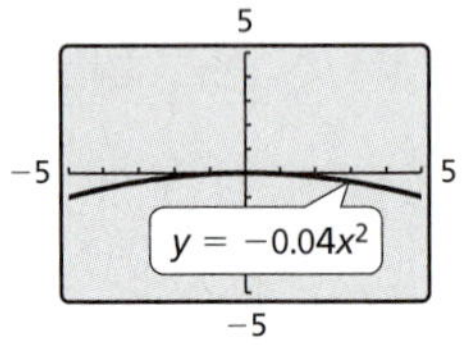

The graph of $y = -0.04x^2$ is a vertical shrink by a factor of $\frac{1}{100}$ of the graph of $y = -4x^2$.

16.

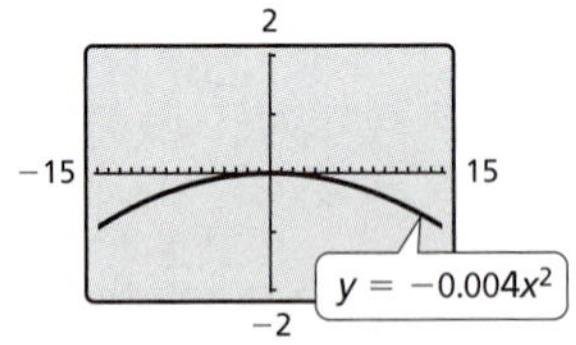

The graph of $y = -0.004x^2$ is a vertical shrink by a factor of $\frac{1}{1000}$ of the graph of $y = -4x^2$.

17. The graph of $y = 0.5x^2$ should be wider than the graph of $y = x^2$;

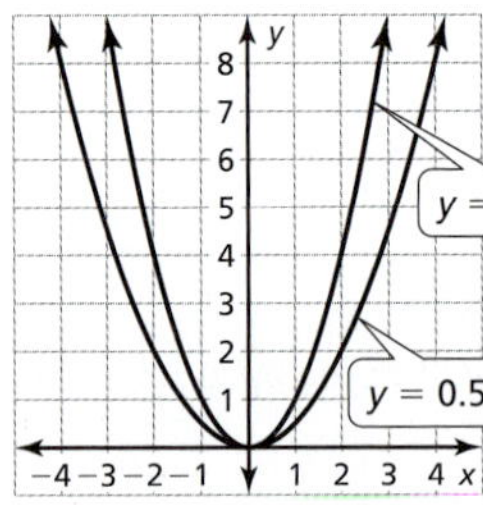

The graphs have the same vertex and the same axis of symmetry. The graph of $y = 0.5x^2$ is wider than the graph of $y = x^2$.

18. 300 feet high, 1000 feet wide

19. **a.** domain: $d \geq 0$, range: $z \geq 0$

b.

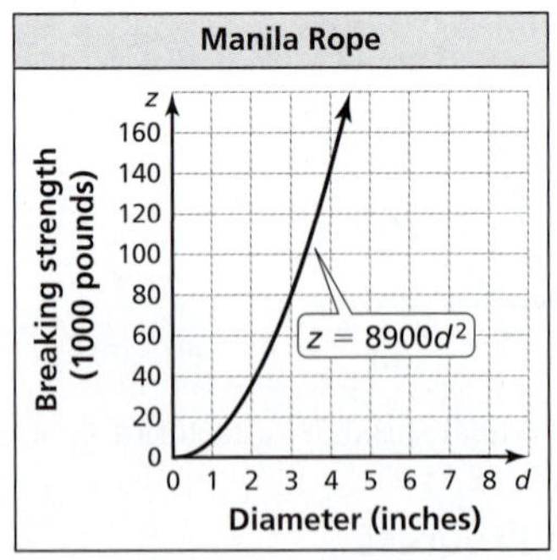

c. no; *Sample answer:* The relationship is quadratic, so a rope with 4 times the diameter will have $4^2 = 16$ times the breaking strength.

22. f is decreasing when $x < 0$. g is decreasing when $x > 0$.

23. f; $a = \frac{3}{4}$

24. yes; *Sample answer:* At an x-intercept, $y = 0$, so $ax^2 = 0$. By the definition of a quadratic function, $a \neq 0$, so by the Zero-Product Property, $x^2 = 0$, which means $x = 0$.

25. *Sample answer:* The vertex of a parabola that opens up is the minimum point, so its y-coordinate is the minimum value of y. The graph passes through $(6, -3)$, so 2 is not the minimum value of y.

26. sometimes; *Sample answer:* The graph of f will be narrower than the graph of g when $a > 1$, but it will be wider when $0 < a < 1$.

27. always; *Sample answer:* When $|a| > 1$, the graph of f will be a vertical stretch of the graph of g, so it will be narrower.

28. always; *Sample answer:* When $0 < |a| < 1$, the graph of f will be a vertical shrink of the graph of g, so it will be wider.

29. never; *Sample answer:* When $|a| > |d|$, the graph of f will be a vertical stretch of the graph of g, so it will be narrower, not wider.

30.

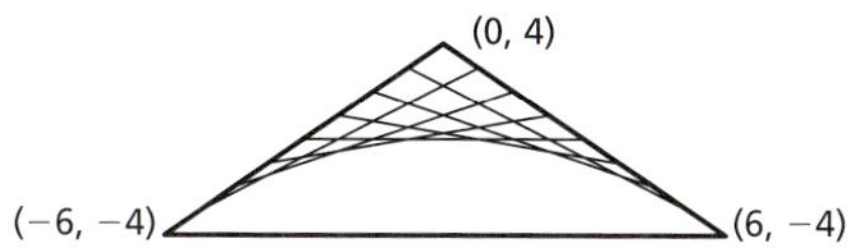

$y = -\frac{1}{9}x^2$

31. a. 8 cm

b. no; *Sample answer:* A faster rotational speed would increase the depth. The diagram shown has a depth of 3.2 centimeters. A model of $y = 0.1x^2$ would only have a depth of 1.6 centimeters, so it would have a slower rotational speed.

8.1 Maintaining Mathematical Proficiency

32. 14

33. 3

34. −25

35. 11

8.1 Mini-Assessment

1. vertex: $(-2, -4)$, axis of symmetry: $x = -2$, domain: all real numbers, range: $y \geq -4$; when $x < -2$, y increases as x decreases, when $x > -2$, y increases as x increases.

2.

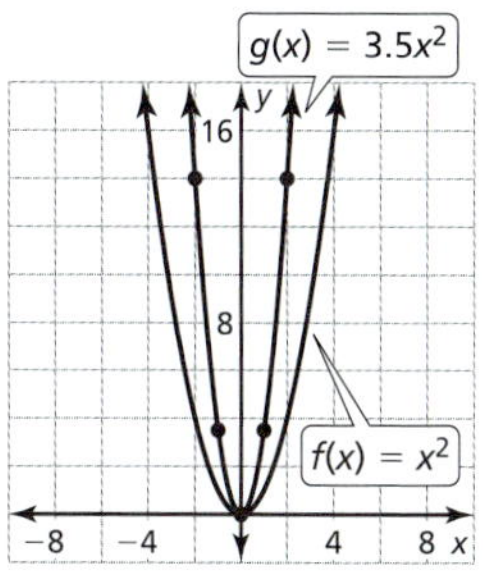

The graph of g is a vertical stretch by a factor of 3.5 of the graph of f.

8.2 Lesson Monitoring Progress

3.

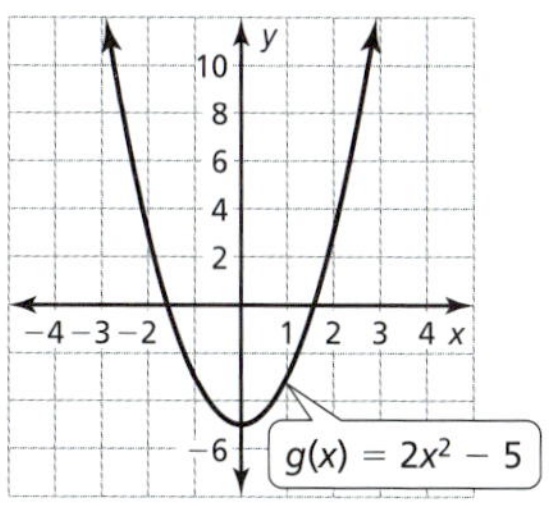

The graph of g is a vertical stretch by a factor of 2 and a vertical translation 5 units down of the graph of f.

4.

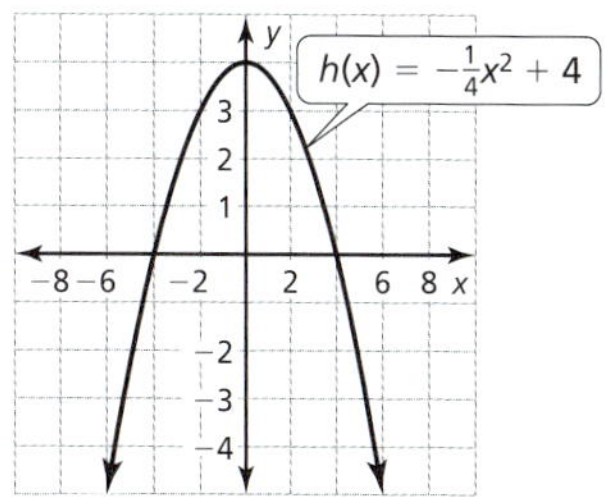

The graph of h is a vertical shrink by a factor of $\frac{1}{4}$, a reflection in the x-axis, and a vertical translation 4 units up of the graph of f.

5. a. The graph of g is a vertical translation 3 units up of the graph of f.

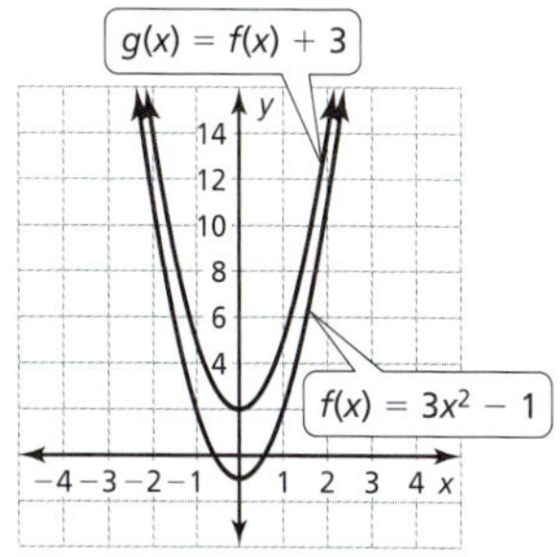

b. $g(x) = 3x^2 + 2$

8.2 Monitoring Progress and Modeling with Mathematics

8.

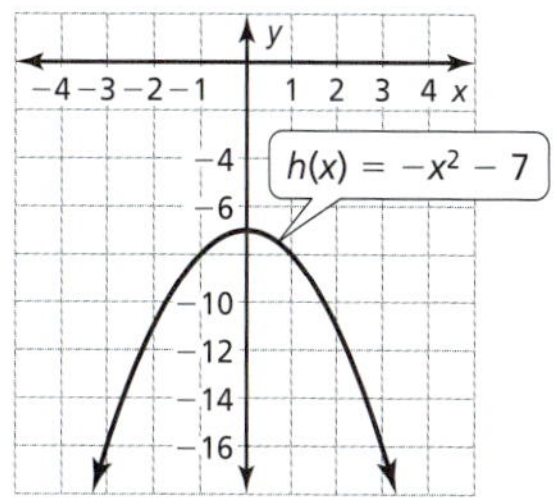

The graph of h is a reflection in the x-axis, and a vertical translation 7 units down of the graph of f.

9.

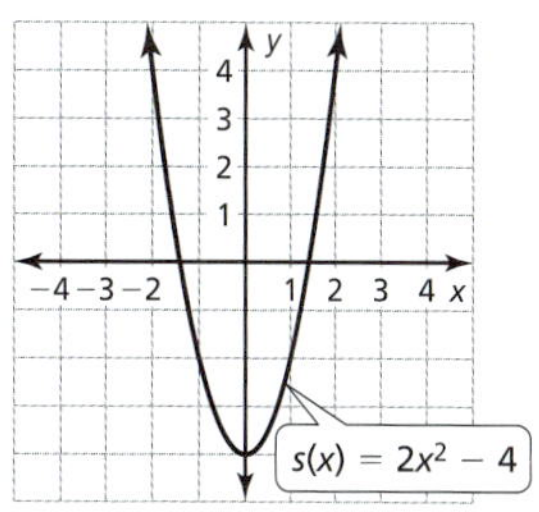

The graph of s is a vertical stretch by a factor of 2 and a vertical translation 4 units down of the graph of f.

10.

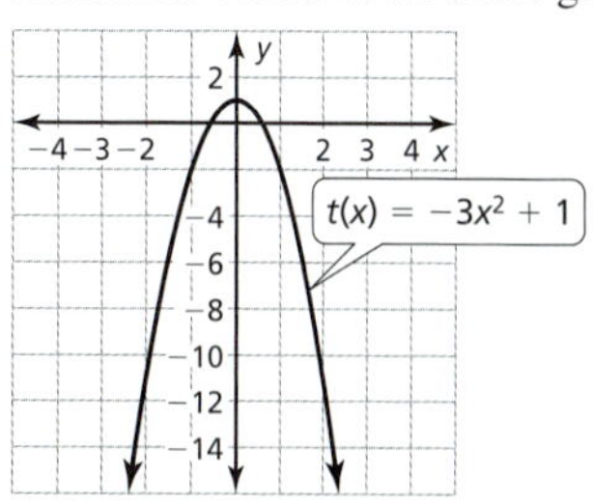

The graph of t is a vertical stretch by a factor of 3, a reflection in the x-axis, and a vertical translation 1 unit up of the graph of f.

11.

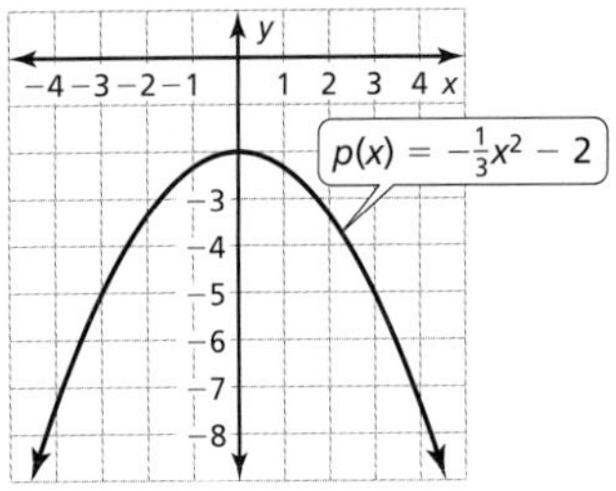

The graph of p is a vertical shrink by a factor of $\frac{1}{3}$, a reflection in the x-axis, and a vertical translation 2 units down of the graph of f.

12.

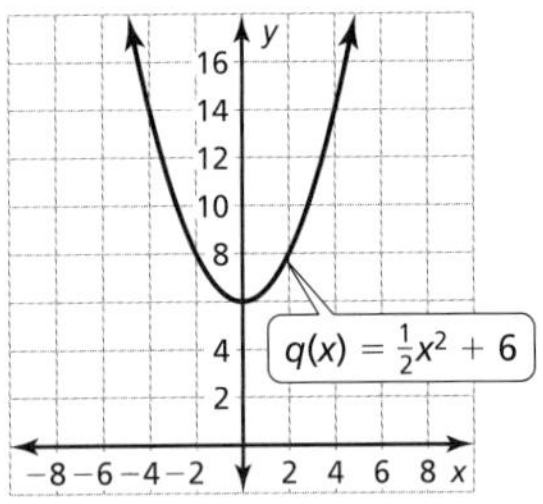

The graph of q is a vertical shrink by a factor of $\frac{1}{2}$ and a vertical translation 6 units up of the graph of f.

13. The graph of g is a vertical translation 2 units up of the graph of f.

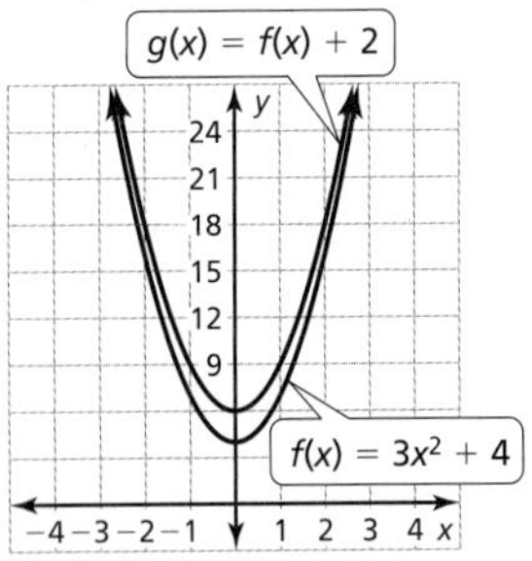

$g(x) = 3x^2 + 6$

14. The graph of g is a vertical translation 4 units down of the graph of f.

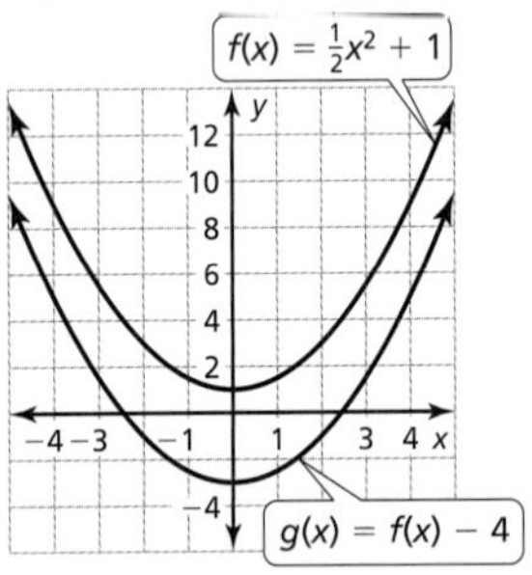

$g(x) = \frac{1}{2}x^2 - 3$

15. The graph of g is a vertical translation 3 units down of the graph of f.

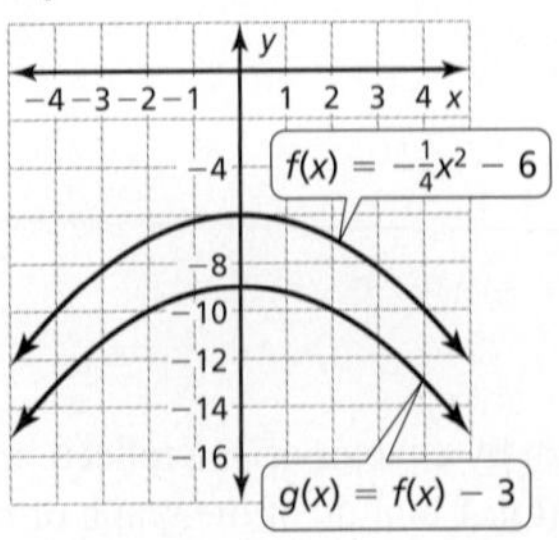

$g(x) = -\frac{1}{4}x^2 - 9$

16. The graph of g is a vertical translation 7 units up of the graph of f.

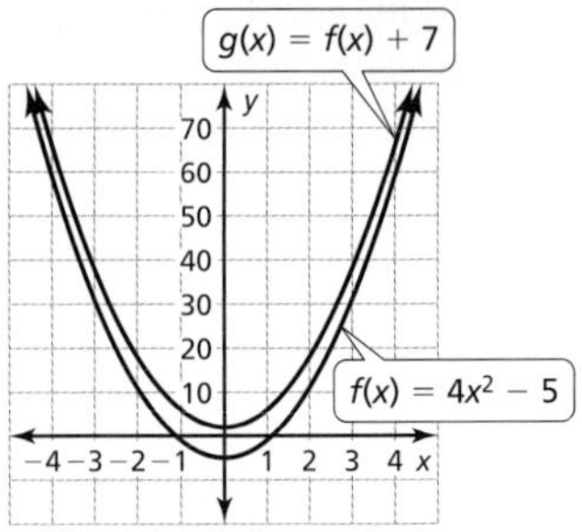

$g(x) = 4x^2 + 2$

17. The graph of $y = 3x^2 + 2$ is narrower, so it should be a stretch not a shrink; The graph of $y = 3x^2 + 2$ is a vertical stretch by a factor of 3 and a translation 2 units up of the graph of $y = x^2$.

18. $g(x)$ is not graphed correctly;

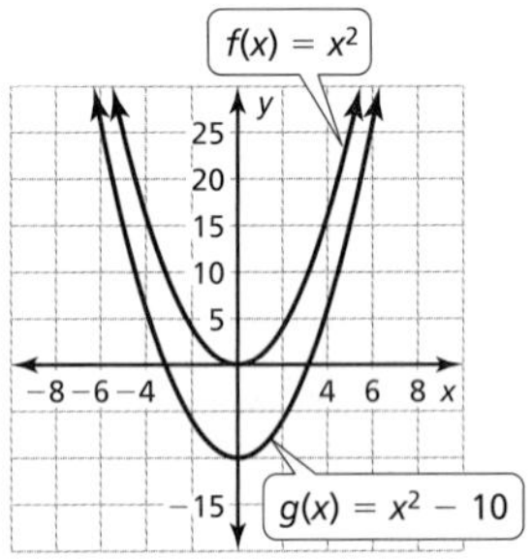

Both graphs open up and have the same axis of symmetry. However, the vertex of the graph of g, $(0, -10)$, is 10 units below the vertex of the graph of f, $(0, 0)$.

19. $x = 1, x = -1$

20. $x = 6, x = -6$

21. $x = 5, x = -5$

22. $x = 7, x = -7$

23. $x = 2, x = -2$

24. $x = 3, x = -3$

25. $x = \frac{1}{2}, x = -\frac{1}{2}$

26. $x = \frac{7}{2}, x = -\frac{7}{2}$

27. **a.** 3 sec

b. When $k > 0$, the water balloon will take more than 3 seconds to hit the ground. When $k < 0$, the water balloon will take less than 3 seconds to hit the ground.

28. The x-intercept is 1.5. This means the apple hits the ground after 1.5 seconds. The y-intercept is 36. This means the apple fell from a height of 36 feet.

29. *Sample answer:*

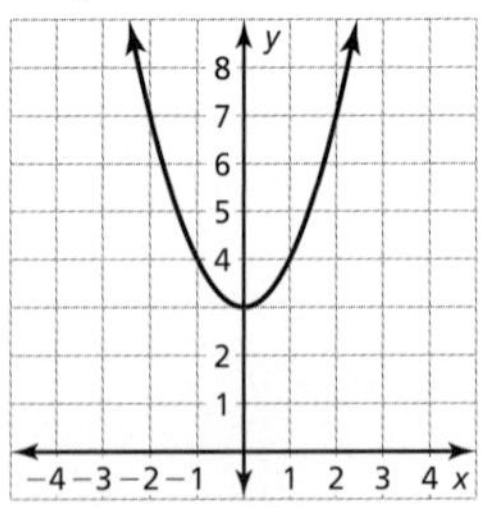

30. *Sample answer:*

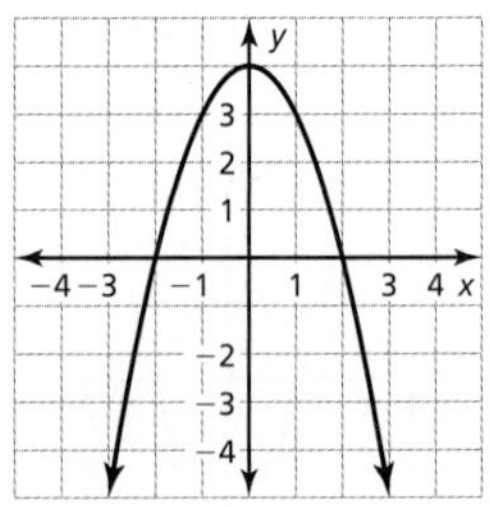

31. *Sample answer:*

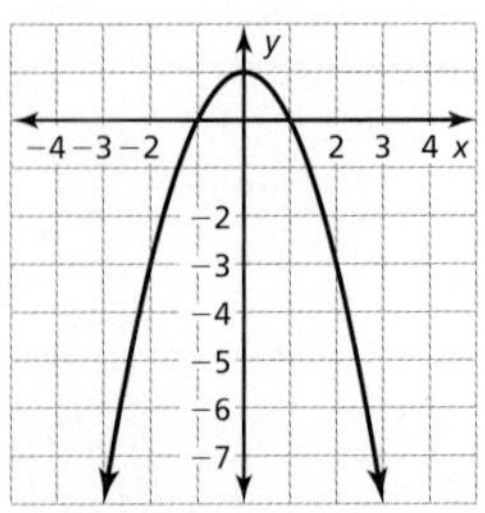

32. *Sample answer:*

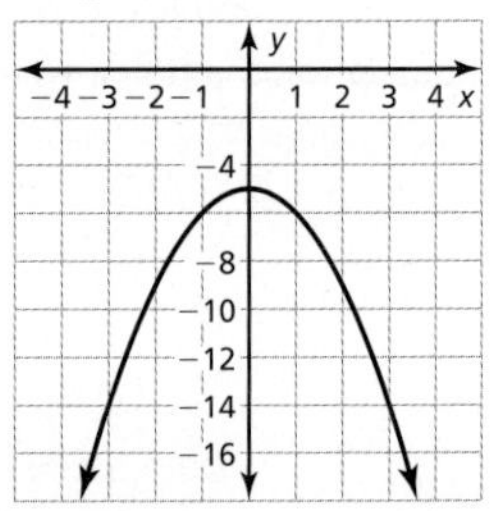

33. **a.** $T(x) = -44$; the distance between the balls

b. 44 ft

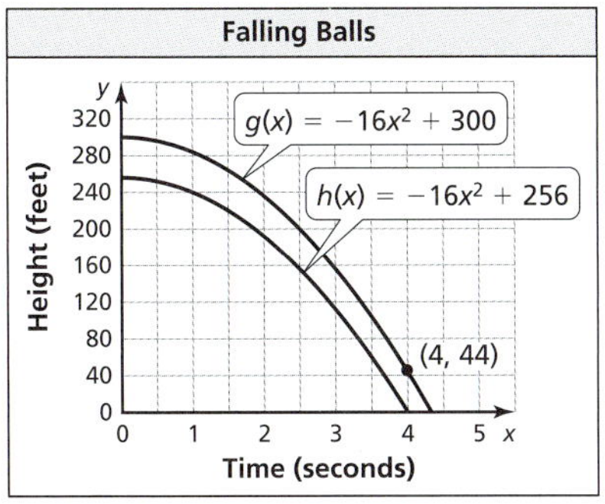

34. no; *Sample answer:* Changing a causes a vertical stretch or shrink, which does not change the vertex.

35. 12 ft by 12 ft

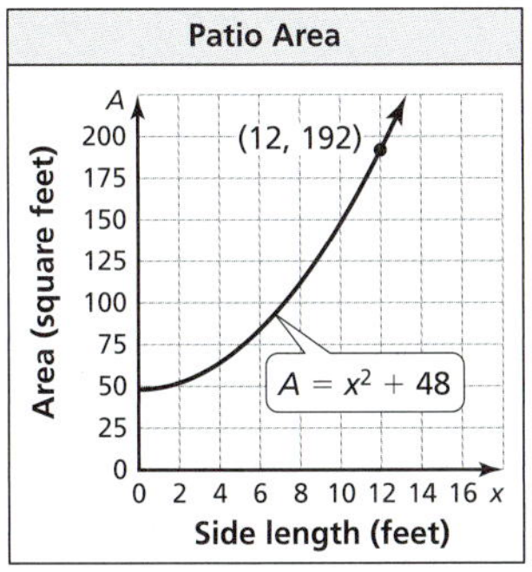

36. A

37. Graph the function and determine the x-intercepts; Set the function equal to 0, factor $-16t^2 + 400$, and apply the Zero-Product Property.

38. **a.** waterfall 1

b. waterfall 2

c. waterfall 1

39. **a.** $h = -16t^2 + 45$; $h = -16t^2 + 32$

b. The graph of $h = -16t^2 + 32$ is a vertical translation 13 units down of the graph of $h = -16t^2 + 45$.

40. *Sample answer:* 8; $A = 2\left(\frac{1}{2}bh\right) = bh = 2(4) = 8$

41. (0, 5.8); *Sample answer:* The outer edges are located 40 feet from the center. Substituting this into $y = 0.012x^2$ indicates they are 19.2 feet above the x-axis. To be 25 feet above the x-axis, they must be vertically translated up 5.8 feet.

8.2 Maintaining Mathematical Proficiency

42. $-\frac{1}{3}$

43. $-\frac{3}{8}$

44. $\frac{7}{9}$

45. $\frac{5}{12}$

8.3 Lesson Monitoring Progress

6.

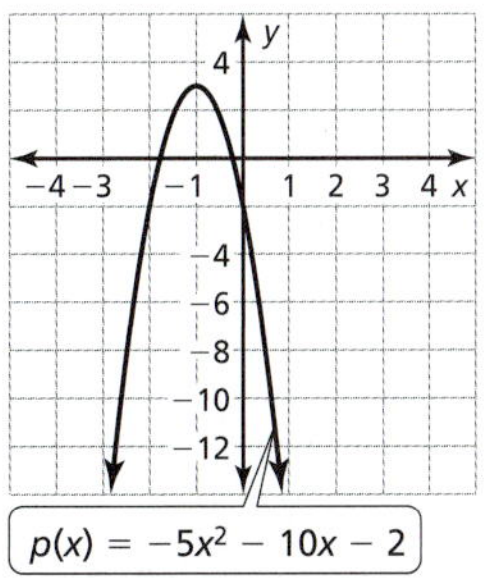

domain: all real numbers, range: $y \le 3$

8.3 Monitoring Progress and Modeling with Mathematics

16.

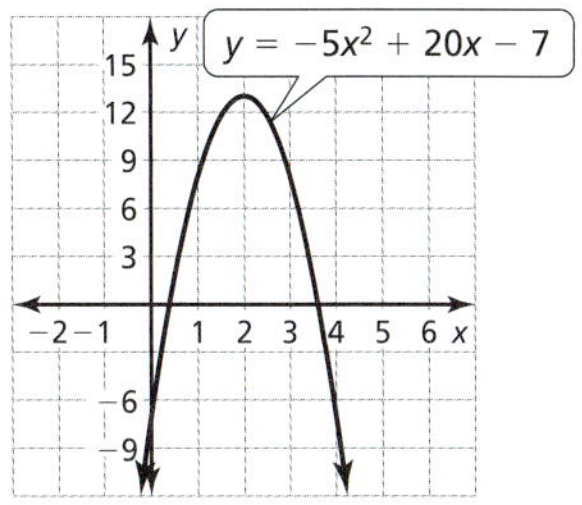

domain: all real numbers, range: $y \le 13$

17.

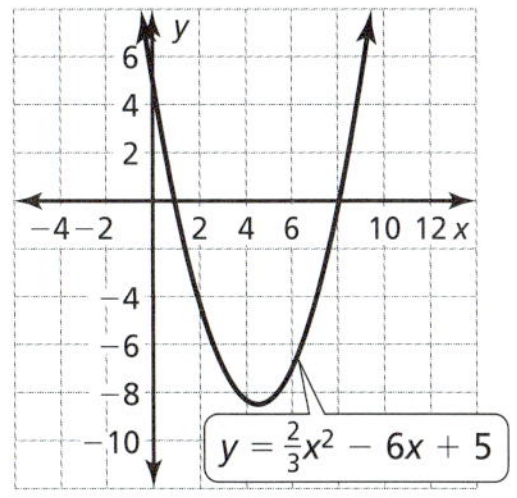

domain: all real numbers, range: $y \ge -\frac{17}{2}$

18.

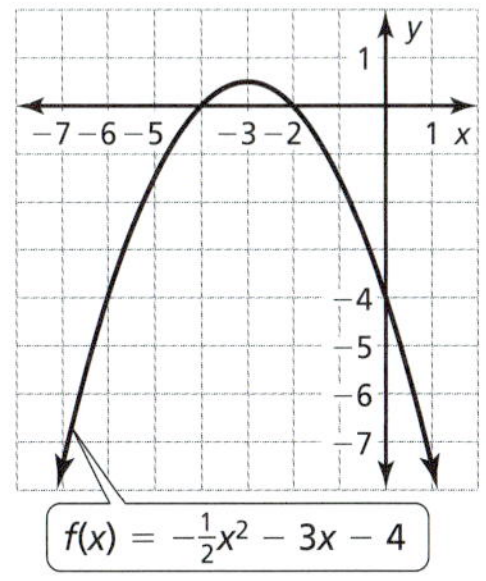

domain: all real numbers, range: $y \le \frac{1}{2}$

19. There should be two negatives in the substitution, one from the formula and one because b is -12; $x = -\dfrac{b}{2a} = -\dfrac{-12}{2(3)} = 2$; The axis of symmetry is $x = 2$.

20. The formula for the axis of symmetry has a negative sign; The axis of symmetry is $x = -\dfrac{b}{2a} = -\dfrac{4}{2(1)} = -2$; $f(-2) = (-2)^2 + 4(-2) + 3 = -1$; So, the vertex is $(-2, -1)$; The y-intercept is 3, so the points (0, 3) and $(-4, 3)$ lie on the graph.

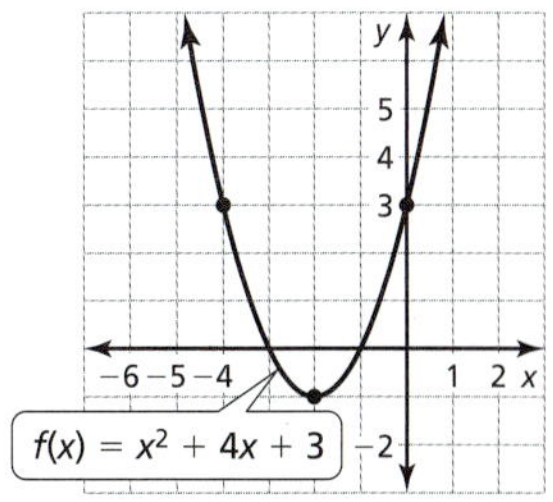

21. minimum value; -12

22. maximum value; 12

23. maximum value; -1

24. minimum value; $\frac{1}{2}$

25. maximum value; $66\frac{1}{2}$

26. minimum value; $-4\frac{1}{4}$

43. down; *Sample answer:* Because (3, 2) and (9, 2) have the same y-coordinate, any point with an x-coordinate between 3 and 9 lies on the part of the parabola between these two points that passes through the vertex. Because 7 is greater than 2, the vertex must be above these 2 points, so the parabola opens down.

44. the maximum value or minimum value of the function; *Sample answer:* Because $-\frac{b}{2a}$ is the x-coordinate of the vertex, $f\left(-\frac{b}{2a}\right)$ is the y-coordinate of the vertex, which is also the maximum value or minimum value of the function.

45. $y = -2x^2 + 8x$

46. It is not possible to identify characteristics of parabola A with only two points; The vertex of parabola B is $(3, -4)$ because 3 is halfway between 1 and 5, which are the x-coordinates of points that have the same y-coordinates. Parabola B opens up because -4, the y-coordinate of the vertex, is the minimum value.

47. 14 ft

48. *Sample answer:* 2; Sketch a graph of $y = x^2$ and a tangent line through $(1, 1)$. Then select another point on the tangent line, $(0, -1)$, and substitute these points into the slope formula to calculate a slope of 2.

49. $\frac{k^2}{8}$ ft^2

8.3 Maintaining Mathematical Proficiency

50. The graph of q is a horizontal translation 6 units left of the graph of f.

51. The graph of h is a vertical shrink by a factor of 0.5 and a reflection in the x-axis of the graph of f.

52. The graph of g is a horizontal translation 2 units right and a vertical translation 5 units up of the graph of f.

53. The graph of p is a vertical stretch by a factor of 3 and a horizontal translation 1 unit left of the graph of f.

8.1–8.3 Quiz

9. The graph of g is a vertical translation 2 units up of the graph of f.

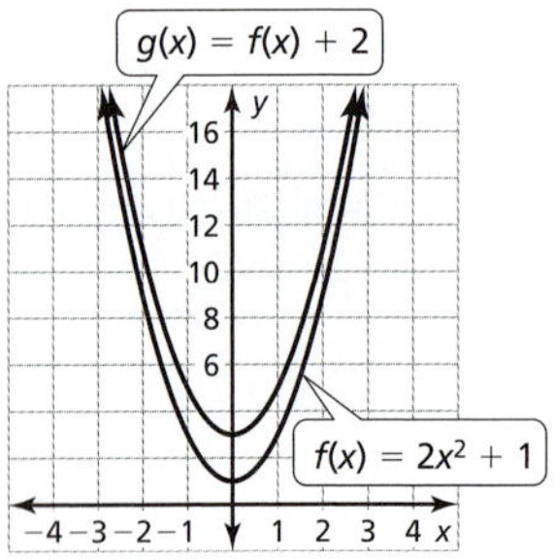

$g(x) = 2x^2 + 3$

10. The graph of g is a vertical translation 9 units down of the graph of f.

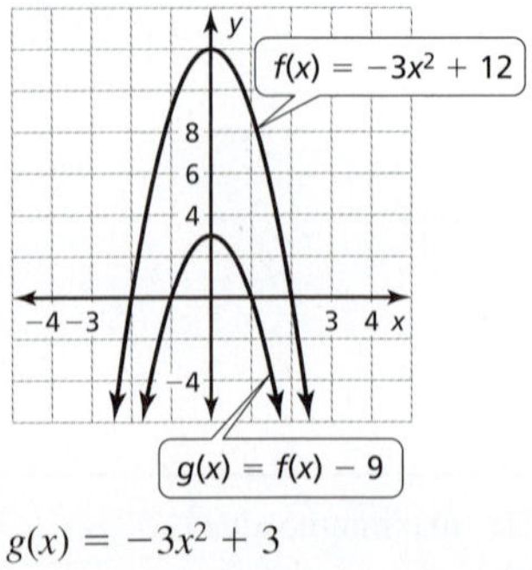

$g(x) = -3x^2 + 3$

11. The graph of g is a vertical translation 6 units down of the graph of f.

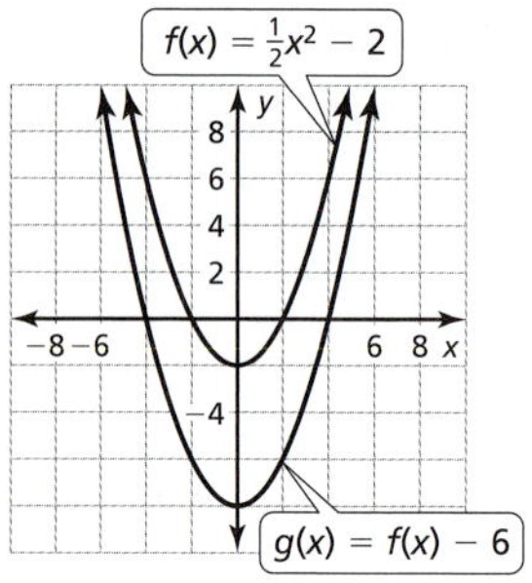

$g(x) = \frac{1}{2}x^2 - 8$

12. The graph of g is a vertical translation 1 unit up of the graph of f.

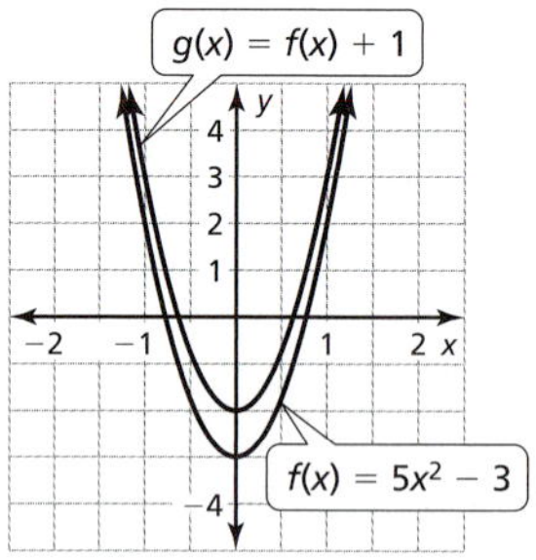

$g(x) = 5x^2 - 2$

13.

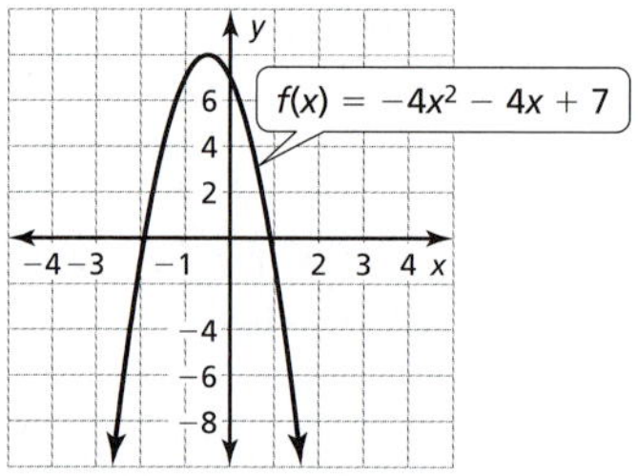

domain: all real numbers, range: $y \leq 8$

14.

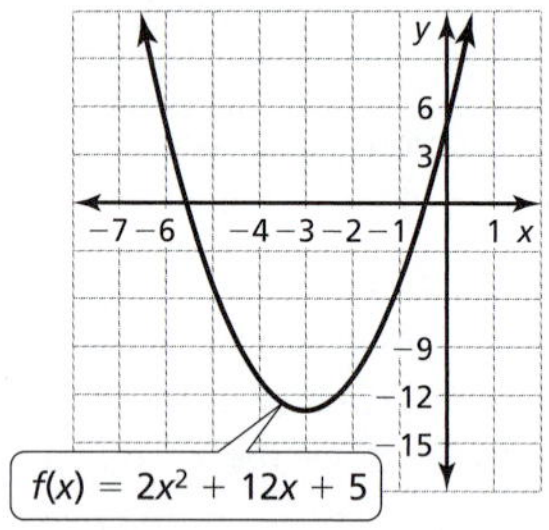

domain: all real numbers, range: $y \geq -13$

15.

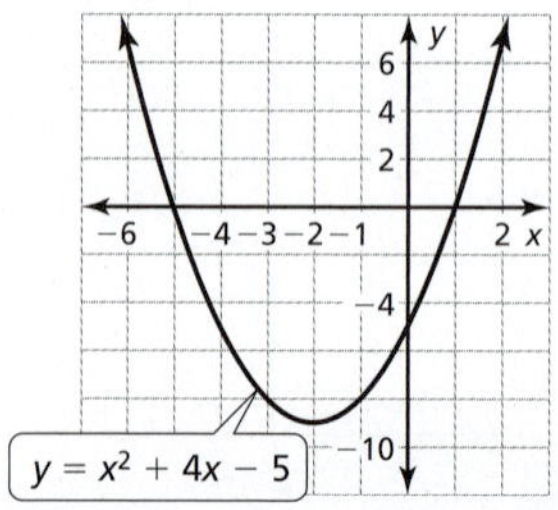

domain: all real numbers, range: $y \geq -9$

16. 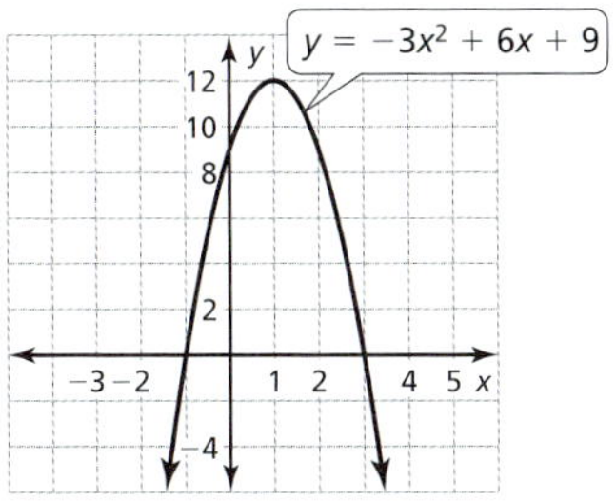

domain: all real numbers, range: $y \leq 12$

17. minimum value; -8

18. maximum value; 18

19. maximum value; 16

20. minimum value; -5

21. 2 sec

22. a. 1.25 sec

b. the first pinecone; *Sample answer:* The second pinecone will take 1.5 seconds to fall, which is longer than the first.

23. domain: $0 \leq t \leq 2$, range: $0 \leq h \leq 18$; 18 ft

8.4 Lesson Monitoring Progress

4. 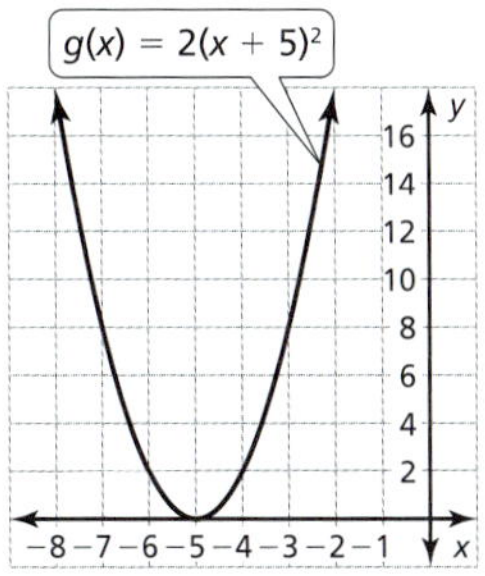

The graph of g is a horizontal translation 5 units left and a vertical stretch by a factor of 2 of the graph of f.

5. 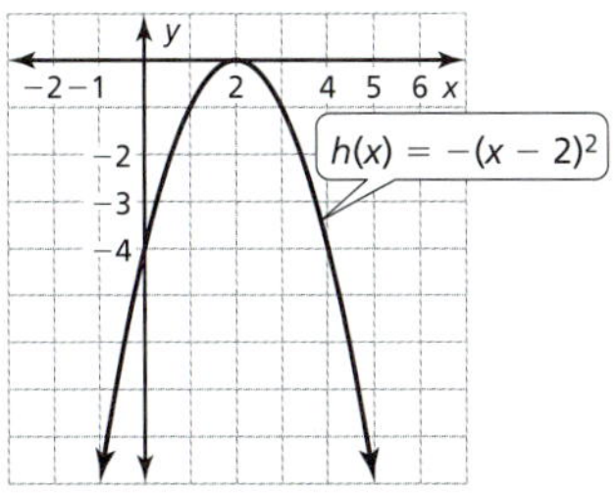

The graph of h is a horizontal translation 2 units right and a reflection in the x-axis of the graph of f.

7. 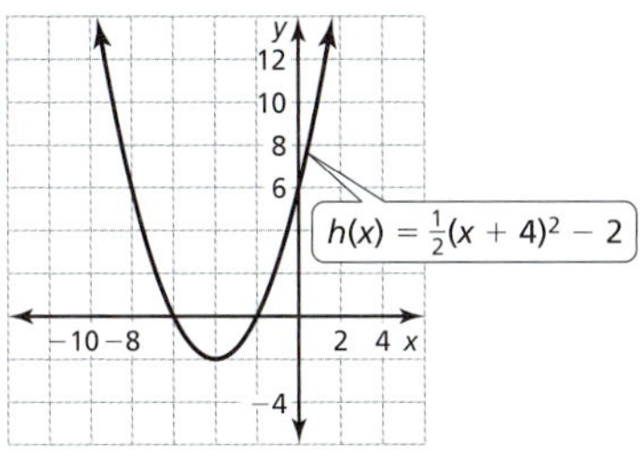

The graph of h is a vertical shrink by a factor of $\frac{1}{2}$, and a translation 4 units left and 2 units down of the graph of f.

8. 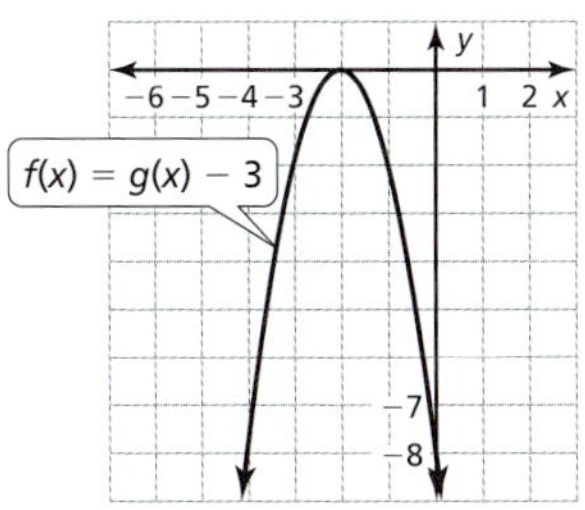

8.4 Monitoring Progress and Modeling with Mathematics

25. 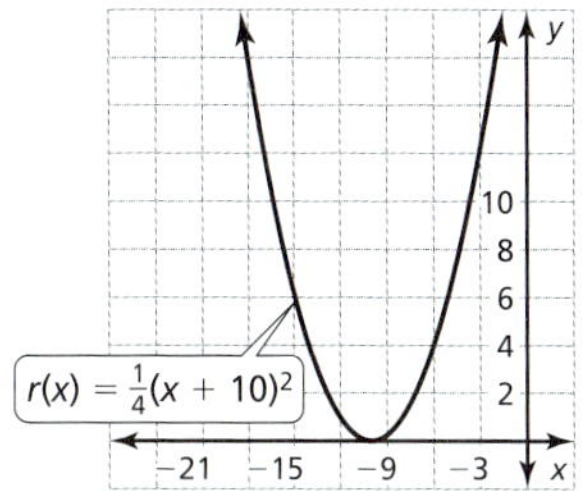

The graph of r is a horizontal translation 10 units left and a vertical shrink by a factor of $\frac{1}{4}$ of the graph of f.

26. 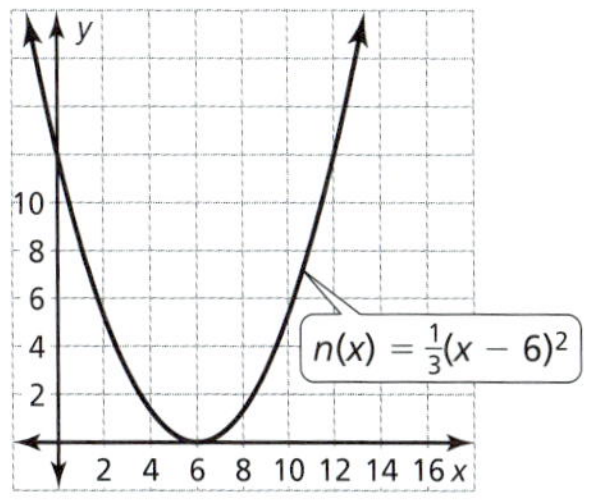

The graph of n is a horizontal translation 6 units right and a vertical shrink by a factor of $\frac{1}{3}$ of the graph of f.

27. 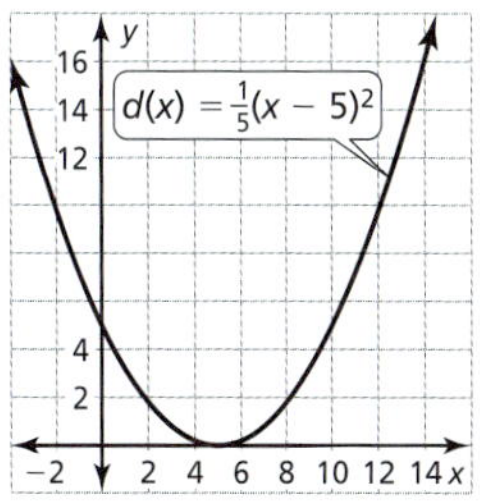

The graph of d is a horizontal translation 5 units right and a vertical shrink by a factor of $\frac{1}{5}$ of the graph of f.

28. 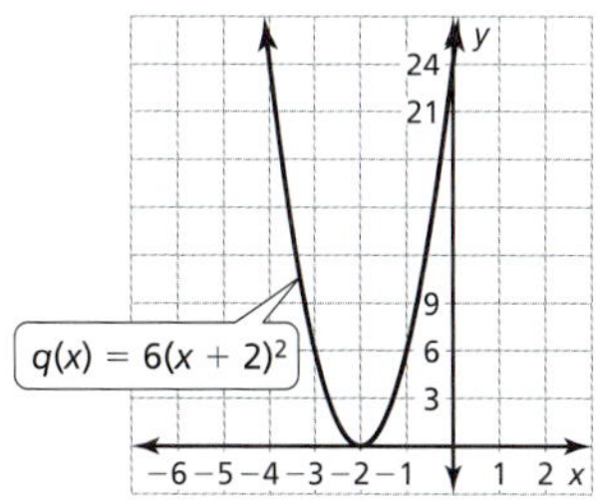

The graph of q is a horizontal translation 2 units left and a vertical stretch by a factor of 6 of the graph of f.

29. If $f(-x) = f(x)$ the function is even; So, $f(x)$ is an even function.

30. h is the x-coordinate of the vertex, not the y-coordinate; Because $h = -8$, the vertex is $(-8, 0)$.

43. 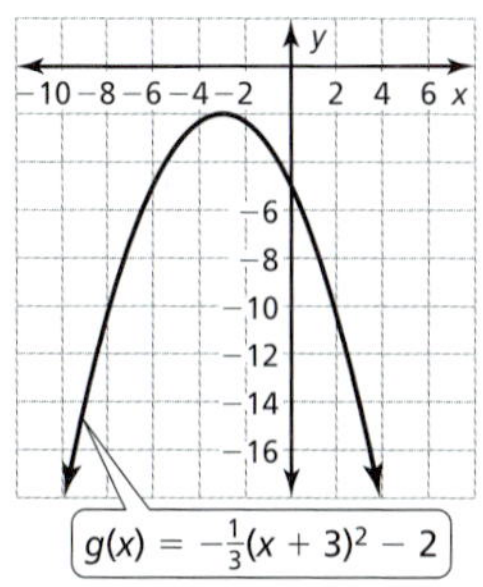

The graph of g is a vertical shrink by a factor of $\frac{1}{3}$, a reflection in the x-axis, and a translation 3 units left and 2 units down of the graph of f.

Additional Answers

44. 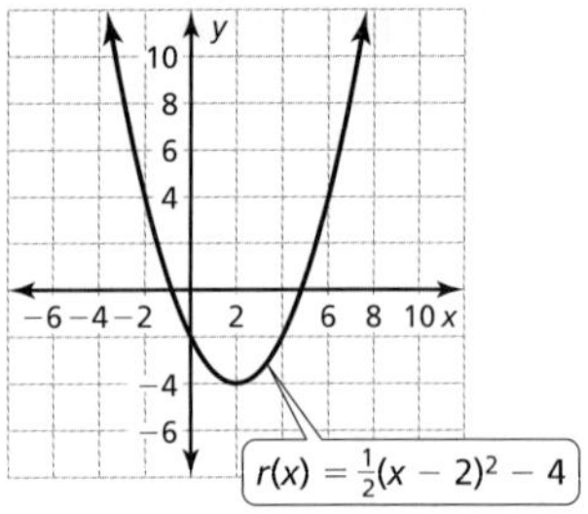

The graph of r is a vertical shrink by a factor of $\frac{1}{2}$, and a translation 2 units right and 4 units down of the graph of f.

49. 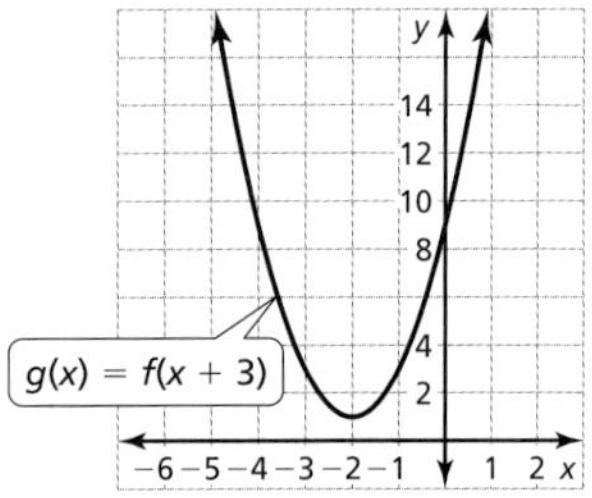

50. 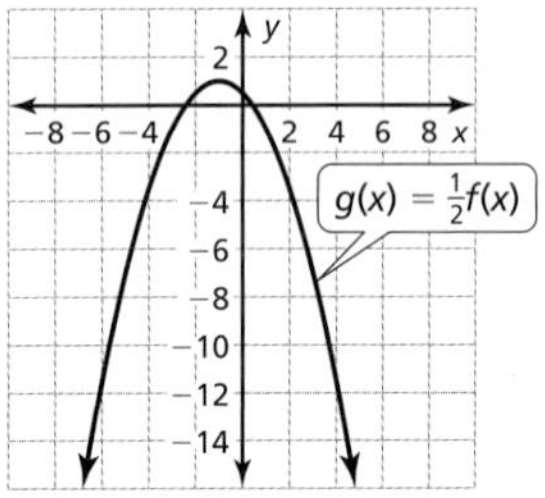

51. 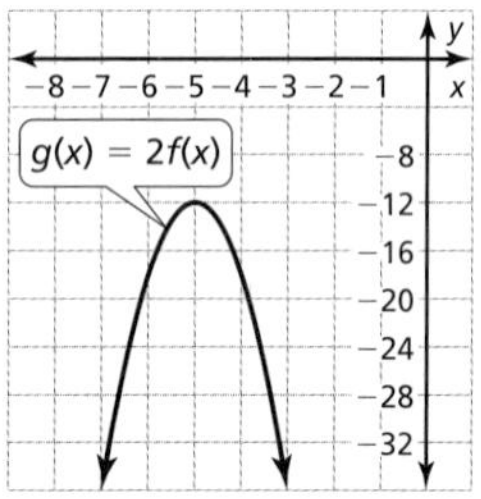

52. 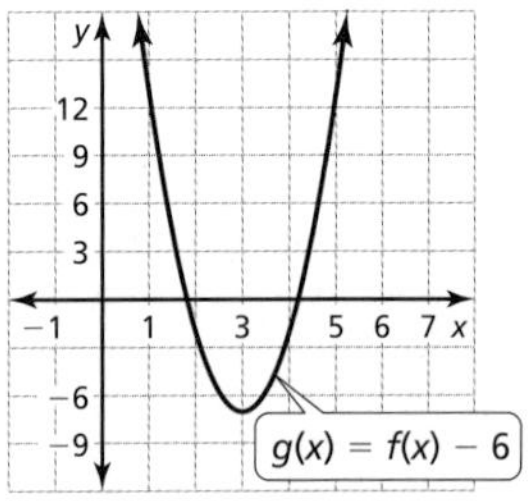

53. 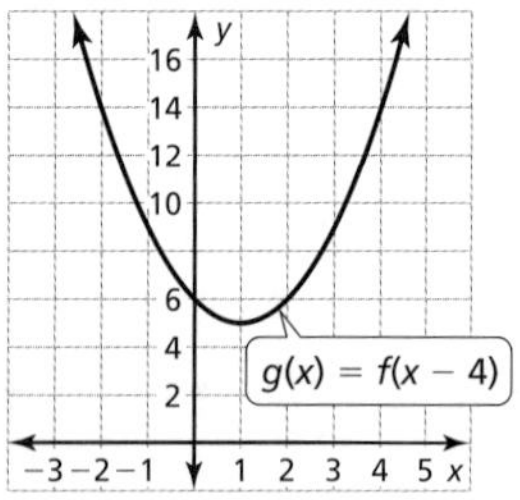

54. 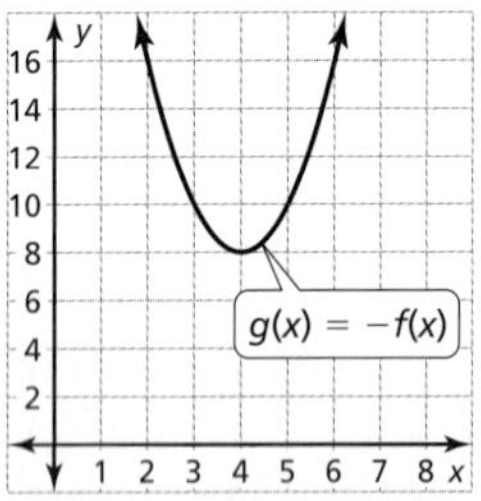

55. a. 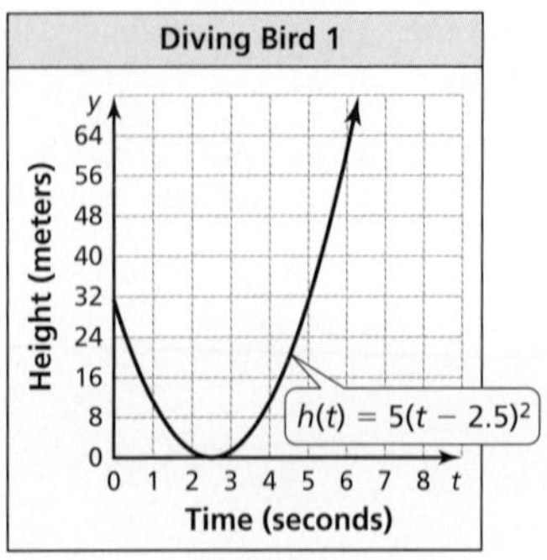

b. 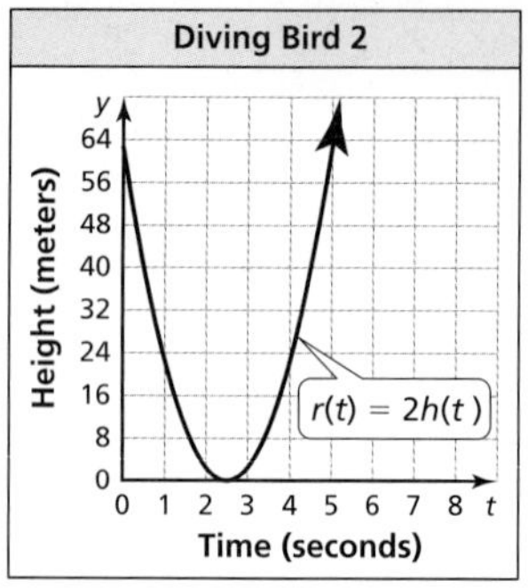

c. The graph of r is a vertical stretch by a factor of 2 of the graph of h; the second bird; *Sample answer:* Because $r(t)$ is twice $h(t)$, the second bird starts at a height twice as high as the first bird.

56. a. 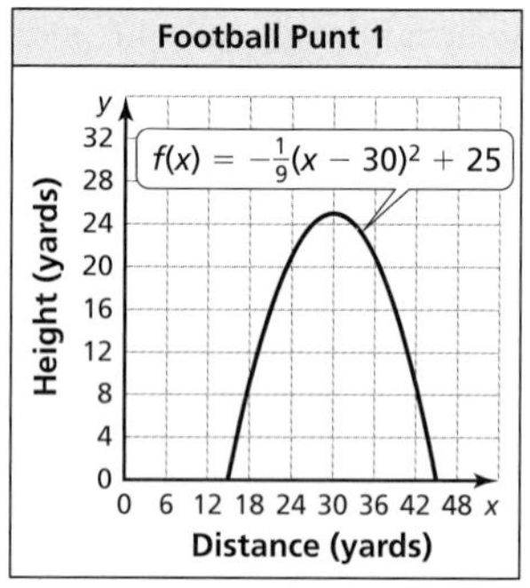

domain: $15 \le x \le 45$, range: $0 \le y \le 25$

b. 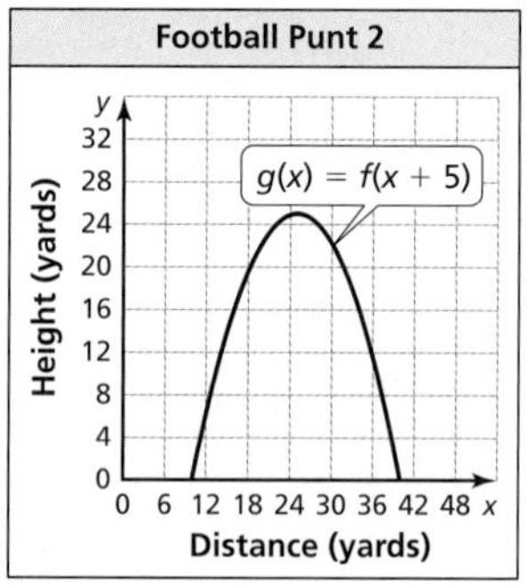

domain: $10 \le x \le 40$, range: $0 \le y \le 25$

c. The graph of g is a horizontal translation 5 units left of the graph of f; second possession; *Sample answer:* On the first possession, the ball is punted 15 yards from the kicker's goal line. On the second possession, the ball is punted 10 yards from the kicker's goal line.

63. $f(x) = -\frac{18}{125}(x-25)^2 + 90$

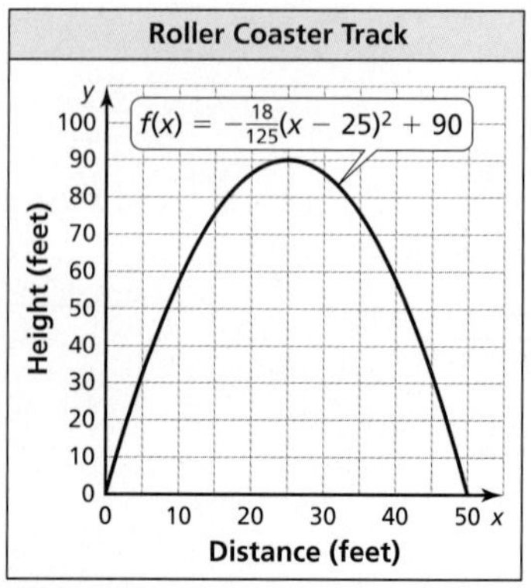

64. $f(x) = -\frac{1}{12}(x - 59)^2 + 300$

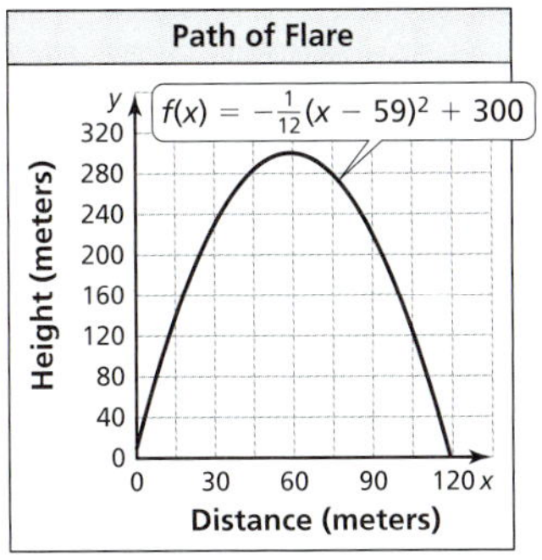

65. $y = 2(x - 2)^2 - 4$

66. $y = 3(x + 1)^2 - 4$

67. $f(x) = -5(x - 1)^2 + 8$

68. $f(x) = -(x + 2)^2 + 6$

69. no; *Sample answer:* The graph would not pass the vertical line test.

70. +; −; *Sample answer:* The vertex is $(-2, -3)$, so in the vertex form $h = -2$ and $k = -3$.

71. The graph of h is a vertical translation 4 units up of the graph of f; $h(x) = -(x + 1)^2 + 2$

72. The graph of h is a horizontal translation 5 units right of the graph of f; $h(x) = 2(x - 6)^2 + 1$

73. The graph of h is a vertical stretch by a factor of 2 of the graph of f; $h(x) = 8(x - 2)^2 + 6$

74. The graph of h is a vertical shrink by a factor of $\frac{1}{3}$ of the graph of f; $h(x) = -\frac{1}{3}(x + 5)^2 - 2$

75. $y = (x - 2)^2 - 5$; $y = x^2 - 4x - 1$; *Sample answer:* The vertex, $(2, -5)$, can be quickly determined from the vertex form; The y-intercept, -1, can be quickly determined from the standard form.

76. **a.** true; *Sample answer:* If $g(x) = af(x)$, then $g(-x) = af(-x)$. Because f is even, $g(-x) = af(x) = g(x)$, so g is even.

b. true; *Sample answer:* If $g(x) = af(x)$, then $g(-x) = af(-x)$. Because f is odd, $g(-x) = a(-f(x)) = -af(x) = -g(x)$, so g is odd.

c. true; *Sample answer:* If $h(x) = f(x) + g(x)$, then $h(-x) = f(-x) + g(-x)$. Because f and g are even, $h(-x) = f(x) + g(x) = h(x)$, so h is even.

d. true; *Sample answer:* If $h(x) = f(x) + g(x)$, then $h(-x) = f(-x) + g(-x)$. Because f and g are odd, $h(-x) = -f(x) - g(x) = -(f(x) + g(x)) = -h(x)$, so h is odd.

e. false; *Sample answer:* A counterexample is illustrated by adding $f(x) = x^2 + 4$, an even function, to $g(x) = 3x$, an odd function. The sum $h(x) = x^2 + 3x + 4$ is neither even nor odd.

77. **a.** the second birdbath; *Sample answer:* The first birdbath has a depth of 4 inches and the second birdbath has a depth of 6 inches.

b. the first birdbath; *Sample answer:* The first birdbath has a width of 36 inches and the second birdbath has a width of 30 inches.

78. The graph of $y = 2x^2 + 8x + 8$ is a vertical stretch by a factor of 2 and a horizontal translation 2 units left of the graph of $y = x^2$; *Sample answer:* Factoring the right side of $y = 2x^2 + 8x + 8$ changes the equation to vertex form, $y = 2(x + 2)^2$, from which the transformations can be easily determined.

8.4 Mini-Assessment

4. $f(x) = -\frac{9}{25}(x - 10)^2 + 36$

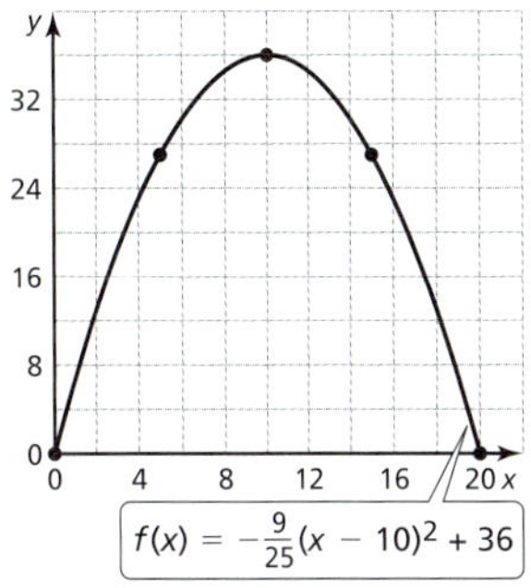

8.5 Lesson Monitoring Progress

3.

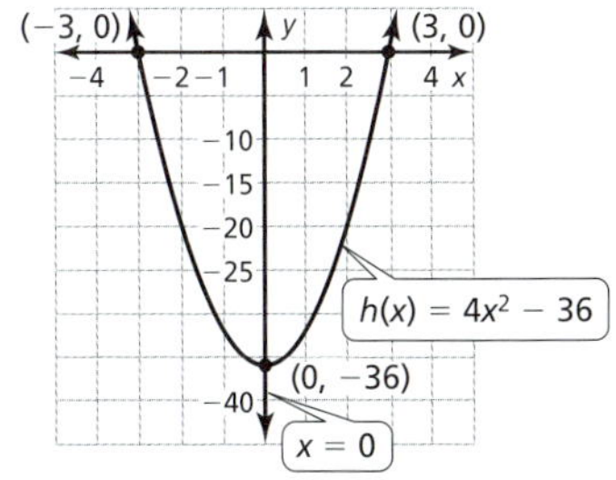

domain: all real numbers, range: $y \geq -36$

8.5 Monitoring Progress and Modeling with Mathematics

11.

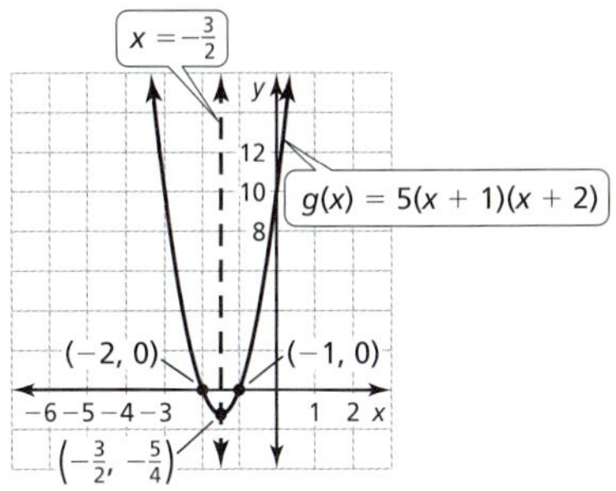

domain: all real numbers, range: $y \geq -\frac{5}{4}$

12.

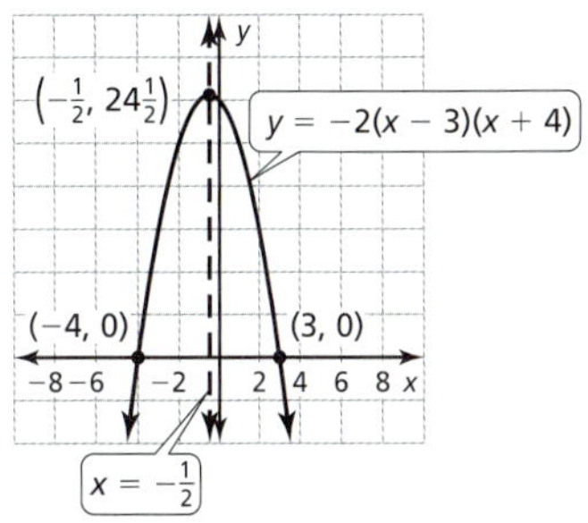

domain: all real numbers, range: $y \leq 24\frac{1}{2}$

13.

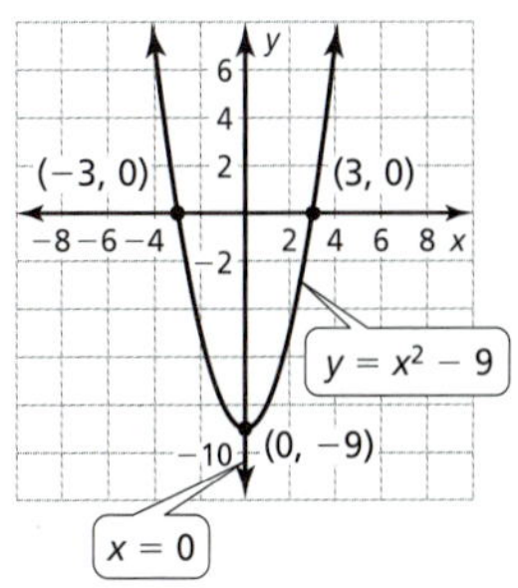

domain: all real numbers, range: $y \geq -9$

Additional Answers

14.

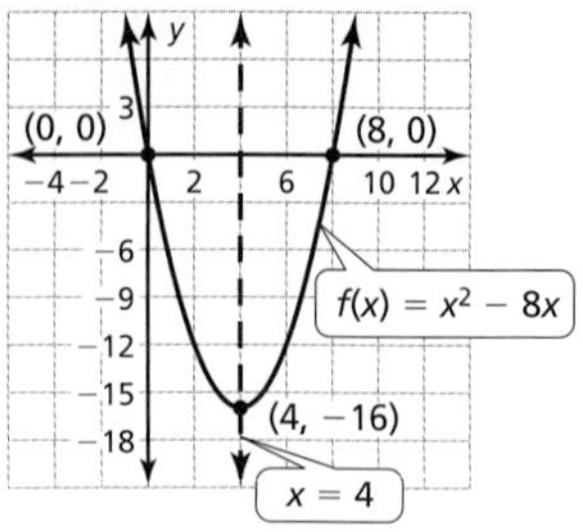

domain: all real numbers, range: $y \geq -16$

15.

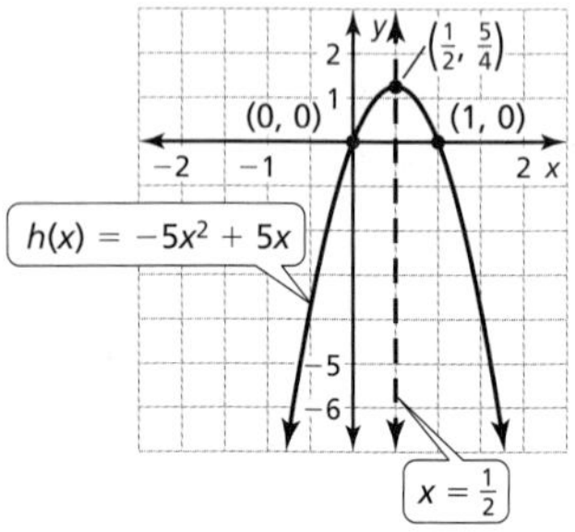

domain: all real numbers, range: $y \leq \frac{5}{4}$

16.

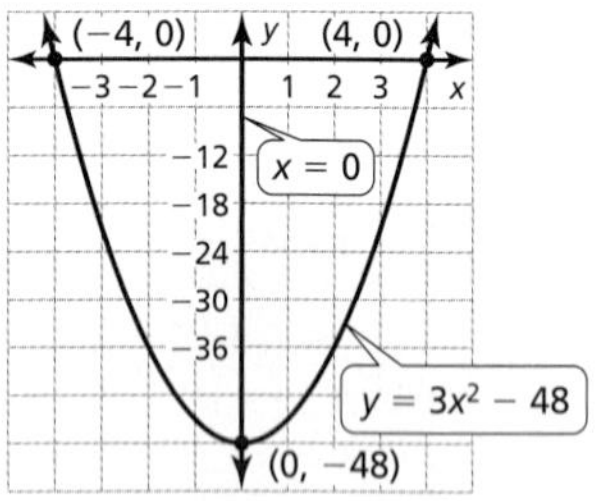

domain: all real numbers, range: $y \geq -48$

17.

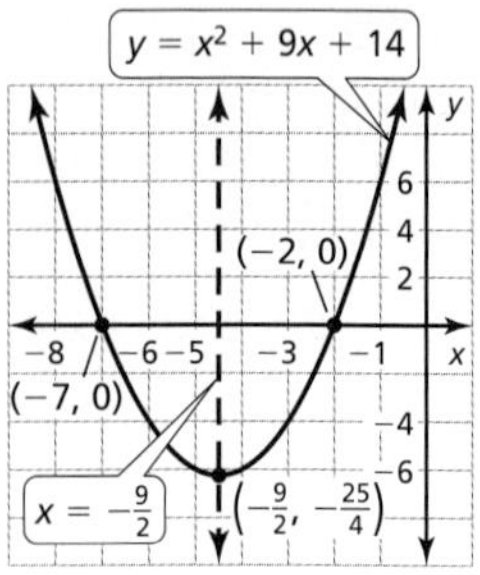

domain: all real numbers, range: $y \geq -\frac{25}{4}$

18.

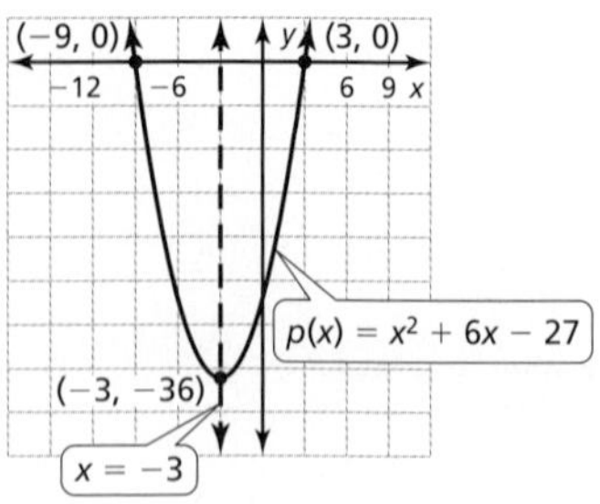

domain: all real numbers, range: $y \geq -36$

19.

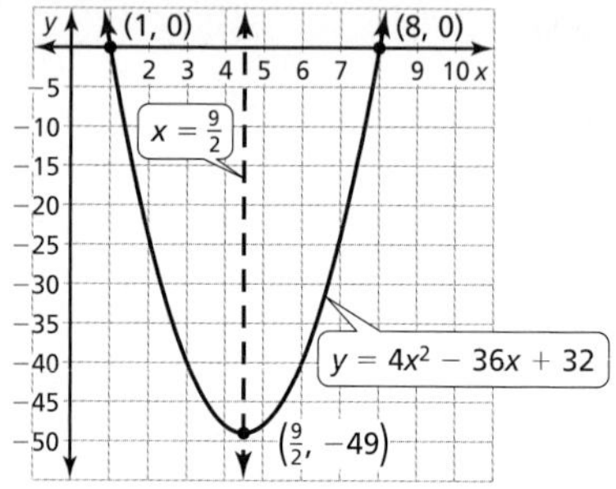

domain: all real numbers, range: $y \geq -49$

20.

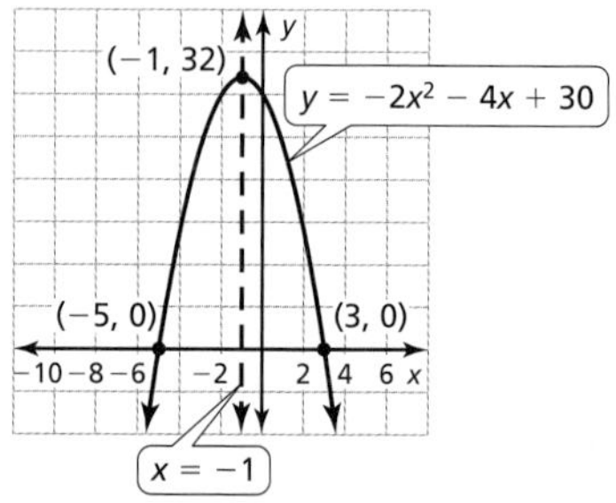

domain: all real numbers, range: $y \leq 32$

21. 2, 10 **22.** −5, 1

23. −8, 3 **24.** 4, 13

25. −2, 7 **26.** −1

27. −5, −2, 2 **28.** −6, 6, 11

29. −7, 0, 7 **30.** −3, 1, 3

31. D **32.** F

33. C **34.** B

35. A **36.** E

61.

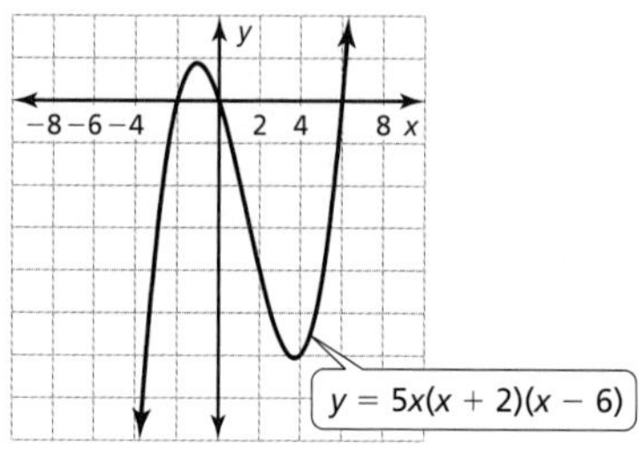

62.

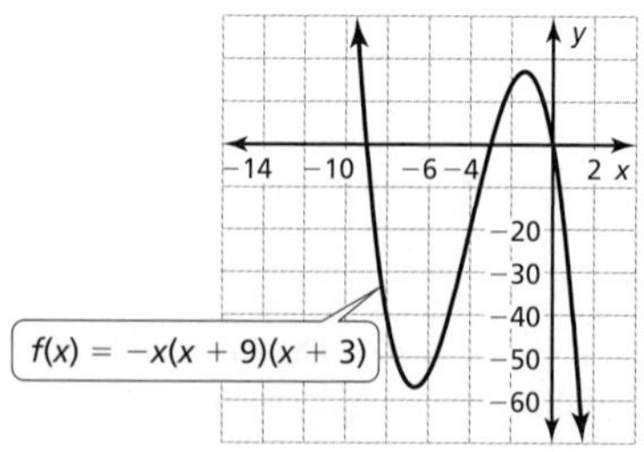

63.

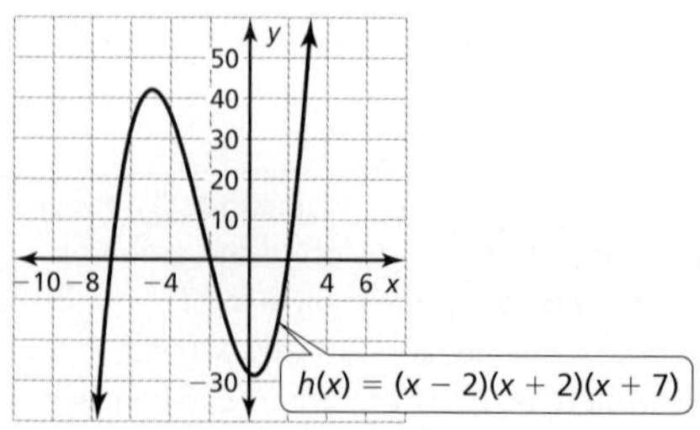

64.

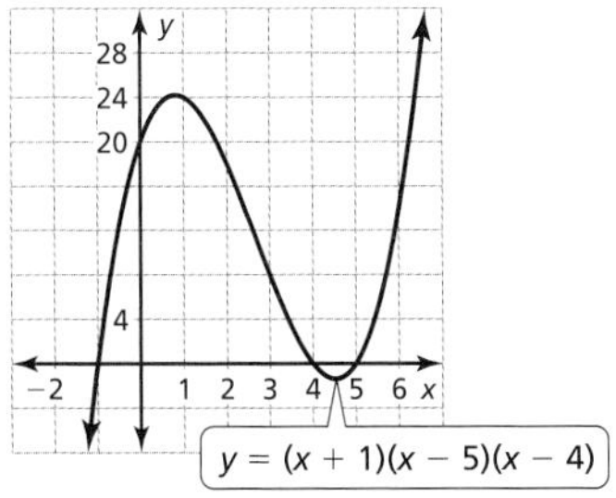

65.

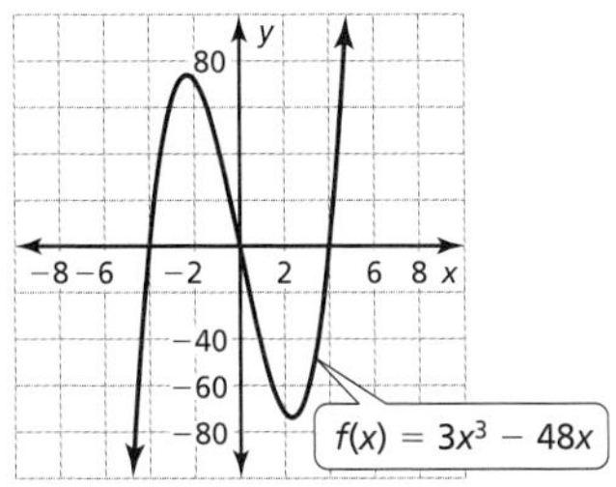

66.

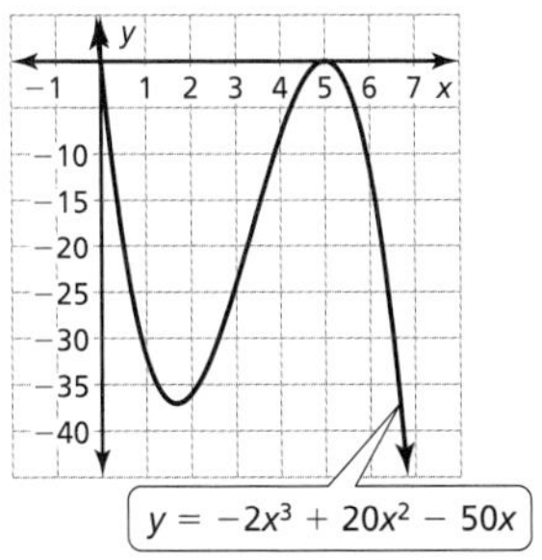

67.

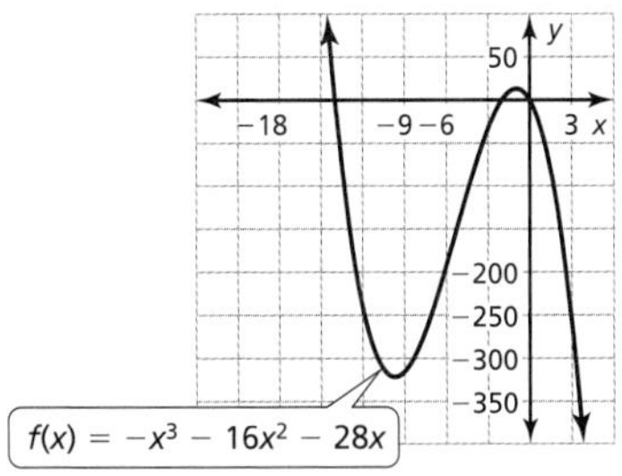

68.

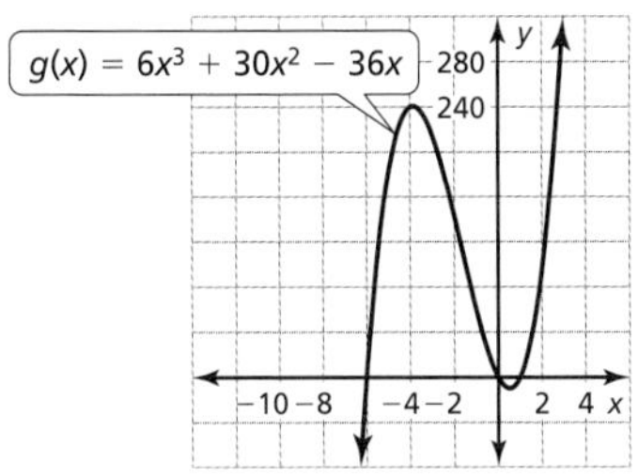

69. $f(x) = 4x^3 - 12x^2 - 16x$ **70.** $f(x) = -6x^3 - 6x^2 + 36x$

71. $f(x) = -2x^3 - 22x^2 - 56x$

72. $f(x) = 5x^3 - 50x^2 + 135x - 90$

73. *Sample answer:* $f(x) = x^3 - 9x^2 + 2x + 48$

74. *Sample answer:* $f(x) = x^3 + 12x^2 + 35x$

75. *Sample answer:* $f(x) = x^3 - 8x^2 + 7x$

76. *Sample answer:* $f(x) = x^3 + 6$

92. **a.** *Sample answer:* $p = 75, q = 425$

b. *Sample answer:* 350 feet wide, 61 feet high; Use the height as the value of $f(x)$ from part (a) along with the values of p and q estimated in part (a). Use the values of p and q to calculate the value of the x-coordinate of the vertex to substitute for x in the function in part (a), and solve the function for a.

93. **a.** $f(x) = -3(x + 4)(x - 2)$

b.

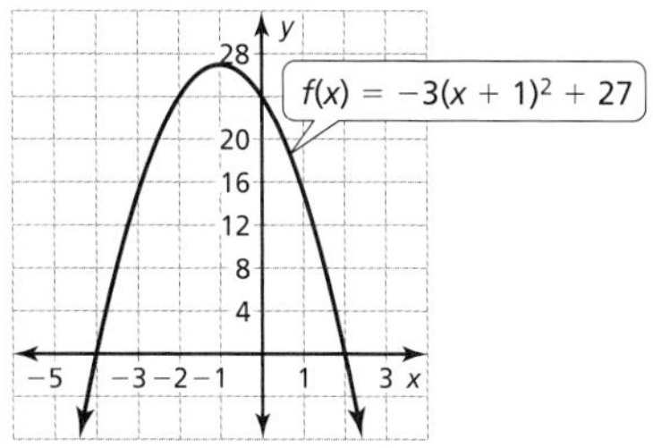

Sample answer: Plot the vertex $(-1, 27)$, which can be determined from the vertex form. Then plot the x-intercepts $(-4, 0)$ and $(2, 0)$, which can be determined from the intercept form. Draw a smooth curve through these points.

94. **a.** $g(x) = 2x(x - 2)$

b.

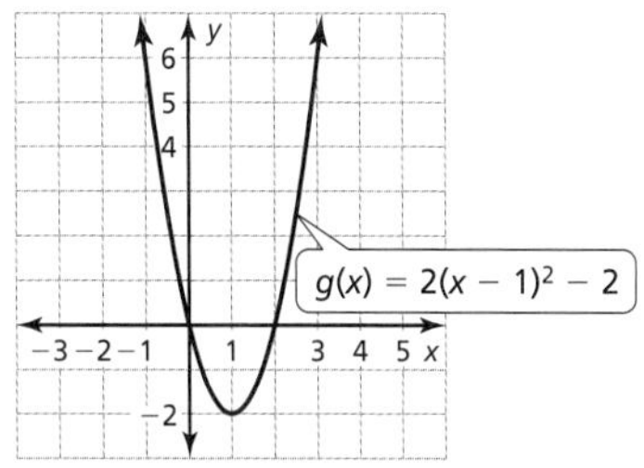

Sample answer: Plot the vertex $(1, -2)$, which can be determined from the vertex form. Then plot the x-intercepts $(0, 0)$ and $(2, 0)$, which can be determined from the intercept form. Draw a smooth curve through these points.

95. yes; *Sample answer:* Let p and q be real zeros of a quadratic function. When $p = q$, the function has exactly one real zero and can be written as $y = a(x - p)(x - p)$.

96. yes; *Sample answer:* In standard form, $f(x) = ax^2 + bx + c$, missing terms can be included by using 0 for b or c. In vertex form, $f(x) = a(x - h)^2 + k$, missing values can be included by using 0 for h or k.

97. $f(x) = \frac{1}{10}(x + 5)(x + 2)(x - 1)(x - 4)(x - 8)$

98. **a.**

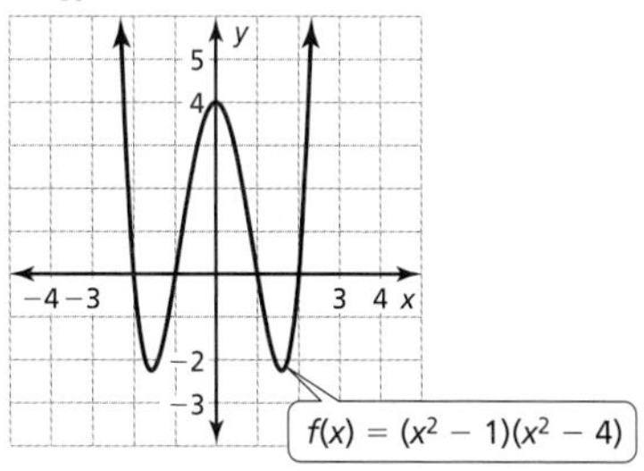

b.

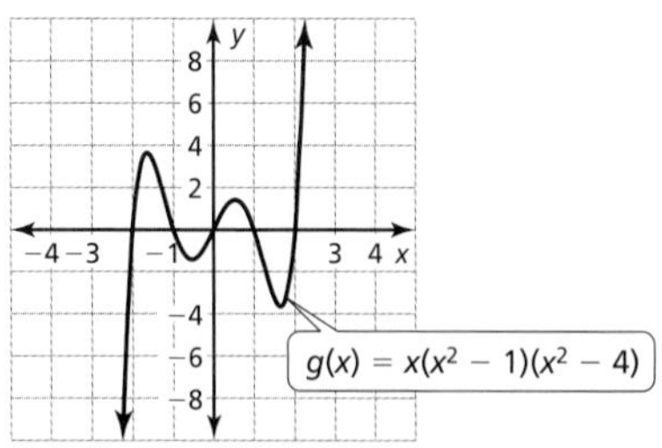

Sample answer: Factor the functions into intercept form. Plot the intercepts. Calculate the values for several additional points, then plot them to get an idea of the general shape of the curve. Draw a smooth curve through the plotted points.

99. $-k, 2k$

100. *Sample answer:* $y = (x + 4)(x - 2)$ and $y = 2(x + 4)(x - 2)$; The two given points are x-intercepts, so the graphs of any two quadratic equations having these two intercepts would intersect at these points.

101. *Sample answer:* $y = (x - 5)^2 + 2$ and $y = -(x - 5)^2 + 10$; The two given points have the same y-coordinate, so the axis of symmetry of any parabola passing through them would be halfway between, which is $x = 5$. The vertex form can be used with $h = 5$ and any selected value of k, where $k \neq 6$, to find quadratic equations that would pass through these points. Any two of these equations would form a system with these two points as solutions.

8.5 Maintaining Mathematical Proficiency

102. 300 calories

103. 29 g

104. increases

105. neither; There is no common difference or common ratio.

106. geometric; There is a common ratio of 3.

107. arithmetic; There is a common difference of -8.

108. neither; There is no common difference or common ratio.

8.6 Lesson Monitoring Progress

1.

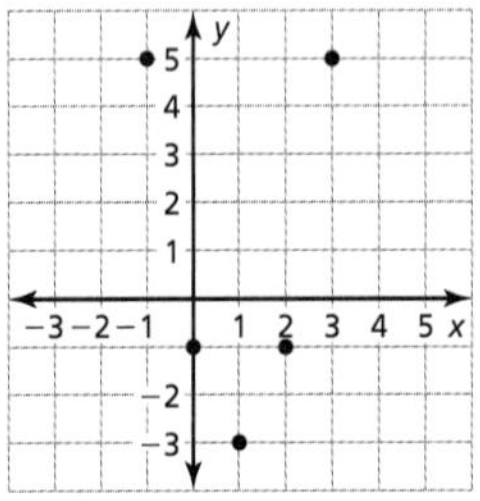

quadratic

2.

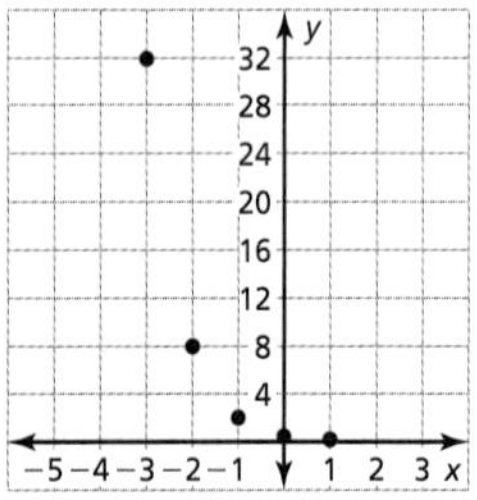

exponential

3.

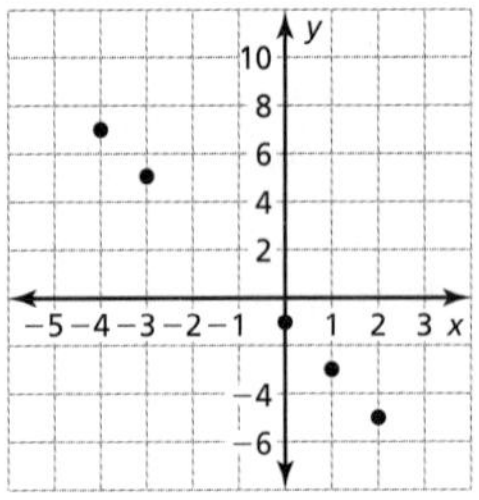

linear

8.6 Monitoring Progress and Modeling with Mathematics

13.

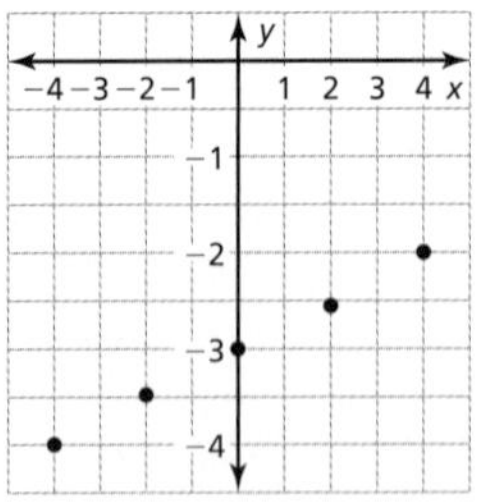

linear

14.

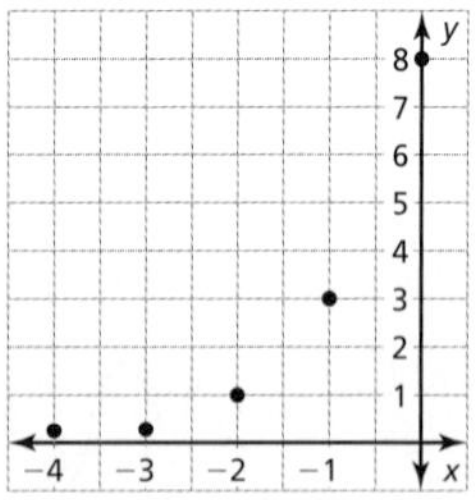

exponential

15. linear

16. exponential

17. exponential

18. quadratic

35. The average rate of change of a linear function is constant because the dependent variable of a linear function increases by the same amount for each constant change in the independent variable. The average rate of change of a quadratic or exponential function is not constant because the dependent variable of a quadratic or exponential function changes by a different amount for each constant change in the independent variable.

36. **a.** D; A line is the graph of a linear function.

b. C; The graph appears to represent an exponential decay function, which has a base between 0 and 1. Choice C is a translation of this type of function.

c. B; The graph appears to represent an exponential growth function, which has a base greater than 1. Choice B is a translation of this type of function.

d. A; A parabola is the graph of a quadratic function.

37. quadratic; The second differences have a constant value of $9n - 5$.

38. *Sample answer:* $y = 1.5x^2$

39. no; *Sample answer:* There may not be enough points to clearly determine the shape of the graph.

40. *Sample answer:* the number of new branches, the length of new branches, the total length of all branches, the total number of branches; exponential, exponential, linear, exponential; $y = (2)^{n-1}, y = \left(\frac{1}{2}\right)^{n-1}, y = n, y = 2^n - 1$

41. no; The graph of an exponential function will always eventually have greater y-values than the graph of a quadratic function.

42. *Sample answer:* $y = -0.414x^2 + 6.17x + 47.0$; Compare the linear, quadratic, and exponential models obtained using the regression features of a graph utility; about \$69.9 billion

8.6 Maintaining Mathematical Proficiency

43. 11

44. 5

45. 8

46. 3

47. $x^2 - 64$

48. $16y^2 - 4$

49. $9a^2 - 25b^2$

50. $4r^2 - 36s^2$

Chapter 8 Review

8.

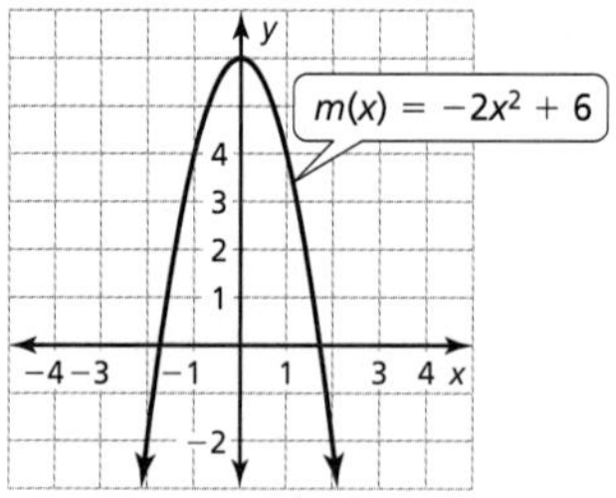

The graph of m is a vertical stretch by a factor of 2, a reflection in the x-axis, and a vertical translation 6 units up of the graph of f.

9.

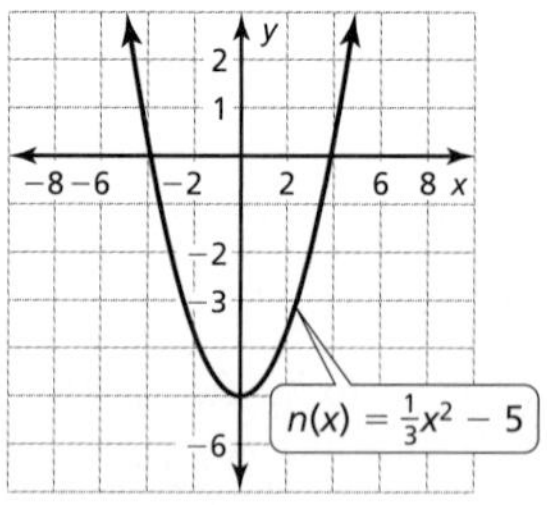

The graph of n is a vertical shrink by a factor of $\frac{1}{3}$ and a vertical translation 5 units down of the graph of f.

Chapter 8 Test

5.

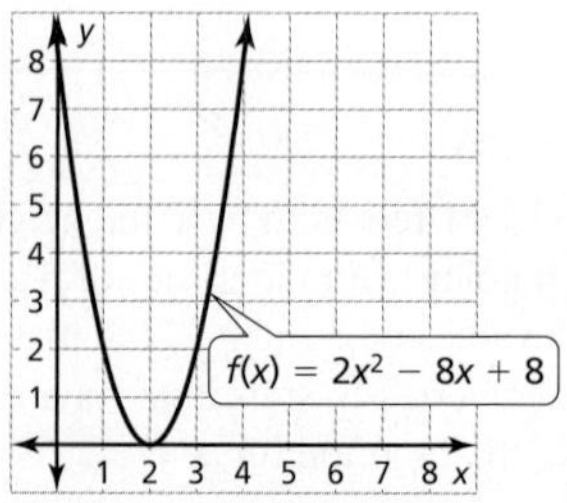

domain: all real numbers, range: $y \geq 0$

6. 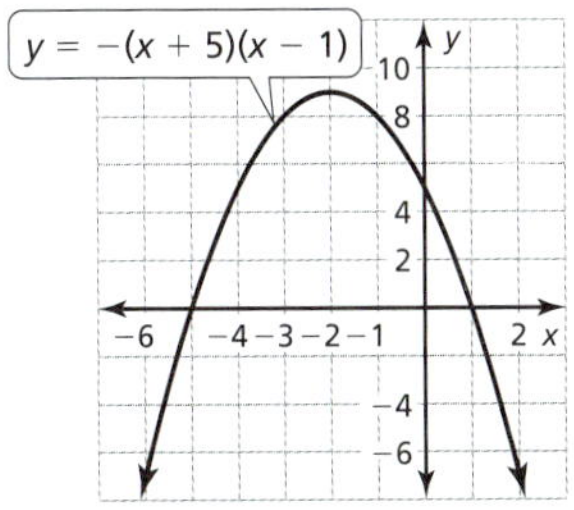

domain: all real numbers, range: $y \le 9$

7. 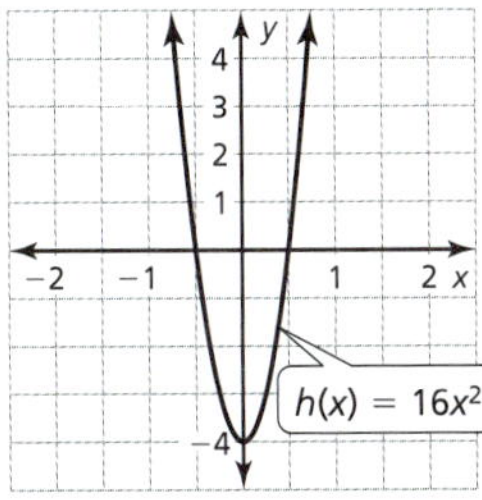

domain: all real numbers, range: $y \ge -4$

8. exponential; The y-values increase by a constant factor; $y = 8(2)^x$

9. quadratic; The second differences increase by a constant amount; $y = -2x^2$

10. $f(x) = -\frac{1}{2}x^2 - 5x - 8$; *Sample answer:* Use the intercepts to write an equation in intercept form. Substitute the third point into the equation to find a. Write the equation in intercept form with the value of a. Simplify the equation to put it in standard form.

11. $f(x) = 3x^2 - 30x$; *Sample answer:* Use the intercepts to write an equation in intercept form. Substitute the third point into the equation to find a. Write the equation in intercept form with the value of a. Simplify the equation to put it in standard form.

12. *Sample answer:* $f(x) = x^2 + 3$; To be even, the vertex needs to be on the y-axis, so it must be in the form $ax^2 + c$. Because the range is $y \ge 3$, a must be positive and $c = 3$. Any positive value of a is acceptable.

13. *Sample answer:* $f(x) = x^2 - 8x + 16$; Use the x-intercept as the vertex. Use the intercept to write an equation in intercept form. Substitute the second point into the equation to find a. Write the equation in intercept form with the value of a. Simplify the equation to put it in standard form.

14. linear; The first differences are constant; $y = 19x$

15. **a.** 4.445 ft

b. yes; *Sample answer:* After traveling a horizontal distance of 30 feet, the height is 3.6 feet, so the ball will be 0.6 feet above the net.

16. **a.** *Sample answer:* $a = 1, b = 0, c = 0$

b. *Sample answer:* $a = 0, b = 1, c = 0$

c. *Sample answer:* $a = 1, b = 2, c = 3$

17. 1, 3, 5; They are increasing by a constant amount.

Chapter 9

9.1 Monitoring Progress and Modeling with Mathematics

104. no; Applying the sum and difference pattern to the denominator results in $16 - \sqrt[3]{5^2}$, which still contains a radical.

105. 377

106. **a.** $\left(\frac{1+\sqrt{5}}{2}\right)^2 - \frac{1+\sqrt{5}}{2} - 1 = 0$, $\left(\frac{1-\sqrt{5}}{2}\right)^2 - \frac{1-\sqrt{5}}{2} - 1 = 0$

b. *Sample answer:* $DF = \frac{1+\sqrt{5}}{2}$

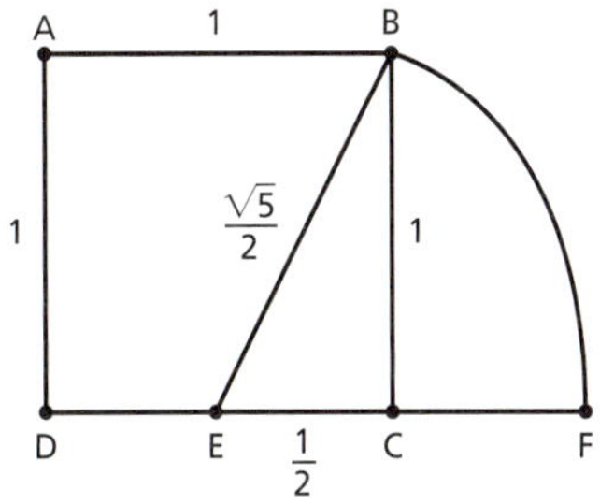

107. $\frac{2\sqrt[3]{x^2} - 2\sqrt[3]{x} + 2}{x+1}$; Multiplying the numerator and denominator by $\sqrt[3]{x^2} - \sqrt[3]{x} + 1$ rationalizes the denominator.

9.1 Maintaining Mathematical Proficiency

108. 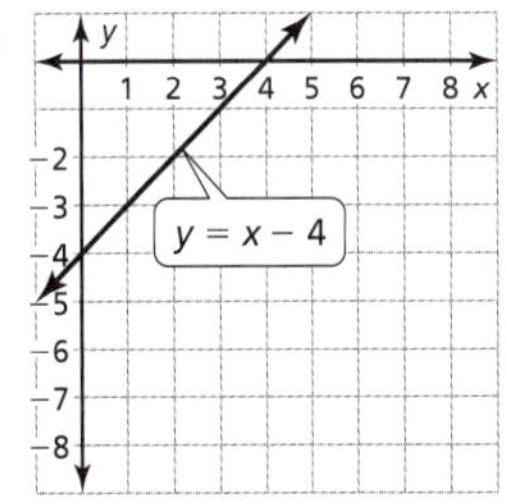

4

109. 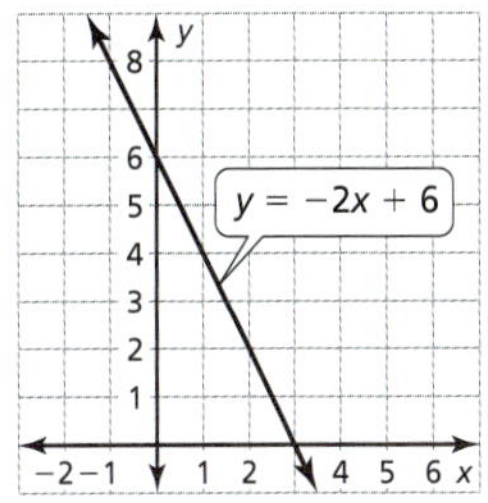

3

110. 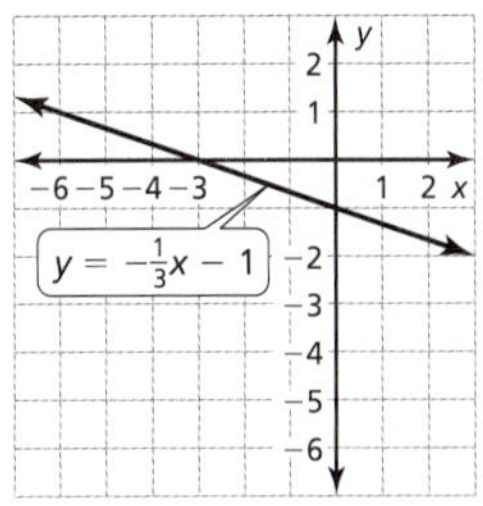

-3

111. 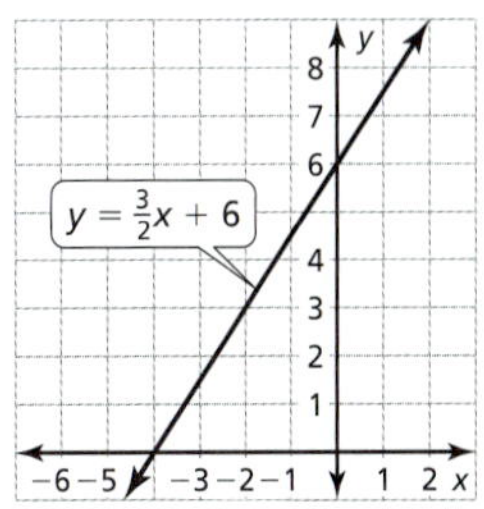

-4

112. $x = 5$

113. $x = -3$

114. $x = 3$

115. $x = -\frac{2}{3}$

9.2 Monitoring Progress and Modeling with Mathematics

51.

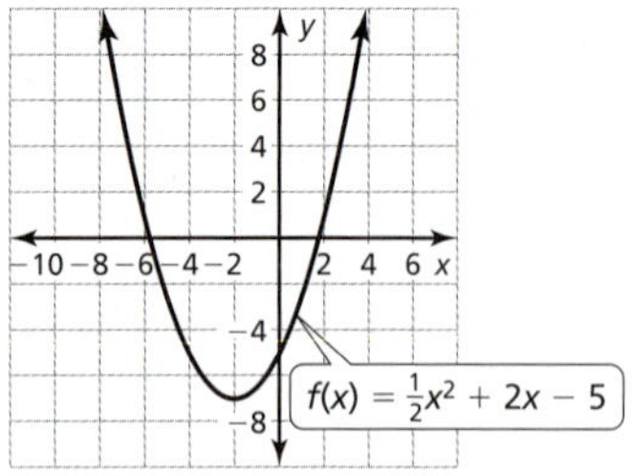

about 1.7, about −5.7

52.

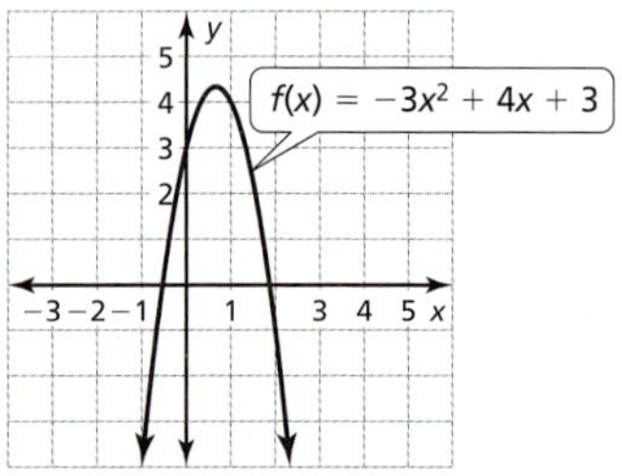

about 1.9, about −0.5

53. **a.** 118 ft, 198 ft, 246 ft, 262 ft, 246 ft, 198 ft, 118 ft, 6 ft

b. about 1.5 sec, about 6.5 sec

c. about 1.4 sec, about 6.6 sec

55. about 3.7 ft

56. about 6.2 m

57. Graph the function to determine which integers the solutions are between. Then make tables using x-values between the integers with an interval of 0.1. Look for a change of sign in the function values, then select the value closest to zero.

58. **a.** 2; The graphs intersect in two places.

b. The x-intercepts are 1 and −4.

59. *Sample answer:* Method 1; Only one graph needs to be drawn.

60. infinitely many; *Sample answer:*

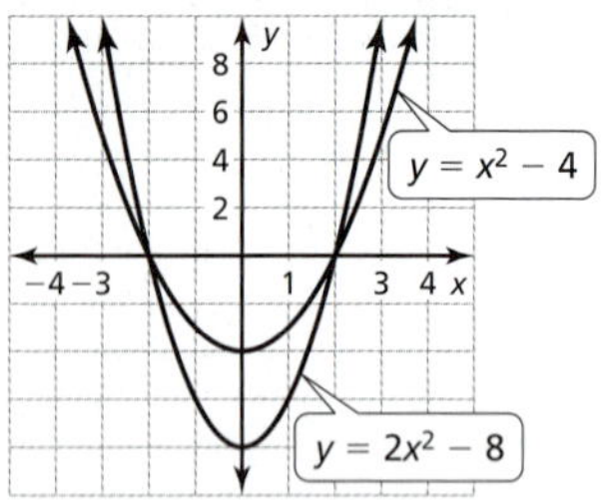

61. about 24.1 ft

62. yes; When $x = 57$, $y = 26.313$.

63. sometimes; $y = -2x^2 + 1$ has two x-intercepts, but $y = -2x^2 + (-1)$ has no x-intercepts.

64. always; The sign of y does not change.

65. never; The graph of $y = ax^2 + bx + c$ has at most two x-intercepts.

9.2 Maintaining Mathematical Proficiency

66. exponential decay function; As x increases by 1, y is multiplied by $\frac{1}{6}$.

67. exponential growth function; As x increases by 1, y is multiplied by 4.

9.3 Monitoring Progress and Modeling with Mathematics

42. $x = \frac{-b}{2a} \pm \sqrt{\frac{b^2 - 4ac}{4a^2}}$ or $x = \frac{-b \pm \sqrt{b^2 - 4ac}}{2a}$

43. $x = 6, x = -2$

44. **a.** $x = 14, x = -2$

b. $x = -3, x = -11$

9.3 Maintaining Mathematical Proficiency

45. $(x + 4)^2$

46. $(x - 2)^2$

47. $(x - 7)^2$

48. $(x + 9)^2$

49. $(x + 6)^2$

50. $(x - 11)^2$

9.4 Monitoring Progress and Modeling with Mathematics

68.

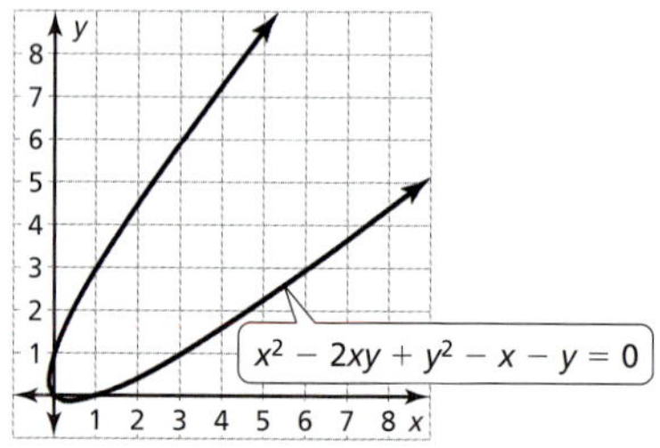

a parabola at an angle

71. yes; Substituting 23.50 for y in the model, the stock is worth \$23.50 ten days and thirty days after the stock is purchased.

72. Factoring does not work for this equation, and graphing does not produce an exact solution.

73. length: 66 in., width: 6 in.

74. 0; If $c < -\left(\frac{b}{2}\right)^2$, then $\left(\frac{b}{2}\right)^2 + c < 0$, so there are no real solutions.

9.4 Maintaining Mathematical Proficiency

75. $a_1 = 10, a_n = a_{n-1} + 5$

76. $a_1 = 3, a_n = 2a_{n-1}$

77. $a_1 = -20, a_n = a_{n-1} + 4$

78. $2\sqrt{3}$

79. $6\sqrt{2}$

80. $2\sqrt{5}$

9.5 Monitoring Progress and Modeling with Mathematics

39. $x \approx 0.74, x \approx -6.74$; *Sample answer:* $a = 1$ and b is even, so solve by completing the square.

40. $x \approx 2.24, x \approx -2.24$; *Sample answer:* The equation can be written in the form $x^2 = d$, so solve using square roots.

41. $x = -4, x = 3$; *Sample answer:* The equation is easily factorable, so solve by factoring.

42. $x \approx 3.73, x \approx 0.27$; *Sample answer:* $a = 1$ and b is even, so solve by completing the square.

43. $x \approx 2.19, x \approx -1.94$; *Sample answer:* The equation cannot be factored and $a \neq 1$, so solve using the quadratic formula.

44. $x = 1, x = -7$; *Sample answer:* The left side is a perfect square trinomial, so solve using square roots.

79. **a.** $(x + 1), (x - 6); x^2 - 5x - 6 = 0; x, (x - 2); x^2 - 2x = 0;$
$\left(x + \frac{1}{2}\right), (x - 5); x^2 - \frac{9}{2}x - \frac{5}{2} = 0$

b.

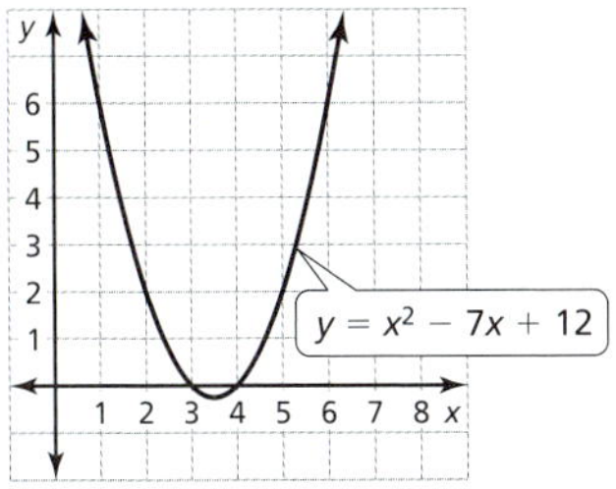

3, 4

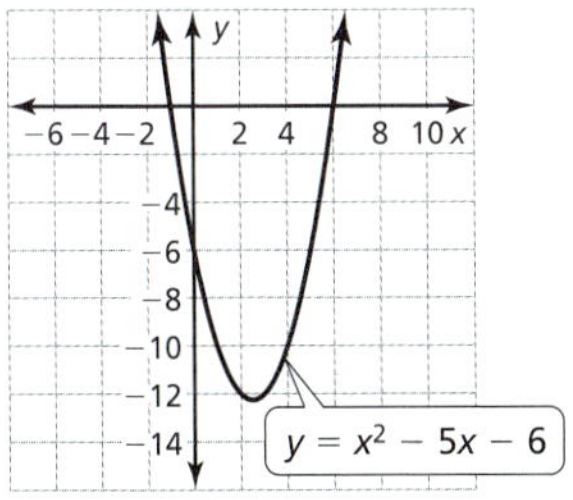

$-1, 6$

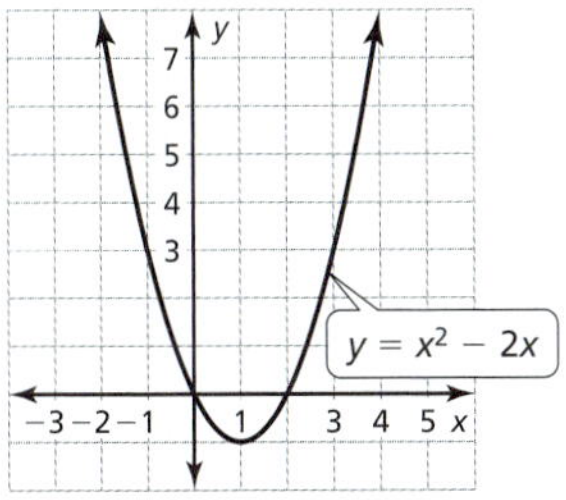

0, 2

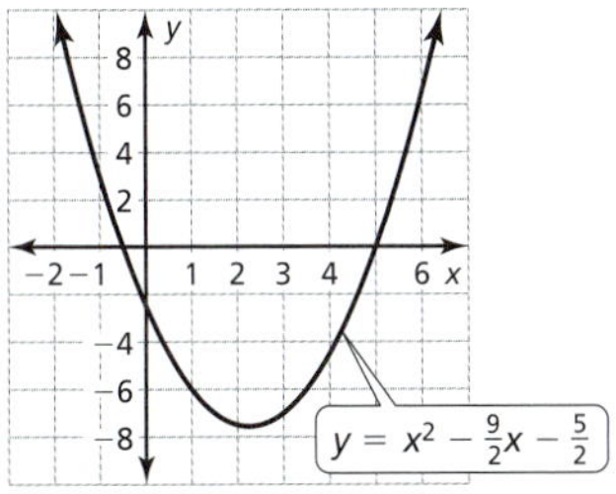

$-\frac{1}{2}, 5$

82. **a.** $k > -\frac{25}{64}$

b. $k = -\frac{25}{64}$

c. $k < -\frac{25}{64}$

9.5 Maintaining Mathematical Proficiency

83. (4, 0); *Sample answer:* substitution because both equations are solved for y

84. $(-4, 5)$; *Sample answer:* substitution because one equation is solved for x

85. (5, 3); *Sample answer:* elimination because one pair of like terms has the same coefficient

86. $(-2, 7)$; *Sample answer:* substitution because one of the variables has a coefficient of 1

9.6 Monitoring Progress and Modeling with Mathematics

57. The possible number of solutions is 0, 1, or 2.

9.6 Maintaining Mathematical Proficiency

63.

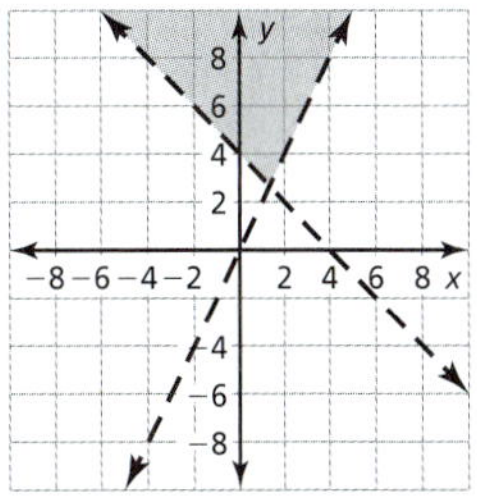

64.

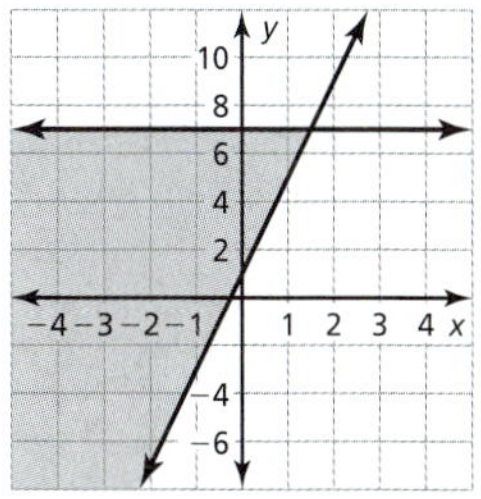

65.

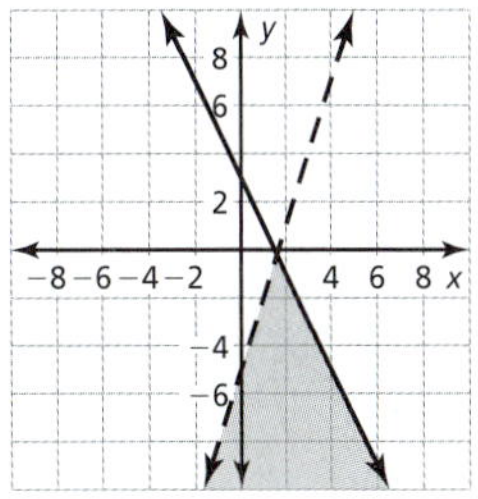

66.

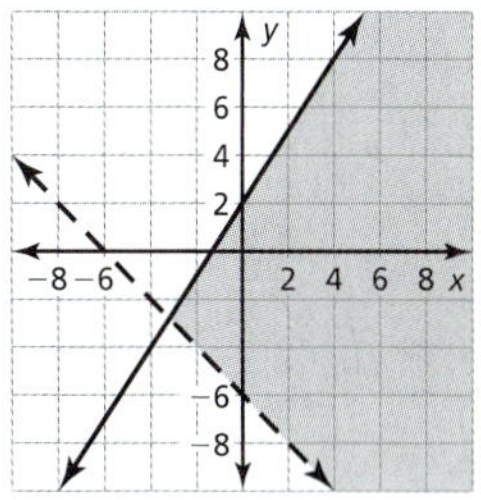

67.

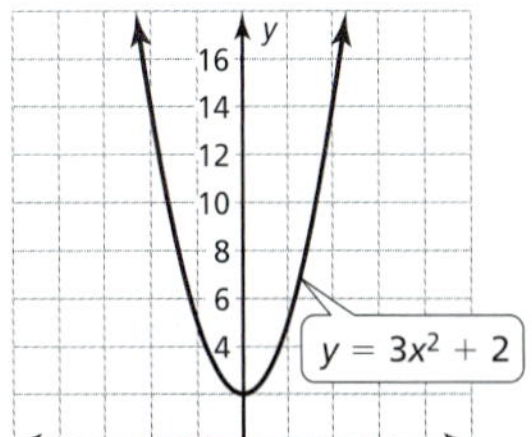

domain: all real numbers; range: $y \geq 2$

68.

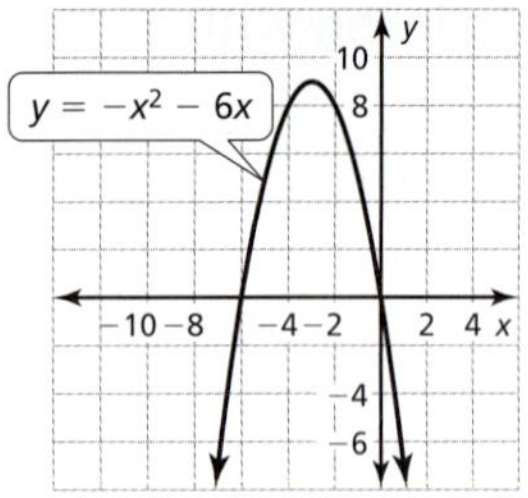

domain: all real numbers; range: $y \leq 9$

69.

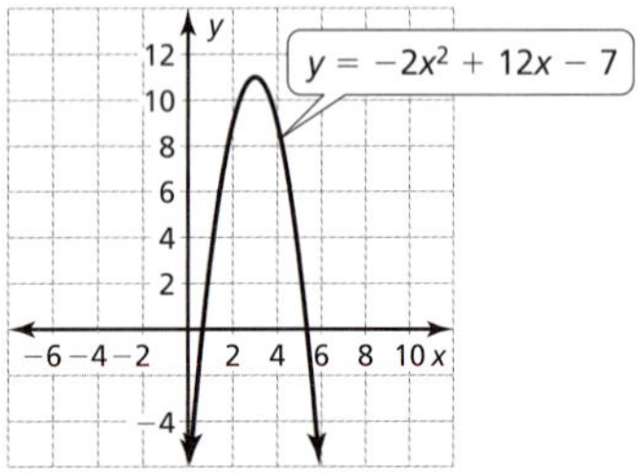

domain: all real numbers; range: $y \leq 11$

70.

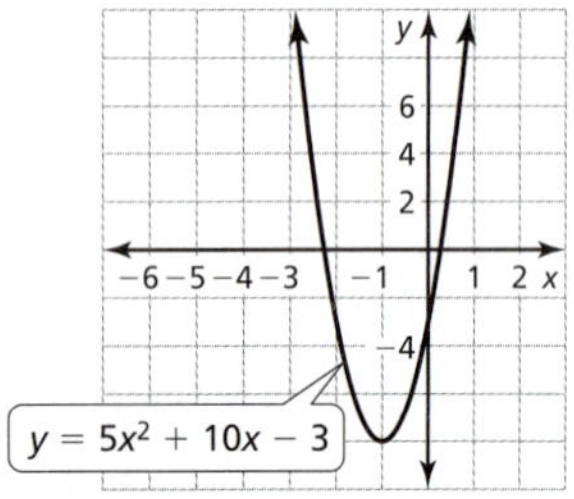

domain: all real numbers; range: $y \geq -8$

Chapter 10

10.1 Explorations

1. a.

x	$f(x)$
0	0
1	1
4	2
9	3

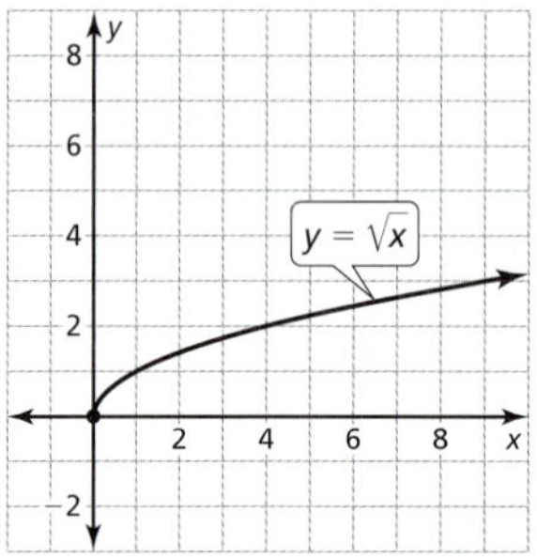

$x \geq 0$; $y \geq 0$

b.

x	$f(x)$
−2	0
−1	1
2	2
7	3

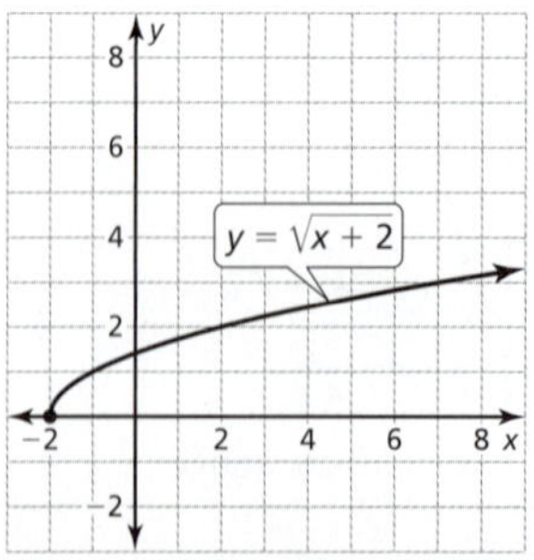

$x \geq -2$; $y \geq 0$

10.1 Lesson Monitoring Progress

6.

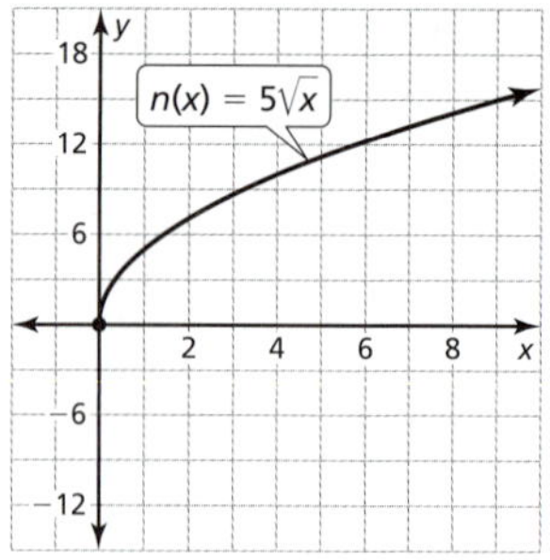

$y \geq 0$

9.

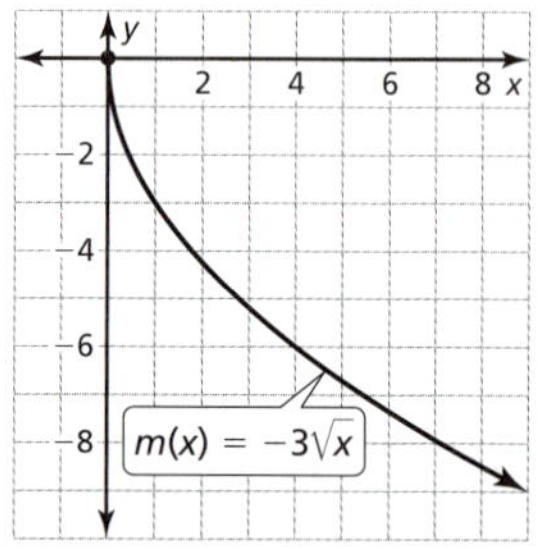

The graph of g is a vertical stretch by a factor of 3 and a reflection in the x-axis of the graph of f.

10.1 Monitoring Progress and Modeling with Mathematics

23.

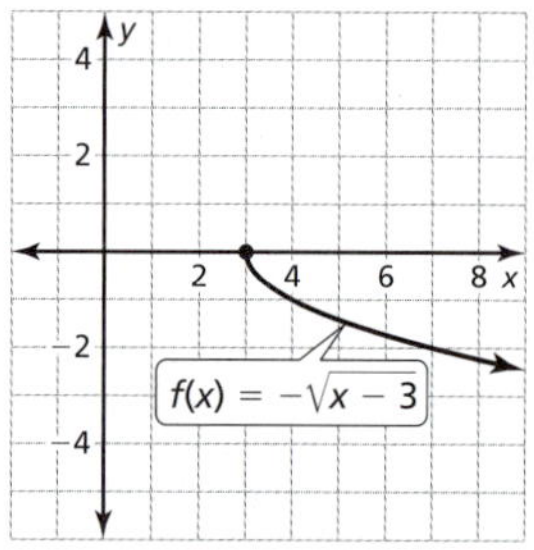

$y \leq 0$

24.

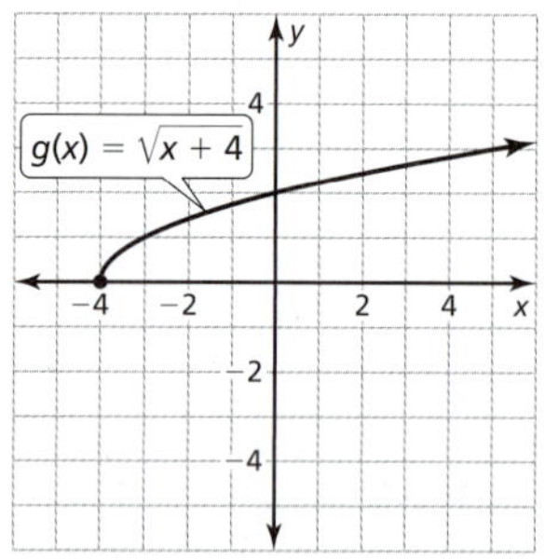

$y \geq 0$

25.

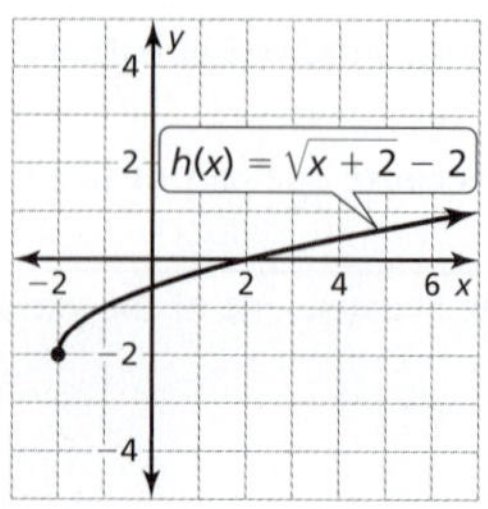

$y \geq -2$

26.

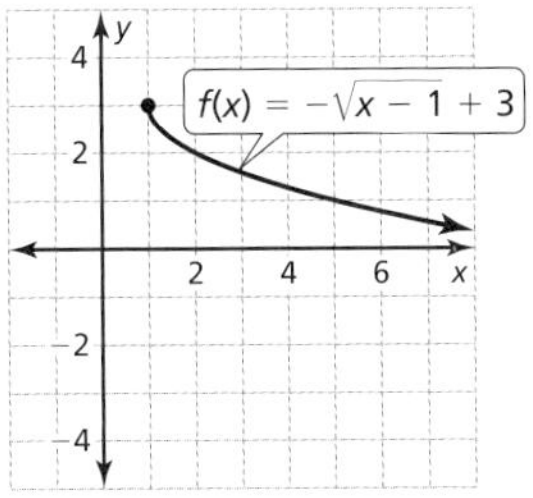

$y \leq 3$

27.

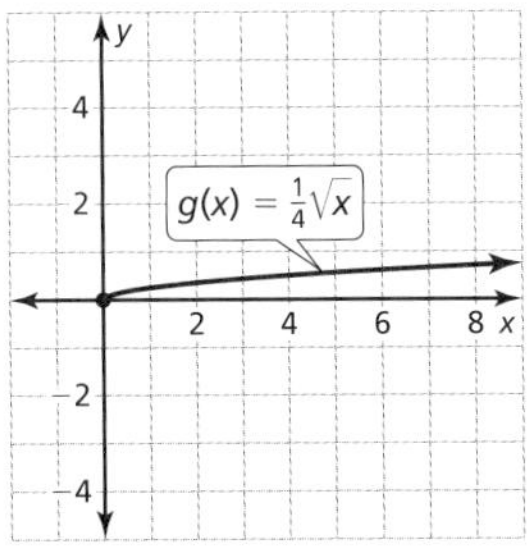

The graph of g is a vertical shrink by a factor of $\frac{1}{4}$ of the graph of f.

28.

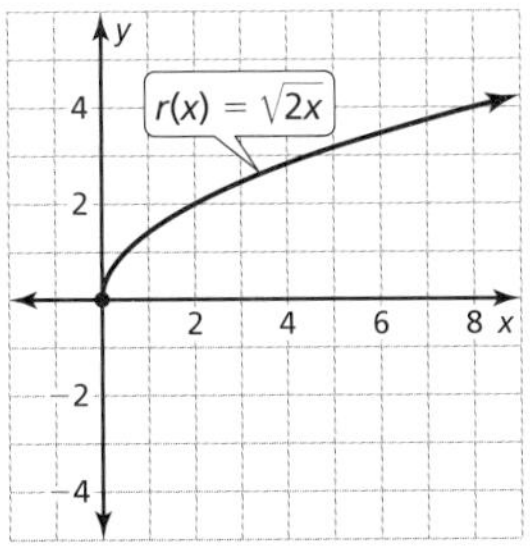

The graph of r is a horizontal shrink by a factor of $\frac{1}{2}$ of the graph of f.

29.

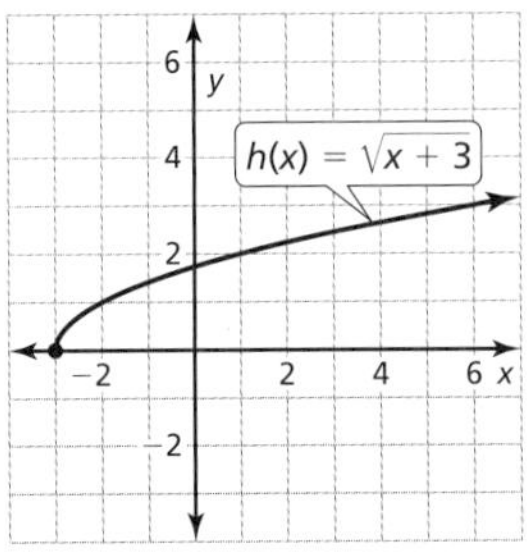

The graph of h is a translation 3 units left of the graph of f.

30.

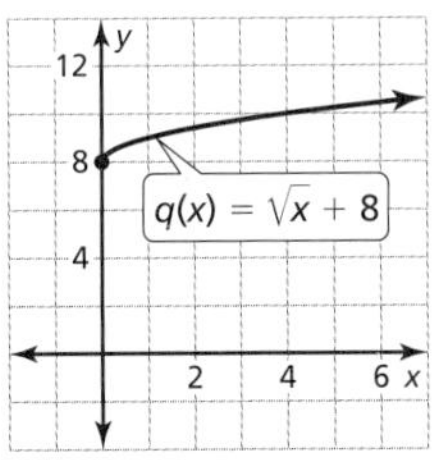

The graph of q is a translation 8 units up of the graph of f.

31.

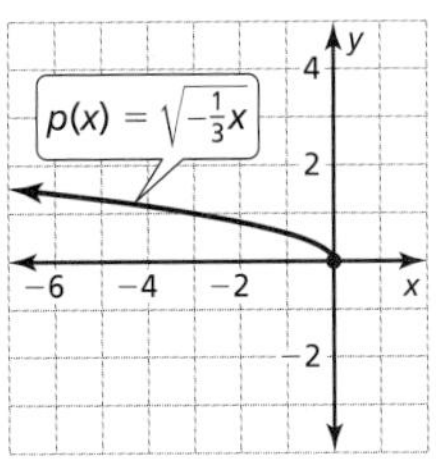

The graph of p is a horizontal stretch by a factor of 3 and a reflection in the y-axis of the graph of f.

32.

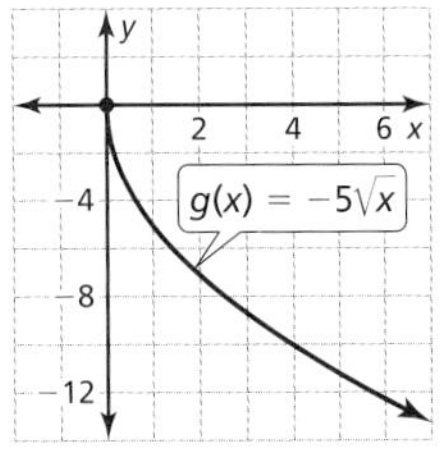

The graph of g is a vertical stretch by a factor of 5 and a reflection in the x-axis of the graph of f.

33.

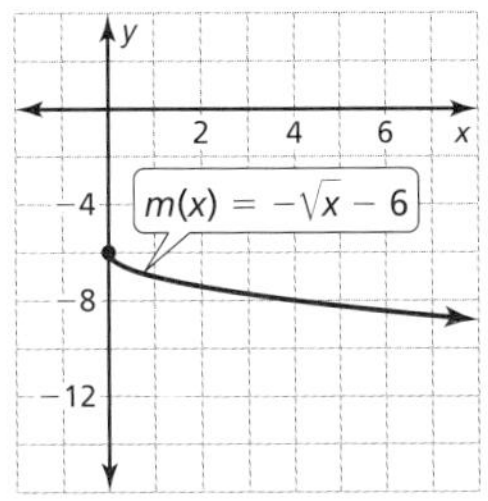

The graph of m is a reflection in the x-axis and a translation 6 units down of the graph of f.

34.

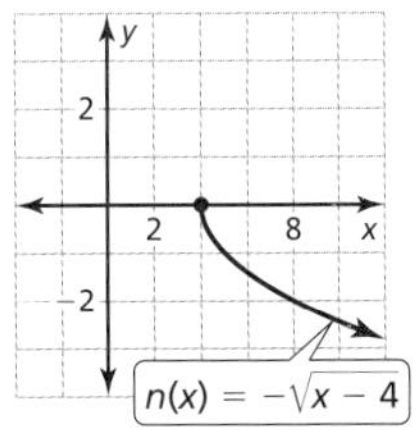

The graph of n is a translation 4 units right and a reflection in the x-axis of the graph of f.

35. The graph of $y = \sqrt{x}$ is translated 1 unit up instead of 1 unit left.

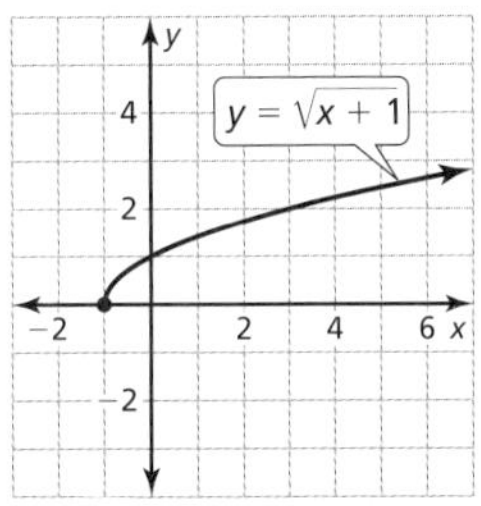

42. Translate 1 unit right, stretch vertically by a factor of 2, and translate 4 units up to obtain the graph of h.

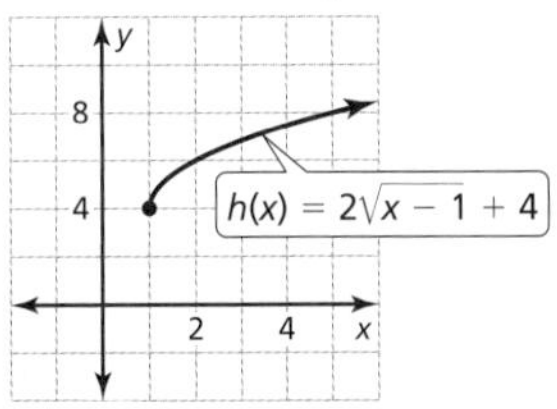

43. Translate 1 unit right, stretch vertically by a factor of 2, reflect in the x-axis, and translate 5 units up to obtain the graph of h.

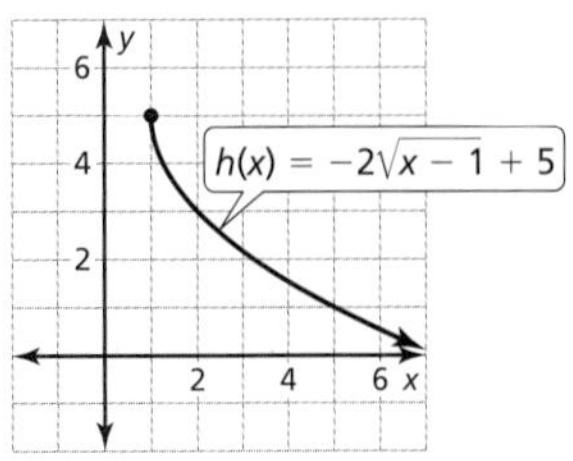

44. Translate 2 units left, stretch vertically by a factor of 5, reflect in the x-axis, and translate 1 unit down to obtain the graph of h.

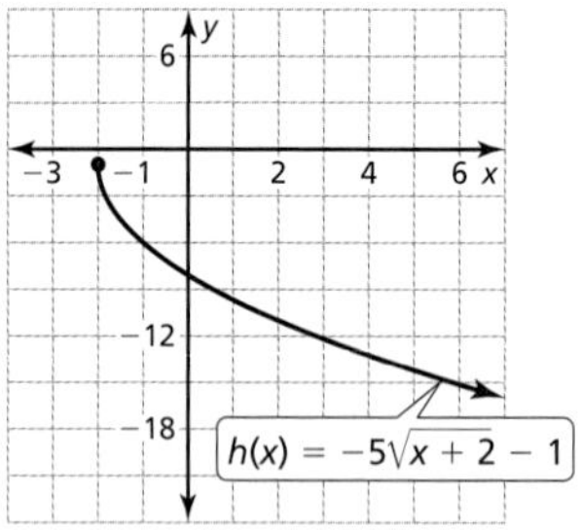

45. As skid mark lengths increase from 0 to 15 feet, the initial van speeds increase at an average rate of about 1.22 miles per hour per foot on Road Surface A and about 1.00 mile per hour per foot on Road Surface B.

46. From 0 to 6 joules, the velocity of Object A increases at an average rate of about 0.29 meter per second per joule and the velocity of Object B increases at an average rate of about 0.33 meter per second per joule.

47. a. *Sample answer:* $y = \sqrt{x} + 1$

b. *Sample answer:* $y = -\sqrt{x}$

48. yes; yes; Transformations of the graph of the parent square root function, $f(x) = \sqrt{x}$, could result in x or y coordinates that are negative.

49. a.

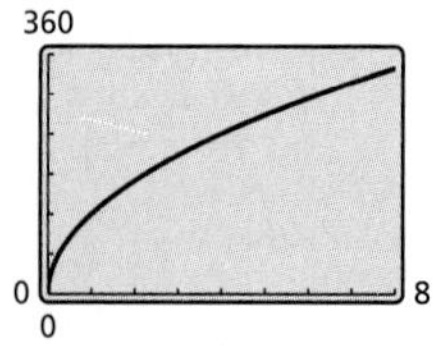

6.25 lb/in.^{2}

b. decreases

51. a. $A > 0$

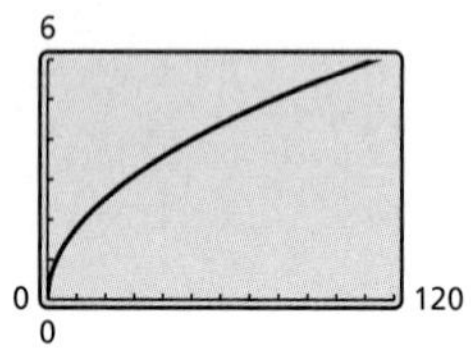

b. about 91.61 in.^{2}

52. a. $a = \frac{1}{8}$

b. $a > \frac{1}{8}$

c. $-\frac{1}{8} < a < 0$

53. a. 0

b. no; $f(x)$ increases as x increases.

c. *Sample answer:* $f(x) = -\sqrt{x}$; no; $f(x)$ decreases as x increases.

d. *Sample answer:* $f(x) = -4 + \sqrt{x}$

54. a. B; This is the graph of the parent square root function, $f(x) = \sqrt{x}$, translated 2 units up.

b. D; This is the graph of $f(x)$ translated 4 units down.

c. C; This is the graph of $f(x)$ reflected in the y-axis.

d. A; This is the graph of $f(x)$ with a horizontal shrink.

55. $f(x) = 5\sqrt{x}$; The graph of $f(x) = 5\sqrt{x}$ is a vertical stretch of the graph of $y = \sqrt{x}$ by a factor of 5, and the graph of $g(x) = \sqrt{5x} = \sqrt{5} \cdot \sqrt{x}$ is a vertical stretch of the graph of $y = \sqrt{x}$ by a factor of $\sqrt{5}$.

56. Graph $y = x - 1$ and $y = \sqrt{5x - 9}$ on the same set of coordinate axes.

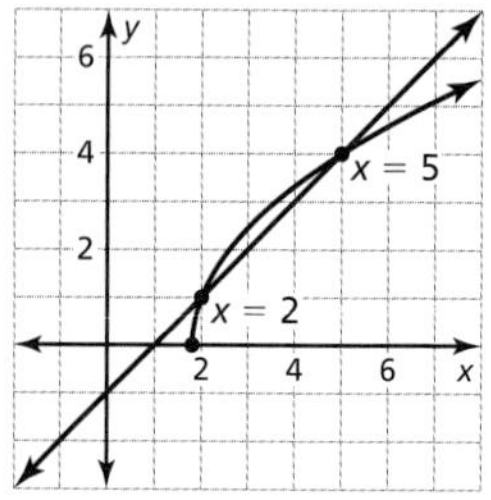

$x = 2, x = 5$

The x-coordinates of the intersection points are the solutions of $x - 1 = \sqrt{5x - 9}$.

57. *Sample answer:* $f(x) = 3 - \sqrt{x + 5}$

10.1 Maintaining Mathematical Proficiency

58. 7

59. −4

60. $\frac{1}{3}$

61. $(x + 1)(x + 6)$

62. $(d - 4)(d - 7)$

63. $(y - 8)(y + 5)$

10.2 Explorations

2. a. $y = \sqrt[3]{x + 4}$

x	$f(x)$
−4	0
−3	1
−2	$\sqrt[3]{2}$
−1	$\sqrt[3]{3}$
0	$\sqrt[3]{4}$
1	$\sqrt[3]{5}$
2	$\sqrt[3]{6}$
3	$\sqrt[3]{7}$
4	2
5	$\sqrt[3]{9}$

b. $y = 1 + \sqrt[3]{x + 4}$

x	$f(x)$
−4	1
−3	2
−2	$1 + \sqrt[3]{2}$
−1	$1 + \sqrt[3]{3}$
0	$1 + \sqrt[3]{4}$
1	$1 + \sqrt[3]{5}$
2	$1 + \sqrt[3]{6}$
3	$1 + \sqrt[3]{7}$
4	3
5	$1 + \sqrt[3]{9}$

4. c.

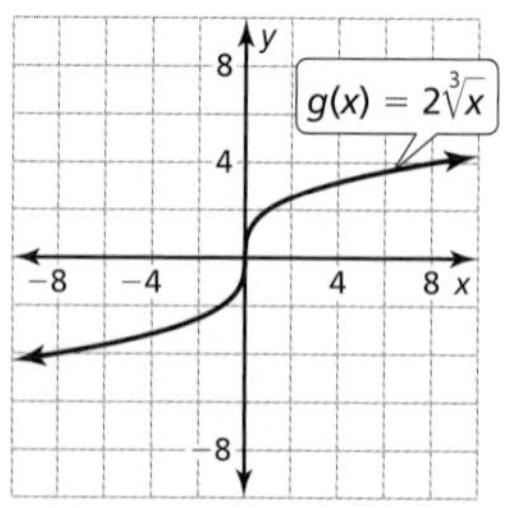

The graph of g is a vertical stretch by a factor of 2 of the graph of f.

d.

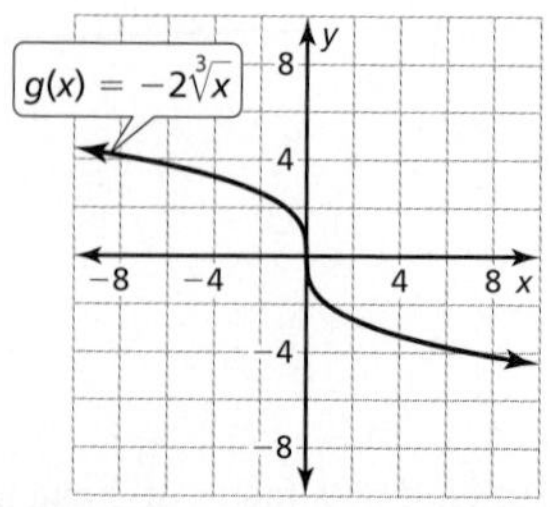

The graph of g is a vertical stretch by a factor of 2 and a reflection in the x-axis of the graph of f.

10.2 Lesson Monitoring Progress

3.

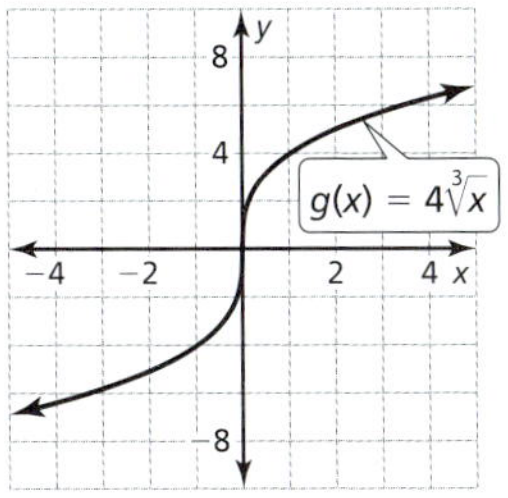

The graph of g is a vertical stretch by a factor of 4 of the graph of f.

5.

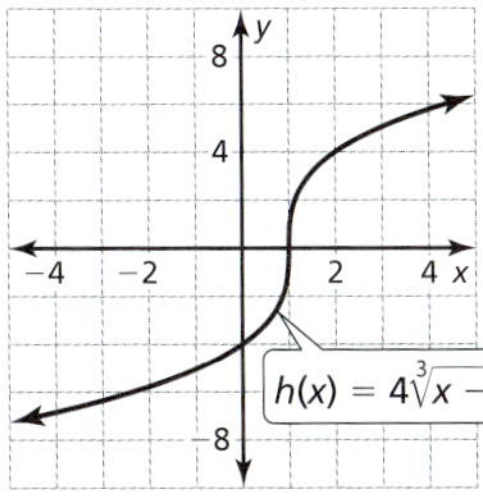

The graph of h is a vertical stretch by a factor of 4 and a translation 1 unit right of the graph of f.

6.

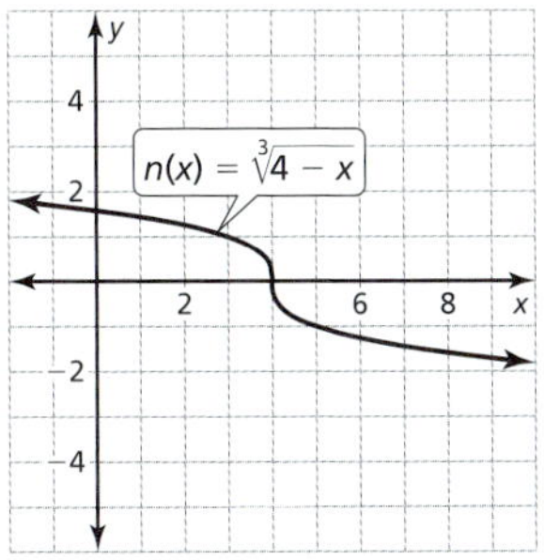

The graph of n is a reflection in the y-axis and a translation 4 units right of the graph of f.

7. Translate 2 units left, shrink vertically by a factor of $\frac{1}{2}$, reflect in the x-axis, and translate 4 units down to obtain the graph of g.

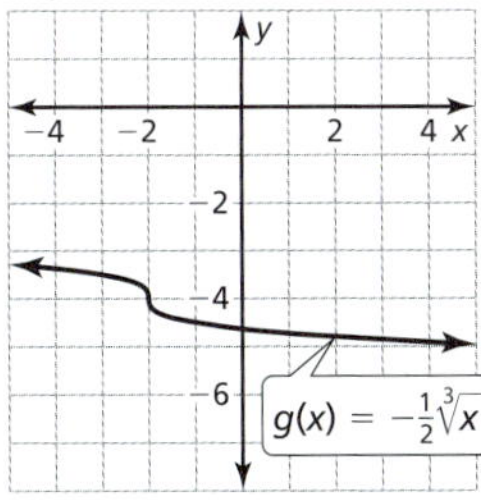

10.2 Monitoring Progress and Modeling with Mathematics

12.

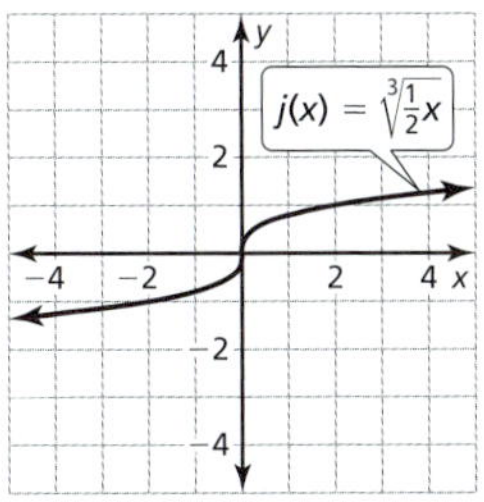

The graph of j is a horizontal stretch by a factor of 2 of the graph of f.

13. The graph of q is a translation 5 units left of the graph of f; $h = -5$

14. The graph of g is a translation 4 units up of the graph of f; $k = 4$

15. The graph of v is a translation 6 units down of the graph of f; $k = -6$

16. The graph of p is a vertical stretch by a factor of 3 of the graph of f; $a = 3$

17.

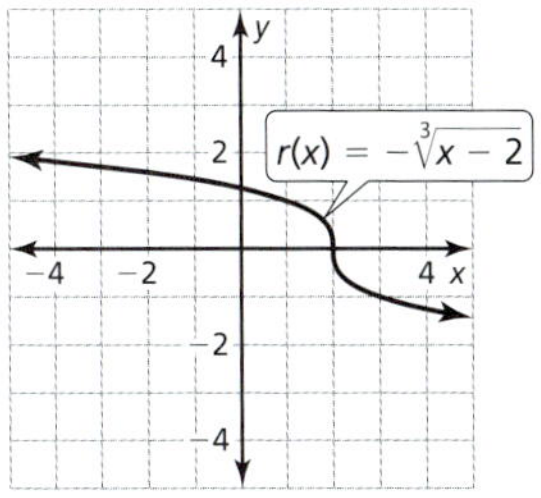

The graph of r is a reflection in the x-axis and a translation 2 units right of the graph of f.

18.

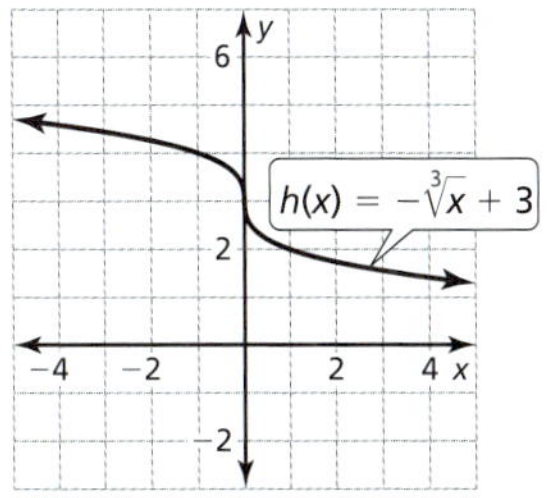

The graph of h is a reflection in the x-axis and a translation 3 units up of the graph of f.

19.

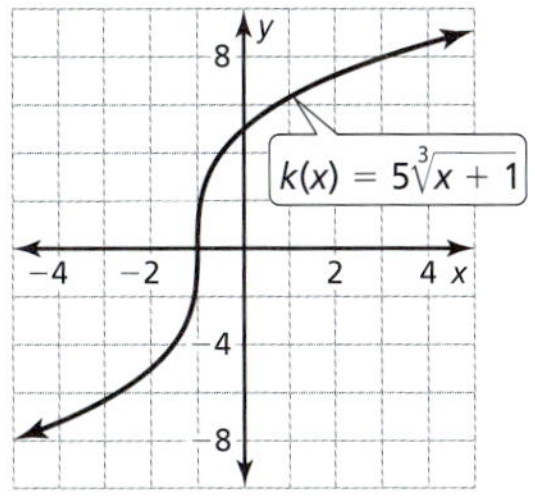

The graph of k is a vertical stretch by a factor of 5 and a translation 1 unit left of the graph of f.

20.

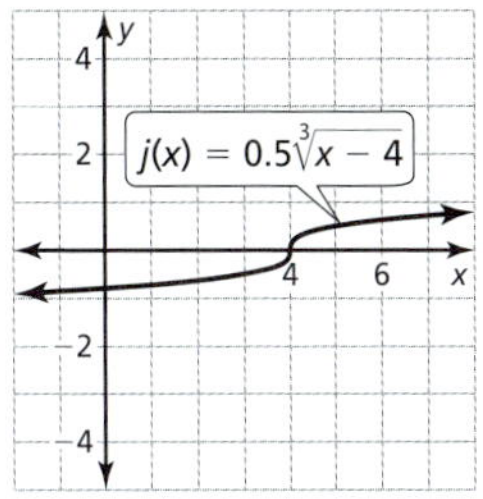

The graph of j is a vertical shrink by a factor of 0.5 and a translation 4 units right of the graph of f.

21.

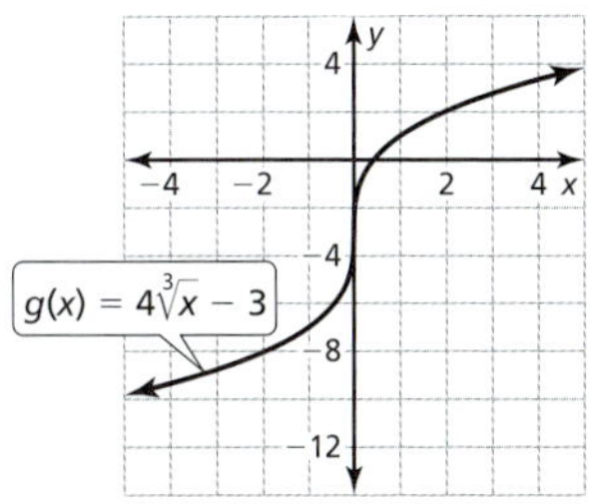

The graph of g is a vertical stretch by a factor of 4 and a translation 3 units down of the graph of f.

22.

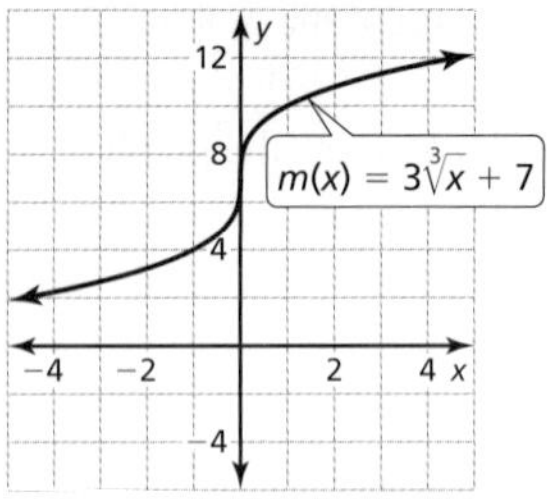

The graph of m is a vertical stretch by a factor of 3 and a translation 7 units up of the graph of f.

23.

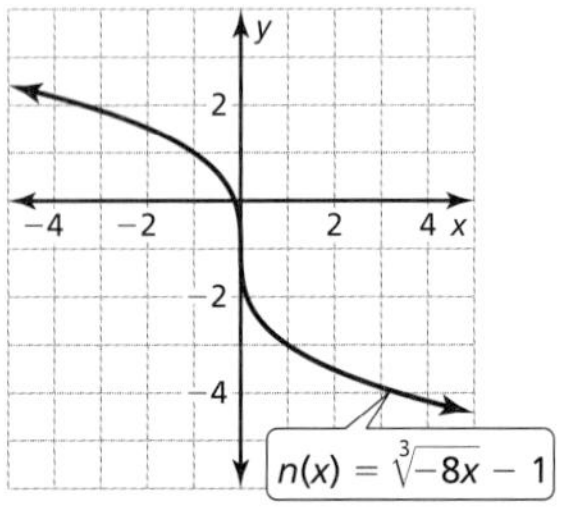

The graph of n is a horizontal shrink by a factor of $\frac{1}{8}$, a reflection in the y-axis, and a translation 1 unit down of the graph of f.

24.

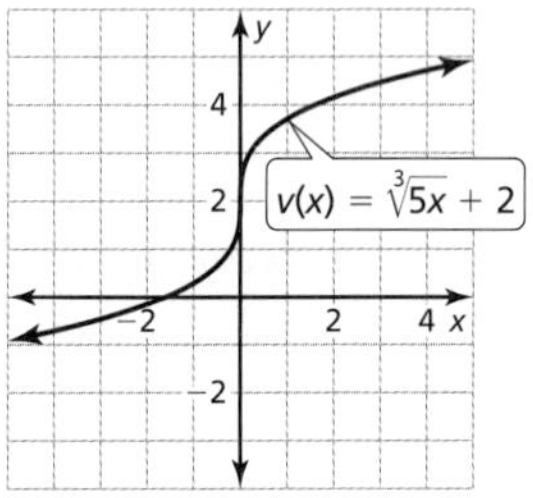

The graph of v is a horizontal shrink by a factor of $\frac{1}{5}$ and a translation 2 units up of the graph of f.

25.

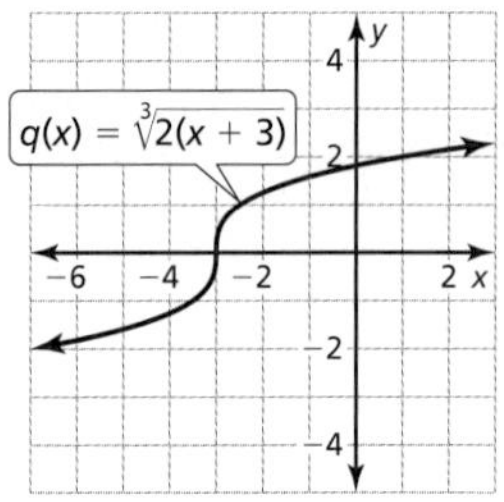

The graph of q is a horizontal shrink by a factor of $\frac{1}{2}$ and a translation 3 units left of the graph of f.

26.

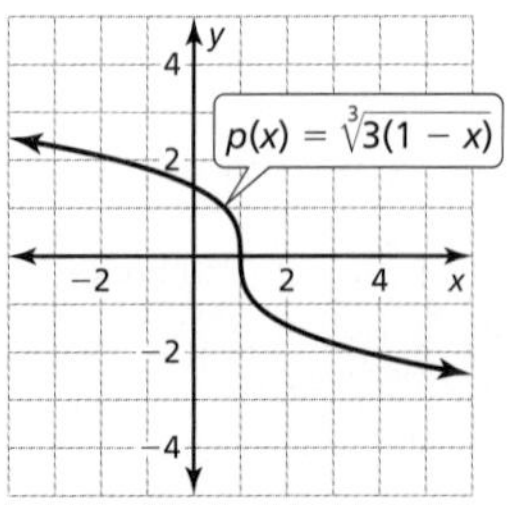

The graph of p is a horizontal shrink by a factor of $\frac{1}{3}$, a reflection in the y-axis, and a translation 1 unit right of the graph of f.

27. Translate 4 units right and 2 units up to obtain the graph of g.

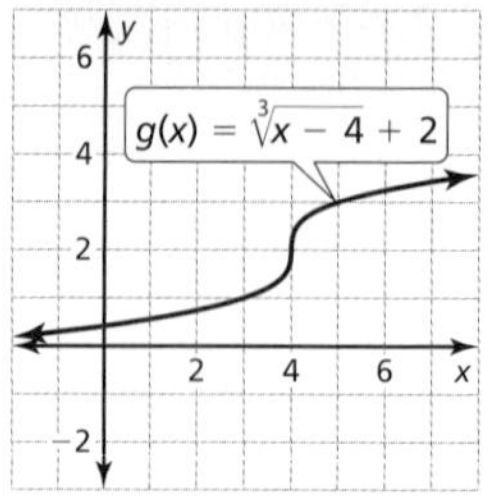

28. Translate 1 unit left and 3 units down to obtain the graph of n.

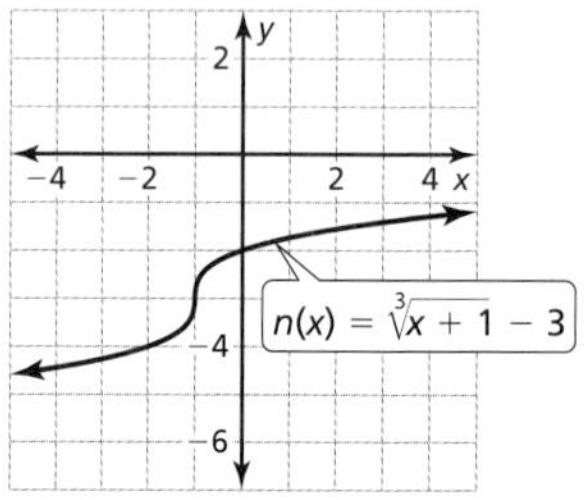

29. Translate 3 units left, stretch vertically by a factor of 5, reflect in the x-axis, and translate 2 units up to obtain the graph of j.

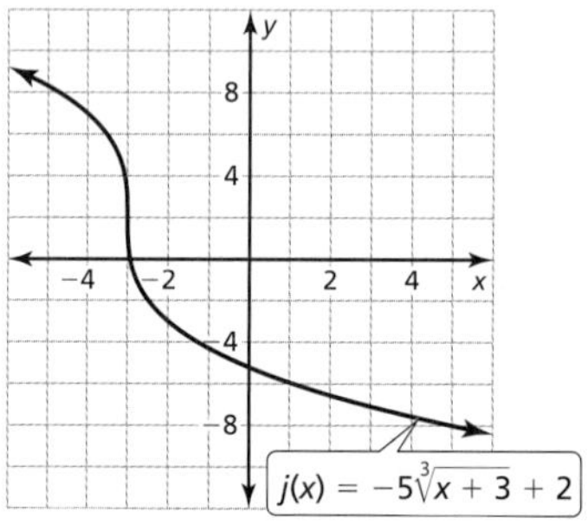

30. Translate 9 units right, stretch vertically by a factor of 6, and translate 5 units down obtain the graph of k.

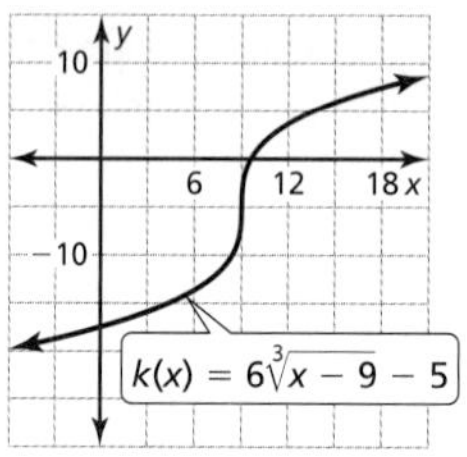

31. Translate 1 unit right, shrink vertically by a factor of $\frac{1}{3}$, and translate 7 units up to obtain the graph of v.

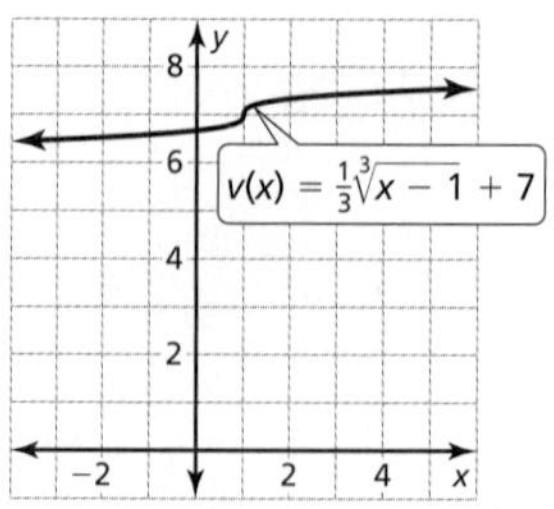

32. Translate 4 units left, stretch vertically by a factor of $\frac{3}{2}$, reflect in the x-axis, and translate 3 units down to obtain the graph of h.

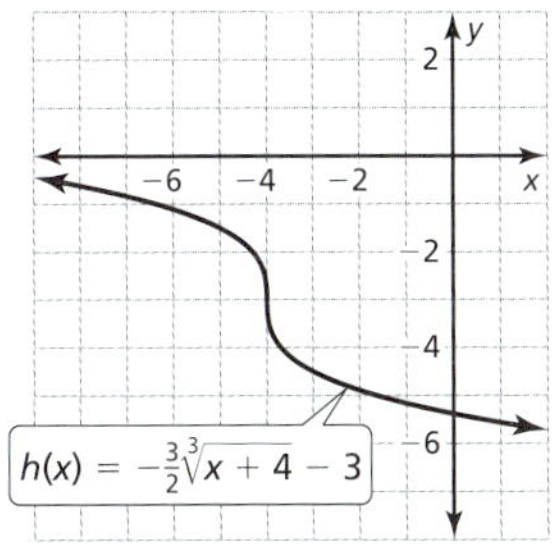

33. The graph should be a translation 3 units right, not a translation 3 units left, of the graph of $y = \sqrt[3]{x}$.

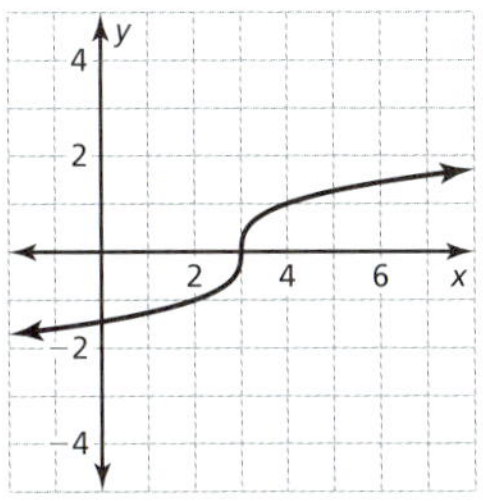

34. Values of $x \leq 0$ are not shown in the graph.

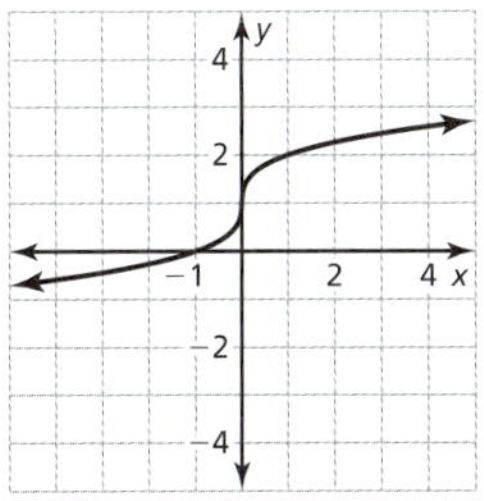

35. $\dfrac{2}{3\sqrt[3]{6}} \approx 0.37$ times greater

36. The average rate of change of g is $\frac{2}{6} \approx 0.33$ times the average rate of change of h.

37.

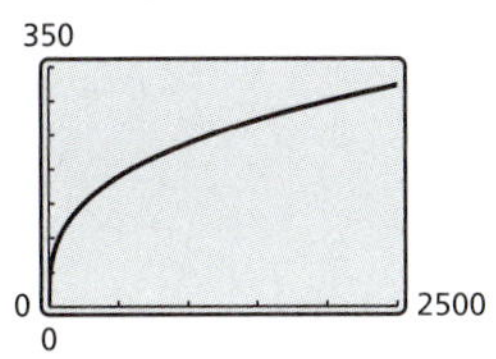

about 790 hp

38.

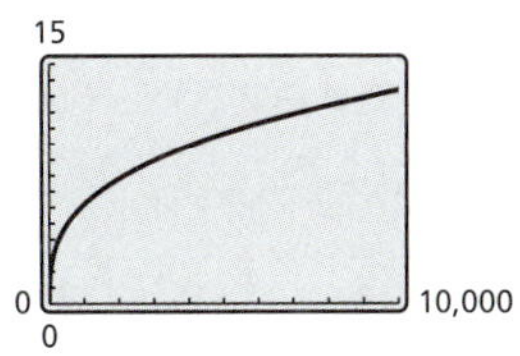

about 9203 in.^{3}

39. no; A function $f(x)$ is an odd function when $f(-x) = -f(x)$ for each x in the domain of f. *Sample answer:* A cube root function, such as $f(x) = \sqrt[3]{x} + 1$, is not an odd function because $f(-x) = \sqrt[3]{-x} + 1 = -\sqrt[3]{x} + 1$ and $-f(x) = -(\sqrt[3]{x} + 1) = -\sqrt[3]{x} - 1$. $f(-x) \neq -f(x)$.

40. **a.** $x < 0$; $x > 0$

b. f is not decreasing on any interval; f is increasing for all real numbers.

c. no; Because f is increasing for all real numbers, it does not have a maximum or a minimum value.

d. 1

41. *Sample answer:* $f(x) = -7\sqrt[3]{x - 3} + 4$

42. *Sample answer:* $f(x) = 2\sqrt[3]{x} - 1$

10.2 Maintaining Mathematical Proficiency

43. $3(x + 6)(x - 2)$ **44.** $(x - 1)(2x - 9)$

45. $(x + 3)(4x - 5)$ **46.** $x = -6, x = 6$

47. no real solutions **48.** $x = -13, x = 5$

49. $x = \frac{7}{5}, x = \frac{13}{5}$

10.2 Mini-Assessment

3.

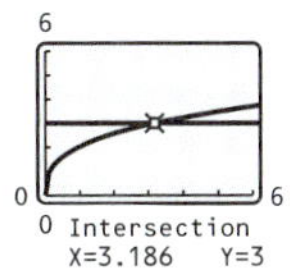

The volume is about 3.2 cubic feet.

10.1–10.2 Quiz

10. The graph of v is a translation 3 units up of the graph of f; $k = 3$

11. The graph of g is a vertical stretch by a factor of 2 and a reflection in the x-axis of the graph of f; $a = -2$

12. The graph of p is a translation 1 unit left of the graph of f; $h = -1$

13. Translate 2 units left, stretch vertically by a factor of 3, reflect in the x-axis, and translate 6 units up to obtain the graph of h.

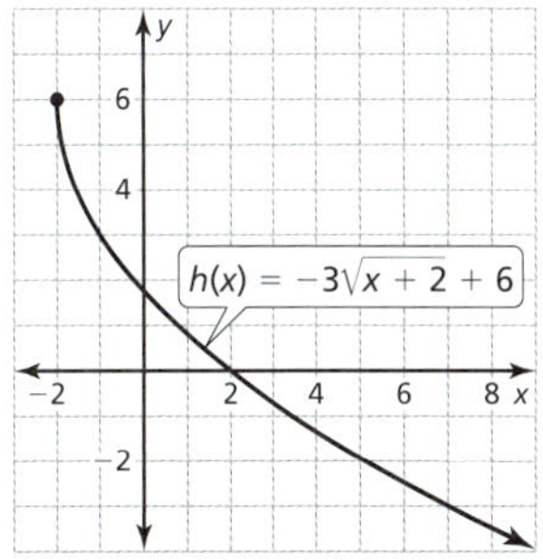

14. Shrink vertically by a factor of $\frac{1}{2}$ and translate 3 units down to obtain the graph of h.

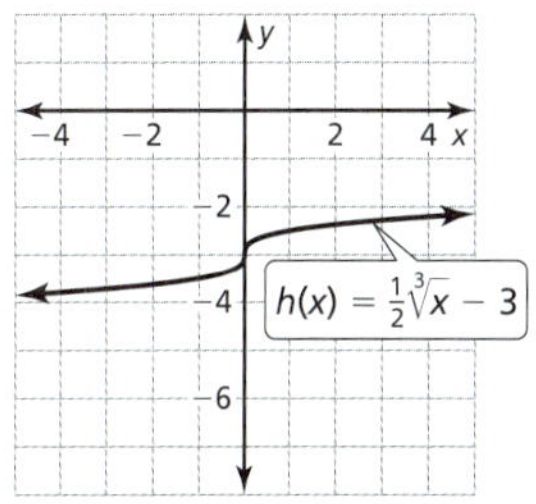

15. **a.**

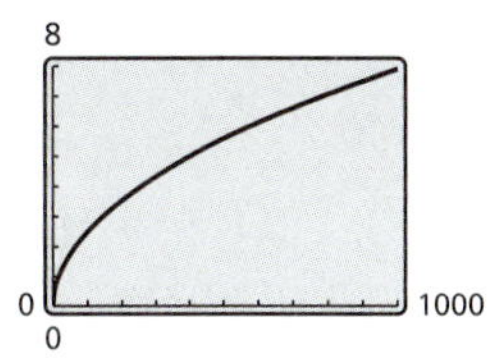

$h \geq 0, t \geq 0$

b. about 876 ft

16.

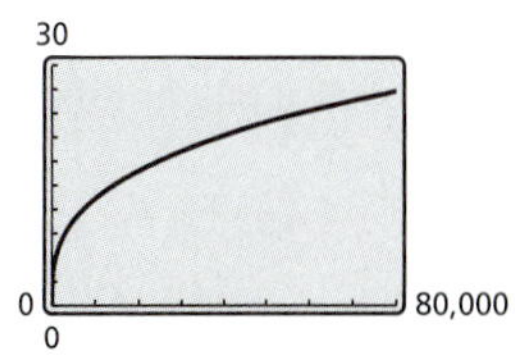

about 65,450 m^3

17. $1 \div \dfrac{\sqrt[3]{9/2}}{3} \approx 1.8$ times greater

10.3 Monitoring Progress and Modeling with Mathematics

81. no; Because $\sqrt{(2x + 5)^2}$ cannot be negative, the equation is true only for $x \geq -\frac{5}{2}$, where $2x + 5$ is nonnegative.

82. *Sample answer:* Graph $y = \sqrt[3]{x} + 1$ and $y = \sqrt{x - 3}$.

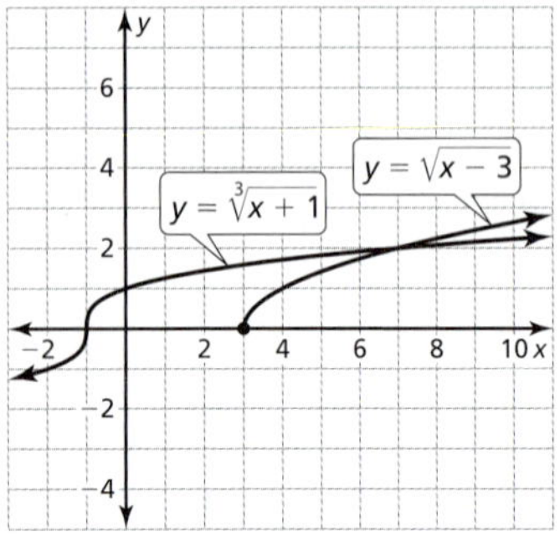

Determine the x-coordinate of the intersection point to be $x = 7$. Check your solution.

83. **a.** about 71.5 N

b. more; To keep the same ratio $\frac{T}{m}$, if m is increased, then T must be increased also.

10.3 Maintaining Mathematical Proficiency

84. $x^2 + 6x - 16$

85. $12p^2 + 11p - 5$

86. $s^3 + 5s^2 + 2s - 8$

87.

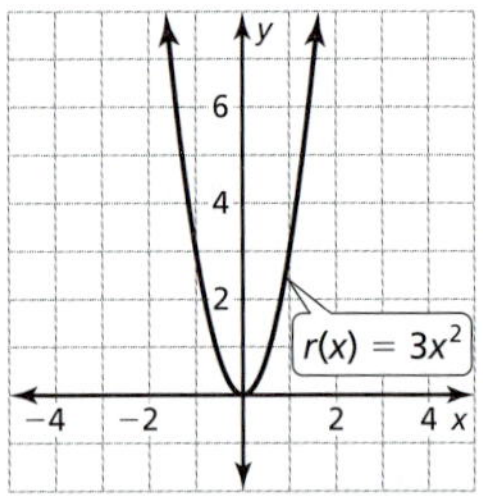

Both graphs open up and have the same vertex, (0, 0), and the same axis of symmetry, $x = 0$. The graph of r is narrower than the graph of f because the graph of r is a vertical stretch by a factor of 3 of the graph of f.

88.

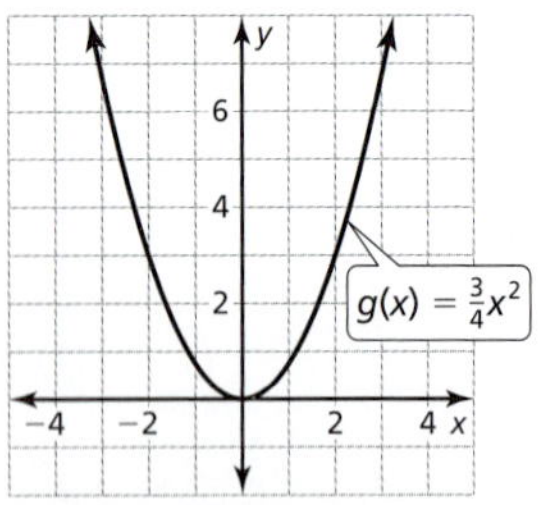

Both graphs open up and have the same vertex, (0, 0), and the same axis of symmetry, $x = 0$. The graph of g is wider than the graph of f because the graph of g is a vertical shrink by a factor of $\frac{3}{4}$ of the graph of f.

89.

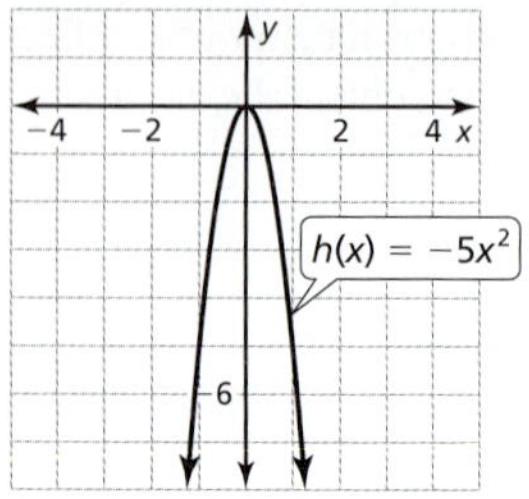

Both the graphs have the same vertex, (0, 0), and the same axis of symmetry, $x = 0$, but the graph of h opens down and is narrower than the graph of f, because the graph of h is a vertical stretch by a factor of 5 and a reflection in the x-axis of the graph of f.

10.4 Monitoring Progress and Modeling with Mathematics

11. $x = 4y + 4$; 12

12. $x = \frac{3y - 12}{2}$; -3

13. $x = \pm\frac{\sqrt{y}}{3}$; $\pm\frac{\sqrt{2}}{3}$

14. $x = \pm\sqrt{2y + 14}$; $\pm 3\sqrt{2}$

15.

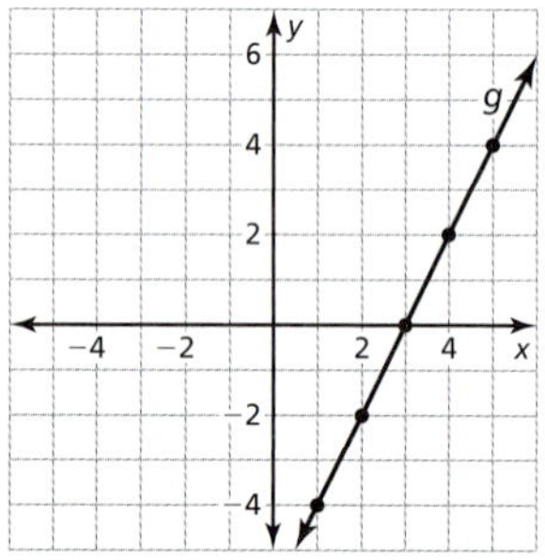

domain of inverse: all real numbers; range of inverse: all real numbers

16.

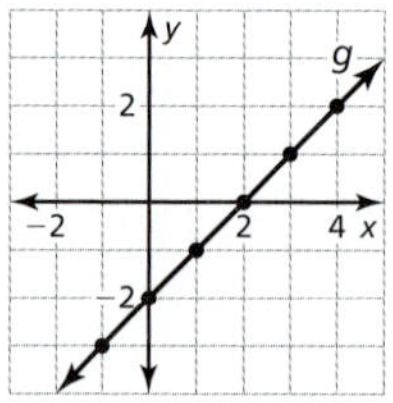

domain of inverse: all real numbers; range of inverse: all real numbers

17. $g(x) = \frac{x + 1}{4}$

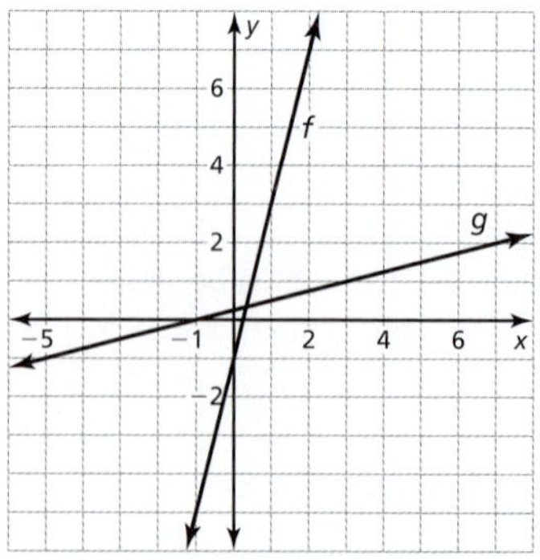

18. $g(x) = \dfrac{5 - x}{2}$

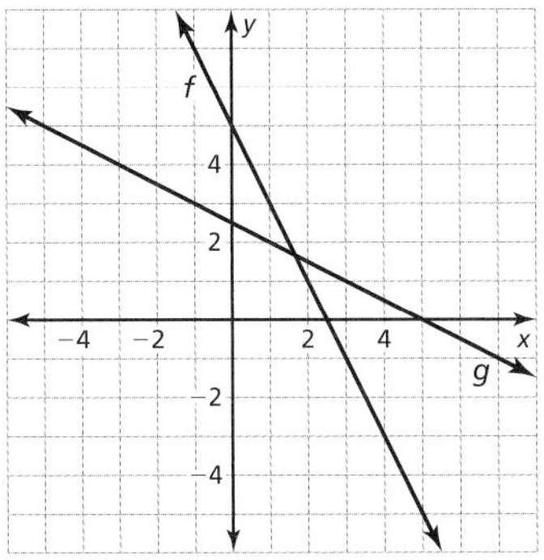

19. $g(x) = -\dfrac{x + 2}{3}$

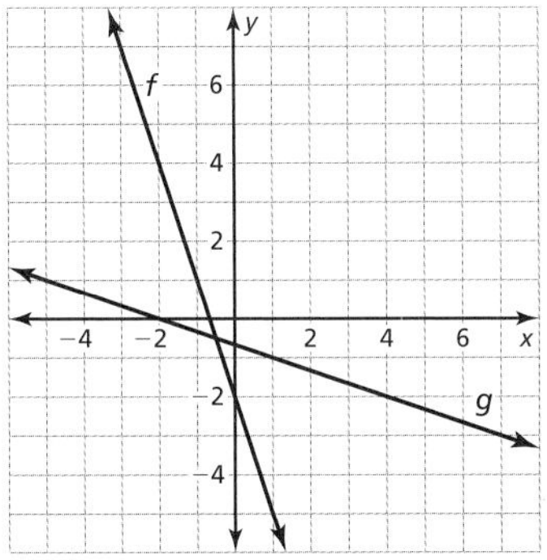

20. $g(x) = \dfrac{x - 3}{2}$

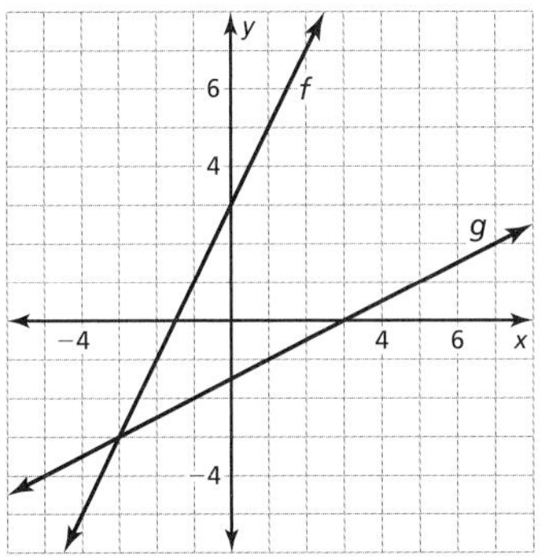

21. $g(x) = 3x - 24$

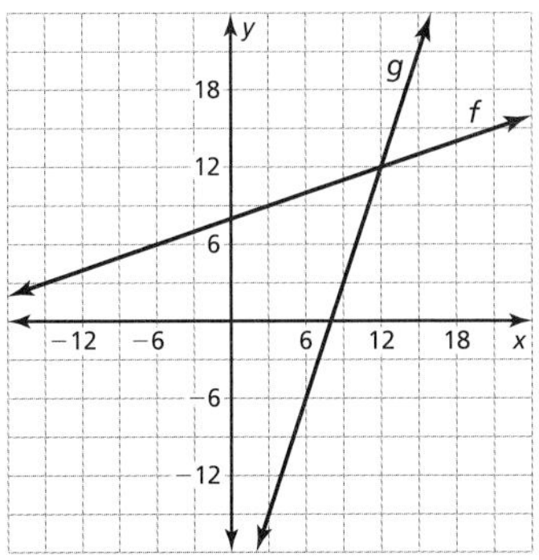

22. $g(x) = \dfrac{7 - 2x}{3}$

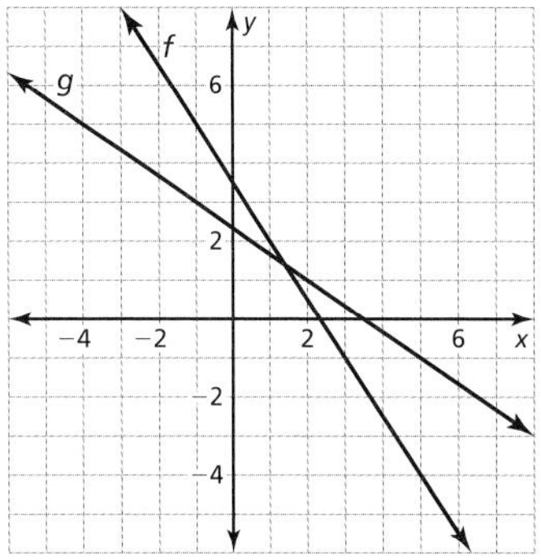

29. not a function

30. function

31. function

32. not a function

33. yes; $g(x) = x^2 - 3, x \geq 0$

34. yes; $g(x) = x^2 + 5, x \geq 0$

35. yes; $g(x) = \dfrac{x^2 + 6}{2}, x \geq 0$

36. yes; $g(x) = \dfrac{x^2 - 1}{4}, x \geq 0$

37. yes; $g(x) = \dfrac{x^2}{9} + 8, x \geq 0$

38. yes; $g(x) = \dfrac{16x^2 - 2}{5}, x \leq 0$

39. yes; $g(x) = \dfrac{(-x - 2)^2 - 5}{3}; x \leq -2$

40. yes; $g(x) = \left(\dfrac{x - 6}{2}\right)^2 + 7, x \geq 6$

41. no; $y = \pm\sqrt{\dfrac{x}{2}}$

42. no; $y = x, y \geq 0$ and $y = -x, y < 0$

43. x and y were not switched in the equation; The inverse of f is $g(x) = \dfrac{x - 5}{3}$ or $g(x) = \dfrac{x}{3} - \dfrac{5}{3}$.

44. A restriction should be placed on the domain of g; The inverse of f is $g(x) = x^2 + 3, x \geq 0$.

45. $D = 1.33899E$; about 334.75 U.S. dollars

46. $t = \frac{1}{4}\sqrt{50 - h}$; about 1.48 seconds

47. $s = \dfrac{2\sqrt{3}}{3}h$; $s \approx 18.48$ in.

48. $s = \dfrac{2(3)^{3/4}}{3}\sqrt{A}$; $s \approx 5.04$ ft

52. $g(x) = \frac{1}{2}\sqrt[3]{x} - 2$

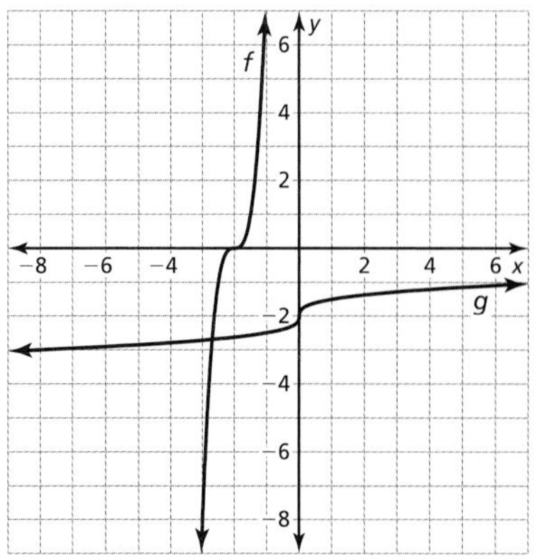

53. $g(x) = \left(\frac{x}{4}\right)^3$

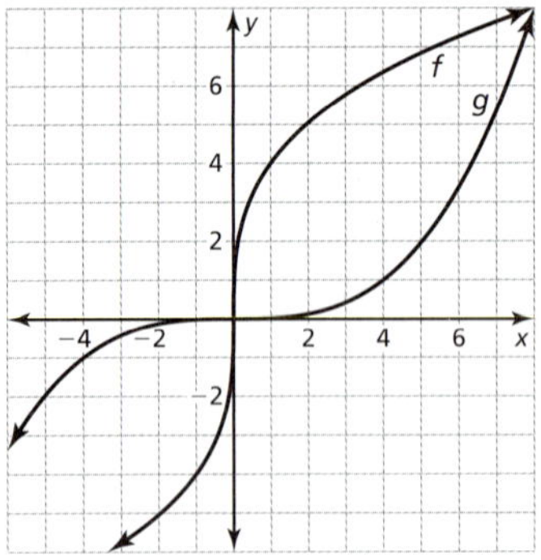

54. $g(x) = -x^3 + 1$

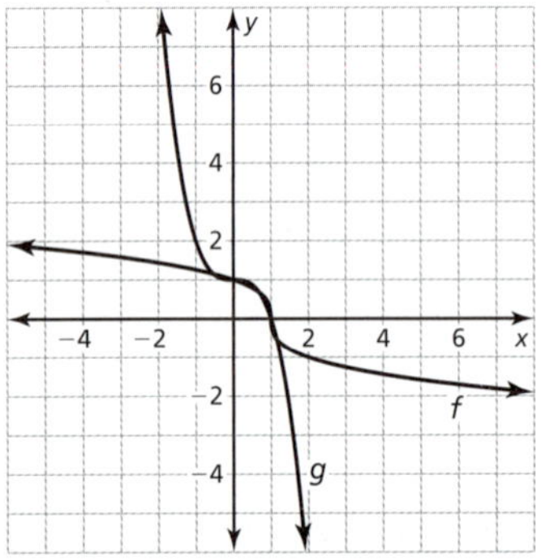

55. no; The inverse of $f(x) = 3$ is the vertical line $x = 3$, which does not pass the Horizontal Line Test and is not a function.

56. A, D; B, F; C, E

57. *Sample answer:* Restrict the domain to nonnegative values of x; new function: $x \geq 0$; $y \geq -5$; inverse function: $x \geq -5$; $y \geq 0$

58. no; The graph of an even function is symmetric about the y-axis, so by the Horizontal Line Test, the inverse of an even function with at least two values in its domain is not a function.

59. *Sample answer:* $f(x) = \frac{1}{4}x$

60. a.

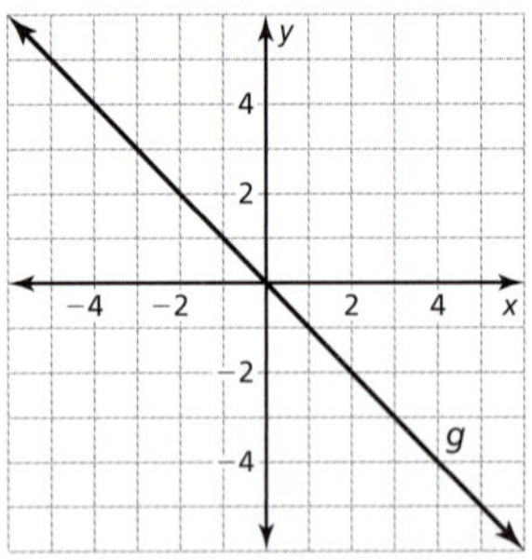

For each point $(a, -a)$ that is on the graph of $g(x) = -x$, the point $(-a, a)$ is also on the graph.

b.

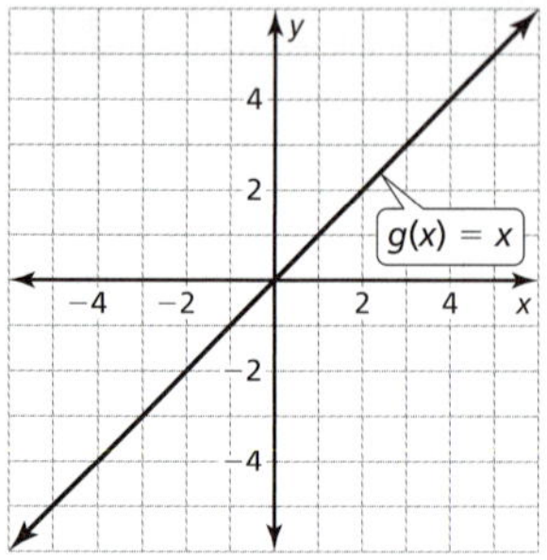

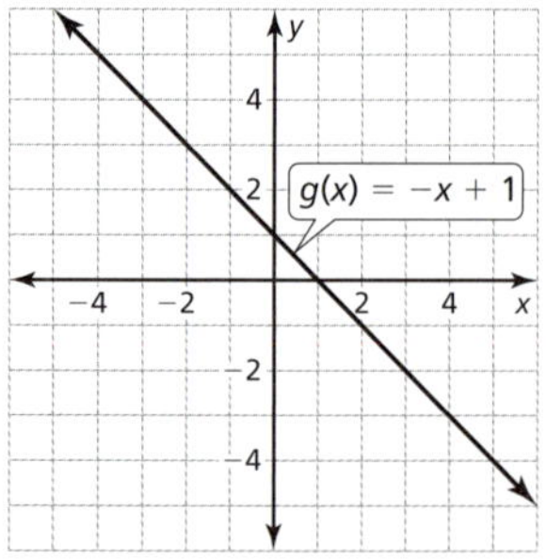

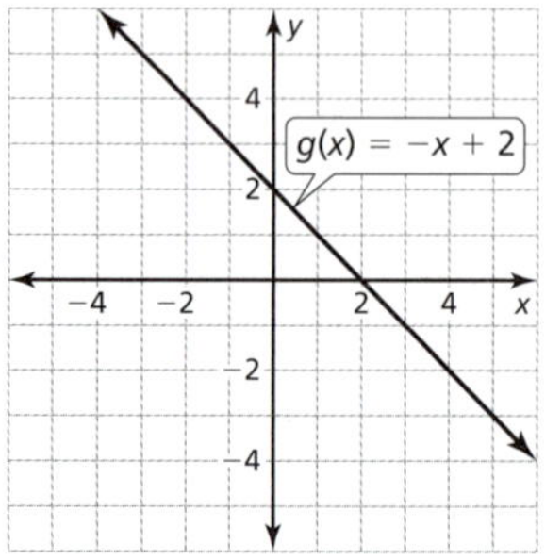

c. The linear function $f(x) = mx + b$ is its own inverse when $m = -1$ and b is any real number or when $m = 1$ and $b = 0$.

61. The inverse of $f(x) = mx + b$, where $b \neq 0$, is $g(x) = \frac{x - b}{m} = \frac{1}{m}x - \frac{b}{m}$, which is a linear function. The graph of the inverse function g has a slope of $\frac{1}{m}$ and a y-intercept of $-\frac{b}{m}$.

62. $-1 \leq x \leq 1$; $x = y^3 - 3y$

63. no; no; For any values of h, k, and a, a horizontal line intersects the graph of f more than once.

10.4 Maintaining Mathematical Proficiency

64. $-4x - 14$

65. $7y - 11$

66. $8t^2 + 8t - 1$

67. $-10d^2 + 11d - 2$

68.

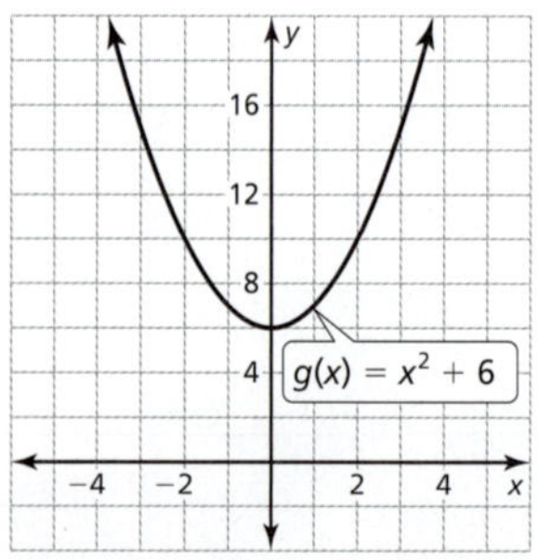

Both the graph of f and the graph of g open up and have the same axis of symmetry, $x = 0$. The vertex of the graph of g, $(0, 6)$, is above the vertex of the graph of f, $(0, 0)$, because the graph of g is a vertical translation 6 units up of the graph of f.

69. 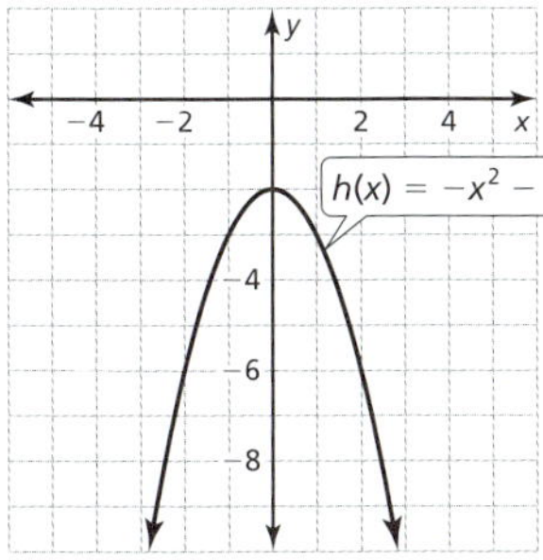

Both the graph of f and the graph of h have the same axis of symmetry, $x = 0$, but the graph of f opens up and the graph of h opens down. The vertex of the graph of f is $(0, 0)$ and the vertex of the graph of h is $(0, -2)$. So, the graph of h is a reflection in the x-axis and a vertical translation 2 units down of the graph of f.

70. 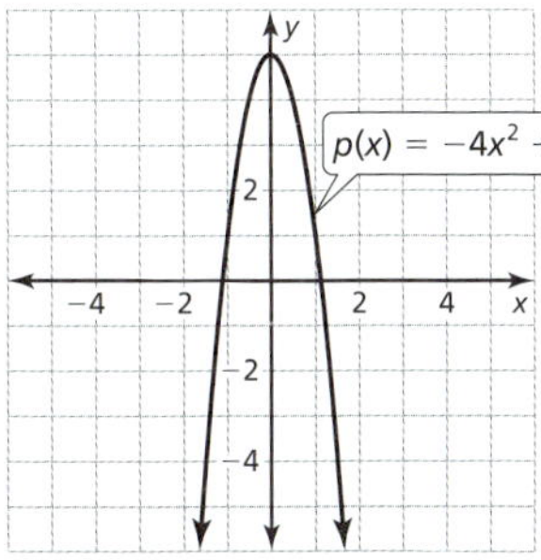

Both the graph of f and the graph of p have the same axis of symmetry, $x = 0$, but the graph of p is narrower than the graph of f. The graph of p opens down but the graph of f opens up, and the vertex of the graph of p is $(0, 5)$ but the vertex of the graph of f is $(0, 0)$. The graph of p is a vertical stretch by a factor of 4, a reflection in the x-axis, and a vertical translation 5 units up of the graph of f.

71. 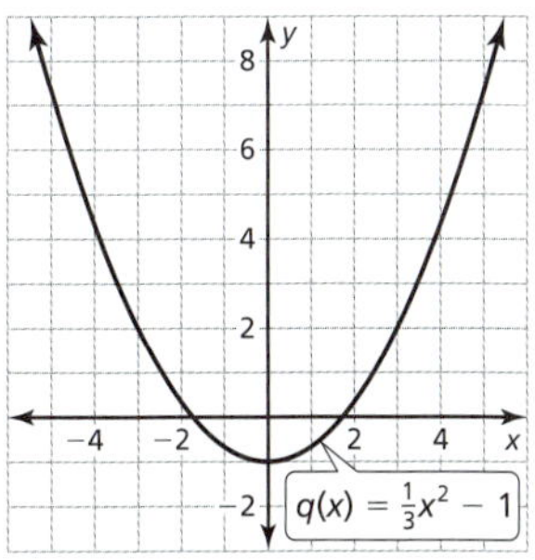

Both the graph of f and the graph of q open up and have the same axis of symmetry, $x = 0$, but the graph of q is wider than the graph of f. Also, the vertex $(0, -1)$ of the graph of q is below the vertex $(0, 0)$ of the graph of f. The graph of q is a vertical shrink by a factor of $\frac{1}{3}$ and a vertical translation 1 unit down of the graph of f.

Chapter 10 Test

12. The solution $x = 2$ is extraneous, because it gives a value of $2 - 5 = -3$ on the left side, which cannot equal the radical expression on the right side of the equation.

13. Translate 1 unit right, stretch vertically by a factor of 4, and translate 5 units up to obtain the graph of h.

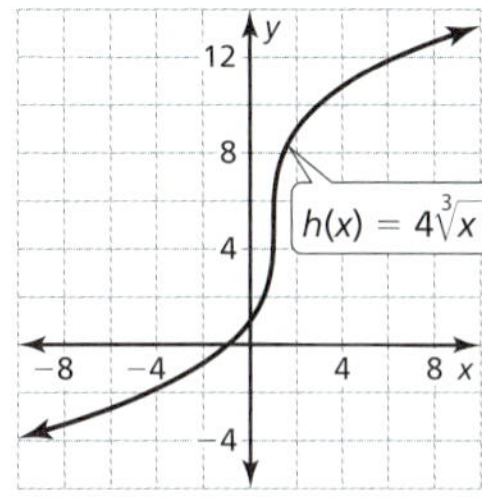

14. Translate 7 units left, reflect in the x-axis, and translate 2 units down to obtain the graph of w.

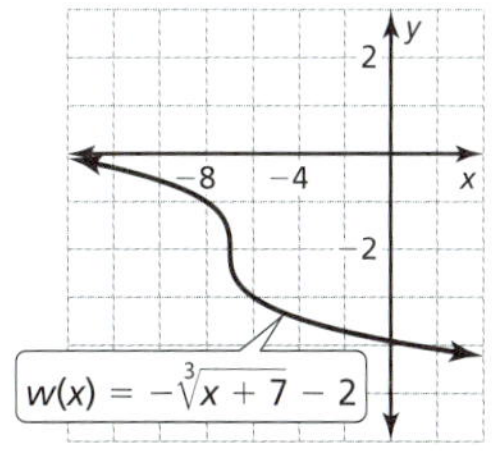

15. a. 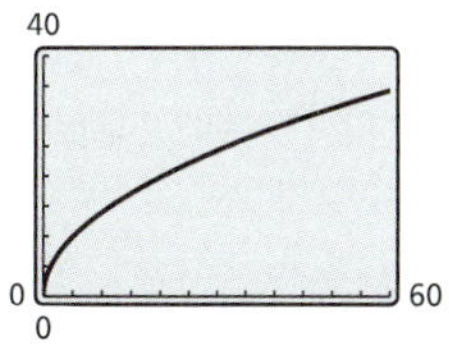

domain: $h \geq 0$; range: $v \geq 0$

b. at least 40 m

c. decreases

16. a. 16°C

b. 5.5 sec

17. *Sample answer:* You can exclude values of x that are less than 3.

18. *Sample answer:* $f(x) = \sqrt{-x} + 9$

Chapter 10 Standards Assessment

7.

No zeros	$f(x) = 3x^2 + 4x + 2$	$f(x) = -6x^2 - 5$
One zero	$f(x) = 4x^2 - 8x + 4$	$f(x) = 7x^2$
Two zeros	$f(x) = -x^2 + 2x$	$f(x) = x^2 - 3x - 21$

Chapter 11

11.1 Explorations

2. 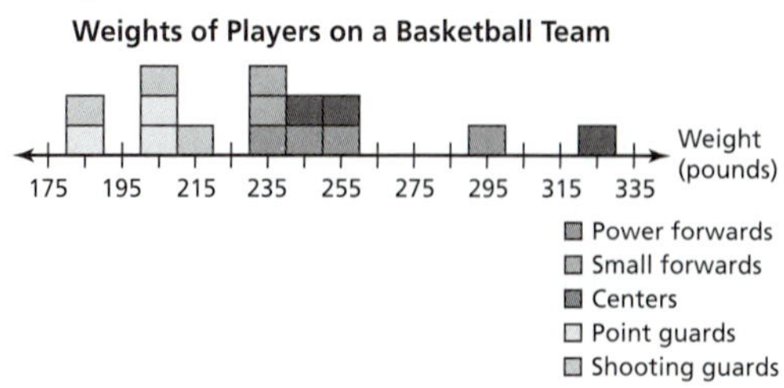

yes; Guards tend to be lighter, forwards tend to be heavier.

11.1 Monitoring Progress and Modeling with Mathematics

39. a. mean: 19.37, median: 19, mode: 18

b. 37, 37; They increase the mean and median but do not affect the mode.

c. College Student Ages

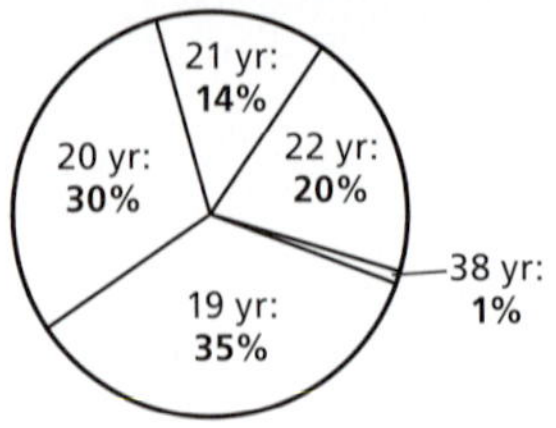

mean: 20.37, median: 20, mode: 19

11.2 Lesson Monitoring Progress

3. 25% of the ages are between 18 and 23, 50% of the ages are between 23 and 29, and 25% of the ages are between 29 and 30.

11.2 Monitoring Progress and Modeling with Mathematics

16. a. 17, 7; The number of runs varies by no more than 17 and the middle half of the number of runs vary by no more than 7 runs.

b. 25% of the number of runs scored are between 0 and 2, 50% of the number of runs scored are between 2 and 9, and 25% of the number of runs scored are between 9 and 17

c. Q2 and Q3; *Sample answer:* Q3 − Q2 = 5 and Q2 − Q1 = 2

17. a. Sales Rep A: symmetric; Sales Rep B: skewed right

b. Sales Rep B; The range and interquartile range are greater.

c. Sales Rep B; The least value is 4.

18. When a distribution is skewed left, most of the data are not on the left side of the plot; Most of the data are on the right side of the plot.

19. range: 36, interquartile range: 12; The range is greater.

20. a. yes; 11 is the greatest value

b. no; *Sample answer:* The data could be 2, 4, 5, 5, 8, 12.

c. yes; The right whisker is longer and most of the data are on the left side of the plot.

d. no; The mean is not always the same as the median.

21. a. Both distributions are skewed right.

b. Brand A: 3.5, Brand B: 2.5

c. The interquartile range of Brand A is greater.

d. Brand A; The range is greater.

e. Brand A; 75% of the battery lives are greater than 3.5 hours.

22. *Sample answer:* 3, 4, 4, 10, 11, 14, 16, 18; Q1 = 4, median = 10.5, Q3 = 15

23. yes; *Sample answer:*

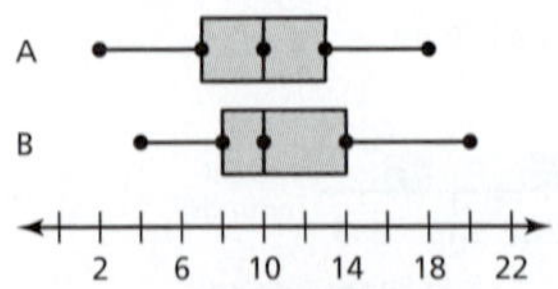

11.2 Maintaining Mathematical Proficiency

24.

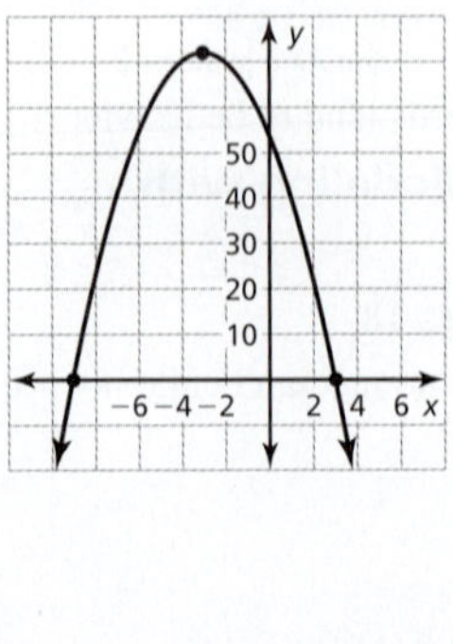

25.

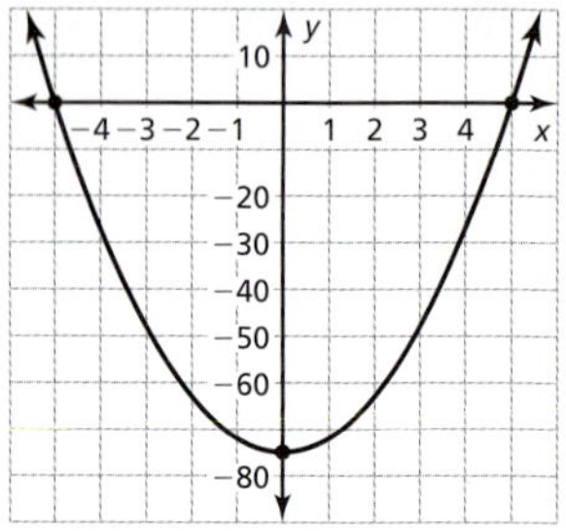

26.

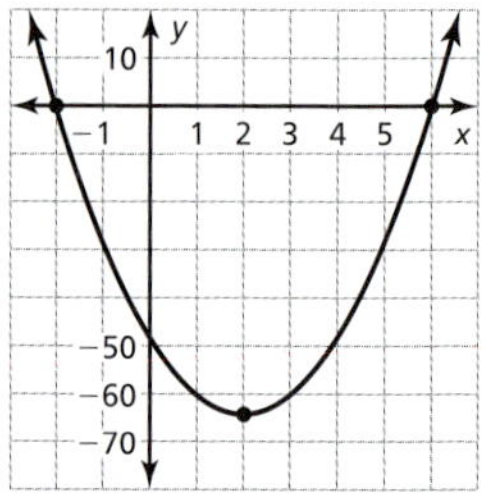

27.

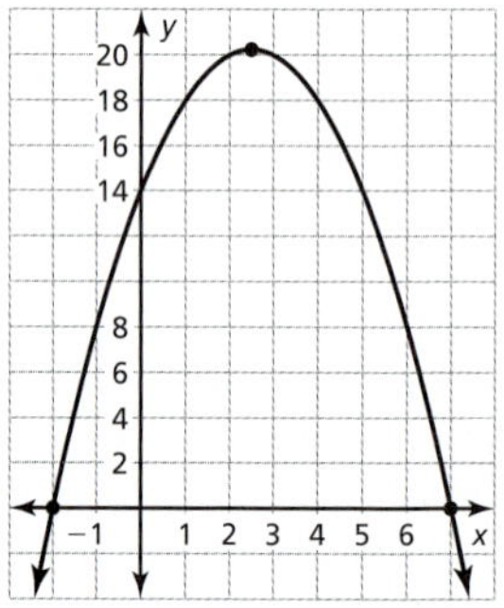

11.3 Lesson Monitoring Progress

1.

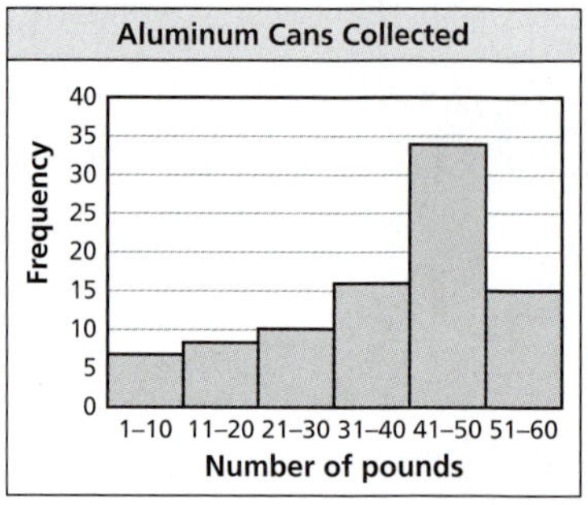

skewed left

2. a.

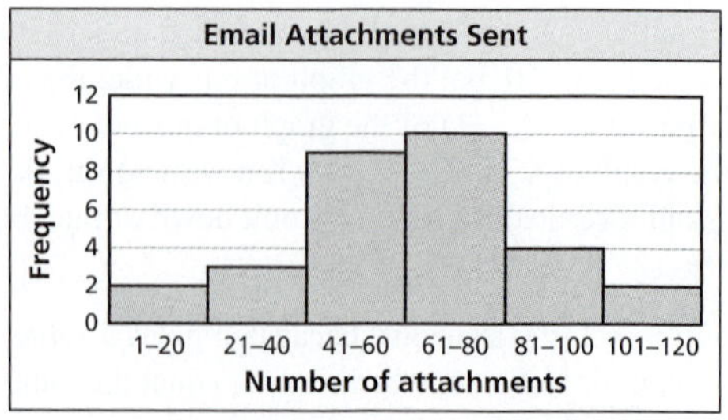

b. mean, standard deviation; The distribution is symmetric.

11.3 Monitoring Progress and Modeling with Mathematics

3.

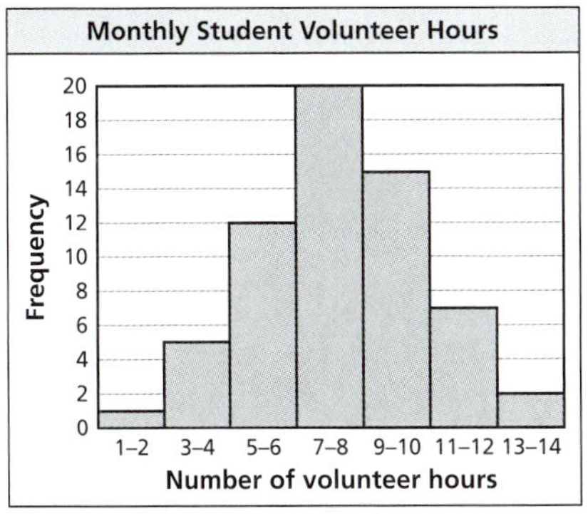

symmetric

4.

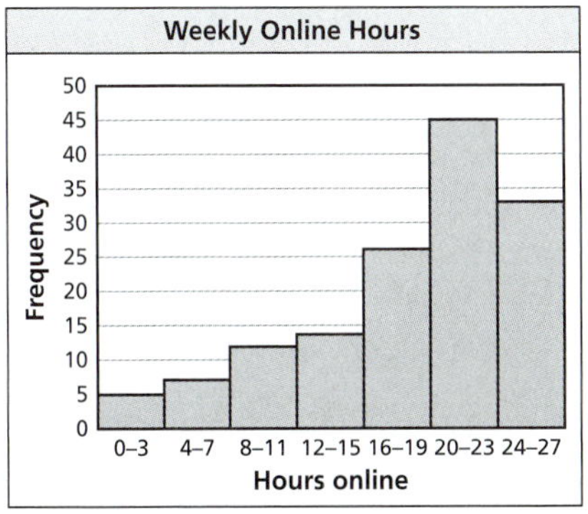

skewed left

16.

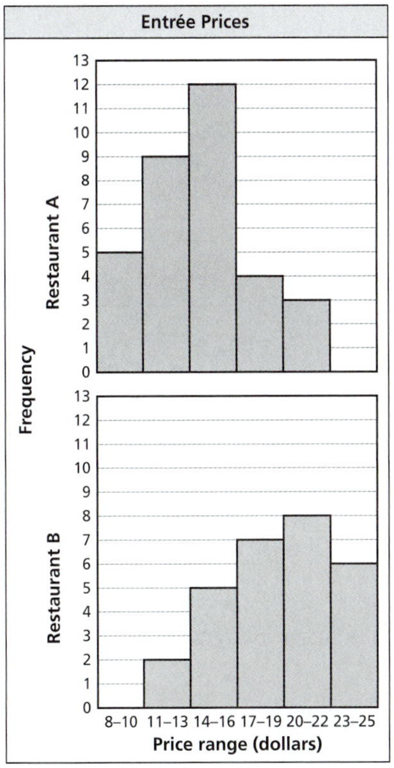

Entrée prices for restaurant B are typically higher, and tend to vary more.

19. **a.**

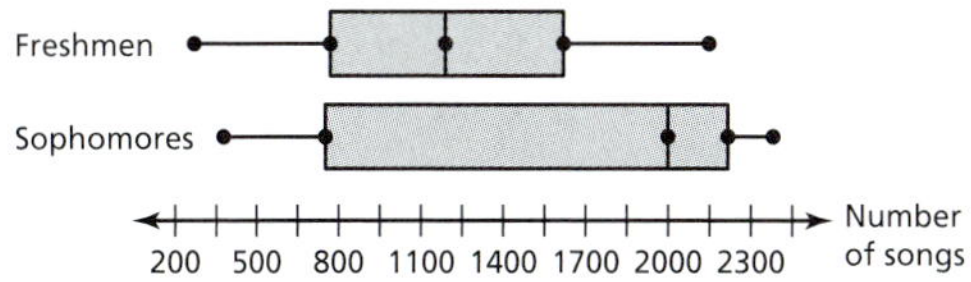

The distribution of data for freshmen is symmetric. The distribution of data for sophomores is skewed left.

b. The number of songs downloaded by freshmen tends to be less than the number of songs downloaded by sophomores.

c. about 31 freshmen

d. about 95 freshmen

20. **a.** The number of songs downloaded by this group of freshmen still tends to be less than the number of songs downloaded by sophomores.

b. The distribution is skewed left.

21. symmetric; yes; The measures of center and variation double.

22. **a.** C

b. A

c. B

23. **a.**

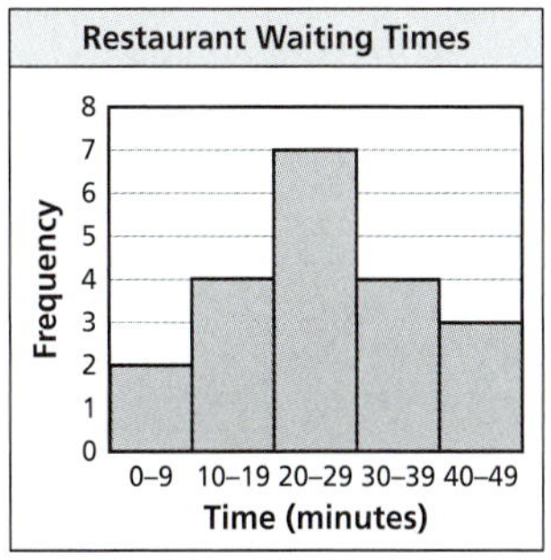

approximately symmetric

b.

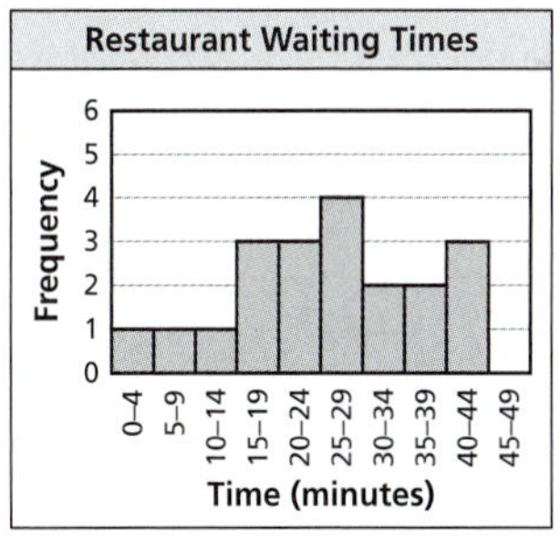

The distribution becomes skewed left.

c. part (b); More intervals show the spread of the data better.

24. *Sample answer:* age at which people receive diplomas or degrees

11.3 Maintaining Mathematical Proficiency

25. $x \geq -6$

26. $x \geq 0$

27. $x \geq 7$

11.3 Mini-Assessment

1. a.

3. Store B typically sells more CDs than Store A, and the number of CDs sold each hour tends to differ more at Store B.

11.4 Lesson Monitoring Progress

3.

		Summer Job		
		Yes	No	Total
Gender	Male	0.43	0.14	0.57
	Female	0.34	0.09	0.43
	Total	0.77	0.23	1

about 23%

4.

		Major in Medical Field	
		Yes	No
Class	Junior	0.36	0.64
	Senior	0.30	0.70

about 30%

11.4 Monitoring Progress and Modeling with Mathematics

15.

		Spanish Class		
		Yes	No	Total
French Class	Yes	45	54	99
	No	64	82	146
	Total	109	136	245

16.

		Condition		
		New	Used	Total
Origin	Foreign	36	28	64
	Domestic	19	15	34
	Total	55	43	98

17.

		Exercise Preference		
		Aerobic	Anaerobic	Total
Gender	Male	0.25	0.30	0.55
	Female	0.27	0.18	0.45
	Total	0.52	0.48	1

18.

		Meat		
		Turkey	Ham	Total
Bread	White	0.34	0.11	0.45
	Wheat	0.24	0.31	0.55
	Total	0.58	0.42	1

24.

		Type		
		Tiger	Hawk	Dragon
Style	Realistic	0.54	0.80	0.68
	Cartoon	0.46	0.20	0.32

about 20%

25. yes; About 71% of students who have a car and about 92% of students who do not have a car live on campus.

26. no; About 71% of students who participate in a sport and about 69% of students who do not participate in a sport watch sports on TV.

27. yes; About 35% of adults ages 21–30 and about 14% of adults ages 61–70 participate in recreational skiing.

28. no; Percentages for each type of degree are similar between genders.

29. Venn diagrams can be used to display the information in a two-way table that has two rows and two columns.

30. a.

		Genre		
		Comedy	Action	Horror
Gender	Male	30	50	20
	Female	40	30	20

b. *Sample answer:* the graph; It displays the data visually.

31. a.

		Seat Location		
		Main Floor	Balcony	Total
Ticket Type	Adult	$2x + y$	$x + 2y$	$3x + 3y$
	Child	$x - 40$	$3x - y - 80$	$4x - y - 120$
	Total	800	1009	1809

b. $x = 249$; $y = 93$

c. about 57%

d. about 73%

32. "One-way tables" display data collected from one source that belong to one category. "Two-way tables" display data collected from one source that belong to two different categories; yes; *Sample answer:*

	Gender and Socio-Economic Status					
	Male			Female		
Type of Program	Low	Middle	High	Low	Middle	High
General	7	10	4	9	10	5
Academic	4	22	21	15	22	21
Vocation	4	15	4	8	16	3

11.4 Maintaining Mathematical Proficiency

33. exponential **34.** quadratic

11.4 Mini-Assessment

1. 110 students have a brother, 182 students have a sister, 140 students do not have a brother, and 68 students do not have a sister.

11.5 Explorations

1. a. *Sample answer:*

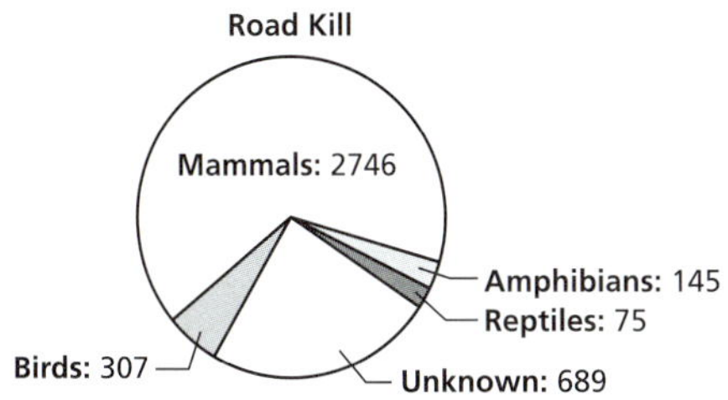

The circle graph shows the types of animals as parts of a whole.

b. *Sample answer:*

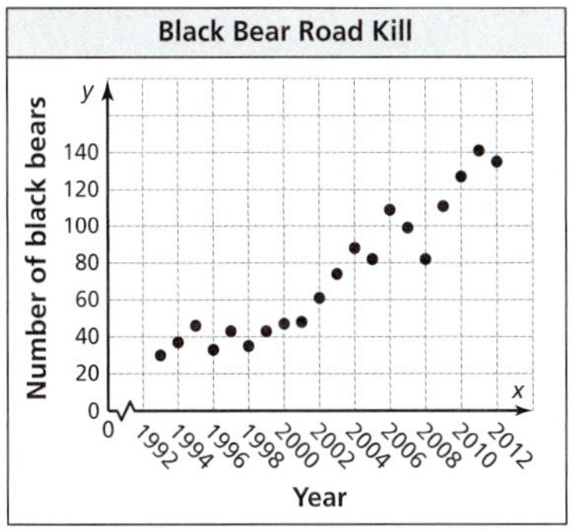

The scatter plot shows the relationship between the data.

c. *Sample answer:*

Raccoon Road Kill Weights

Stem	Leaf
9	4 5
10	
11	0
12	4 9
13	4 6 9
14	0 5 8 8
15	2 7
16	8
17	0 2 3 5
18	5 5 6 7
19	0 1 4
20	4
21	3 5 5 5
22	
23	
24	
25	4

Key: 9 | 4 = 9.4 lb

The stem-and-leaf plot shows how the raccoon weights are distributed.

d. *Sample answer:*

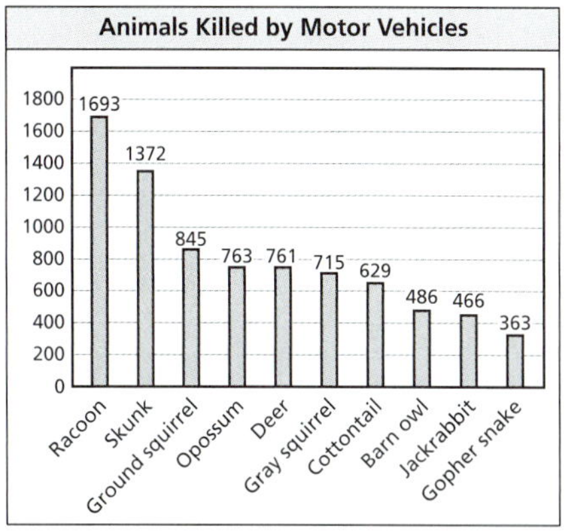

The bar graph shows the data in each specific category.

11.5 Lesson Monitoring Progress

6. *Sample answer:*

Interstate A		Interstate B
9	5	
8 8 8 8 8 7 5 5 5 2	6	5 6 7 9 9
7 5 5 5 2 1 0	7	0 0 1 1 2 3 5 7 8 9 9
2 0	8	0 1 4 4

Key: 2 | 6 | 5 = 62 mi/h on Interstate A, 65 mi/h on Interstate B

7. *Sample answer:*

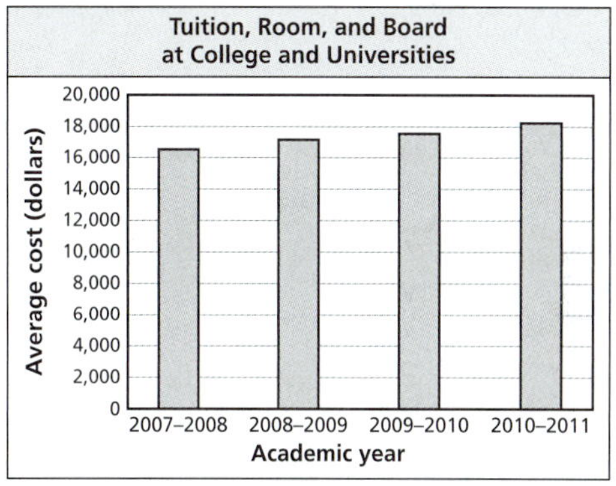

11.5 Monitoring Progress and Modeling with Mathematics

14. *Sample answer:*

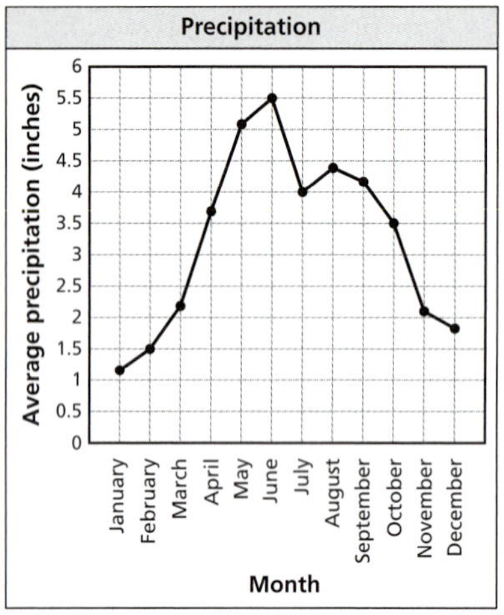

A line graph shows how the data change over time.

15. *Sample answer:*

Grades (out of 100) on a Test

Stem	Leaf
5	2 3 9
6	2 3
7	4 5 7
8	0 1 3 4 5 7 8 9
9	5 6 7
10	0

Key: 5 | 3 = 53 out of 100

A stem-and-leaf plot shows how the grades are distributed.

16. *Sample answer:*

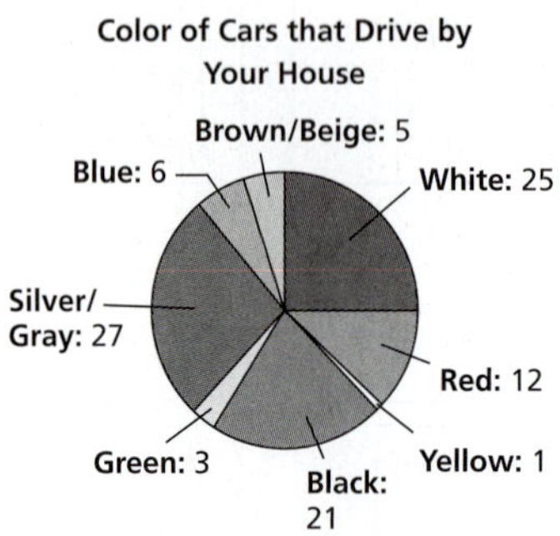

A circle graph shows the data as parts of a whole.

17. *Sample answer:*

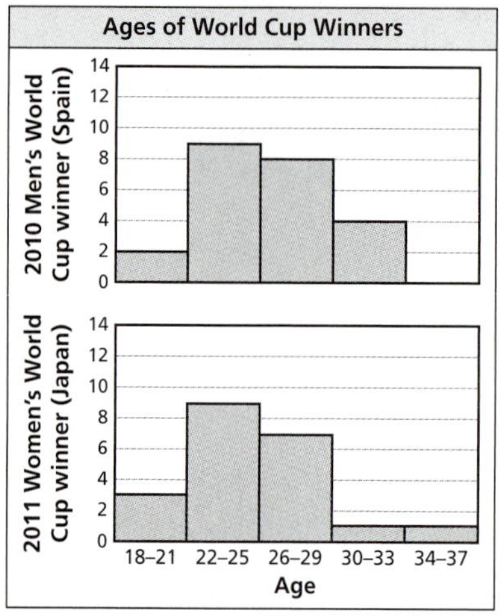

18. *Sample answer:*

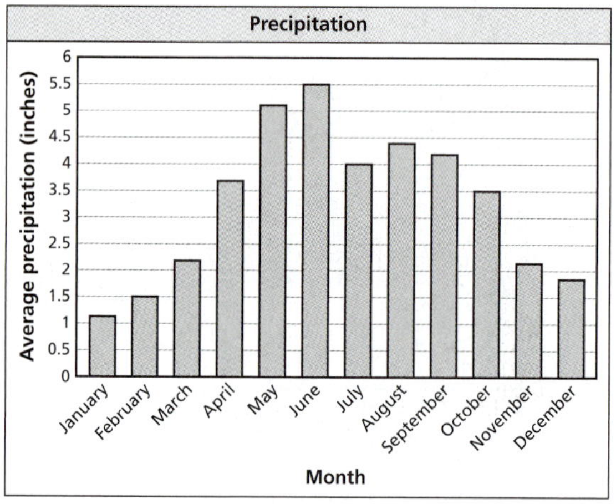

19. *Sample answer:*

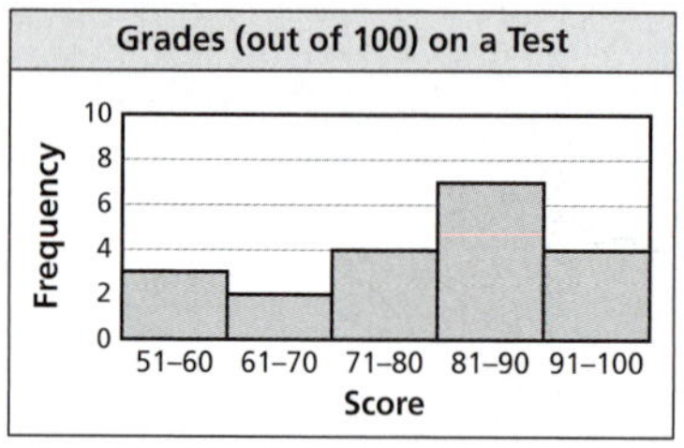

20. *Sample answer:*

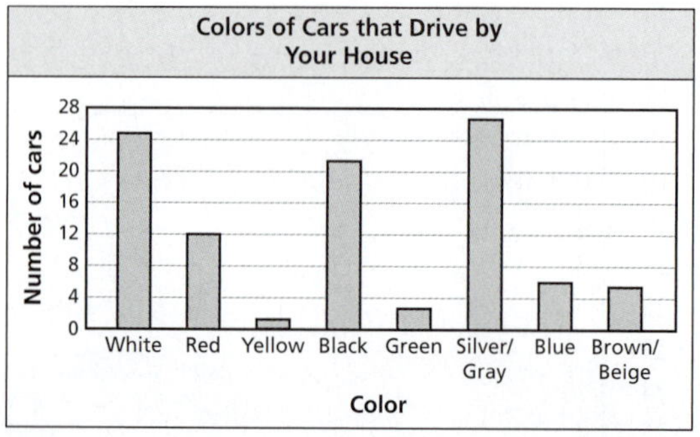

21. The scale on the vertical axis has very small increments; Someone might believe that the annual sales more than tripled from 2010 to 2013.

22. The interval for the third bar is greater than the other two bars and the vertical scale does not have a break. Someone might believe that the frequency for each 30 minute interval is increasing.

23. The increments on the scale are not equal; Someone might believe that the temperatures are evenly distributed.

24. The increments on the vertical axis are not equal; Someone might believe that the decay rate is linear.

25. *Sample answer:*

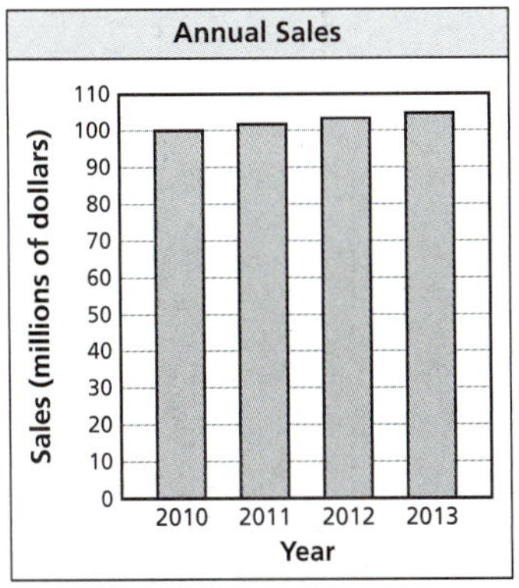

26. *Sample answer:*

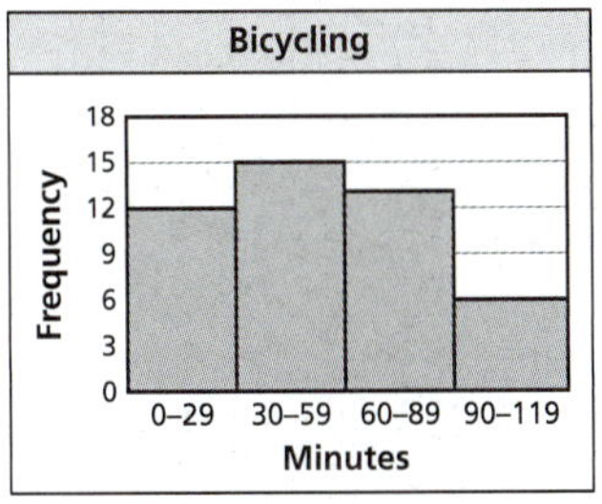

27. Classmate B; The data are quantitative, not qualitative.

28. units on the vertical axis; *Sample answer:* The units may be hundreds of stores opened.

29. **a.** There are 120 responses, so the regions do not represent numbers of students.

b. *Sample answer:* bar graph; A bar graph shows data in categories.

30. *Sample answer:* Doughnut display

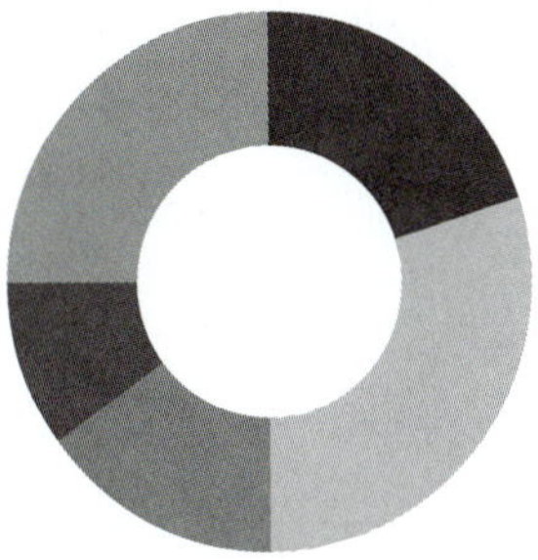

31. *Sample answer:* dot plot

11.5 Maintaining Mathematical Proficiency

32. no; -5 has two outputs.

33. yes; Every input has exactly one output.

11.5 Mini-Assessment

3. The scale on the vertical axis starts at 800 and does not have a break. Someone might believe that sales more than doubled from March to June, when they actually increased by only about 700 games.

Chapter 11 Review

12. **a.**

		Food Court		
		Like	Dislike	Total
Shoppers	Adults	0.10	0.40	0.50
	Teenagers	0.48	0.02	0.50
	Total	0.58	0.42	1

b.

		Food Court	
		Like	Dislike
Shoppers	Adults	0.18	0.95
	Teenagers	0.82	0.05

13. *Sample answer:*

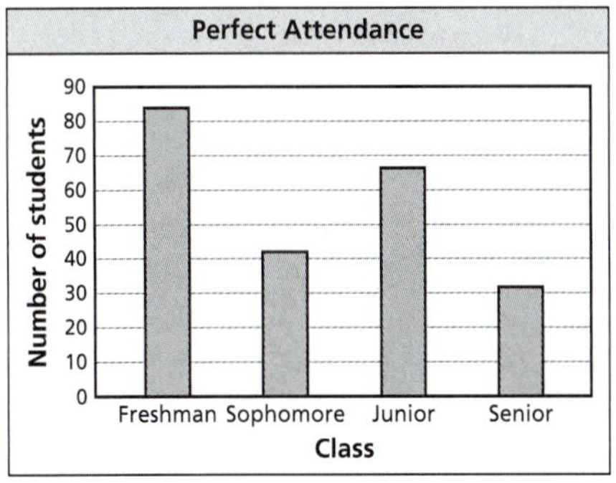

A bar graph shows data in specific categories.

Chapter 11 Test

8. **a.**

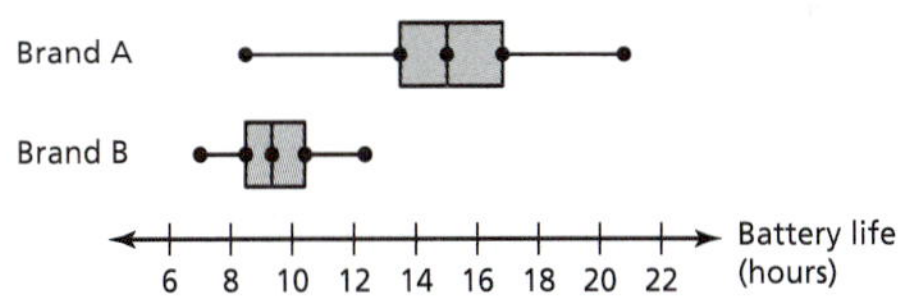

b. Brand A: approximately symmetric; Brand B: skewed right

c. Brand A; The range and interquartile range of the Brand A data are greater.

d. The battery life for brand A is longer and varies more.

9. *Sample answer:*

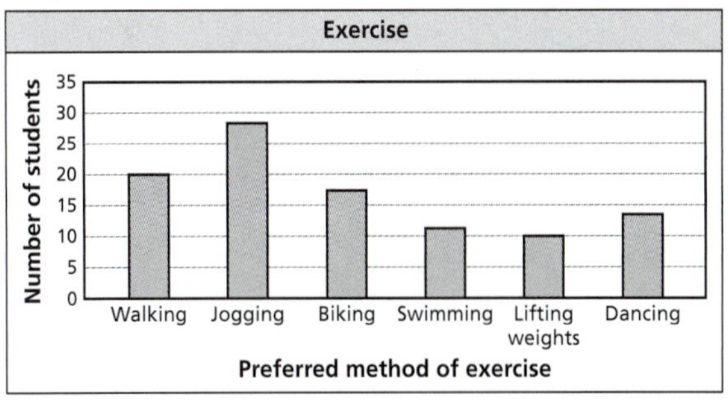

A bar graph shows data in specific categories.

10. a.

		Attending Field Trip		
		Yes	No	Total
Gender	Male	92	29	121
	Female	119	31	150
	Total	211	60	271

211 students are attending the field trip, 60 students are not attending the field trip, 121 males responded, 150 females responded, 271 students were surveyed.

b. about 79.3%

English-Spanish Glossary

English	Spanish
A	
absolute deviation *(p. 90)* The absolute value of the difference of a number x and a given value	**desviación absoluta** *(p. 90)* El valor absoluto de la diferencia de un número x y un valor dado
absolute value equation *(p. 28)* An equation that contains an absolute value expression	**ecuación de valor absoluto** *(p. 28)* Una ecuación que contiene una expresión de valor absoluto
absolute value function *(p. 156)* A function that contains an absolute value expression	**función de valor absoluto** *(p. 156)* Una función que contiene una expresión de valor absoluto
absolute value inequality *(p. 88)* An inequality that contains an absolute value expression	**desigualdad de valor absoluto** *(p. 88)* Una desigualdad que contiene una expresión de valor absoluto
arithmetic sequence *(p. 210)* An ordered list of numbers in which the difference between each pair of consecutive terms is the same	**secuencia aritmética** *(p. 210)* Una lista ordenada en donde la diferencia entre cada par de términos consecutivos es la misma
average rate of change *(p. 462)* The slope of the line through $(a, f(a))$ and $(b, f(b))$ of a function $y = f(x)$ between $x = a$ and $x = b$	**tasa de variación media** *(p. 462)* La pendiente de la línea a través de $(a, f(a))$ y $(b, f(b))$ de una función $y = f(x)$ entre $x = a$ y $x = b$
axis of symmetry *(p. 420)* The vertical line that divides a parabola into two symmetric parts	**eje de simetría** *(p. 420)* La línea vertical que divide una parábola en dos partes simétricas
B	
binomial *(p. 359)* A polynomial with two terms	**binomio** *(p. 359)* Un polinomio con dos términos
box-and-whisker plot *(p. 594)* A graph that shows the variability of a data set along a number line using the least value, the greatest value, and the quartiles of the data	**diagrama de cajas y bigotes** *(p. 594)* Un gráfico que muestra la variabilidad de un conjunto de datos a lo largo de una línea de números usando el valor menor, el valor mayor y los cuartiles de los datos
C	
categorical data *(p. 618)* Data that consist of labels or nonnumerical entries that can be separated into different categories	**datos categóricos** *(p. 618)* Datos que consisten en etiquetas o valores no numéricos que pueden separarse en categorías distintas
causation *(p. 205)* When a change in one variable causes a change in another variable	**causalidad** *(p. 205)* Cuando un cambio en una variable causa un cambio en otra variable
closed *(p. 360)* When an operation performed on any two numbers in the set results in a number that is also in the set	**operación interna** *(p. 360)* Cuando una operación efectuada en dos números cualesquiera del conjunto da como resultado un número que también está en el conjunto
common difference *(p. 210)* The difference between each pair of consecutive terms in an arithmetic sequence	**diferencia común** *(p. 210)* La diferencia entre cada par de términos consecutivos en una secuencia aritmética

common ratio *(p. 332)* The ratio between each pair of consecutive terms in a geometric sequence

razón común *(p. 332)* La razón entre cada par de términos consecutivos en una secuencia geométrica

completing the square *(p. 506)* To add a constant c to an expression of the form $x^2 + bx$ so that $x^2 + bx + c$ is a perfect square trinomial

completando el cuadrado *(p. 506)* Agregar una constante c a una expresión de la forma $x^2 + bx$ para que $x^2 + bx + c$ sea un trinomio de cuadrado perfecto

compound inequality *(p. 82)* An inequality formed by joining two inequalities with the word "and" or the word "or"

desigualdad compuesta *(p. 82)* Una desigualdad formada por dos desigualdades con la palabra "y" o la palabra "o"

compound interest *(p. 317)* The interest earned on the principal and on previously earned interest

interés compuesto *(p. 317)* El interés obtenido en el principal y en intereses previamente obtenidos

conditional relative frequency *(p. 612)* The ratio of a joint relative frequency to the marginal relative frequency

frecuencia relativa condicional *(p. 612)* La razón de una frecuencia relativa conjunta a una frecuencia relativa marginal

conjecture *(p. 3)* An unproven statement about a general mathematical concept

conjetura *(p. 3)* Una afirmación no comprobada sobre un concepto matemático general

conjugates *(p. 482)* Binomials of the form $a\sqrt{b} + c\sqrt{d}$ and $a\sqrt{b} - c\sqrt{d}$, where a, b, c, and d are rational numbers

conjugados *(p. 482)* Binomios de la forma $a\sqrt{b} + c\sqrt{d}$ y $a\sqrt{b} - c\sqrt{d}$, donde a, b, c y d son números racionales

constant function *(p. 138)* A linear equation written in the form $y = 0x + b$ or $y = b$

función constante *(p. 138)* Una ecuación lineal escrita bajo la forma $y = 0x + b$ o $y = b$

continuous domain *(p. 114)* A set of input values that consists of all numbers in an interval

dominio continuo *(p. 114)* Un conjunto de valores que consiste en todos los números en un intervalo

correlation *(p. 197)* A relationship between data sets

correlación *(p. 197)* Una relación entre conjuntos de datos

correlation coefficient *(p. 203)* A value r that tells how closely the equation of the line of best fit models the data

coeficiente de correlación *(p. 203)* Un valor r que indica cuán fielmente la ecuación de la línea de mejor ajuste modela los datos

counterexample *(p. 479)* An example that proves that a general statement is not true

contraejemplo *(p. 479)* Un ejemplo que prueba que una afirmación general no es verdadera

cube root function *(p. 552)* A radical function with an index of 3

función de raíz cúbica *(p. 552)* Una función radical con un índice de 3

D

data transformation *(p. 589)* A procedure that uses a mathematical operation to change a data set into a different data set

transformación de datos *(p. 589)* Un procedimiento que usa una operación matemática para cambiar un conjunto de datos en un conjunto de datos diferentes

degree of a monomial *(p. 358)* The sum of the exponents of the variables in the monomial

grado de un monomio *(p. 358)* La suma de los exponentes de las variables en el monomio

degree of a polynomial *(p. 359)* The greatest degree of the terms in a polynomial

grado de un polinomio *(p. 359)* El grado mayor de los términos en un polinomio

dependent variable *(p. 107)* The variable that represents the output values of a function

variable dependiente *(p. 107)* La variable que representa los resultados de una función

discrete domain *(p. 114)* A set of input values that consists of only certain numbers in an interval

dominio discreto *(p. 114)* Un conjunto de valores que consiste en solamente ciertos números en un intervalo

discriminant *(p. 518)* The expression $b^2 - 4ac$ in the Quadratic Formula

discriminante *(p. 518)* La expresión $b^2 - 4ac$ en la Fórmula Cuadrática

domain *(p. 106)* The set of all possible input values of a function

dominio *(p. 106)* El conjunto de todos los valores posibles de una función

E

equation *(p. 4)* A statement that two expressions are equal

ecuación *(p. 4)* Una afirmación de que dos expresiones son iguales

equivalent equations *(p. 4)* Equations that have the same solution(s)

ecuaciones equivalentes *(p. 4)* Ecuaciones que tiene(n) la(s) misma(s) solución(es)

equivalent inequalities *(p. 62)* Inequalities that have the same solutions

desigualdades equivalentes *(p. 62)* Desigualdades que tiene(n) la(s) misma(s) solución(es)

even function *(p. 442)* A function $y = f(x)$ is even when $f(-x) = f(x)$ for each x in the domain of f.

par función *(p. 442)* Una función $y = f(x)$ es par cuando $f(-x) = f(x)$ para cada x en el dominio de f.

explicit rule *(p. 340)* A rule to define arithmetic and geometric sequences that gives a_n as a function of the term's position number n in the sequence

regla explícita *(p. 340)* Una regla para definir secuencias aritméticas y geométricas que da a_n como función del número de posición del término n en la secuencia

exponential decay *(p. 315)* When a quantity decreases by the same factor over equal intervals of time

decaimiento exponencial *(p. 315)* Cuando una cantidad disminuye por el mismo factor sobre intervalos iguales de tiempo

exponential decay function *(p. 315)* A function of the form $y = a(1 - r)^t$, where $a > 0$ and $0 < r < 1$

función de decaimiento exponencial *(p. 315)* Una función de la forma $y = a(1 - r)^t$, donde $a > 0$ y $0 < r < 1$

exponential equation *(p. 326)* An equation in which variable expressions occur as exponents

ecuación exponencial *(p. 326)* Una ecuación en donde las expresiones de una variable ocurren como exponentes

exponential function *(p. 306)* A nonlinear function of the form $y = ab^x$, where $a \neq 0$, $b \neq 1$, and $b > 0$

función exponencial *(p. 306)* Una función no lineal de la forma $y = ab^x$, donde $a \neq 0$, $b \neq 1$, y $b > 0$

exponential growth *(p. 314)* When a quantity increases by the same factor over equal intervals of time

crecimiento exponencial *(p. 314)* Cuando una cantidad se incrementa por el mismo factor sobre intervalos iguales de tiempo

exponential growth function *(p. 314)* A function of the form $y = a(1 + r)^t$, where $a > 0$ and $r > 0$

función de crecimiento exponencial *(p. 314)* Una función de la forma $y = a(1 + r)^t$, donde $a > 0$ y $r > 0$

extraneous solution *(p. 31)* An apparent solution that must be rejected because it does not satisfy the original equation

solución extraña *(p. 31)* Una solución aparente que debe ser rechazada porque no satisface la ecuación original

extrapolation *(p. 205)* To predict a value outside the range of known values using a graph or its equation

extrapolación *(p. 205)* Predecir un valor fuera del rango de valores conocidos usando un gráfico o su ecuación

F

factored completely *(p. 404)* A polynomial that is written as a product of unfactorable polynomials with integer coefficients

factorizado completamente *(p. 404)* Un polinomio que se escribe como un producto de polinomios no factorizables con coeficientes de números enteros

factored form *(p. 378)* A polynomial that is written as a product of factors

forma factorizada *(p. 378)* Un polinomio que se escribe como un producto de factores

factoring by grouping *(p. 404)* To use the Distributive Property to factor a polynomial with four terms

factorización por agrupamiento *(p. 404)* Uso de la propiedad distributiva para factorizar un polinomio con cuatro términos

family of functions *(p. 146)* A group of functions with similar characteristics

familia de funciones *(p. 146)* Un grupo de funciones con características similares

five-number summary *(p. 594)* The five numbers that make up a box-and-whisker plot

resumen de cinco números *(p. 594)* Los cinco números que componen un diagrama de cajas y bigotes

FOIL Method *(p. 367)* A shortcut for multiplying two binomials by finding the sum of the products of the first terms, outer terms, inner terms, and last terms

método FOIL o de multiplicación de binomios *(p. 367)* Un método rápido para multiplicar dos binomios encontrando la suma de los productos de los primeros términos, los segundos términos, los terceros términos y los últimos términos

formula *(p. 37)* A literal equation that shows how one variable is related to one or more other variables

fórmula *(p. 37)* Una ecuación literal que muestra cómo una variable está relacionada con una o más variables

function *(p. 104)* A relation that pairs each input with exactly one output

función *(p. 104)* Una relación que asocia cada valor exactamente con un resultado

function notation *(p. 122)* Another name for y denoted as $f(x)$ and read as "the value of f at x" or "f of x"

notación de función *(p. 122)* Otro nombre para y que se denota como $f(x)$ y que se lee como "el valor de f en x" o "f de x"

G

geometric sequence *(p. 332)* An ordered list of numbers in which the ratio between each pair of consecutive terms is the same

secuencia geométrica *(p. 332)* Una lista ordenada en donde la razón entre cada par de términos consecutivos es la misma

graph of an inequality *(p. 56)* A graph that shows the solution set of an inequality on a number line

gráfico de una desigualdad *(p. 56)* Un gráfico que muestra el conjunto de soluciones de una desigualdad en una línea numérica

graph of a linear inequality *(p. 268)* The graph in two variables that shows all the solutions of the inequality in a coordinate plane

gráfico de una desigualdad lineal *(p. 268)* El gráfico en dos variables que muestra todas las soluciones de la desigualdad en un plano coordenado

graph of a system of linear inequalities *(p. 275)* The graph of all the solutions of the system of linear inequalities

gráfico de un sistema de desigualdades lineales *(p. 275)* El gráfico de todas las soluciones del sistema de las desigualdades lineales

H

half-planes *(p. 268)* Two regions of the coordinate plane divided by a boundary line

semiplanos *(p. 268)* Dos regiones del plano coordenado divididas por una línea limítrofe

horizontal shrink *(p. 148)* A transformation that causes the graph of a function to shrink toward the y-axis when all the x-coordinates are multiplied by a factor a, where $a > 1$

reducción horizontal *(p. 148)* Una transformación que hace que el gráfico de una función se reduzca hacia el eje y cuando todas las coordenadas x se multiplican por un factor a, donde $a > 1$

horizontal stretch *(p. 148)* A transformation that causes the graph of a function to stretch away from the y-axis when all the x-coordinates are multiplied by a factor a, where $0 < a < 1$

ampliación horizontal *(p. 148)* Una transformación que hace que el gráfico de una función se amplíe desde el eje y cuando todas las coordenadas x se multiplican por un factor a, donde $0 < a < 1$

I

identity *(p. 21)* An equation that is true for all values of the variable

identidad *(p. 21)* Una ecuación que es verdadera para todos los valores de la variable

independent variable *(p. 107)* The variable that represents the input values of a function

variable independiente *(p. 107)* La variable que representa los valores de una función

index of a radical *(p. 300)* The value of n in the radical $\sqrt[n]{a}$

índice de un radical *(p. 300)* El valor de n en el radical $\sqrt[n]{a}$

inequality *(p. 54)* A mathematical sentence that compares expressions

desigualdad *(p. 54)* Una oración matemática que compara expresiones

intercept form *(p. 450)* A quadratic function written in the form $f(x) = a(x - p)(x - q)$, where $a \neq 0$

forma intersección *(p. 450)* Una ecuación cuadrática escrita en la forma $f(x) = a(x - p)(x - q)$, donde $a \neq 0$

interpolation *(p. 205)* To approximate a value between two known values using a graph or its equation

interpolación *(p. 205)* Aproximar un valor entre dos valores conocidos usando un gráfico o su ecuación

interquartile range *(p. 595)* A measure of variation for a data set, which is the difference of the third quartile and the first quartile

rango de intercuartiles *(p. 595)* Una medida de variación para un conjunto de datos, el cual es la diferencia del tercer cuartil y el primer cuartil

inverse function *(p. 569)* Functions that undo each other

función inversa *(p. 569)* Funciones que se anulan entre sí

inverse operations *(p. 4)* Two operations that undo each other, such as addition and subtraction

operaciones inversas *(p. 4)* Dos operaciones que se anulan entre sí, como la suma y la resta

inverse relation *(p. 568)* When the input and output values of the original relation are switched

relación inversa *(p. 568)* Cuando los valores y los resultados de la relación original se intercambian

J

joint frequency *(p. 610)* Each entry in a two-way table

frecuencia conjunta *(p. 610)* Cada valor en una tabla de doble entrada

joint relative frequency *(p. 611)* The ratio of a frequency that is not in the "total" row or the "total" column to the total number of values or observations

frecuencia relativa conjunta *(p. 611)* La razón de una frecuencia que no está en la hilera "total" o columna "total" del número total de valores u observaciones

L

leading coefficient *(p. 359)* The coefficient of the first term of a polynomial written in standard form

coeficiente principal *(p. 359)* El coeficiente del primer término de un polinomio escrito en forma estándar

like radicals *(p. 484)* Radicals with the same index and radicand

radicales semejantes *(p. 484)* Radicales con el mismo índice y radicando

line of best fit *(p. 203)* A line that best models a set of data

línea de mejor ajuste *(p. 203)* Una línea que mejor modela un conjunto de datos

line of fit *(p. 198)* A line drawn on a scatter plot that is close to most of the data points

línea de ajuste *(p. 198)* Una línea dibujada en un diagrama de dispersión que está cerca a la mayoría de los puntos de datos

linear equation in one variable *(p. 4)* An equation that can be written in the form $ax + b = 0$, where a and b are constants and $a \neq 0$

ecuación lineal en una variable *(p. 4)* Una ecuación que puede escribirse en la forma $ax + b = 0$, donde a y b son constantes y $a \neq 0$

linear equation in two variables *(p. 112)* An equation that can be written in the form $y = mx + b$, where m and b are constants

ecuación lineal en dos variables *(p. 112)* Una ecuación que puede escribirse en la forma $y = mx + b$, donde m y b son constantes

linear function *(p. 112)* A function whose graph is a nonvertical line

función lineal *(p. 112)* Una función cuyo gráfico es una línea no vertical

linear inequality in two variables *(p. 268)* An inequality written in the form $ax + by < c$, $ax + by \leq c$, $ax + by > c$, or $ax + by \geq c$, where a, b, and c are real numbers

desigualdad lineal en dos variables *(p. 268)* Una desigualdad escrita en la forma $ax + by < c$, $ax + by \leq c$, $ax + by > c$, o $ax + by \geq c$, donde a, b y c son números reales

linear model *(p. 178)* A linear function that models a real-life situation

modelo lineal *(p. 178)* Una función lineal que modela una situación de la vida real

linear regression *(p. 203)* A method that graphing calculators use to find a precise line of fit that models a set of data

regresión lineal *(p. 203)* Un método que usan las calculadoras gráficas para encontrar una línea precisa de ajuste que modela un conjunto de datos

literal equation *(p. 36)* An equation that has two or more variables

ecuación literal *(p. 36)* Una ecuación que tiene dos o más variables

M

marginal frequency *(p. 610)* The sums of the rows and columns in a two-way table

frecuencia marginales *(p. 610)* Las sumas de las hileras y columnas en una tabla de doble entrada

marginal relative frequency *(p. 611)* The sum of the joint relative frequencies in a row or a column

frecuencia relativa marginal *(p. 611)* La suma de las frecuencias relativas conjuntas en una hilera o columna

maximum value *(p. 433)* The y-coordinate of the vertex of the graph of $f(x) = ax^2 + bx + c$, where $a < 0$

valor máximo *(p. 433)* La coordenada y del vértice del gráfico de $f(x) = ax^2 + bx + c$, donde $a < 0$

mean *(p. 586)* The sum of a numerical data set divided by the number of data values

media *(p. 586)* La suma de un conjunto de datos numéricos divididos entre el número de valores de datos

measure of center *(p. 586)* A measure that represents the center, or typical value, of a data set

medida del centro *(p. 586)* Una medida que representa el centro, o valor típico, de un conjunto de datos

measure of variation *(p. 587)* A measure that describes the spread, or distribution, of a data set

medida de variación *(p. 587)* Una medida que describe la extensión, o distribución, de un conjunto de datos

median *(p. 586)* The middle number of a numerical data set when the values are written in numerical order

mediana *(p. 586)* El número medio de un conjunto de datos numéricos cuando los valores se escriben en orden numérico

minimum value *(p. 433)* The y-coordinate of the vertex of the graph of $f(x) = ax^2 + bx + c$, where $a > 0$

valor mínimo *(p. 433)* La coordenada y del vértice del gráfico de $f(x) = ax^2 + bx + c$, donde $a > 0$

misleading graph *(p. 620)* A statistical graph that is not drawn appropriately

gráfico engañoso *(p. 620)* Un gráfico estadístico que no está dibujado apropiadamente

mode *(p. 586)* The value or values that occur most often in a data set

modo *(p. 586)* El valor o valores que ocurren con mayor frecuencia en un conjunto de datos

monomial *(p. 358)* A number, a variable, or the product of a number and one or more variables with whole number exponents

monomio *(p. 358)* Un número, una variable o el producto de un número y una o más variables con exponentes en números enteros

N

nonlinear function *(p. 112)* A function that does not have a constant rate of change and whose graph is not a line

función no lineal *(p. 112)* Una función que no tiene una tasa constante de cambio y cuyo gráfico no es una línea

n*th root of *a *(p. 300)* For an integer n greater than 1, if $b^n = a$, then b is an nth root of a.

raíz de orden *n* de *a* *(p. 300)* Para un número entero n mayor que 1, si $b^n = a$, entonces b es una raíz de orden n de a.

O

odd function *(p. 442)* A function $y = f(x)$ is odd when $f(-x) = -f(x)$ for each x in the domain of f.

impar funcíon *(p. 442)* Una función $y = f(x)$ es impar cuando $f(-x) = -f(x)$ para cada x en el dominio de f.

outlier *(p. 587)* A data value that is much greater than or much less than the other values in a data set

valor atípico *(p. 587)* Una valor de datos que es mucho mayor o mucho menor que los otros valores en un conjunto de datos

P

parabola *(p. 420)* The U-shaped graph of a quadratic function

parábola *(p. 420)* El gráfico en forma de U de una función cuadrática

parallel lines *(p. 188)* Two lines in the same plane that never intersect

líneas paralelas *(p. 188)* Dos líneas en el mismo plano que nunca se intersectan

parent function *(p. 146)* The most basic function in a family of functions

función principal *(p. 146)* La función más básica en una familia de funciones

perpendicular lines *(p. 189)* Two lines in the same plane that intersect to form right angles

líneas perpendiculares *(p. 189)* Dos líneas en el mismo plano que se intersectan para formar ángulos rectos

piecewise function *(p. 218)* A function defined by two or more equations

función por tramos *(p. 218)* Una función definida por dos o más ecuaciones

point-slope form *(p. 182)* A linear equation written in the form $y - y_1 = m(x - x_1)$

forma punto-pendiente *(p. 182)* Una ecuación lineal escrita en la forma $y - y_1 = m(x - x_1)$

polynomial *(p. 359)* A monomial or a sum of monomials

polinomio *(p. 359)* Un monomio o una suma de monomios

Q

quadratic equation *(p. 490)* A nonlinear equation that can be written in the standard form $ax^2 + bx + c = 0$, where $a \neq 0$

ecuación cuadrática *(p. 490)* Una ecuación no lineal que puede escribirse en la forma estándar $ax^2 + bx + c = 0$, donde $a \neq 0$

Quadratic Formula *(p. 516)* The real solutions of the quadratic equation $ax^2 + bx + c = 0$ are

$x = \frac{-b \pm \sqrt{b^2 - 4ac}}{2a}$, where $a \neq 0$ and $b^2 - 4ac \geq 0$.

fórmula cuadrática *(p. 516)* Las soluciones reales de la ecuación cuadrática $ax^2 + bx + c = 0$ son

$x = \frac{-b \pm \sqrt{b^2 - 4ac}}{2a}$, donde $a \neq 0$ y $b^2 - 4ac \geq 0$.

quadratic function *(p. 420)* A nonlinear function that can be written in the standard form $y = ax^2 + bx + c$, where $a \neq 0$

función cuadrática *(p. 420)* Una ecuación no lineal que puede escribirse en la forma estándar $y = ax^2 + bx + c$, donde $a \neq 0$

qualitative data *(p. 618)* Data that consist of labels or nonnumerical entries that can be separated into different categories

datos cualitativos *(p. 618)* Datos que consisten en etiquetas o valores no numéricos que pueden separarse en categorías distintas

quantitative data *(p. 618)* Data that consist of numbers that represent counts or measurements

datos cuantitativos *(p. 618)* Datos que consisten en números que representan conteos o medidas

quartiles *(p. 594)* Values of a box-and-whisker plot that divide a data set into four equal parts

cuartiles *(p. 594)* Valores de un diagrama de cajas y bigotes que dividen un conjunto de datos en cuatro partes iguales

R

radical *(p. 300)* An expression of the form $\sqrt[n]{a}$

radical *(p. 300)* Una expresión de la forma $\sqrt[n]{a}$

radical equation *(p. 560)* An equation that contains a radical expression with a variable in the radicand

ecuación radical *(p. 560)* Una ecuación que contiene una expresión radical con una variable en el radicando

radical expression *(p. 480)* An expression that contains a radical

expresión radical *(p. 480)* Una expresión que contiene un radical

radical function *(p. 545)* A function that contains a radical expression with the independent variable in the radicand

función radical *(p. 545)* Una ecuación que contiene una expresión radical con la variable independiente en el radicando

range of a data set *(p. 587)* The difference of the greatest value and the least value of a data set

rango de un conjunto de datos *(p. 587)* La diferencia del valor mayor y el valor menor de un conjunto de datos

range of a function *(p. 106)* The set of all possible output values of a function

rango de una función *(p. 106)* El conjunto de todos los resultados posibles de una función

rationalizing the denominator *(p. 482)* To eliminate a radical from the denominator of a fraction by multiplying by an appropriate form of 1

racionalización del denominador *(p. 482)* Eliminar un radical del denominador de una fracción multiplicando por una forma apropiada de 1

recursive rule *(p. 340)* A rule to define arithmetic and geometric sequences that gives the beginning term(s) of a sequence and a recursive equation that tells how a_n is related to one or more preceding terms

regla recursiva *(p. 340)* Una regla para definir las secuencias aritméticas y geométricas que da el(los) primer(os) término(s) de una secuencia y una ecuación recursiva que indica cómo se relaciona a_n a uno o más términos precedentes

reflection *(p. 147)* A transformation that flips a graph over a line called the *line of reflection*

reflexión *(p. 147)* Una transformación que voltea un gráfico sobre una línea llamada la *línea de reflexión*

relation *(p. 104)* A pairing of inputs with outputs

relación *(p. 104)* Una pareja de valores con resultados

repeated roots *(p. 379)* Two or more roots of an equation that are the same number

raíces repetidas *(p. 379)* Dos o más raíces de una ecuación que son el mismo número

residual *(p. 202)* The difference of the y-value of a data point and the corresponding y-value found using the line of fit

residual *(p. 202)* La diferencia del valor y de un punto de datos y el valor y correspondiente usando la línea de ajuste

rise *(p. 136)* The change in y between any two points on a line

desplazamiento vertical *(p. 136)* El cambio en y entre cualquier dos puntos en una recta

roots *(p. 378)* The solutions of a polynomial equation

raíces *(p. 378)* Las soluciones deuna ecuación de polinomios

rule *(p. 3)* A proven statement about a general mathematical concept

regla *(p. 3)* Una afirmación comprobada sobre un concepto matemático general

run *(p. 136)* The change in x between any two points on a line

desplazamiento horizontal *(p. 136)* El cambio en x entre cualquier dos puntos en una recta

S

scatter plot *(p. 196)* A graph that shows the relationship between two data sets

diagrama de dispersión *(p. 196)* Un gráfico que muestra la relación entre dos conjuntos de datos

sequence *(p. 210)* An ordered list of numbers

secuencia *(p. 210)* Una lista ordenada de números

simplest form of a radical *(p. 480)* An expression involving a radical with index n that has no radicands with perfect nth powers as factors other than 1, no radicands that contain fractions, and no radicals that appear in the denominator of a fraction

mínima expresión de un radical *(p. 480)* Una expresión que conlleva un radical con índice n que no tiene radicandos con potencias perfectas de orden n como factores distintos a 1, que no tiene radicandos que contengan fracciones y que no tiene radicales que aparezcan en el denominador de una fracción

slope *(p. 136)* The rate of change between any two points on a line

pendiente *(p. 136)* La tasa de cambio entre dos puntos cualesquiera en una línea

slope-intercept form *(p. 138)* A linear equation written in the form $y = mx + b$

forma intersección-pendiente *(p. 138)* Una ecuación lineal escrita en la forma $y = mx + b$

solution of an equation *(p. 4)* A value that makes an equation true

solución de una ecuación *(p. 4)* Un valor que hace que una ecuación sea verdadera

solution of an inequality *(p. 55)* A value that makes an inequality true

solución de una desigualdad *(p. 55)* Un valor que hace que una desigualdad sea verdadera

solution of a linear equation in two variables *(p. 114)* An ordered pair (x, y) that makes an equation true

solución de una ecuación lineal en dos variables *(p. 114)* Un par ordenado (x, y) que hace que una ecuación sea verdadera

solution of a linear inequality in two variables *(p. 268)* An ordered pair (x, y) that makes the inequality true

solución de una desigualdad lineal en dos variables *(p. 268)* Un par ordenado (x, y) que hace que la desigualdad sea verdadera

solution set *(p. 55)* The set of all solutions of an inequality

conjunto solución *(p. 55)* El conjunto de todas las soluciones de una desigualdad

solution of a system of linear equations *(p. 236)* An ordered pair that is a solution of each equation in the system

solución de un sistema de ecuaciones lineales *(p. 236)* Un par ordenado que es una solución de cada ecuación en el sistema

solution of a system of linear inequalities *(p. 274)* An ordered pair that is a solution of each inequality in the system

solución de un sistema de desigualdades lineales *(p. 274)* Un par ordenado que es una solución de cada desigualdad en el sistema

square root function *(p. 544)* A function that contains a square root with the independent variable in the radicand

función de raíz cuadrada *(p. 544)* Una función que contiene una raíz cuadrada con la variable independiente en el radicando

standard deviation *(p. 588)* A measure of how much a typical value in a numerical data set differs from the mean

desviación estándar *(p. 588)* Una medida de cuánto difiere un valor típico en un conjunto de datos numéricos de la media

standard form of a linear equation *(p. 130)* A linear equation written in the form $Ax + By = C$, where A, B, and C are real numbers and A and B are not both zero

forma estándar de una ecuación lineal *(p. 130)* Una ecuación lineal escrita en la forma $Ax + By = C$, donde A, B y C son números reales y ni A ni B son cero

standard form of a polynomial *(p. 359)* A polynomial in one variable written with the exponents of the terms decreasing from left to right

forma estándar de un polinomio *(p. 359)* Un polinomio en una variable escrita con los exponentes de los términos decreciendo de izquierda a derecha

step function *(p. 220)* A piecewise function defined by a constant value over each part of its domain

función escalón *(p. 220)* Una función por tramos definida por un valor constante sobre cada parte de su dominio

system of linear equations *(p. 236)* A set of two or more linear equations in the same variables

sistema de ecuaciones lineales *(p. 236)* Un conjunto de dos o más ecuaciones lineales en las mismas variables

system of linear inequalities *(p. 274)* A set of two or more linear inequalities in the same variables

sistema de desigualdades lineales *(p. 274)* Un conjunto de dos o más desigualdades lineales en las mismas variables

system of nonlinear equations *(p. 526)* A system in which at least one of the equations is nonlinear

sistema de ecuaciones no lineales *(p. 526)* Un sistema en donde al menos una de las ecuaciones no es lineal

T

term of a sequence *(p. 210)* Each number in a sequence

término de una sucesión *(p. 210)* Cada número en una secuencia

theorem *(p. 3)* A proven statement about a general mathematical concept

teorema *(p. 3)* Una declaración comprobada acerca de un concepto matemático general

transformation *(p. 146)* A change in the size, shape, position, or orientation of a graph

transformación *(p. 146)* Un cambio en el tamaño, forma, posición u orientación de un gráfico

translation *(p. 146)* A transformation that shifts a graph horizontally and/or vertically but does not change the size, shape, or orientation of the graph

traslación *(p. 146)* Una transformación que desplaza un gráfico horizontal y/o verticalmente pero no cambia el tamaño, forma u orientación del gráfico

trinomial *(p. 359)* A polynomial with three terms

trinomio *(p. 359)* Un polinomio con tres términos

two-way table *(p. 610)* A frequency table that displays data collected from one source that belong to two different categories

tabla de doble entrada *(p. 610)* Una tabla de frecuencia que muestra los datos recogidos de una fuente que pertenece a dos categorías distintas

V

vertex *(p. 156)* The point where a graph changes direction

vértice *(p. 156)* El punto en donde un gráfico cambia de dirección

vertex form of an absolute value function *(p. 158)* An absolute value function written in the form $f(x) = a|x - h| + k$, where $a \neq 0$

forma de vértice de una función de valor absoluto *(p. 158)* Una función de valor absoluto escrita en la forma $f(x) = a|x - h| + k$, donde $a \neq 0$

vertex form of a quadratic function *(p. 444)* An quadratic function written in the form $f(x) = a(x - h)^2 + k$, where $a \neq 0$

forma de vértice de una función cuadrática *(p. 444)* Una función cuadrática escrita en la forma $f(x) = a(x - h)^2 + k$, donde $a \neq 0$

vertex of a parabola *(p. 420)* The lowest point on a parabola that opens up or the highest point on a parabola that opens down

vértice de una parábola *(p. 420)* El punto más bajo de una parábola que se abre hacia arriba o el punto más alto de una parábola que se abre hacia abajo

vertical shrink *(p. 148)* A transformation that causes the graph of a function to shrink toward the x-axis when all the y-coordinates are multiplied by a factor a, where $0 < a < 1$

reducción vertical *(p. 148)* Una transformación que hace que el gráfico de una función se reduzca hacia el eje x cuando todas las coordenadas y se multiplican por un factor a, donde $0 < a < 1$

vertical stretch *(p. 148)* A transformation that causes the graph of a function to stretch away from the x-axis when all the y-coordinates are multiplied by a factor a, where $a > 1$

ampliación vertical *(p. 148)* Una transformación que hace que el gráfico de una función se amplíe desde el eje x cuando todas las coordenadas y se multiplican por un factor a, donde $a > 1$

X

***x*-intercept** *(p. 131)* The x-coordinate of a point where the graph crosses the x-axis

intersección *x* *(p. 131)* La coordenada x de un punto donde el gráfico cruza el eje x

Y

***y*-intercept** *(p. 131)* The y-coordinate of a point where the graph crosses the y-axis

intersección *y* *(p. 131)* La coordenada y de un punto donde el gráfico cruza el eje y

Z

zero of a function *(p. 428)* An x-value of a function f for which $f(x) = 0$

cero de una función *(p. 428)* Un valor x de una función f para el cual $f(x) = 0$

Zero-Product Property *(p. 378)* If the product of two real numbers is 0, then at least one of the numbers is 0.

propiedad de producto cero *(p. 378)* Si el producto de dos números reales es 0, entonces al menos uno de los números es 0.

Index

Index

F

G

H

I

Index

Index

S

Reference

Properties

Properties of Equality

Addition Property of Equality
If $a = b$, then $a + c = b + c$.

Subtraction Property of Equality
If $a = b$, then $a - c = b - c$.

Multiplication Property of Equality
If $a = b$, then $a \cdot c = b \cdot c, c \neq 0$.

Division Property of Equality
If $a = b$, then $a \div c = b \div c, c \neq 0$.

Properties of Inequality

Addition Property of Inequality
If $a > b$, then $a + c > b + c$.
If $a < b$, then $a + c < b + c$.

Subtraction Property of Inequality
If $a > b$, then $a - c > b - c$.
If $a < b$, then $a - c < b - c$.

Multiplication Property of Inequality ($c > 0$)

If $a > b$ and $c > 0$, then $ac > bc$.

If $a < b$ and $c > 0$, then $ac < bc$.

Division Property of Inequality ($c > 0$)

If $a > b$ and $c > 0$, then $\frac{a}{c} > \frac{b}{c}$.

If $a < b$ and $c > 0$, then $\frac{a}{c} < \frac{b}{c}$.

Multiplication Property of Inequality ($c < 0$)

If $a > b$ and $c < 0$, then $ac < bc$.
If $a < b$ and $c < 0$, then $ac > bc$.

Division Property of Inequality ($c < 0$)

If $a > b$ and $c < 0$, then $\frac{a}{c} < \frac{b}{c}$.

If $a < b$ and $c < 0$, then $\frac{a}{c} > \frac{b}{c}$.

* The Properties of Inequality are also true for $\geq$ and $\leq$.

Properties of Exponents

Zero Exponent
$a^0 = 1$, where $a \neq 0$

Negative Exponent
$a^{-n} = \frac{1}{a^n}$, where $a \neq 0$

Product of Powers Property
$a^m \cdot a^n = a^{m+n}$

Quotient of Powers Property
$\frac{a^m}{a^n} = a^{m-n}$, where $a \neq 0$

Power of a Power Property
$(a^m)^n = a^{mn}$

Power of a Product Property
$(ab)^m = a^m b^m$

Power of a Quotient Property
$\left(\frac{a}{b}\right)^m = \frac{a^m}{b^m}$, where $b \neq 0$

Rational Exponents
$a^{m/n} = (a^{1/n})^m = (\sqrt[n]{a})^m$

Properties of Absolute Value

$|a| \geq 0$

$|-a| = |a|$

$|ab| = |a||b|$

$\left|\frac{a}{b}\right| = \frac{|a|}{|b|}, b \neq 0$

Properties of Radicals

Product Property of Square Roots
$\sqrt{ab} = \sqrt{a} \cdot \sqrt{b}$, where $a, b \geq 0$

Quotient Property of Square Roots
$\sqrt{\frac{a}{b}} = \frac{\sqrt{a}}{\sqrt{b}}$, where $a \geq 0$ and $b > 0$

Other Properties

Property of Equality for Exponential Equations
If $b > 0$ and $b \neq 1$, then $b^x = b^y$ if and only if $x = y$.

Zero-Product Property
If a and b are real numbers and $ab = 0$, then $a = 0$ or $b = 0$.

Patterns

Square of a Binomial Pattern
$(a + b)^2 = a^2 + 2ab + b^2$
$(a - b)^2 = a^2 - 2ab + b^2$

Sum and Difference Pattern
$(a + b)(a - b) = a^2 - b^2$

Difference of Two Squares Pattern
$a^2 - b^2 = (a + b)(a - b)$

Perfect Square Trinomial Pattern
$a^2 + 2ab + b^2 = (a + b)^2$
$a^2 - 2ab + b^2 = (a - b)^2$

Formulas

Slope
$m = \dfrac{y_2 - y_1}{x_2 - x_1}$

Slope-intercept form
$y = mx + b$

Point-slope form
$y - y_1 = m(x - x_1)$

Standard form of a linear equation
$Ax + By = C$, where A and B are not both 0

Vertex form of an absolute value function
$f(x) = a|x - h| + k$, where $a \neq 0$

Standard form of a quadratic function
$f(x) = ax^2 + bx + c$, where $a \neq 0$

Vertex form of a quadratic function
$f(x) = a(x - h)^2 + k$, where $a \neq 0$

Intercept form of a quadratic function
$f(x) = a(x - p)(x - q)$, where $a \neq 0$

Quadratic Formula
$x = \dfrac{-b \pm \sqrt{b^2 - 4ac}}{2a}$, where $a \neq 0$ and $b^2 - 4ac \geq 0$

Exponential growth
$y = a(1 + r)^t$, where $a > 0$ and $r > 0$

Exponential decay
$y = a(1 - r)^t$, where $a > 0$ and $0 < r < 1$

Explicit rule for an arithmetic sequence
$a_n = a_1 + (n - 1)d$

Explicit rule for a geometric sequence
$a_n = a_1 r^{n-1}$

Recursive equation for an arithmetic sequence
$a_n = a_{n-1} + d$

Recursive equation for a geometric sequence
$a_n = r \cdot a_{n-1}$

Standard deviation

$$\sigma = \sqrt{\frac{(x_1 - \bar{x})^2 + (x_2 - \bar{x})^2 + \cdots + (x_n - \bar{x})^2}{n}}$$

Perimeter, Area, and Volume Formulas

Square

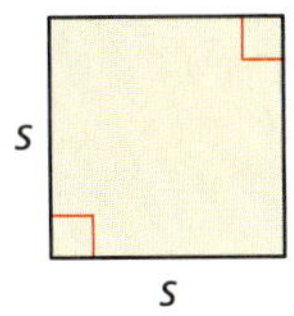

$P = 4s$
$A = s^2$

Rectangle

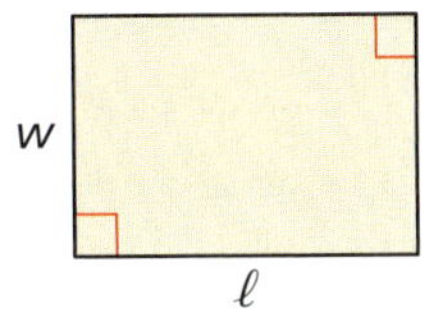

$P = 2\ell + 2w$
$A = \ell w$

Triangle

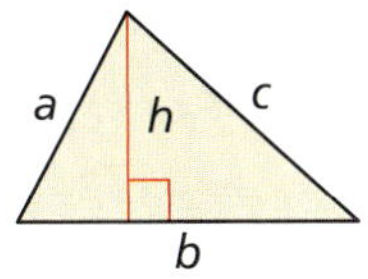

$P = a + b + c$
$A = \frac{1}{2}bh$

Circle

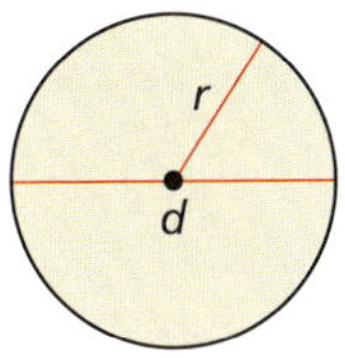

$C = \pi d$ or $C = 2\pi r$
$A = \pi r^2$

Parallelogram

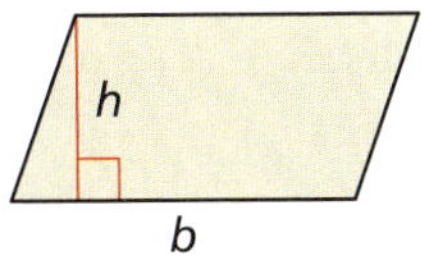

$A = bh$

Trapezoid

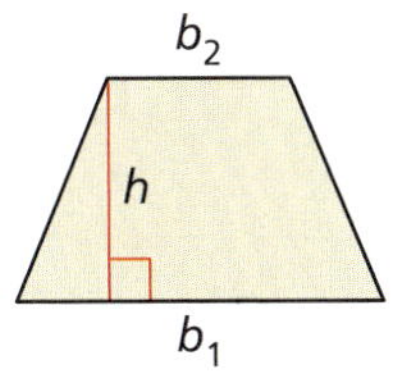

$A = \frac{1}{2}h(b_1 + b_2)$

Rhombus/Kite

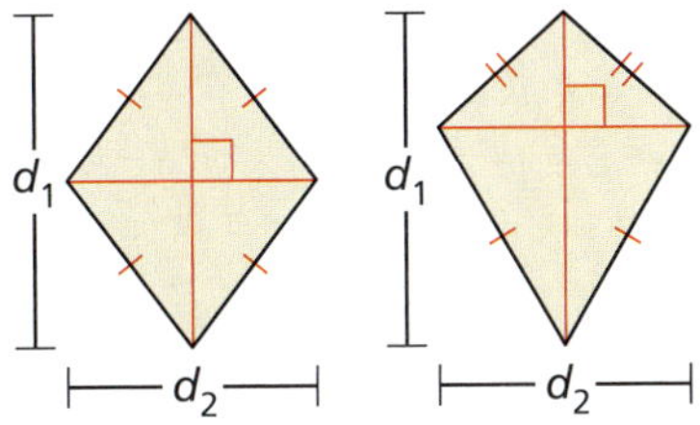

$A = \frac{1}{2}d_1d_2$

Regular *n*-gon

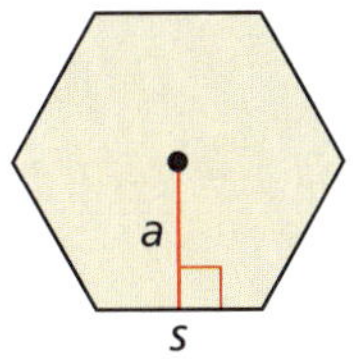

$A = \frac{1}{2}aP$ or $A = \frac{1}{2}a \cdot ns$

Prism

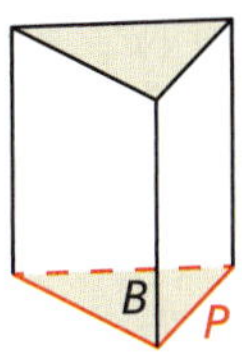

$L = Ph$
$S = 2B + Ph$
$V = Bh$

Cylinder

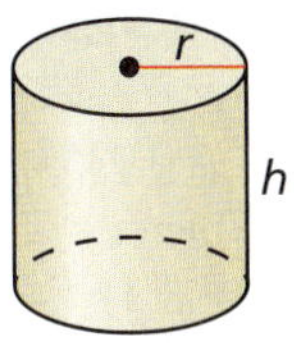

$L = 2\pi rh$
$S = 2\pi r^2 + 2\pi rh$
$V = \pi r^2h$

Pyramid

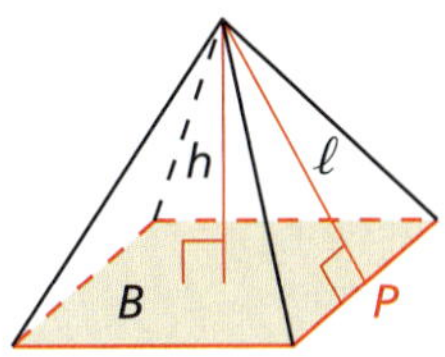

$L = \frac{1}{2}P\ell$
$S = B + \frac{1}{2}P\ell$
$V = \frac{1}{3}Bh$

Cone

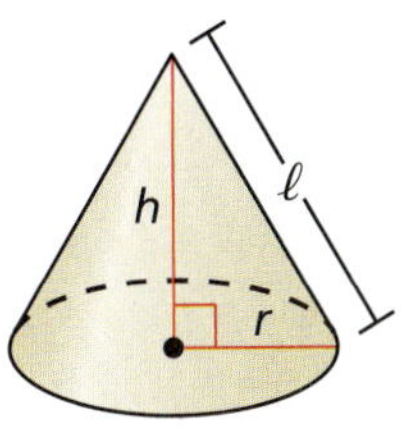

$L = \pi r\ell$

$S = \pi r^2 + \pi r\ell$

$V = \frac{1}{3}\pi r^2h$

Sphere

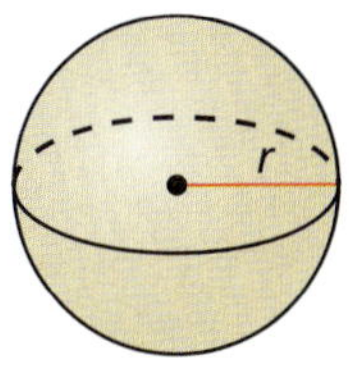

$S = 4\pi r^2$

$V = \frac{4}{3}\pi r^3$

Other Formulas

Pythagorean Theorem
$a^2 + b^2 = c^2$

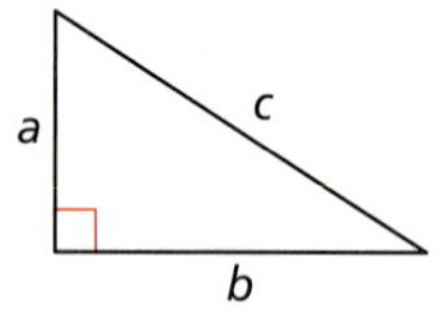

Simple Interest
$I = Prt$

Compound Interest
$y = P\left(1 + \frac{r}{n}\right)^{nt}$

Distance
$d = rt$

Conversions

U.S. Customary
1 foot = 12 inches
1 yard = 3 feet
1 mile = 5280 feet
1 mile = 1760 yards
1 acre = 43,560 square feet
1 cup = 8 fluid ounces
1 pint = 2 cups
1 quart = 2 pints
1 gallon = 4 quarts
1 gallon = 231 cubic inches
1 pound = 16 ounces
1 ton = 2000 pounds

U.S. Customary to Metric
1 inch = 2.54 centimeters
1 foot ≈ 0.3 meter
1 mile ≈ 1.61 kilometers
1 quart ≈ 0.95 liter
1 gallon ≈ 3.79 liters
1 cup ≈ 237 milliliters
1 pound ≈ 0.45 kilogram
1 ounce ≈ 28.3 grams
1 gallon ≈ 3785 cubic centimeters

Time
1 minute = 60 seconds
1 hour = 60 minutes
1 hour = 3600 seconds
1 year = 52 weeks

Temperature
$C = \frac{5}{9}(F - 32)$
$F = \frac{9}{5}C + 32$

Metric
1 centimeter = 10 millimeters
1 meter = 100 centimeters
1 kilometer = 1000 meters
1 liter = 1000 milliliters
1 kiloliter = 1000 liters
1 milliliter = 1 cubic centimeter
1 liter = 1000 cubic centimeters
1 cubic millimeter = 0.001 milliliter
1 gram = 1000 milligrams
1 kilogram = 1000 grams

Metric to U.S. Customary
1 centimeter ≈ 0.39 inch
1 meter ≈ 3.28 feet
1 meter ≈ 39.37 inches
1 kilometer ≈ 0.62 mile
1 liter ≈ 1.06 quarts
1 liter ≈ 0.26 gallon
1 kilogram ≈ 2.2 pounds
1 gram ≈ 0.035 ounce
1 cubic meter ≈ 264 gallons

Credits

Front Matter

viii Stephanie Lupoli/Shutterstock.com; **x** © Anthony Berenyi | Dreamstime.com; **x** zhangyang13576997233/Shutterstock.com; **xi** WDG Photo/Shutterstock.com; **xii** eye-blink/Shutterstock.com; **xiii** Amnartk/Shutterstock.com; **xiv** anyaivanova/Shutterstock.com; **xv** Johnny Adolphson/Shutterstock.com; **xvi** ©iStockphoto.com/MinistryOfJoy; **xvii** Marafona/Shutterstock.com; **xviii** ©iStockphoto.com/Cesar Okada; **xix** Monkey Business Images/Shutterstock.com; **xxiii** Tyler Olson/Shutterstock.com; **xxvii** Goodluz/Shutterstock.com

Chapter 1

0 *top left* Doug Meek/Shutterstock.com; *top right* aceshot1/Shutterstock.com; *center left* Stephanie Lupoli/Shutterstock.com; *bottom right* Stefan Schurr/Shutterstock.com; *bottom left* Ververidis Vasilis/Shutterstock.com; **6** Ververidis Vasilis/Shutterstock.com; **9** cowardlion/Shutterstock.com; **14** Stefan Schurr/Shutterstock.com; **15** Eric Isselee/Shutterstock.com; **22** Stephanie Lupoli/Shutterstock.com; **25** Aleksander Erin/Shutterstock.com; **26** Invisible Studio/Shutterstock.com, avNY/Shutterstock.com, barragan/Shutterstock.com; **29** aceshot1/Shutterstock.com; **32** Jagodka/Shutterstock.com; **38** Petr84/Shutterstock.com; **40** Maxim Tupikov/Shutterstock.com; **41** *bottom left* Vladimir Dokovski/Shutterstock.com; *top right* Potapov Alexander/Shutterstock.com; **42** John Blanton/Shutterstock.com; **43** ollyy/Shutterstock.com

Chapter 2

50 *top left* DDCoral/Shutterstock.com; *top right* Christopher Boswell/Shutterstock.com; *center left* Catalin Petolea/Shutterstock.com; *bottom right* © Anthony Berenyi | Dreamstime.com; *bottom left* © Archhunter.de; **58** Gaieva Tetiana/Shutterstock.com; **59** © Archhunter.de; **64** © Anthony Berenyi | Dreamstime.com; **65** ©iStockphoto.com/NoDerog; **70** Oleksiy Mark/Shutterstock.com; **77** ©iStockphoto.com/fotoVoyager; **78** terekhov igor/Shutterstock.com; **79** Aleksander Erin/Shutterstock.com; **84** *top left* Umberto Shtanzman/Shutterstock.com; *bottom left* marekuliasz/Shutterstock.com; **85** Christopher Boswell/Shutterstock.com; **86** Niyazz/Shutterstock.com; **91** DDCoral/Shutterstock.com; **93** Lucky Business/Shutterstock.com; **97** Dmitry Naumov/Shutterstock.com

Chapter 3

100 *top left* © Sandra Sims | Dreamstime.com; *top right* Yiorgos GR/Shutterstock.com; *center left* zhangyang13576997233/Shutterstock.com; *bottom right* Myotis/Shutterstock.com; *bottom left* guroldinneden/Shutterstock.com; **107** Monika Gniot/Shutterstock.com; **109** *top left* guroldinneden/Shutterstock.com; *bottom right* ©iStockphoto.com/sorbetto; **116** Myotis/Shutterstock.com; **118** © Beaucroft | Dreamstime.com; **119** Vixit/Shutterstock.com; **120** Coprid/Shutterstock.com; **125** zhangyang13576997233/Shutterstock.com; **127** Aleksander Erin/Shutterstock.com; **134** Pertusinas/Shutterstock.com; **140** © Sandra Sims | Dreamstime.com; **142** Beneda Miroslav/Shutterstock.com; **151** Jacek Chabraszewski/Shutterstock.com; **153** ecco/Shutterstock.com, bazzier/Shutterstock.com; **161** Pakhnyushcha/Shutterstock.com; **163** ©iStockphoto.com/NejroN

Chapter 4

172 *top left* © Shannon Fagan | Dreamstime.com; *top right* © Joseph Helfenberger | Dreamstime.com; *center left* Asmus/Shutterstock.com; *bottom right* © Rmarmion | Dreamstime.com; *bottom left* WDG Photo/Shutterstock.com; **178** WDG Photo/Shutterstock.com; **180** bikeriderlondon/Shutterstock.com; **184** © Rmarmion | Dreamstime.com; **190** Asmus/Shutterstock.com; **192** ©iStockphoto.com/mecaleha; **193** Aleksander Erin/Shutterstock.com; **204** © Joseph Helfenberger | Dreamstime.com; **207** Adisa/Shutterstock.com; **215** NatUlrich/Shutterstock.com; **220** © Shannon Fagan | Dreamstime.com; **222** manish mansinh/Shutterstock.com; **225** and **231** Africa Studio/Shutterstock.com

Chapter 5

232 *top left* eye-blink/Shutterstock.com; *top right* Mike Neale/Shutterstock.com; *center left* majeczka/Shutterstock.com; *bottom right* Igor Bulgarin/Shutterstock.com; *bottom left* urbanlight/Shutterstock.com; **238** urbanlight/Shutterstock.com; **240** *top left* Tatuasha/Shutterstock.com; *top right* Rudy Umans/Shutterstock.com; *center left* 7yonov/Shutterstock.com; **244** Igor Bulgarin/Shutterstock.com; **245** BluIz60/Shutterstock.com; **250** VoodooDot/Shutterstock.com, Catherinecml/Shutterstock.com; **251** Pixachi/Shutterstock.com; **258** Viktoria/Shutterstock.com, t_lidiya/Shutterstock.com, ecco/Shutterstock.com, Yeko Photo Studio/Shutterstock.com; **259** Aleksander Erin/Shutterstock.com; **260** *left* Irina Rogova/Shutterstock.com; *right* Kostyantyn Ivanyshen/Shutterstock.com; **264** Ijzendoorn/Shutterstock.com; **266** Mike Neale/Shutterstock.com; **270** Tatyana Vyc/Shutterstock.com; **277** fotohunter/Shutterstock.com; **279** *center left* Marek R. Swadzba/Shutterstock.com; *Exercise 31 left* Chad King / NOAA MBNMS, *right* eye-blink/Shutterstock.com; **281** Melpomene/Shutterstock.com; **285** *left* Sashkin/Shutterstock.com; *right* Boris Sosnovyy/Shutterstock.com

Chapter 6

288 *top left* Amnartk/Shutterstock.com; *top right* ©iStockphoto.com/Sam Burt Photography; *center left* © Ivan Cholakov | Dreamstime.com; *bottom right* U.S. Department of Energy; *bottom left* karl umbriaco/Shutterstock.com; **290** Kasia/Shutterstock.com; **296** *top right* Vitaly Korovin/Shutterstock.com; *center right* Jiri Hera/Shutterstock.com; **297** *center left* Mark III Photonics/Shutterstock.com; *top right* NASA; **302** Glenda M. Powers/Shutterstock.com, siamionau pavel/Shutterstock.com; **304** *top left* ©iStockphoto.com/katkov, ©iStockphoto.com/Andrew Johnson; *center right* DTR; **311** karl umbriaco/Shutterstock.com; **314** locote/Shutterstock.com; **318** risteski goce/Shutterstock.com; **319** Sergey Goruppa/Shutterstock.com; **320** rudall30/Shutterstock.com; **321** *center left* bornholm/Shutterstock.com, Masalski Maksim/Shutterstock.com; *top right* Tupungato/Shutterstock.com; *center right* U.S. Department of Energy; **322** mariematata/Shutterstock.com; **323** Aleksander Erin/Shutterstock.com; **324** © Leerobin | Dreamstime.com; **330** © Ivan Cholakov | Dreamstime.com; **337** Alexander Lukin/Shutterstock.com; **338** ©iStockphoto.com/Sam Burt Photography; **339** Eric Isselee/Shutterstock.com, Vasyl Helevachuk/Shutterstock.com; **341** Steshkin Yevgeniy/Shutterstock.com, mayer kleinostheim/Shutterstock.com, ©iStockphoto.com/Matej Michelizza, Refat/Shutterstock.com; **343** Photo by Alvesgaspar/Joaquim Alves Gaspar, Lisboa, Portugal, modified by RDBury.; **344** *top right* jovan vitanovski/Shutterstock.com; *bottom right* Verdateo/Shutterstock.com; **345** aquatic creature/Shutterstock.com; **347** mangostock/Shutterstock.com

Chapter 7

354 *top left* anyaivanova/Shutterstock.com; *top right* jaroslava V/ Shutterstock.com; *center left* luchschen/Shutterstock.com; *bottom right* Rudy Balasko/Shutterstock.com; *bottom left* Andresr/Shutterstock.com; **361** ©iStockphoto.com/edge69; **362** ecco/Shutterstock.com; **363** *center left* Johanna Goodyear/Shutterstock.com; *Exercise 51* ©iStockphoto.com/ edge69; *Exercise 52* Matt Antonino/Shutterstock.com; *Exercise 53* ©iStockphoto.com/lumpynoodles; **368** Fejas/Shutterstock.com; **370** Andresr/Shutterstock.com; **374** Eric Isselee/Shutterstock.com, Erik Lam/Shutterstock.com; **376** Tamara Kulikova/Shutterstock.com; **382** Rudy Balasko/Shutterstock.com; **383** Aleksander Erin/Shutterstock.com; **384** *center right* Tuja/Shutterstock.com; *bottom right* ©iStockphoto.com/ tirc83; **389** deer boy/Shutterstock.com; **390** *top left* luchschen/ Shutterstock.com; *center left* cobalt88/Shutterstock.com; **394** jaroslava V/ Shutterstock.com; **400** G Tipene/Shutterstock.com; **401** Mark Stout Photography/Shutterstock.com; **402** *top left* MarijaPiliponyte/ Shutterstock.com; *center left* Sergiy Telesh/Shutterstock.com; **406** Dirk Ercken/Shutterstock.com; **407** Anna Breitenberger/Shutterstock.com; **409** Denise Lett/Shutterstock.com

Chapter 8

416 *top left* Johnny Adolphson/Shutterstock.com; *top right* Serg64/ Shutterstock.com; *center left* Susana Ortega/Shutterstock.com; *bottom right* MishAl/Shutterstock.com; *bottom left* Ron Zmiri/Shutterstock.com; **423** Charles Brutlag/Shuttterstock.com; **424** Wolna/Shutterstock.com, ©iStockphoto.com/PLAINVIEW; **428** ©iStockphoto.com/edge69; **430** Ron Zmiri/Shutterstock.com; **437** *top left* James Thew/ Shutterstock.com; *bottom right* James Hoenstine/Shutterstock.com; **439**Aleksander Erin/Shutterstock.com; **440** Rob Byron/Shutterstock.com; **445** dompr/Shutterstock.com; **447** Paul St. Clair/Shutterstock.com; **448** Susana Ortega/Shutterstock.com; **457** Serg64/Shutterstock.com; **464** Johnny Adolphson/Shutterstock.com; **466** Zarja/Shutterstock.com; **467** *top left* homydesign/Shutterstock.com; *bottom right* BlueOrange Studio/ Shutterstock.com, Nikolamirejovska/Shutterstock.com; **469** Alin Brotea/ Shutterstock.com

Chapter 9

476 *top left* © Kjersti Joergensen | Dreamstime.com; *top right* NatalieJean/ Shutterstock.com; *center left* Zhukov Oleg/Shutterstock.com; *bottom right* Richard Paul Kane/Shutterstock.com; *bottom left* ©iStockphoto.com/ MinistryOfJoy; **483** *top left* Radovan Spurny/Shutterstock.com, Worldpics/ Shutterstock.com, wong yu liang/Shutterstock.com; *center left* ©iStockphoto.com/MinistryOfJoy; **486** Jon Le-Bon/Shutterstock.com; **487** Dan Gerber/Shutterstock.com; **493** Richard Paul Kane/ Shutterstock.com; **495** *top left* Ben Haslam/Haslam Photography/ Shutterstock.com; *bottom right* Don Williamson/Shutterstock.com; **496** Steve Cukrov/Shutterstock.com; **501** Zhukov Oleg/Shutterstock.com; **502** Baker Alhashki/Shutterstock.com; **503** Aleksander Erin/ Shutterstock.com; **504** ©iStockphoto.com/mikdam; **513** *center left* graphixmania/Shutterstock.com; *bottom left* Petrovic Igor/Shutterstock.com; *center right* © Kaleb Timberlake | Dreamstime.com; **514** *top right* Ermek/ Shutterstock.com; *center right* © South12th | Dreamstime.com; **521** © Kjersti Joergensen | Dreamstime.com; **522** *center left* ©iStockphoto.com/edhor; *top right* Vereshchagin Dmitry/Shutterstock.com; **523** bikeriderlondon/Shutterstock.com, Aspen Photo/Shutterstock.com, Ruth Peterkin/Shutterstock.com; **531** luigi nifosi/Shutterstock.com; **533** wavebreakmedia/Shutterstock.com; **537** Artazum and Iriana Shiyan/ Shutterstock.com

Chapter 10

540 *top left* ©iStockphoto.com/Andrew_Howe; *top right* © Lostafichuk | Dreamstime.com; *center left* BsChan/Shutterstock.com; *bottom right* Johnny Habell/Shutterstock.com; *bottom left* Marafona/Shutterstock.com; **547** Marafona/Shutterstock.com; **549** *bottom left* Eduard Härkönen/ Shutterstock.com; *bottom right* Johnny Habell/Shutterstock.com; **550** ©iStockphoto.com/technotr; **554** BsChan/Shutterstock.com; **557** Aleksander Erin/Shutterstock.com; **558** Henryk Sadura/ Shutterstock.com; **563** Julia Baturina/Shutterstock.com; **564** Stubbs; **565** *center right* Neamov/Shutterstock.com, Subbotina Anna/ Shutterstock.com, Coprid/Shutterstock.com; *bottom right* © Lostafichuk | Dreamstime.com; **566** mekCar/Shutterstock.com; **573** Eric Isselee/ Shutterstock.com, mycola/Shutterstock.com; **575** Nikola Bilic/ Shutterstock.com; **578** © Andres Rodriguez | Dreamstime.com

Chapter 11

582 *top left* ©iStockphoto.com/David Sucsy; *top right* ©iStockphoto.com/ Cesar Okada; *center left* My Good Images/Shutterstock.com; *bottom right* bikeriderlondon/Shutterstock.com; *bottom left* Iryna Rasko/ Shutterstock.com; **591** bikeriderlondon/Shutterstock.com; **594** My Good Images/Shutterstock.com; **595** Stokkete/Shutterstock.com; **597** Krasowit/ Shutterstock.com; **602** Kostenyukova Nataliya/Shutterstock.com; **603** ©iStockphoto.com/Cesar Okada; **604** Monkey Business Images/ Shutterstock.com; **607** Aleksander Erin/Shutterstock.com; **608** Alexandr79/ Shutterstock.com; **615** Kenneth William Caleno/Shutterstock.com; **616** *center left* ©iStockphoto.com/David Sucsy; *center right* Johann Helgason/Shutterstock.com; **619** Taiga/Shutterstock.com; **623** Lucky Business/Shutterstock.com